Your Guide to Success in Math

Complete Step 0 as soon as you begin your math course.

STEP 0: Plan Your Semester

☐ Register for the online part of the course (if there is one) as soon as possible.

☐ Fill in your Course and Contact information on this pull-out card.

☐ Write important dates from your syllabus on the Semester Organizer section on this pull-out card.

Follow Steps 1–3 during your course. Your instructor will tell you which resources to use—and when—in the textbook or eText, *MyMathGuide* workbook, videos, and MyLab Math. Use these resources for extra help and practice.

STEP 1: PREPARE: Studying the Concepts

☐ Read the textbook or eText, listen to your instructor's lecture, and/or watch the section videos. You can work in *MyMathGuide* as you do this, saving all your work to review throughout the course.

☐ Work the Skill to Review exercises and/or watch the videos in MyLab Math in each section.

☐ Stop and do the Margin Exercises, including the Guided Solution Exercises, as directed.

STEP 2: PARTICIPATE: Making Connections through Active Exploration

☐ Explore the concepts using the Animations in MyLab Math.

☐ Work the Visualizing for Success or Translating for Success exercises in the text and/or in MyLab Math.

☐ Answer the Check Your Understanding exercises in the Section Exercises in the textbook and/or in MyLab Math.

STEP 3: PRACTICE: Reinforcing Understanding

☐ Complete your assigned homework from the textbook and/or in MyLab Math.

  ☐ When doing homework from the textbook, use the answer section to check your work.

  ☐ When doing homework in MyLab Math, use the Learning Aids, such as Help Me Solve This and View an Example, as needed, working toward being able to complete exercises without the aids.

☐ Work the exercises in the Mid-Chapter Review.

☐ Read the Study Guide and work the Review Exercises in the Chapter Summary and Review.

☐ Take the Chapter Test as a practice exam. To watch an instructor solve each problem, go to the Chapter Test Prep Videos in MyLab Math.

Use the ***Studying for Success*** tips in the text and the ***MyLab Math Study Skills*** modules (with videos) to help you develop effective time-management, note-taking, test-prep, and other skills.

Fraction, Decimal, Percent Equivalents

Fraction Notation	$\frac{1}{10}$	$\frac{1}{8}$	$\frac{1}{6}$	$\frac{1}{5}$	$\frac{1}{4}$	$\frac{3}{10}$	$\frac{1}{3}$	$\frac{3}{8}$	$\frac{2}{5}$	$\frac{1}{2}$	$\frac{3}{5}$	$\frac{5}{8}$	$\frac{2}{3}$	$\frac{7}{10}$	$\frac{3}{4}$	$\frac{4}{5}$	$\frac{5}{6}$	$\frac{7}{8}$	$\frac{9}{10}$	$\frac{1}{1}$
Decimal Notation	0.1	0.125	$0.16\overline{6}$	0.2	0.25	0.3	$0.33\overline{3}$	0.375	0.4	0.5	0.6	0.625	$0.66\overline{6}$	0.7	0.75	0.8	$0.83\overline{3}$	0.875	0.9	1
Percent Notation	10%	12.5% or $12\frac{1}{2}$%	$16.\overline{6}$% or $16\frac{2}{3}$%	20%	25%	30%	$33.\overline{3}$% or $33\frac{1}{3}$%	37.5% or $37\frac{1}{2}$%	40%	50%	60%	62.5% or $62\frac{1}{2}$%	$66.\overline{6}$% or $66\frac{2}{3}$%	70%	75%	80%	$83.\overline{3}$% or $83\frac{1}{3}$%	87.5% or $87\frac{1}{2}$%	90%	100%

Percent Notation

$3\% = 3 \times \frac{1}{100} = \frac{3}{100}$ $\quad$ $3\% = 3 \times 0.01 = 0.03$

Percent Increase and Decrease

Price increased from \$480 to \$504:

$$\text{Percent increase} = \frac{\text{Change}}{\text{Original}} = \frac{504 - 480}{480} = \frac{24}{480} = 0.05 = 5\%$$

Price decreased from \$60 to \$45:

$$\text{Percent decrease} = \frac{\text{Change}}{\text{Original}} = \frac{60 - 45}{60} = \frac{15}{60} = 0.25 = 25\%$$

Translating to Percent Equations

What is 20% of 75?
↓ ↓ ↓ ↓ ↓
$a = 20\% \cdot 75$

15 is 20% of what?
↓ ↓ ↓ ↓ ↓
$15 = 20\% \cdot b$

15 is what percent of 75?
↓ ↓ ↓ ↓ ↓
$15 = p \cdot 75$

Interest

Simple: $I = P \cdot r \cdot t$

Compound: $A = P \cdot \left(1 + \frac{r}{n}\right)^{n \cdot t}$

American Units of Length

12 inches (in.) = 1 foot (ft)
36 inches = 1 yard (yd)
3 feet = 1 yard
5280 feet = 1 mile (mi)

Temperature

$F = \frac{9}{5} \cdot C + 32$ $\quad$ $C = \frac{5}{9}(F - 32)$

Metric Units of Length

1 kilometer (km) = 1000 meters (m)
1 hectometer (hm) = 100 meters
1 dekameter (dam) = 10 meters
1 meter
1 decimeter (dm) = $\frac{1}{10}$ meter
1 centimeter (cm) = $\frac{1}{100}$ meter
1 millimeter (mm) = $\frac{1}{1000}$ meter

Formulas

Perimeter of a Square: $P = 4 \cdot s$
Perimeter of a Rectangle: $P = 2 \cdot l + 2 \cdot w$
Area of a Square: $A = s^2$
Area of a Rectangle: $A = l \cdot w$
Area of a Parallelogram: $A = b \cdot h$
Area of a Triangle: $A = \frac{1}{2} \cdot b \cdot h$
Area of a Trapezoid: $A = \frac{1}{2} \cdot h \cdot (a + b)$
Circumference of a Circle: $C = \pi \cdot d$, or $C = 2 \cdot \pi \cdot r$
Area of a Circle: $A = \pi \cdot r^2$
Volume of a Rectangular Solid: $V = l \cdot w \cdot h$
Volume of a Circular Cylinder: $V = \pi \cdot r^2 \cdot h$

Pythagorean Theorem

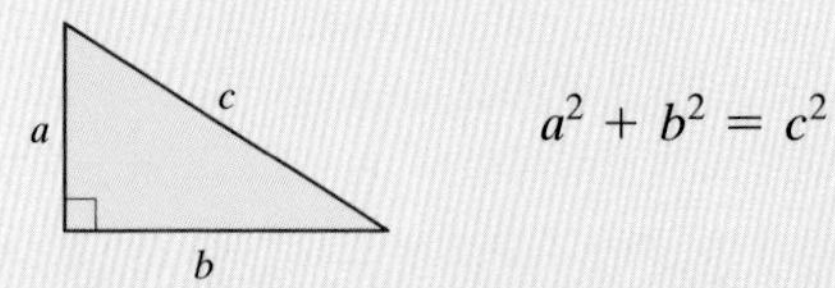

$a^2 + b^2 = c^2$

Exponents and Scientific Notation

$x^0 = 1;\ x^1 = x;\ x^{-3} = \frac{1}{x^3};\ x^2 \cdot x^5 = x^7;$

$\frac{x^5}{x^2} = x^3;\ (x^2)^5 = x^{10};\ (xy)^5 = x^5y^5;\ \left(\frac{x}{y}\right)^5 = \frac{x^5}{y^5}$

$3.26 \times 10^4 = 32{,}600$

$3.26 \times 10^{-4} = 0.000326$

At a Glance: Prealgebra

Place Value

529,630,718,249.70651

Billions			Millions			Thousands			Ones							
5	2	9	6	3	0	7	1	8	2	4	9	7	0	6	5	1
Hundreds	Tens	Ones	Hundreds	Tens	Ones	Hundreds	Tens	Ones	Hundreds	Tens	Ones	Tenths	Hundredths	Thousandths	Ten-thousandths	Hundred-thousandths

Problem-Solving Strategy

1. Familiarize
2. Translate
3. Solve
4. Check
5. State

Exponential Notation

Base → $3^4 = \underbrace{3 \cdot 3 \cdot 3 \cdot 3}_{\text{4 factors of 3}}$ (4 = Exponent)

Order of Operations

1. Do all calculations within grouping symbols before operations outside.
2. Evaluate all exponential expressions.
3. Do all multiplications and divisions in order from left to right.
4. Do all additions and subtractions in order from left to right.

Operations with Integers

$-18 + 3 = -15$

$-6 + (-4) = -10$

$9 - 12 = -3$

$-7 - (-10) = 3$

$-9 \cdot 6 = -54$

$-5 \cdot (-3) = 15$

$18 \div (-3) = -6$

$-10 \div (-2) = 5$

Absolute value: $|-4| = 4$

Prime Factorization

$220 = 2 \cdot 2 \cdot 5 \cdot 11$ All factors are prime.

Least Common Multiple

$12 = 2 \cdot 2 \cdot 3$ $15 = 3 \cdot 5$

The LCM of 12 and 15 $= 2 \cdot 2 \cdot 3 \cdot 5$, or 60.

Fraction Notation

$\frac{8}{1} = 8$ $\frac{8}{8} = 1$ $\frac{0}{8} = 0$ $\frac{8}{0}$ Not defined

$$\frac{5}{6} + \frac{3}{4} = \frac{5}{6} \cdot \frac{2}{2} + \frac{3}{4} \cdot \frac{3}{3} = \frac{10}{12} + \frac{9}{12} = \frac{10 + 9}{12} = \frac{19}{12}$$

$$\frac{5}{6} - \frac{3}{4} = \frac{5}{6} \cdot \frac{2}{2} - \frac{3}{4} \cdot \frac{3}{3} = \frac{10}{12} - \frac{9}{12} = \frac{10 - 9}{12} = \frac{1}{12}$$

$$\frac{5}{6} \cdot \frac{3}{4} = \frac{5 \cdot 3}{6 \cdot 4} = \frac{15}{24} = \frac{3 \cdot 5}{3 \cdot 8} = \frac{3}{3} \cdot \frac{5}{8} = 1 \cdot \frac{5}{8} = \frac{5}{8}$$

$$\frac{5}{6} \div \frac{3}{4} = \frac{5}{6} \cdot \frac{4}{3} = \frac{20}{18} = \frac{2 \cdot 10}{2 \cdot 9} = \frac{2}{2} \cdot \frac{10}{9} = 1 \cdot \frac{10}{9} = \frac{10}{9}$$

Mixed Numerals

$\frac{11}{3} = 3\frac{2}{3}$ $4\frac{1}{2} = \frac{9}{2}$

Graphing Linear Equations

x	$y = 2x - 3$	(x, y)
3	3	(3, 3)
1	−1	(1, −1)
0	−3	(0, −3)

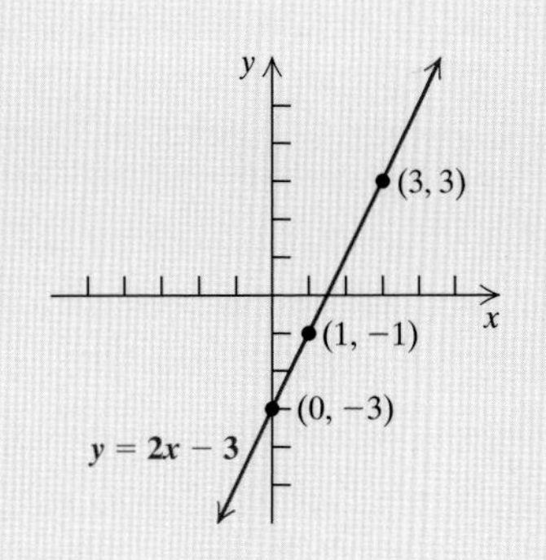

Student Organizer

Course Information

Course Number: ______________ **Name:** ______________________________

Location: ______________________________ **Days/Time:** ______________________

Contact Information

Contact	Name	Email	Phone	Office Hours	Location
Instructor					
Tutor					
Math Lab					
Classmate					
Classmate					

Semester Organizer

Week	Homework	Quizzes and Tests	Other

EDITION 13

Intermediate Algebra

Marvin L. Bittinger
Indiana University Purdue University Indianapolis

Judith A. Beecher

Barbara L. Johnson
Ivy Tech Community College of Indiana

Director, Courseware Portfolio Management:	Michael Hirsch
Courseware Portfolio Manager:	Cathy Cantin
Courseware Portfolio Management Assistants:	Shannon Bushee; Shannon Slocum
Managing Producer:	Karen Wernholm
Content Producer:	Ron Hampton
Producer:	Erin Carreiro
Manager, Courseware QA:	Mary Durnwald
Manager, Content Development:	Eric Gregg
Field Marketing Managers:	Jennifer Crum; Lauren Schur
Product Marketing Manager:	Kyle DiGiannantonio
Product Marketing Assistant:	Brooke Imbornone
Senior Author Support/ Technology Specialist:	Joe Vetere
Manager, Rights and Permissions:	Gina Cheselka
Manufacturing Buyer:	Carol Melville, LSC Communications
Associate Director of Design:	Blair Brown
Program Design Lead:	Barbara T. Atkinson
Text Design:	Geri Davis/The Davis Group, Inc.
Editorial and Production Service:	Martha Morong/Quadrata, Inc.
Composition:	Cenveo® Publisher Services
Illustration:	Network Graphics; William Melvin
Cover Design:	Cenveo® Publisher Services
Cover Image:	Nick Veasey/Untitled X-Ray/Getty Images

Library of Congress Cataloging-in-Publication Data

Names: Bittinger, Marvin L., author. | Beecher, Judith A., author. | Johnson, Barbara L., author.
Title: Intermediate algebra / Marvin L. Bittinger (Indiana University Purdue University Indianapolis), Judith A. Beecher, Barbara L. Johnson (Ivy Tech Community College of Indiana).
Description: 13 edition. | New York, NY : Pearson, [2019] | Includes index.
Identifiers: LCCN 2017060142| ISBN 9780134707365 (pbk.) | ISBN 0134707362 (pbk.)
Subjects: LCSH: Algebra–Textbooks.
Classification: LCC QA154.3 .B578 2019 | DDC 512.9–dc23
LC record available at https://lccn.loc.gov/2017060142

12 2022

ISBN 13: 978-0-13-470736-5
ISBN 10: 0-13-470736-2

Contents

KD 12.19.2022 0831

Index of Activities

Index of Animations

Preface

Math doesn't change, but students' needs—and the way students learn—do.

With this in mind, *Intermediate Algebra,* 13th Edition, continues the Bittinger tradition of objective-based, guided learning, while integrating many updates with the proven pedagogy. These updates are motivated by feedback that we received from students and instructors, as well as our own experience in the classroom. In this edition, our focus is on guided learning and retention: helping each student (and instructor) get the most out of all the available program resources—wherever and whenever they engage with the math.

We believe that student success in math hinges on four key areas: **Foundation, Engagement, Application,** and **Retention**. In the 13th edition, we have added key new program features (highlighted below, for quick reference) in each area to make it easier for each student to personalize his or her learning experience. In addition, you will recognize many proven features and presentations from the previous edition of the program.

FOUNDATION

Studying the Concepts

Students can learn the math concepts by reading the textbook or the eText, participating in class, watching the videos, working in the *MyMathGuide* workbook—or using whatever combination of these course resources works best for them.

In order to understand new math concepts, students must recall and use skills and concepts previously studied. To support student learning, we have integrated two important new features throughout the 13th Edition program:

- ☐ *New!* **Just-in-Time Review** at the beginning of the text and the eText is a set of quick reviews of the key topics from previous courses that are prerequisites for the new material in this course. A note on each Chapter Opener alerts students to the topics they should review for that chapter. In MyLab Math, students will find a concise presentation of each topic in the **Just-in-Time Review Videos.**
- ☐ *New!* **Skill Review,** in nearly every section of the text and the eText, reviews a previously presented skill at the objective level where it is key to learning the new material. This feature offers students two practice exercises with answers. In MyLab Math, new **Skill Review Videos,** created by the Bittinger author team, offer a concise, step-by-step solution for each Skill Review exercise.

Margin Exercises with Guided Solutions, with fill-in blanks at key steps in the problem-solving process, appear in nearly every text section and can be assigned in MyLab Math.

Algebraic–Graphical Connections in the text draw explicit connections between the algebra and the corresponding graphical visualization.

Intermediate Algebra Video Program, our comprehensive program of objective-based, interactive videos, can be used hand-in-hand with our *MyMathGuide* workbook. **Interactive Your Turn exercises** in the videos prompt students to solve problems and receive instant feedback. These videos can be accessed at the section, objective, and example levels.

MyMathGuide offers students a guided, hands-on learning experience. This objective-based workbook (available in print and in MyLab Math) includes vocabulary, skill, and concept review—as well as problem-solving practice with space for students to fill in the answers and stepped-out solutions to problems, to show (and keep) their work, and to write notes. Students can use *MyMathGuide,* while watching the videos, listening to the instructor's lecture, or reading the text or the eText, in order to reinforce and self-assess their learning.

Studying for Success sections are checklists of study skills designed to ensure that students develop the skills they need to succeed in math, school, and life. They are available at the beginning of selected sections.

ENGAGEMENT

Making Connections through Active Exploration

Since understanding the big picture is key to student success, we offer many active learning opportunities for the practice, review, and reinforcement of important concepts and skills.

- ☐ *New!* **Chapter Opener Applications** with graphics use current data and applications to present the math in context. Each application is related to exercises in the text to help students model, visualize, learn, and retain the math.
- ☐ *New!* **Student Activities,** included with each chapter, have been developed as multistep, data-based activities for students to apply the math in the context of an authentic application. Student Activities are available in *MyMathGuide* and in MyLab Math.
- ☐ *New!* **Interactive Animations** can be manipulated by students in MyLab Math through guided and open-ended exploration to further solidify their understanding of important concepts.

Translating for Success offers extra practice with the important first step of the process for solving applied problems. **Visualizing for Success** asks students to match an equation or an inequality with its graph by focusing on characteristics of the equation or the inequality and the corresponding attributes of the graph. Both of these activities are available in the text and in MyLab Math.

Calculator Corner is an optional feature in each chapter that helps students use a calculator to perform calculations and to visualize concepts.

Learning Catalytics uses students' mobile devices for an engagement, assessment, and classroom intelligence system that gives instructors real-time feedback on student learning.

APPLICATION

Reinforcing Understanding

As students explore the math, they have frequent opportunities to apply new concepts, practice, self-assess, and reinforce their understanding.

Margin Exercises, labeled "Do Exercise . . . ," give students frequent opportunities to apply concepts just discussed by solving problems that parallel text examples.

Exercise Sets in each section offer abundant opportunity for practice and review in the text and in MyLab Math. The Section Exercises are grouped by objective for ease of use, and each set includes the following special exercise types:

- ☐ *New!* **Check Your Understanding** with **Reading Check** and **Concept Check** exercises, at the beginning of each exercise set, gives students the opportunity to assess their grasp of the skills and concepts before moving on to the objective-based section exercises. In MyLab Math, many of these exercises use drag & drop functionality.
- ☐ **Skill Maintenance Exercises** offer a thorough review of the math in the preceding sections of the text.
- ☐ **Synthesis Exercises** help students develop critical-thinking skills by requiring them to use what they know in combination with content from the current and previous sections.

RETENTION

Carrying Success Forward

Because continual practice and review is so important to retention, we have integrated both throughout the program in the text and in MyLab Math.

- ☐ *New!* **Skill Builder** adaptive practice, available in MyLab Math, offers each student a personalized learning experience. When a student struggles with the assigned homework, Skill Builder exercises offer just-in-time additional adaptive practice. The adaptive engine tracks student performance and delivers to each individual questions that are appropriate for his or her level of understanding. When the system has determined that the student has a high probability of successfully completing the assigned exercise, it suggests that the student return to the assigned homework.

Mid-Chapter Review offers an opportunity for active review midway through each chapter. This review offers four types of practice problems:

Concept Reinforcement, Guided Solutions, Mixed Review, and Understanding Through Discussion and Writing

Summary and Review is a comprehensive learning and review section at the end of each chapter. Each of the five sections—**Vocabulary Reinforcement** (fill-in-the-blank), **Concept Reinforcement** (true/false), **Study Guide** (examples with stepped-out solutions paired with similar practice problems), **Review Exercises**, and **Understanding Through Discussion and Writing**—includes references to the section in which the material was covered to facilitate review.

Chapter Test offers students the opportunity for comprehensive review and reinforcement prior to taking their instructor's exam. **Chapter Test Prep Videos** in MyLab Math show step-by-step solutions to the questions on the chapter test.

Cumulative Review follows each chapter beginning with Chapter 2. These revisit skills and concepts from all preceding chapters to help students retain previously presented material.

Resources for Success

MyLab Math Online Course for Bittinger, Beecher, and Johnson, *Intermediate Algebra*, 13th Edition (access code required)

MyLab™ Math is available to accompany Pearson's market-leading text offerings. To give students a consistent tone, voice, and teaching method, the pedagogical approach of the text is tightly integrated throughout the accompanying MyLab Math course, making learning the material as seamless as possible.

UPDATED! Learning Path

Structured, yet flexible, the updated learning path highlights author-created, faculty-vetted content—giving students what they need exactly when they need it. The learning path directs students to resources such as two new types of video: **Just-in-Time Review** (concise presentations of key topics from previous courses) and **Skill Review** (author-created exercises with step-by-step solutions that reinforce previously presented skills), both available in the Multimedia Library and assignable in MyLab Math.

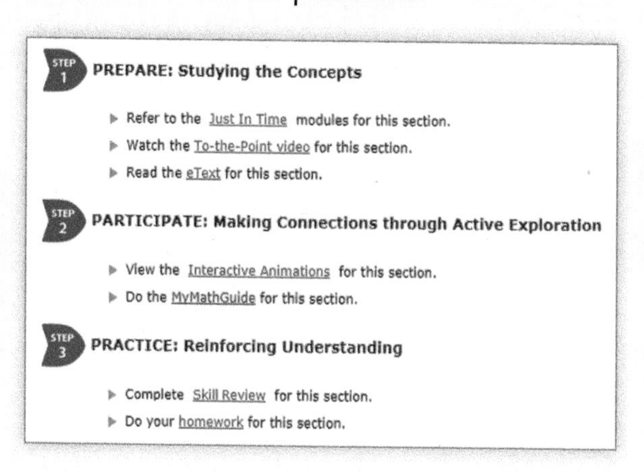

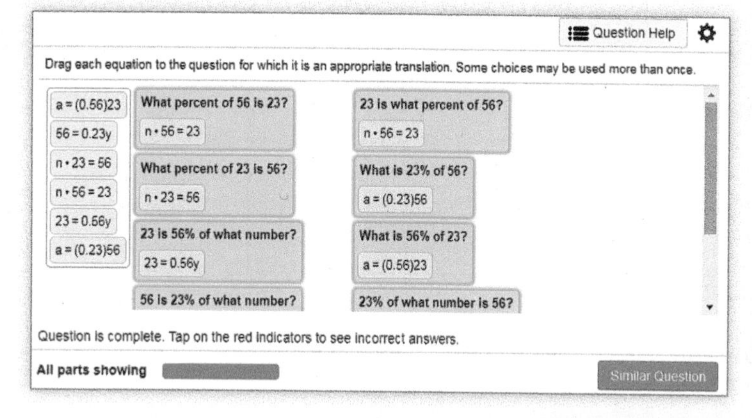

NEW! Drag-and-Drop Exercises

Drag-and-drop exercises are now available in MyLab Math. This new assignment type allows students to drag answers and values within a problem, providing a new and engaging way to test students' concept knowledge.

NEW and UPDATED! Animations

New animations encourage students to learn key concepts through guided and open-ended exploration. Animations are available through the learning path and multimedia library, and they can be assigned within MyLab Math.

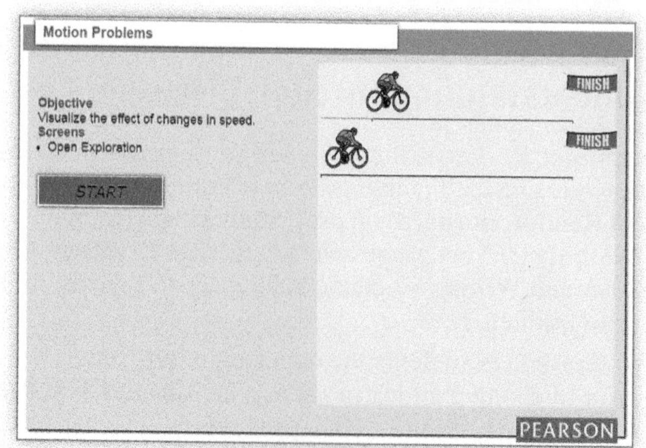

Just-in-Time Review

The common octopus, *Octopus vulgaris*, is most often found in temperate and tropical waters at a depth of 100 to 150 meters. This species can reach a length of 1 meter including its legs. The octopus has three hearts and no bones. Octopuses can learn and are probably the most intelligent of all invertebrates. One-third of its neurons are located in its brain, and the remaining two-thirds of its neurons are in its arms, which are lined with suckers. The table at left shows a comparison of the number of neurons in a human, an octopus, a mouse, and a pond snail.

Comparing Number of Neurons

	Number of neurons
Pond snail	10,000
Mouse	80,000,000
Octopus	500,000,000
Human	86,000,000,000

DATA: *National Geographic*, November 2016, p.70; "The Power of Eight," by Olivia Judson

Data: animaldiversity.org; arkive.org; National Geographic, *November 2016, p. 70, "The Power of Eight," Olivia Judson, Roger Hanlon, Marine Biological Laboratory; Guy Levy and Benny Hochner, Hebrew University of Jerusalem; Cliff Ragsdale, University of Chicago*

We will calculate the number of neurons in the arms of an *Octopus vulgaris* in Example 5 of Just-in-Time 20.

1 THE SET OF REAL NUMBERS

MyLab Math VIDEO

The following figure shows the relationships among various sets of real numbers.

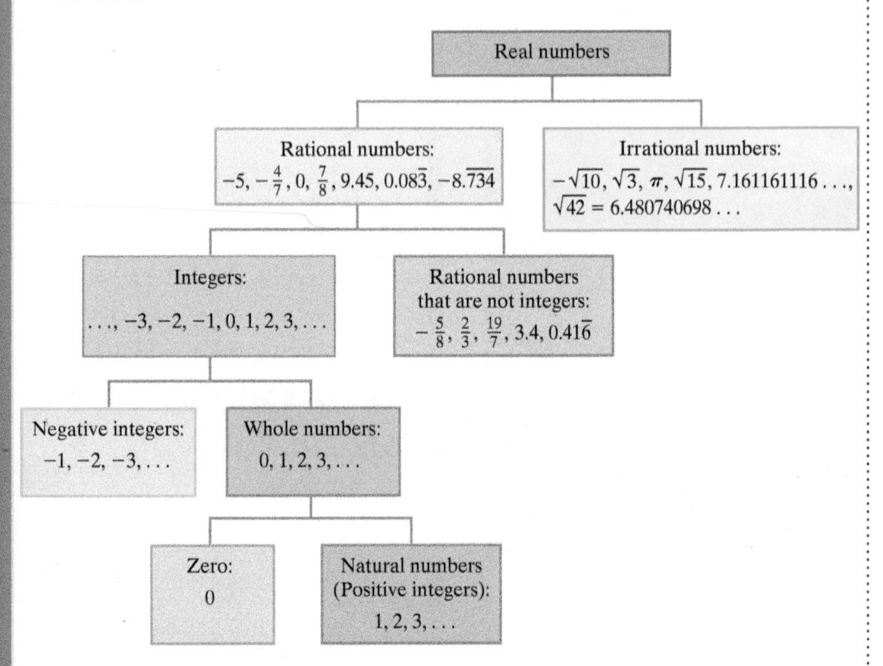

The set containing the numbers -5, 0, and 3 can be named $\{-5, 0, 3\}$. It is described using the **roster method**, which lists all members of a set. We use the roster method to describe three frequently used subsets of real numbers. Note that three dots are used to indicate that the pattern continues without end.

Natural numbers: $\{1, 2, 3, \ldots\}$

Whole numbers: $\{0, 1, 2, 3, \ldots\}$

Integers: $\{\ldots, -3, -2, -1, 0, 1, 2, 3, \ldots\}$

Other subsets of real numbers are described using **set-builder notation**. With this notation, instead of listing all members of a set, we specify conditions under which a number is in a set. For example, the set of all odd natural numbers less than 9 can be described as follows:

$\{x \mid x \text{ is an odd number less than } 9\}$, and read

The set of all x such that x is an odd number less than 9.

Using roster notation, we can write this set as $\{1, 3, 5, 7\}$.

(continued)

EXERCISES

1. Use roster notation to name each set.
   a) The set of all letters in the word "solve"
   b) The set of all even natural numbers

2. Use set-builder notation to name each set.
   a) $\{0, 1, 2, 3, 4, 5\}$
   b) $\{4, 5, 6, 7, 8, 9, 10\}$

The set of **real numbers** is

$$\{x \mid x \text{ is a rational number } or\ x \text{ is an irrational number}\}.$$

Every point on the number line represents some real number, and every real number is represented by some point on the number line.

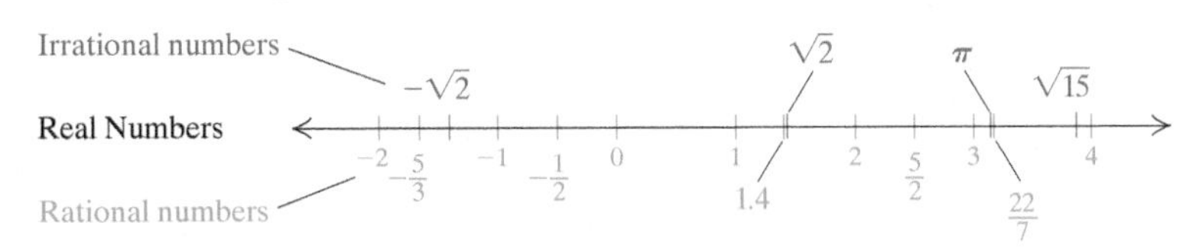

A **rational number** can be expressed as an integer divided by a nonzero integer. The set of rational numbers is

$$\left\{\frac{p}{q} \,\middle|\, p \text{ is an integer, } q \text{ is an integer, and } q \neq 0\right\}.$$

Rational numbers are numbers whose decimal representation either terminates or has a repeating block of digits. The following are examples of rational numbers:

$$\frac{5}{8},\ \frac{12}{-7},\ \frac{-17}{15},\ -\frac{9}{7},\ \frac{39}{1},\ \frac{0}{6}.$$

Note that $\frac{39}{1} = 39$. Thus the set of rational numbers contains the integers. Using long division, we can write a fraction in decimal notation:

$$\frac{5}{8} = \underbrace{0.625}_{\text{Terminating}} \quad \text{and} \quad \frac{6}{11} = \underbrace{0.545454\ldots}_{\text{Repeating}} = 0.\overline{54}.$$

The bar in $0.\overline{54}$ indicates the repeating block of digits in decimal notation.

Irrational numbers are numbers whose decimal representation neither terminates nor has a repeating block of digits. They cannot be represented as the quotient of two integers. Numbers like π, $\sqrt{2}$, $-\sqrt{10}$, $\sqrt{13}$, and $-1.898898889\ldots$ are examples of irrational numbers. The decimal notation for an irrational number *neither* terminates *nor* repeats.

Do Exercises 1–4.

(Exercises 1 and 2 are on the preceding page.)

EXERCISES

3. Given the numbers

$20,\ -10,\ -5.34,\ 18.999,\ 0.58\overline{3},$

$\frac{11}{45},\ \sqrt{7},\ -\sqrt{2},\ \sqrt{16},\ 0,\ -\frac{2}{3},$

$9.34334333433334\ldots$:

a) Name the natural numbers.
b) Name the whole numbers.
c) Name the integers.
d) Name the irrational numbers.
e) Name the rational numbers.
f) Name the real numbers.

4. Given the numbers

$-6,\ 0,\ 1,\ -\frac{1}{2},\ -4,\ \frac{7}{9},$

$12,\ -\frac{6}{5},\ 3.45,\ 5\frac{1}{2},\ \sqrt{3},$

$\sqrt{25},\ -\frac{12}{3},\ 0.131331333133331\ldots$

a) Name the natural numbers.
b) Name the whole numbers.
c) Name the integers.
d) Name the irrational numbers.
e) Name the rational numbers.
f) Name the real numbers.

2 ORDER FOR THE REAL NUMBERS

MyLab Math VIDEO

Real numbers are named in order on the number line. For any two numbers on the line, the one to the left is less than the one to the right.

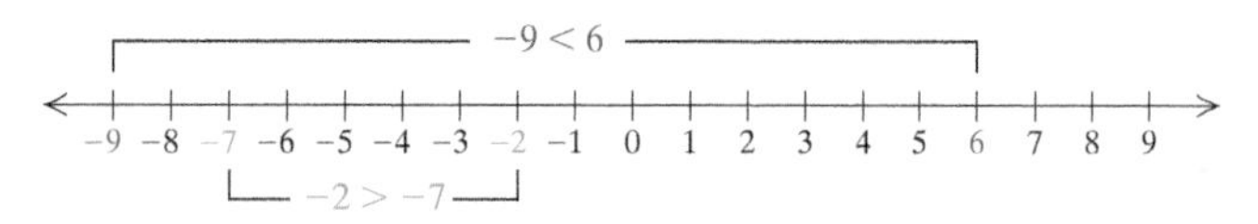

We use the symbol **<** to mean "**is less than.**" The sentence $-9 < 6$ means "-9 is less than 6." The symbol **>** means "**is greater than.**" The sentence $-2 > -7$ means "-2 is greater than -7." Sentences containing $<$ or $>$ are called **inequalities**.

EXAMPLES Use either $<$ or $>$ for $\square$ to write a true sentence.

1. $4 \square 9$ Since 4 is to the left of 9, 4 is less than 9, so $4 < 9$.
2. $-8 \square 3$ Since -8 is to the left of 3, we have $-8 < 3$.
3. $7 \square -12$ Since 7 is to the right of -12, then $7 > -12$.
4. $-21 \square -5$ Since -21 is to the left of -5, we have $-21 < -5$.
5. $4.79 \square 4.97$ Since 4.79 is to the left of 4.97, we have $4.79 < 4.97$.
6. $-2.7 \square -\frac{3}{2}$ Since $-\frac{3}{2} = -1.5$ and -2.7 is to the left of -1.5, we have $-2.7 < -\frac{3}{2}$.
7. $\frac{5}{8} \square \frac{7}{11}$ We convert to decimal notation $\left(\frac{5}{8} = 0.625 \text{ and } \frac{7}{11} = 0.6363\ldots\right)$ and compare. Thus, $\frac{5}{8} < \frac{7}{11}$.

Note that $-8 < 5$ and $5 > -8$ are both true. Every true inequality yields another true inequality if we interchange the numbers or variables and reverse the direction of the inequality sign.

$a < b$ also has the meaning $b > a$.

EXERCISES

Insert $<$ or $>$ for $\square$ to write a true sentence.

1. $-5 \square -4$
2. $-\frac{1}{4} \square -\frac{1}{2}$
3. $87 \square 67$
4. $-9.8 \square -4\frac{2}{3}$
5. $6.78 \square -6.77$
6. $-\frac{4}{5} \square -0.86$
7. $\frac{14}{29} \square \frac{17}{32}$
8. $-\frac{12}{13} \square -\frac{14}{15}$
9. $1.8 \square 1.08$
10. $0 \square -4$

MyLab Math ANIMATION

(continued)

2 ORDER FOR THE REAL NUMBERS (continued)

EXAMPLES Write a different inequality with the same meaning.

8. $a < -5$ The inequality $-5 > a$ has the same meaning.
9. $-3 > -8$ The inequality $-8 < -3$ has the same meaning.

Expressions like $a \leq b$ and $a \geq b$ are also **inequalities**. We read $a \leq b$ as "**a is less than or equal to b**." We read $a \geq b$ as "**a is greater than or equal to b**." If a is nonnegative, then $a \geq 0$.

EXAMPLES Determine whether each of the following is true or false.

10. $-8 \leq 5.7$ True since $-8 < 5.7$ is true.
11. $-8 \leq -8$ True since $-8 = -8$ is true.
12. $-7 \geq 4\frac{1}{3}$ False since neither $-7 > 4\frac{1}{3}$ nor $-7 = 4\frac{1}{3}$ is true.
13. $-\frac{2}{3} \geq -\frac{5}{4}$ True since $-\frac{2}{3} = -0.666\ldots$ and $-\frac{5}{4} = -1.25$ and $-0.666\ldots > -1.25$.

Do Exercises 1–16.
(Exercises 1–10 are on the preceding page.)

EXERCISES

Write a different inequality with the same meaning.

11. $x > 6$

12. $-4 < 7$

Determine whether each of the following is true or false.

13. $6 \geq -9.4$

14. $-18 \leq -18$

15. $-7.6 \leq -10\frac{4}{5}$

16. $-\frac{24}{27} \geq -\frac{25}{28}$

3 GRAPHING INEQUALITIES ON THE NUMBER LINE

MyLab Math VIDEO

A replacement that makes an inequality true is called a **solution**. The set of all solutions is called the **solution set**. A **graph** of an inequality is a drawing that represents its solution set.

EXAMPLE 1 Graph the inequality $x > -3$ on the number line.

The solutions consist of all real numbers greater than -3, so we shade all numbers greater than -3. Since -3 is not a solution, we use a parenthesis at -3 to indicate this. The graph represents the solution set $\{x|x > -3\}$.

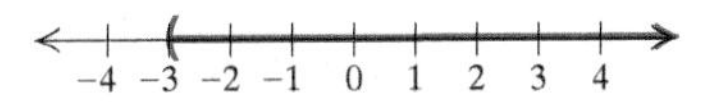

EXAMPLE 2 Graph the inequality $x \leq 2$ on the number line.

We make a drawing that represents the solution set $\{x|x \leq 2\}$. The graph consists of 2 as well as the numbers less than 2. We shade all numbers to the left of 2 and use a bracket at 2 to indicate that it is also a solution.

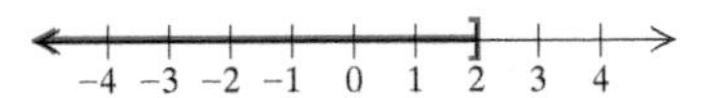

Do Exercises 1–4.

EXERCISES

Graph each inequality.

1. $x > -1$
2. $x \leq 5$
3. $x > 0$
4. $x \leq -\frac{5}{2}$

4 ABSOLUTE VALUE

MyLab Math VIDEO

We call the distance of a number from 0 on the number line the **absolute value** of the number. Since distance is always a nonnegative number, the absolute value of a number is always greater than or equal to 0.

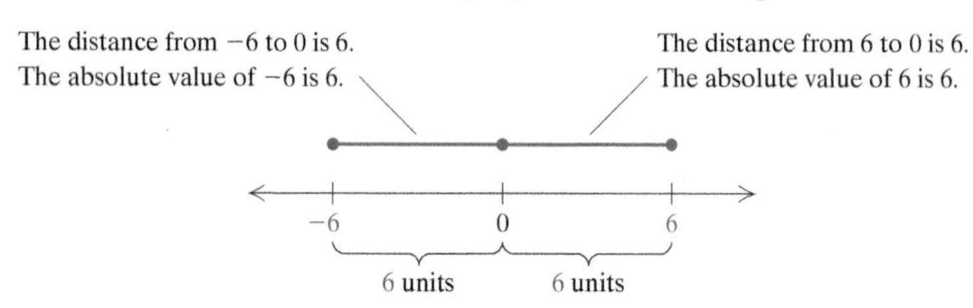

We use the symbol $|x|$ to represent the absolute value of a number x.

EXAMPLES Find the absolute value.

1. $|-7|$ The distance of −7 from 0 is 7, so $|-7|$ is 7.
2. $|12|$ The distance of 12 from 0 is 12, so $|12|$ is 12.
3. $|0|$ The distance of 0 from 0 is 0, so $|0|$ is 0.
4. $\left|\frac{4}{5}\right| = \frac{4}{5}$
5. $|-3.86| = 3.86$

Do Exercises 1–6.

EXERCISES

Find the absolute value.

1. $|-4|$
2. $|0|$
3. $|252.7|$
4. $|31|$
5. $\left|-\frac{1}{2}\right|$
6. $|-0.03|$

5 ADD REAL NUMBERS

MyLab Math VIDEO

RULES FOR ADDITION OF REAL NUMBERS

1. *Positive numbers*: Add the numbers. The result is positive.
2. *Negative numbers*: Add absolute values. Make the answer negative.
3. *A positive number and a negative number*:
   - If the numbers have the same absolute value, the answer is 0.
   - If the numbers have different absolute values, subtract the smaller absolute value from the larger. Then:
     a) If the positive number has the greater absolute value, make the answer positive.
     b) If the negative number has the greater absolute value, make the answer negative.
4. *One number is zero*: The sum is the other number.

IDENTITY PROPERTY OF ZERO

For any real number a, $a + 0 = a$.

(The number 0 is the **additive identity**.)

EXAMPLES

1. $-13 + (-8) = -21$ Two negatives. Add the absolute values: $|-13| + |-8| = 13 + 8 = 21$. Make the answer *negative*: -21.
2. $-2.1 + 8.5 = 6.4$ One negative, one positive. Subtract the smaller absolute value from the larger: $8.5 - 2.1 = 6.4$. The *positive* number, 8.5, has the larger absolute value, so the answer is *positive*, 6.4.
3. $-48 + 31 = -17$ One negative, one positive. Subtract the smaller absolute value from the larger: $48 - 31 = 17$. The *negative* number, -48, has the larger absolute value, so the answer is *negative*, -17.
4. $2.6 + (-2.6) = 0$ One positive, one negative. The numbers have the same absolute value. The sum is 0.
5. $-\frac{5}{9} + 0 = -\frac{5}{9}$ One number is zero. The sum is $-\frac{5}{9}$.
6. $-\frac{2}{3} + \frac{5}{8} = -\frac{16}{24} + \frac{15}{24} = -\frac{1}{24}$

Do Exercises 1–14. ▶

EXERCISES

Add.

1. $8 + (-3)$
2. $-8 + 3$
3. $-8 + (-3)$
4. $8 + 3$
5. $7 + (-2)$
6. $-8 + (-8)$
7. $-13 + 13$
8. $-26 + 0$
9. $-6 + (-15)$
10. $9 + (-2)$
11. $-8.4 + 9.6$
12. $-2.62 + (-6.24)$
13. $-\frac{2}{5} + \frac{3}{4}$
14. $-\frac{5}{6} + \left(-\frac{7}{8}\right)$

6 OPPOSITES, OR ADDITIVE INVERSES

MyLab Math VIDEO

Two numbers whose sum is 0 are called **opposites**, or **additive inverses**.

EXAMPLES Find the opposite, or additive inverse, of each number.

1. 8.6 — The opposite of 8.6 is -8.6 because $8.6 + (-8.6) = 0$.
2. 0 — The opposite of 0 is 0 because $0 + 0 = 0$.
3. $-\frac{7}{9}$ — The opposite of $-\frac{7}{9}$ is $\frac{7}{9}$ because $-\frac{7}{9} + \frac{7}{9} = 0$.

To name the opposite, or additive inverse, we use the symbol $-$, and read the symbolism $-a$ as "the opposite of a" or "the additive inverse of a."

OPPOSITES, OR ADDITIVE INVERSES

For any real number a, the **opposite**, or **additive inverse**, of a, which is $-a$, is such that

$$a + (-a) = (-a) + a = 0.$$

EXAMPLES Evaluate $-x$ and $-(-x)$.

4. If $x = 23$, then $-x = -(23) = -23$ and

$$-(-x) = -(-23) = 23.$$

5. If $x = -5$, then $-x = -(-5) = 5$ and

$$-(-x) = -(-(-5)) = -(5) = -5.$$

Signs of Numbers

We sometimes say that a negative number has a "negative sign," and a positive number has a "positive sign." When we replace a number with its opposite, or additive inverse, we can say that we have "changed its sign."

EXAMPLES Change the sign. (Find the opposite, or additive inverse.)

6. -3 — $-(-3) = 3$
7. $-\frac{3}{8}$ — $-(-\frac{3}{8}) = \frac{3}{8}$
8. 0 — $-0 = 0$
9. 14 — $-(14) = -14$

We can now give a more formal definition of absolute value. For any real number a, the **absolute value** of a, denoted $|a|$, is given by

$$|a| = \begin{cases} a, & \text{if } a \geq 0, \\ -a, & \text{if } a < 0. \end{cases} \quad \begin{array}{l} \text{For example, } |8| = 8 \text{ and } |0| = 0. \\ \text{For example, } |-5| = -(-5) = 5. \end{array}$$

(The absolute value of a is a if a is nonnegative. The absolute value of a is the opposite of a if a is negative.)

Do Exercises 1–6. ▶

EXERCISES

Find the opposite (additive inverse).

1. 10
2. $-\frac{2}{3}$
3. -0.007
4. 0
5. If $x = 9$, evaluate $-x$ and $-(-x)$.
6. If $x = -23$, evaluate $-x$ and $-(-x)$.

Caution!

A symbol such as -8 is usually read "negative 8." It could be read "the opposite of 8," because the opposite of 8 is -8. It could also be read "the additive inverse of 8," because the additive inverse of 8 is -8. When a variable is involved, as in a symbol like $-x$, it can be read "the opposite of x" or "the additive inverse of x" but *not* "negative x," because we do not know whether the symbol represents a positive number, a negative number, or 0. It is never correct to read -8 as "minus 8."

7 SUBTRACT REAL NUMBERS

MyLab Math VIDEO

The difference $a - b$ is the unique number c for which $a = b + c$. That is, $a - b = c$ if c is the number such that $a = b + c$. For example, $3 - 5 = -2$ because $3 = 5 + (-2)$. That is, -2 is the number that when added to 5 gives 3. Although this illustrates the formal definition of subtraction, we generally use the following when we subtract.

SUBTRACTION OF REAL NUMBERS

For any real numbers a and b,

$$a - b = a + (-b).$$

(We can subtract by adding the opposite (additive inverse) of the number being subtracted.)

Compare the following:

$$9 - 4 = 9 + (-4) = 5;$$
$$9 - (-4) = 9 + 4 = 13;$$
$$-9 - 4 = -9 + (-4) = -13; \quad \text{and}$$
$$-9 - (-4) = -9 + 4 = -5.$$

EXAMPLES Subtract.

1. $3 - 5 = 3 + (-5) = -2$
2. $7 - (-3) = 7 + 3 = 10$
3. $0 - 8 = 0 + (-8) = -8$
4. $-19.4 - 5.6 = -19.4 + (-5.6) = -25$
5. $-\frac{4}{3} - \left(-\frac{2}{5}\right) = -\frac{4}{3} + \frac{2}{5} = -\frac{20}{15} + \frac{6}{15} = -\frac{14}{15}$

Do Exercises 1–14.

EXERCISES

Subtract.

1. $5 - 11$
2. $5 - (-11)$
3. $-5 - 11$
4. $-5 - (-11)$
5. $17 - (-17)$
6. $-6 - 14$
7. $-9 - (-9)$
8. $8 - 13$
9. $31 - (-16)$
10. $15.8 - 27.4$
11. $-18.01 - 11.24$
12. $\frac{5}{6} - \left(-\frac{1}{12}\right)$
13. $-\frac{1}{3} - \left(-\frac{1}{12}\right)$
14. $\frac{1}{3} - \frac{4}{5}$

8 MULTIPLY REAL NUMBERS

MyLab Math VIDEO

Look for a pattern:

$3 \cdot 5 = 15$ $-2 \cdot 5 = -10$
$2 \cdot 5 = 10$ $-3 \cdot 5 = -15$
$1 \cdot 5 = 5$ $-4 \cdot 5 = -20$
$0 \cdot 5 = 0$ $-5 \cdot 5 = -25$
$-1 \cdot 5 = -5$ $-6 \cdot 5 = -30$

Look for a pattern:

$4(-5) = -20$ $-1(-5) = 5$
$3(-5) = -15$ $-2(-5) = 10$
$2(-5) = -10$ $-3(-5) = 15$
$1(-5) = -5$ $-4(-5) = 20$
$0(-5) = 0$ $-5(-5) = 25$

RULES FOR MULTIPLICATION OF REAL NUMBERS

To multiply two real numbers:

1. If the signs are the same, multiply the absolute values and make the answer *positive*.
2. If the signs are different, multiply the absolute values and make the answer *negative*.
3. If one number is 0, the product is 0.

EXAMPLES Multiply.

1. $-3 \cdot 5 = -15$ The signs are different. Multiply the absolute values. The answer is negative.
2. $(-8.8)(-3.5) = 30.8$ The signs are the same. Multiply the absolute values. The answer is positive.
3. $3 \cdot \left(-\frac{1}{2}\right) = \frac{3}{1} \cdot \left(-\frac{1}{2}\right) = -\frac{3}{2}$
4. $-3 \cdot (-5) = 15$
5. $(-1.2)(4.5) = -5.4$
6. $-\frac{7}{8} \cdot 0 = 0$
7. $\left(-\frac{3}{4}\right) \cdot \left(-\frac{5}{2}\right) = \frac{15}{8}$

Do Exercises 1–10. ▶

EXERCISES

Multiply.

1. $3(-7)$
2. $(-4.2)(-6.3)$
3. $-5 \cdot 9$
4. $-\frac{13}{7} \cdot \left(-\frac{5}{2}\right)$
5. $0 \cdot (-11)$
6. $-\frac{3}{5} \cdot \frac{4}{7}$
7. $-4(-13)$
8. $-3\left(\frac{2}{3}\right)$
9. $-\frac{9}{11} \cdot \left(-\frac{11}{9}\right)$
10. $-3(-4)(5)$

9 DIVIDE REAL NUMBERS

MyLab Math VIDEO

The quotient $a \div b$, or a/b, where $b \neq 0$, is that unique real number c for which $a = b \cdot c$. Using this definition and the rules for multiplying, we can see how to handle signs when dividing.

EXAMPLES Divide.

1. $\frac{10}{-2} = -5$, because $-2 \cdot (-5) = 10$

2. $\frac{-32}{4} = -8$, because $-8 \cdot (4) = -32$

3. $-25 \div (-5) = 5$, because $5 \cdot (-5) = -25$

RULES FOR DIVISION OF REAL NUMBERS

To divide two real numbers:

1. If the signs are the same, divide the absolute values and make the answer *positive.*
2. If the signs are different, divide the absolute values and make the answer *negative.*

Excluding Division by Zero

We cannot divide a nonzero number n by zero. By the definition of division, $n/0$ would be some number that when multiplied by 0 gives n. But when any number is multiplied by 0, the result is 0. Thus the only possibility for n would be 0.

Consider $0/0$. Using the definition of division, we might say that it is 5 because $5 \cdot 0 = 0$. We might also say that it is -8 because $-8 \cdot 0 = 0$. In fact, $0/0$ could be any number at all. So, division by 0 does not make sense. Division by 0 is not defined and is not possible.

EXAMPLES Divide, if possible.

4. $\frac{7}{0}$ Not defined.

5. $\frac{0}{7} = 0$ The quotient is 0 because $0 \cdot 7 = 0$.

(continued)

9 DIVIDE REAL NUMBERS (continued)

Two numbers whose product is 1 are called **reciprocals** (or **multiplicative inverses**). Every nonzero real number a has a **reciprocal** (or **multiplicative inverse**) $1/a$. The reciprocal of a positive number is positive. The reciprocal of a negative number is negative.

EXAMPLES Find the reciprocal of each number.

6. $\frac{4}{5}$ The reciprocal is $\frac{5}{4}$, because $\frac{4}{5} \cdot \frac{5}{4} = 1$.

7. 8 The reciprocal is $\frac{1}{8}$, because $8 \cdot \frac{1}{8} = 1$.

8. $-\frac{2}{3}$ The reciprocal is $-\frac{3}{2}$, because $-\frac{2}{3} \cdot \left(-\frac{3}{2}\right) = 1$.

RECIPROCALS AND DIVISION

For any real numbers a and b, $b \neq 0$,

$$a \div b = \frac{a}{b} = a \cdot \frac{1}{b}.$$

(To divide, we can multiply by the reciprocal of the divisor.)

EXAMPLES Divide by multiplying by the reciprocal of the divisor.

9. $\frac{1}{4} \div \frac{3}{5} = \frac{1}{4} \cdot \frac{5}{3} = \frac{5}{12}$ "Inverting" the divisor, $\frac{3}{5}$, and multiplying

10. $\frac{2}{3} \div \left(-\frac{4}{9}\right) = \frac{2}{3} \cdot \left(-\frac{9}{4}\right) = -\frac{18}{12}$, or $-\frac{3}{2}$

11. $-\frac{5}{7} \div 3 = -\frac{5}{7} \cdot \frac{1}{3} = -\frac{5}{21}$

For any numbers a and b, $b \neq 0$,

$$\frac{-a}{b} = \frac{a}{-b} = -\frac{a}{b} \quad \text{and} \quad \frac{-a}{-b} = \frac{a}{b}.$$

Do Exercises 1–14. ▶

EXERCISES

Divide, if possible.

1. $\frac{-8}{4}$

2. $\frac{-77}{-11}$

3. $\frac{5}{0}$

4. $\frac{0}{32}$

Find the reciprocal of each number.

5. $\frac{3}{4}$

6. $-\frac{7}{8}$

7. 25

8. 0.2

Divide.

9. $\frac{2}{7} \div \left(-\frac{11}{3}\right)$

10. $-\frac{10}{3} \div \left(-\frac{2}{15}\right)$

11. $-48 \div 0.4$

12. $\frac{8}{19} \div (-2)$

Find the opposite and the reciprocal.

13. $-\frac{3}{8}$

14. 6

10 EXPONENTIAL NOTATION (PART 1)

MyLab Math VIDEO

Exponential notation is a shorthand device. For $3 \cdot 3 \cdot 3 \cdot 3$, we write 3^4. In the **exponential notation** 3^4, the number 3 is called the **base** and the number 4 is called the **exponent**. Exponential notation a^n, where n is an integer greater than 1, means

$$\underbrace{a \cdot a \cdot a \cdots a \cdot a.}_{n \text{ factors}}$$

We read a^n as "a to the nth power," or simply "a to the nth." We can read a^2 as "a-squared" and a^3 as "a-cubed."

EXAMPLES Write exponential notation.

1. $7 \cdot 7 \cdot 7 = 7^3$ **2.** $xxxxx = x^5$ **3.** $\frac{2}{3} \cdot \frac{2}{3} \cdot \frac{2}{3} \cdot \frac{2}{3} = \left(\frac{2}{3}\right)^4$

EXAMPLES Evaluate.

4. $5^3 = 5 \cdot 5 \cdot 5 = 125$ **5.** $\left(\frac{7}{8}\right)^2 = \frac{7}{8} \cdot \frac{7}{8} = \frac{49}{64}$

6. $-(10)^4 = -(10 \cdot 10 \cdot 10 \cdot 10) = -10{,}000$

7. $(-10)^4 = (-10)(-10)(-10)(-10) = 10{,}000$

When an exponent is an integer greater than 1, it tells how many times the base occurs as a factor. What happens when the exponent is 1 or 0? Look for a pattern below. Think of dividing by 10 on the right.

$$10^3 = 10 \cdot 10 \cdot 10 = 1000$$
$$10^2 = 10 \cdot 10 = 100$$
$$10^1 = ?$$
$$10^0 = ?$$

In order for the pattern to continue, 10^1 would have to be 10 and 10^0 would have to be 1.

EXPONENTS OF 0 AND 1

For any number a, we agree that a^1 means a.

For any *nonzero* number a, we agree that a^0 means 1.

Mathematicians agree *not* to define 0^0.

EXAMPLES Rewrite without an exponent.

8. $4^1 = 4$ **9.** $(-97)^1 = -97$

10. $6^0 = 1$ **11.** $(-37.4)^0 = 1$

(continued)

EXERCISES

Write exponential notation.

1. $yyyyyy$

2. $8 \cdot 8 \cdot 8 \cdot 8$

3. $(3.8)(3.8)(3.8)(3.8)(3.8)$

4. $\left(-\frac{4}{5}\right)\left(-\frac{4}{5}\right)\left(-\frac{4}{5}\right)$

Evaluate.

5. $\left(\frac{3}{4}\right)^1$

6. 17^0

7. $(-2)^6$

8. $(-7)^3$

9. $\left(\frac{1}{3}\right)^3$

10. $-(3)^4$

11. $(0.1)^6$

12. $\left(-\frac{5}{7}\right)^2$

10 EXPONENTIAL NOTATION (PART 1) (continued)

Look for a pattern below. Again, think of dividing by 10 on the right.

$10^2 = 100$
$10^1 = 10$
$10^0 = 1$
$10^{-1} = ?$
$10^{-2} = ?$

In order for the pattern to continue, 10^{-1} would have to be $\frac{1}{10}$ and 10^{-2} would have to be $\frac{1}{100}$.

NEGATIVE EXPONENTS

For any real number a that is nonzero and any integer n,

$$a^{-n} = \frac{1}{a^n}.$$

EXAMPLES Rewrite using a positive exponent. Evaluate, if possible.

12. $y^{-5} = \frac{1}{y^5}$

13. $\frac{1}{t^{-4}} = t^4$

14. $(-2)^{-3} = \frac{1}{(-2)^3} = \frac{1}{(-2)(-2)(-2)} = \frac{1}{-8} = -\frac{1}{8}$

15. $\left(\frac{2}{5}\right)^{-2} = \frac{1}{\left(\frac{2}{5}\right)^2} = \frac{1}{\frac{4}{25}} = 1 \cdot \frac{25}{4} = \frac{25}{4}$

Caution!

A negative exponent does *not* necessarily indicate that an answer is negative! For example, 3^{-2} means $1/3^2$, or $1/9$, not -9.

EXAMPLES Rewrite using a negative exponent.

16. $\frac{1}{x^2} = x^{-2}$

17. $\frac{1}{(-7)^4} = (-7)^{-4}$

Do Exercises 1–20.
(Exercises 1–12 are on the preceding page.)

EXERCISES

Rewrite using a positive exponent. Evaluate, if possible.

13. y^{-5}

14. $(-4)^{-3}$

15. $\left(\frac{3}{4}\right)^{-2}$

16. $\frac{1}{a^{-2}}$

17. $\left(\frac{1}{4}\right)^{-2}$

Rewrite using a negative exponent.

18. $\frac{1}{3^4}$

19. $\frac{1}{b^3}$

20. $\frac{1}{(-16)^2}$

MyLab Math
ANIMATION

11 ORDER OF OPERATIONS

MyLab Math VIDEO

RULES FOR ORDER OF OPERATIONS

1. Do all the calculations within grouping symbols, like parentheses, before operations outside.
2. Evaluate all exponential expressions.
3. Do all multiplications and divisions in order from left to right.
4. Do all additions and subtractions in order from left to right.

EXAMPLE 1 Simplify: $8 + 2 \cdot 5^3$.

$8 + 2 \cdot 5^3 = 8 + 2 \cdot 125$	Evaluating the exponential expression
$= 8 + 250$	Doing the multiplication
$= 258$	Adding

EXAMPLE 2 Simplify and compare: $(8 - 10)^2$ and $8^2 - 10^2$.

$(8 - 10)^2 = (-2)^2 = 4;$
$8^2 - 10^2 = 64 - 100 = -36$

We see that $(8 - 10)^2$ and $8^2 - 10^2$ are *not* the same.

EXAMPLE 3 Simplify: $3^4 + 62 \cdot 8 - 2(29 + 33 \cdot 4)$.

$3^4 + 62 \cdot 8 - 2(29 + 33 \cdot 4)$

$= 3^4 + 62 \cdot 8 - 2(29 + 132)$	Carrying out operations inside parentheses first; doing the multiplication
$= 3^4 + 62 \cdot 8 - 2(161)$	Adding inside parentheses
$= 81 + 62 \cdot 8 - 2(161)$	Evaluating the exponential expression
$= 81 + 496 - 2(161)$ $= 81 + 496 - 322$	Doing the multiplications in order from left to right
$= 577 - 322$ $= 255$	Doing all additions and subtractions in order from left to right

(continued)

EXERCISES

Simplify.

1. $9[8 - 7(5 - 2)]$
2. $(9 - 12)^2$
3. $4 \div (8 - 10)^2 + 1$
4. $[64 \div (-4)] \div (-2)$
5. $9^2 - 12^2$
6. $[24 \div (-3)] \div \left(-\frac{1}{2}\right)$
7. $9 \div (-3) + 16 \div 8$
8. $20 + 4^3 \div (-8)$
9. $256 \div (-32) \div (-4)$
10. $9[(8 - 11) - 13]$

11 ORDER OF OPERATIONS (continued)

When parentheses occur within parentheses, we can make them different shapes, such as [ ] (also called "brackets") and { } (usually called "braces"). Parentheses, brackets, and braces all have the same meaning. When parentheses occur within parentheses, **computations in the *innermost* ones are to be done first.**

EXAMPLE 4 Simplify: $5 - \{6 - [3 - (7 + 3)]\}$.

$$\begin{aligned} 5 - \{6 - [3 - (7 + 3)]\} &= 5 - \{6 - [3 - 10]\} && \text{Adding } 7 + 3 \\ &= 5 - \{6 - [-7]\} && \text{Subtracting } 3 - 10 \\ &= 5 - 13 && \text{Subtracting } 6 - [-7] \\ &= -8 \end{aligned}$$

EXAMPLE 5 Simplify: $\dfrac{12|7 - 9| + 8 \cdot 5}{3^2 + 2^3}$.

We do the calculations in the numerator and in the denominator separately, and then divide the results:

$$\begin{aligned} \frac{12|7 - 9| + 8 \cdot 5}{3^2 + 2^3} &= \frac{12|-2| + 8 \cdot 5}{9 + 8} \\ &= \frac{12(2) + 8 \cdot 5}{17} && \text{Subtracting inside the absolute-value signs before taking the absolute value} \\ &= \frac{24 + 40}{17} \\ &= \frac{64}{17}. \end{aligned}$$

Do Exercises 1–20.
(Exercises 1–10 are on the preceding page.)

EXERCISES

Simplify.

11. $18 - 2 \cdot 3 - 9$

12. $(13 \cdot 2 - 8 \cdot 4)^2$

13. $12 - 4(5 - 1)$

14. $[2 \cdot (5 - 3)]^2$

15. $[5(8 - 6) + 12] - [24 - (8 - 4)]$

16. $-32 - 18 \div [4 - (-2)]$

17. $2^5 - [4 \cdot 3 - 2(1 + 4 \cdot 2)]$

18. $10 - 3\{8 \div [4 - (11 - 6)]\}$

19. $\dfrac{4|6 - 7| - 5 \cdot 4}{6 \cdot 7 - 8|4 - 1|}$

20. $\dfrac{(8 - 3)^2 + (7 - 10)^2}{3^2 - 2^3}$

12 TRANSLATE TO AN ALGEBRAIC EXPRESSION

MyLab Math VIDEO

When a letter is used to represent various numbers, it is called a **variable**. Let $t =$ the number of hours that a passenger jet has been flying. Then t is a variable, because t changes as the flight continues. If a letter represents one particular number, it is called a **constant**. Let $d =$ the number of hours in a day. Then d is a constant.

An **algebraic expression** consists of variables, numbers, and operation signs, such as $+, -, \cdot, \div$. When an equals sign, $=$, is placed between two expressions, an **equation** is formed.

(continued)

To translate problems to equations, we need to know that certain words correspond to certain symbols, as shown in the following table.

Key Words

ADDITION	SUBTRACTION	MULTIPLICATION	DIVISION
add	subtract	multiply	divide
sum	difference	product	quotient
plus	minus	times	divided by
total	decreased by	twice	ratio
increased by	less than	of	per
more than			

Expressions like rs represent products and can also be written as $r \cdot s$, $r \times s$, $(r)(s)$, or $r(s)$. The multipliers r and s are also called **factors**. A quotient $m \div 5$ can also be represented as $m/5$ or $\frac{m}{5}$.

EXAMPLE 1 Translate to an algebraic expression:

Eight less than some number.

We can use any variable we wish, such as x, y, t, m, n, and so on. Here we let t represent the number. If we knew the number to be 23, then the translation of "eight less than 23" would be $23 - 8$. If we knew the number to be 345, then the translation of "eight less than 345" would be $345 - 8$. Since we are using a variable for the number, the translation is

$t - 8$. *Caution*! $8 - t$ would be incorrect.

EXAMPLES Translate each of the following to an algebraic expression.

Phrase	*Algebraic Expression*
2. Half *of* a number	$\frac{1}{2}t$, or $\frac{t}{2}$
3. Five *more than* three *times* some number	$3p + 5$, or $5 + 3p$
4. The *difference* of two numbers	$x - y$
5. Seventy-six percent *of* some number	$76\%z$, or $0.76z$, or $\frac{76}{100}z$
6. Eight *less than twice* some number	$2x - 8$

Do Exercises 1–8.

EXERCISES

Translate each phrase to an algebraic expression.

1. x divided by 6

2. Seven more than four times some number

3. 25 less than t

4. One-third of a number

5. The sum of w and twice q

6. -18 multiplied by m

7. m subtracted from s

8. Nineteen percent of some number

13 EVALUATE ALGEBRAIC EXPRESSIONS

MyLab Math VIDEO

When we replace a variable with a number, we say that we are **substituting** for the variable. Carrying out the resulting calculation is called **evaluating the expression**. The result is called the **value** of the expression.

EXAMPLE 1 Evaluate $x - y$ when $x = 83$ and $y = 49$.

We substitute 83 for x and 49 for y and carry out the subtraction:

$$x - y = 83 - 49 = 34.$$

EXAMPLE 2 Evaluate a/b when $a = -63$ and $b = 7$.

We substitute -63 for a and 7 for b and carry out the division:

$$\frac{a}{b} = \frac{-63}{7} = -9.$$

EXAMPLE 3 Evaluate the expression $3xy + z$ when $x = 2$, $y = -5$, and $z = 7$.

We substitute and carry out the calculations according to the rules for order of operations:

$$3xy + z = 3(2)(-5) + 7 = -30 + 7 = -23.$$

EXAMPLE 4 Evaluate $5 + 2(a - 1)^2$ when $a = 4$.

$5 + 2(a - 1)^2 = 5 + 2(4 - 1)^2$	Substituting
$= 5 + 2(3)^2$	Working within parentheses first
$= 5 + 2(9)$	Simplifying 3^2
$= 5 + 18$	Multiplying
$= 23$	Adding

EXAMPLE 5 Evaluate $9 - x^3 + 6 \div 2y^2$ when $x = 2$ and $y = 5$.

$9 - x^3 + 6 \div 2y^2 = 9 - 2^3 + 6 \div 2(5)^2$	Substituting
$= 9 - 8 + 6 \div 2 \cdot 25$	Simplifying 2^3 and 5^2
$= 9 - 8 + 3 \cdot 25$	Dividing
$= 9 - 8 + 75$	Multiplying
$= 1 + 75$	Subtracting
$= 76$	Adding

Do Exercises 1–8. ▶

EXERCISES

Evaluate.

1. $57y$, when $y = -8$

2. $\frac{x}{y}$, when $x = 30$ and $y = -6$

3. $\frac{5}{p + q}$, when $p = 20$ and $q = 30$

4. $\frac{18m}{n}$, when $m = 7$ and $n = 18$

5. $4x - y$, when $x = 3$ and $y = -2$

6. $2c \div 3b$, when $b = 4$ and $c = 6$

7. $25 - r^2 + s \div r^2$, when $r = 3$ and $s = 27$

8. $m + n(5 + n^2)$, when $m = 15$ and $n = 3$

14 EQUIVALENT FRACTION EXPRESSIONS

Two expressions that have the same value for all *allowable* replacements are called **equivalent expressions**.

THE IDENTITY PROPERTY OF 1
For any real number a, $a \cdot 1 = 1 \cdot a = a$.
(The number 1 is the **multiplicative identity**.)

We will often refer to the use of the identity property of 1 as "multiplying by 1."

EXAMPLE 1 Use multiplying by 1 to find an expression equivalent to $\frac{3}{5}$ with a denominator of $10x$.

Because $10x = 5 \cdot 2x$, we multiply by 1, using $2x/(2x)$ as a name for 1:

$$\frac{3}{5} = \frac{3}{5} \cdot 1 = \frac{3}{5} \cdot \frac{2x}{2x} = \frac{3 \cdot 2x}{5 \cdot 2x} = \frac{6x}{10x}.$$

Note that the expressions $3/5$ and $6x/(10x)$ are equivalent. They have the same value for any allowable replacement. Note too that 0 is not an allowable replacement in $6x/(10x)$, but for all nonzero real numbers, the expressions $3/5$ and $6x/(10x)$ have the same value.

In algebra, we consider an expression like $3/5$ to be a "simplified" form of $6x/(10x)$. To find such simplified expressions, we reverse the identity property of 1 in order to "remove a factor of 1."

EXAMPLE 2 Simplify: $\frac{7x}{9x}$.

$$\frac{7x}{9x} = \frac{7 \cdot x}{9 \cdot x} = \frac{7}{9} \cdot \frac{x}{x} = \frac{7}{9} \cdot 1 = \frac{7}{9}$$

EXAMPLE 3 Simplify: $-\frac{24y}{16y}$.

$$-\frac{24y}{16y} = -\frac{3 \cdot 8y}{2 \cdot 8y} = -\frac{3}{2} \cdot \frac{8y}{8y} = -\frac{3}{2} \cdot 1 = -\frac{3}{2}$$

Do Exercises 1–8.

MyLab Math VIDEO

EXERCISES

Use multiplying by 1 to find an equivalent expression with the given denominator.

1. $\frac{7}{8}$; $8x$

2. $\frac{3}{10}$; $50y$

3. $\frac{1}{17}$; $51b$

4. $\frac{300}{49}$; $98w$

Simplify.

5. $\frac{25x}{15x}$

6. $\frac{36y}{18y}$

7. $-\frac{100a}{25a}$

8. $\frac{-625t}{15t}$

15 THE COMMUTATIVE LAWS AND THE ASSOCIATIVE LAWS

MyLab Math VIDEO

EXAMPLE 1 Evaluate $x + y$ and $y + x$ when $x = 5$ and $y = 8$.

$$x + y = 5 + 8 = 13; \qquad y + x = 8 + 5 = 13.$$

EXAMPLE 2 Evaluate xy and yx when $x = 4$ and $y = 3$.

$$xy = 4 \cdot 3 = 12; \qquad yx = 3 \cdot 4 = 12.$$

When only addition is involved, changing the order does not change the answer. Likewise, when only multiplication is involved, changing the order does not change the answer. The expressions $x + y$ and $y + x$ are equivalent, and the expressions xy and yx are equivalent.

THE COMMUTATIVE LAWS

Addition. For any numbers a and b,

$$a + b = b + a.$$

Multiplication. For any numbers a and b,

$$ab = ba.$$

EXAMPLE 3 Evaluate $a + (b + c)$ and $(a + b) + c$ when $a = 4$, $b = 8$, and $c = 5$.

$$a + (b + c) = 4 + (8 + 5) = 4 + 13 = 17;$$
$$(a + b) + c = (4 + 8) + 5 = 12 + 5 = 17$$

EXAMPLE 4 Evaluate $a \cdot (b \cdot c)$ and $(a \cdot b) \cdot c$ when $a = 7$, $b = 4$, and $c = 2$.

$$a \cdot (b \cdot c) = 7 \cdot (4 \cdot 2) = 7 \cdot 8 = 56;$$
$$(a \cdot b) \cdot c = (7 \cdot 4) \cdot 2 = 28 \cdot 2 = 56$$

When only addition is involved, changing the grouping does not change the answer. Likewise, when only multiplication is involved, changing the grouping does not change the answer. The expressions $a + (b + c) = (a + b) + c$ are equivalent, and the expressions $a \cdot (b \cdot c)$ and $(a \cdot b) \cdot c$ are equivalent.

THE ASSOCIATIVE LAWS

Addition. For any numbers a, b, and c,

$$a + (b + c) = (a + b) + c.$$

Multiplication. For any numbers a, b, and c,

$$a \cdot (b \cdot c) = (a \cdot b) \cdot c.$$

(continued)

EXERCISES

Use a commutative law to find an equivalent expression.

1. $w + 3$
2. rt
3. $pq + 14$

Use an associative law to find an equivalent expression.

4. $m + (n + 2)$
5. $(7 \cdot x) \cdot y$

15 THE COMMUTATIVE LAWS AND THE ASSOCIATIVE LAWS (continued)

EXAMPLE 5 Use the commutative laws and the associative laws to write at least three expressions equivalent to $(x + 8) + y$.

a) $(x + 8) + y = x + (8 + y)$ Using the associative law first and then the commutative law
$= x + (y + 8)$

b) $(x + 8) + y = y + (x + 8)$ Using the commutative law and then the commutative law again
$= y + (8 + x)$

c) $(x + 8) + y = (8 + x) + y$ Using the commutative law first and then the associative law
$= 8 + (x + y)$

Do Exercises 1–7.
(Exercises 1–5 are on the preceding page.)

EXERCISES

Use the commutative laws and the associative laws to find three equivalent expressions.

6. $(a + b) + 8$

7. $9 \cdot (x \cdot y)$

16 THE DISTRIBUTIVE LAWS

MyLab Math VIDEO

EXAMPLE 1 Evaluate $8(x + y)$ and $8x + 8y$ when $x = 4$ and $y = 5$.

$$8(x + y) = 8(4 + 5) = 8(9) = 72; \qquad 8x + 8y = 8 \cdot 4 + 8 \cdot 5 = 32 + 40 = 72$$

The expressions $8(x + y)$ and $8x + 8y$ are equivalent.

THE DISTRIBUTIVE LAWS

The Distributive Law of Multiplication Over Addition

For any numbers a, b, and c,

$$a(b + c) = ab + ac, \quad \text{or} \quad (b + c)a = ba + ca.$$

The Distributive Law of Multiplication Over Subtraction

For any real numbers a, b, and c,

$$a(b - c) = ab - ac, \quad \text{or} \quad (b - c)a = ba - ca.$$

EXERCISES

Multiply.

1. $3(c + 1)$

2. $3(x - y)$

3. $-2(3c + 5d)$

4. $5x(y - z + w)$

5. $\frac{1}{2}h(a + b)$

6. $2\pi r(h + 1)$

(continued)

16 THE DISTRIBUTIVE LAWS (continued)

The distributive laws are the basis of multiplication in algebra as well as in arithmetic. In the following examples, note that we multiply each number or letter inside the parentheses by the factor outside.

EXAMPLES Multiply.

2. $4(x - 2) = 4 \cdot x - 4 \cdot 2 = 4x - 8$
3. $b(s - t + f) = bs - bt + bf$
4. $-3(y + 4) = -3 \cdot y + (-3) \cdot 4 = -3y - 12$
5. $-2x(y - 1) = -2x \cdot y - (-2x) \cdot 1 = -2xy + 2x$

The reverse of multiplying is called **factoring**. Factoring an expression involves factoring its *terms*. **Terms** of algebraic expressions are the parts separated by addition signs.

EXAMPLE 6 List the terms of $3x - 4y - 2z$.

We first find an equivalent expression that uses addition signs:

$3x - 4y - 2z = 3x + (-4y) + (-2z)$. Using the property $a - b = a + (-b)$

Thus the terms are $3x$, $-4y$, and $-2z$.

To **factor** an expression is to find an equivalent expression that is a product. If $N = a \cdot b$, then a and b are **factors** of N.

EXAMPLES Factor.

7. $8x + 8y = 8(x + y)$ 8 and $x + y$ are factors.
8. $cx - cy = c(x - y)$ c and $x - y$ are factors.

Generally, we try to factor out the largest factor common to all the terms. In the following example, we might factor out 3, but there is a larger factor common to the terms, 9. So we factor out the 9.

EXAMPLE 9 Factor: $9x + 27y$.

$$9x + 27y = 9 \cdot x + 9 \cdot (3y) = 9(x + 3y)$$

We often must supply a factor of 1 when factoring out a common factor, as in the next example, which is a formula involving simple interest.

EXAMPLE 10 Factor: $P + Prt$.

$P + Prt = P \cdot 1 + P \cdot rt$ Writing P as a product of P and 1

$= P(1 + rt)$ Using a distributive law

Do Exercises 1–14.

(Exercises 1–6 are on the preceding page.)

EXERCISES

List the terms.

7. $4a - 5b + 6$
8. $2x - y - 1$

Factor.

9. $9a + 9b$
10. $7x - 21$
11. $ab + a$
12. $18a - 24b - 48$
13. $8m + 4n - 24$
14. $xy - xz + xw$

17 COLLECTING LIKE TERMS

MyLab Math VIDEO

If two terms have the same letter, or letters, we say that they are **like terms**, or **similar terms**. (If powers, or exponents, are involved, then like terms must have the same letters raised to the same powers.)

If two terms have no letters at all but are just numbers, they are also similar terms. We can simplify by **collecting**, or **combining**, **like terms**, using a distributive law.

EXAMPLES Collect like terms.

1. $3x + 5x = (3 + 5)x = 8x$ Factoring out the x using a distributive law

2. $x - 3x = 1 \cdot x - 3 \cdot x = (1 - 3)x = -2x$ $x = 1 \cdot x$

3. $2x + 3y - 5x - 2y$

$= 2x - 5x + 3y - 2y$ Using a commutative law

$= (2 - 5)x + (3 - 2)y$ Using a distributive law

$= -3x + y$ Simplifying

4. $3x + 2x + 5 + 7 = (3 + 2)x + (5 + 7) = 5x + 12$

5. $4.2x - 6.7y - 5.8x + 23y = (4.2 - 5.8)x + (-6.7 + 23)y$
$= -1.6x + 16.3y$

6. $-\frac{1}{4}a + \frac{1}{2}b - \frac{3}{5}a - \frac{2}{5}b = \left(-\frac{1}{4} - \frac{3}{5}\right)a + \left(\frac{1}{2} - \frac{2}{5}\right)b$
$= \left(-\frac{5}{20} - \frac{12}{20}\right)a + \left(\frac{5}{10} - \frac{4}{10}\right)b$
$= -\frac{17}{20}a + \frac{1}{10}b$

Do Exercises 1–10. ▶

EXERCISES

Collect like terms.

1. $7x + 5x$
2. $8b - 11b$
3. $14y + y$
4. $12a - a$
5. $t - 9t$
6. $5x - 3x + 8x$
7. $3c + 8d - 7c + 4d$
8. $4x - 7 + 18x + 25$
9. $1.3x + 1.4y - 0.11x - 0.47y$
10. $\frac{2}{3}a + \frac{5}{6}b - 27 - \frac{4}{5}a - \frac{7}{6}b$

18 REMOVING PARENTHESES AND COLLECTING LIKE TERMS

MyLab Math VIDEO

THE PROPERTY OF −1

For any number a,

$$-1 \cdot a = -a.$$

(Negative 1 times a is the opposite of a. In other words, changing the sign is the same as multiplying by -1.)

EXAMPLES Find an equivalent expression without parentheses.

1. $-(3x) = -1(3x)$ Replacing − with −1 using the property of −1
   $= (-1 \cdot 3)x$ Using an associative law
   $= -3x$ Multiplying

2. $-(-9y) = -1(-9y) = [-1(-9)]y = 9y$

3. $-(4 + x) = -1(4 + x)$ Replacing − with −1
   $= -1 \cdot 4 + (-1) \cdot x$ Multiplying using the distributive law
   $= -4 + (-x)$ Replacing $-1 \cdot x$ with $-x$
   $= -4 - x$ Adding an opposite is the same as subtracting.

4. $-(a - b) = -1(a - b) = -1 \cdot a - (-1) \cdot b$
   $= -a + [-(-1)b] = -a + b$, or $b - a$

For any real numbers a and b,

$$-(a - b) = b - a.$$

(The opposite of $a - b$ is $b - a$.)

We can find an equivalent expression for an opposite by multiplying every term by -1. We could also say that we change the sign of every term inside the parentheses.

EXAMPLE 5 Find an equivalent expression without parentheses:

$$-\left(-9t + 7z - \tfrac{1}{4}w\right).$$

We have

$$-\left(-9t + 7z - \tfrac{1}{4}w\right) = 9t - 7z + \tfrac{1}{4}w.$$ Changing the sign of every term

(continued)

EXERCISES

Find an equivalent expression without parentheses.

1. $-(-2c)$
2. $-(b + 4)$
3. $-(x - 8)$
4. $-(t - y)$
5. $-(r + s + t)$
6. $-(8x - 6y + 13)$

Some expressions contain parentheses preceded by subtraction signs. These parentheses can be removed by changing the sign of *every* term inside.

EXAMPLES Remove parentheses and simplify.

6. $6x - (4x + 2) = 6x - 4x - 2$ Changing the sign of every term inside parentheses

$= 2x - 2$ Collecting like terms

7. $3y - 4 - (9y - 7) = 3y - 4 - 9y + 7$

$= -6y + 3$, or $3 - 6y$

8. $3y + (3x - 8) - (5 - 12y) = 3y + 3x - 8 - 5 + 12y$

$= 15y + 3x - 13$

9. $x - 3(x + y) = x - 3x - 3y$ Removing parentheses by multiplying $x + y$ by -3

$= -2x - 3y$ Collecting like terms

Caution!

A common error is to forget to change this sign. *Remember*: When multiplying by a negative number, change the sign of *every* term inside the parentheses.

10. $3y - 2(4y - 5) = 3y - 8y + 10 = -5y + 10$

When expressions with grouping symbols contain variables, we still work from the inside out when simplifying, using the rules for order of operations.

EXAMPLE 11 Simplify: $[2(x + 7) - 4^2] - (2 - x)$.

$[2(x + 7) - 4^2] - (2 - x)$

$= [2x + 14 - 4^2] - (2 - x)$ Multiplying to remove the innermost grouping symbols using a distributive law

$= [2x + 14 - 16] - (2 - x)$ Evaluating the exponential expression

$= [2x - 2] - (2 - x)$ Collecting like terms inside the brackets

$= 2x - 2 - 2 + x$ Multiplying by -1 to remove the parentheses

$= 3x - 4$ Collecting like terms

Do Exercises 1–12.

(Exercises 1–6 are on the preceding page.)

EXERCISES

Simplify.

7. $4m - (3m - 1)$

8. $7a - [9 - 3(5a - 2)]$

9. $-2(x + 3) - 5(x - 4)$

10. $8x - (-3y + 7) + (9x - 11)$

11. $[7(x + 5) - 19] - [4(x - 6) + 10]$

12. $5\{-2 + 3[4 - 2(3 + 5)]\}$

19 EXPONENTIAL NOTATION (PART 2)

MyLab Math VIDEO

We often need to find ways to determine *equivalent exponential expressions.* We do this with several rules or properties regarding exponents.

RULES FOR EXPONENTS

For any real numbers a and b and any integers m and n:

Product Rule: $a^m \cdot a^n = a^{m+n}$

Quotient Rule: $\dfrac{a^m}{a^n} = a^{m-n}, a \neq 0$

Power Rule: $(a^m)^n = a^{mn}$

Raising a product to a power: $(ab)^n = a^n b^n$

Raising a quotient to a power: $\left(\dfrac{a}{b}\right)^n = \dfrac{a^n}{b^n}, b \neq 0;$

$\left(\dfrac{a}{b}\right)^{-n} = \left(\dfrac{b}{a}\right)^n, b \neq 0, a \neq 0$

EXAMPLES Simplify.

1. $x^4 \cdot x^3 = x^{4+3} = x^7$ Adding exponents using the product rule

2. $\dfrac{5^7}{5^3} = 5^{7-3} = 5^4$ Subtracting exponents using the quotient rule

3. $\dfrac{5^7}{5^{-3}} = 5^{7-(-3)} = 5^{7+3} = 5^{10}$ Subtracting exponents (adding an opposite)

4. $4^5 \cdot 4^{-3} = 4^{5+(-3)} = 4^2 = 16$ Adding exponents

5. $\dfrac{9^{-2}}{9^5} = 9^{-2-5} = 9^{-7} = \dfrac{1}{9^7}$

Note that we give answers using positive exponents. In some situations, this may not be appropriate, but we do so here.

6. $(8x^n)(6x^{2n}) = 8 \cdot 6 \cdot x^n \cdot x^{2n} = 48 \cdot x^{n+2n} = 48x^{3n}$

7. $\dfrac{16x^4y^7}{-8x^3y^9} = \dfrac{16}{-8} \cdot \dfrac{x^4}{x^3} \cdot \dfrac{y^7}{y^9} = -2x^{4-3}y^{7-9} = -2x^1y^{-2} = -\dfrac{2x}{y^2}$

(continued)

EXERCISES

Simplify.

1. $8^{-2} \cdot 8^{-4}$

2. $\dfrac{a^3}{a^{-2}}$

3. $\dfrac{10^{-3}}{10^6}$

4. $b^2 \cdot b^{-5}$

5. $\dfrac{-24x^6y^7}{18x^{-3}y^9}$

6. $(14m^2n^3)(-2m^3n^2)$

7. $(-3t^{-4a})(-5t^{-a})$

8. $\dfrac{-18x^{-2}y^3}{-12x^{-5}y^5}$

8. $\dfrac{14x^5y^{-3}}{4x^9y^{-5}} = \dfrac{14}{4}\cdot\dfrac{x^5}{x^9}\cdot\dfrac{y^{-3}}{y^{-5}} = \dfrac{7}{2}x^{5-9}y^{-3-(-5)}$

$= \dfrac{7}{2}x^{-4}y^2 = \dfrac{7y^2}{2x^4}$

9. $(8x^4y^{-2})(-3x^{-3}y) = 8\cdot(-3)\cdot x^4\cdot x^{-3}\cdot y^{-2}\cdot y^1$

$= -24x^{4+(-3)}y^{-2+1}$

$= -24xy^{-1} = -\dfrac{24x}{y}$

When an expression inside parentheses is raised to a power, the inside expression is the base. Thus, for $(5^2)^4$, we are raising 5^2 to the fourth power:

$$(5^2)^4 = (5^2)(5^2)(5^2)(5^2) = 5^8.$$

EXAMPLES Simplify.

10. $(x^5)^7 = x^{5\cdot 7} = x^{35}$ Multiplying exponents

11. $\left(\dfrac{3^{-4}}{5^2}\right)^6 = \dfrac{(3^{-4})^6}{(5^2)^6} = \dfrac{3^{-24}}{5^{12}} = \dfrac{1}{3^{24}\cdot 5^{12}}$

12. $(x^4)^{-2t} = x^{4(-2t)} = x^{-8t} = \dfrac{1}{x^{8t}}$

13. $\left(\dfrac{x^2}{y^{-3}}\right)^{-5} = \dfrac{x^{2\cdot(-5)}}{y^{-3\cdot(-5)}} = \dfrac{x^{-10}}{y^{15}} = \dfrac{1}{x^{10}y^{15}}$

14. $\left(\dfrac{-2a^{-5}}{a^{-3}b}\right)^4 = \dfrac{(-2a^{-5})^4}{(a^{-3}b)^4} = \dfrac{(-2)^4(a^{-5})^4}{(a^{-3})^4b^4}$

$= \dfrac{16a^{-20}}{a^{-12}b^4} = \dfrac{16a^{-20-(-12)}}{b^4}$

$= \dfrac{16a^{-8}}{b^4} = \dfrac{16}{a^8b^4}$

15. $(4r^3y^{-2}w)^{-2} = 4^{-2}(r^3)^{-2}(y^{-2})^{-2}w^{-2}$

$= \dfrac{1}{4^2}r^{-6}y^4w^{-2}$

$= \dfrac{y^4}{16r^6w^2}$

Do Exercises 1–16.
(Exercises 1–8 are on the preceding page.)

EXERCISES

Simplify.

9. $(r^9)^5$

10. $\left(\dfrac{4^{-3}}{3^4}\right)^3$

11. $(5a^2b^2)^3$

12. $\left(\dfrac{a^4}{c^7}\right)^2$

13. $(4^3)^2$

14. $\left(\dfrac{2x^3y^{-2}}{3y^{-3}}\right)^3$

15. $(8^x)^{4y}$

16. $\left(\dfrac{-4x^4y^{-2}}{5x^{-1}y^4}\right)^{-4}$

20 SCIENTIFIC NOTATION

MyLab Math VIDEO

Scientific notation is useful when representing very large numbers and very small numbers and when estimating. The following are examples of scientific notation:

- Brazil spent \$9,700,000,000 to stage the Summer 2016 Olympic Games.
  Data: Olympic Organizing Committee, Public Olympic Authority of Brazil, Federal Government of Brazil

  $\$9{,}700{,}000{,}000 = \$9.7 \times 10^9$

- The diameter of a helium atom is about 0.000000022 cm.

  $0.000000022 \text{ cm} = 2.2 \times 10^{-8} \text{ cm}$

Scientific notation for a number is an expression of the type

$$M \times 10^n,$$

where n is an integer, M is greater than or equal to 1 and less than 10 ($1 \le M < 10$), and M is expressed in decimal notation. 10^n is also considered to be scientific notation when $M = 1$.

You should try to make conversions to scientific notation mentally as much as possible. A positive exponent in scientific notation indicates a large number (greater than or equal to 10) and a negative exponent indicates a small number (between 0 and 1).

EXAMPLES Convert mentally to scientific notation.

1. Light travels 9,460,000,000,000 km in one year.

   $9{,}460{,}000{,}000{,}000 = 9.46 \times 10^{12}$ 9.460,000,000,000.
   12 places

   Large number, so the exponent is positive.

2. The mass of a grain of sand is 0.0648 g (grams).

   $0.0648 = 6.48 \times 10^{-2}$ 0.06.48
   2 places

   Small number, so the exponent is negative.

(continued)

EXERCISES

Convert each number to scientific notation.

1. 47,000,000,000
2. 0.000000016
3. 0.000000263
4. 2,600,000,000,000
5. *Insect-Eating Lizard.* A gecko is an insect-eating lizard. Its feet will adhere to virtually any surface because they contain millions of miniscule hairs, or setae, that are 200 billionths of a meter wide. Write 200 billionths in scientific notation.
   Data: *The Proceedings of the National Academy of Sciences*, Dr. Kellar Autumn and Wendy Hansen of Lewis and Clark College, Portland, Oregon

Some expressions contain parentheses preceded by subtraction signs. These parentheses can be removed by changing the sign of *every* term inside.

EXAMPLES Remove parentheses and simplify.

6. $6x - (4x + 2) = 6x - 4x - 2$ Changing the sign of every term inside parentheses
$= 2x - 2$ Collecting like terms

7. $3y - 4 - (9y - 7) = 3y - 4 - 9y + 7$
$= -6y + 3$, or $3 - 6y$

8. $3y + (3x - 8) - (5 - 12y) = 3y + 3x - 8 - 5 + 12y$
$= 15y + 3x - 13$

9. $x - 3(x + y) = x - 3x - 3y$ Removing parentheses by multiplying $x + y$ by -3
$= -2x - 3y$ Collecting like terms

Caution!

A common error is to forget to change this sign. *Remember*: When multiplying by a negative number, change the sign of *every* term inside the parentheses.

10. $3y - 2(4y - 5) = 3y - 8y + 10 = -5y + 10$

When expressions with grouping symbols contain variables, we still work from the inside out when simplifying, using the rules for order of operations.

EXAMPLE 11 Simplify: $[2(x + 7) - 4^2] - (2 - x)$.

$[2(x + 7) - 4^2] - (2 - x)$

$= [2x + 14 - 4^2] - (2 - x)$ Multiplying to remove the innermost grouping symbols using a distributive law

$= [2x + 14 - 16] - (2 - x)$ Evaluating the exponential expression

$= [2x - 2] - (2 - x)$ Collecting like terms inside the brackets

$= 2x - 2 - 2 + x$ Multiplying by -1 to remove the parentheses

$= 3x - 4$ Collecting like terms

Do Exercises 1–12.
(Exercises 1–6 are on the preceding page.)

EXERCISES

Simplify.

7. $4m - (3m - 1)$

8. $7a - [9 - 3(5a - 2)]$

9. $-2(x + 3) - 5(x - 4)$

10. $8x - (-3y + 7) + (9x - 11)$

11. $[7(x + 5) - 19] - [4(x - 6) + 10]$

12. $5\{-2 + 3[4 - 2(3 + 5)]\}$

19 EXPONENTIAL NOTATION (PART 2)

MyLab Math VIDEO

We often need to find ways to determine *equivalent exponential expressions.* We do this with several rules or properties regarding exponents.

RULES FOR EXPONENTS

For any real numbers a and b and any integers m and n:

Product Rule: $a^m \cdot a^n = a^{m+n}$

Quotient Rule: $\frac{a^m}{a^n} = a^{m-n}, a \neq 0$

Power Rule: $(a^m)^n = a^{mn}$

Raising a product to a power: $(ab)^n = a^n b^n$

Raising a quotient to a power: $\left(\frac{a}{b}\right)^n = \frac{a^n}{b^n}, b \neq 0;$

$\left(\frac{a}{b}\right)^{-n} = \left(\frac{b}{a}\right)^n, b \neq 0, a \neq 0$

EXAMPLES Simplify.

1. $x^4 \cdot x^3 = x^{4+3} = x^7$ Adding exponents using the product rule
2. $\frac{5^7}{5^3} = 5^{7-3} = 5^4$ Subtracting exponents using the quotient rule
3. $\frac{5^7}{5^{-3}} = 5^{7-(-3)} = 5^{7+3} = 5^{10}$ Subtracting exponents (adding an opposite)
4. $4^5 \cdot 4^{-3} = 4^{5+(-3)} = 4^2 = 16$ Adding exponents
5. $\frac{9^{-2}}{9^5} = 9^{-2-5} = 9^{-7} = \frac{1}{9^7}$

Note that we give answers using positive exponents. In some situations, this may not be appropriate, but we do so here.

6. $(8x^n)(6x^{2n}) = 8 \cdot 6 \cdot x^n \cdot x^{2n} = 48 \cdot x^{n+2n} = 48x^{3n}$
7. $\frac{16x^4y^7}{-8x^3y^9} = \frac{16}{-8} \cdot \frac{x^4}{x^3} \cdot \frac{y^7}{y^9} = -2x^{4-3}y^{7-9} = -2x^1y^{-2} = -\frac{2x}{y^2}$

(continued)

EXERCISES

Simplify.

1. $8^{-2} \cdot 8^{-4}$
2. $\frac{a^3}{a^{-2}}$
3. $\frac{10^{-3}}{10^6}$
4. $b^2 \cdot b^{-5}$
5. $\frac{-24x^6y^7}{18x^{-3}y^9}$
6. $(14m^2n^3)(-2m^3n^2)$
7. $(-3t^{-4a})(-5t^{-a})$
8. $\frac{-18x^{-2}y^3}{-12x^{-5}y^5}$

8. $\dfrac{14x^5y^{-3}}{4x^9y^{-5}} = \dfrac{14}{4}\cdot\dfrac{x^5}{x^9}\cdot\dfrac{y^{-3}}{y^{-5}} = \dfrac{7}{2}x^{5-9}y^{-3-(-5)}$

$= \dfrac{7}{2}x^{-4}y^2 = \dfrac{7y^2}{2x^4}$

9. $(8x^4y^{-2})(-3x^{-3}y) = 8\cdot(-3)\cdot x^4\cdot x^{-3}\cdot y^{-2}\cdot y^1$

$= -24x^{4+(-3)}y^{-2+1}$

$= -24xy^{-1} = -\dfrac{24x}{y}$

When an expression inside parentheses is raised to a power, the inside expression is the base. Thus, for $(5^2)^4$, we are raising 5^2 to the fourth power:

$$(5^2)^4 = (5^2)(5^2)(5^2)(5^2) = 5^8.$$

EXAMPLES Simplify.

10. $(x^5)^7 = x^{5\cdot7} = x^{35}$ Multiplying exponents

11. $\left(\dfrac{3^{-4}}{5^2}\right)^6 = \dfrac{(3^{-4})^6}{(5^2)^6} = \dfrac{3^{-24}}{5^{12}} = \dfrac{1}{3^{24}\cdot5^{12}}$

12. $(x^4)^{-2t} = x^{4(-2t)} = x^{-8t} = \dfrac{1}{x^{8t}}$

13. $\left(\dfrac{x^2}{y^{-3}}\right)^{-5} = \dfrac{x^{2\cdot(-5)}}{y^{-3\cdot(-5)}} = \dfrac{x^{-10}}{y^{15}} = \dfrac{1}{x^{10}y^{15}}$

14. $\left(\dfrac{-2a^{-5}}{a^{-3}b}\right)^4 = \dfrac{(-2a^{-5})^4}{(a^{-3}b)^4} = \dfrac{(-2)^4(a^{-5})^4}{(a^{-3})^4b^4}$

$= \dfrac{16a^{-20}}{a^{-12}b^4} = \dfrac{16a^{-20-(-12)}}{b^4}$

$= \dfrac{16a^{-8}}{b^4} = \dfrac{16}{a^8b^4}$

15. $(4r^3y^{-2}w)^{-2} = 4^{-2}(r^3)^{-2}(y^{-2})^{-2}w^{-2}$

$= \dfrac{1}{4^2}r^{-6}y^4w^{-2}$

$= \dfrac{y^4}{16r^6w^2}$

Do Exercises 1–16.
(Exercises 1–8 are on the preceding page.)

EXERCISES

Simplify.

9. $(r^9)^5$

10. $\left(\dfrac{4^{-3}}{3^4}\right)^3$

11. $(5a^2b^2)^3$

12. $\left(\dfrac{a^4}{c^7}\right)^2$

13. $(4^3)^2$

14. $\left(\dfrac{2x^3y^{-2}}{3y^{-3}}\right)^3$

15. $(8^x)^{4y}$

16. $\left(\dfrac{-4x^4y^{-2}}{5x^{-1}y^4}\right)^{-4}$

20 SCIENTIFIC NOTATION

MyLab Math VIDEO

Scientific notation is useful when representing very large numbers and very small numbers and when estimating. The following are examples of scientific notation:

- Brazil spent $9,700,000,000 to stage the Summer 2016 Olympic Games.

  Data: Olympic Organizing Committee, Public Olympic Authority of Brazil, Federal Government of Brazil

  $\$9{,}700{,}000{,}000 = \$9.7 \times 10^9$

- The diameter of a helium atom is about 0.000000022 cm.

  $0.000000022 \text{ cm} = 2.2 \times 10^{-8} \text{ cm}$

Scientific notation for a number is an expression of the type

$$M \times 10^n,$$

where n is an integer, M is greater than or equal to 1 and less than 10 $(1 \le M < 10)$, and M is expressed in decimal notation. 10^n is also considered to be scientific notation when $M = 1$.

You should try to make conversions to scientific notation mentally as much as possible. A positive exponent in scientific notation indicates a large number (greater than or equal to 10) and a negative exponent indicates a small number (between 0 and 1).

EXAMPLES Convert mentally to scientific notation.

1. Light travels 9,460,000,000,000 km in one year.

   $9{,}460{,}000{,}000{,}000 = 9.46 \times 10^{12}$ 9.460,000,000,000. (12 places)

   Large number, so the exponent is positive.

2. The mass of a grain of sand is 0.0648 g (grams).

   $0.0648 = 6.48 \times 10^{-2}$ 0.06.48 (2 places)

   Small number, so the exponent is negative.

(continued)

EXERCISES

Convert each number to scientific notation.

1. 47,000,000,000
2. 0.000000016
3. 0.000000263
4. 2,600,000,000,000
5. *Insect-Eating Lizard.* A gecko is an insect-eating lizard. Its feet will adhere to virtually any surface because they contain millions of miniscule hairs, or setae, that are 200 billionths of a meter wide. Write 200 billionths in scientific notation.

   Data: *The Proceedings of the National Academy of Sciences*, Dr. Kellar Autumn and Wendy Hansen of Lewis and Clark College, Portland, Oregon

EXAMPLES Convert mentally to decimal notation.

3. $4.893 \times 10^5 = 489{,}300$ 4.89300. (5 places)

Positive exponent, indicating a large number.

4. $8.7 \times 10^{-8} = 0.000000087$ 0.00000008.7 (8 places)

Negative exponent, indicating a small number.

Each of the following is *not* scientific notation.

13.95×10^{13}, — This number is greater than 10.

0.468×10^{-8} — This number is less than 1.

We can use the properties of exponents when we multiply and divide in scientific notation.

EXAMPLE 5 *Nervous System of an Octopus.* The common octopus, *Octopus vulgaris*, has 5×10^8 neurons in its nervous system, and approximately 67% of its neurons are in its arms. How many neurons are there in the arms of an octopus? Write the answer in scientific notation.

Data: *National Geographic*, November 2016, p. 70, "The Power of Eight," Olivia Judson. Roger Hanlon, Marine Biological Laboratory; Guy Levy and Benny Hochner, Hebrew University of Jerusalem; Cliff Ragsdale, University of Chicago.

We first express 67% in scientific notation:

$$67\% = 0.67 = 6.7 \times 10^{-1}.$$

To determine the number of neurons in the arms of an octopus, we multiply:

$$\begin{aligned}(6.7 \times 10^{-1}) \cdot (5 \times 10^8) &= (6.7 \times 5)(10^{-1} \times 10^8)\\ &= 33.5 \times 10^7\\ &= (3.35 \times 10) \times 10^7\\ &= 3.35 \times 10^8.\end{aligned}$$

There are 3.35×10^8 neurons in the arms of the *Octopus vulgaris*.

Do Exercises 1–10.

(Exercises 1–5 are on the preceding page.)

EXERCISES

Convert each number to decimal notation.

6. 6.73×10^8

7. The mass of an electron is 9.11×10^{-28} g.

8. There are 1.86×10^9 monthly active Facebook users as of February 1, 2017.
Data: Facebook

9. Multiply and write the answer in scientific notation:

$$(6.5 \times 10^3)(5.2 \times 10^{-8}).$$

10. Divide and write the answer in scientific notation:

$$\frac{3.6 \times 10^{-3}}{7.2 \times 10^{-12}}.$$

CHAPTER

1

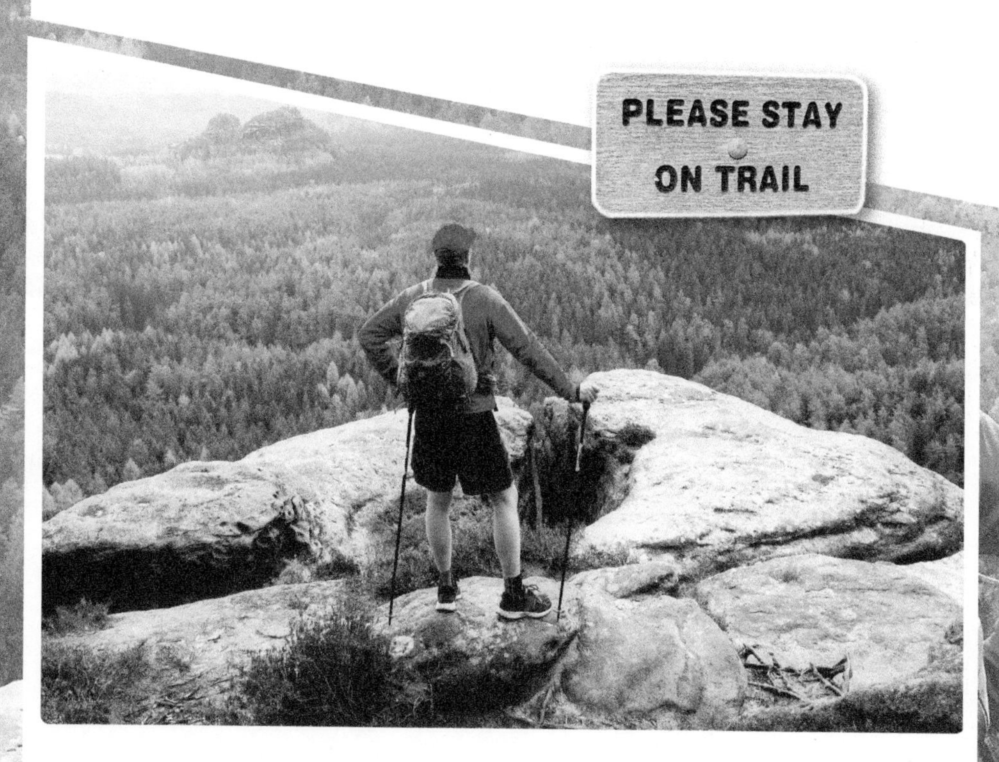

Solving Linear Equations and Inequalities

As a form of recreation, hiking has increased in popularity in recent years, with the number of hikers or backpackers in the United States growing from 30 million in 2009 to 44 million in 2016. Some of the longest trail systems in the United States are illustrated in the accompanying graph. Hikers who complete three of the best-known trails—the Continental Divide National Scenic Trail, the Pacific Coast National Scenic Trail, and the Appalachian National Scenic Trail— are eligible for the Triple Crown Award. As of 2016, there have been 291 Triple Crown Awards given.

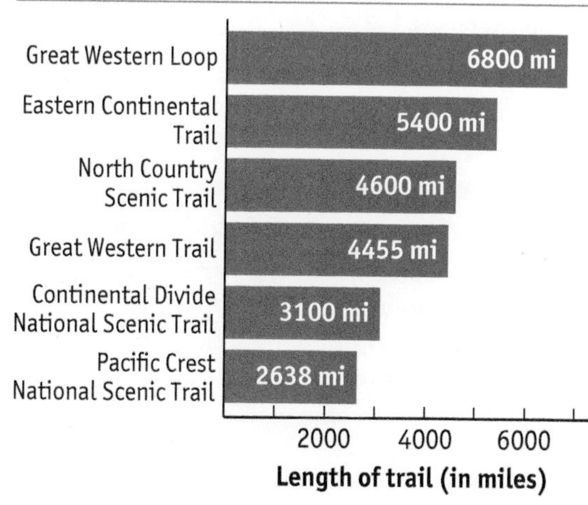

Data: statista.com; American Long Distance Hiking Association—West; adventurestrong.com

In Example 1 of Section 1.3, we will calculate the length of the Appalachian National Scenic Trail, and in Exercise 25 of Section 1.3, we will use a linear equation to estimate the number of young-adult hikers for various years.

JUST IN TIME

Review topics 4, 12, 16, 17, and 18 in the **Just In Time** section at the front of the text. This provides excellent prerequisite skill review for this chapter.

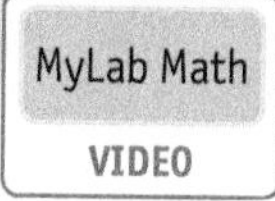

STUDYING FOR SUCCESS *Getting Off to a Good Start*

- ☐ Your syllabus for this course is extremely important. Read it carefully, noting required texts and materials.
- ☐ If there is an online component for your course, register for it as soon as possible.
- ☐ At the front of the text, you will find a Student Organizer card. This foldout card will help you keep track of important dates and useful contact information.

1.1 Solving Equations

OBJECTIVES

- **a** Determine whether a given number is a solution of a given equation.
- **b** Solve equations using the addition principle.
- **c** Solve equations using the multiplication principle.
- **d** Solve equations using the addition principle and the multiplication principle together, removing parentheses where appropriate.

a EQUATIONS AND SOLUTIONS

In order to solve many kinds of problems, we must be able to solve *equations*. Some examples of equations are

$$15 - 10 = 2 + 3, \quad x + 8 = 23, \quad 5x - 2 = 9 - x.$$

EQUATION

An **equation** is a number sentence that says that the expressions on either side of the equals sign, =, represent the same number.

Some equations are true. Some are false. Some are neither true nor false.

EXAMPLES Determine whether the equation is true, false, or neither.

1. $1 + 10 = 11$ Both expressions represent 11. The equation is *true*.

2. $7 - 8 = 9 - 13$ $7 - 8$ represents -1 and $9 - 13$ represents -4. The equation is *false*.

3. $x - 9 = 3$ The equation is *neither* true nor false, because we do not know what number x represents.

◀ **Do Exercises 1–3.**

If an equation contains a variable, then some replacements or values of the variable may make it true and some may make it false.

SOLUTION OF AN EQUATION

The replacements for the variable that make an equation true are called the **solutions** of the equation. The set of all solutions is called the **solution set** of the equation. When we find all the solutions, we say that we have **solved** the equation.

To determine whether a number is a solution of an equation, we evaluate the algebraic expression on each side of the equals sign by substitution. If the values are the same, then the number is a solution of the equation. If they are not, then the number is not a solution.

Consider the following equations.

a) $3 + 4 = 7$
b) $5 - 1 = 2$
c) $21 + 2 = 24$
d) $x - 5 = 12$
e) $9 - x = x$
f) $13 + 2 = 15$

1. Which equations are true?
2. Which equations are false?
3. Which equations are neither true nor false?

Answers

1. (a), (f) **2.** (b), (c) **3.** (d), (e)

EXAMPLE 4 Determine whether 5 is a solution of $x + 6 = 11$.

$$\begin{array}{c|l} x + 6 = 11 & \text{Writing the equation} \\ \hline 5 + 6 \ ? \ 11 & \text{Substituting 5 for } x \\ 11 & \text{TRUE} \end{array}$$

Since the left side and the right side are the same, 5 is a solution of the equation.

EXAMPLE 5 Determine whether 18 is a solution of $2x - 3 = 5$.

$$\begin{array}{c|l} 2x - 3 = 5 & \text{Writing the equation} \\ \hline 2 \cdot 18 - 3 \ ? \ 5 & \text{Substituting 18 for } x \\ 36 - 3 & \\ 33 & \text{FALSE} \end{array}$$

Since the left side and the right side are not the same, 18 is not a solution of the equation.

Do Exercises 4–6.

Determine whether the given number is a solution of the given equation.

4. 8; $x + 5 = 13$

5. -4; $7x = 16$

6. 5; $2x + 3 = 13$

Equivalent Equations

Consider the equation

$$x = 5.$$

The solution of this equation is easily "seen" to be 5. If we replace x with 5, we get

$$5 = 5, \quad \text{which is true.}$$

In Example 4, we saw that the solution of the equation $x + 6 = 11$ is also 5, but the fact that 5 is the solution is not so readily apparent. We now consider principles that allow us to start with one equation and end up with an *equivalent equation*, like $x = 5$, in which the variable is alone on one side, and for which the solution is read directly from the equation.

EQUIVALENT EQUATIONS

Equations with the same solutions are called **equivalent equations**.

Do Exercises 7 and 8.

7. Determine whether

$3x + 2 = 11$ and $x = 3$

are equivalent.

8. Determine whether

$4 - 5x = -11$ and $x = -3$

are equivalent.

b THE ADDITION PRINCIPLE

One of the principles that we use in solving equations involves addition. The equation $a = b$ says that a and b represent the same number. Suppose that $a = b$ is true and we then add a number c to a. We will get the same result if we add c to b, because a and b are the same number.

THE ADDITION PRINCIPLE

For any real numbers a, b, and c,

$$a = b \quad \text{is equivalent to} \quad a + c = b + c.$$

Answers

4. Yes **5.** No **6.** Yes
7. Yes **8.** No

When we use the addition principle, we sometimes say that we "add the same number on both sides of an equation." We can also "subtract the same number on both sides of an equation," because we can express subtraction as the addition of an opposite. That is,

$$a - c = b - c \quad \text{is equivalent to} \quad a + (-c) = b + (-c).$$

EXAMPLE 6 Solve: $x + 6 = 11$.

$$x + 6 = 11$$

$$\left.\begin{aligned} x + 6 + (-6) &= 11 + (-6) \\ x + 6 - 6 &= 11 - 6 \end{aligned}\right\}$$ Using the addition principle: adding -6 on both sides or subtracting 6 on both sides. Note that 6 and -6 are opposites.

$$x + 0 = 5$$ Simplifying

$$x = 5$$ Using the identity property of 0: $x + 0 = x$

Check: $x + 6 = 11$

$5 + 6 \ ? \ 11$ Substituting 5 for x

$11 \mid 11$ TRUE

The solution is 5.

In Example 6, we wanted to get x alone so that we could readily see the solution, so we added the opposite of 6. This eliminated the 6 on the left, giving us the *additive identity* 0, which when added to x is x. We began with $x + 6 = 11$. Using the addition principle, we derived a simpler equation, $x = 5$, that had the same solution. The equations $x + 6 = 11$ and $x = 5$ are *equivalent*.

EXAMPLE 7 Solve: $y - 4.7 = 13.9$.

$$y - 4.7 = 13.9$$

$$y - 4.7 + 4.7 = 13.9 + 4.7$$ Using the addition principle: adding 4.7 on both sides. Note that -4.7 and 4.7 are opposites.

$$y + 0 = 18.6$$ Simplifying

$$y = 18.6$$ Using the identity property of 0: $y + 0 = y$

Check: $y - 4.7 = 13.9$

$18.6 - 4.7 \ ? \ 13.9$ Substituting 18.6 for y

$13.9 \mid 13.9$ TRUE

The solution is 18.6.

EXAMPLE 8 Solve: $-\frac{3}{8} + x = -\frac{5}{7}$.

$$-\frac{3}{8} + x = -\frac{5}{7}$$

$$\frac{3}{8} + \left(-\frac{3}{8}\right) + x = \frac{3}{8} + \left(-\frac{5}{7}\right)$$ Using the addition principle: adding $\frac{3}{8}$

$$0 + x = \frac{3}{8} - \frac{5}{7}$$

$$x = \frac{3}{8} \cdot \frac{7}{7} - \frac{5}{7} \cdot \frac{8}{8}$$ Multiplying by 1 to obtain the least common denominator

$$= \frac{21}{56} - \frac{40}{56}$$

$$= -\frac{19}{56}$$

Check:

$$\begin{array}{c|c}
-\frac{3}{8} + x & = -\frac{5}{7} \\
\hline
-\frac{3}{8} + \left(-\frac{19}{56}\right) & ?\ -\frac{5}{7} \\
-\frac{3}{8}\cdot\frac{7}{7} + \left(-\frac{19}{56}\right) & \\
-\frac{21}{56} + \left(-\frac{19}{56}\right) & \\
-\frac{40}{56} & \\
-\frac{5}{7} & \\
\end{array}$$

Substituting $-\frac{19}{56}$ for x

TRUE

The solution is $-\frac{19}{56}$.

Do Exercises 9–12.

Solve using the addition principle.

9. $x + 9 = 2$

GS **10.** $x + \dfrac{1}{4} = -\dfrac{3}{5}$

$$x + \frac{1}{4} + \left(-\frac{1}{4}\right) = -\frac{3}{5} + (\quad)$$

$$x + \frac{1}{4} - \quad = -\frac{3}{5} - \frac{1}{4}$$

$$x + \quad = -\frac{3}{5}\cdot\frac{4}{4} - \frac{1}{4}\cdot\frac{5}{\quad}$$

$$x = -\frac{12}{\quad} - \frac{5}{20}$$

$$x = -\frac{\quad}{20}$$

11. $13 = -25 + y$

12. $y - 61.4 = 78.9$

C THE MULTIPLICATION PRINCIPLE

A second principle for solving equations involves multiplication. Suppose that $a = b$ is true and we multiply a by a nonzero number c. We get the same result if we multiply b by c, because a and b are the same number.

THE MULTIPLICATION PRINCIPLE

For any real numbers a, b, and c, $c \neq 0$,

$a = b$ is equivalent to $a \cdot c = b \cdot c$.

EXAMPLE 9 Solve: $\frac{4}{5}x = 22$.

$$\frac{4}{5}x = 22$$

$$\frac{5}{4}\cdot\frac{4}{5}x = \frac{5}{4}\cdot 22 \qquad \text{Multiplying by } \tfrac{5}{4}\text{, the reciprocal of } \tfrac{4}{5}$$

$$1\cdot x = \frac{55}{2} \qquad \text{Multiplying and simplifying}$$

$$x = \frac{55}{2} \qquad \text{Using the identity property of 1: } 1\cdot x = x$$

Check:

$$\begin{array}{c|c}
\frac{4}{5}x & = 22 \\
\hline
\frac{4}{5}\cdot\frac{55}{2} & ?\ 22 \\
22 & \\
\end{array}$$

TRUE

The solution is $\frac{55}{2}$.

In Example 9, in order to get x alone, we multiplied by the *multiplicative inverse*, or *reciprocal*, of $\frac{4}{5}$. When we multiplied, we got the *multiplicative identity* 1 times x, or $1 \cdot x$, which simplified to x. This enabled us to eliminate the $\frac{4}{5}$ on the left.

The multiplication principle also tells us that we can "divide by a nonzero number on both sides" because division is the same as multiplying by a reciprocal. That is,

$$\frac{a}{c} = \frac{b}{c} \text{ is equivalent to } a\cdot\frac{1}{c} = b\cdot\frac{1}{c}, \text{ when } c \neq 0.$$

In a product like $\frac{4}{5}x$, the number in front of the variable is called the **coefficient**. When this number is in fraction notation, it is usually most convenient to multiply both sides by its reciprocal. If the coefficient is an integer or is in decimal notation, it is usually more convenient to divide by the coefficient.

Answers

9. -7 **10.** $-\frac{17}{20}$ **11.** 38 **12.** 140.3

Guided Solution:

10. $-\frac{1}{4}, \frac{1}{4}, 0, 5, 20, 17$

Solve using the multiplication principle.

13. $8x = 10$

14. $-\frac{3}{7}y = 21$

15. $-4x = -\frac{6}{7}$

GS

$-\frac{1}{4} \cdot (-4x) = -\frac{1}{\square} \cdot \left(-\frac{6}{7}\right)$

$x = \frac{6}{\square}$

$x = \frac{\square}{14}$

EXAMPLE 10 Solve: $4x = 9$.

$$4x = 9$$

$$\frac{4x}{4} = \frac{9}{4}$$ Using the multiplication principle: multiplying on both sides by $\frac{1}{4}$ or dividing on both sides by the coefficient, 4

$$1 \cdot x = \frac{9}{4}$$ Simplifying

$$x = \frac{9}{4}$$ Using the identity property of 1: $1 \cdot x = x$

Check: $4x = 9$

$4 \cdot \frac{9}{4} \;?\; 9$

9 | TRUE

The solution is $\frac{9}{4}$.

◀ **Do Exercises 13–15.**

EXAMPLE 11 Solve: $5.5 = -0.05y$.

$$5.5 = -0.05y$$

$$\frac{5.5}{-0.05} = \frac{-0.05y}{-0.05}$$ Dividing by -0.05 on both sides

$$-110 = 1 \cdot y$$

$$-110 = y$$

The check is left to the student. The solution is -110.

Note that equations are reversible. That is, $a = b$ is equivalent to $b = a$. Thus, in Example 11, $-110 = y$ and $y = -110$ are equivalent, and the solution of both equations is -110.

◀ **Do Exercise 16.**

16. Solve: $-12.6 = 4.2y$.

EXAMPLE 12 Solve: $-\frac{x}{4} = 10$.

$$-\frac{x}{4} = 10$$

$$-\frac{1}{4}x = 10$$ $-\frac{x}{4} = -\frac{1}{4} \cdot x$

$$-4 \cdot \left(-\frac{1}{4}\right)x = -4 \cdot 10$$ Multiplying by -4 on both sides. Note that -4 is the reciprocal of $-\frac{1}{4}$.

$$1 \cdot x = -40$$ Simplifying

$$x = -40$$

The check is left to the student. The solution is -40.

◀ **Do Exercises 17 and 18.**

Solve.

17. $-\frac{x}{8} = 17$

18. $-x = -5$

Answers

13. $\frac{5}{4}$ **14.** -49 **15.** $\frac{3}{14}$ **16.** -3
17. -136 **18.** 5

Guided Solution:
15. 4, 28, 3

d USING THE PRINCIPLES TOGETHER

Remove parentheses and collect like terms. [J18]

Remove parentheses and simplify.

1. $4(3x - 7) - x + 2$ **2.** $30 - 5(1 - 6x)$

Answers: 1. $11x - 26$ **2.** $30x + 25$

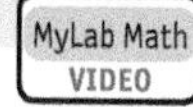

Let's see how we can use the addition and multiplication principles together.

EXAMPLE 13 Solve: $3x - 4 = 13$.

$$3x - 4 = 13$$

$$3x - 4 + 4 = 13 + 4 \qquad \text{Using the addition principle: adding 4}$$

$$3x = 17 \qquad \text{Simplifying}$$

$$\frac{3x}{3} = \frac{17}{3} \qquad \text{Dividing by 3}$$

$$x = \frac{17}{3} \qquad \text{Simplifying}$$

Check:

$3x - 4 = 13$	
$3 \cdot \frac{17}{3} - 4$?	13
$17 - 4$	
13	TRUE

The solution is $\frac{17}{3}$, or $5\frac{2}{3}$.

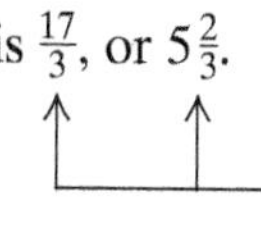

In algebra, "improper" fraction notation, such as $\frac{17}{3}$, is quite "proper." We will generally use such notation rather than $5\frac{2}{3}$.

Do Exercise 19. ▶

19. Solve: $-4 + 9x = 8$.

In a situation such as Example 13, it is easier to first use the addition principle. In a situation in which fractions or decimals are involved, it may be easier to use the multiplication principle first to clear them.

EXAMPLE 14 Clear the fractions and solve: $\frac{3}{16}x + \frac{1}{2} = \frac{11}{8}$.

We multiply on both sides by the least common multiple of the denominators—in this case, 16:

$$\frac{3}{16}x + \frac{1}{2} = \frac{11}{8} \qquad \text{The LCM of the denominators is 16.}$$

$$16\left(\frac{3}{16}x + \frac{1}{2}\right) = 16\left(\frac{11}{8}\right) \qquad \text{Multiplying by 16}$$

$$16 \cdot \frac{3}{16}x + 16 \cdot \frac{1}{2} = 22 \qquad \text{Carrying out the multiplication. We use the distributive law on the left, being careful to multiply } both \text{ terms by 16.}$$

$$3x + 8 = 22. \qquad \text{Simplifying. The fractions are cleared.}$$

Answer

19. $\frac{4}{3}$

20. Clear the fractions and solve:

$$\frac{2}{3} - \frac{5}{6}y = \frac{1}{3}.$$

GS

$$\frac{2}{3} - \frac{5}{6}y = \frac{1}{3}$$

$$\text{LCD} = 6$$

$$6\left(\frac{2}{3} - \frac{5}{6}y\right) = \square \cdot \frac{1}{3}$$

$$6 \cdot \frac{2}{3} - \square \cdot \frac{5}{6}y = 2$$

$$\square - 5y = 2$$

$$\square + 4 - 5y = -4 + 2$$

$$-5y = \square$$

$$-\frac{1}{5} \cdot (-5y) = -\frac{1}{5} \cdot (-2)$$

$$1 \cdot y = \frac{2}{5}$$

$$y = \frac{2}{5}$$

21. Clear the decimals and solve:

$$6.3x - 9.1 = 3x.$$

Answers

20. $4 - 5y = 2; \frac{2}{5}$

21. $63x - 91 = 30x; \frac{91}{33}$

Guided Solution:
20. 6, 6, 4, −4, −2

Then

$$3x + 8 - 8 = 22 - 8 \quad \text{Subtracting 8}$$
$$3x = 14$$
$$\frac{3x}{3} = \frac{14}{3} \quad \text{Dividing by 3}$$
$$x = \frac{14}{3}.$$

The number $\frac{14}{3}$ checks and is the solution.

◀ **Do Exercise 20.**

If there are like terms on opposite sides of an equation, we use the addition principle to get them on the same side of the equation.

EXAMPLE 15 Clear the decimals and solve: $12.4 - 5.12x = 3.14x$.

We multiply on both sides by a power of ten—10, 100, 1000, and so on—to clear the equation of decimals. In this case, we use 10^2, or 100, because the greatest number of decimal places is 2.

$$12.4 - 5.12x = 3.14x$$
$$100(12.4 - 5.12x) = 100(3.14x) \quad \text{Multiplying by 100}$$
$$100(12.4) - 100(5.12x) = 314x \quad \text{Carrying out the multiplication. We use the distributive law on the left.}$$
$$1240 - 512x = 314x \quad \text{Simplifying}$$
$$1240 - 512x + 512x = 314x + 512x \quad \text{Adding } 512x$$
$$1240 = 826x$$
$$\frac{1240}{826} = \frac{826x}{826} \quad \text{Dividing by 826}$$
$$x = \frac{1240}{826}, \text{ or } \frac{620}{413}$$

The solution is $\frac{620}{413}$.

◀ **Do Exercise 21.**

If there are like terms on the same side of an equation, we begin by collecting them.

EXAMPLE 16 Solve: $8x + 6 - 2x = -4x - 14$.

$$8x + 6 - 2x = -4x - 14$$
$$6x + 6 = -4x - 14 \quad \text{Collecting like terms on the left}$$
$$4x + 6x + 6 = 4x - 4x - 14 \quad \text{Adding } 4x$$
$$10x + 6 = -14 \quad \text{Collecting like terms}$$
$$10x + 6 - 6 = -14 - 6 \quad \text{Subtracting 6}$$
$$10x = -20$$
$$\frac{10x}{10} = \frac{-20}{10} \quad \text{Dividing by 10}$$
$$x = -2$$

Check:

$$\begin{array}{r|l} 8x + 6 - 2x & = -4x - 14 \\ \hline 8(-2) + 6 - 2(-2) & ? \; -4(-2) - 14 \\ -16 + 6 + 4 & 8 - 14 \\ -6 & -6 \quad \text{TRUE} \end{array}$$

The solution is -2.

Do Exercises 22–24. ▶

Solve.

22. $\frac{5}{2}x + \frac{9}{2}x = 21$

23. $1.4x - 0.9x + 0.7 = -2.2$

24. $-4x + 2 + 5x = 3x - 15$

Special Cases

Some equations have no solution.

EXAMPLE 17 Solve: $-8x + 5 = 14 - 8x$.

We have

$$-8x + 5 = 14 - 8x$$

$$8x - 8x + 5 = 8x + 14 - 8x \quad \text{Adding } 8x$$

$$5 = 14. \quad \text{We get a false equation.}$$

No matter what number we use for x, we get a false sentence. Thus the equation has *no* solution. ■

There are some equations for which any real number is a solution.

EXAMPLE 18 Solve: $-8x + 5 = 5 - 8x$.

We have

$$-8x + 5 = 5 - 8x$$

$$8x - 8x + 5 = 8x + 5 - 8x \quad \text{Adding } 8x$$

$$5 = 5. \quad \text{We get a true equation.}$$

Replacing x with any real number gives a true sentence. Thus any real number is a solution. The equation has *infinitely* many solutions.

Do Exercises 25 and 26. ▶

Solve.

25. $4 + 7x = 7x + 9$

26. $3 + 9x = 9x + 3$

Equations Containing Parentheses

Equations containing parentheses can often be solved by first multiplying to remove parentheses and then proceeding as before.

EXAMPLE 19 Solve: $30 + 5(x + 3) = -3 + 5x + 48$.

We have

$$30 + 5(x + 3) = -3 + 5x + 48$$

$$30 + 5x + 15 = -3 + 5x + 48 \quad \text{Multiplying, using the distributive law, to remove parentheses}$$

$$45 + 5x = 45 + 5x \quad \text{Collecting like terms on each side}$$

$$45 + 5x - 5x = 45 + 5x - 5x \quad \text{Subtracting } 5x$$

$$45 = 45. \quad \text{Simplifying. We get a true equation.}$$

All real numbers are solutions.

Do Exercises 27–29. ▶

Solve.

27. $7x - 17 = 4 + 7(x - 3)$

28. $3x + 4(x + 2) = 11 + 7x$

29. $3x + 8(x + 2) = 11 + 7x$

Answers

22. 3 **23.** $-\frac{29}{5}$ **24.** $\frac{17}{2}$
25. No solution **26.** All real numbers are solutions. **27.** All real numbers are solutions. **28.** No solution **29.** $-\frac{5}{4}$

EXAMPLE 20 Solve: $3(7 - 2x) = 14 - 8(x - 1)$.

$$3(7 - 2x) = 14 - 8(x - 1)$$

$$21 - 6x = 14 - 8x + 8 \quad \text{Multiplying, using the distributive law, to remove parentheses}$$

$$21 - 6x = 22 - 8x \quad \text{Collecting like terms}$$

$$21 - 6x + 8x = 22 - 8x + 8x \quad \text{Adding } 8x$$

$$21 + 2x = 22 \quad \text{Collecting like terms}$$

$$21 + 2x - 21 = 22 - 21 \quad \text{Subtracting 21}$$

$$2x = 1$$

$$\frac{2x}{2} = \frac{1}{2} \quad \text{Dividing by 2}$$

$$x = \frac{1}{2}$$

Check:

$3(7 - 2x)$	$= 14 - 8(x - 1)$	
$3(7 - 2\cdot\frac{1}{2})$?	$14 - 8(\frac{1}{2} - 1)$	
$3(7 - 1)$	$14 - 8(-\frac{1}{2})$	
$3\cdot 6$	$14 + 4$	
18	18	TRUE

The solution is $\frac{1}{2}$.

Do Exercises 30 and 31.

Solve.

30. $30 + 7(x - 1) = 3(2x + 7)$

31. $3(y - 1) - 1 = 2 - 5(y + 5)$

AN EQUATION-SOLVING PROCEDURE

1. Clear the equation of fractions or decimals if that is needed. (This step is optional, but it can ease calculations.)
2. If parentheses occur, multiply to remove them using the distributive law.
3. Collect like terms on each side of the equation, if necessary.
4. Use the addition principle to get all terms with the variable on one side and all other terms on the other side.
5. Collect like terms on each side again, if necessary.
6. Use the multiplication principle to solve for the variable.

Answers

30. -2 **31.** $-\frac{19}{8}$

CALCULATOR CORNER

Checking Possible Solutions Although a calculator is *not* required for this textbook, the book contains a series of *optional* discussions on using a graphing calculator. The keystrokes for the TI-84 Plus graphing calculator will be shown throughout. For keystrokes for other models of calculators, consult the user's manual for your particular model.

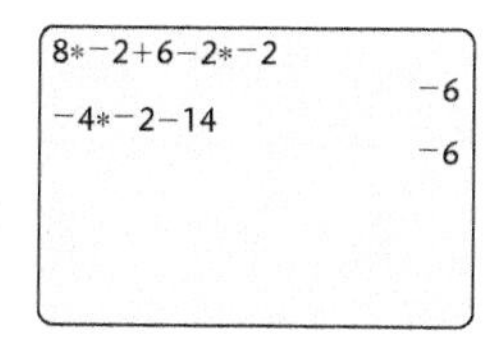

To check possible solutions of an equation on a calculator, we can substitute and carry out the calculations on each side of the equation. If the left and the right sides of the equation have the same value, then the number that was substituted is a solution of the equation. To check the possible solution -2 in the equation $8x + 6 - 2x = -4x - 14$ in Example 16, for instance, we first substitute -2 for x in the expression on the left side of the equation and get -6. Then we substitute -2 for x in the expression on the right side of the equation. Again, we get -6. Since the two sides of the equation have the same value when x is -2, we know that -2 is the solution of the equation.

A table can also be used to check possible solutions of equations. First, we press Y= to display the equation-editor screen. If an expression for Y1 is currently entered, we place the cursor on it and press CLEAR to delete it. We do the same for any other entries that are present. Next, we position the cursor to the right of Y1 = and enter the left side of the equation. Then we position the cursor beside Y2 = and enter the right side of the equation. Now we press 2ND TBLSET to display the Table Setup screen. (TBLSET is the second operation associated with the WINDOW key.) On the INDPNT line, we position the cursor on "ASK" and press ENTER to set up a table in ASK mode. (The settings for TblStart and ΔTbl are irrelevant in ASK mode.)

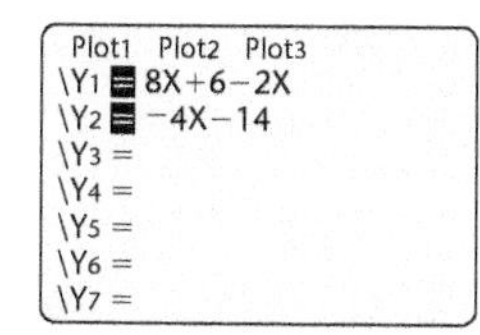

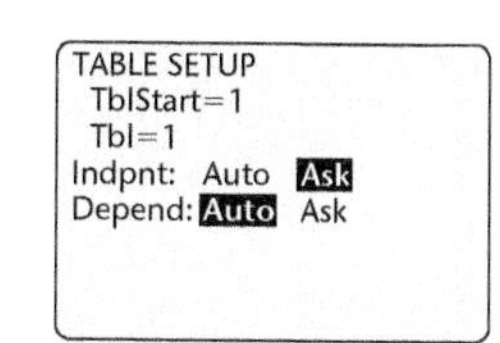

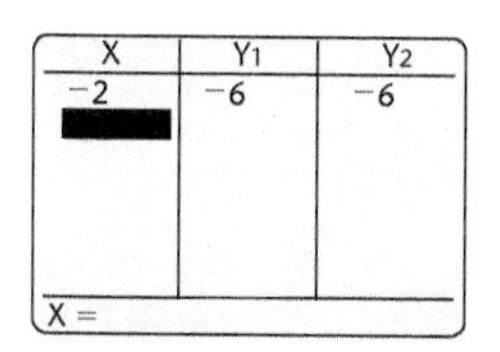

We press 2ND TABLE to display the table. (TABLE is the second operation associated with the GRAPH key.) We enter the possible solution, -2, and see that $Y1 = -6 = Y2$ for this value of x. This confirms that the left and the right sides of the equation have the same value for $x = -2$, so -2 is the solution of the equation.

EXERCISES:

1. Use substitution to check the solutions found in Examples 9, 13, and 15.
2. Use a table set in ASK mode to check the solutions found in Margin Exercises 24, 30, and 31.

1.1 Exercise Set

FOR EXTRA HELP  MyLab Math

✓ Check Your Understanding

Reading Check Choose from the column on the right the most appropriate first step in solving each equation.

RC1. $5 + x = -12$

RC2. $16 = x - 3$

RC3. $3 = -\frac{1}{9}x$

RC4. $5x = -16$

a) Divide by 3 on both sides.

b) Multiply by -9 on both sides.

c) Divide by 5 on both sides.

d) Subtract 5 on both sides.

e) Add $\frac{1}{9}$ on both sides.

f) Add 3 on both sides.

Concept Check For each equation, **(a)** give the smallest positive number by which both sides of the equation can be multiplied to clear fractions, and **(b)** multiply by that number to clear fractions. Do not solve.

CC1. $\frac{2}{5}x - \frac{1}{10} = \frac{7}{10}$

CC2. $4 - \frac{2}{3}x = \frac{5}{2}$

For each equation, **(a)** give the smallest power of 10 by which both sides of the equation can be multiplied to clear decimals, and **(b)** multiply by that number to clear decimals. Do not solve.

CC3. $1.2y - 0.5 = 2.7y$

CC4. $0.06 - 7x = 1.2x$

Remember to review the objectives before doing the exercises.

a Determine whether the given number is a solution of the given equation.

1. $17;\ x + 23 = 40$

2. $24;\ 47 - x = 23$

3. $-8;\ 2x - 3 = -18$

4. $-10;\ 3x + 14 = -27$

5. $45;\ \frac{-x}{9} = -2$

6. $32;\ \frac{-x}{8} = -3$

7. $10;\ 2 - 3x = 21$

8. $-11;\ 4 - 5x = 59$

9. $19;\ 5x + 7 = 102$

10. $9;\ 9y + 5 = 86$

11. $-11;\ 7(y - 1) = 84$

12. $-13;\ x + 5 = 5 + x$

b Solve using the addition principle. Don't forget to check.

13. $y + 6 = 13$

14. $x + 7 = 14$

15. $-20 = x - 12$

16. $-27 = y - 17$

17. $-8 + x = 19$

18. $-8 + r = 17$

19. $-12 + z = -51$

20. $-37 + x = -89$

21. $p - 2.96 = 83.9$

22. $z - 14.9 = -5.73$

23. $-\frac{3}{8} + x = -\frac{5}{24}$

24. $x + \frac{1}{12} = -\frac{5}{6}$

c Solve using the multiplication principle. Don't forget to check.

25. $3x = 18$

26. $5x = 30$

27. $-11y = 44$

28. $-4x = 124$

29. $-\frac{x}{7} = 21$

30. $-\frac{x}{3} = -25$

31. $-96 = -3z$

32. $-120 = -8y$

33. $4.8y = -28.8$

34. $0.39t = -2.73$

35. $\frac{3}{2}t = -\frac{1}{4}$

36. $-\frac{7}{6}y = -\frac{7}{8}$

d Solve using the principles together. Don't forget to check.

37. $6x - 15 = 45$

38. $4x - 7 = 81$

39. $5x - 10 = 45$

40. $6z - 7 = 11$

41. $9t + 4 = -104$

42. $5x + 7 = -108$

43. $-\frac{7}{3}x + \frac{2}{3} = -18$

44. $-\frac{9}{2}y + 4 = -\frac{91}{2}$

45. $x + 9x = 3$

46. $2x + x = -1$

47. $6x - 8x - 1 = 4$

48. $-2x + 6x - 2 = 11$

49. $\frac{6}{5}x + \frac{4}{10}x = \frac{32}{10}$

50. $\frac{9}{5}y + \frac{4}{10}y = \frac{66}{10}$

51. $0.9y - 0.7y = 4.2$

52. $0.8t - 0.3t = 6.5$

53. $8x + 48 = 3x - 12$

54. $15x + 40 = 8x - 9$

55. $7y - 1 = 27 + 7y$

56. $3x - 15 = 15 + 3x$

57. $3x - 4 = 5 + 12x$

58. $9t - 4 = 14 + 15t$

59. $5 - 4a = a - 13$

60. $6 - 7x = x - 14$

61. $3m - 7 = -7 - 4m - m$

62. $5x - 8 = -8 + 3x - x$

63. $5x + 3 = 11 - 4x + x$

64. $6y + 20 = 10 + 3y + y$

65. $-7 + 9x = 9x - 7$

66. $-3t + 4 = 5 - 3t$

67. $6y - 8 = 9 + 6y$

68. $5 - 2y = -2y + 5$

69. $\frac{1}{2}a - \frac{2}{3} = \frac{1}{6} - \frac{1}{3}a$

70. $\frac{3}{4} - \frac{5}{8}m = \frac{1}{2}m - \frac{3}{8}$

71. $1.2x - 3.4x = 6.8 - 2.1x$

72. $0.2t + 1.7 = 5.8 + 0.3t$

73. $2.5 + 0.05a = 6 - 0.7a$

74. $2 - 0.01x = 0.4x + 2.5$

75. $\frac{1}{2}y - \frac{3}{4}y + 2 = \frac{5}{8} + \frac{1}{4}y$

76. $\frac{2}{3}x + 1 - \frac{1}{6} = \frac{1}{2}x - \frac{5}{6}x$

77. $2(x + 7) = 4x$

78. $3(y + 6) = 9y$

79. $80 = 10(3t + 2)$

80. $27 = 9(5y - 2)$

81. $180(n - 2) = 900$

82. $210(x - 3) = 840$

83. $5y - (2y - 10) = 25$

84. $8x - (3x - 5) = 40$

85. $7(3x + 6) = 11 - (x + 2)$

86. $3(4 - 2x) = 4 - (6x - 8)$

87. $2[9 - 3(-2x - 4)] = 12x + 42$

88. $-40x + 45 = 3[7 - 2(7x - 4)]$

89. $\frac{1}{8}(16y + 8) - 17 = -\frac{1}{4}(8y - 16)$

90. $\frac{1}{6}(12t + 48) - 20 = -\frac{1}{8}(24t - 144)$

91. $3[5 - 3(4 - t)] - 2 = 5[3(5t - 4) + 8] - 26$

92. $6[4(8 - y) - 5(9 + 3y)] - 21 = -7[3(7 + 4y) - 4]$

93. $\frac{2}{3}\left(\frac{7}{8} + 4x\right) - \frac{5}{8} = \frac{3}{8}$

94. $\frac{3}{4}\left(3x - \frac{1}{2}\right) + \frac{2}{3} = \frac{1}{3}$

95. $5(4x - 3) - 2(6 - 8x) + 10(-2x + 7) = -4(9 - 12x)$

96. $9(4x + 7) - 3(5x - 8) = 6(\frac{2}{3} - x) - 5(\frac{3}{5} + 2x)$

Skill Maintenance

This heading indicates that the exercises that follow are *Skill Maintenance* exercises, which review any skill previously studied in the text. You will see them in virtually every exercise set. Answers to *all* skill maintenance exercises are found at the back of the book. To review a topic, restudy the objective shown in red.

Multiply or divide, and simplify. [J19]

97. $a^{-9} \cdot a^{23}$

98. $\frac{a^{-9}}{a^{23}}$

99. $(6x^5y^{-4})(-3x^{-3}y^{-7})$

100. $\frac{6x^5y^{-4}}{-3x^{-3}y^{-7}}$

Multiply. [J16]

101. $2(6 - 10x)$

102. $-1(5 - 6x)$

103. $-4(3x - 2y + z)$

104. $5(-2x + 7y - 4)$

Factor. [J16]

105. $2x - 6y$

106. $-4x - 24y$

107. $4x - 10y + 2$

108. $-10x + 35y - 20$

109. Name the set consisting of the positive integers less than 10, using both roster notation and set-builder notation. [J1]

110. Name the set consisting of the negative integers greater than -9 using both roster notation and set-builder notation. [J1]

Synthesis

To the student and the instructor: The *Synthesis* exercises found at the end of every exercise set challenge students to combine concepts or skills studied in that section or in preceding parts of the text.

Solve. (The symbol [calculator icon] indicates an exercise designed to be done using a calculator.)

111. [calculator icon] $4.23x - 17.898 = -1.65x - 42.454$

112. [calculator icon] $-0.00458y + 1.7787 = 13.002y - 1.005$

113. $\frac{3x}{2} + \frac{5x}{3} - \frac{13x}{6} - \frac{2}{3} = \frac{5}{6}$

114. $\frac{2x - 5}{6} + \frac{4 - 7x}{8} = \frac{10 + 6x}{3}$

115. $x - \{3x - [2x - (5x - (7x - 1))]\} = x + 7$

116. $23 - 2\{4 + 3(x - 1)\} + 5\{x - 2(x + 3)\} = 7\{x - 2[5 - (2x + 3)]\}$

1.2

OBJECTIVE

a Evaluate formulas and solve a formula for a specified letter.

Formulas and Applications

a EVALUATING AND SOLVING FORMULAS

A **formula** is an equation that represents or models a relationship between two or more quantities. For example, the relationship between the perimeter P of a square and the length s of its sides is given by the formula $P = 4s$. The formula $A = s^2$ represents the relationship between the area A of a square and the length s of its sides.

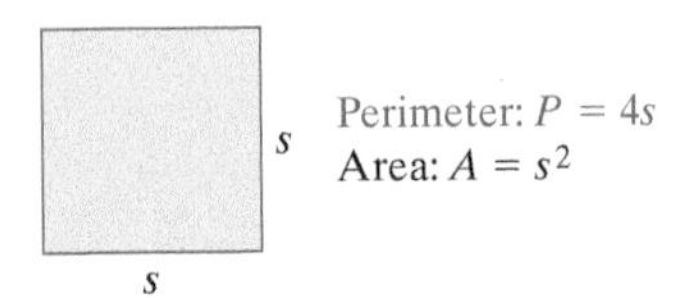

Other important geometric formulas are $A = \pi r^2$ (for the area A of a circle of radius r), $C = \pi d$ (for the circumference C of a circle of diameter d), and $A = b \cdot h$ (for the area A of a parallelogram of height h and base b). A more complete list of geometric formulas appears on the last page of this text.

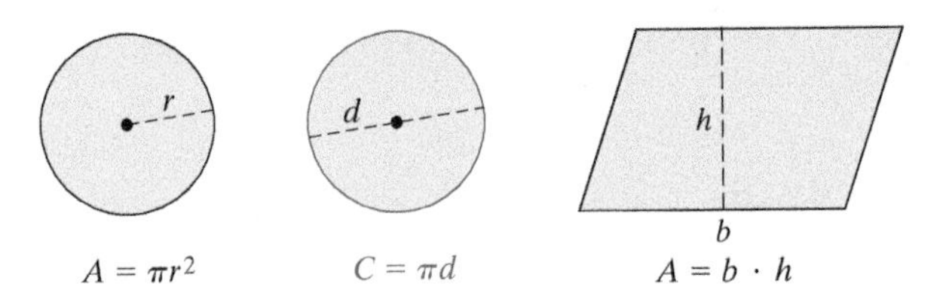

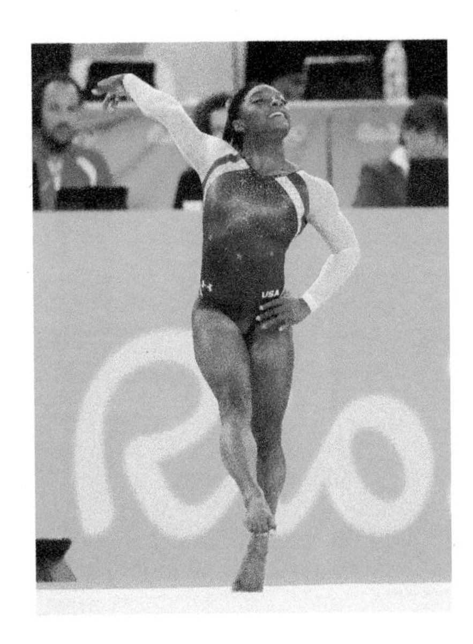

EXAMPLE 1 *Body Mass Index.* **Body mass index** I can be used to determine whether an individual has a healthy weight for his or her height. An index in the range 18.5–24.9 indicates a normal weight. Body mass index is given by the formula, or model,

$$I = \frac{703W}{H^2},$$

where W is weight, in pounds, and H is height, in inches.

Data: Centers for Disease Control and Prevention

a) Simone Biles of the gold-medal-winning 2016 U.S. Olympic gymnastics team is 4 ft 9 in. tall and weighs 103 lb. What is her body mass index?

b) Formula One driver Sebastian Vettel has a body mass index of 20.8 and a height of 5 ft 9 in. What is his weight?

a) We substitute 103 lb for W and 4 ft 9 in., or $4 \cdot 12 + 9 = 57$ in., for H. Then we have

$$I = \frac{703W}{H^2} = \frac{703(103)}{57^2} \approx 22.3.$$

Thus Simone Biles's body mass index is 22.3.

b) We substitute 20.8 for I and 5 ft 9 in., or $5 \cdot 12 + 9 = 69$ in., for H and solve for W using the equation-solving principles introduced in Section 1.1:

$$I = \frac{703W}{H^2}$$

$$20.8 = \frac{703W}{69^2} \qquad \text{Substituting}$$

$$20.8 = \frac{703W}{4761}$$

$$4761 \cdot 20.8 = 4761 \cdot \frac{703W}{4761} \qquad \text{Multiplying by 4761}$$

$$99{,}028.8 = 703W \qquad \text{Simplifying}$$

$$\frac{99{,}028.8}{703} = \frac{703W}{703} \qquad \text{Dividing by 703}$$

$$141 \approx W.$$

Sebastian Vettel weighs about 141 lb.

Do Exercise 1. ▶

If we want to make repeated calculations of W, as in Example 1(b), it might be easier to first solve for W, getting it alone on one side of the equation. We "solve" for W as we did above, using the equation-solving principles of Section 1.1.

EXAMPLE 2 Solve for W: $I = \dfrac{703W}{H^2}$.

$$I = \frac{703W}{H^2} \qquad \text{We want this letter alone.}$$

$$I \cdot H^2 = \frac{703W}{H^2} \cdot H^2 \qquad \text{Multiplying by } H^2 \text{ on both sides to clear the fraction}$$

$$IH^2 = 703W \qquad \text{Simplifying}$$

$$\frac{IH^2}{703} = \frac{703W}{703} \qquad \text{Dividing by 703}$$

$$\frac{IH^2}{703} = W$$

Do Exercise 2. ▶

1. *Body Mass Index.*
   a) Roland is 6 ft 1 in. tall and weighs 195 lb. What is his body mass index?
   b) Keisha has a body mass index of 24.5 and a height of 5 ft 8 in. What is her weight?
   c) Calculate your own body mass index.

2. Solve for m: $F = \dfrac{mv^2}{r}$.
   (This is a physics formula.)

Answers

1. (a) 25.7; **(b)** 161 lb; **(c)** Answers will vary.
2. $m = \dfrac{rF}{v^2}$

EXAMPLE 3 Solve for r: $H = 2r + 3m$.

$$H = 2r + 3m \quad \text{We want this letter alone.}$$
$$H - 3m = 2r \quad \text{Subtracting } 3m$$
$$\frac{H - 3m}{2} = r \quad \text{Dividing by 2}$$

◀ **Do Exercise 3.**

3. Solve for m: $H = 2r + 3m$.

EXAMPLE 4 Solve for b: $A = \frac{5}{2}(b - 20)$.

$$A = \tfrac{5}{2}(b - 20) \quad \text{We want this letter alone.}$$
$$2A = 5(b - 20) \quad \text{Multiplying by 2 to clear the fraction}$$
$$2A = 5b - 100 \quad \text{Removing parentheses}$$
$$2A + 100 = 5b \quad \text{Adding 100}$$
$$\frac{2A + 100}{5} = b, \quad \text{or} \quad b = \frac{2A}{5} + \frac{100}{5} = \frac{2A}{5} + 20 \quad \text{Dividing by 5}$$

◀ **Do Exercise 4.**

4. Solve for c: $P = \frac{3}{5}(c + 10)$. GS

$$P = \frac{3}{5}(c + 10)$$
$$\_\_\_ \cdot P = 5 \cdot \frac{3}{5}(c + 10)$$
$$5P = \_\_\_(c + 10)$$
$$5P = 3c + \_\_\_$$
$$5P - \_\_\_ = 3c$$
$$\frac{5P - 30}{\_\_\_} = c, \text{ or}$$
$$c = \frac{5}{3}P - \_\_\_$$

EXAMPLE 5 *Area of a Trapezoid.* Solve for a: $A = \frac{1}{2}h(a + b)$. (To find the area of a trapezoid, take half the product of the height, h, and the sum of the lengths of the parallel sides, a and b.)

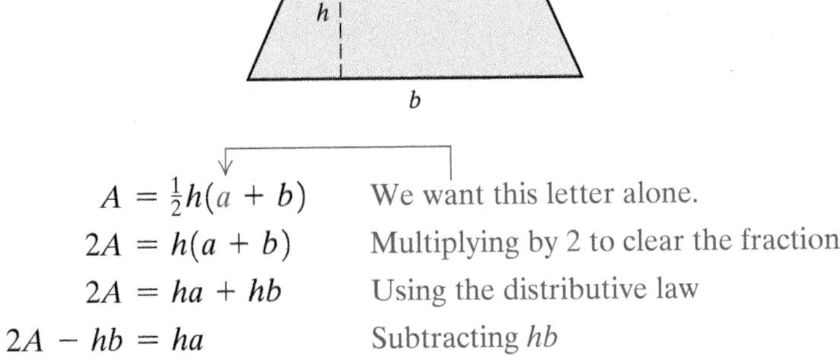

$$A = \tfrac{1}{2}h(a + b) \quad \text{We want this letter alone.}$$
$$2A = h(a + b) \quad \text{Multiplying by 2 to clear the fraction}$$
$$2A = ha + hb \quad \text{Using the distributive law}$$
$$2A - hb = ha \quad \text{Subtracting } hb$$
$$\frac{2A - hb}{h} = a, \quad \text{or} \quad a = \frac{2A}{h} - \frac{hb}{h} = \frac{2A}{h} - b \quad \text{Dividing by } h$$

Note that there is more than one correct form of the answers in Examples 4 and 5. This is a common occurrence when we solve formulas.

◀ **Do Exercise 5.**

5. Solve for b: $A = \frac{1}{2}h(a + b)$.

Answers

3. $m = \frac{H - 2r}{3}$ **4.** $c = \frac{5P - 30}{3}$, or $c = \frac{5}{3}P - 10$ **5.** $b = \frac{2A - ha}{h}$, or $b = \frac{2A}{h} - a$

Guided Solution:

4. 5, 3, 30, 30, 3, 10

We used the addition principle and the multiplication principle to solve equations in Section 1.1. In a similar manner, we use the same principles in this section to solve a formula for a given letter.

To solve a formula for a given letter, identify the letter, and:

1. Multiply on both sides to clear the fractions or decimals, if necessary.
2. If parentheses occur, multiply to remove them using the distributive law.
3. Collect like terms on each side, if necessary. This may require factoring if a variable is in more than one term.
4. Using the addition principle, get all terms with the letter to be solved for on one side of the equation and all other terms on the other side.
5. Collect like terms again, if necessary.
6. Solve for the letter in question using the multiplication principle.

As indicated in step (3) above, sometimes we must factor in order to isolate a letter.

EXAMPLE 6 *Simple Interest.* Solve for P: $A = P + Prt$. (To find the amount A in an account to which the principal P will grow at simple interest rate r, in t years, add the principal P to the interest, Prt.)

$$A = P + Prt \quad \text{We want this letter alone.}$$

$$A = P(1 + rt) \quad \text{Factoring (or collecting like terms)}$$

$$\frac{A}{1 + rt} = P \quad \text{Dividing by } 1 + rt \text{ on both sides}$$

Do Exercise 6. ▶

6. Solve for Q: $T = Q + Qvy$.

EXAMPLE 7 *Chess Ratings.* The formula

$$R = r + \frac{400(W - L)}{N}$$

is used to establish a chess player's rating R, after he or she has played N games, where W is the number of wins, L is the number of losses, and r is the average rating of the opponents.

Data: U.S. Chess Federation

a) Cara plays 8 games in a chess tournament, winning 5 games and losing 3 games. The average rating of her opponents is 1205. Find Cara's chess rating.

b) Solve the formula for L.

a) We substitute 8 for N, 5 for W, 3 for L, and 1205 for r in the formula. Then we calculate R:

$$R = r + \frac{400(W - L)}{N} = 1205 + \frac{400(5 - 3)}{8} = 1305.$$

Answer

6. $Q = \dfrac{T}{1 + vy}$

b) We solve as follows:

$$R = r + \frac{400(W - L)}{N}$$ We want the letter L alone.

$$NR = N\left[r + \frac{400(W - L)}{N}\right]$$ Multiplying by N to clear the fraction

$$NR = N \cdot r + N \cdot \frac{400(W - L)}{N}$$ Multiplying using the distributive law

$$NR = Nr + 400(W - L)$$ Simplifying

$$NR = Nr + 400W - 400L$$ Using the distributive law

$$NR - Nr - 400W = -400L$$ Subtracting Nr and $400W$

$$\frac{NR - Nr - 400W}{-400} = L.$$ Dividing by -400

Another correct form of the answer is $L = \dfrac{Nr + 400W - NR}{400}$.

◀ **Do Exercise 7.**

7. ***Chess Ratings.*** Use the formula given in Example 7.

a) Martin plays 6 games in a tournament, winning 2 games and losing 4. The average rating of his opponents is 1384. Find Martin's chess rating.

b) Solve the formula for W.

Answers

7. (a) About 1251; **(b)** $W = \dfrac{NR - Nr + 400L}{400}$

1.2 Exercise Set

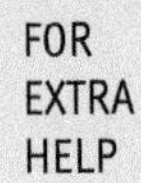

FOR EXTRA HELP

MyLab Math

✓ Check Your Understanding

Reading Check Choose from the list on the right the word that best completes each statement.

RC1. A formula is a(n) ______________ that represents a relationship between two or more quantities.

RC2. Body mass index is calculated using an individual's weight and ______________.

RC3. The formula $A = \pi r^2$ represents the relationship between a circle's radius and its ______________.

RC4. The formula $A = \frac{1}{2}h(a + b)$ represents the relationship between the area of a(n) ______________, its height, and the lengths of its bases.

area
circumference
equation
height
perimeter
trapezoid
triangle

Concept Check Choose from the column on the right the correct solution of each equation.

CC1. Solve $s = t + 4$ for t.

CC2. Solve $qs + 4r = t$ for q.

CC3. Solve $r = \dfrac{1}{4}q - t$ for t.

CC4. Solve $4q = 7r$ for q.

CC5. Solve $\dfrac{1}{4}s = t - q$ for s.

CC6. Solve $7r - t = 4s$ for r.

a) $q = \dfrac{7r}{4}$, or $\dfrac{7}{4}r$

b) $q = \dfrac{t - 4r}{s}$

c) $s = 4(t - q)$

d) $t = s - 4$

e) $r = \dfrac{4s + t}{7}$

f) $t = \dfrac{q - 4r}{4}$

a Solve for the given letter.

1. *Motion Formula:*

$d = rt$, for r

(Distance d, speed r, time t)

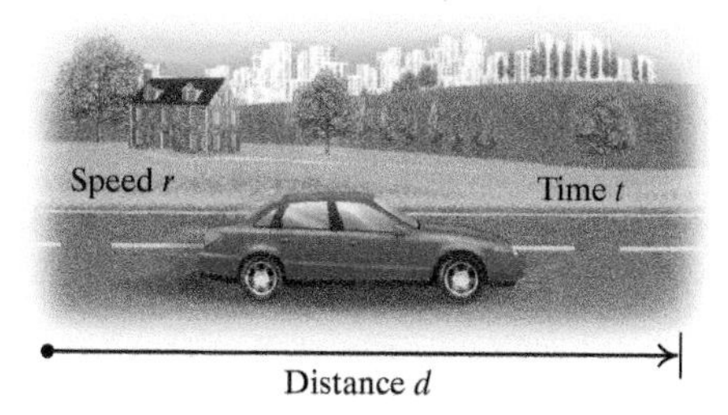

2. *Motion Formula:*

$d = rt$, for t

3. *Area of a Parallelogram:*

$A = bh$, for h

(Area A, base b, height h)

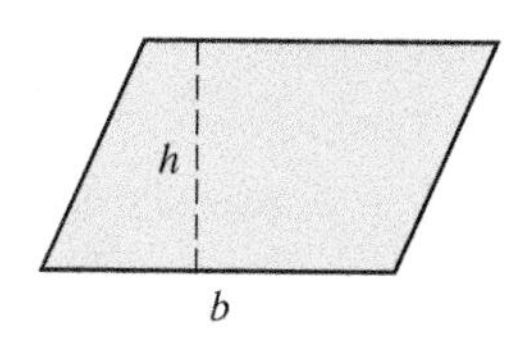

4. *Volume of a Sphere:*

$V = \frac{4}{3}\pi r^3$, for r^3

(Volume V, radius r)

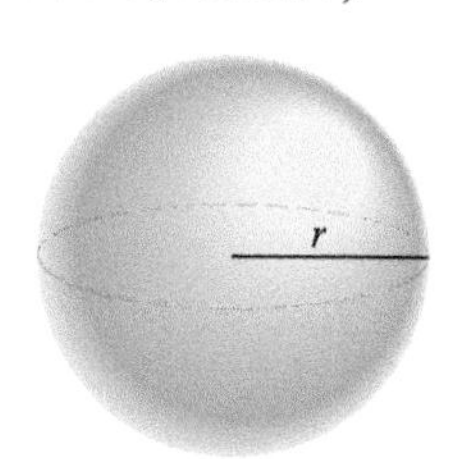

5. *Perimeter of a Rectangle:*

$P = 2l + 2w$, for w

(Perimeter P, length l, and width w)

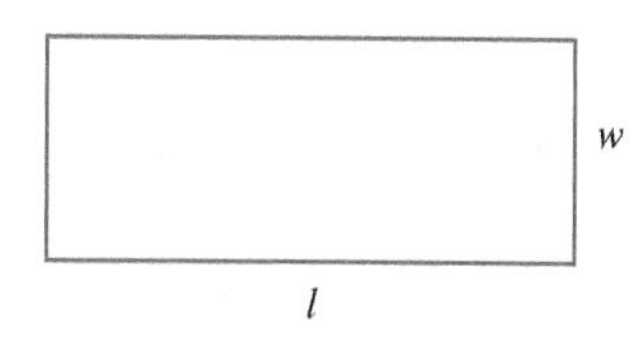

6. *Perimeter of a Rectangle:*

$P = 2l + 2w$, for l

7. *Area of a Triangle:*

$A = \frac{1}{2}bh$, for b

(Area A, base b, height h)

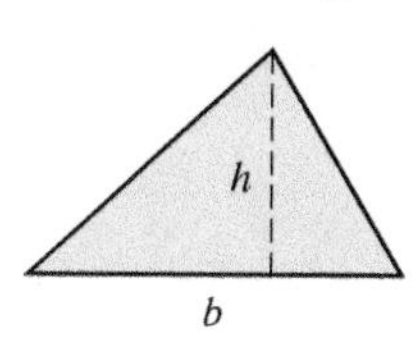

8. *Area of a Triangle:*

$A = \frac{1}{2}bh$, for h

9. *Average of Two Numbers:*

$A = \frac{a + b}{2}$, for a

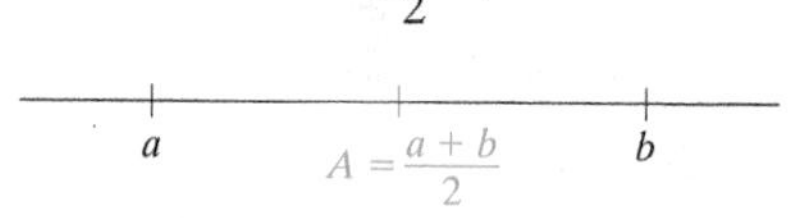

10. *Average of Two Numbers:*

$A = \frac{a + b}{2}$, for b

11. *Force:*

$F = ma$, for m

(Force F, mass m, acceleration a)

12. *Force:*

$F = ma$, for a

13. *Simple Interest:*

$I = Prt$, for t

(Interest I, principal P, interest rate r, time t)

14. *Simple Interest:*

$I = Prt$, for P

15. *Relativity:*

$E = mc^2$, for c^2

(Energy E, mass m, speed of light c)

16. $E = mc^2$, for m

17. $Q = \frac{p - q}{2}$, for p

18. $Q = \frac{p - q}{2}$, for q

19. $Ax + By = c$, for y

20. $Ax + By = c$, for x

21. $I = 1.08\frac{T}{N}$, for N

22. $F = \frac{mv^2}{r}$, for v^2

23. $C = \frac{3}{4}(m + 5)$, for m

24. $N = \frac{1}{3}M(t + w)$, for w

25. $n = \frac{1}{3}(a + b - c)$, for b

26. $t = \frac{1}{6}(x - y + z)$, for z

27. $d = R - Rst$, for R

28. $g = m + mnp$, for m

29. $T = B + Bqt$, for B

30. $Z = Q - Qab$, for Q

Basal Metabolic Rate. An individual's basal metabolic rate is the minimum number of calories required to sustain life when the individual is at rest. It can be thought of as the number of calories burned by an individual who sleeps all day. The Harris–Benedict formula for basal metabolic rate for a man is $R = 66 + 6.23w + 12.7h - 6.8a$. The formula for a woman is $R = 655 + 4.35w + 4.7h - 4.7a$. In each formula, R is in calories, w is weight, in pounds, h is height, in inches, and a is age, in years.

Data: Shapefit

31. **a)** Gary weighs 185 lb, is 5 ft 11 in. tall, and is 28 years old. Use the formula for the basal metabolic rate for a man to find Gary's basal metabolic rate.
b) Solve the formula for w.

32. **a)** Alyssa weighs 145 lb, is 5 ft 6 in. tall, and is 32 years old. Use the formula for the basal metabolic rate for a woman to find Alyssa's basal metabolic rate.
b) Solve the formula for h.

33. *Caloric Requirement.* The number of calories K required each day by a moderately active female who wants to maintain her weight is estimated by the formula

$$K = 1015.25 + 6.74w + 7.29h - 7.29a,$$

where w is weight, in pounds, h is height, in inches, and a is age, in years.

Data: Shapefit

a) Serena is a moderately active 25-year-old woman who weighs 150 lb and is 5 ft 8 in. tall. Find the number of calories she requires each day in order to maintain her weight.

b) Solve the formula for a.

34. *Caloric Requirement.* The number of calories K required each day by a moderately active male who wants to maintain his weight is estimated by the formula

$$K = 102.3 + 9.66w + 19.69h - 10.54a,$$

where w is weight, in pounds, h is height, in inches, and a is age, in years.

Data: Shapefit

a) Dan is a moderately active man who weighs 210 lb, is 6 ft 2 in. tall, and is 34 years old. Find the number of calories he requires each day in order to maintain his weight.

b) Solve the formula for a.

Projecting Birth Weight. Ultrasonic images of 29-week-old fetuses can be used to predict birth weight. One model, or formula, developed by Thurnau, is $P = 9.337da - 299$; a second model, developed by Weiner, is $P = 94.593c + 34.227a - 2134.616$. For the formulas, P is the estimated birth weight, in grams, d is the diameter of the fetal head, in centimeters, c is the circumference of the fetal head, in centimeters, and a is the circumference of the fetal abdomen, in centimeters.

Data: G. R. Thurnau, R. K. Tamura, R. E. Sabbagha, et al. *Am. J. Obstet Gynecol* 1983; **145**:557; C. P. Weiner, R. E. Sabbagha, N. Vaisrub, et al. *Obstet Gynecol* 1985; **65**:812.

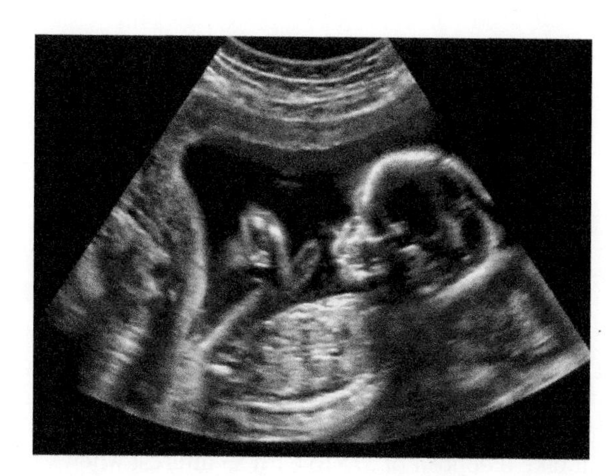

35. **a)** Use Thurnau's model to estimate the birth weight of a 29-week-old fetus when the diameter of the fetal head is 8.5 cm and the circumference of the fetal abdomen is 24.1 cm.

b) Solve the formula for a.

36. **a)** Use Weiner's model to estimate the birth weight of a 29-week-old fetus when the circumference of the fetal head is 26.7 cm and the circumference of the fetal abdomen is 24.1 cm.

b) Solve the formula for c.

37. *Young's Rule in Medicine.* Young's rule for determining the amount of a medicine dosage for a child is given by

$$c = \frac{ad}{a + 12},$$

where a is the child's age, in years, and d is the usual adult dosage, in milligrams. (*Warning*! Do not apply this formula without checking with a physician!)

Data: June Looby Olsen, et al., *Medical Dosage Calculations*, 6th ed. Reading, MA: Addison Wesley Longman, p. A-31

a) The usual adult dosage of a particular medication is 250 mg. Find the dosage for a child of age 3.

b) Solve the formula for d.

38. *Full-Time-Equivalent Students.* Colleges accommodate students who need to take different total-credit-hour loads. They determine the number of "full-time-equivalent" students, F, using the formula

$$F = \frac{n}{15},$$

where n is the total number of credits students enroll in for a given semester.

a) Determine the number of full-time-equivalent students on a campus in which students register for 42,690 credits.

b) Solve the formula for n.

Skill Maintenance

Divide. [J9]

39. $\frac{80}{-16}$

40. $-2000 \div (-8)$

41. $-\frac{1}{2} \div \frac{1}{4}$

42. $120 \div (-4.8)$

43. $-\frac{2}{3} \div \left(-\frac{5}{6}\right)$

44. $\frac{-90}{-15}$

45. $\frac{-90}{15}$

46. $\frac{-80}{16}$

Synthesis

Solve.

47. $A = \pi rs + \pi r^2$, for s

48. $s = v_1 t + \frac{1}{2}at^2$, for a; for v_1

49. $\frac{P_1V_1}{T_1} = \frac{P_2V_2}{T_2}$, for V_1; for P_2

50. $\frac{P_1V_1}{T_1} = \frac{P_2V_2}{T_2}$, for T_2; for P_1

51. In Exercise 13, you solved the formula $I = Prt$ for t. Now use the formula for t to determine how long it will take a deposit of \$75 to earn \$3 interest when invested at 5% simple interest.

52. The area of the shaded triangle ABE is 20 cm². Find the area of the trapezoid. (See Example 5.)

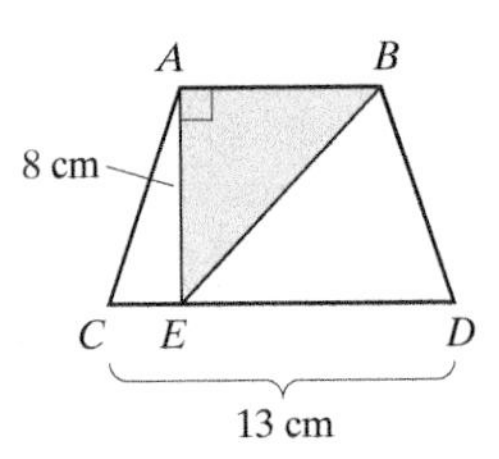

53. *Horsepower of an Engine.* The horsepower of an engine can be calculated by the formula

$$H = W\left(\frac{v}{234}\right)^3,$$

where W is the weight, in pounds, of the car, including the driver, fluids, and fuel, and v is the maximum velocity, or speed, in miles per hour, of the car attained a quarter mile after beginning acceleration.

a) Find the horsepower of a V-6, 2.8-liter engine if $W = 2700$ lb and $v = 83$ mph.

b) Find the horsepower of a 4-cylinder, 2.0-liter engine if $W = 3100$ lb and $v = 73$ mph.

Applications and Problem Solving

1.3

OBJECTIVES

a Solve applied problems by translating to equations.

b Solve basic motion problems.

a FIVE STEPS FOR PROBLEM SOLVING

Translate a phrase to an algebraic expression. [J12]

Translate each phrase to an algebraic expression.

1. Five times x **2.** 2 less than w

Answers: 1. $5x$ **2.** $w - 2$

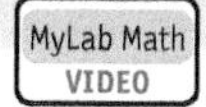

One very important use of algebra is as a tool for problem solving. The following five-step strategy for solving problems will be used throughout this text.

FIVE STEPS FOR PROBLEM SOLVING

1. *Familiarize* yourself with the problem situation.
2. *Translate* the problem to an equation.
3. *Solve* the equation.
4. *Check* the answer in the original problem.
5. *State* the answer to the problem clearly.

Of the five steps, probably the most important is the first one: becoming familiar with the problem situation. Here are some hints for familiarization.

To familiarize yourself with the problem:

- If a problem is given in words, read it carefully.
- List the information given and the question to be answered. Choose a variable (or variables) to represent the unknown(s) and clearly state what the variable represents. Be descriptive! For example, let L = length (in meters), d = distance (in miles), and so on.
- Make a drawing and label it with known information. Also, indicate unknown information, using specific units if given.
- Find further information if necessary. Look up a formula at the back of this book or in a reference book. Look up the topic using an Internet search engine.
- Make a table that lists all the information you have collected. Look for patterns that may help in the translation to an equation.
- Think of a possible answer and check the guess. Note the manner in which the guess is checked. This will help you translate the problem to an equation.

Long Hikes in the United States

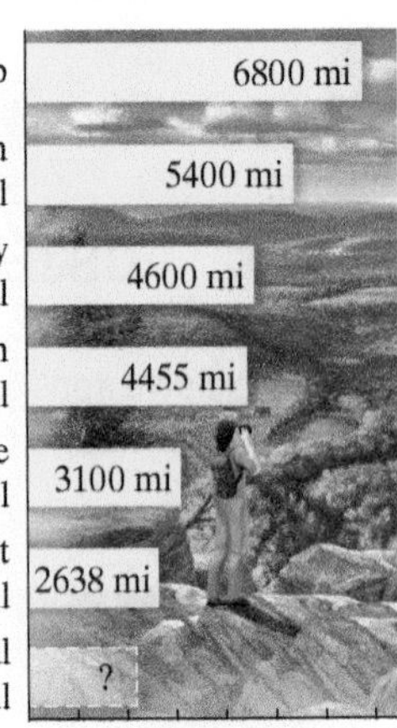

EXAMPLE 1 *Hiking.* At a length of 5400 mi, the Eastern Continental Trail is one of the longest hiking trails in the United States. This trail connects several trails, including the Appalachian National Scenic Trail, often called the Appalachian Trail. The Appalachian Trail alone is one of the 10 longest hiking trails in the United States. The Eastern Continental Trail is 1052 mi longer than twice the length of the Appalachian Trail. How long is the Appalachian Trail?

Data: adventurestrong.com

1. **Familiarize.** We can familiarize ourselves with the problem by estimating the length of the Appalachian Trail and checking our estimate. If the Appalachian Trail were 2000 mi long, then the length of the Eastern Continental Trail would be

$$2(2000) + 1052 = 4000 + 1052 = 5052 \text{ mi}.$$

Since the Eastern Continental Trail is longer than 5052, we see that our estimate of 2000 mi is too low. We let a = the length of the Appalachian Trail, in miles.

2. **Translate.** We translate as follows:

$$\underbrace{\text{Twice the length of the Appalachian Trail}}_{2 \cdot a} \;\; \underbrace{\text{plus}}_{+} \;\; \underbrace{1052 \text{ mi}}_{1052} \;\; \underbrace{\text{is}}_{=} \;\; \underbrace{5400 \text{ mi}}_{5400.}$$

3. **Solve.** We solve the equation as follows:

$$\begin{aligned} 2a + 1052 &= 5400 & \\ 2a + 1052 - 1052 &= 5400 - 1052 & &\text{Subtracting 1052} \\ 2a &= 4348 & &\text{Simplifying} \\ \frac{2a}{2} &= \frac{4348}{2} & &\text{Dividing by 2} \\ a &= 2174. & \end{aligned}$$

4. **Check.** If the Appalachian Trail is 2174 mi long, then the Eastern Continental Trail is $2(2174) + 1052$, or 5400, mi long. The length checks.

5. **State.** The Appalachian Trail is 2174 mi long.

◀ **Do Exercise 1.**

1. *Population Increase.* From 2015 to 2016, Texas gained 432,957 residents. This was 13,451 fewer than four times the number of residents gained by North Carolina during the same time period. How many residents did North Carolina gain?

Data: census.gov

EXAMPLE 2 *Solar Panel Support.* The cross section of a support for a solar energy panel is triangular. The second angle of the triangle is five times as large as the first angle. The third angle is 2° less than the first angle. Find the measures of the angles.

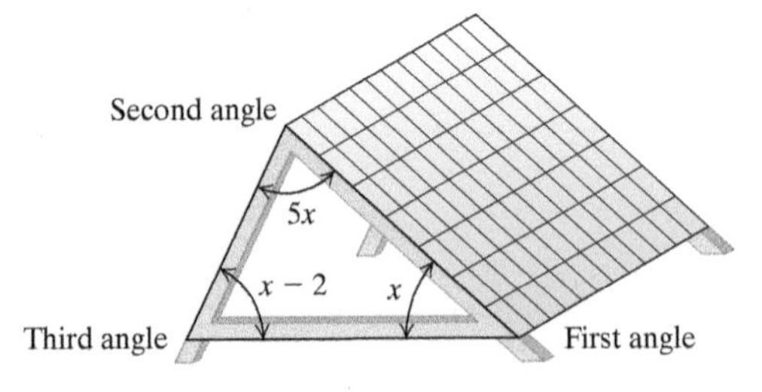

1. **Familiarize.** The second and third angles are described in terms of the first angle so we begin by assigning a variable to the first angle. Then we use that variable to describe the other two angles.

We let x = the measure of the first angle. Then $5x$ = the measure of the second angle and $x - 2$ = the measure of the third angle. Recall that the sum of the measures of the angles of a triangle is 180°.

2. **Translate.** We have the following translation:

$$\underbrace{\text{Measure of first angle}}_{x} \;\; \underbrace{\text{plus}}_{+} \;\; \underbrace{\text{Measure of second angle}}_{5x} \;\; \underbrace{\text{plus}}_{+} \;\; \underbrace{\text{Measure of third angle}}_{(x - 2)} \;\; \underbrace{\text{is}}_{=} \;\; \underbrace{180^\circ}_{180.}$$

Answer

1. 111,602 residents

3. **Solve.** We solve the equation as follows:

$$x + 5x + (x - 2) = 180$$
$$7x - 2 = 180 \quad \text{Collecting like terms}$$
$$7x - 2 + 2 = 180 + 2 \quad \text{Adding 2}$$
$$7x = 182 \quad \text{Simplifying}$$
$$\frac{7x}{7} = \frac{182}{7} \quad \text{Dividing by 7}$$
$$x = 26.$$

Thus the possible measures of the angles are

First angle: $x = 26°$;
Second angle: $5x = 5 \cdot 26 = 130°$;
Third angle: $x - 2 = 26 - 2 = 24°$.

4. **Check.** Do we have a solution of the *original* problem? The sum of the measures of the angles is

$$26° + 130° + 24° = 180°.$$

The measure of the second angle is five times the measure of the first angle: $130° = 5 \cdot 26°$. The measure of the third angle is 2° less than the measure of the first angle: $24° = 26° - 2°$. The answer checks.

5. **State.** The measure of the first angle is 26°, the measure of the second angle is 130°, and the measure of the third angle is 24°.

Do Exercise 2. ▶

2. *Cross Section of a Roof.* In a triangular cross section of a roof, the second angle is twice as large as the first. The third angle is 20° greater than the first angle. Find the measures of the angles.

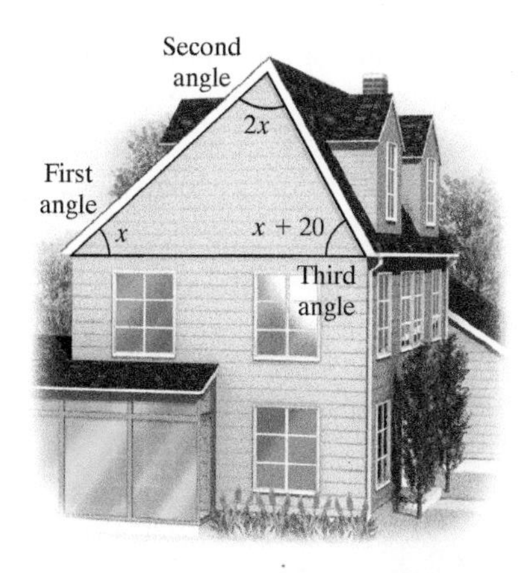

EXAMPLE 3 *Price of Auto Detailing.* Using a coupon from an online coupon service, Edward paid $152.75 for full-service, interior and exterior, detailing of the cab of his truck. He paid 35% less than the original price for this service. What was the original price?

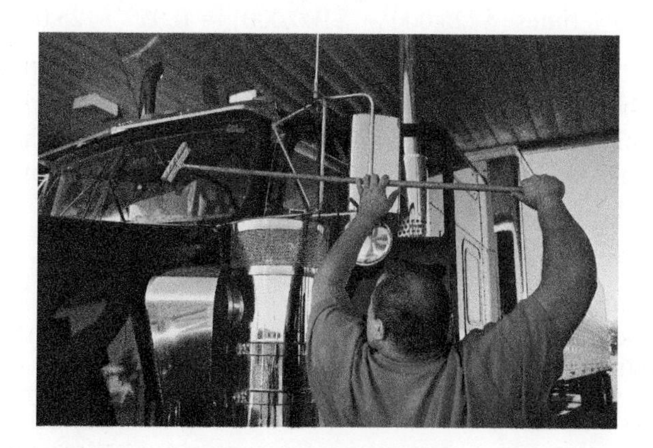

1. **Familiarize.** We let $x =$ the original price.
2. **Translate.**

Original price	minus	35%	of	Original price	is	New price	
↓	↓	↓	↓	↓	↓	↓	
x	$-$	35%	$\cdot$	x	$=$	152.75.	Translation

Answer
2. 40°, 80°, 60°

3. Solve. We solve the equation:

$$x - 35\% \cdot x = 152.75$$
$$1x - 0.35x = 152.75 \qquad x = 1x \text{ and } 35\% \cdot x = 0.35x$$
$$\left.\begin{aligned}(1 - 0.35)x &= 152.75\\ 0.65x &= 152.75\end{aligned}\right\} \qquad \text{Collecting like terms}$$
$$x = 235. \qquad \text{Dividing by } 0.65$$

4. Check. If the original price were \$235, we would have:

Price reduction: $35\% \cdot \$235 = 0.35 \cdot 235 = \$82.25$;

Reduced price: $\$235 - \$82.25 = \$152.75$.

We get the price that Edward paid, so the answer checks.

5. State. The original price of the full-service detailing of the cab of the truck was \$235.

3. *Price of Cross-Training Shoes.* Acton Sporting Goods lowers the price of a pair of cross-training shoes 20% to a sale price of \$68. What was the original price?

Do Exercise 3.

EXAMPLE 4 *Real Estate Commission.* The Mendozas negotiated to pay the realtor who sold their house the following commission:

6% for the first \$100,000 of the selling price, and

5.5% for the amount that exceeded \$100,000.

The realtor received a commission of \$10,565 for selling the house. What was the selling price?

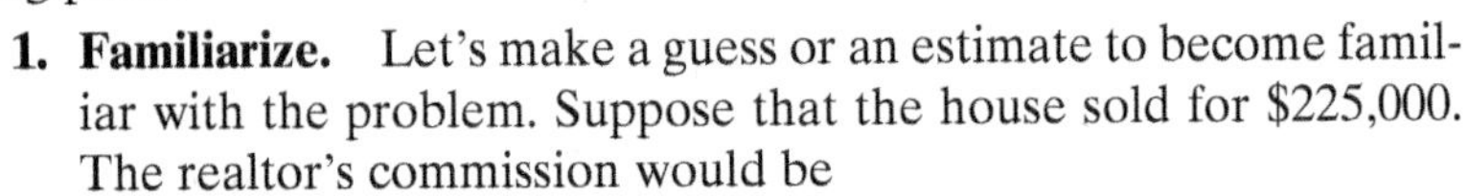

1. Familiarize. Let's make a guess or an estimate to become familiar with the problem. Suppose that the house sold for \$225,000. The realtor's commission would be

$$6\% \text{ of } \$100{,}000 = 0.06(\$100{,}000) = \$6000$$

plus

$$5.5\% \text{ times } (\$225{,}000 - \$100{,}000) = 0.055(\$125{,}000) = \$6875.$$

The total commission would be \$6000 + \$6875, or \$12,875. Although our guess is not correct, the calculation we performed familiarizes us with the problem, and it also tells us that the house sold for less than \$225,000, since \$10,565 is less than \$12,875. We let $S =$ the selling price of the house.

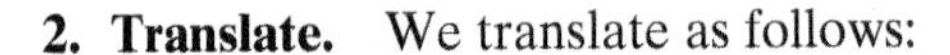

2. Translate. We translate as follows:

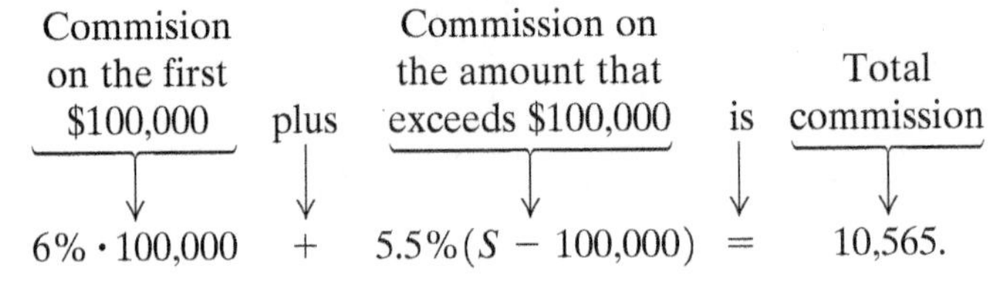

Commision on the first \$100,000	plus	Commission on the amount that exceeds \$100,000	is	Total commission
↓	↓	↓	↓	↓
$6\% \cdot 100{,}000$	$+$	$5.5\%(S - 100{,}000)$	$=$	$10{,}565.$

3. Solve. We solve the equation:

$$6\% \cdot 100{,}000 + 5.5\%(S - 100{,}000) = 10{,}565$$
$$0.06 \cdot 100{,}000 + 0.055(S - 100{,}000) = 10{,}565 \qquad \text{Converting to decimal notation}$$
$$6000 + 0.055S - 0.055 \cdot 100{,}000 = 10{,}565. \qquad \text{Simplifying and using the distributive law}$$

Answer

3. \$85

Then

$$\begin{aligned}
6000 + 0.055S - 5500 &= 10{,}565 && \text{Simplifying} \\
0.055S + 500 &= 10{,}565 && \text{Collecting like terms} \\
0.055S + 500 - 500 &= 10{,}565 - 500 && \text{Subtracting 500} \\
0.055S &= 10{,}065 \\
\frac{0.055S}{0.055} &= \frac{10{,}065}{0.055} && \text{Dividing by 0.055} \\
S &= \$183{,}000.
\end{aligned}$$

4. **Check.** Performing the check is similar to the sample calculation in the *Familiarize* step. The check is left to the student.
5. **State.** The selling price of the house was $183,000.

Do Exercise 4. ▶

4. *Real Estate Commission.* The Currys negotiated to pay the realtor who sold their house the following commission:

   7% for the first $100,000 of the selling price

   and

   5% for the amount that exceeded $100,000.

   The realtor received a commission of $13,400 for selling the house. What was the selling price?

EXAMPLE 5 *Insurance Premiums.* The mortality rate for smokers is much higher than it is for those who do not smoke. With all other factors being equal, the smoker can expect to pay a higher life-insurance premium. The equation

$$y = 8.33x + 32.81$$

can be used to estimate the monthly premium for a $250,000 term life insurance policy for a female smoker age 40 or older, where x is the issue age—that is, $x = 0$ corresponds to issue age 40, $x = 3$ corresponds to issue age 43, and so on.

Data: American General Life Insurance Company

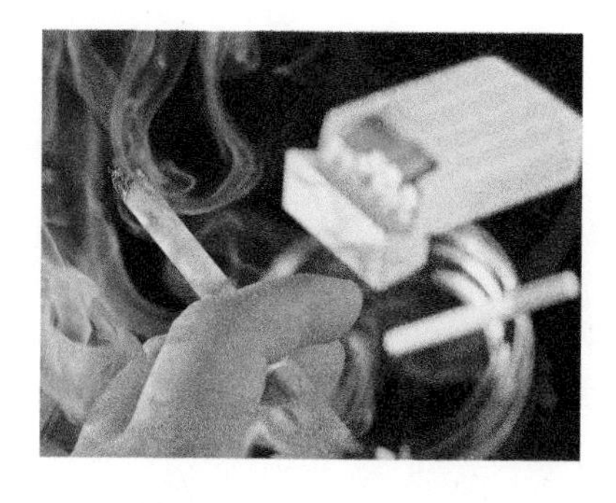

a) Estimate the monthly insurance premium for a female smoker who is 48 years old.

b) At what issue age would the monthly premium be approximately $167?

Since an equation is given, we know that we have a correct translation and thus we will not use the five-step problem-solving strategy in this case.

a) To estimate the monthly premium for a female smoker whose age is 48, we first note that 48 is 8 years beyond 40. We substitute 8 for x in the equation:

$$y = 8.33x + 32.81 = 8.33(8) + 32.81 = 99.45.$$

The estimated monthly insurance premium for a female smoker who is 48 years old when the policy is issued is $99.45.

b) To determine the issue age for a monthly premium of $167, we substitute 167 for y in the equation and solve for x:

$$\begin{aligned}
y &= 8.33x + 32.81 \\
167 &= 8.33x + 32.81 && \text{Substituting 167 for } y \\
167 - 32.81 &= 8.33x + 32.81 - 32.81 && \text{Subtracting 32.81} \\
134.19 &= 8.33x \\
\frac{134.19}{8.33} &= \frac{8.33x}{8.33} && \text{Dividing by 8.33} \\
16 &\approx x.
\end{aligned}$$

It is estimated that about 16 years after age 40, or at issue age 56, the monthly premium for a $250,000 term life insurance policy for a female smoker is $167.

Do Exercise 5. ▶

5. *Insurance Premiums.* Refer to Example 5.

   a) Estimate the monthly insurance premium for a female smoker who is 51 years old.

   b) At what issue age would the monthly premium be approximately $190?

Answers

4. $228,000 **5. (a)** $124.44; **(b)** about issue age 59

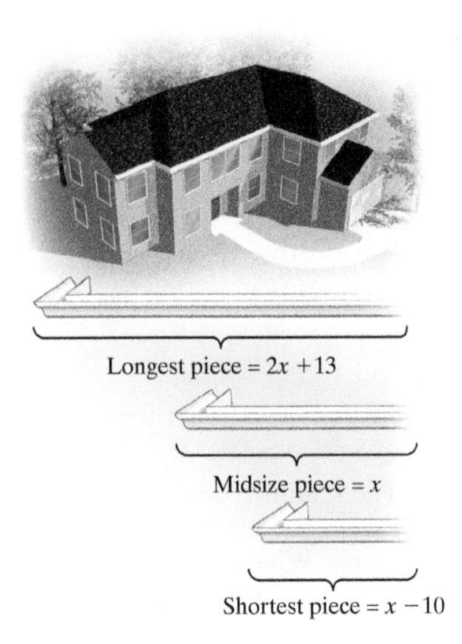

EXAMPLE 6 *Installing Seamless Guttering.* Seamless guttering is delivered on a continuous roll and sections are cut from the roll as needed. The Jordans know that they need 127 ft of guttering for six separate sections of their home and that the four shortest sections will be the same size. The longest piece of guttering is 13 ft more than twice the length of the midsize piece. The shortest piece of guttering is 10 ft less than the midsize piece. How long is each piece of guttering?

1. **Familiarize.** All the pieces are described in terms of the midsize piece, so we begin by assigning a variable to that piece and using that variable to describe the lengths of the other pieces.

   We let x = the length of the midsize piece, in feet, $2x + 13$ = the length of the longest piece, and $x - 10$ = the length of the shortest piece.
2. **Translate.** The sum of the length of the longest piece, plus the length of the midsize piece, plus four times the length of the shortest piece is 127 ft. This gives us the following translation:

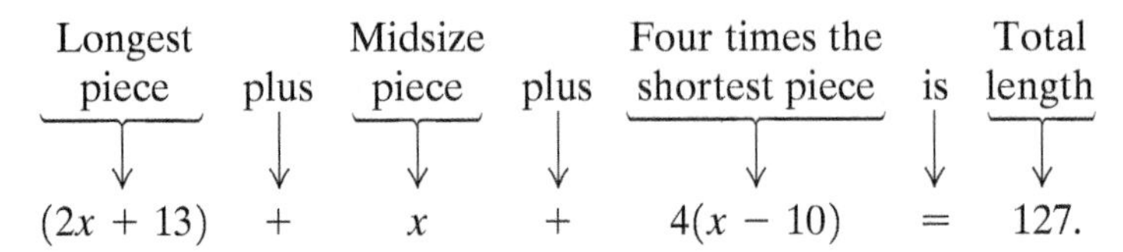

3. **Solve.** We solve the equation, as follows:

$$(2x + 13) + x + 4(x - 10) = 127$$
$$2x + 13 + x + 4x - 40 = 127 \quad \text{Using the distributive law}$$
$$7x - 27 = 127 \quad \text{Collecting like terms}$$
$$7x - 27 + 27 = 127 + 27 \quad \text{Adding 27}$$
$$7x = 154 \quad \text{Collecting like terms}$$
$$\frac{7x}{7} = \frac{154}{7} \quad \text{Dividing by 7}$$
$$x = 22.$$

4. **Check.** Do we have an answer to the *problem*? If the length of the midsize piece is 22 ft, then the length of the longest piece is

   $2 \cdot 22 + 13$, or 57 ft,

   and the length of the shortest piece is

   $22 - 10$, or 12 ft.

   The sum of the lengths of the longest piece, the midsize piece, and four times the shortest piece must be 127 ft:

   57 ft + 22 ft + 4(12 ft) = 127 ft.

   These lengths check.
5. **State.** The length of the longest piece is 57 ft, the length of the midsize piece is 22 ft, and the length of the shortest piece is 12 ft.

Do Exercise 6.

6. *Cutting a Board.* A 106-in. board is cut into three pieces. The shortest length is two-thirds of the midsize length and the longest length is 15 in. less than two times the midsize length. Find the length of each piece.

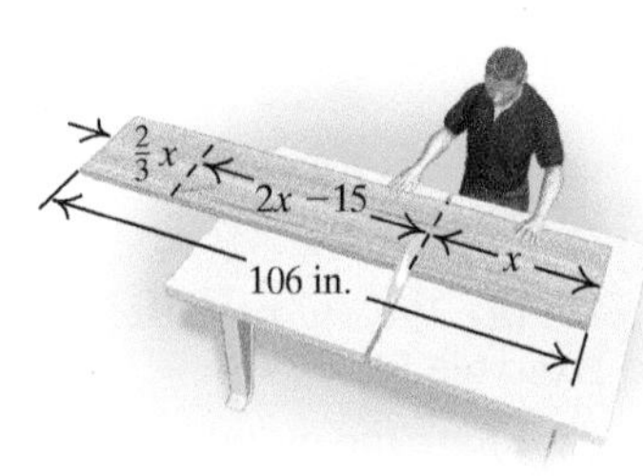

Answer
6. 22 in., 33 in., 51 in.

Sometimes applied problems involve **consecutive integers** like 19, 20, 21, 22 or $-34, -33, -32, -31$. Consecutive integers can be represented in the form $x, x + 1, x + 2, x + 3$, and so on.

Some examples of **consecutive even integers** are 20, 22, 24, 26 and $-34, -32, -30, -28$. Consecutive even integers can be represented in the form $x, x + 2, x + 4, x + 6$, and so on, as can **consecutive odd integers** like 19, 21, 23, 25 and $-33, -31, -29, -27$.

EXAMPLE 7 *Artist's Prints.* Often artists will number in sequence a limited number of prints in order to increase their value. An artist creates 500 prints and saves three for her children. The numbers of those prints are consecutive integers whose sum is 189. Find the number of each of those prints.

1. **Familiarize.** The numbers of the prints are consecutive integers. Thus we let $x =$ the first integer, $x + 1 =$ the second, and $x + 2 =$ the third.
2. **Translate.** We translate as follows:

$$\underbrace{\text{First integer}}_{x} + \underbrace{\text{Second integer}}_{(x+1)} + \underbrace{\text{Third integer}}_{(x+2)} = 189$$

$$x + (x + 1) + (x + 2) = 189.$$

3. **Solve.** We solve the equation:

$$\begin{aligned} x + (x + 1) + (x + 2) &= 189 \\ 3x + 3 &= 189 && \text{Collecting like terms} \\ 3x + 3 - 3 &= 189 - 3 && \text{Subtracting 3} \\ 3x &= 186 \\ \frac{3x}{3} &= \frac{186}{3} && \text{Dividing by 3} \\ x &= 62. \end{aligned}$$

Then $x + 1 = 62 + 1 = 63$ and $x + 2 = 62 + 2 = 64$.

4. **Check.** The numbers 62, 63, and 64 are consecutive integers, and their sum is 189. The numbers check.
5. **State.** The numbers of the prints are 62, 63, and 64.

Do Exercise 7. ▶

7. *Artist's Prints.* Refer to Example 7. The artist saves three other prints for her own archives. These are also numbered consecutively. The sum of the numbers is 1266. Find the numbers of each of those prints.

b BASIC MOTION PROBLEMS

When a problem deals with speed, distance, and time, we can expect to use the following **motion formula**.

THE MOTION FORMULA

Distance = Rate (or speed) · Time

$d = rt$

Answer

7. 421, 422, 423

EXAMPLE 8 *Moving Walkways.* A moving walkway in O'Hare Airport is 300 ft long and moves at a speed of 5 ft/sec. If Kate walks at a speed of 4 ft/sec, how long will it take her to travel the distance of 300 ft using the moving walkway?

1. **Familiarize.** First read the problem carefully. You might want to talk about it with a classmate, reword it, or sketch the situation described as illustrated at left. Organizing the information in a table can also be helpful.

Distance to be traveled	300 ft
Kate's walking speed	4 ft/sec
Speed of the moving walkway	5 ft/sec
Kate's total speed on the walkway	?
Time required	?

Since Kate is walking on the walkway in the same direction in which it is moving, the two speeds can be added to determine Kate's total speed on the walkway. We can then complete the table, letting t = the time, in seconds, required to travel 300 ft on the moving walkway.

Distance to be traveled	300 ft
Kate's walking speed	4 ft/sec
Speed of the moving walkway	5 ft/sec
Kate's total speed on the walkway	9 ft/sec
Time required	t

2. **Translate.** To translate, we use the motion formula $d = rt$, where d = distance, r = speed, or rate, and t = time. We substitute 300 ft for d and 9 ft/sec for r:

$$d = rt$$
$$300 = 9 \cdot t.$$

3. **Solve.** We solve the equation:

$$300 = 9t$$
$$\frac{300}{9} = \frac{9t}{9} \quad \text{Dividing by 9}$$
$$\frac{100}{3} = t.$$

4. **Check.** At a speed of 9 ft/sec and in a time of 100/3, or $33\frac{1}{3}$ sec, Kate would travel $d = 9 \cdot \frac{100}{3} = 300$ ft. This answer checks.

5. **State.** Kate will travel the distance of 300 ft in 100/3, or $33\frac{1}{3}$ sec.

Do Exercise 8.

8. *Marine Travel.* Tim's fishing boat travels 12 km/h in still water. How long will it take him to travel 25 km upstream if the river's current is 3 km/h? 25 km downstream if the river's current is 3 km/h? (*Hint*: To find the boat's speed traveling upstream, subtract the speed of the current from the speed of the boat in still water. To find the boat's speed downstream, add the speed of the current to the speed of the boat in still water.)

Answer

8. $2\frac{7}{9}$ hr; $1\frac{2}{3}$ hr

1.3 Exercise Set

FOR EXTRA HELP 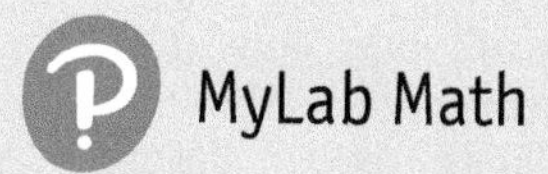

✓ Check Your Understanding

Reading Check Choose from the column on the right the word that completes each step in the five steps for problem solving.

RC1. ____________ yourself with the problem situation.

RC2. ____________ the problem to an equation.

RC3. ____________ the equation.

RC4. ____________ the answer in the original problem.

RC5. ____________ the answer to the problem clearly.

Solve
Familiarize
State
Translate
Check

Concept Check Translate each statement to an equation. Do not solve.

CC1. In a 400-mi car race, the lead car is four times as far from the start as it is from the finish. Let x represent the distance of the lead car from the finish, in miles.

CC2. The second angle of the triangle is 3° more than the first, and the third angle is twice the first. Let a represent the measure of the first angle, in degrees.

CC3. The price of a book, including a 6% sales tax, was $17.49. Let p represent the original marked price of the book, in dollars.

CC4. The perimeter of a rectangle is 100 ft, and the length is 12 ft longer than the width. Let w represent the width of the rectangle, in feet.

a Solve.

1. *Diana Nyad.* On September 2, 2013, Diana Nyad became the first person to swim from Havana, Cuba, to Key West, Florida, across the Florida Straits. She swam the 110-mi distance in approximately 53 hr. At 11.00 P.M. on Sunday, September 1, she was approximately three times as far from Cuba as she was from Florida. How far was she from the Florida coast?

Data: ChicagoTribune.com, September 3, 2013

2. *Ironman Triathlon.* Held annually in Hawaii since 1978, the Ironman Triathlon championship is a series of long-distance races consisting of a 2.4-mi swim, a 112-mi bicycle ride, and a 26.2-mi marathon. At one point, a participant had completed twice as many miles as the number of miles left to complete. How many miles had he completed at that mark?

Data: ironman.com

3. *Landscape Design.* The residents of a downtown neighborhood designed a triangular-shaped park as part of a city beautification program. The park is bound by streets on all sides. The second angle of the triangle is 7° more than the first. The third angle is 7° less than twice the first. Find the measures of the angles.

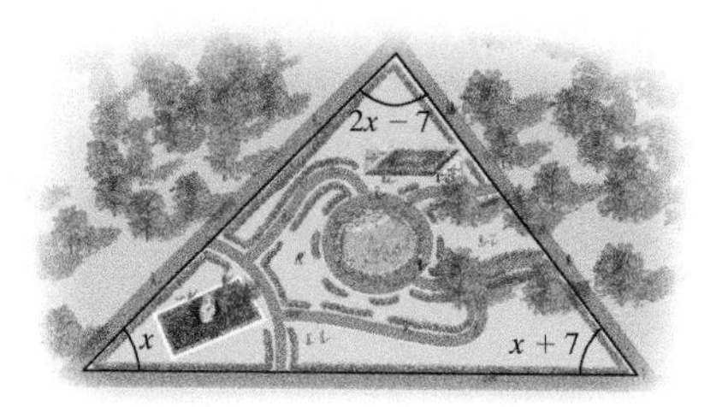

4. *Angles of a Triangle.* The second angle of a triangle is three times as large as the first. The measure of the third angle is 25° greater than that of the first angle. How large are the angles?

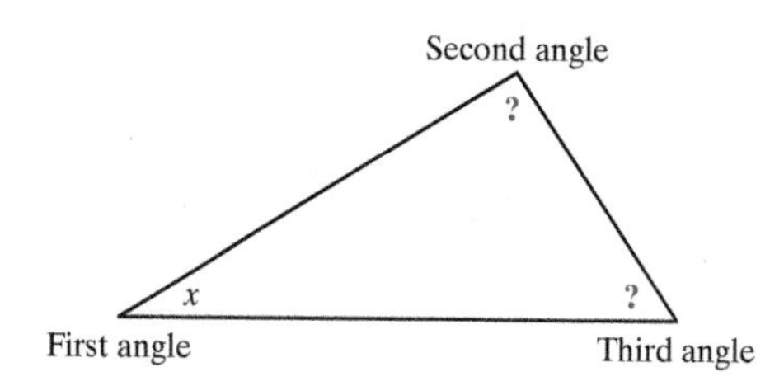

5. *Gorillas.* With a total population in the wild of only a little over 100,000, all species of gorillas are considered endangered. There are approximately 17,000 Eastern gorillas living in the wild. This is 1000 less than two-ninths of the number of Western gorillas living in the Wild. How many Western gorillas are living in the wild?

Data: allaboutwildlife.com

6. *Climber Deaths.* Deaths of Mt. Everest climbers were first recorded in 1921. From that time through June 2016, there have been over 280 deaths. The total number of deaths of native Nepalis is 114. This is 30 more than one-half of the number of foreigners who died. How many foreigners died climbing Mount Everest?

7. *Sale Price.* A premium drawing-pencil set of 144 pieces was on sale online at 30% off for a 24-hr period. Abby purchased the set while it was on sale for $176.40. What was the original price?

8. *Sales Tax.* Gareth purchased a fitness watch online for $64.15. This amount included a 7% sales tax. What was the price of the watch itself?

9. *Perimeter of an NBA Court.* The perimeter of an NBA-sized basketball court is 288 ft. The length is 44 ft longer than the width. Find the dimensions of the court.

Data: National Basketball Association

10. *Perimeter of a Tennis Court.* The width of a standard tennis court used for playing doubles is 42 ft less than the length. The perimeter of the court is 228 ft. Find the dimensions of the court.

Data: *Dunlop Illustrated Encyclopedia of Facts*

11. *Rope Cutting.* A rope that is 168 ft long is to be cut into three pieces such that the second piece is 6 ft less than three times the first, and the third is 2 ft more than two-thirds of the second. Find the length of the longest piece.

12. *Wire Cutting.* A piece of wire that is 100 cm long is to be cut into two pieces, each to be bent to make a square. The length of a side of one square is to be twice the length of a side of the other. How should the wire be cut?

13. *Real Estate Commission.* The Carlsons negotiated the following real estate commission on the selling price of their house:

    7% for the first $100,000, and

    5% for the amount that exceeds $100,000.

    The realtor received a commission of $15,250 for selling the house. What was the selling price?

14. *Real Estate Commission.* The Youngs negotiated the following real estate commission on the selling price of their house:

    8% for the first $100,000, and

    3% for the amount that exceeds $100,000.

    The realtor received a commission of $9200 for selling the house. What was the selling price?

15. *Consecutive Odd Integers.* Find three consecutive odd integers such that the sum of the first, two times the second, and three times the third is 70.

16. *Consecutive Even Integers.* Find three consecutive even integers such that the sum of the first, five times the second, and four times the third is 1226.

17. *Interstate Mile Markers.* U.S. interstate highways post numbered markers every mile to indicate location in case of an accident or break-down. In many states, the numbers increase from west to east. The sum of two consecutive mile markers on I-80 in Iowa is 459. Find the numbers on the markers.

18. *Post-Office Box Numbers.* The sum of the numbers on two adjacent post-office boxes is 697. What are the numbers?

19. *School Photos.* Memory Makers prices its school photos as shown here.

    ☆☆☆ **Memory Makers** ☆☆☆

BASIC PACKAGE	PRICE
1 8×10 2 5×7 12 wallet-size	$14.95
1 sheet of 6 extra wallet-size	$1.35

    The Morris family purchases the basic package for each of its three children, along with extra wallet-size photos. How many wallet-size photos did they buy in all if their total bill for the photos is $57?

20. *Carpet Cleaning.* A1 Carpet Cleaners charges $75 to clean the first 200 sq ft of carpet. There is an additional charge of 25¢ per square foot for any footage that exceeds 200 sq ft and $1.40 per step for any carpeting on a staircase. A customer's cleaning bill was $253.95. This included the cleaning of a staircase with 13 steps. In addition to the staircase, how many square feet of carpet did the customer have cleaned?

21. *Original Salary.* An editorial assistant receives an 8% raise, bringing her salary to $42,066. What was her salary before the raise?

22. *Original Salary.* After a salesman receives a 5% raise, his new salary is $40,530. What was his old salary?

23. *Diabetes.* In 2014, there were 22 million Americans diagnosed with diabetes. This number of cases represents an increase of 300% over the number of cases in 1980. How many cases of diabetes were there in 1980?

Data: Centers for Disease Control and Prevention

24. *College Tuition.* The average cost of tuition for one year at a private university was $38,760 in 2015. This represented a 179% increase over the average tuition for one year at a private university in 1995. How much did tuition cost, on average, at a private university in 1995?

Data: *U.S. News and World Report*

25. *Hiking.* The number of young adult hikers in the United States has steadily increased since 2010. The equation

$$y = 0.25x + 3.68$$

can be used to estimate the number of young adults participating in hiking in the United States, in millions, x years after 2010.

Data: statista.com

a) Estimate the number of young adult hikers in the United States in 2018.

b) In what year will 6.5 million young adults participate in hiking in the United States?

26. *Insurance Premiums.* The mortality rate for nonsmokers is much lower than it is for those who do smoke. With all other factors being equal, the nonsmoker can expect to pay a lower insurance premium. The equation

$$y = 2.06x + 10.08$$

can be used to estimate the monthly premium for a $250,000 term life insurance policy for a female nonsmoker age 40 or older, where x is the issue age—that is, $x = 0$ corresponds to issue age 40, $x = 3$ corresponds to issue age 43, and so on.

Data: American General Life Insurance Company

a) Estimate the monthly insurance premium for a female nonsmoker who is 50 years old.

b) At what issue age would the monthly premium be approximately $52?

b Solve.

27. *Cruising Altitude.* A Boeing 767 has been instructed to climb from its present altitude of 8000 ft to a cruising altitude of 29,000 ft. The plane ascends at a rate of 3500 ft/min. How long will it take the plane to reach the cruising altitude?

28. *Air Travel.* A pilot has been instructed to descend from an altitude of 26,000 ft to 11,000 ft. If the pilot descends at a rate of 2500 ft/min, how long will it take the plane to reach the new altitude?

29. *Boating.* Jen's motorboat travels at a speed of 10 mph in still water. Booth River flows at a speed of 2 mph. How long will it take Jen to travel 15 mi downstream? 15 mi upstream?

30. *Flight into a Headwind.* An airplane traveling 390 mph in still air encounters a 65-mph headwind. How long will it take the plane to travel 725 mi into the wind?

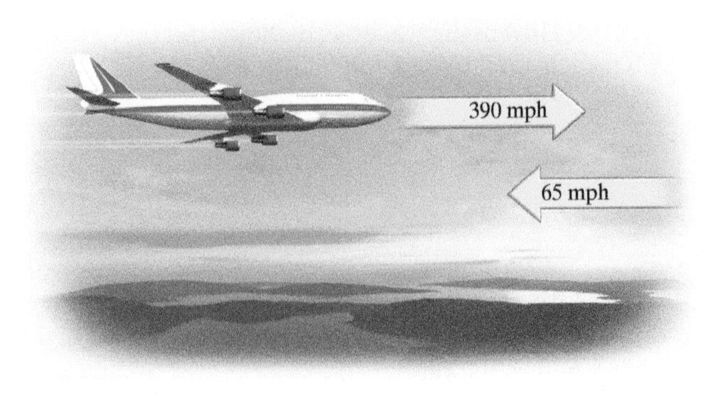

31. *Swimming.* Fran swims at a speed of 5 mph in still water. The Lazy River flows at a speed of 2.3 mph. How long will it take Fran to swim 1.8 mi upstream? 1.8 mi downstream?

32. *River Cruising.* Now being used as a floating hotel and restaurant in Chattanooga, Tennessee, the *Delta Queen* is a sternwheel steamboat that once cruised the Mississippi River system. It was not uncommon for the *Delta Queen* to travel at a speed of 7 mph in still water and for the Mississippi to flow at a speed of 3 mph. At these rates, how long did it take the boat to cruise 2 mi upstream?

Data: *Delta Queen* information; *The Natchez Democrat*, February 12, 2009

Skill Maintenance

Simplify. [J11]

33. $5^2 - 2 \cdot 5 \cdot 12 + 12^2$

34. $16 \cdot 8 + 200 \div 25 \cdot 10$

35. $\dfrac{12|8 - 10| + 9 \cdot 6}{5^4 + 4^5}$

36. $\dfrac{(9 - 4)^2 + (8 - 11)^2}{4^2 + 2^2}$

Synthesis

37. *Real Estate Prices.* Home prices in Panduski increased 1% from 2015 to 2016. Prices dropped 3% from 2016 to 2017 and dropped another 7% from 2017 to 2018. If a house sold for $105,000 in 2018, what was it worth in 2015? (Round to the nearest dollar.)

38. *Adjusted Wages.* Christina's salary is reduced $n\%$ during a period of financial difficulty. By what number should her salary be multiplied in order to bring it back to where it was before the reduction?

39. *Population Change.* The yearly changes in the population census of Poplarville for three consecutive years are, respectively, a 20% increase, a 30% increase, and a 20% decrease. What is the total percent change from the beginning of the first year to the end of the third year, to the nearest percent?

40. *Watch Time.* Your watch loses $1\frac{1}{2}$ sec every hour. You have a friend whose watch gains 1 sec every hour. The watches show the same time now. After how many more seconds will they show the same time again?

41. *Geometry.* Consider the geometric figure below. Suppose that $L \| M$, $m\angle 8 = 5x + 25$, and $m\angle 4 = 8x + 4$. Find $m\angle 2$ and $m\angle 1$. (*Note:* "$m\angle 1$" means "measure of angle 1.")

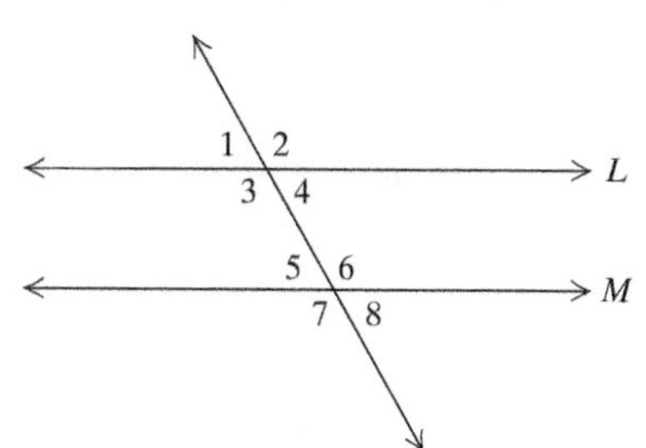

42. *Geometry.* Suppose that the figure $ABCD$ shown below is a square. Point A is folded onto the midpoint of $\overline{AB}$ and point D is folded onto the midpoint of $\overline{DC}$. The perimeter of the smaller figure formed is 25 in. Find the area of the square $ABCD$.

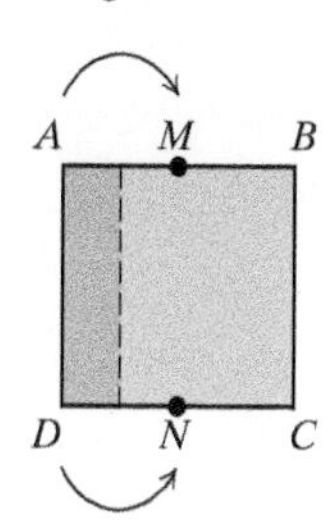

Mid-Chapter Review

Concept Reinforcement

Determine whether each statement is true or false.

______________ **1.** $2x + 3 = 7$ and $x = 2$ are equivalent equations. [1.1a]

______________ **2.** It is possible for an equation to be false. [1.1a]

______________ **3.** Every equation has at least one solution. [1.1d]

______________ **4.** When we solve an applied problem, we check the possible solution in the equation to which the problem was translated. [1.3a]

Guided Solutions

GS Fill in each box with the number or the expression that creates a correct statement or solution.

5. Solve: $2x - 5 = 1 - 4x$. [1.1d]

$$2x - 5 = 1 - 4x$$
$$2x - 5 + 4x = 1 - 4x + \square$$
$$6x - 5 = \square \quad \text{Collecting like terms}$$
$$6x - 5 + \square = 1 + 5$$
$$6x = \square \quad \text{Collecting like terms}$$
$$\frac{6x}{6} = \frac{6}{\square}$$
$$x = \square \quad \text{Simplifying}$$

6. Solve for y: $Mx + Ny = T$. [1.2a]

$$Mx + Ny = T$$
$$Mx + Ny - Mx = T - \square$$
$$\square = T - Mx$$
$$y = \frac{T - Mx}{\square}$$

Mixed Review

Determine whether the given number is a solution of the given equation. [1.1a]

7. 7; $x + 5 = 12$

8. $\frac{1}{3}$; $3x - 4 = 5$

9. -24; $\frac{-x}{8} = -3$

10. 9; $6(x - 3) = 36$

Solve. [1.1b, c, d]

11. $x - 7 = -10$

12. $-7x = 56$

13. $8x - 9 = 23$

14. $1 - x = 3x - 7$

15. $2 - 4y = -4y + 2$

16. $\frac{3}{4}y + 2 = \frac{7}{2}$

17. $5t - 9 = 7t - 4$

18. $4x - 11 = 11 + 4x$

19. $2(y - 4) = 8y$

20. $4y - (y - 1) = 16$

21. $t - 3(t - 4) = 9$

22. $6(2x + 3) = 10 - (4x - 5)$

Solve for the given letter. [1.2a]

23. $P = mn$, for n

24. $z = 3t + 3w$, for t

25. $N = \frac{r + s}{4}$, for s

26. $T = 1.5\frac{A}{B}$, for B

27. $H = \frac{2}{3}(t - 5)$, for t

28. $f = g + ghm$, for g

Solve. [1.3a, b]

29. *Female Medical School Graduates.* The number of female medical school graduates in the United States totaled 8784 in 2016. This number was an increase of 10.8% from the total in 2006. What was the number of female medical school graduates in 2006?

Data: Association of American Medical Colleges

30. *Calories Burned While Walking.* A person weighing 154 lb burns approximately 230 calories while walking 4.5 mph for 30 min. This number of calories is 50 calories less than twice the number burned by a 154-lb person while walking 3.5 mph for 30 min. How many calories would a 154-lb person burn while walking 3.5 mph for 30 min?

Data: Centers for Disease Control and Prevention

31. *Carpet Dimensions.* The width of an Oriental carpet is 2 ft less than the length. The perimeter of the carpet is 24 ft. Find the dimensions of the carpet.

32. *Boating.* Frederick's boat travels at a speed of 9 mph in still water. The Bailey River flows at a speed of 3 mph. How long will it take Frederick to travel 18 mi downstream? 18 mi upstream?

Understanding Through Discussion and Writing

To the student and the instructor: The Discussion and Writing exercises are meant to be answered with one or more sentences. They can be discussed and answered collaboratively by the entire class or by small groups.

33. Explain the difference between equivalent expressions and equivalent equations. [1.1a]

34. Devise an application in which it would be useful to solve the motion formula $d = rt$ for r. [1.2a]

35. The equations

$$P = 2l + 2w \quad \text{and} \quad w = \frac{P}{2} - l$$

are equivalent formulas involving the perimeter P, length l, and width w of a rectangle. Devise a problem for which the second of the two formulas would be more useful. [1.2a]

36. Explain why we can use the addition principle to subtract the same number on both sides of an equation and why we can use the multiplication principle to divide by the same nonzero number on both sides of an equation. [1.1b, c]

37. How can a guess or an estimate help prepare you for the *Translate* step when solving problems? [1.3a]

38. Why is it important to label clearly what a variable represents in an applied problem? [1.3a]

STUDYING FOR SUCCESS *Using the Textbook*

- ☐ Study the step-by-step solutions to the examples, paying attention to the extra explanations shown in red.
- ☐ Stop and do the margin exercises when directed to do so. Answers are at the bottom of the margin, so you can immediately check your work.
- ☐ The objective symbols a, b, c, and so on, allow you to refer to the appropriate place in the text whenever you need to review a topic.

1.4 Sets, Inequalities, and Interval Notation

OBJECTIVES

- **a** Determine whether a given number is a solution of an inequality.
- **b** Write interval notation for the solution set or the graph of an inequality.
- **c** Solve an inequality using the addition principle and the multiplication principle and then graph the inequality.
- **d** Solve applied problems by translating to inequalities.

a INEQUALITIES

We can extend our equation-solving skills to the solving of inequalities.

INEQUALITY

An **inequality** is a sentence containing $<, >, \leq, \geq$, or $\neq$.

Some examples of inequalities are

$-2 < a, \quad x > 4, \quad x + 3 \leq 6,$

$6 - 7y \geq 10y - 4, \quad$ and $\quad 5x \neq 10.$

SOLUTION OF AN INEQUALITY

Any replacement or value for the variable that makes an inequality true is called a **solution** of the inequality. The set of all solutions is called the **solution set**. When all the solutions of an inequality have been found, we say that we have **solved** the inequality.

EXAMPLES Determine whether the given number is a solution of the inequality.

1. $x + 3 < 6;\ 5$

We substitute 5 for x and get $5 + 3 < 6$, or $8 < 6$, a *false* sentence. Therefore, 5 is not a solution.

2. $2x - 3 > -3;\ 1$

We substitute 1 for x and get $2(1) - 3 > -3$, or $-1 > -3$, a *true* sentence. Therefore, 1 is a solution.

3. $4x - 1 \leq 3x + 2;\ -3$

We substitute -3 for x and get $4(-3) - 1 \leq 3(-3) + 2$, or $-13 \leq -7$, a *true* sentence. Therefore, -3 is a solution.

◀ **Do Exercises 1–3.**

Determine whether the given number is a solution of the inequality.

1. $3 - x < 2;\ 8$

2. $3x + 2 > -1;\ -2$

3. $3x + 2 \leq 4x - 3;\ 5$

Answers

1. Yes **2.** No **3.** Yes

b INEQUALITIES AND INTERVAL NOTATION

Graph an inequality on the number line. [J3]

Graph each inequality.

1. $x > -2$
2. $x \leq 1$

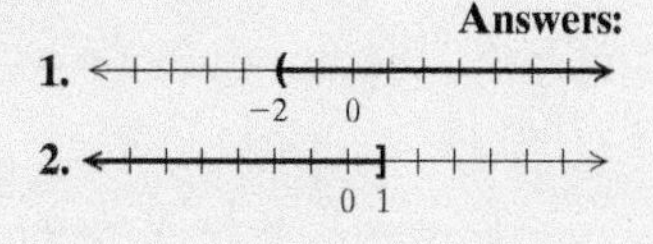

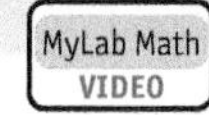

MyLab Math ANIMATION

The solutions of an inequality can be represented graphically. The solution set can be written using both *set-builder notation* and *interval notation.*

The **graph** of an inequality in one variable is a drawing on the number line that represents its solutions. For example, to graph $x < 4$, we note that the solutions are all real numbers less than 4. We shade all numbers less than 4 on the number line and indicate that 4 is not a solution by using a right parenthesis ")" at 4.

$\{x|x < 4\}$, or $(-\infty, 4)$

We write the solution set for $x < 4$ using **set-builder notation** as $\{x|x < 4\}$. This is read "The set of all x such that x is less than 4." We write the solution set for $x < 4$ using **interval notation** as $(-\infty, 4)$. Interval notation uses parentheses ( ) and brackets [ ] and closely parallels the graph of an inequality on the number line.

If a and b are real numbers such that $a < b$, we define the **open interval** (a, b) as the set of all numbers between but not including a and b—that is, the set of all x for which $a < x < b$. Thus,

$$(a, b) = \{x|a < x < b\}.$$

The points a and b are the **endpoints** of the interval. The parentheses indicate that the endpoints are *not* included in the graph.

The **closed interval** $[a, b]$ is defined as the set of all numbers x for which $a \leq x \leq b$. Thus,

$$[a, b] = \{x|a \leq x \leq b\}.$$

The brackets indicate that the endpoints *are* included in the graph.*

The following **half-open intervals** include one endpoint and exclude the other:

$(a, b] = \{x|a < x \leq b\}$. The graph excludes a and includes b.

$[a, b) = \{x|a \leq x < b\}$. The graph includes a and excludes b.

Caution!

The *interval* (a, b) uses the same notation as the *ordered pair* (a, b), which denotes a point in the plane, as we will see in Chapter 2. The context in which the notation appears usually makes the meaning clear.

*Some books use the representations (open dots at a and b) and (solid dots at a and b) instead of, respectively, (a, b) and $[a, b]$.

Some intervals extend without bound in one or both directions. We use the symbols ∞, read "infinity," and $-\infty$, read "negative infinity," to name these intervals. The notation (a, ∞) represents the set of all numbers greater than a—that is,

$$(a, \infty) = \{x|x > a\}.$$

(a, ∞)

a

Similarly, the notation $(-\infty, a)$ represents the set of all numbers less than a—that is,

$$(-\infty, a) = \{x|x < a\}.$$

$(-\infty, a)$

a

The notations $[a, \infty)$ and $(-\infty, a]$ are used when we want to include the endpoint a. The interval $(-\infty, \infty)$ names the set of all real numbers.

$(-\infty, \infty) = \{x \mid x \text{ is a real number}\}$

Interval notation is summarized in the following table.

Intervals: Notation and Graphs

INTERVAL NOTATION	SET NOTATION	GRAPH
(a, b)	$\{x\|a < x < b\}$	a b
$[a, b]$	$\{x\|a \le x \le b\}$	a b
$[a, b)$	$\{x\|a \le x < b\}$	a b
$(a, b]$	$\{x\|a < x \le b\}$	a b
(a, ∞)	$\{x\|x > a\}$	a
$[a, \infty)$	$\{x\|x \ge a\}$	a
$(-\infty, b)$	$\{x\|x < b\}$	b
$(-\infty, b]$	$\{x\|x \le b\}$	b
$(-\infty, \infty)$	$\{x\|x \text{ is a real number}\}$	

Caution!

Whenever the symbol ∞ is included in interval notation, a right parenthesis ")" is used. Similarly, when $-\infty$ is included, a left parenthesis "(" is used.

EXAMPLES Write interval notation for the given set.

4. $\{x|-4 < x < 5\} = (-4, 5)$

5. $\{x|x \ge -2\} = [-2, \infty)$

6. $\{x|7 > x \ge 1\} = \{x|1 \le x < 7\} = [1, 7)$

EXAMPLES Write interval notation for the given graph.

7. −6 −5 −4 −3 −2 −1 0 1 2 3 4 5 6

$(-2, 4]$

8. −6 −5 −4 −3 −2 −1 0 1 2 3 4 5 6

$(-\infty, -1)$

Do Exercises 4–8. ▶

c SOLVING INEQUALITIES

Two inequalities are **equivalent** if they have the same solution set. For example, the inequalities $x > 4$ and $4 < x$ are equivalent. Just as the addition principle for equations gives us equivalent equations, the addition principle for inequalities gives us equivalent inequalities.

THE ADDITION PRINCIPLE FOR INEQUALITIES

For any real numbers a, b, and c:

$a < b$ is equivalent to $a + c < b + c$;

$a > b$ is equivalent to $a + c > b + c$.

Similar statements hold for $\leq$ and $\geq$.

Since subtracting c is the same as adding $-c$, there is no need for a separate subtraction principle.

EXAMPLE 9 Solve and graph: $x + 5 > 1$.

We have

$$x + 5 > 1$$

$$x + 5 - 5 > 1 - 5 \quad \text{Using the addition principle: adding } -5 \text{ or subtracting } 5$$

$$x > -4.$$

We used the addition principle to show that the inequalities $x + 5 > 1$ and $x > -4$ are equivalent. The solution set is $\{x|x > -4\}$ and consists of an infinite number of solutions. We cannot possibly check them all. Instead, we can perform a partial check by substituting one member of the solution set (here we use -1) into the original inequality:

$$\begin{array}{c|l} x + 5 > 1 & \\ \hline -1 + 5 \ ? \ 1 & \\ 4 & \text{TRUE} \end{array}$$

Since $4 > 1$ is true, we have a partial check. The solution set is $\{x|x > -4\}$, or $(-4, \infty)$. The graph is as follows:

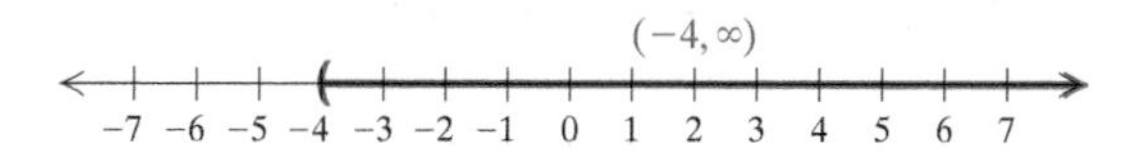

Do Exercises 9 and 10. ▶

Write interval notation for the given set or graph.

4. $\{x|-4 \leq x < 5\}$

5. $\{x|x \leq -2\}$

6. $\{x|6 \geq x > 2\}$

7.

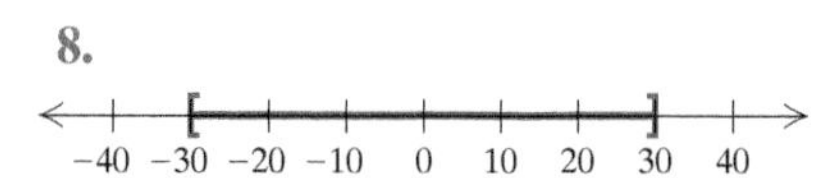

8.

−40 −30 −20 −10 0 10 20 30 40

Solve and graph.

9. $x + 6 > 9$

−5 −4 −3 −2 −1 0 1 2 3 4 5

10. $x + 4 \leq 7$

−5 −4 −3 −2 −1 0 1 2 3 4 5

Answers

4. $[-4, 5)$ **5.** $(-\infty, -2]$ **6.** $(2, 6]$
7. $[10, \infty)$ **8.** $[-30, 30]$
9. $\{x|x > 3\}$, or $(3, \infty)$;

0 3

10. $\{x|x \leq 3\}$, or $(-\infty, 3]$;

0 3

EXAMPLE 10 Solve and graph: $4x - 1 \geq 5x - 2$.

We have

$$
\begin{aligned}
4x - 1 &\geq 5x - 2 \\
4x - 1 + 2 &\geq 5x - 2 + 2 && \text{Adding 2} \\
4x + 1 &\geq 5x && \text{Simplifying} \\
4x + 1 - 4x &\geq 5x - 4x && \text{Subtracting } 4x \\
1 &\geq x. && \text{Simplifying}
\end{aligned}
$$

The inequalities $1 \geq x$ and $x \leq 1$ have the same meaning and the same solutions. The solution set is $\{x | 1 \geq x\}$ or, more commonly, $\{x | x \leq 1\}$. Using interval notation, we write that the solution set is $(-\infty, 1]$. The graph is as follows:

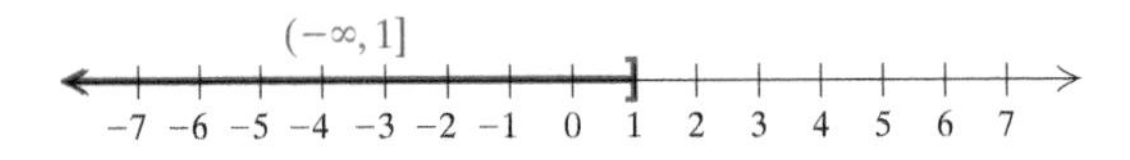

◀ **Do Exercise 11.**

11. Solve and graph:

$2x - 3 \geq 3x - 1$.

GS

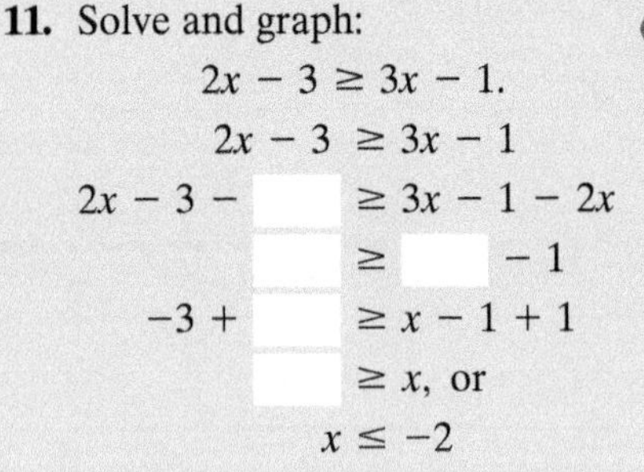

The solution set is $\{x | x \leq -2\}$, or $(-\infty, \quad]$.

−5 −4 −3 −2 −1 0 1 2 3 4 5

The multiplication principle for inequalities differs from the multiplication principle for equations. Consider the true inequality

$-4 < 9$.

If we multiply both numbers by 2, we get another true inequality:

$-4(2) < 9(2)$, or $-8 < 18$. True

If we multiply both numbers by -3, we get a false inequality:

$-4(-3) < 9(-3)$, or $12 < -27$. False

However, if we now *reverse* the inequality symbol above, we get a true inequality:

$12 > -27$. True

THE MULTIPLICATION PRINCIPLE FOR INEQUALITIES

For any real numbers a and b, and any *positive* number c:

$a < b$ is equivalent to $ac < bc$;
$a > b$ is equivalent to $ac > bc$.

For any real numbers a and b, and any *negative* number c:

$a < b$ is equivalent to $ac > bc$;
$a > b$ is equivalent to $ac < bc$.

Similar statements hold for $\leq$ and $\geq$.

Since division by c is the same as multiplication by $1/c$, there is no need for a separate division principle.

The multiplication principle tells us that when we multiply or divide on both sides of an inequality by a negative number, we must reverse the inequality symbol to obtain an equivalent inequality.

Answer

11. $\{x | x \leq -2\}$, or $(-\infty, -2]$;

−2 0

Guided Solution:

11. $2x, -3, x, 1, -2, -2$

EXAMPLE 11 Solve and graph: $3y < \frac{3}{4}$.

We have

$$3y < \frac{3}{4}$$

$$\frac{1}{3} \cdot 3y < \frac{1}{3} \cdot \frac{3}{4} \quad \text{Multiplying by } \tfrac{1}{3}. \text{ Since } \tfrac{1}{3} > 0, \text{ the symbol stays the same.}$$

$$y < \frac{1}{4}. \quad \text{Simplifying}$$

Any number less than $\frac{1}{4}$ is a solution. The solution set is $\{y \mid y < \frac{1}{4}\}$, or $(-\infty, \frac{1}{4})$. The graph is as follows:

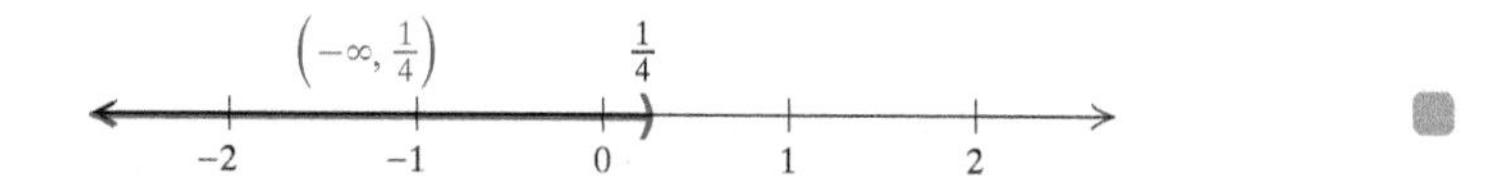

EXAMPLE 12 Solve and graph: $-5x \geq -80$.

We have

$$-5x \geq -80$$

$$\frac{-5x}{-5} \leq \frac{-80}{-5} \quad \text{Dividing by } -5. \text{ Since } -5 < 0, \text{ the inequality symbol must be reversed.}$$

$$x \leq 16.$$

The solution set is $\{x \mid x \leq 16\}$, or $(-\infty, 16]$. The graph is as follows:

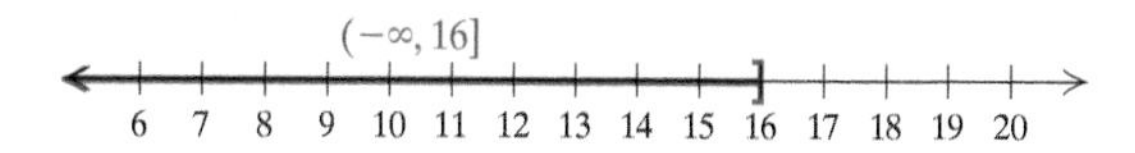

Do Exercises 12–14.

Solve and graph.

12. $5y \leq \frac{3}{2}$

−5 −4 −3 −2 −1 0 1 2 3 4 5

13. $-2y > 10$

−5 −4 −3 −2 −1 0 1 2 3 4 5

14. $-\frac{1}{3}x \leq -4$

0 4 8 12 16 20

We use the addition and multiplication principles together in solving inequalities in much the same way as in solving equations.

EXAMPLE 13 Solve: $16 - 7y \geq 10y - 4$.

We have

$$16 - 7y \geq 10y - 4$$

$$-16 + 16 - 7y \geq -16 + 10y - 4 \quad \text{Adding } -16$$

$$-7y \geq 10y - 20 \quad \text{Collecting like terms}$$

$$-10y + (-7y) \geq -10y + 10y - 20 \quad \text{Adding } -10y$$

$$-17y \geq -20 \quad \text{Collecting like terms}$$

$$\frac{-17y}{-17} \leq \frac{-20}{-17} \quad \text{Dividing by } -17. \text{ The symbol must be reversed.}$$

$$y \leq \frac{20}{17}. \quad \text{Simplifying}$$

The solution set is $\{y \mid y \leq \frac{20}{17}\}$ or $(-\infty, \frac{20}{17}]$.

We can avoid multiplying or dividing by a negative number by using the addition principle in a different way. Let's rework Example 13 by adding $7y$ instead of $-10y$.

Answers

12. $\left\{y \mid y \leq \frac{3}{10}\right\}$, or $\left(-\infty, \frac{3}{10}\right]$

13. $\{y \mid y < -5\}$, or $(-\infty, -5)$

14. $\{x \mid x \geq 12\}$, or $[12, \infty)$

$$16 - 7y \geq 10y - 4$$
$$16 - 7y + 7y \geq 10y - 4 + 7y \quad \text{Adding } 7y. \text{ This makes the coefficient of the } y\text{-term positive.}$$
$$16 \geq 17y - 4 \quad \text{Collecting like terms}$$
$$16 + 4 \geq 17y - 4 + 4 \quad \text{Adding 4}$$
$$20 \geq 17y \quad \text{Collecting like terms}$$
$$\frac{20}{17} \geq \frac{17y}{17} \quad \text{Dividing by 17. The symbol stays the same.}$$
$$\frac{20}{17} \geq y, \text{ or } y \leq \frac{20}{17}$$

Solve.

15. $6 - 5y \geq 7$

$6 - 5y - 6 \geq 7 - \square$

$\square \geq 1$

$\frac{-5y}{-5} \ \square \ \frac{1}{\square}$

$\square \leq -\frac{1}{5}$

The solution set is

$\left\{y \mid y \leq -\frac{1}{5}\right\}$, or $\left(\square, -\frac{1}{5}\right]$.

GS

16. $3x + 5x < 4$

17. $17 - 5(y - 2) \leq 45y + 8(2y - 3) - 39y$

EXAMPLE 14 Solve: $-3(x + 8) - 5x > 4x - 9$.

$$-3(x + 8) - 5x > 4x - 9$$
$$-3x - 24 - 5x > 4x - 9 \quad \text{Using the distributive law}$$
$$-24 - 8x > 4x - 9 \quad \text{Collecting like terms}$$
$$-24 - 8x + 8x > 4x - 9 + 8x \quad \text{Adding } 8x$$
$$-24 > 12x - 9 \quad \text{Collecting like terms}$$
$$-24 + 9 > 12x - 9 + 9 \quad \text{Adding 9}$$
$$-15 > 12x$$
$$\frac{-15}{12} > \frac{12x}{12} \quad \text{Dividing by 12. The symbol stays the same.}$$
$$-\frac{5}{4} > x.$$

The solution set is $\{x \mid -\frac{5}{4} > x\}$, or $\{x \mid x < -\frac{5}{4}\}$, or $(-\infty, -\frac{5}{4})$.

◀ Do Exercises 15–17.

d APPLICATIONS AND PROBLEM SOLVING

Many problem-solving and applied situations translate to inequalities.

IMPORTANT WORDS	SAMPLE SENTENCE	TRANSLATION
is at least	Max is at least 5 years old.	$m \geq 5$
is at most	At most 6 people could fit in the elevator.	$n \leq 6$
cannot exceed	Total weight in the elevator cannot exceed 2000 pounds.	$w \leq 2000$
must exceed	The speed must exceed 15 mph.	$s > 15$
is between	Heather's income is between \$23,000 and \$35,000.	$23{,}000 < h < 35{,}000$
no more than	Bing weighs no more than 90 pounds.	$w \leq 90$
no less than	Saul would accept no less than \$4000 for the piano.	$p \geq 4000$

Answers

15. $\left\{y \mid y \leq -\frac{1}{5}\right\}$, or $\left(-\infty, -\frac{1}{5}\right]$
16. $\left\{x \mid x < \frac{1}{2}\right\}$, or $\left(-\infty, \frac{1}{2}\right)$
17. $\left\{y \mid y \geq \frac{17}{9}\right\}$, or $\left[\frac{17}{9}, \infty\right)$

Guided Solution:
15. $6, -5y, \leq, -5, y, -\infty$

The following phrases deserve special attention.

TRANSLATING "AT LEAST" AND "AT MOST"

A quantity x is **at least** some amount q: $x \geq q$.
(If x is at least q, it cannot be less than q.)

A quantity x is **at most** some amount q: $x \leq q$.
(If x is at most q, it cannot be more than q.)

Do Exercises 18–24. ▶

EXAMPLE 15 *Web Developers.* As a result of the increase in online activity, the employment demand for web developers is expected to increase 27% from 2014 to 2024. The equation

$$P = 3950t + 148{,}500$$

can be used to estimate the number of web developers in the work force, where t is the number of years after 2014. Determine the years for which the number of web developers will be greater than 200,000.
Data: U.S. Department of Labor

1. **Familiarize.** We already have an equation. To become more familiar with it, we might make a substitution for t. Suppose that we want to know the number of web developers 8 years after 2014, or in 2022. We substitute 8 for t:

$$P = 3950(8) + 148{,}500 = 180{,}100.$$

We see that in 2022, the number of web developers will be fewer than 200,000. To find the years in which the number of web developers exceeds 200,000, we could make guesses greater than 8, but it is more efficient to proceed to the next step.

2. **Translate.** The number of web developers is to be more than 200,000. Thus we have

$$P > 200{,}000.$$

We replace P with $3950t + 148{,}500$:

$$3950t + 148{,}500 > 200{,}000.$$

3. **Solve.** We solve the inequality:

$$3950t + 148{,}500 > 200{,}000$$
$$3950t > 51{,}500 \quad \text{Subtracting 148,500}$$
$$t > 13.04. \quad \text{Dividing by 3950 and rounding}$$

4. **Check.** As a partial check, we can substitute a value for t that is greater than 13.04, say 14: $P = 3950(14) + 148{,}500 = 203{,}800$.

5. **State.** The number of web developers will be greater than 200,000 for years more than 13.04 years after 2014, so we have $\{t \mid t > 13.04\}$.

Do Exercise 25. ▶

Translate.

18. Russell will pay at most $250 for that plane ticket.

19. Emma scored at least an 88 on her Spanish test.

20. The time of the test was between 50 min and 60 min.

21. The University of Southern Indiana is more than 25 mi away.

22. Sarah's weight is less than 110 lb.

23. That number is greater than -8.

24. The costs of production of that bar-code scanner cannot exceed $135,000.

25. *Web Developers.* Refer to Example 15. Determine, in terms of an inequality, the years for which the number of web developers is greater than 175,000.

Answers

18. $t \leq 250$ **19.** $s \geq 88$ **20.** $50 < t < 60$ **21.** $d > 25$ **22.** $w < 110$ **23.** $n > -8$ **24.** $c \leq 135{,}000$ **25.** More than 6.71 years after 2014, or $\{t \mid t > 6.71\}$

EXAMPLE 16 *Salary Plans.* On her new job, Rose can be paid in one of two ways: *Plan A* is a salary of \$600 per month, plus a commission of 4% of sales; and *Plan B* is a salary of \$800 per month, plus a commission of 6% of sales in excess of \$10,000. For what amount of monthly sales is plan A better than plan B, if we assume that sales are always more than \$10,000?

1. **Familiarize.** Listing the given information in a table will be helpful.

PLAN A: MONTHLY INCOME	PLAN B: MONTHLY INCOME
\$600 salary 4% of sales *Total*: \$600 + 4% of sales	\$800 salary 6% of sales over \$10,000 *Total*: \$800 + 6% of sales over \$10,000

Next, suppose that Rose had sales of \$12,000 in one month. Which plan would be better? Under plan A, she would earn \$600 plus 4% of \$12,000, or

$$600 + 0.04(12{,}000) = \$1080.$$

Since with plan B commissions are paid only on sales in excess of \$10,000, Rose would earn \$800 plus 6% of (\$12,000 − \$10,000), or

$$800 + 0.06(12{,}000 - 10{,}000) = 800 + 0.06(2000) = \$920.$$

This shows that for monthly sales of \$12,000, plan A is better. Similar calculations will show that for sales of \$30,000 per month, plan B is better. To determine *all* values for which plan A pays more money, we must solve an inequality that is based on the calculations above.

2. **Translate.** We let $S =$ the amount of monthly sales. If we examine the calculations in the *Familiarize* step, we see that the monthly income from plan A is $600 + 0.04S$ and from plan B is $800 + 0.06(S - 10{,}000)$. Thus we want to find all values of S for which

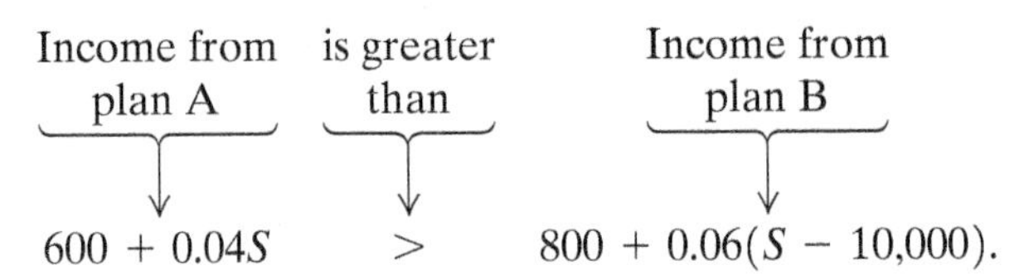

3. **Solve.** We solve the inequality:

$$600 + 0.04S > 800 + 0.06(S - 10{,}000)$$
$$600 + 0.04S > 800 + 0.06S - 600 \quad \text{Using the distributive law}$$
$$600 + 0.04S > 200 + 0.06S \quad \text{Collecting like terms}$$
$$400 > 0.02S \quad \text{Subtracting 200 and } 0.04S$$
$$20{,}000 > S, \text{ or } S < 20{,}000. \quad \text{Dividing by 0.02}$$

4. **Check.** For $S = 20{,}000$, the income from plan A is

$$600 + 4\% \cdot 20{,}000, \text{ or } \$1400.$$

The income from plan B is

$$800 + 6\% \cdot (20{,}000 - 10{,}000), \text{ or } \$1400.$$

This confirms that for sales of \$20,000, Rose's pay is the same under either plan.

In the *Familiarize* step, we saw that for sales of \$12,000, plan A pays more. Since $12{,}000 < 20{,}000$, this is a partial check. Since we cannot check all possible values of S, we will stop here.

5. **State.** For monthly sales of less than \$20,000, plan A is better.

◀ **Do Exercise 26.**

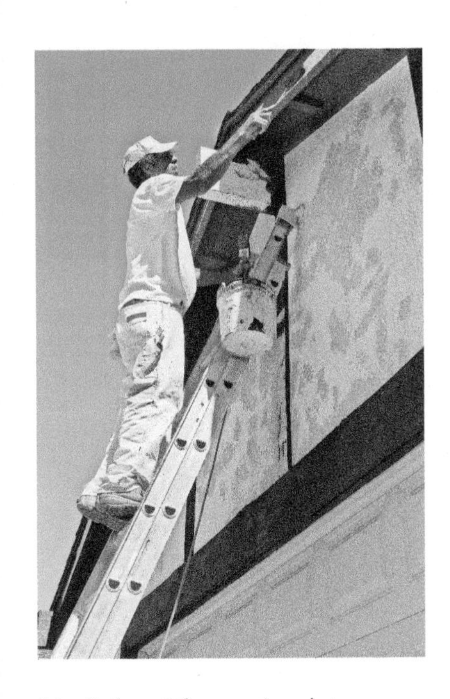

26. *Salary Plans.* A painter can be paid in one of two ways:

Plan A: \$500 plus \$4 per hour;

Plan B: Straight \$9 per hour.

Suppose that the job takes n hours. For what values of n is plan A better for the painter?

Answer

26. For $\{n \mid n < 100\}$, plan A is better.

Translating for Success

1. *Consecutive Integers.* The sum of two consecutive even integers is 102. Find the integers.

2. *Salary Increase.* After Susanna earned a 5% raise, her new salary was \$25,750. What was her former salary?

3. *Dimensions of a Rectangle.* The length of a rectangle is 6 in. more than the width. The perimeter of the rectangle is 102 in. Find the length and the width.

4. *Population.* The population of Doddville is decreasing at a rate of 5% per year. The current population is 25,750. What was the population the previous year?

5. *Reading Assignment.* Quinn has 6 days to complete a 150-page reading assignment. How many pages must he read the first day so that he has no more than 102 pages left to read on the 5 remaining days?

The goal of these matching questions is to practice step (2), Translate, of the five-step problem-solving process. Translate each word problem to an equation or an inequality and select a correct translation from A–O.

A. $0.05(25{,}750) = x$

B. $x + 2x = 102$

C. $2x + 2(x + 6) = 102$

D. $150 - x \leq 102$

E. $x - 0.05x = 25{,}750$

F. $x + (x + 2) = 102$

G. $x + (x + 6) > 102$

H. $x + 5x = 150$

I. $x + 0.05x = 25{,}750$

J. $x + (2x + 6) = 102$

K. $x + (x + 1) = 102$

L. $102 + x > 150$

M. $0.05x = 25{,}750$

N. $102 + 5x > 150$

O. $x + (x + 6) = 102$

Answers on page A-3

6. *Numerical Relationship.* One number is 6 more than twice another. The sum of the numbers is 102. Find the numbers.

7. *Clients.* Together Ella and Ken have 102 clients. If Ken has 6 more clients than Ella, how many does each have?

8. *Sales Commissions.* Will earns a commission of 5% on his sales. One year he earned commissions totaling \$25,750. What were his total sales for the year?

9. *Fencing.* Jess has 102 ft of fencing that he plans to use to enclose two dog runs. The perimeter of one run is to be twice the perimeter of the other. Into what lengths should the fencing be cut?

10. *Quiz Scores.* Lupe has a total of 102 points on the first 6 quizzes in her sociology class. How many total points must she earn on the 5 remaining quizzes in order to have more than 150 points for the semester?

1.4 Exercise Set

FOR EXTRA HELP — MyLab Math

✓ Check Your Understanding

Reading Check Choose from the column on the right the word that best completes each statement. Not every word will be used.

RC1. Because $-5 > -6$ is true, -5 is a(n) ________ of $x > -6$.

RC2. The solution set $\{x|x > 9\}$ is written in ________ notation.

RC3. The interval $[5, 7]$ is a(n) ________ interval.

RC4. The interval $[-3, 10)$ is a(n) ________ interval.

RC5. We reverse the direction of the inequality symbol when we multiply both sides of an inequality by a(n) ________ number.

closed
half-open
interval
negative
open
positive
set-builder
solution

Concept Check For each solution set expressed in set-builder notation, select from the column on the right the equivalent interval notation.

CC1. $\{x|a \le x < b\}$

CC2. $\{x|x < b\}$

CC3. $\{x|x \text{ is a real number}\}$

CC4. $\{x|a < x < b\}$

CC5. $\{x|a \le x \le b\}$

CC6. $\{x|x \ge a\}$

a) (a, b)
b) $[a, b)$
c) $(-\infty, \infty)$
d) $[a, \infty)$
e) $(-\infty, b]$
f) $(a, b]$
g) $[a, b]$
h) $(-\infty, b)$
i) (a, ∞)

a Determine whether the given numbers are solutions of the inequality.

1. $x - 2 \ge 6;\ -4, 0, 4, 8$

2. $3x + 5 \le -10;\ -5, -10, 0, 27$

3. $t - 8 > 2t - 3;\ 0, -8, -9, -3, -\frac{7}{8}$

4. $5y - 7 < 8 - y;\ 2, -3, 0, 3, \frac{2}{3}$

b Write interval notation for the given set or graph.

5. $\{x|x < 5\}$

6. $\{t|t \ge -5\}$

7. $\{x|-3 \le x \le 3\}$

8. $\{t|-10 < t \le 10\}$

9. $\{x|-4 > x > -8\}$

10. $\{x|13 > x \ge 5\}$

11. −6 −5 −4 −3 −2 −1 0 1 2 3 4 5 6

12. −40 −30 −20 −10 0 10 20 30 40

13. $-\sqrt{2}$ −2 −1 0 1 2

14. −12 −8 −4 0 4 8 12

C Solve and graph.

15. $x + 2 > 1$

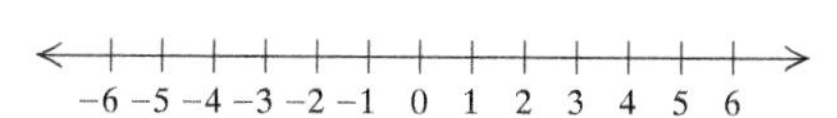

16. $x + 8 > 4$

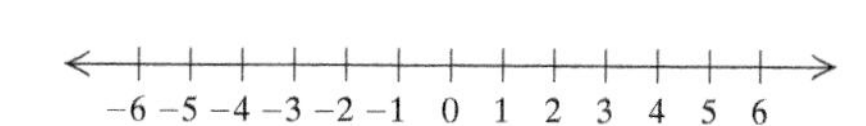

17. $y + 3 < 9$

−6 −5 −4 −3 −2 −1 0 1 2 3 4 5 6

18. $y + 4 < 10$

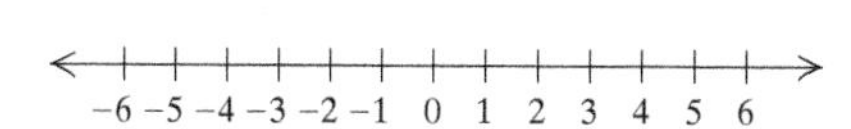

19. $a - 9 \leq -31$

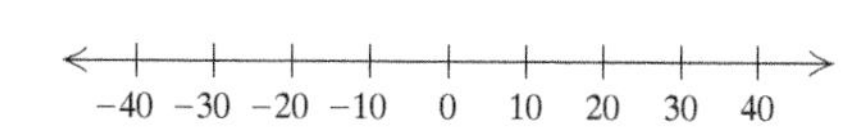

20. $a + 6 \leq -14$

−40 −30 −20 −10 0 10 20 30 40

21. $t + 13 \geq 9$

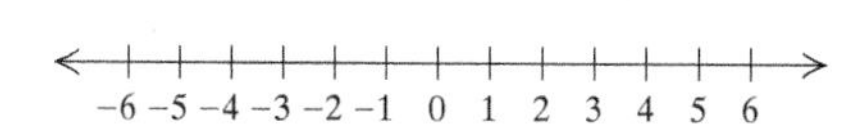

22. $x - 8 \leq 17$

−40 −30 −20 −10 0 10 20 30 40

23. $y - 8 > -14$

−6 −5 −4 −3 −2 −1 0 1 2 3 4 5 6

24. $y - 9 > -18$

−12 −8 −4 0 4 8 12

25. $x - 11 \leq -2$

−12 −8 −4 0 4 8 12

26. $y - 18 \leq -4$

−40 −30 −20 −10 0 10 20 30 40

27. $8x \geq 24$

−6 −5 −4 −3 −2 −1 0 1 2 3 4 5 6

28. $8t < -56$

−12 −8 −4 0 4 8 12

29. $0.3x < -18$

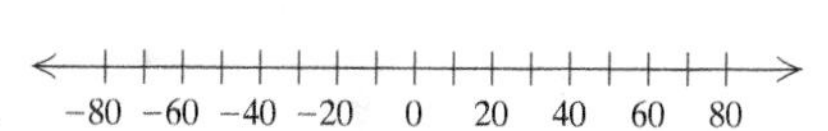

30. $0.6x < 30$

−80 −60 −40 −20 0 20 40 60 80

31. $\frac{2}{3}x > 2$

−6 −5 −4 −3 −2 −1 0 1 2 3 4 5 6

32. $\frac{3}{5}x > -3$

−6 −5 −4 −3 −2 −1 0 1 2 3 4 5 6

Solve.

33. $-9x \geq -8.1$

34. $-5y \leq 3.5$

35. $-\frac{3}{4}x \geq -\frac{5}{8}$

36. $-\frac{1}{8}y \leq -\frac{9}{8}$

37. $2x + 7 < 19$

38. $5y + 13 > 28$

39. $5y + 2y \leq -21$

40. $-9x + 3x \geq -24$

41. $2y - 7 < 5y - 9$

42. $8x - 9 < 3x - 11$

43. $0.4x + 5 \le 1.2x - 4$

44. $0.2y + 1 > 2.4y - 10$

45. $5x - \frac{1}{12} \le \frac{5}{12} + 4x$

46. $2x - 3 < \frac{13}{4}x + 10 - 1.25x$

47. $4(4y - 3) \ge 9(2y + 7)$

48. $2m + 5 \ge 16(m - 4)$

49. $3(2 - 5x) + 2x < 2(4 + 2x)$

50. $2(0.5 - 3y) + y > (4y - 0.2)8$

51. $5[3m - (m + 4)] > -2(m - 4)$

52. $[8x - 3(3x + 2)] - 5 \ge 3(x + 4) - 2x$

53. $3(r - 6) + 2 > 4(r + 2) - 21$

54. $5(t + 3) + 9 < 3(t - 2) + 6$

55. $19 - (2x + 3) \le 2(x + 3) + x$

56. $13 - (2c + 2) \ge 2(c + 2) + 3c$

57. $\frac{1}{4}(8y + 4) - 17 < -\frac{1}{2}(4y - 8)$

58. $\frac{1}{3}(6x + 24) - 20 > -\frac{1}{4}(12x - 72)$

59. $2[4 - 2(3 - x)] - 1 \ge 4[2(4x - 3) + 7] - 25$

60. $5[3(7 - t) - 4(8 + 2t)] - 20 \le -6[2(6 + 3t) - 4]$

61. $\frac{4}{5}(7x - 6) < 40$

62. $\frac{2}{3}(4x - 3) > 30$

63. $\frac{3}{4}(3 + 2x) + 1 \geq 13$

64. $\frac{7}{8}(5 - 4x) - 17 \geq 38$

65. $\frac{3}{4}\left(3x - \frac{1}{2}\right) - \frac{2}{3} < \frac{1}{3}$

66. $\frac{2}{3}\left(\frac{7}{8} - 4x\right) - \frac{5}{8} < \frac{3}{8}$

67. $0.7(3x + 6) \geq 1.1 - (x + 2)$

68. $0.9(2x + 8) < 20 - (x + 5)$

69. $a + (a - 3) \leq (a + 2) - (a + 1)$

70. $0.8 - 4(b - 1) > 0.2 + 3(4 - b)$

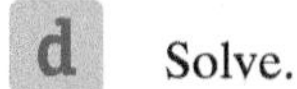
Solve.

Body Mass Index. *Body mass index I* can be used to determine whether an individual has a healthy weight for his or her height. An index in the range 18.5–24.9 indicates a normal weight. Body mass index is given by the formula, or model,

$$I = \frac{703W}{H^2},$$

where W is weight, in pounds, and H is height, in inches. Use this formula for Exercises 71 and 72.

Data: Centers for Disease Control and Prevention

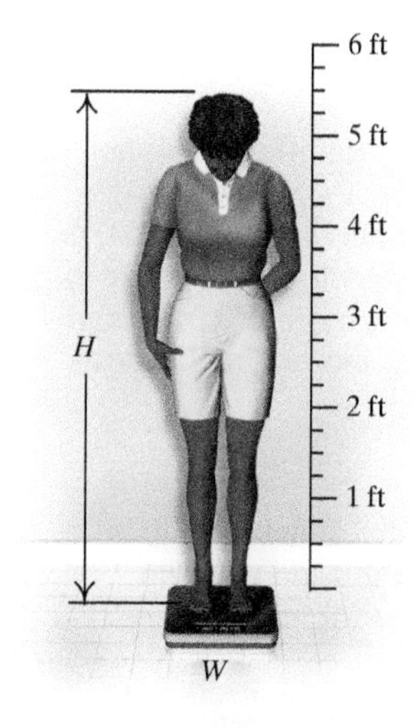

71. *Body Mass Index.* Alexandra's height is 62 in. Determine, in terms of an inequality, those weights W that will keep her body mass index below 25.

72. *Body Mass Index.* Josiah's height is 77 in. Determine, in terms of an inequality, those weights W that will keep his body mass index below 25.

73. *Grades.* David is taking an economics course in which there will be 4 tests, each worth 100 points. He has scores of 89, 92, and 95 on the first three tests. He must make a total of at least 360 in order to get an A. What scores on the last test will give David an A?

74. *Grades.* Elizabeth is taking a mathematics course in which there will be 5 tests, each worth 100 points. She has scores of 94, 90, and 89 on the first three tests. She must make a total of at least 450 in order to get an A. What scores on the fourth test will keep Elizabeth eligible for an A?

75. *Insurance Claims.* After a serious automobile accident, most insurance companies will replace the damaged car with a new one if repair costs exceed 80% of the N.A.D.A., or "blue-book," value of the car. Miguel's car recently sustained \$9200 worth of damage but was not replaced. What was the blue-book value of his car?

76. *Delivery Service.* Jay's Express prices cross-town deliveries at \$15 for the first 10 miles plus \$1.25 for each additional mile. PDQ, Inc., prices its cross-town deliveries at \$25 for the first 10 miles plus \$0.75 for each additional mile. For what number of miles is PDQ less expensive?

77. *Salary Plans.* Imani can be paid in one of two ways:

Plan A: A salary of \$400 per month plus a commission of 8% of gross sales;

Plan B: A salary of \$610 per month, plus a commission of 5% of gross sales.

For what amount of gross sales should Imani select plan A?

78. *Salary Plans.* Aiden can be paid for his masonry work in one of two ways:

Plan A: \$300 plus \$9.00 per hour;

Plan B: Straight \$12.50 per hour.

Suppose that the job takes n hours. For what values of n is plan B better for Aiden?

79. *Prescription Coverage.* Low Med offers two prescription-drug insurance plans. With plan 1, James would pay the first \$150 of his prescription costs and 30% of all costs after that. With plan 2, James would pay the first \$280 of costs, but only 10% of the rest. For what amount of prescription costs will plan 2 save James money? (Assume that his prescription costs exceed \$280.)

80. *Insurance Benefits.* Bayside Insurance offers two plans. Under plan A, Giselle would pay the first \$50 of her medical bills and 20% of all bills after that. Under plan B, Giselle would pay the first \$250 of bills, but only 10% of the rest. For what amount of medical bills will plan B save Giselle money? (Assume that her bills will exceed \$250.)

81. *Wedding Costs.* The Arnold Inn offers two plans for wedding parties. Under plan A, the inn charges \$30 for each person in attendance. Under plan B, the inn charges \$1300 plus \$20 for each person in excess of the first 25 who attend. For what size parties will plan B cost less? (Assume that more than 25 guests will attend.)

82. *Investing.* Matthew is about to invest \$20,000, part at 3% and the rest at 4%. What is the most that he can invest at 3% and still be guaranteed at least \$650 in interest per year?

83. *Renting Office Space.* An investment group is renovating a commercial building and will rent offices to small businesses. The formula

$$R = 2(s + 70)$$

can be used to determine the monthly rent for an office with s square feet. All utilities are included in the monthly payment. For what square footage will the rent be less than \$2100?

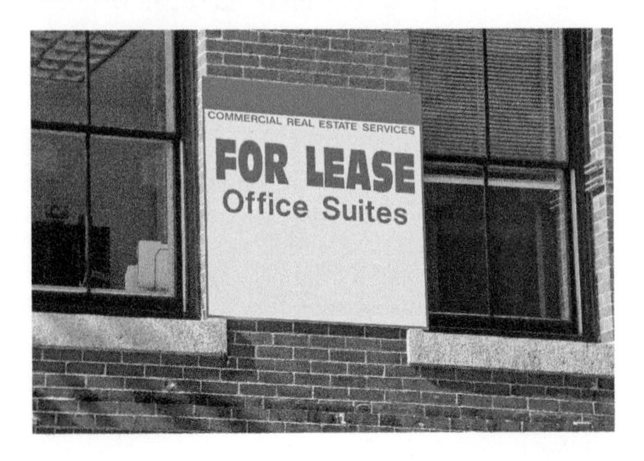

84. *Temperatures of Solids.* The formula

$$C = \tfrac{5}{9}(F - 32)$$

can be used to convert Fahrenheit temperatures F to Celsius temperatures C.

a) Gold is a solid at Celsius temperatures less than 1063°C. Find the Fahrenheit temperatures for which gold is a solid.

b) Silver is a solid at Celsius temperatures less than 960.8°C. Find the Fahrenheit temperatures for which silver is a solid.

85. *Tuition and Fees at Two-Year Colleges.* The equation

$$C = 103t + 1923$$

can be used to estimate the average cost of tuition and fees at two-year public institutions of higher education, where t is the number of years after 2005.

Data: National Center for Education Statistics, U.S. Department of Education

a) What was the average cost of tuition and fees in 2010? in 2016?

b) For what years will the cost of tuition and fees be more than $3500?

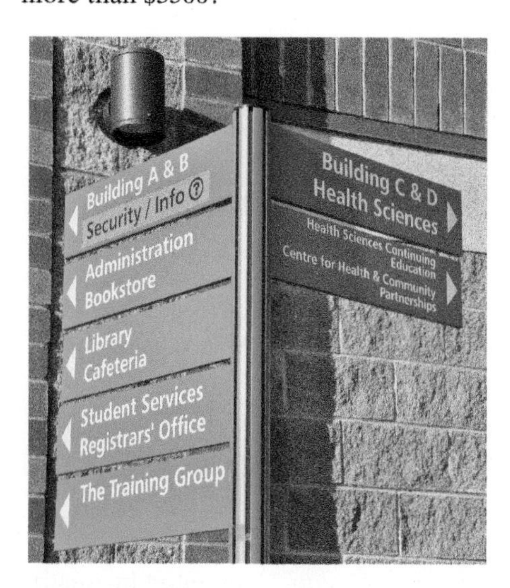

86. *Dewpoint Spread.* Pilots use the **dewpoint spread**, or the difference between the current temperature and the dewpoint (the temperature at which dew occurs), to estimate the height of the cloud cover. Each 3° of dewpoint spread corresponds to an increased height of cloud cover of 1000 ft. A plane, flying with limited instruments, must have a cloud cover higher than 3500 ft. What dewpoint spreads will allow the plane to fly?

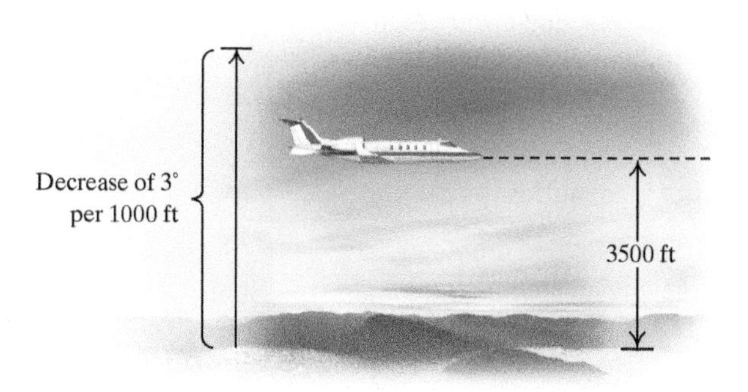

Skill Maintenance

Simplify. [J18]

87. $3a - 6(2a - 5b)$

88. $2(x - y) + 10(3x - 7y)$

89. $4(a - 2b) - 6(2a - 5b)$

90. $-3(2a - 3b) + 8b$

Factor. [J16]

91. $30x - 70y - 40$

92. $-12a + 30ab$

93. $-8x + 24y - 4$

94. $10n - 45mn + 100m$

Add or subtract. [J5], [J7]

95. $-2.3 - 8.9$

96. $-2.3 + 8.9$

97. $-2.3 + (-8.9)$

98. $-2.3 - (-8.9)$

Synthesis

99. *Supply and Demand.* The supply S and demand D for a certain product are given by

$$S = 460 + 94p \quad \text{and} \quad D = 2000 - 60p.$$

a) Find those values of p for which supply exceeds demand.

b) Find those values of p for which supply is less than demand.

Determine whether each statement is true or false. If false, give a counterexample.

100. For any real numbers x and y, if $x < y$, then $x^2 < y^2$.

101. For any real numbers a, b, c, and d, if $a < b$ and $c < d$, then $a + c < b + d$.

102. Determine whether the inequalities

$$x < 3 \quad \text{and} \quad 0 \cdot x < 0 \cdot 3$$

are equivalent. Give reasons to support your answer.

Solve.

103. $x + 5 \leq 5 + x$

104. $x + 8 < 3 + x$

105. $x^2 + 1 > 0$

1.5 Intersections, Unions, and Compound Inequalities

OBJECTIVES

- **a** Find the intersection of two sets. Solve and graph conjunctions of inequalities.
- **b** Find the union of two sets. Solve and graph disjunctions of inequalities.
- **c** Solve applied problems involving conjunctions and disjunctions of inequalities.

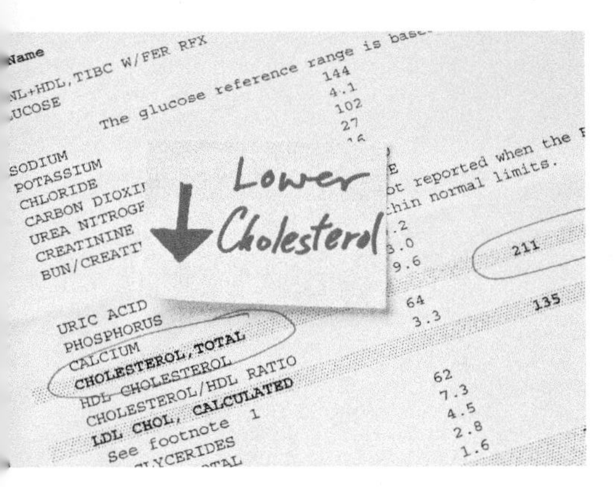

Cholesterol is a substance that is found in every cell of the human body. High levels of cholesterol can cause fatty deposits in the blood vessels that increase the risk of heart attack or stroke. A blood test can be used to measure *total cholesterol*. The following table shows the health risks associated with various cholesterol levels.

TOTAL CHOLESTEROL	RISK LEVEL
Less than 200	Normal
From 200 to 239	Borderline high
240 or higher	High

A total-cholesterol level T from 200 to 239 is considered border-line high. We can express this by the sentence

$$200 \leq T \quad and \quad T \leq 239$$

or more simply by

$$200 \leq T \leq 239.$$

This is an example of a *compound inequality*. **Compound inequalities** consist of two or more inequalities joined by the word *and* or the word *or*. We now "solve" such sentences—that is, we find the set of all solutions.

a INTERSECTIONS OF SETS AND CONJUNCTIONS OF INEQUALITIES

INTERSECTION

The **intersection** of two sets A and B is the set of all members that are common to A and B. We denote the intersection of sets A and B as

$$A \cap B.$$

The intersection of two sets is often illustrated as shown below.

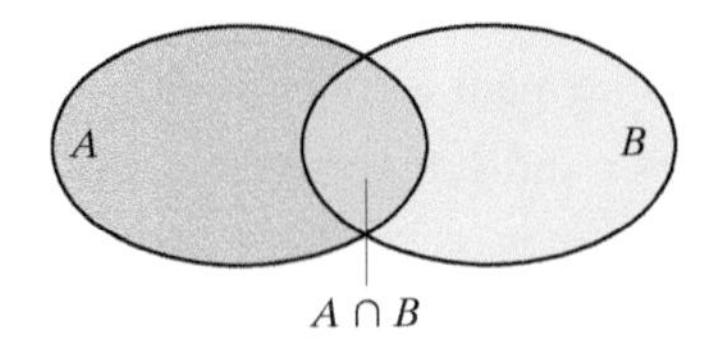

EXAMPLE 1 Find the intersection: $\{1, 2, 3, 4, 5\} \cap \{-2, -1, 0, 1, 2, 3\}$.

Only the numbers 1, 2, and 3 are common to the two sets, so the intersection is $\{1, 2, 3\}$.

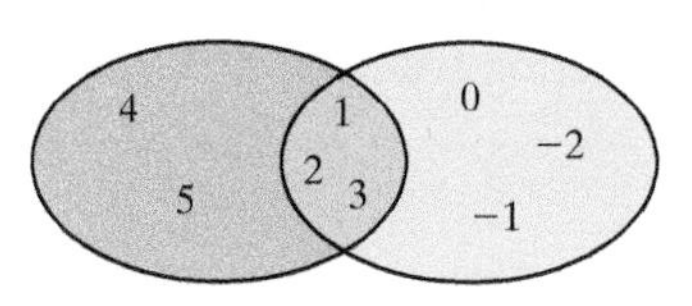

Do Exercises 1 and 2. ▶

1. Find the intersection: $\{0, 3, 5, 7\} \cap \{0, 1, 3, 11\}$.

2. Shade the intersection of sets A and B.

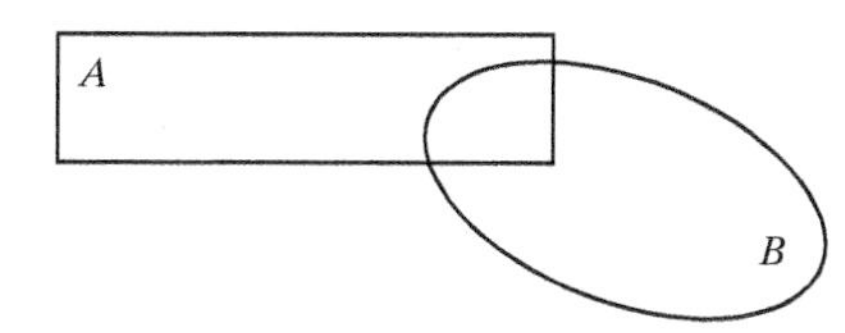

CONJUNCTION

When two or more sentences are joined by the word *and* to make a compound sentence, the new sentence is called a **conjunction** of the sentences.

The following is a conjunction of inequalities:

$$-2 < x \quad \textit{and} \quad x < 1.$$

A number is a solution of a conjunction if it is a solution of *both* inequalities. For example, 0 is a solution of $-2 < x$ *and* $x < 1$ because $-2 < 0$ *and* $0 < 1$. Shown below is the graph of $-2 < x$, followed by the graph of $x < 1$, and then by the graph of the conjunction $-2 < x$ *and* $x < 1$. As the graphs demonstrate, *the solution set of a conjunction is the intersection of the solution sets of the individual inequalities.*

Set	Number line (−7 to 7)	Interval
$\{x \mid -2 < x\}$	−7 −6 −5 −4 −3 −2 −1 0 1 2 3 4 5 6 7	$(-2, \infty)$
$\{x \mid x < 1\}$	−7 −6 −5 −4 −3 −2 −1 0 1 2 3 4 5 6 7	$(-\infty, 1)$
$\{x \mid -2 < x\} \cap \{x \mid x < 1\}$ $= \{x \mid -2 < x \textit{ and } x < 1\}$	−7 −6 −5 −4 −3 −2 −1 0 1 2 3 4 5 6 7	$(-2, 1)$

Because there are numbers that are both greater than -2 and less than 1, the conjunction $-2 < x$ *and* $x < 1$ can be abbreviated as $-2 < x < 1$. Thus the interval $(-2, 1)$ can be represented as $\{x \mid -2 < x < 1\}$, the set of all numbers that are *simultaneously* greater than -2 *and* less than 1. Note that, in general, for $a < b$,

$a < x$ ***and*** $x < b$ **can be abbreviated** $a < x < b$;

and $b > x$ ***and*** $x > a$ **can be abbreviated** $b > x > a$.

Caution!

"$a > x$ *and* $x < b$" cannot be abbreviated as "$a > x < b$".

Answers

1. $\{0, 3\}$

2.

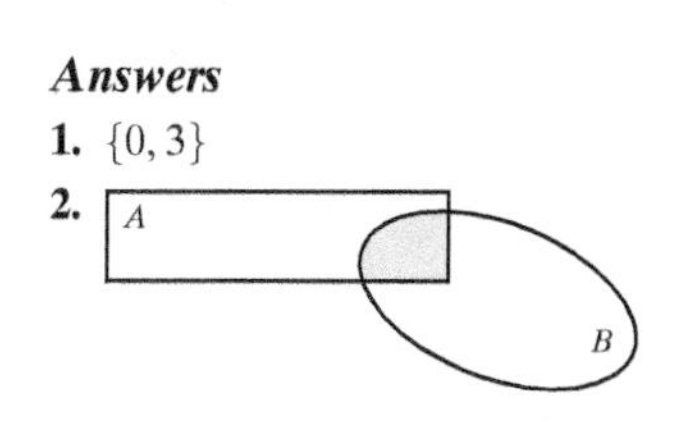

3. Graph and write interval notation:

$-1 < x$ *and* $x < 4$.

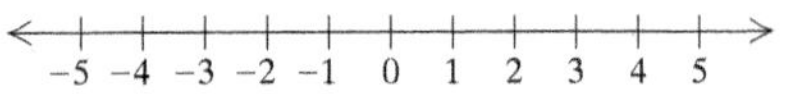

"AND"; "INTERSECTION"

The word **"and"** corresponds to **"intersection"** and to the symbol "∩". In order for a number to be a solution of a conjunction, it must make each part of the conjunction true.

Do Exercise 3.

EXAMPLE 2 Solve and graph: $-1 \le 2x + 5 < 13$.

This inequality is an abbreviation for the conjunction

$$-1 \le 2x + 5 \quad \textit{and} \quad 2x + 5 < 13.$$

The word *and* corresponds to set *intersection*, ∩. To solve the conjunction, we solve each of the two inequalities separately and then find the intersection of the solution sets:

$-1 \le 2x + 5$	*and*	$2x + 5 < 13$	
$-6 \le 2x$	*and*	$2x < 8$	Subtracting 5
$-3 \le x$	*and*	$x < 4$.	Dividing by 2

We now abbreviate the result:

$$-3 \le x < 4.$$

The solution set is $\{x \mid -3 \le x < 4\}$, or, in interval notation, $[-3, 4)$. The graph is the intersection of the two separate solution sets.

$\{x \mid -3 \le x\}$

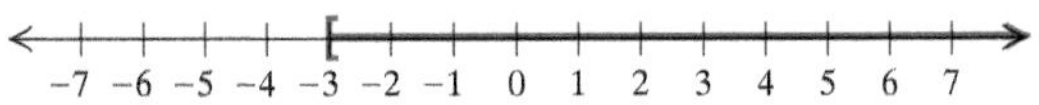

$[-3, \infty)$

$\{x \mid x < 4\}$

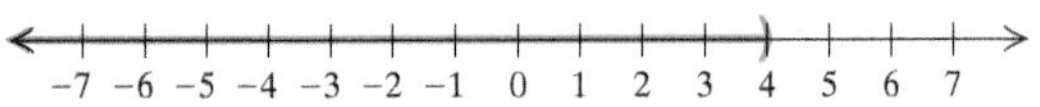

$(-\infty, 4)$

$\{x \mid -3 \le x\} \cap \{x \mid x < 4\}$
$= \{x \mid -3 \le x < 4\}$

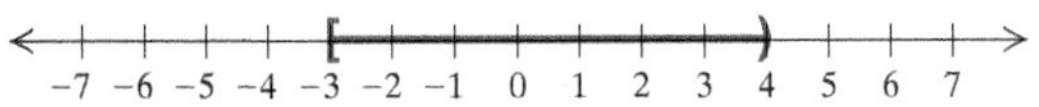

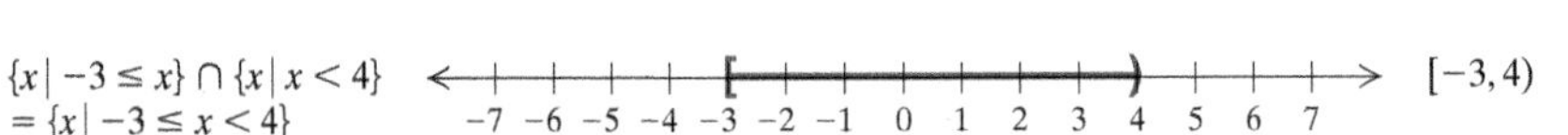

$[-3, 4)$

The steps above are generally combined as follows:

$-1 \le 2x + 5 < 13$	$2x + 5$ appears in both inequalities.
$-6 \le 2x < 8$	Subtracting 5
$-3 \le x < 4$.	Dividing by 2

4. Solve and graph:

$-22 < 3x - 7 \le 23$.

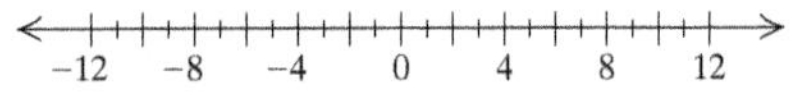

Do Exercise 4.

EXAMPLE 3 Solve and graph: $2x - 5 \ge -3$ *and* $5x + 2 \ge 17$.

We first solve each inequality separately:

$2x - 5 \ge -3$	*and*	$5x + 2 \ge 17$
$2x \ge 2$	*and*	$5x \ge 15$
$x \ge 1$	*and*	$x \ge 3$.

Answers

3. 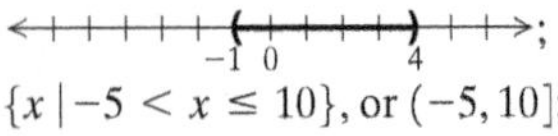

; $(-1, 4)$

4. $\{x \mid -5 < x \le 10\}$, or $(-5, 10]$;

Next, we find the intersection of the two separate solution sets:

$\{x \mid x \geq 1\}$ $[1, \infty)$

$\{x \mid x \geq 3\}$ $[3, \infty)$

$\{x \mid x \geq 1\} \cap \{x \mid x \geq 3\}$
$= \{x \mid x \geq 3\}$ $[3, \infty)$

The numbers common to both sets are those that are greater than or equal to 3. Thus the solution set is $\{x|x \geq 3\}$, or, in interval notation, $[3, \infty)$. You should check that any number in $[3, \infty)$ satisfies the conjunction whereas numbers outside $[3, \infty)$ do not.

Do Exercise 5.

5. Solve and graph:
$3x + 4 < 10$ *and* $2x - 7 < -13$.

EMPTY SET; DISJOINT SETS

Sometimes two sets have no elements in common. In such a case, we say that the intersection of the two sets is the **empty set**, denoted { } or $\varnothing$. Two sets with an empty intersection are said to be **disjoint**.

$A \cap B = \varnothing$

EXAMPLE 4 Solve and graph: $2x - 3 > 1$ *and* $3x - 1 < 2$.

We solve each inequality separately:

$$2x - 3 > 1 \quad and \quad 3x - 1 < 2$$
$$2x > 4 \quad and \quad 3x < 3$$
$$x > 2 \quad and \quad x < 1.$$

The solution set is the intersection of the solution sets of the individual inequalities.

$\{x \mid x > 2\}$ $(2, \infty)$

$\{x \mid x < 1\}$ $(-\infty, 1)$

$\{x \mid x > 2\} \cap \{x \mid x < 1\}$
$= \{x \mid x > 2 \text{ and } x < 1\}$
$= \varnothing$ $\varnothing$

Since no number is both greater than 2 and less than 1, the solution set is the empty set, $\varnothing$.

Do Exercise 6.

6. Solve and graph:
$3x - 7 \leq -13$ *and* $4x + 3 > 8$.

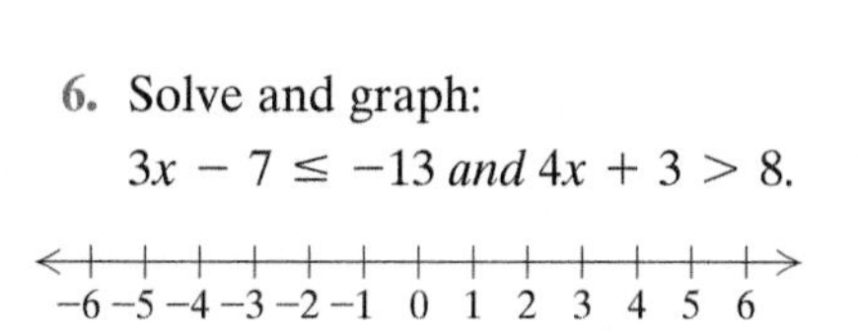

Answers

5. $\{x|x < -3\}$;

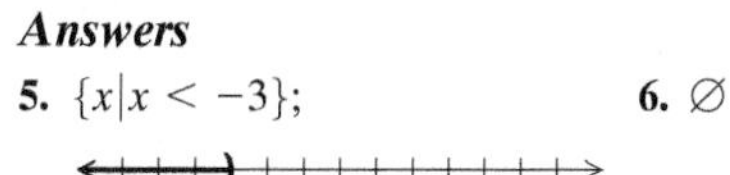

6. $\varnothing$

EXAMPLE 5 Solve: $3 \le 5 - 2x < 7$.

We have

$$3 \le 5 - 2x < 7$$
$$3 - 5 \le 5 - 2x - 5 < 7 - 5 \quad \text{Subtracting 5}$$
$$-2 \le -2x < 2 \quad \text{Simplifying}$$
$$\frac{-2}{-2} \ge \frac{-2x}{-2} > \frac{2}{-2} \quad \text{Dividing by } -2. \text{ The symbols must be reversed.}$$
$$1 \ge x > -1. \quad \text{Simplifying}$$

The solution set is $\{x | 1 \ge x > -1\}$, or $\{x | -1 < x \le 1\}$, since the inequalities $1 \ge x > -1$ and $-1 < x \le 1$ are equivalent. The solution, in interval notation, is $(-1, 1]$.

◀ Do Exercise 7.

7. Solve: $-4 \le 8 - 2x \le 4$. GS

$$-4 \le 8 - 2x \le 4$$
$$-4 - \square \le 8 - 2x - 8 \le 4 - 8$$
$$\square \le -2x \le \square$$
$$\frac{-12}{-2} \;\square\; \frac{-2x}{-2} \ge \frac{-4}{-2}$$
$$6 \ge \square \ge 2, \text{ or}$$
$$2 \;\square\; x \le 6$$

The solution set is $\{x | 2 \le x \le 6\}$, or $[\square, 6]$.

b UNIONS OF SETS AND DISJUNCTIONS OF INEQUALITIES

UNION

The **union** of two sets A and B is the collection of elements belonging to A and/or B. We denote the union of A and B by

$$A \cup B.$$

The union of two sets is often illustrated as shown below.

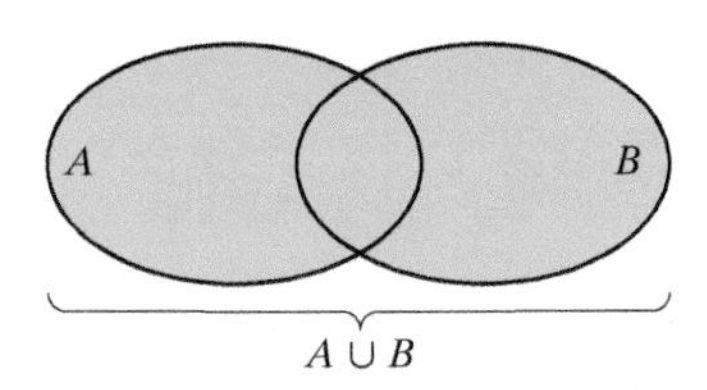

EXAMPLE 6 Find the union: $\{2, 3, 4\} \cup \{3, 5, 7\}$.

The numbers in either or both sets are 2, 3, 4, 5, and 7, so the union is $\{2, 3, 4, 5, 7\}$. We don't list the number 3 twice.

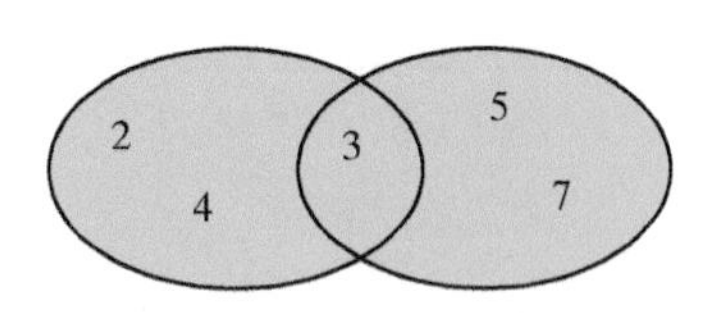

◀ Do Exercises 8 and 9.

8. Find the union:
$\{0, 1, 3, 4\} \cup \{0, 1, 7, 9\}$.

9. Shade the union of sets A and B.

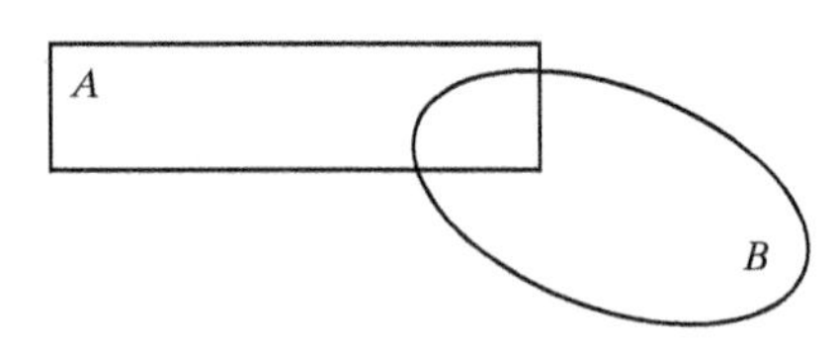

DISJUNCTION

When two or more sentences are joined by the word *or* to make a compound sentence, the new sentence is called a **disjunction** of the sentences.

Answers

7. $\{x | 2 \le x \le 6\}$, or $[2, 6]$
8. $\{0, 1, 3, 4, 7, 9\}$
9.

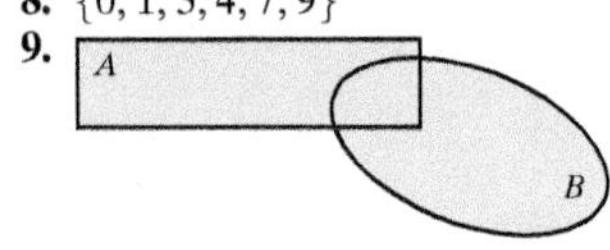

Guided Solution:
7. $8, -12, -4, \ge, x, \le, 2$

The following is an example of a disjunction:

$x < -3 \quad or \quad x > 3.$

A number is a solution of a disjunction if it is a solution of at least one of the individual inequalities. For example, -7 is a solution of $x < -3$ *or* $x > 3$ because $-7 < -3$. Similarly, 5 is also a solution because $5 > 3$.

Shown below is the graph of $x < -3$, followed by the graph of $x > 3$, and then by the graph of the disjunction $x < -3$ *or* $x > 3$. As the graphs demonstrate, *the solution set of a disjunction is the union of the solution sets of the individual sentences.*

$\{x \mid x < -3\}$ — $(-\infty, -3)$

$\{x \mid x > 3\}$ — $(3, \infty)$

$\{x \mid x < -3\} \cup \{x \mid x > 3\} = \{x \mid x < -3 \text{ or } x > 3\}$ — $(-\infty, -3) \cup (3, \infty)$

The solution set of

$x < -3 \quad or \quad x > 3$

is written $\{x \mid x < -3 \text{ or } x > 3\}$, or, in interval notation, $(-\infty, -3) \cup (3, \infty)$. This cannot be written in a more condensed form.

"OR"; "UNION"

The word "**or**" corresponds to "**union**" and the symbol "$\cup$". In order for a number to be in the solution set of a disjunction, it must be in *at least one* of the solution sets of the individual sentences.

Do Exercise 10. ▶

10. Graph and write interval notation:

$x \leq -2$ *or* $x > 4$.

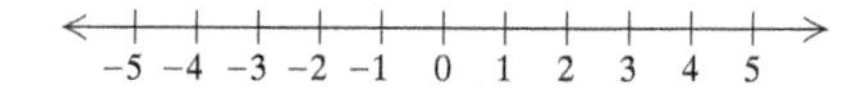

EXAMPLE 7 Solve and graph: $7 + 2x < -1$ *or* $13 - 5x \leq 3$.

We solve each inequality separately, retaining the word *or*:

$$7 + 2x < -1 \quad or \quad 13 - 5x \leq 3$$
$$2x < -8 \quad or \quad -5x \leq -10$$
$$x < -4 \quad or \quad x \geq 2.$$

Dividing by -5. The symbol must be reversed.

To find the solution set of the disjunction, we consider the individual graphs. We graph $x < -4$ and then $x \geq 2$. Then we take the union of the graphs.

$\{x \mid x < -4\}$ — $(-\infty, -4)$

$\{x \mid x \geq 2\}$ — $[2, \infty)$

$\{x \mid x < -4 \text{ or } x \geq 2\}$ — $(-\infty, -4) \cup [2, \infty)$

The solution set is written $\{x \mid x < -4 \text{ or } x \geq 2\}$, or, in interval notation, $(-\infty, -4) \cup [2, \infty)$.

Answer

10. $(-\infty, -2] \cup (4, \infty)$

Solve and graph.

11. $x - 4 < -3$ *or* $x - 3 \geq 3$

−10 −8 −6 −4 −2 0 2 4 6 8 10

12. $-2x + 4 \leq -3$ *or* $x + 5 < 3$

−6 −5 −4 −3 −2 −1 0 1 2 3 4 5 6

Caution!

A compound inequality like

$$x < -4 \quad or \quad x \geq 2,$$

as in Example 7, *cannot* be expressed as $2 \leq x < -4$ because to do so would be to say that x is *simultaneously* less than -4 and greater than or equal to 2. No number is both less than -4 *and* greater than or equal to 2, but many are less than -4 *or* greater than or equal to 2.

◀ **Do Exercises 11 and 12.**

13. Solve:

$-3x - 7 < -1$ *or* $x + 4 < -1$.

$-3x - 7 < -1$ *or* $x + 4 < -1$

$-3x <$ ____ *or* $x <$ ____

$\frac{-3x}{-3}$ ____ $\frac{6}{-3}$ *or* $x < -5$

$x >$ ____ *or* $x < -5$

The solution set is $\{x|x <$ ____ *or* $x > -2\}$, or, in interval notation, $(-\infty, -5) \cup (-2,$ ____ $)$.

EXAMPLE 8 Solve: $-2x - 5 < -2$ *or* $x - 3 < -10$.

We solve the individual inequalities separately, retaining the word *or*:

$$-2x - 5 < -2 \quad or \quad x - 3 < -10$$

$$-2x < 3 \quad or \quad x < -7$$

Reversing the symbol

$$x > -\tfrac{3}{2} \quad or \quad x < -7.$$

Keep the word "or."

The solution set is written $\{x|x < -7 \text{ or } x > -\frac{3}{2}\}$, or, in interval notation, $(-\infty, -7) \cup (-\frac{3}{2}, \infty)$.

◀ **Do Exercise 13.**

EXAMPLE 9 Solve: $3x - 11 < 4$ *or* $4x + 9 \geq 1$.

We solve the individual inequalities separately, retaining the word *or*:

$$3x - 11 < 4 \quad or \quad 4x + 9 \geq 1$$

$$3x < 15 \quad or \quad 4x \geq -8$$

$$x < 5 \quad or \quad x \geq -2.$$

To find the solution set, we first look at the individual graphs.

$\{x|x < 5\}$ −7 −6 −5 −4 −3 −2 −1 0 1 2 3 4 5 6 7 $(-\infty, 5)$

$\{x|x \geq -2\}$ −7 −6 −5 −4 −3 −2 −1 0 1 2 3 4 5 6 7 $[-2, \infty)$

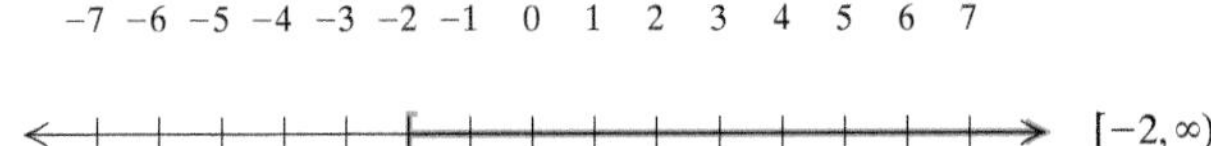

$\{x|x < 5\} \cup \{x|x \geq -2\}$
$= \{x|x < 5 \text{ or } x \geq -2\}$
$= \{x|x \text{ is a real number}\}$

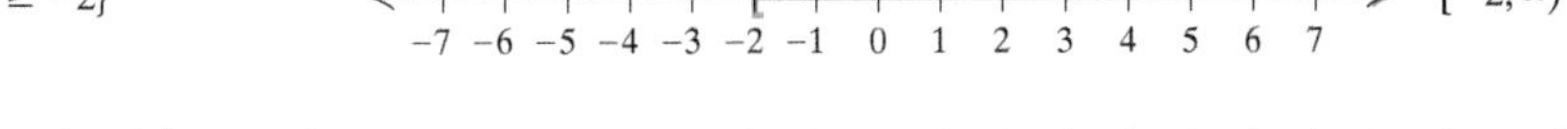

$(-\infty, \infty)$ = The set of all real numbers

Since any number is either less than 5 or greater than or equal to -2, the two sets fill the entire number line. Thus the solution set is the set of all real numbers, $(-\infty, \infty)$.

◀ **Do Exercise 14.**

14. Solve and graph:

$5x - 7 \leq 13$ *or* $2x - 1 \geq -7$.

−6 −5 −4 −3 −2 −1 0 1 2 3 4 5 6

Answers

11. $\{x|x < 1 \text{ or } x \geq 6\}$, or $(-\infty, 1) \cup [6, \infty)$;

0 1 6

12. $\left\{x|x < -2 \text{ or } x \geq \frac{7}{2}\right\}$, or $(-\infty, -2) \cup \left[\frac{7}{2}, \infty\right)$;

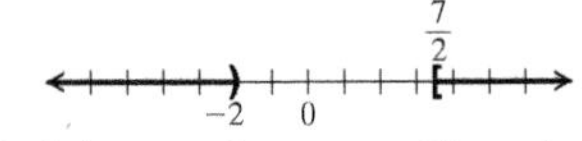

13. $\{x|x < -5 \text{ or } x > -2\}$, or $(-\infty, -5) \cup (-2, \infty)$

14. All real numbers;

0

Guided Solution:

13. 6, −5, >, −2, −5, ∞

C APPLICATIONS AND PROBLEM SOLVING

EXAMPLE 10 *Renting Office Space.* The equation

$$R = 2(s + 70)$$

can be used to determine the monthly rent R for an office with s square footage in a renovated commercial building. All utilities are included in the monthly payment. A florist shop has a monthly rental budget between \$1720 and \$2560. What square footage can be rented and remain within budget?

1. **Familiarize.** We have an equation for calculating the monthly rent. Thus we can substitute a value into the formula. For 720 ft^2, the rent is found as follows:

   $$R = 2(720 + 70) = 2 \cdot 790 = \$1580.$$

   This familiarizes us with the equation and also tells us that the number of square feet that we are looking for must be larger than 720 since \$1580 is less than \$1720.

2. **Translate.** We want the monthly rent to be between \$1720 and \$2560, so we need to find those values of s for which $1720 < R < 2560$. Substituting $2(s + 70)$ for R, we have

   $$1720 < 2(s + 70) < 2560.$$

3. **Solve.** We solve the inequality:

   $$1720 < 2(s + 70) < 2560$$
   $$\frac{1720}{2} < \frac{2(s + 70)}{2} < \frac{2560}{2} \quad \text{Dividing by 2}$$
   $$860 < s + 70 < 1280$$
   $$790 < s < 1210. \quad \text{Subtracting 70}$$

4. **Check.** We substitute some values as we did in the *Familiarize* step.

5. **State.** Square footage between 790 ft^2 and 1210 ft^2 can be rented for a budget between \$1720 and \$2560 per month.

Do Exercise 15. ▶

15. *Renting Office Space.* Refer to Example 10. What square footage can be rented for a budget between \$2000 and \$3200?

Answer

15. Between 930 ft^2 and 1530 ft^2

1.5 Exercise Set

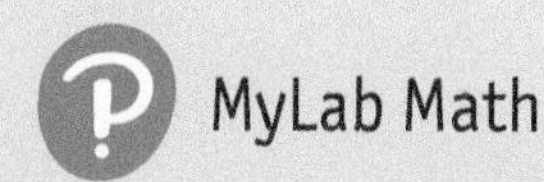

✓ Check Your Understanding

Reading Check Determine whether each statement is true or false.

RC1. A compound inequality like $x < 5$ *and* $x > -2$ can be expressed as $-2 < x < 5$.

RC2. A compound inequality like $x \geq 5$ *and* $x < -2$ can be expressed as $5 \leq x < -2$.

RC3. The solution set of $x < -4$ *and* $x > 4$ can be written as $\varnothing$.

RC4. The solution set of $x \leq 3$ *and* $x \geq 0$ can be written as $[0, 3]$.

Concept Check Determine whether -5 is in the solution set of each compound inequality.

CC1. $-7 < x < 7$

CC2. $-5 \leq x < 3$

CC3. $-5 < x \leq 3$

CC4. $x > -10$ *and* $x < 0$

CC5. $x < -10$ *or* $x > 0$

CC6. $x < -4$ *or* $x > 1$

a, **b** Find the intersection or union.

1. $\{9, 10, 11\} \cap \{9, 11, 13\}$

2. $\{1, 5, 10, 15\} \cap \{5, 15, 20\}$

3. $\{a, b, c, d\} \cap \{b, f, g\}$

4. $\{m, n, o, p\} \cap \{m, o, p\}$

5. $\{9, 10, 11\} \cup \{9, 11, 13\}$

6. $\{1, 5, 10, 15\} \cup \{5, 15, 20\}$

7. $\{a, b, c, d\} \cup \{b, f, g\}$

8. $\{m, n, o, p\} \cup \{m, o, p\}$

9. $\{2, 5, 7, 9\} \cap \{1, 3, 4\}$

10. $\{a, e, i, o, u\} \cap \{m, q, w, s, t\}$

11. $\{3, 5, 7\} \cup \varnothing$

12. $\{3, 5, 7\} \cap \varnothing$

a Graph and write interval notation.

13. $-4 < a$ *and* $a \leq 1$

-6 -5 -4 -3 -2 -1 0 1 2 3 4 5 6

14. $-\frac{5}{2} \leq m$ *and* $m < \frac{3}{2}$

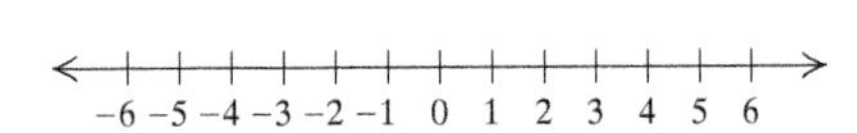

15. $1 < x < 6$

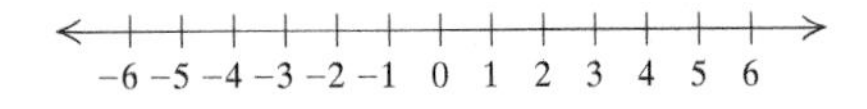

16. $-3 \leq y \leq 4$

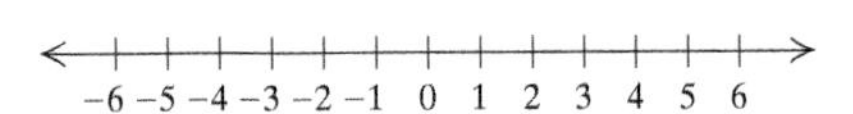

Solve and graph.

17. $-10 \leq 3x + 2$ *and* $3x + 2 < 17$

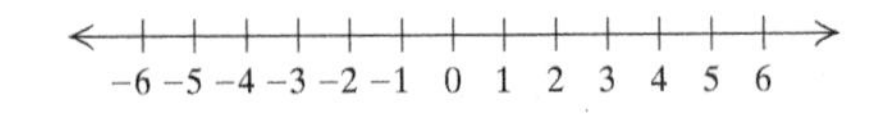

18. $-11 < 4x - 3$ *and* $4x - 3 \leq 13$

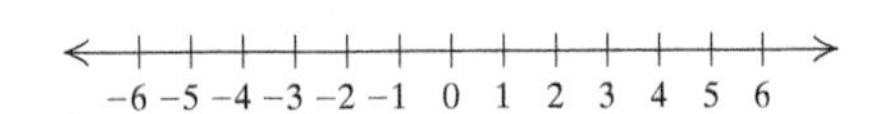

19. $3x + 7 \geq 4$ *and* $2x - 5 \geq -1$

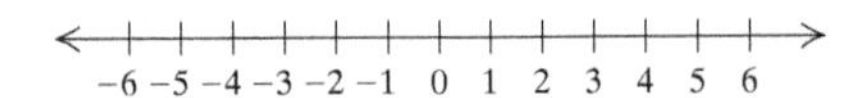

20. $4x - 7 < 1$ *and* $7 - 3x > -8$

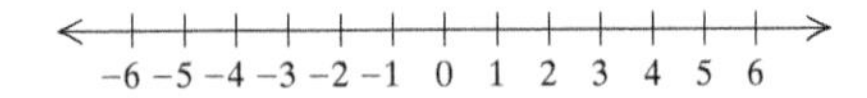

21. $4 - 3x \geq 10$ *and* $5x - 2 > 13$

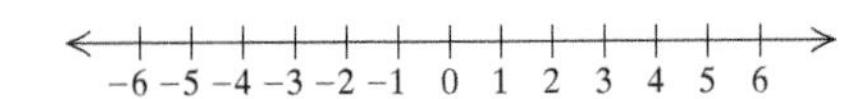

22. $5 - 7x > 19$ *and* $2 - 3x < -4$

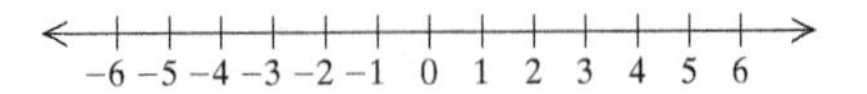

Solve.

23. $-4 < x + 4 < 10$

24. $-6 < x + 6 \le 8$

25. $6 > -x \ge -2$

26. $3 > -x \ge -5$

27. $2 < x + 3 \le 9$

28. $-6 \le x + 1 < 9$

29. $1 < 3y + 4 \le 19$

30. $5 \le 8x + 5 \le 21$

31. $-10 \le 3x - 5 \le -1$

32. $-6 \le 2x - 3 < 6$

33. $-18 \le -2x - 7 < 0$

34. $4 > -3m - 7 \ge 2$

35. $-\frac{1}{2} < \frac{1}{4}x - 3 \le \frac{1}{2}$

36. $-\frac{2}{3} \le 4 - \frac{1}{4}x < \frac{2}{3}$

37. $-4 \le \frac{7 - 3x}{5} \le 4$

38. $-3 < \frac{2x - 5}{4} < 8$

b Graph and write interval notation.

39. $x < -2$ *or* $x > 1$

−6 −5 −4 −3 −2 −1 0 1 2 3 4 5 6

40. $x < -4$ *or* $x > 0$

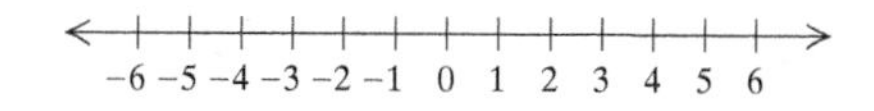

41. $x \le -3$ *or* $x > 1$

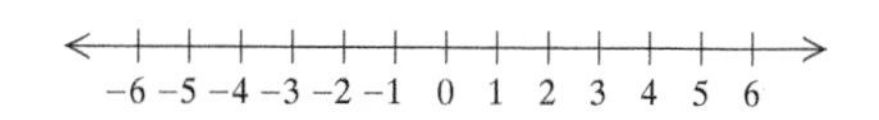

42. $x \le -1$ *or* $x > 3$

−6 −5 −4 −3 −2 −1 0 1 2 3 4 5 6

Solve and graph.

43. $x + 3 < -2$ *or* $x + 3 > 2$

−6 −5 −4 −3 −2 −1 0 1 2 3 4 5 6

44. $x - 2 < -1$ *or* $x - 2 > 3$

−6 −5 −4 −3 −2 −1 0 1 2 3 4 5 6

45. $2x - 8 \le -3$ *or* $x - 1 \ge 3$

−6 −5 −4 −3 −2 −1 0 1 2 3 4 5 6

46. $x - 5 \le -4$ *or* $2x - 7 \ge 3$

−6 −5 −4 −3 −2 −1 0 1 2 3 4 5 6

47. $7x + 4 \ge -17$ *or* $6x + 5 \ge -7$

−6 −5 −4 −3 −2 −1 0 1 2 3 4 5 6

48. $4x - 4 < -8$ *or* $4x - 4 < 12$

−6 −5 −4 −3 −2 −1 0 1 2 3 4 5 6

Solve.

49. $7 > -4x + 5$ *or* $10 \le -4x + 5$

50. $6 > 2x - 1$ *or* $-4 \le 2x - 1$

51. $3x - 7 > -10$ *or* $5x + 2 \leq 22$

52. $3x + 2 < 2$ *or* $4 - 2x < 14$

53. $-2x - 2 < -6$ *or* $-2x - 2 > 6$

54. $-3m - 7 < -5$ *or* $-3m - 7 > 5$

55. $\frac{2}{3}x - 14 < -\frac{5}{6}$ *or* $\frac{2}{3}x - 14 > \frac{5}{6}$

56. $\frac{1}{4} - 3x \leq -3.7$ *or* $\frac{1}{4} - 5x \geq 4.8$

57. $\frac{2x - 5}{6} \leq -3$ *or* $\frac{2x - 5}{6} \geq 4$

58. $\frac{7 - 3x}{5} < -4$ *or* $\frac{7 - 3x}{5} > 4$

C Solve.

59. *Pressure at Sea Depth.* The equation

$$P = 1 + \frac{d}{33}$$

gives the pressure P, in atmospheres (atm), at a depth of d feet in the sea. For what depths d is the pressure at least 1 atm and at most 7 atm?

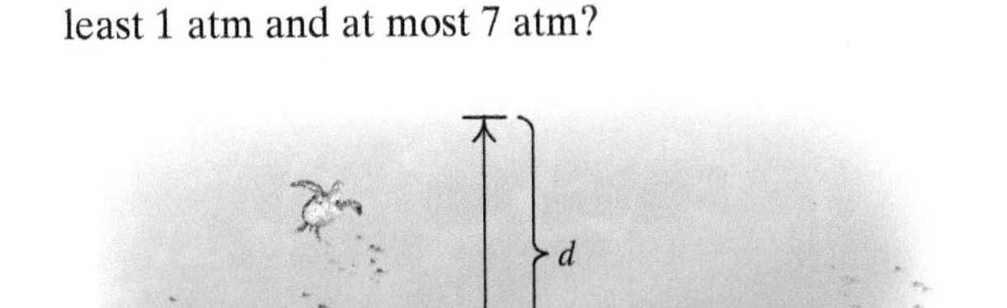

60. *Temperatures of Liquids.* The formula

$$C = \tfrac{5}{9}(F - 32)$$

can be used to convert Fahrenheit temperatures F to Celsius temperatures C.

a) Gold is a liquid for Celsius temperatures C such that $1063^\circ \leq C < 2660^\circ$. Find such an inequality for the corresponding Fahrenheit temperatures.

b) Silver is a liquid for Celsius temperatures C such that $960.8^\circ \leq C < 2180^\circ$. Find such an inequality for the corresponding Fahrenheit temperatures.

61. *Aerobic Exercise.* In order to achieve maximum results from aerobic exercise, one should maintain one's heart rate at a certain level. A 30-year-old woman with a resting heart rate of 60 beats per minute should keep her heart rate between 138 and 162 beats per minute while exercising. She checks her pulse for 10 sec while exercising. What should the number of beats be?

62. *Minimizing Tolls.* A \$6.00 toll is charged to cross the bridge from mainland Florida to Sanibel Island. A six-month pass, costing \$50.00, reduces the toll to \$2.00. A six-month unlimited-trip pass costs \$300 and allows for unlimited crossings. How many crossings in six months does it take, on average, for the reduced-toll pass to be the most economical choice?

Data: leewayinfo.com

Solve.

23. $-4 < x + 4 < 10$

24. $-6 < x + 6 \le 8$

25. $6 > -x \ge -2$

26. $3 > -x \ge -5$

27. $2 < x + 3 \le 9$

28. $-6 \le x + 1 < 9$

29. $1 < 3y + 4 \le 19$

30. $5 \le 8x + 5 \le 21$

31. $-10 \le 3x - 5 \le -1$

32. $-6 \le 2x - 3 < 6$

33. $-18 \le -2x - 7 < 0$

34. $4 > -3m - 7 \ge 2$

35. $-\frac{1}{2} < \frac{1}{4}x - 3 \le \frac{1}{2}$

36. $-\frac{2}{3} \le 4 - \frac{1}{4}x < \frac{2}{3}$

37. $-4 \le \frac{7 - 3x}{5} \le 4$

38. $-3 < \frac{2x - 5}{4} < 8$

b Graph and write interval notation.

39. $x < -2$ *or* $x > 1$

−6 −5 −4 −3 −2 −1 0 1 2 3 4 5 6

40. $x < -4$ *or* $x > 0$

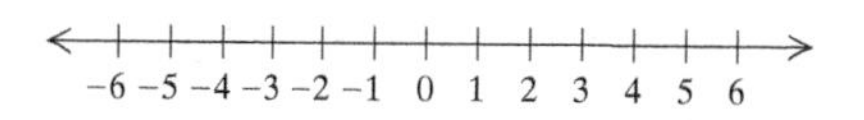

41. $x \le -3$ *or* $x > 1$

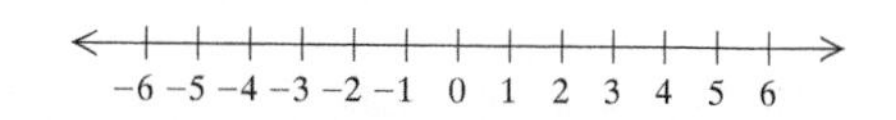

42. $x \le -1$ *or* $x > 3$

−6 −5 −4 −3 −2 −1 0 1 2 3 4 5 6

Solve and graph.

43. $x + 3 < -2$ *or* $x + 3 > 2$

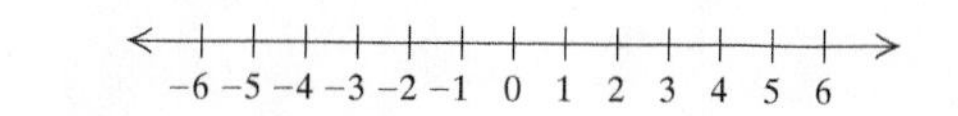

44. $x - 2 < -1$ *or* $x - 2 > 3$

−6 −5 −4 −3 −2 −1 0 1 2 3 4 5 6

45. $2x - 8 \le -3$ *or* $x - 1 \ge 3$

−6 −5 −4 −3 −2 −1 0 1 2 3 4 5 6

46. $x - 5 \le -4$ *or* $2x - 7 \ge 3$

−6 −5 −4 −3 −2 −1 0 1 2 3 4 5 6

47. $7x + 4 \ge -17$ *or* $6x + 5 \ge -7$

−6 −5 −4 −3 −2 −1 0 1 2 3 4 5 6

48. $4x - 4 < -8$ *or* $4x - 4 < 12$

−6 −5 −4 −3 −2 −1 0 1 2 3 4 5 6

Solve.

49. $7 > -4x + 5$ *or* $10 \le -4x + 5$

50. $6 > 2x - 1$ *or* $-4 \le 2x - 1$

51. $3x - 7 > -10$ *or* $5x + 2 \leq 22$

52. $3x + 2 < 2$ *or* $4 - 2x < 14$

53. $-2x - 2 < -6$ *or* $-2x - 2 > 6$

54. $-3m - 7 < -5$ *or* $-3m - 7 > 5$

55. $\frac{2}{3}x - 14 < -\frac{5}{6}$ *or* $\frac{2}{3}x - 14 > \frac{5}{6}$

56. $\frac{1}{4} - 3x \leq -3.7$ *or* $\frac{1}{4} - 5x \geq 4.8$

57. $\frac{2x - 5}{6} \leq -3$ *or* $\frac{2x - 5}{6} \geq 4$

58. $\frac{7 - 3x}{5} < -4$ *or* $\frac{7 - 3x}{5} > 4$

C Solve.

59. *Pressure at Sea Depth.* The equation

$$P = 1 + \frac{d}{33}$$

gives the pressure P, in atmospheres (atm), at a depth of d feet in the sea. For what depths d is the pressure at least 1 atm and at most 7 atm?

60. *Temperatures of Liquids.* The formula

$$C = \tfrac{5}{9}(F - 32)$$

can be used to convert Fahrenheit temperatures F to Celsius temperatures C.

a) Gold is a liquid for Celsius temperatures C such that $1063° \leq C < 2660°$. Find such an inequality for the corresponding Fahrenheit temperatures.

b) Silver is a liquid for Celsius temperatures C such that $960.8° \leq C < 2180°$. Find such an inequality for the corresponding Fahrenheit temperatures.

61. *Aerobic Exercise.* In order to achieve maximum results from aerobic exercise, one should maintain one's heart rate at a certain level. A 30-year-old woman with a resting heart rate of 60 beats per minute should keep her heart rate between 138 and 162 beats per minute while exercising. She checks her pulse for 10 sec while exercising. What should the number of beats be?

62. *Minimizing Tolls.* A \$6.00 toll is charged to cross the bridge from mainland Florida to Sanibel Island. A six-month pass, costing \$50.00, reduces the toll to \$2.00. A six-month unlimited-trip pass costs \$300 and allows for unlimited crossings. How many crossings in six months does it take, on average, for the reduced-toll pass to be the most economical choice?

Data: leewayinfo.com

63. *Body Mass Index.* Refer to Exercises 71 and 72 in Exercise Set 1.4. Alexandra's height is 62 in. What weights W will allow Alexandra to keep her body mass index I in the 18.5–24.9 range?

64. *Body Mass Index.* Refer to Exercises 71 and 72 in Exercise Set 1.4. Josiah's height is 77 in. What weights W will allow Josiah to keep his body mass index in the 18.5–24.9 range?

65. *Young's Rule in Medicine.* Refer to Exercise 37 in Exercise Set 1.2. The dosage of a medication for an 8-year-old child must stay between 100 mg and 200 mg. Find the equivalent adult dosage.

66. *Young's Rule in Medicine.* Refer to Exercise 65. The dosage of a medication for a 5-year-old child must stay between 50 mg and 100 mg. Find the equivalent adult dosage.

Skill Maintenance

Solve. [1.1d]

67. $8y - 3 = 3 + 8y$

68. $-\frac{1}{2}t + 5 = -\frac{7}{2}t$

69. $20 = 4(3y - 7)$

70. $3x - (x - 1) = 19$

71. $-3 + 2x = 2x - 3$

72. $6(x - 5) = 2(x + 3)$

Synthesis

Solve.

73. $x - 10 < 5x + 6 \le x + 10$

74. $4m - 8 > 6m + 5$ *or* $5m - 8 < -2$

75. $-\frac{2}{15} \le \frac{2}{3}x - \frac{2}{5} \le \frac{2}{15}$

76. $2[5(3 - y) - 2(y - 2)] > y + 4$

77. $3x < 4 - 5x < 5 + 3x$

78. $2x - \frac{3}{4} < -\frac{1}{10}$ *or* $2x - \frac{3}{4} > \frac{1}{10}$

79. $x + 4 < 2x - 6 \le x + 12$

80. $2x + 3 \le x - 6$ *or* $3x - 2 \le 4x + 5$

Determine whether each sentence is true or false for all real numbers a, b, and c.

81. If $-b < -a$, then $a < b$.

82. If $a \le c$ and $c \le b$, then $b \ge a$.

83. If $a < c$ and $b < c$, then $a < b$.

84. If $-a < c$ and $-c > b$, then $a > b$.

85. What is the union of the set of all rational numbers with the set of all irrational numbers? the intersection?

1.6 Absolute-Value Equations and Inequalities

OBJECTIVES

- **a** Simplify expressions containing absolute-value symbols.
- **b** Find the distance between two points on the number line.
- **c** Solve equations with absolute-value expressions.
- **d** Solve equations with two absolute-value expressions.
- **e** Solve inequalities with absolute-value expressions.

a PROPERTIES OF ABSOLUTE VALUE

SKILL REVIEW *Find the absolute value of a real number.* [J4]

Find each absolute value.

1. $|-4|$ **2.** $|3.5|$

Answers: 1. 4 **2.** 3.5

MyLab Math VIDEO

We can think of the **absolute value** of a number as the number's distance from zero on the number line.

ABSOLUTE VALUE

The **absolute value** of x, denoted $|x|$, is defined as follows:

$$x \geq 0 \longrightarrow |x| = x; \qquad x < 0 \longrightarrow |x| = -x.$$

This definition tells us that, when x is nonnegative, the absolute value of x is x and, when x is negative, the absolute value of x is the opposite of x. For example, $|3| = 3$ and $|-3| = -(-3) = 3$. We see that absolute value is never negative.

Some simple properties of absolute value allow us to manipulate or simplify algebraic expressions.

PROPERTIES OF ABSOLUTE VALUE

a) $|ab| = |a| \cdot |b|$, for any real numbers a and b.
(The absolute value of a product is the product of the absolute values.)

b) $\left|\frac{a}{b}\right| = \frac{|a|}{|b|}$, for any real numbers a and b and $b \neq 0$.
(The absolute value of a quotient is the quotient of the absolute values.)

c) $|-a| = |a|$, for any real number a.
(The absolute value of the opposite of a number is the same as the absolute value of the number.)

EXAMPLES Simplify, leaving as little as possible inside the absolute-value signs.

1. $|5x| = |5| \cdot |x| = 5|x|$
2. $|-3y| = |-3| \cdot |y| = 3|y|$
3. $|7x^2| = |7| \cdot |x^2| = 7|x^2| = 7x^2$ Since x^2 is never negative for any number x
4. $\left|\frac{6x}{-3x^2}\right| = \left|\frac{-2}{x}\right| = \frac{|-2|}{|x|} = \frac{2}{|x|}$

Do Exercises 1–5. ▶

Simplify, leaving as little as possible inside the absolute-value signs.

1. $|7x|$
2. $|x^8|$
3. $|5a^2b|$
4. $\left|\frac{7a}{b^2}\right|$
5. $|-9x|$

b DISTANCE ON THE NUMBER LINE

The number line at right shows that the distance between -3 and 2 is 5.

Another way to find the distance between two numbers on the number line is to determine the absolute value of the difference, as follows:

$$|-3 - 2| = |-5| = 5, \quad \text{or} \quad |2 - (-3)| = |5| = 5.$$

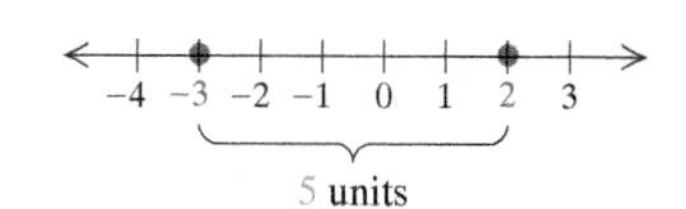

Note that the order in which we subtract does not matter because we are taking the absolute value after we have subtracted.

DISTANCE AND ABSOLUTE VALUE

For any real numbers a and b, the **distance** between them is $|a - b|$.

We should note that the distance is also $|b - a|$, because $a - b$ and $b - a$ are opposites and hence have the same absolute value.

EXAMPLE 5 Find the distance between -8 and -92 on the number line.

$|-8 - (-92)| = |84| = 84$, or $|-92 - (-8)| = |-84| = 84$ ■

EXAMPLE 6 Find the distance between x and 0 on the number line.

$$|x - 0| = |x|$$

Do Exercises 6–8. ▶

Find the distance between the points.

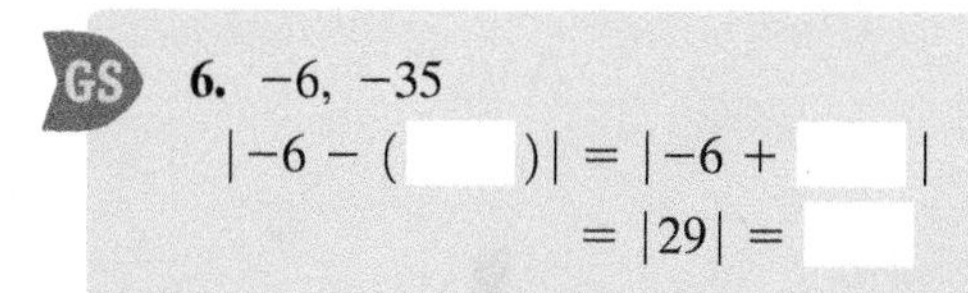

7. 19, 14
8. 0, p

c EQUATIONS WITH ABSOLUTE VALUE

EXAMPLE 7 Solve: $|x| = 4$. Then graph on the number line.

Note that $|x| = |x - 0|$, so that $|x - 0|$ is the distance from x to 0. Thus solutions of the equation $|x| = 4$, or $|x - 0| = 4$ are those numbers x whose distance from 0 is 4. Those numbers are -4 and 4. The solution set is $\{-4, 4\}$. The graph consists of just two points, as shown.

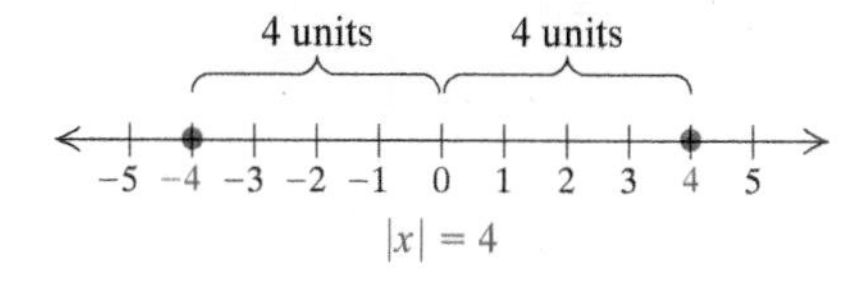

■

Answers

1. $7|x|$ **2.** x^8 **3.** $5a^2|b|$ **4.** $\frac{7|a|}{b^2}$
5. $9|x|$ **6.** 29 **7.** 5 **8.** $|p|$

Guided Solution:
6. −35, 35, 29

EXAMPLE 8 Solve: $|x| = 0$.

The only number whose absolute value is 0 is 0 itself. Thus the solution is 0. The solution set is $\{0\}$.

EXAMPLE 9 Solve: $|x| = -7$.

The absolute value of a number is always nonnegative. Thus there is no number whose absolute value is -7; consequently, the equation has no solution. The solution set is $\varnothing$.

Examples 7–9 lead us to the following principle for solving linear equations with absolute value.

THE ABSOLUTE-VALUE PRINCIPLE

For any positive number p and any algebraic expression X:

a) The solution of $|X| = p$ is those numbers that satisfy $X = -p$ *or* $X = p$.

b) The equation $|X| = 0$ is equivalent to the equation $X = 0$.

c) The equation $|X| = -p$ has no solution.

9. Solve: $|x| = 6$. Then graph on the number line.

10. Solve: $|x| = -6$.

11. Solve: $|p| = 0$.

Do Exercises 9–11.

We can use the absolute-value principle with the addition and multiplication principles to solve equations with absolute value.

EXAMPLE 10 Solve: $2|x| + 5 = 9$.

We first use the addition and multiplication principles to get $|x|$ by itself. Then we use the absolute-value principle.

$$2|x| + 5 = 9$$
$$2|x| = 4 \quad \text{Subtracting 5}$$
$$|x| = 2 \quad \text{Dividing by 2}$$
$$x = -2 \quad or \quad x = 2 \quad \text{Using the absolute-value principle}$$

The solutions are -2 and 2. The solution set is $\{-2, 2\}$.

Do Exercises 12–14.

Solve.

12. $|3x| = 6$

13. $4|x| + 10 = 27$

14. $3|x| - 2 = 10$

EXAMPLE 11 Solve: $|x - 2| = 3$.

We can consider solving this equation in two different ways.

Method 1: This allows us to see the meaning of the solutions graphically. The solution set consists of those numbers that are 3 units from 2 on the number line.

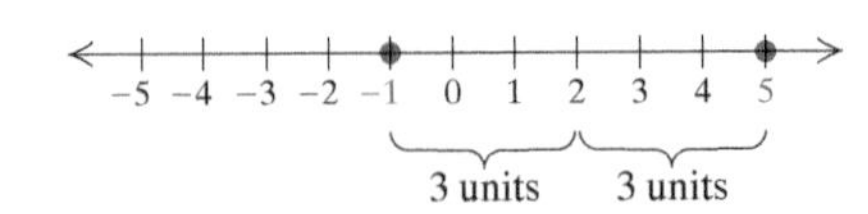

The solutions of $|x - 2| = 3$ are -1 and 5. The solution set is $\{-1, 5\}$.

Answers

9. $\{6, -6\}$;

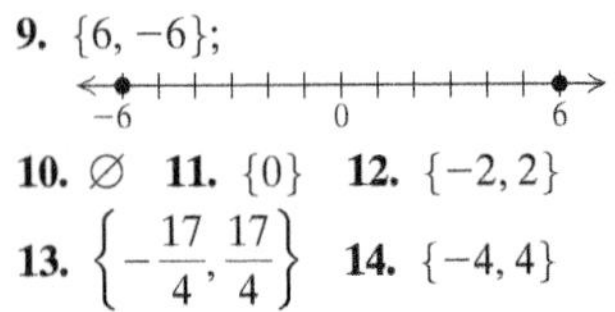

10. $\varnothing$ **11.** $\{0\}$ **12.** $\{-2, 2\}$
13. $\left\{-\frac{17}{4}, \frac{17}{4}\right\}$ **14.** $\{-4, 4\}$

Method 2: This method is more efficient. We use the absolute-value principle, replacing X with $x - 2$ and p with 3. Then we solve each equation separately.

$$|X| = p$$
$$|x - 2| = 3$$
$$x - 2 = -3 \quad or \quad x - 2 = 3 \qquad \text{Absolute-value principle}$$
$$x = -1 \quad or \quad x = 5$$

The solutions are -1 and 5. The solution set is $\{-1, 5\}$.

Do Exercise 15. ▶

15. Solve: $|x - 4| = 1$. Use two methods as in Example 11.

EXAMPLE 12 Solve: $|2x + 5| = 13$.

We use the absolute-value principle, replacing X with $2x + 5$ and p with 13:

$$|X| = p$$
$$|2x + 5| = 13$$
$$2x + 5 = -13 \quad or \quad 2x + 5 = 13 \qquad \text{Absolute-value principle}$$
$$2x = -18 \quad or \quad 2x = 8$$
$$x = -9 \quad or \quad x = 4.$$

The solutions are -9 and 4. The solution set is $\{-9, 4\}$.

Do Exercise 16. ▶

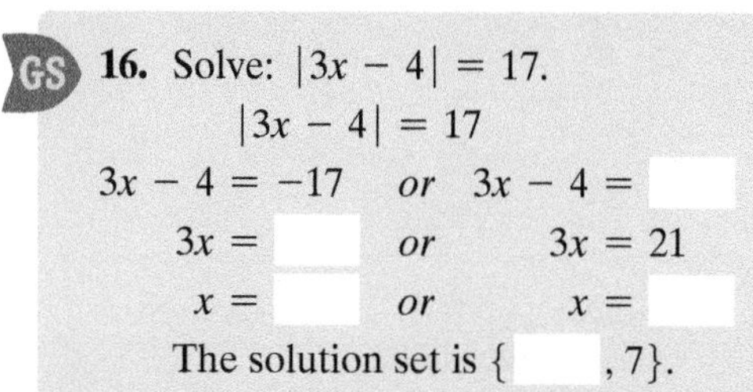

EXAMPLE 13 Solve: $|4 - 7x| = -8$.

Since absolute value is always nonnegative, this equation has no solution. The solution set is $\varnothing$.

Do Exercise 17. ▶

17. Solve: $|6 + 2x| = -3$.

d EQUATIONS WITH TWO ABSOLUTE-VALUE EXPRESSIONS

Sometimes equations have two absolute-value expressions. Consider $|a| = |b|$. This means that a and b are the same distance from 0. If a and b are the same distance from 0, then either they are the same number or they are opposites.

EXAMPLE 14 Solve: $|2x - 3| = |x + 5|$.

Either $2x - 3 = x + 5$ *or* $2x - 3 = -(x + 5)$. We solve each equation:

$$2x - 3 = x + 5 \quad or \quad 2x - 3 = -(x + 5)$$
$$x - 3 = 5 \quad or \quad 2x - 3 = -x - 5$$
$$x = 8 \quad or \quad 3x - 3 = -5$$
$$x = 8 \quad or \quad 3x = -2$$
$$x = 8 \quad or \quad x = -\tfrac{2}{3}.$$

The solutions are 8 and $-\frac{2}{3}$. The solution set is $\{8, -\frac{2}{3}\}$. ■

Answers

15. $\{3, 5\}$ **16.** $\left\{-\frac{13}{3}, 7\right\}$ **17.** $\varnothing$

Guided Solution:

16. $17, -13, -\frac{13}{3}, 7, -\frac{13}{3}$

EXAMPLE 15 Solve: $|x + 8| = |x - 5|$.

$$\begin{aligned} x + 8 = x - 5 \quad &or \quad x + 8 = -(x - 5) \\ 8 = -5 \quad &or \quad x + 8 = -x + 5 \\ 8 = -5 \quad &or \quad 2x = -3 \\ 8 = -5 \quad &or \quad x = -\tfrac{3}{2} \end{aligned}$$

The first equation has no solution. The solution of the second equation is $-\frac{3}{2}$. The solution set is $\{-\frac{3}{2}\}$.

◀ Do Exercises 18 and 19.

Solve.

18. $|5x - 3| = |x + 4|$

19. $|x - 3| = |x + 10|$

MyLab Math

ANIMATION

e INEQUALITIES WITH ABSOLUTE VALUE

We can extend our methods for solving equations with absolute value to those for solving inequalities with absolute value.

EXAMPLE 16 Solve: $|x| = 4$. Then graph on the number line.

From Example 7, we know that the solutions are -4 and 4. The solution set is $\{-4, 4\}$. The graph consists of just two points, as shown here.

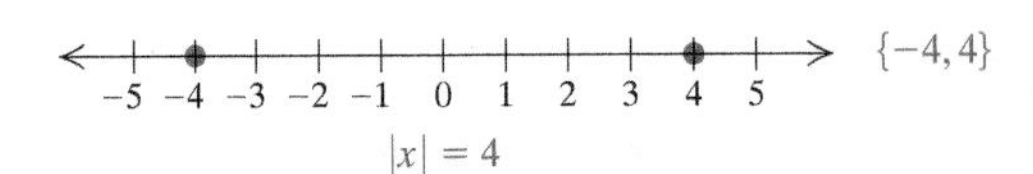

◀ Do Exercise 20.

20. Solve: $|x| = 5$. Then graph on the number line.

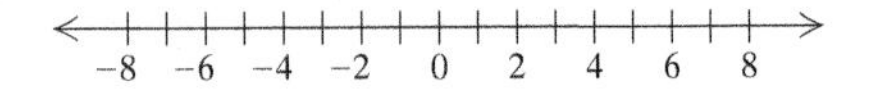

EXAMPLE 17 Solve: $|x| < 4$. Then graph.

The solutions of $|x| < 4$ are the solutions of $|x - 0| < 4$ and are those numbers x whose distance from 0 is less than 4. We can check by substituting or by looking at the number line that numbers like $-3, -2, -1, -\frac{1}{2}, -\frac{1}{4}, 0, \frac{1}{4}, \frac{1}{2}$, 1, 2, and 3 are all solutions. In fact, the solutions are all the real numbers x between -4 and 4. The solution set is $\{x \mid -4 < x < 4\}$ or, in interval notation, $(-4, 4)$. The graph is as follows.

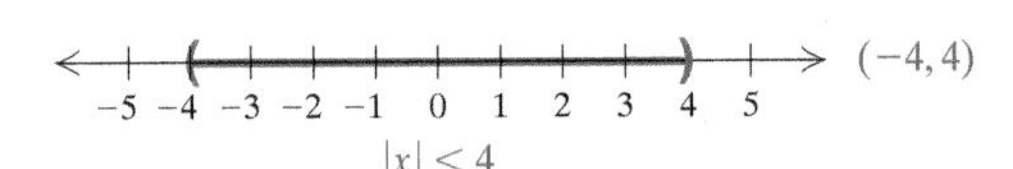

◀ Do Exercise 21.

21. Solve: $|x| < 5$. Then graph.

−8 −6 −4 −2 0 2 4 6 8

EXAMPLE 18 Solve: $|x| \geq 4$. Then graph.

The solutions of $|x| \geq 4$ are solutions of $|x - 0| \geq 4$ and are those numbers whose distance from 0 is greater than or equal to 4—in other words, those numbers x such that $x \leq -4$ *or* $x \geq 4$. The solution set is $\{x \mid x \leq -4 \text{ or } x \geq 4\}$, or $(-\infty, -4] \cup [4, \infty)$. The graph is as follows.

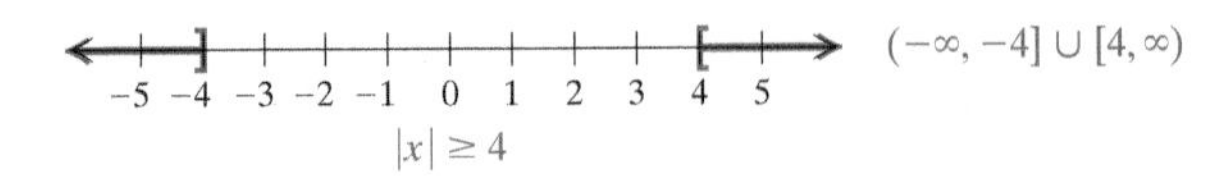

◀ Do Exercise 22.

22. Solve: $|x| \geq 5$. Then graph.

−8 −6 −4 −2 0 2 4 6 8

Examples 16–18 illustrate three cases of solving equations and inequalities with absolute value. The following is a general principle for solving equations and inequalities with absolute value.

Answers

18. $\left\{\frac{7}{4}, -\frac{1}{6}\right\}$ 19. $\left\{-\frac{7}{2}\right\}$

20. $\{-5, 5\}$;

−5 0 5

21. $\{x \mid -5 < x < 5\}$, or $(-5, 5)$;

−5 0 5

22. $\{x \mid x \leq -5 \text{ or } x \geq 5\}$, or $(-\infty, -5] \cup [5, \infty)$;

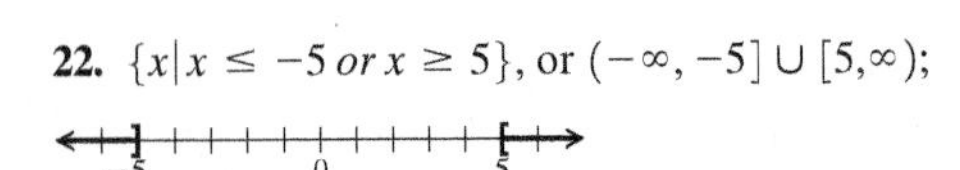

SOLUTIONS OF ABSOLUTE-VALUE EQUATIONS AND INEQUALITIES

For any positive number p and any algebraic expression X:

a) The solutions of $|X| = p$ are those numbers that satisfy $X = -p$ *or* $X = p$.

As an example, replacing X with $5x - 1$ and p with 8, we see that the solutions of $|5x - 1| = 8$ are those numbers x for which

$$5x - 1 = -8 \quad or \quad 5x - 1 = 8$$
$$5x = -7 \quad or \quad 5x = 9$$
$$x = -\tfrac{7}{5} \quad or \quad x = \tfrac{9}{5}.$$

The solution set is $\{-\tfrac{7}{5}, \tfrac{9}{5}\}$.

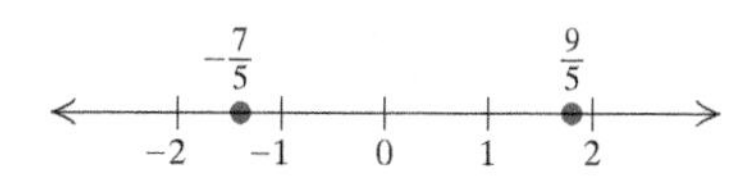

b) The solutions of $|X| < p$ are those numbers that satisfy $-p < X < p$.

As an example, replacing X with $6x + 7$ and p with 5, we see that the solutions of $|6x + 7| < 5$ are those numbers x for which

$$-5 < 6x + 7 < 5$$
$$-12 < 6x < -2$$
$$-2 < x < -\tfrac{1}{3}.$$

The solution set is $\{x \mid -2 < x < -\tfrac{1}{3}\}$, or $(-2, -\tfrac{1}{3})$.

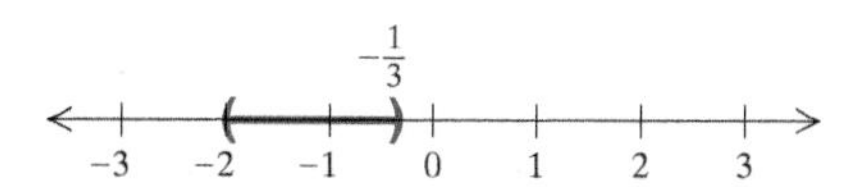

c) The solutions of $|X| > p$ are those numbers that satisfy $X < -p$ *or* $X > p$.

As an example, replacing X with $2x - 9$ and p with 4, we see that the solutions of $|2x - 9| > 4$ are those numbers x for which

$$2x - 9 < -4 \quad or \quad 2x - 9 > 4$$
$$2x < 5 \quad or \quad 2x > 13$$
$$x < \tfrac{5}{2} \quad or \quad x > \tfrac{13}{2}.$$

The solution set is $\{x \mid x < \tfrac{5}{2} \text{ or } x > \tfrac{13}{2}\}$, or $(-\infty, \tfrac{5}{2}) \cup (\tfrac{13}{2}, \infty)$.

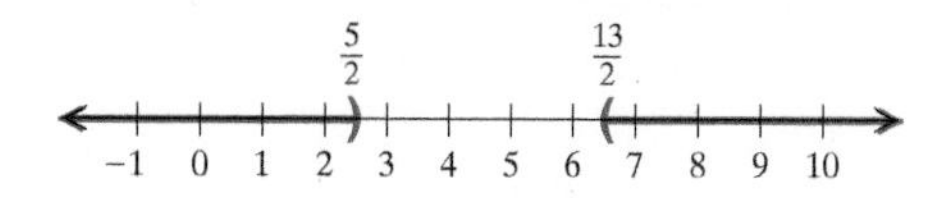

EXAMPLE 19 Solve: $|3x - 2| < 4$. Then graph.

We use part (b). In this case, X is $3x - 2$ and p is 4:

$$|X| < p$$

$$|3x - 2| < 4 \qquad \text{Replacing } X \text{ with } 3x - 2 \text{ and } p \text{ with } 4$$

$$-4 < 3x - 2 < 4$$

$$-2 < 3x < 6$$

$$-\tfrac{2}{3} < x < 2.$$

The solution set is $\{x \mid -\frac{2}{3} < x < 2\}$, or $(-\frac{2}{3}, 2)$. The graph is as follows.

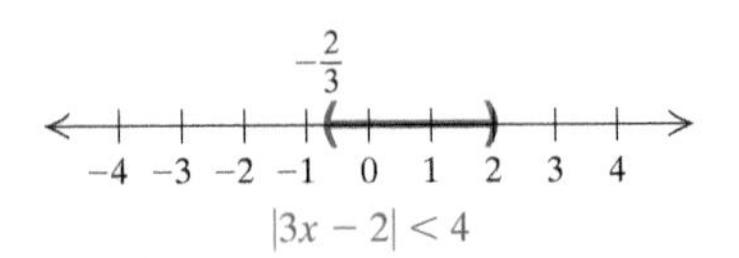

$|3x - 2| < 4$

EXAMPLE 20 Solve: $|8 - 4x| \le 5$. Then graph.

We use part (b). In this case, X is $8 - 4x$ and p is 5:

$$|X| \le p$$

$$|8 - 4x| \le 5 \qquad \text{Replacing } X \text{ with } 8 - 4x \text{ and } p \text{ with } 5$$

$$-5 \le 8 - 4x \le 5$$

$$-13 \le -4x \le -3$$

$$\tfrac{13}{4} \ge x \ge \tfrac{3}{4}. \qquad \text{Dividing by } -4 \text{ and reversing the inequality symbols}$$

The solution set is $\{x \mid \frac{13}{4} \ge x \ge \frac{3}{4}\}$, or $\{x \mid \frac{3}{4} \le x \le \frac{13}{4}\}$, or $[\frac{3}{4}, \frac{13}{4}]$.

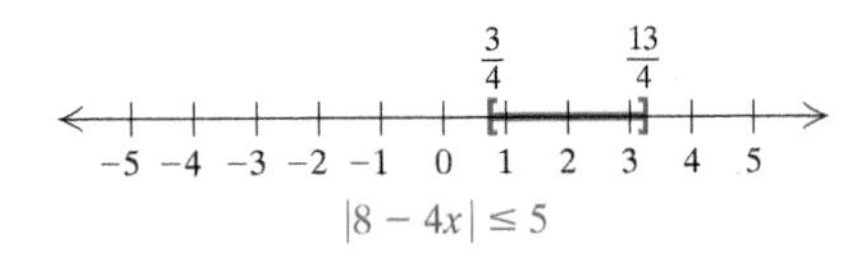

$|8 - 4x| \le 5$

EXAMPLE 21 Solve: $|4x + 2| \ge 6$. Then graph.

We use part (c). In this case, X is $4x + 2$ and p is 6:

$$|X| \ge p$$

$$|4x + 2| \ge 6 \qquad \text{Replacing } X \text{ with } 4x + 2 \text{ and } p \text{ with } 6$$

$$4x + 2 \le -6 \quad \text{or} \quad 4x + 2 \ge 6$$

$$4x \le -8 \quad \text{or} \quad 4x \ge 4$$

$$x \le -2 \quad \text{or} \quad x \ge 1.$$

The solution set is $\{x \mid x \le -2 \text{ or } x \ge 1\}$, or $(-\infty, -2] \cup [1, \infty)$.

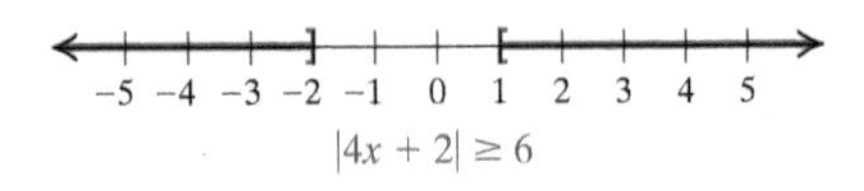

$|4x + 2| \ge 6$

Do Exercises 23–25.

Solve. Then graph.

23. $|2x - 3| < 7$

−8 −6 −4 −2 0 2 4 6 8

GS **24.** $|7 - 3x| \le 4$

$$\_\_\_\_ \le 7 - 3x \le 4$$

$$-11 \le -3x \le -3$$

$$\frac{-11}{-3} \ \_\_\_\_ \ \frac{-3x}{-3} \ \_\_\_\_ \ \frac{-3}{-3}$$

$$\frac{11}{3} \ge \_\_\_\_ \ge 1$$

The solution set is $\{x \mid \_\_\_\_ \le x \le \frac{11}{3}\}$, or $[1, \frac{11}{3}]$.

−8 −6 −4 −2 0 2 4 6 8

25. $|3x + 2| \ge 5$

$$3x + 2 \le \_\_\_\_ \quad \text{or} \quad 3x + 2 \ge 5$$

$$3x \le -7 \quad \text{or} \quad 3x \ge \_\_\_\_$$

$$x \le \_\_\_\_ \quad \text{or} \quad x \ge 1$$

The solution set is $\{x \mid x \le -\frac{7}{3} \text{ or } x \ge \_\_\_\_\}$, or $(\_\_\_\_, -\frac{7}{3}] \cup [\_\_\_\_, \infty)$.

−8 −6 −4 −2 0 2 4 6 8

Answers

23. $\{x \mid -2 < x < 5\}$, or $(-2, 5)$;

24. $\left\{x \mid 1 \le x \le \frac{11}{3}\right\}$, or $\left[1, \frac{11}{3}\right]$;

25. $\left\{x \mid x \le -\frac{7}{3} \text{ or } x \ge 1\right\}$, or $\left(-\infty, -\frac{7}{3}\right] \cup [1, \infty)$;

Guided Solutions:

24. $-4, \ge, \ge, x, 1$

25. $-5, 3, -\frac{7}{3}, 1, -\infty, 1$

1.6 Exercise Set

FOR EXTRA HELP MyLab Math

✓ Check Your Understanding

Reading Check Determine whether each statement is true or false.

RC1. The number a is $|a|$ units from 0 on the number line.

RC2. $|x|$ is never negative.

RC3. $|x|$ is always positive.

RC4. If two numbers are opposites, their absolute values are equal.

RC5. The distance between -53 and 46 is $|-53 - 46|$.

RC6. There are exactly two solutions of $|2x| = 10$.

Concept Check Choose from the column on the right the graph of the solution set of each equation or inequality.

CC1. $|x| > 3$

CC2. $|x| \geq 3$

CC3. $|x| < 3$

CC4. $|x| = 3$

CC5. $|x| \leq 3$

CC6. $|x| > -3$

a) −5 −4 −3 −2 −1 0 1 2 3 4 5

b) −5 −4 −3 −2 −1 0 1 2 3 4 5

c) −5 −4 −3 −2 −1 0 1 2 3 4 5

d) −5 −4 −3 −2 −1 0 1 2 3 4 5

e) −5 −4 −3 −2 −1 0 1 2 3 4 5

f) −5 −4 −3 −2 −1 0 1 2 3 4 5

a Simplify, leaving as little as possible inside absolute-value signs.

1. $|9x|$

2. $|26x|$

3. $|2x^2|$

4. $|8x^2|$

5. $|-2x^2|$

6. $|-20x^2|$

7. $|-6y|$

8. $|-17y|$

9. $\left|\frac{-2}{x}\right|$

10. $\left|\frac{y}{3}\right|$

11. $\left|\frac{x^2}{-y}\right|$

12. $\left|\frac{x^4}{-y}\right|$

13. $\left|\frac{-8x^2}{2x}\right|$

14. $\left|\frac{-9y^2}{3y}\right|$

15. $\left|\frac{4y^3}{-12y}\right|$

16. $\left|\frac{5x^3}{-25x}\right|$

b Find the distance between the points on the number line.

17. $-8, -46$

18. $-7, -32$

19. $36, 17$

20. $52, 18$

21. $-3.9, 2.4$

22. $-1.8, -3.7$

23. $-5, 0$

24. $\frac{2}{3}, -\frac{5}{6}$

c Solve.

25. $|x| = 3$

26. $|x| = 5$

27. $|x| = -3$

28. $|x| = -9$

29. $|q| = 0$

30. $|y| = 7.4$

31. $|x - 3| = 12$

32. $|3x - 2| = 6$

33. $|2x - 3| = 4$

34. $|5x + 2| = 3$

35. $|4x - 9| = 14$

36. $|9y - 2| = 17$

37. $|x| + 7 = 18$

38. $|x| - 2 = 6.3$

39. $574 = 283 + |t|$

40. $-562 = -2000 + |x|$

41. $|5x| = 40$

42. $|2y| = 18$

43. $|3x| - 4 = 17$

44. $|6x| + 8 = 32$

45. $7|w| - 3 = 11$

46. $5|x| + 10 = 26$

47. $\left|\frac{2x - 1}{3}\right| = 5$

48. $\left|\frac{4 - 5x}{6}\right| = 7$

49. $|m + 5| + 9 = 16$

50. $|t - 7| - 5 = 4$

51. $10 - |2x - 1| = 4$

52. $2|2x - 7| + 11 = 25$

53. $|3x - 4| = -2$

54. $|x - 6| = -8$

55. $|\frac{5}{9} + 3x| = \frac{1}{6}$

56. $|\frac{2}{3} - 4x| = \frac{4}{5}$

d Solve.

57. $|3x + 4| = |x - 7|$

58. $|2x - 8| = |x + 3|$

59. $|x + 3| = |x - 6|$

60. $|x - 15| = |x + 8|$

61. $|2a + 4| = |3a - 1|$

62. $|5p + 7| = |4p + 3|$

63. $|y - 3| = |3 - y|$

64. $|m - 7| = |7 - m|$

65. $|5 - p| = |p + 8|$

66. $|8 - q| = |q + 19|$

67. $\left|\dfrac{2x - 3}{6}\right| = \left|\dfrac{4 - 5x}{8}\right|$

68. $\left|\dfrac{6 - 8x}{5}\right| = \left|\dfrac{7 + 3x}{2}\right|$

69. $|\frac{1}{2}x - 5| = |\frac{1}{4}x + 3|$

70. $|2 - \frac{2}{3}x| = |4 + \frac{7}{8}x|$

e Solve.

71. $|x| < 3$

72. $|x| \le 5$

73. $|x| \ge 2$

74. $|y| > 12$

75. $|x - 1| < 1$

76. $|x + 4| \le 9$

77. $5|x + 4| \le 10$

78. $2|x - 2| > 6$

79. $|2x - 3| \le 4$

80. $|5x + 2| \le 3$

81. $|2y - 7| > 10$

82. $|3y - 4| > 8$

83. $|4x - 9| \ge 14$

84. $|9y - 2| \ge 17$

85. $|y - 3| < 12$

86. $|p - 2| < 6$

87. $|2x + 3| \le 4$

88. $|5x + 2| \le 13$

89. $|4 - 3y| > 8$

90. $|7 - 2y| > 5$

91. $|9 - 4x| \ge 14$

92. $|2 - 9p| \ge 17$

93. $|3 - 4x| < 21$

94. $|-5 - 7x| \le 30$

95. $\left|\frac{1}{2} + 3x\right| \geq 12$

96. $\left|\frac{1}{4}y - 6\right| > 24$

97. $\left|\frac{x - 7}{3}\right| < 4$

98. $\left|\frac{x + 5}{4}\right| \leq 2$

99. $\left|\frac{2 - 5x}{4}\right| \geq \frac{2}{3}$

100. $\left|\frac{1 + 3x}{5}\right| > \frac{7}{8}$

101. $|m + 5| + 9 \leq 16$

102. $|t - 7| + 3 \geq 4$

103. $7 - |3 - 2x| \geq 5$

104. $16 \leq |2x - 3| + 9$

105. $\left|\frac{2x - 1}{3}\right| \leq 1$

106. $\left|\frac{3x - 2}{5}\right| \geq 1$

Skill Maintenance

Solve. [1.4c]

107. $-11x + 2x \geq -36$

108. $\frac{7}{9}y < -\frac{7}{10}$

109. $2(r - 1) + 4 < 3(r - 2) - 8$

Solve.

110. $8 > -x \geq 4$ [1.5a]

111. $-3 \leq 2x + 5$ *or* $10 > 2x - 1$ [1.5b]

112. $-2 \leq 6x - 4 < 20$ [1.5a]

Synthesis

113. *Motion of a Spring.* A weighted spring is bouncing up and down so that its distance d above the ground satisfies the inequality $|d - 6 \text{ ft}| \leq \frac{1}{2}$ ft. Find all possible distances d.

114. *Container Sizes.* A container company is manufacturing rectangular boxes of various sizes. The length of any box must exceed the width by at least 3 in., but the perimeter cannot exceed 24 in. What widths are possible?

$$l \geq w + 3,$$
$$2l + 2w \leq 24$$

Solve.

115. $|x + 5| > x$

116. $1 - \left|\frac{1}{4}x + 8\right| = \frac{3}{4}$

117. $|7x - 2| = x + 4$

118. $|x - 1| = x - 1$

119. $|x - 6| \leq -8$

120. $|3x - 4| > -2$

Find an equivalent inequality with absolute value.

121. $-3 < x < 3$

122. $-5 \leq y \leq 5$

123. $x \leq -6$ *or* $x \geq 6$

124. $-5 < x < 1$

125. $x < -8$ *or* $x > 2$

CHAPTER

1 Summary and Review

Vocabulary Reinforcement

Complete each statement with the correct term from the column on the right. Some of the choices may not be used and some may be used more than once.

addition principle
multiplication principle
union
set-builder
empty set
absolute value
disjunction
inequality
intersection
distance
interval
disjoint sets
compound
conjunction

1. A(n) ______________ is a sentence containing $<, \leq, >, \geq$, or $\neq$. [1.4a]

2. Using ______________ notation, we write the solution set for $x < 7$ as $\{x|x < 7\}$. [1.4b]

3. Using ______________ notation, we write the solution set of $-5 \leq y < 16$ as $[-5, 16)$. [1.4b]

4. The ______________ of two sets A and B is the set of all members that are common to A and B. [1.5a]

5. When two or more sentences are joined by the word *and* to make a compound sentence, the new sentence is called a(n) ______________ of the sentences. [1.5a]

6. When two sets have no elements in common, the intersection of the two sets is the ______________. [1.5a]

7. Two sets with an empty intersection are said to be ______________. [1.5a]

8. The ______________ of two sets A and B is the collection of elements belonging to A and/or B. [1.5b]

9. When two or more sentences are joined by the word *or* to make a compound sentence, the new sentence is called a(n) ______________ of the sentences. [1.5b]

10. The ______________ for equations states that for any real numbers a, b, and c, $a = b$ is equivalent to $a + c = b + c$. [1.1b]

11. The ______________ for equations states that for any real numbers a, b, and c, $c \neq 0$, $a = b$ is equivalent to $a \cdot c = b \cdot c$. [1.1c]

12. For any real numbers a and b, the ______________ between them is $|a - b|$. [1.6b]

Concept Reinforcement

Determine whether each statement is true or false.

______________ 1. For any real numbers a, b, and c, $c \neq 0$, $a = b$ is equivalent to $a \cdot c = b \cdot c$. [1.1c]

______________ 2. When we solve $3B = mt + nt$ for t, we get $t = \dfrac{3B - mt}{n}$. [1.2a]

______________ 3. For any real numbers a, b, and c, $c \neq 0$, $a \leq b$ is equivalent to $ac \leq bc$. [1.4c]

______________ 4. The inequalities $x < 2$ and $x \leq 1$ are equivalent. [1.4c]

______________ 5. If x is negative, $|x| = -x$. [1.6a]

______________ 6. $|x|$ is always positive. [1.6a]

______________ 7. $|a - b| = |b - a|$. [1.6b]

Study Guide

Objective 1.1a Determine whether a given number is a solution of a given equation.

Example Determine whether 10 is a solution of $5x - 6 = 44$.

$$\begin{array}{c|c} 5x - 6 & 44 \\ \hline 5 \cdot 10 - 6 \ ? & 44 \\ 50 - 6 & \\ 44 & \end{array} \quad \text{TRUE}$$

The number 10 is a solution of the equation.

Practice Exercise

1. Determine whether -3 is a solution of $28 - 7x = 7$.

Objective 1.1d Solve equations using the addition principle and the multiplication principle together, removing parentheses where appropriate.

Example Solve: $10y - 2(3y + 1) = 6$.

$$10y - 2(3y + 1) = 6$$
$$10y - 6y - 2 = 6 \quad \text{Removing parentheses}$$
$$4y - 2 = 6 \quad \text{Collecting like terms}$$
$$4y = 8 \quad \text{Adding 2}$$
$$y = 2 \quad \text{Dividing by 4}$$

The solution is 2.

Practice Exercise

2. Solve: $2(x + 2) = 5(x - 4)$.

Objective 1.2a Evaluate formulas and solve a formula for a specified letter.

Example Solve for z: $T = \frac{w + z}{3}$.

$$T = \frac{w + z}{3}$$
$$3 \cdot T = 3\left(\frac{w + z}{3}\right) \quad \text{Multiplying by 3 to clear the fraction}$$
$$3T = w + z \quad \text{Simplifying}$$
$$3T - w = z \quad \text{Subtracting } w$$

Practice Exercise

3. Solve for h: $F = \frac{1}{4}gh$.

Objective 1.4a Determine whether a given number is a solution of an inequality.

Example Determine whether -3 and 1 are solutions of the inequality $4 - x \geq 2 - 5x$.

We substitute -3 for x and get

$$4 - (-3) \geq 2 - 5(-3), \text{ or } 7 \geq 17,$$

a *false* sentence. Therefore, -3 is not a solution.

We substitute 1 for x and get

$$4 - 1 \geq 2 - 5 \cdot 1, \text{ or } 3 \geq -3,$$

a *true* sentence. Therefore, 1 is a solution.

Practice Exercise

4. Determine whether -2 and 5 are solutions of the inequality $8 - 3x \leq 3x + 6$.

Objective 1.4b Write interval notation for the solution set of an inequality.

Example Write interval notation for the solution set.

a) $\{x \mid x \leq -12\} = (-\infty, -12]$
b) $\{r \mid r > -1\} = (-1, \infty)$
c) $\{y \mid -8 \leq y < 9\} = [-8, 9)$
d) $\{x \mid 0 \geq x \geq -6\} = [-6, 0]$
e) $\{c \mid -25 < c \leq 25\} = (-25, 25]$

Practice Exercise

5. Write interval notation for the solution set.
a) $\{t \mid t < -8\}$
b) $\{x \mid -7 \leq x < 10\}$
c) $\{b \mid b \geq 3\}$

Objective 1.4c Solve an inequality using the addition principle and the multiplication principle and then graph the inequality.

Example Solve and graph: $6x - 7 \leq 3x + 2$.

$6x - 7 \leq 3x + 2$
$3x - 7 \leq 2$ Subtracting $3x$
$3x \leq 9$ Adding 7
$x \leq 3$ Dividing by 3

The solution set is $\{x \mid x \leq 3\}$, or $(-\infty, 3]$. We graph the solution set.

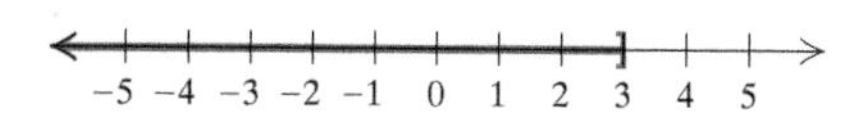

Practice Exercise

6. Solve and graph: $5y + 5 < 2y - 1$.

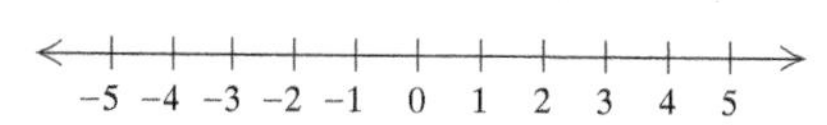

Objective 1.5a Find the intersection of two sets. Solve and graph conjunctions of inequalities.

Example Solve and graph: $-5 < 2x - 3 \leq 3$.

$-5 < 2x - 3 \leq 3$
$-2 < 2x \leq 6$ Adding 3
$-1 < x \leq 3$ Dividing by 2

The solution set is $\{x \mid -1 < x \leq 3\}$, or $(-1, 3]$. We graph the solution set.

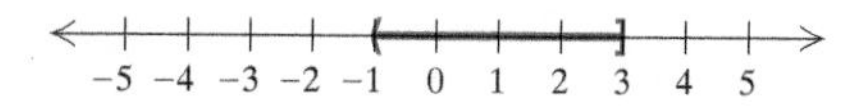

Practice Exercise

7. Solve and graph: $-4 \leq 5z + 6 < 11$.

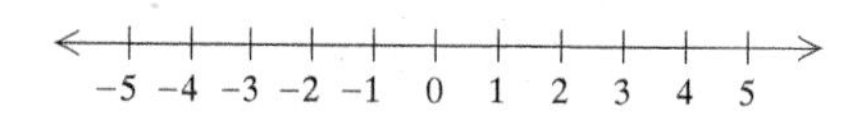

Objective 1.5b Find the union of two sets. Solve and graph disjunctions of inequalities.

Example Solve and graph:
$2x + 1 \leq -5$ *or* $3x + 1 > 7$.

$2x + 1 \leq -5$ *or* $3x + 1 > 7$
$2x \leq -6$ *or* $3x > 6$
$x \leq -3$ *or* $x > 2$

The solution set is $\{x \mid x \leq -3 \text{ or } x > 2\}$, or $(-\infty, -3] \cup (2, \infty)$. We graph the solution set.

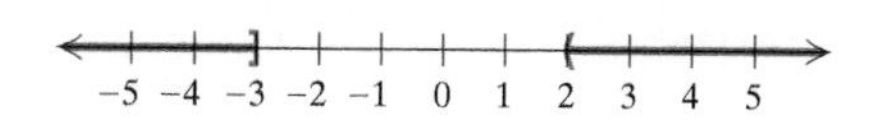

Practice Exercise

8. Solve and graph: $z + 4 < 3$ *or* $4z + 1 \geq 5$.

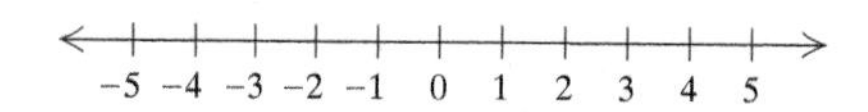

Objective 1.6a Simplify expressions containing absolute-value symbols.

Example Simplify: $|-6c|$.

$$|-6c| = |-6| \cdot |c| = 6|c|$$

Practice Exercise

9. Simplify: $|8y^2|$.

Objective 1.6b Find the distance between two points on the number line.

Example Find the distance between -10 and 3 on the number line.

$$|-10 - 3| = |-13| = 13$$

Practice Exercise

10. Find the distance between 8 and -20 on the number line.

Objective 1.6c Solve equations with absolute-value expressions.

Example Solve: $|y - 2| = 1$.

$$y - 2 = -1 \quad or \quad y - 2 = 1$$
$$y = 1 \quad or \quad y = 3$$

The solution set is $\{1, 3\}$.

Practice Exercise

11. Solve: $|5x - 1| = 9$.

Objective 1.6d Solve equations with two absolute-value expressions.

Example Solve: $|4x - 4| = |2x + 8|$.

$$4x - 4 = 2x + 8 \quad or \quad 4x - 4 = -(2x + 8)$$
$$2x - 4 = 8 \quad or \quad 4x - 4 = -2x - 8$$
$$2x = 12 \quad or \quad 6x - 4 = -8$$
$$x = 6 \quad or \quad 6x = -4$$
$$x = 6 \quad or \quad x = -\frac{2}{3}$$

The solution set is $\left\{6, -\frac{2}{3}\right\}$.

Practice Exercise

12. Solve: $|z + 4| = |3z - 2|$.

Objective 1.6e Solve inequalities with absolute-value expressions.

Example Solve: **(a)** $|5x + 3| < 2$; **(b)** $|x + 3| \geq 1$.

a) $|5x + 3| < 2$

$$-2 < 5x + 3 < 2$$
$$-5 < 5x < -1$$
$$-1 < x < -\frac{1}{5}$$

The solution set is $\left\{x \middle| -1 < x < -\frac{1}{5}\right\}$, or $\left(-1, -\frac{1}{5}\right)$.

b) $|x + 3| \geq 1$

$$x + 3 \leq -1 \quad or \quad x + 3 \geq 1$$
$$x \leq -4 \quad or \quad x \geq -2$$

The solution set is $\{x | x \leq -4 \; or \; x \geq -2\}$, or $(-\infty, -4] \cup [-2, \infty)$.

Practice Exercise

13. Solve: **(a)** $|2x + 3| < 5$; **(b)** $|3x + 2| \geq 8$.

Review Exercises

Solve. [1.1b, c, d]

1. $-11 + y = -3$

2. $-7x = -3$

3. $-\frac{5}{3}x + \frac{7}{3} = -5$

4. $6(2x - 1) = 3 - (x + 10)$

5. $2.4x + 1.5 = 1.02$

6. $2(3 - x) - 4(x + 1) = 7(1 - x)$

Solve for the indicated letter. [1.2a]

7. $C = \frac{4}{11}d + 3$, for d

8. $A = 2a - 3b$, for b

9. *Interstate Mile Markers.* If you are traveling on a U.S. interstate highway, you will notice numbered markers every mile to indicate your location in case of an accident or other emergency. In many states, the numbers on the markers increase from west to east. The sum of two consecutive mile markers on I-70 in Utah is 371. Find the numbers on the markers. [1.3a]

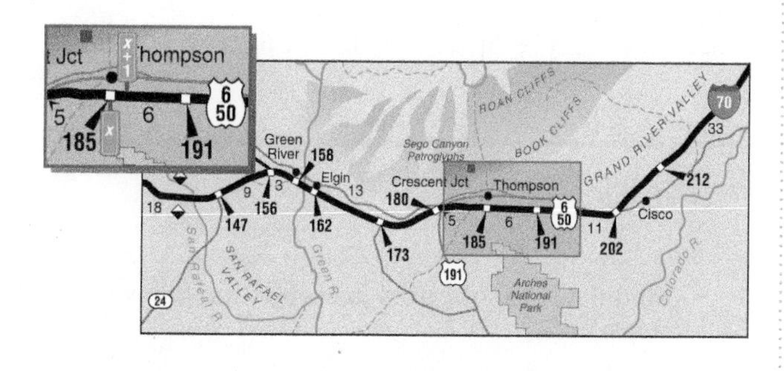

10. *Rope Cutting.* A piece of rope 27 m long is cut into two pieces so that one piece is four-fifths as long as the other. Find the length of each piece. [1.3a]

11. *Population Growth.* The population of Newcastle grew 12% from one year to the next to a total of 179,200. What was the former population? [1.3a]

12. *Moving Walkway.* A moving walkway in an airport is 360 ft long and moves at a speed of 6 ft/sec. If Arnie walks at a speed of 3 ft/sec, how long will it take him to walk the length of the moving walkway? [1.3b]

Write interval notation for the given set or graph. [1.4b]

13. $\{x \mid -8 \le x < 9\}$

14.

−80 −60 −40 −20 0 20 40 60 80

Solve and graph. Write interval notation for the solution set. [1.4c]

15. $x - 2 \le -4$

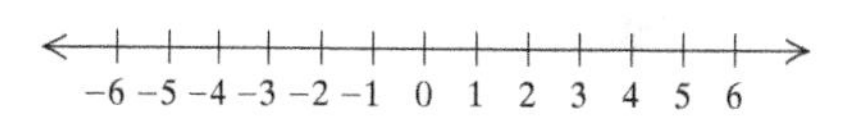

16. $x + 5 > 6$

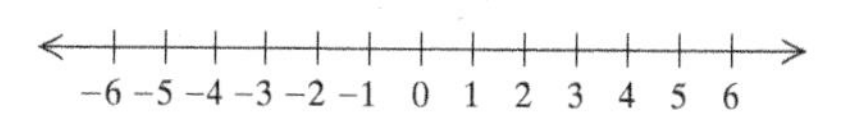

Solve. [1.4c]

17. $a + 7 \le -14$

18. $y - 5 \ge -12$

19. $4y > -16$

20. $-0.3y < 9$

21. $-6x - 5 < 13$

22. $4y + 3 \le -6y - 9$

23. $-\frac{1}{2}x - \frac{1}{4} > \frac{1}{2} - \frac{1}{4}x$

24. $0.3y - 8 < 2.6y + 15$

25. $-2(x - 5) \ge 6(x + 7) - 12$

26. *Moving Costs.* Metro Movers charges $85 plus $40 per hour to move households across town. Champion Moving charges $60 per hour for cross-town moves. For what lengths of time is Champion more expensive? [1.4d]

27. *Investments.* Joe plans to invest $30,000, part at 3% and part at 4%, for one year. What is the most that can be invested at 3% in order to make at least $1100 interest in one year? [1.4d]

Graph and write interval notation. [1.5a, b]

28. $-2 \le x < 5$

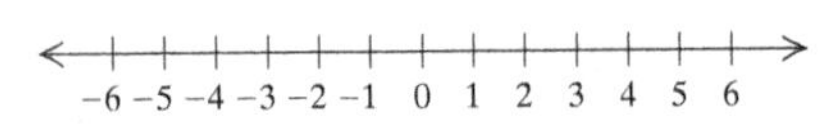

29. $x \le -2$ *or* $x > 5$

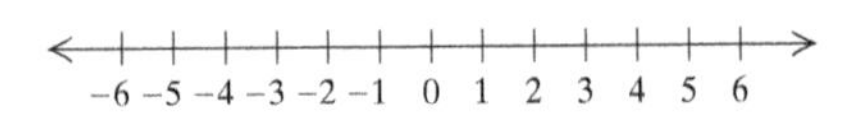

30. Find the intersection:

$\{1, 2, 5, 6, 9\} \cap \{1, 3, 5, 9\}$. [1.5a]

31. Find the union:

$\{1, 2, 5, 6, 9\} \cup \{1, 3, 5, 9\}$. [1.5b]

Solve. [1.5a, b]

32. $2x - 5 < -7$ *and* $3x + 8 \ge 14$

33. $-4 < x + 3 \le 5$

34. $-15 < -4x - 5 < 0$

35. $3x < -9$ *or* $-5x < -5$

36. $2x + 5 < -17$ *or* $-4x + 10 \le 34$

37. $2x + 7 \le -5$ *or* $x + 7 \ge 15$

Simplify. [1.6a]

38. $\left|-\frac{3}{x}\right|$ **39.** $\left|\frac{2x}{y^2}\right|$ **40.** $\left|\frac{12y}{-3y^2}\right|$

41. Find the distance between -23 and 39. [1.6b]

Solve. [1.6c, d]

42. $|x| = 6$ **43.** $|x - 2| = 7$

44. $|2x + 5| = |x - 9|$ **45.** $|5x + 6| = -8$

Solve. [1.6e]

46. $|2x + 5| < 12$ **47.** $|x| \ge 3.5$

48. $|3x - 4| \ge 15$ **49.** $|x| < 0$

Invasive Species. The equation

$$S = 1.75t + 120$$

can be used to estimate the number of invasive species S in the Great Lakes t years after 1970—that is, $t =$ corresponds to 1970, $t = 10$ corresponds to 1980, and so on. Use this equation in Exercises 50 and 51.

Data: wbez.org

50. Estimate the number of invasive species in the Great Lakes in 2010. [1.2a], [1.3a]

A. About 138 species
B. About 190 species
C. About 210 species
D. About 3638 species

51. For what years is the number of invasive species in the Great Lakes predicted to be between 180 and 200? [1.5c]

A. Between 2004 and 2016
B. Between 2002 and 2010
C. Between 2030 and 2050
D. Between 2000 and 2080

Synthesis

52. Solve: $|2x + 5| \le |x + 3|$. [1.6d, e]

Understanding Through Discussion and Writing

1. Explain in your own words why the inequality symbol must be reversed when both sides of an inequality are multiplied or divided by a negative number. [1.4c]

2. Explain in your own words why the solutions of the inequality $|x + 5| \le 2$ can be interpreted as "all those numbers x whose distance from -5 is at most 2 units." [1.6e]

3. Describe the circumstances under which, for intervals, $[a, b] \cup [c, d] = [a, d]$. [1.5b]

4. Explain in your own words why the interval $[6, \infty)$ is only part of the solution set of $|x| \ge 6$. [1.6e]

5. Find the error or errors in each of the following steps:

$7 - 9x + 6x < -9(x + 2) + 10x$

$7 - 9x + 6x < -9x + 2 + 10x$ **(1)**

$7 + 6x > 2 + 10x$ **(2)**

$-4x > 8$ **(3)**

$x > -2$. [1.4c] **(4)**

6. Explain why the conjunction $3 < x$ *and* $x < 5$ is equivalent to $3 < x < 5$, but the disjunction $3 < x$ *or* $x < 5$ is not. [1.5a, b]

CHAPTER 1 Test

For Extra Help: For step-by-step test solutions, access the Chapter Test Prep Videos in MyLab Math.

Solve.

1. $x + 7 = 5$

2. $-12x = -8$

3. $x - \frac{3}{5} = \frac{2}{3}$

4. $3y - 4 = 8$

5. $1.7y - 0.1 = 2.1 - 0.3y$

6. $5(3x + 6) = 6 - (x + 8)$

7. Solve $A = 3B - C$ for B.

8. Solve $m = n - nt$ for n.

Solve.

9. *Room Dimensions.* A rectangular room has a perimeter of 48 ft. The width is two-thirds of the length. What are the dimensions of the room?

10. *Copy Budget.* Copy Solutions rents a copier for \$240 per month plus 1.5¢ per copy. A law firm needs to lease a copy machine for use during a special case that they anticipate will take 3 months. If they allot a budget of \$1500 for copying costs, how many copies can they make?

11. *Population Decrease.* The population of Baytown dropped 12% from one year to the next to a total of 158,400. What was the former population?

12. *Angles in a Triangle.* The measures of the angles of a triangle are three consecutive integers. Find the measures of the angles.

13. *Boating.* A paddleboat moves at a rate of 12 mph in still water. If the river's current moves at a rate of 3 mph, how long will it take the boat to travel 36 mi downstream? 36 mi upstream?

Write interval notation for the given set or graph.

14. $\{x \mid -3 < x \leq 2\}$

15. 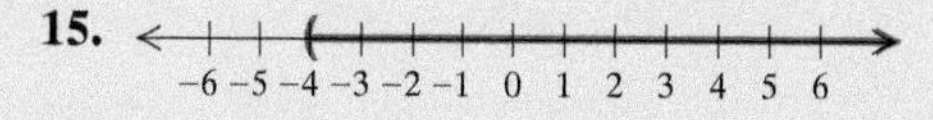

Solve and graph. Write interval notation for the solution set.

16. $x - 2 \leq 4$

17. $-4y - 3 \geq 5$

-6 -5 -4 -3 -2 -1 0 1 2 3 4 5 6

Solve.

18. $x - 4 \geq 6$

19. $-0.6y < 30$

20. $3a - 5 \leq -2a + 6$

21. $-5y - 1 > -9y + 3$

22. $4(5 - x) < 2x + 5$

23. $-8(2x + 3) + 6(4 - 5x) \geq 2(1 - 7x) - 4(4 + 6x)$

Solve.

24. *Moving Costs.* Mitchell Moving Company charges \$105 plus \$30 per hour to move households across town. Quick-Pak Moving charges \$80 per hour for cross-town moves. For what lengths of time is Quick-Pak more expensive?

25. *Pressure at Sea Depth.* The equation

$$P = 1 + \frac{d}{33}$$

gives the pressure P, in atmospheres (atm), at a depth of d feet in the sea. For what depths d is the pressure at least 2 atm and at most 8 atm?

Graph and write interval notation.

26. $-3 \le x \le 4$

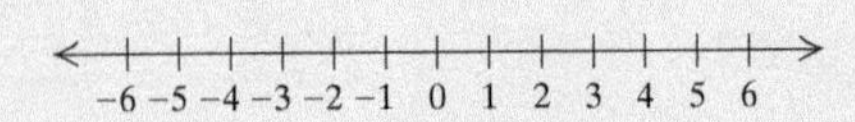

27. $x < -3$ *or* $x > 4$

−6 −5 −4 −3 −2 −1 0 1 2 3 4 5 6

Solve.

28. $5 - 2x \le 1$ *and* $3x + 2 \ge 14$

29. $-3 < x - 2 < 4$

30. $-11 \le -5x - 2 < 0$

31. $-3x > 12$ *or* $4x > -10$

32. $x - 7 \le -5$ *or* $x - 7 \ge -10$

33. $3x - 2 < 7$ *or* $x - 2 > 4$

Simplify.

34. $\left|\frac{7}{x}\right|$

35. $\left|\frac{-6x^2}{3x}\right|$

36. Find the distance between 4.8 and −3.6.

37. Find the intersection:

$$\{1, 3, 5, 7, 9\} \cap \{3, 5, 11, 13\}.$$

38. Find the union:

$$\{1, 3, 5, 7, 9\} \cup \{3, 5, 11, 13\}.$$

Solve.

39. $|x| = 9$

40. $|x - 3| = 9$

41. $|x + 10| = |x - 12|$

42. $|2 - 5x| = -10$

43. $|4x - 1| < 4.5$

44. $|x| > 3$

45. $\left|\frac{6 - x}{7}\right| \le 15$

46. $|-5x - 3| \ge 10$

47. The solution of $2(3x - 6) + 5 = 1 - (x - 6)$ is which of the following?

A. Less than 0
B. Between 0 and 1
C. Between 1 and 3
D. Greater than 3

Synthesis

Solve.

48. $|3x - 4| \le -3$

49. $7x < 8 - 3x < 6 + 7x$

CHAPTER

2

Graphs, Functions, and Applications

With the increasing number of both cars with better gas mileage as well as electric cars, U.S. gas tax revenue is expected to decrease in the next decade. When planning for needed cash to build and repair roads, states are considering increasing the number of tollways in order to generate the revenue lost in declining gas tax revenue.

Data: Joint Commission on Taxation

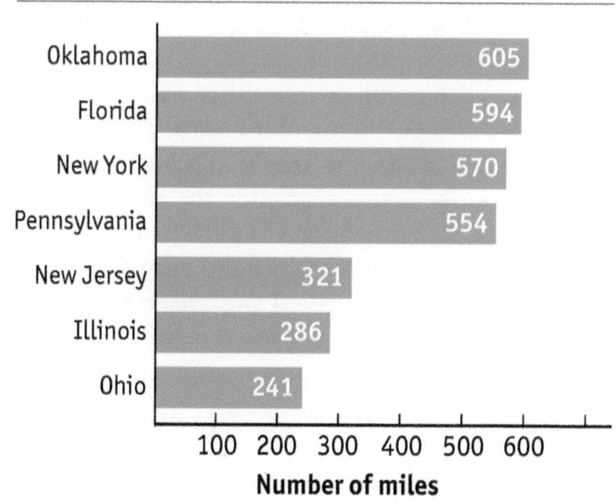

In Exercise 52 of Exercise Set 2.6, we will find a linear function that fits the gas tax revenue data and then use the function to determine when the tax revenue will be $19.8 billion.

JUST IN TIME

Review Topics 7, 9, and 13 in the **Just In Time** section at the front of the text provide excellent prerequisite skill review for this chapter.

MyLab Math
VIDEO

STUDYING FOR SUCCESS *Learning Resources on Campus*

- ☐ There may be a learning lab or a tutoring center for drop-in tutoring.
- ☐ There may be group tutoring sessions for this course.
- ☐ The math department may have a bulletin board or a network that provides contact information for private tutors.

2.1 Graphs of Equations

OBJECTIVES

- **a** Plot points associated with ordered pairs of numbers.
- **b** Determine whether an ordered pair of numbers is a solution of an equation.
- **c** Graph linear equations using tables.
- **d** Graph nonlinear equations using tables.

Graphs display information and can provide a visual approach to problem solving. We often see graphs in newspapers and magazines.

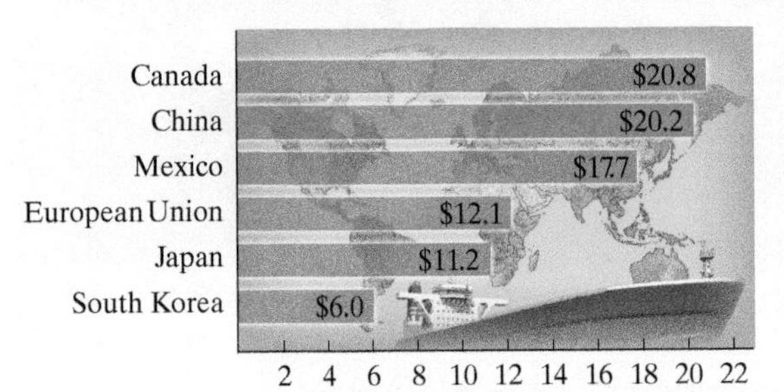

DATA: United States Department of Agriculture

Worldwide Urban Population

Percent of world population that is urban: 70%, 60, 50, 40, 30, 20, 10

(1950, 30%) (2017, 55%)

1950 1970 1990 2010 2030 2050

Year

DATA: United Nations Population Division

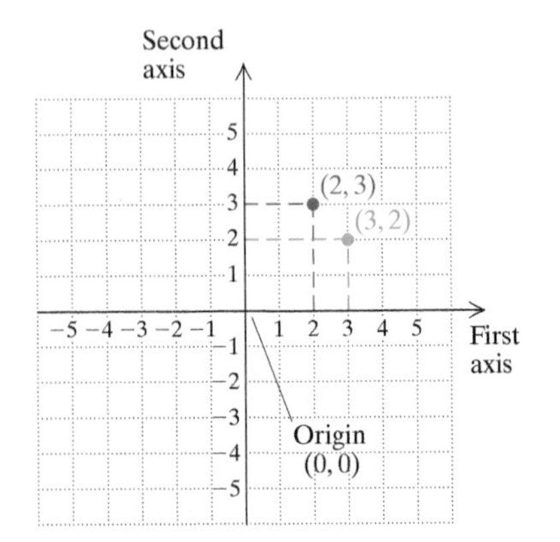

a PLOTTING ORDERED PAIRS

We have already learned to graph numbers and inequalities in one variable on a line. To graph an equation that contains two variables, we graph pairs of numbers on a plane.

On the number line, each point is the graph of a number. On a plane, each point is the graph of a number pair. To locate points on a plane, we use two perpendicular number lines called **axes**. They cross at a point called the **origin**. The arrows show the positive directions on the axes. Consider the **ordered pair** $(2, 3)$. The numbers in an ordered pair are called **coordinates**. In $(2, 3)$, the **first coordinate** is 2 and the **second coordinate** is 3. (The first coordinate is sometimes called the **abscissa** and the second the **ordinate**.) To plot $(2, 3)$, we start at the origin and move 2 units in the positive horizontal direction (2 units to the right). Then we move 3 units in the positive vertical direction (3 units up) and make a dot.

The point $(3, 2)$ is also plotted in the figure. Note that $(3, 2)$ and $(2, 3)$ are different points. The order of the numbers in the pair is indeed important. They are called *ordered pairs* because it makes a difference which number is listed first.

The coordinates of the origin are $(0, 0)$. In general, the first axis is called the x-axis and the second axis is called the y-axis. We call this the **Cartesian coordinate system** in honor of the great French mathematician and philosopher René Descartes (1596–1650).

EXAMPLE 1 Plot the points $(-4, 3)$, $(-5, -3)$, $(0, 4)$, and $(2.5, 0)$.

To plot $(-4, 3)$, we note that the first number, -4, tells us the distance in the first, or horizontal, direction. We move 4 units in the negative direction, *left*. The second number tells us the distance in the second, or vertical, direction. We move 3 units in the positive direction, *up*. The point $(-4, 3)$ is then marked, or plotted.

The points $(-5, -3)$, $(0, 4)$, and $(2.5, 0)$ are plotted in the same manner.

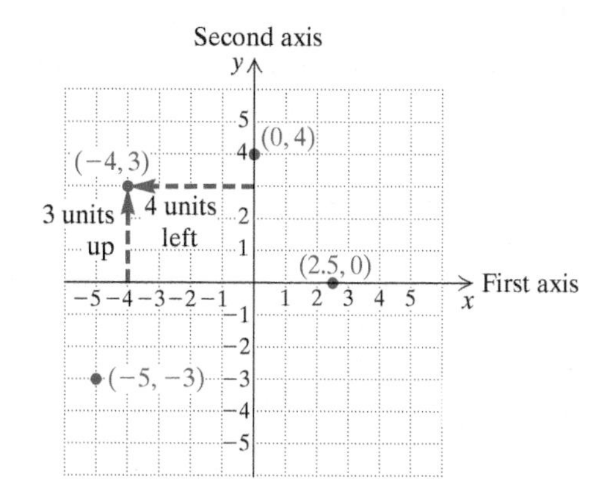

Do Exercises 1–10.

Plot each point on the plane below.

1. $(6, 4)$
2. $(4, 6)$
3. $(-3, 5)$
4. $(5, -3)$
5. $(-4, -3)$
6. $(4, -2)$
7. $(0, 3)$
8. $(3, 0)$
9. $(0, -4)$
10. $(-4, 0)$

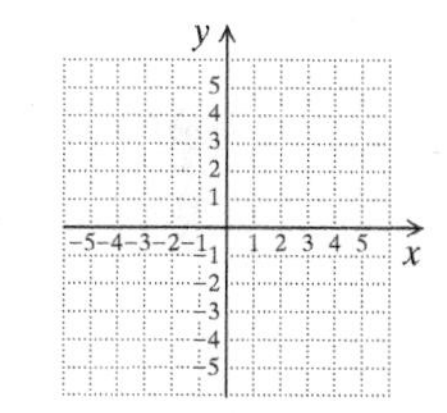

Quadrants

The axes divide the plane into four regions called **quadrants**, denoted by Roman numerals and numbered counterclockwise starting at the upper right. In region I (the *first* quadrant), both coordinates of a point are positive. In region II (the *second* quadrant), the first coordinate is negative and the second coordinate is positive. In the *third* quadrant, both coordinates are negative, and in the *fourth* quadrant, the first coordinate is positive and the second coordinate is negative.

Points with one or more 0's as coordinates, such as $(0, -5)$, $(4, 0)$, and $(0, 0)$ are on axes and *not* in quadrants.

Do Exercises 11 and 12 on the following page.

b SOLUTIONS OF EQUATIONS

SKILL REVIEW

Determine whether a given number is a solution of a given equation. [1.1a]

Determine whether the given number is a solution of the given equation.

1. $5;\ -5(2 - y) = -15$ **2.** $-7;\ 2x - 6 = -20$

Answers: 1. No **2.** Yes

Answers

1.–10.

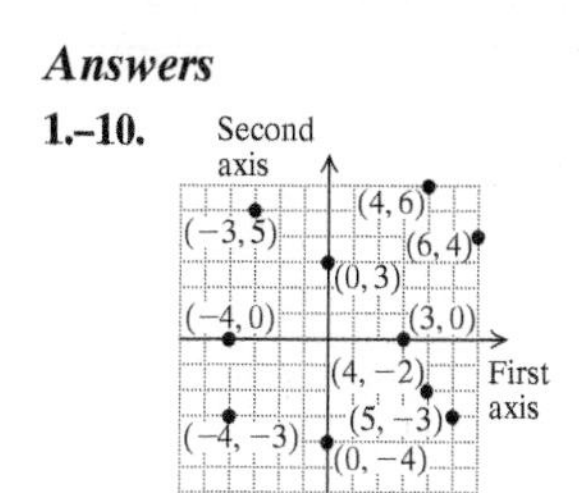

11. What can you say about the coordinates of a point in the third quadrant?

12. What can you say about the coordinates of a point in the fourth quadrant?

If an equation has two variables, its solutions are pairs of numbers. When such a solution is written as an ordered pair, the first number listed in the pair generally replaces the variable that occurs first alphabetically.

EXAMPLE 2 Determine whether each of the following pairs is a solution of $5b - 3a = 34$: $(2, 8)$ and $(-1, 6)$.

For the pair $(2, 8)$, we substitute 2 for a and 8 for b (alphabetical order of variables):

$$\begin{array}{r|l} 5b - 3a & = 34 \\ \hline 5\cdot 8 - 3\cdot 2 & ?\ 34 \\ 40 - 6 & \\ 34 & \qquad \text{TRUE} \end{array}$$

Thus, $(2, 8)$ is a solution of the equation.

For $(-1, 6)$, we substitute -1 for a and 6 for b:

$$\begin{array}{r|l} 5b - 3a & = 34 \\ \hline 5\cdot 6 - 3\cdot(-1) & ?\ 34 \\ 30 + 3 & \\ 33 & \qquad \text{FALSE} \end{array}$$

Thus, $(-1, 6)$ is *not* a solution of the equation.

◀ **Do Exercises 13 and 14.**

13. Determine whether $(2, -4)$ is a solution of $5b - 3a = 34$.

14. Determine whether $(2, -4)$ is a solution of $7p + 5q = -6$.

EXAMPLE 3 Show that the pairs $(-4, 3)$, $(0, 1)$, and $(4, -1)$ are solutions of $y = 1 - \frac{1}{2}x$. Then plot the three points and use them to help determine another pair that is a solution.

We replace x with the first coordinate and y with the second coordinate of each pair:

$$\begin{array}{l|l} y & = 1 - \frac{1}{2}x \\ \hline 3 & ?\ 1 - \frac{1}{2}\cdot(-4) \\ & 1 + 2 \\ & 3 \qquad \text{TRUE} \end{array} \qquad \begin{array}{l|l} y & = 1 - \frac{1}{2}x \\ \hline 1 & ?\ 1 - \frac{1}{2}\cdot(0) \\ & 1 - 0 \\ & 1 \qquad \text{TRUE} \end{array} \qquad \begin{array}{l|l} y & = 1 - \frac{1}{2}x \\ \hline -1 & ?\ 1 - \frac{1}{2}\cdot(4) \\ & 1 - 2 \\ & -1 \qquad \text{TRUE} \end{array}$$

In each case, the substitution results in a true equation. Thus all the pairs are solutions of the equation.

We plot the points as shown at right. Note that the three points appear to "line up." That is, they appear to be on a straight line. We use a ruler and draw a line passing through $(-4, 3)$, $(0, 1)$, and $(4, -1)$.

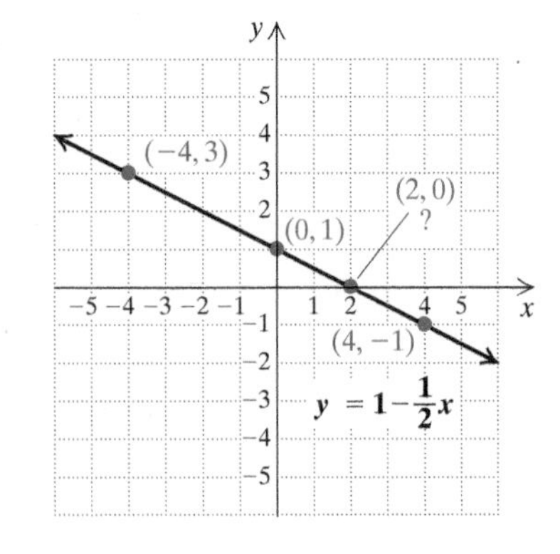

The line appears to pass through $(2, 0)$ as well. Let's see if this pair is a solution of $y = 1 - \frac{1}{2}x$:

$$\begin{array}{l|l} y & = 1 - \frac{1}{2}x \\ \hline 0 & ?\ 1 - \frac{1}{2}\cdot(2) \\ & 1 - 1 \\ & 0 \qquad \text{TRUE} \end{array}$$

We see that $(2, 0)$ is another solution of the equation.

◀ **Do Exercise 15.**

15. Use the line in Example 3 to find at least two more points that are solutions.

Answers

11. Both negative
12. First positive, second negative
13. No **14.** Yes **15.** $(-6, 4)$, $(-2, 2)$; answers may vary

Example 3 leads us to believe that any point on the line that passes through $(-4, 3)$, $(0, 1)$, and $(4, -1)$ represents a solution of $y = 1 - \frac{1}{2}x$. In fact, every solution of $y = 1 - \frac{1}{2}x$ is represented by a point on that line and every point on that line represents a solution. The line is said to be the *graph* of the equation.

GRAPH OF AN EQUATION

The **graph** of an equation is a drawing that represents all of its solutions.

C GRAPHS OF LINEAR EQUATIONS

MyLab Math ANIMATION

Equations like $y = 1 - \frac{1}{2}x$ and $2x + 3y = 6$ are said to be **linear** because the graph of their solutions is a line. In general, a linear equation is any equation equivalent to one of the form $y = mx + b$ or $Ax + By = C$, where m, b, A, B, and C are constants (that is, they are numbers, not variables) and A and B are not both 0.

EXAMPLE 4 Graph: $y = 2x$.

We find some ordered pairs that are solutions. This time we list the pairs in a table. To find an ordered pair, we can choose *any* number for x and then determine y. For example, if we choose 3 for x, then $y = 2 \cdot 3 = 6$ (substituting into the equation $y = 2x$). We choose some negative values for x, as well as some positive ones. If a number takes us off the graph paper, we generally do not use it. Next, we plot these points. If we plotted *many* such points, they would appear to make a solid line. We draw the line with a ruler and label it $y = 2x$.

x	y	(x, y)
0	0	$(0, 0)$
1	2	$(1, 2)$
3	6	$(3, 6)$
-2	-4	$(-2, -4)$
-3	-6	$(-3, -6)$

Choose any x.
Compute y.
Form the pair.
Plot the points.

(3, 6)
$y = 2x$
(0, 0)
(1, 2)
(−2, −4)
(−3, −6)

To graph a linear equation:

1. Select a value for one variable and calculate the corresponding value of the other variable. Form an ordered pair using alphabetical order as indicated by the variables.
2. Repeat step (1) to obtain at least two other ordered pairs. Two ordered pairs are essential. A third serves as a check.
3. Plot the ordered pairs and draw a straight line passing through the points.

Do Exercises 16 and 17.

Graph.

16. $y = -2x$

x	y	(x, y)
-3		
-1		
0		
1		
3		

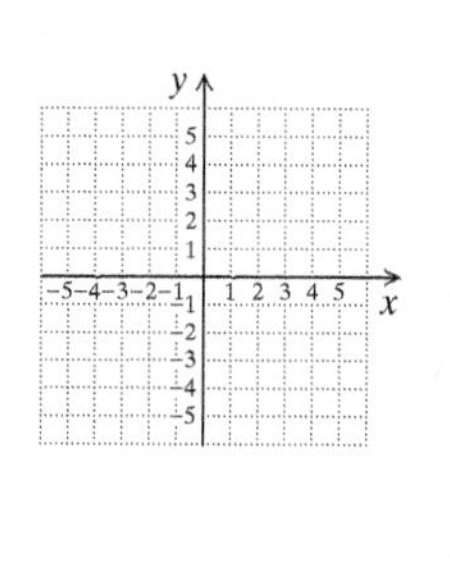

GS **17.** $y = \frac{1}{2}x$

x	y	(x, y)
4	2	$(4, 2)$
2	☐	$(2, ☐)$
0	0	$(0, 0)$
-2	☐	$(☐, -1)$
-4	☐	$(-4, ☐)$

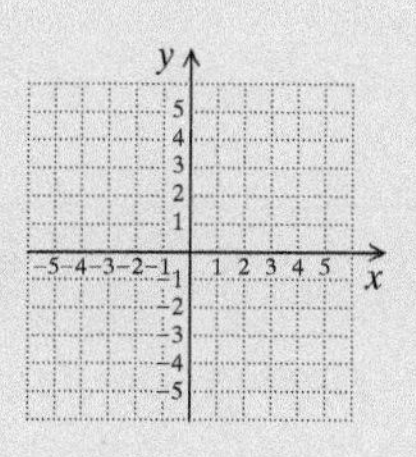

Answers

16.

x	y	(x, y)
-3	6	$(-3, 6)$
-1	2	$(-1, 2)$
0	0	$(0, 0)$
1	-2	$(1, -2)$
3	-6	$(3, -6)$

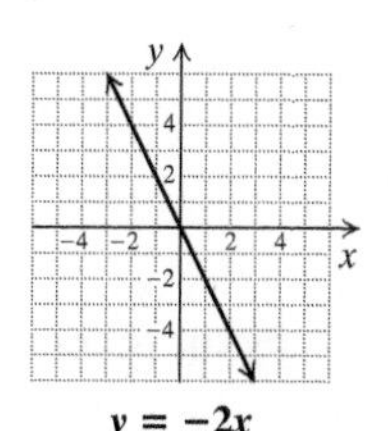

17.

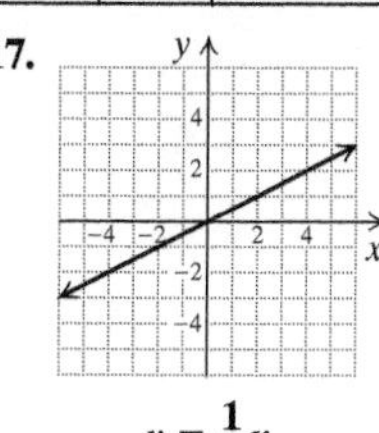

Guided Solution:
17. 1, 1, −1, −2, −2, −2

EXAMPLE 5 Graph: $y = -\frac{1}{2}x + 3$.

By choosing even integers for x, we can avoid fraction values when calculating y. For example, if we choose 4 for x, we get

$$y = -\tfrac{1}{2}x + 3 = -\tfrac{1}{2}(4) + 3 = -2 + 3 = 1.$$

When x is -6, we get

$$y = -\tfrac{1}{2}x + 3 = -\tfrac{1}{2}(-6) + 3 = 3 + 3 = 6,$$

and when x is 0, we get

$$y = -\tfrac{1}{2}x + 3 = -\tfrac{1}{2}(0) + 3 = 0 + 3 = 3.$$

We list the results in a table. Then we plot the points corresponding to each pair.

x	y	(x, y)
4	1	(4, 1)
−6	6	(−6, 6)
0	3	(0, 3)

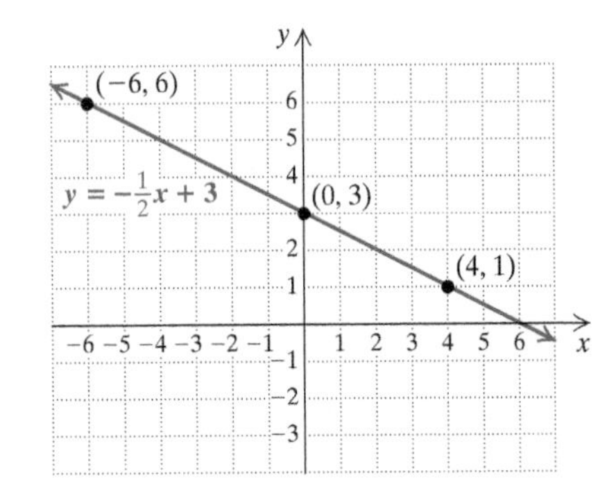

Note that the three points line up. If they did not, we would know that we had made a mistake. When only two points are plotted, an error is harder to detect. We use a ruler or other straightedge to draw a line through the points and then label the graph. Every point on the line represents a solution of $y = -\frac{1}{2}x + 3$.

Do Exercises 18 and 19.

18. Graph: $y = 2x + 3$.

x	y	(x, y)

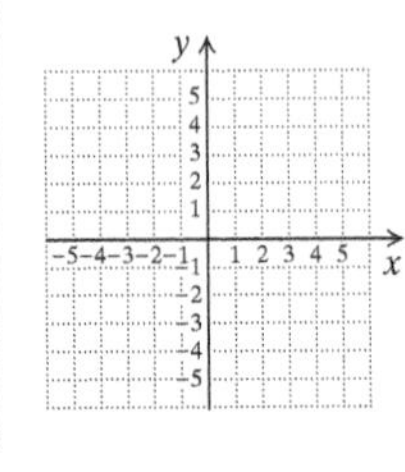

19. Graph: $y = -\dfrac{1}{2}x - 3$.

x	y	(x, y)

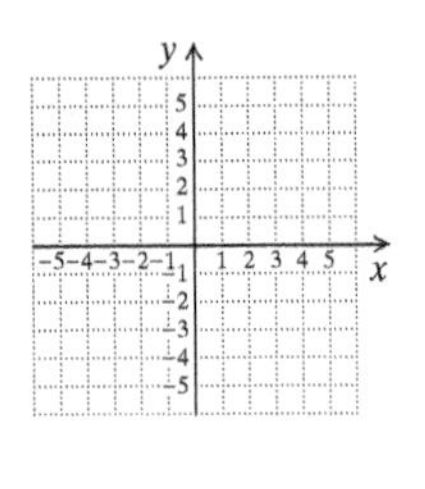

Answers

Answers to Margin Exercises 18 and 19 are on p. 123.

CALCULATOR CORNER

Finding Solutions of Equations A table of values representing ordered pairs that are solutions of an equation can be displayed on a graphing calculator. To do this for the equation in Example 4, $y = 2x$, we first access the equation-editor screen. Then we clear any equations that are present. (See the Calculator Corner on p. 41 for the procedure for doing this.) Next, we enter the equation, display the table set-up screen, and set both INDPNT and DEPEND to AUTO.

We will display a table of values that starts with $x = -2$ (TBLSTART) and adds 1 (ΔTBL) to the preceding x-value.

X	Y1	
−2	−4	
−1	−2	
0	0	
1	2	
2	4	
3	6	
4	8	
X = −2		

EXERCISES: Create a table of ordered pairs that are solutions of the equation.

1. Example 5
2. Example 7

Calculating ordered pairs is usually easiest when y is isolated on one side of the equation, as in $y = 2x$ and $y = -\frac{1}{2}x + 3$. To graph an equation in which y is not isolated, we can use the addition principle and the multiplication principle to first solve for y. (See Sections 1.1 and 1.2.)

EXAMPLE 6 Graph: $3x + 5y = 10$.

We first solve for y:

$$\begin{aligned} 3x + 5y &= 10 \\ 3x + 5y - 3x &= 10 - 3x && \text{Subtracting } 3x \\ 5y &= 10 - 3x && \text{Simplifying} \\ \tfrac{1}{5} \cdot 5y &= \tfrac{1}{5} \cdot (10 - 3x) && \text{Multiplying by } \tfrac{1}{5}\text{, or dividing by 5} \\ y &= \tfrac{1}{5} \cdot (10) - \tfrac{1}{5} \cdot (3x) && \text{Using the distributive law} \\ y &= 2 - \tfrac{3}{5}x, \text{ or } y = -\tfrac{3}{5}x + 2. \end{aligned}$$

Thus the equation $3x + 5y = 10$ is equivalent to $y = -\frac{3}{5}x + 2$. We now find three ordered pairs, using multiples of 5 for x to avoid fractions.

x	y	(x, y)
0	2	(0, 2)
5	−1	(5, −1)
−5	5	(−5, 5)

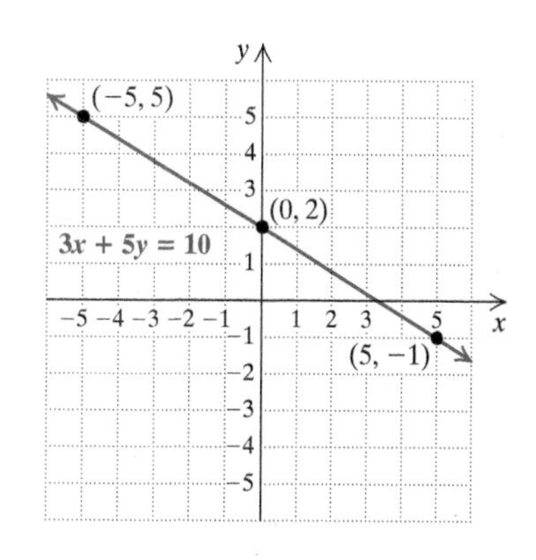

We plot the points, draw the line, and label the graph as shown.

Do Exercises 20 and 21. ▶

d GRAPHING NONLINEAR EQUATIONS

We have seen that equations whose graphs are straight lines are called **linear**. There are many equations whose graphs are not straight lines. Here are some examples.

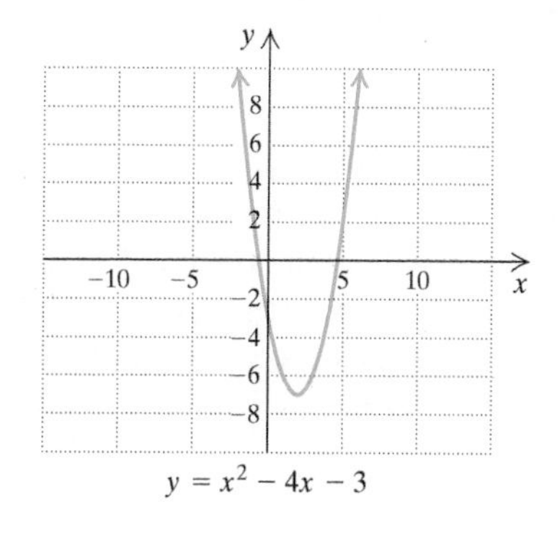

$y = x^2 - 4x - 3$

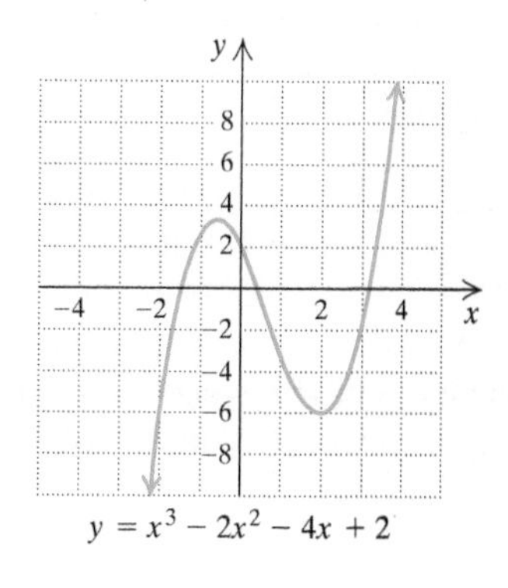

$y = x^3 - 2x^2 - 4x + 2$

20. Graph: $4y - 3x = -8$.

We first solve for y:

$$\begin{aligned} 4y - 3x &= -8 \\ 4y &= \square - 8 \\ y &= \frac{3}{4}x - \square. \end{aligned}$$

x	y	(x, y)
0	−2	(0, −2)
4	☐	(4, ☐)
−4	☐	(☐, −5)

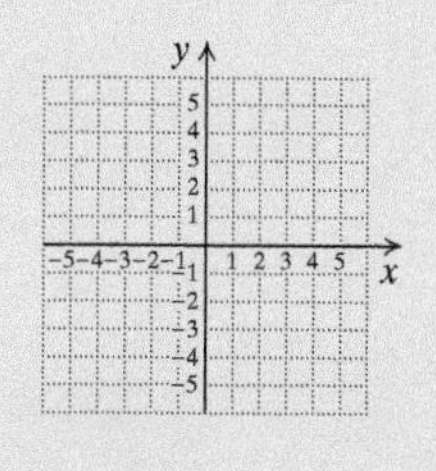

21. Graph: $5x + 2y = 4$.

x	y

Answers

18.

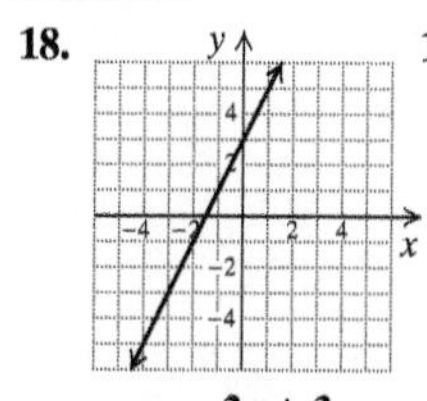

$y = 2x + 3$

19.

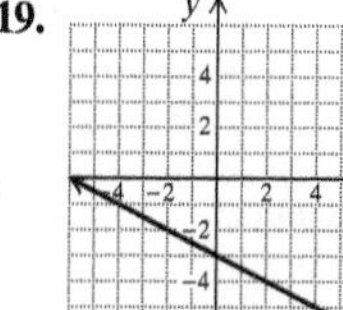

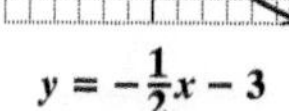

$y = -\frac{1}{2}x - 3$

20.

$4y - 3x = -8$

21.

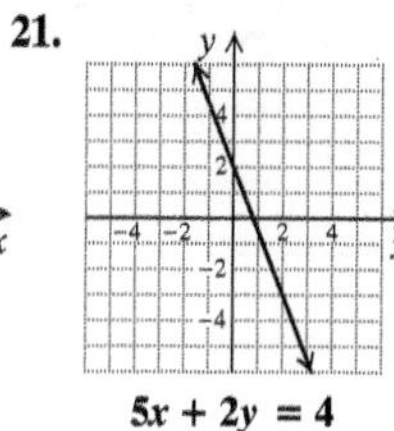

$5x + 2y = 4$

Guided Solution:
20. $3x$, 2, 1, 1, −5, −4

22. Graph: $y = 4 - x^2$.

x	y
0	
1	
-1	
2	
-2	
3	
-3	

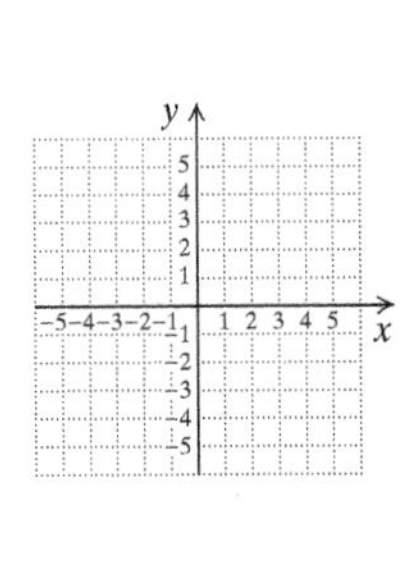

23. Graph: $y = \frac{2}{x}$.

x	y
1	
2	
4	
-1	
-2	
-4	
$\frac{1}{2}$	
$-\frac{1}{2}$	

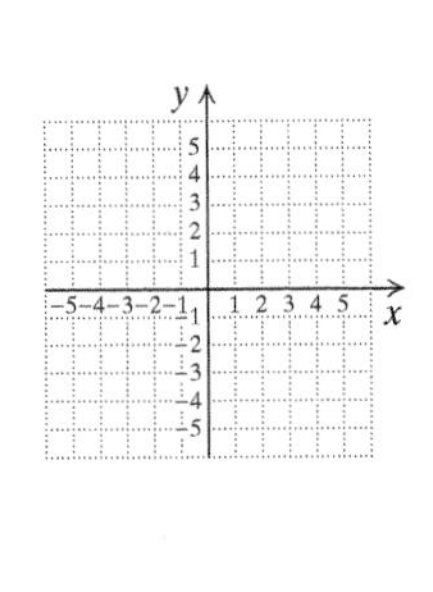

Answers

22.

x	y
0	4
1	3
-1	3
2	0
-2	0
3	-5
-3	-5

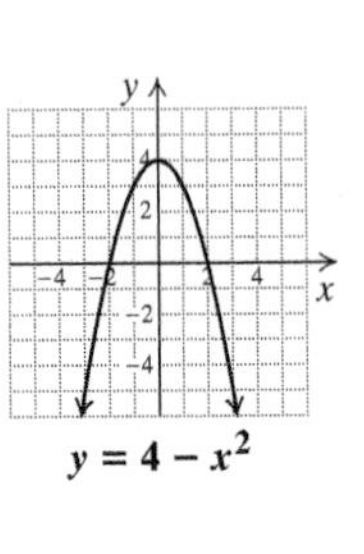

Answer for Margin Exercise 23 is on p. 125.

Let's graph some of these **nonlinear equations**. We usually need to plot more than three points in order to get a good idea of the shape of the graph.

EXAMPLE 7 Graph: $y = x^2 - 5$.

We select numbers for x and find the corresponding values for y. For example, if we choose -2 for x, we get $y = (-2)^2 - 5 = 4 - 5 = -1$. The table lists several ordered pairs.

x	y
0	-5
-1	-4
1	-4
-2	-1
2	-1
-3	4
3	4

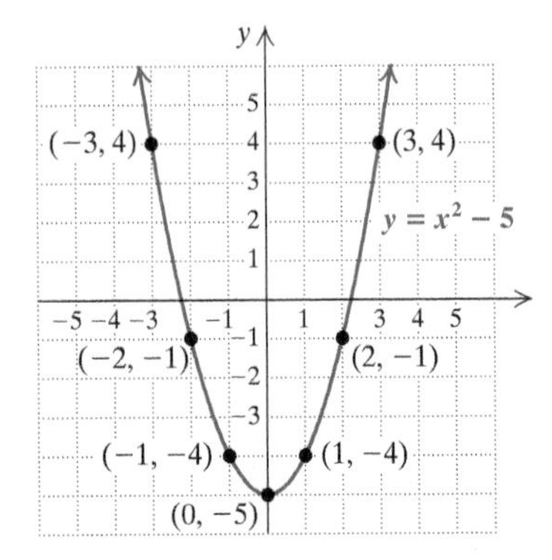

Next, we plot the points. The more points we plot, the more clearly we see the shape of the graph. Since the value of $x^2 - 5$ grows rapidly as x moves away from the origin, the graph rises steeply on either side of the y-axis.

Do Exercise 22.

EXAMPLE 8 Graph: $y = 1/x$.

We select x-values and find the corresponding y-values. The table lists the ordered pairs $(3, \frac{1}{3})$, $(2, \frac{1}{2})$, $(1, 1)$, and so on.

x	y
3	$\frac{1}{3}$
2	$\frac{1}{2}$
1	1
$\frac{1}{2}$	2
$-\frac{1}{2}$	-2
-1	-1
-2	$-\frac{1}{2}$
-3	$-\frac{1}{3}$

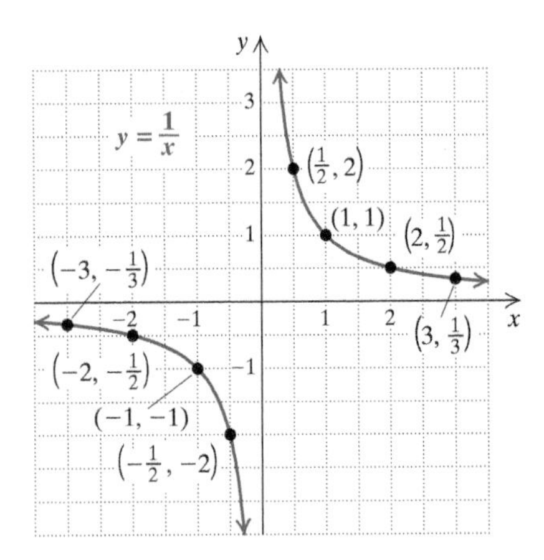

We plot these points, noting that each first coordinate is paired with its reciprocal. Since $1/0$ is undefined, we cannot use 0 as a first coordinate. Thus there are two "branches" to this graph—one on each side of the y-axis. Note that for x-values far to the right or far to the left of 0, the graph approaches, but does not touch, the x-axis; and for x-values close to 0, the graph approaches, but does not touch, the y-axis.

Do Exercise 23.

EXAMPLE 9 Graph: $y = |x|$.

We select numbers for x and find the corresponding values for y. For example, if we choose -1 for x, we get $y = |-1| = 1$. Several ordered pairs are listed in the following table.

x	y
-3	3
-2	2
-1	1
0	0
1	1
2	2
3	3

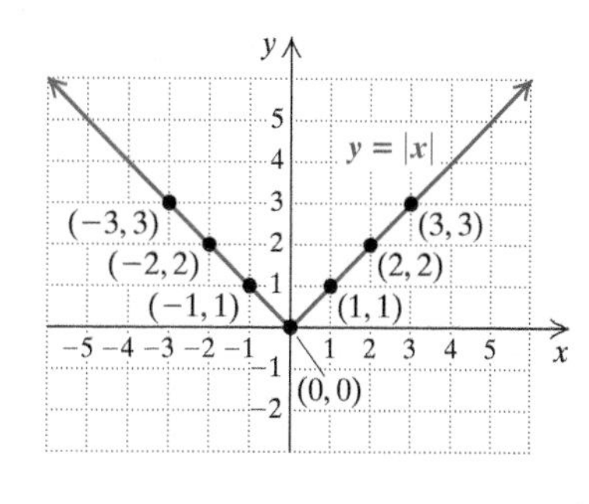

We plot these points, noting that the absolute value of a positive number is the same as the absolute value of its opposite. Thus the x-values 3 and -3 both are paired with the y-value 3. Note that the graph is V-shaped and centered at the origin.

Do Exercise 24. ▶

24. Graph: $y = 4 - |x|$.

x	y
0	
2	
-2	
4	
-4	
5	
-5	

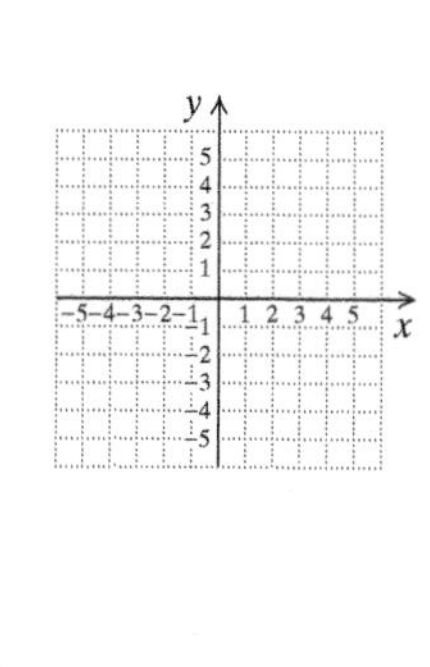

With equations like $y = -\frac{1}{2}x + 3$, $y = x^2 - 5$, and $y = |x|$, which we have graphed in this section, it is understood that y is the **dependent variable** and x is the **independent variable**, since y is expressed in terms of x and consequently y is calculated after first choosing x.

CALCULATOR CORNER

Graphing Equations Equations must be solved for y before they can be graphed on the TI-84 Plus. Consider the equation $3x + 2y = 6$. Solving for y, we enter $y_1 = (6 - 3x)/2$ as described on p. 41. Then we select a window and press GRAPH to see the graph of the equation. (Press ZOOM 6 to see the graph in the standard window as shown on the right below.)

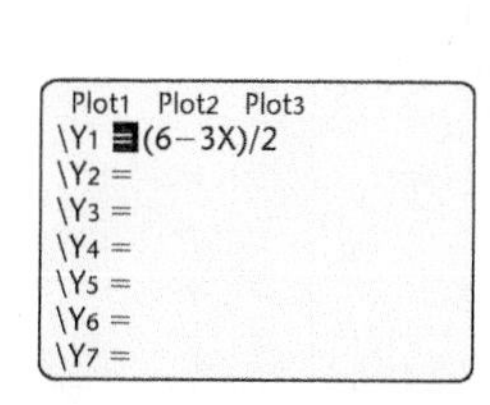

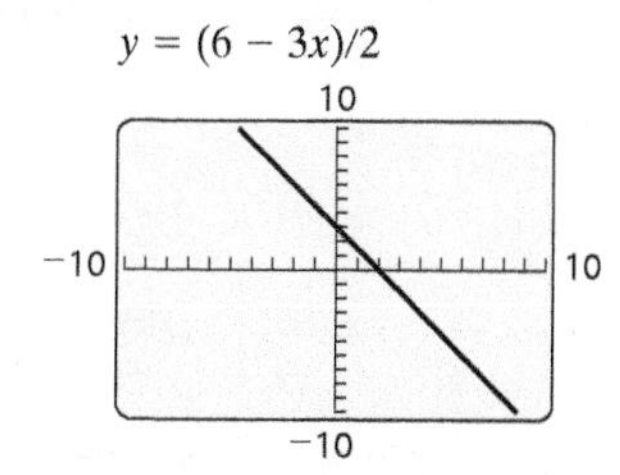

EXERCISES: Graph each equation in the standard viewing window $[-10, 10, -10, 10]$, with Xscl = 1 and Yscl = 1.

1. $y = 2x - 1$

2. $3x + y = 2$

3. $y = 5x - 3$

4. $y = -4x + 5$

5. $y = \frac{2}{3}x - 3$

6. $y = -\frac{3}{4}x + 4$

7. $y = 3.104x - 6.21$

8. $2.98x + y = -1.75$

Answers

23.

x	y
1	2
2	1
4	$\frac{1}{2}$
-1	-2
-2	-1
-4	$-\frac{1}{2}$
$\frac{1}{2}$	4
$-\frac{1}{2}$	-4

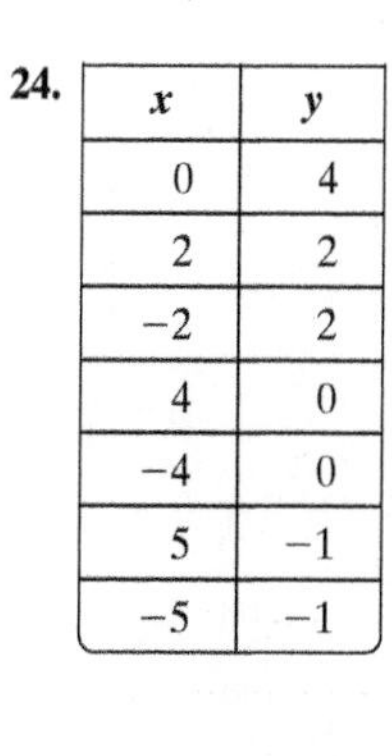

24.

x	y
0	4
2	2
-2	2
4	0
-4	0
5	-1
-5	-1

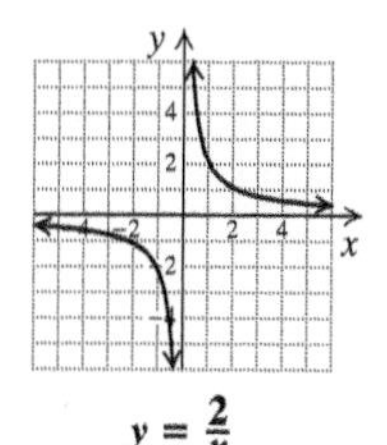

$y = \frac{2}{x}$

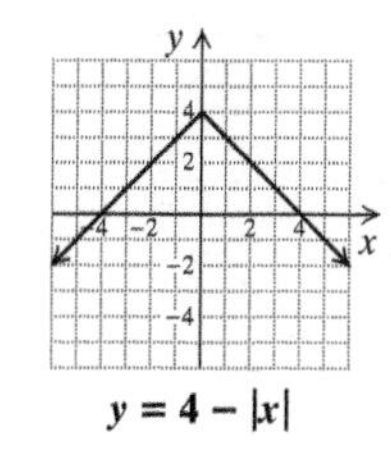

$y = 4 - |x|$

2.1 Exercise Set

FOR EXTRA HELP MyLab Math

✓ Check Your Understanding

Reading Check Determine whether each statement is true or false.

RC1. The point $(5, 0)$ is in quadrant I and in quadrant IV.

RC2. The ordered pairs $(1, -6)$ and $(-6, 1)$ name the same point.

RC3. In the ordered pair $(-8, 3)$, the first coordinate, -8, is also called the abscissa.

RC4. To plot the point $(-2, 7)$, we start at the origin and move horizontally to -2. Then we move up vertically 7 units and make a "dot."

RC5. In the ordered pair $(4, -10)$, the second coordinate, -10, is also called the ordinate.

RC6. The point $(0, -3)$ is on the x-axis.

Concept Check Choose from the column on the right an equivalent equation for each of the following equations.

CC1. $3x + 4y = 0$

CC2. $4y - 3x = 0$

CC3. $4x - 3y = -4$

CC4. $3y + 4x = -12$

a) $y = \frac{4}{3}x + \frac{4}{3}$

b) $y = \frac{3}{4}x$

c) $y = -\frac{4}{3}x - 4$

d) $y = -\frac{3}{4}x$

a Plot the following points.

1. $A(4, 1), B(2, 5), C(0, 3), D(0, -5), E(6, 0), F(-3, 0), G(-2, -4), H(-5, 1), J(-6, 6)$

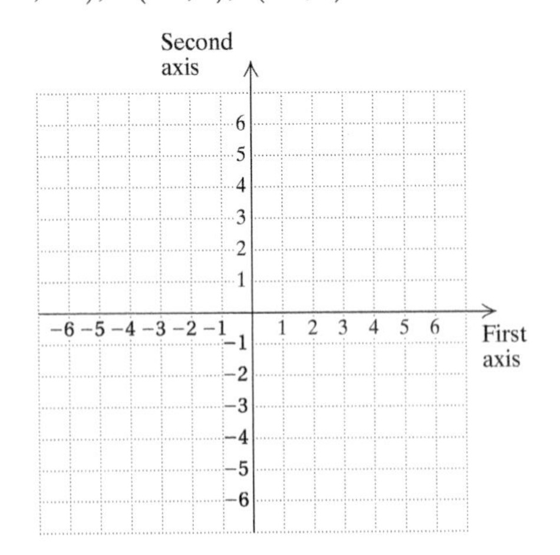

2. $A(-3, -5), B(1, 3), C(0, 7), D(0, -2), E(5, 0), F(-4, 0), G(1, -7), H(-6, 4), J(-3, 3)$

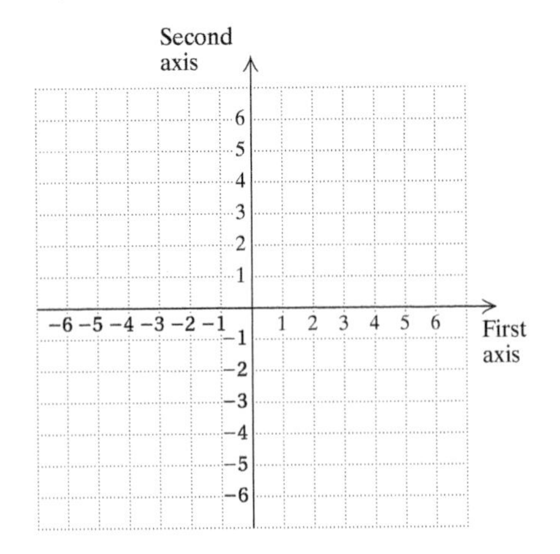

3. Plot the points $M(2, 3)$, $N(5, -3)$, and $P(-2, -3)$. Draw $\overline{MN}$, $\overline{NP}$, and $\overline{MP}$. ($\overline{MN}$ means the line segment from M to N.) What kind of geometric figure is formed? What is its area?

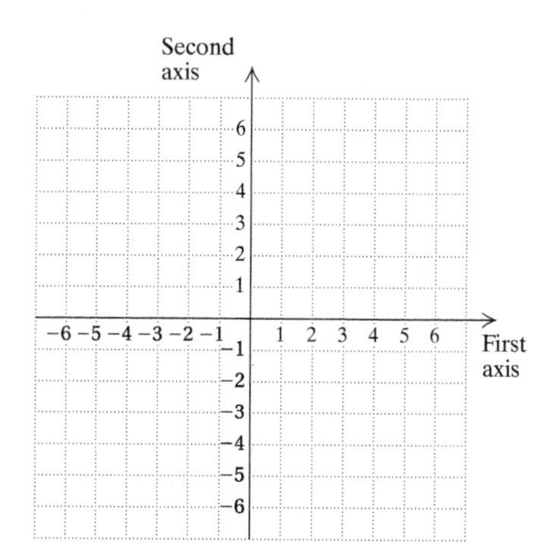

4. Plot the points $Q(-4, 3)$, $R(5, 3)$, $S(2, -1)$, and $T(-7, -1)$. Draw $\overline{QR}$, $\overline{RS}$, $\overline{ST}$, and $\overline{TQ}$. What kind of figure is formed? What is its area?

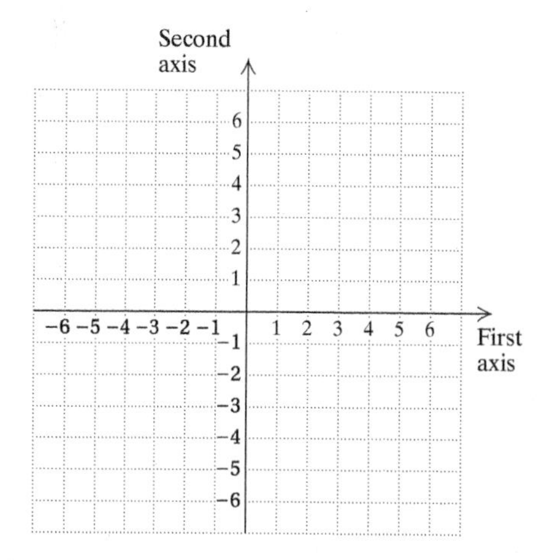

b Determine whether the given point is a solution of the equation.

5. $(1, -1)$; $y = 2x - 3$

6. $(3, 4)$; $t = 4 - 3s$

7. $(3, 5)$; $4x - y = 7$

8. $(2, -1)$; $4r + 3s = 5$

9. $\left(0, \frac{3}{5}\right)$; $2a + 5b = 7$

10. $(-5, 1)$; $2p - 3q = -13$

In Exercises 11–16, an equation and two ordered pairs are given. Show that each pair is a solution of the equation. Then graph the equation and use the graph to determine another solution. Answers for solutions may vary, but the graphs do not.

11. $y = 4 - x$; $(-1, 5), (3, 1)$

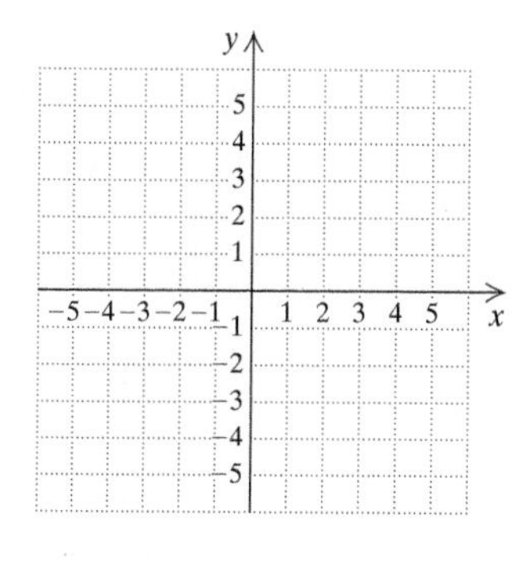

12. $y = x - 3$; $(5, 2), (-1, -4)$

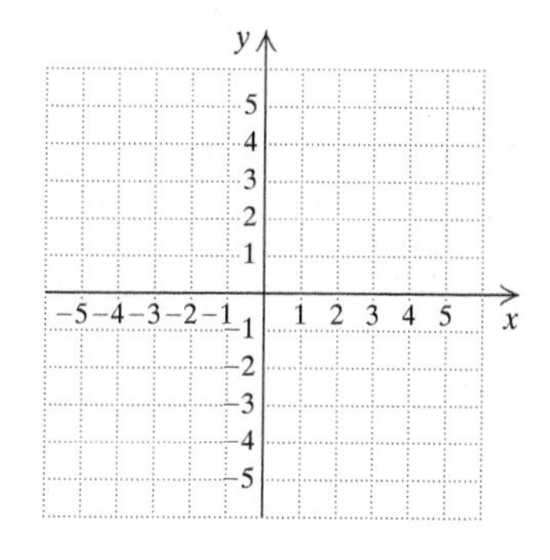

13. $3x + y = 7$; $(2, 1), (4, -5)$

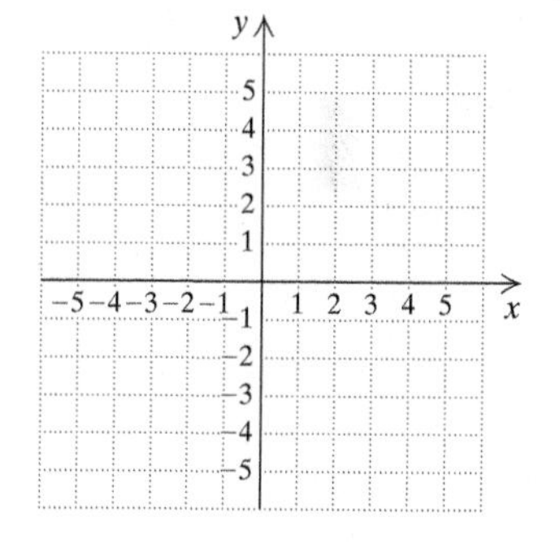

14. $y = \frac{1}{2}x + 3$; $(4, 5), (-2, 2)$

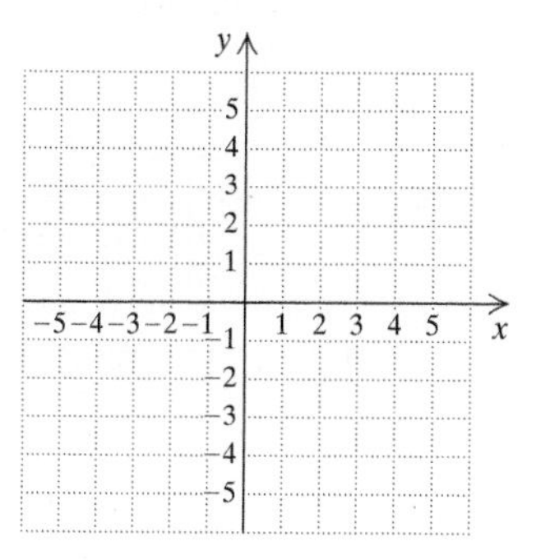

15. $6x - 3y = 3$; $(1, 1), (-1, -3)$

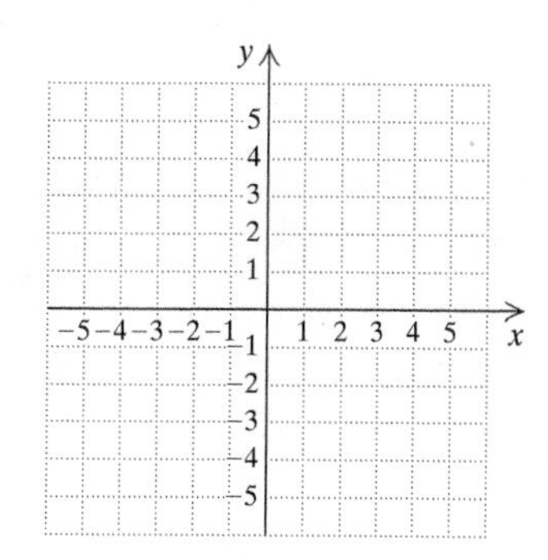

16. $4x - 2y = 10$; $(0, -5), (4, 3)$

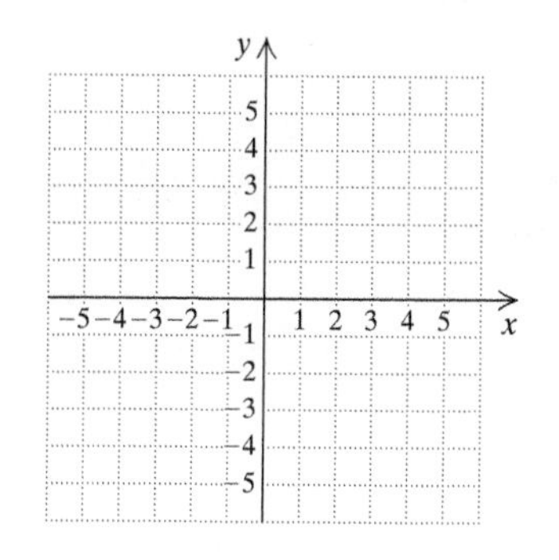

C Graph.

17. $y = x - 1$

x	y

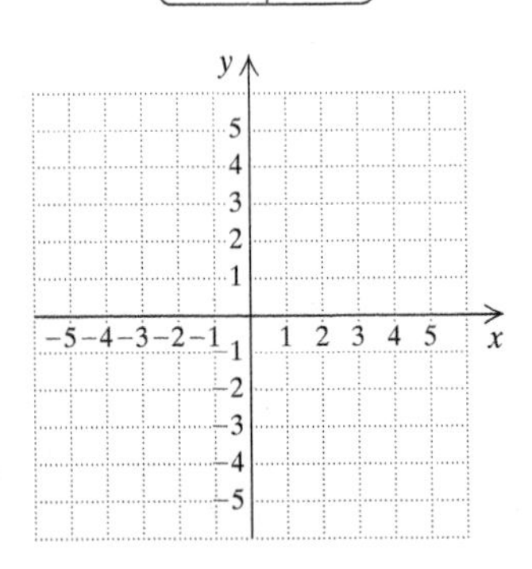

18. $y = x + 1$

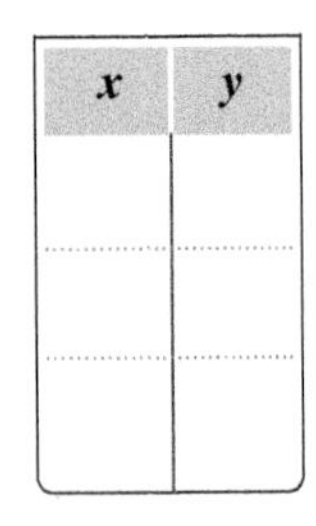

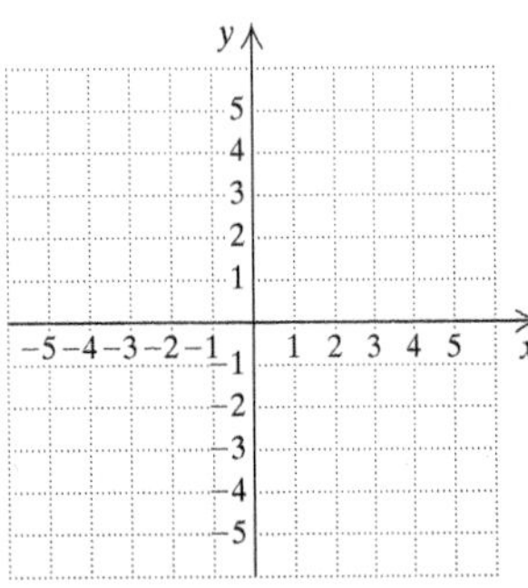

19. $y = x$

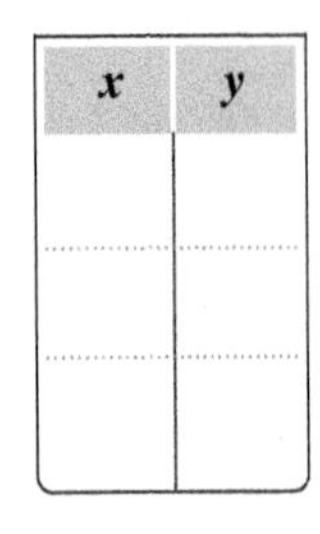

20. $y = -3x$

x	y

21. $y = \frac{1}{4}x$

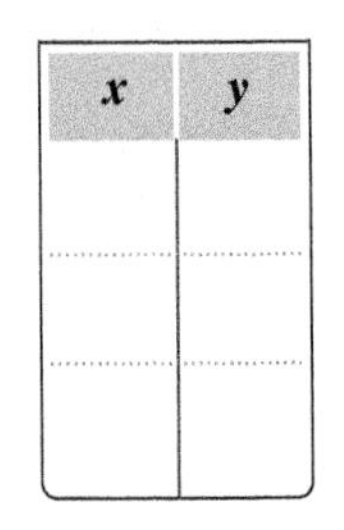

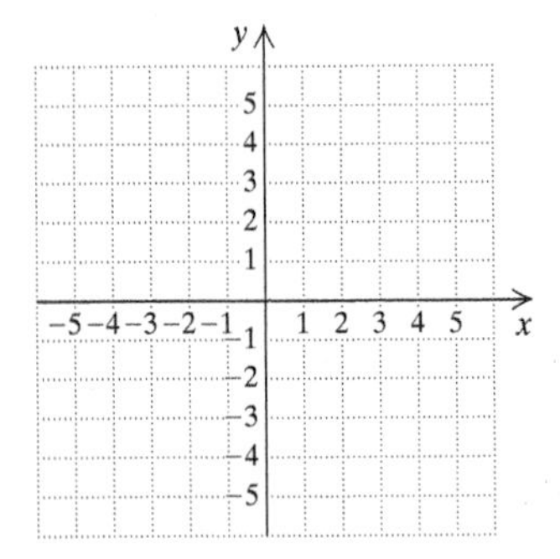

22. $y = \frac{1}{3}x$

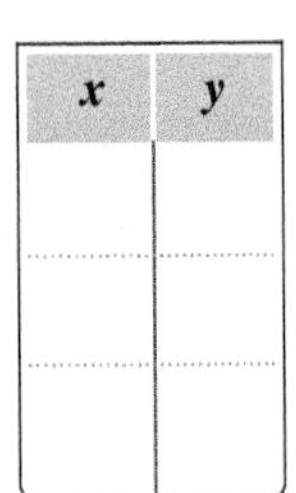

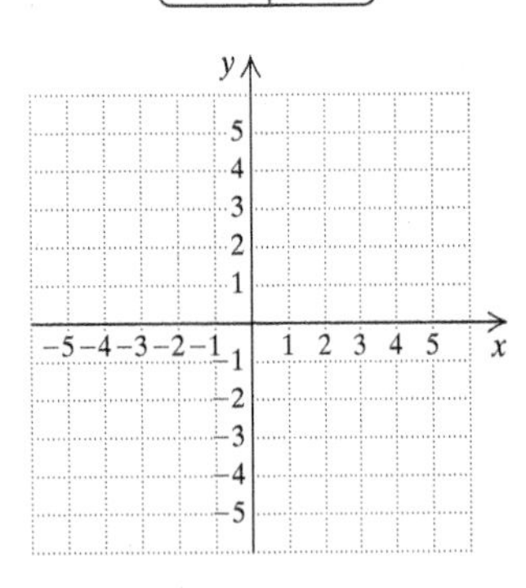

23. $y = 3 - x$

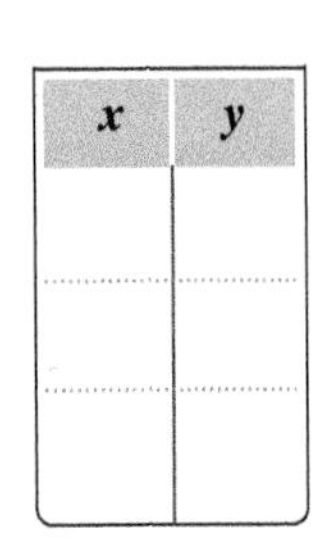

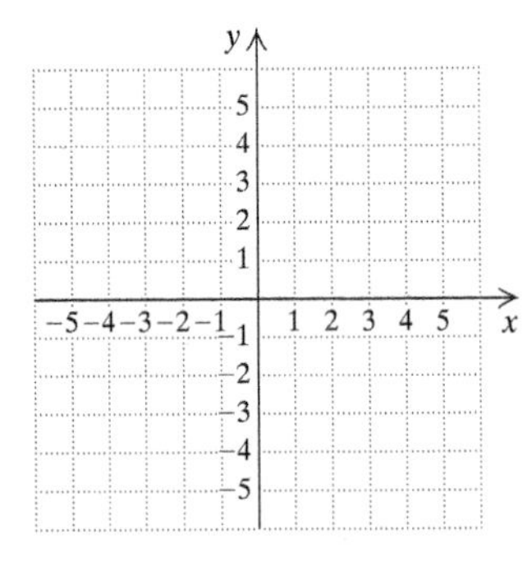

24. $y = x + 3$

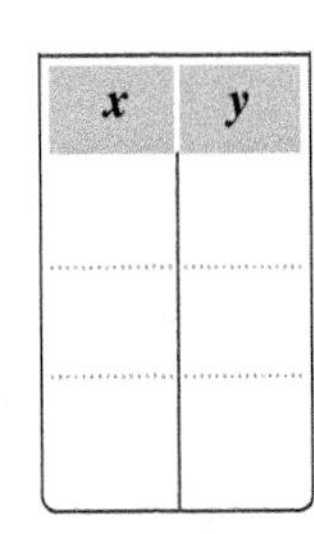

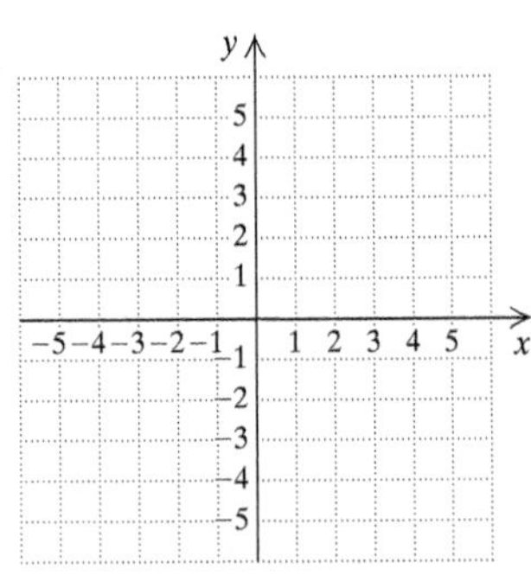

25. $y = 5x - 2$

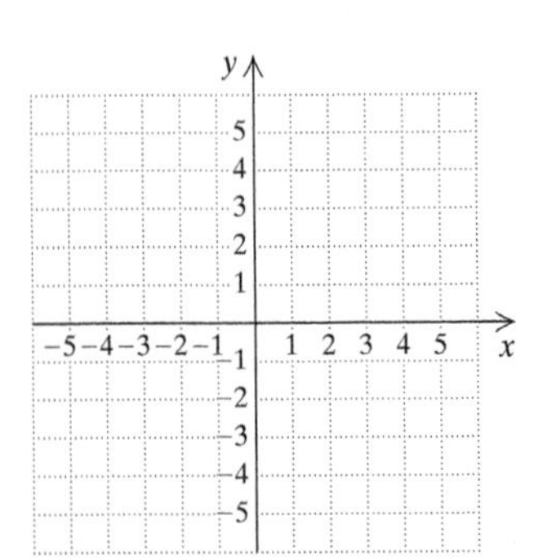

26. $y = \frac{1}{4}x + 2$

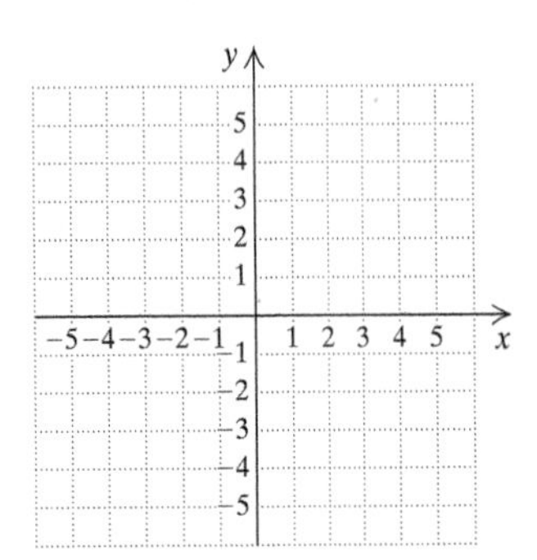

27. $y = \frac{1}{2}x + 1$

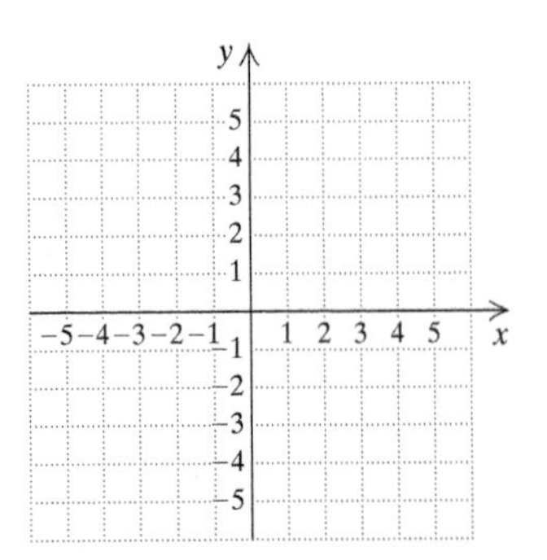

28. $y = \frac{1}{3}x - 4$

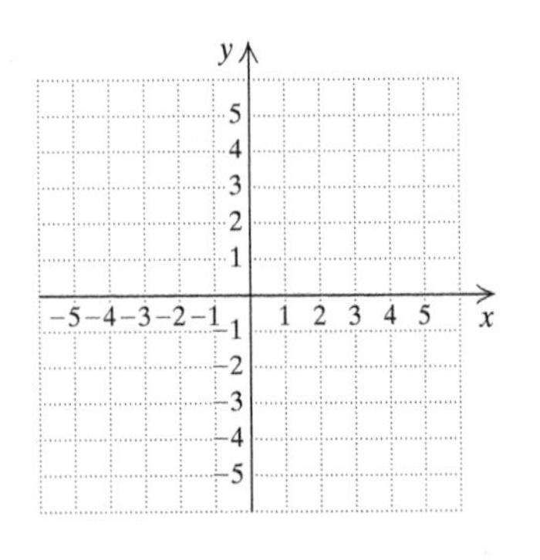

29. $x + y = 5$

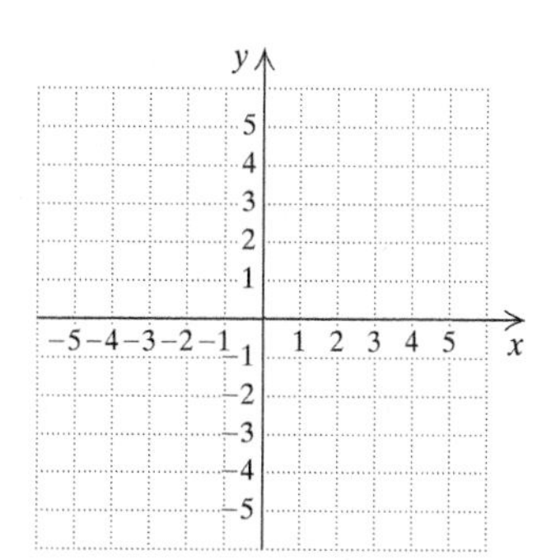

30. $x + y = -4$

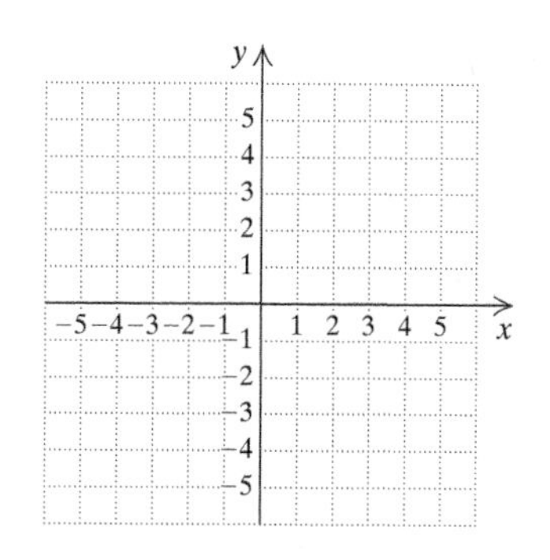

31. $y = -\frac{5}{3}x - 2$

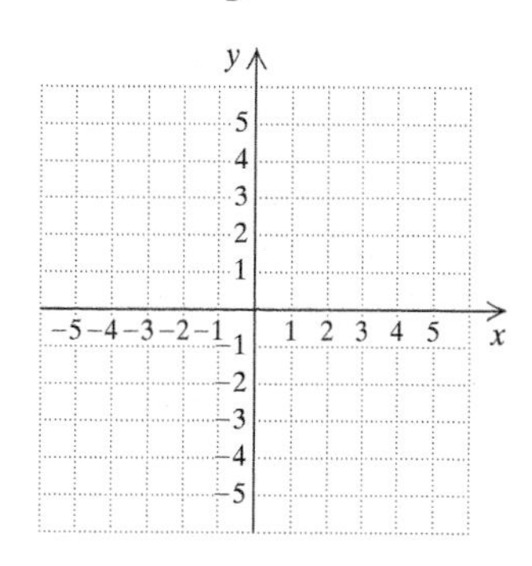

32. $y = -\frac{5}{2}x + 3$

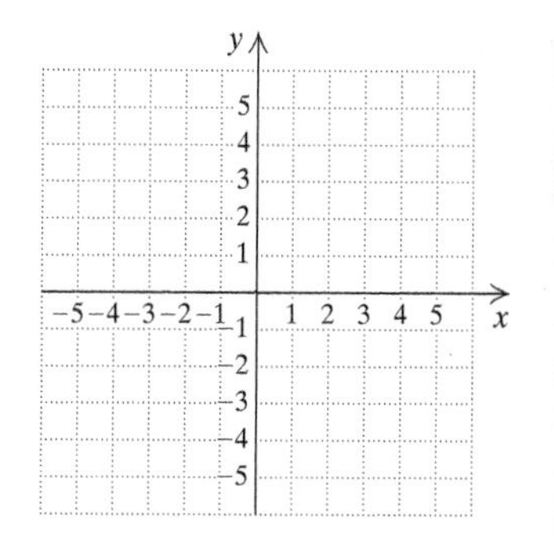

33. $x + 2y = 8$

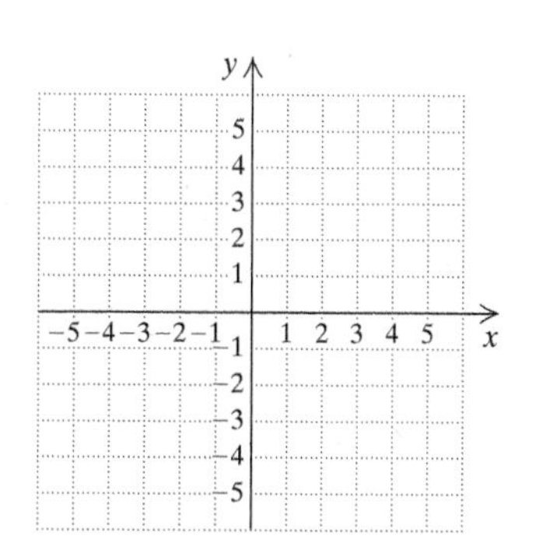

34. $x + 2y = -6$

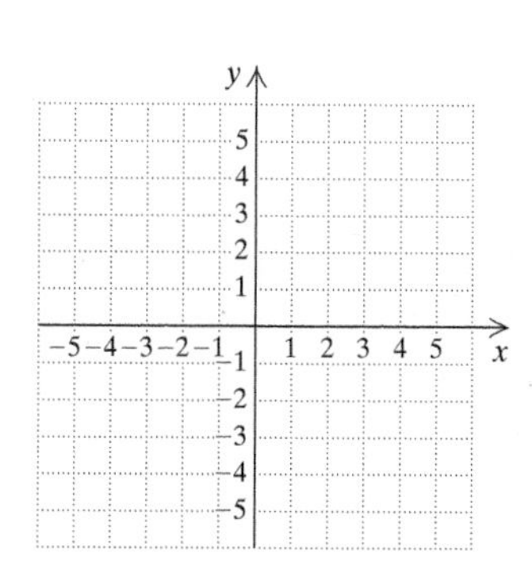

35. $y = \frac{3}{2}x + 1$

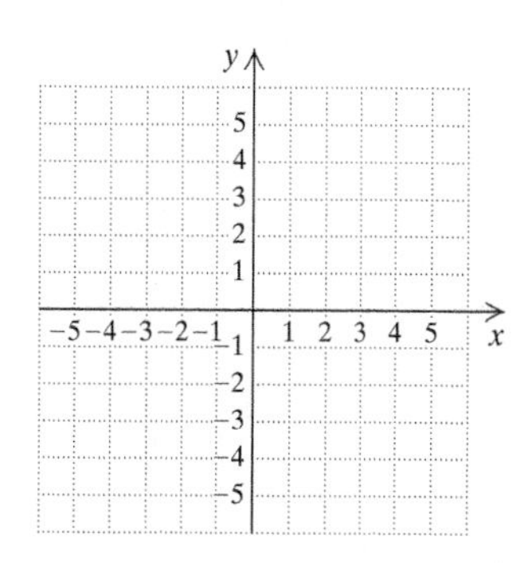

36. $y = -\frac{1}{2}x - 3$

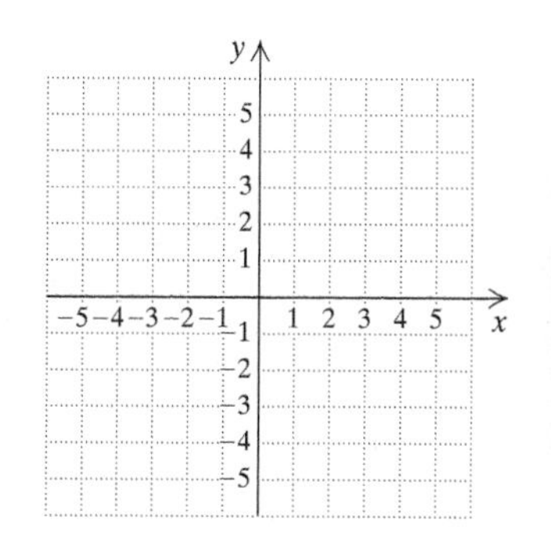

37. $8y + 2x = 4$

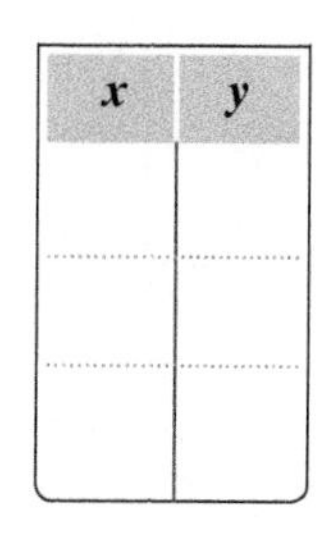

38. $6x - 3y = -9$

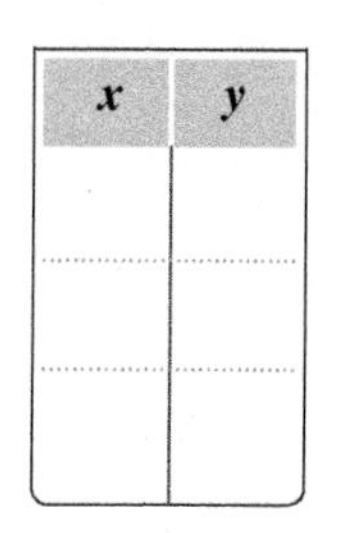

39. $8y + 2x = -4$

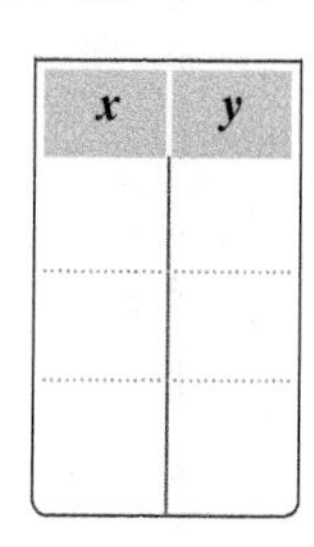

40. $6y + 2x = 8$

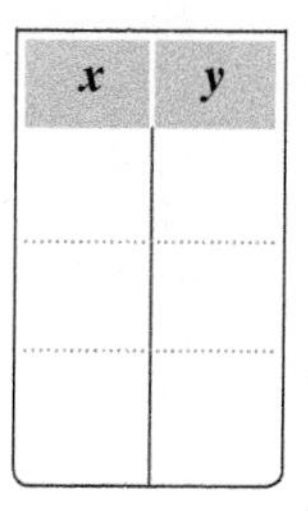

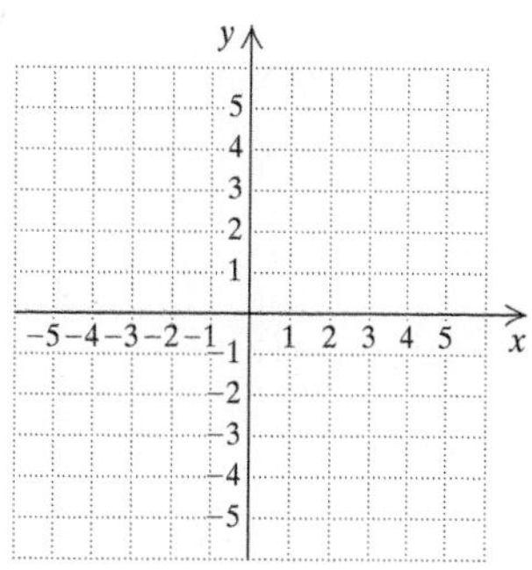

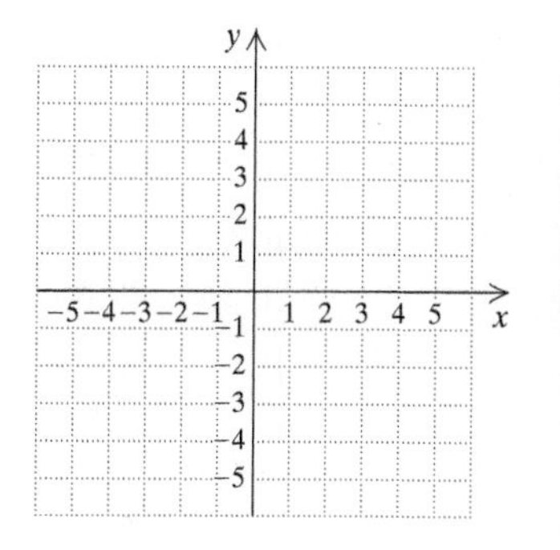

d Graph.

41. $y = x^2$

x	y

y
5
4
3
2
1
−5 −4 −3 −2 −1 1 2 3 4 5 x
−1
−2
−3
−4
−5

42. $y = -x^2$
($Hint$: $-x^2 = -1 \cdot x^2$.)

x	y

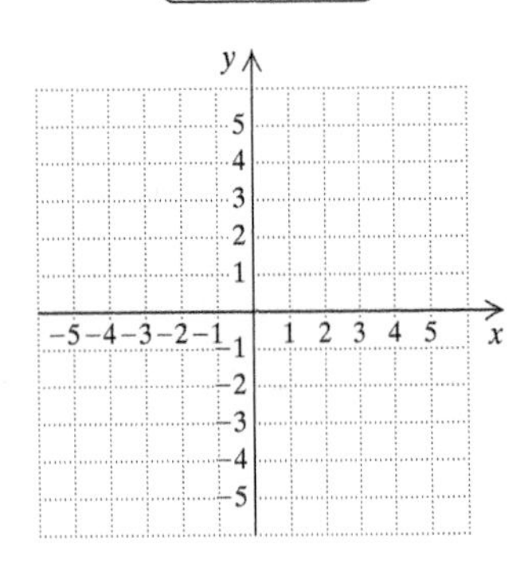

43. $y = x^2 + 2$

x	y

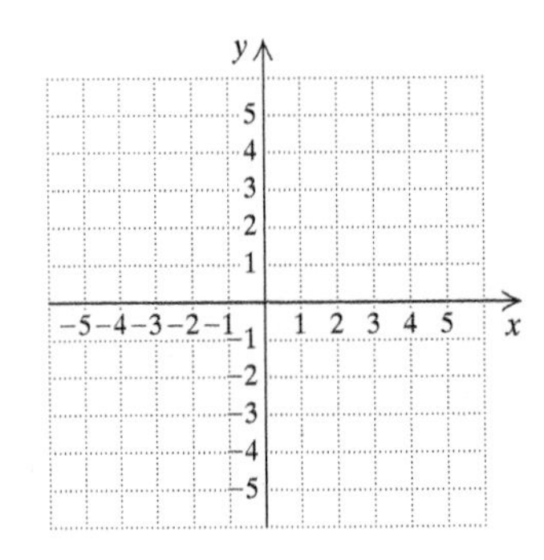

44. $y = 3 - x^2$

x	y

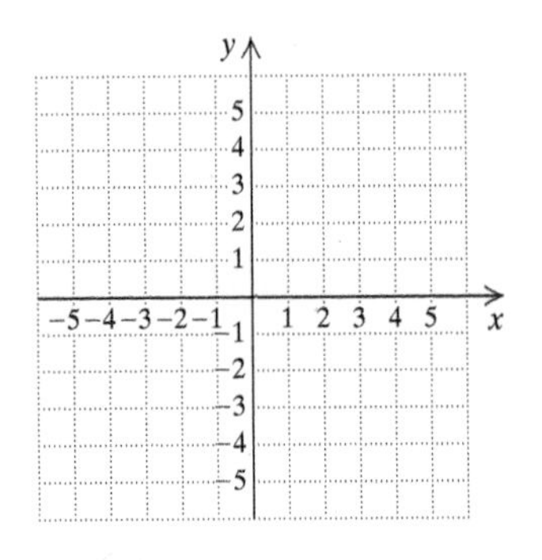

45. $y = x^2 - 3$

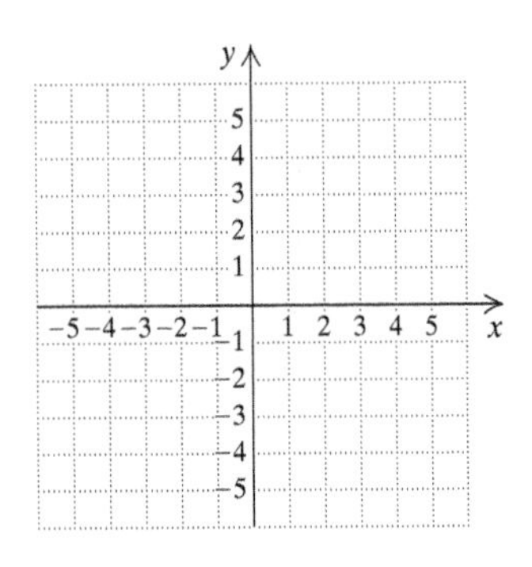

46. $y = x^2 - 3x$

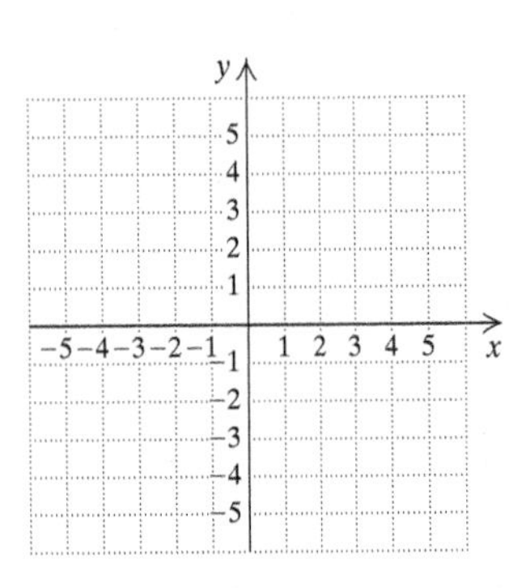

47. $y = -\frac{1}{x}$

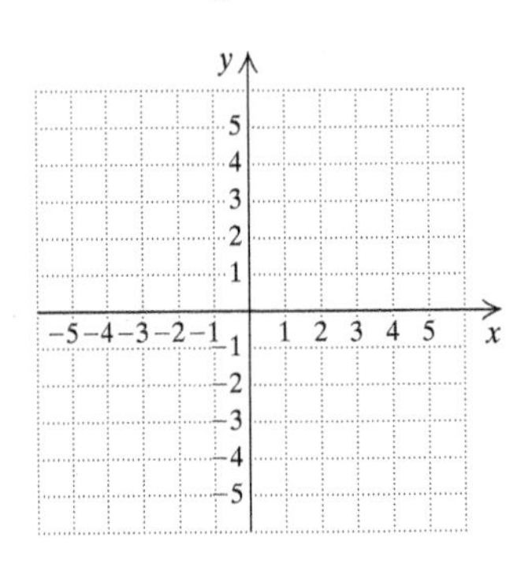

48. $y = \frac{3}{x}$

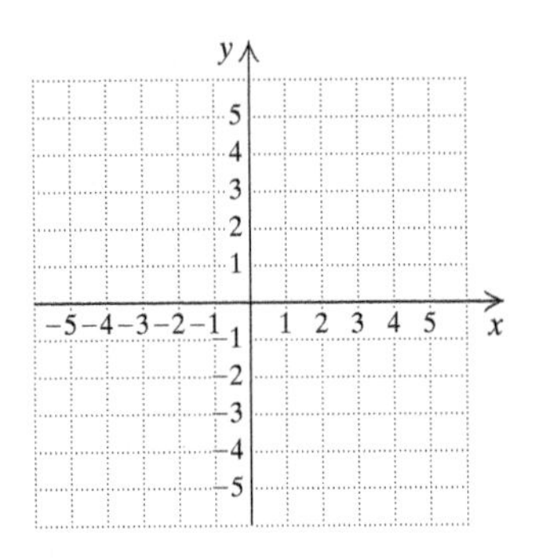

49. $y = |x - 2|$

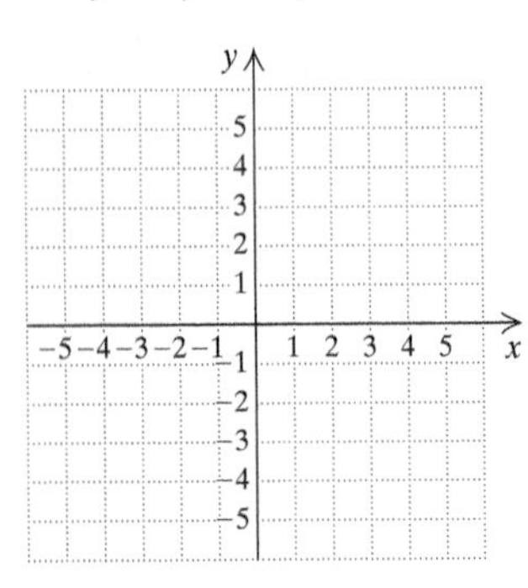

50. $y = |x| + 2$

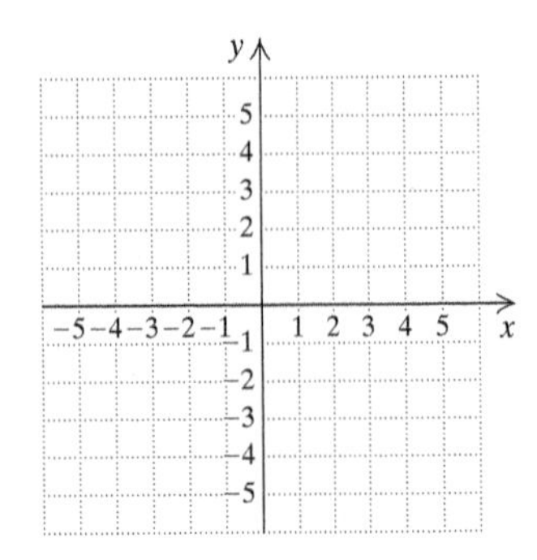

51. $y = x^3$

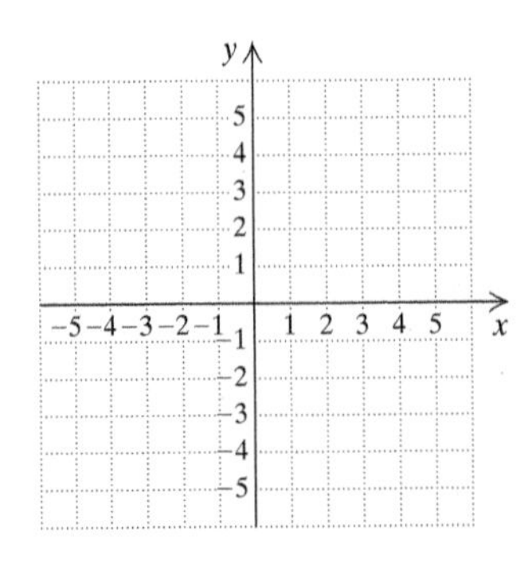

52. $y = x^3 - 2$

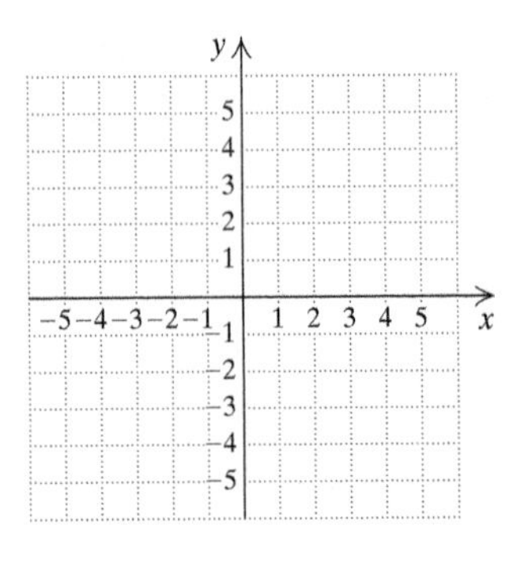

Skill Maintenance

Solve. [1.5a, b]

53. $-3 < 2x - 5 \leq 10$

54. $2x - 5 \geq -10$ *or* $-4x - 2 < 10$

55. $3x - 5 \leq -12$ *or* $3x - 5 \geq 12$

56. $-13 < 3x + 5 < 23$

Solve. [1.3a]

57. *Waiting Lists for Organ Transplants.* In July of 2016, there were more than 120,000 people on waiting lists for organ transplants. There were 114,171 people waiting for a kidney or a liver, and 84,867 fewer were waiting for a liver than for a kidney. How many were on the waiting list for a kidney? for a liver?

Data: Organ Procurement and Transplantation Network

58. *Landscaping.* Grass seed is being spread on a triangular traffic island. If the grass seed can cover an area of 200 ft^2 and the island's base is 16 ft long, how tall a triangle can the seed fill?

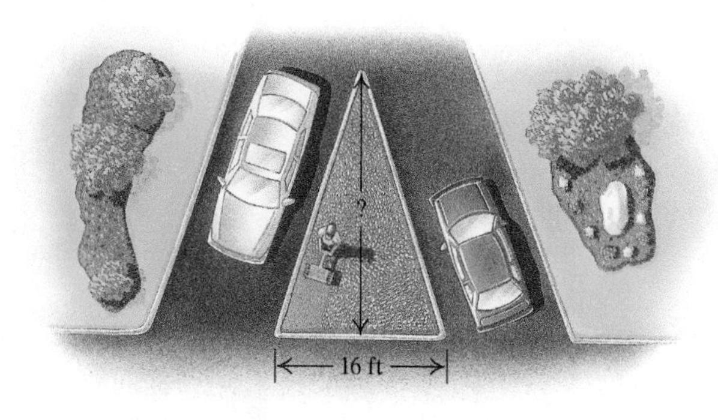

59. *Taxi Fare.* The fare for a taxi ride from Jen's office to the South Bay Health Center is \$19.85. The driver charges \$2.00 for the first $\frac{1}{2}$ mi and \$1.05 for each additional $\frac{1}{4}$ mi. How far is it from Jen's office to the South Bay Health Center?

60. *Real Estate Commission.* The Clines negotiated the following real estate commission on the selling price of their house:

7% for the first \$100,000 and

4% for the amount that exceeds \$100,000.

The realtor received a commission of \$16,200 for selling the house. What was the selling price?

Synthesis

Use a graphing calculator to graph each of the equations in Exercises 61–64. Use a standard viewing window of $[-10, 10, -10, 10]$, with Xscl = 1 and Yscl = 1.

61. $y = x^3 - 3x + 2$

62. $y = x - |x|$

63. $y = \frac{1}{x - 2}$

64. $y = \frac{1}{x^2}$

In Exercises 65–68, find an equation for the given graph.

65.

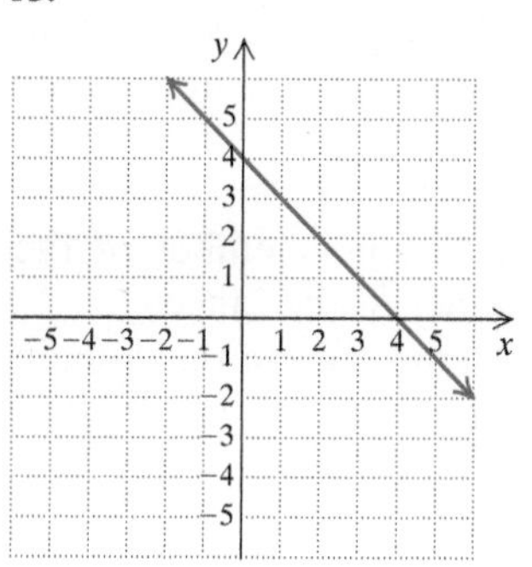

66.

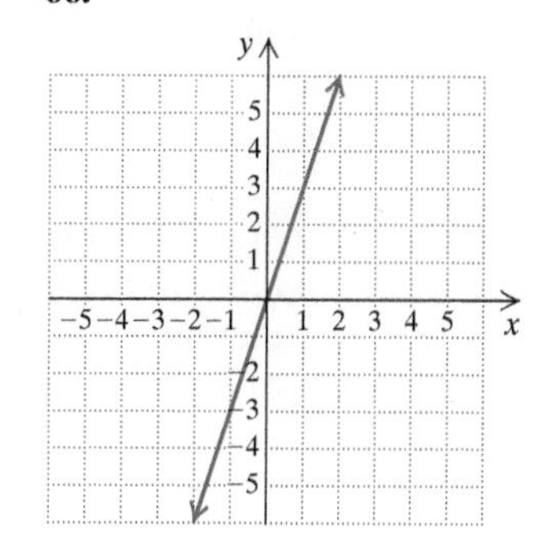

67.

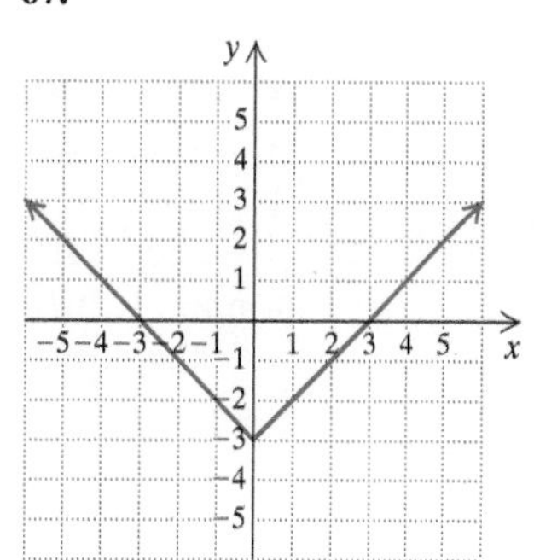

68.

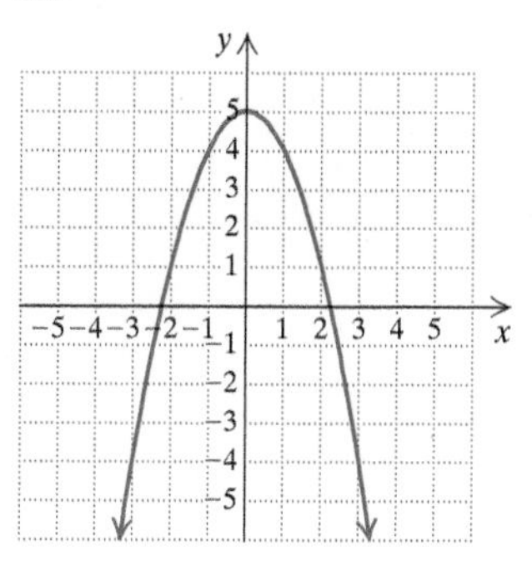

2.2 Functions and Graphs

OBJECTIVES

a Determine whether a correspondence is a function.

b Given a function described by an equation, find function values (outputs) for specified values (inputs).

c Draw the graph of a function.

d Determine whether a graph is that of a function using the vertical-line test.

e Solve applied problems involving functions and their graphs.

a IDENTIFYING FUNCTIONS

Consider the equation $y = 2x - 3$. If we substitute a value for x—say, 5—we get a value for y, 7:

$$y = 2x - 3 = 2(5) - 3 = 10 - 3 = 7.$$

The equation $y = 2x - 3$ is an example of a *function*, one of the most important concepts in mathematics.

In much the same way that ordered pairs form correspondences between first and second coordinates, a *function* is a correspondence from one set to another. For example:

To each student in a college, there corresponds his or her student ID.

To each item in a store, there corresponds its price.

To each real number, there corresponds the cube of that number.

In each case, the first set is called the **domain** and the second set is called the **range**. Each of these correspondences is a **function**, because given a member of the domain, there is *just one* member of the range to which it corresponds. Given a student, there is *just one* ID. Given an item, there is *just one* price. Given a real number, there is *just one* cube.

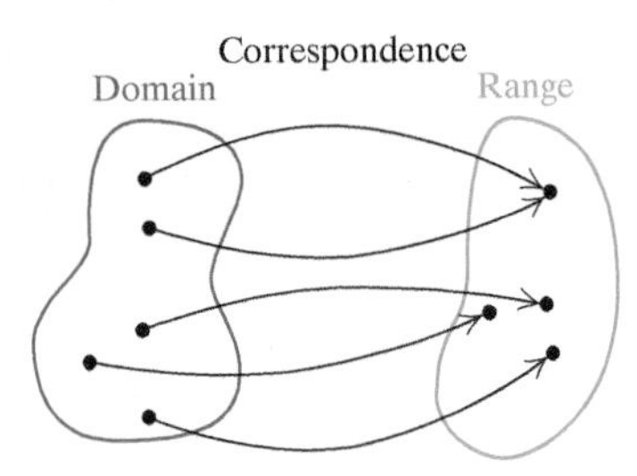

EXAMPLE 1 Determine whether the correspondence is a function.

f:

Domain		Range
1	⟶	\$107.40
2	⟶	\$ 34.10
3	⟶	\$ 29.60
4	⟶	\$ 19.60

g:

Domain		Range
3	⟶	5
4	⟶	9
5	⟶	−7
6	⟶	−7

h:

Domain		Range
Chicago	⟶	Cubs
Chicago	⟶	White Sox
Baltimore	⟶	Orioles
San Diego	⟶	Padres

p:

Domain		Range
Cubs	⟶	Chicago
White Sox	⟶	Chicago
Orioles	⟶	Baltimore
Padres	⟶	San Diego

The correspondence f *is* a function because each member of the domain is matched to *only one* member of the range.

The correspondence g *is* a function because each member of the domain is matched to *only one* member of the range. Note that a function allows two or more members of the domain to correspond to the same member of the range.

The correspondence h *is not* a function because one member of the domain, Chicago, is matched to *more than one* member of the range.

The correspondence p *is* a function because each member of the domain is matched to *only one* member of the range. ■

FUNCTION; DOMAIN; RANGE

A **function** is a correspondence between a first set, called the **domain**, and a second set, called the **range**, such that each member of the domain corresponds to **exactly one** member of the range.

Do Exercises 1–4.

Determine whether each correspondence is a function.

1. *Domain* — *Range*

Cheetah ⟶ 70 mph
Human ⟶ 28 mph
Lion ⟶ 50 mph
Chicken ⟶ 9 mph

2. *Domain* — *Range*

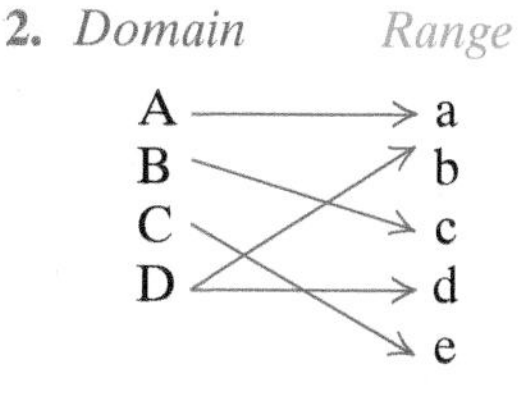

3. *Domain* — *Range*

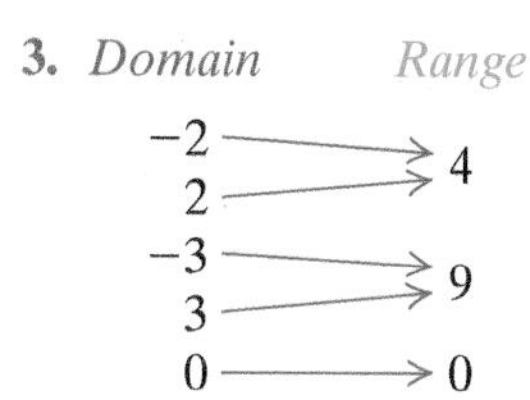

4. *Domain* — *Range*

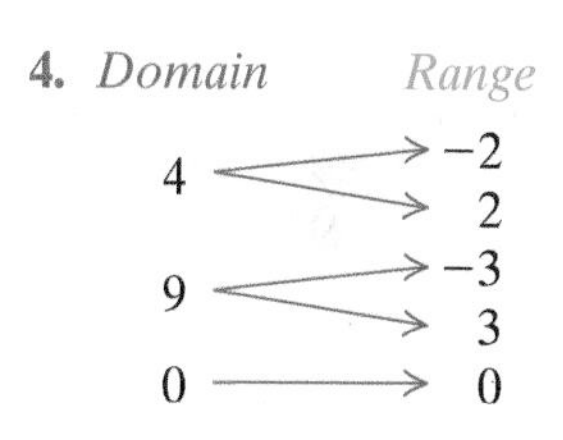

EXAMPLE 2 Determine whether each correspondence is a function.

	Domain	*Correspondence*	*Range*
a)	The integers	Each number's square	A set of nonnegative integers
b)	A set of presidents (listed below)	Each president's appointees to the Supreme Court	A set of Supreme Court Justices (listed below)

APPOINTING PRESIDENT	SUPREME COURT JUSTICE
George H. W. Bush	Samuel A. Alito, Jr.
William Jefferson Clinton	Stephen G. Breyer
	Ruth Bader Ginsburg
	Neil Gorsuch
George W. Bush	Elena Kagan
	John G. Roberts, Jr.
Barack H. Obama	Sonia M. Sotomayor
Donald J. Trump	Clarence Thomas

a) The correspondence *is* a function because each integer has *only one* square.

b) The correspondence *is not* a function because there is at least one member of the domain who is paired with more than one member of the range (William Jefferson Clinton with Stephen G. Breyer and Ruth Bader Ginsburg; George W. Bush with Samuel A. Alito, Jr., and John G. Roberts, Jr.; Barack H. Obama with Elena Kagan and Sonia M. Sotomayor).

Do Exercises 5–7 on the following page.

When a correspondence between two sets is not a function, it is still an example of a **relation**.

RELATION

A **relation** is a correspondence between a first set, called the **domain**, and a second set, called the **range**, such that each member of the domain corresponds to **at least one** member of the range.

Thus, although the correspondences of Examples 1 and 2 are not all functions, they *are* all relations. A function is a special type of relation—one in which each member of the domain is paired with *exactly one* member of the range.

Answers

1. Yes **2.** No **3.** Yes **4.** No

b FINDING FUNCTION VALUES

Evaluate algebraic expressions by substitution. [J13]

Evaluate.

1. $-\frac{1}{4}x$, when $x = 40$

2. $y^2 - 2y + 6$, when $y = -1$

Answers: 1. -10 **2.** 9

Most functions considered in mathematics are described by equations like $y = 2x + 3$ or $y = 4 - x^2$. We graph the function $y = 2x + 3$ by first performing calculations like the following:

for $x = 4, y = 2x + 3 = 2 \cdot 4 + 3 = 8 + 3 = 11$;
for $x = -5, y = 2x + 3 = 2 \cdot (-5) + 3 = -10 + 3 = -7$;
for $x = 0, y = 2x + 3 = 2 \cdot 0 + 3 = 0 + 3 = 3$; and so on.

For $y = 2x + 3$, the **inputs** (members of the domain) are values of x substituted into the equation. The **outputs** (members of the range) are the resulting values of y. If we call the function f, we can use x to represent an arbitrary *input* and $f(x)$—read "f of x," or "f at x," or "the value of f at x"—to represent the corresponding *output*. In this notation, the function given by $y = 2x + 3$ is written as $f(x) = 2x + 3$ and the calculations above can be written more concisely as follows:

$y = f(4) = 2 \cdot 4 + 3 = 8 + 3 = 11$;
$y = f(-5) = 2 \cdot (-5) + 3 = -10 + 3 = -7$;
$y = f(0) = 2 \cdot 0 + 3 = 0 + 3 = 3$; and so on.

Thus instead of writing "when $x = 4$, the value of y is 11," we can simply write "$f(4) = 11$," which can also be read as "f of 4 is 11" or "for the input 4, the output of f is 11."

We can think of a function as a machine. Think of $f(4) = 11$ as putting 4, a member of the domain (an input), into the machine. The machine knows the correspondence $f(x) = 2x + 3$, multiplies 4 by 2 and adds 3, and produces 11, a member of the range (the output).

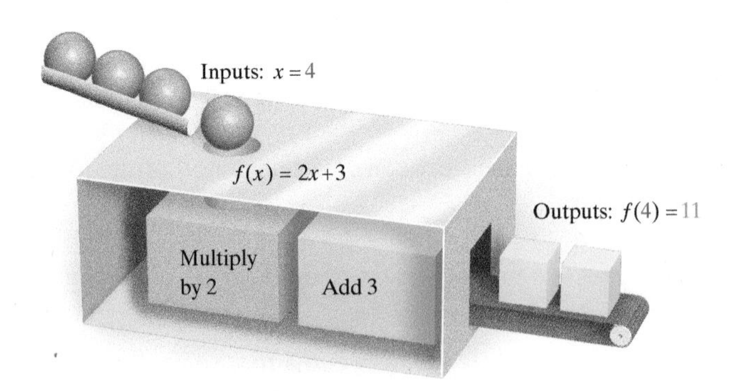

Determine whether each correspondence is a function.

5. *Domain*
A set of numbers
Correspondence
Square each number and subtract 10.
Range
A set of numbers

6. *Domain*
A set of polygons
Correspondence
Find the perimeter of each polygon.
Range
A set of numbers

7. Determine whether the correspondence is a function.
Domain
A set of numbers
Correspondence
The area of a rectangle
Range
A set of rectangles

Caution!

The notation $f(x)$ *does not mean* "f times x" and should not be read that way.

Answers
5. Yes **6.** Yes **7.** No

EXAMPLE 3 A function f is given by $f(x) = 3x^2 - 2x + 8$. Find each of the indicated function values.

a) $f(0)$ **b)** $f(-5)$ **c)** $f(7a)$

One way to find function values when a formula is given is to think of the formula with blanks, or placeholders, replacing the variable as follows:

$$f(\square) = 3\square^2 - 2\square + 8.$$

To find an output for a given input, we think: "Whatever goes in the blank on the left goes in the blank(s) on the right." With this in mind, let's complete the example.

a) $f(0) = 3\cdot 0^2 - 2\cdot 0 + 8 = 8$

b) $f(-5) = 3(-5)^2 - 2\cdot(-5) + 8 = 3\cdot 25 + 10 + 8 = 75 + 10 + 8 = 93$

c) $f(7a) = 3(7a)^2 - 2(7a) + 8 = 3\cdot 49a^2 - 2(7a) + 8 = 147a^2 - 14a + 8$

Do Exercise 8. ▶

EXAMPLE 4 Find the indicated function value.

a) $f(5)$, for $f(x) = 3x + 2$ **b)** $g(-2)$, for $g(x) = 7$

c) $F(a + 1)$, for $F(x) = 5x - 8$ **d)** $f(a + h)$, for $f(x) = -2x + 1$

a) $f(5) = 3\cdot 5 + 2 = 15 + 2 = 17$

b) For the function given by $g(x) = 7$, all inputs share the same output, 7. Thus, $g(-2) = 7$. The function g is an example of a **constant function**.

c) $F(a + 1) = 5(a + 1) - 8 = 5a + 5 - 8 = 5a - 3$

d) $f(a + h) = -2(a + h) + 1 = -2a - 2h + 1$

Do Exercise 9. ▶

8. Find the indicated function values for the function $f(x) = 2x^2 + 3x - 4$.

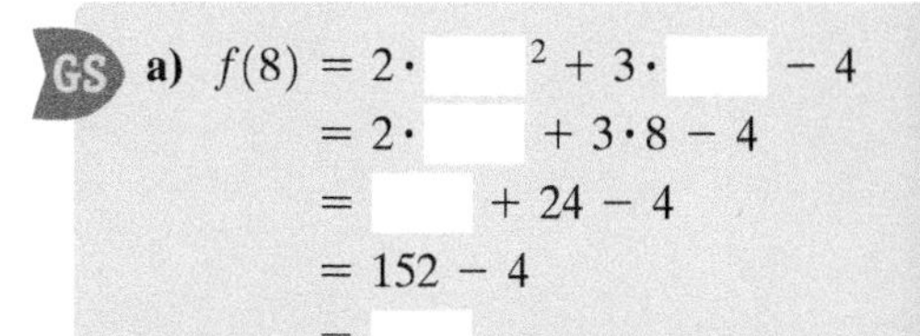
GS **a)** $f(8) = 2\cdot\_\_\_^2 + 3\cdot\_\_\_ - 4$
$= 2\cdot\_\_\_ + 3\cdot 8 - 4$
$= \_\_\_ + 24 - 4$
$= 152 - 4$
$= \_\_\_$

b) $f(0)$

c) $f(-5)$

d) $f(2a)$

9. Find the indicated function value.

a) $f(-6)$, for $f(x) = 5x - 3$

b) $g(55)$, for $g(x) = -3$

c) $F(a + 2)$, for $F(x) = -5x + 8$

d) $f(a - h)$, for $f(x) = 6x - 7$

Answers

8. (a) 148; **(b)** -4; **(c)** 31; **(d)** $8a^2 + 6a - 4$ **9. (a)** -33; **(b)** -3; **(c)** $-5a - 2$; **(d)** $6a - 6h - 7$

Guided Solution:

8. (a) 8, 8, 64, 128, 148

CALCULATOR CORNER

Finding Function Values We can find function values using a graphing calculator. One method is to substitute inputs directly into the formula. Consider the function $f(x) = x^2 + 3x - 4$. We find that $f(-5) = 6$. See Figure 1.

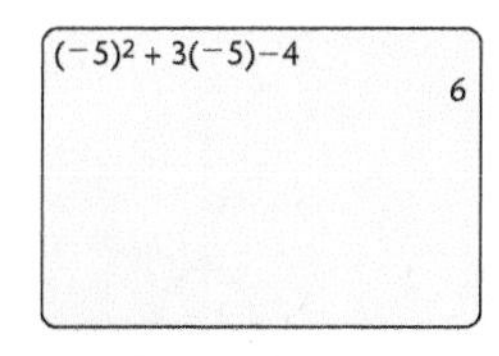

FIGURE 1

X	Y1	
−5	6	

X =

FIGURE 2

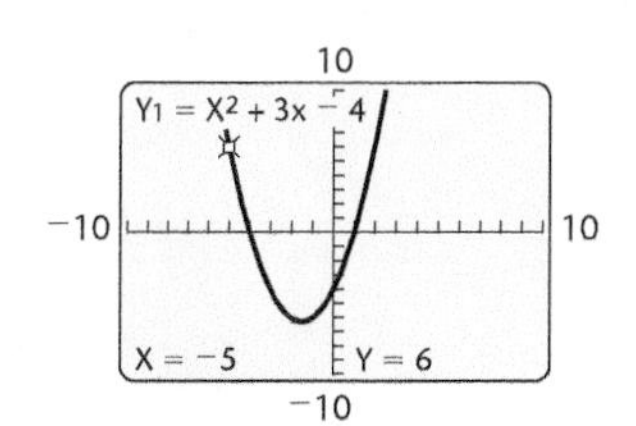

FIGURE 3

After we have entered the function as $y_1 = x^2 + 3x - 4$ on the equation-editor screen, there are other methods that we can use to find function values. We can use a table set in ASK mode and enter $x = -5$. We see that the function value, y_1, is 6. See Figure 2. We can also use the VALUE feature to evaluate the function. To do this, we first graph the function. Then we press 2ND CALC 1 to access the VALUE feature. Next, we supply the desired x-value. Finally, we press ENTER to see $X = -5, Y = 6$ at the bottom of the screen. See Figure 3. Again we see that the function value is 6. Note that when the VALUE feature is used to find a function value, the x-value must be in the viewing window.

EXERCISES: Find each function value.

1. $f(-5.1)$, for $f(x) = -3x + 2$

2. $f(3)$, for $f(x) = 4x^2 + x - 5$

MyLab Math
ANIMATION

C GRAPHS OF FUNCTIONS

To graph a function, we find ordered pairs (x, y) or $(x, f(x))$, plot them, and connect the points. Note that y and $f(x)$ are used interchangeably—that is, $y = f(x)$—when we are working with functions and their graphs.

EXAMPLE 5 Graph: $f(x) = x + 2$.

A list of some function values is shown in the following table. We plot the points and connect them. The graph is a straight line. The "y" on the vertical axis could also be labeled "$f(x)$."

x	$f(x)$
−4	−2
−3	−1
−2	0
−1	1
0	2
1	3
2	4
3	5
4	6

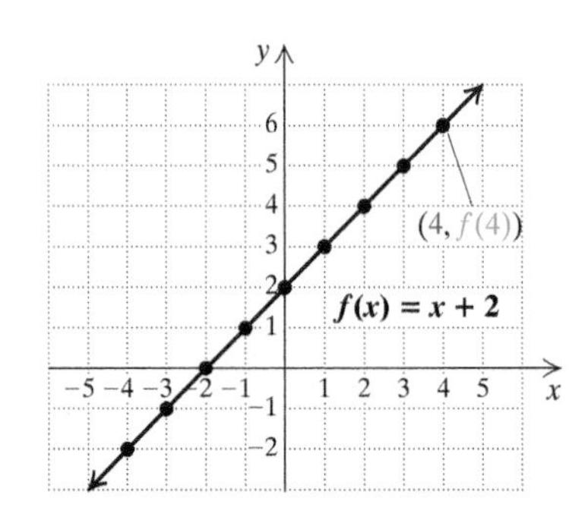

◀ Do Exercise 10.

EXAMPLE 6 Graph: $g(x) = 4 - x^2$.

We calculate some function values, plot the corresponding points, and draw the curve.

$$g(0) = 4 - 0^2 = 4 - 0 = 4,$$
$$g(-1) = 4 - (-1)^2 = 4 - 1 = 3,$$
$$g(1) = 4 - 1^2 = 4 - 1 = 3,$$
$$g(-2) = 4 - (-2)^2 = 4 - 4 = 0,$$
$$g(2) = 4 - 2^2 = 4 - 4 = 0,$$
$$g(-3) = 4 - (-3)^2 = 4 - 9 = -5$$
$$g(3) = 4 - 3^2 = 4 - 9 = -5$$

x	$g(x)$
−3	−5
−2	0
−1	3
0	4
1	3
2	0
3	−5

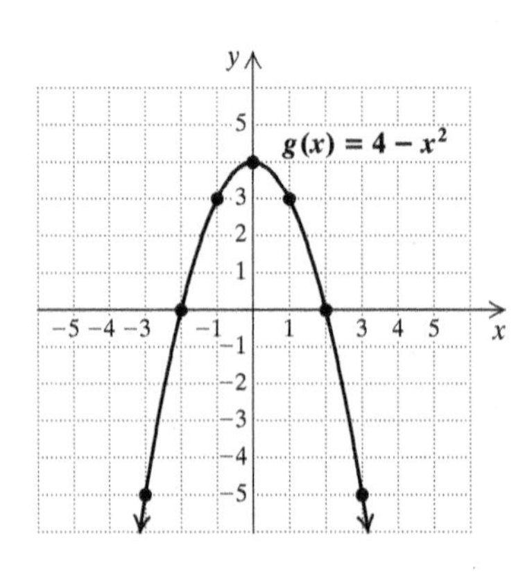

◀ Do Exercise 11.

10. Graph: $f(x) = x - 4$.

x	$f(x)$

11. Graph: $g(x) = 5 - x^2$.

x	$g(x)$

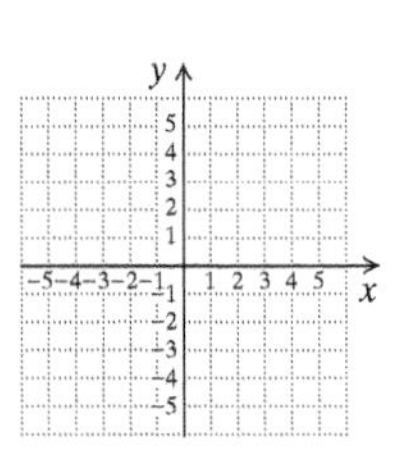

Answers

10.

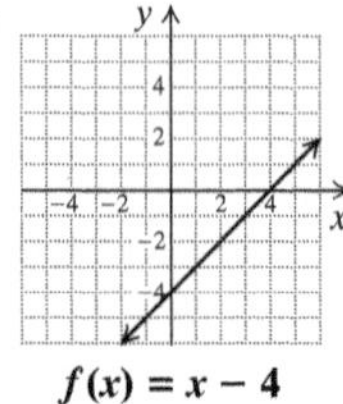

$f(x) = x - 4$

11.

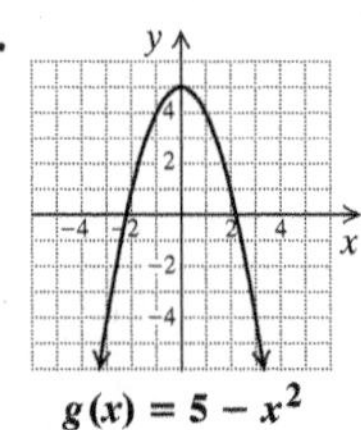

$g(x) = 5 - x^2$

EXAMPLE 7 Graph: $h(x) = |x|$.

A list of some function values is shown in the following table. We plot the points and connect them. The graph is a V-shaped "curve" that rises on either side of the vertical axis.

x	$h(x)$
−3	3
−2	2
−1	1
0	0
1	1
2	2
3	3

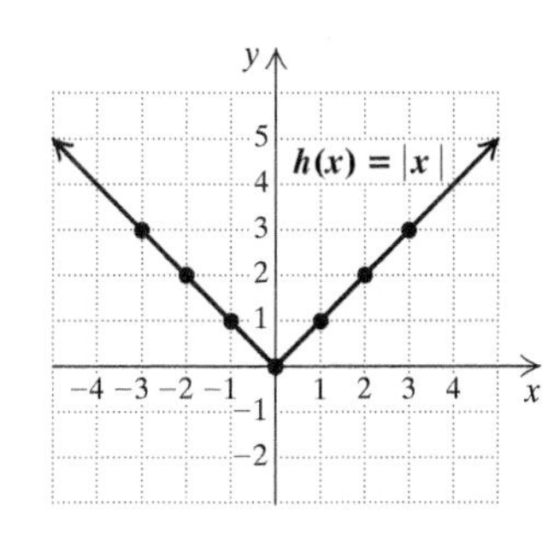

Do Exercise 12. ▶

12. Graph: $t(x) = 3 - |x|$.

x	$t(x)$

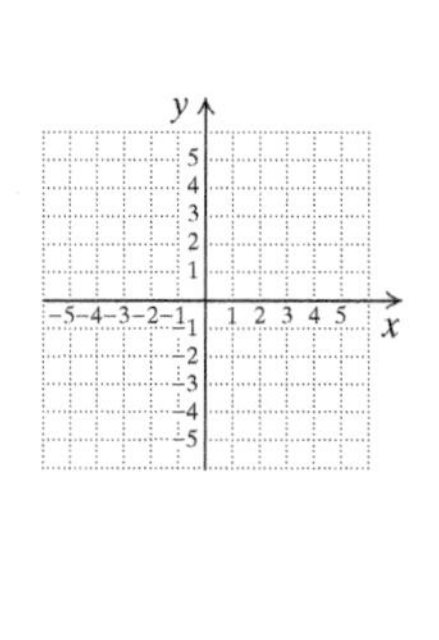

d THE VERTICAL-LINE TEST

Consider the graph of the function f described by $f(x) = x^2 - 5$ shown at right. It is also the graph of the equation $y = x^2 - 5$.

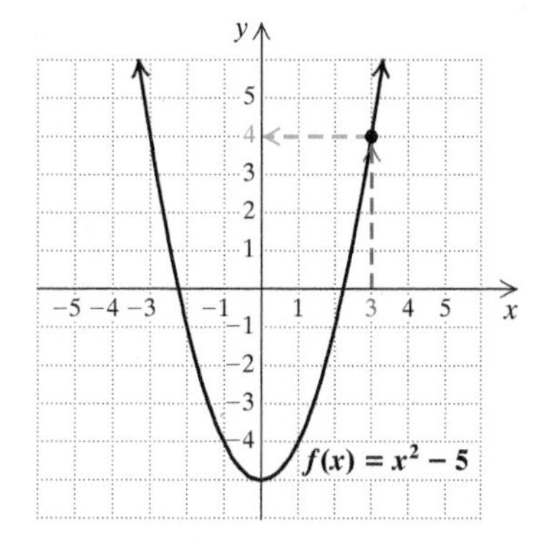

To find a function value, like $f(3)$, from a graph, we locate the input on the horizontal axis, move directly up or down to the graph of the function, and then move left or right to find the output on the vertical axis. Thus, $f(3) = 4$. Keep in mind that members of the domain are found on the horizontal axis, members of the range are found on the vertical axis, and the y on the vertical axis could also be labeled $f(x)$.

When one member of the domain is paired with two or more different members of the range, the correspondence is not a function. Thus, when a graph contains two or more different points with the same first coordinate, the graph cannot represent a function. Points sharing a common first coordinate are vertically above or below each other. (See the following graph.) This observation leads to the *vertical-line test.*

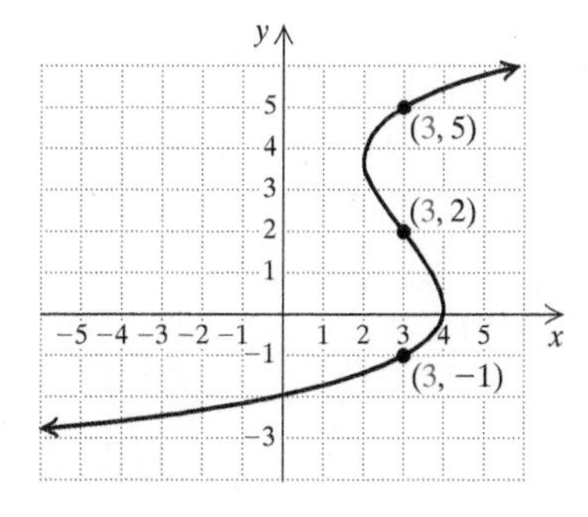

Since 3 is paired with more than one member of the range, the graph does not represent a function.

THE VERTICAL-LINE TEST

If it is possible for a vertical line to cross a graph more than once, then the graph is *not* the graph of a function.

CALCULATOR CORNER

Graphing Functions

To graph a function using a graphing calculator, we replace the function notation with y and proceed as described in the Calculator Corner on p. 125. To graph $f(x) = 2x^2 + x$ in the standard window, for example, we replace $f(x)$ with y and enter $y_1 = 2x^2 + x$ on the Y = screen and then press ZOOM 6.

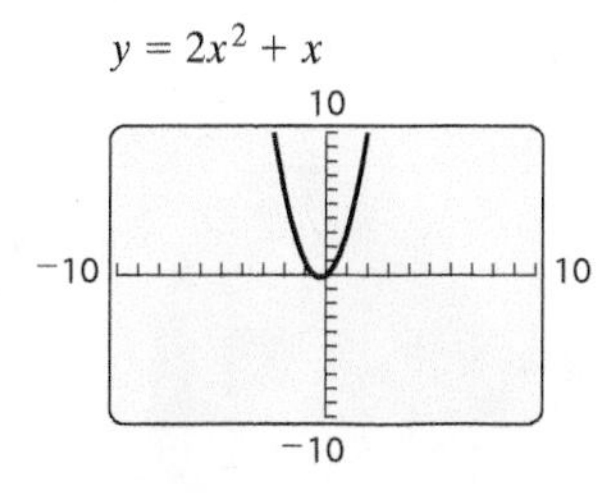

EXERCISES: Graph each function.

1. $f(x) = x - 4$
2. $f(x) = -2x - 3$
3. $h(x) = 1 - x^2$
4. $f(x) = 3x^2 - 4x + 1$
5. $f(x) = x^3$
6. $f(x) = |x + 3|$

Answer

12. $t(x) = 3 - |x|$

Determine whether each of the following is the graph of a function.

13.
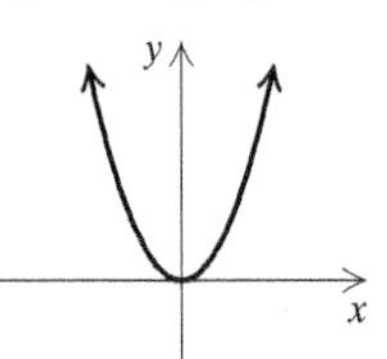

14.
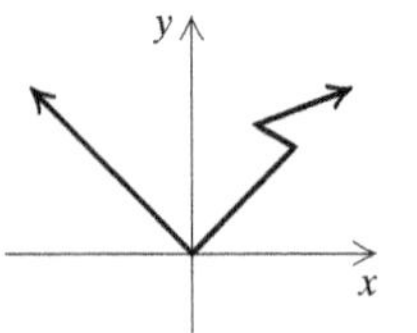

15.
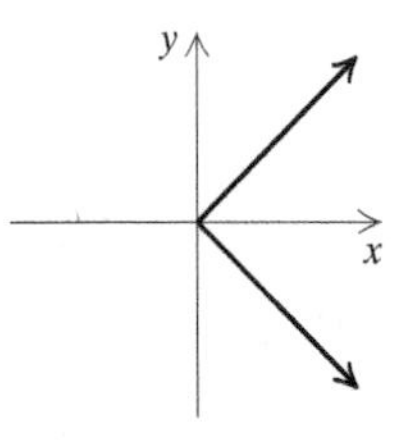

16.
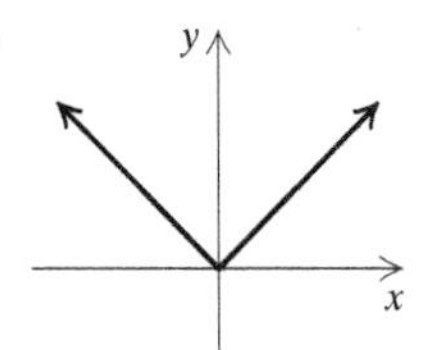

EXAMPLE 8 Determine whether each of the following is the graph of a function.

a)
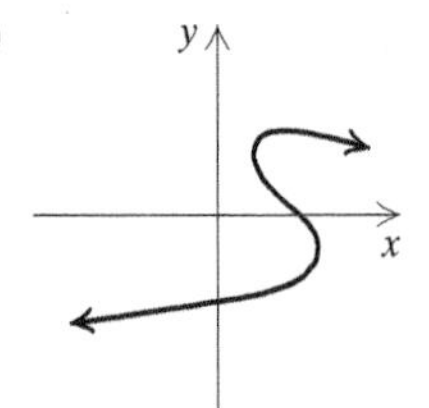

b)
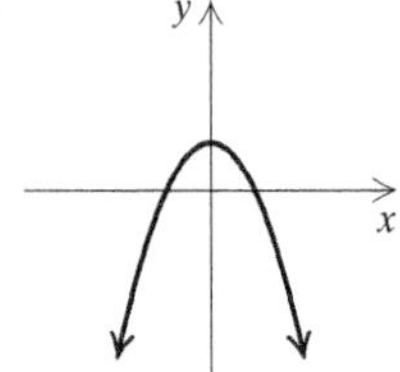

a) The graph *is not* that of a function because a vertical line can cross the graph at more than one point.

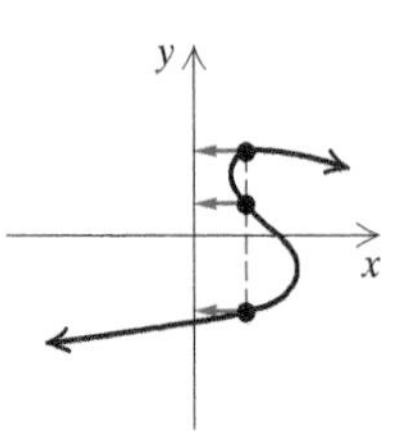

b) The graph *is* that of a function because no vertical line can cross the graph more than once.

Do Exercises 13–16.

e APPLICATIONS OF FUNCTIONS AND THEIR GRAPHS

Functions are often described by graphs, whether or not an equation is given. To use a graph in an application, we note that each point on the graph represents a pair of values.

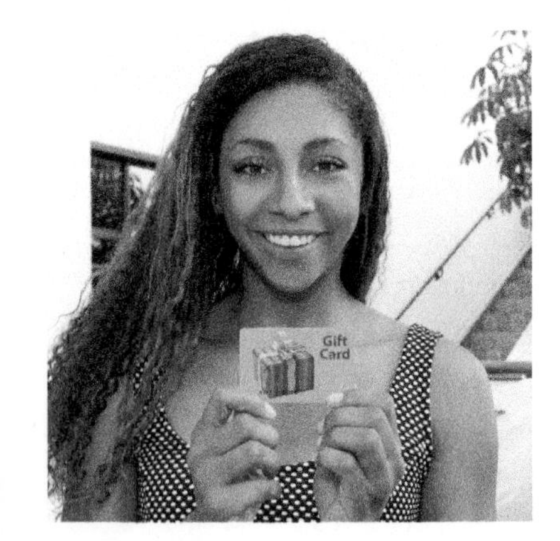

EXAMPLE 9 *Gift-Card Sales.* The following graph represents worldwide gift-card sales from 2008 to 2018. This amount of sales, in billions of dollars, is a function of the year. Note that no equation is given for the function.

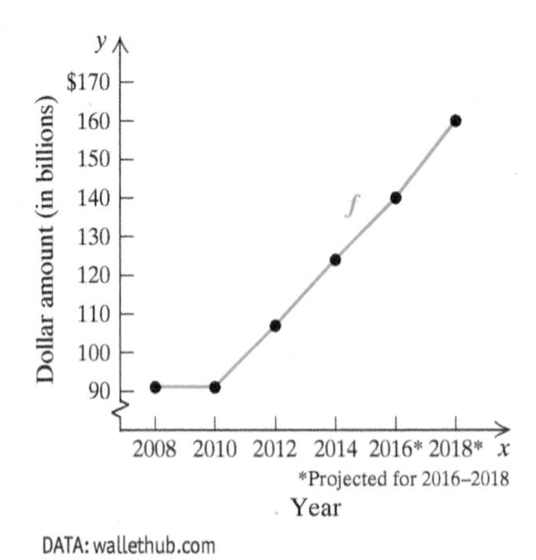

DATA: wallethub.com

a) What was the total amount spent on gift cards in 2012? That is, find $f(2012)$.

b) What is the projected amount spend on gift cards in 2017? That is, find $f(2017)$.

Answers
13. Yes **14.** No **15.** No **16.** Yes

a) To estimate the amount spent on gift cards in 2012, we locate 2012 on the horizontal axis and move directly up until we reach the graph. Then we move across to the vertical axis. We come to a point that is about 107, so we estimate the amount spent on gift cards in 2012 to be $107 billion.

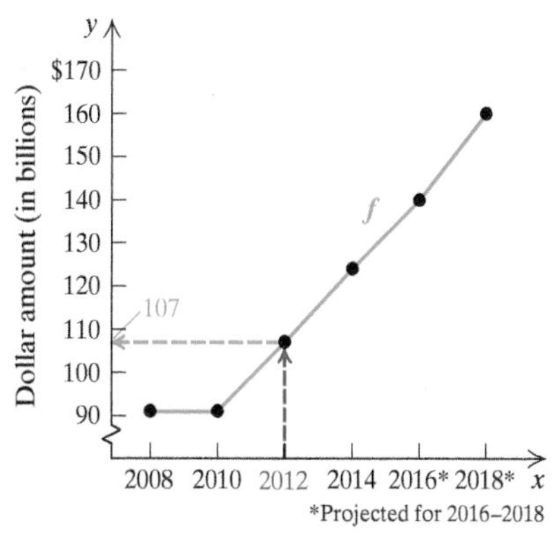

DATA: wallethub.com

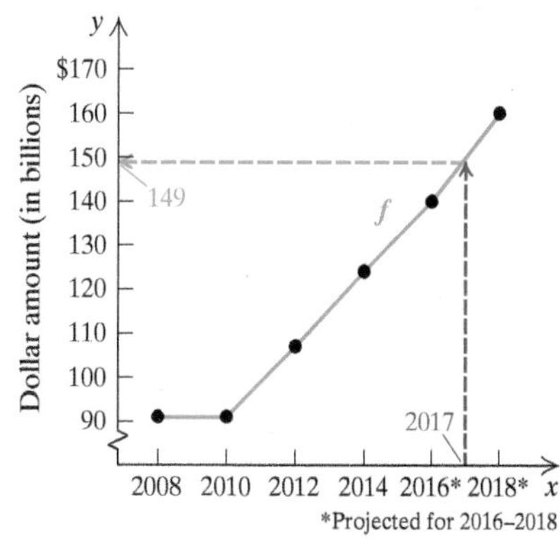

DATA: wallethub.com

b) To estimate the amount spent on gift cards in 2017, we locate 2017 on the horizontal axis and move directly up until we reach the graph. Then we move across to the vertical axis. We come to a point that is about 149, so we estimate the amount spent on gift cards in 2017 to be $149 billion.

Do Exercises 17 and 18.

Refer to the graph in Example 9.

17. What was the total spent on gift cards in 2014?

18. What was the total spent on gift cards in 2011?

Answers

17. $124 billion **18.** $99 billion

2.2 Exercise Set

FOR EXTRA HELP

MyLab Math

✓ Check Your Understanding

Reading Check Choose from the column on the right the word that completes each definition. Some words will be used more than once.

RC1. A function is a correspondence between a first set, called the __________, and a second set, called the __________, such that each member of the __________ corresponds to __________ member of the __________.

RC2. A relation is a correspondence between a first set, called the __________, and a second set, called the __________, such that each member of the __________ corresponds to __________ member of the __________.

at least one
exactly one
domain
range

Concept Check Use the graph at right to find the given function value by locating the input on the horizontal axis, moving directly up or down to the graph of the function, and then moving left or right to find the output on the vertical axis. As an example, finding $f(4) = -5$ is illustrated.

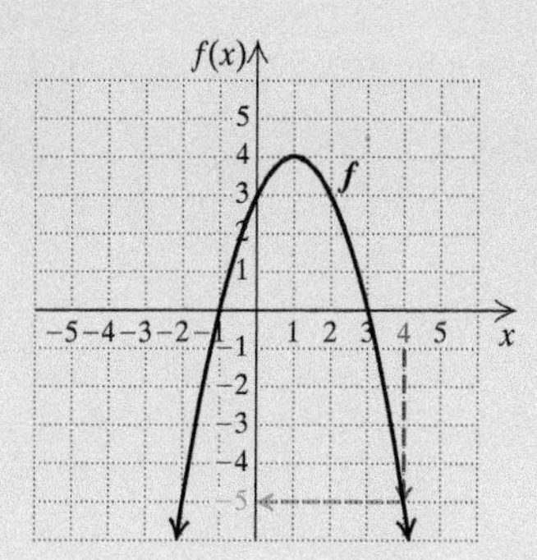

CC1. $f(2)$

CC2. $f(0)$

CC3. $f(-2)$

CC4. $f(3)$

a Determine whether each correspondence is a function.

1.

Domain	*Range*
2	9
5	8
19	

2.

Domain	*Range*
5	3
-3	7
7	
-7	

3.

Domain	*Range*
-5	1
5	
8	

4.

Domain	*Range*
6	-6
7	-7
3	-3

5.

Domain	*Range*
9	3, -3
16	4, -4
25	5, -5

6.

Domain	*Range*
The Color Purple, 1982 (Pulitzer Prize 1983)	Ray Bradbury
East of Eden, 1952	Pearl Buck
Fahrenheit 451, 1953	Ernest Hemingway
The Good Earth, 1931 (Pulitzer Prize 1932)	Harper Lee
For Whom the Bell Tolls, 1940	John Steinbeck
The Grapes of Wrath, 1939 (Pulitzer Prize 1940)	Alice Walker
To Kill a Mockingbird, 1960 (Pulitzer Prize 1961)	
The Old Man and the Sea, 1952 (Pulitzer Prize 1953)	

7.

Domain	*Range*
Florida	Florida State University, University of Florida, University of Miami
Kansas	Baker University, Kansas State University, University of Kansas

8.

Domain	*Range*
Colorado State University, University of Colorado, University of Denver	Colorado
Gonzaga University, University of Washington, Washington State University	Washington

	Domain	*Correspondence*	*Range*
9.	A set of numbers	The area of a triangle	A set of triangles
10.	A family	Each person's height, in inches	A set of positive numbers
11.	The set of U.S. Senators	The state that a Senator represents	The set of all states
12.	The set of all states	Each state's members of the U.S. Senate	The set of U.S. Senators

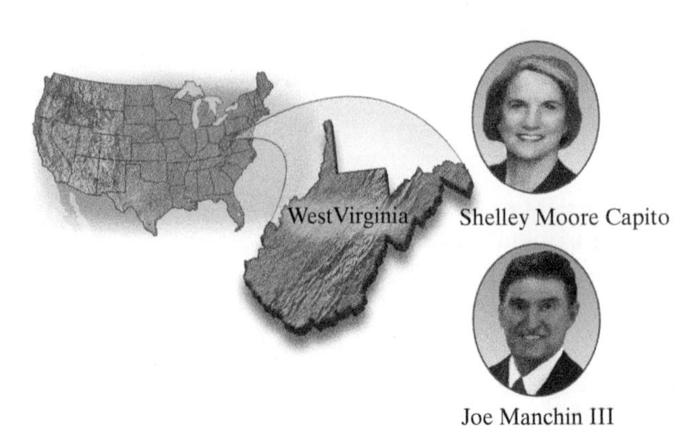

b Find the function values.

13. $f(x) = x + 5$
a) $f(4)$ **b)** $f(7)$
c) $f(-3)$ **d)** $f(0)$
e) $f(2.4)$ **f)** $f\left(\frac{2}{3}\right)$

14. $g(t) = t - 6$
a) $g(0)$ **b)** $g(6)$
c) $g(13)$ **d)** $g(-1)$
e) $g(-1.08)$ **f)** $g\left(\frac{7}{8}\right)$

15. $h(p) = 3p$
a) $h(-7)$ **b)** $h(5)$
c) $h\left(\frac{2}{3}\right)$ **d)** $h(0)$
e) $h(6a)$ **f)** $h(a + 1)$

16. $f(x) = -4x$
a) $f(6)$ **b)** $f\left(-\frac{1}{2}\right)$
c) $f(0)$ **d)** $f(-1)$
e) $f(3a)$ **f)** $f(a - 1)$

17. $g(s) = 3s + 4$
a) $g(1)$ **b)** $g(-7)$
c) $g\left(\frac{2}{3}\right)$ **d)** $g(0)$
e) $g(a - 2)$ **f)** $g(a + h)$

18. $h(x) = 19$, a constant function
a) $h(4)$ **b)** $h(-6)$
c) $h(12.5)$ **d)** $h(0)$
e) $h\left(\frac{2}{3}\right)$ **f)** $h(a + 3)$

19. $f(x) = 2x^2 - 3x$
a) $f(0)$ **b)** $f(-1)$
c) $f(2)$ **d)** $f(10)$
e) $f(-5)$ **f)** $f(4a)$

20. $f(x) = 3x^2 - 2x + 1$
a) $f(0)$ **b)** $f(1)$
c) $f(-1)$ **d)** $f(10)$
e) $f(-3)$ **f)** $f(2a)$

21. $f(x) = |x| + 1$
a) $f(0)$ **b)** $f(-2)$
c) $f(2)$ **d)** $f(-10)$
e) $f(a - 1)$ **f)** $f(a + h)$

22. $g(t) = |t - 1|$
a) $g(4)$ **b)** $g(-2)$
c) $g(-1)$ **d)** $g(100)$
e) $g(5a)$ **f)** $g(a + 1)$

23. $f(x) = x^3$
a) $f(0)$ **b)** $f(-1)$
c) $f(2)$ **d)** $f(10)$
e) $f(-5)$ **f)** $f(-3a)$

24. $f(x) = x^4 - 3$
a) $f(1)$ **b)** $f(-1)$
c) $f(0)$ **d)** $f(2)$
e) $f(-2)$ **f)** $f(-a)$

25. *Gold Production in China.* The function $G(x) = 0.943x + 10.943$ can be used to estimate the production of gold, in thousands of troy ounces, in China from 2010 through 2015. Let $G(x) =$ the amount of gold produced and $x =$ the number of years after 2010. What was the amount of gold produced in China in 2012? in 2015?

Data: Mineral Commodity Summaries 2015, U.S. Geological Survey, U.S. Department of Interior

26. *International Travelers to the United States.* The function $T(x) = 3.72x + 55.54$ can be used to estimate the number of international visitors, in millions, to the United States from 2009 through 2015. Let $T(x) =$ the number of international travelers and $x =$ the number of years after 2009. How many international travelers, to the nearest million, visited the United States in 2011? in 2014?

Data: National Travel and Tourism Office, International Trade Administration, U.S. Department of Commerce, World Tourism Organization

27. ***Pressure at Sea Depth.*** The function $P(d) = 1 + (d/33)$ gives the pressure, in *atmospheres* (atm), at a depth of d feet in the sea. Note that $P(0) = 1$ atm, $P(33) = 2$ atm, and so on. Find the pressure at 20 ft, 30 ft, and 100 ft.

28. ***Temperature as a Function of Depth.*** The function $T(d) = 10d + 20$ gives the temperature, in degrees Celsius, inside the earth as a function of the depth d, in kilometers. Find the temperature at 5 km, 20 km, and 1000 km.

29. ***Melting Snow.*** The function $W(d) = 0.112d$ approximates the amount of water, in centimeters, that results from d centimeters of snow melting. Find the amount of water that results from snow melting from depths of 16 cm, 25 cm, and 100 cm.

30. ***Temperature Conversions.*** The function $C(F) = \frac{5}{9}(F - 32)$ determines the Celsius temperature that corresponds to F degrees Fahrenheit. Find the Celsius temperature that corresponds to 62°F, 77°F, and 23°F.

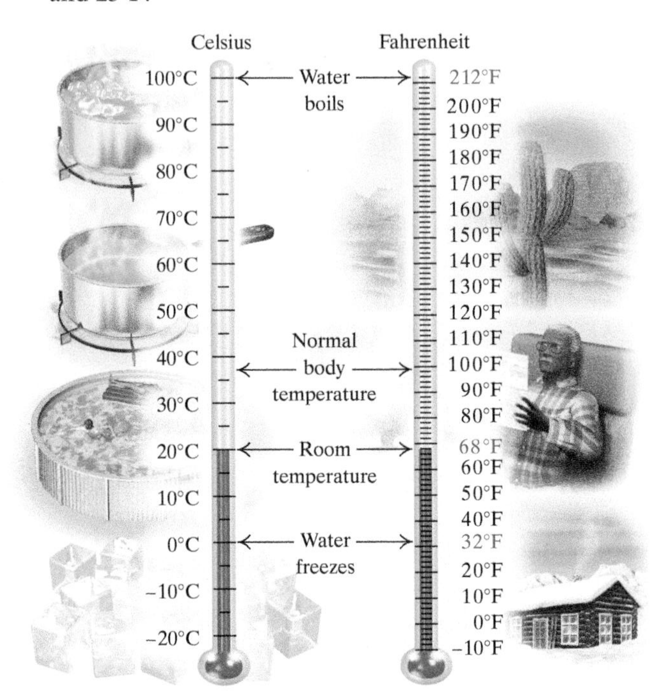

C Graph each function.

31. $f(x) = -2x$

x	$f(x)$

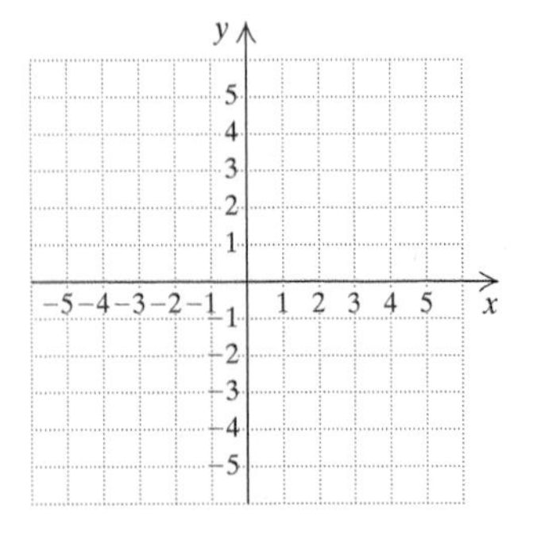

32. $g(x) = 3x$

x	$g(x)$

33. $g(x) = 3x - 1$

x	$g(x)$

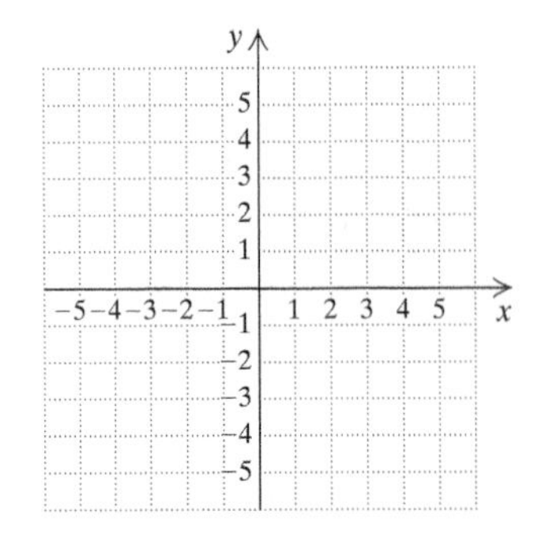

34. $f(x) = 2x + 5$

x	$f(x)$

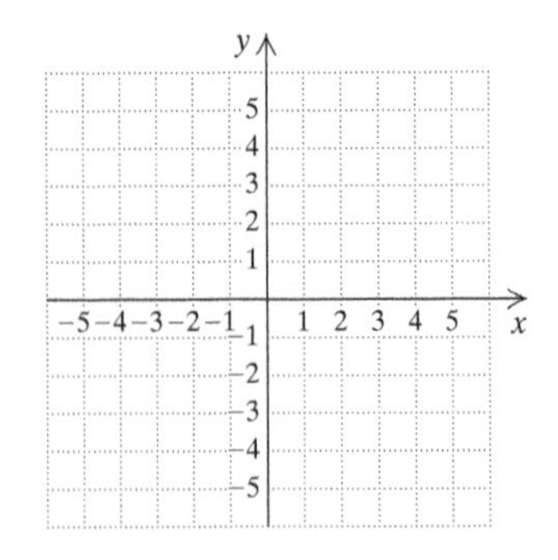

35. $g(x) = -2x + 3$

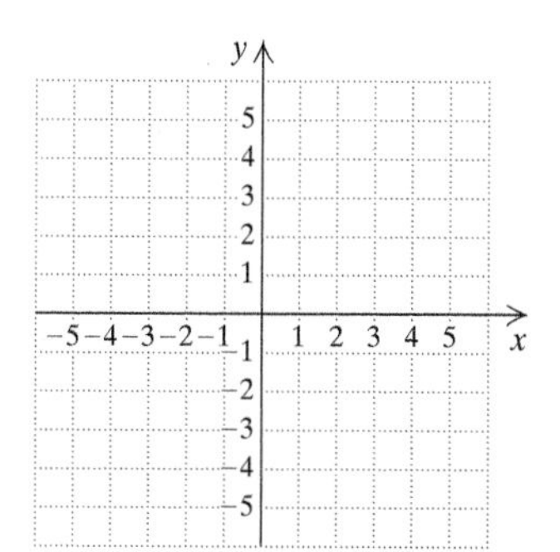

36. $f(x) = -\frac{1}{2}x + 2$

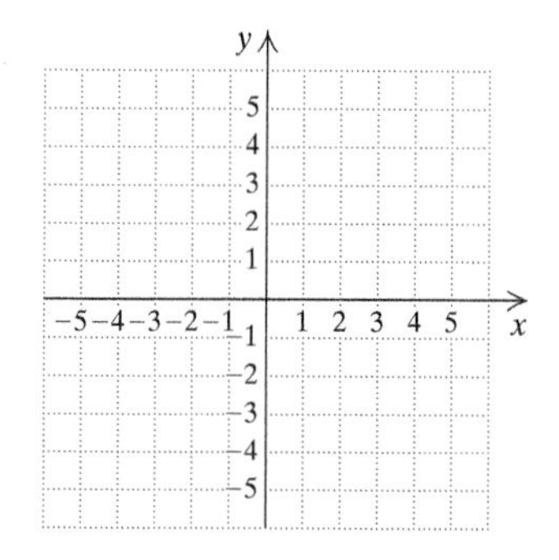

37. $f(x) = \frac{1}{2}x + 1$

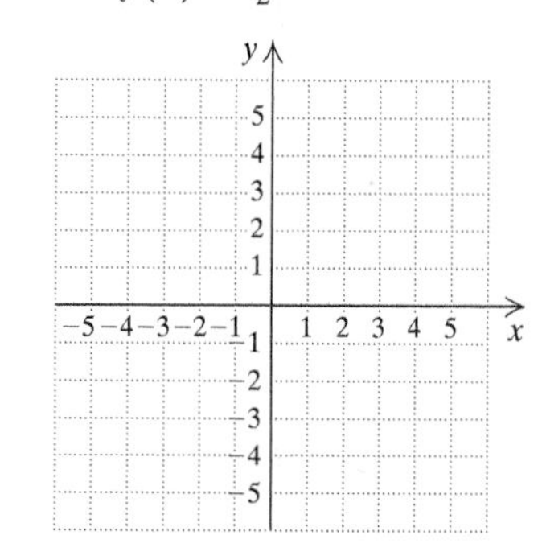

38. $f(x) = -\frac{3}{4}x - 2$

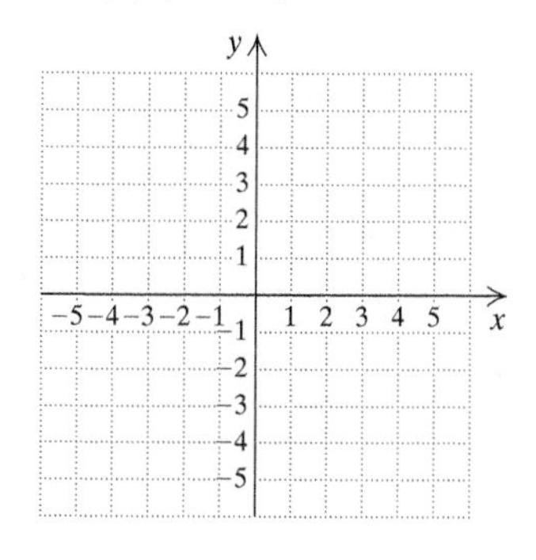

39. $f(x) = 2 - |x|$

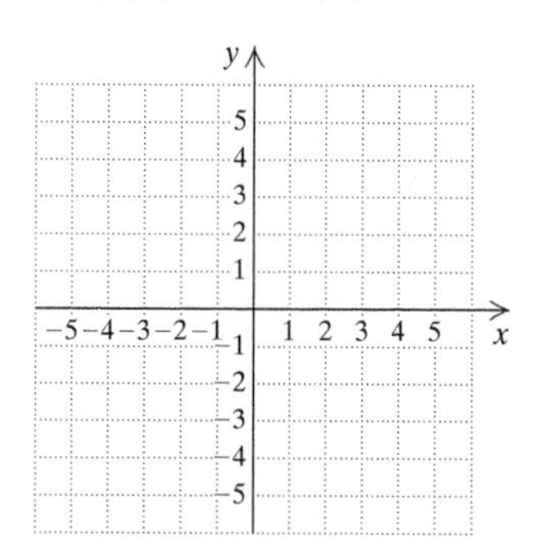

40. $f(x) = |x| - 4$

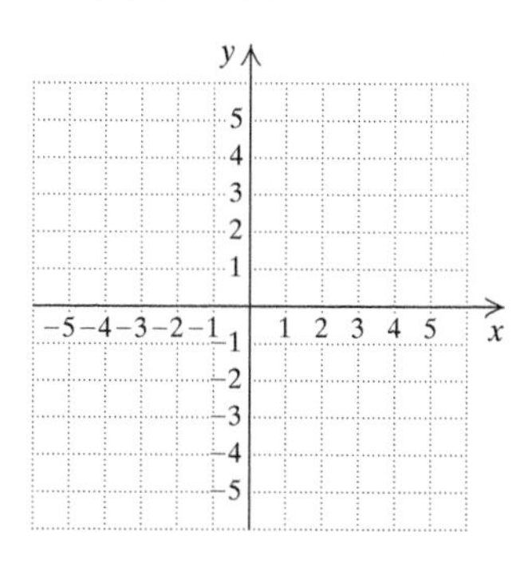

41. $g(x) = |x - 1|$

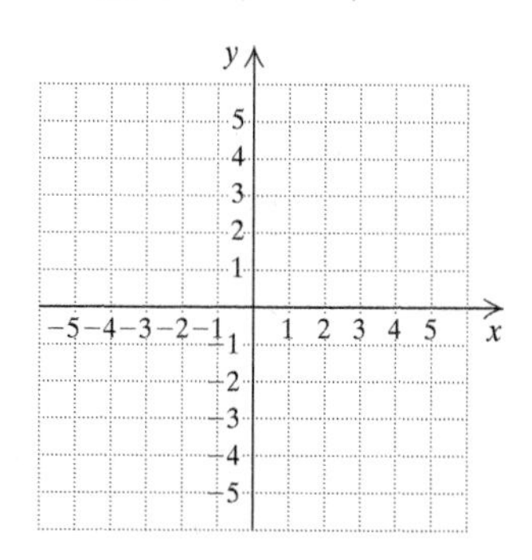

42. $g(x) = |x + 3|$

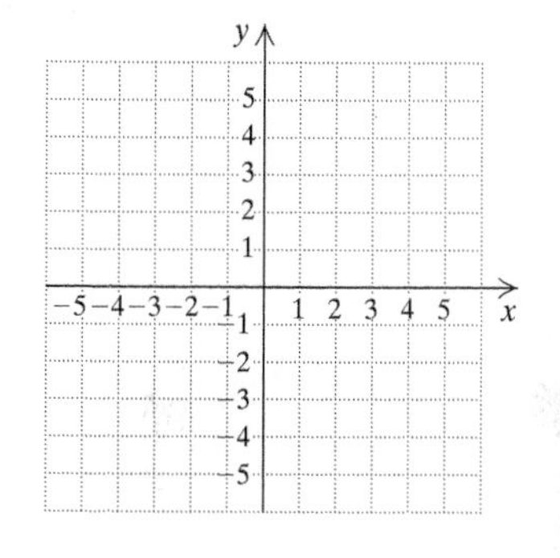

43. $g(x) = x^2 + 2$

x	$g(x)$
-2	
-1	
0	
1	
2	

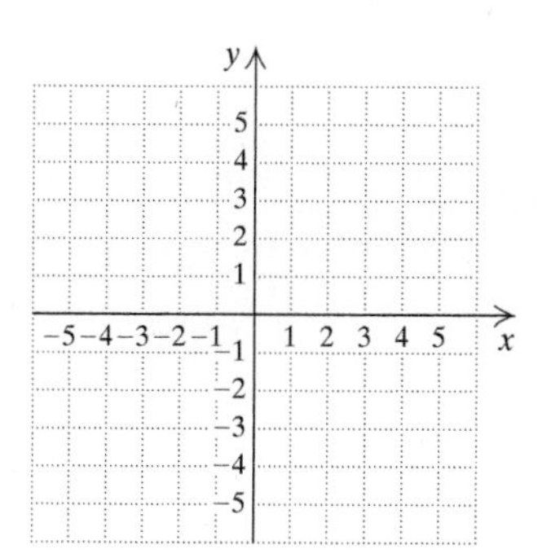

44. $f(x) = x^2 + 1$

x	$f(x)$
-2	
-1	
0	
1	
2	

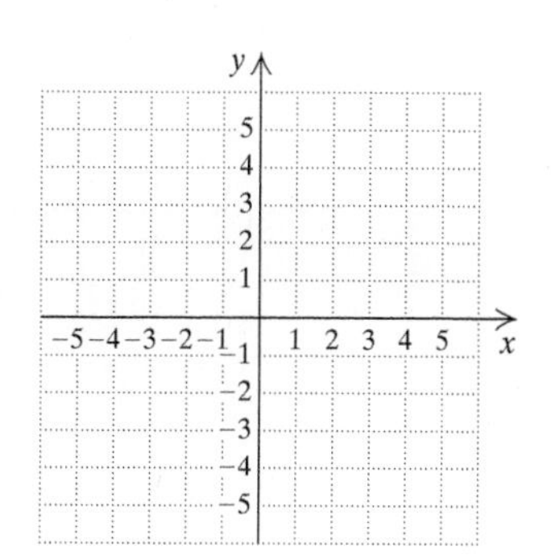

45. $f(x) = x^2 - 2x - 3$

x	$f(x)$
-2	
-1	
0	
1	
2	
3	
4	

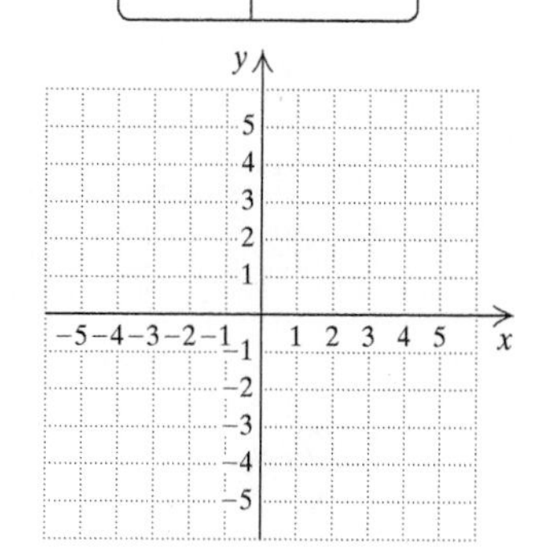

46. $g(x) = x^2 + 6x + 5$

x	$g(x)$
-6	
-5	
-4	
-3	
-2	
-1	
0	

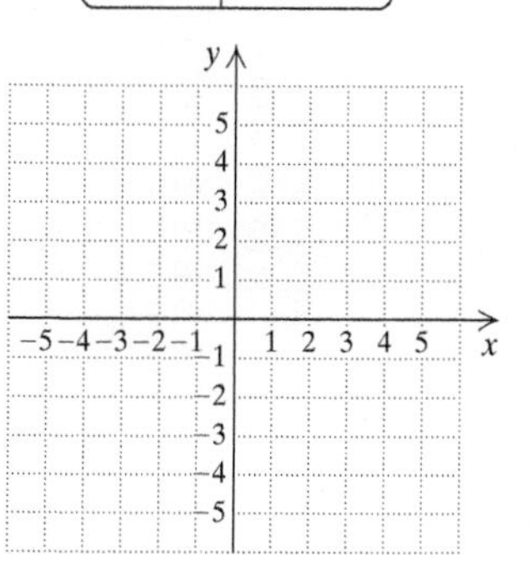

47. $f(x) = -x^2 + 1$

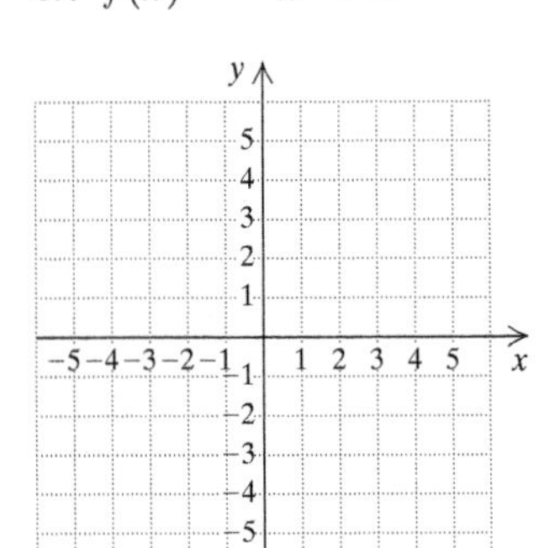

48. $f(x) = -x^2 + 2$

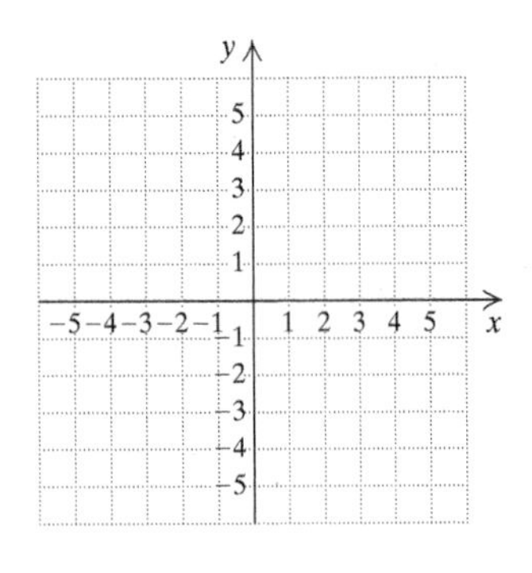

49. $f(x) = x^3 + 1$

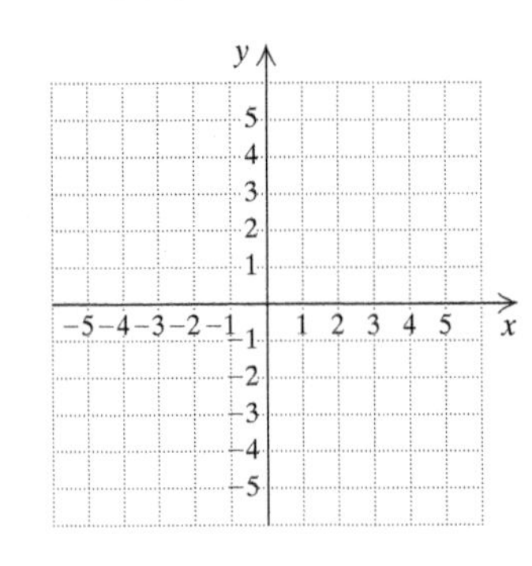

50. $f(x) = x^3 - 2$

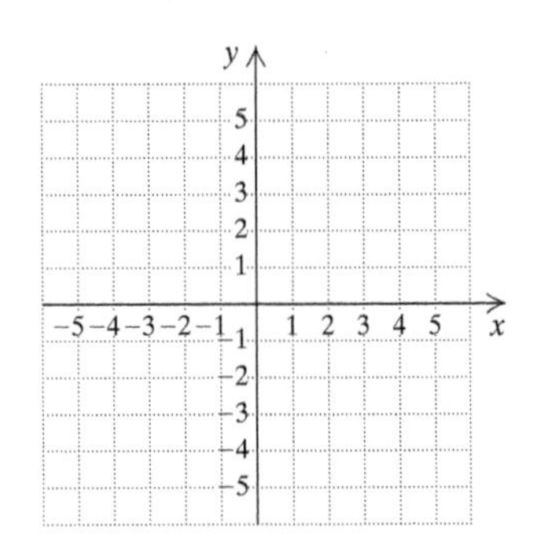

d Determine whether each of the following is the graph of a function.

51.

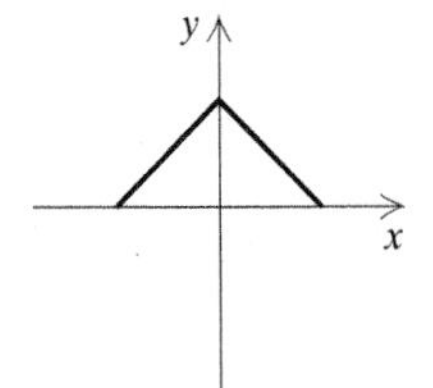

52.

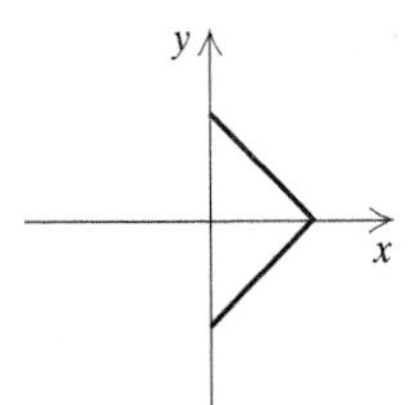

53.

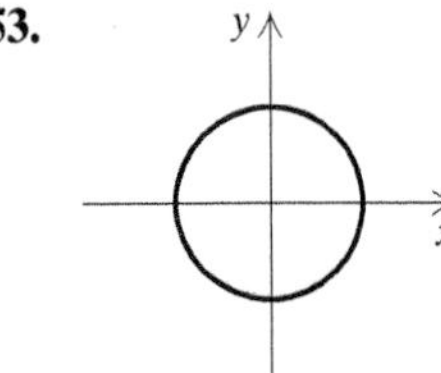

54.

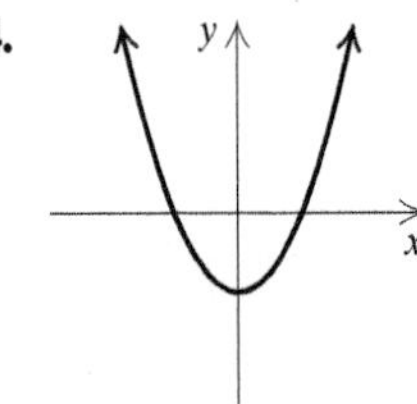

55.

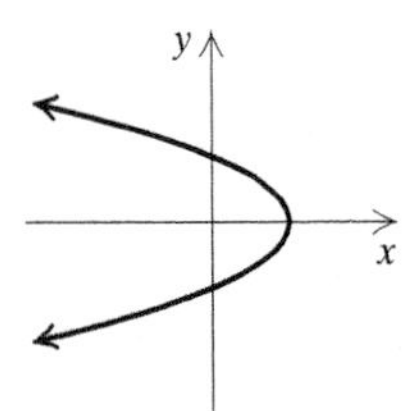

56.

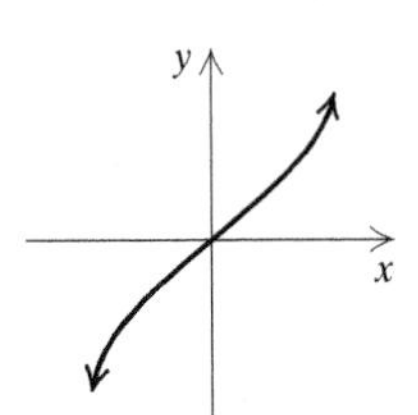

57.

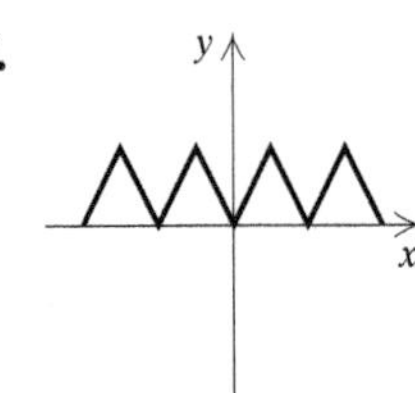

58.

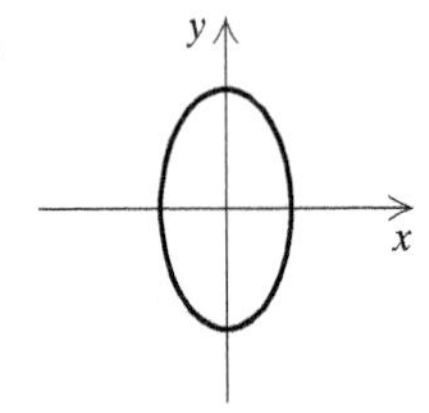

e Solve.

Median Price of a Home. The following graph approximates the median price of a home in the United States from 2011 through 2016. The price is a function f of the year x.

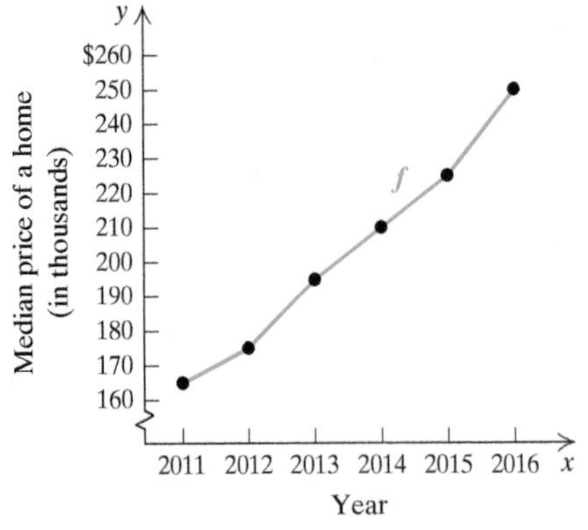

DATA: National Association of REALTORS®

59. Approximate the median price of a home in 2014.

60. Approximate the median price of a home in 2012.

Scheduled Airline Passengers. The following graph approximates the number of scheduled airline passengers in the United States from 2000 through 2016. The number of airline passengers is a function g of the year x.

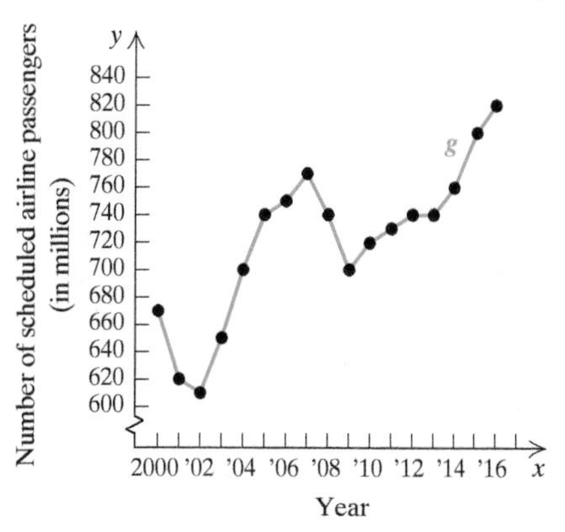

DATA: Airlines for America; airlines.org

61. Approximate the number of passengers in 2010.

62. Approximate the number of passengers in 2016.

63. Approximate the number of passengers in 2003.

64. Approximate the number of passengers in 2011.

Skill Maintenance

Solve.

65. $4 - 7y > 2y - 32$ [1.4c]

66. $\frac{2}{3}(4x - 2) \geq 60$ [1.4c]

67. $7y - 2 = 3 + 7y$ [1.1d]

68. $6x - 31 = 11 + 6(x - 7)$ [1.1d]

69. $13x - 5 - x = 2(x + 5)$ [1.1d]

70. $\frac{1}{16}x + 4 = \frac{5}{8}x - 1$ [1.1d]

Synthesis

71. Suppose that for some function g, $g(x - 6) = 10x - 1$. Find $g(-2)$.

72. Suppose that for some function h, $h(x + 5) = x^2 - 4$. Find $h(3)$.

For Exercises 73 and 74, let $f(x) = 3x^2 - 1$ and $g(x) = 2x + 5$.

73. Find $f(g(-4))$ and $g(f(-4))$.

74. Find $f(g(-1))$ and $g(f(-1))$.

75. Suppose that a function g is such that $g(-1) = -7$ and $g(3) = 8$. Find a formula for g if $g(x)$ is of the form $g(x) = mx + b$, where m and b are constants.

2.3 Finding Domain and Range

OBJECTIVE

a Find the domain and the range of a function.

MyLab Math ANIMATION

SKILL REVIEW *Solve equations using both the addition principle and the multiplication principle.* [1.1d]

Solve.

1. $6x - 3 = 51$
2. $15 - 2x = 0$

Answers: 1. 9 2. $\frac{15}{2}$, or 7.5

MyLab Math VIDEO

1. Find the domain and the range of the function f whose graph is shown below.

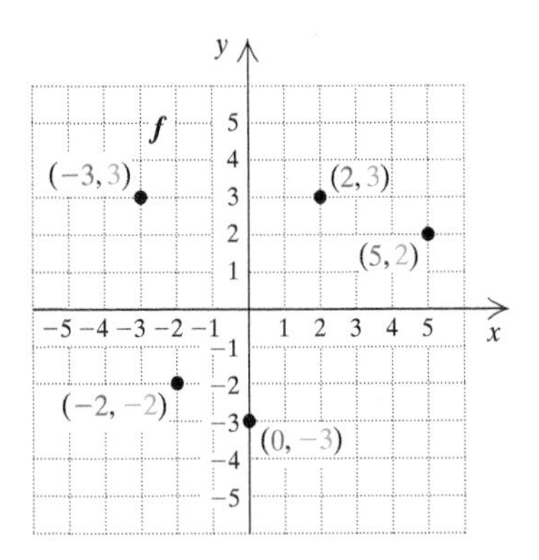

a FINDING DOMAIN AND RANGE

The solutions of an equation in two variables consist of a set of ordered pairs. A set of ordered pairs is called a **relation**. When a set of ordered pairs is such that no two different pairs share a common first coordinate, we have a **function**. The **domain** is the set of all first coordinates, and the **range** is the set of all second coordinates.

EXAMPLE 1 Find the domain and the range of the function f whose graph is shown below.

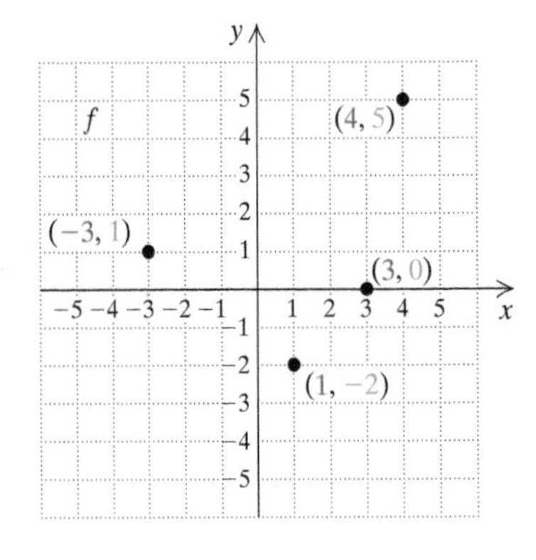

This function contains just four ordered pairs and it can be written as

$$\{(-3,1),(1,-2),(3,0),(4,5)\}.$$

We can determine the domain and the range by reading the x- and y-values directly from the graph.

The domain is the set of all first coordinates, or x-values, $\{-3, 1, 3, 4\}$. The range is the set of all second coordinates, or y-values, $\{1, -2, 0, 5\}$.

◀ **Do Exercise 1.**

EXAMPLE 2 For the function f whose graph is shown at right, determine each of the following.

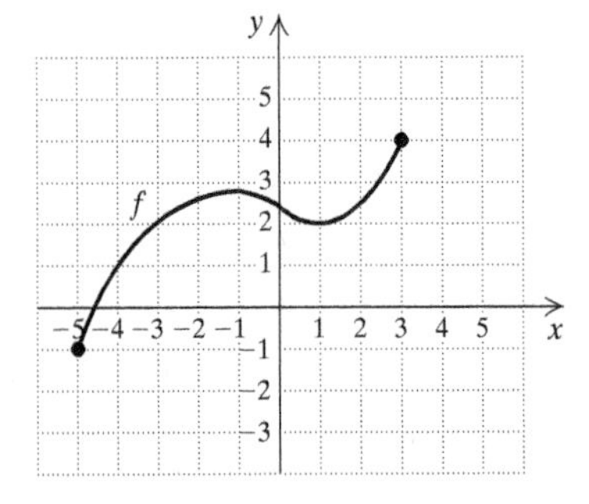

a) The number in the range that is paired with 1 from the domain. That is, find $f(1)$.

b) The domain of f

c) The numbers in the domain that are paired with 1 from the range. That is, find all x such that $f(x) = 1$.

d) The range of f

a) To determine which number in the range is paired with 1 in the domain, we locate 1 on the horizontal axis. Next, we find the point on the graph of f for which 1 is the first coordinate. From that point, we can look to the vertical axis to find the corresponding y-coordinate, 2. The input 1 has the output 2—that is, $f(1) = 2$.

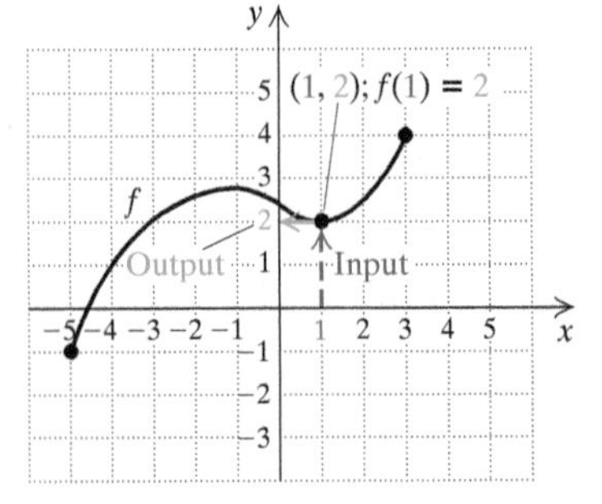

Answer

1. Domain = $\{-3, -2, 0, 2, 5\}$; range = $\{3, -2, -3, 2\}$

b) The domain of the function is the set of all x-values, or inputs, of the points on the graph. These extend from -5 to 3 and can be viewed as the curve's shadow, or projection, onto the x-axis. Thus the domain is $\{x \mid -5 \le x \le 3\}$, or, in interval notation, $[-5, 3]$.

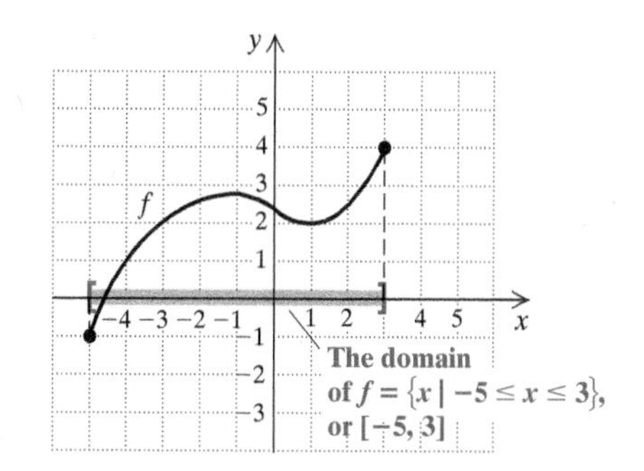

c) To determine which numbers in the domain are paired with 1 in the range, we locate 1 on the vertical axis. From there, we look left and right to the graph of f to find any points for which 1 is the second coordinate (output). One such point exists, $(-4, 1)$. For this function, we note that $x = -4$ is the only member of the domain paired with 1. For other functions, there might be more than one member of the domain paired with a member of the range.

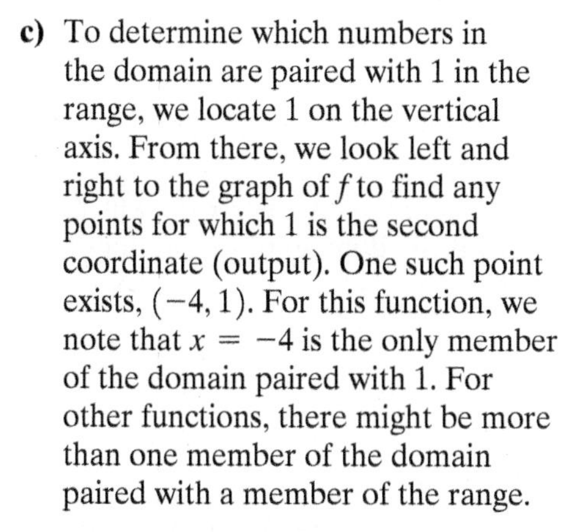

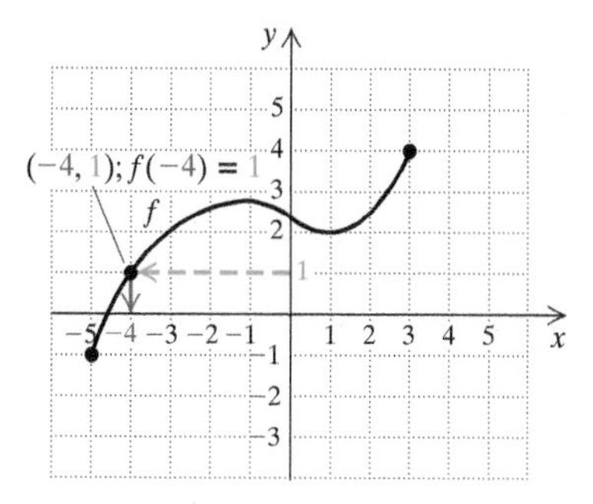

d) The range of the function is the set of all y-values, or outputs, of the points on the graph. These extend from -1 to 4 and can be viewed as the curve's shadow, or projection, onto the y-axis. Thus the range is $\{y \mid -1 \le y \le 4\}$, or, in interval notation, $[-1, 4]$.

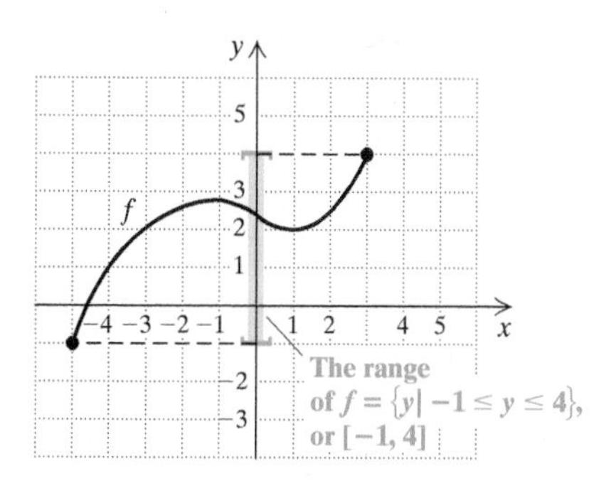

Do Exercise 2.

2. For the function f whose graph is shown below, determine each of the following.

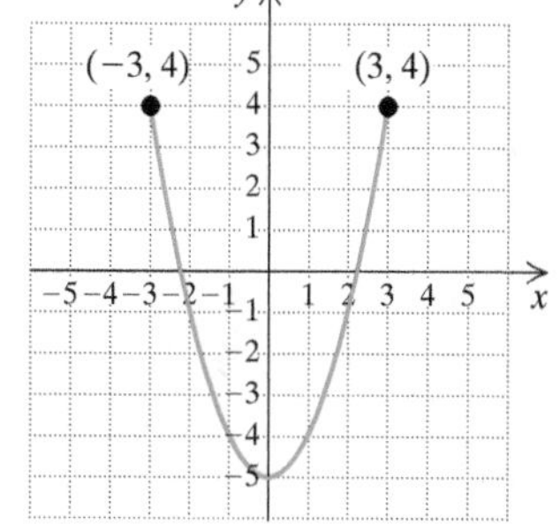

a) The number in the range that is paired with the input 1. That is, find $f(1)$.

b) The domain of f

c) The numbers in the domain that are paired with 4

d) The range of f

EXAMPLE 3 Find the domain and the range of the function h whose graph is shown below.

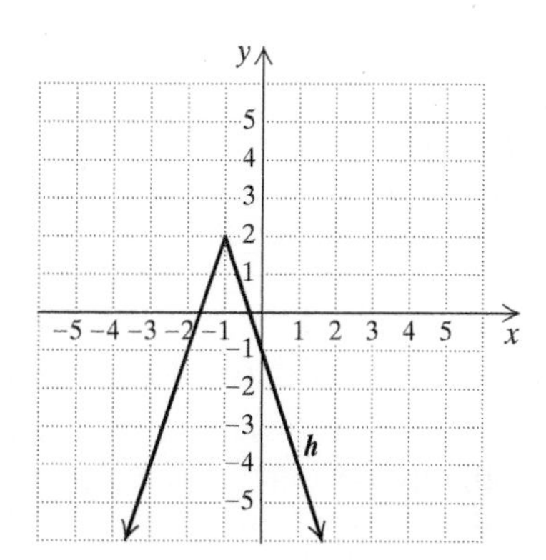

Since no endpoints are indicated, the graph extends indefinitely horizontally. Thus the domain, or the set of inputs, is the set of all real numbers. The range, or the set of outputs, is the set of all y-values of the points on the graph. Thus the range is $\{y \mid y \le 2\}$, or $(-\infty, 2]$.

Do Exercise 3.

3. Find the domain and the range of the function f whose graph is shown below.

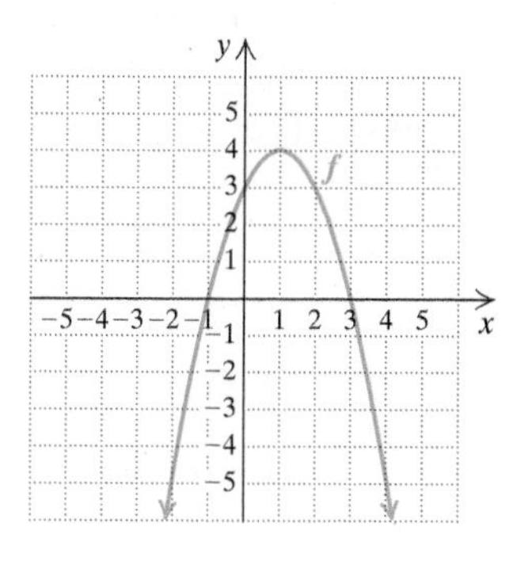

Answers

2. (a) -4; **(b)** $\{x \mid -3 \le x \le 3\}$, or $[-3, 3]$; **(c)** $-3, 3$; **(d)** $\{y \mid -5 \le y \le 4\}$, or $[-5, 4]$ **3.** Domain: all real numbers; range: $\{y \mid y \le 4\}$, or $(-\infty, 4]$

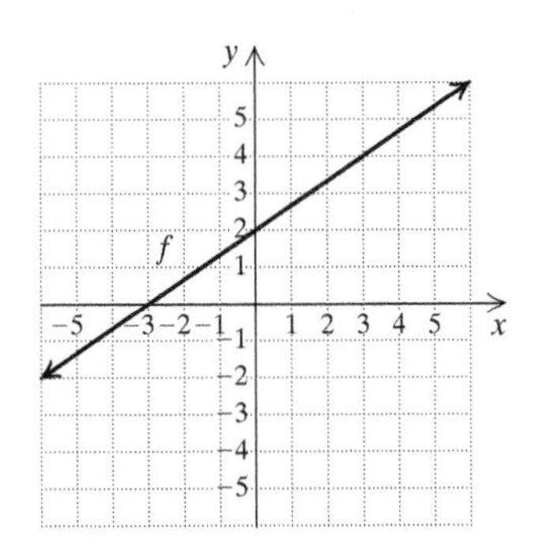

4. Find the domain and the range of the function f whose graph is shown below.

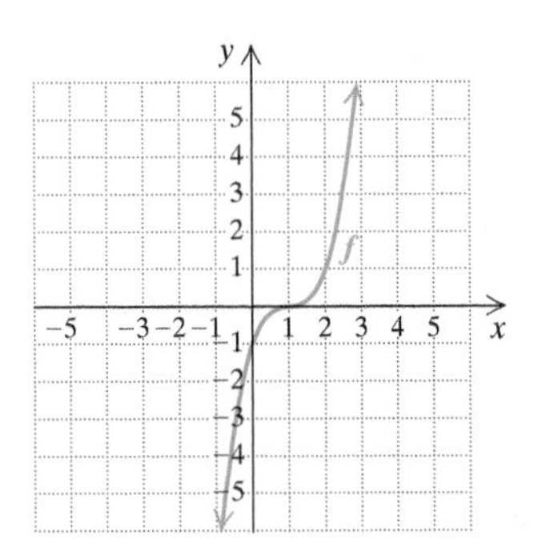

Find the domain.

5. $f(x) = x^3 - |x|$

6. $f(x) = \dfrac{4}{3x + 2}$

Set the denominator equal to 0 and solve for x:

$$3x + 2 = \square$$
$$3x = -2$$
$$x = \square.$$

Thus, $\square$ is not in the domain of $f(x)$; all other real numbers are.

Domain $= \{x \mid x$ is a real number *and* $x \neq \square\}$, or $(\square, -\frac{2}{3}) \cup (-\frac{2}{3}, \infty)$

Answers

4. Domain: all real numbers; range: all real numbers **5.** All real numbers

6. $\left\{x \mid x \text{ is a real number } and\ x \neq -\frac{2}{3}\right\}$, or $\left(-\infty, -\frac{2}{3}\right) \cup \left(-\frac{2}{3}, \infty\right)$

Guided Solution:

6. $0, -\frac{2}{3}, -\frac{2}{3}, -\frac{2}{3}, -\infty$

EXAMPLE 4 Find the domain and the range of the function f whose graph is shown at left.

Since no endpoints are indicated, the graph extends indefinitely both horizontally and vertically. Thus the domain is the set of all real numbers. Likewise, the range is the set of all real numbers.

Do Exercise 4.

When a function is given by an equation or a formula, the domain is understood to be the largest set of real numbers (inputs) for which function values (outputs) can be calculated. That is, the domain is the set of all possible allowable inputs into the formula. To find the domain, think, "What can we substitute?"

EXAMPLE 5 Find the domain: $f(x) = |x|$.

We ask, "What can we substitute?" Is there any number x for which we cannot calculate $|x|$? The answer is no. Thus the domain of f is the set of all real numbers.

EXAMPLE 6 Find the domain: $f(x) = \dfrac{3}{2x - 5}$.

We ask, "What can we substitute?" Is there any number x for which we cannot calculate $3/(2x - 5)$? Since $3/(2x - 5)$ cannot be calculated when the denominator $2x - 5$ is 0, we solve the following equation to find those real numbers that must be excluded from the domain of f:

$$2x - 5 = 0 \quad \text{Setting the denominator equal to 0}$$
$$2x = 5 \quad \text{Adding 5}$$
$$x = \tfrac{5}{2}. \quad \text{Dividing by 2}$$

Thus, $\frac{5}{2}$ is not in the domain, whereas all other real numbers are.

The domain of f is $\{x \mid x \text{ is a real number } and\ x \neq \frac{5}{2}\}$. In interval notation, the domain is $\left(-\infty, \frac{5}{2}\right) \cup \left(\frac{5}{2}, \infty\right)$.

Do Exercises 5 and 6.

Functions: A Review

The following is a review of the function concepts considered in Sections 2.1–2.3. Use the graph below to visualize the concepts.

Function Concepts

- Formula for f: $f(x) = x^2 - 7$
- For every input of f, there is exactly one output.
- When 1 is the input, -6 is the output.
- $f(1) = -6$
- $(1, -6)$ is on the graph.
- Domain = The set of all inputs = The set of all real numbers
- Range = The set of all outputs $= \{y \mid y \geq -7\} = [-7, \infty)$

Graph

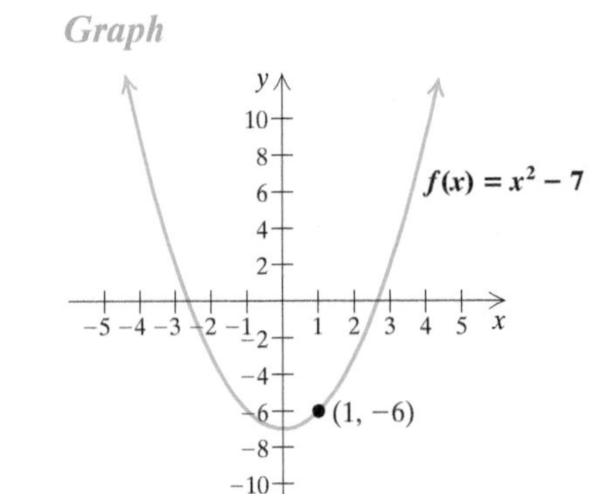

2.3 Exercise Set

FOR EXTRA HELP MyLab Math

✓ Check Your Understanding

Reading Check and Concept Check Choose from the column on the right the domain of the function. Some choices may be used more than once, others not at all.

RC1. $f(x) = 5 - x$

RC2. $f(x) = \dfrac{-5}{5 - x}$

RC3. $f(x) = |5 - x|$

RC4. $f(x) = \dfrac{5}{|x - 5|}$

RC5. $f(x) = 5 - |x|$

RC6. $f(x) = \dfrac{x - 5}{x + 5}$

a) All real numbers

b) $\{x | x \text{ is a real number } and\ x \neq 5\}$

c) $\{x | x \text{ is a real number } and\ x \neq -5\ and\ x \neq 5\}$

d) $\{x | x \text{ is a real number } and\ x \neq -5\}$

a In Exercises 1–8, the graph is that of a function. Determine for each one **(a)** $f(1)$; **(b)** the domain; **(c)** all x-values such that $f(x) = 2$; and **(d)** the range. An open dot indicates that the point does not belong to the graph.

1.

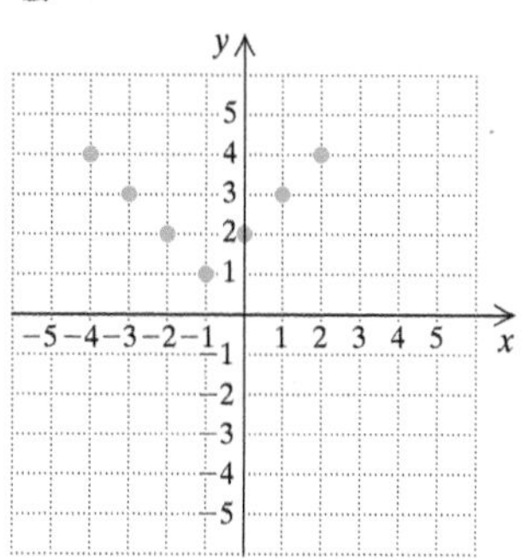

2.

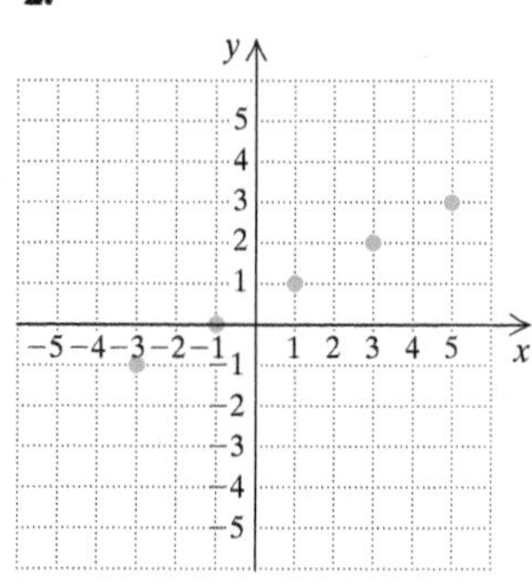

3.

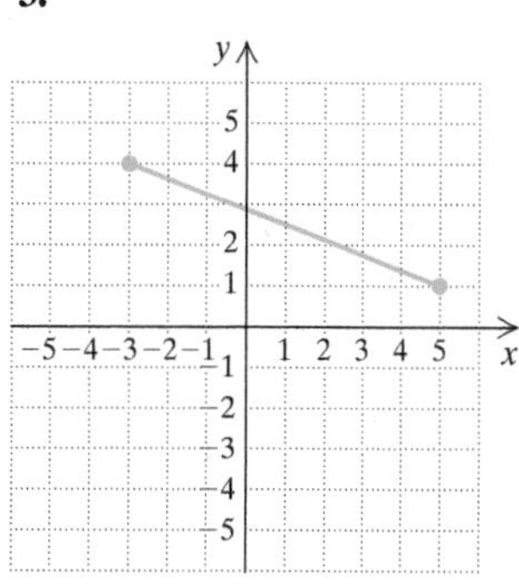

4.

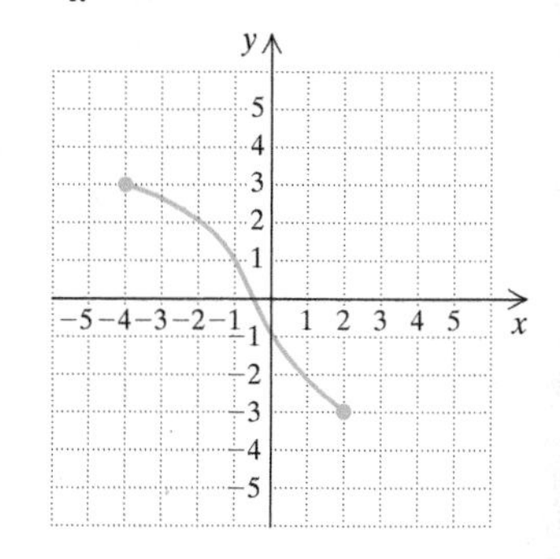

5.

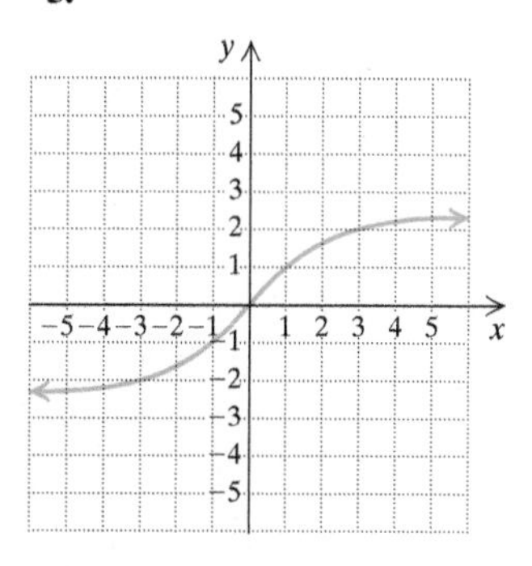

6.

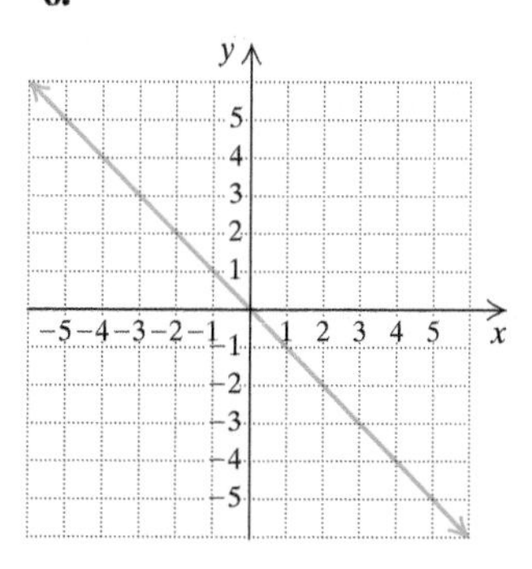

7.

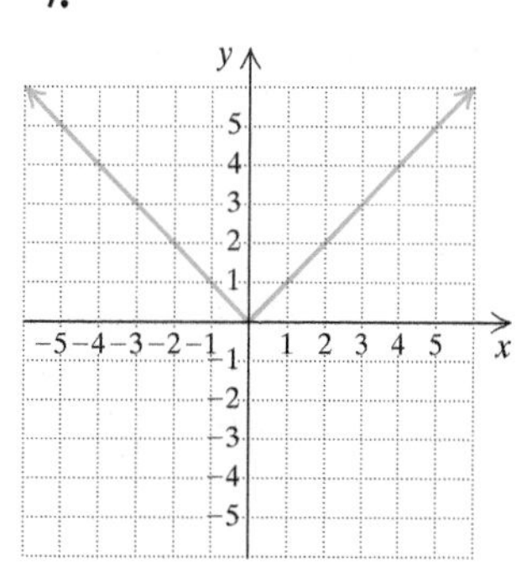

8.

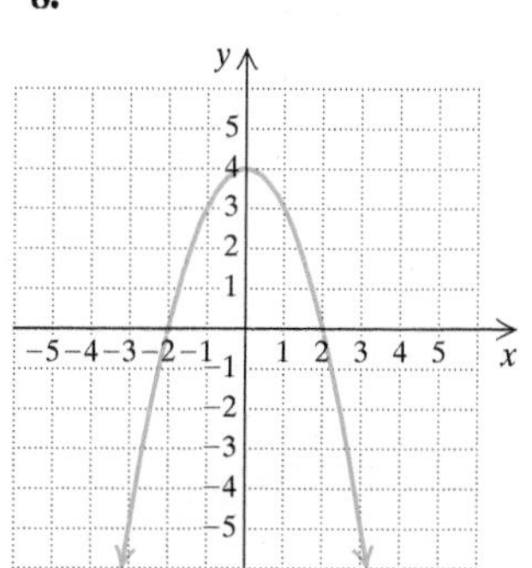

Find the domain.

9. $f(x) = \dfrac{2}{x + 3}$

10. $f(x) = \dfrac{7}{5 - x}$

11. $f(x) = 2x + 1$

12. $f(x) = 4 - 5x$

13. $f(x) = x^2 + 3$

14. $f(x) = x^2 - 2x + 3$

15. $f(x) = \dfrac{8}{5x - 14}$

16. $f(x) = \dfrac{x - 2}{3x + 4}$

17. $f(x) = |x| - 4$

18. $f(x) = |x - 4|$

19. $f(x) = \dfrac{x^2 - 3x}{|4x - 7|}$

20. $f(x) = \dfrac{4}{|2x - 3|}$

21. $g(x) = \dfrac{1}{x - 1}$

22. $g(x) = \dfrac{-11}{4 + x}$

23. $g(x) = x^2 - 2x + 1$

24. $g(x) = 8 - x^2$

25. $g(x) = x^3 - 1$

26. $g(x) = 4x^3 + 5x^2 - 2x$

27. $g(x) = \dfrac{7}{20 - 8x}$

28. $g(x) = \dfrac{2x - 3}{6x - 12}$

29. $g(x) = |x + 7|$

30. $g(x) = |x| + 1$

31. $g(x) = \dfrac{-2}{|4x + 5|}$

32. $g(x) = \dfrac{x^2 + 2x}{|10x - 20|}$

33. For the function f whose graph is shown below, find $f(-1), f(0)$, and $f(1)$.

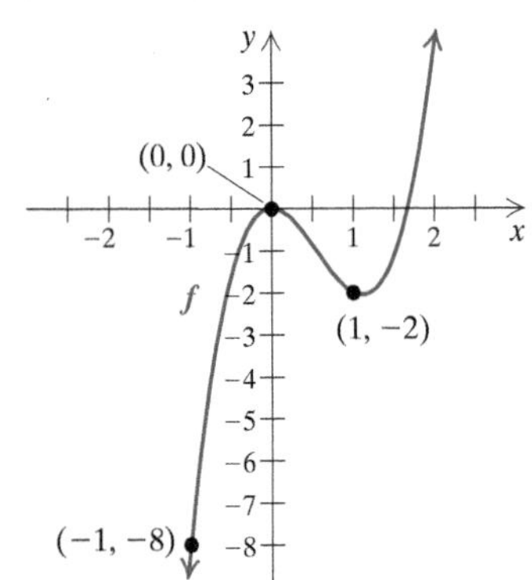

34. For the function g whose graph is shown below, find all the x-values for which $g(x) = 1$.

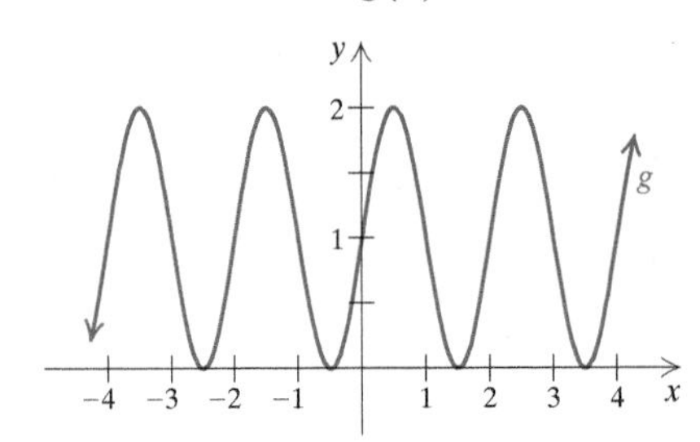

Skill Maintenance

Solve. [1.6c, d]

35. $|x| = 8$

36. $|x| = -8$

37. $|x - 7| = 11$

38. $|2x + 3| = 13$

39. $|3x - 4| = |x + 2|$

40. $|5x - 6| = |3 - 8x|$

41. $|3x - 8| = -11$

42. $|3x - 8| = 0$

Synthesis

43. Determine the range of each of the functions in Exercises 9, 14, 17, and 18.

44. Determine the range of each of the functions in Exercises 22, 23, 24, and 30.

Find the domain of each function.

45. $f(x) = \sqrt[3]{x - 1}$

46. $g(x) = \sqrt{2 - x}$

Mid-Chapter Review

Concept Reinforcement

Determine whether each statement is true or false.

______________ **1.** Every function is a relation. [2.2a]

______________ **2.** It is possible for one input of a function to have two or more outputs. [2.2a]

______________ **3.** It is possible for all the inputs of a function to have the same output. [2.2a]

______________ **4.** If it is possible for a vertical line to cross a graph more than once, the graph is not the graph of a function. [2.2d]

______________ **5.** If the domain of a function is the set of real numbers, then the range is the set of real numbers. [2.3a]

Guided Solutions

Use the graph to complete the table of ordered pairs that name points on the graph.

6. [2.1c]

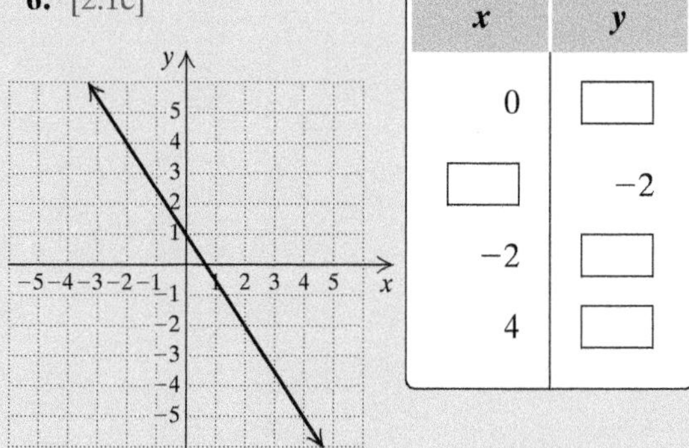

x	y
0	☐
☐	-2
-2	☐
4	☐

7. [2.2c]

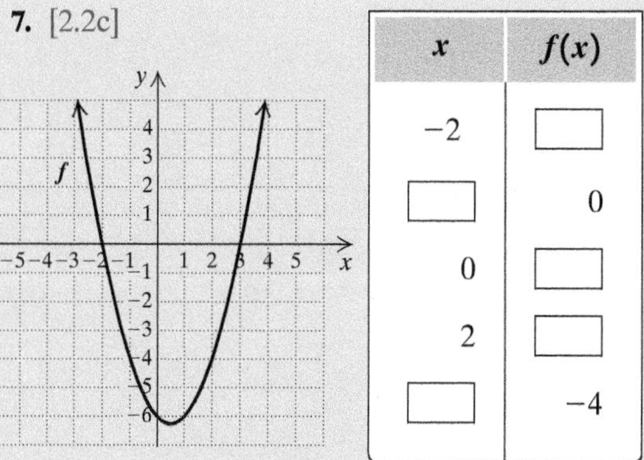

x	$f(x)$
-2	☐
☐	0
0	☐
2	☐
☐	-4

Mixed Review

Determine whether the given point is a solution of the equation. [2.1b]

8. $(-2, -1)$; $5y + 6 = 4x$

9. $\left(\frac{1}{2}, 0\right)$; $8a = 4 - b$

Determine whether the correspondence is a function. [2.2a]

10. *Domain* *Range*

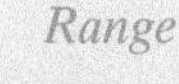

11. *Domain* *Range*

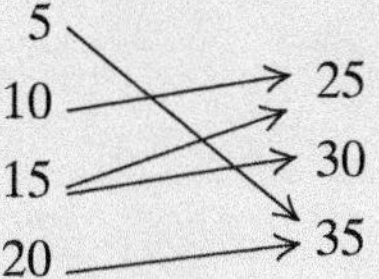

12. Find the domain and the range. [2.3a]

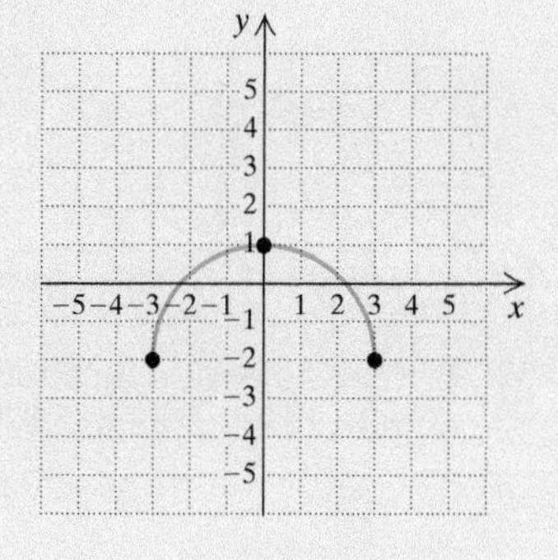

Find the function value. [2.2b]

13. $g(x) = 2 + x$; $g(-5)$

14. $f(x) = x - 7$; $f(0)$

15. $h(x) = 8$; $h\left(\frac{1}{2}\right)$

16. $f(x) = 3x^2 - x + 5$; $f(-1)$

17. $g(p) = p^4 - p^3$; $g(10)$

18. $f(t) = \frac{1}{2}t + 3$; $f(-6)$

Determine whether each of the following is the graph of a function. [2.2d]

19.

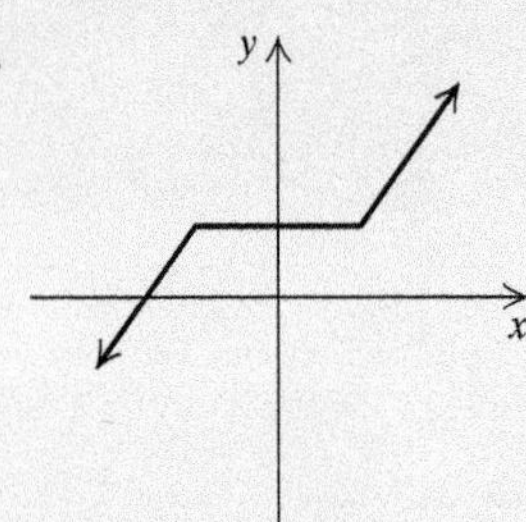

20.

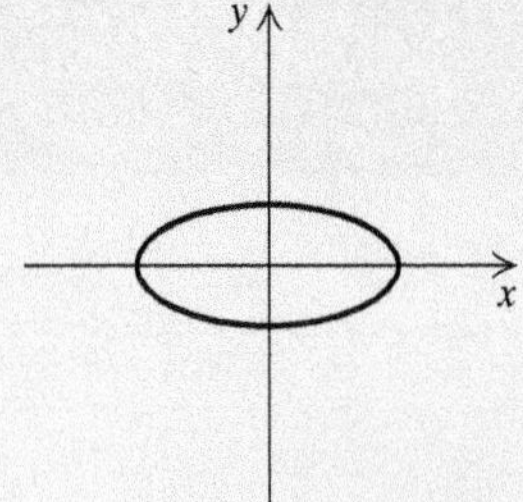

21.

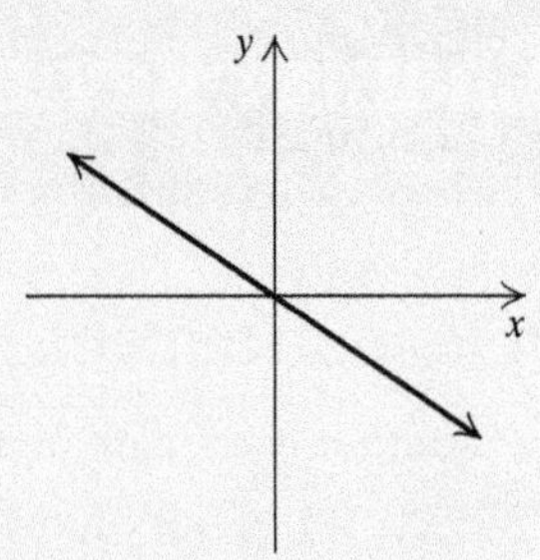

Find the domain. [2.3a]

22. $g(x) = \dfrac{3}{12 - 3x}$

23. $f(x) = x^2 - 10x + 3$

24. $h(x) = \dfrac{x - 2}{x + 2}$

25. $f(x) = |x - 4|$

Graph. [2.1c], [2.2c]

26. $y = -\dfrac{2}{3}x - 2$

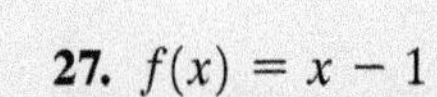

27. $f(x) = x - 1$

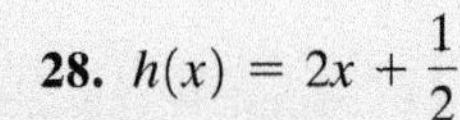

28. $h(x) = 2x + \dfrac{1}{2}$

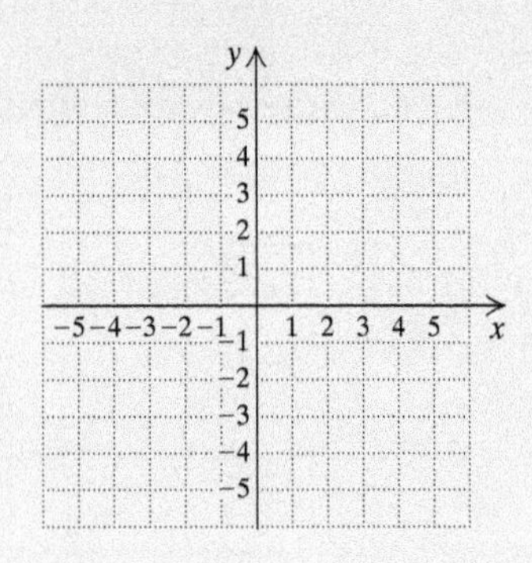

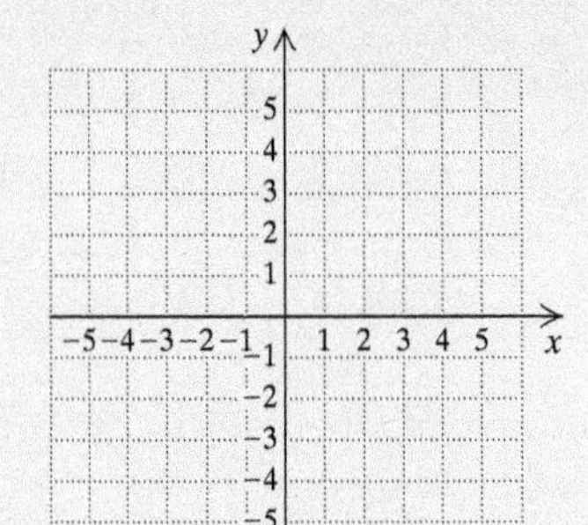

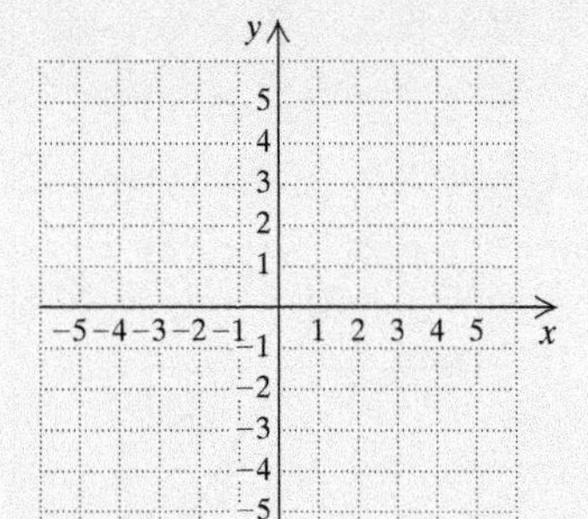

29. $g(x) = |x| - 3$

30. $f(x) = 1 + x^2$

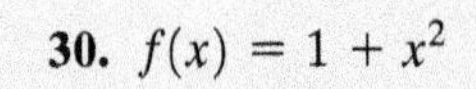

31. $f(x) = -\dfrac{1}{4}x$

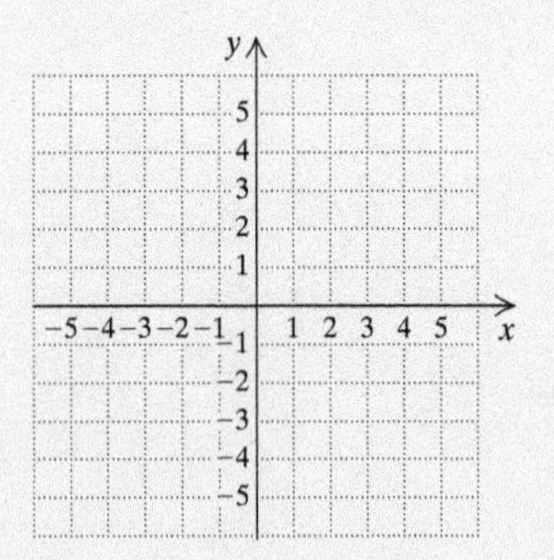

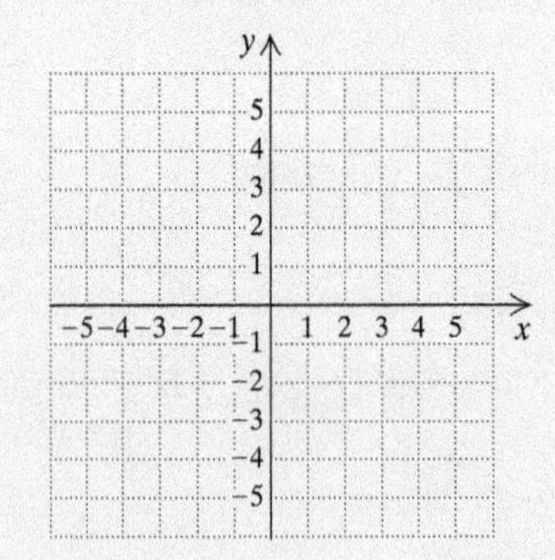

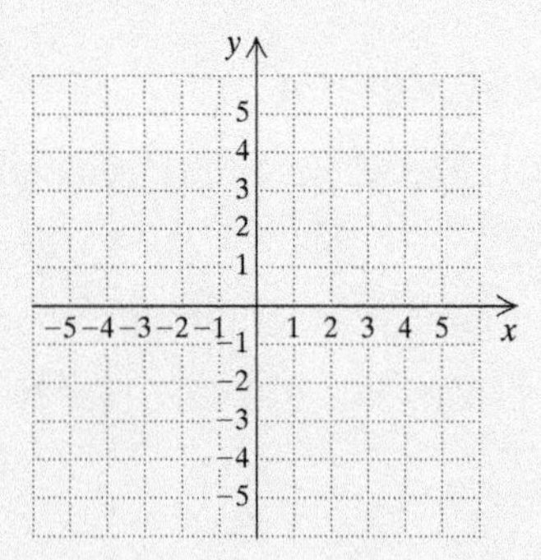

Understanding Through Discussion and Writing

32. Is it possible for a function to have more numbers as outputs than as inputs? Why or why not? [2.2a]

33. Without making a drawing, how can you tell that the graph of $y = x - 30$ passes through three quadrants? [2.1c]

34. For a given function f, it is known that $f(2) = -3$. Give as many interpretations of this fact as you can. [2.2b], [2.3a]

35. Explain the difference between the domain and the range of a function. [2.3a]

STUDYING FOR SUCCESS *A Valuable Resource—Your Instructor*

- ☐ Don't be afraid to ask questions in class. Other students probably have the same questions that you do.
- ☐ Visit your instructor during office hours if you need additional help.
- ☐ Many instructors welcome e-mails from students who have questions.

Linear Functions: Graphs and Slope

2.4

OBJECTIVES

- **a** Find the y-intercept of a line from the equation $y = mx + b$ or $f(x) = mx + b$.
- **b** Given two points on a line, find the slope. Given a linear equation, derive the equivalent slope–intercept equation and determine the slope and the y-intercept.
- **c** Solve applied problems involving slope.

We now turn our attention to functions whose graphs are straight lines. Such functions are called **linear** and can be written in the form $f(x) = mx + b$.

LINEAR FUNCTION

A **linear function** f is any function that can be described by $f(x) = mx + b$.

Compare the two equations $7y + 2x = 11$ and $y = 3x + 5$. Both are linear equations because their graphs are straight lines. Each can be expressed in an equivalent form that is a linear function.

The equation $y = 3x + 5$ can be expressed as $f(x) = mx + b$, where $m = 3$ and $b = 5$.

The equation $7y + 2x = 11$ also has an equivalent form $f(x) = mx + b$. To see this, we solve for y:

$$7y + 2x = 11$$

$$7y + 2x - 2x = -2x + 11 \qquad \text{Subtracting } 2x$$

$$7y = -2x + 11$$

$$\frac{7y}{7} = \frac{-2x + 11}{7} \qquad \text{Dividing by 7}$$

$$y = -\frac{2}{7}x + \frac{11}{7}. \qquad \text{Simplifying}$$

We now have an equivalent function in the form $f(x) = mx + b$:

$$f(x) = -\frac{2}{7}x + \frac{11}{7}, \quad \text{where} \quad m = -\frac{2}{7} \quad \text{and} \quad b = \frac{11}{7}.$$

In this section, we consider the effects of the constants m and b on the graphs of linear functions.

1. Graph $y = 3x$ and $y = 3x - 6$ using the same set of axes. Then compare the graphs.

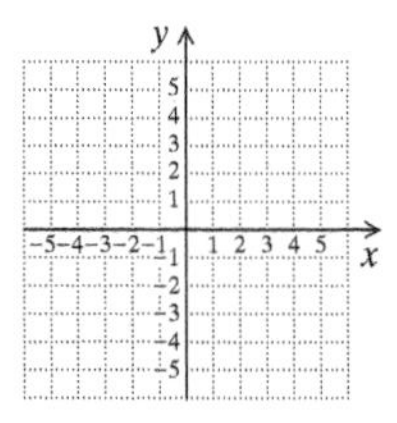

2. Graph $y = -2x$ and $y = -2x + 3$ using the same set of axes. Then compare the graphs.

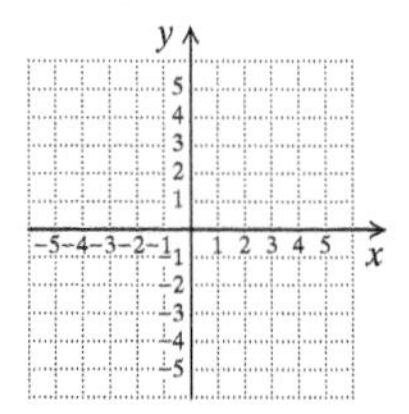

a THE CONSTANT b: THE y-INTERCEPT

Let's first explore the effect of the constant b.

EXAMPLE 1 Graph $y = 2x$ and $y = 2x + 3$ using the same set of axes. Then compare the graphs.

We first make a table of solutions of both equations. Next, we plot these points. Drawing a red line for $y = 2x$ and a blue line for $y = 2x + 3$, we note that the graph of $y = 2x + 3$ is simply the graph of $y = 2x$ shifted, or *translated*, up 3 units. The lines are parallel.

	y	y
x	$y = 2x$	$y = 2x + 3$
0	0	3
1	2	5
−1	−2	1
2	4	7
−2	−4	−1

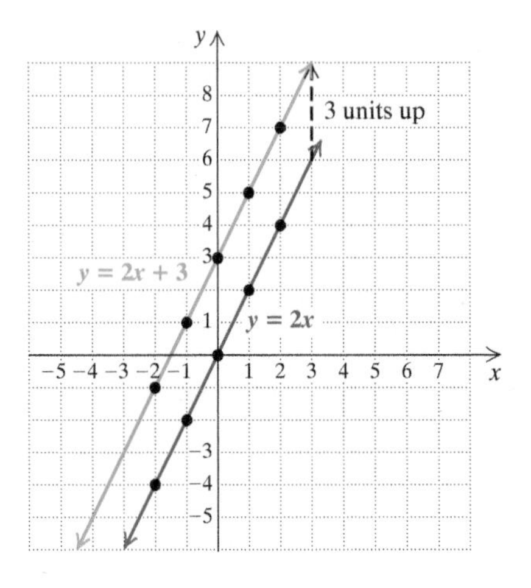

◀ Do Exercises 1 and 2.

EXAMPLE 2 Graph $f(x) = \frac{1}{3}x$ and $g(x) = \frac{1}{3}x - 2$ using the same set of axes. Then compare the graphs.

We first make a table of solutions of both equations. By choosing multiples of 3, we can avoid fractions.

	$f(x)$	$g(x)$
x	$f(x) = \frac{1}{3}x$	$g(x) = \frac{1}{3}x - 2$
0	0	−2
3	1	−1
−3	−1	−3
6	2	0

We then plot these points. Drawing a red line for $f(x) = \frac{1}{3}x$ and a blue line for $g(x) = \frac{1}{3}x - 2$, we see that the graph of $g(x) = \frac{1}{3}x - 2$ is simply the graph of $f(x) = \frac{1}{3}x$ shifted, or translated, down 2 units. The lines are parallel.

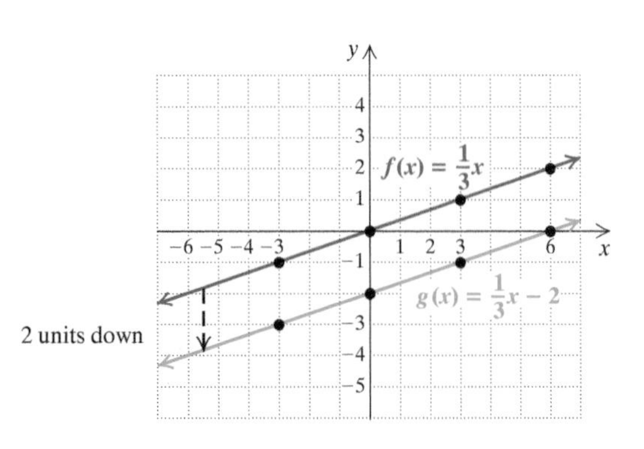

CALCULATOR CORNER

Exploring b We can use a graphing calculator to explore the effect of the constant b on the graph of a function of the form $f(x) = mx + b$. Graph $y_1 = x$ in the standard $[-10, 10, -10, 10]$ viewing window. Then graph $y_2 = x + 4$, followed by $y_3 = x - 3$, in the same viewing window.

EXERCISES:

1. Compare the graph of y_2 with the graph of y_1.
2. Compare the graph of y_3 with the graph of y_1.

Answers

1. 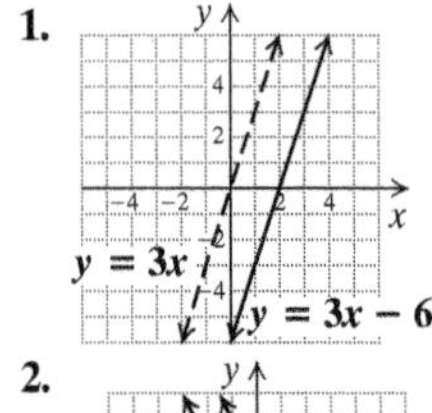

The graph of $y = 3x - 6$ is the graph of $y = 3x$ shifted down 6 units.

2. The graph of $y = -2x + 3$ is the graph of $y = -2x$ shifted up 3 units.

In Example 1, we saw that the graph of $y = 2x + 3$ is parallel to the graph of $y = 2x$ and that it passes through the point $(0, 3)$. Similarly, in Example 2, we saw that the graph of $y = \frac{1}{3}x - 2$ is parallel to the graph of $y = \frac{1}{3}x$ and that it passes through the point $(0, -2)$. In general, the graph of $y = mx + b$ is a line parallel to $y = mx$, passing through the point $(0, b)$. The point $(0, b)$ is called the **y-intercept** because it is the point at which the graph crosses the y-axis. Often it is convenient to refer to the number b as the y-intercept. The constant b has the effect of moving the graph of $y = mx$ up or down $|b|$ units to obtain the graph of $y = mx + b$.

Do Exercise 3.

3. Graph $f(x) = \frac{1}{3}x$ and $g(x) = \frac{1}{3}x + 2$ using the same set of axes. Then compare the graphs.

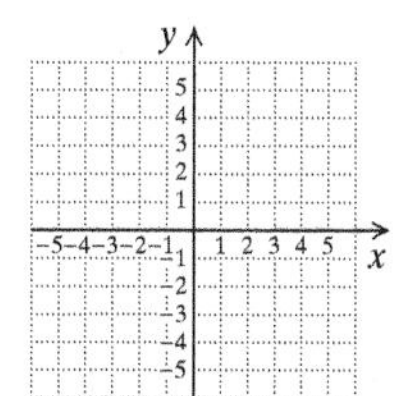

y-INTERCEPT

The y-intercept of the graph of $f(x) = mx + b$ is the point $(0, b)$ or, simply, b.

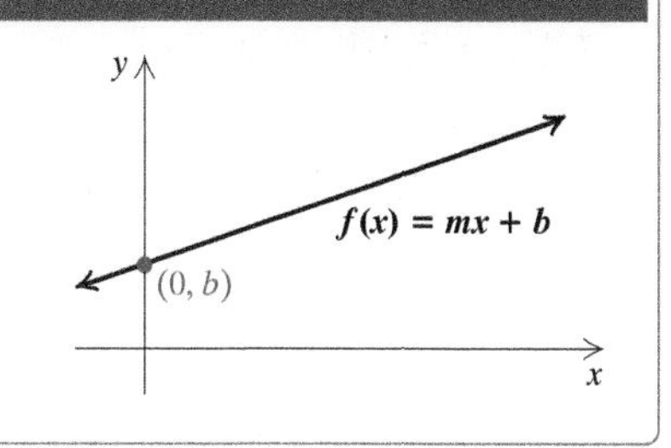

EXAMPLE 3 Find the y-intercept: $y = -5x + 4$.

$y = -5x + 4$

$\downarrow$

$(0, 4)$ is the y-intercept.

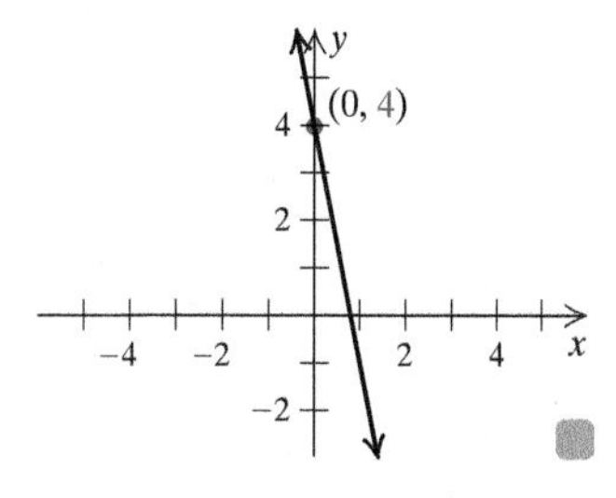

EXAMPLE 4 Find the y-intercept: $f(x) = 6.3x - 7.8$.

$f(x) = 6.3x - 7.8$

$\downarrow$

$(0, -7.8)$ is the y-intercept.

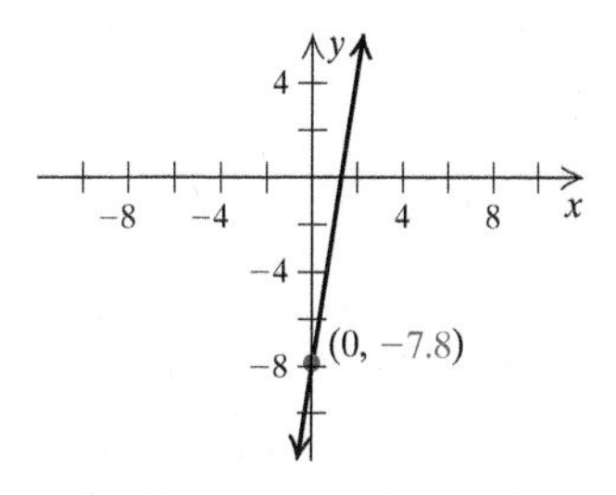

Do Exercises 4 and 5.

Find the y-intercept.

4. $y = 7x + 8$

5. $f(x) = -6x - \frac{2}{3}$

MyLab Math ANIMATION

b THE CONSTANT m: SLOPE

SKILL REVIEW *Subtract real numbers.* [J7]

Subtract.

1. $11 - (-8)$

2. $-6 - (-6)$

Answers: **1.** 19 **2.** 0

Answers

3.

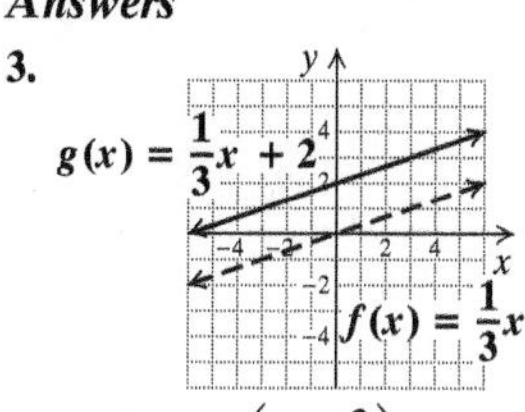

The graph of $g(x)$ is the graph of $f(x)$ shifted up 2 units.

4. $(0, 8)$ 5. $\left(0, -\frac{2}{3}\right)$

Look again at the graphs in Examples 1 and 2. Note that the slant of each red line seems to match the slant of each blue line. This leads us to believe that the number m in the equation $y = mx + b$ is related to the slant of the line. Let's consider some examples.

Graphs with $m < 0$:

$m = -1$

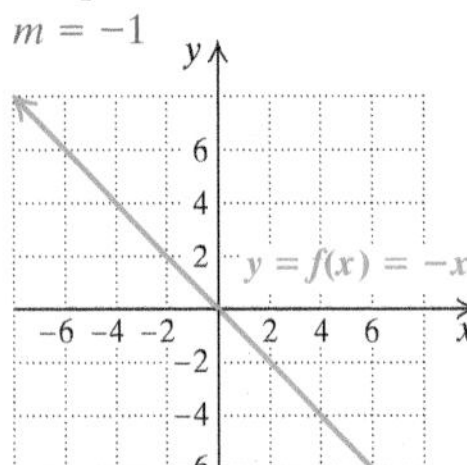

$m = -5$ and $m = -\frac{1}{2}$

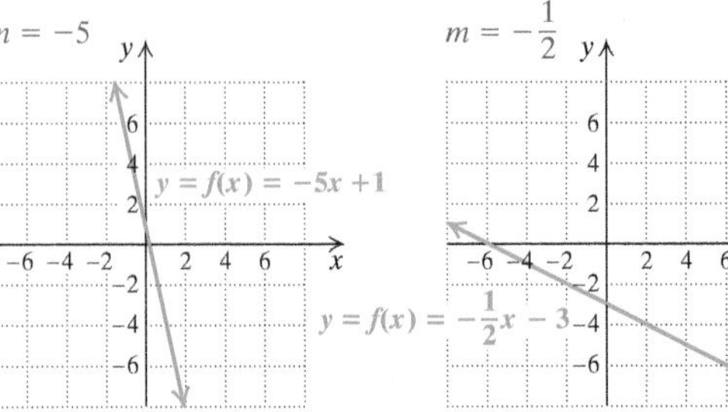

Graphs with $m = 0$:

$m = 0$

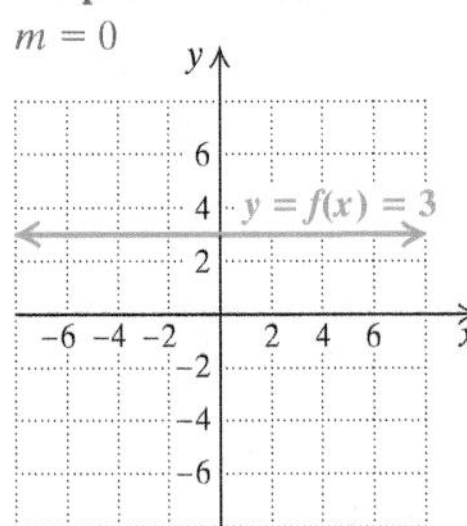

$m = 0$

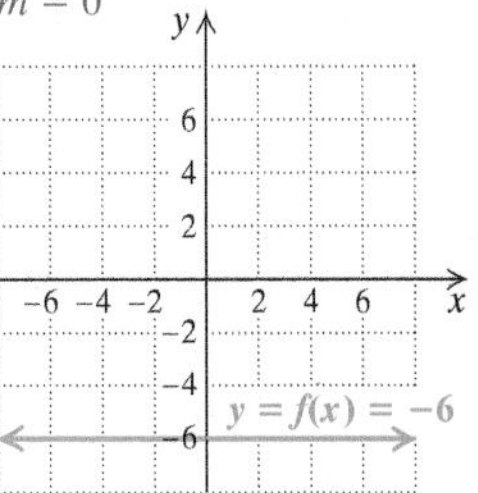

$m = 0$

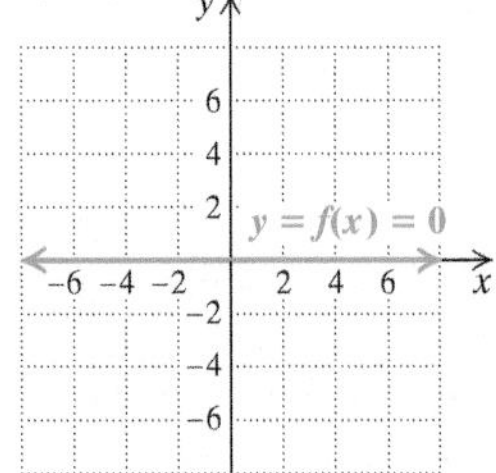

Graphs with $m > 0$:

$m = 1$

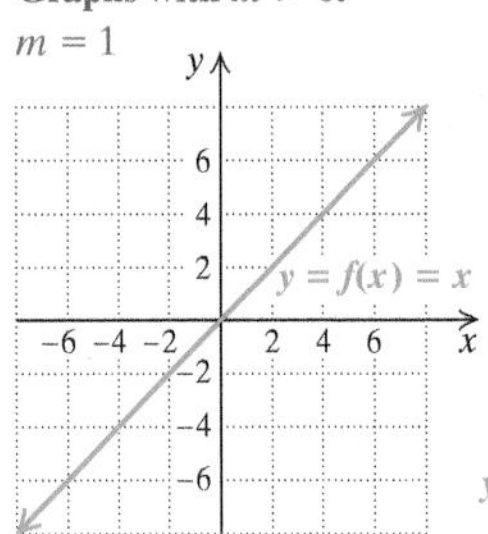

$m = 6$

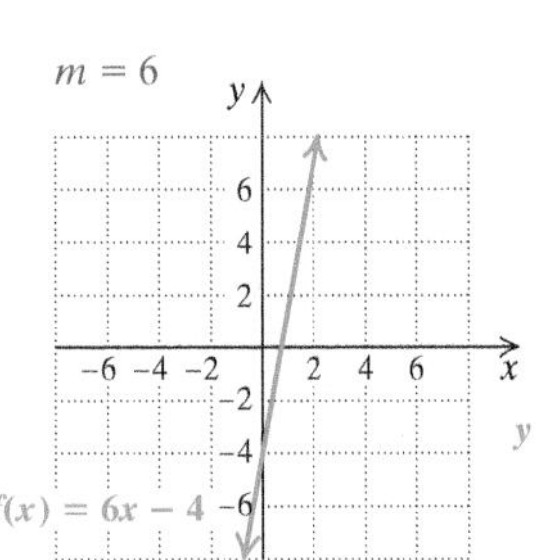

$m = \frac{1}{3}$

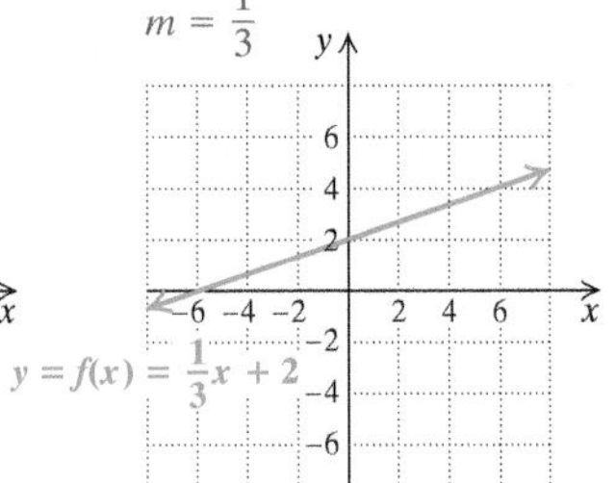

Note that

$m < 0 \longrightarrow$ The graph slants down from left to right;

$m = 0 \longrightarrow$ the graph is horizontal; and

$m > 0 \longrightarrow$ the graph slants up from left to right.

The following definition enables us to visualize the slant and attach a number, a geometric ratio, or *slope*, to the line.

SLOPE

The **slope** m of a line containing points (x_1, y_1) and (x_2, y_2) is given by

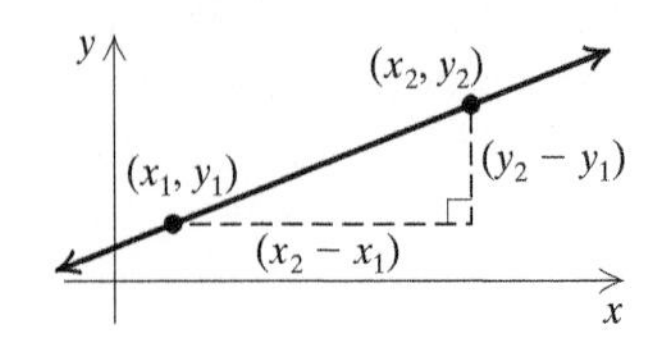

$$m = \frac{\text{rise}}{\text{run}}$$

$$= \frac{\text{change in } y}{\text{change in } x} = \frac{y_2 - y_1}{x_2 - x_1} = \frac{y_1 - y_2}{x_1 - x_2}.$$

Consider a line with two points marked P_1 and P_2, as follows. As we move from P_1 to P_2, the y-coordinate changes from 1 to 3 and the x-coordinate changes from 2 to 7. The change in y is $3 - 1$, or 2. The change in x is $7 - 2$, or 5.

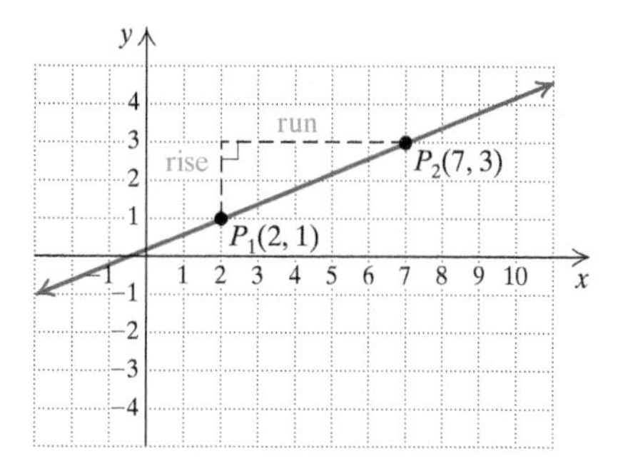

We call the change in y the **rise** and the change in x the **run**. The ratio rise/run is the same for any two points on a line. We call this ratio the **slope**. Slope describes the slant of a line. The slope of the line in the graph above is given by

$$\frac{\text{rise}}{\text{run}}, \quad \text{or} \quad \frac{\text{change in } y}{\text{change in } x}, \quad \text{or} \quad \frac{2}{5}.$$

Whenever x increases by 5 units, y increases by 2 units. Equivalently, whenever x increases by 1 unit, y increases by $\frac{2}{5}$ unit.

EXAMPLE 5 Graph the line containing the points $(-4, 3)$ and $(2, -5)$ and find the slope.

The graph is shown below. Moving from $(-4, 3)$ to $(2, -5)$, we see that the change in y, or the rise, is $-5 - 3$, or -8. The change in x, or the run, is $2 - (-4)$, or 6.

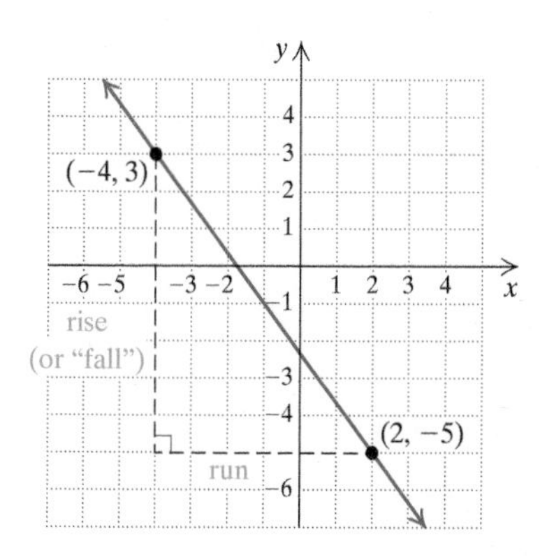

$$\text{Slope} = \frac{\text{rise}}{\text{run}} = \frac{\text{change in } y}{\text{change in } x}$$
$$= \frac{-5 - 3}{2 - (-4)}$$
$$= \frac{-8}{6} = -\frac{8}{6}, \text{ or } -\frac{4}{3}$$

The formula

$$m = \frac{y_2 - y_1}{x_2 - x_1} = \frac{y_1 - y_2}{x_1 - x_2}$$

tells us that we can subtract in two ways. We must remember, however, to subtract the x-coordinates in the same order that we subtract the y-coordinates.

Let's do Example 5 again:

$$\text{Slope} = \frac{\text{change in } y}{\text{change in } x} = \frac{3 - (-5)}{-4 - 2} = \frac{8}{-6} = -\frac{8}{6} = -\frac{4}{3}.$$

We see that both ways give the same value for the slope.

CALCULATOR CORNER

Visualizing Slope

EXERCISES: Use the window settings [−6, 6, −4, 4], with Xscl = 1 and Yscl = 1.

1. Graph $y = x$, $y = 2x$, and $y = 5x$ in the same window. What do you think the graph of $y = 10x$ will look like?
2. Graph $y = x$, $y = 0.5x$, and $y = 0.1x$ in the same window. What do you think the graph of $y = 0.005x$ will look like?
3. Graph $y = -x$, $y = -2x$, and $y = -5x$ in the same window. What do you think the graph of $y = -10x$ will look like?
4. Graph $y = -x$, $y = -0.5x$, and $y = -0.1x$ in the same window. What do you think the graph of $y = -0.005x$ will look like?

Graph the line through the given points and find its slope.

6. $(-1, -1)$ and $(2, -4)$

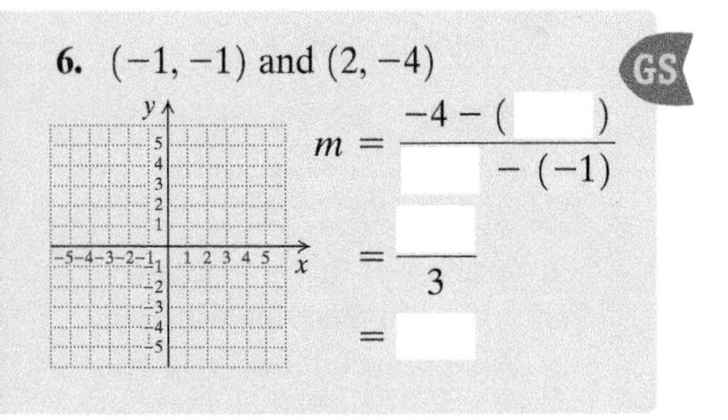

7. $(0, 2)$ and $(3, 1)$

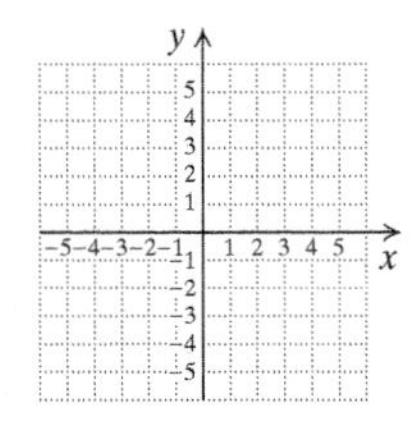

8. Find the slope of the line $f(x) = -\frac{2}{3}x + 1$. Use the points $(9, -5)$ and $(3, -1)$.

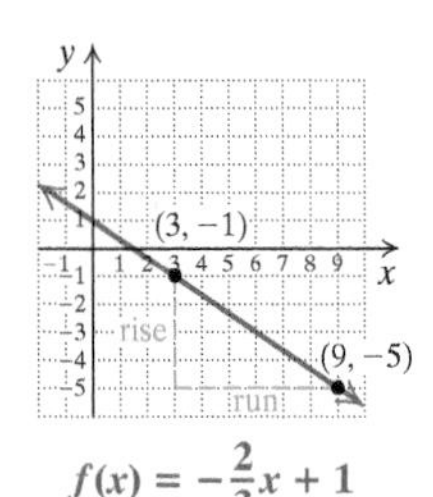

The slope of a line tells how it slants. A line with positive slope slants up from left to right. The larger the positive number, the steeper the slant. A line with negative slope slants down from left to right. The smaller the negative number, the steeper the line.

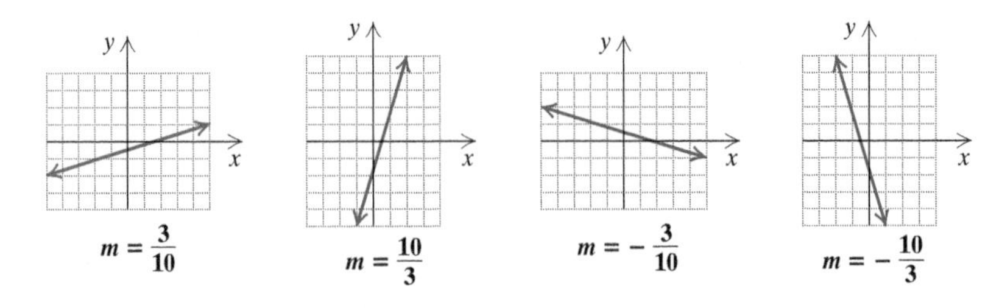

◀ **Do Exercises 6 and 7.**

How can we find the slope from a given equation? Let's consider the equation $y = 2x + 3$, which is in the form $y = mx + b$. We can find two points by choosing convenient values for x—say, 0 and 1—and substituting to find the corresponding y-values.

If $x = 0$, $y = 2 \cdot 0 + 3 = 3$.

If $x = 1$, $y = 2 \cdot 1 + 3 = 5$.

We find two points on the line to be

$(0, 3)$ and $(1, 5)$.

The slope of the line is found as follows, using the definition of slope:

$$m = \frac{\text{change in } y}{\text{change in } x} = \frac{5 - 3}{1 - 0} = \frac{2}{1} = 2.$$

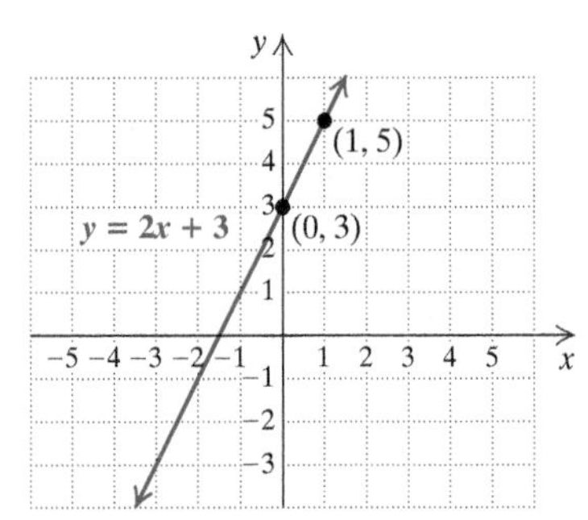

The slope is 2. Note that this is the coefficient of the x-term in the equation $y = 2x + 3$.

If we had chosen different points on the line—say, $(-2, -1)$ and $(4, 11)$—the slope would still be 2, as we see in the following calculation:

$$m = \frac{11 - (-1)}{4 - (-2)} = \frac{11 + 1}{4 + 2} = \frac{12}{6} = 2.$$

◀ **Do Exercise 8.**

We see that the slope of the line $y = mx + b$ is indeed the constant m, the coefficient of x.

SLOPE

The **slope** of the line $y = mx + b$ is m.

Answers

6. 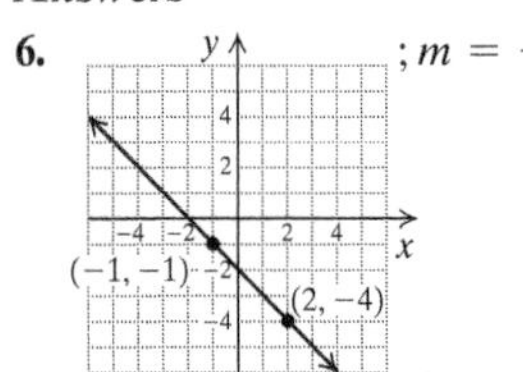

$; m = -1$

7. 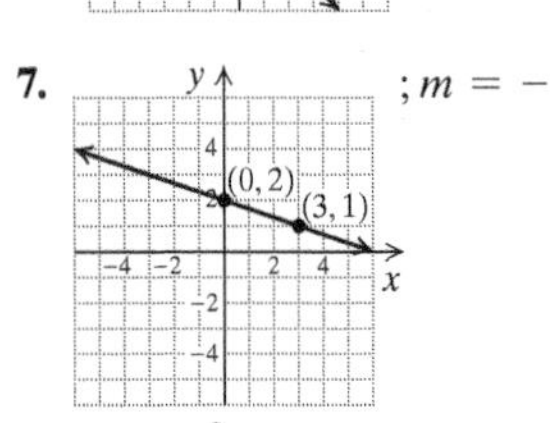

$; m = -\frac{1}{3}$

8. $m = -\frac{2}{3}$

Guided Solution:

6. $-1, 2, -3, -1$

From a linear equation in the form $y = mx + b$, we can read the slope and the y-intercept of the graph directly.

SLOPE–INTERCEPT EQUATION

The equation $y = mx + b$ is called the **slope–intercept equation.** The slope is m and the y-intercept is $(0, b)$.

Note that any graph of an equation $y = mx + b$ passes the vertical-line test and thus represents a function.

EXAMPLE 6 Find the slope and the y-intercept of $y = 5x - 4$.

Since the equation is already in the form $y = mx + b$, we simply read the slope and the y-intercept from the equation:

$$y = 5x - 4.$$

The slope is 5. The y-intercept is $(0, -4)$.

EXAMPLE 7 Find the slope and the y-intercept of $2x + 3y = 8$.

We first solve for y so we can easily read the slope and the y-intercept:

$$2x + 3y = 8$$

$$3y = -2x + 8 \quad \text{Subtracting } 2x$$

$$\frac{3y}{3} = \frac{-2x + 8}{3} \quad \text{Dividing by 3}$$

$$y = -\frac{2}{3}x + \frac{8}{3}. \quad \text{Finding the form } y = mx + b$$

The slope is $-\frac{2}{3}$. The y-intercept is $\left(0, \frac{8}{3}\right)$.

Do Exercises 9 and 10. ▶

Find the slope and the y-intercept.

9. $f(x) = -8x + 23$

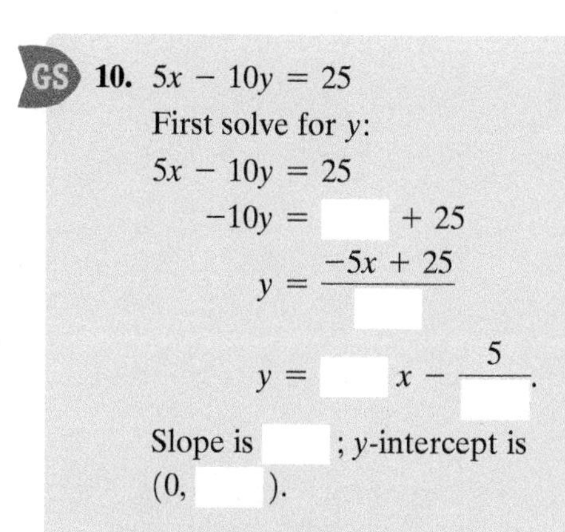

GS 10. $5x - 10y = 25$

First solve for y:

$5x - 10y = 25$

$-10y = \_\_\_ + 25$

$y = \dfrac{-5x + 25}{\_\_\_}$

$y = \_\_\_\,x - \dfrac{5}{\_\_\_}$.

Slope is ___; y-intercept is $(0, \_\_\_)$.

C APPLICATIONS

Slope has many real-world applications. For example, numbers like 2%, 3%, and 6% are often used to represent the *grade* of a road, a measure of how steep a road on a hill or a mountain is. A 3% grade $\left(3\% = \frac{3}{100}\right)$ means that for every horizontal distance of 100 ft that the road runs, the road rises 3 ft, and a -3% grade means that for every horizontal distance of 100 ft, the road drops 3 ft. (Normally, the road-grade signs do not include negative signs, since it is obvious whether you are climbing or descending.)

Road grade $= \frac{a}{b}$ (expressed as a percent)

GreenFoods

y a b x

Answers

9. Slope: -8; y-intercept: $(0, 23)$
10. Slope: $\frac{1}{2}$; y-intercept: $\left(0, -\frac{5}{2}\right)$

Guided Solution:

10. $-5x, -10, \frac{1}{2}, 2, \frac{1}{2}, -\frac{5}{2}$

An athlete might change the grade of a treadmill during a workout. An escape ramp on an airliner might have a slope of about -0.6.

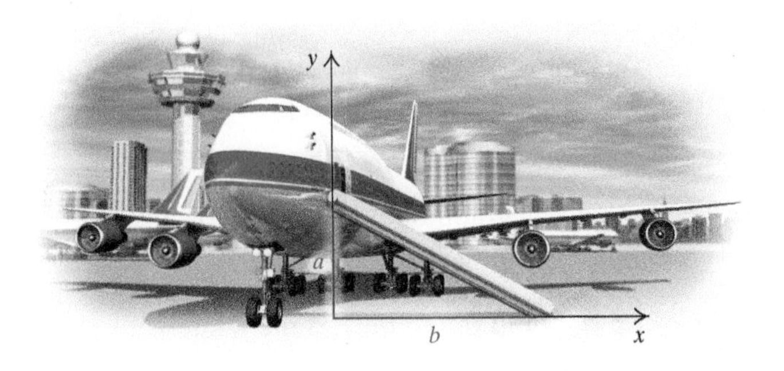

Architects and carpenters use slope when designing and building stairs, ramps, or roof pitches. Another application occurs in hydrology. The strength or force of a river depends on how far the river falls vertically compared to how far it flows horizontally. Slope can also be considered as a **rate of change**.

EXAMPLE 8 *Cancer Center Advertising.* The total spent by U.S. cancer centers on advertising increased from \$54 million in 2005 to \$173 million in 2014. Twenty of the 890 cancer centers who spent \$173 million in 2014 account for 86% of the total amount spent. Find the rate of change in the amount spent on advertising by cancer centers with respect to time, in years.

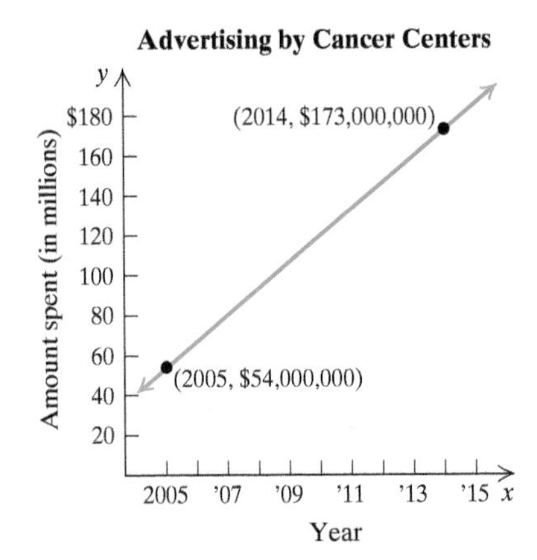

DATA: Indiana University School of Medicine; University of Pittsburgh School of Medicine and Graduate School of Public Health

The rate of change with respect to time, in years, is given by

$$\text{Rate of change} = \frac{\$173{,}000{,}000 - \$54{,}000{,}000}{2014 - 2005}$$
$$= \frac{\$119{,}000{,}000}{9 \text{ years}}$$
$$\approx \$13{,}222{,}222 \text{ per year.}$$

The amount spent on advertising by cancer centers is increasing at a rate of about \$13,222,222 per year.

◀ **Do Exercise 11.**

11. *Holiday Sales.* Retail sales during the holiday season have grown from \$416.4 billion in 2002 to \$655.8 billion in 2016. Find the rate of change in holiday sales with respect to time, in years.

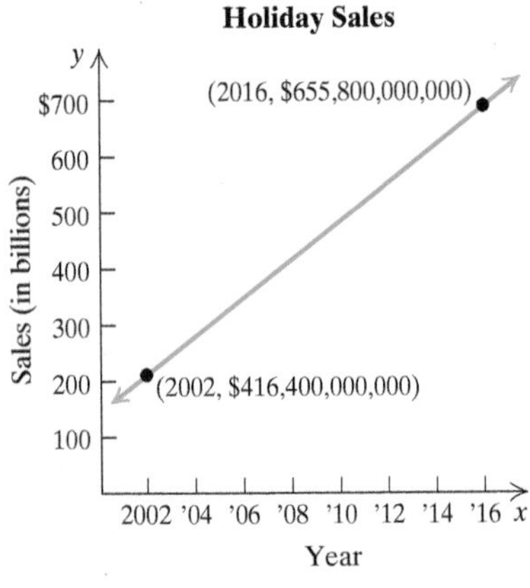

DATA: National Retail Federation

Answer

11. The rate of change in holiday sales is about \$17,100,000,000 per year.

EXAMPLE 9 *Consumption of Potatoes.* The annual per-capita consumption of potatoes has been decreasing since 1970. Find the rate of change of the per-capita consumption of potatoes with respect to time, in years.

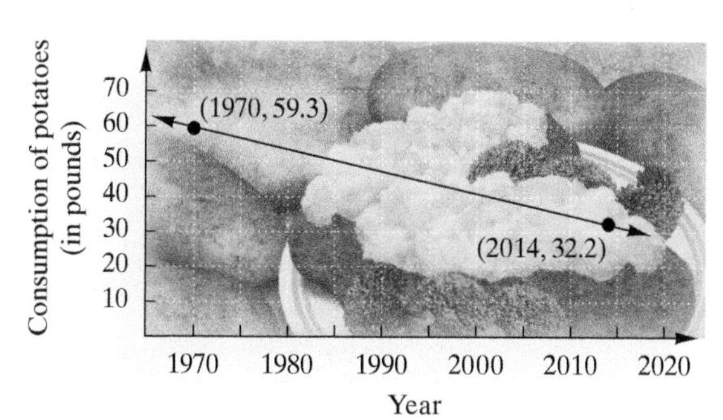

DATA: U.S. Department of Agriculture

The rate of change with respect to time, in years, is given by

$$\text{Rate of change} = \frac{32.2 - 59.3}{2014 - 1970} = \frac{-27.1}{44 \text{ years}} \approx -0.616 \text{ pounds per year.}$$

The per-capita consumption of potatoes is decreasing at a rate of about 0.616 pounds per year.

Do Exercise 12. ▶

12. *Newspaper Circulation.* Daily newspaper circulation has decreased dramatically since 2005. The following graph shows the circulation of daily newspapers, in millions, for 2005 and for 2015. Find the rate of change in the circulation of daily newspapers per year.

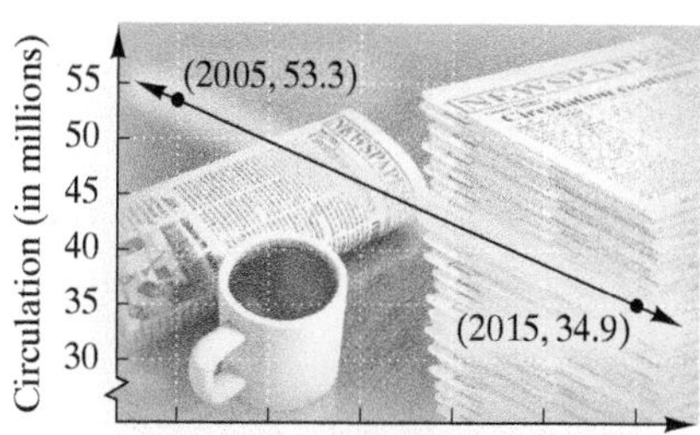

DATA: Editor & Publisher International Data Book

Answer

12. The rate of change is -1.84 million papers per year.

2.4 Exercise Set

FOR EXTRA HELP  MyLab Math

✓ Check Your Understanding

Reading Check From the words below each blank, choose the one that correctly completes the statement.

RC1. If $m > 0$, the graph slants ____________ from left to right.
up/down

RC2. If $m = 0$, the graph is ____________.
horizontal/vertical

RC3. If $m < 0$, the graph slants ____________ from left to right.
up/down

Concept Check Choose from the columns on the right the slope of each line.

CC1.

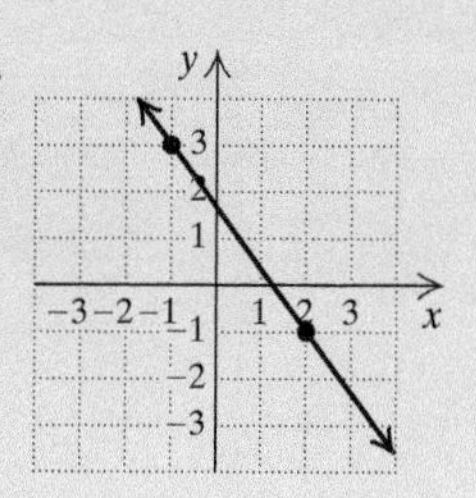

CC2.

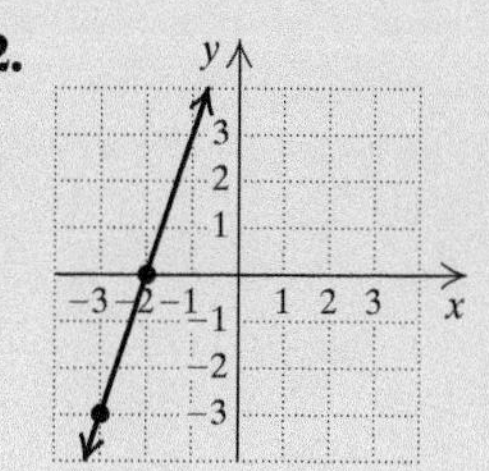

CC3.

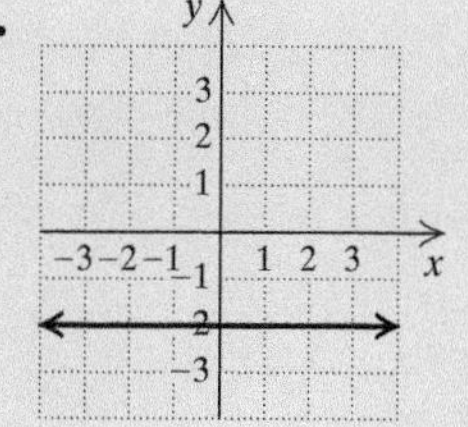

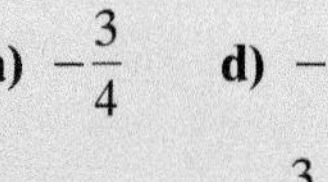

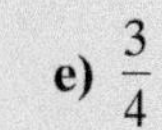

a) $-\frac{3}{4}$ **d)** -4

b) 3 **e)** $\frac{3}{4}$

c) 0 **f)** $-\frac{4}{3}$

a, **b** Find the slope and the y-intercept of each equation.

1. $y = 4x + 5$
2. $y = -5x + 10$
3. $f(x) = -2x - 6$
4. $g(x) = -5x + 7$
5. $y = -\frac{3}{8}x - \frac{1}{5}$
6. $y = \frac{15}{7}x + \frac{16}{5}$
7. $g(x) = 0.5x - 9$
8. $f(x) = -3.1x + 5$
9. $2x - 3y = 8$
10. $-8x - 7y = 24$
11. $9x = 3y + 6$
12. $9y + 36 - 4x = 0$
13. $3 - \frac{1}{4}y = 2x$
14. $5x = \frac{2}{3}y - 10$
15. $17y + 4x + 3 = 7 + 4x$
16. $3y - 2x = 5 + 9y - 2x$

b Find the slope of each line.

17.

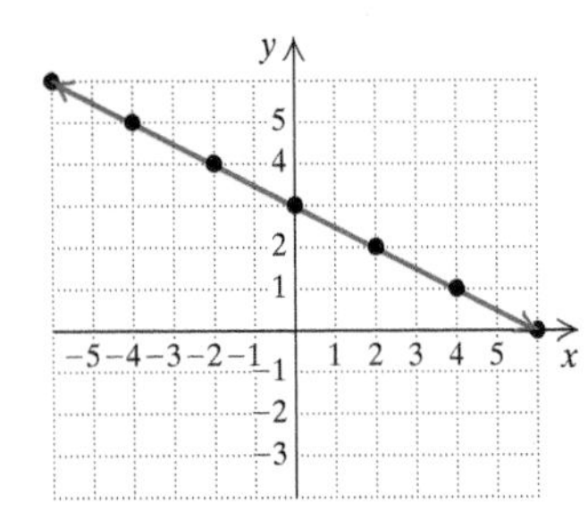

18.

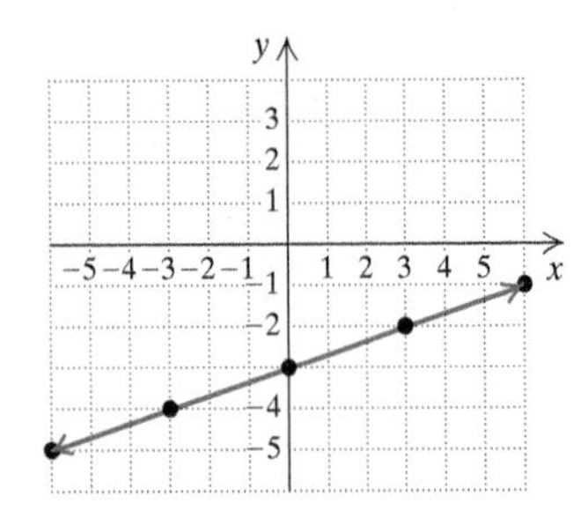

19.

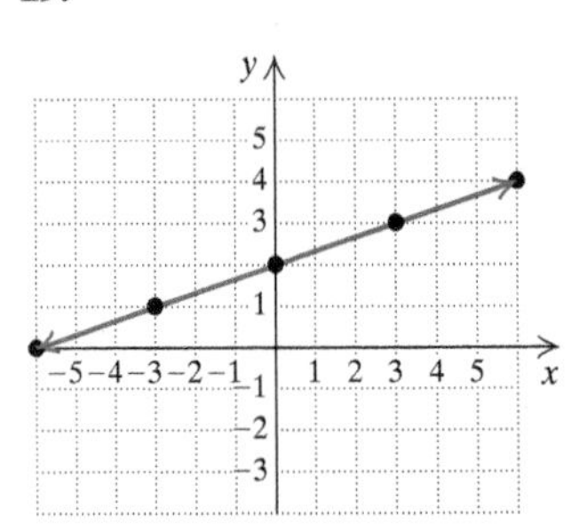

20.

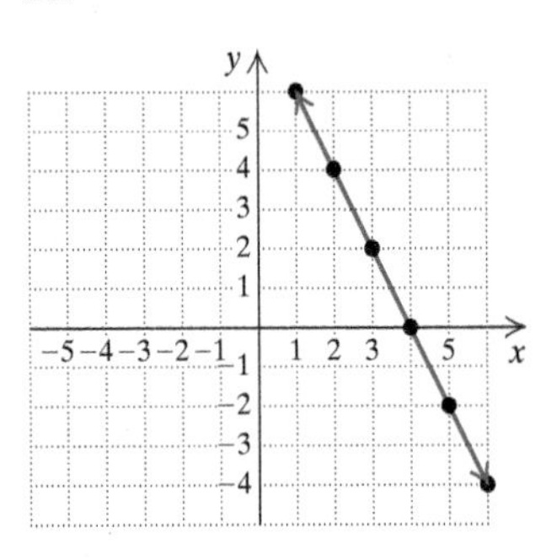

Find the slope of the line containing the given pair of points.

21. $(6, 9)$ and $(4, 5)$
22. $(8, 7)$ and $(2, -1)$
23. $(9, -4)$ and $(3, -8)$
24. $(17, -12)$ and $(-9, -15)$
25. $(-16.3, 12.4)$ and $(-5.2, 8.7)$
26. $(14.4, -7.8)$ and $(-12.5, -17.6)$

c Find the slope (or rate of change).

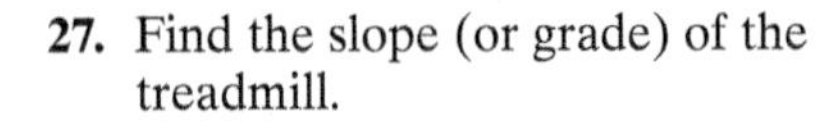

27. Find the slope (or grade) of the treadmill.

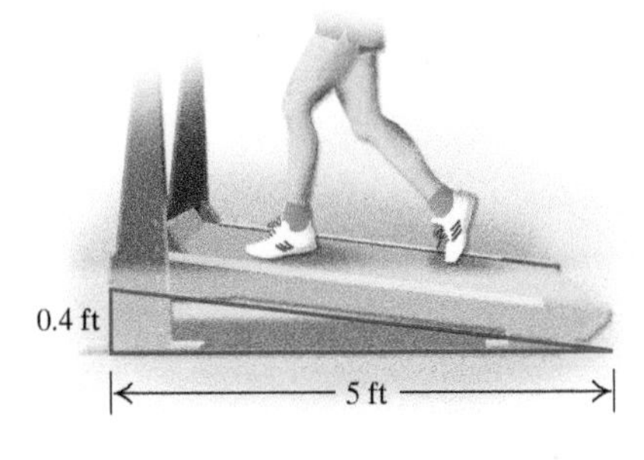

28. Find the slope (or head) of the river.

29. Find the slope (or pitch) of the roof.

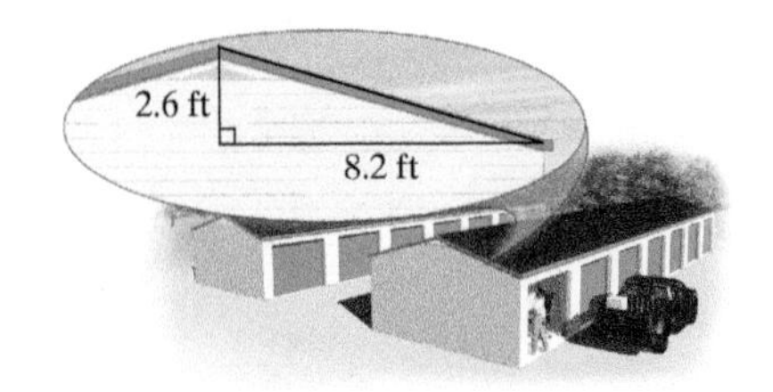

30. Public buildings regularly include steps with 7-in. risers and 11-in. treads. Find the grade of such a stairway.

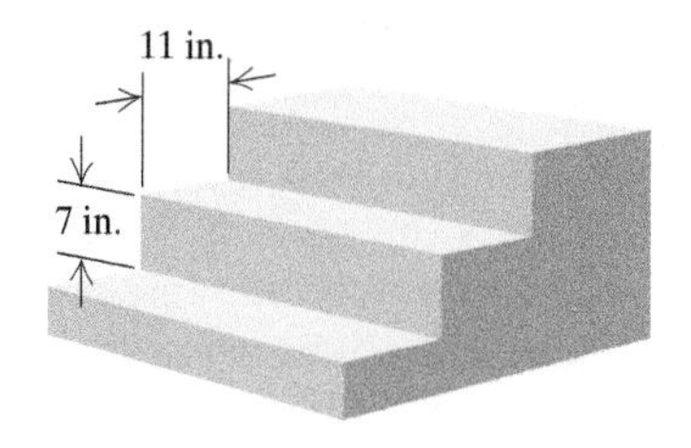

31. *Super Bowl Ad Spending.* Find the rate of change in Super Bowl advertising spending with respect to time, in years.

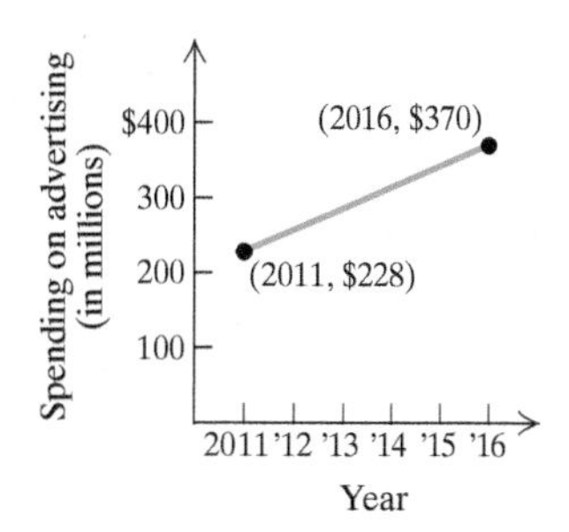

DATA: KANTAR MEDIA

32. *Vehicle Production.* Find the rate of change in U.S. vehicle production with respect to time, in years.

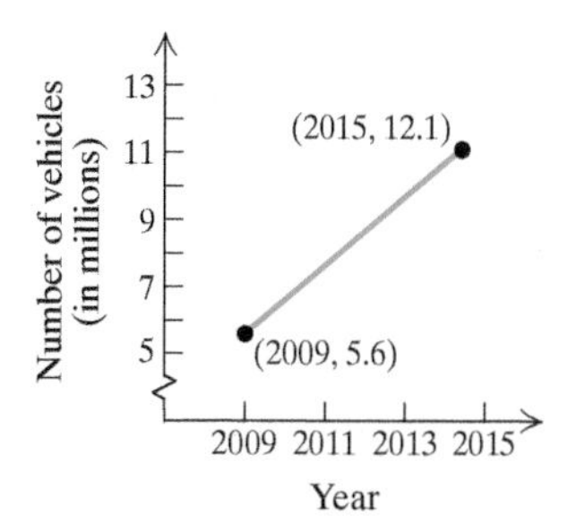

DATA: Automotive News Data Center; R. L. Polk; OICA

Find the rate of change.

33.

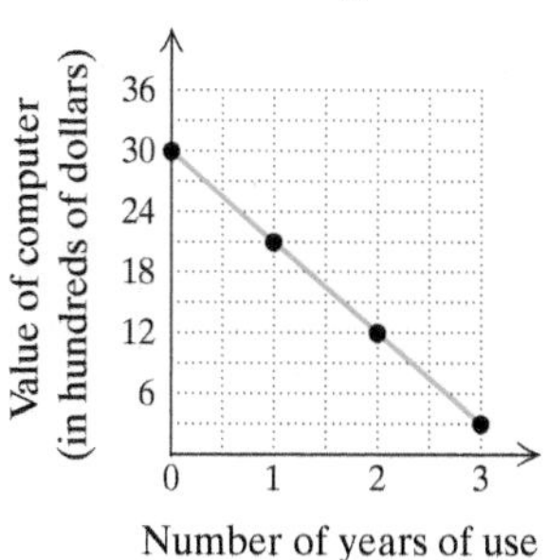

34.

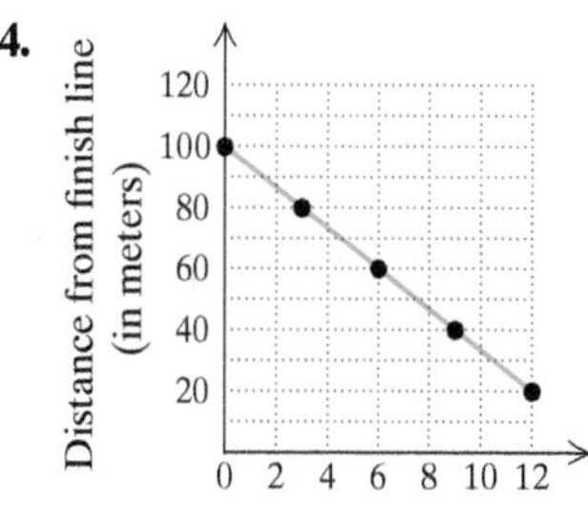

35.

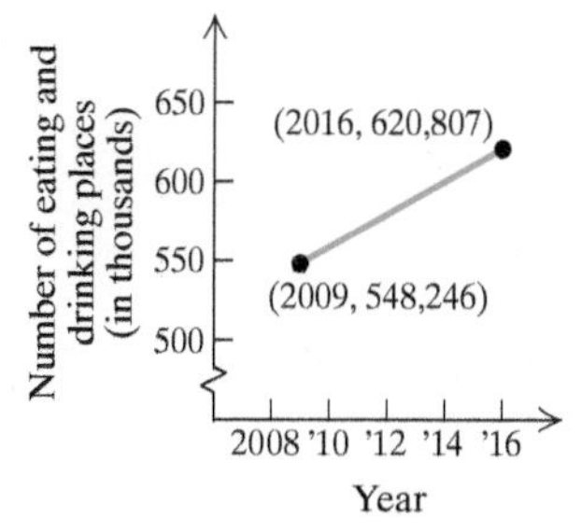

DATA: U.S. Bureau of Labor Statistics

36.

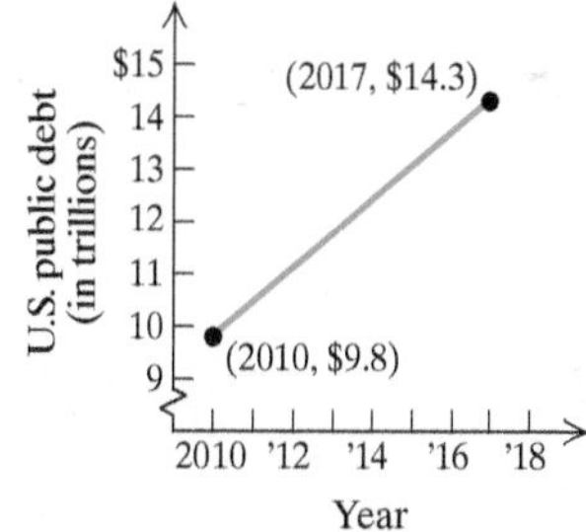

DATA: Budget of the U.S Government; Office of Management and Budget; Executive Office of the President

Skill Maintenance

Simplify. [J11], [J18]

37. $3^2 - 24 \cdot 56 + 144 \div 12$

38. $9\{2x - 3[5x + 2(-3x + y^0 - 2)]\}$

39. $10\{2x + 3[5x - 2(-3x + y^1 - 2)]\}$

40. $5^4 \div 625 \div 5^2 \cdot 5^7 \div 5^3$

Solve. [1.3a]

41. One side of a square is 5 yd less than a side of an equilateral triangle. If the perimeter of the square is the same as the perimeter of the triangle, what is the length of a side of the square? of the triangle?

Solve. [1.6c, e]

42. $|5x - 8| \geq 32$

43. $|5x - 8| < 32$

44. $|5x - 8| = 32$

45. $|5x - 8| = -32$

2.5 More on Graphing Linear Equations

OBJECTIVES

a Graph linear equations using intercepts.

b Given a linear equation in slope–intercept form, use the slope and the y-intercept to graph the line.

c Graph linear equations of the form $x = a$ or $y = b$.

d Given the equations of two lines, determine whether their graphs are parallel or whether they are perpendicular.

SKILL REVIEW *Plot points associated with ordered pairs of numbers.* [2.1a]

1. Plot the following points:
$A(0, 4)$, $B(0, -1)$, $C(0, 0)$, $D(3, 0)$, and $E(-\frac{7}{2}, 0)$.

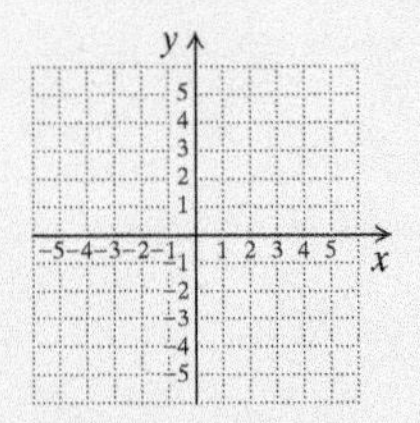

Answer:

1.

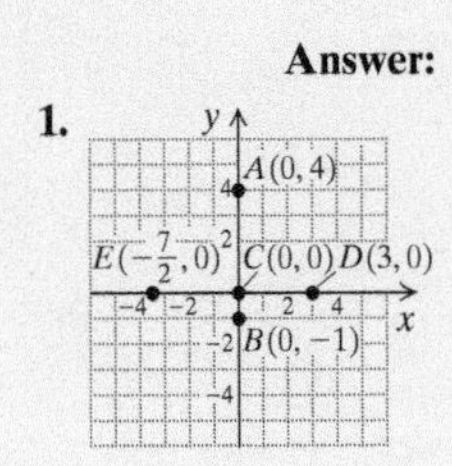

MyLab Math VIDEO

a GRAPHING USING INTERCEPTS

The ***x*-intercept** of the graph of a linear equation or function is the point at which the graph crosses the x-axis. The ***y*-intercept** is the point at which the graph crosses the y-axis. We know from geometry that only one line can be drawn through two given points. Thus, if we know the intercepts, we can graph the line. To ensure that a computation error has not been made, it is a good idea to calculate a third point as a check.

Many equations of the type $Ax + By = C$ can be graphed conveniently using intercepts.

x-INTERCEPTS AND y-INTERCEPTS

A ***y*-intercept** is a point $(0, b)$. To find b, let $x = 0$ and solve for y.
An ***x*-intercept** is a point $(a, 0)$. To find a, let $y = 0$ and solve for x.

EXAMPLE 1 Find the intercepts of $3x + 2y = 12$ and then graph the line.

y-intercept: To find the y-intercept, we let $x = 0$ and solve for y:

$$3x + 2y = 12$$
$$3 \cdot 0 + 2y = 12 \quad \text{Substituting 0 for } x$$
$$2y = 12$$
$$y = 6.$$

The y-intercept is $(0, 6)$.

x-intercept: To find the x-intercept, we let $y = 0$ and solve for x:

$$3x + 2y = 12$$
$$3x + 2 \cdot 0 = 12 \quad \text{Substituting 0 for } y$$
$$3x = 12$$
$$x = 4.$$

The x-intercept is $(4, 0)$.

We plot these points and draw the line, using a third point as a check. We choose $x = 6$ and solve for y:

$$3(6) + 2y = 12$$
$$18 + 2y = 12$$
$$2y = -6$$
$$y = -3.$$

We plot $(6, -3)$ and note that it is on the line so the graph is probably correct.

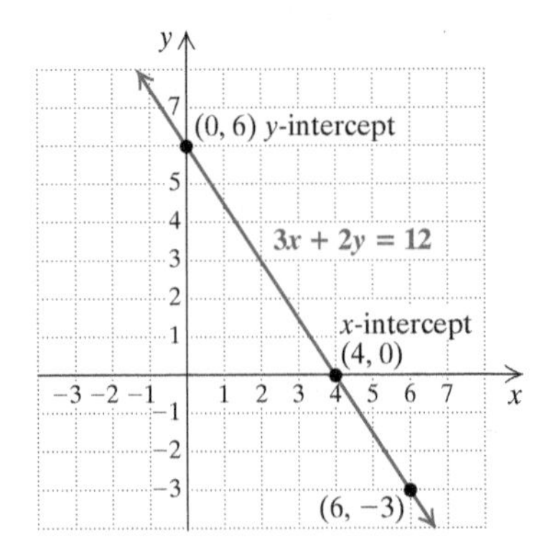

When both the x-intercept and the y-intercept are $(0, 0)$, as is the case with an equation such as $y = 2x$, whose graph passes through the origin, we would need to calculate another point and use a third point as a check.

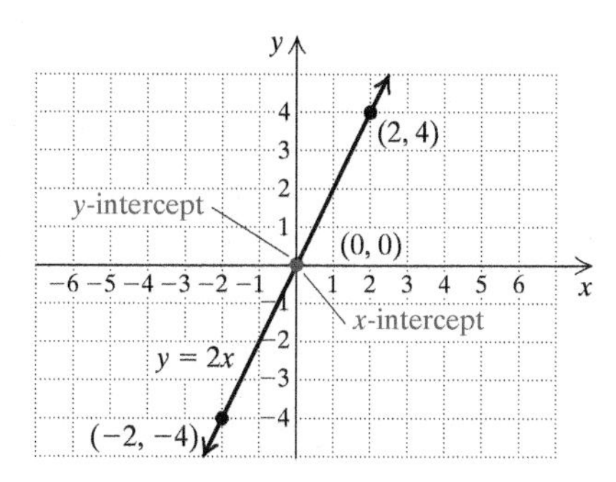

Do Exercise 1. ▶

CALCULATOR CORNER

Viewing the Intercepts Knowing the intercepts of a linear equation helps us determine a good viewing window for the graph of the equation. For example, when we graph the equation $y = -x + 15$ in the standard window, we see only a small portion of the graph in the upper right-hand corner of the screen, as shown on the left below.

$y = -x + 15$

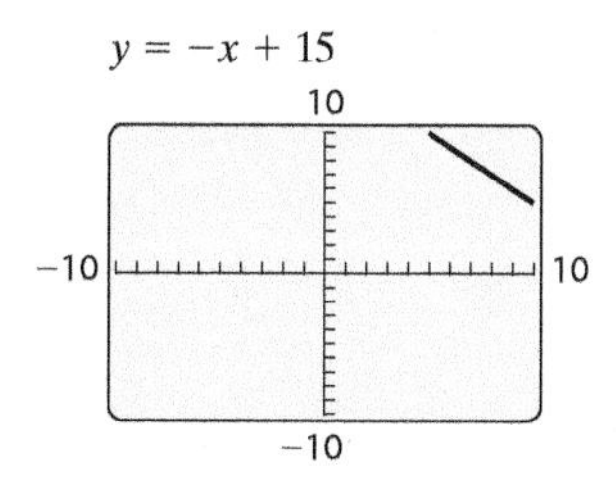

$y = -x + 15$

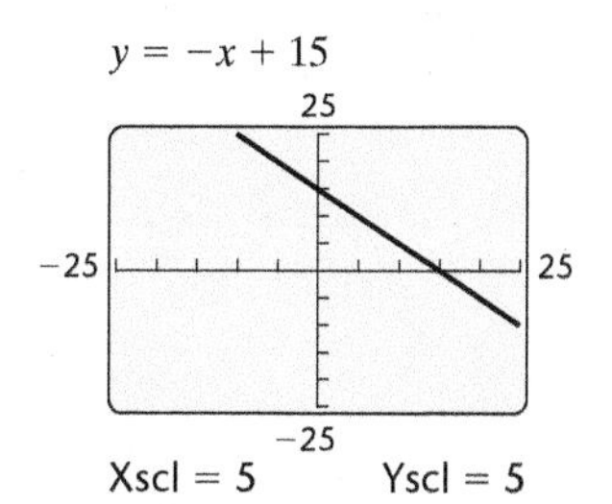

Using algebra, as we did in Example 1, we can find that the intercepts of the graph of this equation are $(0, 15)$ and $(15, 0)$. This tells us that, if we are to see a portion of the graph that includes the intercepts, both Xmax and Ymax should be greater than 15. We can try different window settings until we find one that suits us. One good choice, shown on the right above, is $[-25, 25, -25, 25]$, with Xscl $= 5$ and Yscl $= 5$.

EXERCISES: Find the intercepts of the equation algebraically. Then graph the equation on a graphing calculator, choosing window settings that allow the intercepts to be seen clearly. (Settings may vary.)

1. $y = -3.2x - 16$

2. $y - 4.25x = 85$

3. $6x + 5y = 90$

4. $5x - 6y = 30$

5. $8x + 3y = 9$

6. $y = 0.4x - 5$

7. $y = 1.2x - 12$

8. $4x - 5y = 2$

1. Find the intercepts of $4y - 12 = -6x$ and then graph the line.

To find the y-intercept, set $x = 0$ and solve for y:

$$4y - 12 = -6 \cdot \_\_\_$$
$$4y - 12 = \_\_\_$$
$$4y = \_\_\_$$
$$y = \_\_\_.$$

The y-intercept is $(0, \_\_\_)$.

To find the x-intercept, set $y = 0$ and solve for x:

$$4 \cdot \_\_\_ - 12 = -6x$$
$$\_\_\_ = -6x$$
$$\_\_\_ = x.$$

The x-intercept is $(\_\_\_, 0)$.

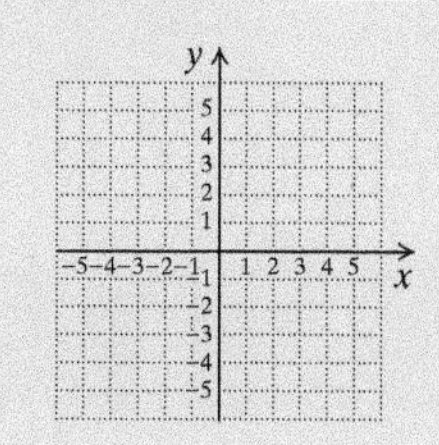

Answer

1.

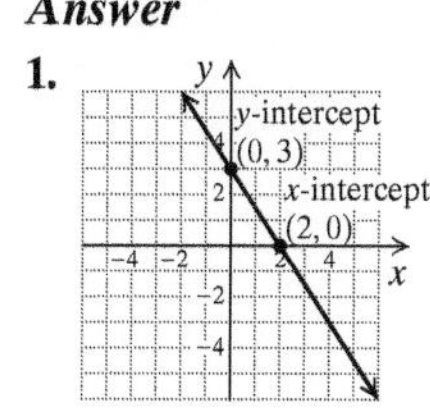

$4y - 12 = -6x$

Guided Solution:

1. 0, 0, 12, 3, 3; 0, −12, 2, 2

b GRAPHING USING THE SLOPE AND THE y-INTERCEPT

We can also graph a line using its slope and y-intercept.

EXAMPLE 2 Graph: $y = -\frac{2}{3}x + 1$.

This equation is in slope–intercept form, $y = mx + b$. The y-intercept is $(0, 1)$. We plot $(0, 1)$. We can think of the slope $(m = -\frac{2}{3})$ as $\frac{-2}{3}$.

$$m = \frac{\text{Rise}}{\text{Run}} = \frac{-2}{3} \qquad \begin{array}{l}\text{Move 2 units down.}\\ \text{Move 3 units right.}\end{array}$$

Starting at the y-intercept and using the slope, we find another point by moving 2 units down (since the numerator is *negative* and corresponds to the change in y) and 3 units to the right (since the denominator is *positive* and corresponds to the change in x). We get to a new point, $(3, -1)$. In a similar manner, we can move from the point $(3, -1)$ to find another point, $(6, -3)$.

We could also think of the slope $(m = -\frac{2}{3})$ as $\frac{2}{-3}$.

$$m = \frac{\text{Rise}}{\text{Run}} = \frac{2}{-3} \qquad \begin{array}{l}\text{Move 2 units up.}\\ \text{Move 3 units left.}\end{array}$$

Then we can start again at $(0, 1)$, but this time we move 2 units up (since the numerator is *positive* and corresponds to the change in y) and 3 units to the left (since the denominator is *negative* and corresponds to the change in x). We get another point on the graph, $(-3, 3)$, and from it we can obtain $(-6, 5)$ and others in a similar manner. We plot the points and draw the line.

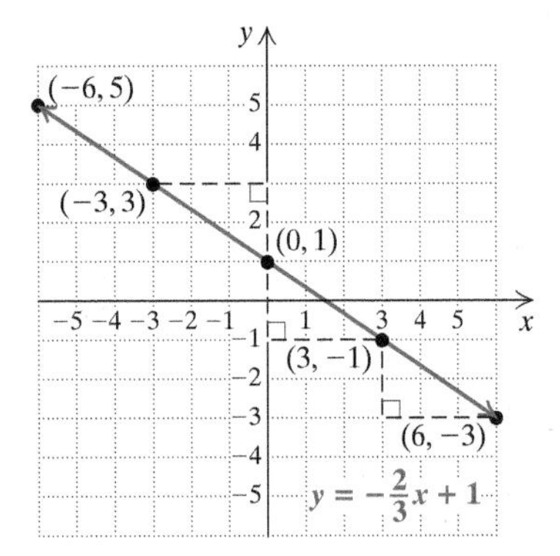

EXAMPLE 3 Graph: $f(x) = \frac{2}{5}x + 4$.

First, we plot the y-intercept, $(0, 4)$. We then consider the slope $\frac{2}{5}$. A slope of $\frac{2}{5}$ tells us that, for every 2 units that the graph rises, it runs 5 units horizontally in the positive direction, or to the right. Thus, starting at the y-intercept and using the slope, we find another point by moving 2 units up (since the numerator is *positive* and corresponds to the change in y) and 5 units to the right (since the denominator is *positive* and corresponds to the change in x). We get to a new point, $(5, 6)$.

Graph using the slope and the y-intercept.

2. $y = \frac{3}{2}x + 1$

3. $f(x) = \frac{3}{4}x - 2$

4. $g(x) = -\frac{3}{5}x + 5$

5. $y = -\frac{5}{3}x - 4$

Answers

2.

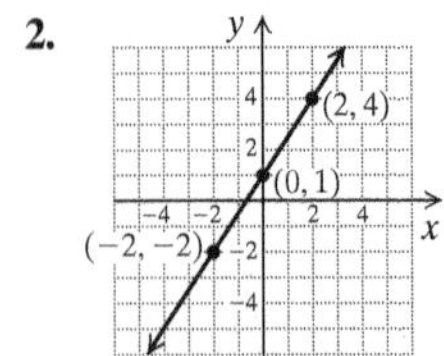

$y = \frac{3}{2}x + 1$

3.

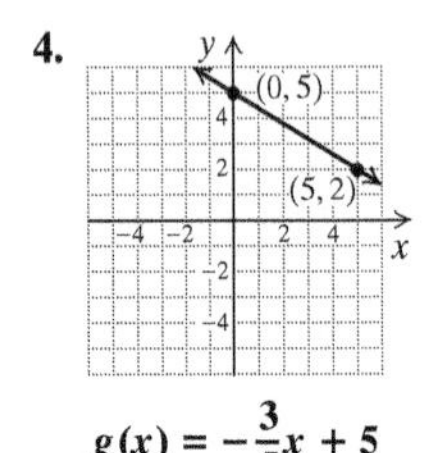

$f(x) = \frac{3}{4}x - 2$

4.

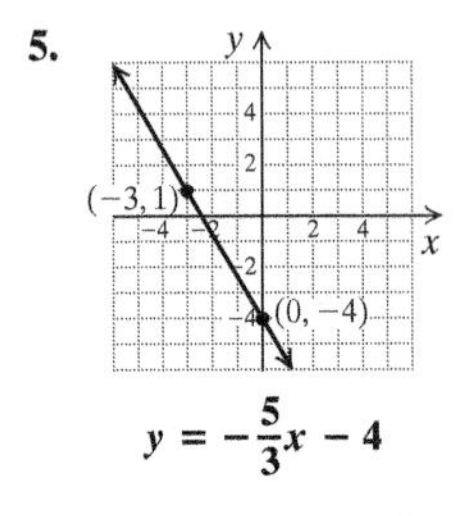

$g(x) = -\frac{3}{5}x + 5$

5. (−3, 1) (0, −4)

$y = -\frac{5}{3}x - 4$

We can also think of the slope $\frac{2}{5}$ as $\frac{-2}{-5}$. A slope of $\frac{-2}{-5}$ tells us that, for every 2 units that the graph drops, it runs 5 units horizontally in the negative direction, or to the left. We again start at the y-intercept, $(0, 4)$. We move 2 units down (since the numerator is *negative* and corresponds to the change in y) and 5 units to the left (since the denominator is *negative* and corresponds to the change in x). We get to another new point, $(-5, 2)$. We plot the points and draw the line.

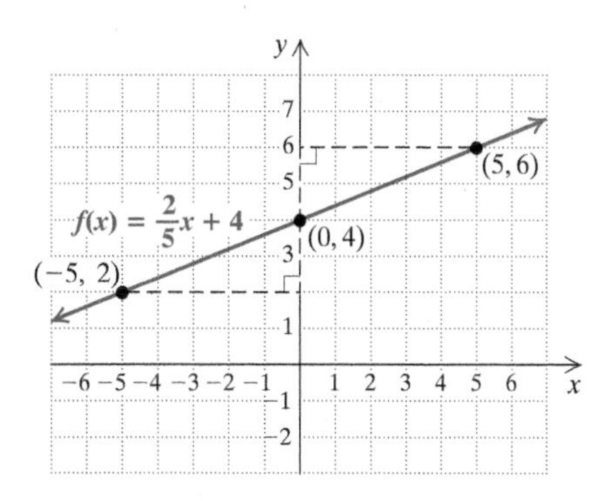

Do Exercises 2–5 on the preceding page.

c HORIZONTAL LINES AND VERTICAL LINES

Some equations have graphs that are parallel to one of the axes. This happens when either A or B is 0 in $Ax + By = C$. These equations have a missing variable; that is, there is only one variable in the equation. In the following example, x is missing.

EXAMPLE 4 Graph: $y = 3$.

Since x is missing, any number for x will do. Thus all ordered pairs $(x, 3)$ are solutions. The graph is a **horizontal line** parallel to the x-axis.

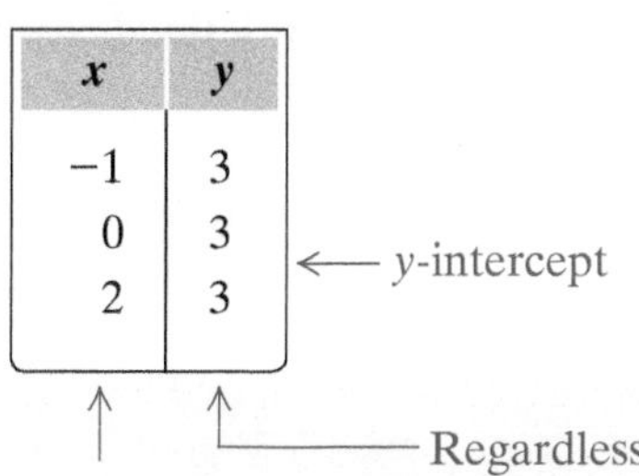

x	y
−1	3
0	3
2	3

← y-intercept

Regardless of x, y must be 3.

Choose *any* number for x.

y = 3
(−1, 3)
(0, 3)
(2, 3)

What about the slope of a horizontal line? In Example 4, consider the points $(-1, 3)$ and $(2, 3)$, which are on the line $y = 3$. The change in y is $3 - 3$, or 0. The change in x is $-1 - 2$, or -3. Thus,

$$m = \frac{3 - 3}{-1 - 2} = \frac{0}{-3} = 0.$$

Any two points on a horizontal line have the same y-coordinate. Thus the change in y is always 0, so the slope is 0.

Do Exercises 6 and 7.

Graph and determine the slope.

6. $f(x) = -4$

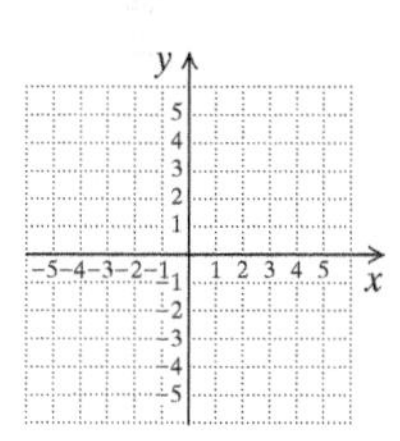

7. $y = 3.6$

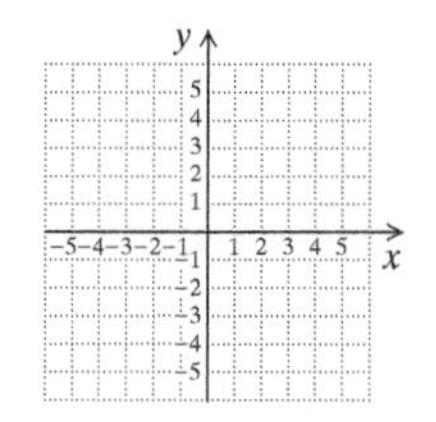

Answers

6. $m = 0$

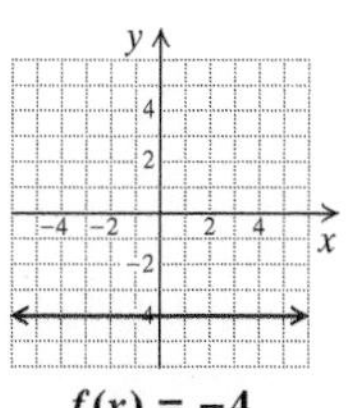

7. $m = 0$

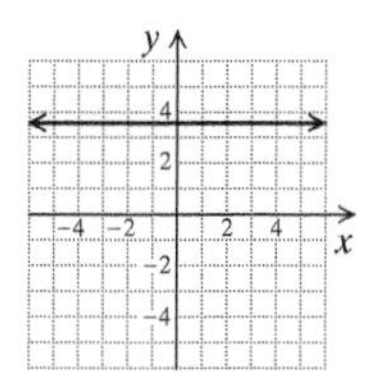

We can also determine the slope by noting that $y = 3$ can be written in slope–intercept form as $y = 0x + 3$, or $f(x) = 0x + 3$. From this equation, we read that the slope is 0. A function of this type is called a **constant function**. We can express it in the form $y = b$, or $f(x) = b$. Its graph is a horizontal line that crosses the y-axis at $(0, b)$.

In the following example, y is missing and the graph is parallel to the y-axis.

EXAMPLE 5 Graph: $x = -2$.

Since y is missing, any number for y will do. Thus all ordered pairs $(-2, y)$ are solutions. The graph is a **vertical line** parallel to the y-axis.

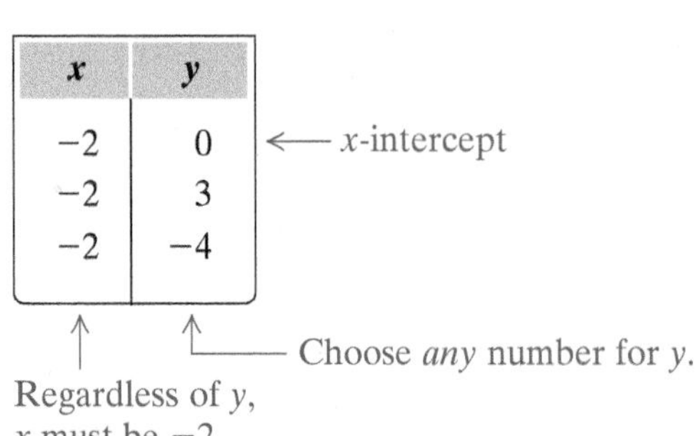

x	y
−2	0
−2	3
−2	−4

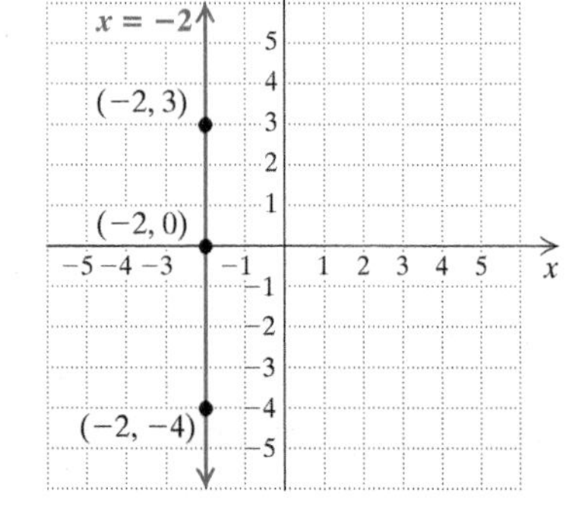

This graph is not the graph of a function because it fails the vertical-line test. The vertical line itself crosses the graph more than once.

◀ **Do Exercises 8 and 9.**

What about the slope of a vertical line? In Example 5, consider the points $(-2, 3)$ and $(-2, -4)$, which are on the line $x = -2$. The change in y is $3 - (-4)$, or 7. The change in x is $-2 - (-2)$, or 0. Thus,

$$m = \frac{3 - (-4)}{-2 - (-2)} = \frac{7}{0}. \quad \text{Not defined}$$

Since division by 0 is not defined, the slope of this line is not defined. Any two points on a vertical line have the same x-coordinate. Thus the change in x is always 0, so the slope of any vertical line is not defined.

The following summarizes the characteristics of horizontal lines and vertical lines and their equations.

HORIZONTAL LINE; VERTICAL LINE

The graph of $y = b$, or $f(x) = b$, is a **horizontal line** with y-intercept $(0, b)$. It is the graph of a constant function with slope 0.

The graph of $x = a$ is a **vertical line** with x-intercept $(a, 0)$. The slope is not defined. It is not the graph of a function.

◀ **Do Exercise 10.**

Graph.

8. $x = -5$

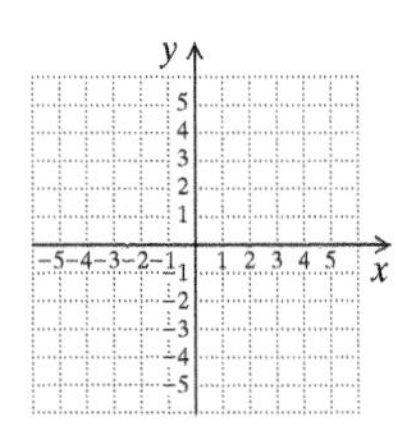

9. $8x - 5 = 19$ (*Hint*: Solve for x.)

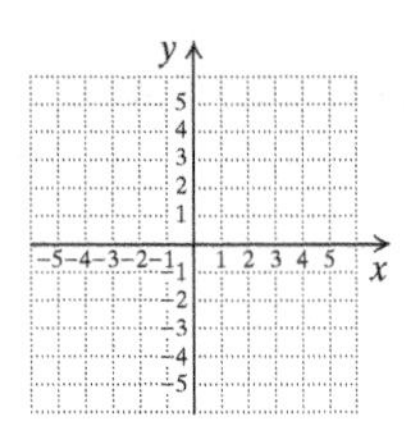

10. Determine, if possible, the slope of each line.
    a) $x = -12$
    b) $y = 6$
    c) $2y + 7 = 11$
    d) $x = 0$
    e) $y = -\frac{3}{4}$
    f) $10 - 5x = 15$

Answers

8.

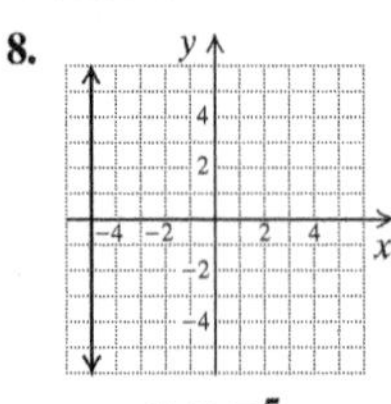

$x = -5$

9.

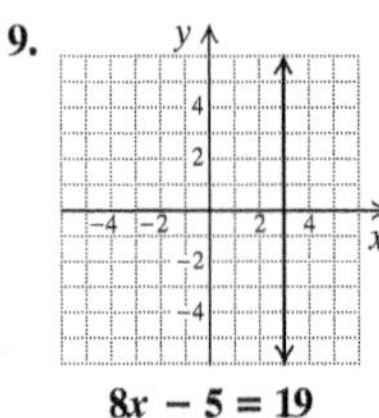

$8x - 5 = 19$

10. **(a)** Not defined; **(b)** 0; **(c)** 0; **(d)** not defined; **(e)** 0; **(f)** not defined

We have graphed linear equations in several ways in this chapter. Although, in general, you can use any method that works best for you, we list some guidelines in the margin at right.

To graph a linear equation:

1. Is the equation of the type $x = a$ or $y = b$? If so, the graph will be a line parallel to an axis; $x = a$ is vertical and $y = b$ is horizontal.
2. If the line is of the type $y = mx$, both intercepts are the origin, $(0, 0)$. Plot $(0, 0)$ and one other point.
3. If the line is of the type $y = mx + b$, plot the y-intercept and one other point.
4. If the equation is of the form $Ax + By = C$, graph using intercepts. If the intercepts are too close together, choose another point farther from the origin.
5. In all cases, use a third point as a check.

d PARALLEL LINES AND PERPENDICULAR LINES

Parallel Lines

Parallel lines extend indefinitely without intersecting. If two lines are vertical, they are parallel. How can we tell whether nonvertical lines are parallel? We examine their slopes and y-intercepts.

PARALLEL LINES

Two nonvertical lines are **parallel** if they have the *same* slope and *different* y-intercepts. Vertical lines are parallel.

EXAMPLE 6 Determine if the graphs of $y - 3x = 1$ and $3x + 2y = -2$ are parallel.

To determine if lines are parallel, we first find their slopes. To do this, we find the slope–intercept form of each equation by solving for y:

$$y - 3x = 1$$
$$y = 3x + 1;$$

$$3x + 2y = -2$$
$$2y = -3x - 2$$
$$y = \tfrac{1}{2}(-3x - 2)$$
$$y = -\tfrac{3}{2}x - 1.$$

The slopes, 3 and $-\frac{3}{2}$, are different. Thus the lines are not parallel, as the graphs at right confirm.

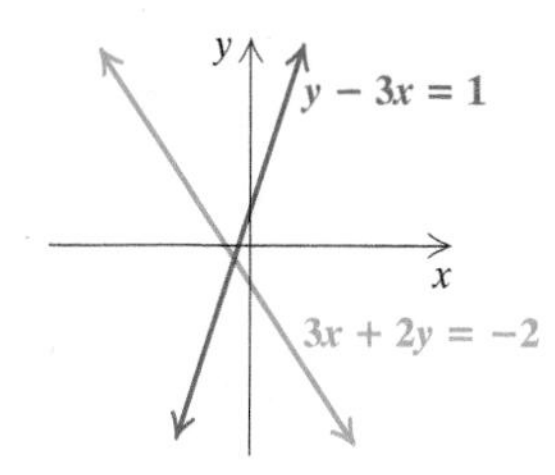

EXAMPLE 7 Determine if the graphs of $3x - y = -5$ and $y - 3x = -2$ are parallel.

We first find the slope–intercept form of each equation by solving for y:

$$3x - y = -5$$
$$-y = -3x - 5$$
$$-1(-y) = -1(-3x - 5)$$
$$y = 3x + 5;$$

$$y - 3x = -2$$
$$y = 3x - 2.$$

The slopes, 3, are the same. The y-intercepts, $(0, 5)$ and $(0, -2)$, are different. Thus the lines are parallel, as the graphs appear to confirm.

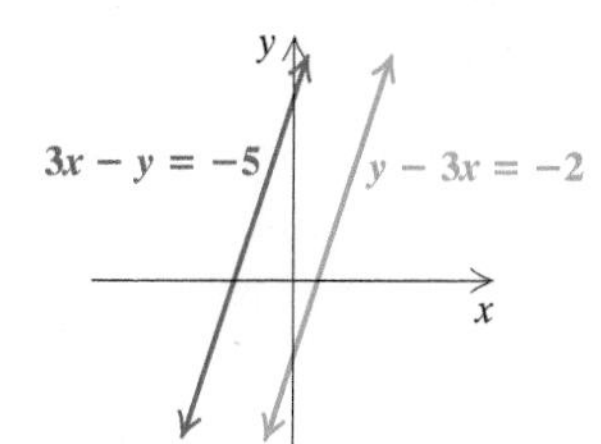

Do Exercises 11–13.

Determine whether the graphs of the given pair of lines are parallel.

GS **11.** $x + 4 = y$,
$y - x = -3$

Write each equation in the form $y = mx + b$:

$x + 4 = y \rightarrow y = x + 4;$

$y - x = -3 \rightarrow y = \_\_\_\_ - 3.$

The slope of each line is ____, and the y-intercepts, $(0, 4)$ and $(0, \_\_\_\_)$, are different. Thus the lines ________ (are/are not) parallel.

12. $y + 4 = 3x$,
$4x - y = -7$

13. $y = 4x + 5$,
$2y = 8x + 10$

Answers

11. Yes **12.** No **13.** No; they are the same line.

Guided Solution:

11. x, 1, -3, are

Perpendicular Lines

If one line is vertical and another is horizontal, they are perpendicular. For example, the lines $x = 5$ and $y = -3$ are perpendicular. Otherwise, how can we tell whether two lines are perpendicular?

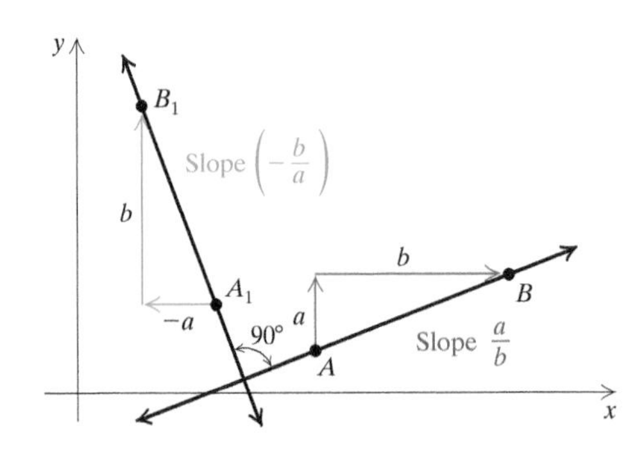

Consider a line $\overleftrightarrow{AB}$, as shown in the figure above, with slope a/b. Then think of rotating the line 90° to get a line $\overleftrightarrow{A_1B_1}$ perpendicular to $\overleftrightarrow{AB}$. For the new line, the rise and the run are interchanged, but the run is now negative. Thus the slope of the new line is $-b/a$, which is the opposite of the reciprocal of the slope of the first line. Also note that when we multiply the slopes, we get

$$\frac{a}{b}\left(-\frac{b}{a}\right) = -1.$$

This is the condition under which lines will be perpendicular.

> **PERPENDICULAR LINES**
>
> Two lines are **perpendicular** if the product of their slopes is -1. (If one line has slope m, then the slope of a line perpendicular to it is $-1/m$. That is, to find the slope of a line perpendicular to a given line, we take the reciprocal of the given slope and change the sign.)
>
> Lines are also perpendicular if one of them is vertical ($x = a$) and one of them is horizontal ($y = b$).

EXAMPLE 8 Determine whether the graphs of $5y = 4x + 10$ and $4y = -5x + 4$ are perpendicular.

In order to determine whether the lines are perpendicular, we determine whether the product of their slopes is -1. We first find the slope–intercept form of each equation by solving for y.

We have

$$\begin{aligned} 5y &= 4x + 10 \\ y &= \tfrac{1}{5}(4x + 10) \\ y &= \tfrac{4}{5}x + 2; \end{aligned} \qquad \begin{aligned} 4y &= -5x + 4 \\ y &= \tfrac{1}{4}(-5x + 4) \\ y &= -\tfrac{5}{4}x + 1. \end{aligned}$$

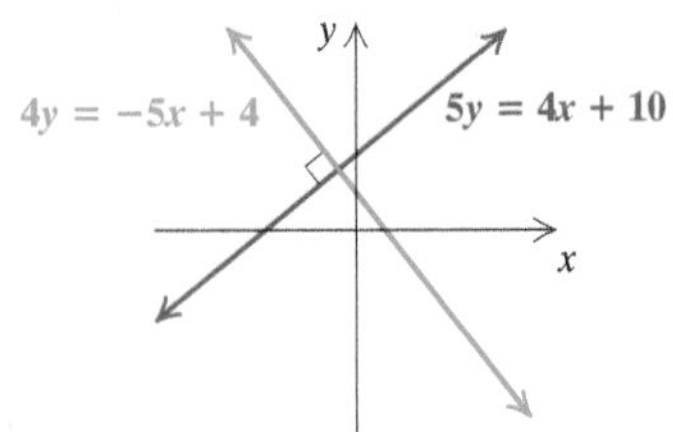

The slope of the first line is $\frac{4}{5}$, and the slope of the second line is $-\frac{5}{4}$. The product of the slopes is $\frac{4}{5}\cdot\left(-\frac{5}{4}\right) = -1$. Thus the lines are perpendicular.

Do Exercises 14 and 15.

Determine whether the graphs of the given pair of lines are perpendicular.

14. $2y - x = 2$,
$y + 2x = 4$ GS

Write each equation in the form $y = mx + b$:

$2y - x = 2 \rightarrow y =$ ____ $x + 1$;

$y + 2x = 4 \rightarrow y =$ ____ $x + 4$.

The slopes of these lines are ____ and -2. The product of the slopes $\frac{1}{2}\cdot(-2) =$ ____.

Thus the lines ____ (are/are not) perpendicular.

15. $3y = 2x + 15$,
$2y = 3x + 10$

Answers

14. Yes **15.** No

Guided Solution:

14. $\frac{1}{2}, -2, \frac{1}{2}, -1$, are

Visualizing for Success

A

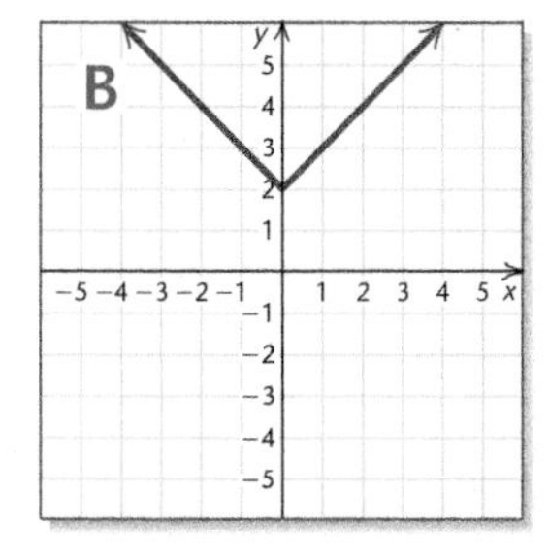

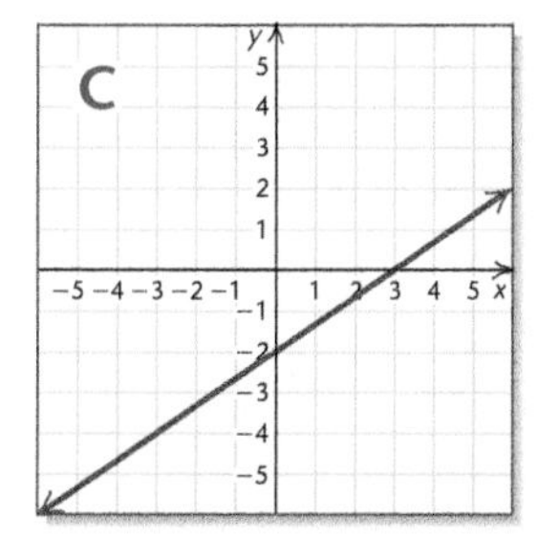

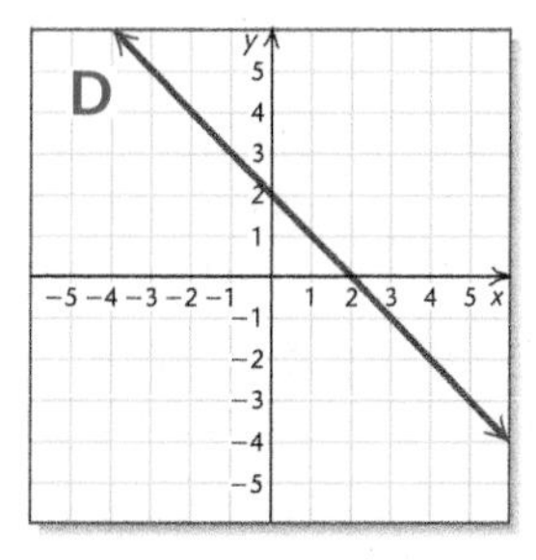

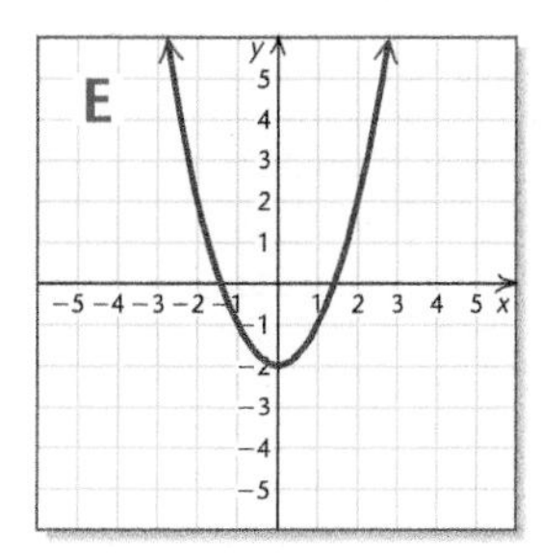

Match each equation with its graph.

1. $y = 2 - x$
2. $x - y = 2$
3. $x + 2y = 2$
4. $2x - 3y = 6$
5. $x = 2$
6. $y = 2$
7. $y = |x + 2|$
8. $y = |x| + 2$
9. $y = x^2 - 2$
10. $y = 2 - x^2$

Answers on page A-7

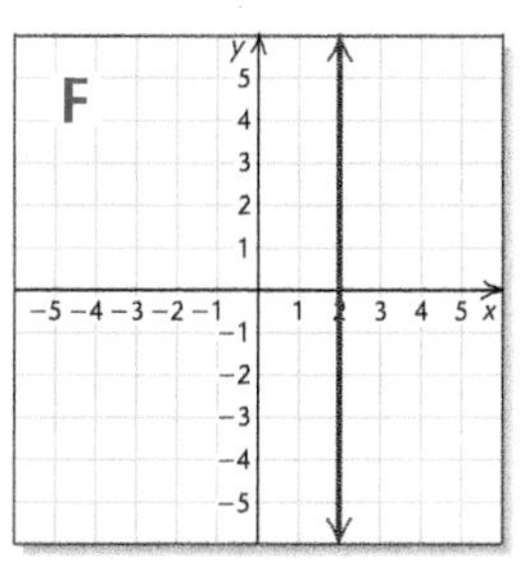

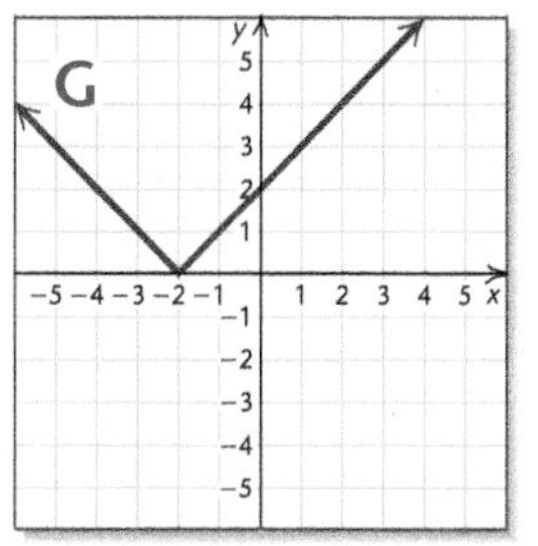

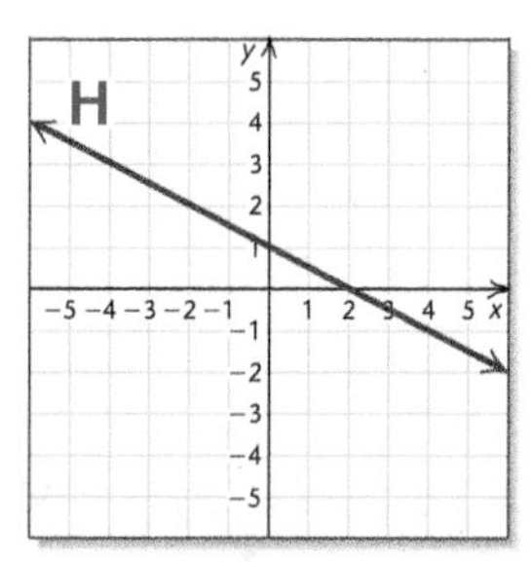

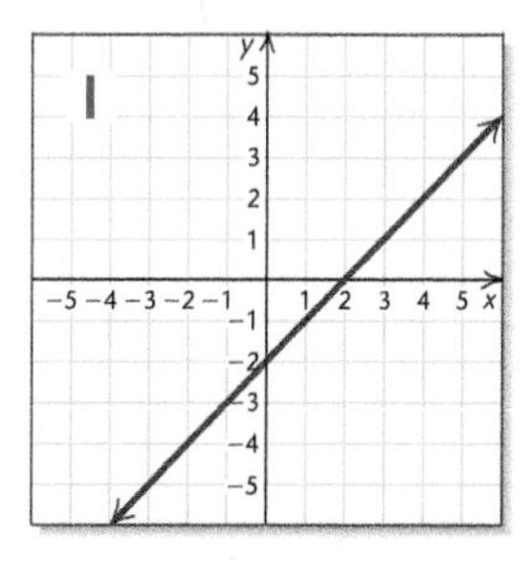

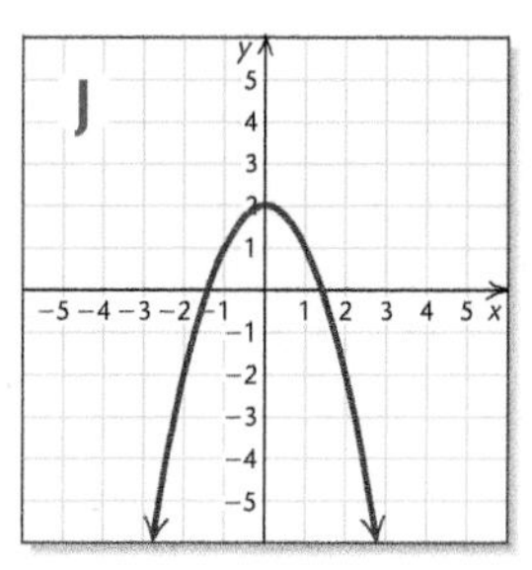

2.5 Exercise Set

FOR EXTRA HELP MyLab Math

✓ Check Your Understanding

Reading Check Determine whether each statement is true or false.

RC1. Two nonvertical lines are parallel if they have the same slope and the same y-intercepts.

RC2. The x-intercept of $x = a$, a vertical line, is $(a, 0)$.

RC3. The slope of a horizontal line is 0.

RC4. Two lines are perpendicular if the product of their slopes is 1.

Concept Check Choose from the columns on the right the ordered pair, equation, or word that completes the statement.

CC1. The x-intercept of $x = -\frac{2}{7}$ is ______________

CC2. The y-intercept of $y = -2x + 7$ is ______________.

CC3. The graphs of the lines $x = -4$ and $y = 5$ are ______________.

CC4. The graphs of the lines $y = -13$ and ______________ are parallel.

$(0, 7)$ | parallel
$(-2, 0)$ | perpendicular
$\left(0, -\frac{2}{7}\right)$ | $y = \frac{2}{7}$
$\left(-\frac{2}{7}, 0\right)$ | $x = -7$

a Find the intercepts and then graph the line.

1. $x - 2 = y$

2. $x + 3 = y$

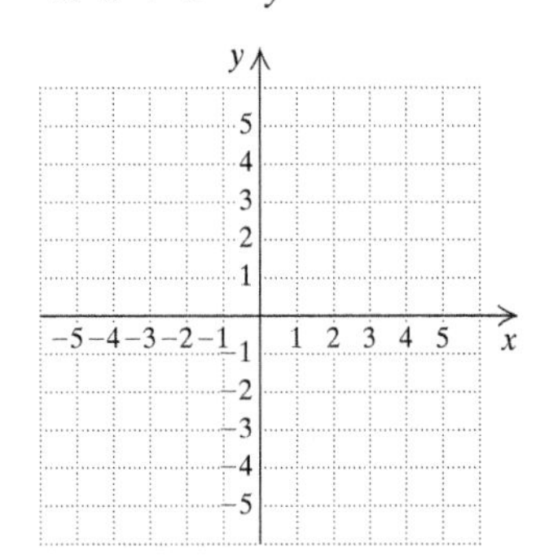

3. $x + 3y = 6$

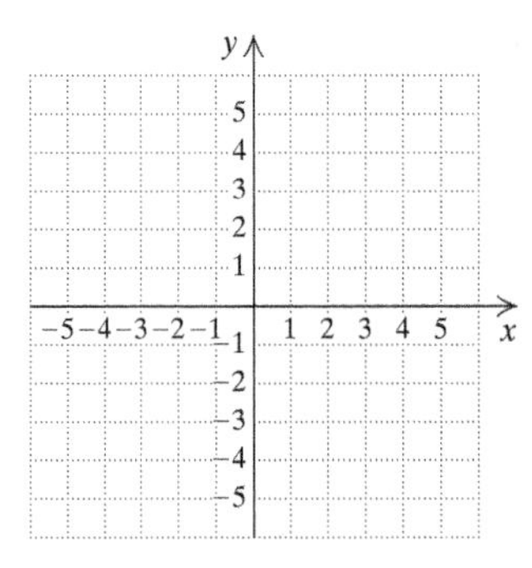

4. $x - 2y = 4$

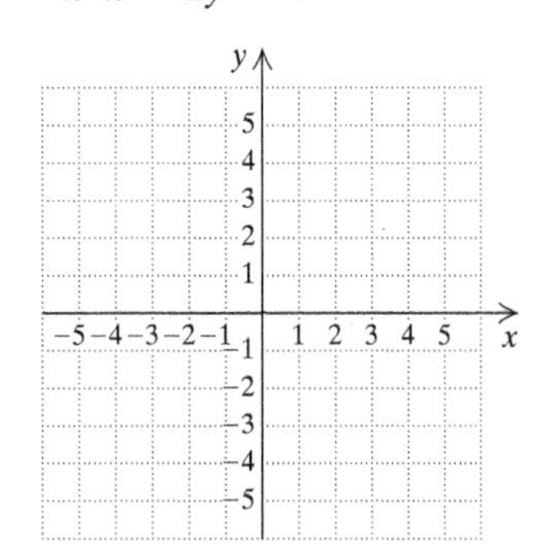

5. $2x + 3y = 6$

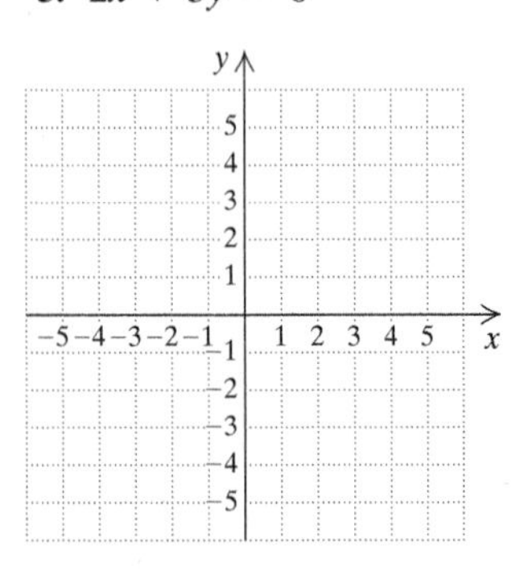

6. $5x - 2y = 10$

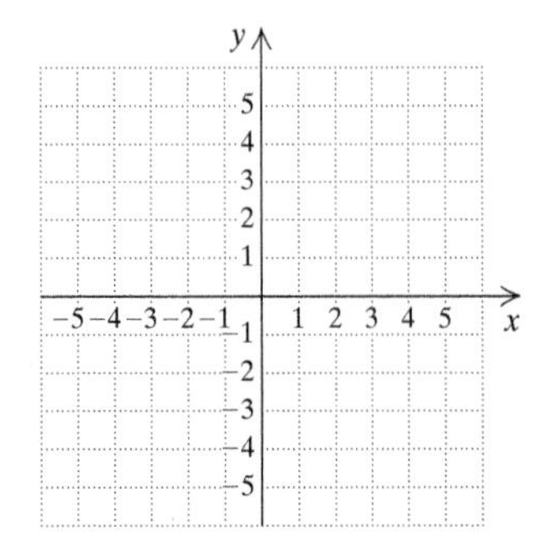

7. $f(x) = -2 - 2x$

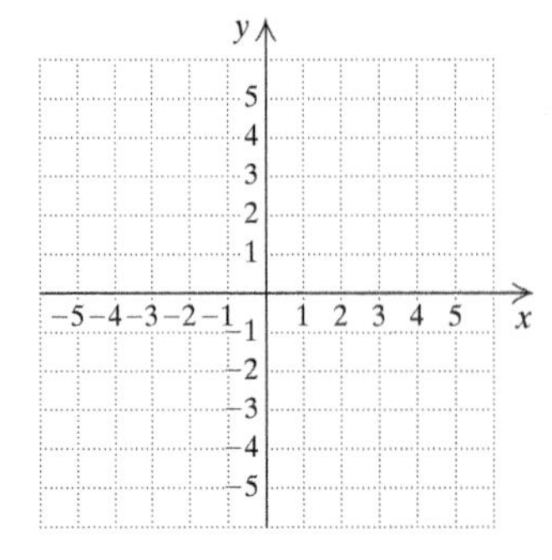

8. $g(x) = 5x - 5$

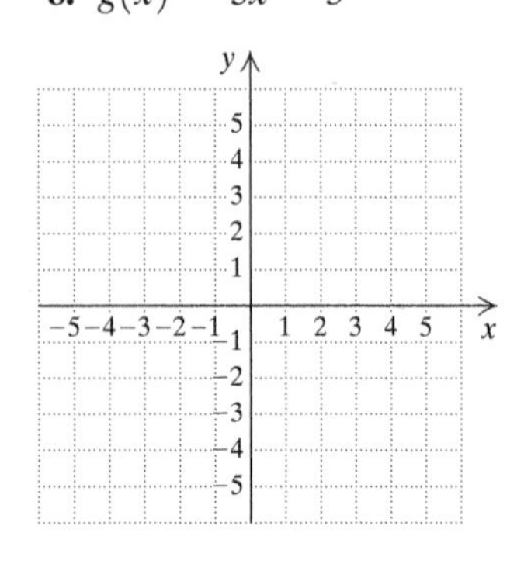

9. $5y = -15 + 3x$

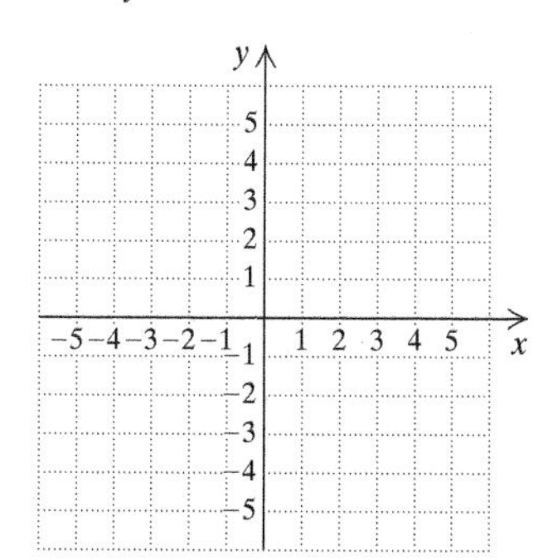

10. $5x - 10 = 5y$

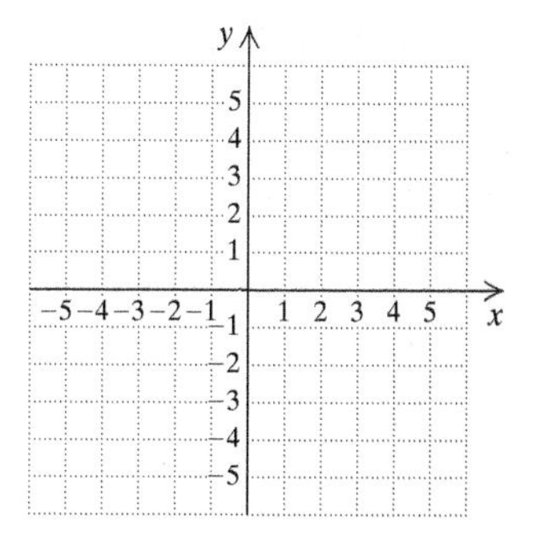

11. $2x - 3y = 6$

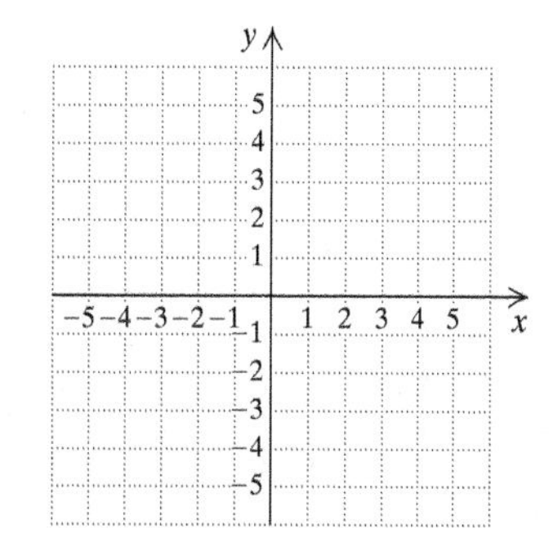

12. $4x + 5y = 20$

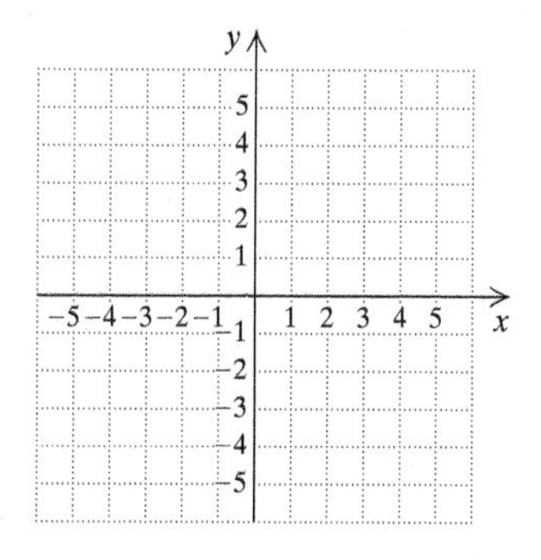

13. $2.8y - 3.5x = -9.8$

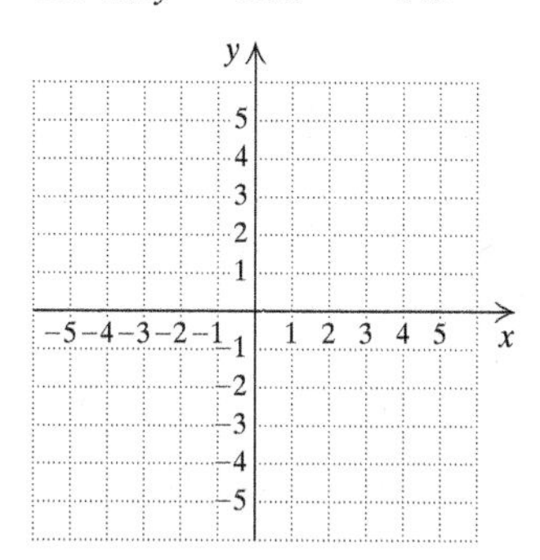

14. $10.8x - 22.68 = 4.2y$

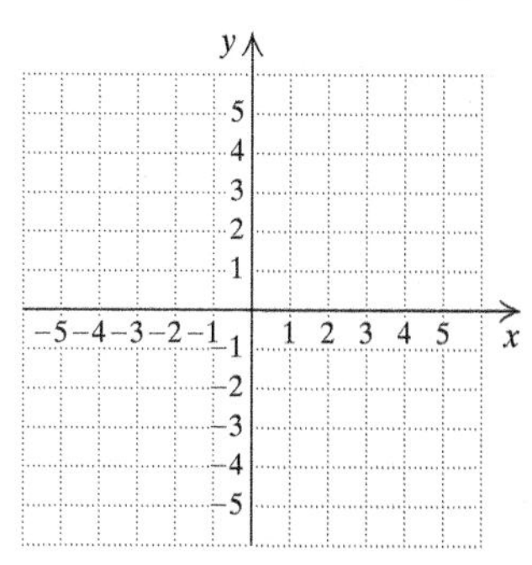

15. $5x + 2y = 7$

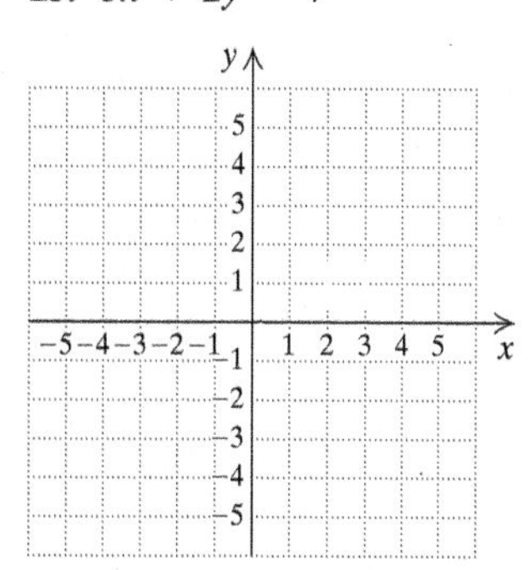

16. $3x - 4y = 10$

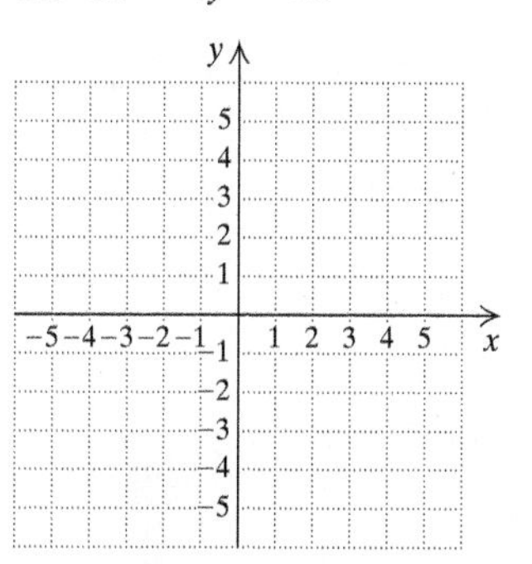

b Graph using the slope and the y-intercept.

17. $y = \frac{5}{2}x + 1$

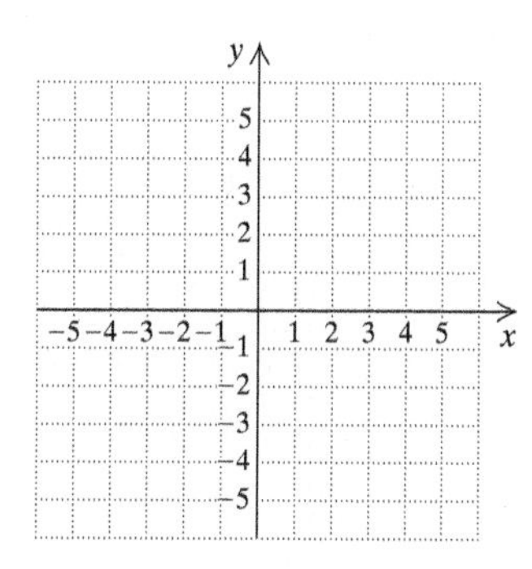

18. $y = \frac{2}{5}x - 4$

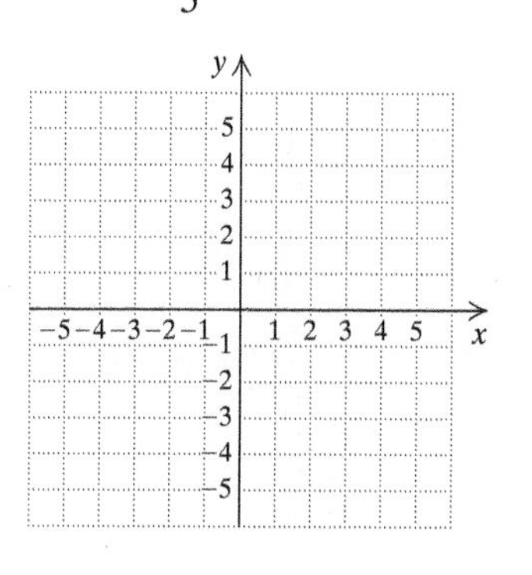

19. $f(x) = -\frac{5}{2}x - 4$

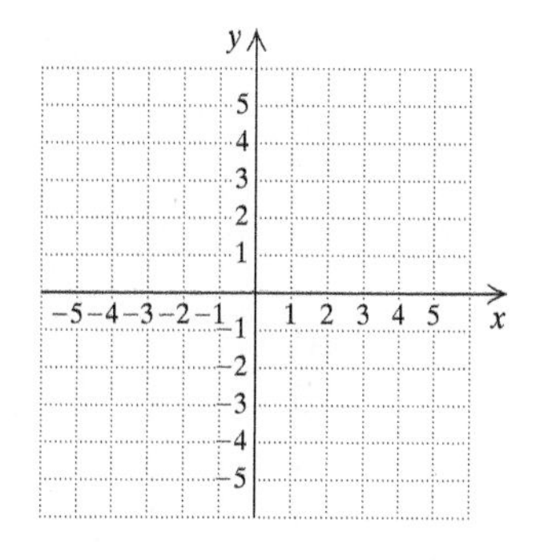

20. $f(x) = \frac{2}{5}x + 3$

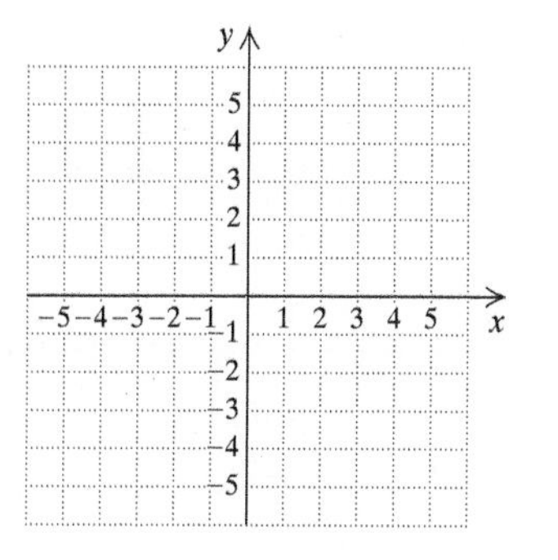

21. $x + 2y = 4$

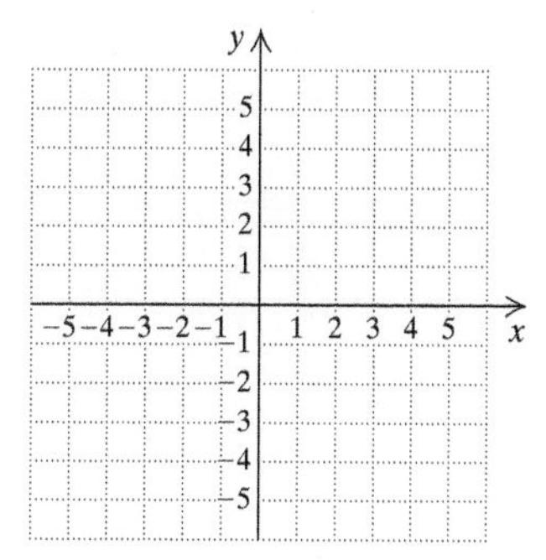

22. $x - 3y = 6$

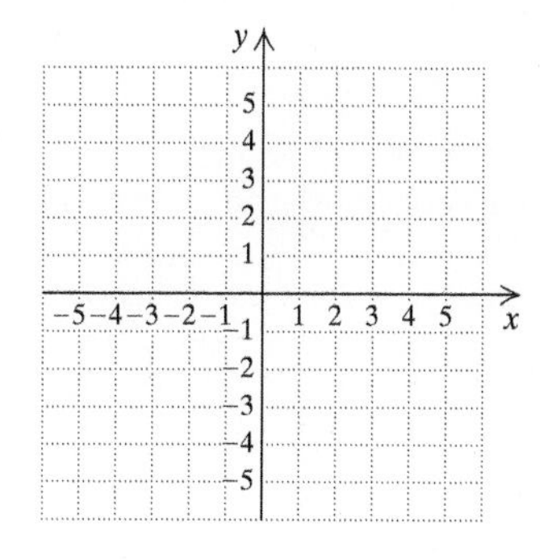

23. $4x - 3y = 12$

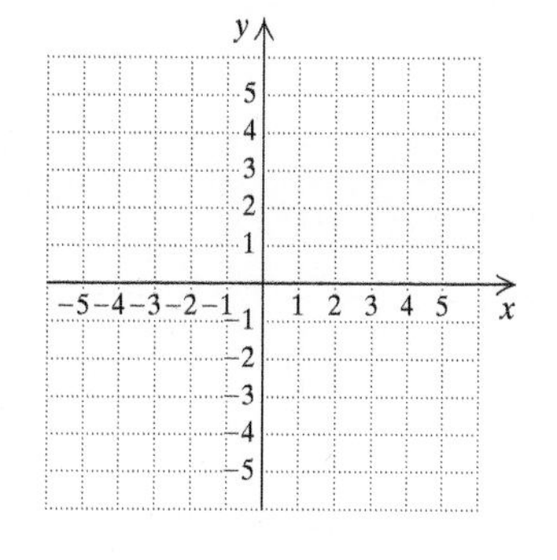

24. $2x + 6y = 12$

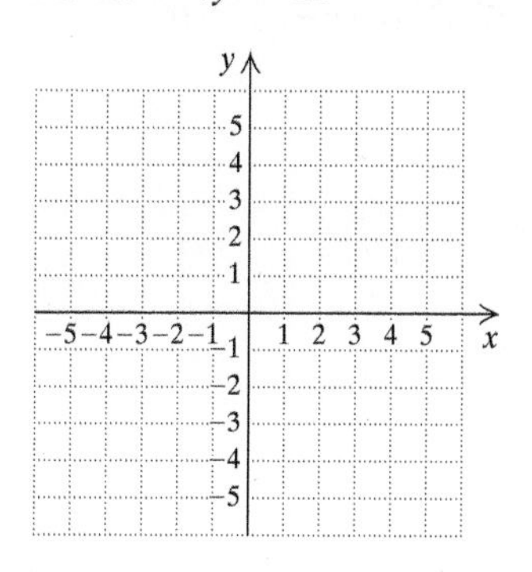

25. $f(x) = \frac{1}{3}x - 4$

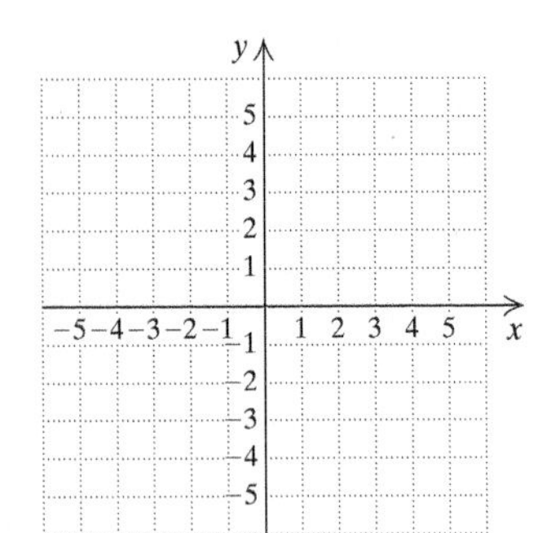

26. $g(x) = -0.25x + 2$

27. $5x + 4 \cdot f(x) = 4$
(*Hint*: Solve for $f(x)$.)

28. $3 \cdot f(x) = 4x + 6$

C Graph and, if possible, determine the slope.

29. $x = 1$

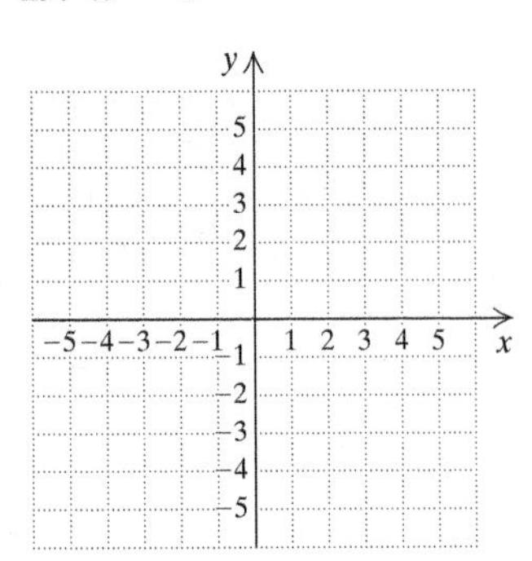

30. $x = -4$

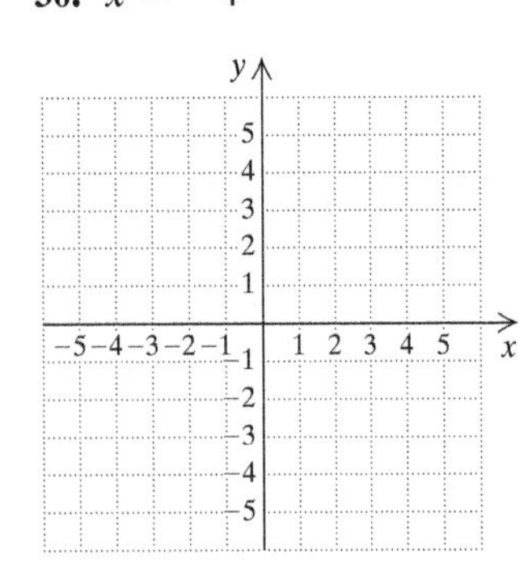

31. $y = -1$

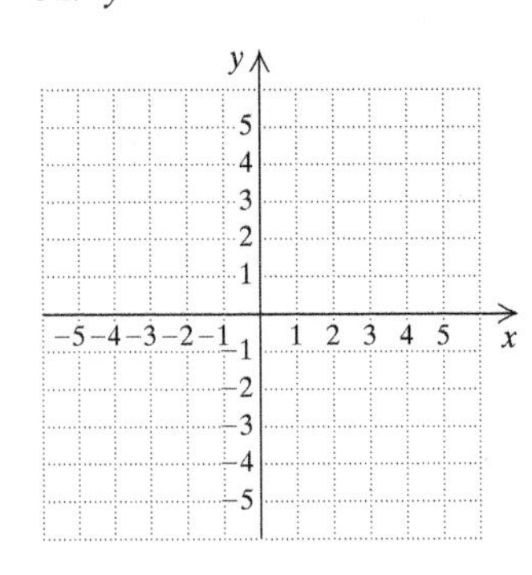

32. $y = \frac{3}{2}$

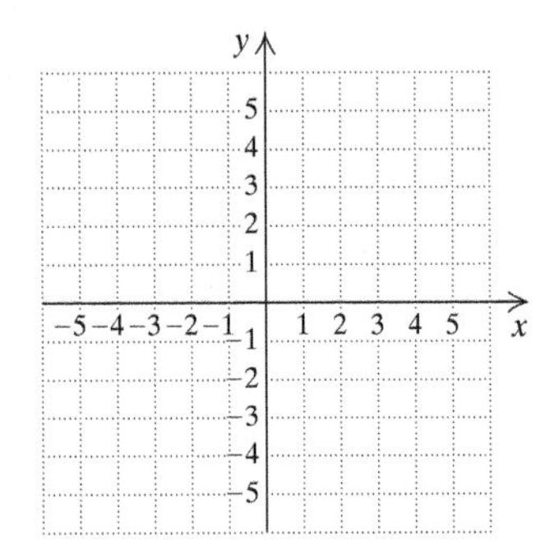

33. $f(x) = -6$

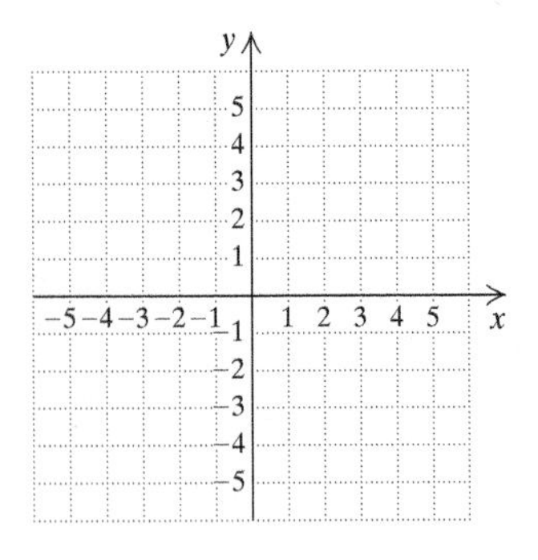

34. $f(x) = 2$

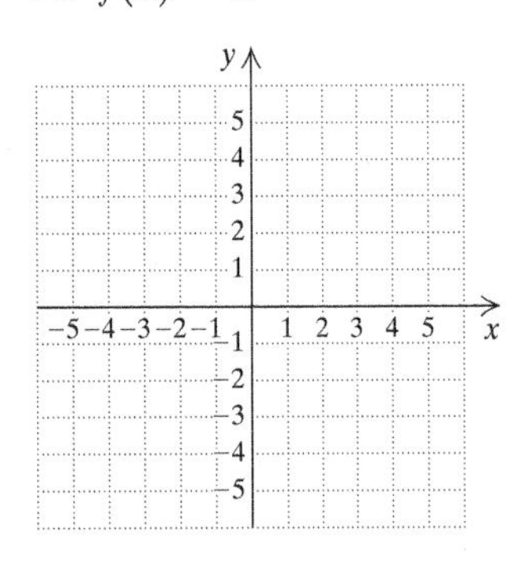

35. $y = 0$

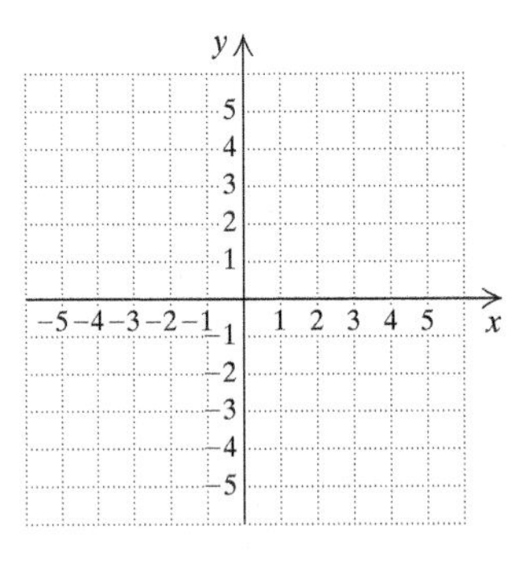

36. $x = 0$

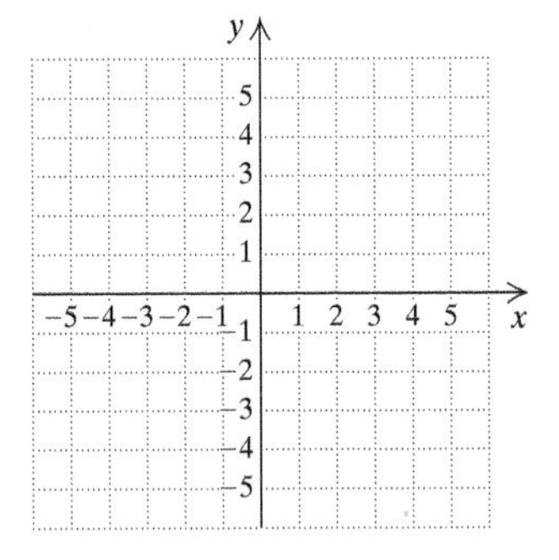

37. $2 \cdot f(x) + 5 = 0$

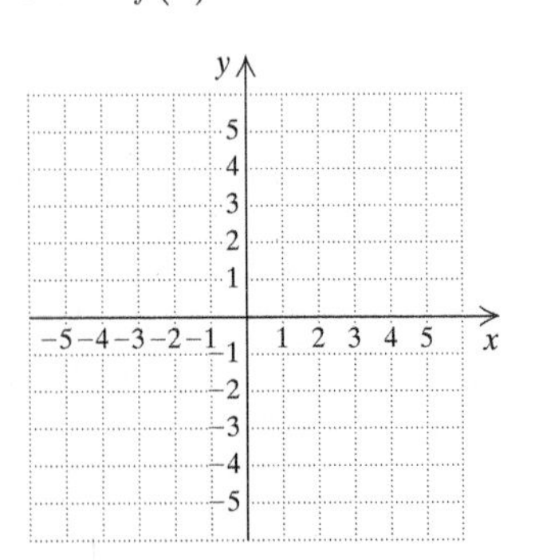

38. $4 \cdot g(x) + 3x = 12 + 3x$

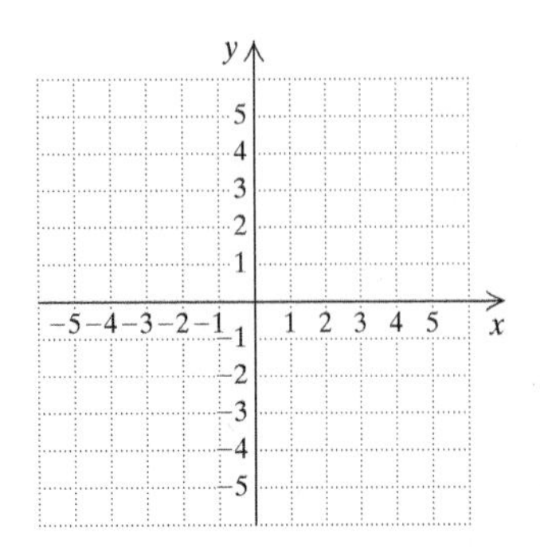

39. $7 - 3x = 4 + 2x$

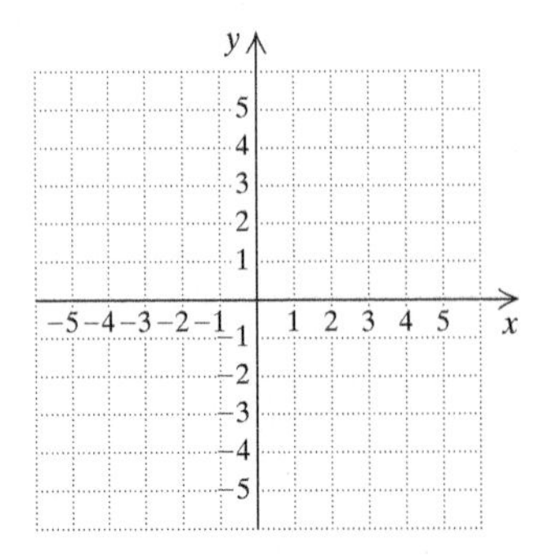

40. $3 - f(x) = 2$

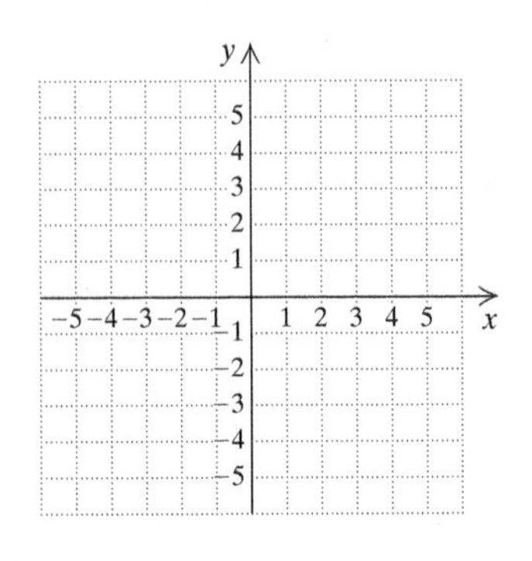

d Determine whether the graphs of the given pair of lines are parallel.

41. $x + 6 = y$,
$y - x = -2$

42. $2x - 7 = y$,
$y - 2x = 8$

43. $y + 3 = 5x$,
$3x - y = -2$

44. $y + 8 = -6x$,
$-2x + y = 5$

45. $y = 3x + 9$,
$2y = 6x - 2$

46. $y + 7x = -9$,
$-3y = 21x + 7$

47. $12x = 3$,
$-7x = 10$

48. $5y = -2$,
$\frac{3}{4}x = 16$

Determine whether the graphs of the given pair of lines are perpendicular.

49. $y = 4x - 5$,
$4y = 8 - x$

50. $2x - 5y = -3$,
$2x + 5y = 4$

51. $x + 2y = 5$,
$2x + 4y = 8$

52. $y = -x + 7$,
$y = x + 3$

53. $2x - 3y = 7$,
$2y - 3x = 10$

54. $x = y$,
$y = -x$

55. $2x = 3$,
$-3y = 6$

56. $-5y = 10$,
$y = -\frac{4}{9}$

Skill Maintenance

Write in scientific notation. [J20]

57. 53,000,000,000

58. 0.000047

59. 0.018

60. 99,902,000

Write in decimal notation. [J20]

61. 2.13×10^{-5}

62. 9.01×10^{8}

63. 2×10^{4}

64. 8.5677×10^{-2}

Factor. [J16]

65. $9x - 15y$

66. $12a + 21ab$

67. $21p - 7pq + 14p$

68. $64x - 128y + 256$

Synthesis

69. Find the value of a such that the graphs of $5y = ax + 5$ and $\frac{1}{4}y = \frac{1}{10}x - 1$ are parallel.

70. Find the value of k such that the graphs of $x + 7y = 70$ and $y + 3 = kx$ are perpendicular.

71. Write an equation of the line that has x-intercept $(-3, 0)$ and y-intercept $(0, \frac{2}{5})$.

72. Find the coordinates of the point of intersection of the graphs of the equations $x = -4$ and $y = 5$.

73. Write an equation for the x-axis. Is this equation a function?

74. Write an equation for the y-axis. Is this equation a function?

75. Find the value of m in $y = mx + 3$ so that the x-intercept of its graph will be $(4, 0)$.

76. Find the value of b in $2y = -7x + 3b$ so that the y-intercept of its graph will be $(0, -13)$.

2.6 Finding Equations of Lines; Applications

OBJECTIVES

a Find an equation of a line when the slope and the y-intercept are given.

b Find an equation of a line when the slope and a point are given.

c Find an equation of a line when two points are given.

d Given a line and a point not on the given line, find an equation of the line parallel to the line and containing the point, and find an equation of the line perpendicular to the line and containing the point.

e Solve applied problems involving linear functions.

In this section, we will learn to find an equation of a line for which we have been given two pieces of information.

a FINDING AN EQUATION OF A LINE WHEN THE SLOPE AND THE y-INTERCEPT ARE GIVEN

If we know the slope and the y-intercept of a line, we can find an equation of the line using the slope–intercept equation $y = mx + b$.

EXAMPLE 1 A line has slope -0.7 and y-intercept $(0, 13)$. Find an equation of the line.

We use the slope–intercept equation and substitute -0.7 for m and 13 for b:

$$y = mx + b$$

$$y = -0.7x + 13.$$

◀ Do Exercise 1.

1. A line has slope 3.4 and y-intercept $(0, -8)$. Find an equation of the line.

b FINDING AN EQUATION OF A LINE WHEN THE SLOPE AND A POINT ARE GIVEN

Suppose that we know the slope of a line and the coordinates of one point on the line. We can use the slope–intercept equation to find an equation of the line. Or, we can use the **point–slope equation**. We first develop a formula for such a line.

Suppose that a line of slope m passes through the point (x_1, y_1). For any other point (x, y) on this line, we must have

$$\frac{y - y_1}{x - x_1} = m.$$

MyLab Math ANIMATION

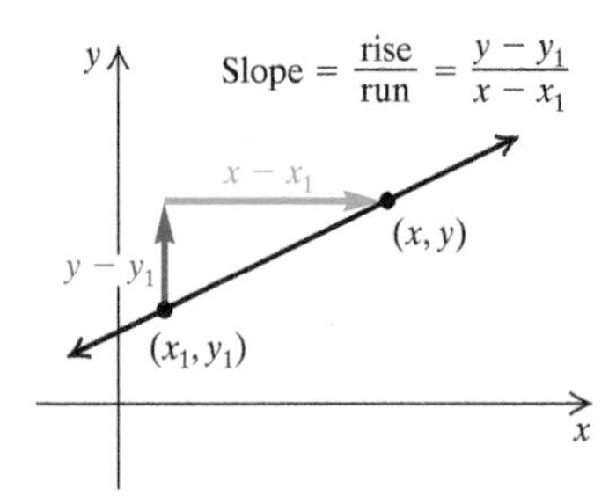

It is tempting to use this last equation as an equation of the line of slope m that passes through (x_1, y_1). The only problem with this form is that when x and y are replaced with x_1 and y_1, we have $\frac{0}{0} = m$, a false equation. To avoid this difficulty, we multiply by $x - x_1$ on both sides and simplify:

$$\frac{y - y_1}{x - x_1}(x - x_1) = m(x - x_1) \qquad \text{Multiplying by } x - x_1 \text{ on both sides}$$

$$y - y_1 = m(x - x_1). \qquad \text{Removing a factor of 1: } \frac{x - x_1}{x - x_1} = 1$$

This is the *point–slope* form of a linear equation.

Answer

1. $y = 3.4x - 8$

POINT–SLOPE EQUATION

The **point–slope equation** of a line with slope m, passing through (x_1, y_1), is

$$y - y_1 = m(x - x_1).$$

If we know the slope of a line and a point on the line, we can find an equation of the line using either the point–slope equation,

$$y - y_1 = m(x - x_1),$$

or the slope–intercept equation,

$$y = mx + b.$$

EXAMPLE 2 Find an equation of the line with slope 5 and containing the point $\left(\frac{1}{2}, -1\right)$.

Using the Point–Slope Equation: We consider $\left(\frac{1}{2}, -1\right)$ to be (x_1, y_1). We substitute $\frac{1}{2}$ for x_1, -1 for y_1, and 5 for m:

$$\begin{aligned} y - y_1 &= m(x - x_1) && \text{Point–slope equation} \\ y - (-1) &= 5\left(x - \tfrac{1}{2}\right) && \text{Substituting} \\ y + 1 &= 5x - \tfrac{5}{2} && \text{Simplifying} \\ y &= 5x - \tfrac{5}{2} - 1 \\ y &= 5x - \tfrac{5}{2} - \tfrac{2}{2} \\ y &= 5x - \tfrac{7}{2}. \end{aligned}$$

Using the Slope–Intercept Equation: The point $\left(\frac{1}{2}, -1\right)$ is on the line, so it is a solution of the equation. Thus we can substitute $\frac{1}{2}$ for x and -1 for y in $y = mx + b$. We also substitute 5 for m, the slope. Then we solve for b:

$$\begin{aligned} y &= mx + b && \text{Slope–intercept equation} \\ -1 &= 5 \cdot \left(\tfrac{1}{2}\right) + b && \text{Substituting} \\ -1 &= \tfrac{5}{2} + b \\ -1 - \tfrac{5}{2} &= b \\ -\tfrac{2}{2} - \tfrac{5}{2} &= b \\ -\tfrac{7}{2} &= b. && \text{Solving for } b \end{aligned}$$

We then use the slope–intercept equation $y = mx + b$ again and substitute 5 for m and $-\frac{7}{2}$ for b:

$$y = 5x - \tfrac{7}{2}.$$

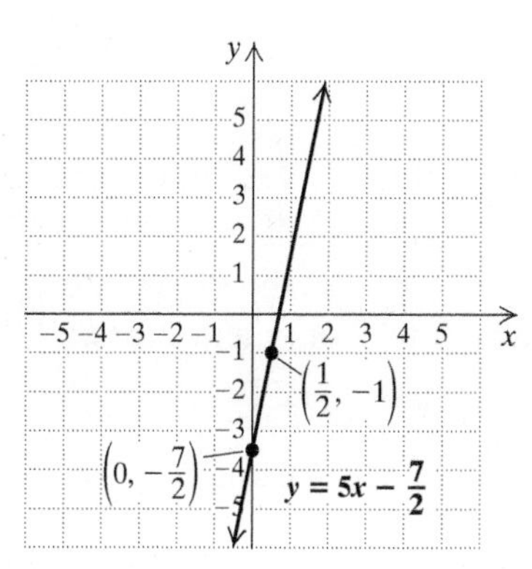

Do Exercises 2–5.

Find an equation of the line with the given slope and containing the given point.

2. $m = -5$, $(-4, 2)$

3. $m = 3$, $(1, -2)$

4. $m = 8$, $(3, 5)$

5. $m = -\frac{2}{3}$, $(1, 4)$

Answers

2. $y = -5x - 18$ **3.** $y = 3x - 5$ **4.** $y = 8x - 19$ **5.** $y = -\frac{2}{3}x + \frac{14}{3}$

C FINDING AN EQUATION OF A LINE WHEN TWO POINTS ARE GIVEN

We can also use the slope–intercept equation or the point–slope equation to find an equation of a line when two points are given.

EXAMPLE 3 Find an equation of the line containing the points $(2, 3)$ and $(-6, 1)$.

First, we find the slope:

$$m = \frac{3 - 1}{2 - (-6)} = \frac{2}{8}, \text{ or } \frac{1}{4}.$$

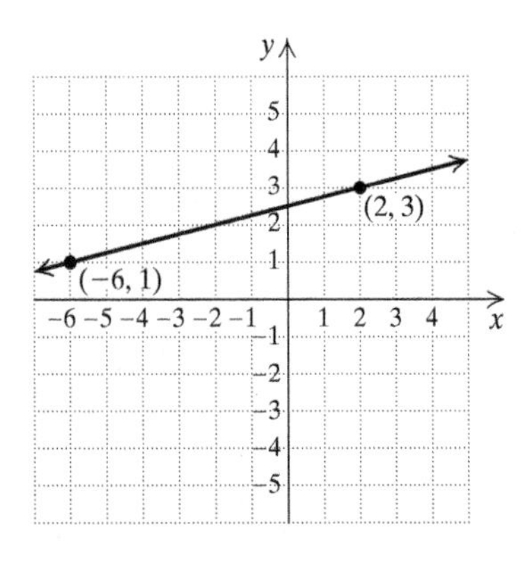

Now we have the slope and two points. We then proceed as we did in Example 2, using either point, and either the point–slope equation or the slope–intercept equation.

Using the Point–Slope Equation: We choose $(2, 3)$ and substitute 2 for x_1, 3 for y_1, and $\frac{1}{4}$ for m:

$$\begin{aligned} y - y_1 &= m(x - x_1) && \text{Point–slope equation} \\ y - 3 &= \tfrac{1}{4}(x - 2) && \text{Substituting} \\ y - 3 &= \tfrac{1}{4}x - \tfrac{1}{2} \\ y &= \tfrac{1}{4}x - \tfrac{1}{2} + 3 \\ y &= \tfrac{1}{4}x - \tfrac{1}{2} + \tfrac{6}{2} \\ y &= \tfrac{1}{4}x + \tfrac{5}{2}. \end{aligned}$$

Using the Slope–Intercept Equation: We choose $(2, 3)$ and substitute 2 for x, 3 for y, and $\frac{1}{4}$ for m:

$$\begin{aligned} y &= mx + b && \text{Slope–intercept equation} \\ 3 &= \tfrac{1}{4} \cdot 2 + b && \text{Substituting} \\ 3 &= \tfrac{1}{2} + b \\ 3 - \tfrac{1}{2} &= \tfrac{1}{2} + b - \tfrac{1}{2} \\ \tfrac{6}{2} - \tfrac{1}{2} &= b \\ \tfrac{5}{2} &= b. && \text{Solving for } b \end{aligned}$$

Finally, we use the slope–intercept equation $y = mx + b$ again and substitute $\frac{1}{4}$ for m and $\frac{5}{2}$ for b:

$$y = \tfrac{1}{4}x + \tfrac{5}{2}.$$

Do Exercises 6 and 7.

6. Find an equation of the line containing the points $(4, -3)$ and $(1, 2)$. **GS**

First, find the slope:

$$m = \frac{\square - (-3)}{1 - 4} = \frac{\square}{-3} = -\frac{5}{\square}.$$

Using the point–slope equation,

$$y - y_1 = m(x - x_1),$$

substitute 4 for x_1, -3 for y_1, and $-\frac{5}{3}$ for m:

$$\begin{aligned} y - (\square) &= \square\,(x - \square) \\ y + \square &= -\frac{5}{3}x + \square \\ y &= -\frac{5}{3}x + \frac{20}{3} - 3 \\ y &= -\frac{5}{3}x + \frac{20}{3} - \frac{\square}{3} \\ y &= -\frac{5}{3}x + \frac{\square}{3}. \end{aligned}$$

7. Find an equation of the line containing the points $(-3, -5)$ and $(-4, 12)$.

Answers

6. $y = -\frac{5}{3}x + \frac{11}{3}$ **7.** $y = -17x - 56$

Guided Solution:

6. $2, 5, 3; -3, -\frac{5}{3}, 4, 3, \frac{20}{3}, 9, 11$

d FINDING AN EQUATION OF A LINE PARALLEL OR PERPENDICULAR TO A GIVEN LINE THROUGH A POINT NOT ON THE LINE

We can also use the methods of Example 2 to find an equation of a line parallel or perpendicular to a given line and containing a point not on the line.

EXAMPLE 4 Find an equation of the line containing the point $(-1, 3)$ and parallel to the line $2x + y = 10$.

A line parallel to the given line $2x + y = 10$ must have the same slope as the given line. To find that slope, we first find the slope–intercept equation by solving for y:

$$2x + y = 10$$
$$y = -2x + 10.$$

Thus the line we want to find through $(-1, 3)$ must also have slope -2.

Using the Point–Slope Equation: We use the point $(-1, 3)$ and the slope -2, substituting -1 for x_1, 3 for y_1, and -2 for m:

$$y - y_1 = m(x - x_1)$$
$$y - 3 = -2(x - (-1)) \quad \text{Substituting}$$
$$y - 3 = -2(x + 1) \quad \text{Simplifying}$$
$$y - 3 = -2x - 2$$
$$y = -2x + 1.$$

Using the Slope–Intercept Equation: We substitute -1 for x, 3 for y, and -2 for m in $y = mx + b$. Then we solve for b:

$$y = mx + b$$
$$3 = -2(-1) + b \quad \text{Substituting}$$
$$3 = 2 + b$$
$$1 = b. \quad \text{Solving for } b$$

We then use the equation $y = mx + b$ again and substitute -2 for m and 1 for b:

$$y = -2x + 1.$$

The given line $2x + y = 10$, or $y = -2x + 10$, and the line $y = -2x + 1$ have the same slope but different y-intercepts. Thus their graphs are parallel.

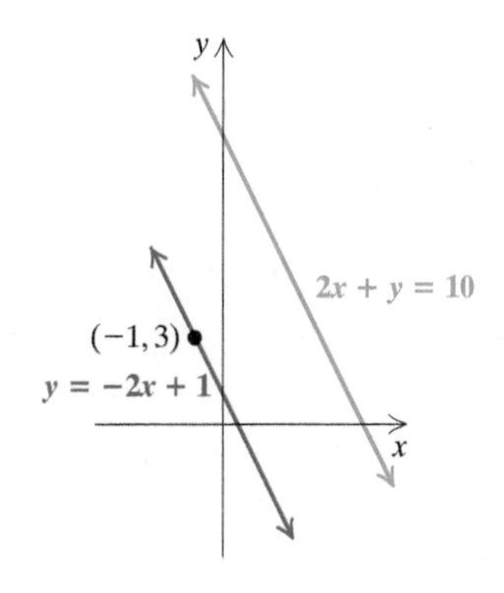

8. Find an equation of the line containing the point $(2, -1)$ and parallel to the line $9x - 3y = 5$.

Find the slope of the given line:

$$9x - 3y = 5$$
$$-3y = \_\_\_\_ + 5$$
$$y = \_\_\_\_\, x - \frac{5}{3}.$$

The slope is ____.

The line parallel to $9x - 3y = 5$ must have slope ____.

Using the slope–intercept equation,

$$y = mx + b,$$

substitute 3 for m, 2 for x, and -1 for y, and solve for b:

$$\_\_\_\_ = 3 \cdot \_\_\_\_ + b$$
$$-1 = 6 + b$$
$$\_\_\_\_ = b.$$

Substitute 3 for m and -7 for b in $y = mx + b$:

$$y = 3x + (\_\_\_\_)$$
$$y = 3x - 7.$$

Do Exercise 8.

Answer

8. $y = 3x - 7$

Guided Solution:

8. $-9x, 3, 3, 3; -1, 2, -7, -7$

EXAMPLE 5 Find an equation of the line containing the point $(2, -3)$ and perpendicular to the line $4y - x = 20$.

To find the slope of the given line, we first find its slope–intercept form by solving for y:

$$4y - x = 20$$
$$4y = x + 20$$
$$\frac{4y}{4} = \frac{x + 20}{4} \quad \text{Dividing by 4}$$
$$y = \tfrac{1}{4}x + 5.$$

We know that the slope of the perpendicular line must be the opposite of the reciprocal of $\frac{1}{4}$. Thus the new line through $(2, -3)$ must have slope -4.

Using the Point–Slope Equation: We use the point $(2, -3)$ and the slope -4, substituting 2 for x_1, -3 for y_1, and -4 for m:

$$y - y_1 = m(x - x_1)$$
$$y - (-3) = -4(x - 2) \quad \text{Substituting}$$
$$y + 3 = -4x + 8$$
$$y = -4x + 5.$$

Using the Slope–Intercept Equation: We now substitute 2 for x and -3 for y in $y = mx + b$. We also substitute -4 for m, the slope. Then we solve for b:

$$y = mx + b$$
$$-3 = -4(2) + b \quad \text{Substituting}$$
$$-3 = -8 + b$$
$$5 = b. \quad \text{Solving for } b$$

Finally, we use the equation $y = mx + b$ again and substitute -4 for m and 5 for b:

$$y = -4x + 5.$$

The product of the slopes of the lines $4y - x = 20$ and $y = -4x + 5$ is $\frac{1}{4} \cdot (-4) = -1$. Thus their graphs are perpendicular.

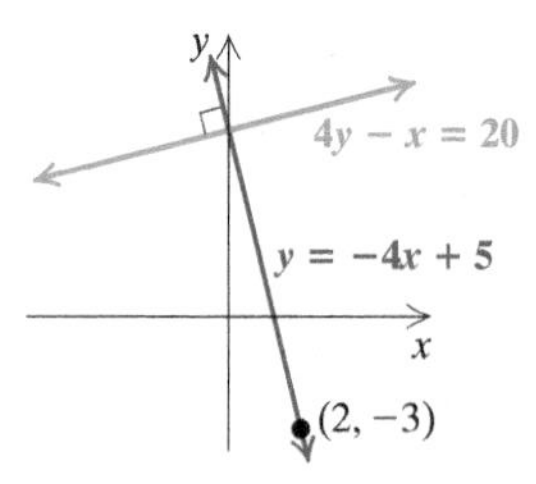

◀ **Do Exercise 9.**

9. Find an equation of the line containing the point $(5, 4)$ and perpendicular to the line $2x - 4y = 9$. GS

Find the slope of the given line:

$2x - 4y = 9$

$-4y = \underline{\quad} + 9$

$y = \underline{\quad}\, x - \frac{9}{4}.$

The slope is $\underline{\quad}$.

The slope of a line perpendicular to $2x - 4y = 9$ is the opposite of the reciprocal of $\frac{1}{2}$, or $\underline{\quad}$.

Using the point–slope equation,

$y - y_1 = m(x - x_1)$,

substitute -2 for m, 5 for x_1, and 4 for y_1:

$y - \underline{\quad} = \underline{\quad}(x - 5)$

$y - 4 = -2x + \underline{\quad}$

$y = -2x + 14.$

Answer

9. $y = -2x + 14$

Guided Solution:

9. $-2x, \frac{1}{2}, \frac{1}{2}, -2; 4, -2, 10$

e APPLICATIONS OF LINEAR FUNCTIONS

When the essential parts of a problem are described in mathematical language, we say that we have a **mathematical model**. We have already studied many kinds of mathematical models in this text—for example, the formulas in Section 1.2 and the functions in Section 2.2. Here we study linear functions as models.

EXAMPLE 6 *Cost of a Necklace.* Amelia's Beads offers a class in designing necklaces. For a necklace made of 6-mm beads, 4.23 beads per inch are needed. The cost of a necklace of 6-mm gemstone beads that sell for 40¢ each is \$7 for the clasp and the crimps and approximately \$1.70 per inch.

a) Formulate a linear function that models the total cost of a necklace $C(n)$, where n is the length of the necklace, in inches.

b) Graph the model.

c) Use the model to determine the cost of a 30-in. necklace.

a) The problem describes a situation in which cost per inch is charged in addition to the fixed cost of the clasp and the crimps. The total cost of a 16-in. necklace is

$$\$7 + \$1.70 \cdot 16 = \$34.20.$$

For a 17-in. necklace, the total cost is

$$\$7 + \$1.70 \cdot 17 = \$35.90.$$

These calculations lead us to generalize that for a necklace that is n inches long, the total cost is given by $C(n) = 7 + 1.7n$, where $n \geq 0$ since the length of the necklace cannot be negative. (Actually most necklaces are at least 14 in. long.) The notation $C(n)$ indicates that the cost C is a function of the length n.

b) Before we draw the graph, we rewrite the model in slope–intercept form:

$$C(n) = 1.7n + 7.$$

The y-intercept is $(0, 7)$ and the slope, or rate of change, is \$1.70, or $\frac{\$17}{10}$ $\left(\frac{1.7}{1} = \frac{17}{10}\right)$, per inch. We first plot $(0, 7)$; from that point, we move 17 units up and 10 units to the right to the point $(10, 24)$. We then draw a line through these points. We also calculate a third value as a check:

$$C(20) = 1.7 \cdot 20 + 7 = 41.$$

The point $(20, 41)$ lines up with the other two points so the graph is correct.

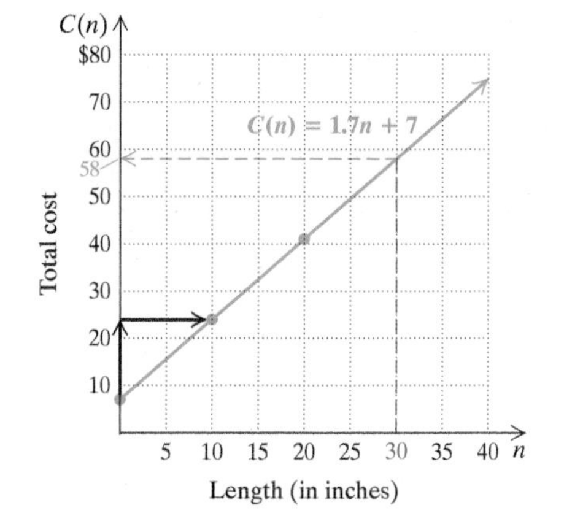

10. *Cost of a Service Call.* For a service call, Belmont Heating and Air Conditioning charges a \$65 trip fee and \$80 per hour for labor.

a) Formulate a linear function for the total cost of the service call $C(t)$, where t is the length of the call, in hours.

b) Graph the model.

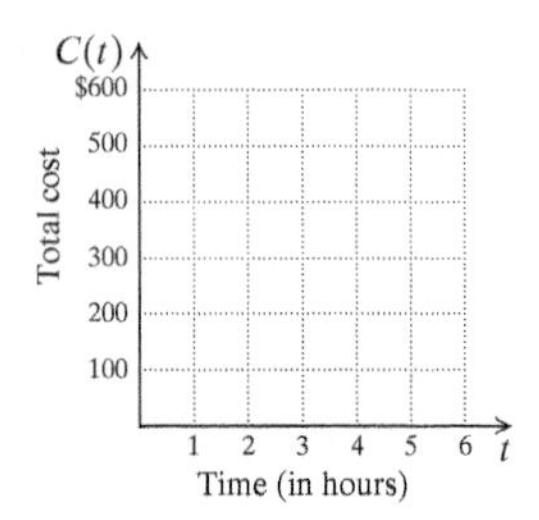

c) Use the model to determine the cost of a $2\frac{1}{2}$-hr service call.

c) To determine the total cost of a 30-in. necklace, we find $C(30)$:

$$C(30) = 1.7 \cdot 30 + 7 = 58.$$

From the graph, we see that the input 30 corresponds to the output 58. Thus we see that a 30-in. necklace costs \$58.

◀ Do Exercise 10.

In the following example, we use two points and find an equation for the linear function through these points. Then we use the equation to estimate.

EXAMPLE 7 *Decreasing Number of Banks.* The number of banks in the United States has decreased steadily in recent years. The following table lists data regarding the correspondence between the year and the number of banks.

YEAR, x (in number of years after 2008)	Number of Banks, N
2008, 0	8441
2016, 8	6068

Data: Federal Deposit Insurance Company

a) Assuming a constant rate of change, use the two data points to find a linear function that fits the data.

b) Use the function to determine the number of banks in 2013.

c) In what year will the number of banks reach 5200?

a) We let x = the number of years after 2008 and N = the number of banks. The table gives us two ordered pairs, $(0, 8441)$ and $(8, 6068)$. We use them to find a linear function that fits the data. First, we find the slope:

$$m = \frac{6068 - 8441}{8 - 0} = \frac{-2373}{8} = -296.625.$$

Next, we find an equation $N = mx + b$ that fits the data. One of the data points, $(0, 8441)$, is the y-intercept. Thus we know the value of b in the slope–intercept equation, $y = mx + b$. We use the equation $N = mx + b$ and substitute -296.625 for m and 8441 for b:

$$N = -296.625x + 8441.$$

Using function notation, we have

$$N(x) = -296.625x + 8441.$$

Answer

10. (a) $C(t) = 80t + 65$;
(b) C(t) $600 500 400 300 200 100 Total cost; 1 2 3 4 5 6 t Time (in hours)
(c) \$265

b) To determine the number of banks in 2013, we substitute 5 for x (2013 is 5 years after 2008) in the function $N(x) = -296.625x + 8441$:

$$\begin{aligned} N(x) &= -296.625x + 8441 \\ N(5) &= -296.625(5) + 8441 && \text{Substituting} \\ &= -1483.125 + 8441 \\ &= 6957.875 \approx 6958. \end{aligned}$$

There were about 6958 banks in 2013.

c) To find the year in which the number of banks will reach 5200, we substitute 5200 for $N(x)$ and solve for x:

$$\begin{aligned} N(x) &= -296.625x + 8441 \\ 5200 &= -296.625x + 8441 && \text{Substituting} \\ -3241 &= -296.625x && \text{Subtracting 8441} \\ 11 &\approx x. && \text{Dividing by } -296.625 \end{aligned}$$

The number of banks will reach 5200 about 11 years after 2008, or in 2019.

Do Exercise 11. ▶

11. ***Hat Size as a Function of Head Circumference.*** The following table lists data relating hat size to head circumference.

HEAD CIRCUMFERENCE, C (in inches)	HAT SIZE, H
21.2	$6\frac{3}{4}$
22	7

Data: Shushan's New Orleans

a) Assuming a constant rate of change, use the two data points to find a linear function that fits the data.

b) Use the function to determine the hat size of a person whose head has a circumference of 24.8 in.

c) Jerome's hat size is 8. What is the circumference of his head?

Answer

11. **(a)** $H(C) = \frac{5}{16}C + \frac{1}{8}$, or $H(C) = 0.3125C + 0.125$; **(b)** $7\frac{7}{8}$, or 7.875; **(c)** 25.2 in.

2.6 Exercise Set

FOR EXTRA HELP  MyLab Math

✓ Check Your Understanding

Reading Check Complete each statement with the correct equation.

RC1. The slope–intercept equation of a line with slope m and passing through (x_1, y_1) is ______________.

RC2. The point–slope equation of a line with slope m and passing through (x_1, y_1) is ______________.

Concept Check For the given equation, determine the slope of the line **(a)** parallel to the given line and **(b)** perpendicular to the given line.

CC1. $y = \frac{4}{11}x - 2$

CC2. $y = -5$

CC3. $2x - y = -4$

CC4. $y - \frac{4}{3} = -\frac{5}{6}x$

CC5. $x = 3$

CC6. $10x + 5y = 14$

a Find an equation of the line having the given slope and y-intercept.

1. Slope: -8; y-intercept: $(0, 4)$
2. Slope: 5; y-intercept: $(0, -3)$
3. Slope: 2.3; y-intercept: $(0, -1)$
4. Slope: -9.1; y-intercept: $(0, 2)$

Find a linear function $f(x) = mx + b$ whose graph has the given slope and y-intercept.

5. Slope: $-\frac{7}{3}$; y-intercept: $(0, -5)$
6. Slope: $\frac{4}{5}$; y-intercept: $(0, 28)$
7. Slope: $\frac{2}{3}$; y-intercept: $\left(0, \frac{5}{8}\right)$
8. Slope: $-\frac{7}{8}$; y-intercept: $\left(0, -\frac{7}{11}\right)$

b Find an equation of the line having the given slope and containing the given point.

9. $m = 5,\ (4, 3)$
10. $m = 4,\ (5, 2)$
11. $m = -3,\ (9, 6)$
12. $m = -2,\ (2, 8)$
13. $m = 1,\ (-1, -7)$
14. $m = 3,\ (-2, -2)$
15. $m = -2,\ (8, 0)$
16. $m = -3,\ (-2, 0)$
17. $m = 0,\ (0, -7)$
18. $m = 0,\ (0, 4)$
19. $m = \frac{2}{3},\ (1, -2)$
20. $m = -\frac{4}{5},\ (2, 3)$

c Find an equation of the line containing the given pair of points.

21. $(1, 4)$ and $(5, 6)$
22. $(2, 5)$ and $(4, 7)$
23. $(-3, -3)$ and $(2, 2)$
24. $(-1, -1)$ and $(9, 9)$
25. $(-4, 0)$ and $(0, 7)$
26. $(0, -5)$ and $(3, 0)$
27. $(-2, -3)$ and $(-4, -6)$
28. $(-4, -7)$ and $(-2, -1)$
29. $(0, 0)$ and $(6, 1)$
30. $(0, 0)$ and $(-4, 7)$
31. $\left(\frac{1}{4}, -\frac{1}{2}\right)$ and $\left(\frac{3}{4}, 6\right)$
32. $\left(\frac{2}{3}, \frac{3}{2}\right)$ and $\left(-3, \frac{5}{6}\right)$

d Write an equation of the line containing the given point and parallel to the given line.

33. $(3, 7)$; $x + 2y = 6$
34. $(0, 3)$; $2x - y = 7$
35. $(2, -1)$; $5x - 7y = 8$
36. $(-4, -5)$; $2x + y = -3$
37. $(-6, 2)$; $3x = 9y + 2$
38. $(-7, 0)$; $2y + 5x = 6$

Write an equation of the line containing the given point and perpendicular to the given line.

39. $(2, 5)$; $2x + y = 3$

40. $(4, 1)$; $x - 3y = 9$

41. $(3, -2)$; $3x + 4y = 5$

42. $(-3, -5)$; $5x - 2y = 4$

43. $(0, 9)$; $2x + 5y = 7$

44. $(-3, -4)$; $6y - 3x = 2$

e Solve.

45. *School Fund-Raiser.* A school club is raising funds by having a "Shred It Day," when residents of the community can bring in their sensitive documents to be shredded. The club is charging \$10 for the first three paper bags full of documents and \$5 for each additional bag.

a) Formulate a linear function that models the total cost $C(x)$ of shredding x additional bags of documents.
b) Graph the model.
c) Use the model to determine the total cost of shredding 7 bags of documents.

46. *Fitness Club Costs.* A fitness club charges an initiation fee of \$165 plus \$24.95 per month.

a) Formulate a linear function that models the total cost $C(t)$ of a club membership for t months.
b) Graph the model.
c) Use the model to determine the total cost of a 14-month membership.

47. *Value of a Lawn Mower.* A landscaping business purchased a ZTR commercial lawn mower for \$9400. The value $V(t)$ of the mower depreciates (declines) at a rate of \$85 per month.

a) Formulate a linear function that models the value $V(t)$ of the mower after t months.
b) Graph the model.
c) Use the model to determine the value of the mower after 18 months.

48. *Value of a Computer.* True Tone Graphics bought a computer for \$3800. The value $V(t)$ of the computer depreciates at a rate of \$50 per month.

a) Formulate a linear function that models the value $V(t)$ of the computer after t months.
b) Graph the model.
c) Use the model to determine the value of the computer after $10\frac{1}{2}$ months.

In Exercises 49–54, assume that a constant rate of change exists for each model formed.

49. *Urban Population.* The following table lists data regarding the percentage of the world population considered urban in 1960 and in 2014.

YEAR, x (in number of years after 1960)	PERCENTAGE OF WORLD POPULATION THAT WAS URBAN
1960, 0	34%, or 0.34
2014, 54	54%, or 0.54

Data: Global Health Observation Data, World Health Organization

a) Use the two data points to find a linear function that fits the data. Let x = the number of years after 1960 and $U(x)$ = the percentage of the world population considered urban. (Use 0.34 and 0.54 when finding the linear function. Then convert answers to percent notation.)
b) Use the function of part (a) to estimate the percentage of the world population considered urban in 1990. Predict the percentage that will be considered urban in 2020.

50. *Disney Revenue.* The following table lists data regarding revenue from the Disney parks and resorts in 2010 and in 2016.

YEAR, x (in number of years after 2010)	REVENUE FROM PARKS AND RESORTS (in billions of dollars)
2010, 0	\$10.667
2016, 6	16.974

Data: The Disney Company

a) Use the two data points to find a linear function that fits the data. Let x = the number of years after 2010 and $R(x)$ = the revenue, in billions of dollars, from the parks and resorts.
b) Use the function of part (a) to estimate the revenue in 2014 and to predict the revenue in 2021.

51. *Bridges Needing Repair.* The number of U.S. bridges in need of repair has been declining recently. In 2000, 89,460 bridges were labeled structurally deficient. By 2016, this number had dropped to 56,000. Let $B(x)$ = the number of bridges in need of repair and x = the number of years after 2000.

Data: U.S. Department of Transportation; *The Wall Street Journal*, "Maligned Bridges Get Needed Repair" by David Harrison, 5/22/17

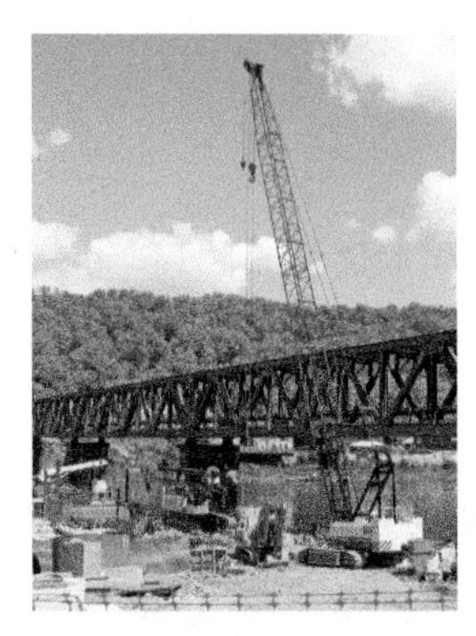

a) Find a linear function that fits the data.
b) Use the function of part (a) to estimate the number of bridges in need of repair in 2008.
c) At this rate of decrease, when will the number of bridges in need of repair reach 48,000?

52. *Revenue from Gas Taxes.* As a result of cars getting better gas mileage, the increased use of electric vehicles, and fewer miles being driven, revenues from U.S. gas taxes are declining. It is projected that the gas-tax revenue will drop from \$24.1 billion in 2017 to \$20.3 billion by 2025. Let $G(x)$ = the gas tax revenue, in billions of dollars, and x = the number of years after 2017.

Data: Joint Commission on Taxation

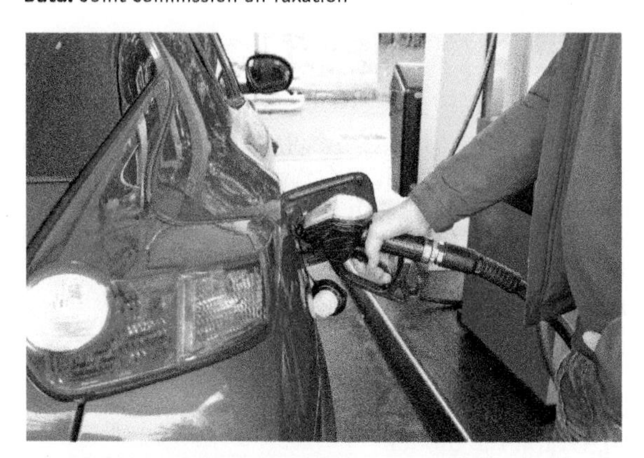

a) Find a linear function that fits the data.
b) Use the function of part (a) to estimate the projected gas tax revenue in 2019 and in 2023.
c) At this rate of decrease, when will the gas tax revenue be \$19.8 billion?

53. *Life Expectancy in South Africa.* In 2003, the life expectancy in South Africa was 46.56 years. In 2015, it was 62.34 years. Let $E(t)$ = life expectancy and t = the number of years after 2003.

Data: *CIA World Factbook* 2003–2012

a) Find a linear function that fits the data.
b) Use the function of part (a) to estimate life expectancy in 2020.

54. *Life Expectancy in Monaco.* In 2003, the life expectancy in Monaco was 79.27 years. In 2015, it was 89.52 years. Let $E(t)$ = life expectancy and t = the number of years since 2003.

Data: *CIA World Factbook* 2003–2017

a) Find a linear function that fits the data.
b) Use the function of part (a) to estimate life expectancy in 2020.

Skill Maintenance

Solve. [1.4c], [1.5a], [1.6c, d, e]

55. $2x + 3 \le 51$

56. $2x + 3 \le 5x - 4$

57. $|2x + 3| \le 13$

58. $|2x + 3| = |x - 4|$

Synthesis

59. Find k such that the line containing the points $(-3, k)$ and $(4, 8)$ is parallel to the line containing the points $(5, 3)$ and $(1, -6)$.

60. Find an equation of the line passing through the point $(4, 5)$ and perpendicular to the line passing through the points $(-1, 3)$ and $(2, 9)$.

Summary and Review

Vocabulary Reinforcement

Complete each statement with the correct term from the column on the right. Some of the choices may be used more than once, and some may not be used at all.

1. The graph of $x = a$ is a(n) ______________ line with x-intercept $(a, 0)$. [2.5c]
2. The ______________ equation of a line with slope m and passing through (x_1, y_1) is $y - y_1 = m(x - x_1)$. [2.6b]
3. A(n) ______________ is a correspondence between a first set, called the ______________, and a second set, called the ______________, such that each member of the ______________ corresponds to ______________ member of the ______________. [2.2a]
4. The ______________ of a line containing points (x_1, y_1) and (x_2, y_2) is given by $m =$ the change in y/the change in x, also described as rise/run. [2.4b]
5. Two lines are ______________ if the product of their slopes is -1. [2.5d]
6. The equation $y = mx + b$ is called the ______________ equation of a line with slope m and y-intercept $(0, b)$. [2.4b]
7. Lines are ______________ if they have the same slope and different y-intercepts. [2.5d]

x-intercept
y-intercept
at least one
exactly one
slope–intercept
point–slope
slope
function
relation
parallel
perpendicular
vertical
horizontal
domain
range

Concept Reinforcement

Determine whether each statement is true or false.

______________ 1. The slope of a vertical line is 0. [2.5c]

______________ 2. A line with slope 1 slants less steeply than a line with slope -5. [2.4b]

______________ 3. Parallel lines have the same slope and y-intercept. [2.5d]

Study Guide

Objective 2.2a Determine whether a correspondence is a function.

Example Determine whether each correspondence is a function.

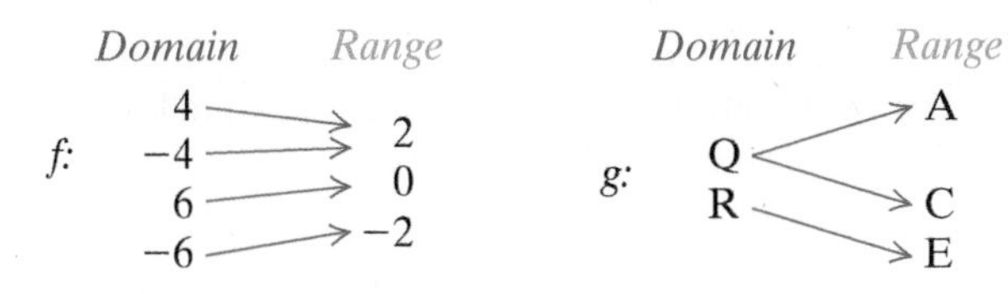

The correspondence f *is* a function because each member of the domain is matched to *only one* member of the range. The correspondence g *is not* a function because one member of the domain, Q, is matched to more than one member of the range.

Practice Exercise

1. Determine whether the correspondence is a function.

Domain *Range*

h: 11 → 20; 15 → 20, 31; 19 → 31

Objective 2.2b Given a function described by an equation, find function values (outputs) for specified values (inputs).

Example Find the indicated function value.

a) $f(0)$, for $f(x) = -x + 6$ **b)** $g(5)$, for $g(x) = -10$

c) $h(-1)$, for $h(x) = 4x^2 + x$

a) $f(x) = -x + 6$: $f(0) = -0 + 6 = 6$

b) $g(x) = -10$: $g(5) = -10$

c) $h(x) = 4x^2 + x$: $h(-1) = 4(-1)^2 + (-1) = 3$

Practice Exercise

2. Find $g(0)$, $g(-2)$, and $g(6)$ for $g(x) = \frac{1}{2}x - 2$.

Objective 2.2c Draw the graph of a function.

Example Graph: $f(x) = -\frac{2}{3}x + 2$.

By choosing multiples of 3 for x, we can avoid fraction values for y. If $x = -3$, then $y = -\frac{2}{3}\cdot(-3) + 2 = 2 + 2 = 4$. We list three ordered pairs in a table, plot the points, draw the line, and label the graph.

x	$f(x)$
3	0
0	2
−3	4

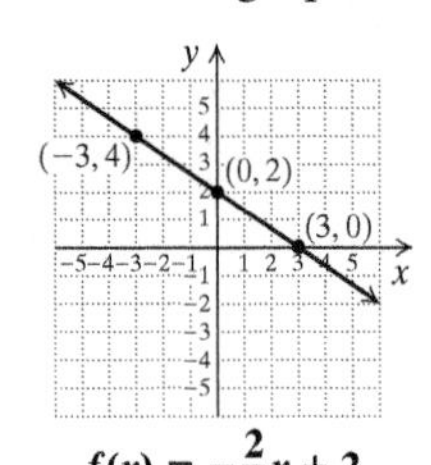

$f(x) = -\frac{2}{3}x + 2$

Practice Exercise

3. Graph: $f(x) = \frac{2}{5}x - 3$.

x	$f(x)$

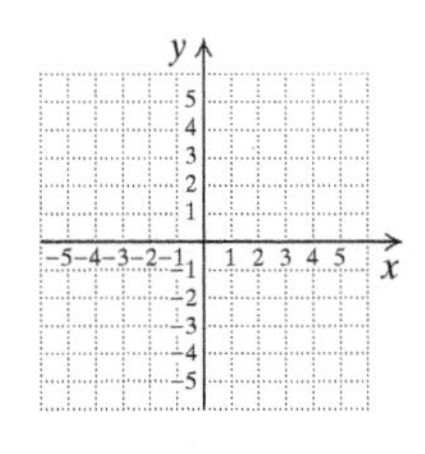

Objective 2.2d Determine whether a graph is that of a function using the vertical-line test.

Example Determine whether each of the following is the graph of a function.

a)

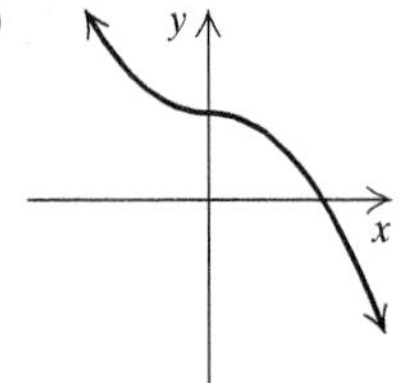

b)

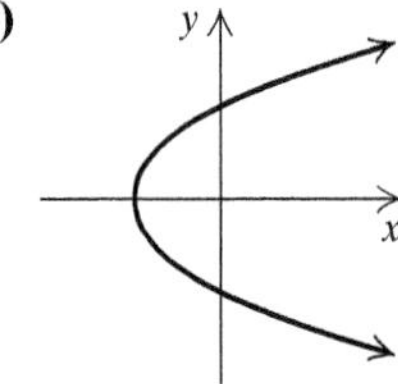

a) The graph is that of a function because no vertical line can cross the graph at more than one point.

b) The graph is not that of a function because a vertical line can cross the graph more than once.

Practice Exercise

4. Determine whether the graph is the graph of a function.

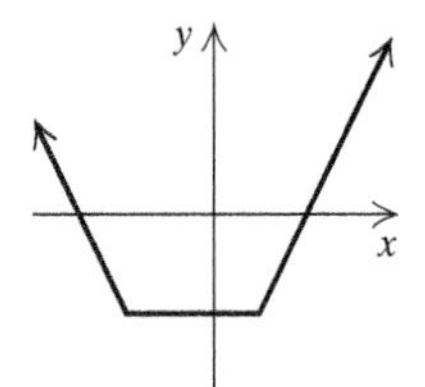

Objective 2.3a Find the domain and the range of a function.

Example For the function f whose graph is shown below, determine the domain and the range.

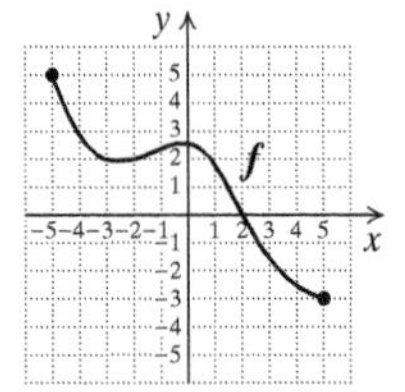

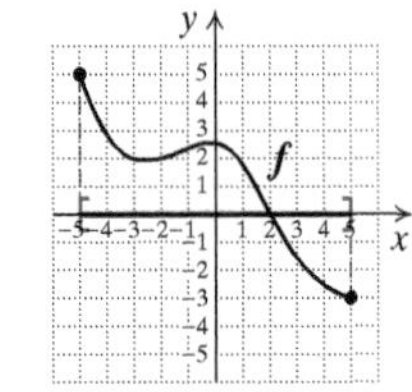

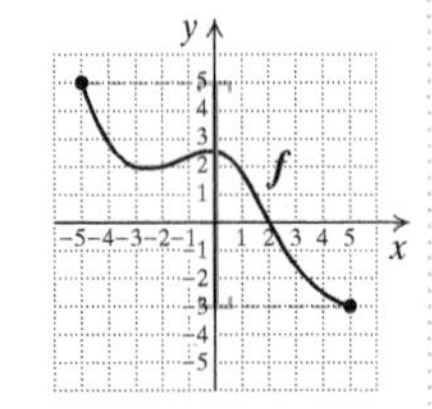

Domain: $[-5, 5]$; range: $[-3, 5]$

Practice Exercises

5. For the function g whose graph is shown below, determine the domain and the range.

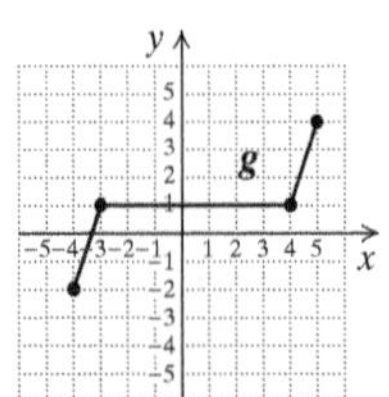

Example Find the domain of $g(x) = \dfrac{x + 1}{2x - 6}$.

Since $(x + 1)/(2x - 6)$ cannot be calculated when the denominator $2x - 6$ is 0, we solve $2x - 6 = 0$ to find the real numbers that must be excluded from the domain of g:

$$2x - 6 = 0$$
$$2x = 6$$
$$x = 3.$$

Thus, 3 is not in the domain. The domain of g is $\{x | x \text{ is a real number } and\ x \neq 3\}$, or $(-\infty, 3) \cup (3, \infty)$.

6. Find the domain of

$$h(x) = \frac{x - 3}{3x + 9}.$$

Objective 2.4b Given two points on a line, find the slope. Given a linear equation, derive the equivalent slope–intercept equation and determine the slope and the y-intercept.

Example Find the slope of the line containing $(-5, 6)$ and $(-1, -4)$.

$$m = \frac{\text{change in } y}{\text{change in } x} = \frac{6 - (-4)}{-5 - (-1)} = \frac{6 + 4}{-5 + 1} = \frac{10}{-4} = -\frac{5}{2}$$

Example Find the slope and the y-intercept of $4x - 2y = 20$.

We first solve for y:

$$4x - 2y = 20$$
$$-2y = -4x + 20 \quad \text{Subtracting } 4x$$
$$y = 2x - 10. \quad \text{Dividing by } -2$$

The slope is 2, and the y-intercept is $(0, -10)$.

Practice Exercises

7. Find the slope of the line containing $(2, -8)$ and $(-3, 2)$.

8. Find the slope and the y-intercept of $3x = -6y + 12$.

Objective 2.5a Graph linear equations using intercepts.

Example Find the intercepts of $x - 2y = 6$ and then graph the line.

To find the y-intercept, we let $x = 0$ and solve for y:

$$0 - 2y = 6 \quad \text{Substituting 0 for } x$$
$$-2y = 6$$
$$y = -3.$$

The y-intercept is $(0, -3)$.

To find the x-intercept, we let $y = 0$ and solve for x:

$$x - 2 \cdot 0 = 6 \quad \text{Substituting 0 for } y$$
$$x = 6.$$

The x-intercept is $(6, 0)$.

We plot these points and draw the line, using a third point as a check. We let $x = -2$ and solve for y:

$$-2 - 2y = 6$$
$$-2y = 8$$
$$y = -4.$$

We plot $(-2, -4)$ and note that it is on the line. Thus the graph is correct.

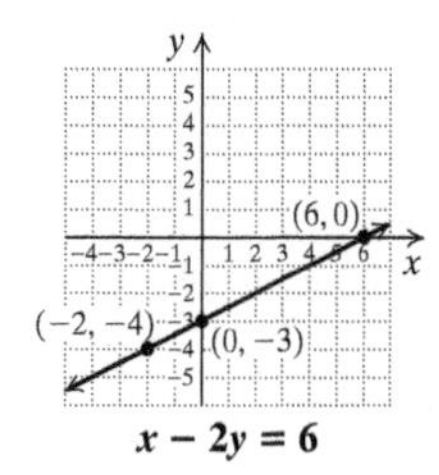

$x - 2y = 6$

Practice Exercise

9. Find the intercepts of $3y - 3 = x$ and then graph the line.

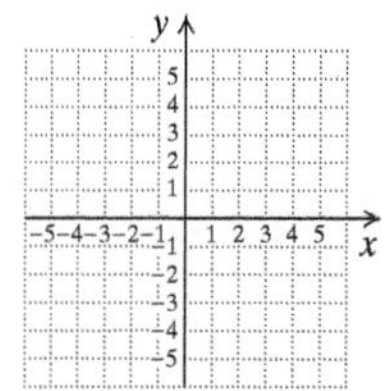

Objective 2.5b Given a linear equation in slope–intercept form, use the slope and the y-intercept to graph the line.

Example Graph using the slope and the y-intercept:

$$y = -\frac{3}{2}x + 5.$$

This equation is in slope–intercept form, $y = mx + b$. The y-intercept is $(0, 5)$. We plot $(0, 5)$. We can think of the slope $(m = -\frac{3}{2})$ as $\frac{-3}{2}$.

Starting at the y-intercept, we use the slope to find another point on the graph. We move 3 units down and 2 units to the right. We get a new point: $(2, 2)$.

To get a third point for a check, we start at $(2, 2)$ and move 3 units down and 2 units to the right to the point $(4, -1)$. We plot the points and draw the line.

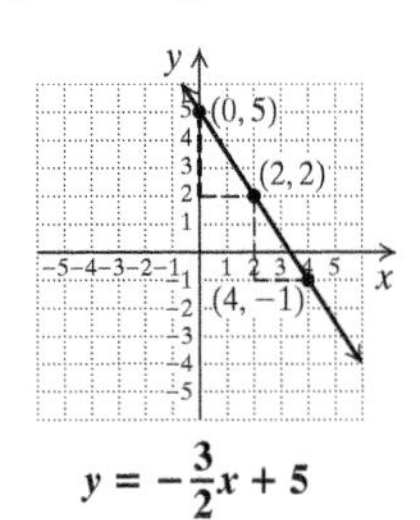

$y = -\frac{3}{2}x + 5$

Practice Exercise

10. Graph using the slope and the y-intercept:

$$y = \frac{1}{4}x - 3.$$

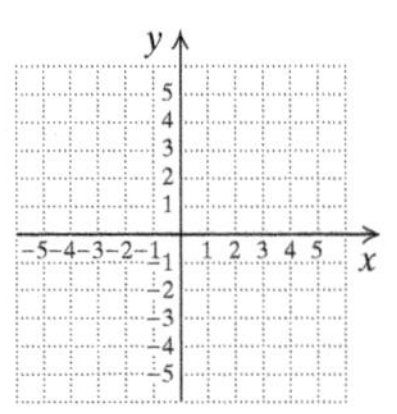

Objective 2.5c Graph linear equations of the form $x = a$ or $y = b$.

Example Graph: $y = -1$.

All ordered pairs $(x, -1)$ are solutions; y is -1 at each point. The graph is a horizontal line that intersects the y-axis at $(0, -1)$.

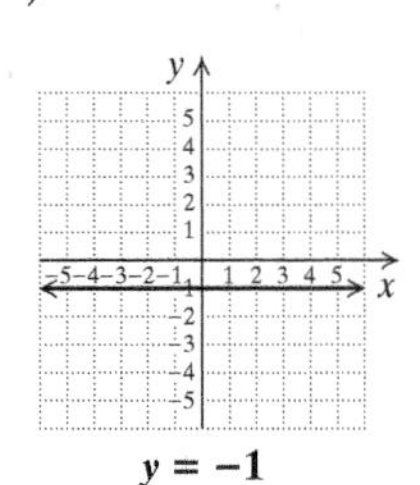

$y = -1$

Example Graph: $x = 2$.

All ordered pairs $(2, y)$ are solutions; x is 2 at each point. The graph is a vertical line that intersects the x-axis at $(2, 0)$.

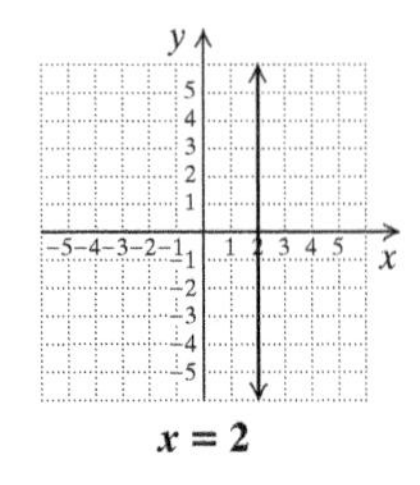

$x = 2$

Practice Exercises

11. Graph: $y = 3$.

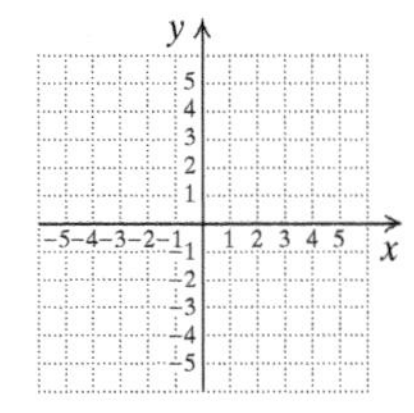

12. Graph: $x = -\frac{5}{2}$.

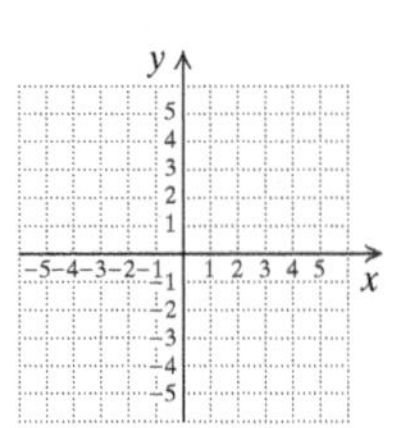

Objective 2.5d Given the equations of two lines, determine whether their graphs are parallel or whether they are perpendicular.

Example Determine whether the graphs of the given pair of lines are parallel, perpendicular, or neither.

a) $2y - x = 16,$
$x + \frac{1}{2}y = 4$

b) $5x - 3 = 2y,$
$2y + 12 = 5x$

a) Writing each equation in slope–intercept form, we have $y = \frac{1}{2}x + 8$ and $y = -2x + 8$. The slopes are $\frac{1}{2}$ and -2. The product of the slopes is -1: $\frac{1}{2}\cdot(-2) = -1$. The graphs are perpendicular.

b) Writing each equation in slope–intercept form, we have $y = \frac{5}{2}x - \frac{3}{2}$ and $y = \frac{5}{2}x - 6$. The slopes are the same, $\frac{5}{2}$, and the y-intercepts are different. The graphs are parallel.

Practice Exercises

Determine whether the graphs of the given pair of lines are parallel, perpendicular, or neither.

13. $-3x + 8y = -8,$
$8y = 3x + 40$

14. $5x - 2y = -8,$
$2x + 5y = 15$

Objective 2.6a Find an equation of a line when the slope and the y-intercept are given.

Example A line has slope 0.8 and y-intercept $(0, -17)$. Find an equation of the line.

We use the slope–intercept equation and substitute 0.8 for m and -17 for b:

$$y = mx + b \quad \text{Slope–intercept equation}$$
$$y = 0.8x - 17.$$

Practice Exercise

15. A line has slope -8 and y-intercept $(0, 0.3)$. Find an equation of the line.

Objective 2.6b Find an equation of a line when the slope and a point are given.

Example Find an equation of the line with slope -2 and containing the point $\left(\frac{1}{3}, -1\right)$.

Using the *point–slope equation*, we substitute -2 for m, $\frac{1}{3}$ for x_1, and -1 for y_1:

$$y - (-1) = -2\left(x - \frac{1}{3}\right) \quad \text{Using } y - y_1 = m(x - x_1)$$
$$y + 1 = -2x + \frac{2}{3}$$
$$y = -2x - \frac{1}{3}.$$

Using the *slope–intercept equation*, we substitute -2 for m, $\frac{1}{3}$ for x, and -1 for y, and then solve for b:

$$-1 = -2\cdot\frac{1}{3} + b \quad \text{Using } y = mx + b$$
$$-1 = -\frac{2}{3} + b$$
$$-\frac{1}{3} = b.$$

Then, substituting -2 for m and $-\frac{1}{3}$ for b in the slope–intercept equation $y = mx + b$, we have $y = -2x - \frac{1}{3}$.

Practice Exercise

16. Find an equation of the line with slope -4 and containing the point $\left(\frac{1}{2}, -3\right)$.

Objective 2.6c Find an equation of a line when two points are given.

Example Find an equation of the line containing the points $(-3, 9)$ and $(1, -2)$.

We first find the slope:

$$\frac{9 - (-2)}{-3 - 1} = \frac{11}{-4} = -\frac{11}{4}.$$

Using the slope–intercept equation and the point $(1, -2)$, we substitute $-\frac{11}{4}$ for m, 1 for x, and -2 for y, and then solve for b. We could also have used the point $(-3, 9)$.

$$y = mx + b$$
$$-2 = -\tfrac{11}{4} \cdot 1 + b$$
$$-\tfrac{8}{4} = -\tfrac{11}{4} + b$$
$$\tfrac{3}{4} = b$$

Then substituting $-\frac{11}{4}$ for m and $\frac{3}{4}$ for b in $y = mx + b$, we have $y = -\frac{11}{4}x + \frac{3}{4}$.

Practice Exercise

17. Find an equation of the line containing the points $(-2, 7)$ and $(4, -3)$.

Objective 2.6d Given a line and a point not on the given line, find an equation of the line parallel to the line and containing the point, and find an equation of the line perpendicular to the line and containing the point.

Example Write an equation of the line containing $(-1, 1)$ and parallel to $3y - 6x = 5$.

Solving $3y - 6x = 5$ for y, we get $y = 2x + \frac{5}{3}$. The slope of the given line is 2.

A line parallel to the given line must have the same slope, 2. We substitute 2 for m, -1 for x_1, and 1 for y_1 in the point–slope equation:

$$y - 1 = 2[x - (-1)] \quad \text{Using } y - y_1 = m(x - x_1)$$
$$y - 1 = 2(x + 1)$$
$$y - 1 = 2x + 2$$
$$y = 2x + 3. \quad \text{Line parallel to the given line and passing through } (-1, 1)$$

Example Write an equation of the line containing the point $(2, -4)$ and perpendicular to $6x + 2y = 13$.

Solving $6x + 2y = 13$ for y, we get $y = -3x + \frac{13}{2}$. The slope of the given line is -3.

The slope of a line perpendicular to the given line is the opposite of the reciprocal of -3, or $\frac{1}{3}$. We substitute $\frac{1}{3}$ for m, 2 for x_1, and -4 for y_1 in the point–slope equation:

$$y - (-4) = \tfrac{1}{3}(x - 2) \quad \text{Using } y - y_1 = m(x - x_1)$$
$$y + 4 = \tfrac{1}{3}x - \tfrac{2}{3}$$
$$y = \tfrac{1}{3}x - \tfrac{14}{3}. \quad \text{Line perpendicular to the given line and passing through } (2, -4)$$

Practice Exercises

18. Write an equation of the line containing the point $(2, -5)$ and parallel to $4x - 3y = 6$.

19. Write an equation of the line containing $(2, -5)$ and perpendicular to $4x - 3y = 6$.

Review Exercises

Determine whether each correspondence is a function. [2.2a]

1.

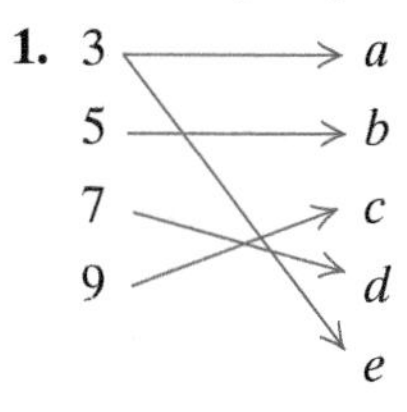

2.

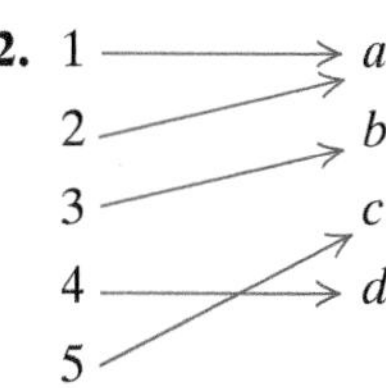

Find the function values. [2.2b]

3. $g(x) = -2x + 5$; $g(0)$ and $g(-1)$

4. $f(x) = 3x^2 - 2x + 7$; $f(0)$ and $f(-1)$

5. *Gift-Card Sales.* The function $G(t) = 6.22t + 76.22$ can be used to approximate the average amount of gift-card sales, in billions of dollars, where t is the number of years after 2006. Estimate the amount of gift-card sales in 2016. That is, find $G(10)$. [2.2b]

Data: National Retail Federation

Graph. [2.1c, d], [2.2c]

6. $y = -3x + 2$

x	y

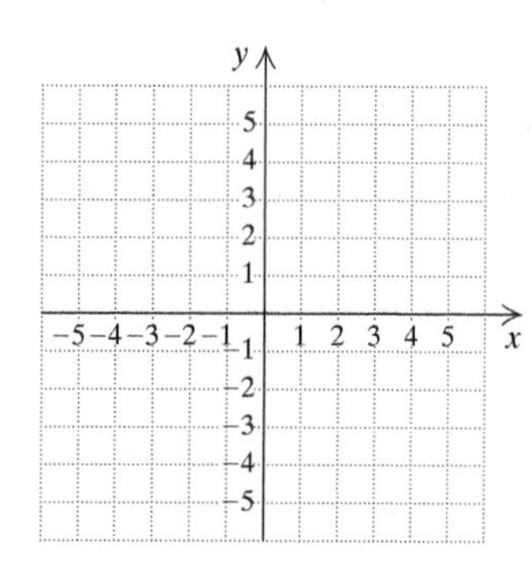

7. $g(x) = \frac{5}{2}x - 3$

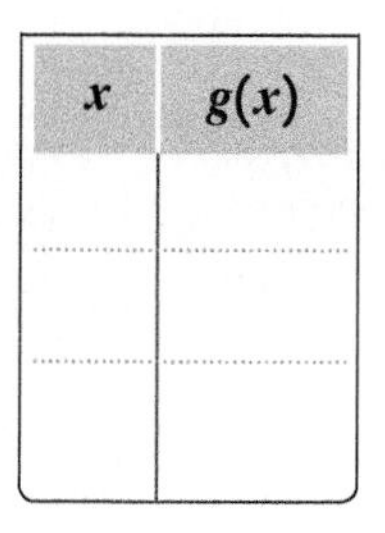

x	$g(x)$

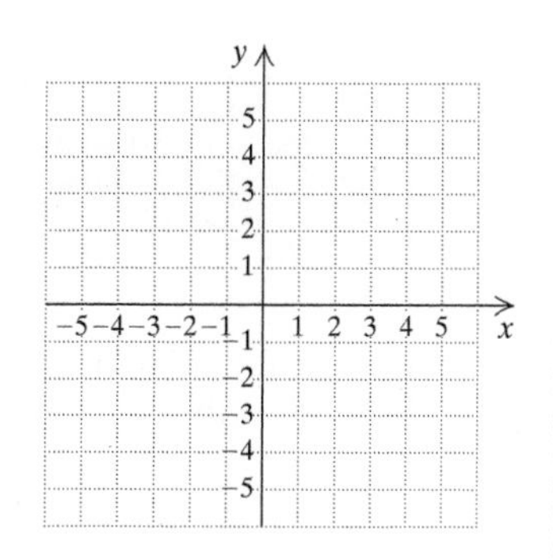

8. $f(x) = |x - 3|$

x	$f(x)$

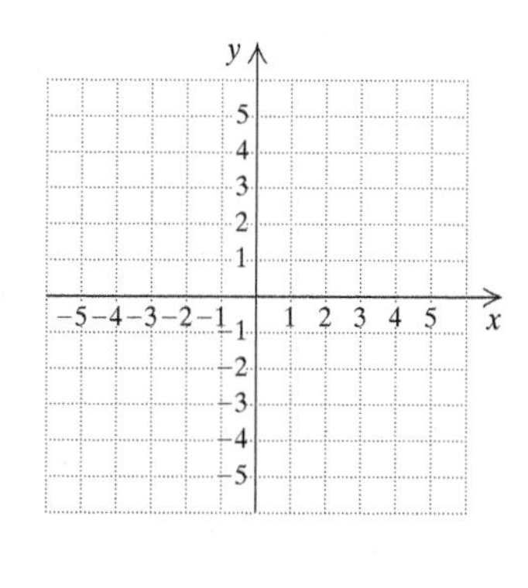

9. $y = 3 - x^2$

x	y

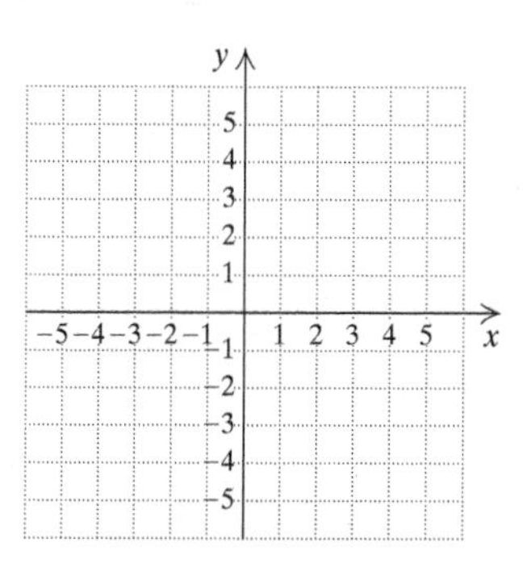

Determine whether each of the following is the graph of a function. [2.2d]

10.

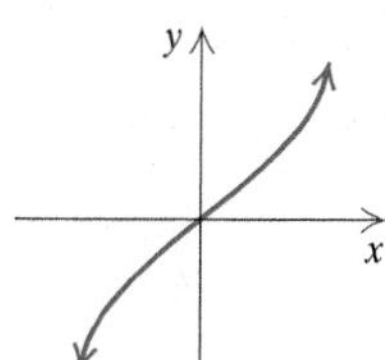

11.

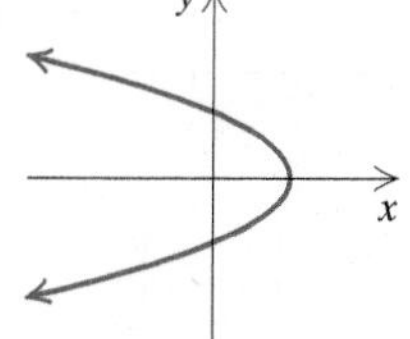

12. For the following graph of a function f, determine **(a)** $f(2)$; **(b)** the domain; **(c)** all x-values such that $f(x) = 2$; and **(d)** the range. [2.3a]

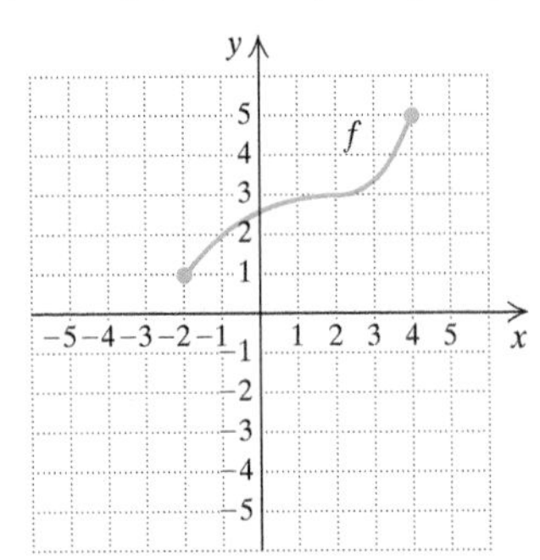

Find the domain. [2.3a]

13. $f(x) = \dfrac{5}{x - 4}$

14. $g(x) = x - x^2$

Find the slope and the y-intercept. [2.4a, b]

15. $y = -3x + 2$

16. $4y + 2x = 8$

17. Find the slope, if it exists, of the line containing the points $(13, 7)$ and $(10, -4)$. [2.4b]

Find the intercepts. Then graph the equation. [2.5a]

18. $2y + x = 4$

x	y	
		← x-intercept
		← y-intercept

19. $2y = 6 - 3x$

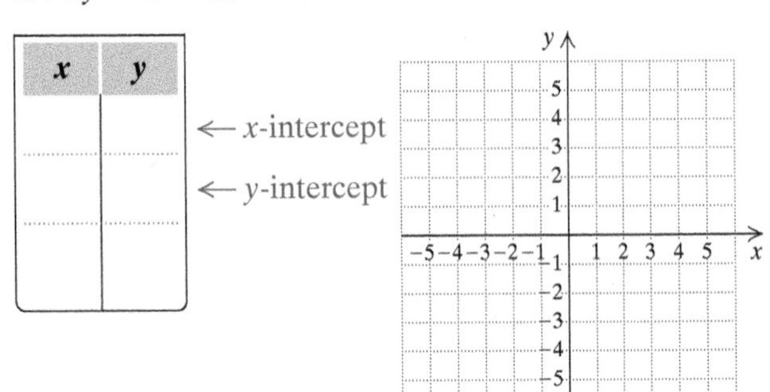

Graph using the slope and the y-intercept. [2.5b]

20. $g(x) = -\frac{2}{3}x - 4$

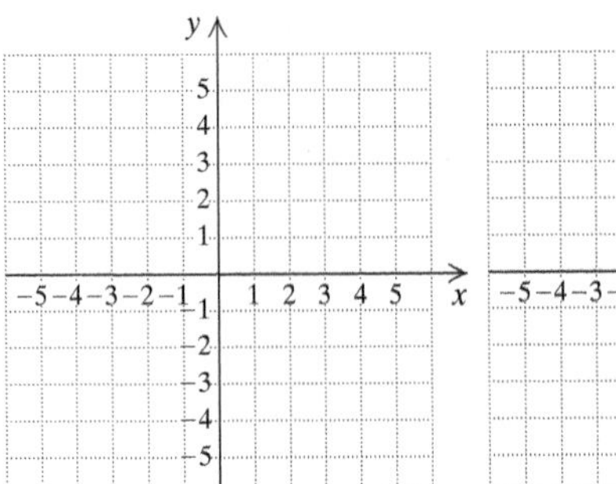

21. $f(x) = \frac{5}{2}x + 3$

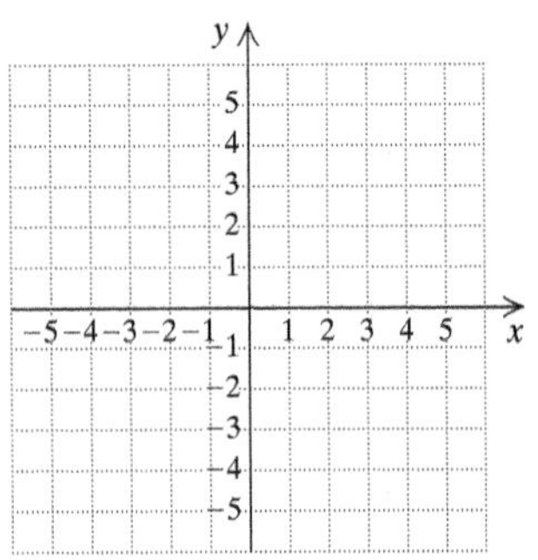

Graph. [2.5c]

22. $x = -3$

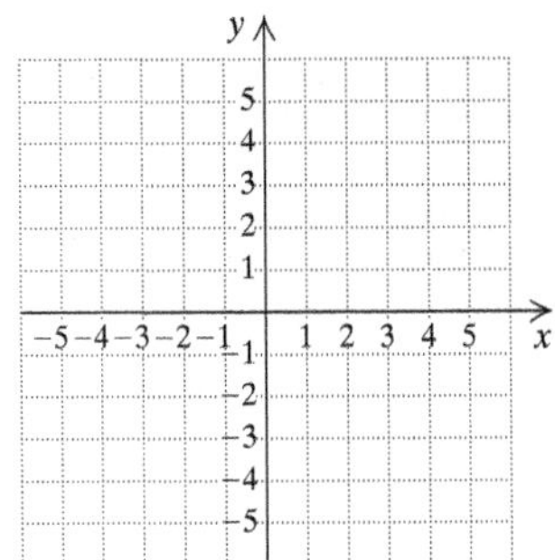

23. $f(x) = 4$

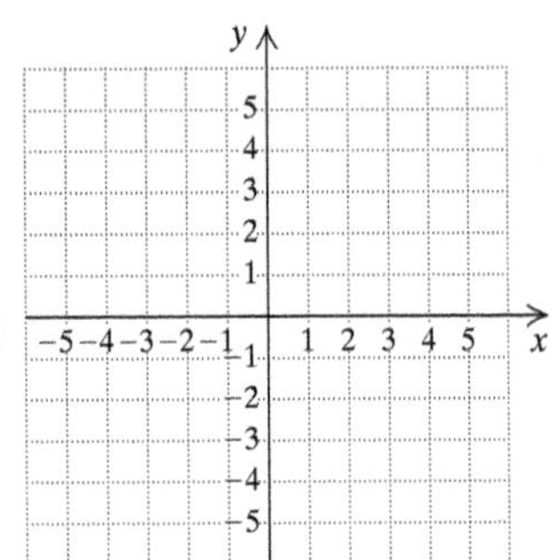

Determine whether the graphs of the given pair of lines are parallel or perpendicular. [2.5d]

24. $y + 5 = -x$,
$x - y = 2$

25. $3x - 5 = 7y$,
$7y - 3x = 7$

26. $4y + x = 3$,
$2x + 8y = 5$

27. $x = 4$,
$y = -3$

28. Find a linear function $f(x) = mx + b$ whose graph has slope 4.7 and y-intercept $(0, -23)$. [2.6a]

29. Find an equation of the line having slope -3 and containing the point $(3, -5)$. [2.6b]

30. Find an equation of the line containing the points $(-2, 3)$ and $(-4, 6)$. [2.6c]

31. Find an equation of the line containing the given point and parallel to the given line:

$$(14, -1);\quad 5x + 7y = 8. \quad [2.6d]$$

32. Find an equation of the line containing the given point and perpendicular to the given line:

$$(5, 2);\quad 3x + y = 5. \quad [2.6d]$$

33. *Records in the 400-Meter Run.* The following table lists data regarding the Summer Olympics winning times in the men's 400-m run. [2.6e]

YEAR	SUMMER OLYMPICS WINNING TIME IN MEN'S 400-M RUN (in seconds)
1972	44.66
2016	43.03

a) Use the two data points to find a linear function that fits the data. Let x = the number of years after 1972 and $R(x)$ = the Summer Olympics winning time x years from 1972.

b) Use the function to estimate the winning time in the men's 400-m run in 2000 and in 2010.

34. What is the domain of $f(x) = \dfrac{x+3}{x-2}$? [2.3a]

A. $\{x \mid x \geq -3\}$
B. $\{x \mid x \text{ is a real number } and\ x \neq -3\ and\ x \neq 2\}$
C. $\{x \mid x \text{ is a real number } and\ x \neq 2\}$
D. $\{x \mid x > -3\}$

35. Find an equation of the line containing the point $(-2, 1)$ and perpendicular to $3y - \frac{1}{2}x = 0$. [2.6d]

A. $6x + y = -11$ **B.** $y = -\dfrac{1}{6}x - 11$

C. $y = -2x - 3$ **D.** $2x + \dfrac{1}{3} = 0$

Synthesis

36. Homespun Jellies charges \$2.49 for each jar of preserves. Shipping charges are \$3.75 for handling, plus \$0.60 per jar. Find a linear function for determining the cost of buying and shipping x jars of preserves. [2.6e]

Understanding Through Discussion and Writing

1. Under what conditions will the x-intercept and the y-intercept of a line be the same? What would the equation for such a line look like? [2.5a]

2. Explain the usefulness of the concept of slope when describing a line. [2.4b, c], [2.5b], [2.6a, b, c, d]

3. A student makes a mistake when using a graphing calculator to draw $4x + 5y = 12$ and the following screen appears. Use algebra to show that a mistake has been made. What do you think the mistake was? [2.4b]

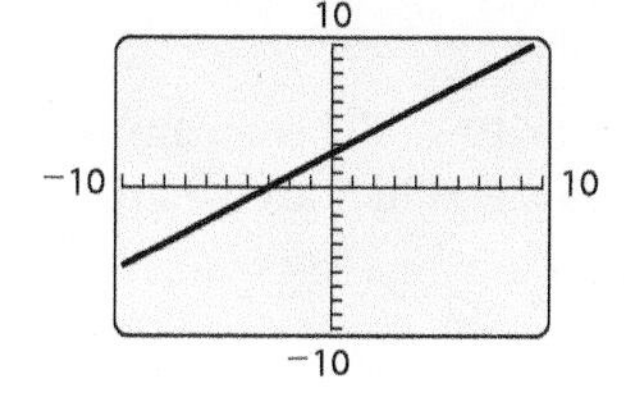

4. *Computer Repair.* The cost $R(t)$, in dollars, of computer repair at PC Pros is given by

$$R(t) = 50t + 35,$$

where t is the number of hours that the repair requires. Determine m and b in this application and explain their meaning. [2.6e]

5. Explain why the slope of a vertical line is not defined but the slope of a horizontal line is 0. [2.5c]

6. A student makes a mistake when using a graphing calculator to draw $5x - 2y = 3$ and the following screen appears. Use algebra to show that a mistake has been made. What do you think the mistake was? [2.4b]

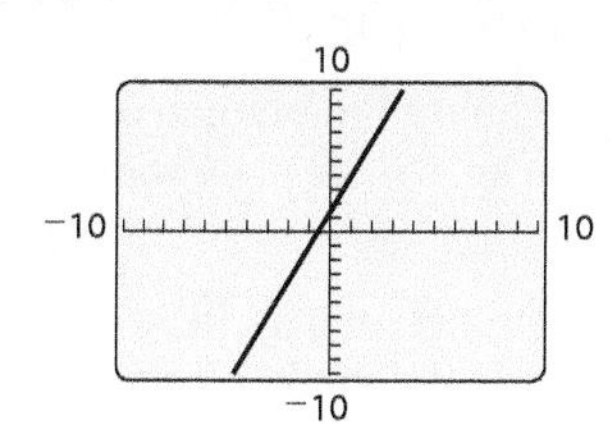

Test

For Extra Help — For step-by-step test solutions, access the Chapter Test Prep Videos in MyLab Math.

Determine whether each correspondence is a function.

1.

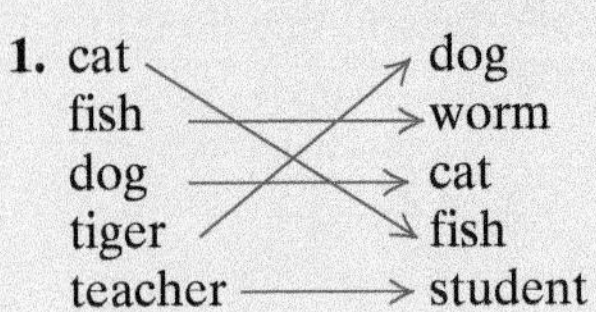

2. Lake Placid, Oslo, Squaw Valley, Innsbruck → 1980, 1976, 1960, 1952, 1932

Find the function values.

3. $f(x) = -3x - 4$; $f(0)$ and $f(-2)$

4. $g(x) = x^2 + 7$; $g(0)$ and $g(-1)$

5. $h(x) = -6$; $h(-4)$ and $h(-6)$

6. $f(x) = |x + 7|$; $f(-10)$ and $f(-7)$

Graph.

7. $y = -2x - 5$

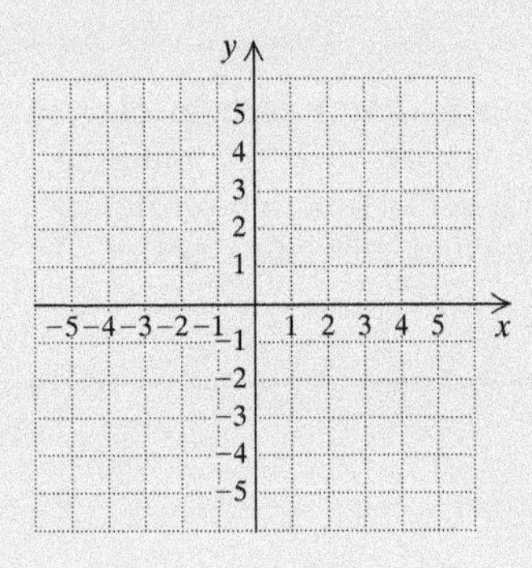

8. $f(x) = -\frac{3}{5}x$

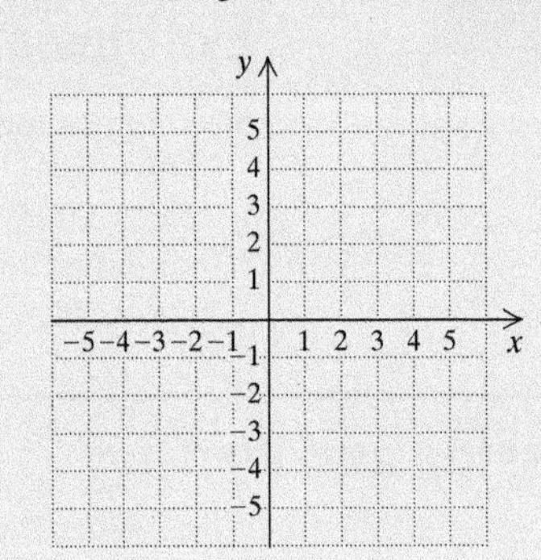

9. $g(x) = 2 - |x|$

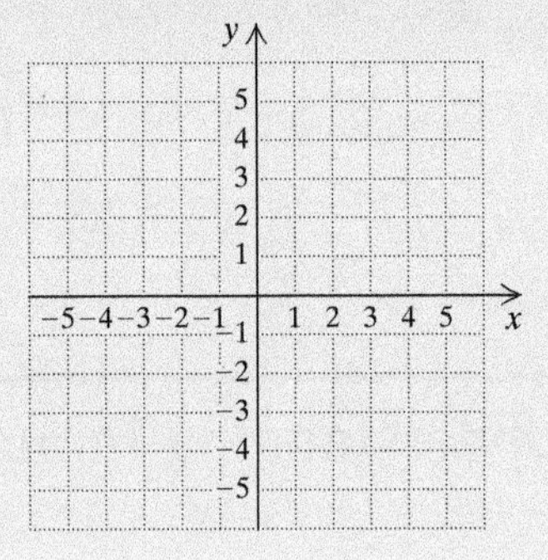

10. $f(x) = x^2 + 2x - 3$

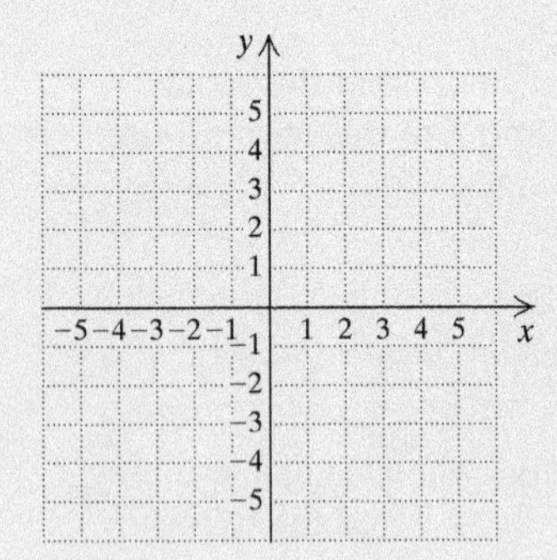

11. $y = f(x) = -3$

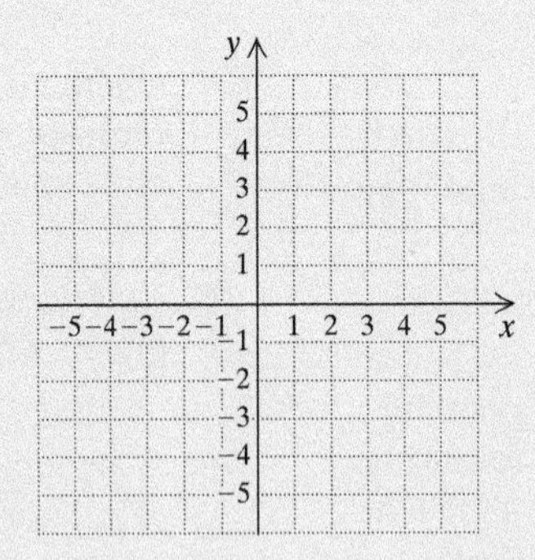

12. $2x = -4$

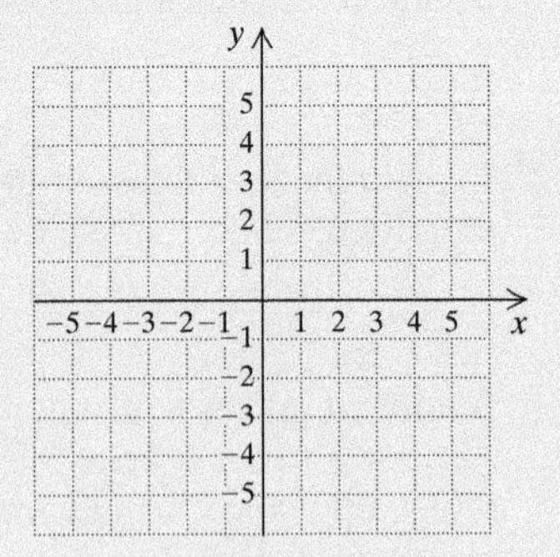

13. *Median Age of Cars.* The function

$$A(t) = 0.22t + 5.87$$

can be used to estimate the median age of cars in the United States t years after 1990. (This means, for example, that if the median age of cars is 4 years, then half the cars are older than 4 years and half are younger.)

Data: The Polk Co; autonews.com

a) Find the median age of cars in 2010.
b) In what year was the median age of cars 11.6 years?

Determine whether each of the following is the graph of a function.

14.

15.

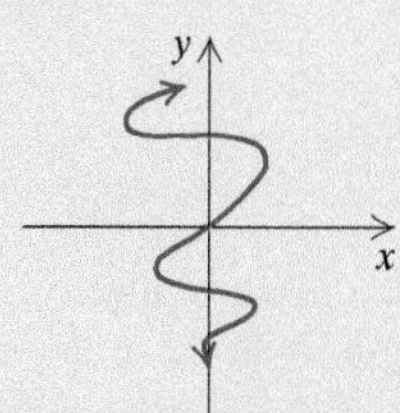

Find the domain.

16. $f(x) = \dfrac{8}{2x + 3}$

17. $g(x) = 5 - x^2$

18. For the following graph of function f, determine **(a)** $f(1)$; **(b)** the domain; **(c)** all x-values such that $f(x) = 2$; and **(d)** the range.

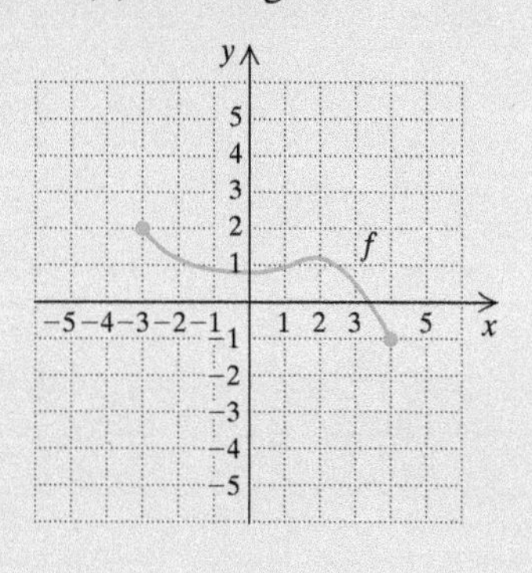

Find the slope and the y-intercept.

19. $f(x) = -\frac{3}{5}x + 12$

20. $-5y - 2x = 7$

Find the slope, if it exists, of the line containing the following points.

21. $(-2, -2)$ and $(6, 3)$

22. $(-3.1, 5.2)$ and $(-4.4, 5.2)$

23. Find the slope, or rate of change, of the graph at right.

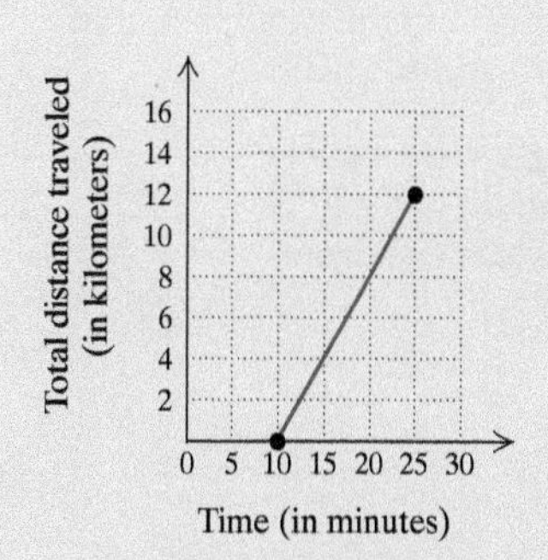

24. Find the intercepts. Then graph the equation.

$2x + 3y = 6$

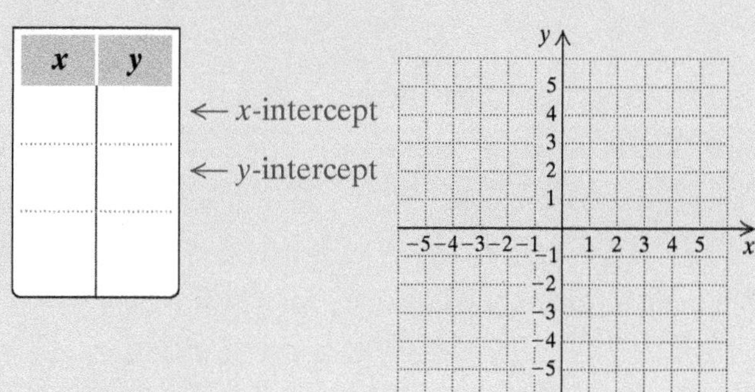

25. Graph using the slope and the y-intercept:

$f(x) = -\dfrac{2}{3}x - 1.$

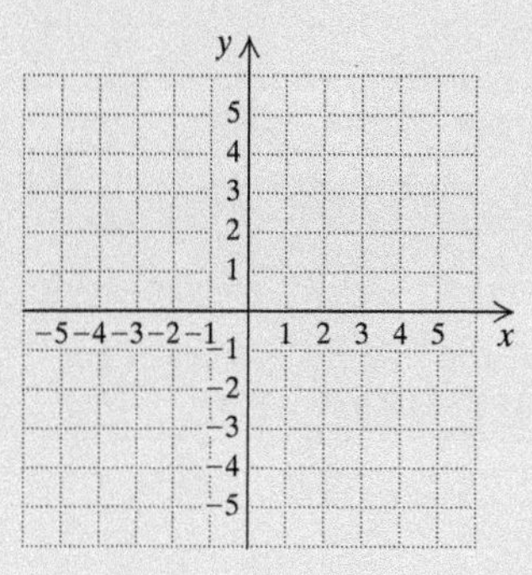

Determine whether the graphs of the given pair of lines are parallel or perpendicular.

26. $4y + 2 = 3x,$
$-3x + 4y = -12$

27. $y = -2x + 5,$
$2y - x = 6$

28. Find an equation of the line that has the given characteristics:

slope: -3; y-intercept: $(0, 4.8)$.

29. Find a linear function $f(x) = mx + b$ whose graph has the given slope and y-intercept:

slope: 5.2; y-intercept: $\left(0, -\frac{5}{8}\right)$.

30. Find an equation of the line having the given slope and containing the given point:

$m = -4$; $(1, -2)$.

31. Find an equation of the line containing the given pair of points:

$(4, -6)$ and $(-10, 15)$.

32. Find an equation of the line containing the given point and parallel to the given line:

$(4, -1)$; $x - 2y = 5$.

33. Find an equation of the line containing the given point and perpendicular to the given line:

$(2, 5)$; $x + 3y = 2$.

34. *Median Age of Men at First Marriage.* The following table lists data regarding the median age of men at first marriage in 1970 and in 2010.

YEAR	MEDIAN AGE OF MEN AT FIRST MARRIAGE
1970	23.2
2015	29.2

Data: U.S. Census Bureau

a) Use the two data points to find a linear function that fits the data. Let x = the number of years after 1970 and A = the median age at first marriage x years after 1970.

b) Use the function to estimate the median age of men at first marriage in 2008 and in 2019.

35. Find an equation of the line having slope -2 and containing the point $(3, 1)$.

A. $y - 1 = 2(x - 3)$
B. $y - 1 = -2(x - 3)$
C. $x - 1 = -2(y - 3)$
D. $x - 1 = 2(y - 3)$

Synthesis

36. Find k such that the line $3x + ky = 17$ is perpendicular to the line $8x - 5y = 26$.

37. Find a formula for a function f for which $f(-2) = 3$.

Cumulative Review

1. *Records in the 1500-Meter Run.* The following table lists data regarding the world indoor records in the men's 1500-m run in 1950 and in 2004.

YEAR	RECORDS IN THE 1500-M RUN (in minutes)
1950	3.85
2004	3.50

a) Use the two data points to find a linear function that fits the data. Let $x =$ the number of years after 1950 and $R(x) =$ the world record x years after 1950.
b) Use the function to estimate the world record in the 1500-m run in 2008 and in 2010.

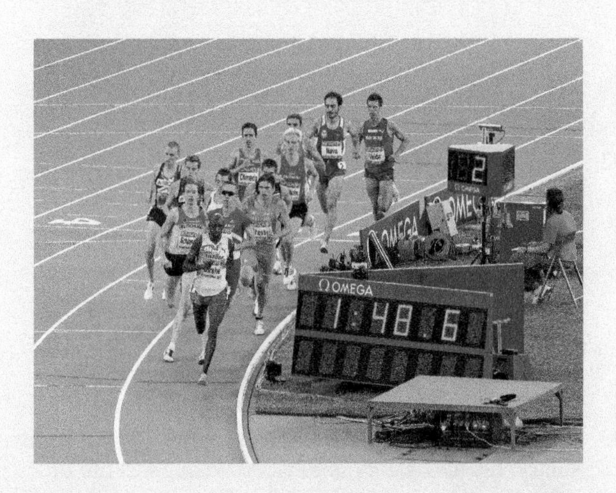

2. For the graph of function f shown below, determine **(a)** $f(15)$; **(b)** the domain; **(c)** all x-values such that $f(x) = 14$; and **(d)** the range.

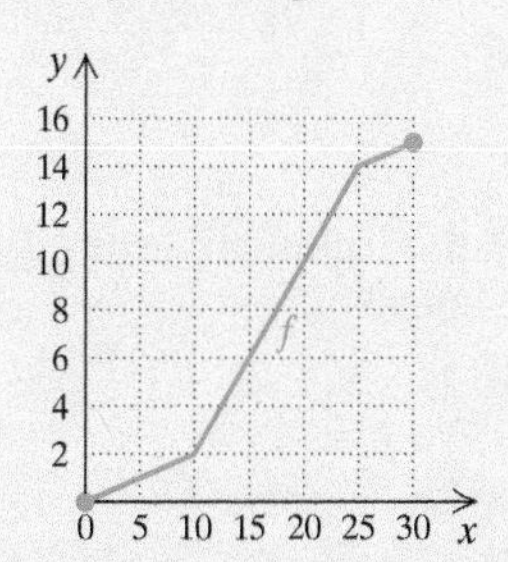

Solve.

3. $x + 9.4 = -12.6$

4. $\frac{2}{3}x - \frac{1}{4} = -\frac{4}{5}x$

5. $-2.4t = -48$

6. $4x + 7 = -14$

7. $3n - (4n - 2) = 7$

8. $5y - 10 = 10 + 5y$

9. Solve $W = Ax + By$ for x.

10. Solve $M = A + 4AB$ for A.

Solve.

11. $y - 12 \leq -5$

12. $6x - 7 < 2x - 13$

13. $5(1 - 2x) + x < 2(3 + x)$

14. $x + 3 < -1$ *or* $x + 9 \geq 1$

15. $-3 < x + 4 \leq 8$

16. $-8 \leq 2x - 4 \leq -1$

17. $|x| = 8$

18. $|y| > 4$

19. $|4x - 1| \leq 7$

20. Find an equation of the line containing the point $(-4, -6)$ and perpendicular to the line whose equation is $4y - x = 3$.

21. Find an equation of the line containing the point $(-4, -6)$ and parallel to the line whose equation is $4y - x = 3$.

Graph on a plane.

22. $y = -2x + 3$

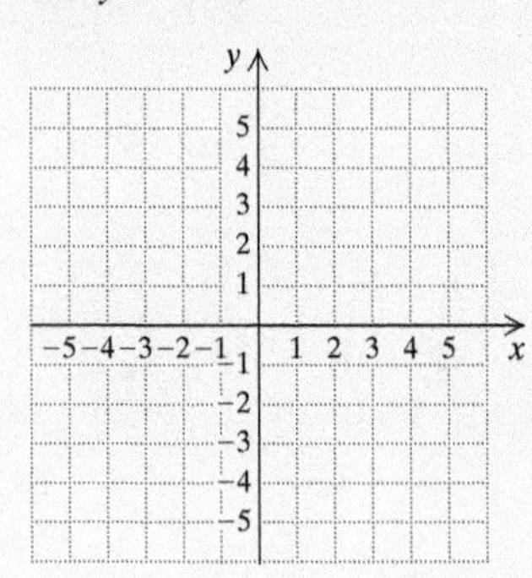

23. $3x = 2y + 6$

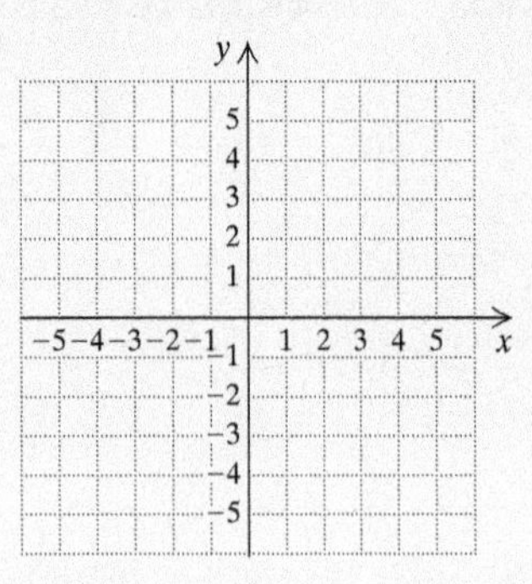

24. $4x + 16 = 0$

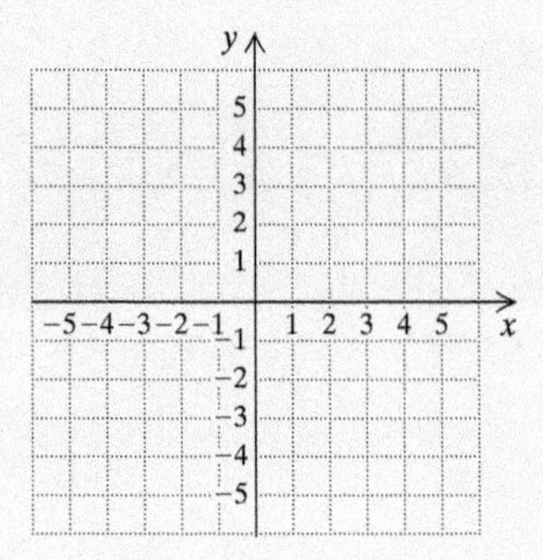

25. $-2y = -6$

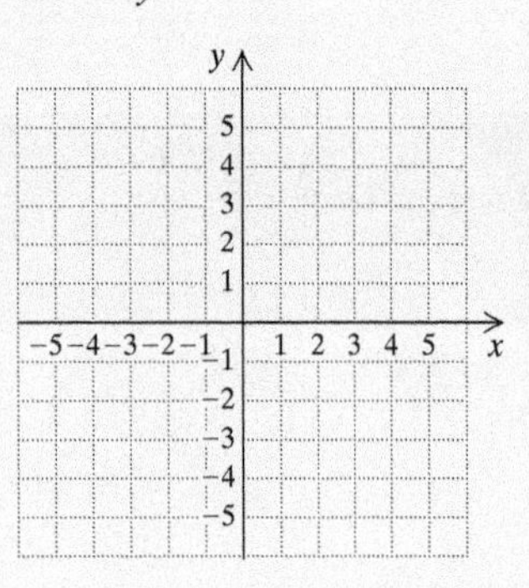

26. $f(x) = \frac{2}{3}x + 1$

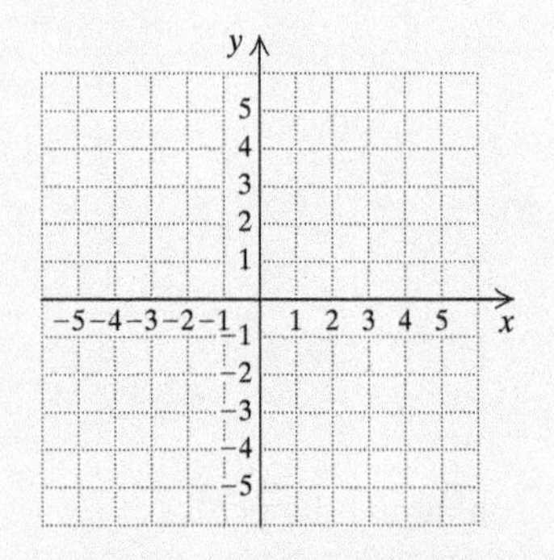

27. $g(x) = 5 - |x|$

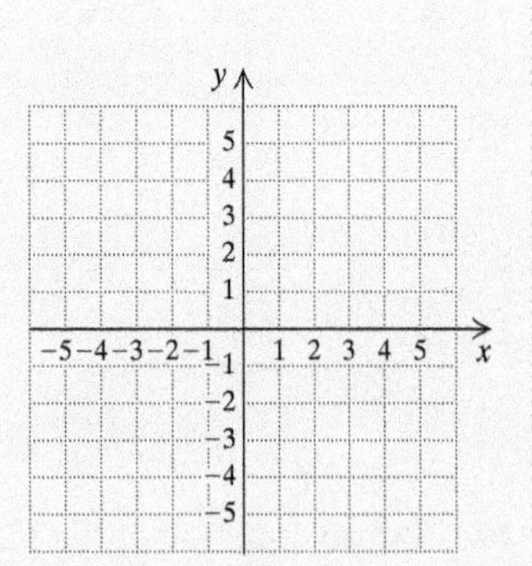

28. Find the slope and the y-intercept of $-4y + 9x = 12$.

29. Find the slope, if it exists, of the line containing the points $(2, 7)$ and $(-1, 3)$.

30. Find an equation of the line with slope -3 and containing the point $(2, -11)$.

31. Find an equation of the line containing the points $(-6, 3)$ and $(4, 2)$.

Solve.

32. *Lot Dimensions.* The perimeter of a lot is 80 m. The length exceeds the width by 6 m. Find the dimensions.

33. *Salary Raise.* After David receives a 20% raise in salary, his new salary is $27,000. What was the old salary?

Synthesis

34. Which pairs of the following four equations represent perpendicular lines?

(1) $7y - 3x = 21$
(2) $-3x - 7y = 12$
(3) $7y + 3x = 21$
(4) $3y + 7x = 12$

35. *Radio Advertising.* Wayside Auto Sales discovers that when $1000 is spent on radio advertising, weekly sales increase by $101,000. When $1250 is spent on radio advertising, weekly sales increase by $126,000. Assuming that sales increase according to a linear equation, by what amount would sales increase when $1500 is spent on radio advertising?

36. Solve: $x + 5 < 3x - 7 \leq x + 13$.

CHAPTER

3

Systems of Equations

Video gamers are present in 65% of U.S. households and are represented in every age group, with about one-fourth of gamers under 18 and one-fourth over 50. In 2016, U.S. consumers spent \$24.5 billion on video game software, an increase of \$7 billion since 2010. The delivery format of video games has changed from mostly physical in 2009 to mostly digital in 2016, as illustrated in the accompanying graph. As the graph indicates, the number of games delivered physically was the same as the number delivered digitally sometime during 2012. The genres of games played is changing as well. In 2016, half of the most popular games were shooting and action games; these genres represented only 32% of the most popular games in 2009.

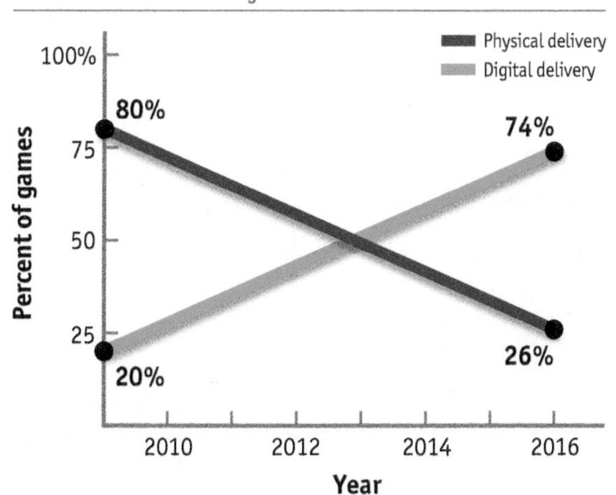

Data: Entertainment Software Association; statista.com

In Exercise 1 of Section 3.4, we will calculate the number of video-game company locations in the United States, and in Exercise 33 of Section 3.6, we will calculate the number of games rated for different audiences.

JUST IN TIME

Review topics 3, 12, 13, and 18 in the **Just In Time** section at the front of the text. This provides excellent prerequisite skill review for this chapter.

MyLab Math
VIDEO

STUDYING FOR SUCCESS *Preparing for a Test*

- ☐ Make up your own test questions as you study.
- ☐ Do an overall review of the chapter, focusing on the objectives and the examples.
- ☐ Do the exercises in the mid-chapter review and in the end-of-chapter review.
- ☐ Take the chapter test at the end of the chapter.

3.1 Systems of Equations in Two Variables

OBJECTIVE

a Solve a system of two linear equations or two functions by graphing and determine whether a system is consistent or inconsistent and whether the equations in a system are dependent or independent.

Sets of equations for which we seek a common solution are called *systems of equations*. We can solve systems of two equations in two variables by graphing.

a SOLVING SYSTEMS OF EQUATIONS GRAPHICALLY

SKILL REVIEW *Graph linear equations.* [2.1c]

Graph.

1. $x + y = 3$

2. $y = x - 2$

Answers: 1.

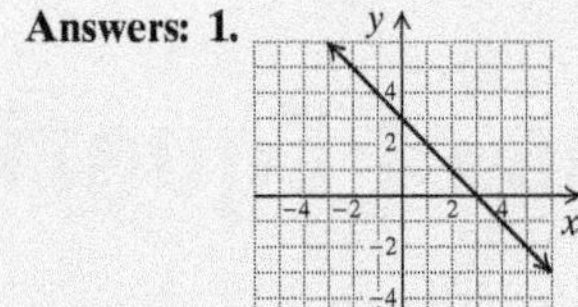

$x + y = 3$

2.

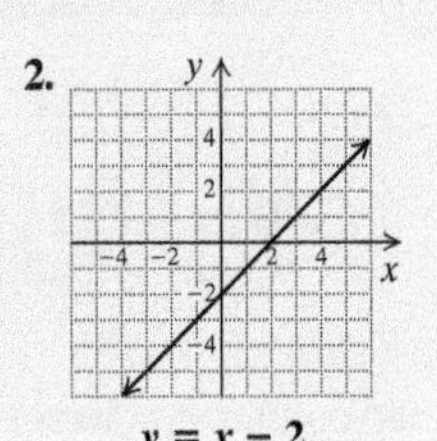

$y = x - 2$

MyLab Math VIDEO

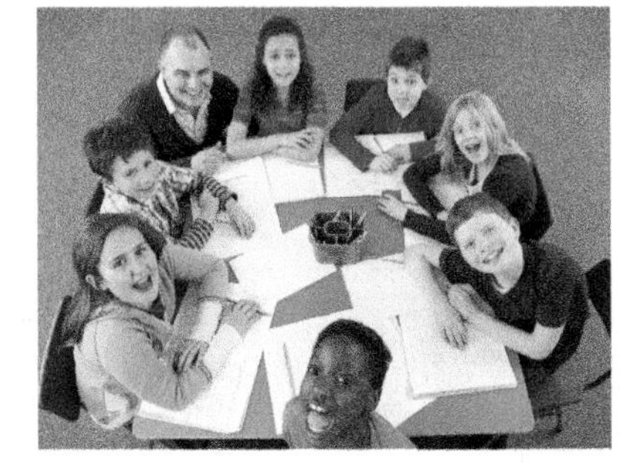

School Enrollment. In 2016, approximately 50 million children were enrolled in public elementary and secondary schools in the United States. There were 20 million more students enrolled in prekindergarten–grade 8 than there were in grades 9–12. How many were enrolled at each level?

Data: National Center for Education Statistics

To solve, we first let

x = the number enrolled in prekindergarten–grade 8, and

y = the number enrolled in grades 9–12,

where x and y are in millions of students. The problem gives us two statements that can be translated to equations.

First, we consider the total number enrolled:

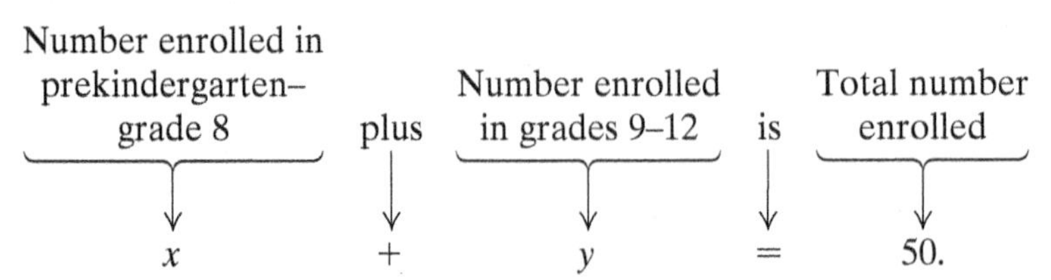

The second statement of the problem compares the enrollment at the two levels:

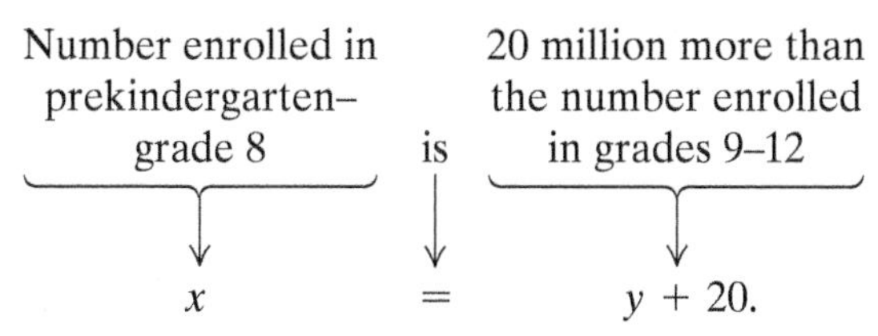

We have now translated the problem to a pair of equations, or a **system of equations**:

$$x + y = 50,$$
$$x = y + 20.$$

A **solution** of a system of two equations in two variables is an ordered pair that makes *both* equations true. If we graph a system of equations, the point at which the graphs intersect will be a solution of *both* equations. To find the solution of the system above, we graph both equations, as shown here.

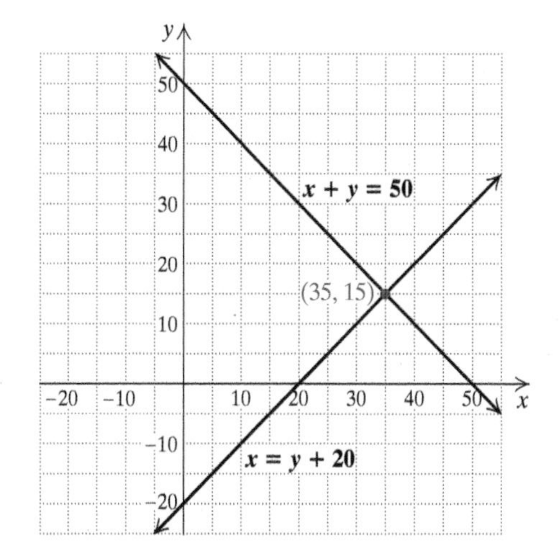

We see that the graphs intersect at the point $(35, 15)$—that is, $x = 35$ and $y = 15$. These numbers check in the statement of the original problem. This tells us that 35 million students were enrolled in prekindergarten–grade 8, and 15 million students were enrolled in grades 9–12.

Systems of two equations may have one solution, no solution, or infinitely many solutions.

One Solution

EXAMPLE 1 Solve this system graphically:

$$y - x = 1,$$
$$y + x = 3.$$

We draw the graph of each equation and find the coordinates of the point of intersection.

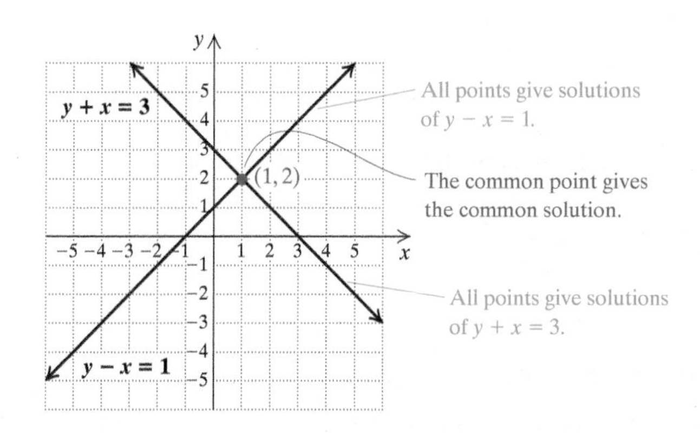

The point of intersection has coordinates that make *both* equations true. The solution seems to be the point $(1, 2)$. However, solving by graphing may give only approximate answers. Thus we check the pair $(1, 2)$ in both equations.

Check:	$y - x = 1$		$y + x = 3$	
	$2 - 1 \; ? \; 1$		$2 + 1 \; ? \; 3$	
	$1 \mid 1$	TRUE	$3 \mid 3$	TRUE

The solution is $(1, 2)$.

Do Exercises 1 and 2. ▶

Solve each system graphically.

1. $-2x + y = 1,$
   $3x + y = 1$

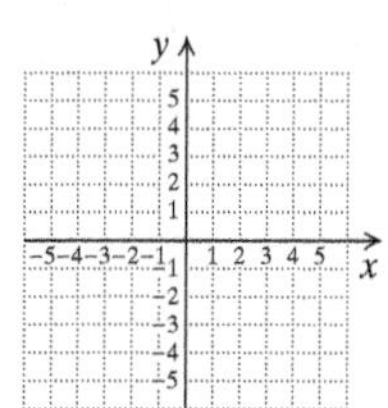

2. $y = \frac{1}{2}x,$
   $y = -\frac{1}{4}x + \frac{3}{2}$

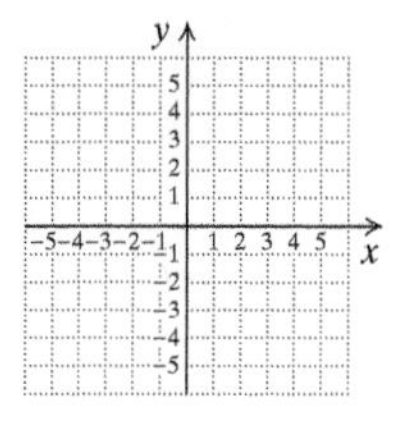

Answers

1. $(0, 1)$ **2.** $(2, 1)$

No Solution

Sometimes the equations in a system have graphs that are parallel lines.

EXAMPLE 2 Solve graphically:

$$f(x) = -3x + 5,$$
$$g(x) = -3x - 2.$$

Note that this system is written using function notation. We graph the functions. The graphs have the same slope, -3, and different y-intercepts, so they are parallel. There is no point at which they cross, so the system has no solution. No matter what point we try, it will *not* check in *both* equations. The solution set is thus the empty set, denoted $\varnothing$, or $\{ \}$.

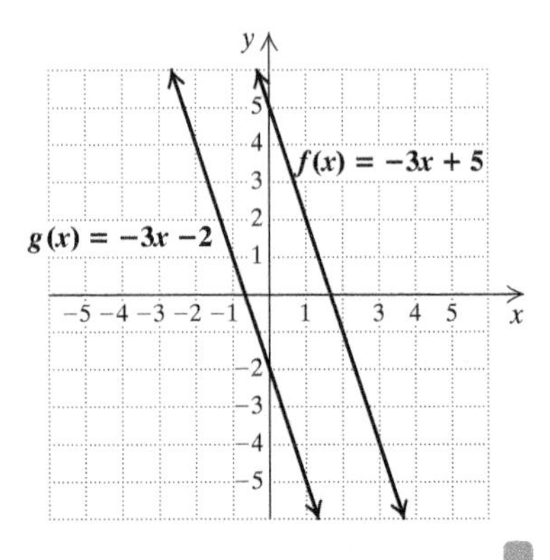

3. Solve graphically:

$$y + 2x = 3,$$
$$y + 2x = -4.$$

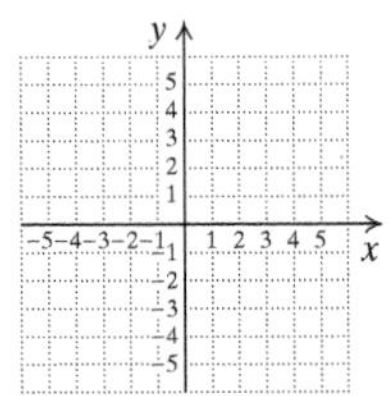

4. Classify each of the systems in Margin Exercises 1–3 as consistent or inconsistent.

The system in Margin Exercise 1 has a solution, so it is ______.

The system in Margin Exercise 2 has a solution, so it is ______.

The system in Margin Exercise 3 does not have a solution, so it is ______.

CONSISTENT SYSTEMS AND INCONSISTENT SYSTEMS

If a system of equations has at least one solution, then it is **consistent**.

If a system of equations has no solution, then it is **inconsistent**.

The system in Example 1 is consistent. The system in Example 2 is inconsistent.

Do Exercises 3 and 4.

Infinitely Many Solutions

Sometimes the equations in a system have the same graph. In such a case, the equations have an *infinite* number of solutions in common.

EXAMPLE 3 Solve graphically:

$$3y - 2x = 6,$$
$$-12y + 8x = -24.$$

We graph the equations and see that the graphs are the same. Thus any solution of one of the equations is a solution of the other. Each equation has an infinite number of solutions, two of which are shown on the graph.

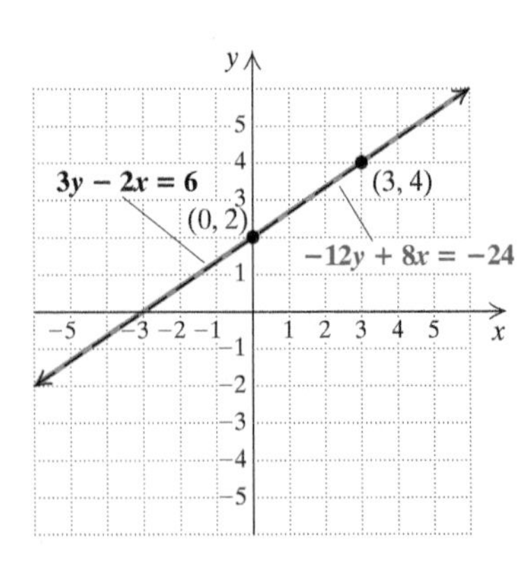

Answers

3. No solution 4. Consistent: Margin Exercises 1 and 2; inconsistent: Margin Exercise 3

Guided Solution:

4. Consistent, consistent, inconsistent

We check one such solution, $(0, 2)$, which is the y-intercept of each equation.

Check:

$$\begin{array}{c|l} 3y - 2x = 6 & \\ \hline 3(2) - 2(0) \ ?\ 6 & \\ 6 - 0 & \\ 6 & \text{TRUE} \end{array} \qquad \begin{array}{c|l} -12y + 8x = -24 & \\ \hline -12(2) + 8(0) \ ?\ -24 & \\ -24 + 0 & \\ -24 & \text{TRUE} \end{array}$$

We leave it to the student to check that $(3, 4)$ is a solution of both equations. If $(0, 2)$ and $(3, 4)$ are solutions, then all points on the line containing them will be solutions. The system has an infinite number of solutions.

DEPENDENT EQUATIONS AND INDEPENDENT EQUATIONS

If for a system of two equations in two variables:

the graphs of the equations are the same line, then the equations are **dependent**.

the graphs of the equations are different lines, then the equations are **independent**.

When we graph a system of two equations, one of the following three situations can occur.

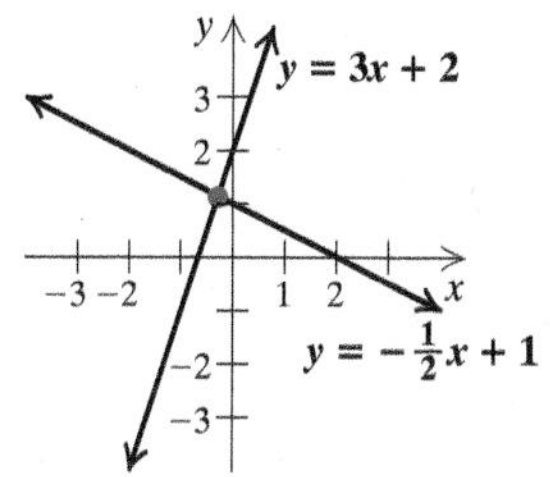

One solution.
Graphs intersect.
The system is consistent and the equations are independent.

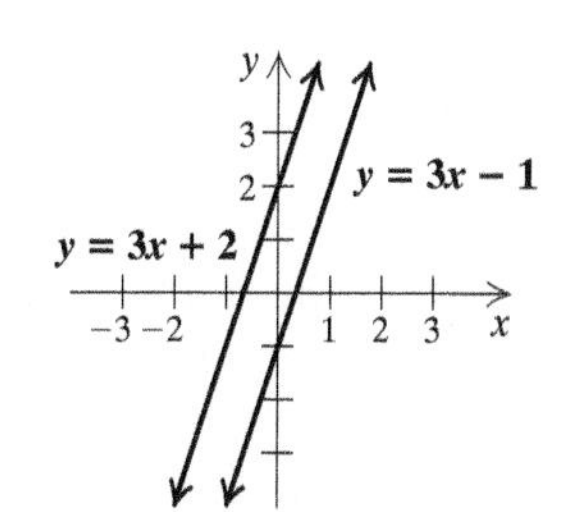

No solution.
Graphs are parallel.
The system is inconsistent and the equations are independent.

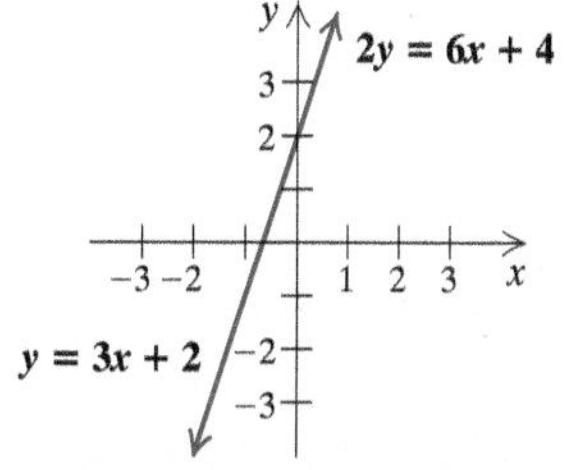

Infinitely many solutions.
Equations have the same graph. *The system is consistent and the equations are dependent.*

Let's summarize what we know about the systems of equations shown in Examples 1–3.

	NUMBER OF SOLUTIONS	GRAPHS OF EQUATIONS
EXAMPLE 1	1 System is consistent.	Different Equations are independent.
EXAMPLE 2	0 System is inconsistent.	Different Equations are independent.
EXAMPLE 3	Infinitely many System is consistent.	Same Equations are dependent.

Do Exercises 5 and 6.

5. Solve graphically:
$$2x - 5y = 10,$$
$$-6x + 15y = -30.$$

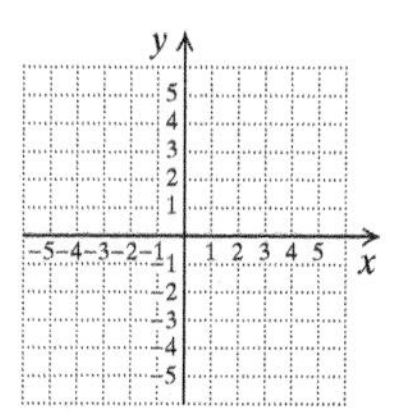

GS 6. Classify the equations in Margin Exercises 1, 2, 3, and 5 as dependent or independent.
In Margin Exercise 1, the graphs are different, so the equations are ______.
In Margin Exercise 2, the graphs are different, so the equations are .
In Margin Exercise 3, the graphs are different, so the equations are ______.
In Margin Exercise 5, the graphs are the same, so the equations are ______.

Answers

5. Infinitely many solutions
6. Independent: Margin Exercises 1, 2, and 3; dependent: Margin Exercise 5

Guided Solution:
6. Independent, independent, independent, dependent

ALGEBRAIC ▶◀ GRAPHICAL CONNECTION

Consider the equation $-2x + 13 = 4x - 17$. Let's solve it algebraically:

$$-2x + 13 = 4x - 17$$
$$13 = 6x - 17 \quad \text{Adding } 2x$$
$$30 = 6x \quad \text{Adding } 17$$
$$5 = x. \quad \text{Dividing by } 6$$

We can also solve the equation graphically, as we see in the following two methods. Using method 1, we graph two functions. The solution of the original equation is the x-coordinate of the point of intersection. Using method 2, we graph one function. The solution of the original equation is the x-coordinate of the x-intercept of the graph.

Method 1: Solve $-2x + 13 = 4x - 17$ graphically.

We let $f(x) = -2x + 13$ and $g(x) = 4x - 17$. Graphing the system of equations, we get the graph shown below.

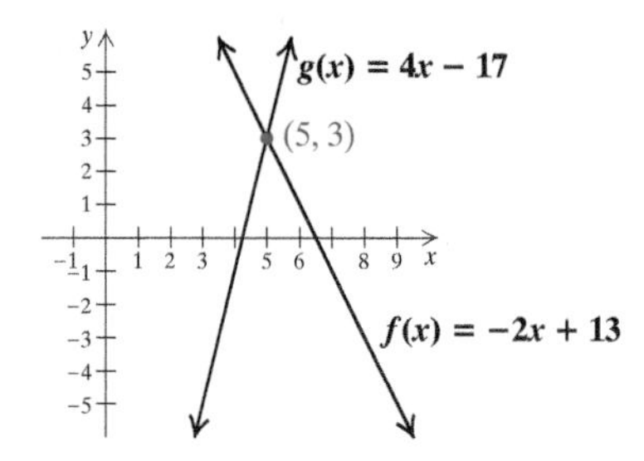

The point of intersection of the two graphs is $(5, 3)$. Note that the x-coordinate of this point is 5. This is the value of x for which $-2x + 13 = 4x - 17$, so it is the solution of the equation.

◀ **Do Exercises 7 and 8.**

7. **a)** Solve $x + 1 = \frac{2}{3}x$ algebraically.
   b) Solve $x + 1 = \frac{2}{3}x$ graphically using method 1.

8. Solve $\frac{1}{2}x + 3 = 2$ graphically using method 1.

Method 2: Solve $-2x + 13 = 4x - 17$ graphically.

Adding $-4x$ and 17 on both sides, we obtain an equation with 0 on one side: $-6x + 30 = 0$. This time we let $f(x) = -6x + 30$ and $g(x) = 0$. Since the graph of $g(x) = 0$, or $y = 0$, is the x-axis, we need only graph $f(x) = -6x + 30$ and see where it crosses the x-axis.

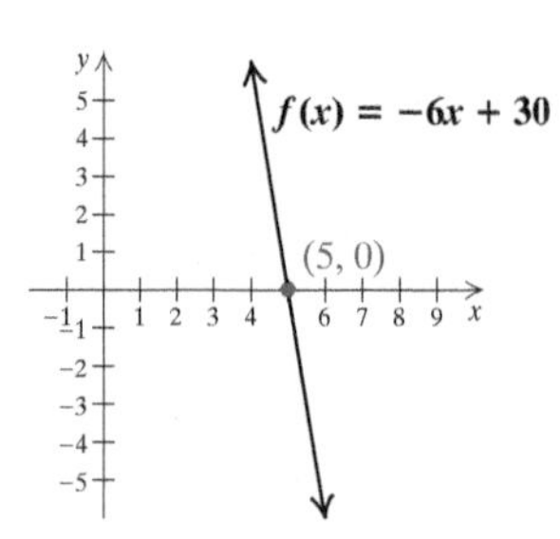

Note that the x-intercept of $f(x) = -6x + 30$ is $(5, 0)$, or just 5. This x-value is the solution of the equation $-2x + 13 = 4x - 17$.

◀ **Do Exercises 9 and 10.**

9. **a)** Solve $x + 1 = \frac{2}{3}x$ graphically using method 2.
   b) Compare your answers to Margin Exercises 7(a), 7(b), and 9(a).

10. Solve $\frac{1}{2}x + 3 = 2$ graphically using method 2.

Answers

7. (a) -3; **(b)** the same: -3 **8.** -2
9. (a) -3; **(b)** All are -3. **10.** -2

CALCULATOR CORNER

Solving Systems of Equations We can solve a system of two equations in two variables using a graphing calculator. Consider the system of equations in Example 1. First, we solve the equations for y, obtaining $y = x + 1$ and $y = -x + 3$. Next, we enter $y_1 = x + 1$ and $y_2 = -x + 3$ on the equation-editor screen and graph the equations. We can use the standard viewing window, $[-10, 10, -10, 10]$.

We will use the INTERSECT feature to find the coordinates of the point of intersection of the lines. To access this feature, we press 2ND CALC 5. (CALC is the second operation associated with the TRACE key.) The query "First curve?" appears on the graph screen. The blinking cursor is positioned on the graph of y_1. We press ENTER to indicate that this is the first curve involved in the intersection. Next, the query "Second curve?" appears and the blinking cursor is positioned on the graph of y_2. We press ENTER to indicate that this is the second curve. Now the query "Guess?" appears. We use the ▷ and ◁ keys to move the cursor close to the point of intersection or we enter an x-value close to the first coordinate of the point of intersection. Then we press ENTER. The coordinates of the point of intersection of the graphs, $x = 1$, $y = 2$, appear at the bottom of the screen. Thus the solution of the system of equations is $(1, 2)$.

$y_1 = x + 1,\ y_2 = -x + 3$

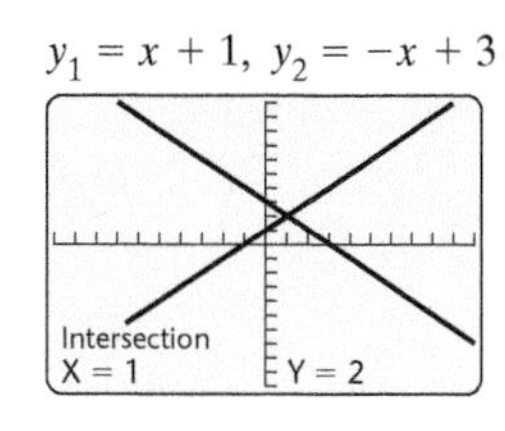

EXERCISES: Use a graphing calculator to solve each system of equations.

1. $x + y = 5,$
$y = x + 1$

2. $y = x + 3,$
$2x - y = -7$

3. $x - y = -6,$
$y = 2x + 7$

4. $x + 4y = -1,$
$x - y = 4$

3.1 Exercise Set

FOR EXTRA HELP   MyLab Math

✓ Check Your Understanding

Reading Check Determine whether each statement is true or false.

RC1. Every system of equations has one solution.

RC2. A solution of a system of equations in two variables is an ordered pair.

RC3. Graphs of two lines may have one point, no points, or an infinite number of points in common.

RC4. If a system of two equations has only one solution, the system is consistent and the equations in the system are independent.

Concept Check Determine whether the given ordered pair is a solution of the system of equations.

CC1. $(-2, -3)$; $x - y = 1,$
$x + y = -5$

CC2. $(-1, 0)$; $b - a = 1,$
$2a = b - 2$

CC3. $(5, -6)$; $c - d = 11,$
$2c = d + 4$

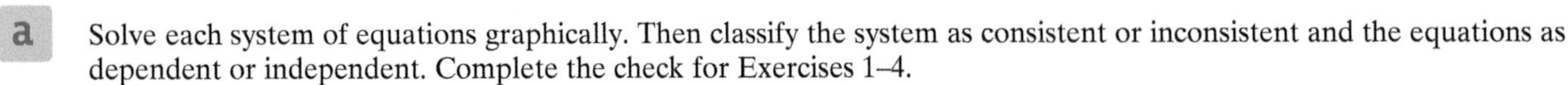

a Solve each system of equations graphically. Then classify the system as consistent or inconsistent and the equations as dependent or independent. Complete the check for Exercises 1–4.

1. $x + y = 4,$
$x - y = 2$

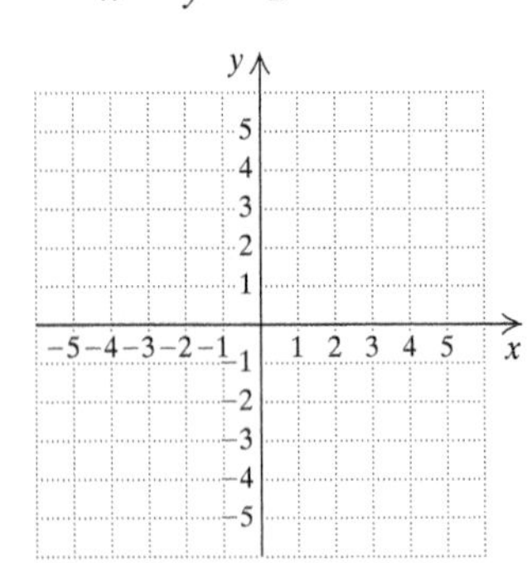

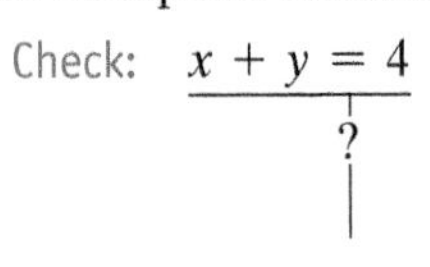

Check: $x + y = 4$ | ?

$x - y = 2$ | ?

2. $x - y = 3,$
$x + y = 5$

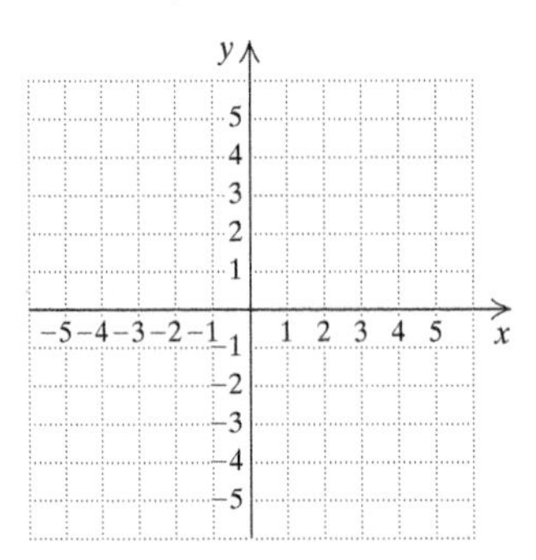

Check: $x - y = 3$ | ?

$x + y = 5$ | ?

3. $2x - y = 4,$
$2x + 3y = -4$

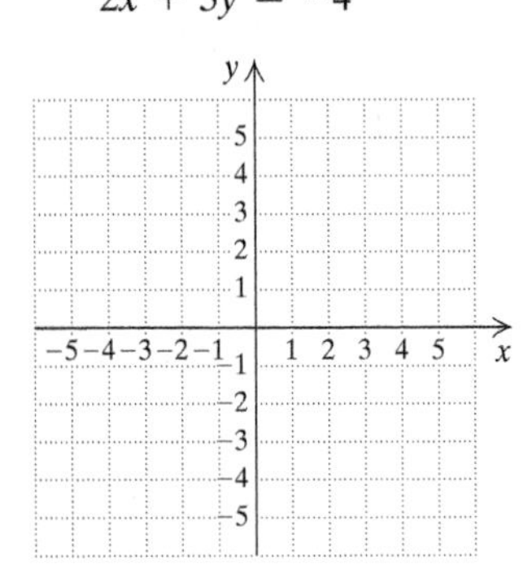

Check: $2x - y = 4$ | ?

$2x + 3y = -4$ | ?

4. $3x + y = 5,$
$x - 2y = 4$

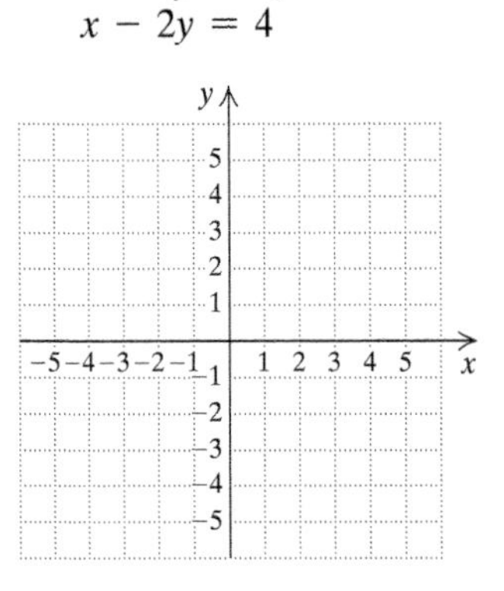

Check: $3x + y = 5$ | ?

$x - 2y = 4$ | ?

5. $2x + y = 6,$
$3x + 4y = 4$

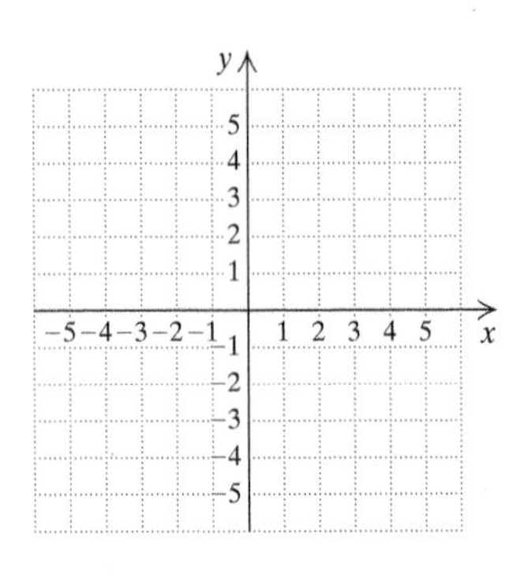

6. $2y = 6 - x,$
$3x - 2y = 6$

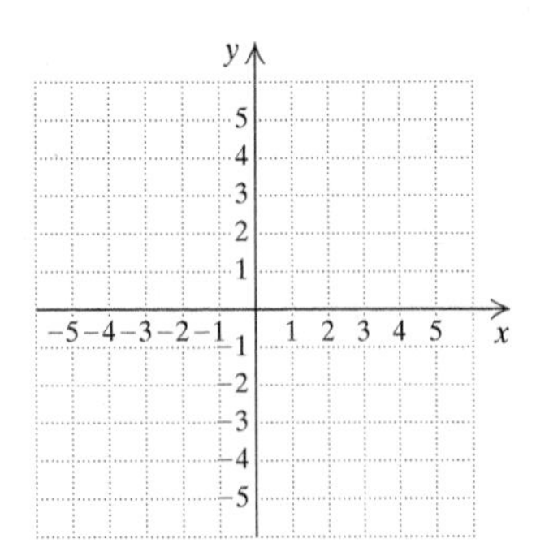

7. $f(x) = x - 1,$
$g(x) = -2x + 5$

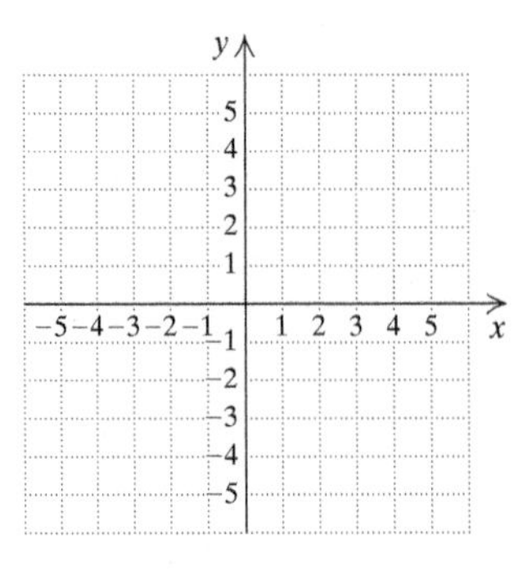

8. $f(x) = x + 1,$
$g(x) = \frac{2}{3}x$

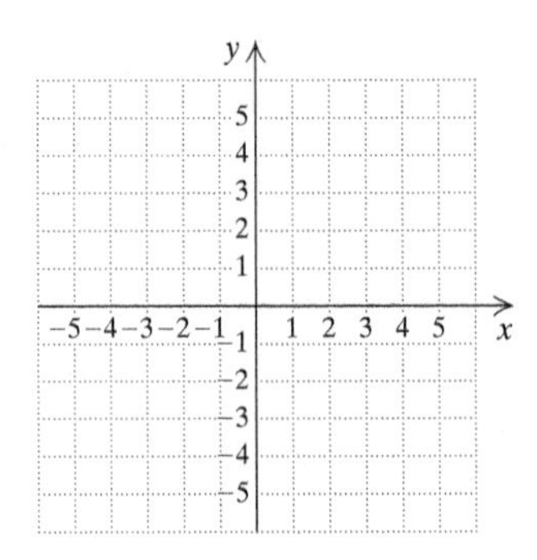

9. $2u + v = 3,$
$2u = v + 7$

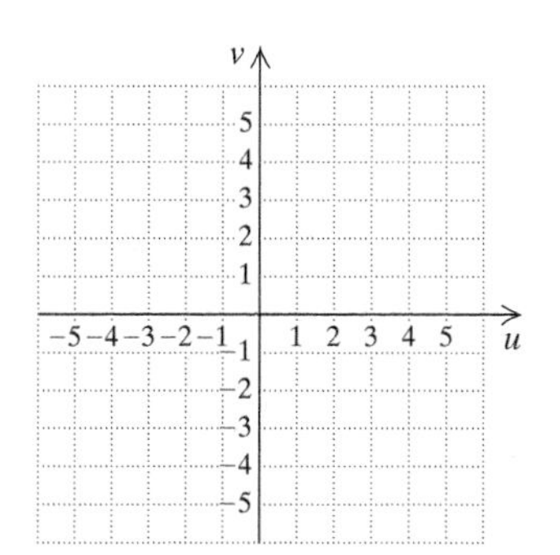

10. $2b + a = 11,$
$a - b = 5$

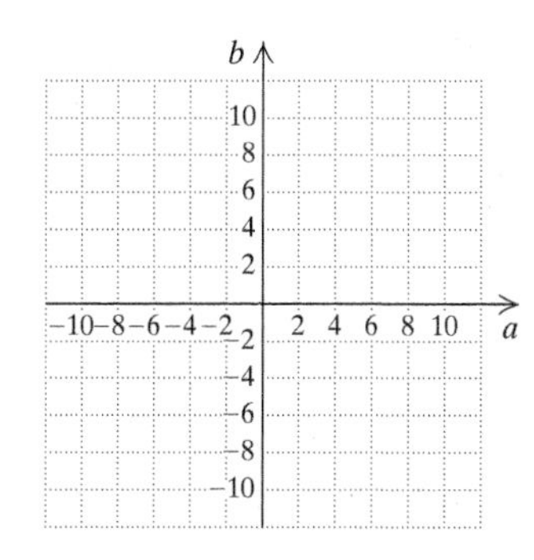

11. $f(x) = -\frac{1}{3}x - 1,$
$g(x) = \frac{4}{3}x - 6$

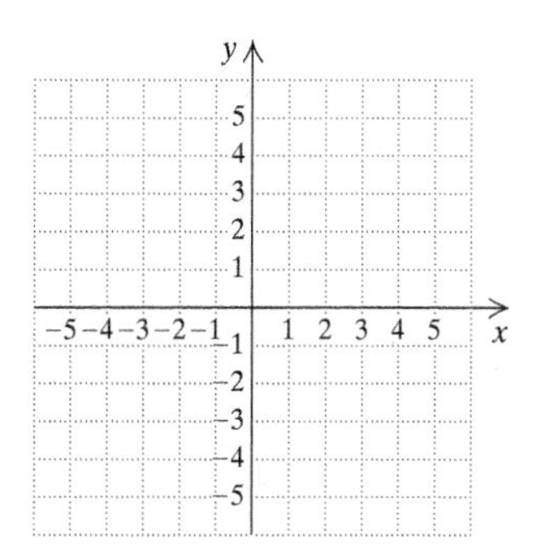

12. $f(x) = -\frac{1}{4}x + 1,$
$g(x) = \frac{1}{2}x - 2$

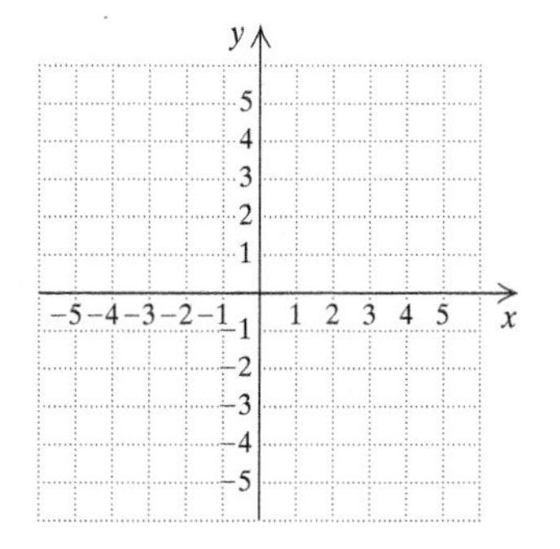

13. $6x - 2y = 2,$
$9x - 3y = 1$

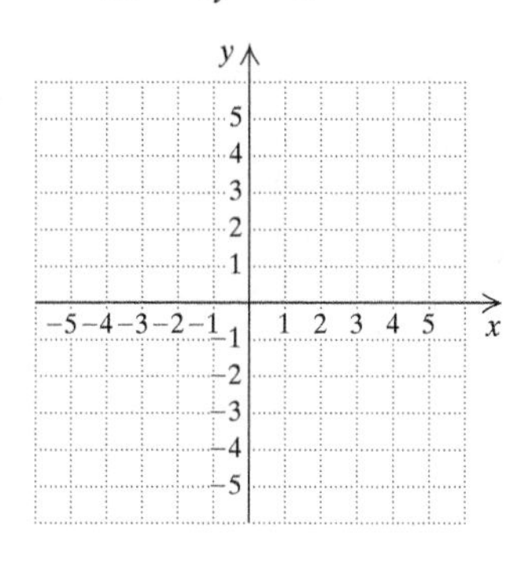

14. $y - x = 5,$
$2x - 2y = 10$

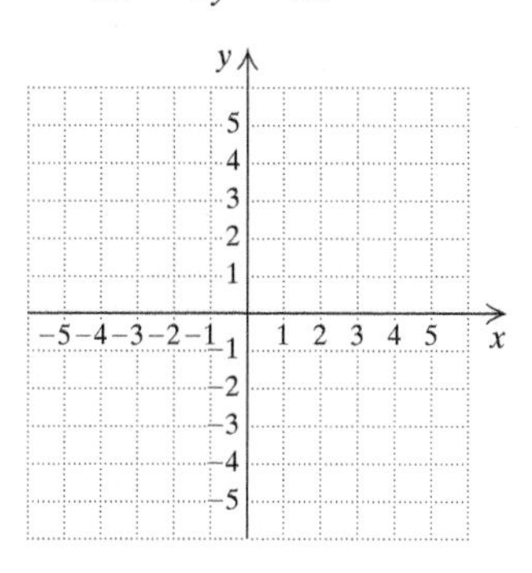

15. $2x - 3y = 6,$
$3y - 2x = -6$

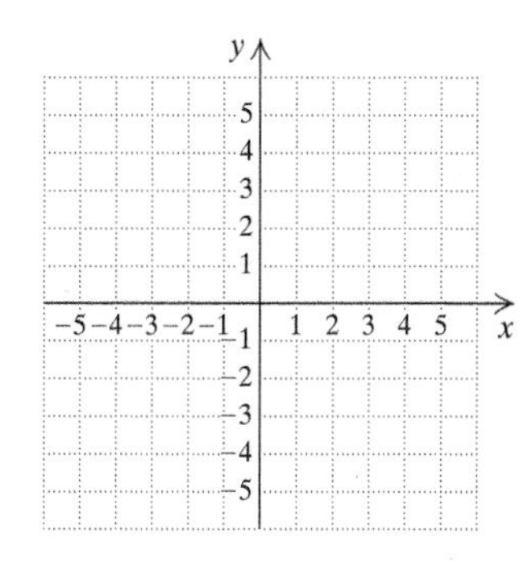

16. $y = 3 - x,$
$2x + 2y = 6$

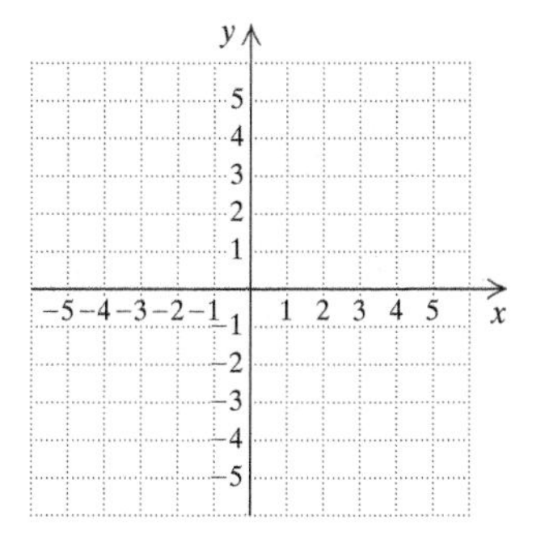

17. $x = 4,$
$y = -5$

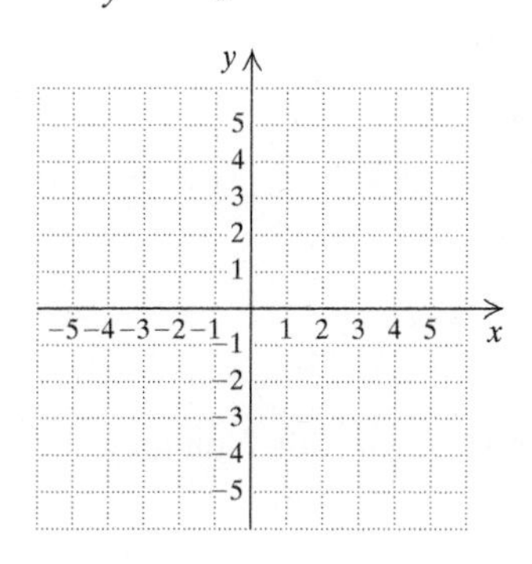

18. $x = -3,$
$y = 2$

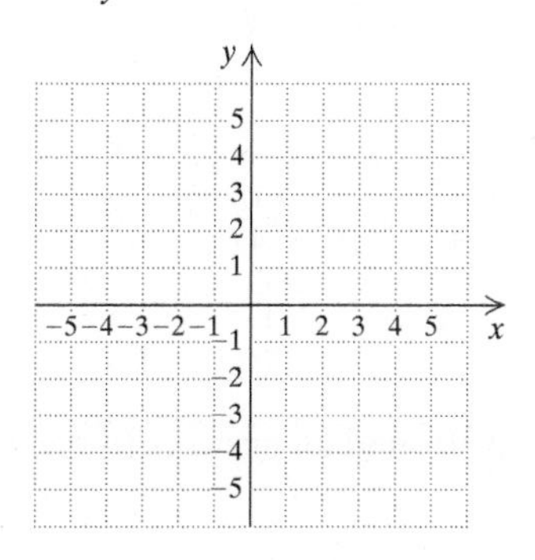

19. $y = -x - 1,$
$x - 2y = 8$

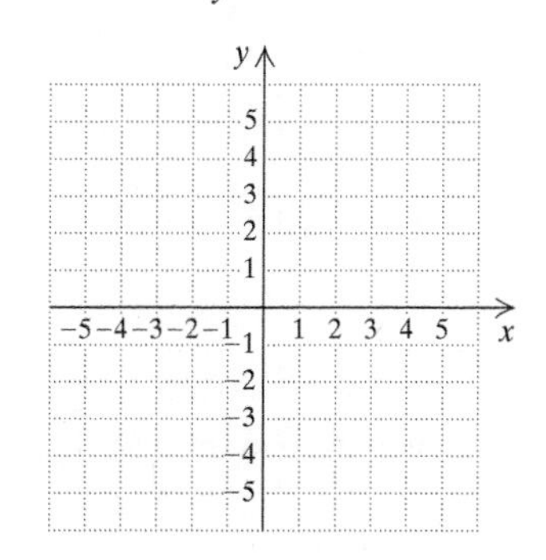

20. $a + 2b = -3,$
$b - a = 6$

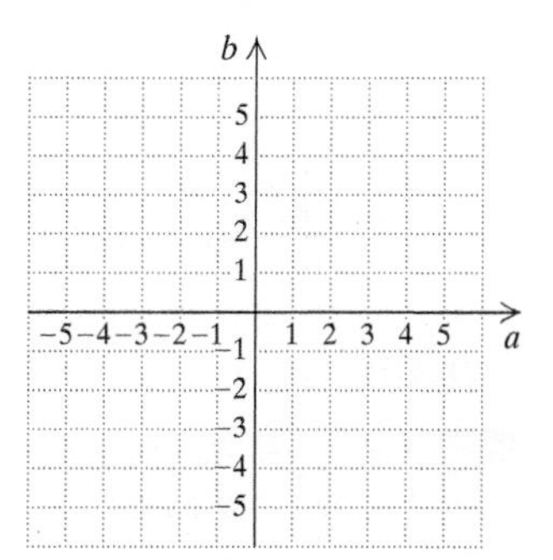

Matching. Each of Exercises 21–26 shows the graph of a system of equations and its solution. First, classify the system as consistent or inconsistent and the equations as dependent or independent. Then match it with one of the appropriate systems of equations (A)–(F), which follow.

21. Solution: $(3, 3)$

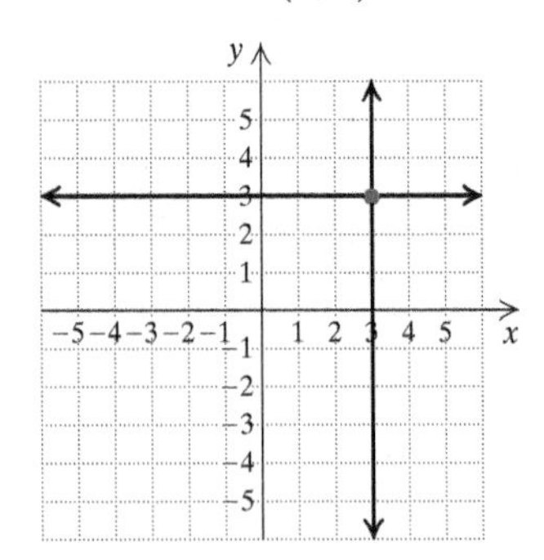

22. Solution: $(1, 1)$

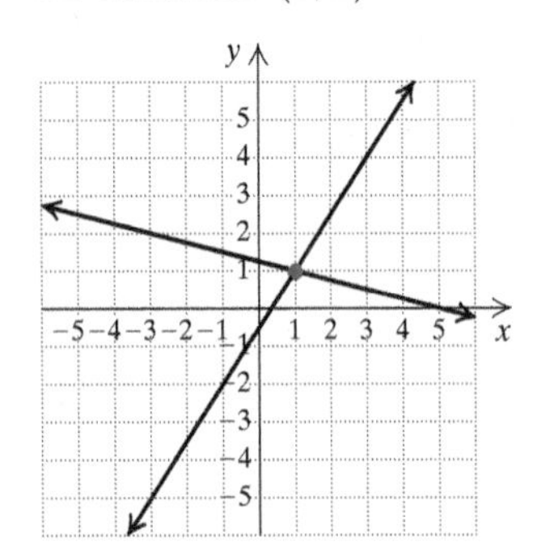

23. Solutions: Infinitely many

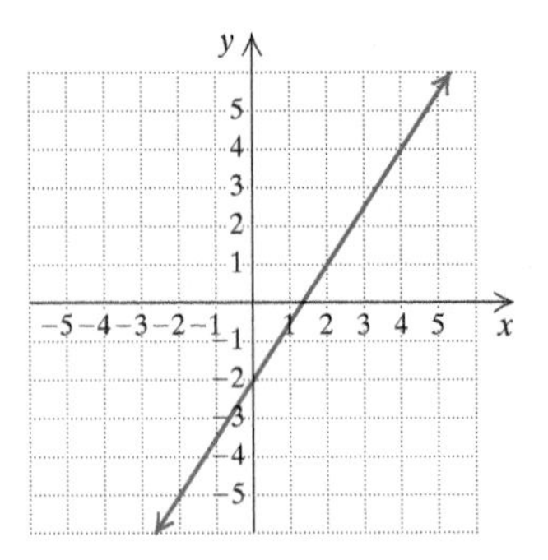

24. Solution: $(4, -3)$

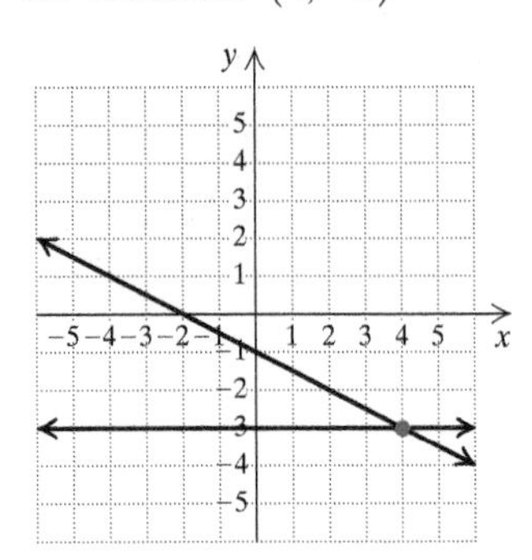

25. Solution: No solution

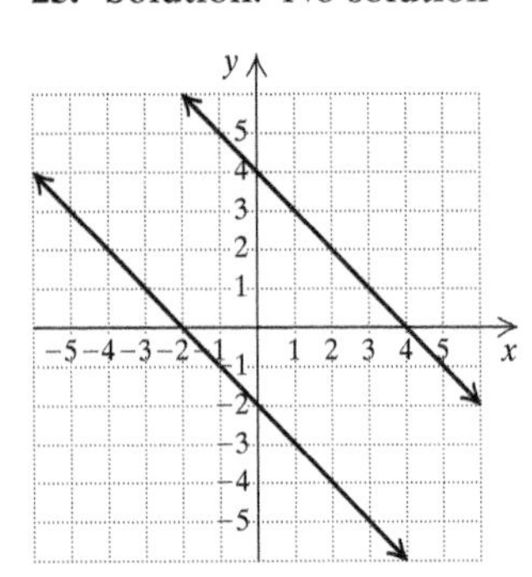

26. Solution: $(-1, 3)$

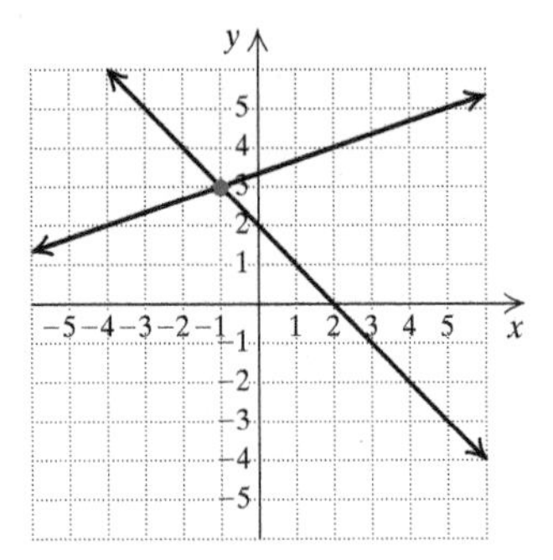

A. $3y - x = 10,$
$x = -y + 2$

B. $9x - 6y = 12,$
$y = \frac{3}{2}x - 2$

C. $2y - 3x = -1,$
$x + 4y = 5$

D. $x + y = 4,$
$y = -x - 2$

E. $\frac{1}{2}x + y = -1,$
$y = -3$

F. $x = 3,$
$y = 3$

Skill Maintenance

Solve. [1.1d]

27. $3x + 4 = x - 2$

28. $\frac{3}{4}x + 2 = \frac{2}{5}x - 5$

29. $4x - 5x = 8x - 9 + 11x$

30. $5(10 - 4x) = -3(7x - 4)$

Synthesis

Use a graphing calculator to solve each system of equations. Round all answers to the nearest hundredth. You may need to solve for y first.

31. $2.18x + 7.81y = 13.78,$
$5.79x - 3.45y = 8.94$

32. $f(x) = 123.52x + 89.32,$
$g(x) = -89.22x + 33.76$

Solve graphically.

33. $y = |x|,$
$x + 4y = 15$

34. $x - y = 0,$
$y = x^2$

35. $2x - 3 = 4 - 5x$

36. $x + 3 = 2x^2$

Solving by Substitution

3.2

OBJECTIVES

a Solve systems of equations in two variables by the substitution method.

b Solve applied problems by solving systems of two equations using substitution.

Consider this system of equations:

$$5x + 9y = 2,$$
$$4x - 9y = 10.$$

What is the solution? It is rather difficult to tell exactly by graphing. It would appear that fractions are involved. It turns out that the solution is

$$\left(\frac{4}{3}, -\frac{14}{27}\right).$$

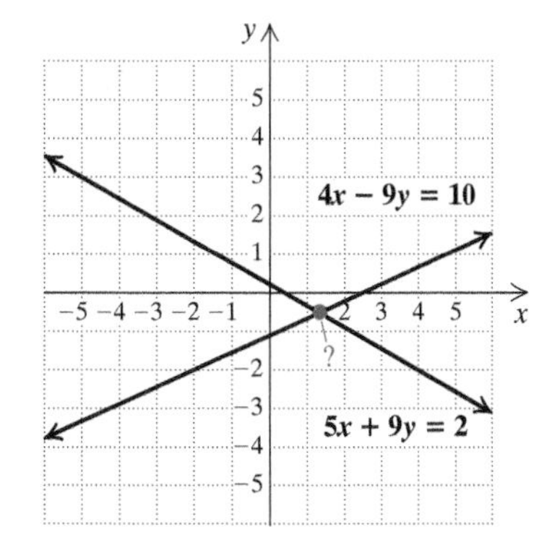

Solving by graphing, though useful in many applied situations, is not always fast or accurate in cases where solutions are not integers. We need techniques involving algebra to determine the solution exactly. Because these techniques use algebra, they are called **algebraic methods**.

a THE SUBSTITUTION METHOD

One nongraphical method for solving systems is known as the **substitution method**.

SKILL REVIEW *Solve equations.* [1.1d]

Solve.

1. $3y - 4 = 2$

2. $2(x + 1) + 5 = 1$

Answers: **1.** 2 **2.** −3

MyLab Math VIDEO

EXAMPLE 1 Solve this system:

$$x + y = 4, \quad (1)$$
$$x = y + 1. \quad (2)$$

Equation (2) says that x and $y + 1$ name the same number. Thus we can substitute $y + 1$ for x in equation (1):

$$x + y = 4 \qquad \text{Equation (1)}$$
$$(y + 1) + y = 4. \qquad \text{Substituting } y + 1 \text{ for } x$$

Since this equation has only one variable, we can solve for y using methods learned earlier:

$$(y + 1) + y = 4$$
$$2y + 1 = 4 \qquad \text{Removing parentheses and collecting like terms}$$
$$2y = 3 \qquad \text{Subtracting 1}$$
$$y = \tfrac{3}{2}. \qquad \text{Dividing by 2}$$

We return to the original pair of equations and substitute $\frac{3}{2}$ for y in *either* equation so that we can solve for x. Calculation will be easier if we choose equation (2) since it is already solved for x:

$$x = y + 1 \qquad \text{Equation (2)}$$
$$= \tfrac{3}{2} + 1 \qquad \text{Substituting } \tfrac{3}{2} \text{ for } y$$
$$= \tfrac{3}{2} + \tfrac{2}{2} = \tfrac{5}{2}.$$

We obtain the ordered pair $\left(\frac{5}{2}, \frac{3}{2}\right)$. Even though we solved for y *first*, it is still the *second* coordinate since x is before y alphabetically. We check to be sure that the ordered pair is a solution.

Check:

$x + y = 4$		$x = y + 1$	
$\frac{5}{2} + \frac{3}{2} \stackrel{?}{=} 4$		$\frac{5}{2} \stackrel{?}{=} \frac{3}{2} + 1$	
$\frac{8}{2}$		$\frac{3}{2} + \frac{2}{2}$	
4	TRUE	$\frac{5}{2}$	TRUE

Since $\left(\frac{5}{2}, \frac{3}{2}\right)$ checks, it is the solution. Even though exact fraction solutions are difficult to determine graphically, a graph can help us to visualize whether the solution is reasonable.

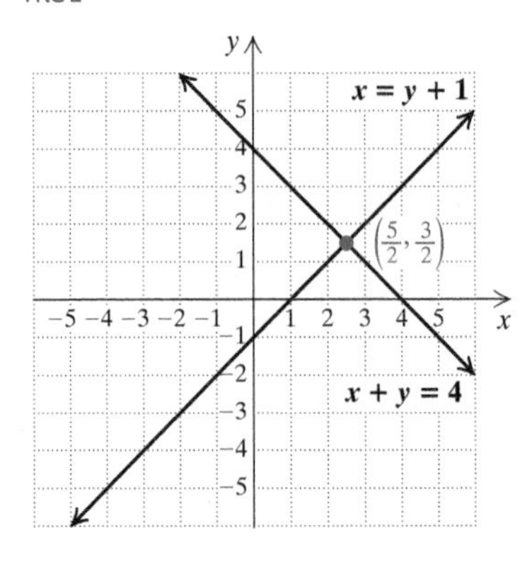

Solve by the substitution method.

1. $x + y = 6,$
   $y = x + 2$

2. $y = 7 - x,$
   $2x - y = 8$

   (*Caution*: Use parentheses when you substitute, being careful about removing them. Remember to solve for both variables.)

◀ **Do Exercises 1 and 2.**

Suppose that neither equation of a pair has a variable alone on one side. We then solve one equation for one of the variables.

EXAMPLE 2 Solve this system:

$$2x + y = 6, \quad (1)$$
$$3x + 4y = 4. \quad (2)$$

First, we solve one equation for one variable. Since the coefficient of y is 1 in equation (1), it is the easier one to solve for y:

$$y = 6 - 2x. \quad (3)$$

Next, we substitute $6 - 2x$ for y in equation (2) and solve for x:

$3x + 4(6 - 2x) = 4$	Substituting $6 - 2x$ for y
$3x + 24 - 8x = 4$	Multiplying to remove parentheses
$24 - 5x = 4$	Collecting like terms
$-5x = -20$	Subtracting 24
$x = 4.$	Dividing by -5

Caution!

Remember to use parentheses when you substitute. Then remove them properly.

In order to find y, we return to either of the original equations, (1) or (2), or equation (3), which we solved for y. It is generally easier to use an equation like (3), where we have solved for the specific variable. We substitute 4 for x in equation (3) and solve for y:

$$y = 6 - 2x = 6 - 2(4) = 6 - 8 = -2.$$

We obtain the ordered pair $(4, -2)$.

Check:

$2x + y = 6$		$3x + 4y = 4$	
$2(4) + (-2) \stackrel{?}{=} 6$		$3(4) + 4(-2) \stackrel{?}{=} 4$	
$8 - 2$		$12 - 8$	
6	TRUE	4	TRUE

Since $(4, -2)$ checks, it is the solution.

◀ **Do Exercises 3 and 4.**

Solve by the substitution method.

3. $2y + x = 1,$
   $y - 2x = 8$

4. GS $8x - 5y = 12, \quad (1)$
   $x - y = 3 \quad (2)$

   Solve for x in equation (2):

   $x - y = 3$

   $x = \_\_\_ + 3. \quad (3)$

   Substitute $y + 3$ for $\_\_\_$ in equation (1) and solve for $\_\_\_$:

   $8x - 5y = 12$

   $8(y + 3) - 5y = 12$

   $8y + \_\_\_ - 5y = 12$

   $3y + 24 = 12$

   $3y = -12$

   $y = \_\_\_.$

   Substitute -4 for y in equation (3) and solve for x:

   $x = y + 3$

   $= -4 + 3$

   $= \_\_\_.$

   The ordered pair checks in both equations. The solution is $(\_\_\_, \_\_\_)$.

Answers

1. $(2, 4)$ **2.** $(5, 2)$ **3.** $(-3, 2)$ **4.** $(-1, -4)$

Guided Solution:

4. $y, x, y, 24, -4, -1, -1, -4$

EXAMPLE 3 Solve this system of equations:

$$y = -3x + 5, \qquad (1)$$
$$y = -3x - 2. \qquad (2)$$

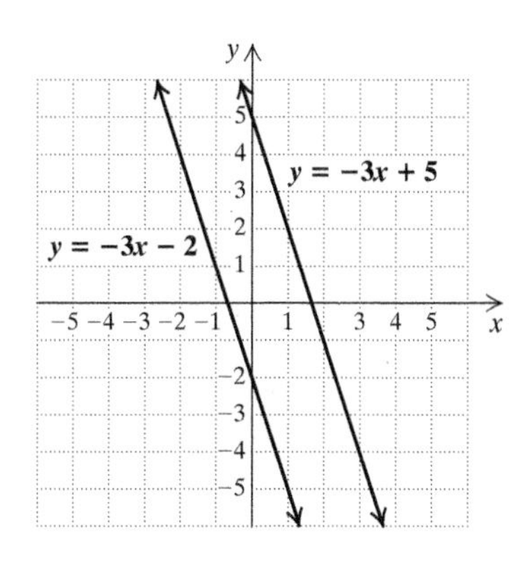

The graphs of the equations in the system are shown at right. Since the graphs are parallel, there is no solution. Let's try to solve this system algebraically using substitution. We substitute $-3x - 2$ for y in equation (1):

$-3x - 2 = -3x + 5$ Substituting $-3x - 2$ for y
$-2 = 5.$ Adding $3x$

We have a false equation. The equation has no solution. This means that the system has **no solution**.

Do Exercise 5. ▶

5. Solve using substitution:
$$y + 2x = 3,$$
$$y + 2x = -4.$$

EXAMPLE 4 Solve this system of equations:

$$x = 2y - 1, \qquad (1)$$
$$4y - 2x = 2. \qquad (2)$$

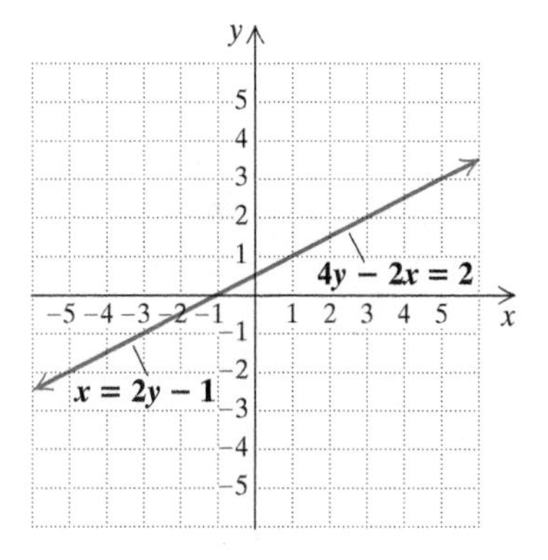

The graphs of the equations in the system are shown at right. Since the graphs are the same, there is an infinite number of solutions.

Let's try to solve this system algebraically using substitution. We substitute $2y - 1$ for x in equation (2):

$4y - 2(2y - 1) = 2$ Substituting $2y - 1$ for x
$4y - 4y + 2 = 2$ Removing parentheses
$2 = 2.$ Simplifying; $4y - 4y = 0$

We have a true equation. Any value of y will make this equation true. This means that the system has **infinitely many solutions**.

Do Exercise 6. ▶

6. Solve using substitution:
$$x - 3y = 1,$$
$$6y - 2x = -2.$$

SPECIAL CASES

When solving a system of two linear equations in two variables:

1. If a false equation is obtained, then the system has no solution.
2. If a true equation is obtained, then the system has an infinite number of solutions.

b SOLVING APPLIED PROBLEMS INVOLVING TWO EQUATIONS

Many applied problems are easier to solve if we first translate to a system of two equations rather than to a single equation.

EXAMPLE 5 *Architecture.* The architects who designed the John Hancock Building in Chicago created a visually appealing building that slants on the sides. The ground floor is in the shape of a rectangle that is larger than the rectangle formed by the top floor. The ground floor has a perimeter of 860 ft. The length is 100 ft more than the width. Find the length and the width.

Answers
5. No solution **6.** Infinitely many solutions

7. *Architecture.* The top floor of the John Hancock Building is also in the shape of a rectangle, but its perimeter is 520 ft. The width is 60 ft less than the length. Find the length and the width.

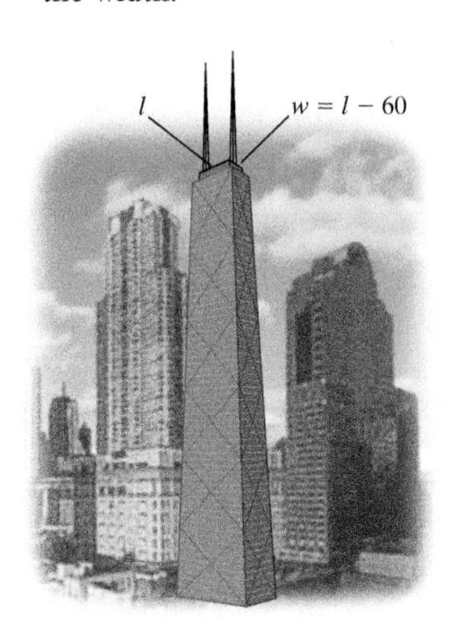

Answer

7. Length: 160 ft; width: 100 ft

1. **Familiarize.** We first make a drawing and label it, using l for length and w for width. We recall or look up the formula for perimeter: $P = 2l + 2w$. This formula can be found at the back of this book.

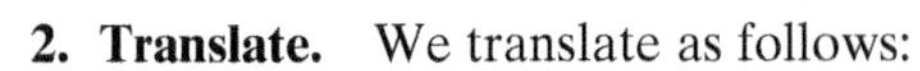

2. **Translate.** We translate as follows:

   The perimeter is 860 ft.

   $$2l + 2w = 860.$$

   We can also write a second equation:

   The length is 100 ft more than the width.

   $$l = w + 100.$$

   We now have a system of equations:

   $$2l + 2w = 860, \quad \textbf{(1)}$$
   $$l = w + 100. \quad \textbf{(2)}$$

3. **Solve.** We substitute $w + 100$ for l in equation (1) and solve for w:

   $$2(w + 100) + 2w = 860 \quad \text{Substituting in equation (1)}$$
   $$2w + 200 + 2w = 860 \quad \text{Multiplying to remove parentheses}$$
   $$4w + 200 = 860 \quad \text{Collecting like terms}$$
   $$\left.\begin{aligned} 4w &= 660 \\ w &= 165. \end{aligned}\right\} \quad \text{Solving for } w$$

   Next, we substitute 165 for w in equation (2) and solve for l:

   $$l = 165 + 100 = 265.$$

4. **Check.** Consider the dimensions 265 ft and 165 ft. The length is 100 ft more than the width. The perimeter is 2(265 ft) + 2(165 ft), or 860 ft. The dimensions 265 ft and 165 ft check in the original problem.
5. **State.** The length is 265 ft, and the width is 165 ft.

◀ **Do Exercise 7.**

3.2 Exercise Set

FOR EXTRA HELP

MyLab Math

✓ Check Your Understanding

Reading Check Determine whether each statement is true or false.

RC1. The substitution method is an algebraic method for solving systems of equations.

RC2. We can find solutions of systems of equations involving fractions using the substitution method.

RC3. When we are writing the solution of a system, the value that we found first is always the first number in the ordered pair.

RC4. When solving using substitution, if we obtain a false equation, then the system has many solutions.

Concept Check For each system of equations, the value of one variable is given. Find the value of the other variable. Then write the solution as an ordered pair, using alphabetical order for the variables.

CC1. $2a + 5b = 36,$
$2a - b = 12$
$a = 8$

CC2. $3x - 2y = 10,$
$2y - 2x = 9$
$y = \frac{47}{2}$

CC3. $5w - p = 7,$
$3w = 8 + p$
$p = -\frac{19}{2}$

a Solve each system of equations by the substitution method.

1. $y = 5 - 4x,$
$2x - 3y = 13$

2. $x = 8 - 4y,$
$3x + 5y = 3$

3. $2y + x = 9,$
$x = 3y - 3$

4. $9x - 2y = 3,$
$3x - 6 = y$

5. $3s - 4t = 14,$
$5s + t = 8$

6. $m - 2n = 3,$
$4m + n = 1$

7. $9x - 2y = -6,$
$7x + 8 = y$

8. $t = 4 - 2s,$
$t + 2s = 6$

9. $-5s + t = 11,$
$4s + 12t = 4$

10. $5x + 6y = 14,$
$-3y + x = 7$

11. $2x - 3 = y,$
$y - 2x = 1$

12. $4p - 2w = 16,$
$5p + 7w = 1$

13. $3a - b = 7,$
$2a + 2b = 5$

14. $3x + y = 4,$
$12 - 3y = 9x$

15. $2x - 6y = 4,$
$3y + 2 = x$

16. $5x + 3y = 4,$
$x - 4y = 3$

17. $2x + 2y = 2,$
$3x - y = 1$

18. $4x + 13y = 5,$
$-6x + y = 13$

b Solve.

19. *Archaeology.* The remains of an ancient ball court in Monte Alban, Mexico, include a rectangular playing alley with a perimeter of about 60 m. The length of the alley is five times the width. Find the length and the width of the playing alley.

20. *Soccer Field.* The perimeter of a soccer field is 340 m. The length exceeds the width by 50 m. Find the length and the width.

21. *Supplementary Angles.* **Supplementary angles** are angles whose sum is 180°. Two supplementary angles are such that one angle is 12° less than three times the other. Find the measures of the angles.

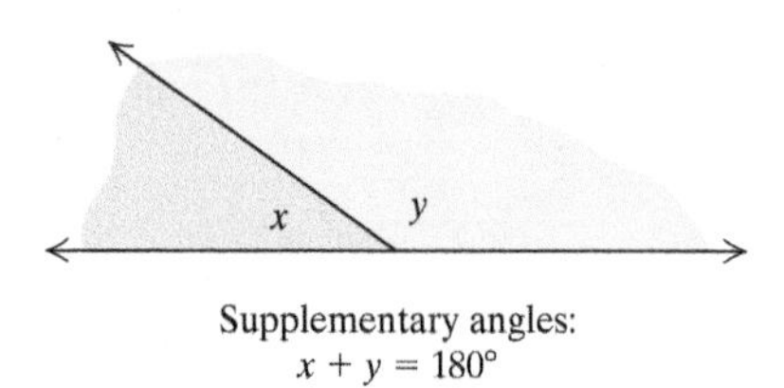

Supplementary angles:
$x + y = 180°$

22. *Complementary Angles.* **Complementary angles** are angles whose sum is 90°. Two complementary angles are such that one angle is 6° more than five times the other. Find the measures of the angles.

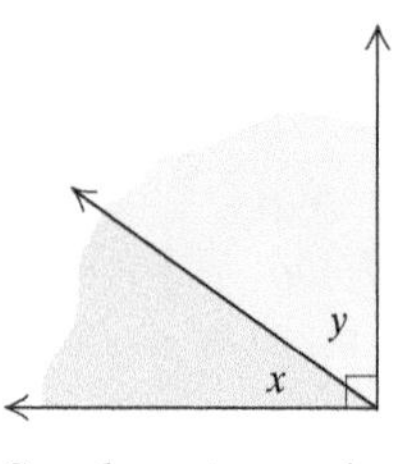

Complementary angles:
$x + y = 90°$

23. *Hockey Points.* At one time, hockey teams received two points when they won a game and one point when they tied. One season, a team won a championship with 60 points. They won 9 more games than they tied. How many wins and how many ties did the team have?

24. *Airplane Seating.* An airplane has a total of 152 seats. The number of coach-class seats is 5 more than six times the number of first-class seats. How many of each type of seat are there on the plane?

Skill Maintenance

25. Find the slope of the line $y = 1.3x - 7$. [2.4b]

26. Simplify: $-9(y + 7) - 6(y - 4)$. [J18]

27. Solve $A = \frac{pq}{7}$ for p. [1.2a]

28. Find the slope of the line containing the points $(-2, 3)$ and $(-5, -4)$. [2.4b]

Solve. [1.1d]

29. $-4x + 5(x - 7) = 8x - 6(x + 2)$

30. $-12(2x - 3) = 16(4x - 5)$

Synthesis

31. Two solutions of $y = mx + b$ are $(1, 2)$ and $(-3, 4)$. Find m and b.

32. Solve for x and y in terms of a and b:

$$5x + 2y = a,$$
$$x - y = b.$$

33. *Design.* A piece of posterboard has a perimeter of 156 in. If you cut 6 in. off the width, the length becomes four times the width. What are the dimensions of the original piece of posterboard?

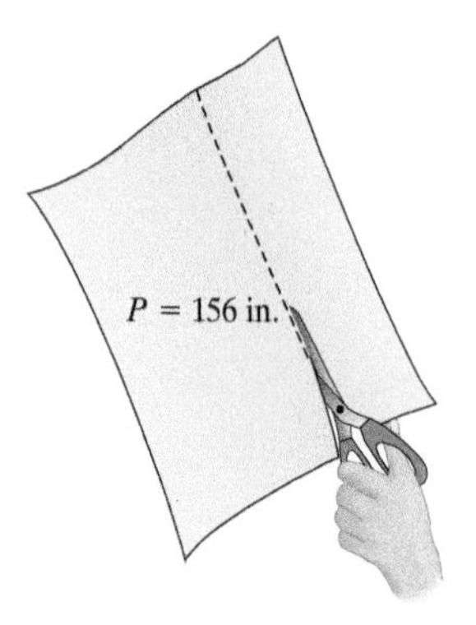

34. *Nontoxic Scouring Powder.* A nontoxic scouring powder is made up of 4 parts baking soda and 1 part vinegar. How much of each ingredient is needed for a 16-oz mixture?

Solving by Elimination

3.3

OBJECTIVES

a Solve systems of equations in two variables by the elimination method.

b Solve applied problems by solving systems of two equations using elimination.

a THE ELIMINATION METHOD

Solve equations. [1.1d]

Solve. Clear the fractions or decimals first.

1. $4.2x - 10.4 = 45.4 - 5.1x$ **2.** $\frac{1}{4}x - \frac{2}{5} + \frac{1}{2}x = \frac{3}{5} + x$

Answers: 1. 6 **2.** −4

MyLab Math VIDEO

The **elimination method*** for solving systems of equations makes use of the *addition principle* for equations.

EXAMPLE 1 Solve this system:

$$2x - 3y = 0, \quad (1)$$
$$-4x + 3y = -1. \quad (2)$$

The elimination method works well to solve this system because the $-3y$ in one equation and the $3y$ in the other are opposites. If we add them, these terms will add to 0 and, in effect, the variable y will have been "eliminated."

We will use the addition principle for equations, adding the same number on both sides of the equation. According to equation (2), $-4x + 3y$ and -1 are the same number. Thus we can use a vertical form and add $-4x + 3y$ on the left side of equation (1) and -1 on the right side:

$$\begin{aligned} 2x - 3y &= 0 && (1)\\ -4x + 3y &= -1 && (2)\\ \hline -2x + 0y &= -1 && \text{Adding}\\ -2x + \ \ 0 &= -1\\ -2x &= -1. \end{aligned}$$

We have eliminated the variable y. We now have an equation with just one variable, which we solve for x:

$$\begin{aligned} -2x &= -1\\ x &= \tfrac{1}{2}. \end{aligned}$$

Next, we substitute $\frac{1}{2}$ for x in either equation and solve for y:

$$\begin{aligned} 2\cdot\tfrac{1}{2} - 3y &= 0 && \text{Substituting in equation (1)}\\ 1 - 3y &= 0\\ -3y &= -1 && \text{Subtracting 1}\\ y &= \tfrac{1}{3}. && \text{Dividing by } -3 \end{aligned}$$

We obtain the ordered pair $\left(\frac{1}{2}, \frac{1}{3}\right)$. We can check by solving graphically, as shown at right. To check algebraically, we substitute $\frac{1}{2}$ for x and $\frac{1}{3}$ for y in both equations.

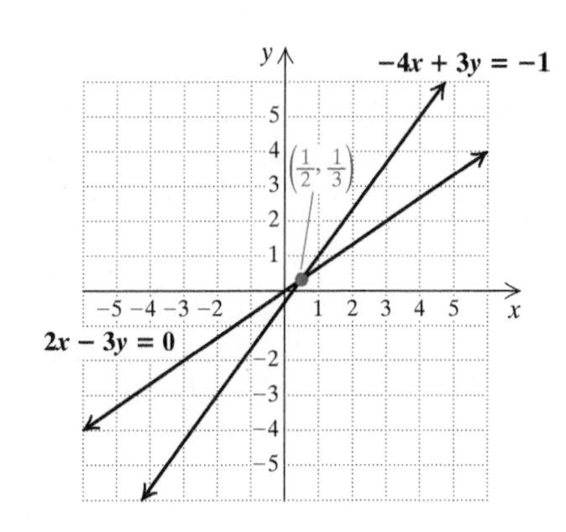

*This method is also called the *addition method*.

Solve by the elimination method.

1. $5x + 3y = 17,$
   $-5x + 2y = 3$

2. $-3a + 2b = 0,$
   $3a - 4b = -1$

Check:

$2x - 3y = 0$		
$2(\frac{1}{2}) - 3(\frac{1}{3})$	?	0
$1 - 1$		
0		TRUE

$-4x + 3y = -1$		
$-4(\frac{1}{2}) + 3(\frac{1}{3})$	?	-1
$-2 + 1$		
-1		TRUE

Since $(\frac{1}{2}, \frac{1}{3})$ checks, it is the solution.

◀ Do Exercises 1 and 2.

In order to eliminate a variable, we sometimes use the multiplication principle to multiply one or both of the equations by a particular number before adding.

EXAMPLE 2 Solve this system:

$$3x + 3y = 15, \quad (1)$$
$$2x + 6y = 22. \quad (2)$$

If we add directly, we will not eliminate a variable. However, note that if the $3y$ in equation (1) were $-6y$, we could eliminate y. Thus we multiply by -2 on both sides of equation (1) and add:

$$-6x - 6y = -30 \quad \text{Multiplying by } -2 \text{ on both sides of equation (1)}$$
$$2x + 6y = 22 \quad \text{Equation (2)}$$
$$-4x + 0 = -8 \quad \text{Adding}$$
$$-4x = -8$$
$$x = 2. \quad \text{Solving for } x$$

Then

$$2 \cdot 2 + 6y = 22 \quad \text{Substituting 2 for } x \text{ in equation (2)}$$
$$\left.\begin{aligned} 4 + 6y &= 22 \\ 6y &= 18 \\ y &= 3. \end{aligned}\right\} \quad \text{Solving for } y$$

We obtain $(2, 3)$, or $x = 2$, $y = 3$.

Check:

$3x + 3y = 15$		
$3(2) + 3(3)$	?	15
$6 + 9$		
15		TRUE

$2x + 6y = 22$		
$2(2) + 6(3)$	?	22
$4 + 18$		
22		TRUE

Since $(2, 3)$ checks, it is the solution. We can also see this in the graph at left.

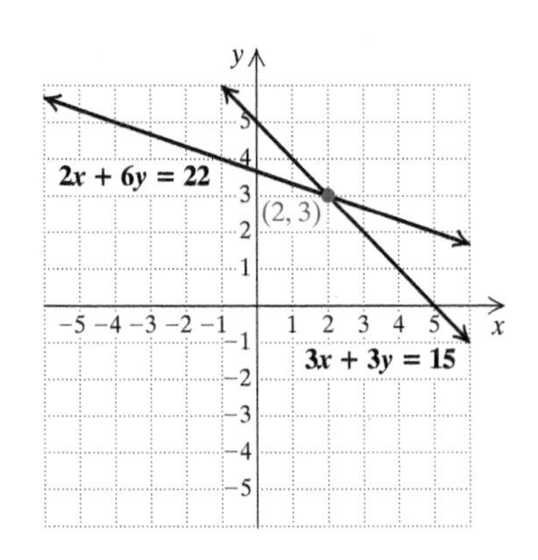

◀ Do Exercise 3.

3. Solve by the elimination method: GS

$2y + 3x = 12, \quad$ **(1)**
$-4y + 5x = -2. \quad$ **(2)**

Multiply by 2 on both sides of equation (1) and add:

$4y + 6x = 24$
$-4y + 5x = -2$
$0 + \underline{\quad} = \underline{\quad}$
$11x = 22$
$x = \underline{\quad}.$

Substitute $\underline{\quad}$ for x in equation (1) and solve for y:

$2y + 3x = 12$
$2y + 3(\underline{\quad}) = 12$
$2y + 6 = 12$
$2y = \underline{\quad}$
$y = \underline{\quad}.$

The ordered pair checks in both equations, so the solution is $(\underline{\quad}, \underline{\quad})$.

EXAMPLE 3 Solve this system:

$$2x + 3y = 17, \quad (1)$$
$$5x + 7y = 29. \quad (2)$$

We must multiply twice in order to make one pair of terms with the same variable opposites. We decide to do this with the x-terms in each equation. We multiply equation (1) by 5 and equation (2) by -2. Then we get $10x$ and $-10x$, which are opposites.

Answers

1. $(1, 4)$ **2.** $\left(\frac{1}{3}, \frac{1}{2}\right)$ **3.** $(2, 3)$

Guided Solution:
3. $11x$, 22, 2, 2, 2, 6, 3, 2, 3

$$\begin{array}{llrl}
\textit{From equation } (1): & 10x + 15y = 85 & & \text{Multiplying by 5} \\
\textit{From equation } (2): & -10x - 14y = -58 & & \text{Multiplying by } -2 \\
\hline
& 0 + y = 27 & & \text{Adding} \\
& y = 27 & & \text{Solving for } y
\end{array}$$

Then

$$2x + 3 \cdot 27 = 17 \quad \text{Substituting 27 for } y \text{ in equation (1)}$$

$$\left.\begin{aligned} 2x + 81 &= 17 \\ 2x &= -64 \\ x &= -32. \end{aligned}\right\} \quad \text{Solving for } x$$

We check the ordered pair $(-32, 27)$.

Check:

$$\begin{array}{c|l}
2x + 3y & = 17 \\
\hline
2(-32) + 3(27) & ?\ 17 \\
-64 + 81 & \\
17 & \quad \text{TRUE}
\end{array}
\qquad
\begin{array}{c|l}
5x + 7y & = 29 \\
\hline
5(-32) + 7(27) & ?\ 29 \\
-160 + 189 & \\
29 & \quad \text{TRUE}
\end{array}$$

We obtain $(-32, 27)$, or $x = -32$, $y = 27$, as the solution.

Do Exercises 4 and 5. ▶

Solve by the elimination method.

4. $4x + 5y = -8,$
   $7x + 9y = 11$

5. $4x - 5y = 38,$
   $7x - 8y = -22$

When solving a system of equations using the elimination method, it helps to first write the equations in the form $Ax + By = C$. When decimals or fractions occur, it also helps to *clear* before solving.

EXAMPLE 4 Solve this system:

$$0.2x + 0.3y = 1.7,$$
$$\tfrac{1}{7}x + \tfrac{1}{5}y = \tfrac{29}{35}.$$

We have

$$0.2x + 0.3y = 1.7, \longrightarrow \begin{array}{l}\text{Multiplying by 10} \\ \text{to clear decimals}\end{array} \longrightarrow 2x + 3y = 17,$$
$$\tfrac{1}{7}x + \tfrac{1}{5}y = \tfrac{29}{35} \longrightarrow \begin{array}{l}\text{Multiplying by 35} \\ \text{to clear fractions}\end{array} \longrightarrow 5x + 7y = 29.$$

We multiplied by 10 to clear the decimals. Multiplication by 35, the least common denominator, clears the fractions. The problem is now identical to Example 3. The solution is $(-32, 27)$, or $x = -32$, $y = 27$.

Do Exercises 6 and 7. ▶

6. Clear the decimals. Then solve.

$$0.02x + 0.03y = 0.01,$$
$$0.3x - 0.1y = 0.7$$

(*Hint*: Multiply the first equation by 100 and the second one by 10.)

7. Clear the fractions. Then solve.

$$\frac{3}{5}x + \frac{2}{3}y = \frac{1}{3},$$
$$\frac{3}{4}x - \frac{1}{3}y = \frac{1}{4}$$

> To use the elimination method to solve systems of two equations:
>
> 1. Write both equations in the form $Ax + By = C$.
> 2. Clear any decimals or fractions.
> 3. Choose a variable to eliminate.
> 4. Make the chosen variable's terms opposites by multiplying one or both equations by appropriate numbers if necessary.
> 5. Eliminate a variable by adding the respective sides of the equations and then solve for the remaining variable.
> 6. Substitute in either of the original equations to find the value of the other variable.

Answers

4. $(-127, 100)$ **5.** $(-138, -118)$
6. $2x + 3y = 1,$
$3x - y = 7;\ (2, -1)$
7. $9x + 10y = 5,$
$9x - 4y = 3;\ \left(\frac{25}{63}, \frac{1}{7}\right)$

Some systems have no solution. How do we recognize such systems if we are solving using elimination?

EXAMPLE 5 Solve this system:

$$y + 3x = 5, \quad (1)$$
$$y + 3x = -2. \quad (2)$$

If we find the slope–intercept equations for this system, we get

$$y = -3x + 5,$$
$$y = -3x - 2.$$

The graphs are parallel lines. The system has no solution.

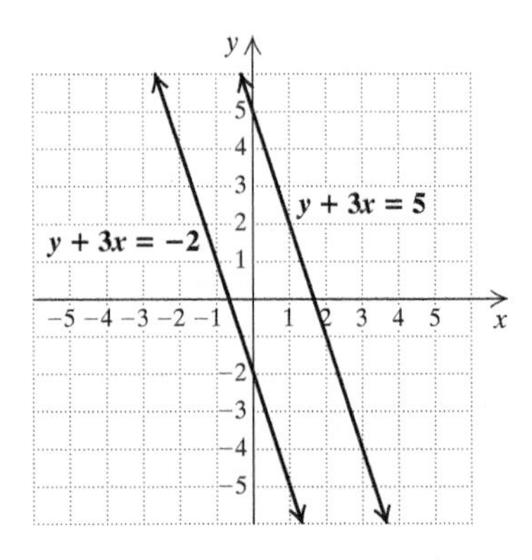

8. Solve by the elimination method:

$$y + 2x = 3,$$
$$y + 2x = -1.$$

Multiply the second equation by -1 and add:

$$y + 2x = 3$$
$$-y - 2x = 1$$
$$0 = \square.$$

The equation is ________ (true/false), so the system has no solution.

Let's attempt to solve the system by the elimination method:

$y + 3x = 5$ Equation (1)
$-y - 3x = 2$ Multiplying equation (2) by -1
$0 = 7.$ Adding, we obtain a false equation.

The x-terms and the y-terms are eliminated and we have a *false* equation. If we obtain a false equation, such as $0 = 7$, when solving algebraically, we know that the system has **no solution**. The system is inconsistent, and the equations are independent.

Do Exercise 8.

Some systems have infinitely many solutions. How can we recognize such a situation when we are solving systems using an algebraic method?

EXAMPLE 6 Solve this system:

$$3y - 2x = 6, \quad (1)$$
$$-12y + 8x = -24. \quad (2)$$

The graphs are the same line. The system has an infinite number of solutions.

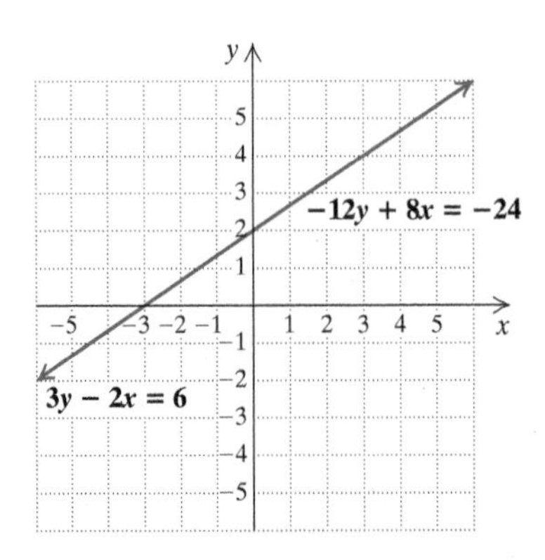

Answer
8. No solution
Guided Solution:
8. 4, false

Suppose that we try to solve this system by the elimination method:

$$\begin{array}{rl} 12y - 8x = 24 & \text{Multiplying equation (1) by 4} \\ -12y + 8x = -24 & \text{Equation (2)} \\ \hline 0 = 0. & \text{Adding, we obtain a true equation.} \end{array}$$

We have eliminated both variables, and what remains is a true equation, $0 = 0$. It can be expressed as $0 \cdot x + 0 \cdot y = 0$, and is true for all numbers x and y. If an ordered pair is a solution of one of the original equations, then it will be a solution of the other. The system has an infinite number of solutions. The system is consistent, and the equations are dependent.

SPECIAL CASES

When solving a system of two linear equations in two variables:

1. If a false equation is obtained, such as $0 = 7$, then the system has no solution. The system is *inconsistent*, and the equations are *independent*.
2. If a true equation is obtained, such as $0 = 0$, then the system has an infinite number of solutions. The system is *consistent*, and the equations are *dependent*.

Do Exercise 9.

9. Solve by the elimination method:

$$2x - 5y = 10,$$
$$-6x + 15y = -30.$$

Comparing Methods

We can solve systems of equations graphically, or we can solve them algebraically using substitution or elimination. When deciding which method to use, consider the information in this table as well as directions from your instructor.

METHOD	STRENGTHS	WEAKNESSES
Graphical	Can "see" solutions.	Inexact when solutions involve numbers that are not integers. Solutions may not appear on the part of the graph drawn.
Substitution	Yields exact solutions. Convenient to use when a variable has a coefficient of 1.	Can introduce extensive computations with fractions. Cannot "see" solutions quickly.
Elimination	Yields exact solutions. Convenient to use when no variable has a coefficient of 1. The preferred method for systems of three or more equations in three or more variables. (See Section 3.5.)	Cannot "see" solutions quickly.

Answer

9. Infinitely many solutions

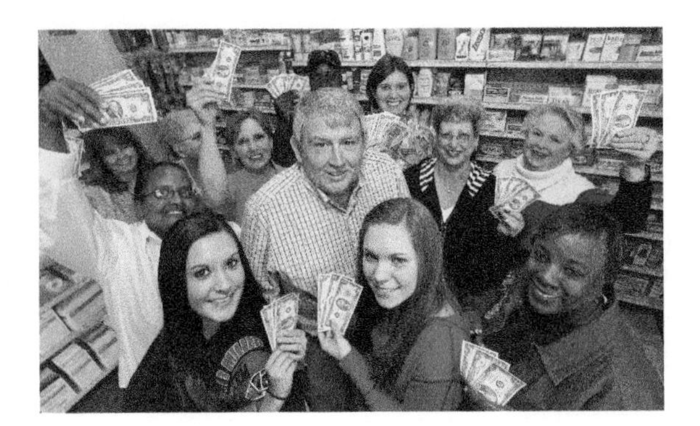

b SOLVING APPLIED PROBLEMS USING ELIMINATION

Let's now solve an applied problem using the elimination method.

EXAMPLE 7 *Stimulating the Hometown Economy.* To stimulate the economy in his town of Brewton, Alabama, in 2009, Danny Cottrell, co-owner of The Medical Center Pharmacy, gave each of his full-time employees \$700 and each part-time employee \$300. He asked that each person donate 15% to a charity of his or her choice and spend the rest locally. The money was paid in \$2 bills, a rarely used currency, so that the business community could easily see how the money circulated. Cottrell gave away a total of \$16,000 to his 24 employees. How many full-time employees and how many part-time employees were there?

Data: *The Press-Register,* March 4, 2009

1. **Familiarize.** We let f = the number of full-time employees and p = the number of part-time employees. Each full-time employee received \$700, so a total of $700f$ was paid to them. Similarly, the part-time employees received a total of $300p$. Thus a total of $700f + 300p$ was given away.

2. **Translate.** We translate to two equations.

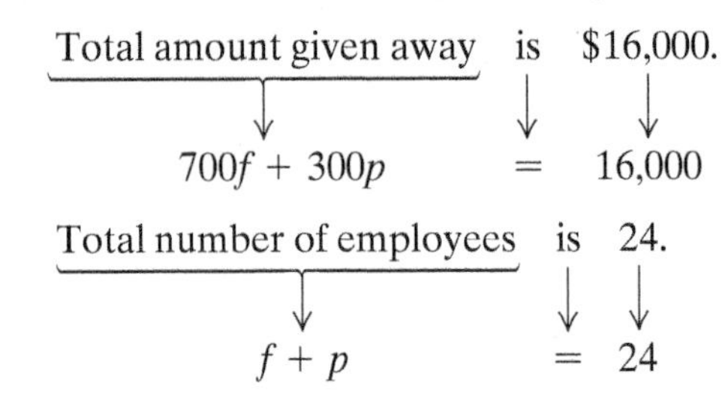

We now have a system of equations:

$$700f + 300p = 16{,}000, \quad (1)$$
$$f + p = 24. \quad (2)$$

3. **Solve.** First, we multiply by -300 on both sides of equation (2) and add:

$700f + 300p = 16{,}000$	Equation (1)
$-300f - 300p = -7200$	Multiplying by -300 on both sides of equation (2)
$400f = 8800$	Adding
$f = 22.$	Solving for f

Next, we substitute 22 for f in equation (2) and solve for p:

$$22 + p = 24$$
$$p = 2.$$

4. **Check.** If there are 22 full-time employees and 2 part-time employees, there is a total of $22 + 2$, or 24, employees. The 22 full-time employees received a total of $\$700 \cdot 22$, or \$15,400, and the 2 part-time employees received a total of $\$300 \cdot 2$, or \$600. Then a total of $\$15{,}400 + \$600$, or \$16,000, was given away. The numbers check in the original problem.

5. **State.** There were 22 full-time employees and 2 part-time employees.

◀ **Do Exercise 10.**

10. *Bonuses.* Monica gave each of the full-time employees in her small business a year-end bonus of \$500 while each part-time employee received \$250. She gave a total of \$4000 in bonuses to her 10 employees. How many full-time employees and how many part-time employees did Monica have?

Answer

10. Full-time: 6; part-time: 4

3.3 Exercise Set

FOR EXTRA HELP MyLab Math

✓ Check Your Understanding

Reading Check Choose from the column on the right the word that best completes each sentence. Words may be used more than once.

RC1. If a system of equations has a solution, then it is ______________.	consistent
RC2. If a system of equations has no solution, then it is ______________.	inconsistent
RC3. If a system of equations has infinitely many solutions, then it is ______________.	dependent
	independent

RC4. If the graphs of the equations in a system of two equations in two variables are the same line, then the equations are ______________.

RC5. If the graphs of the equations in a system of two equations in two variables are parallel, then the system is ______________.

RC6. If the graphs of the equations in a system of two equations in two variables intersect at one point, then the equations are ______________.

Concept Check For each system, determine by what number the second equation should be multiplied in order to eliminate y by adding.

CC1. $3x + 6y = 7,$
$2x + 2y = 1$

CC2. $5x - 6y = 2,$
$2x + 3y = 5$

CC3. $7x + 3y = 2,$
$4x + 3y = 9$

a Solve each system of equations using the elimination method.

1. $x + 3y = 7,$
$-x + 4y = 7$

2. $x + y = 9,$
$2x - y = -3$

3. $9x + 5y = 6,$
$2x - 5y = -17$

4. $2x - 3y = 18,$
$2x + 3y = -6$

5. $5x + 3y = -11,$
$3x - y = -1$

6. $2x + 3y = -9,$
$5x - 6y = -9$

7. $5r - 3s = 19,$
$2r - 6s = -2$

8. $2a + 3b = 11,$
$4a - 5b = -11$

9. $2x + 3y = 1,$
$4x + 6y = 2$

10. $3x - 2y = 1,$
$-6x + 4y = -2$

11. $5x - 9y = 7,$
$7y - 3x = -5$

12. $5x + 4y = 2,$
$2x - 8y = 4$

13. $3x + 2y = 24,$
$2x + 3y = 26$

14. $5x + 3y = 25,$
$3x + 4y = 26$

15. $2x - 4y = 5,$
$2x - 4y = 6$

16. $3x - 5y = -2,$
$5y - 3x = 7$

17. $2a + b = 12,$
$a + 2b = -6$

18. $10x + y = 306,$
$10y + x = 90$

19. $\frac{1}{3}x + \frac{1}{5}y = 7,$
$\frac{1}{6}x - \frac{2}{5}y = -4$

20. $\frac{2}{3}x + \frac{1}{7}y = -11,$
$\frac{1}{7}x - \frac{1}{3}y = -10$

21. $\frac{1}{5}x + \frac{1}{2}y = 6,$
$\frac{2}{5}x - \frac{3}{2}y = -8$

22. $\frac{2}{3}x + \frac{3}{5}y = -17,$
$\frac{1}{2}x - \frac{1}{3}y = -1$

23. $\frac{1}{2}x - \frac{1}{3}y = -4,$
$\frac{1}{4}x + \frac{5}{6}y = 4$

24. $\frac{4}{3}x + \frac{3}{2}y = 4,$
$\frac{5}{6}x - \frac{1}{8}y = -6$

25. $0.3x - 0.2y = 4,$
$0.2x + 0.3y = 0.5$

26. $0.7x - 0.3y = 0.5,$
$-0.4x + 0.7y = 1.3$

27. $0.05x + 0.25y = 22,$
$0.15x + 0.05y = 24$

28. $1.3x - 0.2y = 12,$
$0.4x + 17y = 89$

b Solve. Use the elimination method when solving the translated system.

29. *Finding Numbers.* The sum of two numbers is 63. The larger number minus the smaller number is 9. Find the numbers.

30. *Finding Numbers.* The sum of two numbers is 2. The larger number minus the smaller number is 20. Find the numbers.

31. *Finding Numbers.* The sum of two numbers is 3. Three times the larger number plus two times the smaller number is 24. Find the numbers.

32. *Finding Numbers.* The sum of two numbers is 9. Two times the larger number plus three times the smaller number is 2. Find the numbers.

33. *Complementary Angles.* Two angles are complementary. (**Complementary angles** are angles whose sum is 90°.) Their difference is 6°. Find the angles.

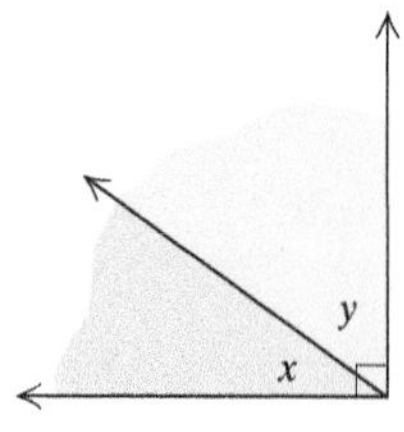

Complementary angles:
$x + y = 90°$

34. *Supplementary Angles.* Two angles are supplementary. (**Supplementary angles** are angles whose sum is 180°.) Their difference is 22°. Find the angles.

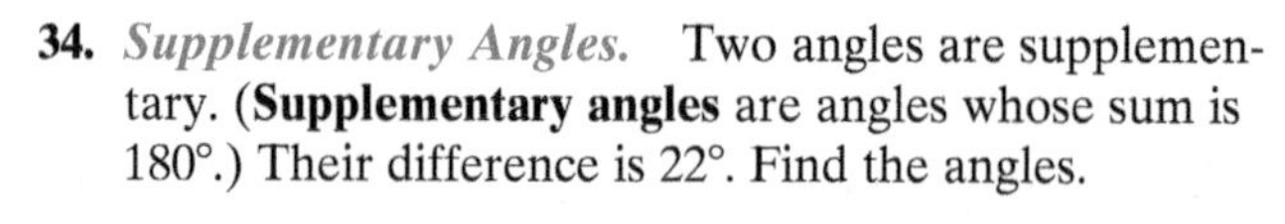

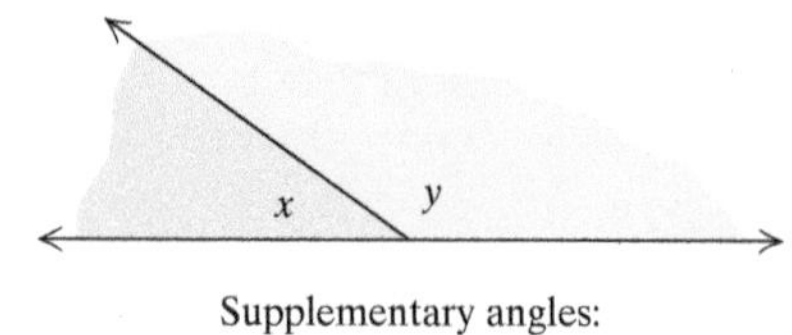

Supplementary angles:
$x + y = 180°$

35. *Basketball Scoring.* In their championship game, the Eastside Golden Eagles scored 60 points on a combination of two-point shots and three-point shots. If they made a total of 27 shots, how many of each kind of shot was made?

36. *Basketball Scoring.* Wilt Chamberlain once scored 100 points, setting a record for points scored in an NBA game. Chamberlain took only two-point shots and (one-point) foul shots and made a total of 64 shots. How many shots of each type did he make?

37. Each course offered during the winter session at New Heights Community College is worth either 3 credits or 4 credits. The members of the Touring Concert Chorale took a total of 33 courses during the winter session, worth a total of 107 credits. How many of each type of class did the chorale members take?

38. Daphne's Lawn and Garden Center offered customers who bought a custom lawn-care package a free ornamental tree, either an Eastern Redbud or a Kousa Dogwood. The center's cost for each Eastern Redbud was \$37, and its cost for each Kousa Dogwood was \$45. A total of 18 customers took advantage of the offer. The center's total cost for the promotional items was \$754. How many patrons chose each type of ornamental tree?

Skill Maintenance

Given the function $f(x) = 3x^2 - x + 1$, find each of the following function values. [2.2b]

39. $f(0)$

40. $f(-1)$

41. $f(-2)$

42. $f(2a)$

43. Find the domain of the function
$$f(x) = \frac{x - 5}{x + 7}. \quad [2.3a]$$

44. Find the domain and the range of the function
$$g(x) = 5 - x^2. \quad [2.3a]$$

45. Find an equation of the line with slope $-\frac{3}{5}$ and y-intercept $(0, -7)$. [2.6a]

46. Find an equation of the line containing the points $(-10, 2)$ and $(-2, 10)$. [2.6c]

Synthesis

47. Use the INTERSECT feature to solve the following system of equations. You may need to first solve for y. Round answers to the nearest hundredth.
$$3.5x - 2.1y = 106.2,$$
$$4.1x + 16.7y = -106.28$$

48. Solve:
$$\frac{x + y}{2} - \frac{x - y}{5} = 1,$$
$$\frac{x - y}{2} + \frac{x + y}{6} = -2.$$

49. The solution of this system is $(-5, -1)$. Find A and B.
$$Ax - 7y = -3,$$
$$x - By = -1$$

50. Find an equation to pair with $6x + 7y = -4$ such that $(-3, 2)$ is a solution of the system.

51. The points $(0, -3)$ and $\left(-\frac{3}{2}, 6\right)$ are two of the solutions of the equation $px - qy = -1$. Find p and q.

52. Determine a and b for which $(-4, -3)$ will be a solution of the system
$$ax + by = -26,$$
$$bx - ay = 7.$$

3.4 Solving Applied Problems: Two Equations

OBJECTIVES

a Solve applied problems involving total value and mixture using systems of two equations.

b Solve applied problems involving motion using systems of two equations.

MyLab Math ANIMATION

a TOTAL-VALUE PROBLEMS AND MIXTURE PROBLEMS

Systems of equations can be a useful tool in solving applied problems. Using systems often makes the *Translate* step easier than using a single equation. The first kind of problem we consider involves quantities of items purchased and the total value, or cost, of the items. We refer to this type of problem as a **total-value problem**.

EXAMPLE 1 *Lunch Orders.* In order to pick up lunch, Cathy collected \$181.50 from her co-workers for a total of 21 salads and sandwiches. When she got to the deli, she forgot how many of each were ordered. If salads cost \$7.50 and sandwiches cost \$9.50, how many of each should she buy?

1. **Familiarize.** Let's begin by guessing that 5 salads were ordered. Since there was a total of 21 orders, this means that 16 sandwiches were ordered. The total cost of the order would then be

$$\underbrace{\$7.50(5)}_{\text{Cost of salads}} + \underbrace{\$9.50(16)}_{\text{Cost of sandwiches}} = \$37.50 + \$152 = \$189.50.$$

The guess is incorrect, but we can use the same process to translate the problem to a system of equations. We also note that our guess resulted in a total that was too high, so there were more salads and fewer sandwiches ordered than we guessed.

We let d = the number of salads and w = the number of sandwiches ordered. The cost of d salads is \$7.50d, and the cost of w sandwiches is \$9.50w. Organizing the information in a table can help us translate the information to a system of equations.

	SALADS	SANDWICHES	TOTAL	
NUMBER OF ORDERS	d	w	21	$\longrightarrow d + w = 21$
COST PER ORDER	\$7.50	\$9.50		
TOTAL COST	\$7.50d	\$9.50w	\$181.50	$\longrightarrow 7.50d + 9.50w = 181.50$

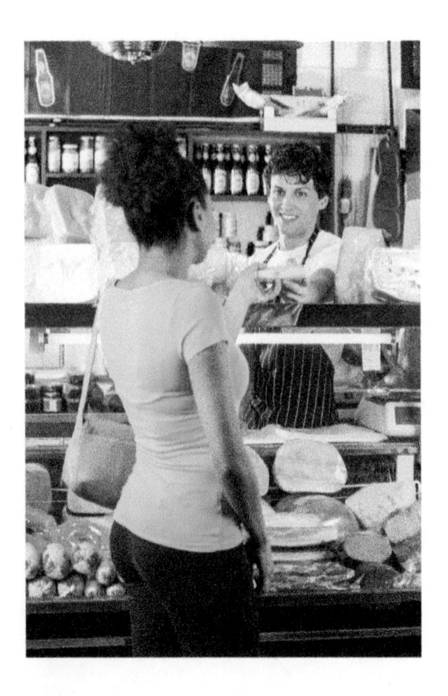

2. **Translate.** The first row of the table gives us one equation:

$$d + w = 21.$$

The last row of the table gives us a second equation:

$$7.50d + 9.50w = 181.50.$$

Clearing decimals in the second equation gives us the following system of equations:

$$d + w = 21, \qquad (1)$$
$$75d + 95w = 1815. \qquad (2)$$

3. Solve. We use the elimination method to solve the system of equations. We eliminate d by multiplying by -75 on both sides of equation (1) and then adding the result to equation (2):

$-75d - 75w = -1575$	Multiplying equation (1) by -75
$75d + 95w = 1815$	Equation (2)
$20w = 240$	Adding
$w = 12.$	Dividing by 20

Next, we substitute 12 for w in equation (1) and solve for d:

$d + w = 21$	Equation (1)
$d + 12 = 21$	Substituting 12 for w
$d = 9.$	Solving for d

We obtain $(9, 12)$, or $d = 9$, $w = 12$.

4. Check. We check in the original problem.

Total number of orders:	$d + w = 9 + 12 = 21$
Cost of salads:	$\$7.50d = \$7.50(9) = \$\ 67.50$
Cost of sandwiches:	$\$9.50w = \$9.50(12) = \$114.00$
	Total $= \$181.50$

The numbers check.

5. State. Cathy should buy 9 salads and 12 sandwiches.

Do Exercise 1. ▶

The following problem, similar to Example 1, is called a **mixture problem**.

EXAMPLE 2 *Blending Spices.* Spice It Up sells ground turmeric for \$1.35 per ounce and ground sumac for \$1.85 per ounce. Ethan wants to make a 20-oz seasoning blend of the two spices that sells for \$1.65 per ounce. How much of each should he use?

1. *Retail Sales of Tee Shirts.* A campus bookstore sells college tee shirts. White tee shirts sell for \$18.95 each and red ones sell for \$19.50 each. If receipts for the sale of 30 tee shirts total \$572.90, how many of each color did the shop sell?

Complete the following table, letting w = the number of white tee shirts and r = the number of red tee shirts.

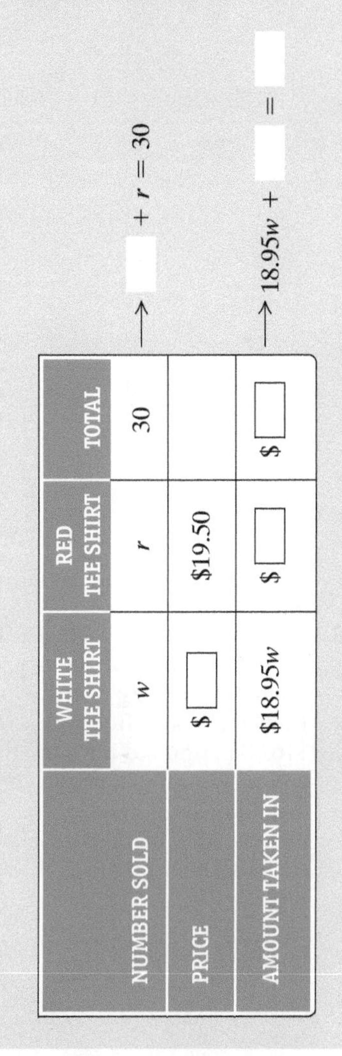

	White Tee Shirt	Red Tee Shirt	Total	
Number Sold	w	r	30	→ ☐ + r = 30
Price	\$☐	\$19.50		
Amount Taken In	\$18.95w	\$☐	\$☐	→ $18.95w$ + ☐ = ☐

Answer

1. White: 22; red: 8

Guided Solution:

1.

White	Red	Total	
w	r	30	→ $w + r = 30$
\$18.95	\$19.50		
\$18.95w	\$19.50r	\$572.90	→ $18.95w + 19.50r = 572.90$

1. **Familiarize.** Suppose that Ethan uses 4 oz of sumac. Since he wants a total of 20 oz, he will need 16 oz of turmeric. We compare the value of the spices separately with the desired value of the blend:

   Spices purchased separately: $\$1.85(4) + \$1.35(16)$, or \$29.

   Blend: $\$1.65(20) = \$33$

   Since these amounts are not the same, our guess is not correct, but these calculations help us to translate the problem.

   We let s = the number of ounces of sumac and t = the number of ounces of turmeric. Then we organize the information in a table as follows.

	SUMAC	TURMERIC	BLEND	
NUMBER OF OUNCES	s	t	20	$\longrightarrow s + t = 20$
PRICE PER OUNCE	\$1.85	\$1.35	\$1.65	
VALUE OF SPICES	\$1.85s	\$1.35t	\$1.65(20), or \$33	$\longrightarrow 1.85s + 1.35t = 33$

2. **Translate.** The total number of ounces in the blend is 20, so we have one equation:

   $$s + t = 20.$$

   The value of the sumac is $1.85s$, and the value of the turmeric is $1.35t$. These amounts are in dollars. Since the total value is to be 1.65(20), or 33, we have

   $$1.85s + 1.35t = 33.$$

   We can multiply by 100 on both sides of the second equation in order to clear the decimals. Thus we have translated to a system of equations:

   $$s + t = 20, \quad (1)$$
   $$185s + 135t = 3300. \quad (2)$$

3. **Solve.** We will solve this system using substitution, but elimination is also an appropriate method to use. When equation (1) is solved for t, we get $t = 20 - s$. We substitute $20 - s$ for t in equation (2) and solve:

   $185s + 135(20 - s) = 3300$ Substituting

   $185s + 2700 - 135s = 3300$ Using the distributive law

   $50s = 600$ Subtracting 2700 and collecting like terms

   $s = 12.$

   We have $s = 12$. Substituting 12 for s in the equation $t = 20 - s$, we obtain $t = 20 - 12$, or 8.

4. **Check.** We check in a manner similar to our guess in the *Familiarize* step.

   Total number of ounces: $12 + 8 = 20$

   Value of the blend: $1.85(12) + 1.35(8) = 33$

   Thus the number of ounces of each spice checks.

5. **State.** Ethan should use 12 oz of sumac and 8 oz of turmeric.

◀ **Do Exercise 2.**

2. *Blending Coffees.* The Coffee Counter charges \$18.00 per pound for organic Kenyan French Roast coffee and \$16.00 per pound for Sumatran coffee. How much of each type should be used in order to make a 20-lb blend that sells for \$16.70 per pound?

Answer

2. Kenyan: 7 lb; Sumatran: 13 lb

EXAMPLE 3 *Student Loans.* Jeron's student loans totaled \$16,200. Part was a Perkins loan made at 5% interest and the rest was a Stafford loan made at 4% interest. After one year, Jeron's loans accumulated \$715 in interest. What was the amount of each loan?

1. **Familiarize.** Listing the given information in a table will help. The columns in the table come from the formula for simple interest: $I = Prt$. We let $x =$ the number of dollars in the Perkins loan and $y =$ the number of dollars in the Stafford loan.

	PERKINS LOAN	STAFFORD LOAN	TOTAL	
PRINCIPAL	x	y	\$16,200	$\longrightarrow x + y = 16{,}200$
RATE OF INTEREST	5%	4%		
TIME	1 year	1 year		
INTEREST	$0.05x$	$0.04y$	\$715	$\longrightarrow 0.05x + 0.04y = 715$

2. **Translate.** The total of the amounts of the loans is found in the first row of the table. This gives us one equation:

$$x + y = 16{,}200.$$

Look at the last row of the table. The interest totals \$715. This gives us a second equation:

$$5\%x + 4\%y = 715, \quad \text{or} \quad 0.05x + 0.04y = 715.$$

After we multiply on both sides to clear the decimals, we have

$$5x + 4y = 71{,}500.$$

3. **Solve.** Using either elimination or substitution, we solve the resulting system:

$$x + y = 16{,}200,$$
$$5x + 4y = 71{,}500.$$

We find that $x = 6700$ and $y = 9500$.

4. **Check.** The sum is \$6700 + \$9500, or \$16,200. The interest from \$6700 at 5% for one year is 5%(\$6700), or \$335. The interest from \$9500 at 4% for one year is 4%(\$9500), or \$380. The total amount of interest is \$335 + \$380, or \$715. The numbers check in the problem.

5. **State.** The Perkins loan was for \$6700, and the Stafford loan was for \$9500.

Do Exercise 3.

3. ***Client Investments.*** Infinite Financial Services invested Jasmine's IRA contribution of \$3700 for one year at simple interest, yielding \$297. Part of the money is invested at 7% and the rest at 9%. How much was invested at each rate?

Do the *Familiarize* and *Translate* steps by completing the following table. Let $x =$ the number of dollars invested at 7% and $y =$ the number of dollars invested at 9%.

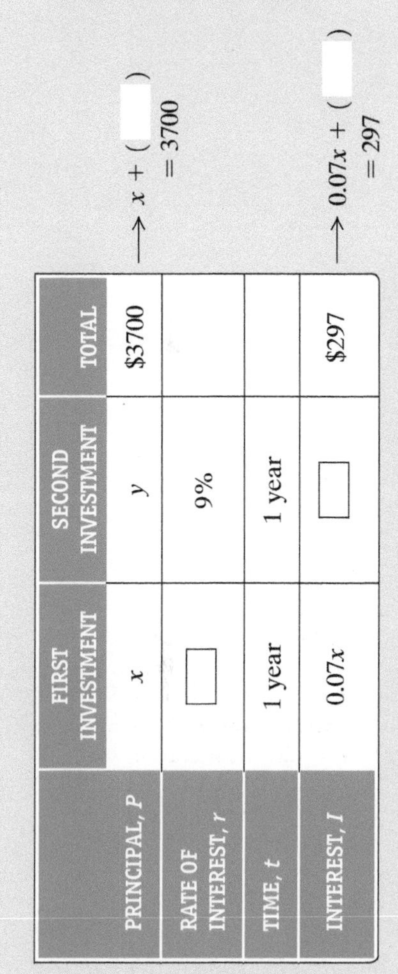

Answer

3. \$1800 at 7%; \$1900 at 9%

Guided Solution:

3.

First Investment	Second Investment	Total	
x	y	\$3700	$\longrightarrow x + y = 3700$
7%	9%		
1 year	1 year		
$0.07x$	$0.09y$	\$297	$\longrightarrow 0.07x + 0.09y = 297$

EXAMPLE 4 *Mixing Fertilizers.* Nature's Landscapes carries two kinds of fertilizer containing nitrogen and water. "Gently Green" is 5% nitrogen and "Sun Saver" is 15% nitrogen. Nature's Landscapes needs to combine the two types of solution in order to make 90 L of a solution that is 12% nitrogen. How much of each brand should be used?

1. **Familiarize.** We first make a drawing and a guess to become familiar with the problem.

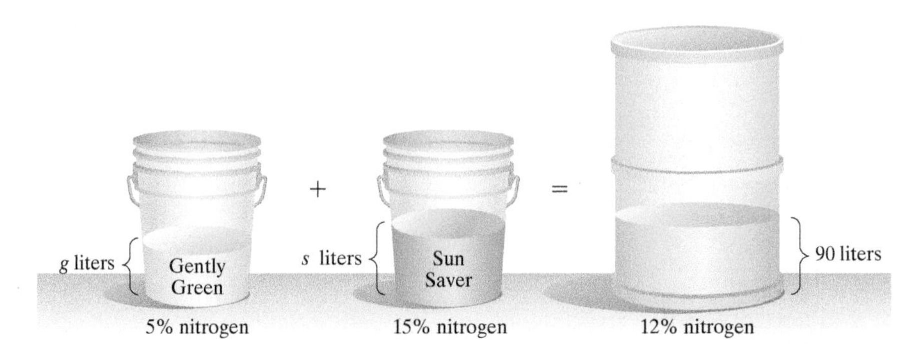

We choose two numbers that total 90 L—say, 40 L of Gently Green and 50 L of Sun Saver—for the amounts of each fertilizer. Will the resulting mixture have the correct percentage of nitrogen? To find out, we multiply as follows:

$$5\%(40 \text{ L}) = 2 \text{ L of nitrogen} \quad \text{and} \quad 15\%(50 \text{ L}) = 7.5 \text{ L of nitrogen.}$$

Thus the total amount of nitrogen in the mixture is 2 L + 7.5 L, or 9.5 L. The final mixture of 90 L is supposed to be 12% nitrogen. Now

$$12\%(90 \text{ L}) = 10.8 \text{ L.}$$

Since 9.5 L and 10.8 L are not the same, our guess is incorrect. But these calculations help us to make the translation.

We let g = the number of liters of Gently Green and s = the number of liters of Sun Saver in the mixture.

	GENTLY GREEN	SUN SAVER	MIXTURE	
NUMBER OF LITERS	g	s	90	$\rightarrow g + s = 90$
PERCENT OF NITROGEN	5%	15%	12%	
AMOUNT OF NITROGEN	$0.05g$	$0.15s$	0.12×90, or 10.8 liters	$\rightarrow 0.05g + 0.15s = 10.8$

2. **Translate.** If we add g and s in the first row, we get 90, and this gives us one equation:

$$g + s = 90.$$

If we add the amounts of nitrogen listed in the third row, we get 10.8, and this gives us another equation:

$$5\%g + 15\%s = 10.8, \quad \text{or} \quad 0.05g + 0.15s = 10.8.$$

After clearing the decimals, we have the following system:

$$g + s = 90, \quad \textbf{(1)}$$
$$5g + 15s = 1080. \quad \textbf{(2)}$$

3. **Solve.** We solve the system using elimination. We multiply equation (1) by -5 and add the result to equation (2):

$$\begin{aligned} -5g - 5s &= -450 && \text{Multiplying equation (1) by } -5 \\ 5g + 15s &= 1080 && \text{Equation (2)} \\ \hline 10s &= 630 && \text{Adding} \\ s &= 63. && \text{Dividing by 10} \end{aligned}$$

Next, we substitute 63 for s in equation (1) and solve for g:

$$\begin{aligned} g + 63 &= 90 && \text{Substituting in equation (1)} \\ g &= 27. && \text{Solving for } g \end{aligned}$$

We obtain $(27, 63)$, or $g = 27, s = 63$.

4. **Check.** Remember that g is the number of liters of Gently Green, with 5% nitrogen, and s is the number of liters of Sun Saver, with 15% nitrogen.

Total number of liters of mixture: $g + s = 27 + 63 = 90$ L

Amount of nitrogen: $5\%(27) + 15\%(63) = 1.35 + 9.45 = 10.8$ L

Percentage of nitrogen in mixture: $\dfrac{10.8}{90} = 0.12 = 12\%$

The numbers check in the original problem.

5. **State.** Nature's Landscapes should mix 27 L of Gently Green and 63 L of Sun Saver.

Do Exercise 4. ▶

4. ***Mixing Cleaning Solutions.*** King's Service Station uses two kinds of cleaning solution containing acid and water. "Attack" is 2% acid and "Blast" is 6% acid. They want to mix the two in order to get 60 qt of a solution that is 5% acid. How many quarts of each should they use?

Do the *Familiarize* and *Translate* steps by completing the following table. Let a = the number of quarts of Attack and b = the number of quarts of Blast.

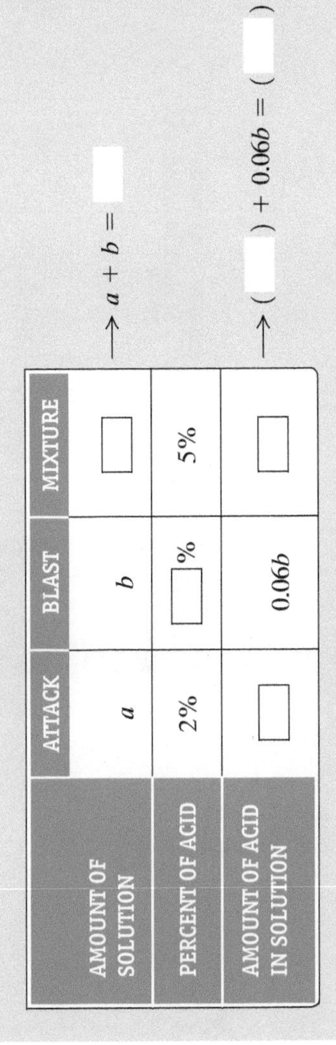

b MOTION PROBLEMS

MyLab Math ANIMATION

Solve basic motion problems. [1.3b]

Solve.

1. Frank's boat travels at a rate of 8 mph in still water. Boynton River flows at a speed of 2 mph. How long will it take Frank to travel 15 mi downstream?
2. Refer to Exercise 1. How long will it take Frank to travel 15 mi upstream?

Answers: 1. 1.5 hr **2.** 2.5 hr

MyLab Math VIDEO

When a problem deals with speed, distance, and time, we can expect to use the following *motion formula*.

THE MOTION FORMULA

Distance = Rate (or speed) · Time

$$d = rt$$

Answer

4. Attack: 15 qt; Blast: 45 qt

Guided Solution:

4.

Attack	Blast	Mixture	
a	b	60	$\longrightarrow a + b = 60$
2%	6%	5%	
$0.02a$	$0.06b$	0.05×60, or 3	$\longrightarrow 0.02a + 0.06b = 3$

EXAMPLE 5 *Auto Travel.* Keri left Monday morning to drive to a seminar that began Monday evening. An hour after she had left the office, her assistant, Matt, realized that she had forgotten to take a large portfolio needed for a presentation. Knowing Keri would not answer her cell phone when driving, Matt left immediately with the portfolio to try to catch up with her. If Keri drove at a speed of 55 mph and Matt drove at a speed of 65 mph, how long did it take Matt to catch up with her? Assume that neither driver stopped to take a break.

1. **Familiarize.** We first make a drawing. From the drawing, we see that when Matt catches up with Keri, the distances from the office are the same. We let d = the distance, in miles. If we let t = the time, in hours, for Matt to catch Keri, then $t + 1$ = the time traveled by Keri at a slower speed.

We organize the information in a table as follows.

$d = r \cdot t$

	DISTANCE	RATE	TIME	
KERI	d	55	$t + 1$	$\rightarrow d = 55(t + 1)$
MATT	d	65	t	$\rightarrow d = 65t$

2. **Translate.** Using $d = rt$ in each row of the table, we get an equation. Thus we have a system of equations:

$$d = 55(t + 1), \quad \textbf{(1)}$$
$$d = 65t. \quad \textbf{(2)}$$

3. **Solve.** We solve the system using the substitution method:

$65t = 55(t + 1)$ Substituting $65t$ for d in equation (1)

$65t = 55t + 55$ Multiplying to remove parentheses on the right

$10t = 55$
$t = 5.5.$ } Solving for t

Matt's time is 5.5 hr, which means that Keri's time is $5.5 + 1$, or 6.5 hr.

4. **Check.** At 65 mph, Matt will travel $65 \cdot 5.5$, or 357.5 mi, in 5.5 hr. At 55 mph, Keri will travel $55 \cdot 6.5$, or the same 357.5 mi, in 6.5 hr. The distances are the same, so the numbers check.

5. **State.** Matt caught up with Keri in 5.5 hr.

◀ **Do Exercise 5.**

5. ***Train Travel.*** A train leaves Barstow traveling east at 35 km/h. One hour later, a faster train leaves Barstow, also traveling east on a parallel track at 40 km/h. How far from Barstow will the faster train catch up with the slower one?

GS

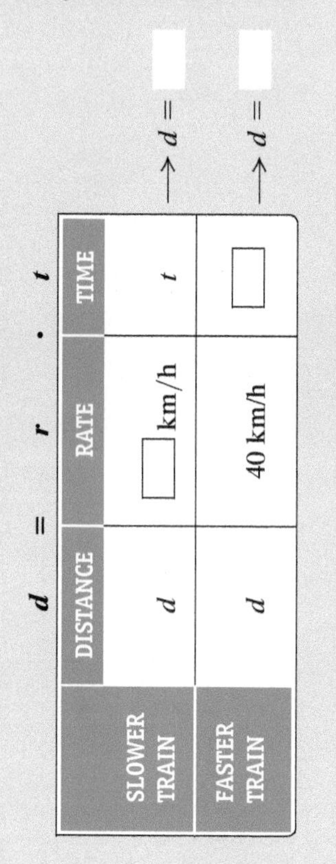

	DISTANCE	RATE	TIME	
SLOWER TRAIN	d	☐ km/h	t	$\rightarrow d =$ ☐
FASTER TRAIN	d	40 km/h	☐	$\rightarrow d =$ ☐

Answer

5. 280 km

Guided Solution:

5.

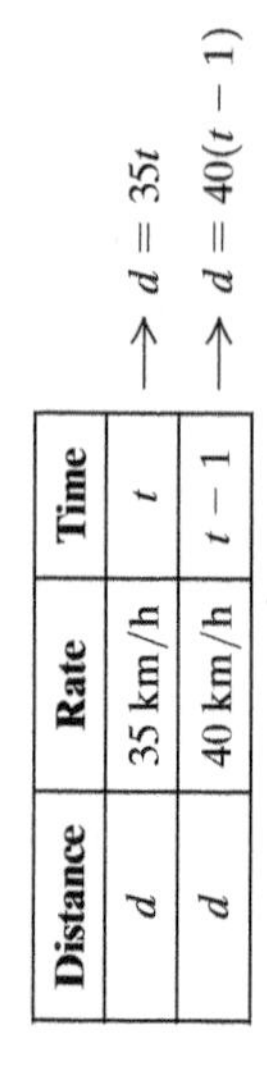

Distance	Rate	Time	
d	35 km/h	t	$\rightarrow d = 35t$
d	40 km/h	$t - 1$	$\rightarrow d = 40(t - 1)$

EXAMPLE 6 *Marine Travel.* A Coast Guard patrol boat travels 4 hr on a trip downstream with a 6-mph current. The return trip against the same current takes 5 hr. Find the speed of the boat in still water.

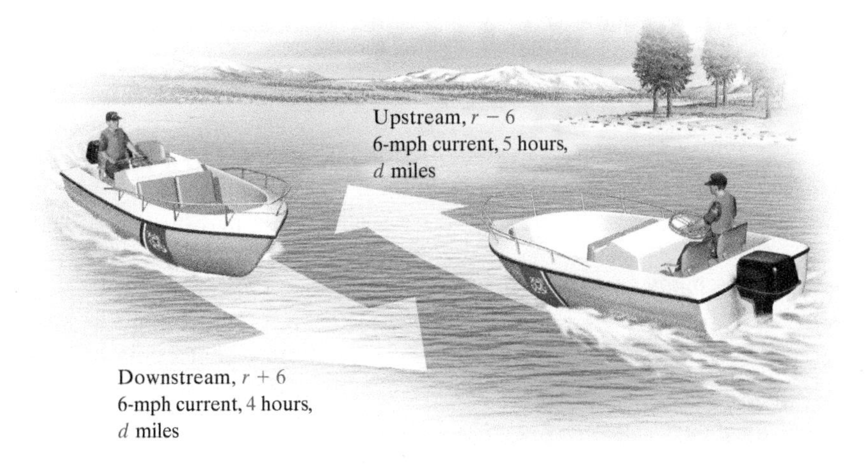

1. **Familiarize.** We first make a drawing. From the drawing, we see that the distances are the same. We let d = the distance, in miles, and r = the speed of the boat in still water, in miles per hour. Then, when the boat is traveling downstream, its speed is $r + 6$. (The current helps the boat along.) When it is traveling upstream, its speed is $r - 6$. (The current holds the boat back.) We can organize the information in a table. We use the formula $d = rt$.

$d = r \cdot t$

	DISTANCE	RATE	TIME	
DOWNSTREAM	d	$r + 6$	4	$\rightarrow d = (r + 6)4$
UPSTREAM	d	$r - 6$	5	$\rightarrow d = (r - 6)5$

2. **Translate.** From each row of the table, we get an equation, $d = rt$:

$$d = 4r + 24, \quad (1)$$
$$d = 5r - 30. \quad (2)$$

3. **Solve.** We solve the system using the substitution method:

$$4r + 24 = 5r - 30 \qquad \text{Substituting } 4r + 24 \text{ for } d \text{ in equation (2)}$$
$$\left.\begin{aligned} 24 &= r - 30 \\ 54 &= r. \end{aligned}\right\} \qquad \text{Solving for } r$$

4. **Check.** If $r = 54$, then $r + 6 = 60$; and $60 \cdot 4 = 240$ mi, the distance traveled downstream. If $r = 54$, then $r - 6 = 48$; and $48 \cdot 5 = 240$ mi, the distance traveled upstream. The distances are the same.

   When checking your answer, always ask, "Have I found what the problem asked for?" We could solve for a certain variable but still have not answered the question of the original problem. For example, we might have found speed when the problem wanted distance. In this problem, we want the speed of the boat in still water, and that is r.

5. **State.** The speed in still water is 54 mph.

Do Exercise 6.

GS **6.** ***Air Travel.*** An airplane flew for 4 hr with a 20-mph tailwind. The return flight against the same wind took 5 hr. Find the speed of the plane in still air.

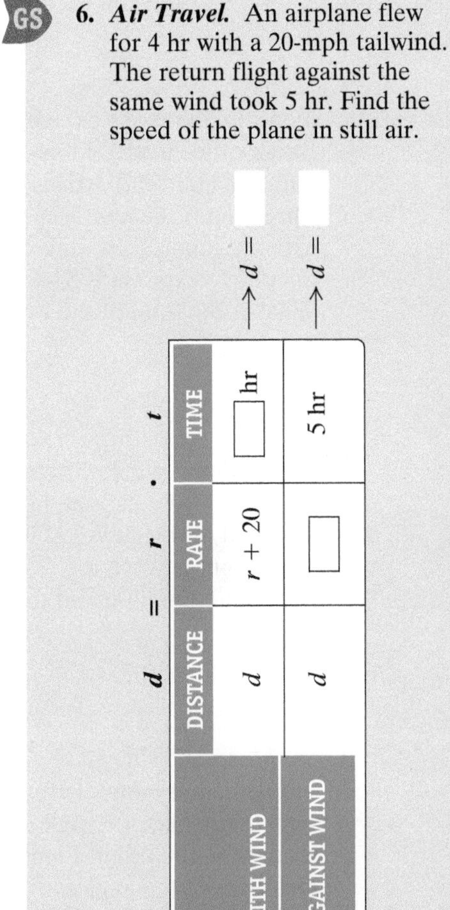

Answer

6. 180 mph

Guided Solution:

6.

Distance	Rate	Time	
d	$r + 20$	4 hr	$\rightarrow d = (r + 20)4$
d	$r - 20$	5 hr	$\rightarrow d = (r - 20)5$

Translating for Success

The goal of these matching questions is to practice step (2), Translate, of the five-step problem-solving process. Translate each word problem to a system of equations and select a correct translation from systems A–J.

1. *Office Expense.* The monthly telephone expense for an office is \$1094 less than the janitorial expense. Three times the janitorial expense minus four times the telephone expense is \$248. What is the total of the two expenses?

2. *Dimensions of a Triangle.* The sum of the base and the height of a triangle is 192 in. The height is twice the base. Find the base and the height.

3. *Supplementary Angles.* Two supplementary angles are such that twice one angle is 7° more than the other. Find the measures of the angles.

4. *SAT Scores.* The total of Megan's writing and math scores on the SAT was 1094. Her math score was 248 points higher than her writing score. What were her math and writing SAT scores?

5. *Sightseeing Boat.* A sightseeing boat travels 3 hr on a trip downstream with a 2.5-mph current. The return trip against the same current takes 3.5 hr. Find the speed of the boat in still water.

A. $x = y + 248,$
$x + y = 1094$

B. $5x = 2y - 3,$
$y = \frac{2}{3}x + 5$

C. $y = \frac{1}{2}x,$
$2x + 2y = 192$

D. $2x = 7 + y,$
$x + y = 180$

E. $x + y = 192,$
$x = 2y$

F. $x + y = 180,$
$x = 2y + 7$

G. $x - 1094 = y,$
$3x - 4y = 248$

H. $3\%x + 2.5\%y = 97.50,$
$x + y = 2500$

I. $2x = 5 + \frac{2}{3}y,$
$3y = 15x - 4$

J. $x = (y + 2.5) \cdot 3,$
$3.5(y - 2.5) = x$

Answers on page A-10

6. *Running Distances.* Each day Tricia runs 5 mi more than two-thirds the distance that Chris runs. Five times the distance that Chris runs is 3 mi less than twice the distance that Tricia runs. How far does Tricia run daily?

7. *Dimensions of a Rectangle.* The perimeter of a rectangle is 192 in. The width is half the length. Find the length and the width.

8. *Mystery Numbers.* Teka asked her students to determine the two numbers that she placed in a sealed envelope. Twice the smaller number is 5 more than two-thirds the larger number. Three times the larger number is 4 less than fifteen times the smaller. Find the numbers.

9. *Supplementary Angles.* Two supplementary angles are such that one angle is 7° more than twice the other. Find the measures of the angles.

10. *Student Loans.* Brandt's student loans totaled \$2500. Part was borrowed at 3% interest and the rest at 2.5%. After one year, Brandt had accumulated \$97.50 in interest. What was the amount of each loan?

3.4 Exercise Set

FOR EXTRA HELP 

✓ Check Your Understanding

Reading Check and Concept Check Consider the following mixture problem and the table used to translate the problem.

Cherry Breeze is 30% fruit juice and Berry Choice is 15% fruit juice. How much of each should be used in order to make 10 gal of a drink that is 20% fruit juice?

Choose from the options below the expression that best fits each numbered space in the table.

2   10   15 $0.15y$

	CHERRY BREEZE	BERRY CHOICE	MIXTURE
GALLONS OF DRINK	x	y	**RC1.** ___
PERCENT OF FRUIT JUICE	30%	**RC2.** ___ %	20%
GALLONS OF FRUIT JUICE IN MIXTURE	$0.3x$	**RC3.** ___	**RC4.** ___

a Solve.

1. *Video-Game Production.* Video games are developed or published at 2848 locations in the United States. There are 1796 more locations at which video games are developed than locations at which video games are published. At how many locations are video games developed and at how many are they published?

   Data: Entertainment Software Association

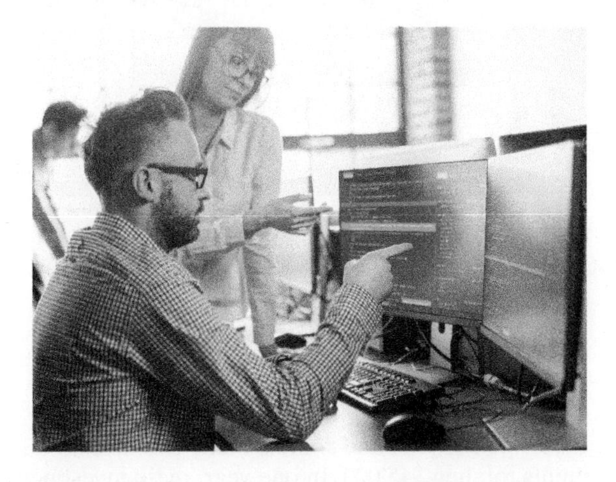

2. *Spoken Languages.* Of a total of 422 indigenous and immigrant languages spoken in the United States, 10 more spoken languages are indigenous than are immigrant. How many indigenous languages and how many immigrant languages are spoken in the United States?

   Data: *USA Today*, 11/8/2015

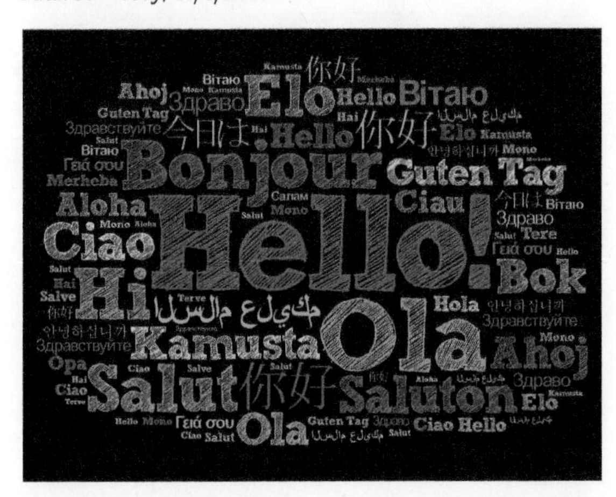

3. *Entertainment.* For her personal-finance class, Laura was required to estimate her annual entertainment expenditures. She discovered that during the previous year, she spent \$225.32 on a total of 68 e-books and game applications. If each book cost \$3.99 and each game cost \$1.99, how many books and how many games did she purchase?

4. *Flowers.* Kevin's Floral Emporium offers two types of sunflowers for sale by the stem. When in season, the small ones sell for \$2.50 per stem, and the large ones sell for \$3.95 per stem. One late summer weekend, Kevin sold a total of 118 stems for \$376.20. How many of each size did he sell?

5. *Furniture Polish.* A nontoxic furniture polish can be made by combining vinegar and olive oil. The amount of oil should be three times the amount of vinegar. How much of each ingredient is needed in order to make 30 oz of furniture polish?

6. *Nontoxic Floor Wax.* A nontoxic floor wax can be made by combining lemon juice and food-grade linseed oil. The amount of oil should be twice the amount of lemon juice. How much of each ingredient is needed in order to make 32 oz of floor wax? (The mix should be spread with a rag and buffed when dry.)

7. *Catering.* Stella's Catering is planning a wedding reception. The bride and groom would like to serve a nut mixture containing 25% peanuts. Stella has available mixtures that are either 40% or 10% peanuts. How much of each type should be mixed in order to get a 10-lb mixture that is 25% peanuts?

8. *Blending Granola.* Deep Thought Granola is 25% nuts and dried fruit. Oat Dream Granola is 10% nuts and dried fruit. How much of Deep Thought and how much of Oat Dream should be mixed in order to form a 20-lb batch of granola that is 19% nuts and dried fruit?

9. *Ink Remover.* Etch Clean Graphics uses one cleanser that is 25% acid and a second that is 50% acid. How many liters of each should be mixed in order to get 10 L of a solution that is 40% acid?

10. *Livestock Feed.* Soybean meal is 16% protein and corn meal is 9% protein. How many pounds of each should be mixed in order to get a 350-lb mixture that is 12% protein?

11. *Vegetable Seeds.* Tara's website, verdantveggies.com, specializes in the sale of rare or unusual vegetable seeds. Tara sells packets of sweet-pepper seeds for \$2.85 each and packets of hot-pepper seeds for \$4.29 each. She also offers a 16-packet mixed-pepper assortment combining packets of both types of seeds at \$3.30 per packet. How many packets of each type of seed are in the assortment?

Sweet peppers Hot peppers Assorted

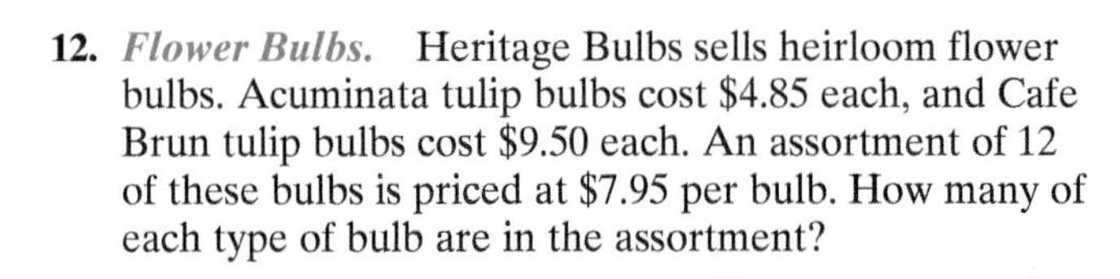

12. *Flower Bulbs.* Heritage Bulbs sells heirloom flower bulbs. Acuminata tulip bulbs cost \$4.85 each, and Cafe Brun tulip bulbs cost \$9.50 each. An assortment of 12 of these bulbs is priced at \$7.95 per bulb. How many of each type of bulb are in the assortment?

13. *Student Loans.* Sarah's two student loans totaled \$12,000. One of her loans was at 6% simple interest and the other at 3%. After one year, Sarah owed \$585 in interest. What was the amount of each loan?

14. *Investments.* Ana and Johnny made two investments totaling \$45,000. In one year, these investments yielded \$2430 in simple interest. Part of the money was invested at 4% and the rest at 6%. How much was invested at each rate?

15. *Food Science.* The following bar graph shows the milk fat percentages in three dairy products. How many pounds each of whole milk and cream should be mixed in order to form 200 lb of milk for cream cheese?

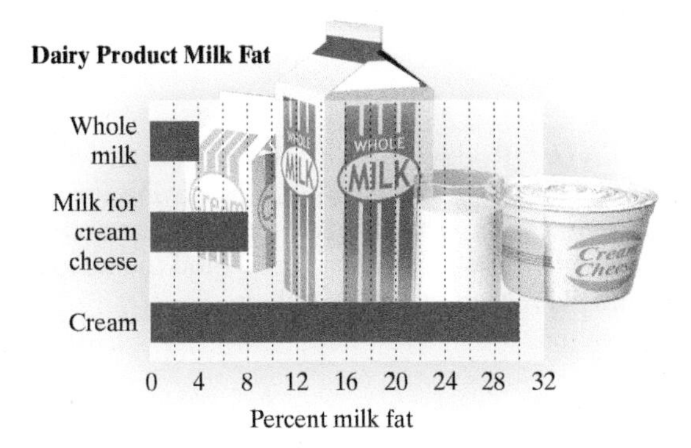

16. *Automotive Maintenance.* Arctic Antifreeze is 18% alcohol and Frost No-More is 10% alcohol. How many liters of Arctic Antifreeze should be mixed with 7.5 L of Frost No-More in order to get a mixture that is 15% alcohol?

17. *Investments.* William opened two investment accounts for his daughter's college fund. The first year, these investments, which totaled $3200, yielded $155 in simple interest. Part of the money was invested at 5.5% and the rest at 4%. How much was invested at each rate?

18. *Student Loans.* Cole's two student loans totaled $31,000. One of his loans was at 2.8% simple interest and the other at 4.5%. After one year, Cole owed $1024.40 in interest. What was the amount of each loan?

19. *Making Change.* Juan goes to a bank and gets change for a $50 bill consisting of all $5 bills and $1 bills. There are 22 bills in all. How many of each kind are there?

20. *Making Change.* Christina makes a $9.25 purchase at a bookstore with a $20 bill. The store has no bills and gives her the change in quarters and dollar coins. There are 19 coins in all. How many of each kind are there?

21. *Super Bowl Advertisements.* For commercials aired during Super Bowl L, advertisers paid an average of $5 million to air a 30-sec commercial. Even at this rate, some commercials were longer than 30 sec. A total of 62 commercials ran for either 30 sec or 1 min. The total running time was $39\frac{1}{2}$ min. How many of each length commercial aired during the Super Bowl?

22. *Typing.* It takes Ashleigh 5 min to type a page of her essay and $1\frac{1}{2}$ min to proofread a page. One afternoon she spent 2 hr typing or proofreading a total of 38 pages. How many pages did she type and how many did she proofread?

b Solve.

23. *Train Travel.* A train leaves Danville Junction and travels north at a speed of 75 mph. Two hours later, a second train leaves on a parallel track and travels north at 125 mph. How far from the station will the second train catch up to the first train?

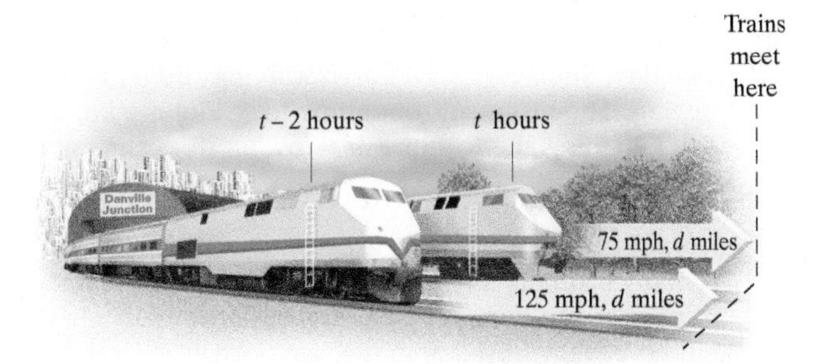

24. *Car Travel.* Max leaves Kansas City and drives east at a speed of 80 km/h. One hour later, Olivia leaves Kansas City traveling in the same direction as Max but at 96 km/h. Assuming that neither driver stops for a break, how far from Kansas City will they be when Olivia catches up with Max?

25. *Canoeing.* Darren paddled for 4 hr with a 6-km/h current to reach a campsite. The return trip against the same current took 10 hr. Find the speed of Darren's canoe in still water.

26. *Boating.* Mia's motorboat took 3 hr to make a trip downstream with a 6-mph current. The return trip against the same current took 5 hr. Find the speed of the boat in still water.

27. *Air Travel.* Christie pilots her Cessna 150 plane for 270 mi against a headwind in 3 hr. The flight would take 1 hr and 48 min with a tailwind of the same speed. Find the headwind and the speed of the plane in still air.

28. *Air Travel.* Rod is a pilot for Crossland Airways. He computes his flight time against a headwind for a trip of 2900 mi at 5 hr. The flight would take 4 hr and 50 min if the headwind were half as great. Find the headwind and the plane's air speed in still air.

29. *Air Travel.* Two airplanes start at the same time and fly toward each other from points 1000 km apart at rates of 420 km/h and 330 km/h. After how many hours will they meet?

30. *Air Travel.* Two planes start at the same time and travel toward each other from cities that are 780 km apart at rates of 190 km/h and 200 km/h. In how many hours will they meet?

31. *Point of No Return.* A plane flying the 3458-mi trip from New York City to London has a 50-mph tailwind. The flight's *point of no return* is the point at which the flight time required to return to New York is the same as the time required to continue to London. If the speed of the plane in still air is 360 mph, how far is New York from the point of no return?

32. *Point of No Return.* A plane is flying the 2553-mi trip from Los Angeles to Honolulu into a 60-mph headwind. If the speed of the plane in still air is 310 mph, how far from Los Angeles is the plane's point of no return? (See Exercise 31.)

Skill Maintenance

33. Find the intersection:

$\{2, 4, 6, 8, 10\} \cap \{6, 7, 8, 9, 10\}$. [1.5a]

34. Find the union:

$\{2, 4, 6, 8, 10\} \cup \{6, 7, 8, 9, 10\}$. [1.5b]

Simplify. [1.6a]

35. $|3a|$

36. $|-7x^2|$

37. $\left|\frac{-3}{y}\right|$

38. $\left|\frac{a^4}{c}\right|$

Synthesis

39. *Automotive Maintenance.* The radiator in Michelle's car contains 16 L of antifreeze and water. This mixture is 30% antifreeze. How much of this mixture should she drain and replace with pure antifreeze so that there will be a mixture of 50% antifreeze?

40. *Physical Exercise.* Natalie jogs and walks to school each day. She averages 4 km/h walking and 8 km/h jogging. The distance from home to school is 6 km and Natalie makes the trip in 1 hr. How far does she jog in a trip?

41. *Fuel Economy.* Ashlee's SUV gets 18 miles per gallon (mpg) in city driving and 24 mpg in highway driving. The SUV is driven 465 mi on 23 gal of gasoline. How many miles were driven in the city and how many were driven on the highway?

42. *Siblings.* Phil and Maria are siblings. Maria has twice as many brothers as she has sisters. Phil has the same number of brothers as he has sisters. How many girls and how many boys are in the family?

Mid-Chapter Review

Concept Reinforcement

Determine whether each statement is true or false.

______________ **1.** If, when solving a system of two linear equations in two variables, a false equation is obtained, the system has infinitely many solutions. [3.2a], [3.3a]

______________ **2.** Every system of equations has at least one solution. [3.1a]

______________ **3.** If the graphs of two linear equations intersect, then the system is consistent. [3.1a]

______________ **4.** The intersection of the graphs of the lines $x = a$ and $y = b$ is (a, b). [3.1a]

Guided Solutions

Fill in each box with the number, the variable, or the expression that creates a correct statement or solution.

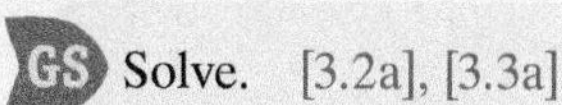
Solve. [3.2a], [3.3a]

5. $x + 2y = 3,$ **(1)**
$y = x - 6$ **(2)**

$x + 2(\square) = 3$ Substituting for y in equation (1)
$x + \square x - \square = 3$ Removing parentheses
$\square x - 12 = 3$ Collecting like terms
$3x = \square$
$x = \square$

$y = \square - 6$ Substituting in equation (2)
$y = \square$ Subtracting
The solution is $(\square, \square)$.

6. $3x - 2y = 5,$ **(1)**
$2x + 4y = 14$ **(2)**

$\square x - \square y = \square$ Multiplying equation (1) by 2
$2x + 4y = 14$ Equation (2)
$\square x = \square$ Adding
$x = \square$

$2 \cdot \square + 4y = 14$ Substituting for x in equation (2)
$\square + 4y = 14$ Multiplying
$4y = \square$
$y = \square$
The solution is $(\square, \square)$.

Mixed Review

Solve each system of equations graphically. Then classify the system as consistent or inconsistent and the equations as dependent or independent. [3.1a]

7. $y = x - 6,$
$y = 4 - x$

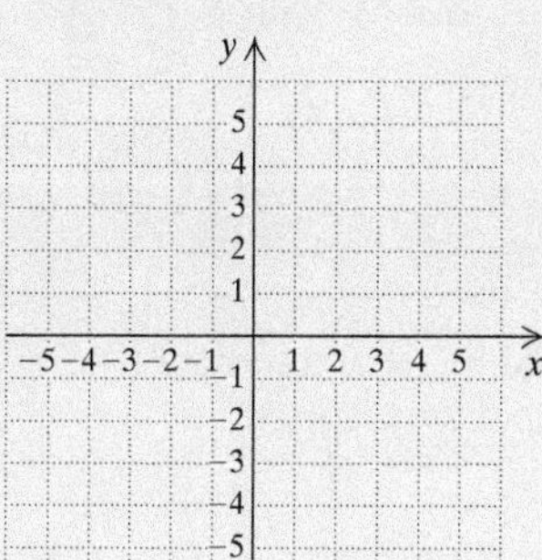

8. $x + y = 3,$
$3x + y = 3$

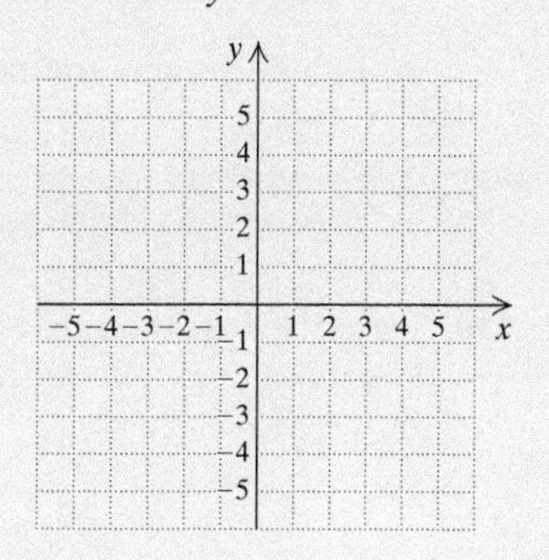

9. $y = 2x - 3,$
$4x - 2y = 6$

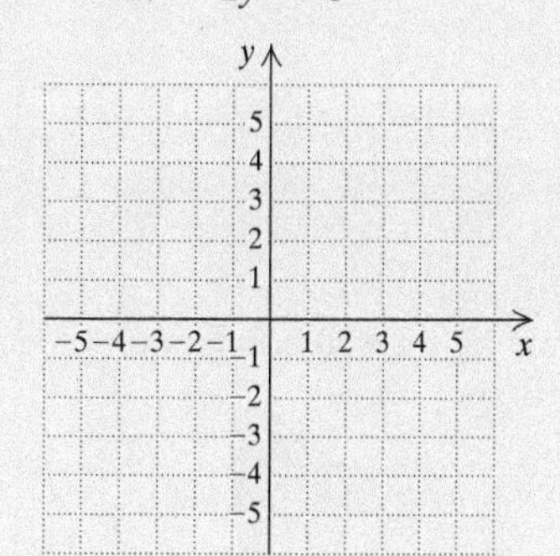

10. $x - y = 3,$
$2y - 2x = 6$

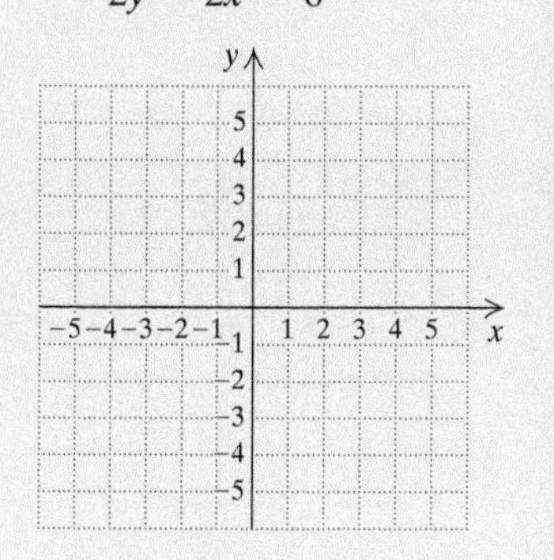

Solve using the substitution method. [3.2a]

11. $x = y + 2,$
$2x - 3y = -2$

12. $y = x - 5,$
$x - 2y = 8$

13. $4x + 3y = 3,$
$y = x + 8$

14. $3x - 2y = 1,$
$x = y + 1$

Solve using the elimination method. [3.3a]

15. $2x + y = 2,$
$x - y = 4$

16. $x - 2y = 13,$
$x + 2y = -3$

17. $3x - 4y = 5,$
$5x - 2y = -1$

18. $3x + 2y = 11,$
$2x + 3y = 9$

19. $x - 2y = 5,$
$3x - 6y = 10$

20. $4x - 6y = 2,$
$-2x + 3y = -1$

21. $\frac{1}{2}x + \frac{1}{3}y = 1,$
$\frac{1}{5}x - \frac{3}{4}y = 11$

22. $0.2x + 0.3y = 0.6,$
$0.1x - 0.2y = -2.5$

Solve.

23. *Garden Dimensions.* A landscape architect designs a garden with a perimeter of 44 ft. The width is 2 ft less than the length. Find the length and the width. [3.2b]

24. *Investments.* Sandy made two investments totaling $5000. Part of the money was invested at 2% and the rest at 3%. In one year, these investments earned $129 in simple interest. How much was invested at each rate? [3.4a]

25. *Mixing Solutions.* A lab technician wants to mix a solution that is 20% acid with a second solution that is 50% acid in order to get 84 L of a solution that is 30% acid. How many liters of each solution should be used? [3.4a]

26. *Boating.* Monica's motorboat took 5 hr to make a trip downstream with a 6-mph current. The return trip against the same current took 8 hr. Find the speed of the boat in still water. [3.4b]

Understanding Through Discussion and Writing

27. Explain how to find the solution of $\frac{3}{4}x + 2 = \frac{2}{5}x - 5$ in two ways graphically and in two ways algebraically. [3.1a], [3.2a], [3.3a]

28. Write a system of equations with the given solution. Answers may vary. [3.1a], [3.2a], [3.3a]

a) $(4, -3)$

b) No solution

c) Infinitely many solutions

29. Describe a method that could be used to create an inconsistent system of equations. [3.1a], [3.2a], [3.3a]

30. Describe a method that could be used to create a system of equations with dependent equations. [3.1a], [3.2a], [3.3a]

STUDYING FOR SUCCESS *Taking a Test*

- ☐ Read each question carefully. Know what the question is before you answer it.
- ☐ Try to work all the questions the first time through, marking those you need to recheck if you have time at the end.
- ☐ Pace yourself.
- ☐ Write your test answers in a neat and orderly manner.

3.5 Systems of Equations in Three Variables

OBJECTIVE

a Solve systems of three equations in three variables.

a SOLVING SYSTEMS IN THREE VARIABLES

SKILL REVIEW *Solve systems of equations in two variables by the elimination method.* [3.3a]

Solve.

1. $3x + y = 1,$
$5x - y = 7$

2. $2x + 3y = 9,$
$3x + 2y = 1$

Answers: 1. $(1, -2)$ **2.** $(-3, 5)$

MyLab Math VIDEO

A **linear equation in three variables** is an equation equivalent to one of the type $Ax + By + Cz = D$. A **solution** of a system of three equations in three variables is an ordered triple (x, y, z) that makes *all three* equations true.

The substitution method can be used to solve systems of three equations, but it is not efficient unless a variable has already been eliminated from one or more of the equations. Therefore, we will use only the elimination method.* The first step is to eliminate a variable and obtain a system of two equations in two variables.

EXAMPLE 1 Solve the following system of equations:

$$\begin{aligned} x + y + z &= 4, & (1)\\ x - 2y - z &= 1, & (2)\\ 2x - y - 2z &= -1. & (3) \end{aligned}$$

a) We first use *any* two of the three equations to get an equation in two variables. In this case, let's use equations (1) and (2) and add to eliminate z:

$$\begin{aligned} x + y + z &= 4 & (1)\\ x - 2y - z &= 1 & (2)\\ \hline 2x - y \quad &= 5. & (4) \end{aligned} \qquad \text{Adding to eliminate } z$$

b) We use a *different* pair of equations and eliminate the **same variable** that we did in part (a). Let's use equations (1) and (3) and again eliminate z.

*Other methods for solving systems of equations are considered in Appendixes B and C.

Caution!

A common error is to eliminate a different variable the second time.

$$x + y + z = 4, \quad (1)$$
$$2x - y - 2z = -1; \quad (3)$$

$$\begin{aligned} 2x + 2y + 2z &= 8 && \text{Multiplying equation (1) by 2} \\ 2x - y - 2z &= -1 && (3) \\ \hline 4x + y &= 7 && (5) \quad \text{Adding to eliminate } z \end{aligned}$$

c) Now we solve the resulting system of equations, (4) and (5). That solution will give us two of the numbers. Note that we now have two equations in two variables. Had we eliminated two *different* variables in parts (a) and (b), this would not be the case.

$$\begin{aligned} 2x - y &= 5 && (4) \\ 4x + y &= 7 && (5) \\ \hline 6x &= 12 && \text{Adding} \\ x &= 2 \end{aligned}$$

We can use either equation (4) or (5) to find y. We choose equation (5):

$$\begin{aligned} 4x + y &= 7 && (5) \\ 4(2) + y &= 7 && \text{Substituting 2 for } x \\ 8 + y &= 7 \\ y &= -1. \end{aligned}$$

d) We now have $x = 2$ and $y = -1$. To find the value for z, we use any of the original three equations, substitute, and solve for z. Let's use equation (1) and substitute our two numbers in it:

$$\begin{aligned} x + y + z &= 4 && (1) \\ 2 + (-1) + z &= 4 && \text{Substituting 2 for } x \text{ and } -1 \text{ for } y \\ 1 + z &= 4 && \text{Solving for } z \\ z &= 3. \end{aligned}$$

We have obtained the ordered triple $(2, -1, 3)$. To check, we substitute $(2, -1, 3)$ into each of the three equations using alphabetical order of the variables.

Check:

$$\begin{array}{c|c} x + y + z & 4 \\ \hline 2 + (-1) + 3 \ ? & 4 \\ 4 & \end{array} \quad \text{TRUE}$$

$$\begin{array}{c|c} x - 2y - z & 1 \\ \hline 2 - 2(-1) - 3 \ ? & 1 \\ 2 + 2 - 3 & \\ 1 & \end{array} \quad \text{TRUE}$$

$$\begin{array}{c|c} 2x - y - 2z & -1 \\ \hline 2(2) - (-1) - 2 \cdot 3 \ ? & -1 \\ 4 + 1 - 6 & \\ -1 & \end{array} \quad \text{TRUE}$$

The triple $(2, -1, 3)$ checks and is the solution.

To use the elimination method to solve systems of three equations:

1. Write all equations in the standard form, $Ax + By + Cz = D$.
2. Clear any decimals or fractions.
3. Choose a variable to eliminate. Then use *any* two of the three equations to eliminate that variable, getting an equation in two variables.
4. Next, use a different pair of equations and get another equation in *the same two variables*. That is, eliminate the same variable that you did in step (3).
5. Solve the resulting system (pair) of equations. That will give two of the numbers.
6. Then use any of the original three equations to find the third number.

Do Exercise 1. ▶

1. Solve. Don't forget to check.

$$\begin{aligned} 4x - y + z &= 6, \\ -3x + 2y - z &= -3, \\ 2x + y + 2z &= 3 \end{aligned}$$

EXAMPLE 2 Solve this system:

$$\begin{aligned} 4x - 2y - 3z &= 5, & (1) \\ -8x - y + z &= -5, & (2) \\ 2x + y + 2z &= 5. & (3) \end{aligned}$$

a) The equations are in standard form and do not contain decimals or fractions.

b) We decide to eliminate the variable y since the y-terms are opposites in equations (2) and (3). We add:

$$\begin{aligned} -8x - y + z &= -5 & (2) \\ 2x + y + 2z &= 5 & (3) \\ \hline -6x \qquad + 3z &= 0. & (4) \quad \text{Adding} \end{aligned}$$

c) We use another pair of equations to get an equation in the same two variables, x and z. We use equations (1) and (3) and eliminate y:

$$\begin{aligned} 4x - 2y - 3z &= 5, & (1) \\ 2x + y + 2z &= 5; & (3) \end{aligned}$$

$$\begin{aligned} 4x - 2y - 3z &= 5 & (1) \\ 4x + 2y + 4z &= 10 & \text{Multiplying equation (3) by 2} \\ \hline 8x \qquad + z &= 15. & (5) \quad \text{Adding} \end{aligned}$$

d) Next, we solve the resulting system of equations (4) and (5). That will give us two of the numbers:

$$\begin{aligned} -6x + 3z &= 0, & (4) \\ 8x + z &= 15. & (5) \end{aligned}$$

We multiply equation (5) by -3 and then add:

$$\begin{aligned} -6x + 3z &= 0 & (4) \\ -24x - 3z &= -45 & \text{Multiplying equation (5) by } -3 \\ \hline -30x \qquad &= -45 & \text{Adding} \\ x &= \tfrac{-45}{-30} = \tfrac{3}{2}. \end{aligned}$$

Answer

1. $(2, 1, -1)$

We now use equation (5) to find z:

$$
\begin{aligned}
8x + z &= 15 && \text{(5)} \\
8\left(\tfrac{3}{2}\right) + z &= 15 && \text{Substituting } \tfrac{3}{2} \text{ for } x \\
12 + z &= 15 && \\
z &= 3. && \text{Solving for } z
\end{aligned}
$$

e) Next, we use any of the original equations and substitute to find the third number, y. We choose equation (3) since the coefficient of y there is 1:

$$
\begin{aligned}
2x + y + 2z &= 5 && \text{(3)} \\
2\left(\tfrac{3}{2}\right) + y + 2(3) &= 5 && \text{Substituting } \tfrac{3}{2} \text{ for } x \text{ and 3 for } z \\
3 + y + 6 &= 5 && \\
y + 9 &= 5 && \text{Solving for } y \\
y &= -4. &&
\end{aligned}
$$

The solution is $\left(\frac{3}{2}, -4, 3\right)$. The check is as follows.

Check:

$$
\begin{array}{r|l}
4x - 2y - 3z & = 5 \\
\hline
4\cdot\frac{3}{2} - 2(-4) - 3(3) & ?\ 5 \\
6 + 8 - 9 & \\
5 & \quad \text{TRUE}
\end{array}
$$

$$
\begin{array}{r|l}
-8x - y + z & = -5 \\
\hline
-8\cdot\frac{3}{2} - (-4) + 3 & ?\ -5 \\
-12 + 4 + 3 & \\
-5 & \quad \text{TRUE}
\end{array}
$$

$$
\begin{array}{r|l}
2x + y + 2z & = 5 \\
\hline
2\cdot\frac{3}{2} + (-4) + 2(3) & ?\ 5 \\
3 - 4 + 6 & \\
5 & \quad \text{TRUE}
\end{array}
$$

Do Exercise 2.

2. Solve. Don't forget to check.

$$
\begin{aligned}
2x + y - 4z &= 0, \\
x - y + 2z &= 5, \\
3x + 2y + 2z &= 3
\end{aligned}
$$

In Example 3, two of the equations have a missing variable.

EXAMPLE 3 Solve this system:

$$
\begin{aligned}
x + y + z &= 180, && \text{(1)} \\
x \qquad - z &= -70, && \text{(2)} \\
2y - z &= 0. && \text{(3)}
\end{aligned}
$$

We note that there is no y in equation (2). In order to have a system of two equations in the variables x and z, we need to find another equation without a y. We use equations (1) and (3) to eliminate y:

$$
\begin{aligned}
x + y + z &= 180, && \text{(1)} \\
2y - z &= 0; && \text{(3)}
\end{aligned}
$$

$$
\begin{aligned}
-2x - 2y - 2z &= -360 && \text{Multiplying equation (1) by } -2 \\
2y - z &= 0 && \text{(3)} \\
\hline
-2x \qquad - 3z &= -360. && \text{(4)} \quad \text{Adding}
\end{aligned}
$$

Answer

2. $\left(2, -2, \frac{1}{2}\right)$

Next, we solve the resulting system of equations (2) and (4):

$$x - z = -70, \quad (2)$$
$$-2x - 3z = -360; \quad (4)$$

$$\begin{aligned} 2x - 2z &= -140 && \text{Multiplying equation (2) by 2} \\ -2x - 3z &= -360 && (4) \\ \hline -5z &= -500 && \text{Adding} \\ z &= 100. \end{aligned}$$

To find x, we substitute 100 for z in equation (2) and solve for x:

$$\begin{aligned} x - z &= -70 \\ x - 100 &= -70 \\ x &= 30. \end{aligned}$$

To find y, we substitute 100 for z in equation (3) and solve for y:

$$\begin{aligned} 2y - z &= 0 \\ 2y - 100 &= 0 \\ 2y &= 100 \\ y &= 50. \end{aligned}$$

The triple (30, 50, 100) is the solution. The check is left to the student.

Do Exercise 3.

3. Solve. Don't forget to check.

$$\begin{aligned} x + y + z &= 100, && (1) \\ x - y &= -10, && (2) \\ x - z &= -30 && (3) \end{aligned}$$

Add equations (1) and (3):

$$\begin{aligned} x + y + z &= 100 && (1) \\ x - z &= -30 && (3) \\ \hline 2x + y &= \_\_\_\_ . && (4) \end{aligned}$$

Add equations (2) and (4) and solve for x:

$$\begin{aligned} x - y &= -10 && (2) \\ 2x + y &= 70 && (4) \\ \hline 3x &= \_\_\_\_ \\ x &= \_\_\_\_ . \end{aligned}$$

Substitute 20 for x in equation (4) and solve for y:

$$\begin{aligned} 2(20) + y &= 70 \\ y &= \_\_\_\_ . \end{aligned}$$

Substitute 20 for x and 30 for y in equation (1) and solve for z:

$$\begin{aligned} 20 + 30 + z &= 100 \\ z &= \_\_\_\_ . \end{aligned}$$

The numbers check. The solution is (20, 30, ____).

It is possible for a system of three equations to have no solution, that is, to be inconsistent. An example is the system

$$\begin{aligned} x + y + z &= 14, \\ x + y + z &= 11, \\ 2x - 3y + 4z &= -3. \end{aligned}$$

Note the first two equations. It is not possible for a sum of three numbers to be both 14 and 11. Thus the system has no solution. We will not consider such systems here, nor will we consider systems with infinitely many solutions, which also exist.

Answer

3. (20, 30, 50)

Guided Solution:

3. 70, 60, 20, 30, 50, 50

3.5 Exercise Set

FOR EXTRA HELP

MyLab Math

✓ Check Your Understanding

Reading Check Choose from the column on the right the option that is an example of each term. Choices may be used more than once.

RC1. A linear equation in three variables

RC2. A system of equations in three variables

RC3. A solution of a linear equation in three variables

RC4. A solution of a system of equations in three variables

a) $(4, -3, 0)$

b) $a + b - c = 1$

c) $\begin{aligned} a + 3b - c &= 1, \\ 2a + 3b - c &= -1, \\ a - 2b + 3c &= 10 \end{aligned}$

Concept Check Determine whether $(2, -1, -3)$ is a solution of each system.

CC1. $x + y - 2z = 7,$
$2x - y - z = 8,$
$-x - 2y - 3z = 9$

CC2. $r - s + t = 0,$
$2r - s + t = 2,$
$r - 2s + t = -3$

CC3. $a - c = 5,$
$3b - c = 0,$
$2a + 5b = -1$

a Solve.

1. $x + y + z = 2,$
$2x - y + 5z = -5,$
$-x + 2y + 2z = 1$

2. $2x - y - 4z = -12,$
$2x + y + z = 1,$
$x + 2y + 4z = 10$

3. $2x - y + z = 5,$
$6x + 3y - 2z = 10,$
$x - 2y + 3z = 5$

4. $x - y + z = 4,$
$3x + 2y + 3z = 7,$
$2x + 9y + 6z = 5$

5. $2x - 3y + z = 5,$
$x + 3y + 8z = 22,$
$3x - y + 2z = 12$

6. $6x - 4y + 5z = 31,$
$5x + 2y + 2z = 13,$
$x + y + z = 2$

7. $3a - 2b + 7c = 13,$
$a + 8b - 6c = -47,$
$7a - 9b - 9c = -3$

8. $x + y + z = 0,$
$2x + 3y + 2z = -3,$
$-x + 2y - 3z = -1$

9. $2x + 3y + z = 17,$
$x - 3y + 2z = -8,$
$5x - 2y + 3z = 5$

10. $2x + y - 3z = -4,$
$4x - 2y + z = 9,$
$3x + 5y - 2z = 5$

11. $2x + y + z = -2,$
$2x - y + 3z = 6,$
$3x - 5y + 4z = 7$

12. $2x + y + 2z = 11,$
$3x + 2y + 2z = 8,$
$x + 4y + 3z = 0$

13. $x - y + z = 4,$
$5x + 2y - 3z = 2,$
$3x - 7y + 4z = 8$

14. $2x + y + 2z = 3,$
$x + 6y + 3z = 4,$
$3x - 2y + z = 0$

15. $4x - y - z = 4,$
$2x + y + z = -1,$
$6x - 3y - 2z = 3$

16. $2r + s + t = 6,$
$3r - 2s - 5t = 7,$
$r + s - 3t = -10$

17. $a - 2b - 5c = -3,$
$3a + b - 2c = -1,$
$2a + 3b + c = 4$

18. $x + 4y - z = 5,$
$2x - y + 3z = -5,$
$4x + 3y + z = 5$

19. $2r + 3s + 12t = 4,$
$4r - 6s + 6t = 1,$
$r + s + t = 1$

20. $10x + 6y + z = 7,$
$5x - 9y - 2z = 3,$
$15x - 12y + 2z = -5$

21. $a + 2b + c = 1,$
$7a + 3b - c = -2,$
$a + 5b + 3c = 2$

22. $3p + 2r = 11,$
$q - 7r = 4,$
$p - 6q = 1$

23. $x + y + z = 57,$
$-2x + y = 3,$
$x - z = 6$

24. $4a + 9b = 8,$
$8a + 6c = -1,$
$6b + 6c = -1$

25. $r + s = 5,$
$3s + 2t = -1,$
$4r + t = 14$

26. $a - 5c = 17,$
$b + 2c = -1,$
$4a - b - 3c = 12$

27. $x + y + z = 105,$
$10y - z = 11,$
$2x - 3y = 7$

Skill Maintenance

Solve for the indicated letter. [1.2a]

28. $F = 3ab$, for a

29. $Q = 4(a + b)$, for a

30. $F = \frac{1}{2}t(c - d)$, for d

31. $F = \frac{1}{2}t(c - d)$, for c

32. $Ax + By = c$, for y

33. $Ax - By = c$, for y

Find the slope and the y-intercept. [2.4b]

34. $y = -\frac{2}{3}x - \frac{5}{4}$

35. $y = 5 - 4x$

36. $2x - 5y = 10$

37. $7x - 6.4y = 20$

Synthesis

Solve.

38. $w + x - y + z = 0,$
$w - 2x - 2y - z = -5,$
$w - 3x - y + z = 4,$
$2w - x - y + 3z = 7$

39. $w + x + y + z = 2,$
$w + 2x + 2y + 4z = 1,$
$w - x + y + z = 6,$
$w - 3x - y + z = 2$

3.6 Solving Applied Problems: Three Equations

OBJECTIVE

a Solve applied problems using systems of three equations.

Solving systems of three or more equations is important in many applications occurring in the natural and social sciences, business, and engineering.

a USING SYSTEMS OF THREE EQUATIONS

SKILL REVIEW *Solve applied problems by translating to equations.* [1.3a]

Solve.

1. The second angle of a triangle is twice as large as the first angle. The third angle is 5° larger than the second angle. Find the measures of the angles.
2. Malika invested her tax refund check in a fund paying 4% interest. After one year, she had earned $50.40 in interest. How much did she invest?

Answers: 1. 35°, 70°, 75° **2.** $1260

MyLab Math VIDEO

EXAMPLE 1 *Jewelry Design.* Kim is designing a triangular-shaped pendant for a client of her custom jewelry business. The largest angle of the triangle is 70° greater than the smallest angle. The largest angle is twice as large as the remaining angle. Find the measure of each angle.

1. **Familiarize.** We first make a drawing. We let $x =$ the smallest angle, $z =$ the largest angle, and $y =$ the remaining angle.

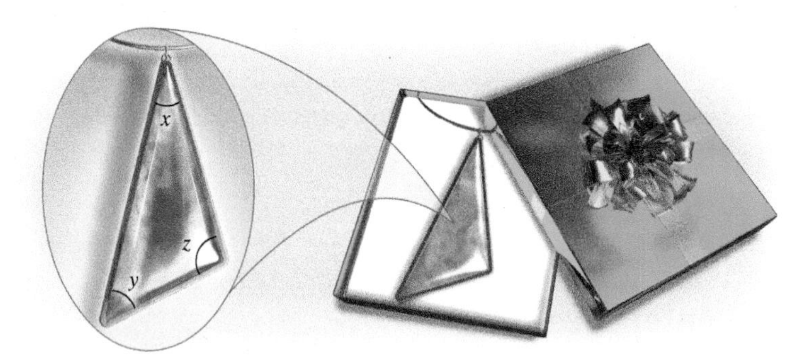

2. **Translate.** In order to translate the problem, we use the fact that the sum of the measures of the angles of a triangle is 180°:

$$x + y + z = 180.$$

There are two statements in the problem that we can translate directly.

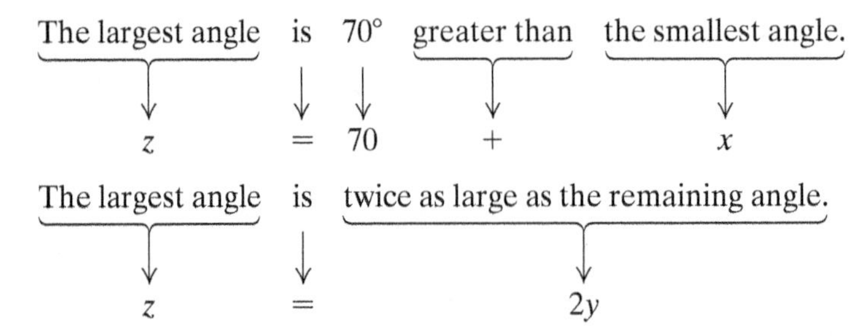

We now have a system of three equations:

$$\begin{aligned} x + y + z &= 180, \qquad & x + y + z &= 180, \quad & (1)\\ x + 70 &= z, \qquad \text{or} & x - z &= -70, \quad & (2)\\ 2y &= z; \qquad & 2y - z &= 0. \quad & (3) \end{aligned}$$

3. **Solve.** We can solve the system using elimination. However, because we can write both x and y in terms of z, we can also use substitution to solve this system.

   We first solve equation (2) for x:

$$\begin{aligned} x - z &= -70\\ x &= z - 70. \qquad (4) \end{aligned}$$

   We then solve equation (3) for y:

$$\begin{aligned} 2y - z &= 0\\ y &= \frac{z}{2}. \qquad (5) \end{aligned}$$

   Finally, we substitute in equation (1) and solve for z:

$$\begin{aligned} x + y + z &= 180\\ (z - 70) + \frac{z}{2} + z &= 180 \qquad \text{Substituting}\\ \frac{5}{2}z - 70 &= 180\\ \frac{5}{2}z &= 250\\ z &= 100. \qquad \text{Solving for } z \end{aligned}$$

   We now substitute 100 for z in equations (4) and (5) and find $x = 30$ and $y = 50$.

4. **Check.** The sum of the numbers is 180. The largest angle measures 100° and the smallest measures 30°, so the largest angle is 70° greater than the smallest. The largest angle is twice as large as 50°, the remaining angle. We have an answer to the problem.

5. **State.** The measures of the angles of the triangle are 30°, 50°, and 100°.

Do Exercise 1. ▶

1. ***Triangle Measures.*** One angle of a triangle is twice as large as a second angle. The remaining angle is 20° greater than the first angle. Find the measure of each angle.

EXAMPLE 2 ***Airline Tickets.*** Some airlines charge fees for baggage that is checked in rather than carried with the passenger. A Spanish class flew to Costa Rica using an airline that charged \$25 for one checked bag and \$60 for two checked bags. The 30 students paid a total of \$1120 in baggage fees for 40 checked bags. How many students checked no bags, how many checked one bag, and how many checked two bags?

1. **Familiarize.** A careful reading of the problem leads us to recognize that there are three relationships given: the total number of students, the total baggage fee, and the total number of bags. These three relationships will give us three equations. We let $x =$ the number of students who checked no bags, $y =$ the number of students who checked one bag, and $z =$ the number of students who checked two bags.

Answer

1. 64°, 32°, 84°

2. **Translate.** We translate the three relationship statements to equations.
   - A total of 30 students went on the trip. $\longrightarrow x + y + z = 30,$
   - The baggage fee was \$1120. $\longrightarrow 0 \cdot x + 25 \cdot y + 60 \cdot z = 1120,$
   - A total of 40 bags were checked. $\longrightarrow 0 \cdot x + 1 \cdot y + 2 \cdot z = 40$
3. **Solve.** We write the equations in standard form:

$$x + y + z = 30, \quad (1)$$
$$25y + 60z = 1120, \quad (2)$$
$$y + 2z = 40. \quad (3)$$

Equations (2) and (3) form a system of two equations in two unknowns that we can solve for y and z.

$$\begin{aligned} 25y + 60z &= 1120, & \\ -25y - 50z &= -1000 & \text{Multiplying equation (3) by } -25 \\ \hline 10z &= 120 & \text{Adding} \\ z &= 12 & \text{Solving for } z \end{aligned}$$

We substitute 12 for z in equation (3) and solve for y:

$$\begin{aligned} y + 2 \cdot 12 &= 40 \\ y + 24 &= 40 \\ y &= 16. \end{aligned}$$

Finally, we find x by substituting 16 for y and 12 for z in equation (1):

$$\begin{aligned} x + 16 + 12 &= 30 \\ x &= 2. \end{aligned}$$

We have $x = 2$, $y = 16$, and $z = 12$.

4. **Check.** We check our answers in each statement of the problem.
   - The total number of students on the trip was $2 + 16 + 12 = 30$.
   - The baggage fee was $0 \cdot 2 + 25 \cdot 16 + 60 \cdot 12 = 1120$.
   - The total number of bags was $0 \cdot 2 + 1 \cdot 16 + 2 \cdot 12 = 40$.
5. **State.** There were 2 students who checked no bags, 16 students who checked 1 bag, and 12 students who checked 2 bags.

◀ **Do Exercise 2.**

2. ***Client Investments.*** Kaufman Financial Corporation makes investments for corporate clients. One year, a client receives \$1120 in simple interest from three investments that total \$25,000. Part is invested at 3%, part at 4%, and part at 5%. There is \$11,000 more invested at 5% than at 4%. How much was invested at each rate? GS

Let x = the amount invested at 3%, y = the amount invested at 4%, and z = the amount invested at 5%. Complete the following table to help in the translation.

	FIRST INVESTMENT	SECOND INVESTMENT	THIRD INVESTMENT	TOTAL
PRINCIPAL, P	x	y	z	\$ [ ]
RATE OF INTEREST, r	3%	4%	[ ] %	
TIME, t	1 year	1 year	1 year	
INTEREST, I	$0.03x$	$0.04y$	$0.05z$	\$ [ ]

Answer

2. \$4000 at 3%; \$5000 at 4%; \$16,000 at 5%

Guided Solution:

2. 25,000, 5, 1120

3.6 Exercise Set

FOR EXTRA HELP MyLab Math

✓ Check Your Understanding

Reading Check and Concept Check Choose from the column on the right an appropriate translation for each statement.

RC1. The sum of three numbers is 60.

RC2. The first number minus the second number plus the third number is 60.

RC3. The first number is 60 more than the sum of the other two numbers.

RC4. The first number is 60 less than the sum of the other two numbers.

a) $x = y + z + 60$

b) $x = y + z - 60$

c) $x + y + z = 60$

d) $x - y + z = 60$

a Solve.

1. *Advanced Placement Exams.* In 2016, approximately 4.7 million Advanced Placement (AP) exams were taken in the United States. A total of 40 thousand exams was taken in Chinese, French, and German Language and Culture. The number of Chinese exams taken was 8 thousand more than the number of German exams taken. The number of French exams taken was 4 thousand more than the sum of the number of Chinese and German exams. Find the number of exams taken in each language.

Data: collegeboard.org

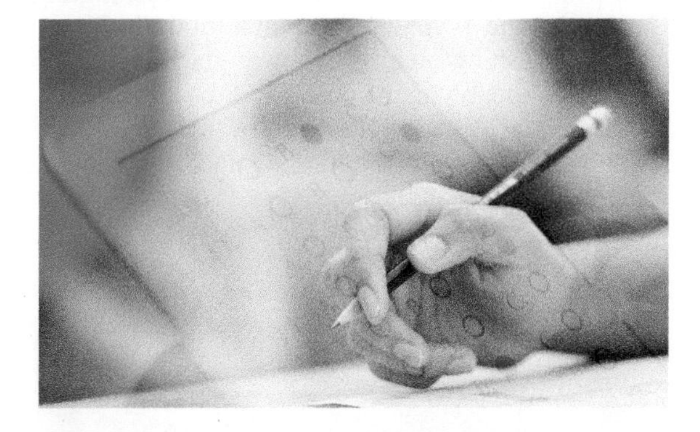

2. *Fat Content of Fast Food.* A meal at McDonald's consisting of a Big Mac, a medium order of fries, and a medium vanilla milkshake contains 61 g of fat. The Big Mac has 11 more grams of fat than the milkshake. The total fat content of the fries and the shake exceeds that of the Big Mac by 5 g. Find the fat content of each food item.

Data: McDonald's

3. *Triangle Measures.* In triangle ABC, the measure of angle B is three times that of angle A. The measure of angle C is 20° more than that of angle A. Find the measure of each angle.

4. *Triangle Measures.* In triangle ABC, the measure of angle B is twice the measure of angle A. The measure of angle C is 80° more than that of angle A. Find the measure of each angle.

5. The sum of three numbers is 55. The difference of the largest and the smallest is 49, and the sum of the two smaller numbers is 13. Find the numbers.

6. *History.* Find the year in which the first U.S. transcontinental railroad was completed. The following are some facts about the number. The sum of the digits in the year is 24. The ones digit is 1 more than the hundreds digit. Both the tens and the ones digits are multiples of 3.

7. *Eggs.* Chicken eggs are sorted by size before they are sold. A dozen small eggs weighs 18 oz, a dozen medium eggs weighs 21 oz, and a dozen large eggs weighs 24 oz. A camp cook needed 480 oz of eggs and purchased 22 dozen eggs at a local convenience store for a total of $43.88. If a dozen small eggs cost $1.79, a dozen medium eggs cost $1.99, and a dozen large eggs cost $2.09, how many dozens of each size egg did the cook purchase?

Data: thekitchn.com

8. *Coffee.* A coffee shop on campus sells coffee in three sizes: a 12-oz small, a 16-oz medium, and a 20-oz large. A small coffee sells for $1.75, a medium coffee for $1.95, and a large coffee for $2.25. One morning, Brandie served 50 coffees for a total of $98.70. She made the coffee in 80-oz batches, and used exactly 10 of the batches during the morning. How many of each size did she sell?

9. *Cholesterol Levels.* Recent studies indicate that a child's intake of cholesterol should be no more than 300 mg per day. By eating 1 egg, 1 cupcake, and 1 slice of pizza, a child consumes 302 mg of cholesterol. If the child eats 2 cupcakes and 3 slices of pizza, he or she takes in 65 mg of cholesterol. By eating 2 eggs and 1 cupcake, a child consumes 567 mg of cholesterol. How much cholesterol is in each item?

10. *Book Sale.* Katie, Rachel, and Logan went together to a library book sale. Katie bought 22 children's books, 10 paperbacks, and 5 hardbacks for a total of $63.50. Rachel bought 12 paperbacks and 15 hardbacks for a total of $52.50. Logan bought 8 children's books and 6 hardbacks for a total of $29.00. How much did each type of book cost?

11. *Automobile Pricing.* A recent basic model of a particular automobile had a price of $14,685. The basic model with the added features of automatic transmission and power door locks was $16,070. The basic model with air conditioning (AC) and power door locks was $15,580. The basic model with AC and automatic transmission was $15,925. What was the individual cost of each of the three options?

12. *Computer Pricing.* Lindsay plans to buy a new desktop computer for gaming. The base price of the computer is $480. If she upgrades the processor and the memory, the price of the computer is $745. If she upgrades the memory and the graphics card, the price of the computer is $690. If she upgrades the processor and the graphics card, the price of the computer is $805. What is the price of each upgrade?

13. *Pet Ownership.* In 2017, a total of 204 million birds, cats, and dogs lived as pets in U.S. homes. The number of cats was 4 million more than the number of dogs, and the number of birds was 3 million less than one-eighth of the sum of the numbers of cats and dogs. Find the number of each type of pet living in U.S. homes.

Data: American Pet Productions Association

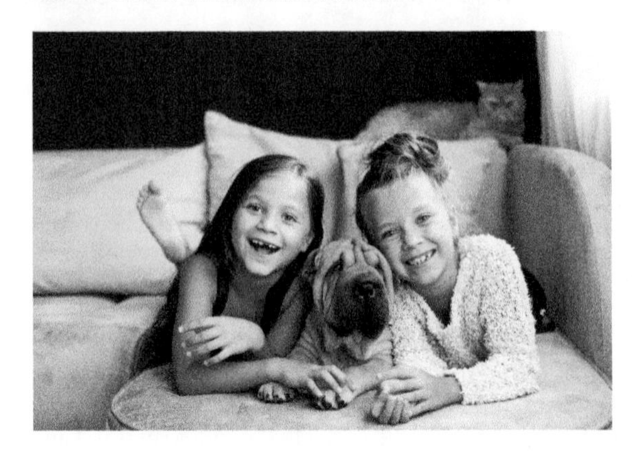

14. *Glaciers.* The western U.S. states of Washington, California, and Wyoming contain a total of 6366 glaciers. Washington has 164 fewer glaciers than do California and Wyoming combined. California has 262 more glaciers than one-third of the sum of the number of glaciers in Washington and Wyoming. Find the number of glaciers in each state.

Data: glaciers.research.pdx.edu

15. *Nutrition.* A dietician in a hospital prepares meals under the guidance of a physician. Suppose that for a particular patient a physician prescribes a meal to have 800 calories, 55 g of protein, and 220 mg of vitamin C. The dietician prepares a meal of roast beef, baked potato, and broccoli according to the data in the following table. How many servings of each food are needed in order to satisfy the doctor's orders?

16. *Nutrition.* Repeat Exercise 15 but replace the broccoli with asparagus, for which one 180-g serving contains 50 calories, 5 g of protein, and 44 mg of vitamin C. Which meal would you prefer eating?

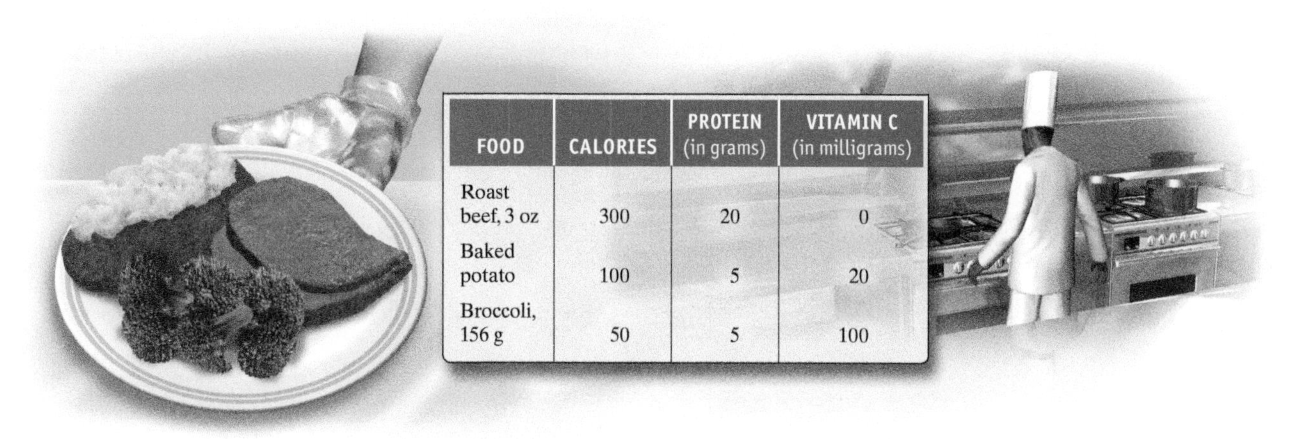

FOOD	CALORIES	PROTEIN (in grams)	VITAMIN C (in milligrams)
Roast beef, 3 oz	300	20	0
Baked potato	100	5	20
Broccoli, 156 g	50	5	100

17. *Investments.* A business class divided an imaginary investment of \$80,000 among three mutual funds. The first fund grew by 2%, the second by 6%, and the third by 3%. Total earnings were \$2250. The earnings from the first fund were \$150 more than the earnings from the third. How much was invested in each fund?

18. *Student Loans.* Terrence owes \$32,000 in student loans. The interest rate on his Perkins loan is 5%, the rate on his Stafford loan is 4%, and the rate on his bank loan is 7%. Interest for one year totaled \$1500. The interest for one year from the Perkins loan is \$220 more than the interest from the bank loan. What is the amount of each loan?

19. *Golf.* On an 18-hole golf course, there are par-3 holes, par-4 holes, and par-5 holes. A golfer who shoots par on every hole has a total of 70. There are twice as many par-4 holes as there are par-5 holes. How many of each type of hole are there on the golf course?

20. *Basketball Scoring.* The New York Knicks once scored a total of 92 points on a combination of 2-point field goals, 3-point field goals, and 1-point foul shots. Altogether, the Knicks made 50 baskets and 19 more 2-pointers than foul shots. How many shots of each kind were made?

21. *Lens Production.* When Sight-Rite's three polishing machines, A, B, and C, are all working, 5700 lenses can be polished in one week. When only A and B are working, 3400 lenses can be polished in one week. When only B and C are working, 4200 lenses can be polished in one week. How many lenses can be polished in a week by each machine alone?

22. *Telemarketing.* Steve, Teri, and Isaiah can process 740 telephone orders per day. Steve and Teri together can process 470 orders, while Teri and Isaiah together can process 520 orders per day. How many orders can each person process alone?

Skill Maintenance

Graph each function. [2.2c]

23. $f(x) = 2x - 3$

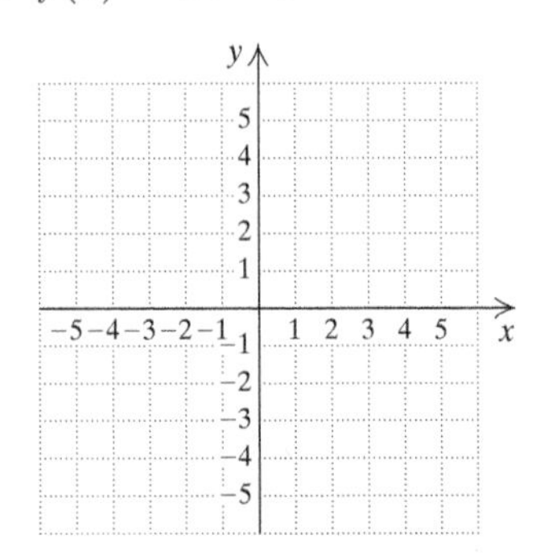

24. $g(x) = |x + 1|$

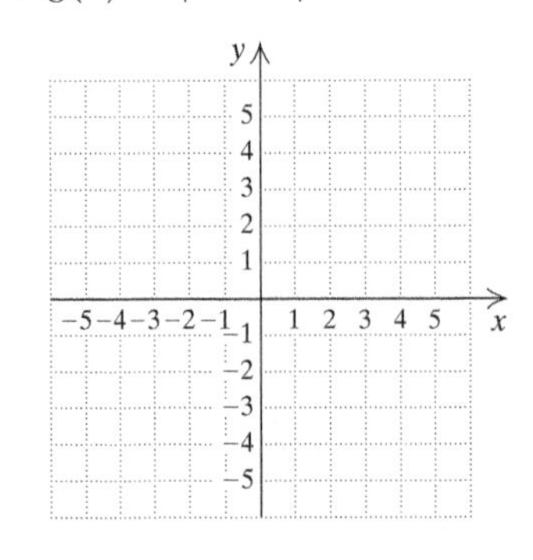

25. $h(x) = x^2 - 2$

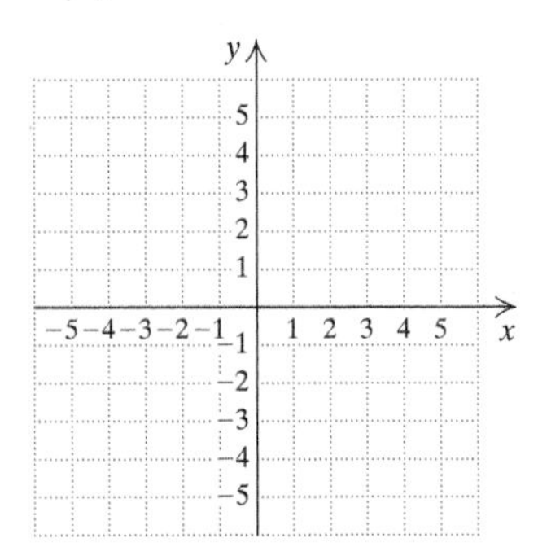

Determine whether each of the following is the graph of a function. [2.2d]

26.

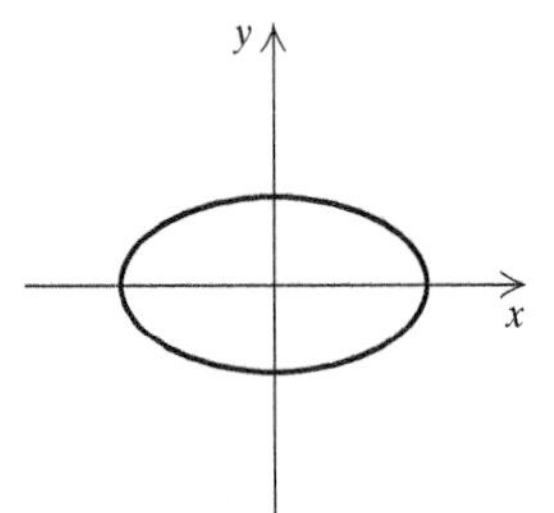

27.

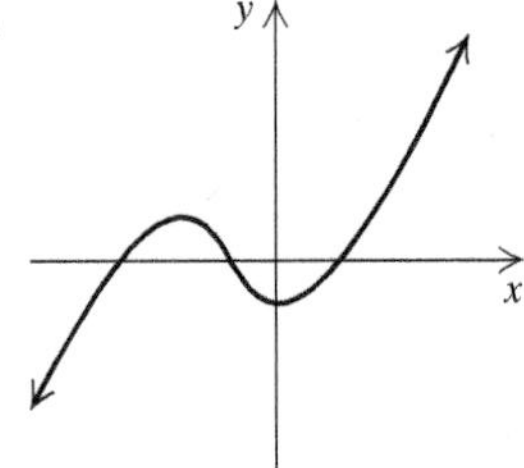

28.

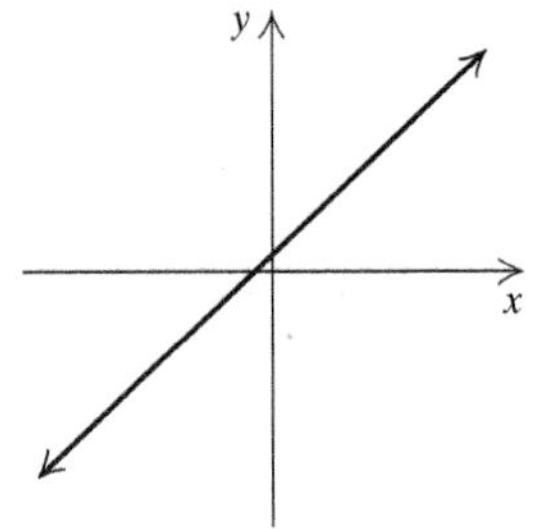

Synthesis

29. Find the sum of the angle measures at the tips of the star in this figure.

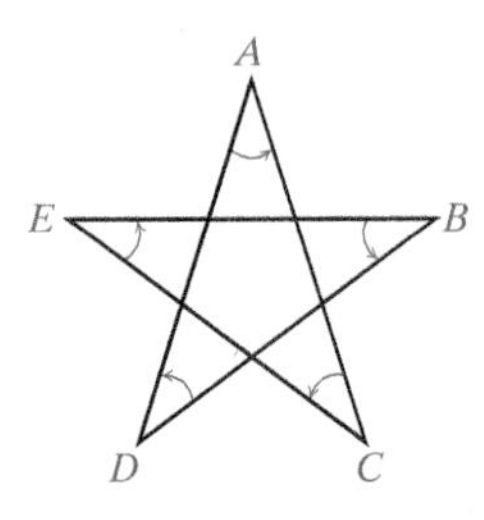

30. *Sharing Raffle Tickets.* Hal gives Tom as many raffle tickets as Tom has and Gary as many as Gary has. In like manner, Tom then gives Hal and Gary as many tickets as each then has. Similarly, Gary gives Hal and Tom as many tickets as each then has. If each finally has 40 tickets, with how many tickets does Tom begin?

31. *Digits.* Find a three-digit positive integer such that the sum of all three digits is 14, the tens digit is 2 more than the ones digit, and if the digits are reversed, the number is unchanged.

32. *Ages.* Tammy's age is the sum of the ages of Carmen and Dennis. Carmen's age is 2 more than the sum of the ages of Dennis and Mark. Dennis's age is four times Mark's age. The sum of all four ages is 42. How old is Tammy?

33. *Video-Game Ratings.* The Entertainment Software Rating Board assigned one of the following ratings to 1476 video games produced in 2016: E (everyone), E10+ (everyone ages 10 and older), T (teens), and M (mature audiences). More games were rated E than any other rating, with only 74 fewer E games than the total number of the other games. The sum of the number of games rated E or E10+ was 45 more than twice the sum of the number of games rated T or M. The number of games rated T was 15 less than twice the number of games rated M. How many games were assigned each rating?

Data: Entertainment Software Association

Systems of Inequalities in Two Variables

3.7

OBJECTIVES

a Determine whether an ordered pair of numbers is a solution of an inequality in two variables.

b Graph linear inequalities in two variables.

c Graph systems of linear inequalities and find coordinates of any vertices.

A **graph** of an inequality is a drawing that represents its solutions. An inequality in one variable can be graphed on the number line. An inequality in two variables can be graphed on a coordinate plane.

A **linear inequality** is one that we can get from a related linear equation by changing the equals symbol to an inequality symbol. The graph of a linear inequality is the half-plane on one side of the graph of the related equation. The graph sometimes includes the graph of the related line at the boundary of the half-plane.

a SOLUTIONS OF INEQUALITIES IN TWO VARIABLES

The solutions of an inequality in two variables are ordered pairs.

EXAMPLES Determine whether the ordered pair is a solution of the inequality $5x - 4y > 13$.

1. $(-3, 2)$

$$5x - 4y > 13$$

$5(-3) - 4 \cdot 2 \; ? \; 13$	We use alphabetical order to replace x with -3 and y with 2.
$-15 - 8$	
-23	FALSE

Since $-23 > 13$ is false, $(-3, 2)$ is not a solution.

2. $(4, -3)$

$$5x - 4y > 13$$

$5(4) - 4(-3) \; ? \; 13$	Replacing x with 4 and y with -3
$20 + 12$	
32	TRUE

Since $32 > 13$ is true, $(4, -3)$ is a solution.

Do Exercises 1 and 2. ▶

1. Determine whether $(1, -4)$ is a solution of $4x - 5y < 12$.

$$4x - 5y < 12$$

?

2. Determine whether $(4, -3)$ is a solution of $3y - 2x \le 6$.

$$3y - 2x \le 6$$

?

b GRAPHING INEQUALITIES IN TWO VARIABLES

MyLab Math ANIMATION

SKILL REVIEW

Graph linear equations using intercepts. [2.5a]

Find the intercepts. Then graph the equation.

1. $3x - 2y = 6$ **2.** $2x + y = 4$

Answers: 1.

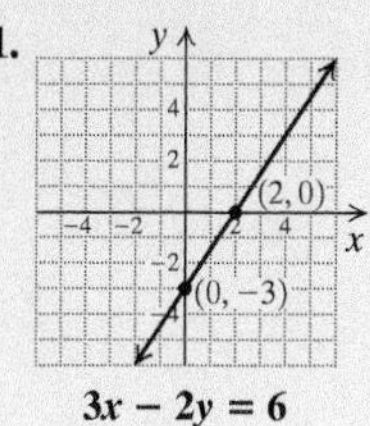

$3x - 2y = 6$

2.

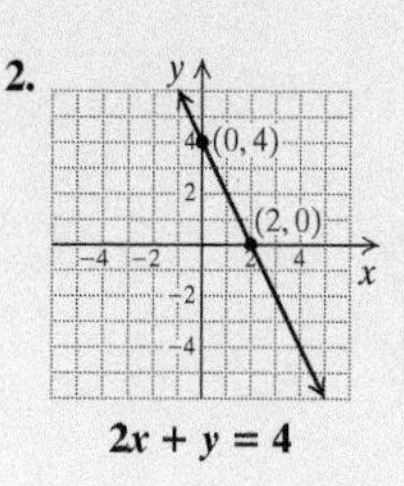

$2x + y = 4$

MyLab Math VIDEO

Answers

1. No **2.** Yes

Let's visualize the results of Examples 1 and 2. The equation $5x - 4y = 13$ is represented by the dashed line in the following graphs.

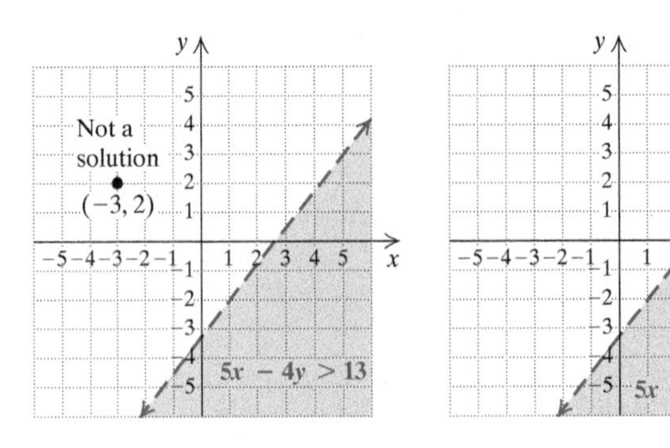

The solutions of the inequality $5x - 4y > 13$ are shaded below that dashed line. As shown in the graph on the left above, the pair $(-3, 2)$ is not a solution of the inequality $5x - 4y > 13$ and is not in the shaded region.

The pair $(4, -3)$ is a solution of the inequality $5x - 4y > 13$ and is in the shaded region. See the graph on the right above.

We now consider how to graph inequalities.

EXAMPLE 3 Graph: $y < x$.

We first graph the line $y = x$. Every solution of $y = x$ is an ordered pair like $(3, 3)$, where the first and second coordinates are the same. The graph of $y = x$ is shown in Figure 1. We draw it dashed because these points are *not* solutions of $y < x$.

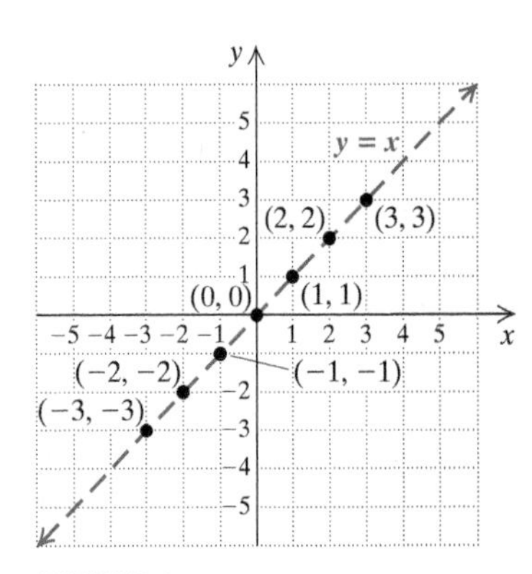

FIGURE 1

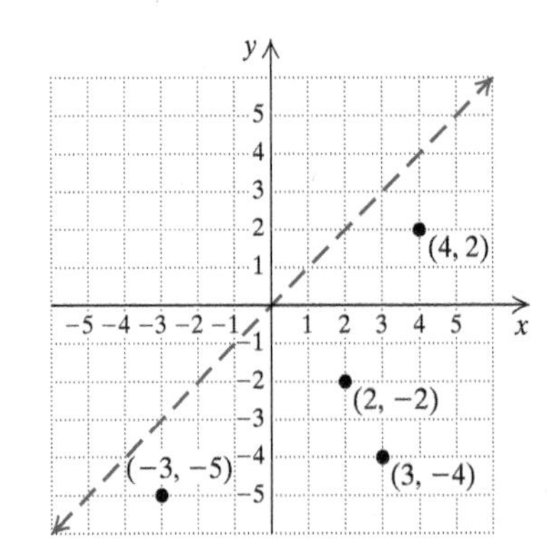

FIGURE 2

Now look at the graph in Figure 2. Several ordered pairs are plotted on the half-plane below $y = x$. Each is a solution of $y < x$. We can check the pair $(4, 2)$ as follows:

$$\begin{array}{c|c} y & < x \\ \hline 2 & ?\ 4 \end{array} \quad \text{TRUE}$$

It turns out that any point on the same side of $y = x$ as $(4, 2)$ is also a solution. Thus, *if you know that one point in a half-plane is a solution of an inequality, then all points in that half-plane are solutions.* In this text, we will usually indicate this by color shading. We shade the half-plane below $y = x$, as shown in Figure 3.

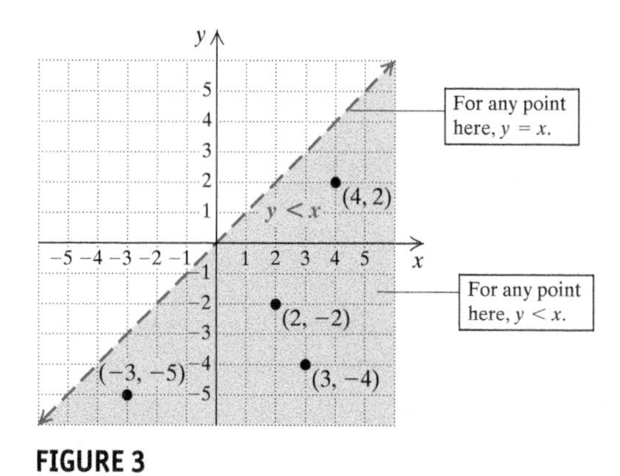

FIGURE 3

EXAMPLE 4 Graph: $8x + 3y \geq 24$.

First, we sketch the line $8x + 3y = 24$. Points on the line $8x + 3y = 24$ are also in the graph of $8x + 3y \geq 24$, so we draw the line solid. This indicates that all points on the line are solutions. The rest of the solutions are in the half-plane either to the left or to the right of the line. To determine which, we select a point that is not on the line and determine whether it is a solution of $8x + 3y \geq 24$. We try $(-3, 4)$ as a test point.

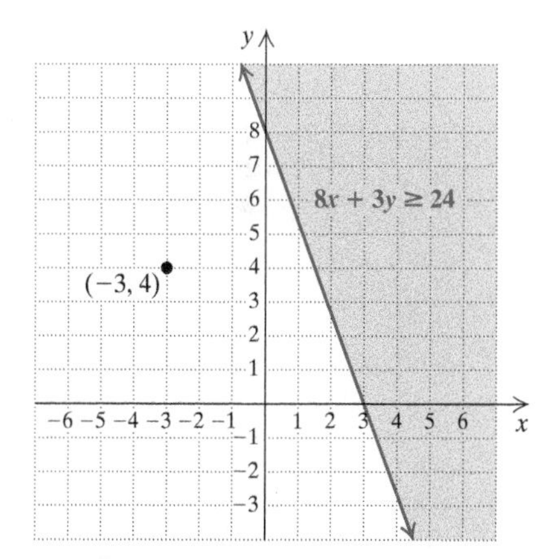

$$8x + 3y \geq 24$$

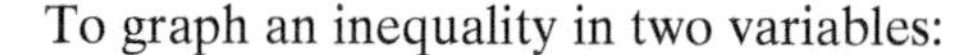

$8(-3) + 3(4)$	$?\ 24$	Using $(-3, 4)$ as a test point
$-24 + 12$		
-12		FALSE

We see that $-12 \geq 24$ is *false*. Since $(-3, 4)$ is not a solution, none of the points in the half-plane containing $(-3, 4)$ is a solution. Thus the points in the opposite half-plane are solutions. We shade that half-plane and obtain the graph shown at right.

To graph an inequality in two variables:

1. Replace the inequality symbol with an equals sign and graph this related equation. This separates points that represent solutions from those that do not.
2. If the inequality symbol is $<$ or $>$, draw the line dashed. If the inequality symbol is $\leq$ or $\geq$, draw the line solid.
3. The graph consists of a half-plane that is either above or below or to the left or to the right of the line and, if the line is solid, the line as well. To determine which half-plane to shade, choose a point not on the line as a test point. Substitute to determine whether that point is a solution. If so, shade the half-plane containing that point. If not, shade the opposite half-plane.

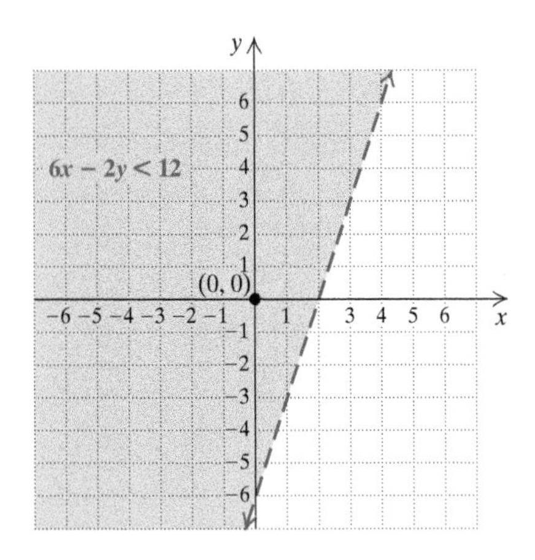

Graph.

3. $6x - 3y < 18$

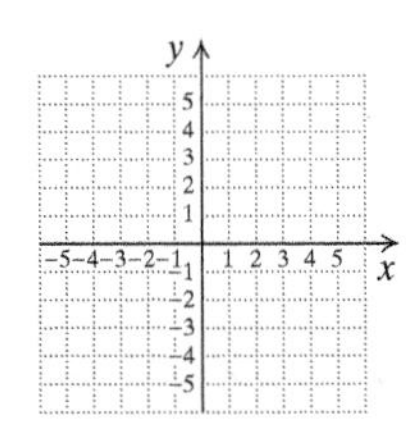

4. $4x + 3y \geq 12$

GS

1. Graph the related equation $4x + 3y$ ______ 12.
2. Since the inequality symbol is $\geq$, draw a ______ line. (dashed/solid)
3. Try the test point $(0, 0)$.

$$\begin{array}{c|c} 4x + 3y \geq 12 & \\ \hline 4(0) + 3(\quad) & ? \\ 0 & \end{array}$$

FALSE

$(0, 0)$ is not a solution, so we shade the opposite half-plane.

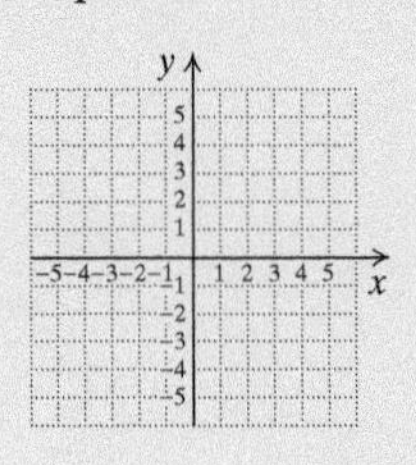

EXAMPLE 5 Graph: $6x - 2y < 12$.

1. We first graph the related equation $6x - 2y = 12$.
2. Since the inequality uses the symbol $<$, points on the line are not solutions of the inequality, so we draw a dashed line.
3. To determine which half-plane to shade, we consider a test point *not* on the line. We try $(0, 0)$ and substitute:

$$\begin{array}{c|c} 6x - 2y < 12 & \\ \hline 6(0) - 2(0) & ?\ 12 \\ 0 - 0 & \\ 0 & \end{array}$$

TRUE

Since the inequality $0 < 12$ is *true*, the point $(0, 0)$ is a solution; each point in the half-plane containing $(0, 0)$ is a solution. Thus each point in the opposite half-plane is *not* a solution. The graph is shown at left.

Do Exercises 3 and 4.

EXAMPLE 6 Graph $x > -3$ on a plane.

There is only one variable in this inequality. If we graph the inequality on the number line, its graph is as follows:

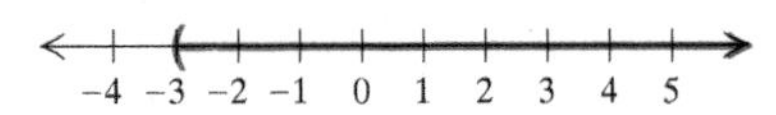

However, we can also write this inequality as $x + 0y > -3$ and consider graphing it in the plane. We first graph the related equation $x = -3$ in the plane. We draw the boundary with a dashed line. The rest of the graph is a half-plane to the right or to the left of the line $x = -3$. To determine which, we consider a test point, $(2, 5)$.

$$\begin{array}{c|c} x + 0y > -3 & \\ \hline 2 + 0(5) & ?\ -3 \\ 2 & \end{array}$$

TRUE

Since $(2, 5)$ is a solution, all the points in the half-plane containing $(2, 5)$ are solutions. We shade that half-plane.

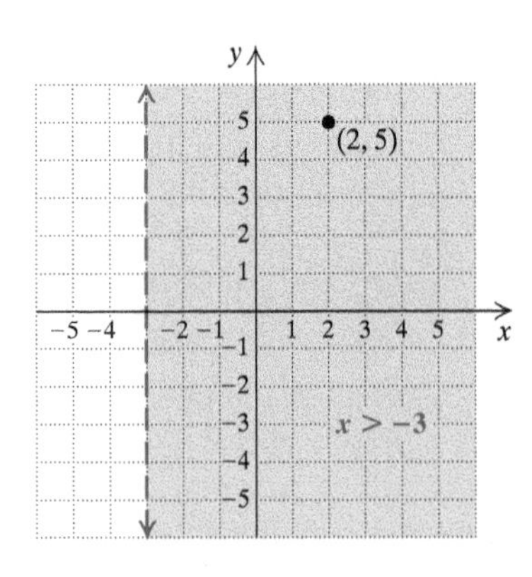

Answers

3. 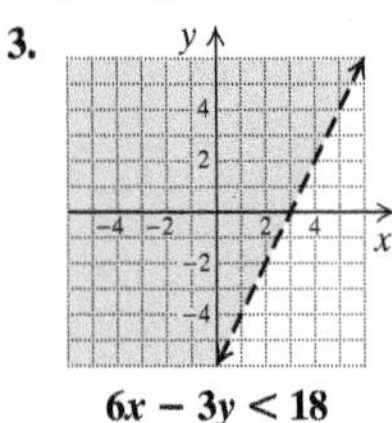

$6x - 3y < 18$

4. 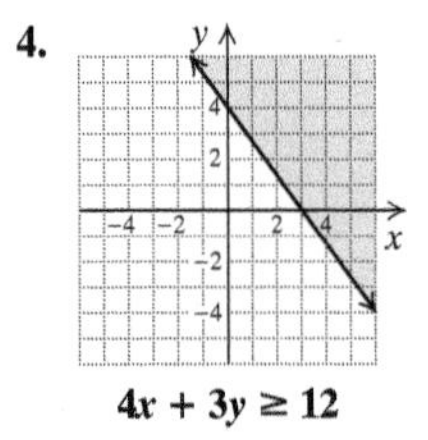

$4x + 3y \geq 12$

Guided Solution:
4. $=$, solid, 0

EXAMPLE 7 Graph $y \leq 4$ on a plane.

We first graph $y = 4$ using a solid line. We then use $(-2, 5)$ as a test point and substitute in $0x + y \leq 4$.

$$\begin{array}{c|l} 0x + y \leq 4 & \\ \hline 0(-2) + 5 \;?\; 4 & \\ 0 + 5 & \\ 5 & \text{FALSE} \end{array}$$

We see that $(-2, 5)$ is *not* a solution, so all the points in the half-plane containing $(-2, 5)$ are not solutions. Thus each point in the opposite half-plane is a solution. The graph of the inequality is shown at right.

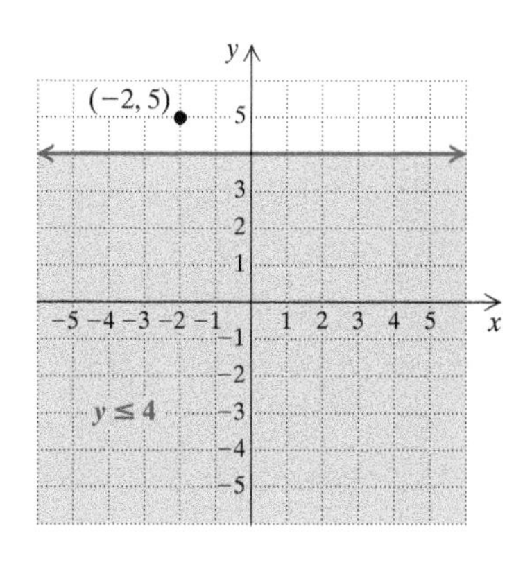

Do Exercises 5 and 6.

Graph on a plane.

5. $x < 3$

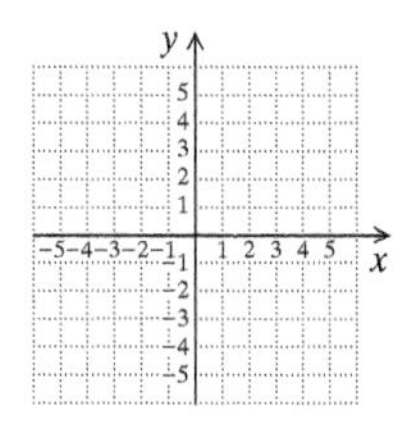

6. $y \geq -4$

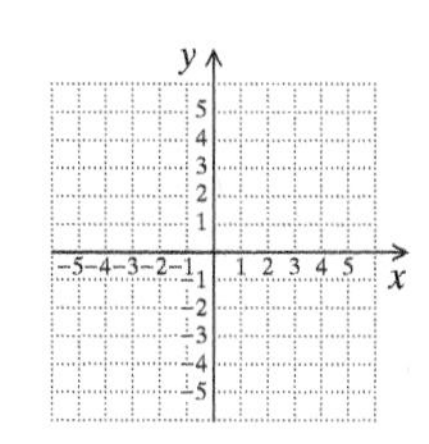

C SYSTEMS OF LINEAR INEQUALITIES

The following is an example of a system of two linear inequalities in two variables:

$$x + y \leq 4,$$
$$x - y < 4.$$

A **solution** of a system of linear inequalities is an ordered pair that is a solution of *both* inequalities. To graph solutions of systems of linear inequalities, we graph each inequality and determine where the graphs overlap, or intersect. That will be a region in which the ordered pairs are solutions of both inequalities.

EXAMPLE 8 Graph the solutions of the system

$$x + y \leq 4,$$
$$x - y < 4.$$

We graph $x + y \leq 4$ by first graphing the equation $x + y = 4$ using a solid red line. We consider $(0, 0)$ as a test point and find that it is a solution, so we shade all points on that side of the line using red shading. (See the graph on the left below.) The arrows near the ends of the line also indicate the half-plane, or region, that contains the solutions.

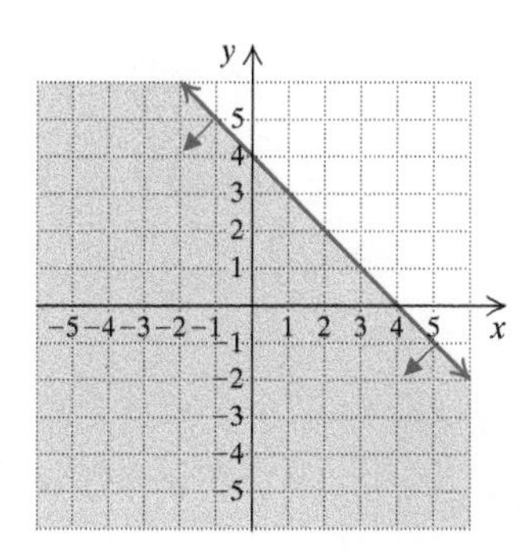

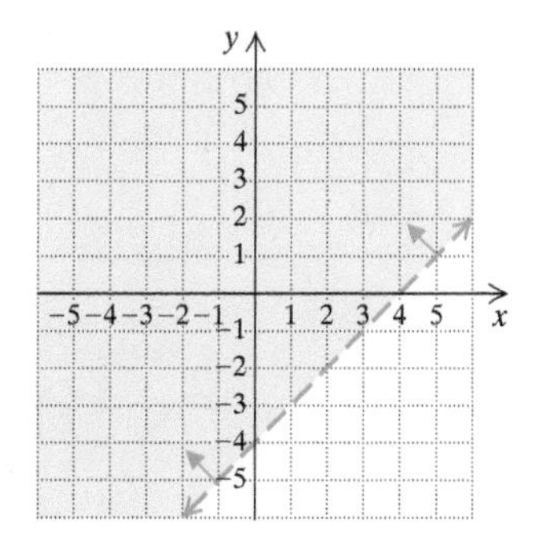

Answers

5.

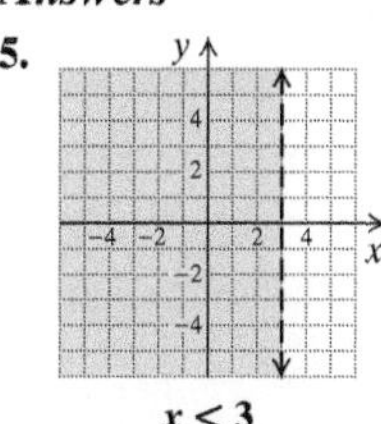

6.

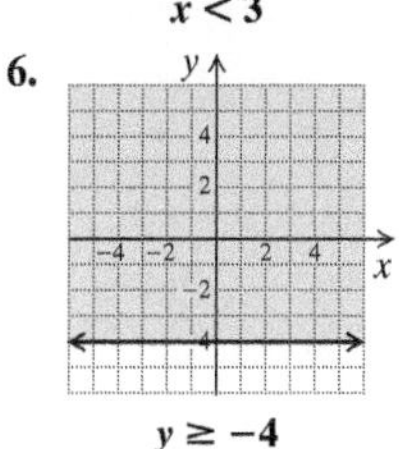

Next, we graph $x - y < 4$. We begin by graphing the equation $x - y = 4$ using a dashed blue line and consider $(0, 0)$ as a test point. Again, $(0, 0)$ is a solution so we shade that side of the line using blue shading. (See the graph on the right at the bottom of the preceding page.) The solution set of the system is the region that is shaded both red and blue and part of the line $x + y = 4$.

7. Graph:

$$x + y \geq 1,$$
$$y - x \geq 2.$$

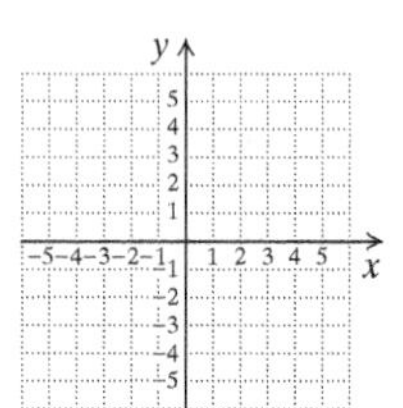

◀ **Do Exercise 7.**

EXAMPLE 9 Graph: $-2 < x \leq 5$.

This is actually a system of inequalities:

$$-2 < x,$$
$$x \leq 5.$$

We graph the equation $-2 = x$ and see that the graph of the first inequality is the half-plane to the right of the line $-2 = x$. (See the graph on the left below.)

Next, we graph the second inequality, starting with the line $x = 5$, and find that its graph is the line as well as the half-plane to the left of it. (See the graph on the right below.) Then we shade the intersection of these graphs.

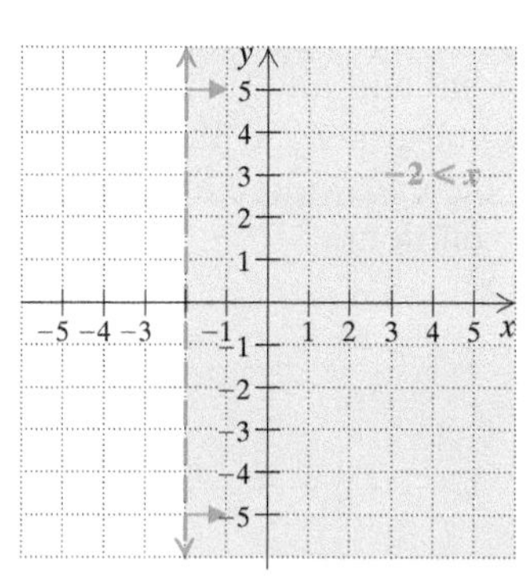

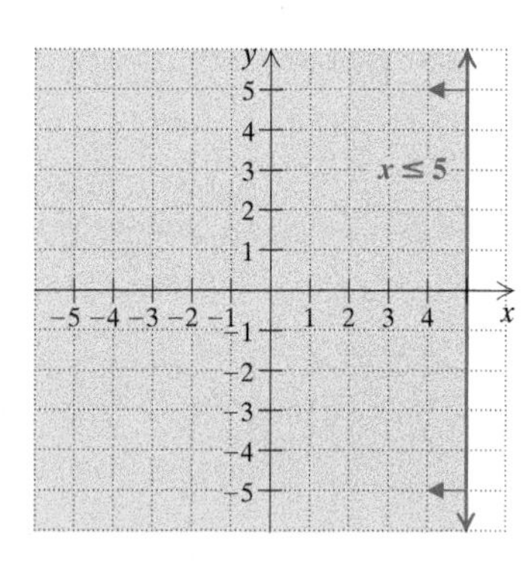

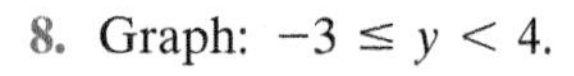
8. Graph: $-3 \leq y < 4$.

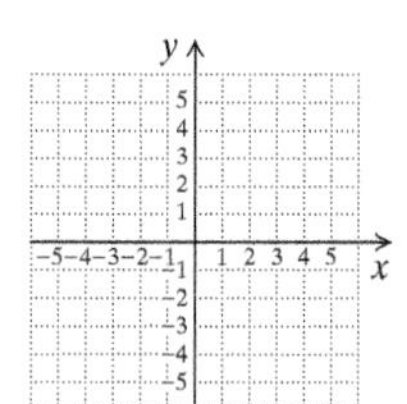

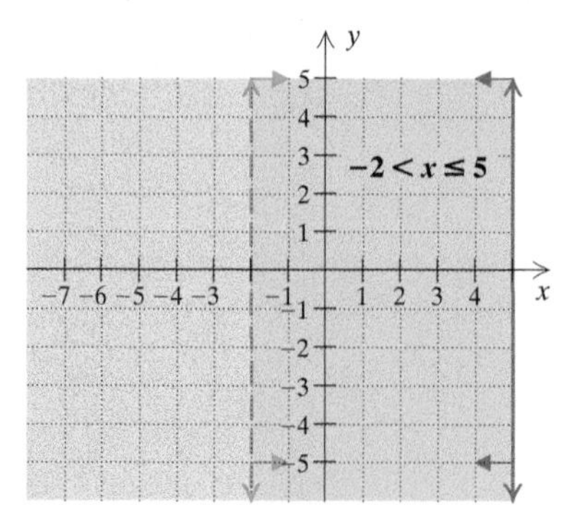

◀ **Do Exercise 8.**

Answers

7.

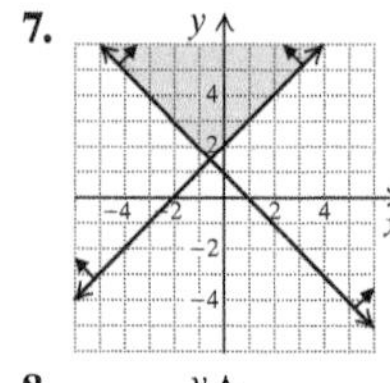

8.

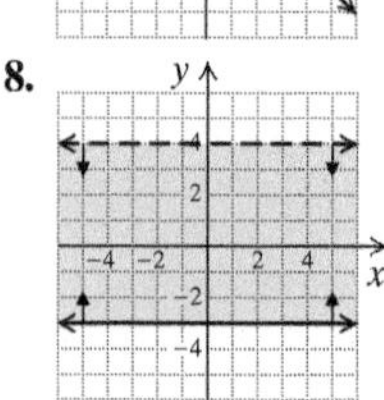

A system of inequalities may have a graph that consists of a polygon and its interior. In *linear programming*, which is a topic rich in application that you may study in a later course, it is important to be able to find the vertices of such a polygon.

EXAMPLE 10 Graph the following system of inequalities. Find the coordinates of any vertices formed.

$$6x - 2y \leq 12, \quad (1)$$
$$y - 3 \leq 0, \quad (2)$$
$$x + y \geq 0 \quad (3)$$

We graph the lines $6x - 2y = 12$, $y - 3 = 0$, and $x + y = 0$ using solid lines. The regions for each inequality are indicated by the arrows at the ends of the lines. We then note where the regions overlap and shade the region of solutions using one color.

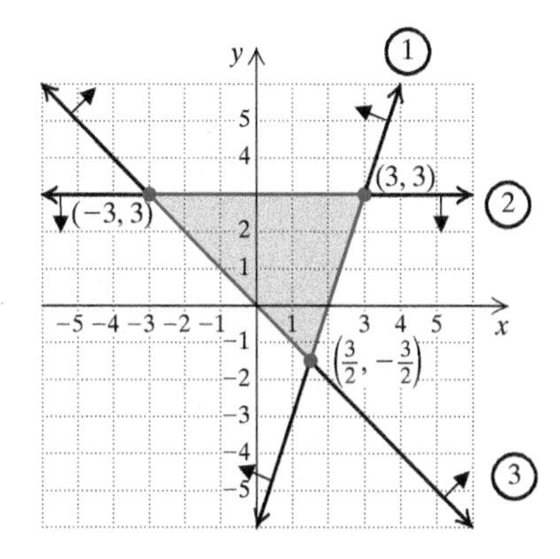

To find the vertices, we solve three different systems of equations. The system of equations from inequalities (1) and (2) is

$$6x - 2y = 12, \quad (1)$$
$$y - 3 = 0. \quad (2)$$

Solving, we obtain the vertex $(3, 3)$.

The system of equations from inequalities (1) and (3) is

$$6x - 2y = 12, \quad (1)$$
$$x + y = 0. \quad (3)$$

Solving, we obtain the vertex $\left(\frac{3}{2}, -\frac{3}{2}\right)$.

The system of equations from inequalities (2) and (3) is

$$y - 3 = 0, \quad (2)$$
$$x + y = 0. \quad (3)$$

Solving, we obtain the vertex $(-3, 3)$.

Do Exercise 9. ▶

9. Graph the system of inequalities. Find the coordinates of any vertices formed.

$$5x + 6y \leq 30,$$
$$0 \leq y \leq 3,$$
$$0 \leq x \leq 4$$

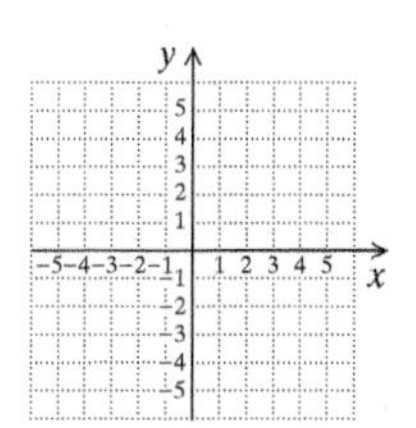

Answer

9.

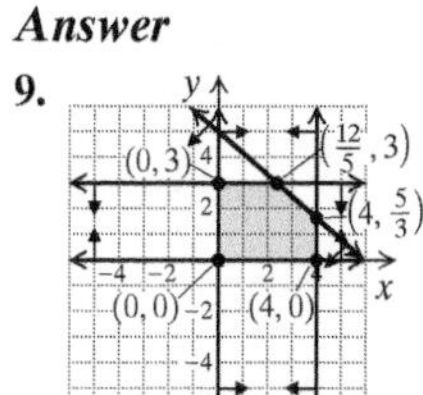

EXAMPLE 11 Graph the following system of inequalities. Find the coordinates of any vertices formed.

$$
\begin{aligned}
x + y &\le 16, && (1)\\
3x + 6y &\le 60, && (2)\\
x &\ge 0, && (3)\\
y &\ge 0 && (4)
\end{aligned}
$$

We graph each inequality using solid lines. The regions for each inequality are indicated by the arrows at the ends of the lines. We then note where the regions overlap and shade the region of solutions using one color.

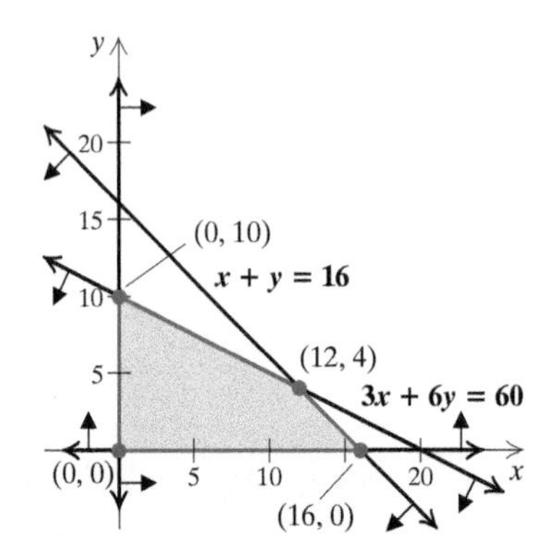

To find the vertices, we solve four different systems of equations. The system of equations from inequalities (1) and (2) is

$$
\begin{aligned}
x + y &= 16, && (1)\\
3x + 6y &= 60. && (2)
\end{aligned}
$$

Solving, we obtain the vertex $(12, 4)$.

The system of equations from inequalities (1) and (4) is

$$
\begin{aligned}
x + y &= 16, && (1)\\
y &= 0. && (4)
\end{aligned}
$$

Solving, we obtain the vertex $(16, 0)$.

The system of equations from inequalities (3) and (4) is

$$
\begin{aligned}
x &= 0, && (3)\\
y &= 0. && (4)
\end{aligned}
$$

The vertex is $(0, 0)$.

The system of equations from inequalities (2) and (3) is

$$
\begin{aligned}
3x + 6y &= 60, && (2)\\
x &= 0. && (3)
\end{aligned}
$$

Solving, we obtain the vertex $(0, 10)$.

Do Exercise 10.

10. Graph the system of inequalities. Find the coordinates of any vertices formed.

$$
\begin{aligned}
2x + 4y &\le 8,\\
x + y &\le 3,\\
x &\ge 0,\\
y &\ge 0
\end{aligned}
$$

Answer

10.

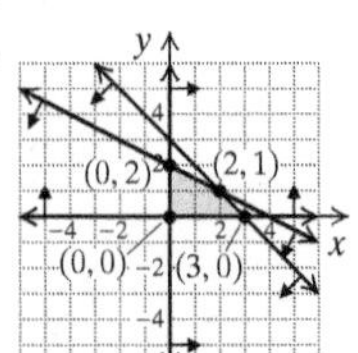

Visualizing for Success

Match the equation, inequality, system of equations, or system of inequalities with its graph.

1. $x + y = -4,$
   $2x + y = -8$

2. $2x + 5y \geq 10$

3. $2x - 2y = 5$

4. $2x - 5y = 10$

5. $-2y < 8$

6. $5x - 2y = 10$

7. $2x = 10$

8. $5x + 2y < 10,$
   $2x - 5y > 10$

9. $5x \geq -10$

10. $y - 2x < 8$

A

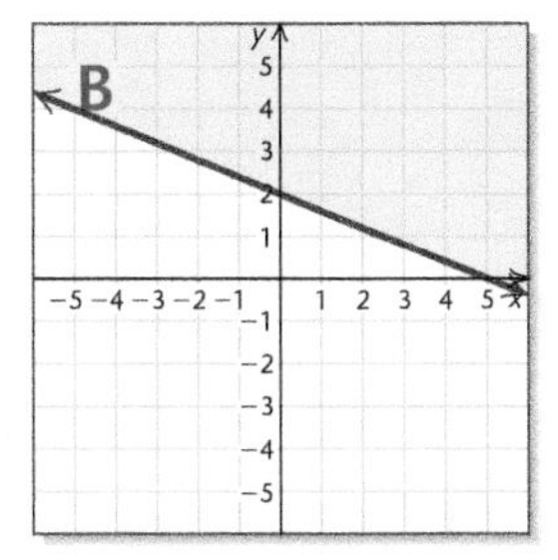

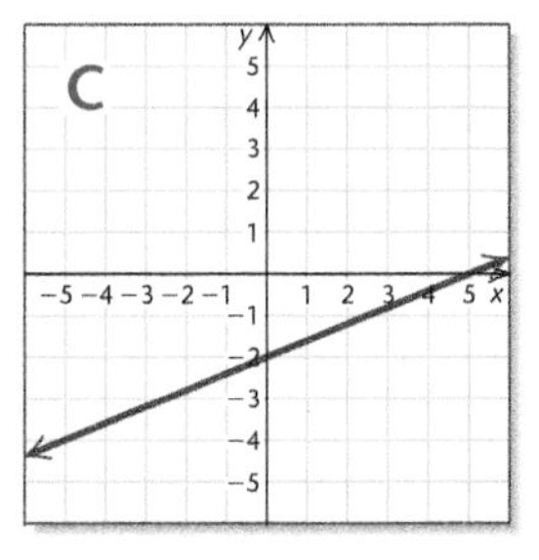

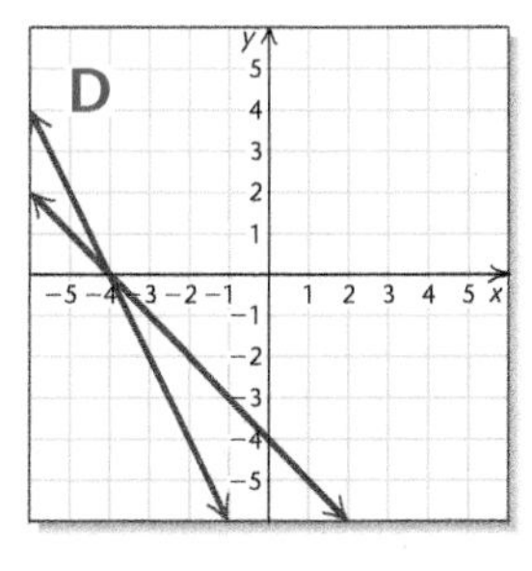

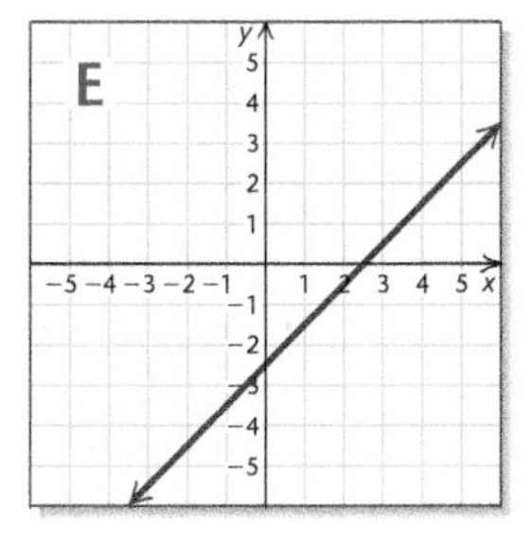

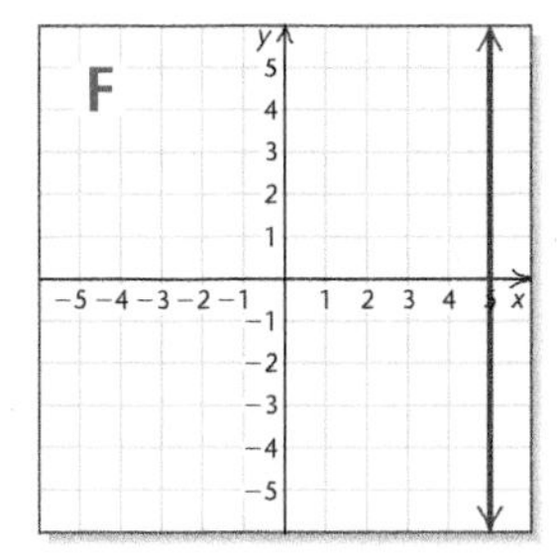

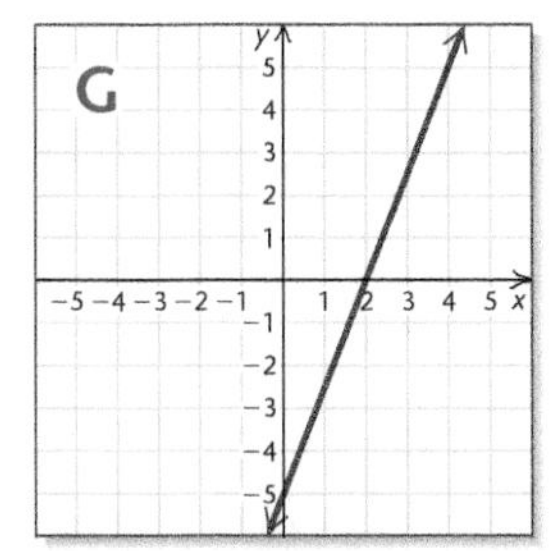

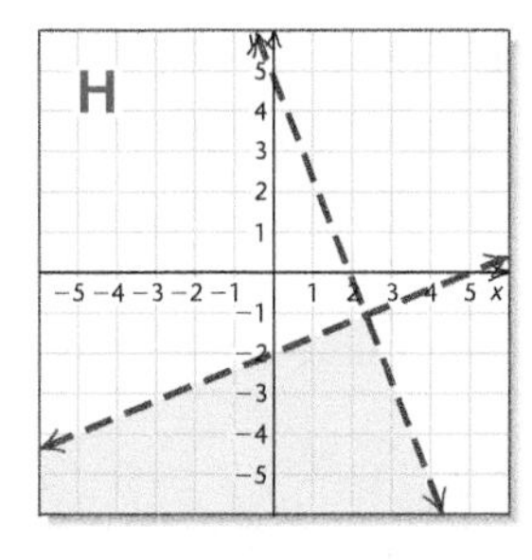

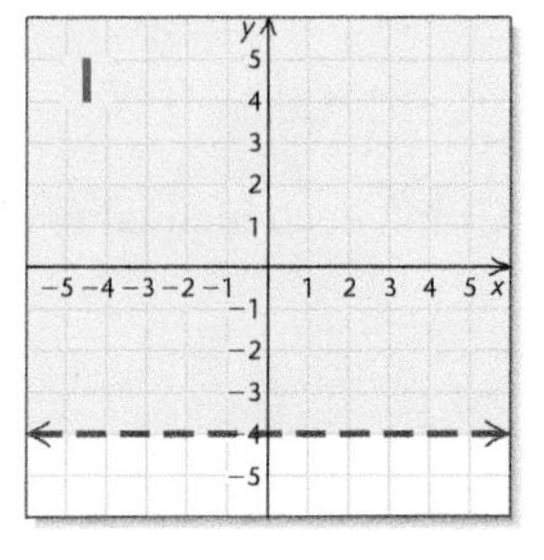

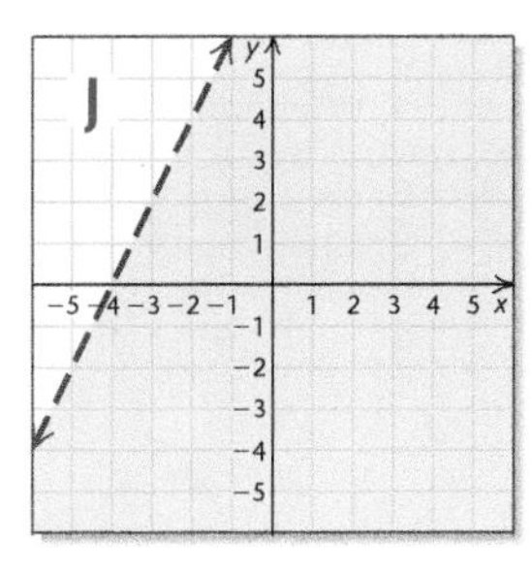

Answers on page A-11

3.7 Exercise Set

FOR EXTRA HELP MyLab Math

✓ Check Your Understanding

Reading Check Choose from the column on the right the word that best completes each statement.

RC1. A(n) __________ of an inequality is a drawing that represents its solutions.

RC2. The sentence $4x - y < 3$ is an example of a linear __________.

RC3. The graph of $4x - y < 3$ is a(n) __________.

RC4. The ordered pair $(1, 6)$ is a(n) __________ of $4x - y < 3$.

RC5. For $4x - y < 3$, the related __________ is $4x - y = 3$.

RC6. To determine which half-plane to shade when graphing an inequality, we can use a(n) __________ point.

equation
graph
half-plane
inequality
solution
test

Concept Check Determine whether the graph of the related equation should be drawn dashed or solid in order to graph the inequality.

CC1. $2x - y < 5$ **CC2.** $x \geq -3$ **CC3.** $4y \leq 2 - x$

Determine whether the point $(0, 0)$ is a solution of each inequality.

CC4. $2x + y < 5$ **CC5.** $y < x - 3$ **CC6.** $x \leq 0.5$

a Determine whether the given ordered pair is a solution of the given inequality.

1. $(-3, 3)$; $3x + y < -5$

2. $(6, -8)$; $4x + 3y \geq 0$

3. $(5, 9)$; $2x - y > -1$

4. $(5, -2)$; $6y - x > 2$

b Graph each inequality on a plane.

5. $y > 2x$

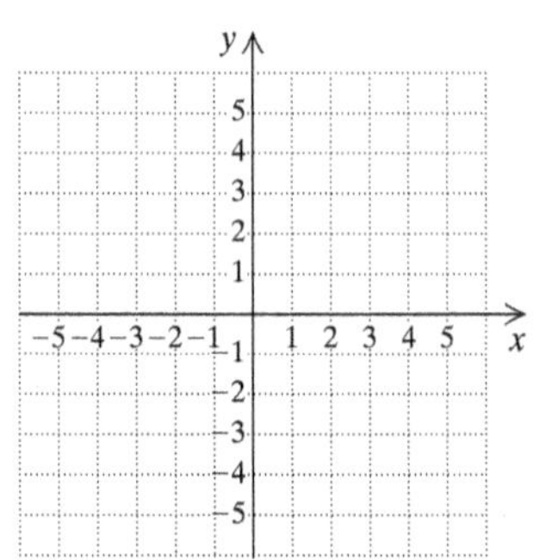

6. $y < 3x$

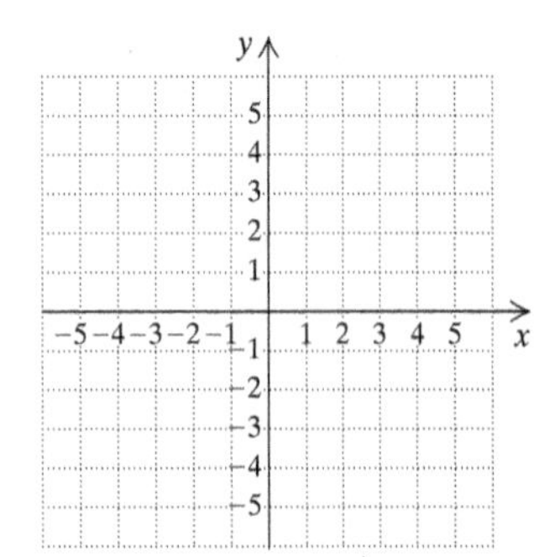

7. $y < x + 1$

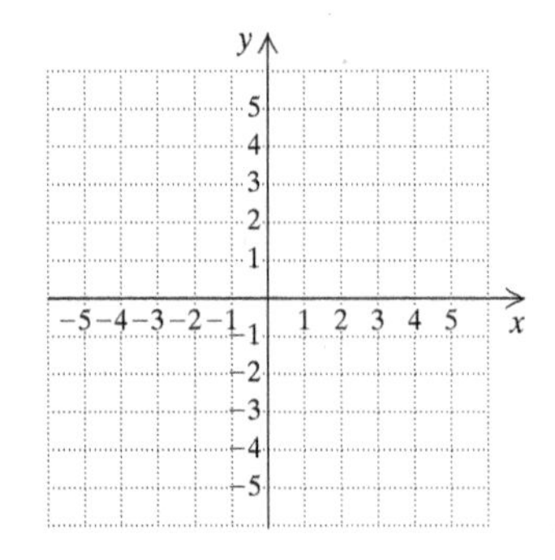

8. $y \leq x - 3$

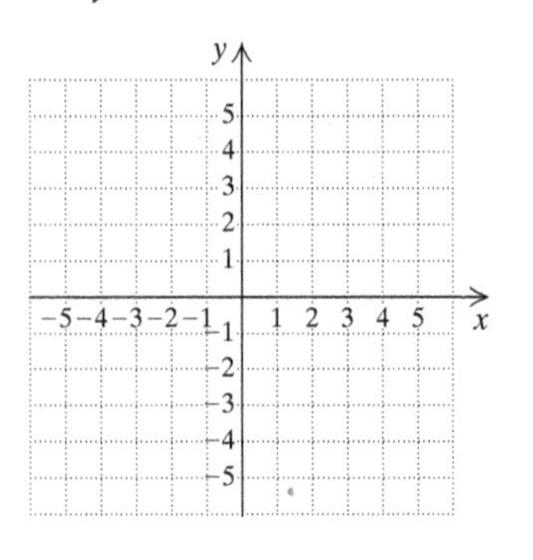

9. $y > x - 2$

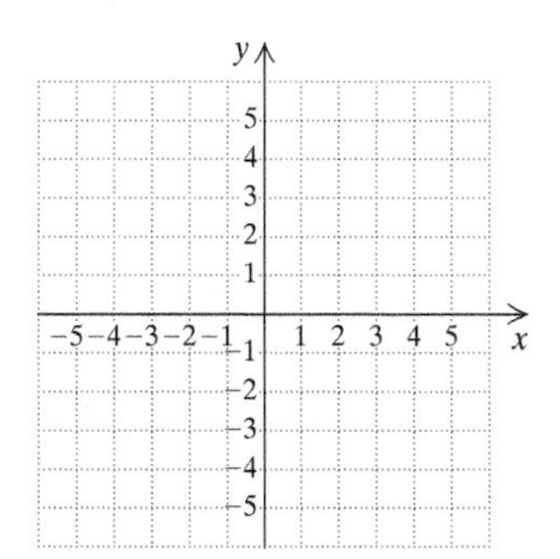

10. $y \geq x + 4$

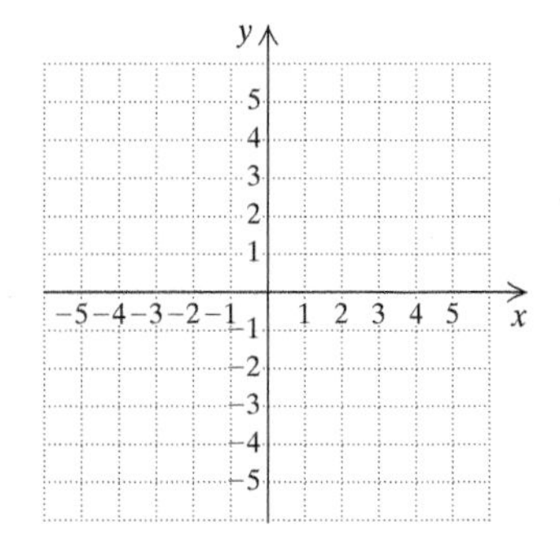

11. $x + y < 4$

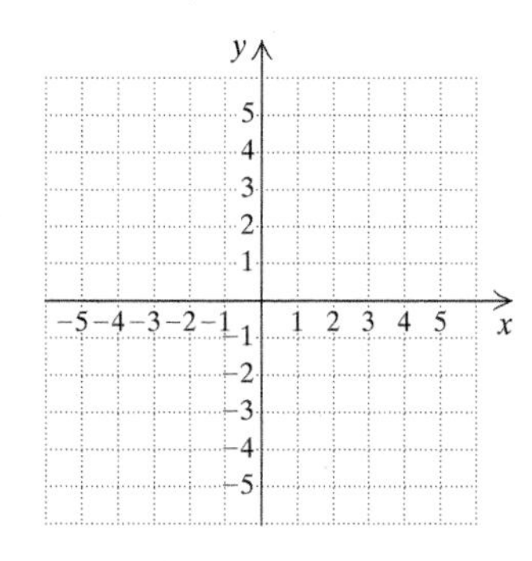

12. $x - y \geq 3$

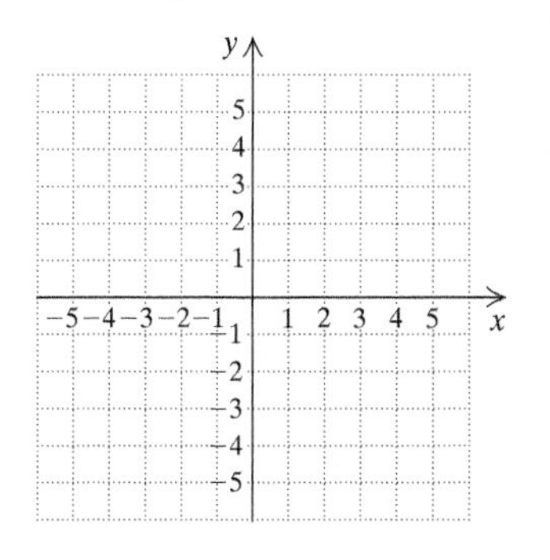

13. $3x + 4y \leq 12$

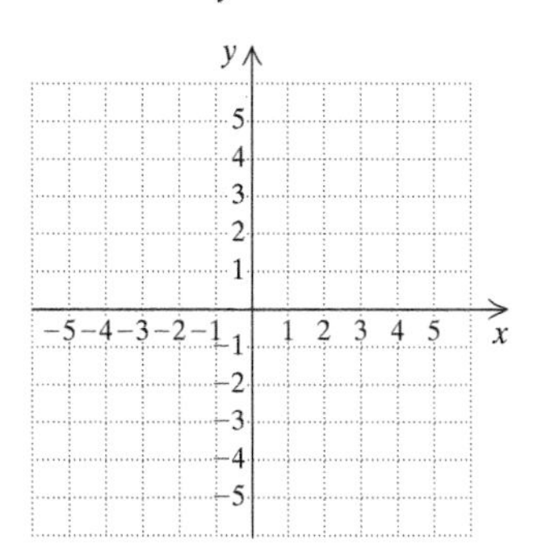

14. $2x + 3y < 6$

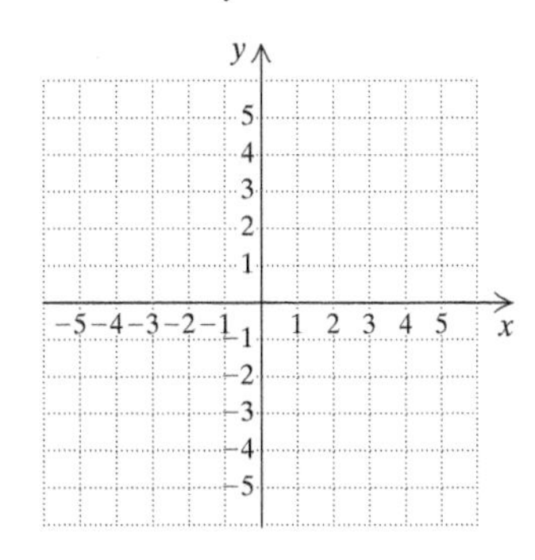

15. $2y - 3x > 6$

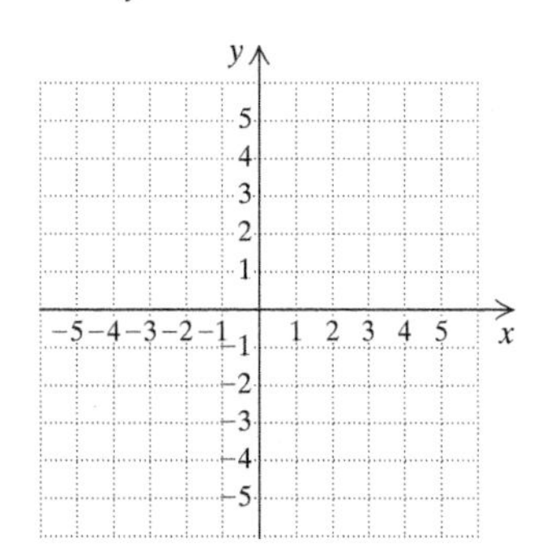

16. $2y - x \leq 4$

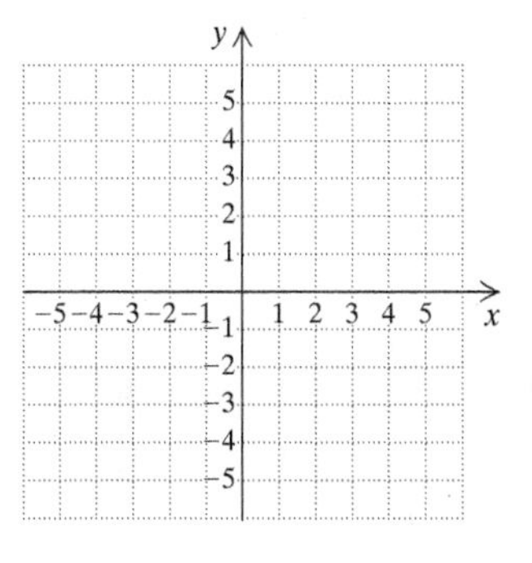

17. $3x - 2 \leq 5x + y$

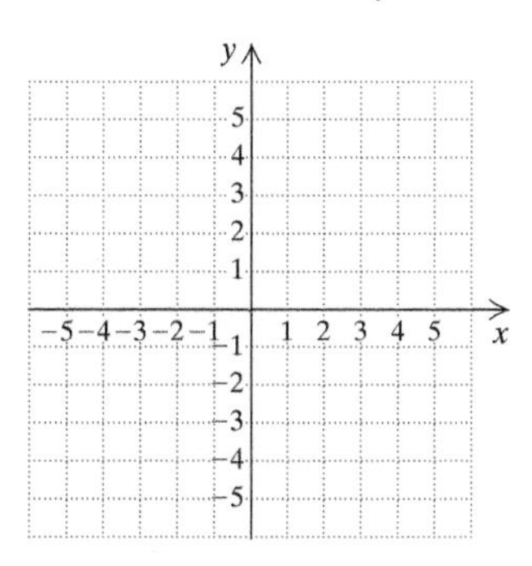

18. $2x - 2y \geq 8 + 2y$

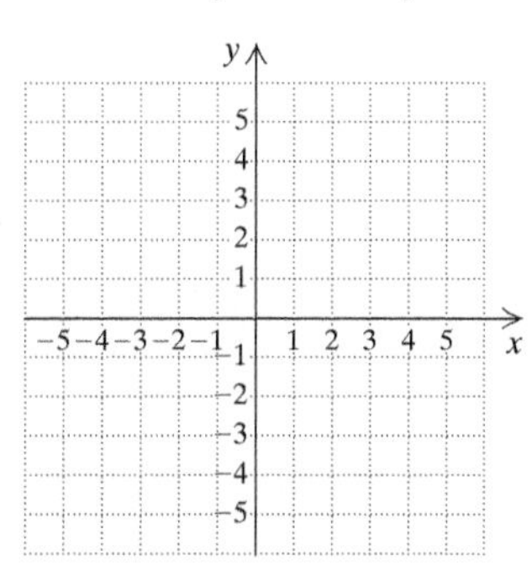

19. $x < 5$

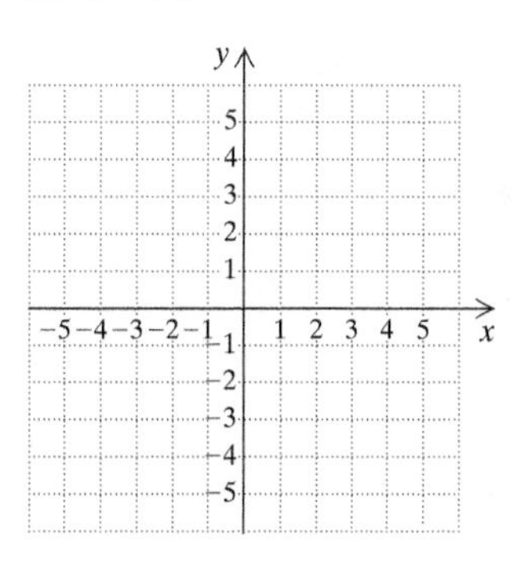

20. $y \geq -2$

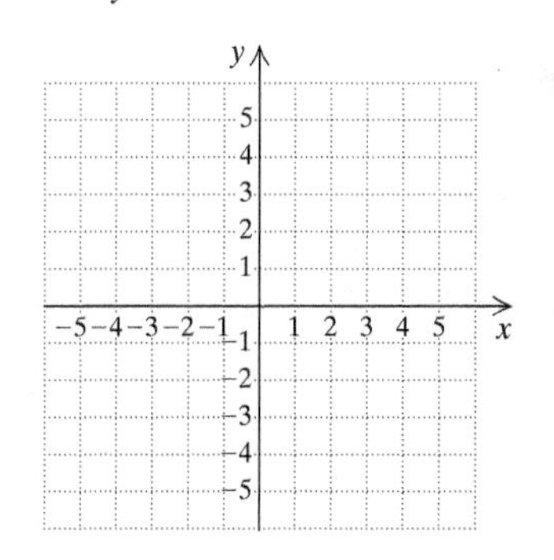

21. $y > 2$

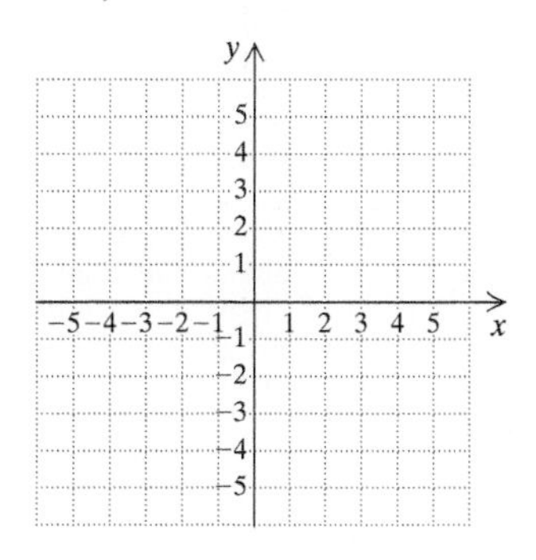

22. $x \leq -4$

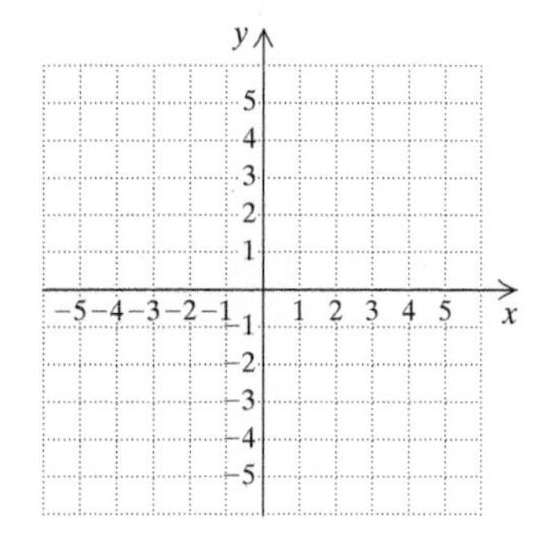

23. $2x + 3y \leq 6$

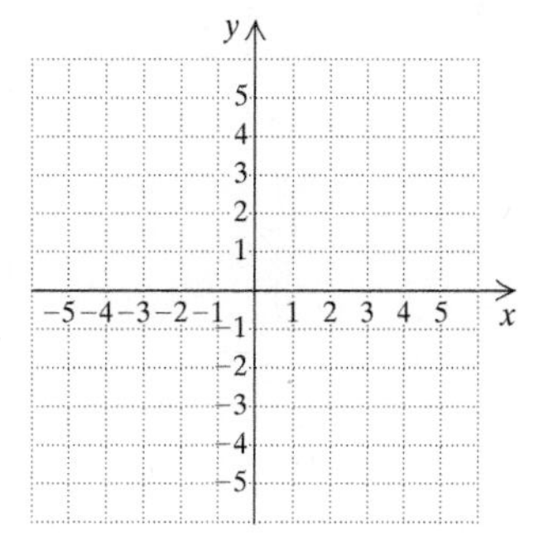

24. $7x + 2y \geq 21$

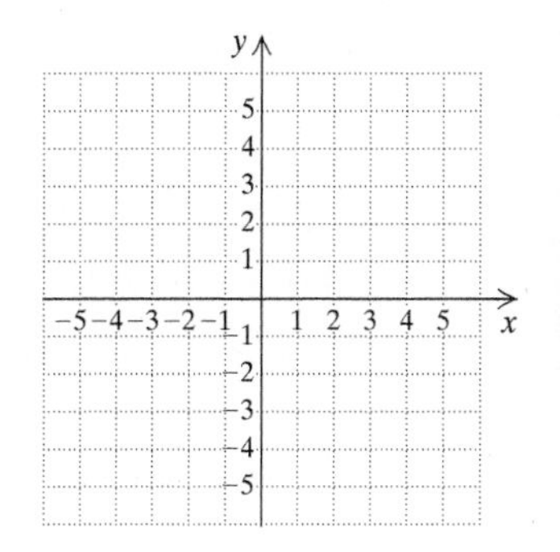

Matching. Each of Exercises 25–30 shows the graph of an inequality. Match the graph with one of the appropriate inequalities (A)–(F), which follow.

25.

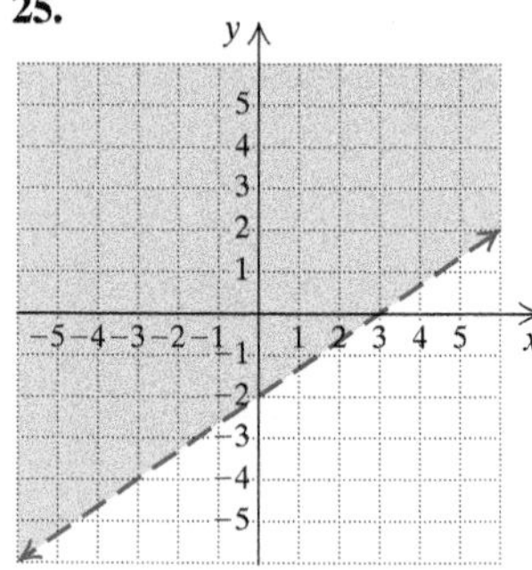

26.

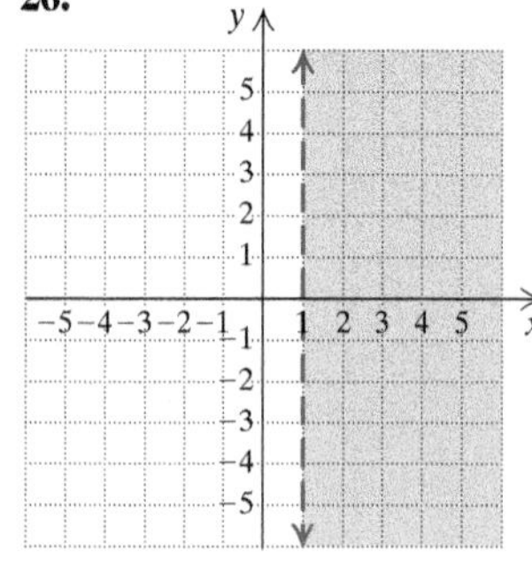

27.

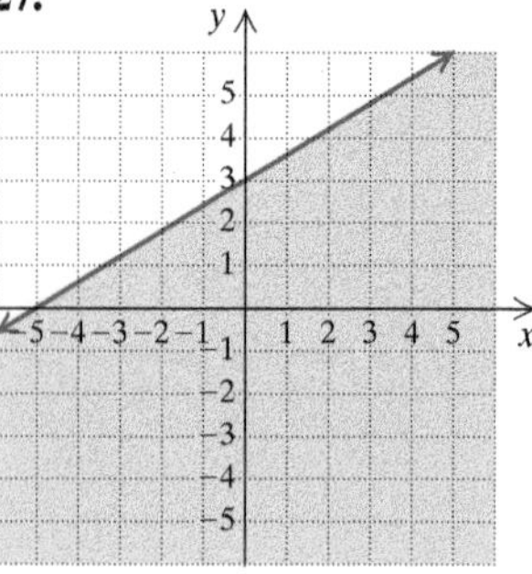

28.

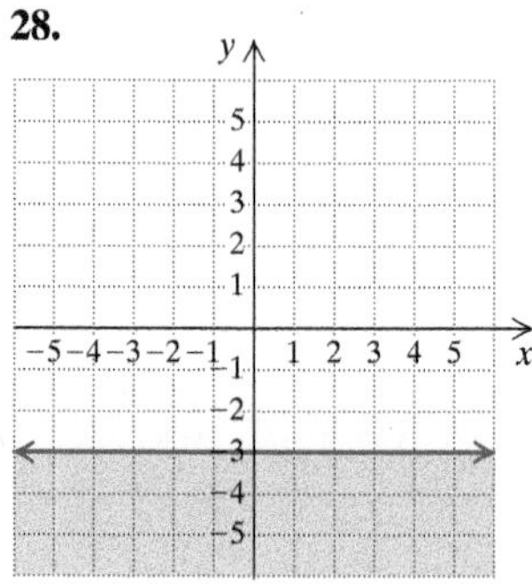

29.

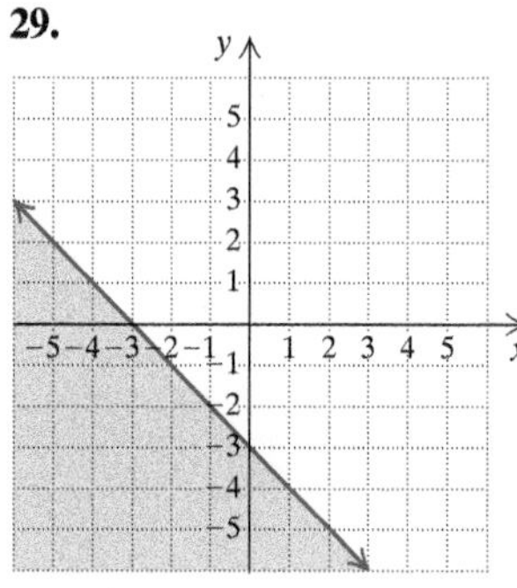

30.

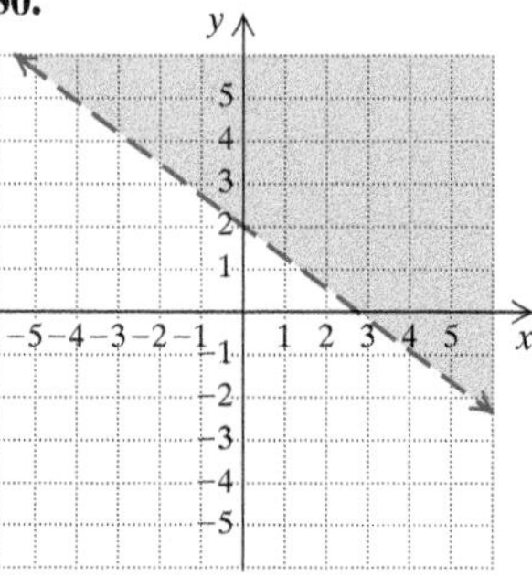

A. $4y > 8 - 3x$ **B.** $3x \geq 5y - 15$ **C.** $y + x \leq -3$ **D.** $x > 1$ **E.** $y \leq -3$ **F.** $2x - 3y < 6$

C Graph each system of inequalities. Find the coordinates of any vertices formed.

31. $y \geq x,$
$y \leq -x + 2$

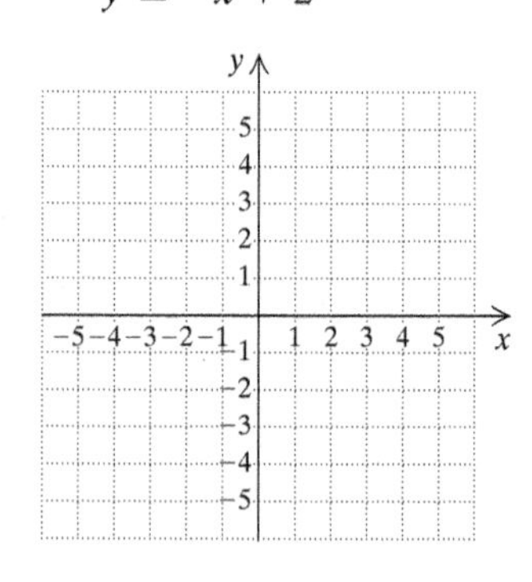

32. $y \geq x,$
$y \leq -x + 4$

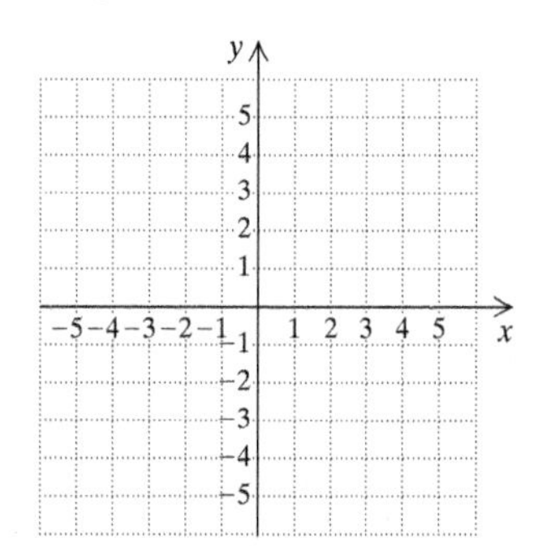

33. $y > x,$
$y < -x + 1$

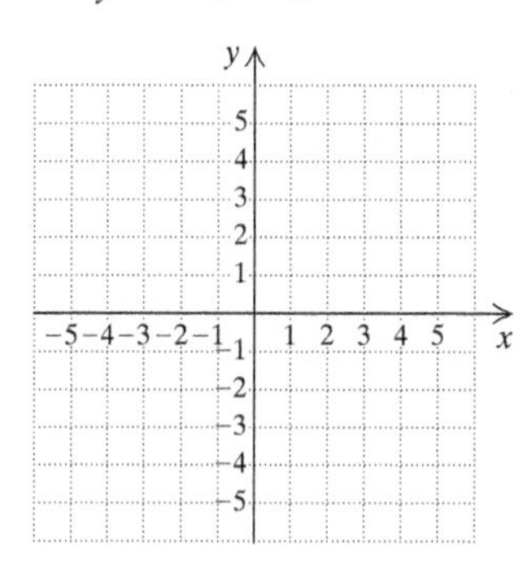

34. $y < x,$
$y > -x + 3$

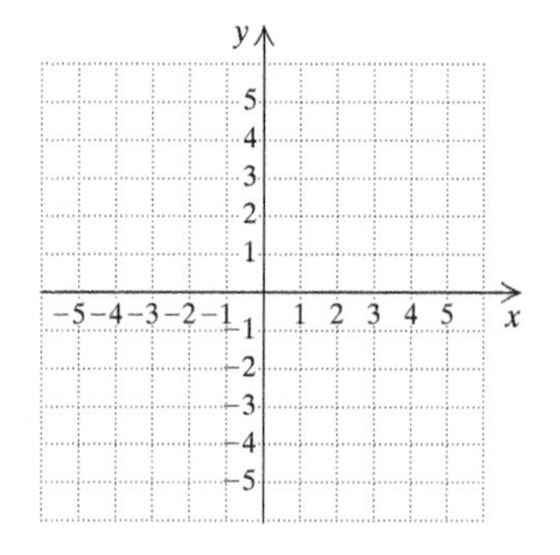

35. $x \leq 3,$
$y \geq -3x + 2$

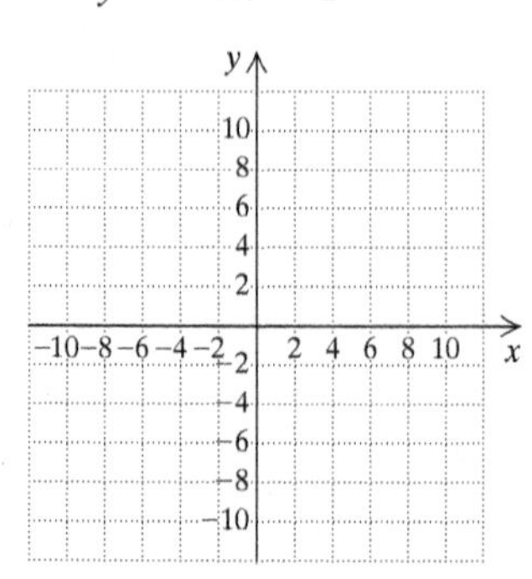

36. $x \geq -2,$
$y \leq -2x + 3$

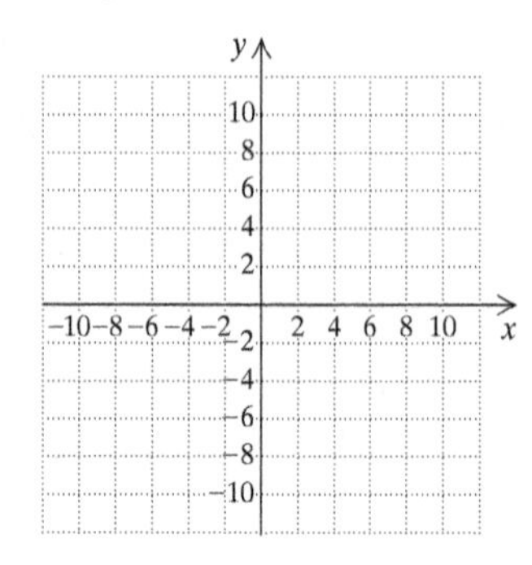

37. $x + y \leq 1,$
$x - y \leq 2$

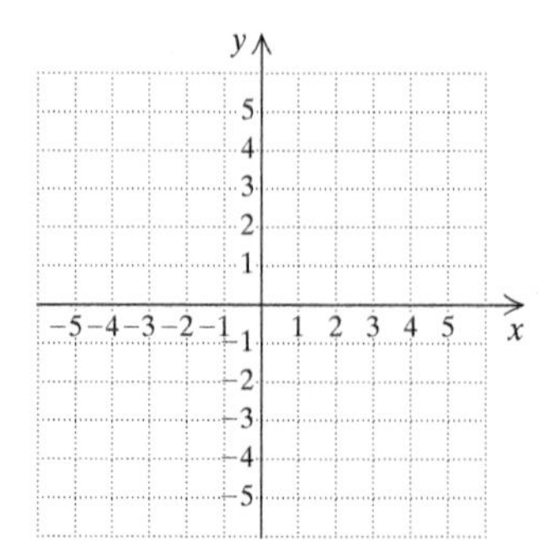

38. $x + y \leq 3,$
$x - y \leq 4$

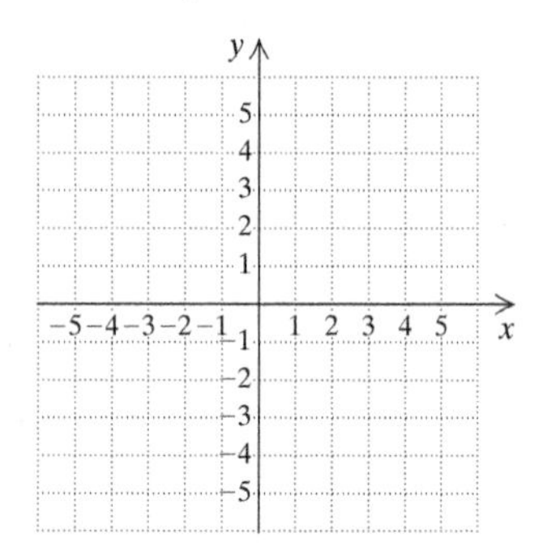

39. $y \leq 2x + 1,$
$y \geq -2x + 1,$
$x \leq 2$

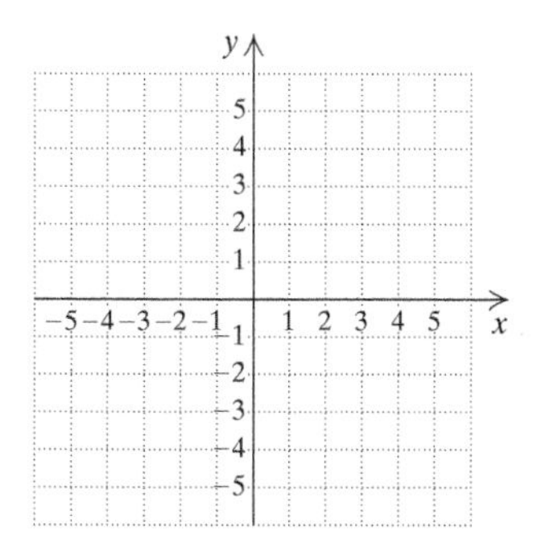

40. $x - y \leq 2,$
$x + 2y \geq 8,$
$y \leq 4$

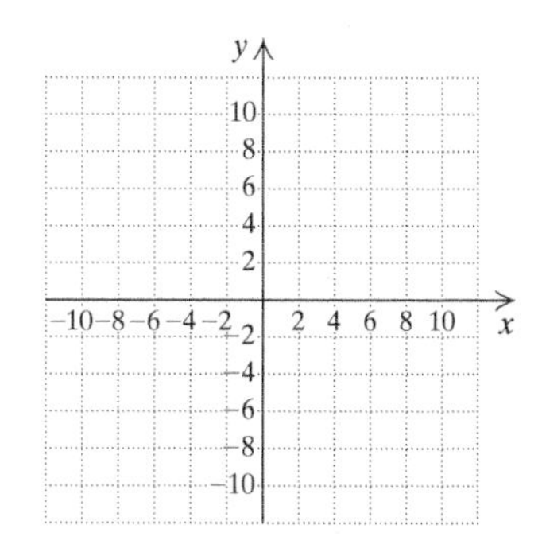

41. $x + 2y \leq 12,$
$2x + y \leq 12,$
$x \geq 0,$
$y \geq 0$

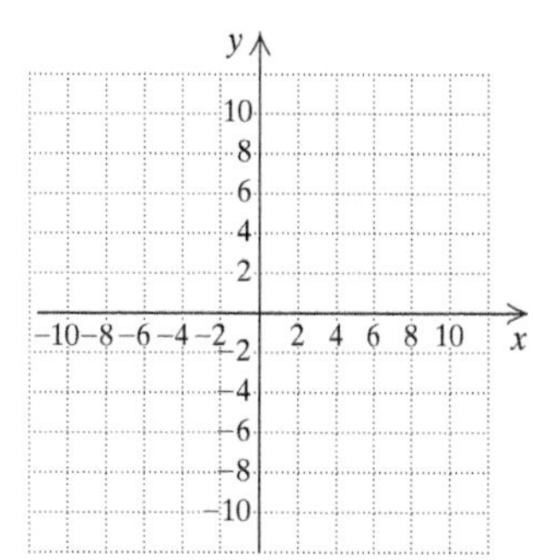

42. $y - x \geq 1,$
$y - x \leq 3,$
$2 \leq x \leq 5$

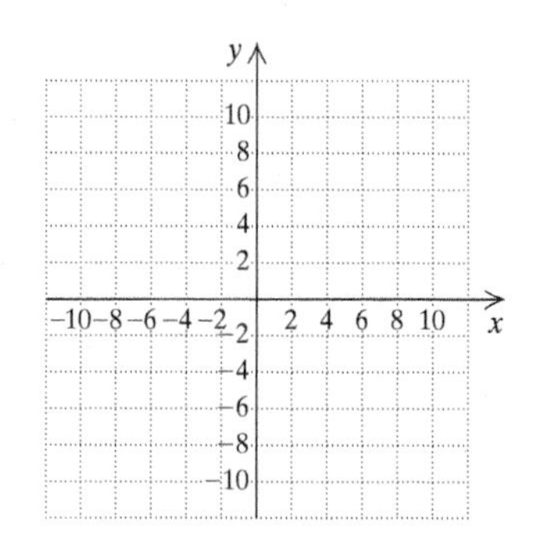

Skill Maintenance

Solve. [1.1d]

43. $5(3x - 4) = -2(x + 5)$

44. $4(3x + 4) = 2 - x$

45. $2(x - 1) + 3(x - 2) - 4(x - 5) = 10$

46. $10x - 8(3x - 7) = 2(4x - 1)$

47. $5x + 7x = -144$

48. $0.5x - 2.34 + 2.4x = 7.8x - 9$

Given the function $f(x) = |2 - x|$, find each of the following function values. [2.2b]

49. $f(0)$

50. $f(-1)$

51. $f(10)$

52. $f(2a)$

Synthesis

53. *Waterfalls.* In order for a waterfall to be classified as a classical waterfall, its height must be less than twice its crest width, and its crest width cannot exceed one-and-a-half times its height. The tallest waterfall in the world is about 3200 ft high. Let h represent a waterfall's height, in feet, and w the crest width, in feet. Write and graph a system of inequalities that represents all possible combinations of heights and crest widths of classical waterfalls.

54. *Exercise Danger Zone.* It is dangerous to exercise when the weather is hot and humid. The solutions of the following system of inequalities give a "danger zone" for which it is dangerous to exercise intensely:

$$4H - 3F < 70,$$
$$F + H > 160,$$
$$2F + 3H > 390,$$

where F is the temperature, in degrees Fahrenheit, and H is the humidity.

a) Draw the danger zone by graphing the system of inequalities.

b) Is it dangerous to exercise when $F = 80°$ and $H = 80\%$?

CHAPTER

3 Summary and Review

Vocabulary Reinforcement

Complete each statement with the correct term from the column on the right. Some of the choices may be used more than once and some may not be used at all.

1. A solution of a system of two equations in two variables is an ordered ______________ that makes both equations true. [3.1a]
2. A(n) ______________ system of equations has at least one solution. [3.1a]
3. A solution of a system of three equations in three variables is an ordered ______________ that makes all three equations true. [3.5a]
4. If, for a system of two equations in two variables, the graphs of the equations are different lines, then the equations are ______________. [3.1a]
5. The graph of an inequality like $x > 2y$ is a(n) ______________. [3.7b]

line
half-plane
independent
dependent
consistent
inconsistent
pair
triple

Concept Reinforcement

Determine whether each statement is true or false.

______________ 1. A system of equations with infinitely many solutions is inconsistent. [3.1a]

______________ 2. It is not possible for the equations in an inconsistent system of two equations to be dependent. [3.1a]

______________ 3. When $(0, b)$ is a solution of each equation in a system of two equations, the graphs of the two equations have the same y-intercept. [3.1a]

______________ 4. The system of equations $x = 4$ and $y = -4$ is inconsistent. [3.1a]

Study Guide

Objective 3.1a Solve a system of two linear equations or two functions by graphing and determine whether a system is consistent or inconsistent and whether the equations in a system are dependent or independent.

Example Solve this system of equations graphically. Then classify the system as consistent or inconsistent and the equations as dependent or independent.

$$x - y = 3,$$
$$y = 2x - 4$$

We graph the equations. The point of intersection appears to be $(1, -2)$. This checks in both equations, so it is the solution. The system has one solution, so it is consistent and the equations are independent.

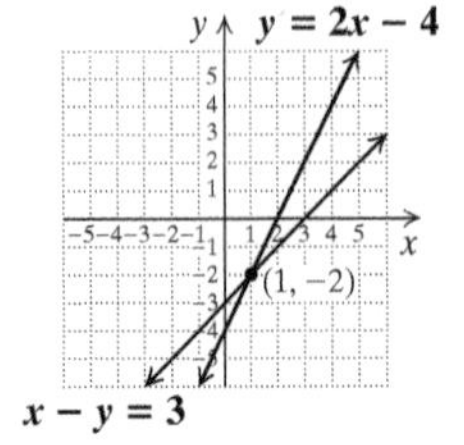

Practice Exercise

1. Solve this system of equations graphically. Then classify the system as consistent or inconsistent and the equations as dependent or independent.

$$x + 3y = 1,$$
$$x + y = 3$$

Objective 3.2a Solve systems of equations in two variables by the substitution method.

Example Solve the system

$$x - 2y = 1, \quad \textbf{(1)}$$
$$2x - 3y = 3. \quad \textbf{(2)}$$

We solve equation (1) for x, since the coefficient of x is 1 in that equation:

$$x - 2y = 1$$
$$x = 2y + 1. \quad \textbf{(3)}$$

Next, we substitute for x in equation (2) and solve for y:

$$2x - 3y = 3$$
$$2(2y + 1) - 3y = 3$$
$$4y + 2 - 3y = 3$$
$$y + 2 = 3$$
$$y = 1.$$

Then we substitute 1 for y in equation (1), (2), or (3) and find x. We choose equation (3) since it is already solved for x:

$$x = 2y + 1 = 2 \cdot 1 + 1 = 2 + 1 = 3$$

Check:

$x - 2y = 1$		$2x - 3y = 3$	
$3 - 2 \cdot 1 \; ? \; 1$		$2 \cdot 3 - 3 \cdot 1 \; ? \; 3$	
$3 - 2$		$6 - 3$	
1	TRUE	3	TRUE

The ordered pair (3, 1) checks in both equations, so it is the solution of the system of equations.

Practice Exercise

2. Solve the system

$$2x + y = 2,$$
$$3x + 2y = 5$$

using the substitution method.

Objective 3.3a Solve systems of equations in two variables by the elimination method.

Example Solve the system

$$2a + 3b = -1, \quad \textbf{(1)}$$
$$3a + 2b = 6. \quad \textbf{(2)}$$

We could eliminate either a or b. In this case, we decide to eliminate the a-terms. We multiply equation (1) by 3 and equation (2) by -2 and then add and solve for b:

$$6a + 9b = -3$$
$$\underline{-6a - 4b = -12}$$
$$5b = -15$$
$$b = -3.$$

Next, we substitute -3 for b in either of the original equations:

$$2a + 3b = -1 \quad \textbf{(1)}$$
$$2a + 3(-3) = -1$$
$$2a - 9 = -1$$
$$2a = 8$$
$$a = 4.$$

The ordered pair $(4, -3)$ checks in both equations, so it is a solution of the system of equations.

Practice Exercise

3. Solve the system

$$2x + 3y = 5,$$
$$3x + 4y = 6$$

using the elimination method.

Objective 3.4a Solve applied problems involving total value and mixture using systems of two equations.

Example To start a small business, Michael took two loans totaling \$18,000. One of the loans was at 7% interest and the other at 8%. After one year, Michael owed \$1365 in interest. What was the amount of each loan?

1. **Familiarize.** We let x and y represent the amounts of the two loans. Next, we organize the information in a table and use the simple interest formula, $I = Prt$.

	LOAN 1	LOAN 2	TOTAL
PRINCIPAL	x	y	\$18,000
RATE OF INTEREST	7%	8%	
TIME	1 year	1 year	
INTEREST	$7\%x$, or $0.07x$	$8\%y$, or $0.08y$	\$1365

2. **Translate.** The total amount of the loans is found in the first row of the table. This gives us one equation:

$$x + y = 18{,}000.$$

From the last row of the table, we see that the interest totals \$1365. This gives us a second equation:

$$0.07x + 0.08y = 1365, \text{ or } 7x + 8y = 136{,}500.$$

3. **Solve.** We solve the resulting system of equations:

$$x + y = 18{,}000, \quad (1)$$
$$7x + 8y = 136{,}500. \quad (2)$$

We multiply by -7 on both sides of equation (1) and add:

$$-7x - 7y = -126{,}000$$
$$7x + 8y = 136{,}500 \quad (2)$$
$$y = 10{,}500. \quad \text{Adding}$$

Then

$$x + 10{,}500 = 18{,}000 \quad \text{Substituting 10,500 for } y \text{ in equation (1)}$$
$$x = 7500. \quad \text{Solving for } x$$

We find that $x = 7500$ and $y = 10{,}500$.

4. **Check.** The sum is \$7500 + \$10,500, or \$18,000. The interest from \$7500 at 7% for one year is 7%(\$7500), or \$525. The interest from \$10,500 at 8% for one year is 8%(\$10,500), or \$840. The total amount of interest is \$525 + \$840, or \$1365. The numbers check in the problem.

5. **State.** Michael took loans of \$7500 at 7% interest and \$10,500 at 8% interest.

Practice Exercise

4. Jaretta made two investments totaling \$23,000. In one year, these investments yielded \$1237 in simple interest. Part of the money was invested at 6% and the rest at 5%. How much was invested at each rate?

Objective 3.5a Solve systems of three equations in three variables.

Example Solve:

$$\begin{aligned} x - y - z &= -2, & \textbf{(1)} \\ 2x + 3y + z &= 3, & \textbf{(2)} \\ 5x - 2y - 2z &= -1. & \textbf{(3)} \end{aligned}$$

The equations are in standard form and do not contain decimals or fractions. We choose to eliminate z since the z-terms in equations (1) and (2) are opposites.

First, we add these two equations:

$$\begin{aligned} x - y - z &= -2 \\ 2x + 3y + z &= 3 \\ \hline 3x + 2y \quad &= 1. \quad \textbf{(4)} \end{aligned}$$

Next, we multiply equation (2) by 2 and add it to equation (3) to eliminate z from another pair of equations:

$$\begin{aligned} 4x + 6y + 2z &= 6 \\ 5x - 2y - 2z &= -1 \\ \hline 9x + 4y \quad &= 5. \quad \textbf{(5)} \end{aligned}$$

Now we solve the system consisting of equations (4) and (5). We multiply equation (4) by -2 and add:

$$\begin{aligned} -6x - 4y &= -2 \\ 9x + 4y &= 5 \\ \hline 3x \quad &= 3 \\ x &= 1. \end{aligned}$$

Then we use either equation (4) or (5) to find y:

$$\begin{aligned} 3x + 2y &= 1 \quad \textbf{(4)} \\ 3 \cdot 1 + 2y &= 1 \\ 3 + 2y &= 1 \\ 2y &= -2 \\ y &= -1. \end{aligned}$$

Finally, we use one of the original equations to find z:

$$\begin{aligned} 2x + 3y + z &= 3 \quad \textbf{(2)} \\ 2 \cdot 1 + 3(-1) + z &= 3 \\ -1 + z &= 3 \\ z &= 4. \end{aligned}$$

Check:

$x - y - z = -2$	
$1 - (-1) - 4$	? -2
$1 + 1 - 4$	
-2	TRUE

$2x + 3y + z = 3$	
$2 \cdot 1 + 3(-1) + 4$	? 3
$2 - 3 + 4$	
3	TRUE

$5x - 2y - 2z = -1$	
$5 \cdot 1 - 2(-1) - 2 \cdot 4$	? -1
$5 + 2 - 8$	
-1	TRUE

The ordered triple $(1, -1, 4)$ checks in all three equations, so it is the solution of the system of equations.

Practice Exercise

5. Solve:

$$\begin{aligned} x - y + z &= 9, \\ 2x + y + 2z &= 3, \\ 4x + 2y - 3z &= -1. \end{aligned}$$

Objective 3.7b Graph linear inequalities in two variables.

Example Graph: $2x + y \le 4$.

First, we graph the line $2x + y = 4$. The intercepts are $(0, 4)$ and $(2, 0)$. We draw the line solid because the inequality symbol is $\le$. Next, we choose a test point not on the line and determine whether it is a solution of the inequality. We choose $(0, 0)$, since it is usually an easy point to use.

$$\begin{array}{c|l} 2x + y \le 4 & \\ \hline 2 \cdot 0 + 0 \ ? \ 4 & \\ 0 & \text{TRUE} \end{array}$$

Since $(0, 0)$ is a solution, we shade the half-plane that contains $(0, 0)$.

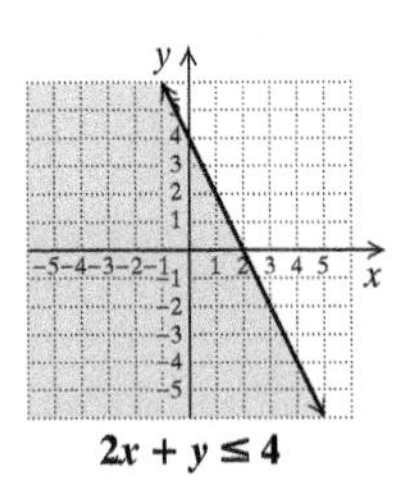

$2x + y \le 4$

Practice Exercise

6. Graph: $3x - 2y > 6$.

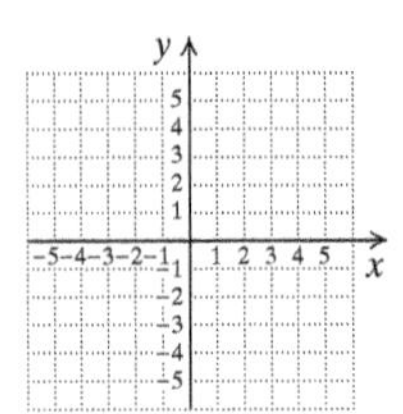

Objective 3.7c Graph systems of linear inequalities and find coordinates of any vertices.

Example Graph this system of inequalities and find the coordinates of any vertices formed:

$$x - 2y \ge -2, \quad \textbf{(1)}$$
$$3x - y \le 4, \quad \textbf{(2)}$$
$$y \ge -1. \quad \textbf{(3)}$$

We graph the related equations using solid lines. Then we indicate the region for each inequality by arrows at the ends of the line. Next, we shade the region of overlap.

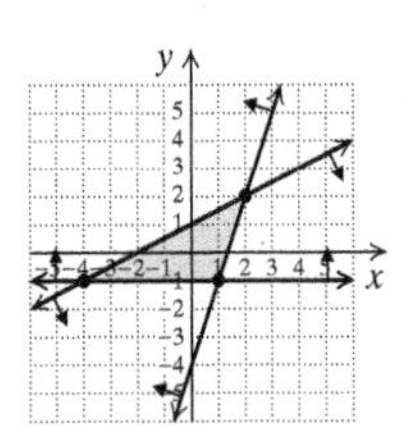

To find the vertices, we solve three different systems of related equations. From (1) and (2), we solve

$$x - 2y = -2,$$
$$3x - y = 4$$

to find the vertex $(2, 2)$. From (1) and (3), we solve

$$x - 2y = -2,$$
$$y = -1$$

to find the vertex $(-4, -1)$. From (2) and (3), we solve

$$3x - y = 4,$$
$$y = -1$$

to find the vertex $(1, -1)$.

Practice Exercise

7. Graph this system of inequalities and find the coordinates of any vertices found:

$$x - 2y \le 4,$$
$$x + y \le 4,$$
$$x - 1 \ge 0.$$

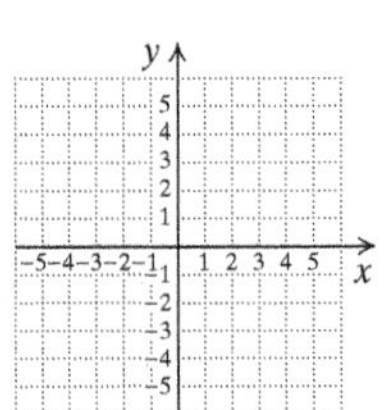

Review Exercises

Solve graphically. Then classify the system as consistent or inconsistent and the equations as dependent or independent. [3.1a]

1. $4x - y = -9,$
$x - y = -3$

2. $15x + 10y = -20,$
$3x + 2y = -4$

3. $y - 2x = 4,$
$y - 2x = 5$

Solve by the substitution method. [3.2a]

4. $2x - 3y = 5,$
$x = 4y + 5$

5. $y = x + 2,$
$y - x = 8$

6. $7x - 4y = 6,$
$y - 3x = -2$

Solve by the elimination method. [3.3a]

7. $x + 3y = -3,$
$2x - 3y = 21$

8. $3x - 5y = -4,$
$5x - 3y = 4$

9. $\frac{1}{3}x + \frac{2}{9}y = 1,$
$\frac{3}{2}x + \frac{1}{2}y = 6$

10. $1.5x - 3 = -2y,$
$3x + 4y = 6$

Solve.

11. *Retail Sales.* Paint Town sold 45 paintbrushes, one kind at $8.50 each and another at $9.75 each. In all, $398.75 was taken in for the brushes. How many of each kind were sold? [3.4a]

12. *Orange Drink Mixtures.* "Orange Thirst" is 15% orange juice and "Quencho" is 5% orange juice. How many liters of each should be combined in order to get 10 L of a mixture that is 10% orange juice? [3.4a]

13. *Train Travel.* A train leaves Watsonville at noon traveling north at 44 mph. One hour later, another train, going 52 mph, travels north on a parallel track. How many hours will the second train travel before it overtakes the first train? [3.4b]

Solve. [3.5a]

14. $x + 2y + z = 10,$
$2x - y + z = 8,$
$3x + y + 4z = 2$

15. $3x + 2y + z = 1,$
$2x - y - 3z = 1,$
$-x + 3y + 2z = 6$

16. $2x - 5y - 2z = -4,$
$7x + 2y - 5z = -6,$
$-2x + 3y + 2z = 4$

17. $x + y + 2z = 1,$
$x - y + z = 1,$
$x + 2y + z = 2$

18. *Triangle Measure.* In triangle ABC, the measure of angle A is four times the measure of angle C, and the measure of angle B is 45° more than the measure of angle C. What are the measures of the angles of the triangle? [3.6a]

19. *Popcorn.* Paul paid a total of $49 for 1 bag of caramel nut crunch popcorn, 1 bag of plain popcorn, and 1 bag of mocha choco latte popcorn. The price of the caramel nut crunch popcorn was six times the price of the plain popcorn and $16 more than the price of mocha choco latte popcorn. What was the price of each type of popcorn? [3.6a]

Graph. [3.7b]

20. $2x + 3y < 12$

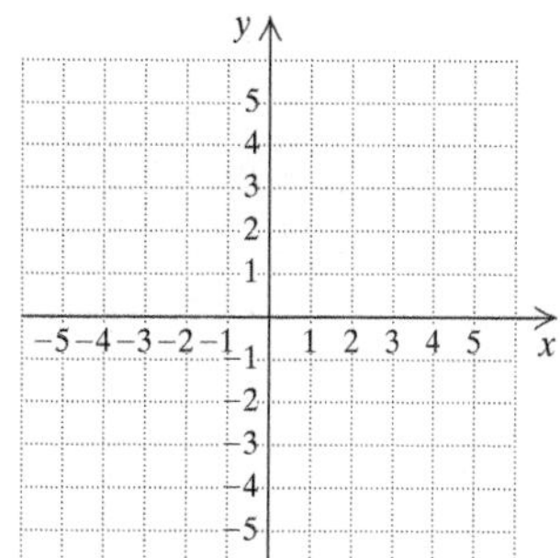

21. $y \leq 0$

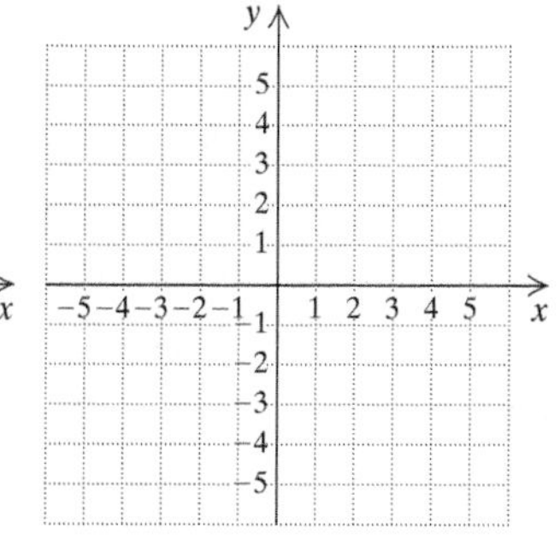

22. $x + y \geq 1$

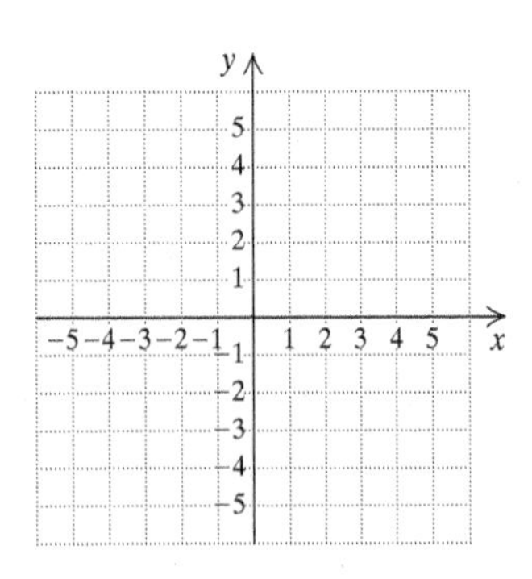

Graph. Find the coordinates of any vertices formed. [3.7c]

23. $y \geq -3,$
$x \geq 2$

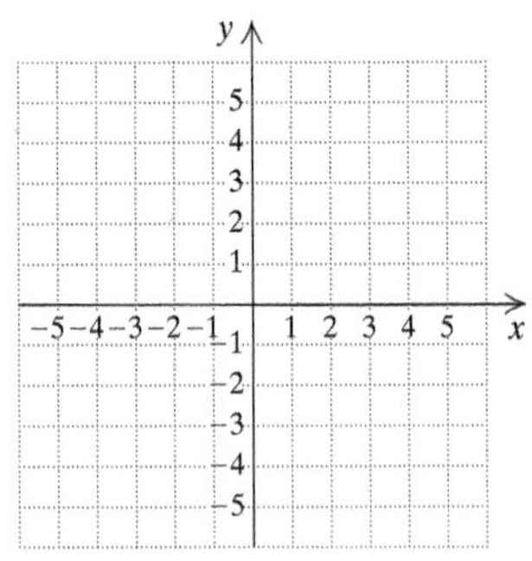

24. $x + 3y \geq -1,$
$x + 3y \leq 4$

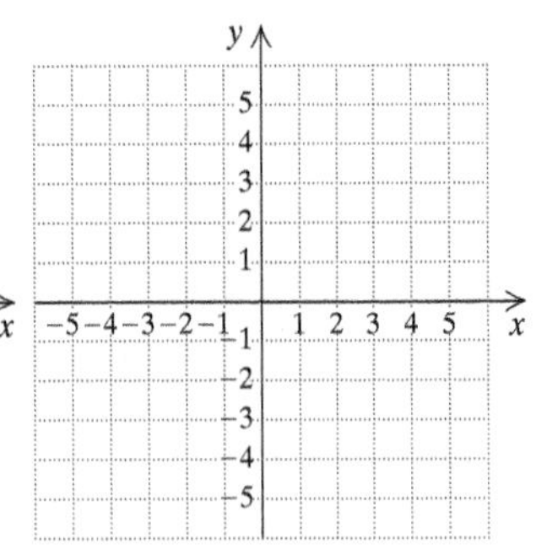

25. $x - y \leq 3,$
$x + y \geq -1,$
$y \leq 2$

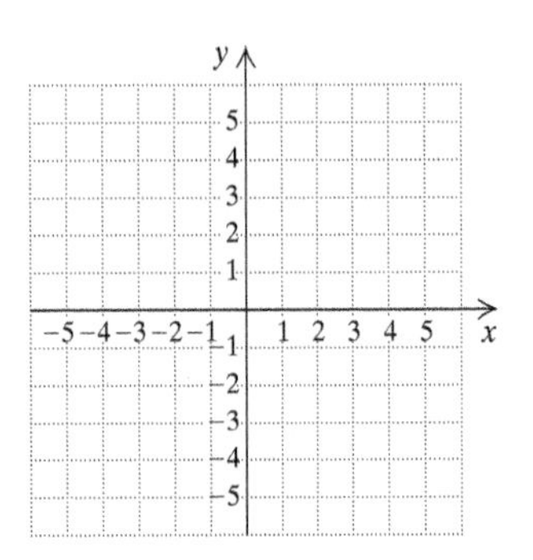

26. The sum of two numbers is -2. The sum of twice one number and the other is 4. One number is which of the following? [3.3b]

A. -6 **B.** 2
C. 6 **D.** 8

27. *Motorcycle Travel.* Sally and Elliot travel on motorcycles toward each other from Chicago and Indianapolis, which are about 350 km apart, and they are biking at rates of 110 km/h and 90 km/h. They started at the same time. In how many hours will they meet? [3.4b]

A. 1.75 hr **B.** 3.9 hr
C. 3.2 hr **D.** 17.5 hr

Synthesis

28. Solve graphically: [2.1d], [3.1a]
$y = x + 2,$
$y = x^2 + 2.$

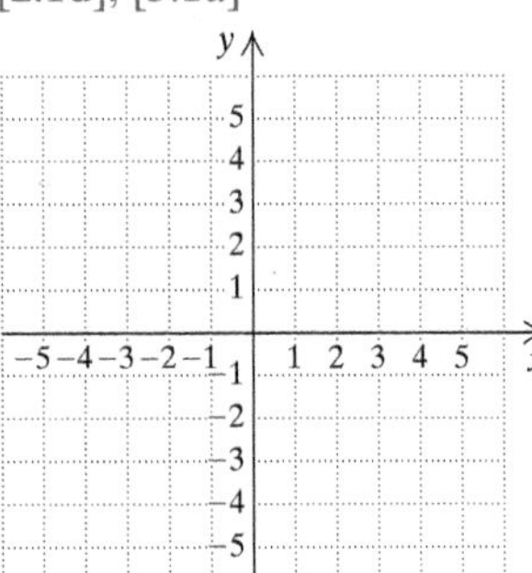

Understanding Through Discussion and Writing

1. Write a problem for a classmate to solve. Design the problem so the answer is "The florist sold 14 hanging baskets and 9 flats of petunias." [3.4a]

2. Exercise 21 in Exercise Set 3.6 can be solved mentally after a careful reading of the problem. Explain how this can be done. [3.6a]

3. *Ticket Revenue.* A pops-concert audience of 100 people consists of adults, senior citizens, and children. The ticket prices are \$10 each for adults, \$3 each for senior citizens, and \$0.50 each for children. The total amount of money taken in is \$100. How many adults, senior citizens, and children are in attendance? Does there seem to be some information missing? Do some careful reasoning and explain. [3.6a]

4. When graphing linear inequalities, Ron always shades above the line when he sees a $\geq$ symbol. Is this wise? Why or why not? [3.7a]

CHAPTER

3 Test

For Extra Help: For step-by-step test solutions, access the Chapter Test Prep Videos in MyLab Math.

Solve graphically. Then classify the system as consistent or inconsistent and the equations as dependent or independent.

1. $y = 3x + 7,$
$3x + 2y = -4$

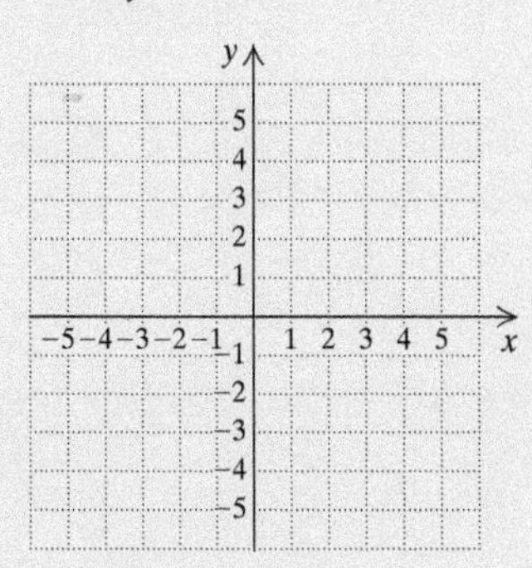

2. $y = 3x + 4,$
$y = 3x - 2$

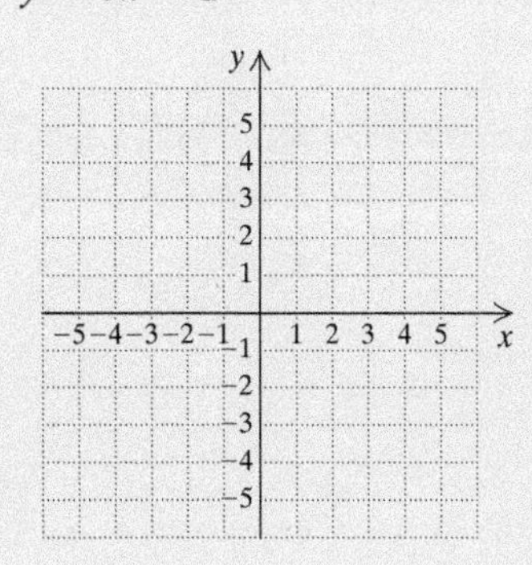

3. $y - 3x = 6,$
$6x - 2y = -12$

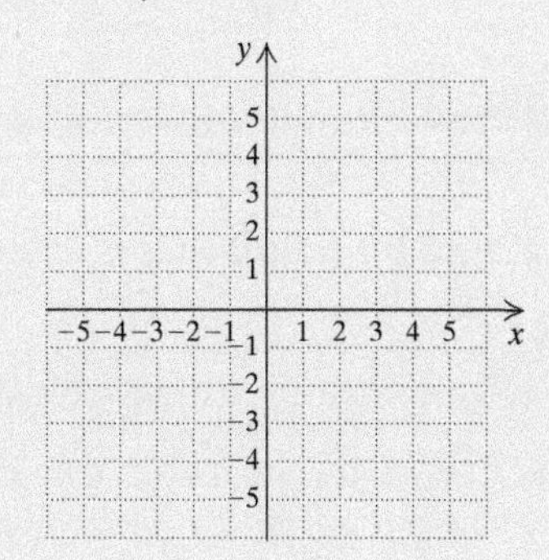

Solve by the substitution method.

4. $4x + 3y = -1,$
$y = 2x - 7$

5. $x = 3y + 2,$
$2x - 6y = 4$

6. $x + 2y = 6,$
$2x + 3y = 7$

Solve by the elimination method.

7. $2x + 5y = 3,$
$-2x + 3y = 5$

8. $x + y = -2,$
$4x - 6y = -3$

9. $\frac{2}{3}x - \frac{4}{5}y = 1,$
$\frac{1}{3}x - \frac{2}{5}y = 2$

Solve.

10. *Tennis Court.* The perimeter of a standard tennis court used for playing doubles is 288 ft. The width of the court is 42 ft less than the length. Find the length and the width.

11. *Air Travel.* An airplane flew for 5 hr with a 20-km/h tailwind and returned in 7 hr against the same wind. Find the speed of the plane in still air.

12. *Chicken Dinners.* High Flyin' Wings charges \$12 for a bucket of chicken wings and \$7 for a chicken dinner. After filling 28 orders for buckets and dinners during a football game, the waiters had collected \$281. How many buckets and how many dinners did they sell?

13. *Mixing Solutions.* A chemist has one solution that is 20% salt and a second solution that is 45% salt. How many liters of each should be used in order to get 20 L of a solution that is 30% salt?

14. Solve:

$$\begin{aligned} 6x + 2y - 4z &= 15, \\ -3x - 4y + 2z &= -6, \\ 4x - 6y + 3z &= 8. \end{aligned}$$

15. *Repair Rates.* An electrician, a carpenter, and a plumber are hired to work on a house. The electrician earns \$21 per hour, the carpenter \$19.50 per hour, and the plumber \$24 per hour. The first day on the job, they worked a total of 21.5 hr and earned a total of \$469.50. If the plumber worked 2 hr more than the carpenter did, how many hours did the electrician work?

Graph. Find the coordinates of any vertices formed.

16. $y \geq x - 2$

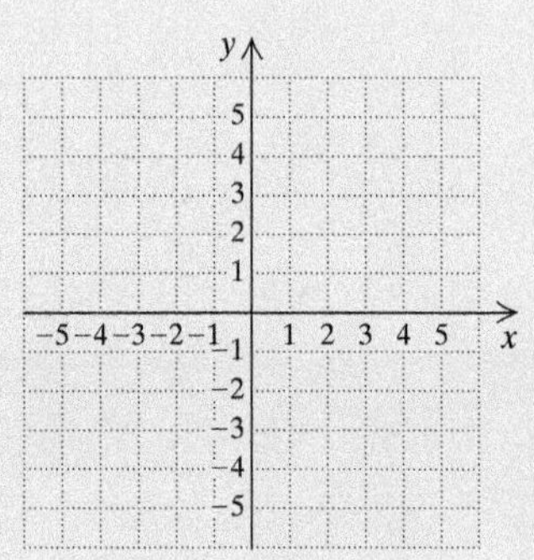

17. $x - 6y < -6$

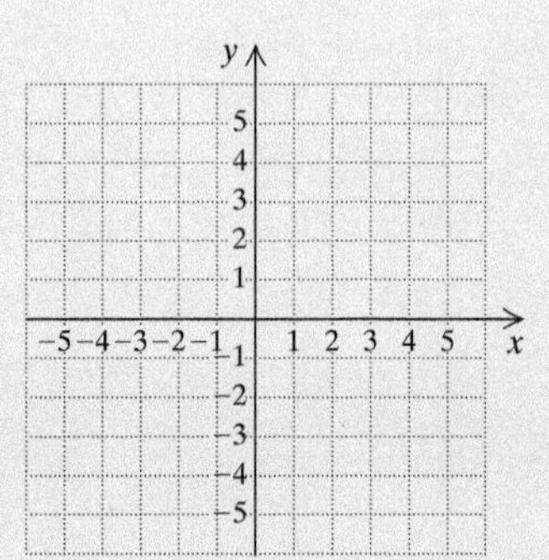

18. $x + y \geq 3,$
$x - y \geq 5$

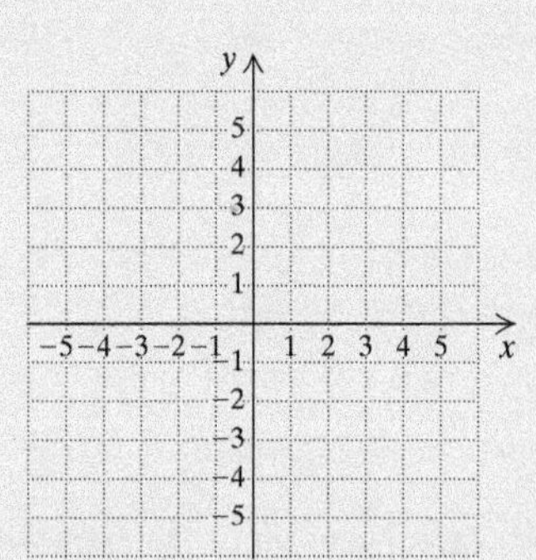

19. $2y - x \geq -4,$
$2y + 3x \leq -6,$
$y \leq 0,$
$x \leq 0$

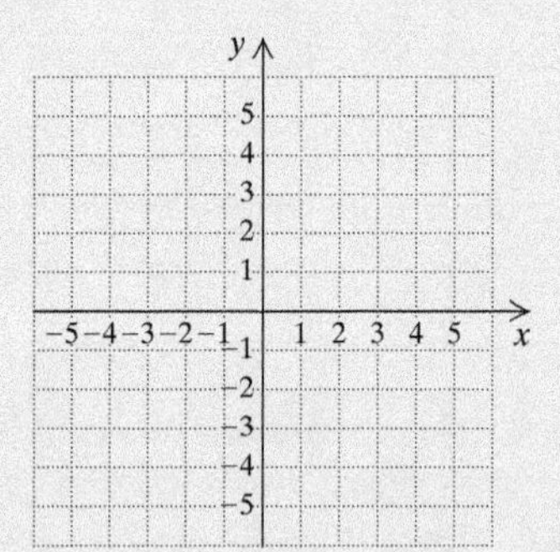

20. A business class divided an imaginary \$30,000 investment among three funds. The first fund grew 2%, the second grew 3%, and the third grew 5%. Total earnings were \$990. The earnings from the third fund were \$280 more than the earnings from the first. How much was invested at 5%?

A. \$9000 **B.** \$10,000 **C.** \$11,000 **D.** \$12,000

Synthesis

21. The graph of the function $f(x) = mx + b$ contains the points $(-1, 3)$ and $(-2, -4)$. Find m and b.

Cumulative Review

Solve.

1. $6y - 5(3y - 4) = 10$

2. $-3 + 5x = 2x + 15$

3. $A = \pi r^2 h$, for h

4. $L = \dfrac{1}{3}m(k + p)$, for p

5. $5x + 8 > 2x + 5$

6. $-12 \leq -3x + 1 < 0$

7. $2x - 10 \leq -4$ *or* $x - 4 \geq 3$

8. $|x + 1| = 4$

9. $|8y - 3| \geq 15$

10. $|2x + 1| = |x - 4|$

11. Find the distance between -18 and -7 on the number line.

Graph on a plane.

12. $3y = 9$

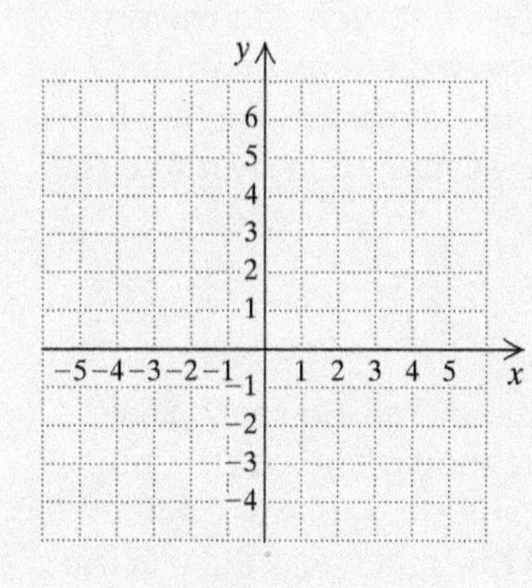

13. $f(x) = -\dfrac{1}{2}x - 3$

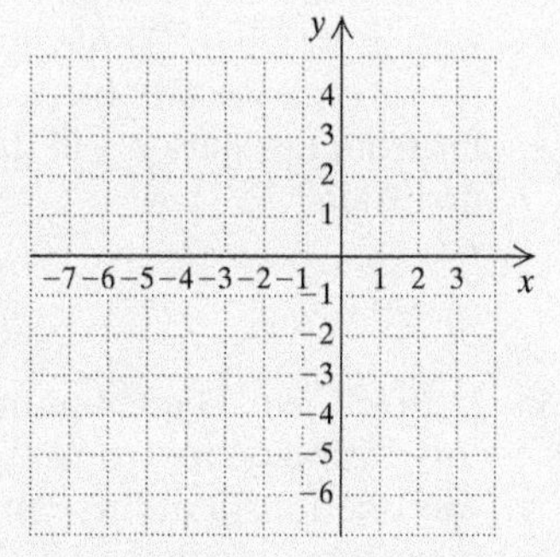

14. $3x - 1 = y$

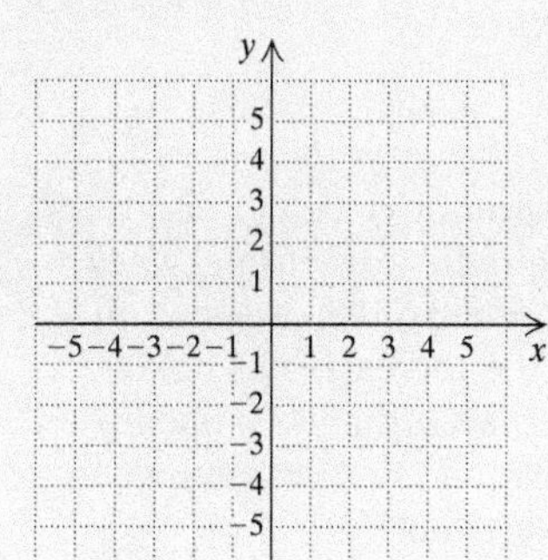

15. $3x + 5y = 15$

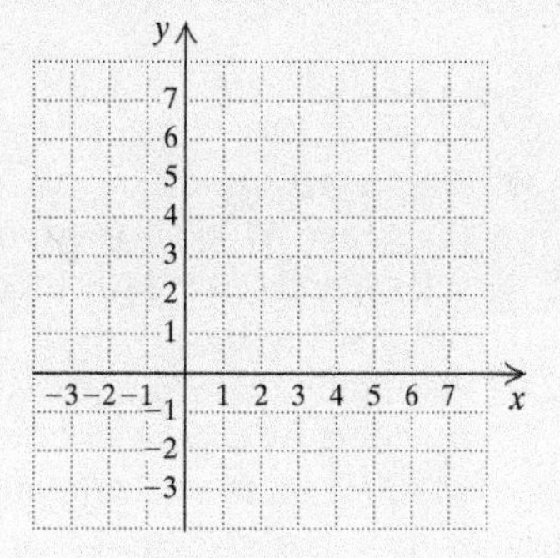

16. $y > 3x - 4$

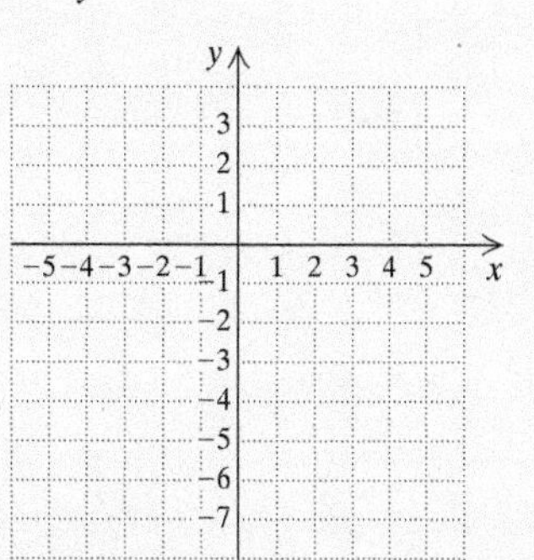

17. $2x - y \leq 6$

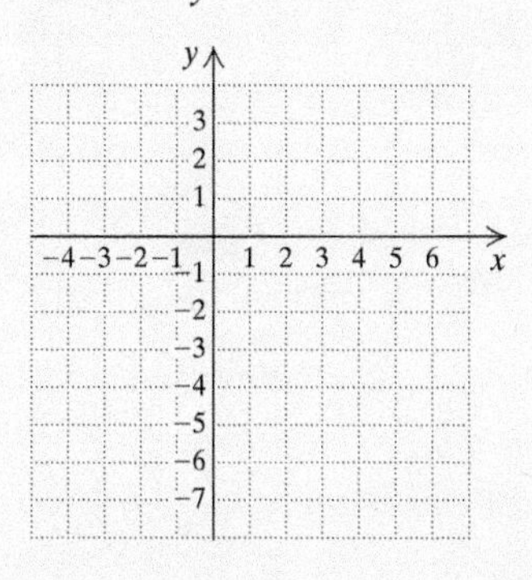

18. Solve graphically. Then classify the system as consistent or inconsistent and the equations as dependent or independent.

$$2x - y = 7,$$
$$x + 3y = 0$$

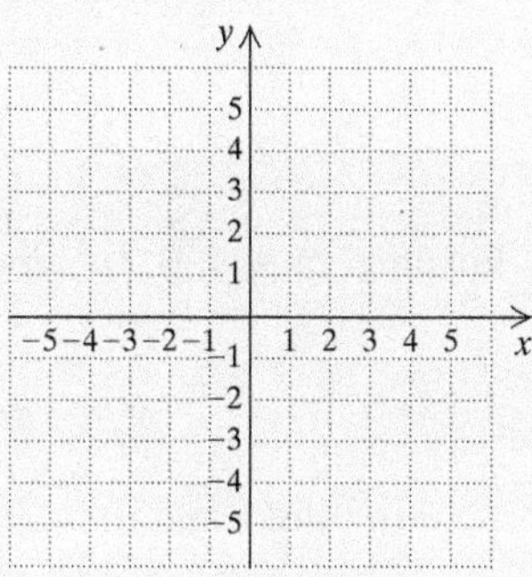

Solve.

19. $3x + 4y = 4,$
$x = 2y + 2$

20. $3x + y = 2,$
$6x - y = 7$

21. $4x + 3y = 5,$
$3x + 2y = 3$

22. $x - y + z = 1,$
$2x + y + z = 3,$
$x + y - 2z = 4$

Graph. Find the coordinates of any vertices formed.

23. $x + y \leq -3,$
$x - y \leq 1$

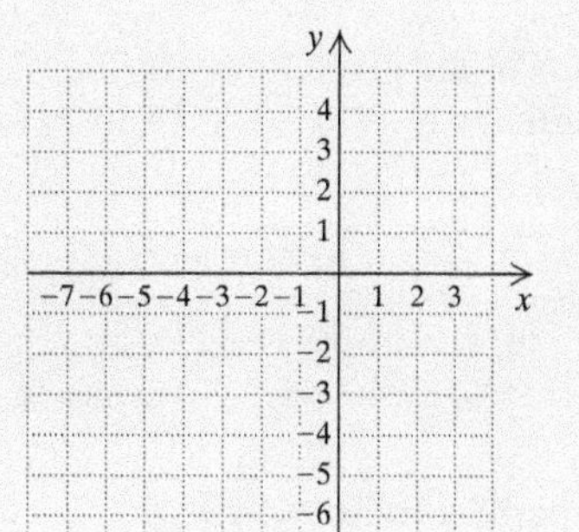

24. $4y - 3x \geq -12,$
$4y + 3x \geq -36,$
$y \leq 0,$
$x \leq 0$

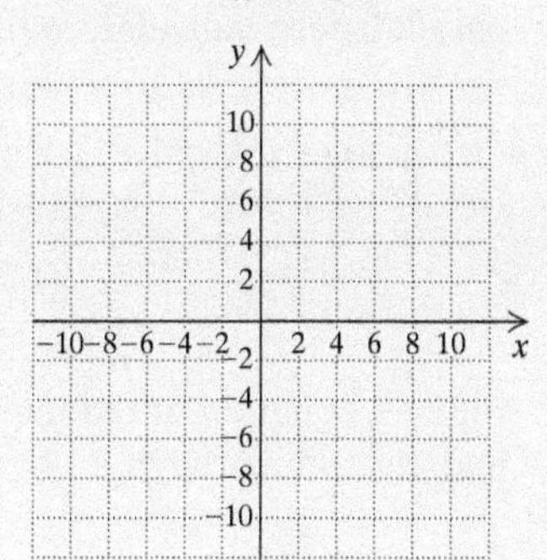

25. For the function f whose graph is shown below, determine **(a)** the domain, **(b)** the range, **(c)** $f(-3)$, and **(d)** any input for which $f(x) = 5$.

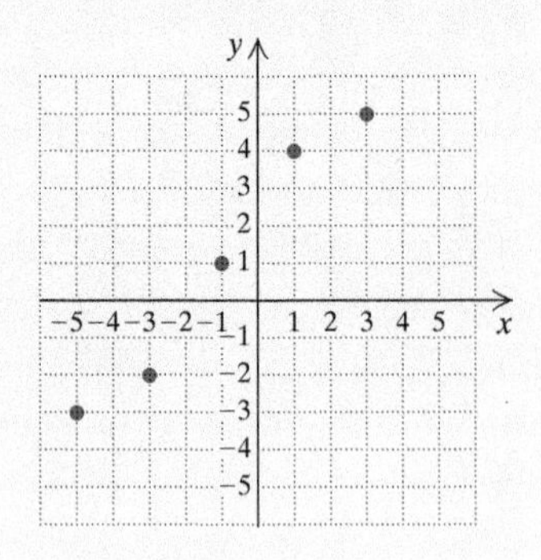

26. Find the domain of the function given by

$$f(x) = \frac{7}{2x - 1}.$$

27. Given $g(x) = 1 - 2x^2$, find $g(-1)$, $g(0)$, and $g(3)$.

28. Find the slope and the y-intercept of $5y - 4x = 20$.

29. Find an equation of the line with slope -3 and containing the point $(5, 2)$.

30. Find an equation of the line containing the points $(-1, -3)$ and $(-3, 5)$.

31. Determine whether the graphs of the given lines are parallel, perpendicular, or neither.

$$x - 2y = 4,$$
$$4x + 2y = 1$$

32. Find an equation of the line parallel to $3x - 9y = 2$ and containing the point $(-6, 2)$.

Solve.

33. *Wire Cutting.* Rolly's Electric wants to cut a piece of copper wire 10 m long into two pieces, one of them two-thirds as long as the other. How should the wire be cut?

34. *Test Scores.* Adam is taking a geology course in which there will be 4 tests, each worth 100 points. He has scores of 87, 94, and 91 on the first three tests. He must have a total of at least 360 in order to get an A. What scores on the last test will give Adam an A?

35. *Inventory.* The Everton College store paid \$2268 for an order of 45 calculators. The store paid \$9 for each scientific calculator. The others, all graphing calculators, cost the store \$78 each. How many of each type of calculator was ordered?

36. *Mixing Solutions.* A technician wants to mix one solution that is 15% alcohol with another solution that is 25% alcohol in order to get 30 L of a solution that is 18% alcohol. How much of each solution should be used?

37. *Train Travel.* A train leaves a station and travels west at 80 km/h. Three hours later, a second train leaves on a parallel track and travels 120 km/h. How far from the station will the second train overtake the first train?

38. *Utility Cost.* One month Ladi and Bo spent \$680 for electricity, rent, and telephone. The electric bill was one-fourth of the rent and the rent was \$400 more than the phone bill. How much was the electric bill?

Synthesis

39. *Radio Advertising.* An automotive dealer discovers that when \$1000 is spent on radio advertising, weekly sales increase by \$101,000. When \$1250 is spent on radio advertising, weekly sales increase by \$126,000. Assuming that sales increase according to a linear function, by what amount would sales increase when \$1500 is spent on radio advertising?

40. Given that $f(x) = mx + b$ and that $f(5) = -3$ when $f(-4) = 2$, find m and b.

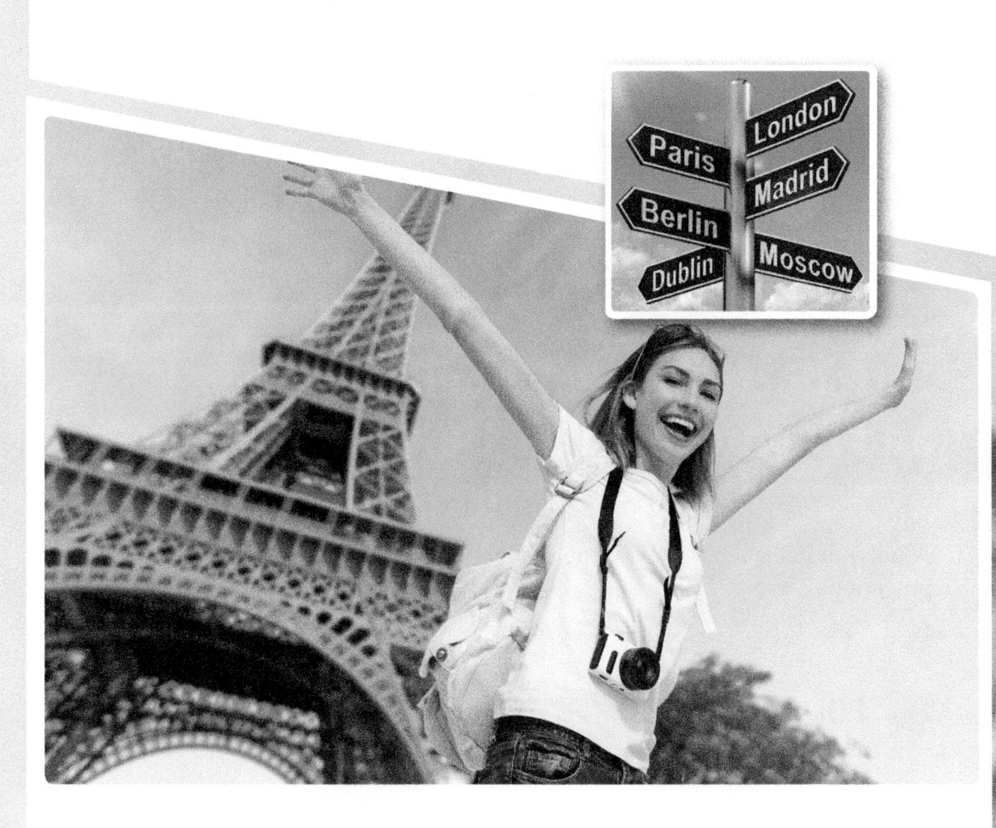

CHAPTER

4

Polynomials and Polynomial Functions

The number of Americans traveling by air out of the country grew rapidly from 33 million flights in 2003 to 74 million flights in 2015. Growth in travel has not been constant over those 15 years. For example, the accompanying graph shows that the number of Americans traveling to Europe has not risen linearly, but instead can be modeled by a higher-order *polynomial function*. Growth in travel has also varied between regions of the world, with travel to Africa rising sharply in 2009 and then decreasing through 2015. Some of these travel patterns can be explained by world economic and political changes.

Travel to Europe and Africa

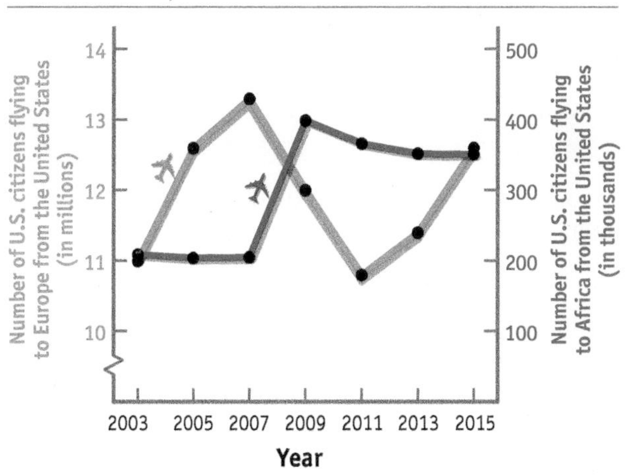

Data: travel.trade.gov

In Exercise 25 of Section 4.1, we will use a graph of a polynomial function to estimate the number of Americans traveling to Europe for various years.

JUST IN TIME

Review topics 10, 13, 16, and 19 in the **Just In Time** section at the front of the text. This provides excellent prerequisite skill review for this chapter.

MyLab Math VIDEO

STUDYING FOR SUCCESS *Time Management*

- ☐ As a general rule, budget two to three hours for homework and study for every hour that you spend in class.
- ☐ Balance work and study carefully. Be honest with yourself about how much time you have available to work, attend class, and study.
- ☐ Make an hour-by-hour schedule of your week, planning time for leisure as well as work and study.
- ☐ Use your syllabus to help you plan your time. Transfer project deadlines and test dates to your calendar.

4.1 Introduction to Polynomials and Polynomial Functions

OBJECTIVES

- a Identify terms, degrees, and coefficients in polynomials; identify types of polynomials; and arrange polynomials in ascending order or descending order.
- b Evaluate a polynomial function for given inputs.
- c Collect like terms in a polynomial and add polynomials.
- d Find the opposite of a polynomial and subtract polynomials.

A **polynomial** is a particular type of algebraic expression. Some examples of polynomials are

$$x + 7, \quad abc, \quad 5t^2 - 6t + 1, \quad \text{and} \quad 7.$$

a POLYNOMIAL EXPRESSIONS

The following are examples of *monomials*:

$$0, \quad -3, \quad z, \quad 8x, \quad -7y^2, \quad 4a^2b^3, \quad 1.3p^4q^5r^7.$$

MONOMIAL

A **monomial** is a constant or a constant times some variable or variables raised to powers that are nonnegative integers.

Expressions like these are called **polynomials in one variable:**

$$5x^2, \quad 2y^2 + 5y - 3, \quad 5a^4 - 3a^2 + \tfrac{1}{4}a - 8.$$

Expressions like these are called **polynomials in several variables:**

$$15x^3y^2, \quad a - b, \quad \frac{1}{2}xy^2z - 4x^3z + y^3 + 9.$$

POLYNOMIAL

A **polynomial** is a monomial or a combination of sums and/or differences of monomials.

The following are algebraic expressions that are not polynomials:

$$(1)\ \frac{y^2 - 3}{y^2 + 4}, \quad (2)\ 8x^4 - 2x^3 + \frac{1}{x}, \quad (3)\ \frac{2xy}{x^3 - y^3}.$$

Expressions (1) and (3) are not polynomials because they represent quotients. In expression (2), although we can write $1/x$ as x^{-1}, this is not a monomial because the exponent is negative.

The **terms** of a polynomial are separated by + signs. The polynomial $5x^3y - 7xy^2 - y^3 + 2$ has four terms:

$$5x^3y, \quad -7xy^2, \quad -y^3, \text{ and } 2.$$

The **coefficients** of the terms are 5, −7, −1, and 2. The term 2 is called a **constant term.**

The **degree of a term** is the sum of the exponents of the variables, if there are variables. For example,

the degree of the term $9x^5$ is 5 and
the degree of the term $0.6a^2b^7$ is $2 + 7$, or 9.

The degree of a nonzero constant term, such as 2, is 0. We can express 2 as $2x^0$.

Because we can express 0 as $0 = 0x^5 = 0x^{12}$, and so on, using any exponent we wish, the term 0 has *no* degree.

The **degree of a polynomial** is the same as the degree of its term of highest degree. For example,

the degree of the polynomial $4 - x^3 + 5x^2 - x^6$ is 6.

The **leading term** of a polynomial is the term of highest degree. Its coefficient is called the **leading coefficient.** For example,

the leading term of $9x^2 - 5x^3 + x - 10$ is $-5x^3$ and
the leading coefficient is −5.

EXAMPLE 1 Identify the terms, the degree of each term, and the degree of the polynomial $2x^3 + 8x^2 - 17x - 3$. Then identify the leading term, the leading coefficient, and the constant term.

TERM	$2x^3$	$8x^2$	$-17x$	-3
DEGREE OF TERM	3	2	1	0
DEGREE OF POLYNOMIAL	3			
LEADING TERM	$2x^3$			
LEADING COEFFICIENT	2			
CONSTANT TERM	−3			

EXAMPLE 2 Identify the terms, the degree of each term, and the degree of the polynomial $6x^2 + 8x^2y^3 - 17xy - 24xy^2z^4 + 2y + 3$. Then identify the leading term, the leading coefficient, and the constant term.

TERM	$6x^2$	$8x^2y^3$	$-17xy$	$-24xy^2z^4$	$2y$	3
DEGREE OF TERM	2	5	2	7	1	0
DEGREE OF POLYNOMIAL	7					
LEADING TERM	$-24xy^2z^4$					
LEADING COEFFICIENT	−24					
CONSTANT TERM	3					

1. Identify the terms and the leading term:
$-92x^5 - 8x^4 + x^2 + 5$.

2. Identify the coefficient of each term and the leading coefficient:
$5x^3y - 4xy^2 - 2x^3 + xy - y - 5$.

3. Identify the terms, the degree of each term, and the degree of the polynomial. Then identify the leading term, the leading coefficient, and the constant term.
a) $6x^2 - 5x^3 + 2x - 7$
b) $2y - 4 - 5x + 9x^2y^3z^2 + 5xy^2$

Answers

1. $-92x^5, -8x^4, x^2, 5; -92x^5$
2. $5, -4, -2, 1, -1, -5; 5$ **3. (a)** $6x^2, -5x^3, 2x, -7; 2, 3, 1, 0; 3; -5x^3; -5; -7$;
(b) $2y, -4, -5x, 9x^2y^3z^2, 5xy^2; 1, 0, 1, 7, 3; 7; 9x^2y^3z^2; 9; -4$

4. Consider the following polynomials.
   a) $3x^2 - 2$
   b) $5x^3 + 9x - 3$
   c) $4x^2$
   d) $-7y$
   e) -3
   f) $8x^3 - 2x^2$
   g) $-4y^2 - 5 - 5y$
   h) $5 - 3x$

   Identify the monomials, the binomials, and the trinomials.

5. a) Arrange in ascending order: $5 - 6x^2 + 7x^3 - x^4 + 10x$.
   b) Arrange in descending order: $5 - 6x^2 + 7x^3 - x^4 + 10x$.

6. a) Arrange in ascending powers of y: $5x^4y - 3y^2 + 3x^2y^3 - 5$.
   b) Arrange in descending powers of y: $5x^4y - 3y^2 + 3x^2y^3 - 5$.

7. For the polynomial function $P(x) = x^2 - 2x + 5$, find $P(0)$, $P(4)$, and $P(-2)$.

Answers

4. Monomials: (c), (d), (e); binomials: (a), (f), (h); trinomials: (b), (g)
5. (a) $5 + 10x - 6x^2 + 7x^3 - x^4$; **(b)** $-x^4 + 7x^3 - 6x^2 + 10x + 5$
6. (a) $-5 + 5x^4y - 3y^2 + 3x^2y^3$; **(b)** $3x^2y^3 - 3y^2 + 5x^4y - 5$
7. 5; 13; 13

Do Exercises 1–3 on the preceding page.

The following are some names for certain types of polynomials.

TYPE	DEFINITION: POLYNOMIAL OF	EXAMPLES
Monomial	One term	4, $-3p$, $-7a^2b^3$, 0, xyz
Binomial	Two terms	$2x + 7$, $a^2 - 3b$, $5x^3 + 8x$
Trinomial	Three terms	$x^2 - 7x + 12$, $4a^2 + 2ab + b^2$

Do Exercise 4.

We generally arrange polynomials in one variable in **descending order** so that the exponents *decrease* from left to right. Sometimes they may be written so that the exponents *increase* from left to right, which is **ascending order**. In general, if an exercise is written in a particular order, we write the answer in that same order.

EXAMPLE 3 Consider $12 + x^2 - 7x$. Arrange in descending order and then in ascending order.

Descending order: $x^2 - 7x + 12$
Ascending order: $12 - 7x + x^2$

EXAMPLE 4 Consider $x^4 + 2 - 5x^2 + 3x^3y + 7xy^2$. Arrange in descending powers of x and then in ascending powers of x.

Descending powers of x: $x^4 + 3x^3y - 5x^2 + 7xy^2 + 2$
Ascending powers of x: $2 + 7xy^2 - 5x^2 + 3x^3y + x^4$

Do Exercises 5 and 6.

b EVALUATING POLYNOMIAL FUNCTIONS

A polynomial function is one like $P(x) = 5x^7 + 3x^5 - 4x^2 - 5$, in which the algebraic expression used to describe the function is a polynomial. To find the outputs of a polynomial function for a given input, we substitute the input for each occurrence of the variable.

EXAMPLE 5 For the polynomial function $P(x) = -x^2 + 4x - 1$, find $P(2)$, $P(10)$, and $P(-10)$.

$$P(2) = -2^2 + 4(2) - 1 = -4 + 8 - 1 = 3;$$
$$P(10) = -10^2 + 4(10) - 1 = -100 + 40 - 1 = -61;$$
$$P(-10) = -(-10)^2 + 4(-10) - 1 = -100 - 40 - 1 = -141$$

Do Exercise 7.

EXAMPLE 6 *Veterinary Medicine.* Gentamicin is an antibiotic frequently used by veterinarians. The concentration C, in micrograms per milliliter (mcg/mL), of Gentamicin in a horse's bloodstream t hours after injection can be approximated by the polynomial function

$$C(t) = -0.005t^4 + 0.003t^3 + 0.35t^2 + 0.5t.$$

a) Evaluate $C(2)$ to find the concentration 2 hr after injection.

b) Use only the following graph to estimate $C(4)$.

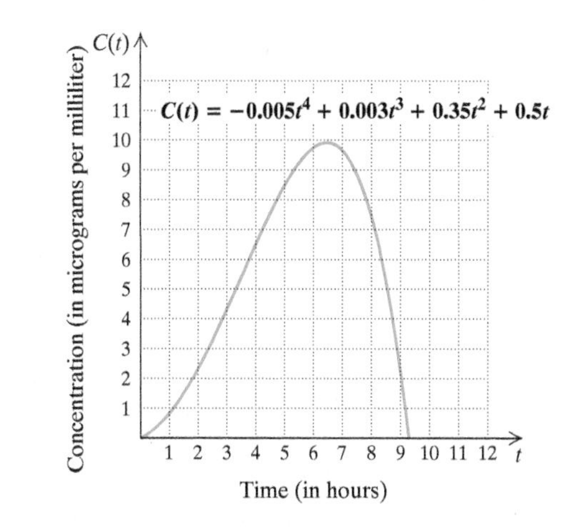

a) We evaluate the function when $t = 2$:

$$\begin{aligned} C(t) &= -0.005t^4 + 0.003t^3 + 0.35t^2 + 0.5t \\ C(2) &= -0.005(2)^4 + 0.003(2)^3 + 0.35(2)^2 + 0.5(2) \\ &= -0.005(16) + 0.003(8) + 0.35(4) + 0.5(2) \\ &= -0.08 + 0.024 + 1.4 + 1 \\ &= 2.344. \end{aligned}$$

We carry out the calculation using the rules for order of operations.

The concentration after 2 hr is about 2.344 mcg/mL.

b) To estimate $C(4)$, the concentration after 4 hr, we locate 4 on the horizontal axis. From there we move vertically to the graph of the function and then horizontally to the $C(t)$-axis. This locates a value of about 6.5. Thus, $C(4) \approx 6.5$. The concentration after 4 hr is about 6.5 mcg/mL.

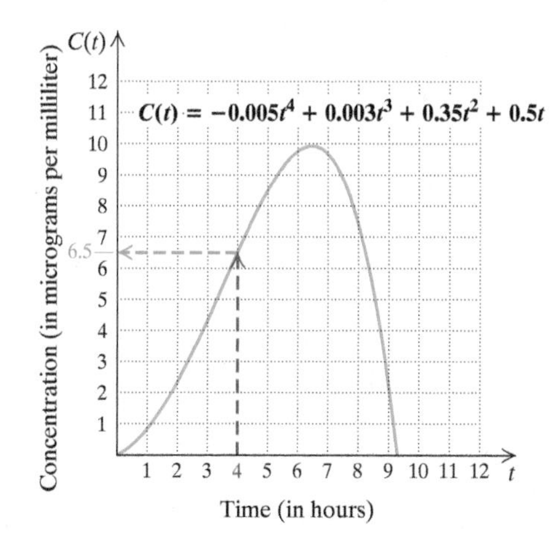

Do Exercise 8. ▶

8. *Veterinary Medicine.* Refer to the function and the graph of Example 6.

a) Evaluate $C(3)$ to find the concentration 3 hr after injection.

b) Use only the graph at left to estimate $C(9)$.

C ADDING POLYNOMIALS

SKILL REVIEW *Collect like terms.* [J17]

Collect like terms.

1. $a + b + 7b$ **2.** $x - 6x$

Answers: **1.** $a + 8b$ **2.** $-5x$

Answers

8. (a) $C(3) = 4.326$ mcg/mL; **(b)** $C(9) \approx 2$ mcg/mL

When two terms have the same variable(s) raised to the same power(s), they are called **like terms**, or **similar terms**, and they can be "collected," or "combined," using the distributive laws.

EXAMPLES Collect like terms.

7. $3x^2 - 4y + 2x^2 = 3x^2 + 2x^2 - 4y$ — Rearranging using the commutative law for addition

GS

$= (3 + 2)x^2 - 4y$ — Using the distributive law

$= 5x^2 - 4y$ — Adding the coefficients of x^2

8. $9x^3 + 5x - 4x^2 - 2x^3 + 5x^2 = 7x^3 + x^2 + 5x$

9. $3x^2y + 5xy^2 - 3x^2y - xy^2 = 4xy^2$

Do Exercises 9 and 10.

Collect like terms.

9. $3y - 4x + 6xy^2 - 2xy^2$

10. $3xy^3 + 2x^3y + 5xy^3 - 8x + 15 - 3x^2y - 6x^2y + 11x - 8$

$3xy^3$ and ____ are like terms.

$-3x^2y$ and ____ are like terms.

$-8x$ and ____ are like terms.

15 and ____ are like terms.

Rearranging:

$2x^3y + 3xy^3 + 5xy^3 - 3x^2y - 6x^2y - 8x + 11x + 15 - 8.$

Collecting like terms:

$2x^3y +$ ____ $xy^3 -$ ____ $x^2y +$ ____ $x +$ ____.

The sum of two polynomials can be found by writing a plus sign between them and then collecting like terms to simplify the expression.

EXAMPLE 10 Add: $(-3x^3 + 2x - 4) + (4x^3 + 3x^2 + 2)$.

$(-3x^3 + 2x - 4) + (4x^3 + 3x^2 + 2) = x^3 + 3x^2 + 2x - 2$ ■

EXAMPLE 11 Add: $13x^3y + 3x^2y - 5y$ and $x^3y + 4x^2y - 3xy$.

$(13x^3y + 3x^2y - 5y) + (x^3y + 4x^2y - 3xy) = 14x^3y + 7x^2y - 3xy - 5y$

Do Exercises 11–13.

Add.

11. $(3x^3 + 4x^2 - 7x - 2) + (-7x^3 - 2x^2 + 3x + 4)$

12. $(7y^5 - 5) + (3y^5 - 4y^2 + 10)$

13. $(5p^2q^4 - 2p^2q^2 - 3q) + (-6p^2q^2 + 3q + 5)$

In order to use columns to add, we write the polynomials one under the other, listing like terms under one another and leaving spaces for missing terms.

EXAMPLE 12 Add: $4ax^2 + 4bx - 5$ and $-6ax^2 + 8$.

$$\begin{array}{r} 4ax^2 + 4bx - 5 \\ \underline{-6ax^2 \qquad\quad + 8} \\ -2ax^2 + 4bx + 3 \end{array}$$ ■

d SUBTRACTING POLYNOMIALS

If the sum of two polynomials is 0, the polynomials are **opposites**, or **additive inverses**, of each other. For example,

$$(3x^2 - 5x + 2) + (-3x^2 + 5x - 2) = 0,$$

so the opposite of $3x^2 - 5x + 2$ is $-3x^2 + 5x - 2$. We can say the same thing using algebraic symbolism, as follows:

The opposite of $(3x^2 - 5x + 2)$ is $(-3x^2 + 5x - 2)$.

$$-\quad(3x^2 - 5x + 2) \quad = \quad -3x^2 + 5x - 2$$

Thus, $-(3x^2 - 5x + 2)$ and $-3x^2 + 5x - 2$ are equivalent.

The *opposite* of a polynomial P can be symbolized by $-P$ or by replacing each term with its opposite. The two expressions for the opposite are equivalent.

Answers

9. $3y - 4x + 4xy^2$
10. $2x^3y + 8xy^3 - 9x^2y + 3x + 7$
11. $-4x^3 + 2x^2 - 4x + 2$
12. $10y^5 - 4y^2 + 5$ **13.** $5p^2q^4 - 8p^2q^2 + 5$

Guided Solution:

10. $5xy^3, -6x^2y, 11x, -8, 8, 9, 3, 7$

EXAMPLE 13 Write two equivalent expressions for the opposite of $7xy^2 - 6xy - 4y + 3$.

First expression: $-(7xy^2 - 6xy - 4y + 3)$ Writing an inverse sign in front

Second expression: $-7xy^2 + 6xy + 4y - 3$ Writing the opposite of each term

Do Exercises 14–16. ▶

To subtract a polynomial, we add its opposite.

EXAMPLE 14 Subtract: $(-5x^2 + 4) - (2x^2 + 3x - 1)$.

We have

$$\begin{aligned}(-5x^2 + 4) &- (2x^2 + 3x - 1)\\ &= (-5x^2 + 4) + [-(2x^2 + 3x - 1)] && \text{Adding the opposite}\\ &= (-5x^2 + 4) + (-2x^2 - 3x + 1) && -2x^2 - 3x + 1 \text{ is equivalent to } -(2x^2 + 3x - 1).\\ &= -7x^2 - 3x + 5. && \text{Adding}\end{aligned}$$

With practice, you may find that you can skip some steps, by mentally taking the opposite of each term and then combining like terms. Eventually, all you will write is the answer.

$$\begin{aligned}(-5x^2 + 4) &- (2x^2 + 3x - 1)\\ &= -7x^2 - 3x + 5\end{aligned}$$

Think:
$-5x^2 - 2x^2 = -5x^2 + (-2x^2) = -7x^2$,
$0x - 3x = 0x + (-3x) = -3x$,
$4 - (-1) = 4 + 1 = 5$.

Do Exercises 17–19. ▶

To use columns for subtraction, we mentally change the signs of the terms being subtracted.

EXAMPLE 15 Subtract:

$$(4x^2y - 6x^3y^2 + x^2y^2) - (4x^2y + x^3y^2 + 3x^2y^3 - 8x^2y^2).$$

Write: (Subtract)

$$\begin{array}{l} 4x^2y - 6x^3y^2 \qquad\qquad\quad + x^2y^2 \\ \underline{-(4x^2y + x^3y^2 + 3x^2y^3 - 8x^2y^2)} \end{array}$$

Think: (Add)

$$\begin{array}{r} 4x^2y - 6x^3y^2 \qquad\qquad\quad + x^2y^2 \\ \underline{-4x^2y - x^3y^2 - 3x^2y^3 + 8x^2y^2} \\ -7x^3y^2 - 3x^2y^3 + 9x^2y^2 \end{array}$$

Take the opposite of each term mentally and add.

Do Exercises 20–22. ▶

Write two equivalent expressions for the opposite, or additive inverse.

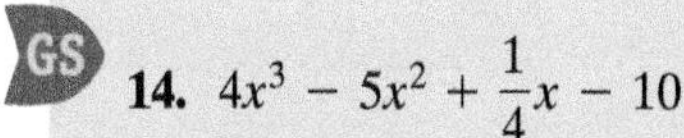

GS **14.** $4x^3 - 5x^2 + \frac{1}{4}x - 10$

One expression uses an inverse sign in front:

$\square\left(4x^3 - 5x^2 + \frac{1}{4}x - 10\right)$.

For a second expression, write the opposite of each term:

$-4x^3 + 5x^2 \ \square\ \frac{1}{4}x \ \square\ 10$.

15. $8xy^2 - 4x^3y^2 - 9x - \frac{1}{5}$

16. $-9y^5 - 8y^4 + \frac{1}{2}y^3 - y^2 + y - 1$

Subtract.

17. $(6x^2 + 4) - (3x^2 - 1)$

18. $(9y^3 - 2y - 4) - (-5y^3 - 8)$

19. $(-3p^2 + 5p - 4) - (-4p^2 + 11p - 2)$

Subtract.

20. $(2y^5 - y^4 + 3y^3 - y^2 - y - 7) - (-y^5 + 2y^4 - 2y^3 + y^2 - y - 4)$

21. $(4p^4q - 5p^3q^2 + p^2q^3 + 2q^4) - (-5p^4q + 5p^3q^2 - 3p^2q^3 - 7q^4)$

22. $\left(\frac{3}{2}y^3 - \frac{1}{2}y^2 + 0.3\right) - \left(\frac{1}{2}y^3 + \frac{1}{2}y^2 - \frac{4}{3}y + 0.2\right)$

Answers

14. $-(4x^3 - 5x^2 + \frac{1}{4}x - 10)$; $-4x^3 + 5x^2 - \frac{1}{4}x + 10$
15. $-(8xy^2 - 4x^3y^2 - 9x - \frac{1}{5})$; $-8xy^2 + 4x^3y^2 + 9x + \frac{1}{5}$
16. $-(-9y^5 - 8y^4 + \frac{1}{2}y^3 - y^2 + y - 1)$; $9y^5 + 8y^4 - \frac{1}{2}y^3 + y^2 - y + 1$
17. $3x^2 + 5$ **18.** $14y^3 - 2y + 4$
19. $p^2 - 6p - 2$ **20.** $3y^5 - 3y^4 + 5y^3 - 2y^2 - 3$
21. $9p^4q - 10p^3q^2 + 4p^2q^3 + 9q^4$
22. $y^3 - y^2 + \frac{4}{3}y + 0.1$

Guided Solution:
14. $-, -, +$

4.1 Exercise Set

FOR EXTRA HELP MyLab Math

✓ Check Your Understanding

Reading Check Choose from the column on the right the expression that best fits each description.

RC1. A binomial

RC2. A trinomial

RC3. A polynomial with more than one term written in ascending order

RC4. A polynomial in several variables

RC5. The coefficient of the term $7x^5$

RC6. The degree of the term $6xy^2z$

RC7. The constant term in the polynomial $3x^9 - 7x + 5$

RC8. The leading coefficient in the polynomial $5x^3 - 6x + x^4 + 7$

a) $8x - 9x^2 + x^4 + 2x^5$

b) $7x^2yz^3$

c) $3a^4 - 9$

d) $4x^2 + 8x + 2$

e) 1

f) 4

g) 5

h) 7

Concept Check List the like terms in each polynomial.

CC1. $4x + 7x^2 + 2x^3 - 3x - x^2 + 5x^3$

CC2. $2x^2y - 3xy^2 + 4xy^2 + 7x^2y$

CC3. $2a^2 - 7 + a^2 + 3 + 8a^2$

CC4. $x^2 + 6xy - 5x + 2y^2 + 8x$

a Identify the terms, the degree of each term, and the degree of the polynomial. Then identify the leading term, the leading coefficient, and the constant term.

1. $-9x^4 - x^3 + 7x^2 + 6x - 8$

2. $y^3 - 5y^2 + y + 1$

3. $t^3 + 4t^7 + s^2t^4 - 2$

4. $a^2 + 9b^5 - a^4b^3 - 11$

5. $u^7 + 8u^2v^6 + 3uv + 4u - 1$

6. $2p^6 + 5p^4w^4 - 13p^3w + 7p^2 - 10$

Arrange in descending powers of y.

7. $23 - 4y^3 + 7y - 6y^2$

8. $5 - 8y + 6y^2 + 11y^3 - 18y^4$

9. $x^2y^2 + x^3y - xy^3 + 1$

10. $x^3y - x^2y^2 + xy^3 + 6$

11. $2by - 9b^5y^5 - 8b^2y^3$

12. $dy^6 - 2d^7y^2 + 3cy^5 - 7y - 2d$

Arrange in ascending powers of x.

13. $12x + 5 + 8x^5 - 4x^3$

14. $-3x^2 + 8x + 2$

15. $-9x^3y + 3xy^3 + x^2y^2 + 2x^4$

16. $5x^2y^2 - 9xy + 8x^3y^2 - 5x^4$

17. $4ax - 7ab + 4x^6 - 7ax^2$

18. $5xy^8 - 3ax^5 + 4ax^3 - 12a + 5x^5$

b Evaluate each polynomial function for the given values of the variable.

19. $P(x) = 3x^2 - 2x + 5;\ P(4), P(-2), P(0)$

20. $f(x) = -7x^3 + 10x^2 - 13;\ f(4), f(-1), f(0)$

21. $p(x) = 9x^3 + 8x^2 - 4x - 9;\ p(-3), p(0), p(1), p\left(\frac{1}{2}\right)$

22. $Q(x) = 6x^3 - 11x - 4;\ Q(-2), Q\left(\frac{1}{3}\right), Q(0), Q(10)$

23. *Wind Energy.* The number of watts of power P generated by a particular home-sized turbine at a wind speed of x miles per hour can be approximated by the polynomial function

$$P(x) = 0.0157x^3 + 0.1163x^2 - 1.3396x + 3.7063.$$

Estimate the power, in watts, generated by a 25-mph wind.

24. *Golf Ball Stacks.* Each stack of golf balls pictured below is formed by square layers of golf balls. The number N of balls in the stack is given by the polynomial function

$$N(x) = \tfrac{1}{3}x^3 + \tfrac{1}{2}x^2 + \tfrac{1}{6}x,$$

where x is the number of layers. How many golf balls are in each of the stacks?

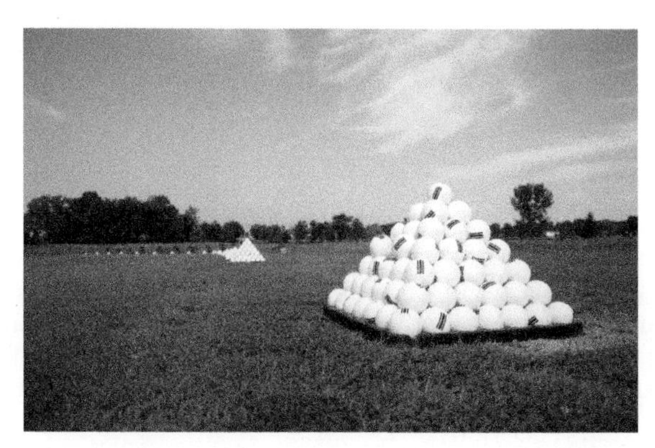

StudioOneNine/Shutterstock

25. *Travel Abroad.* The polynomial function

$$f(t) = 0.0144t^3 - 0.2965t^2 + 1.61t + 10.23$$

can be used to estimate the number of U.S. citizens, in millions, who flew to Europe from the United States t years after 2002.

Data: travel.trade.gov

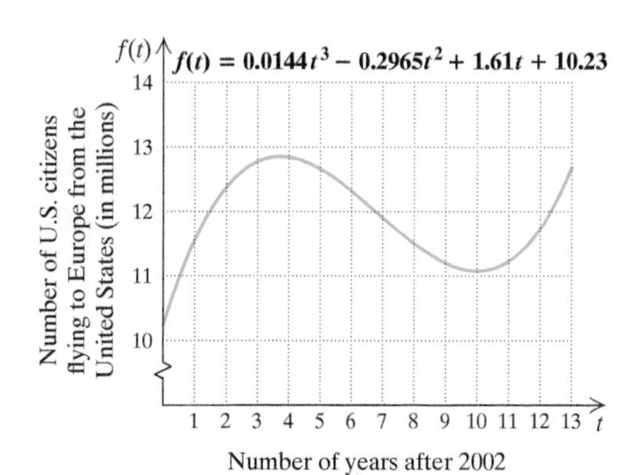

a) Use the graph above to estimate the number of U.S. citizens who flew to Europe in 2007.
b) Use the graph above to estimate the number of U.S. citizens who flew to Europe in 2013.
c) Approximate $f(7)$.
d) Approximate $f(4)$.

26. *Medicine.* Ibuprofen is a medication used to relieve pain. The polynomial function

$$M(t) = 0.5t^4 + 3.45t^3 - 96.65t^2 + 347.7t,$$
$$0 \le t \le 6,$$

can be used to estimate the number of milligrams of ibuprofen in the bloodstream t hours after 400 mg of the medication has been swallowed.

Data: Dr. P. Carey, Burlington, VT

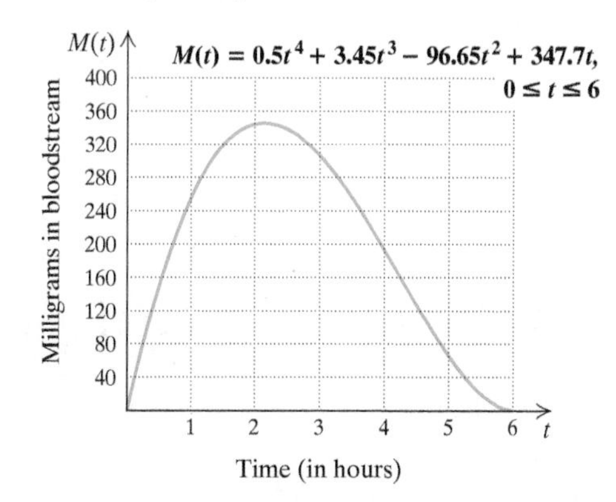

a) Use the graph above to estimate the number of milligrams of ibuprofen in the bloodstream 2 hr after 400 mg has been swallowed.
b) Use the graph above to estimate the number of milligrams of ibuprofen in the bloodstream 4 hr after 400 mg has been swallowed.
c) Approximate $M(5)$.
d) Approximate $M(3)$.

27. *Total Revenue.* A firm is marketing a new style of sunglasses. The firm determines that when it sells x pairs of sunglasses, its total revenue is

$$R(x) = 240x - 0.5x^2 \text{ dollars.}$$

a) What is the total revenue from the sale of 50 pairs of sunglasses?
b) What is the total revenue from the sale of 95 pairs of sunglasses?

28. *Total Cost.* A firm determines that the total cost, in dollars, of producing x pairs of sunglasses is given by

$$C(x) = 5000 + 0.4x^2.$$

a) What is the total cost of producing 50 pairs of sunglasses?
b) What is the total cost of producing 95 pairs of sunglasses?

Total Profit. **Total profit P** is defined as total revenue R minus total cost C, and is given by the function

$$P(x) = R(x) - C(x).$$

For each of the following, find the total profit $P(x)$.

29. $R(x) = 280x - 0.4x^2$, $C(x) = 7000 + 0.6x^2$

30. $R(x) = 280x - 0.7x^2$, $C(x) = 8000 + 0.5x^2$

Magic Number. In a recent Major League Baseball season, the Arizona Diamondbacks were leading the San Francisco Giants for the Western Division championship of the National League. In the following table, the number in parentheses, 18, was the **magic number** for the Diamondbacks. It means that any combination of Diamondbacks wins and Giants losses that totals 18 would ensure the championship for the Diamondbacks. The magic number M is given by the polynomial

$$M = G - W_1 - L_2 + 1,$$

where W_1 is the number of wins for the first-place team, L_2 is the number of losses for the second-place team, and G is the total number of games in the season, which is 162 in the major leagues. When the magic number reaches 1, a tie for the championship is clinched. When the magic number reaches 0, the championship is clinched. For the situation shown below, $G = 162$, $W_1 = 81$, and $L_2 = 64$. Then the magic number is

$$\begin{aligned} M &= G - W_1 - L_2 + 1 \\ &= 162 - 81 - 64 + 1 \\ &= 18. \end{aligned}$$

WEST	W	L	Pct.	GB
Arizona (18)	81	62	.566	–
San Francisco	80	64	.556	$1\frac{1}{2}$
Los Angeles	78	65	.545	3
San Diego	70	73	.490	11
Colorado	62	80	.437	$18\frac{1}{2}$

Magic number in parentheses

31. Compute the magic number for Atlanta.

EAST	W	L	PCT.	GB
Atlanta (?)	78	64	.549	–
Philadelphia	75	68	.524	$3\frac{1}{2}$
New York	71	73	.493	8
Florida	66	77	.462	$12\frac{1}{2}$
Montreal	61	82	.427	$17\frac{1}{2}$

32. Compute the magic number for Houston.

CENTRAL	W	L	PCT.	GB
Houston (?)	84	59	.587	–
St. Louis	78	64	.549	$5\frac{1}{2}$
Chicago	78	65	.545	6
Milwaukee	63	80	.441	21
Cincinnati	58	86	.403	$26\frac{1}{2}$
Pittsburgh	55	88	.385	29

c Collect like terms.

33. $6x^2 - 7x^2 + 3x^2$

34. $-2y^2 - 7y^2 + 5y^2$

35. $7x - 2y - 4x + 6y$

36. $a - 8b - 5a + 7b$

37. $3a + 9 - 2 + 8a - 4a + 7$

38. $13x + 14 - 6 - 7x + 3x + 5$

39. $3a^2b + 4b^2 - 9a^2b - 6b^2$

40. $5x^2y^2 + 4x^3 - 8x^2y^2 - 12x^3$

41. $8x^2 - 3xy + 12y^2 + x^2 - y^2 + 5xy + 4y^2$

42. $a^2 - 2ab + b^2 + 9a^2 + 5ab - 4b^2 + a^2$

43. $4x^2y - 3y + 2xy^2 - 5x^2y + 7y + 7xy^2$

44. $3xy^2 + 4xy - 7xy^2 + 7xy + x^2y$

Add.

45. $(3x^2 + 5y^2 + 6) + (2x^2 - 3y^2 - 1)$

46. $(11y^2 + 6y - 3) + (9y^2 - 2y + 9)$

47. $(2a - c + 3b) + (4a - 2b + 2c)$

48. $(8x + z - 7y) + (5x + 10y - 4z)$

49. $(a^2 - 3b^2 + 4c^2) + (-5a^2 + 2b^2 - c^2)$

50. $(x^2 - 5y^2 - 9z^2) + (-6x^2 + 9y^2 - 2z^2)$

51. $(x^2 + 3x - 2xy - 3) + (-4x^2 - x + 3xy + 2)$

52. $(5a^2 - 3b + ab + 6) + (-a^2 + 8b - 8ab - 4)$

53. $(7x^2y - 3xy^2 + 4xy) + (-2x^2y - xy^2 + xy)$

54. $(7ab - 3ac + 5bc) + (13ab - 15ac - 8bc)$

55. $(2x^2 - y^2 + 3xy) + (7x^2 + 4xy - 9y)$

56. $(ab^2 + 3a^2 - 2a) + (4ab^2 - 7b - a)$

57. $(2r^2 + 12r - 11) + (6r^2 - 2r + 4) + (r^2 - r - 2)$

58. $(5x^2 + 19x - 23) + (-7x^2 - 11x + 12) + (-x^2 - 9x + 8)$

59. $\left(\frac{2}{3}xy + \frac{5}{6}xy^2 + 5.1x^2y\right) + \left(-\frac{4}{5}xy + \frac{3}{4}xy^2 - 3.4x^2y\right)$

60. $\left(\frac{1}{8}xy - \frac{3}{5}x^3y^2 + 4.3y^3\right) + \left(-\frac{1}{3}xy - \frac{3}{4}x^3y^2 - 2.9y^3\right)$

d Write two equivalent expressions for the opposite of the polynomial.

61. $5x^3 - 7x^2 + 3x - 6$

62. $-8y^4 - 18y^3 + 4y - 9$

63. $-13y^2 + 6ay^4 - 5by^2$

64. $9ax^5y^3 - 8by^5 - abx - 16ay$

Subtract.

65. $(7x - 2) - (-4x + 5)$

66. $(8y + 1) - (-5y - 2)$

67. $(-3x^2 + 2x + 9) - (x^2 + 5x - 4)$

68. $(-9y^2 + 4y + 8) - (4y^2 + 2y - 3)$

69. $(5a + c - 2b) - (3a + 2b - 2c)$

70. $(z + 8x - 4y) - (4x + 6y - 3z)$

71. $(3x^2 - 2x - x^3) - (5x^2 - x^3 - 8x)$

72. $(8y^2 - 4y^3 - 3y) - (3y^2 - 9y - 7y^3)$

73. $(5a^2 + 4ab - 3b^2) - (9a^2 - 4ab + 2b^2)$

74. $(9y^2 - 14yz - 8z^2) - (12y^2 - 8yz + 4z^2)$

75. $(6ab - 4a^2b + 6ab^2) - (3ab^2 - 10ab - 12a^2b)$

76. $(10xy - 4x^2y^2 - 3y^3) - (-9x^2y^2 + 4y^3 - 7xy)$

77. $(0.09y^4 - 0.052y^3 + 0.93) - (0.03y^4 - 0.084y^3 + 0.94y^2)$

78. $(1.23x^4 - 3.122x^3 + 1.11x) - (0.79x^4 - 8.734x^3 + 0.04x^2 + 6.71x)$

79. $\left(\frac{5}{8}x^4 - \frac{1}{4}x^2 - \frac{1}{2}\right) - \left(-\frac{3}{8}x^4 + \frac{3}{4}x^2 + \frac{1}{2}\right)$

80. $\left(\frac{5}{6}y^4 - \frac{1}{2}y^2 - 7.8y + \frac{1}{3}\right) - \left(-\frac{3}{8}y^4 + \frac{3}{4}y^2 + 3.4y - \frac{1}{5}\right)$

Skill Maintenance

Solve.

81. $-3x - 7 = x - 5$ [1.1d]

82. $\frac{1}{3}t - \frac{1}{2} = \frac{1}{6}t$ [1.1d]

83. $x - (7 - x) = 2(x + 3)$ [1.1d]

84. $-9y \leq -18$ [1.4c]

Graph using the slope and the y-intercept. [2.5b]

85. $y = \frac{4}{3}x + 2$

86. $y = -0.4x + 1$

87. $y = 0.4x - 3$

88. $y = -\frac{2}{3}x - 4$

Synthesis

89. *Triangular Layers.* The number of spheres in a triangular pyramid with x triangular layers is given by the function

$$N(x) = \frac{1}{6}x^3 + \frac{1}{2}x^2 + \frac{1}{3}x.$$

The volume of a sphere of radius r is given by the function

$$V(r) = \frac{4}{3}\pi r^3,$$

where π can be approximated as 3.14.

Chocolate Heaven has a window display of truffles piled in triangular pyramid formations, each 5 layers deep. If the diameter of each truffle is 3 cm, find the volume of chocolate in each triangular pyramid in the display.

90. *Surface Area.* Find a polynomial function that gives the outside surface area of a box like this one, with dimensions as shown.

91. A student who is trying to graph $f(x) = 0.05x^4 - x^2 + 5$ gets the following screen. How can the student tell at a glance that a mistake has been made?

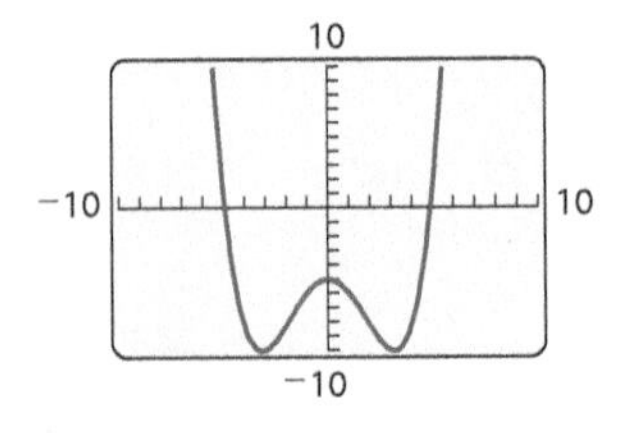

Perform the indicated operations. Assume that the exponents are natural numbers.

92. $(3x^{6a} - 5x^{5a} + 4x^{3a} + 8) - (2x^{6a} + 4x^{4a} + 3x^{3a} + 2x^{2a})$

93. $(47x^{4a} + 3x^{3a} + 22x^{2a} + x^a + 1) + (37x^{3a} + 8x^{2a} + 3)$

Multiplication of Polynomials

4.2

OBJECTIVES

a Multiply any two polynomials.

b Use the FOIL method to multiply two binomials.

c Use a rule to square a binomial.

d Use a rule to multiply a sum and a difference of the same two terms.

e For functions f described by second-degree polynomials, find and simplify notation like $f(a + h)$ and $f(a + h) - f(a)$.

a MULTIPLICATION OF ANY TWO POLYNOMIALS

Use the distributive laws to find equivalent expressions by multiplying. [J16]

Multiply.

1. $3(x - y)$

2. $-\frac{1}{2}(6a - 10b)$

Answers: 1. $3x - 3y$ **2.** $-3a + 5b$

MyLab Math VIDEO

Multiplying Monomials

Monomials are expressions like $10x^2$, $8x^5$, and $-7a^2b^3$. To multiply monomials, we first multiply their coefficients. Then we multiply the variables using the commutative and associative laws and the rules for exponents.

EXAMPLES Multiply and simplify.

1. $(10x^2)(8x^5) = (10 \cdot 8)(x^2 \cdot x^5)$
$= 80x^{2+5}$ Adding exponents
$= 80x^7$

2. $(-8x^4y^7)(5x^3y^2) = (-8 \cdot 5)(x^4 \cdot x^3)(y^7 \cdot y^2)$
$= -40x^{4+3}y^{7+2}$ Adding exponents
$= -40x^7y^9$

Do Exercises 1–3. ▶

Multiply.

1. $(9y^2)(-2y)$

2. $(4x^3y)(6x^5y^2)$

3. $(-5xy^7z^4)(18x^3y^2z^8)$

Multiplying Monomials and Binomials

The distributive law is the basis for multiplying polynomials other than monomials. We first multiply a monomial and a binomial.

EXAMPLE 3 Multiply: $2x(3x - 5)$.

$2x \cdot (3x - 5) = 2x \cdot 3x - 2x \cdot 5$ Using the distributive law
$= 6x^2 - 10x$ Multiplying monomials ■

EXAMPLE 4 Multiply: $3a^2b(a^2 - b^2)$.

$3a^2b \cdot (a^2 - b^2) = 3a^2b \cdot a^2 - 3a^2b \cdot b^2$ Using the distributive law
$= 3a^4b - 3a^2b^3$

Do Exercises 4 and 5. ▶

Multiply.

4. $(-3y)(2y + 6)$

5. $(2xy)(4y^2 - 5)$

Answers

1. $-18y^3$ **2.** $24x^8y^3$ **3.** $-90x^4y^9z^{12}$
4. $-6y^2 - 18y$ **5.** $8xy^3 - 10xy$

Multiplying Binomials

Next, we multiply two binomials. To do so, we use the distributive law twice, first considering one of the binomials as a single expression and multiplying it by each term of the other binomial.

EXAMPLE 5 Multiply: $(3y^2 + 4)(y^2 - 2)$.

$$(3y^2 + 4)(y^2 - 2) = (3y^2 + 4) \cdot y^2 - (3y^2 + 4) \cdot 2 \quad \text{Using the distributive law}$$
$$= [3y^2 \cdot y^2 + 4 \cdot y^2] - [3y^2 \cdot 2 + 4 \cdot 2] \quad \text{Using the distributive law}$$
$$= 3y^2 \cdot y^2 + 4 \cdot y^2 - 3y^2 \cdot 2 - 4 \cdot 2 \quad \text{Removing parentheses}$$
$$= 3y^4 + 4y^2 - 6y^2 - 8 \quad \text{Multiplying the monomials}$$
$$= 3y^4 - 2y^2 - 8 \quad \text{Collecting like terms}$$

Multiply.

6. $(5x^2 - 4)(x + 3)$

7. $(2y + 3)(3y - 4)$

◀ **Do Exercises 6 and 7.**

Multiplying Any Two Polynomials

EXAMPLE 6 Multiply: $(p + 2)(p^4 - 2p^3 + 3)$.

By the distributive law, we have

$$(p + 2)(p^4 - 2p^3 + 3)$$
$$= (p + 2)(p^4) - (p + 2)(2p^3) + (p + 2)(3)$$
$$= p(p^4) + 2(p^4) - p(2p^3) - 2(2p^3) + p(3) + 2(3)$$
$$= p^5 + 2p^4 - 2p^4 - 4p^3 + 3p + 6$$
$$= p^5 - 4p^3 + 3p + 6. \quad \text{Collecting like terms}$$

Multiply.

8. $(p - 3)(p^3 + 4p^2 - 5)$

9. $(2x^3 + 4x - 5)(x - 4)$

◀ **Do Exercises 8 and 9.**

PRODUCT OF TWO POLYNOMIALS

To multiply two polynomials P and Q, select one of the polynomials, say P. Then multiply each term of P by every term of Q and collect like terms.

We can use columns when doing long multiplications. We multiply each term at the top by every term at the bottom, keeping like terms in columns and *adding spaces for missing terms.* Then we add.

EXAMPLE 7 Multiply: $(5x^3 + 3x^2 + x - 4)(-2x^2 + 3x + 6)$.

$$\begin{array}{rrrrrrl}
 & & 5x^3 & +\ 3x^2 & +\ x & -\ 4 & \\
 & & & -2x^2 & +\ 3x & +\ 6 & \\
\hline
 & & 30x^3 & +\ 18x^2 & +\ 6x & -\ 24 & \text{Multiplying by 6} \\
 & 15x^4 & +\ 9x^3 & +\ 3x^2 & -\ 12x & & \text{Multiplying by } 3x \\
-10x^5 & -\ 6x^4 & -\ 2x^3 & +\ 8x^2 & & & \text{Multiplying by } -2x^2 \\
\hline
-10x^5 & +\ 9x^4 & +\ 37x^3 & +\ 29x^2 & -\ 6x & -\ 24 &
\end{array}$$

Answers

6. $5x^3 + 15x^2 - 4x - 12$
7. $6y^2 + y - 12$
8. $p^4 + p^3 - 12p^2 - 5p + 15$
9. $2x^4 - 8x^3 + 4x^2 - 21x + 20$

EXAMPLE 8 Multiply: $(5x^3 + x - 4)(-2x^2 + 3x + 6)$.

$$\begin{array}{rrrrrr} & & 5x^3 & & +\ x & -\ 4 \\ & & & -2x^2 & +\ 3x & +\ 6 \\ \hline & & 30x^3 & & +\ 6x & -\ 24 \\ & 15x^4 & & +\ 3x^2 & -\ 12x & \\ -10x^5 & & -\ 2x^3 & +\ 8x^2 & & \\ \hline -10x^5 & +\ 15x^4 & +\ 28x^3 & +\ 11x^2 & -\ 6x & -\ 24 \end{array}$$

Multiplying by 6
Multiplying by $3x$
Multiplying by $-2x^2$

Do Exercises 10–12.

Multiply. Use columns.

10. $(-4x^3 + 5x^2 - 2x + 1) \times (-2x^2 - 3x + 6)$

11. $(-4x^3 - 2x + 1) \times (-2x^2 - 3x + 6)$

12. $(a^2 - 2ab + b^2) \times (a^3 + 3ab - b^2)$

b PRODUCT OF TWO BINOMIALS USING THE FOIL METHOD

We now consider some **special products**. Let's find a faster special-product rule for the product of two binomials. Consider $(x + 7)(x + 4)$. We multiply each term of $(x + 7)$ by each term of $(x + 4)$:

$$(x + 7)(x + 4) = x \cdot x + x \cdot 4 + 7 \cdot x + 7 \cdot 4.$$

This multiplication illustrates a pattern that occurs whenever two binomials are multiplied:

First terms, Outside terms, Inside terms, Last terms

$$(x + 7)(x + 4) = x \cdot x + 4x + 7x + 7(4) = x^2 + 11x + 28.$$

This special method of multiplying is called the **FOIL method**. Keep in mind that this method is based on the distributive law.

A visualization of $(x + 7)(x + 4)$ using areas

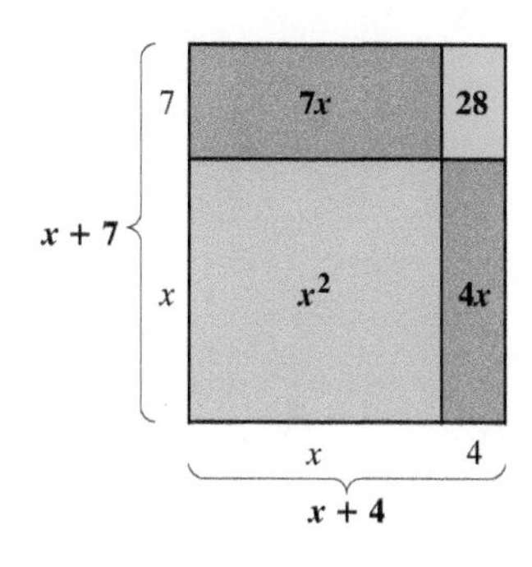

THE FOIL METHOD

To multiply two binomials, $A + B$ and $C + D$, multiply the **F**irst terms AC, the **O**utside terms AD, the **I**nside terms BC, and then the **L**ast terms BD. Then collect like terms, if possible.

$$(A + B)(C + D) = AC + AD + BC + BD$$

1. Multiply First terms: AC.
2. Multiply Outside terms: AD.
3. Multiply Inside terms: BC.
4. Multiply Last terms: BD.

↓
FOIL

F, L, $(A + B)(C + D)$, I, O

EXAMPLE 9 Multiply: $(x + 5)(x - 8)$.

F O I L

$$(x + 5)(x - 8) = x^2 - 8x + 5x - 40$$
$$= x^2 - 3x - 40 \quad \text{Collecting like terms}$$

We write the result in descending order since the original binomials are in descending order.

Answers

10. $8x^5 + 2x^4 - 35x^3 + 34x^2 - 15x + 6$
11. $8x^5 + 12x^4 - 20x^3 + 4x^2 - 15x + 6$
12. $a^5 - 2a^4b + 3a^3b + a^3b^2 - 7a^2b^2 + 5ab^3 - b^4$

EXAMPLES

$$\text{F} \quad \text{O} \quad \text{I} \quad \text{L}$$

10. $(3xy + 2x)(x^2 + 2xy^2) = 3x^3y + 6x^2y^3 + 2x^3 + 4x^2y^2$

11. $(2x - 3)(y + 2) = 2xy + 4x - 3y - 6$

12. $(2x + 3y)(x - 4y) = 2x^2 - 8xy + 3xy - 12y^2$
$= 2x^2 - 5xy - 12y^2$ Collecting like terms

Do Exercises 13–15.

Multiply.

13. $(y - 4)(y + 10)$

14. $(p + 5q)(2p - 3q)$

15. $(x^2y + 2x)(xy^2 + y^2)$

c SQUARES OF BINOMIALS

MyLab Math ANIMATION

A visualization of $(A + B)^2$ using areas

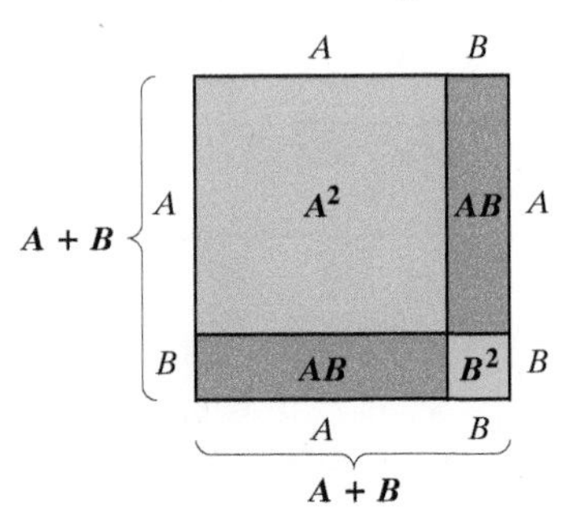

We can use the FOIL method to develop special products for the square of a binomial:

$$(A + B)^2 = (A + B)(A + B) = A^2 + AB + AB + B^2 = A^2 + 2AB + B^2;$$

$$(A - B)^2 = (A - B)(A - B) = A^2 - AB - AB + B^2 = A^2 - 2AB + B^2.$$

SQUARE OF A BINOMIAL

The **square of a binomial** is the square of the first term, plus twice the product of the two terms, plus the square of the last term.

$$(A + B)^2 = A^2 + 2AB + B^2;$$
$$(A - B)^2 = A^2 - 2AB + B^2$$

Caution!

In general,

$$(AB)^2 = A^2B^2, \quad \text{but} \quad (A + B)^2 \neq A^2 + B^2.$$

EXAMPLES Multiply.

$(A - B)^2 = A^2 - 2\,A\,B + B^2$

13. $(y - 5)^2 = y^2 - 2(y)(5) + 5^2$
$= y^2 - 10y + 25$

$(A + B)^2 = A^2 + 2\,A\,B + B^2$

14. $(2x + 3y)^2 = (2x)^2 + 2(2x)(3y) + (3y)^2$
$= 4x^2 + 12xy + 9y^2$

15. $(3x^2 + 5xy^2)^2 = (3x^2)^2 + 2(3x^2)(5xy^2) + (5xy^2)^2$
$= 9x^4 + 30x^3y^2 + 25x^2y^4$

16. $(\frac{1}{2}a^2 - b^3)^2 = (\frac{1}{2}a^2)^2 - 2(\frac{1}{2}a^2)(b^3) + (b^3)^2$
$= \frac{1}{4}a^4 - a^2b^3 + b^6$

Do Exercises 16–19.

Multiply.

16. $(a - b)^2$

17. $(x + 8)^2$

18. $(3x - 7)^2$

19. $\left(m^3 + \frac{1}{4}n\right)^2$ GS

$= (\quad)^2 + 2(\quad)\left(\frac{1}{4}n\right) + \left(\frac{1}{4}n\right)^2$

$= m^{\quad} + \quad m^3n + \quad n^2$

Answers

13. $y^2 + 6y - 40$ **14.** $2p^2 + 7pq - 15q^2$
15. $x^3y^3 + x^2y^3 + 2x^2y^2 + 2xy^2$
16. $a^2 - 2ab + b^2$ **17.** $x^2 + 16x + 64$
18. $9x^2 - 42x + 49$ **19.** $m^6 + \frac{1}{2}m^3n + \frac{1}{16}n^2$

Guided Solution:

19. $m^3, m^3, 6, \frac{1}{2}, \frac{1}{16}$

d PRODUCTS OF SUMS AND DIFFERENCES

Another special case of a product of two binomials is the product of a sum and a difference. Note the following:

$$(A + B)(A - B) = \overset{\text{F}}{A^2} - \overset{\text{O}}{AB} + \overset{\text{I}}{AB} - \overset{\text{L}}{B^2} = A^2 - B^2.$$

PRODUCT OF A SUM AND A DIFFERENCE

The product of the sum and the difference of the same two terms is the square of the first term minus the square of the second term (the difference of their squares).

$$(A + B)(A - B) = A^2 - B^2$$

This is called a **difference of squares.**

EXAMPLES Multiply. (Say the rule as you work.)

$$(A + B)(A - B) = A^2 - B^2$$

17. $(y + 5)(y - 5) = y^2 - 5^2 = y^2 - 25$

18. $(2xy^2 + 3x)(2xy^2 - 3x) = (2xy^2)^2 - (3x)^2 = 4x^2y^4 - 9x^2$

19. $(0.2t - 1.4m)(0.2t + 1.4m) = (0.2t)^2 - (1.4m)^2 = 0.04t^2 - 1.96m^2$

20. $\left(\frac{2}{3}n - m^2\right)\left(\frac{2}{3}n + m^2\right) = \left(\frac{2}{3}n\right)^2 - (m^2)^2 = \frac{4}{9}n^2 - m^4$

Do Exercises 20–23.

Multiply.

20. $(x + 8)(x - 8)$

21. $(4y - 7)(4y + 7)$

22. $(2.8a + 4.1b)(2.8a - 4.1b)$

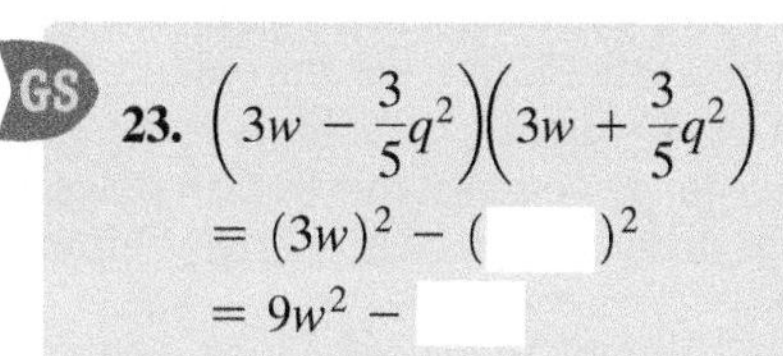

GS **23.** $\left(3w - \frac{3}{5}q^2\right)\left(3w + \frac{3}{5}q^2\right)$

$= (3w)^2 - (\quad)^2$

$= 9w^2 - \quad$

EXAMPLES Multiply.

21. $(5y + 4 + 3x)(5y + 4 - 3x) = (5y + 4)^2 - (3x)^2$

$= 25y^2 + 40y + 16 - 9x^2$

Here we treat the binomial $5y + 4$ as the first expression, A, and $3x$ as the second, B.

22. $(3xy^2 + 4y)(-3xy^2 + 4y) = (4y + 3xy^2)(4y - 3xy^2)$

$= (4y)^2 - (3xy^2)^2$

$= 16y^2 - 9x^2y^4$

Do Exercises 24 and 25.

Multiply.

24. $(2x + 3 - 5y)(2x + 3 + 5y)$

25. $(7x^2y + 2y)(-2y + 7x^2y)$

Try to multiply polynomials mentally. When several types are mixed, first check to see what types of polynomials are to be multiplied. Then use the quickest method. Sometimes we might use more than one method to find a product. Remember that FOIL *always* works for multiplying binomials!

Answers

20. $x^2 - 64$ **21.** $16y^2 - 49$
22. $7.84a^2 - 16.81b^2$ **23.** $9w^2 - \frac{9}{25}q^4$
24. $4x^2 + 12x + 9 - 25y^2$
25. $49x^4y^2 - 4y^2$

Guided Solution:

23. $\frac{3}{5}q^2, \frac{9}{25}q^4$

EXAMPLE 23 Multiply: $(s - 5t)(s + 5t)(s^2 - 25t^2)$.

We first note that $s - 5t$ and $s + 5t$ can be multiplied using the rule $(A - B)(A + B) = A^2 - B^2$. Then we have the product of two identical binomials, so we square, using $(A - B)^2 = A^2 - 2AB + B^2$.

$$
\begin{aligned}
&(s - 5t)(s + 5t)(s^2 - 25t^2) \\
&\quad = (s^2 - 25t^2)(s^2 - 25t^2) && \text{Using } (A - B)(A + B) = A^2 - B^2 \\
&\quad = (s^2 - 25t^2)^2 \\
&\quad = (s^2)^2 - 2(s^2)(25t^2) + (25t^2)^2 && \text{Using } (A - B)^2 = A^2 - 2AB + B^2 \\
&\quad = s^4 - 50s^2t^2 + 625t^4
\end{aligned}
$$

◀ **Do Exercise 26.**

26. Multiply:
$(3x + 2y)(3x - 2y)(9x^2 + 4y^2)$.

Answer

26. $81x^4 - 16y^4$

CALCULATOR CORNER

Checking Using Tables and Graphs A partial check of operations with polynomials in one variable can be done using tables or graphs. For example, a table set in AUTO mode can be used to check the addition $(-3x^3 + 2x - 4) + (4x^3 + 3x^2 + 2) = x^3 + 3x^2 + 2x - 2$. To do so, we enter $y_1 = (-3x^3 + 2x - 4) + (4x^3 + 3x^2 + 2)$ and $y_2 = x^3 + 3x^2 + 2x - 2$. If the addition has been done correctly, the values of y_1 and y_2 will be the same regardless of the table settings used.

X	Y1	Y2
−2	−2	−2
−1	−2	−2
0	−2	−2
1	4	4
2	22	22
3	58	58
4	118	118

X=−2

To check an operation graphically, we compare two graphs. This is often easier to do when two graph styles are used. To check the product $(x + 3)(x - 2) = x^2 + x - 6$, we first use MODE to select the SEQUENTIAL mode. Next, on the Y= screen, we enter $y_1 = (x + 3)(x - 2)$ and $y_2 = x^2 + x - 6$. We will select the line-graph style for y_1 and the path style for y_2. To select these graph styles, we use ◀ to position the cursor over the icon to the left of the equation and press ENTER repeatedly until the desired style of icon appears, as shown below.

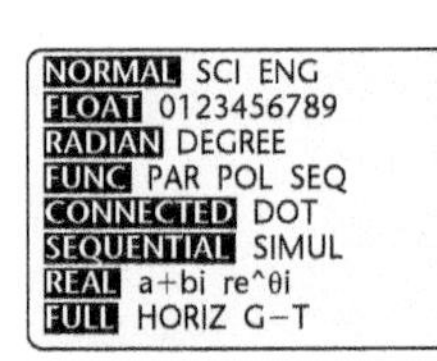

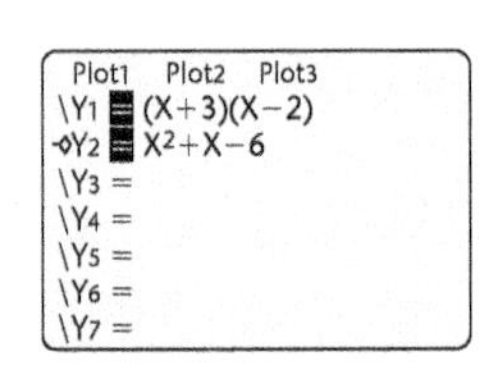

$y_1 = (x + 3)(x - 2)$,
$y_2 = x^2 + x - 6$

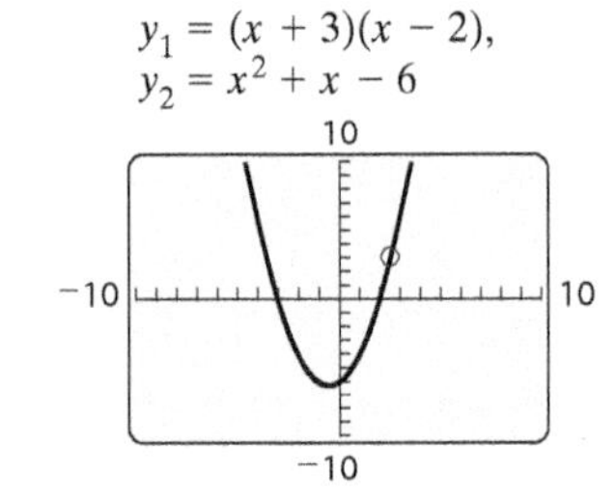

The graphing calculator will graph y_1 first as a solid curve. Then it will graph y_2 as the circular cursor traces the leading edge of the graph, allowing us to determine visually whether the graphs coincide. In this case, the graphs appear to coincide, so the multiplication is probably correct.

EXERCISES: Use a table or graphs to determine whether each sum, difference, or product is correct.

1. $(x + 4)(x + 3) = x^2 + 7x + 12$
2. $(3x + 2)(x - 1) = 3x^2 + x - 2$
3. $(4x - 1)(x - 5) = 4x^2 - 21x + 5$
4. $(2x - 1)(3x - 4) = 6x^2 - 11x - 4$
5. $(x - 1)(x - 1) = x^2 + 1$
6. $(x - 2)(x + 2) = x^2 - 4$
7. $(x^3 - 2x^2 + 3x - 7) + (3x^2 - 4x + 5) = x^3 + x^2 - x - 2$
8. $(2x^2 + 3x - 6) + (5x^2 - 7x + 4) = 7x^2 + 4x - 2$
9. $(7x^5 + 2x^4 - 5x) - (-x^5 - 2x^4 + 3) = 8x^5 + 4x^4 - 5x - 3$
10. $(3x^4 - 2x^2 - 1) - (2x^4 - 3x^2 - 4) = x^4 + x^2 - 5$

e USING FUNCTION NOTATION

ALGEBRAIC ▶◀ GRAPHICAL CONNECTION

Let's stop for a moment and look back at what we have done in this section. We have shown, for example, that

$$(x - 2)(x + 2) = x^2 - 4,$$

that is, $x^2 - 4$ and $(x - 2)(x + 2)$ are equivalent expressions.

From the viewpoint of functions, if

$$f(x) = (x - 2)(x + 2)$$

and

$$g(x) = x^2 - 4,$$

then for any given input x, the outputs $f(x)$ and $g(x)$ are identical. Thus the graphs of these functions are identical and we say that f and g represent the same function. Functions like these are graphed in detail in Chapter 7.

x	$f(x)$	$g(x)$
3	5	5
2	0	0
1	−3	−3
0	−4	−4
−1	−3	−3
−2	0	0
−3	5	5

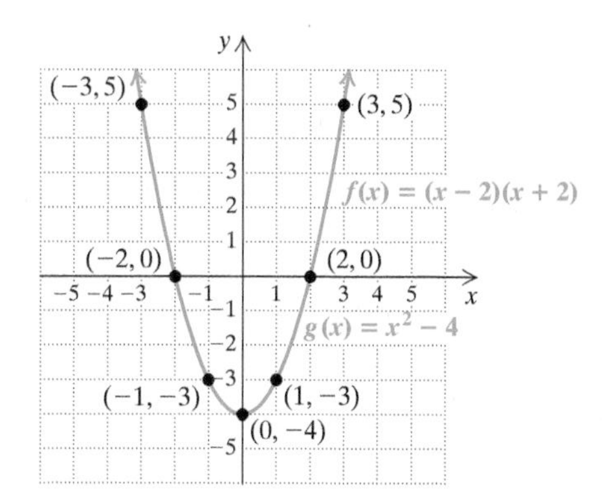

Our work with multiplying can be used when manipulating functions.

EXAMPLE 24 Given $f(x) = x^2 - 4x + 5$, find and simplify $f(a + 3)$ and $f(a + h) - f(a)$.

To find $f(a + 3)$, we replace x with $a + 3$. Then we simplify:

$$\begin{aligned} f(a + 3) &= (a + 3)^2 - 4(a + 3) + 5 \\ &= a^2 + 6a + 9 - 4a - 12 + 5 = a^2 + 2a + 2. \end{aligned}$$

To find $f(a + h) - f(a)$, we replace x with $a + h$ for $f(a + h)$ and x with a for $f(a)$. Then we simplify:

$$\begin{aligned} f(a + h) - f(a) &= [(a + h)^2 - 4(a + h) + 5] - [a^2 - 4a + 5] \\ &= a^2 + 2ah + h^2 - 4a - 4h + 5 - a^2 + 4a - 5 \\ &= 2ah + h^2 - 4h. \end{aligned}$$

Do Exercise 27. ▶

27. Given $f(x) = x^2 + 2x - 7$, find and simplify $f(a + 1)$ and $f(a + h) - f(a)$.

Answer

27. $f(a + 1) = a^2 + 4a - 4;$
$f(a + h) - f(a) = 2ah + h^2 + 2h$

4.2 Exercise Set

FOR EXTRA HELP MyLab Math

✓ Check Your Understanding

Reading Check Determine whether each statement is true or false.

RC1. We use the distributive law to multiply polynomials.

RC2. The square of a binomial is a binomial.

RC3. We can use FOIL to multiply any two binomials.

RC4. The product of two monomials is always a monomial.

Concept Check Choose from the column on the right an equivalent expression for each product. Not every choice will be used.

CC1. $(x + 3)^2$

CC2. $(x - 3)^2$

CC3. $(x + 3)(x - 3)$

CC4. $(a + 10)(a - 10)$

CC5. $(a - 10)^2$

CC6. $(a + 10)^2$

a) $x^2 + 9$
b) $x^2 - 9$
c) $x^2 + 6x + 9$
d) $x^2 - 6x + 9$
e) $a^2 + 100$
f) $a^2 - 100$
g) $a^2 + 20a + 100$
h) $a^2 - 20a + 100$

a Multiply.

1. $8y^2 \cdot 3y$

2. $-5x^2 \cdot 6xy$

3. $2x(-10x^2y)$

4. $-7ab^2(4a^2b^2)$

5. $(5x^5y^4)(-2xy^3)$

6. $(2a^2bc^2)(-3ab^5c^4)$

7. $2z(7 - x)$

8. $4a(a^2 - 3a)$

9. $6ab(a + b)$

10. $2xy(2x - 3y)$

11. $5cd(3c^2d - 5cd^2)$

12. $a^2(2a^2 - 5a^3)$

13. $(5x + 2)(3x - 1)$

14. $(2a - 3b)(4a - b)$

15. $(s + 3t)(s - 3t)$

16. $(y + 4)(y - 4)$

17. $(x - y)(x - y)$

18. $(a + 2b)(a + 2b)$

19. $(x^3 + 8)(x^3 - 5)$

20. $(2x^4 - 7)(3x^3 + 5)$

21. $(a^2 - 2b^2)(a^2 - 3b^2)$

22. $(2m^2 - n^2)(3m^2 - 5n^2)$

23. $(x - 4)(x^2 + 4x + 16)$

24. $(y + 3)(y^2 - 3y + 9)$

25. $(x + y)(x^2 - xy + y^2)$

26. $(a - b)(a^2 + ab + b^2)$

27. $(a^2 + a - 1)(a^2 + 4a - 5)$

28. $(x^2 - 2x + 1)(x^2 + x + 2)$

29. $(4a^2b - 2ab + 3b^2)(ab - 2b + a)$

30. $(2x^2 + y^2 - 2xy)(x^2 - 2y^2 - xy)$

31. $\left(x + \frac{1}{4}\right)\left(x + \frac{1}{4}\right)$

32. $\left(b - \frac{1}{3}\right)\left(b - \frac{1}{3}\right)$

33. $\left(\frac{1}{2}x - \frac{2}{3}\right)\left(\frac{1}{4}x + \frac{1}{3}\right)$

34. $\left(\frac{2}{3}a + \frac{1}{6}b\right)\left(\frac{1}{3}a - \frac{5}{6}b\right)$

35. $(1.3x - 4y)(2.5x + 7y)$

36. $(40a - 0.24b)(0.3a + 10b)$

b, **c** Multiply.

37. $(a + 8)(a + 5)$

38. $(x + 2)(x + 3)$

39. $(y + 7)(y - 4)$

40. $(y - 2)(y + 3)$

41. $\left(3a + \frac{1}{2}\right)^2$

42. $\left(2x - \frac{1}{3}\right)^2$

43. $(x - 2y)^2$

44. $(2s + 3t)^2$

45. $\left(b - \frac{1}{3}\right)\left(b - \frac{1}{2}\right)$

46. $\left(x - \frac{1}{2}\right)\left(x - \frac{1}{4}\right)$

47. $(2x + 9)(x + 2)$

48. $(3b + 2)(2b - 5)$

49. $(20a - 0.16b)^2$

50. $(10p^2 + 2.3y)^2$

51. $(2x - 3y)(2x + y)$

52. $(2a - 3b)(2a - b)$

53. $(x^3 + 2)^2$

54. $(y^4 - 7)^2$

55. $(2x^2 - 3y^2)^2$

56. $(3s^2 + 4t^2)^2$

57. $(a^3b^2 + 1)^2$

58. $(x^2y - xy^3)^2$

59. $(0.1a^2 - 5b)^2$

60. $(6m + 0.45p^2)^2$

61. *Compound Interest.* Suppose that P dollars is invested in a savings account at interest rate i, compounded annually, for 2 years. The amount A in the account after 2 years is given by

$$A = P(1 + i)^2.$$

Find an equivalent expression for A without parentheses.

62. *Compound Interest.* Suppose that P dollars is invested in a savings account at interest rate i, compounded semiannually, for 1 year. The amount A in the account after 1 year is given by

$$A = P\left(1 + \frac{i}{2}\right)^2.$$

Find an equivalent expression for A without parentheses.

d Multiply.

63. $(d + 8)(d - 8)$

64. $(y - 3)(y + 3)$

65. $(2c + 3)(2c - 3)$

66. $(1 - 2x)(1 + 2x)$

67. $(6m - 5n)(6m + 5n)$

68. $(3x + 7y)(3x - 7y)$

69. $(x^2 + yz)(x^2 - yz)$

70. $(2a^2 + 5ab)(2a^2 - 5ab)$

71. $(-mn + m^2)(mn + m^2)$

72. $(1.6 + cw)(-1.6 + cw)$

73. $(-3pt + 4p^2)(4p^2 + 3pt)$

74. $(-10xy + 5x^2)(5x^2 + 10xy)$

75. $\left(\frac{1}{2}p - \frac{2}{3}n\right)\left(\frac{1}{2}p + \frac{2}{3}n\right)$

76. $\left(\frac{3}{5}ab + 4c\right)\left(\frac{3}{5}ab - 4c\right)$

77. $(x + 1)(x - 1)(x^2 + 1)$

78. $(y - 2)(y + 2)(y^2 + 4)$

79. $(a - b)(a + b)(a^2 - b^2)$

80. $(2x - y)(2x + y)(4x^2 - y^2)$

81. $(a + b + 1)(a + b - 1)$

82. $(m + n + 2)(m + n - 2)$

83. $(2x + 3y + 4)(2x + 3y - 4)$

84. $(3a - 2b + c)(3a - 2b - c)$

e For each of the following functions, find $f(t - 1)$, $f(p + 1)$, $f(a + h) - f(a)$, $f(t - 2) + c$, and $f(a) + 5$.

85. $f(x) = 5x + x^2$

86. $f(x) = 4x + 2x^2$

87. $f(x) = 3x^2 - 7x + 8$

88. $f(x) = 3x^2 - 4x + 7$

89. $f(x) = 5x - x^2$

90. $f(x) = 4x - 2x^2$

91. $f(x) = 4 + 3x - x^2$

92. $f(x) = 2 - 4x - 3x^2$

Skill Maintenance

Graph. [2.1c, d], [2.2c]

93. $f(x) = \frac{2}{3}x - 1$

94. $g(x) = |x| - 1$

95. $g(x) = \frac{4}{x - 3}$

96. $f(x) = 1 - x^2$

Solve. [3.4b]

97. *Auto Travel.* Rachel leaves on a business trip, forgetting her laptop computer. Her sister discovers Rachel's laptop 2 hr later, and knowing that Rachel needs it for her sales presentation and that Rachel normally travels at a speed of 55 mph, she decides to follow her at a speed of 75 mph. After how long will Rachel's sister catch up with her?

98. *Air Travel.* An airplane flew for 5 hr against a 20-mph headwind. The return trip with the wind took 4 hr. Find the speed of the plane in still air.

Solve. [3.2a], [3.3a]

99. $5x + 9y = 2,$
$4x - 9y = 10$

100. $x + 4y = 13,$
$5x - 7y = -16$

101. $2x - 3y = 1,$
$4x - 6y = 2$

102. $9x - 8y = -2,$
$3x + 2y = 3$

Synthesis

103. Use the TABLE feature or the GRAPH feature of a graphing calculator to check your answers to Exercises 28, 40, and 77.

104. Use the TABLE feature or the GRAPH feature of a graphing calculator to determine whether each of the following is correct.

a) $(x - 1)^2 = x^2 - 1$
b) $(x - 2)(x + 3) = x^2 + x - 6$
c) $(x - 1)^3 = x^3 - 3x^2 + 3x - 1$
d) $(x + 1)^4 = x^4 + 1$

Multiply. Assume that variables in exponents represent natural numbers.

105. $(z^{n^2})^{n^3}(z^{4n^3})^{n^2}$

106. $y^3z^n(y^{3n}z^3 - 4yz^{2n})$

107. $(r^2 + s^2)^2(r^2 + 2rs + s^2)(r^2 - 2rs + s^2)$

108. $(y - 1)^6(y + 1)^6$

109. $\left(3x^5 - \frac{5}{11}\right)^2$

110. $(4x^2 + 2xy + y^2)(4x^2 - 2xy + y^2)$

111. $(x^a + y^b)(x^a - y^b)(x^{2a} + y^{2b})$

112. $\left(x - \frac{1}{7}\right)\left(x^2 + \frac{1}{7}x + \frac{1}{49}\right)$

113. $(x - 1)(x^2 + x + 1)(x^3 + 1)$

114. $(x^{a-b})^{a+b}$

4.3

OBJECTIVES

a Factor polynomials whose terms have a common factor.

b Factor certain polynomials with four terms by grouping.

SKILL REVIEW *Use the distributive law to factor.* [J16]

Factor.

1. $2y - 2$

2. $15y - 10x + 25$

Answers:
1. $2(y - 1)$
2. $5(3y - 2x + 5)$

MyLab Math VIDEO

Caution!

Be careful not to confuse terms with factors! The terms of $x^2 - 9$ are x^2 and -9. Terms are used to form sums. Factors of $x^2 - 9$ are $x - 3$ and $x + 3$. Factors are used to form products.

Introduction to Factoring

Factoring is the reverse of multiplication. To **factor** an expression is to find an equivalent expression that is a product. For example, we can *factor* $x^2 - 9$: $x^2 - 9 = (x + 3)(x - 3)$. We say that $x + 3$ and $x - 3$ are **factors** of $x^2 - 9$ and that $(x + 3)(x - 3)$ is a **factorization.**

FACTOR AND FACTORIZATION

To **factor** a polynomial is to express it as a product.

A **factor** of a polynomial P is a polynomial that can be used to express P as a product.

A **factorization** of a polynomial P is an expression that names P as a product of factors.

a TERMS WITH COMMON FACTORS

To multiply a monomial and a polynomial with more than one term, we multiply each term by the monomial using the distributive laws. To factor, we do the reverse. We express a polynomial as a product using the distributive laws in reverse. Compare.

Multiply

$$5x(x^2 - 3x + 1) = 5x \cdot x^2 - 5x \cdot 3x + 5x \cdot 1 = 5x^3 - 15x^2 + 5x$$

Factor

$$5x^3 - 15x^2 + 5x = 5x \cdot x^2 - 5x \cdot 3x + 5x \cdot 1 = 5x(x^2 - 3x + 1)$$

EXAMPLE 1 Factor: $4y^2 - 8$.

$4y^2 - 8 = 4 \cdot y^2 - 4 \cdot 2$ 4 is the largest common factor.

$= 4(y^2 - 2)$ Factoring out the common factor 4

In some cases, there is more than one common factor. In Example 2 below, for instance, the common factors are $5, x, x^2, x^3, 5x, 5x^2$, and $5x^3$. If there is more than one common factor, we generally choose the one with the largest coefficient and the largest exponent.

EXAMPLES Factor.

2. $20x^4 - 5x^3 = 5x^3 \cdot 4x - 5x^3 \cdot 1$ The factor of 1 is essential.

$= 5x^3(4x - 1)$ Multiply mentally to check your answer.

3. $12x^2y - 20x^3y = 4x^2y(3 - 5x)$

The polynomials in Examples 1–3 have been **factored completely**. They cannot be factored further. The factors in the resulting factorization are said to be **prime polynomials**.

EXAMPLE 4 Factor: $10a^6b^2 - 4a^5b^3 + 2a^4b^4$.

First, we look for the greatest positive common factor in the coefficients:

$$10, -4, 2 \longrightarrow \text{Greatest common factor} = 2.$$

Second, we look for the greatest common factor in the powers of a:

$$a^6, a^5, a^4 \longrightarrow \text{Greatest common factor} = a^4.$$

Third, we look for the greatest common factor in the powers of b:

$$b^2, b^3, b^4 \longrightarrow \text{Greatest common factor} = b^2.$$

Thus, $2a^4b^2$ is the greatest common factor of the given polynomial. Then

$$10a^6b^2 - 4a^5b^3 + 2a^4b^4 = 2a^4b^2 \cdot 5a^2 - 2a^4b^2 \cdot 2ab + 2a^4b^2 \cdot b^2$$
$$= 2a^4b^2(5a^2 - 2ab + b^2).$$

Do Exercises 1–5.

Factor.

1. $3x^2 - 6$
2. $4x^5 - 8x^3$
3. $9y^4 - 15y^3 + 3y^2$
4. $6x^2y - 21x^3y^2 + 3x^2y^3$
5. $30a^8b^4c^5 - 50a^6b^5c^{11}$

When the leading coefficient is a negative number, we generally factor out a negative coefficient.

EXAMPLES Factor out a common factor with a negative coefficient.

5. $-4x - 24 = -4(x + 6)$ *Check*: $-4(x + 6) = -4x - 24$
6. $-2x^2 + 6x - 10 = -2(x^2 - 3x + 5)$

Do Exercises 6 and 7.

Factor out a common factor with a negative coefficient.

6. $-8x + 32$
7. $-3x^2 - 15x + 9$

EXAMPLE 7 *Height of a Rocket.* A water rocket is launched upward with an initial velocity of 96 ft/sec. Its height h, in feet, after t seconds is given by the function $h(t) = -16t^2 + 96t$.

a) Find an equivalent expression for $h(t)$ by factoring out a common factor with a negative coefficient.

b) Check your factoring by evaluating both expressions for $h(t)$ at $t = 2$.

a) We factor out $-16t$ as follows:

$$h(t) = -16t^2 + 96t = -16t(t - 6).$$

b) We check as follows:

$$h(2) = -16 \cdot 2^2 + 96 \cdot 2 = 128;$$
$$h(2) = -16 \cdot 2(2 - 6) = 128. \quad \text{Using the factorization}$$

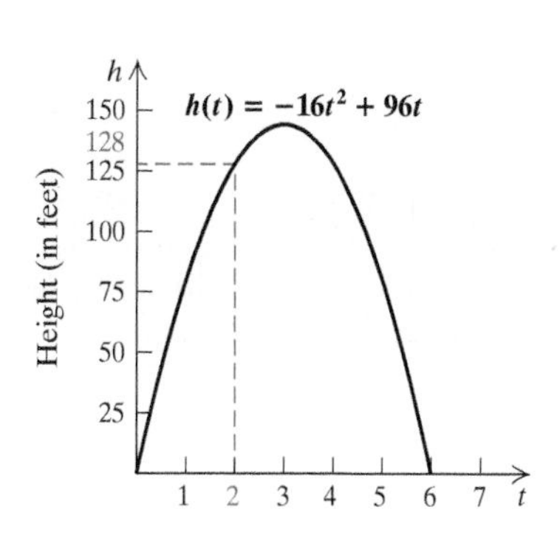

Do Exercise 8.

8. *Height of a Softball.* Suppose that a softball is thrown upward with an initial velocity of 64 ft/sec. Its height h, in feet, after t seconds is given by the function

$$h(t) = -16t^2 + 64t.$$

a) Find an equivalent expression for $h(t)$ by factoring out a common factor with a negative coefficient.

b) Check your factoring by evaluating both expressions for $h(t)$ at $t = 1$.

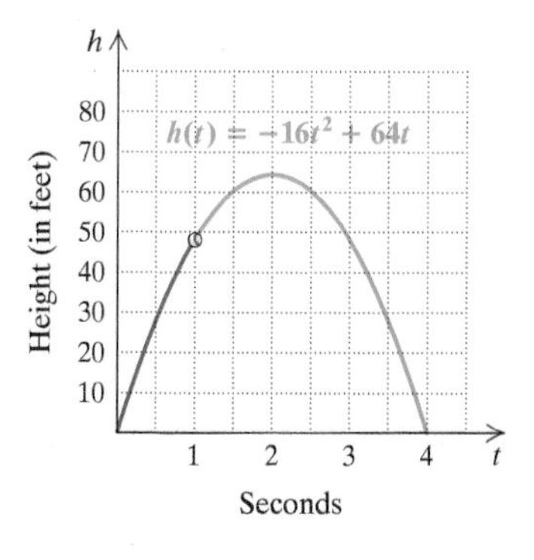

Answers

1. $3(x^2 - 2)$ **2.** $4x^3(x^2 - 2)$
3. $3y^2(3y^2 - 5y + 1)$
4. $3x^2y(2 - 7xy + y^2)$
5. $10a^6b^4c^5(3a^2 - 5bc^6)$
6. $-8(x - 4)$ **7.** $-3(x^2 + 5x - 3)$
8. (a) $h(t) = -16t(t - 4)$;
(b) $h(1) = 48$ in each expression

Factor.

9. $(p + q)(x + 2) + (p + q)(x + y)$

10. $(y + 3)(y - 21) + (y + 3)(y + 10)$ GS

The common binomial factor is ____.

$(y + 3)(y - 21) + (y + 3)(y + 10)$
$= (\quad)(y - 21 + y + 10)$
$= (y + 3)(\quad)$

b FACTORING BY GROUPING

In expressions of four or more terms, there may be a *common binomial factor*. We proceed as in the following examples.

EXAMPLE 8 Factor: $(a - b)(x + 5) + (a - b)(x - y^2)$.

$$(a - b)(x + 5) + (a - b)(x - y^2) = (a - b)[(x + 5) + (x - y^2)]$$
$$= (a - b)(2x + 5 - y^2)$$

Do Exercises 9 and 10.

In Example 9, we group before we factor.

EXAMPLE 9 Factor: $y^3 + 3y^2 + 4y + 12$.

$y^3 + 3y^2 + 4y + 12 = (y^3 + 3y^2) + (4y + 12)$ Grouping
$= y^2(y + 3) + 4(y + 3)$ Factoring each binomial
$= (y + 3)(y^2 + 4)$ Factoring out the common factor $y + 3$

EXAMPLE 10 Factor: $3x^3 - 6x^2 - x + 2$.

First, we factor out the greatest common factor in the first two terms:

$$3x^3 - 6x^2 = 3x^2(x - 2).$$

Next, we look at the third and fourth terms to see if we can factor them in order to have $x - 2$ as a factor. We see that if we factor out -1, we get $x - 2$:

$$-x + 2 = -1 \cdot (x - 2).$$

Finally, we factor out the common factor $x - 2$:

$3x^3 - 6x^2 - x + 2 = (3x^3 - 6x^2) + (-x + 2)$
$= 3x^2(x - 2) + (-x + 2)$
$= 3x^2(x - 2) - 1(x - 2)$ *Check*: $-1(x - 2) = -x + 2$
$= (x - 2)(3x^2 - 1).$ Factoring out the common factor $x - 2$

EXAMPLE 11 Factor: $4x^3 - 15 + 20x^2 - 3x$.

$4x^3 - 15 + 20x^2 - 3x = 4x^3 + 20x^2 - 3x - 15$ Rearranging
$= 4x^2(x + 5) - 3(x + 5)$ *Check*: $-3(x + 5) = -3x - 15$
$= (x + 5)(4x^2 - 3)$ Factoring out $x + 5$

Not all polynomials with four terms can be factored by grouping. An example is $x^3 + x^2 + 3x - 3$. Note that in a grouping like $x^2(x + 1) + 3(x - 1)$, the expressions $x + 1$ and $x - 1$ are not the same. No grouping allows us to factor out a common binomial.

Do Exercises 11 and 12.

Factor by grouping, if possible.

11. $5y^3 + 2y^2 - 10y - 4$ GS
$= y^2(\quad) - 2(\quad)$
$= (\quad)(y^2 - 2)$

12. $x^3 + 5x^2 + 4x - 20$

Answers

9. $(p + q)(2x + y + 2)$ 10. $(y + 3)(2y - 11)$
11. $(5y + 2)(y^2 - 2)$
12. Cannot be factored by grouping

Guided Solutions:
10. $y + 3, y + 3, 2y - 11$
11. $5y + 2, 5y + 2, 5y + 2$

4.3 Exercise Set

FOR EXTRA HELP MyLab Math

✓ Check Your Understanding

Reading Check Choose from the column on the right the word that best completes each statement.

RC1. To factor a polynomial is to express it as a(n) ________.

RC2. In the expression $x(x - 2)$, x and $x - 2$ are ________.

RC3. The expression $x(x - 2)$ is a(n) ________ of $x^2 - 2x$.

RC4. A polynomial that cannot be factored is said to be ________.

RC5. The expression $4x$ is a(n) ________ factor of the terms of the polynomial $8x + 12x^3$.

RC6. When we factor by grouping, we look for a common ________ factor.

binomial
common
factorization
factors
prime
product

Concept Check Determine the greatest positive common factor of the terms in each polynomial. Do not factor.

CC1. $12x + 20y - 48$

CC2. $8x^7 + 9x^{10} - 25x^3$

CC3. $36a^4 + 60a^7 - 6a^3$

CC4. $28a^3b^4c^4 - 40a^5bc^9 + 16a^2b^7c^6$

a Factor.

1. $6a^2 + 3a$

2. $4x^2 + 2x$

3. $x^3 + 9x^2$

4. $y^3 + 8y^2$

5. $8x^2 - 4x^4$

6. $6x^2 + 3x^4$

7. $4x^2y - 12xy^2$

8. $5x^2y^3 + 15x^3y^2$

9. $3y^2 - 3y - 9$

10. $5x^2 - 5x + 15$

11. $4ab - 6ac + 12ad$

12. $8xy + 10xz - 14xw$

13. $10a^4 + 15a^2 - 25a - 30$

14. $12t^5 - 20t^4 + 8t^2 - 16$

15. $15x^2y^5z^3 - 12x^4y^4z^7$

16. $21a^3b^5c^7 - 14a^7b^6c^2$

17. $14a^4b^3c^5 + 21a^3b^5c^4 - 35a^4b^4c^3$

18. $9x^3y^6z^2 - 12x^4y^4z^4 + 15x^2y^5z^3$

Factor out a common factor with a negative coefficient.

19. $-5x - 45$

20. $-3t + 18$

21. $-6a - 84$

22. $-8t + 40$

23. $-2x^2 + 2x - 24$

24. $-2x^2 + 16x - 20$

25. $-3y^2 + 24y$

26. $-7x^2 - 56y$

27. $-a^4 + 2a^3 - 13a^2 - 1$

28. $-m^3 - m^2 + m - 2$

29. $-3y^3 + 12y^2 - 15y + 24$

30. $-4m^4 - 32m^3 + 64m - 12$

31. *Volume of a Propane Gas Tank.* A propane gas tank is shaped like a circular cylinder with half of a sphere at each end. The volume of the tank with length h and radius r of the cylindrical section is given by the polynomial

$$\pi r^2 h + \tfrac{4}{3}\pi r^3.$$

Find an equivalent expression by factoring out a common factor.

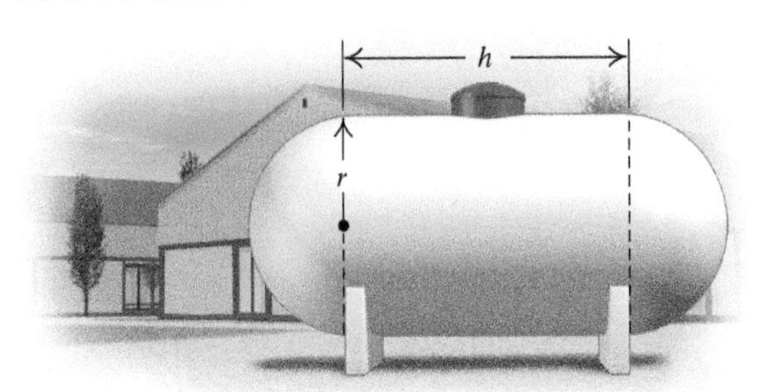

32. *Triangular Layers.* The stack of truffles shown below is formed by triangular layers of truffles. The number N of truffles in the stack is given by the polynomial function

$$N(x) = \tfrac{1}{6}x^3 + \tfrac{1}{2}x^2 + \tfrac{1}{3}x,$$

where x is the number of layers. Find an equivalent expression for $N(x)$ by factoring out a common factor.

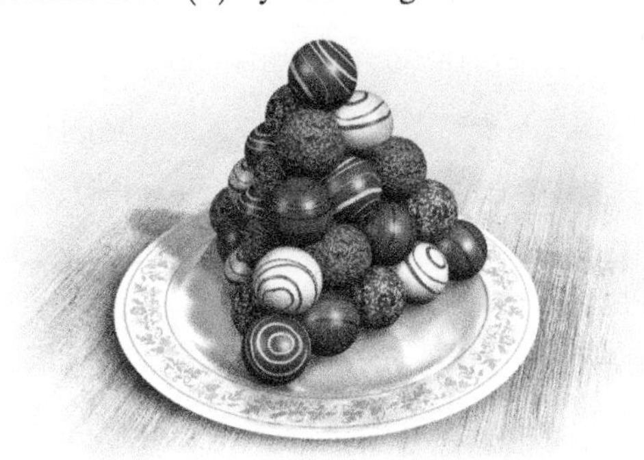

33. *Height of a Baseball.* A baseball is popped up with an upward velocity of 72 ft/sec. Its height h, in feet, after t seconds is given by

$$h(t) = -16t^2 + 72t.$$

a) Find an equivalent expression for $h(t)$ by factoring out a common factor with a negative coefficient.

b) Perform a partial check of part (a) by evaluating both expressions for $h(t)$ at $t = 2$.

34. *Number of Diagonals.* The number of diagonals of a polygon having n sides is given by the polynomial function

$$P(n) = \tfrac{1}{2}n^2 - \tfrac{3}{2}n.$$

Find an equivalent expression for $P(n)$ by factoring out a common factor.

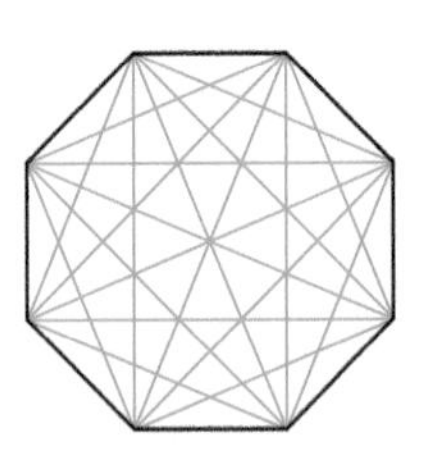

35. *Total Revenue.* Perfect Sound is marketing a new kind of home theater chair. The firm determines that when it sells x chairs, the total revenue R is given by the polynomial function

$$R(x) = 280x + 0.4x^2 \text{ dollars.}$$

Find an equivalent expression for $R(x)$ by factoring out $0.4x$.

36. *Total Cost.* Perfect Sound determines that the total cost C of producing x home theater chairs is given by the polynomial function

$$C(x) = 0.18x + 0.6x^2.$$

Find an equivalent expression for $C(x)$ by factoring out $0.6x$.

b Factor.

37. $a(b - 2) + c(b - 2)$

38. $a(x^2 - 3) - 2(x^2 - 3)$

39. $(x - 2)(x + 5) + (x - 2)(x + 8)$

40. $(m - 4)(m + 3) + (m - 4)(m - 3)$

41. $y^8 - 7y^7 + y - 7$

42. $b^5 - 3b^4 + b - 3$

43. $ac + ad + bc + bd$

44. $xy + xz + wy + wz$

45. $b^3 - b^2 + 2b - 2$

46. $y^3 - y^2 + 3y - 3$

47. $y^3 + 8y^2 - 5y - 40$

48. $t^3 + 6t^2 - 2t - 12$

49. $24x^3 + 72x - 36x^2 - 108$

50. $10a^3 + 50a - 15a^2 - 75$

51. $a^4 - a^3 + a^2 + a$

52. $p^6 + p^5 - p^3 + p^2$

53. $2y^4 + 6y^2 - 5y^2 - 15$

54. $2xy + x^2y - 6 - 3x$

Skill Maintenance

Solve.

55. $|x - 3| = 10$ [1.6c]

56. $|2a - 3| = |3a + 5|$ [1.6d]

57. $|2 - x| \leq 12$ [1.6e]

58. $|3y - 7| + 2 > 8$ [1.6e]

59. $8 \leq x - 7 \leq 10$ [1.5a]

60. $-2 < -3x + 1 < 0$ [1.5a]

61. $2x - 7 > 6$ *or* $3x + 1 < 2$ [1.5b]

62. $-m + 3 \leq 2$ *or* $m + 5 > 5m - 1$ [1.5b]

Synthesis

Complete each of the following.

63. $x^5y^4 + \_\_\_\_ = x^3y(\_\_\_\_ + xy^5)$

64. $a^3b^7 - \_\_\_\_ = \_\_\_\_(ab^4 - c^2)$

Factor.

65. $rx^2 - rx + 5r + sx^2 - sx + 5s$

66. $3a^2 + 6a + 30 + 7a^2b + 14ab + 70b$

67. $a^4x^4 + a^4x^2 + 5a^4 + a^2x^4 + a^2x^2 + 5a^2 + 5x^4 + 5x^2 + 25$ (*Hint*: Use three groups of three.)

Factor out the smallest power of x in each of the following.

68. $x^{1/2} + 5x^{3/2}$

69. $x^{1/3} - 7x^{4/3}$

70. $x^{3/4} + x^{1/2} - x^{1/4}$

71. $x^{1/3} - 5x^{1/2} + 3x^{3/4}$

Factor. Assume that all exponents are natural numbers.

72. $2x^{3a} + 8x^a + 4x^{2a}$

73. $3a^{n+1} + 6a^n - 15a^{n+2}$

74. $4x^{a+b} + 7x^{a-b}$

75. $7y^{2a+b} - 5y^{a+b} + 3y^{a+2b}$

4.4 Factoring Trinomials: $x^2 + bx + c$

OBJECTIVE

a Factor trinomials of the type $x^2 + bx + c$.

a FACTORING TRINOMIALS: $x^2 + bx + c$

We now consider factoring trinomials of the type $x^2 + bx + c$. We use a process that is based on the FOIL method.

Constant Term Positive

Recall the FOIL method of multiplying two binomials:

$$\begin{aligned}(x+3)(x+5) &= \overset{F}{x^2} + \overset{O}{5x} + \overset{I}{3x} + \overset{L}{15} \\ &= x^2 + 8x + 15.\end{aligned}$$

The product is a trinomial. To factor $x^2 + 8x + 15$, we think of FOIL in reverse. Since the first term of the trinomial is x^2, the first term of each binomial factor is x. We want to find numbers p and q such that

$$x^2 + 8x + 15 = (x + p)(x + q).$$

We now look for two numbers whose product is 15 and whose sum is 8. Those numbers are 3 and 5. Thus the factorization is

$$(x + 3)(x + 5), \quad \text{or} \quad (x + 5)(x + 3).$$

Thus we can factor using the following general form in reverse:

$$(x + p)(x + q) = x^2 + (p + q)x + pq.$$

EXAMPLE 1 Factor: $x^2 + 9x + 8$.

Think of FOIL in reverse. The first term of each factor is x. We are looking for numbers p and q such that

$$x^2 + 9x + 8 = (x + p)(x + q) = x^2 + (p + q)x + pq.$$

We look for two numbers p and q whose product is 8 and whose sum is 9. Since both 8 and 9 are positive, we need consider only positive factors.

PAIRS OF FACTORS	SUMS OF FACTORS
2, 4	6
1, 8	9

← The numbers we need are 1 and 8.

The factorization is $(x + 1)(x + 8)$. We can check by multiplying:

$$(x + 1)(x + 8) = x^2 + 9x + 8.$$

Do Exercises 1 and 2.

When the constant term of a trinomial is positive, we look for two factors with the same sign (both positive or both negative). The sign is that of the middle term.

Factor. Check by multiplying.

1. $x^2 + 5x + 6$

2. $y^2 + 7y + 10$ GS

Because both 7 and 10 are positive, we need consider only ____ factors of 10.

PAIRS OF FACTORS	SUMS OF FACTORS
1, 10	☐
2, 5	☐

The numbers we need are 2 and ____. Thus,

$y^2 + 7y + 10 = (y + 2)(\quad)$.

Answers

1. $(x + 2)(x + 3)$ **2.** $(y + 2)(y + 5)$

Guided Solution:

2. positive, 11, 7, 5, $y + 5$

EXAMPLE 2 Factor: $20 - 9y + y^2$.

We begin by writing the trinomial in descending order:

$$y^2 - 9y + 20.$$

Since the constant term, 20, is positive and the coefficient of the middle term, -9, is negative, we look for a factorization of 20 in which both factors are negative. Their sum must be -9.

PAIRS OF FACTORS	SUMS OF FACTORS
−1, −20	−21
−2, −10	−12
−4, −5	−9

← The numbers we need are −4 and −5.

The factorization is $(y - 4)(y - 5)$.

Do Exercises 3 and 4.

Factor.

3. $m^2 - 8m + 12$

4. $24 - 11t + t^2$

Constant Term Negative

When the constant term of a trinomial is negative, we look for two factors whose product is negative. One of them must be positive and the other negative. Their sum must be the coefficient of the middle term.

EXAMPLE 3 Factor: $x^3 - x^2 - 30x$.

Always look first for the largest common factor. This time x is the common factor. We first factor it out:

$$x^3 - x^2 - 30x = x(x^2 - x - 30).$$

Now consider $x^2 - x - 30$. Since the constant term, -30, is negative, the pair of factors we seek will contain one positive factor and one negative factor. The sum of the factors must be -1, the coefficient of the middle term, so the negative factor must have the larger absolute value. Thus we consider only pairs of factors in which the negative factor has the larger absolute value.

PAIRS OF FACTORS	SUMS OF FACTORS
1, −30	−29
2, −15	−13
3, −10	−7
5, −6	−1

← The numbers we need are 5 and −6.

The factorization of $x^2 - x - 30$ is $(x + 5)(x - 6)$. But do not forget the common factor! The factorization of the original trinomial is

$$x(x + 5)(x - 6).$$

Do Exercises 5–7.

5. **a)** Factor: $x^2 - x - 20$.
   b) Explain why you would not consider these pairs of factors in factoring $x^2 - x - 20$.

PAIRS OF FACTORS	PRODUCTS OF FACTORS
1, 20	
2, 10	
4, 5	
−1, −20	
−2, −10	
−4, −5	

Factor.

6. $x^3 - 3x^2 - 54x$

7. $2x^3 - 2x^2 - 84x$

Answers

3. $(m - 2)(m - 6)$ **4.** $(t - 3)(t - 8)$, or $(3 - t)(8 - t)$ **5. (a)** $(x - 5)(x + 4)$; **(b)** The product of each pair is positive. **6.** $x(x - 9)(x + 6)$ **7.** $2x(x - 7)(x + 6)$

EXAMPLE 4 Factor: $x^2 + 17x - 110$.

Since the constant term, -110, is negative, factorizations of -110 will have one positive factor and one negative factor. The sum of the factors must be 17, so the positive factor must have the larger absolute value.

PAIRS OF FACTORS	SUMS OF FACTORS
$-1, 110$	109
$-2, 55$	53
$-5, 22$	17
$-10, 11$	1

We consider only pairs of factors in which the positive term has the larger absolute value.

The numbers we need are -5 and 22.

The factorization is $(x - 5)(x + 22)$.

Do Exercises 8–10.

Factor.

8. $x^3 + 4x^2 - 12x$

9. $y^2 - 4y - 12$

10. $x^2 - 110 - x$

Some trinomials are not factorable.

EXAMPLE 5 Factor: $x^2 - x - 7$.

There are no factors of -7 whose sum is -1. This trinomial is *not* factorable into binomials.

Do Exercise 11.

11. Factor: $x^2 + x - 5$.

To factor $x^2 + bx + c$:

1. First arrange in descending order.
2. Use a trial-and-error procedure that looks for factors of c whose sum is b.
   - If c is positive, then the signs of the factors are the same as the sign of b.
   - If c is negative, then one factor is positive and the other is negative. (If the sum of the two factors is the opposite of b, changing the signs of each factor will give the desired factors whose sum is b.)
3. Check your result by multiplying.

The procedure considered here can also be applied to a trinomial with more than one variable.

EXAMPLE 6 Factor: $x^2 - 2xy - 48y^2$.

We look for numbers p and q such that

$$x^2 - 2xy - 48y^2 = (x + py)(x + qy).$$

Our thinking is much the same as if we were factoring $x^2 - 2x - 48$. We look for factors of -48 whose sum is -2. Those factors are 6 and -8. Then

$$x^2 - 2xy - 48y^2 = (x + 6y)(x - 8y).$$

We can check by multiplying.

Do Exercises 12 and 13.

Factor.

12. $x^2 - 5xy + 6y^2$

13. $p^2 - 6pq - 16q^2$

Answers

8. $x(x + 6)(x - 2)$ **9.** $(y - 6)(y + 2)$
10. $(x + 10)(x - 11)$ **11.** Not factorable
12. $(x - 2y)(x - 3y)$ **13.** $(p - 8q)(p + 2q)$

EXAMPLE 7 Factor: $x^4 + 2x^2 - 15$.

We look for numbers p and q such that

$$x^4 + 2x^2 - 15 = (x^2 + p)(x^2 + q).$$

The constant term is negative and the middle term is positive. Thus we look for pairs of factors of -15, such that the positive factor has the larger absolute value and the sum of the factors is 2. Those factors are -3 and 5. The desired factorization is

$$(x^2 - 3)(x^2 + 5).$$

Do Exercises 14 and 15. ▶

Leading Coefficient of −1

EXAMPLE 8 Factor: $-x^2 + 5x + 14$.

Note that this trinomial has a leading coefficient of -1. In such a case, we begin by factoring out a -1:

$$-x^2 + 5x + 14 = -1(x^2 - 5x - 14)$$
$$= -1(x - 7)(x + 2). \quad \text{Factoring } x^2 - 5x - 14$$

We can also express this answer two other ways by multiplying through either binomial by -1. Thus each of the following is a correct answer:

$$-x^2 + 5x + 14 = -1(x - 7)(x + 2);$$
$$= (-x + 7)(x + 2); \quad \text{Multiplying } x - 7 \text{ by } -1$$
$$= (x - 7)(-x - 2). \quad \text{Multiplying } x + 2 \text{ by } -1$$

Do Exercises 16 and 17. ▶

Factor.

14. $x^4 - 9x^2 + 14$

15. $y^6 + y^3 - 6$

We look for numbers p and q such that

$y^6 + y^3 - 6 = (y^3 + p)(y^3 + q)$.

The product of p and q is ____.

The sum of p and q is ____.

The factors we want are -2 and ____.

The factorization is

$(y^3 - 2)(y^3 +$ ____ $)$.

Factor.

16. $10 - 3x - x^2$

17. $-x^2 + 8x - 16$

Answers

14. $(x^2 - 2)(x^2 - 7)$ **15.** $(y^3 - 2)(y^3 + 3)$
16. $-(x + 5)(x - 2)$, or $(-x - 5)(x - 2)$, or $(x + 5)(-x + 2)$ **17.** $-(x - 4)(x - 4)$, or $(-x + 4)(x - 4)$

Guided Solution:
15. $-6, 1, 3, 3$

4.4 Exercise Set

MyLab Math

✓ Check Your Understanding

Reading Check Choose from the column on the right the phrase that best completes each statement.

RC1. To factor $x^2 + 19x - 20$, we look for a factorization of -20 in which ____.

RC2. To factor $x^2 - 11x - 12$, we look for a factorization of -12 in which ____.

RC3. To factor $x^2 + 7x + 12$, we look for a factorization of 12 in which ____.

RC4. To factor $x^2 - 12x + 20$, we look for a factorization of 20 in which ____.

a) both factors are positive

b) both factors are negative

c) the positive factor has the greater absolute value

d) the negative factor has the greater absolute value

Concept Check List all pairs of positive factors of each number.

CC1. 36

CC2. 18

CC3. 54

CC4. 72

CC5. 80

CC6. 84

a Factor.

1. $x^2 + 13x + 36$

2. $x^2 + 9x + 18$

3. $t^2 - 8t + 15$

4. $y^2 - 10y + 21$

5. $x^2 - 8x - 33$

6. $t^2 - 15 - 2t$

7. $2y^2 - 16y + 32$

8. $2a^2 - 20a + 50$

9. $p^2 + 3p - 54$

10. $m^2 + m - 72$

11. $12x + x^2 + 27$

12. $10y + y^2 + 24$

13. $y^2 - \frac{2}{3}y + \frac{1}{9}$

14. $p^2 + \frac{2}{5}p + \frac{1}{25}$

15. $t^2 - 4t + 3$

16. $y^2 - 14y + 45$

17. $5x + x^2 - 14$

18. $x + x^2 - 90$

19. $x^2 + 5x + 6$

20. $y^2 + 8y + 7$

21. $56 + x - x^2$

22. $32 + 4y - y^2$

23. $32y + 4y^2 - y^3$

24. $56x + x^2 - x^3$

25. $x^4 + 11x^2 - 80$

26. $y^4 + 5y^2 - 84$

27. $x^2 - 3x + 7$

28. $x^2 + 12x + 13$

29. $x^2 + 12xy + 27y^2$

30. $p^2 - 5pq - 24q^2$

31. $2x^2 - 8x - 90$

32. $3x^2 - 21x - 90$

33. $-z^2 + 36 - 9z$

34. $24 - a^2 - 10a$

35. $x^4 + 50x^2 + 49$

36. $p^4 + 80p^2 + 79$

37. $x^6 + 11x^3 + 18$

38. $x^6 - x^3 - 42$

39. $x^8 - 11x^4 + 24$

40. $x^8 - 7x^4 + 10$

41. $y^2 - 0.8y + 0.16$

42. $a^2 + 1.4a + 0.49$

43. $12 - b^{10} - b^{20}$

44. $8 - 7t^{15} - t^{30}$

Skill Maintenance

Solve. [3.4a]

45. *Mixing Rice.* Countryside Rice is 90% white rice and 10% wild rice. Mystic Rice is 50% wild rice. How much of each type should be used in order to create a 25-lb batch of rice that is 35% wild rice?

46. *Wages.* Takako worked a total of 17 days last month at her father's restaurant. She earned $50 per day during the week and $60 per day during the weekend. Last month Takako earned $940. How many weekdays did she work?

Determine whether each of the following is the graph of a function. [2.2d]

47.

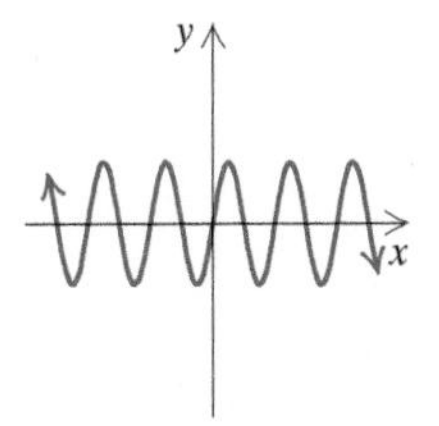

48.

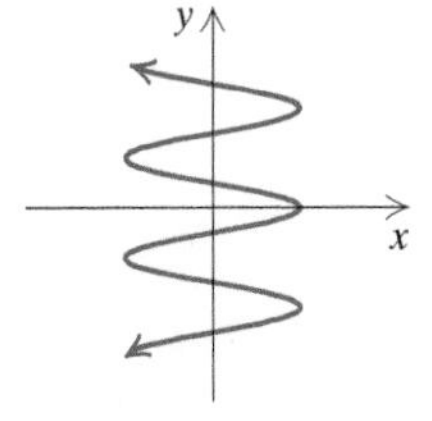

49.

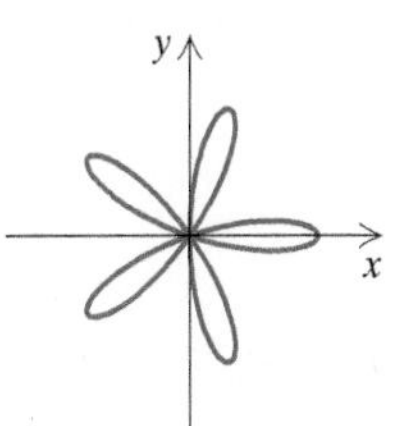

50.

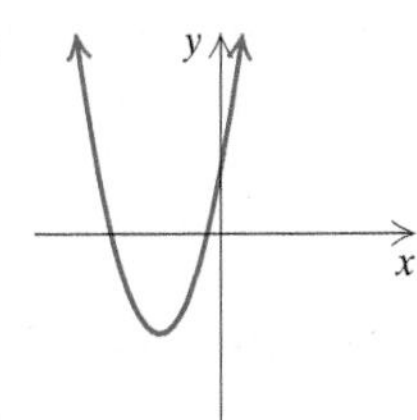

Find the domain of f. [2.3a]

51. $f(x) = x^2 - 2$

52. $f(x) = 3 - 2x$

53. $f(x) = \dfrac{3}{4x - 7}$

54. $f(x) = 3 - |x|$

Synthesis

55. Find all integers m for which $x^2 + mx + 75$ can be factored.

56. Find all integers q for which $x^2 + qx - 32$ can be factored.

57. One of the factors of $x^2 - 345x - 7300$ is $x + 20$. Find the other factor.

58. Use the TABLE feature and the GRAPH feature of a graphing calculator to check your answers to Exercises 1–6.

Mid-Chapter Review

Concept Reinforcement

Determine whether each statement is true or false.

______________ 1. The polynomial $5x + 2x^2 - 4x^3$ can be factored. [4.3a]

______________ 2. The expression $17x^{-2}y^3$ is a monomial. [4.1a]

______________ 3. The degree of a polynomial is the same as the degree of the leading term. [4.1a]

______________ 4. The opposite of $-x^2 + x$ is $x - x^2$. [4.1d]

______________ 5. The binomial $144 - x^2$ is a difference of squares. [4.2d]

Guided Solutions

GS Fill in each blank with the number or expression that creates a correct solution.

6. Multiply: $(8w - 3)(w - 5)$. [4.2b]

$$(8w - 3)(w - 5) = \overset{F}{(8w)(\quad)} + \overset{O}{(8w)(-5)} + \overset{I}{(-3)(w)} + \overset{L}{(-3)(\quad)}$$
$$= 8w^2 - \quad w - 3w + \quad = 8w^2 - \quad w + \quad$$

7. Factor: $c^3 - 8c^2 - 48c$. [4.3a], [4.4a]

$$c^3 - 8c^2 - 48c = c \cdot c^2 - c \cdot \quad - c \cdot 48 = c(c^2 - \quad - 48) = c(c + \quad)(c - \quad)$$

8. Factor: $x^{10} + 8x^5 - 9$. [4.4a]

$$x^{10} + 8x^5 - 9 = (\quad)^2 + 8(\quad) - 9 = (\quad + 9)(\quad - 1)$$

9. Factor by grouping: $5y^3 + 20y^2 - y - 4$. [4.3b]

$$5y^3 + 20y^2 - y - 4 = 5y^2(\quad) - 1(\quad) = (\quad)(5y^2 - 1)$$

Mixed Review

For each polynomial, identify the terms, the degree of each term, and the degree of the polynomial. Then identify the leading term, the leading coefficient, and the constant term. [4.1a]

10. $-a^7 + a^4 - a + 8$

11. $3x^4 + 2x^3w^5 - 12x^2w + 4x^2 - 1$

12. Arrange in ascending powers of y: $-2y + 5 - y^3 + y^9 - 2y^4$. [4.1a]

13. Arrange in descending powers of x: $2qx - 9qr + 2x^5 - 4qx^2$. [4.1a]

Evaluate each polynomial function for the given values of the variable. [4.1b]

14. $h(x) = -x^3 - 4x + 5$; $h(0), h(-2)$, and $h\left(\frac{1}{2}\right)$

15. $f(x) = \frac{1}{2}x^4 - x^3$; $f(-1), f(1)$, and $f(0)$

16. Given $f(x) = x^2 + 2x - 9$, find and simplify $f(a - 2)$ and $f(a + h) - f(a)$. [4.2e]

Add, subtract, or multiply,

17. $(3a^2 - 7b + ab + 2) + (-5a^2 + 4b - 5ab - 3)$ [4.1c]

18. $(x^2 + 10x - 4) + (9x^2 - 2x + 1) + (x^2 - x - 5)$ [4.1c]

19. $(b - 12)(b + 1)$ [4.2b]

20. $c^2(3c^2 - c^3)$ [4.2a]

21. $(y^4 - 6)(y^4 + 3)$ [4.2b]

22. $(7y^2 - 2y^3 - 5y) - (y^2 - 3y - 6y^3)$ [4.1d]

23. $(8x - 11) - (-x + 1)$ [4.1d]

24. $(4x - 5)^2$ [4.2c]

25. $(2x + 5)^2$ [4.2c]

26. $(0.01x - 0.5y) - (2.5y - 0.1x)$ [4.1d]

27. $-13x^2 \cdot 10xy$ [4.2a]

28. $(x + y)(x^2 - 2xy + 3y^2)$ [4.2a]

29. $(5x - 7)(2x + 9)$ [4.2b]

30. $(9x - 4)(9x + 4)$ [4.2d]

Factor.

31. $5h^2 + 7h$ [4.3a]

32. $x^2 + 8x - 20$ [4.4a]

33. $21 - 4b - b^2$ [4.4a]

34. $m^2 + \frac{2}{7}m + \frac{1}{49}$ [4.4a]

35. $2xy - x^2y - 5x + 10$ [4.3b]

36. $3w^2 - 6w + 3$ [4.4a]

37. $t^3 + 3t^2 + t + 3$ [4.3b]

38. $24xy^6z^4 - 16x^4y^3z$ [4.3a]

39. $x^2 + 8x + 6$ [4.4a]

Understanding Through Discussion and Writing

40. Explain in your own words why $-(a - b) = b - a$. [4.1d], [4.3a]

41. Is the sum of two binomials always a binomial? Why or why not? [4.1c]

42. Is it true that if a polynomial's coefficients and exponents are all prime numbers, then the polynomial itself is prime? Why or why not? [4.3a]

43. Under what conditions would it be easier to evaluate a polynomial function after it has been factored? [4.1b], [4.4a]

44. Explain the error in each of the following.

a) $(a + 3)^2 = a^2 + 9$ [4.2c]

b) $(a - b)(a - b) = a^2 - b^2$ [4.2c]

c) $(x + 3)(x - 4) = x^2 - 12$ [4.2b]

d) $(p + 7)(p - 7) = p^2 + 49$ [4.2d]

e) $(t - 3)^2 = t^2 - 9$ [4.2c]

45. Checking the factorization of a second-degree polynomial by making a single replacement is only a *partial* check. Write an *incorrect* factorization and explain how evaluating both the polynomial and the factorization might not catch a possible error. [4.4a]

STUDYING FOR SUCCESS *As You Study*

- ☐ Find a quiet place to study.
- ☐ Be disciplined in your use of electronics such as video games, the Internet, and television. Study first!
- ☐ Pace yourself. It is usually better to study for shorter periods several times per week than to study in one marathon session each week.

4.5 Factoring Trinomials: $ax^2 + bx + c, a \neq 1$

OBJECTIVES

- a Factor trinomials of the type $ax^2 + bx + c, a \neq 1$, by the FOIL method.
- b Factor trinomials of the type $ax^2 + bx + c, a \neq 1$, by the ac-method.

SKILL REVIEW *Multiply two binomials.* [4.2b]

Multiply.

1. $(8x - 7)(2x + 1)$
2. $(6a - b)(3a + 5b)$

Answers:
1. $16x^2 - 6x - 7$
2. $18a^2 + 27ab - 5b^2$

MyLab Math VIDEO

To the instructor: Here we present two ways to factor general trinomials: the FOIL method and the ac-method. You can teach both methods and let the student use the one he or she prefers or you can select just one for the student.

Now we learn to factor trinomials of the type $ax^2 + bx + c, a \neq 1$. We use two methods: the FOIL method and the ac-method.

a THE FOIL METHOD

We first consider the **FOIL method** for factoring trinomials of the type $ax^2 + bx + c, a \neq 1$. Consider the following multiplication.

$$\begin{aligned}(3x + 2)(4x + 5) &= \overset{F}{12x^2} + \overset{O}{15x} + \overset{I}{8x} + \overset{L}{10} \\ &= 12x^2 + 23x + 10\end{aligned}$$

To factor $12x^2 + 23x + 10$, we must reverse what we just did. We look for two binomials whose product is this trinomial. The product of the First terms must be $12x^2$. The product of the Outside terms plus the product of the Inside terms must be $23x$. The product of the Last terms must be 10. In general, finding such an answer involves trial and error. We use the following method.

THE FOIL METHOD FOR FACTORING TRINOMIALS

1. Factor out the largest common factor. The remaining trinomial is $ax^2 + bx + c$.
2. Find two First terms whose product is ax^2:

   $$(\square x + \quad)(\square x + \quad) = ax^2 + bx + c.$$
   FOIL

3. Find two Last terms whose product is c:

   $$(\quad x + \square)(\quad x + \square) = ax^2 + bx + c.$$
   FOIL

4. Repeat steps (2) and (3), if necessary, until a combination is found for which the sum of the Outside and Inside products is bx:

   $$(\square x + \square)(\square x + \square) = ax^2 + bx + c.$$
   I, O — FOIL

5. Check by multiplying.

EXAMPLE 1 Factor: $3x^2 + 10x - 8$.

1. First, we factor out the largest common factor, if any. There is none (other than 1 or -1).
2. Next, we factor the first term, $3x^2$. The only possibility is $3x \cdot x$. The desired factorization is then of the form $(3x + \square)(x + \square)$.
3. We then factor the last term, -8, which is negative. The possibilities are $(-8)(1)$, $8(-1)$, $2(-4)$, and $(-2)(4)$. They can be written in either order.
4. We look for combinations of factors from steps (2) and (3) such that the sum of the outside and the inside products is the middle term, $10x$:

$3x$, $-8x$: $(3x - 8)(x + 1) = 3x^2 - 5x - 8$; Wrong middle term

$-3x$, $8x$: $(3x + 8)(x - 1) = 3x^2 + 5x - 8$; Wrong middle term

$-12x$, $2x$: $(3x + 2)(x - 4) = 3x^2 - 10x - 8$; Wrong middle term

$12x$, $-2x$: $(3x - 2)(x + 4) = 3x^2 + 10x - 8$. Correct middle term!

5. *Check:* $(3x - 2)(x + 4) = 3x^2 + 10x - 8$.

Do Exercises 1 and 2. ▶

Factor by the FOIL method.

1. $3x^2 - 13x - 56$
2. $3x^2 + 5x + 2$

EXAMPLE 2 Factor: $18x^6 - 57x^5 + 30x^4$.

1. First, we factor out the largest common factor, if any. The expression $3x^4$ is common to all terms, so we factor it out: $3x^4(6x^2 - 19x + 10)$.
2. Next, we factor the trinomial $6x^2 - 19x + 10$. We factor the first term, $6x^2$, and get $6x \cdot x$, or $3x \cdot 2x$. We then have these as possibilities for factorizations: $(3x + \square)(2x + \square)$ or $(6x + \square)(x + \square)$.
3. We then factor the last term, 10, which is positive. The possibilities are $(10)(1)$, $(-10)(-1)$, $(5)(2)$, and $(-5)(-2)$. They can be written in either order. Since the middle term, $-19x$, is negative, we consider only $(-10)(-1)$ and $(-5)(-2)$.
4. We look for combinations of factors from steps (2) and (3) such that the sum of the outside and the inside products is the middle term, $-19x$. We begin by using these factors with $(3x + \square)(2x + \square)$. Should we not find the correct factorization, we will consider $(6x + \square)(x + \square)$.

$-3x$, $-20x$: $(3x - 10)(2x - 1) = 6x^2 - 23x + 10$; Wrong middle term

$-30x$, $-2x$: $(3x - 1)(2x - 10) = 6x^2 - 32x + 10$; Wrong middle term

$-6x$, $-10x$: $(3x - 5)(2x - 2) = 6x^2 - 16x + 10$; Wrong middle term

$-15x$, $-4x$: $(3x - 2)(2x - 5) = 6x^2 - 19x + 10$ Correct middle term!

We have a correct answer. We need not consider $(6x + \square)(x + \square)$.

Answers

1. $(x - 7)(3x + 8)$ **2.** $(3x + 2)(x + 1)$

The factorization of $6x^2 - 19x + 10$ is $(3x - 2)(2x - 5)$. But do not forget the common factor! We must include it in order to get a complete factorization of the original trinomial:

$$18x^6 - 57x^5 + 30x^4 = 3x^4(3x - 2)(2x - 5).$$

5. *Check*: $3x^4(3x - 2)(2x - 5) = 3x^4(6x^2 - 19x + 10)$
$= 18x^6 - 57x^5 + 30x^4.$

From Examples 1 and 2, we can make some observations that might speed up the factoring.

- Note in Example 1 that changing the signs in the binomial factors changed the signs of the middle terms.
- In Example 2, we examined the signs of the terms before forming possible products. If a and c are both positive, the signs of the factors of c will match the sign of b.
- In Example 2, look again at the possibility $(3x - 1)(2x - 10)$. Without multiplying, we can reject such a possibility, noting that the expression $2x - 10$ has a common factor, 2. But we removed the largest common factor before we began. If this expression were a factorization, then 2 would have to be a common factor along with $3x^4$.

Given that we factored out the largest common factor at the outset, we can now eliminate factorizations that have a common factor.

TIPS FOR FACTORING $ax^2 + bx + c$ USING THE FOIL METHOD

1. If the largest common factor has been factored out of the original trinomial, then no binomial factor can have a common factor (other than 1 or -1).
2. **a)** If the signs of all the terms are positive, then the signs of all the terms of the binomial factors are positive.
   b) If a and c are positive and b is negative, then the signs of the factors of c are negative.
   c) If a is positive and c is negative, then the factors of c will have opposite signs.
3. Be systematic about your trials. Keep track of those you have tried and those you have not.
4. Changing the signs of the factors of c will change the sign of the middle term.

Keep in mind that this method of factoring trinomials of the type $ax^2 + bx + c$ involves trial and error. As you practice, you will find that you will need fewer trials to arrive at the factorization.

Do Exercises 3–6.

The procedure considered here can also be applied to a trinomial with more than one variable.

EXAMPLE 3 Factor: $30m^2 + 23mn - 11n^2$.

1. First, we factor out the largest common factor, if any. In this polynomial, there is no common factor (other than 1 or -1).

Factor.

3. $24y^2 - 46y + 10$

4. $20x^5 - 46x^4 + 24x^3$ GS

First, factor out the largest common factor:
$\_\_\_\_(10x^2 - 23x + 12)$.
$10x^2$ can be factored as $2x \cdot 5x$ or as $\_\_\_\_ \cdot x$.
Since -23 is negative, the factors of 12 must be negative: 12 can be factored as $(-1)(-12)$, $(-2)(-6)$, or $(-3)(-4)$.
Next, list all possibilities with $2x$ and $5x$ as the first terms. If none of these is correct, we will check $10x$ and x.
$(2x - 1)(5x - 12)$,
$(2x - 12)(5x - 1)$,
$(2x - 2)(5x - 6)$,
$(2x - 6)(5x - \_\_\_\_)$,
$(2x - 3)(5x - 4)$,
$(2x - 4)(5x - \_\_\_\_)$
Only two possibilities have no common factor:
$(2x - 1)(5x - 12)$
$= 10x^2 - \_\_\_\_ + 24$,
$(2x - 3)(5x - 4)$
$= 10x^2 - \_\_\_\_ + 12$.
The factorization of $20x^5 - 46x^4 + 24x^3$ is
$\_\_\_\_(2x - 3)(5x - 4)$.

5. $3x^2 + 19x + 20$

6. $16x^2 - 12 + 16x$

Answers

3. $2(4y - 1)(3y - 5)$ **4.** $2x^3(2x - 3)(5x - 4)$
5. $(3x + 4)(x + 5)$ **6.** $4(2x - 1)(2x + 3)$

Guided Solution:

4. $2x^3, 10x, 2, 3, 29x, 23x, 2x^3$

2. Next, we factor the first term, $30m^2$, and get the following possibilities:

$30m \cdot m, \quad 15m \cdot 2m, \quad 10m \cdot 3m, \quad \text{and} \quad 6m \cdot 5m.$

We then have these as possibilities for factorizations:

$$(30m + \square)(m + \square), \qquad (15m + \square)(2m + \square),$$
$$(10m + \square)(3m + \square), \qquad (6m + \square)(5m + \square).$$

3. We then factor the last term, $-11n^2$, which is negative. The possibilities are $-11n \cdot n$ and $11n \cdot (-n)$.
4. We look for combinations of factors from steps (2) and (3) such that the sum of the outside and the inside products is the middle term, $23mn$. Since the coefficient of the middle term is positive, let's begin our search using $11n \cdot (-n)$. Should we not find the correct factorization, we will consider $-11n \cdot n$.

$$(30m + 11n)(m - n) = 30m^2 - 19mn - 11n^2;$$
$$(30m - n)(m + 11n) = 30m^2 + 329mn - 11n^2;$$

Note that changing the order of $11n$ and $-n$ changes the middle term.

$$(15m + 11n)(2m - n) = 30m^2 + 7mn - 11n^2;$$
$$(15m - n)(2m + 11n) = 30m^2 + 163mn - 11n^2;$$
$$(10m + 11n)(3m - n) = 30m^2 + 23mn - 11n^2$$ ← Correct middle term

We have a correct product: $30m^2 + 23mn - 11n^2$. The factorization of $30m^2 + 23mn - 11n^2$ is $(10m + 11n)(3m - n)$.

5. *Check:* $(10m + 11n)(3m - n) = 30m^2 + 23mn - 11n^2$.

Do Exercises 7 and 8. ▶

Factor.

7. $21x^2 - 5xy - 4y^2$

8. $60a^2 + 123ab - 27b^2$

b THE *ac*-METHOD

The second method of factoring trinomials of the type $ax^2 + bx + c$, $a \neq 1$, is known as the ***ac*-method**, or the **grouping method**.

We can factor $x^2 + 7x + 10$ by "splitting" the middle term, $7x$, and using factoring by grouping:

$$\begin{aligned} x^2 + 7x + 10 &= x^2 + 2x + 5x + 10 \\ &= x(x + 2) + 5(x + 2) \\ &= (x + 2)(x + 5). \end{aligned}$$

If the leading coefficient is not 1, as in $6x^2 + 23x + 20$, we use a method for factoring similar to what we just did with $x^2 + 7x + 10$.

THE *ac*-METHOD FOR FACTORING TRINOMIALS

1. Factor out the largest common factor. The remaining trinomial is $ax^2 + bx + c$.
2. Multiply the leading coefficient a and the constant c.
3. Try to factor the product ac so that the sum of the factors is b. That is, find integers p and q such that $pq = ac$ and $p + q = b$.
4. Split the middle term, writing it as a sum using the factors found in step (3).
5. Factor by grouping.
6. Check by multiplying.

Answers

7. $(7x - 4y)(3x + y)$
8. $3(4a + 9b)(5a - b)$

Factor by the ac-method.

9. $4x^2 + 4x - 3$

10. $4x^2 + 37x + 9$ GS

1. There is no common factor.
2. Multiply the leading coefficient and the constant: $4(9) = \_\_\_$.
3. Look for a pair of factors of 36 whose sum is 37. Both factors will be positive.

PAIRS OF FACTORS	SUMS OF FACTORS
1, 36	☐
2, 18	20
3, 12	15
4, 9	13
6, 6	12

4. Split the middle term, $37x$: $37x = x + \_\_\_$.
5. Factor by grouping: $4x^2 + x + 36x + 9$ $= x(4x + 1) + 9(\quad)$ $= (\quad)(x + 9)$.
6. *Check*: $(4x + 1)(x + 9) = 4x^2 + 37x + 9$.

Answers

9. $(2x + 3)(2x - 1)$ 10. $(4x + 1)(x + 9)$

Guided Solution:

10. $36, 37, 36x, 4x + 1, 4x + 1$

EXAMPLE 4 Factor: $6x^2 + 23x + 20$.

1. First, factor out a common factor, if any. There is none (other than 1 or -1).
2. Multiply the leading coefficient, 6, and the constant, 20: $6 \cdot 20 = 120$.
3. Then look for a factorization of 120 in which the sum of the factors is the coefficient of the middle term, 23. Since both 120 and 23 are positive, we need consider only positive factors of 120.

PAIRS OF FACTORS	SUMS OF FACTORS
1, 120	121
2, 60	62
3, 40	43
4, 30	34

PAIRS OF FACTORS	SUMS OF FACTORS
5, 24	29
6, 20	26
8, 15	23
10, 12	22

4. Split the middle term: $23x = 8x + 15x$.
5. Factor by grouping:

$$6x^2 + 23x + 20 = 6x^2 + 8x + 15x + 20 \qquad \text{Substituting } 8x + 15x \text{ for } 23x$$
$$= 2x(3x + 4) + 5(3x + 4) \qquad \text{Factoring by grouping}$$
$$= (3x + 4)(2x + 5).$$

We could also split the middle term as $15x + 8x$. We still get the same factorization, although the factors are in a different order:

$$6x^2 + 23x + 20 = 6x^2 + 15x + 8x + 20$$
$$= 3x(2x + 5) + 4(2x + 5)$$
$$= (2x + 5)(3x + 4).$$

6. *Check:* $(3x + 4)(2x + 5) = 6x^2 + 23x + 20$.

◀ **Do Exercises 9 and 10.**

EXAMPLE 5 Factor: $6x^4 - 116x^3 - 80x^2$.

1. First, factor out the largest common factor, if any. The expression $2x^2$ is common to all three terms: $2x^2(3x^2 - 58x - 40)$.
2. Now, factor the trinomial $3x^2 - 58x - 40$. Multiply the leading coefficient, 3, and the constant, -40: $3(-40) = -120$.
3. Next, try to factor -120 so that the sum of the factors is -58. Since the coefficient of the middle term, -58, is negative, the negative factor of -120 must have the larger absolute value.

PAIRS OF FACTORS	SUMS OF FACTORS
1, −120	−119
2, −60	−58
3, −40	−37
4, −30	−26

PAIRS OF FACTORS	SUMS OF FACTORS
5, −24	−19
6, −20	−14
8, −15	−7
10, −12	−2

4. Split the middle term, $-58x$, as follows: $-58x = 2x - 60x$.

5. Factor by grouping:

$$3x^2 - 58x - 40 = 3x^2 + 2x - 60x - 40 \quad \text{Substituting } 2x - 60x \text{ for } -58x$$
$$= x(3x + 2) - 20(3x + 2) \quad \text{Factoring by grouping}$$
$$= (3x + 2)(x - 20).$$

The factorization of $3x^2 - 58x - 40$ is $(3x + 2)(x - 20)$. But don't forget the common factor! The factorization of the original polynomial is

$$6x^4 - 116x^3 - 80x^2 = 2x^2(3x + 2)(x - 20).$$

6. *Check*: $2x^2(3x + 2)(x - 20) = 2x^2(3x^2 - 58x - 40)$
$= 6x^4 - 116x^3 - 80x^2$.

Do Exercises 11 and 12. ▶

Factor by the *ac*-method.

11. $10y^4 - 7y^3 - 12y^2$

12. $6a^3 - 7a^2 - 5a$

Answers

11. $y^2(5y + 4)(2y - 3)$
12. $a(3a - 5)(2a + 1)$

4.5 Exercise Set

FOR EXTRA HELP

MyLab Math

✓ Check Your Understanding

Concept Check Choose the word or expression shown under each blank that best completes the description of factoring $2x^2 - 3x + 1$ using the FOIL method.

CC1. The product of the First terms must be ______.
$2x^2/-3x/1$

CC2. The product of the Last terms must be ______.
$2x^2/-3x/1$

CC3. The sum of the Outside and the Inside products must be ______.
$2x^2/-3x/1$

CC4. Both Last terms must be ______.
positive/negative

Choose the word or expression shown under each blank that best completes the description of factoring $10x^2 + 21x + 2$ using the *ac*-method.

CC5. Multiply the ______ (leading/largest) coefficient 10 and the ______ (constant/degree) 2. The product is 20.

CC6. Find two integers whose ______ (sum/product) is 20 and whose ______ (sum/product) is 21. The integers are 20 and 1.

CC7. Split the middle term, ______ ($10x^2/21x/2$), writing it as the sum of $20x$ and x.

CC8. ______ (Factor/Add) by grouping.

Reading Check Determine whether each statement is true or false.

RC1. The first step in factoring any polynomial is to look for a common factor.

RC2. If a trinomial has no common factor, then neither of its binomial factors can have a common factor.

RC3. Trinomials in more than one variable cannot be factored.

RC4. Every trinomial can be factored.

RC5. If a trinomial can be factored using the FOIL method, then it can also be factored using the *ac*-method.

a, **b** Factor.

1. $3x^2 - 14x - 5$

2. $8x^2 - 6x - 9$

3. $10y^3 + y^2 - 21y$

4. $6x^3 + x^2 - 12x$

5. $3c^2 - 20c + 32$

6. $12b^2 - 8b + 1$

7. $35y^2 + 34y + 8$

8. $9a^2 + 18a + 8$

9. $4t + 10t^2 - 6$

10. $8x + 30x^2 - 6$

11. $8x^2 - 16 - 28x$

12. $18x^2 - 24 - 6x$

13. $18a^2 - 51a + 15$

14. $30a^2 - 85a + 25$

15. $30t^2 + 85t + 25$

16. $18y^2 + 51y + 15$

17. $12x^3 - 31x^2 + 20x$

18. $15x^3 - 19x^2 - 10x$

19. $14x^4 - 19x^3 - 3x^2$

20. $70x^4 - 68x^3 + 16x^2$

21. $3a^2 - a - 4$

22. $6a^2 - 7a - 10$

23. $9x^2 + 15x + 4$

24. $6y^2 - y - 2$

25. $3 + 35z - 12z^2$

26. $8 - 6a - 9a^2$

27. $-4t^2 - 4t + 15$

28. $-12a^2 + 7a - 1$

29. $3x^3 - 5x^2 - 2x$

30. $18y^3 - 3y^2 - 10y$

31. $24x^2 - 2 - 47x$

32. $15y^2 - 10 - 15y$

33. $-8t^3 - 8t^2 + 30t$

34. $-36a^3 + 21a^2 - 3a$

35. $-24x^3 + 2x + 47x^2$

36. $-15y^3 + 10y + 47y^2$

37. $21x^2 + 37x + 12$

38. $10y^2 + 23y + 12$

39. $40x^4 + 16x^2 - 12$

40. $24y^4 + 2y^2 - 15$

41. $12a^2 - 17ab + 6b^2$

42. $20p^2 - 23pq + 6q^2$

43. $2x^2 + xy - 6y^2$

44. $8m^2 - 6mn - 9n^2$

45. $12x^2 - 58xy + 56y^2$

46. $30a^2 + 21ab - 36b^2$

47. $9x^2 - 30xy + 25y^2$

48. $4p^2 + 12pq + 9q^2$

49. $3x^6 + 4x^3 - 4$

50. $2p^8 + 11p^4 + 15$

51. $16x^3 + 12x^2 - 6x$

52. $18x^5 + 45x^3 + 6x$

53. *Height of a Thrown Baseball.* Suppose that a baseball is thrown upward with an initial velocity of 80 ft/sec from a height of 224 ft. Its height h, in feet, after t seconds is given by the function

$$h(t) = -16t^2 + 80t + 224.$$

a) What is the height of the ball after 0 sec? 1 sec? 3 sec? 4 sec? 6 sec?

b) Find an equivalent expression for $h(t)$ by factoring.

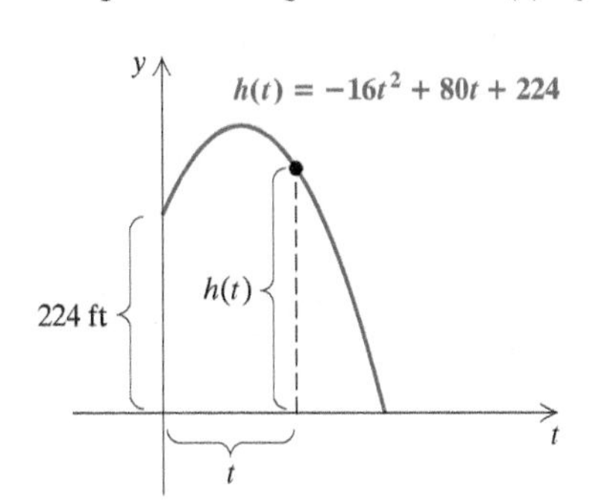

54. *Fireworks.* Suppose that a bottle rocket is launched upward with an initial velocity of 96 ft/sec from a height of 880 ft. Its height h, in feet, after t seconds is given by the function

$$h(t) = -16t^2 + 96t + 880.$$

a) What is the height of the bottle rocket after 0 sec? 1 sec? 3 sec? 8 sec? 10 sec?

b) Find an equivalent expression for $h(t)$ by factoring.

Skill Maintenance

Solve. [3.5a]

55. $x + 2y - z = 0,$ $4x + 2y + 5z = 6,$ $2x - y + z = 5$

56. $2x + y + 2z = 5,$ $4x - 2y - 3z = 5,$ $-8x - y + z = -5$

57. $2x + 9y + 6z = 5,$ $x - y + z = 4,$ $3x + 2y + 3z = 7$

58. $x - 3y + 2z = -8,$ $2x + 3y + z = 17,$ $5x - 2y + 3z = 5$

Determine whether the graphs of the given pairs of lines are parallel, perpendicular, or neither. [2.5d]

59. $y - 2x = 18,$ $2x - 7 = y$

60. $21x + 7 = -3y,$ $y + 7x = -9$

61. $2x + 5y = 4,$ $2x - 5y = -3$

62. $y + x = 7,$ $y - x = 3$

Find an equation of the line containing the given pair of points. [2.6c]

63. $(-2, -3)$ and $(5, -4)$

64. $(2, -3)$ and $(5, -4)$

65. $(-10, 3)$ and $(7, -4)$

66. $\left(-\frac{2}{3}, 1\right)$ and $\left(\frac{4}{3}, -4\right)$

Synthesis

67. Use the TABLE feature and the GRAPH feature of a graphing calculator to check your answers to Exercises 2, 17, and 28.

68. Use the TABLE feature and the GRAPH feature of a graphing calculator to check your answers to Exercises 4, 11, and 32.

Factor. Assume that variables in exponents represent positive integers.

69. $7a^2b^2 + 6 + 13ab$

70. $2x^4y^6 - 3x^2y^3 - 20$

71. $9x^2y^2 - 4 + 5xy$

72. $\frac{1}{4}p^2 - \frac{2}{5}p + \frac{4}{25}$

73. $x^{2a} + 5x^a - 24$

74. $4x^{2a} - 4x^a - 3$

4.6 Special Factoring

OBJECTIVES

a Factor trinomial squares.

b Factor differences of squares.

c Factor certain polynomials with four terms by grouping and possibly using the factoring of a trinomial square or the difference of squares.

d Factor sums and differences of cubes.

In this section, we consider some special factoring methods.

a TRINOMIAL SQUARES

Consider the trinomial $x^2 + 6x + 9$. To factor it, we can look for factors of 9 whose sum is 6. We see that these factors are 3 and 3 and the factorization is

$$x^2 + 6x + 9 = (x + 3)(x + 3) = (x + 3)^2.$$

Note that the result is the square of a binomial. We also call $x^2 + 6x + 9$ a **trinomial square**, or **perfect-square trinomial**. We now develop a faster procedure for factoring trinomial squares.

How can we recognize when an expression is a trinomial square? Look at $A^2 + 2AB + B^2$ and $A^2 - 2AB + B^2$.

How to recognize a **trinomial square**:

a) The two expressions A^2 and B^2 must be squares.

b) There must be no minus sign before either A^2 or B^2.

c) Multiplying A and B (expressions whose squares are A^2 and B^2) and doubling the result gives either the remaining term or its opposite.

EXAMPLES Determine whether the polynomial is a trinomial square.

1. $x^2 + 10x + 25$

a) Two terms are squares: x^2 and 25.

b) There is no minus sign before either x^2 or 25.

c) If we multiply the expressions whose squares are x^2 and 25, x and 5, and double the product, we get $10x$, the remaining term.

Thus this is a trinomial square.

2. $4x + 16 + 3x^2$

a) Only one term, 16, is a square ($3x^2$ is not a square because 3 is not a perfect square and $4x$ is not a square because x is not a square).

Thus this is not a trinomial square.

3. $100y^2 + 81 - 180y$

(It can help to first write this in descending order: $100y^2 - 180y + 81$.)

a) Two of the terms, $100y^2$ and 81, are squares.

b) There is no minus sign before either $100y^2$ or 81.

c) If we multiply the expressions whose squares are $100y^2$ and 81, $10y$ and 9, and double the product, we get $2(10y)(9) = 180y$. This is the opposite of the remaining term, $-180y$.

Thus this is a trinomial square.

◀ **Do Exercise 1.**

1. Which of the following are trinomial squares?

a) $x^2 + 6x + 9$

b) $x^2 - 8x + 16$

c) $x^2 + 6x + 11$

d) $4x^2 + 25 - 20x$

e) $16x^2 - 20x + 25$

f) $16 + 14x + 5x^2$

g) $x^2 + 8x - 16$

h) $x^2 - 8x - 16$

Answer

1. (a), (b), (d)

The factors of a trinomial square are two identical binomials.

TRINOMIAL SQUARES

$A^2 + 2AB + B^2 = (A + B)^2$;
$A^2 - 2AB + B^2 = (A - B)^2$

EXAMPLE 4 Factor: $x^2 - 10x + 25$.

$x^2 - 10x + 25 = (x - 5)^2$ We find the square terms and write their square roots with a minus sign between them.

Note the sign!

EXAMPLE 5 Factor: $16y^2 + 49 + 56y$.

$16y^2 + 49 + 56y = 16y^2 + 56y + 49$ Rewriting in descending order

$= (4y + 7)^2$ We find the square terms and write their square roots with a plus sign between them.

EXAMPLE 6 Factor: $-20xy + 4y^2 + 25x^2$.

$-20xy + 4y^2 + 25x^2 = 4y^2 - 20xy + 25x^2$ Writing descending order in y

$= (2y - 5x)^2$.

This square can also be expressed as $25x^2 - 20xy + 4y^2 = (5x - 2y)^2$.

Do Exercises 2–5.

Factor.

2. $x^2 + 14x + 49$
3. $9y^2 - 30y + 25$
4. $16x^2 + 72xy + 81y^2$

5. $16x^4 - 40x^2y^3 + 25y^6$
Factor as a trinomial square:
$16x^4 - 40x^2y^3 + 25y^6$
$= (4x^2)^2 - 2 \cdot 4x^2 \cdot$ ____
$+ ($ ____ $)^2$
$= (4x^2 -$ ____ $)^2$.

In factoring, we must always remember to look *first* for the largest factor common to all the terms. If the first term is negative, we generally factor out a negative common factor.

EXAMPLE 7 Factor: $-4y^2 - 144y^8 + 48y^5$.

$-4y^2 - 144y^8 + 48y^5$
$= -4y^2(1 + 36y^6 - 12y^3)$ Factoring out the common factor $-4y^2$
$= -4y^2(1 - 12y^3 + 36y^6)$ Changing order
$= -4y^2(1 - 6y^3)^2$ Factoring the trinomial square

Do Exercises 6 and 7.

Factor.

6. $-8a^2 + 24ab - 18b^2$
7. $3a^2 - 30ab + 75b^2$

b DIFFERENCES OF SQUARES

Multiply a sum and a difference of the same two terms. [4.2d]

Multiply.

1. $(y - 3)(y + 3)$
2. $(5x - 7)(5x + 7)$

Answers: 1. $y^2 - 9$ **2.** $25x^2 - 49$

MyLab Math VIDEO

Answers

2. $(x + 7)^2$ **3.** $(3y - 5)^2$ **4.** $(4x + 9y)^2$ **5.** $(4x^2 - 5y^3)^2$ **6.** $-2(2a - 3b)^2$ **7.** $3(a - 5b)^2$

Guided Solution:
5. $5y^3, 5y^3, 5y^3$

The following are *differences of squares:*

$$x^2 - 9, \qquad 49 - 4y^2, \qquad a^2 - 49b^2.$$

To factor a difference of squares, we start with the special product $\mathbf{(A + B)(A - B) = A^2 - B^2}$ and reverse the equation.

FACTORING A DIFFERENCE OF SQUARES

$$A^2 - B^2 = (A + B)(A - B)$$

Factor.

8. $y^2 - 4$

9. $49x^4 - 25y^{10}$

10. $m^2 - \frac{1}{9}$

Factor as a difference of squares:

$m^2 - \frac{1}{9}$

$= m^2 - (\quad)^2$

$= (m + \quad)(m - \quad).$

EXAMPLE 8 Factor: $x^2 - 9$.

$$\begin{array}{c} A^2 - B^2 = (A + B)(A - B) \\ \downarrow \quad \downarrow \quad \downarrow \downarrow \downarrow \downarrow \\ x^2 - 9 = x^2 - 3^2 = (x + 3)(x - 3) \end{array}$$

GS **EXAMPLE 9** Factor: $25y^6 - 49x^2$.

$$\begin{array}{c} A^2 \quad - \quad B^2 \quad = \quad (A + B)(A - B) \\ \downarrow \quad \downarrow \quad \downarrow \downarrow \downarrow \downarrow \\ 25y^6 - 49x^2 = (5y^3)^2 - (7x)^2 = (5y^3 + 7x)(5y^3 - 7x) \end{array}$$

EXAMPLE 10 Factor: $x^2 - \frac{1}{16}$.

$$x^2 - \tfrac{1}{16} = x^2 - \left(\tfrac{1}{4}\right)^2 = \left(x + \tfrac{1}{4}\right)\left(x - \tfrac{1}{4}\right)$$

Do Exercises 8–10.

Common factors should always be factored out first.

EXAMPLE 11 Factor: $5 - 5x^2y^6$.

There is a common factor, 5.

$5 - 5x^2y^6 = 5(1 - x^2y^6)$ Factoring out the common factor 5

$= 5[1^2 - (xy^3)^2]$ Recognizing the difference of squares; $x^2y^6 = (x^1y^3)^2 = (xy^3)^2$

$= 5(1 + xy^3)(1 - xy^3)$ Factoring the difference of squares

EXAMPLE 12 Factor: $2x^4 - 8y^4$.

There is a common factor, 2.

$2x^4 - 8y^4 = 2(x^4 - 4y^4)$ Factoring out the common factor 2

$= 2[(x^2)^2 - (2y^2)^2]$ Recognizing the difference of squares

$= 2(x^2 + 2y^2)(x^2 - 2y^2)$ Factoring the difference of squares

EXAMPLE 13 Factor: $16x^4y - 81y$.

There is a common factor, y.

$16x^4y - 81y = y(16x^4 - 81)$ Factoring out the common factor y

$= y[(4x^2)^2 - 9^2]$

$= y(4x^2 + 9)(4x^2 - 9)$ Factoring the difference of squares

$= y(4x^2 + 9)(2x + 3)(2x - 3)$ Factoring $4x^2 - 9$, which is also a difference of squares

Answers

8. $(y + 2)(y - 2)$
9. $(7x^2 + 5y^5)(7x^2 - 5y^5)$
10. $\left(m + \frac{1}{3}\right)\left(m - \frac{1}{3}\right)$

Guided Solution:

10. $\frac{1}{3}, \frac{1}{3}, \frac{1}{3}$

In Example 13, note that $4x^2 + 9$ is a sum of two squares, and it cannot be factored further. Also note that one of the factors, $4x^2 - 9$, could be factored further. Whenever that is possible, you should do so. That way you will be factoring *completely*.

Caution!

We cannot factor a sum of squares as the square of a binomial. In particular,

$$A^2 + B^2 \neq (A + B)^2.$$

Do Exercises 11–14.

Factor.

11. $25x^2y^2 - 4a^2$

12. $9x^2 - 16y^2$

13. $20x^2 - 5y^2$

14. $81x^4y^2 - 16y^2$

c MORE FACTORING BY GROUPING

Sometimes when factoring a polynomial with four terms, we get a factor that can be factored further using other methods that we have learned.

EXAMPLE 14 Factor completely: $x^3 + 3x^2 - 4x - 12$.

$$\begin{aligned} x^3 + 3x^2 - 4x - 12 &= x^2(x + 3) - 4(x + 3) \\ &= (x + 3)(x^2 - 4) \\ &= (x + 3)(x + 2)(x - 2) \end{aligned}$$

Do Exercise 15.

15. Factor: $a^3 + a^2 - 16a - 16$.

A difference of squares can have more than two terms. For example, one of the squares may be a trinomial. We can factor by a type of grouping.

EXAMPLE 15 Factor completely: $x^2 + 6x + 9 - y^2$.

$$\begin{aligned} x^2 + 6x + 9 - y^2 &= (x^2 + 6x + 9) - y^2 && \text{Grouping as a trinomial minus } y^2 \text{ to show a difference of squares} \\ &= (x + 3)^2 - y^2 \\ &= (x + 3 + y)(x + 3 - y) \end{aligned}$$

Do Exercises 16–19.

Factor completely.

16. $x^2 + 2x + 1 - p^2$

17. $y^2 - 8y + 16 - 9m^2$

18. $x^2 + 8x + 16 - 100t^2$

19. $64p^2 - (x^2 + 8x + 16)$

d SUMS OR DIFFERENCES OF CUBES

We can factor the sum or the difference of two expressions that are cubes.

Consider the following products:

$$\begin{aligned} (A + B)(A^2 - AB + B^2) &= A(A^2 - AB + B^2) + B(A^2 - AB + B^2) \\ &= A^3 - A^2B + AB^2 + A^2B - AB^2 + B^3 \\ &= A^3 + B^3 \end{aligned}$$

and

$$\begin{aligned} (A - B)(A^2 + AB + B^2) &= A(A^2 + AB + B^2) - B(A^2 + AB + B^2) \\ &= A^3 + A^2B + AB^2 - A^2B - AB^2 - B^3 \\ &= A^3 - B^3. \end{aligned}$$

The above equations (reversed) show how we can factor a sum or a difference of two cubes. Each factors as a product of a binomial and a trinomial.

Answers

11. $(5xy + 2a)(5xy - 2a)$
12. $(3x + 4y)(3x - 4y)$
13. $5(2x + y)(2x - y)$
14. $y^2(9x^2 + 4)(3x + 2)(3x - 2)$
15. $(a + 1)(a + 4)(a - 4)$
16. $(x + 1 + p)(x + 1 - p)$
17. $(y - 4 + 3m)(y - 4 - 3m)$
18. $(x + 4 + 10t)(x + 4 - 10t)$
19. $[8p + (x + 4)][8p - (x + 4)]$, or $(8p + x + 4)(8p - x - 4)$

N	N^3
0.2	0.008
0.1	0.001
0	0
1	1
2	8
3	27
4	64
5	125
6	216
7	343
8	512
9	729
10	1000

SUM OR DIFFERENCE OF CUBES

$A^3 + B^3 = (A + B)(A^2 - AB + B^2)$;
$A^3 - B^3 = (A - B)(A^2 + AB + B^2)$

Note that what we are considering here is a sum or a difference of cubes. We are not cubing a binomial. For example, $(A + B)^3$ is *not* the same as $A^3 + B^3$. The table of cubes in the margin is helpful.

EXAMPLE 16 Factor: $x^3 - 27$.

We have

$$x^3 - 27 = x^3 - 3^3.$$

(with $x^3 \to A^3$, $3^3 \to B^3$)

In one set of parentheses, we write the cube root of the first term, x. Then we write the cube root of the second term, -3. This gives us the expression $x - 3$:

$$(x - 3)(\qquad).$$

To get the next factor, we think of $x - 3$ and do the following:

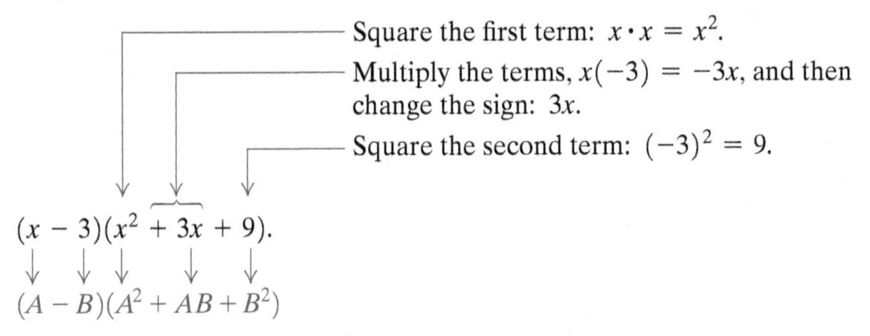

Note that we cannot factor $x^2 + 3x + 9$. It is not a trinomial square nor can it be factored by trial and error. Check this on your own.

Do Exercises 20 and 21.

Factor.

20. $x^3 - 8$

21. $64 - y^3$

EXAMPLE 17 Factor: $125x^3 + y^3$.

We have

$$125x^3 + y^3 = (5x)^3 + y^3.$$

In one set of parentheses, we write the cube root of the first term, $5x$. Then we write the cube root of the second term, y. This gives us the expression $5x + y$:

$$(5x + y)(\qquad).$$

To get the next factor, we think of $5x + y$ and do the following:

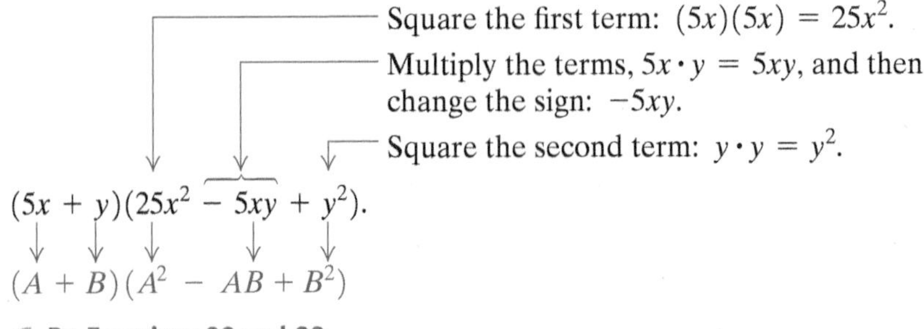

Do Exercises 22 and 23.

Factor.

22. $27x^3 + y^3$

23. $8y^3 + z^3$

Answers

20. $(x - 2)(x^2 + 2x + 4)$
21. $(4 - y)(16 + 4y + y^2)$
22. $(3x + y)(9x^2 - 3xy + y^2)$
23. $(2y + z)(4y^2 - 2yz + z^2)$

EXAMPLE 18 Factor: $128y^7 - 250x^6y$.

We first look for the largest common factor:

$$\begin{aligned}128y^7 - 250x^6y &= 2y(64y^6 - 125x^6)\\ &= 2y[(4y^2)^3 - (5x^2)^3]\\ &= 2y(4y^2 - 5x^2)(16y^4 + 20x^2y^2 + 25x^4).\end{aligned}$$

EXAMPLE 19 Factor: $a^6 - b^6$.

We factor a difference of squares:

$$a^6 - b^6 = (a^3)^2 - (b^3)^2 = (a^3 + b^3)(a^3 - b^3).$$

One factor is a sum of two cubes, and the other factor is a difference of two cubes. We factor them:

$$a^6 - b^6 = (a + b)(a^2 - ab + b^2)(a - b)(a^2 + ab + b^2).$$

We have now factored completely.

In Example 19, had we thought of factoring first as a difference of two cubes, we would have had

$$\begin{aligned}(a^2)^3 - (b^2)^3 &= (a^2 - b^2)(a^4 + a^2b^2 + b^4)\\ &= (a + b)(a - b)(a^4 + a^2b^2 + b^4).\end{aligned}$$

In this case, we might have missed some factors; $a^4 + a^2b^2 + b^4$ can be factored as $(a^2 - ab + b^2)(a^2 + ab + b^2)$, but we probably would not have known to do such factoring.

When you can factor as either a difference of squares or a difference of cubes, factor as a difference of squares first.

EXAMPLE 20 Factor: $64a^6 - 729b^6$.

We have

$$\begin{aligned}64a^6 - 729b^6 &= (8a^3)^2 - (27b^3)^2\\ &= (8a^3 - 27b^3)(8a^3 + 27b^3) \quad \text{Factoring a difference of squares}\\ &= [(2a)^3 - (3b)^3][(2a)^3 + (3b)^3].\end{aligned}$$

Each factor is a sum or a difference of cubes. We factor each:

$$= (2a - 3b)(4a^2 + 6ab + 9b^2)(2a + 3b)(4a^2 - 6ab + 9b^2).$$

FACTORING SUMMARY

Sum of cubes:	$A^3 + B^3 = (A + B)(A^2 - AB + B^2)$;
Difference of cubes:	$A^3 - B^3 = (A - B)(A^2 + AB + B^2)$;
Difference of squares:	$A^2 - B^2 = (A + B)(A - B)$;
Sum of squares:	$A^2 + B^2$ cannot be factored as the square of a binomial: $A^2 + B^2 \neq (A + B)^2$.

Do Exercises 24–27.

Factor.

24. $m^6 - n^6$

25. $16x^7y + 54xy^7$

26. $729x^6 - 64y^6$

27. $x^3 - 0.027$

Answers

24. $(m + n)(m^2 - mn + n^2)(m - n)(m^2 + mn + n^2)$
25. $2xy(2x^2 + 3y^2)(4x^4 - 6x^2y^2 + 9y^4)$
26. $(3x + 2y)(9x^2 - 6xy + 4y^2)(3x - 2y)(9x^2 + 6xy + 4y^2)$
27. $(x - 0.3)(x^2 + 0.3x + 0.09)$

Visualizing for Success

In each of Exercises 1–10, choose from the following list two algebraic expressions for the shaded area of the figure.

A. $(5x + 2)^2$

B. $13x$

C. $400 - 4x^2$

D. $x^2 - (x - 2y)^2$

E. $25x^2 + 20x + 4$

F. $\frac{1}{2}(x^2 - y^2)$

G. $(x + 1)^2$

H. $4y(x - y)$

I. $4(10 - x)(10 + x)$

J. $\frac{1}{2}(x - y)(x + y)$

K. $x^2 + 2x + 1$

L. $6x(14x - 5) - 3x(3x + 5)$

M. $x^2 + 9x + 20$

N. $(x - 2)^2$

O. $(x + 4)(x + 5)$

P. $8(x - 5) + (x - 5)(x - 8) + 5(x - 8)$

Q. $x^2 - 40$

R. $5x + 8x$

S. $15x(5x - 3)$

T. $x^2 - 4x + 4$

Answers on page A-15

1

5x
2
5x
2

2

x − y
x + y

3

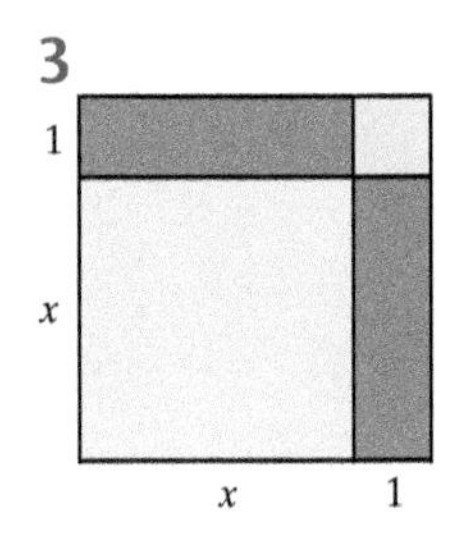

4

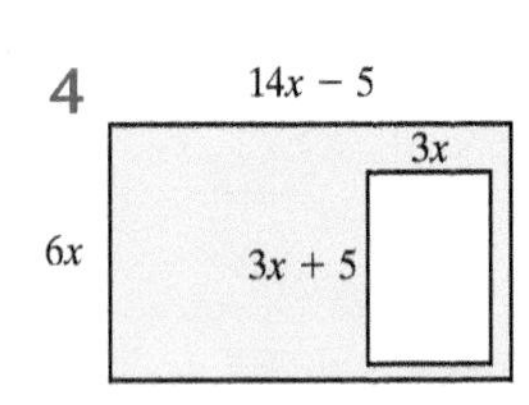

5

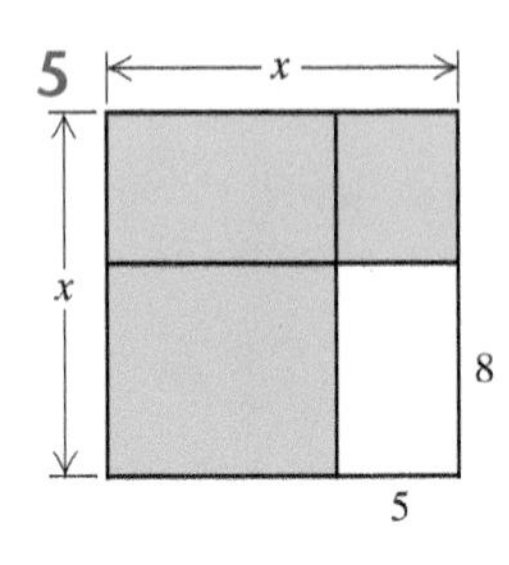

6

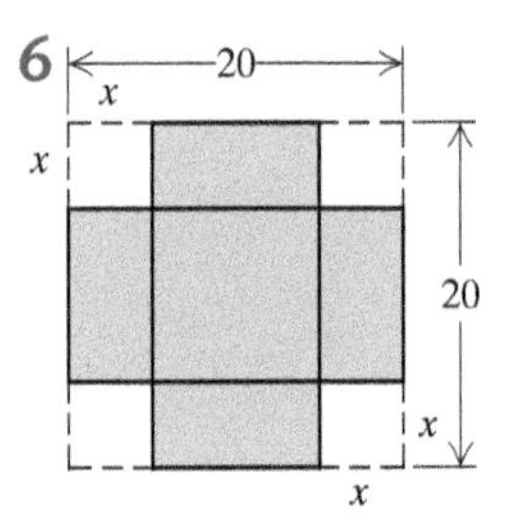

7

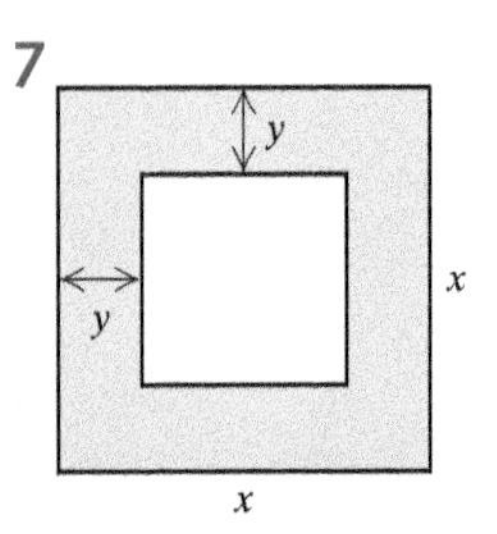

8

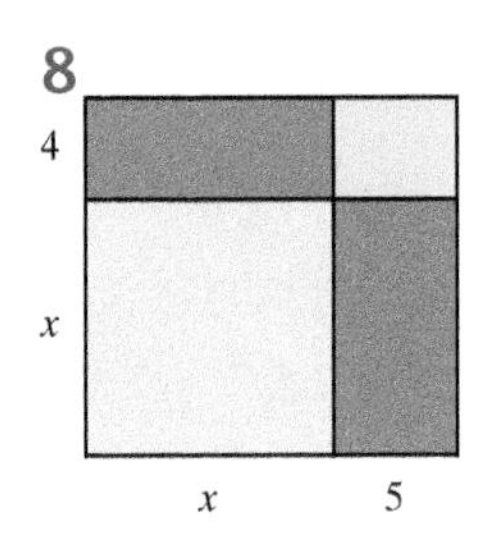

9

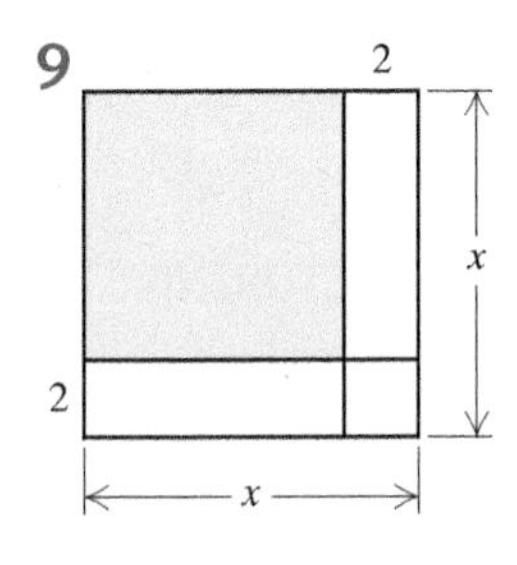

10

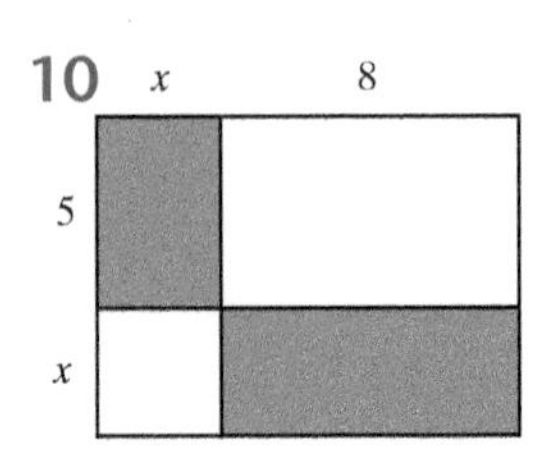

4.6 Exercise Set

FOR EXTRA HELP MyLab Math

✓ Check Your Understanding

Reading Check Classify each of the following as either a trinomial square, a difference of squares, a difference of cubes, a sum of cubes, or none of these.

RC1. $x^2 - 100$

RC2. $x^2 - 20x + 100$

RC3. $x^3 - 1000$

RC4. $4x^2 + 49$

RC5. $16x^2 + 40x + 25$

RC6. $x^2 - 9x + 6$

RC7. $27x^3 + 1$

RC8. $36x^4 - 1$

Concept Check Determine whether each polynomial is a trinomial square.

CC1. $x^2 + 2x + 1$

CC2. $x^2 + x + 1$

CC3. $4x^2 - 2x + 1$

CC4. $4x^2 - 20x + 25$

CC5. $25x^2 + 9 + 30x$

CC6. $100y^2 + 60xy + 9x^2$

a Factor.

1. $x^2 - 4x + 4$

2. $y^2 - 16y + 64$

3. $y^2 + 18y + 81$

4. $x^2 + 8x + 16$

5. $x^2 + 1 + 2x$

6. $x^2 + 1 - 2x$

7. $9y^2 + 12y + 4$

8. $25x^2 - 60x + 36$

9. $-18y^2 + y^3 + 81y$

10. $24a^2 + a^3 + 144a$

11. $12a^2 + 36a + 27$

12. $20y^2 + 100y + 125$

13. $2x^2 - 40x + 200$

14. $32x^2 + 48x + 18$

15. $1 - 8d + 16d^2$

16. $64 + 25y^2 - 80y$

17. $3a^3 - 6a^2 + 3a$

18. $5c^3 + 20c^2 + 20c$

19. $0.25x^2 + 0.30x + 0.09$

20. $0.04x^2 - 0.28x + 0.49$

21. $p^2 - 2pt + t^2$

22. $m^2 + 2mn + n^2$

23. $a^2 + 4ab + 4b^2$

24. $49x^2 - 14xy + y^2$

25. $25a^2 - 30ab + 9b^2$

26. $49p^2 - 84pw + 36w^2$

27. $y^6 + 26y^3 + 169$

28. $p^6 - 10p^3 + 25$

29. $16x^{10} - 8x^5 + 1$

30. $9x^{10} + 12x^5 + 4$

31. $x^4 + 2x^2y^2 + y^4$

32. $a^6 - 2a^3b^4 + b^8$

33. $-20x^2 + 60x - 45$

34. $-75x^2 + 60x - 12$

35. $-18a^4 + 60a^5 - 50a^6$

36. $-3c^2 + 24c^3 - 48c^4$

b Factor.

37. $p^2 - 49$

38. $m^2 - 64$

39. $y^4 - 8y^2 + 16$

40. $y^4 - 18y^2 + 81$

41. $p^2q^2 - 25$

42. $a^2b^2 - 81$

43. $6x^2 - 6y^2$

44. $8x^2 - 8y^2$

45. $4xy^4 - 4xz^4$

46. $25ab^4 - 25az^4$

47. $4a^3 - 49a$

48. $9x^3 - 25x$

49. $3x^8 - 3y^8$

50. $2a^9 - 32a$

51. $9a^4 - 25a^2b^4$

52. $16x^6 - 121x^2y^4$

53. $\frac{1}{36} - z^2$

54. $\frac{1}{100} - y^2$

55. $0.04x^2 - 0.09y^2$

56. $0.01x^2 - 0.04y^2$

c Factor.

57. $m^3 - 7m^2 - 4m + 28$

58. $x^3 + 8x^2 - x - 8$

59. $a^3 - ab^2 - 2a^2 + 2b^2$

60. $p^2q - 25q + 3p^2 - 75$

61. $(a + b)^2 - 100$

62. $(p - 7)^2 - 144$

63. $144 - (p - 8)^2$

64. $100 - (x - 4)^2$

65. $a^2 + 2ab + b^2 - 9$

66. $x^2 - 2xy + y^2 - 25$

67. $r^2 - 2r + 1 - 4s^2$

68. $c^2 + 4cd + 4d^2 - 9p^2$

69. $2m^2 + 4mn + 2n^2 - 50b^2$

70. $12x^2 + 12x + 3 - 3y^2$

71. $9 - (a^2 + 2ab + b^2)$

72. $16 - (x^2 - 2xy + y^2)$

d Factor.

73. $z^3 + 27$

74. $a^3 + 8$

75. $x^3 - 1$

76. $c^3 - 64$

77. $8 - 27b^3$

78. $64 - 125x^3$

79. $8a^3 + 1$

80. $27x^3 + 1$

81. $8x^3 + 27$

82. $27y^3 + 64$

83. $a^3 - b^3$

84. $x^3 - y^3$

85. $a^3 + \frac{1}{8}$

86. $b^3 + \frac{1}{27}$

87. $x^3 + 0.001$

88. $y^3 + 0.125$

89. $2y^3 - 128$

90. $3z^3 - 3$

91. $24a^3 + 3$

92. $54x^3 + 2$

93. $rs^3 + 64r$

94. $ab^3 + 125a$

95. $5x^5 - 40x^2z^3$

96. $2y^3z^4 - 54z^7$

97. $64x^6 - 8t^6$

98. $125c^6 - 8d^6$

99. $z^6 - 1$

100. $p^6 - q^6$

101. $t^6 + 64y^6$

102. $t^6 + 1$

103. $8w^9 - z^9$

104. $a^9 + 64b^9$

105. $x^6 - 64y^6$

106. $64c^6 - d^6$

107. $729w^6 - 64z^{12}$

108. $64x^{12} - 729y^{18}$

109. $\frac{1}{8}c^3 + d^3$

110. $\frac{27}{125}x^3 - y^3$

111. $0.001x^3 - 0.008y^3$

112. $0.125r^3 - 0.216s^3$

Skill Maintenance

Solve. [3.2a], [3.3a]

113. $7x - 2y = -11,$
$2x + 7y = 18$

114. $y = 3x - 8,$
$4x - 6y = 100$

115. $x - y = -12,$
$x + y = 14$

116. $7x - 2y = -11,$
$2y - 7x = -18$

Graph the given system of inequalities and determine coordinates of any vertices formed. [3.7c]

117. $x - y \leq 5,$
$x + y \geq 3$

118. $x - y \leq 5,$
$x + y \geq 3,$
$x \leq 6$

119. $x - y \geq 5,$
$x + y \leq 3,$
$x \geq 1$

120. $x - y \geq 5,$
$x + y \leq 3$

Given the line and a point not on the line, find an equation through the point parallel to the given line, and find an equation through the point perpendicular to the given line. [2.6d]

121. $x - y = 5; (-2, -4)$

122. $2x - 3y = 6; (1, -7)$

123. $y = -\frac{1}{2}x + 3; (4, 5)$

124. $x - 4y = -10; (6, 0)$

Synthesis

125. Given that $P(x) = x^3$, use factoring to simplify $P(a + h) - P(a)$.

126. Given that $P(x) = x^4$, use factoring to simplify $P(a + h) - P(a)$.

127. *Volume of Carpeting.* The volume of a carpet that is rolled up can be estimated by the polynomial $\pi R^2h - \pi r^2h$, where R and r represent the radii of two circles, as shown in the following figure.

a) Factor the polynomial.
b) Use both the original form and the factored form to find the volume of a roll for which $R = 50$ cm, $r = 10$ cm, and $h = 4$ m. Use 3.14 for π.

128. Show how the geometric model below can be used to verify the formula for factoring $a^3 - b^3$.

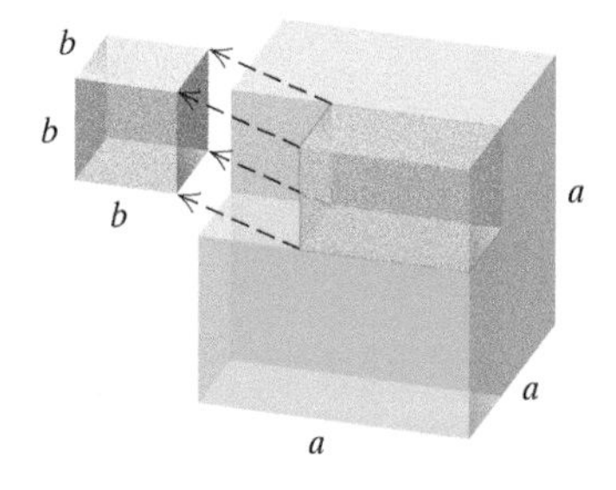

Factor. Assume that variables in exponents represent positive integers.

129. $5c^{100} - 80d^{100}$

130. $9x^{2n} - 6x^n + 1$

131. $x^{6a} + y^{3b}$

132. $a^3x^3 - b^3y^3$

133. $3x^{3a} + 24y^{3b}$

134. $\frac{8}{27}x^3 + \frac{1}{64}y^3$

135. $\frac{1}{24}x^3y^3 + \frac{1}{3}z^3$

136. $7x^3 - \frac{7}{8}$

137. $(x + y)^3 - x^3$

138. $(1 - x)^3 + (x - 1)^6$

139. $(a + 2)^3 - (a - 2)^3$

140. $y^4 - 8y^3 - y + 8$

Factoring: A General Strategy

4.7

OBJECTIVE

a Factor polynomials completely using any of the methods considered in this chapter.

a A GENERAL FACTORING STRATEGY

Identify types of polynomials. [4.1a]
Determine whether the polynomial is either a monomial, a binomial, a trinomial, or none of these.

1. $49 - 64x^2$ **2.** $5x^7 + 3x^2 - 25x$

Answers: 1. Binomial **2.** Trinomial

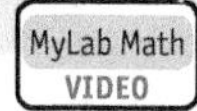

Factoring is an important algebraic skill, used for solving equations and many other manipulations of algebraic symbolism. We now consider polynomials of many types and learn to use a general strategy for factoring. The key is to recognize the type of polynomial to be factored.

A STRATEGY FOR FACTORING

a) Always look for a *common factor* (other than 1 or -1). If there are any, factor out the largest one.

b) Then look at the number of terms.

Two terms: Try factoring as a difference of squares first. Next, try factoring as a sum or a difference of cubes. Do *not* try to factor a *sum* of squares: $A^2 + B^2$.

Three terms: Determine whether the expression is a trinomial square. If it is, you know how to factor. If not, try the trial-and-error method or the *ac*–method.

Four or more terms: Try factoring by grouping and removing a common binomial factor. Next, try grouping into a difference of squares, one of which is a trinomial.

c) Always *factor completely*. If a factor with more than one term can be factored, you should factor it.

d) Always *check* by multiplying.

EXAMPLE 1 Factor: $10a^2x - 40b^2x$.

a) We look first for a common factor:

$10x(a^2 - 4b^2)$. Factoring out the largest common factor

b) The factor $a^2 - 4b^2$ has only two terms. It is a difference of squares. We factor it, keeping the common factor: $10x(a + 2b)(a - 2b)$.

c) Have we factored completely? Yes, because none of the factors with more than one term can be factored further using polynomials of smaller degree.

d) *Check*: $10x(a + 2b)(a - 2b) = 10x(a^2 - 4b^2) = 10xa^2 - 40xb^2$, or $10a^2x - 40b^2x$.

Factor completely.

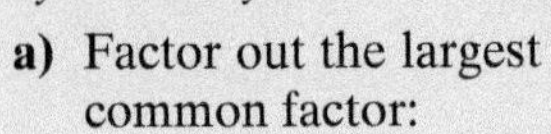

1. $3y^3 - 12x^2y$ **GS**

   a) Factor out the largest common factor:

   $3y^3 - 12x^2y = \underline{\quad\quad}(y^2 - 4x^2)$.

   b) There are two terms inside the parentheses. The expression is a difference of squares. Factor the difference of squares:

   $3y(y^2 - 4x^2)$
   $= 3y(y + \underline{\quad\quad})(y - \underline{\quad\quad})$.

   c) We have factored completely.

   d) *Check:*

   $3y(y + 2x)(y - 2x)$
   $= 3y(y^2 - 4x^2)$
   $= 3y^3 - 12x^2y$.

2. $50x^7 + 40x^4 + 8x$

3. $3x - 6 - bx^2 + 2bx$

EXAMPLE 2 Factor: $2x^2 + 50a^2 - 20ax$.

a) We remove the largest common factor: $2(x^2 + 25a^2 - 10ax)$.

b) In the parentheses, there are three terms. The trinomial is a square. We factor it: $2(x - 5a)^2$.

c) None of the factors with more than one term can be factored further.

d) *Check:* $2(x - 5a)^2 = 2(x^2 - 10ax + 25a^2) = 2x^2 - 20ax + 50a^2$, or $2x^2 + 50a^2 - 20ax$.

EXAMPLE 3 Factor: $3x + 12 + ax^2 + 4ax$.

a) There is no common factor (other than 1 or -1).

b) There are four terms. We try grouping to remove a common binomial factor:

$3(x + 4) + ax(x + 4)$	Factoring two grouped binomials
$= (x + 4)(3 + ax)$.	Factoring out the common binomial factor

c) None of the factors with more than one term can be factored further.

d) *Check:* $(x + 4)(3 + ax) = 3x + ax^2 + 12 + 4ax$, or $3x + 12 + ax^2 + 4ax$.

Do Exercises 1–3.

EXAMPLE 4 Factor: $x^6 - y^6$.

a) We look for a common factor. There isn't one (other than 1 or -1).

b) There are only two terms. The binomial is a difference of squares and also a difference of cubes. We factor first as a difference of squares: $(x^3)^2 - (y^3)^2$. Factoring, we have: $(x^3 + y^3)(x^3 - y^3)$.

c) One factor is a sum of two cubes, and the other factor is a difference of two cubes. We factor them:

$$x^6 - y^6 = (x^3 + y^3)(x^3 - y^3)$$
$$= (x + y)(x^2 - xy + y^2)(x - y)(x^2 + xy + y^2).$$

We have now factored completely because none of the factors can be factored further using polynomials of smaller degree.

d) *Check:* $(x + y)(x^2 - xy + y^2)(x - y)(x^2 + xy + y^2)$
$= (x^3 + y^3)(x^3 - y^3) = (x^3)^2 - (y^3)^2 = x^6 - y^6$.

EXAMPLE 5 Factor: $x^3 - xy^2 + x^2y - y^3$.

a) There is no common factor (other than 1 or -1).

b) There are four terms. We factor by grouping:

$x(x^2 - y^2) + y(x^2 - y^2)$	Factoring two grouped binomials
$= (x^2 - y^2)(x + y)$.	Factoring out the common binomial factor

c) The factor $x^2 - y^2$ can be factored further, giving

$(x + y)(x - y)(x + y)$. Factoring a difference of squares

None of the factors with more than one term can be factored further, so we have factored completely.

d) *Check:* $(x + y)(x - y)(x + y) = (x^2 - y^2)(x + y)$
$= x^3 + x^2y - y^2x - y^3$, or
$x^3 - xy^2 + x^2y - y^3$.

Answers

1. $3y(y + 2x)(y - 2x)$ **2.** $2x(5x^3 + 2)^2$
3. $(x - 2)(3 - bx)$

Guided Solution:
1. $3y, 2x, 2x$

EXAMPLE 6 Factor: $10x^6 + 40y^2$.

a) We remove the largest common factor: $10(x^6 + 4y^2)$.

b) In the parentheses, there are two terms, a sum of squares, which cannot be factored.

c) We have factored $10x^6 + 40y^2$ completely as $10(x^6 + 4y^2)$.

d) *Check*: $10(x^6 + 4y^2) = 10x^6 + 40y^2$.

EXAMPLE 7 Factor: $6x^2 - 20x - 16$.

a) We remove the largest common factor: $2(3x^2 - 10x - 8)$.

b) In the parentheses, there are three terms. The trinomial is not a square. We factor: $2(x - 4)(3x + 2)$.

c) We cannot factor further.

d) *Check*: $2(x - 4)(3x + 2) = 2(3x^2 - 10x - 8) = 6x^2 - 20x - 16$.

EXAMPLE 8 Factor: $y^2 - 9a^2 + 12y + 36$.

a) There is no common factor (other than 1 or −1).

b) There are four terms. We try grouping to remove a common binomial factor, but that is not possible. We try grouping as a difference of squares:

$$(y^2 + 12y + 36) - 9a^2 = (y + 6)^2 - (3a)^2$$
$$= (y + 6 + 3a)(y + 6 - 3a).$$
Factoring the difference of squares

c) No factor with more than one term can be factored further.

d) *Check*: $(y + 6 + 3a)(y + 6 - 3a) = [(y + 6) + 3a][(y + 6) - 3a]$
$= (y + 6)^2 - (3a)^2$
$= y^2 + 12y + 36 - 9a^2$, or
$y^2 - 9a^2 + 12y + 36$.

Do Exercises 4–9.

Factor.

4. $7a^3 - 7$
5. $6x^2 - 3x - 18$
6. $5y^4 + 20x^6$
7. $a^3 - ab^2 - a^2b + b^3$
8. $64x^6 - 729y^6$
9. $2x^2 - 20x + 50 - 18b^2$

Answers

4. $7(a - 1)(a^2 + a + 1)$
5. $3(x - 2)(2x + 3)$ **6.** $5(y^4 + 4x^6)$
7. $(a - b)^2(a + b)$
8. $(2x + 3y)(4x^2 - 6xy + 9y^2) \times (2x - 3y)(4x^2 + 6xy + 9y^2)$
9. $2(x - 5 + 3b)(x - 5 - 3b)$

4.7 Exercise Set

FOR EXTRA HELP

MyLab Math

✓ Check Your Understanding

Reading Check Choose from the column on the right the word that best completes each step in the factoring strategy.

RC1. Always look first for a ______________ factor.

RC2. If there are two terms, determine whether the binomial is a ______________ of squares, a sum of cubes, or a difference of cubes.

RC3. If there are three terms, determine whether the trinomial is a ______________.

RC4. If there are four terms, try factoring by ______________.

RC5. Always factor ______________.

RC6. Always ______________ by multiplying.

check
common
completely
difference
grouping
square

Concept Check Choose from the column on the right the description that best corresponds to the type of polynomial. Choices may be used more than once.

CC1. $16x^2 - 25y^2$

CC2. $27a^6 - 1$

CC3. $9a^6 + 100b^8$

CC4. $300c^4 - 12d^2$

CC5. $8x^3 + y^6z^{12}$

CC6. $4x^2 + 4x + 16$

CC7. $4x^2 - 28x + 49$

CC8. $4x^2 + 20x - 25$

a) Polynomial with a common factor

b) Difference of two squares

c) Sum of two cubes

d) Difference of two cubes

e) Trinomial square

f) None of these

a Factor completely.

1. $y^2 - 225$

2. $x^2 - 400$

3. $2x^2 + 11x + 12$

4. $8a^2 + 18a - 5$

5. $5x^4 - 20$

6. $3xy^2 - 75x$

7. $p^2 + 36 + 12p$

8. $a^2 + 49 + 14a$

9. $2x^2 - 10x - 132$

10. $3y^2 - 15y - 252$

11. $9x^2 - 25y^2$

12. $16a^2 - 81b^2$

13. $4m^4 - 100$

14. $2x^2 - 288$

15. $6w^2 + 12w - 18$

16. $8z^2 - 8z - 16$

17. $2xy^2 - 50x$

18. $3a^3b - 108ab$

19. $225 - (a - 3)^2$

20. $625 - (t - 10)^2$

21. $6x^5 + 54x^3$

22. $36y^2 + 9z^4$

23. $m^6 - 1$

24. $64t^6 - 1$

25. $x^2 + 6x - y^2 + 9$

26. $t^2 + 10t - p^2 + 25$

27. $250x^3 - 128y^3$

28. $27a^3 - 343b^3$

29. $8m^3 + m^6 - 20$

30. $-37x^2 + x^4 + 36$

31. $ac + cd - ab - bd$

32. $xw - yw + xz - yz$

33. $50b^2 - 5ab - a^2$

34. $9c^2 + 12cd - 5d^2$

35. $10a + 10 - 5a^2c - 5ac$

36. $18x - 18 + 3x^2y - 3xy$

37. $-7x^2 + 2x^3 + 4x - 14$

38. $9m^2 + 3m^3 + 8m + 24$

39. $2x^3 + 6x^2 - 8x - 24$

40. $3x^3 + 6x^2 - 27x - 54$

41. $16x^3 + 54y^3$

42. $250a^3 + 54b^3$

43. $6y - 60x^2y - 9xy$

44. $2b - 28a^2b + 10ab$

45. $a^8 - b^8$

46. $2x^4 - 32$

47. $a^3b - 16ab^3$

48. $x^3y - 25xy^3$

49. $\frac{1}{16}x^2 - \frac{1}{6}xy^2 + \frac{1}{9}y^4$

50. $36x^2 + 15x + \frac{25}{16}$

51. $5x^3 - 5x^2y - 5xy^2 + 5y^3$

52. $a^3 - ab^2 + a^2b - b^3$

53. $42ab + 27a^2b^2 + 8$

54. $-23xy + 20x^2y^2 + 6$

55. $8y^4 - 125y$

56. $64p^4 - p$

57. $8x^3 + 44x^2 + 60x$

58. $15a^3 - 115a^2 + 70a$

59. $3a^5b^4 - 75a$

60. $2c - 2c^7d^4$

61. $a^2 - b^2 - 6b - 9$

62. $m^2 - n^2 - 8n - 16$

63. $q^2 - 10q + 25 - r^2$

64. $y^2 - 14y + 49 - z^2$

Skill Maintenance

Solve. [3.2b]

65. *Exam Scores.* There are 75 questions on a college entrance examination. Two points are awarded for each correct answer, and one half point is deducted for each incorrect answer. A score of 100 indicates how many correct and how many incorrect answers, assuming that all questions are answered?

66. *Perimeter.* A pentagon with all five sides the same length has the same perimeter as an octagon with all eight sides the same length. One side of the pentagon is 2 less than three times the length of one side of the octagon. Find the perimeters.

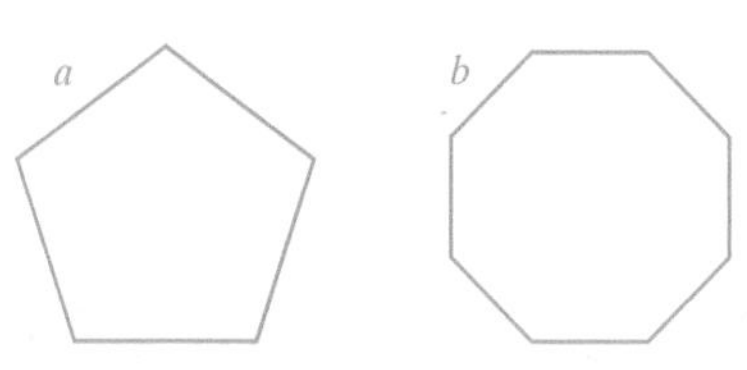

Synthesis

Factor. Assume that variables in exponents represent natural numbers.

67. $30y^4 - 97xy^2 + 60x^2$

68. $3x^2y^2z + 25xyz^2 + 28z^3$

69. $5x^3 - \frac{5}{27}$

70. $9y^3 - \frac{9}{1000}$

71. $(x - p)^2 - p^2$

72. $s^6 - 729t^6$

73. $(y - 1)^4 - (y - 1)^2$

74. $27x^{6s} + 64y^{3t}$

75. $4x^2 + 4xy + y^2 - r^2 + 6rs - 9s^2$

76. $c^4d^4 - a^{16}$

77. $c^{2w+1} + 2c^{w+1} + c$

78. $24x^{2a} - 6$

79. $3(x + 1)^2 + 9(x + 1) - 12$

80. $8(a - 3)^2 - 64(a - 3) + 128$

81. $x^6 - 2x^5 + x^4 - x^2 + 2x - 1$

82. $1 - \dfrac{x^{27}}{1000}$

83. $y^9 - y$

84. $(m - 1)^3 - (m + 1)^3$

4.8 Applications of Polynomial Equations and Functions

OBJECTIVES

a Solve quadratic and other polynomial equations by first factoring and then using the principle of zero products.

b Solve applied problems involving quadratic and other polynomial equations that can be solved by factoring.

SKILL REVIEW *Factor polynomials whose terms have a common factor.* [4.3a]

Factor.

1. $x^2 + 20x$

2. $3y^2 - 6y$

Answers: 1. $x(x + 20)$ **2.** $3y(y - 2)$

MyLab Math VIDEO

Whenever two polynomials are set equal to each other, we have a **polynomial equation**. Some examples of polynomial equations are

$$4x^3 + x^2 + 5x = 6x - 3,$$
$$x^2 - x = 6,$$
and $3y - 7 = 0$

A second-degree polynomial equation in one variable is often called a **quadratic equation**. Of the equations listed above, only $x^2 - x = 6$ is a quadratic equation.

Polynomial equations, and quadratic equations in particular, occur frequently in applications, so the ability to solve them is an important skill. One way of solving certain polynomial equations involves factoring.

a THE PRINCIPLE OF ZERO PRODUCTS

MyLab Math ANIMATION

When we multiply two or more numbers, if either factor is 0, then the product is 0. Conversely, if a product is 0, then at least one of the factors must be 0. This property of 0 gives us a new principle for solving equations.

THE PRINCIPLE OF ZERO PRODUCTS

For any real numbers a and b:

If $ab = 0$, then $a = 0$ *or* $b = 0$ (or both).

If $a = 0$ *or* $b = 0$, then $ab = 0$.

To solve an equation using the principle of zero products, we first write it in *standard form*: with 0 on one side of the equation and the leading coefficient positive.

EXAMPLE 1 Solve: $x^2 - x = 6$.

In order to use the principle of zero products, we must have 0 on one side of the equation, so we subtract 6 on both sides:

$$x^2 - x - 6 = 0. \qquad \text{Getting 0 on one side}$$

We need a factorization on the other side, so we factor the polynomial:

$$(x - 3)(x + 2) = 0. \qquad \text{Factoring}$$

We now have two expressions, $x - 3$ and $x + 2$, whose product is 0. Using the principle of zero products, we set each expression or factor equal to 0:

$$x - 3 = 0 \quad or \quad x + 2 = 0. \qquad \text{Using the principle of zero products}$$

This gives us two simple linear equations. We solve them separately,

$$x = 3 \quad or \quad x = -2,$$

and check in the original equation as follows.

Check:

$$\begin{array}{c|l} x^2 - x = 6 & \\ \hline 3^2 - 3 \ ?\ 6 & \\ 9 - 3 & \\ 6 & \text{TRUE} \end{array} \qquad \begin{array}{c|l} x^2 - x = 6 & \\ \hline (-2)^2 - (-2) \ ?\ 6 & \\ 4 + 2 & \\ 6 & \text{TRUE} \end{array}$$

The numbers 3 and -2 are both solutions.

To solve an equation using the principle of zero products:

1. Obtain a 0 on one side of the equation.
2. Factor the other side.
3. Set each factor equal to 0.
4. Solve the resulting equations.

Do Exercise 1.

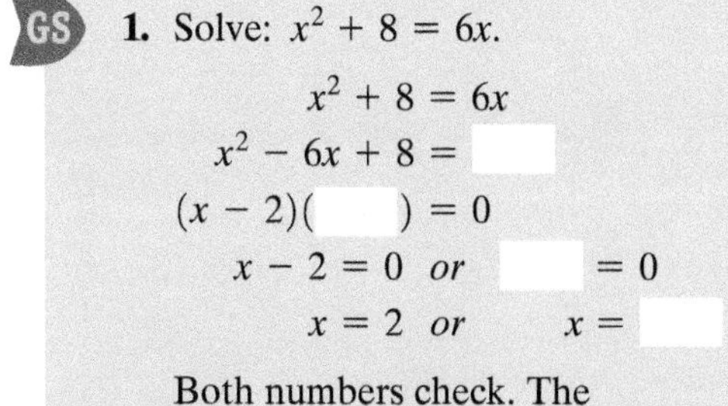

GS **1.** Solve: $x^2 + 8 = 6x$.

$$x^2 + 8 = 6x$$
$$x^2 - 6x + 8 = \underline{\quad}$$
$$(x - 2)(\underline{\quad}) = 0$$
$$x - 2 = 0 \ \ or \ \ \underline{\quad} = 0$$
$$x = 2 \ \ or \ \ x = \underline{\quad}$$

Both numbers check. The solutions are 2 and ___.

When you solve an equation using the principle of zero products, you can always check by substitution as we did in Example 1.

Caution!

When we are using the principle of zero products, it is important to be sure that there is a 0 on one side of the equation. If neither side of the equation is 0, the procedure will not work.

For example, consider $x^2 - x = 6$ in Example 1 as

$$x(x - 1) = 6.$$

Suppose that we reasoned as follows, setting factors equal to 6:

$$x = 6 \quad or \quad x - 1 = 6 \qquad \text{This step is incorrect!}$$
$$x = 7.$$

Neither 6 nor 7 checks, as shown below:

$$\begin{array}{c|l} x(x - 1) = 6 & \\ \hline 6(6 - 1) \ ?\ 6 & \\ 6(5) & \\ 30 & \text{FALSE} \end{array} \qquad \begin{array}{c|l} x(x - 1) = 6 & \\ \hline 7(7 - 1) \ ?\ 6 & \\ 7(6) & \\ 42 & \text{FALSE} \end{array}$$

EXAMPLE 2 Solve: $7y + 3y^2 = -2$.

Since there must be a 0 on one side of the equation, we add 2 to get 0 on the right-hand side and arrange in descending order. Then we factor and use the principle of zero products.

$$7y + 3y^2 = -2$$
$$3y^2 + 7y + 2 = 0 \qquad \text{Getting 0 on one side}$$
$$(3y + 1)(y + 2) = 0 \qquad \text{Factoring}$$
$$3y + 1 = 0 \quad or \quad y + 2 = 0 \qquad \text{Using the principle of zero products}$$
$$y = -\tfrac{1}{3} \quad or \quad y = -2$$

The solutions are $-\frac{1}{3}$ and -2.

Do Exercise 2.

2. Solve: $5y + 2y^2 = 3$.

Answers

1. 4, 2 **2.** $\frac{1}{2}, -3$

Guided Solution:
1. $0, x - 4, x - 4, 4, 4$

EXAMPLE 3 Solve: $5b^2 = 10b$.

$$\begin{aligned} 5b^2 &= 10b \\ 5b^2 - 10b &= 0 && \text{Getting 0 on one side} \\ 5b(b-2) &= 0 && \text{Factoring} \\ 5b = 0 \quad &or \quad b - 2 = 0 && \text{Using the principle of zero products} \\ b = 0 \quad &or \quad b = 2 \end{aligned}$$

The solutions are 0 and 2.

Do Exercise 3.

3. Solve: $8b^2 = 16b$.

EXAMPLE 4 Solve: $x^2 - 6x + 9 = 0$.

$$\begin{aligned} x^2 - 6x + 9 &= 0 && \text{Getting 0 on one side} \\ (x-3)(x-3) &= 0 && \text{Factoring} \\ x - 3 = 0 \quad &or \quad x - 3 = 0 && \text{Using the principle of zero products} \\ x = 3 \quad &or \quad x = 3 \end{aligned}$$

There is only one solution, 3.

Do Exercise 4.

4. Solve: $25 + x^2 = -10x$.

EXAMPLE 5 Solve: $3x^3 - 9x^2 = 30x$.

$$\begin{aligned} 3x^3 - 9x^2 &= 30x \\ 3x^3 - 9x^2 - 30x &= 0 && \text{Getting 0 on one side} \\ 3x(x^2 - 3x - 10) &= 0 && \text{Factoring out a common factor} \\ 3x(x+2)(x-5) &= 0 && \text{Factoring the trinomial} \\ 3x = 0 \quad or \quad x + 2 = 0 \quad &or \quad x - 5 = 0 && \text{Using the principle of zero products} \\ x = 0 \quad or \quad x = -2 \quad &or \quad x = 5 \end{aligned}$$

The solutions are 0, −2, and 5.

Do Exercise 5.

5. Solve: $x^3 + x^2 = 6x$.

EXAMPLE 6 Given that $f(x) = 3x^2 - 4x$, find all values of x for which $f(x) = 4$.

We want all numbers x for which $f(x) = 4$. Since $f(x) = 3x^2 - 4x$, we must have

$$\begin{aligned} 3x^2 - 4x &= 4 && \text{Setting } f(x) \text{ equal to 4} \\ 3x^2 - 4x - 4 &= 0 && \text{Getting 0 on one side} \\ (3x+2)(x-2) &= 0 && \text{Factoring} \\ 3x + 2 = 0 \quad &or \quad x - 2 = 0 \\ x = -\tfrac{2}{3} \quad &or \quad x = 2. \end{aligned}$$

We can check as follows.

$$f\left(-\tfrac{2}{3}\right) = 3\left(-\tfrac{2}{3}\right)^2 - 4\left(-\tfrac{2}{3}\right) = 3 \cdot \tfrac{4}{9} + \tfrac{8}{3} = \tfrac{4}{3} + \tfrac{8}{3} = \tfrac{12}{3} = 4;$$
$$f(2) = 3(2)^2 - 4(2) = 3 \cdot 4 - 8 = 12 - 8 = 4$$

To have $f(x) = 4$, we must have $x = -\frac{2}{3}$ or $x = 2$.

Do Exercise 6.

6. Given that $f(x) = 10x^2 + 13x$, find all values of x for which $f(x) = 3$.

Answers

3. 0, 2 **4.** −5 **5.** 0, 2, −3 **6.** $-\frac{3}{2}, \frac{1}{5}$

EXAMPLE 7 Find the domain of F if $F(x) = \frac{x - 2}{x^2 + 2x - 15}$.

The domain of F is the set of all values for which

$$\frac{x - 2}{x^2 + 2x - 15}$$

is a real number. Since division by 0 is undefined, $F(x)$ cannot be calculated for any x-value for which the denominator, $x^2 + 2x - 15$, is 0. To make sure that these values are *excluded*, we solve:

$$x^2 + 2x - 15 = 0 \quad \text{Setting the denominator equal to 0}$$
$$(x - 3)(x + 5) = 0 \quad \text{Factoring}$$
$$x - 3 = 0 \quad or \quad x + 5 = 0$$
$$x = 3 \quad or \quad x = -5. \quad \text{These are the values to } exclude.$$

The domain of F is $\{x | x \text{ is a real number } and\ x \neq -5\ and\ x \neq 3\}$.

Do Exercise 7. ▶

7. Find the domain of the function G if

$$G(x) = \frac{2x - 9}{x^2 - 3x - 28}.$$

EXAMPLE 8 Use the graph of $f(x) = x^2 + 6x + 8$ to find the x-intercepts of the graph and the solutions of the equation $x^2 + 6x + 8 = 0$.

The graph of the function $f(x) = x^2 + 6x + 8$ and its x-intercepts are shown below.

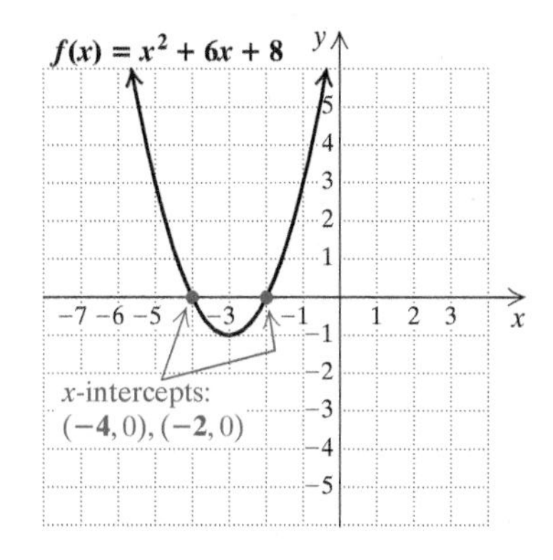

The x-intercepts are $(-4, 0)$ and $(-2, 0)$. These pairs are also the points of intersection of the graphs of $f(x) = x^2 + 6x + 8$ and $g(x) = 0$ (the x-axis). Thus, -4 and -2 are the solutions of the equation $x^2 + 6x + 8 = 0$.

To verify this, let's solve the equation $x^2 + 6x + 8 = 0$.

$$x^2 + 6x + 8 = 0$$
$$(x + 4)(x + 2) = 0 \quad \text{Factoring}$$
$$x + 4 = 0 \quad or \quad x + 2 = 0 \quad \text{Principle of zero products}$$
$$x = -4 \quad or \quad x = -2.$$

We see that the solutions of $x^2 + 6x + 8 = 0$, -4 and -2, are the first coordinates of the x-intercepts, $(-4, 0)$ and $(-2, 0)$, of the graph of $f(x) = x^2 + 6x + 8$.

Do Exercise 8. ▶

8. Consider solving the equation

$$x^2 - 6x + 8 = 0$$

graphically.

a) Below is the graph of $f(x) = x^2 - 6x + 8$. Use *only* the graph to find the x-intercepts of the graph.

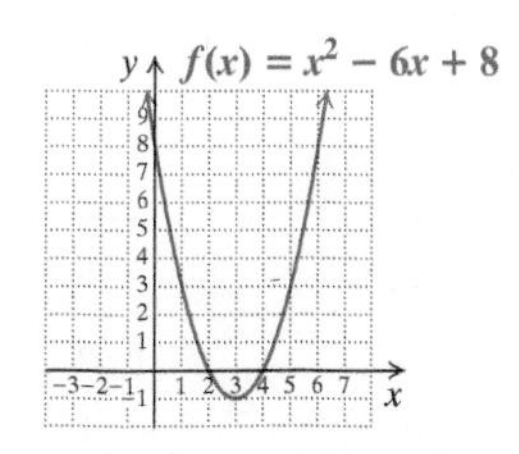

b) Use *only* the graph to find the solutions of $x^2 - 6x + 8 = 0$.

c) Compare your answers to parts (a) and (b).

Answers

7. $\{x | x \text{ is a real number } and\ x \neq -4\ and\ x \neq 7\}$
8. (a) $(2, 0)$ and $(4, 0)$; **(b)** 2, 4; **(c)** The solutions of $x^2 - 6x + 8 = 0$, 2 and 4, are the first coordinates of the x-intercepts, $(2, 0)$ and $(4, 0)$, of the graph of $f(x) = x^2 - 6x + 8$.

CALCULATOR CORNER

Solving Quadratic Equations To solve the equation $x^2 - x = 6$ graphically, we can first write the equation with 0 on one side. We get $x^2 - x - 6 = 0$. Next, we graph $y = x^2 - x - 6$ in a window that shows the x-intercepts. The standard window works well in this case.

The solutions of the equation are the values of x for which $x^2 - x - 6 = 0$. These are also the first coordinates of the x-intercepts of the graph. To find the solution corresponding to the leftmost x-intercept, we first select the ZERO feature from the CALC menu. The prompt "Left Bound?" appears. We use the ◁ key or the ▷ key to move the cursor to the left of the intercept and press ENTER. Now, the prompt "Right Bound?" appears. We move the cursor to the right of the intercept and press ENTER. Next, the prompt "Guess?" appears. We move the cursor close to the intercept and press ENTER again. We now see the cursor positioned at the leftmost x-intercept and the coordinates of that point, $x = -2, y = 0$, are displayed. Thus, $x^2 - x - 6 = 0$ when $x = -2$. This is one solution of the equation.

We repeat this procedure to find the first coordinate of the other x-intercept. We see that $x = 3$ at that point. Thus the solutions of the equation are -2 and 3.

$y = x^2 - x - 6$

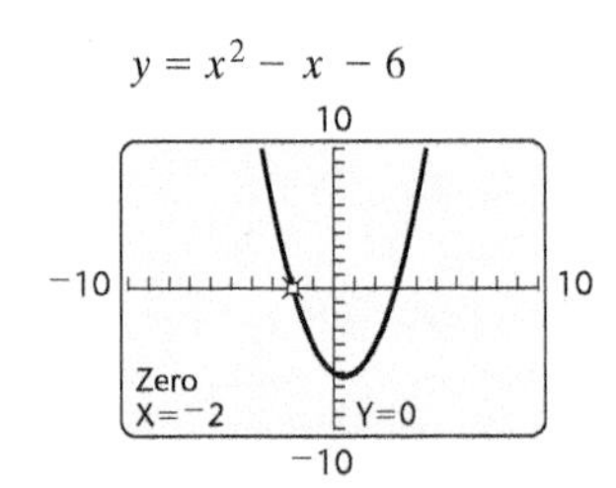

$y = x^2 - x - 6$

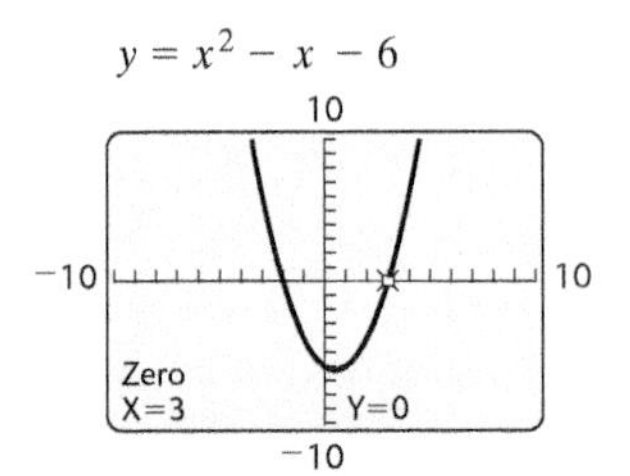

This equation could also be solved by entering $y_1 = x^2 - x$ and $y_2 = 6$ and finding the first coordinate of the points of intersection using the INTERSECT feature.

$y_1 = x^2 - x, \; y_2 = 6$

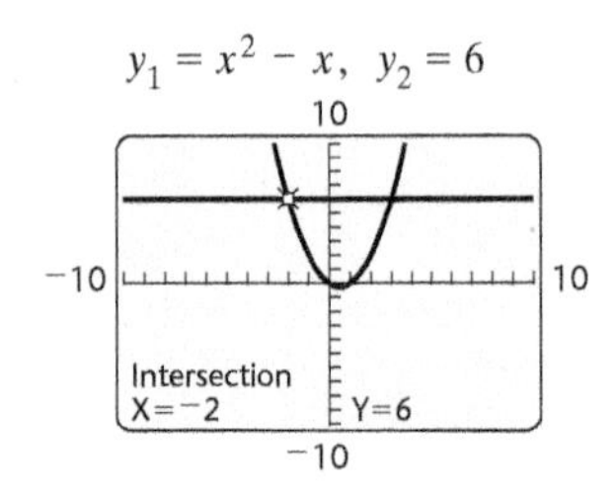

$y_1 = x^2 - x, \; y_2 = 6$

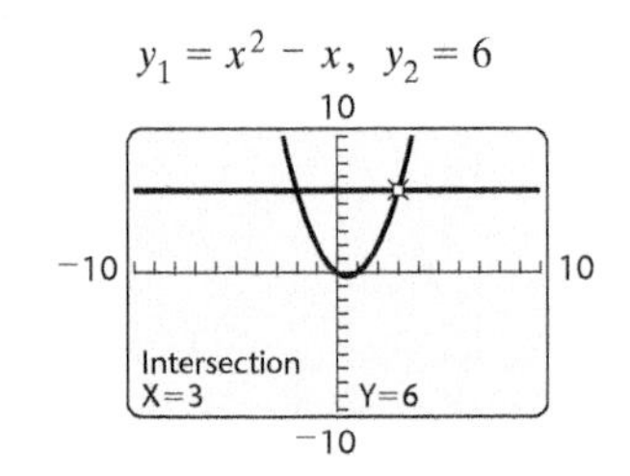

EXERCISE:

1. Solve the equations in Examples 2–5 graphically. Note that, regardless of the variable used in an example, each equation should be entered on the equation-editor screen in terms of x.

b APPLICATIONS AND PROBLEM SOLVING

Some problems can be translated to quadratic equations. The problem-solving process is the same one that we use for other kinds of applied problems.

EXAMPLE 9 *Prize Tee Shirts.* During intermission at sporting events, team mascots commonly use a powerful slingshot to launch tightly rolled tee shirts into the stands. The height $h(t)$, in feet, of an airborne tee shirt t seconds after having been launched can be approximated by

$$h(t) = -15t^2 + 75t + 10.$$

After peaking, a rolled-up tee shirt is caught by a fan 70 ft above ground level. For how long was the tee shirt in the air?

1. **Familiarize.** We make a drawing and label it, using the information provided. (See the figure at right.) We could evaluate $h(t)$ for a few values of t. Note that t cannot be negative, since it represents time from launch.

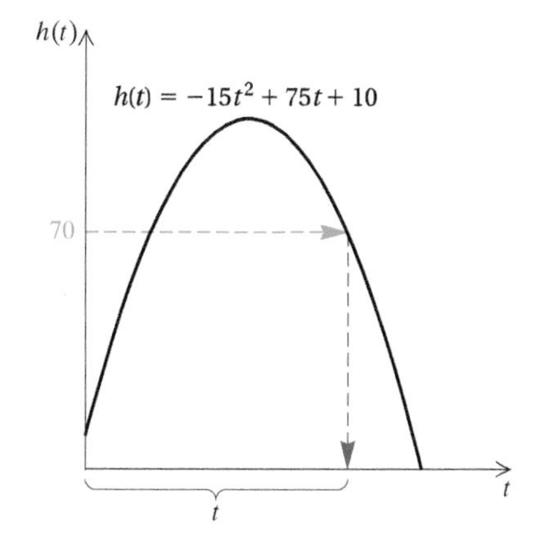

2. **Translate.** The function is given. Since we are asked to determine how long it will take for the shirt to reach someone 70 ft above ground level, we are interested in the value of t for which $h(t) = 70$:

$$-15t^2 + 75t + 10 = 70.$$

3. **Solve.** We solve by factoring:

$$\begin{aligned} -15t^2 + 75t + 10 &= 70 \\ -15t^2 + 75t - 60 &= 0 \quad \text{Subtracting 70} \\ -15(t^2 - 5t + 4) &= 0 \\ -15(t - 4)(t - 1) &= 0 \quad \text{Factoring} \\ t - 4 = 0 \quad &or \quad t - 1 = 0 \\ t = 4 \quad &or \quad t = 1. \end{aligned}$$

The solutions appear to be 4 and 1.

4. **Check.** We have

$$h(4) = -15 \cdot 4^2 + 75 \cdot 4 + 10 = -240 + 300 + 10 = 70 \text{ ft};$$
$$h(1) = -15 \cdot 1^2 + 75 \cdot 1 + 10 = -15 + 75 + 10 = 70 \text{ ft}.$$

Both 1 and 4 check, as we can also see from the following graph.

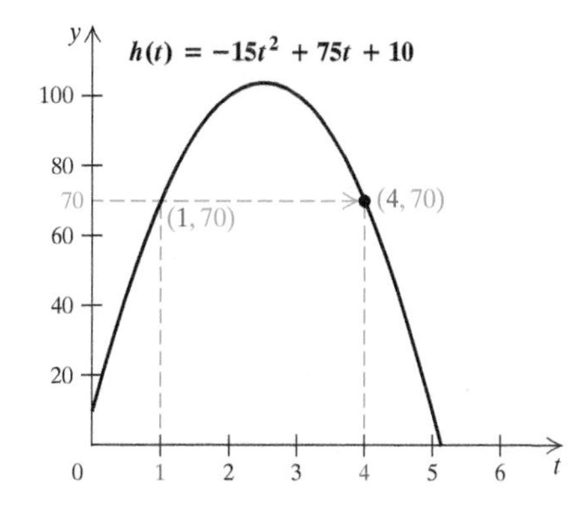

However, the problem states that the tee shirt is caught on the way down from its peak height. Thus we reject the solution 1 since that would indicate when the height of the tee shirt was 70 ft on the way up.

5. **State.** The tee shirt was in the air for 4 sec.

Do Exercise 9. ▶

The following example involves the **Pythagorean theorem**, which relates the lengths of the sides of a right triangle. A **right triangle** has a 90°, or right, angle, which is denoted by a symbol like ⌈. The longest side, opposite the 90° angle, is called the **hypotenuse**. The other sides, called **legs**, form the two sides of the right angle.

9. *Fireworks Displays.* Fireworks are typically launched from a mortar with an upward velocity (initial speed) of about 64 ft/sec. The height $h(t)$, in feet, of a "weeping willow" display, t seconds after having been launched from an 80-ft high rooftop, is given by
$h(t) = -16t^2 + 64t + 80.$
How long will it take the cardboard shell from the fireworks to reach the ground?

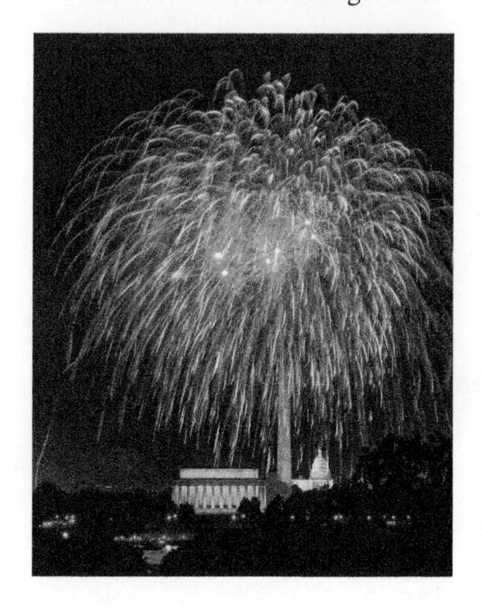

Answer
9. 5 sec

THE PYTHAGOREAN THEOREM

The sum of the squares of the lengths of the legs of a right triangle is equal to the square of the length of the hypotenuse:

$$a^2 + b^2 = c^2.$$

EXAMPLE 10 *Carpentry.* In order to build a deck at a right angle to her house, Geri places a stake in the ground a precise distance from the back wall of her house. This stake will combine with two marks on the house to form a right triangle. From a course in geometry, Geri remembers that there are three consecutive integers that can work as sides of a right triangle. Find the measurements of that triangle.

1. **Familiarize.** Recall that $x, x + 1$, and $x + 2$ can be used to represent three unknown consecutive integers. Since $x + 2$ is the largest number, it must represent the hypotenuse. The legs serve as the sides of the right angle, so one leg must be formed by the marks on the house.

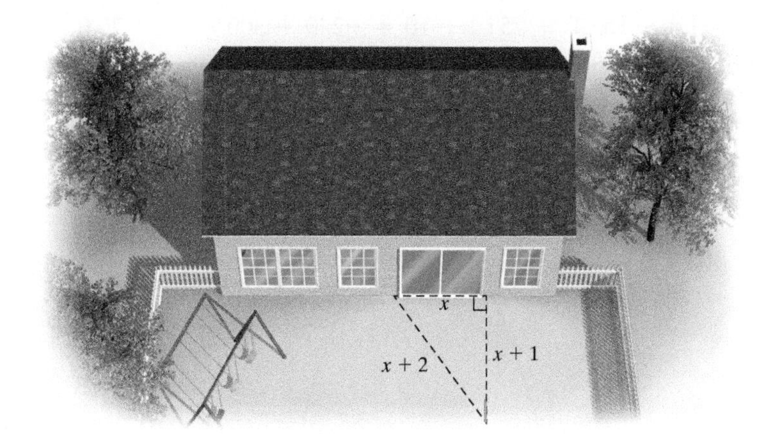

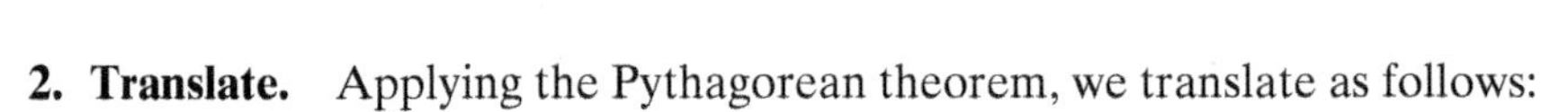

2. **Translate.** Applying the Pythagorean theorem, we translate as follows:

$$a^2 + b^2 = c^2$$
$$x^2 + (x + 1)^2 = (x + 2)^2.$$

3. **Solve.** We solve the equation as follows:

$$\begin{aligned} x^2 + (x^2 + 2x + 1) &= x^2 + 4x + 4 && \text{Squaring the binomials} \\ 2x^2 + 2x + 1 &= x^2 + 4x + 4 && \text{Collecting like terms} \\ x^2 - 2x - 3 &= 0 && \text{Subtracting } x^2 + 4x + 4 \\ (x - 3)(x + 1) &= 0 && \text{Factoring} \\ x - 3 = 0 \quad &or \quad x + 1 = 0 \\ x = 3 \quad &or \quad x = -1. \end{aligned}$$

4. **Check.** The integer -1 cannot be a length of a side because it is negative. For $x = 3$, we have $x + 1 = 4$, and $x + 2 = 5$. Since $3^2 + 4^2 = 5^2$, the lengths 3, 4, and 5 determine a right triangle. Thus, 3, 4, and 5 check.

5. **State.** If the marks on the house are 3 yd apart, Geri should locate the stake at the point in the yard that is precisely 4 yd from one mark and 5 yd from the other mark.

Do Exercise 10.

10. ***Triangle Dimensions.*** One leg of a right triangle has length 10 cm. The other sides have lengths that are consecutive even integers. Find these lengths. GS

1. **Familiarize.** Let x and $x + 2$ represent the lengths of the unknown sides of the triangle.

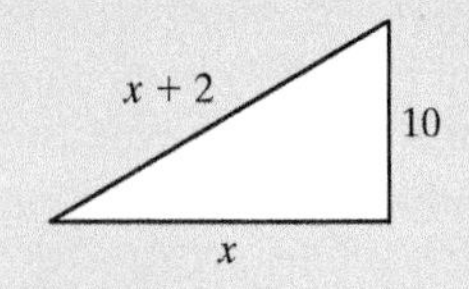

2. **Translate.**
$a^2 + b^2 = c^2$
$x^2 + 10^2 = (\quad)^2$

3. **Solve.**
$x^2 + 100 = x^2 + 4x + 4$
$100 = 4x + 4$
$\quad = 4x$
$\quad = x$

4. **Check.** If $x = 24$, then $x + 2 =$ ____. Then $10^2 + 24^2 = 100 + 576 =$ ____ $= 26^2$.

5. **State.** The lengths are ____ cm and 26 cm.

Answer
10. 24 cm, 26 cm

Guided Solution:
10. $x + 2$, 96, 24, 26, 676, 24

Translating for Success

1. *Car Travel.* Two cars leave town at the same time going in opposite directions. One travels 50 mph and the other travels 55 mph. In how many hours will they be 200 mi apart?

2. *Mixture of Solutions.* Solution A is 27% alcohol and solution B is 55% alcohol. How much of each should be used in order to make 10 L of a solution that is 48% alcohol?

3. *Triangle Dimensions.* The base of a triangle is 3 cm less than the height. The area is 27 cm^2. Find the height and the base.

4. *Three Numbers.* The sum of three numbers is 38. The first number is 3 less than twice the second number. The second number minus the third number is -7. What are the numbers?

5. *Supplementary Angles.* Two angles are supplementary. One angle measures 27° more than three times the measure of the other. Find the measure of each angle.

Translate each word problem to an equation or a system of equations and select a correct translation from choices A–Q.

A. $x + y + z = 38,$
$x = 2y - 3,$
$y - z = -7$

B. $\frac{1}{2}x(x - 3) = 27$

C. $x + y = 180,$
$x = 3y - 27$

D. $x^2 + 36 = (x + 4)^2$

E. $x^2 + (x + 4)^2 = 36$

F. $x + y = 10,$
$0.27x + 0.55y = 4.8$

G. $x + y = 45,$
$10x - 7y = 402$

H. $x + y + z = 180,$
$y - 3x - 38 = 0,$
$x - z = 7$

I. $x + y = 90,$
$x = 3y + 10$

J. $2x + 2(x - 3) = 27$

K. $x + y + z = 38,$
$x - 2y = 3,$
$x - z = -7$

L. $x + y = 10,$
$27x + 55y = 4.8$

M. $55x - 50x = 200$

N. $x(x - 3) = 27$

O. $x + y = 45,$
$7x + 10y = 402$

P. $x + y = 180,$
$x = 3y + 27$

Q. $50x + 55x = 200$

Answers on page A-16

6. *Triangle Dimensions.* The length of one leg of a right triangle is 6 m. The length of the hypotenuse is 4 m longer than the length of the other leg. Find the lengths of the hypotenuse and the other leg.

7. *Pizza Sales.* Todd's fraternity sold 45 pizzas over a football weekend. Small pizzas sold for \$7 each and large pizzas for \$10 each. The total amount of the sales was \$402. How many of each size pizza were sold?

8. *Angle Measures.* The second angle of a triangle measures 38° more than three times the measure of the first. The measure of the third angle is 7° less than the first. Find the measures of each angle of the triangle.

9. *Complementary Angles.* Two angles are complementary. One angle measures 10° more than three times the measure of the other. Find the measure of each angle.

10. *Rectangle Dimensions.* The base of a rectangle is 3 cm less than the height. The area is 27 cm^2. Find the height and the base.

4.8 Exercise Set

FOR EXTRA HELP MyLab Math

✓ Check Your Understanding

Reading Check Determine whether each sentence is true or false.

RC1. If $(x + 2)(x + 3) = 24$, then $x + 2 = 24$ *or* $x + 3 = 24$.

RC2. A quadratic equation always has two different solutions.

RC3. The number 0 is never a solution of a quadratic equation.

RC4. The Pythagorean theorem states that the sum of the lengths of the legs of a right triangle is equal to the length of the hypotenuse.

Concept Check Solve.

CC1. $(x - 3)(x - 1) = 0$

CC2. $(t + 9)(t + 4) = 0$

CC3. $(2x + 3)(5x - 8) = 0$

CC4. $7a(a + 8) = 0$

CC5. $x(3x - 1) = 0$

CC6. $(x + 5)(x - 6)(x - 2) = 0$

a Solve.

1. $x^2 + 3x = 28$

2. $y^2 - 4y = 45$

3. $y^2 + 9 = 6y$

4. $r^2 + 4 = 4r$

5. $x^2 + 20x + 100 = 0$

6. $y^2 + 10y + 25 = 0$

7. $9x + x^2 + 20 = 0$

8. $8y + y^2 + 15 = 0$

9. $x^2 + 8x = 0$

10. $t^2 + 9t = 0$

11. $x^2 - 25 = 0$

12. $p^2 - 49 = 0$

13. $z^2 = 144$

14. $y^2 = 64$

15. $y^2 + 2y = 63$

16. $a^2 + 3a = 40$

17. $32 + 4x - x^2 = 0$

18. $27 + 6t - t^2 = 0$

19. $3b^2 + 8b + 4 = 0$

20. $9y^2 + 15y + 4 = 0$

21. $3 + 8y^2 = 10y$

22. $11x + 4x^2 = -6$

23. $6z - z^2 = 0$

24. $8y - y^2 = 0$

25. $12z^2 + z = 6$

26. $6x^2 - 7x = 10$

27. $7x^2 - 7 = 0$

28. $4y^2 - 36 = 0$

29. $10 - r - 21r^2 = 0$

30. $28 + 5a - 12a^2 = 0$

31. $15y^2 = 3y$

32. $18x^2 = 9x$

33. $14 = x(x - 5)$

34. $x(x - 5) = 24$

35. $2x^3 - 2x^2 = 12x$

36. $50y + 5y^3 = 35y^2$

37. $2x^3 = 128x$

38. $147y = 3y^3$

39. $t^4 - 26t^2 + 25 = 0$

40. $x^4 - 13x^2 + 36 = 0$

41. $(a - 4)(a + 4) = 20$

42. $(t - 6)(t + 6) = 45$

43. $x(5 + 12x) = 28$

44. $a(1 + 21a) = 10$

45. Given that $f(x) = x^2 + 12x + 40$, find all values of x such that $f(x) = 8$.

46. Given that $f(x) = x^2 + 14x + 50$, find all values of x such that $f(x) = 5$.

47. Given that $g(x) = 2x^2 + 5x$, find all values of x such that $g(x) = 12$.

48. Given that $g(x) = 2x^2 - 15x$, find all values of x such that $g(x) = -7$.

49. Given that $h(x) = 12x + x^2$, find all values of x such that $h(x) = -27$.

50. Given that $h(x) = 4x - x^2$, find all values of x such that $h(x) = -32$.

Find the domain of the function f given by each of the following.

51. $f(x) = \dfrac{3}{x^2 - 4x - 5}$

52. $f(x) = \dfrac{2}{x^2 - 7x + 6}$

53. $f(x) = \dfrac{x}{6x^2 - 54}$

54. $f(x) = \dfrac{2x}{5x^2 - 20}$

55. $f(x) = \dfrac{x - 5}{25x^2 - 10x + 1}$

56. $f(x) = \dfrac{1 + x}{9x^2 + 30x + 25}$

57. $f(x) = \dfrac{7}{5x^3 - 35x^2 + 50x}$

58. $f(x) = \dfrac{3}{2x^3 - 2x^2 - 12x}$

In each of Exercises 59–62, an equation $ax^2 + bx + c = 0$ is given. Use *only* the graph of $f(x) = ax^2 + bx + c$ to find the x-intercepts of the graph and the solutions of the equation $ax^2 + bx + c = 0$.

59. $x^2 - 4x - 45 = 0$

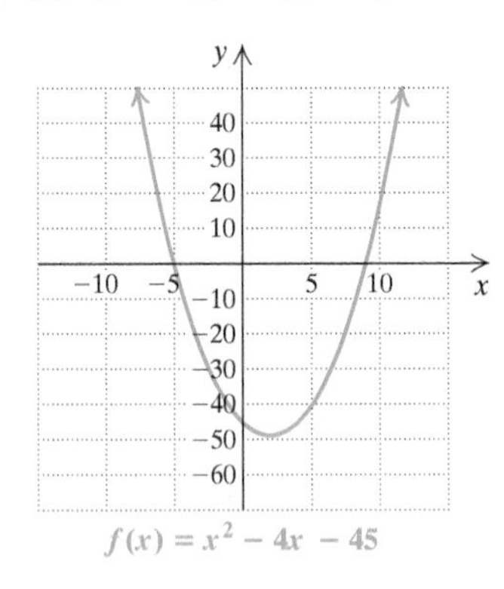

$f(x) = x^2 - 4x - 45$

60. $-x^2 - 3x + 40 = 0$

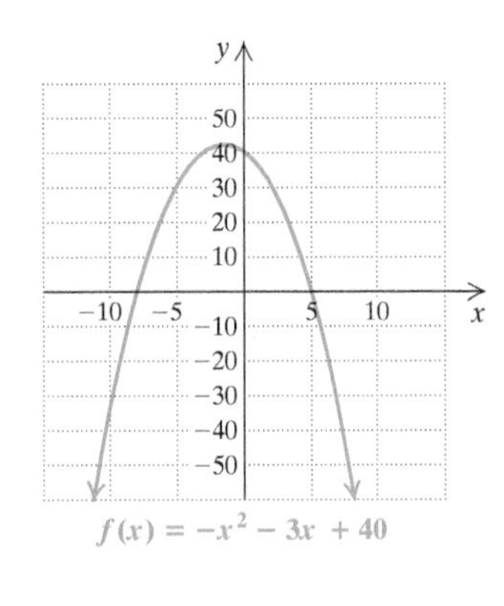

$f(x) = -x^2 - 3x + 40$

61. $32 + 4x - x^2 = 0$

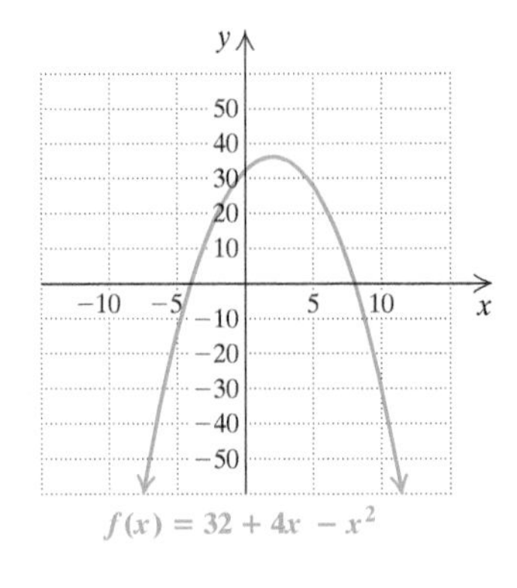

$f(x) = 32 + 4x - x^2$

62. $3x^2 - 12x = 0$

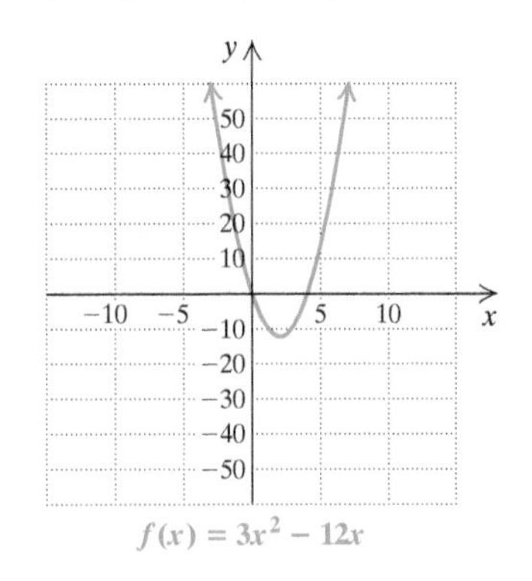

$f(x) = 3x^2 - 12x$

b Solve.

63. ***Book Area.*** A book is 5 cm longer than it is wide. The area is 84 cm^2. Find the length and the width.

64. ***Area of an Envelope.*** An envelope is 4 cm longer than it is wide. The area is 96 cm^2. Find the length and the width.

65. ***Tent Design.*** The triangular entrance to a tent is 2 ft taller than it is wide. The area of the entrance is 12 ft^2. Find the height and the base.

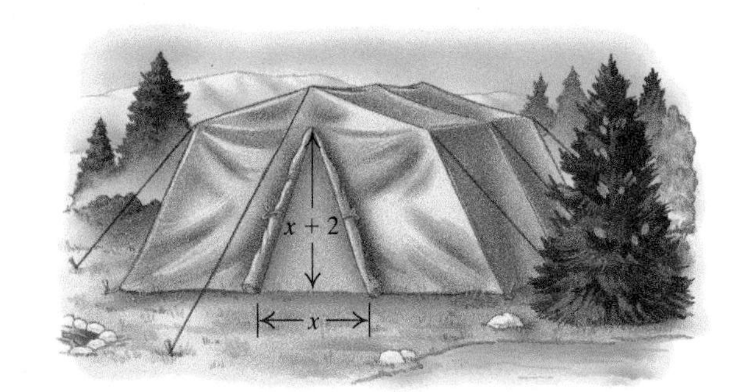

66. ***Sailing.*** A triangular sail is 9 m taller than it is wide. The area is 56 m^2. Find the height and the base of the sail.

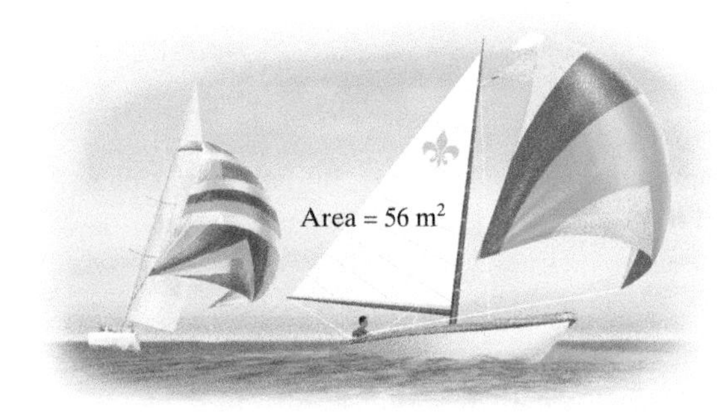

67. ***Geometry.*** If each of the sides of a square is lengthened by 6 cm, the area becomes 144 cm^2. Find the length of a side of the original square.

68. ***Geometry.*** If each of the sides of a square is lengthened by 4 m, the area becomes 49 m^2. Find the length of a side of the original square.

69. ***Consecutive Even Integers.*** Three consecutive even integers are such that the square of the third is 76 more than the square of the second. Find the three integers.

70. ***Consecutive Even Integers.*** Three consecutive even integers are such that the square of the first plus the square of the third is 136. Find the three integers.

71. *Framing a Picture.* A picture frame measures 12 cm by 20 cm, and 84 cm^2 of picture shows. Find the width of the frame.

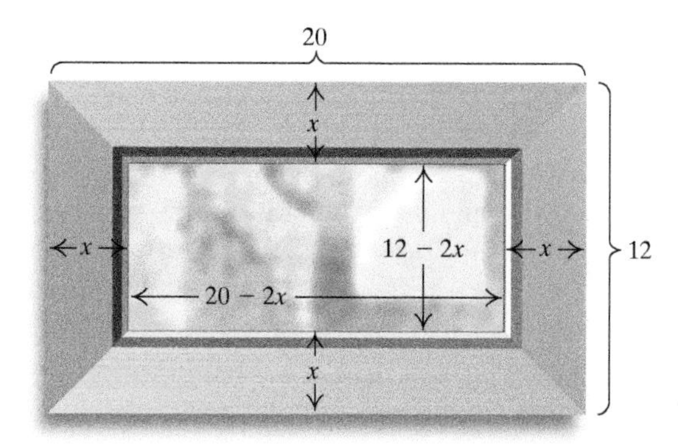

72. *Flower Bed Design.* A rectangular flower bed is to be 3 m longer than it is wide. The flower bed will have an area of 108 m^2. What will its dimensions be?

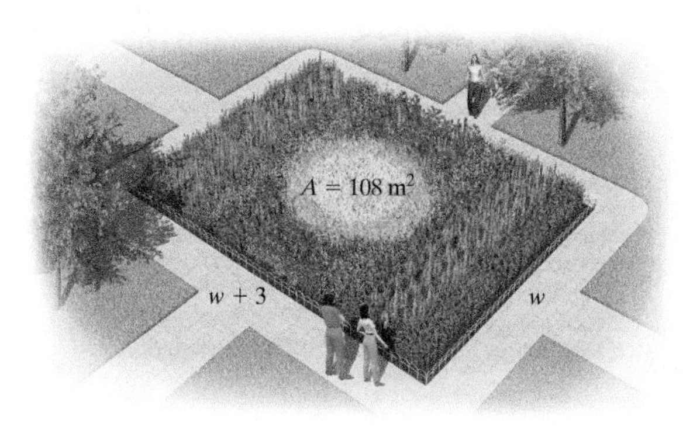

73. *Storage Shed Design.* The length of a storage shed is 4 ft greater than the width. The area is 96 ft^2. Find the length and the width.

74. *Framing a Picture.* A picture frame measures 14 cm by 20 cm, and 160 cm^2 of picture shows. Find the width of the frame.

75. *Antenna Wires.* A wire is stretched from the ground to the top of an antenna tower, as shown. The wire is 20 ft long. The height of the tower is 4 ft greater than the distance d from the tower's base to the end of the wire. Find the distance d and the height of the tower.

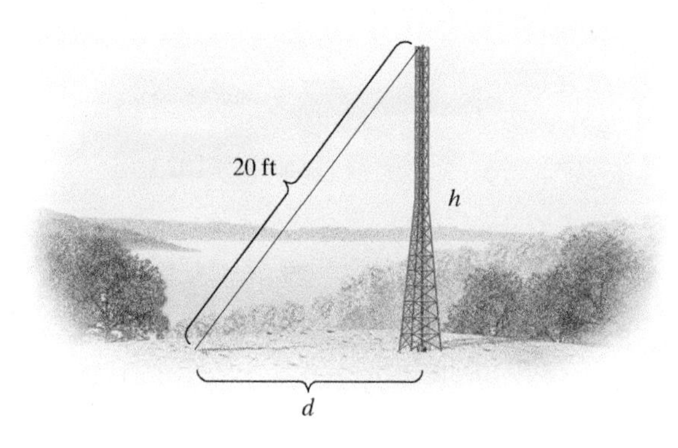

76. *Parking Lot Design.* A rectangular parking lot is 50 ft longer than it is wide. Determine the dimensions of the parking lot if it measures 250 ft diagonally.

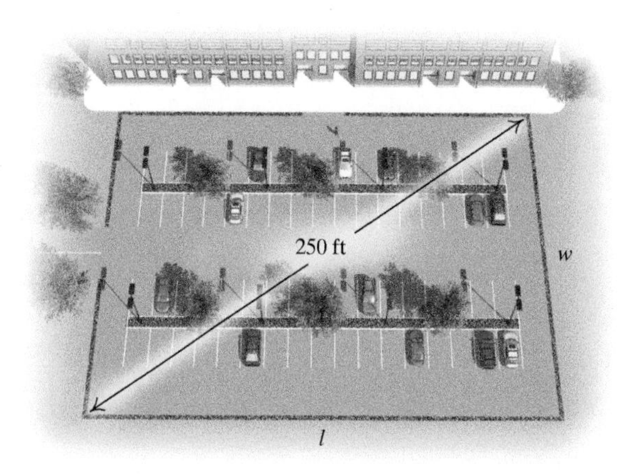

77. *Ladder Location.* The foot of an extension ladder is 9 ft from a wall. The height that the ladder reaches on the wall and the length of the ladder are consecutive integers. How long is the ladder?

78. *Child's Block.* The lengths of the sides of a right triangle formed by a child's wooden block are such that one leg has length 5 cm. The lengths of the other sides are consecutive integers. Find the lengths of the other sides of the triangle.

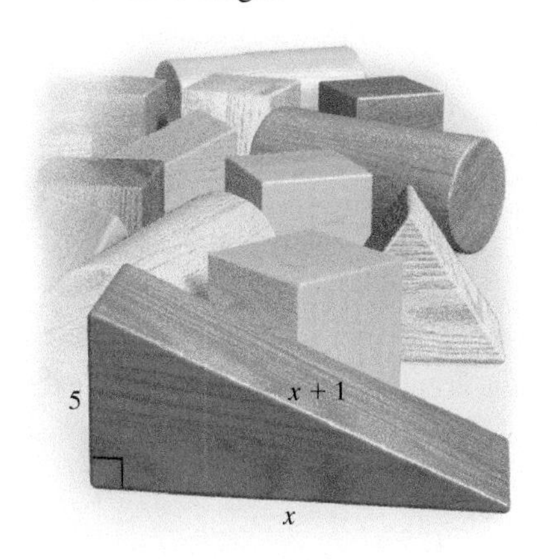

79. *Triangle Dimensions.* The lengths of the sides of a right triangle are consecutive even integers. Find the lengths of the sides.

80. *Triangle Dimensions.* The lengths of the hypotenuse and one leg of a right triangle are consecutive odd integers. The other leg is 9 ft shorter than the hypotenuse. Find the lengths of the sides

81. *Fireworks.* Suppose that a bottle rocket is launched upward with an initial velocity of 96 ft/sec from a height of 880 ft. Its height h, in feet, after t seconds is given by

$$h(t) = -16t^2 + 96t + 880.$$

After how long will the rocket reach the ground?

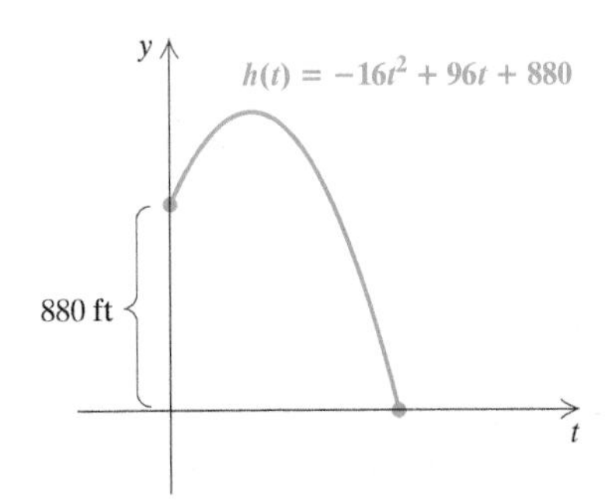

82. *Safety Flares.* Suppose that a flare is launched upward with an initial velocity of 80 ft/sec from a height of 224 ft. Its height h, in feet, after t seconds is given by

$$h(t) = -16t^2 + 80t + 224.$$

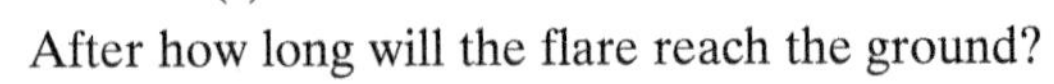

After how long will the flare reach the ground?

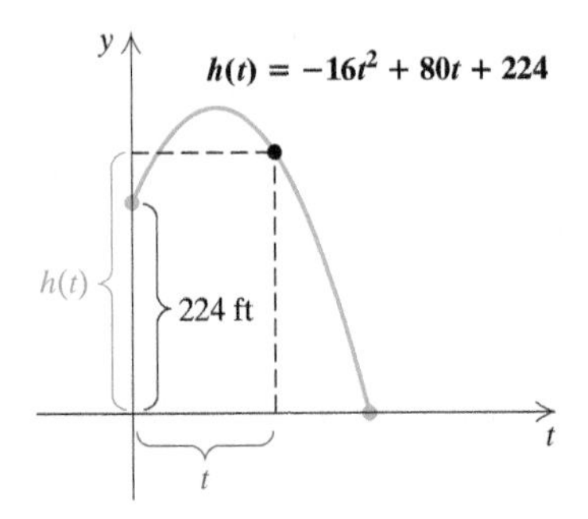

Skill Maintenance

Find the distance between the given pair of points on the number line. [1.6b]

83. $-3, -4$

84. $3.6, 4.9$

85. $-\frac{3}{5}, \frac{2}{3}$

86. $0, -1023$

Find an equation of the line containing the given pair of points. [2.6c]

87. $(-2, 7)$ and $(-8, -4)$

88. $(-2, 7)$ and $(8, -4)$

89. $(-2, 7)$ and $(8, 4)$

90. $(-24, 10)$ and $(-86, -42)$

Synthesis

91. Following is the graph of $f(x) = -x^2 - 2x + 3$. Use *only* the graph to solve $-x^2 - 2x + 3 = 0$ and $-x^2 - 2x + 3 \geq -5$.

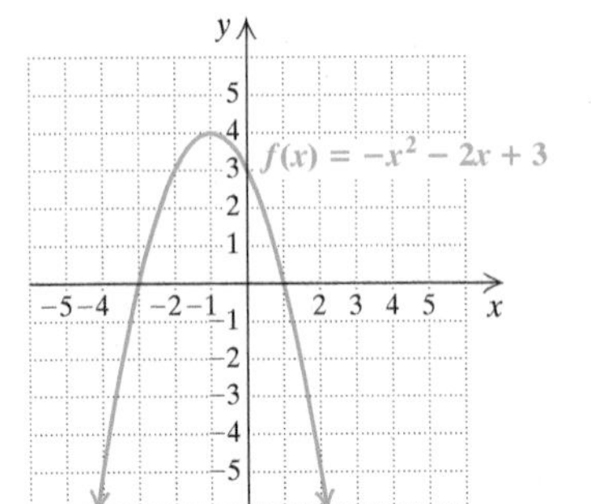

92. Following is the graph of $f(x) = x^4 - 3x^3$. Use *only* the graph to solve $x^4 - 3x^3 = 0$, $x^4 - 3x^3 \leq 0$, and $x^4 - 3x^3 > 0$.

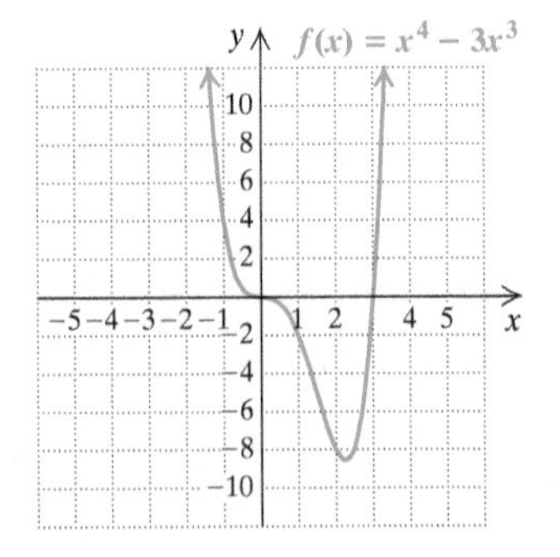

93. Use a graphing calculator to solve each equation.

a) $x^4 - 3x^3 - x^2 + 5 = 0$
b) $x^4 - 3x^3 - x^2 + 5 = 5$
c) $x^4 - 3x^3 - x^2 + 5 = -8$
d) $x^4 = 1 + 3x^3 + x^2$

94. Solve each of the following equations.

a) $(8x + 11)(12x^2 - 5x - 2) = 0$
b) $(3x^2 - 7x - 20)(x - 5) = 0$
c) $3x^3 + 6x^2 - 27x - 54 = 0$
(*Hint*: Factor by grouping.)
d) $2x^3 + 6x^2 = 8x + 24$

CHAPTER

4 Summary and Review

Vocabulary Reinforcement

Complete each statement with the correct term from the column on the right. Some of the choices may be used more than once and some may not be used at all.

product
difference
factor
factorization
grouping
ascending
descending
monomial
binomial
trinomial
zero

1. When the terms of a polynomial are written such that the exponents increase from left to right, we say that the polynomial is written in ______________ order. [4.1a]

2. To ______________ a polynomial is to express it as a product. [4.3a]

3. A(n) ______________ of a polynomial P is a polynomial that can be used to express P as a product. [4.3a]

4. A(n) ______________ of a polynomial is an expression that names that polynomial as a product. [4.3a]

5. When factoring a polynomial with four terms, try factoring by ______________. [4.7a]

6. A trinomial square is the square of a(n) ______________. [4.6a]

7. The principle of ______________ products states that if $ab = 0$, then $a = 0$ *or* $b = 0$. [4.8a]

8. The factorization of a(n) ______________ of squares is the product of the sum and the difference of two terms. [4.6c]

Concept Reinforcement

Determine whether each statement is true or false.

______________ **1.** According to the principle of zero products, if $ab = 0$, then $a = 0$ *and* $b = 0$. [4.8a]

______________ **2.** The binomial $27 - t^3$ is a difference of cubes. [4.6d]

______________ **3.** The expression $5x^2 - 6y^{-1}$ is a binomial. [4.1a]

Study Guide

Objective 4.1a Identify terms, degrees, and coefficients in polynomials; identify types of polynomials; and arrange polynomials in ascending order or descending order.

Example Identify the terms, the degree of each term, and the degree of the polynomial. Then identify the leading term, the leading coefficient, and the constant term:

$$-x^5 + 3x^4 - 7x^3 - 2x^2 + x - 10.$$

Terms: $-x^5, 3x^4, -7x^3, -2x^2, x, -10$
Degree of each term: 5, 4, 3, 2, 1, 0
Degree of polynomial: 5
Leading term: $-x^5$
Leading coefficient: -1
Constant term: -10

Practice Exercise

1. Identify the terms, the degree of each term, and the degree of the polynomial. Then identify the leading term, the leading coefficient, and the constant term:

$$-6x^4 + 5x^3 - x^2 + 10x - 1.$$

Objective 4.1d Find the opposite of a polynomial and subtract polynomials.

Example Subtract: $(4t^2 - t - t^3) - (7t^2 - t^3 - 5t)$.

$$\begin{aligned}(4t^2 - t - t^3) &- (7t^2 - t^3 - 5t)\\ &= (4t^2 - t - t^3) + (-7t^2 + t^3 + 5t)\\ &= 4t^2 - t - t^3 - 7t^2 + t^3 + 5t\\ &= -3t^2 + 4t\end{aligned}$$

Practice Exercise

2. Subtract:
$(3y^2 - 6y^3 + 7y) - (y^2 - 10y - 8y^3 + 8)$.

Objective 4.2b Use the FOIL method to multiply two binomials.

Example Multiply: $(7a - b)(4a + 9b)$.

$$\begin{aligned}(7a - b)(4a + 9b) &= \overset{F}{28a^2} + \overset{O}{63ab} - \overset{I}{4ab} - \overset{L}{9b^2}\\ &= 28a^2 + 59ab - 9b^2\end{aligned}$$

Practice Exercise

3. Multiply: $(3x - 5y)(x + 2y)$.

Objective 4.2c Use a rule to square a binomial.

Example Multiply: $(3q - 4)^2$.

$$(A - B)^2 = A^2 - 2AB + B^2$$

$$\begin{aligned}(3q - 4)^2 &= (3q)^2 - 2(3q)(4) + 4^2\\ &= 9q^2 - 24q + 16\end{aligned}$$

Practice Exercise

4. Multiply: $(2y + 7)^2$.

Objective 4.2d Use a rule to multiply a sum and a difference of the same two terms.

Example Multiply: $(8x + 5)(8x - 5)$.

$$(A + B)(A - B) = A^2 - B^2$$

$$\begin{aligned}(8x + 5)(8x - 5) &= (8x)^2 - 5^2\\ &= 64x^2 - 25\end{aligned}$$

Practice Exercise

5. Multiply: $(5d + 10)(5d - 10)$.

Objective 4.2e For functions f described by second-degree polynomials, find and simplify notation like $f(a + h)$ and $f(a + h) - f(a)$.

Example Given $f(x) = 2x - x^2$, find $f(x - 1)$ and $f(a + h) - f(a)$.

$$\begin{aligned}f(x - 1) &= 2(x - 1) - (x - 1)^2 = 2(x - 1) - (x^2 - 2x + 1)\\ &= 2x - 2 - x^2 + 2x - 1 = -x^2 + 4x - 3;\end{aligned}$$

$$\begin{aligned}f(a + h) - f(a) &= [2(a + h) - (a + h)^2] - [2a - a^2]\\ &= [2(a + h) - (a^2 + 2ah + h^2)] - [2a - a^2]\\ &= 2a + 2h - a^2 - 2ah - h^2 - 2a + a^2\\ &= -h^2 - 2ah + 2h\end{aligned}$$

Practice Exercise

6. Given $f(x) = 3x^2 - x + 2$, find $f(x + 1)$ and $f(a + h) - f(a)$.

Objective 4.3b Factor certain polynomials with four terms by grouping.

Example Factor: $x^3 - 6x^2 + 3x - 18$.

$$\begin{aligned}x^3 - 6x^2 + 3x - 18 &= (x^3 - 6x^2) + (3x - 18)\\ &= x^2(x - 6) + 3(x - 6)\\ &= (x - 6)(x^2 + 3)\end{aligned}$$

Practice Exercise

7. Factor: $y^3 + 3y^2 - 8y - 24$.

Objective 4.5a Factor trinomials of the type $ax^2 + bx + c, a \neq 1$, by the FOIL method.

Example Factor $15x^2 - 4x - 3$ by the FOIL method.

The terms of $15x^2 - 4x - 3$ do not have a common factor. We factor the first term, $15x^2$, and get $15x \cdot x$ and $5x \cdot 3x$. We then have

$$(15x + \square)(x + \square) \text{ and } (5x + \square)(3x + \square)$$

as possible factorizations. We then factor the last term, -3. The possibilities are $(-3)(1)$ and $(3)(-1)$. We look for combinations of factors such that the sum of the outside product and the inside product is the middle term, $-4x$. We list possible factorizations until a correct middle term is found:

$(15x - 3)(x + 1) = 15x^2 + 12x - 3;$

$(5x - 3)(3x + 1) = 15x^2 - 4x - 3. \rightarrow$ Correct middle term

Thus, $15x^2 - 4x - 3 = (5x - 3)(3x + 1)$.

Practice Exercise

8. Factor $3x^2 + 19x - 72$ by the FOIL method.

Objective 4.5b Factor trinomials of the type $ax^2 + bx + c, a \neq 1$, by the *ac*-method.

Example Factor $6x^2 - 5x - 4$ by the *ac*-method.

There are no common factors. We multiply the leading coefficient, 6, and the constant, -4: $6(-4) = -24$. Next, we try to factor -24 so that the sum of the factors is -5. Since -5 is negative, the negative factor of -24 must have the larger absolute value.

PAIRS OF FACTORS	SUMS OF FACTORS
1, −24	−23
2, −12	−10

PAIRS OF FACTORS	SUMS OF FACTORS
3, −8	−5
4, −6	−2

Next, we split the middle term using the factors 3 and -8:

$$\begin{aligned} 6x^2 - 5x - 4 &= 6x^2 + 3x - 8x - 4 \\ &= 3x(2x + 1) - 4(2x + 1) \\ &= (2x + 1)(3x - 4). \end{aligned}$$

Practice Exercise

9. Factor $10x^2 - 33x - 7$ by the *ac*-method.

Objective 4.6a Factor trinomial squares.

Example Factor: $4x^2 - 44x + 121$.

$$A^2 - 2AB + B^2 = (A - B)^2$$

$$\begin{aligned} 4x^2 - 44x + 121 &= (2x)^2 - 2 \cdot (2x) \cdot 11 + 11^2 \\ &= (2x - 11)^2 \end{aligned}$$

Practice Exercise

10. Factor: $81x^2 - 72x + 16$.

Objective 4.6b Factor differences of squares.

Example Factor: $64y^2 - 9$.

$$A^2 - B^2 = (A + B)(A - B)$$

$$64y^2 - 9 = (8y)^2 - 3^2 = (8y + 3)(8y - 3)$$

Practice Exercise

11. Factor: $100t^2 - 1$.

Objective 4.6d Factor sums and differences of cubes.

Example Factor: $8w^3 + 125$.

$A^3 + B^3 = (A + B)(A^2 - AB + B^2)$

$8w^3 + 125 = (2w)^3 + 5^3 = (2w + 5)(4w^2 - 10w + 25)$

Example Factor: $125x^3 - 8$.

$A^3 - B^3 = (A - B)(A^2 + AB + B^2)$

$125x^3 - 8 = (5x)^3 - 2^3 = (5x - 2)(25x^2 + 10x + 4)$

Practice Exercises

12. Factor: $216x^3 + 1$.

13. Factor: $1000y^3 - 27$.

Objective 4.8a Solve quadratic and other polynomial equations by first factoring and then using the principle of zero products.

Example Solve: $5x^2 + 11x = 12$.

$5x^2 + 11x - 12 = 0$ Getting 0 on one side

$(5x - 4)(x + 3) = 0$ Factoring

$5x - 4 = 0$ *or* $x + 3 = 0$ Using the principle of zero products

$5x = 4$ *or* $x = -3$

$x = \frac{4}{5}$ *or* $x = -3$

The solutions are -3 and $\frac{4}{5}$.

Practice Exercise

14. Solve: $3x^2 - x = 14$.

Review Exercises

1. Given the polynomial [4.1a]

$$3x^6y - 7x^8y^3 + 2x^3 - 3y^2:$$

a) Identify the degree of each term and the degree of the polynomial.
b) Identify the leading term and the leading coefficient.
c) Arrange in ascending powers of x.
d) Arrange in descending powers of y.

Evaluate the polynomial function for the given values. [4.1b]

2. $P(x) = x^3 - x^2 + 4x$; $P(0)$ and $P(-1)$

3. $P(x) = 4 - 2x - x^2$; $P(-2)$ and $P(5)$

Collect like terms. [4.1c]

4. $8x + 13y - 15x + 10y$

5. $3ab - 10 + 5ab^2 - 2ab + 7ab^2 + 14$

6. *Median Income by Age.* The polynomial function

$$I(x) = -0.0560x^4 + 7.9980x^3 - 436.1840x^2 + 11{,}627.8376x - 90{,}625.0001, \quad 13 \le x \le 65,$$

can be used to approximate the median income $I(x)$ of a person of age x living in the United States. Use the following graph to estimate the median income of a 40-year-old. [4.1b]

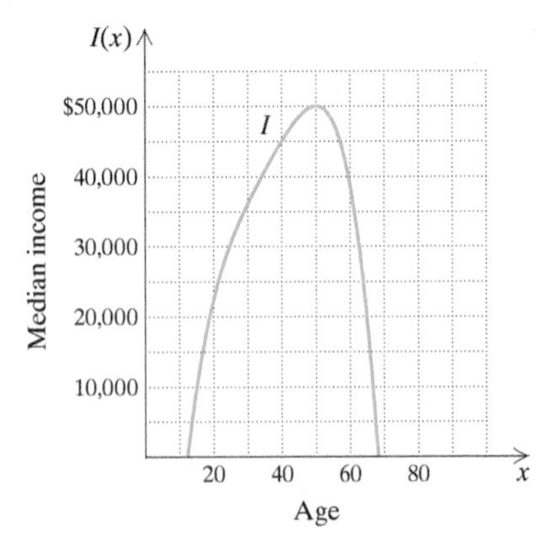

DATA: U.S. Census Bureau; The Conference Board; Simmons Bureau of Labor Statistics

Add, subtract, or multiply. [4.1c, d], [4.2a, b, c, d]

7. $(-6x^3 - 4x^2 + 3x + 1) + (5x^3 + 2x + 6x^2 + 1)$

8. $(4x^3 - 2x^2 - 7x + 5) + (8x^2 - 3x^3 - 9 + 6x)$

9. $(-9xy^2 - xy + 6x^2y) + (-5x^2y - xy + 4xy^2) + (12x^2y - 3xy^2 + 6xy)$

10. $(3x - 5) - (-6x + 2)$

11. $(4a - b + 3c) - (6a - 7b - 4c)$

12. $(9p^2 - 4p + 4) - (-7p^2 + 4p + 4)$

13. $(6x^2 - 4xy + y^2) - (2x^2 + 3xy - 2y^2)$

14. $(3x^2y)(-6xy^3)$

15. $(x^4 - 2x^2 + 3)(x^4 + x^2 - 1)$

16. $(4ab + 3c)(2ab - c)$

17. $(2x + 5y)(2x - 5y)$

18. $(2x - 5y)^2$

19. $(5x^2 - 7x + 3)(4x^2 + 2x - 9)$

20. $(x^2 + 4y^3)^2$

21. $(x - 5)(x^2 + 5x + 25)$

22. $\left(x - \frac{1}{3}\right)\left(x - \frac{1}{6}\right)$

23. Given that $f(x) = x^2 - 2x - 7$, find and simplify $f(a - 1)$ and $f(a + h) - f(a)$. [4.2e]

Factor. [4.3a, b], [4.4a], [4.5a, b], [4.6a, b, c, d], [4.7a]

24. $9y^4 - 3y^2$

25. $15x^4 - 18x^3 + 21x^2 - 9x$

26. $a^2 - 12a + 27$

27. $3m^2 + 14m + 8$

28. $25x^2 + 20x + 4$

29. $4y^2 - 16$

30. $ax + 2bx - ay - 2by$

31. $4x^4 + 4x^2 + 20$

32. $27x^3 - 8$

33. $0.064b^3 - 0.125c^3$

34. $y^5 - y$

35. $2z^8 - 16z^6$

36. $54x^6y - 2y$

37. $1 + a^3$

38. $36x^2 - 120x + 100$

39. $6t^2 + 17pt + 5p^2$

40. $x^3 + 2x^2 - 9x - 18$

41. $a^2 - 2ab + b^2 - 4t^2$

Solve. [4.8a]

42. $x^2 - 20x = -100$

43. $6b^2 - 13b + 6 = 0$

44. $8y^2 = 14y$

45. $r^2 = 16$

46. Given that $f(x) = x^2 - 7x - 40$, find all values of x such that $f(x) = 4$. [4.8a]

47. Find the domain of the function f given by

$$f(x) = \frac{x - 3}{3x^2 + 19x - 14}. \quad [4.8a]$$

Solve. [4.8b]

48. *Photograph Dimensions.* A photograph is 3 in. longer than it is wide. When a 2-in. matte border is placed around the photograph, the total area of the photograph and the border is 108 in^2. Find the dimensions of the photograph.

49. The sum of the squares of three consecutive odd integers is 83. Find the integers.

50. *Area.* The number of *square units* in the area of a square is 7 more than six times the number of *units* in the length of a side. What is the length of a side of the square?

51. Which of the following is a factor of $t^3 - 64$? [4.6d]

A. $t - 4$ **B.** $t^2 - 4t + 16$
C. $t^2 + 8t + 16$ **D.** $t + 4$

52. Which of the following is a factor of

$$hm + 5hn - gm - 5gn? \quad [4.3b]$$

A. $m - n$ **B.** $h + g$
C. $m + 5n$ **D.** $m - 5n$

Synthesis

Factor. [4.6d]

53. $128x^6 - 2y^6$

54. $(x + 1)^3 - (x - 1)^3$

55. Multiply: $[a - (b - 1)][(b - 1)^2 + a(b - 1) + a^2]$. [4.6d]

56. Solve: $64x^3 = x$. [4.8a]

Understanding Through Discussion and Writing

1. Under what conditions, if any, can the sum of two squares be factored? Explain. [4.3a], [4.6b]

2. Explain how to use the *ac*-method to factor trinomials of the type $ax^2 + bx + c, a \neq 1$. [4.5b]

3. Annie claims that she can add any two polynomials but finds subtraction difficult. What advice would you offer her? [4.1d]

4. Suppose that you are given a detailed graph of $y = P(x)$, where $P(x)$ is a polynomial. How could you use the graph to solve the equation $P(x) = 0$? $P(x) = 4$? [4.8a]

5. Explain how you could use factoring or graphing to explain why $x^3 - 8 \neq (x - 2)^3$. [4.6d]

6. Emily has factored a particular polynomial as $(a - b)(x - y)$. George factors the same polynomial and gets $(b - a)(y - x)$. Who is correct and why? [4.3a], [4.7a]

7. Explain how one could write a quadratic equation that has 5 and -3 as solutions. Can the number of solutions of a quadratic equation exceed two? Why or why not? [4.8a]

8. In this chapter, we learned to solve equations that we could not have solved before. Describe these new equations and the way we go about solving them. How is the procedure different from those we have used before now? [4.8a]

CHAPTER

4 Test

For Extra Help

For step-by-step test solutions, access the Chapter Test Prep Videos in MyLab Math.

1. Given the polynomial

$$3xy^3 - 4x^2y + 5x^5y^4 - 2x^4y:$$

a) Identify the degree of each term and the degree of the polynomial.
b) Identify the leading term and the leading coefficient.
c) Arrange in ascending powers of x.
d) Arrange in descending powers of y.

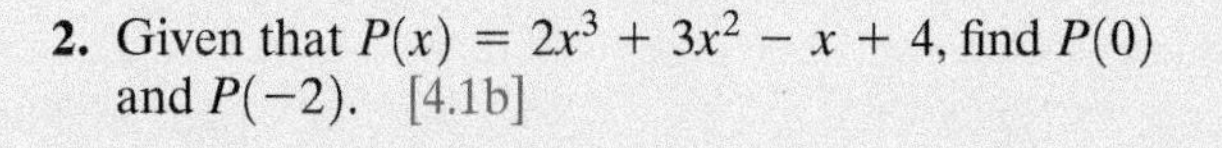

2. Given that $P(x) = 2x^3 + 3x^2 - x + 4$, find $P(0)$ and $P(-2)$. [4.1b]

3. *Age of College Students.* The percentage of Canadian college students working toward a bachelor's degree who are x years old can be estimated using the polynomial function

$$c(x) = -0.005382x^4 + 0.567608x^3 - 22.05744x^2 + 372.8x - 2297, \quad 17 \le x \le 33.$$

Use the graph to estimate the percentage of Canadian college students working toward a bachelor's degree who are 22 years old.

Data: statcan.gc.ca

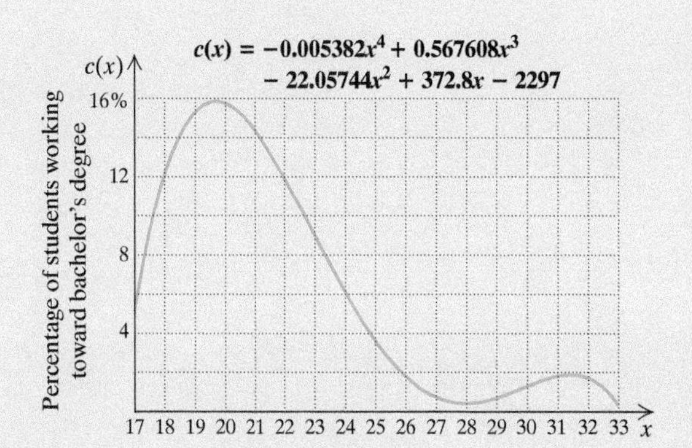

4. Collect like terms: $5xy - 2xy^2 - 2xy + 5xy^2$.

Add, subtract, or multiply.

5. $(-6x^3 + 3x^2 - 4y) + (3x^3 - 2y - 7y^2)$

6. $(4a^3 - 2a^2 + 6a - 5) + (3a^3 - 3a + 2 - 4a^2)$

7. $(5m^3 - 4m^2n - 6mn^2 - 3n^3) + (9mn^2 - 4n^3 + 2m^3 + 6m^2n)$

8. $(9a - 4b) - (3a + 4b)$

9. $(4x^2 - 3x + 7) - (-3x^2 + 4x - 6)$

10. $(6y^2 - 2y - 5y^3) - (4y^2 - 7y - 6y^3)$

11. $(-4x^2y)(-16xy^2)$

12. $(6a - 5b)(2a + b)$

13. $(x - y)(x^2 - xy - y^2)$

14. $(3m^2 + 4m - 2)(-m^2 - 3m + 5)$

15. $(4y - 9)^2$

16. $(x - 2y)(x + 2y)$

17. Given that $f(x) = x^2 - 5x$, find and simplify $f(a + 10)$ and $f(a + h) - f(a)$.

Factor.

18. $9x^2 + 7x$

19. $24y^3 + 16y^2$

20. $y^3 + 5y^2 - 4y - 20$

21. $p^2 - 12p - 28$

22. $12m^2 + 20m + 3$

23. $9y^2 - 25$

24. $3r^3 - 3$

25. $9x^2 + 25 - 30x$

26. $(z + 1)^2 - b^2$

27. $x^8 - y^8$

28. $y^2 + 8y + 16 - 100t^2$

29. $20a^2 - 5b^2$

30. $24x^2 - 46x + 10$

31. $16a^7b + 54ab^7$

Solve.

32. $x^2 - 18 = 3x$

33. $5y^2 - 125 = 0$

34. $2x^2 + 21 = -17x$

35. Given that $f(x) = 3x^2 - 15x + 11$, find all values of x such that $f(x) = 11$.

36. Find the domain of the function f given by

$$f(x) = \frac{3 - x}{x^2 + 2x + 1}.$$

Solve.

37. *Photograph Dimensions.* A photograph is 3 cm longer than it is wide. Its area is 40 cm^2. Find its length and its width.

38. *Ladder Location.* An extension ladder leaning against a wall is 1 ft longer than the distance that it reaches up the wall and 8 ft longer than the distance from the foot of the ladder to the wall. How long is the ladder?

39. *Number of Games in a League.* If there are n teams in a league and each team plays every other team once, the total number of games played is given by the polynomial function $f(n) = \frac{1}{2}n^2 - \frac{1}{2}n$. Find an equivalent expression for $f(n)$ by factoring completely.

40. Factor: $8x^3 - 1$.

A. $(2x - 1)(2x - 1)(2x - 1)$
B. $(2x - 1)(2x + 1)$
C. $(2x - 1)(4x^2 + 2x + 1)$
D. $(2x + 1)(4x^2 - 2x + 1)$

Synthesis

41. Factor: $6x^{2n} - 7x^n - 20$.

42. If $pq = 5$ and $(p + q)^2 = 29$, find the value of $p^2 + q^2$.

CHAPTERS

1–4 Cumulative Review

Simplify.

1. $(x^2 + 4x - xy - 9) + (-3x^2 - 3x + 8)$

2. $(6x^2 - 3x + 2x^3) - (8x^2 - 9x + 2x^3)$

3. $(a^2 - a - 3) \cdot (a^2 + 2a - 3)$

4. $(x + 4)(x + 9)$

Solve.

5. $8 - 3x = 6x - 10$

6. $\frac{1}{2}x - 3 = \frac{7}{2}$

7. $A = \frac{1}{2}h(a + b)$, for b

8. $6x - 1 \leq 3(5x + 2)$

9. $4x - 3 < 2$ *or* $x - 3 > 1$

10. $|2x - 3| < 7$

11. $x + y + z = -5,$ $x - z = 10,$ $y - z = 12$

12. $2x + 5y = -2,$ $5x + 3y = 14$

13. $3x - y = 7,$ $2x + 2y = 5$

14. $x + 2y - z = 0,$ $3x + y - 2z = -1,$ $x - 4y + z = -2$

15. $11x + x^2 + 24 = 0$

16. $2x^2 - 15x = -7$

17. Given that $f(x) = 3x^2 + 4x$, find all values of x such that $f(x) = 4$.

18. Find the domain of the function F given by

$$F(x) = \frac{x + 7}{x^2 - 2x - 15}.$$

Factor.

19. $3x^3 - 12x^2$

20. $2x^4 + x^3 + 2x + 1$

21. $x^2 + 5x - 14$

22. $20a^2 - 23a + 6$

23. $4x^2 - 25$

24. $2x^2 - 28x + 98$

25. $a^3 + 1000$

26. $64x^3 - 1$

27. $4a^3 + a^6 - 12$

28. $4x^4y^2 - x^2y^4$

29. *Photo Prices.* Greg ordered two 8×10 prints and one 11×14 print from the college photo services department. The total cost was \$41.40. Sara ordered one 8×10 print, one 11×14 print, and one 24×36 print for a total cost of \$101.20. Austin ordered two 11×14 prints and two 24×36 prints for a total cost of \$184. How much did each size print cost?

Graph.

30. $x < 1$ *or* $x \geq 2$

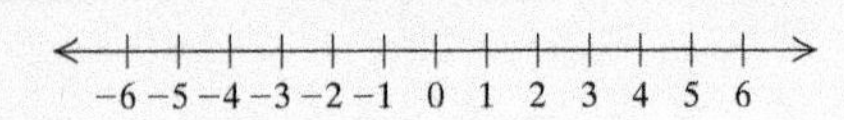

31. $y = -2x$

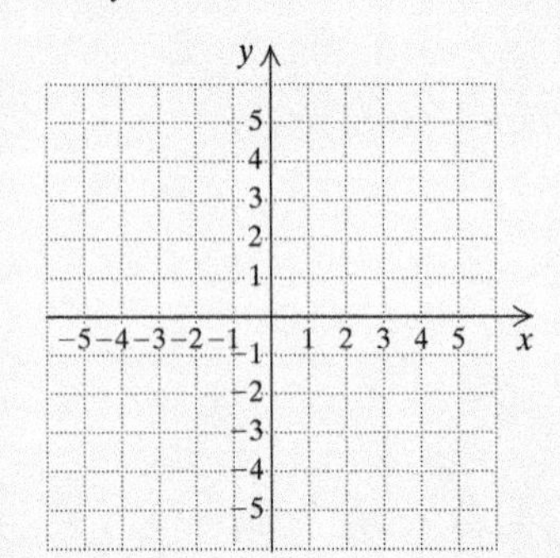

32. $6y + 24 = 0$

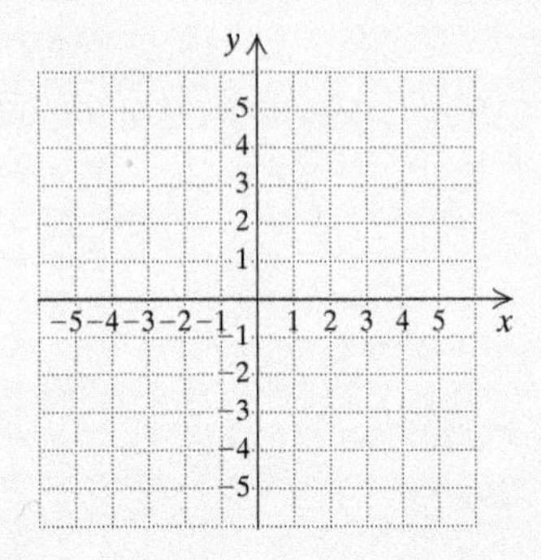

33. $y > x + 6$

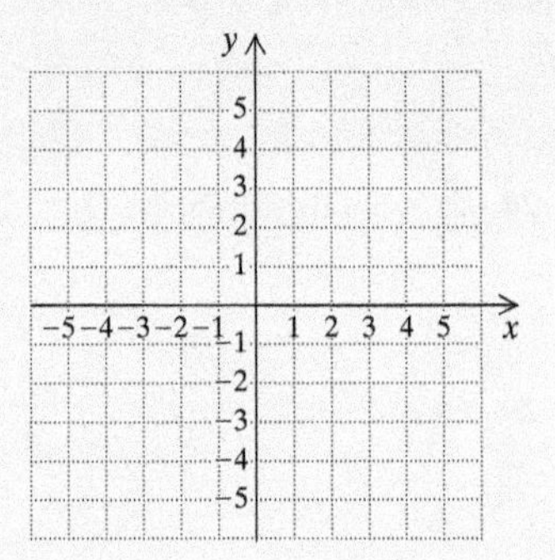

34. $f(x) = x^2 - 3$

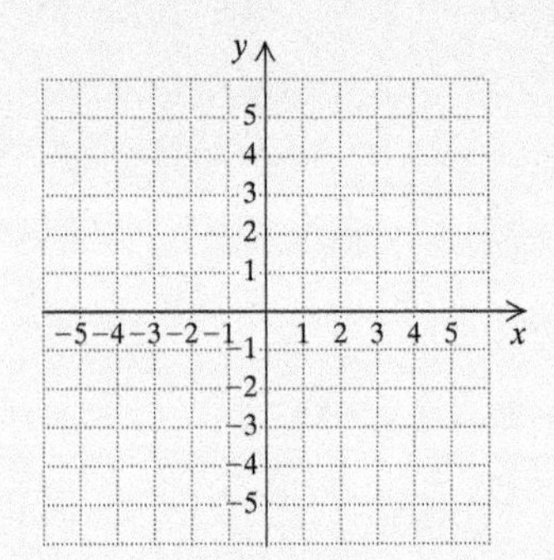

35. $g(x) = 4 - |x|$

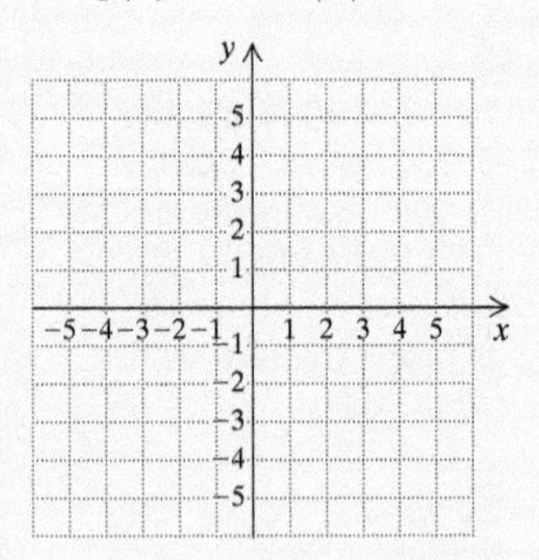

36. $2x + 3y \leq 6,$
$5x - 5y \leq 15,$
$x \geq 0$
Label the vertices.

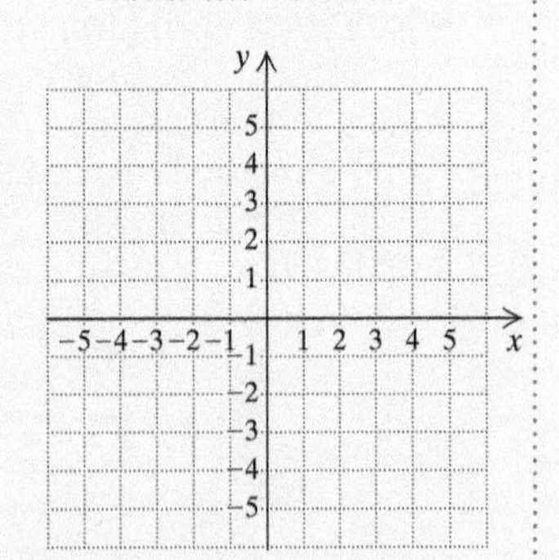

37. Find an equation of the line containing the point $(3, 7)$ and parallel to the line $x + 2y = 6$.

38. Find an equation of the line containing the point $(3, -2)$ and perpendicular to the line $3x + 4y = 5$.

39. Find an equation of the line containing the points $(-1, 4)$ and $(-2, 0)$.

40. Find an equation of the line with slope -3 and through the point $(2, 1)$.

41. *Maple Syrup Production.* Farmers in North America produced 7.6 million gal of maple syrup in 2006 and 18.8 million gal in 2016. Find the rate of change in production of maple syrup with respect to time, in years.

Data: U.S. Department of Agriculture; statcan.gc.ca; *The Wall Street Journal*, 3/6/2017

42. *Games in a Sports League.* In a sports league of n teams in which each team plays every other team twice, the total number N of games to be played is given by the function

$$N(n) = n^2 - n.$$

a) A women's college volleyball league has 6 teams. If we assume that each team plays every other team twice, what is the total number of games to be played?

b) Another volleyball league plays a total of 72 games. If we assume that each team plays every other team twice, how many teams are in the league?

Synthesis

43. *Display of a Sports Card.* A valuable sports card is 4 cm wide and 5 cm long. The card is to be sandwiched by two pieces of Lucite, each of which is $5\frac{1}{2}$ times the area of the card. Determine the dimensions of the Lucite that will ensure a uniform border.

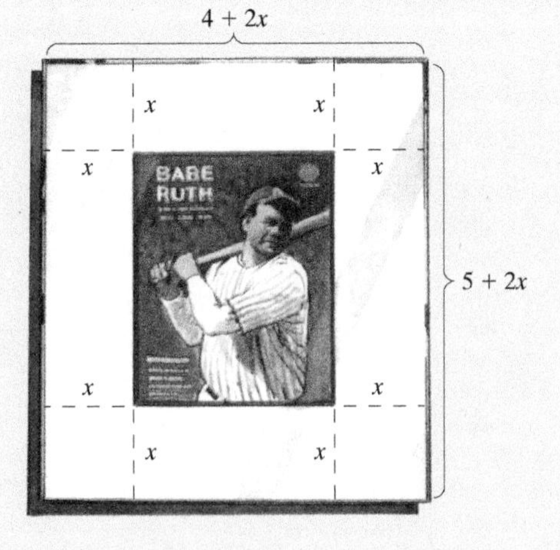

44. Solve: $|x + 1| \leq |x - 3|$.

CHAPTER

5

Rational Expressions, Equations, and Functions

Over 106 million bales of cotton were produced worldwide from August 1, 2016, through August 1, 2017. The graph below shows the percentage produced by the top six countries. The United States supplied approximately 17 million bales. U.S. textile manufacturers use about 8 million bales annually, of which 57% goes into apparel, 36% into home furnishings, and 7% into industrial products.

Data: Foreign Agricultural Service / USDA, Office of Global Analysis; cotton.org

Leaders in Cotton Production

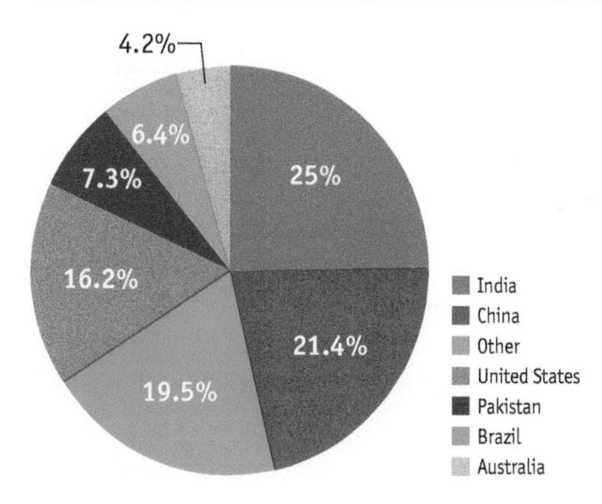

In Example 4 of Section 5.8, we will discuss the description of knitted cotton fabric in terms of wales per inch (WPI) and courses per inch (CPI). We will also calculate the number of courses per inch given the stitch length.

JUST IN TIME

Review topics 6, 13, 14, and 19 in the **Just In Time** section at the front of the text provide excellent prerequisite skill review for this chapter.

MyLab Math

VIDEO

STUDYING FOR SUCCESS *Working Exercises*

- ☐ Don't begin solving a homework problem by working backward from the answer given at the back of the text. Remember: Quizzes and tests have no answer section!
- ☐ Check answers to odd-numbered exercises at the back of the book.
- ☐ Work some even-numbered exercises to practice doing exercises without answers. Check your answers later with a friend or your instructor.

5.1 Rational Expressions and Functions: Multiplying, Dividing, and Simplifying

OBJECTIVES

- **a** Find all numbers for which a rational expression is not defined or that are not in the domain of a rational function, and state the domain of the function.
- **b** Multiply a rational expression by 1, using an expression like A/A.
- **c** Simplify rational expressions.
- **d** Multiply rational expressions and simplify.
- **e** Divide rational expressions and simplify.

a RATIONAL EXPRESSIONS AND FUNCTIONS

SKILL REVIEW *Find the domain and the range of a function.* [2.3a]

Find the domain.

1. $f(x) = 3x + 7$ **2.** $f(x) = \dfrac{x - 7}{2x + 3}$

Answers: 1. All real numbers **2.** $\left\{x \mid x \text{ is a real number } \textit{and } x \neq -\frac{3}{2}\right\}$, or $\left(-\infty, -\frac{3}{2}\right) \cup \left(-\frac{3}{2}, \infty\right)$

MyLab Math VIDEO

An expression that consists of the quotient of two polynomials, where the polynomial in the denominator is nonzero, is called a **rational expression**. The following are examples of rational expressions:

$$\frac{7}{8}, \quad \frac{z}{-6}, \quad \frac{a}{b}, \quad \frac{8}{y+5}, \quad \frac{t^4 - 5t}{t^2 - 3t - 28}, \quad \frac{x^2 + 7xy - 4}{x^3 - y^3}.$$

Note that every rational number is a rational expression.

Rational expressions indicate division. Thus we cannot make a replacement of the variable that allows a denominator to be 0. (For a discussion of why we exclude division by 0, see Section J9.)

EXAMPLE 1 Find all numbers for which the rational expression

$$\frac{2x + 1}{x - 3}$$

is not defined.

When x is replaced with 3, the denominator is 0, and the rational expression is not defined:

$$\frac{2x + 1}{x - 3} = \frac{2 \cdot 3 + 1}{3 - 3} = \frac{7}{0}. \longleftarrow \text{Division by 0 is not defined.}$$

You can check some replacements other than 3 to see that it appears that 3 is the only replacement that is not allowable. Thus the rational expression is not defined for the number 3. ■

You may have noticed that the procedure in Example 1 is similar to one that we have performed when finding the domain of a function.

EXAMPLE 2 Find the domain of f if $f(x) = \frac{2x + 1}{x - 3}$.

The domain is the set of all replacements for which the rational expression is defined (see Section 2.3). We must exclude the replacements that make the denominator 0. We find them by first setting the denominator equal to 0. Solving $x - 3 = 0$ for x, we get $x = 3$. The domain of f is $\{x|x$ is a real number *and* $x \neq 3\}$, or, in interval notation, $(-\infty, 3) \cup (3, \infty)$.

Do Exercises 1 and 2. ▶

ALGEBRAIC ▶◀ GRAPHICAL CONNECTION

Let's make a visual check of Example 2 by looking at the following graph.

Note that the graph consists of two unconnected "branches." If a vertical line were drawn at $x = 3$, shown dashed here, it would not touch the graph of f. Thus, 3 is not in the domain of f.

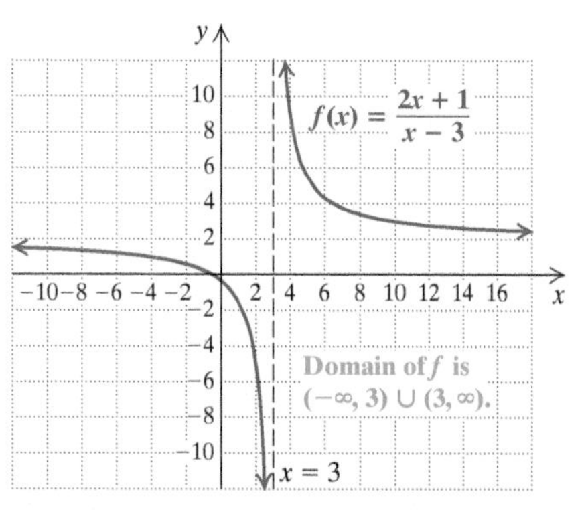

EXAMPLE 3 Find all numbers for which the rational expression

$$\frac{t^4 - 5t}{t^2 - 3t - 28}$$

is not defined.

The rational expression is not defined for a replacement that makes the denominator 0. To determine those replacements to exclude, we set the denominator equal to 0 and solve:

$$t^2 - 3t - 28 = 0 \quad \text{Setting the denominator equal to 0}$$
$$(t - 7)(t + 4) = 0 \quad \text{Factoring}$$
$$t - 7 = 0 \quad or \quad t + 4 = 0 \quad \text{Using the principle of zero products}$$
$$t = 7 \quad or \quad t = -4.$$

Thus the expression is not defined for the replacements 7 and -4.

EXAMPLE 4 Find the domain of g if

$$g(t) = \frac{t^4 - 5t}{t^2 - 3t - 28}.$$

We proceed as we did in Example 3. The expression is not defined for the replacements 7 and -4. Thus the domain is $\{t|t$ is a real number *and* $t \neq 7$ *and* $t \neq -4\}$, or, in interval notation, $(-\infty, -4) \cup (-4, 7) \cup (7, \infty)$.

Do Exercises 3 and 4. ▶

1. Find all numbers for which the rational expression

$$\frac{x^2 - 4x + 9}{2x + 5}$$

is not defined.

2. Find the domain of f if

$$f(x) = \frac{x^2 - 4x + 9}{2x + 5}.$$

Write both set-builder notation and interval notation for the answer.

3. Find all numbers for which the rational expression

$$\frac{t^2 - 9}{t^2 - 7t + 10}$$

is not defined.

4. Find the domain of g if

$$g(t) = \frac{t^2 - 9}{t^2 - 7t + 10}.$$

Write both set-builder notation and interval notation for the answer.

Answers

1. $-\frac{5}{2}$

2. $\left\{x \middle| x \text{ is a real number } and\ x \neq -\frac{5}{2}\right\}$; $\left(-\infty, -\frac{5}{2}\right) \cup \left(-\frac{5}{2}, \infty\right)$

3. 2, 5

4. $\{t|t$ is a real number *and* $t \neq 2$ *and* $t \neq 5\}$; $(-\infty, 2) \cup (2, 5) \cup (5, \infty)$

b FINDING EQUIVALENT RATIONAL EXPRESSIONS

Calculations with rational expressions are similar to those with rational numbers.

> **MULTIPLYING RATIONAL EXPRESSIONS**
>
> To multiply rational expressions, multiply numerators and multiply denominators:
>
> $$\frac{A}{B} \cdot \frac{C}{D} = \frac{AC}{BD}.$$

For example, we have the following:

$$\frac{3}{5} \cdot \frac{2}{7} = \frac{3 \cdot 2}{5 \cdot 7} = \frac{6}{35}, \qquad \frac{3x}{4} \cdot \frac{5x}{7} = \frac{(3x)(5x)}{4 \cdot 7} = \frac{15x^2}{28},$$

and $\dfrac{x+3}{y-4} \cdot \dfrac{x^3}{y+5} = \dfrac{(x+3)x^3}{(y-4)(y+5)}$. Multiplying numerators and multiplying denominators

For purposes of our work in this chapter, it is better in the example above to leave the numerator $(x + 3)x^3$ and the denominator $(y - 4)(y + 5)$ in factored form because it is easier to simplify if we do not multiply.

Before discussing simplifying rational expressions, we first consider multiplying by 1.

> Any rational expression with the same numerator and denominator is a symbol for 1:
>
> $$\frac{73}{73} = 1, \quad \frac{x-y}{x-y} = 1, \quad \frac{4x^2-5}{4x^2-5} = 1, \quad \frac{-1}{-1} = 1, \quad \frac{x+5}{x+5} = 1.$$

We can multiply by 1 to get equivalent expressions—for example,

$$\frac{7}{9} = \frac{7}{9} \cdot \frac{4}{4} = \frac{7 \cdot 4}{9 \cdot 4} = \frac{28}{36} \quad \text{and} \quad \frac{5}{6} = \frac{5}{6} \cdot \frac{x}{x} = \frac{5 \cdot x}{6 \cdot x} = \frac{5x}{6x}.$$

As another example, let's multiply $(x + y)/5$ by 1, using the symbol $(x - y)/(x - y)$:

$$\frac{x+y}{5} \cdot \frac{x-y}{x-y} = \frac{(x+y)(x-y)}{5(x-y)}.$$ Multiplying by $\dfrac{x-y}{x-y}$, which is 1

We know that the expressions

$$\frac{x+y}{5} \quad \text{and} \quad \frac{(x+y)(x-y)}{5(x-y)}$$

are equivalent. This means that they will name the same number for all replacements that do not make a denominator 0.

EXAMPLES Multiply to obtain an equivalent expression.

5. $\dfrac{x^2+3}{x-1}\cdot 1 = \dfrac{x^2+3}{x-1}\cdot\dfrac{x+1}{x+1} = \dfrac{(x^2+3)(x+1)}{(x-1)(x+1)}$ Using $\dfrac{x+1}{x+1}$ for 1

6. $1\cdot\dfrac{x-4}{x-y} = \dfrac{-1}{-1}\cdot\dfrac{x-4}{x-y} = \dfrac{-1\cdot(x-4)}{-1\cdot(x-y)}$ Using $\dfrac{-1}{-1}$ for 1

$= \dfrac{-x+4}{-x+y} = \dfrac{4-x}{y-x}$

Do Exercises 5–7. ▶

Multiply.

5. $\dfrac{3x+2y}{5x+4y}\cdot\dfrac{x}{x}$

6. $\dfrac{2x^2-y}{3x+4}\cdot\dfrac{3x+2}{3x+2}$

7. $\dfrac{-1}{-1}\cdot\dfrac{2a-5}{a-b}$

c SIMPLIFYING RATIONAL EXPRESSIONS

We simplify rational expressions using the identity property of 1 (see Section J14) in reverse. That is, we "remove" factors that are equal to 1. We first factor the numerator and the denominator and then factor the rational expression, so that a factor is equal to 1. We also say, accordingly, that we "remove a factor of 1."

EXAMPLE 7 Simplify: $\dfrac{120}{320}$.

$\dfrac{120}{320} = \dfrac{40\cdot 3}{40\cdot 8}$ Factoring the numerator and the denominator, looking for common factors

$= \dfrac{40}{40}\cdot\dfrac{3}{8}$ Factoring the rational expression; $\dfrac{40}{40}$ is a factor of 1

$= 1\cdot\dfrac{3}{8}$ $\dfrac{40}{40} = 1$

$= \dfrac{3}{8}$ Removing a factor of 1

Do Exercise 8. ▶

8. Simplify: $\dfrac{128}{160}$.

EXAMPLES Simplify.

8. $\dfrac{5x^2}{x} = \dfrac{5x\cdot x}{1\cdot x}$ Factoring the numerator and the denominator

$= \dfrac{5x}{1}\cdot\dfrac{x}{x}$ Factoring the rational expression; $\dfrac{x}{x}$ is a factor of 1

$= 5x\cdot 1$ $\dfrac{5x}{1} = 5x;\ \dfrac{x}{x} = 1$

$= 5x$ Removing a factor of 1

In this example, we supplied a 1 in the denominator. This can always be done, but it is not necessary.

9. $\dfrac{4a+8}{2} = \dfrac{2(2a+4)}{2\cdot 1}$ Factoring the numerator and the denominator

$= \dfrac{2}{2}\cdot\dfrac{2a+4}{1}$ Factoring the rational expression; $\dfrac{2}{2}$ is a factor of 1

$= \dfrac{2a+4}{1}$ Removing a factor of 1

$= 2a+4$

Do Exercises 9 and 10. ▶

Simplify.

9. $\dfrac{7x^2}{x}$

10. $\dfrac{6a+9}{3}$

Answers

5. $\dfrac{(3x+2y)x}{(5x+4y)x}$ **6.** $\dfrac{(2x^2-y)(3x+2)}{(3x+4)(3x+2)}$
7. $\dfrac{-2a+5}{-a+b}$, or $\dfrac{5-2a}{b-a}$ **8.** $\dfrac{4}{5}$ **9.** $7x$
10. $2a+3$

CALCULATOR CORNER

Checking Multiplication and Simplification We can use the TABLE feature as a partial check that rational expressions have been multiplied and/or simplified correctly. To check the simplification in Example 11,

$$\frac{x^2 - 1}{2x^2 - x - 1} = \frac{x + 1}{2x + 1},$$

we first enter

$y_1 = (x^2 - 1)/(2x^2 - x - 1)$ and
$y_2 = (x + 1)/(2x + 1)$.

Then, using AUTO mode, we look at a table of values of y_1 and y_2. (See p. 122.) If the simplification is correct, the values should be the same for all replacements for which the rational expression is defined. The ERROR messages indicate that -0.5 and 1 are replacements in the first rational expression for which the expression is not defined and -0.5 is a replacement in the second rational expression for which the expression is not defined. For all other numbers, we see that y_1 and y_2 are the same, so the simplification appears to be correct. Remember, this is only a partial check since we cannot check all possible values of x.

X	Y1	Y2
−1.5	.25	.25
−1	0	0
−.5	ERROR	ERROR
0	1	1
.5	.75	.75
1	ERROR	.66667
1.5	.625	.625

X=−1.5

EXERCISES: Use the TABLE feature to determine whether each of the following is correct.

1. $\dfrac{5x^2}{x} = 5x$
2. $\dfrac{2x^2 + 4x}{6x^2 + 2x} = \dfrac{x + 2}{3x + 1}$
3. $\dfrac{x^2 - 3x + 2}{x^2 - 1} = \dfrac{x + 2}{x - 1}$
4. $\dfrac{x^2 - 16}{x^2 - 4} = 4$
5. $\dfrac{x^2 - 5x}{x^2} \cdot \dfrac{4}{x^2 - 25} = \dfrac{4}{x(x + 5)}$

EXAMPLES Simplify.

10. $\dfrac{2x^2 + 4x}{6x^2 + 2x} = \dfrac{2x(x + 2)}{2x(3x + 1)}$ Factoring the numerator and the denominator

$= \dfrac{2x}{2x} \cdot \dfrac{x + 2}{3x + 1}$ Factoring the rational expression

$= \dfrac{x + 2}{3x + 1}$ Removing a factor of 1

11. $\dfrac{x^2 - 1}{2x^2 - x - 1} = \dfrac{(x - 1)(x + 1)}{(2x + 1)(x - 1)}$ Factoring the numerator and the denominator

$= \dfrac{x + 1}{2x + 1} \cdot \dfrac{x - 1}{x - 1}$ Factoring the rational expression

$= \dfrac{x + 1}{2x + 1}$ Removing a factor of 1

12. $\dfrac{9x^2 + 6xy - 3y^2}{12x^2 - 12y^2} = \dfrac{3(x + y)(3x - y)}{3(4)(x + y)(x - y)}$ Factoring the numerator and the denominator

$= \dfrac{3(x + y)}{3(x + y)} \cdot \dfrac{3x - y}{4(x - y)}$ Factoring the rational expression

$= \dfrac{3x - y}{4(x - y)}$ Removing a factor of 1

For purposes of later work, we generally do not multiply out the numerator and the denominator after simplifying rational expressions.

Canceling

Canceling is a shortcut that you may have used for removing a factor of 1 when working with fraction notation or rational expressions. With great concern, we mention it here as a possible way to speed up your work. **Canceling can be done for removing factors of 1 only in products.** It *cannot* be done in sums or when adding expressions together. Our concern is that canceling be done with care and understanding. Example 12 might have been done faster as follows:

$\dfrac{9x^2 + 6xy - 3y^2}{12x^2 - 12y^2} = \dfrac{\cancel{3}\cancel{(x + y)}(3x - y)}{\cancel{3}(4)\cancel{(x + y)}(x - y)}$ When a factor of 1 is noted, it is "canceled" as shown.

$= \dfrac{3x - y}{4(x - y)}.$ Removing a factor of 1: $\dfrac{3(x + y)}{3(x + y)} = 1$

Caution!

The difficulty with canceling is that it can be applied incorrectly in situations such as the following:

$\dfrac{\cancel{2} + 3}{\cancel{2}} = 3,$ Wrong! $\dfrac{\cancel{4} + 1}{\cancel{4} + 2} = \dfrac{1}{2},$ Wrong! $\dfrac{1\cancel{5}}{\cancel{5}4} = \dfrac{1}{4}.$ Wrong!

In each of these situations, the expressions canceled are *not* factors of 1. Factors are parts of products. For example, in $2 \cdot 3$, 2 and 3 are factors, but in $2 + 3$, 2 and 3 are *not* factors. **If you can't factor, you can't cancel! If in doubt, don't cancel!**

Do Exercises 11–13.

Simplify.

11. $\frac{6x^2 + 4x}{4x^2 + 8x}$

12. $\frac{2y^2 + 6y + 4}{y^2 - 1}$

13. $\frac{20a^2 - 80b^2}{16a^2 - 64ab + 64b^2}$

Opposites in Rational Expressions

Expressions of the form $a - b$ and $b - a$ are opposites, or additive inverses, of each other. When either of these binomials is multiplied by -1, the result is the other binomial:

$$\left.\begin{aligned} -1(a - b) &= -a + b = b - a; \\ -1(b - a) &= -b + a = a - b. \end{aligned}\right\}$$ Multiplication by -1 reverses the order in which subtraction occurs.

Consider

$$\frac{x - 8}{8 - x}.$$

At first glance, the numerator and the denominator do not appear to have any common factors other than 1. But $x - 8$ and $8 - x$ are opposites of each other. Therefore, we can rewrite one as the opposite of the other by factoring out a -1.

EXAMPLE 13 Simplify: $\frac{x - 8}{8 - x}$.

$$\frac{x - 8}{8 - x} = \frac{x - 8}{-(x - 8)}$$ Rewriting $8 - x$ as $-(x - 8)$. See Section J18.

$$= \frac{1(x - 8)}{-1(x - 8)}$$

$$= \frac{1}{-1} \cdot \frac{x - 8}{x - 8}$$

$$= -1 \cdot 1$$ Note that $\frac{1}{-1} = -1$, not 1.

$$= -1$$

Do Exercises 14–16.

Simplify.

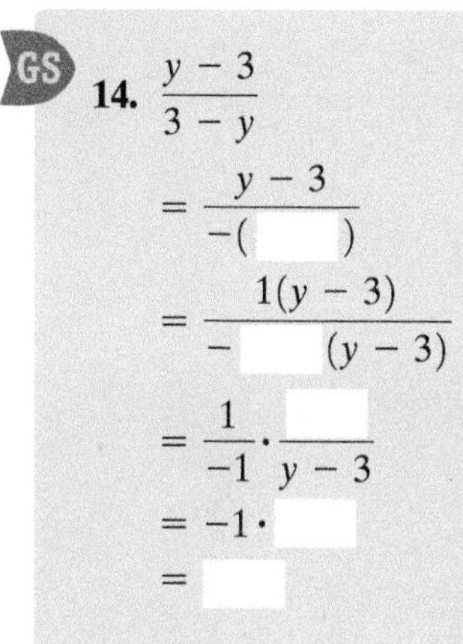

GS 14. $\frac{y - 3}{3 - y}$

$= \frac{y - 3}{-(\quad)}$

$= \frac{1(y - 3)}{-\quad(y - 3)}$

$= \frac{1}{-1} \cdot \frac{\quad}{y - 3}$

$= -1 \cdot \quad$

$= \quad$

15. $\frac{p - q}{q - p}$

16. $\frac{t + 8}{-t - 8}$

d MULTIPLYING AND SIMPLIFYING

After multiplying, we generally simplify, if possible. That is one reason why we leave the numerator and the denominator in factored form. Even so, we might need to factor them further in order to simplify.

EXAMPLES Multiply and simplify.

14. $\frac{x + 2}{x - 3} \cdot \frac{x^2 - 4}{x^2 + x - 2} = \frac{(x + 2)(x^2 - 4)}{(x - 3)(x^2 + x - 2)}$ Multiplying the numerators and the denominators

$= \frac{(x + 2)(x + 2)(x - 2)}{(x - 3)(x + 2)(x - 1)}$ Factoring the numerator and the denominator

$= \frac{(x + 2)\cancel{(x + 2)}(x - 2)}{(x - 3)\cancel{(x + 2)}(x - 1)}$ Removing a factor of 1: $\frac{x + 2}{x + 2} = 1$

$= \frac{(x + 2)(x - 2)}{(x - 3)(x - 1)}$ Simplifying

Answers

11. $\frac{3x + 2}{2(x + 2)}$ **12.** $\frac{2(y + 2)}{y - 1}$ **13.** $\frac{5(a + 2b)}{4(a - 2b)}$ **14.** -1 **15.** -1 **16.** -1

Guided Solution:
14. $y - 3, 1, y - 3, 1, -1$

Multiply and simplify.

17. $\dfrac{(x-y)^3}{x+y} \cdot \dfrac{3x+3y}{x^2-y^2}$

18. $\dfrac{a^3+b^3}{a^2-b^2} \cdot \dfrac{a^2-2ab+b^2}{a^2-ab+b^2}$

Find the reciprocal.

19. $\dfrac{x+3}{x-5}$

20. $x+7$

21. $\dfrac{1}{y^3-9}$

15. $\dfrac{a^3-b^3}{a^2-b^2} \cdot \dfrac{a^2+2ab+b^2}{a^2+ab+b^2}$

$= \dfrac{(a^3-b^3)(a^2+2ab+b^2)}{(a^2-b^2)(a^2+ab+b^2)}$

$= \dfrac{(a-b)(a^2+ab+b^2)(a+b)(a+b)}{(a-b)(a+b)(a^2+ab+b^2)\cdot 1}$ Factoring the numerator and the denominator

$= \dfrac{\cancel{(a-b)}\cancel{(a^2+ab+b^2)}\cancel{(a+b)}(a+b)}{\cancel{(a-b)}\cancel{(a+b)}\cancel{(a^2+ab+b^2)}\cdot 1}$

Removing a factor of 1: $\dfrac{(a-b)(a^2+ab+b^2)(a+b)}{(a-b)(a^2+ab+b^2)(a+b)} = 1$

$= \dfrac{a+b}{1} = a+b$ Simplifying

Do Exercises 17 and 18.

e DIVIDING AND SIMPLIFYING

Two expressions are reciprocals (or multiplicative inverses) of each other if their product is 1. To find the reciprocal of a rational expression, we interchange the numerator and the denominator.

The reciprocal of $\dfrac{3}{7}$ is $\dfrac{7}{3}$. The reciprocal of $y-8$ is $\dfrac{1}{y-8}$.

The reciprocal of $\dfrac{x+2y}{x+y-1}$ is $\dfrac{x+y-1}{x+2y}$.

Do Exercises 19–21.

We divide rational expressions in the same way that we divide fraction notation in arithmetic. For a review, see Section J9.

DIVIDING RATIONAL EXPRESSIONS

To divide by a rational expression, multiply by its reciprocal:

$$\frac{A}{B} \div \frac{C}{D} = \frac{A}{B}\cdot\frac{D}{C} = \frac{AD}{BC}.$$

Then factor and simplify if possible.

For example,

$$\frac{2}{3} \div \frac{4}{5} = \frac{2}{3}\cdot\frac{5}{4} = \frac{2\cdot 5}{3\cdot 2\cdot 2} = \frac{5}{3\cdot 2}\cdot\frac{2}{2} = \frac{5}{6}\cdot 1 = \frac{5}{6}.$$

EXAMPLE 16 Divide and simplify.

$$\frac{x-2}{x+1} \div \frac{x+5}{x-3} = \frac{x-2}{x+1}\cdot\frac{x-3}{x+5} = \frac{(x-2)(x-3)}{(x+1)(x+5)}$$

Answers

17. $\dfrac{3(x-y)(x-y)}{x+y}$ 18. $a-b$
19. $\dfrac{x-5}{x+3}$ 20. $\dfrac{1}{x+7}$ 21. y^3-9

EXAMPLE 17 Divide and simplify.

$$\frac{a^2 - 1}{a - 1} \div \frac{a^2 - 2a + 1}{a + 1}$$

$$= \frac{a^2 - 1}{a - 1} \cdot \frac{a + 1}{a^2 - 2a + 1}$$ Multiplying by the reciprocal of the divisor

$$= \frac{(a^2 - 1)(a + 1)}{(a - 1)(a^2 - 2a + 1)}$$ Multiplying the numerators and the denominators

$$= \frac{(a + 1)(a - 1)(a + 1)}{(a - 1)(a - 1)(a - 1)}$$ Factoring the numerator and the denominator

$$= \frac{(a + 1)\cancel{(a - 1)}(a + 1)}{(a - 1)\cancel{(a - 1)}(a - 1)}$$ Removing a factor of 1: $\frac{a - 1}{a - 1} = 1$

$$= \frac{(a + 1)(a + 1)}{(a - 1)(a - 1)}$$ Simplifying

Do Exercises 22 and 23. ▶

EXAMPLE 18 Perform the indicated operations and simplify:

$$\frac{c^3 - d^3}{(c + d)^2} \div (c - d) \cdot (c + d).$$

Using the rules for order of operations, we do the division first:

$$\frac{c^3 - d^3}{(c + d)^2} \div (c - d) \cdot (c + d) = \frac{c^3 - d^3}{(c + d)^2} \cdot \frac{1}{c - d} \cdot (c + d)$$

$$= \frac{(c^3 - d^3)(c + d)}{(c + d)^2(c - d)} = \frac{(c - d)(c^2 + cd + d^2)(c + d)}{(c + d)(c + d)(c - d)}$$

$$= \frac{\cancel{(c - d)}(c^2 + cd + d^2)\cancel{(c + d)}}{(c + d)\cancel{(c + d)}\cancel{(c - d)}} = \frac{c^2 + cd + d^2}{c + d}.$$

Do Exercise 24. ▶

Divide and simplify.

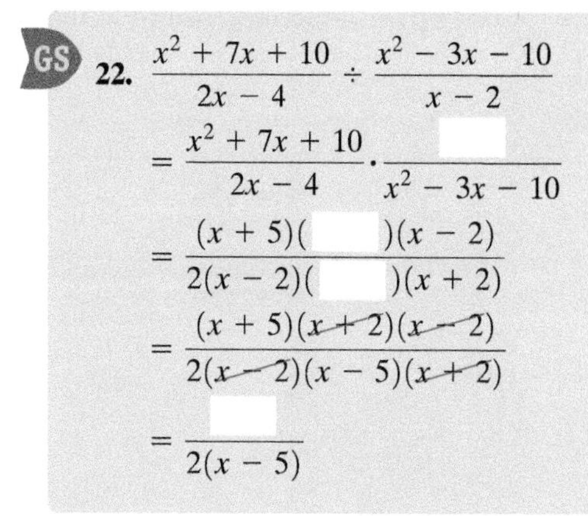

GS **22.** $\frac{x^2 + 7x + 10}{2x - 4} \div \frac{x^2 - 3x - 10}{x - 2}$

$= \frac{x^2 + 7x + 10}{2x - 4} \cdot \frac{\square}{x^2 - 3x - 10}$

$= \frac{(x + 5)(\quad)(x - 2)}{2(x - 2)(\quad)(x + 2)}$

$= \frac{(x + 5)\cancel{(x + 2)}\cancel{(x - 2)}}{2\cancel{(x - 2)}(x - 5)\cancel{(x + 2)}}$

$= \frac{\square}{2(x - 5)}$

23. $\frac{a^2 - b^2}{ab} \div \frac{a^2 - 2ab + b^2}{2a^2b^2}$

24. Perform the indicated operations and simplify:

$$\frac{a^3 + 8}{a - 2} \div (a^2 - 2a + 4) \cdot (a - 2)^2.$$

Answers

22. $\frac{x + 5}{2(x - 5)}$ **23.** $\frac{2ab(a + b)}{a - b}$
24. $(a + 2)(a - 2)$

Guided Solution:
22. $x - 2, x + 2, x - 5, x + 5$

5.1 Exercise Set

FOR EXTRA HELP  MyLab Math

✓ Check Your Understanding

Reading Check Choose the word below each blank that best completes the statement.

RC1. A rational expression can be written as a ______________ of two polynomials.
product/quotient

RC2. A rational expression is simplified when the numerator and the denominator have no ______________ (other than 1) in common.
factors/terms

RC3. Expressions that have the same value for all allowable replacements are called ______________ expressions.
rational/equivalent

RC4. A rational expression is undefined when the ______________ is zero.
denominator/numerator

Concept Check Choose from the column on the right an expression that is equivalent to the given expression.

CC1. $\dfrac{1}{x} \div \dfrac{1}{8}$

CC2. The opposite of $x - 8$

CC3. The reciprocal of $\dfrac{8}{x - 8}$

CC4. $\dfrac{1}{x} \div \dfrac{8}{x}$

CC5. $\dfrac{1}{x} \div 8$

CC6. The reciprocal of $x - 8$

CC7. $\dfrac{x}{8} \cdot \dfrac{8}{x}$

CC8. $\dfrac{x}{8} \cdot \dfrac{8}{x + 1}$

a) $\dfrac{1}{x} \cdot \dfrac{x}{8}$

b) $\dfrac{1}{8x}$

c) $\dfrac{8}{x}$

d) $\dfrac{x}{8} \div \dfrac{x + 1}{8}$

e) $\dfrac{x - 8}{8}$

f) 1

g) $8 - x$

h) $\dfrac{1}{x - 8}$

a Find all numbers for which the rational expression is not defined.

1. $\dfrac{5t^2 - 64}{3t + 17}$

2. $\dfrac{x^2 + x + 105}{5x - 45}$

3. $\dfrac{x^3 - x^2 + x + 2}{x^2 + 12x + 35}$

4. $\dfrac{x^2 - 3x - 4}{x^2 - 18x + 77}$

Find the domain. Write both set-builder notation and interval notation for the answer.

5. $f(x) = \dfrac{4x - 5}{x + 7}$

6. $f(r) = \dfrac{5r + 3}{r - 6}$

7. $g(x) = \dfrac{7}{3x - x^2}$

8. $g(x) = \dfrac{9}{8x + x^2}$

9. $f(t) = \dfrac{5t^2 - 64}{3t + 17}$

10. $f(x) = \dfrac{x^2 + x + 105}{5x - 45}$

11. $f(x) = \dfrac{x^3 - x^2 + x + 2}{x^2 + 12x + 35}$

12. $f(x) = \dfrac{x^2 - 3x - 4}{x^2 - 18x + 77}$

b Multiply the expression by the given form of 1 to obtain an equivalent expression. Do not simplify.

13. $\dfrac{x + 2}{x + 8}; \dfrac{7x}{7x}$

14. $\dfrac{2 - y^2}{8 - y}; \dfrac{-1}{-1}$

15. $\dfrac{q - 5}{q + 3}; \dfrac{q + 5}{q + 5}$

16. $\dfrac{p + 1}{p + 4}; \dfrac{p - 4}{p - 4}$

c Simplify.

17. $\dfrac{15y^5}{5y^4}$

18. $\dfrac{7w^3}{28w^2}$

19. $\dfrac{16p^3}{24p^7}$

20. $\dfrac{48t^5}{56t^{11}}$

21. $\dfrac{9a - 27}{9}$

22. $\dfrac{6a - 30}{6}$

23. $\dfrac{12x - 15}{21}$

24. $\dfrac{18a - 2}{22}$

25. $\dfrac{4y - 12}{4y + 12}$

26. $\dfrac{8x + 16}{8x - 16}$

27. $\dfrac{t^2 - 16}{t^2 - 8t + 16}$

28. $\dfrac{p^2 - 25}{p^2 + 10p + 25}$

29. $\dfrac{x^2 - 9x + 8}{x^2 + 3x - 4}$

30. $\dfrac{y^2 + 8y - 9}{y^2 - 5y + 4}$

31. $\dfrac{w^3 - z^3}{w^2 - z^2}$

32. $\dfrac{a^2 - b^2}{a^3 + b^3}$

d Multiply and simplify.

33. $\dfrac{x^4}{3x + 6} \cdot \dfrac{5x + 10}{5x^7}$

34. $\dfrac{10t}{6t - 12} \cdot \dfrac{20t - 40}{30t^3}$

35. $\dfrac{x^2 - 16}{x^2} \cdot \dfrac{x^2 - 4x}{x^2 - x - 12}$

36. $\dfrac{y^2 + 10y + 25}{y^2 - 9} \cdot \dfrac{y^2 - 3y}{y + 5}$

37. $\dfrac{y^2 - 16}{2y + 6} \cdot \dfrac{y + 3}{y - 4}$

38. $\dfrac{m^2 - n^2}{4m + 4n} \cdot \dfrac{m + n}{m - n}$

39. $\dfrac{x^2 - 2x - 35}{2x^3 - 3x^2} \cdot \dfrac{4x^3 - 9x}{7x - 49}$

40. $\dfrac{y^2 - 10y + 9}{y^2 - 1} \cdot \dfrac{y + 4}{y^2 - 5y - 36}$

41. $\dfrac{c^3 + 8}{c^2 - 4} \cdot \dfrac{c^2 - 4c + 4}{c^2 - 2c + 4}$

42. $\dfrac{x^3 - 27}{x^2 - 9} \cdot \dfrac{x^2 - 6x + 9}{x^2 + 3x + 9}$

43. $\dfrac{x^2 - y^2}{x^3 - y^3} \cdot \dfrac{x^2 + xy + y^2}{x^2 + 2xy + y^2}$

44. $\dfrac{4x^2 - 9y^2}{8x^3 - 27y^3} \cdot \dfrac{4x^2 + 6xy + 9y^2}{4x^2 + 12xy + 9y^2}$

e Divide and simplify.

45. $\dfrac{12x^8}{3y^4} \div \dfrac{16x^3}{6y}$

46. $\dfrac{9a^7}{8b^2} \div \dfrac{12a^2}{24b^7}$

47. $\dfrac{3y + 15}{y} \div \dfrac{y + 5}{y}$

48. $\dfrac{6x + 12}{x} \div \dfrac{x + 2}{x^3}$

49. $\dfrac{y^2 - 9}{y} \div \dfrac{y + 3}{y + 2}$

50. $\dfrac{x^2 - 4}{x} \div \dfrac{x - 2}{x + 4}$

51. $\dfrac{4a^2 - 1}{a^2 - 4} \div \dfrac{2a - 1}{a - 2}$

52. $\dfrac{25x^2 - 4}{x^2 - 9} \div \dfrac{5x - 2}{x + 3}$

53. $\dfrac{x^2 - 16}{x^2 - 10x + 25} \div \dfrac{3x - 12}{x^2 - 3x - 10}$

54. $\dfrac{y^2 - 36}{y^2 - 8y + 16} \div \dfrac{3y - 18}{y^2 - y - 12}$

55. $\dfrac{y^3 + 3y}{y^2 - 9} \div \dfrac{y^2 + 5y - 14}{y^2 + 4y - 21}$

56. $\dfrac{a^3 + 4a}{a^2 - 16} \div \dfrac{a^2 + 8a + 15}{a^2 + a - 20}$

57. $\dfrac{x^3 - 64}{x^3 + 64} \div \dfrac{x^2 - 16}{x^2 - 4x + 16}$

58. $\dfrac{8y^3 + 27}{64y^3 - 1} \div \dfrac{4y^2 - 9}{16y^2 + 4y + 1}$

59. $\dfrac{8x^3y^3 + 27x^3}{64x^3y^3 - x^3} \div \dfrac{4x^2y^2 - 9x^2}{16x^2y^2 + 4x^2y + x^2}$

60. $\dfrac{x^3y - 64y}{x^3y + 64y} \div \dfrac{x^2y^2 - 16y^2}{x^2y^2 - 4xy^2 + 16y^2}$

Perform the indicated operations and simplify.

61. $\dfrac{r^2 - 4s^2}{r + 2s} \div (r + 2s) \cdot \dfrac{2s}{r - 2s}$

62. $\dfrac{d^2 - d}{d^2 - 6d + 8} \cdot \dfrac{d - 2}{d^2 + 5d} \div \dfrac{5d}{d^2 - 9d + 20}$

63. $\dfrac{y^2 - 2y}{y^2 + y - 2} \cdot \dfrac{y - 1}{y^2 + 4y + 4} \div \dfrac{y^2 + 2y - 8}{y^4}$

64. $\dfrac{9x^2}{x^2 - 16y^2} \div \dfrac{1}{x^2 + 4xy} \cdot \dfrac{x - 4y}{3x}$

Skill Maintenance

In Exercises 65–68, the graph is that of a function. Determine the domain and the range. [2.3a]

65.

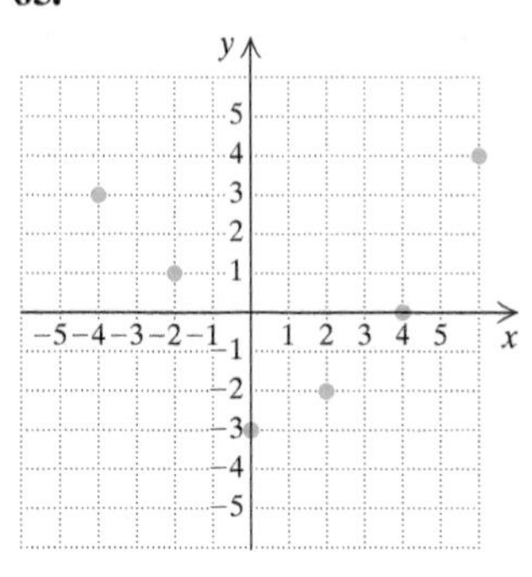

66.

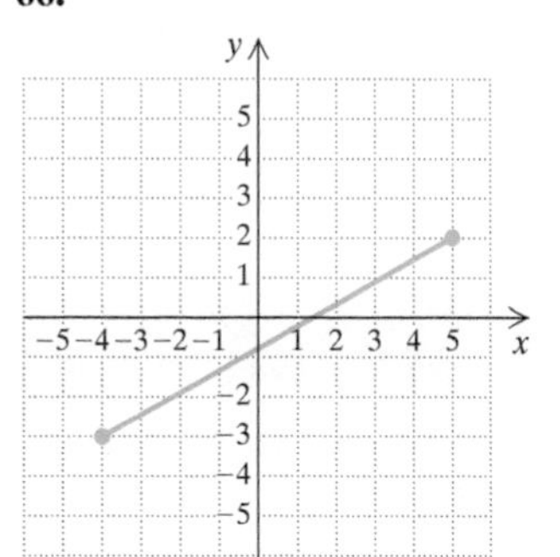

67.

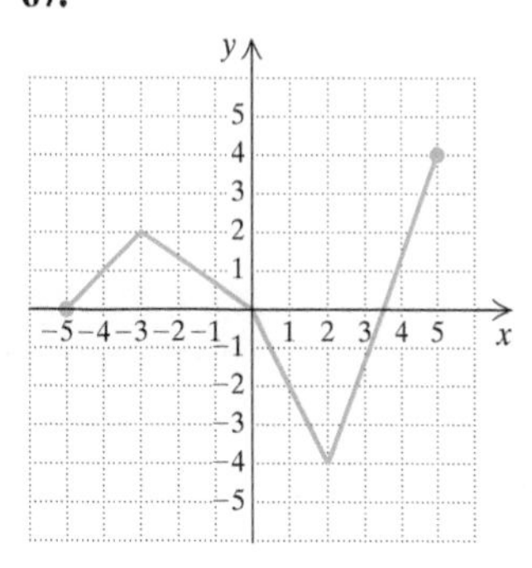

68.

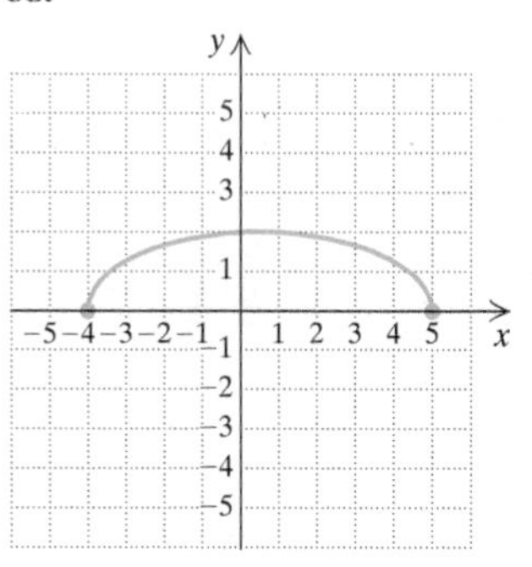

Factor. [4.7a]

69. $10x^2 - 80x + 70$

70. $10x^2 - 13x + 4$

71. $2x^3 - 16x^2 - 66x$

72. $10y^2 + 80y - 650$

73. Find an equation of the line with slope $-\frac{2}{3}$ and y-intercept $(0, -5)$. [2.6a]

74. Find an equation of the line having slope $-\frac{2}{7}$ and containing the point $(-4, 8)$. [2.6b]

Synthesis

Simplify.

75. $\dfrac{m^2 - t^2}{m^2 + t^2 + m + t + 2mt}$

76. $\dfrac{2x - 5(x + 2) - (x - 2)}{x^2 - 4}$

77. Let

$$g(x) = \frac{2x + 3}{4x - 1}.$$

Find $g(5)$, $g(0)$, $g(\frac{1}{4})$, and $g(a + h)$.

5.2 LCMs, LCDs, Addition, and Subtraction

OBJECTIVES

a Find the LCM of several algebraic expressions by factoring.

b Add and subtract rational expressions.

c Simplify combined additions and subtractions of rational expressions.

a FINDING LCMs BY FACTORING

SKILL REVIEW *Find the opposite of a polynomial and subtract polynomials.* [4.1d]

Subtract.

1. $(2y - 5) - (y - 6)$

2. $(3x^2 + x - 4) - (5x^2 - x + 10)$

Answers: 1. $y + 1$ **2.** $-2x^2 + 2x - 14$

MyLab Math VIDEO

To add rational expressions when denominators are different, we first find a common denominator. Let's review the procedure used in arithmetic first. To do the addition

$$\frac{5}{42} + \frac{7}{12},$$

we find a common denominator. We look for the least common multiple (LCM) of 42 and 12. That number becomes the least common denominator (LCD).

To find the LCM, we factor both numbers completely (into primes).

$42 = 2 \cdot 3 \cdot 7 \leftarrow$ Any multiple of 42 has these factors.

$12 = 2 \cdot 2 \cdot 3 \leftarrow$ Any multiple of 12 has these factors.

The LCM is the number that has 2 as a factor twice, 3 as a factor once, and 7 as a factor once: LCM $= 2 \cdot 2 \cdot 3 \cdot 7$, or 84.

FINDING LCMs

To find the LCM, use each factor the greatest number of times that it occurs in any one prime factorization.

EXAMPLE 1 Find the LCM of 18 and 24.

$18 = 2 \cdot 3 \cdot 3$

$24 = 2 \cdot 2 \cdot 2 \cdot 3$

The LCM is $2 \cdot 2 \cdot 2 \cdot 3 \cdot 3$, or 72.

◀ Do Exercises 1 and 2.

Find the LCM by factoring.

1. 18, 30

2. 12, 18, 24

Now let's return to adding $\frac{5}{42}$ and $\frac{7}{12}$:

$$\frac{5}{42} + \frac{7}{12} = \frac{5}{2 \cdot 3 \cdot 7} + \frac{7}{2 \cdot 2 \cdot 3}. \quad \text{Factoring the denominators}$$

Answers

1. 90 **2.** 72

The LCD is the LCM of the denominators, $2 \cdot 2 \cdot 3 \cdot 7$. To get this LCD in the first denominator, we need a factor of 2. In the second denominator, we need a factor of 7. We multiply by 1, as follows:

$$\frac{5}{2 \cdot 3 \cdot 7} \cdot \frac{2}{2} + \frac{7}{2 \cdot 2 \cdot 3} \cdot \frac{7}{7} = \frac{10}{2 \cdot 2 \cdot 3 \cdot 7} + \frac{49}{2 \cdot 2 \cdot 3 \cdot 7}$$
$$= \frac{59}{2 \cdot 2 \cdot 3 \cdot 7} = \frac{59}{84}.$$

Multiplying the first fraction by $\frac{2}{2}$ gave us an equivalent fraction with a denominator that is the LCD. Multiplying the second fraction by $\frac{7}{7}$ also gave us an equivalent fraction with a denominator that is the LCD. Once we had a common denominator, we added the numerators and kept the common denominator.

Do Exercises 3 and 4. ▶

Add, first finding the LCD of the denominators.

3. $\frac{5}{12} + \frac{11}{30}$

4. $\frac{7}{12} + \frac{13}{18} + \frac{1}{24}$

We find the LCM of algebraic expressions in the same way that we find the LCM of natural numbers.

Our reasoning for learning how to find LCMs is so that we will be able to add rational expressions. For example, to do the addition

$$\frac{7}{12xy^2} + \frac{8}{15x^3y},$$

we first need to find the LCM of $12xy^2$ and $15x^3y$, which is $60x^3y^2$.

EXAMPLES

2. Find the LCM of $12xy^2$ and $15x^3y$.

We factor each expression completely. To find the LCM, we use each factor the greatest number of times that it occurs in any one prime factorization.

$$\left.\begin{aligned} 12xy^2 &= 2 \cdot 2 \cdot 3 \cdot x \cdot y \cdot y; \\ 15x^3y &= 3 \cdot 5 \cdot x \cdot x \cdot x \cdot y \end{aligned}\right\} \quad \text{Factoring}$$

$$\text{LCM} = 2 \cdot 2 \cdot 3 \cdot 5 \cdot x \cdot x \cdot x \cdot y \cdot y = 60x^3y^2$$

$12xy^2$ is a factor.
$15x^3y$ is a factor.

The LCM of $12xy^2$ and $15x^3y$ is $60x^3y^2$.

3. Find the LCM of $x^2 + 2x + 1$, $5x^2 - 5x$, and $x^2 - 1$.

$$\left.\begin{aligned} x^2 + 2x + 1 &= (x + 1)(x + 1); \\ 5x^2 - 5x &= 5x(x - 1); \\ x^2 - 1 &= (x + 1)(x - 1) \end{aligned}\right\} \quad \text{Factoring}$$

Both factors of $x^2 - 1$ are already present in the previous factorizations.

$$\text{LCM} = 5x(x + 1)(x + 1)(x - 1)$$

4. Find the LCM of $x^2 - y^2$, $x^3 + y^3$, and $x^2 + 2xy + y^2$.

$$\left.\begin{aligned} x^2 - y^2 &= (x - y)(x + y); \\ x^3 + y^3 &= (x + y)(x^2 - xy + y^2); \\ x^2 + 2xy + y^2 &= (x + y)(x + y) = (x + y)^2 \end{aligned}\right\} \quad \text{Factoring}$$

$$\text{LCM} = (x - y)(x + y)^2(x^2 - xy + y^2)$$

Answers

3. $\frac{47}{60}$ **4.** $\frac{97}{72}$

Recall that $-(x - 3) = -1(x - 3) = 3 - x$. If $(x - 3)(x + 2)$ is an LCM, then $-1(x - 3)(x + 2) = (3 - x)(x + 2)$ is also an LCM.

If, when we are finding LCMs, factors that are opposites occur, we do not use both of them. For example, if $a - b$ occurs in one factorization and $b - a$ occurs in another, we do not use both, since they are opposites.

Find the LCM.

5. $a^2b^2,\ 5a^3b$

6. $y^2 + 7y + 12,\ y^2 + 8y + 16,\ y + 4$

7. $x^2 - 9,\ x^3 - x^2 - 6x,\ 2x^2$

8. $a^2 - b^2,\ 2b - 2a$

EXAMPLE 5 Find the LCM of $x^2 - y^2$ and $3y - 3x$.

$$x^2 - y^2 = (x + y)(x - y)$$
$$3y - 3x = 3(y - x), \text{ or } -3(x - y)$$

We can use $(x - y)$ or $(y - x)$, but we do not use both.

$$\text{LCM} = 3(x + y)(x - y), \text{ or } 3(x + y)(y - x), \text{ or } -3(x + y)(x - y)$$

In most cases, we would use the form $3(x + y)(x - y)$.

Do Exercises 5–8.

b ADDING AND SUBTRACTING RATIONAL EXPRESSIONS

ADDITION AND SUBTRACTION WITH LIKE DENOMINATORS

To add or subtract when denominators are the same, add or subtract the numerators and keep the same denominator.

$$\frac{A}{C} + \frac{B}{C} = \frac{A + B}{C} \quad \text{and} \quad \frac{A}{C} - \frac{B}{C} = \frac{A - B}{C}, \quad \text{where } C \neq 0.$$

Then factor and simplify if possible.

EXAMPLE 6 Add: $\dfrac{3 + x}{x} + \dfrac{4}{x}$.

$$\frac{3 + x}{x} + \frac{4}{x} = \frac{3 + x + 4}{x}$$ Adding numerators and keeping the same denominator

$$= \frac{7 + x}{x}$$

Caution!

This expression does *not* simplify to 7: $\dfrac{7 + x}{x} \neq 7$.

Example 6 shows that

$$\frac{3 + x}{x} + \frac{4}{x} \quad \text{and} \quad \frac{7 + x}{x}$$

are equivalent expressions. They name the same number for all replacements for which the rational expressions are defined.

Answers

5. $5a^3b^2$ **6.** $(y + 3)(y + 4)(y + 4)$
7. $2x^2(x - 3)(x + 3)(x + 2)$
8. $2(a + b)(a - b)$, or $2(a + b)(b - a)$, or $-2(a + b)(a - b)$

EXAMPLE 7 Add: $\dfrac{4x^2 - 5xy}{x^2 - y^2} + \dfrac{2xy - y^2}{x^2 - y^2}$.

$$\frac{4x^2 - 5xy}{x^2 - y^2} + \frac{2xy - y^2}{x^2 - y^2} = \frac{4x^2 - 3xy - y^2}{x^2 - y^2} \quad \text{Adding the numerators}$$

$$= \frac{(4x + y)(x - y)}{(x + y)(x - y)} \quad \text{Factoring the numerator and the denominator}$$

$$= \frac{(4x + y)\cancel{(x - y)}}{(x + y)\cancel{(x - y)}} \quad \text{Removing a factor of 1: } \frac{x - y}{x - y} = 1$$

$$= \frac{4x + y}{x + y}$$

Do Exercises 9 and 10. ▶

EXAMPLE 8 Subtract: $\dfrac{4x + 5}{x + 3} - \dfrac{x - 2}{x + 3}$.

$$\frac{4x + 5}{x + 3} - \frac{x - 2}{x + 3} = \frac{4x + 5 - (x - 2)}{x + 3} \quad \text{Subtracting numerators}$$

$$= \frac{4x + 5 - x + 2}{x + 3}$$

$$= \frac{3x + 7}{x + 3}$$

A common error: forgetting these parentheses. If you forget them, you will be subtracting only *part* of the second numerator.

Do Exercises 11 and 12. ▶

When denominators are different, we find the least common denominator, LCD. The procedure we will use is as follows.

ADDITION AND SUBTRACTION WITH DIFFERENT DENOMINATORS

To add or subtract rational expressions with different denominators:

1. Find the LCM of the denominators. This is the least common denominator (LCD).
2. For each rational expression, find an equivalent expression with the LCD. To do so, multiply by 1 using an expression for 1 made up of factors of the LCD that are missing from the original denominator.
3. Add or subtract the numerators. Write the result over the LCD.
4. Simplify, if possible.

EXAMPLE 9 Add: $\dfrac{2a}{5} + \dfrac{3b}{2a}$.

We first find the LCD: $\left.\begin{matrix} 5 \\ 2a \end{matrix}\right\}$ LCD $= 5 \cdot 2a$, or $10a$.

Now we multiply each expression by 1. We choose symbols for 1 that will give us the LCD in each denominator. In this case, we use $2a/(2a)$ and $5/5$:

$$\frac{2a}{5} \cdot \frac{2a}{2a} + \frac{3b}{2a} \cdot \frac{5}{5} = \frac{4a^2}{10a} + \frac{15b}{10a} = \frac{4a^2 + 15b}{10a}.$$

Multiplying the first term by $2a/(2a)$ gave us a denominator of $10a$. Multiplying the second term by $\frac{5}{5}$ also gave us a denominator of $10a$. ■

Add.

9. $\dfrac{5 + y}{y} + \dfrac{7}{y}$

10. $\dfrac{2x^2 + 5x - 9}{x - 5} + \dfrac{x^2 - 19x + 4}{x - 5}$

Subtract.

11. $\dfrac{a}{b + 2} - \dfrac{b}{b + 2}$

12. $\dfrac{4y + 7}{x^2 + y^2} - \dfrac{3y - 5}{x^2 + y^2}$

Answers

9. $\dfrac{12 + y}{y}$ 10. $3x + 1$ 11. $\dfrac{a - b}{b + 2}$ 12. $\dfrac{y + 12}{x^2 + y^2}$

EXAMPLE 10 Add: $\dfrac{3x^2 + 3xy}{x^2 - y^2} + \dfrac{2 - 3x}{x - y}$.

We first find the LCD of the denominators. To do so, we first factor:

$$\left.\begin{aligned} x^2 - y^2 &= (x + y)(x - y) \\ x - y &= x - y \end{aligned}\right\} \quad \text{LCD} = (x + y)(x - y).$$

The first expression already has the LCD. We multiply by 1 to get the LCD in the second expression. Then we add and simplify if possible.

$$\frac{3x^2 + 3xy}{(x + y)(x - y)} + \frac{2 - 3x}{x - y} \cdot \frac{x + y}{x + y} \qquad \text{Multiplying by 1 to get the LCD}$$

$$= \frac{3x^2 + 3xy}{(x + y)(x - y)} + \frac{(2 - 3x)(x + y)}{(x - y)(x + y)}$$

$$= \frac{3x^2 + 3xy}{(x + y)(x - y)} + \frac{2x + 2y - 3x^2 - 3xy}{(x - y)(x + y)} \qquad \text{Multiplying in the numerator}$$

$$= \frac{3x^2 + 3xy + 2x + 2y - 3x^2 - 3xy}{(x + y)(x - y)} \qquad \text{Adding the numerators}$$

$$= \frac{2x + 2y}{(x + y)(x - y)} \qquad \text{Combining like terms}$$

$$= \frac{2(x + y)}{(x + y)(x - y)} \qquad \text{Factoring the numerator}$$

$$= \frac{2\cancel{(x + y)}}{\cancel{(x + y)}(x - y)} \qquad \text{Removing a factor of 1: } \frac{x + y}{x + y} = 1$$

$$= \frac{2}{x - y}$$

Do Exercises 13 and 14.

Add.

13. $\dfrac{3x}{7} + \dfrac{4y}{3x}$

GS

$= \dfrac{3x}{7} \cdot \dfrac{\square}{3x} + \dfrac{4y}{3x} \cdot \dfrac{7}{\square}$

$= \dfrac{9x^2}{21\square} + \dfrac{\square}{21x}$

$= \dfrac{\square + 28y}{\square}$

14. $\dfrac{2xy - 2x^2}{x^2 - y^2} + \dfrac{2x + 3}{x + y}$

EXAMPLE 11 Subtract: $\dfrac{2y + 1}{y^2 - 7y + 6} - \dfrac{y + 3}{y^2 - 5y - 6}$.

$$\frac{2y + 1}{y^2 - 7y + 6} - \frac{y + 3}{y^2 - 5y - 6}$$

$$= \frac{2y + 1}{(y - 6)(y - 1)} - \frac{y + 3}{(y - 6)(y + 1)} \qquad \text{LCD} = (y - 6)(y - 1)(y + 1)$$

$$= \frac{2y + 1}{(y - 6)(y - 1)} \cdot \frac{y + 1}{y + 1} - \frac{y + 3}{(y - 6)(y + 1)} \cdot \frac{y - 1}{y - 1} \qquad \text{Multiplying by 1 to get the LCD}$$

$$= \frac{(2y + 1)(y + 1) - (y + 3)(y - 1)}{(y - 6)(y - 1)(y + 1)} \qquad \text{Subtracting the numerators}$$

$$= \frac{(2y^2 + 3y + 1) - (y^2 + 2y - 3)}{(y - 6)(y - 1)(y + 1)} \qquad \text{Multiplying. Note the use of parentheses.}$$

$$= \frac{2y^2 + 3y + 1 - y^2 - 2y + 3}{(y - 6)(y - 1)(y + 1)}$$

$$= \frac{y^2 + y + 4}{(y - 6)(y - 1)(y + 1)} \qquad \text{The numerator cannot be factored. The rational expression is simplified.}$$

Answers

13. $\dfrac{9x^2 + 28y}{21x}$ **14.** $\dfrac{3}{x + y}$

Guided Solution:

13. $3x, 7, x, 28y, 9x^2, 21x$

We generally do not multiply out a numerator or a denominator if it has two or more factors (other than monomials). This will be helpful when we solve equations.

Do Exercises 15 and 16.

Subtract.

15. $\dfrac{a}{a+3} - \dfrac{a-4}{a}$

16. $\dfrac{4y-5}{y^2-7y+12} - \dfrac{y+7}{y^2+2y-15}$

Denominators That Are Opposites

When one denominator is the opposite of the other, we can first multiply either expression by 1 using $-1/-1$.

EXAMPLE 12 Add: $\dfrac{a}{2a} + \dfrac{a^3}{-2a}$.

$$\frac{a}{2a} + \frac{a^3}{-2a} = \frac{a}{2a} + \frac{a^3}{-2a}\cdot\frac{-1}{-1} \qquad \text{Multiplying by 1, using } \frac{-1}{-1}$$

This is equal to 1 (not -1).

$$= \frac{a}{2a} + \frac{-a^3}{2a}$$

$$= \frac{a-a^3}{2a} \qquad \text{Adding numerators}$$

$$= \frac{a(1+a)(1-a)}{2a} \qquad \text{Factoring}$$

$$= \frac{\cancel{a}(1+a)(1-a)}{2\cancel{a}} \qquad \text{Removing a factor of 1: } \frac{a}{a}=1$$

$$= \frac{(1+a)(1-a)}{2}$$

Do Exercises 17 and 18.

Add.

17. $\dfrac{b}{3b} + \dfrac{b^3}{-3b}$

18. $\dfrac{3x^2+4}{x-5} + \dfrac{x^2-7}{5-x}$

EXAMPLE 13 Subtract: $\dfrac{x^2}{5y} - \dfrac{x^3}{-5y}$.

$$\frac{x^2}{5y} - \frac{x^3}{-5y} = \frac{x^2}{5y} - \frac{x^3}{-5y}\cdot\frac{-1}{-1} \qquad \text{Multiplying by } \frac{-1}{-1}$$

$$= \frac{x^2}{5y} - \frac{-x^3}{5y}$$

$$= \frac{x^2-(-x^3)}{5y}$$

Don't forget these parentheses!

$$= \frac{x^2+x^3}{5y}, \text{ or } \frac{x^2(1+x)}{5y}$$

EXAMPLE 14 Subtract: $\dfrac{5x}{x-2y} - \dfrac{3y-7}{2y-x}$.

$$\frac{5x}{x-2y} - \frac{3y-7}{2y-x} = \frac{5x}{x-2y} - \frac{3y-7}{2y-x}\cdot\frac{-1}{-1}$$

$$= \frac{5x}{x-2y} - \frac{-3y+7}{x-2y}$$

Remember: $(2y-x)(-1) = -2y + x = x - 2y$.

$$= \frac{5x-(-3y+7)}{x-2y} \qquad \text{Subtracting numerators}$$

$$= \frac{5x+3y-7}{x-2y}$$

Do Exercises 19 and 20.

Subtract.

19. $\dfrac{3}{4y} - \dfrac{7x}{-4y}$

GS 20. $\dfrac{4x^2}{2x-y} - \dfrac{7x^2}{y-2x}$

$= \dfrac{4x^2}{2x-y} - \dfrac{7x^2}{y-2x}\cdot\dfrac{\square}{-1}$

$= \dfrac{4x^2}{2x-y} - \dfrac{-7x^2}{\square}$

$= \dfrac{4x^2-(\square)}{2x-y}$

$= \dfrac{\square\, x^2}{2x-y}$

Answers

15. $\dfrac{a+12}{a(a+3)}$ 16. $\dfrac{3(y^2+4y+1)}{(y-4)(y-3)(y+5)}$

17. $\dfrac{(1+b)(1-b)}{3}$ 18. $\dfrac{2x^2+11}{x-5}$

19. $\dfrac{3+7x}{4y}$ 20. $\dfrac{11x^2}{2x-y}$

Guided Solution:

20. $-1, 2x-y, -7x^2, 11$

CALCULATOR CORNER

Checking Addition and Subtraction Use the TABLE feature, as described on p. 122, to check the sums and differences in Examples 6, 8, 11, and 15. Then check your answers to Margin Exercises 15 and 21.

c COMBINED ADDITIONS AND SUBTRACTIONS

EXAMPLE 15 Perform the indicated operations and simplify.

$$\frac{2x}{x^2-4}+\frac{5}{2-x}-\frac{1}{2+x}$$

$$=\frac{2x}{(x-2)(x+2)}+\frac{5}{2-x}-\frac{1}{2+x}$$

$$=\frac{2x}{(x-2)(x+2)}+\frac{5}{2-x}\cdot\frac{-1}{-1}-\frac{1}{x+2}\qquad \text{Multiplying by } \frac{-1}{-1}$$

$$=\frac{2x}{(x-2)(x+2)}+\frac{-5}{x-2}-\frac{1}{x+2}\qquad \text{LCD} = (x-2)(x+2)$$

$$=\frac{2x}{(x-2)(x+2)}+\frac{-5}{x-2}\cdot\frac{x+2}{x+2}-\frac{1}{x+2}\cdot\frac{x-2}{x-2}\qquad \text{Multiplying by 1 to get the LCD}$$

$$=\frac{2x-5(x+2)-(x-2)}{(x-2)(x+2)}\qquad \text{Adding and subtracting the numerators}$$

$$=\frac{2x-5x-10-x+2}{(x-2)(x+2)}\qquad \text{Removing parentheses}$$

$$=\frac{-4x-8}{(x-2)(x+2)}=\frac{-4(x+2)}{(x-2)(x+2)}\qquad \text{Removing a factor of 1: } \frac{x+2}{x+2}=1$$

$$=\frac{-4}{x-2},\ \text{or}\ -\frac{4}{x-2}$$

Another correct form of the answer is $\frac{4}{2-x}$. It is found by multiplying by $-1/-1$.

◀ **Do Exercise 21.**

21. Perform the indicated operations and simplify:

$$\frac{8x}{x^2-1}+\frac{2}{1-x}-\frac{4}{x+1}.$$

Answer

21. $\frac{2}{x-1}$

5.2 Exercise Set

FOR EXTRA HELP MyLab Math

✓ Check Your Understanding

Reading Check Choose from the column on the right the best option to complete each statement. Some choices will not be used.

To add $\frac{7}{90}+\frac{5}{81}$, we begin by finding a **RC1.** ______________ denominator. We first look for the least common **RC2.** ______________ of 90 and 81. That number becomes the least common **RC3.** ______________ of the two fractions. We factor 90 and 81: $90 = 2\cdot3\cdot3\cdot5$ and $81 = 3\cdot3\cdot3\cdot3$. Then to find the LCM of 90 and 81, we use each factor the **RC4.** ______________ number of times that it appears in any one factorization. The LCM is **RC5.** ______________.

$2\cdot3\cdot3\cdot5$
$2\cdot3\cdot3\cdot3\cdot3\cdot5$
common
multiple
numerator
denominator
greatest
least

Concept Check When we are subtracting rational expressions, it is important to use parentheses to make sure that we subtract the entire numerator. In Exercises CC1 and CC2, complete each numerator by **(a)** filling in the parentheses, **(b)** removing the parentheses, and **(c)** collecting like terms.

CC1. $\frac{10x}{x-7} - \frac{3x+5}{x-7} = \frac{10x - (\quad)}{x-7} = \frac{\qquad}{x-7} = \frac{\qquad}{x-7}$
(a) (b) (c)

CC2. $\frac{7}{4+a} - \frac{4-9a}{4+a} = \frac{7 - (\quad)}{4+a} = \frac{\qquad}{4+a} = \frac{\qquad}{4+a}$
(a) (b) (c)

a Find the LCM by factoring.

1. 15, 40

2. 12, 32

3. 18, 48

4. 45, 54

5. 30, 105

6. 24, 60

7. 9, 15, 5

8. 27, 35, 63

Add. Find the LCD first.

9. $\frac{5}{6} + \frac{4}{15}$

10. $\frac{5}{12} + \frac{13}{18}$

11. $\frac{7}{36} + \frac{1}{24}$

12. $\frac{11}{30} + \frac{19}{75}$

13. $\frac{3}{4} + \frac{7}{30} + \frac{1}{16}$

14. $\frac{5}{8} + \frac{7}{12} + \frac{11}{40}$

Find the LCM.

15. $21x^2y,\ 7xy$

16. $18a^2b,\ 50ab^3$

17. $y^2 - 100,\ 10y + 100$

18. $r^2 - s^2,\ rs + s^2$

19. $15ab^2,\ 3ab,\ 10a^3b$

20. $6x^2y^2,\ 9x^3y,\ 15y^3$

21. $5y - 15,\ y^2 - 6y + 9$

22. $x^2 + 10x + 25,\ x^2 + 2x - 15$

23. $y^2 - 25,\ 5 - y$

24. $x^2 - 36,\ 6 - x$

25. $2r^2 - 5r - 12,\ 3r^2 - 13r + 4,\ r^2 - 16$

26. $2x^2 - 5x - 3,\ 2x^2 - x - 1,\ x^2 - 6x + 9$

27. $x^5 + 4x^3,\ x^3 - 4x^2 + 4x$

28. $9x^3 + 9x^2 - 18x,\ 6x^5 + 24x^4 + 24x^3$

29. $x^5 - 2x^4 + x^3,\ 2x^3 + 2x,\ 5x + 5$

30. $x^5 - 4x^4 + 4x^3,\ 3x^2 - 12,\ 2x + 4$

b Add or subtract. Then simplify. If a denominator has two or more factors (other than monomials), leave it in factored form.

31. $\frac{x - 2y}{x + y} + \frac{x + 9y}{x + y}$

32. $\frac{a - 8b}{a + b} + \frac{a + 13b}{a + b}$

33. $\frac{4y + 3}{y - 2} - \frac{y - 2}{y - 2}$

34. $\frac{3t + 2}{t - 4} - \frac{t - 4}{t - 4}$

35. $\frac{a^2}{a - b} + \frac{b^2}{b - a}$

36. $\frac{r^2}{r - s} + \frac{s^2}{s - r}$

37. $\frac{6}{y} - \frac{7}{-y}$

38. $\frac{4}{x} - \frac{9}{-x}$

39. $\frac{4a - 2}{a^2 - 49} + \frac{5 + 3a}{49 - a^2}$

40. $\frac{2y - 3}{y^2 - 1} - \frac{4 - y}{1 - y^2}$

41. $\frac{a^3}{a - b} + \frac{b^3}{b - a}$

42. $\frac{x^3}{x^2 - y^2} + \frac{y^3}{y^2 - x^2}$

43. $\frac{y - 2}{y + 4} + \frac{y + 3}{y - 5}$

44. $\frac{x - 2}{x + 3} + \frac{x + 2}{x - 4}$

45. $\frac{4xy}{x^2 - y^2} + \frac{x - y}{x + y}$

46. $\frac{5ab}{a^2 - b^2} + \frac{a + b}{a - b}$

47. $\dfrac{9x + 2}{3x^2 - 2x - 8} + \dfrac{7}{3x^2 + x - 4}$

48. $\dfrac{3y + 2}{2y^2 - y - 10} + \dfrac{8}{2y^2 - 7y + 5}$

49. $\dfrac{4}{x + 1} + \dfrac{x + 2}{x^2 - 1} + \dfrac{3}{x - 1}$

50. $\dfrac{-2}{y + 2} + \dfrac{5}{y - 2} + \dfrac{y + 3}{y^2 - 4}$

51. $\dfrac{x - 1}{3x + 15} - \dfrac{x + 3}{5x + 25}$

52. $\dfrac{y - 2}{4y + 8} - \dfrac{y + 6}{5y + 10}$

53. $\dfrac{5ab}{a^2 - b^2} - \dfrac{a - b}{a + b}$

54. $\dfrac{6xy}{x^2 - y^2} - \dfrac{x + y}{x - y}$

55. $\dfrac{3y}{y^2 - 7y + 10} - \dfrac{2y}{y^2 - 8y + 15}$

56. $\dfrac{5x}{x^2 - 6x + 8} - \dfrac{3x}{x^2 - x - 12}$

57. $\dfrac{y}{y^2 - y - 20} + \dfrac{2}{y + 4}$

58. $\dfrac{6}{y^2 + 6y + 9} + \dfrac{5}{y^2 - 9}$

59. $\dfrac{3y + 2}{y^2 + 5y - 24} + \dfrac{7}{y^2 + 4y - 32}$

60. $\dfrac{3y + 2}{y^2 - 7y + 10} + \dfrac{2y}{y^2 - 8y + 15}$

61. $\dfrac{3x - 1}{x^2 + 2x - 3} - \dfrac{x + 4}{x^2 - 9}$

62. $\dfrac{3p - 2}{p^2 + 2p - 24} - \dfrac{p - 3}{p^2 - 16}$

C Perform the indicated operations and simplify.

63. $\dfrac{1}{x + 1} - \dfrac{x}{x - 2} + \dfrac{x^2 + 2}{x^2 - x - 2}$

64. $\dfrac{2}{y + 3} - \dfrac{y}{y - 1} + \dfrac{y^2 + 2}{y^2 + 2y - 3}$

65. $\dfrac{y - 3}{y - 4} - \dfrac{y + 2}{y + 4} + \dfrac{y - 7}{y^2 - 16}$

66. $\dfrac{x - 1}{x - 2} - \dfrac{x + 1}{x + 2} + \dfrac{x - 6}{x^2 - 4}$

67. $\frac{y+2}{y+4}+\frac{y-7}{y^2-16}-\frac{y-3}{y-4}$

68. $\frac{x-6}{x^2-4}-\frac{x-1}{x-2}-\frac{x+1}{x+2}$

69. $\frac{4x}{x^2-1}+\frac{3x}{1-x}-\frac{4}{x-1}$

70. $\frac{5y}{1-2y}-\frac{2y}{2y+1}+\frac{3}{4y^2-1}$

71. $\frac{1}{x+y}+\frac{1}{y-x}-\frac{2x}{x^2-y^2}$

72. $\frac{1}{b-a}+\frac{1}{a+b}-\frac{2b}{a^2-b^2}$

73. $\frac{x+5}{x-3}-\frac{x+2}{x+1}-\frac{6x+10}{x^2-2x-3}$

74. $\frac{13x+2}{x^2+3x-10}-\frac{x+2}{x-2}+\frac{x-3}{x+5}$

Skill Maintenance

Graph. [3.7b]

75. $2x - 3y > 6$

76. $5x - 3y \leq 15$

Factor. [4.6d]

77. $23x^4 + 23x$

78. $64a^3 - 27b^3$

79. Find an equation of the line that passes through the point $(4, -6)$ and is parallel to the line $3y + 8x = 10$. [2.6d]

80. Find an equation of the line that passes through the point $(-2, 3)$ and is perpendicular to the line $5y + 4x = 7$. [2.6d]

Synthesis

81. Determine the domain and the range of the function graphed below.

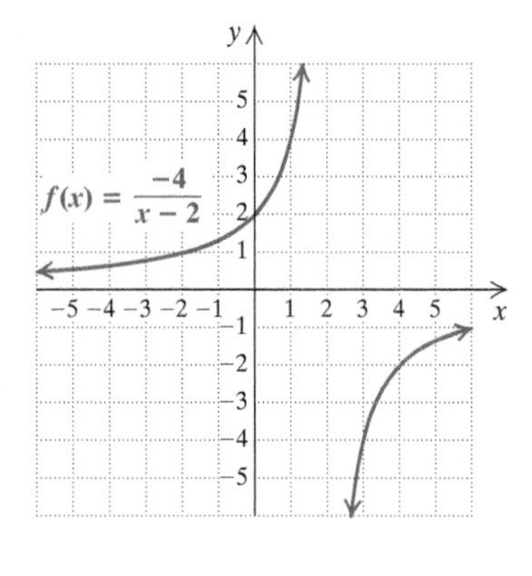

Find the LCM.

82. 18, 42, 82, 120, 300, 700

83. $x^8 - x^4,\ x^5 - x^2,\ x^5 - x^3,\ x^5 + x^2$

84. The LCM of two expressions is $8a^4b^7$. One of the expressions is $2a^3b^7$. List all possibilities for the other expression.

Perform the indicated operations and simplify.

85. $\frac{x+y+1}{y-(x+1)}+\frac{x+y-1}{x-(y-1)}-\frac{x-y-1}{1-(y-x)}$

86. $\frac{b-c}{a-(b-c)}-\frac{b-a}{(b-a)-c}$

Division of Polynomials

5.3

OBJECTIVES

a Divide a polynomial by a monomial.

b Divide a polynomial by a divisor that is not a monomial, and if there is a remainder, express the result in two ways.

c Use synthetic division to divide a polynomial by a binomial of the type $x - a$.

A rational expression represents division. "Long" division of polynomials, like division of real numbers, relies on our multiplication and subtraction skills.

a DIVISOR A MONOMIAL

We first consider division by a monomial (a term like $45x^{10}$ or $48a^2b^5$). When we are dividing a monomial by a monomial, we can use the rules of exponents and subtract exponents when the bases are the same. (We studied this in Section J19.) For example,

$$\frac{45x^{10}}{3x^4} = \frac{45}{3}x^{10-4} = 15x^6 \quad \text{and} \quad \frac{48a^2b^5}{-3ab^2} = \frac{48}{-3}a^{2-1}b^{5-2} = -16ab^3.$$

When we divide a polynomial by a monomial, we break up the division into a sum of quotients of monomials. To do so, we reverse the rule for adding fractions. That is, since

$$\frac{A}{C} + \frac{B}{C} = \frac{A+B}{C}, \quad \text{we know that} \quad \frac{A+B}{C} = \frac{A}{C} + \frac{B}{C}.$$

SKILL REVIEW *Use exponential notation in division.* [J19]

Divide and simplify.

1. $\dfrac{15w^6}{3w^4}$
2. $\dfrac{36x^8y^{15}}{-4x^3y^{10}}$

Answers: 1. $5w^2$ 2. $-9x^5y^5$

MyLab Math VIDEO

EXAMPLE 1 Divide $12x^3 + 8x^2 + x + 4$ by $4x$.

$\dfrac{12x^3 + 8x^2 + x + 4}{4x}$ — Writing a fraction expression

$= \dfrac{12x^3}{4x} + \dfrac{8x^2}{4x} + \dfrac{x}{4x} + \dfrac{4}{4x}$ — Dividing each term of the numerator by the monomial

$= 3x^2 + 2x + \dfrac{1}{4} + \dfrac{1}{x}$ — Doing the four indicated divisions ■

EXAMPLE 2 Divide: $(8x^4y^5 - 3x^3y^4 + 5x^2y^3) \div (x^2y^3)$.

$$\frac{8x^4y^5 - 3x^3y^4 + 5x^2y^3}{x^2y^3} = \frac{8x^4y^5}{x^2y^3} - \frac{3x^3y^4}{x^2y^3} + \frac{5x^2y^3}{x^2y^3}$$
$$= 8x^2y^2 - 3xy + 5$$ ■

DIVIDING BY A MONOMIAL

To divide a polynomial by a monomial, divide each term by the monomial.

Do Exercises 1–3. ▶

Divide.

1. $\dfrac{x^3 + 16x^2 + 6x}{2x}$
2. $(15y^5 - 6y^4 + 18y^3) \div (3y^2)$
3. $(x^4y^3 + 10x^3y^2 + 16x^2y) \div (2x^2y)$

b DIVISOR NOT A MONOMIAL

When the divisor is not a monomial, we use a procedure very much like long division in arithmetic.

Answers

1. $\dfrac{x^2}{2} + 8x + 3$ 2. $5y^3 - 2y^2 + 6y$
3. $\dfrac{x^2y^2}{2} + 5xy + 8$

EXAMPLE 3 Divide $x^2 + 5x + 8$ by $x + 3$.

We have

$$\begin{array}{l} x \leftarrow \text{Divide the first term by the first term: } x^2/x = x. \\ x + 3\overline{)x^2 + 5x + 8} \\ \quad x^2 + 3x \leftarrow \text{Multiply } x \text{ above by the divisor, } x + 3. \\ \qquad 2x \leftarrow \text{Subtract: } (x^2 + 5x) - (x^2 + 3x) = x^2 + 5x - x^2 - 3x = 2x. \end{array}$$

We now "bring down" the next term of the dividend—in this case, 8.

$$\begin{array}{l} x + 2 \leftarrow \text{Divide the first term by the first term: } 2x/x = 2. \\ x + 3\overline{)x^2 + 5x + 8} \\ \quad x^2 + 3x \\ \qquad 2x + 8 \leftarrow \text{The 8 has been "brought down."} \\ \qquad 2x + 6 \leftarrow \text{Multiply 2 above by the divisor, } x + 3. \\ \qquad\quad 2 \leftarrow \text{Subtract: } (2x + 8) - (2x + 6) = 2x + 8 - 2x - 6 = 2. \end{array}$$

The answer is $x + 2$, R 2; or

$$x + 2 + \frac{2}{x + 3}.$$

This expression is the remainder over the divisor.

Note that the answer is not a polynomial unless the remainder is 0.

To check, we multiply the quotient by the divisor and add the remainder to see if we get the dividend:

$$\underbrace{(x + 3)}_{\text{Divisor}} \cdot \underbrace{(x + 2)}_{\text{Quotient}} + \underbrace{2}_{\text{Remainder}} = \underbrace{(x^2 + 5x + 6) + 2}_{\text{Dividend}} = x^2 + 5x + 8.$$

The answer checks.

Do Exercise 4.

4. Divide $x^2 + 7x + 9$ by $x + 2$. GS

$$\begin{array}{l} \quad\quad x + \square \\ x + 2\overline{)x^2 + 7x + 9} \\ \quad x^2 + \square x \\ \quad \square x + 9 \\ \quad\quad 5x + 10 \\ \quad\quad\quad \square \end{array}$$

The answer is $x + \square$, R -1; or $x + 5 + \frac{\square}{x + 2}$.

EXAMPLE 4 Divide: $(5x^4 + x^3 - 3x^2 - 6x - 8) \div (x - 1)$.

$$\begin{array}{l} \quad 5x^3 + 6x^2 + 3x - 3 \\ x - 1\overline{)5x^4 + x^3 - 3x^2 - 6x - 8} \\ \quad 5x^4 - 5x^3 \leftarrow \text{Subtract: } (5x^4 + x^3) - (5x^4 - 5x^3) = 6x^3. \\ \qquad 6x^3 - 3x^2 \\ \qquad 6x^3 - 6x^2 \leftarrow \text{Subtract: } (6x^3 - 3x^2) - (6x^3 - 6x^2) = 3x^2. \\ \qquad\quad 3x^2 - 6x \\ \qquad\quad 3x^2 - 3x \leftarrow \text{Subtract: } (3x^2 - 6x) - (3x^2 - 3x) = -3x. \\ \qquad\qquad -3x - 8 \\ \qquad\qquad -3x + 3 \leftarrow \text{Subtract: } (-3x - 8) - (-3x + 3) = -11. \\ \qquad\qquad\quad -11 \end{array}$$

The answer is $5x^3 + 6x^2 + 3x - 3$, R -11; or

$$5x^3 + 6x^2 + 3x - 3 + \frac{-11}{x - 1}.$$

Do Exercises 5 and 6.

5. Divide and check: $x - 2\overline{)x^2 + 3x - 10}$.

6. Divide and check: $(2x^4 + 3x^3 - x^2 - 7x + 9) \div (x + 4)$.

Answers

4. $x + 5$, R -1; or $x + 5 + \frac{-1}{x + 2}$ 5. $x + 5$
6. $2x^3 - 5x^2 + 19x - 83$, R 341; or $2x^3 - 5x^2 + 19x - 83 + \frac{341}{x + 4}$

Guided Solution:
4. 5, 2, 5, -1, 5, -1

When dividing polynomials, remember to always arrange the polynomials in descending order. In a polynomial division, if there are *missing* terms in the dividend, either write them with 0 coefficients or leave space for them. For example, in $125y^3 - 8$, we say that "the y^2-term and the y-term are **missing**." We could write them in as follows: $125y^3 + 0y^2 + 0y - 8$.

EXAMPLE 5 Divide: $(125y^3 - 8) \div (5y - 2)$.

$$
\begin{array}{r}
25y^2 + 10y + 4 \\
5y - 2 \overline{)\, 125y^3 + 0y^2 + 0y - 8} \\
\underline{125y^3 - 50y^2} \\
50y^2 + 0y \\
\underline{50y^2 - 20y} \\
20y - 8 \\
\underline{20y - 8} \\
0
\end{array}
$$

When there are missing terms, we can write them in.

Subtract: $125y^3 - (125y^3 - 50y^2) = 50y^2$.

Subtract: $50y^2 - (50y^2 - 20y) = 20y$.

Subtract: $(20y - 8) - (20y - 8) = 0$.

The answer is $25y^2 + 10y + 4$.

Do Exercise 7.

7. Divide and check:
$(9y^4 + 14y^2 - 8) \div (3y + 2)$.

Another way to deal with missing terms is to leave space for them, as we see in Example 6.

EXAMPLE 6 Divide: $(x^4 - 9x^2 - 5) \div (x - 2)$.

Note that the x^3-term and the x-term are missing in the dividend.

$$
\begin{array}{r}
x^3 + 2x^2 - 5x - 10 \\
x - 2 \overline{)\, x^4 - 9x^2 - 5} \\
\underline{x^4 - 2x^3} \\
2x^3 - 9x^2 \\
\underline{2x^3 - 4x^2} \\
-5x^2 \\
\underline{-5x^2 + 10x} \\
-10x - 5 \\
\underline{-10x + 20} \\
-25
\end{array}
$$

We leave spaces for missing terms.

Subtract: $x^4 - (x^4 - 2x^3) = 2x^3$.

Subtract: $(2x^3 - 9x^2) - (2x^3 - 4x^2) = -5x^2$.

Subtract: $-5x^2 - (-5x^2 + 10x) = -10x$.

Subtract: $(-10x - 5) - (-10x + 20) = -25$.

The answer is $x^3 + 2x^2 - 5x - 10$, R -25; or

$$x^3 + 2x^2 - 5x - 10 + \frac{-25}{x - 2}.$$

Do Exercises 8 and 9.

Divide and check.

8. $(y^3 - 11y^2 + 6) \div (y - 3)$

9. $(x^3 + 9x^2 - 5) \div (x - 1)$

Answers

7. $3y^3 - 2y^2 + 6y - 4$
8. $y^2 - 8y - 24$, R -66; or $y^2 - 8y - 24 + \frac{-66}{y - 3}$
9. $x^2 + 10x + 10$, R 5; or $x^2 + 10x + 10 + \frac{5}{x - 1}$

When dividing, we may "come out even" (have a remainder of 0) or we may not. If not, how long should we keep working? We continue until the degree of the remainder is less than the degree of the divisor, as in the next example.

EXAMPLE 7 Divide: $(6x^3 + 9x^2 - 5) \div (x^2 - 2x)$.

$$\begin{array}{r|l} x^2 - 2x & 6x^3 + 9x^2 + 0x - 5 \end{array}$$

$6x + 21$	
$x^2 - 2x \overline{)6x^3 + 9x^2 + 0x - 5}$	We have a missing term. We can write it in.
$\underline{6x^3 - 12x^2}$	
$21x^2 + 0x$	
$\underline{21x^2 - 42x}$	
$42x - 5$	The degree of the remainder, 1, is less than the degree of the divisor, 2, so we are finished.

The answer is $6x + 21$, R $(42x - 5)$; or

$$6x + 21 + \frac{42x - 5}{x^2 - 2x}.$$

10. Divide and check:

$(y^3 - 11y^2 + 6) \div (y^2 - 3)$.

◀ **Do Exercise 10.**

C SYNTHETIC DIVISION

To divide a polynomial by a binomial of the type $x - a$, we can streamline the general procedure by a process called **synthetic division**.

Compare the following. In **A**, we perform a division. In **B**, we also divide but we do not write the variables. In **B**, we eliminate some duplication of writing.

A.

$$\begin{array}{r} 4x^2 + 5x + 11 \\ x - 2 \overline{)4x^3 - 3x^2 + x + 7} \\ \underline{4x^3 - 8x^2} \\ 5x^2 + x \\ \underline{5x^2 - 10x} \\ 11x + 7 \\ \underline{11x - 22} \\ 29 \end{array}$$

B.

$$\begin{array}{r} 4 + 5 + 11 \\ 1 - 2 \overline{)4 - 3 + 1 + 7} \\ \underline{4 - 8} \\ 5 + 1 \\ \underline{5 - 10} \\ 11 + 7 \\ \underline{11 - 22} \\ 29 \end{array}$$

In **B**, there is still some duplication of writing. Also, since we can subtract by adding the opposite, we can use 2 instead of -2 and then add instead of subtracting.

Answer

10. $y - 11$, R $(3y - 27)$; or $y - 11 + \dfrac{3y - 27}{y^2 - 3}$

C. *Synthetic Division*

a)
$$\begin{array}{r|rrrr} 2 & 4 & -3 & 1 & 7 \\ & & & & \\ \hline & 4 & & & \end{array}$$

Write 2, the number that is subtracted in the divisor $x - 2$, and the coefficients of the dividend.

Bring down the first coefficient.

b)
$$\begin{array}{r|rrrr} 2 & 4 & -3 & 1 & 7 \\ & & 8 & & \\ \hline & 4 & 5 & & \end{array}$$

Multiply 4 by 2 to get 8. Add 8 and −3.

c)
$$\begin{array}{r|rrrr} 2 & 4 & -3 & 1 & 7 \\ & & 8 & 10 & \\ \hline & 4 & 5 & 11 & \end{array}$$

Multiply 5 by 2 to get 10. Add 10 and 1.

d)
$$\begin{array}{r|rrr|r} 2 & 4 & -3 & 1 & 7 \\ & & 8 & 10 & 22 \\ \hline & 4 & 5 & 11 & 29 \end{array}$$

Multiply 11 by 2 to get 22. Add 22 and 7.

Quotient Remainder

The last number, 29, is the remainder. The other numbers are the coefficients of the quotient with that of the term of highest degree first. Note that the degree of the term of highest degree is 1 less than the degree of the dividend.

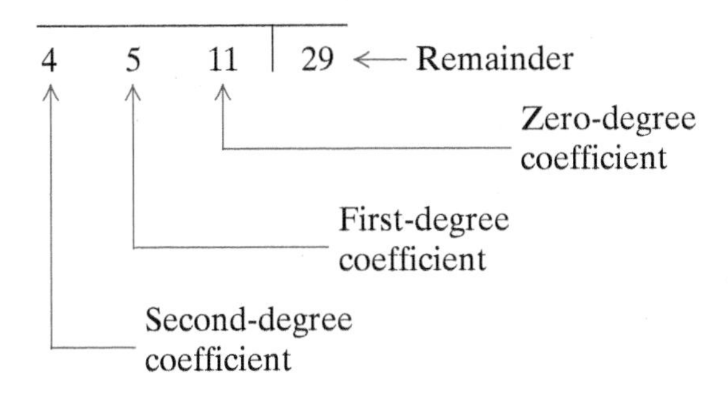

The answer is $4x^2 + 5x + 11$, R 29; or $4x^2 + 5x + 11 + \frac{29}{x - 2}$.

> It is important to remember that in order for synthetic division to work, the divisor must be of the form $x - a$, that is, a variable minus a constant. The coefficient of the variable must be 1.

Do Exercise 11. ▶

11. Use synthetic division to divide:
$(2x^3 - 4x^2 + 8x - 8) \div (x - 3)$.

$$\begin{array}{r|rrr|r} \square & 2 & -4 & 8 & -8 \\ & & 6 & \square & 42 \\ \hline & 2 & \square & 14 & \square \end{array}$$

The answer is $2x^2 + \square x + 14$, R $\square$;

or $2x^2 + 2x + 14 + \frac{34}{\square}$.

EXAMPLE 8 Use synthetic division to divide:

$$(x^3 + 6x^2 - x - 30) \div (x - 2).$$

We have

$$\begin{array}{r|rrr|r} 2 & 1 & 6 & -1 & -30 \\ & & 2 & 16 & 30 \\ \hline & 1 & 8 & 15 & 0 \end{array}$$

The answer is $x^2 + 8x + 15$, R 0; or just $x^2 + 8x + 15$.

Do Exercise 12. ▶

12. Use synthetic division to divide:
$x^3 + 6x^2 - 19x - 24 \div (x - 3)$.

Answers

11. $2x^2 + 2x + 14$, R 34; or $2x^2 + 2x + 14 + \frac{34}{x - 3}$
12. $x^2 + 9x + 8$, R 0; or $x^2 + 9x + 8$

Guided Solution:

11. 3, 6, 2, 34, 2, 34, $x - 3$

When there are missing terms, be sure to write 0's for their coefficients.

EXAMPLES Use synthetic division to divide.

9. $(2x^3 + 7x^2 - 5) \div (x + 3)$

There is no x-term, so we must write a 0 for its coefficient. Note that $x + 3 = x - (-3)$, so we write -3 at the left.

$$\begin{array}{r|rrrr} -3 & 2 & 7 & 0 & -5 \\ & & -6 & -3 & 9 \\ \hline & 2 & 1 & -3 & \big| \; 4 \end{array}$$

The answer is $2x^2 + x - 3$, R 4; or $2x^2 + x - 3 + \frac{4}{x + 3}$.

10. $(x^3 + 4x^2 - x - 4) \div (x + 4)$

Note that $x + 4 = x - (-4)$, so we write -4 at the left.

$$\begin{array}{r|rrrr} -4 & 1 & 4 & -1 & -4 \\ & & -4 & 0 & 4 \\ \hline & 1 & 0 & -1 & \big| \; 0 \end{array}$$

The answer is $x^2 - 1$.

11. $(x^4 - 1) \div (x - 1)$

The divisor is $x - 1$, so we write 1 at the left.

$$\begin{array}{r|rrrrr} 1 & 1 & 0 & 0 & 0 & -1 \\ & & 1 & 1 & 1 & 1 \\ \hline & 1 & 1 & 1 & 1 & \big| \; 0 \end{array}$$

There are 3 missing terms.

The answer is $x^3 + x^2 + x + 1$.

Do Exercises 13 and 14.

Use synthetic division to divide.

13. $(x^3 - 2x^2 + 5x - 4) \div (x + 2)$

14. $(y^3 + 1) \div (y + 1)$

Answers

13. $x^2 - 4x + 13$, R -30; or $x^2 - 4x + 13 + \frac{-30}{x + 2}$

14. $y^2 - y + 1$

5.3 Exercise Set

FOR EXTRA HELP  MyLab Math

✓ Check Your Understanding

Reading Check and Concept Check In order for synthetic division to work, the divisor must be of the form $x - a$, that is, a variable minus a constant. The coefficient of the variable must be 1. For each divisor in Exercises RC1–RC6, determine the constant a.

RC1. $(x^2 - x + 3) \div (x - 4)$

RC2. $(x^3 + 2x^2 + 5) \div (x + 6)$

RC3. $(2x^2 + 4x - 7) \div (x - 7)$

RC4. $(4x^4 - x^3 + x^2 - x) \div (x - 6)$

RC5. $(x^4 - 6x^2 - x + 4) \div (x + 4)$

RC6. $(10x^2 - 6) \div (x + 7)$

a Divide and check.

1. $\dfrac{24x^6 + 18x^5 - 36x^2}{6x^2}$

2. $\dfrac{30y^8 - 15y^6 + 40y^4}{5y^4}$

3. $\dfrac{45y^7 - 20y^4 + 15y^2}{5y^2}$

4. $\dfrac{60x^8 + 44x^5 - 28x^3}{4x^3}$

5. $(32a^4b^3 + 14a^3b^2 - 22a^2b) \div (2a^2b)$

6. $(7x^3y^4 - 21x^2y^3 + 28xy^2) \div (7xy)$

b Divide.

7. $(x^2 + 10x + 21) \div (x + 3)$

8. $(y^2 - 8y + 16) \div (y - 4)$

9. $(a^2 - 8a - 16) \div (a + 4)$

10. $(y^2 - 10y - 25) \div (y - 5)$

11. $(x^2 + 7x + 14) \div (x + 5)$

12. $(t^2 - 7t - 9) \div (t - 3)$

13. $(4y^3 + 6y^2 + 14) \div (2y + 4)$

14. $(6x^3 - x^2 - 10) \div (3x + 4)$

15. $(10y^3 + 6y^2 - 9y + 10) \div (5y - 2)$

16. $(6x^3 - 11x^2 + 11x - 2) \div (2x - 3)$

17. $(2x^4 - x^3 - 5x^2 + x - 6) \div (x^2 + 2)$

18. $(3x^4 + 2x^3 - 11x^2 - 2x + 5) \div (x^2 - 2)$

19. $(2x^5 - x^4 + 2x^3 - x) \div (x^2 - 3x)$

20. $(2x^5 + 3x^3 + x^2 - 4) \div (x^2 + x)$

C Use synthetic division to divide.

21. $(x^3 - 2x^2 + 2x - 5) \div (x - 1)$

22. $(x^3 - 2x^2 + 2x - 5) \div (x + 1)$

23. $(a^2 + 11a - 19) \div (a + 4)$

24. $(a^2 + 11a - 19) \div (a - 4)$

25. $(x^3 - 7x^2 - 13x + 3) \div (x - 2)$

26. $(x^3 - 7x^2 - 13x + 3) \div (x + 2)$

27. $(3x^3 + 7x^2 - 4x + 3) \div (x + 3)$

28. $(3x^3 + 7x^2 - 4x + 3) \div (x - 3)$

29. $(y^3 - 3y + 10) \div (y - 2)$

30. $(x^3 - 2x^2 + 8) \div (x + 2)$

31. $(3x^4 - 25x^2 - 18) \div (x - 3)$

32. $(6y^4 + 15y^3 + 28y + 6) \div (y + 3)$

33. $(x^3 - 8) \div (x - 2)$

34. $(y^3 + 125) \div (y + 5)$

35. $(y^4 - 16) \div (y - 2)$

36. $(x^5 - 32) \div (x - 2)$

37. $(y^8 - 1) \div (y + 1)$

38. $(y^6 - 2) \div (y - 1)$

Skill Maintenance

Graph. [2.2c], [3.7b]

39. $2x - 3y < 6$

40. $5x + 3y \leq 15$

41. $y > 4$

42. $x \leq -2$

43. $f(x) = x^2$

44. $g(x) = x^2 - 3$

45. $f(x) = 3 - x^2$

46. $f(x) = x^2 + 6x + 6$

Solve. [4.8a]

47. $x^2 - 5x = 0$

48. $25y^2 = 64$

49. $12x^2 = 17x + 5$

50. $12x^2 + 11x + 2 = 0$

Synthesis

51. Let $f(x) = 4x^3 + 16x^2 - 3x - 45$. Find $f(-3)$ and then solve $f(x) = 0$.

52. Let $f(x) = 6x^3 - 13x^2 - 79x + 140$. Find $f(4)$ and then solve $f(x) = 0$.

53. When $x^2 - 3x + 2k$ is divided by $x + 2$, the remainder is 7. Find k.

54. Find k such that when $x^3 - kx^2 + 3x + 7k$ is divided by $x + 2$, the remainder is 0.

Divide.

55. $(4a^3b + 5a^2b^2 + a^4 + 2ab^3) \div (a^2 + 2b^2 + 3ab)$

56. $(a^7 + b^7) \div (a + b)$

Complex Rational Expressions

5.4

OBJECTIVE

a Simplify complex rational expressions.

a COMPLEX RATIONAL EXPRESSIONS

SKILL REVIEW

Divide rational expressions and simplify. [5.1e]

Divide and simplify.

1. $\dfrac{5x-10}{x} \div \dfrac{x-2}{x^5}$ **2.** $\dfrac{a^2-49}{a+3} \div \dfrac{a+7}{a+3}$

Answers: 1. $5x^4$ **2.** $a-7$

MyLab Math VIDEO

A **complex rational expression** is a rational expression that contains rational expressions within its numerator and/or its denominator. Here are some examples:

$$\frac{\frac{2}{3}}{\frac{4}{5}}, \quad \frac{1+\frac{5}{x}}{4x}, \quad \frac{\frac{x-y}{x+y}}{\frac{2x-y}{3x+y}}, \quad \frac{\frac{3x}{5}-\frac{2}{x}}{\frac{4x}{3}+\frac{7}{6x}}.$$

The rational expressions within each complex rational expression are red.

There are two methods that can be used to simplify complex rational expressions. We will consider both of them.

To the instructor and the student: Students can be instructed to try both methods and then choose the one that works best for them, or one method can be chosen by the instructor.

Method 1: Multiplying by the LCM of All the Denominators

Method 1. To simplify a complex rational expression:

1. First, find the LCM of all the denominators of all the rational expressions occurring within both the numerator and the denominator of the (original) complex rational expression.
2. Multiply by 1 using LCM/LCM.
3. If possible, simplify.

EXAMPLE 1 Simplify: $\dfrac{x+\frac{1}{5}}{x-\frac{1}{3}}$.

We first find the LCM of all the denominators of all the rational expressions occurring in both the numerator and the denominator of the complex rational expression. The denominators are 3 and 5. The LCM of these denominators is $3 \cdot 5$, or 15. We multiply by 15/15.

$$\frac{x+\frac{1}{5}}{x-\frac{1}{3}} = \left(\frac{x+\frac{1}{5}}{x-\frac{1}{3}}\right)\cdot\frac{15}{15} \qquad \text{Multiplying by 1}$$

$$= \frac{\left(x+\frac{1}{5}\right)\cdot 15}{\left(x-\frac{1}{3}\right)\cdot 15} \qquad \text{Multiplying the numerators and the denominators}$$

Then

$$= \frac{15x + \frac{1}{5}\cdot 15}{15x - \frac{1}{3}\cdot 15}$$ Carrying out the multiplications using the distributive laws

$$= \frac{15x + 3}{15x - 5}, \text{ or } \frac{3(5x + 1)}{5(3x - 1)}.$$ No further simplification is possible.

In Example 1, if you feel more comfortable doing so, you can always write denominators of 1 where there are no denominators written. In this case, you could start out by writing

$$\frac{\frac{x}{1} + \frac{1}{5}}{\frac{x}{1} - \frac{1}{3}}.$$

Do Exercise 1.

1. Simplify. Use method 1.

$$\frac{y + \frac{1}{2}}{y - \frac{1}{7}}$$

EXAMPLE 2 Simplify: $\dfrac{1 + \frac{1}{x}}{1 - \frac{1}{x^2}}$.

We first find the LCM of all the denominators of all the rational expressions occurring in both the numerator and the denominator of the complex rational expression. The denominators are x and x^2. The LCM of these denominators is x^2. We multiply by x^2/x^2.

$$\frac{1 + \frac{1}{x}}{1 - \frac{1}{x^2}} = \left(\frac{1 + \frac{1}{x}}{1 - \frac{1}{x^2}}\right) \cdot \frac{x^2}{x^2}$$ The LCM of the denominators is x^2. We multiply by 1: $\frac{x^2}{x^2}$.

$$= \frac{\left(1 + \frac{1}{x}\right) \cdot x^2}{\left(1 - \frac{1}{x^2}\right) \cdot x^2}$$ Multiplying the numerators and the denominators

$$= \frac{x^2 + \frac{1}{x} \cdot x^2}{x^2 - \frac{1}{x^2} \cdot x^2}$$ Carrying out the multiplications using the distributive laws

$$= \frac{x^2 + x}{x^2 - 1}$$

$$= \frac{x(x + 1)}{(x + 1)(x - 1)}$$ Factoring

$$= \frac{x\cancel{(x + 1)}}{\cancel{(x + 1)}(x - 1)}$$ Removing a factor of 1: $\frac{x + 1}{x + 1} = 1$

$$= \frac{x}{x - 1}$$

Do Exercise 2.

2. Simplify. Use method 1.

$$\frac{1 - \frac{1}{x}}{1 - \frac{1}{x^2}}$$

Answers

1. $\frac{14y + 7}{14y - 2}$, or $\frac{7(2y + 1)}{2(7y - 1)}$ **2.** $\frac{x}{x + 1}$

EXAMPLE 3 Simplify: $\dfrac{\dfrac{1}{a} + \dfrac{1}{b}}{\dfrac{1}{a^3} + \dfrac{1}{b^3}}$.

The denominators are a, b, a^3, and b^3. The LCM of these denominators is a^3b^3. We multiply by $a^3b^3/(a^3b^3)$.

$$\frac{\frac{1}{a} + \frac{1}{b}}{\frac{1}{a^3} + \frac{1}{b^3}} = \left(\frac{\frac{1}{a} + \frac{1}{b}}{\frac{1}{a^3} + \frac{1}{b^3}}\right) \cdot \frac{a^3b^3}{a^3b^3}$$

The LCM of the denominators is a^3b^3. We multiply by 1: $\dfrac{a^3b^3}{a^3b^3}$.

$$= \frac{\left(\frac{1}{a} + \frac{1}{b}\right) \cdot a^3b^3}{\left(\frac{1}{a^3} + \frac{1}{b^3}\right) \cdot a^3b^3}$$

Multiplying the numerators and the denominators

$$= \frac{\frac{1}{a} \cdot a^3b^3 + \frac{1}{b} \cdot a^3b^3}{\frac{1}{a^3} \cdot a^3b^3 + \frac{1}{b^3} \cdot a^3b^3}$$

Carrying out the multiplications using a distributive law

$$= \frac{a^2b^3 + a^3b^2}{b^3 + a^3} = \frac{a^2b^2(b + a)}{(b + a)(b^2 - ba + a^2)}$$

Factoring

$$= \frac{a^2b^2\cancel{(b + a)}}{\cancel{(b + a)}(b^2 - ba + a^2)}$$

Removing a factor of 1: $\dfrac{b + a}{b + a} = 1$

$$= \frac{a^2b^2}{b^2 - ba + a^2}.$$

Do Exercises 3 and 4. ▶

Simplify. Use method 1.

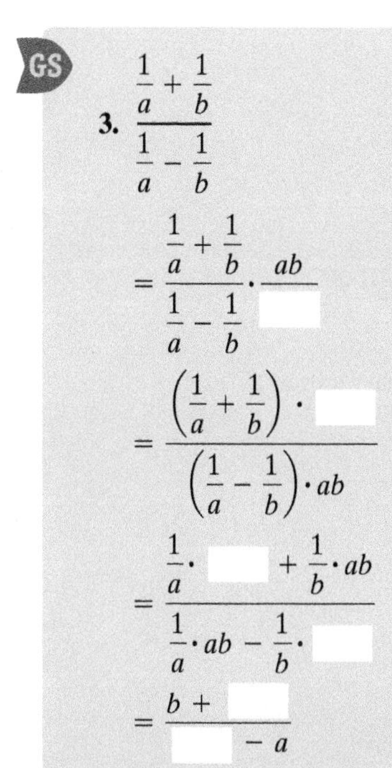

4. $\dfrac{\dfrac{1}{a} - \dfrac{1}{b}}{\dfrac{1}{a^3} - \dfrac{1}{b^3}}$

Method 2: Adding or Subtracting in the Numerator and the Denominator

Method 2. To simplify a complex rational expression:

1. Add or subtract, as necessary, to get a single rational expression in the numerator.
2. Add or subtract, as necessary, to get a single rational expression in the denominator.
3. Divide the numerator by the denominator.
4. If possible, simplify.

We will redo Examples 1–3 using this method.

Answers

3. $\dfrac{b + a}{b - a}$ **4.** $\dfrac{a^2b^2}{b^2 + ba + a^2}$

Guided Solution:
3. ab, ab, ab, ab, a, b

Simplify. Use method 2.

5. $\dfrac{y+\frac{1}{2}}{y-\frac{1}{7}}$

6. $\dfrac{1-\frac{1}{x}}{1-\frac{1}{x^2}}$

$= \dfrac{1\cdot\frac{x}{\square}-\frac{1}{x}}{1\cdot\frac{\square}{x^2}-\frac{1}{x^2}}$

$= \dfrac{\frac{x}{x}-\frac{1}{x}}{\frac{x^2}{x^2}-\frac{1}{x^2}}$

$= \dfrac{\frac{x-1}{x}}{\frac{\square}{x^2}}$

$= \dfrac{x-1}{x}\cdot\dfrac{\square}{x^2-1}$

$= \dfrac{(x-1)\cdot x\cdot x}{x(\square)(x-1)}$

$= \dfrac{\cancel{(x-1)}\cdot \cancel{x}\cdot x}{\cancel{x}(x+1)\cancel{(x-1)}}$

$= \dfrac{\square}{x+1}$

EXAMPLE 4 Simplify: $\dfrac{x+\frac{1}{5}}{x-\frac{1}{3}}$.

$$\frac{x+\frac{1}{5}}{x-\frac{1}{3}} = \frac{x\cdot\frac{5}{5}+\frac{1}{5}}{x-\frac{1}{3}} = \frac{\frac{5x+1}{5}}{x-\frac{1}{3}}$$

To get a single rational expression in the numerator, we note that the LCM in the numerator is 5. We multiply by 1 and add.

$$= \frac{\frac{5x+1}{5}}{x\cdot\frac{3}{3}-\frac{1}{3}} = \frac{\frac{5x+1}{5}}{\frac{3x-1}{3}}$$

To get a single rational expression in the denominator, we note that the LCM in the denominator is 3. We multiply by 1 and subtract.

$$= \frac{5x+1}{5}\cdot\frac{3}{3x-1}$$

Multiplying by the reciprocal of the denominator

$$= \frac{3(5x+1)}{5(3x-1)},\text{ or }\frac{15x+3}{15x-5}$$

No further simplification is possible.

EXAMPLE 5 Simplify: $\dfrac{1+\frac{1}{x}}{1-\frac{1}{x^2}}$.

$$\frac{1+\frac{1}{x}}{1-\frac{1}{x^2}} = \frac{1\cdot\frac{x}{x}+\frac{1}{x}}{1\cdot\frac{x^2}{x^2}-\frac{1}{x^2}}$$

Finding the LCM in the numerator and multiplying by 1

Finding the LCM in the denominator and multiplying by 1

$$= \frac{\frac{x}{x}+\frac{1}{x}}{\frac{x^2}{x^2}-\frac{1}{x^2}}$$

$$= \frac{\frac{x+1}{x}}{\frac{x^2-1}{x^2}}$$

Adding in the numerator and subtracting in the denominator

$$= \frac{x+1}{x}\cdot\frac{x^2}{x^2-1}$$

Multiplying by the reciprocal of the denominator

$$= \frac{\cancel{(x+1)}\cdot\cancel{x}\cdot x}{\cancel{x}(x-1)\cancel{(x+1)}}$$

Factoring and removing a factor of 1: $\dfrac{x(x+1)}{x(x+1)}=1$

$$= \frac{x}{x-1}$$

Do Exercises 5 and 6.

Answers

5. $\dfrac{14y+7}{14y-2}$, or $\dfrac{7(2y+1)}{2(7y-1)}$ 6. $\dfrac{x}{x+1}$

Guided Solution:

6. $x, x^2, x^2-1, x^2, x+1, x$

EXAMPLE 6 Simplify: $\dfrac{\dfrac{1}{a}+\dfrac{1}{b}}{\dfrac{1}{a^3}+\dfrac{1}{b^3}}$.

The LCM in the numerator is ab, and the LCM in the denominator is a^3b^3.

$$\frac{\frac{1}{a}+\frac{1}{b}}{\frac{1}{a^3}+\frac{1}{b^3}}=\frac{\frac{1}{a}\cdot\frac{b}{b}+\frac{1}{b}\cdot\frac{a}{a}}{\frac{1}{a^3}\cdot\frac{b^3}{b^3}+\frac{1}{b^3}\cdot\frac{a^3}{a^3}}=\frac{\frac{b}{ab}+\frac{a}{ab}}{\frac{b^3}{a^3b^3}+\frac{a^3}{a^3b^3}}$$

$$=\frac{\frac{b+a}{ab}}{\frac{b^3+a^3}{a^3b^3}}$$ Adding in the numerator and the denominator

$$=\frac{b+a}{ab}\cdot\frac{a^3b^3}{b^3+a^3}$$ Multiplying by the reciprocal of the denominator

$$=\frac{(b+a)a^3b^3}{ab(b^3+a^3)}=\frac{\cancel{(b+a)}\cdot\cancel{ab}\cdot a^2b^2}{\cancel{ab}\cancel{(b+a)}(b^2-ba+a^2)}$$

$$=\frac{a^2b^2}{b^2-ba+a^2}$$

Do Exercises 7 and 8. ▶

Simplify. Use method 2.

7. $\dfrac{\dfrac{1}{a}+\dfrac{1}{b}}{\dfrac{1}{a}-\dfrac{1}{b}}$

8. $\dfrac{\dfrac{1}{a}-\dfrac{1}{b}}{\dfrac{1}{a^3}-\dfrac{1}{b^3}}$

Answers

7. $\dfrac{b+a}{b-a}$ 8. $\dfrac{a^2b^2}{b^2+ba+a^2}$

5.4 Exercise Set

FOR EXTRA HELP

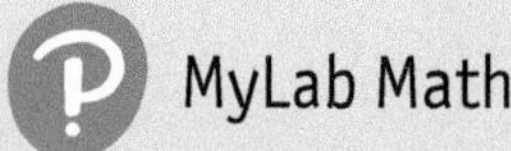

✓ Check Your Understanding

Reading Check Consider the expression $\dfrac{\dfrac{7}{x}-\dfrac{2}{3}}{\dfrac{4}{x}}$. Choose from the column on the right the term that best completes the statement.

RC1. The expression given above is a(n) ______________ rational expression.

RC2. The expression $\dfrac{7}{x}-\dfrac{2}{3}$ is the ______________ of the above expression.

RC3. The ______________ of the rational expressions $\dfrac{7}{x}, \dfrac{2}{3}$, and $\dfrac{4}{x}$ above is $3x$.

RC4. After subtracting in the numerator, we multiply the numerator by the ______________ of the denominator, $\dfrac{4}{x}$.

numerator
denominator
opposite
reciprocal
complex
least common denominator

Concept Check For each complex rational expression, find the LCM of the denominators in the numerator, the LCM of the denominators in the denominator, and the LCM of all the denominators of all the rational expressions within both the numerator and the denominator of the complex rational expression.

CC1. $\dfrac{\dfrac{7}{8} - \dfrac{3}{20}}{\dfrac{9}{10} + \dfrac{5}{14}}$

CC2. $\dfrac{\dfrac{3}{a} + \dfrac{4}{15a}}{\dfrac{1}{12a} - \dfrac{3}{8a}}$

a Simplify.

1. $\dfrac{2 + \dfrac{3}{5}}{4 - \dfrac{1}{2}}$

2. $\dfrac{\dfrac{3}{8} - 5}{\dfrac{2}{3} + 6}$

3. $\dfrac{\dfrac{2}{3} + \dfrac{4}{5}}{\dfrac{3}{4} - \dfrac{1}{2}}$

4. $\dfrac{\dfrac{5}{8} - \dfrac{2}{3}}{\dfrac{3}{4} + \dfrac{5}{6}}$

5. $\dfrac{\dfrac{x}{y^2}}{\dfrac{y^3}{x^2}}$

6. $\dfrac{\dfrac{a^3}{b^5}}{\dfrac{a^4}{b^2}}$

7. $\dfrac{\dfrac{9x^2 - y^2}{xy}}{\dfrac{3x - y}{y}}$

8. $\dfrac{\dfrac{a^2 - 16b^2}{ab}}{\dfrac{a + 4b}{b}}$

9. $\dfrac{\dfrac{1}{a} + 2}{\dfrac{1}{a} - 1}$

10. $\dfrac{\dfrac{1}{t} + 6}{\dfrac{1}{t} - 5}$

11. $\dfrac{x - \dfrac{1}{x}}{x + \dfrac{1}{x}}$

12. $\dfrac{y + \dfrac{1}{y}}{y - \dfrac{1}{y}}$

13. $\dfrac{\dfrac{3}{x} + \dfrac{4}{y}}{\dfrac{4}{x} - \dfrac{3}{y}}$

14. $\dfrac{\dfrac{2}{y} + \dfrac{5}{z}}{\dfrac{1}{y} - \dfrac{4}{z}}$

15. $\dfrac{a - \dfrac{3a}{b}}{b - \dfrac{b}{a}}$

16. $\dfrac{1 - \dfrac{2}{3x}}{x - \dfrac{4}{9x}}$

17. $\dfrac{\dfrac{1}{a} + \dfrac{1}{b}}{\dfrac{a^2 - b^2}{ab}}$

18. $\dfrac{\dfrac{1}{x} - \dfrac{1}{y}}{\dfrac{x^2 - y^2}{xy}}$

19. $\dfrac{\dfrac{1}{x + h} - \dfrac{1}{x}}{h}$

20. $\dfrac{\dfrac{1}{a - h} - \dfrac{1}{a}}{h}$

It may help you to write h as $\dfrac{h}{1}$.

21. $\dfrac{\dfrac{x^2 - x - 12}{x^2 - 2x - 15}}{\dfrac{x^2 + 8x + 12}{x^2 - 5x - 14}}$

22. $\dfrac{\dfrac{y^2 - y - 6}{y^2 - 5y - 14}}{\dfrac{y^2 + 6y + 5}{y^2 - 6y - 7}}$

23. $\dfrac{\dfrac{1}{x + 2} + \dfrac{4}{x - 3}}{\dfrac{2}{x - 3} - \dfrac{7}{x + 2}}$

24. $\dfrac{\dfrac{1}{y - 4} + \dfrac{1}{y + 5}}{\dfrac{6}{y + 5} + \dfrac{2}{y - 4}}$

25. $\dfrac{\dfrac{6}{x^2 - 4} - \dfrac{5}{x + 2}}{\dfrac{7}{x^2 - 4} - \dfrac{4}{x - 2}}$

26. $\dfrac{\dfrac{1}{x^2 - 1} + \dfrac{5}{x^2 - 5x + 4}}{\dfrac{1}{x^2 - 1} + \dfrac{2}{x^2 + 3x + 2}}$

27. $\dfrac{\dfrac{1}{z^2} - \dfrac{1}{w^2}}{\dfrac{1}{z^3} + \dfrac{1}{w^3}}$

28. $\dfrac{\dfrac{1}{b^2} - \dfrac{1}{c^2}}{\dfrac{1}{b^3} - \dfrac{1}{c^3}}$

29. $\dfrac{\dfrac{3}{x^2 + 2x - 3} - \dfrac{1}{x^2 - 3x - 10}}{\dfrac{3}{x^2 - 6x + 5} - \dfrac{1}{x^2 + 5x + 6}}$

30. $\dfrac{\dfrac{1}{a^2 + 7a + 12} + \dfrac{1}{a^2 + a - 6}}{\dfrac{1}{a^2 + 2a - 8} + \dfrac{1}{a^2 + 5a + 4}}$

Skill Maintenance

Factor. [4.3a], [4.4a], [4.6d]

31. $4x^3 + 20x^2 + 6x$

32. $y^3 + 8$

33. $y^3 - 8$

34. $2x^3 - 32x^2 + 126x$

35. $1000x^3 + 1$

36. $1 - 1000a^3$

37. $y^3 - 64x^3$

38. $\frac{1}{8}a^3 - 343$

39. Solve for s: $T = \dfrac{r + s}{3}$. [1.2a]

40. Graph: $f(x) = -3x + 2$. [2.2c]

41. Given that $f(x) = x^2 - 3$, find $f(-5)$. [2.2b]

42. Solve: $|2x - 5| = 7$. [1.6c]

Synthesis

For each function in Exercises 43–46, find and simplify $\dfrac{f(a + h) - f(a)}{h}$.

43. $f(x) = \dfrac{3}{x^2}$

44. $f(x) = \dfrac{5}{x}$

45. $f(x) = \dfrac{1}{1 - x}$

46. $f(x) = \dfrac{x}{1 + x}$

Simplify.

47. $\dfrac{5x^{-1} - 5y^{-1} + 10x^{-1}y^{-1}}{6x^{-1} - 6y^{-1} + 12x^{-1}y^{-1}}$

48. $\left[\dfrac{\dfrac{x + 3}{x - 3} + 1}{\dfrac{x + 3}{x - 3} - 1}\right]^8$

Mid-Chapter Review

Concept Reinforcement

Determine whether each statement is true or false.

______________ **1.** For synthetic division, the divisor must be in the form $x - a$. [5.3c]

______________ **2.** The sum of two rational expressions is the sum of the numerators over the sum of the denominators. [5.2b]

______________ **3.** The domain of $f(x) = \dfrac{(x-5)(x+4)}{x-4}$ is $\{x \mid x \neq 5 \text{ and } x \neq -4 \text{ and } x \neq 4\}$. [5.1a]

Guided Solutions

GS Fill in each blank with the number or the expression that creates a correct solution.

4. Subtract: $\dfrac{7x-2}{x-4} - \dfrac{x+1}{x+3}$. [5.2b]

$$\frac{7x-2}{x-4} - \frac{x+1}{x+3} = \frac{7x-2}{x-4} \cdot \frac{\square}{\square} - \frac{x+1}{x+3} \cdot \frac{\square}{\square}$$

$$= \frac{7x^2 + \square x - \square}{(\square)(x+3)} - \frac{x^2 - \square x - \square}{(\square)(x-4)}$$

$$= \frac{\square x^2 + 19x - 6 - \square + \square + 4}{(\square)(x+3)}$$

$$= \frac{\square x^2 + \square x - \square}{(x-4)(\square)}$$

5. Simplify: $\dfrac{\frac{1}{m} + 3}{\frac{1}{m} - 5}$. [5.4a]

$$\frac{\frac{1}{m} + 3}{\frac{1}{m} - 5} = \frac{\frac{1}{m} + 3}{\frac{1}{m} - 5} \cdot \frac{\square}{\square} = \frac{\square + 3\square}{\square - 5\square}$$

Mixed Review

Find the domain of each function. [5.1a]

6. $f(x) = \dfrac{x+5}{x^2 - 100}$

7. $g(x) = \dfrac{-3}{x-7}$

8. $h(x) = \dfrac{x^2 - 9}{x^2 + 8x - 9}$

Simplify. [5.1c]

9. $\dfrac{24p^2}{36p^9}$

10. $\dfrac{42y - 3}{33}$

11. $\dfrac{x^2 - y^2}{x^3 + y^3}$

12. $\dfrac{x^2 - x - 30}{x^2 - 4x - 12}$

13. $\dfrac{9a - 18}{9a + 18}$

14. $\dfrac{3 - t}{t^2 - t - 6}$

Find the LCM. [5.2a]

15. $x^3,\ 14x^2y,\ 35x^4y^5$

16. $x^2 - 25,\ x^2 - 10x + 25,\ x^2 + 3x - 40$

Perform the indicated operations and simplify.

17. $\frac{45}{x^2 - 1} \div \frac{x + 1}{x - 1}$ [5.1e]

18. $\frac{3x - 1}{x + 6} + \frac{x}{x - 2}$ [5.2b]

19. $\frac{q}{q + 2} - \frac{q + 1}{q}$ [5.2b]

20. $\frac{2y}{y^2 + 2y - 3} - \frac{3y + 1}{y^2 + y - 2}$ [5.2b]

21. $\frac{\frac{1}{b} - 1}{\frac{1}{b^2} - 1}$ [5.4a]

22. $\frac{w^2 - z^2}{5w - 5z} \cdot \frac{w - z}{w + z}$ [5.1d]

23. $\frac{t^3 - 8}{2t + 3} \cdot \frac{2t^2 + t - 3}{t - 2}$ [5.1d]

24. $\frac{5c}{3} + \frac{2a}{5c}$ [5.2b]

25. $\frac{x^2 - 4x}{x^2 + 2x} \div \frac{x^2 - 8x + 16}{x^2 + 4x + 4}$ [5.1e]

Divide and if there is a remainder, express it in two ways. Use synthetic division in Exercises 28–30. [5.3b, c]

26. $(6x^2 - 5x + 11) \div (2x - 3)$

27. $(x^4 - 1) \div (x + 1)$

28. $(2x^3 - x^2 + 5x - 4) \div (x + 2)$

29. $(x^2 - 4x - 12) \div (x - 6)$

30. $(x^4 - 3x^2 + 2) \div (x + 3)$

31. $(15x^2 - 2x + 6) \div (5x + 1)$

Understanding Through Discussion and Writing

32. Explain how synthetic division can be useful when factoring a polynomial. [5.3c]

33. Do addition, subtraction, multiplication, and division of polynomials always result in a polynomial? Why or why not? [5.1d], [5.2b], [5.3b]

34. Is it possible to understand how to simplify rational expressions without first understanding how to multiply? Why or why not? [5.1c]

35. Janine found that the sum of two rational expressions was $(3 - x)/(x - 5)$, but the answer at the back of the book was $(x - 3)/(5 - x)$. Was Janine's answer correct? Why or why not? [5.2b]

36. Nancy *incorrectly* simplifies $(x + 2)/x$ as follows:

$$\frac{x + 2}{x} = \frac{\cancel{x} + 2}{\cancel{x}} = 1 + 2 = 3.$$

She insists that this is correct because when x is replaced with 1, her answer checks. Explain her error. [5.1c]

37. Explain why it is easier to use method 1 than method 2 to simplify the following expression. [5.4a]

$$\frac{\frac{a}{b} + \frac{c}{d}}{\frac{a}{b} - \frac{c}{d}}$$

STUDYING FOR SUCCESS *Applications*

- ☐ If you find applied problems challenging, don't give up! Your skill will improve with each problem that you solve.
- ☐ Make applications real! Look in newspapers, in magazines, and online for applications of the math that you are studying.

5.5 Solving Rational Equations

OBJECTIVE

a Solve rational equations.

a RATIONAL EQUATIONS

SKILL REVIEW *Solve quadratic and other polynomial equations by first factoring and then using the principle of zero products.* [4.8a]

Solve.

1. $x^2 - 3x - 18 = 0$ **2.** $3x^2 - 12x = 0$

Answers: 1. $-3, 6$ **2.** $0, 4$

MyLab Math VIDEO

In Sections 5.1–5.4, we studied operations with *rational expressions.* These expressions do not have equals signs. Although we can perform operations on and simplify rational expressions, we cannot solve them. Note the following examples:

$$\frac{x^2 - 6x + 9}{x^2 - 4} \cdot \frac{x - 2}{x - 3}, \quad \frac{x + y}{x - y} \div \frac{x^2 + y}{x^2 - y^2}, \quad \text{and} \quad \frac{a + 7}{a^2 - 16} + \frac{5}{5a - 15}.$$

Operation signs occur. There are no equals signs!

Most often, the result of our manipulation is another rational expression that is not cleared of fractions.

Equations *do have* equals signs, and we can clear them of fractions as we did in Section 1.1. A **rational**, or **fraction, equation** is an equation containing one or more rational expressions. Here are some examples:

$$\frac{2}{3} - \frac{5}{6} = \frac{1}{x}, \quad x + \frac{6}{x} = 5, \quad \text{and} \quad \frac{2x}{x - 3} - \frac{6}{x} = \frac{18}{x^2 - 3x}.$$

There are equals signs as well as operation signs.

SOLVING RATIONAL EQUATIONS

To solve a rational equation, the first step is to clear the equation of fractions. To do this, multiply all terms on both sides of the equation by the LCM of all the denominators. Then carry out the equation-solving process as discussed in Chapters 1 and 4.

EXAMPLE 1 Solve: $\frac{2}{3} - \frac{5}{6} = \frac{1}{x}$.

Multiplying all terms on both sides of the equation by the LCM, we have

$$(2 \cdot 3 \cdot x) \cdot \left(\frac{2}{3} - \frac{5}{6}\right) = (2 \cdot 3 \cdot x) \cdot \frac{1}{x}$$ Multiplying both sides by the LCM, $2 \cdot 3 \cdot x$

$$2 \cdot 3 \cdot x \cdot \frac{2}{3} - 2 \cdot 3 \cdot x \cdot \frac{5}{6} = 2 \cdot 3 \cdot x \cdot \frac{1}{x}$$ Multiplying to remove parentheses

When clearing fractions, be sure to multiply *every* term in the equation by the LCM.

$$\begin{aligned} 2 \cdot x \cdot 2 - x \cdot 5 &= 2 \cdot 3 \\ 4x - 5x &= 6 \\ -x &= 6 \\ -1 \cdot x &= 6 \\ x &= -6 \end{aligned}$$

Check: $\frac{2}{3} - \frac{5}{6} = \frac{1}{x}$

$\frac{2}{3} - \frac{5}{6} \;?\; \frac{1}{-6}$

$\frac{4}{6} - \frac{5}{6} \;\Big|\; -\frac{1}{6}$

$-\frac{1}{6}$ TRUE

The solution is -6.

Do Exercise 1. ▶

EXAMPLE 2 Solve: $\frac{x+1}{2} - \frac{x-3}{3} = 3$.

The LCM of all the denominators is $2 \cdot 3$, or 6.

$$2 \cdot 3 \cdot \left(\frac{x+1}{2} - \frac{x-3}{3}\right) = 2 \cdot 3 \cdot 3$$ Multiplying both sides by the LCM

$$2 \cdot 3 \cdot \frac{x+1}{2} - 2 \cdot 3 \cdot \frac{x-3}{3} = 2 \cdot 3 \cdot 3$$ Multiplying to remove parentheses

$$3(x+1) - 2(x-3) = 18$$ Simplifying

$$\begin{aligned} 3x + 3 - 2x + 6 &= 18 \\ x + 9 &= 18 \end{aligned}$$ Multiplying and collecting like terms

$$x = 9$$

Check: $\frac{x+1}{2} - \frac{x-3}{3} = 3$

$\frac{9+1}{2} - \frac{9-3}{3} \;?\; 3$

$5 - 2$

3 TRUE

The solution is 9.

Caution!

Clearing fractions is a valid procedure only when solving equations, *not* when adding, subtracting, multiplying, or dividing rational expressions.

Do Exercise 2. ▶

GS **1.** Solve: $\frac{2}{3} + \frac{5}{6} = \frac{1}{x}$.

LCM = $6x$, or $2 \cdot 3 \cdot x$

$$2 \cdot 3 \cdot x\left(\frac{2}{3} + \frac{5}{6}\right) = 2 \cdot \underline{\quad} \cdot x \cdot \frac{1}{x}$$

$$2 \cdot 3 \cdot x \cdot \frac{\underline{\quad}}{3} + 2 \cdot 3 \cdot x \cdot \frac{5}{6} = 2 \cdot 3 \cdot x \cdot \frac{1}{x}$$

$$\begin{aligned} 2 \cdot x \cdot \underline{\quad} + x \cdot 5 &= 2 \cdot 3 \\ 4x + 5x &= 6 \\ \underline{\quad}\, x &= 6 \\ x &= \frac{6}{9} \\ x &= \frac{\underline{\quad}}{3} \end{aligned}$$

The number $\frac{2}{3}$ checks and is the solution.

2. Solve: $\frac{y-4}{5} - \frac{y+7}{2} = 5$.

Answers

1. $\frac{2}{3}$ **2.** -31

Guided Solution:
1. 3, 2, 2, 9, 2

CHECKING POSSIBLE SOLUTIONS

When we multiply all terms on both sides of an equation by the LCM, the resulting equation might yield numbers that are *not* solutions of the original equation. Thus we must *always* check possible solutions in the original equation.

The next example illustrates the importance of checking all possible solutions.

EXAMPLE 3 Solve: $\frac{2x}{x-3} - \frac{6}{x} = \frac{18}{x^2 - 3x}$.

The LCM of the denominators is $x(x-3)$. We multiply all terms on both sides by $x(x-3)$.

$$x(x-3)\left(\frac{2x}{x-3} - \frac{6}{x}\right) = x(x-3)\left(\frac{18}{x^2-3x}\right) \quad \text{Multiplying both sides by the LCM}$$

$$x(x-3)\cdot\frac{2x}{x-3} - x(x-3)\cdot\frac{6}{x} = x(x-3)\cdot\frac{18}{x^2-3x} \quad \text{Multiplying to remove parentheses}$$

$$2x^2 - 6(x-3) = 18 \quad \text{Simplifying}$$

$$2x^2 - 6x + 18 = 18$$

$$2x^2 - 6x = 0$$

$$2x(x-3) = 0 \quad \text{Factoring}$$

$$2x = 0 \quad or \quad x - 3 = 0 \quad \text{Using the principle of zero products}$$

$$x = 0 \quad or \quad x = 3$$

The numbers 0 and 3 are possible solutions. We look at the original equation and see that each makes a denominator 0, so neither is a solution. We can carry out a check, as follows.

Check:

For 0:

$$\frac{2x}{x-3} - \frac{6}{x} = \frac{18}{x^2-3x}$$

$$\frac{2(0)}{0-3} - \frac{6}{0} \overset{?}{=} \frac{18}{0^2 - 3(0)}$$

$$0 - \frac{6}{0} \;\Big|\; \frac{18}{0} \qquad \text{NOT DEFINED}$$

For 3:

$$\frac{2x}{x-3} - \frac{6}{x} = \frac{18}{x^2-3x}$$

$$\frac{2(3)}{3-3} - \frac{6}{3} \overset{?}{=} \frac{18}{3^2 - 3(3)}$$

$$\frac{6}{0} - 2 \;\Big|\; \frac{18}{0} \qquad \text{NOT DEFINED}$$

The equation has *no solution*.

Do Exercise 3.

3. Solve:
$\frac{4x}{x+5} + \frac{20}{x} = \frac{100}{x^2+5x}$.

Answer
3. No solution

EXAMPLE 4 Solve: $\frac{x^2}{x-2} = \frac{4}{x-2}$.

The LCM of the denominators is $x - 2$. We multiply all terms on both sides by $x - 2$.

$$(x-2)\cdot\frac{x^2}{x-2} = (x-2)\cdot\frac{4}{x-2}$$

$$x^2 = 4 \qquad \text{Simplifying}$$

$$x^2 - 4 = 0$$

$$(x+2)(x-2) = 0$$

$$x + 2 = 0 \quad or \quad x - 2 = 0 \qquad \text{Using the principle of zero products}$$

$$x = -2 \quad or \quad x = 2$$

Check: For 2:

$$\frac{x^2}{x-2} = \frac{4}{x-2}$$

$$\frac{2^2}{2-2} \stackrel{?}{=} \frac{4}{2-2}$$

$$\frac{4}{0} \;\Big|\; \frac{4}{0} \qquad \text{NOT DEFINED}$$

For −2:

$$\frac{x^2}{x-2} = \frac{4}{x-2}$$

$$\frac{(-2)^2}{-2-2} \stackrel{?}{=} \frac{4}{-2-2}$$

$$\frac{4}{-4} \;\Big|\; \frac{4}{-4}$$

$$-1 \;\Big|\; -1 \qquad \text{TRUE}$$

The number −2 is a solution, but 2 is not (it results in division by 0).

Do Exercise 4.

4. Solve: $\frac{x^2}{x-3} = \frac{9}{x-3}$.

EXAMPLE 5 Solve: $\frac{2}{x-1} = \frac{3}{x+1}$.

The LCM of the denominators is $(x - 1)(x + 1)$. We multiply all terms on both sides by $(x - 1)(x + 1)$.

$$(x-1)(x+1)\cdot\frac{2}{x-1} = (x-1)(x+1)\cdot\frac{3}{x+1} \qquad \text{Multiplying}$$

$$2(x+1) = 3(x-1) \qquad \text{Simplifying}$$

$$2x + 2 = 3x - 3$$

$$5 = x$$

The check is left to the student. The number 5 checks and is the solution.

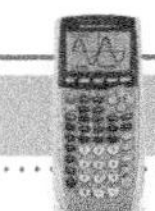

CALCULATOR CORNER

Checking Solutions of Rational Equations We can use a table to check possible solutions of rational equations. Consider the equation in Example 4,

$$\frac{x^2}{x-2} = \frac{4}{x-2},$$

and the possible solutions that were found, −2 and 2. To check these solutions, we enter $y_1 = x^2/(x-2)$ and $y_2 = 4/(x-2)$ on the equation-editor screen. Then, with a table set in ASK mode, we enter $x = -2$. (See p. 41.) Since y_1 and y_2 have the same value, we know that the equation is true when $x = -2$, and thus −2 is a solution. Now we enter $x = 2$. The ERROR messages indicate that 2 is not a solution because it is not an allowable replacement for x in the equation.

X	Y1	Y2
−2	−1	−1
2	ERROR	ERROR

X=

EXERCISES:

1. Use a graphing calculator to check the possible solutions found in Examples 1, 2, and 3.
2. Use a graphing calculator to check the possible solutions you found in Margin Exercises 1–4.

Answer

4. −3

Solve.

5. $\dfrac{2}{x-1} = \dfrac{3}{x+2}$

LCM $= (x-1)(\quad)$

$(x-1)(x+2) \cdot \dfrac{2}{x-1}$

$= (\quad)(x+2) \cdot \dfrac{3}{x+2}$

$2(x+2) = 3(\quad)$

$\quad + 4 = 3x - \quad$

$7 = x$

The number 7 checks and is the solution.

6. $\dfrac{2}{x^2-9} + \dfrac{5}{x-3} = \dfrac{3}{x+3}$

ALGEBRAIC ▶◀ GRAPHICAL CONNECTION

Let's make a visual check of Example 7 by looking at a graph. We can think of the equation

$$x + \frac{6}{x} = 5$$

as the intersection of the graphs of

$$f(x) = x + \frac{6}{x} \quad \text{and} \quad g(x) = 5.$$

We see in the graph that there are two points of intersection, at $x = 2$ and at $x = 3$.

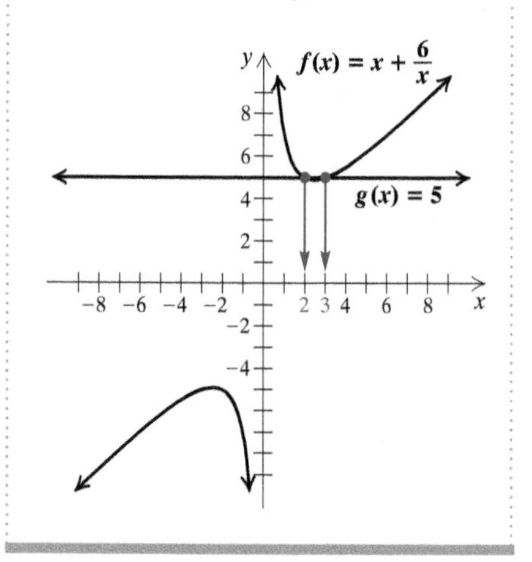

Answer

Answers for Margin Exercises 5 and 6 are on p. 409.

EXAMPLE 6 Solve: $\dfrac{2}{x+5} + \dfrac{1}{x-5} = \dfrac{16}{x^2-25}$.

The LCM of the denominators is $(x+5)(x-5)$. We multiply all terms on both sides by $(x+5)(x-5)$.

$$(x+5)(x-5) \cdot \left[\frac{2}{x+5} + \frac{1}{x-5}\right] = (x+5)(x-5) \cdot \frac{16}{x^2-25}$$

$$(x+5)(x-5) \cdot \frac{2}{x+5} + (x+5)(x-5) \cdot \frac{1}{x-5} = (x+5)(x-5) \cdot \frac{16}{x^2-25}$$

$$2(x-5) + (x+5) = 16$$
$$2x - 10 + x + 5 = 16$$
$$3x - 5 = 16$$
$$3x = 21$$
$$x = 7$$

Check: $\dfrac{2}{x+5} + \dfrac{1}{x-5} = \dfrac{16}{x^2-25}$

$$\frac{2}{7+5} + \frac{1}{7-5} \ ? \ \frac{16}{7^2-25}$$

$\frac{2}{12} + \frac{1}{2}$	$\frac{16}{49-25}$
$\frac{8}{12}$	$\frac{16}{24}$
$\frac{2}{3}$	$\frac{2}{3}$ TRUE

The solution is 7.

◀ **Do Exercises 5 and 6.**

EXAMPLE 7 Given that $f(x) = x + 6/x$, find all values of x for which $f(x) = 5$.

Since $f(x) = x + 6/x$, we want to find all values of x for which

$$x + \frac{6}{x} = 5.$$

The LCM of the denominators is x. We multiply all terms on both sides by x:

$x\left(x + \frac{6}{x}\right) = x \cdot 5$	Multiplying by x on both sides
$x \cdot x + x \cdot \frac{6}{x} = 5x$	
$x^2 + 6 = 5x$	Simplifying
$x^2 - 5x + 6 = 0$	Getting 0 on one side
$(x-3)(x-2) = 0$	Factoring
$x - 3 = 0$ *or* $x - 2 = 0$	Using the principle of zero products
$x = 3$ *or* $x = 2$.	

Check: For $x = 3$, $f(3) = 3 + \dfrac{6}{3} = 3 + 2 = 5$.

$$\text{For } x = 2,\ f(2) = 2 + \frac{6}{2} = 2 + 3 = 5.$$

The solutions are 2 and 3.

Do Exercise 7. ▶

7. Given that $f(x) = x - 12/x$, find all values of x for which $f(x) = 1$.

Caution!

In this section, we have introduced a new use of the LCM. Before, you used the LCM in adding or subtracting rational expressions. Now we are working with equations. There are equals signs. We clear the fractions by multiplying all terms on both sides of the equation by the LCM. This eliminates the denominators. *Do not* make the mistake of trying to "clear the fractions" when you do not have an equation!

Answers

5. 7 **6.** -13 **7.** 4, -3

Guided Solution:

5. $x + 2, x - 1, x - 1, 2x, 3$

5.5 Exercise Set

FOR EXTRA HELP

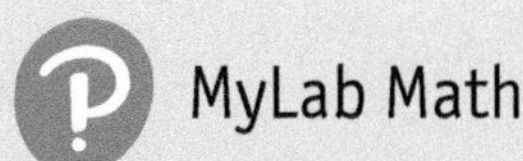

Check Your Understanding

Reading Check One of the common difficulties with studying the material in this chapter is being sure about the task at hand. Are you combining expressions using operations to get another *rational expression*, or are you solving equations for which the results are numbers that are *solutions* of an equation? To practice making these decisions, determine for each of the following exercises the type of answer you should get: "Rational expression" or "Solutions." You need not complete the mathematical operations.

RC1. Add: $\dfrac{6w}{w^2 - 1} + \dfrac{w}{w^2 - w}$.

RC2. Solve: $\dfrac{5}{y - 3} - \dfrac{30}{y^2 - 9} = 1$.

RC3. Subtract: $\dfrac{2}{a - 2} - \dfrac{1}{a + 2}$.

RC4. Divide: $\dfrac{x + 4}{x - 2} \div \dfrac{6x}{x^2 - 4}$.

RC5. Solve: $\dfrac{x^2}{x - 1} = \dfrac{1}{x - 1}$.

RC6. Solve: $\dfrac{10}{y} + y = -2$.

RC7. Multiply: $\dfrac{2t^2}{t^2 - 25} \cdot \dfrac{t^2 + 10t + 25}{t^8}$.

RC8. Solve: $\dfrac{7}{x - 4} - \dfrac{2}{x + 4} = \dfrac{1}{x^2 - 16}$.

Concept Check Determine whether each statement is true or false.

CC1. A rational equation is an equation containing one or more rational expressions.

CC2. When we multiply by the LCM on both sides of a rational equation, the resulting equation might have solutions that are not solutions of the original equation.

a Solve. Don't forget to check!

1. $\frac{y}{10} = \frac{2}{5} + \frac{3}{8}$

2. $\frac{3}{8} + \frac{1}{3} = \frac{t}{12}$

3. $\frac{1}{4} - \frac{5}{6} = \frac{1}{a}$

4. $\frac{5}{8} - \frac{2}{5} = \frac{1}{y}$

5. $\frac{x}{3} - \frac{x}{4} = 12$

6. $\frac{y}{5} - \frac{y}{3} = 15$

7. $x + \frac{8}{x} = -9$

8. $y + \frac{22}{y} = -13$

9. $\frac{3}{y} + \frac{7}{y} = 5$

10. $\frac{4}{3y} - \frac{3}{y} = \frac{10}{3}$

11. $\frac{1}{2} = \frac{z-5}{z+1}$

12. $\frac{x-6}{x+9} = \frac{2}{7}$

13. $\frac{3}{y+1} = \frac{2}{y-3}$

14. $\frac{4}{x-1} = \frac{3}{x+2}$

15. $\frac{y-1}{y-3} = \frac{2}{y-3}$

16. $\frac{x-2}{x-4} = \frac{2}{x-4}$

17. $\frac{x+1}{x} = \frac{3}{2}$

18. $\frac{y+2}{y} = \frac{5}{3}$

19. $\frac{1}{2} - \frac{4}{9x} = \frac{4}{9} - \frac{1}{6x}$

20. $-\frac{1}{3} - \frac{5}{4y} = \frac{3}{4} - \frac{1}{6y}$

21. $\frac{60}{x} - \frac{60}{x-5} = \frac{2}{x}$

22. $\frac{50}{y} - \frac{50}{y-2} = \frac{4}{y}$

23. $\frac{7}{5x-2} = \frac{5}{4x}$

24. $\frac{5}{y+4} = \frac{3}{y-2}$

25. $\frac{x}{x-2} + \frac{x}{x^2-4} = \frac{x+3}{x+2}$

26. $\frac{3}{y-2} + \frac{2y}{4-y^2} = \frac{5}{y+2}$

27. $\frac{6}{x^2-4x+3} - \frac{1}{x-3} = \frac{1}{4x-4}$

28. $\frac{1}{2x+10} = \frac{8}{x^2-25} - \frac{2}{x-5}$

29. $\frac{5}{y+3} = \frac{1}{4y^2-36} + \frac{2}{y-3}$

30. $\frac{7}{x-2} - \frac{8}{x+5} = \frac{1}{2x^2+6x-20}$

31. $\frac{a}{2a-6} - \frac{3}{a^2-6a+9} = \frac{a-2}{3a-9}$

32. $\frac{1}{x-2} = \frac{2}{x+4} + \frac{2x-1}{x^2+2x-8}$

33. $\frac{2x+3}{x-1} = \frac{10}{x^2-1} + \frac{2x-3}{x+1}$

34. $\frac{y}{y+1} + \frac{3y+5}{y^2+4y+3} = \frac{2}{y+3}$

35. $\frac{3x}{x+2} + \frac{72}{x^3+8} = \frac{24}{x^2-2x+4}$

36. $\frac{4}{x+3} + \frac{7}{x^2-3x+9} = \frac{108}{x^3+27}$

37. $\frac{5x}{x-7} - \frac{35}{x+7} = \frac{490}{x^2-49}$

38. $\frac{3x}{x+2} + \frac{6}{x} + 4 = \frac{12}{x^2+2x}$

39. $\frac{x^2}{x^2-4} = \frac{x}{x+2} - \frac{2x}{2-x}$

For the given rational function f, find all values of x for which $f(x)$ has the indicated value.

40. $f(x) = 2x - \frac{15}{x};\ f(x) = 1$

41. $f(x) = 2x - \frac{6}{x};\ f(x) = 1$

42. $f(x) = \frac{x-5}{x+1};\ f(x) = \frac{3}{5}$

43. $f(x) = \frac{x-3}{x+2};\ f(x) = \frac{1}{5}$

44. $f(x) = \frac{12}{x} - \frac{12}{2x};\ f(x) = 8$

45. $f(x) = \frac{6}{x} - \frac{6}{2x};\ f(x) = 5$

Skill Maintenance

Factor. [4.6d]

46. $4t^3 + 500$

47. $1 - t^6$

48. $a^3 + 8b^3$

49. $a^3 - 8b^3$

Solve. [4.8a]

50. $x^2 - 6x + 9 = 0$

51. $(x-3)(x+4) = 0$

52. $x^2 - 49 = 0$

53. $12x^2 - 11x + 2 = 0$

Solve. [2.4c]

54. *Student Loan Debt.* After mortgage debt, student loan debt is the second highest consumer debt category. In 1999, the total student loan debt was $90 billion. This debt had increased to $1.3 trillion in 2017. Find the average rate of change in the student loan with respect to time, in years.

Data: "Student Loan Debt in 2017," Zach Friedman, *Forbes*, February 21, 2017; "Chart of the Day: "Student Loans Have Grown 511% Since 1999," Daniel Indiviglio, *The Atlantic*, August 18, 2011

55. *Mail Volume.* First-class mail volume has steadily decreased in the last decade. The volume in 2007 was 42.3 billion pieces of mail. By 2016, this amount had decreased to only 19.7 billion pieces. Find the average rate of change in the volume of first-class mail with respect to time, in years.

Data: about.usps.com

Synthesis

56.

a) Use the INTERSECT feature of a graphing calculator to find the points of intersection of the graphs of

$$f(x) = \frac{1}{1+x} + \frac{x}{1-x} \quad \text{and} \quad g(x) = \frac{1}{1-x} - \frac{x}{1+x}.$$

b) Use the algebraic methods of this section to check your answers to part (a).

c) Explain which procedure you prefer.

57.

a) Use the INTERSECT feature of a graphing calculator to find the points of intersection of the graphs of

$$f(x) = \frac{x+3}{x+2} - \frac{x+4}{x+3} \quad \text{and} \quad g(x) = \frac{x+5}{x+4} - \frac{x+6}{x+5}.$$

b) Use the algebraic methods of this section to check your answers to part (a).

c) Explain which procedure you prefer.

5.6 Applications and Proportions

OBJECTIVES

a Solve work problems and certain basic problems using rational equations.

b Solve applied problems involving proportions.

c Solve motion problems using rational equations.

SKILL REVIEW *Solve equations using the addition principle and the multiplication principle together, removing parentheses where appropriate.* [1.1d]

Solve.

1. $-\frac{4}{3}b + \frac{2}{3} = -6$

2. $\frac{2}{5}x - \frac{3}{10}x = \frac{6}{5}$

Answers: 1. 5 **2.** 12

MyLab Math VIDEO

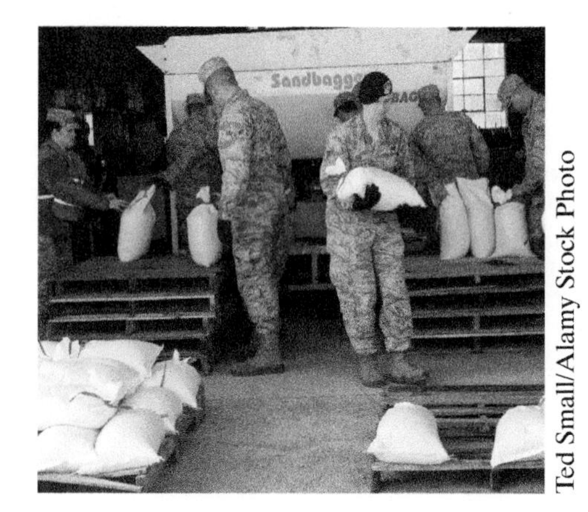

Ted Small/Alamy Stock Photo

a WORK PROBLEMS

EXAMPLE 1 *Filling Sandbags.* The Sandbagger Corporation sells machines that fill sandbags at a job site. The Sandbagger™ can fill an order of 8000 sandbags in 5 hr. The Multibagger™ can fill the same order in 8 hr. If both machines are used together, how long would it take to fill an order of 8000 sandbags?

1. **Familiarize.** We familiarize ourselves with the problem by considering two incorrect ways of translating the problem to mathematical language.

   a) A common *incorrect* way to translate the problem is to average the two times:

   $$\frac{5+8}{2}\text{ hr} = \frac{13}{2}\text{ hr, or } 6\frac{1}{2}\text{ hr.}$$

   Let's think about this. Using only the Sandbagger, the job is completed in 5 hr. If the two baggers are used together, then the time it takes to complete the order should be less than 5 hr. Thus we reject $6\frac{1}{2}$ hr as a solution, but we do have a partial check on any answer we get. The answer should be less than 5 hr.

   b) Another *incorrect* way to translate the problem is as follows. Suppose that the two machines are used in such a way that half of the job is done by the Sandbagger and the other half by the Multibagger. Then

   the Sandbagger fills $\frac{1}{2}$ of the bags in $\frac{1}{2}(5\text{ hr})$, or $2\frac{1}{2}$ hr,

   and

   the Multibagger fills $\frac{1}{2}$ of the bags in $\frac{1}{2}(8\text{ hr})$, or 4 hr.

   But time is wasted since the Sandbagger completed its part $1\frac{1}{2}$ hr earlier than the Multibagger. In effect, the machines were not used together to complete the job as fast as possible. If the Sandbagger is used in addition to the Multibagger after completing its half, the entire job could be finished in a time somewhere between $2\frac{1}{2}$ hr and 4 hr.

We proceed to a translation by considering how much of the job is finished in 1 hr, 2 hr, 3 hr, and so on. It takes the Sandbagger 5 hr to fill 8000 bags alone. Then in 1 hr, it can do $\frac{1}{5}$ of the job. It takes the Multibagger 8 hr to complete the job alone. Then in 1 hr, it can do $\frac{1}{8}$ of the job. Both baggers together can complete

$$\frac{1}{5} + \frac{1}{8}, \text{ or } \frac{13}{40} \text{ of the job in 1 hr.}$$

In 2 hr, the Sandbagger can do $2\left(\frac{1}{5}\right)$ of the job and the Multibagger can do $2\left(\frac{1}{8}\right)$ of the job. Both baggers together can complete

$$2\left(\frac{1}{5}\right) + 2\left(\frac{1}{8}\right), \text{ or } \frac{13}{20} \text{ of the job in 2 hr.}$$

Continuing this reasoning, we can form a table like the one at right. From the table, we see that if the baggers work together for 4 hr, the fraction of the job that will be completed is $1\frac{3}{10}$, which is more of the job than needs to be done. We also see that the answer is somewhere between 3 hr and 4 hr. What we are looking for is a number of hours t such that the fraction of the job that is completed is 1; that is, the job is just completed—not more $\left(1\frac{3}{10}\right)$ and not less $\left(\frac{39}{40}\right)$.

	FRACTION OF JOB COMPLETED		
TIME	SANDBAGGER	MULTIBAGGER	TOGETHER
1 hr	$\frac{1}{5}$	$\frac{1}{8}$	$\frac{1}{5}+\frac{1}{8}$, or $\frac{13}{40}$
2 hr	$2\left(\frac{1}{5}\right)$	$2\left(\frac{1}{8}\right)$	$2\left(\frac{1}{5}\right)+2\left(\frac{1}{8}\right)$, or $\frac{13}{20}$
3 hr	$3\left(\frac{1}{5}\right)$	$3\left(\frac{1}{8}\right)$	$3\left(\frac{1}{5}\right)+3\left(\frac{1}{8}\right)$, or $\frac{39}{40}$
4 hr	$4\left(\frac{1}{5}\right)$	$4\left(\frac{1}{8}\right)$	$4\left(\frac{1}{5}\right)+4\left(\frac{1}{8}\right)$, or $1\frac{3}{10}$
t hr	$t\left(\frac{1}{5}\right)$	$t\left(\frac{1}{8}\right)$	$t\left(\frac{1}{5}\right)+t\left(\frac{1}{8}\right)$

2. **Translate.** From the table, we see that the time we are looking for is some number t for which

$$t\left(\frac{1}{5}\right)+t\left(\frac{1}{8}\right)=1, \quad \text{or} \quad \frac{t}{5}+\frac{t}{8}=1,$$

where 1 represents the idea that the entire job is completed in time t.

3. **Solve.** We solve the equation:

$$\frac{t}{5}+\frac{t}{8}=1$$

$$40\left(\frac{t}{5}+\frac{t}{8}\right)=40\cdot 1 \qquad \text{The LCM is } 5\cdot2\cdot2\cdot2\text{, or 40. We multiply by 40.}$$

$$40\cdot\frac{t}{5}+40\cdot\frac{t}{8}=40 \qquad \text{Using the distributive law}$$

$$8t+5t=40 \qquad \text{Simplifying}$$

$$13t=40$$

$$t=\frac{40}{13}, \text{ or } 3\frac{1}{13}\text{ hr.}$$

4. **Check.** The check can be done by using $\frac{40}{13}$ for t and substituting into the original equation:

$$\frac{40}{13}\left(\frac{1}{5}\right)+\frac{40}{13}\left(\frac{1}{8}\right)=\frac{8}{13}+\frac{5}{13}=\frac{13}{13}=1.$$

We also have a partial check in what we learned from the *Familiarize* step. The answer, $3\frac{1}{13}$ hr, is between 3 hr and 4 hr (see the table), and it is less than 5 hr, the time it takes the Sandbagger to do the job alone.

5. **State.** It takes $3\frac{1}{13}$ hr for the two baggers to complete the job working together.

THE WORK PRINCIPLE

Suppose that a is the time it takes A to do a job, b is the time it takes B to do the same job, and t is the time it takes them to do the job working together. Then

$$\frac{t}{a}+\frac{t}{b}=1.$$

Do Exercise 1. ▶

1. *Building a Playground.* In a recent year, a city development group set a goal of building playgrounds within walking distance of every child in the city. The city provides the plans and assigns the construction to volunteer groups. Volunteer group A can build a playground in 20 hr. Volunteer group B can build the same playground in 25 hr. How long will it take them, working together, to build the playground?

Answer

1. $11\frac{1}{9}$ hr

Austin, Texas, has the fastest growing food truck industry in the United States.

EXAMPLE 2 *Catering a Business Luncheon.* A convention planner is organizing a business luncheon for 2300 guests in Austin, Texas. The chosen menu will include tacos and tamales. To complete the order, two food caterers are needed. Since Austin has the second highest number of food trucks per capita in the United States, the organizer was able to hire two food trucks as caterers. Together they can complete the order in 12 hr. If the second caterer would take 10 hr longer than the first caterer to complete the order alone, how long would it take each, working alone, to prepare the luncheon?

Data: "Another No. 1 for Austin: The city where food trucks pop up the fastest," Michael Theis *Austin Business Journal*, May 5, 2017; *The Economist*; IBISWorld

1. **Familiarize.** Comparing this problem to Example 1, we note that we do not know the times required by each caterer to complete the task had each worked alone. We let

   $h =$ the amount of time, in hours, that it would take the first caterer working alone.

   Then

   $h + 10 =$ the amount of time, in hours, that it would take the second caterer working alone.

   We also know that $t = 12$ hr $=$ total time. Thus,

   $\frac{12}{h} =$ the fraction of the job that the first caterer could finish in 12 hr

   and

   $\frac{12}{h+10} =$ the fraction of the job that the second caterer could finish in 12 hr.

2. **Translate.** Using the work principle, we know that

$$\frac{t}{a} + \frac{t}{b} = 1 \qquad \text{Using the work principle}$$

$$\frac{12}{h} + \frac{12}{h+10} = 1. \qquad \text{Substituting } \frac{12}{h} \text{ for } \frac{t}{a} \text{ and } \frac{12}{h+10} \text{ for } \frac{t}{b}$$

3. **Solve.** We solve the equation:

$$\frac{12}{h} + \frac{12}{h+10} = 1$$

$$h(h+10)\left(\frac{12}{h} + \frac{12}{h+10}\right) = h(h+10)\cdot 1 \qquad \text{We multiply by the LCM, which is } h(h+10).$$

$$h(h+10)\cdot\frac{12}{h} + h(h+10)\cdot\frac{12}{h+10} = h(h+10) \qquad \text{Using the distributive law}$$

$$(h+10)\cdot 12 + h\cdot 12 = h^2 + 10h \qquad \text{Simplifying}$$

$$12h + 120 + 12h = h^2 + 10h$$

$$0 = h^2 - 14h - 120 \qquad \text{Getting 0 on one side}$$

$$0 = (h-20)(h+6) \qquad \text{Factoring}$$

$$h - 20 = 0 \quad or \quad h + 6 = 0 \qquad \text{Using the principle of zero products}$$

$$h = 20 \quad or \quad h = -6.$$

4. **Check.** Since negative time has no meaning in the problem, we reject -6 as a solution to the original problem. The number 20 checks since if the first caterer takes 20 hr alone and the second caterer takes $20 + 10$, or 30 hr alone, in 12 hr, working together, they would have completed

$$\frac{12}{20} + \frac{12}{30} = \frac{3}{5} + \frac{2}{5} = \frac{5}{5}, \text{ or 1 job.}$$

5. **State.** It would take the first caterer 20 hr and the second caterer 30 hr to complete the task alone.

Do Exercise 2. ▶

2. *Filling a Water Tank.* Two pipes carry water to the same tank. Pipe A, working alone, can fill the tank three times as fast as pipe B. Together, the pipes can fill the tank in 24 hr. Find the time that it would take each pipe to fill the tank alone.

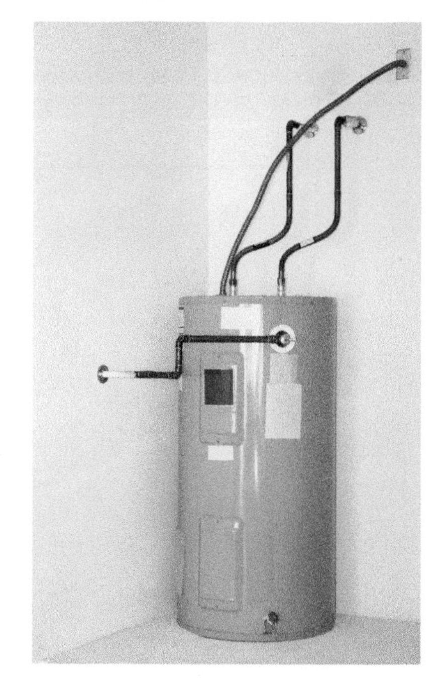

b APPLICATIONS INVOLVING PROPORTIONS

Any rational expression a/b represents a **ratio**. Percent can be considered a ratio. For example, 67% is the ratio of 67 to 100, or 67/100. The ratio of two different kinds of measure is called a **rate**. Speed is an example of a rate. For example, Usain Bolt, from Jamaica, completed the 100-m run in the 2012 Summer Olympics in London with a time of 9.63 sec. His speed, or rate, was

$$\frac{100\text{ m}}{9.63\text{ sec}}, \quad \text{or} \quad 10.4\,\frac{\text{m}}{\text{sec}}. \qquad \text{Rounded to the nearest tenth}$$

He also holds the world record for this event with a time of 9.58 sec, which he raced on August 16, 2009, in Berlin, Germany.

PROPORTION

An equality of ratios, $A/B = C/D$, read "A is to B as C is to D," is called a **proportion**. The numbers named in a true proportion are said to be **proportional** to each other.

We can use proportions to solve applied problems by expressing a ratio in two ways, as shown below. For example, suppose that it takes 8 gal of gas to drive for 120 mi, and we want to determine how much gas will be required to drive for 550 mi. If we assume that the car uses gas at the same rate throughout the trip, then the ratios are the same, and we can write a proportion. We let x represent the number of gallons it takes to drive 550 mi.

$$\begin{matrix}\text{Miles} \longrightarrow \\ \text{Gallons} \longrightarrow\end{matrix} \frac{120}{8} = \frac{550}{x} \begin{matrix}\longleftarrow \text{Miles} \\ \longleftarrow \text{Gallons}\end{matrix}$$

To solve this proportion, we note that the LCM is $8x$. Thus we multiply by $8x$.

$$\begin{aligned}
8x \cdot \frac{120}{8} &= 8x \cdot \frac{550}{x} && \text{Multiplying by } 8x \\
x \cdot 120 &= 8 \cdot 550 && \text{Simplifying} \\
120x &= 8 \cdot 550 \\
x &= \frac{8 \cdot 550}{120} && \text{Dividing by 120} \\
x &\approx 36.67
\end{aligned}$$

Thus, 36.67 gal will be required to drive 550 mi.

Answer

2. Pipe A: 32 hr; pipe B: 96 hr

It is common to use **cross products** to solve proportions, as follows:

$$\frac{120}{8} = \frac{550}{x} \qquad \text{If } \frac{A}{B} = \frac{C}{D}, \text{ then } AD = BC.$$

$$120 \cdot x = 8 \cdot 550$$ $120 \cdot x$ and $8 \cdot 550$ are called *cross products.* Note that this is the equation that results from clearing fractions above.

$$x = \frac{8 \cdot 550}{120}$$

$$x \approx 36.67.$$

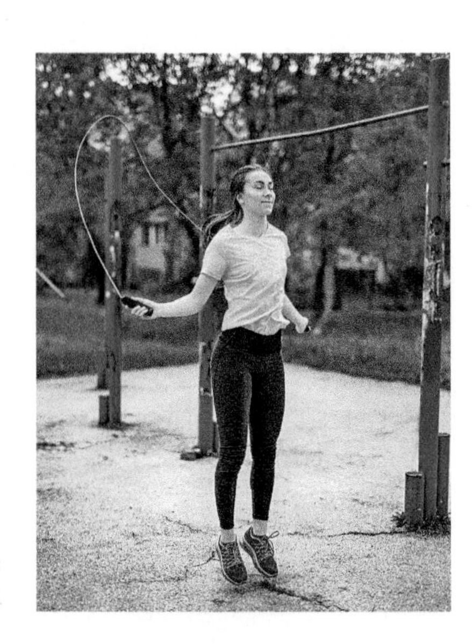

EXAMPLE 3 *Calories Burned.* Jumping rope is an excellent cardiovascular exercise. A person of average weight can burn about 125 calories while jumping rope for only 10 min. The recommended speed is about 2 rope-jumps per second. How many calories will Carly burn if she jumps rope for 38 min?

Data: Livestrong.com, "How Many Calories Will Be Burned Jumping Rope for Ten Minutes?" by Max Roman Dilthey; The Jump Rope Institute; Calorie Lab

1. **Familiarize.** We let $c =$ the number of calories burned in 38 min.
2. **Translate.** Next, we translate to a proportion. We make each side the ratio of number of minutes to number of calories, with number of minutes in the numerator and number of calories in the denominator.

$$\begin{matrix}\text{Minutes} \longrightarrow \\ \text{Calories} \longrightarrow\end{matrix} \frac{10}{125} = \frac{38}{c} \begin{matrix}\longleftarrow \text{Minutes} \\ \longleftarrow \text{Calories}\end{matrix}$$

3. **Solve.** We solve the proportion:

$$\frac{10}{125} = \frac{38}{c}$$

$10c = 125 \cdot 38$ — Equating cross products

$c = \frac{125 \cdot 38}{10}$ — Dividing by 10

$c = 475.$ — Multiplying and dividing

4. **Check.** We substitute into the proportion and check cross products:

$$\frac{10}{125} = \frac{38}{475};$$

$$10 \cdot 475 = 4750; \quad 125 \cdot 38 = 4750.$$

Since the cross products are the same, the answer checks.

5. **State.** In 38 min, Carly will burn 475 calories.

Do Exercise 3.

3. *Calories Burned.* Mia, who weighs 120 lb, will burn 110 calories in 20 min during an aerobics class. How many calories will she burn in 35 min? Let $c =$ the number of calories burned in 35 min. GS

$$\frac{20}{\square} = \frac{35}{c}$$

$$20c = 110 \cdot \square$$

$$20c = 3850$$

$$c = \square$$

EXAMPLE 4 *Estimating Wild Burro Population.* Wild burros still exist in at least five U.S. states. There are more wild burros in Arizona than in any other state. To estimate the number in Arizona, a ranger catches 383 wild burros, tags them, and releases them. Later, 103 burros are caught, and it is found that 10 of them are tagged. Estimate how many wild burros there are in Arizona.

Data: U.S. Department of the Interior Bureau of Land Management, The Wild Horse and Burro Program; "Growing Arizona wild burro population threatening native wildlife," Ben Margiott, *Cronkite News*, February 16, 2016

Answer

3. 192.5 calories

Guided Solution:

3. 110, 35, 192.5

1. **Familiarize.** We let B = the number of wild burros in Arizona. For the purposes of this example, we assume that the tagged burros mix freely with others in the state. We also assume that when some burros have been captured, the ratio of those tagged to the total number captured is the same as the ratio of burros originally tagged to the total number of wild burros in the state. For example, if 1 of every 3 burros captured later is tagged, we would assume that 1 of every 3 burros in the state was originally tagged.
2. **Translate.** We translate to a proportion, as follows:

$$\begin{array}{r} \text{Burros tagged originally} \longrightarrow \\ \text{Burros in Arizona} \longrightarrow \end{array} \frac{383}{B} = \frac{10}{103}. \begin{array}{l} \longleftarrow \text{Tagged burros caught} \\ \longleftarrow \text{Burros caught} \end{array}$$

3. **Solve.** We solve the proportion:

$$383 \cdot 103 = B \cdot 10 \qquad \text{Equating cross products}$$
$$\frac{383 \cdot 103}{10} = B \qquad \text{Dividing by 10}$$
$$3945 \approx B. \qquad \text{Multiplying and dividing and rounding}$$

4. **Check.** We substitute into the proportion and check cross products:

$$\frac{383}{3945} = \frac{10}{103}; \quad 383 \cdot 103 = 39{,}449; \quad 3945 \cdot 10 = 39{,}450.$$

The cross products are close but not exact because we rounded the total.

5. **State.** We estimate that there are about 3945 wild burros in Arizona.

Do Exercise 4. ▶

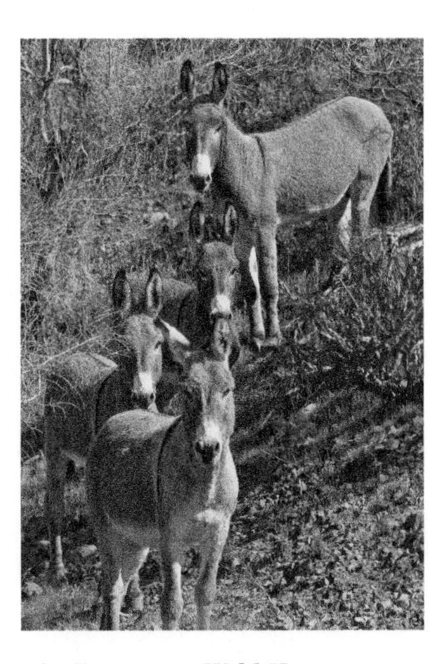

4. ***Estimating Wild Horse Population.*** To estimate the number of wild horses in Nevada, a ranger catches 560 wild horses, tags them, and releases them. Later, 133 horses are caught, and it is found that 4 of them are tagged. Estimate how many wild horses there are in Nevada.

Data: U.S. Department of the Interior Bureau of Land Management, The Wild Horse and Burro Program

C MOTION PROBLEMS

We considered motion problems earlier in Sections 1.3 and 3.4. To translate them, we know that we can use either the basic motion formula, $d = rt$, or either of two formulas, $r = d/t$ or $t = d/r$, which can be derived from $d = rt$.

MOTION FORMULAS

The following are the formulas for motion problems:

$d = rt \longrightarrow \text{Distance} = \text{Rate} \cdot \text{Time};$

$$r = \frac{d}{t} \longrightarrow \text{Rate} = \frac{\text{Distance}}{\text{Time}}; \qquad t = \frac{d}{r} \longrightarrow \text{Time} = \frac{\text{Distance}}{\text{Rate}}.$$

EXAMPLE 5 *Bicycling.* A racer is bicycling 15 km/h faster than a person on a mountain bike. In the time that it takes the racer to travel 80 km, the person on the mountain bike has gone 50 km. Find the speed of each bicyclist.

1. **Familiarize.** Let's guess that the person on the mountain bike is going 10 km/h. The racer would then be traveling 10 + 15, or 25 km/h. At 25 km/h, the racer will travel 80 km in $\frac{80}{25} = 3.2$ hr. Going 10 km/h, the mountain bike will cover 50 km in $\frac{50}{10} = 5$ hr. Since $3.2 \neq 5$, our guess is wrong, but we can see that if r is the rate, in kilometers per hour, of the slower bike, then the rate of the racer, who is traveling 15 km/h faster, is $r + 15$.

Answer

4. About 18,620 wild horses

Making a drawing and organizing the facts in a chart can be helpful.

	DISTANCE	SPEED	TIME
MOUNTAIN BIKE	50	r	t
RACING BIKE	80	$r + 15$	t

$\longrightarrow 50 = rt \longrightarrow t = \dfrac{50}{r}$

$\longrightarrow 80 = (r + 15)t \longrightarrow t = \dfrac{80}{r + 15}$

2. **Translate.** The time is the same for both bikes. Using the formula $d = rt$ and then $t = d/r$ across both rows of the table, we find two expressions for time and can equate them as

$$\frac{50}{r} = \frac{80}{r + 15}.$$

3. **Solve.** We solve the equation:

$$\frac{50}{r} = \frac{80}{r + 15}$$

$$r(r + 15) \cdot \frac{50}{r} = r(r + 15) \cdot \frac{80}{r + 15}$$ The LCM is $r(r + 15)$. We multiply by $r(r + 15)$.

$$(r + 15) \cdot 50 = r \cdot 80$$ Simplifying. We can also obtain this by equating cross products.

$$50r + 750 = 80r$$ Using the distributive law

$$750 = 30r$$ Subtracting $50r$

$$\frac{750}{30} = r$$ Dividing by 30

$$25 = r.$$

4. **Check.** If our answer checks, the mountain bike is going 25 km/h and the racing bike is going 25 + 15, or 40 km/h.

Traveling 80 km at 40 km/h, the racer is riding for $\frac{80}{40} = 2$ hr. Traveling 50 km at 25 km/h, the person on the mountain bike is riding for $\frac{50}{25} = 2$ hr. Our answer checks since the two times are the same.

5. **State.** The speed of the racer is 40 km/h, and the speed of the person on the mountain bike is 25 km/h.

◀ **Do Exercise 5.**

5. *Four-Wheeler Travel.* Olivia's four-wheeler travels 8 km/h faster than Emma's. Olivia travels 69 km in the same time that it takes Emma to travel 45 km. Find the speed of each person's four-wheeler.

Answer

5. Olivia: 23 km/h; Emma: 15 km/h

EXAMPLE 6 *Transporting by Barge.* A river barge travels 98 mi downstream in the same time that it takes to travel 52 mi upstream. The speed of the current in the river is 2.3 mph. Find the speed of the barge in still water.

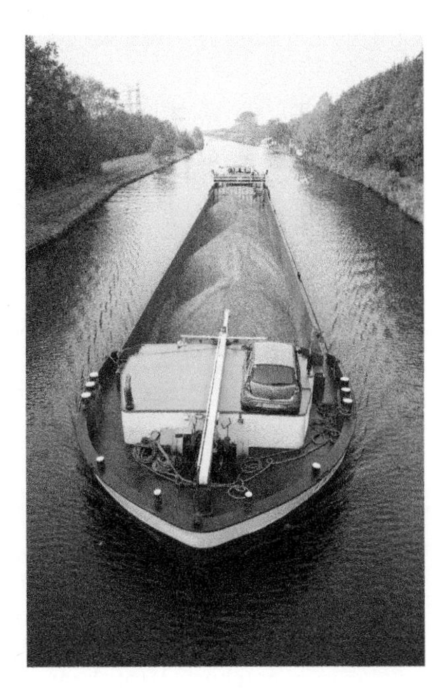

1. **Familiarize.** We first make a drawing. We let s = the speed of the barge in still water and t = the time, and then organize the facts in a table.

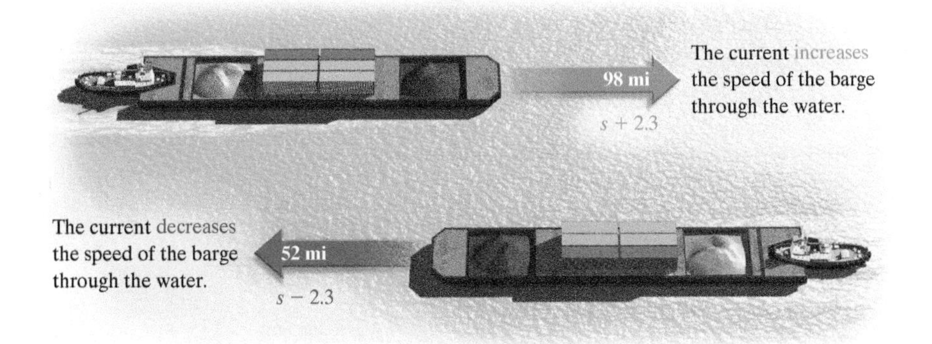

	DISTANCE	SPEED	TIME
DOWNSTREAM	98	$s + 2.3$	t
UPSTREAM	52	$s - 2.3$	t

$\rightarrow 98 = (s + 2.3)t \rightarrow t = \frac{98}{s + 2.3}$

$\rightarrow 52 = (s - 2.3)t \rightarrow t = \frac{52}{s - 2.3}$

2. **Translate.** Using the formula $t = d/r$ across both rows of the table, we find two expressions for time and equate them as

$$\frac{98}{s + 2.3} = \frac{52}{s - 2.3}.$$

3. **Solve.** We solve the equation:

$$\frac{98}{s + 2.3} = \frac{52}{s - 2.3}$$
$$(s + 2.3)(s - 2.3)\left(\frac{98}{s + 2.3}\right) = (s + 2.3)(s - 2.3)\left(\frac{52}{s - 2.3}\right)$$
$$(s - 2.3)98 = (s + 2.3)52$$
$$98s - 225.4 = 52s + 119.6$$
$$46s = 345$$
$$s = 7.5.$$

4. **Check.** Downstream, the speed of the barge is $7.5 + 2.3$, or 9.8 mph. Dividing the distance, 98 mi, by the speed, 9.8 mph, we get 10 hr. Upstream, the speed of the barge is $7.5 - 2.3$, or 5.2 mph. Dividing the distance, 52 mi, by the speed, 5.2 mph, we get 10 hr. Since the times are the same, the answer checks.
5. **State.** The speed of the barge in still water is 7.5 mph.

Do Exercise 6.

6. *Riverboat Speed.* A riverboat cruise line has a boat that can travel 76 mi downstream in the same time that it takes to travel 52 mi upstream. The speed of the current in the river is 1.5 mph. Find the speed of the boat in still water.

Answer
6. 8 mph

Translating for Success

1. *Sums of Squares.* The sum of the squares of two consecutive odd integers is 650. Find the integers.

2. *Estimating Fish Population.* To determine the number of fish in a lake, a conservationist catches 225 of them, tags them, and releases them back into the lake. Later, 108 fish are caught, and it is found that 15 of them are tagged. Estimate how many fish are in the lake.

3. *Consecutive Integers.* The sum of two consecutive even integers is 650. Find the integers.

4. *Sums of Squares.* The sum of the squares of two consecutive integers is 685. Find the integers.

5. *Hockey Results.* A hockey team played 81 games in a season. They won 1 fewer game than three times the number of ties and lost 8 fewer games than they won. How many games did they win? lose? tie?

Translate each word problem to an equation or a system of equations and select a correct translation from choices A–O.

A. $x + (x + 2) = 650$

B. $\frac{225}{x} = \frac{15}{108}$

C. $x^2 + (x + 1)^2 = 685$

D. $\frac{30}{x + 3} = \frac{40}{x}$

E. $x + y + z = 81,$
$x = 3y - 1,$
$z = x - 8$

F. $x + y + z = 81,$
$x - 1 = 3y,$
$z = x - 8$

G. $x^2 + (x + 5)^2 = 650$

H. $x + y + z = 650,$
$x + y = 480,$
$y + z = 685$

I. $\frac{40}{x + 3} = \frac{30}{x}$

J. $\frac{15}{x} = \frac{108}{225}$

K. $x + y + z = 685,$
$x + y = 480,$
$y + z = 650$

L. $x^2 + (x + 2)^2 = 685$

M. $\frac{x}{3} + \frac{x}{8} = 1$

N. $x^2 + (x + 2)^2 = 650$

O. $x = y + 3,$
$2x + 2y = 81$

Answers on page A-19

6. *Sides of a Square.* If each side of a square is increased by 5 ft, the area of the original square plus the area of the enlarged square is 650 ft^2. Find the length of a side of the original square.

7. *Bicycling.* The speed of one mountain biker is 3 km/h faster than the speed of another biker. The first biker travels 40 km in the same amount of time that it takes the second to travel 30 km. Find the speed of each biker.

8. *PDQ Shopping Network.* Sarah, Claire, and Maggie can process 685 telephone orders per day for PDQ shopping network. Sarah and Claire together can process 480 orders per day, while Claire and Maggie can process 650 orders. How many orders can each process alone?

9. *Filling Time.* A spa can be filled in 3 hr by hose A alone and in 8 hr by hose B alone. How long would it take to fill the spa if both hoses are working?

10. *Rectangle Dimensions.* The length of a rectangle is 3 ft longer than its width. Find the dimensions of the rectangle such that the perimeter of the rectangle is 81 ft.

5.6 Exercise Set

FOR EXTRA HELP 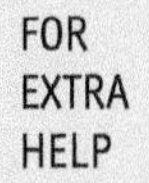 

✓ Check Your Understanding

Reading Check and Concept Check Choose from the column on the right the word that best completes the statement. Some words may be used more than once, and some words may not be used.

RC1. A ratio of two quantities is their ______________.

RC2. An equality of ratios, $\frac{A}{B} = \frac{C}{D}$, is called a(n) ______________.

RC3. Distance equals ______________ times time.

RC4. Rate equals ______________ divided by time.

RC5. Time equals ______________ divided by rate.

RC6. The numbers named in a true proportion are said to be ______________ to each other.

product
quotient
distance
rate
proportion
proportional
similar

a Solve.

1. *Painting a House.* Jose can paint a house in 28 hr. His brother, Miguel, can paint the same house in 36 hr. Working together, how long will it take them to paint the house?

2. *Filling a Pool.* An in-ground backyard pool can be filled in 12 hr if water enters through a pipe alone, or in 30 hr if water enters through a hose alone. If water is entering through both the pipe and the hose, how long will it take to fill the pool?

3. *Printing Tee Shirts.* In 30 hr, one machine can print tee shirts honoring the winning team in a national championship sporting event. Another machine can complete the same order in only 24 hr. If both machines are used, how long will it take to print the order?

4. *Washing Elephants.* Leah can wash the zoo's elephants in 3 hr. Ian, who is less experienced, needs 4 hr to do the same job. Working together, how long will it take them to wash the elephants?

5. *Placing Wrappers on Canned Goods.* Machine A can place wrappers on a batch of canned goods in 4 fewer hours than machine B. Together, they can complete the job in 1.5 hr. How long would it take each machine working alone?

6. *Cutting Firewood.* Tom can cut and split a cord of firewood in 6 fewer hours than Henry can. When they work together, it takes them 4 hr. How long would it take each of them to do the job alone?

7. *Clearing a Lot.* A commercial contractor needs to clear a plot of land for a new bank. Ryan can clear the lot in 7.5 hr. Ethan can do the same job in 10.5 hr. How long will it take them, working together, to clear the land? (*Hint*: You may find that multiplying by $\frac{1}{10}$ on both sides of the equation will clear the decimals.)

8. *Sorting Donations.* At the neighborhood food pantry, Grace can sort a morning's donations in 4.5 hr. Caleb can do the same job in 3 hr. Working together, how long would it take them to sort the food donations?

9. *Skimming a Swimming Pool.* Cole can skim a swimming pool with a leaf net three times as fast as Jim can. Together they can skim the pool in 6 min. How long would it take each to skim the pool alone?

10. *Pressing Shirts.* Karla and William press shirts for Perfection Laundry. Each week an amusement park drops off 320 shirts to be laundered and pressed. Karla can press shirts twice as fast as William. Together they can press 320 shirts in 11 hr. How long would it take each to press this order alone?

b Solve.

11. *Three-Point Field Goals.* After 15 games of an NBA season, a player had scored 9 three-point field goals. Assuming that he would continue to score at the same rate, how many three-point field goals would he score in the entire 82-game season?

12. *Touchdown Pace.* After 5 games of a recent NFL season, a quarterback had passed for 12 touchdowns. Assuming that he would continue to pass for touchdowns at the same rate, how many touchdown passes would he throw in the entire 16-game season?

13. *Models of Indy Cars.* Mattel, Inc., recently added some made-to-scale models of Indy Cars to their Hot Wheels® product line. The length of an IRL car is 15 ft. Its width is 7 ft. The width of the die-cast replica is 3.5 in. Find the length of the model.

Data: Mattel, IndyCar.com

14. *USS Constitution.* For a woodworking class, Alexis is building a scale model of the USS Constitution, known as "Old Ironsides." The length of the ship at the water line is 175 ft; the beam (or width) is 43.5 ft. The width of the model is 6.75 in. Find the length of the model.

15. *Weight on Moon.* The ratio of the weight of an object on the moon to the weight of an object on Earth is 0.16 to 1. How much will a 180-lb astronaut weigh on the moon?

16. *Weight on Mars.* The ratio of the weight of an object on Mars to the weight of an object on Earth is 0.378 to 1. How much will a 120-lb astronaut weigh on Mars?

17. *Coffee Consumption.* Coffee beans from 14 trees are required to produce 7.7 kg of coffee. (This is the average amount that each person in the United States drinks each year.) The beans from how many trees are required to produce 638 kg of coffee?

18. *Human Blood.* 10 cm^3 of a normal specimen of human blood contains 1.2 g of hemoglobin. How many grams does 32 cm^3 of the same blood contain?

19. *Wine Production.* In a recent year, a winery produced 4320 bottles of wine from 8 tons of grapes. They expect the demand to reach 7850 bottles next year. How many tons of grapes will they need?

Data: napanow.com

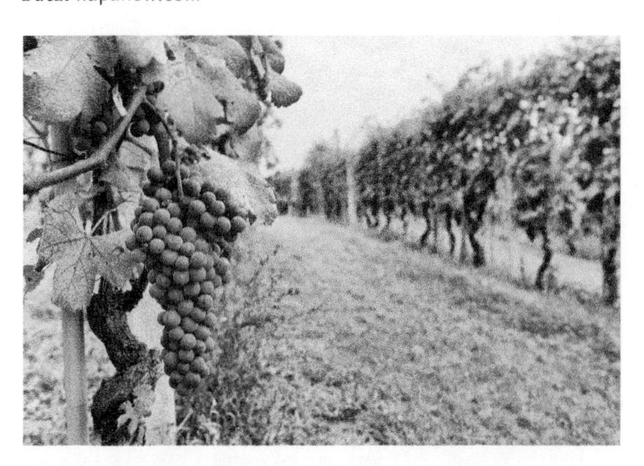

20. *Wind Turbines.* Data show that when operating at full capacity, fifteen 300-ft tall wind turbines can power 5250 homes. How many homes can be powered by 923 wind turbines?

Data: Indiana Office of Energy Development, E.ON Climate & Renewables

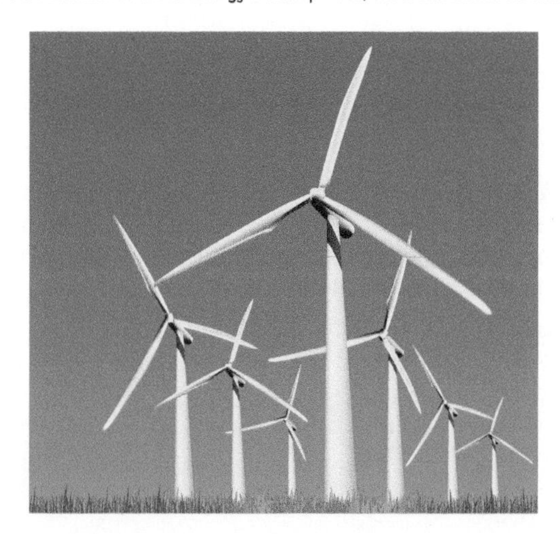

21. *Estimating Wildlife Populations.* To determine the number of trout in a lake, a conservationist catches 112 trout, tags them, and releases them back into the lake. Later, 82 trout are caught; 32 of them are tagged. How many trout are in the lake?

22. *Estimating Wildlife Populations.* To determine the number of deer in a game preserve, a conservationist catches 318 deer, tags them, and lets them loose. Later, 168 deer are caught; 56 of them are tagged. How many deer are in the preserve?

23. *Rope Cutting.* A rope is 28 ft long. How can the rope be cut in such a way that the ratio of the resulting two segments is 3 to 5?

24. Consider the numbers 1, 2, 3, and 5. If the same number is added to each of the numbers, it is found that the ratio of the first new number to the second is the same as the ratio of the third new number to the fourth. Find the number.

25. *Retaining Wall.* On average, a retaining wall requires approximately 1017 kg of stone for each 3.2 m^2. The area includes the face of the wall and its upper surface. How many kilograms of stone are needed for 65 m^2? Round the answer to the nearest kilogram.

Data: Reed, David, *The Art and Craft of Stonescaping* (Sterling Publishing Co., Inc.: New York, 2000)

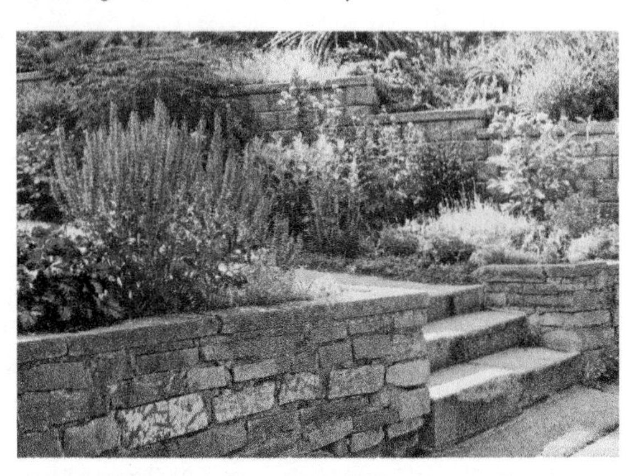

26. *Corona Arch.* The following photograph shows Corona Arch in Moab, Utah, one of the favorite hiking places of one of your authors, Marv Bittinger. He appears at the bottom of the photograph. Assume that an $8\frac{1}{2}$-in. by 11-in. photograph has been printed from a digital file and that in that photo Marv is $\frac{11}{32}$, or 0.34375 in. tall, and the height of the arch in the photo is $7\frac{5}{8}$, or 7.625 in. Given that Marv is 73 in. tall, find the actual height H of the arch.

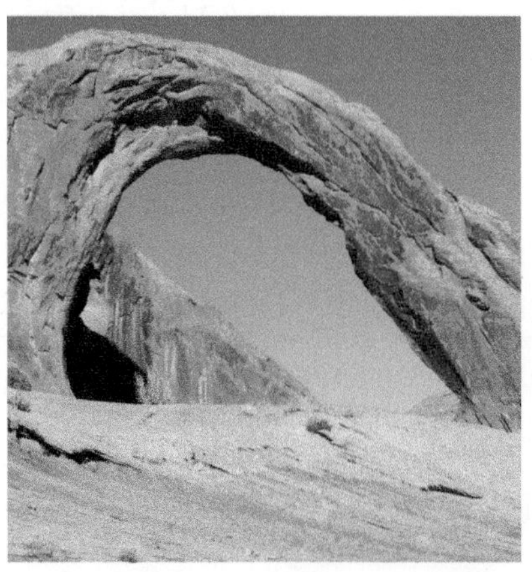

Gary Whitton/Shutterstock

C Solve.

27. *Jet Travel.* A Boeing 747 flies 2420 mi with the wind. In the same amount of time, it can fly 2140 mi against the wind. The cruising speed (in still air) is 570 mph. Find the speed of the wind.

Data: Boeing

28. *Transporting Cargo.* Boeing has a jumbo jet that is used to transport cargo. This jet flies 1430 mi with the wind. In the same amount of time, it can fly 1320 mi against the wind. The cruising speed (in still air) is 550 mph. Find the speed of the wind.

Data: Boeing

29. *Kayaking.* The speed of the current in the Wabash River is 3 mph. Brooke's kayak can travel 4 mi upstream in the same time that it takes to travel 10 mi downstream. What is the speed of Brooke's kayak in still water?

30. *Boating.* The current in the Animas River moves at a rate of 4 mph. Sydney's dinghy motors 6 mi upstream in the same time that it takes to motor 12 mi downstream. What is the speed of the dinghy in still water?

31. *Tour Travel.* Adventure Tours has 6 leisure-tour trolleys that travel 15 mph slower than their 3 express-tour buses. The bus travels 132 mi in the time that it takes the trolley to travel 99 mi. Find the speed of each mode of transportation.

32. *Hiking.* Vanessa hikes 2 mph slower than Xavier. In the time that it takes Xavier to hike 8 mi, Vanessa hikes 5 mi. Find the speed of each person.

33. *Moving Sidewalks.* A moving sidewalk at an airport moves at a rate of 1.8 ft/sec. Walking on the moving sidewalk, Thomas travels 105 ft forward in the time that it takes to travel 51 ft in the opposite direction. How fast would Thomas be walking on a nonmoving sidewalk?

34. *Moving Sidewalks.* A moving sidewalk moves at a rate of 1.7 ft/sec. Walking on the moving sidewalk, Hunter can travel 120 ft forward in the same time that it takes to travel 52 ft in the opposite direction. How fast would Hunter be walking on a nonmoving sidewalk?

Skill Maintenance

In Exercises 35–38, the graph is that of a function. Determine the domain and the range. [2.3a]

35.

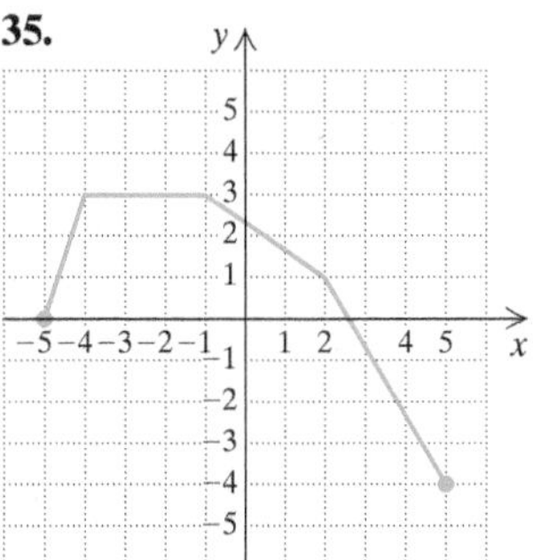

36.

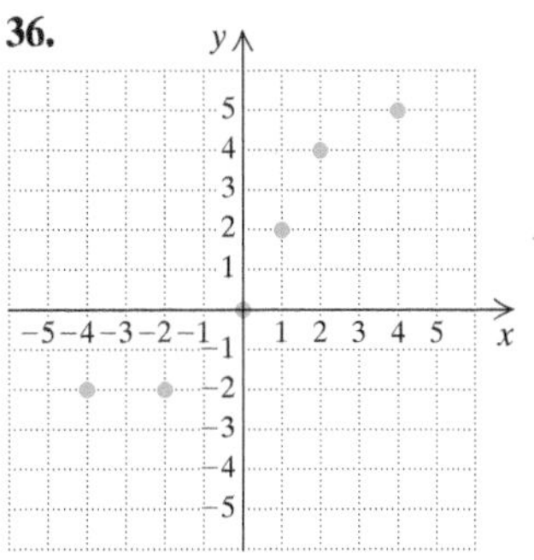

37.

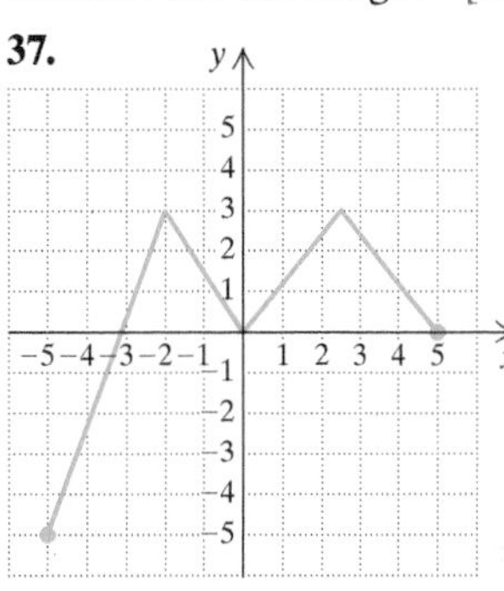

38.

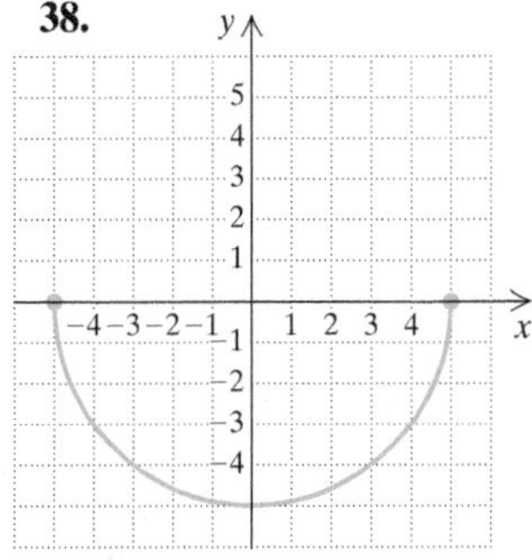

Graph. [3.7b]

39. $x - 4y \geq 4$

40. $x \geq 3$

Graph. [2.2c]

41. $f(x) = |x + 3|$

42. $f(x) = 5 - |x|$

Synthesis

43. Three trucks, A, B, and C, working together, can move a load of mulch in t hours. When working alone, it takes A 1 extra hour to move the mulch, B 6 extra hours, and C t extra hours. Find t.

44. *Escalators.* Together, a 100-cm wide escalator and a 60-cm wide escalator can empty a 1575-person auditorium in 14 min. The wider escalator moves twice as many people as the narrower one does. How many people per hour does the 60-cm wide escalator move?

45. *Gas Mileage.* An automobile gets 22.5 miles per gallon (mpg) in city driving and 30 mpg in highway driving. The car is driven 465 mi on a full tank of 18.4 gal of gasoline. How many miles were driven in the city and how many were driven on the highway?

46. *Travel by Car.* Mackenzie drives to work at 50 mph and arrives 1 min late. She drives to work at 60 mph and arrives 5 min early. How far does Mackenzie live from work?

5.7 Formulas and Applications

OBJECTIVE

a Solve a formula for a letter.

a FORMULAS

SKILL REVIEW *Evaluate formulas and solve a formula for a specified letter.* [1.2a]

Solve for the given letter.

1. $Dx + Ey = z$, for y

2. $C = \frac{1}{2}a(x + y + z)$, for z

Answers: 1. $y = \frac{z - Dx}{E}$ **2.** $z = \frac{2C - ax - ay}{a}$

MyLab Math VIDEO

To solve a rational formula for a given letter, identify the letter, and:

1. Multiply on both sides to clear fractions or decimals, if that is needed.
2. Multiply if necessary to remove parentheses.
3. Get all terms with the letter to be solved for on one side of the equation and all other terms on the other side, using the addition principle.
4. Factor out the letter if it appears in more than one term.
5. Solve for the letter in question, using the multiplication principle.

Formulas occur frequently as mathematical models.

EXAMPLE 1 *Optics.* The formula $f = L/d$ defines a camera's "f-stop," where L is the *focal length* (the distance from the lens to the film) and d is the *aperture* (the diameter of the lens). Solve the formula for d.

We solve this equation as we did the rational equations in Section 5.5:

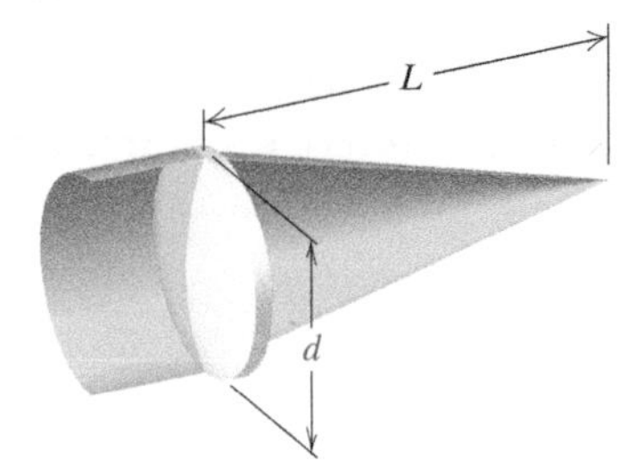

$$f = \frac{L}{d}$$ We want the letter d alone.

$$d \cdot f = d \cdot \frac{L}{d}$$ The LCM is d. We multiply by d.

$$df = L$$ Simplifying

$$d = \frac{L}{f}.$$ Dividing by f

The formula $d = L/f$ can now be used to find the aperture if we know the focal length and the f-stop.

Do Exercise 1.

1. *Combined Gas Law.* The formula

$$\frac{PV}{T} = k$$

relates the pressure P, the volume V, and the temperature T of a gas. Solve the formula for T. (*Hint*: Begin by clearing the fraction.)

EXAMPLE 2 *Astronomy.* The formula $L = \frac{dR}{D - d}$, where D is the diameter of the sun, d is the diameter of the earth, R is the earth's distance from the sun, and L is some fixed distance, is used to calculate when lunar eclipses occur. Solve the formula for D.

$$L = \frac{dR}{D - d}$$ We want the letter D alone.

$$(D - d) \cdot L = (D - d) \cdot \frac{dR}{D - d}$$ The LCM is $D - d$. We multiply by $D - d$.

$$(D - d)L = dR$$ Simplifying

$$DL - dL = dR$$

$$DL = dR + dL$$ Adding dL

$$D = \frac{dR + dL}{L}$$ Dividing by L

Answer

1. $T = \frac{PV}{k}$

Since D appears by itself on one side and not on the other, we have solved for D.

Do Exercise 2.

EXAMPLE 3 Solve the formula $L = \dfrac{dR}{D - d}$ for d.

We proceed as we did in Example 2 until we reach the equation

$DL - dL = dR.$ We want d alone.

We must get all terms containing d alone on one side:

$$DL - dL = dR$$
$$DL = dR + dL \quad \text{Adding } dL$$
$$DL = d(R + L) \quad \text{Factoring out the letter } d$$
$$\frac{DL}{R + L} = d. \quad \text{Dividing by } R + L$$

We now have d alone on one side, so we have solved the formula for d.

Caution!

If, when you are solving an equation for a letter, the letter appears on both sides of the equation, you know that the answer is wrong. The letter must be alone on one side and *not* occur on the other.

Do Exercise 3.

EXAMPLE 4 *Resistance.* The formula

$$\frac{1}{R} = \frac{1}{r_1} + \frac{1}{r_2}$$

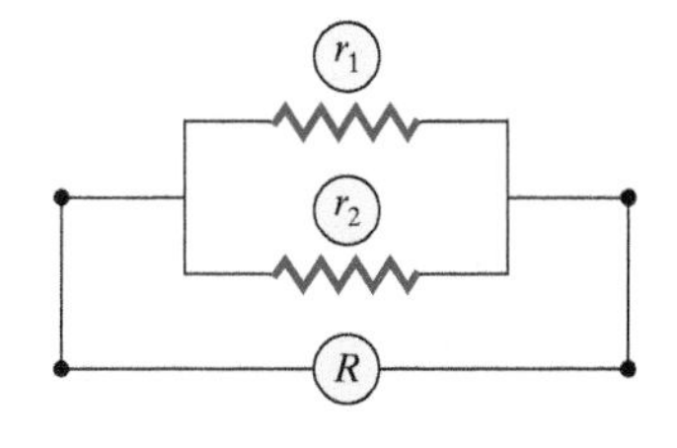

involves the resistance R of two resistors r_1 and r_2 connected in parallel.* Solve the formula for r_1.

We multiply by the LCM, which is Rr_1r_2:

$$Rr_1r_2 \cdot \frac{1}{R} = Rr_1r_2 \cdot \left(\frac{1}{r_1} + \frac{1}{r_2}\right) \quad \text{Multiplying by the LCM}$$
$$Rr_1r_2 \cdot \frac{1}{R} = Rr_1r_2 \cdot \frac{1}{r_1} + Rr_1r_2 \cdot \frac{1}{r_2} \quad \text{Multiplying to remove parentheses}$$
$$r_1r_2 = Rr_2 + Rr_1. \quad \text{Simplifying by removing factors of 1}$$

We might be tempted at this point to multiply by $1/r_2$ to get r_1 alone on the left, *but* note that there is an r_1 on the right. We must get all the terms involving r_1 on the *same side* of the equation.

$$r_1r_2 - Rr_1 = Rr_2 \quad \text{Subtracting } Rr_1$$
$$r_1(r_2 - R) = Rr_2 \quad \text{Factoring out } r_1$$
$$r_1 = \frac{Rr_2}{r_2 - R} \quad \text{Dividing by } r_2 - R \text{ to get } r_1 \text{ alone}$$

Do Exercise 4.

*Note that R, r_1, and r_2 are all different variables. It is common to use subscripts, as in r_1 (read "r sub 1") and r_2, to distinguish variables.

2. Solve the formula

$$I = \frac{pT}{M + pn}$$

for n.

$$I = \frac{pT}{M + pn}$$
$$I(\underline{\quad} + pn) = pT$$
$$IM + \underline{\quad} = pT$$
$$Ipn = pT - \underline{\quad}$$
$$n = \frac{pT - IM}{\underline{\quad}}$$

3. *The Doppler Effect.* The formula

$$F = \frac{sg}{s + v}$$

is used to determine the frequency F of a sound that is moving at velocity v toward a listener who hears the sound as frequency g. Here s is the speed of sound in a particular medium. Solve the formula for s.

4. *Work Formula.* The formula

$$\frac{t}{a} + \frac{t}{b} = 1$$

involves the total time t for some work to be done by two workers whose individual times are a and b. Solve the formula for t.

Answers

2. $n = \dfrac{pT - IM}{Ip}$ **3.** $s = \dfrac{Fv}{g - F}$

4. $t = \dfrac{ab}{b + a}$

Guided Solution:

2. M, Ipn, IM, Ip

5.7 Exercise Set

FOR EXTRA HELP MyLab Math

✓ Check Your Understanding

Reading Check and Concept Check Solve each equation for y. Choose from the column on the right an equivalent equation.

RC1. $\frac{x}{y} = \frac{w}{t}$

RC2. $w = \frac{(x + y)t}{4}$

RC3. $\frac{1}{w} = \frac{1}{x} + \frac{1}{y}$

RC4. $w = \frac{(x + t)y}{4}$

a) $y = \frac{4w - xt}{t}$

b) $y = \frac{xw}{t}$

c) $y = \frac{xt}{w}$

d) $y = \frac{4w}{x + t}$

e) $y = \frac{wx}{x - w}$

f) $y = \frac{xw}{y - x}$

a Solve.

1. $\frac{W_1}{W_2} = \frac{d_1}{d_2}$, for W_2

2. $\frac{W_1}{W_2} = \frac{d_1}{d_2}$, for d_1

3. $\frac{1}{R} = \frac{1}{r_1} + \frac{1}{r_2}$, for r_2
(Electricity formula)

4. $\frac{1}{R} = \frac{1}{r_1} + \frac{1}{r_2}$, for R
(Electricity formula)

5. $s = \frac{(v_1 + v_2)t}{2}$, for t

6. $s = \frac{(v_1 + v_2)t}{2}$, for v_1

7. $R = \frac{gs}{g + s}$, for s

8. $I = \frac{2V}{V + 2r}$, for V

9. $\frac{1}{p} + \frac{1}{q} = \frac{1}{f}$, for p
(An optics formula)

10. $\frac{1}{p} + \frac{1}{q} = \frac{1}{f}$, for f
(Optics formula)

11. $\frac{t}{a} + \frac{t}{b} = 1$, for a
(Work formula)

12. $\frac{t}{a} + \frac{t}{b} = 1$, for b
(Work formula)

13. $I = \frac{nE}{E + nr}$, for E

14. $I = \frac{nE}{E + nr}$, for n

15. $I = \frac{704.5W}{H^2}$, for H^2

16. $S = \frac{H}{m(t_1 - t_2)}$, for t_1

17. $\frac{E}{e} = \frac{R + r}{r}$, for r

18. $\frac{E}{e} = \frac{R + r}{r}$, for e

19. $V = \frac{1}{3}\pi h^2(3R - h)$, for R

20. $A = P(1 + rt)$, for r
(Interest formula)

21. $S = 2\pi rh + 2\pi r^2$, for h
(Surface area of a cylinder)

22. *Escape Velocity.* The formula

$$\frac{V^2}{R^2} = \frac{2g}{R + h}$$

is used to find a satellite's *escape velocity* V, where R is a planet's radius, h is the satellite's height above the planet, and g is the planet's acceleration due to gravity. Solve the formula for h.

23. *Average Speed.* The formula

$$v = \frac{d_2 - d_1}{t_2 - t_1}$$

gives an object's average speed v when that object has traveled d_1 miles in t_1 hours and d_2 miles in t_2 hours. Solve the formula for t_2.

24. *Earned Run Average.* The formula

$$A = 9 \cdot \frac{R}{I}$$

gives a pitcher's *earned run average* A, where R is the number of earned runs, and I is the number of innings pitched. How many earned runs were given up if a pitcher's earned run average is 2.4 after 45 innings? Solve the formula for I.

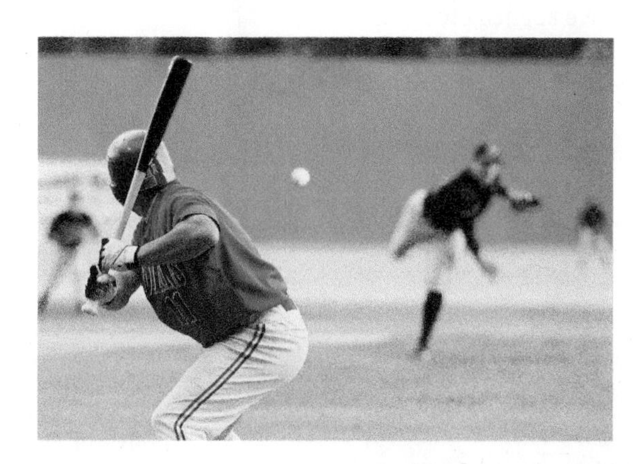

Skill Maintenance

Solve. [3.6a]

25. *Coin Value.* There are 50 dimes in a roll of dimes, 40 nickels in a roll of nickels, and 40 quarters in a roll of quarters. Rob has 12 rolls of coins with a total value of \$70.00. He has 3 more rolls of nickels than dimes. How many rolls of each type does he have?

Given that $f(x) = x^3 - x$, find each of the following. [2.2b]

26. $f(-2)$

27. $f(2)$

28. $f(0)$

29. $f(2a)$

30. Find the slope of the line containing the points $(-2, 5)$ and $(8, -3)$. [2.4b]

31. Find an equation of the line containing the points $(-2, 5)$ and $(8, -3)$. [2.6c]

Synthesis

32. *Escape Velocity.* (Refer to Exercise 22.) A satellite's escape velocity is 6.5 mi/sec, the radius of the earth is 3960 mi, and the acceleration due to gravity is 32.2 ft/sec^2. How far is the satellite from the surface of the earth?

5.8 Variation and Applications

OBJECTIVES

- **a** Find an equation of direct variation given a pair of values of the variables.
- **b** Solve applied problems involving direct variation.
- **c** Find an equation of inverse variation given a pair of values of the variables.
- **d** Solve applied problems involving inverse variation.
- **e** Find equations of other kinds of variation given values of the variables.
- **f** Solve applied problems involving other kinds of variation.

We now extend our study of formulas and functions by considering applications involving variation.

a EQUATIONS OF DIRECT VARIATION

A substitute teacher earns \$65 per day. For 1 day, \$65 is earned; for 2 days, \$130 is earned; for 3 days, \$195 is earned; and so on. We plot this information on a graph, using the number of hours as the first coordinate and the amount earned as the second coordinate to form a set of ordered pairs:

$$(1, 65), \quad (2, 130),$$
$$(3, 195), \quad (4, 260),$$

and so on.

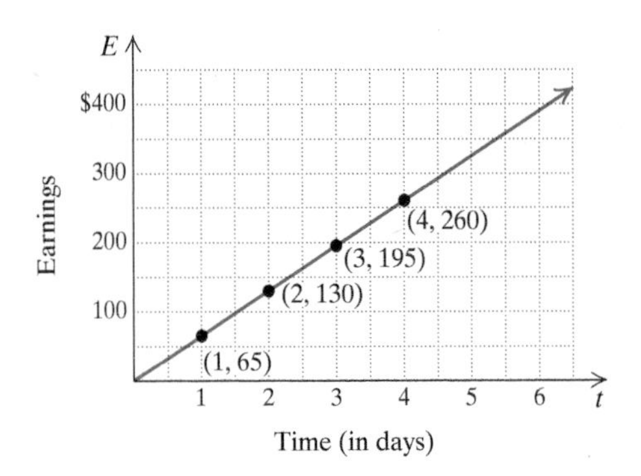

Note that the ratio of the second coordinate to the first coordinate is the same number for each point:

$$\frac{65}{1} = 65, \quad \frac{130}{2} = 65, \quad \frac{195}{3} = 65, \quad \frac{260}{4} = 65, \quad \text{and so on.}$$

Whenever a situation produces pairs of numbers in which the *ratio is constant*, we say that there is **direct variation**. Here the amount earned varies directly as the time:

$$\frac{E}{t} = 65 \text{ (a constant)}, \quad \text{or} \quad E = 65t,$$

or, using function notation, $E(t) = 65t$. The equation is an **equation of direct variation**. The coefficient, 65 in the situation above, is called the **variation constant**. In this case, it is the rate of change of earnings with respect to time.

DIRECT VARIATION

If a situation gives rise to a linear function $f(x) = kx$, or $y = kx$, where k is a positive constant, we say that we have **direct variation**, or that ***y* varies directly as *x***, or that ***y* is directly proportional to *x***. The number k is called the **variation constant**, or the **constant of proportionality**.

EXAMPLE 1 Find the variation constant and an equation of variation in which y varies directly as x, and $y = 32$ when $x = 2$.

We know that (2, 32) is a solution of $y = kx$. Thus,

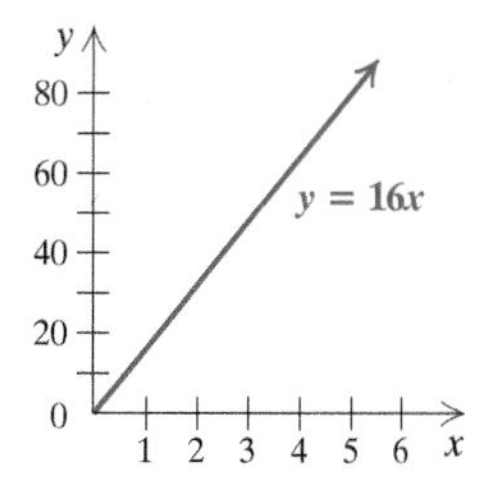

$$y = kx$$

$$32 = k \cdot 2 \quad \text{Substituting}$$

$$\frac{32}{2} = k, \text{ or } k = 16. \quad \text{Solving for } k$$

The variation constant, 16, is the rate of change of y with respect to x. The equation of variation is $y = 16x$.

The graph of $y = kx$, $k > 0$, always goes through the origin and rises from left to right. Note that as x increases, y increases. The constant k is also the slope of the line.

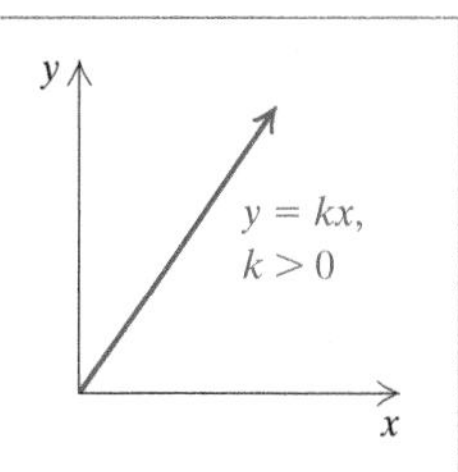

Do Exercises 1 and 2.

b APPLICATIONS OF DIRECT VARIATION

EXAMPLE 2 ***Water from Melting Snow.*** The number of centimeters W of water produced from melting snow varies directly as S, the number of centimeters of snow. Meteorologists have found that, under certain conditions, 150 cm of snow will melt to 16.8 cm of water. To how many centimeters of water will 200 cm of snow melt?

We first find the variation constant using the data and then find an equation of variation:

$$W = kS \quad W \text{ varies directly as } S.$$

$$16.8 = k \cdot 150 \quad \text{Substituting}$$

$$\frac{16.8}{150} = k \quad \text{Solving for } k$$

$$0.112 = k. \quad \text{This is the variation constant.}$$

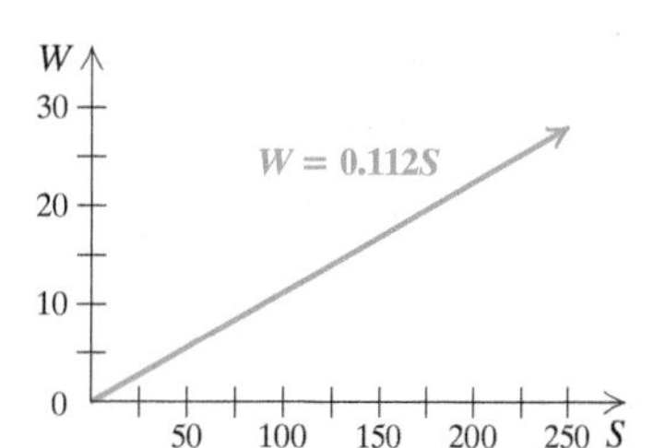

The equation of variation is $W = 0.112S$.

Next, we use the equation to find how many centimeters of water will result from melting 200 cm of snow:

$$W = 0.112S$$

$$= 0.112(200) \quad \text{Substituting}$$

$$= 22.4.$$

Thus, 200 cm of snow will melt to 22.4 cm of water.

Do Exercises 3 and 4. (Exercise 4 is on the following page.)

GS **1.** Find the variation constant and an equation of variation in which y varies directly as x, and $y = 8$ when $x = 20$.

$$y = kx$$

$$8 = k \cdot \square$$

$$\frac{\square}{20} = k$$

$$\frac{2}{\square} = k$$

The variation constant is $\square$.
The equation of variation is $y = \square \cdot x$.

2. Find the variation constant and an equation of variation in which y varies directly as x, and $y = 5.6$ when $x = 8$.

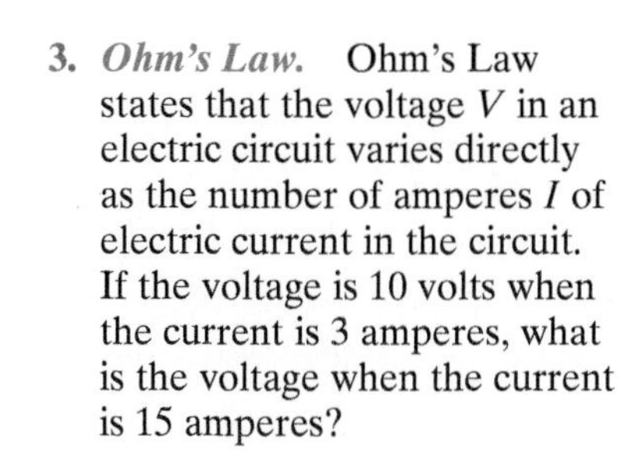

3. ***Ohm's Law.*** Ohm's Law states that the voltage V in an electric circuit varies directly as the number of amperes I of electric current in the circuit. If the voltage is 10 volts when the current is 3 amperes, what is the voltage when the current is 15 amperes?

Answers

1. $\frac{2}{5}$; $y = \frac{2}{5}x$ **2.** 0.7; $y = 0.7x$ **3.** 50 volts

Guided Solution:

1. 20, 8, 5, $\frac{2}{5}$, $\frac{2}{5}$

4. *Bees and Honey.* The amount of honey H produced varies directly as the number of bees who produce the honey. It takes 15,000 bees to produce 25 lb of honey. How much honey is produced by 40,000 bees?

C EQUATIONS OF INVERSE VARIATION

A bus is traveling a distance of 20 mi. At a speed of 5 mph, the trip will take 4 hr; at 20 mph, it will take 1 hr; at 40 mph, it will take $\frac{1}{2}$ hr; and so on. We plot this information on a graph, using speed as the first coordinate and time as the second coordinate to determine a set of ordered pairs:

$(5, 4)$, $(10, 2)$,
$(20, 1)$, $(40, \frac{1}{2})$,

and so on.

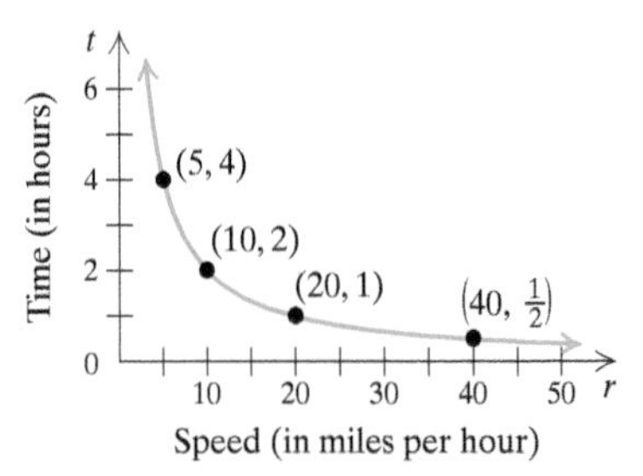

Note that the products of the coordinates are all the same number:

$$5 \cdot 4 = 20, \quad 20 \cdot 1 = 20, \quad 40 \cdot \tfrac{1}{2} = 20, \quad \text{and so on.}$$

Whenever a situation produces pairs of numbers in which the *product is constant*, we say that there is **inverse variation**. Here the time varies inversely as the speed:

$$rt = 20 \text{ (a constant)}, \quad \text{or} \quad t = \frac{20}{r}.$$

The equation is an **equation of inverse variation**. The constant, 20 in the situation above, is called the **variation constant**. Note that as the first number (speed) increases, the second number (time) decreases.

INVERSE VARIATION

If a situation gives rise to a function $f(x) = k/x$, or $y = k/x$, where k is a positive constant, we say that we have **inverse variation**, or that ***y* varies inversely as *x***, or that ***y* is inversely proportional to *x***. The number k is called the **variation constant**, or the **constant of proportionality**.

EXAMPLE 3 Find the variation constant and an equation of variation in which y varies inversely as x, and $y = 32$ when $x = 0.2$.

We know that $(0.2, 32)$ is a solution of $y = k/x$. We substitute:

$$y = \frac{k}{x}$$

$$32 = \frac{k}{0.2} \qquad \text{Substituting}$$

$$(0.2)32 = k \qquad \text{Solving for } k$$

$$6.4 = k.$$

The variation constant is 6.4. The equation of variation is $y = \dfrac{6.4}{x}$.

Answer

4. $66\frac{2}{3}$ lb

It is helpful to look at the graph of $y = k/x$, $k > 0$. The graph is like the one shown at right for positive values of x. Note that as x increases, y decreases.

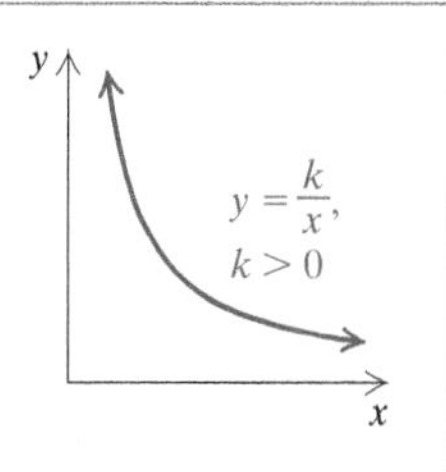

Do Exercise 5.

5. Find the variation constant and an equation of variation in which y varies inversely as x, and $y = 0.012$ when $x = 50$.

$$y = \frac{k}{x}$$

$$0.012 = \frac{k}{\square}$$

$$\square \cdot 50 = k$$

$$\square = k$$

The variation constant is $\square$.
The equation of variation is

$$y = \frac{\square}{x}.$$

d APPLICATIONS OF INVERSE VARIATION

EXAMPLE 4 *Fabric Manufacturing.* Knitted cotton fabric is described in terms of wales per inch (WPI) (for the fabric width) and courses per inch (CPI) (for the fabric length). The CPI, C, is inversely proportional to the stitch length l. For a specific fabric with a stitch length of 0.162 in., the CPI is 35.71. What would the CPI be if the stitch length were increased to 0.171 in.?
Data: Cotton Incorporated, "Guidelines for Engineering Cotton Knits"

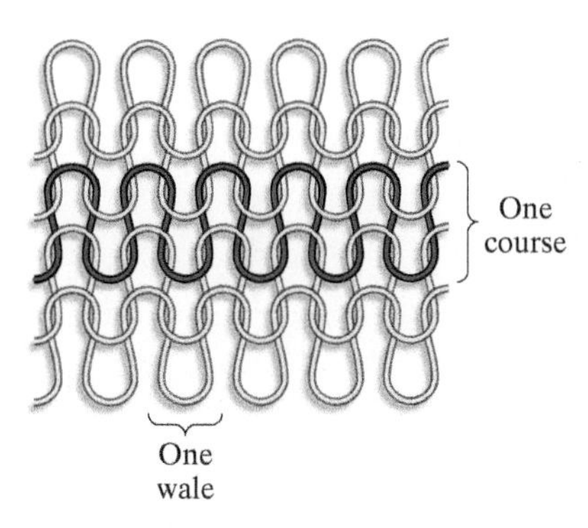

We first find the variation constant using the data given and then find an equation of variation:

$C = \frac{k}{l}$ — C varies inversely as l.

$35.71 = \frac{k}{0.162}$ — Substituting

$5.785 \approx k.$ — Solving for k, the variation constant

The equation of variation is $C = \frac{5.785}{l}$.

Next, we use the equation to find the CPI for a fabric that has a stitch length of 0.171 in.:

$C = \frac{5.785}{l}$ — Equation of variation

$= \frac{5.785}{0.171}$ — Substituting

$\approx 33.83.$

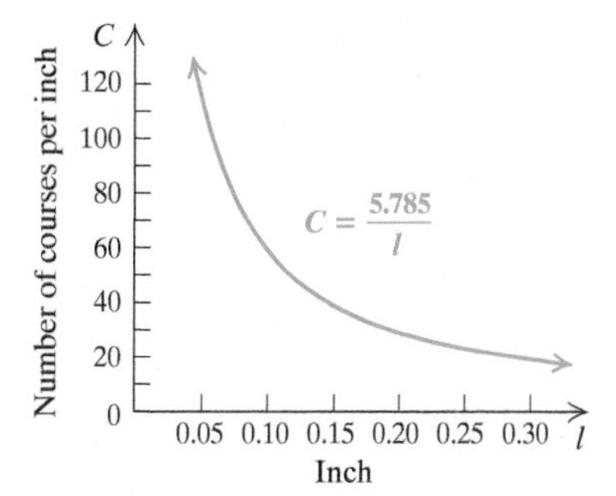

The CPI of a fabric that has a stitch length of 0.171 in. is 33.83.

Do Exercise 6.

6. *Cleaning Bleachers.* The time t to do a job varies inversely as the number of people P who work on the job (assuming that all work at the same rate). It takes 4.5 hr for 12 people to clean a section of bleachers after a NASCAR race. How long would it take 15 people to complete the same job?

e OTHER KINDS OF VARIATION

We now look at other kinds of variation. Consider the equation for the area of a circle, in which A and r are variables and π is a constant:

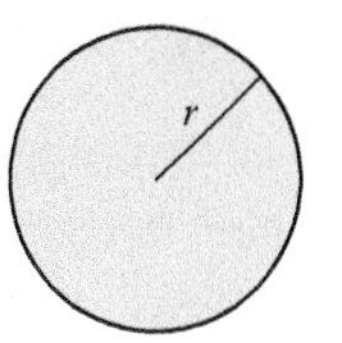

$$A = \pi r^2, \quad \text{or, as a function,} \quad A(r) = \pi r^2.$$

We say that the area *varies directly* as the square of the radius.

Answers

5. 0.6; $y = \frac{0.6}{x}$ **6.** 3.6 hr

Guided Solution:
5. 50, 0.012, 0.6, 0.6, 0.6

y varies directly as the *n*th power of *x* if there is some positive constant *k* such that

$$y = kx^n.$$

EXAMPLE 5 Find an equation of variation in which *y* varies directly as the square of *x*, and $y = 12$ when $x = 2$.

We write an equation of variation and find *k*:

$$y = kx^2$$
$$12 = k \cdot 2^2$$
$$12 = k \cdot 4$$
$$3 = k.$$

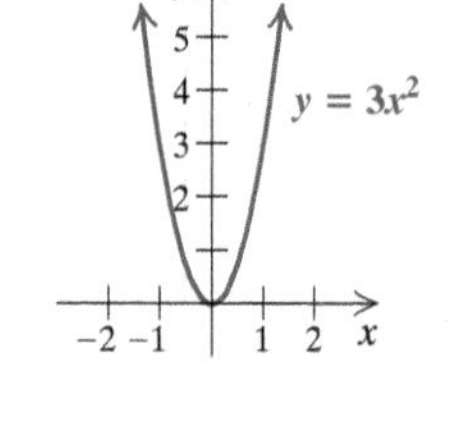

Thus, $y = 3x^2$.

◀ Do Exercise 7.

7. Find an equation of variation in which *y* varies directly as the square of *x*, and $y = 175$ when $x = 5$.

$$y = kx^2$$
$$\square = k \cdot 5^2$$
$$175 = k \cdot 25$$
$$\square = k$$

The equation of variation is $y = \square\, x^2$.

From the law of gravity, we know that the weight *W* of an object *varies inversely* as the square of its distance *d* from the center of the earth:

$$W = \frac{k}{d^2}.$$

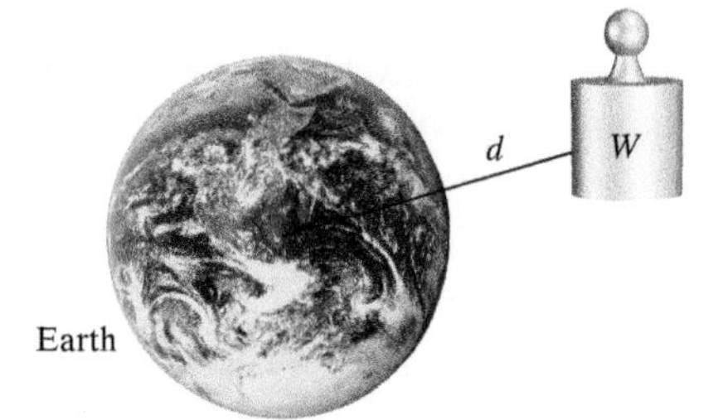

y varies inversely as the *n*th power of *x* if there is some positive constant *k* such that

$$y = \frac{k}{x^n}.$$

EXAMPLE 6 Find an equation of variation in which *W* varies inversely as the square of *d*, and $W = 3$ when $d = 5$.

$$W = \frac{k}{d^2}$$
$$3 = \frac{k}{5^2} \quad \text{Substituting}$$
$$3 = \frac{k}{25}$$
$$75 = k$$

Thus, $W = \frac{75}{d^2}$.

◀ Do Exercise 8.

8. Find an equation of variation in which *y* varies inversely as the square of *x*, and $y = \frac{1}{4}$ when $x = 6$.

Consider the equation for the area *A* of a triangle with height *h* and base *b*: $A = \frac{1}{2}bh$. We say that the area **varies jointly** as the height and the base.

Answers

7. $y = 7x^2$ **8.** $y = \frac{9}{x^2}$

Guided Solution:
7. 175, 7, 7

y varies jointly as x and z if there is some positive constant k such that

$$y = kxz.$$

EXAMPLE 7 Find an equation of variation in which y varies jointly as x and z, and $y = 42$ when $x = 2$ and $z = 3$.

$y = kxz$

$42 = k \cdot 2 \cdot 3$ Substituting

$42 = k \cdot 6$

$7 = k$

Thus, $y = 7xz$.

Do Exercise 9. ▶

9. Find an equation of variation in which y varies jointly as x and z, and $y = 65$ when $x = 10$ and $z = 13$.

$y = kxz$

$65 = k \cdot \underline{\quad} \cdot 13$

$65 = k \cdot 130$

$\frac{65}{\underline{\quad}} = k$

$\underline{\quad} = k$

The equation of variation is $y = \underline{\quad}\, xz$.

Different types of variation can be combined. For example, the equation

$$y = k \cdot \frac{xz^2}{w}$$

asserts that y varies jointly as x and the square of z, and inversely as w.

EXAMPLE 8 Find an equation of variation in which y varies jointly as x and z and inversely as the square of w, and $y = 105$ when $x = 3$, $z = 20$, and $w = 2$.

$y = k \cdot \frac{xz}{w^2}$

$105 = k \cdot \frac{3 \cdot 20}{2^2}$ Substituting

$105 = k \cdot 15$

$7 = k$

Thus, $y = 7 \cdot \frac{xz}{w^2}$.

Do Exercise 10. ▶

10. Find an equation of variation in which y varies jointly as x and the square of z and inversely as w, and $y = 80$ when $x = 4$, $z = 10$, and $w = 25$.

f OTHER APPLICATIONS OF VARIATION

EXAMPLE 9 *Volume of a Tree.* The volume of wood V in a tree varies jointly as the height h and the square of the girth g (girth is distance around). If the volume of a redwood tree is 216 m^3 when the height is 30 m and the girth is 1.5 m, what is the height of a tree whose volume is 960 m^3 and girth is 2 m?

Answers

9. $y = \frac{1}{2}xz$ **10.** $y = \frac{5xz^2}{w}$

Guided Solution:

9. 10, 130, $\frac{1}{2}, \frac{1}{2}$

We first find k using the first set of data. Then we solve for h using the second set of data.

$$V = khg^2$$
$$216 = k \cdot 30 \cdot 1.5^2$$
$$3.2 = k$$

Then the equation of variation is $V = 3.2hg^2$. We substitute the second set of data into the equation:

$$960 = 3.2 \cdot h \cdot 2^2$$
$$75 = h.$$

Thus the height of the tree is 75 m.

EXAMPLE 10 *Intensity of a Light.* The intensity of light I from a light bulb varies inversely as the square of the distance d from the bulb. Suppose that I is 90 W/m^2 (watts per square meter) when the distance is 5 m. How much *further* would it be to a point at which the intensity is 40 W/m^2?

We first find k using the first set of data. Then we solve for I using the second set of data.

$$I = \frac{k}{d^2}$$
$$90 = \frac{k}{5^2}$$
$$2250 = k$$

Then the equation of variation is $I = 2250/d^2$. We substitute the second intensity into the equation and solve for d:

$$I = \frac{2250}{d^2}$$
$$40 = \frac{2250}{d^2}$$
$$40d^2 = 2250$$
$$d^2 = 56.25$$
$$d = 7.5.$$

When the intensity is 40 W/m^2, the distance is 7.5 m. Thus the point at which the intensity is 40 W/m^2 is $7.5 - 5$, or 2.5, m further.

Do Exercises 11 and 12.

11. *Distance of a Dropped Object.* The distance s that an object falls when dropped from some point above the ground varies directly as the square of the time t that it falls. If the object falls 19.6 m in 2 sec, how far will the object fall in 10 sec?

12. *Electrical Resistance.* At a fixed temperature, the resistance R of a wire varies directly as the length l and inversely as the square of its diameter d. If the resistance is 0.1 ohm when the diameter is 1 mm and the length is 50 cm, what is the resistance when the length is 2000 cm and the diameter is 2 mm?

Answers

11. 490 m **12.** 1 ohm

5.8 Exercise Set

FOR EXTRA HELP MyLab Math

✓ Check Your Understanding

Reading Check Choose the word from below each blank that best completes the statement.

RC1. If a situation gives rise to a function $f(x) = k/x$, or $y = k/x$, when k is a positive constant, we say that we have ____________ variation.
direct/inverse

RC2. If a situation gives rise to a linear function $f(x) = kx$, or $y = kx$, where k is a positive constant, we say that we have ____________ variation.
direct/inverse

Concept Check Match each description of variation with the appropriate equation of variation listed on the right.

CC1. a varies directly as z.

CC2. y is inversely proportional to b.

CC3. y varies jointly as z and b and inversely as c.

CC4. x is directly proportional to c.

CC5. b varies directly as z.

CC6. a varies inversely as z.

CC7. c varies inversely as x.

CC8. y varies jointly as c and x.

a) $a = \frac{k}{z}$ **b)** $y = kcx$

c) $b = kz$ **d)** $y = \frac{k}{b}$

e) $x = ky$ **f)** $a = kz$

g) $c = \frac{k}{x}$ **h)** $y = \frac{kzb}{c}$

i) $x = kc$ **j)** $y = \frac{kbc}{z}$

a Find the variation constant and an equation of variation in which y varies directly as x and the following are true.

1. $y = 40$ when $x = 8$

2. $y = 54$ when $x = 12$

3. $y = 4$ when $x = 30$

4. $y = 3$ when $x = 33$

5. $y = 0.9$ when $x = 0.4$

6. $y = 0.8$ when $x = 0.2$

b Solve.

7. ***Shipping by Semi Truck.*** The number of semi trucks T needed to ship metal varies directly as the weight W of the metal. It takes 75 semi trucks to ship 1500 tons of metal. How many trucks are needed for 3500 tons of metal?

Data: scrappy.com

8. ***Shipping by Rail Cars.*** The number of rail cars R needed to ship metal varies directly as the weight W of the metal. It takes approximately 21 rail cars to ship 1500 tons of metal. How many rail cars are needed for 3500 tons of metal?

Data: scrappy.com

9. *Fat Intake.* The maximum number of grams of fat that should be in a diet varies directly as a person's weight. A person weighing 120 lb should have no more than 60 g of fat per day. What is the maximum daily fat intake for a person weighing 180 lb?

10. *Relative Aperture.* The relative aperture, or f-stop, of a 23.5-mm diameter lens is directly proportional to the focal length F of the lens. If a 150-mm focal length has an f-stop of 6.3, find the f-stop of a 23.5-mm diameter lens with a focal length of 80 mm.

11. *Mass of Water in Human Body.* The number of kilograms W of water in a human body varies directly as the mass of the body. A 96-kg person contains 64 kg of water. How many kilograms of water are in a 60-kg person?

12. *Weight on Mars.* The weight M of an object on Mars varies directly as its weight E on Earth. A person who weighs 95 lb on Earth weighs 38 lb on Mars. How much would a 100-lb person weigh on Mars?

13. *Aluminum Usage.* The number of aluminum cans N used each year varies directly as the number of people using them. If 250 people use 60,000 cans in one year, how many cans are used each year in St. Louis, Missouri, which has a population of 318,172?

14. *Hooke's Law.* Hooke's law states that the distance d that a spring is stretched by a hanging object varies directly as the weight w of the object. If a spring is stretched 40 cm by a 3-kg barbell, what is the distance stretched by a 5-kg barbell?

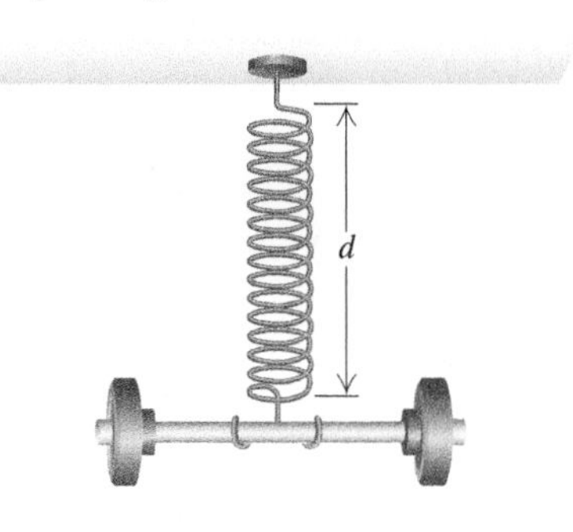

c Find the variation constant and an equation of variation in which y varies inversely as x and the following are true.

15. $y = 14$ when $x = 7$

16. $y = 1$ when $x = 8$

17. $y = 3$ when $x = 12$

18. $y = 12$ when $x = 5$

19. $y = 0.1$ when $x = 0.5$

20. $y = 1.8$ when $x = 0.3$

d Solve.

21. *Work Rate.* The time T required to do a job varies inversely as the number of people P working. It takes 5 hr for 7 bricklayers to build a park wall. How long will it take 10 bricklayers to complete the job?

22. *Pumping Rate.* The time t required to empty a tank varies inversely as the rate r of pumping. If a pump can empty a tank in 45 min at the rate of 600 kL/min, how long will it take the pump to empty the same tank at the rate of 1000 kL/min?

23. *Current and Resistance.* The current I in an electrical conductor varies inversely as the resistance R of the conductor. If the current is $\frac{1}{2}$ ampere when the resistance is 240 ohms, what is the current when the resistance is 540 ohms?

24. *Wavelength and Frequency.* The wavelength W of a radio wave varies inversely as its frequency F. A wave with a frequency of 1200 kilohertz has a length of 300 meters. What is the length of a wave with a frequency of 800 kilohertz?

25. *Beam Weight.* The weight W that a horizontal beam can support varies inversely as the length L of the beam. Suppose that a 12-ft beam can support 1200 lb. How many pounds can a 15-ft beam support?

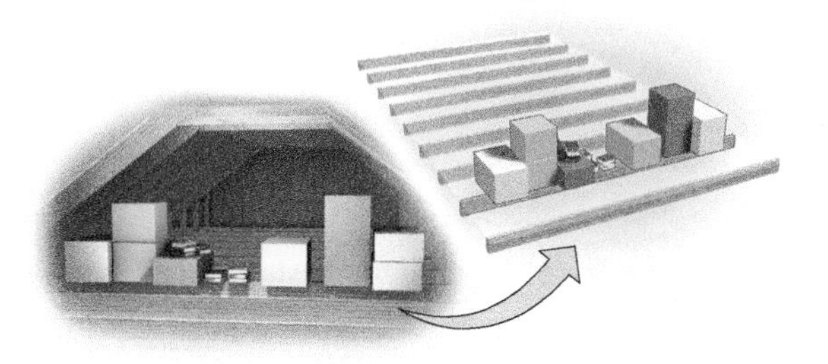

26. *Musical Pitch.* The pitch P of a musical tone varies inversely as its wavelength W. One tone has a pitch of 440 vibrations per second and a wavelength of 2.4 ft. Find the wavelength of another tone that has a pitch of 275 vibrations per second.

27. *Rate of Travel.* The time t required to drive a fixed distance varies inversely as the speed r. It takes 5 hr at a speed of 80 km/h to drive a fixed distance. How long will it take to drive the same distance at a speed of 70 km/h?

28. *Volume and Pressure.* The volume V of a gas varies inversely as the pressure P upon it. The volume of a gas is 200 cm^3 under a pressure of 32 kg/cm^2. What will its volume be under a pressure of 40 kg/cm^2?

e Find an equation of variation in which the following are true.

29. y varies directly as the square of x, and $y = 0.15$ when $x = 0.1$

30. y varies directly as the square of x, and $y = 6$ when $x = 3$

31. y varies inversely as the square of x, and $y = 0.15$ when $x = 0.1$

32. y varies inversely as the square of x, and $y = 6$ when $x = 3$

33. y varies jointly as x and z, and $y = 56$ when $x = 7$ and $z = 8$

34. y varies directly as x and inversely as z, and $y = 4$ when $x = 12$ and $z = 15$

35. y varies jointly as x and the square of z, and $y = 105$ when $x = 14$ and $z = 5$

36. y varies jointly as x and z and inversely as w, and $y = \frac{3}{2}$ when $x = 2$, $z = 3$, and $w = 4$

37. y varies jointly as x and z and inversely as the product of w and p, and $y = \frac{3}{28}$ when $x = 3$, $z = 10$, $w = 7$, and $p = 8$

38. y varies jointly as x and z and inversely as the square of w, and $y = \frac{12}{5}$ when $x = 16$, $z = 3$, and $w = 5$

f Solve.

39. *Stopping Distance of a Car.* The stopping distance d of a car after the brakes have been applied varies directly as the square of the speed r. If a car traveling 60 mph can stop in 200 ft, how fast can a car travel and still stop in 72 ft?

40. *TV Signal.* The intensity I of a TV signal varies inversely as the square of the distance d from the transmitter. If the intensity is 23 W/m^2 at a distance of 2 km, what is the intensity at a distance of 6 km?

41. *Weight of a Sphere.* The weight W of a sphere of a given material varies directly as its volume V, and its volume V varies directly as the cube of its diameter.

a) Find an equation of variation relating the weight W to the diameter d.

b) An iron ball that is 5 in. in diameter is known to weigh 25 lb. Find the weight of an iron ball that is 8 in. in diameter.

42. *Combined Gas Law.* The volume V of a given mass of a gas varies directly as the temperature T and inversely as the pressure P. If $V = 231\text{ cm}^3$ when $T = 42°$ and $P = 20\text{ kg/cm}^2$, what is the volume when $T = 30°$ and $P = 15\text{ kg/cm}^2$?

43. *Earned-Run Average.* A pitcher's earned-run average E varies directly as the number R of earned runs allowed and inversely as the number I of innings pitched. In 2016, Clayton Kershaw of the Los Angeles Dodgers had an earned-run average of 1.69. He gave up 28 earned runs in 149.0 innings. How many earned runs would he have given up had he pitched 210 innings with the same average?

Data: Major League Baseball

44. *Atmospheric Drag.* Wind resistance, or atmospheric drag, tends to slow down moving objects. Atmospheric drag varies jointly as an object's surface area A and velocity v. If a car traveling at a speed of 40 mph with a surface area of 37.8 ft^2 experiences a drag of 222 N (Newtons), how fast must a car with 51 ft^2 of surface area travel in order to experience a drag force of 430 N?

45. *Water Flow.* The amount Q of water emptied by a pipe varies directly as the square of the diameter d. A pipe 5 in. in diameter will empty 225 gal of water over a fixed time period. If we assume the same kind of flow, how many gallons of water are emptied in the same amount of time by a pipe that is 9 in. in diameter?

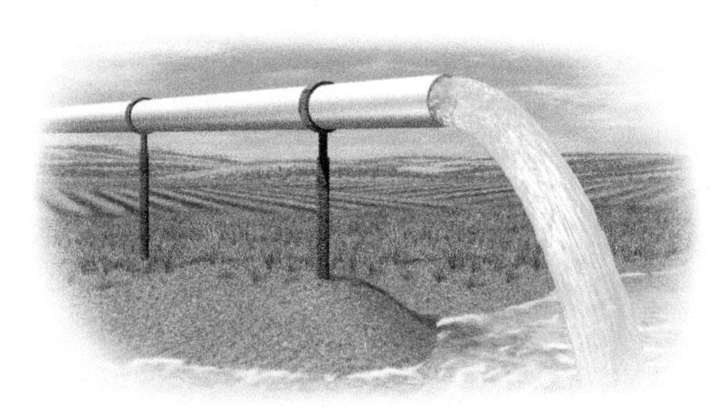

46. *Wind Power.* Wind power P from a turbine varies directly as the square of the length r of one of its blades. Two common blade lengths for commercial wind turbines are 35 m and 50 m. When the blade length is 35 m, about 1.5 MW (megawatts) of power is produced under favorable conditions. How much power would be produced, under favorable conditions, by a turbine with 50-m blades?

Data: aweo.org

Skill Maintenance

Write interval notation for the given set.

47. $\{y \mid y \geq -8\}$ [1.4b]

48. $\{x \mid x \text{ is a real number}\}$ [1.4b]

49. $\{t \mid -5 \leq t < 15\}$ [1.4b]

50. $\{a \mid a < -4 \text{ or } a \geq 0\}$ [1.5b]

51. $\{q \mid q < -\frac{1}{2}\}$ [1.4b]

52. $\{x \mid x > 2 \text{ or } x \leq -\frac{5}{8}\}$ [1.5b]

Solve.

53. $3x - y = 7,$
$y = 1 - x$ [3.2a]

54. $6x - 2y = 12,$
$3x + 2y = 24$ [3.3a]

Synthesis

55. In each of the following equations, state whether y varies directly as x, inversely as x, or neither directly nor inversely as x.

a) $7xy = 14$

b) $x - 2y = 12$

c) $-2x + 3y = 0$

d) $x = \frac{3}{4}y$

56. *Area of a Circle.* The area of a circle varies directly as the square of the length of a diameter. What is the variation constant?

CHAPTER 5

Summary and Review

Vocabulary Reinforcement

Complete each statement with the correct term from the column on the right. Some of the choices may be used more than once, and some may not be used.

positive
negative
proportional
proportion
rational
complex
constant
inverse
direct
joint

1. An equality of ratios, $A/B = C/D$, is called a(n) ____________. [5.6b]
2. An expression that consists of the quotient of two polynomials, where the polynomial in the denominator is nonzero, is called a(n) ____________ expression. [5.1a]
3. A(n) ____________ equation is an equation containing one or more rational expressions. [5.5a]
4. If a situation gives rise to a function $f(x) = k/x$, or $y = k/x$, where k is a(n) ____________ constant, we say that we have ____________ variation. [5.8c]
5. A(n) ____________ rational expression is a rational expression that contains rational expressions within its numerator and/or its denominator. [5.4a]
6. If a situation gives rise to a linear function $f(x) = kx$, or $y = kx$, where k is a(n) ____________ constant, we say that we have ____________ variation. The number k is called the variation ____________. [5.8a]

Concept Reinforcement

Determine whether each statement is true or false.

____________ 1. If y is inversely proportional to x, then the rational function $f(x) = k/x$ can model the situation. [5.8c]

____________ 2. Clearing fractions is a valid procedure only when solving equations, not when adding, subtracting, multiplying, or dividing rational expressions. [5.5a]

Study Guide

Objective 5.1a Find all numbers for which a rational expression is not defined or that are not in the domain of a rational function, and state the domain of the function.

Example Find the domain of $f(x) = \dfrac{x^2 - 12x + 27}{x^2 + 6x - 16}$.

The rational expression is not defined for a replacement that makes the denominator 0. We set the denominator equal to 0 and solve for x.

$$x^2 + 6x - 16 = 0$$
$$(x + 8)(x - 2) = 0$$
$$x + 8 = 0 \quad or \quad x - 2 = 0$$
$$x = -8 \quad or \quad x = 2$$

The expression is not defined for replacements -8 and 2. Thus the domain is

$\{x | x \text{ is a real number } and\ x \neq -8\ and\ x \neq 2\}$, or $(-\infty, -8) \cup (-8, 2) \cup (2, \infty)$.

Practice Exercise

1. Find the domain of

$$f(x) = \frac{x^2 + 3x - 28}{x^2 + 3x - 54}.$$

Objective 5.1c Simplify rational expressions.

Example Simplify: $\frac{a^2 - 1}{a^2 + 7a - 8}$.

$$\frac{a^2 - 1}{a^2 + 7a - 8} = \frac{(a + 1)(a - 1)}{(a + 8)(a - 1)} = \frac{a + 1}{a + 8} \cdot \frac{a - 1}{a - 1} = \frac{a + 1}{a + 8}$$

Practice Exercise

2. Simplify:

$$\frac{b^2 - 9}{b^2 - 5b - 24}.$$

Objective 5.1e Divide rational expressions and simplify.

Example Divide and simplify: $\frac{t^2 + 2t + 4}{3t^2 + 6t} \div \frac{t^3 - 8}{t^3 + 2t^2}$.

$$\frac{t^2 + 2t + 4}{3t^2 + 6t} \div \frac{t^3 - 8}{t^3 + 2t^2} = \frac{t^2 + 2t + 4}{3t^2 + 6t} \cdot \frac{t^3 + 2t^2}{t^3 - 8}$$

$$= \frac{(t^2 + 2t + 4)(t)(t)(t + 2)}{3t(t + 2)(t - 2)(t^2 + 2t + 4)}$$

$$= \frac{t}{3(t - 2)}$$

Practice Exercise

3. Divide and simplify:

$$\frac{w^3 - 125}{w^3 + 8w^2 + 15w} \div \frac{w - 5}{w^3 - 25w}.$$

Objective 5.2a Find the LCM of several algebraic expressions by factoring.

Example Find the LCM of x^2, $16x^2 - 25$, and $4x^3 - 15x^2 - 25x$.

We factor each expression completely:

$x^2 = x \cdot x;$

$16x^2 - 25 = (4x + 5)(4x - 5);$

$4x^3 - 15x^2 - 25x = x(4x + 5)(x - 5).$

$$\text{LCM} = x \cdot x \cdot (4x + 5)(4x - 5)(x - 5)$$
$$= x^2(4x + 5)(4x - 5)(x - 5)$$

Practice Exercise

4. Find the LCM of x^4, $x^5 - 9x^3$, and $2x^2 + 11x + 15$.

Objective 5.2b Add and subtract rational expressions.

Example Subtract:

$$\frac{x - y}{x^2 + 3xy + 2y^2} - \frac{3y}{x^2 + 6xy + 5y^2}.$$

First, we factor the denominators.

$$\frac{x - y}{(x + 2y)(x + y)} - \frac{3y}{(x + 5y)(x + y)}$$

The LCM is $(x + 2y)(x + y)(x + 5y)$.

$$= \frac{x - y}{(x + 2y)(x + y)} \cdot \frac{x + 5y}{x + 5y} - \frac{3y}{(x + 5y)(x + y)} \cdot \frac{x + 2y}{x + 2y}$$

$$= \frac{(x - y)(x + 5y)}{(x + 2y)(x + y)(x + 5y)} - \frac{3y(x + 2y)}{(x + 5y)(x + y)(x + 2y)}$$

$$= \frac{(x^2 + 4xy - 5y^2) - (3xy + 6y^2)}{(x + 2y)(x + y)(x + 5y)}$$

$$= \frac{x^2 + 4xy - 5y^2 - 3xy - 6y^2}{(x + 2y)(x + y)(x + 5y)} = \frac{x^2 + xy - 11y^2}{(x + 2y)(x + y)(x + 5y)}$$

Practice Exercise

5. Subtract:

$$\frac{r + s}{r^2 + rs - 2s^2} - \frac{5s}{r^2 - s^2}.$$

Objective 5.3b Divide a polynomial by a divisor that is not a monomial, and if there is a remainder, express the result in two ways.

Example Divide: $(y^2 - 2y + 13) \div (y + 2)$.

$$\begin{array}{r} y - 4 \\ y + 2\overline{)y^2 - 2y + 13} \\ \underline{y^2 + 2y} \qquad \\ -4y + 13 \\ \underline{-4y - 8} \\ 21 \end{array}$$

The answer is $y - 4$, R 21, or $y - 4 + \dfrac{21}{y + 2}$.

Practice Exercise

6. Divide:

$$(y^2 - 5y + 9) \div (y - 1).$$

Objective 5.3c Use synthetic division to divide a polynomial by a binomial of the type $x - a$.

Example Use synthetic division to divide:

$$(x^3 - 2x^2 - 6) \div (x + 2).$$

There is no x-term, so we write 0 for its coefficient. Note that $x + 2 = x - (-2)$, so we write -2 on the left.

$$\begin{array}{r|rrrr} -2 & 1 & -2 & 0 & -6 \\ & & -2 & 8 & -16 \\ \hline & 1 & -4 & 8 & -22 \end{array}$$

The answer is $x^2 - 4x + 8$, R -22, or $x^2 - 4x + 8 + \dfrac{-22}{x + 2}$.

Practice Exercise

7. Use synthetic division to divide:

$$(x^3 - 5x^2 - 1) \div (x + 3).$$

Objective 5.4a Simplify complex rational expressions.

Example Simplify: $\dfrac{\frac{2}{x} - \frac{5}{y}}{\frac{5}{x} + \frac{2}{y}}$.

The LCM of all denominators is xy.

$$\frac{\frac{2}{x} - \frac{5}{y}}{\frac{5}{x} + \frac{2}{y}} = \frac{\frac{2}{x} - \frac{5}{y}}{\frac{5}{x} + \frac{2}{y}} \cdot \frac{xy}{xy} = \frac{\frac{2}{x} \cdot xy - \frac{5}{y} \cdot xy}{\frac{5}{x} \cdot xy + \frac{2}{y} \cdot xy} = \frac{2y - 5x}{5y + 2x}$$

Practice Exercise

8. Simplify:

$$\frac{\frac{2}{a} + \frac{8}{b}}{\frac{8}{a} - \frac{2}{b}}.$$

Objective 5.5a Solve rational equations.

Example Solve:

$$\frac{12}{x^2 - 6x - 7} - \frac{3}{x - 7} = \frac{1}{x + 1}.$$

The LCM of the denominators is $(x - 7)(x + 1)$. We multiply all terms on both sides by $(x - 7)(x + 1)$.

$$(x - 7)(x + 1)\left(\frac{12}{x^2 - 6x - 7} - \frac{3}{x - 7}\right) = (x - 7)(x + 1) \cdot \frac{1}{x + 1}$$

$$12 - 3(x + 1) = x - 7$$

$$12 - 3x - 3 = x - 7$$

$$9 - 3x = x - 7$$

$$-4x = -16$$

$$x = 4$$

We must always check possible solutions. The number 4 checks in the original equation. The solution is 4.

Practice Exercise

9. Solve:

$$\frac{5}{x - 4} - \frac{3}{x + 5} = \frac{4}{x^2 + x - 20}.$$

Objective 5.8a Find an equation of direct variation given a pair of values of the variables.

Example Find the variation constant and an equation of variation in which y varies directly as x, and $y = 44$ when $x = \frac{11}{5}$.

$$y = kx \quad \text{Direct variation}$$
$$44 = k \cdot \frac{11}{5} \quad \text{Substituting}$$
$$\frac{5}{11} \cdot 44 = k$$
$$20 = k$$

The variation constant is 20. The equation of variation is $y = 20x$.

Practice Exercise

10. Find the variation constant and an equation of variation in which y varies directly as x, and $y = 62$ when $x = \frac{2}{3}$.

Objective 5.8c Find an equation of inverse variation given a pair of values of the variables.

Example Find the variation constant and an equation of variation in which y varies inversely as x, and $y = \frac{5}{18}$ when $x = 2$.

$$y = \frac{k}{x} \quad \text{Inverse variation}$$
$$\frac{5}{18} = \frac{k}{2} \quad \text{Substituting}$$
$$2 \cdot \frac{5}{18} = k$$
$$\frac{5}{9} = k$$

The variation constant is $\frac{5}{9}$. The equation of variation is $y = \frac{\frac{5}{9}}{x}$, or $y = \frac{5}{9x}$.

Practice Exercise

11. Find the variation constant and an equation of variation in which y varies inversely as x, and $y = \frac{3}{10}$ when $x = 15$.

Review Exercises

1. Find all numbers for which the rational expression
$$\frac{x^2 - 3x + 2}{x^2 - 9}$$
is not defined. [5.1a]

2. Find the domain of f where
$$f(x) = \frac{x^2 - 3x + 2}{x^2 - 9}. \quad [5.1a]$$

Simplify. [5.1c]

3. $\dfrac{4x^2 - 7x - 2}{12x^2 + 11x + 2}$

4. $\dfrac{a^2 + 2a + 4}{a^3 - 8}$

Find the LCM. [5.2a]

5. $6x^3, 16x^2$

6. $x^2 - 49, 3x + 1$

7. $x^2 + x - 20, x^2 + 3x - 10$

Perform the indicated operations and simplify. [5.1d, e], [5.2b, c]

8. $\dfrac{y^2 - 64}{2y + 10} \cdot \dfrac{y + 5}{y + 8}$

9. $\dfrac{x^3 - 8}{x^2 - 25} \cdot \dfrac{x^2 + 10x + 25}{x^2 + 2x + 4}$

10. $\dfrac{9a^2 - 1}{a^2 - 9} \div \dfrac{3a + 1}{a + 3}$

11. $\dfrac{x^3 - 64}{x^2 - 16} \div \dfrac{x^2 + 5x + 6}{x^2 - 3x - 18}$

12. $\dfrac{x}{x^2 + 5x + 6} - \dfrac{2}{x^2 + 3x + 2}$

13. $\dfrac{2x^2}{x-y} + \dfrac{2y^2}{x+y}$

14. $\dfrac{3}{y+4} - \dfrac{y}{y-1} + \dfrac{y^2+3}{y^2+3y-4}$

Divide.

15. $(16ab^3c - 10ab^2c^2 + 12a^2b^2c) \div (4ab)$ [5.3a]

16. $(y^2 - 20y + 64) \div (y - 6)$ [5.3b]

17. $(6x^4 + 3x^2 + 5x + 4) \div (x^2 + 2)$ [5.3b]

Divide using synthetic division. Show your work. [5.3c]

18. $(x^3 + 5x^2 + 4x - 7) \div (x - 4)$

19. $(3x^4 - 5x^3 + 2x - 7) \div (x + 1)$

Simplify. [5.4a]

20. $\dfrac{3 + \dfrac{3}{y}}{4 + \dfrac{4}{y}}$

21. $\dfrac{\dfrac{2}{a} + \dfrac{2}{b}}{\dfrac{4}{a^3} + \dfrac{4}{b^3}}$

22. $\dfrac{\dfrac{x^2-5x-36}{x^2-36}}{\dfrac{x^2+x-12}{x^2-12x+36}}$

23. $\dfrac{\dfrac{4}{x+3} - \dfrac{2}{x^2-3x+2}}{\dfrac{3}{x-2} + \dfrac{1}{x^2+2x-3}}$

Solve. [5.5a]

24. $\dfrac{x}{4} + \dfrac{x}{7} = 1$

25. $\dfrac{5}{3x+2} = \dfrac{3}{2x}$

26. $\dfrac{4x}{x+1} + \dfrac{4}{x} + 9 = \dfrac{4}{x^2+x}$

27. $\dfrac{90}{x^2-3x+9} - \dfrac{5x}{x+3} = \dfrac{405}{x^3+27}$

28. $\dfrac{2}{x-3} + \dfrac{1}{4x+20} = \dfrac{1}{x^2+2x-15}$

29. Given that

$$f(x) = \frac{6}{x} + \frac{4}{x},$$

find all x for which $f(x) = 5$.

30. *House Painting.* David can paint the outside of a house in 12 hr. Bill can paint the same house in 9 hr. How long would it take them working together to paint the house? [5.6a]

31. *Boat Travel.* The current of the Gold River is 6 mph. A boat travels 50 mi downstream in the same time that it takes to travel 30 mi upstream. Complete the following table and then find the speed of the boat in still water. [5.6c]

	DISTANCE	SPEED	TIME
DOWNSTREAM			
UPSTREAM			

32. *Travel Distance.* Fred operates a potato-chip delivery route. He drives 800 mi in 3 days. How far will he travel in 15 work days? [5.6b]

Solve for the indicated letter. [5.7a]

33. $W = \frac{cd}{c + d}$, for d; for c

34. $S = \frac{p}{a} + \frac{t}{b}$, for b; for t

35. Find an equation of variation in which y varies directly as x, and $y = 100$ when $x = 25$. [5.8a]

36. Find an equation of variation in which y varies inversely as x, and $y = 100$ when $x = 25$. [5.8c]

37. *Pumping Time.* The time t required to empty a tank varies inversely as the rate r of pumping. If a pump can empty a tank in 35 min at the rate of 800 kL per minute, how long will it take the pump to empty the same tank at the rate of 1400 kL per minute? [5.8d]

38. *Test Score.* The score N on a test varies directly as the number of correct responses a. Ellen answers 29 questions correctly and earns a score of 87. What would Ellen's score have been if she had answered 25 questions correctly? [5.8b]

39. *Power of Electric Current.* The power P expended by heat in an electric circuit of fixed resistance varies directly as the square of the current C in the circuit. A circuit expends 180 watts when a current of 6 amperes is flowing. What is the amount of heat expended when the current is 10 amperes? [5.8f]

40. Find the domain of $f(x) = \frac{x^2 - x}{x^2 - 2x - 35}$. [5.1a]

A. $(0, 1)$
B. $(-\infty, -5) \cup (-5, 7) \cup (7, \infty)$
C. $(-5, 7)$
D. $(-\infty, 0) \cup (0, 1) \cup (1, \infty)$

41. Find the LCM of x^5, $x - 4$, $x^2 - 4$, and $x^2 - 4x$. [5.2a]

A. $x(x - 4)^2$
B. $(x - 4)(x + 4)$
C. $x^5(x - 4)(x - 2)(x + 2)$
D. $x^5(x - 4)^2$

Synthesis

42. Find the reciprocal and simplify: $\frac{a - b}{a^3 - b^3}$. [5.1c, e]

43. Solve: $\frac{5}{x - 13} - \frac{5}{x} = \frac{65}{x^2 - 13x}$. [5.5a]

Understanding Through Discussion and Writing

1. Discuss at least three different uses of the LCM studied in this chapter. [5.2b], [5.4a], [5.5a]

2. You have learned to solve a new kind of equation in this chapter. Explain how this type differs from those you have studied previously and how the equation-solving process differs. [5.5a]

3. Explain why it is sufficient, when checking a possible solution of a rational equation, to verify that the number in question does not make a denominator 0. [5.5a]

4. If y varies directly as x and x varies inversely as z, how does y vary with regard to z? Why? [5.8a, c, e]

5. Explain how you might easily create rational equations for which only one of the two possible solutions is a solution. (See Example 4 of Section 5.5 for a hint.) [5.5a]

6. Which is easier to solve for x? Explain why. [5.7a]

$$\frac{1}{38} + \frac{1}{47} = \frac{1}{x} \quad \text{or} \quad \frac{1}{a} + \frac{1}{b} = \frac{1}{x}$$

CHAPTER 5 Test

For Extra Help — For step-by-step test solutions, access the Chapter Test Prep Videos in MyLab Math.

1. Find all numbers for which the rational expression
$$\frac{x^2 - 16}{x^2 - 3x + 2}$$
is not defined.

2. Find the domain of f where
$$f(x) = \frac{x^2 - 16}{x^2 - 3x + 2}.$$

Simplify.

3. $\dfrac{12x^2 + 11x + 2}{4x^2 - 7x - 2}$

4. $\dfrac{p^3 + 1}{p^2 - p - 2}$

5. Find the LCM of $x^2 + x - 6$ and $x^2 + 8x + 15$.

Perform the indicated operations and simplify.

6. $\dfrac{2x^2 + 20x + 50}{x^2 - 4} \cdot \dfrac{x + 2}{x + 5}$

7. $\dfrac{x}{x^2 + 11x + 30} - \dfrac{5}{x^2 + 9x + 20}$

8. $\dfrac{y^2 - 16}{2y + 6} \div \dfrac{y - 4}{y + 3}$

9. $\dfrac{x^2}{x - y} + \dfrac{y^2}{y - x}$

10. $\dfrac{1}{x + 1} - \dfrac{x + 2}{x^2 - 1} + \dfrac{3}{x - 1}$

11. $\dfrac{a}{a - b} + \dfrac{b}{a^2 + ab + b^2} - \dfrac{2}{a^3 - b^3}$

Divide.

12. $(20r^2s^3 + 15r^2s^2 - 10r^3s^3) \div (5r^2s)$

13. $(y^3 + 125) \div (y + 5)$

14. $(4x^4 + 3x^3 - 5x - 2) \div (x^2 + 1)$

Divide using synthetic division. Show your work.

15. $(x^3 + 3x^2 + 2x - 6) \div (x - 3)$

16. $(3x^3 + 22x^2 - 160) \div (x + 4)$

Simplify.

17. $\dfrac{1 - \dfrac{1}{x^2}}{1 - \dfrac{1}{x}}$

18. $\dfrac{\dfrac{1}{a^3} + \dfrac{1}{b^3}}{\dfrac{1}{a} + \dfrac{1}{b}}$

19. Given that
$$f(x) = \frac{2}{x - 1} + \frac{2}{x + 2},$$
find all x for which $f(x) = 1$.

Solve.

20. $\frac{2}{x - 1} = \frac{3}{x + 3}$

21. $\frac{7x}{x + 3} + \frac{21}{x - 3} = \frac{126}{x^2 - 9}$

22. $\frac{2x}{x + 7} = \frac{5}{x + 1}$

23. $\frac{1}{3x - 6} - \frac{1}{x^2 - 4} = \frac{3}{x + 2}$

24. *Completing a Puzzle.* Working together, Rachel and Jessie can complete a jigsaw puzzle in 1.5 hr. Rachel takes 4 hr longer than Jessie does when working alone. How long would it take Jessie to complete the puzzle?

25. *Bicycle Travel.* David can bicycle at a rate of 12 mph when there is no wind. Against the wind, David bikes 8 mi in the same time that it takes to bike 14 mi with the wind. What is the speed of the wind?

26. *Predicting Paint Needs.* Logan and Noah run a summer painting company to defray their college expenses. They need 4 gal of paint to paint 1700 ft^2 of clapboard. How much paint would they need for a building with 6000 ft^2 of clapboard?

Solve for the indicated letter.

27. $T = \frac{ab}{a - b}$, for a; for b

28. $Q = \frac{2}{a} - \frac{t}{b}$, for a

29. Find an equation of variation in which Q varies jointly as x and y, and $Q = 25$ when $x = 2$ and $y = 5$.

30. Find an equation of variation in which y varies inversely as x, and $y = 10$ when $x = 25$.

31. *Income vs. Time.* Kaylee's income I varies directly as the time t worked. She gets a job that pays \$550 for 40 hr of work. What is she paid for working 72 hr, assuming that there is no change in pay scale for overtime?

32. *Time and Speed.* The time t required to drive a fixed distance varies inversely as the speed r. It takes 5 hr at 60 km/h to drive a fixed distance. How long would it take to drive that same distance at 40 km/h?

33. *Area of a Balloon.* The surface area of a balloon varies directly as the square of its radius. The area is 314 cm^2 when the radius is 5 cm. What is the area when the radius is 7 cm?

34. Find the LCM of $6x^2$, $3x^2 - 3y^2$, and $x^2 - 2xy - 3y^2$.

A. $3x^2(2x + y)(x - 3y)$
B. $6x(x + y)(x - y)(x - 3y)$
C. $3x^2(x + y)^2(x - y)$
D. $6x^2(x + y)(x - y)(x - 3y)$

Synthesis

35. Solve: $\frac{6}{x - 15} - \frac{6}{x} = \frac{90}{x^2 - 15x}$.

36. Find the x- and y-intercepts of the function f given by

$$f(x) = \frac{\frac{5}{x + 4} - \frac{3}{x - 2}}{\frac{2}{x - 3} + \frac{1}{x + 4}}.$$

1–5 Cumulative Review

Graph.

1. $y = -5x + 4$

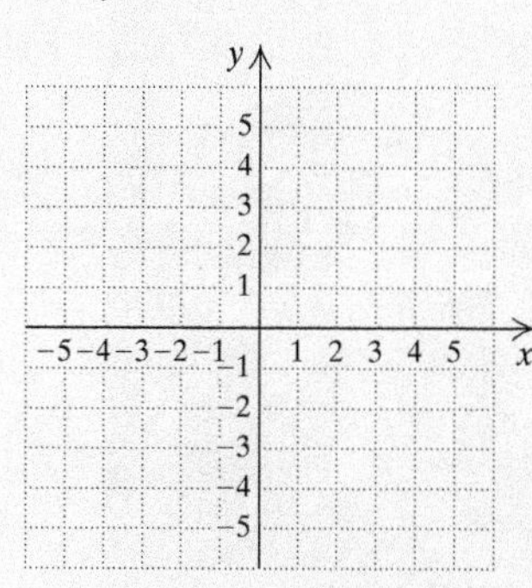

2. $3x - 18 = 0$

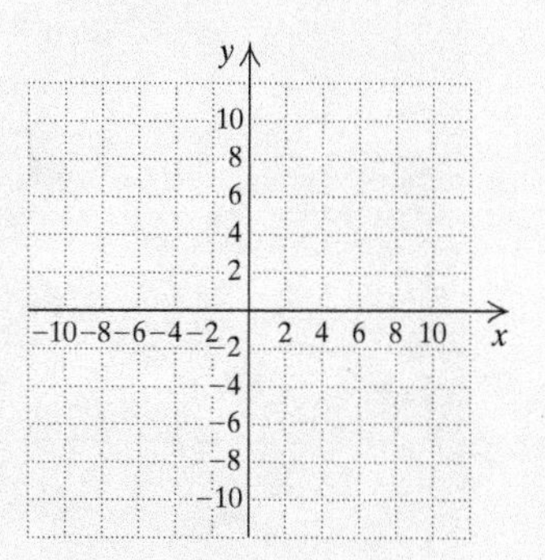

3. $x + 3y < 4$

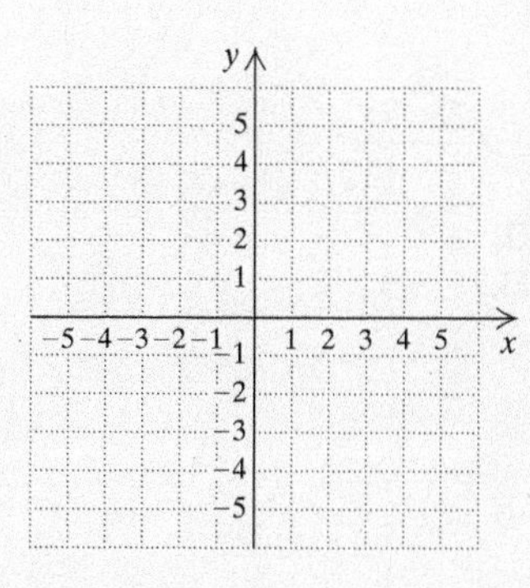

4. $x + y \geq 4,$
   $x - y > 1$

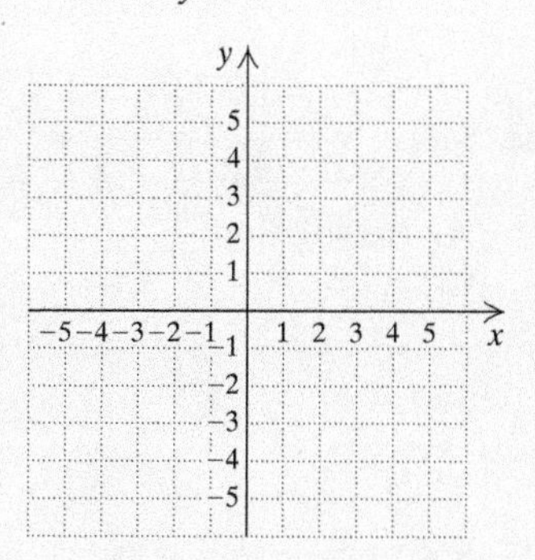

5. Given that $g(x) = |x - 4| + 5$, find $g(-2)$.

6. Given that
$$f(x) = \frac{x - 2}{x^2 - 25},$$
find the domain.

7. Find the domain and the range of the function graphed below.

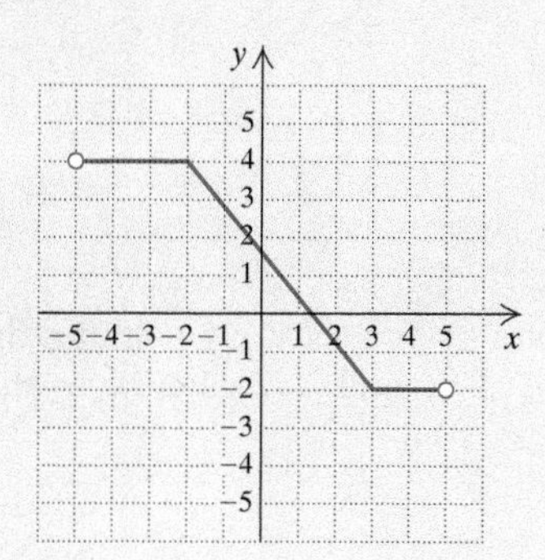

Perform the indicated operation and simplify.

8. $(6m - n)^2$

9. $(3a - 4b)(5a + 2b)$

10. $\frac{y^2 - 4}{3y + 33} \cdot \frac{y + 11}{y + 2}$

11. $\frac{9x^2 - 25}{x^2 - 16} \div \frac{3x + 5}{x - 4}$

12. $\frac{2x + 1}{4x - 12} - \frac{x - 2}{5x - 15}$

13. $\frac{1 - \frac{2}{y^2}}{1 - \frac{1}{y^3}}$

14. $(6p^2 - 2p + 5) - (-10p^2 + 6p + 5)$

15. $\frac{2}{x + 2} + \frac{3}{x - 2} - \frac{x + 1}{x^2 - 4}$

16. $(2x^3 - 7x^2 + x - 3) \div (x + 2)$

Solve.

17. $9y - (5y - 3) = 33$

18. $-3 < -2x - 6 < 0$

19. $\frac{3x}{x - 2} - \frac{6}{x + 2} = \frac{24}{x^2 - 4}$

20. $P = \frac{3a}{a + b}$, for a

21. $F = \frac{9}{5}C + 32$, for C

22. $|x| \geq 2.1$

23. $\frac{6}{x - 5} = \frac{2}{2x}$

24. $8x = 1 + 16x^2$

25. $14 + 3x = 2x^2$

Solve.

26. $4x - 2y = 6,$
$6x - 3y = 9$

27. $4x + 5y = -3,$
$x = 1 - 3y$

28. $x + 2y - 2z = 9,$
$2x - 3y + 4z = -4,$
$5x - 4y + 2z = 5$

29. $x + 6y + 4z = -2,$
$4x + 4y + z = 2,$
$3x + 2y - 4z = 5$

Factor.

30. $4x^3 + 18x^2$

31. $8a^3 - 4a^2 - 6a + 3$

32. $x^2 + 8x - 84$

33. $6x^2 + 11x - 10$

34. $16y^2 - 81$

35. $t^2 - 16t + 64$

36. $64x^3 + 8$

37. $0.027b^3 - 0.008c^3$

38. $x^6 - x^2$

39. $20x^2 + 7x - 3$

40. Find an equation of the line with slope $-\frac{1}{2}$ passing through the point $(2, -2)$.

41. Find an equation of the line that is perpendicular to the line $2x + y = 5$ and passes through the point $(3, -1)$.

42. *Lacrosse Results.* A lacrosse team played 63 games in a season. They won 1 fewer game than nine times the number of ties and lost 30 fewer games than they won. How many games did they win? lose? tie?

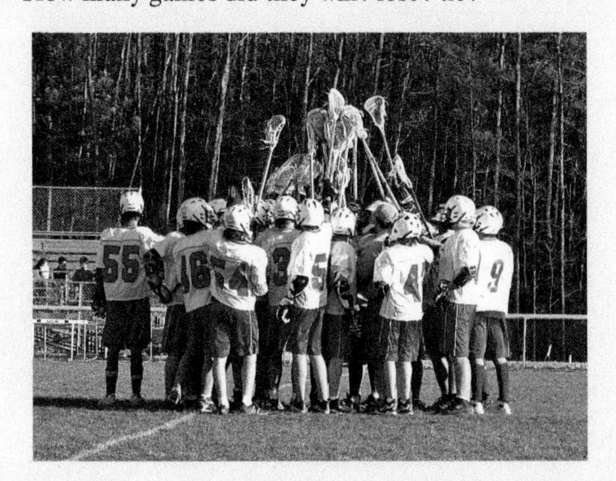

43. *Waste Generation.* The amount of waste generated by a fast-food restaurant varies directly as the number of customers served. A typical restaurant that serves 2000 customers per day generates 238 lb of waste daily. How many pounds of waste would be generated daily by a restaurant that serves 1700 customers per day?

44. Solve: $\frac{x}{x - 4} - \frac{4}{x + 3} = \frac{28}{x^2 - x - 12}$.

A. No solution
B. 0
C. $-4, 3$
D. $4, -3$

45. Solve: $x^2 - x - 6 = 6$.

A. 4, 9
B. 3, 8
C. $4, -3$
D. 0, 1

46. *Tank Filling.* An oil storage tank can be filled in 10 hr by ship A working alone and in 15 hr by ship B working alone. How many hours would it take to fill the oil storage tank if both ships A and B are working?

A. 8 hr
B. 6 hr
C. $12\frac{1}{2}$ hr
D. 25 hr

Synthesis

47. The graph of $y = ax^2 + bx + c$ contains the three points $(4, 2)$, $(2, 0)$, and $(1, 2)$. Find a, b, and c.

Solve.

48. $16x^3 = x$

49. $\frac{18}{x - 9} + \frac{10}{x + 5} = \frac{28x}{x^2 - 4x - 45}$

CHAPTER

6

Radical Expressions, Equations, and Functions

By 2017, smartphone ownership had increased to approximately 44% of the world population and is expected to grow to 59% by 2022. In the first quarter of 2017, worldwide sales of smartphones totaled 380 million units—an increase of 9.1% over the first quarter of 2016. The graph at right shows the percent of market share for five top vendors. The combined market share of three Chinese manufacturers, Huawei, Oppo, and Vivo, made up nearly 24% of the first-quarter sales.

Worldwide Smartphone Sales by Vendor

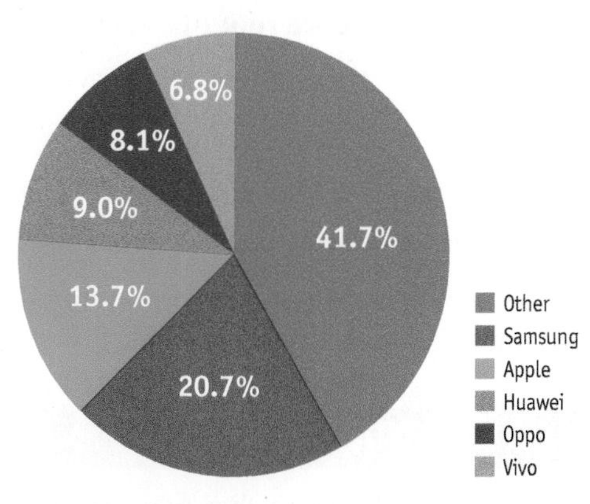

DATA: www.gartner.com

Data: gartner.com; strategyanalytics.com, "44% of Population Will Own Smartphones in 2017," by Linda Sui, December 21, 2016

In Exercise 15 of Exercise Set 6.7, we will calculate the length of the Samsung Galaxy S8 infinity screen.

JUST IN TIME

Review topics 10, 17, and 19 in the **Just In Time** section at the front of the text provide excellent prerequisite skill review for this chapter.

MyLab Math
VIDEO

STUDYING FOR SUCCESS *Quiz and Test Follow-Up*

- ☐ Immediately after completing a chapter quiz or a test, write out a step-by-step solution of each question that you missed.
- ☐ Visit your instructor or tutor for help with problems that are still giving you trouble.
- ☐ Keep your tests and quizzes, along with their corrections, to use as a study guide for the final examination.

6.1 Radical Expressions and Functions

OBJECTIVES

- **a** Find principal square roots and their opposites, approximate square roots, identify radicands, find outputs of square-root functions, graph square-root functions, and find the domains of square-root functions.
- **b** Simplify radical expressions with perfect-square radicands.
- **c** Find cube roots, simplifying certain expressions, and find outputs of cube-root functions.
- **d** Simplify expressions involving odd roots and even roots.

In this section, we consider roots, such as square roots and cube roots. We define the symbolism associated with roots and consider methods of manipulating symbols.

a SQUARE ROOTS AND SQUARE-ROOT FUNCTIONS

MyLab Math ANIMATION

SKILL REVIEW *Find the domain of a function.* [2.3a]

Find the domain.

1. $f(x) = \dfrac{6}{x - 8}$ **2.** $f(x) = x^2 + \dfrac{1}{3}$

Answers: 1. $\{x|x \text{ is a real number } and\ x \neq 8\}$
2. All real numbers

MyLab Math VIDEO

When we raise a number to the second power, we say that we have **squared** the number. Sometimes we may need to find the number that was squared. We call this process **finding a square root** of a number.

SQUARE ROOT

The number c is a **square root** of a if $c^2 = a$.

For example:

5 is a *square root* of 25 because $5^2 = 5 \cdot 5 = 25$;
-5 is a *square root* of 25 because $(-5)^2 = (-5)(-5) = 25$.

The number -4 does not have a real-number square root because there is no real number c such that $c^2 = -4$.

PROPERTIES OF SQUARE ROOTS

Every positive real number has two real-number square roots.
The number 0 has just one square root, 0 itself.
Negative numbers do not have real-number square roots.*

*In Section 6.8, we will consider a number system in which negative numbers do have square roots.

EXAMPLE 1 Find the two square roots of 64.

The square roots of 64 are 8 and -8 because $8^2 = 64$ and $(-8)^2 = 64$.

Do Exercises 1–3.

Find the square roots.

1. 9 2. 36 3. 121

> **PRINCIPAL SQUARE ROOT**
>
> The **principal square root** of a nonnegative number is its nonnegative square root. The symbol $\sqrt{a}$ represents the principal square root of a. To name the negative square root of a, we can write $-\sqrt{a}$.

EXAMPLES Simplify.

2. $\sqrt{25} = 5$

   Remember: $\sqrt{\ }$ indicates the principal (nonnegative) square root.

3. $-\sqrt{25} = -5$
4. $\sqrt{\frac{81}{64}} = \frac{9}{8}$ because $\left(\frac{9}{8}\right)^2 = \frac{9}{8} \cdot \frac{9}{8} = \frac{81}{64}$.
5. $\sqrt{0.0049} = 0.07$ because $(0.07)^2 = (0.07)(0.07) = 0.0049$.
6. $-\sqrt{0.000001} = -0.001$
7. $\sqrt{0} = 0$
8. $\sqrt{-25}$ Does not exist as a real number. Negative numbers do not have real-number square roots.

Do Exercises 4–13.

Simplify.

4. a) $\sqrt{16}$ b) $-\sqrt{16}$ c) $\sqrt{-16}$
5. a) $\sqrt{49}$ b) $-\sqrt{49}$ c) $\sqrt{-49}$
6. $\sqrt{1}$
7. $-\sqrt{36}$
8. $\sqrt{\frac{81}{100}}$
9. $\sqrt{0.0064}$
10. $-\sqrt{\frac{25}{64}}$
11. $\sqrt{\frac{16}{9}}$
12. $-\sqrt{0.81}$
13. $\sqrt{1.44}$

We found exact square roots in Examples 1–8. It would be helpful to memorize the table of exact square roots at right. We often need to use rational numbers to *approximate* square roots that are irrational. Such expressions can be found using a calculator with a square-root key.

TABLE OF SQUARE ROOTS

$\sqrt{1} = 1$	$\sqrt{196} = 14$
$\sqrt{4} = 2$	$\sqrt{225} = 15$
$\sqrt{9} = 3$	$\sqrt{256} = 16$
$\sqrt{16} = 4$	$\sqrt{289} = 17$
$\sqrt{25} = 5$	$\sqrt{324} = 18$
$\sqrt{36} = 6$	$\sqrt{361} = 19$
$\sqrt{49} = 7$	$\sqrt{400} = 20$
$\sqrt{64} = 8$	$\sqrt{441} = 21$
$\sqrt{81} = 9$	$\sqrt{484} = 22$
$\sqrt{100} = 10$	$\sqrt{529} = 23$
$\sqrt{121} = 11$	$\sqrt{576} = 24$
$\sqrt{144} = 12$	$\sqrt{625} = 25$
$\sqrt{169} = 13$	

EXAMPLES Use a calculator to approximate each of the following.

Number	*Using a calculator with a 10-digit readout*	*Rounded to three decimal places*
9. $\sqrt{11}$	3.316624790	3.317
10. $\sqrt{487}$	22.06807649	22.068
11. $-\sqrt{7297.8}$	−85.42716196	−85.427
12. $\sqrt{\frac{463}{557}}$	.9117229728	0.912

Do Exercises 14–17.

Use a calculator to approximate each square root to three decimal places.

14. $\sqrt{17}$
15. $\sqrt{1138}$
16. $-\sqrt{867.6}$
17. $\sqrt{\frac{22}{35}}$

Answers

1. 3, −3 **2.** 6, −6 **3.** 11, −11
4. (a) 4; **(b)** −4; **(c)** does not exist as a real number
5. (a) 7; **(b)** −7; **(c)** does not exist as a real number
6. 1 **7.** −6 **8.** $\frac{9}{10}$ **9.** 0.08 **10.** $-\frac{5}{8}$
11. $\frac{4}{3}$ **12.** −0.9 **13.** 1.2 **14.** 4.123
15. 33.734 **16.** −29.455 **17.** 0.793

RADICAL; RADICAL EXPRESSION; RADICAND

The symbol $\sqrt{}$ is called a **radical**.

An expression written with a radical is called a **radical expression**.

The expression written under the radical is called the **radicand**.

These are radical expressions:

$$\sqrt{5}, \quad \sqrt{a}, \quad -\sqrt{5x}, \quad \sqrt{y^2 + 7}.$$

The radicands in these expressions are 5, a, $5x$, and $y^2 + 7$, respectively.

EXAMPLE 13 Identify the radicand in $x\sqrt{x^2 - 9}$.

The radicand is the expression under the radical, $x^2 - 9$.

◀ **Do Exercises 18 and 19.**

Identify the radicand.

18. $5\sqrt{28 + x}$

19. $\sqrt{\dfrac{y}{y + 3}}$

Since each nonnegative real number x has exactly one principal square root, the symbol $\sqrt{x}$ represents exactly one real number and thus can be used to define a square-root function:

$$f(x) = \sqrt{x}.$$

The domain of this function is the set of nonnegative real numbers. In interval notation, the domain is $[0, \infty)$.

EXAMPLE 14 For the given function, find the indicated function values:

$$f(x) = \sqrt{3x - 2}; \quad f(1), f(5), \text{and } f(0).$$

We have

$f(1) = \sqrt{3 \cdot 1 - 2}$ Substituting 1 for x

$= \sqrt{3 - 2} = \sqrt{1} = 1;$ Simplifying and taking the square root

$f(5) = \sqrt{3 \cdot 5 - 2}$ Substituting 5 for x

$= \sqrt{13} \approx 3.606;$ Simplifying and approximating

$f(0) = \sqrt{3 \cdot 0 - 2}$ Substituting 0 for x

$= \sqrt{-2}.$ Negative radicand. No real-number function value exists; 0 is not in the domain of f.

◀ **Do Exercises 20 and 21.**

For the given function, find the indicated function values.

20. $g(x) = \sqrt{6x + 4}$; $g(0), g(3)$, and $g(-5)$ GS

$g(0) = \sqrt{6 \cdot \_\_\_ + 4}$

$= \sqrt{\_\_\_ + 4}$

$= \sqrt{4} = \_\_\_$

$g(3) = \sqrt{6 \cdot \_\_\_ + 4}$

$= \sqrt{\_\_\_ + 4}$

$= \sqrt{22}$

$g(-5) = \sqrt{6(\_\_\_) + 4}$

$= \sqrt{\_\_\_ + 4}$

$= \sqrt{-26}$

-26 is a ________ (negative/positive) radicand. No real-number function value exists.

21. $f(x) = -\sqrt{x}$; $f(4), f(7)$, and $f(-3)$

EXAMPLE 15 Find the domain of $g(x) = \sqrt{x + 2}$.

The expression $\sqrt{x + 2}$ is a real number only when $x + 2$ is nonnegative. Thus the domain of $g(x) = \sqrt{x + 2}$ is the set of all x-values for which $x + 2 \geq 0$. We solve as follows:

$x + 2 \geq 0$

$x \geq -2.$ Adding -2

The domain of $g = \{x \mid x \geq -2\}$, or $[-2, \infty)$. ■

Answers

18. $28 + x$ **19.** $\dfrac{y}{y + 3}$ **20.** 2; $\sqrt{22}$; does not exist as a real number
21. -2; $-\sqrt{7} \approx -2.646$; does not exist as a real number

Guided Solution:
20. 0, 0, 2; 3, 18; -5, -30, negative

EXAMPLE 16 Graph: **(a)** $f(x) = \sqrt{x}$; **(b)** $g(x) = \sqrt{x + 2}$.

We first find outputs as we did in Example 14. We can either select inputs that have exact outputs or use a calculator to make approximations. Once ordered pairs have been calculated, a smooth curve can be drawn.

a)

x	$f(x) = \sqrt{x}$	$(x, f(x))$
0	0	(0, 0)
1	1	(1, 1)
3	1.7	(3, 1.7)
4	2	(4, 2)
7	2.6	(7, 2.6)
9	3	(9, 3)

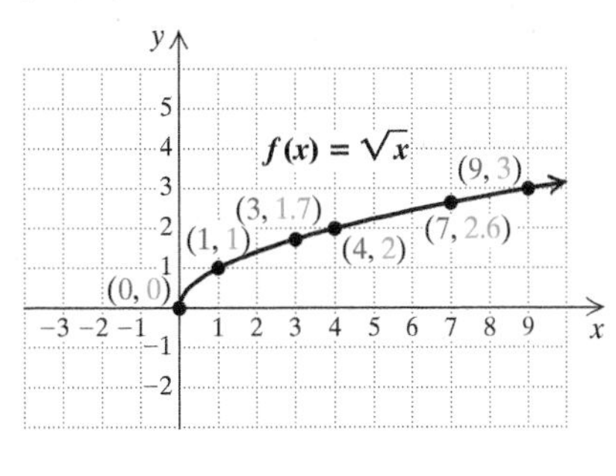

We can see from the table and the graph that the domain of f is $[0, \infty)$. The range is also the set of nonnegative real numbers $[0, \infty)$.

b)

x	$g(x) = \sqrt{x + 2}$	$(x, g(x))$
−2	0	(−2, 0)
−1	1	(−1, 1)
0	1.4	(0, 1.4)
3	2.2	(3, 2.2)
5	2.6	(5, 2.6)
10	3.5	(10, 3.5)

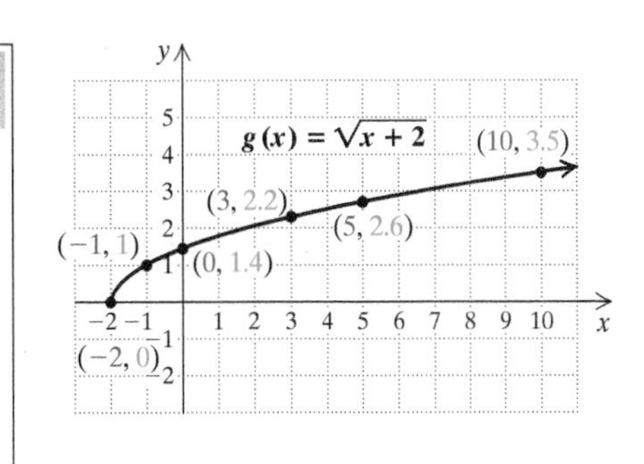

We can see from the table, the graph, and Example 15 that the domain of g is $[-2, \infty)$. The range is the set of nonnegative real numbers $[0, \infty)$.

Do Exercises 22–25.

Find the domain of each function.

22. $f(x) = \sqrt{x - 5}$

23. $g(x) = \sqrt{2x + 3}$

Graph.

24. $g(x) = -\sqrt{x}$

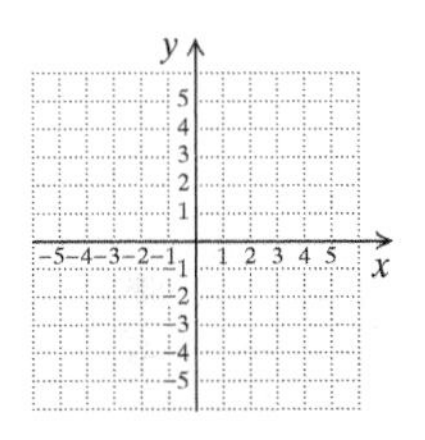

25. $f(x) = 2\sqrt{x + 3}$

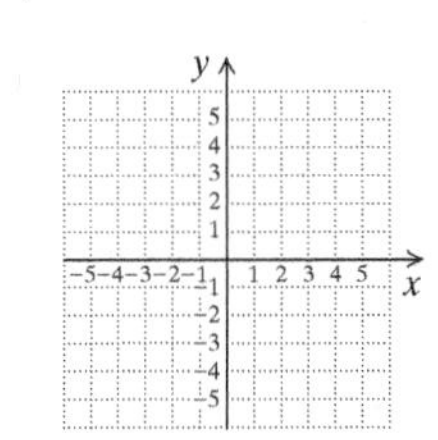

b FINDING $\sqrt{a^2}$

In the expression $\sqrt{a^2}$, the radicand is a perfect square. It is tempting to think that $\sqrt{a^2} = a$, but we see below that this is not always the case.

Suppose $a = 5$. Then we have $\sqrt{5^2}$, which is $\sqrt{25}$, or 5.
Suppose $a = -5$. Then we have $\sqrt{(-5)^2}$, which is $\sqrt{25}$, or 5.
Suppose $a = 0$. Then we have $\sqrt{0^2}$, which is $\sqrt{0}$, or 0.

The symbol $\sqrt{a^2}$ never represents a negative number. It represents the principal square root of a^2. Note the following.

SIMPLIFYING $\sqrt{a^2}$

$a \geq 0 \longrightarrow \sqrt{a^2} = a$

If a is positive or 0, the principal square root of a^2 is a.

$a < 0 \longrightarrow \sqrt{a^2} = -a$

If a is negative, the principal square root of a^2 is the opposite of a.

Answers

22. $\{x | x \geq 5\}$, or $[5, \infty)$
23. $\{x | x \geq -\frac{3}{2}\}$, or $[-\frac{3}{2}, \infty)$
24.

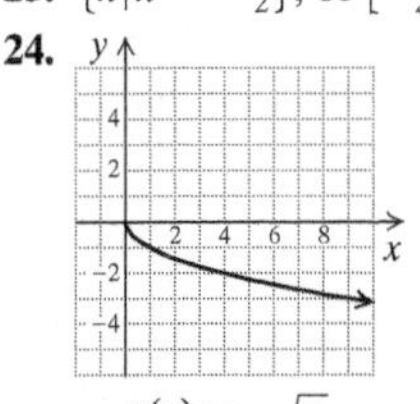

25.

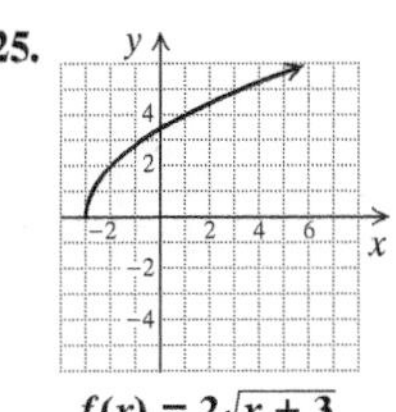

In all cases, the radical expression $\sqrt{a^2}$ represents the absolute value of a.

PRINCIPAL SQUARE ROOT OF a^2

For any real number a, $\sqrt{a^2} = |a|$. The principal (nonnegative) square root of a^2 is the absolute value of a.

The absolute value is used to ensure that the principal square root is nonnegative, which is as it is defined.

Find each of the following. Assume that letters can represent *any* real number.

26. $\sqrt{y^2}$
27. $\sqrt{(-24)^2}$
28. $\sqrt{(5y)^2}$
29. $\sqrt{16y^2}$
30. $\sqrt{(x+7)^2}$
31. $\sqrt{4(x-2)^2}$
32. $\sqrt{49(y+5)^2}$

33. $\sqrt{x^2 - 6x + 9}$
$= \sqrt{(x - \quad)^2}$
$= |\quad - 3|$

GS

EXAMPLES Find each of the following. Assume that letters can represent any real number.

17. $\sqrt{(-16)^2} = |-16|$, or 16

18. $\sqrt{(3b)^2} = |3b| = |3| \cdot |b| = 3|b|$

$|3b|$ can be simplified to $3|b|$ because the absolute value of any product is the product of the absolute values. That is, $|a \cdot b| = |a| \cdot |b|$.

19. $\sqrt{(x-1)^2} = |x - 1|$

20. $\sqrt{x^2 + 8x + 16} = \sqrt{(x+4)^2}$
$= |x + 4|$

Caution!

$|x + 4|$ is *not* the same as $|x| + 4$.

Do Exercises 26–33.

c CUBE ROOTS

CUBE ROOT

The number c is the **cube root** of a, written $\sqrt[3]{a}$, if the third power of c is a—that is, if $c^3 = a$, then $\sqrt[3]{a} = c$.

For example:

2 is the *cube root* of 8 because $2^3 = 2 \cdot 2 \cdot 2 = 8$;

-4 is the *cube root* of -64 because $(-4)^3 = (-4)(-4)(-4) = -64$.

We talk about *the* cube root of a number rather than *a* cube root because of the following.

Every real number has exactly one cube root in the system of real numbers. The symbol $\sqrt[3]{a}$ represents *the* cube root of a.

Answers

26. $|y|$ **27.** 24 **28.** $5|y|$ **29.** $4|y|$
30. $|x+7|$ **31.** $2|x-2|$ **32.** $7|y+5|$
33. $|x-3|$

Guided Solution:
33. 3, x

EXAMPLES Find each of the following.

21. $\sqrt[3]{8} = 2$ because $2^3 = 8$.

22. $\sqrt[3]{-27} = -3$

23. $\sqrt[3]{-\frac{216}{125}} = -\frac{6}{5}$

24. $\sqrt[3]{0.001} = 0.1$

25. $\sqrt[3]{x^3} = x$

26. $\sqrt[3]{-8} = -2$

27. $\sqrt[3]{0} = 0$

28. $\sqrt[3]{-8y^3} = \sqrt[3]{(-2y)^3} = -2y$

When we are determining a cube root, no absolute-value signs are needed because a real number has just one cube root. The real-number cube root of a positive number is positive. The real-number cube root of a negative number is negative. The cube root of 0 is 0. That is, $\sqrt[3]{a^3} = a$ whether $a > 0$, $a < 0$, or $a = 0$.

Do Exercises 34–37.

Find each of the following.

34. $\sqrt[3]{-64}$

35. $\sqrt[3]{27y^3}$

36. $\sqrt[3]{8(x+2)^3}$

37. $\sqrt[3]{-\frac{343}{64}}$

Since the symbol $\sqrt[3]{x}$ represents exactly one real number, it can be used to define a cube-root function: $f(x) = \sqrt[3]{x}$.

EXAMPLE 29 For the given function, find the indicated function values:

$$f(x) = \sqrt[3]{x};\quad f(125), f(0), f(-8), \text{and } f(-10).$$

We have

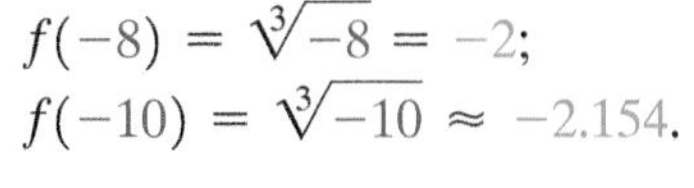

$f(125) = \sqrt[3]{125} = 5;$

$f(0) = \sqrt[3]{0} = 0;$

$f(-8) = \sqrt[3]{-8} = -2;$

$f(-10) = \sqrt[3]{-10} \approx -2.154.$

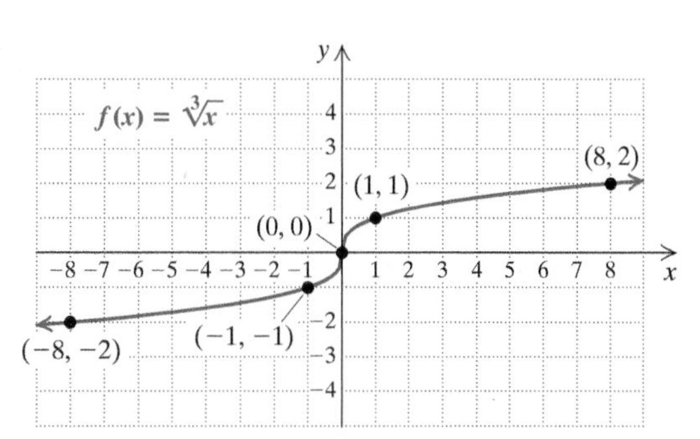

The graph of $f(x) = \sqrt[3]{x}$ is shown above for reference. Note that both the domain and the range consist of the entire set of real numbers, $(-\infty, \infty)$.

Do Exercise 38.

38. For the given function, find the indicated function values:
$g(x) = \sqrt[3]{x-4};\ g(-23),$
$g(4), g(-1), \text{and } g(11).$

d ODD AND EVEN *k*TH ROOTS

In the expression $\sqrt[k]{a}$, we call k the **index** and assume $k \geq 2$.

Odd Roots

The 5th root of a number a is the number c for which $c^5 = a$. There are also 7th roots, 9th roots, and so on. Whenever the number k in $\sqrt[k]{\ }$ is an odd number, we say that we are taking an **odd root**.

Every number has just one real-number odd root. If the number is positive, then the root is positive. If the number is negative, then the root is negative. If the number is 0, then the root is 0. For example, $\sqrt[3]{8} = 2$, $\sqrt[3]{-8} = -2$, and $\sqrt[3]{0} = 0$. Absolute-value signs are *not* needed when we are finding odd roots.

Answers

34. -4 **35.** $3y$ **36.** $2(x+2)$
37. $-\frac{7}{4}$ **38.** -3; 0; $\sqrt[3]{-5} \approx -1.710$;
$\sqrt[3]{7} \approx 1.913$

If k is an *odd* natural number, then for any real number a,

$$\sqrt[k]{a^k} = a.$$

EXAMPLES Find each of the following.

30. $\sqrt[5]{32} = 2$

31. $\sqrt[5]{-32} = -2$

32. $-\sqrt[5]{32} = -2$

33. $-\sqrt[5]{-32} = -(-2) = 2$

34. $\sqrt[7]{x^7} = x$

35. $\sqrt[7]{128} = 2$

36. $\sqrt[7]{-128} = -2$

37. $\sqrt[7]{0} = 0$

38. $\sqrt[5]{a^5} = a$

39. $\sqrt[9]{(x-1)^9} = x - 1$

Do Exercises 39–45.

Find each of the following.

39. $\sqrt[5]{243}$

40. $\sqrt[5]{-243}$

41. $\sqrt[5]{x^5}$

42. $\sqrt[7]{y^7}$

43. $\sqrt[5]{0}$

44. $\sqrt[5]{-32x^5}$

45. $\sqrt[7]{(3x+2)^7}$

Even Roots

When the index k in $\sqrt[k]{}$ is an even number, we say that we are taking an **even root**. When the index is 2, we do not write it. Every positive real number has two real-number kth roots when k is even. One of those roots is positive and one is negative. Negative real numbers do not have real-number kth roots when k is even. When we are finding even kth roots, absolute-value signs are sometimes necessary, as we have seen with square roots. For example,

$$\sqrt{64} = 8, \quad \sqrt[6]{64} = 2, \quad -\sqrt[6]{64} = -2, \quad \sqrt[6]{64x^6} = \sqrt[6]{(2x)^6} = |2x| = 2|x|.$$

Note that in $\sqrt[6]{64x^6}$, we need absolute-value signs because a variable is involved.

EXAMPLES Find each of the following. Assume that variables can represent any real number.

40. $\sqrt[4]{16} = 2$

41. $-\sqrt[4]{16} = -2$

42. $\sqrt[4]{-16}$
Does not exist as a real number.

43. $\sqrt[4]{81x^4} = \sqrt[4]{(3x)^4} = |3x| = 3|x|$

44. $\sqrt[6]{(y+7)^6} = |y+7|$

45. $\sqrt{81y^2} = \sqrt{(9y)^2} = |9y| = 9|y|$

The following is a summary of how absolute value is used when we are taking even roots or odd roots.

SIMPLIFYING

For any real number a:

a) $\sqrt[k]{a^k} = |a|$ when k is an *even* natural number. We use absolute value when k is even unless a is nonnegative.

b) $\sqrt[k]{a^k} = a$ when k is an *odd* natural number greater than 1. We do not use absolute value when k is odd.

Do Exercises 46–54.

Find each of the following. Assume that letters can represent any real number.

46. $\sqrt[4]{81}$

47. $-\sqrt[4]{81}$

48. $\sqrt[4]{-81}$

49. $\sqrt[4]{0}$

50. $\sqrt[4]{16(x-2)^4}$

51. $\sqrt[6]{x^6}$

52. $\sqrt[8]{(x+3)^8}$

53. $\sqrt[7]{(x+3)^7}$

54. $\sqrt[5]{243x^5}$

Answers

39. 3 **40.** −3 **41.** x **42.** y **43.** 0 **44.** $-2x$ **45.** $3x+2$ **46.** 3 **47.** −3 **48.** Does not exist as a real number **49.** 0 **50.** $2|x-2|$ **51.** $|x|$ **52.** $|x+3|$ **53.** $x+3$ **54.** $3x$

6.1 Exercise Set

FOR EXTRA HELP MyLab Math

✓ Check Your Understanding

Reading Check Determine whether each statement is true or false.

RC1. $-\sqrt{16} = -4$

RC2. $\sqrt{-16} = -4$

RC3. $\sqrt{49}$ represents 7 or -7.

RC4. 36 has two square roots.

RC5. There are no real numbers that when squared yield negative numbers.

RC6. The radicand in the expression $\sqrt{x} - 10$ is $x - 10$.

Concept Check Choose from the columns on the right the domain of the given function. Some of the choices may not be used, and some may be used more than once.

CC1. $f(x) = \sqrt{9 - x}$

CC2. $f(x) = \sqrt{x + 9} + 3$

CC3. $g(x) = \sqrt{x - 3}$

CC4. $h(x) = x + 9$

CC5. $f(x) = 3 - x$

CC6. $g(x) = 3 - \sqrt{3 - x}$

a) $[-3, \infty)$
b) $[-9, \infty)$
c) $(3, \infty)$
d) $(-\infty, -9)$
e) $(-\infty, -3]$
f) $[9, \infty)$
g) $(-\infty, 3]$
h) $[3, \infty)$
i) $(-\infty, \infty)$
j) $(-\infty, 9]$

a Find the square roots.

1. 16 **2.** 225 **3.** 144 **4.** 9 **5.** 400 **6.** 81

Simplify.

7. $-\sqrt{\frac{49}{36}}$

8. $-\sqrt{\frac{361}{9}}$

9. $\sqrt{196}$

10. $\sqrt{441}$

11. $\sqrt{0.0036}$

12. $\sqrt{0.04}$

13. $\sqrt{-225}$

14. $\sqrt{-64}$

Use a calculator to approximate to three decimal places.

15. $\sqrt{347}$

16. $-\sqrt{1839.2}$

17. $\sqrt{\frac{285}{74}}$

18. $\sqrt{\frac{839.4}{19.7}}$

Identify the radicand.

19. $9\sqrt{y^2 + 16}$

20. $-3\sqrt{p^2 - 10}$

21. $x^4y^5\sqrt{\frac{x}{y - 1}}$

22. $a^2b^2\sqrt{\frac{a^2 - b}{b}}$

For the given function, find the indicated function values.

23. $f(x) = \sqrt{5x - 10}$; $f(6), f(2), f(1)$, and $f(-1)$

24. $t(x) = -\sqrt{2x + 1}$; $t(4), t(0), t(-1)$, and $t\left(-\frac{1}{2}\right)$

25. $g(x) = \sqrt{x^2 - 25}$; $g(-6), g(3), g(6)$, and $g(13)$

26. $F(x) = \sqrt{x^2 + 1}$; $F(0), F(-1)$, and $F(-10)$

27. Find the domain of the function f in Exercise 23.

28. Find the domain of the function t in Exercise 24.

29. ***Parking-Lot Arrival Spaces.*** The attendants at a parking lot park cars in temporary spaces before the cars are taken to long-term parking stalls. The number N of such spaces needed is approximated by the function

$$N(a) = 2.5\sqrt{a},$$

where a is the average number of arrivals in peak hours. What is the number of spaces needed when the average number of arrivals is 66? 100?

30. ***Body Surface Area.*** Body surface area B can be estimated using the Mosteller formula

$$B = \sqrt{\frac{h \times w}{3600}},$$

where B is in square meters, h is height, in centimeters, and w is weight, in kilograms. Estimate the body surface area of **(a)** a woman whose height is 165 cm and whose weight is 63 kg and **(b)** a man whose height is 183 cm and whose weight is 100 kg. Round to the nearest tenth.

Graph.

31. $f(x) = 2\sqrt{x}$

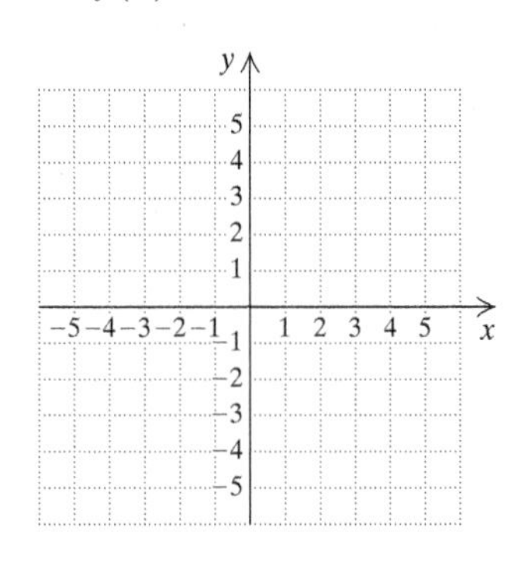

32. $g(x) = 3 - \sqrt{x}$

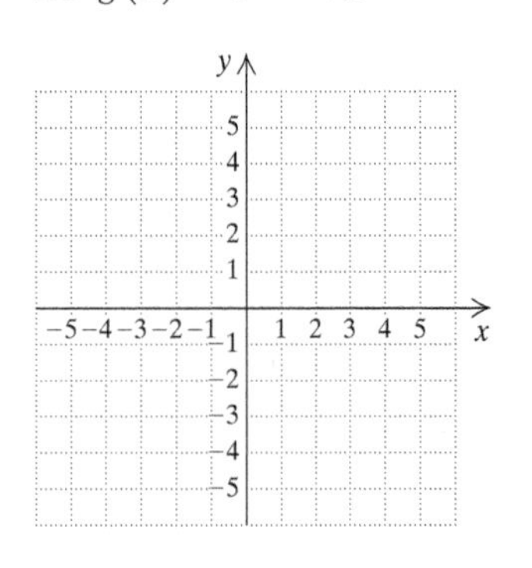

33. $f(x) = -3\sqrt{x}$

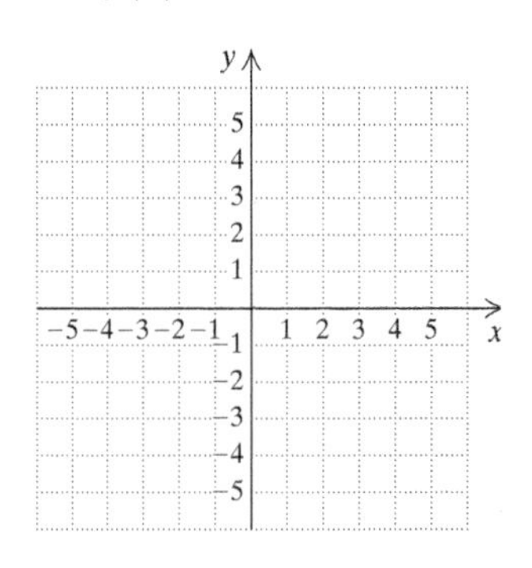

34. $f(x) = 2 + \sqrt{x - 1}$

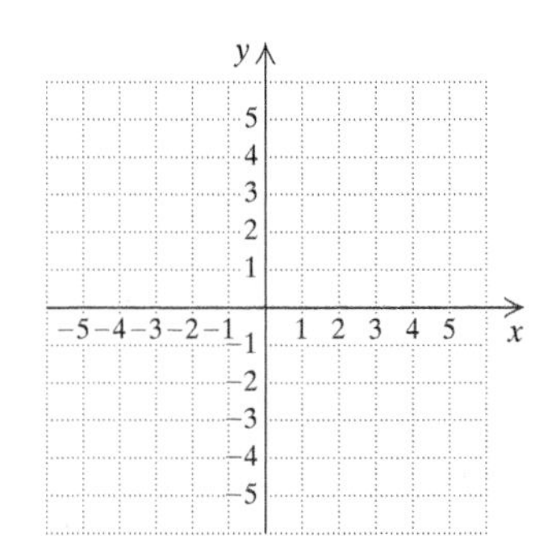

35. $f(x) = \sqrt{x}$

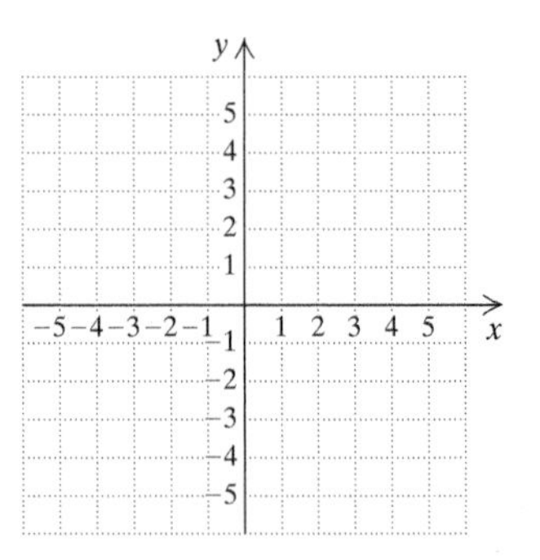

36. $g(x) = -\sqrt{x}$

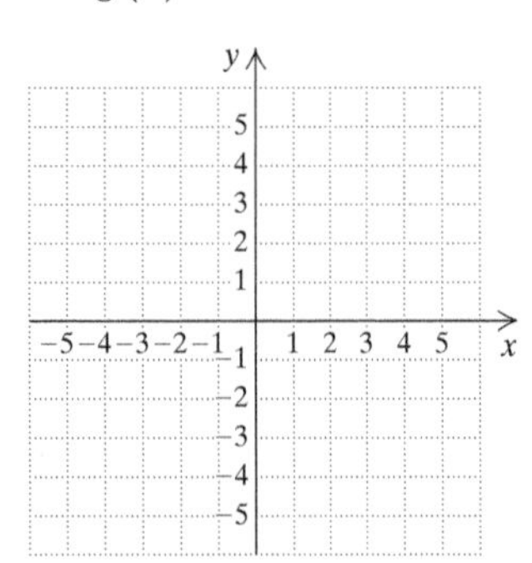

37. $f(x) = \sqrt{x - 2}$

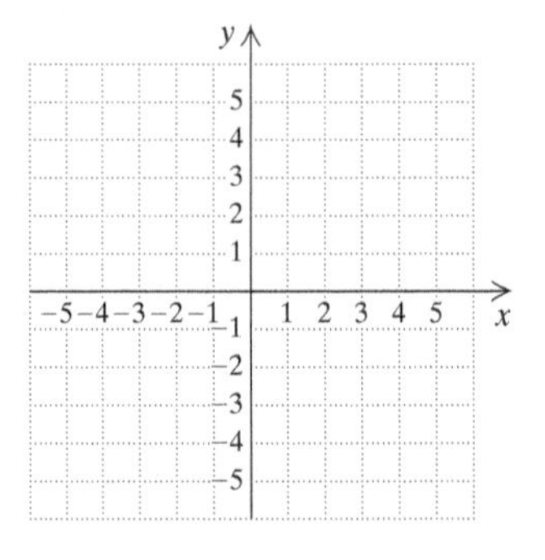

38. $g(x) = \sqrt{x + 3}$

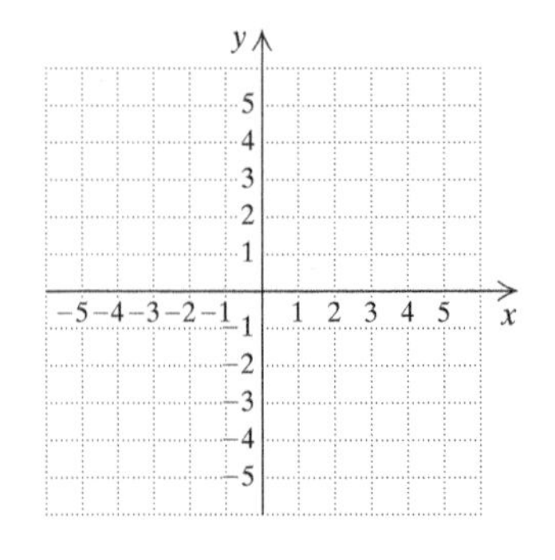

39. $f(x) = \sqrt{12 - 3x}$

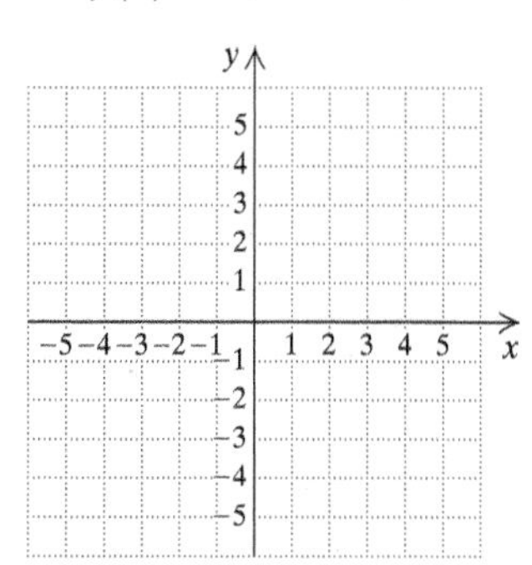

40. $g(x) = \sqrt{8 - 4x}$

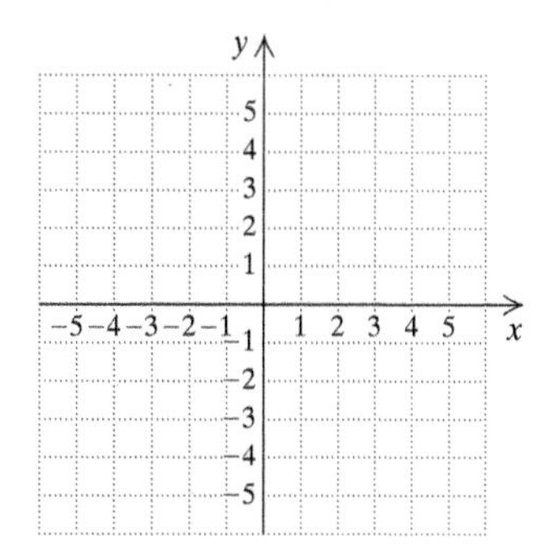

41. $g(x) = \sqrt{3x + 9}$

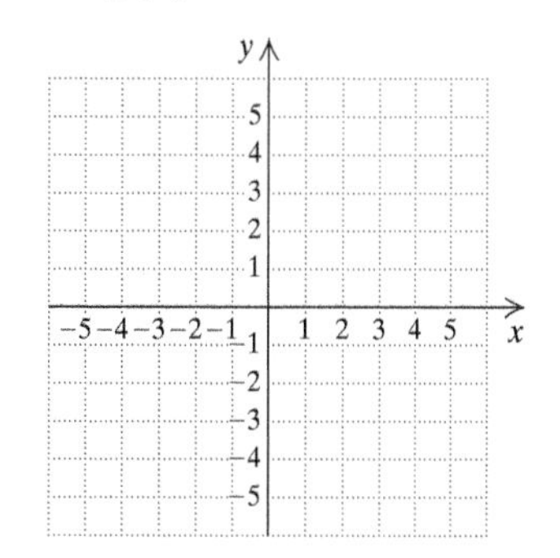

42. $f(x) = \sqrt{3x - 6}$

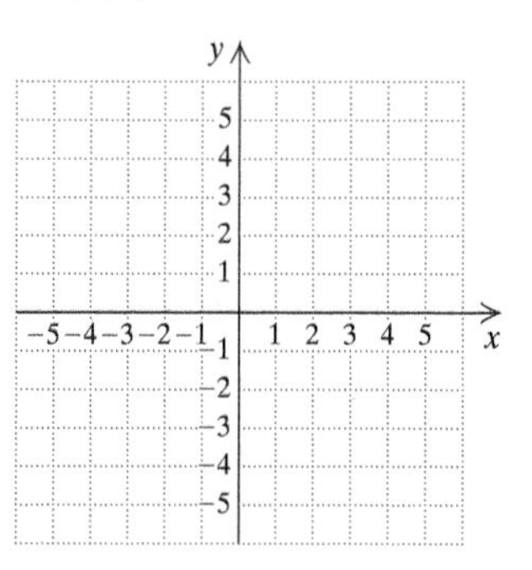

b Find each of the following. Assume that letters can represent *any* real number.

43. $\sqrt{16x^2}$

44. $\sqrt{25t^2}$

45. $\sqrt{(-12c)^2}$

46. $\sqrt{(-9d)^2}$

47. $\sqrt{(p + 3)^2}$

48. $\sqrt{(2 - x)^2}$

49. $\sqrt{x^2 - 4x + 4}$

50. $\sqrt{9t^2 - 30t + 25}$

c Simplify.

51. $\sqrt[3]{27}$

52. $-\sqrt[3]{64}$

53. $\sqrt[3]{-64x^3}$

54. $\sqrt[3]{-125y^3}$

55. $\sqrt[3]{-216}$

56. $-\sqrt[3]{-1000}$

57. $\sqrt[3]{0.343(x + 1)^3}$

58. $\sqrt[3]{0.000008(y - 2)^3}$

For the given function, find the indicated function values.

59. $f(x) = \sqrt[3]{x + 1}$; $f(7), f(26), f(-9)$, and $f(-65)$

60. $g(x) = -\sqrt[3]{2x - 1}$; $g(-62), g(0), g(-13)$, and $g(63)$

61. $f(x) = -\sqrt[3]{3x + 1}$; $f(0), f(-7), f(21)$, and $f(333)$

62. $g(t) = \sqrt[3]{t - 3}$; $g(30), g(-5), g(1)$, and $g(67)$

d Find each of the following. Assume that letters can represent *any* real number.

63. $-\sqrt[4]{625}$

64. $-\sqrt[4]{256}$

65. $\sqrt[5]{-1}$

66. $\sqrt[5]{-32}$

67. $\sqrt[5]{-\frac{32}{243}}$

68. $\sqrt[5]{-\frac{1}{32}}$

69. $\sqrt[6]{x^6}$

70. $\sqrt[8]{y^8}$

71. $\sqrt[4]{(5a)^4}$

72. $\sqrt[4]{(7b)^4}$

73. $\sqrt[10]{(-6)^{10}}$

74. $\sqrt[12]{(-10)^{12}}$

75. $\sqrt[414]{(a+b)^{414}}$

76. $\sqrt[1999]{(2a+b)^{1999}}$

77. $\sqrt[7]{y^7}$

78. $\sqrt[3]{(-6)^3}$

79. $\sqrt[5]{(x-2)^5}$

80. $\sqrt[9]{(2xy)^9}$

Skill Maintenance

Solve. [4.8a]

81. $4x^2 - 49 = 0$

82. $2x^2 - 26x + 72 = 0$

83. $4x^3 - 20x^2 + 25x = 0$

84. $4x^2 - 20x + 25 = 0$

Simplify. [J19]

85. $(a^3b^2c^5)^3$

86. $(5a^7b^8)(2a^3b)$

Synthesis

87. Find the domain of

$$f(x) = \frac{\sqrt{x+3}}{\sqrt{2-x}}.$$

88. Use only the graph of $f(x) = \sqrt[3]{x}$, shown below, to approximate $\sqrt[3]{4}$, $\sqrt[3]{6}$, and $\sqrt[3]{-5}$. Answers may vary.

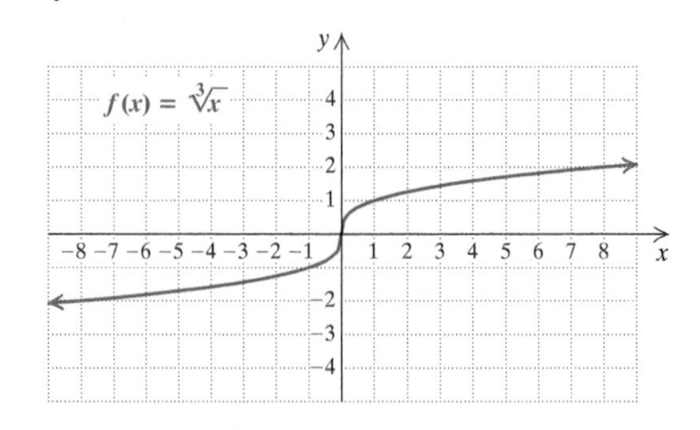

89. Use the TABLE, TRACE, and GRAPH features of a graphing calculator to find the domain and the range of each of the following functions.

a) $f(x) = \sqrt[3]{x}$

b) $g(x) = \sqrt[3]{4x - 5}$

c) $q(x) = 2 - \sqrt{x+3}$

d) $h(x) = \sqrt[4]{x}$

e) $t(x) = \sqrt[4]{x - 3}$

Rational Numbers as Exponents

6.2

OBJECTIVES

a Write expressions with or without rational exponents, and simplify, if possible.

b Write expressions without negative exponents, and simplify, if possible.

c Use the laws of exponents with rational exponents.

d Use rational exponents to simplify radical expressions.

In this section, we give meaning to expressions such as $a^{1/3}, 7^{-1/2}$, and $(3x)^{0.84}$, which have rational numbers as exponents. We will see that using such notation can help simplify certain radical expressions.

a RATIONAL EXPONENTS

Expressions like $a^{1/2}, 5^{-1/4}$, and $(2y)^{4/5}$ have not yet been defined. We will define such expressions so that the general properties of exponents hold.

Consider $a^{1/2} \cdot a^{1/2}$. If we want to multiply by adding exponents, it must follow that $a^{1/2} \cdot a^{1/2} = a^{1/2+1/2}$, or a^1. Thus we should define $a^{1/2}$ to be a square root of a. Similarly, $a^{1/3} \cdot a^{1/3} \cdot a^{1/3} = a^{1/3+1/3+1/3}$, or a^1, so $a^{1/3}$ should be defined to mean $\sqrt[3]{a}$.

$a^{1/n}$

For any *nonnegative* real number a and any natural number index n $(n \neq 1)$,

$a^{1/n}$ means $\sqrt[n]{a}$ (the nonnegative nth root of a).

With rational exponents, we assume that the bases are nonnegative.

EXAMPLES Rewrite without rational exponents, and simplify, if possible.

1. $27^{1/3} = \sqrt[3]{27} = 3$

2. $(abc)^{1/5} = \sqrt[5]{abc}$

3. $x^{1/2} = \sqrt{x}$ An index of 2 is not written.

Do Exercises 1–5. ▶

Rewrite without rational exponents, and simplify, if possible.

1. $y^{1/4}$
2. $(3a)^{1/2}$
3. $16^{1/4}$
4. $(125)^{1/3}$
5. $(a^3b^2c)^{1/5}$

EXAMPLES Rewrite with rational exponents.

4. $\sqrt[5]{7xy} = (7xy)^{1/5}$ We need parentheses around the radicand.

5. $8\sqrt[3]{xy} = 8(xy)^{1/3}$

6. $\sqrt[7]{\dfrac{x^3y}{9}} = \left(\dfrac{x^3y}{9}\right)^{1/7}$

Do Exercises 6–9. ▶

Rewrite with rational exponents.

6. $\sqrt[3]{19ab}$
7. $19\sqrt[3]{ab}$
8. $\sqrt[5]{\dfrac{x^2y}{16}}$
9. $7\sqrt[4]{2ab}$

How should we define $a^{2/3}$? If the general properties of exponents are to hold, we have $a^{2/3} = (a^{1/3})^2$, or $(a^2)^{1/3}$, or $(\sqrt[3]{a})^2$, or $\sqrt[3]{a^2}$. We define this accordingly.

$a^{m/n}$

For any natural numbers m and n $(n \neq 1)$ and any nonnegative real number a,

$a^{m/n}$ means $\sqrt[n]{a^m}$, or $(\sqrt[n]{a})^m$.

Answers

1. $\sqrt[4]{y}$ **2.** $\sqrt{3a}$ **3.** 2 **4.** 5 **5.** $\sqrt[5]{a^3b^2c}$ **6.** $(19ab)^{1/3}$ **7.** $19(ab)^{1/3}$ **8.** $\left(\dfrac{x^2y}{16}\right)^{1/5}$ **9.** $7(2ab)^{1/4}$

EXAMPLES Rewrite without rational exponents, and simplify, if possible.

7. $(27)^{2/3} = \sqrt[3]{27^2} = (\sqrt[3]{27})^2 = 3^2 = 9$

8. $4^{3/2} = \sqrt[2]{4^3} = (\sqrt[2]{4})^3 = 2^3 = 8$

◀ Do Exercises 10–12.

Rewrite without rational exponents, and simplify, if possible.

10. $x^{3/5}$ **11.** $8^{2/3}$

12. $4^{5/2}$

EXAMPLES Rewrite with rational exponents.

The index becomes the denominator of the rational exponent.

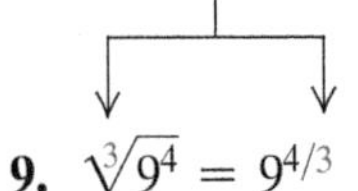

9. $\sqrt[3]{9^4} = 9^{4/3}$

10. $(\sqrt[4]{7xy})^5 = (7xy)^{5/4}$

◀ Do Exercises 13 and 14.

Rewrite with rational exponents.

13. $(\sqrt[3]{7abc})^4$ **14.** $\sqrt[5]{6^7}$

b NEGATIVE RATIONAL EXPONENTS

SKILL REVIEW *Rewrite expressions with or without negative integers as exponents.* [J10]

Express with positive exponents.

1. $3x^{-2}$ **2.** cd^{-5}

Answers: 1. $\frac{3}{x^2}$ **2.** $\frac{c}{d^5}$

Negative rational exponents have a meaning similar to that of negative integer exponents.

$a^{-m/n}$

For any rational number m/n and any positive real number a,

$$a^{-m/n} \text{ means } \frac{1}{a^{m/n}};$$

that is, $a^{m/n}$ and $a^{-m/n}$ are reciprocals.

EXAMPLES Rewrite with positive exponents, and simplify, if possible.

11. $9^{-1/2} = \frac{1}{9^{1/2}} = \frac{1}{\sqrt{9}} = \frac{1}{3}$

12. $(5xy)^{-4/5} = \frac{1}{(5xy)^{4/5}}$

13. $64^{-2/3} = \frac{1}{64^{2/3}} = \frac{1}{(\sqrt[3]{64})^2} = \frac{1}{4^2} = \frac{1}{16}$

14. $4x^{-2/3}y^{1/5} = 4 \cdot \frac{1}{x^{2/3}} \cdot y^{1/5} = \frac{4y^{1/5}}{x^{2/3}}$

15. $\left(\frac{3r}{7s}\right)^{-5/2} = \left(\frac{7s}{3r}\right)^{5/2}$ Since $\left(\frac{a}{b}\right)^{-n} = \left(\frac{b}{a}\right)^{n}$

◀ Do Exercises 15–19.

Rewrite with positive exponents, and simplify, if possible.

15. $16^{-1/4}$ **16.** $(3xy)^{-7/8}$

GS **17.** $81^{-3/4} = \frac{1}{81^{3/4}} = \frac{1}{(\sqrt[4]{81})^{\square}} = \frac{1}{\square^3} = \frac{1}{\square}$

18. $7p^{3/4}q^{-6/5}$ **19.** $\left(\frac{11m}{7n}\right)^{-2/3}$

Answers

10. $\sqrt[5]{x^3}$ **11.** 4 **12.** 32 **13.** $(7abc)^{4/3}$ **14.** $6^{7/5}$ **15.** $\frac{1}{2}$ **16.** $\frac{1}{(3xy)^{7/8}}$ **17.** $\frac{1}{27}$ **18.** $\frac{7p^{3/4}}{q^{6/5}}$ **19.** $\left(\frac{7n}{11m}\right)^{2/3}$

Guided Solution:
17. 3, 3, 27

CALCULATOR CORNER

Rational Exponents We can use a graphing calculator to approximate rational roots of real numbers, as shown at right. To approximate $7^{2/3}$, note that the parentheses around the exponent are necessary. If they are not used, the calculator will read the expression as $7^2 \div 3$. To approximate $14^{-1.9}$, note that parentheses are not required when the exponent is expressed in a single decimal number. The display indicates that $7^{2/3} \approx 3.659$ and $14^{-1.9} \approx 0.007$.

$7^{2/3}$	3.65930571
$14^{-1.9}$	.006642885

EXERCISES: Approximate each of the following.

1. $5^{3/4}$ **2.** $8^{4/7}$ **3.** $29^{-3/8}$

4. $73^{0.56}$ **5.** $34^{-2.78}$ **6.** $32^{0.2}$

c LAWS OF EXPONENTS

The same laws hold for rational-number exponents as for integer exponents. We list them for review.

For any real number a and any rational exponents m and n:

1. $a^m \cdot a^n = a^{m+n}$ In multiplying, we add exponents if the bases are the same.
2. $\dfrac{a^m}{a^n} = a^{m-n}$ In dividing, we subtract exponents if the bases are the same.
3. $(a^m)^n = a^{m \cdot n}$ To raise a power to a power, we multiply the exponents.
4. $(ab)^m = a^m b^m$ To raise a product to a power, we raise each factor to the power.
5. $\left(\dfrac{a}{b}\right)^n = \dfrac{a^n}{b^n}$ To raise a quotient to a power, we raise both the numerator and the denominator to the power.

EXAMPLES Use the laws of exponents to simplify.

16. $3^{1/5} \cdot 3^{3/5} = 3^{1/5+3/5} = 3^{4/5}$ Adding exponents

17. $\dfrac{7^{1/4}}{7^{1/2}} = 7^{1/4-1/2} = 7^{1/4-2/4} = 7^{-1/4} = \dfrac{1}{7^{1/4}}$ Subtracting exponents

18. $(7.2^{2/3})^{3/4} = 7.2^{2/3 \cdot 3/4} = 7.2^{6/12} = 7.2^{1/2}$ Multiplying exponents

19. $(a^{-1/3}b^{2/5})^{1/2} = a^{-1/3 \cdot 1/2} \cdot b^{2/5 \cdot 1/2}$ Raising a product to a power and multiplying exponents

$= a^{-1/6}b^{1/5} = \dfrac{b^{1/5}}{a^{1/6}}$

Do Exercises 20–23.

Use the laws of exponents to simplify.

20. $7^{1/3} \cdot 7^{3/5}$

21. $\dfrac{5^{7/6}}{5^{5/6}}$

22. $(9^{3/5})^{2/3}$

23. $(p^{-2/3}q^{1/4})^{1/2}$

d SIMPLIFYING RADICAL EXPRESSIONS

Rational exponents can be used to simplify some radical expressions. The procedure is as follows.

Answers

20. $7^{14/15}$ **21.** $5^{1/3}$ **22.** $9^{2/5}$ **23.** $\dfrac{q^{1/8}}{p^{1/3}}$

SIMPLIFYING RADICAL EXPRESSIONS

1. Convert radical expressions to exponential expressions.
2. Use arithmetic and the laws of exponents to simplify.
3. Convert back to radical notation when appropriate.

Important: This procedure works only when we assume that a negative number has not been raised to an even power in the radicand. With this assumption, no absolute-value signs will be needed.

EXAMPLES Use rational exponents to simplify.

20. $\sqrt[6]{x^3} = x^{3/6}$ Converting to an exponential expression

$= x^{1/2}$ Simplifying the exponent

$= \sqrt{x}$ Converting back to radical notation

21. $\sqrt[6]{4} = 4^{1/6}$ Converting to exponential notation

$= (2^2)^{1/6}$ Renaming 4 as 2^2

$= 2^{2/6}$ Using $(a^m)^n = a^{mn}$; multiplying exponents

$= 2^{1/3}$ Simplifying the exponent

$= \sqrt[3]{2}$ Converting back to radical notation

22. $\sqrt[8]{a^2b^4} = (a^2b^4)^{1/8}$ Converting to exponential notation

$= a^{2/8} \cdot b^{4/8}$ Using $(ab)^n = a^nb^n$

$= a^{1/4} \cdot b^{1/2}$ Simplifying the exponents

$= a^{1/4} \cdot b^{2/4}$ Rewriting $\frac{1}{2}$ with a denominator of 4

$= (ab^2)^{1/4}$ Using $a^nb^n = (ab)^n$

$= \sqrt[4]{ab^2}$ Converting back to radical notation

◀ **Do Exercises 24–29.**

Use rational exponents to simplify.

24. $\sqrt[4]{a^2}$ **25.** $\sqrt[4]{x^4}$

26. $\sqrt[6]{8}$ **27.** $\sqrt[12]{x^3y^6}$

28. $\sqrt[6]{a^{12}b^3}$ **29.** $\sqrt[5]{a^5b^{10}}$

We can use properties of rational exponents to write a single radical expression for a product or a quotient.

EXAMPLE 23 Use rational exponents to write a single radical expression for $\sqrt[3]{5} \cdot \sqrt{2}$.

$\sqrt[3]{5} \cdot \sqrt{2} = 5^{1/3} \cdot 2^{1/2}$ Converting to exponential notation

$= 5^{2/6} \cdot 2^{3/6}$ Rewriting so that the exponents have a common denominator

$= (5^2 \cdot 2^3)^{1/6}$ Using $a^nb^n = (ab)^n$

$= \sqrt[6]{5^2 \cdot 2^3}$ Converting back to radical notation

$= \sqrt[6]{25 \cdot 8}$

$= \sqrt[6]{200}$ Multiplying under the radical

◀ **Do Exercise 30.**

30. Use rational exponents to write a single radical expression for $\sqrt[4]{7} \cdot \sqrt{3}$.

EXAMPLE 24 Write a single radical expression for $a^{1/2}b^{-1/2}c^{5/6}$.

$a^{1/2}b^{-1/2}c^{5/6} = a^{3/6}b^{-3/6}c^{5/6}$ Rewriting so that the exponents have a common denominator

$= (a^3b^{-3}c^5)^{1/6}$ Using $a^nb^n = (ab)^n$

$= \sqrt[6]{a^3b^{-3}c^5}$ Converting to radical notation ■

Answers

24. $\sqrt{a}$ **25.** x **26.** $\sqrt{2}$ **27.** $\sqrt[4]{xy^2}$
28. $a^2\sqrt{b}$ **29.** ab^2 **30.** $\sqrt[4]{63}$

EXAMPLE 25 Write a single radical expression for $\dfrac{x^{5/6} \cdot y^{3/8}}{x^{4/9} \cdot y^{1/4}}$.

$$\frac{x^{5/6} \cdot y^{3/8}}{x^{4/9} \cdot y^{1/4}} = x^{5/6-4/9} \cdot y^{3/8-1/4}$$ Subtracting exponents

$$= x^{15/18-8/18} \cdot y^{3/8-2/8}$$ Finding common denominators so that exponents can be subtracted

$$= x^{7/18} \cdot y^{1/8}$$ Carrying out the subtraction of exponents

$$= x^{28/72} \cdot y^{9/72}$$ Rewriting so that the exponents have a common denominator

$$= (x^{28}y^{9})^{1/72}$$ Using $a^n b^n = (ab)^n$

$$= \sqrt[72]{x^{28}y^{9}}$$ Converting to radical notation

Do Exercises 31 and 32.

Write a single radical expression

31. $x^{2/3}y^{1/2}z^{5/6}$

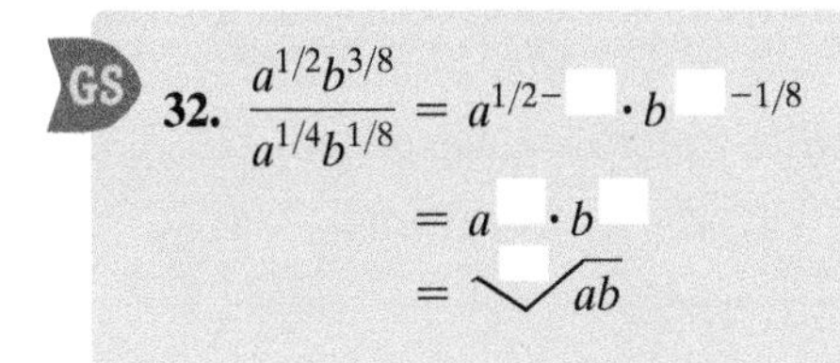

GS **32.** $\dfrac{a^{1/2}b^{3/8}}{a^{1/4}b^{1/8}} = a^{1/2-\square} \cdot b^{\square-1/8}$

$= a^{\square} \cdot b^{\square}$

$= \sqrt[\square]{ab}$

EXAMPLES Use rational exponents to simplify.

26. $\sqrt[6]{(5x)^3} = (5x)^{3/6}$ Converting to exponential notation

$= (5x)^{1/2}$ Simplifying the exponent

$= \sqrt{5x}$ Converting back to radical notation

27. $\sqrt[5]{t^{20}} = t^{20/5}$ Converting to exponential notation

$= t^4$ Simplifying the exponent

28. $(\sqrt[3]{pq^2c})^{12} = (pq^2c)^{12/3}$ Converting to exponential notation

$= (pq^2c)^4$ Simplifying the exponent

$= p^4q^8c^4$ Using $(ab)^n = a^n b^n$

29. $\sqrt{\sqrt[3]{x}} = \sqrt{x^{1/3}}$ Converting the radicand to exponential notation

$= (x^{1/3})^{1/2}$ Try to go directly to this step.

$= x^{1/6}$ Multiplying exponents

$= \sqrt[6]{x}$ Converting back to radical notation

Do Exercises 33–36.

Use rational exponents to simplify.

33. $\sqrt[14]{(5m)^2}$ **34.** $\sqrt[18]{m^3}$

35. $(\sqrt[6]{a^5b^3c})^{24}$ **36.** $\sqrt[5]{\sqrt{x}}$

Answers

31. $\sqrt[6]{x^4y^3z^5}$ **32.** $\sqrt[4]{ab}$ **33.** $\sqrt[7]{5m}$
34. $\sqrt[6]{m}$ **35.** $a^{20}b^{12}c^4$ **36.** $\sqrt[10]{x}$

Guided Solution:
32. 1/4, 3/8, 1/4, 1/4, 4

6.2 Exercise Set

FOR EXTRA HELP

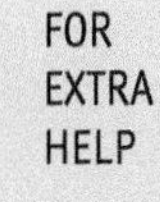

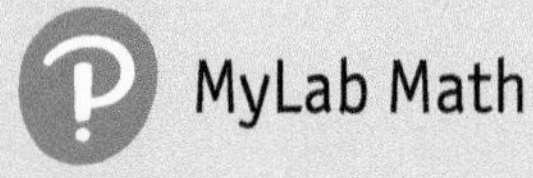

MyLab Math

✓ Check Your Understanding

Reading Check and Concept Check Match the expression with an equivalent expression from the columns on the right.

RC1. $\dfrac{c^2}{c^5}$ **RC2.** $c^{-2/5}$

RC3. $\sqrt{c}$ **RC4.** $c^{2/5}$

RC5. $\sqrt{c^5}$ **RC6.** $(c^2)^5$

RC7. $-c^5 \cdot c^2$ **RC8.** c^2c^5

a) c^{2+5} **b)** $\dfrac{1}{c^{2/5}}$

c) $c^{1/2}$ **d)** $c^{2 \cdot 5}$

e) $c^{5/2}$ **f)** $-c^{5+2}$

g) $(\sqrt[5]{c})^2$ **h)** c^{2-5}

a Rewrite without rational exponents, and simplify, if possible.

1. $y^{1/7}$
2. $x^{1/6}$
3. $8^{1/3}$
4. $16^{1/2}$
5. $(a^3b^3)^{1/5}$
6. $(x^2y^2)^{1/3}$
7. $16^{3/4}$
8. $4^{7/2}$
9. $49^{3/2}$
10. $27^{4/3}$

Rewrite with rational exponents.

11. $\sqrt{17}$
12. $\sqrt{x^3}$
13. $\sqrt[3]{18}$
14. $\sqrt[3]{23}$
15. $\sqrt[5]{xy^2z}$
16. $\sqrt[7]{x^3y^2z^2}$
17. $(\sqrt{3mn})^3$
18. $(\sqrt[3]{7xy})^4$
19. $(\sqrt[7]{8x^2y})^5$
20. $(\sqrt[6]{2a^5b})^7$

b Rewrite with positive exponents, and simplify, if possible.

21. $27^{-1/3}$
22. $100^{-1/2}$
23. $100^{-3/2}$
24. $16^{-3/4}$
25. $3x^{-1/4}$
26. $8y^{-1/7}$
27. $(2rs)^{-3/4}$
28. $(5xy)^{-5/6}$
29. $2a^{3/4}b^{-1/2}c^{2/3}$
30. $5x^{-2/3}y^{4/5}z$
31. $\left(\frac{7x}{8yz}\right)^{-3/5}$
32. $\left(\frac{2ab}{3c}\right)^{-5/6}$
33. $\frac{1}{x^{-2/3}}$
34. $\frac{1}{a^{-7/8}}$
35. $2^{-1/3}x^4y^{-2/7}$
36. $3^{-5/2}a^3b^{-7/3}$
37. $\frac{7x}{\sqrt[3]{z}}$
38. $\frac{6a}{\sqrt[4]{b}}$
39. $\frac{5a}{3c^{-1/2}}$
40. $\frac{2z}{5x^{-1/3}}$

c Use the laws of exponents to simplify. Write the answers with positive exponents.

41. $5^{3/4} \cdot 5^{1/8}$
42. $11^{2/3} \cdot 11^{1/2}$
43. $\frac{7^{5/8}}{7^{3/8}}$
44. $\frac{3^{5/8}}{3^{-1/8}}$
45. $\frac{4.9^{-1/6}}{4.9^{-2/3}}$
46. $\frac{2.3^{-3/10}}{2.3^{-1/5}}$
47. $(6^{3/8})^{2/7}$
48. $(3^{2/9})^{3/5}$
49. $a^{2/3} \cdot a^{5/4}$
50. $x^{3/4} \cdot x^{2/3}$

51. $(a^{2/3} \cdot b^{5/8})^4$

52. $(x^{-1/3} \cdot y^{-2/5})^{-15}$

53. $(x^{2/3})^{-3/7}$

54. $(a^{-3/2})^{2/9}$

55. $\left(\frac{x^{3/4}}{y^{1/2}}\right)^{-2/3}$

56. $\left(\frac{a^{-3/2}}{b^{-5/3}}\right)^{1/3}$

57. $(m^{-1/4} \cdot n^{-5/6})^{-12/5}$

58. $(x^{3/8} \cdot y^{5/2})^{4/3}$

d Use rational exponents to simplify. Write the answer in radical notation if appropriate.

59. $\sqrt[6]{a^2}$

60. $\sqrt[6]{t^4}$

61. $\sqrt[3]{x^{15}}$

62. $\sqrt[4]{a^{12}}$

63. $\sqrt[6]{x^{-18}}$

64. $\sqrt[5]{a^{-10}}$

65. $(\sqrt[3]{ab})^{15}$

66. $(\sqrt[7]{cd})^{14}$

67. $\sqrt[14]{128}$

68. $\sqrt[6]{81}$

69. $\sqrt[6]{4x^2}$

70. $\sqrt[3]{8y^6}$

71. $\sqrt{x^4y^6}$

72. $\sqrt[4]{16x^4y^2}$

73. $\sqrt[5]{32c^{10}d^{15}}$

Use rational exponents to write a single radical expression.

74. $\sqrt[3]{3}\sqrt{3}$

75. $\sqrt[3]{7} \cdot \sqrt[4]{5}$

76. $\sqrt{11} \cdot \sqrt[6]{13}$

77. $\sqrt[4]{5} \cdot \sqrt[5]{7}$

78. $\sqrt[3]{y}\sqrt[5]{3y}$

79. $\sqrt{x}\sqrt[3]{2x}$

80. $(\sqrt[3]{x^2y^5})^{12}$

81. $(\sqrt[5]{a^2b^4})^{15}$

82. $\sqrt[4]{\sqrt{x}}$

83. $\sqrt[3]{\sqrt[6]{m}}$

84. $a^{2/3} \cdot b^{3/4}$

85. $x^{1/3} \cdot y^{1/4} \cdot z^{1/6}$

86. $\frac{x^{8/15} \cdot y^{7/5}}{x^{1/3} \cdot y^{-1/5}}$

87. $\left(\frac{c^{-4/5}d^{5/9}}{c^{3/10}d^{1/6}}\right)^3$

88. $\sqrt[3]{\sqrt[4]{xy}}$

Skill Maintenance

Solve. [1.6c]

89. $|7x - 5| = 9$

90. $|3x| = 120$

91. $8 - |2x + 5| = -2$

92. $\left|\frac{1}{2} + x\right| = \frac{7}{8}$

Synthesis

93. Use the SIMULTANEOUS mode to graph
$$y_1 = x^{1/2}, \quad y_2 = 3x^{2/5}, \quad y_3 = x^{4/7}, \quad y_4 = \tfrac{1}{5}x^{3/4}.$$
Then, looking only at coordinates, match each graph with its equation.

94. Simplify:
$$(\sqrt[10]{\sqrt[5]{x^{15}}})^5(\sqrt[5]{\sqrt[10]{x^{15}}})^5.$$

6.3 Simplifying Radical Expressions

OBJECTIVES

a. Multiply and simplify radical expressions.

b. Divide and simplify radical expressions.

SKILL REVIEW *Find principal square roots and their opposites.* [6.1a]

Find the square roots.

1. 81 **2.** 100

Answers: 1. 9, −9 **2.** 10, −10

MyLab Math VIDEO

a MULTIPLYING AND SIMPLIFYING RADICAL EXPRESSIONS

Note that $\sqrt{4}\sqrt{25} = 2 \cdot 5 = 10$. Also $\sqrt{4 \cdot 25} = \sqrt{100} = 10$. Likewise,

$$\sqrt[3]{27}\sqrt[3]{8} = 3 \cdot 2 = 6 \quad \text{and} \quad \sqrt[3]{27 \cdot 8} = \sqrt[3]{216} = 6.$$

These examples suggest the following.

THE PRODUCT RULE FOR RADICALS

For any nonnegative real numbers a and b and any index k,

$$\sqrt[k]{a} \cdot \sqrt[k]{b} = \sqrt[k]{a \cdot b}, \quad \text{or} \quad a^{1/k} \cdot b^{1/k} = (ab)^{1/k}.$$

The index must be the same throughout.

(To multiply, multiply the radicands.)

EXAMPLES Multiply.

1. $\sqrt{3} \cdot \sqrt{5} = \sqrt{3 \cdot 5} = \sqrt{15}$

2. $\sqrt{5a}\sqrt{2b} = \sqrt{5a \cdot 2b} = \sqrt{10ab}$

3. $\sqrt[3]{4}\sqrt[3]{5} = \sqrt[3]{4 \cdot 5} = \sqrt[3]{20}$

4. $\sqrt[4]{\frac{y}{5}}\sqrt[4]{\frac{7}{x}} = \sqrt[4]{\frac{y}{5} \cdot \frac{7}{x}} = \sqrt[4]{\frac{7y}{5x}}$

Caution! A common error is to omit the index in the answer.

◀ **Do Exercises 1–4.**

Multiply.

1. $\sqrt{19}\sqrt{7}$
2. $\sqrt{3p}\sqrt{7q}$
3. $\sqrt[4]{403}\sqrt[4]{7}$
4. $\sqrt[3]{\frac{5}{p}} \cdot \sqrt[3]{\frac{2}{q}}$

Keep in mind that the product rule can be used only when the indexes are the same. When indexes differ, we can use rational exponents as we did in Examples 23 and 24 of Section 6.2.

EXAMPLE 5 Multiply: $\sqrt{5x} \cdot \sqrt[4]{3y}$.

$\sqrt{5x} \cdot \sqrt[4]{3y} = (5x)^{1/2}(3y)^{1/4}$ — Converting to exponential notation

$= (5x)^{2/4}(3y)^{1/4}$ — Rewriting so that the exponents have a common denominator

$= [(5x)^2(3y)]^{1/4}$ — Using $a^n b^n = (ab)^n$

$= [(25x^2)(3y)]^{1/4}$ — Squaring $5x$

$= \sqrt[4]{(25x^2)(3y)}$ — Converting back to radical notation

$= \sqrt[4]{75x^2y}$ — Multiplying under the radical

◀ **Do Exercises 5 and 6.**

Multiply.

5. $\sqrt{5}\sqrt[3]{2}$
6. $\sqrt[4]{x}\sqrt[3]{2y}$

Answers

1. $\sqrt{133}$ **2.** $\sqrt{21pq}$ **3.** $\sqrt[4]{2821}$ **4.** $\sqrt[3]{\frac{10}{pq}}$ **5.** $\sqrt[6]{500}$ **6.** $\sqrt[12]{16x^3y^4}$

We can reverse the product rule to simplify a product. We simplify the root of a product by taking the root of each factor separately.

FACTORING RADICAL EXPRESSIONS

For any nonnegative real numbers a and b and any index k,

$$\sqrt[k]{ab} = \sqrt[k]{a} \cdot \sqrt[k]{b}, \quad \text{or} \quad (ab)^{1/k} = a^{1/k} \cdot b^{1/k}.$$

(Take the kth root of each factor separately.)

Compare the following:

$$\sqrt{50} = \sqrt{10 \cdot 5} = \sqrt{10}\sqrt{5};$$
$$\sqrt{50} = \sqrt{25 \cdot 2} = \sqrt{25}\sqrt{2} = 5\sqrt{2}.$$

In the second case, the radicand is written with the perfect-square factor 25. If you do not recognize any perfect-square factors, try factoring the radicand into its prime factors. For example,

$$\sqrt{50} = \sqrt{2 \cdot \underbrace{5 \cdot 5}} = 5\sqrt{2}.$$

Perfect square (a pair of the same numbers)

Square-root radical expressions in which the radicand has no perfect-square factors, such as $5\sqrt{2}$, are considered to be in simplest form. A procedure for simplifying kth roots follows.

SIMPLIFYING kTH ROOTS

To simplify a radical expression by factoring:

1. Look for the largest factors of the radicand that are perfect kth powers (where k is the index).
2. Then take the kth root of the resulting factors.
3. A radical expression, with index k, is *simplified* when its radicand has no factors that are perfect kth powers.

EXAMPLES Simplify by factoring.

6. $\sqrt{50} = \sqrt{25 \cdot 2} = \sqrt{25} \cdot \sqrt{2} = \sqrt{5 \cdot 5} \cdot \sqrt{2} = 5\sqrt{2}$

This factor is a perfect square.

7. $\sqrt[3]{32} = \sqrt[3]{8 \cdot 4} = \sqrt[3]{8} \cdot \sqrt[3]{4} = \sqrt[3]{2 \cdot 2 \cdot 2} \cdot \sqrt[3]{2 \cdot 2} = 2\sqrt[3]{4}$

This factor is a perfect cube (third power).

8. $\sqrt[4]{48} = \sqrt[4]{16 \cdot 3} = \sqrt[4]{16} \cdot \sqrt[4]{3} = \sqrt[4]{2 \cdot 2 \cdot 2 \cdot 2} \cdot \sqrt[4]{3} = 2\sqrt[4]{3}$

This factor is a perfect fourth power.

Do Exercises 7 and 8. ▶

Simplify by factoring.

7. $\sqrt{32}$

8. $\sqrt[3]{80}$

Answers

7. $4\sqrt{2}$ **8.** $2\sqrt[3]{10}$

Frequently, expressions under radicals do not contain negative numbers raised to even powers. In such cases, absolute-value notation is not necessary. **For this reason, we will no longer use absolute-value notation.**

EXAMPLES Simplify by factoring. Assume that no radicands were formed by raising negative numbers to even powers.

9. $\sqrt{5x^2} = \sqrt{5 \cdot x^2}$ Factoring the radicand

$= \sqrt{5} \cdot \sqrt{x^2}$ Factoring into two radicals

$= \sqrt{5} \cdot x$ Taking the square root of x^2

Absolute-value notation is not needed because we assume that x is not negative.

10. $\sqrt{18x^2y} = \sqrt{9 \cdot 2 \cdot x^2 \cdot y}$ Looking for perfect-square factors and factoring the radicand

$= \sqrt{9 \cdot x^2 \cdot 2 \cdot y}$

$= \sqrt{9} \cdot \sqrt{x^2} \cdot \sqrt{2 \cdot y}$ Factoring into several radicals

$= 3x\sqrt{2y}$ Taking square roots

11. $\sqrt{216x^5y^3} = \sqrt{36 \cdot 6 \cdot x^4 \cdot x \cdot y^2 \cdot y}$ Looking for perfect-square factors and factoring the radicand

$= \sqrt{36 \cdot x^4 \cdot y^2 \cdot 6 \cdot x \cdot y}$

$= \sqrt{36}\sqrt{x^4}\sqrt{y^2}\sqrt{6xy}$ Factoring into several radicals

$= 6x^2y\sqrt{6xy}$ Taking square roots

Let's look at this example another way. We do a complete factorization and look for pairs of factors. Each pair of factors makes a square:

$\sqrt{216x^5y^3} = \sqrt{2 \cdot 2 \cdot 2 \cdot 3 \cdot 3 \cdot 3 \cdot x \cdot x \cdot x \cdot x \cdot x \cdot y \cdot y \cdot y}$ Each pair of factors makes a perfect square.

$= 2 \cdot 3 \cdot x \cdot x \cdot y \cdot \sqrt{2 \cdot 3 \cdot x \cdot y}$

$= 6x^2y\sqrt{6xy}.$

12. $\sqrt[3]{16a^7b^{11}} = \sqrt[3]{8 \cdot 2 \cdot a^6 \cdot a \cdot b^9 \cdot b^2}$ Factoring the radicand. The index is 3, so we look for the largest powers that are multiples of 3 because these are perfect cubes.

$= \sqrt[3]{8} \cdot \sqrt[3]{a^6} \cdot \sqrt[3]{b^9} \cdot \sqrt[3]{2ab^2}$ Factoring into radicals

$= 2a^2b^3\sqrt[3]{2ab^2}$ Taking cube roots

Let's look at this example another way. We do a complete factorization and look for triples of factors. Each triple of factors makes a cube:

$\sqrt[3]{16a^7b^{11}}$

$= \sqrt[3]{2 \cdot 2 \cdot 2 \cdot 2 \cdot a \cdot a \cdot a \cdot a \cdot a \cdot a \cdot a \cdot b \cdot b \cdot b \cdot b \cdot b \cdot b \cdot b \cdot b \cdot b \cdot b \cdot b}$

Each triple of factors makes a cube.

$= 2 \cdot a \cdot a \cdot b \cdot b \cdot b \cdot \sqrt[3]{2 \cdot a \cdot b \cdot b}$

$= 2a^2b^3\sqrt[3]{2ab^2}.$

Do Exercises 9–14.

Simplify by factoring. Assume that no radicands were formed by raising negative numbers to even powers.

9. $\sqrt{300}$ **10.** $\sqrt{36y^2}$

11. $\sqrt{27zw^2}$ **12.** $\sqrt[3]{16}$

13. $\sqrt{12ab^3c^2}$

$= \sqrt{\boxed{} \cdot 3 \cdot a \cdot b^2 \cdot \boxed{} \cdot c^2}$

$= \sqrt{4} \cdot \sqrt{b^2} \cdot \sqrt{c^2} \cdot \sqrt{\boxed{}}$

$= 2 \cdot \boxed{} \cdot c\sqrt{3ab}$

GS

14. $\sqrt[3]{81x^4y^8}$

Answers

9. $10\sqrt{3}$ **10.** $6y$ **11.** $3w\sqrt{3z}$
12. $2\sqrt[3]{2}$ **13.** $2bc\sqrt{3ab}$ **14.** $3xy^2\sqrt[3]{3xy^2}$

Guided Solution:
13. 4, b, $3ab$, b

Sometimes after we have multiplied, we can simplify by factoring.

EXAMPLES Multiply and simplify. Assume that no radicands were formed by raising negative numbers to even powers.

13. $\sqrt{20}\sqrt{8} = \sqrt{20 \cdot 8} = \sqrt{\underline{4} \cdot 5 \cdot \underline{4} \cdot 2} = 4\sqrt{10}$

14. $3\sqrt[3]{25} \cdot 2\sqrt[3]{5} = 3 \cdot 2 \cdot \sqrt[3]{25} \cdot \sqrt[3]{5} = 6 \cdot \sqrt[3]{25 \cdot 5}$
$= 6 \cdot \sqrt[3]{5 \cdot 5 \cdot 5}$
$= 6 \cdot 5 = 30$

15. $\sqrt[3]{18y^3}\sqrt[3]{4x^2} = \sqrt[3]{18y^3 \cdot 4x^2}$ Multiplying radicands
$= \sqrt[3]{\underline{2} \cdot 3 \cdot 3 \cdot \underline{y \cdot y \cdot y} \cdot \underline{2 \cdot 2} \cdot x \cdot x}$
$= 2 \cdot y \cdot \sqrt[3]{3 \cdot 3 \cdot x \cdot x}$
$= 2y\sqrt[3]{9x^2}$

Do Exercises 15–18. ▶

Multiply and simplify. Assume that no radicands were formed by raising negative numbers to even powers.

15. $\sqrt{3}\sqrt{6}$ **16.** $\sqrt{18y}\sqrt{14y}$

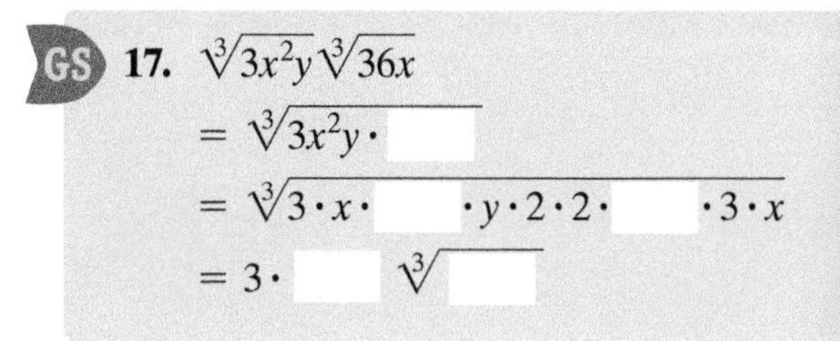

18. $\sqrt{7a}\sqrt{21b}$

b DIVIDING AND SIMPLIFYING RADICAL EXPRESSIONS

Note that $\dfrac{\sqrt[3]{27}}{\sqrt[3]{8}} = \dfrac{3}{2}$ and that $\sqrt[3]{\dfrac{27}{8}} = \dfrac{3}{2}$. This example suggests the following.

THE QUOTIENT RULE FOR RADICALS

For any nonnegative number a, any positive number b, and any index k,

$$\frac{\sqrt[k]{a}}{\sqrt[k]{b}} = \sqrt[k]{\frac{a}{b}}, \quad \text{or} \quad \frac{a^{1/k}}{b^{1/k}} = \left(\frac{a}{b}\right)^{1/k}.$$

(To divide, divide the radicands. After doing this, you can sometimes simplify by taking roots.)

EXAMPLES Divide and simplify. Assume that no radicands were formed by raising negative numbers to even powers.

16. $\dfrac{\sqrt{80}}{\sqrt{5}} = \sqrt{\dfrac{80}{5}} = \sqrt{16} = 4$ We divide the radicands.

17. $\dfrac{5\sqrt[3]{32}}{\sqrt[3]{2}} = 5\sqrt[3]{\dfrac{32}{2}} = 5\sqrt[3]{16} = 5\sqrt[3]{8 \cdot 2} = 5\sqrt[3]{8}\sqrt[3]{2} = 5 \cdot 2\sqrt[3]{2} = 10\sqrt[3]{2}$

18. $\dfrac{\sqrt{72xy}}{2\sqrt{2}} = \dfrac{1}{2} \cdot \dfrac{\sqrt{72xy}}{\sqrt{2}} = \dfrac{1}{2}\sqrt{\dfrac{72xy}{2}} = \dfrac{1}{2}\sqrt{36xy} = \dfrac{1}{2}\sqrt{36}\sqrt{xy}$
$= \dfrac{1}{2} \cdot 6\sqrt{xy} = 3\sqrt{xy}$

Do Exercises 19–22. ▶

Divide and simplify. Assume that no radicands were formed by raising negative numbers to even powers.

19. $\dfrac{\sqrt{75}}{\sqrt{3}}$ **20.** $\dfrac{14\sqrt{128xy}}{2\sqrt{2}}$

21. $\dfrac{\sqrt{50a^3}}{\sqrt{2a}}$ **22.** $\dfrac{4\sqrt[3]{250}}{7\sqrt[3]{2}}$

Answers

15. $3\sqrt{2}$ **16.** $6y\sqrt{7}$ **17.** $3x\sqrt[3]{4y}$ **18.** $7\sqrt{3ab}$ **19.** 5 **20.** $56\sqrt{xy}$ **21.** $5a$ **22.** $\dfrac{20}{7}$

Guided Solution:
17. $36x, x, 3, x, 4y$

We can simplify the root of a quotient by taking the roots of the numerator and of the denominator separately.

kTH ROOTS OF QUOTIENTS

For any nonnegative number a, any positive number b, and any index k,

$$\sqrt[k]{\frac{a}{b}} = \frac{\sqrt[k]{a}}{\sqrt[k]{b}}, \quad \text{or} \quad \left(\frac{a}{b}\right)^{1/k} = \frac{a^{1/k}}{b^{1/k}}.$$

(Take the kth roots of the numerator and of the denominator separately.)

EXAMPLES Simplify by taking the roots of the numerator and of the denominator. Assume that no radicands were formed by raising negative numbers to even powers.

19. $\sqrt[3]{\frac{27}{125}} = \frac{\sqrt[3]{27}}{\sqrt[3]{125}} = \frac{3}{5}$ — We take the cube root of the numerator and of the denominator.

20. $\sqrt{\frac{25}{y^2}} = \frac{\sqrt{25}}{\sqrt{y^2}} = \frac{5}{y}$ — We take the square root of the numerator and of the denominator.

21. $\sqrt{\frac{16x^3}{y^4}} = \frac{\sqrt{16x^3}}{\sqrt{y^4}} = \frac{\sqrt{16x^2 \cdot x}}{\sqrt{y^4}} = \frac{\sqrt{16x^2} \cdot \sqrt{x}}{\sqrt{y^4}} = \frac{4x\sqrt{x}}{y^2}$

22. $\sqrt[3]{\frac{27y^5}{343x^3}} = \frac{\sqrt[3]{27y^5}}{\sqrt[3]{343x^3}} = \frac{\sqrt[3]{27y^3 \cdot y^2}}{\sqrt[3]{343x^3}} = \frac{\sqrt[3]{27y^3} \cdot \sqrt[3]{y^2}}{\sqrt[3]{343x^3}} = \frac{3y\sqrt[3]{y^2}}{7x}$

We are assuming here that no variable represents 0 or a negative number. Thus we need not be concerned about zero denominators or absolute value.

Do Exercises 23–25.

When indexes differ, we can use rational exponents.

EXAMPLE 23 Divide and simplify: $\frac{\sqrt[3]{a^2b^4}}{\sqrt{ab}}$.

$$\frac{\sqrt[3]{a^2b^4}}{\sqrt{ab}} = \frac{(a^2b^4)^{1/3}}{(ab)^{1/2}}$$ Converting to exponential notation

$$= \frac{a^{2/3}b^{4/3}}{a^{1/2}b^{1/2}}$$ Using the product and power rules

$$= a^{2/3-1/2}b^{4/3-1/2}$$ Subtracting exponents

$$= a^{4/6-3/6}b^{8/6-3/6}$$ Finding common denominators so exponents can be subtracted

$$= a^{1/6}b^{5/6}$$

$$= (ab^5)^{1/6}$$ Using $a^n b^n = (ab)^n$

$$= \sqrt[6]{ab^5}$$ Converting back to radical notation

Do Exercise 26.

Simplify by taking the roots of the numerator and of the denominator. Assume that no radicands were formed by raising negative numbers to even powers.

23. $\sqrt{\frac{25}{36}}$

24. $\sqrt{\frac{x^2}{100}}$

25. $\sqrt[3]{\frac{54x^5}{125}}$

26. Divide and simplify:

$$\frac{\sqrt[4]{x^3y^2}}{\sqrt[3]{x^2y}}.$$

Answers

23. $\frac{5}{6}$ **24.** $\frac{x}{10}$ **25.** $\frac{3x\sqrt[3]{2x^2}}{5}$ **26.** $\sqrt[12]{xy^2}$

6.3 Exercise Set

FOR EXTRA HELP MyLab Math

✓ Check Your Understanding

Reading Check Determine whether each statement is true or false.

RC1. For any nonnegative real numbers a and b and any index k, $\sqrt[k]{a} \cdot \sqrt[k]{b} = \sqrt[k]{ab}$.

RC2. For $q > 0$, $\sqrt{q^2 - 100} = q + 10$.

RC3. The expression $\sqrt{Y}$ is not simplified if Y contains a factor that contains a perfect square.

RC4. For any nonnegative number a, any positive number b, and any index k, $\dfrac{\sqrt[k]{a}}{\sqrt[k]{b}} = \sqrt[k]{\dfrac{a}{b}}$.

Concept Check Determine whether each radical expression is in simplest form.

CC1. $\sqrt[4]{162x^2}$

CC2. $3\sqrt{121}$

CC3. $\sqrt[3]{15y^2}$

CC4. $\sqrt{320}$

CC5. $\sqrt{210xy}$

CC6. $\sqrt[3]{6y^6}$

a Simplify by factoring. Assume that no radicands were formed by raising negative numbers to even powers.

1. $\sqrt{24}$

2. $\sqrt{20}$

3. $\sqrt{90}$

4. $\sqrt{18}$

5. $\sqrt[3]{250}$

6. $\sqrt[3]{108}$

7. $\sqrt{180x^4}$

8. $\sqrt{175y^6}$

9. $\sqrt[3]{54x^8}$

10. $\sqrt[3]{40y^3}$

11. $\sqrt[3]{80t^8}$

12. $\sqrt[3]{108x^5}$

13. $\sqrt[4]{80}$

14. $\sqrt[4]{32}$

15. $\sqrt{32a^2b}$

16. $\sqrt{75p^3q^4}$

17. $\sqrt[4]{243x^8y^{10}}$

18. $\sqrt[4]{162c^4d^6}$

19. $\sqrt[5]{96x^7y^{15}}$

20. $\sqrt[5]{p^{14}q^9r^{23}}$

Multiply and simplify. Assume that no radicands were formed by raising negative numbers to even powers.

21. $\sqrt{10}\sqrt{5}$

22. $\sqrt{6}\sqrt{3}$

23. $\sqrt{15}\sqrt{6}$

24. $\sqrt{2}\sqrt{32}$

25. $\sqrt[3]{2}\sqrt[3]{4}$

26. $\sqrt[3]{9}\sqrt[3]{3}$

27. $\sqrt{45}\sqrt{60}$

28. $\sqrt{24}\sqrt{75}$

29. $\sqrt{3x^3}\sqrt{6x^5}$

30. $\sqrt{5a^7}\sqrt{15a^3}$

31. $\sqrt{5b^3}\sqrt{10c^4}$

32. $\sqrt{2x^3y}\sqrt{12xy}$

33. $\sqrt[3]{5a^2}\sqrt[3]{2a}$

34. $\sqrt[3]{7x}\sqrt[3]{3x^2}$

35. $\sqrt[3]{y^4}\sqrt[3]{16y^5}$

36. $\sqrt[3]{s^2t^4}\sqrt[3]{s^4t^6}$

37. $\sqrt[4]{16}\sqrt[4]{64}$

38. $\sqrt[5]{64}\sqrt[5]{16}$

39. $\sqrt{12a^3b}\sqrt{8a^4b^2}$

40. $\sqrt{30x^3y^4}\sqrt{18x^2y^5}$

41. $\sqrt{2}\sqrt[3]{5}$

42. $\sqrt{6}\sqrt[3]{5}$

43. $\sqrt[4]{3}\sqrt{2}$

44. $\sqrt[3]{5}\sqrt[4]{2}$

45. $\sqrt{a}\sqrt[4]{a^3}$

46. $\sqrt[3]{x^2}\sqrt[6]{x^5}$

47. $\sqrt[5]{b^2}\sqrt{b^3}$

48. $\sqrt[4]{a^3}\sqrt[3]{a^2}$

49. $\sqrt{xy^3}\sqrt[3]{x^2y}$

50. $\sqrt{y^5z}\sqrt[3]{yz^4}$

51. $\sqrt{2a^3b}\sqrt[4]{8ab^2}$

52. $\sqrt[4]{9ab^3}\sqrt{3a^4b}$

b Divide and simplify. Assume that all expressions under radicals represent positive numbers.

53. $\dfrac{\sqrt{90}}{\sqrt{5}}$

54. $\dfrac{\sqrt{98}}{\sqrt{2}}$

55. $\dfrac{\sqrt{35q}}{\sqrt{7q}}$

56. $\dfrac{\sqrt{30x}}{\sqrt{10x}}$

57. $\dfrac{\sqrt[3]{54}}{\sqrt[3]{2}}$

58. $\dfrac{\sqrt[3]{40}}{\sqrt[3]{5}}$

59. $\dfrac{\sqrt{56xy^3}}{\sqrt{8x}}$

60. $\dfrac{\sqrt{52ab^3}}{\sqrt{13a}}$

61. $\dfrac{\sqrt[3]{96a^4b^2}}{\sqrt[3]{12a^2b}}$

62. $\dfrac{\sqrt[3]{189x^5y^7}}{\sqrt[3]{7x^2y^2}}$

63. $\dfrac{\sqrt{128xy}}{2\sqrt{2}}$

64. $\dfrac{\sqrt{48ab}}{2\sqrt{3}}$

65. $\dfrac{\sqrt[4]{48x^9y^{13}}}{\sqrt[4]{3xy^5}}$

66. $\dfrac{\sqrt[5]{64a^{11}b^{28}}}{\sqrt[5]{2ab^2}}$

67. $\dfrac{\sqrt[3]{a}}{\sqrt{a}}$

68. $\dfrac{\sqrt{x}}{\sqrt[4]{x}}$

69. $\dfrac{\sqrt[3]{a^2}}{\sqrt[4]{a}}$

70. $\dfrac{\sqrt[3]{x^2}}{\sqrt[5]{x}}$

71. $\dfrac{\sqrt[4]{x^2y^3}}{\sqrt[3]{xy}}$

72. $\dfrac{\sqrt[5]{a^4b^2}}{\sqrt[3]{ab^2}}$

Simplify.

73. $\sqrt{\dfrac{25}{36}}$

74. $\sqrt{\dfrac{49}{64}}$

75. $\sqrt{\dfrac{16}{49}}$

76. $\sqrt{\dfrac{100}{81}}$

77. $\sqrt[3]{\dfrac{125}{27}}$

78. $\sqrt[3]{\dfrac{343}{1000}}$

79. $\sqrt{\dfrac{49}{y^2}}$

80. $\sqrt{\dfrac{121}{x^2}}$

81. $\sqrt{\dfrac{25y^3}{x^4}}$

82. $\sqrt{\dfrac{36a^5}{b^6}}$

83. $\sqrt[3]{\dfrac{81y^5}{64}}$

84. $\sqrt[3]{\dfrac{8z^7}{125}}$

85. $\sqrt[3]{\dfrac{27a^4}{8b^3}}$

86. $\sqrt[3]{\dfrac{64x^7}{216y^6}}$

87. $\sqrt[4]{\dfrac{81x^4}{16}}$

88. $\sqrt[4]{\dfrac{256}{81x^8}}$

89. $\sqrt[4]{\dfrac{16a^{12}}{b^4c^{16}}}$

90. $\sqrt[4]{\dfrac{81x^4}{y^8z^4}}$

91. $\sqrt[5]{\dfrac{32x^8}{y^{10}}}$

92. $\sqrt[5]{\dfrac{32b^{10}}{243a^{20}}}$

93. $\sqrt[5]{\dfrac{w^7}{z^{10}}}$

94. $\sqrt[5]{\dfrac{z^{11}}{w^{20}}}$

95. $\sqrt[6]{\dfrac{x^{13}}{y^6z^{12}}}$

96. $\sqrt[6]{\dfrac{p^9q^{24}}{r^{18}}}$

Skill Maintenance

Solve. [4.8b]

97. The sum of a number and its square is 90. Find the number.

98. *Triangle Dimensions.* The base of a triangle is 2 in. longer than the height. The area is 12 in^2. Find the height and the base.

Solve. [5.5a]

99. $\dfrac{12x}{x-4} - \dfrac{3x^2}{x+4} = \dfrac{384}{x^2-16}$

100. $\dfrac{4x}{x+5} + \dfrac{20}{x} = \dfrac{100}{x^2+5x}$

Synthesis

101. *Pendulums.* The **period** of a pendulum is the time it takes to complete one cycle, swinging to and fro. For a pendulum that is L centimeters long, the period T is given by the function

$$T(L) = 2\pi\sqrt{\frac{L}{980}},$$

where T is in seconds. Find, to the nearest hundredth of a second, the period of a pendulum of length **(a)** 65 cm; **(b)** 98 cm; **(c)** 120 cm. Use a calculator's π key if possible.

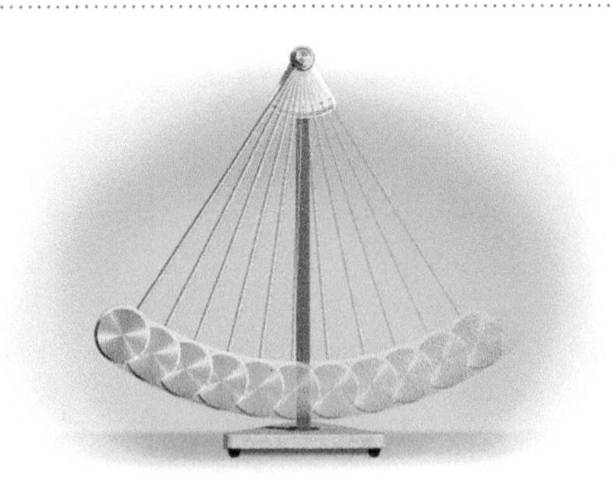

Simplify.

102. $\dfrac{\sqrt[3]{x^3-y^3}}{\sqrt[3]{x-y}}$

103. $\dfrac{\sqrt{44x^2y^9z}\sqrt{22y^9z^6}}{(\sqrt{11xy^8z^2})^2}$

Addition, Subtraction, and More Multiplication

6.4

OBJECTIVES

a Add or subtract with radical notation and simplify.

b Multiply expressions involving radicals in which some factors contain more than one term.

a ADDITION AND SUBTRACTION

Any two real numbers can be added. For example, the sum of 7 and $\sqrt{3}$ can be expressed as $7 + \sqrt{3}$. We cannot simplify this sum. However, when we have **like radicals** (radicals having the same index and radicand), we can use the distributive laws to simplify by collecting like radical terms. For example,

$$7\sqrt{3} + \sqrt{3} = 7\sqrt{3} + 1\sqrt{3} = (7 + 1)\sqrt{3} = 8\sqrt{3}.$$

SKILL REVIEW *Simplify an expression by collecting like terms.* [J17]

Collect like terms.

1. $2x + 5x$
2. $y + 3 - 4y + 1$

Answers: 1. $7x$ **2.** $-3y + 4$

MyLab Math VIDEO

EXAMPLES Add or subtract.

1. $6\sqrt{7} + 4\sqrt{7} = (6 + 4)\sqrt{7}$ Using a distributive law
$= 10\sqrt{7}$

2. $8\sqrt[3]{2} - 7x\sqrt[3]{2} + 5\sqrt[3]{2} = (8 - 7x + 5)\sqrt[3]{2}$ Factoring out $\sqrt[3]{2}$
$= (13 - 7x)\sqrt[3]{2}$

These parentheses are necessary!

3. $3\sqrt[5]{4x} + 7\sqrt[5]{4x} - \sqrt[3]{4x} = (3 + 7)\sqrt[5]{4x} - \sqrt[3]{4x}$
$= 10\sqrt[5]{4x} - \sqrt[3]{4x}$

Note that these expressions have the same *radicand,* but they are not like radicals because they do not have the same *index.*

Do Exercises 1 and 2. ▶

Add or subtract.

1. $5\sqrt{2} + 8\sqrt{2}$
2. $7\sqrt[4]{5x} + 3\sqrt[4]{5x} - \sqrt{7}$

Sometimes, in order to determine whether there are like radical terms, we need to simplify radicals by factoring.

EXAMPLES Add or subtract.

4. $3\sqrt{8} - 5\sqrt{2} = 3\sqrt{4 \cdot 2} - 5\sqrt{2}$ Factoring 8
$= 3\sqrt{4} \cdot \sqrt{2} - 5\sqrt{2}$ Factoring $\sqrt{4 \cdot 2}$ into two radicals
$= 3 \cdot 2\sqrt{2} - 5\sqrt{2}$ Taking the square root of 4
$= 6\sqrt{2} - 5\sqrt{2}$
$= (6 - 5)\sqrt{2}$ Collecting like radical terms
$= \sqrt{2}$

5. $5\sqrt{2} - 4\sqrt{3}$ No simplification is possible.

6. $5\sqrt[3]{16y^4} + 7\sqrt[3]{2y} = 5\sqrt[3]{8y^3 \cdot 2y} + 7\sqrt[3]{2y}$
$= 5\sqrt[3]{8y^3} \cdot \sqrt[3]{2y} + 7\sqrt[3]{2y}$ } Factoring the first radical
$= 5 \cdot 2y \cdot \sqrt[3]{2y} + 7\sqrt[3]{2y}$
$= 10y\sqrt[3]{2y} + 7\sqrt[3]{2y}$ Taking the cube root of $8y^3$
$= (10y + 7)\sqrt[3]{2y}$ Collecting like radical terms

Do Exercises 3–5. ▶

Add or subtract.

3. $7\sqrt{45} - 2\sqrt{5}$
4. $3\sqrt[3]{y^5} + 4\sqrt[3]{y^2} + \sqrt[3]{8y^6}$

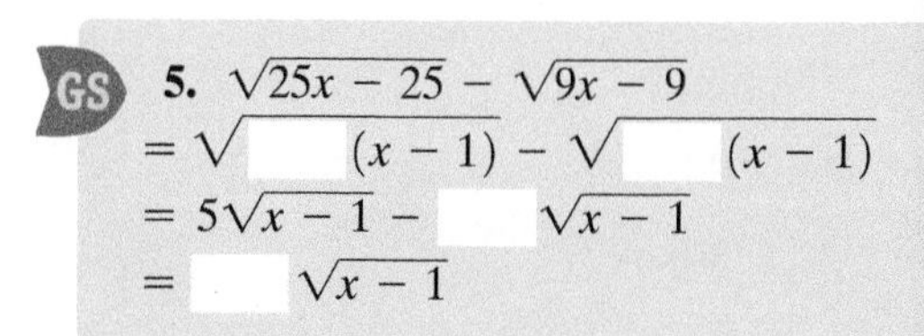

Answers

1. $13\sqrt{2}$ **2.** $10\sqrt[4]{5x} - \sqrt{7}$ **3.** $19\sqrt{5}$
4. $(3y + 4)\sqrt[3]{y^2} + 2y^2$

Guided Solution:
5. 25, 9, 3, 2

b MORE MULTIPLICATION

To multiply expressions in which some factors contain more than one term, we use the procedures for multiplying polynomials.

EXAMPLES Multiply.

7. $\sqrt{3}(x - \sqrt{5}) = \sqrt{3} \cdot x - \sqrt{3} \cdot \sqrt{5}$ Using a distributive law
$= x\sqrt{3} - \sqrt{15}$ Multiplying radicals

8. $\sqrt[3]{y}(\sqrt[3]{y^2} + \sqrt[3]{2}) = \sqrt[3]{y} \cdot \sqrt[3]{y^2} + \sqrt[3]{y} \cdot \sqrt[3]{2}$ Using a distributive law
$= \sqrt[3]{y^3} + \sqrt[3]{2y}$ Multiplying radicals
$= y + \sqrt[3]{2y}$ Simplifying $\sqrt[3]{y^3}$

Do Exercises 6 and 7.

Multiply. Assume that no radicands were formed by raising negative numbers to even powers.

6. $\sqrt{2}(5\sqrt{3} + 3\sqrt{7})$

7. $\sqrt[3]{a^2}(\sqrt[3]{3a} - \sqrt[3]{2})$

EXAMPLE 9 Multiply: $(4\sqrt{3} + \sqrt{2})(\sqrt{3} - 5\sqrt{2})$.

F O I L

$$(4\sqrt{3} + \sqrt{2})(\sqrt{3} - 5\sqrt{2}) = 4(\sqrt{3})^2 - 20\sqrt{3} \cdot \sqrt{2} + \sqrt{2} \cdot \sqrt{3} - 5(\sqrt{2})^2$$
$$= 4 \cdot 3 - 20\sqrt{6} + \sqrt{6} - 5 \cdot 2$$
$$= 12 - 20\sqrt{6} + \sqrt{6} - 10$$
$$= 2 - 19\sqrt{6}$$ Collecting like terms

EXAMPLE 10 Multiply: $(\sqrt{a} + \sqrt{3})(\sqrt{b} + \sqrt{3})$. Assume that all expressions under radicals represent nonnegative numbers.

$$(\sqrt{a} + \sqrt{3})(\sqrt{b} + \sqrt{3}) = \sqrt{a}\sqrt{b} + \sqrt{a}\sqrt{3} + \sqrt{3}\sqrt{b} + \sqrt{3}\sqrt{3}$$
$$= \sqrt{ab} + \sqrt{3a} + \sqrt{3b} + 3$$

EXAMPLE 11 Multiply: $(\sqrt{5} + \sqrt{7})(\sqrt{5} - \sqrt{7})$.

$(\sqrt{5} + \sqrt{7})(\sqrt{5} - \sqrt{7}) = (\sqrt{5})^2 - (\sqrt{7})^2$ This is now a difference of two squares: $(A - B)(A + B) = A^2 - B^2$.
$= 5 - 7 = -2$

EXAMPLE 12 Multiply: $(\sqrt{a} + \sqrt{b})(\sqrt{a} - \sqrt{b})$. Assume that no radicands were formed by raising negative numbers to even powers.

$(\sqrt{a} + \sqrt{b})(\sqrt{a} - \sqrt{b}) = (\sqrt{a})^2 - (\sqrt{b})^2$
$= a - b$ ← No radicals

Expressions of the form $\sqrt{a} + \sqrt{b}$ and $\sqrt{a} - \sqrt{b}$ are called **conjugates**. Their product is always an expression that has no radicals.

Do Exercises 8–11.

Multiply. Assume that no radicands were formed by raising negative numbers to even powers.

8. $(\sqrt{3} - 5\sqrt{2})(2\sqrt{3} + \sqrt{2})$

9. $(\sqrt{a} + 2\sqrt{3})(3\sqrt{b} - 4\sqrt{3})$

GS 10. $(\sqrt{2} + \sqrt{5})(\sqrt{2} - \sqrt{5})$
$= (\quad)^2 - (\sqrt{5})^2$
$= 2 - \quad$
$= \quad$

11. $(\sqrt{p} - \sqrt{q})(\sqrt{p} + \sqrt{q})$

EXAMPLE 13 Multiply: $(\sqrt{3} + x)^2$.

$(\sqrt{3} + x)^2 = (\sqrt{3})^2 + 2 \cdot \sqrt{3} \cdot x + x^2$ Squaring a binomial
$= 3 + 2x\sqrt{3} + x^2$

Do Exercises 12 and 13.

Multiply.

12. $(2\sqrt{5} - y)^2$

13. $(3\sqrt{6} + 2)^2$

Answers

5. $2\sqrt{x - 1}$ **6.** $5\sqrt{6} + 3\sqrt{14}$
7. $a\sqrt[3]{3} - \sqrt[3]{2a^2}$ **8.** $-4 - 9\sqrt{6}$
9. $3\sqrt{ab} - 4\sqrt{3a} + 6\sqrt{3b} - 24$
10. -3 **11.** $p - q$ **12.** $20 - 4y\sqrt{5} + y^2$
13. $58 + 12\sqrt{6}$

Guided Solution:
10. $\sqrt{2}, 5, -3$

6.4 Exercise Set

FOR EXTRA HELP MyLab Math

✓ Check Your Understanding

Reading Check Determine whether each statement is true or false.

RC1. The expression $3\sqrt{7y}$ and $\sqrt[3]{7y}$ are like radicals.

RC2. The expression $3 - \sqrt{3}$ cannot be simplified.

RC3. The expression $\sqrt{x} - \sqrt{y}$ and $\sqrt{x} + \sqrt{y}$ are conjugates.

RC4. The product of expressions that are conjugates is always an expression that has no radicals.

Concept Check Like radical terms have the *same* index and the *same* radicand. Determine whether the given pair of terms are like radicals. Answer "yes" or "no."

CC1. $4\sqrt[3]{5y},\ 2\sqrt[3]{5y}$

CC2. $5,\ 5\sqrt{2}$

CC3. $\sqrt[7]{x^2y^3},\ \sqrt[7]{x^2y^2}$

CC4. $q\sqrt[4]{q^3},\ 2\sqrt[4]{q^3}$

CC5. $-4\sqrt{3},\ \sqrt{3}$

CC6. $x\sqrt[3]{y},\ y\sqrt[3]{x}$

CC7. $3\sqrt[5]{a-b},\ 3\sqrt[4]{a-b}$

CC8. $\frac{1}{4}\sqrt[3]{\frac{x^2}{y}},\ 4\sqrt[3]{\frac{x^2}{y}}$

a Add or subtract. Assume that no radicands were formed by raising negative numbers to even powers.

1. $7\sqrt{5} + 4\sqrt{5}$

2. $2\sqrt{3} + 9\sqrt{3}$

3. $6\sqrt[3]{7} - 5\sqrt[3]{7}$

4. $13\sqrt[5]{3} - 8\sqrt[5]{3}$

5. $4\sqrt[3]{y} + 9\sqrt[3]{y}$

6. $6\sqrt[4]{t} - 3\sqrt[4]{t}$

7. $5\sqrt{6} - 9\sqrt{6} - 4\sqrt{6}$

8. $3\sqrt{10} - 8\sqrt{10} + 7\sqrt{10}$

9. $4\sqrt[3]{3} - \sqrt{5} + 2\sqrt[3]{3} + \sqrt{5}$

10. $5\sqrt{7} - 8\sqrt[4]{11} + \sqrt{7} + 9\sqrt[4]{11}$

11. $8\sqrt{27} - 3\sqrt{3}$

12. $9\sqrt{50} - 4\sqrt{2}$

13. $8\sqrt{45} + 7\sqrt{20}$

14. $9\sqrt{12} + 16\sqrt{27}$

15. $18\sqrt{72} + 2\sqrt{98}$

16. $12\sqrt{45} - 8\sqrt{80}$

17. $3\sqrt[3]{16} + \sqrt[3]{54}$

18. $\sqrt[3]{27} - 5\sqrt[3]{8}$

19. $2\sqrt{128} - \sqrt{18} + 4\sqrt{32}$

20. $5\sqrt{50} - 2\sqrt{18} + 9\sqrt{32}$

21. $\sqrt{5a} + 2\sqrt{45a^3}$

22. $4\sqrt{3x^3} - \sqrt{12x}$

23. $\sqrt[3]{24x} - \sqrt[3]{3x^4}$

24. $\sqrt[3]{54x} - \sqrt[3]{2x^4}$

25. $7\sqrt{27x^3} + \sqrt{3x}$

26. $2\sqrt{45x^3} - \sqrt{5x}$

27. $\sqrt{4} + \sqrt{18}$

28. $\sqrt[3]{8} - \sqrt[3]{24}$

29. $5\sqrt[3]{32} - \sqrt[3]{108} + 2\sqrt[3]{256}$

30. $3\sqrt[3]{8x} - 4\sqrt[3]{27x} + 2\sqrt[3]{64x}$

31. $\sqrt[3]{6x^4} + \sqrt[3]{48x} - \sqrt[3]{6x}$

32. $\sqrt[4]{80x^5} - \sqrt[4]{405x^9} + \sqrt[4]{5x}$

33. $\sqrt{4a - 4} + \sqrt{a - 1}$

34. $\sqrt{9y + 27} + \sqrt{y + 3}$

35. $\sqrt{x^3 - x^2} + \sqrt{9x - 9}$

36. $\sqrt{4x - 4} + \sqrt{x^3 - x^2}$

b Multiply. Assume that no radicands were formed by raising negative numbers to even powers.

37. $\sqrt{5}(4 - 2\sqrt{5})$

38. $\sqrt{6}(2 + \sqrt{6})$

39. $\sqrt{3}(\sqrt{2} - \sqrt{7})$

40. $\sqrt{2}(\sqrt{5} - \sqrt{2})$

41. $\sqrt{3}(-4\sqrt{3} + 6)$

42. $\sqrt{2}(-5\sqrt{2} - 7)$

43. $\sqrt{3}(2\sqrt{5} - 3\sqrt{4})$

44. $\sqrt{2}(3\sqrt{10} - 2\sqrt{2})$

45. $\sqrt[3]{2}(\sqrt[3]{4} - 2\sqrt[3]{32})$

46. $\sqrt[3]{3}(\sqrt[3]{9} - 4\sqrt[3]{21})$

47. $3\sqrt[3]{y}\,(2\sqrt[3]{y^2} - 4\sqrt[3]{y})$

48. $2\sqrt[3]{y^2}(5\sqrt[3]{y} + 4\sqrt[3]{y^2})$

49. $\sqrt[3]{a}\,(\sqrt[3]{2a^2} + \sqrt[3]{16a^2})$

50. $\sqrt[3]{x}\,(\sqrt[3]{3x^2} - \sqrt[3]{81x^2})$

51. $(\sqrt{3} - \sqrt{2})(\sqrt{3} + \sqrt{2})$

52. $(\sqrt{5} + \sqrt{6})(\sqrt{5} - \sqrt{6})$

53. $(\sqrt{8} + 2\sqrt{5})(\sqrt{8} - 2\sqrt{5})$

54. $(\sqrt{18} + 3\sqrt{7})(\sqrt{18} - 3\sqrt{7})$

55. $(7 + \sqrt{5})(7 - \sqrt{5})$

56. $(4 - \sqrt{3})(4 + \sqrt{3})$

57. $(2 - \sqrt{3})(2 + \sqrt{3})$

58. $(11 - \sqrt{2})(11 + \sqrt{2})$

59. $(\sqrt{8} + \sqrt{5})(\sqrt{8} - \sqrt{5})$

60. $(\sqrt{6} - \sqrt{7})(\sqrt{6} + \sqrt{7})$

61. $(3 + 2\sqrt{7})(3 - 2\sqrt{7})$

62. $(6 - 3\sqrt{2})(6 + 3\sqrt{2})$

63. $(\sqrt{c} + \sqrt{d})(\sqrt{c} - \sqrt{d})$

64. $(\sqrt{x} - \sqrt{y})(\sqrt{x} + \sqrt{y})$

65. $(3 - \sqrt{5})(2 + \sqrt{5})$

66. $(2 + \sqrt{6})(4 - \sqrt{6})$

67. $(\sqrt{3} + 1)(2\sqrt{3} + 1)$

68. $(4\sqrt{3} + 5)(\sqrt{3} - 2)$

69. $(2\sqrt{7} - 4\sqrt{2})(3\sqrt{7} + 6\sqrt{2})$

70. $(4\sqrt{5} + 3\sqrt{3})(3\sqrt{5} - 4\sqrt{3})$

71. $(\sqrt{a} + \sqrt{2})(\sqrt{a} + \sqrt{3})$

72. $(2 - \sqrt{x})(1 - \sqrt{x})$

73. $(2\sqrt[3]{3} + \sqrt[3]{2})(\sqrt[3]{3} - 2\sqrt[3]{2})$

74. $(3\sqrt[3]{7} + \sqrt[3]{6})(2\sqrt[3]{7} - 3\sqrt[3]{6})$

75. $(2 + \sqrt{3})^2$

76. $(\sqrt{5} + 1)^2$

77. $(\sqrt[5]{9} - \sqrt[5]{3})(\sqrt[5]{8} + \sqrt[5]{27})$

78. $(\sqrt[3]{8x} - \sqrt[3]{5y})^2$

Skill Maintenance

Multiply or divide and simplify.

79. $\frac{a^3 + 8}{a^2 - 4} \cdot \frac{a^2 - 4a + 4}{a^2 - 2a + 4}$ [5.1d]

80. $\frac{a^2 - 4}{a} \div \frac{a - 2}{a + 4}$ [5.1e]

Simplify. [5.4a]

81. $\dfrac{x - \frac{1}{3}}{x + \frac{1}{4}}$

82. $\dfrac{1 - \frac{1}{x}}{1 - \frac{1}{x^2}}$

83. $\dfrac{\frac{1}{p} - \frac{1}{q}}{\frac{1}{p^2} - \frac{1}{q^2}}$

84. $\dfrac{\frac{1}{a} + \frac{1}{b}}{\frac{1}{a^3} + \frac{1}{b^3}}$

Solve. [1.6c, d, e]

85. $|3x + 7| = 22$

86. $|3x + 7| < 22$

87. $|3x + 7| \geq 22$

88. $|3x + 7| = |2x - 5|$

Synthesis

89. Graph the function $f(x) = \sqrt{(x - 2)^2}$. What is the domain?

90. Use a graphing calculator to check your answers to Exercises 5, 22, and 72.

Multiply and simplify.

91. $\sqrt{9 + 3\sqrt{5}}\sqrt{9 - 3\sqrt{5}}$

92. $(\sqrt{x + 2} - \sqrt{x - 2})^2$

93. $(\sqrt{3} + \sqrt{5} - \sqrt{6})^2$

94. $\sqrt[3]{y}(1 - \sqrt[3]{y})(1 + \sqrt[3]{y})$

95. $(\sqrt[3]{9} - 2)(\sqrt[3]{9} + 4)$

96. $\left[\sqrt{3 + \sqrt{2 + \sqrt{1}}}\right]^4$

Mid-Chapter Review

Concept Reinforcement

Determine whether each statement is true or false.

__________ **1.** Every real number has two real-number square roots. [6.1a]

__________ **2.** If $\sqrt[3]{q}$ is negative, then q is negative. [6.1c]

__________ **3.** $a^{m/n}$ and $a^{n/m}$ are reciprocals. [6.2b]

__________ **4.** To multiply radicals with the same index, we multiply the radicands. [6.3a]

Guided Solutions

Fill in each blank with the number that creates a correct statement or solution.

Perform the indicated operations and simplify. [6.3a], [6.4a]

5. $\sqrt{6}\sqrt{10} = \sqrt{6\cdot\square} = \sqrt{2\cdot\square\cdot 2\cdot\square} = \square\sqrt{\square}$

6. $5\sqrt{32} - 3\sqrt{18} = 5\sqrt{\square\cdot 2} - 3\sqrt{\square\cdot 2}$

$= 5\cdot 4\sqrt{2} - 3\cdot\square\sqrt{2}$

$= \square\sqrt{2} - \square\sqrt{2}$

$= \square\sqrt{2}$

Mixed Review

Simplify. [6.1a]

7. $\sqrt{81}$

8. $-\sqrt{144}$

9. $\sqrt{\frac{16}{25}}$

10. $\sqrt{-9}$

11. For $f(x) = \sqrt{2x+3}$, find $f(3)$ and $f(-2)$. [6.1a]

12. Find the domain of $f(x) = \sqrt{4-x}$. [6.1a]

Graph. [6.1a]

13. $f(x) = -2\sqrt{x}$

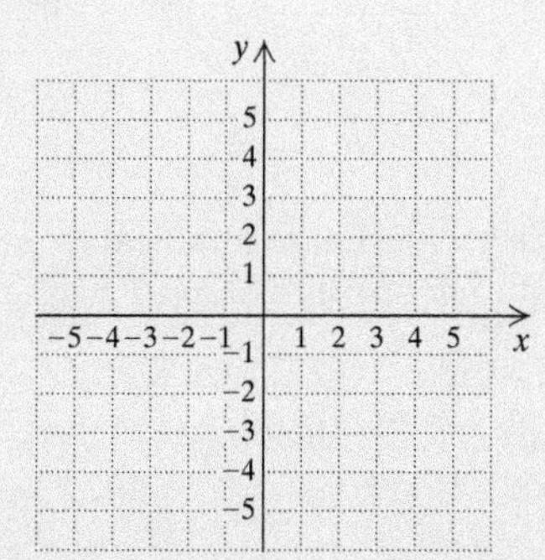

14. $g(x) = \sqrt{x+1}$

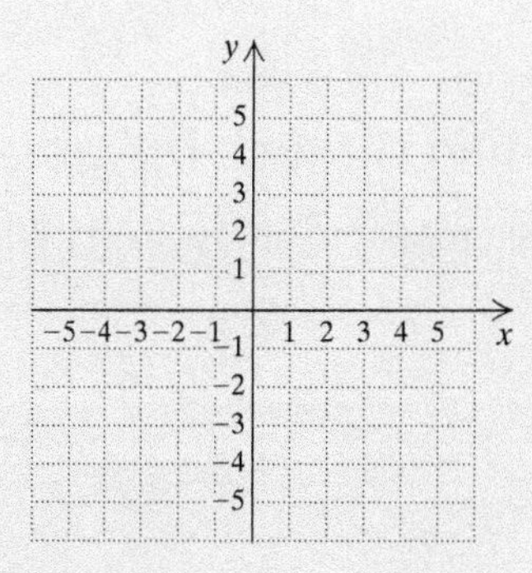

Find each of the following. Assume that letters can represent *any* real number. [6.1b, c, d]

15. $\sqrt{36z^2}$

16. $\sqrt{x^2 - 8x + 16}$

17. $\sqrt[3]{-64}$

18. $-\sqrt[3]{27a^3}$

19. $\sqrt[5]{32}$

20. $\sqrt[10]{y^{10}}$

Rewrite without rational exponents and simplify, if possible. [6.2a]

21. $125^{1/3}$

22. $(a^3b)^{1/4}$

Rewrite with rational exponents. [6.2a]

23. $\sqrt[5]{16}$

24. $\sqrt[3]{6m^2n}$

Simplify. Write the answer with positive exponents. [6.2b, c]

25. $3^{1/4} \cdot 3^{-5/8}$

26. $\dfrac{7^{6/5}}{7^{2/5}}$

27. $(x^{3/4}y^{-2/3})^2$

28. $(n^{-3/5})^{5/4}$

Use rational exponents to simplify. Write the answer in radical notation. [6.2d]

29. $\sqrt[6]{16}$

30. $(\sqrt[10]{ab})^5$

Use rational exponents to write a single radical expression. [6.2d]

31. $\sqrt{y}\,\sqrt[3]{y}$

32. $a^{2/3}b^{3/5}$

Perform the indicated operation and simplify. Assume that no radicands were formed by raising negative numbers to even powers. [6.3a, b], [6.4a, b]

33. $\sqrt{5}\sqrt{15}$

34. $\sqrt[3]{4x^2y}\,\sqrt[3]{6xy^4}$

35. $\dfrac{\sqrt[3]{80}}{\sqrt[3]{2}}$

36. $\sqrt{\dfrac{49a^5}{b^8}}$

37. $5\sqrt{7} + 6\sqrt{7}$

38. $3\sqrt{18x^3} - 6\sqrt{32x}$

39. $\sqrt{3}(2 - 5\sqrt{3})$

40. $(1 - \sqrt{x})(3 - \sqrt{x})$

41. $(\sqrt{m} - \sqrt{n})(\sqrt{m} + \sqrt{n})$

42. $(\sqrt{7} + 2)^2$

43. $(2\sqrt{3} + 3\sqrt{5})(3\sqrt{3} - 4\sqrt{5})$

Understanding Through Discussion and Writing

44. Does the *n*th root of x^2 always exist? Why or why not? [6.1a]

45. Explain how to formulate a radical expression that can be used to define a function f with a domain of $\{x|x \leq 5\}$. [6.1a]

46. Explain why $\sqrt[3]{x^6} = x^2$ for any value of x, but $\sqrt{x^6} = x^3$ only when $x \geq 0$. [6.2d]

47. Is the quotient of two irrational numbers always an irrational number? Why or why not? [6.3b]

STUDYING FOR SUCCESS *Make the Most of Your Time in Class*

- ☐ Before class, try to glance at the next section in your text to be discussed.
- ☐ Get a great seat. Sitting near the front will help you both hear the instruction more clearly and avoid distractions.
- ☐ Let your instructor know in advance if you must miss class, and do your best to keep up with any work.

6.5 More on Division of Radical Expressions

OBJECTIVES

- **a** Rationalize the denominator of a radical expression having one term in the denominator.
- **b** Rationalize the denominator of a radical expression having two terms in the denominator.

SKILL REVIEW *Use a rule to multiply a sum and a difference of the same two terms.* [4.2d]

Multiply.

1. $(x + 3)(x - 3)$
2. $(2y + 5)(2y - 5)$

Answers: **1.** $x^2 - 9$ **2.** $4y^2 - 25$

MyLab Math VIDEO

a RATIONALIZING DENOMINATORS

Sometimes in mathematics it is useful to find an equivalent expression without a radical in the denominator. This provides a standard notation for expressing results. The procedure for finding such an expression is called **rationalizing the denominator**. We carry this out by multiplying by 1.

EXAMPLE 1 Rationalize the denominator: $\sqrt{\frac{7}{3}}$.

We multiply by 1, using $\sqrt{3}/\sqrt{3}$. We do this so that the denominator of the radicand will be a perfect square.

$$\sqrt{\frac{7}{3}} = \frac{\sqrt{7}}{\sqrt{3}} \cdot \frac{\sqrt{3}}{\sqrt{3}} = \frac{\sqrt{7} \cdot \sqrt{3}}{\sqrt{3} \cdot \sqrt{3}}$$

$$= \frac{\sqrt{21}}{\sqrt{3^2}} = \frac{\sqrt{21}}{3}$$

The radicand is a perfect square.

Do Exercise 1.

EXAMPLE 2 Rationalize the denominator: $\sqrt[3]{\frac{7}{25}}$.

We first factor the denominator:

$$\sqrt[3]{\frac{7}{25}} = \sqrt[3]{\frac{7}{5 \cdot 5}}.$$

To eliminate the radical in the denominator, we consider the index 3 and the factors of the denominators. We have 2 factors of 5, and we need 3 factors of 5 in order to have a perfect cube. We achieve this by multiplying by 1, using $\sqrt[3]{5}/\sqrt[3]{5}$.

$$\sqrt[3]{\frac{7}{25}} = \frac{\sqrt[3]{7}}{\sqrt[3]{25}} = \frac{\sqrt[3]{7}}{\sqrt[3]{5 \cdot 5}} \cdot \frac{\sqrt[3]{5}}{\sqrt[3]{5}}$$

$$= \frac{\sqrt[3]{7} \cdot \sqrt[3]{5}}{\sqrt[3]{5 \cdot 5} \cdot \sqrt[3]{5}}$$

$$= \frac{\sqrt[3]{35}}{\sqrt[3]{5^3}} = \frac{\sqrt[3]{35}}{5}$$

The radicand is a perfect cube.

Do Exercise 2.

1. Rationalize the denominator: $\sqrt{\frac{2}{5}}$.

2. Rationalize the denominator: $\sqrt[3]{\frac{5}{4}}$.

Answers

1. $\frac{\sqrt{10}}{5}$ **2.** $\frac{\sqrt[3]{10}}{2}$

EXAMPLE 3 Rationalize the denominator: $\sqrt{\dfrac{2a}{5b}}$. Assume that no radicands were formed by raising negative numbers to even powers.

$$\sqrt{\frac{2a}{5b}} = \frac{\sqrt{2a}}{\sqrt{5b}} \quad \text{Converting to a quotient of radicals}$$

$$= \frac{\sqrt{2a}}{\sqrt{5b}} \cdot \frac{\sqrt{5b}}{\sqrt{5b}} \quad \text{Multiplying by 1}$$

$$= \frac{\sqrt{10ab}}{\sqrt{5^2b^2}} \quad \text{The radicand in the denominator is a perfect square.}$$

$$= \frac{\sqrt{10ab}}{5b}$$

◀ Do Exercise 3.

3. Rationalize the denominator: $\sqrt{\dfrac{4a}{3b}}$.

GS

$$\sqrt{\frac{4a}{3b}} = \frac{\sqrt{4a}}{\sqrt{3b}} \cdot \frac{\boxed{}}{\sqrt{3b}}$$

$$= \frac{\sqrt{\boxed{}\,ab}}{\sqrt{3^2 \cdot \boxed{}}}$$

$$= \frac{\boxed{}\sqrt{3ab}}{\boxed{}}$$

EXAMPLE 4 Rationalize the denominator: $\dfrac{\sqrt[3]{a}}{\sqrt[3]{9x}}$.

We factor the denominator:

$$\frac{\sqrt[3]{a}}{\sqrt[3]{9x}} = \frac{\sqrt[3]{a}}{\sqrt[3]{3 \cdot 3 \cdot x}}.$$

To choose the symbol for 1, we look at $3 \cdot 3 \cdot x$. To make it a cube, we need another 3 and two more x's. Thus we multiply by 1, using $\sqrt[3]{3x^2}/\sqrt[3]{3x^2}$:

$$\frac{\sqrt[3]{a}}{\sqrt[3]{9x}} = \frac{\sqrt[3]{a}}{\sqrt[3]{3 \cdot 3 \cdot x}} \cdot \frac{\sqrt[3]{3x^2}}{\sqrt[3]{3x^2}} \quad \text{Multiplying by 1}$$

$$= \frac{\sqrt[3]{3ax^2}}{\sqrt[3]{3^3x^3}} \quad \text{The radicand in the denominator is a perfect cube.}$$

$$= \frac{\sqrt[3]{3ax^2}}{3x}.$$

◀ Do Exercises 4 and 5.

Rationalize the denominator.

4. $\dfrac{\sqrt[4]{7}}{\sqrt[4]{2}}$

5. $\sqrt[3]{\dfrac{3x^5}{2y}}$

EXAMPLE 5 Rationalize the denominator: $\dfrac{3x}{\sqrt[5]{2x^2y^3}}$.

$$\frac{3x}{\sqrt[5]{2x^2y^3}} = \frac{3x}{\sqrt[5]{2 \cdot x \cdot x \cdot y \cdot y \cdot y}}$$

$$= \frac{3x}{\sqrt[5]{2x^2y^3}} \cdot \frac{\sqrt[5]{2^4x^3y^2}}{\sqrt[5]{2^4x^3y^2}}$$

$$= \frac{3x\sqrt[5]{16x^3y^2}}{\sqrt[5]{2^5x^5y^5}} \quad \text{The radicand in the denominator is a perfect fifth power.}$$

$$= \frac{3x\sqrt[5]{16x^3y^2}}{2xy}$$

$$= \frac{x}{x} \cdot \frac{3\sqrt[5]{16x^3y^2}}{2y}$$

$$= \frac{3\sqrt[5]{16x^3y^2}}{2y}$$

◀ Do Exercise 6.

6. Rationalize the denominator:

$$\frac{7x}{\sqrt[3]{4xy^5}}.$$

Answers

3. $\dfrac{2\sqrt{3ab}}{3b}$ 4. $\dfrac{\sqrt[4]{56}}{2}$ 5. $\dfrac{x\sqrt[3]{12x^2y^2}}{2y}$
6. $\dfrac{7\sqrt[3]{2x^2y}}{2y^2}$

Guided Solution:
3. $\sqrt{3b}$, 12, b^2, 2, $3b$

b RATIONALIZING WHEN THERE ARE TWO TERMS

Certain pairs of expressions containing square roots, such as $c - \sqrt{b}$, $c + \sqrt{b}$ and $\sqrt{a} - \sqrt{b}$, $\sqrt{a} + \sqrt{b}$, are called **conjugates**. The product of such a pair of conjugates has no radicals in it. (See Example 12 of Section 6.4.) Thus when we wish to rationalize a denominator that has two terms and one or more of them involves a square-root radical, we multiply by 1 using the conjugate of the denominator to write a symbol for 1.

Do Exercises 7 and 8. ▶

Multiply.

7. $(c - \sqrt{b})(c + \sqrt{b})$

8. $(\sqrt{a} + \sqrt{b})(\sqrt{a} - \sqrt{b})$

EXAMPLES In each of the following, what symbol for 1 would you use to rationalize the denominator?

Expression *Symbol for 1*

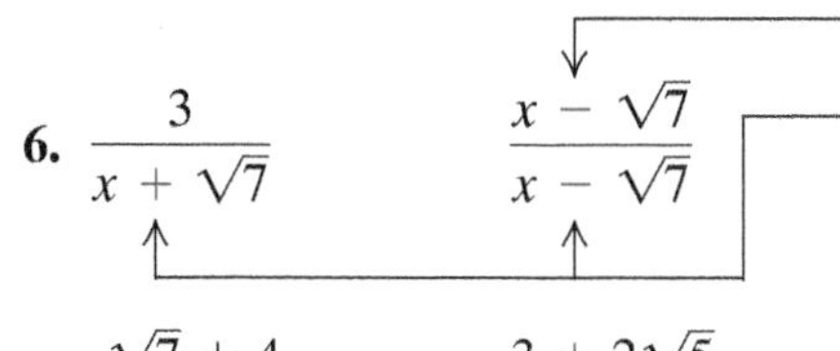

6. $\dfrac{3}{x + \sqrt{7}}$ $\dfrac{x - \sqrt{7}}{x - \sqrt{7}}$

7. $\dfrac{\sqrt{7} + 4}{3 - 2\sqrt{5}}$ $\dfrac{3 + 2\sqrt{5}}{3 + 2\sqrt{5}}$

Change the operation sign in the denominator to obtain the conjugate. Use the conjugate for the numerator and denominator of the symbol for 1.

Do Exercises 9 and 10. ▶

What symbol for 1 would you use to rationalize the denominator?

9. $\dfrac{\sqrt{5} + 1}{\sqrt{3} - y}$ 10. $\dfrac{1}{\sqrt{2} + \sqrt{3}}$

EXAMPLE 8 Rationalize the denominator: $\dfrac{4}{\sqrt{3} + x}$.

$$\frac{4}{\sqrt{3} + x} = \frac{4}{\sqrt{3} + x} \cdot \frac{\sqrt{3} - x}{\sqrt{3} - x}$$
$$= \frac{4(\sqrt{3} - x)}{(\sqrt{3} + x)(\sqrt{3} - x)}$$
$$= \frac{4\sqrt{3} - 4x}{3 - x^2}$$

EXAMPLE 9 Rationalize the denominator: $\dfrac{4 + \sqrt{2}}{\sqrt{5} - \sqrt{2}}$.

$$\frac{4 + \sqrt{2}}{\sqrt{5} - \sqrt{2}} = \frac{4 + \sqrt{2}}{\sqrt{5} - \sqrt{2}} \cdot \frac{\sqrt{5} + \sqrt{2}}{\sqrt{5} + \sqrt{2}}$$ Multiplying by 1, using the conjugate of $\sqrt{5} - \sqrt{2}$, which is $\sqrt{5} + \sqrt{2}$

$$= \frac{(4 + \sqrt{2})(\sqrt{5} + \sqrt{2})}{(\sqrt{5} - \sqrt{2})(\sqrt{5} + \sqrt{2})}$$ Multiplying numerators and denominators

$$= \frac{4\sqrt{5} + 4\sqrt{2} + \sqrt{2}\sqrt{5} + (\sqrt{2})^2}{(\sqrt{5})^2 - (\sqrt{2})^2}$$ Using $(A - B)(A + B) = A^2 - B^2$ in the denominator

$$= \frac{4\sqrt{5} + 4\sqrt{2} + \sqrt{10} + 2}{5 - 2}$$
$$= \frac{4\sqrt{5} + 4\sqrt{2} + \sqrt{10} + 2}{3}$$

Do Exercises 11 and 12. ▶

Rationalize the denominator.

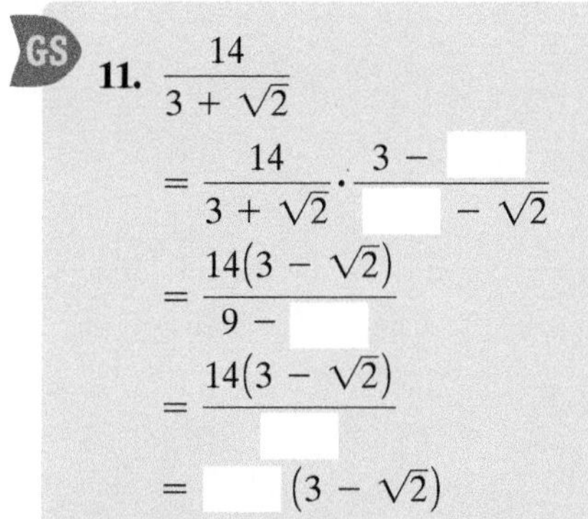

GS 11. $\dfrac{14}{3 + \sqrt{2}}$

$= \dfrac{14}{3 + \sqrt{2}} \cdot \dfrac{3 - \square}{\square - \sqrt{2}}$

$= \dfrac{14(3 - \sqrt{2})}{9 - \square}$

$= \dfrac{14(3 - \sqrt{2})}{\square}$

$= \square(3 - \sqrt{2})$

$= 6 - 2\sqrt{2}$

12. $\dfrac{5 + \sqrt{2}}{1 - \sqrt{2}}$

Answers

7. $c^2 - b$ **8.** $a - b$ **9.** $\dfrac{\sqrt{3} + y}{\sqrt{3} + y}$
10. $\dfrac{\sqrt{2} - \sqrt{3}}{\sqrt{2} - \sqrt{3}}$ **11.** $6 - 2\sqrt{2}$
12. $-7 - 6\sqrt{2}$

Guided Solution:
11. $\sqrt{2}$, 3, 2, 7, 2

6.5 Exercise Set

FOR EXTRA HELP MyLab Math

✓ Check Your Understanding

Reading Check Determine whether the two given expressions are conjugates. Answer "yes" or "no."

RC1. $8 - \sqrt{10},\ \sqrt{10} - 8$

RC2. $\sqrt{4} - \sqrt{15},\ \sqrt{4} + \sqrt{15}$

RC3. $-\sqrt{2} + 9,\ \sqrt{2} + 9$

RC4. $\frac{\sqrt{11}}{\sqrt{3}},\ \frac{\sqrt{3}}{\sqrt{11}}$

Concept Check Choose from the columns on the right the symbol for 1 that you would use to rationalize the denominator. Some choices may be used more than once and others may not be used.

CC1. $\frac{2}{x - \sqrt{3}}$

CC2. $\frac{\sqrt{5}}{\sqrt{3}}$

CC3. $\frac{\sqrt[4]{9x^2}}{\sqrt[4]{x^3}}$

CC4. $\frac{\sqrt[3]{3y}}{\sqrt[3]{9y^2}}$

CC5. $\frac{x + \sqrt{3}}{\sqrt{3}}$

CC6. $\frac{1}{3\sqrt{x}}$

a) $\frac{\sqrt{3x}}{\sqrt{3x}}$

b) $\frac{x - \sqrt{3}}{x - \sqrt{3}}$

c) $\frac{\sqrt{3}}{\sqrt{3}}$

d) $\frac{\sqrt{x}}{\sqrt{x}}$

e) $\frac{\sqrt[4]{x}}{\sqrt[4]{x}}$

f) $\frac{\sqrt{9}}{\sqrt{9}}$

g) $\frac{x + \sqrt{3}}{x + \sqrt{3}}$

h) $\frac{\sqrt[3]{3y}}{\sqrt[3]{3y}}$

a Rationalize the denominator. Assume that no radicands were formed by raising negative numbers to even powers.

1. $\sqrt{\frac{5}{3}}$

2. $\sqrt{\frac{8}{7}}$

3. $\sqrt{\frac{11}{2}}$

4. $\sqrt{\frac{17}{6}}$

5. $\frac{2\sqrt{3}}{7\sqrt{5}}$

6. $\frac{3\sqrt{5}}{8\sqrt{2}}$

7. $\sqrt[3]{\frac{16}{9}}$

8. $\sqrt[3]{\frac{1}{3}}$

9. $\frac{\sqrt[3]{3a}}{\sqrt[3]{5c}}$

10. $\frac{\sqrt[3]{7x}}{\sqrt[3]{3y}}$

11. $\frac{\sqrt[3]{2y^4}}{\sqrt[3]{6x^4}}$

12. $\frac{\sqrt[3]{3a^4}}{\sqrt[3]{7b^2}}$

13. $\frac{1}{\sqrt[4]{st}}$

14. $\frac{1}{\sqrt[3]{yz}}$

15. $\sqrt{\frac{3x}{20}}$

16. $\sqrt{\frac{7a}{32}}$

17. $\sqrt[3]{\dfrac{4}{5x^5y^2}}$

18. $\sqrt[3]{\dfrac{7c}{100ab^5}}$

19. $\sqrt[4]{\dfrac{1}{8x^7y^3}}$

20. $\dfrac{2x}{\sqrt[5]{18x^8y^6}}$

b Rationalize the denominator. Assume that no radicands were formed by raising negative numbers to even powers.

21. $\dfrac{9}{6 - \sqrt{10}}$

22. $\dfrac{3}{8 + \sqrt{5}}$

23. $\dfrac{-4\sqrt{7}}{\sqrt{5} + \sqrt{3}}$

24. $\dfrac{-5\sqrt{2}}{\sqrt{7} - \sqrt{5}}$

25. $\dfrac{6\sqrt{3}}{3\sqrt{2} - \sqrt{5}}$

26. $\dfrac{34\sqrt{5}}{2\sqrt{5} - \sqrt{3}}$

27. $\dfrac{3 + \sqrt{5}}{\sqrt{2} + \sqrt{5}}$

28. $\dfrac{2 + \sqrt{3}}{\sqrt{3} + \sqrt{5}}$

29. $\dfrac{\sqrt{3} - \sqrt{2}}{\sqrt{3} - \sqrt{7}}$

30. $\dfrac{\sqrt{5} - \sqrt{3}}{\sqrt{5} - \sqrt{2}}$

31. $\dfrac{\sqrt{5} - 2\sqrt{6}}{\sqrt{3} - 4\sqrt{5}}$

32. $\dfrac{\sqrt{6} - 3\sqrt{5}}{\sqrt{3} - 2\sqrt{7}}$

33. $\dfrac{2 - \sqrt{a}}{3 + \sqrt{a}}$

34. $\dfrac{5 + \sqrt{x}}{8 - \sqrt{x}}$

35. $\dfrac{2 + 3\sqrt{x}}{3 + 2\sqrt{x}}$

36. $\dfrac{5 + 2\sqrt{y}}{4 + 3\sqrt{y}}$

37. $\dfrac{5\sqrt{3} - 3\sqrt{2}}{3\sqrt{2} - 2\sqrt{3}}$

38. $\dfrac{7\sqrt{2} + 4\sqrt{3}}{4\sqrt{3} - 3\sqrt{2}}$

39. $\dfrac{\sqrt{x} - \sqrt{y}}{\sqrt{x} + \sqrt{y}}$

40. $\dfrac{\sqrt{a} + \sqrt{b}}{\sqrt{a} - \sqrt{b}}$

Skill Maintenance

Solve. [5.5a]

41. $\dfrac{1}{2} - \dfrac{1}{3} = \dfrac{5}{t}$

42. $\dfrac{5}{x - 1} + \dfrac{9}{x^2 + x + 1} = \dfrac{15}{x^3 - 1}$

Divide and simplify. [5.1e]

43. $\dfrac{1}{x^3 - y^3} \div \dfrac{1}{(x - y)(x^2 + xy + y^2)}$

44. $\dfrac{2x^2 - x - 6}{x^2 + 4x + 3} \div \dfrac{2x^2 + x - 3}{x^2 - 1}$

Synthesis

45. Simplify. (*Hint*: Rationalize the denominator.)

$$\sqrt{a^2 - 3} - \frac{a^2}{\sqrt{a^2 - 3}}$$

46. Express each of the following as the product of two radical expressions.

a) $x - 5$

b) $x - a$

6.6

Solving Radical Equations

OBJECTIVES

a Solve radical equations with one radical term.

b Solve radical equations with two radical terms.

c Solve applied problems involving radical equations.

a THE PRINCIPLE OF POWERS

SKILL REVIEW *Solve quadratic equations by first factoring and then using the principle of zero products.* [4.8a]

Solve.

1. $x^2 - x = 6$

2. $x^2 - x = 2x + 4$

Answers: 1. $-2, 3$ **2.** $-1, 4$

MyLab Math VIDEO

A **radical equation** has variables in one or more radicands. For example,

$$\sqrt[3]{2x} + 1 = 5 \quad \text{and} \quad \sqrt{x} + \sqrt{4x - 2} = 7$$

are radical equations. To solve such an equation, we need a new equation-solving principle. Suppose that an equation $a = b$ is true. If we square both sides, we get another true equation: $a^2 = b^2$. This can be generalized.

THE PRINCIPLE OF POWERS

For any natural number n, if an equation $a = b$ is true, then $a^n = b^n$ is true.

However, if an equation $a^n = b^n$ is true, it *may not* be true that $a = b$, if n is even. For example, $3^2 = (-3)^2$ is true, but $3 = -3$ is not true. Thus we *must check* the possible solutions when we solve an equation using the principle of powers.

To solve an equation with a radical term, we first isolate the radical term on one side of the equation. Then we use the principle of powers.

EXAMPLE 1 Solve: $\sqrt{x} - 3 = 4$.

We have

$$\sqrt{x} - 3 = 4$$

$$\sqrt{x} = 7 \qquad \text{Adding 3 to isolate the radical}$$

$$(\sqrt{x})^2 = 7^2 \qquad \text{Using the principle of powers (squaring)}$$

$$x = 49. \qquad \sqrt{x} \cdot \sqrt{x} = x$$

The number 49 is a possible solution. But we *must* check in order to be sure!

Check:

$$\sqrt{x} - 3 = 4$$

$$\sqrt{49} - 3 \ ? \ 4$$

$$7 - 3$$

$$4 \qquad \text{TRUE}$$

Caution!

The principle of powers does not always give equivalent equations. For this reason, a check is a must!

The solution is 49.

EXAMPLE 2 Solve: $\sqrt{x} = -3$.

We might note at the outset that this equation has no solution because the principal square root of a number is never negative. Let's continue as above for comparison.

$$\sqrt{x} = -3$$
$$(\sqrt{x})^2 = (-3)^2$$
$$x = 9$$

Check: $\sqrt{x} = -3$

$\sqrt{9}$	?	-3
3		FALSE

The number 9 does *not* check. Thus the equation $\sqrt{x} = -3$ has no real-number solution. Note that the solution of the equation $x = 9$ is 9, but the equation $\sqrt{x} = -3$ has *no* solution. Thus the equations $x = 9$ and $\sqrt{x} = -3$ are *not* equivalent equations.

Do Exercises 1 and 2.

Solve.

1. $\sqrt{x} - 7 = 3$

2. $\sqrt{x} = -2$

EXAMPLE 3 Solve: $x - 7 = 2\sqrt{x + 1}$.

The radical term is already isolated. We proceed with the principle of powers:

$$x - 7 = 2\sqrt{x + 1}$$
$$(x - 7)^2 = \left(2\sqrt{x + 1}\right)^2 \quad \text{Using the principle of powers (squaring)}$$
$$(x - 7)(x - 7) = \left(2\sqrt{x + 1}\right)\left(2\sqrt{x + 1}\right)$$
$$x^2 - 14x + 49 = 2^2\left(\sqrt{x + 1}\right)^2$$
$$x^2 - 14x + 49 = 4(x + 1)$$
$$x^2 - 14x + 49 = 4x + 4$$
$$x^2 - 18x + 45 = 0$$
$$(x - 3)(x - 15) = 0 \quad \text{Factoring}$$
$$x - 3 = 0 \quad or \quad x - 15 = 0 \quad \text{Using the principle of zero products}$$
$$x = 3 \quad or \quad x = 15.$$

The possible solutions are 3 and 15.

Check: For 3:

$x - 7 = 2\sqrt{x + 1}$

$3 - 7$	?	$2\sqrt{3 + 1}$
-4		$2\sqrt{4}$
		$2(2)$
		4 FALSE

For 15:

$x - 7 = 2\sqrt{x + 1}$

$15 - 7$	?	$2\sqrt{15 + 1}$
8		$2\sqrt{16}$
		$2(4)$
		8 TRUE

The number 3 does *not* check, but the number 15 does check. The solution is 15.

The number 3 in Example 3 is what is sometimes called an *extraneous solution*, but such terminology is risky to use at best because the number 3 is in *no way* a *solution* of the original equation.

Do Exercises 3 and 4.

Solve.

GS 3. $x + 2 = \sqrt{2x + 7}$

$(x + 2)^{\square} = (\sqrt{2x + 7})^2$

$x^2 + \square x + 4 = \square + 7$

$x^2 + 2x - \square = 0$

$(x + \square)(x - 1) = 0$

$x + 3 = 0 \quad or \quad x - 1 = 0$

$x = -3 \quad or \quad x = \square$

The number $\square$ does not check, but the number 1 does check. The solution is $\square$.

4. $x + 1 = 3\sqrt{x - 1}$

Answers

1. 100 **2.** No solution **3.** 1 **4.** 2, 5

Guided Solution:

3. 2, 4, 2x, 3, 3, 1, -3, 1

ALGEBRAIC GRAPHICAL CONNECTION

We can visualize or check the solutions of a radical equation graphically. Consider the equation of Example 3: $x - 7 = 2\sqrt{x + 1}$. We can examine the solutions by graphing the equations

$$y = x - 7 \quad \text{and} \quad y = 2\sqrt{x + 1}$$

using the same set of axes. A hand-drawn graph of $y = 2\sqrt{x + 1}$ would involve approximating square roots on a calculator.

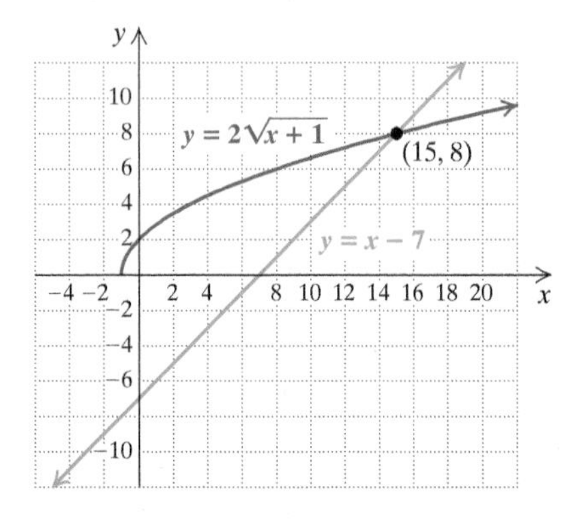

It appears from the graph that when $x = 15$, the values of $y = x - 7$ and $y = 2\sqrt{x + 1}$ are the same, 8. We can check this as we did in Example 3. Note too that the graphs *do not* intersect at $x = 3$, the possible solution that did not check.

CALCULATOR CORNER

Solving Radical Equations We can solve radical equations graphically. Consider the equation in Example 3, $x - 7 = 2\sqrt{x + 1}$.

We first graph each side of the equation. We enter $y_1 = x - 7$ and $y_2 = 2\sqrt{x + 1}$ on the equation-editor screen and graph the equations using the window $[-5, 20, -10, 10]$. Note that there is one point of intersection. We use the INTERSECT feature to find its coordinates. (See the Calculator Corner on p. 207 for the procedure.) The first coordinate, 15, is the value of x for which $y_1 = y_2$, or $x - 7 = 2\sqrt{x + 1}$. It is the solution of the equation. Note that the graph shows a single solution whereas the algebraic solution in Example 3 yields two possible solutions, 3 and 15, that must be checked. The algebraic check shows that 15 is the only solution.

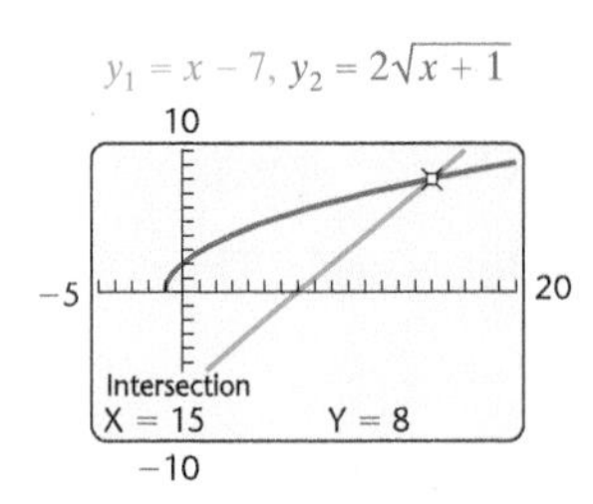

EXERCISES:

1. Solve the equations in Examples 1 and 4 graphically.
2. Solve the equations in Margin Exercises 1, 3, and 4 graphically.

EXAMPLE 2 Solve: $\sqrt{x} = -3$.

We might note at the outset that this equation has no solution because the principal square root of a number is never negative. Let's continue as above for comparison.

$$\sqrt{x} = -3$$
$$(\sqrt{x})^2 = (-3)^2$$
$$x = 9$$

Check: $\sqrt{x} = -3$

$\sqrt{9} \ ? \ -3$

$3 \quad$ FALSE

The number 9 does *not* check. Thus the equation $\sqrt{x} = -3$ has no real-number solution. Note that the solution of the equation $x = 9$ is 9, but the equation $\sqrt{x} = -3$ has *no* solution. Thus the equations $x = 9$ and $\sqrt{x} = -3$ are *not* equivalent equations.

Do Exercises 1 and 2. ▶

Solve.

1. $\sqrt{x} - 7 = 3$

2. $\sqrt{x} = -2$

EXAMPLE 3 Solve: $x - 7 = 2\sqrt{x + 1}$.

The radical term is already isolated. We proceed with the principle of powers:

$$x - 7 = 2\sqrt{x+1}$$
$$(x-7)^2 = \left(2\sqrt{x+1}\right)^2 \qquad \text{Using the principle of powers (squaring)}$$
$$(x-7)(x-7) = \left(2\sqrt{x+1}\right)\left(2\sqrt{x+1}\right)$$
$$x^2 - 14x + 49 = 2^2\left(\sqrt{x+1}\right)^2$$
$$x^2 - 14x + 49 = 4(x+1)$$
$$x^2 - 14x + 49 = 4x + 4$$
$$x^2 - 18x + 45 = 0$$
$$(x-3)(x-15) = 0 \qquad \text{Factoring}$$
$$x - 3 = 0 \quad or \quad x - 15 = 0 \qquad \text{Using the principle of zero products}$$
$$x = 3 \quad or \quad x = 15.$$

The possible solutions are 3 and 15.

Check: For 3:

$x - 7 = 2\sqrt{x+1}$

$3 - 7 \ ? \ 2\sqrt{3+1}$

$-4 \mid 2\sqrt{4}$

$\quad 2(2)$

$\quad 4 \quad$ FALSE

For 15:

$x - 7 = 2\sqrt{x+1}$

$15 - 7 \ ? \ 2\sqrt{15+1}$

$8 \mid 2\sqrt{16}$

$\quad 2(4)$

$\quad 8 \quad$ TRUE

The number 3 does *not* check, but the number 15 does check. The solution is 15. ■

The number 3 in Example 3 is what is sometimes called an *extraneous solution*, but such terminology is risky to use at best because the number 3 is in *no way* a *solution* of the original equation.

Do Exercises 3 and 4. ▶

Solve.

GS 3. $x + 2 = \sqrt{2x + 7}$

$(x+2)^{\square} = \left(\sqrt{2x+7}\right)^2$

$x^2 + \square x + 4 = \square + 7$

$x^2 + 2x - \square = 0$

$(x + \square)(x - 1) = 0$

$x + 3 = 0 \quad or \quad x - 1 = 0$

$x = -3 \quad or \quad x = \square$

The number $\square$ does not check, but the number 1 does check. The solution is $\square$.

4. $x + 1 = 3\sqrt{x - 1}$

Answers

1. 100 **2.** No solution **3.** 1 **4.** 2, 5

Guided Solution:

3. 2, 4, 2*x*, 3, 3, 1, −3, 1

ALGEBRAIC ▶◀ GRAPHICAL CONNECTION

We can visualize or check the solutions of a radical equation graphically. Consider the equation of Example 3: $x - 7 = 2\sqrt{x + 1}$. We can examine the solutions by graphing the equations

$$y = x - 7 \quad \text{and} \quad y = 2\sqrt{x + 1}$$

using the same set of axes. A hand-drawn graph of $y = 2\sqrt{x + 1}$ would involve approximating square roots on a calculator.

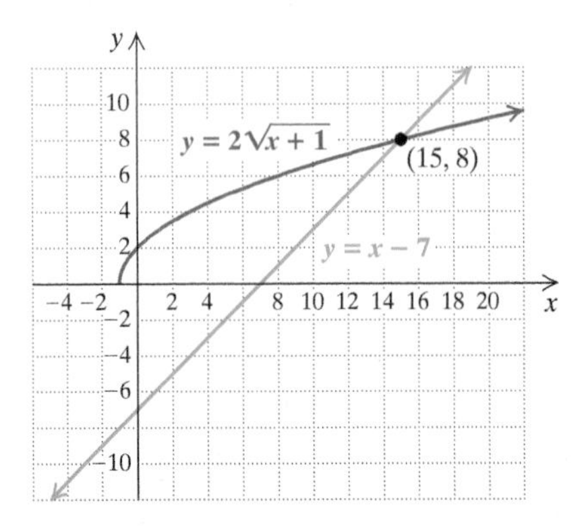

It appears from the graph that when $x = 15$, the values of $y = x - 7$ and $y = 2\sqrt{x + 1}$ are the same, 8. We can check this as we did in Example 3. Note too that the graphs *do not* intersect at $x = 3$, the possible solution that did not check.

CALCULATOR CORNER

Solving Radical Equations We can solve radical equations graphically. Consider the equation in Example 3, $x - 7 = 2\sqrt{x + 1}$.

We first graph each side of the equation. We enter $y_1 = x - 7$ and $y_2 = 2\sqrt{x + 1}$ on the equation-editor screen and graph the equations using the window $[-5, 20, -10, 10]$. Note that there is one point of intersection. We use the INTERSECT feature to find its coordinates. (See the Calculator Corner on p. 207 for the procedure.) The first coordinate, 15, is the value of x for which $y_1 = y_2$, or $x - 7 = 2\sqrt{x + 1}$. It is the solution of the equation. Note that the graph shows a single solution whereas the algebraic solution in Example 3 yields two possible solutions, 3 and 15, that must be checked. The algebraic check shows that 15 is the only solution.

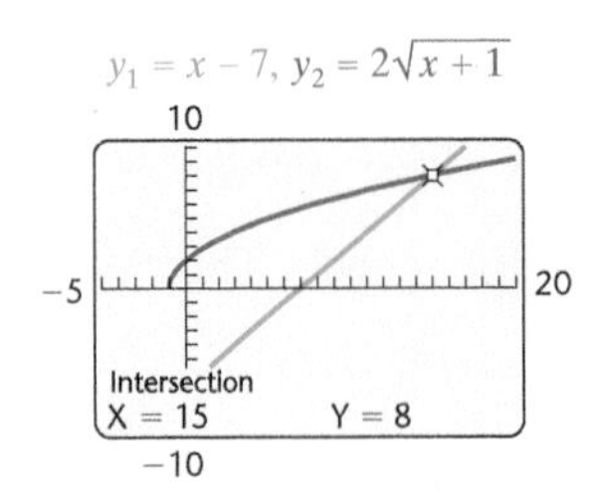

EXERCISES:

1. Solve the equations in Examples 1 and 4 graphically.
2. Solve the equations in Margin Exercises 1, 3, and 4 graphically.

EXAMPLE 4 Solve: $x = \sqrt{x+7} + 5$.

We have

$$x = \sqrt{x+7} + 5$$

$$x - 5 = \sqrt{x+7} \qquad \text{Subtracting 5 to isolate the radical term}$$

$$(x-5)^2 = (\sqrt{x+7})^2 \qquad \text{Using the principle of powers (squaring both sides)}$$

$$x^2 - 10x + 25 = x + 7$$

$$x^2 - 11x + 18 = 0$$

$$(x-9)(x-2) = 0 \qquad \text{Factoring}$$

$$x = 9 \quad or \quad x = 2. \qquad \text{Using the principle of zero products}$$

The possible solutions are 9 and 2.

Check: For 9:

$$\begin{array}{l|l} \multicolumn{2}{l}{x = \sqrt{x+7} + 5} \\ \hline 9 \ ? & \sqrt{9+7} + 5 \\ & \sqrt{16} + 5 \\ & 4 + 5 \\ & 9 \end{array} \qquad \text{TRUE}$$

For 2:

$$\begin{array}{l|l} \multicolumn{2}{l}{x = \sqrt{x+7} + 5} \\ \hline 2 \ ? & \sqrt{2+7} + 5 \\ & \sqrt{9} + 5 \\ & 3 + 5 \\ & 8 \end{array} \qquad \text{FALSE}$$

Since 9 checks but 2 does not, the solution is 9.

EXAMPLE 5 Solve: $\sqrt[3]{2x+1} + 5 = 0$.

We have

$$\sqrt[3]{2x+1} + 5 = 0$$

$$\sqrt[3]{2x+1} = -5 \qquad \text{Subtracting 5. This isolates the radical term.}$$

$$(\sqrt[3]{2x+1})^3 = (-5)^3 \qquad \text{Using the principle of powers (raising to the third power)}$$

$$2x + 1 = -125$$

$$2x = -126 \qquad \text{Subtracting 1}$$

$$x = -63.$$

Check:

$$\begin{array}{r|l} \multicolumn{2}{c}{\sqrt[3]{2x+1} + 5 = 0} \\ \hline \sqrt[3]{2\cdot(-63)+1} + 5 & ?\ 0 \\ \sqrt[3]{-126+1} + 5 & \\ \sqrt[3]{-125} + 5 & \\ -5 + 5 & \\ 0 & \end{array} \qquad \text{TRUE}$$

The solution is -63.

Do Exercises 5 and 6. ▶

Solve.

5. $x = \sqrt{x+5} + 1$

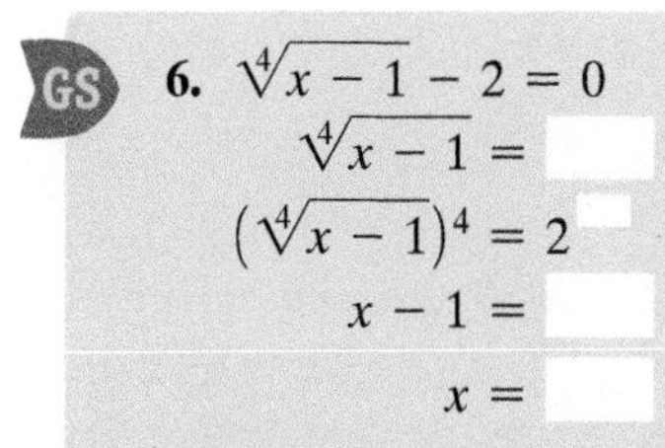

GS 6. $\sqrt[4]{x-1} - 2 = 0$

$\sqrt[4]{x-1} = \square$

$(\sqrt[4]{x-1})^4 = 2^{\square}$

$x - 1 = \square$

$x = \square$

Answers

5. 4 **6.** 17

Guided Solution:

6. 2, 4, 16, 17

b EQUATIONS WITH TWO RADICAL TERMS

A general strategy for solving radical equations, including those with two radical terms, is as follows.

> **SOLVING RADICAL EQUATIONS**
>
> To solve radical equations:
>
> 1. Isolate one of the radical terms.
> 2. Use the principle of powers.
> 3. If a radical remains, perform steps (1) and (2) again.
> 4. Check possible solutions.

EXAMPLE 6 Solve: $\sqrt{x-3} + \sqrt{x+5} = 4$.

$\sqrt{x-3} + \sqrt{x+5} = 4$

$\sqrt{x-3} = 4 - \sqrt{x+5}$ Subtracting $\sqrt{x+5}$. This isolates one of the radical terms.

$(\sqrt{x-3})^2 = (4 - \sqrt{x+5})^2$ Using the principle of powers (squaring both sides)

$x - 3 = 16 - 8\sqrt{x+5} + (x+5)$ Using $(A - B)^2 = A^2 - 2AB + B^2$. See this rule in Section 4.6.

$x - 3 = -8\sqrt{x+5} + x + 21$ Collecting like terms

$-3 = 21 - 8\sqrt{x+5}$ Subtracting x

$-24 = -8\sqrt{x+5}$ Subtracting 21 to isolate the radical term

$3 = \sqrt{x+5}$ Dividing by -8

$3^2 = (\sqrt{x+5})^2$ Squaring

$9 = x + 5$

$4 = x$

The number 4 checks and is the solution.

EXAMPLE 7 Solve: $\sqrt{2x-5} = 1 + \sqrt{x-3}$.

$\sqrt{2x-5} = 1 + \sqrt{x-3}$

$(\sqrt{2x-5})^2 = (1 + \sqrt{x-3})^2$ One radical is already isolated. We square both sides.

$2x - 5 = 1 + 2\sqrt{x-3} + (x-3)$

$2x - 5 = 2\sqrt{x-3} + x - 2$

$x - 3 = 2\sqrt{x-3}$ Isolating the remaining radical term

$(x-3)^2 = (2\sqrt{x-3})^2$ Squaring both sides

$x^2 - 6x + 9 = 4(x-3)$

$x^2 - 6x + 9 = 4x - 12$

$x^2 - 10x + 21 = 0$

$(x-7)(x-3) = 0$ Factoring

$x = 7$ *or* $x = 3$ Using the principle of zero products

The possible solutions are 7 and 3.

Check: For 7:

$$\begin{array}{r|l} \sqrt{2x-5} & = 1+\sqrt{x-3} \\ \hline \sqrt{2(7)-5} & ?\ 1+\sqrt{7-3} \\ \sqrt{14-5} & 1+\sqrt{4} \\ \sqrt{9} & 1+2 \\ 3 & 3 \quad \text{TRUE} \end{array}$$

For 3:

$$\begin{array}{r|l} \sqrt{2x-5} & = 1+\sqrt{x-3} \\ \hline \sqrt{2(3)-5} & ?\ 1+\sqrt{3-3} \\ \sqrt{6-5} & 1+\sqrt{0} \\ \sqrt{1} & 1+0 \\ 1 & 1 \quad \text{TRUE} \end{array}$$

The numbers 7 and 3 check and are the solutions.

Do Exercises 7 and 8. ▶

Solve.

7. $\sqrt{x} - \sqrt{x-5} = 1$

8. $\sqrt{2x-5} - 2 = \sqrt{x-2}$

EXAMPLE 8 Solve: $\sqrt{x+2} - \sqrt{2x+2} + 1 = 0$.

We first isolate one radical.

$$\sqrt{x+2} - \sqrt{2x+2} + 1 = 0$$

$$\sqrt{x+2} + 1 = \sqrt{2x+2} \qquad \text{Adding } \sqrt{2x+2} \text{ to isolate a radical term}$$

$$\left(\sqrt{x+2} + 1\right)^2 = \left(\sqrt{2x+2}\right)^2 \qquad \text{Squaring both sides}$$

$$x + 2 + 2\sqrt{x+2} + 1 = 2x + 2$$

$$2\sqrt{x+2} = x - 1$$

$$\left(2\sqrt{x+2}\right)^2 = (x-1)^2$$

$$4(x+2) = x^2 - 2x + 1$$

$$4x + 8 = x^2 - 2x + 1$$

$$0 = x^2 - 6x - 7$$

$$0 = (x-7)(x+1) \qquad \text{Factoring}$$

$$x - 7 = 0 \quad or \quad x + 1 = 0 \qquad \text{Using the principle of zero products}$$

$$x = 7 \quad or \quad x = -1$$

The possible solutions are 7 and -1.

Check: For 7:

$$\begin{array}{r|l} \sqrt{x+2} - \sqrt{2x+2} + 1 & = 0 \\ \hline \sqrt{7+2} - \sqrt{2\cdot 7+2} + 1 & ?\ 0 \\ \sqrt{9} - \sqrt{16} + 1 & \\ 3 - 4 + 1 & \\ 0 & \quad \text{TRUE} \end{array}$$

For -1:

$$\begin{array}{r|l} \sqrt{x+2} - \sqrt{2x+2} + 1 & = 0 \\ \hline \sqrt{-1+2} - \sqrt{2\cdot(-1)+2} + 1 & ?\ 0 \\ \sqrt{1} - \sqrt{0} + 1 & \\ 1 - 0 + 1 & \\ 2 & \quad \text{FALSE} \end{array}$$

The number 7 checks, but -1 does not. The solution is 7.

Do Exercise 9. ▶

9. Solve:
$\sqrt{3x+1} - 1 - \sqrt{x+4} = 0$.

Answers

7. 9 **8.** 27 **9.** 5

C APPLICATIONS

Speed of Sound. Many applications translate to radical equations. For example, at a temperature of t degrees Fahrenheit, sound travels at a rate of S feet per second, where

$$S = 21.9\sqrt{5t + 2457}.$$

EXAMPLE 9 *Capitol Concert.* Each year, a free concert, know as "A Capitol Fourth," on the west lawn of the Capitol in Washington, D.C., precedes the July 4th fireworks display on the National Mall. A few of the performers at the 2017 concert included the Beach Boys, the Blues Brothers, Trace Adkins, Yolanda Adams, and the *Voice* winner Chris Blue. A scientific instrument determined that the sound of the music was traveling at a rate of 1170 ft/sec. What was the air temperature at the concert?

We substitute 1170 for S in the formula $S = 21.9\sqrt{5t + 2457}$:

$$1170 = 21.9\sqrt{5t + 2457}.$$

Then we solve the equation for t:

$$1170 = 21.9\sqrt{5t + 2457}$$

$$\frac{1170}{21.9} = \sqrt{5t + 2457} \quad \text{Dividing by 21.9}$$

$$\left(\frac{1170}{21.9}\right)^2 = \left(\sqrt{5t + 2457}\right)^2 \quad \text{Squaring both sides}$$

$$2854.2 \approx 5t + 2457 \quad \text{Simplifying}$$

$$397.2 \approx 5t \quad \text{Subtracting 2457}$$

$$79 \approx t. \quad \text{Dividing by 5}$$

The temperature at the concert was about 79°F.

◀ Do Exercise 10.

10. *Marching Band Performance.* When the Fulton High School marching band performed at half-time of a football game, the speed of sound of the music was measured by a scientific instrument to be 1162 ft/sec. What was the air temperature?

Answer

10. About 72°F

6.6 Exercise Set

FOR EXTRA HELP  MyLab Math

✓ Check Your Understanding

Reading Check Choose from the column on the right the term that best completes each statement. Not every word will be used.

RC1. The equation $\sqrt{4 - 11x} = 3$ is a(n) ____________ equation.

RC2. When we square both sides of an equation, we are using the principle of ____________.

RC3. To solve an equation with a radical term, we first ____________ the radical term on one side of the equation.

RC4. A radical equation has variables in one or more ____________.

RC5. A check is essential when we raise both sides of an equation to a(n) ____________ power.

even
radical
isolate
odd
radicands
square roots
powers
raise
rational
principle

The possible solutions are 7 and 3.

Check: For 7:

$$\begin{array}{c|c} \sqrt{2x-5} & = 1+\sqrt{x-3} \\ \hline \sqrt{2(7)-5} & ?\ 1+\sqrt{7-3} \\ \sqrt{14-5} & 1+\sqrt{4} \\ \sqrt{9} & 1+2 \\ 3 & 3 \end{array} \quad \text{TRUE}$$

For 3:

$$\begin{array}{c|c} \sqrt{2x-5} & = 1+\sqrt{x-3} \\ \hline \sqrt{2(3)-5} & ?\ 1+\sqrt{3-3} \\ \sqrt{6-5} & 1+\sqrt{0} \\ \sqrt{1} & 1+0 \\ 1 & 1 \end{array} \quad \text{TRUE}$$

The numbers 7 and 3 check and are the solutions.

Do Exercises 7 and 8. ▶

Solve.

7. $\sqrt{x} - \sqrt{x-5} = 1$

8. $\sqrt{2x-5} - 2 = \sqrt{x-2}$

EXAMPLE 8 Solve: $\sqrt{x+2} - \sqrt{2x+2} + 1 = 0$.

We first isolate one radical.

$$\sqrt{x+2} - \sqrt{2x+2} + 1 = 0$$

$$\sqrt{x+2} + 1 = \sqrt{2x+2}$$ Adding $\sqrt{2x+2}$ to isolate a radical term

$$\left(\sqrt{x+2}+1\right)^2 = \left(\sqrt{2x+2}\right)^2$$ Squaring both sides

$$x + 2 + 2\sqrt{x+2} + 1 = 2x + 2$$

$$2\sqrt{x+2} = x - 1$$

$$\left(2\sqrt{x+2}\right)^2 = (x-1)^2$$

$$4(x+2) = x^2 - 2x + 1$$

$$4x + 8 = x^2 - 2x + 1$$

$$0 = x^2 - 6x - 7$$

$$0 = (x-7)(x+1)$$ Factoring

$$x - 7 = 0 \quad or \quad x + 1 = 0$$ Using the principle of zero products

$$x = 7 \quad or \quad x = -1$$

The possible solutions are 7 and -1.

Check: For 7:

$$\begin{array}{c|c} \sqrt{x+2}-\sqrt{2x+2}+1 & = 0 \\ \hline \sqrt{7+2}-\sqrt{2\cdot 7+2}+1 & ?\ 0 \\ \sqrt{9}-\sqrt{16}+1 & \\ 3-4+1 & \\ 0 & \end{array} \quad \text{TRUE}$$

For -1:

$$\begin{array}{c|c} \sqrt{x+2}-\sqrt{2x+2}+1 & = 0 \\ \hline \sqrt{-1+2}-\sqrt{2\cdot(-1)+2}+1 & ?\ 0 \\ \sqrt{1}-\sqrt{0}+1 & \\ 1-0+1 & \\ 2 & \end{array} \quad \text{FALSE}$$

The number 7 checks, but -1 does not. The solution is 7.

Do Exercise 9. ▶

9. Solve: $\sqrt{3x+1} - 1 - \sqrt{x+4} = 0$.

Answers

7. 9 **8.** 27 **9.** 5

C APPLICATIONS

Speed of Sound. Many applications translate to radical equations. For example, at a temperature of t degrees Fahrenheit, sound travels at a rate of S feet per second, where

$$S = 21.9\sqrt{5t + 2457}.$$

EXAMPLE 9 *Capitol Concert.* Each year, a free concert, know as "A Capitol Fourth," on the west lawn of the Capitol in Washington, D.C., precedes the July 4th fireworks display on the National Mall. A few of the performers at the 2017 concert included the Beach Boys, the Blues Brothers, Trace Adkins, Yolanda Adams, and the *Voice* winner Chris Blue. A scientific instrument determined that the sound of the music was traveling at a rate of 1170 ft/sec. What was the air temperature at the concert?

We substitute 1170 for S in the formula $S = 21.9\sqrt{5t + 2457}$:

$$1170 = 21.9\sqrt{5t + 2457}.$$

Then we solve the equation for t:

$$1170 = 21.9\sqrt{5t + 2457}$$

$$\frac{1170}{21.9} = \sqrt{5t + 2457} \quad \text{Dividing by 21.9}$$

$$\left(\frac{1170}{21.9}\right)^2 = \left(\sqrt{5t + 2457}\right)^2 \quad \text{Squaring both sides}$$

$$2854.2 \approx 5t + 2457 \quad \text{Simplifying}$$

$$397.2 \approx 5t \quad \text{Subtracting 2457}$$

$$79 \approx t. \quad \text{Dividing by 5}$$

The temperature at the concert was about 79°F.

 **Do Exercise 10.**

10. *Marching Band Performance.* When the Fulton High School marching band performed at half-time of a football game, the speed of sound of the music was measured by a scientific instrument to be 1162 ft/sec. What was the air temperature?

Answer

10. About 72°F

6.6 Exercise Set

FOR EXTRA HELP MyLab Math

✓ Check Your Understanding

Reading Check Choose from the column on the right the term that best completes each statement. Not every word will be used.

RC1. The equation $\sqrt{4 - 11x} = 3$ is a(n) ______ equation.

RC2. When we square both sides of an equation, we are using the principle of ______.

RC3. To solve an equation with a radical term, we first ______ the radical term on one side of the equation.

RC4. A radical equation has variables in one or more ______.

RC5. A check is essential when we raise both sides of an equation to a(n) ______ power.

even
radical
isolate
odd
radicands
square roots
powers
raise
rational
principle

Concept Check Determine the number of times that the principle of squaring must be used in order to solve the equation. Do not solve the equation.

CC1. $\sqrt{x - 2} = x + 4$

CC2. $\sqrt{12y - 5} = \sqrt{y + 24}$

CC3. $3 - \sqrt{x + 3} = \sqrt{x - 3}$

CC4. $\sqrt{c} + 2 = \sqrt{c + 7}$

a Solve.

1. $\sqrt{2x - 3} = 4$

2. $\sqrt{5x + 2} = 7$

3. $\sqrt{6x} + 1 = 8$

4. $\sqrt{3x} - 4 = 6$

5. $\sqrt{y + 7} - 4 = 4$

6. $\sqrt{x - 1} - 3 = 9$

7. $\sqrt{5y + 8} = 10$

8. $\sqrt{2y + 9} = 5$

9. $\sqrt[3]{x} = -1$

10. $\sqrt[3]{y} = -2$

11. $\sqrt{x + 2} = -4$

12. $\sqrt{y - 3} = -2$

13. $\sqrt[3]{x + 5} = 2$

14. $\sqrt[3]{x - 2} = 3$

15. $\sqrt[4]{y - 3} = 2$

16. $\sqrt[4]{x + 3} = 3$

17. $\sqrt[3]{6x + 9} + 8 = 5$

18. $\sqrt[3]{3y + 6} + 2 = 3$

19. $8 = \dfrac{1}{\sqrt{x}}$

20. $\dfrac{1}{\sqrt{y}} = 3$

21. $x - 7 = \sqrt{x - 5}$

22. $x - 5 = \sqrt{x + 7}$

23. $2\sqrt{x + 1} + 7 = x$

24. $\sqrt{2x + 7} - 2 = x$

25. $3\sqrt{x - 1} - 1 = x$

26. $x - 1 = \sqrt{x + 5}$

27. $x - 3 = \sqrt{27 - 3x}$

28. $x - 1 = \sqrt{1 - x}$

b Solve.

29. $\sqrt{3y+1} = \sqrt{2y+6}$

30. $\sqrt{5x-3} = \sqrt{2x+3}$

31. $\sqrt{y-5} + \sqrt{y} = 5$

32. $\sqrt{x-9} + \sqrt{x} = 1$

33. $3 + \sqrt{z-6} = \sqrt{z+9}$

34. $\sqrt{4x-3} = 2 + \sqrt{2x-5}$

35. $\sqrt{20-x} + 8 = \sqrt{9-x} + 11$

36. $4 + \sqrt{10-x} = 6 + \sqrt{4-x}$

37. $\sqrt{4y+1} - \sqrt{y-2} = 3$

38. $\sqrt{y+15} - \sqrt{2y+7} = 1$

39. $\sqrt{x+2} + \sqrt{3x+4} = 2$

40. $\sqrt{6x+7} - \sqrt{3x+3} = 1$

41. $\sqrt{3x-5} + \sqrt{2x+3} + 1 = 0$

42. $\sqrt{2m-3} + 2 - \sqrt{m+7} = 0$

43. $2\sqrt{t-1} - \sqrt{3t-1} = 0$

44. $3\sqrt{2y+3} - \sqrt{y+10} = 0$

C Solve.

Sightings to the Horizon. How far can you see to the horizon from a given height? The function $D = 1.2\sqrt{h}$ can be used to approximate the distance D, in miles, that a person can see to the horizon from a height h, in feet.

45. *Observation Deck at Jeddah Tower.* Jeddah Tower, soon to be the tallest building in the world, in Jeddah, Saudi Arabia, is projected to be completed in 2020. An observation deck with a 98-ft diameter outdoor balcony near the top of the tower is 2080 ft high. How far can a tourist see to the horizon from the observation deck?

Data: Council on Tall Buildings and Urban Habitat (CTBUH); *Global Construction News*

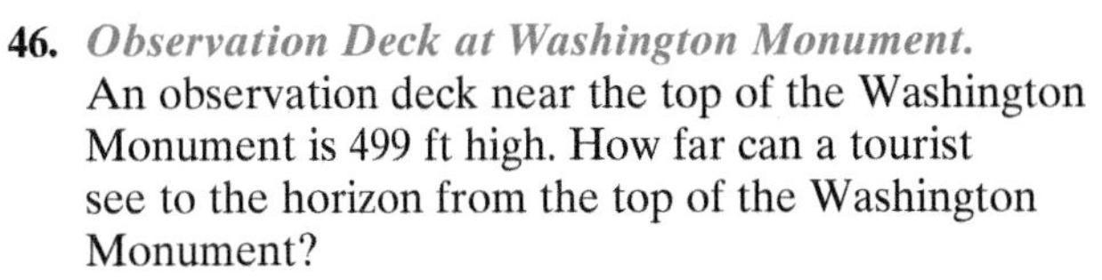

46. *Observation Deck at Washington Monument.* An observation deck near the top of the Washington Monument is 499 ft high. How far can a tourist see to the horizon from the top of the Washington Monument?

Data: Council on Tall Buildings and Urban Habitat (CTBUH); *Global Construction News*

47. Sarah can see 31.3 mi to the horizon from the top of a cliff. What is the height of Sarah's eyes?

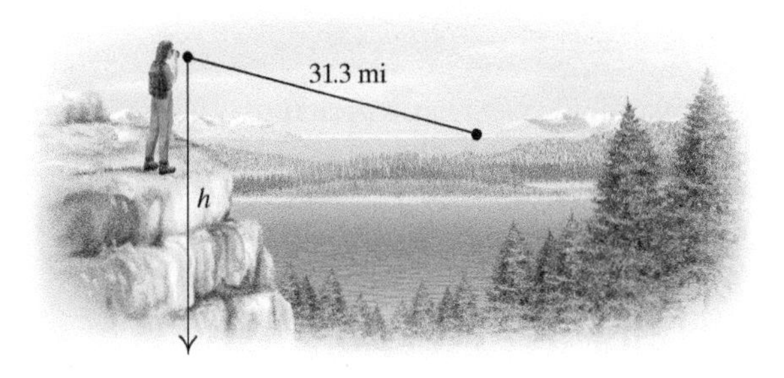

48. A steeplejack can see 13 mi to the horizon from the top of a building. What is the height of the steeplejack's eyes?

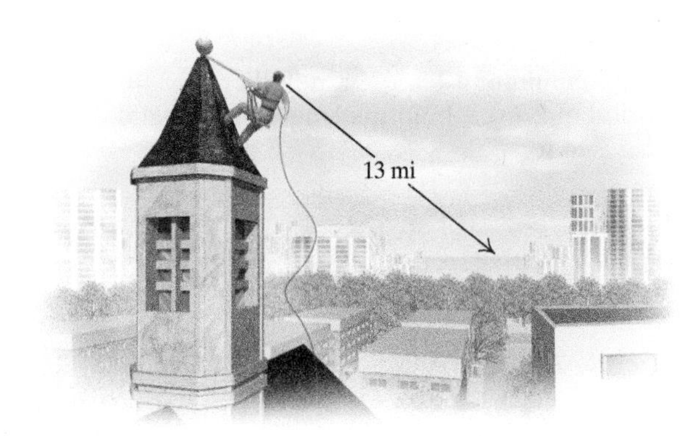

49. A technician can see 30.4 mi to the horizon from the top of a radio tower. How high is the tower?

50. A person can see 230 mi to the horizon from an airplane window. How high is the airplane?

Speed of a Skidding Car. After an automobile accident, how do police determine the speed at which a car had been traveling? The formula

$$r = 2\sqrt{5L}$$

can be used to approximate the speed r, in miles per hour, of a car that has left a skid mark of length L, in feet. Use this formula for Exercises 51 and 52.

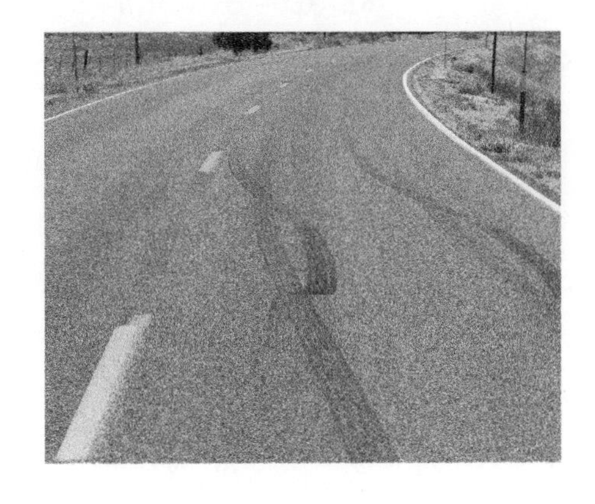

51. How far will a car skid at 55 mph? at 75 mph?

52. How far will a car skid at 65 mph? at 100 mph?

53. *Water Flow and Nozzle Pressure.* For accuracy and safety, firefighters rely on nozzle pressure and water flow. They can calculate water flow using the formula

$$F = 29.72D^2\sqrt{P},$$

where F is the water flow, in gallons per minute, D is the nozzle diameter, in inches, and P is the nozzle pressure, in pounds per square inch. What is the nozzle pressure when the nozzle diameter is $1\frac{3}{4}$ in. and the water flow is 815 gallons per minute?

Data: Manchaca Fire Rescue (mvfd.org); Truckee Meadow Fire Protection District (tmfire.us)

54. *Temperature and the Speed of Sound.* Avalanche blasting creates controlled slides. Squaw Valley, a Tahoe Ski resort, is using the Gazex Inertia Explorer, a new blasting technology that is safer and creates a larger blast. At a temperature of t degrees Fahrenheit, sound travels at a rate of S feet per second, where

$$S = 21.9\sqrt{5t + 2457}.$$

If at a recent blast the sound traveled at a rate of 1115 ft/sec, what was the temperature at that time?

Data: kolotv.com; Squaw-Unveils-New-Avalanche-Control-Technology-351977181.html, by Colin Lygren, November 19, 2015

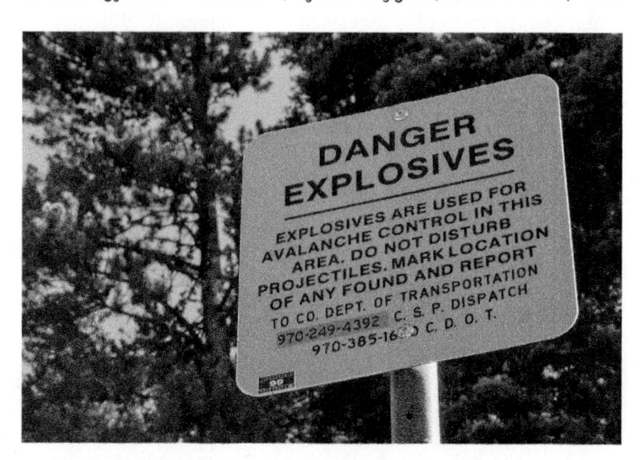

Period of a Swinging Pendulum. The formula $T = 2\pi\sqrt{L/32}$ can be used to find the period T, in seconds, of a pendulum of length L, in feet.

55. What is the length of a pendulum that has a period of 1.0 sec? Use 3.14 for π.

56. What is the length of a pendulum that has a period of 2.0 sec? Use 3.14 for π.

57. The pendulum in Jean's grandfather clock has a period of 2.2 sec. Find the length of the pendulum. Use 3.14 for π.

58. A playground swing has a period of 3.1 sec. Find the length of the swing's chain. Use 3.14 for π.

Skill Maintenance

Solve. [5.6a, b]

59. *Bicycle Travel.* A cyclist traveled 702 mi in 14 days. At this same rate, how far would the cyclist have traveled in 56 days?

60. *Delivering Leaflets.* Jeff can distribute leaflets to homes three times as fast as Grace can. If they work together, it takes them 1 hr to complete the job. How long would it take each to deliver the leaflets alone?

Solve. [4.8a]

61. $x^2 + 2.8x = 0$

62. $3x^2 - 5x = 0$

63. $x^2 - 64 = 0$

64. $2x^2 = x + 21$

For each of the following functions, find and simplify $f(a + h) - f(a)$. [4.2e]

65. $f(x) = x^2$

66. $f(x) = x^2 - x$

67. $f(x) = 2x^2 - 3x$

68. $f(x) = 2x^2 + 3x - 7$

Synthesis

Solve.

69. $\sqrt{\sqrt{y + 49} - \sqrt{y}} = \sqrt{7}$

70. $\sqrt[3]{x^2 + x + 15} - 3 = 0$

71. $\sqrt{\sqrt{x^2 + 9x + 34}} = 2$

72. $6\sqrt{y} + 6y^{-1/2} = 37$

73. $\sqrt{a^2 + 30a} = a + \sqrt{5a}$

74. $\sqrt{\sqrt{x + 4}} = \sqrt{x} - 2$

75. $\sqrt{y + 1} - \sqrt{2y - 5} = \sqrt{y - 2}$

76. $\sqrt{x + 1} - \dfrac{2}{\sqrt{x + 1}} = 1$

Applications Involving Powers and Roots

6.7

OBJECTIVE

a Solve applied problems involving the Pythagorean theorem and powers and roots.

a APPLICATIONS

Approximate square roots of numbers using a calculator. [8.1a]

Use a calculator to approximate each square root. Round to three decimal places.

1. $\sqrt{43}$ **2.** $\sqrt{310}$

Answers: 1. 6.557 **2.** 17.607

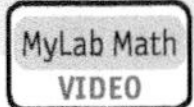

There are many kinds of applied problems that involve powers and roots. Many also make use of right triangles and the Pythagorean theorem: $a^2 + b^2 = c^2$.

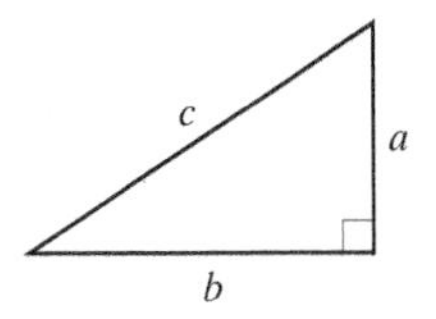

EXAMPLE 1 *Computer Screen Size.* The viewable image size of a wide-screen computer measures 21 in. diagonally and has a width of 18.3 in. What is its height?

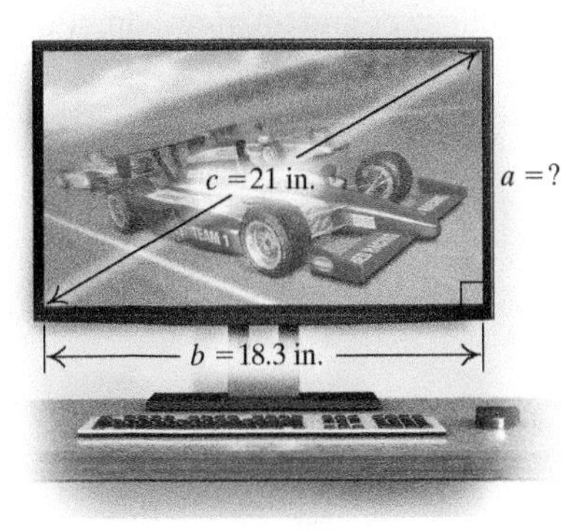

Using the Pythagorean theorem, $a^2 + b^2 = c^2$, we substitute 18.3 for b and 21 for c and then solve for a:

$$a^2 + b^2 = c^2$$

$$a^2 + 18.3^2 = 21^2 \quad \text{Substituting}$$

$$a^2 + 334.89 = 441$$

$$a^2 = 106.11$$

Exact answer: $a = \sqrt{106.11}$ We consider only the positive root since length cannot be negative.

Approximation: $a \approx 10.3$.

The height of the viewable image is about 10.3 in.

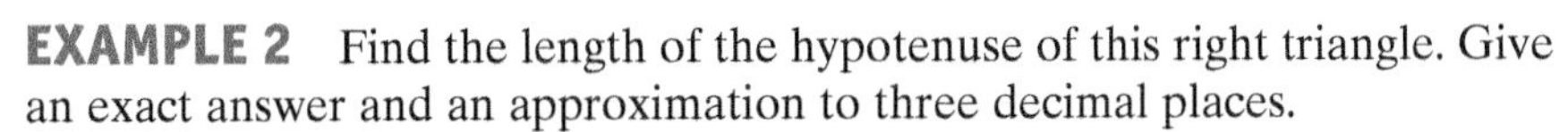

EXAMPLE 2 Find the length of the hypotenuse of this right triangle. Give an exact answer and an approximation to three decimal places.

$$7^2 + 4^2 = c^2 \quad \text{Substituting}$$

$$49 + 16 = c^2$$

$$65 = c^2$$

Exact answer: $c = \sqrt{65}$

Approximation: $c \approx 8.062$ Using a calculator

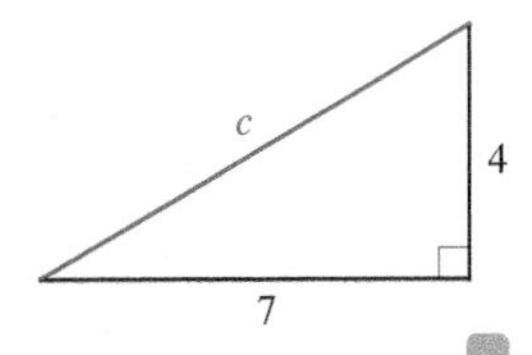

EXAMPLE 3 Find the missing length b in this right triangle. Give an exact answer and an approximation to three decimal places.

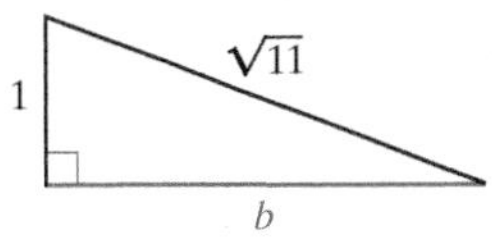

$$1^2 + b^2 = (\sqrt{11})^2 \quad \text{Substituting}$$

$$1 + b^2 = 11$$

$$b^2 = 10$$

Exact answer: $b = \sqrt{10}$

Approximation: $b \approx 3.162$ Using a calculator

(Triangle: legs 1 and b, hypotenuse $\sqrt{11}$)

Do Exercises 1 and 2. ▶

1. Find the length of the hypotenuse of this right triangle. Give an exact answer and an approximation to three decimal places.

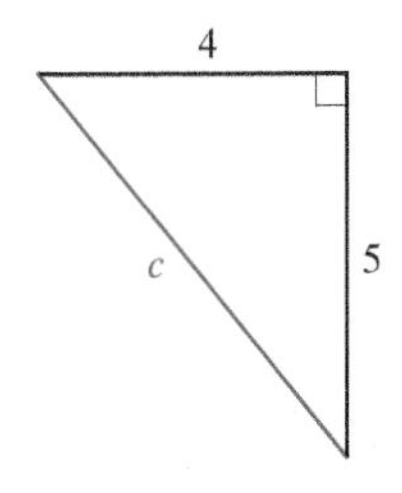

2. Find the length of the leg of this right triangle. Give an exact answer and an approximation to three decimal places.

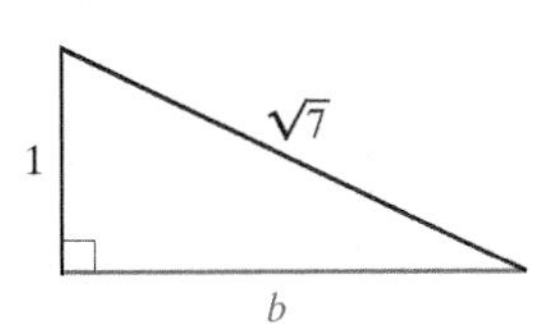

Answers

1. $\sqrt{41}$; 6.403 **2.** $\sqrt{6}$; 2.449

EXAMPLE 4 *Construction.* Darla is laying out the footing of a house. To see if the corner is square, she measures 16 ft from the corner along one wall and 12 ft from the corner along the other wall. How long should the diagonal be between those two points if the corner is a right angle?

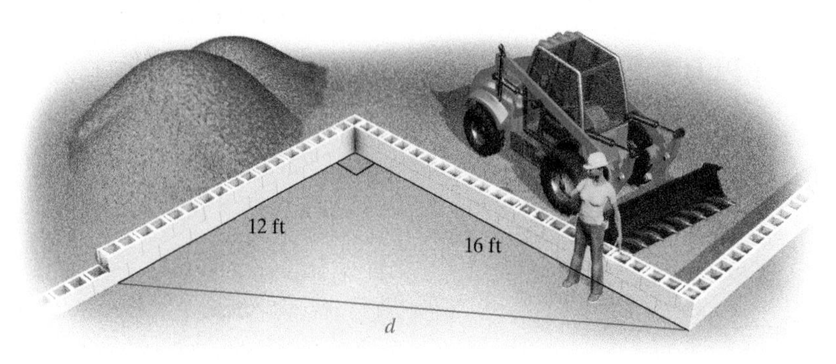

3. *Baseball Diamond.* A baseball diamond is actually in the shape of a square 90 ft on a side. Suppose that a catcher fields a bunt along the third-base line 10 ft from home plate. How far is the catcher's throw to first base? Give an exact answer and an approximation to three decimal places.

We make a drawing and let $d =$ the length of the diagonal. It is the length of the hypotenuse of a right triangle whose legs are 12 ft and 16 ft. We substitute these values in the Pythagorean theorem to find d:

$$d^2 = 12^2 + 16^2$$
$$d^2 = 144 + 256$$
$$d^2 = 400$$
$$d = \sqrt{400} = 20.$$

The length of the diagonal should be 20 ft.

Do Exercise 3.

EXAMPLE 5 *Road-Pavement Messages.* In a psychological study, it was determined that the ideal length L of the letters of a word painted on pavement is given by

$$L = \frac{0.000169d^{2.27}}{h},$$

where d is the distance of a car from the lettering and h is the height of the eye above the road. All units are in feet. For a person h feet above the road, a message d feet away will be the most readable if the length of the letters is L. Find L, given that $h = 4$ ft and $d = 180$ ft.

4. Refer to Example 5. Find L given that $h = 3$ ft and $d = 180$ ft. You will need a calculator with an exponentiation key $\boxed{y^x}$, or ^.

We substitute 4 for h and 180 for d and calculate L using a calculator with an exponentiation key $\boxed{y^x}$, or ^:

$$L = \frac{0.000169(180)^{2.27}}{4} \approx 5.6 \text{ ft}.$$

Do Exercise 4.

Answers

3. $\sqrt{8200}$ ft; 90.554 ft 4. About 7.4 ft

Translating for Success

1. *Angles of a Triangle.* The second angle of a triangle is four times as large as the first. The third is 27° less than the sum of the other angles. Find the measures of the angles.

2. *Lengths of a Rectangle.* The area of a rectangle is 180 ft^2. The length is 26 ft greater than the width. Find the length and the width.

3. *Boat Travel.* The speed of a river is 3 mph. A boat can go 72 mi upstream and 24 mi downstream in a total time of 16 hr. Find the speed of the boat in still water.

4. *Coin Mixture.* A collection of nickels and quarters is worth \$13.85. There are 85 coins in all. How many of each coin are there?

5. *Perimeter.* The perimeter of a rectangle is 180 ft. The length is 26 ft greater than the width. Find the length and the width.

Translate each word problem to an equation or a system of equations and select a correct translation from equations A–O.

A. $12^2 + 12^2 = x^2$

B. $x(x + 26) = 180$

C. $10{,}311 + 5\%x = x$

D. $x + y = 85,$
$5x + 25y = 13.85$

E. $x^2 + 4^2 = 12^2$

F. $\frac{240}{x - 18} = \frac{384}{x}$

G. $x + 5\%x = 10{,}311$

H. $\frac{x}{65} + 1 = \frac{x}{85}$

I. $\frac{x}{65} + \frac{x}{85} = 1$

J. $x + y + z = 180,$
$y = 4x,$
$z = x + y - 27$

K. $2x + 2(x + 26) = 180$

L. $\frac{384}{x - 18} = \frac{240}{x}$

M. $x + y = 85,$
$0.05x + 0.25y = 13.85$

N. $2x + 2(x + 24) = 240$

O. $\frac{72}{x - 3} + \frac{24}{x + 3} = 16$

Answers on page A-23

6. *Shoveling Time.* It takes Marv 65 min to shovel 4 in. of snow from his driveway. It takes Elaine 85 min to do the same job. How long would it take if they worked together?

7. *Money Borrowed.* Claire borrows some money at 5% simple interest. After 1 year, \$10,311 pays off her loan. How much did she originally borrow?

8. *Plank Height.* A 12-ft plank is leaning against a shed. The bottom of the plank is 4 ft from the building. How high up the side of the shed is the top of the plank?

9. *Train Speeds.* The speed of train A is 18 mph slower than the speed of train B. Train A travels 240 mi in the same time that it takes train B to travel 384 mi. Find the speed of train A.

10. *Diagonal of a Square.* Find the length of a diagonal of a square swimming pool whose sides are 12 ft long.

6.7 Exercise Set

FOR EXTRA HELP — MyLab Math

✓ Check Your Understanding

Reading Check Complete each of the following statements.

RC1. The ______________ theorem states that if a and b are the lengths of the legs of a right triangle and c is the length of the third side of a triangle, then $a^2 + b^2 = c^2$.

RC2. In any right triangle, the two shortest sides are called the ______________.

RC3. In any right triangle, the longest side is opposite the right angle and is called the ______________.

Concept Check For each right triangle, choose from the column on the right the equation that can be used to find the missing side.

CC1.

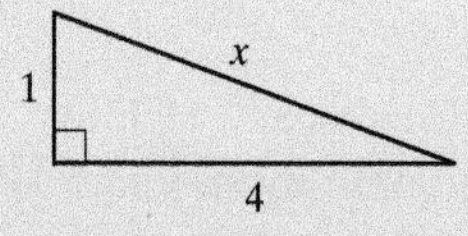

CC2.

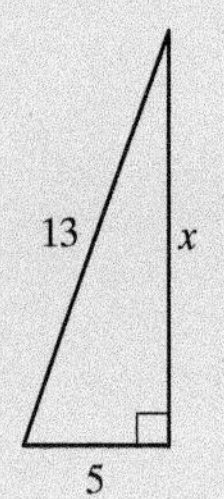

CC3.

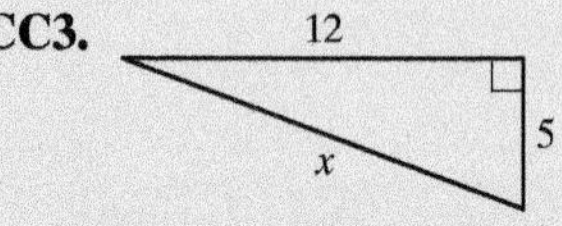

CC4.

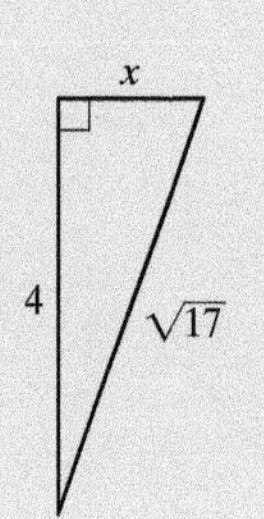

a) $12^2 + x^2 = 5^2$
b) $12^2 + 13^2 = x^2$
c) $5^2 + x^2 = 13^2$
d) $x^2 + 4^2 = (\sqrt{17})^2$
e) $1^2 + 4^2 = x^2$
f) $5^2 + 13^2 = x^2$
g) $5^2 + 12^2 = x^2$
h) $(\sqrt{17})^2 + x^2 = 4^2$

a In a right triangle, find the length of the side not given. Give an exact answer and, where appropriate, an approximation to three decimal places.

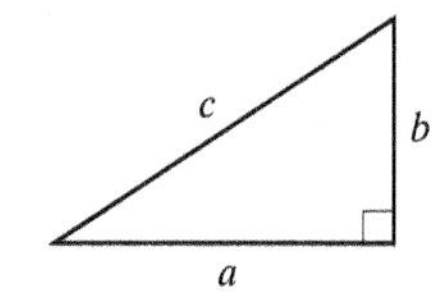

1. $a = 3,\ b = 5$

2. $a = 8,\ b = 10$

3. $a = 15,\ b = 15$

4. $a = 8,\ b = 8$

5. $b = 12,\ c = 13$

6. $a = 5,\ c = 12$

7. $c = 7,\ a = \sqrt{6}$

8. $c = 10,\ a = 4\sqrt{5}$

9. $b = 1,\ c = \sqrt{13}$

10. $a = 1,\ c = \sqrt{12}$

11. $a = 1,\ c = \sqrt{n}$

12. $c = 2,\ a = \sqrt{n}$

In the following problems, give an exact answer and, where appropriate, an approximation to three decimal places.

13. *Road-Pavement Messages.* Using the formula of Example 5, find the length L of a road-pavement message when $h = 4$ ft and $d = 200$ ft.

14. *Road-Pavement Messages.* Using the formula of Example 5, find the length L of a road-pavement message when $h = 8$ ft and $d = 300$ ft.

15. *Length of a Smartphone Screen.* The Samsung Galaxy S8 phone has less bezel at the top and the bottom and a longer screen than previous models. The new infinity screen measures 5.8 in. diagonally and is 2.7 in. wide. Find the length of the display screen.

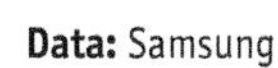

Data: Samsung

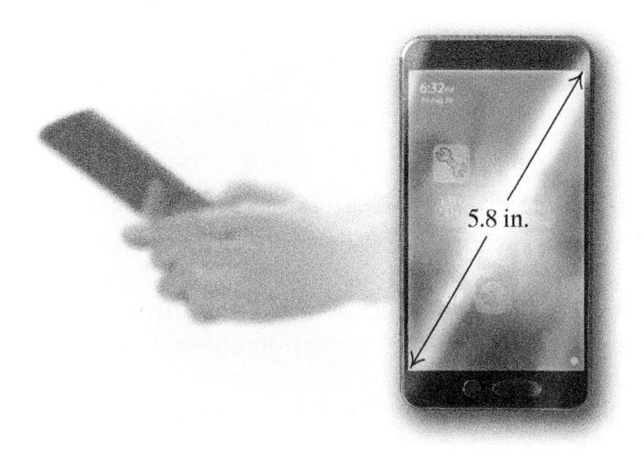

16. *Guy Wire.* How long is a guy wire reaching from the top of a 10-ft pole to a point on the ground 4 ft from the pole?

17. *Pyramide du Louvre.* A large glass and metal pyramid designed by I. M. Pei attracts visitors to the entrance of the Louvre Museum in Paris, France. If a plumb line from the highest point of the pyramid to the center of the base is 71 ft long and each side of the four equilateral triangles measures 100 ft, how far is it from the center of the base to a corner of the base?

Data: glassonweb.com

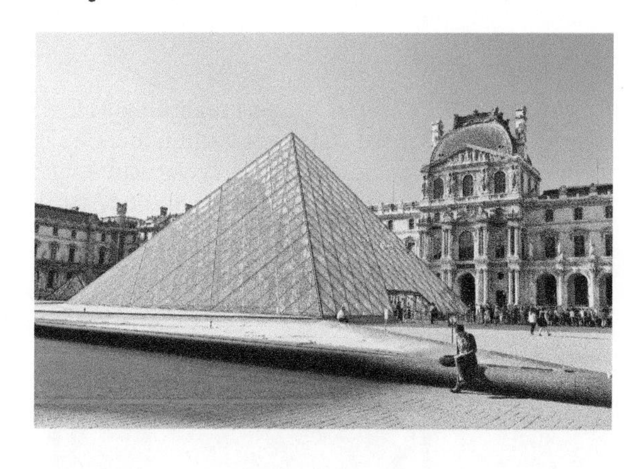

18. *Central Park.* New York City's rectangular Central Park in Manhattan runs 13,725 ft from 59th Street to 110th Street. A diagonal of the park is 13,977 ft. Find the width of the park.

19. Find all ordered pairs on the x-axis of a Cartesian coordinate system that are 5 units from the point $(0, 4)$.

20. Find all ordered pairs on the y-axis of a Cartesian coordinate system that are 5 units from the point $(3, 0)$.

21. *Bridge Expansion.* During the summer heat, a 2-mi bridge expands 2 ft in length. If we assume that the bulge occurs straight up the middle, how high is the bulge? (The answer may surprise you. In reality, bridges are built with expansion spaces to avoid such buckling.)

22. *Triangle Areas.* Triangle ABC has sides of lengths 25 ft, 25 ft, and 30 ft. Triangle PQR has sides of lengths 25 ft, 25 ft, and 40 ft. Which triangle has the greater area and by how much?

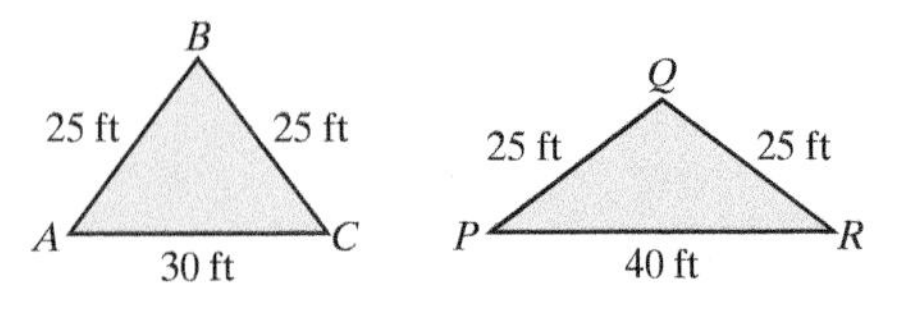

23. *A Baseball Throw.* A baseball diamond is actually in the shape of a square 90 ft on a side. Suppose that a third baseman fields a ball while standing on the third-base line 8 ft from third base, as shown. How far is the third baseman's throw to first base?

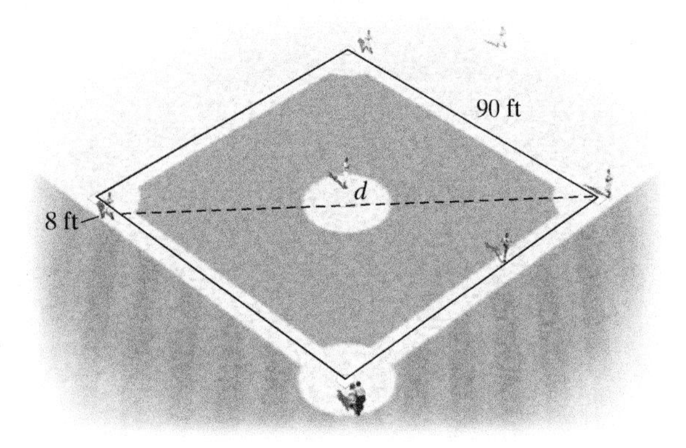

24. *Distance Over Water.* To determine the distance between two points on opposite sides of a pond, a surveyor locates two stakes at either end of the pond and uses instrumentation to place a third stake so that the distance across the pond is the length of a hypotenuse. If the third stake is 90 m from one stake and 70 m from the other, how wide is the pond?

25. *Plumbing.* Plumbers use the Pythagorean theorem to calculate pipe length. If a pipe is to be offset, as shown in the figure, the *travel*, or length, of the pipe, is calculated using the lengths of the *advance* and the *offset*. Find the travel if the offset is 17.75 in. and the advance is 10.25 in.

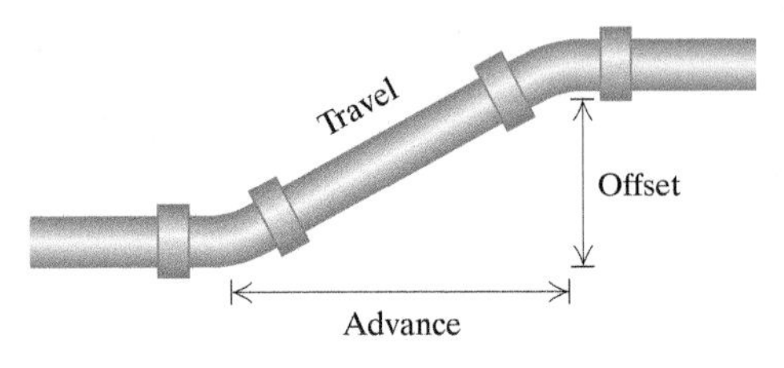

26. *Ramps for the Disabled.* Laws regarding access ramps for the disabled state that a ramp must be in the form of a right triangle, where every vertical length (leg) of 1 ft has a horizontal length (leg) of 12 ft. What is the length of a ramp with a 12-ft horizontal leg and a 1-ft vertical leg?

27. The length and the width of a rectangle are given by consecutive integers. The area of the rectangle is 90 cm^2. Find the length of a diagonal of the rectangle.

28. The diagonal of a square has length $8\sqrt{2}$ ft. Find the length of a side of the square.

29. Each side of a regular octagon has length s. Find a formula for the distance d between the parallel sides of the octagon.

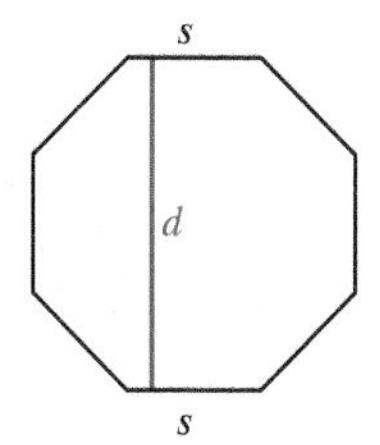

30. The two equal sides of an isosceles right triangle are of length s. Find a formula for the length of the hypotenuse.

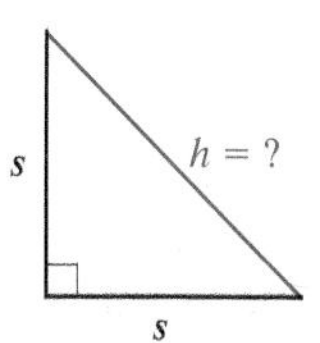

Skill Maintenance

Solve. [5.6c]

31. *Commuter Travel.* The speed of the Zionsville Flash commuter train is 14 mph faster than that of the Carmel Crawler. The Flash travels 290 mi in the same time that it takes the Crawler to travel 230 mi. Find the speed of each train.

32. *Marine Travel.* A motor boat travels three times as fast as the current in the Saskatee River. A trip up the river and back takes 10 hr, and the total distance of the trip is 100 mi. Find the speed of the current.

Solve.

33. $2x^2 + 11x - 21 = 0$ [4.8a]

34. $x^2 + 24 = 11x$ [4.8a]

35. $\frac{x + 2}{x + 3} = \frac{x - 4}{x - 5}$ [5.5a]

36. $3x^2 - 12 = 0$ [4.8a]

37. $\frac{x - 5}{x - 7} = \frac{4}{3}$ [5.5a]

38. $\frac{x - 1}{x - 3} = \frac{6}{x - 3}$ [5.5a]

Synthesis

39. *Roofing.* Kit's cottage, which is 24 ft wide and 32 ft long, needs a new roof. By counting clapboards that are 4 in. apart, Kit determines that the peak of the roof is 6 ft higher than the sides. If one packet of shingles covers $33\frac{1}{3}$ sq ft, how many packets will the job require?

40. *Wind Chill Temperature.* Because wind enhances the loss of heat from the skin, we feel colder when there is wind than when there is not. The *wind chill temperature* is what the temperature would have to be with no wind in order to give the same chilling effect as with the wind. A formula for finding the wind chill temperature, T_W, is

$$T_W = 35.74 + 0.6215T - 35.75V^{0.16} + 0.4275TV^{0.16},$$

where T is the actual temperature given by a thermometer, in degrees Fahrenheit, and V is the wind speed, in miles per hour. This formula can be used only when the wind speed is *above* 3 mph. Use a calculator to find the wind chill temperature in each case. Round to the nearest degree.

Data: National Weather Service

a) $T = 40°\text{F}$, $V = 25$ mph
b) $T = 20°\text{F}$, $V = 25$ mph
c) $T = 10°\text{F}$, $V = 20$ mph
d) $T = 10°\text{F}$, $V = 40$ mph
e) $T = -5°\text{F}$, $V = 35$ mph
f) $T = -15°\text{F}$, $V = 35$ mph

6.8 The Complex Numbers

OBJECTIVES

a Express imaginary numbers as bi, where b is a nonzero real number, and complex numbers as $a + bi$, where a and b are real numbers.

b Add and subtract complex numbers.

c Multiply complex numbers.

d Write expressions involving powers of i in the form $a + bi$.

e Find conjugates of complex numbers and divide complex numbers.

f Determine whether a given complex number is a solution of an equation.

a IMAGINARY NUMBERS AND COMPLEX NUMBERS

Negative numbers do not have square roots in the real-number system. However, mathematicians have described a larger number system that contains the real-number system and in which negative numbers have square roots. That system is called the **complex-number system**. We begin by defining a number that is a square root of -1. We call this new number i.

THE COMPLEX NUMBER i

We define the number i to be $\sqrt{-1}$. That is,

$$i = \sqrt{-1} \quad \text{and} \quad i^2 = -1.$$

To express roots of negative numbers in terms of i, we can use the fact that in the complex-number system, $\sqrt{-p} = \sqrt{-1 \cdot p} = \sqrt{-1}\,\sqrt{p}$ when p is a positive real number.

EXAMPLES Express in terms of i.

1. $\sqrt{-7} = \sqrt{-1 \cdot 7} = \sqrt{-1} \cdot \sqrt{7} = i\sqrt{7}$, or $\sqrt{7}i$

   i is *not* under the radical.

2. $\sqrt{-16} = \sqrt{-1 \cdot 16} = \sqrt{-1} \cdot \sqrt{16} = i \cdot 4 = 4i$
3. $-\sqrt{-13} = -\sqrt{-1 \cdot 13} = -\sqrt{-1} \cdot \sqrt{13} = -i\sqrt{13}$, or $-\sqrt{13}i$
4. $-\sqrt{-64} = -\sqrt{-1 \cdot 64} = -\sqrt{-1} \cdot \sqrt{64} = -i \cdot 8 = -8i$
5. $\sqrt{-48} = \sqrt{-1 \cdot 48} = \sqrt{-1} \cdot \sqrt{48} = i\sqrt{48} = i\sqrt{16 \cdot 3}$
   $= i \cdot 4\sqrt{3} = 4i\sqrt{3}$, or $4\sqrt{3}i$

◀ Do Exercises 1–5.

Express in terms of i.

1. $\sqrt{-5}$
2. $\sqrt{-25}$
3. $-\sqrt{-11}$
4. $-\sqrt{-36}$
5. $\sqrt{-54}$
   $= \sqrt{\quad \cdot 54}$
   $= \sqrt{-1} \cdot \sqrt{54} = i\sqrt{9 \cdot 6}$
   $= i \cdot \quad \sqrt{6}$
   $= 3 \cdot \quad \sqrt{6}$, or $3\sqrt{6}$

GS

IMAGINARY NUMBER

An **imaginary*** **number** is a number that can be named

$$bi,$$

where b is some real number and $b \neq 0$.

To form the system of **complex numbers**, we take the imaginary numbers and the real numbers and all possible sums of real and imaginary numbers. These are complex numbers:

$$7 - 4i, \qquad -\pi + 19i, \qquad 37, \qquad i\sqrt{6}.$$

Answers

1. $i\sqrt{5}$, or $\sqrt{5}i$ **2.** $5i$ **3.** $-i\sqrt{11}$, or $-\sqrt{11}i$ **4.** $-6i$ **5.** $3i\sqrt{6}$, or $3\sqrt{6}i$

Guided Solution:

5. $-1, 3, i, i$

*Don't let the name "imaginary" fool you. The imaginary numbers are very important in such fields as engineering and the physical sciences.

COMPLEX NUMBER

A **complex number** is any number that can be named

$$a + bi,$$

where a and b are any real numbers. (Note that either a or b or both can be 0.)

Since $0 + bi = bi$, every imaginary number is a complex number. Similarly, $a + 0i = a$, so every real number is a complex number. The relationships among various real and complex numbers are shown in the following diagram.

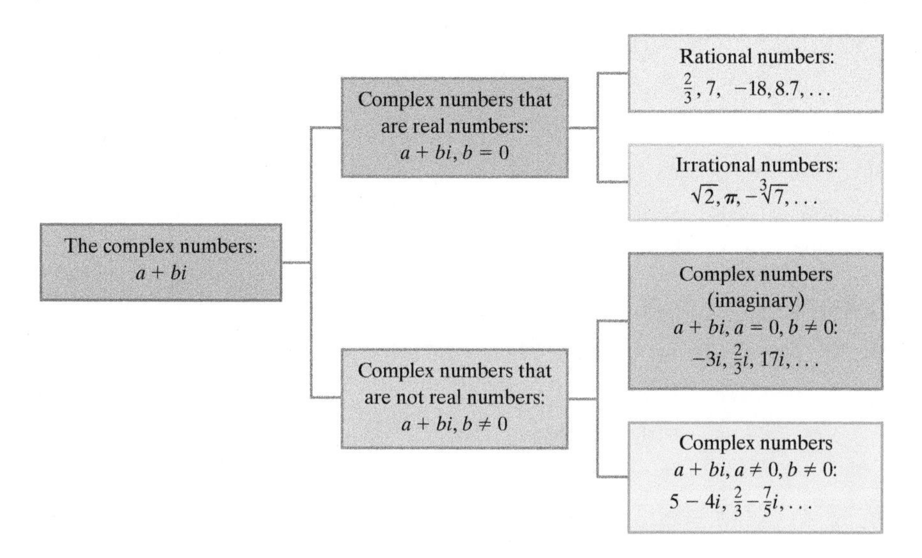

It is important to keep in mind some comparisons between numbers that have real-number roots and those that have complex-number roots that are not real. For example, $\sqrt{-48}$ is a complex number that is not a real number because we are taking the square root of a negative number. *But*, $\sqrt[3]{-125}$ is a real number because we are taking the cube root of a negative number and *any* real number has a cube root that is a real number.

b ADDITION AND SUBTRACTION

The complex numbers follow the commutative and associative laws of addition. Thus we can add and subtract them as we do binomials with real-number coefficients; that is, we collect like terms.

EXAMPLES Add or subtract.

6. $(8 + 6i) + (3 + 2i) = (8 + 3) + (6 + 2)i = 11 + 8i$

7. $(3 + 2i) - (5 - 2i) = (3 - 5) + [2 - (-2)]i = -2 + 4i$

Do Exercises 6–9.

Add or subtract.

6. $(7 + 4i) + (8 - 7i)$

7. $(-5 - 6i) + (-7 + 12i)$

8. $(8 + 3i) - (5 + 8i)$

9. $(5 - 4i) - (-7 + 3i)$

Answers

6. $15 - 3i$ **7.** $-12 + 6i$
8. $3 - 5i$ **9.** $12 - 7i$

C MULTIPLICATION

Use the FOIL method to multiply two binomials. [4.2b]

Multiply.

1. $(w + 4)(w - 6)$

2. $(2x + 3y)(3x - 5y)$

Answers: **1.** $w^2 - 2w - 24$ **2.** $6x^2 - xy - 15y^2$

MyLab Math VIDEO

The complex numbers obey the commutative, associative, and distributive laws. But although the property $\sqrt{a}\sqrt{b} = \sqrt{ab}$ does *not* hold for complex numbers in general, it does hold when $a = -1$ and b is a positive real number.

To multiply square roots of negative real numbers, we first express them in terms of i. For example,

$$\sqrt{-2} \cdot \sqrt{-5} = \sqrt{-1} \cdot \sqrt{2} \cdot \sqrt{-1} \cdot \sqrt{5} = i\sqrt{2} \cdot i\sqrt{5}$$
$$= i^2\sqrt{10} = -\sqrt{10} \quad \text{is correct!}$$

But $\sqrt{-2} \cdot \sqrt{-5} = \sqrt{(-2)(-5)} = \sqrt{10}$ is wrong!

Caution!

The rule $\sqrt{a}\sqrt{b} = \sqrt{ab}$ holds only for nonnegative real numbers.

Keeping this and the fact that $i^2 = -1$ in mind, we multiply in much the same way that we do with real numbers.

EXAMPLES Multiply.

8. $\sqrt{-49} \cdot \sqrt{-16} = \sqrt{-1} \cdot \sqrt{49} \cdot \sqrt{-1} \cdot \sqrt{16}$
$= i \cdot 7 \cdot i \cdot 4 = i^2(28)$
$= (-1)(28) \qquad i^2 = -1$
$= -28$

9. $\sqrt{-3} \cdot \sqrt{-7} = \sqrt{-1} \cdot \sqrt{3} \cdot \sqrt{-1} \cdot \sqrt{7}$
$= i \cdot \sqrt{3} \cdot i \cdot \sqrt{7} = i^2(\sqrt{21})$
$= (-1)\sqrt{21} \qquad i^2 = -1$
$= -\sqrt{21}$

10. $-2i \cdot 5i = -10 \cdot i^2$
$= (-10)(-1) \qquad i^2 = -1$
$= 10$

11. $(-4i)(3 - 5i) = (-4i) \cdot 3 - (-4i)(5i)$ Using a distributive law
$= -12i + 20i^2 = -12i + 20(-1)$
$= -12i - 20 = -20 - 12i$

12. $(1 + 2i)(1 + 3i) = 1 + 3i + 2i + 6i^2$ Using FOIL
$= 1 + 3i + 2i + 6(-1) \qquad i^2 = -1$
$= 1 + 3i + 2i - 6$
$= -5 + 5i$ Collecting like terms

13. $(3 - 2i)^2 = 3^2 - 2(3)(2i) + (2i)^2$ Squaring the binomial
$= 9 - 12i + 4i^2 = 9 - 12i + 4(-1)$
$= 9 - 12i - 4 = 5 - 12i$

Do Exercises 10–17.

Multiply.

10. $\sqrt{-25} \cdot \sqrt{-4}$

11. $\sqrt{-2} \cdot \sqrt{-17}$

12. $-6i \cdot 7i$

13. $-3i(4 - 3i)$

14. $5i(-5 + 7i)$

15. $(1 + 3i)(1 + 5i)$ GS
$= 1 + \_\_\_ + 3i + \_\_\_ i^2$
$= 1 + \_\_\_ i + 15(-1)$
$= 1 + 8i - 15$
$= \_\_\_ + 8i$

16. $(3 - 2i)(1 + 4i)$

17. $(3 + 2i)^2$

Answers

10. -10 **11.** $-\sqrt{34}$ **12.** 42 **13.** $-9 - 12i$ **14.** $-35 - 25i$ **15.** $-14 + 8i$ **16.** $11 + 10i$ **17.** $5 + 12i$

Guided Solution:
15. $5i, 15, 8, -14$

d POWERS OF i

We now want to simplify certain expressions involving powers of i. To do so, we first see how to simplify powers of i. Simplifying powers of i can be done by using the fact that $i^2 = -1$ and expressing the given power of i in terms of even powers, and then in terms of powers of i^2. Consider the following:

$$i,$$
$$i^2 = -1,$$
$$i^3 = i^2 \cdot i = (-1)i = -i,$$
$$i^4 = (i^2)^2 = (-1)^2 = 1,$$
$$i^5 = i^4 \cdot i = (i^2)^2 \cdot i = (-1)^2 \cdot i = i,$$
$$i^6 = (i^2)^3 = (-1)^3 = -1.$$

Note that the powers of i cycle through the values i, -1, $-i$, and 1.

EXAMPLES Simplify.

14. $i^{37} = i^{36} \cdot i = (i^2)^{18} \cdot i = (-1)^{18} \cdot i = 1 \cdot i = i$

15. $i^{58} = (i^2)^{29} = (-1)^{29} = -1$

16. $i^{75} = i^{74} \cdot i = (i^2)^{37} \cdot i = (-1)^{37} \cdot i = -1 \cdot i = -i$

17. $i^{80} = (i^2)^{40} = (-1)^{40} = 1$

Do Exercises 18–21. ▶

Simplify.

18. i^{47}

19. i^{68}

20. i^{85}

21. i^{90}

EXAMPLES Simplify to the form $a + bi$.

18. $8 - i^2 = 8 - (-1) = 8 + 1 = 9$

19. $17 + 6i^3 = 17 + 6 \cdot i^2 \cdot i = 17 + 6(-1)i = 17 - 6i$

20. $i^{22} - 67i^2 = (i^2)^{11} - 67(-1) = (-1)^{11} + 67 = -1 + 67 = 66$

21. $i^{23} + i^{48} = (i^{22}) \cdot i + (i^2)^{24} = (i^2)^{11} \cdot i + (-1)^{24} = (-1)^{11} \cdot i + (-1)^{24}$
$= -i + 1 = 1 - i$

Do Exercises 22–25. ▶

Simplify.

22. $8 - i^5$

23. $7 + 4i^2$

24. $i^{34} - i^{55}$

GS **25.** $6i^{11} + 7i^{14}$
$= 6 \cdot i^{\square} \cdot i + 7 \cdot i^{14}$
$= 6(i^2)^{\square} \cdot i + 7(i^2)^7$
$= 6(\square)^5 \cdot i + 7(-1)^7$
$= 6(\square)i + 7(\square)$
$= -6i - \square$
$= -7 - 6i$

e CONJUGATES AND DIVISION

Conjugates of complex numbers are defined as follows.

CONJUGATE

The **conjugate** of a complex number $a + bi$ is $a - bi$, and the **conjugate** of $a - bi$ is $a + bi$.

EXAMPLES Find the conjugate.

22. $5 + 7i$ The conjugate is $5 - 7i$.

23. $-3 - 9i$ The conjugate is $-3 + 9i$.

24. $4i$ The conjugate is $-4i$.

Do Exercises 26–28. ▶

Find the conjugate.

26. $6 + 3i$

27. $-9 - 5i$

28. $-\frac{1}{4}i$

Answers

18. $-i$ **19.** 1 **20.** i **21.** -1 **22.** $8 - i$ **23.** 3 **24.** $-1 + i$ **25.** $-7 - 6i$ **26.** $6 - 3i$ **27.** $-9 + 5i$ **28.** $\frac{1}{4}i$

Guided Solution:
25. 10, 5, −1, −1, −1, 7

When we multiply a complex number by its conjugate, we get a real number.

EXAMPLES Multiply.

25. $(5 + 7i)(5 - 7i) = 5^2 - (7i)^2$ Using $(A + B)(A - B) = A^2 - B^2$

$= 25 - 49i^2$

$= 25 - 49(-1)$ $i^2 = -1$

$= 25 + 49$

$= 74$

26. $(2 - 3i)(2 + 3i) = 2^2 - (3i)^2$

$= 4 - 9i^2$

$= 4 - 9(-1)$ $i^2 = -1$

$= 4 + 9$

$= 13$

Multiply.

29. $(7 - 2i)(7 + 2i)$

30. $(-3 - i)(-3 + i)$

Do Exercises 29 and 30.

We use conjugates when dividing complex numbers.

EXAMPLE 27 Divide and simplify to the form $a + bi$: $\frac{-5 + 9i}{1 - 2i}$.

$$\frac{-5 + 9i}{1 - 2i} \cdot \frac{1 + 2i}{1 + 2i} = \frac{(-5 + 9i)(1 + 2i)}{(1 - 2i)(1 + 2i)}$$ Multiplying by 1 using the conjugate of the denominator in the symbol for 1

$$= \frac{-5 - 10i + 9i + 18i^2}{1^2 - 4i^2}$$

$$= \frac{-5 - i + 18(-1)}{1 - 4(-1)}$$ $i^2 = -1$

$$= \frac{-5 - i - 18}{1 + 4}$$

$$= \frac{-23 - i}{5}$$

$$= -\frac{23}{5} - \frac{1}{5}i$$

Note the similarity between the preceding example and rationalizing denominators. In both cases, we used the conjugate of the denominator to write another name for 1. In Example 27, the symbol for the number 1 was chosen using the conjugate of the divisor, $1 - 2i$.

EXAMPLE 28 What symbol for 1 would you use to divide?

Division to be done	*Symbol for 1*
$\frac{3 + 5i}{4 + 3i}$	$\frac{4 - 3i}{4 - 3i}$

Answers
29. 53 **30.** 10

EXAMPLE 29 Divide and simplify to the form $a + bi$: $\frac{3 + 5i}{4 + 3i}$.

$$\frac{3 + 5i}{4 + 3i} \cdot \frac{4 - 3i}{4 - 3i} = \frac{(3 + 5i)(4 - 3i)}{(4 + 3i)(4 - 3i)} \quad \text{Multiplying by 1}$$

$$= \frac{12 - 9i + 20i - 15i^2}{4^2 - 9i^2}$$

$$= \frac{12 + 11i - 15(-1)}{16 - 9(-1)} \quad i^2 = -1$$

$$= \frac{12 + 11i + 15}{16 + 9}$$

$$= \frac{27 + 11i}{25} = \frac{27}{25} + \frac{11}{25}i$$

Do Exercises 31 and 32. ▶

Divide and simplify to the form $a + bi$.

31. $\frac{6 + 2i}{1 - 3i}$

32. $\frac{2 + 3i}{-1 + 4i}$

CALCULATOR CORNER

Complex Numbers We can perform operations on complex numbers on a graphing calculator. To do so, we first set the calculator in complex, or $a + bi$, mode by pressing MODE, using the ▽ and ▷ keys to position the blinking cursor over $a + bi$, and then pressing ENTER. We press 2ND QUIT to go to the home screen. Now we can add, subtract, multiply, and divide complex numbers; i is the second operation associated with the (.) key.

To find $(3 + 4i) - (7 - i)$, we note that the parentheses around $3 + 4i$ are optional, but those around $7 - i$ are necessary. To find $\frac{5 - 2i}{-1 + 3i}$, we see that the parentheses must be used to group the numerator and the denominator. To find $\sqrt{-4} \cdot \sqrt{-9}$, note that the calculator supplies the left parenthesis in each radicand and we supply the right parenthesis. The results of these operations are shown at right.

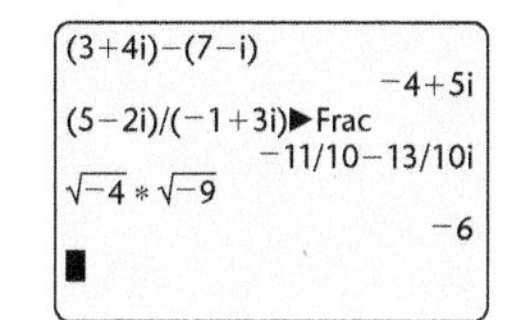

EXERCISES: Carry out each operation.

1. $(9 + 4i) + (-11 - 13i)$
2. $(9 + 4i) - (-11 - 13i)$
3. $(9 + 4i) \cdot (-11 - 13i)$
4. $(9 + 4i) \div (-11 - 13i)$
5. $\sqrt{-16} \cdot \sqrt{-25}$
6. $\sqrt{-23} \cdot \sqrt{-35}$
7. $\frac{4 - 5i}{-6 + 8i}$
8. $(-3i)^4$
9. $(1 - i)^3 - (2 + 3i)^4$

f SOLUTIONS OF EQUATIONS

The equation $x^2 + 1 = 0$ has no real-number solution, but it has *two* nonreal complex solutions.

EXAMPLE 30 Determine whether i is a solution of the equation $x^2 + 1 = 0$.

We substitute i for x in the equation.

$$\begin{array}{c|c} x^2 + 1 & 0 \\ \hline i^2 + 1 \; ? & 0 \\ -1 + 1 & \\ 0 & \text{TRUE} \end{array}$$

The number i is a solution.

Do Exercise 33. ▶

33. Determine whether $-i$ is a solution of $x^2 + 1 = 0$.

$$\begin{array}{c|c} x^2 + 1 & 0 \\ \hline & ? \\ & \end{array}$$

Answers

31. $2i$ **32.** $\frac{10}{17} - \frac{11}{17}i$ **33.** Yes

Any equation consisting of a polynomial in one variable on one side and 0 on the other has complex-number solutions. (Some may be real.) It is not always easy to find the solutions, but they always exist.

EXAMPLE 31 Determine whether $1 + i$ is a solution of the equation $x^2 - 2x + 2 = 0$.

We substitute $1 + i$ for x in the equation.

$$\begin{array}{r|l} x^2 - 2x + 2 & = 0 \\ \hline (1 + i)^2 - 2(1 + i) + 2 & ?\ 0 \\ 1 + 2i + i^2 - 2 - 2i + 2 & \\ 1 + 2i - 1 - 2 - 2i + 2 & \\ (1 - 1 - 2 + 2) + (2 - 2)i & \\ 0 + 0i & \\ 0 & \quad \text{TRUE} \end{array}$$

The number $1 + i$ is a solution.

EXAMPLE 32 Determine whether $2i$ is a solution of $x^2 + 3x - 4 = 0$.

$$\begin{array}{r|l} x^2 + 3x - 4 & = 0 \\ \hline (2i)^2 + 3(2i) - 4 & ?\ 0 \\ 4i^2 + 6i - 4 & \\ -4 + 6i - 4 & \\ -8 + 6i & \quad \text{FALSE} \end{array}$$

The number $2i$ is not a solution.

◀ **Do Exercise 34.**

34. Determine whether $1 - i$ is a solution of $x^2 - 2x + 2 = 0$.

$$\begin{array}{r|l} x^2 - 2x + 2 & = 0 \\ \hline & ? \end{array}$$

Answer

34. Yes

6.8 Exercise Set

FOR EXTRA HELP  MyLab Math

✓ Check Your Understanding

Reading Check and Concept Check Determine whether each statement is true or false.

RC1. Every real number is a complex number, but not every complex number is a real number.

RC2. The conjugate of the complex number $3 - 7i$ is $3 + 7i$.

RC3. All complex numbers are imaginary.

RC4. The square of a complex number is always a real number.

RC5. The product of a complex number and its conjugate is always a real number.

RC6. The imaginary number i raised to an even power is always 1.

RC7. The quotient of two complex numbers is always a complex number.

RC8. We add complex numbers by combining the real parts and combining the imaginary parts.

a Express in terms of i.

1. $\sqrt{-35}$

2. $\sqrt{-21}$

3. $\sqrt{-16}$

4. $\sqrt{-36}$

5. $-\sqrt{-12}$

6. $-\sqrt{-20}$

7. $\sqrt{-3}$

8. $\sqrt{-4}$

9. $\sqrt{-81}$

10. $\sqrt{-27}$

11. $\sqrt{-98}$

12. $-\sqrt{-18}$

13. $-\sqrt{-49}$

14. $-\sqrt{-125}$

15. $4 - \sqrt{-60}$

16. $6 - \sqrt{-84}$

17. $\sqrt{-4} + \sqrt{-12}$

18. $-\sqrt{-76} + \sqrt{-125}$

b Add or subtract and simplify.

19. $(7 + 2i) + (5 - 6i)$

20. $(-4 + 5i) + (7 + 3i)$

21. $(4 - 3i) + (5 - 2i)$

22. $(-2 - 5i) + (1 - 3i)$

23. $(9 - i) + (-2 + 5i)$

24. $(6 + 4i) + (2 - 3i)$

25. $(6 - i) - (10 + 3i)$

26. $(-4 + 3i) - (7 + 4i)$

27. $(4 - 2i) - (5 - 3i)$

28. $(-2 - 3i) - (1 - 5i)$

29. $(9 + 5i) - (-2 - i)$

30. $(6 - 3i) - (2 + 4i)$

c Multiply.

31. $\sqrt{-36} \cdot \sqrt{-9}$

32. $\sqrt{-16} \cdot \sqrt{-64}$

33. $\sqrt{-7} \cdot \sqrt{-2}$

34. $\sqrt{-11} \cdot \sqrt{-3}$

35. $-3i \cdot 7i$

36. $8i \cdot 5i$

37. $-3i(-8 - 2i)$

38. $4i(5 - 7i)$

39. $(3 + 2i)(1 + i)$

40. $(4 + 3i)(2 + 5i)$

41. $(2 + 3i)(6 - 2i)$

42. $(5 + 6i)(2 - i)$

43. $(6 - 5i)(3 + 4i)$

44. $(5 - 6i)(2 + 5i)$

45. $(7 - 2i)(2 - 6i)$

46. $(-4 + 5i)(3 - 4i)$

47. $(3 - 2i)^2$

48. $(5 - 2i)^2$

49. $(1 + 5i)^2$

50. $(6 + 2i)^2$

51. $(-2 + 3i)^2$

52. $(-5 - 2i)^2$

d Simplify.

53. i^7

54. i^{11}

55. i^{24}

56. i^{35}

57. i^{42}

58. i^{64}

59. i^9

60. $(-i)^{71}$

61. i^6

62. $(-i)^4$

63. $(5i)^3$

64. $(-3i)^5$

Simplify to the form $a + bi$.

65. $7 + i^4$

66. $-18 + i^3$

67. $i^{28} - 23i$

68. $i^{29} + 33i$

69. $i^2 + i^4$

70. $5i^5 + 4i^3$

71. $i^5 + i^7$

72. $i^{84} - i^{100}$

73. $1 + i + i^2 + i^3 + i^4$

74. $i - i^2 + i^3 - i^4 + i^5$

75. $5 - \sqrt{-64}$

76. $\sqrt{-12} + 36i$

77. $\frac{8 - \sqrt{-24}}{4}$

78. $\frac{9 + \sqrt{-9}}{3}$

e Divide and simplify to the form $a + bi$.

79. $\frac{4 + 3i}{3 - i}$

80. $\frac{5 + 2i}{2 + i}$

81. $\frac{3 - 2i}{2 + 3i}$

82. $\frac{6 - 2i}{7 + 3i}$

83. $\frac{8 - 3i}{7i}$

84. $\frac{3 + 8i}{5i}$

85. $\frac{4}{3 + i}$

86. $\frac{6}{2 - i}$

87. $\frac{2i}{5 - 4i}$

88. $\frac{8i}{6 + 3i}$

89. $\frac{4}{3i}$

90. $\frac{5}{6i}$

91. $\frac{2 - 4i}{8i}$

92. $\frac{5 + 3i}{i}$

93. $\frac{6 + 3i}{6 - 3i}$

94. $\frac{4 - 5i}{4 + 5i}$

f Determine whether the complex number is a solution of the equation.

95. $1 - 2i$;
$x^2 - 2x + 5 = 0$

96. $1 + 2i$;
$x^2 - 2x + 5 = 0$

97. $2 + i$;
$x^2 - 4x - 5 = 0$

98. $1 - i$;
$x^2 + 2x + 2 = 0$

Skill Maintenance

For each equation, find the intercepts. [2.5a]

99. $2x - y = -30$

100. $4y - 3x = 72$

101. $5x = 10 - 2y$

Find the slope of each line.

102. $y = -\frac{7}{3}x$ [2.4b]

103. $x - 3y = 15$ [2.4b]

104. $6x = -12$ [2.5c]

Multiply or divide and write scientific notation for the result. [J20]

105. $\frac{3.6 \times 10^{-5}}{1.2 \times 10^{-8}}$

106. $(4.1 \times 10^{-2})(6.5 \times 10^{6})$

Synthesis

107. A complex function g is given by

$$g(z) = \frac{z^4 - z^2}{z - 1}.$$

Find $g(2i)$, $g(1 + i)$, and $g(-1 + 2i)$.

108. Evaluate $\frac{1}{w - w^2}$ when $w = \frac{1 - i}{10}$.

Express in terms of i.

109. $\frac{1}{8}(-24 - \sqrt{-1024})$

110. $12\sqrt{-\frac{1}{32}}$

111. $7\sqrt{-64} - 9\sqrt{-256}$

Simplify.

112. $\frac{i^5 + i^6 + i^7 + i^8}{(1 - i)^4}$

113. $(1 - i)^3(1 + i)^3$

114. $\frac{5 - \sqrt{5}i}{\sqrt{5}i}$

115. $\frac{6}{1 + \frac{3}{i}}$

116. $\left(\frac{1}{2} - \frac{1}{3}i\right)^2 - \left(\frac{1}{2} + \frac{1}{3}i\right)^2$

117. $\frac{i - i^{38}}{1 + i}$

118. Find all numbers a for which the opposite of a is the same as the reciprocal of a.

CHAPTER

6 Summary and Review

Vocabulary Reinforcement

Complete each statement with the correct term from the column on the right. Some of the choices may not be used and some may be used more than once.

1. The number c is the ________________ root of a, written $\sqrt[3]{a}$, if the third power of c is a—that is, if $c^3 = a$, then $\sqrt[3]{a} = c$. [6.1c]
2. A(n) ________________ number is any number that can be named $a + bi$, where a and b are any real numbers. [6.8a]
3. For any real number a, $\sqrt{a^2} = |a|$. The ________________ (nonnegative) square root of a^2 is the absolute value of a. [6.1b]
4. To find an equivalent expression without a radical in the denominator is called ________________ the denominator. [6.5a]
5. The symbol $\sqrt{}$ is called a(n) ________________. [6.1a]
6. A(n) ________________ number is a number that can be named bi, where b is some real number and $b \neq 0$. [6.8a]
7. The number c is a(n) ________________ root of a if $c^2 = a$. [6.1a]
8. The expression written under the radical is called the ________________. [6.1a]
9. The ________________ of a complex number $a + bi$ is $a - bi$. [6.8e]
10. In the expression $\sqrt[k]{a}$, we call the k the ________________. [6.1d]

i
radicand
index
square
complex
cube
conjugate
radical
principal
imaginary
rationalizing
odd

Concept Reinforcement

Determine whether each statement is true or false.

________________ 1. For any negative number a, we have $\sqrt{a^2} = -a$. [6.1a]

________________ 2. For any real numbers $\sqrt[m]{a}$ and $\sqrt[n]{b}$, $\sqrt[m]{a} \cdot \sqrt[n]{b} = \sqrt[mn]{ab}$. [6.3a]

________________ 3. For any real numbers $\sqrt[n]{a}$ and $\sqrt[n]{b}$, $\sqrt[n]{a} + \sqrt[n]{b} = \sqrt[n]{a + b}$. [6.4a]

________________ 4. If $x^2 = 4$, then $x = 2$. [6.6a]

________________ 5. All real numbers are complex numbers, but not every complex number is a real number. [6.8a]

________________ 6. The product of a complex number and its conjugate is always a real number. [6.8e]

Study Guide

Objective 6.1b Simplify radical expressions with perfect-square radicands.

Example Simplify: $\sqrt{16x^2}$.

$$\sqrt{16x^2} = \sqrt{(4x)^2} = |4x| = |4| \cdot |x| = 4|x|$$

Example Simplify: $\sqrt{x^2 - 6x + 9}$.

$$\sqrt{x^2 - 6x + 9} = \sqrt{(x-3)^2} = |x - 3|$$

Practice Exercises

1. Simplify: $\sqrt{36y^2}$.
2. Simplify: $\sqrt{a^2 + 4a + 4}$.

Objective 6.2a Write expressions with or without rational exponents, and simplify, if possible.

Example Rewrite $x^{1/4}$ without a rational exponent.

Recall that $a^{1/n}$ means $\sqrt[n]{a}$. Then

$x^{1/4} = \sqrt[4]{x}$.

Example Rewrite $(\sqrt[3]{4xy^2})^4$ with a rational exponent.

Recall that $(\sqrt[n]{a})^m$ means $a^{m/n}$. Then

$(\sqrt[3]{4xy^2})^4 = (4xy^2)^{4/3}$.

Practice Exercises

3. Rewrite $z^{3/5}$ without a rational exponent.

4. Rewrite $(\sqrt{6ab})^5$ with a rational exponent.

Objective 6.2b Write expressions without negative exponents, and simplify, if possible.

Example Rewrite $8^{-2/3}$ with a positive exponent, and simplify, if possible.

Recall that $a^{-m/n}$ means $\frac{1}{a^{m/n}}$. Then

$$8^{-2/3} = \frac{1}{8^{2/3}} = \frac{1}{(\sqrt[3]{8})^2} = \frac{1}{2^2} = \frac{1}{4}.$$

Practice Exercise

5. Rewrite $9^{-3/2}$ with a positive exponent, and simplify, if possible.

Objective 6.2d Use rational exponents to simplify radical expressions.

Example Use rational exponents to simplify: $\sqrt[6]{x^2y^4}$.

$$\begin{aligned}\sqrt[6]{x^2y^4} &= (x^2y^4)^{1/6}\\ &= x^{2/6}y^{4/6}\\ &= x^{1/3}y^{2/3}\\ &= (xy^2)^{1/3}\\ &= \sqrt[3]{xy^2}\end{aligned}$$

Practice Exercise

6. Use rational exponents to simplify: $\sqrt[8]{a^6b^2}$.

Objective 6.3a Multiply and simplify radical expressions.

Example Multiply and simplify: $\sqrt[3]{6xy^2}\sqrt[3]{9y}$.

$$\begin{aligned}\sqrt[3]{6xy^2}\sqrt[3]{9y} &= \sqrt[3]{6xy^2 \cdot 9y}\\ &= \sqrt[3]{54xy^3}\\ &= \sqrt[3]{27y^3 \cdot 2x}\\ &= \sqrt[3]{27y^3}\sqrt[3]{2x}\\ &= 3y\sqrt[3]{2x}\end{aligned}$$

Practice Exercise

7. Multiply and simplify. Assume that all expressions under radicals represent nonnegative numbers.

$\sqrt{5y}\sqrt{30y}$

Objective 6.3b Divide and simplify radical expressions.

Example Divide and simplify: $\frac{\sqrt{24x^5}}{\sqrt{6x}}$.

$$\frac{\sqrt{24x^5}}{\sqrt{6x}} = \sqrt{\frac{24x^5}{6x}} = \sqrt{4x^4} = 2x^2$$

Practice Exercise

8. Divide and simplify: $\frac{\sqrt{20a}}{\sqrt{5}}$.

Objective 6.4a Add or subtract with radical notation and simplify.

Example Subtract: $5\sqrt{2} - 4\sqrt{8}$.

$$\begin{aligned} 5\sqrt{2} - 4\sqrt{8} &= 5\sqrt{2} - 4\sqrt{4 \cdot 2} \\ &= 5\sqrt{2} - 4\sqrt{4}\sqrt{2} \\ &= 5\sqrt{2} - 4 \cdot 2\sqrt{2} = 5\sqrt{2} - 8\sqrt{2} \\ &= (5 - 8)\sqrt{2} = -3\sqrt{2} \end{aligned}$$

Practice Exercise

9. Subtract: $\sqrt{48} - 2\sqrt{3}$.

Objective 6.4b Multiply expressions involving radicals in which some factors contain more than one term.

Example Multiply: $(3 - \sqrt{6})(2 + 4\sqrt{6})$.

We use FOIL:

$$\begin{aligned} &(3 - \sqrt{6})(2 + 4\sqrt{6}) \\ &= 3 \cdot 2 + 3 \cdot 4\sqrt{6} - \sqrt{6} \cdot 2 - \sqrt{6} \cdot 4\sqrt{6} \\ &= 6 + 12\sqrt{6} - 2\sqrt{6} - 4 \cdot 6 \\ &= 6 + 12\sqrt{6} - 2\sqrt{6} - 24 \\ &= -18 + 10\sqrt{6}. \end{aligned}$$

Practice Exercise

10. Multiply: $(5 - \sqrt{x})^2$.

Objective 6.6a Solve radical equations with one radical term.

Example Solve: $x = \sqrt{x - 2} + 4$.

First, we subtract 4 on both sides to isolate the radical. Then we square both sides of the equation.

$$\begin{aligned} x &= \sqrt{x - 2} + 4 \\ x - 4 &= \sqrt{x - 2} \\ (x - 4)^2 &= (\sqrt{x - 2})^2 \\ x^2 - 8x + 16 &= x - 2 \\ x^2 - 9x + 18 &= 0 \\ (x - 3)(x - 6) &= 0 \\ x - 3 = 0 \quad &or \quad x - 6 = 0 \\ x = 3 \quad &or \quad x = 6 \end{aligned}$$

We must check both possible solutions. When we do, we find that 6 checks, but 3 does not. Thus the solution is 6.

Practice Exercise

11. Solve: $3 + \sqrt{x - 1} = x$.

Objective 6.6b Solve radical equations with two radical terms.

Example Solve: $1 = \sqrt{x + 9} - \sqrt{x}$.

$$\begin{aligned} 1 &= \sqrt{x + 9} - \sqrt{x} \\ \sqrt{x} + 1 &= \sqrt{x + 9} && \text{Isolating one radical} \\ (\sqrt{x} + 1)^2 &= (\sqrt{x + 9})^2 && \text{Squaring both sides} \\ x + 2\sqrt{x} + 1 &= x + 9 \\ 2\sqrt{x} &= 8 && \text{Isolating the remaining radical} \\ \sqrt{x} &= 4 \\ (\sqrt{x})^2 &= 4^2 \\ x &= 16 \end{aligned}$$

The number 16 checks. It is the solution.

Practice Exercise

12. Solve: $\sqrt{x + 3} - \sqrt{x - 2} = 1$.

Objective 6.8c Multiply complex numbers.

Example Multiply: $(3 - 2i)(4 + i)$.

$$\begin{aligned}(3 - 2i)(4 + i) &= 12 + 3i - 8i - 2i^2 \quad \text{Using FOIL}\\ &= 12 + 3i - 8i - 2(-1)\\ &= 12 + 3i - 8i + 2\\ &= 14 - 5i\end{aligned}$$

Practice Exercise

13. Multiply: $(2 - 5i)^2$.

Objective 6.8e Find conjugates of complex numbers and divide complex numbers.

Example Divide and simplify to the form $a + bi$:

$$\frac{5 - i}{4 + 3i}.$$

The conjugate of the denominator is $4 - 3i$, so we multiply by 1 using $\frac{4 - 3i}{4 - 3i}$:

$$\begin{aligned}\frac{5 - i}{4 + 3i} &= \frac{5 - i}{4 + 3i} \cdot \frac{4 - 3i}{4 - 3i}\\ &= \frac{20 - 15i - 4i + 3i^2}{16 - 9i^2}\\ &= \frac{20 - 19i + 3(-1)}{16 - 9(-1)}\\ &= \frac{20 - 19i - 3}{16 + 9}\\ &= \frac{17 - 19i}{25} = \frac{17}{25} - \frac{19}{25}i.\end{aligned}$$

Practice Exercise

14. Divide and simplify to the form $a + bi$: $\frac{3 - 2i}{2 + i}$.

Review Exercises

Use a calculator to approximate to three decimal places. [6.1a]

1. $\sqrt{778}$

2. $\sqrt{\frac{963.2}{23.68}}$

3. For the given function, find the indicated function values. [6.1a]

$f(x) = \sqrt{3x - 16}$; $f(0), f(-1), f(1)$, and $f\left(\frac{41}{3}\right)$

4. Find the domain of the function f in Exercise 3. [6.1a]

Simplify. Assume that letters represent any real number. [6.1b]

5. $\sqrt{81a^2}$

6. $\sqrt{(-7z)^2}$

7. $\sqrt{(6 - b)^2}$

8. $\sqrt{x^2 + 6x + 9}$

Simplify. [6.1c]

9. $\sqrt[3]{-1000}$

10. $\sqrt[3]{-\frac{1}{27}}$

11. For the given function, find the indicated function values. [6.1c]

$f(x) = \sqrt[3]{x + 2}$; $f(6), f(-10)$, and $f(25)$

Simplify. Assume that letters represent any real number. [6.1d]

12. $\sqrt[10]{x^{10}}$

13. $-\sqrt[13]{(-3)^{13}}$

Rewrite without rational exponents, and simplify, if possible. [6.2a]

14. $a^{1/5}$

15. $64^{3/2}$

Rewrite with rational exponents. [6.2a]

16. $\sqrt{31}$

17. $\sqrt[5]{a^2b^3}$

Rewrite with positive exponents, and simplify, if possible. [6.2b]

18. $49^{-1/2}$

19. $(8xy)^{-2/3}$

20. $5a^{-3/4}b^{1/2}c^{-2/3}$

21. $\dfrac{3a}{\sqrt[4]{t}}$

Use the laws of exponents to simplify. Write answers with positive exponents. [6.2c]

22. $(x^{-2/3})^{3/5}$

23. $\dfrac{7^{-1/3}}{7^{-1/2}}$

Use rational exponents to simplify. Write the answer in radical notation if appropriate. [6.2d]

24. $\sqrt[3]{x^{21}}$

25. $\sqrt[3]{27x^6}$

Use rational exponents to write a single radical expression. [6.2d]

26. $x^{1/3}y^{1/4}$

27. $\sqrt[4]{x}\sqrt[3]{x}$

Simplify by factoring. Assume that all expressions under radicals represent nonnegative numbers. [6.3a]

28. $\sqrt{245}$

29. $\sqrt[3]{-108}$

30. $\sqrt[3]{250a^2b^6}$

Simplify. Assume that no radicands were formed by raising negative numbers to even powers. [6.3b]

31. $\sqrt{\dfrac{49}{36}}$

32. $\sqrt[3]{\dfrac{64x^6}{27}}$

33. $\sqrt[4]{\dfrac{16x^8}{81y^{12}}}$

Perform the indicated operations and simplify. Assume that no radicands were formed by raising negative numbers to even powers. [6.3a, b], [6.4a]

34. $\sqrt{5x}\sqrt{3y}$

35. $\sqrt[3]{a^5b}\sqrt[3]{27b}$

36. $\sqrt[3]{a}\sqrt[5]{b^3}$

37. $\dfrac{\sqrt[3]{60xy^3}}{\sqrt[3]{10x}}$

38. $\dfrac{\sqrt{75x}}{2\sqrt{3}}$

39. $\dfrac{\sqrt[3]{x^2}}{\sqrt[4]{x}}$

40. $5\sqrt[3]{x} + 2\sqrt[3]{x}$

41. $2\sqrt{75} - 7\sqrt{3}$

42. $\sqrt{50} + 2\sqrt{18} + \sqrt{32}$

43. $\sqrt[3]{8x^4} + \sqrt[3]{xy^6}$

Multiply. [6.4b]

44. $(\sqrt{5} - 3\sqrt{8})(\sqrt{5} + 2\sqrt{8})$

45. $(1 - \sqrt{7})^2$

46. $(\sqrt[3]{27} - \sqrt[3]{2})(\sqrt[3]{27} + \sqrt[3]{2})$

Rationalize the denominator. [6.5a, b]

47. $\sqrt{\dfrac{8}{3}}$

48. $\dfrac{2}{\sqrt{a} + \sqrt{b}}$

Solve. [6.6a, b]

49. $x - 3 = \sqrt{5 - x}$

50. $\sqrt[4]{x + 3} = 2$

51. $\sqrt{x + 8} - \sqrt{3x + 1} = 1$

Automotive Repair. For an engine with a displacement of 2.8 L, the function given by

$$d(n) = 0.75\sqrt{2.8n}$$

can be used to determine the diameter of the carburetor's opening, $d(n)$, in millimeters, where n is the number of rpm at which the engine achieves peak performance. [6.6c]

Data: macdizzy.com

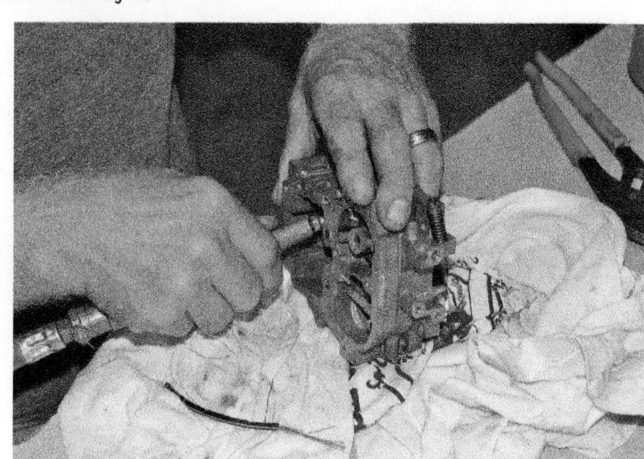

52. If a carburetor's opening is 81 mm, for what number of rpm will the engine produce peak power?

53. If a carburetor's opening is 84 mm, for what number of rpm will the engine produce peak power?

54. *Length of a Side of a Square.* The diagonal of a square has length $9\sqrt{2}$ cm. Find the length of a side of the square. [6.7a]

55. *Bookcase Width.* A bookcase is 5 ft tall and has a 7-ft diagonal brace. How wide is the bookcase? [6.7a]

In a right triangle, find the length of the side not given. Give an exact answer and an answer to three decimal places. [6.7a]

56. $a = 7,\ b = 24$

57. $a = 2,\ c = 5\sqrt{2}$

58. Express in terms of i: $\sqrt{-25} + \sqrt{-8}$. [6.8a]

Add or subtract. [6.8b]

59. $(-4 + 3i) + (2 - 12i)$

60. $(4 - 7i) - (3 - 8i)$

Multiply. [6.8c, d]

61. $(2 + 5i)(2 - 5i)$

62. i^{13}

63. $(6 - 3i)(2 - i)$

Divide. [6.8e]

64. $\dfrac{-3 + 2i}{5i}$

65. $\dfrac{1 - 2i}{3 + i}$

66. Graph: $f(x) = \sqrt{x}$. [6.1a]

67. Which of the following is a solution of $x^2 + 4x + 5 = 0$? [6.8f]

A. $1 - i$
B. $1 + i$
C. $2 + i$
D. $-2 + i$

Synthesis

68. Simplify: $i \cdot i^2 \cdot i^3 \cdots i^{99} \cdot i^{100}$. [6.8c, d]

69. Solve: $\sqrt{11x + \sqrt{6 + x}} = 6$. [6.6a]

Understanding Through Discussion and Writing

1. Find the domain of

$$f(x) = (x + 5)^{1/2}(x + 7)^{-1/2}$$

and explain how you found your answer. [6.1a], [6.2b]

2. Ron is puzzled. When he uses a graphing calculator to graph $y = \sqrt{x} \cdot \sqrt{x}$, he gets the following screen. Explain why Ron did not get the complete line $y = x$. [6.1a], [6.3a]

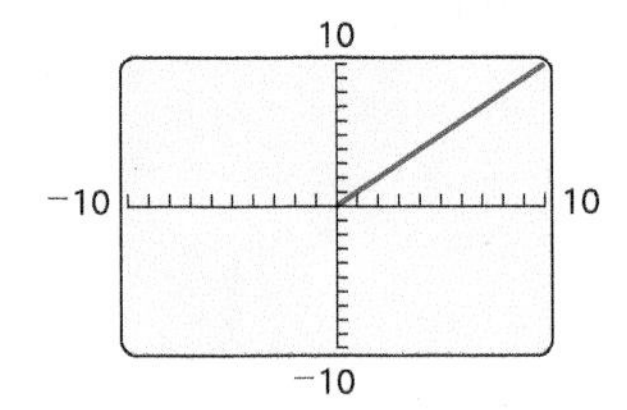

3. In what way(s) is collecting like radical terms the same as collecting like monomial terms? [6.4a]

4. Is checking solutions of equations necessary when the principle of powers is used with an odd power n? Why or why not? [6.1d], [6.6a, b]

5. A student *incorrectly* claims that

$$\frac{5 + \sqrt{2}}{\sqrt{18}} = \frac{5 + \sqrt{1}}{\sqrt{9}} = \frac{5 + 1}{3} = 2.$$

How could you convince the student that a mistake has been made? How would you explain the correct way of rationalizing the denominator? [6.5a]

6. How are conjugates of complex numbers similar to the conjugates used in Section 6.5? [6.8e]

CHAPTER 6 Test

For Extra Help: For step-by-step test solutions, access the Chapter Test Prep Videos in MyLab Math.

1. Use a calculator to approximate $\sqrt{148}$ to three decimal places.

2. For the given function, find the indicated function values.

$$f(x) = \sqrt{8 - 4x}; \quad f(1) \text{ and } f(3)$$

3. Find the domain of the function f in Exercise 2.

Simplify. Assume that letters represent *any* real number.

4. $\sqrt{(-3q)^2}$

5. $\sqrt{x^2 + 10x + 25}$

6. $\sqrt[3]{-\frac{1}{1000}}$

7. $\sqrt[5]{x^5}$

8. $\sqrt[10]{(-4)^{10}}$

Rewrite without rational exponents, and simplify, if possible.

9. $a^{2/3}$

10. $32^{3/5}$

Rewrite with rational exponents.

11. $\sqrt{37}$

12. $(\sqrt{5xy^2})^5$

Rewrite with positive exponents, and simplify, if possible.

13. $1000^{-1/3}$

14. $8a^{3/4}b^{-3/2}c^{-2/5}$

Use the laws of exponents to simplify. Write answers with positive exponents.

15. $(x^{2/3}y^{-3/4})^{12/5}$

16. $\dfrac{2.9^{-5/8}}{2.9^{2/3}}$

Use rational exponents to simplify. Write the answer in radical notation if appropriate. Assume that no radicands were formed by raising negative numbers to even powers.

17. $\sqrt[8]{x^2}$

18. $\sqrt[4]{16x^6}$

Use rational exponents to write a single radical expression.

19. $a^{2/5}b^{1/3}$

20. $\sqrt[4]{2y}\sqrt[3]{y}$

Simplify by factoring. Assume that no radicands were formed by raising negative numbers to even powers.

21. $\sqrt{148}$

22. $\sqrt[4]{80}$

23. $\sqrt[3]{24a^{11}b^{13}}$

Simplify. Assume that no radicands were formed by raising negative numbers to even powers.

24. $\sqrt[3]{\dfrac{16x^5}{y^6}}$

25. $\sqrt{\dfrac{25x^2}{36y^4}}$

Perform the indicated operations and simplify. Assume that no radicands were formed by raising negative numbers to even powers.

26. $\sqrt[3]{2x}\sqrt[3]{5y^2}$

27. $\sqrt[4]{x^3y^2}\sqrt{xy}$

28. $\dfrac{\sqrt[5]{x^3y^4}}{\sqrt[5]{xy^2}}$

29. $\dfrac{\sqrt{300a}}{5\sqrt{3}}$

30. Add: $3\sqrt{128} + 2\sqrt{18} + 2\sqrt{32}$.

Multiply.

31. $(\sqrt{20} + 2\sqrt{5})(\sqrt{20} - 3\sqrt{5})$

32. $(3 + \sqrt{x})^2$

33. Rationalize the denominator: $\dfrac{1 + \sqrt{2}}{3 - 5\sqrt{2}}$.

Solve.

34. $\sqrt[5]{x - 3} = 2$

35. $\sqrt{x - 6} = \sqrt{x + 9} - 3$

36. $\sqrt{x - 1} + 3 = x$

37. *Length of a Side of a Square.* The diagonal of a square has length $7\sqrt{2}$ ft. Find the length of a side of the square.

38. *Sighting to the Horizon.* A person can see 72 mi to the horizon from an airplane window. How high is the airplane? Use the formula $D = 1.2\sqrt{h}$, where D is in miles and h is in feet.

In a right triangle, find the length of the side not given. Give an exact answer and an approximation to three decimal places.

39. $a = 7,\ b = 7$

40. $a = 1,\ c = \sqrt{5}$

41. Express in terms of i: $\sqrt{-9} + \sqrt{-64}$.

42. Subtract: $(5 + 8i) - (-2 + 3i)$.

Multiply.

43. $(3 - 4i)(3 + 7i)$

44. i^{95}

45. Divide: $\dfrac{-7 + 14i}{6 - 8i}$.

46. Determine whether $1 + 2i$ is a solution of $x^2 + 2x + 5 = 0$.

47. Which of the following describes the solution(s) of the equation $x - 4 = \sqrt{x - 2}$?

A. There is exactly one solution, and it is positive.

B. There are one positive solution and one negative solution.

C. There are two positive solutions.

D. There is no solution.

Synthesis

48. Simplify: $\dfrac{1 - 4i}{4i(1 + 4i)^{-1}}$.

49. Solve: $\sqrt{2x - 2} + \sqrt{7x + 4} = \sqrt{13x + 10}$.

CHAPTERS

1–6 Cumulative Review

Simplify. Assume that no radicands were formed by raising negative numbers to even powers.

1. $(2x^2 - 3x + 1) + (6x - 3x^3 + 7x^2 - 4)$

2. $(2x^2 - y)^2$

3. $(5x^2 - 2x + 1)(3x^2 + x - 2)$

4. $\dfrac{x^3 + 64}{x^2 - 49} \cdot \dfrac{x^2 - 14x + 49}{x^2 - 4x + 16}$

5. $\dfrac{\dfrac{y^2 - 5y - 6}{y^2 - 7y - 18}}{\dfrac{y^2 + 3y + 2}{y^2 + 4y + 4}}$

6. $\dfrac{x}{x + 2} + \dfrac{1}{x - 3} - \dfrac{x^2 - 2}{x^2 - x - 6}$

7. $(y^3 + 3y^2 - 5) \div (y + 2)$

8. $\sqrt[3]{-8x^3}$

9. $\sqrt{16x^2 - 32x + 16}$

10. $9\sqrt{75} + 6\sqrt{12}$

11. $\sqrt{2xy^2} \cdot \sqrt{8xy^3}$

12. $\dfrac{3\sqrt{5}}{\sqrt{6} - \sqrt{3}}$

13. $\sqrt[6]{\dfrac{m^{12}n^{24}}{64}}$

14. $6^{2/9} \cdot 6^{2/3}$

15. $(6 + i) - (3 - 4i)$

16. $\dfrac{2 - i}{6 + 5i}$

Solve.

17. $\dfrac{1}{5} + \dfrac{3}{10}x = \dfrac{4}{5}$

18. $M = \dfrac{1}{8}(c - 3)$, for c

19. $3a - 4 < 10 + 5a$

20. $-8 < x + 2 < 15$

21. $|3x - 6| = 2$

22. $625 = 49y^2$

23. $3x + 5y = 30,$
$5x + 3y = 34$

24. $3x + 2y - z = -7,$
$-x + y + 2z = 9,$
$5x + 5y + z = -1$

25. $\dfrac{6x}{x - 5} - \dfrac{300}{x^2 + 5x + 25} = \dfrac{2250}{x^3 - 125}$

26. $\dfrac{3x^2}{x + 2} + \dfrac{5x - 22}{x - 2} = \dfrac{-48}{x^2 - 4}$

27. $I = \dfrac{nE}{R + nr}$, for R

28. $\sqrt{4x + 1} - 2 = 3$

29. $2\sqrt{1 - x} = \sqrt{5}$

30. $13 - x = 5 + \sqrt{x + 4}$

Graph.

31. $f(x) = -\dfrac{2}{3}x + 2$

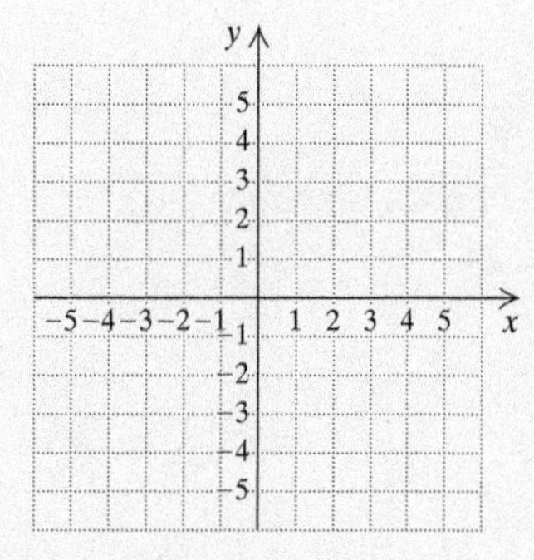

32. $4x - 2y = 8$

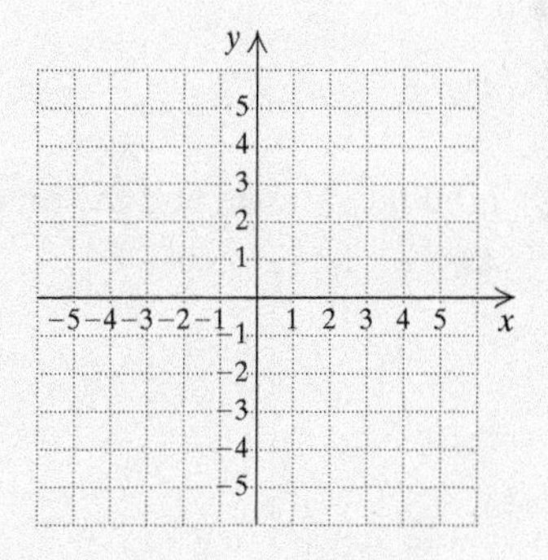

33. $4x \geq 5y + 20$

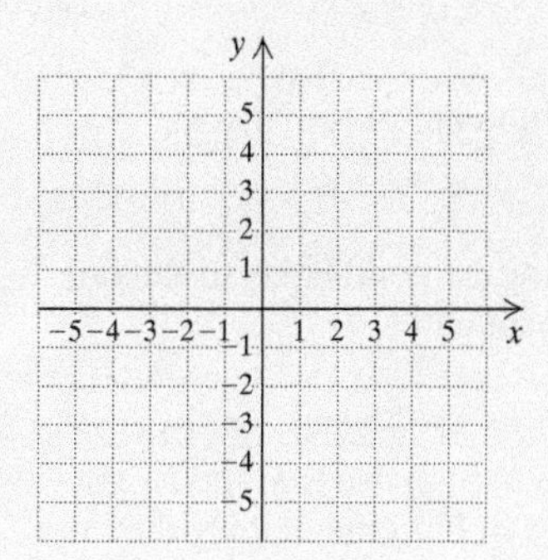

34. $y \geq -3,$
$y \leq 2x + 3$

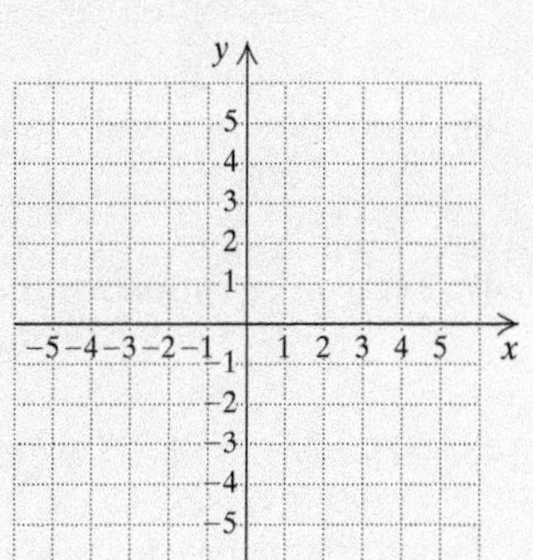

35. $g(x) = x^2 - x - 2$

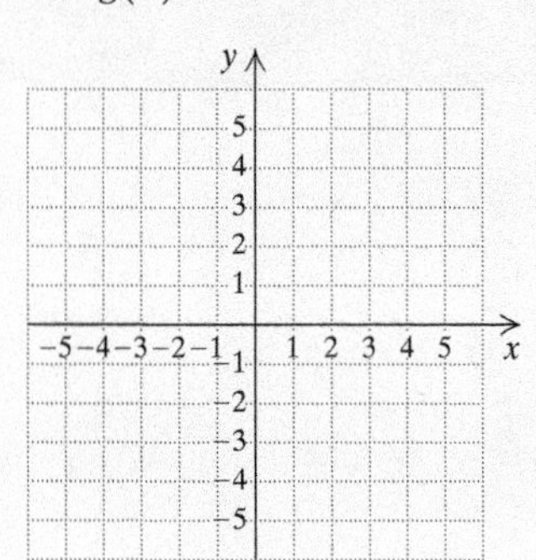

36. $f(x) = |x + 4|$

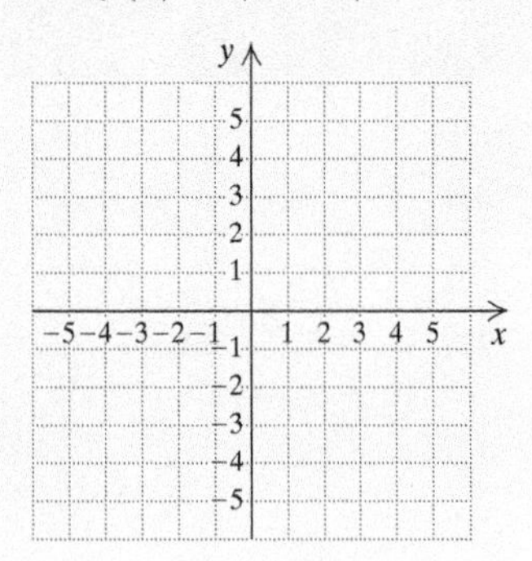

37. $g(x) = \dfrac{4}{x - 3}$

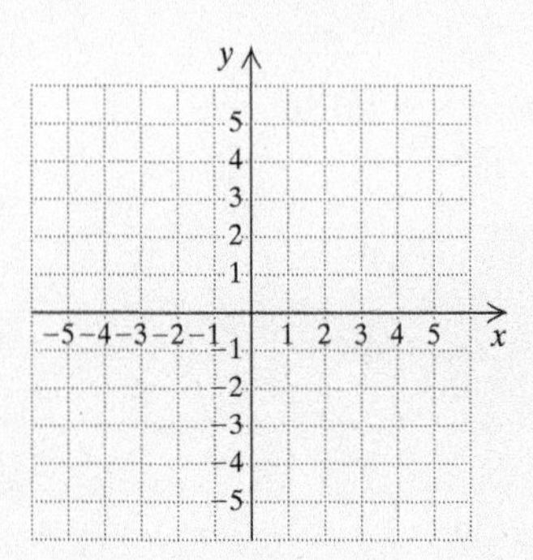

38. $f(x) = 2 - \sqrt{x}$

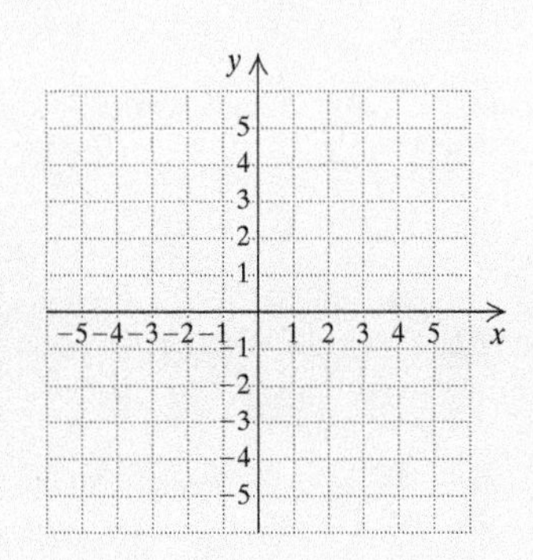

Factor.

39. $12x^2y^2 - 30xy^3$

40. $3x^2 - 17x - 28$

41. $y^2 - y - 132$

42. $27y^3 + 8$

43. $4x^2 - 625$

Find the domain and the range of each function.

44.

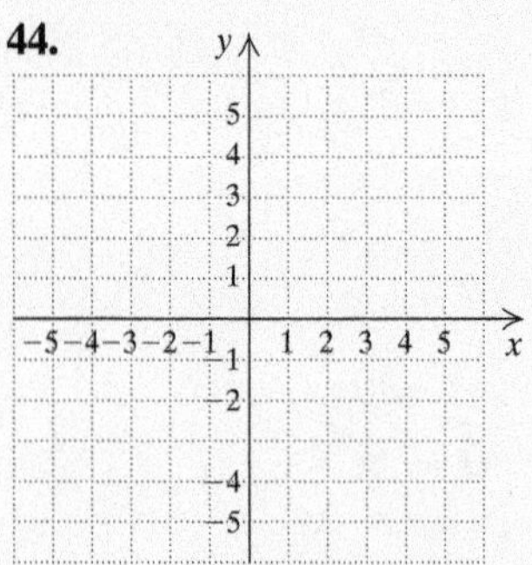

45.

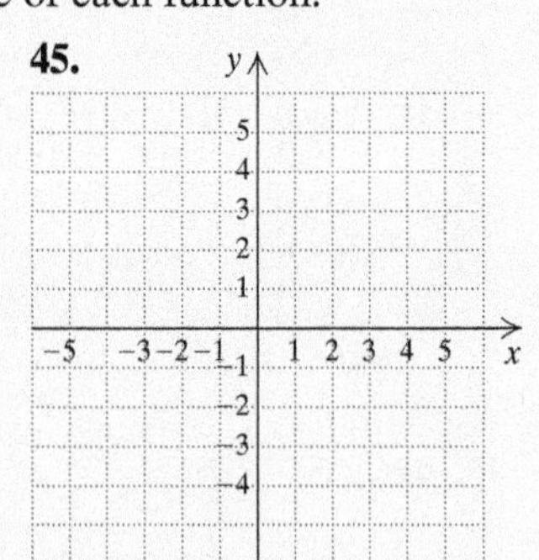

46. Find the slope and the y-intercept of the line $3x - 2y = 8$.

47. Find an equation for the line perpendicular to the line $3x - y = 5$ and passing through $(1, 4)$.

48. *Triangle Area.* The height h of triangles of fixed area varies inversely as the base b. Suppose that the height is 100 ft when the base is 20 ft. Find the height when the base is 16 ft. What is the fixed area?

Solve.

49. *Harvesting Time.* One combine can harvest a field in 3 hr. Another combine can harvest the same field in 1.5 hr. How long should it take them to harvest the field together?

50. *Warning Dye.* A warning dye is used by people in lifeboats to aid search planes. The volume V of the dye used varies directly as the square of the diameter d of the circular area formed by the dye in the water. If 4 L of dye is required for a 10-m wide circle, how much dye is needed for a 40-m wide circle?

51. Rewrite with rational exponents: $\sqrt[5]{xy^4}$.

A. $\dfrac{1}{(xy^4)^5}$ **B.** $(xy^4)^5$
C. $(xy)^{4/5}$ **D.** $(xy^4)^{1/5}$

52. A grain bin can be filled in 3 hr if the grain enters through spout A alone or in 15 hr if the grain enters through spout B alone. If grain is entering through both spouts at the same time, how many hours will it take to fill the bin?

A. $\frac{5}{2}$ hr **B.** 9 hr
C. $22\frac{1}{2}$ hr **D.** $10\frac{1}{2}$ hr

53. Divide: $(x^3 - x^2 + 2x + 4) \div (x - 3)$.

A. $x^2 + 2x + 8$, R 28 **B.** $x^2 + 2x - 4$, R -8
C. $x^2 - 4x - 10$, R -26 **D.** $x^2 - 4x + 14$, R 46

54. Solve: $2x + 6 = 8 + \sqrt{5x + 1}$.

A. $\frac{1}{4}$ **B.** 3
C. $3, \frac{1}{4}$ **D.** 4, 3

Synthesis

55. Solve: $\dfrac{x + \sqrt{x + 1}}{x - \sqrt{x + 1}} = \dfrac{5}{11}$.

CHAPTER

7

Quadratic Equations and Functions

On average, Americans spend about one-fourth of their income on housing. Since 1995, average annual rent has risen from \$6000 to over \$10,000, and the median price for a new home has risen from \$135,000 to over \$300,000. Income has not kept pace with rising home prices; while both median income and average annual rent have risen about 65% since 1995, new-home prices have risen over 120%. Rising home prices relative to income may be, in part, responsible for the decrease in homeownership rate from 2005 to 2016, as illustrated in the accompanying graph.

Homeownership Rates

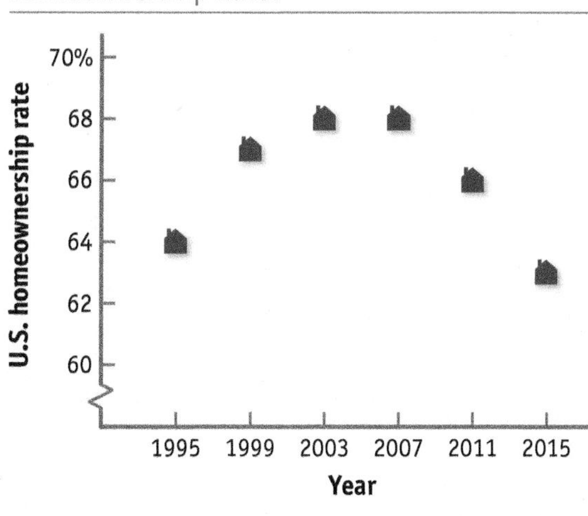

DATA: U.S Census Bureau

Data: pewtrusts.org; areavibes.com; Federal Reserve Bank of St. Louis; U.S. Census Bureau

In Exercise 35 of Section 7.7, we will model the U.S. homeownership rate using a *quadratic function*.

JUST IN TIME

Review topics 1, 3, 11, and 13 in the **Just In Time** section at the front of the text. This provides excellent prerequisite skill review for this chapter.

MyLab Math VIDEO

STUDYING FOR SUCCESS *Doing Your Homework*

- ☐ Prepare for your homework by reading explanations of concepts and by following the step-by-step solutions of examples in the text.
- ☐ Include all the steps. This will keep your work organized, help you to avoid computational errors, and give you a study guide for an exam.
- ☐ Try to do your homework as soon as possible after each class. Avoid waiting until a deadline is near to begin work on an assignment.

7.1 The Basics of Solving Quadratic Equations

OBJECTIVES

- **a** Solve quadratic equations using the principle of square roots, and find the x-intercepts of the graph of a related function.
- **b** Solve quadratic equations by completing the square.
- **c** Solve applied problems using quadratic equations.

ALGEBRAIC GRAPHICAL CONNECTION

The graph of the function $f(x) = x^2 + 6x + 8$ and its x-intercepts are shown below.

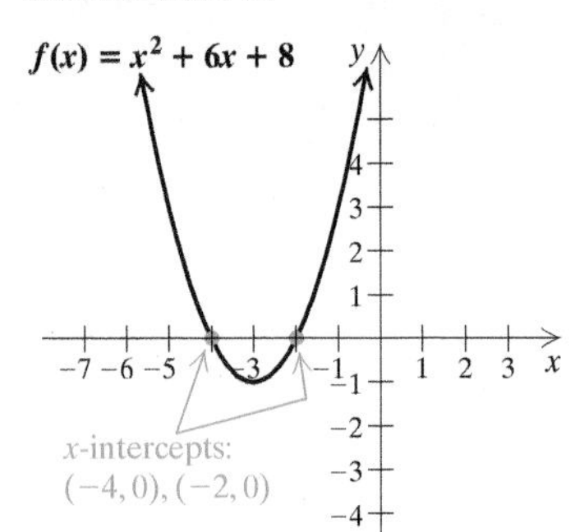

The x-intercepts are $(-4, 0)$ and $(-2, 0)$. These pairs are also the points of intersection of the graphs of $f(x) = x^2 + 6x + 8$ and $g(x) = 0$ (the x-axis). We will analyze the graphs of quadratic functions in greater detail in Sections 7.5–7.7.

We can solve $x^2 + 6x + 8 = 0$ using factoring:

$$x^2 + 6x + 8 = 0$$

$$(x + 4)(x + 2) = 0 \qquad \text{Factoring}$$

$$x + 4 = 0 \quad or \quad x + 2 = 0 \qquad \text{Using the principle of zero products}$$

$$x = -4 \quad or \quad x = -2.$$

We see that the solutions of $x^2 + 6x + 8 = 0$, -4 and -2, are the first coordinates of the x-intercepts, $(-4, 0)$ and $(-2, 0)$, of the graph of $f(x) = x^2 + 6x + 8$.

a THE PRINCIPLE OF SQUARE ROOTS

SKILL REVIEW *Solve quadratic equations using the principle of zero products.* [4.8a]

Solve.

1. $x^2 + 6x - 16 = 0$

2. $6x^2 - 13x - 5 = 0$

Answers: **1.** $-8, 2$ **2.** $-\frac{1}{3}, \frac{5}{2}$

The quadratic equation $5x^2 + 8x - 2 = 0$ is said to be written in **standard form**.

QUADRATIC EQUATION

An equation of the type $ax^2 + bx + c = 0$, where a, b, and c are real-number constants and $a > 0$, is called the **standard form of a quadratic equation**.

To write standard form for the quadratic equation $-5x^2 + 4x - 7 = 0$, we find an equivalent equation by multiplying by -1 on both sides:

$$-1(-5x^2 + 4x - 7) = -1(0)$$
$$5x^2 - 4x + 7 = 0. \quad \text{Writing in standard form}$$

To solve a quadratic equation using the principle of zero products, we first write the equation in standard form and then factor.

EXAMPLE 1

a) Solve: $x^2 = 25$.

b) Find the x-intercepts of $f(x) = x^2 - 25$.

a) We first find standard form and then factor:

$$x^2 - 25 = 0 \quad \text{Subtracting 25}$$
$$(x - 5)(x + 5) = 0 \quad \text{Factoring}$$
$$x - 5 = 0 \quad or \quad x + 5 = 0 \quad \text{Using the principle of zero products}$$
$$x = 5 \quad or \quad x = -5.$$

The solutions are 5 and -5.

b) The x-intercepts of $f(x) = x^2 - 25$ are $(-5, 0)$ and $(5, 0)$. The solutions of the equation $x^2 = 25$ are the first coordinates of the x-intercepts of the graph of $f(x) = x^2 - 25$.

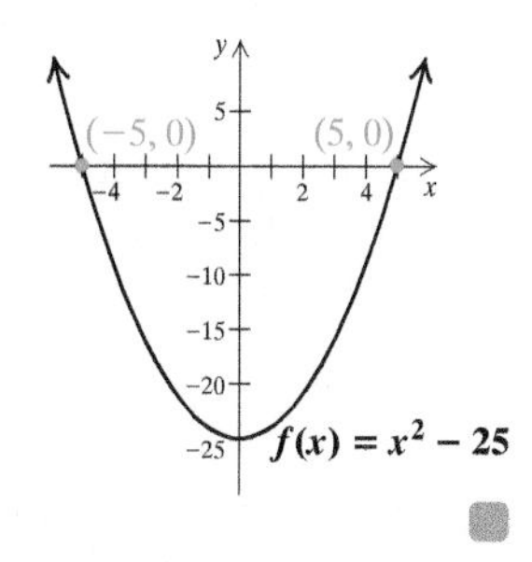

EXAMPLE 2 Solve: $6x^2 - 15x = 0$.

We factor and use the principle of zero products:

$$6x^2 - 15x = 0$$
$$3x(2x - 5) = 0$$
$$3x = 0 \quad or \quad 2x - 5 = 0$$
$$x = 0 \quad or \quad 2x = 5$$
$$x = 0 \quad or \quad x = \tfrac{5}{2}.$$

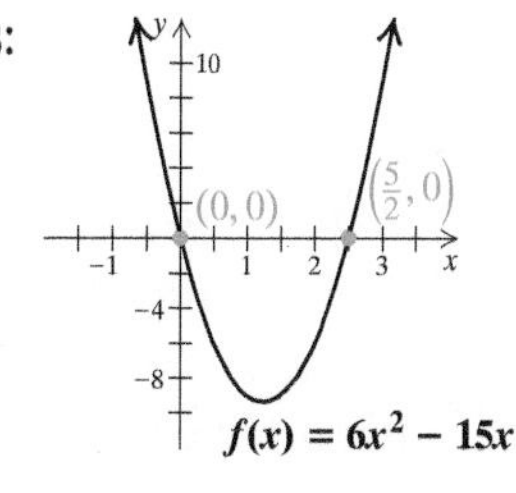

The solutions are 0 and $\frac{5}{2}$. The check is left to the student.

Do Exercises 1–3.

1. Below is the graph of $f(x) = x^2 - 6x + 8$.

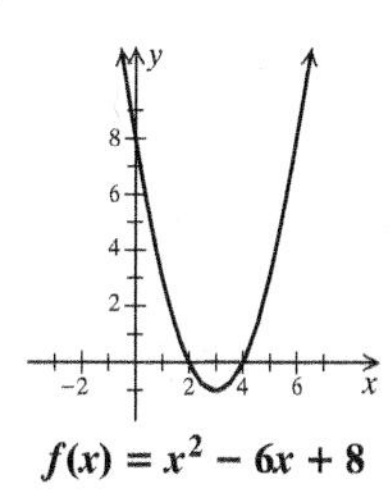

a) What are the x-intercepts of the graph?

b) What are the solutions of $x^2 - 6x + 8 = 0$?

c) What relationship exists between the answers to parts (a) and (b)?

2. **a)** Solve: $x^2 = 16$.

b) Find the x-intercepts of $f(x) = x^2 - 16$.

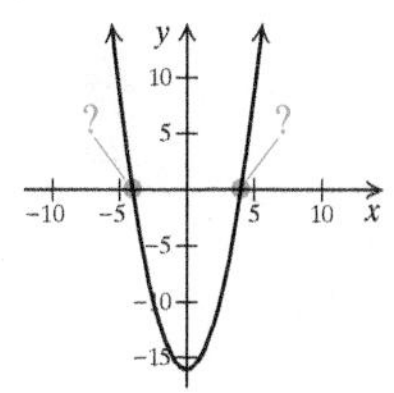

3. **a)** Solve: $4x^2 + 14x = 0$.

b) Find the x-intercepts of $f(x) = 4x^2 + 14x$.

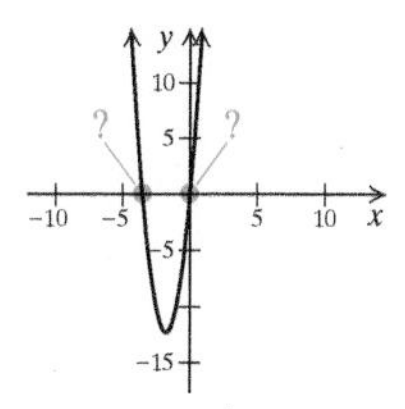

Answers

1. (a) $(2, 0), (4, 0)$; **(b)** 2, 4; **(c)** The solutions of $x^2 - 6x + 8 = 0$, 2 and 4, are the first coordinates of the x-intercepts, $(2, 0)$ and $(4, 0)$, of the graph of $f(x) = x^2 - 6x + 8$.
2. (a) $-4, 4$; **(b)** $(-4, 0), (4, 0)$
3. (a) $0, -\frac{7}{2}$; **(b)** $\left(-\frac{7}{2}, 0\right), (0, 0)$

EXAMPLE 3

a) Solve: $3x^2 = 2 - x$.

b) Find the x-intercepts of $f(x) = 3x^2 + x - 2$.

a) We first find standard form. Then we factor and use the principle of zero products.

$$3x^2 = 2 - x$$

$$3x^2 + x - 2 = 0 \quad \text{Adding } x \text{ and subtracting 2}$$

$$(x + 1)(3x - 2) = 0 \quad \text{Factoring}$$

$$x + 1 = 0 \quad or \quad 3x - 2 = 0 \quad \text{Using the principle of zero products}$$

$$x = -1 \quad or \quad 3x = 2$$

$$x = -1 \quad or \quad x = \tfrac{2}{3}$$

Check: For -1:

$3x^2$	$= 2 - x$	
$3(-1)^2$	? $2 - (-1)$	
$3 \cdot 1$	$2 + 1$	
3	3	TRUE

For $\frac{2}{3}$:

$3x^2$	$= 2 - x$	
$3(\frac{2}{3})^2$	? $2 - (\frac{2}{3})$	
$3 \cdot \frac{4}{9}$	$\frac{6}{3} - \frac{2}{3}$	
$\frac{4}{3}$	$\frac{4}{3}$	TRUE

The solutions are -1 and $\frac{2}{3}$.

b) The x-intercepts of $f(x) = 3x^2 + x - 2$ are $(-1, 0)$ and $(\frac{2}{3}, 0)$. The solutions of the equation $3x^2 = 2 - x$ are the first coordinates of the x-intercepts of the graph of $f(x) = 3x^2 + x - 2$.

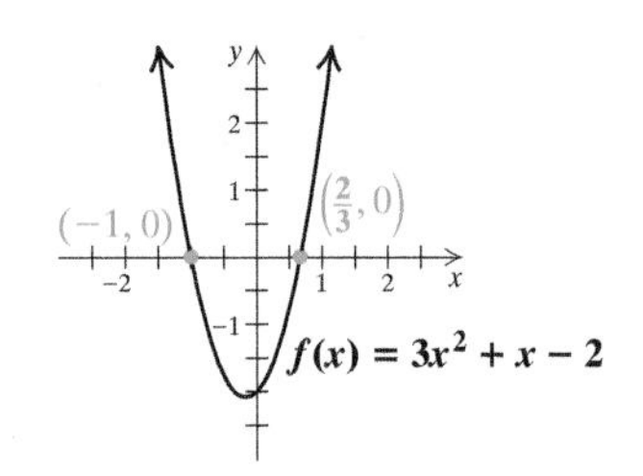

4. a) Solve: $5x^2 = 8x - 3$.

b) Find the x-intercepts of $f(x) = 5x^2 - 8x + 3$.

$f(x) = 5x^2 - 8x + 3$

Do Exercise 4.

Solving Equations of the Type $x^2 = d$

Consider the equation $x^2 = 25$ again. The number 25 has two real-number square roots, namely, 5 and -5. Note that these are the solutions of the equation in Example 1. This illustrates the principle of square roots, which provides a quick method for solving equations of the type $x^2 = d$.

THE PRINCIPLE OF SQUARE ROOTS

The solutions of the equation $x^2 = d$ are $\sqrt{d}$ and $-\sqrt{d}$.

When $d > 0$, the solutions are two real numbers.

When $d = 0$, the only solution is 0.

When $d < 0$, the solutions are two imaginary numbers.

We often use the notation $\pm\sqrt{d}$ to represent both $\sqrt{d}$ and $-\sqrt{d}$.

Answer

4. (a) $\frac{3}{5}$, 1; (b) $\left(\frac{3}{5}, 0\right)$, $(1, 0)$

The quadratic equation $5x^2 + 8x - 2 = 0$ is said to be written in **standard form**.

QUADRATIC EQUATION

An equation of the type $ax^2 + bx + c = 0$, where a, b, and c are real-number constants and $a > 0$, is called the **standard form of a quadratic equation**.

To write standard form for the quadratic equation $-5x^2 + 4x - 7 = 0$, we find an equivalent equation by multiplying by -1 on both sides:

$$-1(-5x^2 + 4x - 7) = -1(0)$$
$$5x^2 - 4x + 7 = 0. \quad \text{Writing in standard form}$$

To solve a quadratic equation using the principle of zero products, we first write the equation in standard form and then factor.

EXAMPLE 1

a) Solve: $x^2 = 25$.

b) Find the x-intercepts of $f(x) = x^2 - 25$.

a) We first find standard form and then factor:

$$x^2 - 25 = 0 \quad \text{Subtracting 25}$$
$$(x - 5)(x + 5) = 0 \quad \text{Factoring}$$
$$x - 5 = 0 \quad or \quad x + 5 = 0 \quad \text{Using the principle of zero products}$$
$$x = 5 \quad or \quad x = -5.$$

The solutions are 5 and -5.

b) The x-intercepts of $f(x) = x^2 - 25$ are $(-5, 0)$ and $(5, 0)$. The solutions of the equation $x^2 = 25$ are the first coordinates of the x-intercepts of the graph of $f(x) = x^2 - 25$.

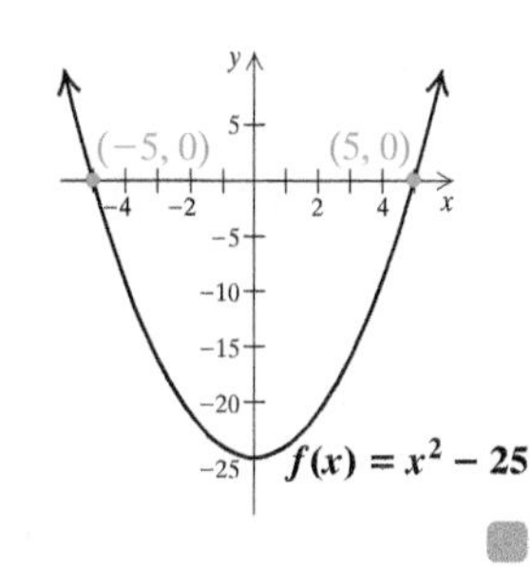

EXAMPLE 2

Solve: $6x^2 - 15x = 0$.

We factor and use the principle of zero products:

$$6x^2 - 15x = 0$$
$$3x(2x - 5) = 0$$
$$3x = 0 \quad or \quad 2x - 5 = 0$$
$$x = 0 \quad or \quad 2x = 5$$
$$x = 0 \quad or \quad x = \tfrac{5}{2}.$$

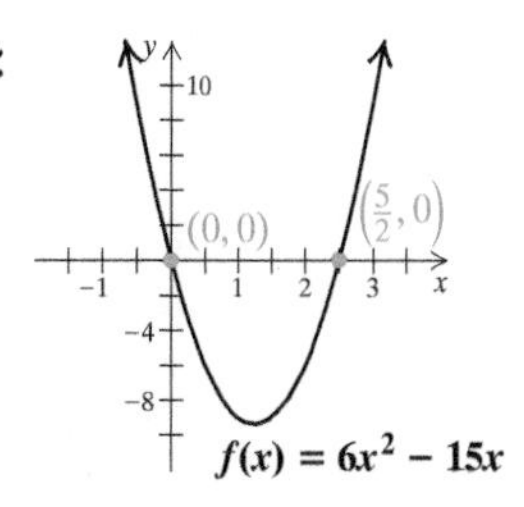

The solutions are 0 and $\frac{5}{2}$. The check is left to the student.

Do Exercises 1–3.

1. Below is the graph of $f(x) = x^2 - 6x + 8$.

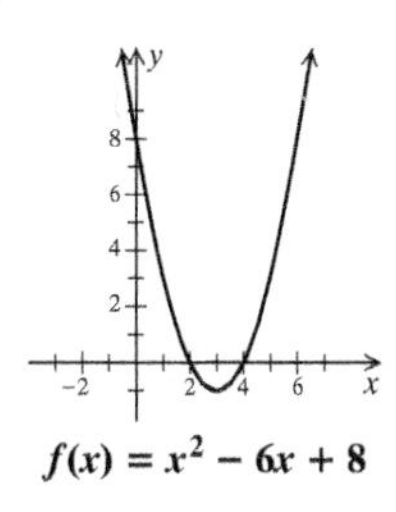

   a) What are the x-intercepts of the graph?

   b) What are the solutions of $x^2 - 6x + 8 = 0$?

   c) What relationship exists between the answers to parts (a) and (b)?

2. a) Solve: $x^2 = 16$.

   b) Find the x-intercepts of $f(x) = x^2 - 16$.

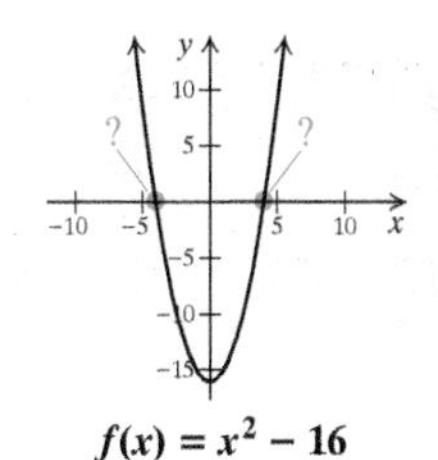

3. a) Solve: $4x^2 + 14x = 0$.

   b) Find the x-intercepts of $f(x) = 4x^2 + 14x$.

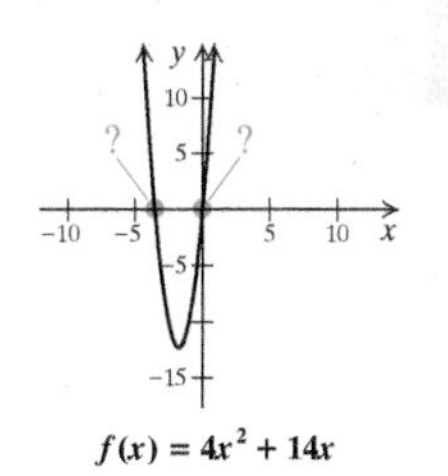

Answers

1. (a) $(2, 0), (4, 0)$; (b) 2, 4; (c) The solutions of $x^2 - 6x + 8 = 0$, 2 and 4, are the first coordinates of the x-intercepts, $(2, 0)$ and $(4, 0)$, of the graph of $f(x) = x^2 - 6x + 8$.
2. (a) $-4, 4$; (b) $(-4, 0), (4, 0)$
3. (a) $0, -\frac{7}{2}$; (b) $\left(-\frac{7}{2}, 0\right), (0, 0)$

EXAMPLE 3

a) Solve: $3x^2 = 2 - x$.

b) Find the x-intercepts of $f(x) = 3x^2 + x - 2$.

a) We first find standard form. Then we factor and use the principle of zero products.

$$3x^2 = 2 - x$$

$$3x^2 + x - 2 = 0 \qquad \text{Adding } x \text{ and subtracting 2}$$

$$(x + 1)(3x - 2) = 0 \qquad \text{Factoring}$$

$$x + 1 = 0 \quad or \quad 3x - 2 = 0 \qquad \text{Using the principle of zero products}$$

$$x = -1 \quad or \quad 3x = 2$$

$$x = -1 \quad or \quad x = \tfrac{2}{3}$$

Check: For -1:

$3x^2$	$= 2 - x$	
$3(-1)^2$	? $2 - (-1)$	
$3 \cdot 1$	$2 + 1$	
3	3	TRUE

For $\frac{2}{3}$:

$3x^2$	$= 2 - x$	
$3(\frac{2}{3})^2$	? $2 - (\frac{2}{3})$	
$3 \cdot \frac{4}{9}$	$\frac{6}{3} - \frac{2}{3}$	
$\frac{4}{3}$	$\frac{4}{3}$	TRUE

The solutions are -1 and $\frac{2}{3}$.

b) The x-intercepts of $f(x) = 3x^2 + x - 2$ are $(-1, 0)$ and $(\frac{2}{3}, 0)$. The solutions of the equation $3x^2 = 2 - x$ are the first coordinates of the x-intercepts of the graph of $f(x) = 3x^2 + x - 2$.

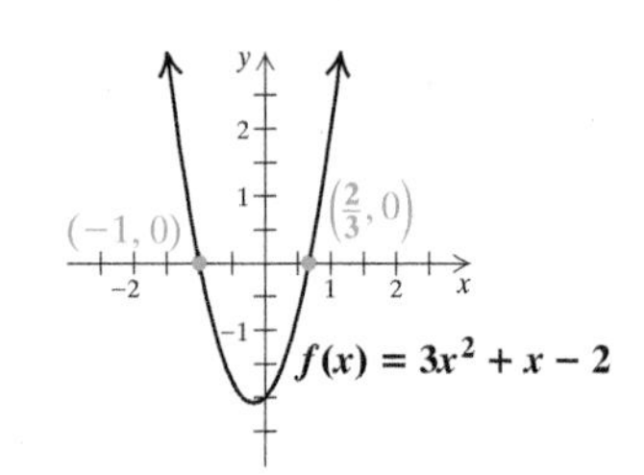

◀ **Do Exercise 4.**

4. a) Solve: $5x^2 = 8x - 3$.

b) Find the x-intercepts of $f(x) = 5x^2 - 8x + 3$.

$f(x) = 5x^2 - 8x + 3$

Solving Equations of the Type $x^2 = d$

Consider the equation $x^2 = 25$ again. The number 25 has two real-number square roots, namely, 5 and -5. Note that these are the solutions of the equation in Example 1. This illustrates the principle of square roots, which provides a quick method for solving equations of the type $x^2 = d$.

THE PRINCIPLE OF SQUARE ROOTS

The solutions of the equation $x^2 = d$ are $\sqrt{d}$ and $-\sqrt{d}$.

When $d > 0$, the solutions are two real numbers.

When $d = 0$, the only solution is 0.

When $d < 0$, the solutions are two imaginary numbers.

We often use the notation $\pm\sqrt{d}$ to represent both $\sqrt{d}$ and $-\sqrt{d}$.

Answer

4. (a) $\frac{3}{5}$, 1; **(b)** $\left(\frac{3}{5}, 0\right)$, $(1, 0)$

EXAMPLE 4 Solve: $3x^2 = 6$. Give the exact solutions and approximate the solutions to three decimal places.

We have

$$3x^2 = 6$$
$$x^2 = 2$$
$$x = \sqrt{2} \quad or \quad x = -\sqrt{2}.$$

Check: For $\sqrt{2}$:

$$\begin{array}{c|l} 3x^2 = 6 & \\ \hline 3(\sqrt{2})^2 \; ? \; 6 & \\ 3 \cdot 2 & \\ 6 & \text{TRUE} \end{array}$$

For $-\sqrt{2}$:

$$\begin{array}{c|l} 3x^2 = 6 & \\ \hline 3(-\sqrt{2})^2 \; ? \; 6 & \\ 3 \cdot 2 & \\ 6 & \text{TRUE} \end{array}$$

$(-\sqrt{2}, 0)$ $(\sqrt{2}, 0)$ $f(x) = 3x^2 - 6$

The solutions are $\sqrt{2}$ and $-\sqrt{2}$, or $\pm\sqrt{2}$, which are about 1.414 and −1.414, or ±1.414, when rounded to three decimal places.

Do Exercise 5. ▶

GS **5.** Solve: $5x^2 = 15$. Give the exact solutions and approximate the solutions to three decimal places.

$$5x^2 = 15$$
$$x^2 = \_\_\_$$
$$x = \sqrt{\_\_\_} \quad or \quad x = -\sqrt{\_\_\_}$$

The solutions can also be written $\pm\sqrt{3}$. If we round to three decimal places, the solutions are ± ___.

Sometimes we rationalize denominators when writing solutions.

EXAMPLE 5 Solve: $-5x^2 + 2 = 0$. Give the exact solutions and approximate the solutions to three decimal places.

$$-5x^2 + 2 = 0$$

$$x^2 = \frac{2}{5} \qquad \text{Subtracting 2 and dividing by } -5$$

$$x = \sqrt{\frac{2}{5}} \quad or \quad x = -\sqrt{\frac{2}{5}} \qquad \text{Using the principle of square roots}$$

$$x = \sqrt{\frac{2}{5} \cdot \frac{5}{5}} \quad or \quad x = -\sqrt{\frac{2}{5} \cdot \frac{5}{5}} \qquad \text{Rationalizing the denominators}$$

$$x = \frac{\sqrt{10}}{5} \quad or \quad x = -\frac{\sqrt{10}}{5}$$

Check: Since there is no x-term in the equation, we can check both numbers at once.

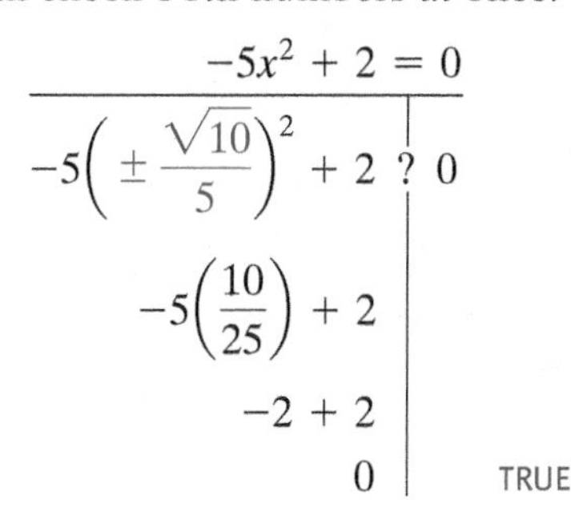

$$\begin{array}{c|l} -5x^2 + 2 = 0 & \\ \hline -5\left(\pm\frac{\sqrt{10}}{5}\right)^2 + 2 \; ? \; 0 & \\ -5\left(\frac{10}{25}\right) + 2 & \\ -2 + 2 & \\ 0 & \text{TRUE} \end{array}$$

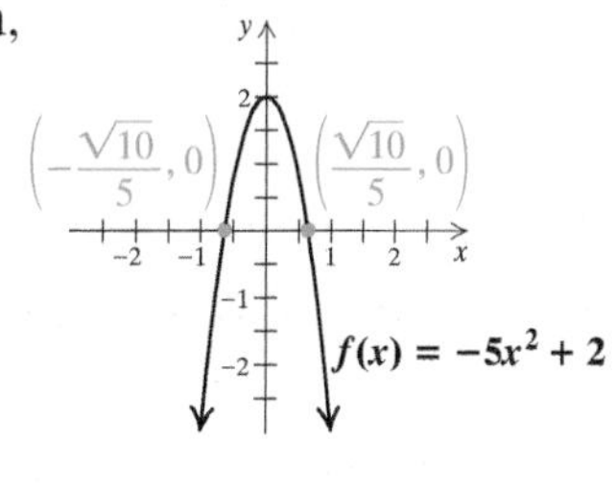

The solutions are $\frac{\sqrt{10}}{5}$ and $-\frac{\sqrt{10}}{5}$, or $\pm\frac{\sqrt{10}}{5}$. The approximate solutions, rounded to three decimal places, are 0.632 and −0.632, or ±0.632.

Do Exercise 6. ▶

6. Solve: $-3x^2 + 8 = 0$. Give the exact solutions and approximate the solutions to three decimal places.

Answers

5. $\sqrt{3}$ and $-\sqrt{3}$, or $\pm\sqrt{3}$; 1.732 and −1.732, or ±1.732

6. $\frac{2\sqrt{6}}{3}$ and $-\frac{2\sqrt{6}}{3}$, or $\pm\frac{2\sqrt{6}}{3}$; 1.633 and −1.633, or ±1.633

Guided Solution:

5. 3, 3, 3, 1.732

Sometimes we get solutions that are imaginary numbers.

EXAMPLE 6 Solve: $4x^2 + 9 = 0$.

$$4x^2 + 9 = 0$$

$$x^2 = -\frac{9}{4} \qquad \text{Subtracting 9 and dividing by 4}$$

$$x = \sqrt{-\frac{9}{4}} \quad or \quad x = -\sqrt{-\frac{9}{4}} \qquad \text{Using the principle of square roots}$$

$$x = \frac{3}{2}i \quad or \quad x = -\frac{3}{2}i \qquad \text{Simplifying; recall that } \sqrt{-1} = i.$$

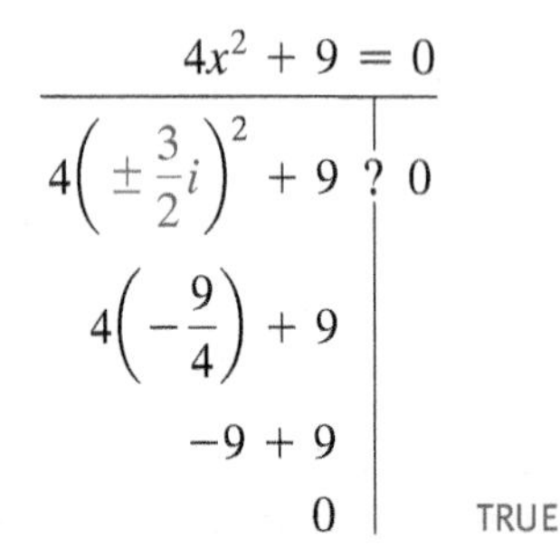

Check:

$$\begin{array}{c|c} 4x^2 + 9 = 0 & \\ \hline 4\left(\pm\frac{3}{2}i\right)^2 + 9 & ?\ 0 \\ 4\left(-\frac{9}{4}\right) + 9 & \\ -9 + 9 & \\ 0 & \text{TRUE} \end{array}$$

$f(x) = 4x^2 + 9$

No x-intercepts

The solutions are $\frac{3}{2}i$ and $-\frac{3}{2}i$, or $\pm\frac{3}{2}i$.

We see that the graph of $f(x) = 4x^2 + 9$ does not cross the x-axis. This is true because the equation $4x^2 + 9 = 0$ has *imaginary* complex-number solutions. Only real-number solutions correspond to x-intercepts.

◀ **Do Exercise 7.**

CALCULATOR CORNER

Imaginary Solutions of Quadratic Equations What happens when you use the ZERO feature to solve the equation in Example 6? Explain why this happens.

7. Solve: $2x^2 + 1 = 0$.

Solving Equations of the Type $(x + c)^2 = d$

Equations like $(x - 2)^2 = 7$ can also be solved using the principle of square roots.

EXAMPLE 7

a) Solve: $(x - 2)^2 = 7$.

b) Find the x-intercepts of $f(x) = (x - 2)^2 - 7$.

a) We have

$$(x - 2)^2 = 7$$

$$x - 2 = \sqrt{7} \quad or \quad x - 2 = -\sqrt{7} \qquad \text{Using the principle of square roots}$$

$$x = 2 + \sqrt{7} \quad or \quad x = 2 - \sqrt{7}.$$

The solutions are $2 + \sqrt{7}$ and $2 - \sqrt{7}$, or $2 \pm \sqrt{7}$.

b) The x-intercepts of $f(x) = (x - 2)^2 - 7$ are $(2 - \sqrt{7}, 0)$ and $(2 + \sqrt{7}, 0)$.

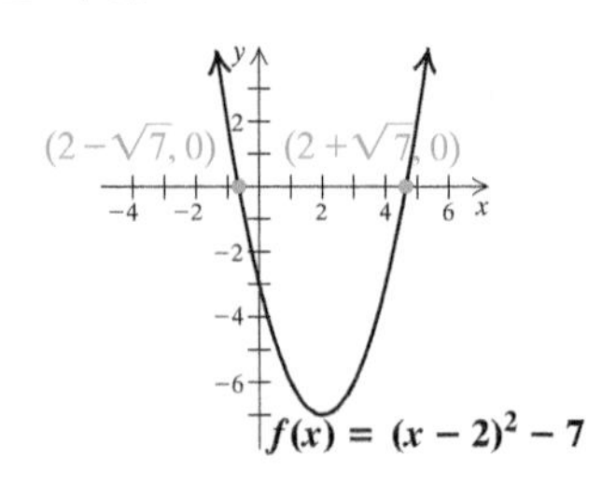

◀ **Do Exercise 8.**

8. a) Solve: $(x - 1)^2 = 5$.

b) Find the x-intercepts of $f(x) = (x - 1)^2 - 5$.

$f(x) = (x - 1)^2 - 5$

Answers

7. $\frac{\sqrt{2}}{2}i$ and $-\frac{\sqrt{2}}{2}i$, or $\pm\frac{\sqrt{2}}{2}i$

8. (a) $1 \pm \sqrt{5}$; **(b)** $(1 - \sqrt{5}, 0)$, $(1 + \sqrt{5}, 0)$

If we can express the left side of an equation as the square of a binomial, we can proceed as we did in Example 7.

EXAMPLE 8 Solve: $x^2 + 6x + 9 = 2$.

We have

$$x^2 + 6x + 9 = 2 \qquad \text{The left side is the square of a binomial.}$$
$$(x + 3)^2 = 2$$
$$x + 3 = \sqrt{2} \quad or \quad x + 3 = -\sqrt{2} \qquad \text{Using the principle of square roots}$$
$$x = -3 + \sqrt{2} \quad or \quad x = -3 - \sqrt{2}.$$

The solutions are $-3 + \sqrt{2}$ and $-3 - \sqrt{2}$, or $-3 \pm \sqrt{2}$.

Do Exercise 9. ▶

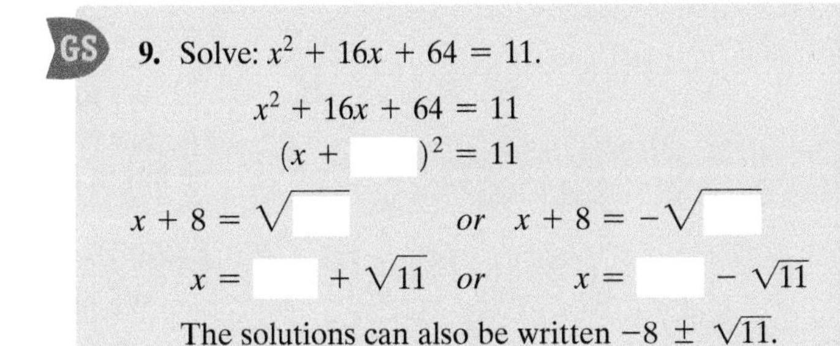

GS **9.** Solve: $x^2 + 16x + 64 = 11$.

$$x^2 + 16x + 64 = 11$$
$$(x + \square)^2 = 11$$
$$x + 8 = \sqrt{\square} \quad or \quad x + 8 = -\sqrt{\square}$$
$$x = \square + \sqrt{11} \quad or \quad x = \square - \sqrt{11}$$

The solutions can also be written $-8 \pm \sqrt{11}$.

b COMPLETING THE SQUARE

Consider the equation $x^2 + 14x - 4 = 0$. Since $x^2 + 14x - 4$ cannot be factored and since it is not in the form $(x + c)^2 = d$, we need a new procedure to solve this equation. This procedure is called **completing the square**. *It can be used to solve any quadratic equation.*

Suppose that we have the quadratic equation

$$x^2 + 14x = 4.$$

If we could add on both sides of the equation a constant that would make the expression on the left the square of a binomial, we could then solve the equation using the principle of square roots.

How can we determine what to add to $x^2 + 14x$ in order to construct the square of a binomial? We want to find a number a such that the following equation is satisfied:

$$x^2 + 14x + a^2 = (x + a)(x + a) = x^2 + 2ax + a^2.$$

Thus, $2a = 14$. Solving, we get $a = 7$. Since $a^2 = 7^2 = 49$, we add 49 to our original expression:

$$x^2 + 14x + 49 = (x + 7)^2. \qquad x^2 + 14x + 49 \text{ is the square of } x + 7.$$

Note that $7 = \frac{14}{2}$. Thus, a is half of the coefficient of x in $x^2 + 14x$.

Returning to solving our original equation, we first add 49 on *both* sides to *complete the square* on the left. Then we solve:

$$x^2 + 14x = 4 \qquad \text{Original equation}$$
$$x^2 + 14x + 49 = 4 + 49 \qquad \text{Adding 49: } \left(\tfrac{14}{2}\right)^2 = 7^2 = 49$$
$$(x + 7)^2 = 53$$
$$x + 7 = \sqrt{53} \quad or \quad x + 7 = -\sqrt{53} \qquad \text{Using the principle of square roots}$$
$$x = -7 + \sqrt{53} \quad or \quad x = -7 - \sqrt{53}.$$

The solutions are $-7 \pm \sqrt{53}$.

Answer

9. $-8 \pm \sqrt{11}$

Guided Solution:

9. 8, 11, 11, −8, −8

COMPLETING THE SQUARE

When solving an equation, to **complete the square** of an expression like $x^2 + bx$, we take half the x-coefficient, which is $b/2$, and square it. Then we add that number, $(b/2)^2$, on both sides of the equation.

We have seen that a quadratic equation $(x + c)^2 = d$ can be solved using the principle of square roots. Any equation, such as $x^2 - 6x + 8 = 0$, can be put in this form by completing the square. Then we can solve as before.

EXAMPLE 9 Solve: $x^2 - 6x + 8 = 0$.

We have

$$x^2 - 6x + 8 = 0$$

$$x^2 - 6x = -8. \quad \text{Subtracting 8}$$

We take half of -6 and square it, to get 9. Then we add 9 on *both* sides of the equation. This makes the left side the square of a binomial, $x - 3$. We have now *completed the square.*

$$x^2 - 6x + 9 = -8 + 9 \quad \text{Adding 9: } \left(\tfrac{-6}{2}\right)^2 = (-3)^2 = 9$$

$$(x - 3)^2 = 1$$

$$x - 3 = 1 \quad or \quad x - 3 = -1 \quad \text{Using the principle of square roots}$$

$$x = 4 \quad or \quad x = 2$$

The solutions are 2 and 4.

◀ **Do Exercises 10 and 11.**

EXAMPLE 10 Solve $x^2 + 4x - 7 = 0$ by completing the square.

We have

$$x^2 + 4x - 7 = 0$$

$$x^2 + 4x = 7 \quad \text{Adding 7}$$

$$x^2 + 4x + 4 = 7 + 4 \quad \text{Adding 4: } \left(\tfrac{4}{2}\right)^2 = (2)^2 = 4$$

$$(x + 2)^2 = 11$$

$$x + 2 = \sqrt{11} \quad or \quad x + 2 = -\sqrt{11} \quad \text{Using the principle of square roots}$$

$$x = -2 + \sqrt{11} \quad or \quad x = -2 - \sqrt{11}.$$

The solutions are $-2 \pm \sqrt{11}$.

◀ **Do Exercise 12.**

Solve.

10. $x^2 + 6x + 8 = 0$

11. $x^2 - 8x - 20 = 0$ GS

$x^2 - 8x =$ ___

$x^2 - 8x +$ ___ $= 20 +$ ___

$(x -$ ___ $)^2 = 36$

$x - 4 = 6 \quad or \quad x - 4 =$ ___

$x = 10 \quad or \quad x =$ ___

12. Solve by completing the square:

$x^2 + 6x - 1 = 0.$

Answers

10. $-2, -4$ **11.** $10, -2$ **12.** $-3 \pm \sqrt{10}$

Guided Solution:

11. 20, 16, 16, 4, -6, -2

When the coefficient of x^2 is not 1, we can multiply to make it 1.

EXAMPLE 11 Solve $3x^2 + 7x = 2$ by completing the square.

We have

$$3x^2 + 7x = 2$$

$$\frac{1}{3}(3x^2 + 7x) = \frac{1}{3}\cdot 2 \quad \text{Multiplying by } \tfrac{1}{3} \text{ to make the } x^2\text{-coefficient 1}$$

$$x^2 + \frac{7}{3}x = \frac{2}{3} \quad \text{Multiplying and simplifying}$$

$$x^2 + \frac{7}{3}x + \frac{49}{36} = \frac{2}{3} + \frac{49}{36} \quad \text{Adding } \frac{49}{36}\text{: } \left[\frac{1}{2}\cdot\frac{7}{3}\right]^2 = \frac{49}{36}$$

$$\left(x + \frac{7}{6}\right)^2 = \frac{24}{36} + \frac{49}{36} \quad \text{Finding a common denominator}$$

$$\left(x + \frac{7}{6}\right)^2 = \frac{73}{36}$$

$$x + \frac{7}{6} = \sqrt{\frac{73}{36}} \quad or \quad x + \frac{7}{6} = -\sqrt{\frac{73}{36}} \quad \text{Using the principle of square roots}$$

$$x + \frac{7}{6} = \frac{\sqrt{73}}{6} \quad or \quad x + \frac{7}{6} = -\frac{\sqrt{73}}{6}$$

$$x = -\frac{7}{6} + \frac{\sqrt{73}}{6} \quad or \quad x = -\frac{7}{6} - \frac{\sqrt{73}}{6}.$$

The solutions are $-\frac{7}{6} \pm \frac{\sqrt{73}}{6}$.

The graph at right shows the x-intercepts of the graph of the related function $f(x) = 3x^2 + 7x - 2$.

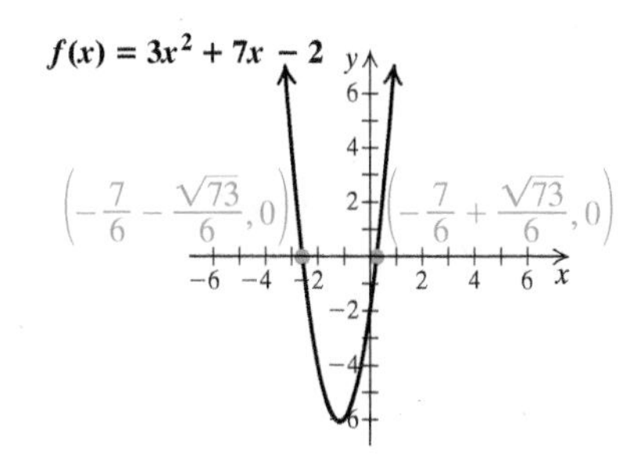

Do Exercises 13 and 14. ▶

Solve by completing the square.

13. $2x^2 + 6x = 5$

14. $3x^2 - 2x = 7$

SOLVING BY COMPLETING THE SQUARE

To solve an equation $ax^2 + bx + c = 0$ by completing the square:

1. If $a \neq 1$, multiply by $1/a$ so that the x^2-coefficient is 1.
2. Once the x^2-coefficient is 1, add or subtract so that the equation is in the form

   $$x^2 + bx = -c, \quad \text{or} \quad x^2 + \frac{b}{a}x = -\frac{c}{a} \text{ if step (1) has been applied.}$$

3. Take half of the x-coefficient and square it. Add the result on both sides of the equation.
4. Express the side with the variables as the square of a binomial.
5. Use the principle of square roots and complete the solution.

Answers

13. $-\frac{3}{2} \pm \frac{\sqrt{19}}{2}$ **14.** $\frac{1}{3} \pm \frac{\sqrt{22}}{3}$

EXAMPLE 12 Solve $2x^2 = 3x - 7$ by completing the square.

We have

$$2x^2 = 3x - 7$$

$$2x^2 - 3x = -7 \quad \text{Subtracting } 3x$$

$$\frac{1}{2}(2x^2 - 3x) = \frac{1}{2}\cdot(-7) \quad \text{Multiplying by } \tfrac{1}{2} \text{ to make the } x^2\text{-coefficient 1}$$

$$x^2 - \frac{3}{2}x = -\frac{7}{2} \quad \text{Multiplying and simplifying}$$

$$x^2 - \frac{3}{2}x + \frac{9}{16} = -\frac{7}{2} + \frac{9}{16} \quad \text{Adding } \frac{9}{16}\text{: } \left[\frac{1}{2}\left(-\frac{3}{2}\right)\right]^2 = \left[-\frac{3}{4}\right]^2 = \frac{9}{16}$$

$$\left(x - \frac{3}{4}\right)^2 = -\frac{56}{16} + \frac{9}{16} \quad \text{Finding a common denominator}$$

$$\left(x - \frac{3}{4}\right)^2 = -\frac{47}{16}$$

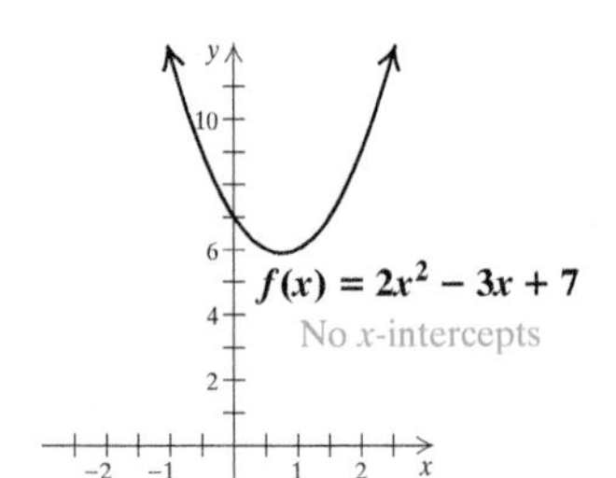

$$x - \frac{3}{4} = \sqrt{-\frac{47}{16}} \quad or \quad x - \frac{3}{4} = -\sqrt{-\frac{47}{16}} \quad \text{Using the principle of square roots}$$

$$x - \frac{3}{4} = \frac{\sqrt{47}}{4}i \quad or \quad x - \frac{3}{4} = -\frac{\sqrt{47}}{4}i \quad \sqrt{-1} = i$$

$$x = \frac{3}{4} + \frac{\sqrt{47}}{4}i \quad or \quad x = \frac{3}{4} - \frac{\sqrt{47}}{4}i.$$

The solutions are $\frac{3}{4} \pm \frac{\sqrt{47}}{4}i$.

We see at left that the graph of $f(x) = 2x^2 - 3x + 7$ does not cross the x-axis. This is true because the equation $2x^2 = 3x - 7$ has nonreal complex-number solutions.

15. Solve by completing the square:

$$3x^2 = 2x - 1.$$

◀ **Do Exercise 15.**

c APPLICATIONS AND PROBLEM SOLVING

EXAMPLE 13 *Hang Time.* One of the most exciting plays in basketball is the dunk shot. The amount of time T that passes from the moment a player leaves the ground, goes up, makes the shot, and arrives back on the ground is called *hang time*. A function relating an athlete's vertical leap V, in inches, to hang time T, in seconds, is given by

$$V(T) = 48T^2.$$

Answer

15. $\frac{1}{3} \pm \frac{\sqrt{2}}{3}i$

a) Hall-of-Famer Michael Jordan had a hang time of about 0.889 sec. What was his vertical leap?

b) Although his height is only 5 ft 7 in., Spud Webb, formerly of the Sacramento Kings, had a vertical leap of about 44 in. What was his hang time?

a) To find Jordan's vertical leap, we substitute 0.889 for T in the function and compute V:

$$V(0.889) = 48(0.889)^2 \approx 37.9 \text{ in.}$$

Jordan's vertical leap was about 37.9 in.

b) To find Webb's hang time, we substitute 44 for V and solve for T:

$$44 = 48T^2 \quad \text{Substituting 44 for } V$$

$$\frac{44}{48} = T^2 \quad \text{Solving for } T^2$$

$$0.91\overline{6} = T^2$$

$$\sqrt{0.91\overline{6}} = T \quad \text{Hang time is positive.}$$

$$0.957 \approx T. \quad \text{Using a calculator}$$

Webb's hang time was about 0.957 sec. Note that his hang time was greater than Jordan's.

Do Exercises 16 and 17. ▶

16. *Vertical Leap.* Blake Griffin of the Los Angeles Clippers has a hang time of about 0.878 sec. What is his vertical leap?

17. *Hang Time.* Russell Westbrook of the Oklahoma Thunder has a vertical leap of 40 in. What is his hang time?

Answers

16. About 37.0 in. **17.** About 0.913 sec

7.1 Exercise Set

FOR EXTRA HELP  MyLab Math

✔ Check Your Understanding

Reading Check Determine whether each statement is true or false.

RC1. The quadratic equation $8x^2 - 11x + 50 = 0$ is in standard form.

RC2. Any quadratic equation can be solved by completing the square.

RC3. A quadratic equation may have solutions that are imaginary numbers.

RC4. The notation $\pm\sqrt{7}$ represents two real numbers.

RC5. To solve $5x^2 = 2x$, we can divide by x on both sides.

RC6. If $(x - 6)^2 = \sqrt{7}$, then $x = \sqrt{7}$ or $x = -\sqrt{7}$.

Concept Check Replace the blanks in each equation with constants in order to complete the square and form a true equation.

CC1. $x^2 + 10x + \_\_\_\_ = (x + \_\_\_\_)^2$

CC2. $t^2 - 12t + \_\_\_\_ = (t - \_\_\_\_)^2$

CC3. $m^2 - 5m + \_\_\_\_ = (m - \_\_\_\_)^2$

CC4. $x^2 + x + \_\_\_\_ = (x + \_\_\_\_)^2$

CC5. $r^2 + \frac{2}{5}r + \_\_\_\_ = (r + \_\_\_\_)^2$

CC6. $y^2 - \frac{3}{2}y + \_\_\_\_ = (y - \_\_\_\_)^2$

a

1. a) Solve:
   $6x^2 = 30.$
   b) Find the x-intercepts of $f(x) = 6x^2 - 30.$

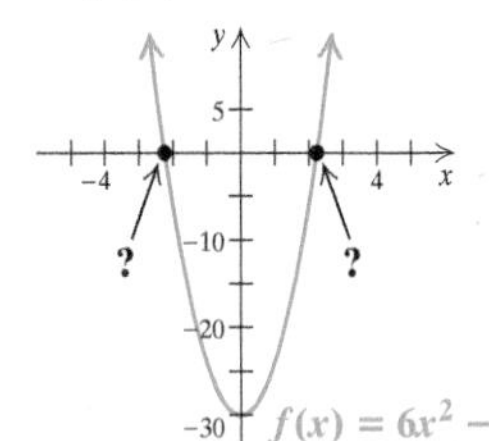

2. a) Solve:
   $5x^2 = 35.$
   b) Find the x-intercepts of $f(x) = 5x^2 - 35.$

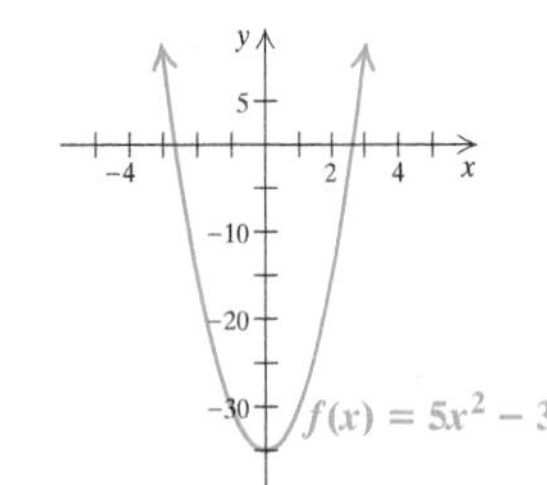

3. a) Solve:
   $9x^2 + 25 = 0.$
   b) Find the x-intercepts of $f(x) = 9x^2 + 25.$

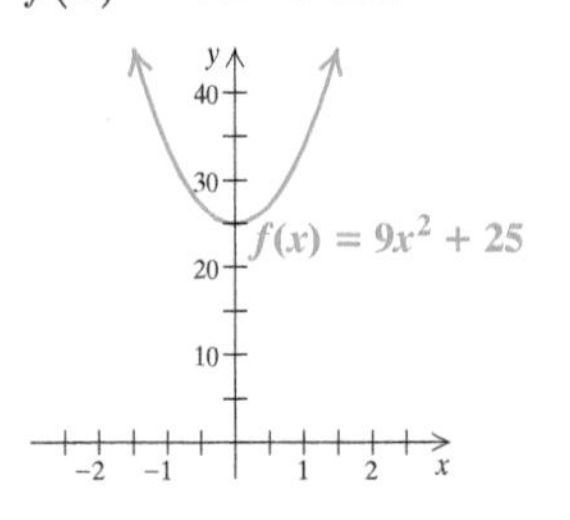

4. a) Solve:
   $36x^2 + 49 = 0.$
   b) Find the x-intercepts of $f(x) = 36x^2 + 49.$

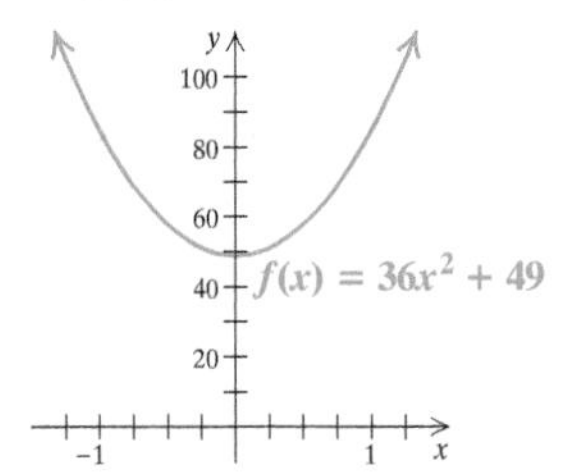

Solve. Give the exact solutions and approximate solutions to three decimal places, when appropriate.

5. $2x^2 - 3 = 0$
6. $3x^2 - 7 = 0$
7. $(x + 2)^2 = 49$
8. $(x - 1)^2 = 6$
9. $(x - 4)^2 = 16$
10. $(x + 3)^2 = 9$
11. $(x - 11)^2 = 7$
12. $(x - 9)^2 = 34$
13. $(x - 7)^2 = -4$
14. $(x + 1)^2 = -9$
15. $(x - 9)^2 = 81$
16. $(t - 2)^2 = 25$
17. $\left(x - \frac{3}{2}\right)^2 = \frac{7}{2}$
18. $\left(y + \frac{3}{4}\right)^2 = \frac{17}{16}$
19. $x^2 + 6x + 9 = 64$
20. $x^2 + 10x + 25 = 100$
21. $y^2 - 14y + 49 = 4$
22. $p^2 - 8p + 16 = 1$

b

Solve by completing the square. Show your work.

23. $x^2 + 4x = 2$
24. $x^2 + 2x = 5$
25. $x^2 - 22x = 11$
26. $x^2 - 18x = 10$
27. $x^2 + x = 1$
28. $x^2 - x = 3$
29. $t^2 - 5t = 7$
30. $y^2 + 9y = 8$
31. $x^2 + \frac{3}{2}x = 3$
32. $x^2 - \frac{4}{3}x = \frac{2}{3}$
33. $m^2 - \frac{9}{2}m = \frac{3}{2}$
34. $r^2 + \frac{2}{5}r = \frac{4}{5}$
35. $x^2 + 6x - 16 = 0$
36. $x^2 - 8x + 15 = 0$
37. $x^2 + 22x + 102 = 0$
38. $x^2 + 18x + 74 = 0$
39. $x^2 - 10x - 4 = 0$
40. $x^2 + 10x - 4 = 0$

41. **a)** Solve:
$x^2 + 7x - 2 = 0.$

b) Find the x-intercepts of $f(x) = x^2 + 7x - 2.$

42. **a)** Solve:
$x^2 - 7x - 2 = 0.$

b) Find the x-intercepts of $f(x) = x^2 - 7x - 2$

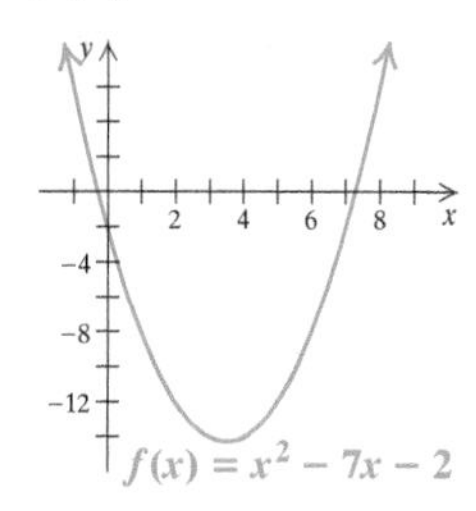

43. **a)** Solve:
$2x^2 - 5x + 8 = 0.$

b) Find the x-intercepts of $f(x) = 2x^2 - 5x + 8.$

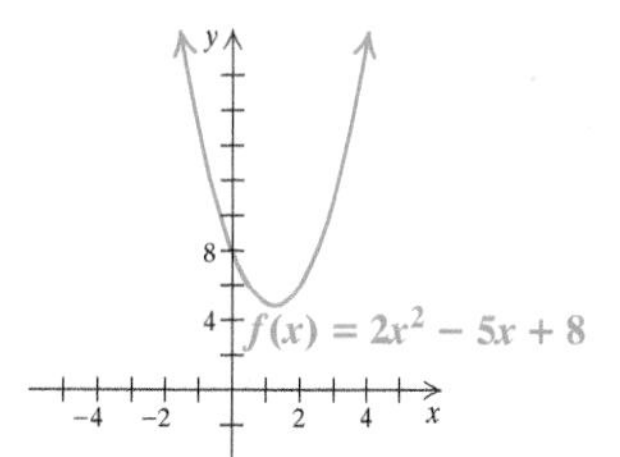

44. **a)** Solve:
$2x^2 - 3x + 9 = 0.$

b) Find the x-intercepts of $f(x) = 2x^2 - 3x + 9.$

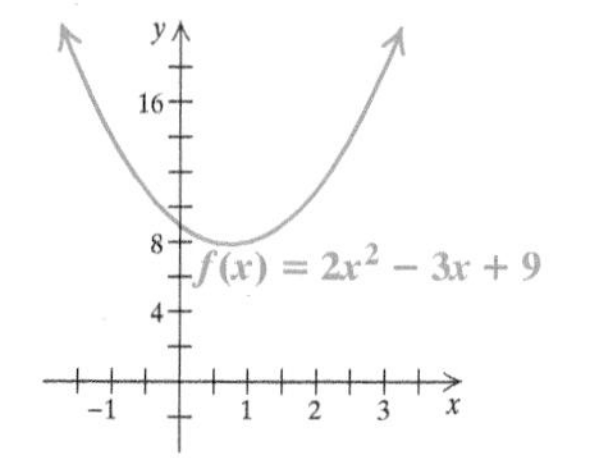

Solve by completing the square. Show your work.

45. $x^2 - \frac{3}{2}x - \frac{1}{2} = 0$

46. $x^2 + \frac{3}{2}x - 2 = 0$

47. $2x^2 - 3x - 17 = 0$

48. $2x^2 + 3x - 1 = 0$

49. $3x^2 - 4x - 1 = 0$

50. $3x^2 + 4x - 3 = 0$

51. $x^2 + x + 2 = 0$

52. $x^2 - x + 1 = 0$

53. $x^2 - 4x + 13 = 0$

54. $x^2 - 6x + 13 = 0$

C *Hang Time.* For Exercises 55 and 56, use the hang-time function $V(T) = 48T^2$, relating vertical leap to hang time.

55. The NBA's Kobe Bryant of the Los Angeles Lakers has a vertical leap of about 38 in. What is his hang time?

56. The NBA's Darrell Griffith had a record vertical leap of 48 in. What was his hang time?

Free-Falling Objects. The function $s(t) = 16t^2$ is used to approximate the distance s, in feet, that an object falls freely from rest in t seconds. Use the formula for Exercises 57–62.

57. The tallest roller coaster in the world is the Kingda Ka, located at Six Flags Great Adventure amusement park, in Jackson, New Jersey. It is 456 ft high. How long would it take an object to fall freely from the top?

58. The Gateway Arch in St. Louis, Missouri, is 630 ft high. How long would it take an object to fall freely from the top?

59. The Bunda Cliffs in Australia extend 62 mi along the coastline and vary in height from 200 ft to 390 ft. How long would it take an object to fall freely from the highest point on these cliffs?

60. Suspended 1854 ft above the water, the Beipanjiang River Bridge in China is the world's highest bridge. How long would it take an object to fall freely from the bridge?

61. The Washington Monument, near the west end of the National Mall in Washington, D.C., is the world's tallest stone structure and the world's tallest obelisk. It is 555.427 ft tall. How long would it take an object to fall freely from that height?

62. Completed in 2010, the Burj Khalifa, in downtown Dubai, is the tallest building in the world. It is 2720 ft tall. How long would it take an object to fall freely from that height?

Skill Maintenance

63. *Marathon Times.* The following table lists the record marathon times in 1981 and in 2011. [2.6e]

NUMBER OF YEARS AFTER 1981	RECORD MARATHON TIME (in minutes)
0	128
30	124

DATA: marathonguide.com

a) Use the two data points in the table to find a linear function $R(t) = mt + b$ that fits the data.
b) Use the function to estimate the record marathon time in 2020.
c) In what year will the marathon record be 122 min?

Graph. [2.2c], [2.5a]

64. $f(x) = 5 - 2x^2$

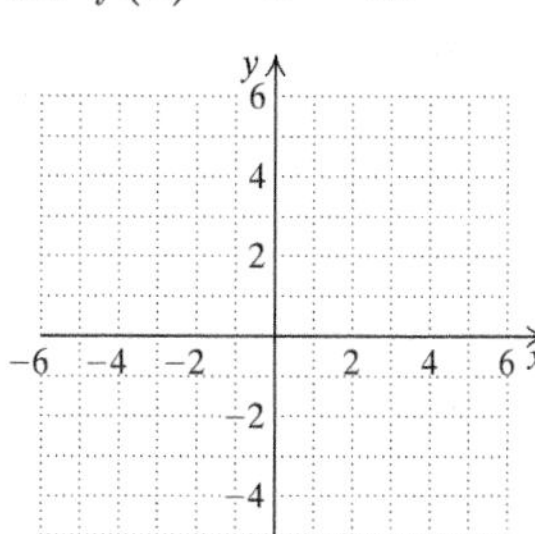

65. $f(x) = 5 - 2x$

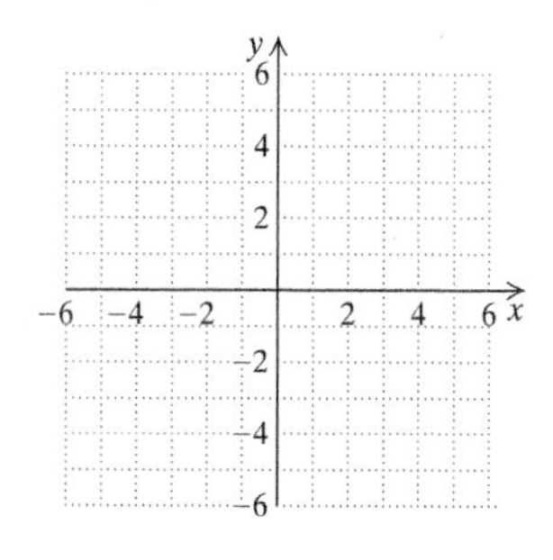

66. $2x - 5y = 10$

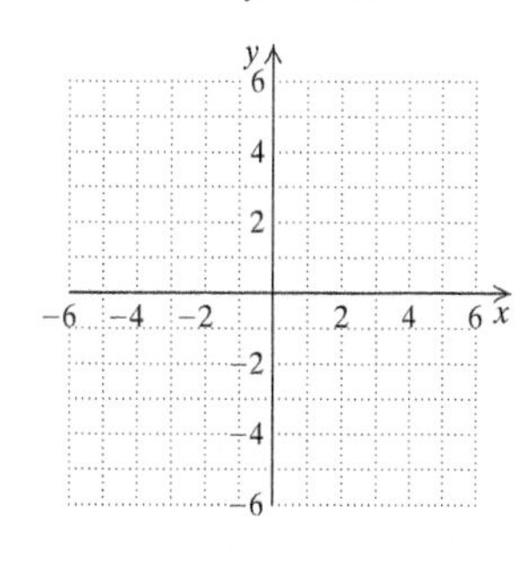

67. $f(x) = |5 - 2x|$

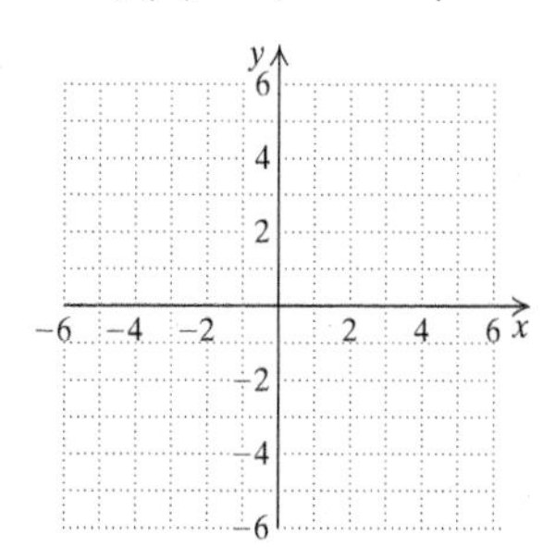

Simplify.

68. $\sqrt{88}$ [6.3a]

69. $\sqrt[5]{32x^5}$ [6.1d]

70. $\dfrac{t^3 - 8}{t^2 - 5t + 6}$ [5.1c]

71. $\dfrac{4x^3 - 6x^2 - 10x}{3x^3 - 3x}$ [5.1c]

72. $\dfrac{\frac{1}{x}}{\frac{1}{2x} - \frac{1}{3x}}$ [5.4a]

73. $\dfrac{\frac{t}{t+1}}{t - \frac{1}{t}}$ [5.4a]

Synthesis

74. Use a graphing calculator to solve each of the following equations.

a) $25.55x^2 - 1635.2 = 0$

b) $-0.0644x^2 + 0.0936x + 4.56 = 0$

c) $2.101x + 3.121 = 0.97x^2$

75. Problems such as those in Exercises 17, 21, and 25 can be solved without first finding standard form by using the INTERSECT feature on a graphing calculator. We let y_1 = the left side of the equation and y_2 = the right side. Use a graphing calculator to solve Exercises 17, 21, and 25 in this manner.

Find b such that the trinomial is a square.

76. $x^2 + bx + 75$

77. $x^2 + bx + 64$

Solve.

78. $\left(x - \frac{1}{3}\right)\left(x - \frac{1}{3}\right) + \left(x - \frac{1}{3}\right)\left(x + \frac{2}{9}\right) = 0$

79. $x(2x^2 + 9x - 56)(3x + 10) = 0$

80. *Boating.* A barge and a fishing boat leave a dock at the same time, traveling at right angles to each other. The barge travels 7 km/h slower than the fishing boat. After 4 hr, the boats are 68 km apart. Find the speed of each vessel.

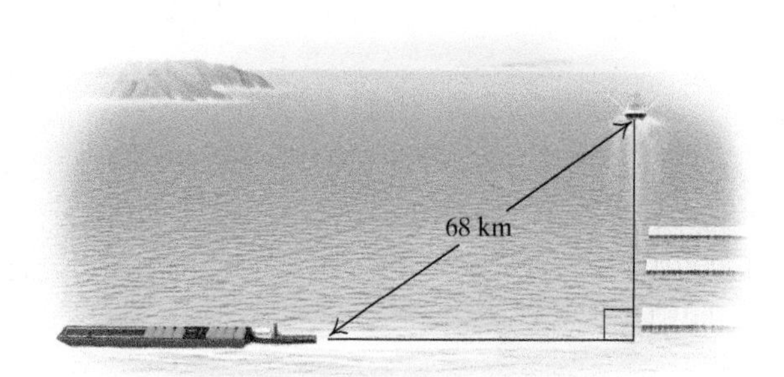

7.2

OBJECTIVE

a Solve quadratic equations using the quadratic formula, and approximate solutions using a calculator.

The Quadratic Formula

a SOLVING USING THE QUADRATIC FORMULA

SKILL REVIEW *Express imaginary numbers as bi and complex numbers as a + bi.* [6.8a]

Express in terms of i.

1. $\sqrt{-100}$

2. $10 - \sqrt{-68}$

Answers: 1. $10i$ **2.** $10 - 2\sqrt{17}i$

Each time you solve by completing the square, the procedure is the same. When we do the same kind of procedure many times, we look for a formula to speed up our work. Consider

$$ax^2 + bx + c = 0, \quad a > 0.$$

Note that if $a < 0$, we can get an equivalent form with $a > 0$ by first multiplying by -1.

Let's solve by *completing the square*. As we carry out the steps, compare them with Example 12 in the preceding section.

$$x^2 + \frac{b}{a}x + \frac{c}{a} = 0 \qquad \text{Multiplying by } \frac{1}{a}$$

$$x^2 + \frac{b}{a}x = -\frac{c}{a} \qquad \text{Subtracting } \frac{c}{a}$$

Half of $\frac{b}{a}$ is $\frac{b}{2a}$. The square is $\frac{b^2}{4a^2}$. We add $\frac{b^2}{4a^2}$ on both sides:

$$x^2 + \frac{b}{a}x + \frac{b^2}{4a^2} = -\frac{c}{a} + \frac{b^2}{4a^2} \qquad \text{Adding } \frac{b^2}{4a^2}$$

$$\left(x + \frac{b}{2a}\right)^2 = -\frac{4ac}{4a^2} + \frac{b^2}{4a^2} \qquad \text{Factoring the left side and finding a common denominator on the right}$$

$$\left(x + \frac{b}{2a}\right)^2 = \frac{b^2 - 4ac}{4a^2}$$

$$x + \frac{b}{2a} = \sqrt{\frac{b^2 - 4ac}{4a^2}} \quad or \quad x + \frac{b}{2a} = -\sqrt{\frac{b^2 - 4ac}{4a^2}}. \qquad \text{Using the principle of square roots}$$

Since $a > 0$, $\sqrt{4a^2} = 2a$, so we can simplify as follows:

$$x + \frac{b}{2a} = \frac{\sqrt{b^2 - 4ac}}{2a} \quad or \quad x + \frac{b}{2a} = -\frac{\sqrt{b^2 - 4ac}}{2a}.$$

Thus,

$$x = -\frac{b}{2a} \pm \frac{\sqrt{b^2 - 4ac}}{2a}, \quad \text{or} \quad x = \frac{-b \pm \sqrt{b^2 - 4ac}}{2a}.$$

This result is called the **quadratic formula.**

THE QUADRATIC FORMULA

The solutions of $ax^2 + bx + c = 0$ are given by

$$x = \frac{-b \pm \sqrt{b^2 - 4ac}}{2a}.$$

A similar proof would show that the formula also holds when $a < 0$.

ALGEBRAIC / GRAPHICAL CONNECTION

The Quadratic Formula (Algebraic). The solutions of $ax^2 + bx + c = 0$, $a \neq 0$, are given by

$$x = \frac{-b \pm \sqrt{b^2 - 4ac}}{2a}.$$

The Quadratic Formula (Graphical). The x-intercepts of the graph of the function $f(x) = ax^2 + bx + c$, $a \neq 0$, if they exist, are given by

$$\left(\frac{-b \pm \sqrt{b^2 - 4ac}}{2a}, 0\right).$$

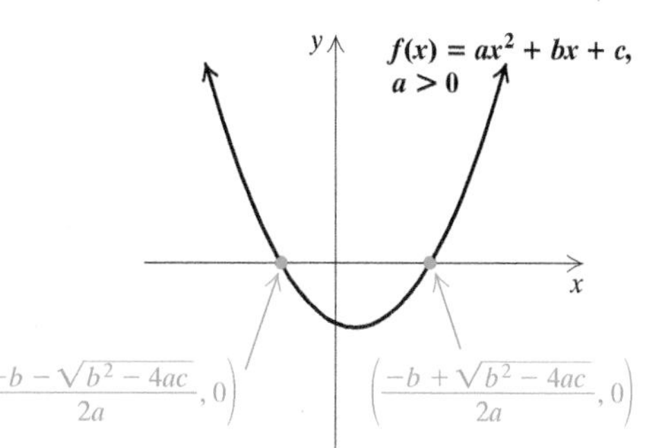

EXAMPLE 1 Solve $5x^2 + 8x = -3$ using the quadratic formula.

We first find standard form and determine a, b, and c:

$$5x^2 + 8x + 3 = 0;$$
$$a = 5, \quad b = 8, \quad c = 3.$$

We then use the quadratic formula, $x = \dfrac{-b \pm \sqrt{b^2 - 4ac}}{2a}$:

$$x = \frac{-8 \pm \sqrt{8^2 - 4 \cdot 5 \cdot 3}}{2 \cdot 5} \quad \text{Substituting}$$

Be sure to write the fraction bar all the way across.

$$x = \frac{-8 \pm \sqrt{64 - 60}}{10}$$
$$x = \frac{-8 \pm \sqrt{4}}{10}$$
$$x = \frac{-8 \pm 2}{10}$$
$$x = \frac{-8 + 2}{10} \quad or \quad x = \frac{-8 - 2}{10}$$
$$x = \frac{-6}{10} \quad or \quad x = \frac{-10}{10}$$
$$x = -\frac{3}{5} \quad or \quad x = -1.$$

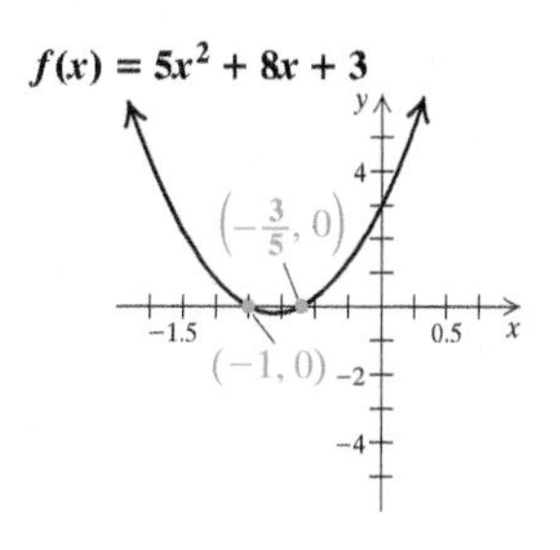

The solutions are $-\frac{3}{5}$ and -1.

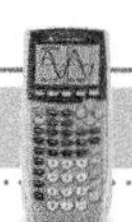

CALCULATOR CORNER

Solving Quadratic Equations We can use the INTERSECT feature to solve a quadratic equation. Consider the equation

$4x(x - 2) - 5x(x - 1) = 2.$

First, we enter the equations $y_1 = 4x(x - 2) - 5x(x - 1)$ and $y_2 = 2$. We then graph the equations in a window that shows the point(s) of intersection of the graphs. Next, we use the INTERSECT feature to find the coordinates of the left-hand point of intersection.

The first coordinate of this point, -2, is one solution of the equation. We use the INTERSECT feature again to find the other solution, -1.

$y_1 = 4x(x - 2) - 5x(x - 1)$, $y_2 = 2$

4, −5, 3, −2
Intersection
X = −2 Y = 2

$y_1 = 4x(x - 2) - 5x(x - 1)$, $y_2 = 2$

4, −5, 3, −2
Intersection
X = −1 Y = 2

EXERCISES: Solve.

1. $5x^2 = -11x + 12$
2. $2x^2 - 15 = 7x$
3. $6(x - 3) = (x - 3)(x - 2)$
4. $(x + 1)(x - 4) = 3(x - 4)$

1. Consider the equation $2x^2 = 4 + 7x$.
   a) Solve using the quadratic formula.
   b) Solve by factoring.

It turns out that we could have solved the equation in Example 1 more easily by factoring, as follows:

$$\begin{aligned} 5x^2 + 8x + 3 &= 0 \\ (5x + 3)(x + 1) &= 0 \\ 5x + 3 = 0 \quad &or \quad x + 1 = 0 \\ 5x = -3 \quad &or \quad x = -1 \\ x = -\tfrac{3}{5} \quad &or \quad x = -1. \end{aligned}$$

To solve a quadratic equation:

1. Check for the form $x^2 = d$ or $(x + c)^2 = d$. If it is in this form, use the principle of square roots as in Section 7.1.
2. If it is not in the form of step (1), write it in standard form $ax^2 + bx + c = 0$ with a and b nonzero.
3. Then try factoring.
4. If it is not possible to factor or if factoring seems difficult, use the quadratic formula.

Although the solutions of a quadratic equation cannot always be found by factoring, they can *always* be found using the quadratic formula.

CALCULATOR CORNER

Approximating Solutions of Quadratic Equations In Example 2, we find that the solutions of the equation $5x^2 - 8x = 3$ are $\frac{4 + \sqrt{31}}{5}$ and $\frac{4 - \sqrt{31}}{5}$. We can use a calculator to approximate these solutions. Parentheses must be used carefully. For example, to approximate $\frac{4 + \sqrt{31}}{5}$, we press

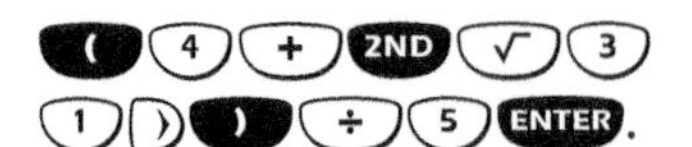

The solutions are approximately 1.914 and −0.314.

(4+√31)/5	
	1.913552873
(4−√31)/5	
	−.3135528726

EXERCISES: Use a calculator to approximate the solutions in each of the following. Round to three decimal places.

1. Example 4
2. Margin Exercise 2
3. Margin Exercise 4

The solutions to all the exercises in this section could also be found by completing the square. However, the quadratic formula is the preferred method because it is faster.

◀ **Do Exercise 1.**

We will see in Example 2 that we cannot always rely on factoring.

EXAMPLE 2 Solve: $5x^2 - 8x = 3$. Give the exact solutions and approximate the solutions to three decimal places.

We first find standard form and determine a, b and c:

$$5x^2 - 8x - 3 = 0;$$
$$a = 5, \quad b = -8, \quad c = -3.$$

We then use the quadratic formula, $x = \frac{-b \pm \sqrt{b^2 - 4ac}}{2a}$:

$$x = \frac{-(-8) \pm \sqrt{(-8)^2 - 4 \cdot 5 \cdot (-3)}}{2 \cdot 5} \qquad \text{Substituting}$$

$$= \frac{8 \pm \sqrt{64 + 60}}{10} = \frac{8 \pm \sqrt{124}}{10} = \frac{8 \pm \sqrt{4 \cdot 31}}{10}$$

$$= \frac{8 \pm 2\sqrt{31}}{10} = \frac{2(4 \pm \sqrt{31})}{2 \cdot 5} = \frac{2}{2} \cdot \frac{4 \pm \sqrt{31}}{5} = \frac{4 \pm \sqrt{31}}{5}.$$

↑

Caution!

To avoid a common error in simplifying, remember to *factor the numerator and the denominator* and then remove a factor of 1.

Answer

1. (a) $-\frac{1}{2}, 4$; (b) $-\frac{1}{2}, 4$

We can use a calculator to approximate the solutions:

$$\frac{4+\sqrt{31}}{5} \approx 1.914; \qquad \frac{4-\sqrt{31}}{5} \approx -0.314.$$

Check: Checking the exact solutions $(4 \pm \sqrt{31})/5$ can be quite cumbersome. It could be done on a calculator or by using the approximations. Here we check 1.914; the check for -0.314 is left to the student.

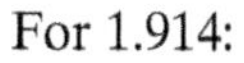
For 1.914:

$$\begin{array}{l|l} 5x^2 - 8x = 3 & \\ \hline 5(1.914)^2 - 8(1.914) \ ?\ 3 & \\ 5(3.663396) - 15.312 & \\ 3.00498 & \end{array}$$

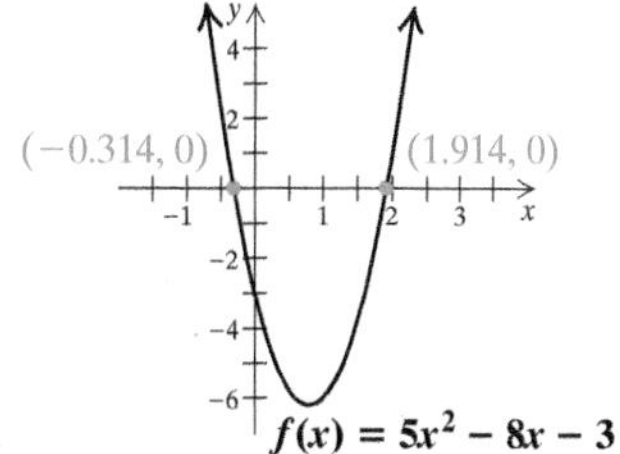

We do not have a perfect check due to the rounding error, but our check seems to confirm the solutions.

Do Exercise 2.

Some quadratic equations have solutions that are nonreal complex numbers.

EXAMPLE 3 Solve: $x^2 + x + 1 = 0$.

We have $a = 1$, $b = 1$, $c = 1$. We use the quadratic formula:

$$x = \frac{-1 \pm \sqrt{1^2 - 4 \cdot 1 \cdot 1}}{2 \cdot 1}$$
$$= \frac{-1 \pm \sqrt{1-4}}{2}$$
$$= \frac{-1 \pm \sqrt{-3}}{2}$$
$$= \frac{-1 \pm \sqrt{3}i}{2}.$$

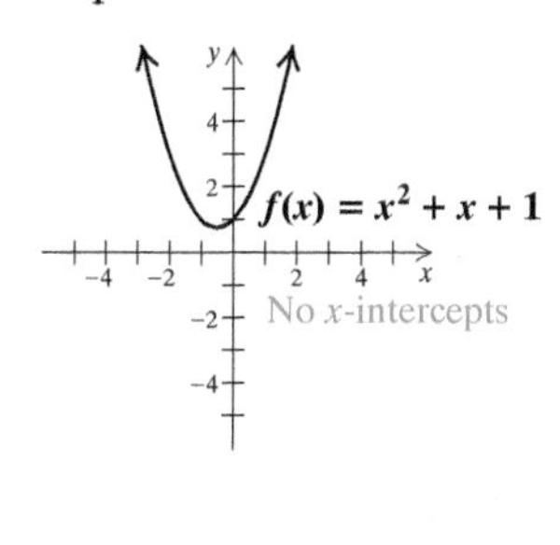

The solutions are

$$\frac{-1+\sqrt{3}i}{2} \text{ and } \frac{-1-\sqrt{3}i}{2}, \text{ or } -\frac{1}{2} + \frac{\sqrt{3}}{2}i \text{ and } -\frac{1}{2} - \frac{\sqrt{3}}{2}i.$$

Do Exercise 3.

EXAMPLE 4 Solve: $2 + \frac{7}{x} = \frac{5}{x^2}$. Give the exact solutions and approximate solutions to three decimal places.

We first find an equivalent quadratic equation in standard form:

$$x^2\left(2 + \frac{7}{x}\right) = x^2 \cdot \frac{5}{x^2} \qquad \text{Multiplying by } x^2 \text{ to clear fractions, noting that } x \neq 0$$
$$2x^2 + 7x = 5$$
$$2x^2 + 7x - 5 = 0. \qquad \text{Subtracting 5}$$

2. Solve using the quadratic formula:

$$3x^2 + 2x = 7.$$

Give the exact solutions and approximate solutions to three decimal places.

Write the equation in standard form.

$$3x^2 + 2x - 7 = \square$$

$a = \square$, $b = \square$, $c = \square$;

$$x = \frac{-\square \pm \sqrt{2^2 - 4 \cdot 3 \cdot (\square)}}{2 \cdot 3}$$
$$= \frac{-2 \pm \sqrt{\square}}{6}$$
$$= \frac{-2 \pm 2\sqrt{\square}}{6}$$
$$= \frac{2(-1 \pm \sqrt{\square})}{2 \cdot 3}$$
$$= \frac{-1 \pm \sqrt{22}}{\square}$$

Approximate the solutions and round to three decimal places.

$$\frac{-1+\sqrt{22}}{3} \approx \square$$
$$\frac{-1-\sqrt{22}}{3} \approx \square$$

3. Solve: $x^2 - x + 2 = 0$.

Answers

2. $\frac{-1 \pm \sqrt{22}}{3}$; 1.230, −1.897

3. $\frac{1 \pm \sqrt{7}i}{2}$, or $\frac{1}{2} \pm \frac{\sqrt{7}}{2}i$

Guided Solution:
2. 0, 3, 2, −7, 2, −7, 88, 22, 22, 3, 1.230, −1.897

Then

$$a = 2, \quad b = 7, \quad c = -5;$$

$$x = \frac{-7 \pm \sqrt{7^2 - 4 \cdot 2 \cdot (-5)}}{2 \cdot 2} \qquad \text{Substituting}$$

$$x = \frac{-7 \pm \sqrt{49 + 40}}{4}$$

$$x = \frac{-7 \pm \sqrt{89}}{4}$$

$$x = \frac{-7 + \sqrt{89}}{4} \quad or \quad x = \frac{-7 - \sqrt{89}}{4}.$$

$f(x) = 2 + \frac{7}{x} - \frac{5}{x^2}$

(−4.108, 0)

(0.608, 0)

Since we began with a rational equation, we need to check. We cleared the fractions before obtaining a quadratic equation in standard form, and this step could introduce numbers that do not check in the original rational equation. We need to show that neither of the numbers makes a denominator 0. Since neither of them does, the solutions are

$$\frac{-7 + \sqrt{89}}{4} \quad \text{and} \quad \frac{-7 - \sqrt{89}}{4}.$$

We can use a calculator to approximate the solutions:

$$\frac{-7 + \sqrt{89}}{4} \approx 0.608;$$

$$\frac{-7 - \sqrt{89}}{4} \approx -4.108.$$

4. Solve:

$$3 = \frac{5}{x} + \frac{4}{x^2}.$$

Give the exact solutions and approximate solutions to three decimal places.

Answer

4. $\frac{5 \pm \sqrt{73}}{6}$; 2.257, −0.591

◀ **Do Exercise 4.**

7.2 Exercise Set

FOR EXTRA HELP

MyLab Math

✔ Check Your Understanding

Reading Check Determine whether each statement is true or false.

RC1. The quadratic formula can be used to solve any quadratic equation.

RC2. The quadratic formula does not work if the solutions are imaginary numbers.

RC3. Using the quadratic formula is always the fastest way to solve a quadratic equation.

RC4. In the quadratic formula, if $b^2 - 4ac$ is positive but not a perfect square, the solutions will be irrational.

Concept Check Complete each statement with the correct number or expression.

CC1. When we are using the quadratic formula to solve $3x^2 - x - 8 = 0$, the value of a is ______.

CC2. When we are using the quadratic formula to solve $3x^2 - x - 8 = 0$, the value of b is ______.

CC3. Standard form for the quadratic equation $5x^2 = 9 - x$ is ______ $= 0$.

CC4. When we are using the quadratic formula to solve $3x^2 = 10x$, the value of c is ______.

a Solve.

1. $x^2 + 8x + 2 = 0$

2. $x^2 - 6x - 4 = 0$

3. $3p^2 = -8p - 1$

4. $3u^2 = 18u - 6$

5. $x^2 - x + 1 = 0$

6. $x^2 + x + 2 = 0$

7. $x^2 + 13 = 4x$

8. $x^2 + 13 = 6x$

9. $r^2 + 3r = 8$

10. $h^2 + 4 = 6h$

11. $1 + \frac{2}{x} + \frac{5}{x^2} = 0$

12. $1 + \frac{5}{x^2} = \frac{2}{x}$

13. a) Solve: $3x + x(x - 2) = 0$.
    b) Find the x-intercepts of $f(x) = 3x + x(x - 2)$.

14. a) Solve: $4x + x(x - 3) = 0$.
    b) Find the x-intercepts of $f(x) = 4x + x(x - 3)$.

15. a) Solve: $11x^2 - 3x - 5 = 0$.
    b) Find the x-intercepts of $f(x) = 11x^2 - 3x - 5$.

16. a) Solve: $7x^2 + 8x = -2$.
    b) Find the x-intercepts of $f(x) = 7x^2 + 8x + 2$.

17. a) Solve: $25x^2 = 20x - 4$.
    b) Find the x-intercepts of $f(x) = 25x^2 - 20x + 4$.

18. a) Solve: $49x^2 - 14x + 1 = 0$.
    b) Find the x-intercepts of $f(x) = 49x^2 - 14x + 1$.

Solve.

19. $4x(x - 2) - 5x(x - 1) = 2$

20. $3x(x + 1) - 7x(x + 2) = 6$

21. $14(x - 4) - (x + 2) = (x + 2)(x - 4)$

22. $11(x - 2) + (x - 5) = (x + 2)(x - 6)$

23. $5x^2 = 17x - 2$

24. $15x = 2x^2 + 16$

25. $x^2 + 5 = 4x$

26. $x^2 + 5 = 2x$

27. $x + \frac{1}{x} = \frac{13}{6}$

28. $\frac{3}{x} + \frac{x}{3} = \frac{5}{2}$

29. $\frac{1}{y} + \frac{1}{y + 2} = \frac{1}{3}$

30. $\frac{1}{x} + \frac{1}{x + 4} = \frac{1}{7}$

31. $(2t - 3)^2 + 17t = 15$

32. $2y^2 - (y + 2)(y - 3) = 12$

33. $(x - 2)^2 + (x + 1)^2 = 0$

34. $(x + 3)^2 + (x - 1)^2 = 0$

35. $x^3 - 1 = 0$
(*Hint*: Factor the difference of cubes. Then use the quadratic formula.)

36. $x^3 + 27 = 0$

Solve. Give the exact solutions and approximate solutions to three decimal places.

37. $x^2 + 6x + 4 = 0$

38. $x^2 + 4x - 7 = 0$

39. $x^2 - 6x + 4 = 0$

40. $x^2 - 4x + 1 = 0$

41. $2x^2 - 3x - 7 = 0$

42. $3x^2 - 3x - 2 = 0$

43. $5x^2 = 3 + 8x$

44. $2y^2 + 2y - 3 = 0$

Skill Maintenance

Solve.

45. $x = \sqrt{x + 2}$ [6.6a]

46. $\sqrt{x + 1} + 2 = \sqrt{3x + 1}$ [6.6b]

47. $\sqrt{2x - 6} + 11 = 2$ [6.6a]

48. $\sqrt[3]{4x - 7} = 2$ [6.6a]

49. $2x^2 = x + 3$ [4.8a]

50. $100x^2 + 1 = 20x$ [4.8a]

51. $\frac{3}{x} - \frac{1}{4} = \frac{7}{2x}$ [5.5a]

52. $\frac{3}{x - 2} = \frac{5}{6x}$ [5.5a]

Synthesis

53. Use a graphing calculator to solve the equations in Exercises 3, 16, 17, and 43 using the INTERSECT feature, letting $y_1 =$ the left side and $y_2 =$ the right side. Then solve $2.2x^2 + 0.5x - 1 = 0$.

54. Use a graphing calculator to solve the equations in Exercises 9, 27, and 30. Then solve $5.33x^2 = 8.23x + 3.24$.

Solve.

55. $2x^2 - x - \sqrt{5} = 0$

56. $\frac{5}{x} + \frac{x}{4} = \frac{11}{7}$

57. $ix^2 - x - 1 = 0$

58. $\sqrt{3}x^2 + 6x + \sqrt{3} = 0$

59. $\frac{x}{x + 1} = 4 + \frac{1}{3x^2 - 3}$

60. $(1 + \sqrt{3})x^2 - (3 + 2\sqrt{3})x + 3 = 0$

61. Let $f(x) = (x - 3)^2$. Find all inputs x such that $f(x) = 13$.

62. Let $f(x) = x^2 + 14x + 49$. Find all inputs x such that $f(x) = 36$.

Applications Involving Quadratic Equations

7.3

OBJECTIVES

a Solve applied problems involving quadratic equations.

b Solve a formula for a given letter.

a APPLICATIONS AND PROBLEM SOLVING

Sometimes when we translate a problem to mathematical language, the result is a quadratic equation.

EXAMPLE 1 *Beach Volleyball.* The beach volleyball court at Lake Jean State Park measures 24 m by 16 m. The playing area is surrounded by a free zone of uniform width. The area of the playing area is one-third of the area of the entire court. How wide is the free zone?

1. **Familiarize.** We let x = the width of the free zone and make a drawing.

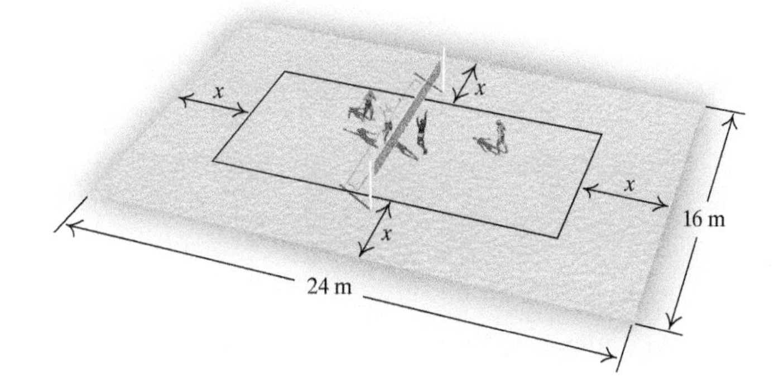

2. **Translate.** The area of a rectangle is lw (length times width). Then:

$$\text{Area of entire court} = 24 \cdot 16;$$
$$\text{Area of playing area} = (24 - 2x)(16 - 2x).$$

Since the playing area is one-third of the area of the court, we have

$$(24 - 2x)(16 - 2x) = \frac{1}{3} \cdot 24 \cdot 16.$$

3. **Solve.** We solve the equation:

$$384 - 80x + 4x^2 = 128 \quad \text{Using FOIL on the left}$$
$$4x^2 - 80x + 256 = 0 \quad \text{Finding standard form}$$
$$x^2 - 20x + 64 = 0 \quad \text{Dividing by 4}$$
$$(x - 4)(x - 16) = 0 \quad \text{Factoring}$$
$$x = 4 \quad or \quad x = 16. \quad \text{Using the principle of zero products}$$

4. **Check.** We check in the original problem. We see that 16 is not a solution because a 24-m by 16-m court cannot have a 16-m free zone.

If the free zone is 4 m wide, then the playing area will have length $24 - 2 \cdot 4$, or 16 m. The width will be $16 - 2 \cdot 4$, or 8 m. The area of the playing area is thus $16 \cdot 8$, or 128 m^2. The area of the entire court is $24 \cdot 16$, or 384 m^2. The area of the playing area is one-third of 384 m^2, so the number 4 checks.

5. **State.** The free zone is 4 m wide.

Do Exercise 1. ▶

MyLab Math
ANIMATION

1. *Landscaping.* A rectangular garden is 60 ft by 80 ft. Part of the garden is torn up to install a sidewalk of uniform width around it. The area of the new garden is one-half of the old area. How wide is the sidewalk?

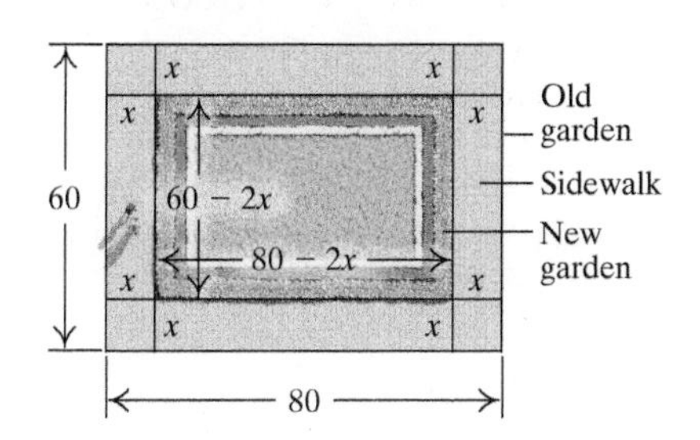

Answer

1. 10 ft

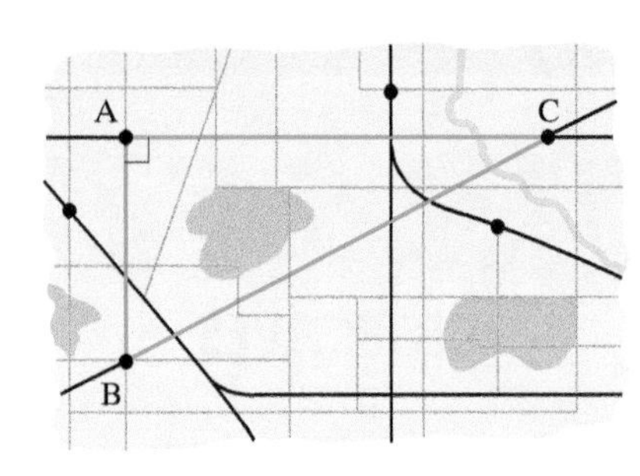

EXAMPLE 2 *Town Planning.* Three towns A, B, and C are situated as shown in the figure at left. The roads at A form a right angle. The distance from A to B is 2 mi less than the distance from A to C. The distance from B to C is 10 mi. Find the distance from A to B and the distance from A to C.

1. **Familiarize.** We first make a drawing and label it. We let $d =$ the distance from A to C. Then the distance from A to B is $d - 2$.

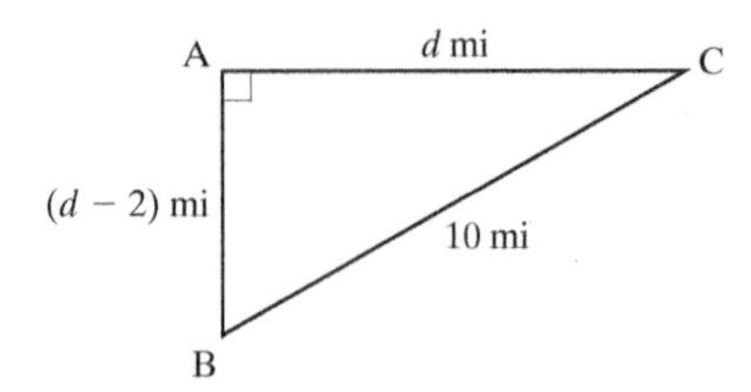

2. **Translate.** We see that a right triangle is formed. We can use the Pythagorean equation, $c^2 = a^2 + b^2$:

$$10^2 = d^2 + (d - 2)^2.$$

3. **Solve.** We solve the equation:

$$\begin{aligned} 10^2 &= d^2 + (d - 2)^2 \\ 100 &= d^2 + d^2 - 4d + 4 && \text{Squaring} \\ 2d^2 - 4d - 96 &= 0 && \text{Finding standard form} \\ d^2 - 2d - 48 &= 0 && \text{Dividing by 2} \\ (d - 8)(d + 6) &= 0 && \text{Factoring} \\ d = 8 \quad or \quad d &= -6. && \text{Using the principle of zero products} \end{aligned}$$

4. **Check.** We know that -6 cannot be a solution because distances are not negative. If $d = 8$, then $d - 2 = 6$, and $8^2 + 6^2 = 64 + 36 = 100$. Since $10^2 = 100$, the distance 8 mi checks.

5. **State.** The distance from A to C is 8 mi, and the distance from A to B is 6 mi.

Do Exercise 2.

2. *Ladder Location.* A ladder leans against a building, as shown below. The ladder is 20 ft long. The distance to the top of the ladder is 4 ft greater than the distance d from the building. Find the distance d and the distance to the top of the ladder.

EXAMPLE 3 *Landscape Design.* Melanie plans to build a fire pit in her backyard at a safe distance from both her gardening shed and her daughter's playset. The three structures will form a right triangle, with the distance of the fire pit from the gardening shed 2 m less than the distance from the fire pit to the playset. The distance from the playset to the gardening shed is 20 m and forms the longest side of the triangle. How far is the fire pit from the gardening shed and from the playset?

1. **Familiarize.** We make a drawing and label it. We let $d =$ the distance from the fire pit to the playset. Then the distance from the fire pit to the gardening shed is $d - 2$.

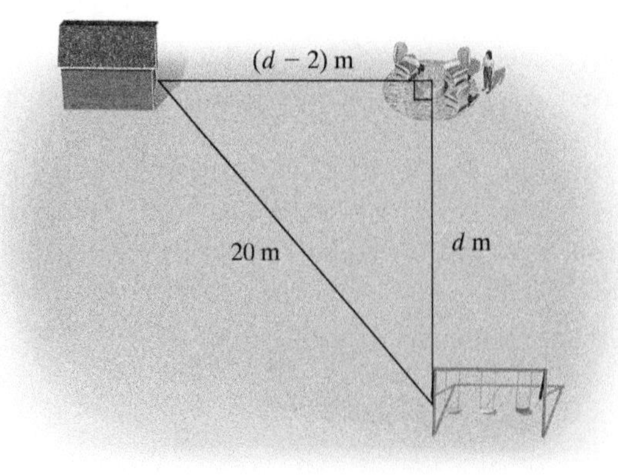

Answer

2. The distance d is 12 ft; the distance to the top of the ladder is 16 ft.

2. **Translate.** We use the Pythagorean equation as in Example 2:

$$20^2 = d^2 + (d - 2)^2.$$

3. **Solve.** We solve the equation using the quadratic formula:

$$\begin{aligned} 400 &= d^2 + d^2 - 4d + 4 && \text{Squaring} \\ 0 &= 2d^2 - 4d - 396 && \text{Finding standard form} \\ 0 &= d^2 - 2d - 198. && \text{Dividing by 2 (or multiplying by } \tfrac{1}{2}) \end{aligned}$$

Then

$$\begin{aligned} d &= \frac{-b \pm \sqrt{b^2 - 4ac}}{2a} \\ &= \frac{-(-2) \pm \sqrt{(-2)^2 - 4(1)(-198)}}{2(1)} \qquad a = 1, b = -2, c = -198 \\ &= \frac{2 \pm \sqrt{796}}{2} = \frac{2 \pm \sqrt{4 \cdot 199}}{2} = \frac{2 \pm 2\sqrt{199}}{2} = 1 \pm \sqrt{199}. \end{aligned}$$

4. **Check.** Since $1 - \sqrt{199}$ is negative, it cannot be the distance between the structures. Using a calculator, we find that $1 + \sqrt{199} \approx 15.1$. Thus we have $d \approx 15.1$ and $d - 2 \approx 13.1$. The square of the hypotenuse in the triangle is 20^2, or 400. Since $(15.1)^2 + (13.1)^2 = 399.62 \approx 400$, the numbers check.

5. **State.** The fire pit is 13.1 m from the gardening shed and 15.1 m from the playset.

Do Exercise 3. ▶

3. *Ladder Location.* Refer to Margin Exercise 2. Suppose that the ladder has length 10 ft. Find the distance d and the distance $d + 4$.

EXAMPLE 4 *Motorcycle Travel.* Karin's motorcycle traveled 300 mi at a certain speed. Had she gone 10 mph faster, she could have made the trip in 1 hr less time. Find her speed.

1. **Familiarize.** We make a drawing, labeling it with known and unknown information, and organize the information in a table. We let r = the speed, in miles per hour, and t = the time, in hours.

DISTANCE	SPEED	TIME	
300	r	t	$\longrightarrow r = \frac{300}{t}$
300	$r + 10$	$t - 1$	$\longrightarrow r + 10 = \frac{300}{t - 1}$

Recalling the motion formula $d = rt$ and solving for r, we get $r = d/t$. From the rows of the table, we obtain

$$r = \frac{300}{t} \quad \text{and} \quad r + 10 = \frac{300}{t - 1}.$$

Answer

3. The distance d is about 4.8 ft; the distance to the top of the ladder is about 8.8 ft.

4. *Marine Travel.* Two ships make the same voyage of 3000 nautical miles. The faster ship travels 10 knots faster than the slower one. (A *knot* is 1 nautical mile per hour.) The faster ship makes the voyage in 50 hr less time than the slower one. Find the speeds of the two ships.

GS

Complete this table to help with the familiarization.

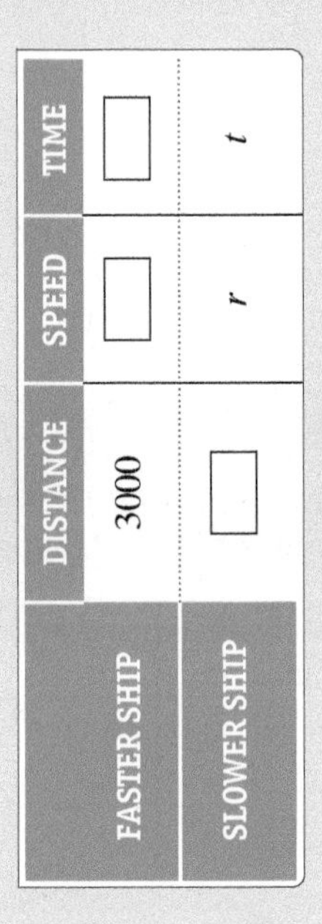

	DISTANCE	SPEED	TIME
FASTER SHIP	3000	[ ]	[ ]
SLOWER SHIP	[ ]	r	t

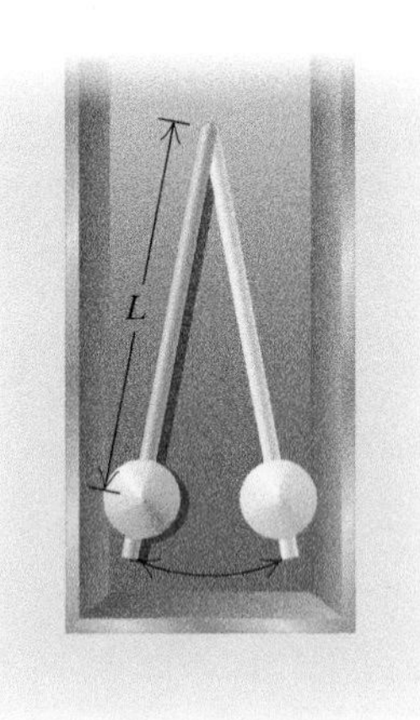

Answer

4. 20 knots, 30 knots

Guided Solution:

4.

	Distance	Speed	Time
Faster ship	3000	$r + 10$	$t - 50$
Slower ship	3000	r	t

2. **Translate.** We substitute for r from the first equation into the second and get a translation:

$$\frac{300}{t} + 10 = \frac{300}{t - 1}.$$

3. **Solve.** We solve as follows:

$$\frac{300}{t} + 10 = \frac{300}{t - 1}$$

$$t(t-1)\left[\frac{300}{t} + 10\right] = t(t-1)\cdot\frac{300}{t-1} \quad \text{Multiplying by the LCM}$$

$$t(t-1)\cdot\frac{300}{t} + t(t-1)\cdot 10 = t(t-1)\cdot\frac{300}{t-1}$$

$$300(t-1) + 10(t^2 - t) = 300t$$

$$300t - 300 + 10t^2 - 10t = 300t$$

$$10t^2 - 10t - 300 = 0 \quad \text{Standard form}$$

$$t^2 - t - 30 = 0 \quad \text{Dividing by 10}$$

$$(t-6)(t+5) = 0 \quad \text{Factoring}$$

$$t = 6 \quad or \quad t = -5. \quad \text{Using the principle of zero products}$$

4. **Check.** Since negative time has no meaning in this problem, we try 6 hr. Remembering that $r = d/t$, we get $r = 300/6 = 50$ mph.

To check, we take the speed 10 mph faster, which is 60 mph, and see how long the trip would have taken at that speed:

$$t = \frac{d}{r} = \frac{300}{60} = 5 \text{ hr}.$$

This is 1 hr less than the trip actually took, so we have an answer.

5. **State.** Karin's speed was 50 mph.

◀ **Do Exercise 4.**

b SOLVING FORMULAS

Recall that to solve a formula for a certain letter, we use the principles for solving equations to get that letter alone on one side.

EXAMPLE 5 *Period of a Pendulum.* The time T required for a pendulum of length L to swing back and forth (complete one period) is given by the formula $T = 2\pi\sqrt{L/g}$, where g is the gravitational constant. Solve for L.

$$T = 2\pi\sqrt{\frac{L}{g}} \quad \text{This is a radical equation.}$$

$$T^2 = \left(2\pi\sqrt{\frac{L}{g}}\right)^2 \quad \text{Principle of powers (squaring)}$$

$$T^2 = 2^2\pi^2\frac{L}{g}$$

$$gT^2 = 4\pi^2 L \quad \text{Clearing fractions and simplifying}$$

$$\frac{gT^2}{4\pi^2} = L \quad \text{Multiplying by } \frac{1}{4\pi^2}$$

We now have L alone on one side and L does not appear on the other side, so the formula is solved for L.

Do Exercise 5.

5. Solve $A = \sqrt{\frac{w_1}{w_2}}$ for w_2.

In most formulas, variables represent nonnegative numbers, so we need only the positive root when taking square roots.

EXAMPLE 6 *Hang Time.* An athlete's *hang time* is the amount of time that the athlete can remain airborne when jumping. A formula relating an athlete's vertical leap V, in inches, to hang time T, in seconds, is $V = 48T^2$. Solve for T.

We have

$$48T^2 = V$$

$$T^2 = \frac{V}{48} \quad \text{Multiplying by } \tfrac{1}{48} \text{ to get } T^2 \text{ alone}$$

$$T = \sqrt{\frac{V}{48}} \quad \text{Using the principle of square roots; note that } T \geq 0.$$

$$= \sqrt{\frac{V}{2\cdot 2\cdot 2\cdot 2\cdot 3}\cdot\frac{3}{3}}$$

$$= \frac{\sqrt{3V}}{2\cdot 2\cdot 3}$$

$$= \frac{\sqrt{3V}}{12}.$$

Do Exercise 6.

EXAMPLE 7 *Falling Distance.* An object that is tossed downward with an initial speed (velocity) of v_0 will travel a distance of s meters, where $s = 4.9t^2 + v_0t$ and t is measured in seconds. Solve for t.

GS 6. Solve $V = \pi r^2 h$ for r. (Volume of a right circular cylinder)

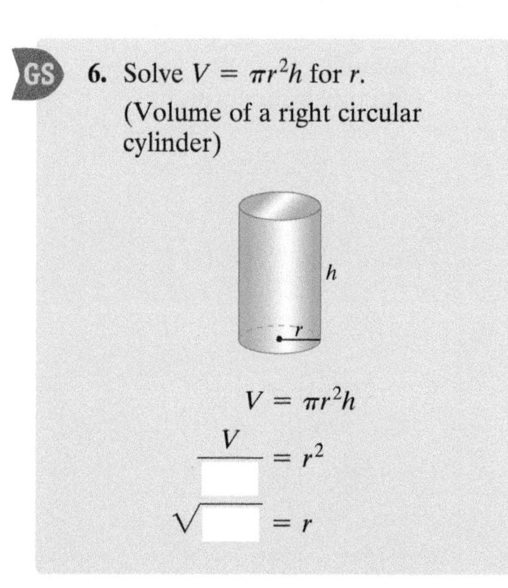

$$V = \pi r^2 h$$

$$\frac{V}{\square} = r^2$$

$$\sqrt{\square} = r$$

Answers

5. $w_2 = \frac{w_1}{A^2}$ **6.** $r = \sqrt{\frac{V}{\pi h}}$

Guided Solution:

6. $\pi h, \frac{V}{\pi h}$

To solve a formula for a letter, say, t:

1. Clear the fractions and use the principle of powers, as needed, until t does not appear in any radicand or denominator. (In some cases, you may clear the fractions first, and in some cases, you may use the principle of powers first.)
2. Collect all terms with t^2 in them. Also collect all terms with t in them.
3. If t^2 does not appear, you can finish by using just the addition and multiplication principles.
4. If t^2 appears but t does not, solve the equation for t^2. Then take square roots on both sides.
5. If there are terms containing both t and t^2, write the equation in standard form and use the quadratic formula.

Since t is squared in one term and raised to the first power in the other term, the equation is quadratic in t. The variable is t; v_0 and s are treated as constants.

We have

$$4.9t^2 + v_0t = s$$

$$4.9t^2 + v_0t - s = 0 \quad \text{Writing standard form}$$

$$a = 4.9, \quad b = v_0, \quad c = -s$$

$$t = \frac{-v_0 \pm \sqrt{(v_0)^2 - 4(4.9)(-s)}}{2(4.9)} \quad \text{Using the quadratic formula: } t = \frac{-b \pm \sqrt{b^2 - 4ac}}{2a}$$

$$= \frac{-v_0 \pm \sqrt{(v_0)^2 + 19.6s}}{9.8}.$$

Since the negative square root would yield a negative value for t, we use only the positive root:

$$t = \frac{-v_0 + \sqrt{(v_0)^2 + 19.6s}}{9.8}.$$

Do Exercise 7.

The steps listed in the margin should help you when solving formulas for a given letter. Try to remember that, when solving a formula, you do the same things you would do to solve an equation.

EXAMPLE 8 Solve $t = \dfrac{a}{\sqrt{a^2 + b^2}}$ for a.

In this case, we could either clear the fractions first or use the principle of powers first. Let's clear the fractions. Multiplying by $\sqrt{a^2 + b^2}$, we have

$$t\sqrt{a^2 + b^2} = a.$$

Now we square both sides and then continue:

$$\left(t\sqrt{a^2 + b^2}\right)^2 = a^2 \quad \text{Squaring}$$

Caution!

Don't forget to square both t and $\sqrt{a^2 + b^2}$.

$$t^2\left(\sqrt{a^2 + b^2}\right)^2 = a^2$$

$$t^2(a^2 + b^2) = a^2$$

$$t^2a^2 + t^2b^2 = a^2$$

$$t^2b^2 = a^2 - t^2a^2 \quad \text{Getting all } a^2\text{-terms together}$$

$$t^2b^2 = a^2(1 - t^2) \quad \text{Factoring out } a^2$$

$$\frac{t^2b^2}{1 - t^2} = a^2 \quad \text{Dividing by } 1 - t^2$$

$$\sqrt{\frac{t^2b^2}{1 - t^2}} = a \quad \text{Taking the square root}$$

$$\frac{tb}{\sqrt{1 - t^2}} = a. \quad \text{Simplifying}$$

You need not rationalize denominators in situations such as this.

Do Exercise 8.

7. Solve $s = gt + 16t^2$ for t.

8. Solve $\dfrac{b}{\sqrt{a^2 - b^2}} = t$ for b.

Answers

7. $t = \dfrac{-g + \sqrt{g^2 + 64s}}{32}$

8. $b = \dfrac{ta}{\sqrt{1 + t^2}}$

Translating for Success

Translate each word problem to an equation or a system of equations and select a correct translation from equations A–O.

1. *Car Travel.* Sarah drove her car 800 mi to see her friend. The return trip was 2 hr faster at a speed that was 10 mph more. Find her return speed.

2. *Coin Mixture.* A collection of dimes and quarters is worth \$26.95. There are 117 coins in all. How many of each coin are there?

3. *Wire Cutting.* A 537-in. wire is cut into three pieces. The second piece is 7 in. shorter than the first. The third is half as long as the first. How long is each piece?

4. *Marine Travel.* The Columbia River flows at a rate of 2 mph for the length of a popular boating route. In order for a motorized dinghy to travel 3 mi upriver and return in a total of 4 hr, how fast must the boat be able to travel in still water?

5. *Locker Numbers.* The numbers on three adjoining lockers are consecutive integers whose sum is 537. Find the integers.

A. $(80 - 2x)(100 - 2x) = \frac{1}{3} \cdot 80 \cdot 100$

B. $\dfrac{800}{x} + 10 = \dfrac{800}{x - 2}$

C. $x + 18\% \cdot x = 3.24$

D. $x + 25y = 26.95,$ $x + y = 117$

E. $2x + 2(x - 7) = 537$

F. $x + (x - 7) + \frac{1}{2}x = 537$

G. $0.10x + 0.25y = 26.95,$ $x + y = 117$

H. $3.24 - 18\% \cdot 3.24 = x$

I. $\dfrac{4}{x + 2} + \dfrac{4}{x - 2} = 3$

J. $x^2 + (x + 1)^2 = 7^2$

K. $75^2 + x^2 = 78^2$

L. $\dfrac{3}{x + 2} + \dfrac{3}{x - 2} = 4$

M. $75^2 + 78^2 = x^2$

N. $x + (x + 1) + (x + 2) = 537$

O. $\dfrac{800}{x} + \dfrac{800}{x - 2} = 10$

Answers on page A-25

6. *Gasoline Prices.* One day the price of gasoline was increased 18% to a new price of \$3.24 per gallon. What was the original price?

7. *Triangle Dimensions.* The hypotenuse of a right triangle is 7 ft. The length of one leg is 1 ft longer than the other. Find the lengths of the legs.

8. *Rectangle Dimensions.* The perimeter of a rectangle is 537 ft. The width of the rectangle is 7 ft shorter than the length. Find the length and the width.

9. *Guy Wire.* A guy wire is 78 ft long. It is attached to the top of a 75-ft cell-phone tower. How far is it from the base of the pole to the point where the wire is attached to the ground?

10. *Landscaping.* A rectangular garden is 80 ft by 100 ft. Part of the garden is torn up to install a sidewalk of uniform width around it. The area of the new garden is $\frac{1}{3}$ of the old area. How wide is the sidewalk?

7.3 Exercise Set

FOR EXTRA HELP MyLab Math

✔ Check Your Understanding

Reading Check and Concept Check Match each formula with the appropriate description from the column on the right.

RC1. ____ $s = 4.9t^2 + v_0t$

RC2. ____ $V = 48T^2$

RC3. ____ $a^2 + b^2 = c^2$

RC4. ____ $A = lw$

RC5. ____ $T = 2\pi\sqrt{\frac{l}{g}}$

RC6. ____ $t = \frac{d}{r}$

a) Area of a rectangle
b) Pythagorean theorem
c) Motion
d) Period of a pendulum
e) Vertical leap
f) Distance

a Solve.

1. *Flower Bed.* The width of a rectangular flower bed is 7 ft less than the length. The area is 18 ft^2. Find the length and the width.

2. *Feed Lot.* The width of a rectangular feed lot is 8 m less than the length. The area is 20 m^2. Find the length and the width.

3. *Parking Lot.* The length of a rectangular parking lot is twice the width. The area is 162 yd^2. Find the length and the width.

4. *Flag Dimensions.* The length of an American flag that is displayed at a government office is 3 in. less than twice its width. The area is 1710 in^2. Find the length and the width of the flag.

5. *Easter Island.* Easter Island is roughly triangular in shape. The height of the triangle is 7 mi less than the base. The area is 60 mi^2. Find the base and the height of the triangular-shaped island.

6. *Sailing.* The base of a triangular sail is 9 m less than the height. The area is 56 m^2. Find the base and the height of the sail.

7. *Parking Lot.* The width of a rectangular parking lot is 51 ft less than its length. Determine the dimensions of the parking lot if it measures 250 ft diagonally.

8. *Sailing.* The base of a triangular sail is 8 ft less than the height. The area is 56 ft^2. Find the base and the height of the sail.

9. *Mirror Framing.* The outside of a mosaic mirror frame measures 14 in. by 20 in., and 160 in^2 of mirror shows. Find the width of the frame.

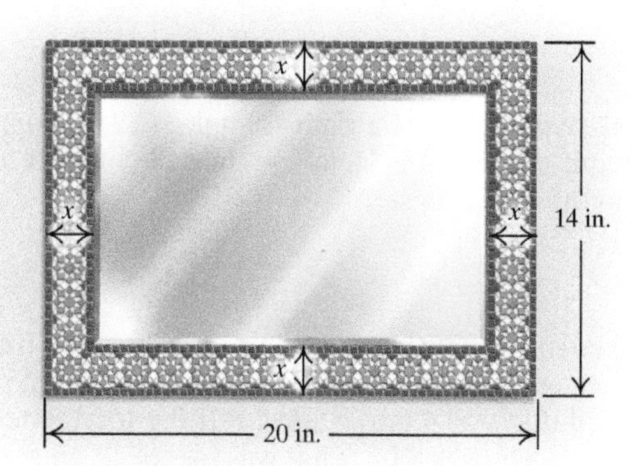

10. *Quilt Dimensions.* Michelle is making a quilt for a wall hanging at the entrance to a state museum. The finished quilt will measure 8 ft by 6 ft. The quilt has a border of uniform width around it. The area of the interior rectangular section is one-half of the area of the entire quilt. How wide is the border?

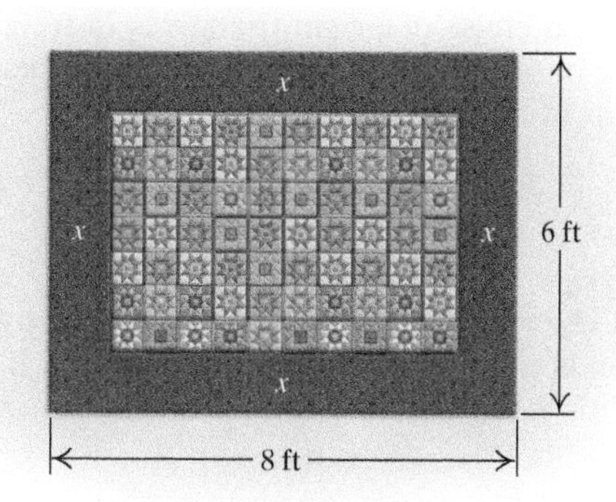

11. *Landscaping.* A landscaper is designing a flower garden in the shape of a right triangle. She wants 10 ft of a perennial border to form the hypotenuse of the triangle, and one leg is to be 2 ft longer than the other. Find the lengths of the legs.

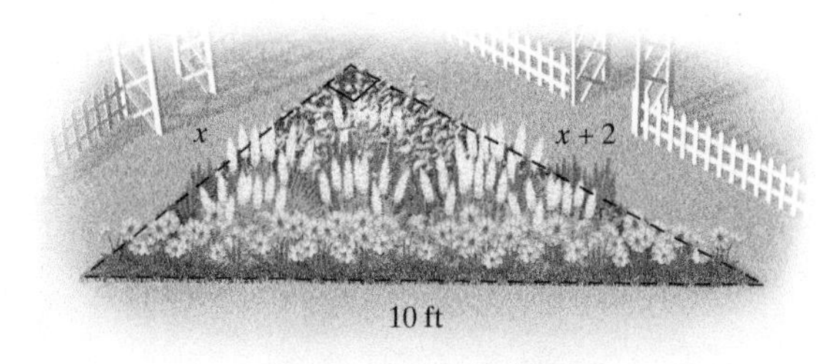

12. *Flag Dimensions.* The diagonal of a Papua New Guinea flag displayed in a school is 60 in. The length of the flag is 12 in. longer than the width. Find the dimensions of the flag.

13. *Raffle Tickets.* Margaret and Zane purchased consecutively numbered raffle tickets at a charity auction. The product of the ticket numbers was 552. Find the ticket numbers.

14. *Box Construction.* An open box is to be made from a 10-ft by 20-ft rectangular piece of cardboard by cutting a square from each corner. The area of the bottom of the box is to be 96 ft^2. What is the length of the sides of the squares that are cut from the corners?

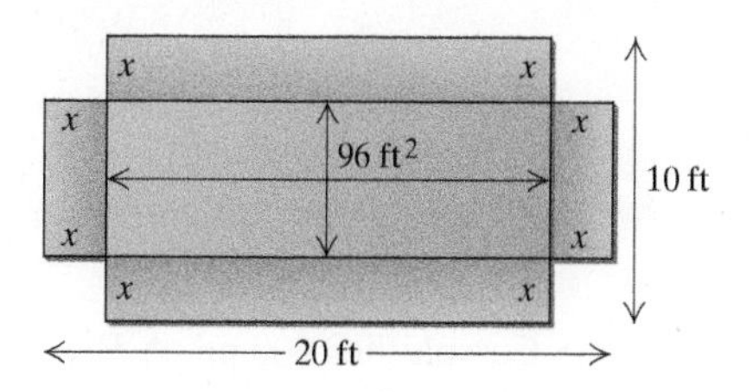

Solve. Find exact answers and approximate answers rounded to three decimal places.

15. The width of a rectangle is 4 ft less than the length. The area is 10 ft^2. Find the length and the width.

16. The length of a rectangle is twice the width. The area is 328 cm^2. Find the length and the width.

17. *Page Dimensions.* The outside of an oversized book page measures 14 in. by 20 in., and 100 in^2 of printed text shows. Find the width of the margin.

18. *Picture Framing.* The outside of a picture frame measures 13 cm by 20 cm, and 80 cm^2 of picture shows. Find the width of the frame.

19. The hypotenuse of a right triangle is 24 ft long. The length of one leg is 14 ft more than the other. Find the lengths of the legs.

20. The hypotenuse of a right triangle is 22 m long. The length of one leg is 10 m less than the other. Find the lengths of the legs.

21. *Car Trips.* During the first part of a trip, Sam's Toyota Prius Hybrid traveled 120 mi. Sam then drove another 100 mi at a speed that was 10 mph slower. If the total time for Sam's trip was 4 hr, what was his speed on each part of the trip?

DISTANCE	SPEED	TIME

22. *Canoeing.* During the first part of a canoe trip, Doug covered 60 km. He then traveled 24 km at a speed that was 4 km/h slower. If the total time for Doug's trip was 8 hr, what was his speed on each part of the trip?

DISTANCE	SPEED	TIME

23. *Skiing.* Kingdom Trails is a nonprofit conservation organization in Vermont, working with private landowners to manage outdoor recreation opportunities and to preserve and protect trails. In January, Colleen skied 24 km along part of the trails. If she had gone 2 km/h faster, the trip would have taken 1 hr less. Find Colleen's speed.

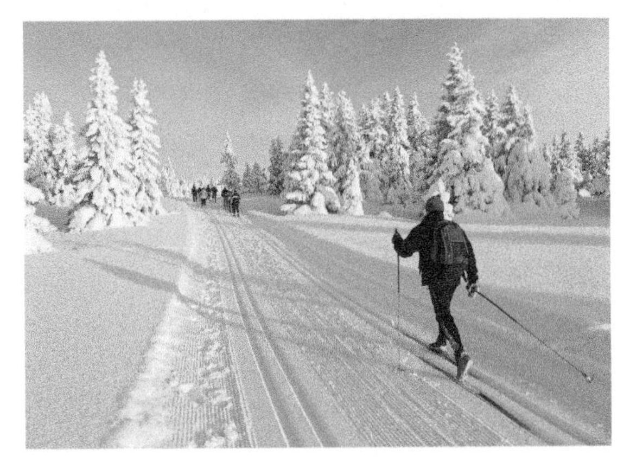

24. *Bicycling.* In July, Art bicycled 120 mi in Vermont. If he had gone 5 mph faster, the trip would have taken 4 hr less. Find Art's speed.

25. *Air Travel.* A Cessna flies 600 mi. A Beechcraft flies 1000 mi at a speed that is 50 mph faster, but takes 1 hr longer. Find the speed of each plane.

26. *Air Travel.* A turbo-jet flies 50 mph faster than a super-prop plane. If a turbo-jet goes 2000 mi in 3 hr less time than it takes the super-prop to go 2800 mi, find the speed of each plane.

27. *Bicycling.* Naoki bikes 40 mi to Hillsboro. The return trip is made at a speed that is 6 mph slower. Total travel time for the round trip is 14 hr. Find Naoki's speed on each part of the trip.

28. *Car Speed.* On a sales trip, Gail drives 600 mi to Richmond. The return trip is made at a speed that is 10 mph slower. Total travel time for the round trip is 22 hr. How fast did Gail travel on each part of the trip?

29. *Navigation.* The current in a typical Mississippi River shipping route flows at a rate of 4 mph. In order for a barge to travel 24 mi upriver and then return in a total of 5 hr, approximately how fast must the barge be able to travel in still water?

30. *Navigation.* The Hudson River flows at a rate of 3 mph. A patrol boat travels 60 mi upriver and returns in a total time of 9 hr. What is the speed of the boat in still water?

b Solve each formula for the given letter. Assume that all variables represent nonnegative numbers.

31. $A = 6s^2$, for s
(Surface area of a cube)

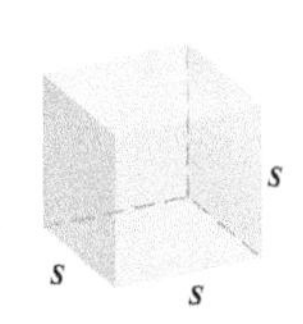

32. $A = 4\pi r^2$, for r
(Surface area of a sphere)

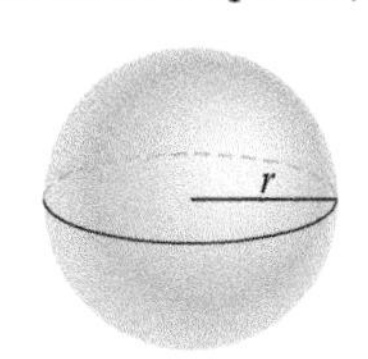

33. $F = \dfrac{Gm_1m_2}{r^2}$, for r

34. $N = \dfrac{kQ_1Q_2}{s^2}$, for s
(Number of phone calls between two cities)

35. $E = mc^2$, for c
(Einstein's energy–mass relationship)

36. $V = \frac{1}{3}s^2h$, for s
(Volume of a pyramid)

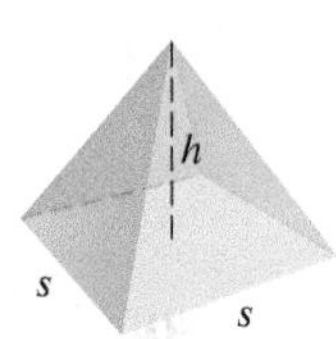

37. $a^2 + b^2 = c^2$, for b
(Pythagorean formula in two dimensions)

38. $a^2 + b^2 + c^2 = d^2$, for c
(Pythagorean formula in three dimensions)

39. $N = \dfrac{k^2 - 3k}{2}$, for k
(Number of diagonals of a polygon of k sides)

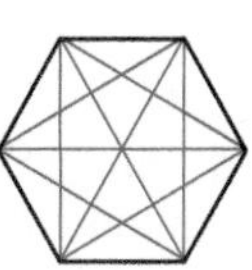

40. $s = v_0t + \dfrac{gt^2}{2}$, for t
(A motion formula)

41. $A = 2\pi r^2 + 2\pi rh$, for r
(Surface area of a cylinder)

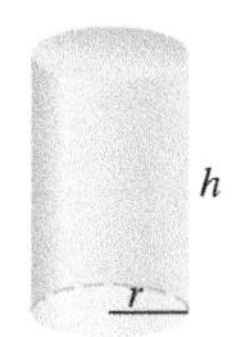

42. $A = \pi r^2 + \pi rs$, for r
(Surface area of a cone)

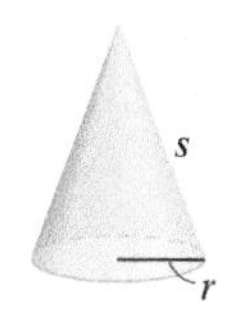

43. $T = 2\pi\sqrt{\frac{L}{g}}$, for g
(A pendulum formula)

44. $W = \sqrt{\frac{1}{LC}}$, for L
(An electricity formula)

45. $I = \frac{703W}{H^2}$, for H
(Body mass index)

46. $N + p = \frac{6.2A^2}{pR^2}$, for R

47. $m = \frac{m_0}{\sqrt{1 - \frac{v^2}{c^2}}}$, for v
(A relativity formula)

48. Solve the formula given in Exercise 47 for c.

Skill Maintenance

Add or subtract. [5.2b, c]

49. $\frac{1}{x - 1} + \frac{1}{x^2 - 3x + 2}$

50. $\frac{x + 1}{x - 1} - \frac{x + 1}{x^2 + x + 1}$

51. $\frac{2}{x + 3} - \frac{x}{x - 1} + \frac{x^2 + 2}{x^2 + 2x - 3}$

52. Multiply and simplify: $\sqrt{3x^2}\sqrt{3x^3}$. [6.3a]

53. Express in terms of i: $\sqrt{-20}$. [6.8a]

Synthesis

54. Solve: $\frac{4}{2x + i} - \frac{1}{x - i} = \frac{2}{x + i}$.

55. Find a when the reciprocal of $a - 1$ is $a + 1$.

56. *Pizza Crusts.* At Pizza Perfect, Ron can make 100 large pizza crusts in 1.2 hr less than Chad. Together they can do the job in 1.8 hr. How long does it take each to do the job alone?

57. *Surface Area.* A sphere is inscribed in a cube as shown in the following figure. Express the surface area of the sphere as a function of the surface area S of the cube.

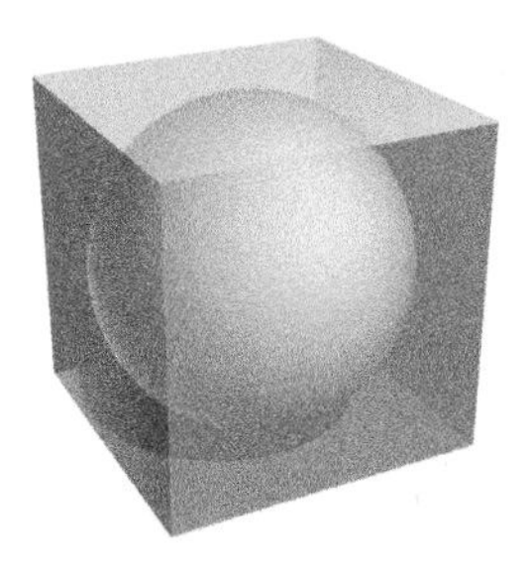

58. *Bungee Jumping.* Jesse is tied to one end of a 40-m elasticized (bungee) cord. The other end of the cord is tied to the middle of a train trestle. If Jesse steps off the bridge, for how long will he fall before the cord begins to stretch? (See Example 7 and let $v_0 = 0$.)

59. *The Golden Rectangle.* For over 2000 years, the proportions of a "golden" rectangle have been considered visually appealing. A rectangle of width w and length l is considered "golden" if

$$\frac{w}{l} = \frac{l}{w + l}.$$

Solve for l.

More on Quadratic Equations

7.4

OBJECTIVES

a Determine the nature of the solutions of a quadratic equation.

b Write a quadratic equation having two given numbers as solutions.

c Solve equations that are quadratic in form.

a THE DISCRIMINANT

From the quadratic formula, we know that the solutions x_1 and x_2 of a quadratic equation are given by

$$x_1 = \frac{-b + \sqrt{b^2 - 4ac}}{2a} \quad \text{and} \quad x_2 = \frac{-b - \sqrt{b^2 - 4ac}}{2a}.$$

The expression $b^2 - 4ac$ is called the **discriminant**. When we are using the quadratic formula, it is helpful to compute the discriminant first. If it is 0, there will be just one real solution. If it is positive, there will be two real solutions. If it is negative, we will be taking the square root of a negative number; hence there will be two nonreal complex-number solutions, and they will be complex conjugates.

DISCRIMINANT $b^2 - 4ac$	NATURE OF SOLUTIONS	x-INTERCEPTS
0	Only one solution; it is a real number	Only one
Positive	Two different real-number solutions	Two different
Negative	Two different nonreal complex-number solutions (complex conjugates)	None

If the discriminant is a perfect square, we can solve the equation by factoring, not needing the quadratic formula.

EXAMPLE 1 Determine the nature of the solutions of $9x^2 - 12x + 4 = 0$.

We have

$$a = 9, \quad b = -12, \quad c = 4.$$

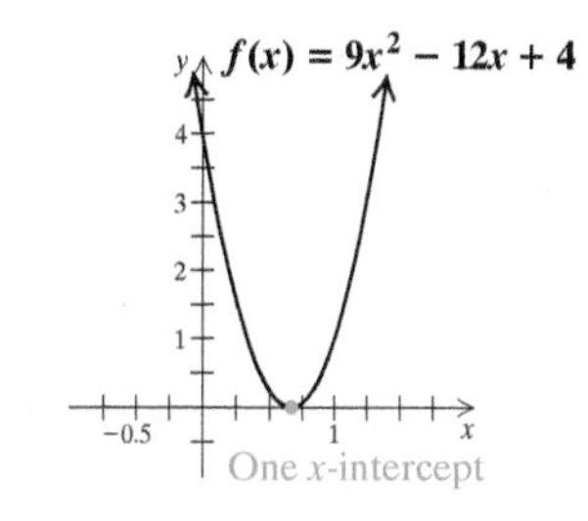

We compute the discriminant:

$$\begin{aligned} b^2 - 4ac &= (-12)^2 - 4 \cdot 9 \cdot 4 \\ &= 144 - 144 \\ &= 0. \end{aligned}$$

There is just one solution, and it is a real number. Since 0 is a perfect square, the equation can be solved by factoring.

EXAMPLE 2 Determine the nature of the solutions of $x^2 + 5x + 8 = 0$.

We have

$$a = 1, \quad b = 5, \quad c = 8.$$

We compute the discriminant:

$$\begin{aligned} b^2 - 4ac &= 5^2 - 4 \cdot 1 \cdot 8 \\ &= 25 - 32 \\ &= -7. \end{aligned}$$

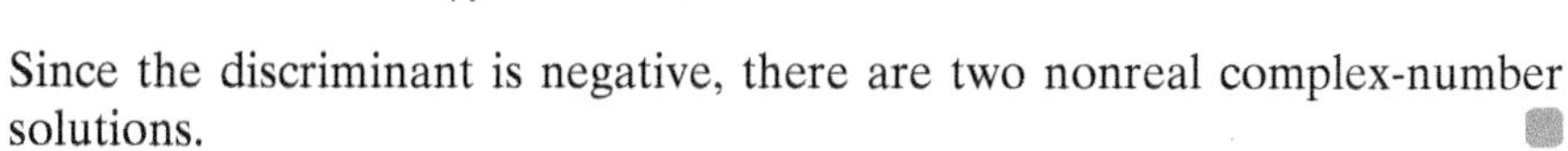

Since the discriminant is negative, there are two nonreal complex-number solutions.

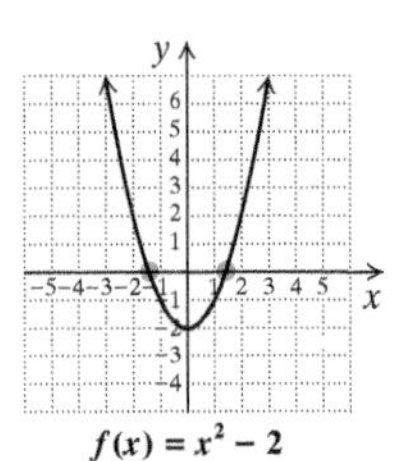

$f(x) = x^2 - 2$
$b^2 - 4ac = 8 > 0$
Two real solutions
Two x-intercepts

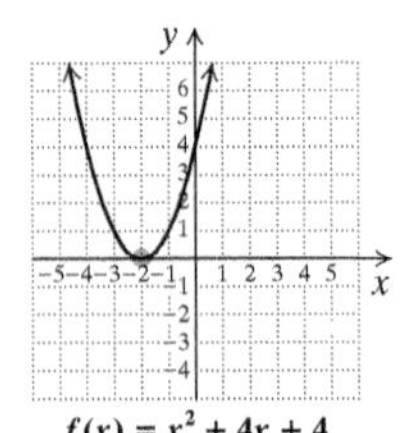

$f(x) = x^2 + 4x + 4$
$b^2 - 4ac = 0$
One real solution
One x-intercept

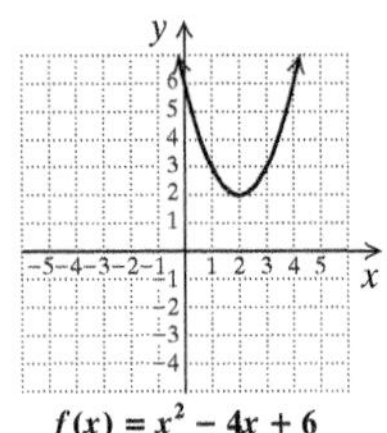

$f(x) = x^2 - 4x + 6$
$b^2 - 4ac = -8 < 0$
No real solutions
No x-intercept

EXAMPLE 3 Determine the nature of the solutions of $x^2 + 5x + 6 = 0$.

We have

$$a = 1, \quad b = 5, \quad c = 6;$$
$$b^2 - 4ac = 5^2 - 4 \cdot 1 \cdot 6 = 1.$$

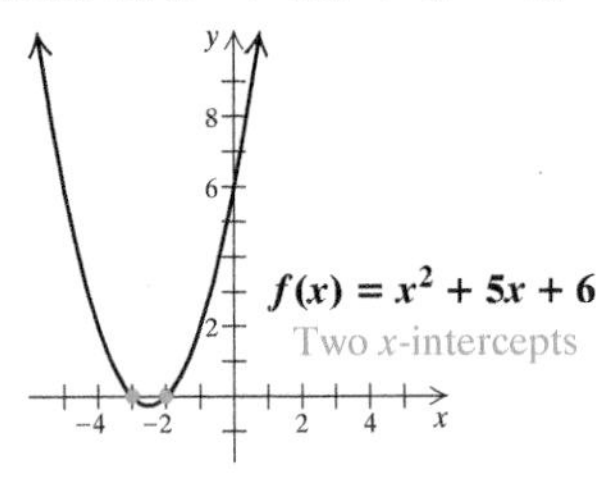

Since the discriminant is positive, there are two solutions, and they are real numbers. The equation can be solved by factoring since the discriminant is a perfect square.

EXAMPLE 4 Determine the nature of the solutions of $5x^2 + x - 3 = 0$.

We have

$$a = 5, \quad b = 1, \quad c = -3;$$
$$b^2 - 4ac = 1^2 - 4 \cdot 5 \cdot (-3) = 1 + 60 = 61.$$

Since the discriminant is positive, there are two solutions, and they are real numbers. The equation cannot be solved by factoring because 61 is not a perfect square.

The discriminant, $b^2 - 4ac$, tells us how many real-number solutions the equation $ax^2 + bx + c = 0$ has, so it also indicates how many x-intercepts the graph of $f(x) = ax^2 + bx + c$ has. Compare the graphs at left.

Do Exercises 1–3.

Determine the nature of the solutions without solving.

1. $x^2 + 5x - 3 = 0$

2. $9x^2 - 6x + 1 = 0$

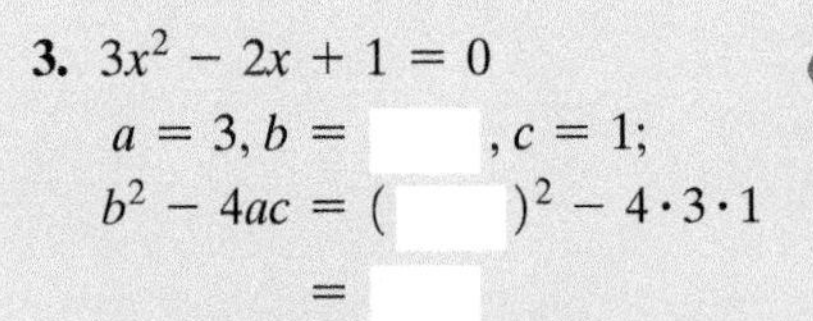

3. $3x^2 - 2x + 1 = 0$ GS
$a = 3, b = \_\_\_, c = 1;$
$b^2 - 4ac = (\_\_\_)^2 - 4 \cdot 3 \cdot 1$
$= \_\_\_$
Since the discriminant is negative, there are ___ nonreal solutions.

b WRITING EQUATIONS FROM SOLUTIONS

We know by the principle of zero products that $(x - 2)(x + 3) = 0$ has solutions 2 and -3. If we know the solutions of an equation, we can write the equation, using this principle in reverse.

EXAMPLE 5 Find a quadratic equation whose solutions are 3 and $-\frac{2}{5}$.

We have

$$x = 3 \quad or \quad x = -\tfrac{2}{5}$$
$$x - 3 = 0 \quad or \quad x + \tfrac{2}{5} = 0 \qquad \text{Getting the 0's on one side}$$
$$x - 3 = 0 \quad or \quad 5x + 2 = 0 \qquad \text{Clearing the fraction}$$
$$(x - 3)(5x + 2) = 0 \qquad \text{Using the principle of zero products in reverse}$$
$$5x^2 - 13x - 6 = 0. \qquad \text{Using FOIL}$$

EXAMPLE 6 Write a quadratic equation whose solutions are $2i$ and $-2i$.

We have

$$x = 2i \quad or \quad x = -2i$$
$$x - 2i = 0 \quad or \quad x + 2i = 0 \qquad \text{Getting the 0's on one side}$$
$$(x - 2i)(x + 2i) = 0 \qquad \text{Using the principle of zero products in reverse}$$
$$x^2 - (2i)^2 = 0 \qquad \text{Using } (A - B)(A + B) = A^2 - B^2$$
$$x^2 - 4i^2 = 0$$
$$x^2 - 4(-1) = 0$$
$$x^2 + 4 = 0.$$

Answers

1. Two real **2.** One real **3.** Two nonreal

Guided Solution:

3. $-2, -2, -8, 2$

EXAMPLE 7 Write a quadratic equation whose solutions are $\sqrt{3}$ and $-2\sqrt{3}$.

We have

$$x = \sqrt{3} \quad or \quad x = -2\sqrt{3}$$

$$x - \sqrt{3} = 0 \quad or \quad x + 2\sqrt{3} = 0 \qquad \text{Getting the 0's on one side}$$

$$(x - \sqrt{3})(x + 2\sqrt{3}) = 0 \qquad \text{Using the principle of zero products}$$

$$x^2 + 2\sqrt{3}x - \sqrt{3}x - 2(\sqrt{3})^2 = 0 \qquad \text{Using FOIL}$$

$$x^2 + \sqrt{3}x - 6 = 0. \qquad \text{Collecting like terms}$$

EXAMPLE 8 Write a quadratic equation whose solutions are $-12i$ and $12i$.

We have

$$x = -12i \quad or \quad x = 12i$$

$$x + 12i = 0 \quad or \quad x - 12i = 0 \qquad \text{Getting the 0's on one side}$$

$$(x + 12i)(x - 12i) = 0 \qquad \text{Using the principle of zero products}$$

$$x^2 - 12ix + 12ix - 144i^2 = 0 \qquad \text{Using FOIL}$$

$$x^2 - 144(-1) = 0 \qquad \text{Collecting like terms; substituting } -1 \text{ for } i^2$$

$$x^2 + 144 = 0.$$

Do Exercises 4–8.

Find a quadratic equation having the following solutions.

4. 7 and -2

5. -4 and $\frac{5}{3}$

6. $5i$ and $-5i$

7. $-2\sqrt{2}$ and $\sqrt{2}$

8. $-7i$ and $7i$

c EQUATIONS QUADRATIC IN FORM

SKILL REVIEW

Solve radical equations. [6.6a]

Solve.

1. $\sqrt{x} = 10$ **2.** $\sqrt{x} = -10$

Answers: 1. 100 **2.** No solution

Certain equations that are not really quadratic can still be solved as quadratic. Consider this fourth-degree equation.

$$x^4 - 9x^2 + 8 = 0$$

$$(x^2)^2 - 9(x^2) + 8 = 0 \qquad \text{Thinking of } x^4 \text{ as } (x^2)^2$$

$$u^2 - 9u + 8 = 0 \qquad \text{To make this clearer, write } u \text{ instead of } x^2.$$

The equation $u^2 - 9u + 8 = 0$ can be solved by factoring or by the quadratic formula. After that, we can find x by remembering that $x^2 = u$. Equations that can be solved like this are said to be **quadratic in form**, or **reducible to quadratic**.

Answers

4. $x^2 - 5x - 14 = 0$
5. $3x^2 + 7x - 20 = 0$
6. $x^2 + 25 = 0$ **7.** $x^2 + \sqrt{2}x - 4 = 0$
8. $x^2 + 49 = 0$

EXAMPLE 9 Solve: $x^4 - 9x^2 + 8 = 0$.

Let $u = x^2$. Then we solve the equation found by substituting u for x^2:

$$\begin{aligned} u^2 - 9u + 8 &= 0 \\ (u - 8)(u - 1) &= 0 \quad \text{Factoring} \\ u - 8 = 0 \quad &or \quad u - 1 = 0 \quad \text{Using the principle of zero products} \\ u = 8 \quad &or \quad u = 1. \end{aligned}$$

Next, we substitute x^2 for u and solve these equations:

$$\begin{aligned} x^2 = 8 \quad &or \quad x^2 = 1 \\ x = \pm\sqrt{8} \quad &or \quad x = \pm 1 \\ x = \pm 2\sqrt{2} \quad &or \quad x = \pm 1. \end{aligned}$$

Note that when a number and its opposite are raised to an even power, the results are the same. Thus we can make one check for $\pm 2\sqrt{2}$ and one for ± 1.

Check:

For $\pm 2\sqrt{2}$:

$$\begin{array}{c|l} x^4 - 9x^2 + 8 = 0 & \\ \hline (\pm 2\sqrt{2})^4 - 9(\pm 2\sqrt{2})^2 + 8 \;?\; 0 & \\ 64 - 9 \cdot 8 + 8 & \\ 0 & \text{TRUE} \end{array}$$

For ± 1:

$$\begin{array}{c|l} x^4 - 9x^2 + 8 = 0 & \\ \hline (\pm 1)^4 - 9(\pm 1)^2 + 8 \;?\; 0 & \\ 1 - 9 + 8 & \\ 0 & \text{TRUE} \end{array}$$

The solutions are 1, -1, $2\sqrt{2}$, and $-2\sqrt{2}$.

Caution!

A common error is to solve for u and then forget to solve for x. Remember that you *must* find values for the *original* variable!

Do Exercise 9.

9. Solve: $x^4 - 10x^2 + 9 = 0$. GS

Let $u =$ ____ . Then $u^2 =$ ____ .

$$\begin{aligned} x^4 - 10x^2 + 9 &= 0 \\ u^2 - 10u + 9 &= 0 \\ (u - 1)(\ \_\_\_\_\ ) &= 0 \\ u - 1 = 0 \quad &or \quad \_\_\_\_ = 0 \\ u = 1 \quad &or \quad u = \_\_\_\_ \\ x^2 = 1 \quad &or \quad x^2 = \_\_\_\_ \\ x = \pm 1 \quad &or \quad x = \_\_\_\_ \end{aligned}$$

All four numbers check. The solutions are -1, 1, -3, and ____ .

Solving equations quadratic in form can sometimes introduce numbers that are not solutions of the original equation. Thus a check by substitution in the original equation is necessary.

EXAMPLE 10 Solve: $x - 3\sqrt{x} - 4 = 0$.

Let $u = \sqrt{x}$. Then we solve the equation found by substituting u for $\sqrt{x}$ and u^2 for x:

$$\begin{aligned} u^2 - 3u - 4 &= 0 \\ (u - 4)(u + 1) &= 0 \\ u = 4 \quad &or \quad u = -1. \end{aligned}$$

Next, we substitute $\sqrt{x}$ for u and solve these equations:

$$\sqrt{x} = 4 \quad or \quad \sqrt{x} = -1.$$

Squaring, we get $x = 16$ or $x = 1$. We check both possible solutions.

Answers

9. $\pm 3, \pm 1$

Guided Solution:

9. $x^2, x^4, u - 9, u - 9, 9, 9, \pm 3, 3$

Check:

For 16:

$$\begin{array}{r|l} x - 3\sqrt{x} - 4 = 0 & \\ \hline 16 - 3\sqrt{16} - 4 \ ? \ 0 & \\ 16 - 3 \cdot 4 - 4 & \\ 16 - 12 - 4 & \\ 0 & \text{TRUE} \end{array}$$

For 1:

$$\begin{array}{r|l} x - 3\sqrt{x} - 4 = 0 & \\ \hline 1 - 3\sqrt{1} - 4 \ ? \ 0 & \\ 1 - 3 \cdot 1 - 4 & \\ 1 - 3 - 4 & \\ -6 & \text{FALSE} \end{array}$$

Since 16 checks but 1 does not, the solution is 16. Had we noted that 1 is not a solution of $\sqrt{x} = -1$, we could have eliminated 1 at that point as a possible solution.

Do Exercise 10. ▶

10. Solve: $x + 3\sqrt{x} - 10 = 0$. Be sure to check.

EXAMPLE 11 Solve: $y^{-2} - y^{-1} - 2 = 0$.

Let $u = y^{-1}$. Then we solve the equation found by substituting u for y^{-1} and u^2 for y^{-2}:

$$u^2 - u - 2 = 0$$
$$(u - 2)(u + 1) = 0$$
$$u = 2 \quad or \quad u = -1.$$

Next, we substitute y^{-1} or $1/y$ for u and solve these equations:

$$\frac{1}{y} = 2 \quad or \quad \frac{1}{y} = -1$$
$$y = \frac{1}{2} \quad or \quad y = \frac{1}{(-1)} = -1.$$

The numbers $\frac{1}{2}$ and -1 both check. They are the solutions.

Do Exercise 11. ▶

11. Solve: $x^{-2} + x^{-1} - 6 = 0$.

EXAMPLE 12 Find the x-intercepts of the graph of

$$f(x) = (x^2 - 1)^2 - (x^2 - 1) - 2.$$

The x-intercepts occur where $f(x) = 0$, so we must have

$$(x^2 - 1)^2 - (x^2 - 1) - 2 = 0.$$

Let $u = x^2 - 1$. Then we solve the equation found by substituting u for $x^2 - 1$:

$$u^2 - u - 2 = 0$$
$$(u - 2)(u + 1) = 0$$
$$u = 2 \quad or \quad u = -1.$$

Next, we substitute $x^2 - 1$ for u and solve these equations:

$$x^2 - 1 = 2 \quad or \quad x^2 - 1 = -1$$
$$x^2 = 3 \quad or \quad x^2 = 0$$
$$x = \pm\sqrt{3} \quad or \quad x = 0.$$

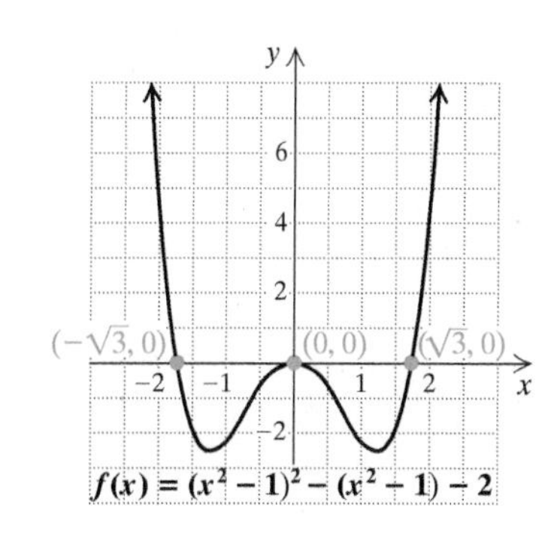

Since the numbers $\sqrt{3}, -\sqrt{3}$, and 0 check, they are the solutions of $(x^2 - 1)^2 - (x^2 - 1) - 2 = 0$. Thus the x-intercepts of the graph of $f(x)$ are $(-\sqrt{3}, 0)$, $(0, 0)$, and $(\sqrt{3}, 0)$.

Do Exercise 12. ▶

12. Find the x-intercepts of $f(x) = (x^2 - x)^2 - 14(x^2 - x) + 24$.

Answers

10. 4 **11.** $-\frac{1}{3}, \frac{1}{2}$
12. $(-3, 0), (-1, 0), (2, 0), (4, 0)$

7.4 Exercise Set

FOR EXTRA HELP MyLab Math

✔ Check Your Understanding

Reading Check Match each discriminant with the appropriate description of the solution(s) from the column on the right. Answers may be used more than once.

RC1. ____ $b^2 - 4ac = 9$

RC2. ____ $b^2 - 4ac = 0$

RC3. ____ $b^2 - 4ac = -1$

RC4. ____ $b^2 - 4ac = 1$

RC5. ____ $b^2 - 4ac = 8$

RC6. ____ $b^2 - 4ac = -12$

a) One real-number solution

b) Two different real-number solutions

c) Two different nonreal complex-number solutions

Concept Check Determine the substitution that could be used to make each equation quadratic in u.

CC1. For $3x - 4\sqrt{x} + 1 = 0$, use $u =$ ____________.

CC2. For $p^4 - p^2 - 6 = 0$, use $u =$ ____________.

CC3. For $x^{-2} + 5x^{-1} - 14 = 0$, use $u =$ ____________.

CC4. For $(2y + 7)^2 - 3(2y + 7) - 40 = 0$, use $u =$ ____________.

a Determine the nature of the solutions of each equation.

1. $x^2 - 8x + 16 = 0$

2. $x^2 + 12x + 36 = 0$

3. $x^2 + 1 = 0$

4. $x^2 + 6 = 0$

5. $x^2 - 6 = 0$

6. $x^2 - 3 = 0$

7. $4x^2 - 12x + 9 = 0$

8. $4x^2 + 8x - 5 = 0$

9. $x^2 - 2x + 4 = 0$

10. $x^2 + 3x + 4 = 0$

11. $9t^2 - 3t = 0$

12. $4m^2 + 7m = 0$

13. $y^2 = \frac{1}{2}y + \frac{3}{5}$

14. $y^2 + \frac{9}{4} = 4y$

15. $4x^2 - 4\sqrt{3}x + 3 = 0$

16. $6y^2 - 2\sqrt{3}y - 1 = 0$

b Write a quadratic equation having the given numbers as solutions.

17. -4 and 4

18. -11 and 9

19. $-4i$ and $4i$

20. $-i$ and i

21. 8, only solution
[*Hint*: It must be a double solution, that is, $(x - 8)(x - 8) = 0$.]

22. -3, only solution

23. $-\frac{2}{5}$ and $\frac{6}{5}$

24. $-\frac{1}{4}$ and $-\frac{1}{2}$

25. $\frac{k}{3}$ and $\frac{m}{4}$

26. $\frac{c}{2}$ and $\frac{d}{2}$

27. $-\sqrt{3}$ and $2\sqrt{3}$

28. $\sqrt{2}$ and $3\sqrt{2}$

29. $6i$ and $-6i$

30. $8i$ and $-8i$

C Solve.

31. $x^4 - 6x^2 + 9 = 0$

32. $x^4 - 7x^2 + 12 = 0$

33. $x - 10\sqrt{x} + 9 = 0$

34. $2x - 9\sqrt{x} + 4 = 0$

35. $(x^2 - 6x)^2 - 2(x^2 - 6x) - 35 = 0$

36. $(x^2 + 5x)^2 + 2(x^2 + 5x) - 24 = 0$

37. $x^{-2} - 5x^{-1} - 36 = 0$

38. $3x^{-2} - x^{-1} - 14 = 0$

39. $(1 + \sqrt{x})^2 + (1 + \sqrt{x}) - 6 = 0$

40. $(2 + \sqrt{x})^2 - 3(2 + \sqrt{x}) - 10 = 0$

41. $(y^2 - 5y)^2 - 2(y^2 - 5y) - 24 = 0$

42. $(2t^2 + t)^2 - 4(2t^2 + t) + 3 = 0$

43. $w^4 - 29w^2 + 100 = 0$

44. $t^4 - 10t^2 + 9 = 0$

45. $2x^{-2} + x^{-1} - 1 = 0$

46. $m^{-2} + 9m^{-1} - 10 = 0$

47. $6x^4 - 19x^2 + 15 = 0$

48. $6x^4 - 17x^2 + 5 = 0$

49. $x^{2/3} - 4x^{1/3} - 5 = 0$

50. $x^{2/3} + 2x^{1/3} - 8 = 0$

51. $\left(\frac{x-4}{x+1}\right)^2 - 2\left(\frac{x-4}{x+1}\right) - 35 = 0$

52. $\left(\frac{x+3}{x-3}\right)^2 - \left(\frac{x+3}{x-3}\right) - 6 = 0$

53. $9\left(\frac{x+2}{x+3}\right)^2 - 6\left(\frac{x+2}{x+3}\right) + 1 = 0$

54. $16\left(\frac{x-1}{x-8}\right)^2 + 8\left(\frac{x-1}{x-8}\right) + 1 = 0$

55. $\left(\frac{x^2-2}{x}\right)^2 - 7\left(\frac{x^2-2}{x}\right) - 18 = 0$

56. $\left(\frac{y^2-1}{y}\right)^2 - 4\left(\frac{y^2-1}{y}\right) - 12 = 0$

Find the x-intercepts of the graph of each function.

57. $f(x) = 5x + 13\sqrt{x} - 6$

58. $f(x) = 3x + 10\sqrt{x} - 8$

59. $f(x) = (x^2 - 3x)^2 - 10(x^2 - 3x) + 24$

60. $f(x) = (x^2 - x)^2 - 8(x^2 - x) + 12$

61. $f(x) = x^{2/3} + x^{1/3} - 2$

62. $f(x) = x^{2/5} + x^{1/5} - 6$

Skill Maintenance

Solve. [3.4a]

63. *Coffee Beans.* Twin Cities Roasters sells Kenyan coffee worth \$9.75 per pound and Peruvian coffee worth \$13.25 per pound. How many pounds of each kind should be mixed in order to obtain a 50-lb mixture that is worth \$11.15 per pound?

64. *Solution Mixtures.* Solution A is 18% alcohol and solution B is 45% alcohol. How many liters of each should be mixed in order to get 12 L of a solution that is 36% alcohol?

Multiply and simplify. Assume that no radicands were formed by raising negative numbers to even powers. [6.3a]

65. $\sqrt{8x}\,\sqrt{2x}$

66. $\sqrt[3]{x^2}\,\sqrt[3]{27x^4}$

67. $\sqrt[4]{9a^2}\,\sqrt[4]{18a^3}$

68. $\sqrt[5]{16}\,\sqrt[5]{64}$

Graph. [2.2c], [2.5a, c]

69. $f(x) = -\frac{3}{5}x + 4$

70. $5x - 2y = 8$

71. $y = 4$

72. $f(x) = -x - 3$

Synthesis

73. Use a graphing calculator to check your answers to Exercises 32, 34, 36, and 39.

74. Use a graphing calculator to solve each of the following equations.

a) $6.75x - 35\sqrt{x} - 5.26 = 0$
b) $\pi x^4 - \pi^2 x^2 = \sqrt{99.3}$
c) $x^4 - x^3 - 13x^2 + x + 12 = 0$

For each equation under the given condition, **(a)** find k and **(b)** find the other solution.

75. $kx^2 - 2x + k = 0$; one solution is -3.

76. $kx^2 - 17x + 33 = 0$; one solution is 3.

77. Find a quadratic equation for which the sum of the solutions is $\sqrt{3}$ and the product is 8.

78. Find k given that $kx^2 - 4x + (2k - 1) = 0$ and the product of the solutions is 3.

79. The graph of a function of the form

$$f(x) = ax^2 + bx + c$$

is a curve similar to the one shown below. Determine a, b, and c from the information given.

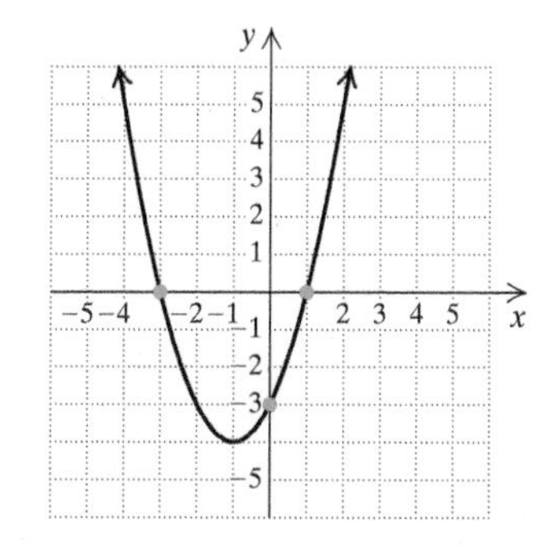

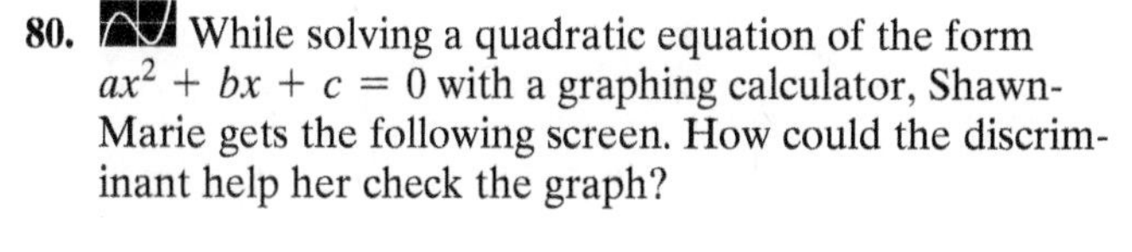

80. While solving a quadratic equation of the form $ax^2 + bx + c = 0$ with a graphing calculator, Shawn-Marie gets the following screen. How could the discriminant help her check the graph?

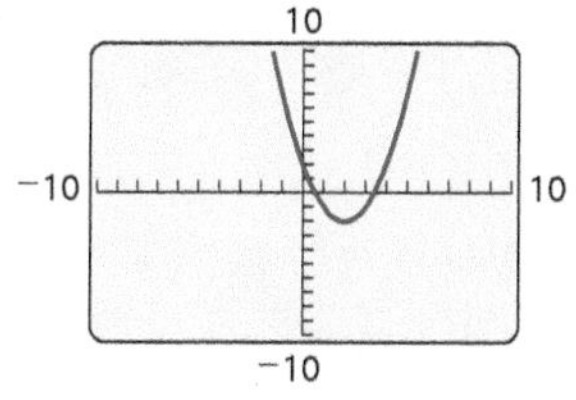

Solve.

81. $\dfrac{x}{x-1} - 6\sqrt{\dfrac{x}{x-1}} - 40 = 0$

82. $\dfrac{x}{x-3} - 24 = 10\sqrt{\dfrac{x}{x-3}}$

83. $\sqrt{x-3} - \sqrt[4]{x-3} = 12$

84. $a^3 - 26a^{3/2} - 27 = 0$

85. $x^6 - 28x^3 + 27 = 0$

86. $x^6 + 7x^3 - 8 = 0$

Mid-Chapter Review

Concept Reinforcement

Determine whether each statement is true or false.

________ **1.** Every quadratic equation has exactly two real-number solutions. [7.4a]

________ **2.** The quadratic formula can be used to find all the solutions of any quadratic equation. [7.2a]

________ **3.** If the graph of a quadratic equation crosses the x-axis, then it has exactly two real-number solutions. [7.4a]

________ **4.** The x-intercepts of $f(x) = x^2 - t$ are $(0, \sqrt{t})$ and $(0, -\sqrt{t})$. [7.1a]

Guided Solutions

GS Fill in each blank with the number that creates a correct solution.

5. Solve $5x^2 + 3x = 4$ by completing the square. [7.1b]

$$5x^2 + 3x = 4$$

$$\square(5x^2 + 3x) = \square \cdot 4$$

$$x^2 + \frac{3}{\square}x = \frac{4}{\square}$$

$$x^2 + \frac{3}{5}x + \square = \frac{4}{5} + \square$$

$$(x + \square)^2 = \frac{\square}{100}$$

$$x + \frac{3}{10} = \sqrt{\square} \quad or \quad x + \frac{3}{10} = -\sqrt{\square}$$

$$x + \frac{3}{10} = \frac{\sqrt{\square}}{10} \quad or \quad x + \frac{3}{10} = -\frac{\sqrt{\square}}{10}$$

$$x = -\frac{\square}{10} + \frac{\sqrt{\square}}{10} \quad or \quad x = -\frac{\square}{10} - \frac{\sqrt{\square}}{10}.$$

The solutions are $-\frac{\square}{10} \pm \frac{\sqrt{\square}}{10}$.

6. Use the quadratic formula to solve $5x^2 + 3x = 4$. [7.2a]

$$5x^2 + 3x = 4$$

$$5x^2 + 3x - \square = 0$$

$$a = \square, \quad b = \square, \quad c = \square$$

$$x = \frac{-b \pm \sqrt{b^2 - 4ac}}{2a}$$

$$x = \frac{-\square \pm \sqrt{\square^2 - 4 \cdot \square \cdot \square}}{2 \cdot \square}$$

$$x = \frac{-3 \pm \sqrt{9 + \square}}{\square}$$

$$x = \frac{-3 \pm \sqrt{\square}}{10}$$

$$x = -\frac{3}{10} \pm \frac{\sqrt{\square}}{10}$$

Mixed Review

Solve by completing the square. [7.1b]

7. $x^2 + 1 = -4x$

8. $2x^2 + 5x - 3 = 0$

9. $x^2 + 10x - 6 = 0$

10. $x^2 - x = 5$

Determine the nature of the solutions of each equation $ax^2 + bx + c = 0$ and the number of x-intercepts of the graph of the function $f(x) = ax^2 + bx + c$. [7.4a]

11. $x^2 - 10x + 25 = 0$

12. $x^2 - 11 = 0$

13. $y^2 = \frac{1}{3}y - \frac{4}{7}$

14. $x^2 + 5x + 9 = 0$

15. $x^2 - 4 = 2x$

16. $x^2 - 8x = 0$

Write a quadratic equation having the given numbers as solutions. [7.4b]

17. -1 and 10

18. -13 and 13

19. $-\sqrt{5}$ and $3\sqrt{5}$

20. $-4i$ and $4i$

21. -6, only solution

22. $-\frac{4}{3}$ and $\frac{2}{7}$

Solve.

23. Jacob traveled 780 mi by car. Had he gone 5 mph faster, he could have made the trip in 1 hr less time. Find his speed. [7.3a]

24. $R = as^2$, for s [7.3b]

Solve. [7.1a], [7.2a], [7.4c]

25. $3x^2 + x = 4$

26. $x^4 - 8x^2 + 15 = 0$

27. $4x^2 = 15x - 5$

28. $7x^2 + 2 = -9x$

29. $2x + x(x - 1) = 0$

30. $(x + 3)^2 = 64$

31. $49x^2 + 16 = 0$

32. $(x^2 - 2)^2 + 2(x^2 - 2) - 24 = 0$

33. $r^2 + 5r = 12$

34. $s^2 + 12s + 37 = 0$

35. $\left(x - \frac{5}{2}\right)^2 = \frac{11}{4}$

36. $x + \frac{1}{x} = \frac{7}{3}$

37. $4x + 1 = 4x^2$

38. $(x - 3)^2 + (x + 5)^2 = 0$

39. $b^2 - 16b + 64 = 3$

40. $(x - 3)^2 = -10$

41. $\frac{1}{x} + \frac{1}{x + 2} = \frac{1}{5}$

42. $x - \sqrt{x} - 6 = 0$

Understanding Through Discussion and Writing

43. Given the solutions of a quadratic equation, is it possible to reconstruct the original equation? Why or why not? [7.4b]

44. Explain how the quadratic formula can be used to factor a quadratic polynomial into two binomials. Use it to factor $5x^2 + 8x - 3$. [7.2a]

45. Describe a procedure that could be used to write an equation having the first seven natural numbers as solutions. [7.4b]

46. Describe a procedure that could be used to write an equation that is quadratic in $3x^2 + 1$ and has real-number solutions. [7.4c]

STUDYING FOR SUCCESS *Working with Others*

- ☐ Try being a tutor for a fellow student. You may find that you understand some concepts better after you have explained them to someone else.
- ☐ Consider forming a study group.
- ☐ Verbalize the math. Often simply talking about a concept with a classmate can help clarify the material.

7.5 Graphing $f(x) = a(x - h)^2 + k$

OBJECTIVES

a Graph quadratic functions of the type $f(x) = ax^2$ and then label the vertex and the line of symmetry.

b Graph quadratic functions of the type $f(x) = a(x - h)^2$ and then label the vertex and the line of symmetry.

c Graph quadratic functions of the type $f(x) = a(x - h)^2 + k$, finding the vertex, the line of symmetry, and the maximum or minimum function value, or y-value.

a GRAPHS OF $f(x) = ax^2$

The most basic quadratic function is $f(x) = x^2$.

EXAMPLE 1 Graph: $f(x) = x^2$.

We choose some values for x and compute $f(x)$ for each. Then we plot the ordered pairs and connect them with a smooth curve.

x	$f(x) = x^2$	$(x, f(x))$
-3	9	$(-3, 9)$
-2	4	$(-2, 4)$
-1	1	$(-1, 1)$
0	0	$(0, 0)$
1	1	$(1, 1)$
2	4	$(2, 4)$
3	9	$(3, 9)$

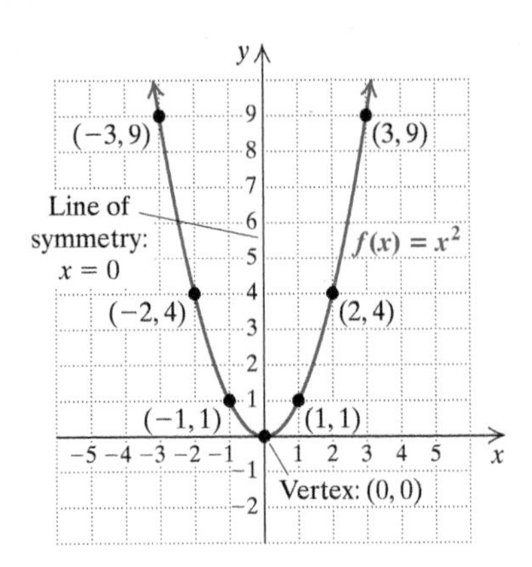

Let's compare the graphs of $g(x) = \frac{1}{2}x^2$ and $h(x) = 2x^2$ with the graph of $f(x) = x^2$. We choose x-values and plot points for both functions.

x	$g(x) = \frac{1}{2}x^2$
-3	$\frac{9}{2}$
-2	2
-1	$\frac{1}{2}$
0	0
1	$\frac{1}{2}$
2	2
3	$\frac{9}{2}$

x	$h(x) = 2x^2$
-3	18
-2	8
-1	2
0	0
1	2
2	8
3	18

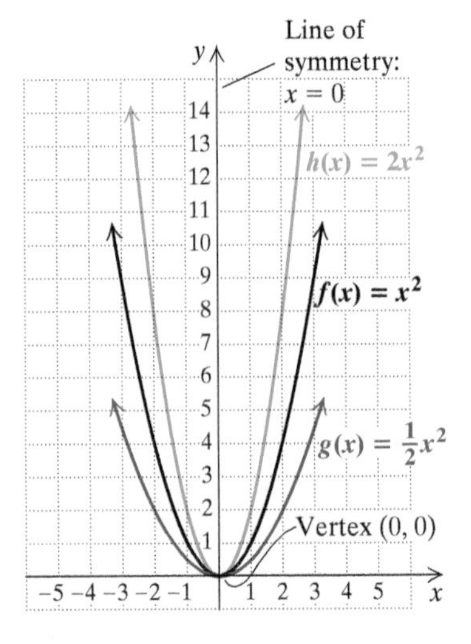

Note the symmetry: For equal increments to the left and right of the vertex, the y-values are the same.

Graphs of quadratic functions are called **parabolas**. They are cup-shaped curves that are symmetric with respect to a vertical line known as the parabola's **line of symmetry**, or **axis of symmetry**. In the graphs shown on the preceding page, the y-axis (or the line $x = 0$) is the line of symmetry. If the paper were to be folded on this line, the two halves of the curve would coincide. The point $(0, 0)$ is the **vertex** of each of the parabolas shown on the preceding page.

Note that the graph of $g(x) = \frac{1}{2}x^2$ is a wider parabola than the graph of $f(x) = x^2$, and the graph of $h(x) = 2x^2$ is narrower.

When we consider the graph of $k(x) = -\frac{1}{2}x^2$, we see that the parabola opens down and has the same shape as the graph of $g(x) = \frac{1}{2}x^2$.

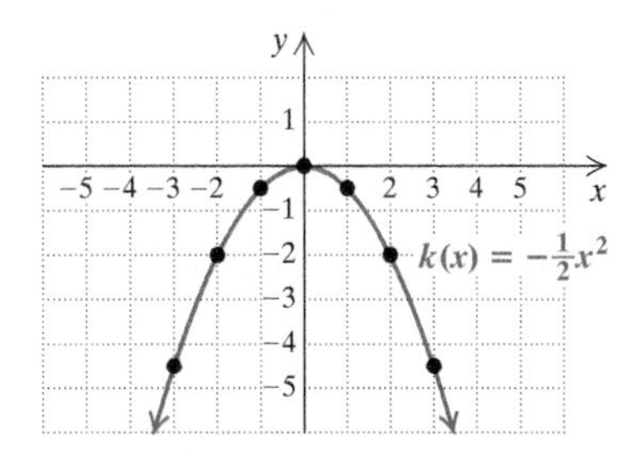

GRAPHS OF $f(x) = ax^2$

The graph of $f(x) = ax^2$, or $y = ax^2$, is a parabola with $x = 0$ as its line of symmetry; its vertex is the origin.

For $a > 0$, the parabola opens up; for $a < 0$, the parabola opens down.

If $|a|$ is greater than 1, the parabola is narrower than $y = x^2$.

If $|a|$ is between 0 and 1, the parabola is wider than $y = x^2$.

Do Exercises 1–3.

b GRAPHS OF $f(x) = a(x - h)^2$

EXAMPLE 2 Graph: $g(x) = (x - 3)^2$.

We choose some values for x and compute $g(x)$. Then we plot the points and draw the curve.

x	$g(x) = (x - 3)^2$
3	0
4	1
5	4
6	9
2	1
1	4
0	9

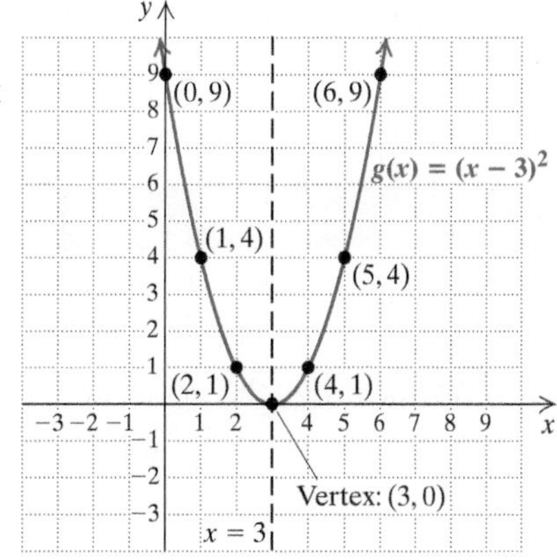

For an x-value of 3, $g(3) = (3 - 3)^2 = 0$. As we increase x-values from 3, the corresponding y-values increase. Then as we decrease x-values from 3, the corresponding y-values increase again. The line $x = 3$ is the line of symmetry.

Graph.

1. $f(x) = -\frac{1}{3}x^2$

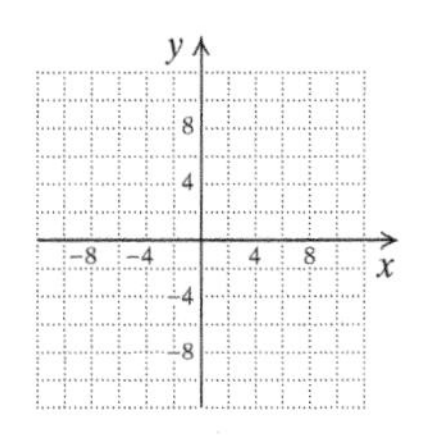

2. $f(x) = 3x^2$

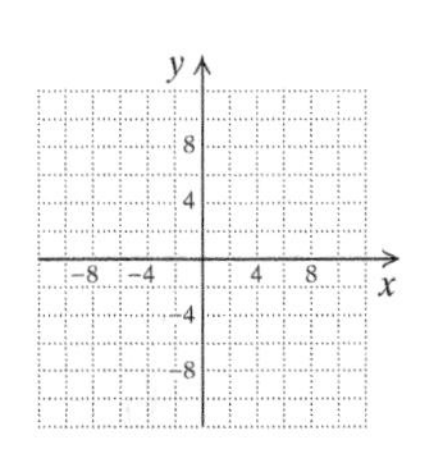

3. $f(x) = -2x^2$

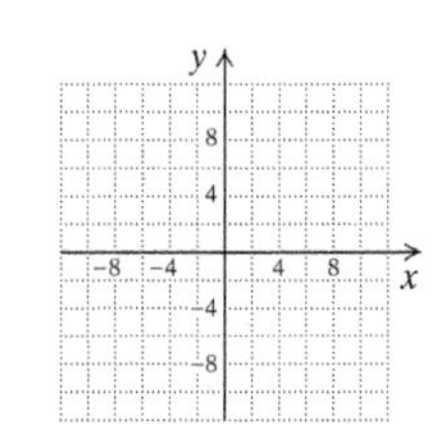

Answers

1.

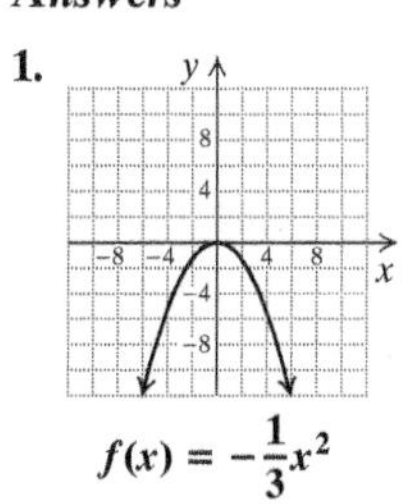

$f(x) = -\frac{1}{3}x^2$

2.

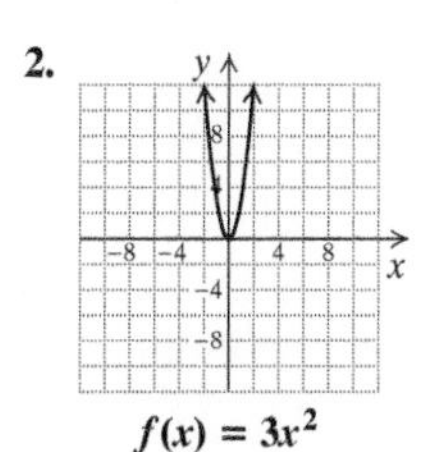

$f(x) = 3x^2$

3.

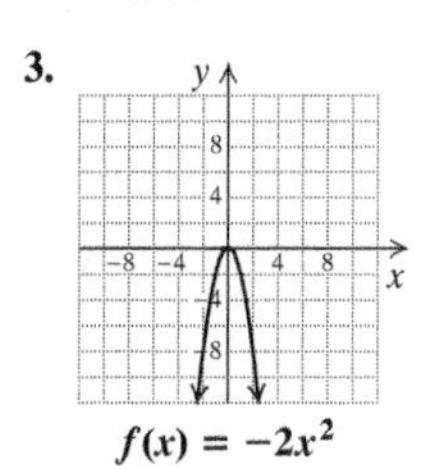

$f(x) = -2x^2$

4. Graph: $g(x) = (x + 2)^2$.

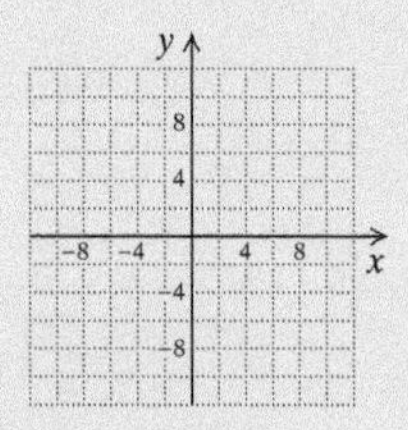

Compute $g(x)$ for each value of x shown.

x	$g(x) = (x + 2)^2$
−2	☐
−1	1
0	4
1	☐
−3	1
−4	4
−5	☐

Plot the points and draw the curve.

EXAMPLE 3 Graph: $t(x) = (x + 3)^2$.

We choose some values for x and compute $t(x)$. Then we plot the points and draw the curve.

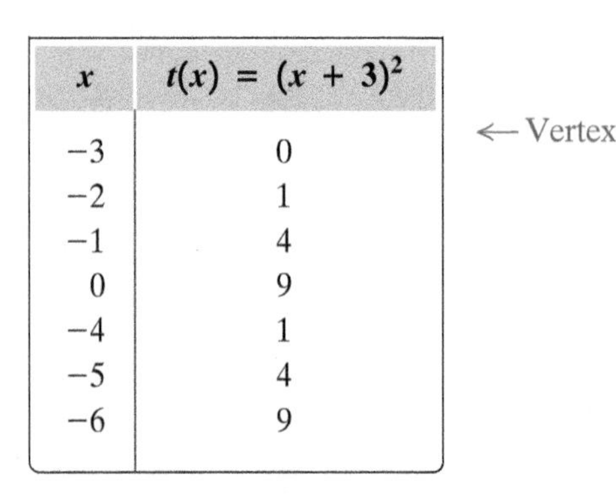

x	$t(x) = (x + 3)^2$
−3	0 ← Vertex
−2	1
−1	4
0	9
−4	1
−5	4
−6	9

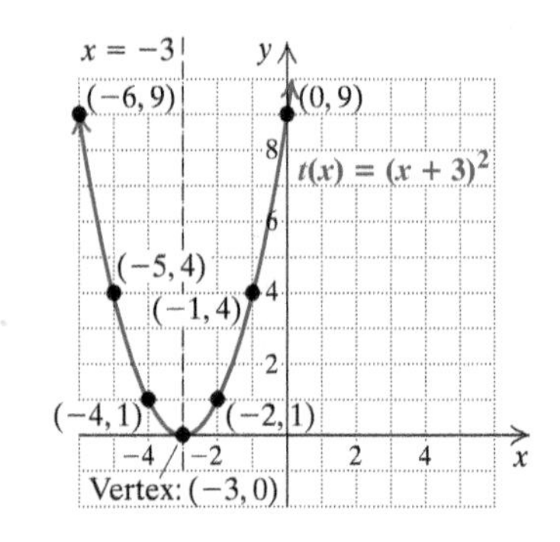

For an x-value of -3, $t(-3) = (-3 + 3)^2 = 0$. As we increase x-values from -3, the corresponding y-values increase. Then as we decrease x-values from -3, the y-values increase again. The line $x = -3$ is the line of symmetry.

◀ Do Exercise 4.

The graph of $g(x) = (x - 3)^2$ in Example 2 looks just like the graph of $f(x) = x^2$ in Example 1, except that it is moved, or translated, 3 units to the right. Comparing the pairs for $g(x)$ with those for $f(x)$, we see that when an input for $g(x)$ is 3 more than an input for $f(x)$, the outputs are the same.

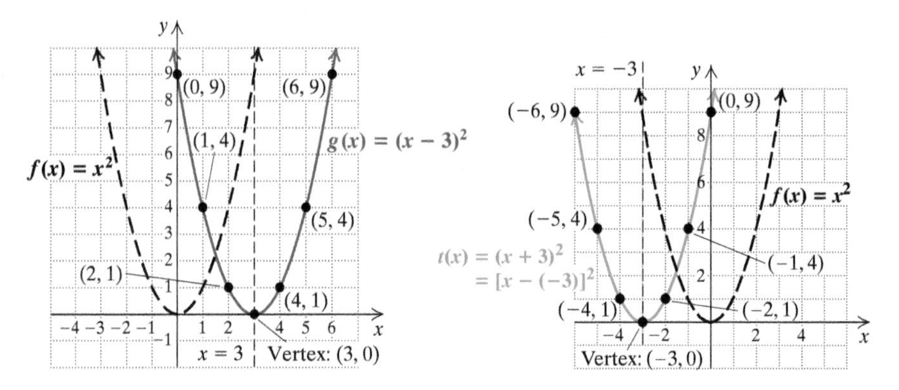

The graph of $t(x) = (x + 3)^2 = [x - (-3)]^2$ in Example 3 looks just like the graph of $f(x) = x^2$ in Example 1, except that it is moved, or translated, 3 units to the left. Comparing the pairs for $t(x)$ with those for $f(x)$, we see that when an input for $t(x)$ is 3 less than an input for $f(x)$, the outputs are the same.

GRAPHS OF $f(x) = a(x - h)^2$

The graph of $f(x) = a(x - h)^2$ has the same shape as the graph of $y = ax^2$.

If h is positive, the graph of $y = ax^2$ is shifted h units to the right.

If h is negative, the graph of $y = ax^2$ is shifted $|h|$ units to the left.

The vertex is $(h, 0)$, and the line of symmetry is $x = h$.

Answer

4.

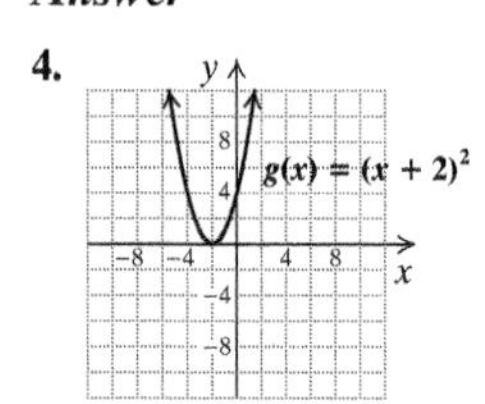

Guided Solution:
4. 0, 9, 9

EXAMPLE 4 Graph: $f(x) = -2(x + 3)^2$.

We first rewrite the equation as $f(x) = -2[x - (-3)]^2$. In this case, $a = -2$ and $h = -3$, so the graph looks like that of $g(x) = 2x^2$ translated 3 units to the left and, since $-2 < 0$, the graph opens down. The vertex is $(-3, 0)$, and the line of symmetry is $x = -3$.

x	$f(x) = -2(x + 3)^2$
-3	0 ← Vertex
-2	-2
-1	-8
-4	-2
-5	-8

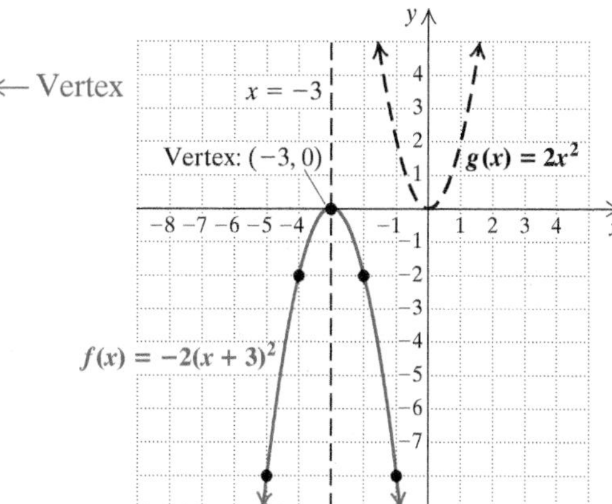

Do Exercises 5 and 6. ▶

Graph. Find and label the vertex and the line of symmetry.

5. $f(x) = \frac{1}{2}(x - 4)^2$

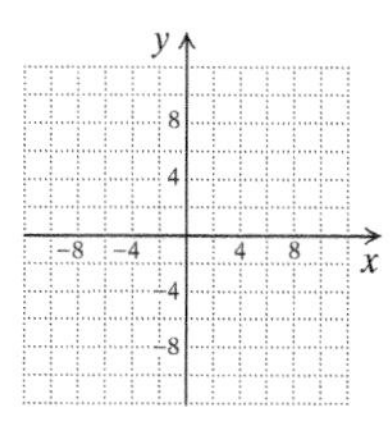

6. $f(x) = -\frac{1}{2}(x - 4)^2$

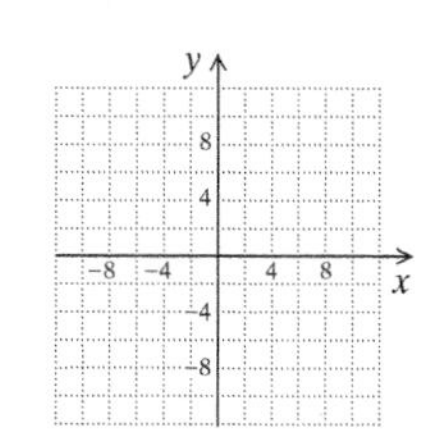

c GRAPHS OF $f(x) = a(x - h)^2 + k$

MyLab Math ANIMATION

Given a graph of $f(x) = a(x - h)^2$, if we add a positive constant k, each function value $f(x)$ is increased by k, so the curve is moved up. If k is negative, the curve is moved down. The line of symmetry for the parabola remains $x = h$, but the vertex will be at (h, k).

Note that if a parabola opens up $(a > 0)$, the function value, or y-value, at the vertex is a least, or **minimum**, value. That is, it is less than the y-value at any other point on the graph. If the parabola opens down $(a < 0)$, the function value at the vertex is a greatest, or **maximum**, value.

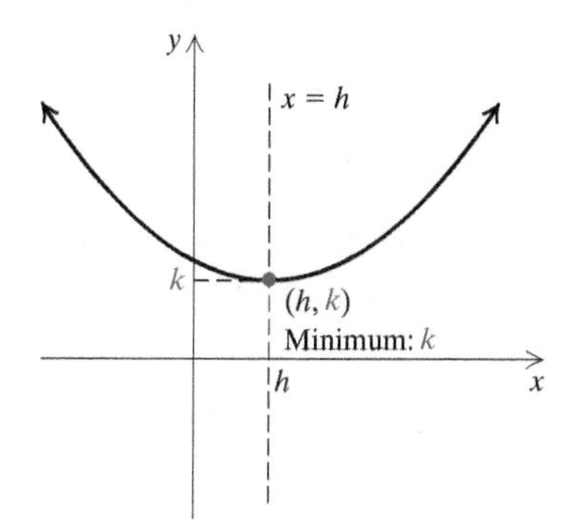

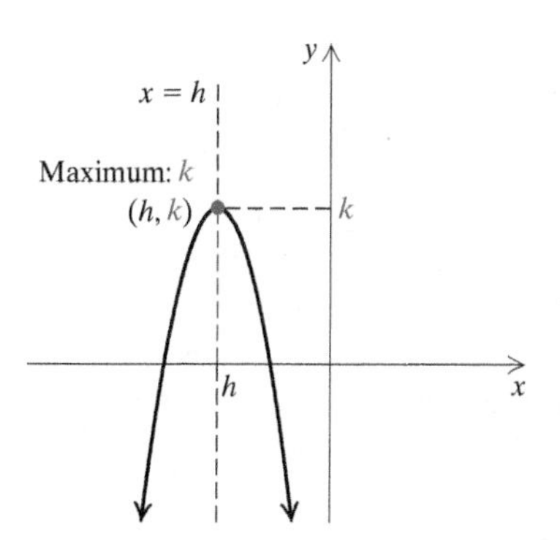

GRAPHS OF $f(x) = a(x - h)^2 + k$

The graph of $f(x) = a(x - h)^2 + k$ has the same shape as the graph of $y = a(x - h)^2$.

If k is positive, the graph of $y = a(x - h)^2$ is shifted k units up.

If k is negative, the graph of $y = a(x - h)^2$ is shifted $|k|$ units down.

The vertex is (h, k), and the line of symmetry is $x = h$.

For $a > 0$, k is the minimum function value. For $a < 0$, k is the maximum function value.

Answers

5.

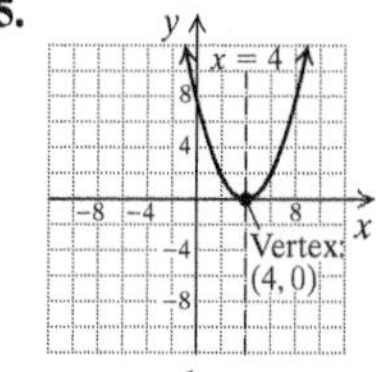

$f(x) = \frac{1}{2}(x - 4)^2$

6.

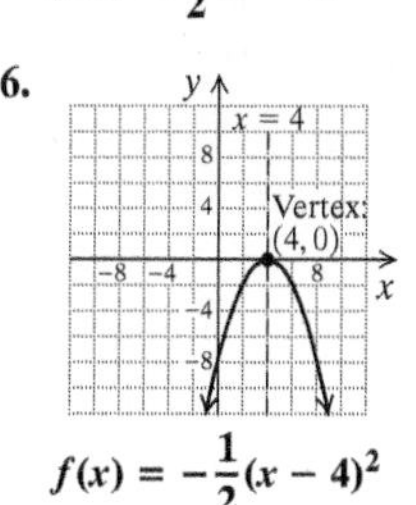

$f(x) = -\frac{1}{2}(x - 4)^2$

7. Graph $g(x) = (x - 1)^2 - 3$, and find the minimum function value.

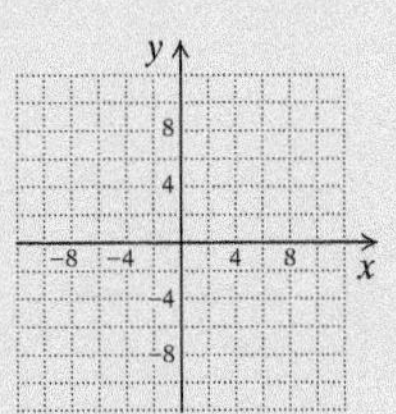

Compute $g(x)$ for each value of x shown.

x	$g(x) = (x - 1)^2 - 3$
1	-3
2	☐
3	1
4	6
0	☐
-1	1
-2	6

The vertex is $(1, -3)$. The line of symmetry is $x =$ ____, and the minimum function value is ____.

Plot the points and draw the curve.

EXAMPLE 5 Graph $f(x) = (x - 3)^2 - 5$, and find the minimum function value.

The graph will look like that of $g(x) = (x - 3)^2$ (see Example 2) but translated 5 units down. You can confirm this by plotting some points. For instance,

$$f(4) = (4 - 3)^2 - 5 = -4,$$

whereas in Example 2,

$$g(4) = (4 - 3)^2 = 1.$$

Note that the vertex is $(h, k) = (3, -5)$, so we begin calculating points on both sides of $x = 3$. The line of symmetry is $x = 3$, and the minimum function value is -5.

x	$f(x) = (x - 3)^2 - 5$	
3	-5	← Vertex
4	-4	
5	-1	
6	4	
2	-4	
1	-1	
0	4	

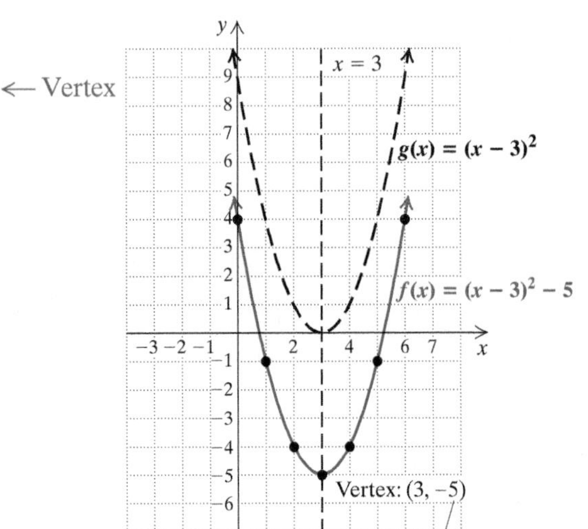

Do Exercise 7.

EXAMPLE 6 Graph $t(x) = \frac{1}{2}(x - 3)^2 + 5$, and find the minimum function value.

The graph looks just like that of $f(x) = \frac{1}{2}x^2$ but moved 3 units to the right and 5 units up. The vertex is $(3, 5)$, and the line of symmetry is $x = 3$. We draw $f(x) = \frac{1}{2}x^2$ and then shift the curve over and up. The minimum function value is 5. By plotting some points, we have a check.

x	$t(x) = \frac{1}{2}(x - 3)^2 + 5$	
3	5	← Vertex
4	$5\frac{1}{2}$	
5	7	
6	$9\frac{1}{2}$	
2	$5\frac{1}{2}$	
1	7	
0	$9\frac{1}{2}$	

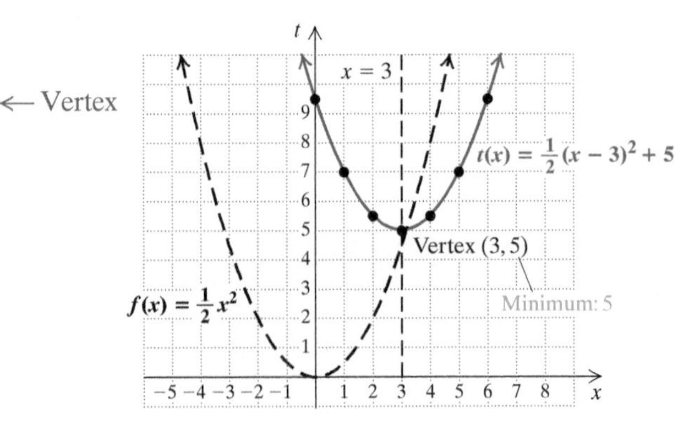

Answer

7.

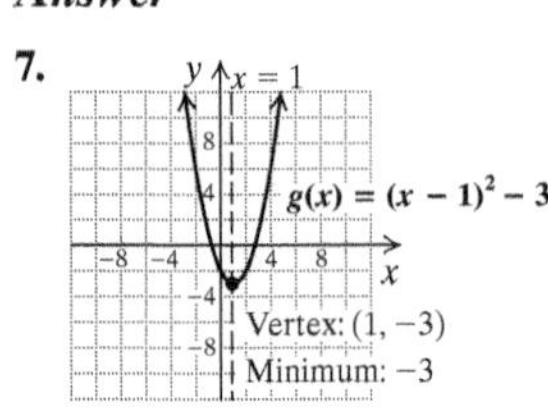

Guided Solution:

7. $-2, -2, 1, -3$

EXAMPLE 7 Graph $f(x) = -2(x + 3)^2 + 5$. Find the vertex, the line of symmetry, and the maximum or minimum value.

We first express the equation in the equivalent form

$$f(x) = -2[x - (-3)]^2 + 5.$$

The graph looks like that of $g(x) = -2x^2$ translated 3 units to the left and 5 units up. The vertex is $(-3, 5)$, and the line of symmetry is $x = -3$. Since $-2 < 0$, we know that the graph opens down so 5, the second coordinate of the vertex, is the maximum y-value. We compute points and draw the graph.

x	$f(x) = -2(x + 3)^2 + 5$	
-3	5	← Vertex
-2	3	
-1	-3	
-4	3	
-5	-3	

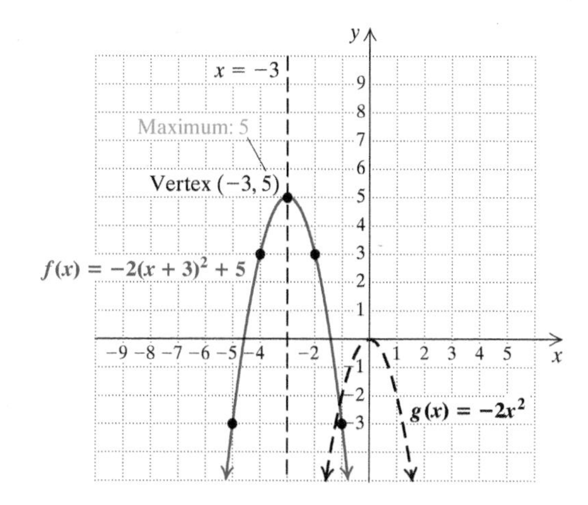

Do Exercises 8 and 9. ▶

Graph. Find the vertex, the line of symmetry, and the maximum or minimum y-value.

8. $f(x) = \frac{1}{2}(x + 2)^2 - 4$

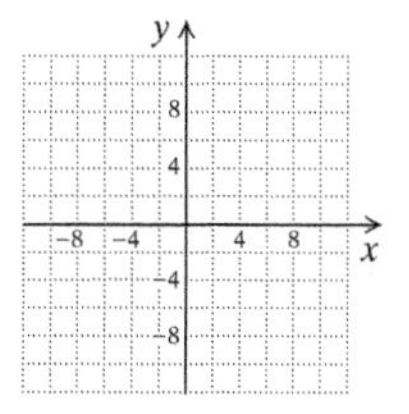

9. $f(x) = -2(x - 5)^2 + 3$

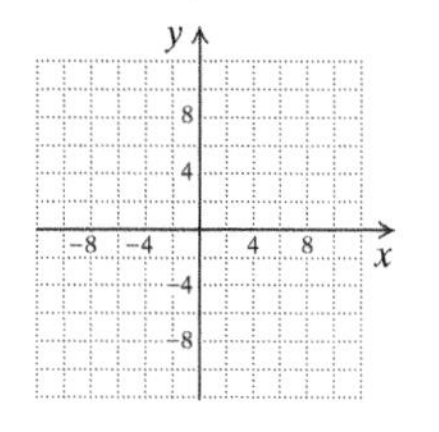

Answers

8.

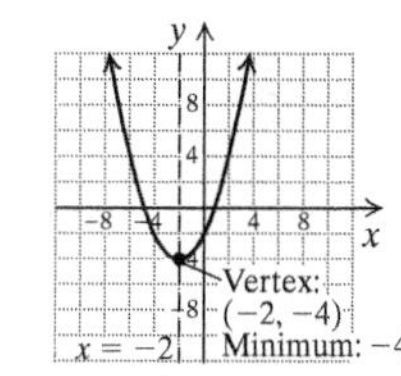

$f(x) = \frac{1}{2}(x + 2)^2 - 4$

9.

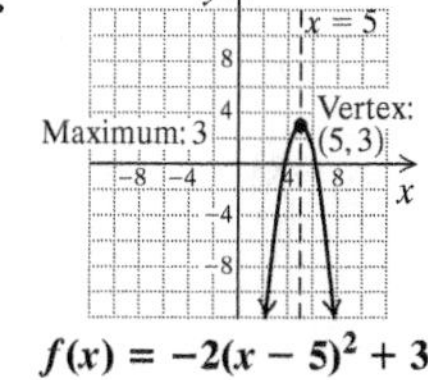

$f(x) = -2(x - 5)^2 + 3$

7.5 Exercise Set

FOR EXTRA HELP

MyLab Math

✔ Check Your Understanding

Reading Check Determine whether each statement is true or false.

RC1. The graph of a quadratic function may be a straight line or a parabola.

RC2. The graph of every quadratic function is symmetric with respect to a vertical line.

RC3. Every quadratic function has either a maximum value or a minimum value.

RC4. The graph of $f(x) = 5x^2$ is wider than the graph of $f(x) = 3x^2$.

Concept Check Determine whether the graph of each function opens up or down.

CC1. $f(x) = -6x^2$

CC2. $g(x) = 8(x - 5)^2$

CC3. $f(x) = \frac{1}{3}(x - 2)^2 - 9$

CC4. $g(x) = -\frac{1}{2}(x + 11)^2 + 24$

a, **b** Graph. Find and label the vertex and the line of symmetry.

1. $f(x) = 4x^2$

x	$f(x)$
0	
1	
2	
−1	
−2	

Vertex: (____, ____)
Line of symmetry: $x =$ ____

2. $f(x) = 5x^2$

x	$f(x)$
0	
1	
2	
−1	
−2	

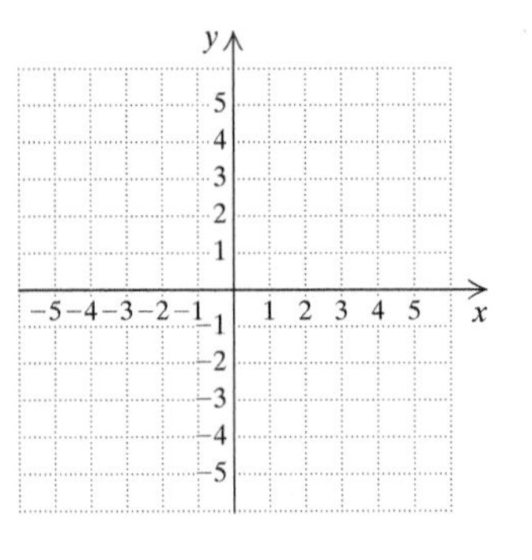

Vertex: (____, ____)
Line of symmetry: $x =$ ____

3. $f(x) = \frac{1}{3}x^2$

x	$f(x)$
0	
1	
2	
−1	
−2	

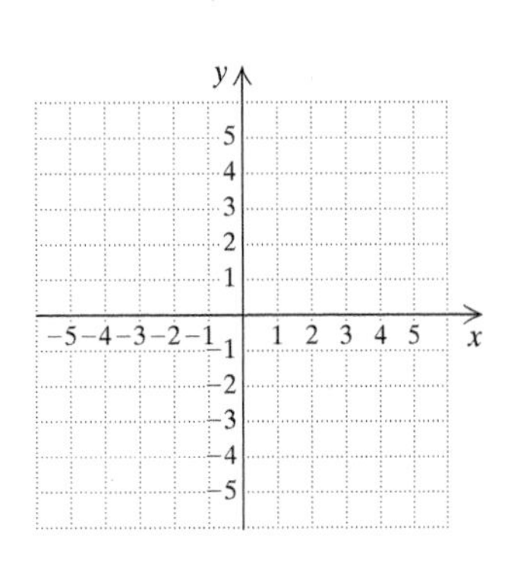

Vertex: (____, ____)
Line of symmetry: $x =$ ____

4. $f(x) = \frac{1}{4}x^2$

x	$f(x)$
0	
1	
2	
−1	
−2	

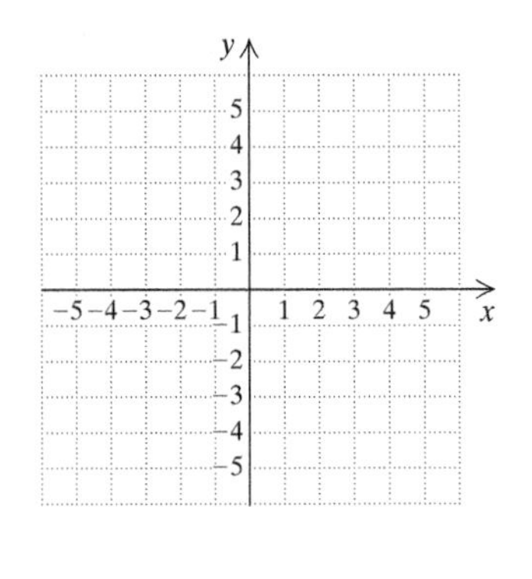

Vertex: (____, ____)
Line of symmetry: $x =$ ____

5. $f(x) = (x + 3)^2$

x	$f(x)$
−3	
−2	
−1	
−4	
−5	

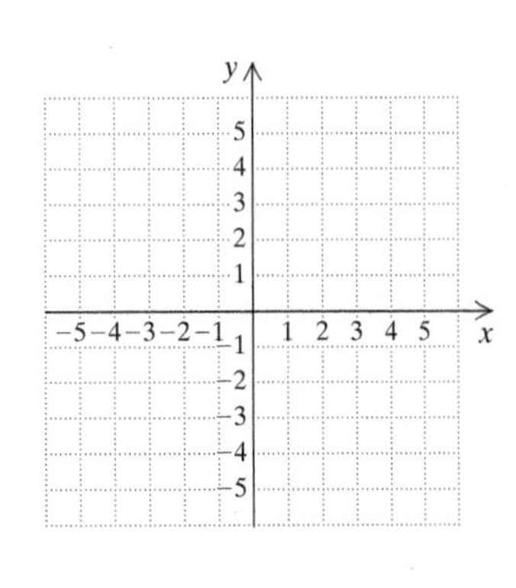

Vertex: (____, ____)
Line of symmetry: $x =$ ____

6. $f(x) = (x + 1)^2$

x	$f(x)$
−1	
0	
1	
−2	
−3	

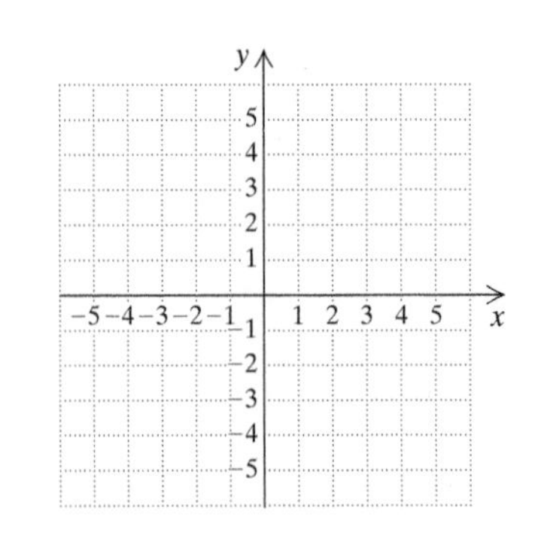

Vertex: (____, ____)
Line of symmetry: $x =$ ____

7. $f(x) = -4x^2$

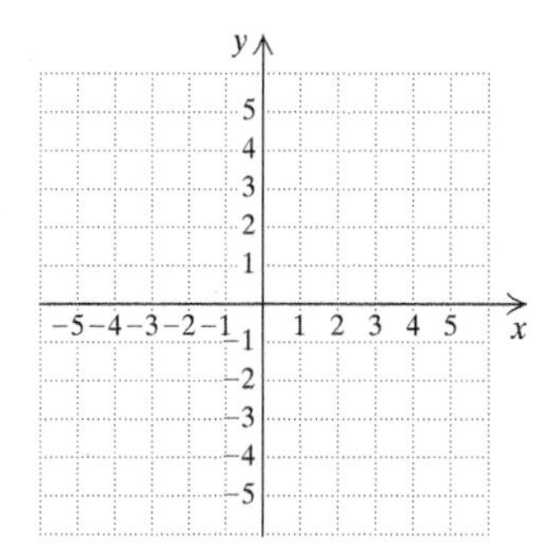

Vertex: (____, ____)
Line of symmetry: $x =$ ____

8. $f(x) = -3x^2$

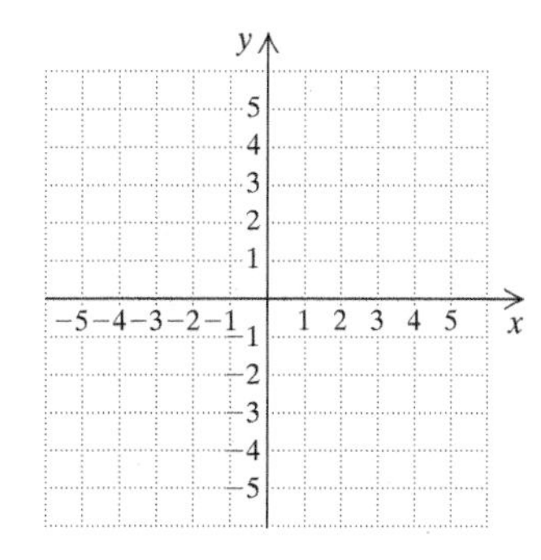

Vertex: (____, ____)
Line of symmetry: $x =$ ____

9. $f(x) = -\frac{1}{2}x^2$

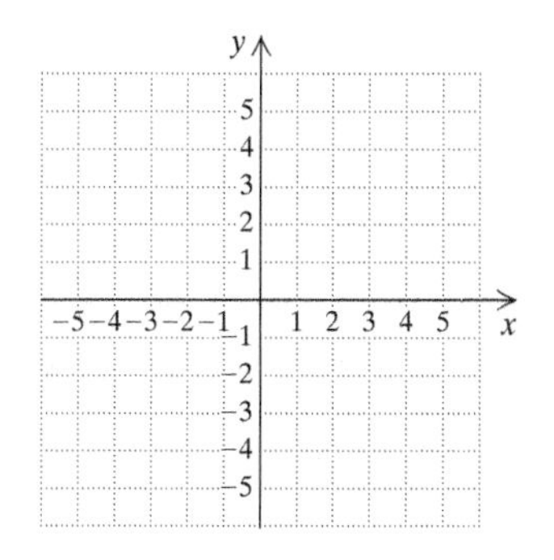

Vertex: (____, ____)
Line of symmetry: $x =$ ____

10. $f(x) = -\frac{1}{4}x^2$

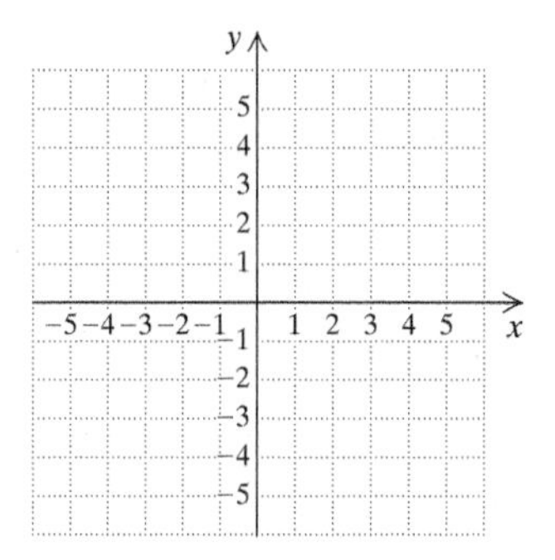

Vertex: (____, ____)
Line of symmetry: $x =$ ____

11. $f(x) = 2(x - 4)^2$

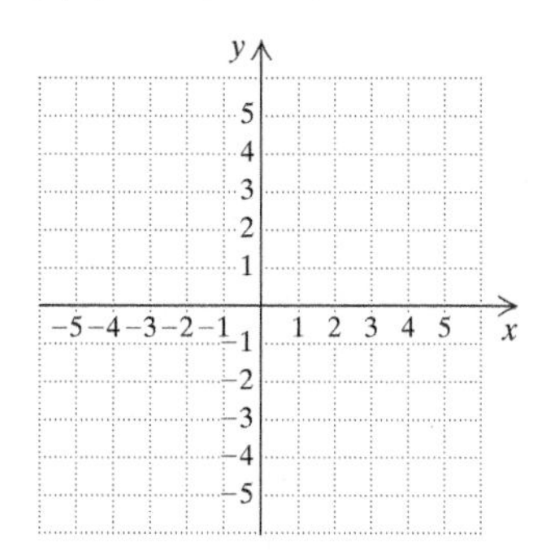

Vertex: (____, ____)
Line of symmetry: $x =$ ____

12. $f(x) = 4(x - 1)^2$

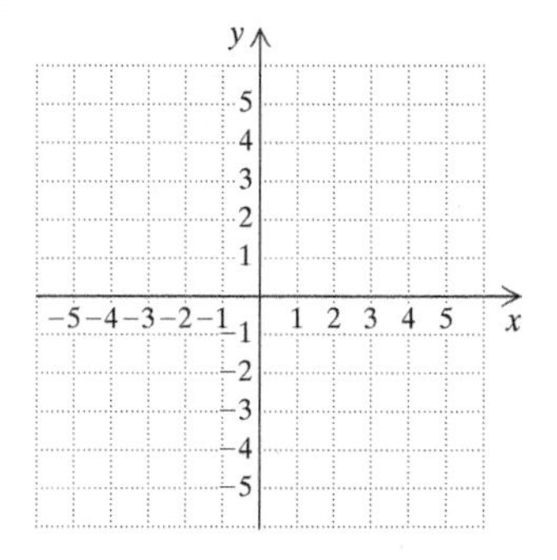

Vertex: (____, ____)
Line of symmetry: $x =$ ____

13. $f(x) = -2(x + 2)^2$

x	$f(x)$
-2	
-3	
-1	
-4	
0	

Vertex: (____, ____)
Line of symmetry: $x =$ ____

14. $f(x) = -2(x + 4)^2$

x	$f(x)$
-4	
-5	
-3	
-6	
-2	

Vertex: (____, ____)
Line of symmetry: $x =$ ____

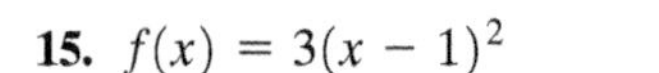

15. $f(x) = 3(x - 1)^2$

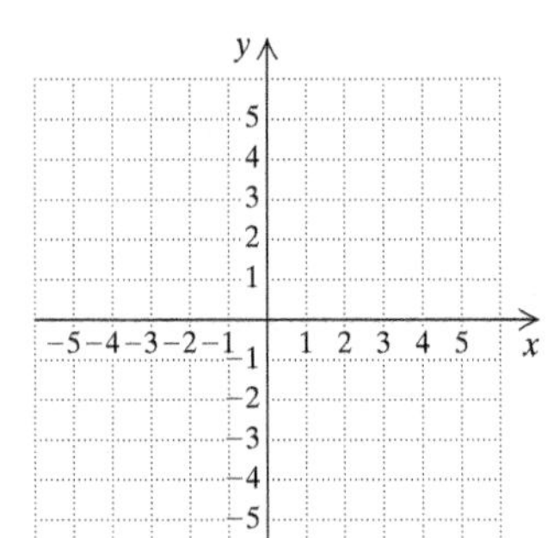

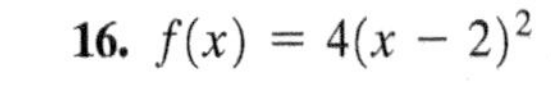

16. $f(x) = 4(x - 2)^2$

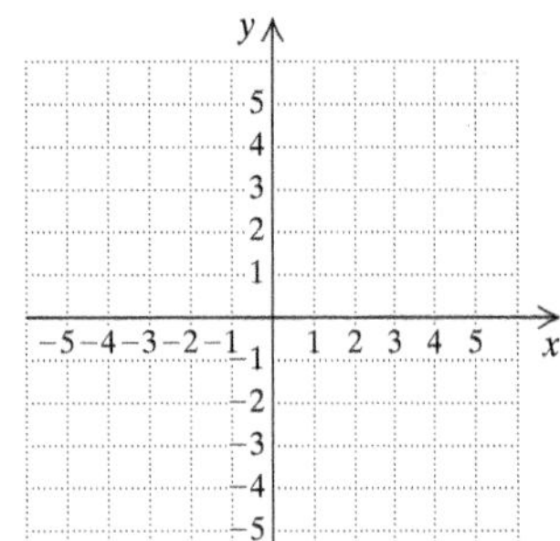

17. $f(x) = -\frac{3}{2}(x + 2)^2$

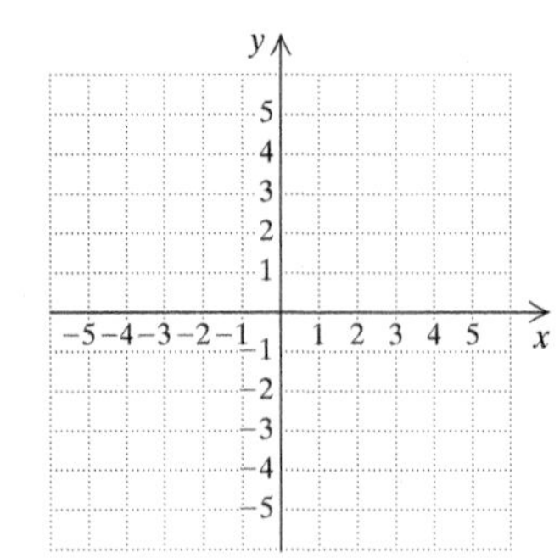

18. $f(x) = -\frac{5}{2}(x + 3)^2$

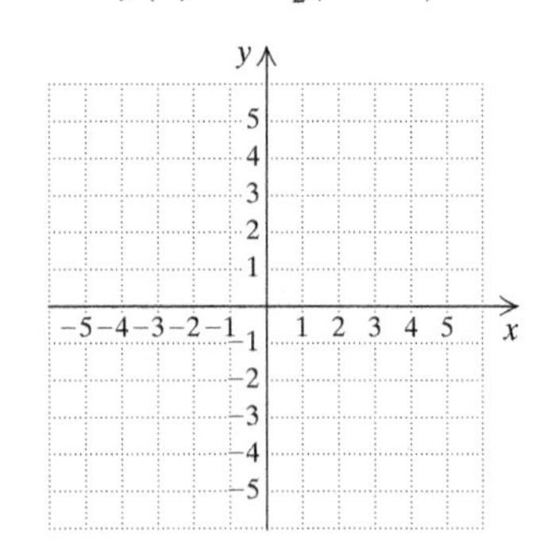

C Graph. Find and label the vertex and the line of symmetry. Find the maximum or minimum value.

19. $f(x) = (x - 3)^2 + 1$

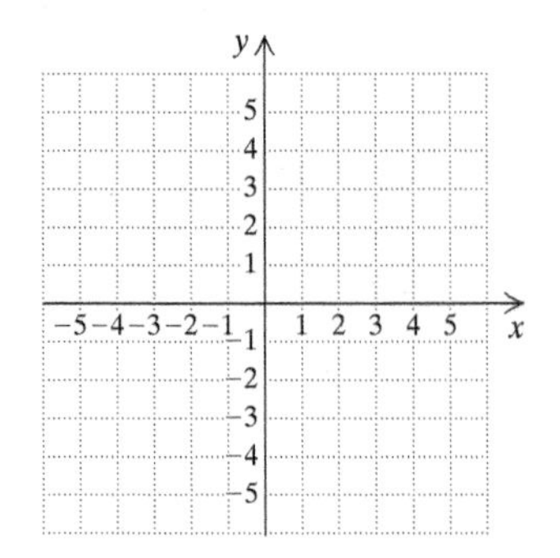

Vertex: (____, ____)
Line of symmetry: $x =$ ____
Minimum value: ____

20. $f(x) = (x + 2)^2 - 3$

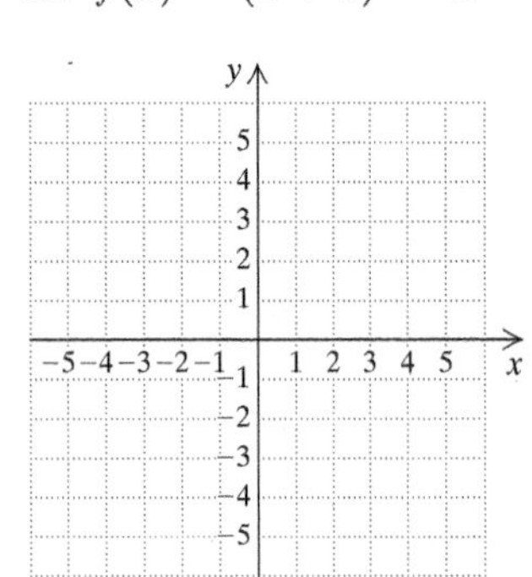

Vertex: (____, ____)
Line of symmetry: $x =$ ____
Minimum value: ____

21. $f(x) = -3(x + 4)^2 + 1$

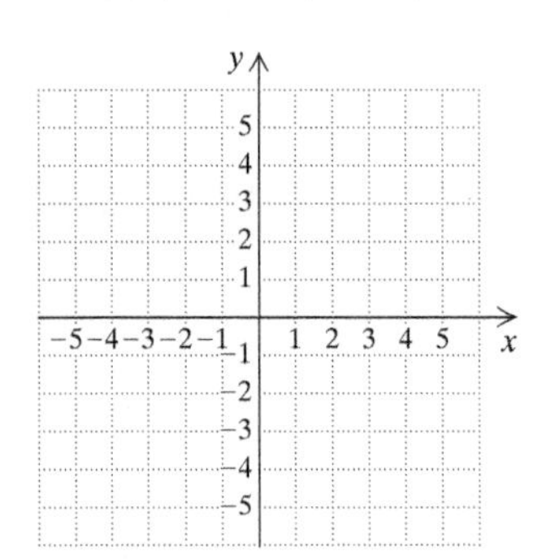

Vertex: (____, ____)
Line of symmetry: $x =$ ____
Maximum value: ____

22. $f(x) = -\frac{1}{2}(x - 1)^2 - 3$

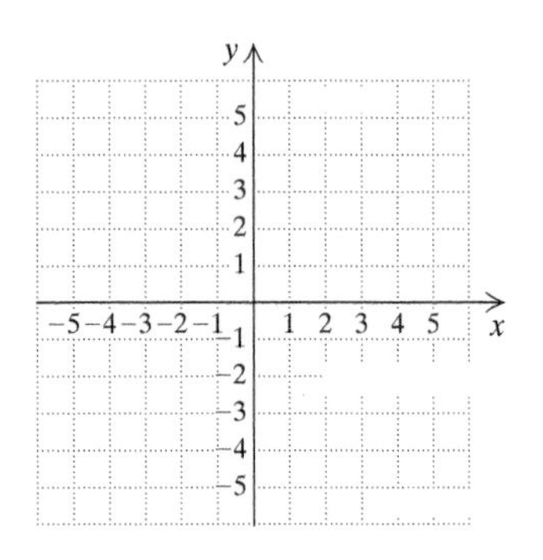

Vertex: (____, ____)
Line of symmetry: $x =$ ____
Maximum value: ____

23. $f(x) = \frac{1}{2}(x + 1)^2 + 4$

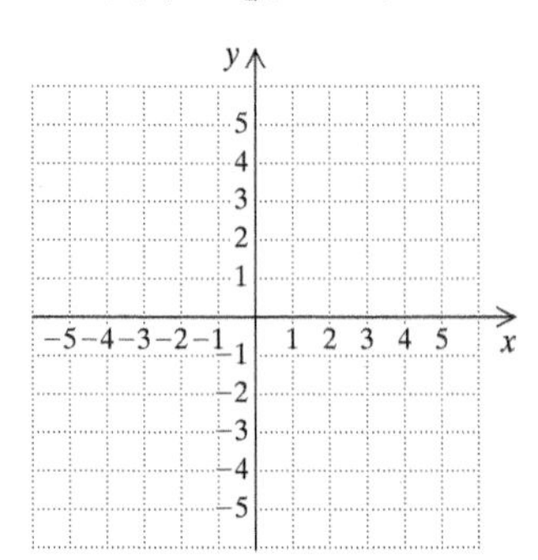

Vertex: (____, ____)
Line of symmetry: $x =$ ____
________ value: ____

24. $f(x) = -2(x - 5)^2 - 3$

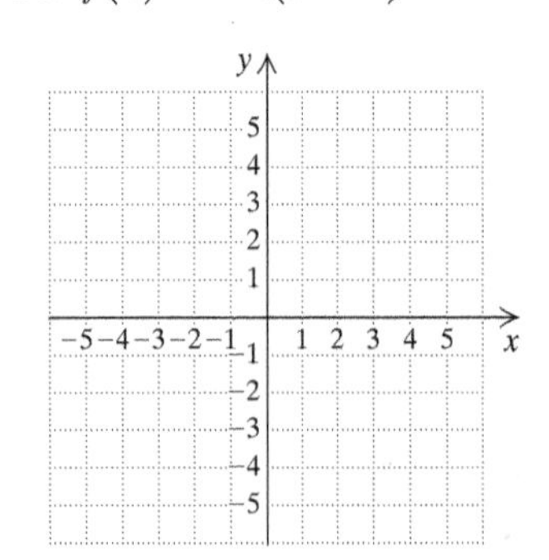

Vertex: (____, ____)
Line of symmetry: $x =$ ____
________ value: ____

25. $f(x) = -(x + 1)^2 - 2$

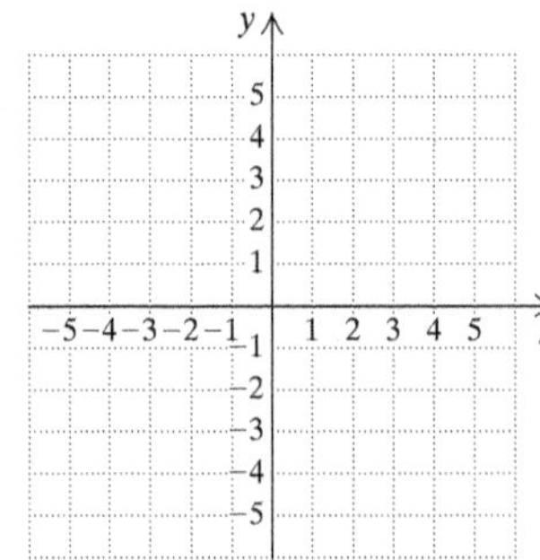

Vertex: (____, ____)
Line of symmetry: $x =$ ____
________ value: ____

26. $f(x) = 3(x - 4)^2 + 2$

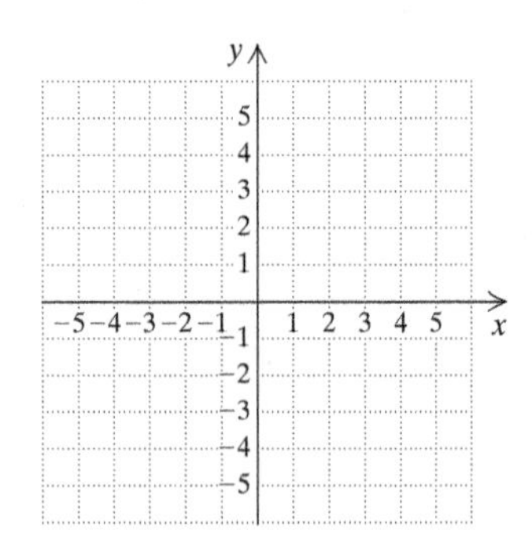

Vertex: (____, ____)
Line of symmetry: $x =$ ____
________ value: ____

Skill Maintenance

Find the slope, if it exists, of the line containing the following points. [2.4b]

27. $(-10, 0)$ and $(5, -6)$

28. $(7, -3)$ and $(15, -3)$

Graphing $f(x) = ax^2 + bx + c$

7.6

OBJECTIVES

a For a quadratic function, find the vertex, the line of symmetry, and the maximum or the minimum value, and then graph the function.

b Find the intercepts of a quadratic function.

a ANALYZING AND GRAPHING $f(x) = ax^2 + bx + c$

By *completing the square,* we can begin with any quadratic polynomial $ax^2 + bx + c$ and find an equivalent expression $a(x - h)^2 + k$. This allows us to analyze and graph any quadratic function $f(x) = ax^2 + bx + c$.

EXAMPLE 1 For $f(x) = x^2 - 6x + 4$, find the vertex, the line of symmetry, and the maximum or the minimum value. Then graph.

We first find the vertex and the line of symmetry. To do so, we find the equivalent form $a(x - h)^2 + k$ by completing the square, beginning as follows:

$$f(x) = x^2 - 6x + 4 = (x^2 - 6x \qquad) + 4.$$

We complete the square inside the parentheses, but in a different manner than we did before. We take half the x-coefficient, $-6/2 = -3$, and square it: $(-3)^2 = 9$. Then we add 0, or $9 - 9$, inside the parentheses. (Instead of adding $(b/2)^2$ on both sides of an equation, we add and subtract it on the same side, effectively adding 0 and not changing the value of the expression.)

$f(x) = (x^2 - 6x + 0) + 4$ — Adding 0

$= (x^2 - 6x + 9 - 9) + 4$ — Substituting $9 - 9$ for 0

$= (x^2 - 6x + 9) + (-9 + 4)$ — Using the associative law of addition to regroup

$= (x - 3)^2 - 5$ — Factoring and simplifying

The vertex is $(3, -5)$, and the line of symmetry is $x = 3$. The coefficient of x^2 is 1, which is positive, so the graph opens up. This tells us that -5 is the minimum value. We plot the vertex and draw the line of symmetry. We choose some x-values on both sides of the vertex and graph the parabola. Suppose we compute the pair $(5, -1)$:

$$f(5) = 5^2 - 6(5) + 4 = 25 - 30 + 4 = -1.$$

We note that it is 2 units to the right of the line of symmetry. There will also be a pair with the same y-coordinate on the graph 2 units *to the left* of the line of symmetry. Thus we get a second point, $(1, -1)$, without making another calculation.

x	$f(x)$
3	-5
4	-4
5	-1
6	4
2	-4
1	-1
0	4

← Vertex

x = 3
(0, 4)
(6, 4)
(1, −1)
(5, −1)
$f(x) = x^2 - 6x + 4$
$= (x - 3)^2 - 5$
(4, −4)
(2, −4)
(3, −5)
Minimum: −5

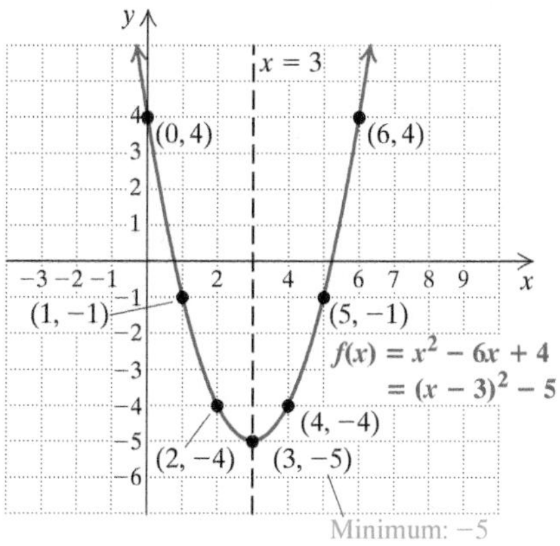

1. For $f(x) = x^2 - 4x + 7$, find the vertex, the line of symmetry, and the maximum or the minimum value. Then graph.

x	$f(x)$

Vertex: (____, ____)
Line of symmetry:
$x =$ ____
Minimum value: ____

◀ **Do Exercise 1.**

EXAMPLE 2 For $f(x) = 3x^2 + 12x + 13$, find the vertex, the line of symmetry, and the maximum or the minimum value. Then graph.

Since the coefficient of x^2 is not 1, we factor out 3 from only the *first two* terms of the expression. Remember that we want to write the function in the form $f(x) = a(x - h)^2 + k$:

$$f(x) = 3x^2 + 12x + 13$$
$$= 3(x^2 + 4x) + 13. \quad \text{Factoring 3 out of the first two terms}$$

Next, we complete the square inside the parentheses:

$$f(x) = 3(x^2 + 4x \qquad) + 13.$$

We take half the x-coefficient, $\frac{1}{2} \cdot 4 = 2$, and square it: $2^2 = 4$. Then we add 0, or $4 - 4$, inside the parentheses:

$$f(x) = 3(x^2 + 4x + 0) + 13 \quad \text{Adding 0}$$
$$= 3(x^2 + 4x + 4 - 4) + 13 \quad \text{Substituting } 4 - 4 \text{ for } 0$$
$$= 3(x^2 + 4x + 4 - 4) + 13$$
$$= 3(x^2 + 4x + 4) + 3(-4) + 13 \quad \text{Using the distributive law to separate } -4 \text{ from the trinomial}$$
$$= 3(x^2 + 4x + 4) - 12 + 13$$
$$= 3(x + 2)^2 + 1 \quad \text{Factoring and simplifying}$$
$$= 3[x - (-2)]^2 + 1.$$

The vertex is $(-2, 1)$, and the line of symmetry is $x = -2$. The coefficient of x^2 is 3, so the graph is narrow and opens up. This tells us that 1 is the minimum value of the function. We choose a few x-values on one side of the line of symmetry, compute y-values, and use the resulting coordinates to find more points on the other side of the line of symmetry. We plot points and graph the parabola.

2. For $f(x) = 3x^2 - 24x + 43$, find the vertex, the line of symmetry, and the maximum or the minimum value. Then graph.

x	$f(x)$

Vertex: (____, ____)
Line of symmetry:
$x =$ ____
Minimum value: ____

x	$f(x)$	
−2	1	← Vertex
−1	4	
−3	4	
0	13	
−4	13	

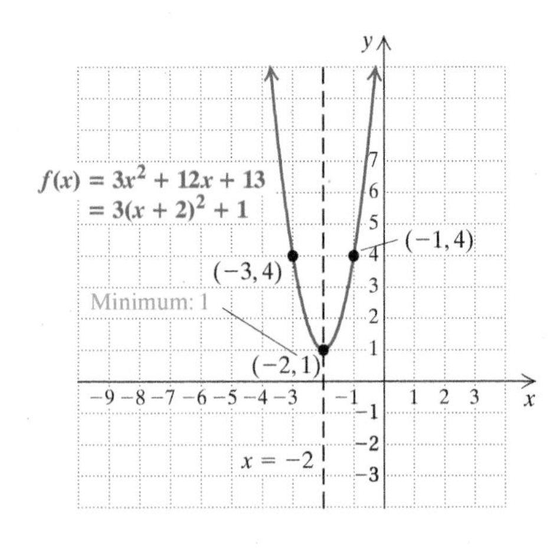

◀ **Do Exercise 2.**

Answers

1.

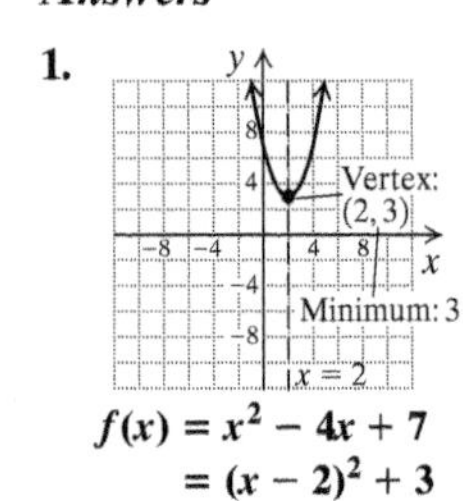

$f(x) = x^2 - 4x + 7$
$= (x - 2)^2 + 3$

2.

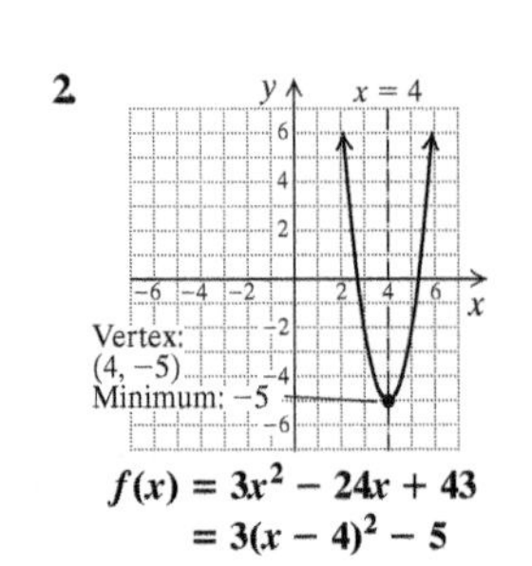

$f(x) = 3x^2 - 24x + 43$
$= 3(x - 4)^2 - 5$

EXAMPLE 3 For $f(x) = -2x^2 + 10x - 7$, find the vertex, the line of symmetry, and the maximum or the minimum value. Then graph.

Again, the coefficient of x^2 is not 1. We factor out -2 from only the *first two* terms of the expression. This makes the coefficient of x^2 inside the parentheses 1:

$$f(x) = -2x^2 + 10x - 7$$
$$= -2(x^2 - 5x) - 7.$$

Next, we complete the square as before:

$$f(x) = -2(x^2 - 5x \qquad) - 7.$$

We take half the x-coefficient, $\frac{1}{2}(-5) = -\frac{5}{2}$, and square it: $\left(-\frac{5}{2}\right)^2 = \frac{25}{4}$. Then we add 0, or $\frac{25}{4} - \frac{25}{4}$, inside the parentheses:

$$f(x) = -2\left(x^2 - 5x + \tfrac{25}{4} - \tfrac{25}{4}\right) - 7 \qquad \text{Adding 0, or } \tfrac{25}{4} - \tfrac{25}{4}$$
$$= -2\left(x^2 - 5x + \tfrac{25}{4} - \tfrac{25}{4}\right) - 7$$
$$= -2\left(x^2 - 5x + \tfrac{25}{4}\right) + (-2)\left(-\tfrac{25}{4}\right) - 7 \qquad \text{Using the distributive law to separate } -\tfrac{25}{4} \text{ from the trinomial}$$
$$= -2\left(x^2 - 5x + \tfrac{25}{4}\right) + \tfrac{25}{2} - 7$$
$$= -2\left(x - \tfrac{5}{2}\right)^2 + \tfrac{11}{2}. \qquad \text{Factoring and simplifying}$$

The vertex is $\left(\frac{5}{2}, \frac{11}{2}\right)$, and the line of symmetry is $x = \frac{5}{2}$. The coefficient of x^2 is -2, so the graph is narrow and opens down. This tells us that $\frac{11}{2}$ is the maximum value of the function. We choose a few x-values on one side of the line of symmetry, compute y-values, and use the resulting coordinates to find more points on the other side of the line of symmetry. We plot points and graph the parabola.

x	$f(x)$	
$\frac{5}{2}$	$\frac{11}{2}$, or $5\frac{1}{2}$	← Vertex
3	5	
4	1	
5	-7	
2	5	
1	1	
0	-7	

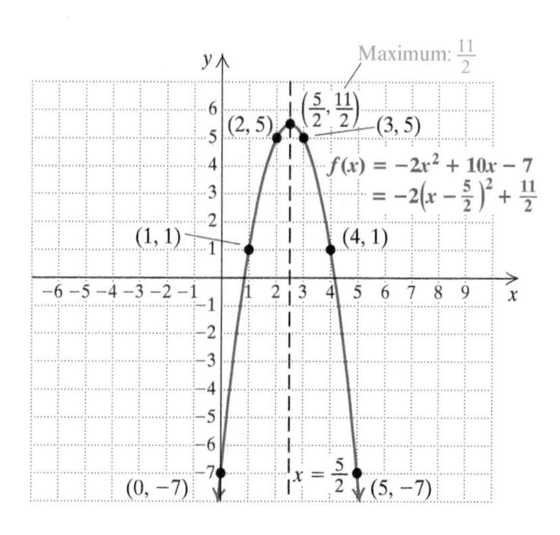

Do Exercise 3. ▶

3. For $f(x) = -4x^2 + 12x - 5$, find the vertex, the line of symmetry, and the maximum or the minimum value. Then graph.

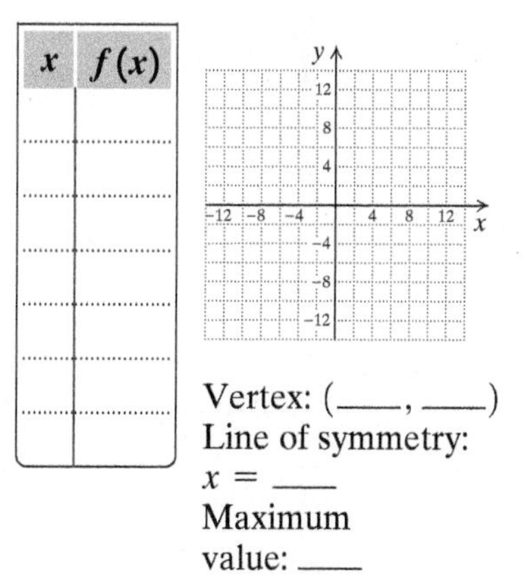

Vertex: (____, ____)
Line of symmetry:
$x =$ ____
Maximum value: ____

Answer

3. Vertex: $\left(\frac{3}{2}, 4\right)$; $x = \frac{3}{2}$; Maximum: 4

$f(x) = -4x^2 + 12x - 5$
$= -4\left(x - \frac{3}{2}\right)^2 + 4$

The method used in Examples 1–3 can be generalized to find a formula for locating the vertex. We complete the square as follows:

$$f(x) = ax^2 + bx + c$$

$$= a\left(x^2 + \frac{b}{a}x\right) + c. \qquad \text{Factoring } a \text{ out of the first two terms. Check by multiplying.}$$

Half of the x-coefficient, $\frac{b}{a}$, is $\frac{b}{2a}$. We square it to get $\frac{b^2}{4a^2}$ and add $\frac{b^2}{4a^2} - \frac{b^2}{4a^2}$ inside the parentheses. Then we distribute a:

$$f(x) = a\left(x^2 + \frac{b}{a}x + \frac{b^2}{4a^2} - \frac{b^2}{4a^2}\right) + c$$

$$= a\left(x^2 + \frac{b}{a}x + \frac{b^2}{4a^2}\right) + a\left(-\frac{b^2}{4a^2}\right) + c \qquad \text{Using the distributive law}$$

$$= a\left(x + \frac{b}{2a}\right)^2 + \frac{-b^2}{4a} + \frac{4ac}{4a} \qquad \text{Factoring and finding a common denominator}$$

$$= a\left[x - \left(-\frac{b}{2a}\right)\right]^2 + \frac{4ac - b^2}{4a}.$$

Thus we have the following.

VERTEX; LINE OF SYMMETRY

The **vertex** of a parabola given by $f(x) = ax^2 + bx + c$ is

$$\left(-\frac{b}{2a}, \frac{4ac - b^2}{4a}\right), \quad \text{or} \quad \left(-\frac{b}{2a}, f\left(-\frac{b}{2a}\right)\right).$$

The x-coordinate of the vertex is $-b/(2a)$. The **line of symmetry** is $x = -b/(2a)$. The second coordinate of the vertex is easiest to find by computing $f\left(-\frac{b}{2a}\right)$.

Let's reexamine Example 3 to see how we could have found the vertex directly. From the formula above,

$$\text{the } x\text{-coordinate of the vertex is } -\frac{b}{2a} = -\frac{10}{2(-2)} = \frac{5}{2}.$$

Substituting $\frac{5}{2}$ into $f(x) = -2x^2 + 10x - 7$, we find the second coordinate of the vertex:

$$f\left(\tfrac{5}{2}\right) = -2\left(\tfrac{5}{2}\right)^2 + 10\left(\tfrac{5}{2}\right) - 7$$

$$= -2\left(\tfrac{25}{4}\right) + 25 - 7$$

$$= -\tfrac{25}{2} + 25 - 7 = -\tfrac{25}{2} + \tfrac{50}{2} - \tfrac{14}{2} = \tfrac{11}{2}.$$

The vertex is $\left(\frac{5}{2}, \frac{11}{2}\right)$. The line of symmetry is $x = \frac{5}{2}$.

We have developed two methods for finding the vertex. One is by completing the square and the other is by using a formula. You should check with your instructor about which method to use.

◀ **Do Exercises 4–6.**

Find the vertex of each parabola using the formula.

4. $f(x) = x^2 - 6x + 4$

5. $f(x) = 3x^2 - 24x + 43$

6. $f(x) = -4x^2 + 12x - 5$ GS

The x-coordinate of the vertex is

$$-\frac{b}{2a} = -\frac{\_\_\_}{2(-4)} = \frac{\_\_\_}{2}.$$

The second coordinate of the vertex is $f\left(\frac{3}{2}\right)$:

$$f\left(\frac{3}{2}\right) = -4(\quad)^2 + 12(\quad) - 5$$

$$= -4(\quad) + \_\_\_ - 5$$

$$= \_\_\_ + 18 - 5$$

$$= \_\_\_.$$

The vertex is (___ , ___).

Answers

4. $(3, -5)$ **5.** $(4, -5)$ **6.** $\left(\frac{3}{2}, 4\right)$

Guided Solution:

6. $12, 3, \frac{3}{2}, \frac{3}{2}, \frac{9}{4}, 18, -9, 4, \frac{3}{2}, 4$

b FINDING THE INTERCEPTS OF A QUADRATIC FUNCTION

MyLab Math ANIMATION

SKILL REVIEW *Graph linear equations using intercepts.* [2.5a]

Find the intercepts and then graph the line.

1. $3x - y = 3$ **2.** $2x + 4y = -8$

Answers: **1.** y-intercept: $(0, -3)$; x-intercept: $(1, 0)$ **2.** y-intercept: $(0, -2)$; x-intercept: $(-4, 0)$

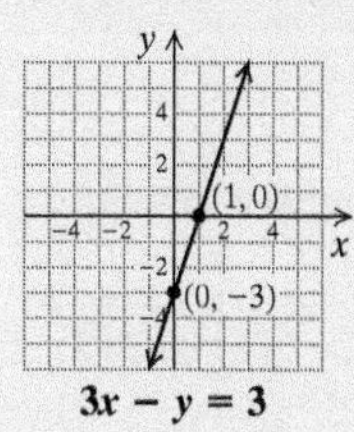

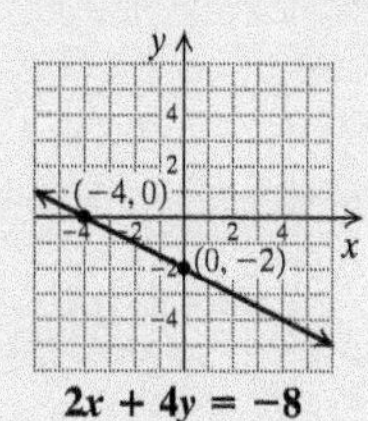

MyLab Math VIDEO

The points at which a graph crosses an axis are called **intercepts**. We determine the y-intercept by finding $f(0)$. For $f(x) = ax^2 + bx + c$, $f(0) = a \cdot 0^2 + b \cdot 0 + c = c$, so the y-intercept is $(0, c)$.

To find the x-intercepts, we look for values of x for which $f(x) = 0$. For $f(x) = ax^2 + bx + c$, we solve

$$0 = ax^2 + bx + c.$$

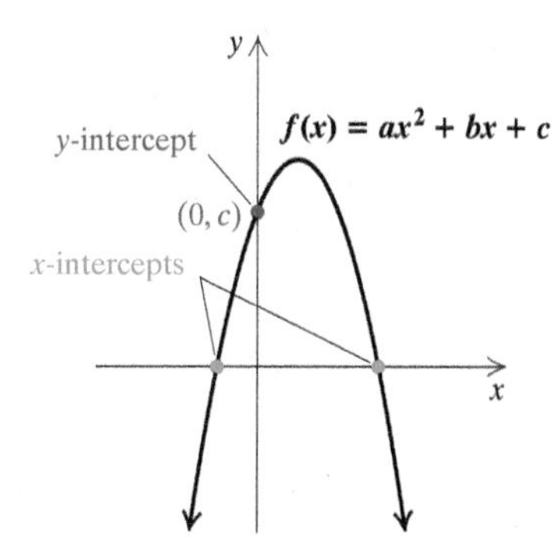

EXAMPLE 4 Find the intercepts of $f(x) = x^2 - 2x - 2$.

The y-intercept is $(0, f(0))$. Since $f(0) = 0^2 - 2 \cdot 0 - 2 = -2$, the y-intercept is $(0, -2)$. To find the x-intercepts, we solve

$$0 = x^2 - 2x - 2.$$

Using the quadratic formula, we have $x = 1 \pm \sqrt{3}$. Thus the x-intercepts are $(1 - \sqrt{3}, 0)$ and $(1 + \sqrt{3}, 0)$, or, approximately, $(-0.732, 0)$ and $(2.732, 0)$.

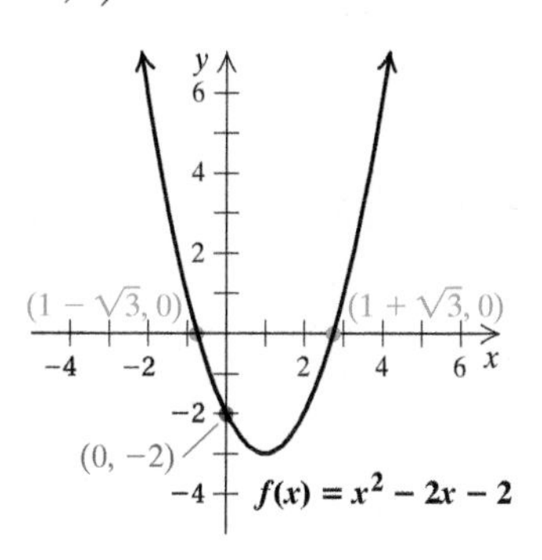

Do Exercises 7–9. ▶

Find the intercepts.

GS **7.** $f(x) = x^2 + 2x - 3$

The y-intercept is $(0, f(0))$.

$f(0) = \_\_\_^2 + 2 \cdot \_\_\_ - 3 = -3$

The y-intercept is $(0, \_\_\_)$. To find the x-intercepts, we solve $0 = x^2 + 2x - 3$:

$0 = x^2 + 2x - 3$

$0 = (x - 1)(\_\_\_)$

$x - 1 = 0$ *or* $\_\_\_ = 0$

$x = 1$ *or* $x = \_\_\_$.

The x-intercepts are $(1, 0)$ and $(\_\_\_, 0)$.

8. $f(x) = x^2 + 8x + 16$

9. $f(x) = x^2 - 4x + 1$

Answers

7. y-intercept: $(0, -3)$; x-intercepts: $(-3, 0)$, $(1, 0)$
8. y-intercept: $(0, 16)$; x-intercept: $(-4, 0)$
9. y-intercept: $(0, 1)$; x-intercepts: $(2 - \sqrt{3}, 0)$, $(2 + \sqrt{3}, 0)$, or $(0.268, 0)$, $(3.732, 0)$

Guided Solution:
7. $0, 0, -3, x + 3, x + 3, -3, -3$

Visualizing for Success

Match each equation or inequality with its graph.

1. $y = -(x - 5)^2 + 2$

2. $2x + 5y = 10$

3. $5x - 2y = 10$

4. $2x - 5y = 10$

5. $y = (x - 5)^2 - 2$

6. $y = x^2 - 5$

7. $5x + 2y \geq 10$

8. $5x + 2y = -10$

9. $y < 5x$

10. $y = -(x + 5)^2 + 2$

A

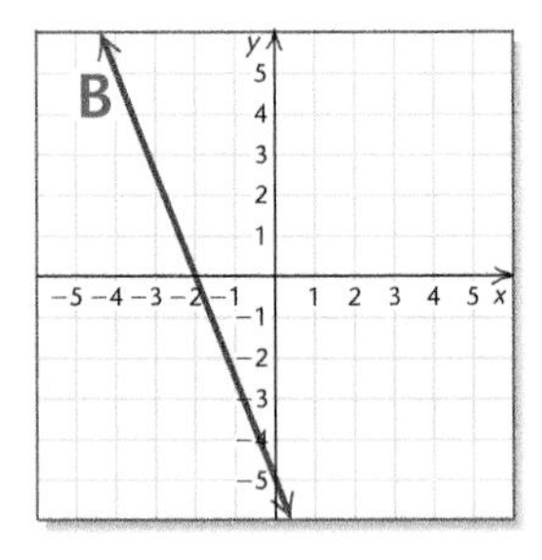

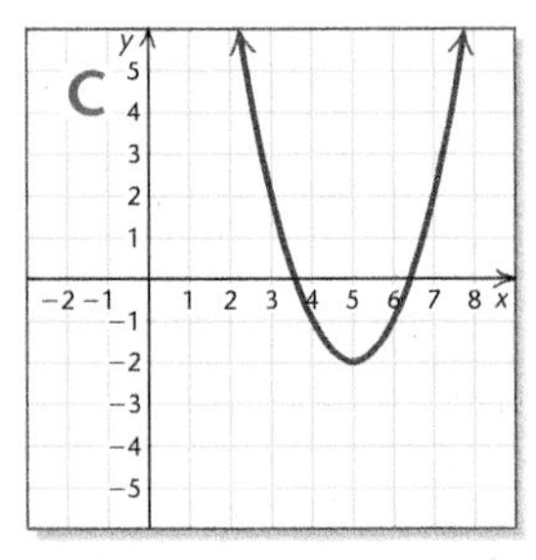

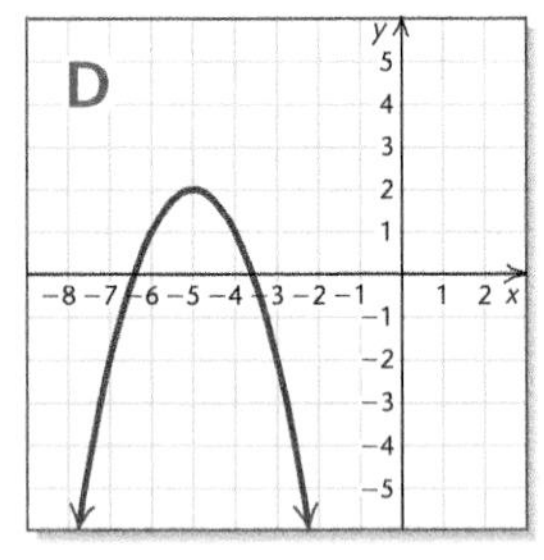

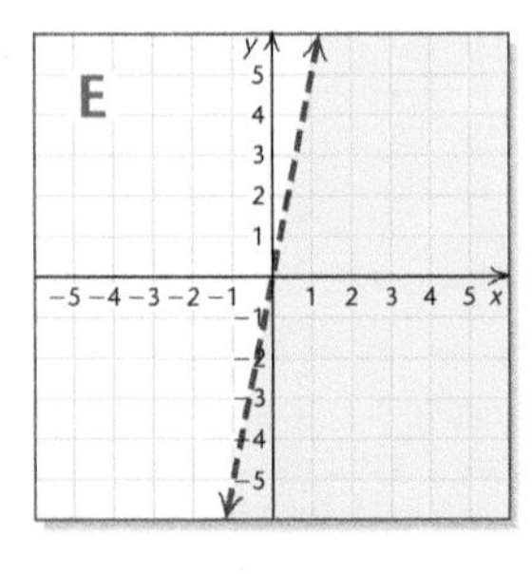

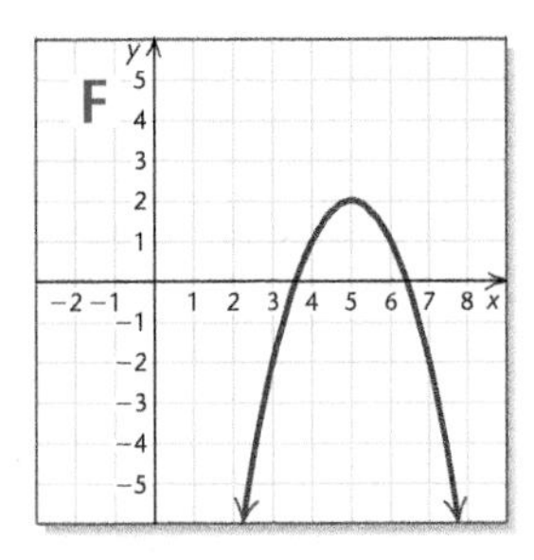

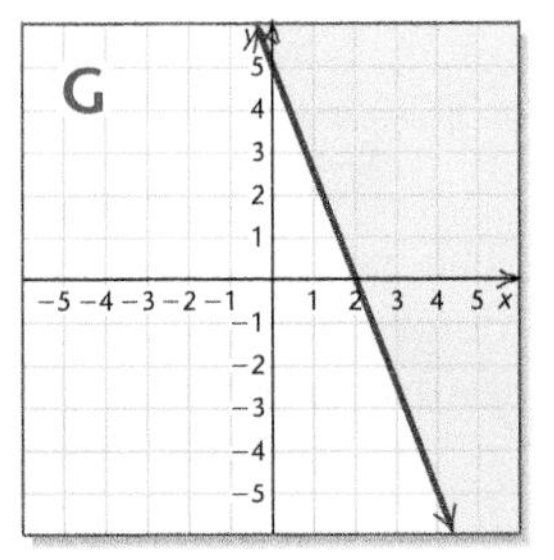

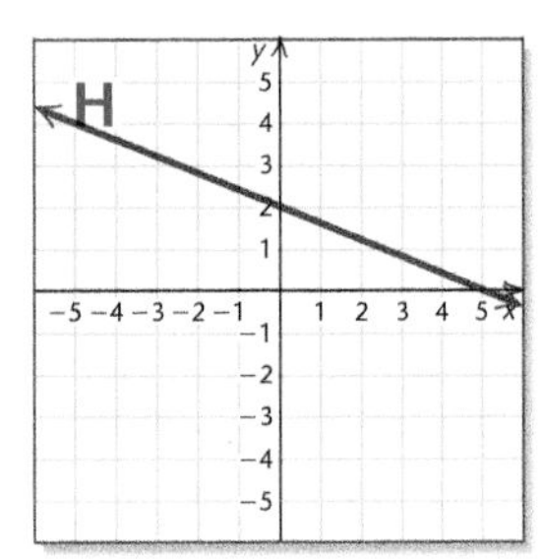

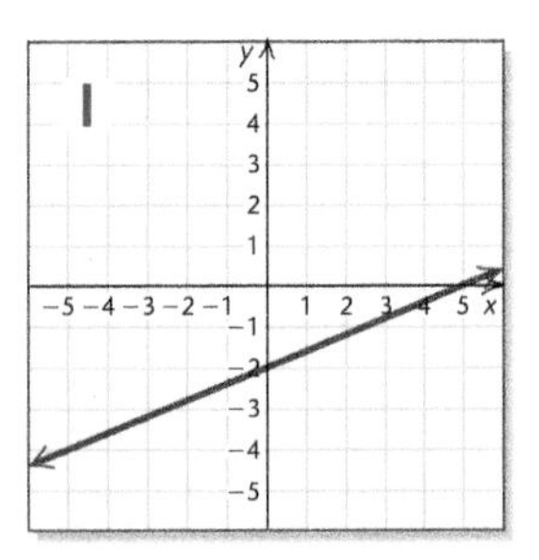

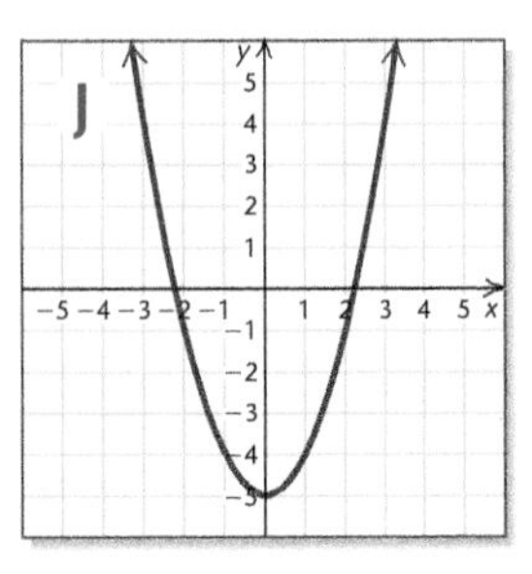

Answers on page A-27

7.6 Exercise Set

FOR EXTRA HELP MyLab Math

✔ Check Your Understanding

Reading Check and Concept Check Choose from beneath the blank the word that best completes each statement.

RC1. The graph of $f(x) = 5x^2 - 10x - 3$ opens ________________.
downward/upward

RC2. The function given by $f(x) = 2x^2 - x - 7$ has a ________________ value.
maximum/minimum

RC3. The graph of $g(x) = (x + 3)^2 - 2$ has its ________________ at $(-3, -2)$.
vertex/x-intercept

RC4. The ________________ of the graph of $g(x) = -x^2 + 6x + 9$ is $(0, 9)$.
x-intercept/y-intercept

a For each quadratic function, find **(a)** the vertex, **(b)** the line of symmetry, and **(c)** the maximum or the minimum value. Then **(d)** graph the function.

1. $f(x) = x^2 - 2x - 3$

x	$f(x)$

a) Vertex: (_____, _____)
b) Line of symmetry: $x =$ _____
c) _________ value: _____
d)

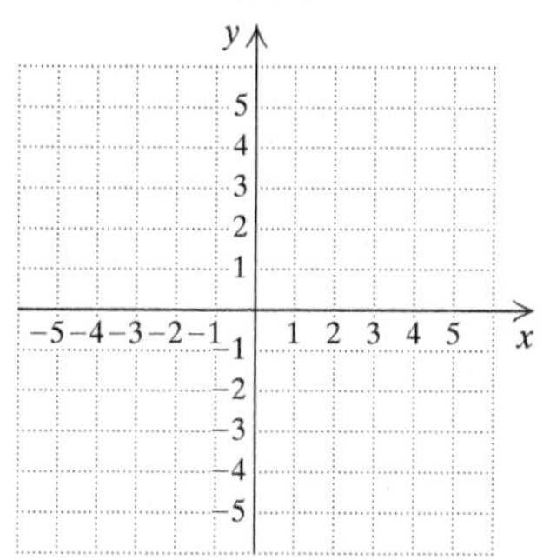

2. $f(x) = x^2 + 2x - 5$

x	$f(x)$

a) Vertex: (_____, _____)
b) Line of symmetry: $x =$ _____
c) _________ value: _____
d)

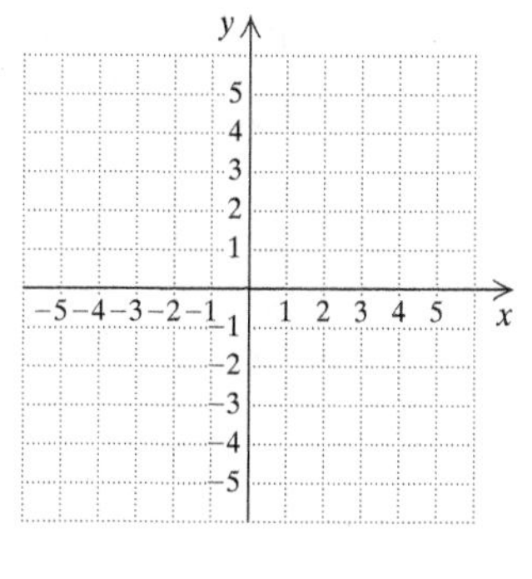

3. $f(x) = -x^2 - 4x - 2$

x	$f(x)$

a) Vertex: (_____, _____)
b) Line of symmetry: $x =$ _____
c) _________ value: _____
d)

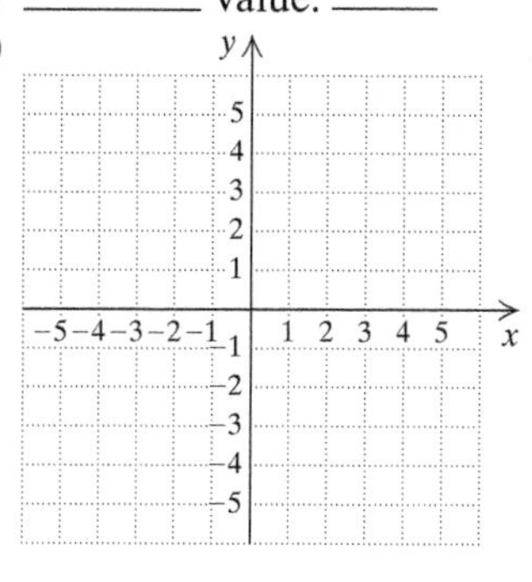

4. $f(x) = -x^2 + 4x + 1$

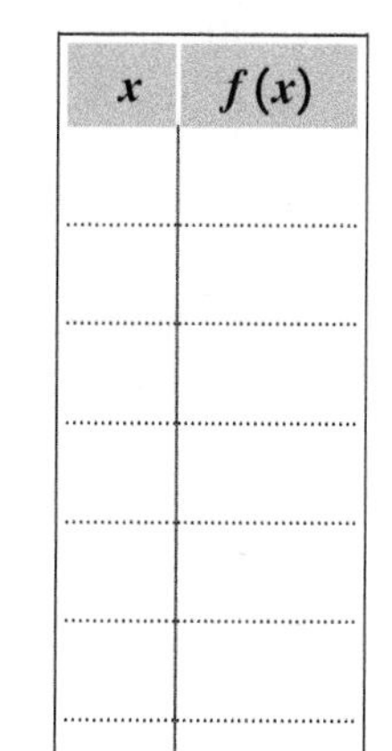

x	$f(x)$

a) Vertex: (_____, _____)
b) Line of symmetry: $x =$ _____
c) _________ value: _____
d)

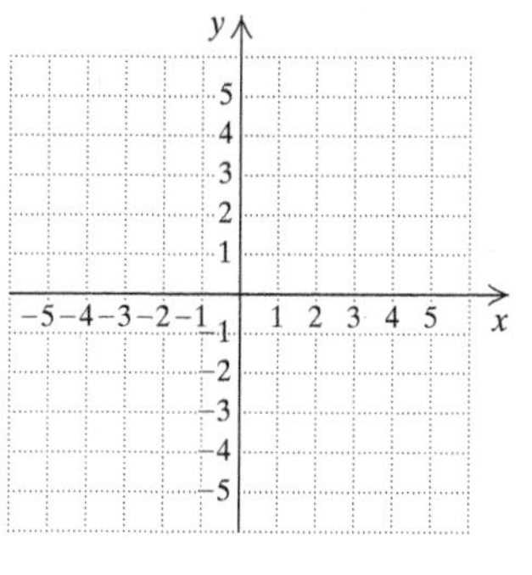

5. $f(x) = 3x^2 - 24x + 50$

a) Vertex: (_____, _____)
b) Line of symmetry: $x =$ _____
c) ________ value: _____
d)

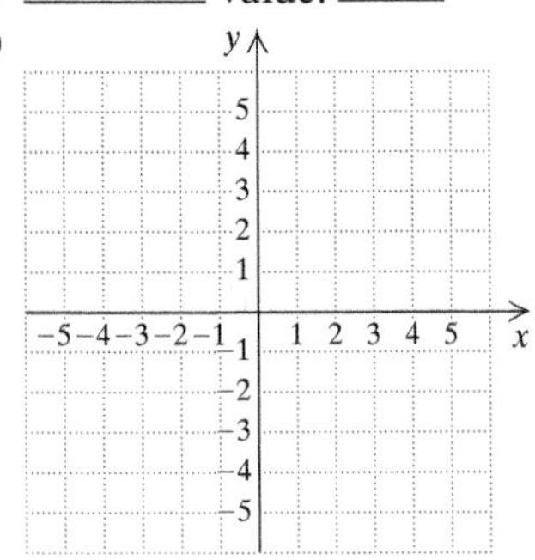

6. $f(x) = 4x^2 + 8x + 1$

a) Vertex: (_____, _____)
b) Line of symmetry: $x =$ _____
c) ________ value: _____
d)

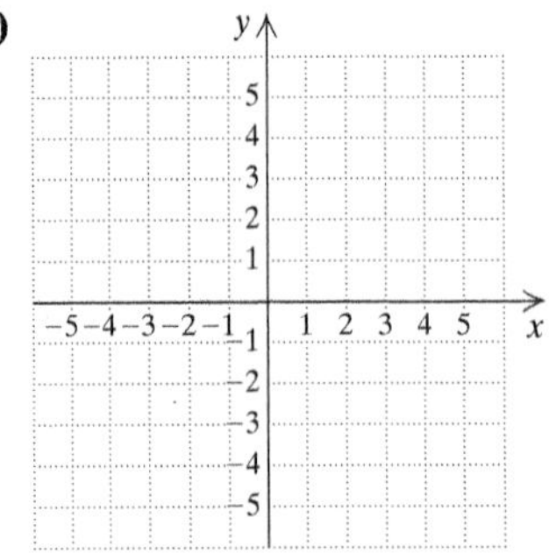

7. $f(x) = -2x^2 - 2x + 3$

a) Vertex: (_____, _____)
b) Line of symmetry: $x =$ _____
c) ________ value: _____
d)

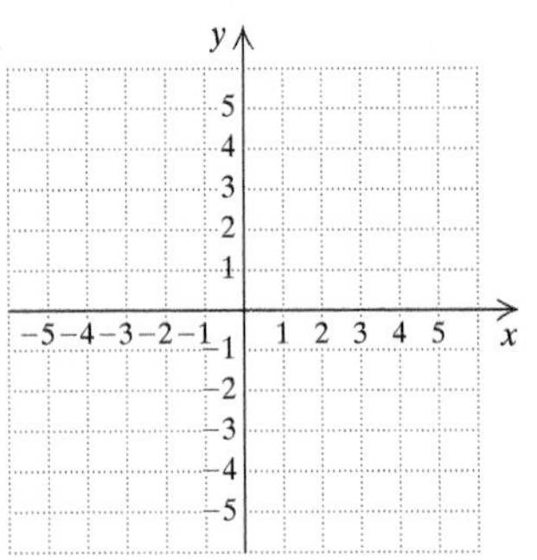

8. $f(x) = -2x^2 + 2x + 1$

a) Vertex: (_____, _____)
b) Line of symmetry: $x =$ _____
c) ________ value: _____
d)

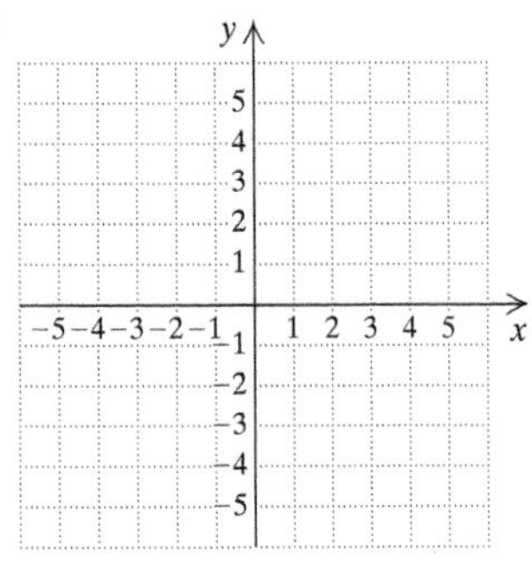

9. $f(x) = 5 - x^2$

a) Vertex: (_____, _____)
b) Line of symmetry: $x =$ _____
c) ________ value: _____
d)

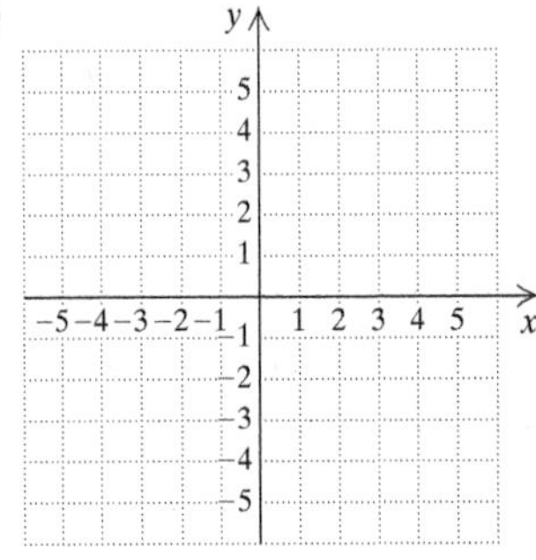

10. $f(x) = x^2 - 3x$

a) Vertex: (_____, _____)
b) Line of symmetry: $x =$ _____
c) ________ value: _____
d)

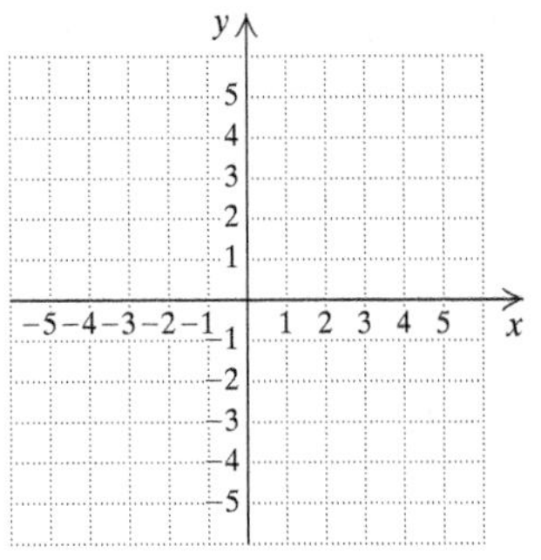

11. $f(x) = 2x^2 + 5x - 2$

a) Vertex: (_____, _____)
b) Line of symmetry: $x =$ _____
c) ________ value: _____
d)

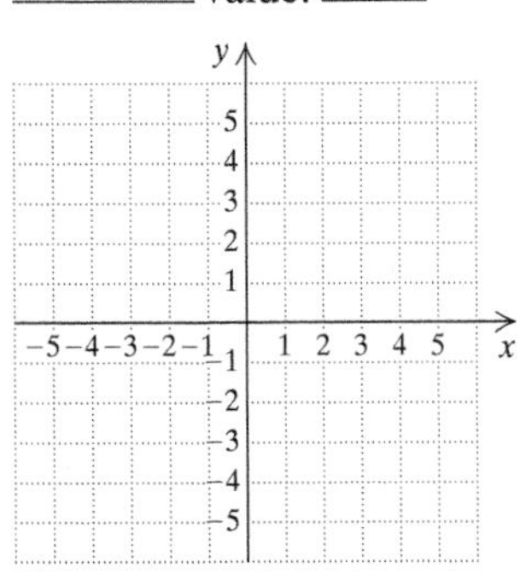

12. $f(x) = -4x^2 - 7x + 2$

a) Vertex: (_____, _____)
b) Line of symmetry: $x =$ _____
c) ________ value: _____
d)

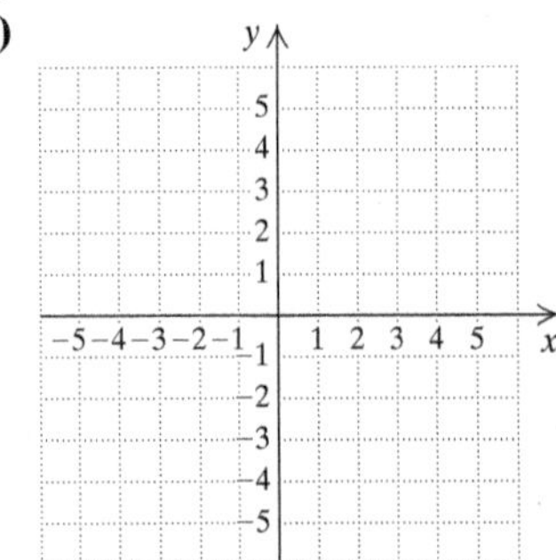

b Find the x- and y-intercepts.

13. $f(x) = x^2 - 6x + 1$
14. $f(x) = x^2 + 2x + 12$
15. $f(x) = -x^2 + x + 20$
16. $f(x) = -x^2 + 5x + 24$
17. $f(x) = 4x^2 + 12x + 9$
18. $f(x) = 3x^2 - 6x + 1$
19. $f(x) = 4x^2 - x + 8$
20. $f(x) = 2x^2 + 4x - 1$

Skill Maintenance

Solve. [5.8b]

21. *Determining Medication Dosage.* A child's dosage D, in milligrams, of a medication varies directly as the child's weight w, in kilograms. To control a fever, a doctor suggests that a child who weighs 28 kg be given 420 mg of analgesic medication. Find an equation of variation.

22. *Calories Burned.* The number C of calories burned while exercising varies directly as the time t, in minutes, spent exercising. Harold exercises for 24 min on a StairMaster and burns 356 calories. Find an equation of variation.

Find the variation constant and an equation of variation in which y varies inversely as x and the following are true. [5.8c]

23. $y = 125$ when $x = 2$
24. $y = 2$ when $x = 125$

Find the variation constant and an equation of variation in which y varies directly as x and the following are true. [5.8a]

25. $y = 125$ when $x = 2$
26. $y = 2$ when $x = 125$

Synthesis

27. Use the TRACE and/or TABLE features of a graphing calculator to estimate the maximum or minimum values of the following functions.
    a) $f(x) = 2.31x^2 - 3.135x - 5.89$
    b) $f(x) = -18.8x^2 + 7.92x + 6.18$

28. Use the INTERSECT feature of a graphing calculator to find the points of intersection of the graphs of the functions.
    $f(x) = x^2 + 2x + 1, \quad g(x) = -2x^2 - 4x + 1$

Graph.

29. $f(x) = |x^2 - 1|$
30. $f(x) = |x^2 + 6x + 4|$
31. $f(x) = |x^2 - 3x - 4|$
32. $f(x) = |2(x - 3)^2 - 5|$

33. A quadratic function has $(-1, 0)$ as one of its intercepts and $(3, -5)$ as its vertex. Find an equation for the function.

34. A quadratic function has $(4, 0)$ as one of its intercepts and $(-1, 7)$ as its vertex. Find an equation for the function.

35. Consider

$$f(x) = \frac{x^2}{8} + \frac{x}{4} - \frac{3}{8}.$$

Find the vertex, the line of symmetry, and the maximum or the minimum value. Then draw the graph.

36. Use only the graph in Exercise 35 to approximate the solutions of each of the following equations.
    a) $\frac{x^2}{8} + \frac{x}{4} - \frac{3}{8} = 0$
    b) $\frac{x^2}{8} + \frac{x}{4} - \frac{3}{8} = 1$
    c) $\frac{x^2}{8} + \frac{x}{4} - \frac{3}{8} = 2$

7.7

Mathematical Modeling with Quadratic Functions

OBJECTIVES

a Solve maximum–minimum problems involving quadratic functions.

b Fit a quadratic function to a set of data to form a mathematical model, and solve related applied problems.

MyLab Math ANIMATION

We now consider some of the many situations in which quadratic functions can serve as mathematical models.

a MAXIMUM–MINIMUM PROBLEMS

We have seen that for any quadratic function $f(x) = ax^2 + bx + c$, the value of $f(x)$ at the vertex is either a maximum or a minimum, meaning that either all outputs are smaller than that value for a maximum or larger than that value for a minimum.

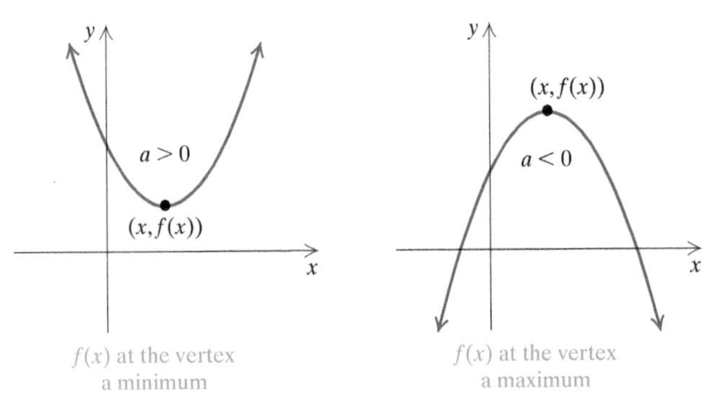

There are many applied problems in which we want to find a maximum or a minimum value. If a quadratic function can be used as a model, we can find such maximums or minimums by finding coordinates of the vertex.

EXAMPLE 1 *Landscape Design.* As part of his backyard design, Christopher plans to create an outdoor dining area by planting boxwood shrubs around the perimeter of a rectangle. If he has enough shrubs to form a perimeter of 64 ft (after allowing for an opening for entry into the area), what are the dimensions of the largest rectangle that Christopher can enclose?

1. **Familiarize.** We first make a drawing and label it. We let l = the length of the dining area and w = the width. Recall the following formulas:

   Perimeter: $2l + 2w$;

   Area: $l \cdot w$.

To become familiar with the problem, let's choose some dimensions (shown at right) for which $2l + 2w = 64$ and then calculate the corresponding areas. What choice of l and w will maximize A?

l	w	A
22	10	220
20	12	240
18	14	252
18.5	13.5	249.75
12.4	19.6	243.04
15	17	255

2. Translate. We have two equations, one for perimeter and one for area:

$$2l + 2w = 64,$$
$$A = l \cdot w.$$

Let's use them to express A as a function of l or w, but not both. To express A in terms of w, for example, we solve for l in the first equation:

$$2l + 2w = 64$$
$$2l = 64 - 2w$$
$$l = \frac{64 - 2w}{2}$$
$$= 32 - w.$$

Substituting $32 - w$ for l, we get a quadratic function $A(w)$, or just A:

$$A = lw = (32 - w)w = 32w - w^2 = -w^2 + 32w.$$

3. Solve. To find the vertex, we complete the square:

$A = -w^2 + 32w$ — This is a parabola opening down, so a maximum exists.

$= -1(w^2 - 32w)$ — Factoring out -1

$= -1(w^2 - 32w + 256 - 256)$ — $\frac{1}{2}(-32) = -16$; $(-16)^2 = 256$. We add 0, or $256 - 256$.

$= -1(w^2 - 32w + 256) + (-1)(-256)$ — Using the distributive law

$= -(w - 16)^2 + 256.$

The vertex is $(16, 256)$. Thus the maximum value is 256. It occurs when $w = 16$ and $l = 32 - w = 32 - 16 = 16$.

4. Check. We note that 256 is larger than any of the values found in the *Familiarize* step. To be more certain, we could make more calculations. We can also use the graph of the function to check the maximum value.

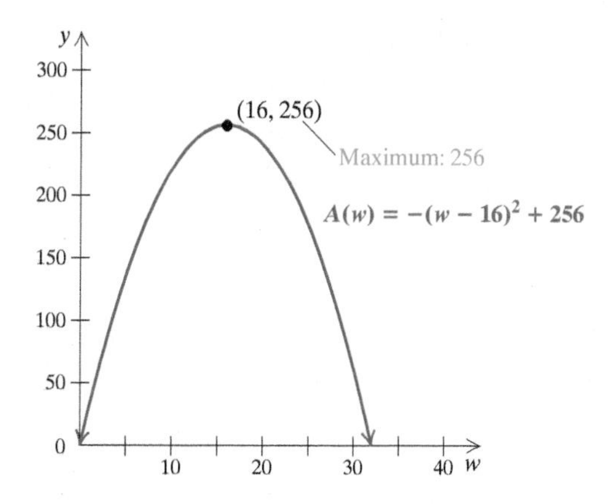

5. State. The largest rectangular dining area that can be enclosed is 16 ft by 16 ft; that is, it is a square with sides of 16 ft.

Do Exercise 1. ▶

1. ***Stained-Glass Window Design.*** An artist is designing a rectangular stained-glass window with a perimeter of 84 in. What dimensions will yield the maximum area?

To familiarize yourself with the problem, complete the following table.

l	w	A
10	32	320
12	30	
15	27	
20	22	
20.5	21.5	

Answer

1. 21 in. by 21 in.

CALCULATOR CORNER

Maximum and Minimum Values We can use a graphing calculator to find the maximum or the minimum value of a quadratic function. Consider the quadratic function in Example 1, $A = -w^2 + 32w$. First, we replace w with x and A with y and graph the function in a window that displays the vertex of the graph. We choose [0, 40, 0, 300], with Xscl = 5 and Yscl = 20. Now, we select the MAXIMUM feature from the CALC menu. We are prompted to select a left bound for the maximum point. This means that we must choose an x-value that is to the left of the x-value of the point where the maximum occurs. This can be done by using the left- and right-arrow keys to move the cursor to a point to the left of the maximum point or by keying in an appropriate value. Once this is done, we press ENTER. Now, we are prompted to select a right bound. We move the cursor to a point to the right of the maximum point or key in an appropriate value.

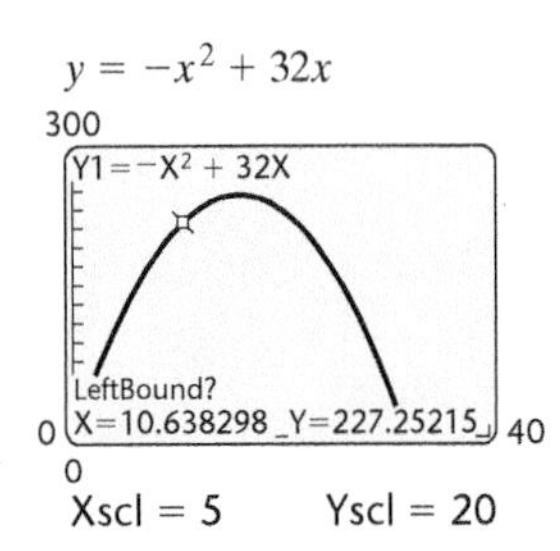

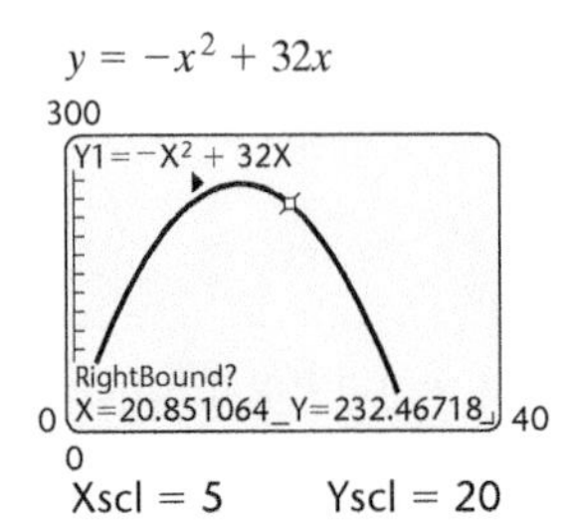

We press ENTER again. Finally, we are prompted to guess the x-value at which the maximum occurs. We move the cursor close to the maximum or key in an x-value. We press ENTER a third time and see that the maximum function value of 256 occurs when $x = 16$. (One or both coordinates of the maximum point might be approximations of the actual values, as shown with the x-value below.)

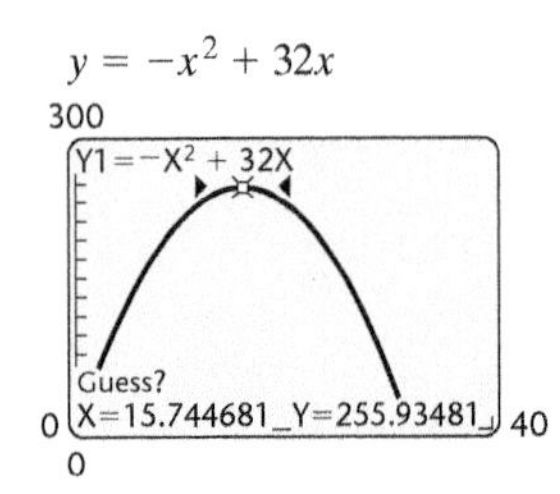

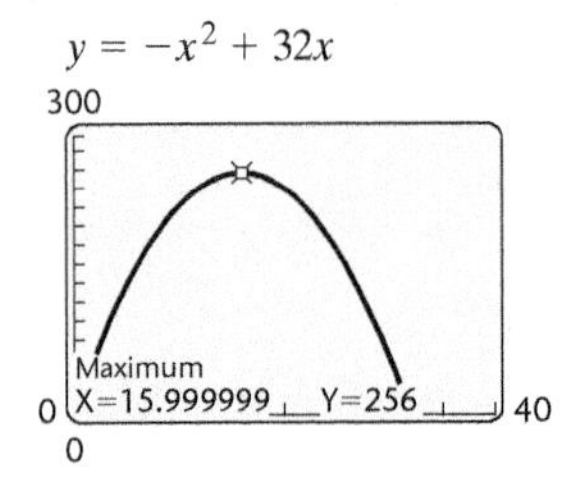

EXERCISES: Use the maximum or minimum feature on a graphing calculator to find the maximum or minimum value of each function.

1. $y = 3x^2 - 6x + 4$

2. $y = 2x^2 + x + 5$

3. $y = -x^2 + 4x + 2$

4. $y = -4x^2 + 5x - 1$

b FITTING QUADRATIC FUNCTIONS TO DATA

Find an equation of a line when two points are given. [2.6c]

Find an equation of the line containing the given points.

1. (2, 9) and (0, 8)

2. (−3, −1) and (−5, 7)

Answers: 1. $y = \frac{1}{2}x + 8$ **2.** $y = -4x - 13$

MyLab Math VIDEO

As we move through our study of mathematics, we develop a library of functions. These functions can serve as models for many applications. Some of them are graphed below. We have not considered the cubic or quartic functions in detail (we leave that discussion to a later course), but we show them here for reference.

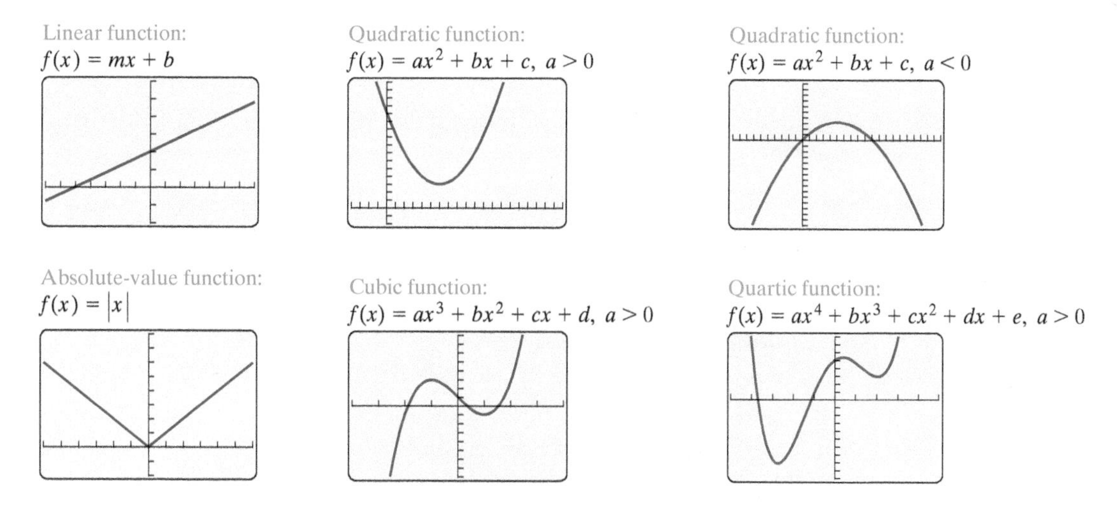

Now let's consider some real-world data. How can we decide which type of function might fit the data of a particular application? One simple way is to graph the data and look for a pattern resembling one of the graphs above. For example, data might be modeled by a linear function if the graph resembles a straight line. The data might be modeled by a quadratic function if the graph rises and then falls, or falls and then rises, in a curved manner resembling a parabola. For a quadratic, it might also just rise or fall in a curved manner as if following only one part of the parabola.

Let's now use our library of functions to see which, if any, might fit certain data situations.

EXAMPLES *Choosing Models.* For the scatterplots and graphs below, determine which, if any, of the following functions might be used as a model for the data.

Linear, $f(x) = mx + b$;

Quadratic, $f(x) = ax^2 + bx + c, a > 0$;

Quadratic, $f(x) = ax^2 + bx + c, a < 0$;

Polynomial, neither quadratic nor linear

2.

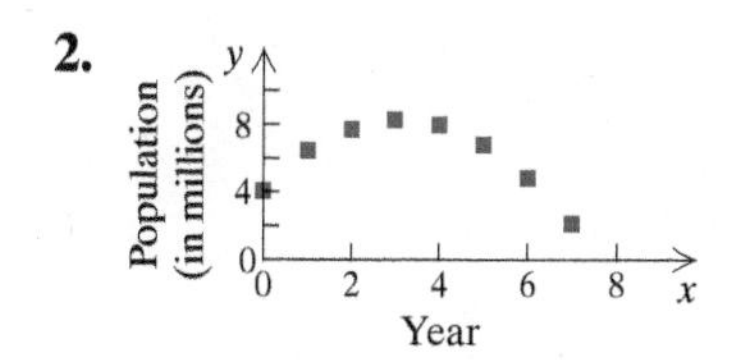

The data rise and then fall in a curved manner fitting a quadratic function $f(x) = ax^2 + bx + c, a < 0$.

3.

The data seem to fit a linear function $f(x) = mx + b$.

4.

The data rise in a manner fitting the right side of a quadratic function $f(x) = ax^2 + bx + c, a > 0$.

5. **Precipitation in Sonoma, California**

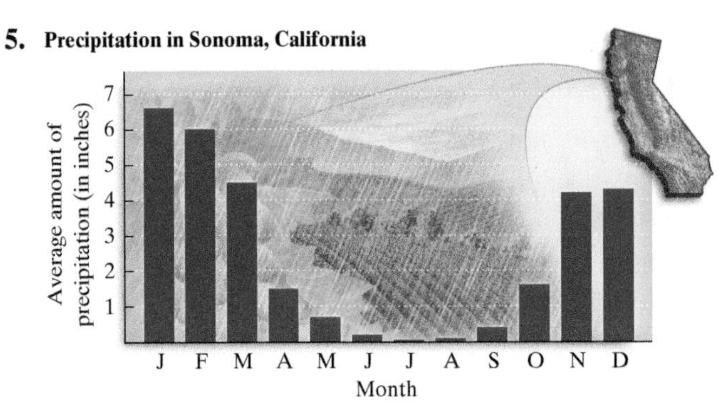

The data fall and then rise in a curved manner fitting a quadratic function $f(x) = ax^2 + bx + c, a > 0$.

6. **Average Number of Motorcyclists Killed per Hour on the Weekend**

DATA: FARS and GES

The data fall, then rise, then fall again. They do not appear to fit a linear function or a quadratic function but might fit a polynomial function that is neither quadratic nor linear.

Do Exercises 2–5.

Choosing Models. For the scatterplots in Margin Exercises 2–5, determine which, if any, of the following functions might be used as a model for the data:

Linear, $f(x) = mx + b$;

Quadratic, $f(x) = ax^2 + bx + c$, $a > 0$;

Quadratic, $f(x) = ax^2 + bx + c$, $a < 0$;

Polynomial, neither quadratic nor linear.

2.

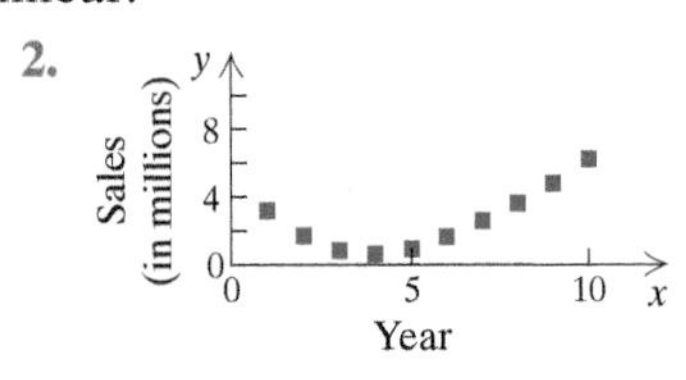

3.

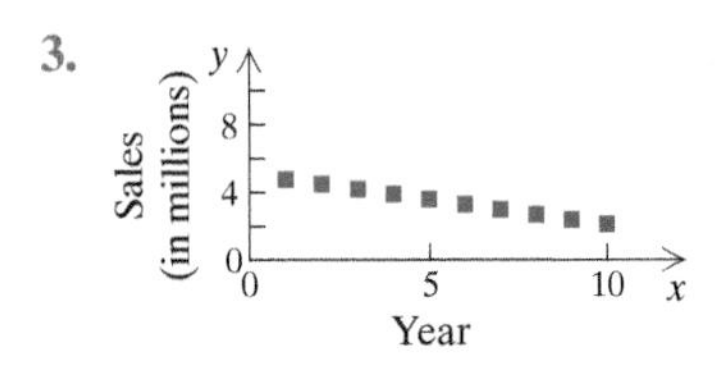

4.

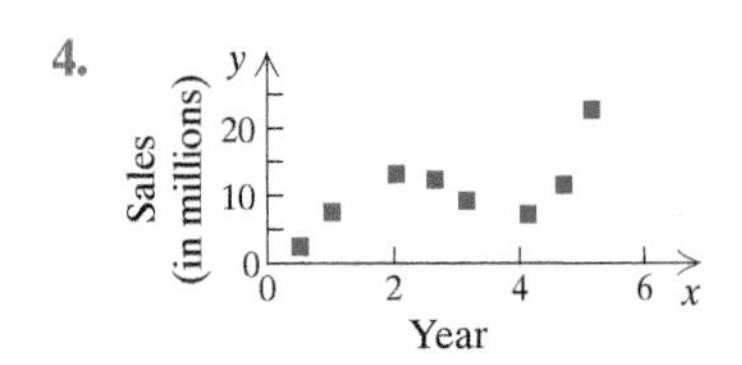

5.

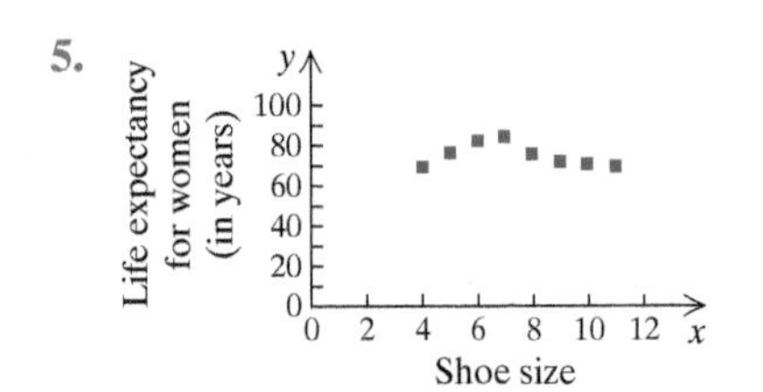

DATA: *Orthopedic Quarterly*

Answers

2. $f(x) = ax^2 + bx + c, a > 0$
3. $f(x) = mx + b$ **4.** Polynomial, neither quadratic nor linear **5.** Polynomial, neither quadratic nor linear

As we move through our study of mathematics, we develop a library of functions. These functions can serve as models for many applications. Some of them are graphed below. We have not considered the cubic or quartic functions in detail (we leave that discussion to a later course), but we show them here for reference.

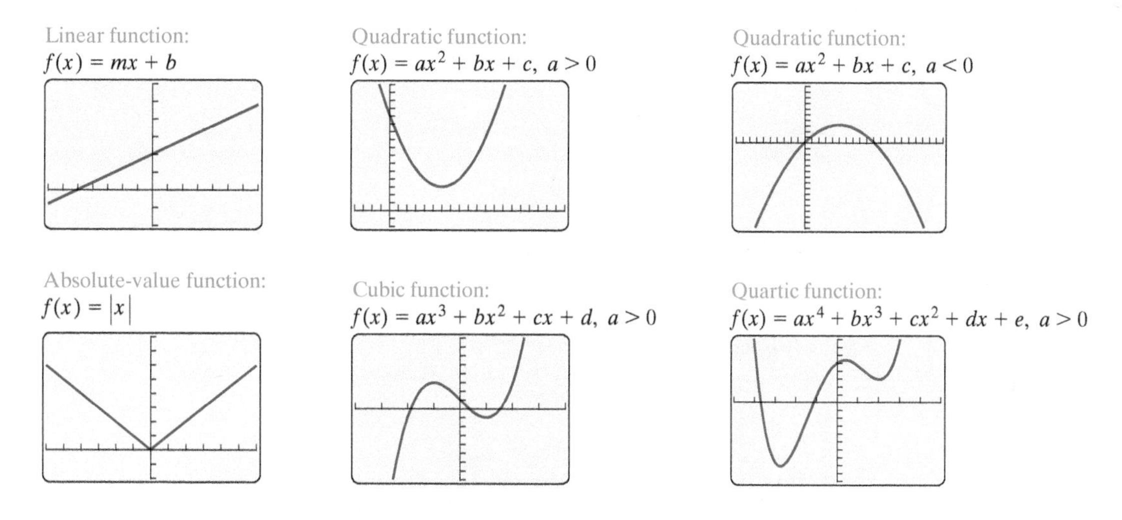

Now let's consider some real-world data. How can we decide which type of function might fit the data of a particular application? One simple way is to graph the data and look for a pattern resembling one of the graphs above. For example, data might be modeled by a linear function if the graph resembles a straight line. The data might be modeled by a quadratic function if the graph rises and then falls, or falls and then rises, in a curved manner resembling a parabola. For a quadratic, it might also just rise or fall in a curved manner as if following only one part of the parabola.

Let's now use our library of functions to see which, if any, might fit certain data situations.

EXAMPLES *Choosing Models.* For the scatterplots and graphs below, determine which, if any, of the following functions might be used as a model for the data.

Linear, $f(x) = mx + b$;

Quadratic, $f(x) = ax^2 + bx + c, a > 0$;

Quadratic, $f(x) = ax^2 + bx + c, a < 0$;

Polynomial, neither quadratic nor linear

2.

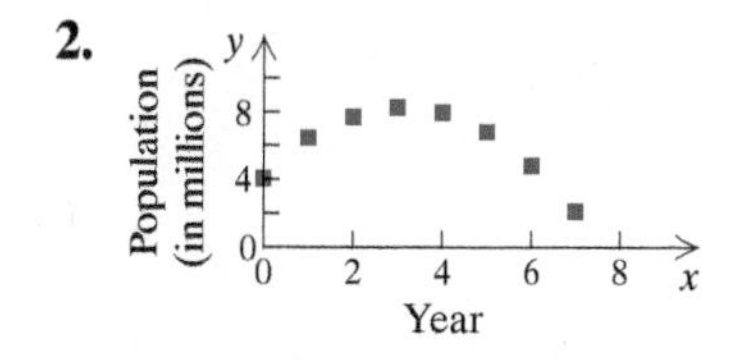

The data rise and then fall in a curved manner fitting a quadratic function $f(x) = ax^2 + bx + c, a < 0$.

3.

The data seem to fit a linear function $f(x) = mx + b$.

4.

The data rise in a manner fitting the right side of a quadratic function $f(x) = ax^2 + bx + c, a > 0$.

5. **Precipitation in Sonoma, California**

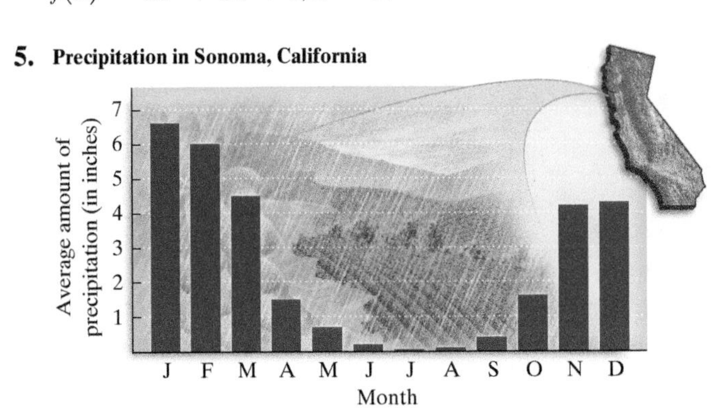

The data fall and then rise in a curved manner fitting a quadratic function $f(x) = ax^2 + bx + c, a > 0$.

6. **Average Number of Motorcyclists Killed per Hour on the Weekend**

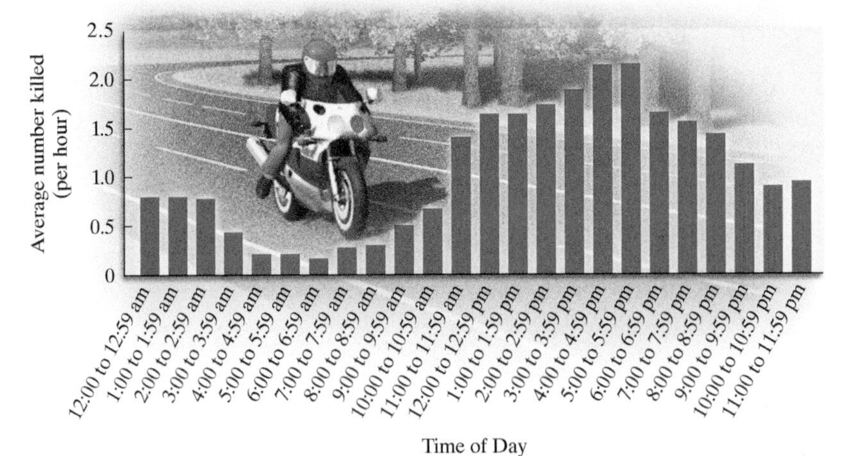

DATA: FARS and GES

The data fall, then rise, then fall again. They do not appear to fit a linear function or a quadratic function but might fit a polynomial function that is neither quadratic nor linear.

◀ **Do Exercises 2–5.**

Choosing Models. For the scatterplots in Margin Exercises 2–5, determine which, if any, of the following functions might be used as a model for the data:

Linear, $f(x) = mx + b$;

Quadratic, $f(x) = ax^2 + bx + c$, $a > 0$;

Quadratic, $f(x) = ax^2 + bx + c$, $a < 0$;

Polynomial, neither quadratic nor linear.

2.

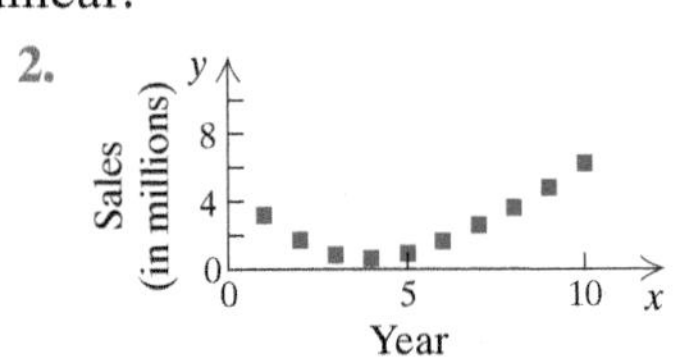

3.

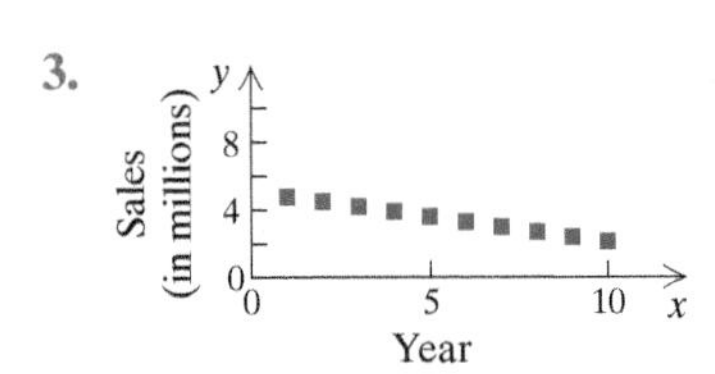

4.

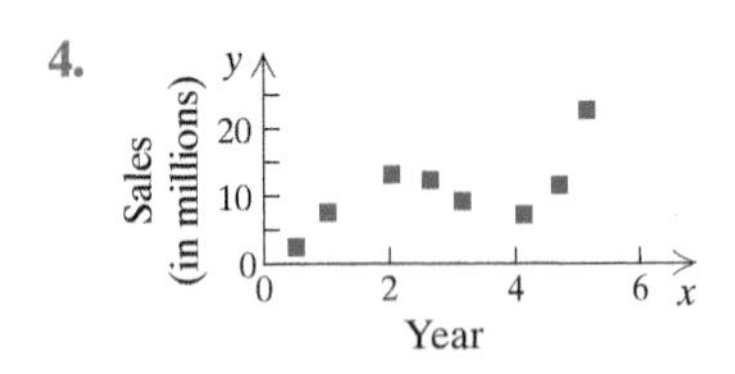

5.

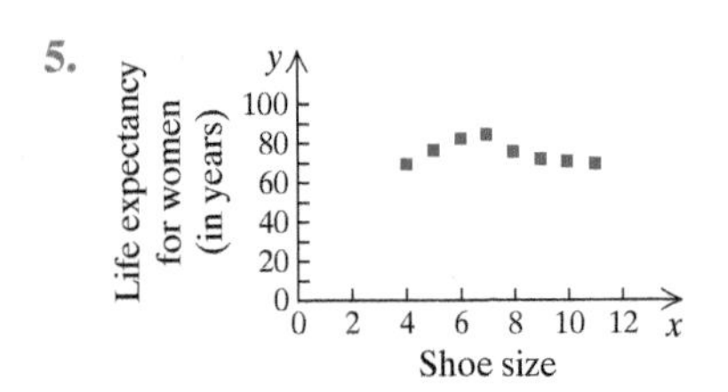

DATA: *Orthopedic Quarterly*

Answers

2. $f(x) = ax^2 + bx + c, a > 0$
3. $f(x) = mx + b$ **4.** Polynomial, neither quadratic nor linear **5.** Polynomial, neither quadratic nor linear

Whenever a quadratic function seems to fit a data situation, that function can be determined if at least three inputs and their outputs are known.

EXAMPLE 7 *Suspension Bridge.* The Clifton Suspension Bridge in Bristol, UK, completed in 1864, is one of the longest bridges suspended by chains. The data in the table at right show the height of the chains above the road surface at different points along the bridge. Here, 0 represents the center of the bridge, negative values of x represent distances in one direction from the center, and positive values of x represent distances in the other direction.

DATA: cliftonbridge.org. uk

LOCATION ON BRIDGE (in meters from center)	HEIGHT OF CHAIN (in meters)
0	1
20	1.7
40	4.0
−40	4.0
−60	7.6

a) Make a scatterplot of the data.

b) Decide whether the data seem to fit a quadratic function.

c) Use the data points (40, 4), (0, 1), and (−40, 4) to find a quadratic function that fits the data.

d) Use the function to estimate the height of the chains at the towers, which are each 107 m from the center of the bridge.

a) The scatterplot is shown at right.

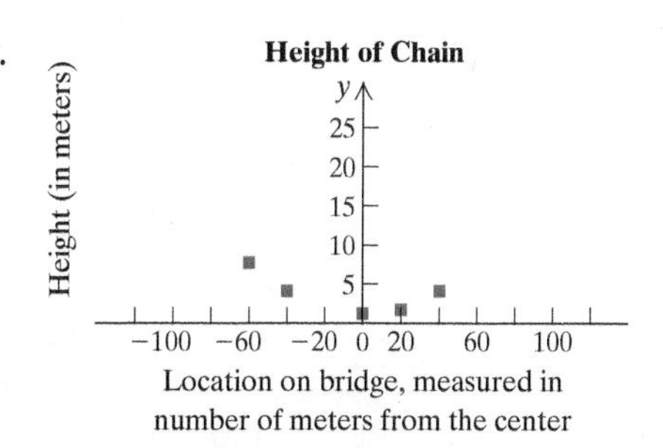

b) The data seem to fall and rise in a manner similar to a quadratic function. Note that there may not be a function that exactly fits all the data.

c) We are looking for a quadratic function

$$H(x) = ax^2 + bx + c.$$

We need to determine the constants a, b, and c. We use the three data points (40, 4), (0, 1), and (−40, 4) and substitute as follows:

$$4 = a \cdot 40^2 + b \cdot 40 + c,$$
$$1 = a \cdot 0^2 + b \cdot 0 + c,$$
$$4 = a \cdot (-40)^2 + b \cdot (-40) + c.$$

After simplifying, we see that we need to solve the system

$$4 = 1600a + 40b + c,$$
$$1 = c,$$
$$4 = 1600a - 40b + c.$$

Since $c = 1$, we need to solve a system of two equations in two variables:

$$4 = 1600a + 40b + 1 \qquad 3 = 1600a + 40b \qquad (1)$$
or
$$4 = 1600a - 40b + 1, \qquad 3 = 1600a - 40b. \qquad (2)$$

We add equations (1) and (2) and solve for a:

$$3 = 1600a + 40b$$
$$\underline{3 = 1600a - 40b}$$
$$6 = 3200a \qquad \text{Adding}$$
$$\frac{3}{1600} = a. \qquad \text{Solving for } a$$

Next, we substitute $\frac{3}{1600}$ for a in equation (1) and solve for b:

$$3 = 1600\left(\frac{3}{1600}\right) + 40b$$
$$3 = 3 + 40b$$
$$0 = 40b$$
$$0 = b.$$

This gives us the quadratic function:

$$H(x) = \frac{3}{1600}x^2 + 1, \quad \text{or} \quad H(x) = 0.001875x^2 + 1.$$

d) The towers are 107 m from the center of the bridge. Since the heights of the towers are the same, we need find only $H(107)$:

$$H(107) = 0.001875(107)^2 + 1 \approx 22.5.$$

At a distance of 107 m from the center of the bridge, the chains are about 22.5 m long.

◀ Do Exercise 6.

6. *Ticket Profits.* Valley Community College is presenting a series of plays. The profit P, in dollars, after x days is listed in the following table.

DAYS x	PROFIT P
0	$-100
90	560
180	872
270	870
360	548
450	−100

a) Make a scatterplot of the data.

b) Decide whether the data can be modeled by a quadratic function.

c) Use the data points $(0, -100)$, $(180, 872)$, and $(360, 548)$ to find a quadratic function that fits the data.

d) Use the function to estimate the profit after 225 days.

Answer

6. (a)

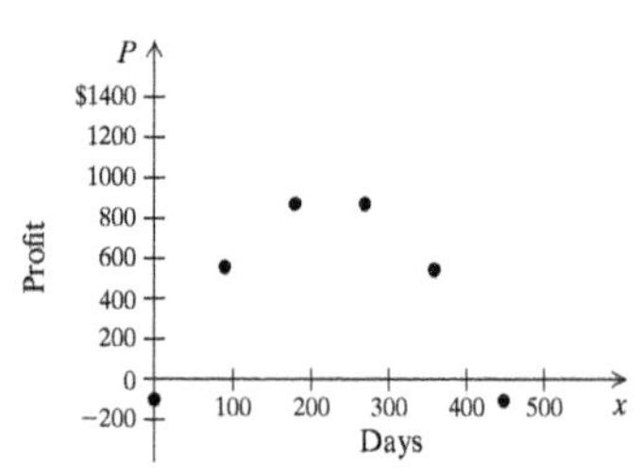

(b) yes; **(c)** $P(x) = -0.02x^2 + 9x - 100$; **(d)** $912.50

7.7 Exercise Set

FOR EXTRA HELP  MyLab Math

✓ Check Your Understanding

Reading Check Determine whether each statement is true or false.

RC1. Sometimes we can solve a problem without solving an equation or an inequality.

RC2. Every quadratic function has a maximum value.

RC3. A scatterplot can help us decide what type of function might fit a set of data.

RC4. When fitting a quadratic function to a set of data, the function must go through all the points on the scatterplot.

Concept Check Write an equation indicating that the ordered pair (x, y) is a solution of $y = ax^2 + bx + c$. (See Example 7.)

CC1. $(2, 4)$

CC2. $(5, 0)$

CC3. $(-6, 10)$

a Solve.

1. *Architecture.* An architect is designing a hotel with a central atrium. Each floor is to be rectangular and is allotted 720 ft of security piping around walls outside the rooms. What dimensions will allow the atrium to have the maximum area?

2. *Fenced-In Land.* A farmer has 100 yd of fencing. What are the dimensions of the largest rectangular pen that the farmer can enclose?

3. *Molding Plastics.* Economite Plastics plans to produce a one-compartment vertical file by bending the long side of an 8-in. by 14-in. sheet of plastic along two lines to form a U shape. How tall should the file be in order to maximize the volume that the file can hold?

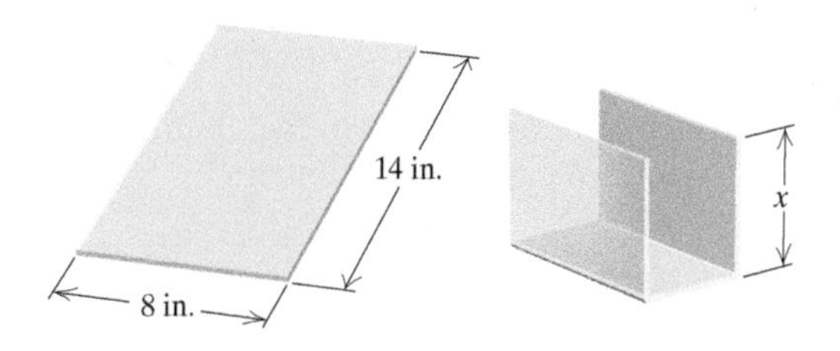

4. *Patio Design.* A stone mason has enough stones to enclose a rectangular patio with a perimeter of 60 ft, assuming that the attached house forms one side of the rectangle. What is the maximum area that the mason can enclose? What should the dimensions of the patio be in order to yield this area?

5. *Minimizing Cost.* Aki's Bicycle Designs has determined that when x hundred bicycles are built, the average cost per bicycle is given by

$$C(x) = 0.1x^2 - 0.7x + 2.425,$$

where $C(x)$ is in hundreds of dollars. How many bicycles should the shop build in order to minimize the average cost per bicycle?

6. *Corral Design.* A rancher needs to enclose two adjacent rectangular corrals, one for sheep and one for cattle. If a river forms one side of the corrals and 180 yd of fencing is available, what is the largest total area that can be enclosed?

7. *Garden Design.* A farmer decides to enclose a rectangular garden, using the side of a barn as one side of the rectangle. What is the maximum area that the farmer can enclose with 40 ft of fence? What should the dimensions of the garden be in order to yield this area?

8. *Composting.* A rectangular compost container is to be formed in a corner of a fenced yard, with 8 ft of chicken wire completing the other two sides of the rectangle. If the chicken wire is 3 ft high, what dimensions of the base will maximize the volume of the container?

9. *Ticket Sales.* The number of tickets sold each day for an upcoming performance of Handel's *Messiah* is given by

$$N(x) = -0.4x^2 + 9x + 11,$$

where x is the number of days after the concert was first announced. When will daily ticket sales peak and how many tickets will be sold that day?

10. *Stock Prices.* The value of a share of a particular stock, in dollars, can be represented by $V(x) = x^2 - 6x + 13$, where x is the number of months after January 2018. What is the lowest value that $V(x)$ will reach, and when will that occur?

Maximizing Profit. Total profit P is the difference between total revenue R and total cost C. Given the following total-revenue and total-cost functions, find the total profit, the maximum value of the total profit, and the value of x at which it occurs.

11. $R(x) = 1000x - x^2$,
$C(x) = 3000 + 20x$

12. $R(x) = 200x - x^2$,
$C(x) = 5000 + 8x$

13. What is the maximum product of two numbers whose sum is 22? What numbers yield this product?

14. What is the maximum product of two numbers whose sum is 45? What numbers yield this product?

15. What is the minimum product of two numbers whose difference is 4? What are the numbers?

16. What is the minimum product of two numbers whose difference is 6? What are the numbers?

17. What is the maximum product of two numbers that add to -12? What numbers yield this product?

18. What is the minimum product of two numbers that differ by 9? What are the numbers?

b *Choosing Models.* For the scatterplots and graphs in Exercises 19–26, determine which, if any, of the following functions might be used as a model for the data: Linear, $f(x) = mx + b$; quadratic, $f(x) = ax^2 + bx + c, a > 0$; quadratic, $f(x) = ax^2 + bx + c, a < 0$; polynomial, neither quadratic nor linear.

19.

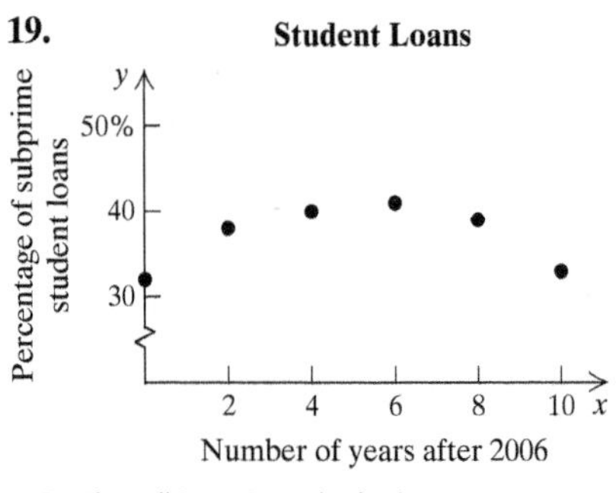

DATA: *The Wall Street Journal*, 4/25/17

20.

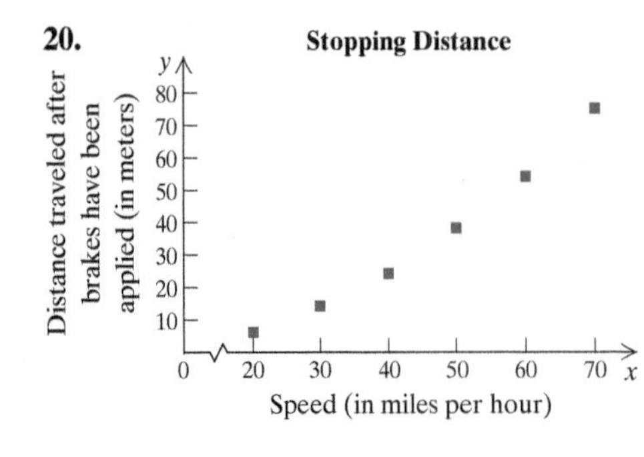

21.

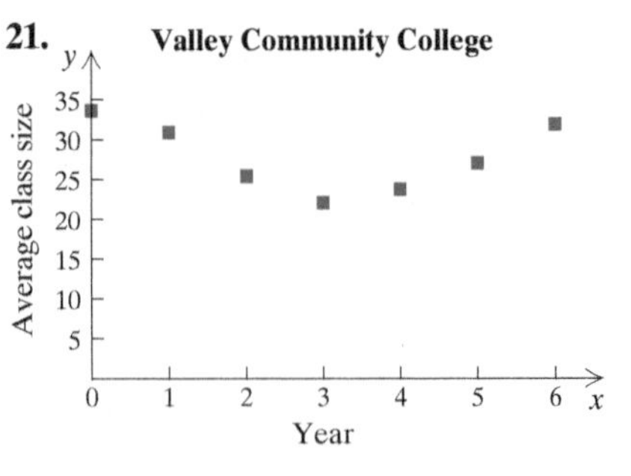

22.

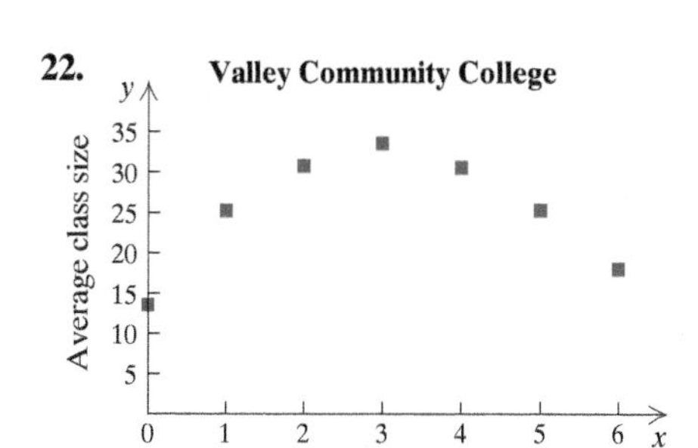

23.

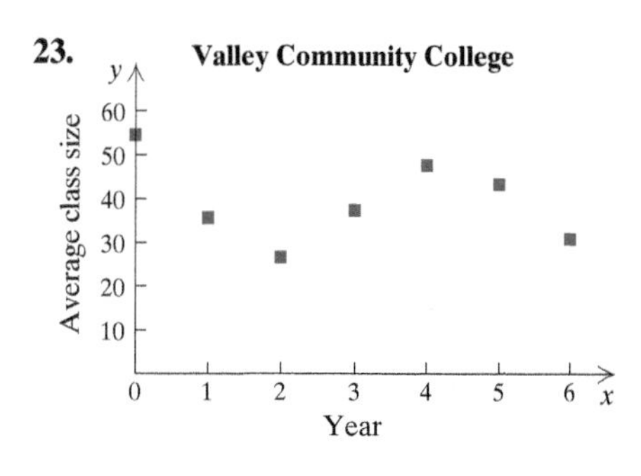

24.

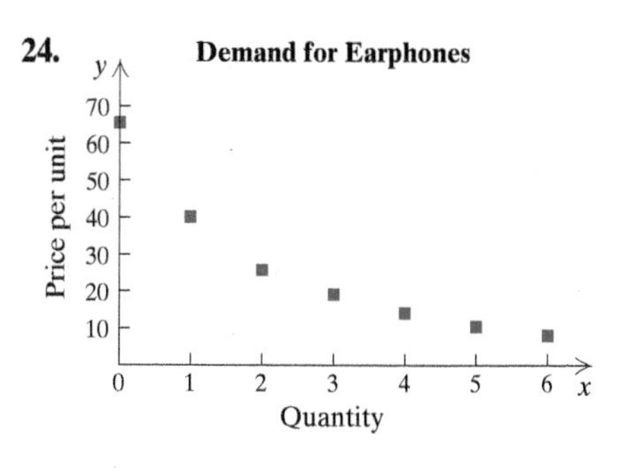

25. **High School Graduation Rate**

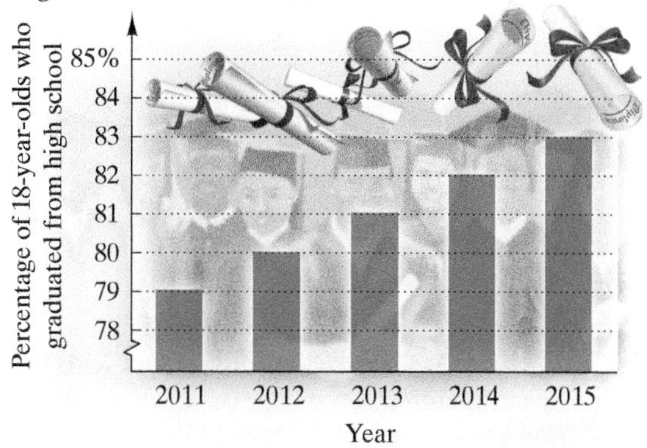

DATA: National Center for Education Statistics

26. **Hurricanes in the Atlantic Basin**

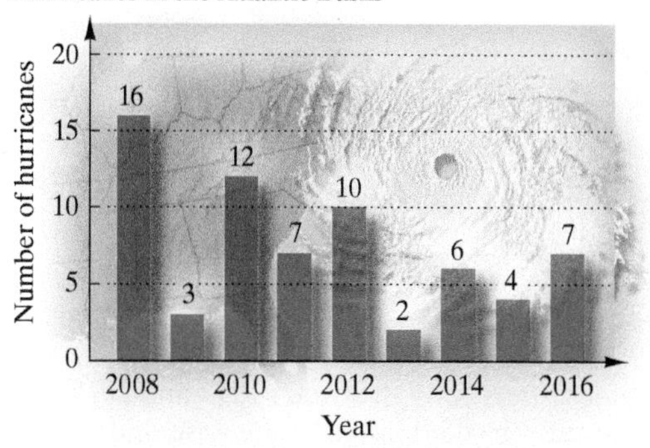

DATA: National Oceanic and Atmospheric Administration/ Hurricane Research Division

Find a quadratic function that fits the set of data points.

27. $(1, 4), (-1, -2), (2, 13)$

28. $(1, 4), (-1, 6), (-2, 16)$

29. $(2, 0), (4, 3), (12, -5)$

30. $(-3, -30), (3, 0), (6, 6)$

31. *Nighttime Accidents.*

a) Find a quadratic function that fits the following data.

TRAVEL SPEED (in kilometers per hour)	NUMBER OF NIGHTTIME ACCIDENTS (for every 200 million kilometers driven)
60	400
80	250
100	250

b) Use the function to estimate the number of nighttime accidents that occur at 50 km/h.

32. *Daytime Accidents.*

a) Find a quadratic function that fits the following data.

TRAVEL SPEED (in kilometers per hour)	NUMBER OF DAYTIME ACCIDENTS (for every 200 million kilometers driven)
60	100
80	130
100	200

b) Use the function to estimate the number of daytime accidents that occur at 50 km/h.

33. *River Depth.* Typically, rivers are deepest in the middle, with the depth decreasing to zero at the edges. A hydrologist measures the depths D, in feet, of a river at distances x, in feet, from one bank. The results are listed in the following table. Use the data points $(0, 0)$, $(50, 20)$, and $(100, 0)$ to find a quadratic function that fits the data. Then use the function to estimate the depth of the river at 75 ft from the bank.

DISTANCE x FROM THE RIVERBANK (in feet)	DEPTH D OF THE RIVER (in feet)
0	0
15	10.2
25	17
50	20
90	7.2
100	0

x = distance from left bank (in feet)

$D(x)$ = depth of river (in feet)

34. *Canoe Depth.* The following figure shows the cross section of a canoe. Canoes are deepest at the middle of the center line, with the depth decreasing to zero at the edges. The following table lists measures of the depths D, in inches, along the center line of one canoe at distances x, in inches, from the edge. Use the data points $(0, 0)$, $(18, 14)$, and $(36, 0)$ to find a quadratic function that fits the data. Then use the function to estimate the depth of the canoe 10 in. from the edge along the center line.

DISTANCE x FROM THE EDGE OF THE CANOE ALONG THE CENTER LINE (in inches)	DEPTH D OF THE CANOE (in inches)
0	0
9	10.5
18	14
30	7.75
36	0

x = distance from edge of canoe (in inches)

$D(x)$ = depth of canoe (in inches)

35. *U.S. Homeownership.* Between 1995 and 2015, homeownership rates in the United States rose and then fell in a pattern that can be modeled by a quadratic function, as shown in the following graph. Use the data points $(4, 67)$, $(8, 68)$, and $(12, 68)$ to find a quadratic function that fits the data. Then use the function to estimate the homeownership rate in 2005.

Homeownership Rates

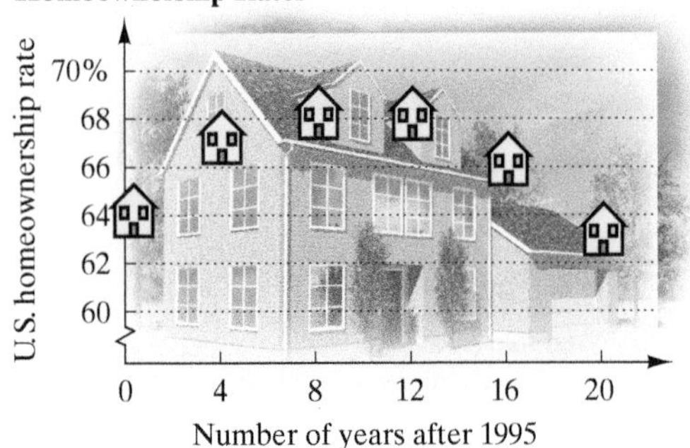

DATA: U.S. Census Bureau

36. *Cake Servings.* Tyler provides customers with the following chart showing the number of servings for various sizes of round cakes.

DIAMETER	NUMBER OF SERVINGS
6 in.	12
8 in.	24
10 in.	35
12 in.	56
14 in.	78

The number of servings should probably be a function of diameter because it should be proportional to the area, and the area is a quadratic function of the diameter. (The area of a circular region is given by $A = \pi r^2$ or $(\pi/4)d^2$.)

a) Express the number of servings as a quadratic function of diameter using the data points $(6, 12)$, $(8, 24)$, and $(12, 56)$.

b) Use the function to find the number of servings in a 9-in. cake.

Skill Maintenance

Add, subtract, or multiply.

37. $(-3x^2 - x - 2) + (x^2 + 3x - 7)$ [4.1c]

38. $(2mn^2 - n^2 - 3m^2n) - (m^2n + 2mn^2 - n^2)$ [4.1d]

39. $(c^2d + 2y)(c^2d - 2y)$ [4.2d]

Factor.

40. $100t^2 - 81$ [4.6b]

41. $12x^3 - 60x^2 + 75x$ [4.6a]

42. $6y^2 + y - 12$ [4.5a, b]

Polynomial Inequalities and Rational Inequalities

7.8

OBJECTIVES

a Solve quadratic inequalities and other polynomial inequalities.

b Solve rational inequalities.

a QUADRATIC AND OTHER POLYNOMIAL INEQUALITIES

Inequalities like the following are called **quadratic inequalities:**

$$x^2 + 3x - 10 < 0, \qquad 5x^2 - 3x + 2 \geq 0.$$

In each case, we have a polynomial of degree 2 on the left. We will solve such inequalities in two ways. The first method provides understanding and the second yields the more efficient method.

We can solve a quadratic inequality, such as $ax^2 + bx + c > 0$, by considering the graph of a related function, $f(x) = ax^2 + bx + c$.

MyLab Math ANIMATION

SKILL REVIEW *Write interval notation for the solution set or graph of an inequality.* [1.4b]

Write interval notation for the given set.

1. $\{x \mid -3 < x \leq 10\}$
2. $\{y \mid y > -\frac{1}{2}\}$

Answers: 1. $(-3, 10]$ 2. $\left(-\frac{1}{2}, \infty\right)$

MyLab Math VIDEO

EXAMPLE 1 Solve: $x^2 + 3x - 10 > 0$.

Consider the function $f(x) = x^2 + 3x - 10$ and its graph. The graph opens up since the leading coefficient $(a = 1)$ is positive. We find the x-intercepts by setting the polynomial equal to 0 and solving:

$$x^2 + 3x - 10 = 0$$
$$(x + 5)(x - 2) = 0$$
$$x + 5 = 0 \quad or \quad x - 2 = 0$$
$$x = -5 \quad or \quad x = 2.$$

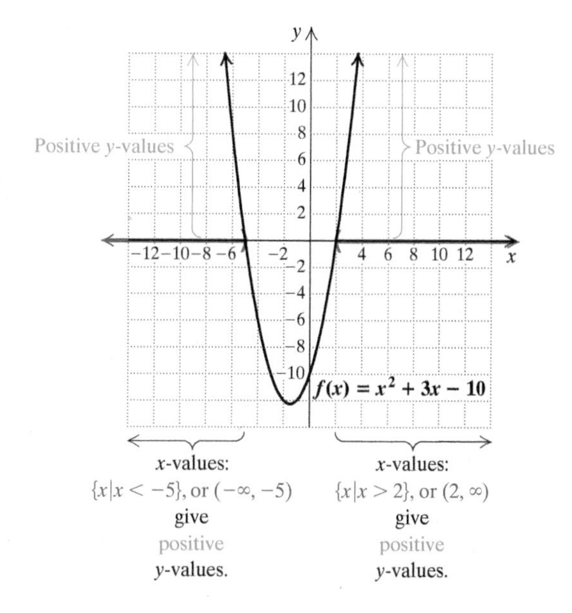

Values of y will be positive to the left and right of the intercepts, as shown. Thus the solution set of the inequality is

$$\{x \mid x < -5 \text{ or } x > 2\}, \quad \text{or} \quad (-\infty, -5) \cup (2, \infty).$$

Do Exercise 1.

1. Solve by graphing:
$x^2 + 2x - 3 > 0$.

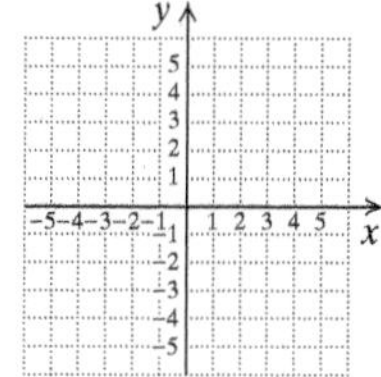

We can solve any inequality by considering the graph of a related function and finding x-intercepts, as in Example 1. In some cases, we may need to use the quadratic formula to find the intercepts.

Answer

1. $\{x \mid x < -3 \text{ or } x > 1\}$, or $(-\infty, -3) \cup (1, \infty)$

EXAMPLE 2 Solve: $x^2 + 3x - 10 < 0$.

Looking again at the graph of $f(x) = x^2 + 3x - 10$ or at least visualizing it tells us that y-values are negative for those x-values between -5 and 2.

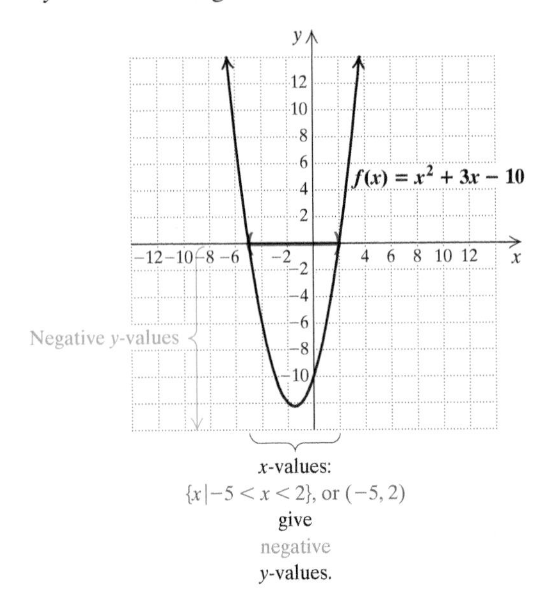

Solve by graphing.

2. $x^2 + 2x - 3 < 0$

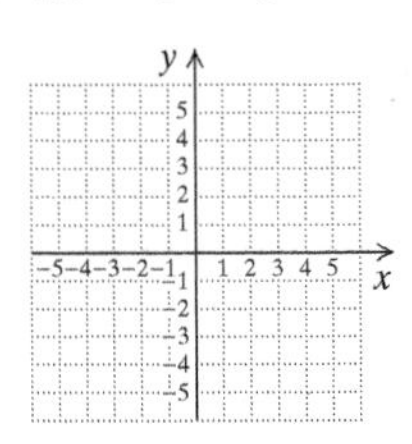

3. $x^2 + 2x - 3 \leq 0$

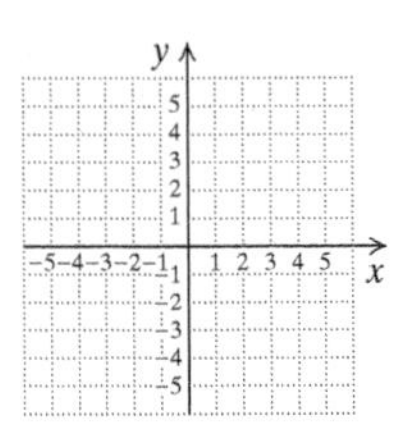

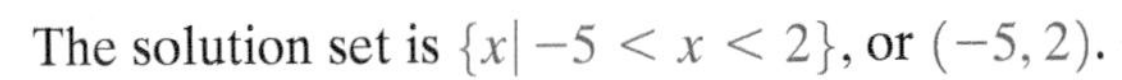

The solution set is $\{x|-5 < x < 2\}$, or $(-5, 2)$.

When an inequality contains $\leq$ or $\geq$, the x-values of the x-intercepts must be included. Thus the solution set of the inequality $x^2 + 3x - 10 \leq 0$ is $\{x|-5 \leq x \leq 2\}$, or $[-5, 2]$.

Do Exercises 2 and 3.

We now consider a more efficient method for solving polynomial inequalities. The preceding discussion provides the understanding for this method. In Examples 1 and 2, we see that the x-intercepts divide the number line into intervals.

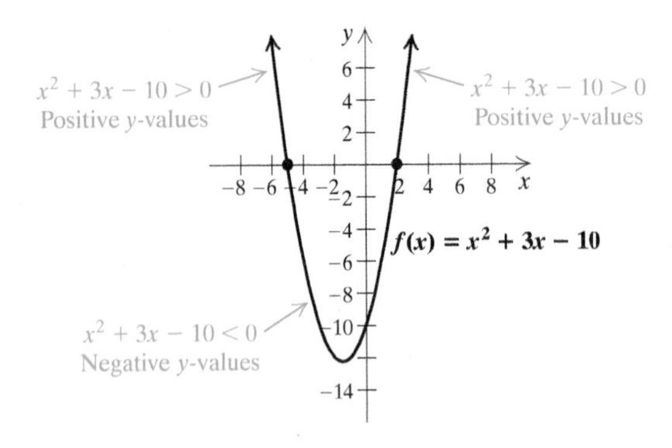

If a function has a positive output for one number in an interval, it will be positive for all the numbers in the interval. The same is true for negative outputs. Thus we can merely make a test substitution in each interval to solve the inequality. This is very similar to our method of using test points to graph a linear inequality in a plane.

Answers

2. $\{x|-3 < x < 1\}$, or $(-3, 1)$
3. $\{x|-3 \leq x \leq 1\}$, or $[-3, 1]$

EXAMPLE 3 Solve: $x^2 + 3x - 10 < 0$.

We set the polynomial equal to 0 and solve. The solutions of $x^2 + 3x - 10 = 0$, or $(x + 5)(x - 2) = 0$, are -5 and 2. We locate the solutions on the number line as follows. Note that the numbers divide the number line into three intervals, which we will call A, B, and C. Within each interval, the values of the function $f(x) = x^2 + 3x - 10$ will be all positive or will be all negative.

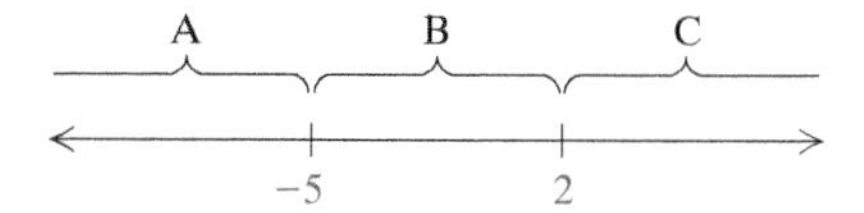

We choose a test number in interval A, say -7, and substitute -7 for x in the function $f(x) = x^2 + 3x - 10$:

$$f(-7) = (-7)^2 + 3(-7) - 10 = 49 - 21 - 10 = 18.$$

Since $f(-7) = 18$ and $18 > 0$, the function values will be positive for any number in interval A.

Next, we try a test number in interval B, say 1, and find the corresponding function value:

$$f(1) = 1^2 + 3(1) - 10 = 1 + 3 - 10 = -6.$$

Since $f(1) = -6$ and $-6 < 0$, the function values will be negative for any number in interval B.

Next, we try a test number in interval C, say 4, and find the corresponding function value:

$$f(4) = 4^2 + 3(4) - 10 = 16 + 12 - 10 = 18.$$

Since $f(4) = 18$ and $18 > 0$, the function values will be positive for any number in interval C.

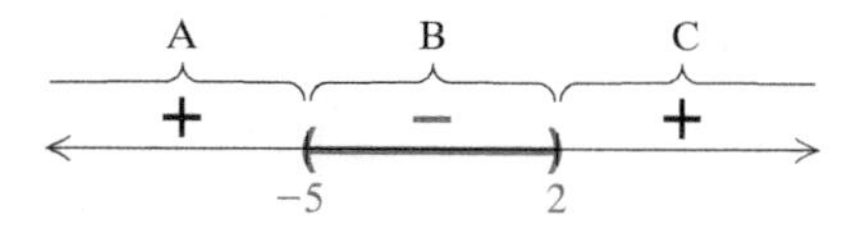

We are looking for numbers x for which $f(x) = x^2 + 3x - 10 < 0$. Thus any number x in interval B is a solution. The solution set is $\{x | -5 < x < 2\}$, or the interval $(-5, 2)$. If the inequality had been $\leq$, it would have been necessary to include the endpoints -5 and 2 in the solution set as well.

Do Exercises 4 and 5. ▶

Solve using the method of Example 3.

4. $x^2 + 3x > 4$

5. $x^2 + 3x \leq 4$

To solve a polynomial inequality:

1. Get 0 on one side, set the expression on the other side equal to 0, and solve to find the x-intercepts.
2. Use the numbers found in step (1) to divide the number line into intervals.
3. Substitute a number from each interval into the related function. If the function value is positive, then the expression will be positive for all numbers in the interval. If the function value is negative, then the expression will be negative for all numbers in the interval.
4. Select the intervals for which the inequality is satisfied and write set-builder notation or interval notation for the solution set.

Answers

4. $\{x | x < -4 \text{ or } x > 1\}$, or $(-\infty, -4) \cup (1, \infty)$
5. $\{x | -4 \leq x \leq 1\}$, or $[-4, 1]$

EXAMPLE 4 Solve: $5x(x + 3)(x - 2) \geq 0$.

The solutions of $f(x) = 0$, or $5x(x + 3)(x - 2) = 0$, are 0, -3, and 2. They divide the number line into four intervals, as shown below.

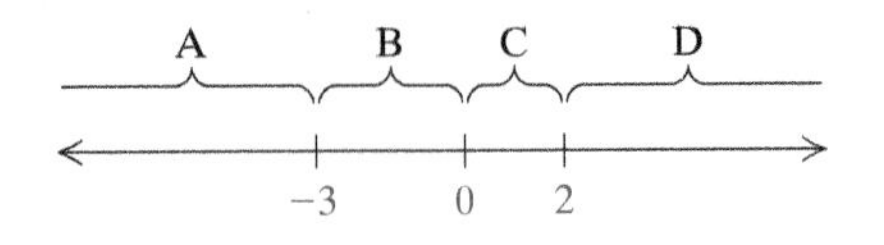

We try test numbers in each interval:

A: Test -5, $f(-5) = 5(-5)(-5 + 3)(-5 - 2) = -350 < 0$.
B: Test -2, $f(-2) = 5(-2)(-2 + 3)(-2 - 2) = 40 > 0$.
C: Test 1, $f(1) = 5(1)(1 + 3)(1 - 2) = -20 < 0$.
D: Test 3, $f(3) = 5(3)(3 + 3)(3 - 2) = 90 > 0$.

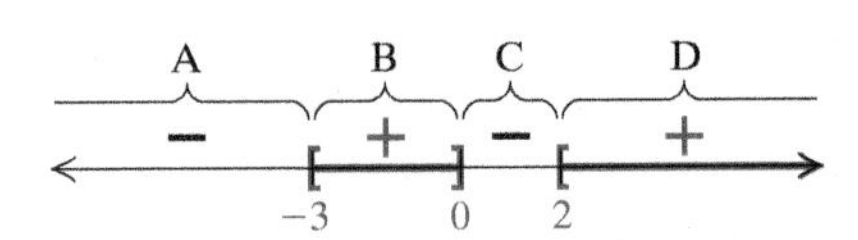

The expression is positive for values of x in intervals B and D. Since the inequality symbol is $\geq$, we must include the x-intercepts. The solution set of the inequality is

$$\{x \mid -3 \leq x \leq 0 \text{ or } x \geq 2\}, \quad \text{or} \quad [-3, 0] \cup [2, \infty).$$

We visualize this with the following graph.

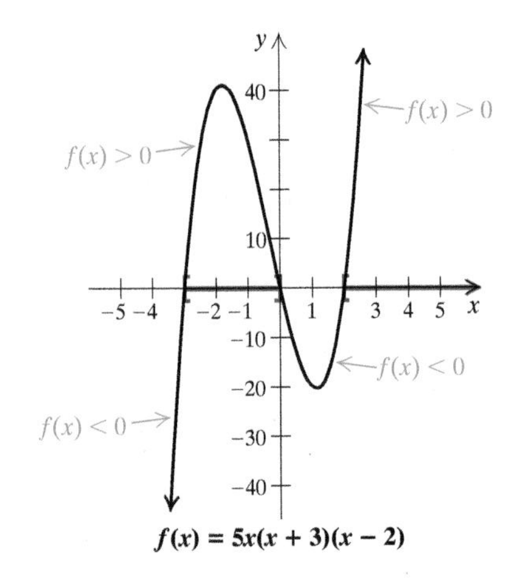

6. Solve: $6x(x + 1)(x - 1) < 0$.

The solutions of $6x(x + 1)(x - 1) = 0$ are 0, -1, and ______. Divide the number line into four intervals and test values of $f(x) = 6x(x + 1)(x - 1)$.

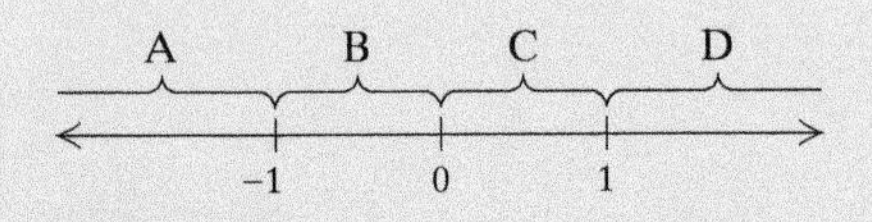

A: Test -2.
$f(-2) = 6(\quad)(-2 + 1)(-2 - 1)$
$= \quad$

B: Test $-\frac{1}{2}$.
$f\left(-\frac{1}{2}\right) = 6(\quad)\left(-\frac{1}{2} + 1\right)\left(-\frac{1}{2} - 1\right)$
$= \quad$

C: Test $\frac{1}{2}$.
$f\left(\frac{1}{2}\right) = 6(\quad)\left(\frac{1}{2} + 1\right)\left(\frac{1}{2} - 1\right)$
$= \quad$

D: Test 2.
$f(2) = 6(\quad)(2 + 1)(2 - 1)$
$= \quad$

The expression is negative for values of x in intervals A and ______. The solution set is $\{x \mid x < -1 \text{ or } 0 < x < 1\}$, or $(-\infty, \quad) \cup (0, \quad)$.

Do Exercise 6.

b RATIONAL INEQUALITIES

We adapt the preceding method for inequalities that involve rational expressions. We call these **rational inequalities**.

Answer

6. $\{x \mid x < -1 \text{ or } 0 < x < 1\}$, or $(-\infty, -1) \cup (0, 1)$

Guided Solution:

6. $1, -2, -36, -\frac{1}{2}, \frac{9}{4}, \frac{1}{2}, -\frac{9}{4}, 2, 36$, C, -1, 1

EXAMPLE 5 Solve: $\dfrac{x-3}{x+4} \geq 2$.

We write a related equation by changing the $\geq$ symbol to $=$:

$$\frac{x-3}{x+4} = 2.$$

Then we solve this related equation. First, we multiply on both sides of the equation by the LCM, which is $x + 4$:

$$(x+4) \cdot \frac{x-3}{x+4} = (x+4) \cdot 2$$
$$x - 3 = 2x + 8$$
$$-11 = x.$$

With rational inequalities, we must also determine those numbers for which the rational expression is not defined—that is, those numbers that make the denominator 0. We set the denominator equal to 0 and solve: $x + 4 = 0$, or $x = -4$. Next, we use the numbers -11 and -4 to divide the number line into intervals, as shown below.

A B C

-11 -4

We try test numbers in each interval to see if each satisfies the original inequality.

A: Test -15,

$$\frac{x-3}{x+4} \geq 2$$
$$\frac{-15-3}{-15+4} \;?\; 2$$
$$\frac{18}{11} \qquad \text{FALSE}$$

Since the inequality is false for $x = -15$, the number -15 is not a solution of the inequality. Interval A *is not* part of the solution set.

B: Test -8,

$$\frac{x-3}{x+4} \geq 2$$
$$\frac{-8-3}{-8+4} \;?\; 2$$
$$\frac{11}{4} \qquad \text{TRUE}$$

Since the inequality is true for $x = -8$, the number -8 is a solution of the inequality. Interval B *is* part of the solution set.

C: Test 1,

$$\frac{x-3}{x+4} \geq 2$$
$$\frac{1-3}{1+4} \;?\; 2$$
$$-\frac{2}{5} \qquad \text{FALSE}$$

Since the inequality is false for $x = 1$, the number 1 is not a solution of the inequality. Interval C *is not* part of the solution set.

The solution set includes interval B. The number -11 is also included since the inequality symbol is $\geq$ and -11 is a solution of the related equation. The number -4 is not included; it is not an allowable replacement because it results in division by 0. Thus the solution set of the original inequality is

$$\{x \mid -11 \leq x < -4\}, \quad \text{or} \quad [-11, -4).$$

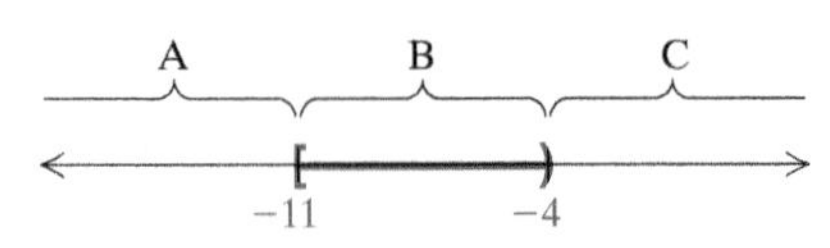

◀ **Do Exercises 7 and 8.**

Solve.

7. $\dfrac{x+1}{x-2} \geq 3$

8. $\dfrac{x}{x-5} < 2$

To solve a rational inequality:

1. Change the inequality symbol to an equals sign and solve the related equation.
2. Find the numbers for which any denominator in the inequality is not defined.
3. Use the numbers found in steps (1) and (2) to divide the number line into intervals.
4. Substitute a number from each interval into the inequality. If the number is a solution, then the interval to which it belongs is part of the solution set.
5. Select the intervals for which the inequality is satisfied and write set-builder notation or interval notation for the solution set.

Answers

7. $\left\{x \mid 2 < x \leq \frac{7}{2}\right\}$, or $\left(2, \frac{7}{2}\right]$

8. $\{x \mid x < 5 \text{ or } x > 10\}$, or $(-\infty, 5) \cup (10, \infty)$

7.8 Exercise Set

FOR EXTRA HELP  MyLab Math

✓ Check Your Understanding

Reading Check Complete each statement using either "positive" or "negative."

RC1. To solve $x^2 - 2 < 0$, we look for intervals for which $f(x) = x^2 - 2$ is ______________.

RC2. To solve $\dfrac{x}{x+1} > 0$, we look for intervals for which $f(x) = \dfrac{x}{x+1}$ is ______________.

RC3. To solve $3x < 5 + x^2$, we look for intervals for which $f(x) = x^2 - 3x + 5$ is ______________.

RC4. To solve $(x-1)(x+2) > 3x^2$, we look for intervals for which $f(x) = 2x^2 - x + 2$ is ______________.

Concept Check Determine whether -1 is a solution of each inequality.

CC1. $x^2 + 3x < 0$

CC2. $x^2 + x < 0$

CC3. $\dfrac{2x}{x-2} \geq 1$

CC4. $\dfrac{x+1}{x-1} \leq 0$

a Solve algebraically and verify results from the graph.

1. $(x - 6)(x + 2) > 0$

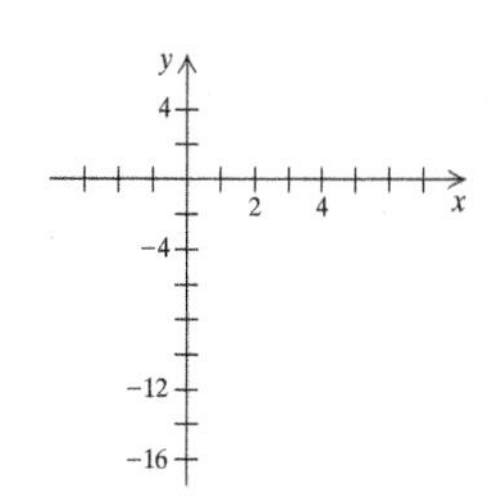

2. $(x - 5)(x + 1) > 0$

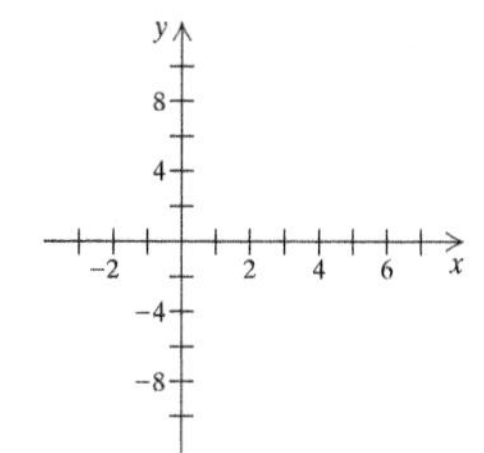

3. $4 - x^2 \geq 0$

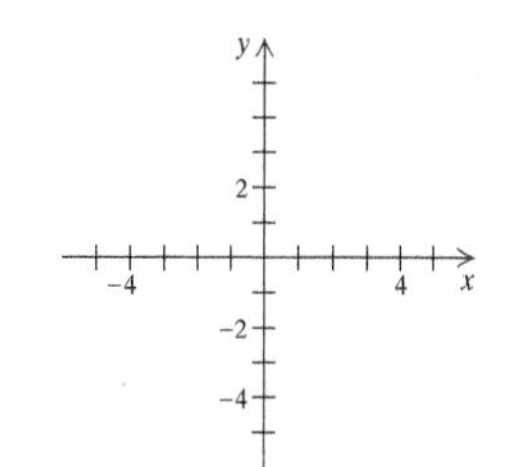

4. $9 - x^2 \leq 0$

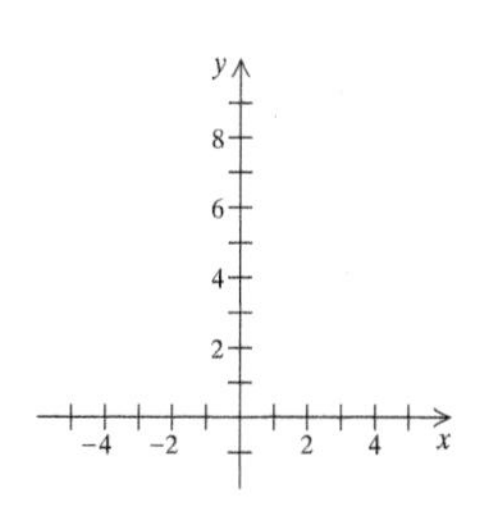

Solve.

5. $3(x + 1)(x - 4) \leq 0$

6. $(x - 7)(x + 3) \leq 0$

7. $x^2 - x - 2 < 0$

8. $x^2 + x - 2 < 0$

9. $x^2 - 2x + 1 \geq 0$

10. $x^2 + 6x + 9 < 0$

11. $x^2 + 8 < 6x$

12. $x^2 - 12 > 4x$

13. $3x(x + 2)(x - 2) < 0$

14. $5x(x + 1)(x - 1) > 0$

15. $(x + 9)(x - 4)(x + 1) > 0$

16. $(x - 1)(x + 8)(x - 2) < 0$

17. $(x + 3)(x + 2)(x - 1) < 0$

18. $(x - 2)(x - 3)(x + 1) < 0$

b Solve.

19. $\frac{1}{x - 6} < 0$

20. $\frac{1}{x + 4} > 0$

21. $\frac{x + 1}{x - 3} > 0$

22. $\frac{x - 2}{x + 5} < 0$

23. $\frac{3x + 2}{x - 3} \leq 0$

24. $\frac{5 - 2x}{4x + 3} \leq 0$

25. $\frac{x - 1}{x - 2} > 3$

26. $\frac{x + 1}{2x - 3} < 1$

27. $\dfrac{(x-2)(x+1)}{x-5} < 0$

28. $\dfrac{(x+4)(x-1)}{x+3} > 0$

29. $\dfrac{x+3}{x} \leq 0$

30. $\dfrac{x}{x-2} \geq 0$

31. $\dfrac{x}{x-1} > 2$

32. $\dfrac{x-5}{x} < 1$

33. $\dfrac{x-1}{(x-3)(x+4)} < 0$

34. $\dfrac{x+2}{(x-2)(x+7)} > 0$

35. $3 < \dfrac{1}{x}$

36. $\dfrac{1}{x} \leq 2$

37. $\dfrac{x^2+x-2}{x^2-x-12} > 0$

38. $\dfrac{x^2-11x+30}{x^2-8x-9} \geq 0$

Skill Maintenance

Simplify. [6.3b]

39. $\sqrt[3]{\dfrac{125}{27}}$

40. $\sqrt{\dfrac{25}{4a^2}}$

41. $\sqrt{\dfrac{16a^3}{b^4}}$

42. $\sqrt[3]{\dfrac{27c^5}{343d^3}}$

Add or subtract. [6.4a]

43. $3\sqrt{8} - 5\sqrt{2}$

44. $7\sqrt{45} - 2\sqrt{20}$

45. $5\sqrt[3]{16a^4} + 7\sqrt[3]{2a}$

46. $3\sqrt{10} + 8\sqrt{20} - 5\sqrt{80}$

Synthesis

47. Use a graphing calculator to solve Exercises 11, 22, and 25 by graphing two curves, one for each side of the inequality.

48. Use a graphing calculator to solve each of the following.

a) $x + \dfrac{1}{x} < 0$

b) $x - \sqrt{x} \geq 0$

c) $\frac{1}{3}x^3 - x + \frac{2}{3} \leq 0$

Solve.

49. $x^2 - 2x \leq 2$

50. $x^2 + 2x > 4$

51. $x^4 + 2x^2 > 0$

52. $x^4 + 3x^2 \leq 0$

53. $\left|\dfrac{x+2}{x-1}\right| < 3$

54. ***Total Profit.*** A company determines that its total profit from the production and sale of x units of a product is given by

$$P(x) = -x^2 + 812x - 9600.$$

a) A company makes a profit for those nonnegative values of x for which $P(x) > 0$. Find the values of x for which the company makes a profit.

b) A company loses money for those nonnegative values of x for which $P(x) < 0$. Find the values of x for which the company loses money.

55. ***Height of a Thrown Object.*** The function

$$H(t) = -16t^2 + 32t + 1920$$

gives the height H of an object thrown from a cliff 1920 ft high, after time t seconds.

a) For what times is the height greater than 1920 ft?

b) For what times is the height less than 640 ft?

CHAPTER

7 Summary and Review

Formulas and Principles

Principle of Square Roots: $x^2 = d$ has solutions $\sqrt{d}$ and $-\sqrt{d}$.

Quadratic Formula: $x = \dfrac{-b \pm \sqrt{b^2 - 4ac}}{2a}$

Discriminant: $b^2 - 4ac$

The *vertex* of the graph of $f(x) = ax^2 + bx + c$ is $\left(-\dfrac{b}{2a}, \dfrac{4ac - b^2}{4a}\right)$, or $\left(-\dfrac{b}{2a}, f\left(-\dfrac{b}{2a}\right)\right)$.

The *line of symmetry* of the graph of $f(x) = ax^2 + bx + c$ is $x = -\dfrac{b}{2a}$.

Vocabulary Reinforcement

Complete each statement with the correct term from the column on the right. Some of the choices may be used more than once and some may not be used at all.

1. The equation $x^2 = 2x - 8$ is an example of a(n) _______________ equation. [7.1a]
2. The inequality $\dfrac{1}{x} < 7$ is an example of a(n) _______________ inequality. [7.8b]
3. We can _______________ the square for $y^2 - 8y$ by adding 16. [7.1b]
4. The expression $b^2 - 4ac$ in the quadratic formula is called the _______________. [7.4a]
5. The equation $m^6 - m^3 - 12 = 0$ is _______________ in form. [7.4c]
6. The graph of a quadratic function is a(n) _______________. [7.5a]
7. The vertical line $x = 0$ is the line of _______________ for the graph of $y = x^2$. [7.5a]
8. The maximum or minimum value of a quadratic function is the y-coordinate of the _______________. [7.5c]

complete
parabola
line
discriminant
symmetry
linear
polynomial
quadratic
rational
vertex
x-intercept
y-intercept

Concept Reinforcement

Determine whether each statement is true or false.

_______________ 1. The graph of $f(x) = -(-x^2 - 8x - 3)$ opens downward. [7.5a]

_______________ 2. If $(-5, 7)$ is the vertex of a parabola, then $x = -5$ is the line of symmetry. [7.6a]

_______________ 3. The graph of $f(x) = -3(x + 2)^2 - 5$ is a translation to the right of the graph of $f(x) = -3x^2 - 5$. [7.5b]

Study Guide

Objective 7.1a Solve quadratic equations using the principle of square roots.

Example Solve: $(x - 3)^2 = -36$.

$$x - 3 = \sqrt{-36} \quad or \quad x - 3 = -\sqrt{-36}$$
$$x - 3 = 6i \quad or \quad x - 3 = -6i$$
$$x = 3 + 6i \quad or \quad x = 3 - 6i$$

The solutions are $3 \pm 6i$.

Practice Exercise

1. Solve: $(x - 2)^2 = -9$.

Objective 7.1b Solve quadratic equations by completing the square.

Example Solve by completing the square:

$$x^2 - 8x + 13 = 0.$$
$$x^2 - 8x = -13$$
$$x^2 - 8x + 16 = -13 + 16$$
$$(x - 4)^2 = 3$$
$$x - 4 = \sqrt{3} \quad or \quad x - 4 = -\sqrt{3}$$
$$x = 4 + \sqrt{3} \quad or \quad x = 4 - \sqrt{3}$$

The solutions are $4 \pm \sqrt{3}$.

Practice Exercise

2. Solve by completing the square:

$$x^2 - 12x + 31 = 0.$$

Objective 7.2a Solve quadratic equations using the quadratic formula, and approximate solutions using a calculator.

Example Solve: $x^2 - 2x = 2$. Give the exact solutions and approximate solutions to three decimal places.

$x^2 - 2x - 2 = 0$ Standard form

$$a = 1, \quad b = -2, \quad c = -2$$

$$x = \frac{-(-2) \pm \sqrt{(-2)^2 - 4 \cdot 1 \cdot (-2)}}{2 \cdot 1} \quad \text{Using the quadratic formula}$$
$$= \frac{2 \pm \sqrt{4 + 8}}{2}$$
$$= \frac{2 \pm \sqrt{12}}{2}$$
$$= \frac{2 \pm 2\sqrt{3}}{2}$$
$$= 1 \pm \sqrt{3}, \text{ or } 2.732 \text{ and } -0.732$$

Practice Exercise

3. Solve: $x^2 - 10x = -23$. Give the exact solutions and approximate solutions to three decimal places.

Objective 7.4a Determine the nature of the solutions of a quadratic equation.

Example Determine the nature of the solutions of the quadratic equation $x^2 - 7x = 1$.

In standard form, we have $x^2 - 7x - 1 = 0$. Thus, $a = 1$, $b = -7$, and $c = -1$. The discriminant, $b^2 - 4ac$, is $(-7)^2 - 4 \cdot 1 \cdot (-1)$, or 53. Since the discriminant is positive, there are two real solutions.

Practice Exercise

4. Determine the nature of the solutions of each quadratic equation.

a) $x^2 - 3x = 7$

b) $2x^2 - 5x + 5 = 0$

Objective 7.4b Write a quadratic equation having two given numbers as solutions.

Example Write a quadratic equation whose solutions are 7 and $-\frac{1}{4}$.

$$x = 7 \quad or \quad x = -\tfrac{1}{4}$$
$$x - 7 = 0 \quad or \quad x + \tfrac{1}{4} = 0$$
$$x - 7 = 0 \quad or \quad 4x + 1 = 0 \quad \text{Clearing the fraction}$$
$$(x - 7)(4x + 1) = 0 \quad \text{Using the principle of zero products in reverse}$$
$$4x^2 - 27x - 7 = 0 \quad \text{Using FOIL}$$

Practice Exercise

5. Write a quadratic equation whose solutions are $-\frac{2}{5}$ and 3.

Objective 7.4c Solve equations that are quadratic in form.

Example Solve: $x - 8\sqrt{x} - 9 = 0$.

Let $u = \sqrt{x}$. Then we substitute u for $\sqrt{x}$ and u^2 for x and solve for u:

$$u^2 - 8u - 9 = 0$$
$$(u - 9)(u + 1) = 0$$
$$u = 9 \quad or \quad u = -1.$$

Next, we substitute $\sqrt{x}$ for u and solve for x:

$$\sqrt{x} = 9 \quad or \quad \sqrt{x} = -1.$$

Squaring each equation, we get

$$x = 81 \quad or \quad x = 1.$$

Checking both 81 and 1 in $x - 8\sqrt{x} - 9 = 0$, we find that 81 checks but 1 does not. The solution is 81.

Practice Exercise

6. Solve:

$$(x^2 - 3)^2 - 5(x^2 - 3) - 6 = 0.$$

Objective 7.6a For a quadratic function, find the vertex, the line of symmetry, and the maximum or minimum value, and then graph the function.

Example For $f(x) = -2x^2 + 4x + 1$, find the vertex, the line of symmetry, and the maximum or minimum value. Then graph.

We factor out -2 from only the first two terms:

$$f(x) = -2(x^2 - 2x) + 1.$$

Next, we complete the square, factor, and simplify:

$$f(x) = -2(x^2 - 2x \qquad) + 1.$$
$$= -2(x^2 - 2x + 1 - 1) + 1$$
$$= -2(x^2 - 2x + 1) + (-2)(-1) + 1$$
$$= -2(x - 1)^2 + 3.$$

The vertex is $(1, 3)$. The line of symmetry is $x = 1$. The coefficient of x^2 is negative, so the graph opens down. Thus, 3 is the maximum value of the function.

We plot points and graph the parabola.

x	y
1	3
2	1
0	1
3	−5
−1	−5

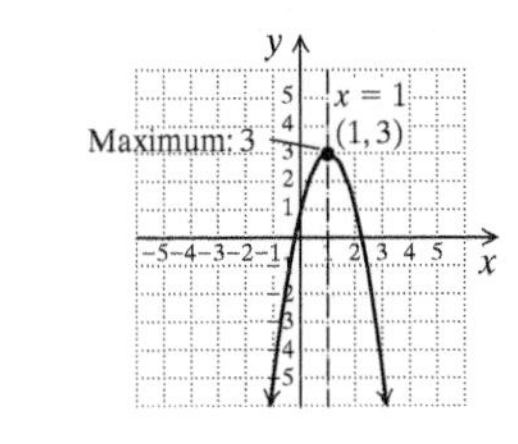

Practice Exercise

7. For $f(x) = -x^2 - 2x - 3$, find the vertex, the line of symmetry, and the maximum or minimum value. Then graph.

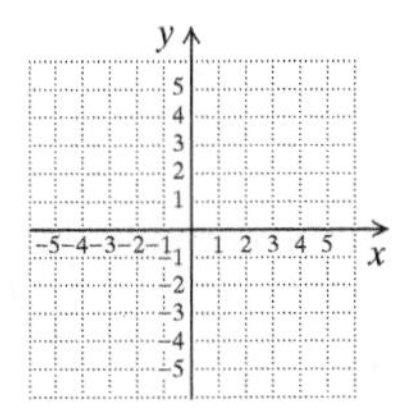

Objective 7.6b Find the intercepts of a quadratic function.

Example Find the intercepts of $f(x) = x^2 - 8x + 14$.

Since $f(0) = 0^2 - 8 \cdot 0 + 14 = 14$, the y-intercept is $(0, 14)$. To find the x-intercepts, we solve $0 = x^2 - 8x + 14$. Using the quadratic formula, we have $x = 4 \pm \sqrt{2}$. Thus the x-intercepts are $(4 - \sqrt{2}, 0)$ and $(4 + \sqrt{2}, 0)$.

Practice Exercise

8. Find the intercepts of $f(x) = x^2 - 6x + 4$.

Objective 7.8a Solve quadratic inequalities and other polynomial inequalities.

Example Solve: $x^2 - 15 > 2x$.

$$x^2 - 2x - 15 > 0 \quad \text{Subtracting } 2x$$

We set the polynomial equal to 0 and solve. The solutions of $x^2 - 2x - 15 = 0$, or $(x + 3)(x - 5) = 0$, are -3 and 5. They divide the number line into three intervals.

−6 −5 −4 −3 −2 −1 0 1 2 3 4 5 6

We try a test point in each interval:

Test -5: $(-5)^2 - 2(-5) - 15 = 20 > 0$;

Test 0: $(0)^2 - 2 \cdot 0 - 15 = -15 < 0$;

Test 6: $(6)^2 - 2 \cdot 6 - 15 = 9 > 0$.

The expression $x^2 - 2x - 15$ is positive for values of x in the intervals $(-\infty, -3)$ and $(5, \infty)$. The inequality symbol is $>$, so -3 and 5 are not solutions. The solution set is $\{x | x < -3 \text{ or } x > 5\}$, or $(-\infty, -3) \cup (5, \infty)$.

Practice Exercise

9. Solve: $x^2 + 40 > 14x$.

Objective 7.8b Solve rational inequalities.

Example Solve: $\frac{x + 3}{x - 6} \geq 2$.

We first solve the related equation $\frac{x + 3}{x - 6} = 2$. The solution is 15. We must also determine those numbers for which the rational expression is not defined. We set the denominator equal to 0 and solve: $x - 6 = 0$, or $x = 6$. The numbers 6 and 15 divide the number line into three intervals. We test a point in each interval.

4 5 6 7 8 9 10 11 12 13 14 15 16

Test 5: $\frac{5 + 3}{5 - 6} \geq 2$, or $-8 \geq 2$, which is false.

Test 9: $\frac{9 + 3}{9 - 6} \geq 2$, or $4 \geq 2$, which is true.

Test 17: $\frac{17 + 3}{17 - 6} \geq 2$, or $\frac{20}{11} \geq 2$, which is false.

The solution set includes the interval $(6, 15)$ and the number 15, the solution of the related equation. The number 6 is not included. It is not an allowable replacement because it results in division by 0. The solution set is $\{x | 6 < x \leq 15\}$, or $(6, 15]$.

Practice Exercise

10. Solve: $\frac{x + 7}{x - 5} \geq 3$.

Review Exercises

1. **a)** Solve: $2x^2 - 7 = 0$. [7.1a]
   b) Find the x-intercepts of $f(x) = 2x^2 - 7$. [7.1a]

Solve. [7.2a]

2. $14x^2 + 5x = 0$

3. $x^2 - 12x + 27 = 0$

4. $x^2 - 7x + 13 = 0$

5. $4x^2 + 6x = 1$

6. $4x(x - 1) + 15 = x(3x + 4)$

7. $x^2 + 4x + 1 = 0$. Give exact solutions and approximate solutions to three decimal places.

8. $\dfrac{x}{x - 2} + \dfrac{4}{x - 6} = 0$

9. $\dfrac{x}{4} - \dfrac{4}{x} = 2$

10. $15 = \dfrac{8}{x + 2} - \dfrac{6}{x - 2}$

11. Solve $x^2 + 6x + 2 = 0$ by completing the square. Show your work. [7.1b]

12. *Hang Time.* A basketball player has a vertical leap of 39 in. What is his hang time? Use the function $V(T) = 48T^2$. [7.1c]

13. *Notebook.* The width of a pocket-sized rectangular notebook is 5 cm less than the length. The area is 126 cm^2. Find the length and the width. [7.3a]

14. *Picture Matting.* A picture mat measures 12 in. by 16 in., and 140 in^2 of picture shows. Find the width of the mat. [7.3a]

15. *Motorcycle Travel.* During the first part of a trip, a motorcyclist travels 50 mi. The rider travels 80 mi on the second part of the trip at a speed that is 10 mph slower. The total time for the trip is 3 hr. What is the speed on each part of the trip? [7.3a]

Determine the nature of the solutions of each equation. [7.4a]

16. $x^2 + 3x - 6 = 0$

17. $x^2 + 2x + 5 = 0$

Write a quadratic equation having the given solutions. [7.4b]

18. $\frac{1}{5}, -\frac{3}{5}$

19. -4, only solution

Solve for the indicated letter. [7.3b]

20. $N = 3\pi\sqrt{\dfrac{1}{p}}$, for p

21. $2A = \dfrac{3B}{T^2}$, for T

Solve. [7.4c]

22. $x^4 - 13x^2 + 36 = 0$

23. $15x^{-2} - 2x^{-1} - 1 = 0$

24. $(x^2 - 4)^2 - (x^2 - 4) - 6 = 0$

25. $x - 13\sqrt{x} + 36 = 0$

For each quadratic function in Exercises 26–28, find and label **(a)** the vertex, **(b)** the line of symmetry, and **(c)** the maximum or minimum value. Then **(d)** graph the function. [7.5c], [7.6a]

26. $f(x) = -\frac{1}{2}(x - 1)^2 + 3$

x	$f(x)$

a) Vertex: (____, ____)
b) Line of symmetry: $x =$ ____
c) __________ value: ____
d)

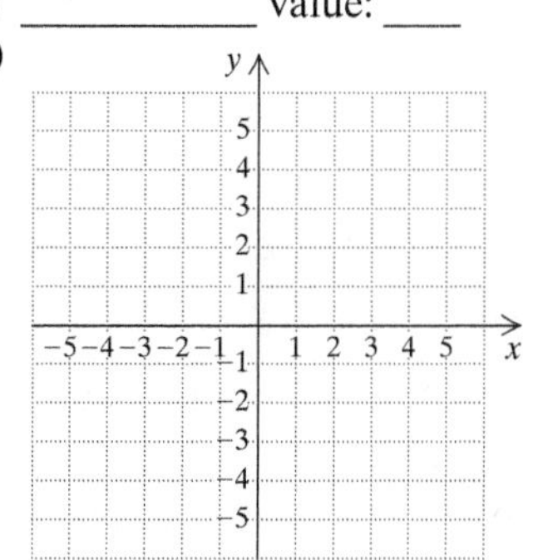

27. $f(x) = x^2 - x + 6$

x	$f(x)$

a) Vertex: (____, ____)
b) Line of symmetry: $x =$ ____
c) __________ value: ____
d)

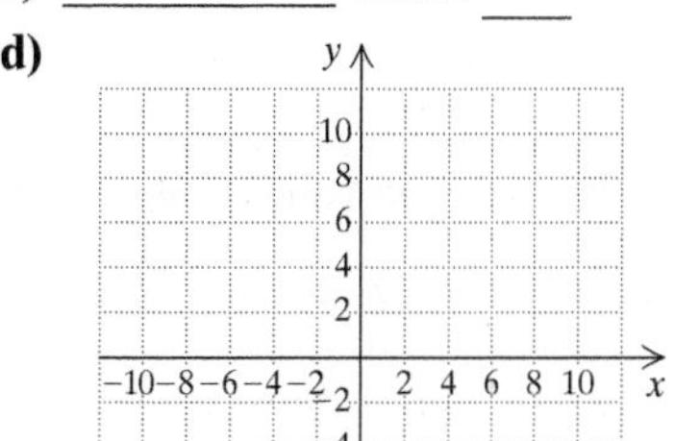

28. $f(x) = -3x^2 - 12x - 8$

x	$f(x)$

a) Vertex: (____, ____)
b) Line of symmetry: $x =$ ____
c) ____________ value: ____
d)

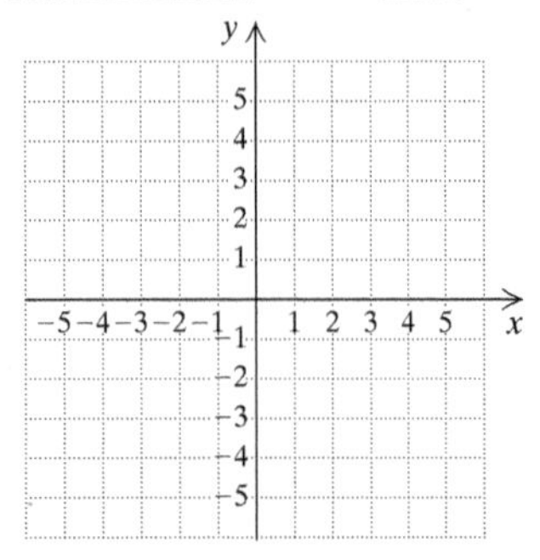

Find the x- and y-intercepts. [7.6b]

29. $f(x) = x^2 - 9x + 14$

30. $g(x) = x^2 - 4x - 3$

31. What is the minimum product of two numbers whose difference is 22? What numbers yield this product? [7.7a]

32. Find a quadratic function that fits the data points $(0, -2)$, $(1, 3)$, and $(3, 7)$. [7.7b]

Solve. [7.8a, b]

33. $(x + 2)(x - 1)(x - 2) > 0$

34. $\dfrac{(x + 4)(x - 1)}{(x + 2)} < 0$

35. ***Online Sales Growth.*** Sales of products and services through the Internet have increased quadratically since 2010, as illustrated in the following graph.

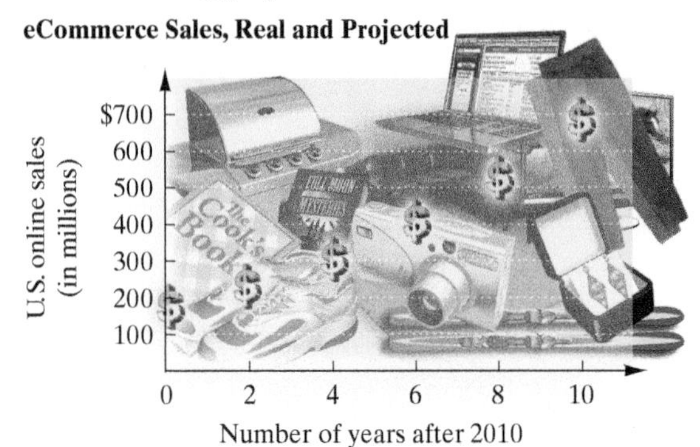

DATA: *USA Today*

a) Use the data points $(0, 170)$, $(4, 300)$, and $(10, 680)$ to fit a quadratic function to the data.
b) Use the quadratic function to estimate the amount of online sales in 2017.

36. Determine the nature of the solutions:

$$x^2 - 10x + 25 = 0. \quad [7.4a]$$

A. Infinite number of solutions
B. One real solution
C. Two real solutions
D. No real solutions

37. Solve: $2x^2 - 6x + 5 = 0$. [7.2a]

A. $\dfrac{3}{2} \pm \dfrac{\sqrt{19}}{2}$ **B.** $3 \pm i$

C. $3 \pm \sqrt{19}$ **D.** $\dfrac{3}{2} \pm \dfrac{1}{2}i$

Synthesis

38. The sum of the base and the height of a triangle is 38 cm. Find the dimensions for which the area is a maximum, and find the maximum area. [7.7a]

39. The average of two numbers is 171. One of the numbers is the square root of the other. Find the numbers. [7.3a]

Understanding Through Discussion and Writing

1. Does the graph of every quadratic function have a y-intercept? Why or why not? [7.6b]

2. Explain how the leading coefficient of a quadratic function can be used to determine whether a maximum or minimum function value exists. [7.7a]

3. Explain, without plotting points, why the graph of $f(x) = (x + 3)^2 - 4$ looks like the graph of $f(x) = x^2$ translated 3 units to the left and 4 units down. [7.5c]

4. Describe a method that could be used to create quadratic inequalities that have no solution. [7.8a]

5. Is it possible for the graph of a quadratic function to have only one x-intercept if the vertex is off the x-axis? Why or why not? [7.6b]

6. Explain how the x-intercepts of a quadratic function can be used to help find the vertex of the function. What piece of information would still be missing? [7.6a, b]

CHAPTER

7 Test

For Extra Help

For step-by-step test solutions, access the Chapter Test Prep Videos in MyLab Math.

1. a) Solve: $3x^2 - 4 = 0$.
b) Find the x-intercepts of $f(x) = 3x^2 - 4$.

Solve.

2. $x^2 + x + 1 = 0$

3. $x - 8\sqrt{x} + 7 = 0$

4. $4x(x - 2) - 3x(x + 1) = -18$

5. $4x^4 - 17x^2 + 15 = 0$

6. $x^2 + 4x = 2$. Give exact solutions and approximate solutions to three decimal places.

7. $\frac{1}{4 - x} + \frac{1}{2 + x} = \frac{3}{4}$

8. Solve $x^2 - 4x + 1 = 0$ by completing the square. Show your work.

9. *Free-Falling Objects.* The Peachtree Plaza in Atlanta, Georgia, is 723 ft tall. Use the function $s(t) = 16t^2$ to approximate how long it would take an object to fall from the top.

10. *Marine Travel.* The Columbia River flows at a rate of 2 mph for the length of a popular boating route. In order for a motorized dinghy to travel 3 mi upriver and then return in a total of 4 hr, how fast must the boat be able to travel in still water?

11. *Memory Board.* A computer-parts company wants to make a rectangular memory board that has a perimeter of 28 cm. What dimensions will allow the board to have the maximum area?

12. *Hang Time.* Professional basketball player Nate Robinson has a vertical leap of 43 in. What is his hang time? Use the function $V(T) = 48T^2$.

13. Determine the nature of the solutions of the equation $x^2 + 5x + 17 = 0$.

14. Write a quadratic equation having the solutions $\sqrt{3}$ and $3\sqrt{3}$.

15. Solve $V = 48T^2$ for T.

For the quadratic functions in Exercises 16 and 17, find and label **(a)** the vertex, **(b)** the line of symmetry, and **(c)** the maximum or minimum value. Then **(d)** graph the function.

16. $f(x) = -x^2 - 2x$

a) Vertex: (____, ____)
b) Line of symmetry: $x =$ ____
c) __________ value: ____

x	$f(x)$

d)

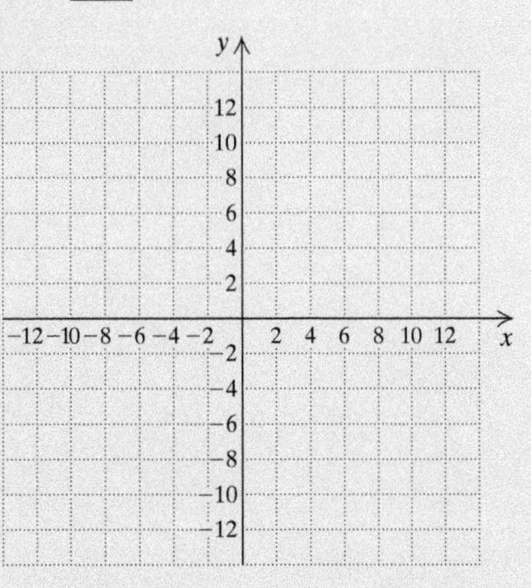

17. $f(x) = 4x^2 - 24x + 41$

a) Vertex: (____, ____)
b) Line of symmetry: $x =$ ____
c) __________ value: ____

x	$f(x)$

d)

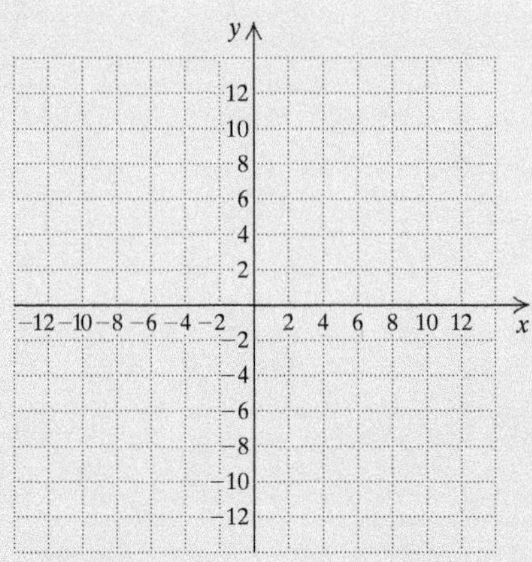

18. Find the x- and y-intercepts:

$$f(x) = -x^2 + 4x - 1.$$

19. What is the minimum product of two numbers whose difference is 8? What numbers yield this product?

20. Find the quadratic function that fits the data points $(0, 0)$, $(3, 0)$, and $(5, 2)$.

21. *Pizza Prices.* Pizza Unlimited has the following prices for pizzas.

DIAMETER	PRICE
8 in.	\$10.00
12 in.	\$12.50
16 in.	\$15.50

Price should probably be a quadratic function of diameter because it should be proportional to the area, and the area is a quadratic function of the diameter. (The area of a circular region is given by $A = \pi r^2$ or $(\pi/4)d^2$.)

a) Express price as a quadratic function of diameter using the data points $(8, 10)$, $(12, 12.50)$, and $(16, 15.50)$.
b) Use the function to find the price of a 14-in. pizza.

Solve.

22. $x^2 < 6x + 7$

23. $\dfrac{x - 5}{x + 3} < 0$

24. $\dfrac{x - 2}{(x + 3)(x - 1)} \geq 0$

25. Write a quadratic equation whose solutions are $\frac{1}{2}i$ and $-\frac{1}{2}i$.

A. $4x^2 - 4ix - 1 = 0$

B. $x^2 - \dfrac{1}{4} = 0$

C. $4x^2 + 1 = 0$

D. $x^2 - ix + 1 = 0$

Synthesis

26. A quadratic function has x-intercepts $(-2, 0)$ and $(7, 0)$ and y-intercept $(0, 8)$. Find an equation for the function. What is its maximum or minimum value?

27. One solution of $kx^2 + 3x - k = 0$ is -2. Find the other solution.

CHAPTERS

1–7 Cumulative Review

1. *Golf Courses.* Most golf courses have a hole such as the one shown here, where the safe way to the hole is to hit straight out on a first shot (the distance a) and then make subsequent shots at a right angle to cover the distance b. Golfers are often lured, however, into taking a shortcut over trees, houses, or lakes. If a golfer makes a hole in one on this hole, how long is the shot?

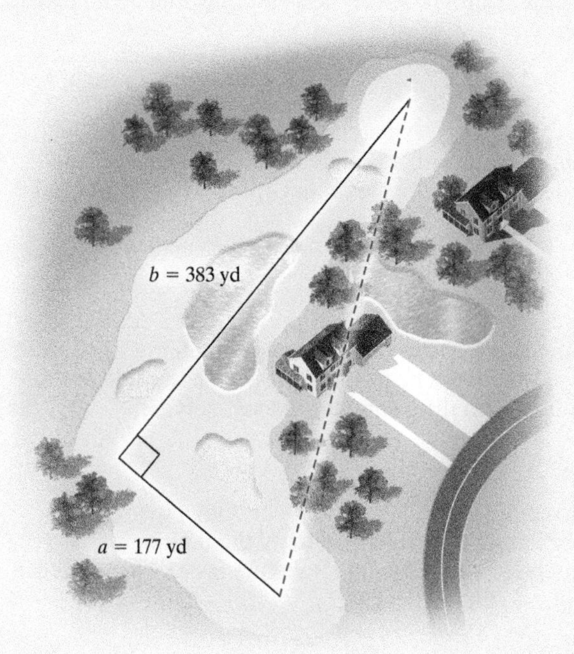

Simplify.

2. $(4 + 8x^2 - 5x) - (-2x^2 + 3x - 2)$

3. $(2x^2 - x + 3)(x - 4)$

4. $\dfrac{a^2 - 16}{5a - 15} \cdot \dfrac{2a - 6}{a + 4}$

5. $\dfrac{y}{y^2 - y - 42} \div \dfrac{y^2}{y - 7}$

6. $\dfrac{2}{m + 1} + \dfrac{3}{m - 5} - \dfrac{m^2 - 1}{m^2 - 4m - 5}$

7. $(9x^3 + 5x^2 + 2) \div (x + 2)$

8. $\dfrac{\frac{1}{x} - \frac{1}{y}}{x + y}$

9. $\sqrt{0.36}$

10. $\sqrt{9x^2 - 36x + 36}$

11. $6\sqrt{45} - 3\sqrt{20}$

12. $\dfrac{2\sqrt{3} - 4\sqrt{2}}{\sqrt{2} - 3\sqrt{6}}$

13. $(8^{2/3})^4$

14. $(3 + 2i)(5 - i)$

15. $\dfrac{6 - 2i}{3i}$

Factor.

16. $2t^2 - 7t - 30$

17. $a^2 + 3a - 54$

18. $-3a^3 + 12a^2$

19. $64a^2 - 9b^2$

20. $3a^2 - 36a + 108$

21. $\dfrac{1}{27}a^3 - 1$

22. $24a^3 + 18a^2 - 20a - 15$

23. $(x + 1)(x - 1) + (x + 1)(x + 2)$

Solve.

24. $3(4x - 5) + 6 = 3 - (x + 1)$

25. $F = \dfrac{mv^2}{r}$, for r

26. $5 - 3(2x + 1) \leq 8x - 3$

27. $3x - 2 < -6$ *or* $x + 3 > 9$

28. $|4x - 1| \leq 14$

29. $5x + 10y = -10,$
$-2x - 3y = 5$

30. $2x + y - z = 9,$
$4x - 2y + z = -9,$
$2x - y + 2z = -12$

31. $10x^2 + 28x - 6 = 0$

32. $\frac{2}{n} - \frac{7}{n} = 3$

33. $\frac{1}{2x - 1} = \frac{3}{5x}$

34. $A = \frac{mh}{m + a}$, for m

35. $\sqrt{2x - 1} = 6$

36. $\sqrt{x - 2} + 1 = \sqrt{2x - 6}$

37. $16(t - 1) = t(t + 8)$

38. $x^2 - 3x + 16 = 0$

39. $\frac{18}{x + 1} - \frac{12}{x} = \frac{1}{3}$

40. $P = \sqrt{a^2 - b^2}$, for a

41. $\frac{(x + 3)(x + 2)}{(x - 1)(x + 1)} < 0$

42. Solve: $4x^2 - 25 > 0$.

Graph.

43. $x + y = 2$

44. $y \geq 6x - 5$

45. $x < -3$

46. $3x - y > 6,$
$4x + y \leq 3$

47. $f(x) = x^2 - 1$

48. $f(x) = -2x^2 + 3$

49. Find an equation of the line with slope $\frac{1}{2}$ and through the point $(-4, 2)$.

50. Find an equation of the line parallel to the line $3x + y = 4$ and through the point $(0, 1)$.

51. *Marine Travel.* The Connecticut River flows at a rate of 4 km/h for the length of a popular scenic route. In order for a cruiser to travel 60 km upriver and then return in a total of 8 hr, how fast must the boat be able to travel in still water?

52. *Architecture.* An architect is designing a rectangular family room with a perimeter of 56 ft. What dimensions will yield the maximum area? What is the maximum area?

53. The perimeter of a hexagon with all six sides the same length is the same as the perimeter of a square. One side of the hexagon is 3 less than the side of the square. Find the perimeter of each polygon.

54. Two pipes can fill a tank in $1\frac{1}{2}$ hr. One pipe requires 4 hr longer running alone to fill the tank than the other. How long would it take the faster pipe, working alone, to fill the tank?

55. Complete the square: $f(x) = 5x^2 - 20x + 15$.
A. $f(x) = 5(x - 2)^2 - 5$
B. $f(x) = 5(x + 2)^2 + 15$
C. $f(x) = 5(x + 2)^2 + 6$
D. $f(x) = 5(x + 2)^2 + 11$

56. How many times does the graph of $f(x) = x^4 - 6x^2 - 16$ cross the x-axis?
A. 1 **B.** 2
C. 3 **D.** 4

Synthesis

57. Solve: $\frac{2x + 1}{x} = 3 + 7\sqrt{\frac{2x + 1}{x}}$.

58. Factor: $\frac{a^3}{8} + \frac{8b^3}{729}$.

CHAPTER

8

Exponential Functions and Logarithmic Functions

In 2016, over 600 million megawatt hours of electricity was generated in the United States from sources considered renewable, such as the sun, the earth, and waste products. This amount accounted for about 17% of the electricity used in 2016, up from 9% in 2007. As the accompanying graph indicates, the amounts of solar power, geothermal power, and power generated from waste products are all increasing. Geothermal power is increasing linearly at a slow rate, power from waste is increasing linearly at a more rapid rate, and solar power is increasing *exponentially*.

Power Generated from Renewable Sources

U.S. power generation (in thousands of megawatt hours)

40,000
35,000
30,000
25,000
20,000
15,000
10,000
5,000
0

Solar
Waste
Geothermal

2008 2010 2012 2014 2016

Year

DATA: *U.S. Energy Information Administration*

In Example 6 of Section 8.7, we will develop an *exponential function* to model solar power generation.

JUST IN TIME

Review topics 10, 11, and 19 in the **Just In Time** section at the front of the text. This provides excellent prerequisite skill review for this chapter.

MyLab Math
VIDEO

STUDYING FOR SUCCESS ***Beginning to Study for the Final Exam***

- ☐ Take a few minutes each week to review highlighted information.
- ☐ Prepare a few pages of notes for the course and then try to condense the notes to just one page.
- ☐ Use the Mid-Chapter Reviews, Summary and Reviews, Chapter Tests, and Cumulative Reviews.

8.1 Exponential Functions

OBJECTIVES

a Graph exponential equations and functions.

b Graph exponential equations in which x and y have been interchanged.

c Solve applied problems involving applications of exponential functions and their graphs.

The following graph approximates the graph of an *exponential function.* We will consider such functions and some of their applications.

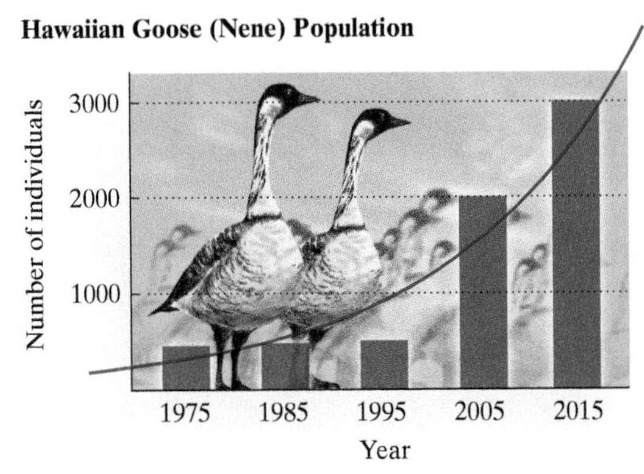

DATA: esasuccess.org

MyLab Math ANIMATION

a GRAPHING EXPONENTIAL FUNCTIONS

SKILL REVIEW *Evaluate expressions containing negative exponents.* [J19]

Evaluate.

1. 2^{-3}

2. $\left(\frac{1}{2}\right)^{-3}$

Answers: 1. $\frac{1}{8}$ **2.** 8

MyLab Math VIDEO

We have defined exponential expressions with rational-number exponents, such as

$$8^{1/4}, \quad 3^{-3/4}, \quad 7^{2.34}, \quad 5^{1.73}.$$

For example, $5^{1.73}$, or $5^{173/100}$, or $\sqrt[100]{5^{173}}$, means to raise 5 to the 173rd power and then take the 100th root. We now develop the meaning of exponential expressions with irrational exponents, such as

$$5^{\sqrt{3}}, \quad 7^{\pi}, \quad 9^{-\sqrt{2}}.$$

Since we can approximate irrational numbers with decimal approximations, we can also approximate expressions with irrational exponents. For example, consider $5^{\sqrt{3}}$. As rational values of r get close to $\sqrt{3}$, 5^r gets close to some real number.

r closes in on $\sqrt{3}$.	5^r closes in on some real number p.
r	5^r
$1 < \sqrt{3} < 2$	$5 = 5^1 < p < 5^2 = 25$
$1.7 < \sqrt{3} < 1.8$	$15.426 \approx 5^{1.7} < p < 5^{1.8} \approx 18.119$
$1.73 < \sqrt{3} < 1.74$	$16.189 \approx 5^{1.73} < p < 5^{1.74} \approx 16.452$
$1.732 < \sqrt{3} < 1.733$	$16.241 \approx 5^{1.732} < p < 5^{1.733} \approx 16.267$

As r closes in on $\sqrt{3}$, 5^r closes in on some real number p. We define $5^{\sqrt{3}}$ to be that number p. To seven decimal places, we have $5^{\sqrt{3}} \approx 16.2424508$.

Any positive irrational exponent can be defined in a similar way. Negative irrational exponents are then defined in the same way as negative integer exponents. Thus the expression a^x has meaning for any real number x. The general laws of exponents still hold, but we will not prove that here.

We now define exponential functions.

EXPONENTIAL FUNCTION

The function $f(x) = a^x$, where a is a positive constant different from 1, is called an **exponential function**, base a.

We restrict the base a to being positive to avoid the possibility of taking even roots of negative numbers such as the square root of -1, $(-1)^{1/2}$, which is not a real number. We restrict the base from being 1 because for $a = 1$, $t(x) = 1^x = 1$, which is a constant. The following are examples of exponential functions:

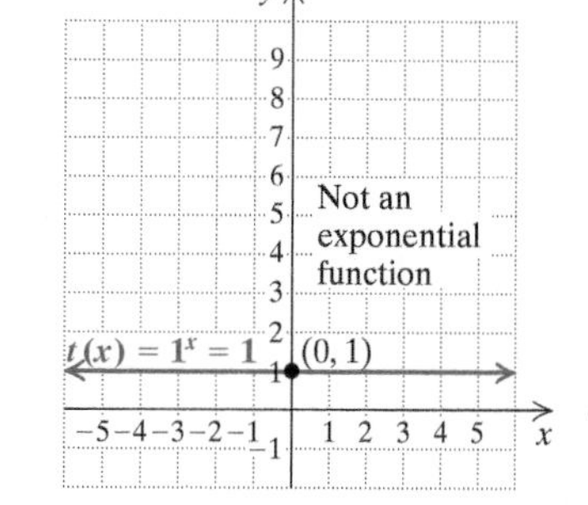

$$f(x) = 2^x, \quad f(x) = \left(\tfrac{1}{2}\right)^x, \quad f(x) = (0.4)^x.$$

Note that in contrast to polynomial functions like $f(x) = x^2$ and $f(x) = x^3$, the variable is *in the exponent.* Let's consider graphs of exponential functions.

EXAMPLE 1 Graph the exponential function $f(x) = 2^x$.

We compute some function values and list the results in a table. It is a good idea to begin by letting $x = 0$.

$f(0) = 2^0 = 1;$
$f(1) = 2^1 = 2;$
$f(2) = 2^2 = 4;$
$f(3) = 2^3 = 8;$
$f(-1) = 2^{-1} = \dfrac{1}{2^1} = \dfrac{1}{2};$
$f(-2) = 2^{-2} = \dfrac{1}{2^2} = \dfrac{1}{4};$
$f(-3) = 2^{-3} = \dfrac{1}{2^3} = \dfrac{1}{8}$

x	$f(x)$
0	1
1	2
2	4
3	8
−1	$\frac{1}{2}$
−2	$\frac{1}{4}$
−3	$\frac{1}{8}$

1. Graph: $f(x) = 3^x$. Complete this table. Then plot the points from the table and connect them with a smooth curve.

x	$f(x)$
0	
1	
2	
3	
-1	
-2	
-3	

2. Graph: $f(x) = \left(\frac{1}{3}\right)^x$.

Complete this table. Then plot the points from the table and connect them with a smooth curve.

x	$f(x)$
0	
1	
2	
3	
-1	
-2	
-3	

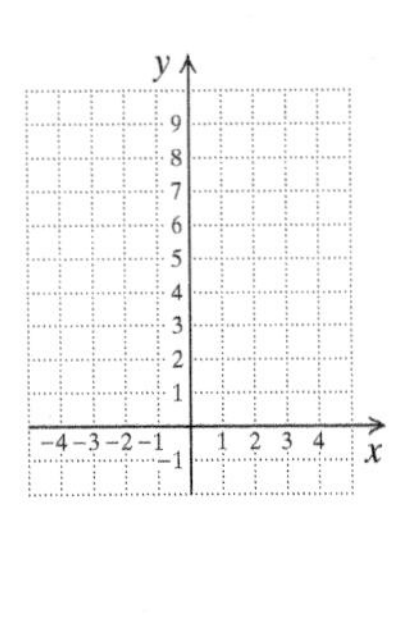

Answers

1.

x	$f(x)$
0	1
1	3
2	9
3	27
-1	$\frac{1}{3}$
-2	$\frac{1}{9}$
-3	$\frac{1}{27}$

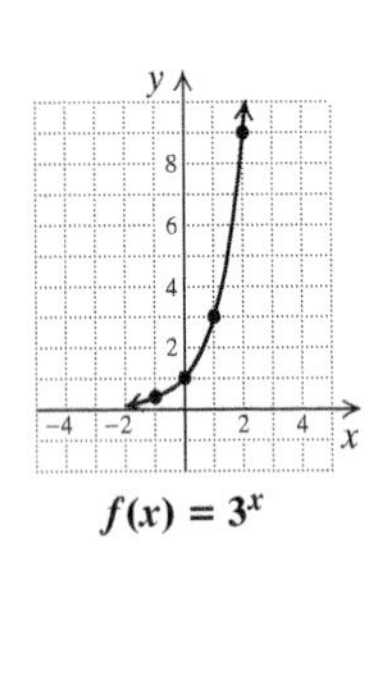

$f(x) = 3^x$

2.

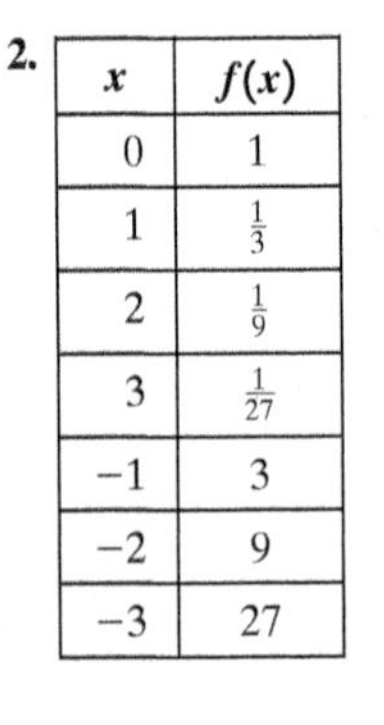

x	$f(x)$
0	1
1	$\frac{1}{3}$
2	$\frac{1}{9}$
3	$\frac{1}{27}$
-1	3
-2	9
-3	27

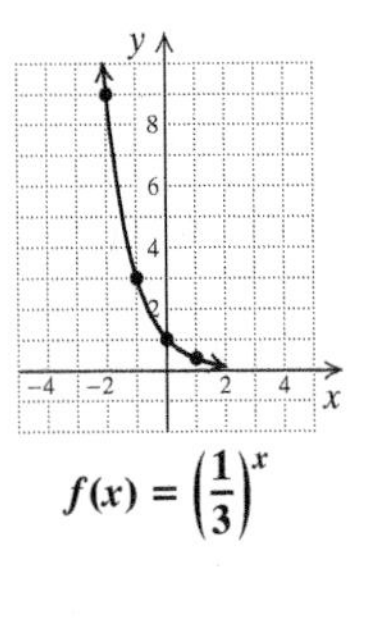

$f(x) = \left(\frac{1}{3}\right)^x$

Next, we plot these points and connect them with a smooth curve.

In graphing, be sure to plot enough points to determine how steeply the curve rises.

The curve comes very close to the x-axis, but does not touch or cross it.

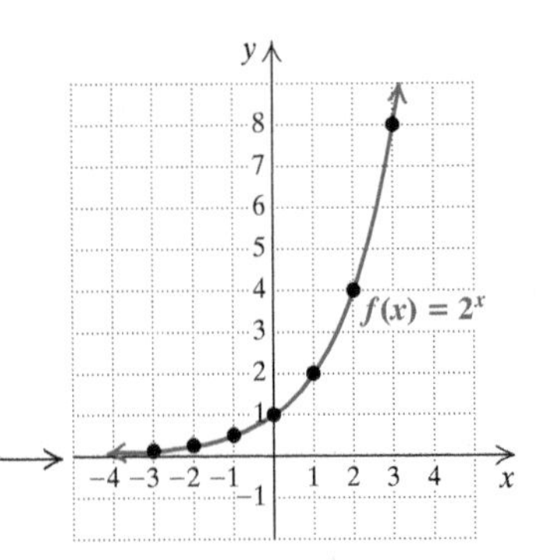

Note that as x increases, the function values increase indefinitely. As x decreases, the function values decrease, getting very close to 0. The x-axis, or the line $y = 0$, is an *asymptote*, meaning here that as x gets very small, the curve comes very close to but never touches the axis.

◀ **Do Exercise 1.**

EXAMPLE 2 Graph the exponential function $f(x) = (\frac{1}{2})^x$.

We compute some function values and list the results in a table. Before we do so, note that

$$f(x) = (\tfrac{1}{2})^x = (2^{-1})^x = 2^{-x}.$$

Then we have

$$f(0) = 2^{-0} = 1;$$
$$f(1) = 2^{-1} = \frac{1}{2^1} = \frac{1}{2};$$
$$f(2) = 2^{-2} = \frac{1}{2^2} = \frac{1}{4};$$
$$f(3) = 2^{-3} = \frac{1}{2^3} = \frac{1}{8};$$
$$f(-1) = 2^{-(-1)} = 2^1 = 2;$$
$$f(-2) = 2^{-(-2)} = 2^2 = 4;$$
$$f(-3) = 2^{-(-3)} = 2^3 = 8.$$

x	$f(x)$
0	1
1	$\frac{1}{2}$
2	$\frac{1}{4}$
3	$\frac{1}{8}$
-1	2
-2	4
-3	8

Next, we plot these points and draw the curve. Note that this graph is a reflection across the y-axis of the graph in Example 1. The line $y = 0$ is again an asymptote.

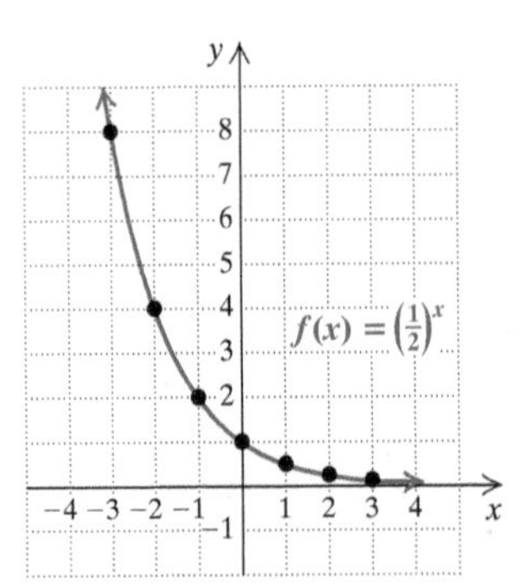

◀ **Do Exercise 2.**

The preceding examples illustrate exponential functions with various bases. Let's list some of their characteristics. Keep in mind that the definition of an exponential function, $f(x) = a^x$, requires that the base be positive and different from 1.

When $a > 1$, the function $f(x) = a^x$ increases from left to right. The greater the value of a, the steeper the curve. As x gets smaller and smaller, the curve gets closer to the line $y = 0$: It is an asymptote.

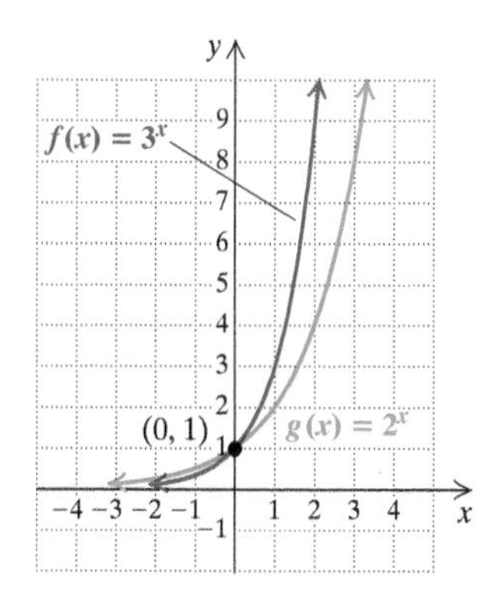

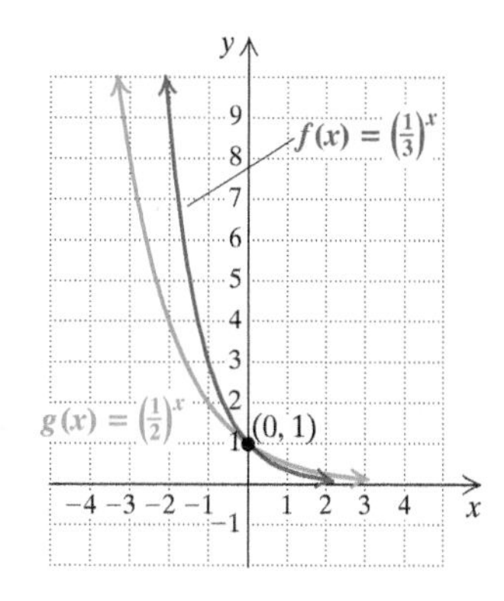

When $0 < a < 1$, the function $f(x) = a^x$ decreases from left to right. As a approaches 1, the curve becomes less steep. As x gets larger and larger, the curve gets closer to the line $y = 0$: It is an asymptote.

> **y-INTERCEPT OF AN EXPONENTIAL FUNCTION**
>
> All functions $f(x) = a^x$ go through the point $(0, 1)$. That is, the y-intercept is $(0, 1)$.

Do Exercises 3 and 4.

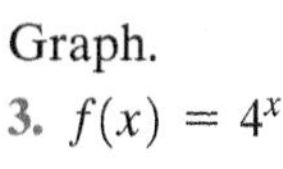

Graph.

3. $f(x) = 4^x$

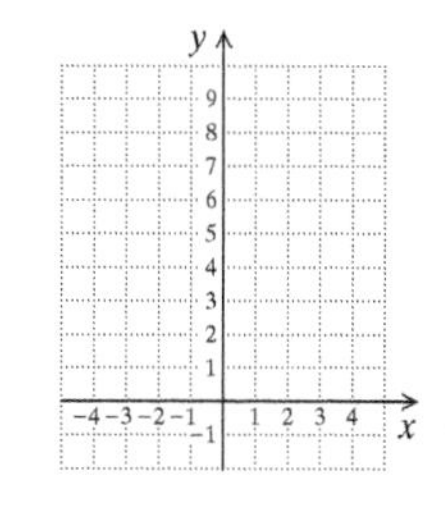

4. $f(x) = \left(\dfrac{1}{4}\right)^x$

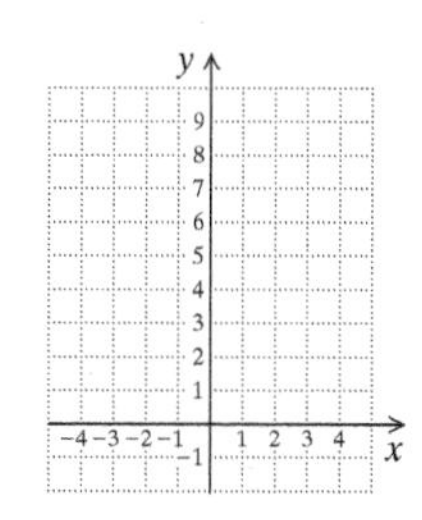

EXAMPLE 3 Graph: $f(x) = 2^{x-2}$.

We construct a table of values. Then we plot the points and connect them with a smooth curve. Be sure to note that $x - 2$ is the *exponent.*

$$f(0) = 2^{0-2} = 2^{-2} = \frac{1}{2^2} = \frac{1}{4};$$

$$f(1) = 2^{1-2} = 2^{-1} = \frac{1}{2^1} = \frac{1}{2};$$

$$f(2) = 2^{2-2} = 2^0 = 1;$$

$$f(3) = 2^{3-2} = 2^1 = 2;$$

$$f(4) = 2^{4-2} = 2^2 = 4;$$

$$f(-1) = 2^{-1-2} = 2^{-3} = \frac{1}{2^3} = \frac{1}{8};$$

$$f(-2) = 2^{-2-2} = 2^{-4} = \frac{1}{2^4} = \frac{1}{16}$$

x	$f(x)$
0	$\frac{1}{4}$
1	$\frac{1}{2}$
2	1
3	2
4	4
−1	$\frac{1}{8}$
−2	$\frac{1}{16}$

Answers

3.

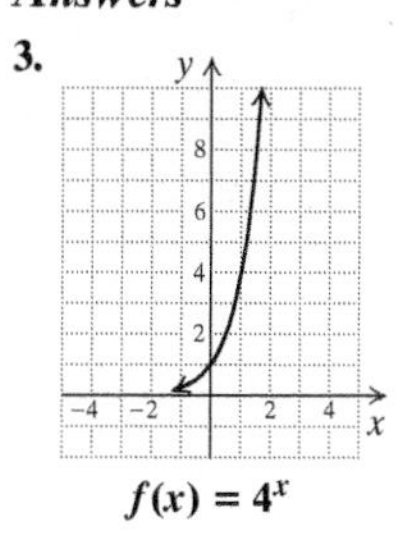

4.

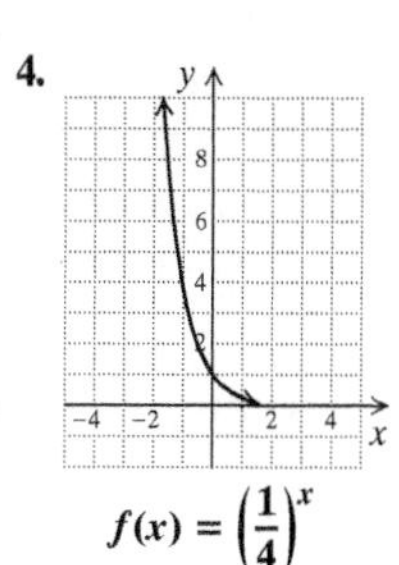

5. Graph: $f(x) = 2^{x+2}$.

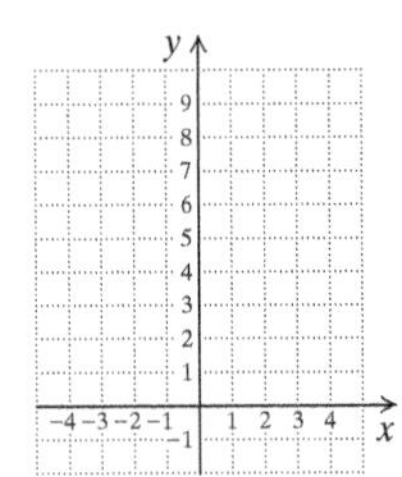

The graph has the same shape as the graph of $g(x) = 2^x$, but it is translated 2 units to the right.

The y-intercept of $g(x) = 2^x$ is $(0, 1)$. The y-intercept of $f(x) = 2^{x-2}$ is $\left(0, \frac{1}{4}\right)$. The line $y = 0$ is still an asymptote.

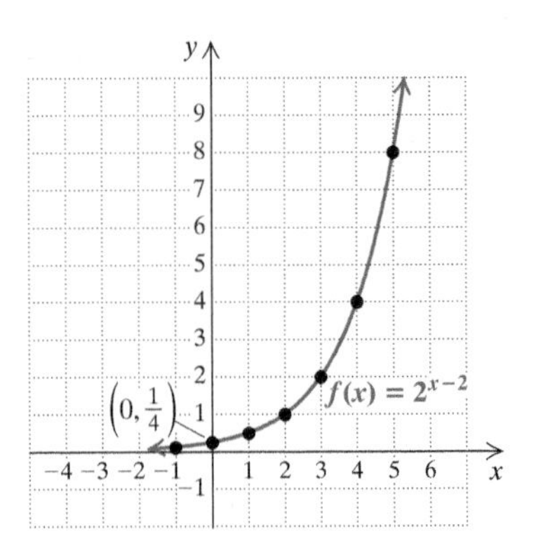

◀ **Do Exercise 5.**

EXAMPLE 4 Graph: $f(x) = 2^x - 3$.

We construct a table of values. Then we plot the points and connect them with a smooth curve. Note that the only expression in the exponent is x.

$$f(0) = 2^0 - 3 = 1 - 3 = -2;$$
$$f(1) = 2^1 - 3 = 2 - 3 = -1;$$
$$f(2) = 2^2 - 3 = 4 - 3 = 1;$$
$$f(3) = 2^3 - 3 = 8 - 3 = 5;$$
$$f(4) = 2^4 - 3 = 16 - 3 = 13;$$
$$f(-1) = 2^{-1} - 3 = \tfrac{1}{2} - 3 = -\tfrac{5}{2};$$
$$f(-2) = 2^{-2} - 3 = \tfrac{1}{4} - 3 = -\tfrac{11}{4}$$

x	$f(x)$
0	-2
1	-1
2	1
3	5
4	13
-1	$-\frac{5}{2}$
-2	$-\frac{11}{4}$

6. Graph: $f(x) = 2^x - 4$.

x	$f(x)$
0	
1	
2	
3	
4	
-1	
-2	

The graph has the same shape as the graph of $g(x) = 2^x$, but it is translated 3 units down. The y-intercept is $(0, -2)$. The line $y = -3$ is an asymptote. The curve gets closer to this line as x gets smaller and smaller.

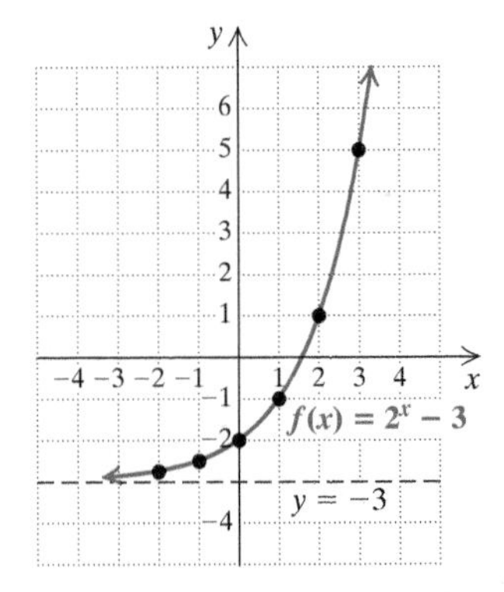

◀ **Do Exercise 6.**

Answers

Answers to Margin Exercises 5 and 6 are on p. 629.

CALCULATOR CORNER

Graphing Exponential Functions We can use a graphing calculator to graph exponential functions. It might be necessary to try several sets of window dimensions in order to find the ones that give a good view of the curve.

To graph $f(x) = 3^x - 1$, we enter the equation as y_1. We can begin graphing with the standard window $[-10, 10, -10, 10]$. Although this window gives a good view of the curve, we might want to adjust it to show more of the curve in the first quadrant. Changing the dimensions to $[-10, 10, -5, 15]$ accomplishes this.

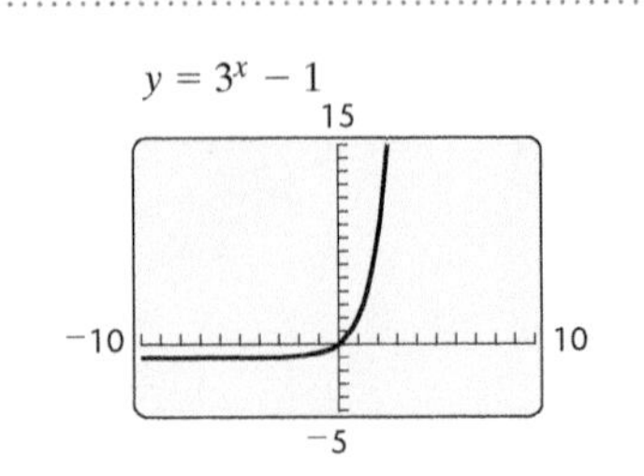

EXERCISE:

1. Use a graphing calculator to graph the functions in Examples 1–4.

b EQUATIONS WITH x AND y INTERCHANGED

It will be helpful in later work to be able to graph an equation in which the x and the y in $y = a^x$ are interchanged.

EXAMPLE 5 Graph: $x = 2^y$.

Note that x is alone on one side of the equation. We can find ordered pairs that are solutions more easily by choosing values for y and then computing the x-values.

For $y = 0, x = 2^0 = 1$.
For $y = 1, x = 2^1 = 2$.
For $y = 2, x = 2^2 = 4$.
For $y = 3, x = 2^3 = 8$.

For $y = -1, x = 2^{-1} = \frac{1}{2^1} = \frac{1}{2}$.

For $y = -2, x = 2^{-2} = \frac{1}{2^2} = \frac{1}{4}$.

For $y = -3, x = 2^{-3} = \frac{1}{2^3} = \frac{1}{8}$.

x	y
1	0
2	1
4	2
8	3
$\frac{1}{2}$	−1
$\frac{1}{4}$	−2
$\frac{1}{8}$	−3

(1) Choose values for y.
(2) Compute values for x.

We plot the points and connect them with a smooth curve. What happens as y-values become smaller?

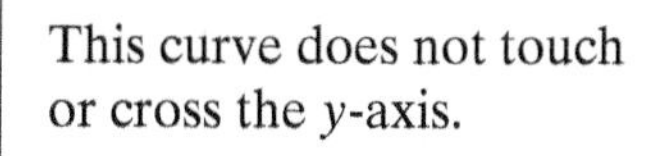
This curve does not touch or cross the y-axis.

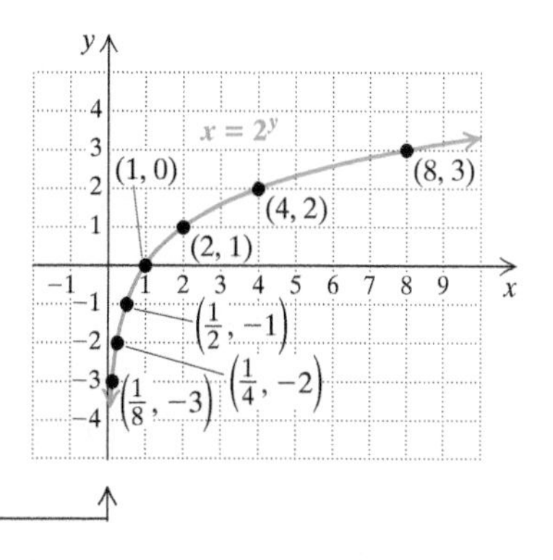

Note that the curve $x = 2^y$ has the same shape as the graph of $y = 2^x$, except that it is reflected, or flipped, across the line $y = x$, as shown below.

$y = 2^x$	
x	y
0	1
1	2
2	4
3	8
−1	$\frac{1}{2}$
−2	$\frac{1}{4}$
−3	$\frac{1}{8}$

$x = 2^y$	
x	y
1	0
2	1
4	2
8	3
$\frac{1}{2}$	−1
$\frac{1}{4}$	−2
$\frac{1}{8}$	−3

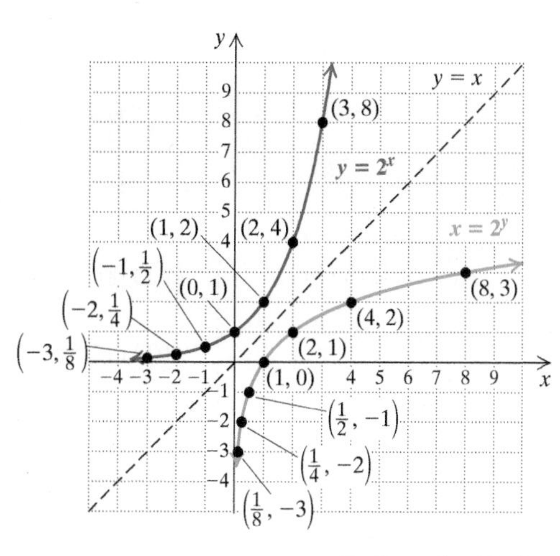

Do Exercise 7. ▶

7. Graph: $x = 3^y$.

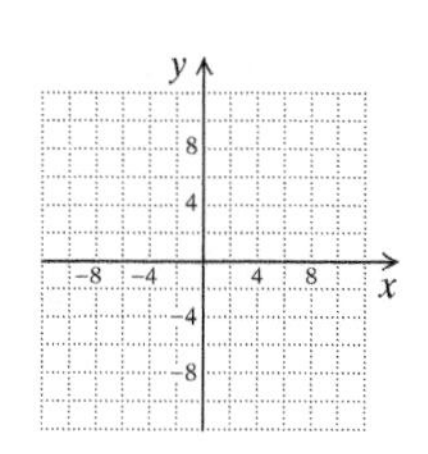

Answers

5.

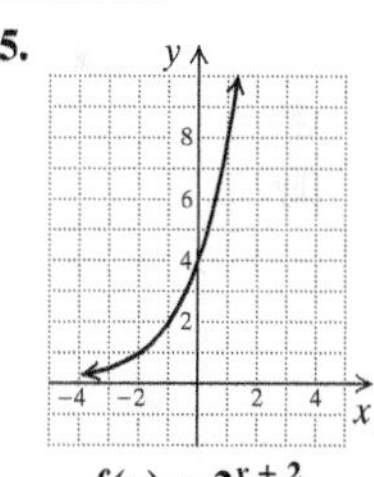
$f(x) = 2^{x+2}$

6.

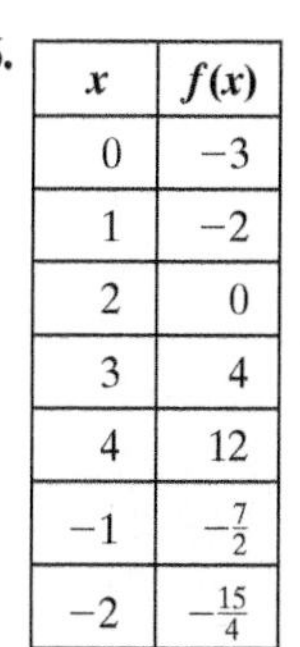

x	$f(x)$
0	−3
1	−2
2	0
3	4
4	12
−1	$-\frac{7}{2}$
−2	$-\frac{15}{4}$

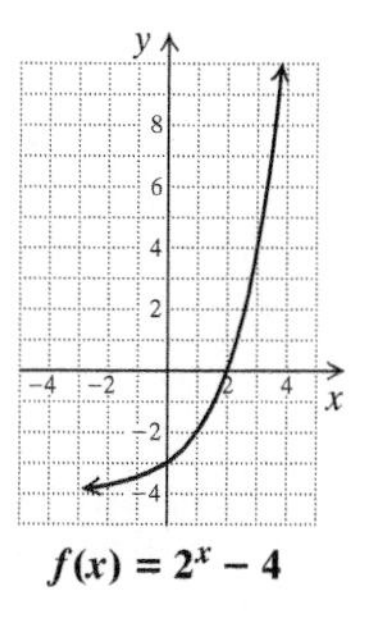
$f(x) = 2^x - 4$

7.

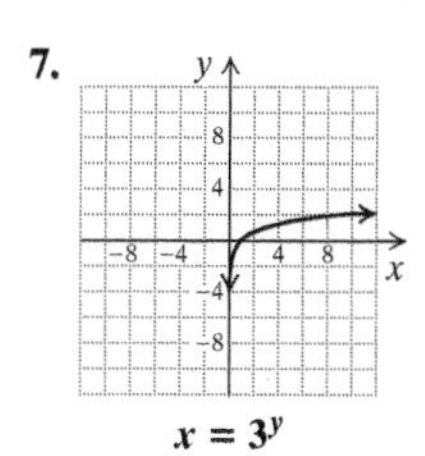
$x = 3^y$

C APPLICATIONS OF EXPONENTIAL FUNCTIONS

When interest is paid on interest, we call it **compound interest**. This is the type of interest paid on investments and loans. Suppose that you have \$100,000 in a savings account at an annual interest rate of 4%. This means that in 1 year, the account will contain the original \$100,000 plus 4% of \$100,000. Thus the total in the account after 1 year will be

$$\$100{,}000 \quad \text{plus} \quad \$100{,}000 \times 0.04.$$

This can also be expressed as

$$\begin{aligned}\$100{,}000 + \$100{,}000 \times 0.04 &= \$100{,}000 \times 1 + \$100{,}000 \times 0.04\\ &= \$100{,}000(1 + 0.04) \quad \text{Factoring out \$100,000 using the distributive law}\\ &= \$100{,}000(1.04)\\ &= \$104{,}000.\end{aligned}$$

Now suppose that the total of \$104,000 remains in the account for another year. At the end of the second year, the account will contain the \$104,000 plus 4% of \$104,000. The total in the account will be

$$\$104{,}000 \quad \text{plus} \quad \$104{,}000 \times 0.04,$$

or

$$\begin{aligned}\$104{,}000(1.04) = [\$100{,}000(1.04)](1.04) &= \$100{,}000(1.04)^2\\ &= \$108{,}160.\end{aligned}$$

Note that in the second year, interest is earned on the first year's interest as well as the original amount. When this happens, we say that the interest is **compounded annually**. If the original amount of \$100,000 earned only simple interest for 2 years, the interest would be

$$\$100{,}000 \times 0.04 \times 2, \quad \text{or} \quad \$8000,$$

and the amount in the account would be

$$\$100{,}000 + \$8000 = \$108{,}000,$$

less than the \$108,160 when interest is compounded annually.

Do Exercise 8.

8. ***Interest Compounded Annually.*** Find the amount in an account after 1 year and after 2 years if \$40,000 is invested at 2%, compounded annually.

GS

Amount after 1 year:

$\$40{,}000 + \$40{,}000 \times$ ____

$= \$40{,}000(1.02)$

$=$ ____

Amount after 2 years:

$\$40{,}800 + \$40{,}800 \times$ ____

$= \$40{,}800(1.02)$

$=$ ____

The following table shows how the computation continues over 4 years.

\$100,000 in an Account

YEAR	WITH INTEREST COMPOUNDED ANNUALLY	WITH SIMPLE INTEREST
Beginning of 1st year	\$100,000	
End of 1st year	$\$100{,}000(1.04)^1 = \$104{,}000$	\$104,000
Beginning of 2nd year	\$104,000	
End of 2nd year	$\$100{,}000(1.04)^2 = \$108{,}160$	\$108,000
Beginning of 3rd year	\$108,160	
End of 3rd year	$\$100{,}000(1.04)^3 = \$112{,}486.40$	\$112,000
Beginning of 4th year	\$112,486.40	
End of 4th year	$\$100{,}000(1.04)^4 \approx \$116{,}985.86$	\$116,000

Answers

8. \$40,800; \$41,616

Guided Solution:

8. 0.02, \$40,800, 0.02, \$41,616

We can express interest compounded annually using an exponential function.

EXAMPLE 6 *Interest Compounded Annually.* The amount of money A that a principal P will grow to after t years at interest rate r, compounded annually, is given by the formula

$$A = P(1 + r)^t.$$

Suppose that \$100,000 is invested at 4% interest, compounded annually.

a) Find a function for the amount in the account after t years.

b) Find the amount of money in the account at $t = 0$, $t = 4$, $t = 8$, and $t = 10$.

c) Graph the function.

a) If $P = \$100{,}000$ and $r = 4\% = 0.04$, we can substitute these values and form the following function:

$$A(t) = \$100{,}000(1 + 0.04)^t = \$100{,}000(1.04)^t.$$

b) To find the function values, you might find a calculator with a power key helpful.

$$A(0) = \$100{,}000(1.04)^0 = \$100{,}000;$$
$$A(4) = \$100{,}000(1.04)^4 \approx \$116{,}985.86;$$
$$A(8) = \$100{,}000(1.04)^8 \approx \$136{,}856.91;$$
$$A(10) = \$100{,}000(1.04)^{10} \approx \$148{,}024.43$$

c) We use the function values computed in (b) with others, if we wish, to draw the graph as follows. Note that the axes are scaled differently because of the large values of A.

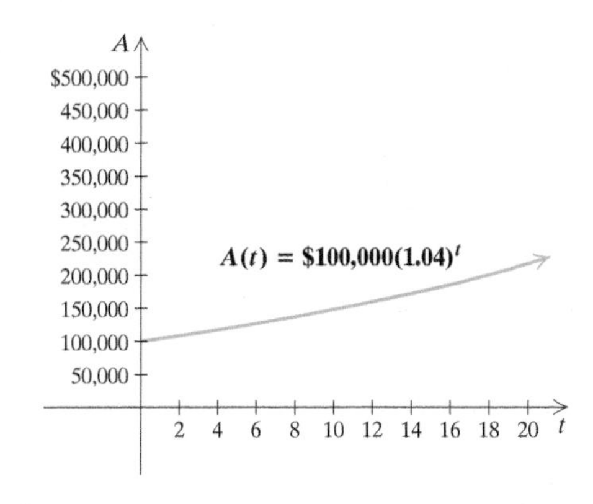

Do Exercise 9.

Suppose that the principal of \$100,000 that we just considered were **compounded semiannually**—that is, every half year. Interest would then be calculated twice a year at a rate of 4% ÷ 2, or 2%, each time. The computations are as follows:

After the first $\frac{1}{2}$ year, the account will contain 102% of \$100,000:

$$\$100{,}000 \times 1.02 = \$102{,}000.$$

After a second $\frac{1}{2}$ year (1 full year), the account will contain 102% of \$102,000:

$$\$102{,}000 \times 1.02 = \$100{,}000 \times (1.02)^2 = \$104{,}040.$$

After a third $\frac{1}{2}$ year $\left(1\frac{1}{2}\text{ full years}\right)$, the account will contain 102% of \$104,040:

$$\$104{,}040 \times 1.02 = \$100{,}000 \times (1.02)^3 = \$106{,}120.80.$$

9. *Interest Compounded Annually.* Suppose that \$40,000 is invested at 5% interest, compounded annually.

a) Find a function for the amount in the account after t years.

b) Find the amount of money in the account at $t = 0$, $t = 4$, $t = 8$, and $t = 10$.

c) Graph the function.

Answers

9. (a) $A(t) = \$40{,}000(1.05)^t$; **(b)** \$40,000; \$48,620.25; \$59,098.22; \$65,155.79

(c)

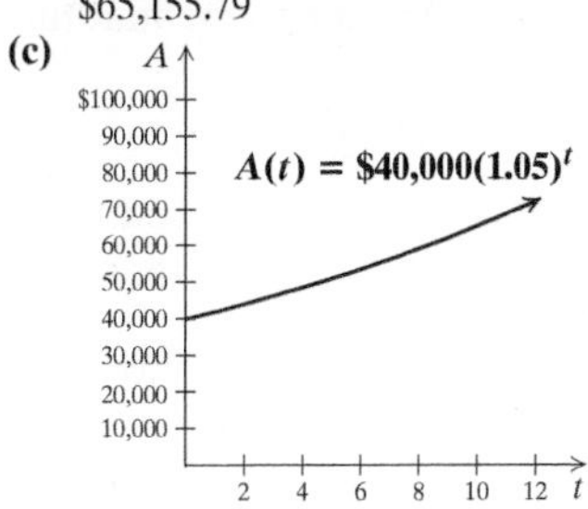

After a fourth $\frac{1}{2}$ year (2 full years), the account will contain 102% of \$106,120.80:

$$\$106{,}120.80 \times 1.02 = \$100{,}000 \times (1.02)^4$$
$$\approx \$108{,}243.22. \quad \text{Rounded to the nearest cent}$$

Comparing these results with those in the table on p. 630, we can see that by having more compounding periods, we increase the amount in the account at the end of the second year. We have illustrated the following result.

COMPOUND-INTEREST FORMULA

If a principal P has been invested at interest rate r, compounded n times per year, in t years it will grow to an amount A given by

$$A = P \cdot \left(1 + \frac{r}{n}\right)^{n \cdot t}.$$

10. A couple invests \$7000 in an account paying 3.4%, compounded quarterly. Find the amount in the account after $5\frac{1}{2}$ years.

GS

$$A = P \cdot \left(1 + \frac{r}{n}\right)^{n \cdot t}$$
$$= 7000 \cdot \left(1 + \frac{3.4\%}{\square}\right)^{\square \cdot \frac{11}{2}}$$
$$= 7000 \cdot \left(1 + \frac{0.034}{4}\right)^{\square}$$
$$= 7000 \cdot (\square)^{22}$$
$$\approx \square$$

EXAMPLE 7 The Ibsens invest \$4000 in an account paying $2\frac{5}{8}\%$, compounded quarterly. Find the amount in the account after $2\frac{1}{2}$ years.

The compounding is quarterly—that is, four times per year—so in $2\frac{1}{2}$ years, there are ten $\frac{1}{4}$-year periods. We substitute \$4000 for P, $2\frac{5}{8}\%$, or 0.02625, for r, 4 for n, and $2\frac{1}{2}$, or $\frac{5}{2}$, for t and compute A:

$$A = P \cdot \left(1 + \frac{r}{n}\right)^{n \cdot t}$$
$$= 4000 \cdot \left(1 + \frac{2\frac{5}{8}\%}{4}\right)^{4 \cdot \frac{5}{2}}$$
$$= 4000 \cdot \left(1 + \frac{0.02625}{4}\right)^{10}$$
$$= 4000(1.0065625)^{10} \quad \text{Using a calculator}$$
$$\approx \$4270.39.$$

The amount in the account after $2\frac{1}{2}$ years is \$4270.39.

Do Exercise 10.

Answer

10. \$8432.72

Guided Solution:

10. 4, 4, 22, 1.0085, \$8432.72

8.1 Exercise Set

FOR EXTRA HELP

MyLab Math

✓ Check Your Understanding

Reading Check Determine whether each statement is true or false.

RC1. In an exponential function, the variable is in the exponent.

RC2. The graph of $f(x) = 3^x$ goes through the point (1, 0).

RC3. If x and y are interchanged in an equation, then the graph of the new equation will be the reflection of the graph of the original equation across the y-axis.

RC4. For a given interest rate, the amount of interest earned in a given period of time will be greater with semiannual compounding than with annual compounding.

Concept Check Each of the following exercises shows the graph of a function $f(x) = a^x$. Determine from the graph whether $a > 1$ or $0 < a < 1$.

CC1.

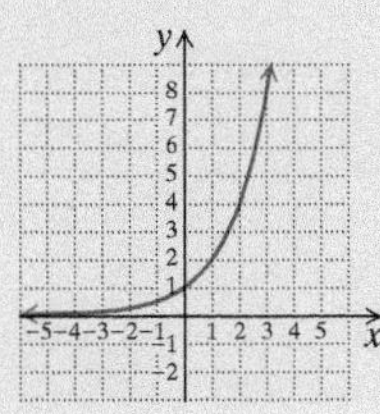

CC2.

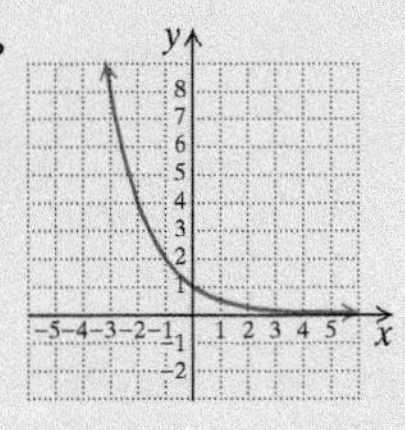

CC3.

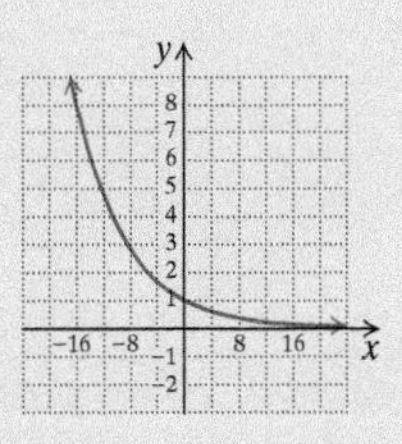

CC4.

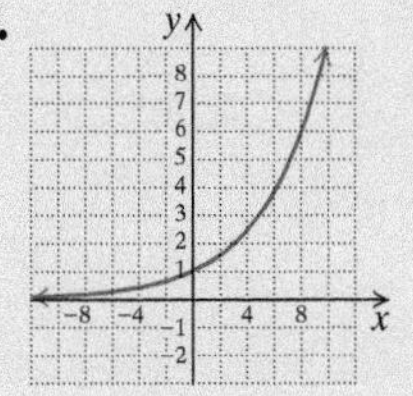

a Graph.

1. $f(x) = 2^x$

x	$f(x)$
0	
1	
2	
3	
−1	
−2	
−3	

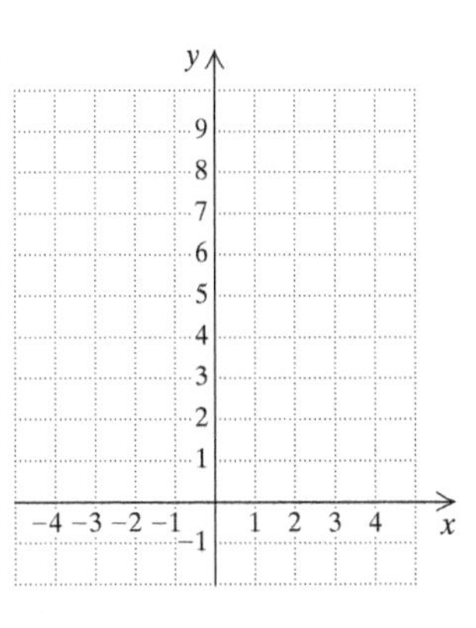

2. $f(x) = 3^x$

x	$f(x)$
0	
1	
2	
3	
−1	
−2	
−3	

3. $f(x) = 5^x$

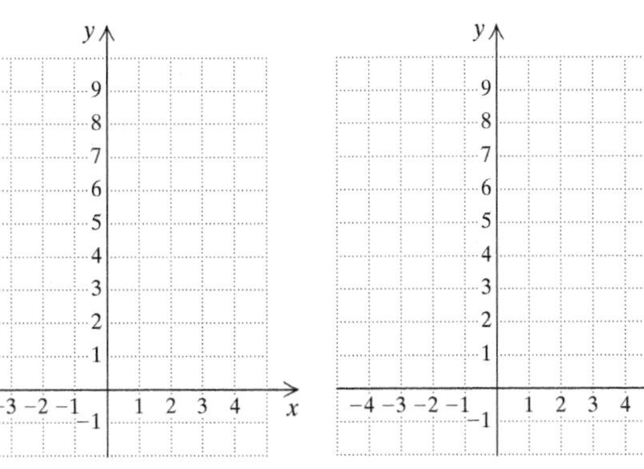

4. $f(x) = 6^x$

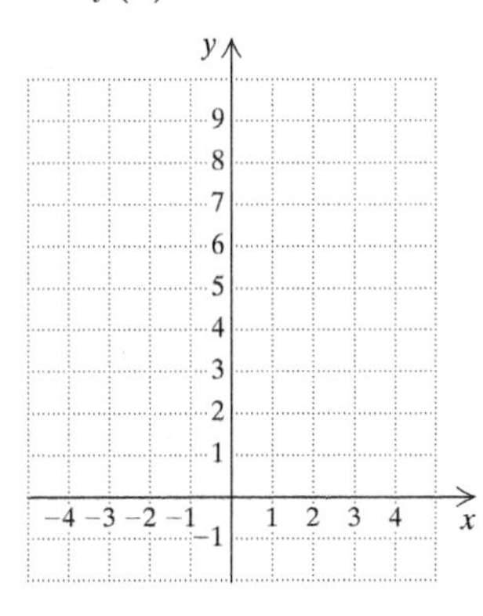

5. $f(x) = 2^{x+1}$

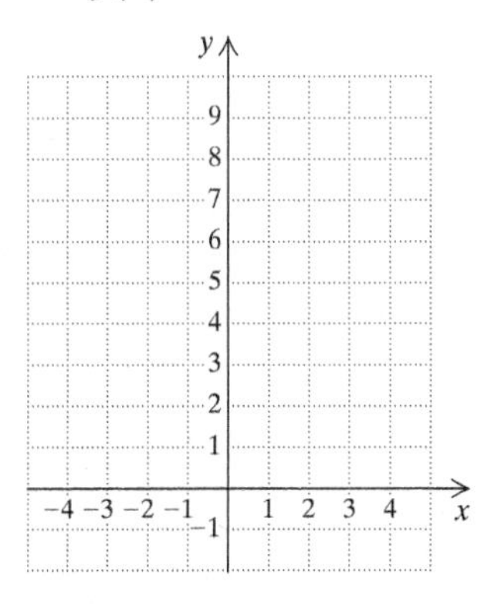

6. $f(x) = 2^{x-1}$

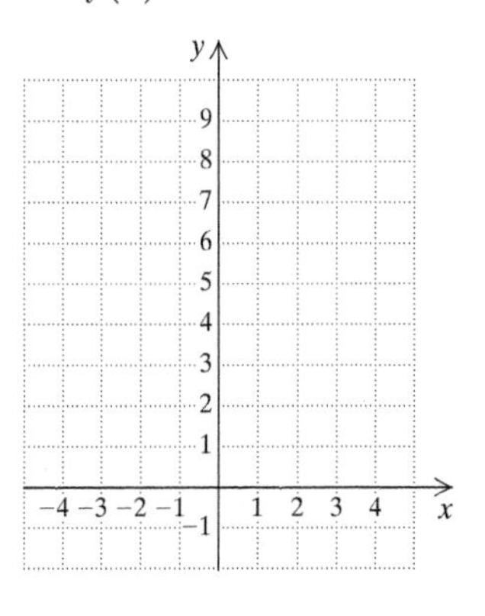

7. $f(x) = 3^{x-2}$

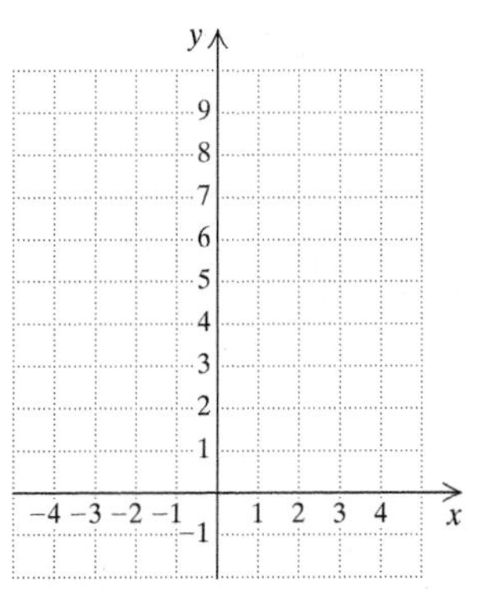

8. $f(x) = 3^{x+2}$

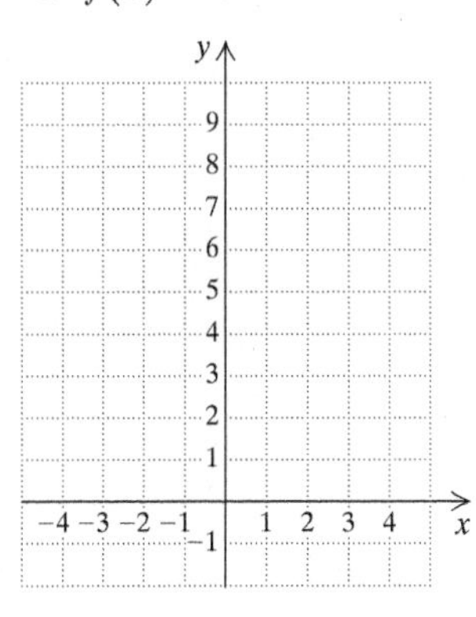

9. $f(x) = 2^x - 3$

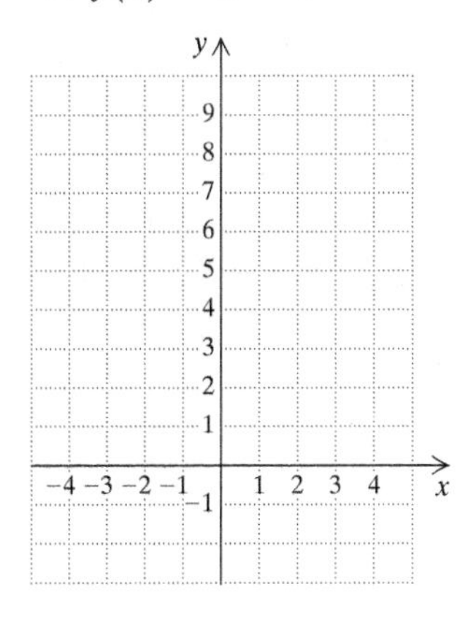

10. $f(x) = 2^x + 1$

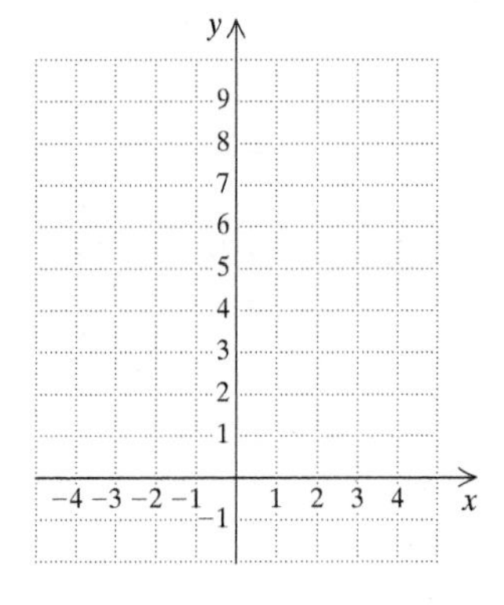

11. $f(x) = 5^{x+3}$

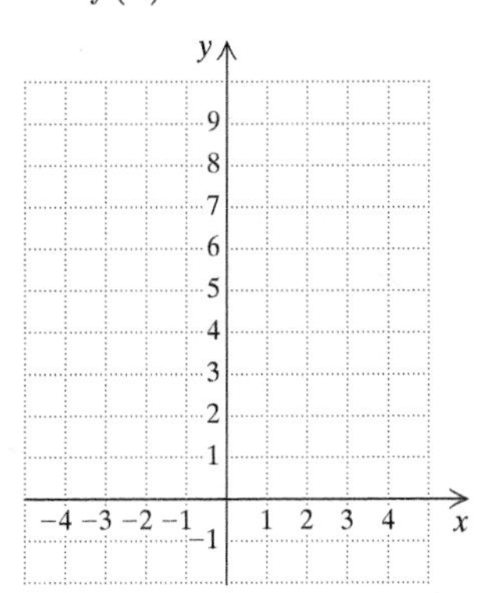

12. $f(x) = 6^{x-4}$

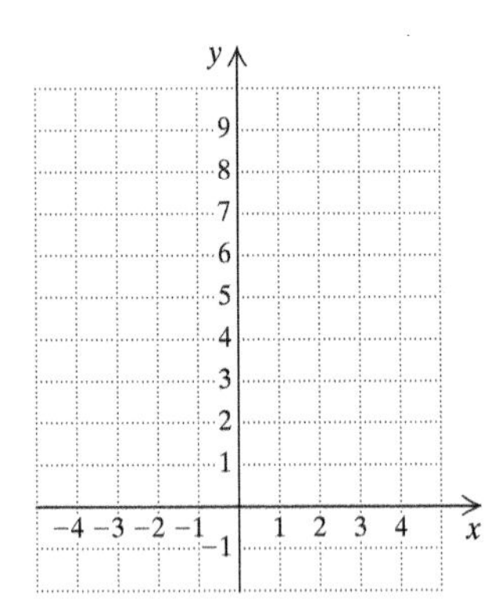

13. $f(x) = \left(\frac{1}{2}\right)^x$

x	$f(x)$
0	
1	
2	
3	
−1	
−2	
−3	

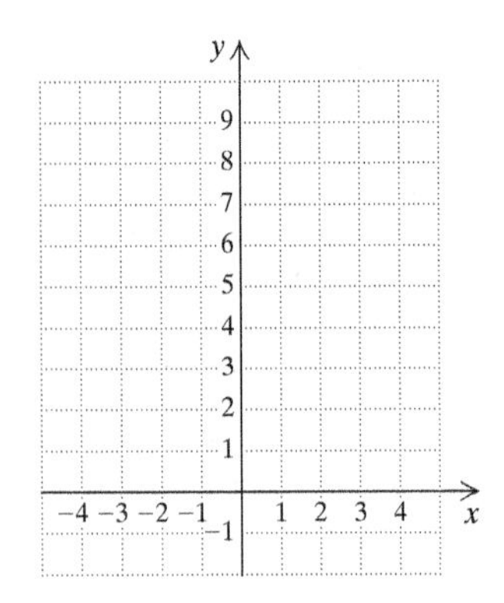

14. $f(x) = \left(\frac{1}{3}\right)^x$

x	$f(x)$
0	
1	
2	
3	
−1	
−2	
−3	

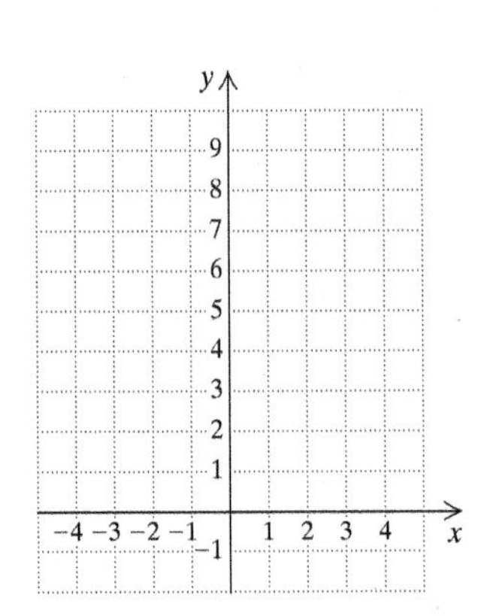

15. $f(x) = \left(\frac{1}{5}\right)^x$

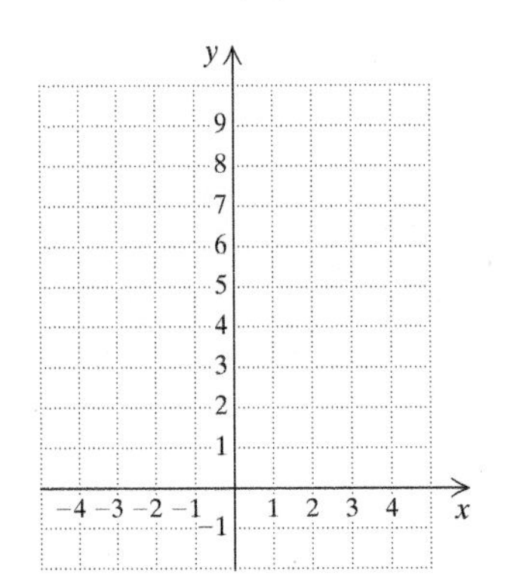

16. $f(x) = \left(\frac{1}{4}\right)^x$

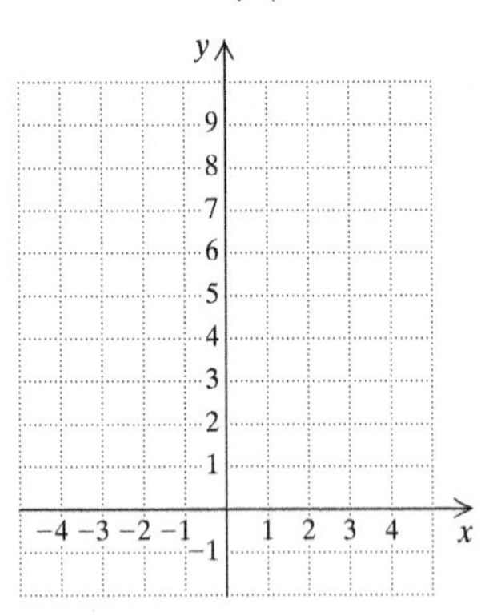

17. $f(x) = 2^{2x-1}$

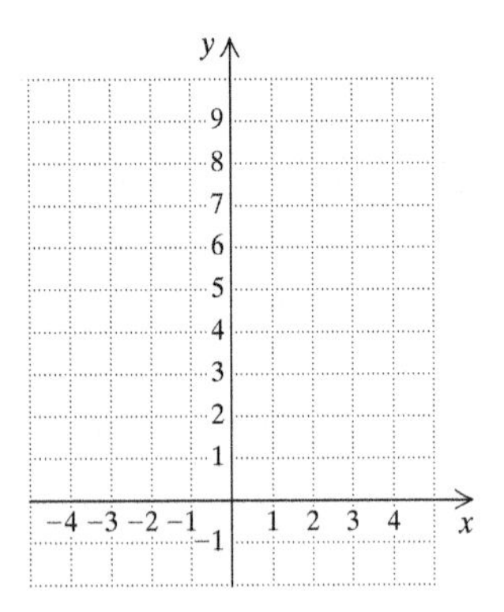

18. $f(x) = 3^{3-x}$

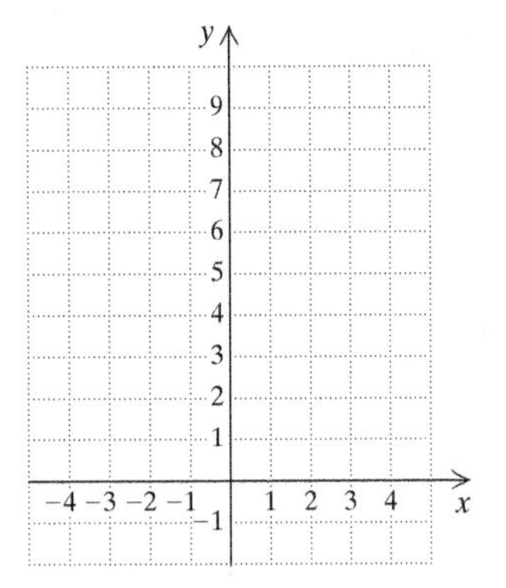

b Graph.

19. $x = 2^y$

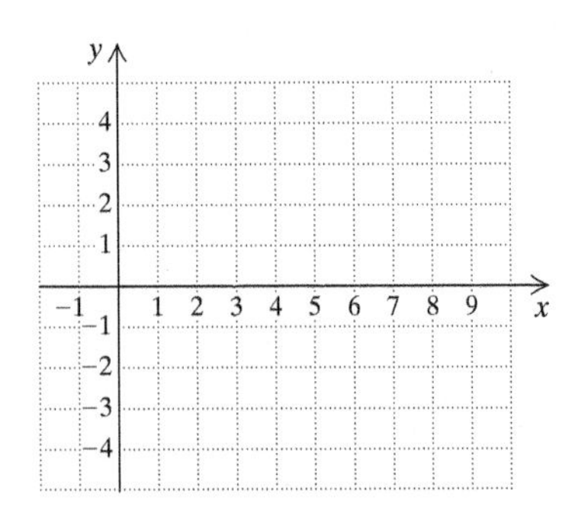

20. $x = 6^y$

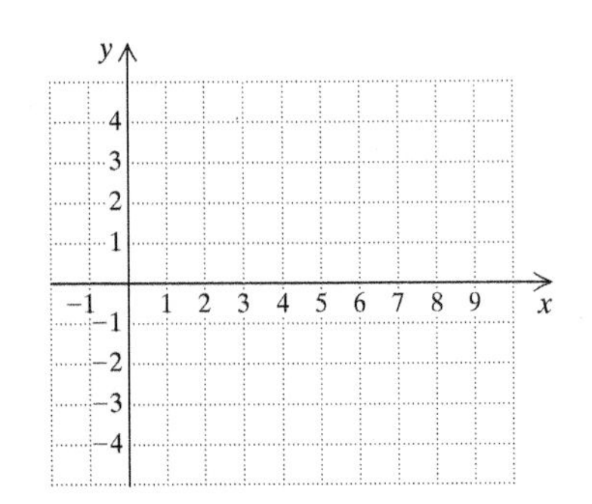

21. $x = \left(\frac{1}{2}\right)^y$

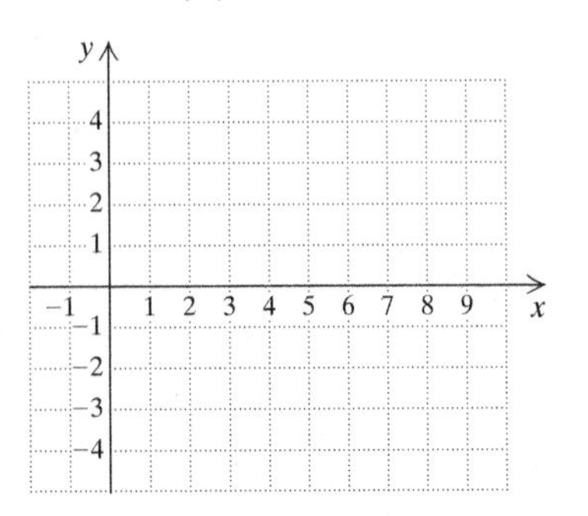

22. $x = \left(\frac{1}{3}\right)^y$

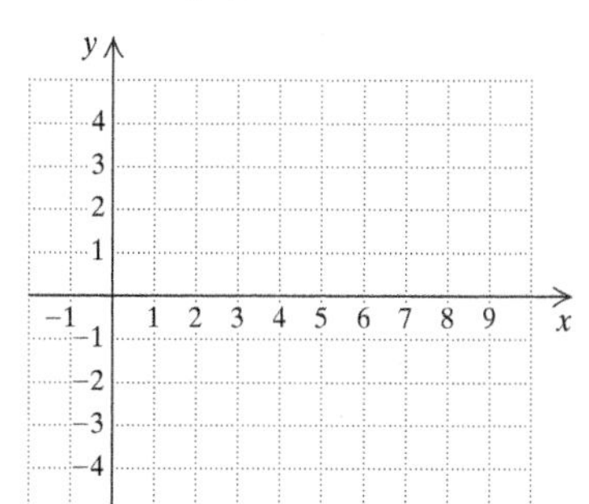

23. $x = 5^y$

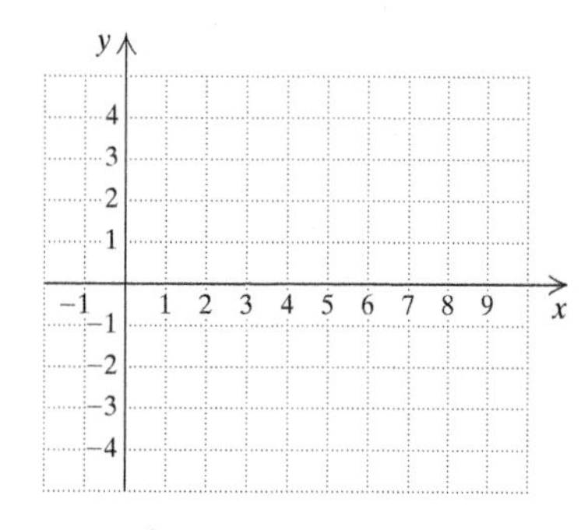

24. $x = \left(\frac{2}{3}\right)^y$

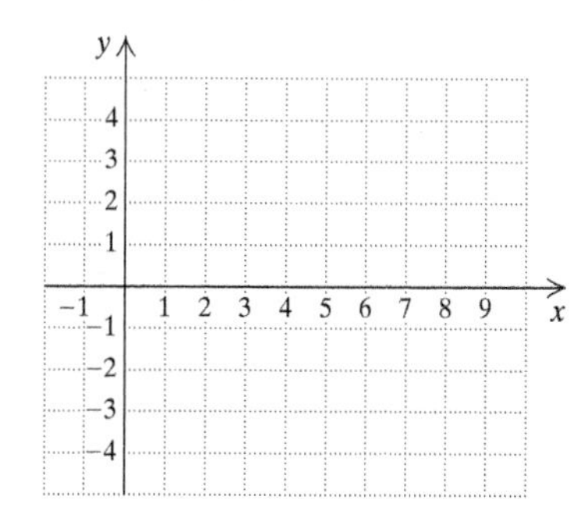

Graph both equations using the same set of axes.

25. $y = 2^x,\ x = 2^y$

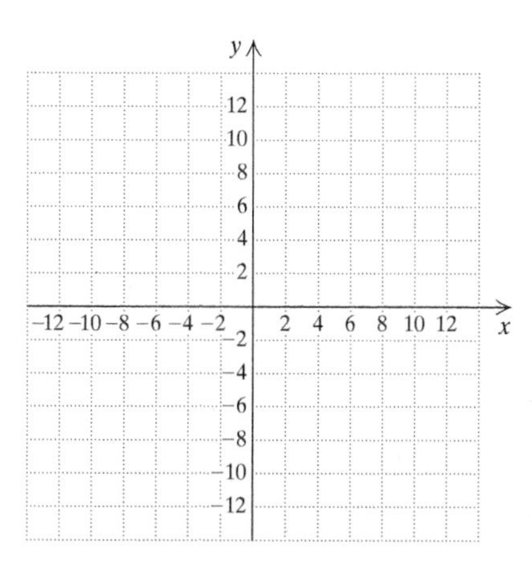

26. $y = \left(\frac{1}{2}\right)^x,\ x = \left(\frac{1}{2}\right)^y$

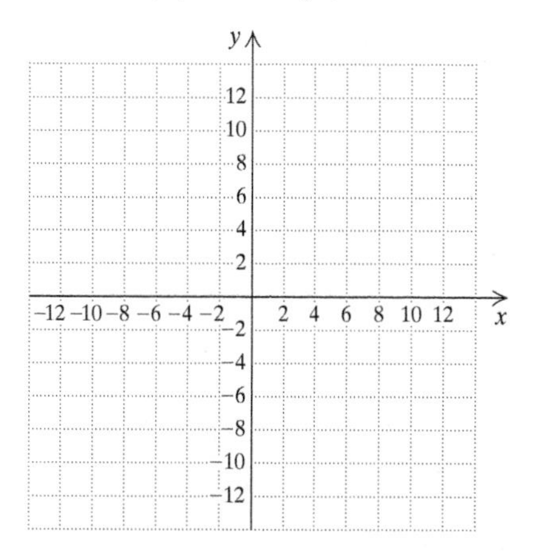

C Solve.

27. *Interest Compounded Annually.* Suppose that \$50,000 is invested at 2% interest, compounded annually.

a) Find a function A for the amount in the account after t years.

b) Complete the following table of function values.

t	$A(t)$
0	
1	
2	
4	
8	
10	
20	

c) Graph the function.

28. *Interest Compounded Annually.* Suppose that \$50,000 is invested at 3% interest, compounded annually.

a) Find a function A for the amount in the account after t years.

b) Complete the following table of function values.

t	$A(t)$
0	
1	
2	
4	
8	
10	
20	

c) Graph the function.

29. *Interest Compounded Semiannually.* Jesse deposits \$2000 in an account paying 2.6%, compounded semiannually. Find the amount in the account after 3 years.

30. *Interest Compounded Semiannually.* Rory deposits \$3500 in an account paying 3.2%, compounded semiannually. Find the amount in the account after 2 years.

31. *Interest Compounded Quarterly.* The Jansens invest \$4500 in an account paying 3.6%, compounded quarterly. Find the amount in the account after $4\frac{1}{2}$ years.

32. *Interest Compounded Quarterly.* The Gemmers invest \$4000 in an account paying 2.8%, compounded quarterly. Find the amount in the account after $3\frac{1}{2}$ years.

33. *Endangered Species.* After the Hawaiian goose, or nene, was placed on the endangered species list, its population began to increase from a critically low number. The number of nene $N(t)$ in Hawaii t years after 1975 can be approximated by

$$N(t) = 374(1.054)^t,$$

where $t = 0$ corresponds to 1975.

Data: esasuccess.org

a) How many nene were in Hawaii in 1975? in 2005? in 2015?
b) Graph the function.

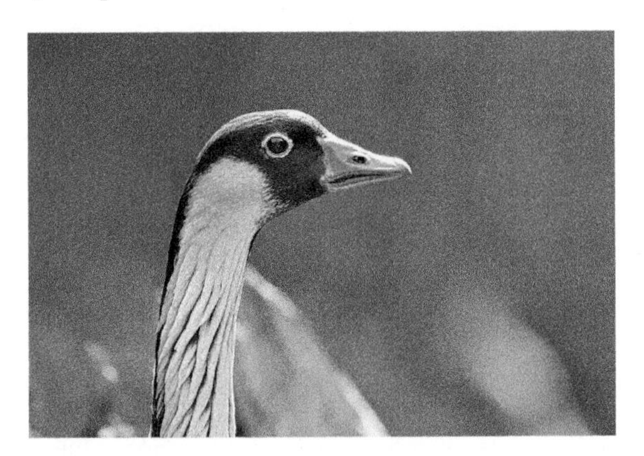

34. *Snapchat.* The popularity of the social media platform Snapchat has increased exponentially since 2014. The average number of daily active Snapchat users $S(t)$, in millions, t years after January 2014 can be approximated by

$$S(t) = 48(1.58)^t,$$

where $t = 0$ corresponds to January 2014.

Data: *The Wall Street Journal*

a) How many daily active users were there, on average, in January 2014? in January 2015? in January 2016?
b) Graph the function.

35. *TV prices.* The average price of an OLED panel for a TV can be approximated by

$$P(t) = 6553(0.612)^t,$$

where t is the number of years after 2013.

Data: smarttvradar.com

a) What was the average price of an OLED panel in 2013? in 2015? in 2017?
b) Graph the function.

36. *Salvage Value.* An office machine is purchased for \$5200. Its value each year is about 80% of the value the preceding year. Its value after t years is given by the exponential function

$$V(t) = \$5200(0.8)^t.$$

a) Find the value of the machine after 0 year, 1 year, 2 years, 5 years, and 10 years.
b) Graph the function.

37. *Recycling Aluminum Cans.* Although Americans discard $1 billion worth of aluminum cans every year, 67% of the aluminum is recycled. If a beverage company distributes 500,000 cans, the amount of aluminum still in use after t years can be made into $N(t)$ cans, where

$$N(t) = 500{,}000(0.67)^t.$$

Data: The Aluminum Association

a) How many cans can be made from the original 500,000 cans after 1 year? after 3 years? after 7 years?
b) Graph the function.

38. *Growth of Bacteria.* Bladder infections are often caused when the bacteria *Escherichia coli* reach the human bladder. Suppose that 3000 of the bacteria are present at time $t = 0$. Then t minutes later, the number of bacteria present will be

$$N(t) = 3000(2)^{t/20}.$$

Data: Hayes, Chris, "Detecting a Human Health Risk: *E. coli*," *Laboratory Medicine* **29**, no. 6, pp. 347–355, June 1998

a) How many bacteria will be present after 10 min? 20 min? 30 min? 40 min? 60 min?
b) Graph the function.

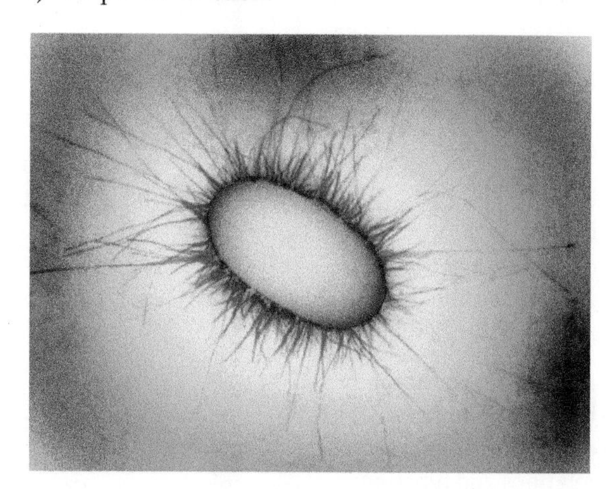

Skill Maintenance

39. Multiply and simplify: $x^{-5} \cdot x^3$. [J19]

40. Simplify: $(x^{-3})^4$. [J19]

Simplify. [J10]

41. 9^0

42. $\left(\frac{2}{3}\right)^0$

43. $\left(\frac{2}{3}\right)^1$

44. 2.7^1

Divide and simplify. [J19]

45. $\dfrac{x^{-3}}{x^4}$

46. $\dfrac{x}{x^{11}}$

47. $\dfrac{x}{x^0}$

48. $\dfrac{x^{-3}}{x^{-4}}$

Synthesis

49. Simplify: $(5^{\sqrt{2}})^{2\sqrt{2}}$.

50. Which is larger: $\pi^{\sqrt{2}}$ or $(\sqrt{2})^{\pi}$?

Graph.

51. $y = 2^x + 2^{-x}$

52. $y = |2^x - 2|$

53. $y = \left|\left(\frac{1}{2}\right)^x - 1\right|$

54. $y = 2^{-x^2}$

Graph both equations using the same set of axes.

55. $y = 3^{-(x-1)}$, $x = 3^{-(y-1)}$

56. $y = 1^x$, $x = 1^y$

57. Use a graphing calculator to graph each of the equations in Exercises 51–54.

8.2 Composite Functions and Inverse Functions

OBJECTIVES

a Find the composition of functions and express certain functions as a composition of functions.

b Find the inverse of a relation if it is described as a set of ordered pairs or as an equation.

c Given a function, determine whether it is one-to-one and has an inverse that is a function.

d Find a formula for the inverse of a function, if it exists, and graph inverse relations and functions.

e Determine whether a function is an inverse by checking its composition with the original function.

Tea Mug Collective's Shane Kimberlin, Alaskan artist

Later in this chapter, we discuss two closely related types of functions: exponential functions and logarithmic functions. In order to properly understand the link between these functions, we must first understand composite functions and inverse functions.

a COMPOSITE FUNCTIONS

Functions frequently occur in which some quantity depends on a variable that, in turn, depends on another variable. For instance, a firm's profits may be a function of the number of items the firm produces, which may in turn be a function of the number of employees hired. In this case, the firm's profits may be considered a **composite function.**

Let's consider an example of a profit function. Tea Mug Collective sells hand-painted tee shirts. Suppose that the monthly profit p, in dollars, from the sale of m shirts is given by $p = 15m - 1200$, and the number of shirts m produced in a month by x employees is given by $m = 40x$.

If Tea Mug Collective employs 10 people, then in one month they can produce $m = 40(10) = 400$ shirts. The profit from selling these 400 shirts would be $p = 15(400) - 1200 = 4800$ dollars. Can we find an equation that would allow us to calculate the monthly profit on the basis of the number of employees? We begin with the profit equation and substitute:

$$\begin{aligned} p &= 15m - 1200 \\ &= 15(40x) - 1200 \qquad \text{Substituting } 40x \text{ for } m \\ &= 600x - 1200. \end{aligned}$$

The equation $p = 600x - 1200$ gives the monthly profit when Tea Mug Collective has x employees.

To find a composition of functions, we follow the same reasoning described above using function notation:

$$p(m) = 15m - 1200, \qquad \text{Profit as a function of the number of shirts produced}$$

$$m(x) = 40x; \qquad \text{Number of shirts as a function of the number of employees}$$

$$\begin{aligned} p(m(x)) &= p(40x) \\ &= 15(40x) - 1200 \\ &= 600x - 1200. \end{aligned}$$

If we call this new function P, then $P(x) = 600x - 1200$. This gives profit as a function of the number of employees.

We call P the *composition* of p and m. In general, the composition of f and g is written $f \circ g$ and is read "the composition of f and g," "f composed with g," or "f circle g."

It is not uncommon to use the same variable to represent the input in more than one function.

Throughout this chapter, keep in mind that equations such as $m(x) = 40x$ and $m(t) = 40t$ describe the same function. Both equations tell us to find a function value by multiplying the input by 40.

COMPOSITE FUNCTION

The **composite function** $f \circ g$, the **composition** of f and g, is defined as

$$(f \circ g)(x) = f(g(x)).$$

We can visualize the composition of functions as follows.

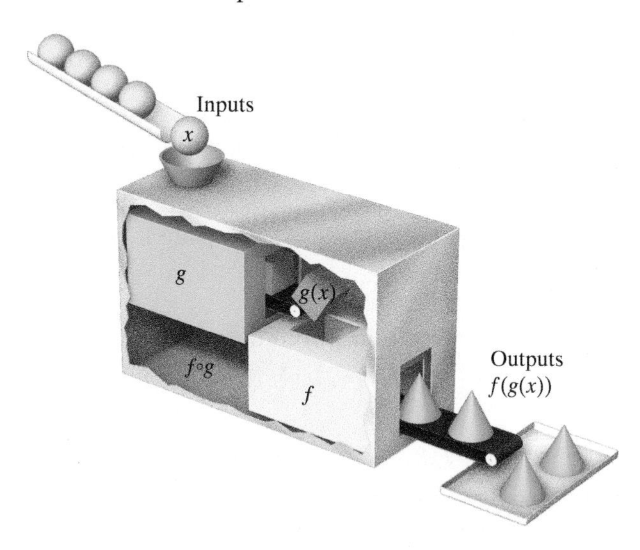

EXAMPLE 1 Given $f(x) = 3x$ and $g(x) = 1 + x^2$:

a) Find $(f \circ g)(5)$ and $(g \circ f)(5)$.

b) Find $(f \circ g)(x)$ and $(g \circ f)(x)$.

We consider each function separately:

$f(x) = 3x$ This function multiplies each input by 3.

and $g(x) = 1 + x^2$. This function adds 1 to the square of each input.

a) $(f \circ g)(5) = f(g(5)) = f(1 + 5^2) = f(26) = 3(26) = 78$;

$(g \circ f)(5) = g(f(5)) = g(3 \cdot 5) = g(15) = 1 + 15^2 = 1 + 225 = 226$

b) $(f \circ g)(x) = f(g(x))$

$= f(1 + x^2)$ Substituting $1 + x^2$ for $g(x)$

$= 3(1 + x^2)$

$= 3 + 3x^2$;

$(g \circ f)(x) = g(f(x))$

$= g(3x)$ Substituting $3x$ for $f(x)$

$= 1 + (3x)^2$

$= 1 + 9x^2$

We can check the values in part (a) with the formulas found in part (b):

$(f \circ g)(x) = 3 + 3x^2$	$(g \circ f)(x) = 1 + 9x^2$
$(f \circ g)(5) = 3 + 3 \cdot 5^2$	$(g \circ f)(5) = 1 + 9 \cdot 5^2$
$= 3 + 3 \cdot 25$	$= 1 + 9 \cdot 25$
$= 3 + 75$	$= 1 + 225$
$= 78$;	$= 226$.

Do Exercise 1. ▶

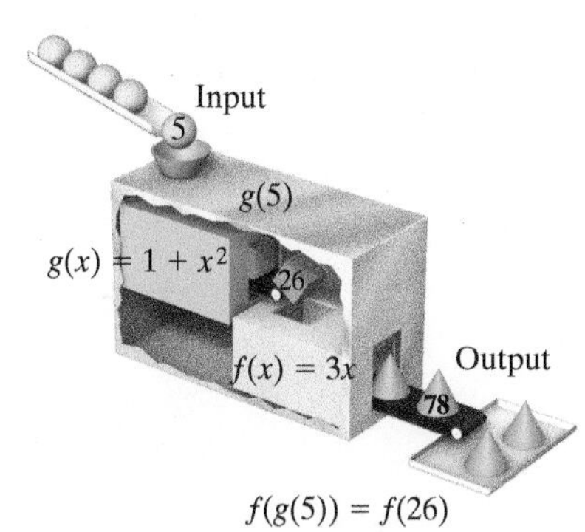

A composition machine for Example 1

1. Given $f(x) = x + 5$ and $g(x) = x^2 - 1$, find $(f \circ g)(x)$ and $(g \circ f)(x)$.

Answer

1. $x^2 + 4$; $x^2 + 10x + 24$

2. Given $f(x) = 4x + 5$ and $g(x) = \sqrt[3]{x}$, find $(f \circ g)(x)$ and $(g \circ f)(x)$.

GS

$(f \circ g)(x) = f(g(x))$
$= f(\sqrt[3]{\quad})$
$= 4(\quad) + 5;$

$(g \circ f)(x) = g(f(x))$
$= g(4x + \quad)$
$= \sqrt[3]{\quad}$

3. Find $f(x)$ and $g(x)$ such that $h(x) = (f \circ g)(x)$. Answers may vary.

a) $h(x) = \sqrt[3]{x^2 + 1}$

b) $h(x) = \dfrac{1}{(x + 5)^4}$

Answers

2. $4\sqrt[3]{x} + 5$; $\sqrt[3]{4x + 5}$
3. (a) $f(x) = \sqrt[3]{x}$; $g(x) = x^2 + 1$;
(b) $f(x) = \dfrac{1}{x^4}$; $g(x) = x + 5$

Guided Solution:
2. x, $\sqrt[3]{x}$, 5, $4x + 5$

Example 1 shows that $(f \circ g)(5) \neq (g \circ f)(5)$ and, in general,

$$(f \circ g)(x) \neq (g \circ f)(x).$$

EXAMPLE 2 Given $f(x) = \sqrt{x}$ and $g(x) = x - 1$, find $(f \circ g)(x)$ and $(g \circ f)(x)$.

$$(f \circ g)(x) = f(g(x)) = f(x - 1) = \sqrt{x - 1};$$
$$(g \circ f)(x) = g(f(x)) = g(\sqrt{x}) = \sqrt{x} - 1$$

◀ Do Exercise 2.

It is important to be able to recognize how a function can be expressed, or "broken down," as a composition. Such a situation can occur in a study of calculus.

EXAMPLE 3 Find $f(x)$ and $g(x)$ such that $h(x) = (f \circ g)(x)$:

$$h(x) = (7x + 3)^2.$$

This is $7x + 3$ to the 2nd power. Two functions that can be used for the composition are $f(x) = x^2$ and $g(x) = 7x + 3$. We can check by forming the composition:

$$h(x) = (f \circ g)(x) = f(g(x)) = f(7x + 3) = (7x + 3)^2.$$

This is the most "obvious" answer to the question. There can be other less obvious answers. For example, if

$$f(x) = (x - 1)^2 \quad \text{and} \quad g(x) = 7x + 4,$$

then

$$h(x) = (f \circ g)(x) = f(g(x)) = f(7x + 4) = (7x + 4 - 1)^2 = (7x + 3)^2.$$

◀ Do Exercise 3.

b INVERSES

A set of ordered pairs is called a **relation**. A function is a special kind of relation in which to each first coordinate there corresponds one and only one second coordinate.

Consider the relation h given as follows:

$$h = \{(-7, 4), (3, -1), (-6, 5), (0, 2)\}.$$

Suppose that we *interchange* the first and second coordinates. The relation we obtain is called the **inverse** of the relation h and is given as follows:

$$\text{Inverse of } h = \{(4, -7), (-1, 3), (5, -6), (2, 0)\}.$$

INVERSE RELATION (ORDERED PAIRS)

Interchanging the coordinates of the ordered pairs in a relation produces the **inverse relation**.

EXAMPLE 4 Consider the relation g given by

$$g = \{(2, 4), (-1, 3), (-2, 0)\}.$$

In the following figure, the relation g is shown in red. The inverse of the relation is

$$\{(4, 2), (3, -1), (0, -2)\}$$

and is shown in blue.

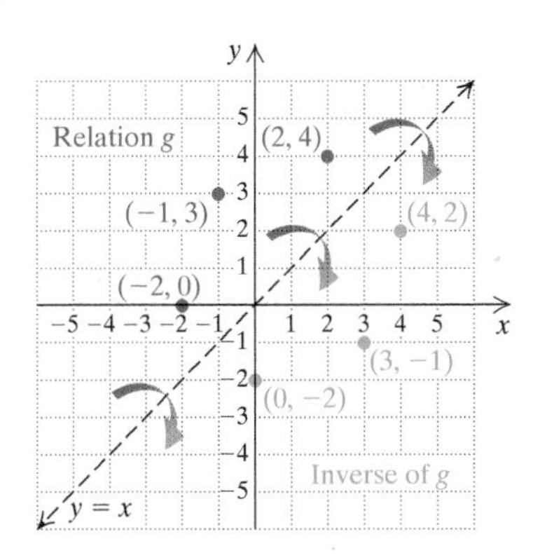

Do Exercise 4. ▶

INVERSE RELATION (EQUATION)

If a relation is defined by an equation, interchanging the variables produces an equation of the **inverse relation.**

EXAMPLE 5 Find an equation of the inverse of $y = 3x - 4$. Then graph both the relation and its inverse.

We interchange x and y and obtain an equation of the inverse:

$$x = 3y - 4.$$

Relation: $y = 3x - 4$ ⟶

x	y
0	−4
1	−1
2	2
3	5

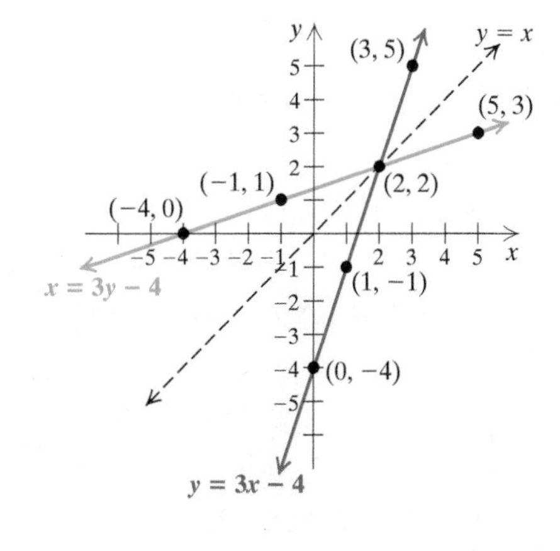

Inverse: $x = 3y - 4$

x	y
−4	0
−1	1
2	2
5	3

Do Exercise 5. ▶

Note in Example 5 that the relation $y = 3x - 4$ is a function and its inverse relation $x = 3y - 4$ is also a function. Each graph passes the vertical-line test. (See Section 2.2.)

4. Consider the relation g given by
$g = \{(2, 5), (-1, 4), (-2, 1)\}$.
The graph of the relation is shown below in red. Find the inverse and draw its graph.

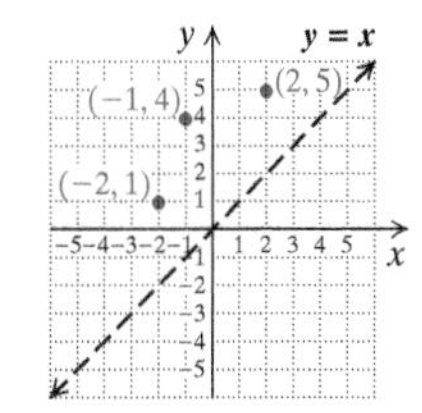

5. Find an equation of the inverse relation. Then complete the table and graph both the original relation and its inverse.

Relation:
$y = 6 - 2x$

x	y
0	6
2	2
3	0
5	−4

Inverse:
____

x	y
	0
	2
	3
	5

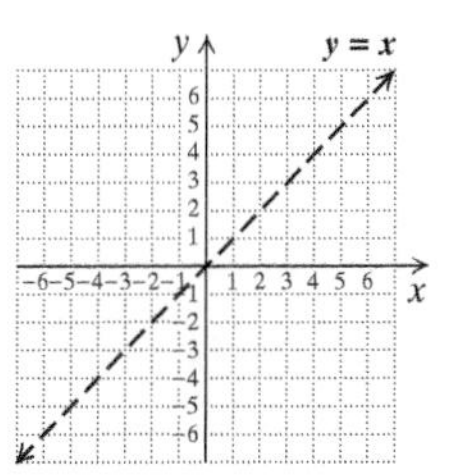

Answer

4. Inverse of $g = \{(5, 2), (4, -1), (1, -2)\}$

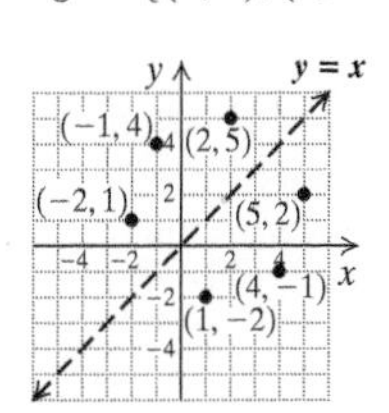

Answer to Margin Exercise 5 is on p. 642.

6. Find an equation of the inverse relation. Then complete the table and graph both the original relation and its inverse.

Relation:

$y = x^2 - 4x + 7$

x	y
0	7
1	4
2	3
3	4
4	7

Inverse:

______

x	y
	0
	1
	2
	3
	4

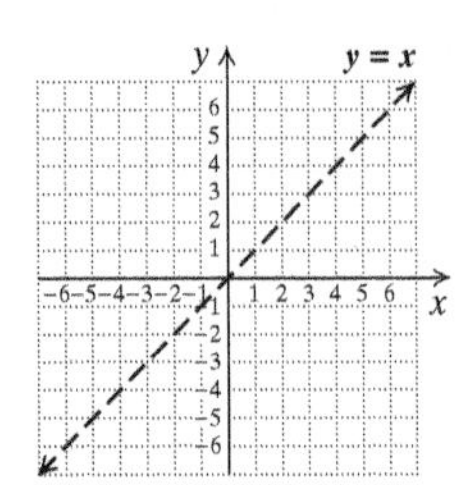

EXAMPLE 6 Find an equation of the inverse of $y = 6x - x^2$. Then graph both the original relation and its inverse.

We interchange x and y and obtain an equation of the inverse:

$$x = 6y - y^2.$$

Relation: $y = 6x - x^2$

x	y
−1	−7
0	0
1	5
3	9
5	5

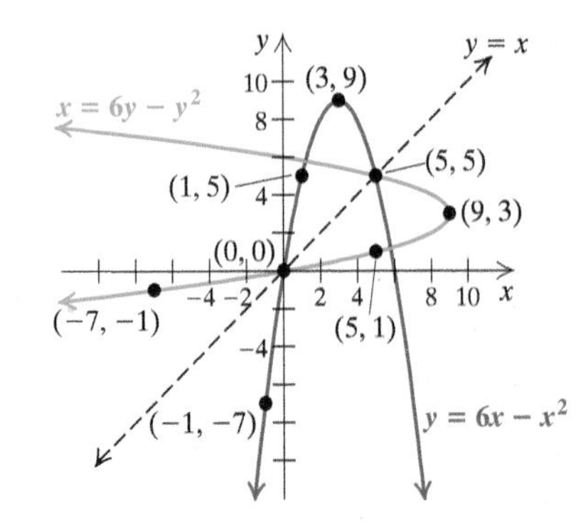

Inverse: $x = 6y - y^2$

x	y
−7	−1
0	0
5	1
9	3
5	5

Note in Example 6 that the relation $y = 6x - x^2$ is a function because it passes the vertical-line test. However, its inverse relation $x = 6y - y^2$ is not a function because its graph fails the vertical-line test. Therefore, the inverse of a function is *not* always a function.

◀ **Do Exercise 6.**

C INVERSES AND ONE-TO-ONE FUNCTIONS

Let's consider the following two functions.

NUMBER (Domain)	CUBE (Range)
−3 →	−27
−2 →	−8
−1 →	−1
0 →	0
1 →	1
2 →	8
3 →	27

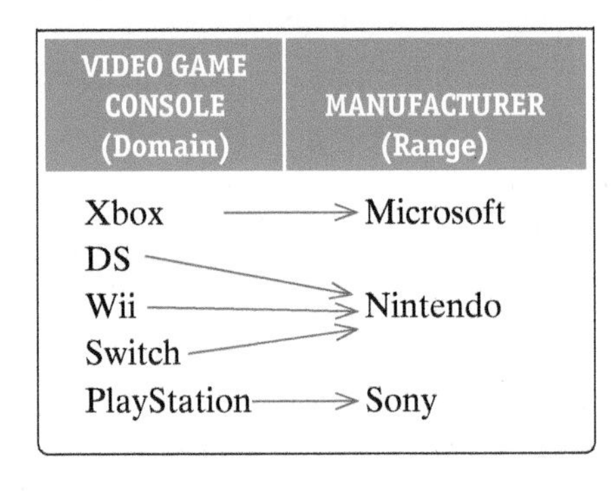

VIDEO GAME CONSOLE (Domain)	MANUFACTURER (Range)
Xbox →	Microsoft
DS →	Nintendo
Wii →	Nintendo
Switch →	Nintendo
PlayStation →	Sony

Suppose that we reverse the arrows. Are these inverse relations functions?

CUBE ROOT (Range)	NUMBER (Domain)
−3	← −27
−2	← −8
−1	← −1
0	← 0
1	← 1
2	← 8
3	← 27

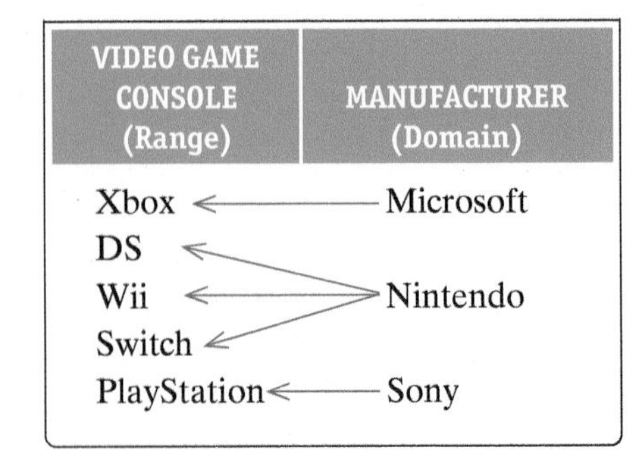

VIDEO GAME CONSOLE (Range)	MANUFACTURER (Domain)
Xbox	← Microsoft
DS	← Nintendo
Wii	← Nintendo
Switch	← Nintendo
PlayStation	← Sony

Answers

5. *Inverse:* $x = 6 - 2y$

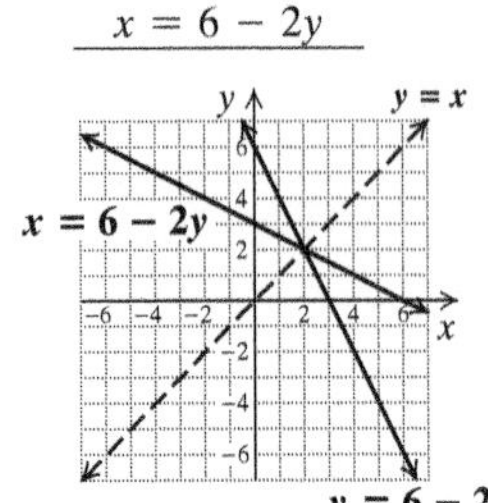

x	y
6	0
2	2
0	3
−4	5

6. *Inverse:* $x = y^2 - 4y + 7$

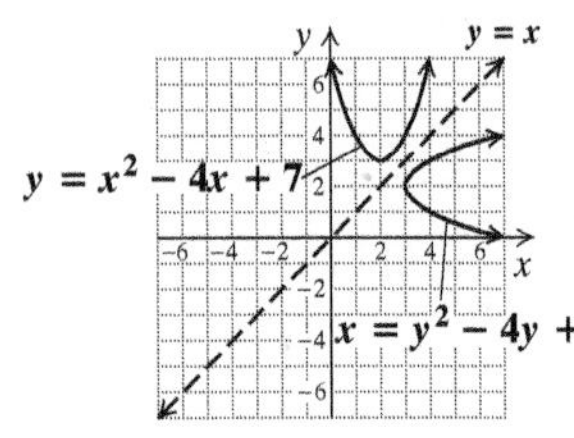

x	y
7	0
4	1
3	2
4	3
7	4

We see that the inverse of the cubing function is a function. The inverse of the game platform function is not a function, however, because the input Nintendo has *three* outputs: DS, Wii, and Switch. Recall that for a function, each input has exactly one output. However, it can happen that the same output comes from two or more different inputs. If this is the case, the inverse cannot be a function. When this possibility is excluded, the inverse is also a function.

In the cubing function, different inputs have different outputs. Thus its inverse is also a function. The cubing function is what is called a **one-to-one function.**

ONE-TO-ONE FUNCTION AND INVERSES

A function f is **one-to-one** if different inputs have different outputs—that is,

if $a \neq b$, then $f(a) \neq f(b)$. Or,

A function f is **one-to-one** if when the outputs are the same, the inputs are the same—that is,

if $f(a) = f(b)$, then $a = b$.

If a function is one-to-one, then its inverse is a function.

How can we tell graphically whether a function is one-to-one and thus has an inverse that is a function?

EXAMPLE 7 The graph of the exponential function $f(x) = 2^x$, or $y = 2^x$, is shown on the left below. The graph of the inverse $x = 2^y$ is shown on the right. How can we tell by examining only the graph on the left whether it has an inverse that is a function?

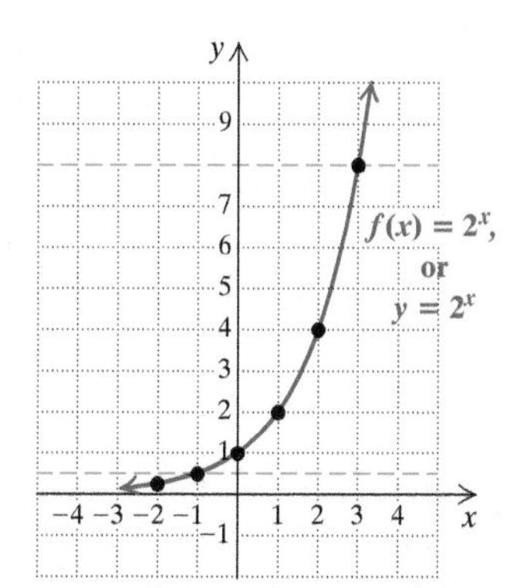

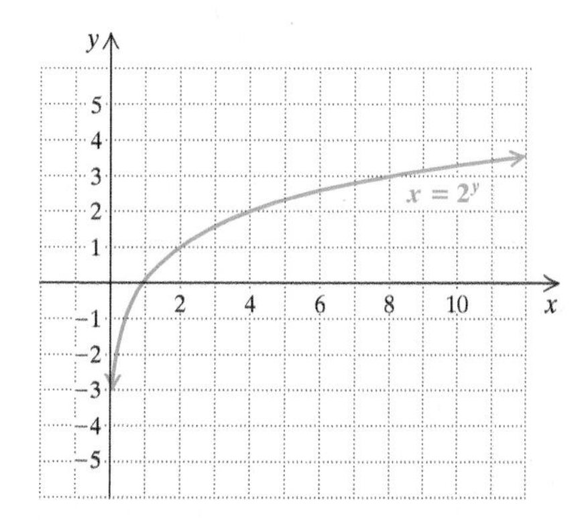

We see that the graph on the right passes the vertical-line test, so we know that it is the graph of a function. However, if we look only at the graph on the left, we think as follows:

A function is one-to-one if different inputs have different outputs. That is, no two x-values will have the same y-value. For this function, no horizontal line can be drawn that will cross the graph more than once. The function is thus one-to-one and its inverse is a function.

THE HORIZONTAL-LINE TEST

If it is possible for a horizontal line to intersect the graph of a function more than once, then the function is not one-to-one and therefore its inverse is not a function.

A graph is that of a function if no vertical line crosses the graph more than once. A function has an inverse that is also a function if no horizontal line crosses the graph more than once.

EXAMPLE 8 Determine whether the function $f(x) = x^2$ is one-to-one and has an inverse that is also a function.

The graph of $f(x) = x^2$, or $y = x^2$, is shown on the left below. There are many horizontal lines that cross the graph more than once, so this function is not one-to-one and does not have an inverse that is a function.

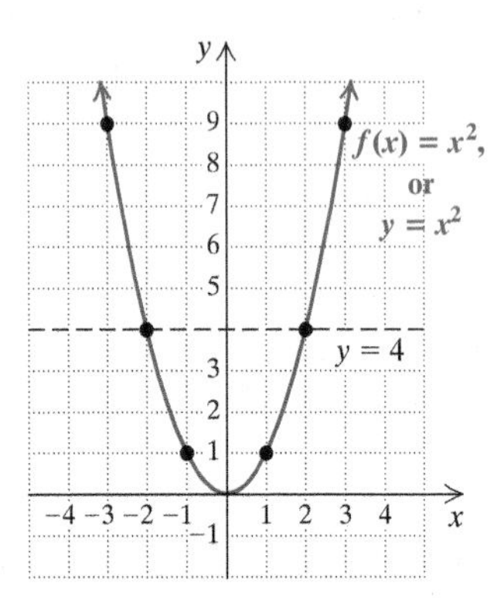

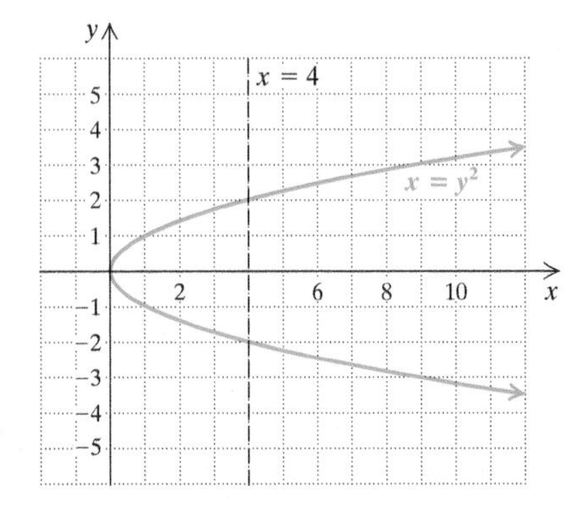

The inverse of the function $y = x^2$ is the relation $x = y^2$. The graph of $x = y^2$ is shown on the right above. It fails the vertical-line test and is not a function.

◀ Do Exercises 7–10.

Determine whether the function is one-to-one and thus has an inverse that is also a function.

7. $f(x) = 4 - x$

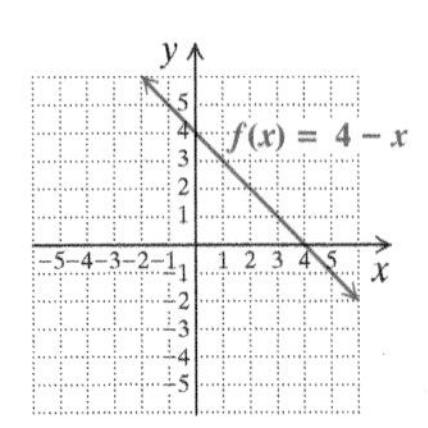

8. $f(x) = x^2 - 1$

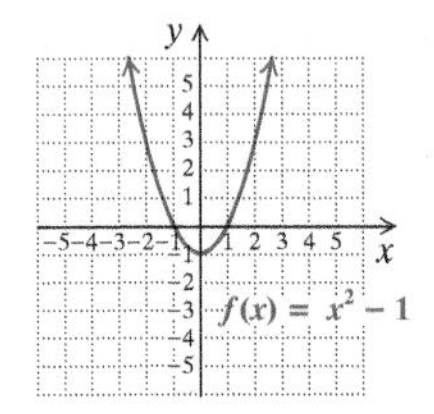

9. $f(x) = 4^x$
(Sketch this graph yourself.)

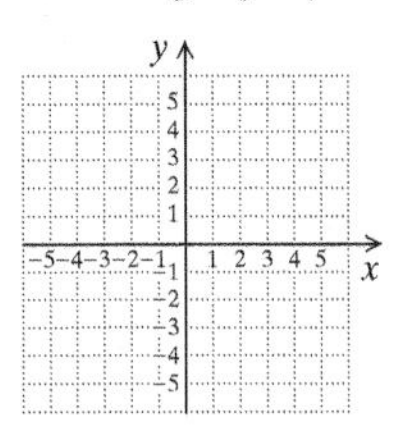

10. $f(x) = |x| - 3$
(Sketch this graph yourself.)

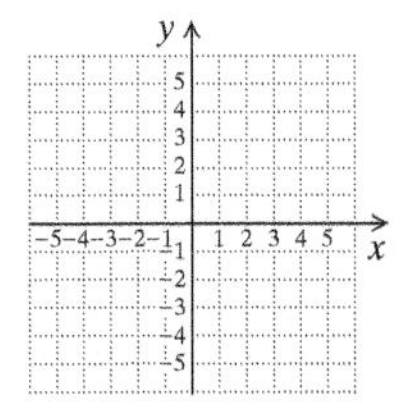

d INVERSE FORMULAS AND GRAPHS

If the inverse of a function f is also a function, then it is named f^{-1} (read "f-inverse").

Caution!

The -1 in f^{-1} is *not* an exponent and f^{-1} does *not* represent a reciprocal!

MyLab Math
ANIMATION

Suppose that a function is described by a formula. If it has an inverse that is a function, how do we find a formula for the inverse function? If for any equation with two variables such as x and y we interchange the variables, we obtain an equation of the inverse relation. We proceed as follows to find a formula for f^{-1}.

If a function f is one-to-one, a formula for its inverse f^{-1} can be found as follows:

1. Replace $f(x)$ with y.
2. Interchange x and y. (This gives the inverse relation.)
3. Solve for y.
4. Replace y with $f^{-1}(x)$.

Answers
7. Yes 8. No 9. Yes 10. No

EXAMPLE 9 Given $f(x) = x + 1$:

a) Determine whether the function is one-to-one.

b) If it is one-to-one, find a formula for $f^{-1}(x)$.

c) Graph the inverse function, if it exists.

a) The graph of $f(x) = x + 1$ is shown below. It passes the horizontal-line test, so it is one-to-one. Thus its inverse is a function.

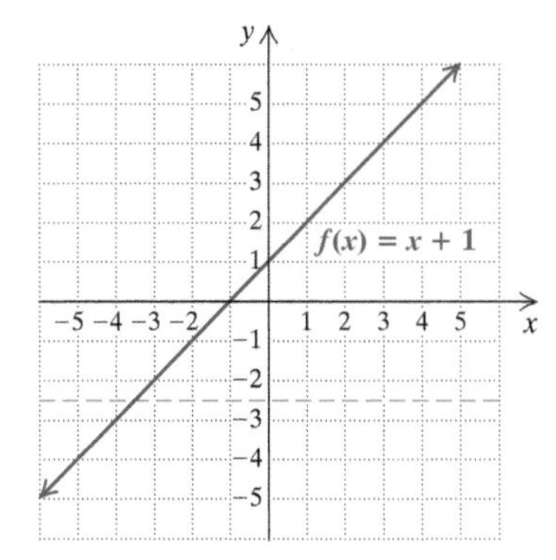

b) 1. Replace $f(x)$ with y: $\quad y = x + 1.$

2. Interchange x and y: $\quad x = y + 1.$ This gives the inverse relation.

3. Solve for y: $\quad x - 1 = y.$

4. Replace y with $f^{-1}(x)$: $\quad f^{-1}(x) = x - 1.$

c) We graph $f^{-1}(x) = x - 1$, or $y = x - 1$. The graph is shown below.

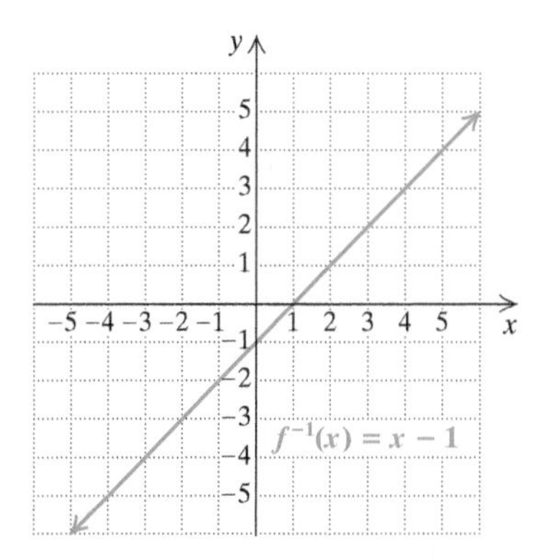

EXAMPLE 10 Given $f(x) = 2x - 3$:

a) Determine whether the function is one-to-one.

b) If it is one-to-one, find a formula for $f^{-1}(x)$.

c) Graph the inverse function, if it exists.

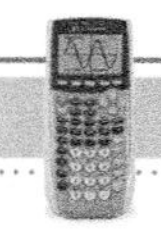

CALCULATOR CORNER

Graphing an Inverse Function The DRAWINV operation can be used to graph a function and its inverse on the same screen. A formula for the inverse function need not be found in order to do this. The graphing calculator must be set in FUNC mode when this operation is used.

To graph $f(x) = 2x - 3$ and $f^{-1}(x)$ using the same set of axes, we first clear any existing equations on the equation-editor screen and then enter $y_1 = 2x - 3$. We graph the function and then use the DRAWINV operation to graph its inverse. The graphs are shown here in a squared window.

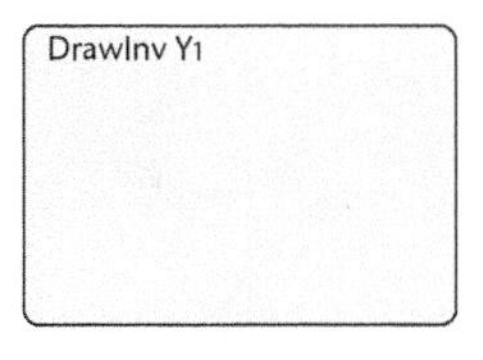

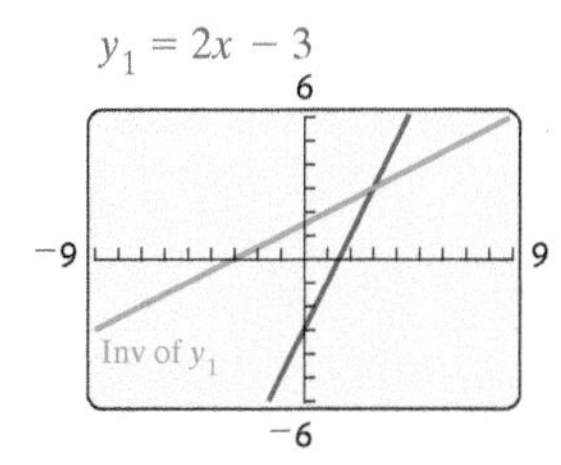

EXERCISES: Use the DRAWINV operation on a graphing calculator to graph each function with its inverse on the same screen.

1. $f(x) = x - 5$
2. $f(x) = \frac{2}{3}x$
3. $f(x) = x^2 + 2$
4. $f(x) = x^3 - 3$

a) The graph of $f(x) = 2x - 3$ is shown below. It passes the horizontal-line test and is one-to-one.

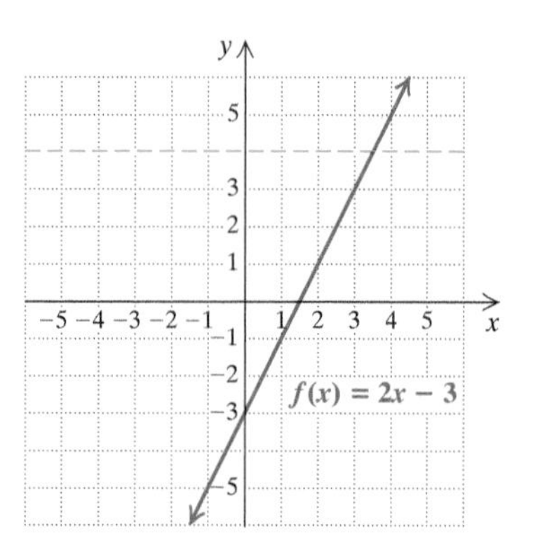

b) 1. Replace $f(x)$ with y: $\quad y = 2x - 3.$

2. Interchange x and y: $\quad x = 2y - 3.$

3. Solve for y:

$$x + 3 = 2y$$

$$\frac{x+3}{2} = y.$$

4. Replace y with $f^{-1}(x)$: $\quad f^{-1}(x) = \dfrac{x+3}{2}.$

c) We graph

$$f^{-1}(x) = \frac{x+3}{2}, \quad \text{or}$$

$$y = \frac{1}{2}x + \frac{3}{2}.$$

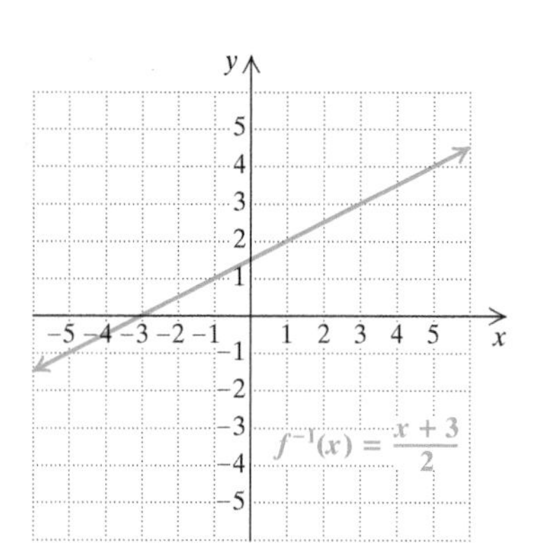

◀ **Do Exercises 11 and 12.**

Let's consider inverses of functions in terms of a function machine. Suppose that a one-to-one function f is programmed into a machine. If the machine is run in reverse, it will perform the inverse function f^{-1}. Inputs then enter at the opposite end, and the entire process is reversed.

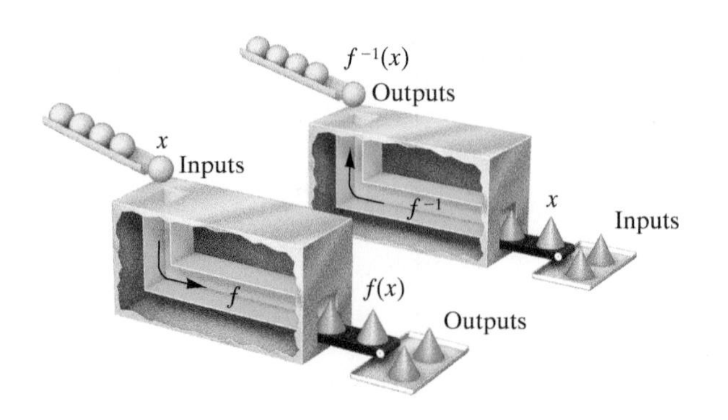

Given each function:

a) Determine whether it is one-to-one.

b) If it is one-to-one, find a formula for the inverse.

c) Graph the inverse function, if it exists.

11. $f(x) = 3 - x$

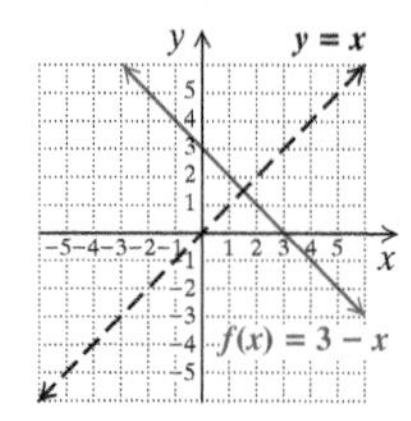

12. $g(x) = 3x - 2$

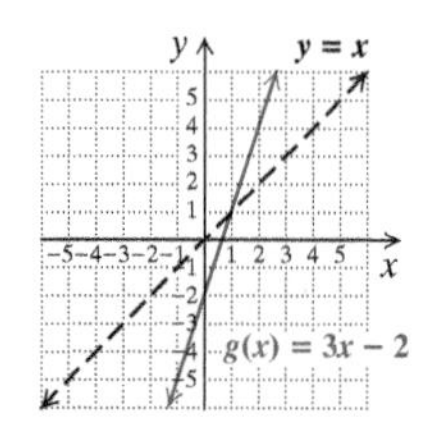

Answers

11. (a) Yes; **(b)** $f^{-1}(x) = 3 - x$;
(c)

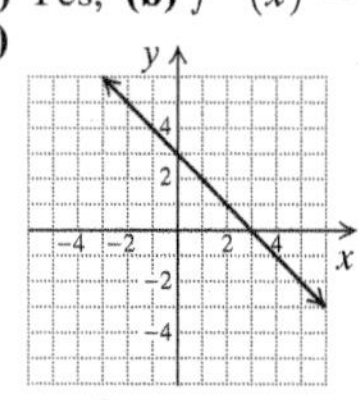

$f^{-1}(x) = 3 - x$

12. (a) Yes; **(b)** $g^{-1}(x) = \dfrac{x+2}{3}$;
(c)

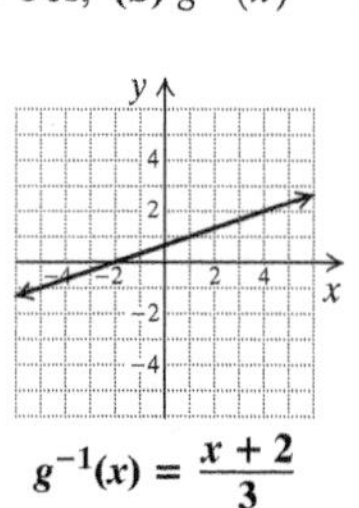

$g^{-1}(x) = \dfrac{x+2}{3}$

Consider $f(x) = 2x - 3$ and $f^{-1}(x) = (x + 3)/2$ from Example 10. For the input 5,

$$f(5) = 2 \cdot 5 - 3 = 10 - 3 = 7.$$

The output is 7. Now we use 7 for the input in the inverse:

$$f^{-1}(7) = \frac{7 + 3}{2} = \frac{10}{2} = 5.$$

The function f takes 5 to 7. The inverse function f^{-1} takes the number 7 back to 5.

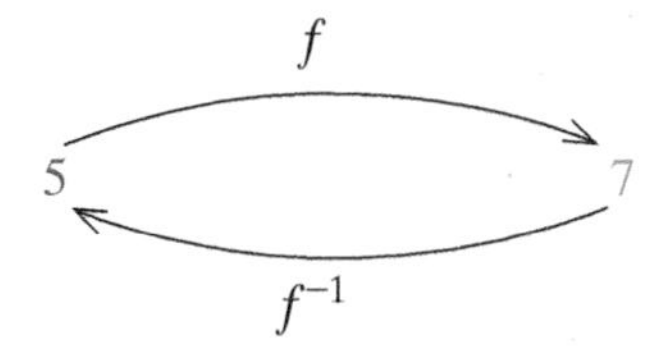

How do the graphs of a function and its inverse compare?

EXAMPLE 11 Graph $f(x) = 2x - 3$ and $f^{-1}(x) = (x + 3)/2$ using the same set of axes. Then compare.

The graph of each function follows. Note that the graph of f^{-1} can be drawn by reflecting the graph of f across the line $y = x$. That is, if we graph $f(x) = 2x - 3$ in wet ink and fold the paper along the line $y = x$, the graph of $f^{-1}(x) = (x + 3)/2$ will appear as the impression made by f.

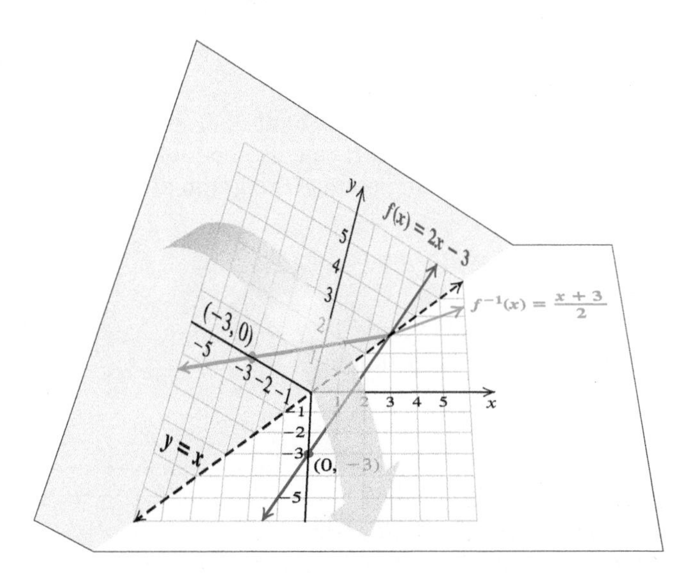

When x and y are interchanged to find a formula for the inverse, we are, in effect, flipping the graph of $f(x) = 2x - 3$ over the line $y = x$. For example, when the coordinates of the y-intercept of the graph of f, $(0, -3)$, are reversed, we get the x-intercept of the graph of f^{-1}, $(-3, 0)$.

The graph of f^{-1} is a reflection of the graph of f across the line $y = x$.

Do Exercise 13. 

13. Graph $g(x) = 3x - 2$ and $g^{-1}(x) = (x + 2)/3$ using the same set of axes.

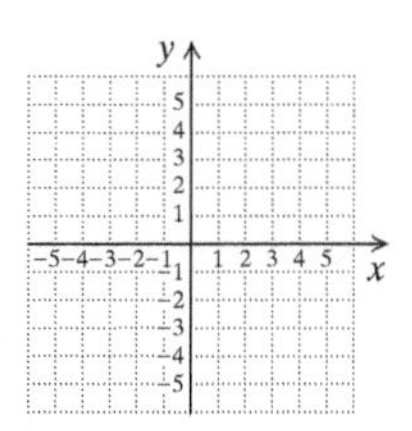

Answer

13.

$g^{-1}(x) = \dfrac{x + 2}{3}$, $y = x$, $g(x) = 3x - 2$

EXAMPLE 12 Consider $g(x) = x^3 + 2$.

a) Determine whether the function is one-to-one.

b) If it is one-to-one, find a formula for its inverse.

c) Graph the inverse, if it exists.

a) The graph of $g(x) = x^3 + 2$ is shown at right in red. It passes the horizontal-line test and thus is one-to-one.

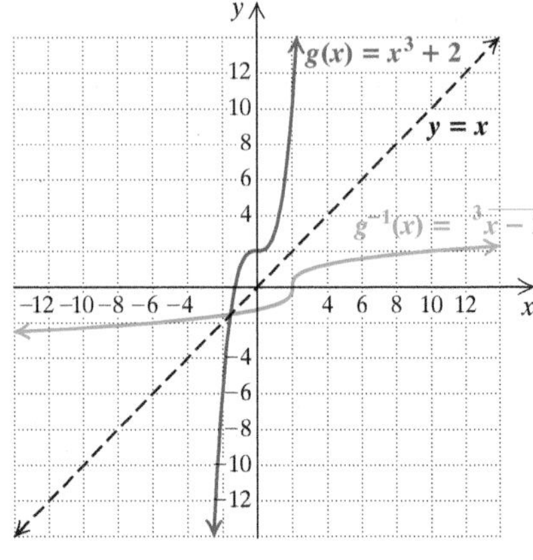

b) 1. Replace $g(x)$ with y: $\quad y = x^3 + 2.$

2. Interchange x and y: $\quad x = y^3 + 2.$

3. Solve for y: $\quad x - 2 = y^3$

$$\sqrt[3]{x - 2} = y.$$ Since a number has only one cube root, we can solve for y.

4. Replace y with $g^{-1}(x)$: $\quad g^{-1}(x) = \sqrt[3]{x - 2}.$

c) To find the graph, we reflect the graph of $g(x) = x^3 + 2$ across the line $y = x$, as we did in Example 11. It can also be found by substituting into $g^{-1}(x) = \sqrt[3]{x - 2}$ and plotting points. The graph of $g^{-1}(x) = \sqrt[3]{x - 2}$ is shown above in blue.

◀ **Do Exercise 14.**

14. Given $f(x) = x^3 + 1$:

a) Determine whether the function is one-to-one.

b) If it is one-to-one, find a formula for its inverse.

c) Graph the function and its inverse using the same set of axes.

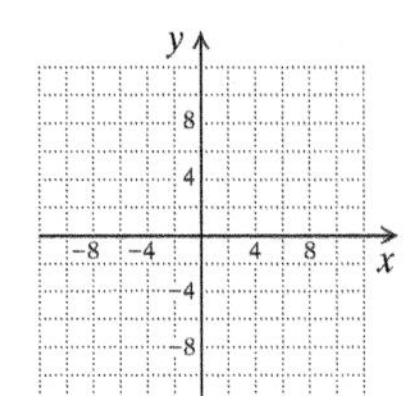

We can now see why we exclude 1 as a base for an exponential function. Consider

$$f(x) = a^x = 1^x = 1.$$

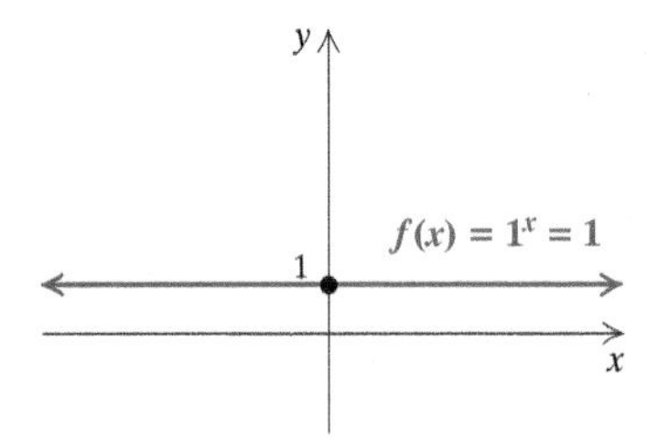

The graph of f is the horizontal line $y = 1$. The graph is not one-to-one. The function does not have an inverse that is a function. All other positive bases yield exponential functions that are one-to-one.

If a function f is one-to-one, then the domain of f is the range of f^{-1}, and the range of f is the domain of f^{-1}.

DOMAIN AND RANGE OF INVERSE FUNCTIONS

The domain of a one-to-one function f is the range of the inverse f^{-1}.
The range of a one-to-one function f is the domain of the inverse f^{-1}.

Answer

14. (a) Yes; **(b)** $f^{-1}(x) = \sqrt[3]{x - 1}$;
(c)

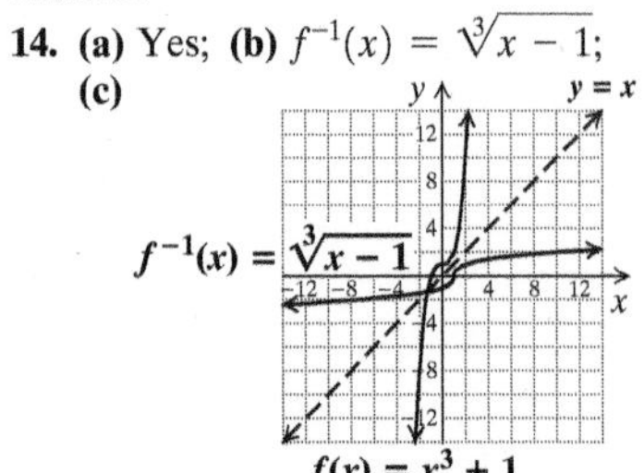

e INVERSE FUNCTIONS AND COMPOSITION

Suppose that we used some input x for the function f and found its output, $f(x)$. The function f^{-1} would then take that output back to x. Similarly, if we began with an input x for the function f^{-1} and found its output, $f^{-1}(x)$, the original function f would then take that output back to x.

> If a function f is one-to-one, then f^{-1} is the unique function for which
>
> $$(f^{-1} \circ f)(x) = x \quad \text{and} \quad (f \circ f^{-1})(x) = x.$$

EXAMPLE 13 Let $f(x) = 2x - 3$. Use composition to show that

$$f^{-1}(x) = \frac{x+3}{2}. \qquad \text{(See Example 10.)}$$

We find $(f^{-1} \circ f)(x)$ and $(f \circ f^{-1})(x)$ and check to see that each is x.

$$\begin{aligned}(f^{-1} \circ f)(x) &= f^{-1}(f(x)) \\ &= f^{-1}(2x-3) \\ &= \frac{(2x-3)+3}{2} \\ &= \frac{2x}{2} \\ &= x;\end{aligned}$$

$$\begin{aligned}(f \circ f^{-1})(x) &= f(f^{-1}(x)) \\ &= f\left(\frac{x+3}{2}\right) \\ &= 2 \cdot \frac{x+3}{2} - 3 \\ &= x + 3 - 3 \\ &= x\end{aligned}$$

Do Exercise 15. ▶

15. Let $f(x) = \frac{2}{3}x - 4$.
Use composition to show that

$$f^{-1}(x) = \frac{3x+12}{2}.$$

Answer

15. $$\begin{aligned}(f^{-1} \circ f)(x) &= f^{-1}(f(x)) = f^{-1}\left(\tfrac{2}{3}x - 4\right) \\ &= \frac{3\left(\frac{2}{3}x - 4\right) + 12}{2} \\ &= \frac{2x - 12 + 12}{2} \\ &= \frac{2x}{2} = x; \\ (f \circ f^{-1})(x) &= f(f^{-1}(x)) = f\left(\frac{3x+12}{2}\right) \\ &= \frac{2}{3}\left(\frac{3x+12}{2}\right) - 4 \\ &= \frac{6x+24}{6} - 4 \\ &= x + 4 - 4 = x\end{aligned}$$

8.2 Exercise Set

FOR EXTRA HELP MyLab Math

✓ Check Your Understanding

Reading Check Choose from the column on the right the word that best completes each statement. Words may be used more than once or not at all.

RC1. Any set of ordered pairs is a(n) ________.

RC2. The relation $\{(3, 6), (0, -1), (2, 5)\}$ is the ________ of $\{(6, 3), (-1, 0), (5, 2)\}$.

RC3. If the graph of a function passes the ________-line test, the inverse of the function is also a function.

RC4. A function whose inverse is also a function is called a(n) ________ function.

RC5. The function $g(x) = x - 10$ is the ________ of $f(x) = x + 10$.

RC6. The function $g(x) = (x-1)^2$ is the ________ $(f \circ h)(x)$ of the functions given by $f(x) = x^2$ and $h(x) = x - 1$.

composition
inverse
function
horizontal
one-to-one
relation
vertical

Concept Check Find each of the following, if possible, given $f(x) = 2x + 3$, $g(x) = x^2 - 1$, and $h(x) = \frac{2}{x}$.

CC1. $(f \circ g)(1)$

CC2. $(g \circ f)(1)$

CC3. $(f \circ h)(-2)$

CC4. $(h \circ f)(-2)$

CC5. $(g \circ h)(-1)$

CC6. $(h \circ g)(-1)$

a

Find $(f \circ g)(x)$ and $(g \circ f)(x)$.

1. $f(x) = 2x - 3$, $g(x) = 6 - 4x$

2. $f(x) = 9 - 6x$, $g(x) = 0.37x + 4$

3. $f(x) = 3x^2 + 2$, $g(x) = 2x - 1$

4. $f(x) = 4x + 3$, $g(x) = 2x^2 - 5$

5. $f(x) = 4x^2 - 1$, $g(x) = \frac{2}{x}$

6. $f(x) = \frac{3}{x}$, $g(x) = 2x^2 + 3$

7. $f(x) = x^2 + 5$, $g(x) = x^2 - 5$

8. $f(x) = \frac{1}{x^2}$, $g(x) = x - 1$

Find $f(x)$ and $g(x)$ such that $h(x) = (f \circ g)(x)$. Answers may vary.

9. $h(x) = (5 - 3x)^2$

10. $h(x) = 4(3x - 1)^2 + 9$

11. $h(x) = \sqrt{5x + 2}$

12. $h(x) = (3x^2 - 7)^5$

13. $h(x) = \frac{1}{x - 1}$

14. $h(x) = \frac{3}{x} + 4$

15. $h(x) = \frac{1}{\sqrt{7x + 2}}$

16. $h(x) = \sqrt{x - 7} - 3$

17. $h(x) = (\sqrt{x} + 5)^4$

18. $h(x) = \frac{x^3 + 1}{x^3 - 1}$

b

Find the inverse of each relation. Graph the original relation in red and then graph the inverse relation in blue.

19. $\{(1, 2), (6, -3), (-3, -5)\}$

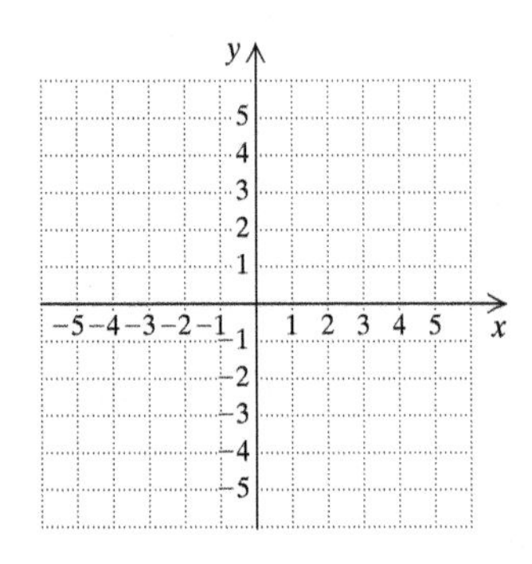

20. $\{(3, -1), (5, 2), (5, -3), (2, 0)\}$

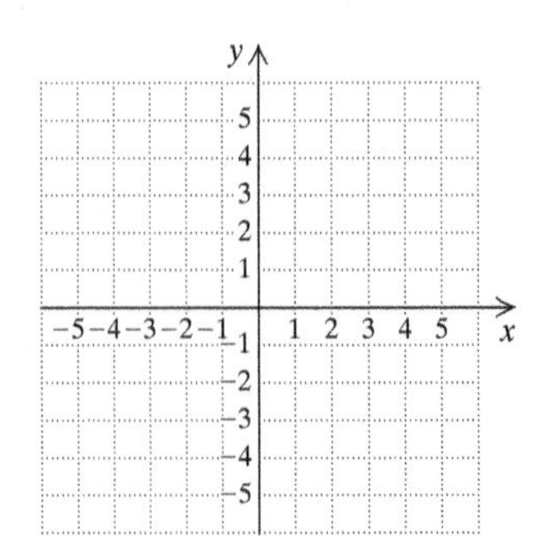

Find an equation of the inverse of the relation. Then complete the second table and graph both the original relation and its inverse.

21. $y = 2x + 6$

x	y
-1	4
0	6
1	8
2	10
3	12

x	y
4	
6	
8	
10	
12	

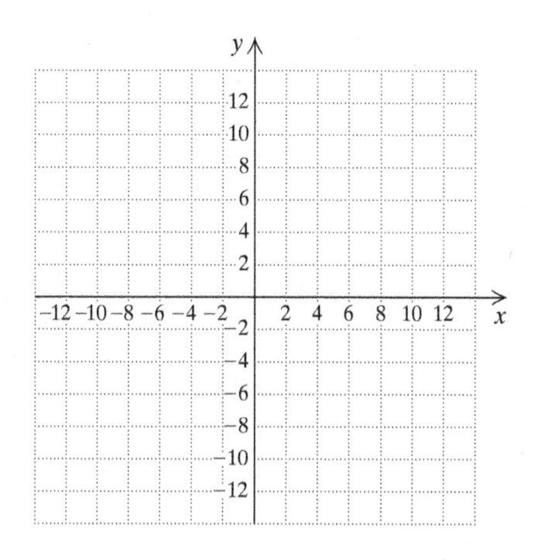

22. $y = \frac{1}{2}x^2 - 8$

x	y
-4	0
-2	-6
0	-8
2	-6
4	0

x	y
0	
-6	
-8	
-6	
0	

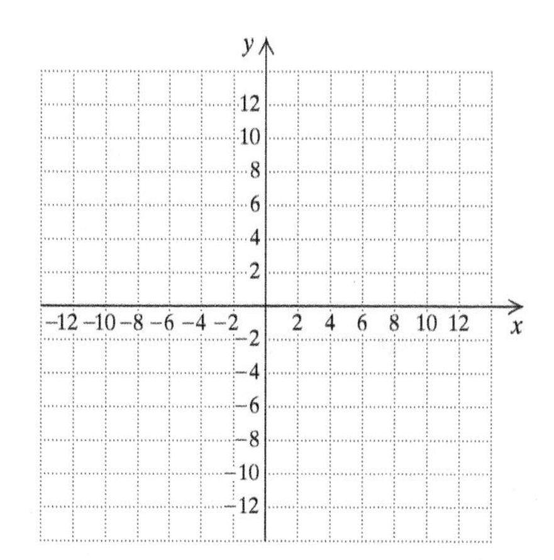

c Determine whether each function is one-to-one.

23. $f(x) = x - 5$

24. $f(x) = 3 - 6x$

25. $f(x) = x^2 - 2$

26. $f(x) = 4 - x^2$

27. $f(x) = |x| - 3$

28. $f(x) = |x - 2|$

29. $f(x) = 3^x$

30. $f(x) = (\frac{1}{2})^x$

d Determine whether each function is one-to-one. If it is, find a formula for its inverse.

31. $f(x) = 5x - 2$

32. $f(x) = 4 + 7x$

33. $f(x) = \dfrac{-2}{x}$

34. $f(x) = \dfrac{1}{x}$

35. $f(x) = \frac{4}{3}x + 7$

36. $f(x) = -\frac{7}{8}x + 2$

37. $f(x) = \dfrac{2}{x + 5}$

38. $f(x) = \dfrac{1}{x - 8}$

39. $f(x) = 5$

40. $f(x) = -2$

41. $f(x) = \dfrac{2x + 1}{5x + 3}$

42. $f(x) = \dfrac{2x - 1}{5x + 3}$

43. $f(x) = x^3 - 1$

44. $f(x) = x^3 + 5$

45. $f(x) = \sqrt[3]{x}$

46. $f(x) = \sqrt[3]{x - 4}$

Graph each function and its inverse using the same set of axes.

47. $f(x) = \frac{1}{2}x - 3$,
$f^{-1}(x) =$ ________

x	$f(x)$
-4	
0	
2	
4	

x	$f^{-1}(x)$
-5	
-3	
-2	
-1	

48. $g(x) = x + 4$,
$g^{-1}(x) =$ ________

x	$g(x)$
-4	
-2	
0	
1	

x	$g^{-1}(x)$
0	
2	
4	
5	

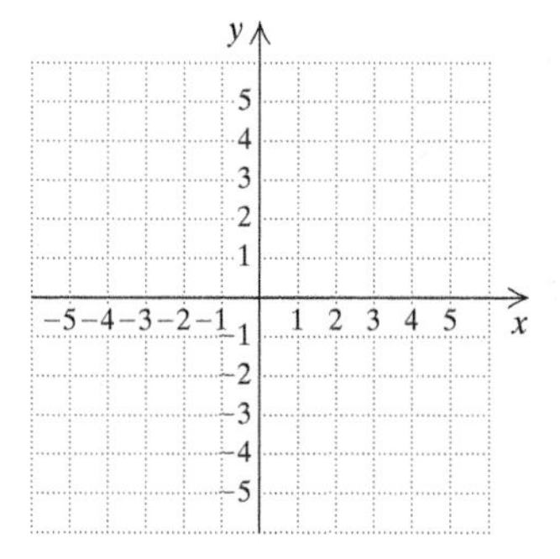

49. $f(x) = x^3$,
$f^{-1}(x) =$ ________

x	$f(x)$
0	
1	
2	
3	
-1	
-2	
-3	

x	$f^{-1}(x)$

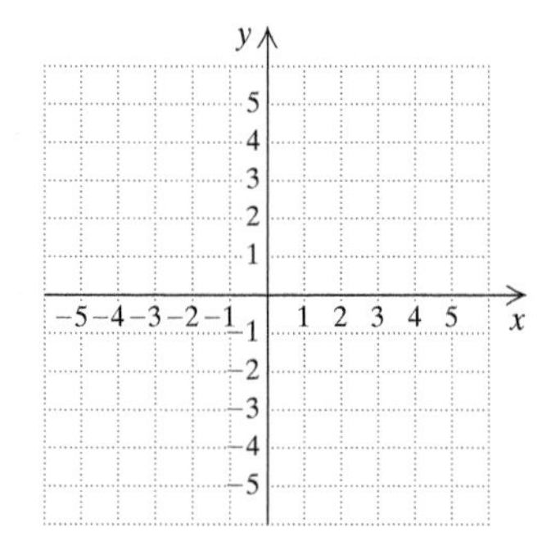

50. $f(x) = x^3 - 1$,
$f^{-1}(x) =$ ________

x	$f(x)$
0	
1	
2	
3	
-1	
-2	
-3	

x	$f^{-1}(x)$

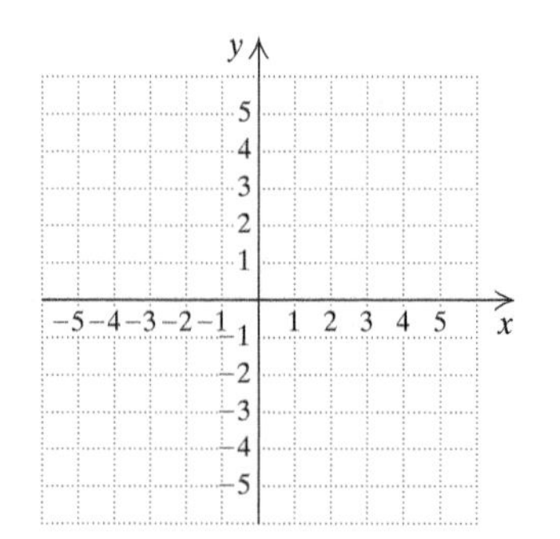

e For each function, use composition to show that the inverse is correct.

51. $f(x) = \frac{4}{5}x$,
$f^{-1}(x) = \frac{5}{4}x$

52. $f(x) = x - 3$,
$f^{-1}(x) = x + 3$

53. $f(x) = \dfrac{x+7}{2}$,
$f^{-1}(x) = 2x - 7$

54. $f(x) = \dfrac{3}{4}x - 1$,
$f^{-1}(x) = \dfrac{4x+4}{3}$

55. $f(x) = \dfrac{1-x}{x}$,
$f^{-1}(x) = \dfrac{1}{x+1}$

56. $f(x) = x^3 - 5$,
$f^{-1}(x) = \sqrt[3]{x+5}$

Find the inverse of the given function by thinking about the operations of the function and then reversing, or undoing, them. You can check your answer using composition.

	Function	*Inverse*
57.	$f(x) = 3x$	$f^{-1}(x) =$ ________
58.	$f(x) = \frac{1}{4}x + 7$	$f^{-1}(x) =$ ________
59.	$f(x) = -x$	$f^{-1}(x) =$ ________
60.	$f(x) = \sqrt[3]{x} - 5$	$f^{-1}(x) =$ ________
61.	$f(x) = \sqrt[3]{x-5}$	$f^{-1}(x) =$ ________
62.	$f(x) = x^{-1}$	$f^{-1}(x) =$ ________

63. *Dress Sizes in the United States and France.* A size-6 dress in the United States is size 38 in France. A function that converts dress sizes in the United States to those in France is

$$f(x) = x + 32.$$

a) Find the dress sizes in France that correspond to sizes of 8, 10, 14, and 18 in the United States.

b) Determine whether this function has an inverse that is a function. If so, find a formula for the inverse.

c) Use the inverse function to find dress sizes in the United States that correspond to sizes of 40, 42, 46, and 50 in France.

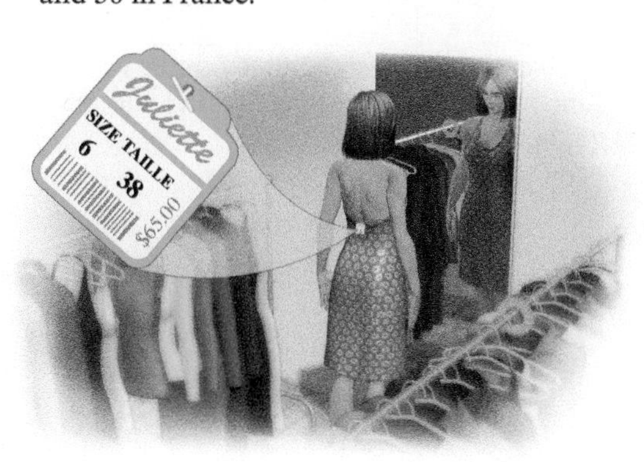

64. *Dress Sizes in the United States and Italy.* A size-6 dress in the United States is size 36 in Italy. A function that converts dress sizes in the United States to those in Italy is

$$f(x) = 2(x + 12).$$

a) Find the dress sizes in Italy that correspond to sizes of 8, 10, 14, and 18 in the United States.

b) Determine whether this function has an inverse that is a function. If so, find a formula for the inverse.

c) Use the inverse function to find dress sizes in the United States that correspond to sizes of 40, 44, 52, and 60 in Italy.

Skill Maintenance

Use rational exponents to simplify. [6.2d]

65. $\sqrt[6]{a^2}$

66. $\sqrt[8]{81}$

67. $\sqrt[12]{64x^6y^6}$

68. $\sqrt[4]{81a^8b^8}$

Simplify.

69. i^{79} [6.8d]

70. $(125x^3y^{-2}z^6)^{-2/3}$ [6.2c]

71. $\sqrt{2400}$ [6.3a]

Multiply.

72. $(x + 1)(x^2 - 7)$ [4.2b]

73. $(2y - 5)^2$ [4.2c]

74. $(7a^2 + c)(7a^2 - c)$ [4.2d]

Synthesis

In Exercises 75–78, use a graphing calculator to help determine whether or not the given functions are inverses of each other.

75. $f(x) = 0.75x^2 + 2;\ g(x) = \sqrt{\dfrac{4(x - 2)}{3}}$

76. $f(x) = 1.4x^3 + 3.2;\ g(x) = \sqrt[3]{\dfrac{x - 3.2}{1.4}}$

77. $f(x) = \sqrt{2.5x + 9.25};\ g(x) = 0.4x^2 - 3.7,\ x \geq 0$

78. $f(x) = 0.8x^{1/2} + 5.23;\ g(x) = 1.25(x^2 - 5.23),\ x \geq 0$

79. Use a graphing calculator to help match each function in Column A with its inverse from Column B.

Column A	*Column B*
(1) $y = 5x^3 + 10$	**A.** $y = \dfrac{\sqrt[3]{x} - 10}{5}$
(2) $y = (5x + 10)^3$	**B.** $y = \sqrt[3]{\dfrac{x}{5}} - 10$
(3) $y = 5(x + 10)^3$	**C.** $y = \sqrt[3]{\dfrac{x - 10}{5}}$
(4) $y = (5x)^3 + 10$	**D.** $y = \dfrac{\sqrt[3]{x - 10}}{5}$

In Exercises 80 and 81, graph the inverse of f.

80.

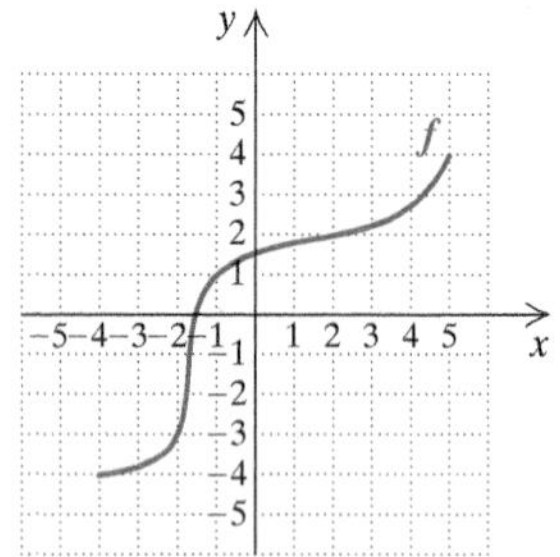

81.

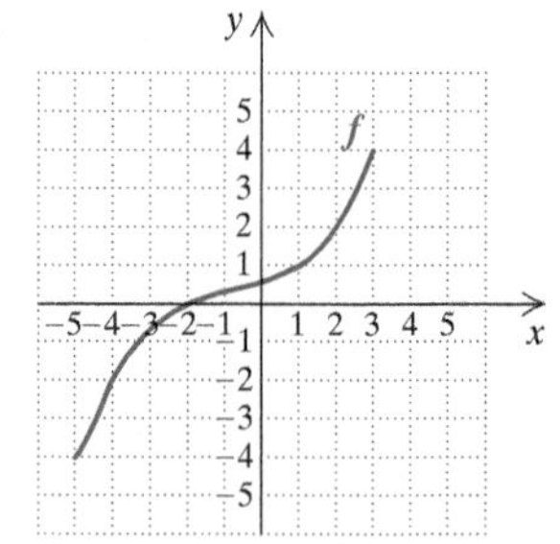

82. Examine the following table. Does it appear that f and g could be inverses of each other? Why or why not?

x	$f(x)$	$g(x)$
6	6	6
7	6.5	8
8	7	10
9	7.5	12
10	8	14
11	8.5	16
12	9	18

83. Refer to Exercise 82. Assume that f and g are both linear functions. Find equations for $f(x)$ and $g(x)$. Are f and g inverses of each other?

Logarithmic Functions

8.3

OBJECTIVES

a Graph logarithmic functions.

b Convert from exponential equations to logarithmic equations and from logarithmic equations to exponential equations.

c Solve logarithmic equations.

d Find common logarithms on a calculator.

We are now ready to study inverses of exponential functions. These functions have many applications and are referred to as *logarithm*, or *logarithmic, functions*.

a GRAPHING LOGARITHMIC FUNCTIONS

Consider the exponential function $f(x) = 2^x$. Like all exponential functions, f is one-to-one. Can a formula for f^{-1} be found? To answer this, we use the method of Section 8.2:

1. Replace $f(x)$ with y: $\quad y = 2^x$.
2. Interchange x and y: $\quad x = 2^y$.
3. Solve for y: $\quad y =$ the power to which we raise 2 to get x.
4. Replace y with $f^{-1}(x)$: $\quad f^{-1}(x) =$ the power to which we raise 2 to get x.

We now define a new symbol to replace the words "the power to which we raise 2 to get x."

MyLab Math ANIMATION

MEANING OF LOGARITHMS

$\log_2 x$, read "the logarithm, base 2, of x," or "log, base 2, of x," means "the power to which we raise 2 to get x."

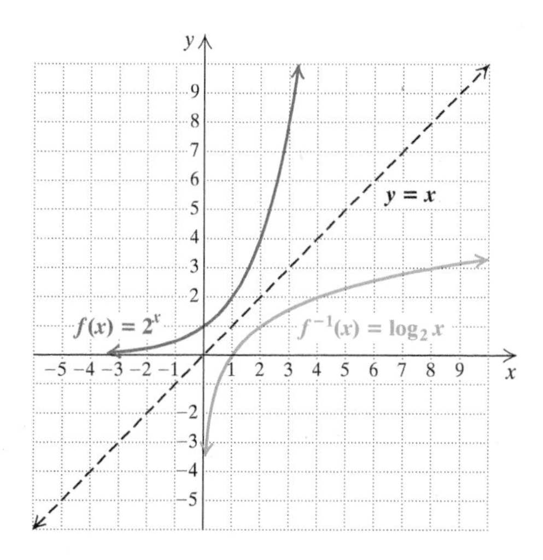

Thus if $f(x) = 2^x$, then $f^{-1}(x) = \log_2 x$. Note that $f^{-1}(8) = \log_2 8 = 3$, because 3 is the *power to which we raise* 2 *to get* 8; that is, $2^3 = 8$.

Although expressions like $\log_2 13$ can be only approximated, remember that $\log_2 13$ represents the ***power*** *to which we raise* 2 *to get* 13. That is, $2^{\log_2 13} = 13$.

Do Exercise 1. ▶

1. Write the meaning of $\log_2 64$. Then find $\log_2 64$.

Answer

1. $\log_2 64$ is the power to which we raise 2 to get 64; 6

For any exponential function $f(x) = a^x$, the inverse is called a **logarithmic function, base *a*.** The graph of the inverse can, of course, be drawn by reflecting the graph of $f(x) = a^x$ across the line $y = x$. It will be helpful to remember that the inverse of $f(x) = a^x$ is given by $f^{-1}(x) = \log_a x$. Normally, we use a number a that is greater than 1 for the logarithm base.

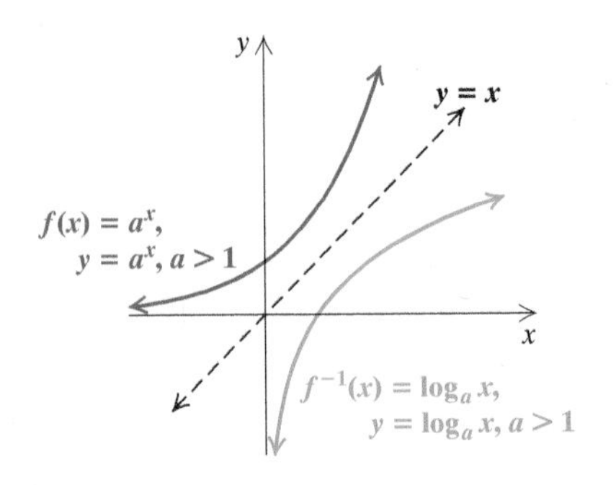

LOGARITHMS

The inverse of $f(x) = a^x$ is given by

$$f^{-1}(x) = \log_a x.$$

We read "$\log_a x$" as "the logarithm, base a, of x." We define $y = \log_a x$ as that number y such that $a^y = x$, where $x > 0$ and a is a positive constant other than 1.

It is helpful in dealing with logarithmic functions to remember that the logarithm of a number is an **exponent**. For instance, $\log_a x$ is the exponent y in $x = a^y$. Keep thinking, "The logarithm, base a, of a number x is the power to which a must be raised in order to get x."

EXPONENTIAL FUNCTION	LOGARITHMIC FUNCTION
$y = a^x$	$x = a^y$
$f(x) = a^x$	$f^{-1}(x) = \log_a x$
$a > 0, a \neq 1$	$a > 0, a \neq 1$
Domain = The set of real numbers	Range = The set of real numbers
Range = The set of positive numbers	Domain = The set of positive numbers

Why do we exclude 1 from being a logarithm base? See the following graph. If we allow 1 as a logarithm base, the graph of the relation $y = \log_1 x$, or $x = 1^y = 1$, is a vertical line, which is not a function and therefore not a logarithmic function.

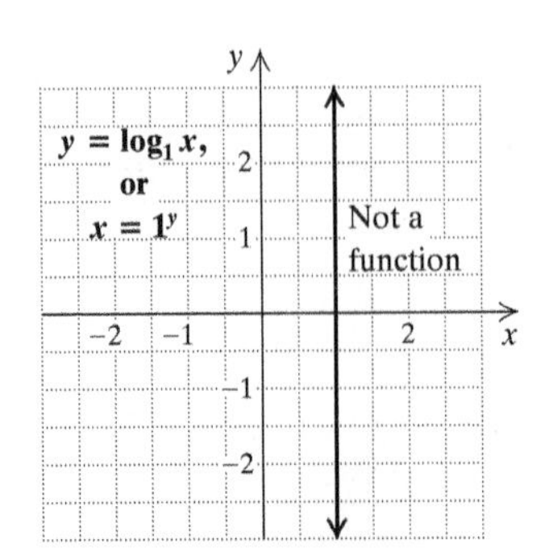

EXAMPLE 1 Graph: $y = f(x) = \log_5 x$.

The equation $y = \log_5 x$ is equivalent to $5^y = x$. We can find ordered pairs that are solutions by choosing values for y and computing the corresponding x-values.

For $y = 0$, $x = 5^0 = 1$.
For $y = 1$, $x = 5^1 = 5$.
For $y = 2$, $x = 5^2 = 25$.
For $y = 3$, $x = 5^3 = 125$.
For $y = -1$, $x = 5^{-1} = \frac{1}{5}$.
For $y = -2$, $x = 5^{-2} = \frac{1}{25}$.

x, or 5^y	y
1	0
5	1
25	2
125	3
$\frac{1}{5}$	−1
$\frac{1}{25}$	−2

(1) Select y.
(2) Compute x.

The table shows the following:

$\log_5 1 = 0;$
$\log_5 5 = 1;$
$\log_5 25 = 2;$
$\log_5 125 = 3;$
$\log_5 \frac{1}{5} = -1;$
$\log_5 \frac{1}{25} = -2.$

These can all be checked using the equations above.

We plot the ordered pairs and connect them with a smooth curve. The graph of $y = 5^x$ has been shown only for reference.

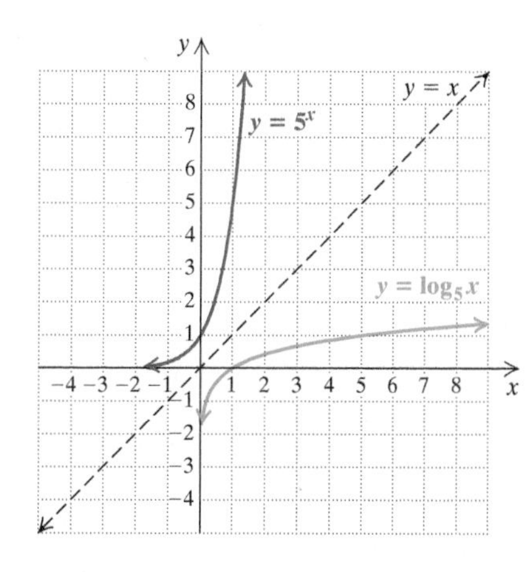

Do Exercise 2.

2. Graph: $y = f(x) = \log_3 x$.

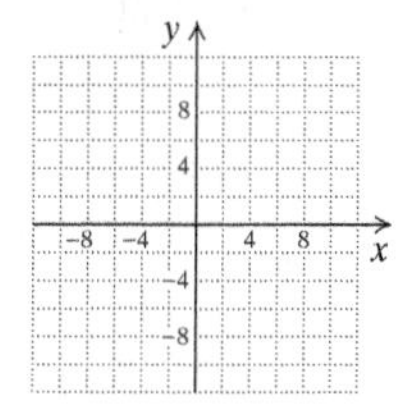

b CONVERTING BETWEEN EXPONENTIAL EQUATIONS AND LOGARITHMIC EQUATIONS

We use the definition of logarithms to convert from exponential equations to logarithmic equations.

CONVERTING BETWEEN EXPONENTIAL EQUATIONS AND LOGARITHMIC EQUATIONS

$y = \log_a x \longrightarrow a^y = x;\qquad a^y = x \longrightarrow y = \log_a x$

Be sure to memorize this relationship! It is probably the most important definition in the chapter. Often this definition will be a justification for a proof or a procedure that we are considering.

Answer

2.

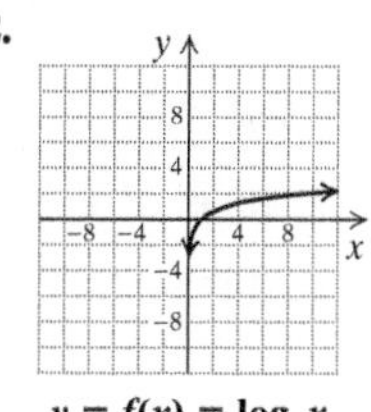

$y = f(x) = \log_3 x$

EXAMPLES Convert to a logarithmic equation.

2. $8 = 2^x \longrightarrow x = \log_2 8$ The exponent is the logarithm. The base remains the same.

3. $y^{-1} = 4 \longrightarrow -1 = \log_y 4$

4. $a^b = c \longrightarrow b = \log_a c$

Do Exercises 3–6.

Convert to a logarithmic equation.

3. $6^0 = 1$

4. $10^{-3} = 0.001$

5. $16^{0.25} = 2$

6. $m^T = P$

We also use the definition of logarithms to convert from logarithmic equations to exponential equations.

EXAMPLES Convert to an exponential equation.

5. $y = \log_3 5 \longrightarrow 3^y = 5$ The logarithm is the exponent. The base does not change.

6. $-2 = \log_a 7 \longrightarrow a^{-2} = 7$

7. $a = \log_b d \longrightarrow b^a = d$

Do Exercises 7–10.

Convert to an exponential equation.

7. $\log_2 32 = 5$

8. $\log_{10} 1000 = 3$

9. $\log_a Q = 7$

10. $\log_t M = x$

C SOLVING CERTAIN LOGARITHMIC EQUATIONS

Certain equations involving logarithms can be solved by first converting to exponential equations. We will solve more complicated equations later.

EXAMPLE 8 Solve: $\log_2 x = -3$.

$$\log_2 x = -3$$

$$2^{-3} = x \quad \text{Converting to an exponential equation}$$

$$\frac{1}{2^3} = x$$

$$\frac{1}{8} = x$$

Check: $\log_2 \frac{1}{8}$ is the exponent to which we raise 2 to get $\frac{1}{8}$. Since $2^{-3} = \frac{1}{8}$, we know that $\frac{1}{8}$ checks and is the solution.

EXAMPLE 9 Solve: $\log_x 16 = 2$.

$$\log_x 16 = 2$$

$$x^2 = 16 \quad \text{Converting to an exponential equation}$$

$$x = 4 \quad or \quad x = -4 \quad \text{Using the principle of square roots}$$

Check: $\log_4 16 = 2$ because $4^2 = 16$. Thus, 4 is a solution. Since all logarithm bases must be positive, $\log_{-4} 16$ is not defined. Therefore, -4 is not a solution.

Do Exercises 11–13.

Solve.

11. $\log_{10} x = 4$

GS

$10^{\square} = x$

$\square = x$

12. $\log_x 81 = 4$

13. $\log_2 x = -2$

Answers

3. $0 = \log_6 1$ **4.** $-3 = \log_{10} 0.001$
5. $0.25 = \log_{16} 2$ **6.** $T = \log_m P$
7. $2^5 = 32$ **8.** $10^3 = 1000$ **9.** $a^7 = Q$
10. $t^x = M$ **11.** 10,000 **12.** 3 **13.** $\frac{1}{4}$

Guided Solution:
11. 4, 10,000

To think of finding logarithms as solving equations may help in some cases.

EXAMPLE 10 Find $\log_{10} 1000$.

Method 1: Let $\log_{10} 1000 = x$. Then

$$10^x = 1000 \quad \text{Converting to an exponential equation}$$
$$10^x = 10^3$$
$$x = 3. \quad \text{The exponents are the same.}$$

Therefore, $\log_{10} 1000 = 3$.

Method 2: Think of the meaning of $\log_{10} 1000$. It is the exponent to which we raise 10 to get 1000. That exponent is 3. Therefore, $\log_{10} 1000 = 3$.

EXAMPLE 11 Find $\log_{10} 0.01$.

Method 1: Let $\log_{10} 0.01 = x$. Then

$$10^x = 0.01 \quad \text{Converting to an exponential equation}$$
$$10^x = \frac{1}{100}$$
$$10^x = 10^{-2}$$
$$x = -2. \quad \text{The exponents are the same.}$$

Therefore, $\log_{10} 0.01 = -2$.

Method 2: $\log_{10} 0.01$ is the exponent to which we raise 10 to get 0.01. Noting that

$$0.01 = \frac{1}{100} = \frac{1}{10^2} = 10^{-2},$$

we see that the exponent is -2. Therefore, $\log_{10} 0.01 = -2$.

EXAMPLE 12 Find $\log_5 1$.

Method 1: Let $\log_5 1 = x$. Then

$$5^x = 1 \quad \text{Converting to an exponential equation}$$
$$5^x = 5^0$$
$$x = 0. \quad \text{The exponents are the same.}$$

Therefore, $\log_5 1 = 0$.

Method 2: $\log_5 1$ is the exponent to which we raise 5 to get 1. That exponent is 0. Therefore, $\log_5 1 = 0$.

Do Exercises 14–16. ▶

THE LOGARITHM OF 1

For any base a,

$$\log_a 1 = 0.$$

The logarithm, base a, of 1 is always 0.

Find each of the following.

14. $\log_{10} 10{,}000$

Let $\log_{10} 10{,}000 = x$.

$\square^x = 10{,}000$

$10^x = \square^4$

$x = \square$

15. $\log_{10} 0.0001$

16. $\log_7 1$

Answers

14. 4 **15.** -4 **16.** 0

Guided Solution:

14. 10, 10, 4

The proof follows from the fact that $a^0 = 1$. This is equivalent to the logarithmic equation $\log_a 1 = 0$.

Another property follows similarly. We know that $a^1 = a$ for any real number a. In particular, it holds for any positive number a. This is equivalent to the logarithmic equation $\log_a a = 1$.

THE LOGARITHM, BASE *a*, OF *a*

For any base a,

$$\log_a a = 1.$$

EXAMPLE 13 Simplify: $\log_m 1$ and $\log_t t$.

$\log_m 1 = 0; \qquad \log_t t = 1$

◀ **Do Exercises 17–20.**

Simplify.

17. $\log_3 1$

18. $\log_3 3$

19. $\log_c c$

20. $\log_c 1$

d FINDING COMMON LOGARITHMS ON A CALCULATOR

Base-10 logarithms are called **common logarithms**. Before calculators became so widely available, common logarithms were used extensively to do complicated calculations. The abbreviation **log**, with no base written, is used for the common logarithm, base-10. Thus,

$$\log 29 \quad \text{means} \quad \log_{10} 29.$$

Be sure to memorize $\log a = \log_{10} a$.

We can approximate log 29. Note the following:

$$\left.\begin{aligned} \log 100 &= \log_{10} 100 = 2; \\ \log 29 &= ?; \\ \log 10 &= \log_{10} 10 = 1. \end{aligned}\right\}$$

It seems reasonable to conclude that log 29 is between 1 and 2.

The calculator key for common logarithms is generally marked LOG. We find that

$$\log 29 \approx 1.462397998 \approx 1.4624,$$

rounded to four decimal places. This also tells us that $10^{1.4624} \approx 29$.

On some scientific calculators, the keystrokes for doing such a calculation might be

2 9 LOG =. The display would then read 1.462398.

If we are using a graphing calculator, the keystrokes might be

LOG 2 9 ENTER. The display would then read 1.462397998.

EXAMPLES Find the common logarithm, to four decimal places, on a scientific calculator or a graphing calculator.

	Function Value	*Readout*	*Rounded*
14.	log 287,523	5.458672591	5.4587
15.	log 0.000486	−3.313363731	−3.3134
16.	log (−5)	NONREAL ANS	Does not exist as a real number

Find the common logarithm, to four decimal places, on a scientific calculator or a graphing calculator.

21. log 78,235.4

22. log 0.0000309

23. $\log(-3)$

24. Find

log 1000 and log 10,000

without using a calculator. Between what two whole numbers is log 9874? Then on a calculator, approximate log 9874, rounded to four decimal places.

Answers

17. 0 **18.** 1 **19.** 1 **20.** 0 **21.** 4.8934 **22.** −4.5100 **23.** Does not exist as a real number **24.** log 1000 = 3, log 10,000 = 4; between 3 and 4; 3.9945

In Example 16, $\log(-5)$ does not exist as a real number because there is no real-number power to which we can raise 10 to get -5. The number 10 raised to any power is nonnegative. The logarithm of a negative number does not exist as a real number (though it can be defined as a complex number).

Do Exercises 21–24 on the preceding page. ▶

We can use common logarithms to express any positive number as a power of 10. Considering very large or very small numbers as powers of 10 might be a helpful way to compare those numbers.

EXAMPLE 17 Complete the following table to express each number in the first column as a power of 10. Round each exponent to the nearest ten-thousandth.

We simply find the common logarithm of the number using a calculator.

NUMBER	EXPRESSED AS A POWER OF 10
4	$4 \approx 10^{0.6021}$
625	$625 \approx 10^{2.7959}$
134,567	$134{,}567 \approx 10^{5.1289}$
0.00567	$0.00567 \approx 10^{-2.2464}$
0.000374859	$0.000374859 \approx 10^{-3.4261}$
186,000	$186{,}000 \approx 10^{5.2695}$
186,000,000	$186{,}000{,}000 \approx 10^{8.2695}$

Do Exercise 25. ▶

25. Complete the following table to express each number in the first column as a power of 10. Round each exponent to the nearest ten-thousandth.

NUMBER	EXPRESSED AS A POWER OF 10
8	
947	
634,567	
0.00708	
0.000778899	
18,600,000	
1860	

The inverse of a logarithmic function is an exponential function. Thus, if $f(x) = \log x$, then $f^{-1}(x) = 10^x$. Because of this, on many calculators, the LOG key doubles as the 10^x key after a 2ND or SHIFT key has been pressed. To find $10^{5.4587}$ on a scientific calculator, we might enter 5.4587 and press 10^x. On many graphing calculators, we press 2ND 10^x, followed by 5.4587. In either case, we get the approximation $10^{5.4587} \approx 287{,}541.1465$. Compare this computation to Example 14. Note that, apart from the rounding error, $10^{5.4587}$ takes us back to about 287,523.

Do Exercise 26. ▶

26. Find $10^{4.8934}$ using a calculator. (Compare your computation to that of Margin Exercise 21.)

Using the scientific keys on a calculator would allow us to construct a graph of $f(x) = \log_{10} x = \log x$ by finding function values directly, rather than converting to exponential form as we did in Example 1.

x	$f(x)$
0.5	-0.3010
1	0
2	0.3010
3	0.4771
5	0.6990
9	0.9542
10	1

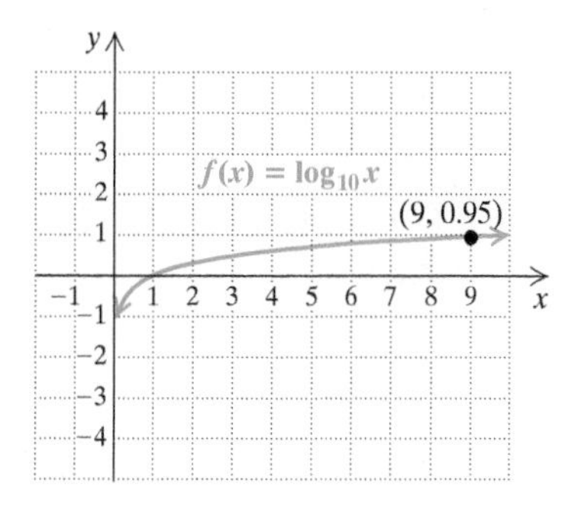

Answers

25. $10^{0.9031}$; $10^{2.9763}$; $10^{5.8025}$; $10^{-2.1500}$; $10^{-3.1085}$; $10^{7.2695}$; $10^{3.2695}$ **26.** 78,234.8042

8.3 Exercise Set

FOR EXTRA HELP

✓ Check Your Understanding

Reading Check Determine whether each statement is true or false.

RC1. The expression $\log_5 12$ represents the power to which we raise 12 to get 5.

RC2. A logarithm base cannot be negative.

RC3. The logarithm of a negative number does not exist.

RC4. The number 1 is not a valid logarithm base.

RC5. If the base of a logarithm is not written, it is assumed to be 2.

RC6. A LOG key on a calculator represents a base-10 logarithm.

RC7. Any positive number can be written as a power of 10.

RC8. Any positive number can be written as a power of 5.

Concept Check Use the powers in the column on the right to find each logarithm.

CC1. $\log_{10} 100 =$ _____

CC2. $\log_3 9 =$ _____

CC3. $\log_2 8 =$ _____

CC4. $\log_2 1024 =$ _____

$2^3 = 8$

$3^2 = 9$

$10^2 = 100$

$2^{10} = 1024$

a Graph.

1. $f(x) = \log_2 x$, or $y = \log_2 x$

$y = \log_2 x \longrightarrow x =$ _____

x, or 2^y	y
	0
	1
	2
	3
	−1
	−2
	−3

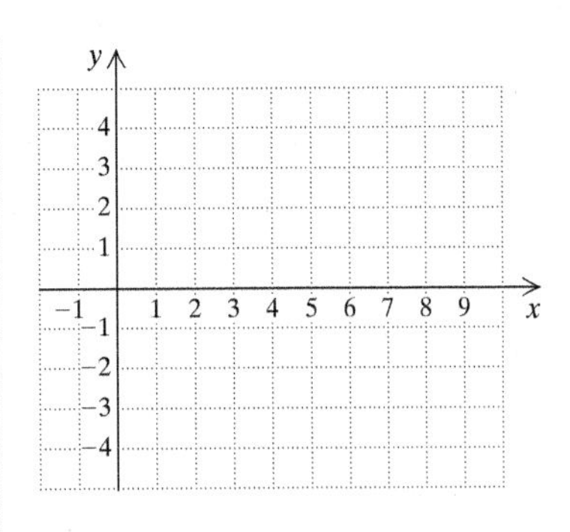

2. $f(x) = \log_{10} x$, or $y = \log_{10} x$

$y = \log_{10} x \longrightarrow x =$ _____

x, or 10^y	y
	0
	1
	2
	3
	−1
	−2
	−3

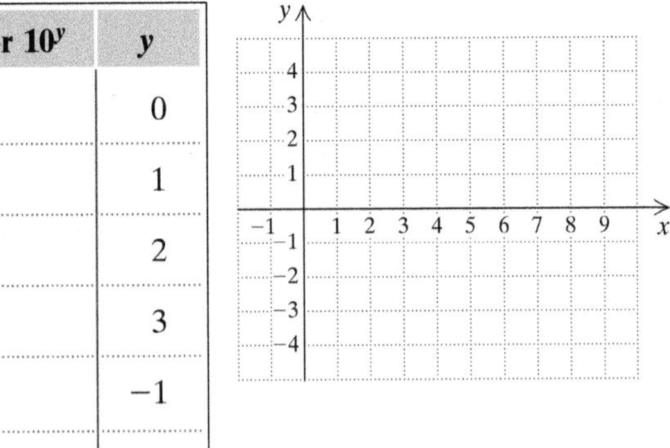

3. $f(x) = \log_{1/3} x$

x	y

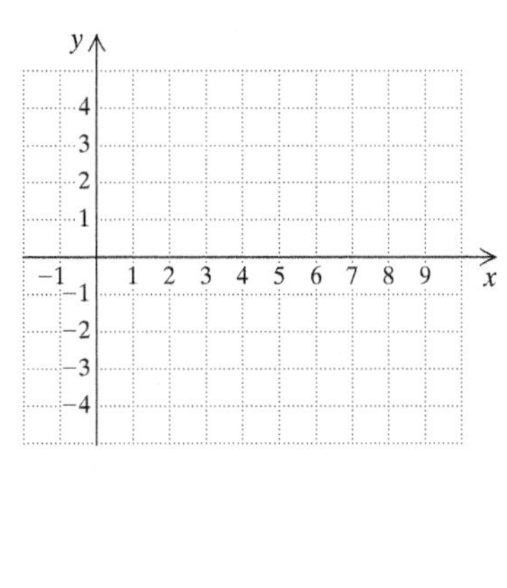

4. $f(x) = \log_{1/2} x$

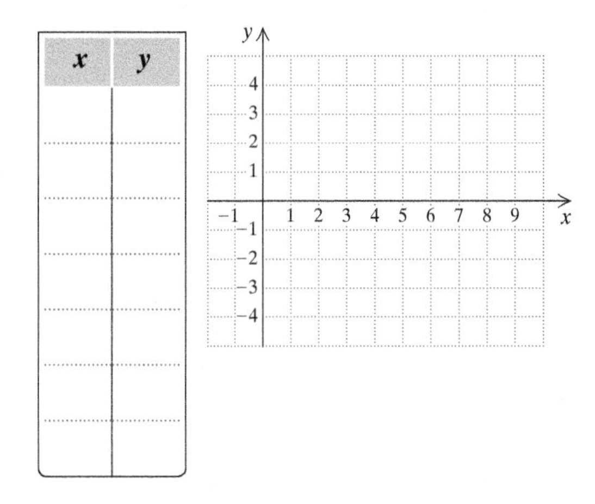

Graph both functions using the same set of axes.

5. $f(x) = 3^x,\ f^{-1}(x) = \log_3 x$

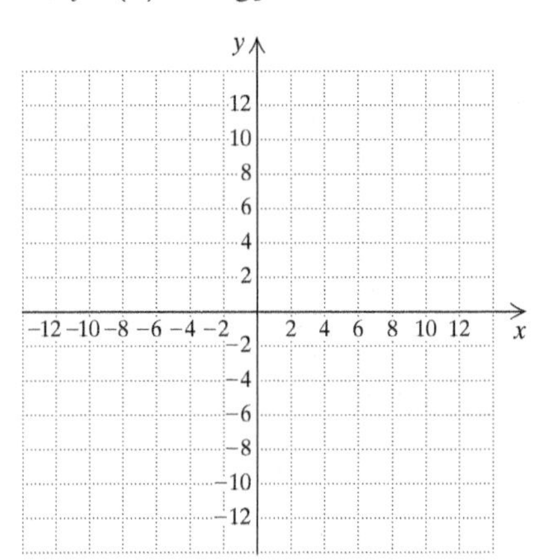

6. $f(x) = 4^x,\ f^{-1}(x) = \log_4 x$

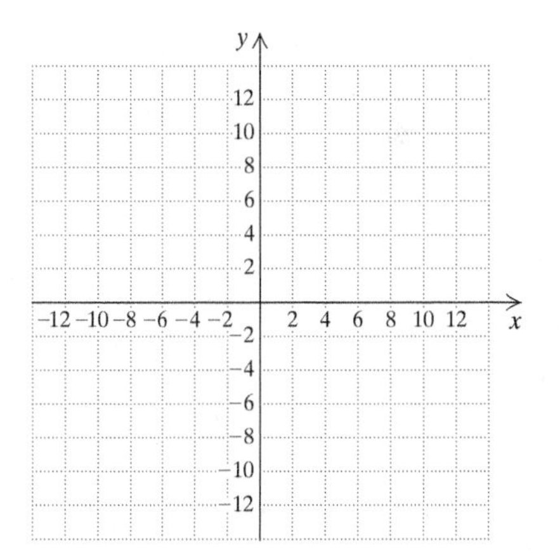

b Convert to a logarithmic equation.

7. $10^3 = 1000$

8. $10^2 = 100$

9. $5^{-3} = \dfrac{1}{125}$

10. $4^{-5} = \dfrac{1}{1024}$

11. $8^{1/3} = 2$

12. $16^{1/4} = 2$

13. $10^{0.3010} = 2$

14. $10^{0.4771} = 3$

15. $e^2 = t$

16. $p^k = 3$

17. $Q^t = x$

18. $P^m = V$

19. $e^2 = 7.3891$

20. $e^3 = 20.0855$

21. $e^{-2} = 0.1353$

22. $e^{-4} = 0.0183$

Convert to an exponential equation.

23. $w = \log_4 10$

24. $t = \log_5 9$

25. $\log_6 36 = 2$

26. $\log_7 7 = 1$

27. $\log_{10} 0.01 = -2$

28. $\log_{10} 0.001 = -3$

29. $\log_{10} 8 = 0.9031$

30. $\log_{10} 2 = 0.3010$

31. $\log_e 100 = 4.6052$

32. $\log_e 10 = 2.3026$

33. $\log_t Q = k$

34. $\log_m P = a$

C Solve.

35. $\log_3 x = 2$

36. $\log_4 x = 3$

37. $\log_x 16 = 2$

38. $\log_x 64 = 3$

39. $\log_2 16 = x$

40. $\log_5 25 = x$

41. $\log_3 27 = x$

42. $\log_4 16 = x$

43. $\log_x 25 = 1$

44. $\log_x 9 = 1$

45. $\log_3 x = 0$

46. $\log_2 x = 0$

47. $\log_2 x = -1$

48. $\log_3 x = -2$

49. $\log_8 x = \frac{1}{3}$

50. $\log_{32} x = \frac{1}{5}$

Find each of the following.

51. $\log_{10} 100$

52. $\log_{10} 100{,}000$

53. $\log_{10} 0.1$

54. $\log_{10} 0.001$

55. $\log_{10} 1$

56. $\log_{10} 10$

57. $\log_5 625$

58. $\log_2 64$

59. $\log_7 49$

60. $\log_5 125$

61. $\log_2 8$

62. $\log_8 64$

63. $\log_9 \frac{1}{81}$

64. $\log_5 \frac{1}{125}$

65. $\log_8 1$

66. $\log_6 6$

67. $\log_e e$

68. $\log_e 1$

69. $\log_{27} 9$

70. $\log_8 2$

d Find the common logarithm, to four decimal places, on a calculator.

71. $\log 78{,}889.2$

72. $\log 9{,}043{,}788$

73. $\log 0.67$

74. $\log 0.0067$

75. $\log(-97)$

76. $\log 0$

77. $\log\left(\frac{289}{32.7}\right)$

78. $\log\left(\frac{23}{86.2}\right)$

79. Complete the following table to express each number in the first column as a power of 10. Round each exponent to the nearest ten-thousandth.

NUMBER	EXPRESSED AS A POWER OF 10
6	
84	
987,606	
0.00987606	
98,760.6	
70,000,000	
7000	

80. Complete the following table to express each number in the first column as a power of 10. Round each exponent to the nearest ten-thousandth.

NUMBER	EXPRESSED AS A POWER OF 10
7	
314	
31.4	
31,400,000	
0.000314	
3.14	
0.0314	

Skill Maintenance

Perform the indicated operations and simplify.

81. $\frac{t^2 - 9}{t^2 - 4t + 4} \cdot \frac{3t - 6}{3t + 9}$ [5.1d]

82. $\frac{c^2}{c - p} + \frac{p^2}{p - c}$ [5.2b]

83. $\frac{3}{x^2 - x - 2} - \frac{1}{2x^2 + 3x + 1}$ [5.2b]

Factor completely.

84. $100x^2 - 9$ [4.6b]

85. $3x^3 - 24x^2 + 48x$ [4.6a]

86. $64a^3 - y^3$ [4.6d]

87. $2x^3 - 3x^2 + 2x - 3$ [4.3b]

88. $10t^6 + 19t^3 - 15$ [4.5a, b]

Synthesis

Graph.

89. $f(x) = \log_3 |x + 1|$

90. $f(x) = \log_2 (x - 1)$

Solve.

91. $\log_{125} x = \frac{2}{3}$

92. $|\log_3 x| = 3$

93. $\log_{128} x = \frac{5}{7}$

94. $\log_4 (3x - 2) = 2$

95. $\log_8 (2x + 1) = -1$

96. $\log_{10} (x^2 + 21x) = 2$

Simplify.

97. $\log_{1/4} \frac{1}{64}$

98. $\log_{81} 3 \cdot \log_3 81$

99. $\log_{10} (\log_4 (\log_3 81))$

100. $\log_2 (\log_2 (\log_4 256))$

101. $\log_{1/5} 25$

8.4 Properties of Logarithmic Functions

OBJECTIVES

a Express the logarithm of a product as a sum of logarithms, and conversely.

b Express the logarithm of a power as a product.

c Express the logarithm of a quotient as a difference of logarithms, and conversely.

d Convert from logarithms of products, quotients, and powers to expressions in terms of individual logarithms, and conversely.

e Simplify expressions of the type $\log_a a^k$.

We now establish some basic properties that are useful in manipulating logarithmic expressions.

a LOGARITHMS OF PRODUCTS

PROPERTY 1: THE PRODUCT RULE

For any positive numbers M and N and any logarithm base a,

$$\log_a (M \cdot N) = \log_a M + \log_a N.$$

(The logarithm of a product is the sum of the logarithms of the factors.)

EXAMPLE 1 Express as a sum of logarithms: $\log_2 (4 \cdot 16)$.

$$\log_2 (4 \cdot 16) = \log_2 4 + \log_2 16 \qquad \text{By Property 1}$$

EXAMPLE 2 Express as a single logarithm: $\log_{10} 0.01 + \log_{10} 1000$.

$$\log_{10} 0.01 + \log_{10} 1000 = \log_{10} (0.01 \times 1000) \qquad \text{By Property 1}$$
$$= \log_{10} 10$$

Do Exercises 1–4.

Express as a sum of logarithms.

1. $\log_5 (25 \cdot 5)$
2. $\log_b (PQ)$

Express as a single logarithm.

3. $\log_3 7 + \log_3 5$
4. $\log_a J + \log_a A + \log_a M$

A Proof of Property 1 (*Optional*): We let $\log_a M = x$ and $\log_a N = y$. Converting to exponential equations, we have $a^x = M$ and $a^y = N$. Then we multiply to obtain

$$M \cdot N = a^x \cdot a^y = a^{x+y}.$$

Converting $M \cdot N = a^{x+y}$ back to a logarithmic equation, we get

$$\log_a (M \cdot N) = x + y.$$

Remembering what x and y represent, we get

$$\log_a (M \cdot N) = \log_a M + \log_a N.$$

b LOGARITHMS OF POWERS

PROPERTY 2: THE POWER RULE

For any positive number M, any real number k, and any logarithm base a,

$$\log_a M^k = k \cdot \log_a M.$$

(The logarithm of a power of M is the exponent times the logarithm of M.)

Answers

1. $\log_5 25 + \log_5 5$ 2. $\log_b P + \log_b Q$
3. $\log_3 35$ 4. $\log_a (JAM)$

EXAMPLES Express as a product.

3. $\log_a 9^{-5} = -5 \log_a 9$ By Property 2

4. $\log_a \sqrt[4]{5} = \log_a 5^{1/4}$ Writing exponential notation

$= \frac{1}{4} \log_a 5$ By Property 2

Do Exercises 5 and 6.

Express as a product.

5. $\log_7 4^5$

6. $\log_a \sqrt{5}$

A Proof of Property 2 (*Optional*): We let $x = \log_a M$. Then we convert to an exponential equation to get $a^x = M$. Raising both sides to the kth power, we obtain

$$(a^x)^k = M^k, \quad \text{or} \quad a^{xk} = M^k.$$

Converting back to a logarithmic equation with base a, we get $\log_a M^k = xk$. But $x = \log_a M$, so

$$\log_a M^k = (\log_a M)k = k \cdot \log_a M.$$

CALCULATOR CORNER

Properties of Logarithms Use a table or a graph to determine whether each of the following is correct.

1. $\log(5x) = \log 5 \cdot \log x$
2. $\log(5x) = \log 5 + \log x$
3. $\log x^2 = \log x \cdot \log x$
4. $\log x^2 = 2 \log x$
5. $\log\left(\frac{x}{3}\right) = \frac{\log x}{\log 3}$
6. $\log\left(\frac{x}{3}\right) = \log x - \log 3$
7. $\log(x + 2) = \log x + \log 2$
8. $\log(x + 2) = \log x \cdot \log 2$

c LOGARITHMS OF QUOTIENTS

PROPERTY 3: THE QUOTIENT RULE

For any positive numbers M and N and any logarithm base a,

$$\log_a \frac{M}{N} = \log_a M - \log_a N.$$

(The logarithm of a quotient is the logarithm of the numerator minus the logarithm of the denominator.)

EXAMPLE 5 Express as a difference of logarithms: $\log_t \frac{6}{U}$.

$\log_t \frac{6}{U} = \log_t 6 - \log_t U$ By Property 3

EXAMPLE 6 Express as a single logarithm: $\log_b 17 - \log_b 27$.

$\log_b 17 - \log_b 27 = \log_b \frac{17}{27}$ By Property 3

EXAMPLE 7 Express as a single logarithm: $\log_{10} 10{,}000 - \log_{10} 100$.

$$\log_{10} 10{,}000 - \log_{10} 100 = \log_{10} \frac{10{,}000}{100} = \log_{10} 100$$

Do Exercises 7 and 8.

7. Express as a difference of logarithms:

$$\log_b \frac{P}{Q}.$$

8. Express as a single logarithm:

$$\log_2 125 - \log_2 25.$$

A Proof of Property 3 (*Optional*): The proof makes use of Property 1 and Property 2.

$\log_a \frac{M}{N} = \log_a M \cdot \frac{1}{N} = \log_a MN^{-1}$ $\quad \frac{1}{N} = N^{-1}$

$= \log_a M + \log_a N^{-1}$ By Property 1

$= \log_a M + (-1) \log_a N$ By Property 2

$= \log_a M - \log_a N$

Answers

5. $5 \log_7 4$ **6.** $\frac{1}{2} \log_a 5$ **7.** $\log_b P - \log_b Q$ **8.** $\log_2 5$

d USING THE PROPERTIES TOGETHER

EXAMPLES Express in terms of logarithms of w, x, y, and z.

8. $\log_a \dfrac{x^2y^3}{z^4} = \log_a (x^2y^3) - \log_a z^4$ Using Property 3

$= \log_a x^2 + \log_a y^3 - \log_a z^4$ Using Property 1

$= 2\log_a x + 3\log_a y - 4\log_a z$ Using Property 2

9. $\log_a \sqrt[4]{\dfrac{xy}{z^3}} = \log_a \left(\dfrac{xy}{z^3}\right)^{1/4}$ Writing exponential notation

$= \frac{1}{4}\log_a \dfrac{xy}{z^3}$ Using Property 2

$= \frac{1}{4}(\log_a (xy) - \log_a z^3)$ Using Property 3 (note the parentheses)

$= \frac{1}{4}(\log_a x + \log_a y - 3\log_a z)$ Using Properties 1 and 2

$= \frac{1}{4}\log_a x + \frac{1}{4}\log_a y - \frac{3}{4}\log_a z$ Distributive law

10. $\log_b \dfrac{xy}{w^3z^4} = \log_b (xy) - \log_b (w^3z^4)$ Using Property 3

$= (\log_b x + \log_b y) - (\log_b w^3 + \log_b z^4)$ Using Property 1

$= \log_b x + \log_b y - \log_b w^3 - \log_b z^4$ Removing parentheses

$= \log_b x + \log_b y - 3\log_b w - 4\log_b z$ Using Property 2

◀ **Do Exercises 9–11.**

Express in terms of logarithms of w, x, y, and z.

9. $\log_a \sqrt{\dfrac{z^3}{xy}}$

10. $\log_a \dfrac{x^2}{y^3z}$

GS **11.** $\log_a \dfrac{x^3y^4}{z^5w^9}$

$= \log_a x^3y^4 - \log_a$ ____

$= (\log_a x^3 + \log_a$ ____ $) -$

$(\log_a z^5 + \log_a$ ____ $)$

$= \log_a x^3 + \log_a y^4 -$

$\log_a z^5 -$ ____

$= 3\log_a x + 4\log_a y -$

____ $\log_a z -$ ____ $\log_a w$

EXAMPLES Express as a single logarithm.

11. $\dfrac{1}{2}\log_a x - 7\log_a y + \log_a z$

$= \log_a x^{1/2} - \log_a y^7 + \log_a z$ Using Property 2

$= \log_a \dfrac{\sqrt{x}}{y^7} + \log_a z$ Using Property 3

$= \log_a \dfrac{z\sqrt{x}}{y^7}$ Using Property 1

12. $\log_a \dfrac{b}{\sqrt{x}} + \log_a \sqrt{bx}$

$= \log_a b - \log_a \sqrt{x} + \log_a \sqrt{bx}$ Using Property 3

$= \log_a b - \frac{1}{2}\log_a x + \frac{1}{2}\log_a (bx)$ Using Property 2

$= \log_a b - \frac{1}{2}\log_a x + \frac{1}{2}(\log_a b + \log_a x)$ Using Property 1

$= \log_a b - \frac{1}{2}\log_a x + \frac{1}{2}\log_a b + \frac{1}{2}\log_a x$

$= \frac{3}{2}\log_a b$ Collecting like terms

$= \log_a b^{3/2}$ Using Property 2

Example 12 could also be done as follows:

$\log_a \dfrac{b}{\sqrt{x}} + \log_a \sqrt{bx} = \log_a \left(\dfrac{b}{\sqrt{x}}\sqrt{bx}\right)$ Using Property 1

$= \log_a \left(\dfrac{b}{\sqrt{x}} \cdot \sqrt{b} \cdot \sqrt{x}\right)$

$= \log_a (b\sqrt{b})$, or $\log_a b^{3/2}$.

Express as a single logarithm.

12. $5\log_a x - \log_a y + \dfrac{1}{4}\log_a z$

13. $\log_a \dfrac{\sqrt{x}}{b} - \log_a \sqrt{bx}$

Answers

9. $\frac{3}{2}\log_a z - \frac{1}{2}\log_a x - \frac{1}{2}\log_a y$
10. $2\log_a x - 3\log_a y - \log_a z$
11. $3\log_a x + 4\log_a y - 5\log_a z - 9\log_a w$
12. $\log_a \dfrac{x^5z^{1/4}}{y}$, or $\log_a \dfrac{x^5\sqrt[4]{z}}{y}$
13. $\log_a \dfrac{1}{b\sqrt{b}}$, or $\log_a b^{-3/2}$

Guided Solution:
11. z^5w^9, y^4, w^9, $\log_a w^9$, 5, 9

Do Exercises 12 and 13 on the preceding page.

EXAMPLES Given $\log_a 2 = 0.301$ and $\log_a 3 = 0.477$, find each of the following, if possible.

13. $\log_a 6 = \log_a (2 \cdot 3) = \log_a 2 + \log_a 3$ Property 1
$= 0.301 + 0.477 = 0.778$

14. $\log_a \frac{2}{3} = \log_a 2 - \log_a 3$ Property 3
$= 0.301 - 0.477 = -0.176$

15. $\log_a 81 = \log_a 3^4 = 4 \log_a 3$ Property 2
$= 4(0.477) = 1.908$

16. $\log_a \frac{1}{3} = \log_a 1 - \log_a 3$ Property 3
$= 0 - 0.477 = -0.477$

17. $\log_a \sqrt{a} = \log_a a^{1/2} = \frac{1}{2} \log_a a = \frac{1}{2} \cdot 1 = \frac{1}{2}$ Property 2

18. $\log_a 2a = \log_a 2 + \log_a a$ Property 1
$= 0.301 + 1 = 1.301$

19. $\log_a 5$ There is no way to find this using these properties ($\log_a 5 \neq \log_a 2 + \log_a 3$).

20. $\frac{\log_a 3}{\log_a 2} = \frac{0.477}{0.301} \approx 1.58$ We simply divide the logarithms, not using any property of logarithms.

Do Exercises 14–22.

e THE LOGARITHM OF THE BASE TO A POWER

PROPERTY 4

For any base a,

$$\log_a a^k = k.$$

(The logarithm, base a, of a to a power is the power.)

A Proof of Property 4 (*Optional*): The proof involves Property 2 and the fact that $\log_a a = 1$:

$\log_a a^k = k(\log_a a)$ Using Property 2
$= k \cdot 1$ Using $\log_a a = 1$
$= k.$

EXAMPLES Simplify.

21. $\log_3 3^7 = 7$

22. $\log_{10} 10^{5.6} = 5.6$

23. $\log_e e^{-t} = -t$

Do Exercises 23–25.

Caution!

Keep in mind that, in general,

$\log_a (M + N) \neq \log_a M + \log_a N$,

$\log_a (M - N) \neq \log_a M - \log_a N$,

$\log_a (MN) \neq (\log_a M)(\log_a N)$,

and

$\log_a (M/N) \neq (\log_a M) \div (\log_a N)$.

Given

$\log_a 2 = 0.301$,
$\log_a 5 = 0.699$,

find each of the following, if possible.

14. $\log_a 4$

15. $\log_a 10$

16. $\log_a \frac{2}{5}$

17. $\log_a \frac{5}{2}$

18. $\log_a \frac{1}{5}$

19. $\log_a \sqrt{a^3}$

20. $\log_a 5a$

21. $\log_a 7$

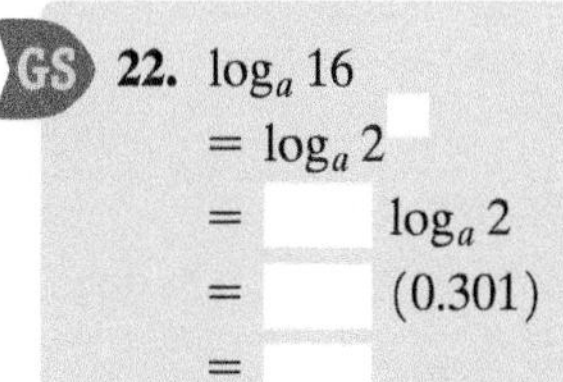

GS **22.** $\log_a 16$
$= \log_a 2^{\square}$
$= \square \log_a 2$
$= \square (0.301)$
$= \square$

Simplify.

23. $\log_2 2^6$

24. $\log_{10} 10^{3.2}$

25. $\log_e e^{12}$

Answers

14. 0.602 **15.** 1 **16.** −0.398 **17.** 0.398 **18.** −0.699 **19.** $\frac{3}{2}$ **20.** 1.699 **21.** Cannot be found using the properties of logarithms **22.** 1.204 **23.** 6 **24.** 3.2 **25.** 12

Guided Solution:
22. 4, 4, 4, 1.204

8.4 Exercise Set

FOR EXTRA HELP MyLab Math

✓ Check Your Understanding

Reading Check Choose from the column on the right the option that best completes each statement.

RC1. $\log_2 (8 \cdot 4) =$ ____

RC2. $\log_2 \left(\frac{8}{4}\right) =$ ____

RC3. $\log_2 4^8 =$ ____

RC4. $\log_2 (3 \cdot 4) =$ ____

RC5. $\log_2 4^3 =$ ____

RC6. $\log_2 3^4 =$ ____

a) $\log_2 3 + \log_2 4$
b) $\log_2 8 + \log_2 4$
c) $\log_2 8 - \log_2 4$
d) $8 \log_2 4$
e) $3 \log_2 4$
f) $4 \log_2 3$

Concept Check Choose from the column on the right the option that makes each statement true. Not all choices will be used.

CC1. $\log_5 12 + \log_5 2 = \log_5$ ____

CC2. $\log_5 12 - \log_5 2 = \log_5$ ____

CC3. $\log_5 5 =$ ____

CC4. $\log_5 5^{10} =$ ____

a) 0
b) 1
c) 5
d) 6
e) 10
f) 24

a Express as a sum of logarithms.

1. $\log_2 (32 \cdot 8)$

2. $\log_3 (27 \cdot 81)$

3. $\log_4 (64 \cdot 16)$

4. $\log_5 (25 \cdot 125)$

5. $\log_a Qx$

6. $\log_r 8Z$

Express as a single logarithm.

7. $\log_b 3 + \log_b 84$

8. $\log_a 75 + \log_a 5$

9. $\log_c K + \log_c y$

10. $\log_t H + \log_t M$

b Express as a product.

11. $\log_c y^4$

12. $\log_a x^3$

13. $\log_b t^6$

14. $\log_{10} y^7$

15. $\log_b C^{-3}$

16. $\log_c M^{-5}$

c Express as a difference of logarithms.

17. $\log_a \frac{67}{5}$

18. $\log_t \frac{T}{7}$

19. $\log_b \frac{2}{5}$

20. $\log_a \frac{z}{y}$

Express as a single logarithm.

21. $\log_c 22 - \log_c 3$

22. $\log_d 54 - \log_d 9$

d Express in terms of logarithms of a single variable or a number.

23. $\log_a x^2y^3z$

24. $\log_a 5xy^4z^3$

25. $\log_b \frac{xy^2}{z^3}$

26. $\log_b \frac{p^2q^5}{m^4n^7}$

27. $\log_c \sqrt[3]{\frac{x^4}{y^3z^2}}$

28. $\log_a \sqrt{\frac{x^6}{p^5q^8}}$

29. $\log_a \sqrt[4]{\frac{m^8n^{12}}{a^3b^5}}$

30. $\log_a \sqrt{\frac{a^6b^8}{a^2b^5}}$

Express as a single logarithm and, if possible, simplify.

31. $\frac{2}{3}\log_a x - \frac{1}{2}\log_a y$

32. $\frac{1}{2}\log_a x + 3\log_a y - 2\log_a x$

33. $\log_a 2x + 3(\log_a x - \log_a y)$

34. $\log_a x^2 - 2\log_a \sqrt{x}$

35. $\log_a \frac{a}{\sqrt{x}} - \log_a \sqrt{ax}$

36. $\log_a (x^2 - 4) - \log_a (x - 2)$

Given $\log_b 3 = 1.099$ and $\log_b 5 = 1.609$, find each of the following.

37. $\log_b 15$

38. $\log_b 8$

39. $\log_b \frac{5}{3}$

40. $\log_b \frac{3}{5}$

41. $\log_b \frac{1}{5}$

42. $\log_b \frac{1}{3}$

43. $\log_b \sqrt{b}$

44. $\log_b \sqrt{b^3}$

45. $\log_b 5b$

46. $\log_b 3b$

47. $\log_b 2$

48. $\log_b 75$

e Simplify.

49. $\log_e e^t$

50. $\log_w w^8$

51. $\log_p p^5$

52. $\log_Y Y^{-4}$

Solve for x.

53. $\log_2 2^7 = x$

54. $\log_9 9^4 = x$

55. $\log_e e^x = -7$

56. $\log_a a^x = 2.7$

Skill Maintenance

Compute and simplify. Express answers in the form $a + bi$, where $i^2 = -1$. [6.8b, c, d, e]

57. i^{29}

58. i^{34}

59. $(2 + i)(2 - i)$

60. $\frac{2 + i}{2 - i}$

61. $(7 - 8i) - (-16 + 10i)$

62. $2i^2 \cdot 5i^3$

63. $(8 + 3i)(-5 - 2i)$

64. $(2 - i)^2$

Synthesis

Determine whether each is true or false.

65. $\frac{\log_a P}{\log_a Q} = \log_a \frac{P}{Q}$

66. $\frac{\log_a P}{\log_a Q} = \log_a P - \log_a Q$

67. $\log_a 3x = \log_a 3 + \log_a x$

68. $\log_a 3x = 3\log_a x$

69. $\log_a (P + Q) = \log_a P + \log_a Q$

70. $\log_a x^2 = 2\log_a x$

Mid-Chapter Review

Concept Reinforcement

Determine whether each statement is true or false.

__________ **1.** The graph of an exponential function never crosses the x-axis. [8.1a]

__________ **2.** A function f is one-to-one if different inputs have different outputs. [8.2c]

__________ **3.** $\log_a 0 = 1$ [8.3c]

__________ **4.** $\log_a \frac{m}{n} = \log_a m - \log_a n$ [8.4c]

Guided Solutions

GS Fill in each box with the number and/or symbol that creates a correct statement or solution.

5. Solve: $\log_5 x = 3$. [8.3c]

$\log_5 x = 3$

$\square^{\square} = x$ Converting to an exponential equation

$\square = x$ Simplifying

6. Given $\log_a 2 = 0.648$ and $\log_a 9 = 2.046$, find $\log_a 18$. [8.4d]

$\log_a 18 = \log_a (\square \cdot \square)$

$= \log_a 2 \;\square\; \log_a \square$

$= 0.648 + \square = \square$

Mixed Review

Graph. [8.1a], [8.3a]

7. $f(x) = 3^{x-1}$

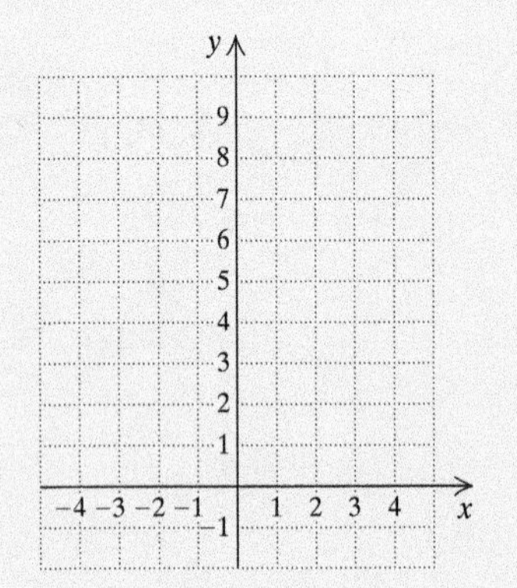

8. $f(x) = \left(\frac{3}{4}\right)^x$

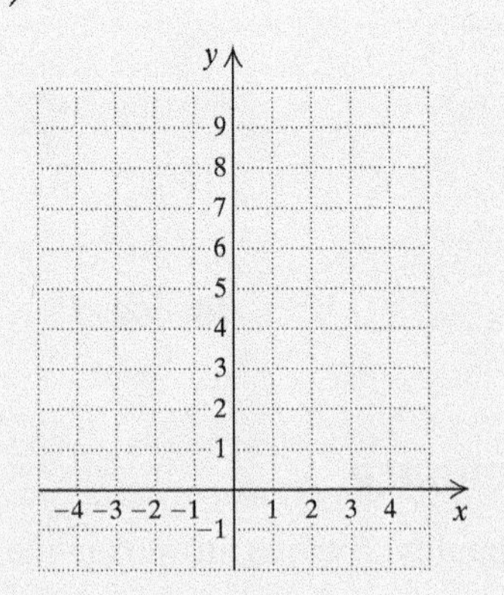

9. $f(x) = \log_4 x$

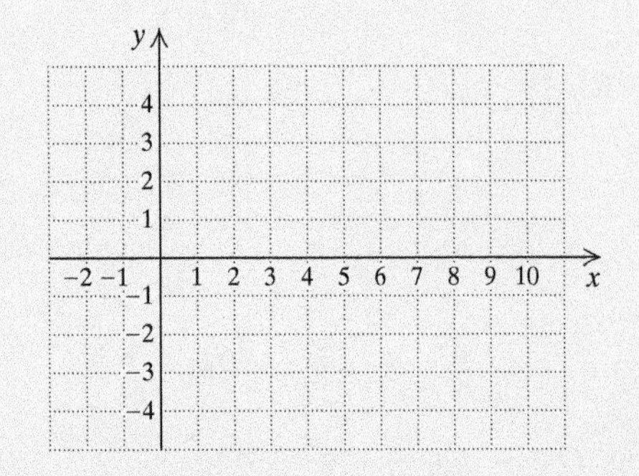

10. $f(x) = \log_{1/4} x$

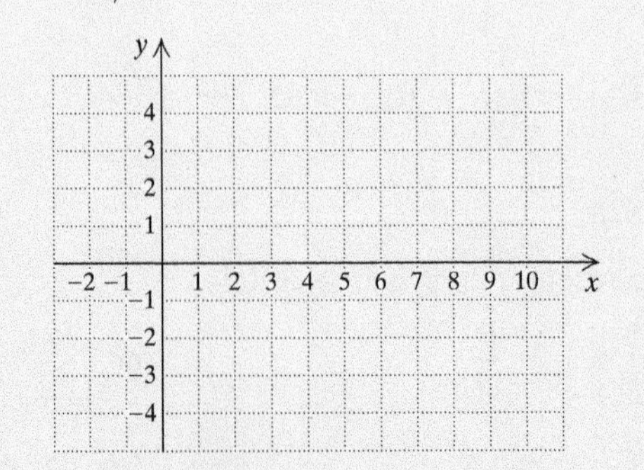

11. *Interest Compounded Annually.* Lucas invests \$500 at 4% interest, compounded annually. [8.1c]

a) Find a function A for the amount in the account after t years.

b) Find the amount in the account at $t = 0$, at $t = 4$, and at $t = 10$.

12. *Interest Compounded Quarterly.* The Currys invest \$1500 in an account paying 3.5% interest, compounded quarterly. Find the amount in the account after $1\frac{1}{2}$ years. [8.1c]

Determine whether each function is one-to-one. If it is, find a formula for its inverse. [8.2c]

13. $f(x) = 3x + 1$

14. $f(x) = x^3 + 2$

Find $(f \circ g)(x)$ and $(g \circ f)(x)$. [8.2a]

15. $f(x) = 2x - 5$, $g(x) = 3 - x$

16. $f(x) = x^2 + 1$, $g(x) = 3x - 1$

Find $f(x)$ and $g(x)$ such that $h(x) = (f \circ g)(x)$. Answers may vary. [8.2a]

17. $h(x) = \dfrac{3}{x + 4}$

18. $h(x) = \sqrt{6x - 7}$

For each function, use composition to show that the inverse is correct. [8.2e]

19. $f(x) = \dfrac{x}{3}$, $f^{-1}(x) = 3x$

20. $f(x) = \sqrt[3]{x + 4}$, $f^{-1}(x) = x^3 - 4$

Convert to a logarithmic equation. [8.3b]

21. $7^3 = 343$

22. $3^{-4} = \dfrac{1}{81}$

Convert to an exponential equation. [8.3b]

23. $\log_6 12 = t$

24. $\log_n T = m$

Solve. [8.3c]

25. $\log_4 64 = x$

26. $\log_x \dfrac{1}{4} = -2$

Find each of the following. [8.3c]

27. $\log_7 49$

28. $\log_2 32$

Use a calculator to find the logarithm, to four decimal places. [8.3d]

29. $\log 243.7$

30. $\log 0.23$

Express in terms of logarithms of a single variable or number. [8.4d]

31. $\log_b \dfrac{2xy^2}{z^3}$

32. $\log_a \sqrt[3]{\dfrac{x^2y^5}{z^4}}$

Express as a single logarithm and, if possible, simplify. [8.4d]

33. $\log_a x - 2\log_a y + \dfrac{1}{2}\log_a z$

34. $\log_m (b^2 - 16) - \log_m (b + 4)$

Simplify. [8.3c], [8.4e]

35. $\log_8 1$

36. $\log_3 3$

37. $\log_a a^{-3}$

38. $\log_c c^5$

Understanding Through Discussion and Writing

39. The function $V(t) = 750(1.2)^t$ is used to predict the value V of a certain rare stamp t years after 2010. Do not calculate $V^{-1}(t)$ but explain how V^{-1} could be used. [8.2c]

40. Explain in your own words what is meant by $\log_a b = c$. [8.3b]

41. Find a way to express $\log_a (x/5)$ as a difference of logarithms without using the quotient rule. Explain your work. [8.4a, b]

42. A student incorrectly reasons that

$$\log_b \frac{1}{x} = \log_b \frac{x}{x \cdot x}$$
$$= \log_b x - \log_b x + \log_b x$$
$$= \log_b x.$$

What mistake has the student made? Explain what the answer should be. [8.4a, c]

STUDYING FOR SUCCESS *Take Good Care of Yourself*

- ☐ Get plenty of rest, especially the night before a test.
- ☐ Try an exercise break when studying. Often a walk or a bike ride will improve your concentration.
- ☐ Plan leisure time in your schedule. Rest and a change of pace will help you avoid burn-out.

8.5 Natural Logarithmic Functions

OBJECTIVES

- **a** Find logarithms or powers, base *e*, using a calculator.
- **b** Use the change-of-base formula to find logarithms with bases other than *e* or 10.
- **c** Graph exponential functions and logarithmic functions, base *e*.

Any positive number other than 1 can serve as the base of a logarithmic function. Common, or base-10, logarithms, which were introduced in Section 8.3, are useful because they have the same base as our "commonly" used decimal system of naming numbers.

Today, another base is widely used. It is an irrational number named *e*. We now consider *e* and **natural logarithms**, or logarithms base *e*.

a THE BASE *e* AND NATURAL LOGARITHMS

When interest is computed n times per year, the compound-interest formula is

$$A = P\left(1 + \frac{r}{n}\right)^{nt},$$

where A is the amount that an initial investment P will grow to after t years at interest rate r. Suppose that \$1 could be invested at 100% interest for 1 year. (In reality, no financial institution would pay such an interest rate.) The preceding formula becomes a function A defined in terms of the number of compounding periods n:

$$A(n) = \left(1 + \frac{1}{n}\right)^{n}.$$

Let's find some function values, using a calculator and rounding to six decimal places. The numbers in the table shown here approach a very important number called e. It is an irrational number, so its decimal representation neither terminates nor repeats.

n	$A(n) = \left(1 + \frac{1}{n}\right)^{n}$
1 (compounded annually)	\$2.00
2 (compounded semiannually)	\$2.25
3	\$2.370370
4 (compounded quarterly)	\$2.441406
5	\$2.488320
100	\$2.704814
365 (compounded daily)	\$2.714567
8760 (compounded hourly)	\$2.718127

THE NUMBER *e*

$e \approx 2.7182818284\ldots$

Logarithms, base e, are called **natural logarithms**, or **Naperian logarithms**, in honor of John Napier (1550–1617), a Scotsman who invented logarithms.

The abbreviation **ln** is commonly used with natural logarithms. Thus,

$$\ln 29 \quad \text{means} \quad \log_e 29.$$

We generally read "ln 29" as "the natural log of 29," or simply "el en of 29."

On a calculator, the key for natural logarithms is generally marked LN. Using that key, we find that

$$\ln 29 \approx 3.36729583 \approx 3.3673, \qquad \text{Be sure to memorize } \ln a = \log_e a.$$

rounded to four decimal places. This also tells us that $e^{3.3673} \approx 29$.

On some scientific calculators, the keystrokes for doing such a calculation might be 2 9 LN =. If we were to use a graphing calculator, the keystrokes might be LN 2 9 ENTER.

EXAMPLES Find the natural logarithm, to four decimal places, on a calculator.

	Function Value	*Readout*	*Rounded*
1.	ln 287,523	12.56905814	12.5691
2.	ln 0.000486	−7.629301934	−7.6293
3.	ln −5	NONREAL ANS	Does not exist as a real number
4.	ln e	1	1
5.	ln 1	0	0

Do Exercises 1–5. ▶

Find the natural logarithm, to four decimal places, on a calculator.

1. ln 78,235.4
2. ln 0.0000309
3. ln (−3)
4. ln 0
5. ln 10

$\ln 1 = 0$ and $\ln e = 1$, for the logarithmic base e.

The inverse of a logarithmic function is an exponential function. Thus, if $f(x) = \ln x$, then $f^{-1}(x) = e^x$. Because of this, on many calculators, the LN key doubles as the e^x key after a 2ND or SHIFT key has been pressed.

EXAMPLE 6 Find $e^{12.5691}$ using a calculator.

On a scientific calculator, we might enter 12.5691 and press e^x. On a graphing calculator, we might press 2ND e^x, followed by 12.5691 ENTER. In either case, we get the approximation

$$e^{12.5691} \approx 287{,}535.0371.$$

Do Exercises 6 and 7. ▶

6. Find $e^{11.2675}$ using a calculator. (Compare this computation to that of Margin Exercise 1.)
7. Find e^{-2} using a calculator.

b CHANGING LOGARITHM BASES

Find common logarithms on a calculator. [8.3d]

Find each of the following using a calculator. Round to four decimal places.

1. $\log \dfrac{8}{3}$
2. $\dfrac{\log 8}{\log 3}$

Answers: 1. 0.4260 **2.** 1.8928

MyLab Math VIDEO

Answers

1. 11.2675 **2.** −10.3848 **3.** Does not exist as a real number **4.** Does not exist **5.** 2.3026 **6.** 78,237.1596 **7.** 0.1353

Most calculators give the values of both common logarithms and natural logarithms. To find a logarithm with some other base, we can use the following conversion formula.

THE CHANGE-OF-BASE FORMULA

For any logarithm bases a and b and any positive number M,

$$\log_b M = \frac{\log_a M}{\log_a b}.$$

CALCULATOR CORNER

The Change-of-Base Formula To find a logarithm with a base other than 10 or e, we can use the change-of-base formula. For example, we can find $\log_5 8$ using common logarithms.

We let $a = 10$, $b = 5$, and $M = 8$ and substitute in the change-of-base formula. We press LOG 8) ÷ LOG 5) ENTER. Note that the parentheses must be closed in the numerator in order to enter the expression correctly. We also close the parentheses in the denominator for completeness. The result is about 1.2920. We could have let $a = e$ and used natural logarithms to find $\log_5 8$ as well.

```
log(8)/log(5)
        1.292029674
```

Some calculators allow us to find a logarithm with any base directly.

```
log5(8)
        1.292029674
```

A Proof of the Change-of-Base Formula (*Optional*): We let $x = \log_b M$. Then, writing an equivalent exponential equation, we have $b^x = M$. Next, we take the logarithm base a on both sides. This gives us

$$\log_a b^x = \log_a M.$$

By Property 2, the Power Rule,

$$x \log_a b = \log_a M,$$

and solving for x, we obtain

$$x = \frac{\log_a M}{\log_a b}.$$

But $x = \log_b M$, so we have

$$\log_b M = \frac{\log_a M}{\log_a b},$$

which is the change-of-base formula.

EXAMPLE 7 Find $\log_4 7$ using common logarithms.

We let $a = 10$, $b = 4$, and $M = 7$. Then we substitute into the change-of-base formula:

$$\log_b M = \frac{\log_a M}{\log_a b}$$

$$\log_4 7 = \frac{\log_{10} 7}{\log_{10} 4} \qquad \text{Substituting 10 for } a, \text{ 4 for } b, \text{ and 7 for } M$$

$$= \frac{\log 7}{\log 4}$$

$$\approx 1.4037.$$

To check, we use a calculator with a power key $\boxed{y^x}$ or ^ to verify that

$$4^{1.4037} \approx 7.$$

We can also use base e for a conversion.

EXAMPLE 8 Find $\log_4 7$ using natural logarithms.

$$\log_b M = \frac{\log_a M}{\log_a b}$$

$$\log_4 7 = \frac{\log_e 7}{\log_e 4} \qquad \text{Substituting } e \text{ for } a, \text{ 4 for } b, \text{ and 7 for } M$$

$$= \frac{\ln 7}{\ln 4}$$

$$\approx 1.4037 \qquad \text{Note that this is the same answer as that for Example 7.}$$

EXAMPLE 9 Find $\log_5 29$ using natural logarithms.

Substituting e for a, 5 for b, and 29 for M, we have

$$\log_5 29 = \frac{\log_e 29}{\log_e 5} \quad \text{Using the change-of-base formula}$$

$$= \frac{\ln 29}{\ln 5} \approx 2.0922.$$

Do Exercises 8 and 9. ▶

C GRAPHS OF EXPONENTIAL FUNCTIONS AND LOGARITHMIC FUNCTIONS, BASE e

EXAMPLE 10 Graph $f(x) = e^x$ and $g(x) = e^{-x}$.

We use a calculator with an e^x key to find approximate values of e^x and e^{-x}. Using these values, we can graph the functions.

x	e^x	e^{-x}
0	1	1
1	2.7	0.4
2	7.4	0.1
−1	0.4	2.7
−2	0.1	7.4

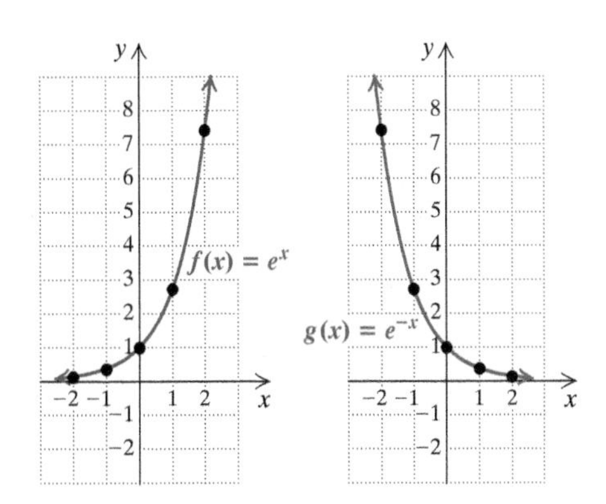

Note that each graph is the image of the other reflected across the y-axis. ■

EXAMPLE 11 Graph: $f(x) = e^{-0.5x}$.

We find some solutions with a calculator, plot them, and then draw the graph. For example, $f(2) = e^{-0.5(2)} = e^{-1} \approx 0.4$.

x	$e^{-0.5x}$
0	1
1	0.6
2	0.4
3	0.2
−1	1.6
−2	2.7
−3	4.5

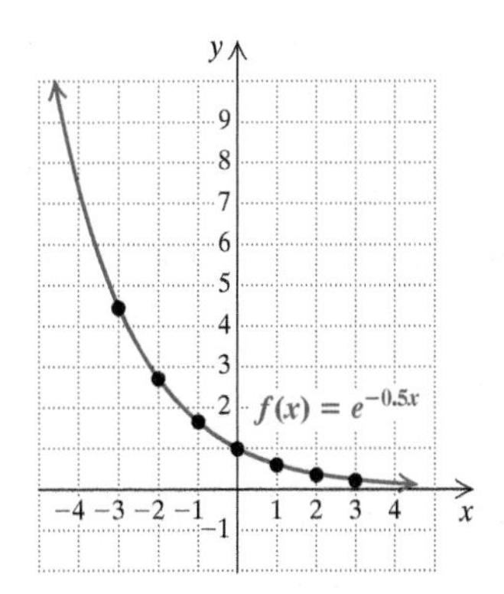

Do Exercises 10 and 11. ▶

8. a) Find $\log_6 7$ using common logarithms.

$$\log_6 7 = \frac{\log \square}{\log \square} \approx \square$$

b) Find $\log_6 7$ using natural logarithms.

$$\log_6 7 = \frac{\ln \square}{\ln \square} \approx \square$$

9. Find $\log_2 46$ using natural logarithms.

Graph.

10. $f(x) = e^{2x}$

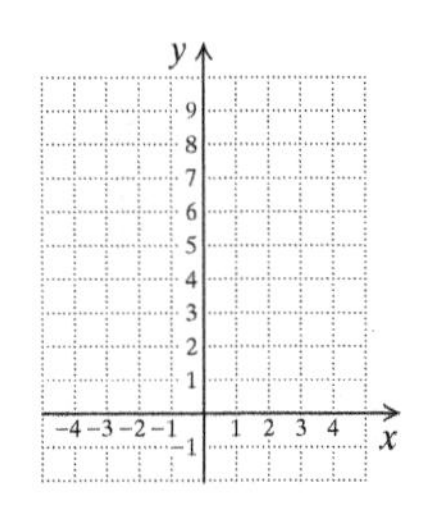

11. $g(x) = \frac{1}{2}e^{-x}$

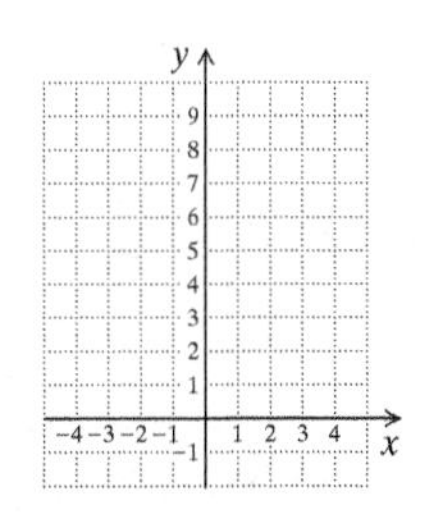

Answers

8. (a) 1.0860; **(b)** 1.0860 **9.** 5.5236

10.

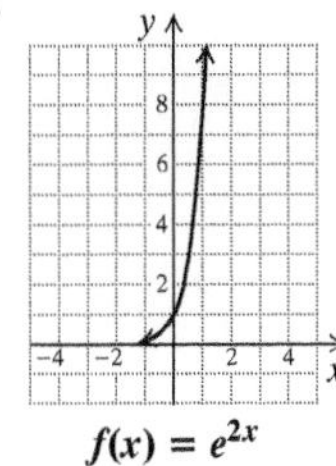

$f(x) = e^{2x}$

11.

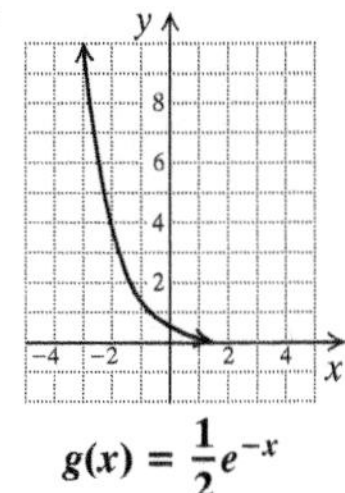

$g(x) = \frac{1}{2}e^{-x}$

Guided Solution:

8. (a) $\frac{7}{6}$, 1.0860; **(b)** $\frac{7}{6}$, 1.0860

Graph.

12. $f(x) = 2 \ln x$

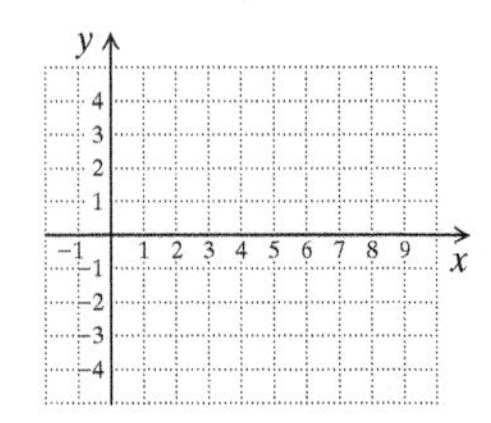

13. $g(x) = \ln(x - 2)$

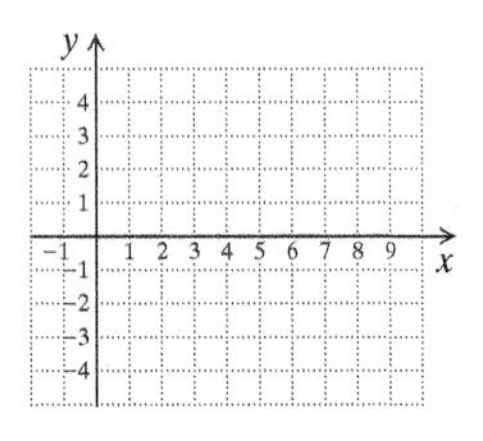

EXAMPLE 12 Graph: $g(x) = \ln x$.

We find some solutions with a calculator and then draw the graph. As expected, the graph is a reflection across the line $y = x$ of the graph of $y = e^x$.

x	$\ln x$
1	0
4	1.4
7	1.9
0.5	−0.7

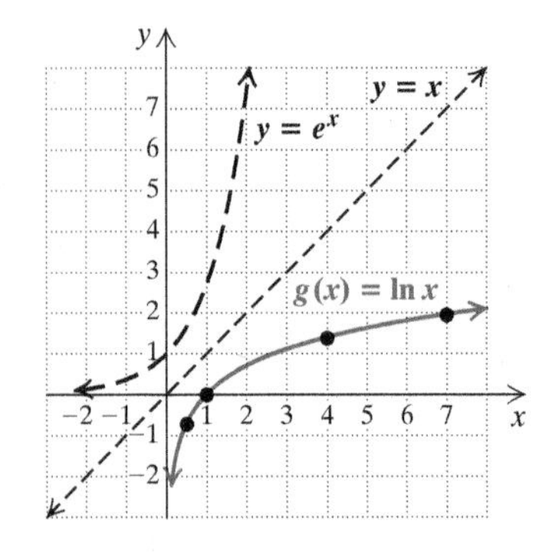

EXAMPLE 13 Graph: $f(x) = \ln(x + 3)$.

We find some solutions with a calculator, and then draw the graph. The graph of $y = \ln(x + 3)$ is the graph of $y = \ln x$ translated 3 units to the left.

x	$\ln(x + 3)$
0	1.1
1	1.4
2	1.6
3	1.8
4	1.9
−1	0.7
−2	0
−2.5	−0.7

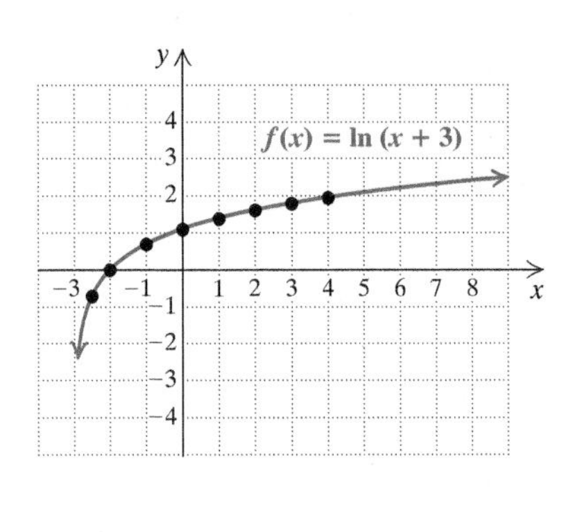

Do Exercises 12 and 13.

Answers

12.

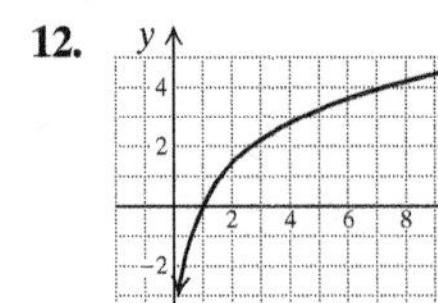

$f(x) = 2 \ln x$

13.

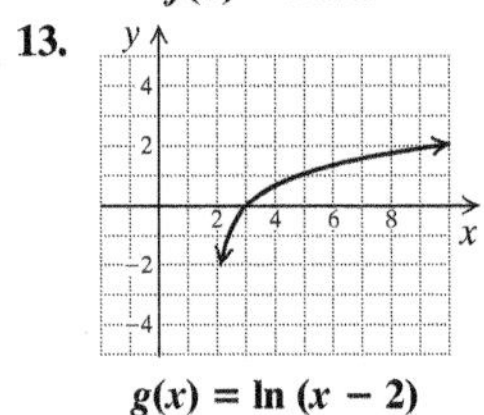

$g(x) = \ln(x - 2)$

CALCULATOR CORNER

Graphing Logarithmic Functions We can graph logarithmic functions with base 10 or base e by entering the function on the equation-editor screen using the LOG or LN key. To graph a logarithmic function with a base other than 10 or e, we must first use the change-of-base formula to change the base to 10 or e, unless our calculator can find other logarithms directly.

We can graph the function $y = \log_5 x$ on a graphing calculator by first changing the base to e. We let $a = e$, $b = 5$, and $M = x$ and substitute in the change-of-base formula. We enter $y_1 = \ln(x)/\ln(5)$ on the equation-editor screen, select a window, and press GRAPH. We could have let $a = 10$ and used base-10 logarithms to graph this function.

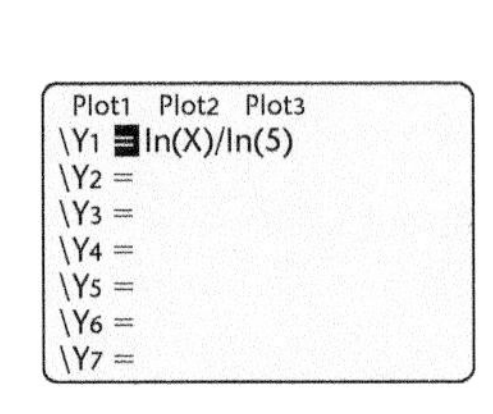

$y = \ln(x)/\ln(5)$

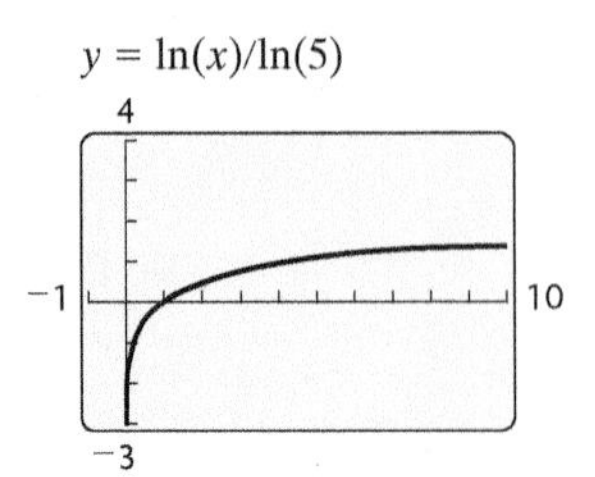

EXERCISES: Graph each of the following on a graphing calculator.

1. $y = \log_2 x$
2. $y = \log_3 x$
3. $y = \log_{1/2} x$
4. $y = \log_{2/3} x$

Visualizing for Success

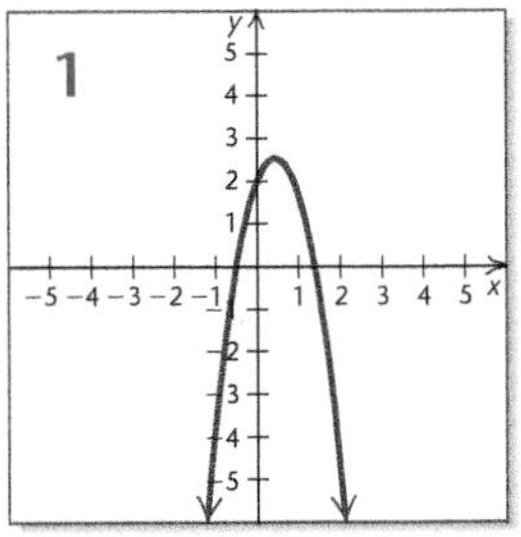

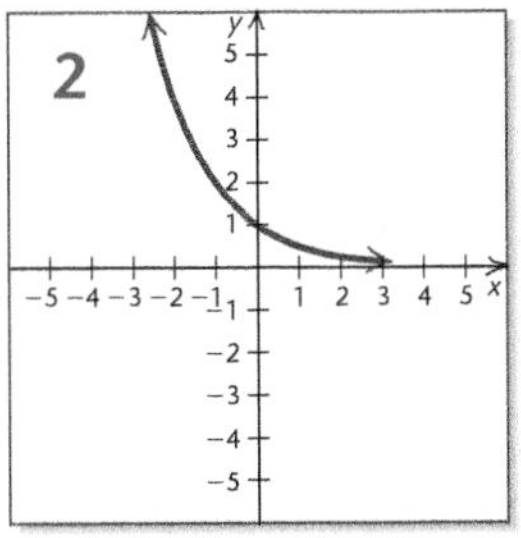

Match each graph with its function.

A. $f(x) = ax^2 + bx + c,$
$a < 0, c < 0$

B. $f(x) = a^x, 0 < a < 1$

C. $f(x) = a^x, a < 0$

D. $f(x) = \log_a x, 0 < a < 1$

E. $f(x) = \log_a x, a < 0$

F. $f(x) = mx + b, m > 0, b < 0$

G. $f(x) = mx + b, m < 0, b > 0$

H. $f(x) = mx + b, m < 0, b < 0$

I. $f(x) = ax^2 + bx + c,$
$a > 0, c < 0$

J. $f(x) = ax^2 + bx + c,$
$a < 0, c > 0$

K. $f(x) = \log_a x, a > 1$

L. $f(x) = ax^2 + bx + c,$
$a > 0, c > 0$

M. $f(x) = mx + b, m < 0, b = 0$

N. $f(x) = mx + b, m = 0, b > 0$

O. $f(x) = a^x, a > 1$

Answers on page A-33

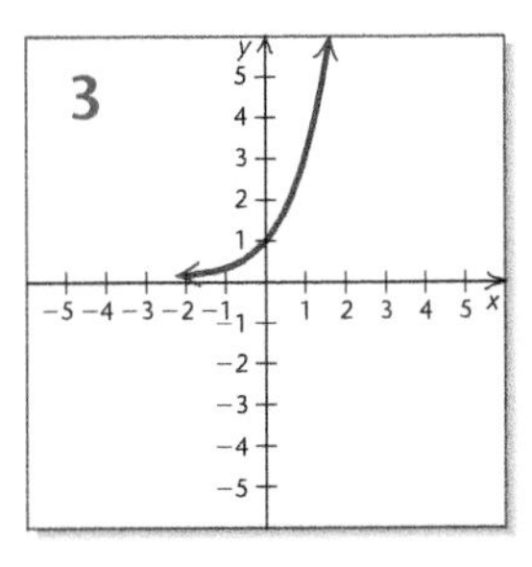

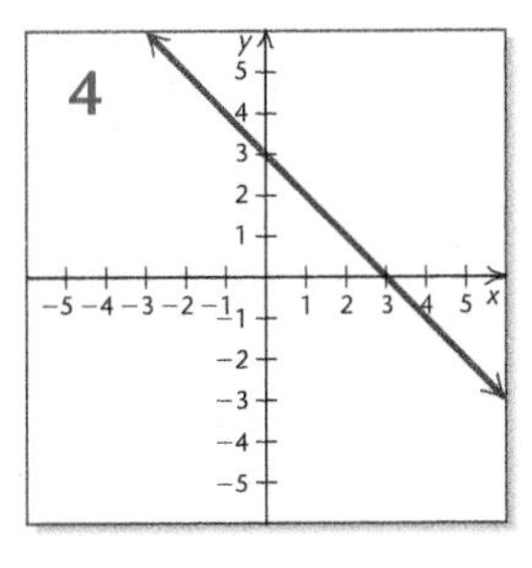

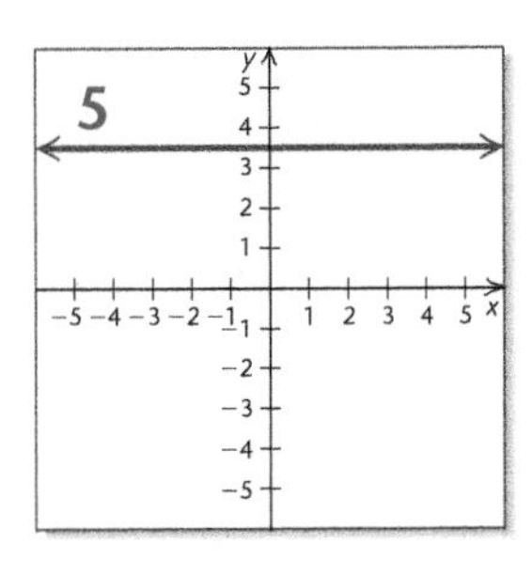

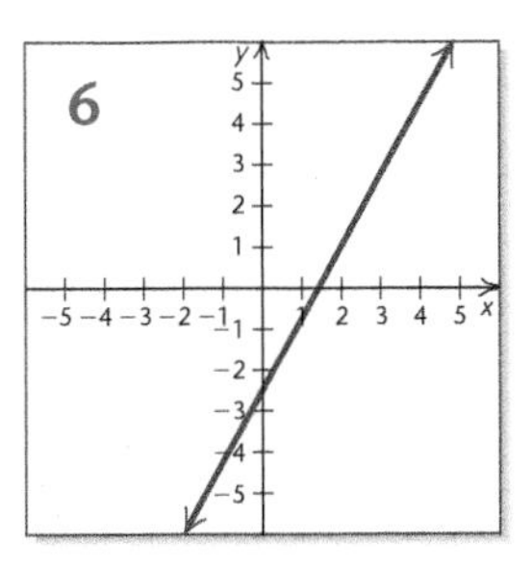

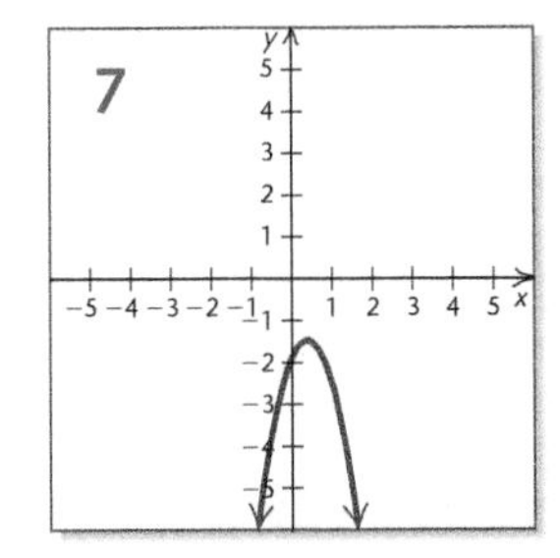

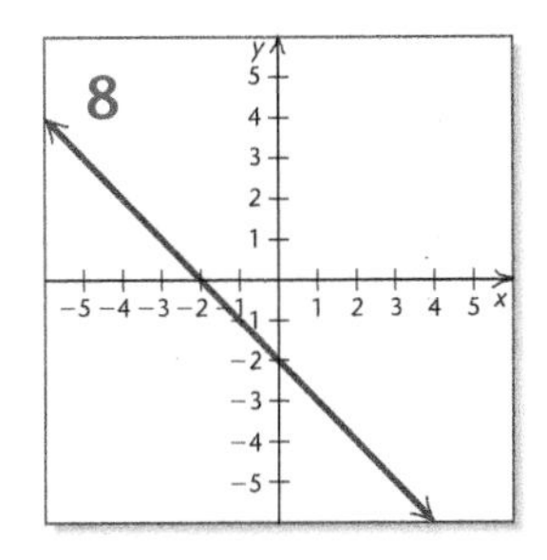

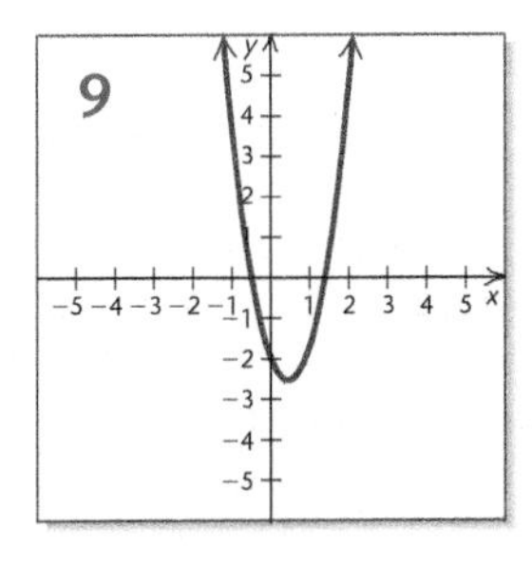

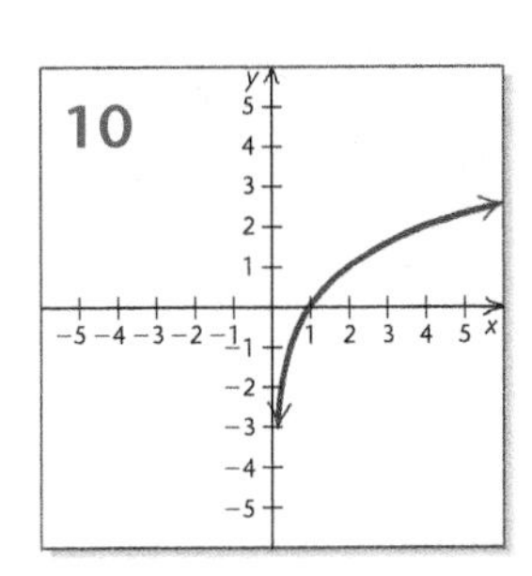

8.5 Exercise Set

FOR EXTRA HELP MyLab Math

✓ Check Your Understanding

Reading Check Determine whether each phrase describes either a base-10 logarithm or a base-e logarithm.

RC1. A common logarithm

RC2. A natural logarithm

RC3. $\log x$

RC4. $\ln x$

Concept Check Choose from the list below the option that best completes each statement. Choices may be used more than once.

a) $\log_5 7$

b) $\log_7 5$

CC1. $\dfrac{\log 5}{\log 7} =$ ____

CC2. $\dfrac{\log 7}{\log 5} =$ ____

CC3. $\dfrac{\ln 5}{\ln 7} =$ ____

CC4. $\dfrac{\ln 7}{\ln 5} =$ ____

a Find each of the following logarithms or powers, base e, using a calculator. Round answers to four decimal places.

1. $\ln 2$

2. $\ln 5$

3. $\ln 62$

4. $\ln 30$

5. $\ln 4365$

6. $\ln 901.2$

7. $\ln 0.0062$

8. $\ln 0.00073$

9. $\ln 0.2$

10. $\ln 0.04$

11. $\ln 0$

12. $\ln(-4)$

13. $\ln\left(\dfrac{97.4}{558}\right)$

14. $\ln\left(\dfrac{786.2}{77.2}\right)$

15. $\ln e$

16. $\ln e^2$

17. $e^{2.71}$

18. $e^{3.06}$

19. $e^{-3.49}$

20. $e^{-2.64}$

21. $e^{4.7}$

22. $e^{1.23}$

23. $\ln e^5$

24. $e^{\ln 7}$

b Find each of the following logarithms using the change-of-base formula.

25. $\log_6 100$

26. $\log_3 100$

27. $\log_2 100$

28. $\log_7 100$

29. $\log_7 65$

30. $\log_5 42$

31. $\log_{0.5} 5$

32. $\log_{0.1} 3$

33. $\log_2 0.2$

34. $\log_2 0.08$

35. $\log_\pi 200$

36. $\log_\pi \pi$

c Graph.

37. $f(x) = e^x$

x	$f(x)$
0	
1	
2	
−1	
−2	

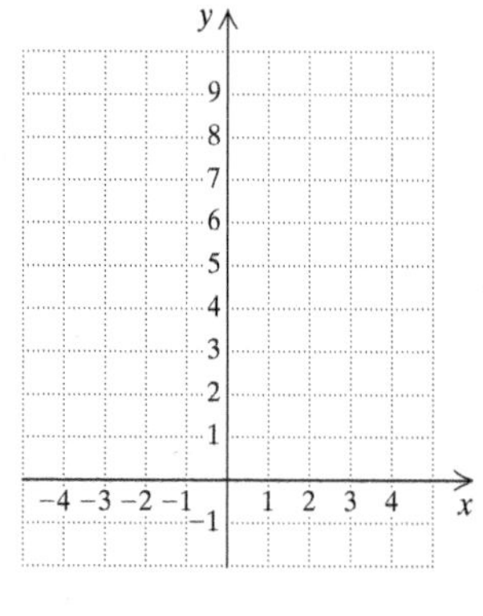

38. $f(x) = e^{0.5x}$

x	$f(x)$
0	
1	
2	
−1	
−2	

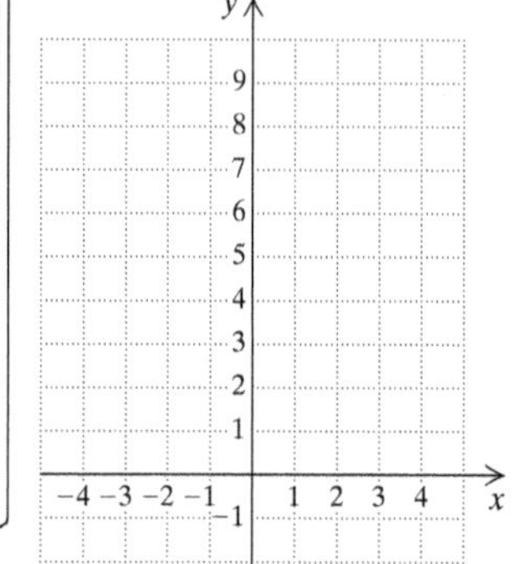

39. $f(x) = e^{-0.5x}$

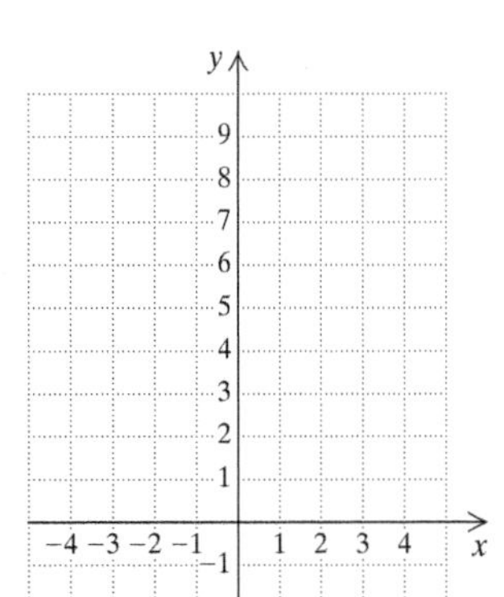

40. $f(x) = e^{-x}$

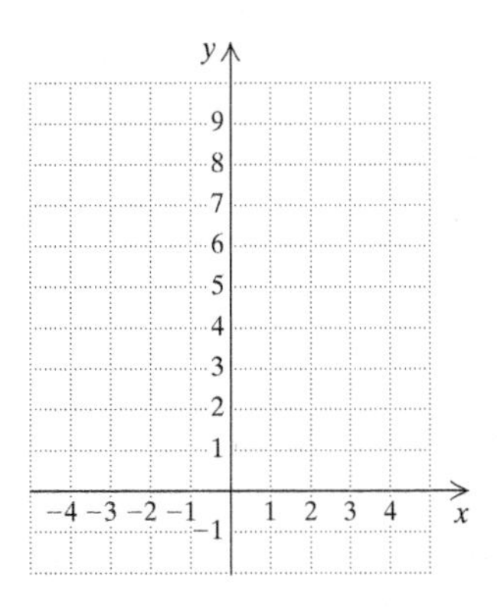

41. $f(x) = e^{x-1}$

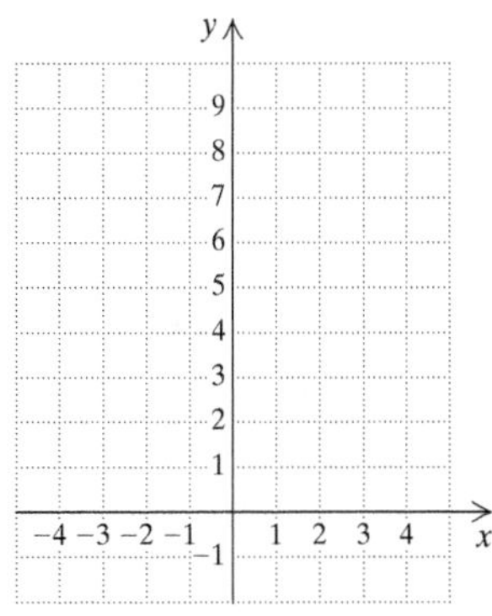

42. $f(x) = e^{-x} + 3$

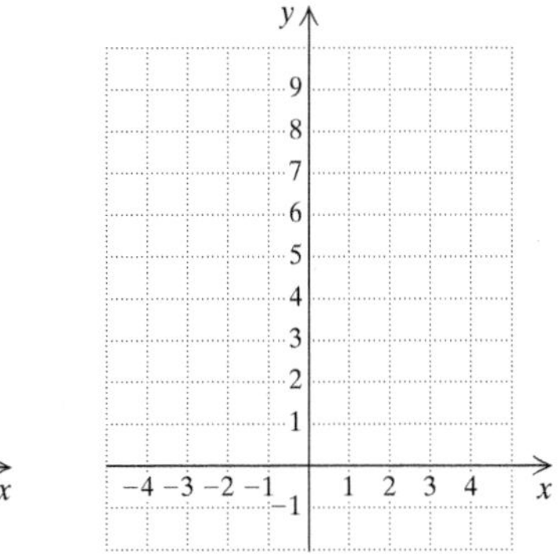

43. $f(x) = e^{x+2}$

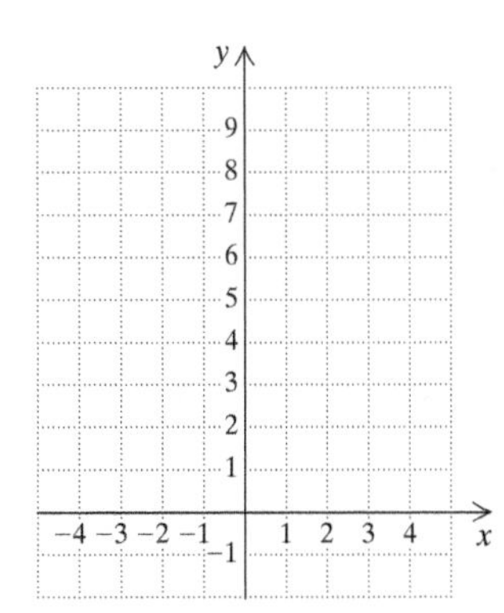

44. $f(x) = e^{x-2}$

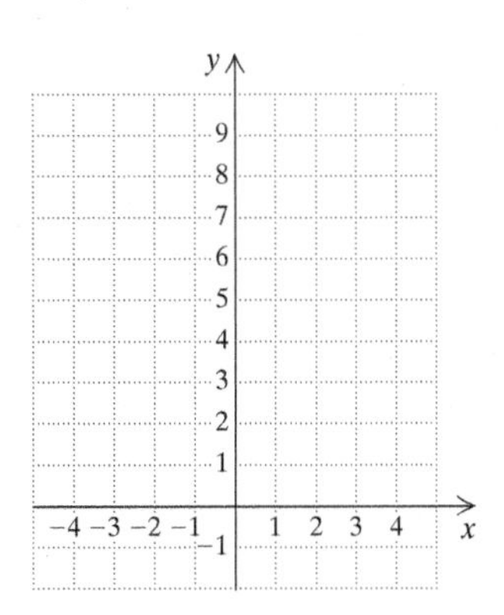

45. $f(x) = e^{x} - 1$

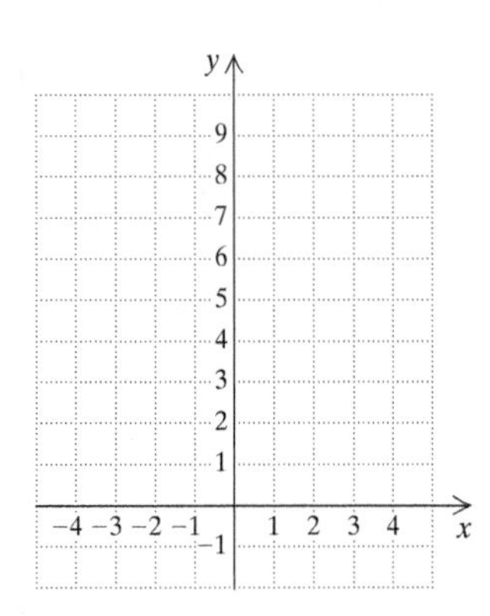

46. $f(x) = 2e^{0.5x}$

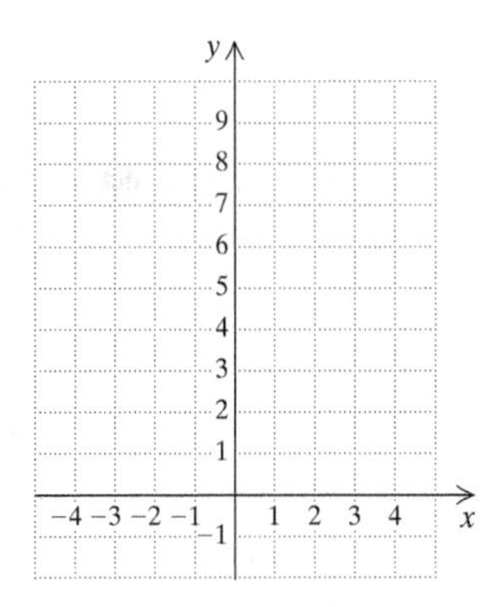

47. $f(x) = \ln(x + 2)$

x	$f(x)$
0	
1	
2	
3	
−0.5	
−1	
−1.5	

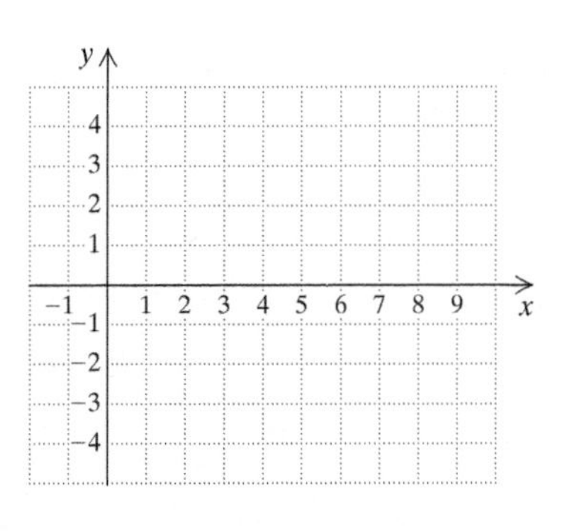

48. $f(x) = \ln(x + 1)$

x	$f(x)$
0	
1	
2	
3	
4	
−0.5	
−0.75	

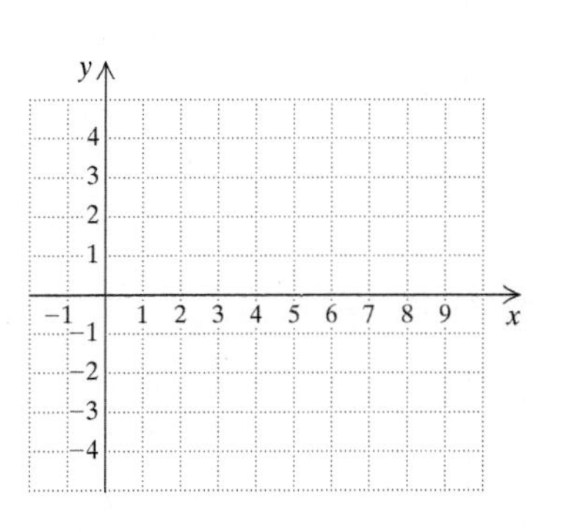

49. $f(x) = \ln(x - 3)$

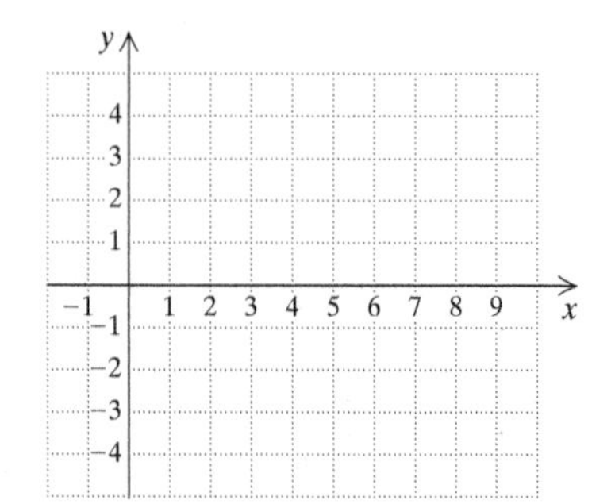

50. $f(x) = 2\ln(x - 2)$

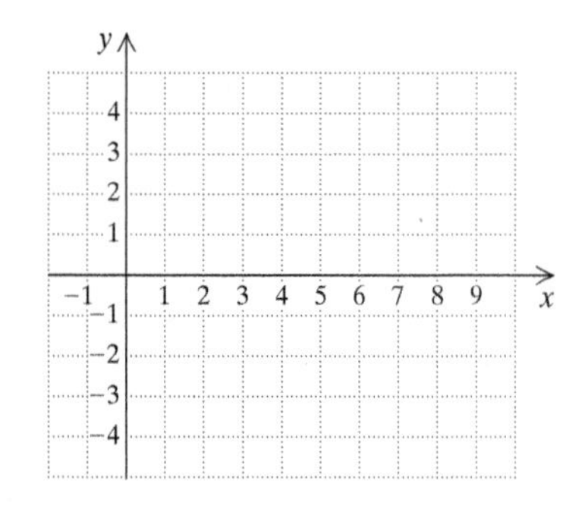

51. $f(x) = 2\ln x$

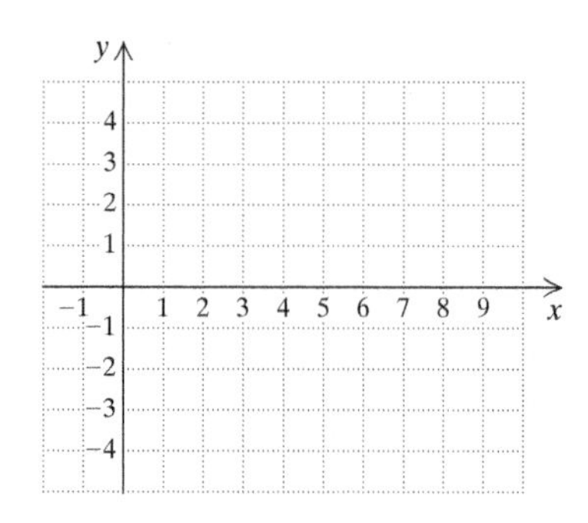

52. $f(x) = \ln x - 3$

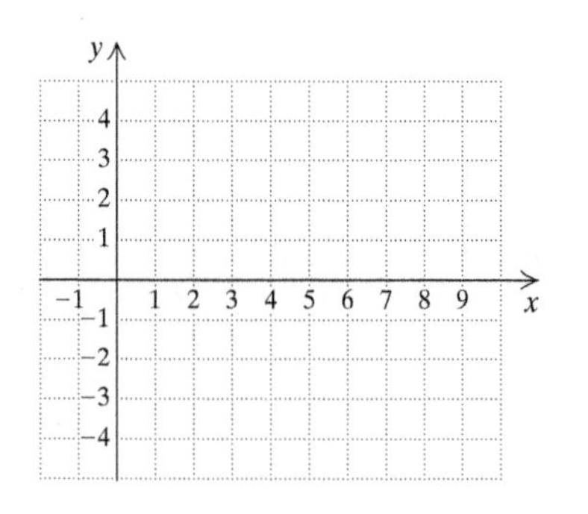

53. $f(x) = \frac{1}{2}\ln x + 1$

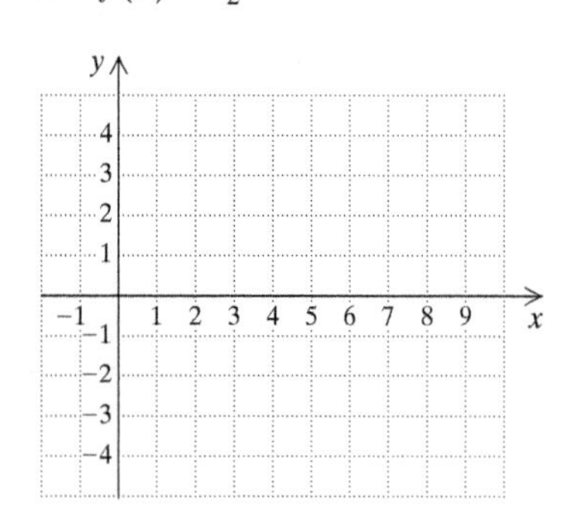

54. $f(x) = \ln x^2$

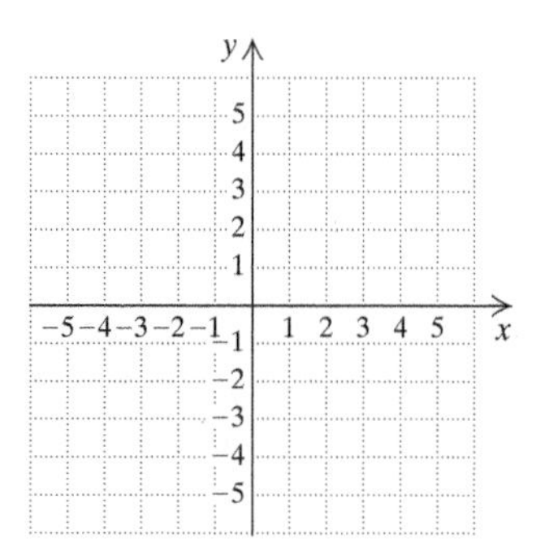

55. $f(x) = |\ln x|$

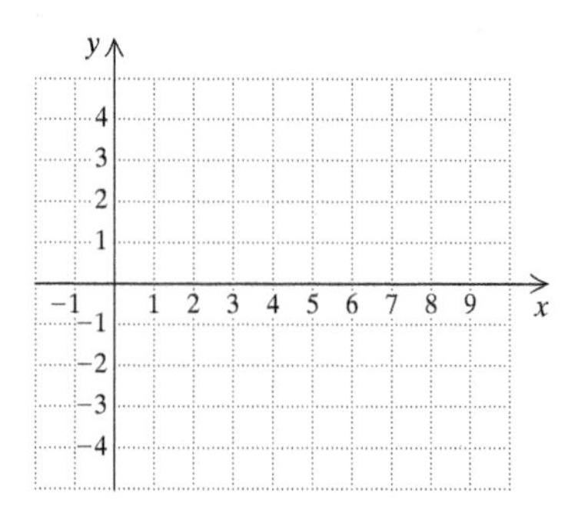

56. $f(x) = \ln |x|$

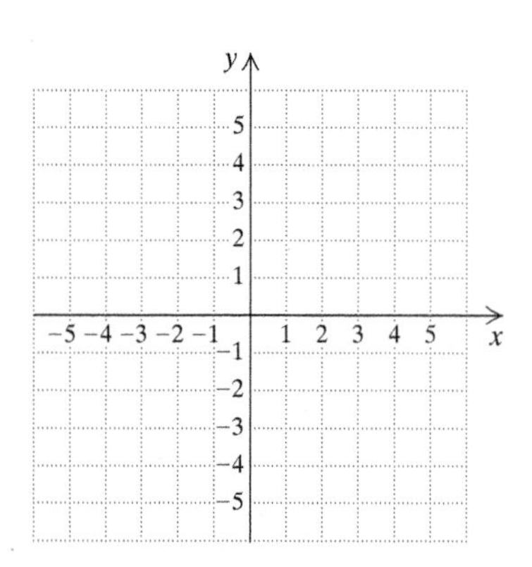

Skill Maintenance

Solve. [7.4c]

57. $x^{1/2} - 6x^{1/4} + 8 = 0$

58. $2y - 7\sqrt{y} + 3 = 0$

59. $x - 18\sqrt{x} + 77 = 0$

60. $x^4 - 25x^2 + 144 = 0$

Synthesis

Use the graph of the function to find the domain and the range.

61. $f(x) = 10x^2e^{-x}$

62. $f(x) = 7.4e^x \ln x$

63. $f(x) = 100(1 - e^{-0.3x})$

Find the domain.

64. $f(x) = \log_3 x^2$

65. $f(x) = \log(2x - 5)$

Solving Exponential Equations and Logarithmic Equations

8.6

OBJECTIVES

a Solve exponential equations.

b Solve logarithmic equations.

a SOLVING EXPONENTIAL EQUATIONS

Equations with variables in exponents, such as $5^x = 12$ and $2^{7x} = 64$, are called **exponential equations**. Sometimes we can solve an exponential equation by writing each side as a power of the same number. We use the following property, which is true because exponential functions are one-to-one.

THE PRINCIPLE OF EXPONENTIAL EQUALITY

For any $a > 0$, $a \neq 1$,

$$a^x = a^y \longrightarrow x = y.$$

(When powers are equal, the exponents are equal.)

EXAMPLE 1 Solve: $2^{3x-5} = 16$.

Because $16 = 2^4$, we can write each side as a power of the same number:

$$2^{3x-5} = 2^4.$$

Since the base is the same, 2, the exponents must be the same. Thus,

$$3x - 5 = 4$$
$$3x = 9$$
$$x = 3.$$

Check: $2^{3x-5} = 16$

$2^{3\cdot 3-5}$? 16

2^{9-5}

2^4

16 | TRUE

The solution is 3.

Do Exercises 1 and 2. ▶

Solve.

1. $3^{2x} = 9$

2. $4^{2x-3} = 64$

ALGEBRAIC ▶◀ GRAPHICAL CONNECTION

The solution, 3, of the equation $2^{3x-5} = 16$ in Example 1 is the x-coordinate of the point of intersection of the graphs of $y = 2^{3x-5}$ and $y = 16$.

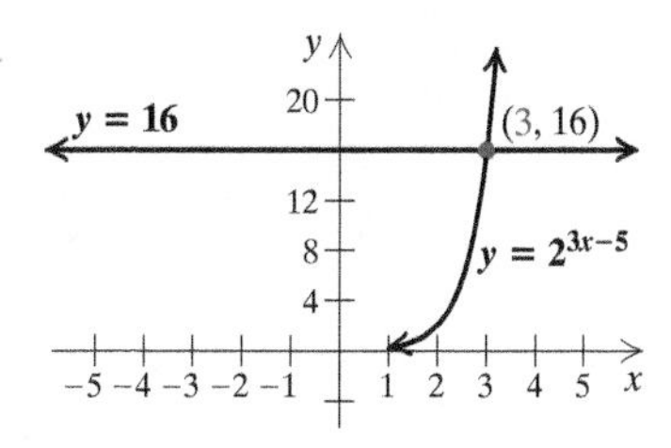

Answers

1. 1 **2.** 3

When it does not seem possible to write both sides of an equation as powers of the same base, we can use the following principle along with the properties developed in Section 8.4.

THE PRINCIPLE OF LOGARITHMIC EQUALITY

For any logarithm base a, and for $x, y > 0$,

$$\log_a x = \log_a y \longrightarrow x = y.$$

(If the logarithms, base a, of two expressions are the same, then the expressions are the same.)

Because calculators can generally find only common or natural logarithms (without resorting to the change-of-base formula), we usually take the common or natural logarithm on both sides of the equation.

The principle of logarithmic equality is useful anytime a variable appears as an exponent.

EXAMPLE 2 Solve: $5^x = 12$.

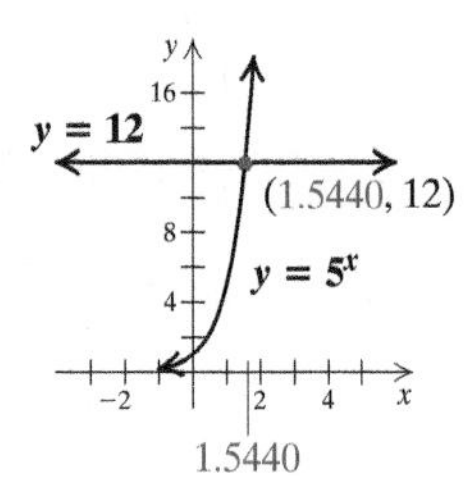

$5^x = 12$

$\log 5^x = \log 12$ Taking the common logarithm on both sides

$x \log 5 = \log 12$ Property 2

$x = \dfrac{\log 12}{\log 5}$ ← **Caution!** This is not $\log \frac{12}{5}$!

This is an exact answer. We cannot simplify further, but we can approximate using a calculator:

$$x = \frac{\log 12}{\log 5} \approx 1.5440.$$

We can also partially check this answer by finding $5^{1.5440}$ using a calculator:

$$5^{1.5440} \approx 12.00078587.$$

We get an answer close to 12, due to the rounding. This checks.

◀ **Do Exercise 3.**

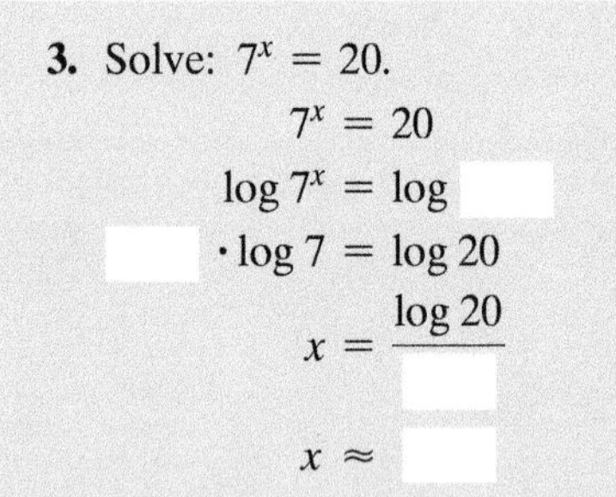

3. Solve: $7^x = 20$. GS

$7^x = 20$

$\log 7^x = \log$ ____

____ $\cdot \log 7 = \log 20$

$x = \dfrac{\log 20}{\text{\_\_\_\_}}$

$x \approx$ ____

If the base is e, we can make our work easier by taking the logarithm, base e, on both sides.

Answer

3. 1.5395

Guided Solution:

3. 20, x, log 7, 1.5395

EXAMPLE 3 Solve: $e^{0.06t} = 1500$.

We take the natural logarithm on both sides:

$$e^{0.06t} = 1500$$
$$\ln e^{0.06t} = \ln 1500 \quad \text{Taking ln on both sides}$$
$$\log_e e^{0.06t} = \ln 1500 \quad \text{Definition of natural logarithms}$$
$$0.06t = \ln 1500 \quad \text{Here we use Property 4: } \log_a a^k = k.$$
$$t = \frac{\ln 1500}{0.06}.$$

We can approximate using a calculator:

$$t = \frac{\ln 1500}{0.06} \approx \frac{7.3132}{0.06} \approx 121.89.$$

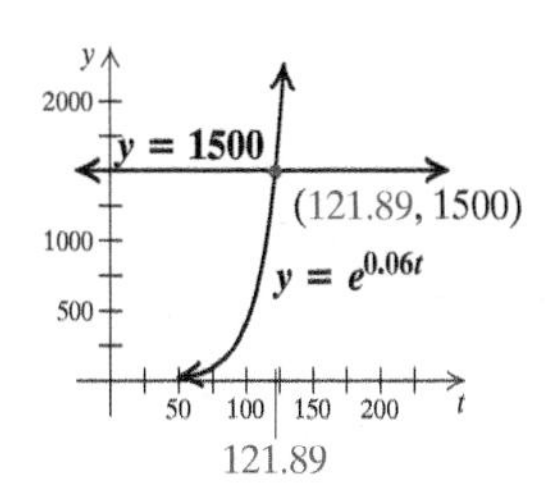

We can also partially check this answer using a calculator.

$e^{0.06t} = 1500$	
$e^{0.06(121.89)}$? 1500	
$e^{7.3134}$	
1500.269444	TRUE

The solution is $\frac{\ln 1500}{0.06}$, or about 121.89.

Do Exercise 4. ▶

4. Solve: $e^{0.3t} = 80$.

b SOLVING LOGARITHMIC EQUATIONS

Solve quadratic equations using the principle of zero products. [4.8a]

Solve.

1. $y^2 - y - 6 = 0$
2. $x^2 - 3x = 4$

Answers: **1.** $-2, 3$ **2.** $-1, 4$

MyLab Math VIDEO

Equations containing logarithmic expressions are called **logarithmic equations**. We can solve logarithmic equations by converting to equivalent exponential equations.

EXAMPLE 4 Solve: $\log_2 x = 3$.

We obtain an equivalent exponential equation:

$$x = 2^3$$
$$x = 8.$$

The solution is 8.

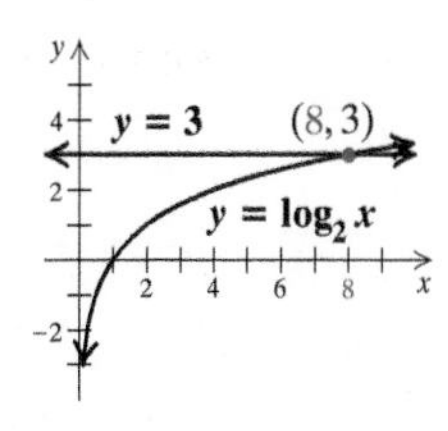

Do Exercise 5. ▶

5. Solve: $\log_5 x = 2$.

Answers

4. 14.6068 **5.** 25

To solve a logarithmic equation, first try to obtain a single logarithmic expression on one side and then write an equivalent exponential equation.

EXAMPLE 5 Solve: $\log_4(8x - 6) = 3$.

We already have a single logarithmic expression, so we write an equivalent exponential equation:

$$8x - 6 = 4^3 \quad \text{Writing an equivalent exponential equation}$$
$$8x - 6 = 64$$
$$8x = 70$$
$$x = \tfrac{70}{8}, \text{ or } \tfrac{35}{4}.$$

Check:

$$\log_4(8x - 6) = 3$$
$$\log_4\left(8 \cdot \tfrac{35}{4} - 6\right) \ ? \ 3$$
$$\log_4(70 - 6)$$
$$\log_4 64$$
$$3 \qquad \text{TRUE}$$

The solution is $\frac{35}{4}$.

6. Solve: $\log_3(5x + 7) = 2$.

Do Exercise 6.

EXAMPLE 6 Solve: $\log x + \log(x - 3) = 1$.

Here we have common logarithms. It helps to first write in the 10's before we obtain a single logarithmic expression on the left.

$$\log_{10} x + \log_{10}(x - 3) = 1$$
$$\log_{10}[x(x - 3)] = 1 \quad \text{Using Property 1 to obtain a single logarithm}$$
$$x(x - 3) = 10^1 \quad \text{Writing an equivalent exponential expression}$$
$$x^2 - 3x = 10$$
$$x^2 - 3x - 10 = 0$$
$$(x + 2)(x - 5) = 0 \quad \text{Factoring}$$
$$x + 2 = 0 \quad or \quad x - 5 = 0 \quad \text{Using the principle of zero products}$$
$$x = -2 \quad or \quad x = 5$$

Check: For −2:

$$\log x + \log(x - 3) = 1$$
$$\log(-2) + \log(-2 - 3) \ ? \ 1$$

The number −2 does *not* check because negative numbers do not have logarithms.

For 5:

$$\log x + \log(x - 3) = 1$$
$$\log 5 + \log(5 - 3) \ ? \ 1$$
$$\log 5 + \log 2$$
$$\log(5 \cdot 2)$$
$$\log 10$$
$$1 \qquad \text{TRUE}$$

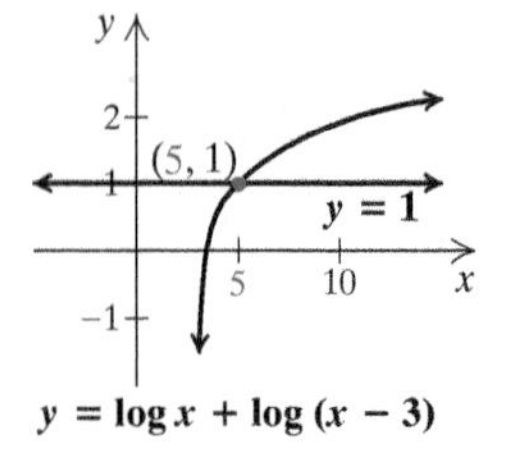

$y = \log x + \log(x - 3)$

7. Solve: $\log x + \log(x + 3) = 1$.

The solution is 5.

Do Exercise 7.

Answers

6. $\frac{2}{5}$ 7. 2

EXAMPLE 7 Solve: $\log_2 (x + 7) - \log_2 (x - 7) = 3$.

$$\log_2 (x + 7) - \log_2 (x - 7) = 3$$

$$\log_2 \frac{x + 7}{x - 7} = 3 \quad \text{Using Property 3 to obtain a single logarithm}$$

$$\frac{x + 7}{x - 7} = 2^3 \quad \text{Writing an equivalent exponential expression}$$

$$\frac{x + 7}{x - 7} = 8$$

$$x + 7 = 8(x - 7) \quad \text{Multiplying by the LCM, } x - 7$$

$$x + 7 = 8x - 56 \quad \text{Using a distributive law}$$

$$63 = 7x$$

$$\frac{63}{7} = x$$

$$9 = x$$

Check:

$\log_2 (x + 7) - \log_2 (x - 7) = 3$	
$\log_2 (9 + 7) - \log_2 (9 - 7)$	? 3
$\log_2 16 - \log_2 2$	
$\log_2 \frac{16}{2}$	
$\log_2 8$	
3	TRUE

The solution is 9.

Do Exercise 8. ▶

GS **8.** Solve:

$\log_3 (2x - 1) - \log_3 (x - 4) = 2$.

$$\log_3 (2x - 1) - \log_3 (x - 4) = 2$$

$$\log_3 \frac{2x - 1}{\_\_\_} = 2$$

$$\frac{2x - 1}{x - 4} = \_\_\_^{2}$$

$$\frac{2x - 1}{x - 4} = \_\_\_$$

$$2x - 1 = \_\_\_(x - 4)$$

$$2x - 1 = 9x - \_\_\_$$

$$\_\_\_ = 7x$$

$$\_\_\_ = x$$

Answer

8. 5

Guided Solution:

8. $x - 4$, 3, 9, 9, 36, 35, 5

8.6 Exercise Set

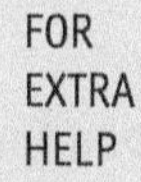

MyLab Math

Check Your Understanding

Reading Check Determine whether each statement is true or false.

RC1. To solve an exponential equation, we can take the common logarithm on both sides of the equation.

RC2. To solve an exponential equation, we can take the natural logarithm on both sides of the equation.

RC3. We cannot calculate the logarithm of a negative number.

RC4. The solution of a logarithmic equation is never a negative number.

Concept Check Determine whether each statement is true or false.

CC1. The solution of $2^x = 6$ is 3.

CC2. The solution of $2^x = 16$ is 4.

CC3. The solution of $\log_2 4 = x$ is 16.

CC4. The solution of $\log_x 8 = 3$ is 2.

CC5. The solution of $\log_8 x = 1$ is 8.

CC6. The solution of $\log_2 1 = x$ is 0.

a Solve.

1. $2^x = 8$
2. $3^x = 81$
3. $4^x = 256$
4. $5^x = 125$
5. $2^{2x} = 32$
6. $4^{3x} = 64$
7. $3^{5x} = 27$
8. $5^{7x} = 625$
9. $2^x = 11$
10. $2^x = 20$
11. $2^x = 43$
12. $2^x = 55$
13. $5^{4x-7} = 125$
14. $4^{3x+5} = 16$
15. $3^{x^2} \cdot 3^{4x} = \frac{1}{27}$
16. $3^{5x} \cdot 9^{x^2} = 27$
17. $4^x = 8$
18. $6^x = 10$
19. $e^t = 100$
20. $e^t = 1000$
21. $e^{-t} = 0.1$
22. $e^{-t} = 0.01$
23. $e^{-0.02t} = 0.06$
24. $e^{0.07t} = 2$
25. $2^x = 3^{x-1}$
26. $3^{x+2} = 5^{x-1}$
27. $(3.6)^x = 62$
28. $(5.2)^x = 70$

b Solve.

29. $\log_4 x = 4$
30. $\log_7 x = 3$
31. $\log_2 x = -5$
32. $\log_9 x = \frac{1}{2}$
33. $\log x = 1$
34. $\log x = 3$
35. $\log x = -2$
36. $\log x = -3$

37. $\ln x = 2$

38. $\ln x = 1$

39. $\ln x = -1$

40. $\ln x = -3$

41. $\log_3 (2x + 1) = 5$

42. $\log_2 (8 - 2x) = 6$

43. $\log x + \log (x - 9) = 1$

44. $\log x + \log (x + 9) = 1$

45. $\log x - \log (x + 3) = -1$

46. $\log (x + 9) - \log x = 1$

47. $\log_2 (x + 1) + \log_2 (x - 1) = 3$

48. $\log_2 x + \log_2 (x - 2) = 3$

49. $\log_4 (x + 6) - \log_4 x = 2$

50. $\log_4 (x + 3) - \log_4 (x - 5) = 2$

51. $\log_4 (x + 3) + \log_4 (x - 3) = 2$

52. $\log_5 (x + 4) + \log_5 (x - 4) = 2$

53. $\log_3 (2x - 6) - \log_3 (x + 4) = 2$

54. $\log_4 (2 + x) - \log_4 (3 - 5x) = 3$

Skill Maintenance

Solve.

55. $-3 \leq x - 12 < 4$ [1.5a]

56. $|2x - 5| > 3$ [1.6e]

57. $x^2 - x = 12$ [4.8a]

58. $x^2 - x \leq 12$ [7.8a]

59. $\sqrt{n - 1} = 8$ [6.6a]

60. $\sqrt{y - 3} = y - 5$ [6.6a]

Synthesis

61. Find the value of x for which the natural logarithm is the same as the common logarithm.

62. Use a graphing calculator to check your answers to Exercises 4, 20, 36, and 54.

63. Use a graphing calculator to solve each of the following equations.

a) $e^{7x} = 14$
b) $8e^{0.5x} = 3$
c) $xe^{3x-1} = 5$
d) $4 \ln (x + 3.4) = 2.5$

64. Use the INTERSECT feature of a graphing calculator to find the points of intersection of the graphs of each pair of functions.

a) $f(x) = e^{0.5x-7}, g(x) = 2x + 6$
b) $f(x) = \ln 3x, g(x) = 3x - 8$
c) $f(x) = \ln x^2, g(x) = -x^2$

Solve.

65. $2^{2x} + 128 = 24 \cdot 2^x$

66. $27^x = 81^{2x-3}$

67. $8^x = 16^{3x+9}$

68. $\log_x (\log_3 27) = 3$

69. $\log_6 (\log_2 x) = 0$

70. $x \log \frac{1}{8} = \log 8$

71. $\log_5 \sqrt{x^2 - 9} = 1$

72. $2^{x^2+4x} = \frac{1}{8}$

73. $\log (\log x) = 5$

74. $\log_5 |x| = 4$

75. $\log x^2 = (\log x)^2$

76. $\log_3 |5x - 7| = 2$

77. $\log_a a^{x^2+4x} = 21$

78. $\sqrt{x} \cdot \sqrt[3]{x} \cdot \sqrt[4]{x} \cdot \sqrt[5]{x} = 146$

79. $3^{2x} - 8 \cdot 3^x + 15 = 0$

80. If $x = (\log_{125} 5)^{\log_5 125}$, what is the value of $\log_3 x$?

8.7 Mathematical Modeling with Exponential Functions and Logarithmic Functions

OBJECTIVES

a Solve applied problems involving logarithmic functions.

b Solve applied problems involving exponential functions.

Exponential functions and logarithmic functions can now be added to our library of functions that can serve as models for many kinds of applications. Let's review some of their graphs.

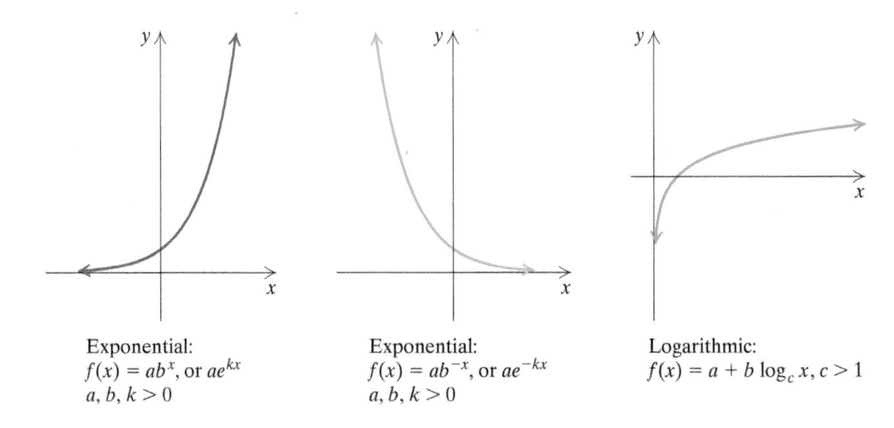

a APPLICATIONS OF LOGARITHMIC FUNCTIONS

EXAMPLE 1 *Sound Levels.* To measure the "loudness" of any particular sound, the decibel scale is used. The loudness L, in decibels (dB), of a sound is given by

$$L = 10 \cdot \log \frac{I}{I_0},$$

where I is the intensity of the sound, in watts per square meter (W/m^2), and $I_0 = 10^{-12}\ W/m^2$. (I_0 is approximately the intensity of the softest sound that can be heard.)

a) An iPod can produce sounds of more than $10^{-0.5}\ W/m^2$, a volume that can damage the hearing of a person exposed to the sound for more than 28 sec. How loud, in decibels, is this sound level?

b) Audiologists and physicians recommend that earplugs be worn when one is exposed to sounds in excess of 90 dB. What is the intensity of such sounds?

Data: American Speech-Language-Hearing Association

a) To find the loudness, in decibels, we use the above formula:

$$\begin{aligned} L &= 10 \cdot \log \frac{I}{I_0} \\ &= 10 \cdot \log \frac{10^{-0.5}}{10^{-12}} && \text{Substituting} \\ &= 10 \cdot \log 10^{11.5} && \text{Subtracting exponents} \\ &= 10 \cdot 11.5 && \log 10^a = a \\ &= 115. \end{aligned}$$

The sound level is 115 decibels.

b) We substitute and solve for I:

$$L = 10 \cdot \log \frac{I}{I_0}$$

$$90 = 10 \cdot \log \frac{I}{10^{-12}} \quad \text{Substituting}$$

$$9 = \log \frac{I}{10^{-12}} \quad \text{Dividing by 10}$$

$$9 = \log I - \log 10^{-12} \quad \text{Using Property 3}$$

$$9 = \log I - (-12) \quad \log 10^a = a$$

$$-3 = \log I \quad \text{Adding } -12$$

$$10^{-3} = I. \quad \text{Converting to an exponential equation}$$

Earplugs are recommended for sounds with intensities that exceed $10^{-3}\ \mathrm{W/m^2}$.

Do Exercises 1 and 2. ▶

EXAMPLE 2 *pH.* In chemistry, the pH of a liquid is defined as

$$\mathrm{pH} = -\log\,[\mathrm{H^+}],$$

where $[\mathrm{H^+}]$ is the hydrogen ion concentration in moles per liter.

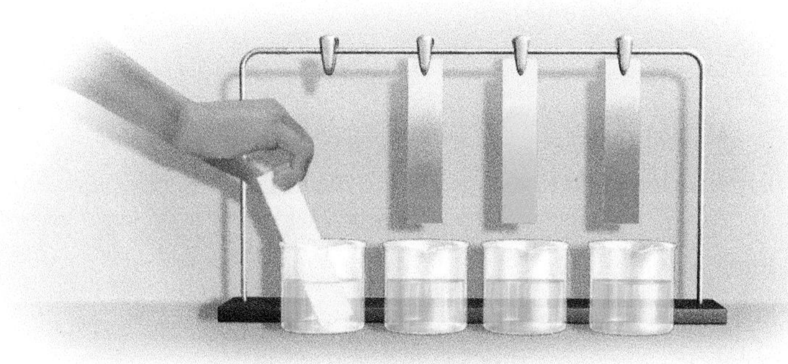

a) The hydrogen ion concentration of human blood is normally about 3.98×10^{-8} moles per liter. Find the pH.

b) The pH of seawater is about 8.3. Find the hydrogen ion concentration.

a) To find the pH of human blood, we use the above formula:

$$\mathrm{pH} = -\log\,[\mathrm{H^+}] = -\log\,[3.98 \times 10^{-8}]$$

$$\approx -(-7.400117) \approx 7.4. \quad \text{Using a calculator and rounding}$$

The pH of human blood is normally about 7.4.

b) We substitute and solve for $[\mathrm{H^+}]$:

$$8.3 = -\log\,[\mathrm{H^+}] \quad \text{Using pH} = -\log\,[\mathrm{H^+}]$$

$$-8.3 = \log\,[\mathrm{H^+}] \quad \text{Dividing by } -1$$

$$10^{-8.3} = [\mathrm{H^+}] \quad \text{Converting to an exponential equation}$$

$$5.01 \times 10^{-9} \approx [\mathrm{H^+}]. \quad \text{Using a calculator; writing scientific notation}$$

The hydrogen ion concentration of seawater is about 5.01×10^{-9} moles per liter.

Do Exercises 3 and 4. ▶

1. *Acoustics.* The intensity of sound in normal conversation is about $3.2 \times 10^{-6}\ \mathrm{W/m^2}$. How high is this sound level in decibels?

$$L = 10 \cdot \log \frac{I}{I_0}$$

$$= 10 \cdot \log \frac{\quad}{10^{-12}}$$

$$= 10 \cdot \log\,(3.2 \times 10^{\quad})$$

$$= 10 \cdot (\log 3.2 + \log 10^{6})$$

$$\approx 10 \cdot (0.5051 + \quad)$$

$$\approx 65$$

The sound level is about ______ decibels.

2. *Audiology.* Overexposure to excessive sound levels can diminish one's hearing to the point where the softest sound that is audible is 28 dB. What is the intensity of such a sound?

3. *Coffee.* The hydrogen ion concentration of freshly brewed coffee is about 1.3×10^{-5} moles per liter. Find the pH.

4. *Acidosis.* When the pH of a patient's blood drops below 7.4, a condition called *acidosis* sets in. Acidosis can be fatal at a pH level of 7. What would the hydrogen ion concentration of the patient's blood be at that point?

$$\mathrm{pH} = -\log\,[\mathrm{H^+}]$$

$$\quad = -\log\,[\mathrm{H^+}]$$

$$\quad = \log\,[\mathrm{H^+}]$$

$$10^{\quad} = [\mathrm{H^+}]$$

Answers

1. About 65 decibels **2.** $10^{-9.2}\ \mathrm{W/m^2}$
3. About 4.9 **4.** 10^{-7} moles per liter

Guided Solutions:

1. 3.2×10^{-6}, 6, 6, 65 **4.** 7, −7, −7

APPLICATIONS OF EXPONENTIAL FUNCTIONS

SKILL REVIEW

Solve logarithmic equations. [8.6b]

Solve.

1. $\log x = 2$

2. $\ln x = -2$

Answers: 1. 100 **2.** $e^{-2} \approx 0.1353$

EXAMPLE 3 *Interest Compounded Annually.* Suppose that $30,000 is invested at 4% interest, compounded annually. In t years, it will grow to the amount A given by the function

$$A(t) = 30{,}000(1.04)^t.$$

(See Example 6 in Section 8.1.)

a) How long will it take to accumulate $150,000 in the account?

b) Let $T =$ the amount of time it takes for the $30,000 to double itself; T is called the **doubling time**. Find the doubling time.

a) We set $A(t) = 150{,}000$ and solve for t:

$$150{,}000 = 30{,}000(1.04)^t$$

$$\frac{150{,}000}{30{,}000} = (1.04)^t \qquad \text{Dividing by 30,000}$$

$$5 = (1.04)^t$$

$$\log 5 = \log (1.04)^t \qquad \text{Taking the common logarithm on both sides}$$

$$\log 5 = t \log 1.04 \qquad \text{Using Property 2}$$

$$\frac{\log 5}{\log 1.04} = t \qquad \text{Dividing by log 1.04}$$

$$41.04 \approx t. \qquad \text{Using a calculator}$$

It will take about 41 years for the $30,000 to grow to $150,000.

b) To find the *doubling time* T, we replace $A(t)$ with 60,000 and t with T and solve for T:

$$60{,}000 = 30{,}000(1.04)^T$$

$$2 = (1.04)^T \qquad \text{Dividing by 30,000}$$

$$\log 2 = \log (1.04)^T \qquad \text{Taking the common logarithm on both sides}$$

$$\log 2 = T \log 1.04 \qquad \text{Using Property 2}$$

$$T = \frac{\log 2}{\log 1.04} \approx 17.7. \qquad \text{Using a calculator}$$

The doubling time is about 17.7 years.

Do Exercise 5.

5. *Interest Compounded Annually.* Suppose that $40,000 is invested at 4.3% interest, compounded annually.

a) After what amount of time will there be $250,000 in the account?

b) Find the doubling time.

Answers

5. (a) 43.5 years; **(b)** 16.5 years

The function in Example 3 illustrates exponential growth. Populations often grow exponentially according to the following model.

EXPONENTIAL GROWTH MODEL

An **exponential growth model** is a function of the form

$$P(t) = P_0e^{kt}, \quad k > 0,$$

where P_0 is the population at time 0, $P(t)$ is the population at time t, and k is the **exponential growth rate** for the situation. The **doubling time** is the amount of time necessary for the population to double in size.

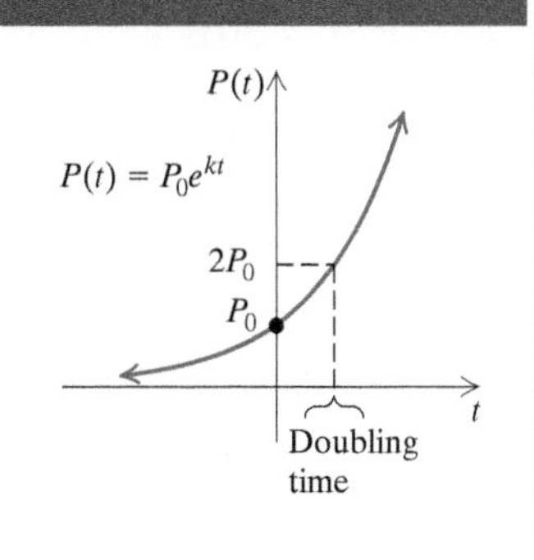

The exponential growth rate is the rate of growth of a population at any *instant* in time. Since the population is continually growing, the percent of total growth after one year will exceed the exponential growth rate.

EXAMPLE 4 *Population Growth.* In 2016, Luxembourg's population was 0.6 million, and the exponential growth rate was 2.05% per year.

Data: Central Intelligence Agency

a) Find the exponential growth function.

b) What will the population be in 2030?

a) The given information allows us to create a model. At $t = 0$ (2016), the population was 0.6 million. We substitute 0.6 for P_0 and 2.05%, or 0.0205, for k to obtain the exponential growth function:

$$P(t) = P_0e^{kt}$$
$$= 0.6e^{0.0205t}.$$

Here $P(t)$ is in millions and t is the number of years after 2016.

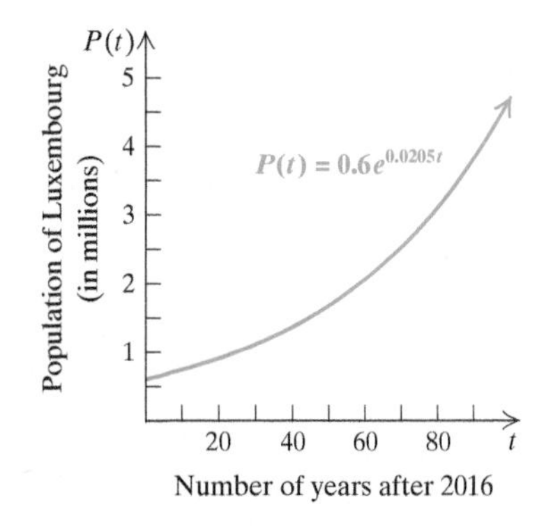

b) In 2030, we have $t = 14$. That is, 14 years have passed since 2016. To find the population in 2030, we substitute 14 for t:

$$P(14) = 0.6e^{0.0205(14)} \quad \text{Substituting 14 for } t$$
$$\approx 0.8. \quad \text{Using a calculator}$$

The population of Luxembourg will be about 0.8 million in 2030.

Do Exercise 6. ▶

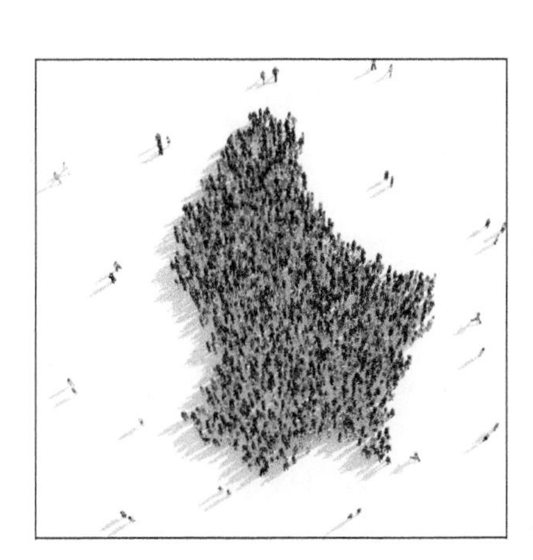

6. *Population Growth.* Refer to Example 4. What will the population of Luxembourg be in 2040?

EXAMPLE 5 *Interest Compounded Continuously.* Suppose that an amount of money P_0 is invested in a savings account at interest rate k, compounded continuously. That is, suppose that interest is computed every "instant" and added to the amount in the account. The balance $P(t)$, after t years, is given by the exponential growth model

$$P(t) = P_0e^{kt}.$$

a) Suppose that $30,000 is invested and grows to $34,855.03 in 5 years. Find the interest rate and then the exponential growth function.

Answer

6. About 1 million

b) What is the balance after 10 years?

c) What is the doubling time?

a) We have $P_0 = 30{,}000$. Thus the exponential growth function is

$$P(t) = 30{,}000e^{kt},$$

where k must still be determined. We know that $P(5) = 34{,}855.03$. We substitute and solve for k:

$$34{,}855.03 = 30{,}000e^{k(5)} = 30{,}000e^{5k}$$

$$\frac{34{,}855.03}{30{,}000} = e^{5k} \qquad \text{Dividing by 30,000}$$

$$1.161834 \approx e^{5k}$$

$$\ln 1.161834 = \ln e^{5k} \qquad \text{Taking the natural logarithm on both sides}$$

$$0.15 \approx 5k \qquad \text{Finding ln 1.161834 on a calculator and simplifying } \ln e^{5k}$$

$$\frac{0.15}{5} = 0.03 \approx k.$$

The interest rate is about 0.03, or 3%, compounded continuously. Note that since interest is being compounded continuously, the interest earned each year is more than 3%. The exponential growth function is

$$P(t) = 30{,}000e^{0.03t}.$$

b) We substitute 10 for t:

$$P(10) = 30{,}000e^{0.03(10)} \approx \$40{,}495.76.$$

The balance in the account after 10 years will be $40,495.76.

c) To find the doubling time T, we replace $P(t)$ with 60,000 and solve for T:

$$60{,}000 = 30{,}000e^{0.03T}$$

$$2 = e^{0.03T} \qquad \text{Dividing by 30,000}$$

$$\ln 2 = \ln e^{0.03T} \qquad \text{Taking the natural logarithm on both sides}$$

$$\ln 2 = 0.03T$$

$$\frac{\ln 2}{0.03} = T \qquad \text{Dividing}$$

$$23.1 \approx T.$$

Thus the original investment of $30,000 will double in about 23.1 years, as shown in the following graph of the growth function.

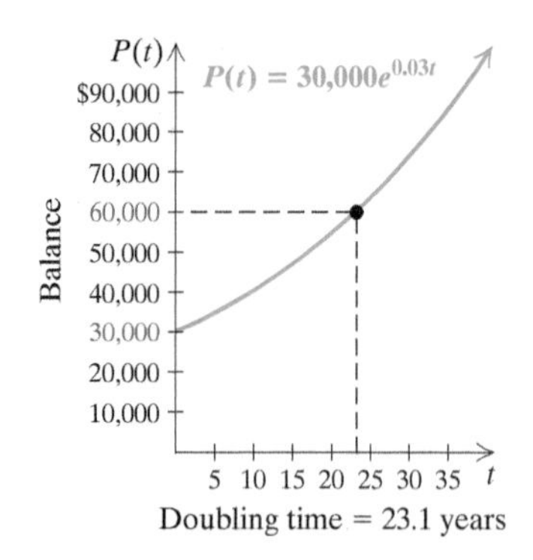

Doubling time = 23.1 years

Do Exercise 7.

7. *Interest Compounded Continuously.*

a) Suppose that $5000 is invested and grows to $6356.25 in 4 years. Find the interest rate and then the exponential growth function.

b) What is the balance after 1 year? 2 years? 10 years?

c) What is the doubling time?

Answers

7. (a) $k = 6\%$, $P(t) = 5000e^{0.06t}$;
(b) $5309.18; $5637.48; $9110.59;
(c) about 11.6 years

EXAMPLE 6 *Solar Power.* The amount of electricity generated from the sun by utilities in the United States has increased exponentially from 864 thousand MGH (megawatt hours) in 2008 to 36,755 thousand MGH in 2016, as shown in the following graph.

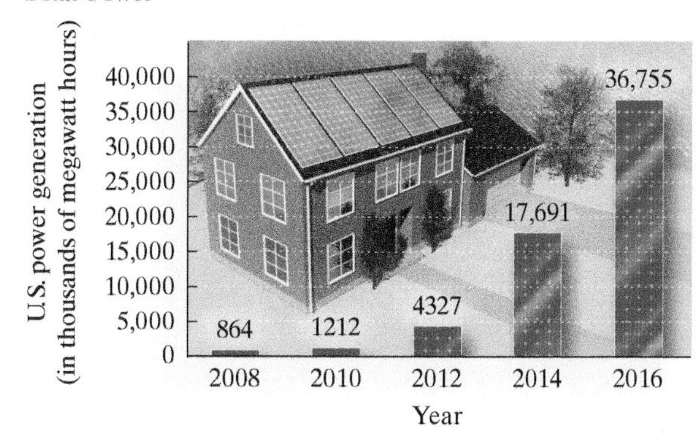

DATA: U.S. Energy Information Administration

a) Let $t =$ the number of years after 2008. Then $t = 0$ corresponds to 2008 and $t = 8$ corresponds to 2016. Use the data points $(0, 864)$ and $(8, 36{,}755)$ to find the exponential growth rate and then the exponential growth function.

b) Use the function found in part (a) to predict the amount of solar power generated in the United States in 2020.

c) Use the function to determine the year in which there will be about 1 billion MGH of solar power generated.

a) We use the equation $P(t) = P_0e^{kt}$, where $P(t)$ is the amount of solar power generated t years after 2008, in thousands of MGH. In 2008, at $t = 0$, there was 864 thousand MGH of solar power generated. Thus we substitute 864 for P_0:

$$P(t) = 864e^{kt}.$$

To find the exponential growth rate k, we note that 8 years later, in 2016, there was 36,755 thousand MGH of solar power generated. We substitute and solve for k:

$$\begin{aligned} P(8) &= 864e^{k(8)} && \text{Substituting} \\ 36{,}755 &= 864e^{k(8)} \\ \frac{36{,}755}{864} &= e^{8k} && \text{Dividing by 864} \\ \ln \frac{36{,}755}{864} &= \ln e^{8k} && \text{Taking the natural logarithm on both sides} \\ 3.750456781 &= 8k && \ln e^a = a \\ 0.469 &\approx k. \end{aligned}$$

The exponential growth rate is 0.469, or 46.9%, and the exponential growth function is $P(t) = 864e^{0.469t}$, where t is the number of years after 2008.

b) Since 2020 is 12 years after 2008, we substitute 12 for t:

$$P(12) = 864e^{0.469(12)} \approx 240{,}283.$$

The amount of solar power generated in 2020 will be about 240,283 thousand MGH.

c) To determine the year in which 1 billion MGH of solar power will be generated, we first note that 1 billion is 1 million • 1 thousand. Thus we can substitute 1,000,000 for $P(t)$ and solve for t:

$$1{,}000{,}000 = 864e^{0.469t}$$

$$\frac{1{,}000{,}000}{864} = e^{0.469t} \quad \text{Dividing by 864}$$

$$\ln \frac{1{,}000{,}000}{864} = \ln e^{0.469t} \quad \text{Taking the natural logarithm on both sides}$$

$$7.053937789 \approx 0.469t \quad \ln e^a = a$$

$$15 \approx t.$$

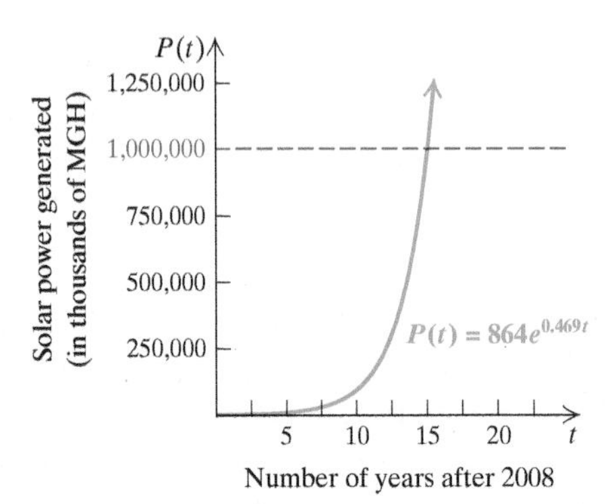

We see that, according to this model, 15 years after 2008, or in 2023, there will be about 1,000,000 thousand, or 1 billion, MGH of solar power generated in the United States.

Do Exercise 8.

In some real-life situations, a quantity or a population is *decreasing* or *decaying* exponentially.

EXPONENTIAL DECAY MODEL

An **exponential decay model** is a function of the form

$$P(t) = P_0 e^{-kt}, \quad k > 0,$$

where P_0 is the quantity present at time 0, $P(t)$ is the amount present at time t, and k is the **decay rate**. The **half-life** is the amount of time necessary for half of the quantity to decay.

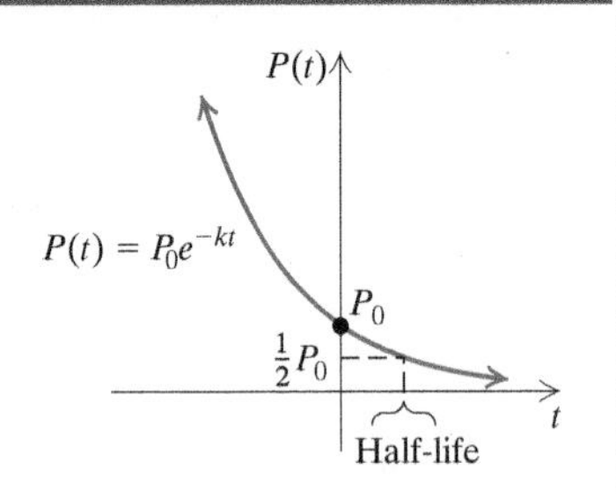

8. *Global Mobile Data Traffic.* The amount of data transferred using mobile devices is expected to increase exponentially from 7.2 exabytes per month in 2016 to 49 exabytes per month in 2021.

Data: Cisco

a) Let $P(t) = P_0 e^{kt}$, where $P(t)$ is the global mobile data traffic, in exabytes per month, t years after 2016. Then $t = 0$ corresponds to 2016 and $t = 5$ corresponds to 2021. Use the data points $(0, 7.2)$ and $(5, 49)$ to find the exponential growth rate and then the exponential growth function.

b) Use the function found in part (a) to estimate the global mobile data traffic per month in 2019.

c) Assuming exponential growth continues at the same rate, predict the year in which the global mobile data traffic will be 100 exabytes per month.

Answers

8. **(a)** $k \approx 0.384$; $P(t) = 7.2e^{0.384t}$; **(b)** about 22.8 exabytes per month; **(c)** 2023

EXAMPLE 7 *Carbon Dating.* The radioactive element carbon-14 has a half-life of 5750 years. The percentage of carbon-14 in the remains of organic matter can be used to determine the age of that material. Recently, near Patuxent River, Maryland, archaeologists discovered charcoal that had lost 8.1% of its carbon-14. The age of this charcoal was evidence that this is the oldest dwelling ever discovered in Maryland. What was the age of the charcoal?

Data: Roylance, Frank D., "Digging Where Indians Camped Before Columbus," *The Baltimore Sun*, July 2, 2009

Reprinted with permission of The Baltimore Sun Media Group. All rights reserved.

We first find k. To do so, we use the concept of half-life. When $t = 5750$ (the half-life), $P(t)$ will be half of P_0. Then

$$0.5P_0 = P_0 e^{-k(5750)}$$

$$0.5 = e^{-5750k} \quad \text{Dividing by } P_0 \text{ on both sides}$$

$$\ln 0.5 = \ln e^{-5750k} \quad \text{Taking the natural logarithm on both sides}$$

$$\ln 0.5 = -5750k$$

$$\frac{\ln 0.5}{-5750} = k$$

$$0.00012 \approx k.$$

Now we have a function for the decay of carbon-14:

$$P(t) = P_0 e^{-0.00012t}. \quad \text{This completes the first part of our solution.}$$

(*Note:* This equation can be used for any subsequent carbon-dating problem.)

If the charcoal has lost 8.1% of its carbon-14 from an initial amount P_0, then 100% − 8.1%, or 91.9%, of P_0 is still present. To find the age t of the charcoal, we solve this equation for t:

$$0.919P_0 = P_0 e^{-0.00012t} \quad \text{We want to find } t \text{ for which } P(t) = 0.919P_0.$$

$$0.919 = e^{-0.00012t} \quad \text{Dividing by } P_0 \text{ on both sides}$$

$$\ln 0.919 = \ln e^{-0.00012t} \quad \text{Taking the natural logarithm on both sides}$$

$$\ln 0.919 = -0.00012t \quad \ln e^a = a$$

$$\frac{\ln 0.919}{-0.00012} = t \quad \text{Dividing by } -0.00012 \text{ on both sides}$$

$$700 \approx t. \quad \text{Using a calculator and rounding}$$

The charcoal is about 700 years old.

Do Exercise 9. ▶

9. *Carbon Dating.* In Chaco Canyon, New Mexico, archaeologists found corn pollen that had lost 38.1% of its carbon-14. What was the age of the pollen?

Answer

9. About 4000 years

Translating for Success

1. *Grain Flow.* Grain flows through spout A four times as fast as through spout B. When grain flows through both spouts, a grain bin is filled in 8 hr. How many hours would it take to fill the bin if grain flows through spout B alone?

2. *Rectangle Dimensions.* The perimeter of a rectangle is 50 ft. The width of the rectangle is 10 ft shorter than the length. Find the length and the width.

3. *Wire Cutting.* A 1086-in. wire is cut into three pieces. The second piece is 8 in. longer than the first. The third is four-fifths as long as the first. How long is each piece?

4. *Movies.* In 2015, there were 30,000 movie screens in China, and this number was increasing exponentially at a rate of 20.7% per year. Write an exponential growth function M for which $M(t)$ approximates the number of movie screens in China t years after 2015.

5. *Charitable Contributions.* In 2017, Kylie donated \$500 to charities. This was an 8% increase over her donations in 2015. How much did Kylie donate to charities in 2015?

The goal of these matching questions is to practice step (2), Translate, *of the five-step problem-solving process. Translate each word problem to an equation or a system of equations and select a correct translation from equations A–O.*

A. $M(t) = 30{,}000e^{0.207t}$

B. $40x = 50(x - 3)$

C. $x^2 + (x - 10)^2 = 50^2$

D. $\dfrac{8}{x} + \dfrac{8}{4x} = 1$

E. $x + 8\%x = 500$

F. $\dfrac{500}{x} + \dfrac{500}{x - 2} = 8$

G. $x + y = 90,$
$0.1x + 0.25y = 16.50$

H. $x + (x + 1) + (x + 2) = 39$

I. $x + (x + 8) + \dfrac{4}{5}x = 1086$

J. $x + (x + 2) + (x + 4) = 39$

K. $M(t) = 20.7e^{30{,}000t}$

L. $x^2 + (x + 8)^2 = 1086$

M. $2x + 2(x - 10) = 50$

N. $\dfrac{500}{x} = \dfrac{500}{x + 2} + 8$

O. $x + y = 90,$
$0.1x + 0.25y = 1650$

Answers on page A-34

6. *Uniform Numbers.* The numbers on three baseball uniforms are consecutive integers whose sum is 39. Find the integers.

7. *Triangle Dimensions.* The hypotenuse of a right triangle is 50 ft. The length of one leg is 10 ft shorter than the other. Find the lengths of the legs.

8. *Coin Mixture.* A collection of dimes and quarters is worth \$16.50. There are 90 coins in all. How many of each coin are there?

9. *Car Travel.* Emma drove her car 500 mi to see her friend. The return trip was 2 hr faster at a speed that was 8 mph more. Find her return speed.

10. *Train Travel.* An Amtrak train leaves a station and travels east at 40 mph. Three hours later, a second train leaves on a parallel track traveling east at 50 mph. After what amount of time will the second train overtake the first?

8.7 Exercise Set

FOR EXTRA HELP MyLab Math

 Check Your Understanding

Reading Check For the exponential growth model $P(t) = P_0 e^{kt}$, $k > 0$, match each variable with its description.

RC1. ___ k

RC2. ___ $P(t)$

RC3. ___ P_0

RC4. ___ T, where $2P_0 = P_0 e^{kT}$

a) Doubling time

b) Exponential growth rate

c) Population at time 0

d) Population at time t

Concept Check A population numbering 550 in 2018 grew at an annual exponential growth rate of 13%. Use the model $P(t) = P_0 e^{kt}$ to find the population in 2020.

CC1. What value should you substitute for P_0?

CC2. What value should you substitute for k?

CC3. What value should you substitute for t?

CC4. What is the population in 2020?

a Solve.

Sound Levels. Use the decibel formula from Example 1 for Exercises 1–4.

1. *Blue Whale.* The blue whale is not only the largest animal on earth, it is also the loudest. The call of a blue whale can reach an intensity of $10^{6.8}$ W/m². What is this sound level, in decibels?

2. *Jackhammer Noise.* A jackhammer can generate sound measurements of 130 dB. What is the intensity of such sounds?

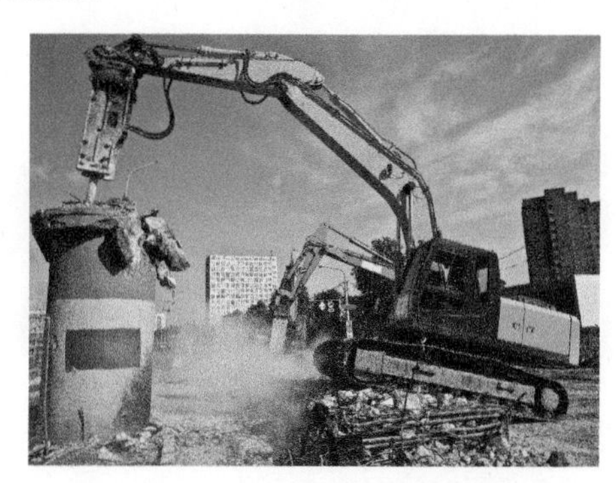

3. *Dishwasher Noise.* A top-of-the-line dishwasher, built to muffle noise, has a sound measurement of 45 dB. A less-expensive dishwasher can have a sound measurement of 60 dB. What is the intensity of each sound?

4. *Sound of an Alarm Clock.* The intensity of sound of an alarm clock is 10^{-4} W/m². What is this sound level, in decibels?

pH. Use the pH formula from Example 2 for Exercises 5–8.

5. *Milk.* The hydrogen ion concentration of milk is about 1.6×10^{-7} moles per liter. Find the pH.

6. *Mouthwash.* The hydrogen ion concentration of mouthwash is about 6.3×10^{-7} moles per liter. Find the pH.

7. *Alkalosis.* When the pH of a person's blood rises above 7.4, a condition called *alkalosis* sets in. Alkalosis can be fatal at a pH level above 7.8. What would the hydrogen ion concentration of the person's blood be at that point?

8. *Orange Juice.* The pH of orange juice is 3.2. What is its hydrogen ion concentration?

Walking Speed. In a study by psychologists Bornstein and Bornstein, it was found that the average walking speed w, in feet per second, of a person living in a city of population P, in thousands, is given by the function

$$w(P) = 0.37 \ln P + 0.05.$$

Data: *International Journal of Psychology*

In Exercises 9–12, various cities and their populations are given. Find the walking speed of people in each city.

9. Minneapolis, Minnesota: 382,600

10. Tallahassee, Florida: 189,900

11. New York City, New York: 8,537,700

12. Phoenix, Arizona: 1,615,100

b Solve.

13. *Otter Population.* Due primarily to the irresponsible use of chemicals, England's otter population declined dramatically during the twentieth century until otters could be found on only a small percentage of riverbanks. After chemical bans were put in place, the number of English riverside sites $R(t)$ occupied by otters rose significantly and can be approximated by the exponential function

$$R(t) = 294(1.074)^t,$$

where t is the number of years since 1985.

Data: Nicolson, Adam, "The Sultans of Streams," *National Geographic* **223**(2), February 2013, pp. 124–134.

a) How many riverbanks were occupied by otters in 2010?
b) In what year were there otters on 1000 riverbanks?
c) What is the doubling time for the number of English riverbanks occupied by otters?

14. *Code-School Graduates.* A need for software developers has led to the rise of code schools, which typically offer 12-week training courses in coding skills. The annual number of code-school graduates $G(t)$ in the United States is increasing exponentially and can be approximated by the exponential function

$$G(t) = 9000(2)^t,$$

where t is the number of years after 2015.

Data: *The Wall Street Journal*, 2/26/2017

a) How many code-school graduates were there in 2017?
b) In what year will there be 144,000 code-school graduates?
c) What is the doubling time for the annual number of code-school graduates?

15. *Drug Overdose.* Although primarily used for relief of pain, opioids have addictive properties that often make withdrawal difficult. Deaths through unintentional opioid overdose have risen sharply across the United States since the early 2000s. Ohio, in particular, has an exceptionally high rate of opioid overdose death. The number of deaths in Ohio in year t, where t is the number of years after 2003, can be approximated by

$$D(t) = 318(1.19)^t.$$

Data: *The Wall Street Journal*, 6/7/2017

a) How many unintentional opioid overdose deaths were there in Ohio in 2013?
b) In what year were there about 3000 unintentional opioid overdose deaths in Ohio?
c) What is the doubling time of unintentional opioid overdose deaths in Ohio?

16. *Salvage Value.* A color photocopier is purchased for \$4800. Its value each year is about 70% of its value in the preceding year. Its salvage value, in dollars, after t years is given by the exponential function

$$V(t) = 4800(0.7)^t.$$

a) Find the salvage value of the copier after 3 years.
b) After what amount of time will the salvage value be \$1200?
c) After what amount of time will the salvage value be half the original value?

Growth. Use the exponential growth model $P(t) = P_0e^{kt}$ for Exercises 17–22.

17. *Interest Compounded Continuously.* Suppose that P_0 is invested in a savings account in which interest is compounded continuously at 3% per year.

a) Express $P(t)$ in terms of P_0 and 0.03.
b) Suppose that $5000 is invested. What is the balance after 1 year? 2 years? 10 years?
c) When will the investment of $5000 double itself?

18. *Interest Compounded Continuously.* Suppose that P_0 is invested in a savings account in which interest is compounded continuously at 5.4% per year.

a) Express $P(t)$ in terms of P_0 and 0.054.
b) Suppose that $10,000 is invested. What is the balance after 1 year? 2 years? 10 years?
c) When will the investment of $10,000 double itself?

19. *World Population Growth.* In 2017, the population of the world reached 7.5 billion, and the exponential growth rate was 1.06% per year.

Data: *CIA World Factbook*

a) Find the exponential growth function.
b) What will the world population be in 2025?
c) In what year will the world population reach 15 billion?
d) What is the doubling time of the world population?

20. *Population Growth.* In 2017, the population of the United States was 324 million, and the exponential growth rate was 0.81% per year.

Data: *CIA World Factbook*

a) Find the exponential growth function.
b) What will the U.S. population be in 2025?
c) In what year will the U.S. population reach 350 million?
d) What is the doubling time of the U.S. population?

21. *Employment.* The number of full-time and part-time employees of the online retailer Amazon increased exponentially from 56 thousand in 2011 to 306.8 thousand in 2016.

Data: *USA Today*, 1/17/2017

a) Let $t = 0$ correspond to 2011 and $t = 5$ correspond to 2016. Then t is the number of years after 2011. Use the data points (0, 56) and (5, 306.8) to find the exponential growth rate and fit an exponential growth function $P(t) = P_0e^{kt}$ to the data, where $P(t)$ is the number of Amazon employees, in thousands, t years after 2011.
b) Use the function found in part (a) to estimate the number of Amazon employees in 2018.
c) According to this model, when will there be one million Amazon employees?

22. *U.S. Tax Code.* Over the years, the U.S. tax code has become increasingly complex. For example, the length of the instruction book for the 1040 tax form increased exponentially from 2 pages in 1935 to 211 pages in 2015.

Data: National Taxpayers Union

a) Let $t = 0$ correspond to 1935 and $t = 80$ correspond to 2015. Then t is the number of years after 1935. Use the data points (0, 2) and (80, 211) to find the exponential growth rate and fit an exponential growth function $C(t) = C_0e^{kt}$ to the data, where $C(t)$ is the number of pages in the instruction book.
b) Use the function found in part (a) to estimate the total number of pages in the instruction book in 2020.
c) When will there be 300 pages in the instruction book?

Carbon Dating. Use the carbon-14 decay function $P(t) = P_0e^{-0.00012t}$ for Exercises 23 and 24.

23. *Carbon Dating.* When archaeologists found the Dead Sea scrolls, they determined that the linen wrapping had lost 22.3% of its carbon-14. How old was the linen wrapping?

24. *Carbon Dating.* In 2005, researchers were able to start a date tree from a seed found at Masada. The seed had lost 21.3% of its carbon-14. How old was the seed?

Decay. Use the exponential decay function $P(t) = P_0e^{-kt}$ for Exercises 25 and 26.

25. *Chemistry.* The decay rate of iodine-131 is 9.6% per day. What is the half-life?

26. *Chemistry.* The decay rate of krypton-85 is 6.3% per day. What is the half-life?

27. *Home Construction.* The chemical urea formaldehyde was found in some insulation used in houses built during the mid to late 1960s. Unknown at the time was the fact that urea formaldehyde emitted toxic fumes as it decayed. The half-life of urea formaldehyde is 1 year. What is its decay rate?

28. *Plumbing.* Lead pipes and solder are often found in older buildings. Unfortunately, as lead decays, toxic chemicals can get in the water resting in the pipes. The half-life of lead is 22 years. What is its decay rate?

29. *Physical Music Sales.* The number of albums sold in a physical format (such as a compact disc) has decreased exponentially from 451 million albums in 2007 to 118 million albums in 2016.

Data: Nielsen Music Industry Reports

a) Find the exponential decay rate, and write an exponential function that represents the number of albums $A(t)$, in millions, sold in a physical format t years after 2007.

b) Estimate the number of albums sold in a physical format in 2020.

c) In what year were there 300 million albums sold in a physical format?

30. *Covered Bridges.* There were as many as 15,000 covered bridges in the United States in the 1800s. Now their number is decreasing exponentially, partly as a result of vandalism. In 1965, there were 1156 covered bridges, but by 2017, only 804 covered bridges remained.

Data: National Society for the Preservation of Covered Bridges; woodcenter.org

a) Find the exponential decay rate, and write an exponential function B that represents the number of covered bridges t years after 1965.

b) Estimate the number of covered bridges in 2002.

c) In what year were there 1000 covered bridges?

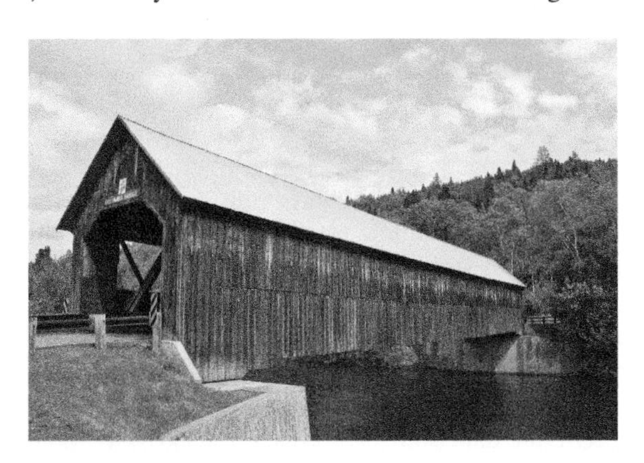

31. *Population Decline.* The population of Cleveland, Ohio, declined from 396,000 in 2010 to 390,000 in 2017. Assume that the population decreases according to the exponential decay model.

Data: worldpopulationreview.com

a) Find the exponential decay rate, and write an exponential function that represents the population of Cleveland t years after 2010.

b) Estimate the population of Cleveland in 2022.

c) In what year will the population of Cleveland reach 380,000?

32. *Online Shopping.* Revenue from online grocery shopping in the United States is expected to increase exponentially from \$35 billion in 2014 to \$175 billion in 2021.

Data: *The Wall Street Journal*, 6/19/2017

a) Find the exponential growth rate, and write a function that represents revenue from online grocery shopping t years after 2014.

b) Estimate the revenue from online grocery shopping in 2018.

c) In what year will revenue from online grocery shopping reach \$300 billion?

33. *Contemporary Art.* In 2017, Jean-Michel Basquiat's "Untitled" was sold at auction for \$110.5 million. The painting had previously been sold for \$19,000 in 1984. Assume that the painting's value increases exponentially.

Data: news.artnet.com, 4/19/2017; *The New York Times*, 5/18/2017

a) Find the exponential growth rate k, and determine the exponential growth function that can be used to estimate the painting's value $V(t)$, in dollars, t years after 1984.
b) Estimate the value of the painting in 2020.
c) What is the doubling time for the value of the painting?
d) How long after 1984 will the value of the painting be \$1 billion?

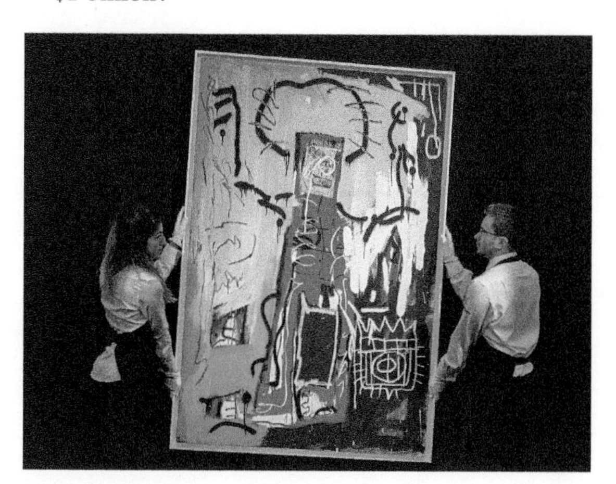

34. *Classic Cars.* In 2014, a 1962 Ferrari 250 GTO (chassis 3851GT) was sold at auction for a record \$38 million. The car had previously been sold for \$4000 in 1965. Assume that the car's value increases exponentially.

Data: *Los Angeles Times*, 8/14/2014

a) Find the exponential growth rate k, and determine the exponential growth function that can be used to estimate the car's value $V(t)$, in dollars, t years after 1965.
b) Estimate the value of the car in 2020.
c) What is the doubling time for the value of the car?
d) How many years after 1965 will the value of the car be \$150 million?

Skill Maintenance

Solve.

35. $5x + 6y = -2,$
$3x + 10y = 2$ [3.2a], [3.3a]

36. $x + y - z = 0,$
$3x + y + z = 6,$
$x - y + 2z = 5$ [3.5a]

37. $x^2 + 2x + 3 = 0$ [7.2a]

38. $\frac{6}{x} + \frac{6}{x+2} = \frac{5}{2}$ [5.5a]

39. $\frac{7}{x^2 - 5x} - \frac{2}{x - 5} = \frac{4}{x}$ [5.5a]

40. $15x^2 + 45 = 0$ [7.1a]

Synthesis

Use a graphing calculator to solve each of the following equations.

41. $2^x = x^{10}$

42. $(\ln 2)x = 10 \ln x$

43. $x^2 = 2^x$

44. $x^3 = e^x$

45. *Nuclear Energy.* Plutonium-239 (Pu-239) is used in nuclear energy plants. The half-life of Pu-239 is 24,360 years. How long will it take for a fuel rod of Pu-239 to lose 90% of its radioactivity?

Data: *Microsoft Encarta 97 Encyclopedia*

46. *Population Growth.* In 2016, the population of China was 1.4 billion, and the exponential growth rate was 0.43%. In the same year, the population of India was 1.3 billion, and the exponential growth rate was 1.19%.

Data: *CIA World Factbook*

a) Find an exponential growth function $C(t)$ that can be used to model the population of China t years after 2016.
b) Find an exponential growth function $I(t)$ that can be used to model the population of India t years after 2016.
c) Use a graphing calculator to estimate the year in which the population of China and the population of India will be the same.

CHAPTER

8 Summary and Review

Key Formulas

Exponential Growth: $P(t) = P_0e^{kt}$

Exponential Decay: $P(t) = P_0e^{-kt}$

Carbon Dating: $P(t) = P_0e^{-0.00012t}$

Interest Compounded Annually: $A = P(1 + r)^t$

Interest Compounded n Times per Year: $A = P\left(1 + \frac{r}{n}\right)^{nt}$

Interest Compounded Continuously: $P(t) = P_0e^{kt}$, where P_0 dollars are invested for t years at interest rate k

Vocabulary Reinforcement

Complete each statement with the correct term from the column on the right. Some of the choices may not be used.

1. The function given by $f(x) = 6^x$ is an example of a(n) ______________ function. [8.1a]
2. The ______________ of a function given by a set of ordered pairs is found by interchanging the first and second coordinates in each ordered pair. [8.2a]
3. When interest is paid on interest previously earned, it is called ______________ interest. [8.1c]
4. Base-10 logarithms are called ______________ logarithms. [8.3d]
5. The logarithm of a number is a(n) ______________. [8.3a]
6. A quantity's ______________ is the amount of time necessary for half of the quantity to decay. [8.7b]

exponential
logarithmic
common
natural
composition
compound
inverse
double time
half-life
base
exponent

Concept Reinforcement

Determine whether each statement is true or false.

______________ 1. The y-intercept of a function $f(x) = a^x$ is $(0, 1)$. [8.1a]

______________ 2. If it is possible for a horizontal line to intersect the graph of a function more than once, its inverse is a function. [8.2c]

______________ 3. $\log_a 1 = 0, \quad a > 0$ [8.3c]

______________ 4. If we find that $\log (78) \approx 1.8921$ on a calculator, we also know that $10^{1.8921} \approx 78$. [8.3d]

______________ 5. $\ln (35) = \ln 7 \cdot \ln 5$ [8.4a]

______________ 6. The functions $f(x) = e^x$ and $g(x) = \ln x$ are inverses of each other. [8.5a]

Study Guide

Objective 8.1a Graph exponential equations and functions.

Example Graph: $f(x) = 4^x$.

We compute some function values and list the results in a table:

$$f(-2) = 4^{-2} = \frac{1}{4^2} = \frac{1}{16};$$

$$f(-1) = 4^{-1} = \frac{1}{4};$$

$$f(0) = 4^0 = 1;$$

$$f(1) = 4^1 = 4;$$

$$f(2) = 4^2 = 16.$$

x	$f(x)$
-2	$\frac{1}{16}$
-1	$\frac{1}{4}$
0	1
1	4
2	16

Now we plot the points $(x, f(x))$ and connect them with a smooth curve.

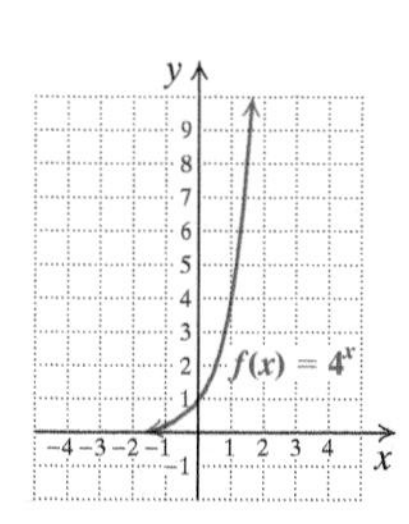

Practice Exercise

1. Graph: $f(x) = 2^x$.

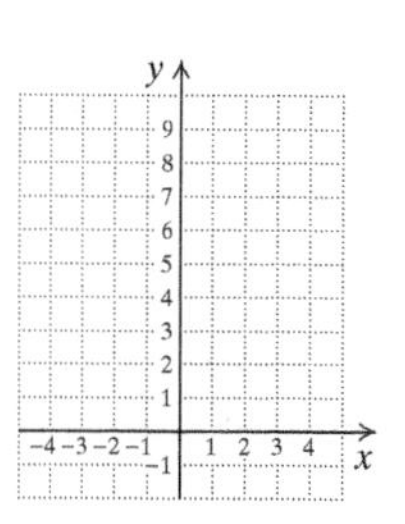

Objective 8.2a Find the composition of functions and express certain functions as a composition of functions.

Example Given $f(x) = x - 2$ and $g(x) = x^2$, find $(f \circ g)(x)$ and $(g \circ f)(x)$.

$$(f \circ g)(x) = f(g(x)) = f(x^2) = x^2 - 2;$$

$$(g \circ f)(x) = g(f(x)) = g(x - 2) = (x - 2)^2 = x^2 - 4x + 4$$

Example Find $f(x)$ and $g(x)$ such that $h(x) = (f \circ g)(x)$:

$$h(x) = \sqrt[3]{x - 5}.$$

Two functions that can be used are $f(x) = \sqrt[3]{x}$ and $g(x) = x - 5$. There are other correct answers.

Practice Exercises

2. Given $f(x) = 2x$ and $g(x) = 4x + 1$, find $(f \circ g)(x)$ and $(g \circ f)(x)$.

3. Find $f(x)$ and $g(x)$ such that $h(x) = (f \circ g)(x)$:

$$h(x) = \frac{1}{3x + 2}.$$

Objective 8.2c Given a function, determine whether it is one-to-one and has an inverse that is a function.

Example Determine whether the function $f(x) = x + 5$ is one-to-one and thus has an inverse that is also a function.

The graph of $f(x) = x + 5$ is shown below. No horizontal line crosses the graph more than once, so the function is one-to-one and has an inverse that is a function.

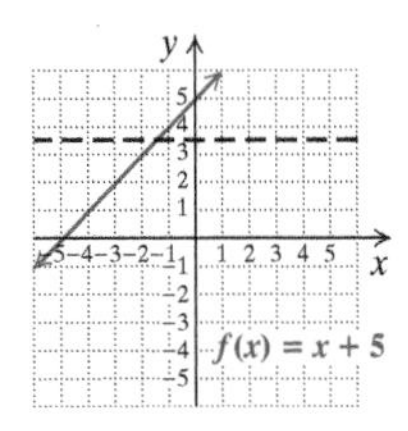

If there is a horizontal line that crosses the graph of a function more than once, the function is not one-to-one and does not have an inverse that is a function.

Practice Exercise

4. Determine whether the function $f(x) = 3^x$ is one-to-one.

Objective 8.2d Find a formula for the inverse of a function, if it exists, and graph inverse relations and functions.

Example Determine whether the function $f(x) = 3x - 1$ is one-to-one. If it is, find a formula for its inverse.

The graph of $f(x) = 3x - 1$ passes the horizontal-line test, so it is one-to-one. Now we find a formula for $f^{-1}(x)$.

1. Replace $f(x)$ with y: $y = 3x - 1$.

2. Interchange x and y: $x = 3y - 1$.

3. Solve for y: $x + 1 = 3y$

$$\frac{x+1}{3} = y.$$

4. Replace y with $f^{-1}(x)$: $f^{-1}(x) = \frac{x+1}{3}$.

Example Graph the one-to-one function $g(x) = x - 3$ and its inverse using the same set of axes.

We graph $g(x) = x - 3$ and then draw its reflection across the line $y = x$.

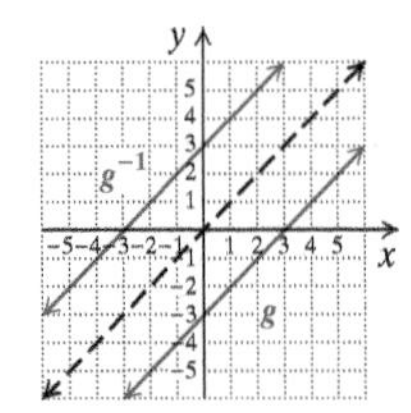

Practice Exercises

5. Determine whether the function $g(x) = 4 - x$ is one-to-one. If it is, find a formula for its inverse.

6. Graph the one-to-one function $f(x) = 2x + 1$ and its inverse using the same set of axes.

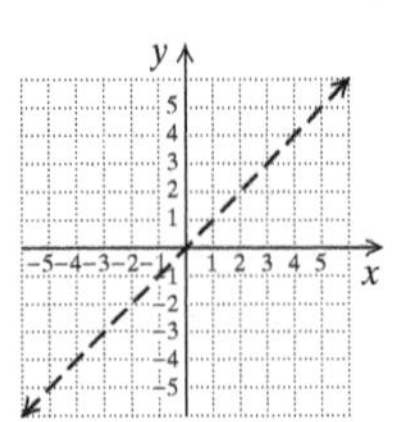

Objective 8.3a Graph logarithmic functions.

Example Graph: $y = f(x) = \log_4 x$.

The equation $y = \log_4 x$ is equivalent to $4^y = x$.

For $y = -2, x = 4^{-2} = \frac{1}{4^2} = \frac{1}{16}$.

For $y = -1, x = 4^{-1} = \frac{1}{4}$.

For $y = 0, x = 4^0 = 1$.

For $y = 1, x = 4^1 = 4$.

For $y = 2, x = 4^2 = 16$.

x	y
$\frac{1}{16}$	-2
$\frac{1}{4}$	-1
1	0
4	1
16	2

Now we plot these points and connect them with a smooth curve.

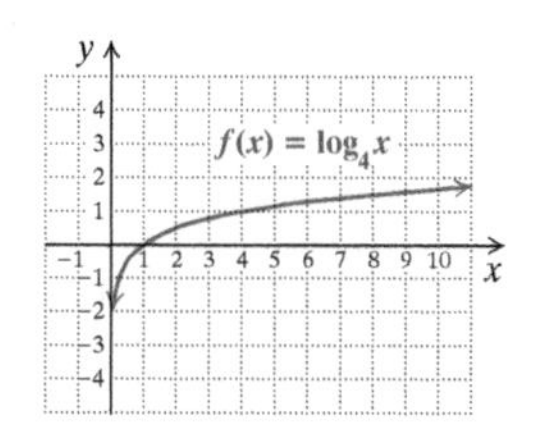

Practice Exercise

7. Graph: $y = \log_5 x$.

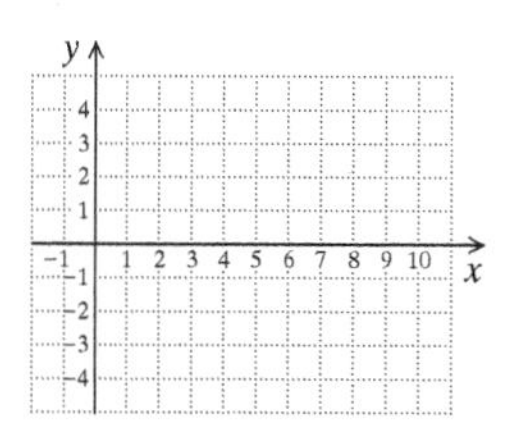

Objective 8.4d Convert from logarithms of products, quotients, and powers to expressions in terms of individual logarithms, and conversely.

Example Express

$$\log_a \frac{x^2y}{z^3}$$

in terms of logarithms of x, y, and z.

$$\log_a \frac{x^2y}{z^3} = \log_a (x^2y) - \log_a z^3$$
$$= \log_a x^2 + \log_a y - \log_a z^3$$
$$= 2 \log_a x + \log_a y - 3 \log_a z$$

Example Express

$$4 \log_a x - \frac{1}{2} \log_a y$$

as a single logarithm.

$$4 \log_a x - \frac{1}{2} \log_a y = \log_a x^4 - \log_a y^{1/2}$$
$$= \log_a \frac{x^4}{y^{1/2}}, \text{ or } \log_a \frac{x^4}{\sqrt{y}}$$

Practice Exercises

8. Express $\log_a \sqrt[5]{\frac{x^3}{y^2}}$ in terms of logarithms of x and y.

9. Express $\frac{1}{2} \log_a x - 3 \log_a y$ as a single logarithm.

Objective 8.5c Graph exponential functions and logarithmic functions, base e.

Example Graph: $f(x) = e^{x-1}$.

x	$f(x)$
-1	0.1
0	0.4
1	1
2	2.7

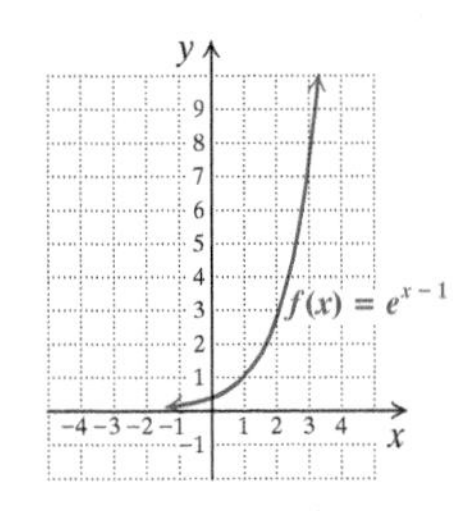

Practice Exercises

10. Graph: $f(x) = e^x - 1$.

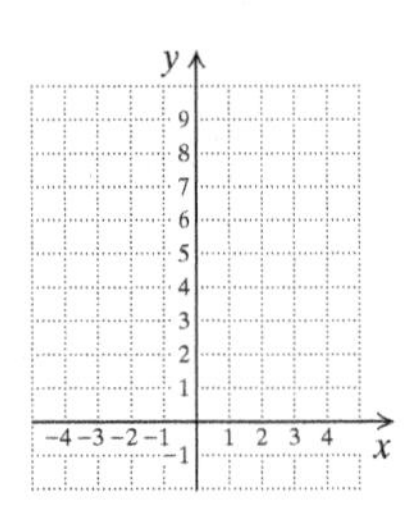

Example Graph: $g(x) = \ln x + 3$.

x	$g(x)$
0.5	2.3
1	3
3	4.1
5	4.6
8	5.1
10	5.3

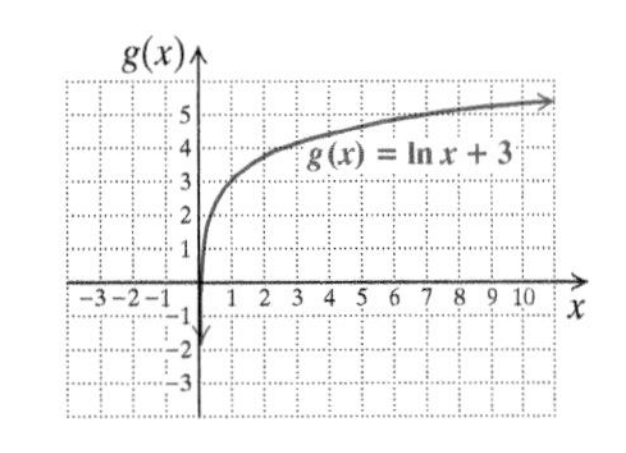

11. Graph: $f(x) = \ln(x + 3)$.

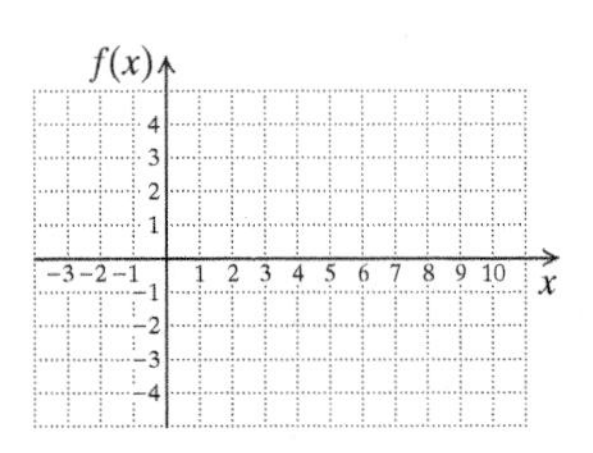

Objective 8.6a Solve exponential equations.

Example Solve: $3^{x-1} = 81$.

$$3^{x-1} = 81$$
$$3^{x-1} = 3^4$$

Since the bases are the same, the exponents must be the same:

$$x - 1 = 4$$
$$x = 5.$$

The solution is 5.

Practice Exercise

12. Solve: $2^{3x} = 16$.

Objective 8.6b Solve logarithmic equations.

Example Solve: $\log x + \log(x + 3) = 1$.

$$\log x + \log(x + 3) = 1$$
$$\log_{10}[x(x + 3)] = 1$$
$$x(x + 3) = 10^1$$
$$x^2 + 3x = 10$$
$$x^2 + 3x - 10 = 0$$
$$(x + 5)(x - 2) = 0$$
$$x + 5 = 0 \quad or \quad x - 2 = 0$$
$$x = -5 \quad or \quad x = 2$$

The number -5 does not check, but 2 does. The solution is 2.

Practice Exercise

13. Solve: $\log_3(2x + 3) = 2$.

Review Exercises

1. Find the inverse of the relation
$\{(-4, 2), (5, -7), (-1, -2), (10, 11)\}$. [8.2b]

Determine whether each function is one-to-one. If it is, find a formula for its inverse. [8.2c, d]

2. $f(x) = 4 - x^2$

3. $g(x) = \dfrac{2x - 3}{7}$

4. $f(x) = 8x^3$

5. $f(x) = \dfrac{4}{3 - 2x}$

6. Graph the function $f(x) = x^3 + 1$ and its inverse using the same set of axes. [8.2d]

Graph.

7. $f(x) = 3^{x-1}$ [8.1a]

x	$f(x)$
0	
1	
2	
3	
−1	
−2	
−3	

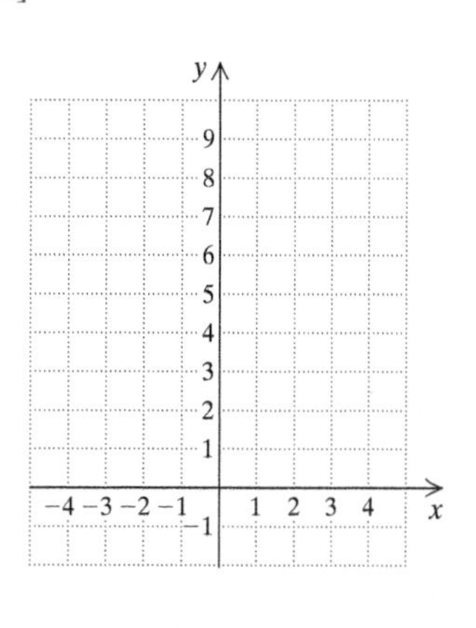

8. $f(x) = \log_3 x$, or $y = \log_3 x$ [8.3a]
$y = \log_3 x \rightarrow x =$ ____

x, or 3^y	y
	0
	1
	2
	3
	−1
	−2
	−3

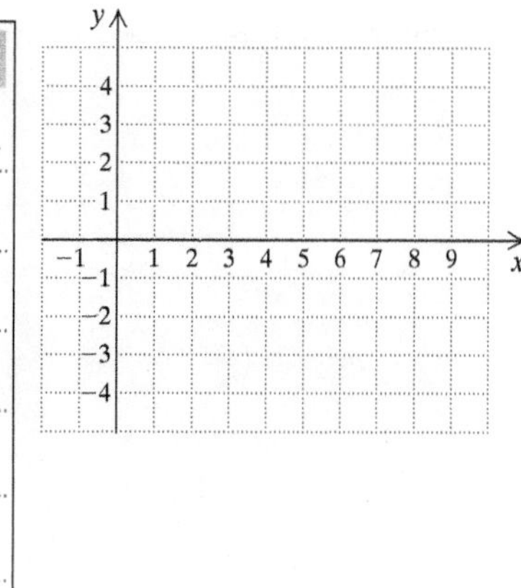

9. $f(x) = e^{x+1}$ [8.5c]

x	$f(x)$
0	
1	
2	
3	
−1	
−2	
−3	

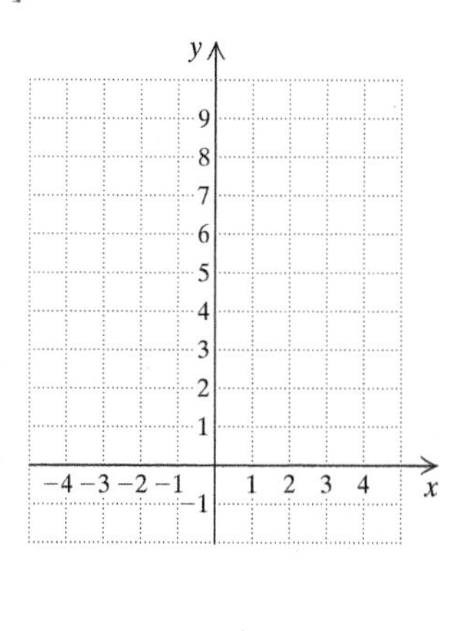

10. $f(x) = \ln (x - 1)$ [8.5c]

x	$f(x)$

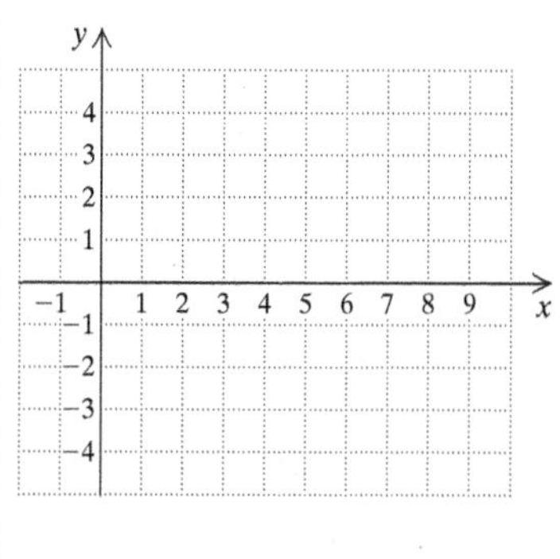

11. Find $(f \circ g)(x)$ and $(g \circ f)(x)$ if $f(x) = x^2$ and $g(x) = 3x - 5$. [8.2a]

12. If $h(x) = \sqrt{4 - 7x}$, find $f(x)$ and $g(x)$ such that $h(x) = (f \circ g)(x)$. [8.2a]

Convert to a logarithmic equation. [8.3b]

13. $10^4 = 10{,}000$

14. $25^{1/2} = 5$

Convert to an exponential equation. [8.3b]

15. $\log_4 16 = x$

16. $\log_{1/2} 8 = -3$

Find each of the following. [8.3c]

17. $\log_3 9$

18. $\log_{10} \frac{1}{10}$

19. $\log_m m$

20. $\log_m 1$

Find the common logarithm, to four decimal places, using a calculator. [8.3d]

21. $\log\left(\frac{78}{43{,}112}\right)$

22. $\log(-4)$

Express in terms of logarithms of x, y, and z. [8.4d]

23. $\log_a x^4y^2z^3$

24. $\log \sqrt[4]{\frac{z^2}{x^3y}}$

Express as a single logarithm. [8.4d]

25. $\log_a 8 + \log_a 15$

26. $\frac{1}{2}\log a - \log b - 2\log c$

Simplify. [8.4e]

27. $\log_m m^{17}$

28. $\log_m m^{-7}$

Given $\log_a 2 = 1.8301$ and $\log_a 7 = 5.0999$, find each of the following. [8.4d]

29. $\log_a 28$

30. $\log_a 3.5$

31. $\log_a \sqrt{7}$

32. $\log_a \frac{1}{4}$

Find each of the following, to four decimal places, using a calculator. [8.5a]

33. $\ln 0.06774$

34. $e^{-0.98}$

35. $e^{2.91}$

36. $\ln 1$

37. $\ln 0$

38. $\ln e$

Find each logarithm using the change-of-base formula. [8.5b]

39. $\log_5 2$

40. $\log_{12} 70$

Solve. Where appropriate, give approximations to four decimal places. [8.6a, b]

41. $\log_3 x = -2$

42. $\log_x 32 = 5$

43. $\log x = -4$

44. $3\ln x = -6$

45. $4^{2x-5} = 16$

46. $2^{x^2} \cdot 2^{4x} = 32$

47. $4^x = 8.3$

48. $e^{-0.1t} = 0.03$

49. $\log_4 16 = x$

50. $\log_4 x + \log_4(x-6) = 2$

51. $\log_2(x+3) - \log_2(x-3) = 4$

52. $\log_3(x-4) = 3 - \log_3(x+4)$

Solve. [8.7a, b]

53. *Sound Level.* The intensity of sound of a symphony orchestra playing at its peak can reach $10^{1.7}\ W/m^2$. How high is this sound level, in decibels? (Use $L = 10 \cdot \log(I/I_0)$ and $I_0 = 10^{-12}\ W/m^2$.)

54. *Internet Traffic.* Internet traffic is increasing exponentially. The number of petabytes (PB) of global consumer Internet usage per month $U(t)$ can be approximated by the exponential function

$$U(t) = 5204(1.47)^t,$$

where t is the number of years after 2010.

Data: Cisco

a) Estimate the global consumer Internet usage per month in 2010, in 2015, and in 2017.
b) In what year will the global consumer Internet usage per month be 250,000 PB?
c) What is the doubling time for global consumer Interner usage per month?
d) Graph the function.

55. *Investment.* In 2015, Lucy invested $40,000 in a mutual fund. By 2018, the value of her investment was $53,000. Assume that the value of her investment increased exponentially.

a) Find the value of k, and write an exponential function that describes the value of Lucy's investment t years after 2015.
b) Predict the value of her investment in 2025.
c) In what year will the value of her investment first reach $85,000?

56. The population of a colony of bacteria doubled in 3 days. What was the exponential growth rate?

57. How long will it take $7600 to double itself if it is invested at 3.4%, compounded continuously?

58. How old is a skeleton that has lost 34% of its carbon-14? (Use $P(t) = P_0e^{-0.00012t}$.)

59. What is the inverse of the function $f(x) = 5^x$, if it exists? [8.3a]

A. $f^{-1}(x) = x^5$ **B.** $f^{-1}(x) = \log_x 5$
C. $f^{-1}(x) = \log_5 x$ **D.** Does not exist

60. Solve: $\log(x^2 - 9) - \log(x + 3) = 1.$ [8.6b]

A. 4 **B.** 5
C. 7 **D.** 13

Synthesis

Solve. [8.6a, b]

61. $\ln(\ln x) = 3$

62. $5^{x+y} = 25, \quad 2^{2x-y} = 64$

Understanding Through Discussion and Writing

1. Explain how the graph of $f(x) = e^x$ could be used to obtain the graph of $g(x) = 1 + \ln x$. [8.2d], [8.5a]

2. Christina first determines that the solution of $\log_3(x + 4) = 1$ is -1, but then rejects it. What mistake do you think she might be making? [8.6b]

3. An organization determines that the cost per person of chartering a bus is given by the function

$$C(x) = \frac{100 + 5x}{x},$$

where x is the number of people in the group and $C(x)$ is in dollars. Determine $C^{-1}(x)$ and explain how this inverse function could be used. [8.2d]

4. Explain how the equation $\ln x = 3$ could be solved using the graph of $f(x) = \ln x$. [8.6b]

5. Explain why you cannot take the logarithm of a negative number. [8.3a]

6. Write a problem for a classmate to solve in which data that seem to fit an exponential growth function are provided. Try to find data in a newspaper to make the problem as realistic as possible. [8.7b]

Test

For Extra Help — For step-by-step test solutions, access the Chapter Test Prep Videos in MyLab Math.

Graph.

1. $f(x) = 2^{x+1}$

x	$f(x)$

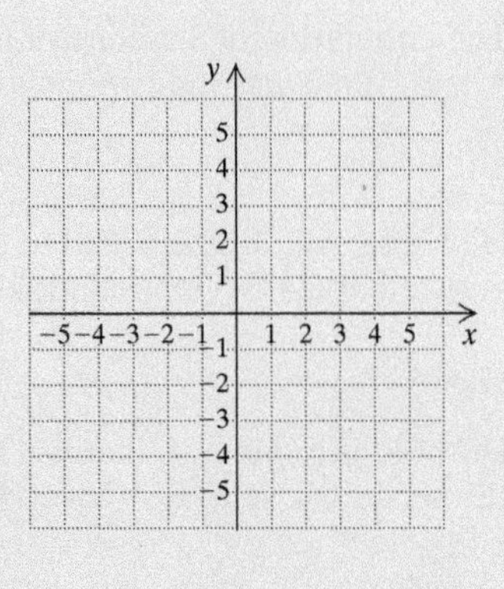

2. $y = \log_2 x$

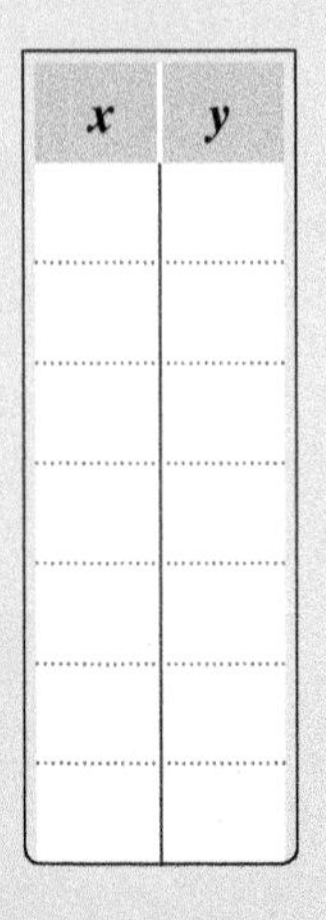

x	y

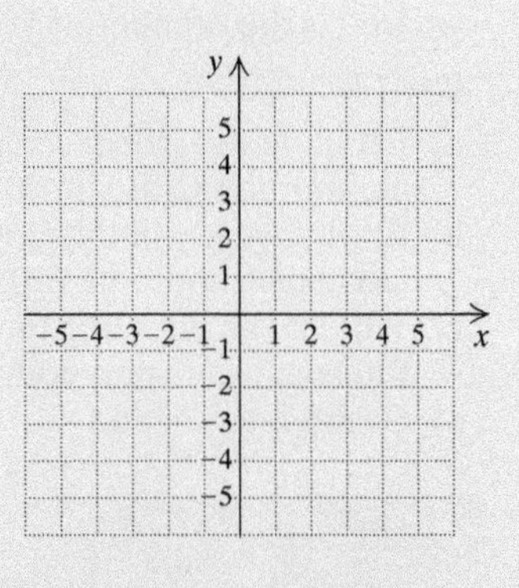

3. $f(x) = e^{x-2}$

x	$f(x)$

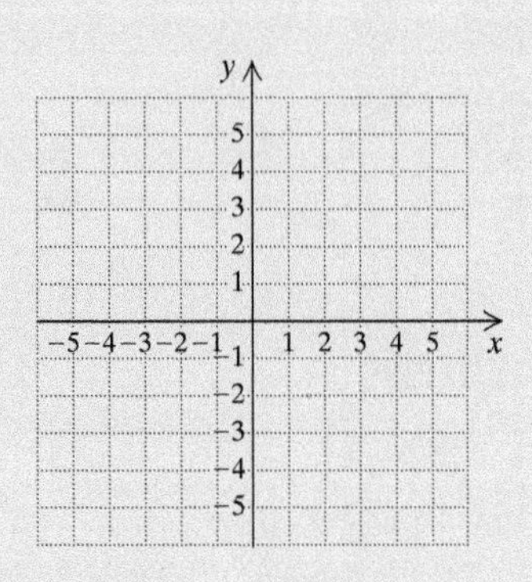

4. $f(x) = \ln(x - 4)$

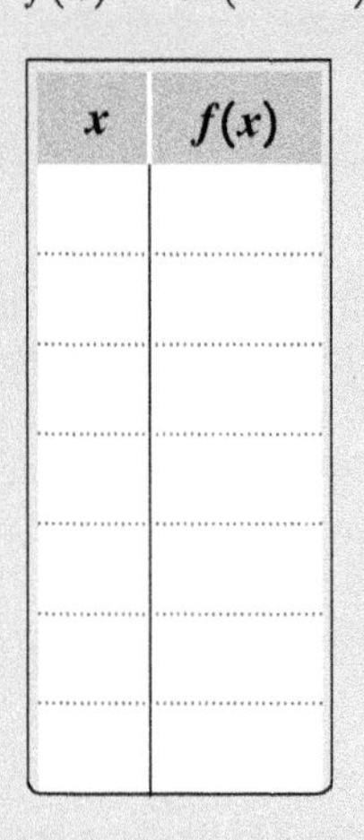

x	$f(x)$

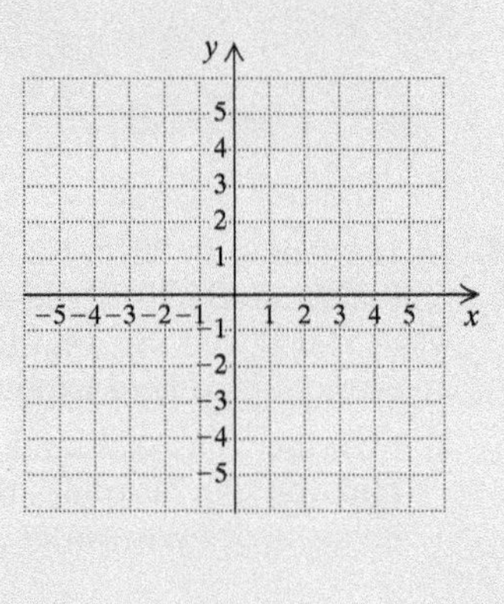

5. Find $(f \circ g)(x)$ and $(g \circ f)(x)$ if $f(x) = x + x^2$ and $g(x) = 5x - 2$.

6. Find the inverse of the relation $\{(-4, 3), (5, -8), (-1, -3), (10, 12)\}$.

Determine whether each function is one-to-one. If it is, find a formula for its inverse.

7. $f(x) = 4x - 3$

8. $f(x) = (x + 1)^3$

9. $f(x) = 2 - |x|$

10. Convert to a logarithmic equation:
$256^{1/2} = 16.$

11. Convert to an exponential equation:
$m = \log_7 49.$

Find each of the following.

12. $\log_5 125$

13. $\log_t t^{23}$

14. $\log_p 1$

Find the common logarithm, to four decimal places, using a calculator.

15. $\log 0.0123$

16. $\log(-5)$

17. Express in terms of logarithms of a, b, and c:
$$\log \frac{a^3 b^{1/2}}{c^2}.$$

18. Express as a single logarithm:
$$\frac{1}{3}\log_a x - 3\log_a y + 2\log_a z.$$

Given $\log_a 2 = 0.301$, $\log_a 6 = 0.778$, and $\log_a 7 = 0.845$, find each of the following.

19. $\log_a \frac{2}{7}$

20. $\log_a 12$

Find each of the following, to four decimal places, using a calculator.

21. $\ln 807.39$

22. $e^{4.68}$

23. $\ln 1$

24. Find $\log_{18} 31$ using the change-of-base formula.

Solve. Where appropriate, give approximations to four decimal places.

25. $\log_x 25 = 2$

26. $\log_4 x = \frac{1}{2}$

27. $\log x = 4$

28. $\ln x = \frac{1}{4}$

29. $7^x = 1.2$

30. $\log(x^2 - 1) - \log(x - 1) = 1$

31. $\log_5 x + \log_5 (x + 4) = 1$

32. *Tomatoes.* What is the pH of tomatoes if the hydrogen ion concentration is 6.3×10^{-5} moles per liter? (Use $\text{pH} = -\log [H^+]$.)

33. *Spread of a Rumor.* The number of people who hear a rumor increases exponentially. If 20 people start a rumor and if each person who hears the rumor repeats it to two people per day, the number of people N who have heard the rumor after t days is given by the function

$$N(t) = 20(3)^t,$$

a) How many people have heard the rumor after 5 days?
b) After what amount of time will 1000 people have heard the rumor?
c) What is the doubling time for the number of people who have heard the rumor?

34. *Interest Compounded Continuously.* Suppose that a \$1000 investment, compounded continuously, grows to \$1150.27 in 5 years.

a) Find the interest rate and the exponential growth function.
b) What is the balance after 8 years?
c) When will the balance be \$1439?
d) What is the doubling time?

35. The population of Masonville grew exponentially and doubled in 23 years. What was the exponential growth rate?

36. How old is an animal bone that has lost 43% of its carbon-14? (Use $P(t) = P_0 e^{-0.00012t}$.)

37. Solve: $\log (3x - 1) + \log x = 1$.

A. There are one positive solution and one negative solution.
B. There is exactly one solution, and it is positive.
C. There is exactly one solution, and it is negative.
D. There is no solution.

Synthesis

38. Solve: $\log_3 |2x - 7| = 4$.

39. If $\log_a x = 2$, $\log_a y = 3$, and $\log_a z = 4$, find

$$\log_a \frac{\sqrt[3]{x^2 z}}{\sqrt[3]{y^2 z^{-1}}}.$$

CHAPTERS

1–8 Cumulative Review

Solve.

1. $8(2x - 3) = 6 - 4(2 - 3x)$

2. $x(x - 3) = 10$

3. $4x - 3y = 15,$
$3x + 5y = 4$

4. $x + y - 3z = -1,$
$2x - y + z = 4,$
$-x - y + z = 1$

5. $\frac{7}{x^2 - 5x} - \frac{2}{x - 5} = \frac{4}{x}$

6. $\sqrt{x - 1} = \sqrt{x + 4} - 1$

7. $x - 8\sqrt{x} + 15 = 0$

8. $x^4 - 13x^2 + 36 = 0$

9. $\log_8 x = 1$

10. $3^{5x} = 7$

11. $\log x - \log(x - 8) = 1$

12. $x^2 + 4x > 5$

13. $|2x - 3| \geq 9$

14. If $f(x) = x^2 + 6x$, find a such that $f(a) = 11$.

15. Solve $D = \frac{ab}{b + a}$ for a.

16. Solve $\frac{1}{p} + \frac{1}{q} = \frac{1}{f}$ for q.

17. Find the domain of the function f given by

$$f(x) = \frac{-4}{3x^2 - 5x - 2}.$$

18. *Chocolate Making.* Greene Brothers' Chocolates are made by hand. It takes Anne 10 min to coat a tray of truffles in chocolate. It takes Clay 12 min to coat a tray of truffles. How long would it take Anne and Clay, working together, to coat the truffles?

19. *Forgetting.* Students in a biology class took a final examination. A forgetting formula for determining what the average exam score would be on a retest t months later is

$$S(t) = 78 - 15 \log(t + 1).$$

a) The average score when the students first took the test occurs when $t = 0$. Find the students' average score on the final exam.

b) What would the average score be on a retest after 4 months?

20. *Acid Mixtures.* Swim Clean is 30% muriatic acid. Pure Swim is 80% muriatic acid. How many liters of each should be mixed together in order to get 100 L of a solution that is 50% muriatic acid?

21. *Marine Travel.* A fishing boat with a trolling motor can move at a speed of 5 km/h in still water. The boat travels 42 km downstream in the same time that it takes to travel 12 km upstream. What is the speed of the stream?

22. *Population Growth.* The population of Spain was approximately 49 million in 2016, and the exponential growth rate was 0.84% per year.

a) Write an exponential function describing the growth of the population of Spain.

b) Estimate the population in 2020 and in 2025.

c) What is the doubling time of the population?

23. *Landscaping.* A rectangular lawn measures 60 ft by 80 ft. Part of the lawn is torn up to install a sidewalk of uniform width around it. The area of the new lawn is 2400 ft^2. How wide is the sidewalk?

24. Given that y varies directly as the square of x and inversely as z, and $y = 2$ when $x = 5$ and $z = 100$, what is y when $x = 3$ and $z = 4$?

Graph.

25. $5x = 15 + 3y$

26. $-2x - 3y \leq 6$

27. $f(x) = 2x^2 - 4x - 1$

28. $f(x) = 3^x$

29. $y = \log_3 x$

Perform the indicated operations and simplify.

30. $(11x^2 - 6x - 3) - (3x^2 + 5x - 2)$

31. $(3x^2 - 2y)^2$

32. $(5a + 3b)(2a - 3b)$

33. $\dfrac{x^2 + 8x + 16}{2x + 6} \div \dfrac{x^2 + 3x - 4}{x^2 - 9}$

34. $\dfrac{1 + \dfrac{3}{x}}{x - 1 - \dfrac{12}{x}}$

35. $\dfrac{3}{x + 6} - \dfrac{2}{x^2 - 36} + \dfrac{4}{x - 6}$

Factor.

36. $1 - 125x^3$

37. $6x^2 + 8xy - 8y^2$

38. $x^4 - 4x^3 + 7x - 28$

39. $2m^2 + 12mn + 18n^2$

40. $x^4 - 16y^4$

41. For the function described by
$$h(x) = -3x^2 + 4x + 8,$$
find $h(-2)$.

42. Divide: $(x^4 - 5x^3 + 2x^2 - 6) \div (x - 3)$.

For the radical expressions that follow, assume that all variables represent positive numbers.

43. Multiply and simplify: $\sqrt{7xy^3} \cdot \sqrt{28x^2y}$.

44. Divide and simplify: $\dfrac{\sqrt[3]{40xy^8}}{\sqrt[3]{5xy}}$.

45. Rationalize the denominator: $\dfrac{3 - \sqrt{y}}{2 - \sqrt{y}}$.

46. Multiply these complex numbers:
$$(1 + i\sqrt{3})(6 - 2i\sqrt{3}).$$

47. Find the inverse of f if $f(x) = 7 - 2x$.

48. Find an equation of the line containing the point $(-3, 5)$ and perpendicular to the line whose equation is $2x + y = 6$.

49. Express as a single logarithm:
$$3 \log x - \tfrac{1}{2} \log y - 2 \log z.$$

50. Convert to an exponential equation:
$$\log_a 5 = x.$$

Find each of the following using a calculator. Round answers to four decimal places.

51. $\log 0.05566$

52. $10^{2.89}$

53. $\ln 12.78$

54. $e^{-1.4}$

55. Complete the square: $f(x) = -2x^2 + 28x - 9$.

A. $f(x) = 2(x - 7)^2 - 89$
B. $f(x) = -2(x - 7)^2 - 58$
C. $f(x) = -2(x - 7)^2 - 107$
D. $f(x) = -2(x - 7)^2 + 89$

56. Solve $B = 2a(b^2 - c^2)$ for c.

A. $c = \sqrt{\dfrac{c}{2b} - 3}$

B. $c = -\sqrt{\dfrac{B}{2a} - b^2}$

C. $c = 2a(b^2 - B^2)$

D. $c = \sqrt{\dfrac{2ab^2 - B}{2a}}$

Synthesis

Solve.

57. $\dfrac{5}{3x - 3} + \dfrac{10}{3x + 6} = \dfrac{5x}{x^2 + x - 2}$

58. $\log \sqrt{3x} = \sqrt{\log 3x}$

59. *Train Travel.* A train travels 280 mi at a certain speed. If the speed had been increased by 5 mph, the trip could have been made in 1 hr less time. Find the actual speed.

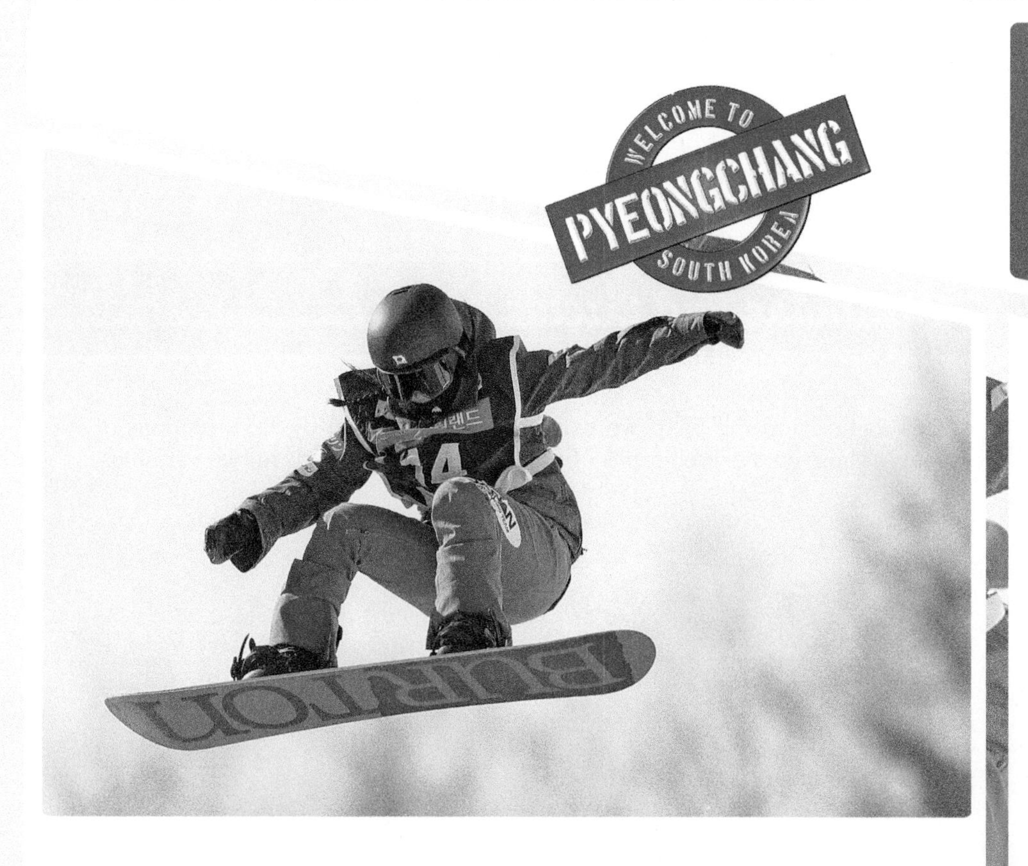

CHAPTER

9

Conic Sections

At the 2018 Winter Olympic Games in PyeongChang, medals in snowboarding will be awarded to both men and women in five events listed in the table below. There will be a maximum of 258 snowboarders (142 male and 116 female) competing in the 2018 Olympics. All participants must have been born before January 21, 2003, and each nation is limited to no more than 26 snowboarders.

Data: nbcolympics.com, "Snowboarding 101: Qualifying for the 2018 Olympic Games"

Winter Olympic Snowboarding Events
Halfpipe
Slopestyle
Big Air
Snowboard cross
Parallel giant slalom

We will calculate the radius used for the edge of each end of a snowboard in Exercise 71 in Exercise Set 9.1.

JUST IN TIME

Review topics 7, 10, and 13 in the **Just In Time** section at the front of the text. This provides excellent prerequisite skill review for this chapter.

MyLab Math VIDEO

STUDYING FOR SUCCESS *Preparing for the Final Exam*

- ☐ Browse through each chapter, reviewing highlighted or boxed information and noting important formulas.
- ☐ Retake the chapter tests that you took in class, or take the chapter tests in the text.
- ☐ Work through the Cumulative Review for Chapters 1–9 as a sample final exam.

9.1 Parabolas and Circles

OBJECTIVES

- **a** Graph parabolas.
- **b** Use the distance formula to find the distance between two points whose coordinates are known.
- **c** Use the midpoint formula to find the midpoint of a segment when the coordinates of its endpoints are known.
- **d** Given an equation of a circle, find its center and radius and graph it. Given the center and the radius of a circle, write an equation of the circle and graph the circle.

This section and the next two examine curves formed by cross sections of cones. These curves are graphs of second-degree equations in two variables.

CONIC SECTIONS IN THREE DIMENSIONS

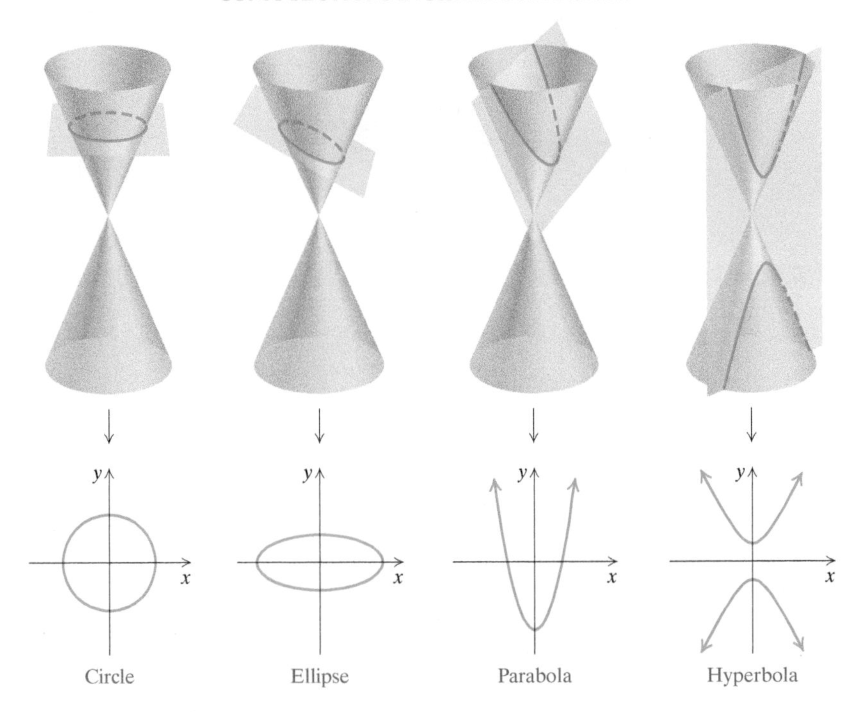

CONIC SECTIONS GRAPHED IN A PLANE

a PARABOLAS

SKILL REVIEW *For a quadratic function, find the vertex, the line of symmetry, and the maximum or minimum value.* [7.6a]

For each quadratic function, find **(a)** the vertex, **(b)** the line of symmetry, and **(c)** the maximum or minimum value.

1. $f(x) = 2x^2 + 4x - 1$ **2.** $f(x) = -x^2 + 4x - 3$

Answers: **1. (a)** $(-1, -3)$; **(b)** $x = -1$; **(c)** minimum: -3 when $x = -1$
2. (a) $(2, 1)$; **(b)** $x = 2$; **(c)** maximum: 1 when $x = 2$

When a cone is cut by a plane parallel to a side of the cone, as shown in the third figure on the preceding page, the conic section formed is a **parabola**. General equations of parabolas are quadratic. Parabolas have many applications in electricity, mechanics, and optics. A cross section of a satellite dish is a parabola, and arches that support certain bridges are parabolas. (Free-hanging cables have a different shape, called a *catenary*.) An arc of a spray can have part of the shape of a parabola.

The arch supporting the Gateshead Millennium Bridge is in the shape of a parabola. This bridge is a pedestrian and cyclist tilt bridge spanning the River Tyne in England.

EQUATIONS OF PARABOLAS

$y = ax^2 + bx + c$ (Line of symmetry is parallel to the y-axis.)

$x = ay^2 + by + c$ (Line of symmetry is parallel to the x-axis.)

Recall from Chapter 7 that the graph of $f(x) = ax^2 + bx + c$ (with $a \neq 0$) is a parabola.

EXAMPLE 1 Graph: $y = x^2 - 4x + 9$.

First, we must locate the vertex. To do so, we can use either of two approaches. One way is to complete the square:

$$\begin{aligned} y &= (x^2 - 4x) + 9 \\ &= (x^2 - 4x + 0) + 9 && \text{Adding 0} \\ &= (x^2 - 4x + 4 - 4) + 9 && \tfrac{1}{2}(-4) = -2;\ (-2)^2 = 4; \text{ substituting } 4 - 4 \text{ for } 0 \\ &= (x^2 - 4x + 4) + (-4 + 9) && \text{Regrouping} \\ &= (x - 2)^2 + 5. && \text{Factoring and simplifying} \end{aligned}$$

The vertex is $(2, 5)$.

A second way to find the vertex is to recall that the x-coordinate of the vertex of the parabola given by $y = ax^2 + bx + c$ is $-b/(2a)$:

$$x = -\frac{b}{2a} = -\frac{-4}{2(1)} = 2.$$

To find the y-coordinate of the vertex, we substitute 2 for x:

$$\begin{aligned} y &= x^2 - 4x + 9 \\ &= 2^2 - 4(2) + 9 = 5. \end{aligned}$$

Either way, the vertex is $(2, 5)$. Next, we calculate and plot some points on each side of the vertex. Since the x^2-coefficient, 1, is positive, the graph opens up.

x	y	
2	5	← Vertex
0	9	← y-intercept
1	6	
3	6	
4	9	

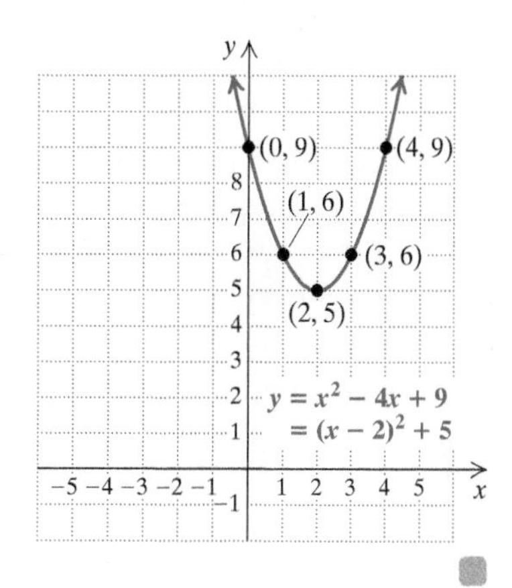

GRAPHING $y = ax^2 + bx + c$

To graph an equation of the type $y = ax^2 + bx + c$ (see Section 7.6):

1. Find the vertex (h, k) either by completing the square to find an equivalent equation $y = a(x - h)^2 + k$, or by using $x = -b/(2a)$ for the x-coordinate and substituting to find the y-coordinate.
2. Choose other values for x on each side of the vertex, and compute the corresponding y-values.
3. The graph opens up for $a > 0$ and opens down for $a < 0$.

1. Graph: $y = x^2 + 4x + 7$.

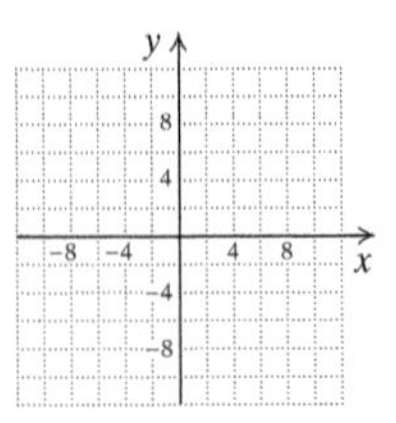

◀ **Do Exercise 1.**

Equations of the form $x = ay^2 + by + c$ represent horizontal parabolas. These parabolas open to the right for $a > 0$ and open to the left for $a < 0$ and have lines of symmetry parallel to the x-axis.

EXAMPLE 2 Graph: $x = y^2 - 4y + 9$.

This equation is like that in Example 1 except that x and y are interchanged. That is, the equations are *inverses* of each other. The vertex is $(5, 2)$ instead of $(2, 5)$. To find ordered pairs, we choose values for y on each side of the vertex. Then we compute values for x. Note that the x- and y-values of the table in Example 1 are interchanged. The graph in Example 2 is the reflection of the graph in Example 1 across the line $y = x$. You should confirm that, by completing the square, we get $x = (y - 2)^2 + 5$.

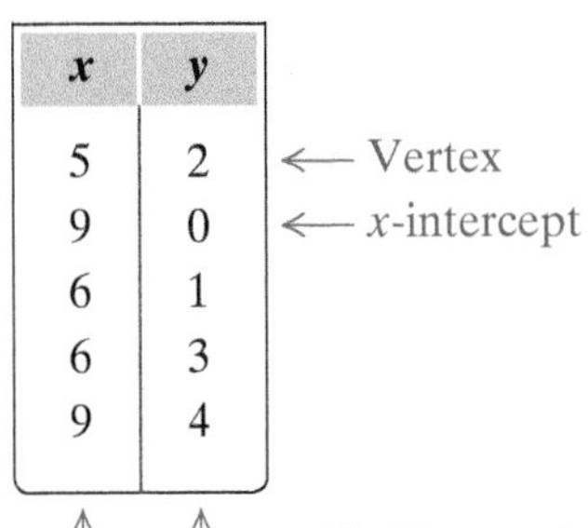

x	y	
5	2	← Vertex
9	0	← x-intercept
6	1	
6	3	
9	4	

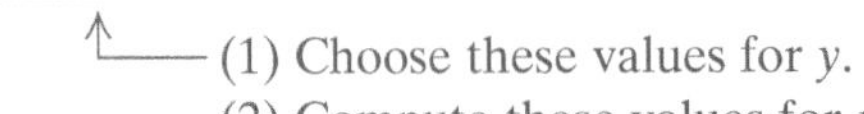
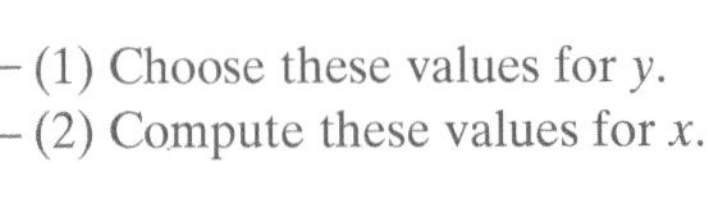

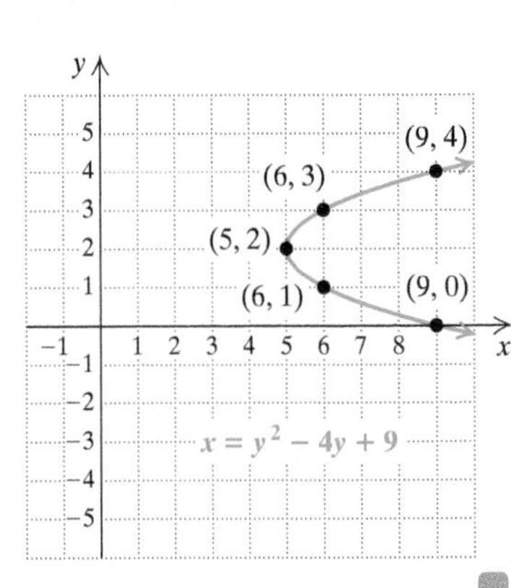

2. Graph: $x = y^2 + 4y + 7$.

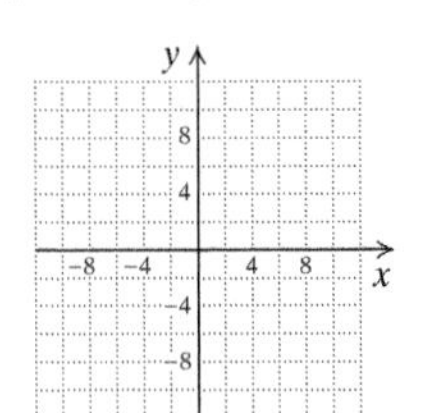

GRAPHING $x = ay^2 + by + c$

To graph an equation of the type $x = ay^2 + by + c$:

1. Find the vertex (h, k) either by completing the square to find an equivalent equation $x = a(y - k)^2 + h$, or by using $y = -b/(2a)$ for the y-coordinate and substituting to find the x-coordinate.
2. Choose other values for y that are above and below the vertex, and compute the corresponding x-values.
3. The graph opens to the right if $a > 0$ and opens to the left if $a < 0$.

◀ **Do Exercise 2.**

Answers

1.

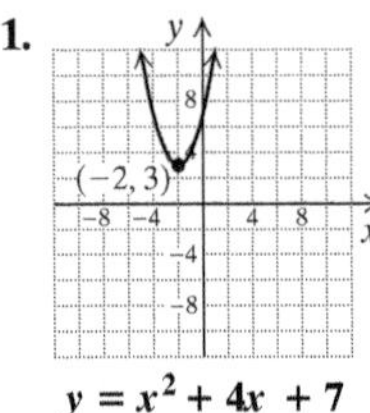

$y = x^2 + 4x + 7$

2.

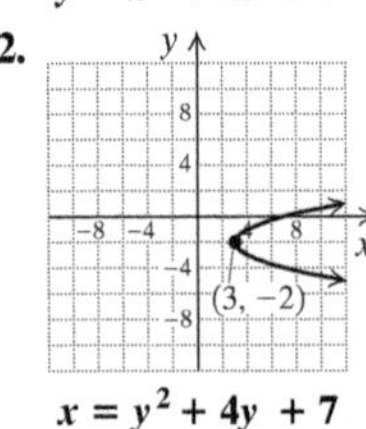

$x = y^2 + 4y + 7$

EXAMPLE 3 Graph: $x = -2y^2 + 10y - 7$.

We use the method of completing the square to find the vertex:

$$\begin{aligned} x &= -2y^2 + 10y - 7 \\ &= -2(y^2 - 5y) - 7 \\ &= -2(y^2 - 5y + 0) - 7 && \text{Adding 0} \\ &= -2\left(y^2 - 5y + \tfrac{25}{4} - \tfrac{25}{4}\right) - 7 && \tfrac{1}{2}(-5) = -\tfrac{5}{2};\ \left(-\tfrac{5}{2}\right)^2 = \tfrac{25}{4}; \text{ substituting } \tfrac{25}{4} - \tfrac{25}{4} \text{ for 0} \\ &= -2\left(y^2 - 5y + \tfrac{25}{4}\right) + (-2)\left(-\tfrac{25}{4}\right) - 7 && \text{Using the distributive law} \\ &= -2\left(y^2 - 5y + \tfrac{25}{4}\right) + \tfrac{25}{2} - 7 \\ &= -2\left(y - \tfrac{5}{2}\right)^2 + \tfrac{11}{2}. && \text{Factoring and simplifying} \end{aligned}$$

The vertex is $\left(\frac{11}{2}, \frac{5}{2}\right)$.

For practice, we also find the vertex by first computing its y-coordinate, $-b/(2a)$, and then substituting to find the x-coordinate:

$$y = -\frac{b}{2a} = -\frac{10}{2(-2)} = \frac{5}{2}$$

$$x = -2y^2 + 10y - 7 = -2\left(\tfrac{5}{2}\right)^2 + 10\left(\tfrac{5}{2}\right) - 7 = \tfrac{11}{2}.$$

To find ordered pairs, we first choose values for y and then compute values for x. A table is shown below, together with the graph. The graph opens to the left because the y^2-coefficient, -2, is negative.

x	y	
$\frac{11}{2}$	$\frac{5}{2}$	← Vertex
-7	0	← x-intercept
5	2	
5	3	
1	1	
1	4	
-7	5	

(1) Choose these values for y.
(2) Compute these values for x.

Do Exercise 3.

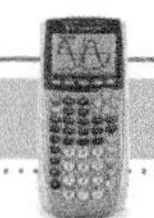

CALCULATOR CORNER

Graphing Parabolas as Inverses Suppose we want to use a graphing calculator to graph the equation

$x = y^2 - 4y + 9$. (1)

One way to do this is to note that this equation is the inverse of the equation

$y = x^2 - 4x + 9$. (2)

We enter $y_1 = x^2 - 4x + 9$. Next, we position the cursor over the equals sign and press ENTER. This deselects that equation, so its graph will not appear in the window. Then we use the DRAWINV feature to graph equation (1). Since y_1 has been deselected, only the graph of equation (1) will appear in the window.

EXERCISES: Use the DRAWINV feature to graph each equation on a graphing calculator.

1. $x = y^2 + 4y + 7$
2. $x = -2y^2 + 10y - 7$
3. $x = 4y^2 - 12y + 5$

3. Graph: $x = 4y^2 - 12y + 5$.

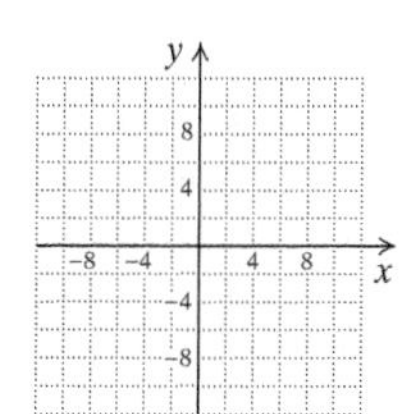

Answer

3.

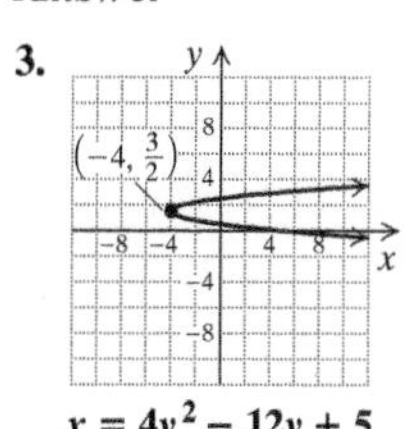

MyLab Math
ANIMATION

b THE DISTANCE FORMULA

Suppose that two points are on a horizontal line, and thus have the same second coordinate. We can find the distance between them by subtracting their first coordinates. This difference may be negative, depending on the order in which we subtract. So, to make sure that we get a positive number, we take the absolute value of this difference. The distance between two points on a horizontal line (x_1, y_1) and (x_2, y_1) is thus $|x_2 - x_1|$. Similarly, the distance between two points on a vertical line (x_2, y_1) and (x_2, y_2) is $|y_2 - y_1|$.

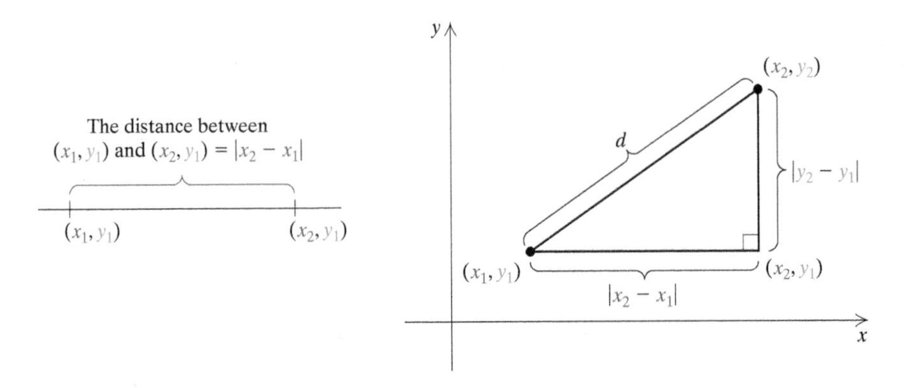

Now consider *any* two points (x_1, y_1) and (x_2, y_2). If $x_1 \neq x_2$ and $y_1 \neq y_2$, these points are vertices of a right triangle, as shown. The other vertex is then (x_2, y_1). The lengths of the legs are $|x_2 - x_1|$ and $|y_2 - y_1|$. We find d, the length of the hypotenuse, by using the Pythagorean equation:

$$d^2 = |x_2 - x_1|^2 + |y_2 - y_1|^2.$$

Since the square of a number is the same as the square of its opposite, we don't need these absolute-value signs. Thus,

$$d^2 = (x_2 - x_1)^2 + (y_2 - y_1)^2.$$

Taking the principal square root, we obtain the formula for the distance between two points.

THE DISTANCE FORMULA

The distance between any two points (x_1, y_1) and (x_2, y_2) is given by

$$d = \sqrt{(x_2 - x_1)^2 + (y_2 - y_1)^2}.$$

This formula holds even when the two points *are* on a vertical line or a horizontal line.

EXAMPLE 4 Find the distance between $(4, -3)$ and $(-5, 4)$. Give an exact answer and an approximation to three decimal places.

We substitute into the distance formula:

$$\begin{aligned} d &= \sqrt{(-5-4)^2 + [4-(-3)]^2} && \text{Substituting} \\ &= \sqrt{(-9)^2 + 7^2} \\ &= \sqrt{130} \approx 11.402. && \text{Using a calculator} \end{aligned}$$

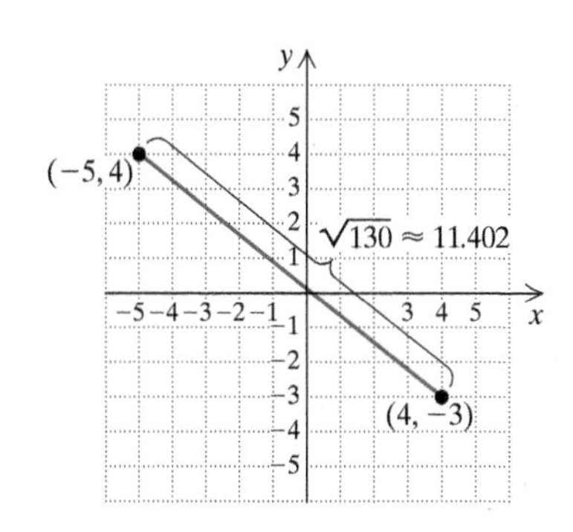

Do Exercises 4 and 5. ▶

Find the distance between each pair of points. Where appropriate, give an approximation to three decimal places.

4. $(2, 6)$ and $(-4, -2)$

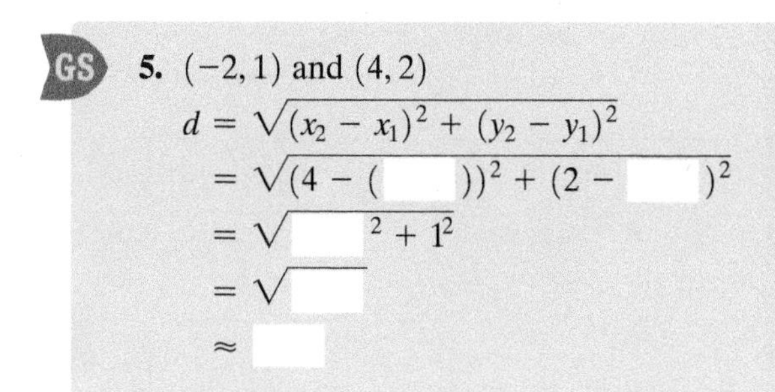

GS 5. $(-2, 1)$ and $(4, 2)$

$$\begin{aligned} d &= \sqrt{(x_2 - x_1)^2 + (y_2 - y_1)^2} \\ &= \sqrt{(4 - (\quad))^2 + (2 - \quad)^2} \\ &= \sqrt{\quad^2 + 1^2} \\ &= \sqrt{\quad} \\ &\approx \quad \end{aligned}$$

c MIDPOINTS OF SEGMENTS

The distance formula can be used to derive a formula for finding the midpoint of a segment when the coordinates of the endpoints are known.

THE MIDPOINT FORMULA

If the endpoints of a segment are (x_1, y_1) and (x_2, y_2), then the coordinates of the midpoint are

$$\left(\frac{x_1 + x_2}{2}, \frac{y_1 + y_2}{2}\right).$$

(To locate the midpoint, determine the average of the x-coordinates and the average of the y-coordinates.)

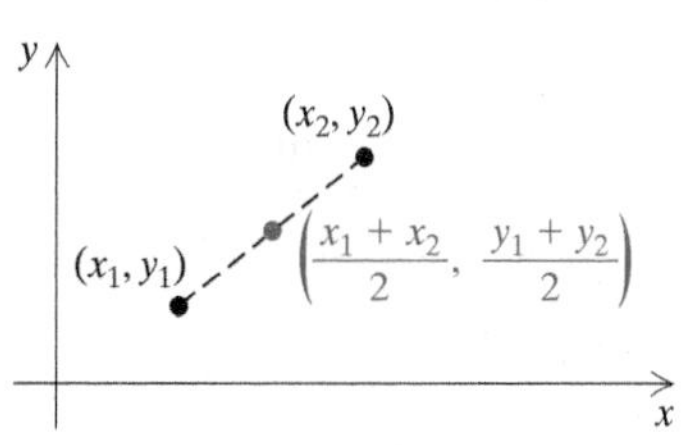

EXAMPLE 5 Find the midpoint of the segment with endpoints $(-2, 3)$ and $(4, -6)$.

Using the midpoint formula, we obtain

$$\left(\frac{-2+4}{2}, \frac{3+(-6)}{2}\right), \text{ or } \left(\frac{2}{2}, \frac{-3}{2}\right), \text{ or } \left(1, -\frac{3}{2}\right).$$

Do Exercises 6 and 7. ▶

Find the midpoint of the segment with the given endpoints.

6. $(-3, 1)$ and $(6, -7)$

7. $(10, -7)$ and $(8, -3)$

Answers

4. 10 **5.** $\sqrt{37} \approx 6.083$ **6.** $\left(\frac{3}{2}, -3\right)$
7. $(9, -5)$

Guided Solution:
5. −2, 1, 6, 37, 6.083

MyLab Math
ANIMATION

d CIRCLES

Another conic section, or curve, shown in the figure at the beginning of this section is a *circle*. A **circle** is defined as the set of all points in a plane that are a fixed distance from a point in that plane.

Let's find an equation for a circle. We call the center (h, k) and let the radius have length r. Suppose that (x, y) is any point on the circle. By the distance formula, we have

$$\sqrt{(x-h)^2 + (y-k)^2} = r.$$

Squaring both sides gives an equation of the circle in standard form:

$$(x-h)^2 + (y-k)^2 = r^2.$$

When $h = 0$ and $k = 0$, the circle is centered at the origin. Otherwise, we can think of that circle being translated $|h|$ units horizontally and $|k|$ units vertically from the origin.

EQUATIONS OF CIRCLES

A circle centered at the origin with radius r has equation

$$x^2 + y^2 = r^2.$$

A circle with center (h, k) and radius r has equation

$$(x-h)^2 + (y-k)^2 = r^2. \qquad \text{(Standard form)}$$

EXAMPLE 6 Find the center and the radius and graph this circle:

$$(x+2)^2 + (y-3)^2 = 16.$$

First, we find an equivalent equation in standard form:

$$[x-(-2)]^2 + (y-3)^2 = 4^2.$$

Thus the center is $(-2, 3)$ and the radius is 4. We draw the graph, shown below, by locating the center and then using a compass, setting its radius at 4, to draw the circle.

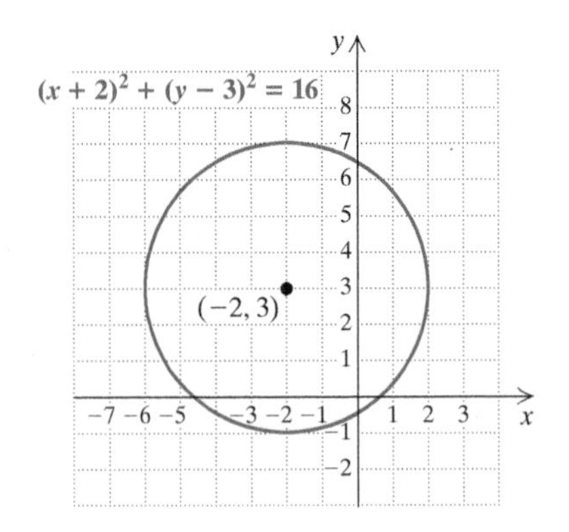

Do Exercises 8 and 9.

8. Find the center and the radius of the circle
$(x-5)^2 + \left(y+\frac{1}{2}\right)^2 = 9.$
Then graph the circle.

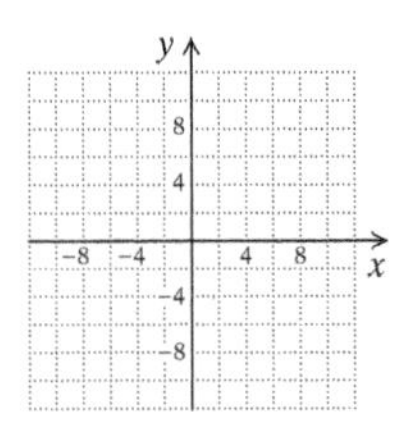

9. Find the center and the radius of the circle $x^2 + y^2 = 64$.

Answers

8.
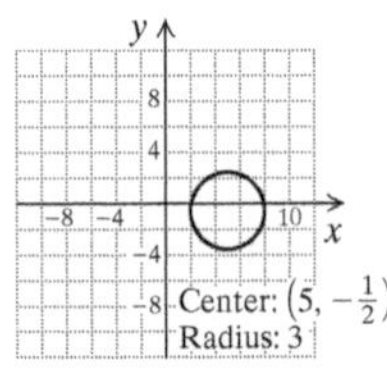

9. $(0, 0)$; $r = 8$

EXAMPLE 7 Write an equation of a circle with center $(9, -5)$ and radius $\sqrt{2}$.

We use standard form $(x - h)^2 + (y - k)^2 = r^2$ and substitute:

$$(x - 9)^2 + [y - (-5)]^2 = (\sqrt{2})^2 \quad \text{Substituting}$$
$$(x - 9)^2 + (y + 5)^2 = 2. \quad \text{Simplifying}$$

Do Exercise 10. ▶

10. Find an equation of a circle with center $(-3, 1)$ and radius 6.

$(x - h)^2 + (y - k)^2 = r^2$

$(x - (\quad))^2 + (y - \quad)^2 = \quad ^2$

$(x + 3)^2 + (y - 1)^2 = \quad$

With certain equations not in standard form, we can complete the square to show that the equations are equations of circles.

EXAMPLE 8 Find the center and the radius and graph this circle:

$$x^2 + y^2 + 8x - 2y + 15 = 0.$$

First, we regroup the terms and then complete the square twice, once with $x^2 + 8x$ and once with $y^2 - 2y$:

$$x^2 + y^2 + 8x - 2y + 15 = 0$$
$$(x^2 + 8x) + (y^2 - 2y) = -15 \quad \text{Regrouping and subtracting 15}$$
$$(x^2 + 8x + 0) + (y^2 - 2y + 0) = -15 \quad \text{Adding 0}$$
$$(x^2 + 8x + 16 - 16) + (y^2 - 2y + 1 - 1) = -15 \quad (\tfrac{8}{2})^2 = 4^2 = 16;\ (\tfrac{-2}{2})^2 = (-1)^2 = 1;\ \text{substituting } 16 - 16 \text{ and } 1 - 1 \text{ for } 0$$
$$(x^2 + 8x + 16) + (y^2 - 2y + 1) - 16 - 1 = -15 \quad \text{Regrouping}$$
$$(x^2 + 8x + 16) + (y^2 - 2y + 1) = -15 + 16 + 1 \quad \text{Adding 16 and 1 on both sides}$$
$$(x + 4)^2 + (y - 1)^2 = 2 \quad \text{Factoring and simplifying}$$
$$[x - (-4)]^2 + (y - 1)^2 = (\sqrt{2})^2. \quad \text{Writing standard form}$$

The center is $(-4, 1)$ and the radius is $\sqrt{2}$.

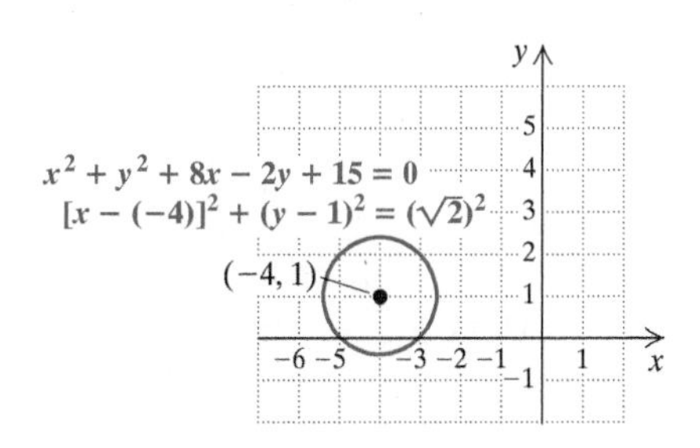

Do Exercise 11. ▶

11. Find the center and the radius of the circle

$$x^2 + 2x + y^2 - 4y + 2 = 0.$$

Then graph the circle.

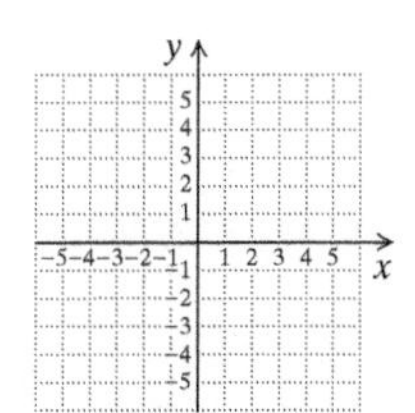

Answers

10. $(x + 3)^2 + (y - 1)^2 = 36$
11. Center: $(-1, 2)$; radius: $\sqrt{3}$;

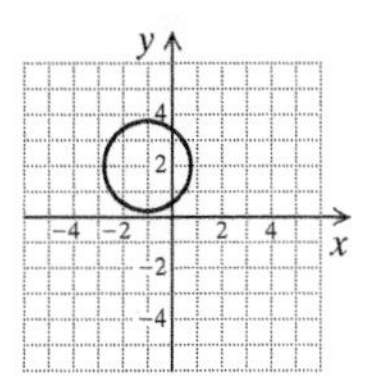

$x^2 + 2x + y^2 - 4y + 2 = 0$

Guided Solution:
10. $-3, 1, 6, 36$

CALCULATOR CORNER

Graphing Circles Equations of circles are not functions, so they cannot be entered directly in "$y =$" form on a graphing calculator. Nevertheless, there are three methods for graphing circles.

Suppose we want to graph the circle $(x - 3)^2 + (y + 1)^2 = 16$. One way to graph this circle is to use the CIRCLE feature from the DRAW menu. The center of the circle is $(3, -1)$ and its radius is 4. To graph it using the CIRCLE feature from the DRAW menu, we first press Y= and clear all previously entered equations. Then we select a square window. We will use $[-3, 9, -5, 3]$. We press 2ND QUIT to go to the home screen and then 2ND DRAW 9 to display "Circle(." We enter the coordinates of the center and the radius, separating the entries by commas, and close the parentheses: 3 , (−) 1 , 4) ENTER.

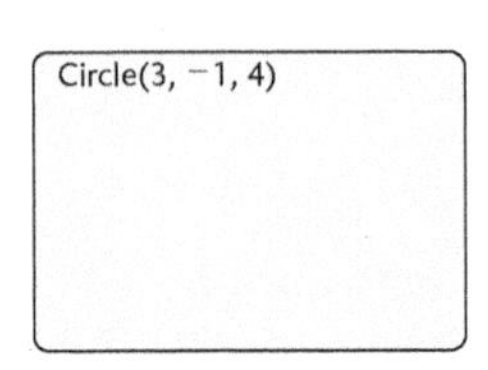

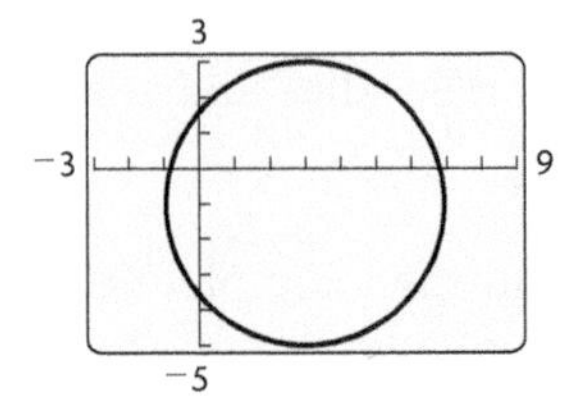

When the Graph screen is displayed, we can use the CLRDRAW operation from the DRAW menu to clear this circle before we graph another circle. To do this, we press 2ND DRAW 1.

Another way to graph a circle is to solve the equation for y first. Consider the equation above:

$$\begin{aligned}(x - 3)^2 + (y + 1)^2 &= 16\\ (y + 1)^2 &= 16 - (x - 3)^2\\ y + 1 &= \pm\sqrt{16 - (x - 3)^2}\\ y &= -1 \pm \sqrt{16 - (x - 3)^2}.\end{aligned}$$

Now we can write two functions, $y_1 = -1 + \sqrt{16 - (x - 3)^2}$ and $y_2 = -1 - \sqrt{16 - (x - 3)^2}$. When we graph these functions in the same window, we have the graph of the circle. The first equation produces the top half of the circle and the second produces the lower half. The graphing calculator does not connect the two parts of the graph because of approximations made near the endpoints of each graph.

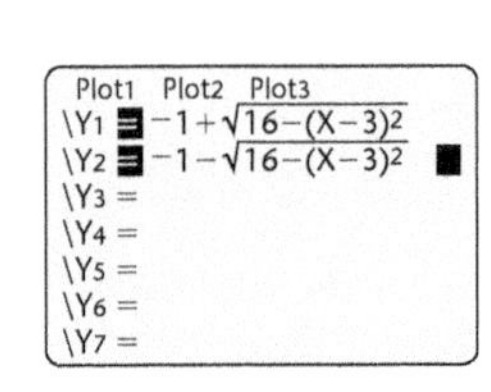

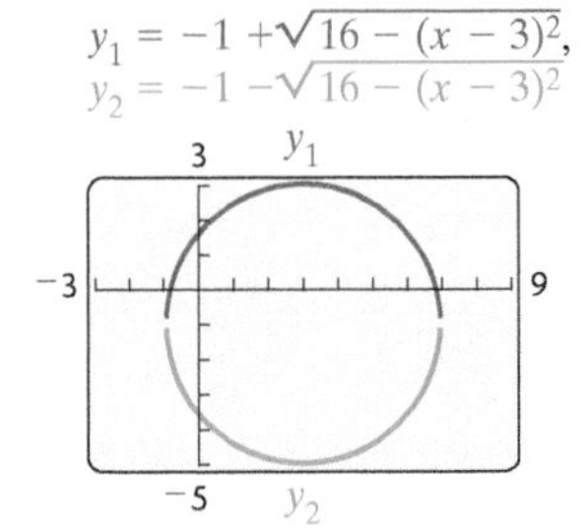

Circles can also be graphed using the CONICS app from the APPS menu.

EXERCISES: Graph each circle using both methods above.

1. $(x - 1)^2 + (y + 2)^2 = 4$
2. $(x + 2)^2 + (y - 2)^2 = 25$
3. $x^2 + y^2 - 16 = 0$
4. $4x^2 + 4y^2 = 100$
5. $x^2 + y^2 - 10x - 11 = 0$

9.1 Exercise Set

FOR EXTRA HELP MyLab Math

✓ Check Your Understanding

Reading Check and Concept Check Determine whether each statement is true or false.

RC1. Parabolas and circles are examples of conic sections.

RC2. The graph of $y = x^2 - 3x + 4$ is a parabola that opens down.

RC3. The distance formula can be used when two points are on a vertical line.

RC4. The x-coordinate of the midpoint of two points is the average of the x-coordinates of the points.

RC5. The center of the circle given by $x^2 + y^2 = 10$ is $(0, 0)$.

RC6. The radius of the circle given by $x^2 + y^2 = 10$ is 10.

RC7. For the parabola $y = 4(x + 7)^2 - 10$, the vertex is $(7, -10)$.

RC8. For the parabola $x = -3y^2 + 5y + 2$, the axis of symmetry is parallel to the x-axis.

a Graph each equation.

1. $y = x^2$

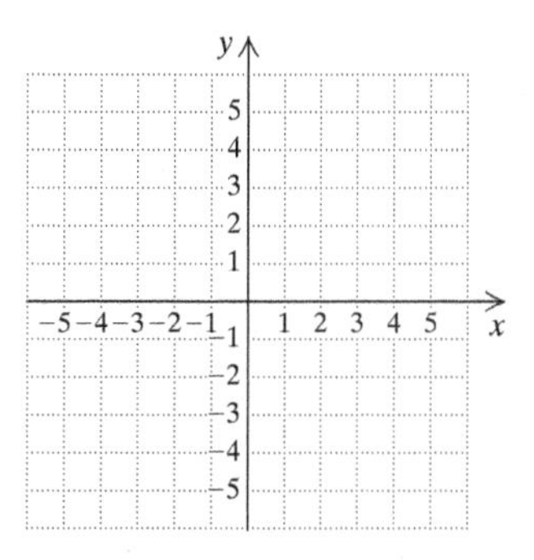

2. $x = y^2$

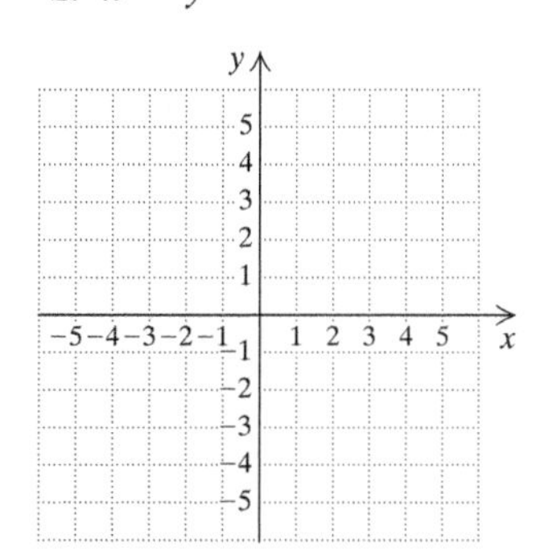

3. $x = y^2 + 4y + 1$

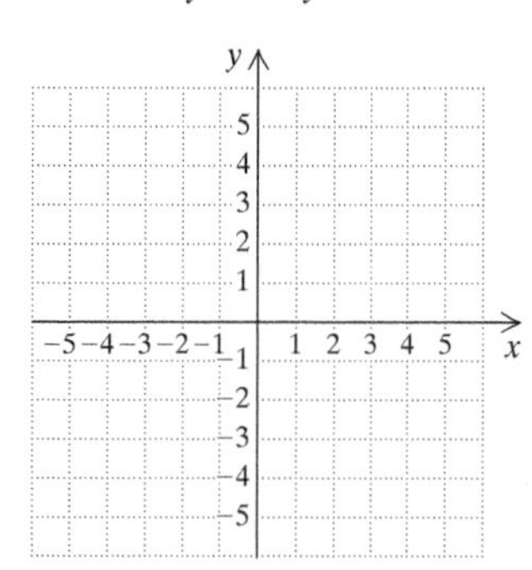

4. $y = x^2 - 2x + 3$

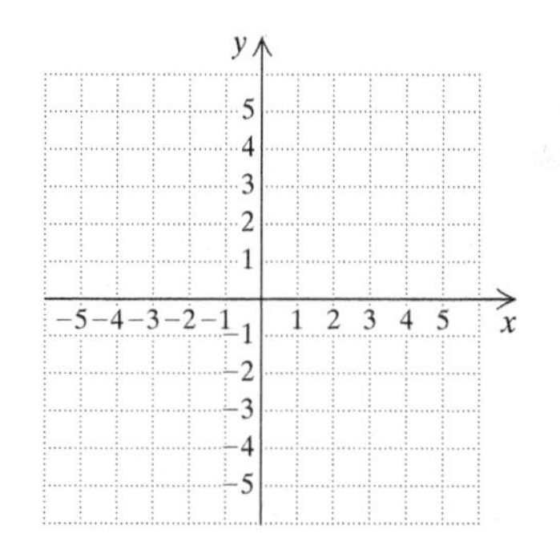

5. $y = -x^2 + 4x - 5$

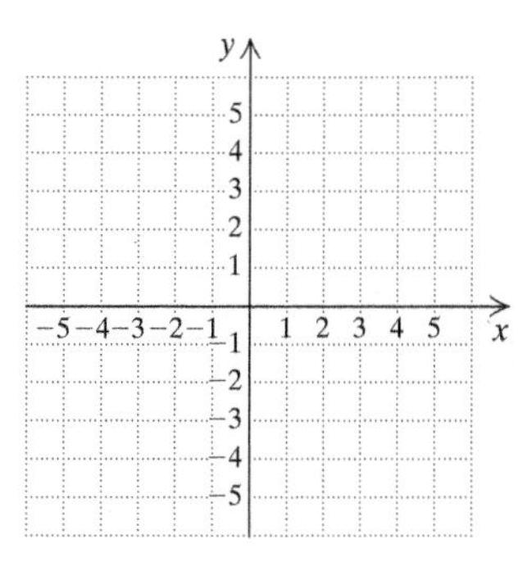

6. $x = 4 - 3y - y^2$

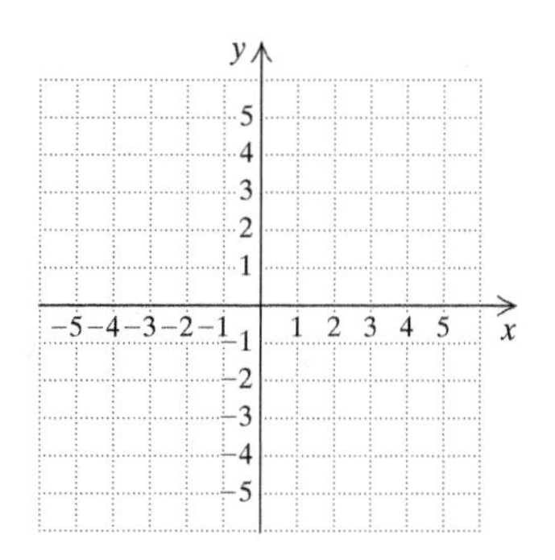

7. $x = -3y^2 - 6y - 1$

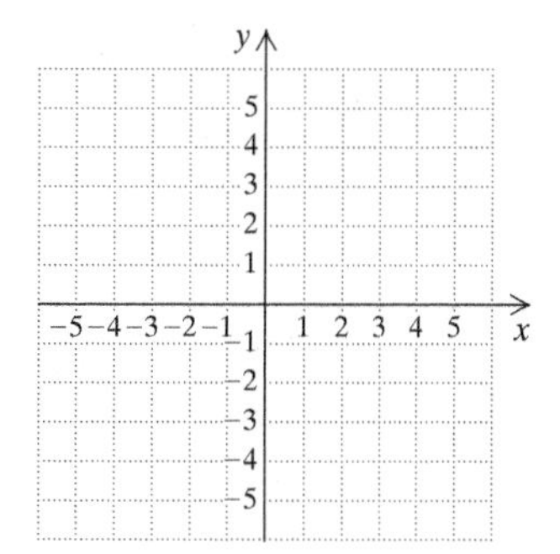

8. $y = -5 - 8x - 2x^2$

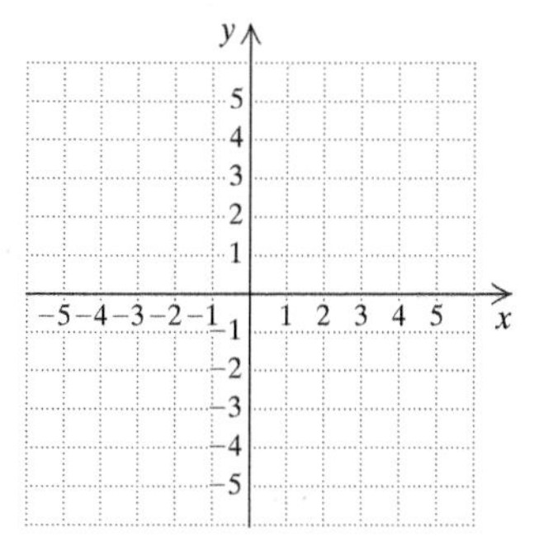

b Find the distance between each pair of points. Where appropriate, give an approximation to three decimal places.

9. $(6, -4)$ and $(2, -7)$

10. $(1, 2)$ and $(-4, 14)$

11. $(0, -4)$ and $(5, -6)$

12. $(8, 3)$ and $(8, -3)$

13. $(9, 9)$ and $(-9, -9)$

14. $(2, 22)$ and $(-8, 1)$

15. $(2.8, -3.5)$ and $(-4.3, -3.5)$

16. $(6.1, 2)$ and $(5.6, -4.4)$

17. $\left(\frac{5}{7}, \frac{1}{14}\right)$ and $\left(\frac{1}{7}, \frac{11}{14}\right)$

18. $(0, \sqrt{7})$ and $(\sqrt{6}, 0)$

19. $(-23, 10)$ and $(56, -17)$

20. $(34, -18)$ and $(-46, -38)$

21. (a, b) and $(0, 0)$

22. $(0, 0)$ and (p, q)

23. $(\sqrt{2}, -\sqrt{3})$ and $(-\sqrt{7}, \sqrt{5})$

24. $(\sqrt{8}, \sqrt{3})$ and $(-\sqrt{5}, -\sqrt{6})$

25. $(1000, -240)$ and $(-2000, 580)$

26. $(-3000, 560)$ and $(-430, -640)$

c Find the midpoint of the segment with the given endpoints.

27. $(-1, 9)$ and $(4, -2)$

28. $(5, 10)$ and $(2, -4)$

29. $(3, 5)$ and $(-3, 6)$

30. $(7, -3)$ and $(4, 11)$

31. $(-10, -13)$ and $(8, -4)$

32. $(6, -2)$ and $(-5, 12)$

33. $(-3.4, 8.1)$ and $(2.9, -8.7)$

34. $(4.1, 6.9)$ and $(5.2, -6.9)$

35. $\left(\frac{1}{6}, -\frac{3}{4}\right)$ and $\left(-\frac{1}{3}, \frac{5}{6}\right)$

36. $\left(-\frac{4}{5}, -\frac{2}{3}\right)$ and $\left(\frac{1}{8}, \frac{3}{4}\right)$

37. $(\sqrt{2}, -1)$ and $(\sqrt{3}, 4)$

38. $(9, 2\sqrt{3})$ and $(-4, 5\sqrt{3})$

d Find the center and the radius of each circle. Then graph the circle.

39. $(x + 1)^2 + (y + 3)^2 = 4$

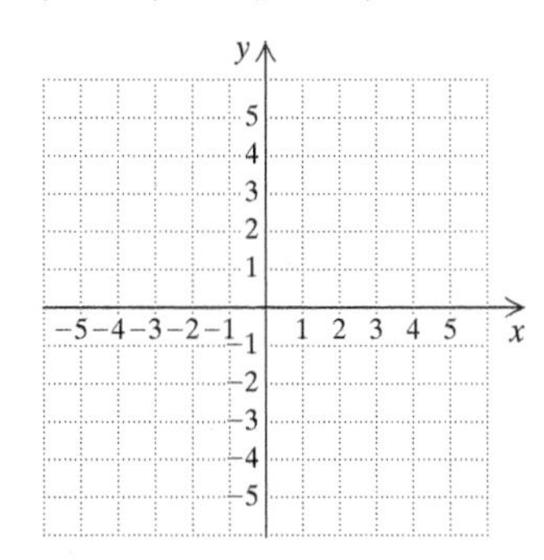

40. $(x - 2)^2 + (y + 3)^2 = 1$

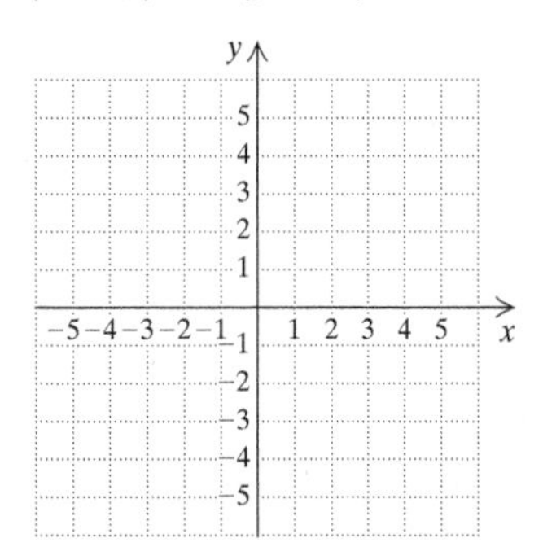

41. $(x - 3)^2 + y^2 = 2$

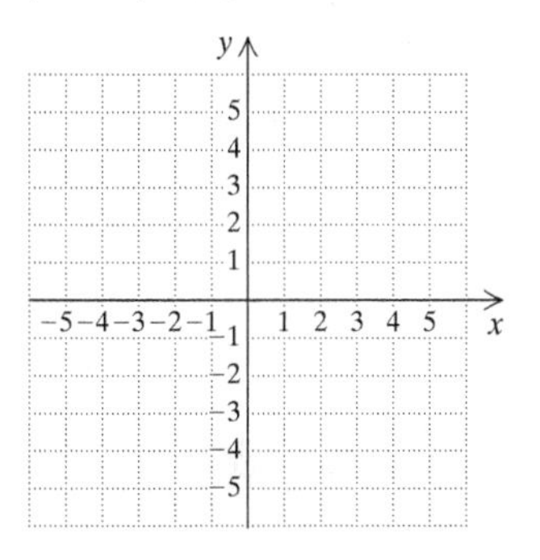

42. $x^2 + (y - 1)^2 = 3$

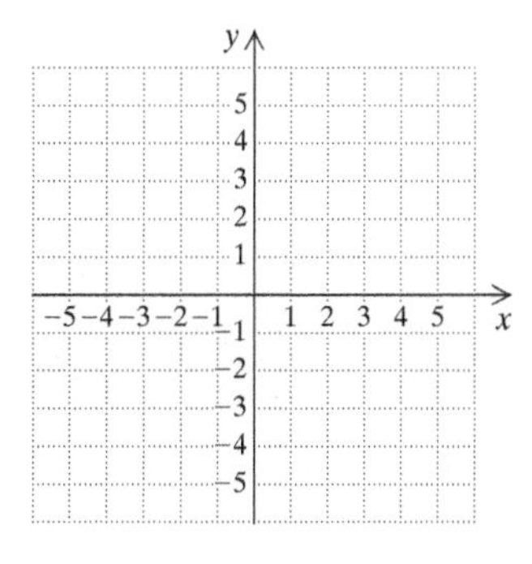

43. $x^2 + y^2 = 25$

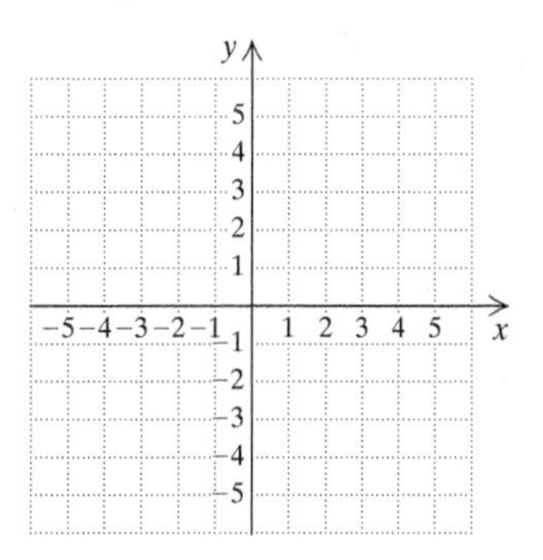

44. $x^2 + y^2 = 9$

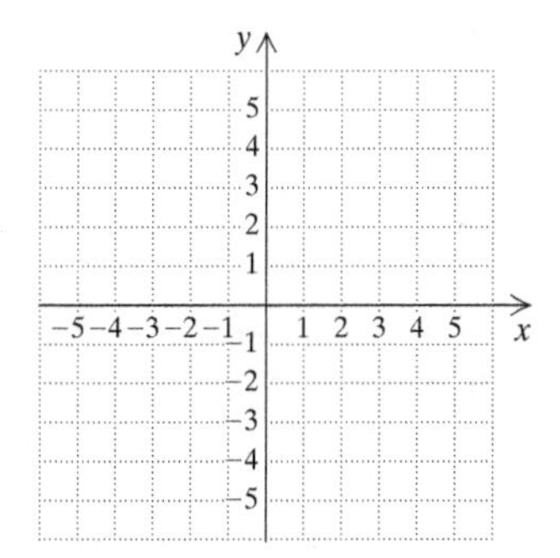

Find an equation of the circle having the given center and radius.

45. Center $(0, 0)$, radius 7

46. Center $(0, 0)$, radius 4

47. Center $(-5, 3)$, radius $\sqrt{7}$

48. Center $(4, 1)$, radius $3\sqrt{2}$

Find the center and the radius of each circle.

49. $x^2 + y^2 + 8x - 6y - 15 = 0$

50. $x^2 + y^2 + 6x - 4y - 15 = 0$

51. $x^2 + y^2 - 8x + 2y + 13 = 0$

52. $x^2 + y^2 + 6x + 4y + 12 = 0$

53. $x^2 + y^2 - 4x = 0$

54. $x^2 + y^2 + 10y - 75 = 0$

Skill Maintenance

Solve. [3.2a], [3.3a]

55. $x - y = 7,$
$x + y = 11$

56. $x + y = 8,$
$x - y = -24$

57. $y = 3x - 2,$
$2x - 4y = 50$

58. $2x + 3y = 8,$
$x - 2y = -3$

59. $-4x + 12y = -9,$
$x - 3y = 2$

Factor. [4.6b]

60. $4a^2 - b^2$

61. $x^2 - 16$

62. $a^2 - 9b^2$

63. $64p^2 - 81q^2$

64. $400c^2d^2 - 225$

Synthesis

Find an equation of a circle satisfying the given conditions.

65. Center $(-3, -2)$, and tangent to the y-axis

66. The endpoints of a diameter are $(7, 3)$ and $(-1, -3)$.

Find the distance between the given points.

67. $(6m, -7n)$ and $(-2m, n)$

68. $\left(-3\sqrt{3}, 1 - \sqrt{6}\right)$ and $\left(\sqrt{3}, 1 + \sqrt{6}\right)$

69. If the sides of a triangle have lengths a, b, and c and $a^2 + b^2 = c^2$, then the triangle is a right triangle. Determine whether the given points are vertices of a right triangle.

$(-8, -5)$, $(6, 1)$, and $(-4, 5)$

70. Find the point on the y-axis that is equidistant from $(2, 10)$ and $(6, 2)$.

71. *Snowboarding.* Each side edge of a snowboard is an arc of a circle. The snowboard shown below has a "running length" of 1150 mm and a "sidecut depth" of 19.5 mm.

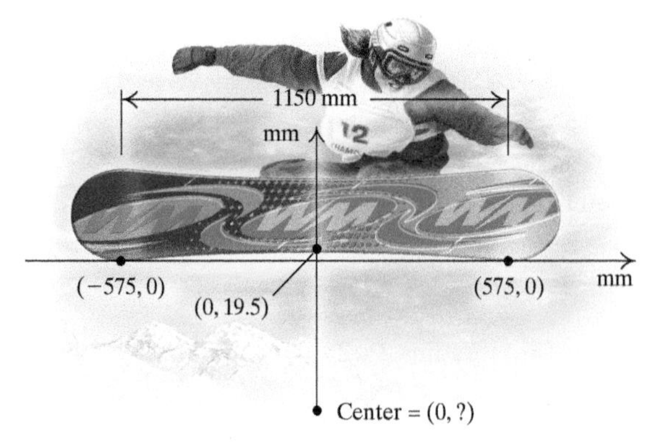

a) Using the coordinates shown, locate the center of the circle. (*Hint*: Equate distances.)
b) What radius is used for the edge of the board?

72. *Doorway Construction.* Ace Carpentry is to cut an arch for the top of an entranceway. The arch needs to be 8 ft wide and 2 ft high. To draw the arch, the carpenters will use a stretched string with chalk attached at an end as a compass.

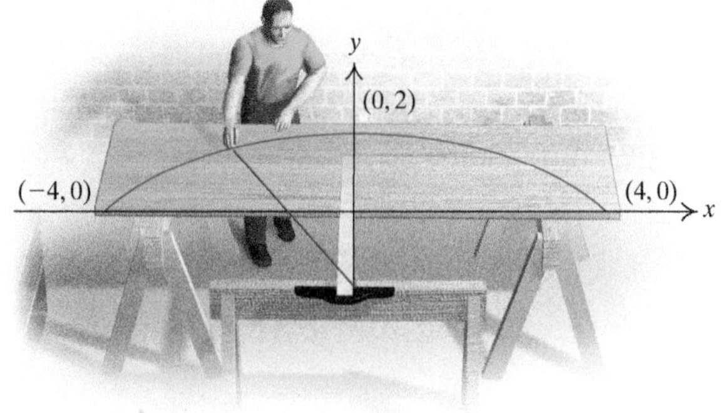

a) Using a coordinate system, locate the center of the circle.
b) What radius should the carpenters use to draw the arch?

Ellipses

9.2

OBJECTIVE

a Graph the standard form of the equation of an ellipse.

a ELLIPSES

When a cone is cut at an angle, as shown below, the conic section formed is an *ellipse*.

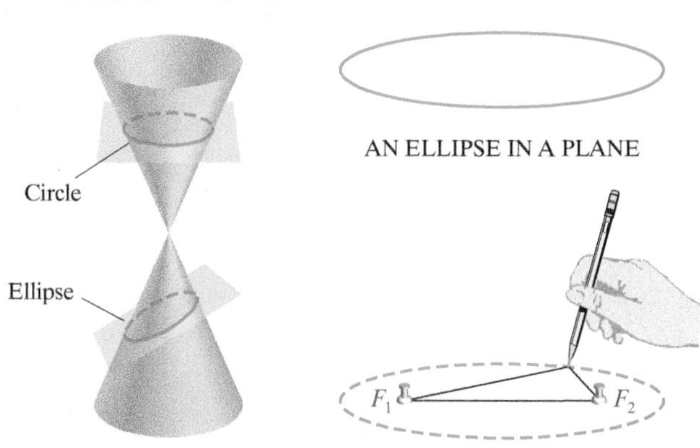

We can draw an ellipse by securing two tacks in a piece of cardboard, tying a string around them, placing a pencil as shown, and drawing with the string kept taut. The formal mathematical definition is related to this method of drawing.

An **ellipse** is defined as the set of all points in a plane such that the *sum* of the distances from two fixed points F_1 and F_2 (called the **foci**; singular, **focus**) is constant. In the preceding drawing, the tacks are at the foci. Ellipses have equations as follows.

EQUATION OF AN ELLIPSE

An ellipse with its center at the origin has equation

$$\frac{x^2}{a^2} + \frac{y^2}{b^2} = 1, \quad a, b > 0, \quad a \neq b. \qquad \text{(Standard form)}$$

We can almost think of a circle as a special kind of ellipse. A circle is formed when $a = b$ and the cutting plane is perpendicular to the axis of the cone. It is also formed when the foci, F_1 and F_2, are the same point. An ellipse with its foci close together is very nearly a circle.

When graphing ellipses, it helps to first find the intercepts. If we replace x with 0 in the standard form of the equation, we can find the y-intercepts:

$$\frac{0^2}{a^2} + \frac{y^2}{b^2} = 1$$

$$\frac{y^2}{b^2} = 1$$

$$y^2 = b^2$$

$$y = \pm b.$$

SKILL REVIEW *Graph linear equations using intercepts.* [2.5a]

Find the intercepts and then graph the line.

1. $x - y = -3$
2. $f(x) = -2x - 4$

Answers:

1. x-intercept: $(-3, 0)$; y-intercept: $(0, 3)$;

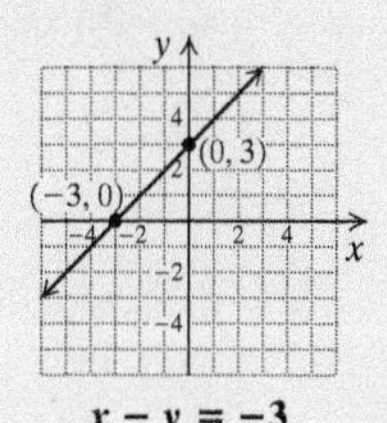

2. x-intercept: $(-2, 0)$; y-intercept: $(0, -4)$;

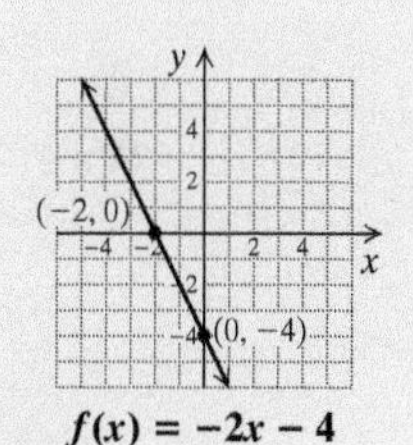

MyLab Math VIDEO

Thus the y-intercepts are $(0, b)$ and $(0, -b)$. Similarly, the x-intercepts are $(a, 0)$ and $(-a, 0)$. If $a > b$, the ellipse is horizontal and $(-a, 0)$ and $(a, 0)$ are **vertices** (singular, **vertex**). If $a < b$, the ellipse is vertical and $(0, -b)$ and $(0, b)$ are the vertices.

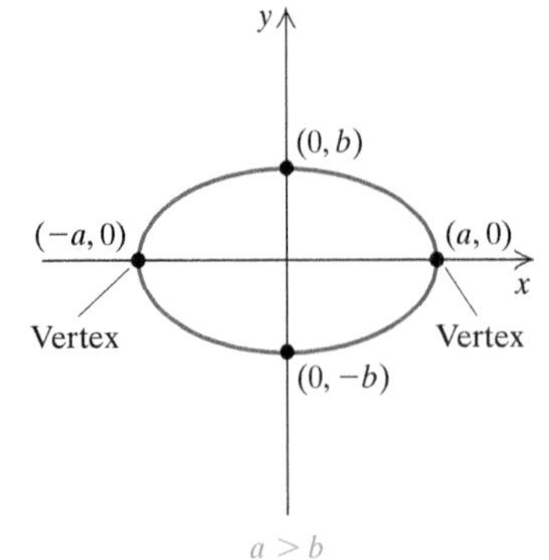

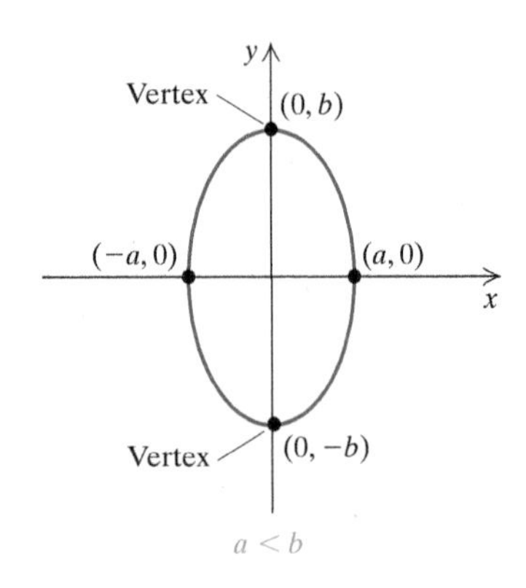

Plotting these points and filling in an oval-shaped curve, we get a graph of the ellipse. If a more precise graph is desired, we can plot more points.

> **INTERCEPTS AND VERTICES OF AN ELLIPSE**
>
> For the ellipse
>
> $$\frac{x^2}{a^2} + \frac{y^2}{b^2} = 1,$$
>
> the ***x*-intercepts** are $(-a, 0)$ and $(a, 0)$, and the ***y*-intercepts** are $(0, -b)$ and $(0, b)$. If $a > b$, then $(-a, 0)$ and $(a, 0)$ are the vertices. If $a < b$, then $(0, -b)$ and $(0, b)$ are the vertices.

EXAMPLE 1 Graph: $\frac{x^2}{4} + \frac{y^2}{9} = 1$.

Note that

$$\frac{x^2}{4} + \frac{y^2}{9} = \frac{x^2}{2^2} + \frac{y^2}{3^2}. \qquad a = 2, b = 3$$

Thus the x-intercepts are $(-2, 0)$ and $(2, 0)$, and the y-intercepts are $(0, -3)$ and $(0, 3)$. The vertices are $(0, -3)$ and $(0, 3)$. We plot these points and connect them with an oval-shaped curve. To be accurate, we might find some other points on the curve. We let $x = 1$ and solve for y:

$$\frac{1^2}{4} + \frac{y^2}{9} = 1$$

$$36\left(\frac{1}{4} + \frac{y^2}{9}\right) = 36 \cdot 1$$

$$36 \cdot \frac{1}{4} + 36 \cdot \frac{y^2}{9} = 36$$

$$9 + 4y^2 = 36$$

$$4y^2 = 27$$

$$y^2 = \frac{27}{4}$$

$$y = \pm\sqrt{\frac{27}{4}} \approx \pm 2.6.$$

Graph each ellipse.

GS **1.** $\frac{x^2}{9} + \frac{y^2}{4} = 1$

$\frac{x^2}{9} + \frac{y^2}{4} = \frac{x^2}{3^2} + \frac{y^2}{\square^2}$

$a = 3$, $b = \square$

The x-intercepts are $(-3, 0)$ and $(\square, 0)$.

The y-intercepts are $(0, \square)$ and $(0, 2)$.

We plot the intercepts and draw the ellipse.

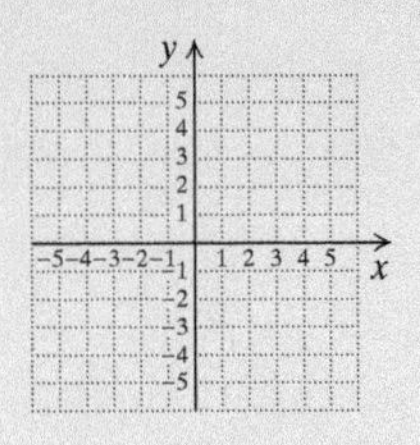

2. $\frac{x^2}{9} + \frac{y^2}{25} = 1$

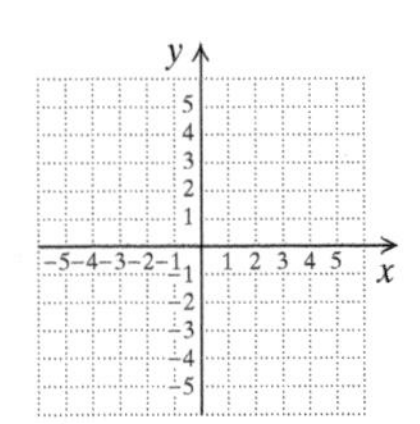

Answers

1.

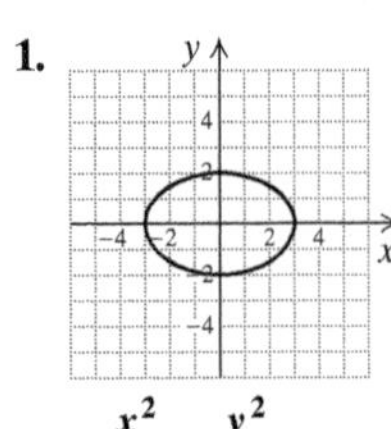

$\frac{x^2}{9} + \frac{y^2}{4} = 1$

2.

$\frac{x^2}{9} + \frac{y^2}{25} = 1$

Guided Solution:

1. 2, 2, 3, −2

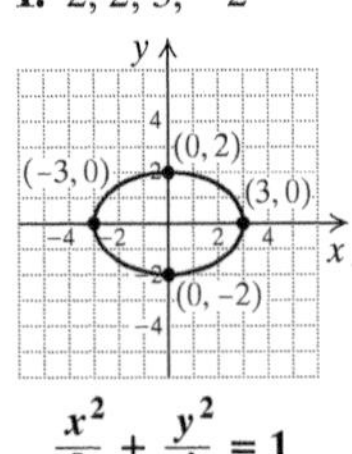

$\frac{x^2}{9} + \frac{y^2}{4} = 1$

Thus, $(1, 2.6)$ and $(1, -2.6)$ can also be plotted and used to draw the graph. Similarly, the points $(-1, -2.6)$ and $(-1, 2.6)$ can also be computed and plotted.

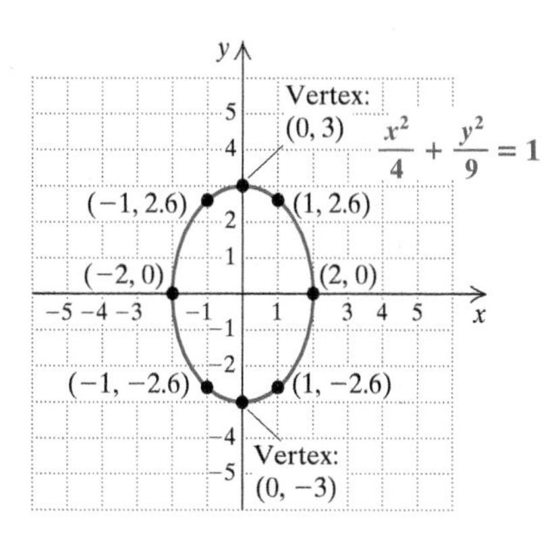

Do Exercises 1 and 2 on the preceding page.

EXAMPLE 2 Graph: $4x^2 + 25y^2 = 100$.

To write the equation in standard form, we multiply both sides by $\frac{1}{100}$:

$$\frac{1}{100}(4x^2 + 25y^2) = \frac{1}{100}(100) \qquad \text{Multiplying by } \tfrac{1}{100} \text{ to get 1 on the right side}$$

$$\left.\begin{aligned} \frac{1}{100}(4x^2) + \frac{1}{100}(25y^2) &= 1 \\ \frac{x^2}{25} + \frac{y^2}{4} &= 1 \end{aligned}\right\} \qquad \text{Simplifying}$$

$$\frac{x^2}{5^2} + \frac{y^2}{2^2} = 1. \qquad a = 5, b = 2$$

The x-intercepts are $(-5, 0)$ and $(5, 0)$, and the y-intercepts are $(0, -2)$ and $(0, 2)$. The vertices are $(-5, 0)$ and $(5, 0)$. We plot the intercepts and connect them with an oval-shaped curve. Other points can also be computed and plotted.

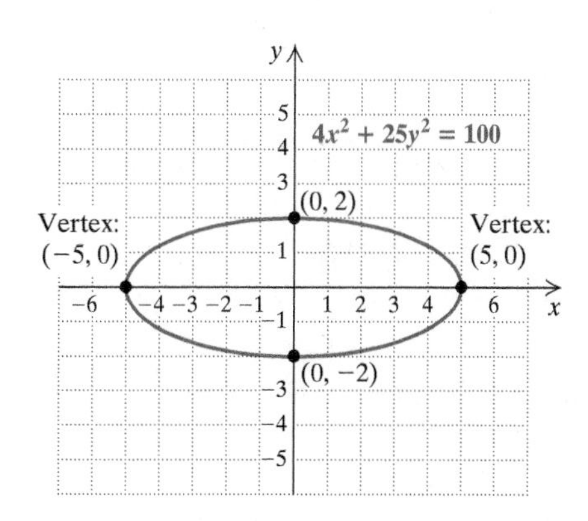

Do Exercise 3.

3. Graph: $16x^2 + 9y^2 = 144$.

$$16x^2 + 9y^2 = 144$$
$$\frac{1}{144}(16x^2 + 9y^2) = \frac{1}{144}(144)$$
$$\frac{x^2}{9} + \frac{y^2}{\square} = \square$$
$$\frac{x^2}{3^2} + \frac{y^2}{\square^2} = 1$$

$a = 3$, $b = \square$

The x-intercepts are $(-3, 0)$ and $(\square, 0)$.

The y-intercepts are $(0, \square)$ and $(0, 4)$.

We draw the ellipse.

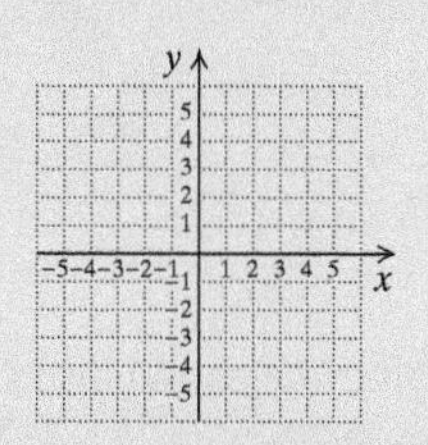

4. Graph:

$$\frac{(x + 2)^2}{16} + \frac{(y - 3)^2}{9} = 1.$$

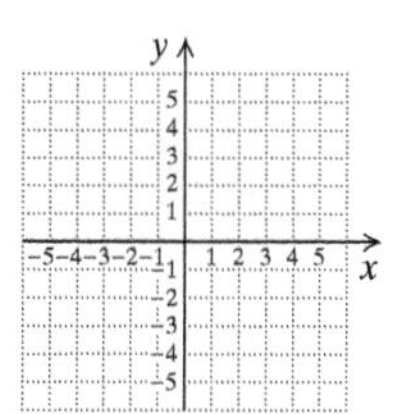

Answers

3.

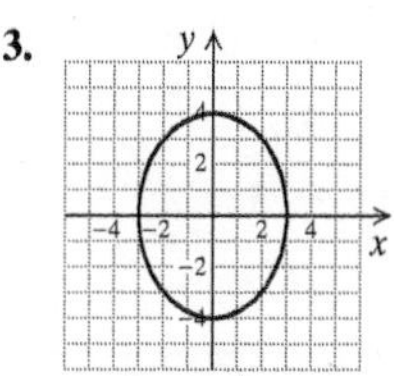

4.

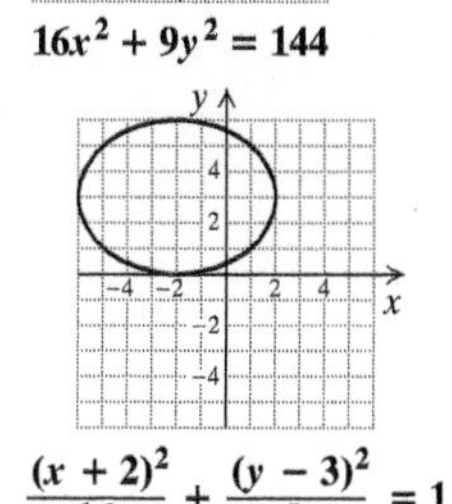

Guided Solution:
3. 16, 1, 4, 4, 3, −4

CALCULATOR CORNER

Graphing Ellipses In a Calculator Corner on p. 726, we graphed a circle by first solving the equation of the circle for y. We can graph an ellipse in the same way. Consider the ellipse in Example 2:

$$4x^2 + 25y^2 = 100$$
$$25y^2 = 100 - 4x^2$$
$$y^2 = \frac{100 - 4x^2}{25}$$
$$y = \pm\sqrt{\frac{100 - 4x^2}{25}}.$$

Now we enter

$$y_1 = \sqrt{\frac{100 - 4x^2}{25}} \quad \text{and}$$
$$y_2 = -\sqrt{\frac{100 - 4x^2}{25}}$$

and graph these equations in a square window.

Ellipses can also be graphed using the CONICS app in the APPS menu.

$y_1 = \sqrt{\frac{100 - 4x^2}{25}}$, $y_2 = -\sqrt{\frac{100 - 4x^2}{25}}$

6, −9, 9, −6

EXERCISES: Graph each ellipse.

1. $\frac{x^2}{9} + \frac{y^2}{4} = 1$
2. $16x^2 + 9y^2 = 144$
3. $\frac{(x-1)^2}{4} + \frac{(y+2)^2}{9} = 1$
4. $\frac{(x+2)^2}{16} + \frac{(y-3)^2}{9} = 1$

Horizontal translations and vertical translations, similar to those used in Chapter 7, can be used to graph ellipses that are not centered at the origin.

STANDARD FORM OF AN ELLIPSE

The standard form of a horizontal ellipse or a vertical ellipse centered at (h, k) is

$$\frac{(x-h)^2}{a^2} + \frac{(y-k)^2}{b^2} = 1.$$

The vertices are $(h + a, k)$ and $(h - a, k)$ if horizontal; $(h, k + b)$ and $(h, k - b)$ if vertical.

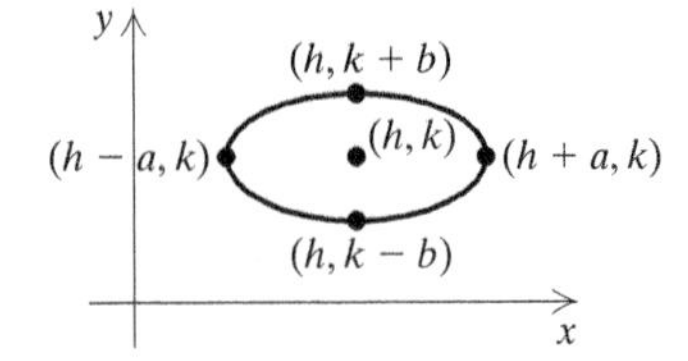

EXAMPLE 3 Graph: $\frac{(x-1)^2}{4} + \frac{(y+5)^2}{9} = 1.$

Note that

$$\frac{(x-1)^2}{4} + \frac{(y+5)^2}{9} = \frac{(x-1)^2}{2^2} + \frac{(y+5)^2}{3^2}.$$

Thus, $a = 2$ and $b = 3$. To determine the center of the ellipse, (h, k), note that

$$\frac{(x-1)^2}{2^2} + \frac{(y+5)^2}{3^2} = \frac{(x-1)^2}{2^2} + \frac{(y-(-5))^2}{3^2}.$$

Thus the center is $(1, -5)$. We locate $(1, -5)$ and then plot $(1 + 2, -5)$, $(1 - 2, -5)$, $(1, -5 + 3)$, and $(1, -5 - 3)$. These are the points $(3, -5)$, $(-1, -5)$, $(1, -2)$, and $(1, -8)$. The vertices are $(1, -8)$ and $(1, -2)$.

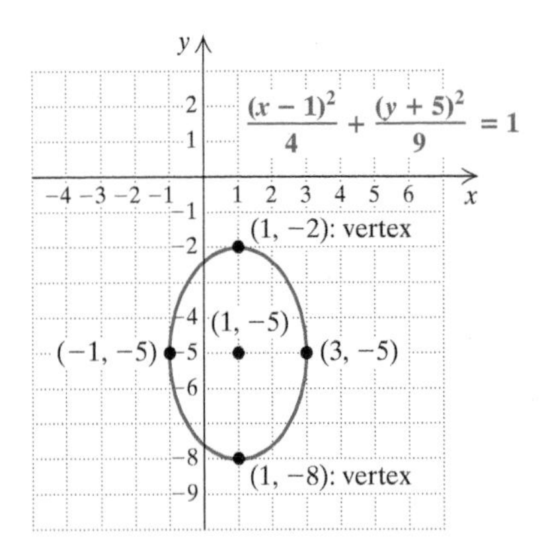

Note that this ellipse is the same as the ellipse in Example 1 but translated 1 unit to the right and 5 units down.

◀ **Do Exercise 4 on the preceding page.**

Ellipses have many applications. The orbits of planets and some comets around the sun are ellipses. The sun is located at one focus. Whispering galleries are also ellipses. A person standing at one focus will be able to hear the whisper of a person standing at the other focus. One example of a whispering gallery is found in the rotunda of the Capitol building in Washington, D.C.

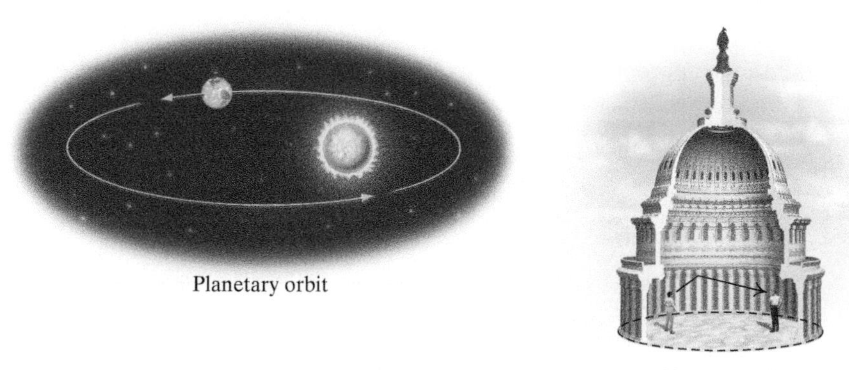
Planetary orbit

Whispering gallery

A medical instrument, the lithotripter, uses shock waves originating at one focus to crush a kidney stone located at the other focus.

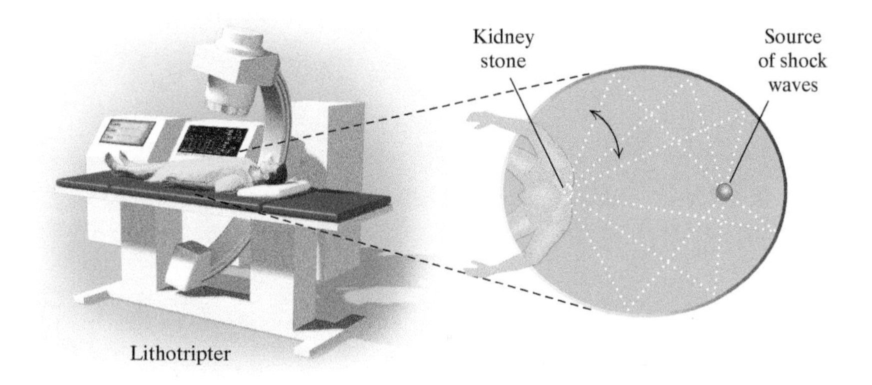

Lithotripter

Wind-driven forest fires can be roughly approximated as the union of "half"-ellipses. Shown below is an illustration of such a forest fire. The smaller half-ellipse on the left moves into the wind, and the elongated half-ellipse on the right moves out in the direction of the wind. The wind tends to spread the fire to the right, but part of the fire will still spread into the wind.

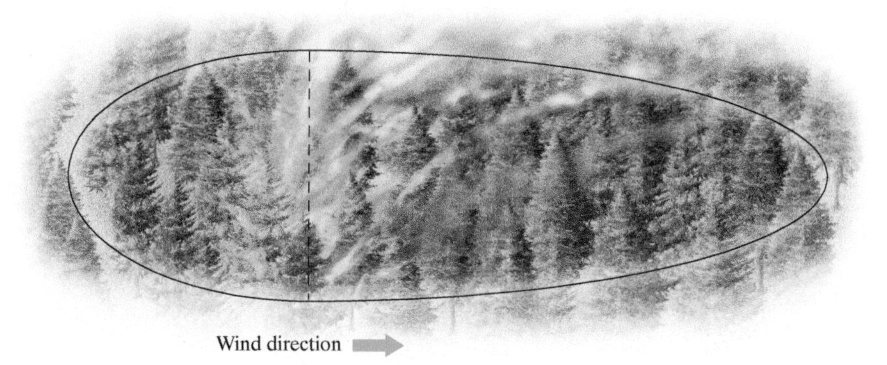

DATA: "Predicting Wind-Driven Wild Land Fire Size and Shape," Hal Anderson, Research Paper INT-305, U.S. Department of Agriculture, Forest Service, February 1983.

9.2 Exercise Set

FOR EXTRA HELP MyLab Math

✓ Check Your Understanding

Reading Check For each term, write the letter of the appropriate labeled part of the drawing.

RC1. ____ Ellipse **RC2.** ____ Focus

RC3. ____ Center **RC4.** ____ Vertex

Concept Check Fill in the blank with the correct ordered pair.

CC1. For the ellipse given by $\frac{x^2}{64} + \frac{y^2}{49} = 1$, the x-intercepts are ________ and ________.

CC2. For the ellipse given by $\frac{(x + 5)^2}{25} + \frac{(y - 2)^2}{100} = 1$, the center is ________ and the vertices are ________ and ________.

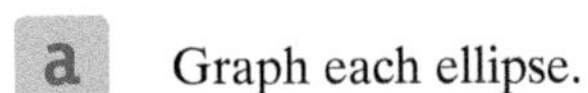

a Graph each ellipse.

1. $\frac{x^2}{9} + \frac{y^2}{36} = 1$

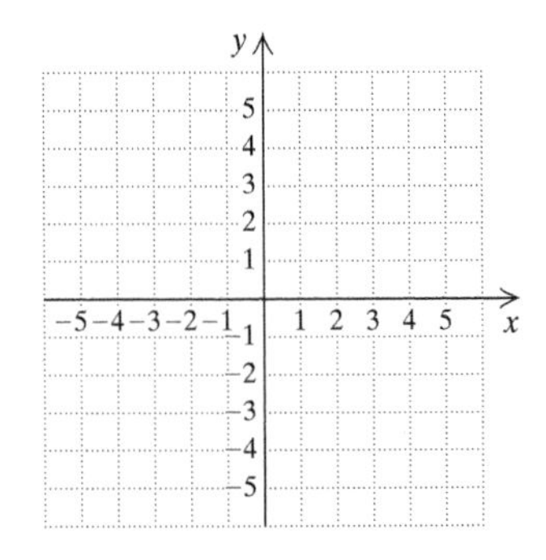

2. $\frac{x^2}{16} + \frac{y^2}{25} = 1$

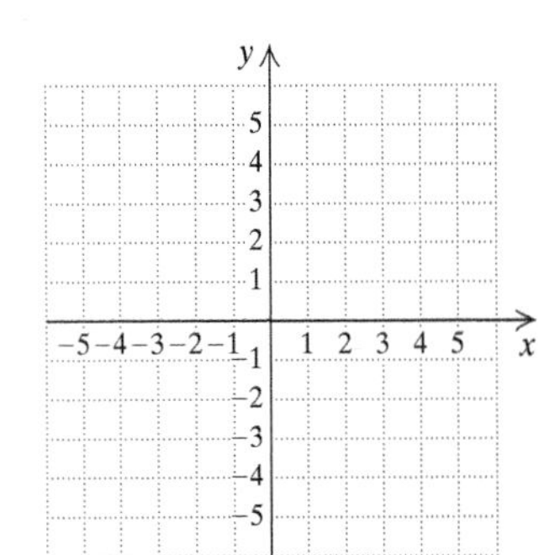

3. $\frac{x^2}{1} + \frac{y^2}{4} = 1$

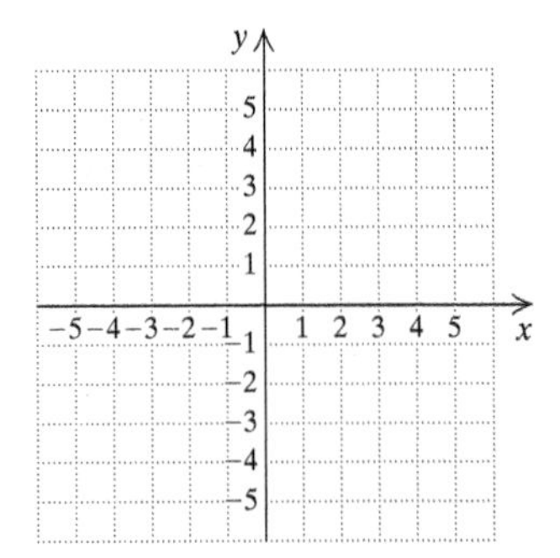

4. $\frac{x^2}{4} + \frac{y^2}{1} = 1$

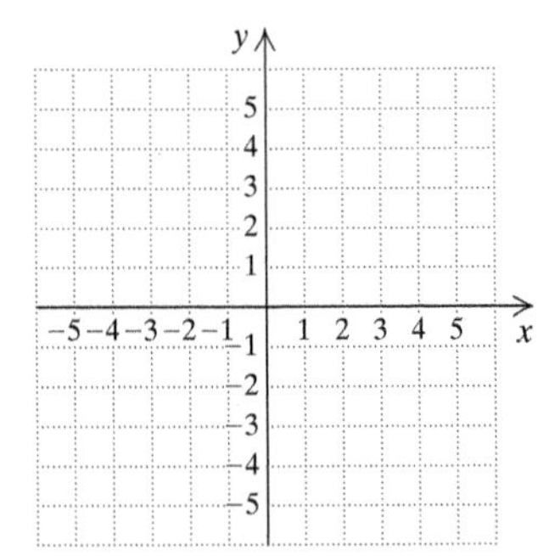

5. $4x^2 + 9y^2 = 36$
(*Hint*: Divide by 36.)

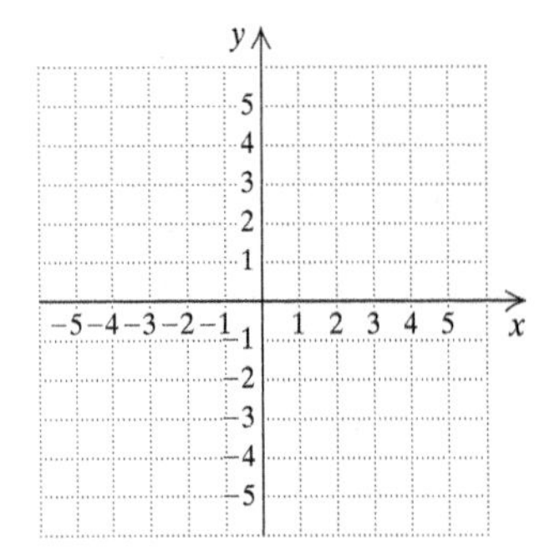

6. $9x^2 + 4y^2 = 36$

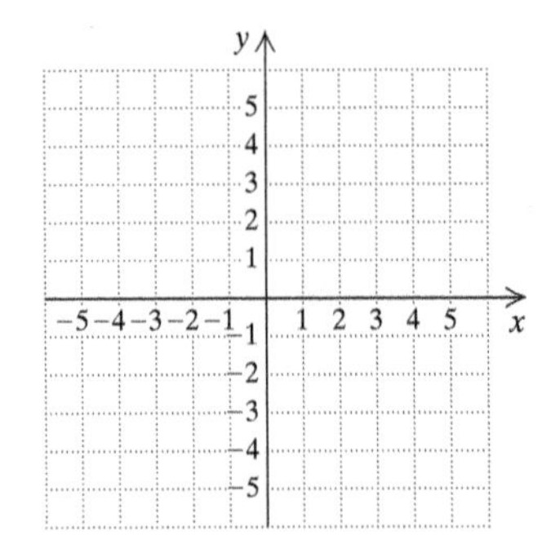

7. $x^2 + 4y^2 = 4$

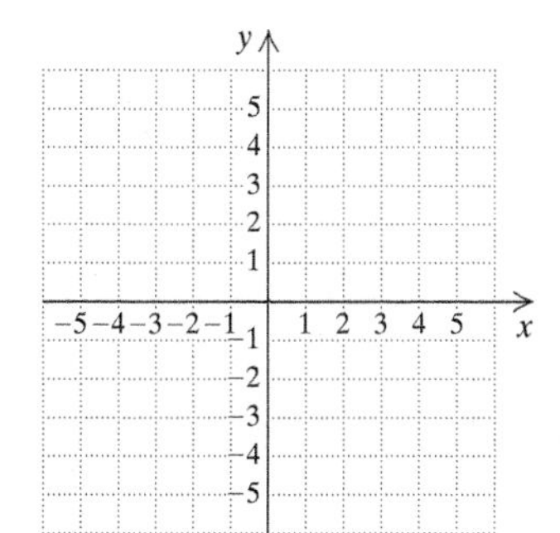

8. $9x^2 + 16y^2 = 144$

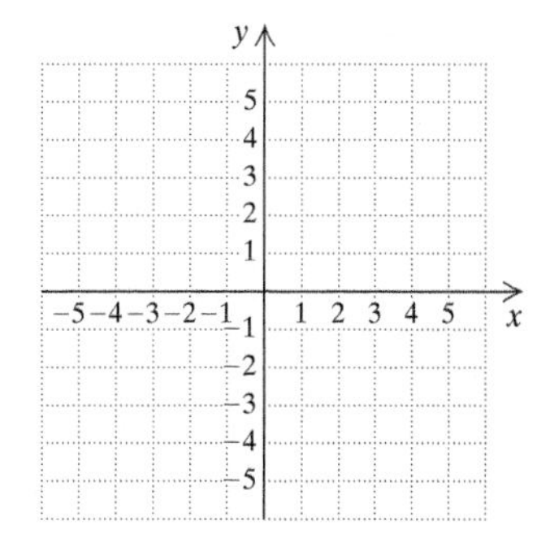

9. $2x^2 + 3y^2 = 6$

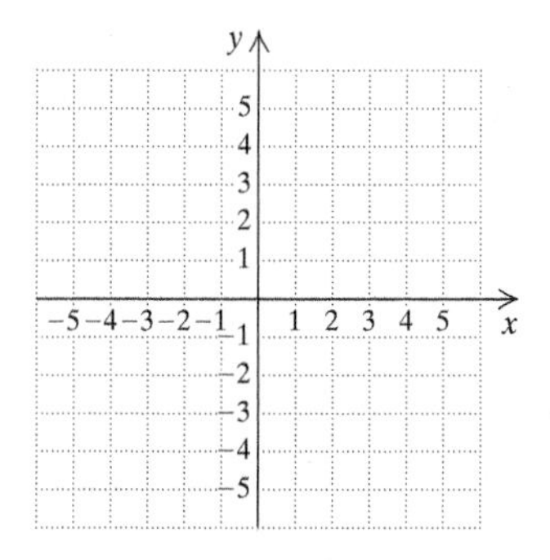

10. $5x^2 + 7y^2 = 35$

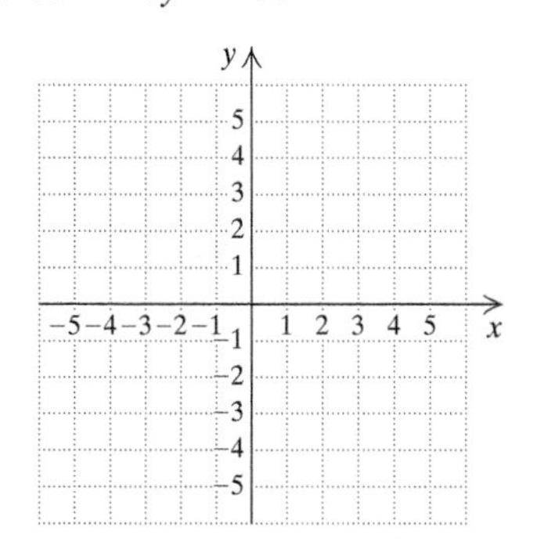

11. $12x^2 + 5y^2 - 120 = 0$

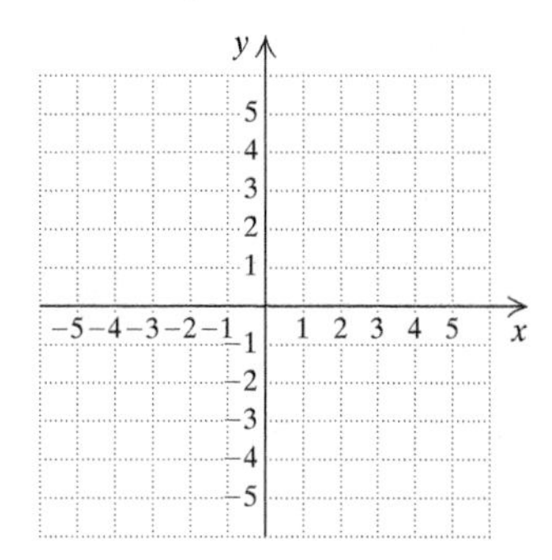

12. $3x^2 + 7y^2 - 63 = 0$

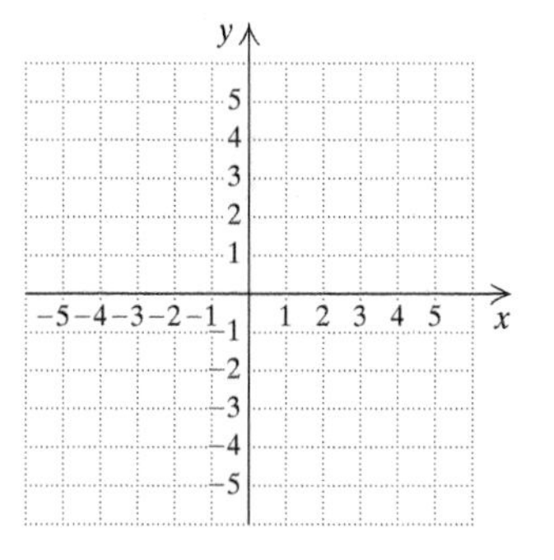

13. $\dfrac{(x-2)^2}{9} + \dfrac{(y-1)^2}{25} = 1$

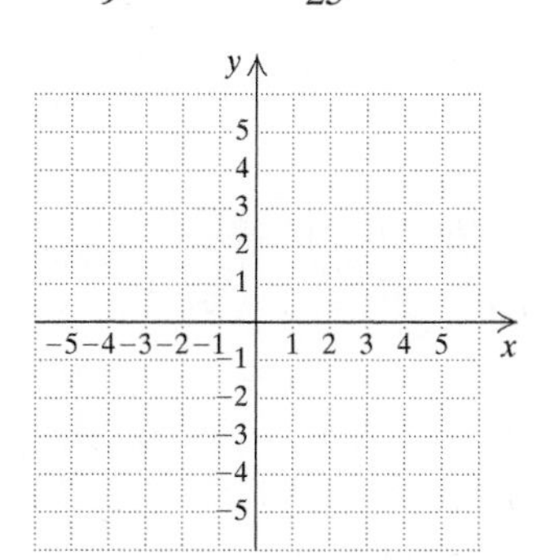

14. $\dfrac{(x-3)^2}{4} + \dfrac{(y-4)^2}{9} = 1$

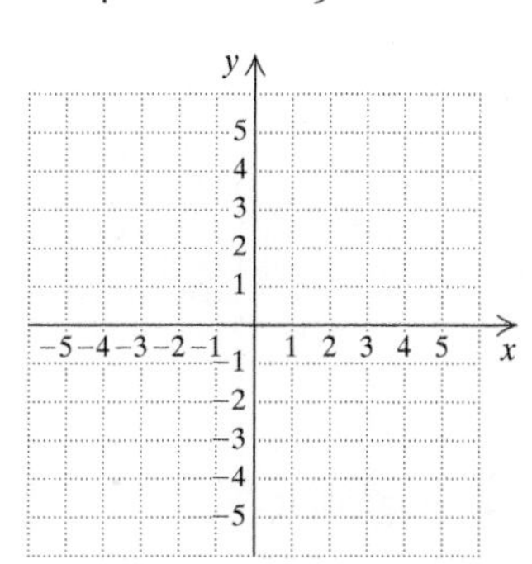

15. $\dfrac{(x+1)^2}{16} + \dfrac{(y+2)^2}{25} = 1$

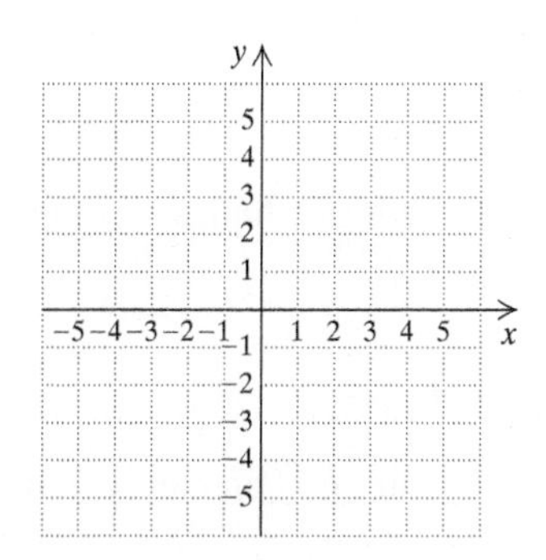

16. $\dfrac{(x+3)^2}{4} + \dfrac{(y-2)^2}{36} = 1$

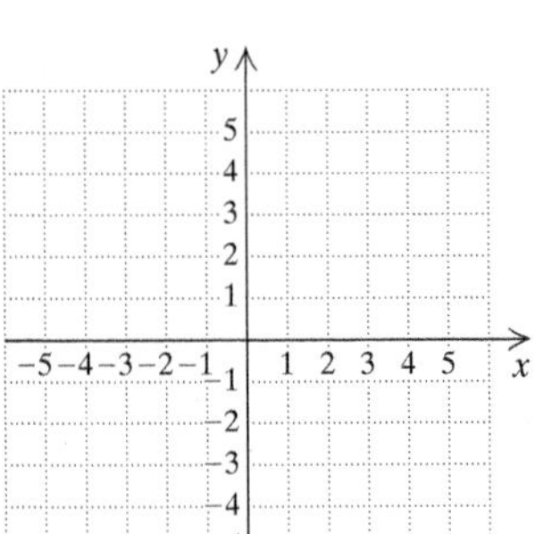

17. $12(x-1)^2 + 3(y+2)^2 = 48$

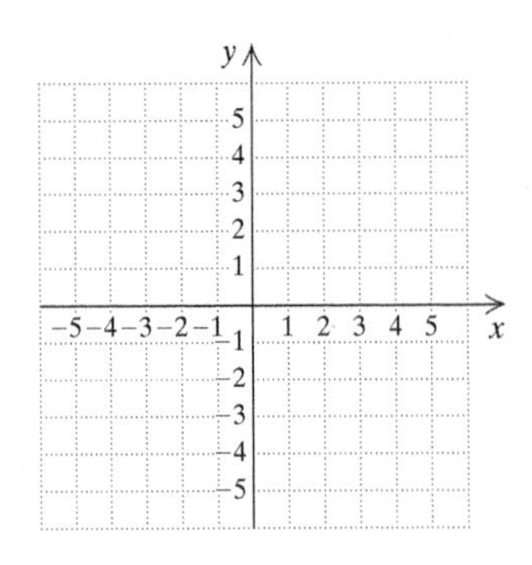

18. $4(x-2)^2 + 9(y+2)^2 = 36$

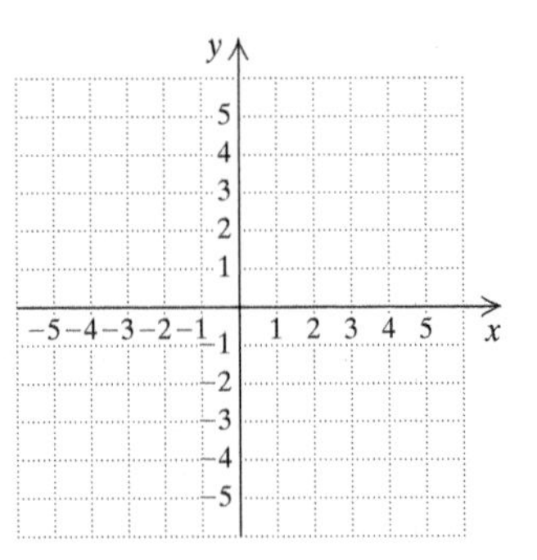

19. $(x+3)^2 + 4(y+1)^2 - 10 = 6$

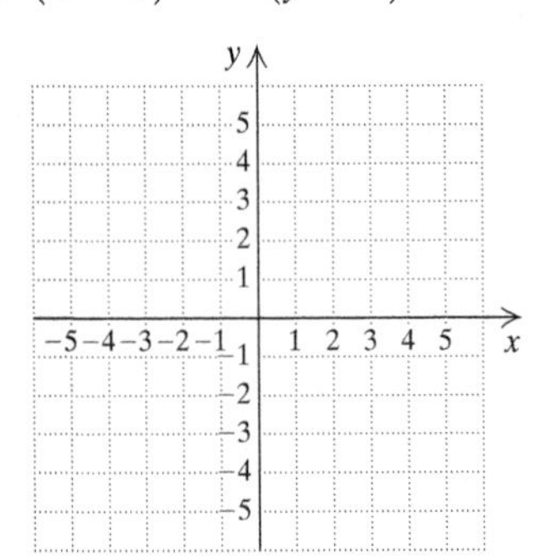

20. $8(x+1)^2 + (y+1)^2 - 12 = 4$

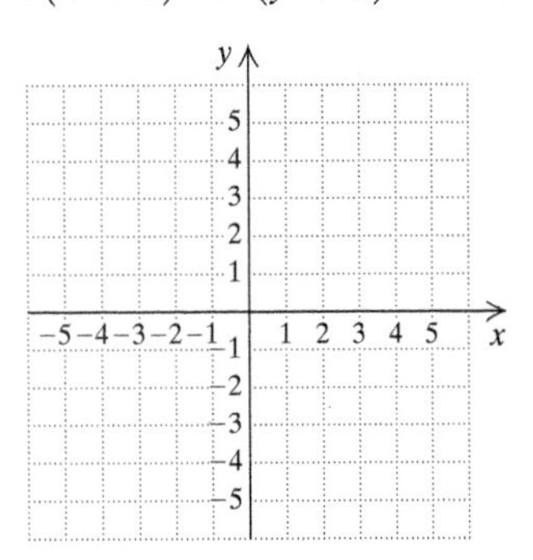

Skill Maintenance

Solve. Give exact solutions. [7.2a]

21. $3x^2 - 2x + 7 = 0$

22. $3x^2 - 12x + 7 = 0$

23. $x^2 + x + 2 = 0$

24. $x^2 + 2x = 10$

Solve. Give both exact and approximate solutions to the nearest tenth. [7.2a]

25. $x^2 + 2x - 17 = 0$

26. $x^2 - 2x = 10$

27. $3x^2 - 12x + 7 = 10 - x^2 + 5x$

28. $2x^2 + 3x - 4 = 0$

Convert to a logarithmic equation. [8.3b]

29. $a^{-t} = b$

30. $8^a = 17$

Convert to an exponential equation. [8.3b]

31. $\ln 24 = 3.1781$

32. $p = \log_e W$

Synthesis

Find an equation of an ellipse that contains the following points.

33. $(-7, 0)$, $(7, 0)$, $(0, -5)$, and $(0, 5)$

34. $(-2, -1)$, $(6, -1)$, $(2, -4)$, and $(2, 2)$

35. *Theatrical Lighting.* The spotlight on a pair of ice skaters casts an ellipse of light on the floor below them that is 6 ft wide and 10 ft long. Find an equation of that ellipse if the performers are in its center, x is the distance from the performers to the side of the ellipse, and y is the distance from the performers to the top of the ellipse.

36. Complete the square as needed and find an equivalent equation in standard form:

$$x^2 - 4x + 4y^2 + 8y - 8 = 0.$$

Mid-Chapter Review

Concept Reinforcement

Determine whether each statement is true or false.

__________ **1.** The graph of $y - x^2 = 5$ is a parabola opening up. [9.1a]

__________ **2.** The graph of $\frac{x^2}{10} + \frac{y^2}{12} = 1$ is an ellipse with its center at the origin. [9.2a]

__________ **3.** The graph of $\frac{(x-1)^2}{10} + \frac{(y-4)^2}{8} = 1$ is an ellipse with its center not at the origin. [9.2a]

__________ **4.** The graph of $(x-1)^2 + (y-4)^2 = 16$ is a circle with its center at $(-1, -4)$. [9.1d]

Guided Solutions

Fill in each blank with the number or the expression that creates a correct solution.

5. For the points $(-6, 2)$ and $(4, -1)$:

a) Find the distance between the points. [9.1b]

b) Find the midpoint of the segment with the given endpoints. [9.1c]

We let $(x_1, y_1) = (-6, 2)$ and $(x_2, y_2) = (4, -1)$.

a) $d = \sqrt{(x_2 - x_1)^2 + (y_2 - y_1)^2} = \sqrt{(\square - \square)^2 + (\square - \square)^2} = \sqrt{(\square)^2 + (\square)^2}$

$= \sqrt{\square + \square} = \sqrt{\square} \approx \square$

b) $\left(\frac{x_1 + x_2}{2}, \frac{y_1 + y_2}{2}\right) = \left(\frac{\square + 4}{2}, \frac{2 + \square}{2}\right) = \left(\frac{\square}{2}, \frac{\square}{2}\right) = (\square, \square)$

6. Find the center and the radius of the circle

$x^2 + y^2 - 20x + 4y + 79 = 0.$ [9.1d]

$$x^2 - 20x + y^2 + 4y = \square$$

$$x^2 - 20x + \square + y^2 + 4y + \square = -79 + \square + \square$$

$$(x - \square)^2 + (y + \square)^2 = \square$$

$$(x - \square)^2 + (y - \square)^2 = \square^2$$

Center: $(\square, \square)$; radius: $\square$.

Mixed Review

Find the distance between each pair of points. Where appropriate, give an approximation to three decimal places. [9.1b]

7. $(5, -6)$ and $(2, -9)$

8. $(2.3, 8)$ and $(-8, 4.2)$

9. $(0, \sqrt{6})$ and $(-\sqrt{5}, 0)$

Find the midpoint of the segment with the given endpoints. [9.1c]

10. $(-11, 3)$ and $(-8, 12)$

11. $\left(-\frac{5}{6}, \frac{1}{3}\right)$ and $\left(\frac{1}{2}, \frac{5}{12}\right)$

12. $(7.2, -4.6)$ and $(-10.2, -3.2)$

Find the center and the radius of each circle. [9.1d]

13. $x^2 + y^2 = 121$

14. $(x - 13)^2 + (y + 9)^2 = 109$

15. $x^2 + (y - 5)^2 = 14$

16. $x^2 + y^2 + 6x - 14y + 42 = 0$

Find an equation of the circle having the given center and radius. [9.1d]

17. Center $(0, 0)$, radius 1

18. Center $\left(-\frac{1}{2}, \frac{3}{4}\right)$, radius $\frac{9}{2}$

19. Center $(-8, 6)$, radius $\sqrt{17}$

20. Center $(3, -5)$, radius $2\sqrt{5}$

Graph. [9.1a], [9.1d], [9.2a]

21. $\frac{x^2}{4} + \frac{y^2}{36} = 1$

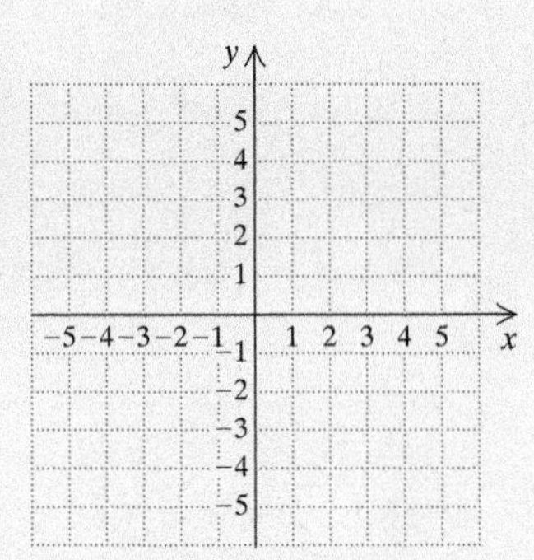

22. $y = x^2 + 2x - 1$

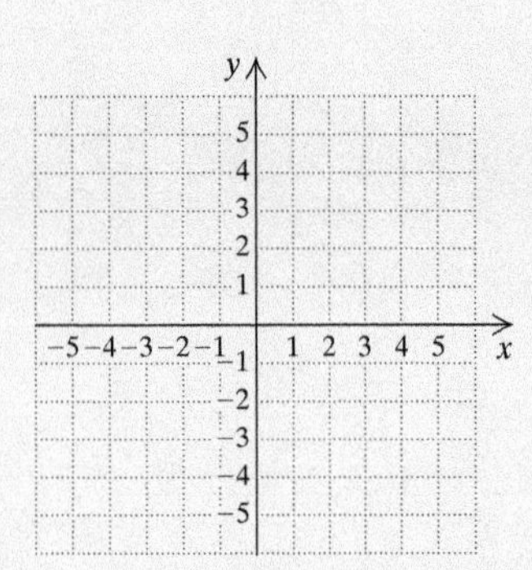

23. $(x - 1)^2 + (y + 2)^2 = 9$

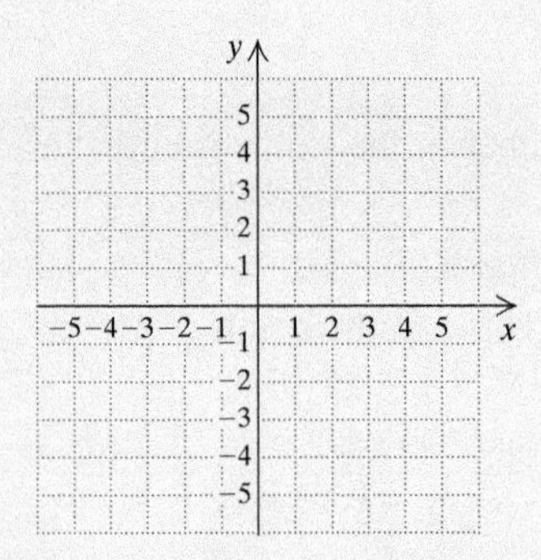

24. $x = y^2 - 2$

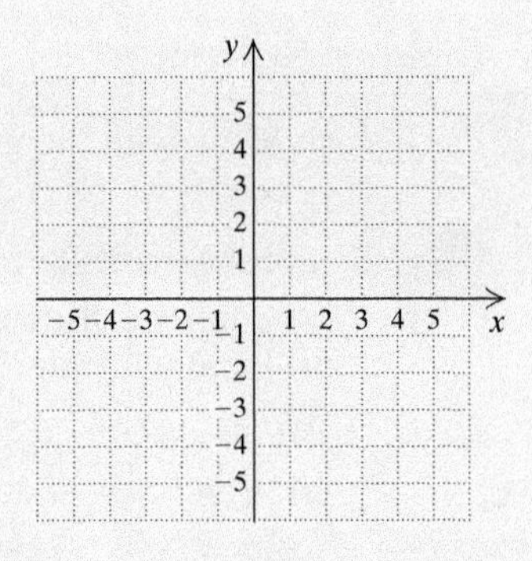

25. $x^2 + y^2 = \frac{9}{4}$

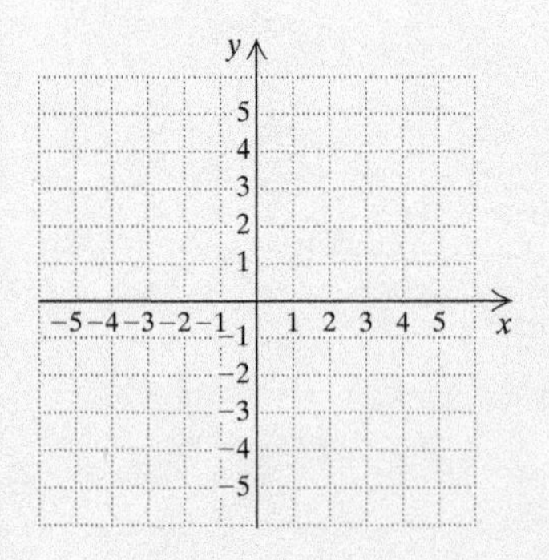

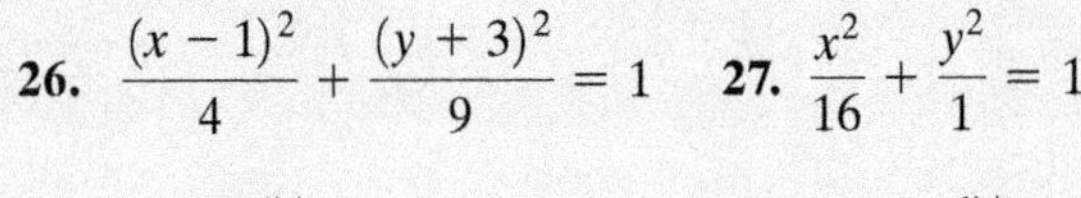

26. $\frac{(x - 1)^2}{4} + \frac{(y + 3)^2}{9} = 1$

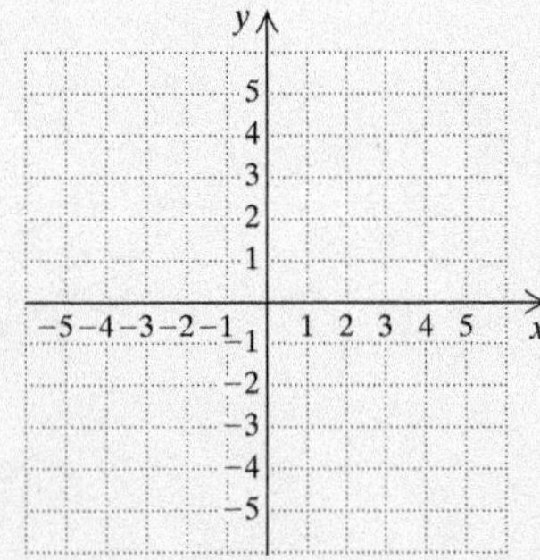

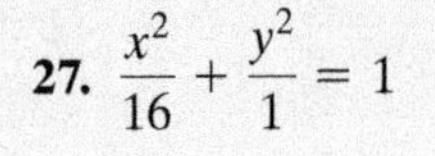

27. $\frac{x^2}{16} + \frac{y^2}{1} = 1$

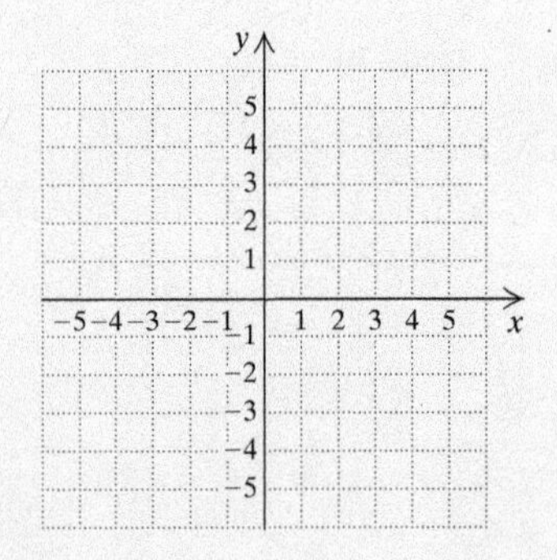

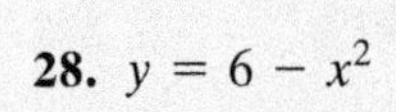

28. $y = 6 - x^2$

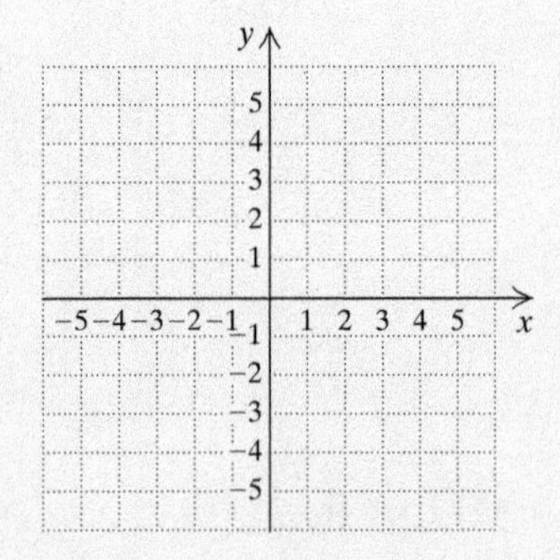

Understanding Through Discussion and Writing

29. How could a graphing calculator be used to graph an equation of the form $x = ay^2 + by + c$? [9.1a]

30. Is the center of a circle part of the circle? Why or why not? [9.1d]

31. An eccentric person builds a pool table in the shape of an ellipse with a hole at one focus and a tiny dot at the other. Guests are amazed at how many bank shots the owner of the pool table makes. Explain how this can happen. [9.2a]

32. *Wind-Driven Forest Fires.*

a) Graph the wind-driven fire formed as the union of the following two curves:

$$\frac{x^2}{10.3^2} + \frac{y^2}{4.8^2} = 1, \quad x \geq 0; \qquad \frac{x^2}{3.6^2} + \frac{y^2}{4.8^2} = 1, \quad x \leq 0.$$

b) What other factors do you think affect the shape of forest fires? [9.2a]

STUDYING FOR SUCCESS *Looking Ahead*

- ☐ As you register for next semester's courses, evaluate both your work and your family commitments.
- ☐ If you are registering for another math course, consider keeping your notes, tests, and text from this course as resources.

Hyperbolas

9.3

OBJECTIVES

a Graph the standard form of the equation of a hyperbola.

b Graph equations (nonstandard form) of hyperbolas.

a HYPERBOLAS

MyLab Math ANIMATION

A **hyperbola** looks like a pair of parabolas, but the actual shapes are different. A hyperbola has two **vertices** and the line through the vertices is known as an **axis**. The point halfway between the vertices is called the **center**.

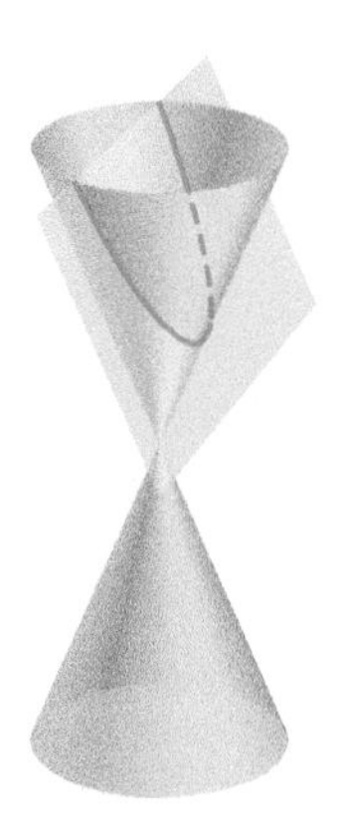

Parabola

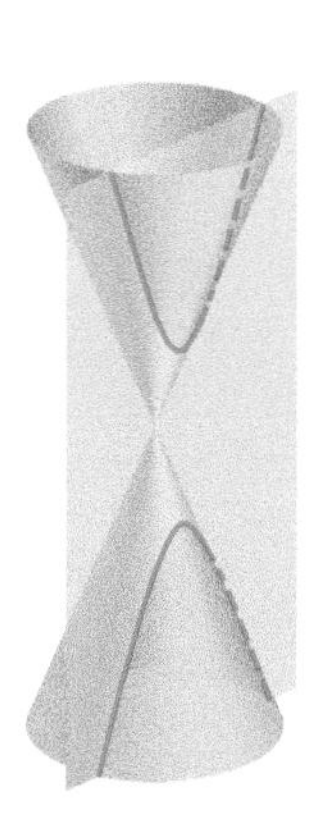

Hyperbola in three dimensions

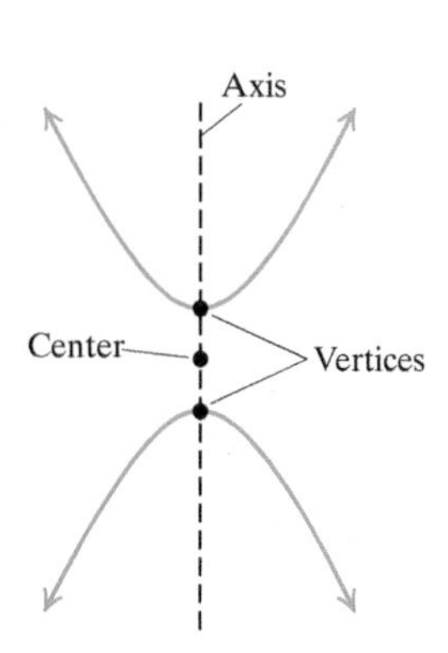

Hyperbola in a plane

SKILL REVIEW *Solve a formula for a specified letter.* [1.2a]

Solve each of the following.

1. $xy = -7$ for x
2. $16x^2 - 9y^2 = 144$ for x^2

Answers: 1. $x = -\frac{7}{y}$ **2.** $x^2 = \frac{144 + 9y^2}{16}$, or $x^2 = 9 + \frac{9}{16}y^2$

MyLab Math VIDEO

EQUATIONS OF HYPERBOLAS

Hyperbolas with their centers at the origin have equations as follows:

Axis horizontal: $\frac{x^2}{a^2} - \frac{y^2}{b^2} = 1$;   Axis vertical: $\frac{y^2}{b^2} - \frac{x^2}{a^2} = 1$.

To graph a hyperbola, it helps to begin by graphing two lines called **asymptotes**. Although the asymptotes themselves are not part of the graph, they serve as guidelines for an accurate sketch.

ASYMPTOTES OF A HYPERBOLA

For hyperbolas with equations as given above, the **asymptotes** are the lines

$$y = \frac{b}{a}x \quad \text{and} \quad y = -\frac{b}{a}x.$$

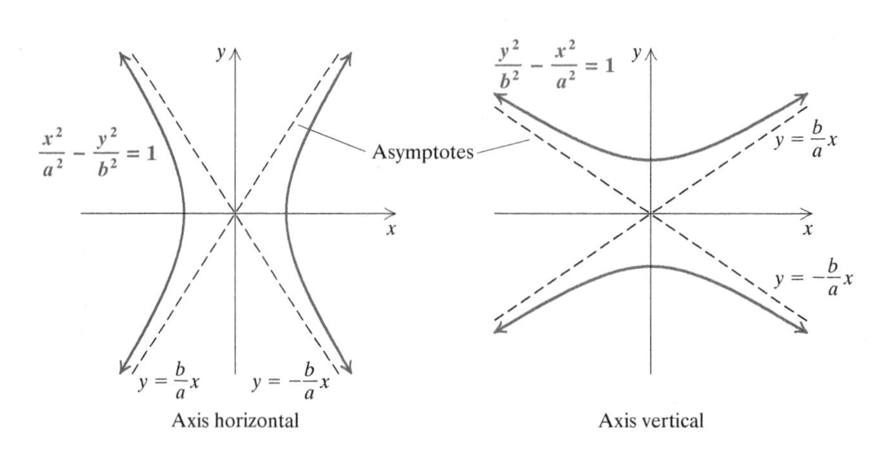

As a hyperbola gets further away from the origin, it gets closer and closer to its asymptotes. The larger $|x|$ gets, the closer the graph gets to an asymptote. The asymptotes act to "constrain" the graph of a hyperbola. On the other hand, parabolas are *not* constrained by any asymptotes.

The next thing to do after sketching asymptotes is to plot vertices. Then it is easy to sketch the curve.

1. Graph: $\frac{x^2}{16} - \frac{y^2}{25} = 1$.

$\frac{x^2}{16} - \frac{y^2}{25} = \frac{x^2}{4^2} - \frac{y^2}{\quad^2}$

$a = 4,\ b = \quad$

The asymptotes are $y = \frac{5}{4}x$ and $y = \quad$.

The hyperbola is horizontal.

The x-intercepts are $(-4, 0)$ and $(\quad, 0)$.

We sketch the asymptotes, plot the intercepts, and draw the hyperbola.

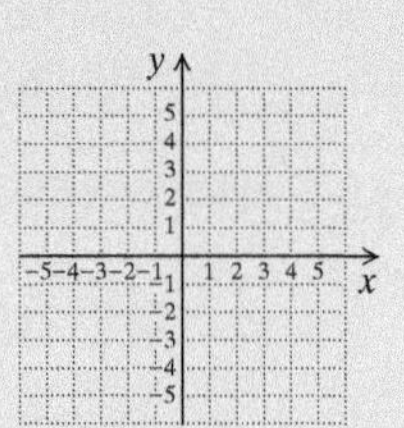

EXAMPLE 1 Graph: $\frac{x^2}{4} - \frac{y^2}{9} = 1$.

Note that

$$\frac{x^2}{4} - \frac{y^2}{9} = \frac{x^2}{2^2} - \frac{y^2}{3^2},\quad \text{Identifying } a \text{ and } b$$

so $a = 2$ and $b = 3$. The asymptotes are thus

$$y = \frac{3}{2}x \quad \text{and} \quad y = -\frac{3}{2}x.$$

We sketch them, as shown in the graph on the left below.

For horizontal hyperbolas or vertical hyperbolas centered at the origin, the vertices also serve as intercepts. Since this hyperbola is horizontal, we replace y with 0 and solve for x. We see that $x^2/2^2 = 1$ when $x = \pm 2$. The intercepts are $(2, 0)$ and $(-2, 0)$. You can check that no y-intercepts exist. The vertices are $(-2, 0)$ and $(2, 0)$.

Finally, we plot the intercepts and sketch the graph. Through each intercept, we draw a smooth curve that approaches the asymptotes closely, as shown.

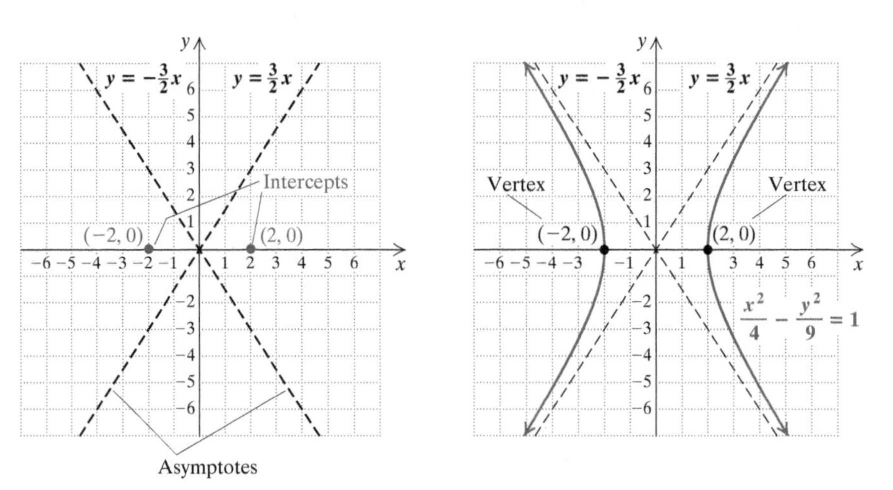

◀ **Do Exercise 1.**

Answer

1.

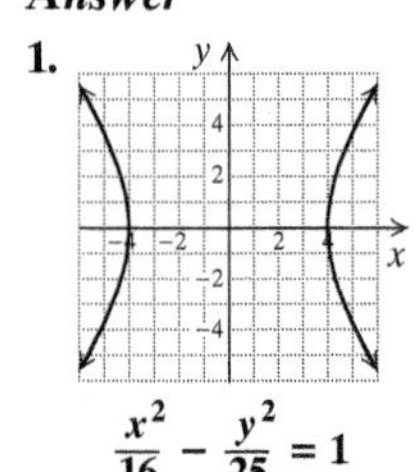

Guided Solution:

1. 5, 5, $-\frac{5}{4}x$, 4

$y = -\frac{5}{4}x$ $y = \frac{5}{4}x$ (4, 0) (−4, 0)

$\frac{x^2}{16} - \frac{y^2}{25} = 1$

EXAMPLE 2 Graph: $\dfrac{y^2}{36} - \dfrac{x^2}{4} = 1$.

Note that

$$\frac{y^2}{36} - \frac{x^2}{4} = \frac{y^2}{6^2} - \frac{x^2}{2^2} = 1.$$

> The intercept distance is found in the term without the minus sign. Here there is a y in this term, so the intercepts are on the y-axis.

The asymptotes are thus $y = \frac{6}{2}x$ and $y = -\frac{6}{2}x$, or $y = 3x$ and $y = -3x$.

The numbers 6 and 2 can be used to sketch a rectangle that helps with graphing. Using ± 2 as x-coordinates and ± 6 as y-coordinates, we form all possible ordered pairs: $(2, 6)$, $(2, -6)$, $(-2, 6)$, and $(-2, -6)$. We plot these pairs and lightly sketch a rectangle through them. The asymptotes pass through the corners (see the figure on the left below). Since the hyperbola is vertical, we plot its y-intercepts, $(0, 6)$ and $(0, -6)$. The vertices are $(0, -6)$ and $(0, 6)$. Finally, we draw curves through the intercepts toward the asymptotes, as shown below.

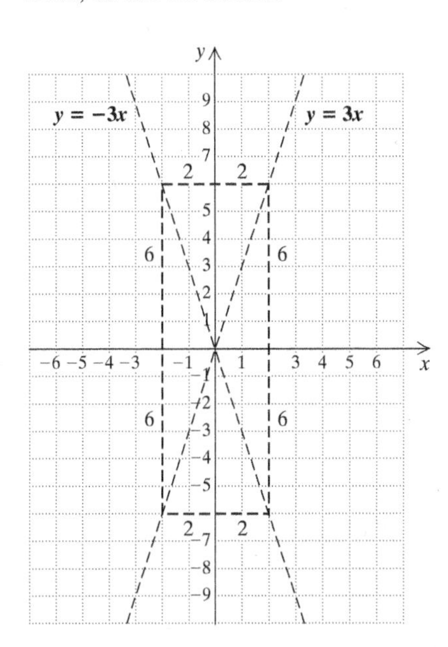

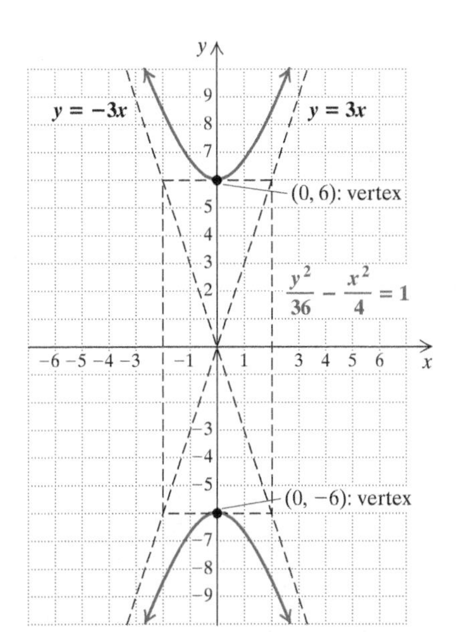

Do Exercise 2. ▶

Although we will not consider these equations here, hyperbolas with center at (h, k) are given by

$$\frac{(x-h)^2}{a^2} - \frac{(y-k)^2}{b^2} = 1 \quad \text{or} \quad \frac{(y-k)^2}{b^2} - \frac{(x-h)^2}{a^2} = 1.$$

Hyperbolas have many applications. A jet breaking the sound barrier creates a sonic boom with a wave front the shape of a cone. The intersection of the cone with the ground is one branch of a hyperbola. Some comets travel in hyperbolic orbits, and a cross section of certain lenses may be hyperbolic in shape.

2. Graph.

a) $\dfrac{y^2}{9} - \dfrac{x^2}{49} = 1$

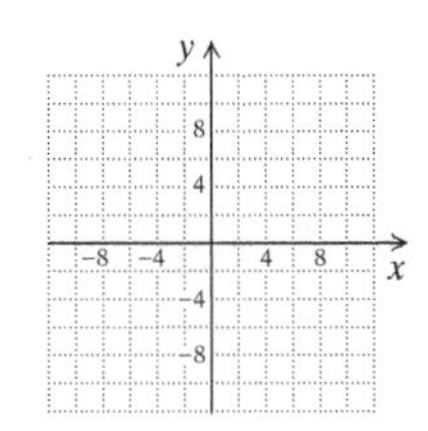

b) $\dfrac{x^2}{49} - \dfrac{y^2}{9} = 1$

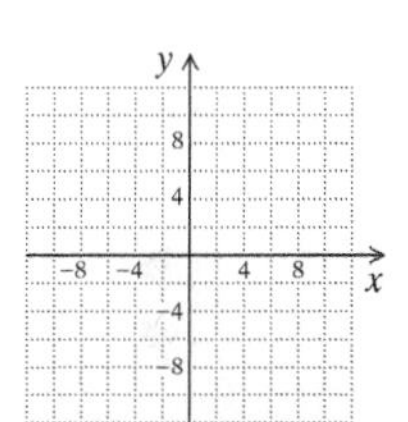

Answers

2. (a)

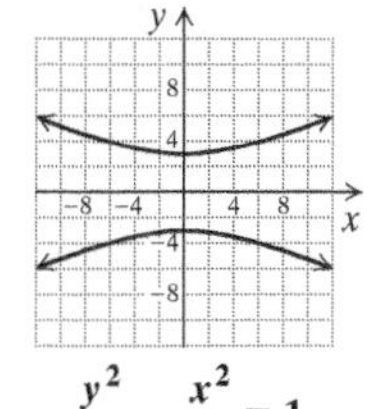

$\dfrac{y^2}{9} - \dfrac{x^2}{49} = 1$

(b)

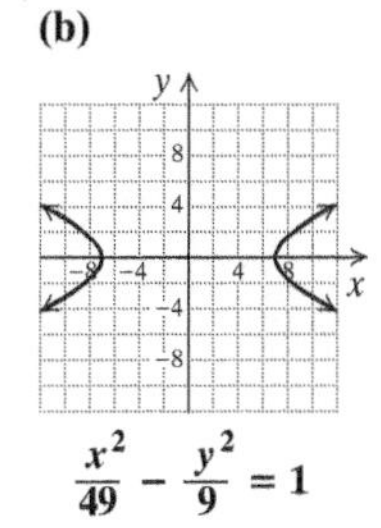

$\dfrac{x^2}{49} - \dfrac{y^2}{9} = 1$

b HYPERBOLAS (NONSTANDARD FORM)

The equations for hyperbolas just examined are the standard ones, but there are other hyperbolas. We consider some of them.

> Hyperbolas having the x- and y-axes as asymptotes have equations as follows:
>
> $xy = c$, where c is a nonzero constant.

EXAMPLE 3 Graph: $xy = -8$.

We first solve for y:

$$y = -\frac{8}{x}.$$ Dividing by x on both sides. Note that $x \neq 0$.

Next, we find some solutions, keeping the results in a table. Note that x cannot be 0 and that for large values of $|x|$, y will be close to 0. Thus the x- and y-axes serve as asymptotes. We plot the points and draw the hyperbola.

x	y
2	−4
−2	4
4	−2
−4	2
1	−8
−1	8
8	−1
−8	1

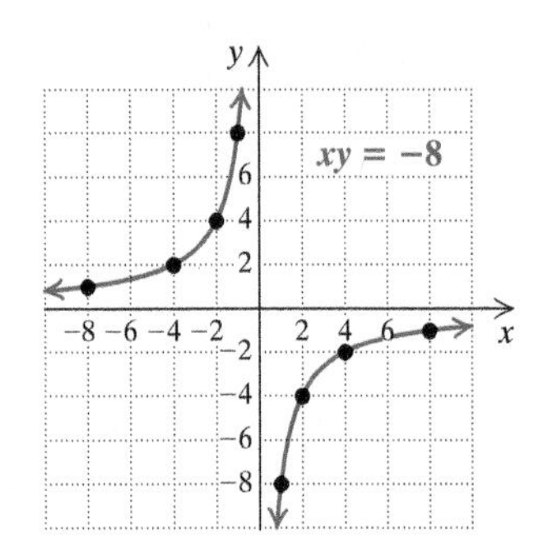

3. Graph: $xy = 8$.

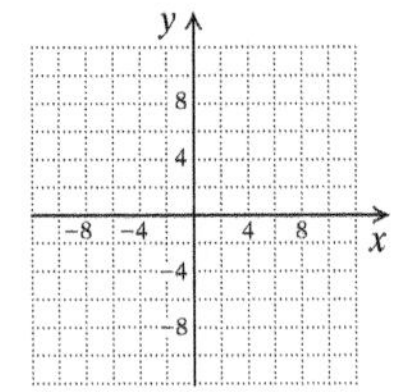

◀ Do Exercise 3.

CALCULATOR CORNER

Graphing Hyperbolas Graphing hyperbolas is similar to graphing circles and ellipses. First, we solve the equation of the hyperbola for y and then graph the two resulting functions. Consider the hyperbola

$$\frac{x^2}{25} - \frac{y^2}{49} = 1.$$

Solving for y, we get

$$y_1 = \frac{7}{5}\sqrt{x^2 - 25} \quad \text{and} \quad y_2 = -\frac{7}{5}\sqrt{x^2 - 25}.$$

Now, we graph these equations in a square viewing window. Hyperbolas can also be graphed using the CONICS app from the APPS menu.

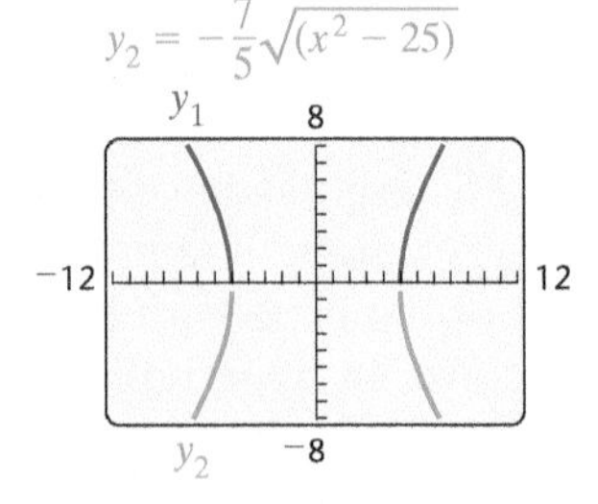

EXERCISES: Graph each hyperbola.

1. $\frac{x^2}{16} - \frac{y^2}{60} = 1$
2. $\frac{x^2}{20} - \frac{y^2}{64} = 1$
3. $16x^2 - 3y^2 = 48$
4. $45x^2 - 9y^2 = 405$

Answer

3.

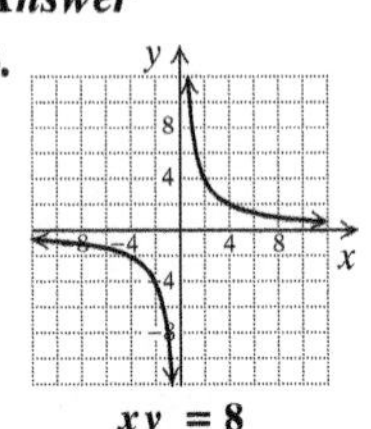

Visualizing for Success

Match each equation with its graph.

1. $4y^2 + x^2 = 16$

2. $y = \left(\frac{1}{2}\right)^x$

3. $y - x^2 = 2$

4. $y^2 - 4x^2 = 16$

5. $4x - 5y = 20$

6. $x^2 + y^2 + 2x - 6y + 6 = 0$

7. $x^2 - y^2 = 16$

8. $4x^2 + y^2 = 16$

9. $y = \log_2 x$

10. $x - y^2 = 2$

Answers on page A-38

A

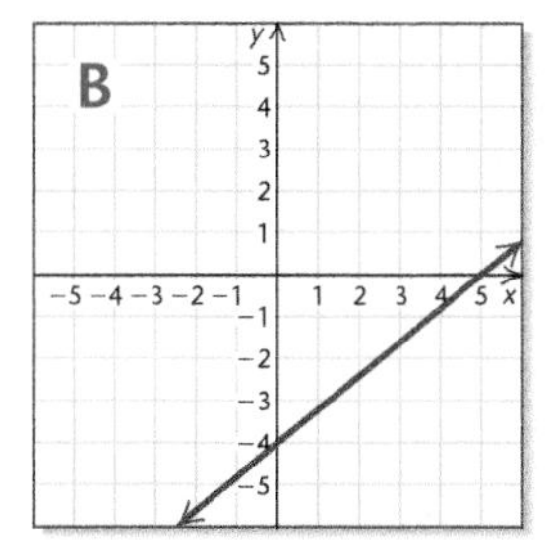

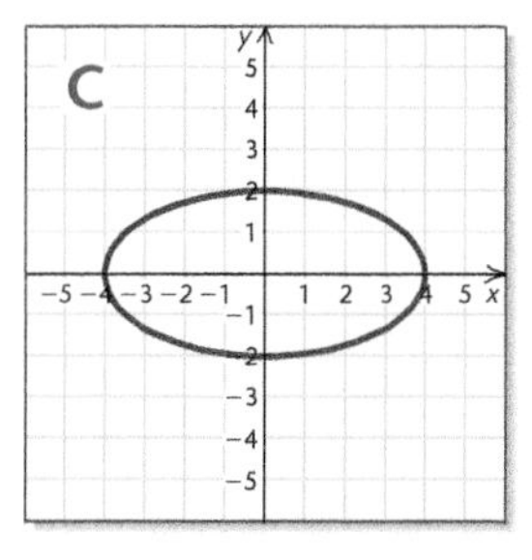

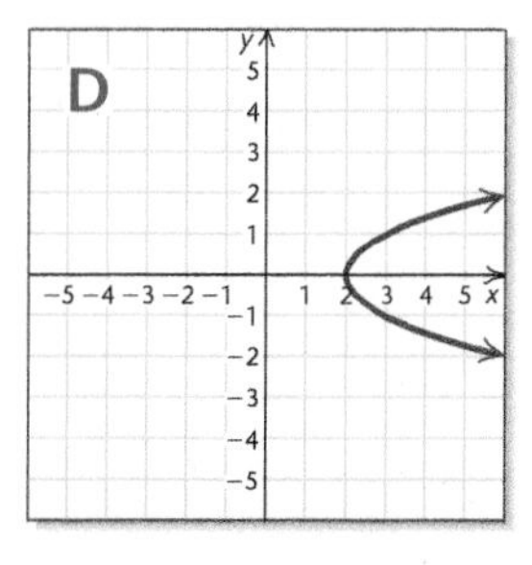

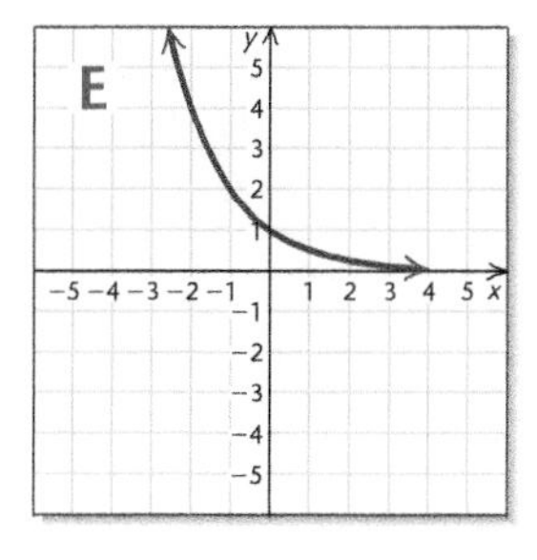

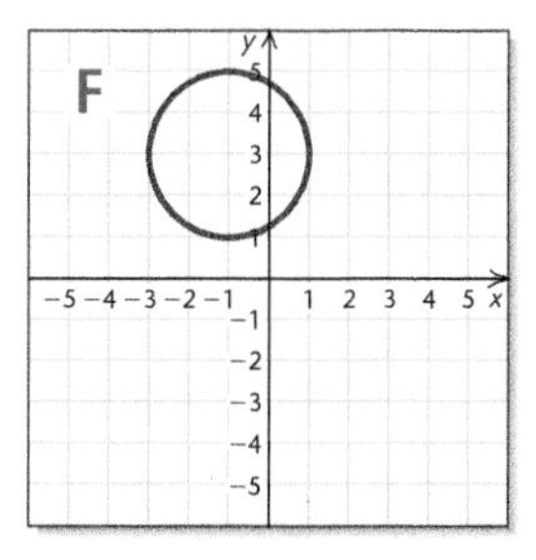

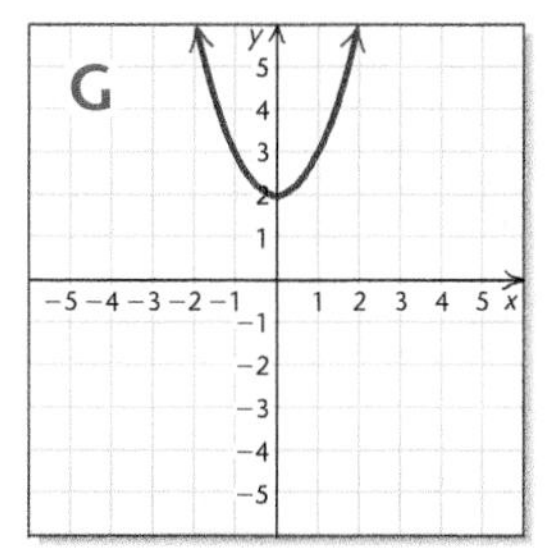

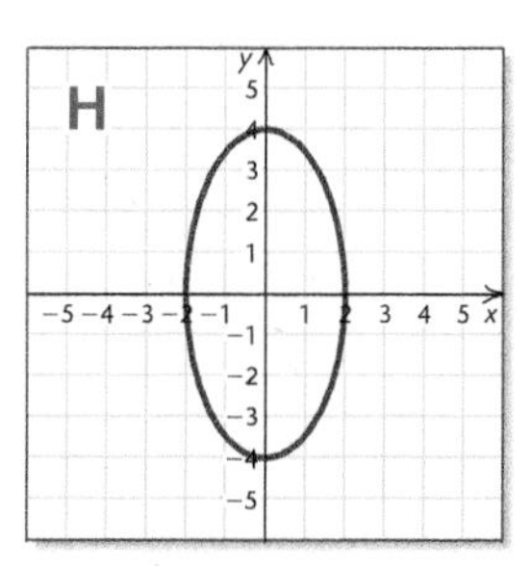

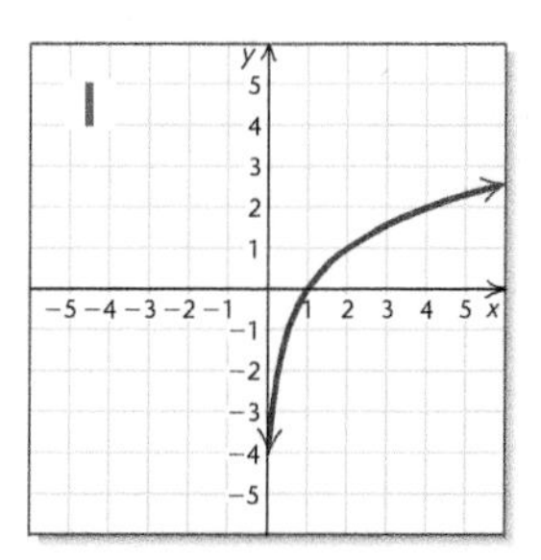

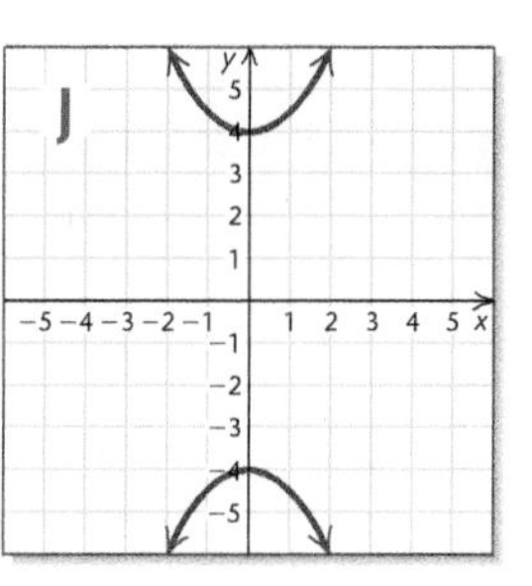

9.3 Exercise Set

FOR EXTRA HELP MyLab Math

✓ Check Your Understanding

Reading Check For each term, write the letter of the appropriate labeled part of the drawing.

RC1. ___ Asymptote

RC2. ___ Axis

RC3. ___ Branch

RC4. ___ Center

RC5. ___ Hyperbola

RC6. ___ Vertex

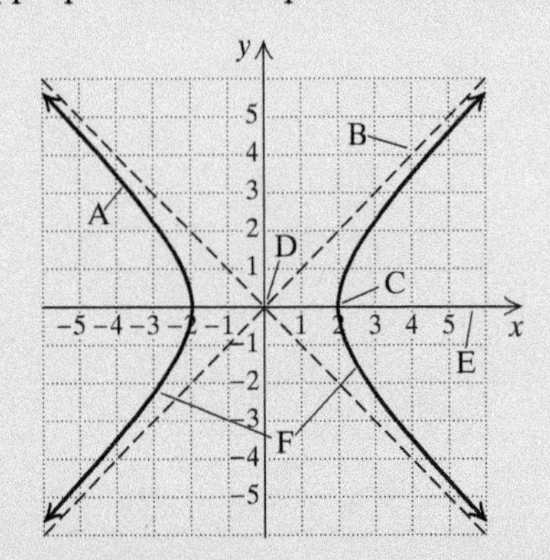

Concept Check Determine whether each statement is true or false.

CC1. For the hyperbola $\frac{x^2}{81} - \frac{y^2}{100} = 1$, the vertices are $(-9, 0)$ and $(9, 0)$.

CC2. For the hyperbola $\frac{x^2}{36} - \frac{y^2}{16} = 1$, no y-intercepts exist.

CC3. For the hyperbola $xy = -3$, the asymptotes are the x- and y-axes.

CC4. For the hyperbola $\frac{y^2}{25} - \frac{x^2}{4} = 1$, the asymptotes are $y = \frac{2}{5}x$ and $y = -\frac{2}{5}x$.

CC5. For the hyperbola $\frac{y^2}{9} - \frac{x^2}{25} = 1$, the vertices are $(3, 0)$ and $(-3, 0)$.

a Graph each hyperbola.

1. $\frac{y^2}{9} - \frac{x^2}{9} = 1$

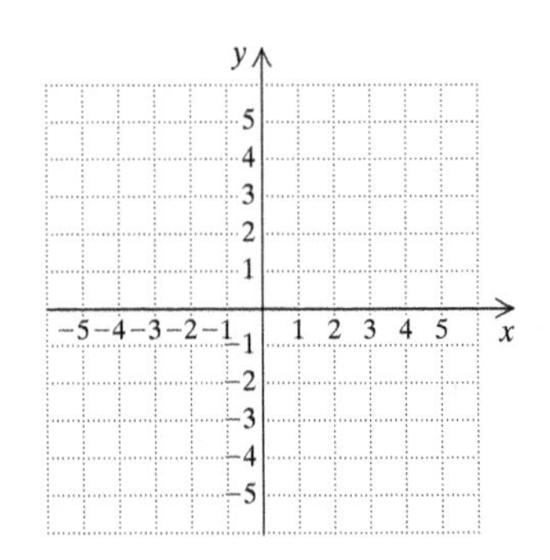

2. $\frac{x^2}{16} - \frac{y^2}{16} = 1$

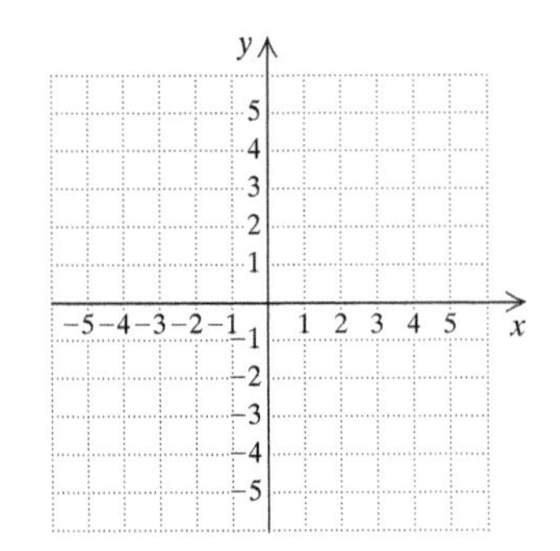

3. $\frac{x^2}{4} - \frac{y^2}{25} = 1$

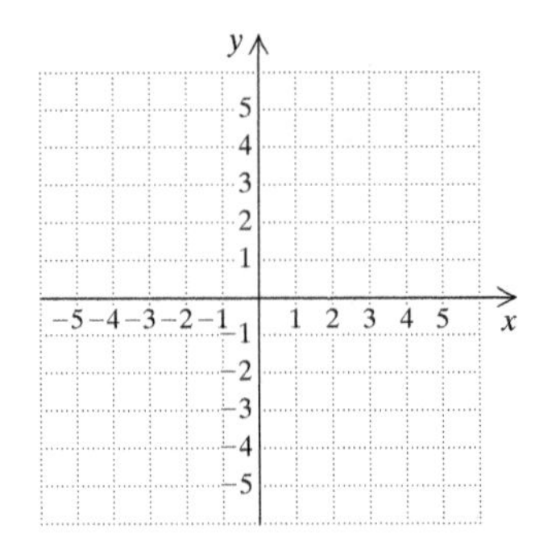

4. $\frac{y^2}{16} - \frac{x^2}{9} = 1$

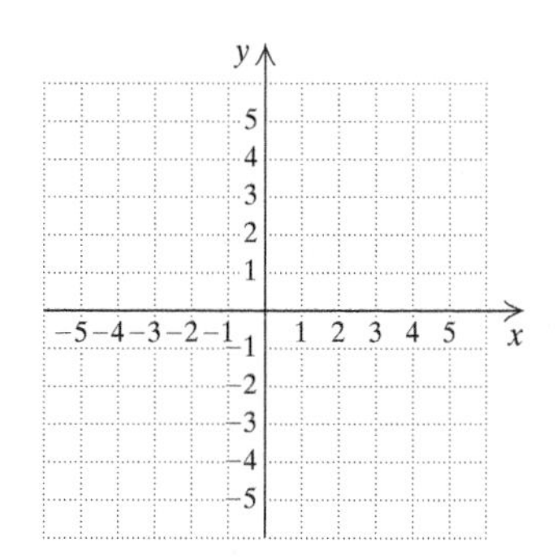

5. $\frac{y^2}{36} - \frac{x^2}{9} = 1$

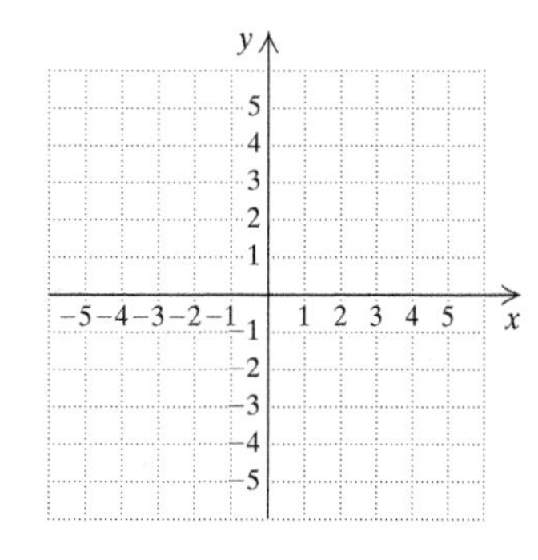

6. $\frac{x^2}{25} - \frac{y^2}{36} = 1$

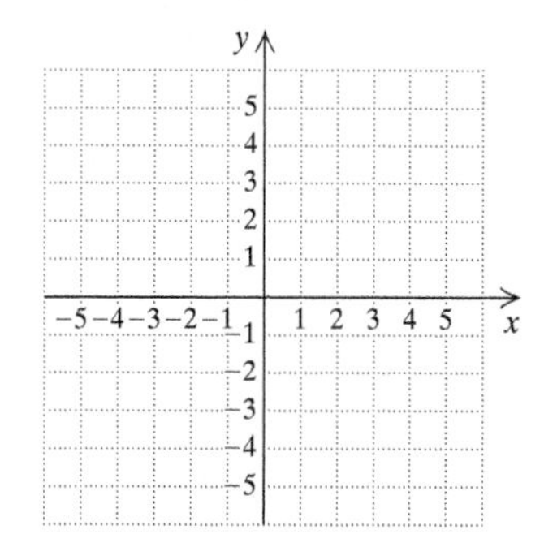

7. $y^2 - x^2 = 25$

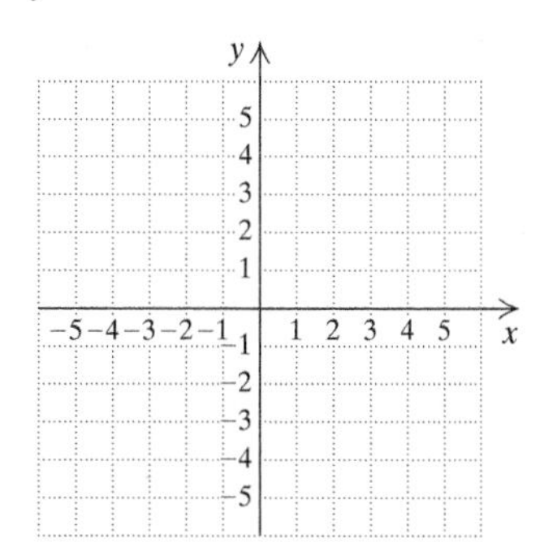

8. $x^2 - y^2 = 4$

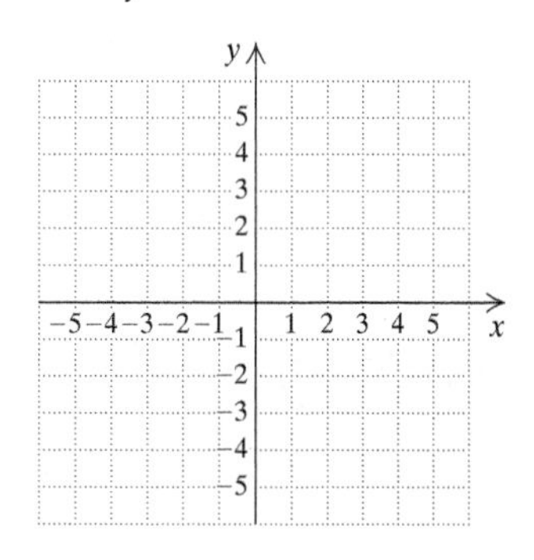

9. $x^2 = 1 + y^2$

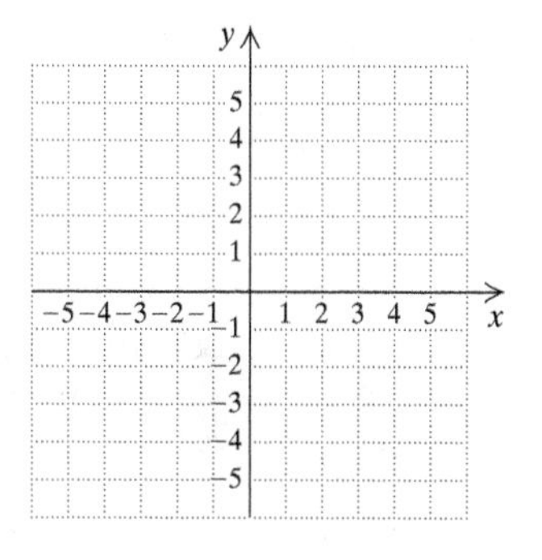

10. $9y^2 = 36 + 4x^2$

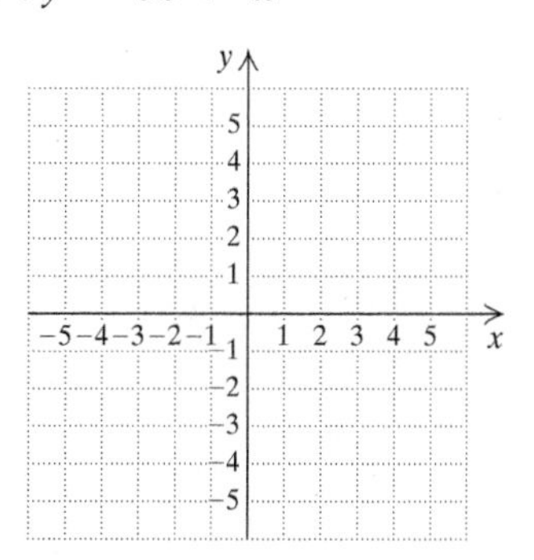

11. $25x^2 - 16y^2 = 400$

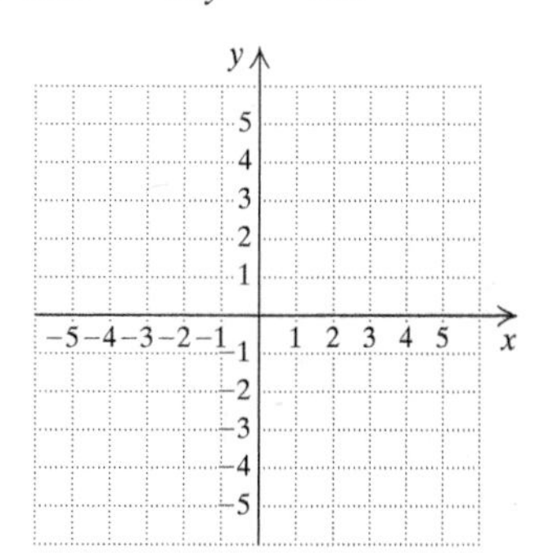

12. $4y^2 - 9x^2 = 36$

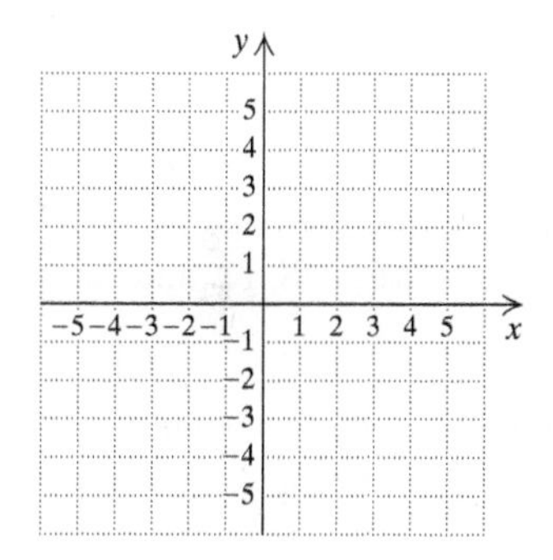

b Graph each hyperbola.

13. $xy = -4$

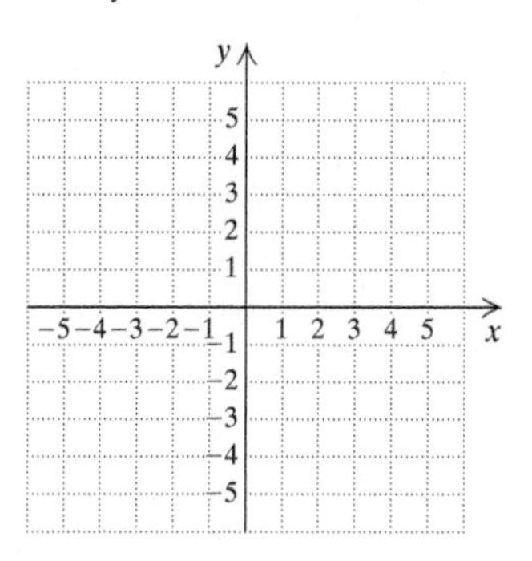

14. $xy = 6$

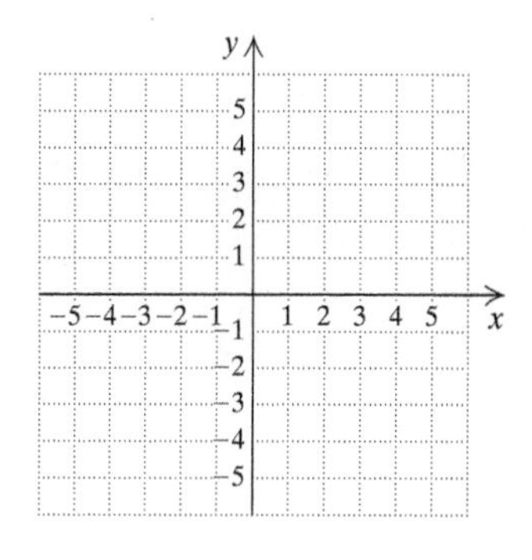

15. $xy = 3$

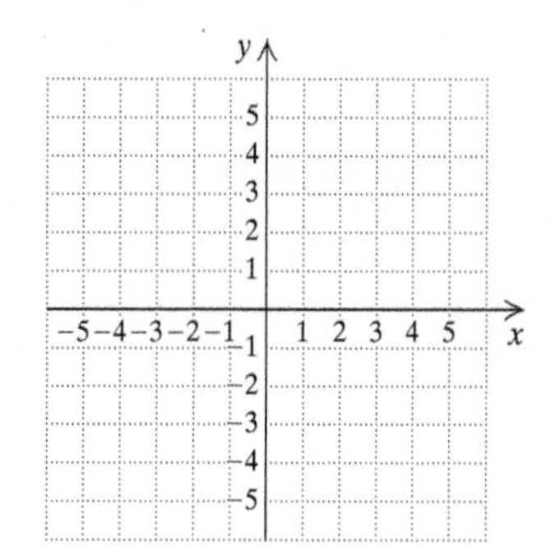

16. $xy = -9$

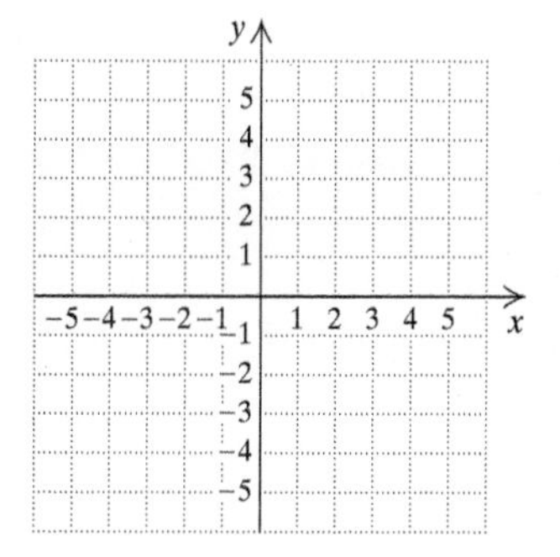

17. $xy = -2$

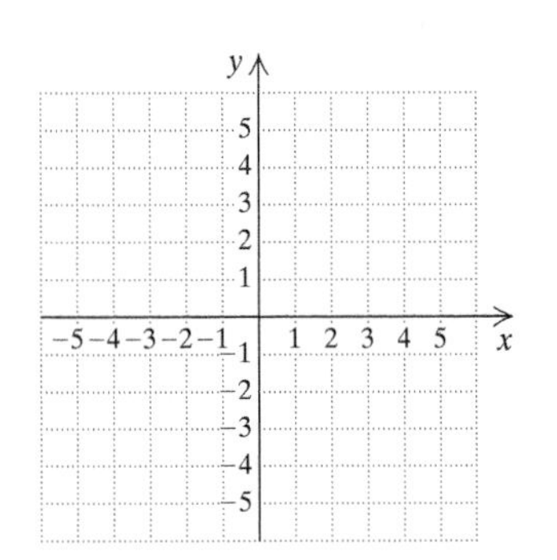

18. $xy = -1$

19. $xy = \frac{1}{2}$

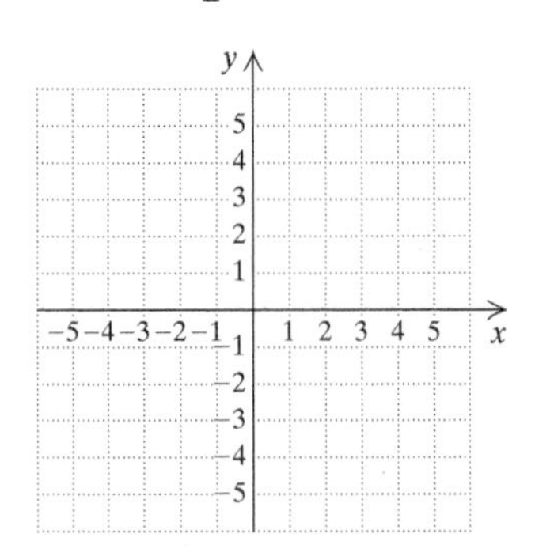

20. $xy = \frac{3}{4}$

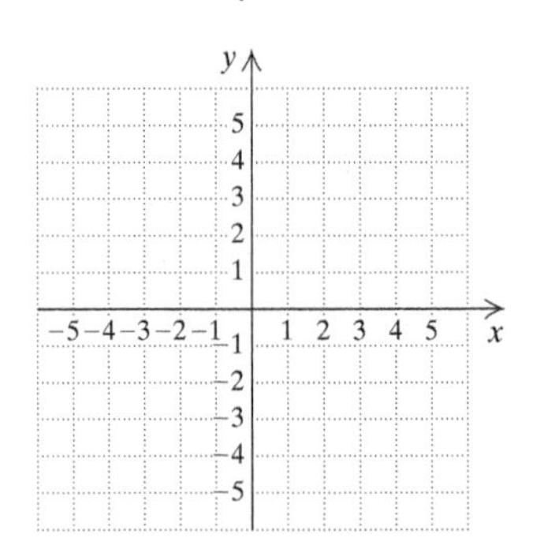

Skill Maintenance

Solve.

21. $3(x - 1) - 4(x - 2) \leq 5 + x$ [1.4c]

22. $0.2a - 0.7 = 0.65$ [1.1d]

23. $-4 < 5 - y < 2$ [1.5a]

24. $|x| > 7$ [1.6e]

25. $2x^2 + 5x = 3$ [4.8a]

26. $\frac{3}{x - 1} = \frac{2}{5 - x}$ [5.5a]

27. $\sqrt{x + 5} = x - 1$ [6.6a]

28. $\log_9 x = \frac{1}{2}$ [8.6b]

Synthesis

29. Use a graphing calculator to check your answers to Exercises 1, 8, 12, and 20.

30. Graph: $\frac{(x - 2)^2}{16} - \frac{(y - 2)^2}{9} = 1.$

Classify the graph of each of the following equations as either a circle, an ellipse, a parabola, or a hyperbola.

31. $x^2 + y^2 - 10x + 8y - 40 = 0$

32. $y + 1 = 2x^2$

33. $1 - 3y = 2y^2 - x$

34. $9x^2 - 4y^2 - 36x + 24y - 36 = 0$

35. $4x^2 + 25y^2 - 8x - 100y + 4 = 0$

36. $\frac{x^2}{7} + \frac{y^2}{7} = 1$

37. $x^2 + y^2 = 8$

38. $y = \frac{2}{x}$

39. $x - \frac{3}{y} = 0$

40. $y + 6x = x^2 + 5$

41. $3x^2 + 5y^2 + x^2 = y^2 + 49$

42. $56x^2 - 17y^2 = 234 - 13x^2 - 38y^2$

Nonlinear Systems of Equations

9.4

OBJECTIVES

a Solve systems of equations in which at least one equation is nonlinear.

b Solve applied problems involving nonlinear systems.

All the systems of equations that we studied in Chapter 3 were linear. We now consider systems of two equations in two variables in which at least one equation is nonlinear.

a ALGEBRAIC SOLUTIONS

Solve systems of equations in two variables by the substitution method. [3.2a]

Solve each system by the substitution method.

1. $3x - 4y = 8,$
$x - 3y = 1$

2. $x = 3 - 5y,$
$2x + y = -6$

Answers: 1. $(4, 1)$ **2.** $\left(-\frac{11}{3}, \frac{4}{3}\right)$

MyLab Math VIDEO

We first consider systems of one first-degree equation and one second-degree equation. As illustrated in the margin, the graphs may be a circle and a line. If so, there are three possibilities for solutions.

For L_1, there is no point of intersection of the line and the circle, hence no solution of the system in the set of real numbers. For L_2, there is one point of intersection, hence one real-number solution. For L_3, there are two points of intersection, hence two real-number solutions.

These systems can be solved graphically by finding the points of intersection. In solving algebraically, we use the substitution method.

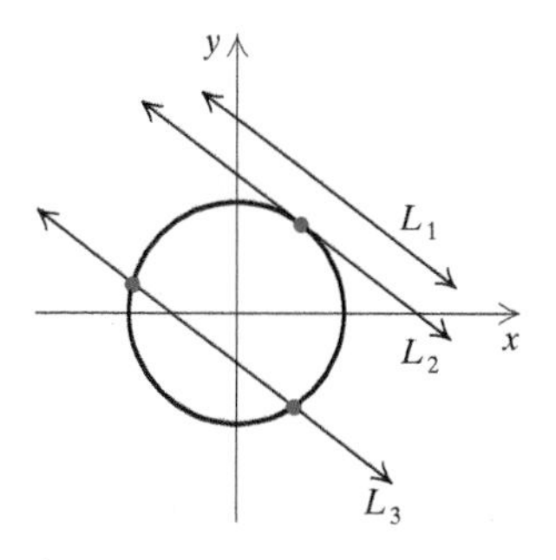

EXAMPLE 1 Solve this system:

$$x^2 + y^2 = 25, \quad (1) \quad \text{(The graph is a circle.)}$$
$$3x - 4y = 0. \quad (2) \quad \text{(The graph is a line.)}$$

We first solve the linear equation (2) for x:

$$x = \tfrac{4}{3}y. \quad (3)$$

We then substitute $\frac{4}{3}y$ for x in equation (1) and solve for y:

$$\left(\tfrac{4}{3}y\right)^2 + y^2 = 25$$
$$\tfrac{16}{9}y^2 + y^2 = 25$$
$$\tfrac{25}{9}y^2 = 25$$
$$y^2 = 9$$
$$y = \pm 3.$$

CALCULATOR CORNER

Solving Nonlinear Systems of Equations

EXERCISES:

1. Use the INTERSECT feature to solve the systems of equations in Examples 1 and 2. Remember that each equation must be solved for y before it is entered in the calculator.
2. Use the INTERSECT feature to solve the systems of equations in Margin Exercises 2 and 3.

Now we substitute these numbers for y in equation (3) and solve for x:

$$x = \tfrac{4}{3}(3) = 4; \qquad x = \tfrac{4}{3}(-3) = -4.$$

Check: For $(4, 3)$:

$$\begin{array}{c|l} x^2 + y^2 = 25 & \\ \hline 4^2 + 3^2 \ ? \ 25 & \\ 16 + 9 & \\ 25 & \text{TRUE} \end{array} \qquad \begin{array}{c|l} 3x - 4y = 0 & \\ \hline 3(4) - 4(3) \ ? \ 0 & \\ 12 - 12 & \\ 0 & \text{TRUE} \end{array}$$

For $(-4, -3)$:

$$\begin{array}{c|l} x^2 + y^2 = 25 & \\ \hline (-4)^2 + (-3)^2 \ ? \ 25 & \\ 16 + 9 & \\ 25 & \text{TRUE} \end{array} \qquad \begin{array}{c|l} 3x - 4y = 0 & \\ \hline 3(-4) - 4(-3) \ ? \ 0 & \\ -12 + 12 & \\ 0 & \text{TRUE} \end{array}$$

The pairs $(4, 3)$ and $(-4, -3)$ check, so they are solutions. We can see the solutions in the graph. The graph of equation (1) is a circle, and the graph of equation (2) is a line. The graphs intersect at the points $(4, 3)$ and $(-4, -3)$.

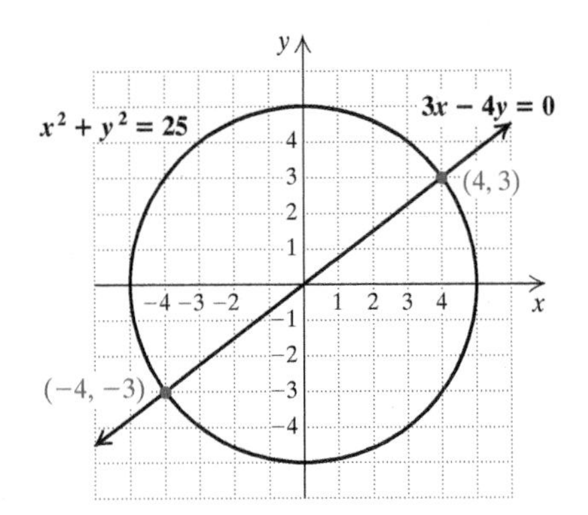

Do Exercises 1 and 2.

Solve. Sketch the graphs to confirm the solutions.

1. $x^2 + y^2 = 25,$
   $y - x = -1$

2. $y = x^2 - 2x - 1,$
   $y = x + 3$

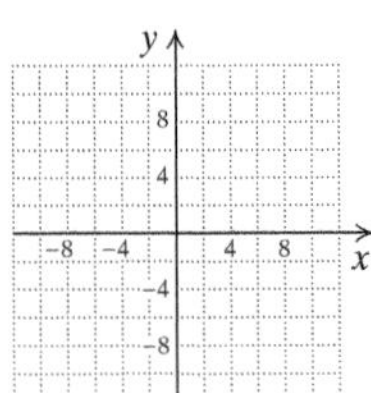

EXAMPLE 2 Solve this system:

$$y + 3 = 2x, \qquad (1)$$
$$x^2 + 2xy = -1. \qquad (2)$$

We first solve the linear equation (1) for y:

$$y = 2x - 3. \qquad (3)$$

We then substitute $2x - 3$ for y in equation (2) and solve for x:

$$x^2 + 2x(2x - 3) = -1$$
$$x^2 + 4x^2 - 6x = -1$$
$$5x^2 - 6x + 1 = 0$$
$$(5x - 1)(x - 1) = 0 \qquad \text{Factoring}$$
$$5x - 1 = 0 \quad or \quad x - 1 = 0 \qquad \text{Using the principle of zero products}$$
$$x = \tfrac{1}{5} \quad or \quad x = 1.$$

Now we substitute these numbers for x in equation (3) and solve for y:

$$y = 2\left(\tfrac{1}{5}\right) - 3 = -\tfrac{13}{5}; \qquad y = 2(1) - 3 = -1.$$

The check is left to the student. The pairs $\left(\frac{1}{5}, -\frac{13}{5}\right)$ and $(1, -1)$ are solutions.

Do Exercise 3.

3. Solve:
   $y + 3x = 1,$
   $x^2 - 2xy = 5.$

Answers

1.

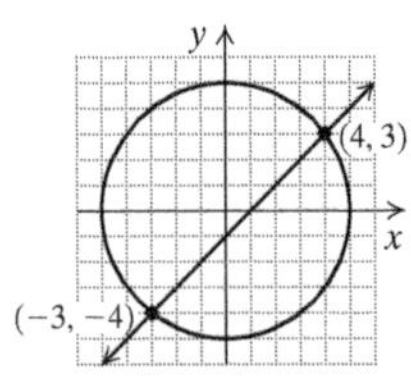

$x^2 + y^2 = 25,$
$y - x = -1$

2.

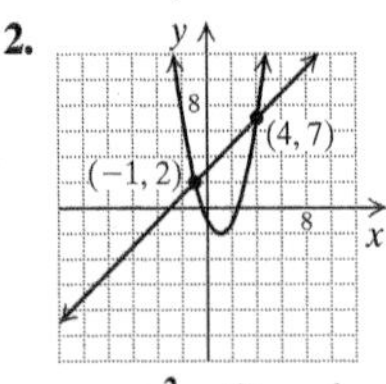

$y = x^2 - 2x - 1,$
$y = x + 3$

3. $\left(-\frac{5}{7}, \frac{22}{7}\right), (1, -2)$

EXAMPLE 3 Solve this system:

$$x + y = 5, \quad \text{(The graph is a line.)}$$
$$y = 3 - x^2. \quad \text{(The graph is a parabola.)}$$

We substitute $3 - x^2$ for y in the first equation:

$$x + 3 - x^2 = 5$$
$$-x^2 + x - 2 = 0$$
$$x^2 - x + 2 = 0. \quad \text{Multiplying by } -1$$

To solve this equation, we need the quadratic formula:

$$x = \frac{-b \pm \sqrt{b^2 - 4ac}}{2a} = \frac{-(-1) \pm \sqrt{(-1)^2 - 4(1)(2)}}{2(1)}$$
$$= \frac{1 \pm \sqrt{1 - 8}}{2} = \frac{1 \pm \sqrt{-7}}{2} = \frac{1}{2} \pm \frac{\sqrt{7}}{2}i.$$

Then solving the first equation for y, we obtain $y = 5 - x$. Substituting values for x gives us

$$y = 5 - \left(\frac{1}{2} + \frac{\sqrt{7}}{2}i\right) = \frac{9}{2} - \frac{\sqrt{7}}{2}i$$

and

$$y = 5 - \left(\frac{1}{2} - \frac{\sqrt{7}}{2}i\right) = \frac{9}{2} + \frac{\sqrt{7}}{2}i.$$

The solutions are

$$\left(\frac{1}{2} + \frac{\sqrt{7}}{2}i, \frac{9}{2} - \frac{\sqrt{7}}{2}i\right)$$

and

$$\left(\frac{1}{2} - \frac{\sqrt{7}}{2}i, \frac{9}{2} + \frac{\sqrt{7}}{2}i\right).$$

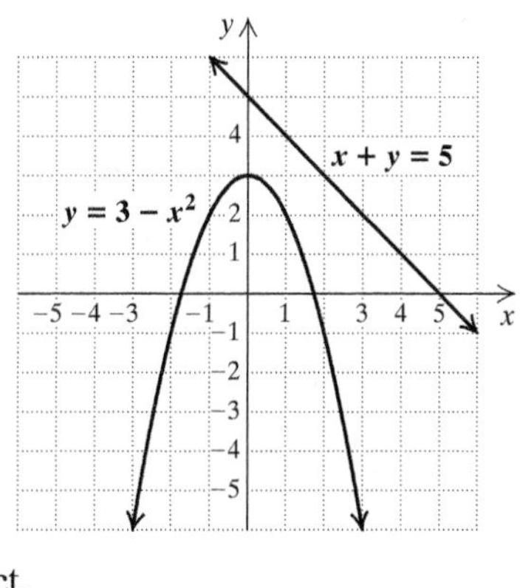

There are no real-number solutions. Note in the figure at right that the graphs do not intersect. Getting only nonreal complex-number solutions tells us that the graphs do not intersect.

Do Exercise 4.

GS **4.** Solve:

$$9x^2 - 4y^2 = 36,$$
$$5x + 2y = 0.$$

Solve $5x + 2y = 0$ for x: $x = -\frac{2}{5}y$.

Substitute $-\frac{2}{5}y$ for x in the first equation and solve for y:

$$9(\quad)^2 - 4y^2 = 36$$
$$9\left(\frac{4}{25}y^2\right) - 4y^2 = 36$$
$$\frac{36}{\quad}y^2 - \frac{100}{25}y^2 = 36$$
$$-\frac{\quad}{25}y^2 = 36$$
$$y^2 = -\frac{225}{\quad}$$
$$y = \pm\sqrt{-\frac{225}{16}} = \pm\frac{15}{\quad}i.$$

If $y = \frac{15}{4}i$, then

$$x = -\frac{2}{5}y = -\frac{2}{5}\cdot\frac{15}{4}i = -\frac{3}{\quad}i.$$

If $y = -\frac{15}{4}i$, then

$$x = -\frac{2}{5}y = -\frac{2}{5}\cdot\left(-\frac{15}{4}i\right) = \frac{3}{\quad}i.$$

The solutions are $\left(-\frac{3}{2}i, \quad\right)$ and $\left(\frac{3}{2}i, \quad\right)$.

Two second-degree equations can have common solutions in various ways. If the graphs happen to be a circle and a hyperbola, for example, there are six possibilities, as shown below.

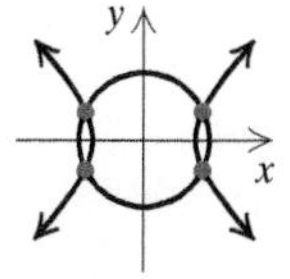

4 real solutions

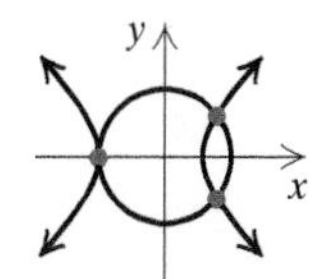

3 real solutions

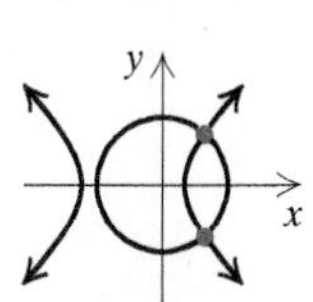

2 real solutions

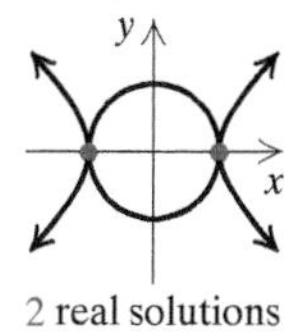

2 real solutions

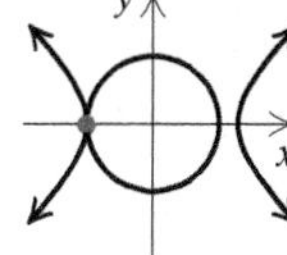
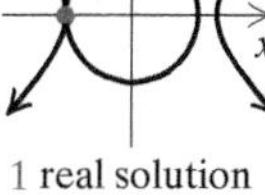

1 real solution

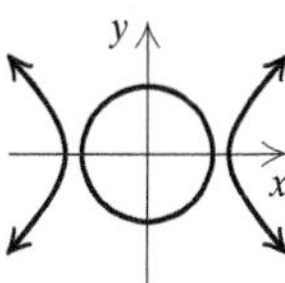

0 real solutions

Answer

4. $\left(-\frac{3}{2}i, \frac{15}{4}i\right), \left(\frac{3}{2}i, -\frac{15}{4}i\right)$

Guided Solution:

4. $-\frac{2}{5}y$, 25, 64, 16, 4, 2, 2, $\frac{15}{4}i$, $-\frac{15}{4}i$

To solve systems of two second-degree equations, we can use either the substitution method or the elimination method. The elimination method is generally used when each equation is of the form $Ax^2 + By^2 = C$. Then we can eliminate an x^2- or a y^2-term.

EXAMPLE 4 Solve:

$$2x^2 + 5y^2 = 22, \quad (1)$$
$$3x^2 - y^2 = -1. \quad (2)$$

In this case, we use the elimination method:

$$\begin{aligned} 2x^2 + 5y^2 &= 22 \\ 15x^2 - 5y^2 &= -5 && \text{Multiplying by 5 on both sides of equation (2)} \\ 17x^2 &= 17 && \text{Adding} \\ x^2 &= 1 \\ x &= \pm 1. \end{aligned}$$

If $x = 1, x^2 = 1$, and if $x = -1, x^2 = 1$, so substituting either 1 or -1 for x in equation (2) gives us

$$\begin{aligned} 3x^2 - y^2 &= -1 \\ 3 \cdot 1 - y^2 &= -1 && \text{Substituting 1 for } x^2 \\ 3 - y^2 &= -1 \\ -y^2 &= -4 \\ y^2 &= 4 \\ y &= \pm 2. \end{aligned}$$

Thus if $x = 1, y = 2$ or $y = -2$, yielding the pairs $(1, 2)$ and $(1, -2)$. If $x = -1, y = 2$ or $y = -2$, yielding the pairs $(-1, 2)$ and $(-1, -2)$.

Check: Since $(2)^2 = 4, (-2)^2 = 4, (1)^2 = 1$, and $(-1)^2 = 1$, we can check all four pairs at one time.

$$\begin{array}{c|c} 2x^2 + 5y^2 & = 22 \\ \hline 2(\pm 1)^2 + 5(\pm 2)^2 & ?\ 22 \\ 2 + 20 & \\ 22 & \text{TRUE} \end{array} \qquad \begin{array}{c|c} 3x^2 - y^2 & = -1 \\ \hline 3(\pm 1)^2 - (\pm 2)^2 & ?\ -1 \\ 3 - 4 & \\ -1 & \text{TRUE} \end{array}$$

The solutions are $(1, 2), (1, -2), (-1, 2)$, and $(-1, -2)$.

Do Exercise 5.

5. Solve:

$$2y^2 - 3x^2 = 6,$$
$$5y^2 + 2x^2 = 53.$$

GS To eliminate the x^2-terms, multiply the first equation by 2 and the second equation by ____:

$$\begin{aligned} 4y^2 - 6x^2 &= \_\_\_\_ \\ 15y^2 + 6x^2 &= \_\_\_\_ \\ 19y^2 &= \_\_\_\_ \\ y^2 &= \_\_\_\_ \\ y &= \pm \_\_\_\_. \end{aligned}$$

Substitute 3 for y in $5y^2 + 2x^2 = 53$:

$$\begin{aligned} 5(3)^2 + 2x^2 &= 53 \\ \_\_\_\_ + 2x^2 &= 53 \\ 2x^2 &= \_\_\_\_ \\ x^2 &= \_\_\_\_ \\ x &= \pm \_\_\_\_. \end{aligned}$$

We get the same result when $y = -3$. The solutions are $(2, 3), (2, \_\_\_\_), (\_\_\_\_, 3)$, and $(-2, \_\_\_\_)$.

When one equation contains a product of two different variables and the other equation is of the form $Ax^2 + By^2 = C$, we often solve for one of the variables in the equation with the product and then substitute in the other.

EXAMPLE 5 Solve:

$$x^2 + 4y^2 = 20, \quad (1)$$
$$xy = 4. \quad (2)$$

Here we use the substitution method. First, we solve equation (2) for y:

$$y = \frac{4}{x}.$$

Answer

5. $(2, 3), (2, -3), (-2, 3), (-2, -3)$

Guided Solution:

5. $3, 12, 159, 171, 9, 3, 45, 8, 4, 2, -3, -2, -3$

Then we substitute $4/x$ for y in equation (1) and solve for x:

$$x^2 + 4\left(\frac{4}{x}\right)^2 = 20$$

$$x^2 + \frac{64}{x^2} = 20$$

$$x^4 + 64 = 20x^2 \qquad \text{Multiplying by } x^2$$

$$x^4 - 20x^2 + 64 = 0 \qquad \text{Obtaining standard form. This equation is quadratic in form.}$$

$$u^2 - 20u + 64 = 0 \qquad \text{Letting } u = x^2$$

$$(u - 16)(u - 4) = 0 \qquad \text{Factoring}$$

$$u = 16 \quad or \quad u = 4. \qquad \text{Using the principle of zero products}$$

Next, we substitute x^2 for u and solve these equations:

$$x^2 = 16 \quad or \quad x^2 = 4$$

$$x = \pm 4 \quad or \quad x = \pm 2.$$

Then $x = 4$ or $x = -4$ or $x = 2$ or $x = -2$. Since $y = 4/x$, if $x = 4, y = 1$; if $x = -4, y = -1$; if $x = 2, y = 2$; and if $x = -2, y = -2$. The ordered pairs $(4, 1)$, $(-4, -1)$, $(2, 2)$, and $(-2, -2)$ check. They are the solutions.

Do Exercise 6.

6. Solve:
$x^2 + xy + y^2 = 19,$
$xy = 6.$

b SOLVING APPLIED PROBLEMS

We now consider applications in which the translation is to a system of equations in which at least one equation is nonlinear.

EXAMPLE 6 *Architecture.* For a college fitness center, an architect plans to lay out a rectangular piece of land that has a perimeter of 204 m and an area of 2565 m^2. Find the dimensions of the piece of land.

1. **Familiarize.** We make a drawing of the area, labeling it using l for the length and w for the width.

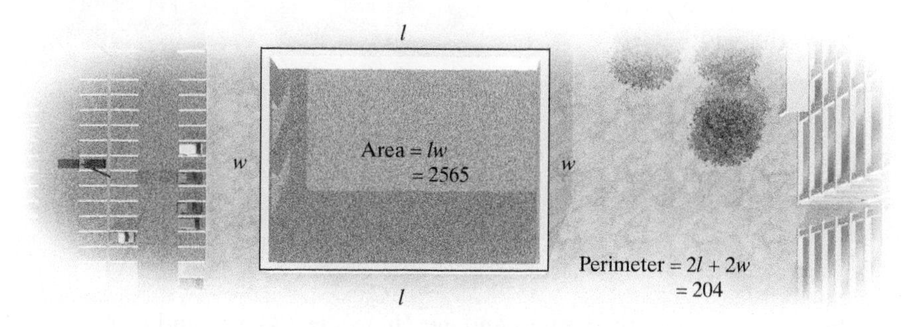

2. **Translate.** We then have the following translation:

Perimeter: $2l + 2w = 204;$

Area: $lw = 2565.$

3. **Solve.** We solve the system

$2l + 2w = 204,$ (The graph is a line.)

$lw = 2565.$ (The graph is a hyperbola.)

Answer

6. $(3, 2), (-3, -2), (2, 3), (-2, -3)$

7. The perimeter of a rectangular mural is 34 m, and the length of a diagonal of the mural is 13 m. Find the dimensions of the mural.

We solve the second equation for l and get $l = 2565/w$. Then we substitute $2565/w$ for l in the first equation and solve for w:

$$2\left(\frac{2565}{w}\right) + 2w = 204$$

$$2(2565) + 2w^2 = 204w \quad \text{Multiplying by } w$$

$$2w^2 - 204w + 2(2565) = 0 \quad \text{Standard form}$$

$$w^2 - 102w + 2565 = 0 \quad \text{Dividing by 2}$$

$$w = \frac{-(-102) \pm \sqrt{(-102)^2 - 4 \cdot 1 \cdot 2565}}{2 \cdot 1}$$

Quadratic formula. Factoring could also be used, but the numbers are quite large.

$$w = \frac{102 \pm \sqrt{144}}{2} = \frac{102 \pm 12}{2}$$

$$w = 57 \quad or \quad w = 45.$$

If $w = 57$, then $l = 2565/w = 2565/57 = 45$. If $w = 45$, then $l = 2565/w = 2565/45 = 57$. Since length is generally considered to be greater than width, we have the solution $l = 57$ and $w = 45$, or $(57, 45)$.

4. **Check.** If $l = 57$ and $w = 45$, the perimeter is $2 \cdot 57 + 2 \cdot 45$, or 204. The area is $57 \cdot 45$, or 2565. The numbers check.
5. **State.** The length is 57 m and the width is 45 m.

Do Exercise 7.

EXAMPLE 7 *HDTV Dimensions.* The ratio of the width to the height of the screen of an HDTV (high-definition television) is 16 to 9. Suppose that a large-screen HDTV has a 70-in. diagonal screen. Find the height and the width of the screen.

1. **Familiarize.** We first make a drawing and label it. Note that there is a right triangle in the figure. We let h = the height and w = the width.

2. **Translate.** Next, we translate to a system of equations:

$$h^2 + w^2 = 70^2, \text{ or } 4900, \quad (1)$$

$$\frac{w}{h} = \frac{16}{9}. \quad (2)$$

3. **Solve.** We solve the system and get $(h, w) \approx (34, 61)$ and $(-34, -61)$.
4. **Check.** Widths cannot be negative, so we need check only $(34, 61)$. In the right triangle, $34^2 + 61^2 = 1156 + 3721 = 4877 \approx 4900 = 70^2$. Also, $\frac{61}{34} \approx \frac{16}{9}$.
5. **State.** The height is about 34 in., and the width is about 61 in.

Do Exercise 8.

8. *HDTV Dimensions.* The ratio of the width to the height of the screen of an HDTV is 16 to 9. Suppose that an HDTV screen has a 46-in. diagonal screen. Find the width and the height of the screen.

Answers

7. 12 m by 5 m **8.** Width: about 40 in.; height: about 22.5 in.

9.4 Exercise Set

FOR EXTRA HELP MyLab Math

✓ Check Your Understanding

Reading Check and Concept Check Determine whether each statement is true or false.

RC1. The solutions of systems of two equations with two variables are ordered pairs.

RC2. If the graphs of the equations in a system are a line and a circle, then the system has at least one real solution.

RC3. Systems of equations that contain at least one nonlinear equation can be solved only by using the substitution method.

RC4. We may need to use the quadratic equation when solving a system of equations.

RC5. If the graphs of the equations in a system are a line and a parabola, then the system has at most two real solutions.

a Solve.

1. $x^2 + y^2 = 100,$
$y - x = 2$

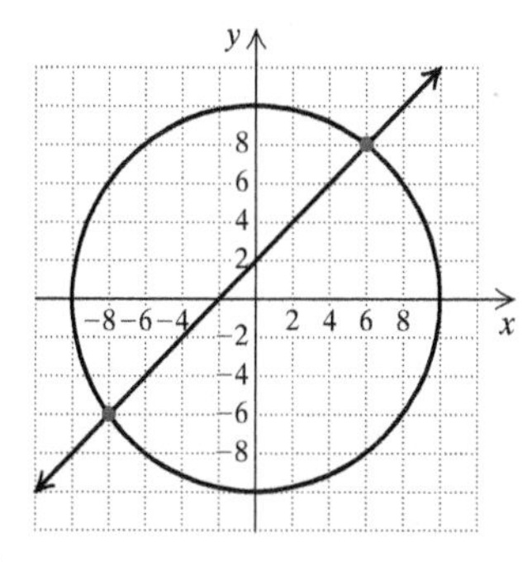

2. $x^2 + y^2 = 25,$
$y - x = 1$

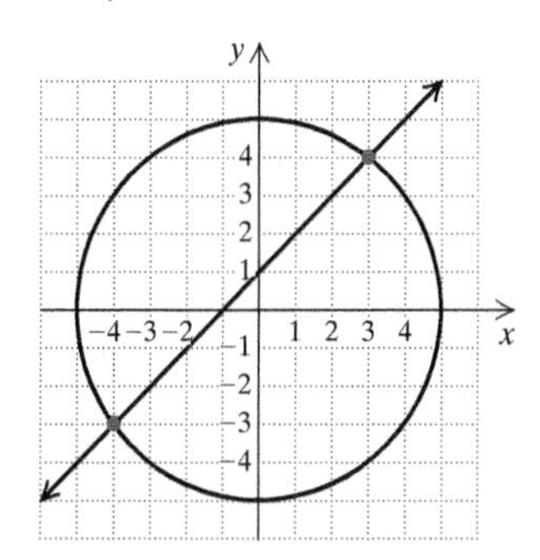

3. $9x^2 + 4y^2 = 36,$
$3x + 2y = 6$

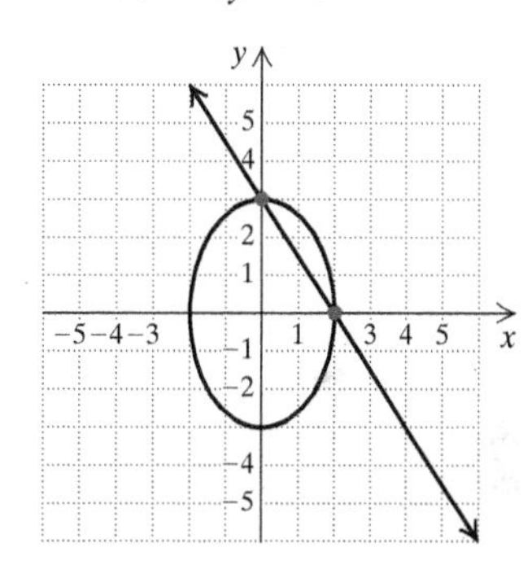

4. $4x^2 + 9y^2 = 36,$
$3y + 2x = 6$

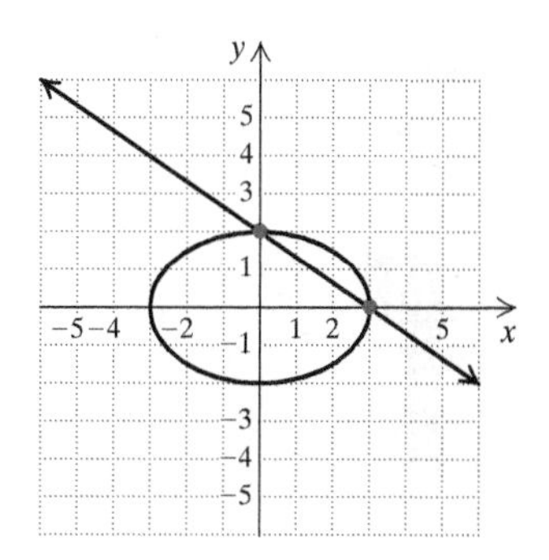

5. $y^2 = x + 3,$
$2y = x + 4$

6. $y = x^2,$
$3x = y + 2$

7. $x^2 - xy + 3y^2 = 27,$
$x - y = 2$

8. $2y^2 + xy + x^2 = 7,$
$x - 2y = 5$

9. $x^2 - xy + 3y^2 = 5,$
$x - y = 2$

10. $a^2 + 3b^2 = 10,$
$a - b = 2$

11. $a + b = -6,$
$ab = -7$

12. $2y^2 + xy = 5,$
$4y + x = 7$

13. $2a + b = 1,$
$b = 4 - a^2$

14. $4x^2 + 9y^2 = 36,$
$x + 3y = 3$

15. $x^2 + y^2 = 5,$
$x - y = 8$

16. $4x^2 + 9y^2 = 36,$
$y - x = 8$

17. $x^2 + y^2 = 25,$
$y^2 = x + 5$

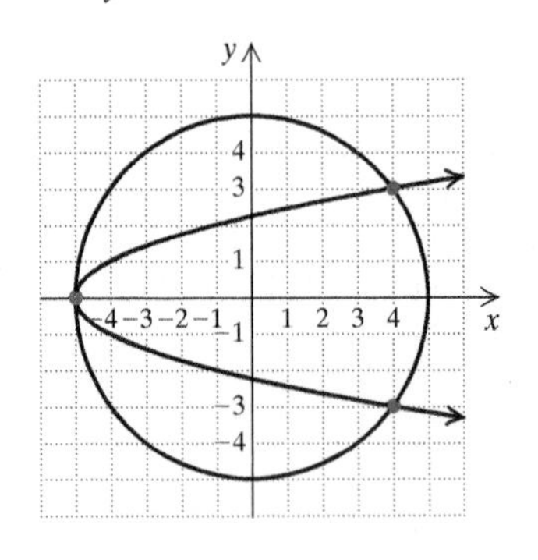

18. $y = x^2,$
$x = y^2$

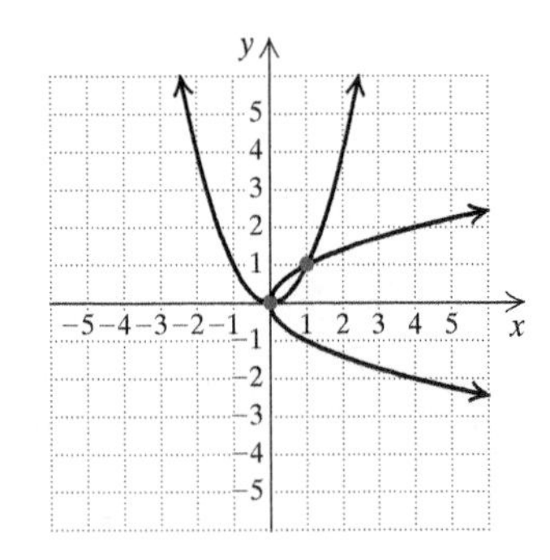

19. $x^2 + y^2 = 9,$
$x^2 - y^2 = 9$

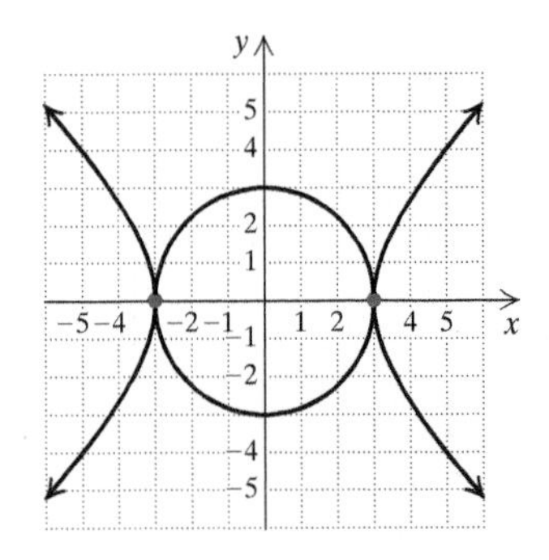

20. $y^2 - 4x^2 = 4,$
$4x^2 + y^2 = 4$

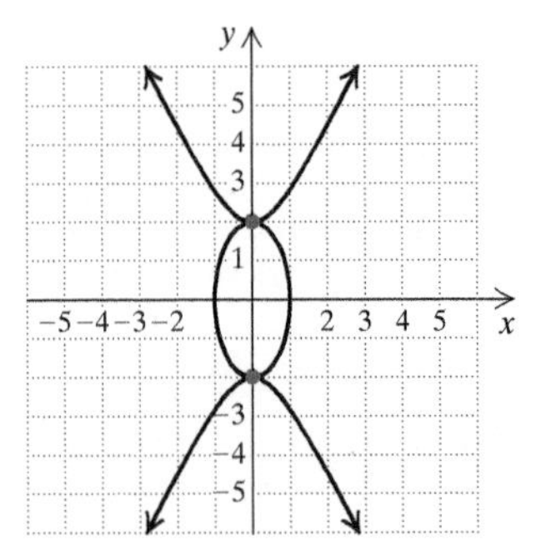

21. $x^2 + y^2 = 20,$
$xy = 8$

22. $x^2 + y^2 = 5,$
$xy = 2$

23. $x^2 + y^2 = 13,$
$xy = 6$

24. $x^2 + y^2 + 6y + 5 = 0,$
$x^2 + y^2 - 2x - 8 = 0$

25. $2xy + 3y^2 = 7,$
$3xy - 2y^2 = 4$

26. $xy - y^2 = 2,$
$2xy - 3y^2 = 0$

27. $4a^2 - 25b^2 = 0,$
$2a^2 - 10b^2 = 3b + 4$

28. $m^2 - 3mn + n^2 + 1 = 0,$
$3m^2 - mn + 3n^2 = 13$

29. $ab - b^2 = -4,$
$ab - 2b^2 = -6$

30. $a^2 + b^2 = 14,$
$ab = 3\sqrt{5}$

31. $x^2 + y^2 = 25,$
$9x^2 + 4y^2 = 36$

32. $x^2 + y^2 = 1,$
$9x^2 - 16y^2 = 144$

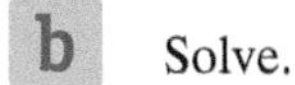

Solve.

33. *Art.* Elliot is designing a rectangular stained glass miniature that has a perimeter of 28 cm and a diagonal of length 10 cm. What should the dimensions of the glass be?

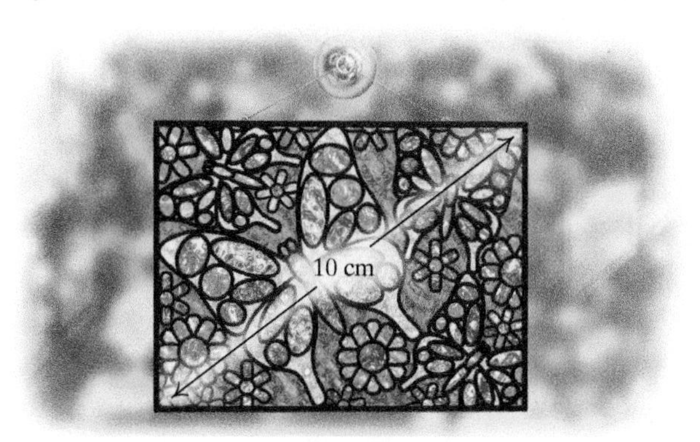

34. *Dimensions of a Van.* The cargo area of a delivery truck must be 108 ft^2, and the length of a diagonal of the truck must accommodate a roll of carpet that is 15 ft wide. Find the dimensions of the cargo area.

35. A rectangle has an area of 14 in^2 and a perimeter of 18 in. Find its dimensions.

36. A rectangle has an area of 40 yd^2 and a perimeter of 26 yd. Find its dimensions.

37. The diagonal of a rectangle is 1 ft longer than the length of the rectangle and 3 ft longer than twice the width. Find the dimensions of the rectangle.

38. It will take 210 yd of fencing to enclose a rectangular field. The area of the field is 2250 yd^2. What are the dimensions of the field?

39. The area of a rectangle is $\sqrt{2}$ m^2, and the length of a diagonal is $\sqrt{3}$ m. Find the dimensions.

40. The area of a rectangle is $\sqrt{3}$ m^2, and the length of a diagonal is 2 m. Find the dimensions.

41. *Garden Design.* A garden contains two square peanut beds. Find the length of each bed if the sum of their areas is 832 ft^2 and the difference of their areas is 320 ft^2.

42. *HDTV Screens.* The ratio of the length to the height of an HDTV screen (see Example 7) is 16 to 9. The Remton Lounge has an HDTV screen with a 42-in. diagonal screen. Find the dimensions of the screen.

43. *Computer Screens.* The ratio of the length to the height of the screen on a computer monitor is 4 to 3. A laptop has a 31-cm diagonal screen. Find the dimensions of the screen.

44. *Investments.* At a local bank, an amount of money invested for 1 year at a certain interest rate yielded \$225 in interest. The bank officer said that if \$750 more had been invested and the rate had been 1% less, the interest would have been the same. Find the principal and the rate.

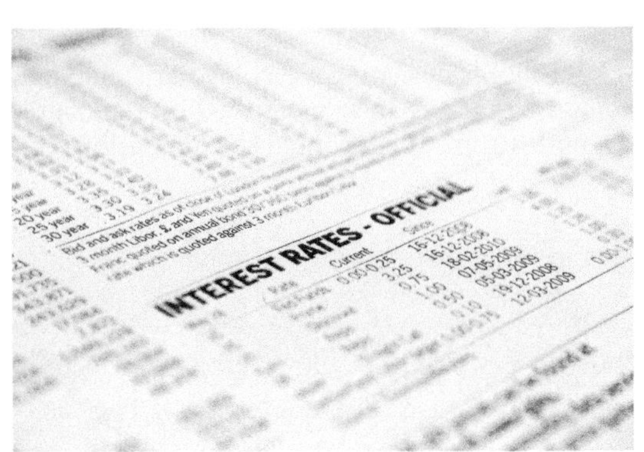

Skill Maintenance

Find a formula for the inverse of each function, if it exists. [8.2d], [8.3a], [8.5a]

45. $f(x) = 10^x$

46. $f(x) = \sqrt[3]{x + 2}$

47. $f(x) = |x|$

48. $f(x) = \dfrac{3}{2x - 7}$

49. $f(x) = \dfrac{x - 2}{x + 3}$

50. $f(x) = \ln x$

Synthesis

51. Solve: $a + b = \dfrac{5}{6}$,

$\dfrac{a}{b} + \dfrac{b}{a} = \dfrac{13}{6}$.

52. Find the equation of an ellipse centered at the origin that passes through the points $(2, -3)$ and $(1, \sqrt{13})$.

53. A piece of wire 100 cm long is to be cut into two pieces and those pieces are each to be bent to make a square. The area of one square is to be 144 cm^2 greater than that of the other. How should the wire be cut?

54. Find the equation of a circle that passes through $(-2, 3)$ and $(-4, 1)$ and whose center is on the line $5x + 8y = -2$.

55. *Railing Sales.* Fireside Castings finds that the total revenue R from the sale of x units of railing is given by the function

$$R(x) = 100x + x^2.$$

Fireside also finds that the total cost C of producing x units of the same product is given by the function

$$C(x) = 80x + 1500.$$

A break-even point is a value of x for which total revenue is the same as total cost; that is, $R(x) = C(x)$. How many units must be sold in order to break even?

Summary and Review

Vocabulary Reinforcement

Complete each statement with the correct term from the column on the right. Words may be used more than once or not at all.

1. A(n) ______________ is the set of all points in a plane that are a fixed distance from a point in that plane. [9.1d]
2. The graph of $xy = 9$ is a(n) ______________. [9.3b]
3. A(n) ______________ parabola opens to the left or to the right. [9.1a]
4. In the equation of a parabola, the point (h, k) represents the ______________ of the parabola. [9.1a]
5. In the equation of a circle, the point (h, k) represents the ______________ of the circle. [9.1d]
6. The point halfway between the vertices of a hyperbola is the ______________ of the hyperbola. [9.3a]

center
intercept
vertex
parabola
circle
ellipse
hyperbola
horizontal
vertical

Concept Reinforcement

Determine whether each statement is true or false.

______________ 1. The graph of $x - 2y^2 = 3$ is a parabola opening to the left. [9.1a]

______________ 2. A system of equations that represent a parabola and a circle can have up to four real solutions. [9.4a]

______________ 3. The graph of $\dfrac{x^2}{10} - \dfrac{y^2}{12} = 1$ is a hyperbola with a vertical axis. [9.3a]

______________ 4. The radius of the circle given by $x^2 + y^2 = 7$ is $\sqrt{7}$. [9.1d]

Study Guide

Objective 9.1a Graph parabolas.

Example Graph: $y = -x^2 - 2x + 5$.

$$\begin{aligned} y &= -(x^2 + 2x + 0) + 5 \\ &= -(x^2 + 2x + (1 - 1)) + 5 \\ &= -(x^2 + 2x + 1) + (-1)(-1) + 5 \\ &= -(x^2 + 2x + 1) + 1 + 5 \\ &= -(x + 1)^2 + 6, \text{ or} \\ &\quad -[x - (-1)]^2 + 6 \end{aligned}$$

The vertex is $(-1, 6)$. Next, we plot some points on each side of the vertex.

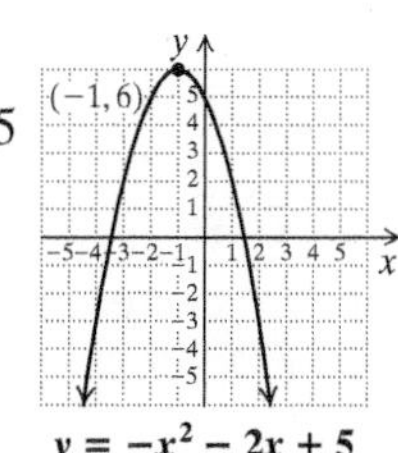

Practice Exercise

1. Graph: $y = -x^2 - 4x - 1$.

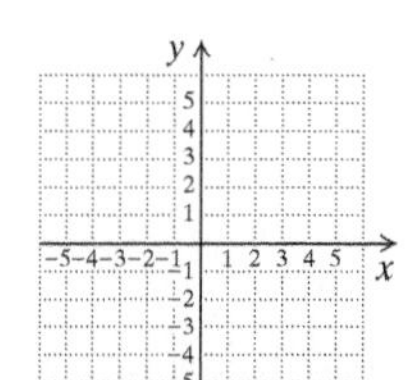

Objective 9.1b Use the distance formula to find the distance between two points whose coordinates are known.

Example Find the distance between $(2, -5)$ and $(-3, 4)$. Give an exact answer and an approximation to three decimal places.

Let $(x_1, y_1) = (2, -5)$ and $(x_2, y_2) = (-3, 4)$. Then

$$d = \sqrt{(x_2 - x_1)^2 + (y_2 - y_1)^2}$$
$$= \sqrt{(-3 - 2)^2 + [4 - (-5)]^2} = \sqrt{(-5)^2 + 9^2}$$
$$= \sqrt{25 + 81} = \sqrt{106} \approx 10.296.$$

Practice Exercise

2. Find the distance between $(-2, 10)$ and $(-1, 7)$. Give an exact answer and an approximation to three decimal places.

Objective 9.1c Use the midpoint formula to find the midpoint of a segment when the coordinates of its endpoints are known.

Example Find the midpoint of the segment with endpoints $(15, -6)$ and $(-3, -20)$.

Let $(x_1, y_1) = (15, -6)$ and $(x_2, y_2) = (-3, -20)$. Then

$$\left(\frac{x_1 + x_2}{2}, \frac{y_1 + y_2}{2}\right) = \left(\frac{15 + (-3)}{2}, \frac{-6 + (-20)}{2}\right)$$
$$= \left(\frac{12}{2}, \frac{-26}{2}\right) = (6, -13).$$

The midpoint is $(6, -13)$.

Practice Exercise

3. Find the midpoint of the segment with endpoints $(17, -14)$ and $(-9, -2)$.

Objective 9.1d Given an equation of a circle, find its center and radius and graph it. Given the center and the radius of a circle, write an equation of the circle and graph the circle.

Example Find the center and the radius of this circle. Then graph the circle.

$$(x + 1)^2 + (y - 3)^2 = 9$$

We first write the equation in standard form:

$$[x - (-1)]^2 + (y - 3)^2 = 3^2.$$

The center is $(-1, 3)$ and the radius is 3.

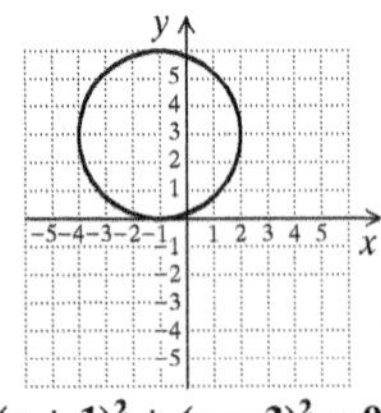

$(x + 1)^2 + (y - 3)^2 = 9$

Practice Exercises

4. Find the center and the radius of this circle. Then graph the circle.

$$(x - 2)^2 + (y + 1)^2 = 16$$

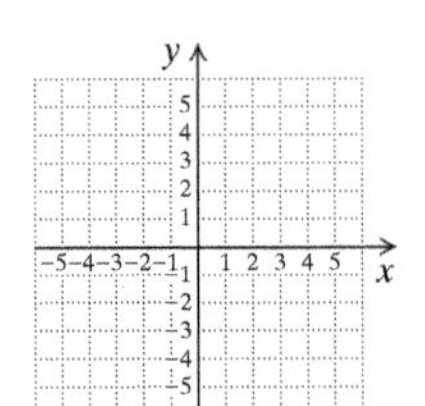

Example Find an equation of the circle having the given center and radius:

Center: $(-6, 0)$; radius: $\sqrt{5}$.

$$(x - h)^2 + (y - k)^2 = r^2 \qquad \text{Standard form}$$
$$[x - (-6)]^2 + (y - 0)^2 = (\sqrt{5})^2$$
$$(x + 6)^2 + y^2 = 5$$

5. Find an equation of the circle having the given center and radius:

Center: $(0, 3)$; radius: 6.

Objective 9.2a Graph the standard form of the equation of an ellipse.

Example Graph: $25x^2 + 9y^2 = 225$.

$$\frac{1}{225}(25x^2 + 9y^2) = \frac{1}{225}\cdot 225$$

$$\frac{x^2}{9} + \frac{y^2}{25} = 1$$

$$\frac{x^2}{3^2} + \frac{y^2}{5^2} = 1$$

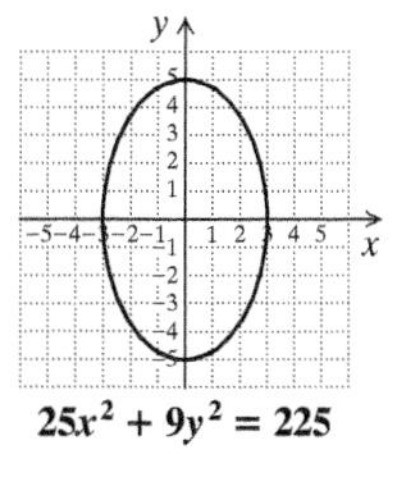

The x-intercepts are $(-3, 0)$ and $(3, 0)$, and the y-intercepts are $(0, -5)$ and $(0, 5)$. The vertices are $(0, -5)$ and $(0, 5)$.

Practice Exercise

6. Graph: $25x^2 + 4y^2 = 100$.

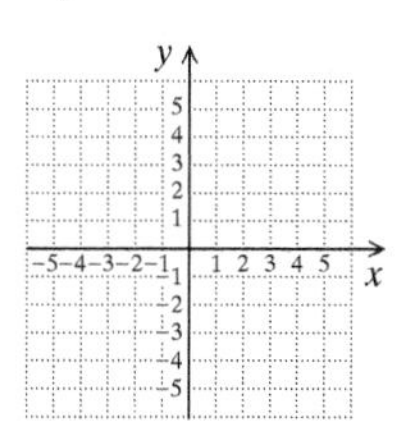

Objective 9.3a Graph the standard form of the equation of a hyperbola.

Example Graph: $\frac{x^2}{1} - \frac{y^2}{4} = 1$.

$$\frac{x^2}{1} - \frac{y^2}{4} = \frac{x^2}{1^2} - \frac{y^2}{2^2}$$

$a = 1, b = 2$; the asymptotes are $y = 2x$ and $y = -2x$.

Since the hyperbola is horizontal, the vertices are the x-intercepts. We replace y with 0 and solve for x:

$$\frac{x^2}{1} - \frac{0^2}{4} = 1$$

$$x^2 = 1$$

$$x = \pm 1.$$

The x-intercepts are $(-1, 0)$ and $(1, 0)$. There are no y-intercepts.

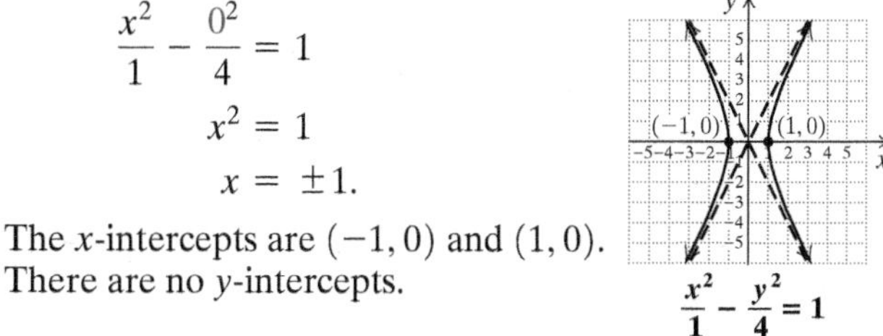

Practice Exercise

7. Graph: $\frac{x^2}{9} - \frac{y^2}{25} = 1$.

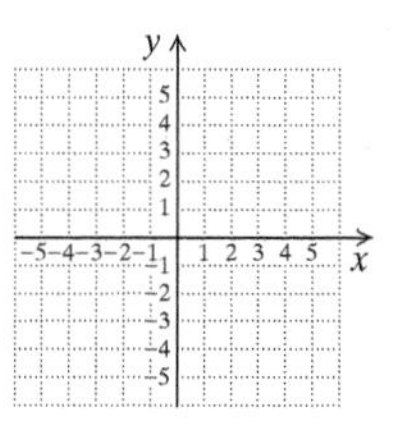

Objective 9.4a Solve systems of equations in which at least one equation is nonlinear.

Example Solve: $\frac{x^2}{4} + \frac{y^2}{16} = 1$,

$2x + y = 4$.

We solve the linear equation for y: $y = -2x + 4$. We then substitute $-2x + 4$ for y in the other equation and solve for x after we clear fractions:

$4x^2 + y^2 = 16$ Clearing fractions

$4x^2 + (-2x + 4)^2 = 16$ Substituting

$4x^2 + (4x^2 - 16x + 16) = 16$

$8x^2 - 16x + 16 = 16$

$8x^2 - 16x = 0$

$8x(x - 2) = 0$

$8x = 0$ *or* $x - 2 = 0$

$x = 0$ *or* $x = 2$.

Next, we substitute these numbers for x in $2x + y = 4$, or $y = -2x + 4$, and solve for y:

$y = -2\cdot 0 + 4 = 4$; $y = -2\cdot 2 + 4 = 0$.

The pairs $(0, 4)$ and $(2, 0)$ check and are the solutions.

Practice Exercise

8. Solve:

$\frac{x^2}{36} + \frac{y^2}{4} = 1$,

$3y - x = 6$.

Review Exercises

Find the distance between each pair of points. Where appropriate, give an approximation to three decimal places. [9.1b]

1. $(2, 6)$ and $(6, 6)$

2. $(-1, 1)$ and $(-5, 4)$

3. $(1.4, 3.6)$ and $(4.7, -5.3)$

4. $(2, 3a)$ and $(-1, a)$

Find the midpoint of the segment with the given endpoints. [9.1c]

5. $(1, 6)$ and $(7, 6)$

6. $(-1, 1)$ and $(-5, 4)$

7. $(1, \sqrt{3})$ and $\left(\frac{1}{2}, -\sqrt{2}\right)$

8. $(2, 3a)$ and $(-1, a)$

Find the center and the radius of each circle. [9.1d]

9. $(x + 2)^2 + (y - 3)^2 = 2$

10. $(x - 5)^2 + y^2 = 49$

11. $x^2 + y^2 - 6x - 2y + 1 = 0$

12. $x^2 + y^2 + 8x - 6y - 10 = 0$

13. Find an equation of the circle with center $(-4, 3)$ and radius $4\sqrt{3}$. [9.1d]

14. Find an equation of the circle with center $(7, -2)$ and radius $2\sqrt{5}$. [9.1d]

Graph.

15. $\dfrac{x^2}{16} + \dfrac{y^2}{4} = 1$ [9.2a]

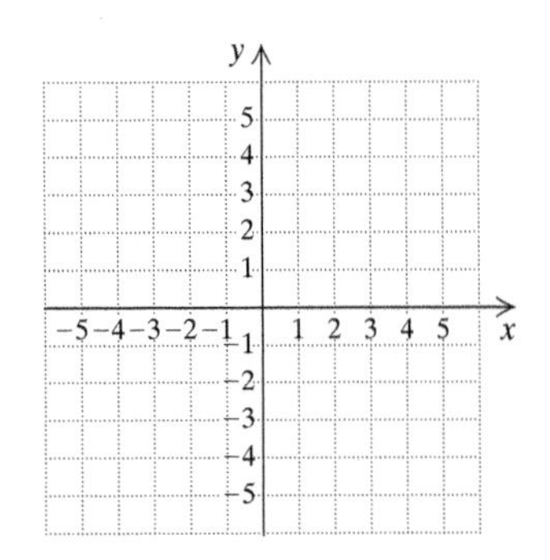

16. $\dfrac{y^2}{9} - \dfrac{x^2}{4} = 1$ [9.3a]

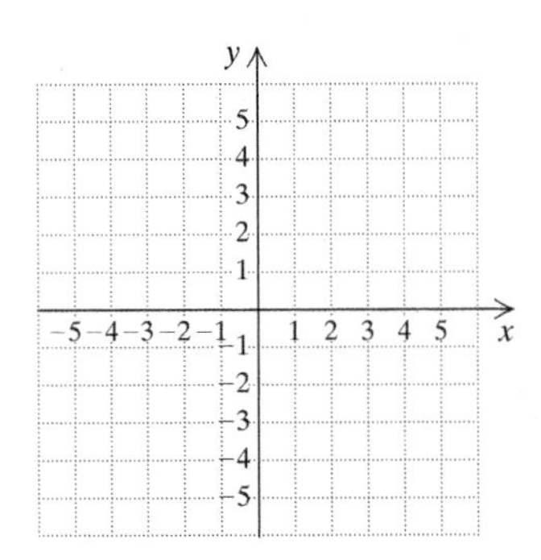

17. $x^2 + y^2 = 16$ [9.1d]

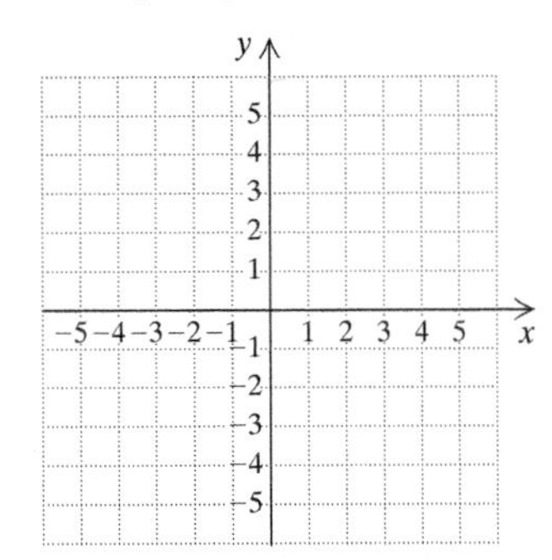

18. $x = y^2 + 2y - 2$ [9.1a]

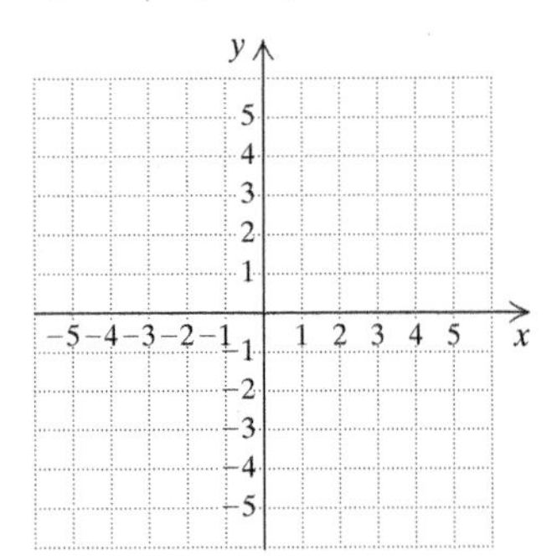

19. $y = -2x^2 - 2x + 3$ [9.1a]

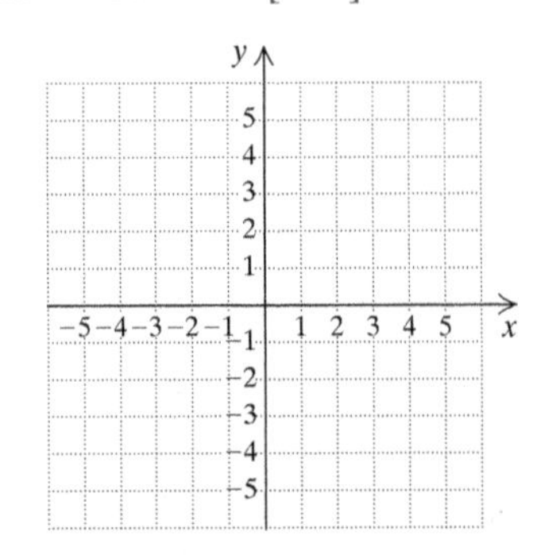

20. $x^2 + y^2 + 2x - 4y - 4 = 0$ [9.1d]

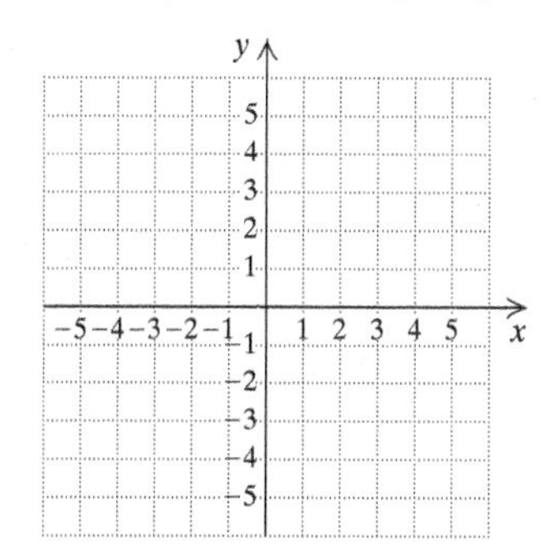

21. $\dfrac{(x - 3)^2}{9} + \dfrac{(y + 4)^2}{4} = 1$ [9.2a]

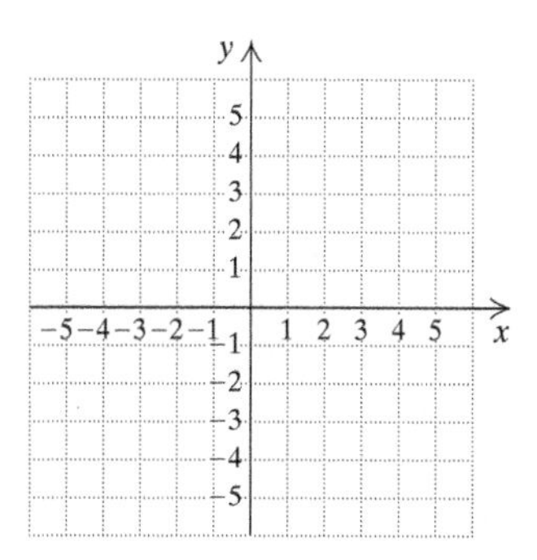

22. $xy = 9$ [9.3b]

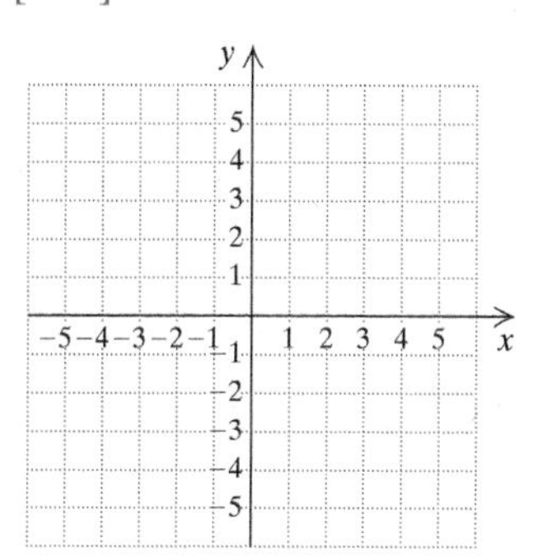

23. $x + y^2 = 2y + 1$ [9.1a]

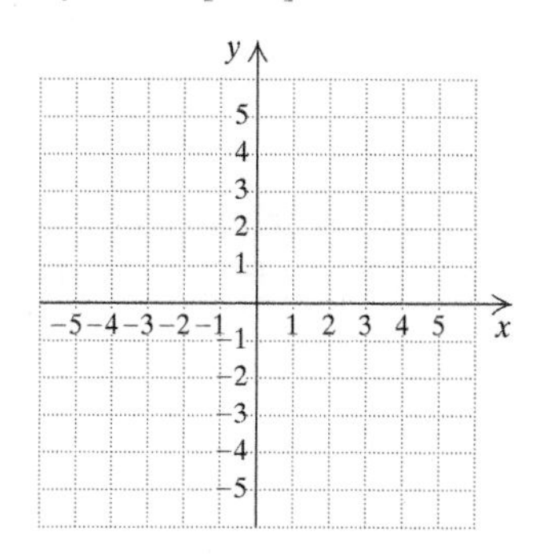

24. $\dfrac{x^2}{4} - \dfrac{y^2}{4} = 1$ [9.3a]

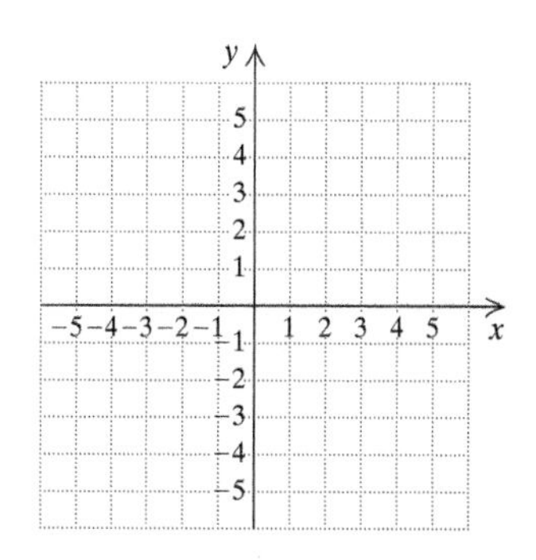

Solve. [9.4a]

25. $x^2 - y^2 = 33,$
$x + y = 11$

26. $x^2 - 2x + 2y^2 = 8,$
$2x + y = 6$

27. $x^2 - y = 3,$
$2x - y = 3$

28. $x^2 + y^2 = 25,$
$x^2 - y^2 = 7$

29. $x^2 - y^2 = 3,$
$y = x^2 - 3$

30. $x^2 + y^2 = 18,$
$2x + y = 3$

31. $x^2 + y^2 = 100,$
$2x^2 - 3y^2 = -120$

32. $x^2 + 2y^2 = 12,$
$xy = 4$

Solve. [9.4b]

33. *Carton Dimensions.* One type of carton used by a manufacturer of stationery products exactly fits both a notecard of area 12 in^2 and a pen of length 5 in., laid diagonally on top of the notecards. What are the dimensions of the carton?

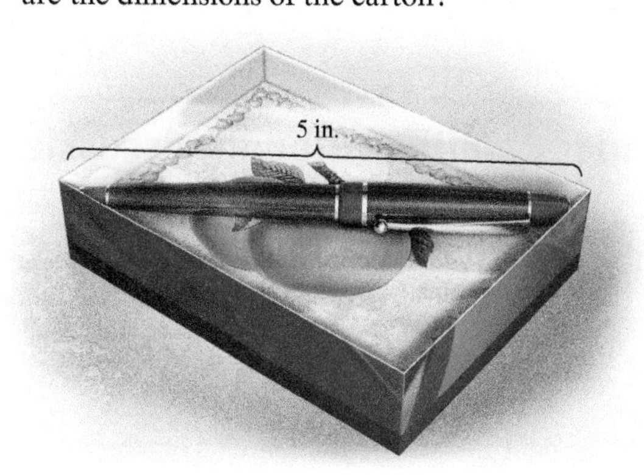

34. *Flower Beds.* The sum of the areas of two circular flower beds is 130π ft^2. The difference of the circumferences is 16π ft. Find the radius of each flower bed.

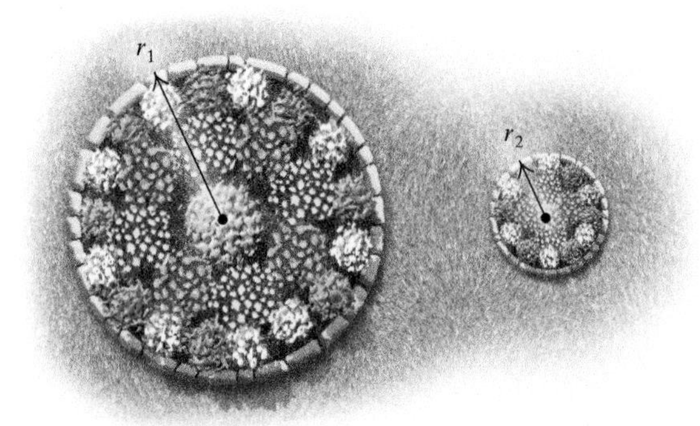

35. Find two positive integers whose sum is 12 and the sum of whose reciprocals is $\frac{3}{8}$.

36. *Vegetable Garden.* A rectangular vegetable garden has a perimeter of 38 m and an area of 84 m^2. What are the dimensions of the garden?

37. From the selections below, choose a graphical representation of the solution set of the system of equations

$$y = \frac{1}{2}x^2 + 1,$$
$$2x - 3y = -6. \quad [9.4a]$$

A.
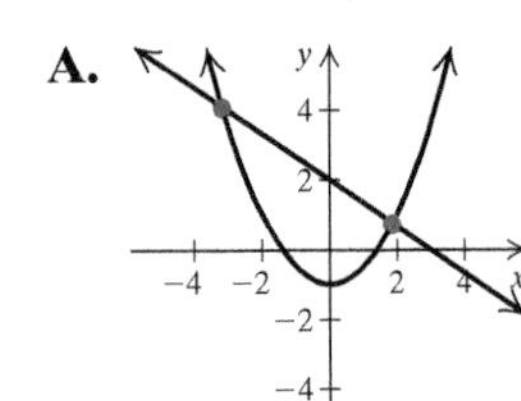

B.
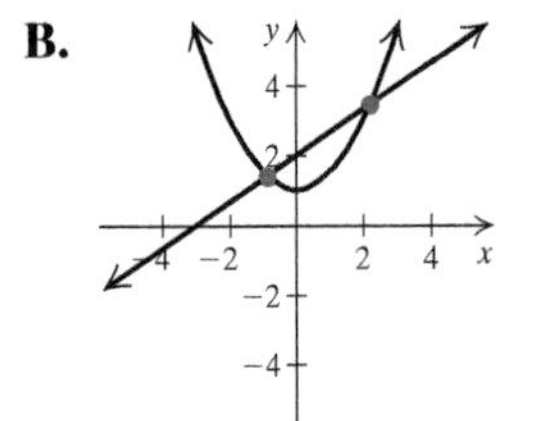

C.
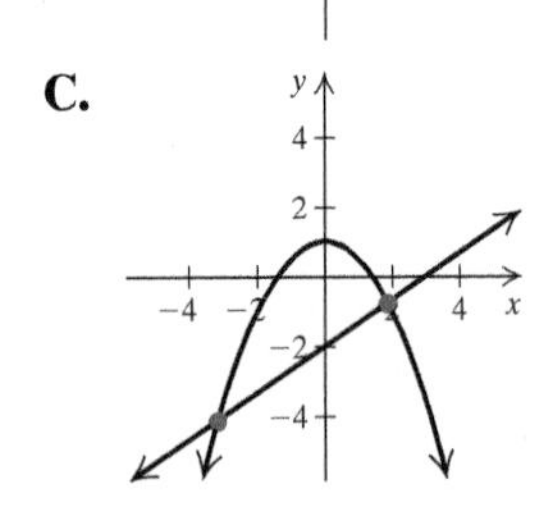

D.
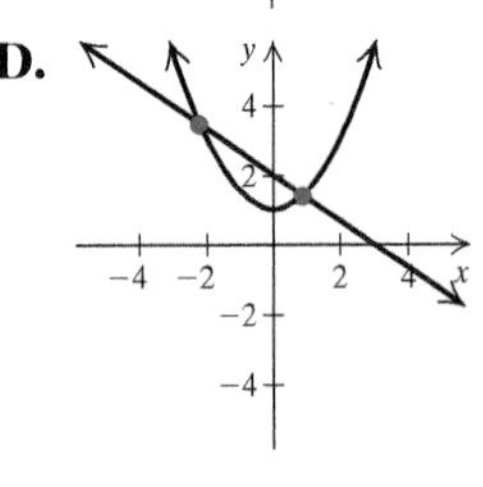

38. Find the center and the radius of the circle

$$x^2 + y^2 + 6x - 16y + 66 = 0. \quad [9.1d]$$

A. Center: $(-3, 8)$; radius: 7
B. Center: $(-6, -8)$; radius: $\sqrt{7}$
C. Center: $(6, -8)$; radius: 7
D. Center: $(-3, 8)$; radius: $\sqrt{7}$

Synthesis

39. Solve:

$$4x^2 - x - 3y^2 = 9,$$
$$-x^2 + x + y^2 = 2. \quad [9.4a]$$

40. Find an equation of the circle that passes through $(-2, -4)$, $(5, -5)$, and $(6, 2)$. [9.1d]

41. Find an equation of the ellipse with the intercepts $(-7, 0)$, $(7, 0)$, $(0, -3)$, and $(0, 3)$. [9.2a]

42. Find the point on the x-axis that is equidistant from $(-3, 4)$ and $(5, 6)$. [9.1b]

Classify each graph as either a circle, an ellipse, a parabola, or a hyperbola.

43. $-y + 4x^2 = 5 - 2x$ [9.1a]

44. $xy = -6$ [9.3b]

45. $\dfrac{x^2}{23} + \dfrac{y^2}{23} = 1$ [9.1d]

46. $43 - 12x^2 + y^2 = 21x^2 + 2y^2$ [9.2a]

47. $3x^2 + 3y^2 = 170$ [9.1d]

48. $\dfrac{x^2}{8} - \dfrac{y^2}{2} = 1$ [9.3a]

Understanding Through Discussion and Writing

1. We have studied techniques for solving systems of equations in this chapter. How do the equations differ from those systems that we studied earlier in the text? [9.4a]

2. Consider the standard equations of a circle, a parabola, an ellipse, and a hyperbola. Which, if any, are functions? Explain. [9.1a], [9.2a], [9.3a]

3. How does the graph of a hyperbola differ from the graph of a parabola? [9.1a], [9.3a]

4. If, in

$$\frac{x^2}{a^2} - \frac{y^2}{b^2} = 1,$$

$a = b$, what are the asymptotes of the graph? Explain. [9.3a]

Test

For Extra Help — For step-by-step test solutions, access the Chapter Test Prep Videos in MyLab Math.

Find the distance between each pair of points. Where appropriate, give an approximation to three decimal places.

1. $(-6, 2)$ and $(6, 8)$

2. $(3, -a)$ and $(-3, a)$

Find the midpoint of the segment with the given endpoints.

3. $(-6, 2)$ and $(6, 8)$

4. $(3, -a)$ and $(-3, a)$

Find the center and the radius of each circle.

5. $(x + 2)^2 + (y - 3)^2 = 64$

6. $x^2 + y^2 + 4x - 6y + 4 = 0$

7. Find an equation of the circle with center $(-2, -5)$ and radius $3\sqrt{2}$.

Graph.

8. $y = x^2 - 4x - 1$

9. $x^2 + y^2 = 36$

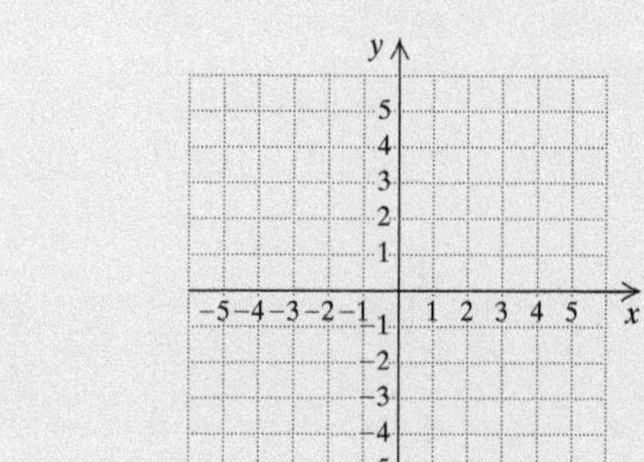

10. $\dfrac{x^2}{9} - \dfrac{y^2}{4} = 1$

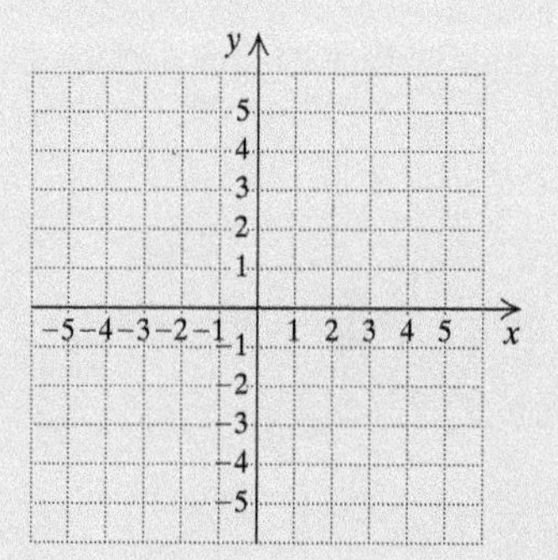

11. $\dfrac{(x + 2)^2}{16} + \dfrac{(y - 3)^2}{9} = 1$

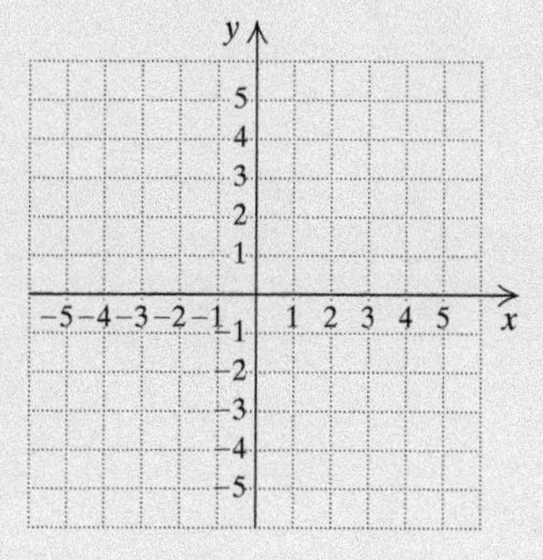

12. $x^2 + y^2 - 4x + 6y + 4 = 0$

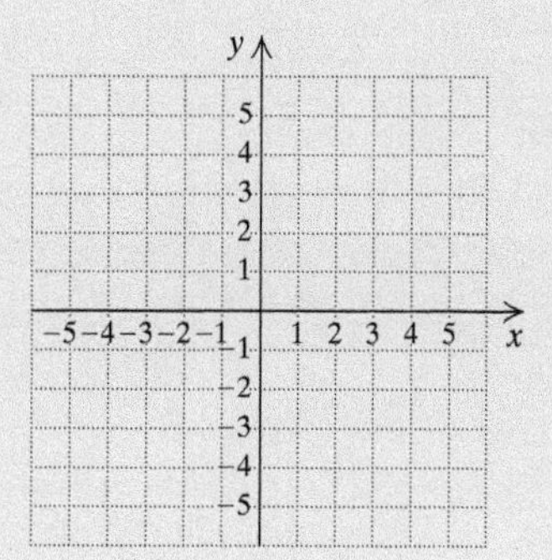

13. $9x^2 + y^2 = 36$

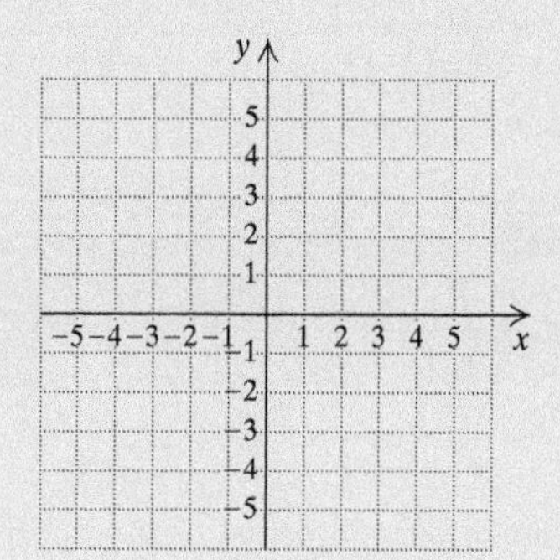

14. $xy = 4$

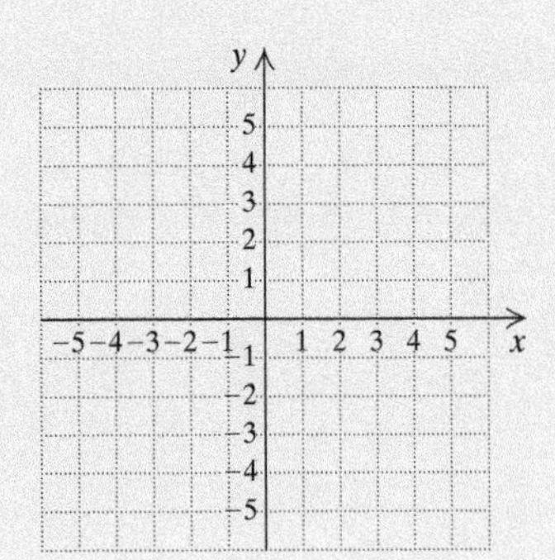

15. $x = -y^2 + 4y$

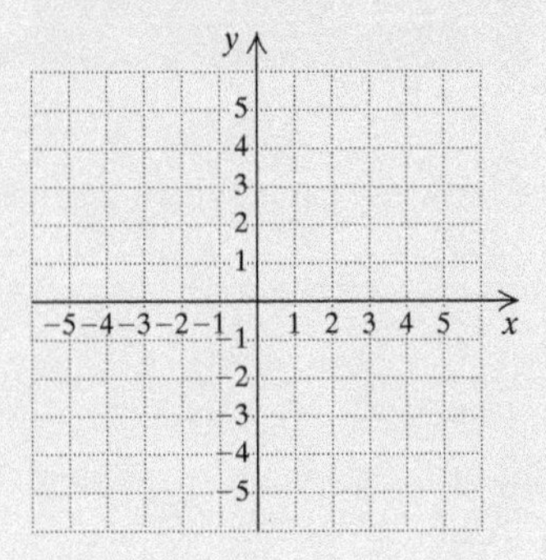

Solve.

16. $\dfrac{x^2}{16} + \dfrac{y^2}{9} = 1,$

$3x + 4y = 12$

17. $x^2 + y^2 = 16,$

$\dfrac{x^2}{16} - \dfrac{y^2}{9} = 1$

18. *Home Office.* A rectangular home office has a diagonal of 20 ft and a perimeter of 56 ft. What are the dimensions of the office?

19. *Investments.* Peggyann invested a certain amount of money for 1 year and earned \$72 in interest. Sally Jean invested \$240 more than Peggyann at an interest rate that was $\frac{5}{6}$ of the rate given to Peggyann, but she earned the same amount of interest. Find the principal and the interest rate of Peggyann's investment.

20. A rectangle with a diagonal of length $5\sqrt{5}$ yd has an area of 22 yd^2. Find the dimensions of the rectangle.

21. *Water Fountains.* The sum of the areas of two square water fountains is 8 m^2, and the difference of their areas is 2 m^2. Find the length of a side of each square.

22. From the selections below, choose a graphical representation of the solution set of the system of equations

$$\frac{x^2}{9} - \frac{y^2}{4} = 1,$$

$$x^2 + y^2 = 16.$$

A.

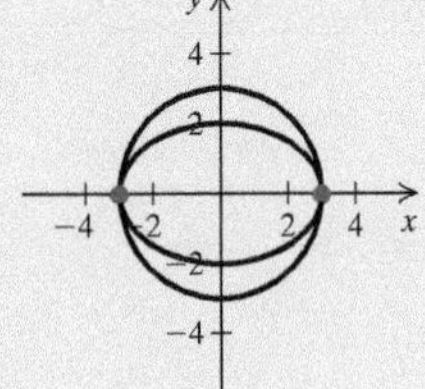

B.

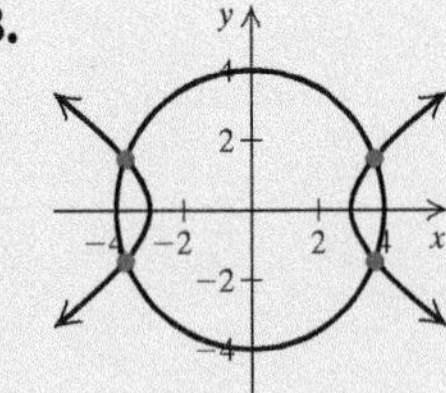

C.

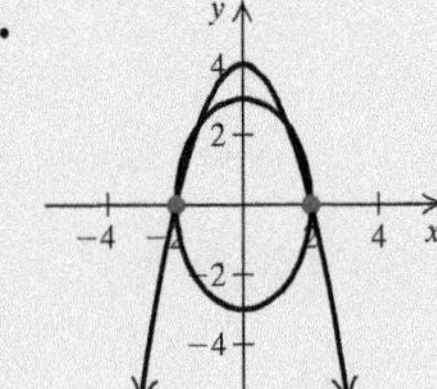

D.

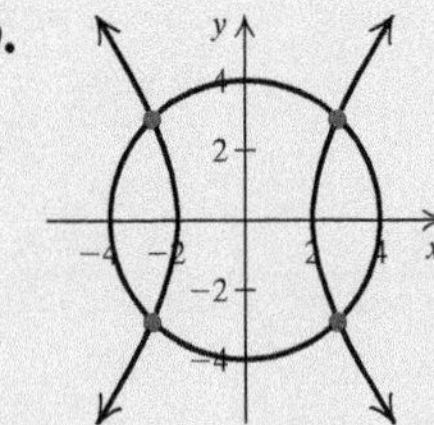

Synthesis

23. Find an equation of the ellipse passing through $(6, 0)$ and $(6, 6)$ with vertices at $(1, 3)$ and $(11, 3)$.

24. Find the points whose distance from $(8, 0)$ is 10.

25. The sum of two numbers is 36, and the product is 4. Find the sum of the reciprocals of the numbers.

26. Find the point on the y-axis that is equidistant from $(-3, -5)$ and $(4, -7)$.

CHAPTERS

1–9 Cumulative Review

Solve.

1. $\frac{1}{3}x - \frac{1}{5} \geq \frac{1}{5}x - \frac{1}{3}$

2. $|x| > 6.4$

3. $3 \leq 4x + 7 < 31$

4. $3x + y = 4,$
$-6x - y = -3$

5. $x^4 - 13x^2 + 36 = 0$

6. $2x^2 = x + 3$

7. $3x - \frac{6}{x} = 7$

8. $\sqrt{x + 5} = x - 1$

9. $x(x + 10) = -21$

10. $2x^2 + x + 1 = 0$

11. $7^x = 30$

12. $\frac{x + 1}{x - 2} > 0$

13. $\log_3 x = 2$

14. $x^2 - 1 \geq 0$

15. $\log_2 x + \log_2 (x + 7) = 3$

16. $\frac{1}{p} + \frac{1}{q} = \frac{1}{f}$, for p

17. $x - y + 2z = 3,$
$-x + z = 4,$
$2x + y - z = -3$

18. $-x^2 + 2y^2 = 7,$
$x^2 + y^2 = 5$

19. $\frac{3}{x - 3} - \frac{x + 2}{x^2 + 2x - 15} = \frac{1}{x + 5}$

20. $P = \frac{3}{4}(M + 2N)$, for N

Solve.

21. *Oil.* Worldwide demand for oil is expected to grow exponentially. The amount of oil $N(t)$, in millions of barrels per day, demanded t years after 2000, can be approximated by

$$N(t) = 77(1.019)^t,$$

where $t = 0$ corresponds to 2000.

Data: euractiv.com

a) How much oil is projected to be demanded in 2016? in 2025?
b) What is the doubling time?
c) Graph the function.

22. *Interest Compounded Annually.* Suppose that \$50,000 is invested at 4% interest, compounded annually.

a) Find a function A for the amount in the account after t years.

b) Find the amount of money in the account at $t = 0$, $t = 4$, $t = 8$, and $t = 10$.

c) Graph the function.

Simplify.

23. $(2x + 3)(x^2 - 2x - 1)$

24. $(3x^2 + x^3 - 1) - (2x^3 + x + 5)$

25. $\dfrac{2m^2 + 11m - 6}{m^3 + 1} \cdot \dfrac{m^2 - m + 1}{m + 6}$

26. $\dfrac{x}{x - 1} + \dfrac{2}{x + 1} - \dfrac{2x}{x^2 - 1}$

27. $\dfrac{1 - \dfrac{5}{x}}{x - 4 - \dfrac{5}{x}}$

28. $(x^4 + 3x^3 - x + 4) \div (x + 1)$

29. $\dfrac{\sqrt{75x^5y^2}}{\sqrt{3xy}}$

30. $4\sqrt{50} - 3\sqrt{18}$

31. $(16^{3/2})^{1/2}$

32. $(2 - i\sqrt{2})(5 + 3i\sqrt{2})$

33. $\dfrac{5 + i}{2 - 4i}$

34. *Medicaid Enrollment.* The projected Medicaid enrollment, in millions, can be modeled by the linear function

$$M(t) = 1.05t + 72.65,$$

where t is the number of years after 2016.

Data: Department of Health and Human Services, 2016 Actuarial Report of the Financial Outlook of Medicaid

a) Find the projected Medicaid enrollment in 2018, in 2021, and in 2025.

b) Graph the function.

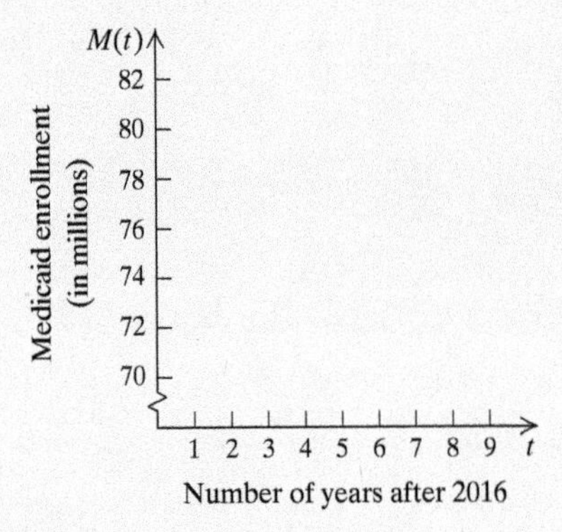

c) Find the y-intercept.

d) Find the slope.

e) Find the rate of change.

35. Find an equation of the line containing the points $(1, 4)$ and $(-1, 0)$.

36. Find an equation of the line containing the point $(1, 2)$ and perpendicular to the line whose equation is $2x - y = 3$.

Graph.

37. $4y - 3x = 12$

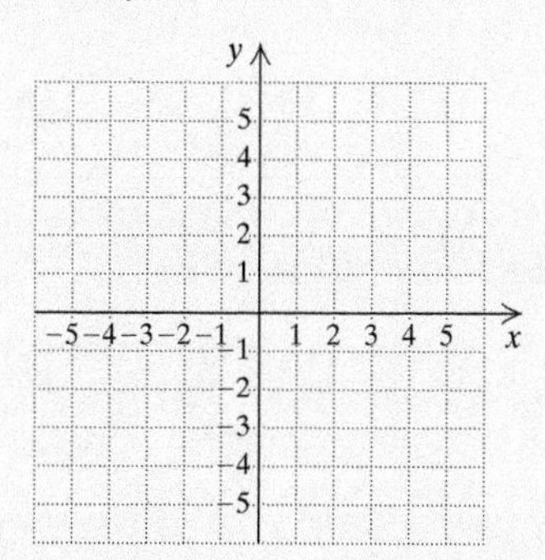

38. $y < -2$

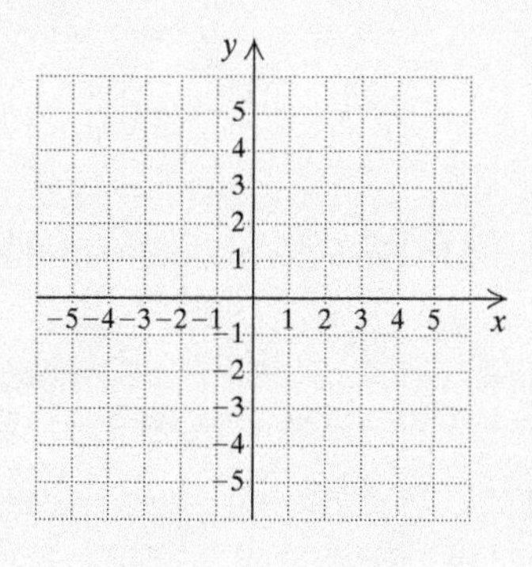

39. $\frac{x^2}{9} + \frac{y^2}{25} = 1$

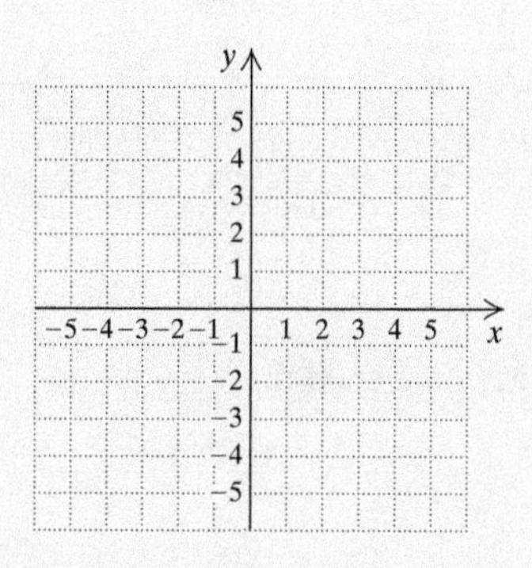

40. $x^2 + y^2 = 2.25$

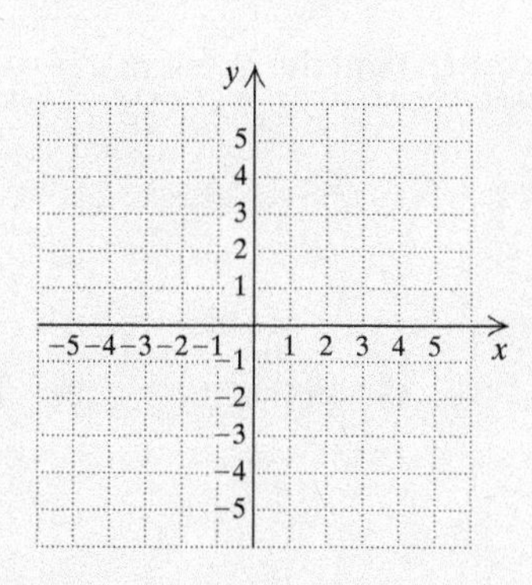

41. $x + y \leq 0,$
$x \geq -4,$
$y \geq -1$

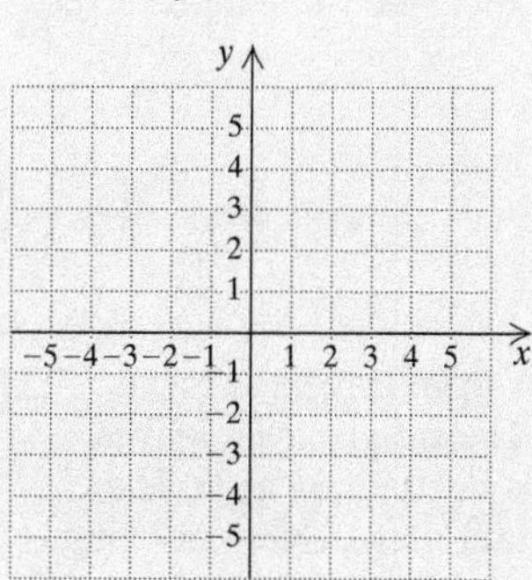

42. $\frac{x^2}{25} - \frac{y^2}{16} = 1$

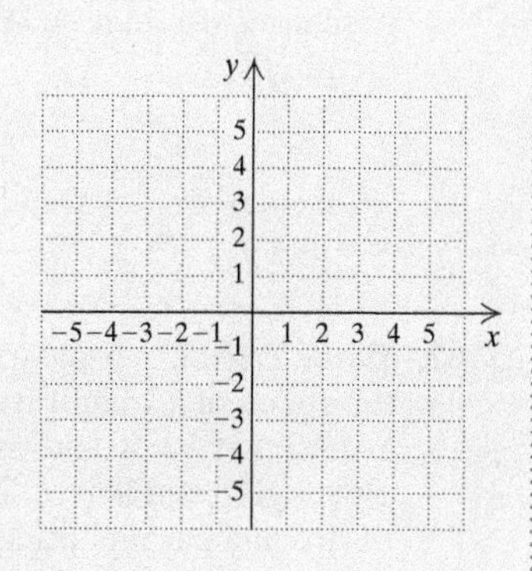

43. $(x - 1)^2 + (y + 1)^2 = 9$

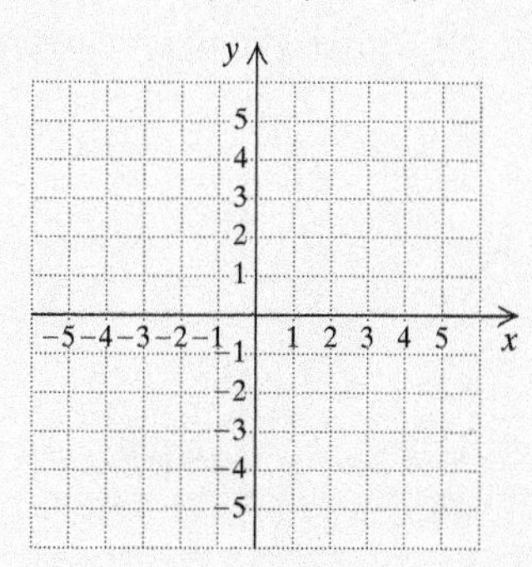

44. $f(x) = 2x^2 - 8x + 9$

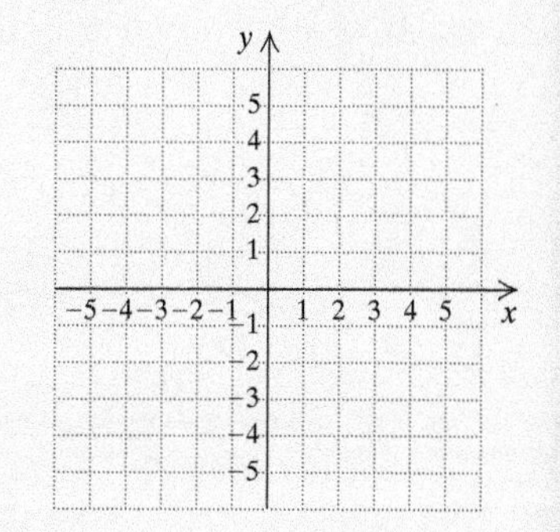

45. $x = 3.5$

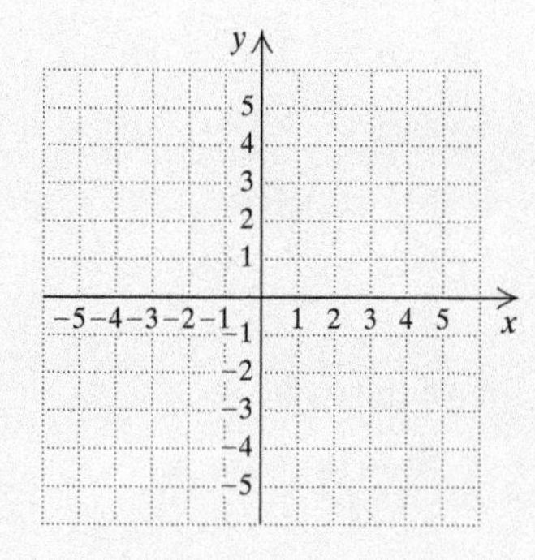

46. $x = y^2 + 1$

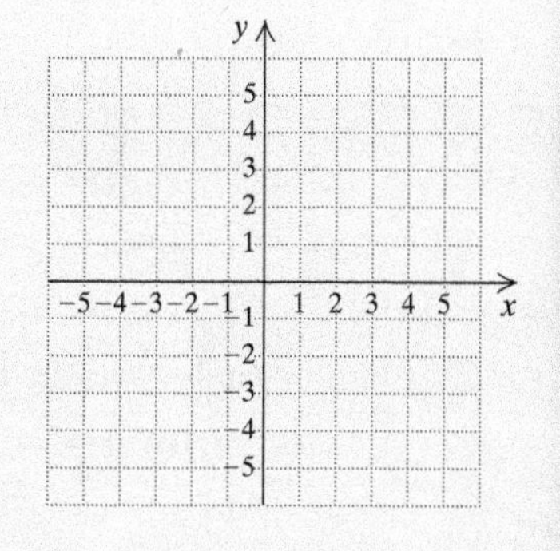

47. $f(x) = e^{-x}$

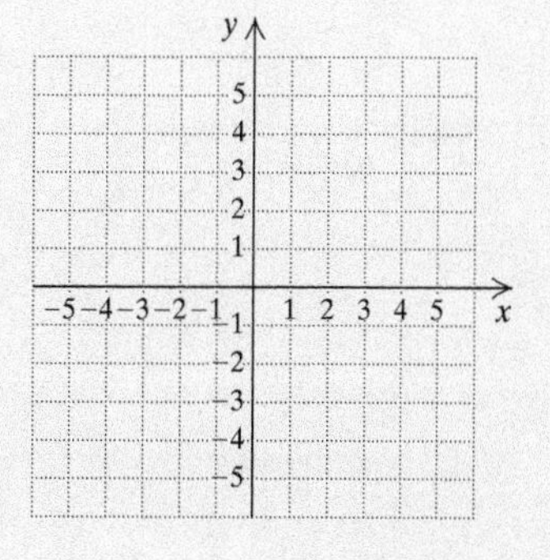

48. $f(x) = \log_2 x$

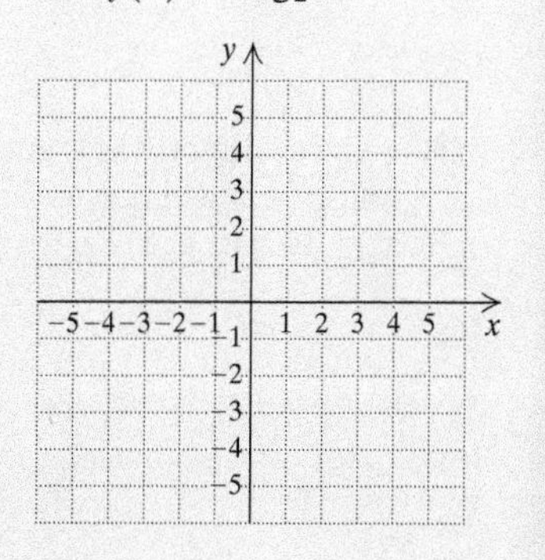

Factor.

49. $2x^4 - 12x^3 + x - 6$

50. $3a^2 - 12ab - 135b^2$

51. $x^2 - 17x + 72$

52. $81m^4 - n^4$

53. $16x^2 - 16x + 4$

54. $81a^3 - 24$

55. $10x^2 + 66x - 28$

56. $6x^3 + 27x^2 - 15x$

57. Find the center and the radius of the circle
$$x^2 - 16x + y^2 + 6y + 68 = 0.$$

58. Find $f^{-1}(x)$ when $f(x) = 2x - 3$.

59. z varies directly as x and inversely as the cube of y, and $z = 5$ when $x = 4$ and $y = 2$. What is z when $x = 10$ and $y = 5$?

60. Given the function f described by $f(x) = x^3 - 2$, find $f(-2)$.

61. Find the distance between the points $(2, 1)$ and $(8, 9)$.

62. Find the midpoint of the segment with endpoints $(-1, -3)$ and $(3, 0)$.

63. Rationalize the denominator: $\frac{5 + \sqrt{a}}{3 - \sqrt{a}}$.

64. Find the domain: $f(x) = \frac{4x - 3}{3x^2 + x}$.

65. Given that $f(x) = 3x^2 + x$, find a such that $f(a) = 2$.

Solve.

66. *Book Club.* A book club offers two types of membership. Limited members pay a fee of \$20 per year and can buy books for \$20 each. Preferred members pay \$40 per year and can buy books for \$15 each. For what numbers of annual book purchases would it be less expensive to be a preferred member?

67. *Train Travel.* A passenger train travels at twice the speed of a freight train. The freight train leaves a station at 2 A.M. and travels north at 34 mph. The passenger train leaves the station at 11 A.M, traveling north on a parallel track. How far from the station will the passenger train overtake the freight train?

68. *Perimeters of Polygons.* A pentagon with all five sides the same size has a perimeter equal to that of an octagon in which all eight sides are the same size. One side of the pentagon is 2 less than three times one side of the octagon. What is the perimeter of each figure?

69. *Ammonia Solutions.* A chemist has two solutions of ammonia and water. Solution A is 6% ammonia and solution B is 2% ammonia. How many liters of each solution are needed in order to obtain 80 L of a solution that is 3.2% ammonia?

70. *Air Travel.* An airplane can fly 190 mi with the wind in the same time that it takes to fly 160 mi against the wind. The speed of the wind is 30 mph. How fast can the plane fly in still air?

71. *Work.* Christy can do a certain job in 21 min. Madeline can do the same job in 14 min. How long would it take to do the job if the two worked together?

72. *Centripetal Force.* The centripetal force F of an object moving in a circle varies directly as the square of the velocity v and inversely as the radius r of the circle. If $F = 8$ when $v = 1$ and $r = 10$, what is F when $v = 2$ and $r = 16$?

73. *Rectangle Dimensions.* The perimeter of a rectangle is 34 ft. The length of a diagonal is 13 ft. Find the dimensions of the rectangle.

74. *Dimensions of a Rug.* The diagonal of a Persian rug is 25 ft. The area of the rug is 300 ft^2. Find the length and the width of the rug.

75. *Maximizing Area.* A farmer wants to fence in a rectangular area next to a river. (Note that no fence will be needed along the river.) What is the area of the largest region that can be fenced in with 100 ft of fencing?

76. *Carbon Dating.* Use the function $P(t) = P_0e^{-0.00012t}$ to find the age of a bone that has lost 25% of its carbon-14.

77. *Beam Load.* The weight W that a horizontal beam can support varies inversely as the length L of the beam. If a 14-m beam can support 1440 kg, what weight can a 6-m beam support?

78. Fit a linear function to the data points $(2, -3)$ and $(5, -4)$.

79. Fit a quadratic function to the data points $(-2, 4)$, $(-5, -6)$, and $(1, -3)$.

80. Convert to a logarithmic equation: $10^6 = r$.

81. Convert to an exponential equation: $\log_3 Q = x$.

82. Express as a single logarithm:

$$\frac{1}{5}(7\log_b x - \log_b y - 8\log_b z).$$

83. Express in terms of logarithms of x, y, and z:

$$\log_b\left(\frac{xy^5}{z}\right)^{-6}.$$

84. What is the maximum product of two numbers whose sum is 26?

85. Determine whether the function $f(x) = 4 - x^2$ is one-to-one.

86. For the graph of function f shown here, determine **(a)** $f(2)$; **(b)** the domain; **(c)** all x-values such that $f(x) = -5$; and **(d)** the range.

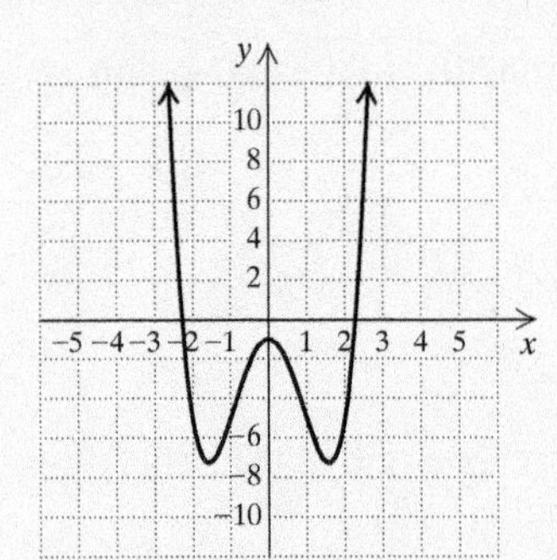

87. *Population Growth of Nevada.* Of the 50 states, Nevada had the greatest percent change in population from 2000 to 2015. In 2015, the population of Nevada was 2,890,845. It had grown from a population of 1,998,257 in 2000. Assume that the population growth increases according to an exponential growth function.

Data: U.S. Census Bureau

a) Let $t = 0$ correspond to 2000 and $t = 15$ correspond to 2015. Then t is the number of years after 2000. Use the data points $(0, 1{,}998{,}257)$ and $(15, 2{,}890{,}845)$ to find the exponential growth rate and fit an exponential growth function $P(t) = P_0e^{kt}$ to the data, where $P(t)$ is the population of Nevada t years after 2000.

b) Use the function found in part (a) to predict the population of Nevada in 2020.

c) If growth continues at this rate, when will the population reach 4.0 million?

Synthesis

88. Solve: $\dfrac{9}{x} - \dfrac{9}{x + 12} = \dfrac{108}{x^2 + 12x}$.

89. Solve: $\log_2(\log_3 x) = 2$.

90. Describe the graph of

$$\frac{x^2}{a^2} + \frac{y^2}{b^2} = 1$$

when $a^2 = b^2$.

91. Diaphantos, a famous mathematician, spent $\frac{1}{6}$ of his life as a child, $\frac{1}{12}$ as a young man, and $\frac{1}{7}$ as a bachelor. Five years after he was married, he had a son who died 4 years before his father at half his father's final age. How long did Diaphantos live?

Appendixes

A

OBJECTIVES

a Find equivalent fraction expressions by multiplying by 1.

b Simplify fraction notation.

c Add, subtract, multiply, and divide using fraction notation.

Fraction Notation

a EQUIVALENT EXPRESSIONS AND FRACTION NOTATION

An example of **fraction notation** for a number is

$$\frac{2}{3} \begin{matrix}\leftarrow \text{Numerator} \\ \leftarrow \text{Denominator}\end{matrix}$$

The top number is called the **numerator**, and the bottom number is called the **denominator**.

The **whole numbers** consist of the natural numbers and 0:

$$0, \quad 1, \quad 2, \quad 3, \quad 4, \quad 5, \ldots.$$

The **arithmetic numbers**, also called the **nonnegative rational numbers**, consist of the whole numbers and the fractions, such as $\frac{2}{3}$ and $\frac{9}{5}$.

ARITHMETIC NUMBERS

The **arithmetic numbers** are the whole numbers and the fractions, such as 8, $\frac{3}{4}$, and $\frac{6}{5}$. All these numbers can be named with fraction notation $\frac{a}{b}$, where a and b are whole numbers and $b \neq 0$.

Note that all whole numbers can be named with fraction notation. For example, we can name the whole number 8 as $\frac{8}{1}$. We call 8 and $\frac{8}{1}$ **equivalent expressions**.

Being able to find an equivalent expression is critical to a study of algebra. Two simple but powerful properties of numbers that allow us to find equivalent expressions are the identity properties of 0 and 1.

THE IDENTITY PROPERTY OF 0 (ADDITIVE IDENTITY)

For any number a,

$$a + 0 = a.$$

(Adding 0 to any number gives that same number—for example, $12 + 0 = 12$.)

THE IDENTITY PROPERTY OF 1 (MULTIPLICATIVE IDENTITY)

For any number a,

$$a \cdot 1 = a.$$

(Multiplying any number by 1 gives that same number—for example, $\frac{3}{5} \cdot 1 = \frac{3}{5}$.)

Here are some ways to name the number 1:

$$\frac{5}{5}, \quad \frac{3}{3}, \quad \text{and} \quad \frac{26}{26}.$$

The following property allows us to find equivalent fraction expressions.

EQUIVALENT EXPRESSIONS FOR 1

For any number a, $a \neq 0$,

$$\frac{a}{a} = 1.$$

We can use the identity property of 1 and the preceding result to find equivalent fraction expressions.

EXAMPLE 1 Write a fraction expression equivalent to $\frac{2}{3}$ with a denominator of 15.

Note that $15 = 3 \cdot 5$. We want fraction notation for $\frac{2}{3}$ that has a denominator of 15, but the denominator 3 is missing a factor of 5. We multiply by 1, using $\frac{5}{5}$ as an equivalent expression for 1. Recall from arithmetic that to multiply with fraction notation, we multiply numerators and we multiply denominators:

$$\frac{2}{3} = \frac{2}{3} \cdot 1 \qquad \text{Using the identity property of 1}$$

$$= \frac{2}{3} \cdot \frac{5}{5} \qquad \text{Using } \frac{5}{5} \text{ for 1}$$

$$= \frac{10}{15}. \qquad \text{Multiplying numerators and denominators}$$

Do Exercises 1–3.

1. Write a fraction expression equivalent to $\frac{2}{3}$ with a denominator of 12.

2. Write a fraction expression equivalent to $\frac{3}{4}$ with a denominator of 28.

3. Multiply by 1 to find three different fraction expressions for $\frac{7}{8}$.

b SIMPLIFYING EXPRESSIONS

We know that $\frac{1}{2}, \frac{2}{4}, \frac{4}{8}$, and so on, all name the same number. Any arithmetic number can be named in many ways. The **simplest fraction notation** is the notation that has the smallest numerator and denominator. We call the process of finding the simplest fraction notation **simplifying**. We reverse the process of Example 1 by first factoring the numerator and the denominator. Then we factor the fraction expression and remove a factor of 1 using the identity property of 1.

EXAMPLE 2 Simplify: $\frac{10}{15}$.

$$\frac{10}{15} = \frac{2 \cdot 5}{3 \cdot 5} \qquad \text{Factoring the numerator and the denominator. In this case, each is the prime factorization.}$$

$$= \frac{2}{3} \cdot \frac{5}{5} \qquad \text{Factoring the fraction expression}$$

$$= \frac{2}{3} \cdot 1$$

$$= \frac{2}{3} \qquad \text{Using the identity property of 1 (removing a factor of 1)}$$

Answers

1. $\frac{8}{12}$ **2.** $\frac{21}{28}$ **3.** $\frac{14}{16}, \frac{21}{24}, \frac{28}{32}$; answers may vary

EXAMPLE 3 Simplify: $\frac{36}{24}$.

$$\frac{36}{24} = \frac{2 \cdot 3 \cdot 2 \cdot 3}{2 \cdot 2 \cdot 3 \cdot 2} \quad \text{Factoring the numerator and the denominator}$$

$$= \frac{2 \cdot 3 \cdot 2}{2 \cdot 3 \cdot 2} \cdot \frac{3}{2} \quad \text{Factoring the fraction expression}$$

$$= 1 \cdot \frac{3}{2}$$

$$= \frac{3}{2} \quad \text{Removing a factor of 1}$$

It is always a good idea to check at the end to see if you have indeed factored out all the common factors of the numerator and the denominator.

Canceling

Canceling is a shortcut that you may have used to remove a factor of 1 when working with fraction notation. With *great* concern, we mention it as a possible way to speed up your work. You should use canceling only when removing common factors in numerators and denominators. Each common factor allows us to remove a factor of 1 in a product. **Canceling *cannot* be done when adding.** Our concern is that "canceling" be performed with care and understanding. Example 3 might have been done faster as follows:

$$\frac{36}{24} = \frac{\cancel{2} \cdot \cancel{3} \cdot \cancel{2} \cdot 3}{\cancel{2} \cdot \cancel{2} \cdot \cancel{3} \cdot 2} = \frac{3}{2}, \quad \text{or} \quad \frac{36}{24} = \frac{3 \cdot \cancel{12}}{2 \cdot \cancel{12}} = \frac{3}{2}, \quad \text{or} \quad \frac{\overset{\overset{3}{\cancel{18}}}{\cancel{36}}}{\underset{\underset{2}{\cancel{12}}}{\cancel{24}}} = \frac{3}{2}.$$

Caution!

The difficulty with canceling is that it is often applied incorrectly in situations like the following:

$$\frac{\cancel{2} + 3}{\cancel{2}} = 3; \quad \frac{\cancel{4} + 1}{\cancel{4} + 2} = \frac{1}{2}; \quad \frac{1\cancel{5}}{\cancel{5}4} = \frac{1}{4}.$$

Wrong! Wrong! Wrong!

The correct answers are

$$\frac{2 + 3}{2} = \frac{5}{2}; \quad \frac{4 + 1}{4 + 2} = \frac{5}{6};$$

$$\frac{15}{54} = \frac{3 \cdot 5}{3 \cdot 18} = \frac{3}{3} \cdot \frac{5}{18} = \frac{5}{18}.$$

In each situation, the number canceled was not a factor of 1. Factors are parts of products. For example, in $2 \cdot 3$, 2 and 3 are factors, but in $2 + 3$, 2 and 3 are *not* factors.

Do Exercises 4–7.

Simplify.

4. $\frac{18}{45}$

5. $\frac{38}{18}$

6. $\frac{72}{27}$

7. $\frac{32}{56}$

Answers

4. $\frac{2}{5}$ 5. $\frac{19}{9}$ 6. $\frac{8}{3}$ 7. $\frac{4}{7}$

We can always insert the number 1 as a factor. The identity property of 1 allows us to do that.

EXAMPLE 4 Simplify: $\frac{18}{72}$.

$$\frac{18}{72} = \frac{2 \cdot \cancel{9}}{8 \cdot \cancel{9}} = \frac{2}{8} = \frac{\cancel{2} \cdot 1}{\cancel{2} \cdot 4} = \frac{1}{4}, \quad \text{or} \quad \frac{18}{72} = \frac{1 \cdot \cancel{18}}{4 \cdot \cancel{18}} = \frac{1}{4}$$

EXAMPLE 5 Simplify: $\frac{72}{9}$.

$\frac{72}{9} = \frac{8 \cdot 9}{1 \cdot 9}$ Factoring and inserting a factor of 1 in the denominator

$= \frac{8 \cdot \cancel{9}}{1 \cdot \cancel{9}}$ Removing a factor of 1: $\frac{9}{9} = 1$

$= \frac{8}{1} = 8$ Simplifying

Do Exercises 8 and 9.

Simplify.

8. $\frac{27}{54}$

9. $\frac{48}{12}$

C MULTIPLICATION, ADDITION, SUBTRACTION, AND DIVISION

After we have performed an operation of multiplication, addition, subtraction, or division, the answer may not be in simplified form. We simplify, if at all possible.

Multiplication

To multiply using fraction notation, we multiply the numerators to get the new numerator, and we multiply the denominators to get the new denominator.

MULTIPLYING FRACTIONS

To multiply fractions, multiply the numerators and multiply the denominators:

$$\frac{a}{b} \cdot \frac{c}{d} = \frac{a \cdot c}{b \cdot d}.$$

EXAMPLE 6 Multiply and simplify: $\frac{5}{6} \cdot \frac{9}{25}$.

$\frac{5}{6} \cdot \frac{9}{25} = \frac{5 \cdot 9}{6 \cdot 25}$ Multiplying numerators and denominators

$= \frac{5 \cdot 3 \cdot 3}{2 \cdot 3 \cdot 5 \cdot 5}$ Factoring the numerator and the denominator

$= \frac{\cancel{5} \cdot \cancel{3} \cdot 3}{2 \cdot \cancel{3} \cdot \cancel{5} \cdot 5}$ Removing a factor of 1: $\frac{3 \cdot 5}{3 \cdot 5} = 1$

$= \frac{3}{10}$ Simplifying

Do Exercises 10 and 11.

Multiply and simplify.

10. $\frac{6}{5} \cdot \frac{25}{12}$

11. $\frac{3}{8} \cdot \frac{5}{3} \cdot \frac{7}{2}$

Answers

8. $\frac{1}{2}$ **9.** 4 **10.** $\frac{5}{2}$ **11.** $\frac{35}{16}$

Addition

When denominators are the same, we can add by adding the numerators and keeping the same denominator.

ADDING FRACTIONS WITH LIKE DENOMINATORS

To add fractions when denominators are the same, add the numerators and keep the same denominator:

$$\frac{a}{c} + \frac{b}{c} = \frac{a+b}{c}.$$

EXAMPLE 7 Add: $\frac{4}{8} + \frac{5}{8}$.

The common denominator is 8. We add the numerators and keep the common denominator:

$$\frac{4}{8} + \frac{5}{8} = \frac{4+5}{8} = \frac{9}{8}.$$

To convert from a mixed numeral to fraction notation:

$$3\frac{5}{8} = \frac{29}{8}$$

ⓐ Multiply the whole number by the denominator:

$3 \cdot 8 = 24.$

ⓑ Add the result to the numerator:

$24 + 5 = 29.$

ⓒ Keep the denominator.

In arithmetic, we generally write $\frac{9}{8}$ as $1\frac{1}{8}$. (See a review of converting from a mixed numeral to fraction notation at left.) In algebra, you will find that *improper fraction* symbols such as $\frac{9}{8}$ are more useful and are quite *proper* for our purposes.

What do we do when denominators are different? We find a common denominator. We can do this by multiplying by 1. Consider adding $\frac{1}{6}$ and $\frac{3}{4}$. There are several common denominators that can be obtained. Let's look at two possibilities.

A.
$$\begin{aligned}\frac{1}{6} + \frac{3}{4} &= \frac{1}{6}\cdot 1 + \frac{3}{4}\cdot 1 \\ &= \frac{1}{6}\cdot\frac{4}{4} + \frac{3}{4}\cdot\frac{6}{6} \\ &= \frac{4}{24} + \frac{18}{24} \\ &= \frac{22}{24} \\ &= \frac{11}{12} \quad \text{Simplifying}\end{aligned}$$

B.
$$\begin{aligned}\frac{1}{6} + \frac{3}{4} &= \frac{1}{6}\cdot 1 + \frac{3}{4}\cdot 1 \\ &= \frac{1}{6}\cdot\frac{2}{2} + \frac{3}{4}\cdot\frac{3}{3} \\ &= \frac{2}{12} + \frac{9}{12} \\ &= \frac{11}{12}\end{aligned}$$

We had to simplify in **A**. We didn't have to simplify in **B**. In **B**, we used the least common multiple of the denominators, 12. That number is called the **least common denominator**, or **LCD**. Using the LCD allows us to add fractions using the smallest numbers possible.

ADDING FRACTIONS WITH DIFFERENT DENOMINATORS

To add fractions when denominators are different:

a) Find the least common multiple of the denominators. That number is the least common denominator, LCD.

b) Multiply by 1, using the appropriate notation n/n for each fraction to express fractions in terms of the LCD.

c) Add the numerators, keeping the same denominator.

d) Simplify, if possible.

EXAMPLE 8 Add and simplify: $\frac{3}{8} + \frac{5}{12}$.

The LCM of the denominators, 8 and 12, is 24. Thus the LCD is 24. We multiply each fraction by 1 to obtain the LCD:

$$\frac{3}{8} + \frac{5}{12} = \frac{3}{8}\cdot\frac{3}{3} + \frac{5}{12}\cdot\frac{2}{2}$$ Multiplying by 1. Since $3 \cdot 8 = 24$, we multiply the first number by $\frac{3}{3}$. Since $2 \cdot 12 = 24$, we multiply the second number by $\frac{2}{2}$.

$$= \frac{9}{24} + \frac{10}{24}$$

$$= \frac{9 + 10}{24}$$ Adding the numerators and keeping the same denominator

$$= \frac{19}{24}.$$

EXAMPLE 9 Add and simplify: $\frac{11}{30} + \frac{5}{18}$.

We first look for the LCM of 30 and 18. That number is then the LCD. We find the prime factorization of each denominator:

$$\frac{11}{30} + \frac{5}{18} = \frac{11}{5\cdot 2\cdot 3} + \frac{5}{2\cdot 3\cdot 3}.$$

The LCD is $5 \cdot 2 \cdot 3 \cdot 3$, or 90. To get the LCD in the first denominator, we need a factor of 3. To get the LCD in the second denominator, we need a factor of 5. We get these numbers by multiplying by 1:

$$\frac{11}{30} + \frac{5}{18} = \frac{11}{5\cdot 2\cdot 3}\cdot\frac{3}{3} + \frac{5}{2\cdot 3\cdot 3}\cdot\frac{5}{5}$$ Multiplying by 1

$$= \frac{33}{5\cdot 2\cdot 3\cdot 3} + \frac{25}{2\cdot 3\cdot 3\cdot 5}$$ The denominators are now the LCD.

$$= \frac{58}{5\cdot 2\cdot 3\cdot 3}$$ Adding the numerators and keeping the LCD

$$= \frac{\cancel{2}\cdot 29}{5\cdot \cancel{2}\cdot 3\cdot 3}$$ Factoring the numerator and removing a factor of 1

$$= \frac{29}{45}.$$ Simplifying

Do Exercises 12–15.

Add and simplify.

12. $\frac{4}{5} + \frac{3}{5}$

13. $\frac{5}{6} + \frac{7}{6}$

14. $\frac{5}{6} + \frac{7}{10}$

15. $\frac{13}{24} + \frac{7}{40}$

Subtraction

When subtracting, we also multiply by 1 to obtain the LCD. After we have made the denominators the same, we can subtract by subtracting the numerators and keeping the same denominator.

EXAMPLE 10 Subtract and simplify: $\frac{9}{8} - \frac{4}{5}$.

$$\frac{9}{8} - \frac{4}{5} = \frac{9}{8}\cdot\frac{5}{5} - \frac{4}{5}\cdot\frac{8}{8}$$ The LCD is 40.

$$= \frac{45}{40} - \frac{32}{40}$$

$$= \frac{45 - 32}{40} = \frac{13}{40}$$ Subtracting the numerators and keeping the same denominator

Answers

12. $\frac{7}{5}$ 13. 2 14. $\frac{23}{15}$ 15. $\frac{43}{60}$

Subtract and simplify.

16. $\frac{7}{8} - \frac{2}{5}$

17. $\frac{5}{12} - \frac{2}{9}$

Find each reciprocal.

18. $\frac{4}{11}$

19. $\frac{15}{7}$

20. 5

21. $\frac{1}{3}$

22. Divide by multiplying by 1:

$$\frac{\frac{3}{5}}{\frac{4}{7}}.$$

Answers

16. $\frac{19}{40}$ **17.** $\frac{7}{36}$ **18.** $\frac{11}{4}$ **19.** $\frac{7}{15}$
20. $\frac{1}{5}$ **21.** 3 **22.** $\frac{21}{20}$

EXAMPLE 11 Subtract and simplify: $\frac{7}{10} - \frac{1}{5}$.

$$\frac{7}{10} - \frac{1}{5} = \frac{7}{10} - \frac{1}{5}\cdot\frac{2}{2} \qquad \text{The LCD is 10; } \frac{7}{10} \text{ already has the LCD.}$$

$$= \frac{7}{10} - \frac{2}{10} = \frac{7-2}{10}$$

$$= \frac{5}{10}$$

$$= \frac{1\cdot \cancel{5}}{2\cdot \cancel{5}} = \frac{1}{2} \qquad \text{Removing a factor of 1: } \frac{5}{5} = 1$$

Do Exercises 16 and 17.

Reciprocals

Two numbers whose product is 1 are called **reciprocals**, or **multiplicative inverses**, of each other. All the arithmetic numbers, except zero, have reciprocals.

EXAMPLES

12. The reciprocal of $\frac{2}{3}$ is $\frac{3}{2}$ because $\frac{2}{3}\cdot\frac{3}{2} = \frac{6}{6} = 1$.

13. The reciprocal of 9 is $\frac{1}{9}$ because $9\cdot\frac{1}{9} = \frac{9}{9} = 1$.

14. The reciprocal of $\frac{1}{4}$ is 4 because $\frac{1}{4}\cdot 4 = \frac{4}{4} = 1$.

Do Exercises 18–21.

Reciprocals and Division

Reciprocals and the number 1 can be used to justify a fast way to divide arithmetic numbers. We multiply by 1, carefully choosing the expression for 1.

EXAMPLE 15 Divide $\frac{2}{3}$ by $\frac{7}{5}$.

This is a symbol for 1.

$$\frac{2}{3} \div \frac{7}{5} = \frac{\frac{2}{3}}{\frac{7}{5}} = \frac{\frac{2}{3}}{\frac{7}{5}}\cdot\frac{\frac{5}{7}}{\frac{5}{7}} \qquad \text{Multiplying by } \frac{\frac{5}{7}}{\frac{5}{7}}\text{. We use } \frac{5}{7} \text{ because it is the reciprocal of } \frac{7}{5}.$$

$$= \frac{\frac{2}{3}\cdot\frac{5}{7}}{\frac{7}{5}\cdot\frac{5}{7}} \qquad \text{Multiplying numerators and denominators}$$

$$= \frac{\frac{10}{21}}{\frac{35}{35}} = \frac{\frac{10}{21}}{1} \qquad \frac{35}{35} = 1$$

$$= \frac{10}{21} \qquad \text{Simplifying}$$

After multiplying in Example 15, we had a denominator of $\frac{35}{35}$, or 1. That was because we used $\frac{5}{7}$, the reciprocal of the divisor, for both the numerator and the denominator of the symbol for 1.

Do Exercise 22.

When multiplying by 1 to divide, we get a denominator of 1. What do we get in the numerator? In Example 15, we got $\frac{2}{3} \cdot \frac{5}{7}$. This is the product of $\frac{2}{3}$, the dividend, and $\frac{5}{7}$, the reciprocal of the divisor. This gives us a procedure for dividing fractions.

DIVIDING FRACTIONS

To divide fractions, multiply by the reciprocal of the divisor:

$$\frac{a}{b} \div \frac{c}{d} = \frac{a}{b} \cdot \frac{d}{c}.$$

EXAMPLE 16 Divide by multiplying by the reciprocal of the divisor: $\frac{1}{2} \div \frac{3}{5}$.

$$\frac{1}{2} \div \frac{3}{5} = \frac{1}{2} \cdot \frac{5}{3} \qquad \frac{5}{3} \text{ is the reciprocal of } \frac{3}{5}$$

$$= \frac{5}{6} \qquad \text{Multiplying}$$

After dividing, always simplify if possible.

EXAMPLE 17 Divide and simplify: $\frac{2}{3} \div \frac{4}{9}$.

$$\frac{2}{3} \div \frac{4}{9} = \frac{2}{3} \cdot \frac{9}{4} \qquad \frac{9}{4} \text{ is the reciprocal of } \frac{4}{9}$$

$$= \frac{2 \cdot 9}{3 \cdot 4} \qquad \text{Multiplying numerators and denominators}$$

$$= \frac{\cancel{2} \cdot \cancel{3} \cdot 3}{\cancel{3} \cdot \cancel{2} \cdot 2} \qquad \text{Removing a factor of 1: } \frac{2 \cdot 3}{2 \cdot 3} = 1$$

$$= \frac{3}{2}$$

Do Exercises 23–26.

Divide by multiplying by the reciprocal of the divisor. Then simplify.

23. $\frac{4}{3} \div \frac{7}{2}$

24. $\frac{5}{4} \div \frac{3}{2}$

25. $\frac{\frac{2}{9}}{\frac{5}{12}}$

26. $\frac{\frac{5}{6}}{\frac{45}{22}}$

EXAMPLE 18 Divide and simplify: $\frac{5}{6} \div 30$.

$$\frac{5}{6} \div 30 = \frac{5}{6} \div \frac{30}{1} = \frac{5}{6} \cdot \frac{1}{30} = \frac{5 \cdot 1}{6 \cdot 30} = \frac{\cancel{5} \cdot 1}{6 \cdot \cancel{5} \cdot 6} = \frac{1}{6 \cdot 6} = \frac{1}{36}$$

Removing a factor of 1: $\frac{5}{5} = 1$

EXAMPLE 19 Divide and simplify: $24 \div \frac{3}{8}$.

$$24 \div \frac{3}{8} = \frac{24}{1} \div \frac{3}{8} = \frac{24}{1} \cdot \frac{8}{3} = \frac{24 \cdot 8}{1 \cdot 3} = \frac{\cancel{3} \cdot 8 \cdot 8}{1 \cdot \cancel{3}} = \frac{8 \cdot 8}{1} = 64$$

Removing a factor of 1: $\frac{3}{3} = 1$

Do Exercises 27 and 28.

Divide and simplify.

27. $\frac{7}{8} \div 56$

28. $36 \div \frac{4}{9}$

Answers

23. $\frac{8}{21}$ 24. $\frac{5}{6}$ 25. $\frac{8}{15}$
26. $\frac{11}{27}$ 27. $\frac{1}{64}$ 28. 81

A Exercise Set

FOR EXTRA HELP

a Write an equivalent expression for each of the following. Use the indicated name for 1.

1. $\frac{3}{4}\left(\text{Use } \frac{3}{3} \text{ for 1.}\right)$

2. $\frac{5}{6}\left(\text{Use } \frac{10}{10} \text{ for 1.}\right)$

3. $\frac{3}{5}\left(\text{Use } \frac{20}{20} \text{ for 1.}\right)$

4. $\frac{8}{9}\left(\text{Use } \frac{4}{4} \text{ for 1.}\right)$

Write an equivalent expression with the given denominator.

5. $\frac{7}{8}$ (Denominator: 24)

6. $\frac{2}{9}$ (Denominator: 54)

b Simplify.

7. $\frac{18}{27}$

8. $\frac{49}{56}$

9. $\frac{56}{14}$

10. $\frac{48}{27}$

11. $\frac{6}{42}$

12. $\frac{13}{104}$

13. $\frac{56}{7}$

14. $\frac{132}{11}$

15. $\frac{19}{76}$

16. $\frac{17}{51}$

17. $\frac{100}{20}$

18. $\frac{150}{25}$

19. $\frac{425}{525}$

20. $\frac{625}{325}$

21. $\frac{2600}{1400}$

22. $\frac{4800}{1600}$

23. $\frac{8 \cdot x}{6 \cdot x}$

24. $\frac{13 \cdot v}{39 \cdot v}$

c Compute and simplify.

25. $\frac{1}{3} \cdot \frac{1}{4}$

26. $\frac{15}{16} \cdot \frac{8}{5}$

27. $\frac{15}{4} \cdot \frac{3}{4}$

28. $\frac{10}{11} \cdot \frac{11}{10}$

29. $\frac{4}{9} + \frac{13}{18}$

30. $\frac{4}{5} + \frac{8}{15}$

31. $\frac{3}{10} + \frac{8}{15}$

32. $\frac{9}{8} + \frac{7}{12}$

33. $\frac{5}{4} - \frac{3}{4}$

34. $\frac{12}{5} - \frac{2}{5}$

35. $\frac{11}{12} - \frac{3}{8}$

36. $\frac{15}{16} - \frac{5}{12}$

37. $\frac{7}{6} \div \frac{3}{5}$

38. $\frac{7}{5} \div \frac{3}{4}$

39. $\frac{8}{9} \div \frac{4}{15}$

40. $\frac{3}{4} \div \frac{3}{7}$

41. $\frac{\frac{13}{12}}{\frac{39}{5}}$

42. $\frac{\frac{17}{6}}{\frac{3}{8}}$

43. $100 \div \frac{1}{5}$

44. $78 \div \frac{1}{6}$

45. $\frac{3}{4} \div 10$

46. $\frac{5}{6} \div 15$

47. $1000 - \frac{1}{100}$

48. $\frac{147}{50} - 2$

Determinants and Cramer's Rule

B

OBJECTIVES

a Evaluate second-order determinants.

b Evaluate third-order determinants.

c Solve systems of equations using Cramer's rule.

The elimination method concerns itself primarily with the coefficients and constants of the equations. We now introduce a method for solving a system of equations using just the coefficients and constants. This method involves *determinants*.

a EVALUATING DETERMINANTS

The following symbolism represents a **second-order determinant**:

$$\begin{vmatrix} a_1 & b_1 \\ a_2 & b_2 \end{vmatrix}.$$

To evaluate a determinant, we do two multiplications and subtract.

EXAMPLE 1 Evaluate:

$$\begin{vmatrix} 2 & -5 \\ 6 & 7 \end{vmatrix}.$$

We multiply and subtract as follows:

$$\begin{vmatrix} 2 & -5 \\ 6 & 7 \end{vmatrix} = 2 \cdot 7 - 6 \cdot (-5) = 14 + 30 = 44.$$

Determinants are defined according to the pattern shown in Example 1.

SECOND-ORDER DETERMINANT

The determinant $\begin{vmatrix} a_1 & b_1 \\ a_2 & b_2 \end{vmatrix}$ is defined to mean $a_1b_2 - a_2b_1$.

The value of a determinant is a *number*. In Example 1, the value is 44.

Do Exercises 1 and 2. ▶

Evaluate.

1. $\begin{vmatrix} 3 & 2 \\ 4 & 1 \end{vmatrix}$

2. $\begin{vmatrix} 5 & -2 \\ -1 & -1 \end{vmatrix}$

b THIRD-ORDER DETERMINANTS

A **third-order determinant** is defined as follows.

$$\begin{vmatrix} a_1 & b_1 & c_1 \\ a_2 & b_2 & c_2 \\ a_3 & b_3 & c_3 \end{vmatrix} = a_1 \begin{vmatrix} b_2 & c_2 \\ b_3 & c_3 \end{vmatrix} - a_2 \begin{vmatrix} b_1 & c_1 \\ b_3 & c_3 \end{vmatrix} + a_3 \begin{vmatrix} b_1 & c_1 \\ b_2 & c_2 \end{vmatrix}$$

Note the minus sign here.

Note that the a's come from the first column.

Answers

1. −5 2. −7

Note that the second-order determinants on the right can be obtained by crossing out the row and the column in which each a occurs.

For a_1: $\begin{vmatrix} \cancel{a_1} & \cancel{b_1} & \cancel{c_1} \\ \cancel{a_2} & b_2 & c_2 \\ \cancel{a_3} & b_3 & c_3 \end{vmatrix}$ For a_2: $\begin{vmatrix} \cancel{a_1} & b_1 & c_1 \\ \cancel{a_2} & \cancel{b_2} & \cancel{c_2} \\ \cancel{a_3} & b_3 & c_3 \end{vmatrix}$

For a_3: $\begin{vmatrix} \cancel{a_1} & b_1 & c_1 \\ \cancel{a_2} & b_2 & c_2 \\ \cancel{a_3} & \cancel{b_3} & \cancel{c_3} \end{vmatrix}$

EXAMPLE 2 Evaluate this third-order determinant:

$$\begin{vmatrix} -1 & 0 & 1 \\ -5 & 1 & -1 \\ 4 & 8 & 1 \end{vmatrix} = -1\begin{vmatrix} 1 & -1 \\ 8 & 1 \end{vmatrix} - (-5)\begin{vmatrix} 0 & 1 \\ 8 & 1 \end{vmatrix} + 4\begin{vmatrix} 0 & 1 \\ 1 & -1 \end{vmatrix}.$$

We calculate as follows:

$$\begin{aligned} &-1\begin{vmatrix} 1 & -1 \\ 8 & 1 \end{vmatrix} - (-5)\begin{vmatrix} 0 & 1 \\ 8 & 1 \end{vmatrix} + 4\begin{vmatrix} 0 & 1 \\ 1 & -1 \end{vmatrix} \\ &= -1[1 \cdot 1 - 8(-1)] + 5(0 \cdot 1 - 8 \cdot 1) + 4[0 \cdot (-1) - 1 \cdot 1] \\ &= -1(9) + 5(-8) + 4(-1) \\ &= -9 - 40 - 4 \\ &= -53. \end{aligned}$$

Do Exercises 3 and 4.

Evaluate.

3. $\begin{vmatrix} 2 & -1 & 1 \\ 1 & 2 & -1 \\ 3 & 4 & -3 \end{vmatrix}$

4. $\begin{vmatrix} 3 & 2 & 2 \\ -2 & 1 & 4 \\ 4 & -3 & 3 \end{vmatrix}$

C SOLVING SYSTEMS USING DETERMINANTS

Here is a system of two equations in two variables:

$$a_1x + b_1y = c_1,$$
$$a_2x + b_2y = c_2.$$

We form three determinants, which we call D, D_x, and D_y.

$D = \begin{vmatrix} a_1 & b_1 \\ a_2 & b_2 \end{vmatrix}$ In D, we have the coefficients of x and y.

$D_x = \begin{vmatrix} c_1 & b_1 \\ c_2 & b_2 \end{vmatrix}$ To form D_x, we replace the x-coefficients in D with the constants on the right side of the equations.

$D_y = \begin{vmatrix} a_1 & c_1 \\ a_2 & c_2 \end{vmatrix}$ To form D_y, we replace the y-coefficients in D with the constants on the right.

It is important that the replacement be done *without changing the order of the columns*. Then the solution of the system can be found as follows. This is known as **Cramer's rule**.

Answers

3. −6 **4.** 93

CRAMER'S RULE

$$x = \frac{D_x}{D}, \quad y = \frac{D_y}{D}$$

EXAMPLE 3 Solve using Cramer's rule:

$$3x - 2y = 7,$$
$$3x + 2y = 9.$$

We compute D, D_x, and D_y:

$$D = \begin{vmatrix} 3 & -2 \\ 3 & 2 \end{vmatrix} = 3 \cdot 2 - 3 \cdot (-2) = 6 + 6 = 12;$$

$$D_x = \begin{vmatrix} 7 & -2 \\ 9 & 2 \end{vmatrix} = 7 \cdot 2 - 9(-2) = 14 + 18 = 32;$$

$$D_y = \begin{vmatrix} 3 & 7 \\ 3 & 9 \end{vmatrix} = 3 \cdot 9 - 3 \cdot 7 = 27 - 21 = 6.$$

Then

$$x = \frac{D_x}{D} = \frac{32}{12}, \text{ or } \frac{8}{3} \quad \text{and} \quad y = \frac{D_y}{D} = \frac{6}{12} = \frac{1}{2}.$$

The solution is $\left(\frac{8}{3}, \frac{1}{2}\right)$.

Do Exercise 5. ▶

5. Solve using Cramer's rule:

$$4x - 3y = 15,$$
$$x + 3y = 0.$$

Cramer's rule for three equations is very similar to that for two.

$$a_1x + b_1y + c_1z = d_1,$$
$$a_2x + b_2y + c_2z = d_2,$$
$$a_3x + b_3y + c_3z = d_3$$

$$D = \begin{vmatrix} a_1 & b_1 & c_1 \\ a_2 & b_2 & c_2 \\ a_3 & b_3 & c_3 \end{vmatrix} \qquad D_x = \begin{vmatrix} d_1 & b_1 & c_1 \\ d_2 & b_2 & c_2 \\ d_3 & b_3 & c_3 \end{vmatrix}$$

$$D_y = \begin{vmatrix} a_1 & d_1 & c_1 \\ a_2 & d_2 & c_2 \\ a_3 & d_3 & c_3 \end{vmatrix}$$

D is again the determinant of the coefficients of x, y, and z. This time we have one more determinant, D_z. We get it by replacing the z-coefficients in D with the constants on the right:

$$D_z = \begin{vmatrix} a_1 & b_1 & d_1 \\ a_2 & b_2 & d_2 \\ a_3 & b_3 & d_3 \end{vmatrix}.$$

Answer

5. $(3, -1)$

The solution of the system is given by the following.

CRAMER'S RULE

$$x = \frac{D_x}{D}, \quad y = \frac{D_y}{D}, \quad z = \frac{D_z}{D}$$

EXAMPLE 4 Solve using Cramer's rule:

$$\begin{aligned} x - 3y + 7z &= 13, \\ x + y + z &= 1, \\ x - 2y + 3z &= 4. \end{aligned}$$

We compute D, D_x, D_y, and D_z:

$$D = \begin{vmatrix} 1 & -3 & 7 \\ 1 & 1 & 1 \\ 1 & -2 & 3 \end{vmatrix} = -10; \qquad D_x = \begin{vmatrix} 13 & -3 & 7 \\ 1 & 1 & 1 \\ 4 & -2 & 3 \end{vmatrix} = 20;$$

$$D_y = \begin{vmatrix} 1 & 13 & 7 \\ 1 & 1 & 1 \\ 1 & 4 & 3 \end{vmatrix} = -6; \qquad D_z = \begin{vmatrix} 1 & -3 & 13 \\ 1 & 1 & 1 \\ 1 & -2 & 4 \end{vmatrix} = -24.$$

Then

$$x = \frac{D_x}{D} = \frac{20}{-10} = -2;$$

$$y = \frac{D_y}{D} = \frac{-6}{-10} = \frac{3}{5};$$

$$z = \frac{D_z}{D} = \frac{-24}{-10} = \frac{12}{5}.$$

The solution is $\left(-2, \frac{3}{5}, \frac{12}{5}\right)$.

6. Solve using Cramer's rule:

$$\begin{aligned} x - 3y - 7z &= 6, \\ 2x + 3y + z &= 9, \\ 4x + y &= 7. \end{aligned}$$

In Example 4, we would not have needed to evaluate D_z. Once we found x and y, we could have substituted them into one of the equations to find z. In practice, it is faster to use determinants to find only two of the numbers; then we find the third by substitution into an equation.

Do Exercise 6.

In using Cramer's rule, we divide by D. If D were 0, we could not do so.

INCONSISTENT SYSTEMS; DEPENDENT EQUATIONS

If $D = 0$ and at least one of the other determinants is not 0, then the system does not have a solution, and we say that it is *inconsistent*.

If $D = 0$ and all the other determinants are also 0, then there is an infinite set of solutions. In that case, we say that the equations in the system are *dependent*.

Answer

6. $(1, 3, -2)$

B Exercise Set

FOR EXTRA HELP MyLab Math

a Evaluate.

1. $\begin{vmatrix} 3 & 7 \\ 2 & 8 \end{vmatrix}$

2. $\begin{vmatrix} 5 & 4 \\ 4 & -5 \end{vmatrix}$

3. $\begin{vmatrix} -3 & -6 \\ -5 & -10 \end{vmatrix}$

4. $\begin{vmatrix} 4 & 5 \\ -7 & 9 \end{vmatrix}$

5. $\begin{vmatrix} 8 & 2 \\ 12 & -3 \end{vmatrix}$

6. $\begin{vmatrix} 1 & 1 \\ 9 & 8 \end{vmatrix}$

7. $\begin{vmatrix} 2 & -7 \\ 0 & 0 \end{vmatrix}$

8. $\begin{vmatrix} 0 & -4 \\ 0 & -6 \end{vmatrix}$

b Evaluate.

9. $\begin{vmatrix} 0 & 2 & 0 \\ 3 & -1 & 1 \\ 1 & -2 & 2 \end{vmatrix}$

10. $\begin{vmatrix} 3 & 0 & -2 \\ 5 & 1 & 2 \\ 2 & 0 & -1 \end{vmatrix}$

11. $\begin{vmatrix} -1 & -2 & -3 \\ 3 & 4 & 2 \\ 0 & 1 & 2 \end{vmatrix}$

12. $\begin{vmatrix} 1 & 2 & 2 \\ 2 & 1 & 0 \\ 3 & 3 & 1 \end{vmatrix}$

13. $\begin{vmatrix} 3 & 2 & -2 \\ -2 & 1 & 4 \\ -4 & -3 & 3 \end{vmatrix}$

14. $\begin{vmatrix} 2 & -1 & 1 \\ 1 & 2 & -1 \\ 3 & 4 & -3 \end{vmatrix}$

15. $\begin{vmatrix} 3 & 2 & 4 \\ 1 & 1 & 1 \\ 1 & 1 & 1 \end{vmatrix}$

16. $\begin{vmatrix} -1 & 6 & -5 \\ 2 & 4 & 4 \\ 5 & 3 & 10 \end{vmatrix}$

c Solve using Cramer's rule.

17. $3x - 4y = 6,$
$5x + 9y = 10$

18. $5x + 8y = 1,$
$3x + 7y = 5$

19. $-2x + 4y = 3,$
$3x - 7y = 1$

20. $5x - 4y = -3,$
$7x + 2y = 6$

21. $4x + 2y = 11,$
$3x - y = 2$

22. $3x - 3y = 11,$
$9x - 2y = 5$

23. $x + 4y = 8,$
$3x + 5y = 3$

24. $x + 4y = 5,$
$-3x + 2y = 13$

25. $2x - 3y + 5z = 27,$
$x + 2y - z = -4,$
$5x - y + 4z = 27$

26. $x - y + 2z = -3,$
$x + 2y + 3z = 4,$
$2x + y + z = -3$

27. $r - 2s + 3t = 6,$
$2r - s - t = -3,$
$r + s + t = 6$

28. $a - 3c = 6,$
$b + 2c = 2,$
$7a - 3b - 5c = 14$

29. $4x - y - 3z = 1,$
$8x + y - z = 5,$
$2x + y + 2z = 5$

30. $3x + 2y + 2z = 3,$
$x + 2y - z = 5,$
$2x - 4y + z = 0$

31. $p + q + r = 1,$
$p - 2q - 3r = 3,$
$4p + 5q + 6r = 4$

32. $x + 2y - 3z = 9,$
$2x - y + 2z = -8,$
$3x - y - 4z = 3$

C

OBJECTIVE

a Solve systems of two or three equations using matrices.

Elimination Using Matrices

The elimination method concerns itself primarily with the coefficients and constants of the equations. In what follows, we learn a method for solving systems using just the coefficients and the constants. This procedure involves what are called *matrices*.

a In solving systems of equations, we perform computations with the constants. The variables play no important role until the end. Thus we can simplify writing a system by omitting the variables. For example, the system

$$\begin{aligned} 3x + 4y &= 5, \\ x - 2y &= 1 \end{aligned} \quad \text{simplifies to} \quad \begin{matrix} 3 & 4 & 5 \\ 1 & -2 & 1 \end{matrix}$$

if we omit the variables, the operation of addition, and the equals signs. The result is a rectangular array of numbers. Such an array is called a **matrix** (plural, **matrices**). We ordinarily write brackets around matrices. The following are matrices.

$$\begin{bmatrix} 4 & 1 & 3 & 5 \\ 1 & 0 & 1 & 2 \\ 6 & 3 & -2 & 0 \end{bmatrix}, \quad \begin{bmatrix} 6 & 2 & 1 & 4 & 7 \\ 1 & 2 & 1 & 3 & 1 \\ 4 & 0 & -2 & 0 & -3 \end{bmatrix}, \quad \begin{bmatrix} 1 & 2 \\ 145 & 0 \\ -7 & 9 \\ 8 & 1 \\ 0 & 0 \end{bmatrix}.$$

The **rows** of a matrix are horizontal, and the **columns** are vertical.

$$\begin{bmatrix} 5 & -2 & 2 \\ 1 & 0 & 1 \\ 0 & 1 & 2 \end{bmatrix} \begin{matrix} \leftarrow \text{row 1} \\ \leftarrow \text{row 2} \\ \leftarrow \text{row 3} \end{matrix}$$

column 1 ↑ column 2 ↑ column 3 ↑

Let's now use matrices to solve systems of linear equations.

EXAMPLE 1 Solve the system

$$\begin{aligned} 5x - 4y &= -1, \\ -2x + 3y &= 2. \end{aligned}$$

We write a matrix using only the coefficients and the constants, keeping in mind that x corresponds to the first column and y to the second. A dashed line separates the coefficients from the constants at the end of each equation:

$$\left[\begin{array}{cc:c} 5 & -4 & -1 \\ -2 & 3 & 2 \end{array}\right].$$

The individual numbers are called *elements*, or *entries*.

Our goal is to transform this matrix into one of the form

$$\left[\begin{array}{cc:c} a & b & c \\ 0 & d & e \end{array}\right].$$

The variables can then be reinserted to form equations from which we can complete the solution.

We do calculations that are similar to those that we would do if we wrote the entire equations. The first step, if possible, is to multiply and/or interchange the rows so that each number in the first column below the first number is a multiple of that number. In this case, we do so by multiplying Row 2 by 5. This corresponds to multiplying the second equation by 5.

$$\left[\begin{array}{rr|r} 5 & -4 & -1 \\ -10 & 15 & 10 \end{array}\right] \qquad \text{New Row 2} = 5(\text{Row 2})$$

Next, we multiply the first row by 2 and add the result to the second row. This corresponds to multiplying the first equation by 2 and adding the result to the second equation. Although we write the calculations out here, we generally try to do them mentally:

$$2 \cdot 5 + (-10) = 0; \qquad 2(-4) + 15 = 7; \qquad 2(-1) + 10 = 8.$$

$$\left[\begin{array}{rr|r} 5 & -4 & -1 \\ 0 & 7 & 8 \end{array}\right] \qquad \text{New Row 2} = 2(\text{Row 1}) + (\text{Row 2})$$

If we now reinsert the variables, we have

$$5x - 4y = -1, \qquad \textbf{(1)}$$
$$7y = 8. \qquad \textbf{(2)}$$

We can now proceed as before, solving equation (2) for y:

$$7y = 8 \qquad \textbf{(2)}$$
$$y = \tfrac{8}{7}.$$

Next, we substitute $\frac{8}{7}$ for y back in equation (1). This procedure is called *back-substitution.*

$$5x - 4y = -1 \qquad \textbf{(1)}$$
$$5x - 4 \cdot \tfrac{8}{7} = -1 \qquad \text{Substituting } \tfrac{8}{7} \text{ for } y \text{ in equation (1)}$$
$$x = \tfrac{5}{7} \qquad \text{Solving for } x$$

The solution is $\left(\frac{5}{7}, \frac{8}{7}\right)$.

Do Exercise 1. ▶

1. Solve using matrices:

$$5x - 2y = -44,$$
$$2x + 5y = -6.$$

EXAMPLE 2 Solve the system

$$2x - y + 4z = -3,$$
$$x \quad\quad - 4z = 5,$$
$$6x - y + 2z = 10.$$

We first write a matrix, using only the coefficients and the constants. Where there are missing terms, we must write 0's:

$$\left[\begin{array}{rrr|r} 2 & -1 & 4 & -3 \\ 1 & 0 & -4 & 5 \\ 6 & -1 & 2 & 10 \end{array}\right]. \qquad \begin{array}{l} \textbf{(P1)} \\ \textbf{(P2)} \\ \textbf{(P3)} \end{array}$$

(P1), (P2), and (P3) designate the equations that are in the first, second, and third position, respectively.

Our goal is to find an equivalent matrix of the form

$$\left[\begin{array}{ccc|c} a & b & c & d \\ 0 & e & f & g \\ 0 & 0 & h & i \end{array}\right].$$

A matrix of this form can be rewritten as a system of equations from which a solution can be found easily.

Answer

1. $(-8, 2)$

The first step, if possible, is to interchange the rows so that each number in the first column below the first number is a multiple of that number. In this case, we do so by interchanging Rows 1 and 2:

$$\left[\begin{array}{rrr|r} 1 & 0 & -4 & 5 \\ 2 & -1 & 4 & -3 \\ 6 & -1 & 2 & 10 \end{array}\right].$$

This corresponds to interchanging the first two equations.

Next, we multiply the first row by -2 and add it to the second row:

$$\left[\begin{array}{rrr|r} 1 & 0 & -4 & 5 \\ 0 & -1 & 12 & -13 \\ 6 & -1 & 2 & 10 \end{array}\right].$$

This corresponds to multiplying new equation (P1) by -2 and adding it to new equation (P2). The result replaces the former (P2). We perform the calculations mentally.

Now we multiply the first row by -6 and add it to the third row:

$$\left[\begin{array}{rrr|r} 1 & 0 & -4 & 5 \\ 0 & -1 & 12 & -13 \\ 0 & -1 & 26 & -20 \end{array}\right].$$

This corresponds to multiplying equation (P1) by -6 and adding it to equation (P3).

Next, we multiply Row 2 by -1 and add it to the third row:

$$\left[\begin{array}{rrr|r} 1 & 0 & -4 & 5 \\ 0 & -1 & 12 & -13 \\ 0 & 0 & 14 & -7 \end{array}\right].$$

This corresponds to multiplying equation (P2) by -1 and adding it to equation (P3).

Reinserting the variables gives us

$$\begin{aligned} x \qquad - 4z &= 5, && \textbf{(P1)} \\ -y + 12z &= -13, && \textbf{(P2)} \\ 14z &= -7. && \textbf{(P3)} \end{aligned}$$

We now solve (P3) for z:

$$\begin{aligned} 14z &= -7 && \textbf{(P3)} \\ z &= -\tfrac{7}{14} && \text{Solving for } z \\ z &= -\tfrac{1}{2}. \end{aligned}$$

Next, we back-substitute $-\frac{1}{2}$ for z in (P2) and solve for y:

$$\begin{aligned} -y + 12z &= -13 && \textbf{(P2)} \\ -y + 12\left(-\tfrac{1}{2}\right) &= -13 && \text{Substituting } -\tfrac{1}{2} \text{ for } z \text{ in equation (P2)} \\ -y - 6 &= -13 \\ -y &= -7 \\ y &= 7. && \text{Solving for } y \end{aligned}$$

Since there is no y-term in (P1), we need only substitute $-\frac{1}{2}$ for z in (P1) and solve for x:

$$\begin{aligned} x - 4z &= 5 && \textbf{(P1)} \\ x - 4\left(-\tfrac{1}{2}\right) &= 5 && \text{Substituting } -\tfrac{1}{2} \text{ for } z \text{ in equation (P1)} \\ x + 2 &= 5 \\ x &= 3. && \text{Solving for } x \end{aligned}$$

The solution is $\left(3, 7, -\frac{1}{2}\right)$.

Do Exercise 2.

2. Solve using matrices:

$$\begin{aligned} x - 2y + 3z &= 4, \\ 2x - y + z &= -1, \\ 4x + y + z &= 1. \end{aligned}$$

Answer

2. $(-1, 2, 3)$

All the operations used in the preceding example correspond to operations with the equations and produce equivalent systems of equations. We call the matrices **row-equivalent** and the operations that produce them **row-equivalent operations**.

ROW-EQUIVALENT OPERATIONS

Each of the following row-equivalent operations produces an equivalent matrix:

a) Interchanging any two rows.

b) Multiplying each element of a row by the same nonzero number.

c) Multiplying each element of a row by a nonzero number and adding the result to another row.

The best overall method of solving systems of equations is by row-equivalent matrices; graphing calculators and computers are programmed to use them. Matrices are part of a branch of mathematics known as linear algebra. They are also studied in more detail in many courses in finite mathematics.

C Exercise Set

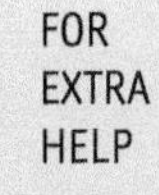

FOR EXTRA HELP MyLab Math

a Solve using matrices.

1. $4x + 2y = 11,$
$3x - y = 2$

2. $3x - 3y = 11,$
$9x - 2y = 5$

3. $x + 4y = 8,$
$3x + 5y = 3$

4. $x + 4y = 5,$
$-3x + 2y = 13$

5. $5x - 3y = -2,$
$4x + 2y = 5$

6. $3x + 4y = 7,$
$-5x + 2y = 10$

7. $2x - 3y = 50,$
$5x + y = 40$

8. $4x + 5y = -8,$
$7x + 9y = 11$

9. $4x - y - 3z = 1,$
$8x + y - z = 5,$
$2x + y + 2z = 5$

10. $3x + 2y + 2z = 3,$
$x + 2y - z = 5,$
$2x - 4y + z = 0$

11. $p + q + r = 1,$
$p - 2q - 3r = 3,$
$4p + 5q + 6r = 4$

12. $x + 2y - 3z = 9,$
$2x - y + 2z = -8,$
$3x - y - 4z = 3$

13. $x - y + 2z = 0,$
$x - 2y + 3z = -1,$
$2x - 2y + z = -3$

14. $4a + 9b = 8,$
$8a + 6c = -1,$
$6b + 6c = -1$

15. $3p + 2r = 11,$
$q - 7r = 4,$
$p - 6q = 1$

16. $m + n + t = 6,$
$m - n - t = 0,$
$m + 2n + t = 5$

D

OBJECTIVE

a Given two functions f and g, find their sum, difference, product, and quotient.

MyLab Math
ANIMATION

The Algebra of Functions

a THE SUM, DIFFERENCE, PRODUCT, AND QUOTIENT OF FUNCTIONS

Suppose that a is in the domain of two functions, f and g. The input a is paired with $f(a)$ by f and with $g(a)$ by g. The outputs can then be added to get $f(a) + g(a)$.

EXAMPLE 1 Let $f(x) = x + 4$ and $g(x) = x^2 + 1$. Find $f(2) + g(2)$.

We visualize two function machines. Because 2 is in the domain of each function, we can compute $f(2)$ and $g(2)$.

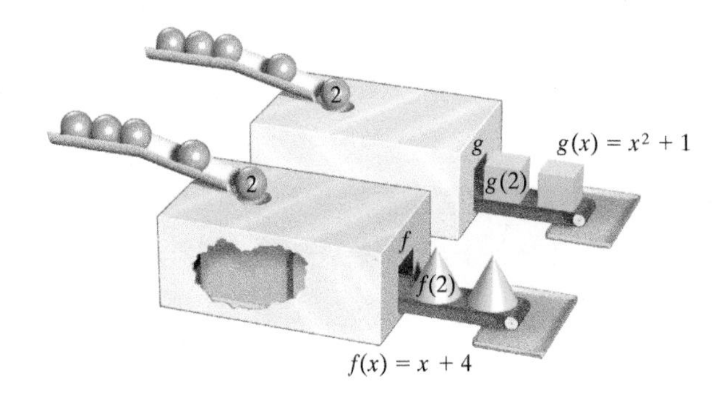

Since

$$f(2) = 2 + 4 = 6 \quad \text{and} \quad g(2) = 2^2 + 1 = 5,$$

we have

$$f(2) + g(2) = 6 + 5 = 11.$$

In Example 1, suppose that we were to write $f(x) + g(x)$ as $(x + 4) + (x^2 + 1)$, or $f(x) + g(x) = x^2 + x + 5$. This could then be regarded as a "new" function: $(f + g)(x) = x^2 + x + 5$. We can alternatively find $f(2) + g(2)$ with $(f + g)(x)$:

$$(f + g)(x) = x^2 + x + 5$$
$$(f + g)(2) = 2^2 + 2 + 5 \qquad \text{Substituting 2 for } x$$
$$= 4 + 2 + 5$$
$$= 11.$$

Similar notations exist for subtraction, multiplication, and division of functions.

THE SUM, DIFFERENCE, PRODUCT, AND QUOTIENT OF FUNCTIONS

For any functions f and g, we can form new functions defined as:

1. The **sum** $f + g$: $(f + g)(x) = f(x) + g(x)$;
2. The **difference** $f - g$: $(f - g)(x) = f(x) - g(x)$;
3. The **product** fg: $(f \cdot g)(x) = f(x) \cdot g(x)$;
4. The **quotient** f/g: $(f/g)(x) = f(x)/g(x)$, where $g(x) \neq 0$.

EXAMPLE 2 Given f and g described by $f(x) = x^2 - 5$ and $g(x) = x + 7$, find $(f + g)(x)$, $(f - g)(x)$, $(f \cdot g)(x)$, $(f/g)(x)$, and $(g \cdot g)(x)$.

$$(f + g)(x) = f(x) + g(x) = (x^2 - 5) + (x + 7) = x^2 + x + 2;$$
$$(f - g)(x) = f(x) - g(x) = (x^2 - 5) - (x + 7) = x^2 - x - 12;$$
$$(f \cdot g)(x) = f(x) \cdot g(x) = (x^2 - 5)(x + 7) = x^3 + 7x^2 - 5x - 35;$$
$$(f/g)(x) = f(x)/g(x) = \frac{x^2 - 5}{x + 7};$$
$$(g \cdot g)(x) = g(x) \cdot g(x) = (x + 7)(x + 7) = x^2 + 14x + 49$$

Note that the sum, the difference, and the product of polynomials are also polynomial functions, but the quotient may not be.

Do Exercise 1.

1. Given $f(x) = x^2 + 3$ and $g(x) = x^2 - 3$, find each of the following.
   a) $(f + g)(x)$
   b) $(f - g)(x)$
   c) $(f \cdot g)(x)$
   d) $(f/g)(x)$
   e) $(f \cdot f)(x)$

EXAMPLE 3 For $f(x) = x^2 - x$ and $g(x) = x + 2$, find $(f + g)(3)$, $(f - g)(-1)$, $(f \cdot g)(5)$, and $(f/g)(-4)$.

We first find $(f + g)(x)$, $(f - g)(x)$, $(f \cdot g)(x)$, and $(f/g)(x)$.

$$(f + g)(x) = f(x) + g(x) = x^2 - x + x + 2$$
$$= x^2 + 2;$$
$$(f - g)(x) = f(x) - g(x) = x^2 - x - (x + 2)$$
$$= x^2 - x - x - 2$$
$$= x^2 - 2x - 2;$$
$$(f \cdot g)(x) = f(x) \cdot g(x) = (x^2 - x)(x + 2)$$
$$= x^3 + 2x^2 - x^2 - 2x$$
$$= x^3 + x^2 - 2x;$$
$$(f/g)(x) = \frac{f(x)}{g(x)} = \frac{x^2 - x}{x + 2}.$$

Then we substitute.

$(f + g)(3) = 3^2 + 2$ Using $(f + g)(x) = x^2 + 2$
$= 9 + 2 = 11;$

$(f - g)(-1) = (-1)^2 - 2(-1) - 2$ Using $(f - g)(x) = x^2 - 2x - 2$
$= 1 + 2 - 2 = 1;$

$(f \cdot g)(5) = 5^3 + 5^2 - 2 \cdot 5$ Using $(f \cdot g)(x) = x^3 + x^2 - 2x$
$= 125 + 25 - 10 = 140;$

$(f/g)(-4) = \frac{(-4)^2 - (-4)}{-4 + 2}$ Using $(f/g)(x) = (x^2 - x)/(x + 2)$
$= \frac{16 + 4}{-2} = \frac{20}{-2} = -10$

Do Exercise 2.

2. Given $f(x) = x^2 + x$ and $g(x) = 2x - 3$, find each of the following.
   a) $(f + g)(-2)$
   b) $(f - g)(4)$
   c) $(f \cdot g)(-3)$
   d) $(f/g)(2)$

Answers

1. (a) $2x^2$; **(b)** 6; **(c)** $x^4 - 9$; **(d)** $\frac{x^2 + 3}{x^2 - 3}$; **(e)** $x^4 + 6x^2 + 9$ **2. (a)** -5; **(b)** 15; **(c)** -54; **(d)** 6

D Exercise Set

FOR EXTRA HELP MyLab Math

a Let $f(x) = -3x + 1$ and $g(x) = x^2 + 2$. Find each of the following.

1. $f(2) + g(2)$

2. $f(-1) + g(-1)$

3. $f(5) - g(5)$

4. $f(4) - g(4)$

5. $f(-1) \cdot g(-1)$

6. $f(-2) \cdot g(-2)$

7. $f(-4)/g(-4)$

8. $f(3)/g(3)$

9. $g(1) - f(1)$

10. $g(2)/f(2)$

11. $g(0)/f(0)$

12. $g(6) - f(6)$

Let $f(x) = x^2 - 3$ and $g(x) = 4 - x$. Find each of the following.

13. $(f + g)(x)$

14. $(f - g)(x)$

15. $(f + g)(-4)$

16. $(f + g)(-5)$

17. $(f - g)(3)$

18. $(f - g)(2)$

19. $(f \cdot g)(x)$

20. $(f/g)(x)$

21. $(f \cdot g)(-3)$

22. $(f \cdot g)(-4)$

23. $(f/g)(0)$

24. $(f/g)(1)$

25. $(f/g)(-2)$

26. $(f/g)(-1)$

For each pair of functions f and g, find $(f + g)(x)$, $(f - g)(x)$, $(f \cdot g)(x)$, and $(f/g)(x)$.

27. $f(x) = x^2$,
$g(x) = 3x - 4$

28. $f(x) = 5x - 1$,
$g(x) = 2x^2$

29. $f(x) = \dfrac{1}{x - 2}$,
$g(x) = 4x^3$

30. $f(x) = 3x^2$,
$g(x) = \dfrac{1}{x - 4}$

31. $f(x) = \dfrac{3}{x - 2}$,
$g(x) = \dfrac{5}{4 - x}$

32. $f(x) = \dfrac{5}{x - 3}$,
$g(x) = \dfrac{1}{x - 2}$

Answers

JUST-IN-TIME REVIEW

1 The Set of Real Numbers

1. (a) $\{s, o, l, v, e\}$; **(b)** $\{2, 4, 6, 8, \ldots\}$
2. (a) $\{x \mid x$ is a whole number less than or equal to $5\}$, or $\{x \mid x$ is a whole number less than $6\}$;
(b) $\{x \mid x$ is an integer greater than 3 and less than $11\}$
3. (a) $20, \sqrt{16}$; **(b)** $20, \sqrt{16}, 0$; **(c)** $20, -10, \sqrt{16}, 0$;
(d) $\sqrt{7}, -\sqrt{2}, 9.34334333433334\ldots$;
(e) $20, -10, -5.34, 18.999, 0.58\overline{3}, \frac{11}{45}, \sqrt{16}, 0, -\frac{2}{3}$;
(f) $20, -10, -5.34, 18.999, 0.58\overline{3}, \frac{11}{45}, \sqrt{7}, -\sqrt{2}, \sqrt{16}, 0, -\frac{2}{3}$, $9.34334333433334\ldots$ **4. (a)** $1, 12, \sqrt{25}$;
(b) $0, 1, 12, \sqrt{25}$; **(c)** $-6, 0, 1, -4, 12, \sqrt{25}, -\frac{12}{3}$;
(d) $\sqrt{3}, 0.131331333133331\ldots$;
(e) $-6, 0, 1, -\frac{1}{2}, -4, \frac{7}{9}, 12, -\frac{6}{5}, 3.45, 5\frac{1}{2}, \sqrt{25}, -\frac{12}{3}$;
(f) $-6, 0, 1, -\frac{1}{2}, -4, \frac{7}{9}, 12, -\frac{6}{5}, 3.45, 5\frac{1}{2}, \sqrt{3}, \sqrt{25}, -\frac{12}{3}$, $0.131331333133331\ldots$

2 Order for the Real Numbers

1. $<$ **2.** $>$ **3.** $>$ **4.** $<$ **5.** $>$ **6.** $>$ **7.** $<$ **8.** $>$
9. $>$ **10.** $>$ **11.** $6 < x$ **12.** $7 > -4$ **13.** True
14. True **15.** False **16.** True

3 Graphing Inequalities on the Number Line

1. 

2. 

3. 

4. 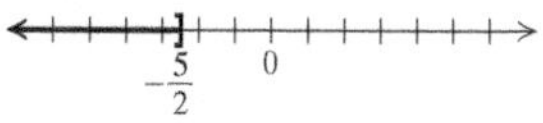

4 Absolute Value

1. 4 **2.** 0 **3.** 252.7 **4.** 31 **5.** $\frac{1}{2}$ **6.** 0.03

5 Add Real Numbers

1. 5 **2.** -5 **3.** -11 **4.** 11 **5.** 5 **6.** -16 **7.** 0
8. -26 **9.** -21 **10.** 7 **11.** 1.2 **12.** -8.86 **13.** $\frac{7}{20}$
14. $-\frac{41}{24}$

6 Opposites, or Additive Inverses

1. -10 **2.** $\frac{2}{3}$ **3.** 0.007 **4.** 0 **5.** $-9; 9$ **6.** $23; -23$

7 Subtract Real Numbers

1. -6 **2.** 16 **3.** -16 **4.** 6 **5.** 34 **6.** -20 **7.** 0
8. -5 **9.** 47 **10.** -11.6 **11.** -29.25 **12.** $\frac{11}{12}$ **13.** $-\frac{1}{4}$
14. $-\frac{7}{15}$

8 Multiply Real Numbers

1. -21 **2.** 26.46 **3.** -45 **4.** $\frac{65}{14}$ **5.** 0 **6.** $-\frac{12}{35}$ **7.** 52
8. -2 **9.** 1 **10.** 60

9 Divide Real Numbers

1. -2 **2.** 7 **3.** Not defined **4.** 0 **5.** $\frac{4}{3}$ **6.** $-\frac{8}{7}$
7. $\frac{1}{25}$ **8.** $\frac{1}{0.2}$, or 5 **9.** $-\frac{6}{77}$ **10.** 25 **11.** -120 **12.** $-\frac{4}{19}$
13. $\frac{3}{8}; -\frac{8}{3}$ **14.** $-6; \frac{1}{6}$

10 Exponential Notation (Part 1)

1. y^6 **2.** 8^4 **3.** $(3.8)^5$ **4.** $(-\frac{4}{5})^3$ **5.** $\frac{3}{4}$ **6.** 1 **7.** 64
8. -343 **9.** $\frac{1}{27}$ **10.** -81 **11.** 0.000001 **12.** $\frac{25}{49}$ **13.** $\frac{1}{y^5}$
14. $-\frac{1}{64}$ **15.** $\frac{16}{9}$ **16.** a^2 **17.** 16 **18.** 3^{-4} **19.** b^{-3}
20. $(-16)^{-2}$

11 Order of Operations

1. -117 **2.** 9 **3.** 2 **4.** 8 **5.** -63 **6.** 16 **7.** -1
8. 12 **9.** 2 **10.** -144 **11.** 3 **12.** 36 **13.** -4 **14.** 16
15. 2 **16.** -35 **17.** 38 **18.** 34 **19.** $-\frac{8}{9}$ **20.** 34

12 Translate to an Algebraic Expression

1. $x \div 6$, or $\frac{x}{6}$, or $x/6$ **2.** $7 + 4y$, or $4y + 7$ **3.** $t - 25$
4. $\frac{1}{3}b$, or $\frac{b}{3}$ **5.** $w + 2q$, or $2q + w$ **6.** $-18m$, or $m(-18)$
7. $s - m$ **8.** $19\%x$, or $0.19x$, or $\frac{19}{100}x$

13 Evaluate Algebraic Expressions

1. -456 **2.** -5 **3.** $\frac{1}{10}$ **4.** 7 **5.** 14 **6.** 16 **7.** 19 **8.** 57

14 Equivalent Fraction Expressions

1. $\frac{7x}{8x}$ **2.** $\frac{15y}{50y}$ **3.** $\frac{3b}{51b}$ **4.** $\frac{600w}{98w}$ **5.** $\frac{5}{3}$ **6.** 2 **7.** -4
8. $-\frac{125}{3}$

15 The Commutative Laws and the Associative Laws

1. $3 + w$ **2.** tr **3.** $14 + pq$, $14 + qp$, or $qp + 14$ **4.** $(m + n) + 2$ **5.** $7 \cdot (x \cdot y)$
6. $a + (8 + b)$, $(a + 8) + b$, $b + (a + 8)$; others are possible
7. $x \cdot (9 \cdot y)$, $y \cdot (x \cdot 9)$, $(y \cdot x) \cdot 9$; others are possible

16 The Distributive Laws

1. $3c + 3$ **2.** $3x - 3y$ **3.** $-6c - 10d$
4. $5xy - 5xz + 5xw$ **5.** $\frac{1}{2}ha + \frac{1}{2}hb$ **6.** $2\pi rh + 2\pi r$
7. $4a, -5b, 6$ **8.** $2x, -y, -1$ **9.** $9(a + b)$

10. $7(x - 3)$ 11. $a(b + 1)$ 12. $6(3a - 4b - 8)$
13. $4(2m + n - 6)$ 14. $x(y - z + w)$

17 Collecting Like Terms

1. $12x$ 2. $-3b$ 3. $15y$ 4. $11a$ 5. $-8t$ 6. $10x$
7. $-4c + 12d$ 8. $22x + 18$ 9. $1.19x + 0.93y$
10. $-\frac{2}{15}a - \frac{1}{3}b - 27$

18 Removing Parentheses and Collecting Like Terms

1. $2c$ 2. $-b - 4$ 3. $-x + 8$, or $8 - x$ 4. $-t + y$, or $y - t$
5. $-r - s - t$ 6. $-8x + 6y - 13$ 7. $m + 1$ 8. $22a - 15$
9. $-7x + 14$ 10. $17x + 3y - 18$ 11. $3x + 30$ 12. -190

19 Exponential Notation (Part 2)

1. $\frac{1}{8^6}$ 2. a^5 3. $\frac{1}{10^9}$ 4. $\frac{1}{b^3}$ 5. $-\frac{4x^9}{3y^2}$ 6. $-28m^5n^5$
7. $\frac{15}{t^{5a}}$ 8. $\frac{3x^3}{2y^2}$ 9. r^{45} 10. $\frac{1}{4^9 \cdot 3^{12}}$ 11. $125a^6b^6$ 12. $\frac{a^8}{c^{14}}$
13. 4^6 14. $\frac{8x^9y^3}{27}$ 15. 8^{4xy} 16. $\frac{5^4y^{24}}{4^4x^{20}}$

20 Scientific Notation

1. 4.7×10^{10} 2. 1.6×10^{-8} 3. 2.63×10^{-7}
4. 2.6×10^{12} 5. 2×10^{-7} 6. 673,000,000
7. 0.000000000000000000000000000000911 g
8. 1,860,000,000 monthly active Facebook users
9. 3.38×10^{-4} 10. 5×10^8

CHAPTER 1

Calculator Corner, p. 41

1. Left to the student 2. Left to the student

Exercise Set 1.1, p. 41

RC1. (d) RC2. (f) RC3. (b) RC4. (c)
CC1. (a) 10; (b) $4x - 1 = 7$ CC2. (a) 6; (b) $24 - 4x = 15$
CC3. (a) 10; (b) $12y - 5 = 27y$ CC4. (a) 100; (b) $6 - 700x = 120x$
1. Yes 3. No 5. No 7. No 9. Yes 11. No
13. 7 15. -8 17. 27 19. -39 21. 86.86 23. $\frac{1}{6}$
25. 6 27. -4 29. -147 31. 32 33. -6 35. $-\frac{1}{6}$
37. 10 39. 11 41. -12 43. 8 45. $\frac{3}{10}$ 47. $-\frac{5}{2}$
49. 2 51. 21 53. -12 55. No solution 57. -1
59. $\frac{18}{5}$ 61. 0 63. 1 65. All real numbers
67. No solution 69. 1 71. -68 73. $\frac{14}{3}$ 75. $\frac{11}{4}$
77. 7 79. 2 81. 7 83. 5 85. $-\frac{3}{2}$
87. All real numbers 89. 5 91. $\frac{23}{66}$ 93. $\frac{5}{32}$ 95. $\frac{79}{32}$
97. a^{14} 98. $\frac{1}{a^{32}}$ 99. $-\frac{18x^2}{y^{11}}$ 100. $-2x^8y^3$
101. $12 - 20x$ 102. $-5 + 6x$ 103. $-12x + 8y - 4z$
104. $-10x + 35y - 20$ 105. $2(x - 3y)$ 106. $-4(x + 6y)$
107. $2(2x - 5y + 1)$ 108. $-5(2x - 7y + 4)$
109. $\{1, 2, 3, 4, 5, 6, 7, 8, 9\}$; $\{x \mid x$ is a positive integer less than $10\}$
110. $\{-8, -7, -6, -5, -4, -3, -2, -1\}$; $\{x \mid x$ is a negative integer greater than $-9\}$ 111. Approximately -4.176 113. $\frac{3}{2}$ 115. 8

Exercise Set 1.2, p. 50

RC1. Equation RC2. Height RC3. Area
RC4. Trapezoid CC1. (d) CC2. (b) CC3. (f)
CC4. (a) CC5. (c) CC6. (e)
1. $r = \frac{d}{t}$ 3. $h = \frac{A}{b}$ 5. $w = \frac{P - 2l}{2}$, or $\frac{P}{2} - l$
7. $b = \frac{2A}{h}$ 9. $a = 2A - b$ 11. $m = \frac{F}{a}$ 13. $t = \frac{I}{Pr}$
15. $c^2 = \frac{E}{m}$ 17. $p = 2Q + q$ 19. $y = \frac{c - Ax}{B}$
21. $N = \frac{1.08T}{I}$ 23. $m = \frac{4}{3}C - 5$, or $\frac{4C - 15}{3}$
25. $b = 3n - a + c$ 27. $R = \frac{d}{1 - st}$ 29. $B = \frac{T}{1 + qt}$
31. (a) About 1930 calories; (b) $w = \frac{R - 66 - 12.7h + 6.8a}{6.23}$
33. (a) About 2340 calories;
(b) $a = \frac{1015.25 + 6.74w + 7.29h - K}{7.29}$
35. (a) 1614 g; (b) $a = \frac{P + 299}{9.337d}$ 37. (a) 50 mg;
(b) $d = \frac{c(a + 12)}{a}$, or $c + \frac{12c}{a}$ 39. -5 40. 250
41. -2 42. -25 43. $\frac{4}{5}$ 44. 6 45. -6 46. -5
47. $s = \frac{A - \pi r^2}{\pi r}$, or $\frac{A}{\pi r} - r$ 49. $V_1 = \frac{P_2V_2T_1}{P_1T_2}$; $P_2 = \frac{P_1V_1T_2}{T_1V_2}$
51. 0.8 year 53. (a) Approximately 120.5 horsepower;
(b) approximately 94.1 horsepower

Exercise Set 1.3, p. 63

RC1. Familiarize RC2. Translate RC3. Solve
RC4. Check RC5. State CC1. $x + 4x = 400$
CC2. $a + (a + 3) + (2a) = 180$ CC3. $p + 0.06p = 17.49$
CC4. $2(w + 12) + 2w = 100$
1. About 27.5 mi 3. 45°, 52°, 83° 5. 81,000 gorillas
7. \$252 9. Length: 94 ft; width: 50 ft 11. 82 ft
13. \$265,000 15. 9, 11, 13 17. 229 and 230 19. 90 photos
21. \$38,950 23. 5.5 million cases 25. (a) 5.68 million hikers; (b) about 11 years after 2010, or in 2021 27. 6 min
29. Downstream: 1.25 hr; upstream: 1.875 hr 31. Upstream: $\frac{2}{3}$ hr; downstream: $\frac{18}{73}$ hr 33. 49 34. 208 35. $\frac{78}{1649}$ 36. $\frac{17}{10}$
37. \$115,243 39. 25% increase 41. $m\angle 2 = 120°$; $m\angle 1 = 60°$

Mid-Chapter Review: Chapter 1, p. 68

1. True 2. True 3. False 4. False
5.
$$\begin{aligned} 2x - 5 &= 1 - 4x \\ 2x - 5 + 4x &= 1 - 4x + 4x \\ 6x - 5 &= 1 \\ 6x - 5 + 5 &= 1 + 5 \\ 6x &= 6 \\ \frac{6x}{6} &= \frac{6}{6} \\ x &= 1 \end{aligned}$$
6.
$$\begin{aligned} Mx + Ny &= T \\ Mx + Ny - Mx &= T - Mx \\ Ny &= T - Mx \\ y &= \frac{T - Mx}{N} \end{aligned}$$
7. Yes 8. No 9. No 10. Yes 11. -3 12. -8
13. 4 14. 2 15. All real numbers 16. 2 17. $-\frac{5}{2}$
18. No solution 19. $-\frac{4}{3}$ 20. 5 21. $\frac{3}{2}$ 22. $-\frac{3}{16}$
23. $n = \frac{P}{m}$ 24. $t = \frac{z - 3w}{3}$, or $\frac{z}{3} - w$ 25. $s = 4N - r$
26. $B = 1.5\frac{A}{T}$ 27. $t = \frac{3H + 10}{2}$, or $\frac{3H}{2} + 5$
28. $g = \frac{f}{1 + hm}$ 29. 7928 female graduates 30. 140 calories
31. Length: 7 ft; width: 5 ft 32. 1.5 hr; 3 hr
33. Equivalent expressions have the same value for all possible replacements. Any replacement that does not make any of the expressions undefined can be substituted for the variable. Equivalent equations have the same solution(s).

34. Answers may vary. A walker who knows how far and how long she walks each day wants to know her average speed each day. **35.** Answers may vary. A decorator wants to have a carpet cut for a bedroom. The perimeter of the room is 54 ft and its length is 15 ft. How wide should the carpet be? **36.** We can subtract by adding an opposite, so we can use the addition principle to subtract the same number on both sides of an equation. Similarly, we can divide by multiplying by a reciprocal, so we can use the multiplication principle to divide both sides of an equation by the same number. **37.** The manner in which a guess or an estimate is manipulated can give insight into the form of the equation to which the problem will be translated. **38.** Labeling the variable clearly makes the *Translate* step more accurate. It also allows us to determine whether the solution of the equation we translated to provides the information asked for in the original problem.

Translating for Success, p. 79

1. F **2.** I **3.** C **4.** E **5.** D **6.** J **7.** O **8.** M **9.** B **10.** L

Exercise Set 1.4, p. 80

RC1. Solution **RC2.** Set-builder **RC3.** Closed **RC4.** Half-open **RC5.** Negative **CC1.** (b) **CC2.** (h) **CC3.** (c) **CC4.** (a) **CC5.** (g) **CC6.** (d)

1. No, no, no, yes **3.** No, yes, yes, no, no **5.** $(-\infty, 5)$ **7.** $[-3, 3]$ **9.** $(-8, -4)$ **11.** $(-2, 5)$ **13.** $(-\sqrt{2}, \infty)$ **15.** $\{x \mid x > -1\}$, or $(-1, \infty)$ **17.** $\{y \mid y < 6\}$, or $(-\infty, 6)$ **19.** $\{a \mid a \le -22\}$, or $(-\infty, -22]$ **21.** $\{t \mid t \ge -4\}$, or $[-4, \infty)$ **23.** $\{y \mid y > -6\}$, or $(-6, \infty)$ **25.** $\{x \mid x \le 9\}$, or $(-\infty, 9]$ **27.** $\{x \mid x \ge 3\}$, or $[3, \infty)$ **29.** $\{x \mid x < -60\}$, or $(-\infty, -60)$ **31.** $\{x \mid x > 3\}$, or $(3, \infty)$ **33.** $\{x \mid x \le 0.9\}$, or $(-\infty, 0.9]$ **35.** $\{x \mid x \le \frac{5}{6}\}$, or $(-\infty, \frac{5}{6}]$ **37.** $\{x \mid x < 6\}$, or $(-\infty, 6)$ **39.** $\{y \mid y \le -3\}$, or $(-\infty, -3]$ **41.** $\{y \mid y > \frac{2}{3}\}$, or $(\frac{2}{3}, \infty)$ **43.** $\{x \mid x \ge 11.25\}$, or $[11.25, \infty)$ **45.** $\{x \mid x \le \frac{1}{2}\}$, or $(-\infty, \frac{1}{2}]$ **47.** $\{y \mid y \le -\frac{75}{2}\}$, or $(-\infty, -\frac{75}{2}]$ **49.** $\{x \mid x > -\frac{2}{17}\}$, or $(-\frac{2}{17}, \infty)$ **51.** $\{m \mid m > \frac{7}{3}\}$, or $(\frac{7}{3}, \infty)$ **53.** $\{r \mid r < -3\}$, or $(-\infty, -3)$ **55.** $\{x \mid x \ge 2\}$, or $[2, \infty)$ **57.** $\{y \mid y < 5\}$, or $(-\infty, 5)$ **59.** $\{x \mid x \le \frac{4}{7}\}$, or $(-\infty, \frac{4}{7}]$ **61.** $\{x \mid x < 8\}$, or $(-\infty, 8)$ **63.** $\{x \mid x \ge \frac{13}{2}\}$, or $[\frac{13}{2}, \infty)$ **65.** $\{x \mid x < \frac{11}{18}\}$, or $(-\infty, \frac{11}{18})$ **67.** $\{x \mid x \ge -\frac{51}{31}\}$, or $[-\frac{51}{31}, \infty)$ **69.** $\{a \mid a \le 2\}$, or $(-\infty, 2]$ **71.** $\{W \mid W <$ (approximately) $136.7 \text{ lb}\}$ **73.** $\{S \mid S \ge 84\}$ **75.** $\{B \mid B \ge \$11{,}500\}$ **77.** $\{S \mid S > \$7000\}$ **79.** $\{c \mid c > \$735\}$ **81.** $\{p \mid p > 80\}$ **83.** $\{s \mid s < 980 \text{ ft}^2\}$ **85. (a)** 2010: \$2438; 2016: \$3056; **(b)** More than 15.31 years after 2005, or $\{t \mid t > 15.31\}$ **87.** $-9a + 30b$ **88.** $32x - 72y$ **89.** $-8a + 22b$ **90.** $-6a + 17b$ **91.** $10(3x - 7y - 4)$ **92.** $-6a(2 - 5b)$ **93.** $-4(2x - 6y + 1)$ **94.** $5(2n - 9mn + 20m)$ **95.** -11.2 **96.** 6.6 **97.** -11.2 **98.** 6.6 **99. (a)** $\{p \mid p > 10\}$; **(b)** $\{p \mid p < 10\}$ **101.** True **103.** All real numbers **105.** All real numbers

Exercise Set 1.5, p. 93

RC1. True **RC2.** False **RC3.** True **RC4.** True **CC1.** Yes **CC2.** Yes **CC3.** No **CC4.** Yes **CC5.** No **CC6.** Yes

1. $\{9, 11\}$ **3.** $\{b\}$ **5.** $\{9, 10, 11, 13\}$ **7.** $\{a, b, c, d, f, g\}$ **9.** $\varnothing$ **11.** $\{3, 5, 7\}$ **13.** ; $(-4, 1]$ **15.** ; $(1, 6)$ **17.** $\{x \mid -4 \le x < 5\}$, or $[-4, 5)$; **19.** $\{x \mid x \ge 2\}$, or $[2, \infty)$; **21.** $\varnothing$ **23.** $\{x \mid -8 < x < 6\}$, or $(-8, 6)$ **25.** $\{x \mid -6 < x \le 2\}$, or $(-6, 2]$ **27.** $\{x \mid -1 < x \le 6\}$, or $(-1, 6]$ **29.** $\{y \mid -1 < y \le 5\}$, or $(-1, 5]$ **31.** $\{x \mid -\frac{5}{3} \le x \le \frac{4}{3}\}$, or $[-\frac{5}{3}, \frac{4}{3}]$ **33.** $\{x \mid -\frac{7}{2} < x \le \frac{11}{2}\}$, or $(-\frac{7}{2}, \frac{11}{2}]$ **35.** $\{x \mid 10 < x \le 14\}$, or $(10, 14]$ **37.** $\{x \mid -\frac{13}{3} \le x \le 9\}$, or $[-\frac{13}{3}, 9]$ **39.** ; $(-\infty, -2) \cup (1, \infty)$ **41.** ; $(-\infty, -3] \cup (1, \infty)$ **43.** $\{x \mid x < -5 \textit{ or } x > -1\}$, or $(-\infty, -5) \cup (-1, \infty)$; **45.** $\{x \mid x \le \frac{5}{2} \textit{ or } x \ge 4\}$, or $(-\infty, \frac{5}{2}] \cup [4, \infty)$; **47.** $\{x \mid x \ge -3\}$, or $[-3, \infty)$; **49.** $\{x \mid x \le -\frac{5}{4} \textit{ or } x > -\frac{1}{2}\}$, or $(-\infty, -\frac{5}{4}] \cup (-\frac{1}{2}, \infty)$ **51.** All real numbers, or $(-\infty, \infty)$ **53.** $\{x \mid x < -4 \textit{ or } x > 2\}$, or $(-\infty, -4) \cup (2, \infty)$ **55.** $\{x \mid x < \frac{79}{4} \textit{ or } x > \frac{89}{4}\}$, or $(-\infty, \frac{79}{4}) \cup (\frac{89}{4}, \infty)$ **57.** $\{x \mid x \le -\frac{13}{2} \textit{ or } x \ge \frac{29}{2}\}$, or $(-\infty, -\frac{13}{2}] \cup [\frac{29}{2}, \infty)$ **59.** $\{d \mid 0 \text{ ft} \le d \le 198 \text{ ft}\}$ **61.** Between 23 beats and 27 beats **63.** $\{W \mid 101.2 \text{ lb} \le W \le 136.2 \text{ lb}\}$ **65.** $\{d \mid 250 \text{ mg} < d < 500 \text{ mg}\}$ **67.** No solution **68.** $-\frac{5}{3}$ **69.** 4 **70.** 9 **71.** All real numbers **72.** 9 **73.** $\{x \mid -4 < x \le 1\}$, or $(-4, 1]$ **75.** $\{x \mid \frac{2}{5} \le x \le \frac{4}{5}\}$, or $[\frac{2}{5}, \frac{4}{5}]$ **77.** $\{x \mid -\frac{1}{8} < x < \frac{1}{2}\}$, or $(-\frac{1}{8}, \frac{1}{2})$ **79.** $\{x \mid 10 < x \le 18\}$, or $(10, 18]$ **81.** True **83.** False **85.** All real numbers; $\varnothing$

Exercise Set 1.6, p. 105

RC1. True **RC2.** True **RC3.** False **RC4.** True **RC5.** True **RC6.** True **CC1.** (f) **CC2.** (b) **CC3.** (e) **CC4.** (c) **CC5.** (a) **CC6.** (d)

1. $9|x|$ **3.** $2x^2$ **5.** $2x^2$ **7.** $6|y|$ **9.** $\dfrac{2}{|x|}$ **11.** $\dfrac{x^2}{|y|}$ **13.** $4|x|$ **15.** $\dfrac{y^2}{3}$ **17.** 38 **19.** 19 **21.** 6.3 **23.** 5 **25.** $\{-3, 3\}$ **27.** $\varnothing$ **29.** $\{0\}$ **31.** $\{-9, 15\}$ **33.** $\{-\frac{1}{2}, \frac{7}{2}\}$ **35.** $\{-\frac{5}{4}, \frac{23}{4}\}$ **37.** $\{-11, 11\}$ **39.** $\{-291, 291\}$ **41.** $\{-8, 8\}$ **43.** $\{-7, 7\}$ **45.** $\{-2, 2\}$ **47.** $\{-7, 8\}$ **49.** $\{-12, 2\}$ **51.** $\{-\frac{5}{2}, \frac{7}{2}\}$ **53.** $\varnothing$ **55.** $\{-\frac{13}{54}, -\frac{7}{54}\}$ **57.** $\{-\frac{11}{2}, \frac{3}{4}\}$ **59.** $\{\frac{3}{2}\}$ **61.** $\{5, -\frac{3}{5}\}$ **63.** All real numbers **65.** $\{-\frac{3}{2}\}$ **67.** $\{\frac{24}{23}, 0\}$ **69.** $\{32, \frac{8}{3}\}$ **71.** $\{x \mid -3 < x < 3\}$, or $(-3, 3)$ **73.** $\{x \mid x \le -2 \textit{ or } x \ge 2\}$, or $(-\infty, -2] \cup [2, \infty)$ **75.** $\{x \mid 0 < x < 2\}$, or $(0, 2)$

77. $\{x \mid -6 \le x \le -2\}$, or $[-6, -2]$
79. $\{x \mid -\frac{1}{2} \le x \le \frac{7}{2}\}$, or $[-\frac{1}{2}, \frac{7}{2}]$
81. $\{y \mid y < -\frac{3}{2} \text{ or } y > \frac{17}{2}\}$, or $(-\infty, -\frac{3}{2}) \cup (\frac{17}{2}, \infty)$
83. $\{x \mid x \le -\frac{5}{4} \text{ or } x \ge \frac{23}{4}\}$, or $(-\infty, -\frac{5}{4}] \cup [\frac{23}{4}, \infty)$
85. $\{y \mid -9 < y < 15\}$, or $(-9, 15)$ **87.** $\{x \mid -\frac{7}{2} \le x \le \frac{1}{2}\}$, or $[-\frac{7}{2}, \frac{1}{2}]$ **89.** $\{y \mid y < -\frac{4}{3} \text{ or } y > 4\}$, or $(-\infty, -\frac{4}{3}) \cup (4, \infty)$
91. $\{x \mid x \le -\frac{5}{4} \text{ or } x \ge \frac{23}{4}\}$, or $(-\infty, -\frac{5}{4}] \cup [\frac{23}{4}, \infty)$
93. $\{x \mid -\frac{9}{2} < x < 6\}$, or $(-\frac{9}{2}, 6)$ **95.** $\{x \mid x \le -\frac{25}{6} \text{ or } x \ge \frac{23}{6}\}$, or $(-\infty, -\frac{25}{6}] \cup [\frac{23}{6}, \infty)$ **97.** $\{x \mid -5 < x < 19\}$, or $(-5, 19)$
99. $\{x \mid x \le -\frac{2}{15} \text{ or } x \ge \frac{14}{15}\}$, or $(-\infty, -\frac{2}{15}] \cup [\frac{14}{15}, \infty)$
101. $\{m \mid -12 \le m \le 2\}$, or $[-12, 2]$ **103.** $\{x \mid \frac{1}{2} \le x \le \frac{5}{2}\}$, or $[\frac{1}{2}, \frac{5}{2}]$ **105.** $\{x \mid -1 \le x \le 2\}$, or $[-1, 2]$
107. $\{x \mid x \le 4\}$, or $(-\infty, -4]$ **108.** $\{y \mid y < -\frac{9}{10}\}$, or $(-\infty, -\frac{9}{10})$
109. $\{r \mid r > 16\}$, or $(16, \infty)$ **110.** $\{x \mid -8 < x \le 4\}$, or $(-8, 4]$
111. All real numbers, or $(-\infty, \infty)$
112. $\{x \mid \frac{1}{3} \le x < 4\}$, or $[\frac{1}{3}, 4)$ **113.** $\{d \mid 5\frac{1}{2} \text{ ft} \le d \le 6\frac{1}{2} \text{ ft}\}$
115. All real numbers **117.** $\{1, -\frac{1}{4}\}$ **119.** $\varnothing$
121. $|x| < 3$ **123.** $|x| \ge 6$ **125.** $|x + 3| > 5$

Summary and Review: Chapter 1, p. 109

Vocabulary Reinforcement

1. Inequality **2.** Set-builder **3.** Interval **4.** Intersection **5.** Conjunction **6.** Empty set **7.** Disjoint sets **8.** Union **9.** Disjunction **10.** Addition principle **11.** Multiplication principle **12.** Distance

Concept Reinforcement

1. True **2.** False **3.** False **4.** False **5.** True **6.** False **7.** True

Study Guide

1. No **2.** 8 **3.** $h = \frac{4F}{g}$ **4.** -2 is not a solution; 5 is a solution.
5. (a) $(-\infty, -8)$; **(b)** $[-7, 10)$; **(c)** $[3, \infty)$ **6.** $\{y \mid y < -2\}$, or $(-\infty, -2)$;
7. $\{z \mid -2 \le z < 1\}$, or $[-2, 1)$;
8. $\{z \mid z < -1 \text{ or } z \ge 1\}$, or $(-\infty, -1) \cup [1, \infty)$;
9. $8y^2$ **10.** 28 **11.** $\{-\frac{8}{5}, 2\}$
12. $\{3, -\frac{1}{3}\}$ **13. (a)** $\{x \mid -4 < x < 1\}$, or $(-4, 1)$; **(b)** $\{x \mid x \le -\frac{10}{3} \text{ or } x \ge 2\}$, or $(-\infty, -\frac{10}{3}] \cup [2, \infty)$

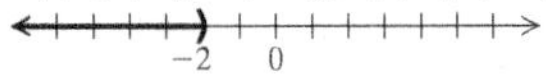

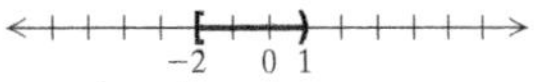

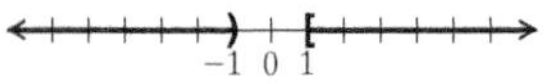

Review Exercises

1. 8 **2.** $\frac{3}{7}$ **3.** $\frac{22}{5}$ **4.** $-\frac{1}{13}$ **5.** -0.2 **6.** 5
7. $d = \frac{11}{4}(C - 3)$ **8.** $b = \frac{A - 2a}{-3}$, or $\frac{2a - A}{3}$
9. 185 and 186 **10.** 15 m, 12 m **11.** 160,000
12. 40 sec **13.** $[-8, 9)$ **14.** $(-\infty, 40]$
15. 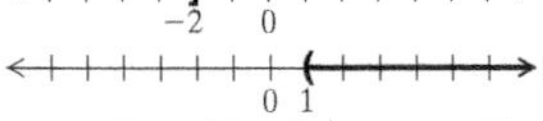 ; $(-\infty, -2]$
16. 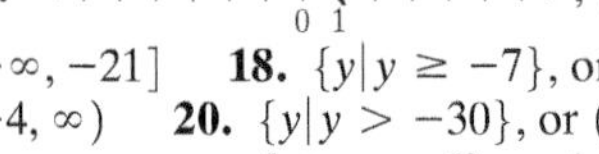 ; $(1, \infty)$ **17.** $\{a \mid a \le -21\}$, or $(-\infty, -21]$ **18.** $\{y \mid y \ge -7\}$, or $[-7, \infty)$ **19.** $\{y \mid y > -4\}$, or $(-4, \infty)$ **20.** $\{y \mid y > -30\}$, or $(-30, \infty)$ **21.** $\{x \mid x > -3\}$, or $(-3, \infty)$ **22.** $\{y \mid y \le -\frac{6}{5}\}$, or $(-\infty, -\frac{6}{5}]$ **23.** $\{x \mid x < -3\}$, or $(-\infty, -3)$ **24.** $\{y \mid y > -10\}$, or $(-10, \infty)$ **25.** $\{x \mid x \le -\frac{5}{2}\}$, or $(-\infty, -\frac{5}{2}]$ **26.** $\{t \mid t > 4\frac{1}{4} \text{ hr}\}$ **27.** \$10,000
28. 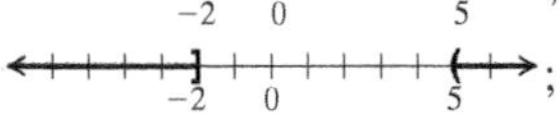 ; $[-2, 5)$
29. ; $(-\infty, -2] \cup (5, \infty)$
30. $\{1, 5, 9\}$ **31.** $\{1, 2, 3, 5, 6, 9\}$ **32.** $\varnothing$
33. $\{x \mid -7 < x \le 2\}$, or $(-7, 2]$ **34.** $\{x \mid -\frac{5}{4} < x < \frac{5}{2}\}$, or $(-\frac{5}{4}, \frac{5}{2})$ **35.** $\{x \mid x < -3 \text{ or } x > 1\}$, or $(-\infty, -3) \cup (1, \infty)$
36. $\{x \mid x < -11 \text{ or } x \ge -6\}$, or $(-\infty, -11) \cup [-6, \infty)$
37. $\{x \mid x \le -6 \text{ or } x \ge 8\}$, or $(-\infty, -6] \cup [8, \infty)$
38. $\frac{3}{|x|}$ **39.** $\frac{2|x|}{y^2}$ **40.** $\frac{4}{|y|}$ **41.** 62 **42.** $\{-6, 6\}$
43. $\{-5, 9\}$ **44.** $\{-14, \frac{4}{3}\}$ **45.** $\varnothing$ **46.** $\{x \mid -\frac{17}{2} < x < \frac{7}{2}\}$, or $(-\frac{17}{2}, \frac{7}{2})$ **47.** $\{x \mid x \le -3.5 \text{ or } x \ge 3.5\}$, or $(-\infty, -3.5] \cup [3.5, \infty)$ **48.** $\{x \mid x \le -\frac{11}{3} \text{ or } x \ge \frac{19}{3}\}$, or $(-\infty, -\frac{11}{3}] \cup [\frac{19}{3}, \infty)$ **49.** $\varnothing$ **50.** B **51.** A
52. $\{x \mid -\frac{8}{3} \le x \le -2\}$, or $[-\frac{8}{3}, -2]$

Understanding Through Discussion and Writing

1. When the signs of the quantities on either side of the inequality symbol are changed, their relative positions on the number line are reversed. **2.** The distance between x and -5 is $|x - (-5)|$, or $|x + 5|$. Then the solutions of the inequality $|x + 5| \le 2$ can be interpreted as "all those numbers x whose distance from -5 is at most 2 units." **3.** When $b \ge c$, then the intervals overlap and $[a, b] \cup [c, d] = [a, d]$. **4.** The solutions of $|x| \ge 6$ are those numbers whose distance from 0 is greater than or equal to 6. In addition to the numbers in $[6, \infty)$, the distance of the numbers in $(-\infty, -6]$ from 0 is also greater than or equal to 6. Thus, $[6, \infty)$ is only part of the solution of the inequality. **5. (1)** $-9(x + 2) = -9x - 18$, not $-9x + 2$. **(2)** This would be correct if (1) were correct except that the inequality symbol should not have been reversed. **(3)** If (2) were correct, the right-hand side would be -5, not 8. **(4)** The inequality symbol should be reversed. The correct solution is

$$\begin{aligned} 7 - 9x + 6x &< -9(x + 2) + 10x \\ 7 - 9x + 6x &< -9x - 18 + 10x \\ 7 - 3x &< x - 18 \\ -4x &< -25 \\ x &> \tfrac{25}{4}. \end{aligned}$$

6. By definition, the notation $3 < x < 5$ indicates that $3 < x$ *and* $x < 5$. A solution of the disjunction $3 < x$ *or* $x < 5$ must be in at least one of these sets but not necessarily in both, so the disjunction cannot be written as $3 < x < 5$.

Test: Chapter 1, p. 115

1. [1.1b] -2 **2.** [1.1c] $\frac{2}{3}$ **3.** [1.1b] $\frac{19}{15}$ **4.** [1.1d] 4
5. [1.1d] 1.1 **6.** [1.1d] -2 **7.** [1.2a] $B = \frac{A + C}{3}$
8. [1.2a] $n = \frac{m}{1 - t}$ **9.** [1.3a] Length: $14\frac{2}{5}$ ft; width: $9\frac{3}{5}$ ft
10. [1.3a] 52,000 copies **11.** [1.3a] 180,000
12. [1.3a] 59°, 60°, 61° **13.** [1.3b] $2\frac{2}{5}$ hr; 4 hr **14.** [1.4b] $(-3, 2]$
15. [1.4b] $(-4, \infty)$ **16.** [1.4c] 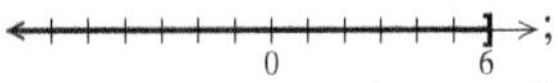 ; $(-\infty, 6]$ **17.** [1.4c] 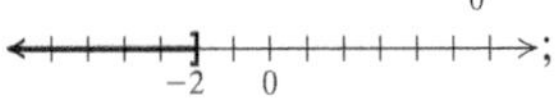 ; $(-\infty, -2]$
18. [1.4c] $\{x \mid x \ge 10\}$, or $[10, \infty)$ **19.** [1.4c] $\{y \mid y > -50\}$, or $(-50, \infty)$ **20.** [1.4c] $\{a \mid a \le \frac{11}{5}\}$, or $(-\infty, \frac{11}{5}]$
21. [1.4c] $\{y \mid y > 1\}$, or $(1, \infty)$ **22.** [1.4c] $\{x \mid x > \frac{5}{2}\}$, or $(\frac{5}{2}, \infty)$ **23.** [1.4c] $\{x \mid x \le \frac{7}{4}\}$, or $(-\infty, \frac{7}{4}]$
24. [1.4d] $\{h \mid h > 2\frac{1}{10} \text{ hr}\}$ **25.** [1.5c] $\{d \mid 33 \text{ ft} \le d \le 231 \text{ ft}\}$
26. [1.5a] 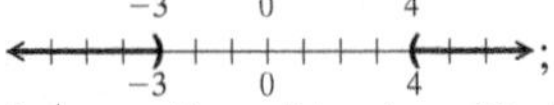 ; $[-3, 4]$
27. [1.5b] 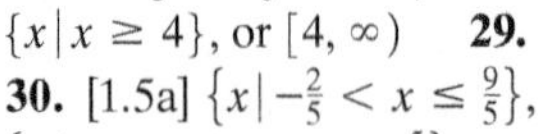 ; $(-\infty, -3) \cup (4, \infty)$
28. [1.5a] $\{x \mid x \ge 4\}$, or $[4, \infty)$ **29.** [1.5a] $\{x \mid -1 < x < 6\}$, or $(-1, 6)$ **30.** [1.5a] $\{x \mid -\frac{2}{5} < x \le \frac{9}{5}\}$, or $(-\frac{2}{5}, \frac{9}{5}]$
31. [1.5b] $\{x \mid x < -4 \text{ or } x > -\frac{5}{2}\}$, or $(-\infty, -4) \cup (-\frac{5}{2}, \infty)$
32. [1.5b] All real numbers, or $(-\infty, \infty)$

33. [1.5b] $\{x | x < 3 \text{ or } x > 6\}$, or $(-\infty, 3) \cup (6, \infty)$

34. [1.6a] $\frac{7}{|x|}$ **35.** [1.6a] $2|x|$ **36.** [1.6b] 8.4

37. [1.5a] $\{3, 5\}$ **38.** [1.5b] $\{1, 3, 5, 7, 9, 11, 13\}$

39. [1.6c] $\{-9, 9\}$ **40.** [1.6c] $\{-6, 12\}$ **41.** [1.6d] $\{1\}$

42. [1.6c] $\varnothing$ **43.** [1.6e] $\{x | -0.875 < x < 1.375\}$, or $(-0.875, 1.375)$ **44.** [1.6e] $\{x | x < -3 \text{ or } x > 3\}$, or $(-\infty, -3) \cup (3, \infty)$ **45.** [1.6e] $\{x | -99 \le x \le 111\}$, or $[-99, 111]$ **46.** [1.6e] $\{x | x \le -\frac{13}{5} \text{ or } x \ge \frac{7}{5}\}$, or $(-\infty, -\frac{13}{5}] \cup [\frac{7}{5}, \infty)$ **47.** [1.1d] C **48.** [1.6e] $\varnothing$

49. [1.5a] $\{x | \frac{1}{5} < x < \frac{4}{5}\}$, or $(\frac{1}{5}, \frac{4}{5})$

CHAPTER 2

Calculator Corner, p. 122

1.

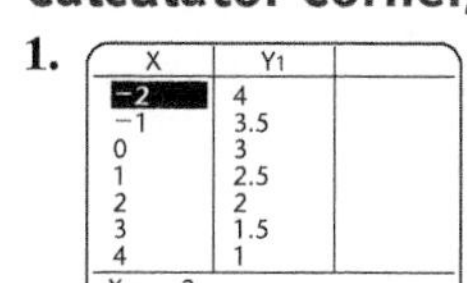

X	Y1
−2	4
−1	3.5
0	3
1	2.5
2	2
3	1.5
4	1

X = −2

2.

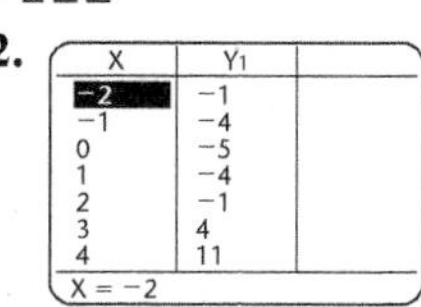

X	Y1
−2	−1
−1	−4
0	−5
1	−4
2	−1
3	4
4	11

X = −2

Calculator Corner, p. 125

1. 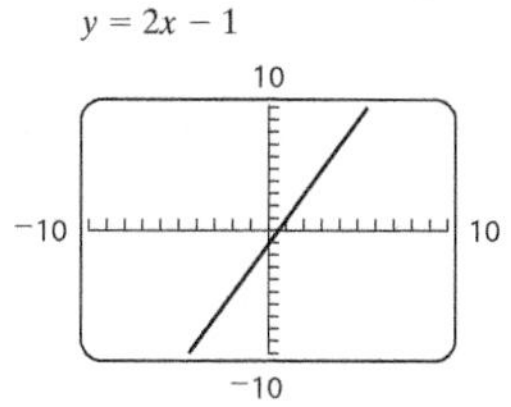 $y = 2x - 1$ **2.** $y = -3x + 2$

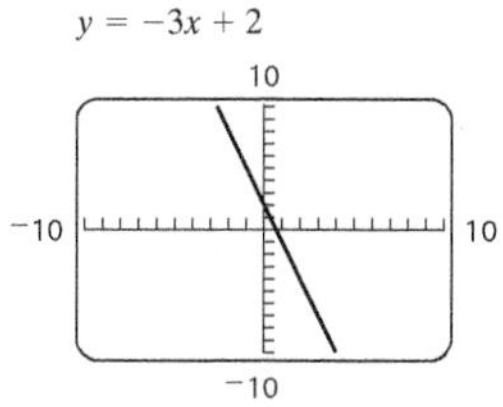

3. 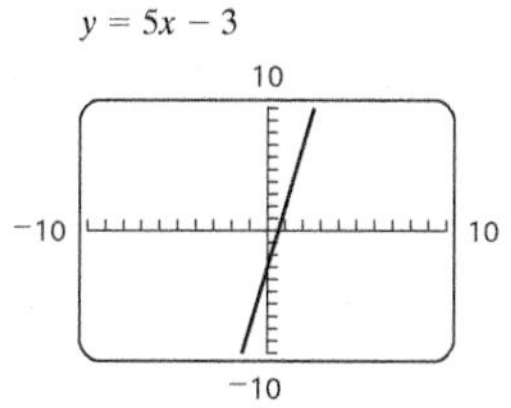 $y = 5x - 3$ **4.** 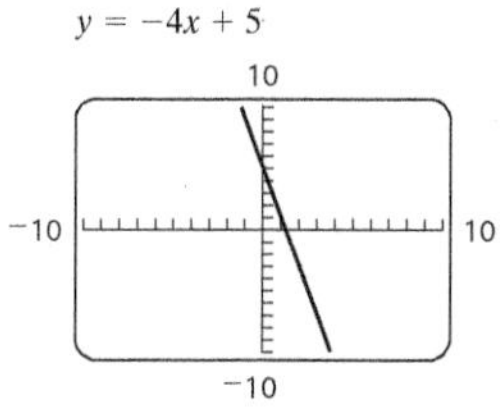 $y = -4x + 5$

5. 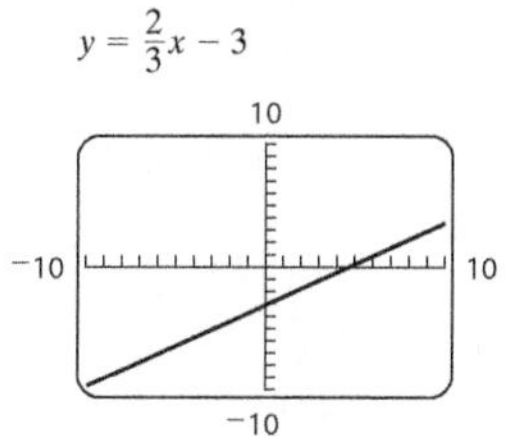 $y = \frac{2}{3}x - 3$ **6.** 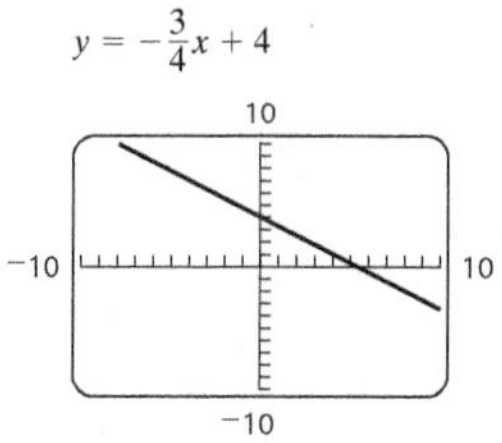 $y = -\frac{3}{4}x + 4$

7. 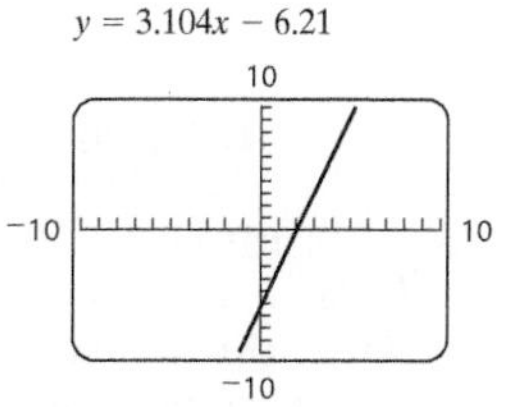 $y = 3.104x - 6.21$ **8.** 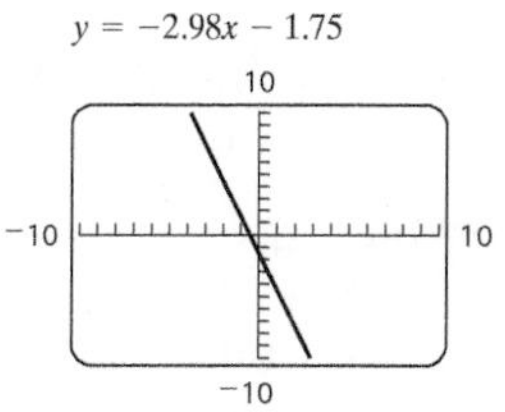 $y = -2.98x - 1.75$

Exercise Set 2.1, p. 126

RC1. False **RC2.** False **RC3.** True **RC4.** True **RC5.** True **RC6.** False **CC1.** (d) **CC2.** (b) **CC3.** (a) **CC4.** (c)

1.

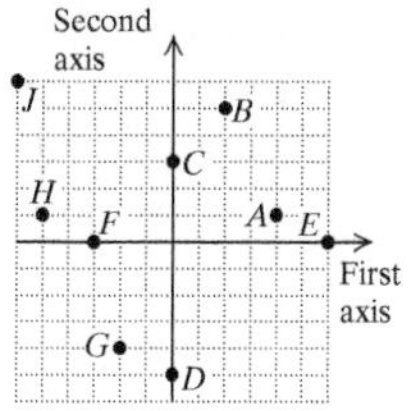

3.

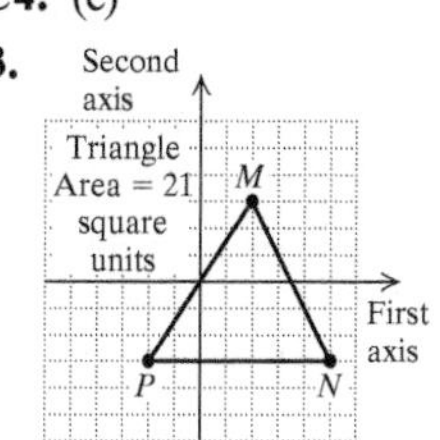

5. Yes **7.** Yes **9.** No

11.

$$\begin{array}{c|l} y = 4 - x & \\ \hline 5 \;?\; 4 - (-1) & \\ & 4 + 1 \\ & 5 \quad \text{TRUE} \end{array}$$

$$\begin{array}{c|l} y = 4 - x & \\ \hline 1 \;?\; 4 - 3 & \\ & 1 \quad \text{TRUE} \end{array}$$

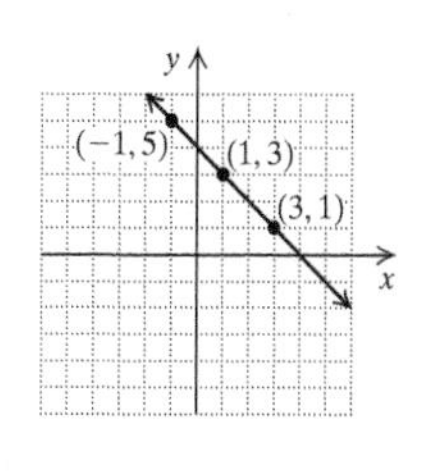

13.

$$\begin{array}{r|l} 3x + y = 7 & \\ \hline 3 \cdot 2 + 1 \;?\; 7 & \\ 6 + 1 & \\ 7 & \text{TRUE} \end{array}$$

$$\begin{array}{r|l} 3x + y = 7 & \\ \hline 3 \cdot 4 + (-5) \;?\; 7 & \\ 12 - 5 & \\ 7 & \text{TRUE} \end{array}$$

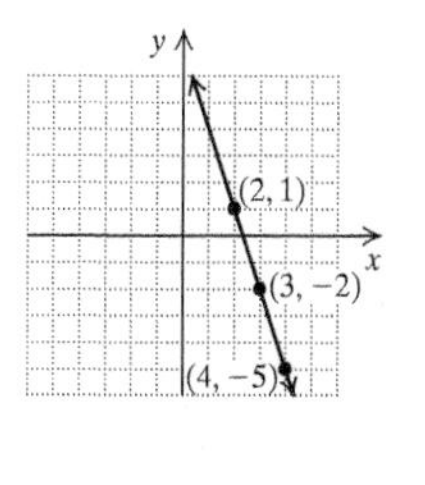

15.

$$\begin{array}{r|l} 6x - 3y = 3 & \\ \hline 6 \cdot 1 - 3 \cdot 1 \;?\; 3 & \\ 6 - 3 & \\ 3 & \text{TRUE} \end{array}$$

$$\begin{array}{r|l} 6x - 3y = 3 & \\ \hline 6(-1) - 3(-3) \;?\; 3 & \\ -6 + 9 & \\ 3 & \text{TRUE} \end{array}$$

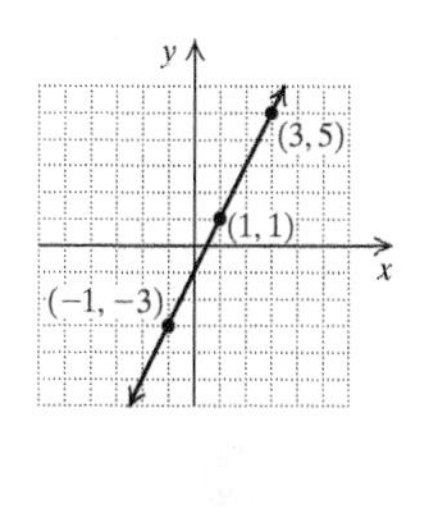

17. 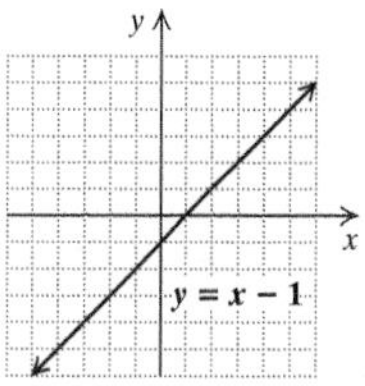 $y = x - 1$ **19.**

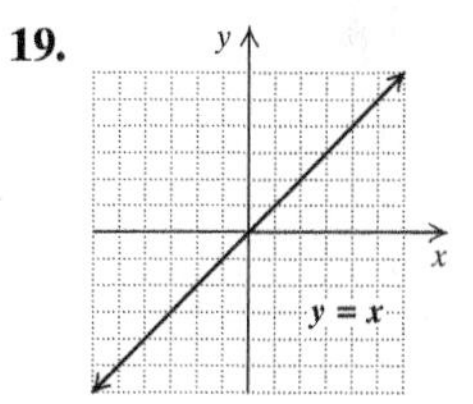

21. 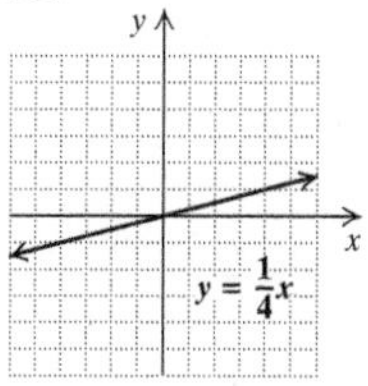 $y = \frac{1}{4}x$ **23.**

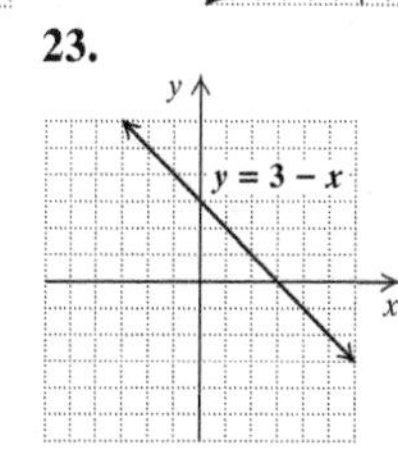

25.

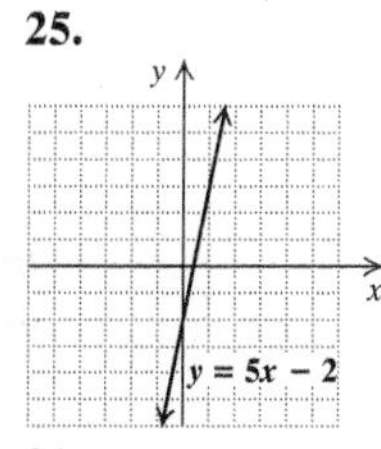

27.

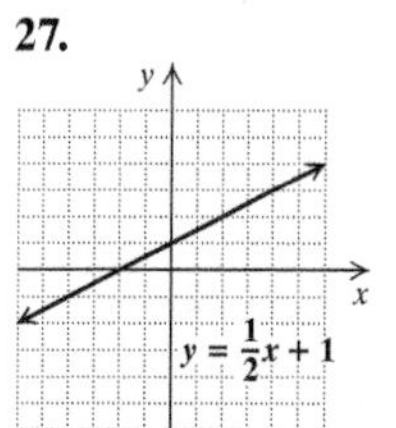

29.

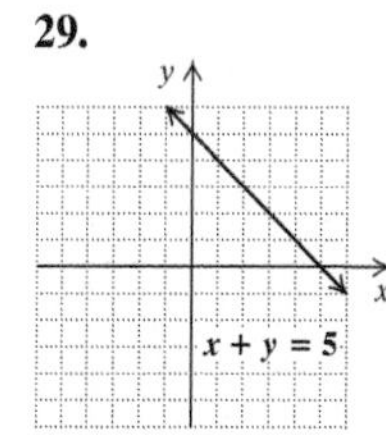

31.

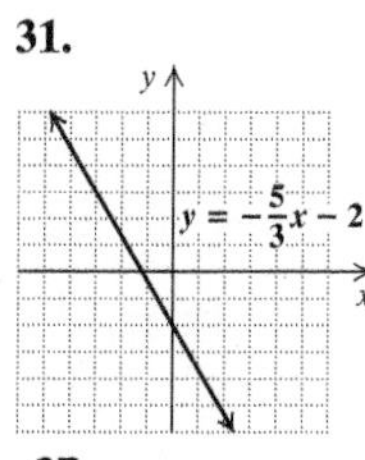

33.

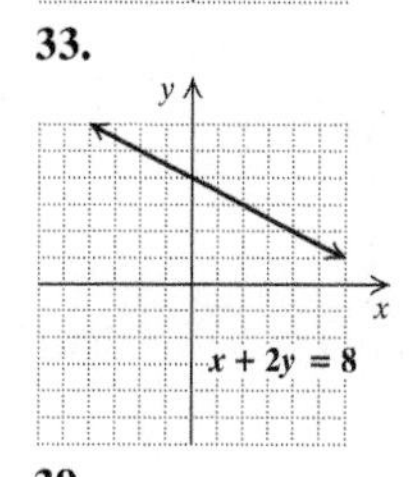

35.

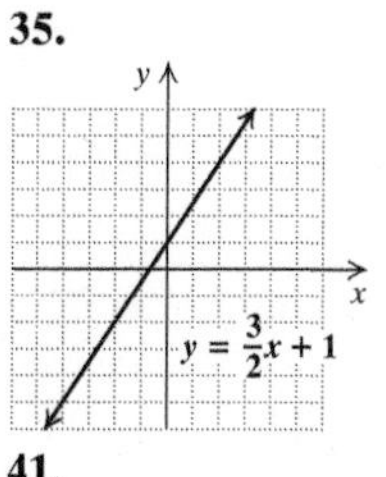

37.

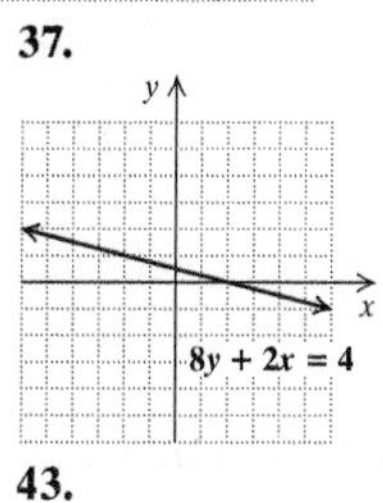

39.

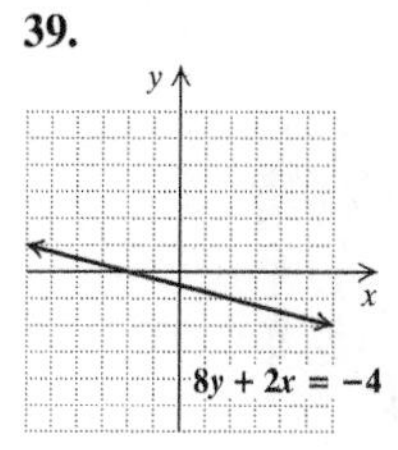

41.

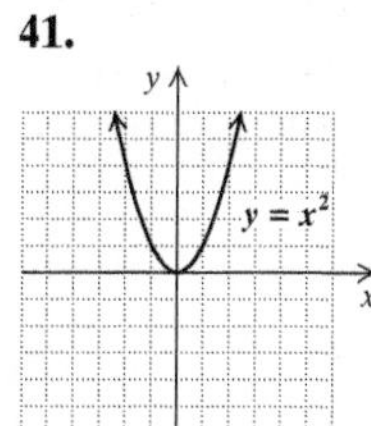

43.

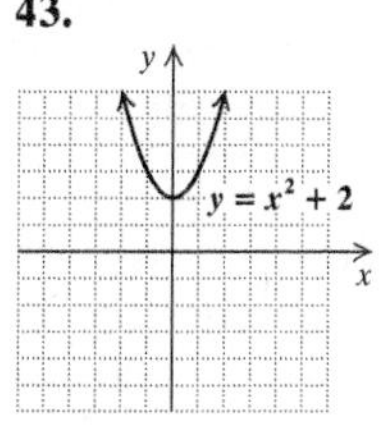

45.

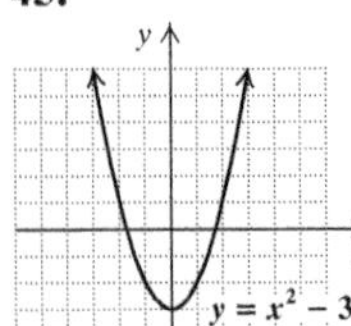

47.

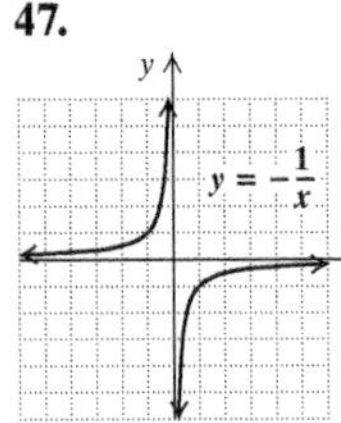

49.

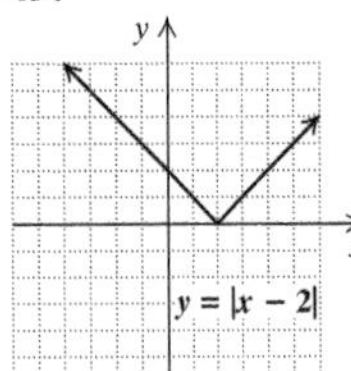

51.

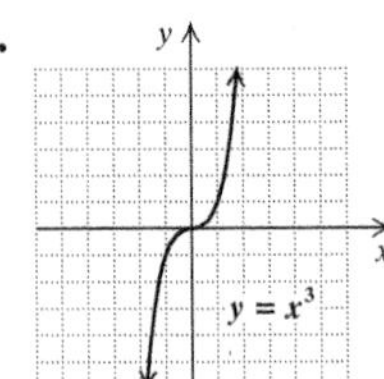

53. $\{x \mid 1 < x \leq \frac{15}{2}\}$, or $\left(1, \frac{15}{2}\right]$

54. $\{x \mid x > -3\}$, or $(-3, \infty)$ **55.** $\{x \mid x \leq -\frac{7}{3}$ *or* $x \geq \frac{17}{3}\}$, or $\left(-\infty, -\frac{7}{3}\right] \cup \left[\frac{17}{3}, \infty\right)$ **56.** $\{x \mid -6 < x < 6\}$, or $(-6, 6)$
57. Kidney: 99,519 people; liver: 14,652 people
58. 25 ft **59.** $4\frac{3}{4}$ mi **60.** \$330,000

61.

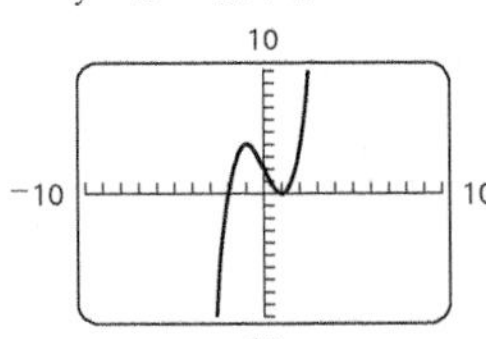

63.

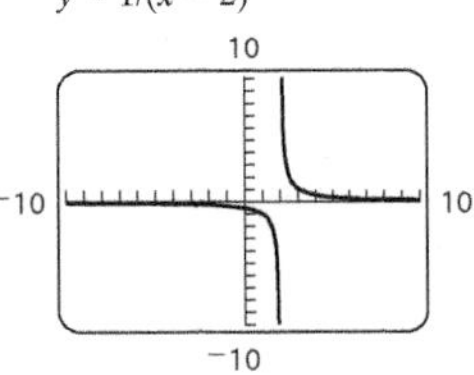

65. $y = -x + 4$ **67.** $y = |x| - 3$

Calculator Corner, p. 135

1. 17.3 **2.** 34

Calculator Corner, p. 137

1.

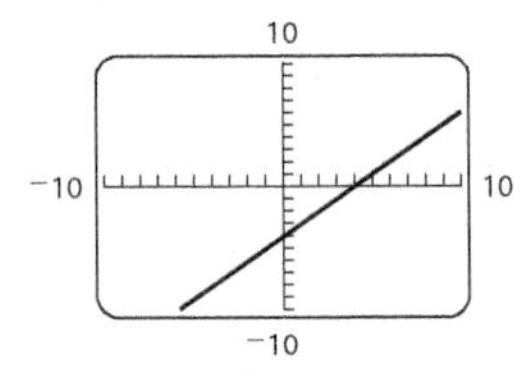

2.

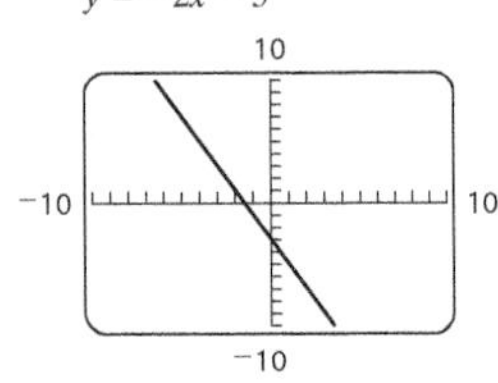

3.

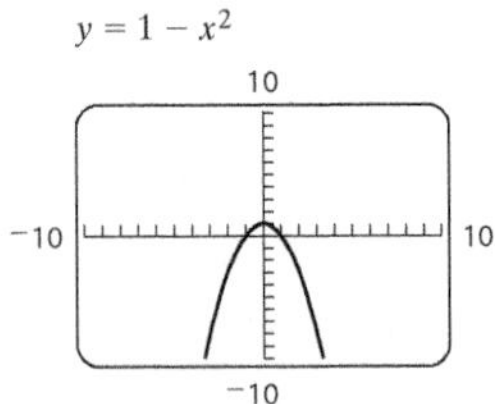

4.

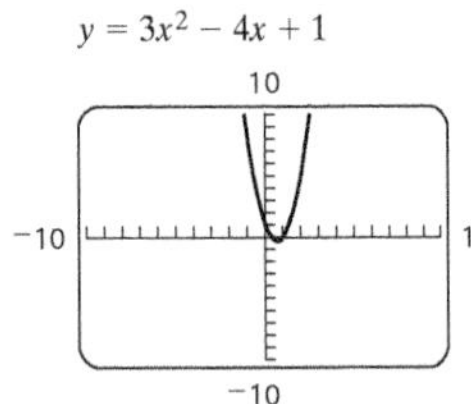

5.

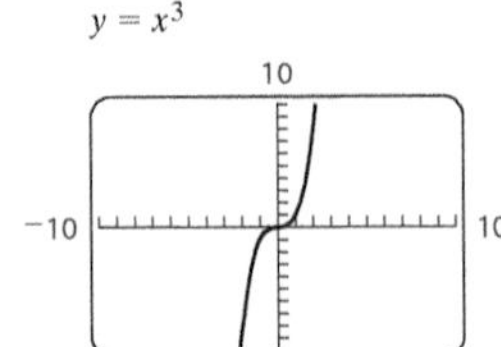

6.

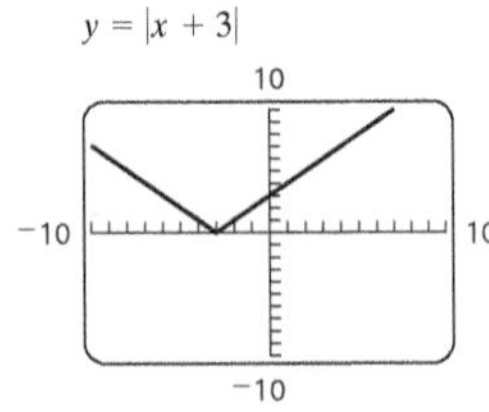

Exercise Set 2.2, p. 139

RC1. Domain, range, domain, exactly one, range
RC2. Domain, range, domain, at least one, range
CC1. $f(2) = 3$ **CC2.** $f(0) = 3$ **CC3.** $f(-2) = -5$
CC4. $f(3) = 0$
1. Yes **3.** Yes **5.** No **7.** No **9.** No **11.** Yes
13. (a) 9; **(b)** 12; **(c)** 2; **(d)** 5; **(e)** 7.4; **(f)** $5\frac{2}{3}$ **15. (a)** −21; **(b)** 15; **(c)** 2; **(d)** 0; **(e)** $18a$; **(f)** $3a + 3$ **17. (a)** 7; **(b)** −17; **(c)** 6; **(d)** 4; **(e)** $3a - 2$; **(f)** $3a + 3h + 4$
19. (a) 0; **(b)** 5; **(c)** 2; **(d)** 170; **(e)** 65; **(f)** $32a^2 - 12a$
21. (a) 1; **(b)** 3; **(c)** 3; **(d)** 11; **(e)** $|a - 1| + 1$; **(f)** $|a + h| + 1$
23. (a) 0; **(b)** −1; **(c)** 8; **(d)** 1000; **(e)** −125; **(f)** $-27a^3$
25. 2012: about 12.8 million troy ounces; 2015: about 15.7 million troy ounces **27.** $1\frac{20}{33}$ atm; $1\frac{10}{11}$ atm; $4\frac{1}{33}$ atm
29. 1.792 cm; 2.8 cm; 11.2 cm

31.

33.

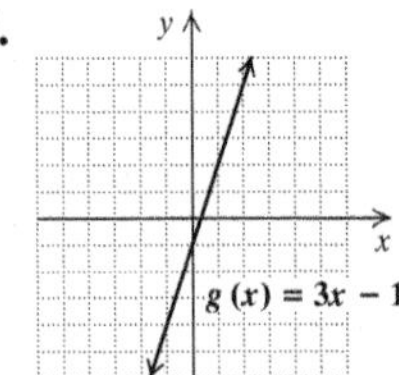

35.

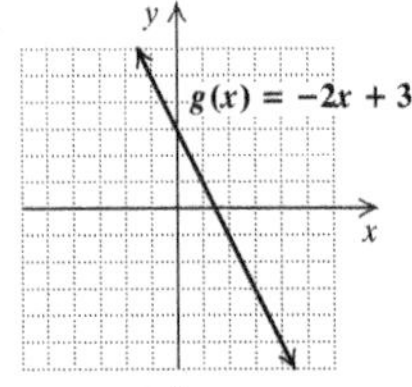

37.

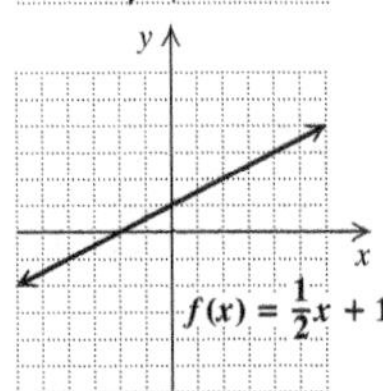

39.

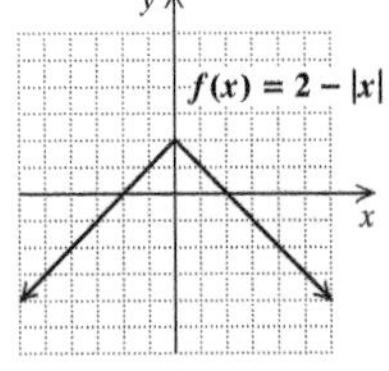

41.

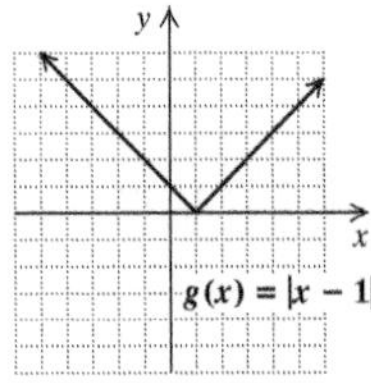

43.

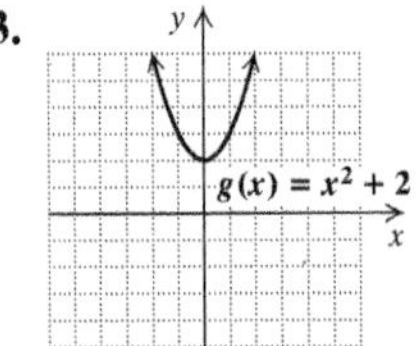

45.

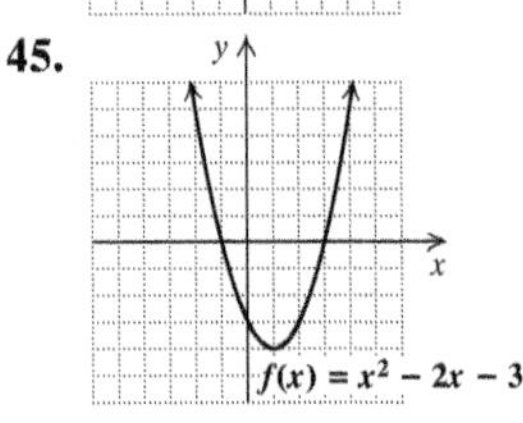

47.

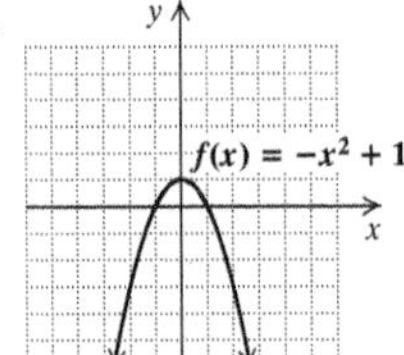

49.

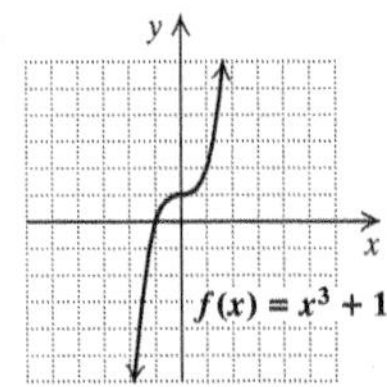

51. Yes **53.** No **55.** No **57.** Yes **59.** \$210,000
61. 720 million passengers **63.** 650 million passengers
65. $\{y \mid y < 4\}$, or $(-\infty, 4)$ **66.** $\{x \mid x \geq 23\}$, or $[23, \infty)$
67. No solution **68.** All real numbers **69.** $\frac{3}{2}$
70. $\frac{80}{9}$ **71.** $g(-2) = 39$ **73.** 26; 99 **75.** $g(x) = \frac{15}{4}x - \frac{13}{4}$

Exercise Set 2.3, p. 149

RC1. (a) **RC2.** (b) **RC3.** (a) **RC4.** (b) **RC5.** (a)
RC6. (d)
1. (a) 3; **(b)** $\{-4, -3, -2, -1, 0, 1, 2\}$; **(c)** −2, 0; **(d)** $\{1, 2, 3, 4\}$
3. (a) $2\frac{1}{2}$; **(b)** $[-3, 5]$; **(c)** $2\frac{1}{4}$; **(d)** $[1, 4]$ **5. (a)** 1; **(b)** all real numbers; **(c)** 3; **(d)** all real numbers **7. (a)** 1; **(b)** all real numbers; **(c)** −2, 2; **(d)** $[0, \infty)$ **9.** $\{x \mid x$ is a real number *and* $x \neq -3\}$, or $(-\infty, -3) \cup (-3, \infty)$ **11.** All real numbers
13. All real numbers **15.** $\{x \mid x$ is a real number *and* $x \neq \frac{14}{5}\}$, or $\left(-\infty, \frac{14}{5}\right) \cup \left(\frac{14}{5}, \infty\right)$ **17.** All real numbers **19.** $\{x \mid x$ is a real number *and* $x \neq \frac{7}{4}\}$, or $\left(-\infty, \frac{7}{4}\right) \cup \left(\frac{7}{4}, \infty\right)$ **21.** $\{x \mid x$ is a real number *and* $x \neq 1\}$, or $(-\infty, 1) \cup (1, \infty)$ **23.** All real numbers
25. All real numbers **27.** $\{x \mid x$ is a real number *and* $x \neq \frac{5}{2}\}$, or $\left(-\infty, \frac{5}{2}\right) \cup \left(\frac{5}{2}, \infty\right)$ **29.** All real numbers **31.** $\{x \mid x$ is a real number *and* $x \neq -\frac{5}{4}\}$, or $\left(-\infty, -\frac{5}{4}\right) \cup \left(-\frac{5}{4}, \infty\right)$ **33.** −8; 0; −2
35. $\{-8, 8\}$ **36.** $\{\ \}$, or $\varnothing$ **37.** $\{-4, 18\}$ **38.** $\{-8, 5\}$
39. $\{\frac{1}{2}, 3\}$ **40.** $\{-1, \frac{9}{13}\}$ **41.** $\{\ \}$, or $\varnothing$ **42.** $\{\frac{8}{3}\}$
43. $(-\infty, 0) \cup (0, \infty)$; $[2, \infty)$; $[-4, \infty)$; $[0, \infty)$
45. All real numbers

Mid-Chapter Review: Chapter 2, p. 151

1. True **2.** False **3.** True **4.** True **5.** False

6.

x	y
0	1
2	−2
−2	4
4	−5

7.

x	$f(x)$
−2	0
3	0
0	−6
2	−4
−1	−4

8. No **9.** Yes **10.** Yes **11.** No **12.** Domain: $\{x \mid -3 \le x \le 3\}$, or $[-3, 3]$; range: $\{y \mid -2 \le y \le 1\}$ **13.** −3 **14.** −7 **15.** 8 **16.** 9 **17.** 9000 **18.** 0 **19.** Yes **20.** No **21.** Yes **22.** $\{x \mid x$ is a real number *and* $x \neq 4\}$, or $(-\infty, 4) \cup (4, \infty)$ **23.** All real numbers **24.** $\{x \mid x$ is a real number *and* $x \neq -2\}$, or $(-\infty, -2) \cup (-2, \infty)$ **25.** All real numbers

26.

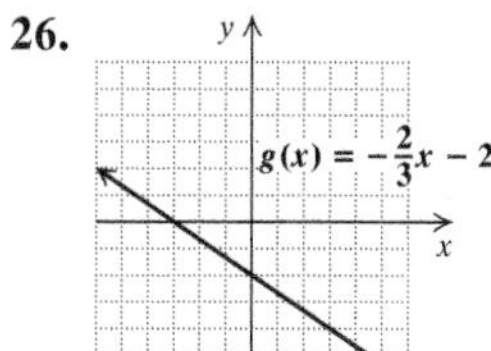
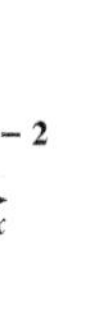

27.

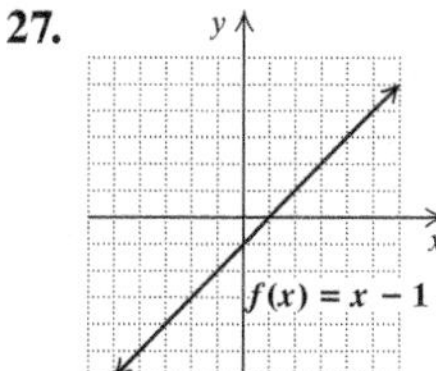

28.

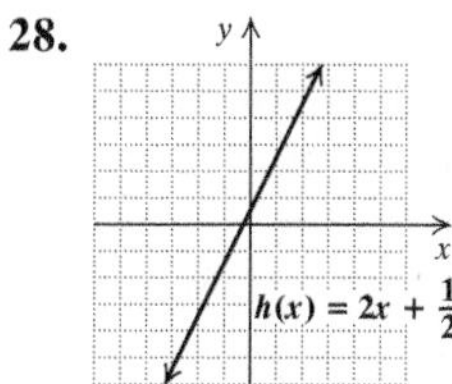

29.

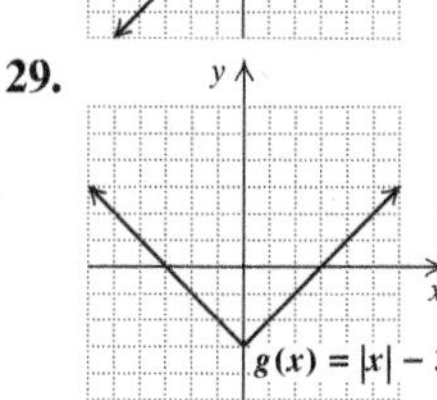

30.

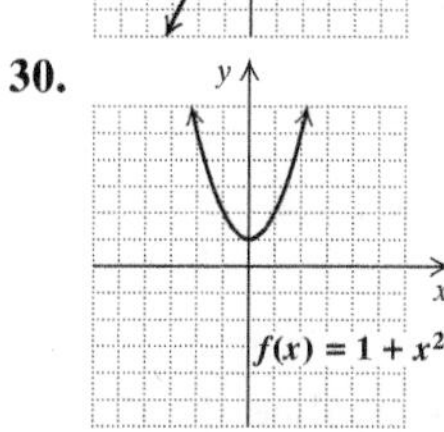

31.

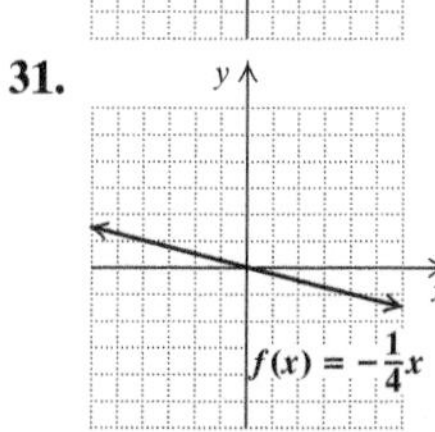

32. No; since each input has exactly one output, the number of outputs cannot exceed the number of inputs. **33.** When $x < 0$, then $y < 0$, and the graph contains points in quadrant III. When $0 < x < 30$, then $y < 0$, and the graph contains points in quadrant IV. When $x > 30$, then $y > 0$, and the graph contains points in quadrant I. Thus the graph passes through three quadrants. **34.** The output −3 corresponds to the input 2. The number −3 in the range is paired with the number 2 in the domain. The point $(2, -3)$ is on the graph of the function. **35.** The domain of a function is the set of all inputs, and the range is the set of all outputs.

Calculator Corner, p. 154

1. The graph of $y_2 = x + 4$ is the graph of $y_1 = x$ shifted up 4 units. **2.** The graph of $y_3 = x - 3$ is the graph of $y_1 = x$ shifted down 3 units.

Calculator Corner, p. 157

1. The graph of $y = 10x$ will slant up from left to right. It will be steeper than the other graphs. **2.** The graph of $y = 0.005x$ will slant up from left to right. It will be less steep than the other graphs. **3.** The graph of $y = -10x$ will slant down from left to right. It will be steeper than the other graphs. **4.** The graph of $y = -0.005x$ will slant down from left to right. It will be less steep than the other graphs.

Exercise Set 2.4, p. 161

RC1. Up **RC2.** Horizontal **RC3.** Down **CC1.** (f) **CC2.** (b) **CC3.** (c) **1.** $m = 4$; y-intercept: $(0, 5)$ **3.** $m = -2$; y-intercept: $(0, -6)$ **5.** $m = -\frac{3}{8}$; y-intercept: $\left(0, -\frac{1}{5}\right)$ **7.** $m = 0.5$; y-intercept: $(0, -9)$ **9.** $m = \frac{2}{3}$; y-intercept: $\left(0, -\frac{8}{3}\right)$ **11.** $m = 3$; y-intercept: $(0, -2)$ **13.** $m = -8$; y-intercept: $(0, 12)$ **15.** $m = 0$; y-intercept: $\left(0, \frac{4}{17}\right)$ **17.** $m = -\frac{1}{2}$ **19.** $m = \frac{1}{3}$ **21.** $m = 2$ **23.** $m = \frac{2}{3}$ **25.** $m = -\frac{1}{3}$ **27.** $\frac{2}{25}$, or 8% **29.** $\frac{13}{41}$, or about 31.7% **31.** The rate of change is about \$28.4 million per year. **33.** The rate of change is −\$900 per year. **35.** The rate of change is about 10,366 eating and drinking places per year. **37.** −1323 **38.** $45x + 54$ **39.** $350x - 60y + 120$ **40.** 25 **41.** Square: 15 yd; triangle: 20 yd **42.** $\{x \mid x \le -\frac{24}{5}$ *or* $x \ge 8\}$, or $\left(-\infty, -\frac{24}{5}\right] \cup [8, \infty)$ **43.** $\{x \mid -\frac{24}{5} < x < 8\}$, or $\left(-\frac{24}{5}, 8\right)$ **44.** $\left\{-\frac{24}{5}, 8\right\}$ **45.** $\{\ \}$, or $\varnothing$

Calculator Corner, p. 165

1.

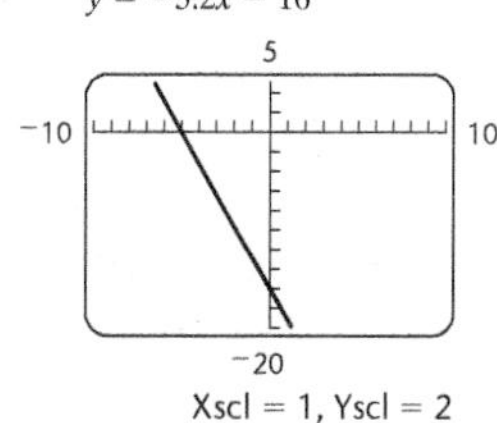

2.

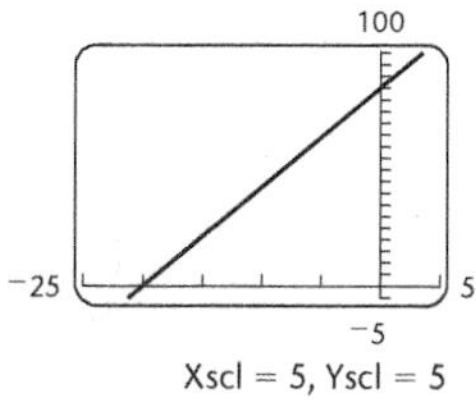

3.

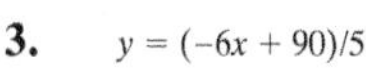
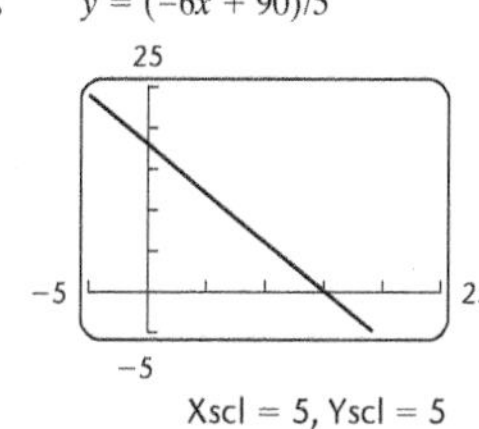

4.

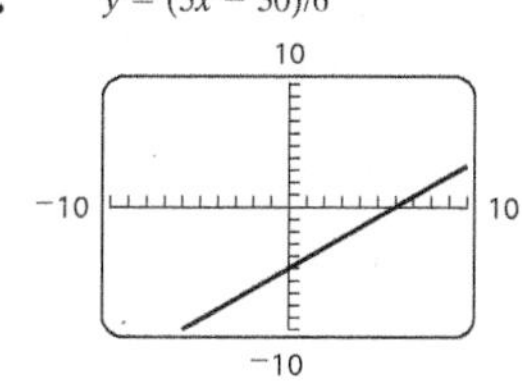

5.

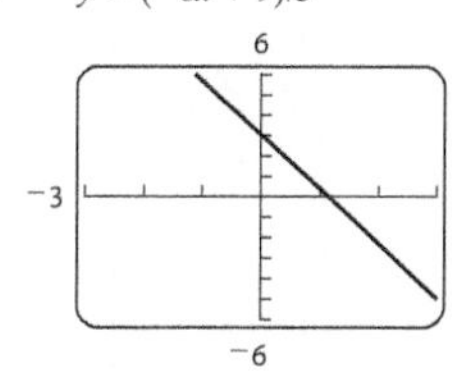

6.

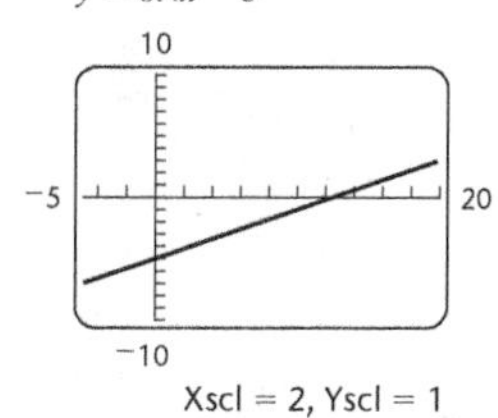

7.

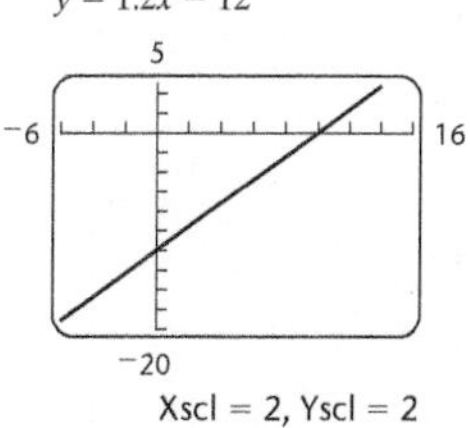

8.

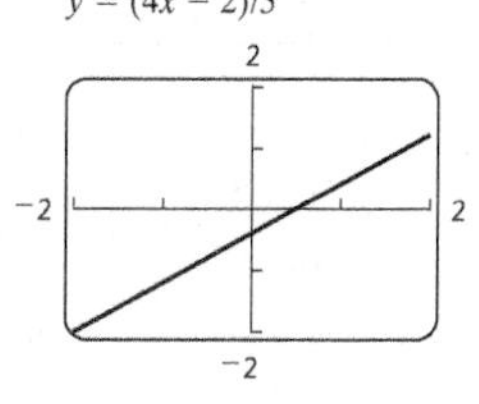

Visualizing for Success, p. 171

1. D **2.** I **3.** H **4.** C **5.** F **6.** A **7.** G **8.** B **9.** E **10.** J

Exercise Set 2.5, p. 172

RC1. False **RC2.** True **RC3.** True **RC4.** False **CC1.** $\left(-\frac{2}{7}, 0\right)$ **CC2.** $(0, 7)$ **CC3.** Perpendicular **CC4.** $y = \frac{2}{7}$

1.

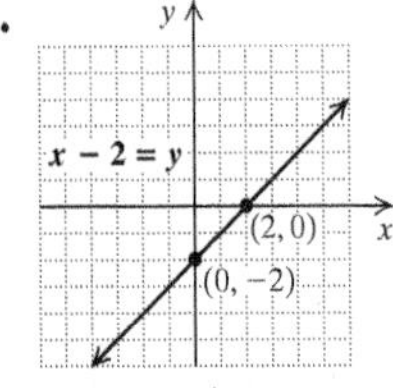

3.

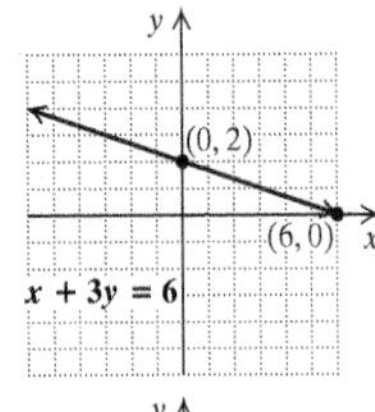

5.

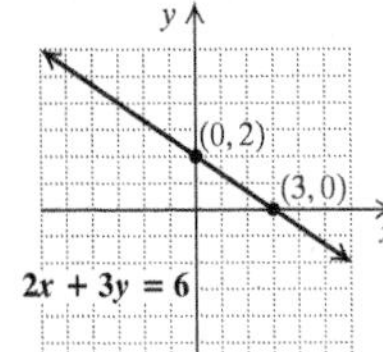

7.

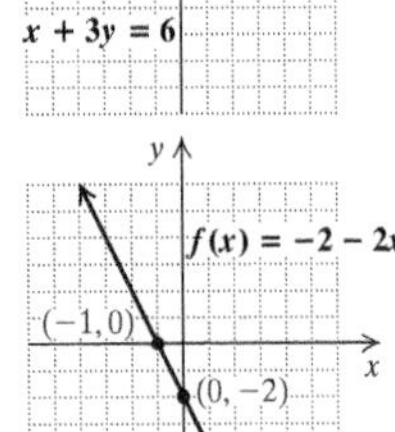

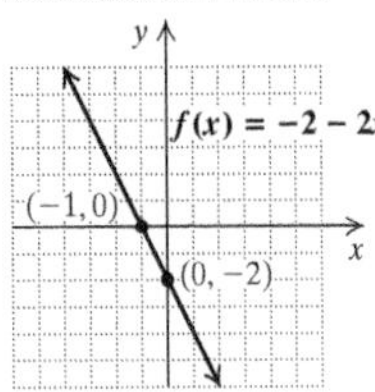

9.

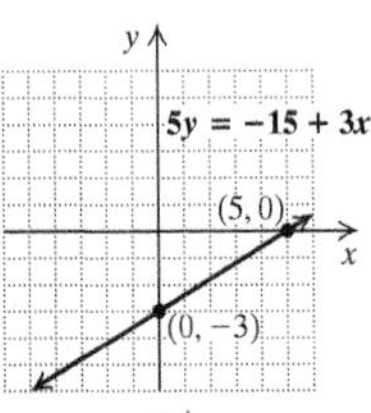

11.

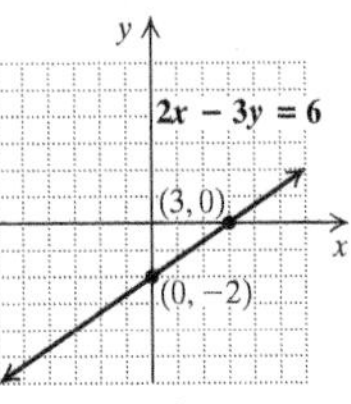

13.

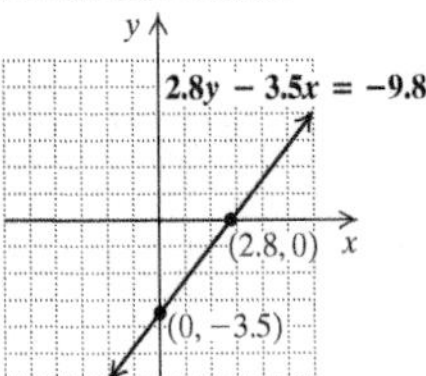

15.

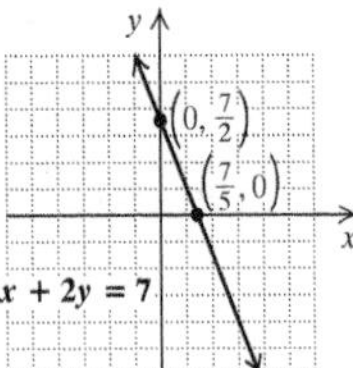

17.

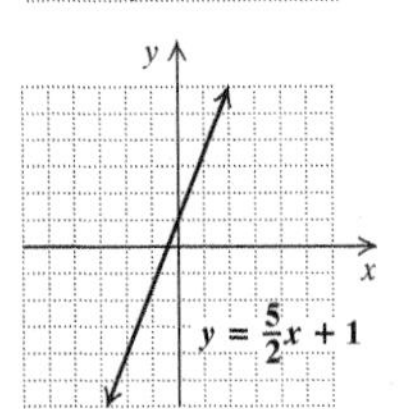

19.

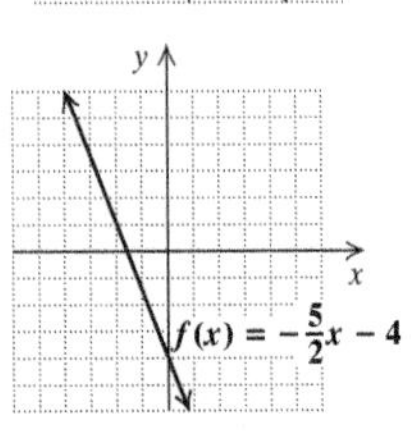

21.

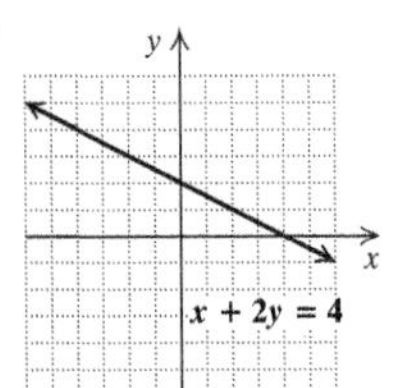

23.

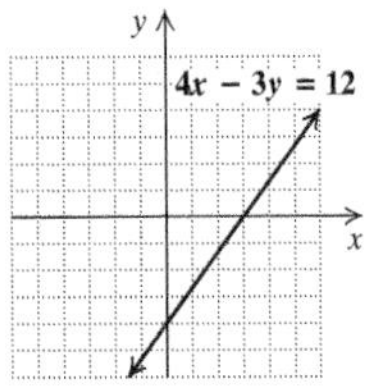

25.

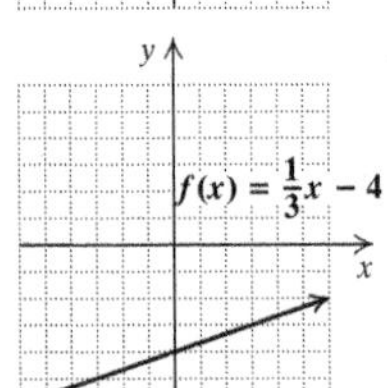

27.

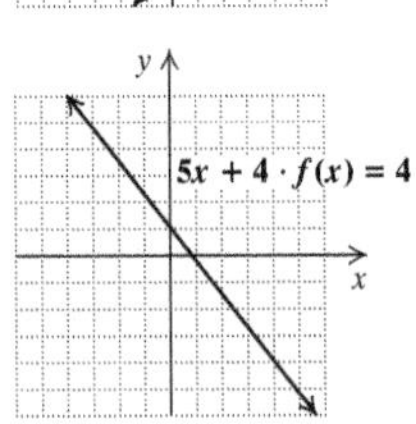

29. Not defined

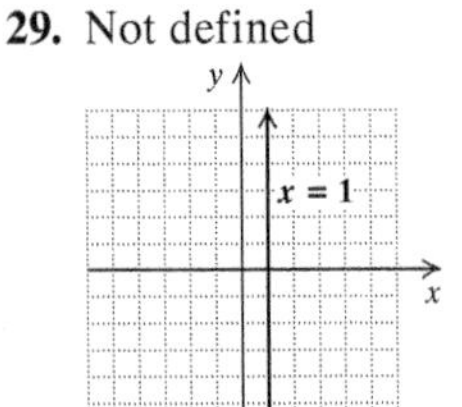

31. $m = 0$

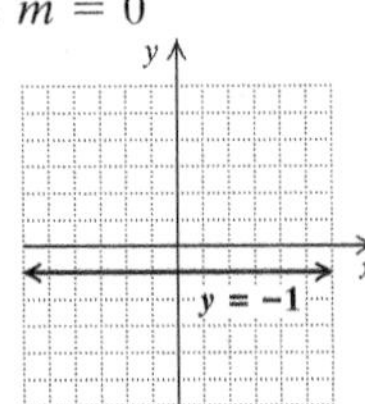

33. $m = 0$

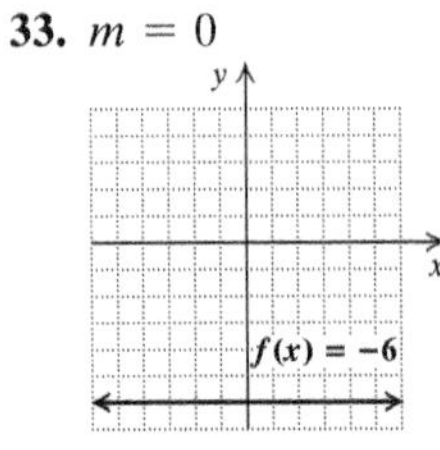

35. $m = 0$

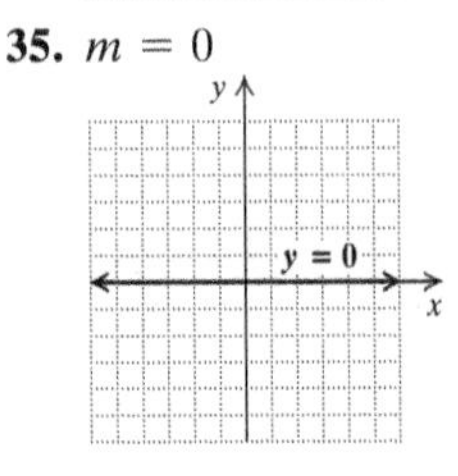

37. $m = 0$

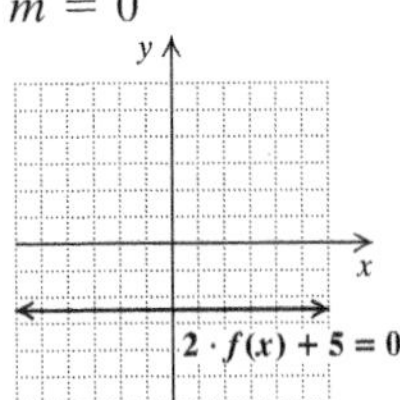

39. Not defined

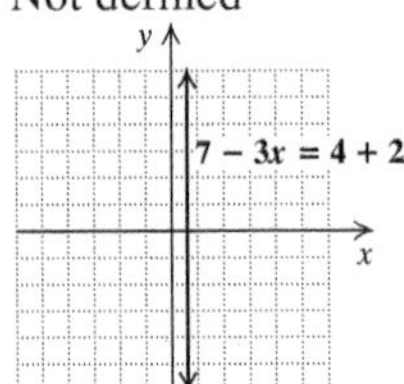

41. Yes **43.** No **45.** Yes **47.** Yes **49.** Yes **51.** No **53.** No **55.** Yes **57.** 5.3×10^{10} **58.** 4.7×10^{-5} **59.** 1.8×10^{-2} **60.** 9.9902×10^{7} **61.** 0.0000213 **62.** 901,000,000 **63.** 20,000 **64.** 0.085677 **65.** $3(3x - 5y)$ **66.** $3a(4 + 7b)$ **67.** $7p(3 - q + 2)$ **68.** $64(x - 2y + 4)$ **69.** $a = 2$ **71.** $y = \frac{2}{15}x + \frac{2}{5}$ **73.** $y = 0$; yes **75.** $m = -\frac{3}{4}$

Exercise Set 2.6, p. 183

RC1. $y = mx + b$ **RC2.** $y - y_1 = m(x - x_1)$ **CC1. (a)** $\frac{4}{11}$; **(b)** $-\frac{11}{4}$ **CC2. (a)** 0; **(b)** not defined **CC3. (a)** 2; **(b)** $-\frac{1}{2}$ **CC4. (a)** $-\frac{5}{6}$; **(b)** $\frac{6}{5}$ **CC5. (a)** Not defined; **(b)** 0 **CC6. (a)** -2; **(b)** $\frac{1}{2}$
1. $y = -8x + 4$ **3.** $y = 2.3x - 1$ **5.** $f(x) = -\frac{7}{3}x - 5$ **7.** $f(x) = \frac{2}{3}x + \frac{5}{8}$ **9.** $y = 5x - 17$ **11.** $y = -3x + 33$ **13.** $y = x - 6$ **15.** $y = -2x + 16$ **17.** $y = -7$ **19.** $y = \frac{2}{3}x - \frac{8}{3}$ **21.** $y = \frac{1}{2}x + \frac{7}{2}$ **23.** $y = x$ **25.** $y = \frac{7}{4}x + 7$ **27.** $y = \frac{3}{2}x$ **29.** $y = \frac{1}{6}x$ **31.** $y = 13x - \frac{15}{4}$ **33.** $y = -\frac{1}{2}x + \frac{17}{2}$ **35.** $y = \frac{5}{7}x - \frac{17}{7}$ **37.** $y = \frac{1}{3}x + 4$ **39.** $y = \frac{1}{2}x + 4$ **41.** $y = \frac{4}{3}x - 6$ **43.** $y = \frac{5}{2}x + 9$
45. (a) $C(x) = 5x + 10$; **(b)**

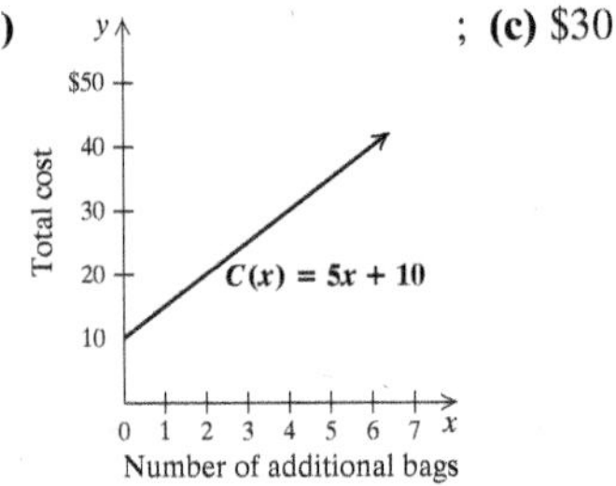

; **(c)** \$30

47. (a) $V(t) = 9400 - 85t$; **(b)**

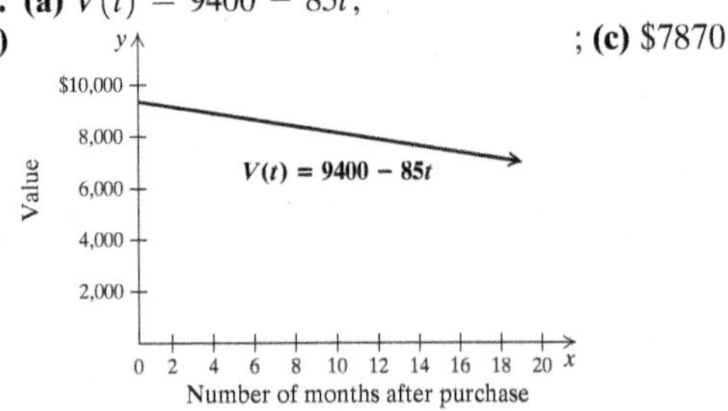

; **(c)** \$7870

49. (a) $U(x) = 0.0037x + 0.34$; **(b)** 1990: 45.1%; 2020: 56.2% **51. (a)** $B(x) = -2091.25x + 89{,}460$; **(b)** 72,730 bridges; **(c)** about 20 years after 2000, or in 2020 **53. (a)** $E(t) = 1.315t + 46.56$; **(b)** about 68.92 years **55.** $\{x \mid x \leq 24\}$, or $(-\infty, 24]$ **56.** $\{x \mid x \geq \frac{7}{3}\}$, or $[\frac{7}{3}, \infty)$ **57.** $\{x \mid -8 \leq x \leq 5\}$, or $[-8, 5]$ **58.** $\{-7, \frac{1}{3}\}$ **59.** -7.75

Summary and Review: Chapter 2, p. 187

Vocabulary Reinforcement

1. Vertical **2.** Point–slope **3.** Function, domain, range, domain, exactly one, range **4.** Slope **5.** Perpendicular **6.** Slope–intercept **7.** Parallel

Concept Reinforcement

1. False **2.** True **3.** False

Study Guide

1. No **2.** $g(0) = -2; g(-2) = -3; g(6) = 1$

3.

4. Yes **5.** Domain: $[-4, 5]$; range: $[-2, 4]$ **6.** $\{x \mid x$ is a real number *and* $x \neq -3\}$, or $(-\infty, -3) \cup (-3, \infty)$ **7.** -2 **8.** Slope: $-\frac{1}{2}$; y-intercept: $(0, 2)$

9.

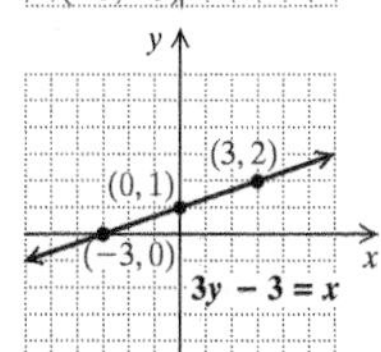

10.

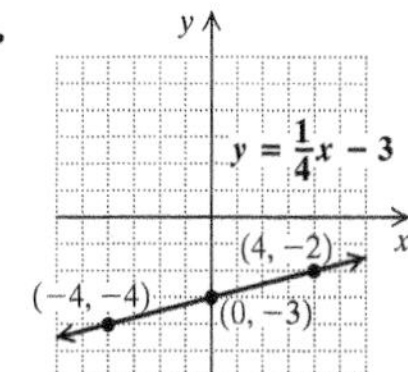

11.

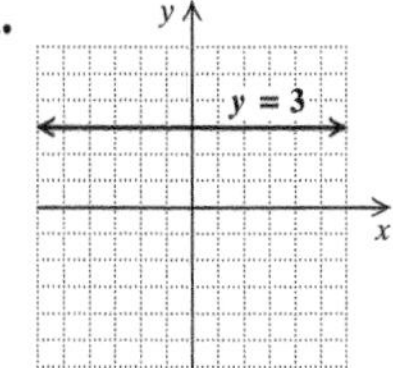

12.

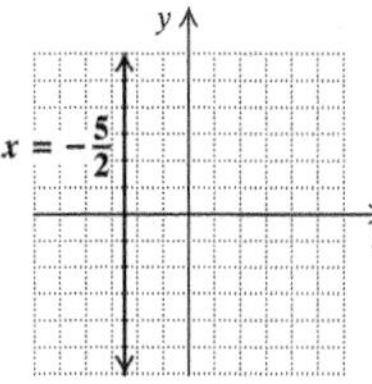

13. Parallel **14.** Perpendicular **15.** $y = -8x + 0.3$ **16.** $y = -4x - 1$ **17.** $y = -\frac{5}{3}x + \frac{11}{3}$ **18.** $y = \frac{4}{3}x - \frac{23}{3}$ **19.** $y = -\frac{3}{4}x - \frac{7}{2}$

Review Exercises

1. No **2.** Yes **3.** $g(0) = 5; g(-1) = 7$ **4.** $f(0) = 7; f(-1) = 12$ **5.** About $138 billion

6.

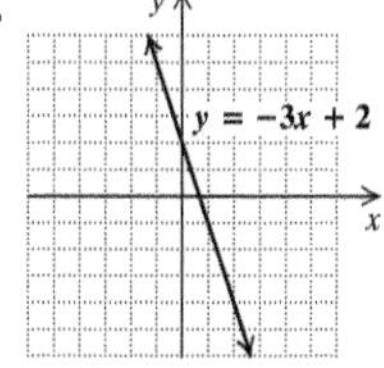

7.

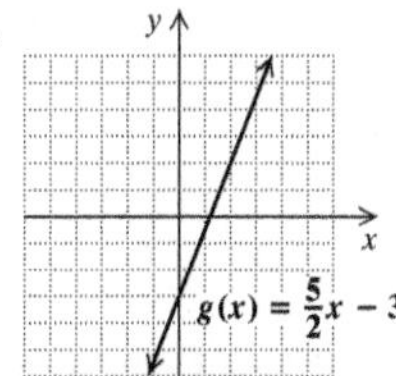

8.

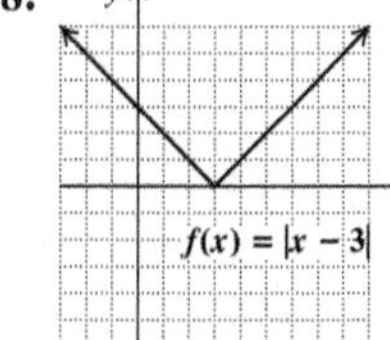

9.

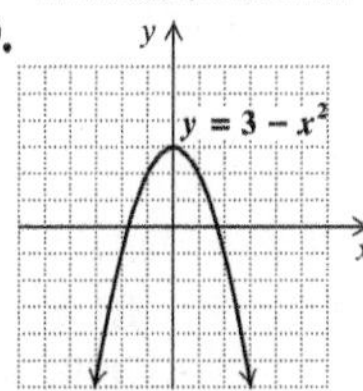

10. Yes **11.** No **12. (a)** $f(2) = 3$; **(b)** $[-2, 4]$; **(c)** -1; **(d)** $[1, 5]$ **13.** $\{x \mid x$ is a real number *and* $x \neq 4\}$, or $(-\infty, 4) \cup (4, \infty)$ **14.** All real numbers **15.** Slope: -3; y-intercept: $(0, 2)$ **16.** Slope: $-\frac{1}{2}$; y-intercept: $(0, 2)$ **17.** $\frac{11}{3}$

18.

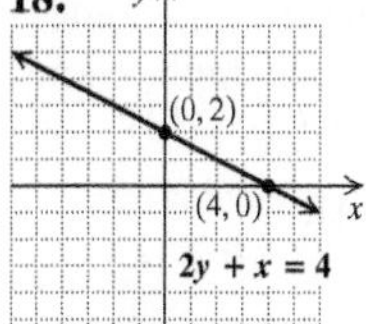

19.

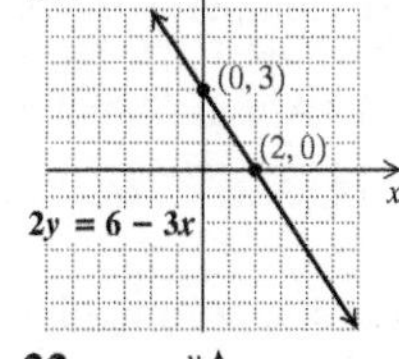

20.

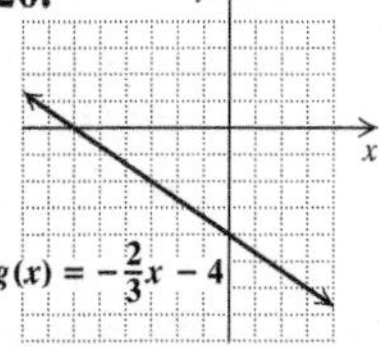

21.

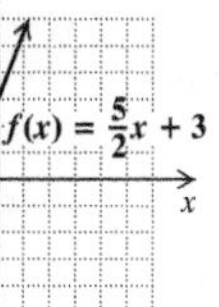

22.

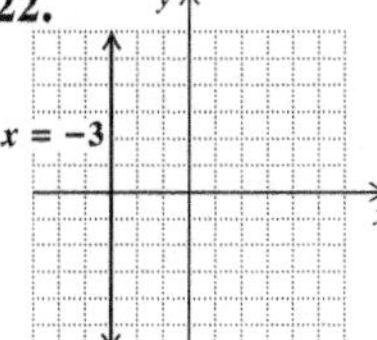

23.

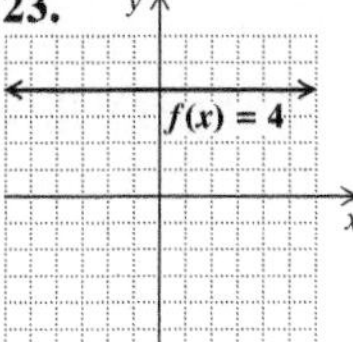

24. Perpendicular **25.** Parallel **26.** Parallel **27.** Perpendicular **28.** $f(x) = 4.7x - 23$ **29.** $y = -3x + 4$ **30.** $y = -\frac{3}{2}x$ **31.** $y = -\frac{5}{7}x + 9$ **32.** $y = \frac{1}{3}x + \frac{1}{3}$ **33. (a)** $R(x) = -0.037x + 44.66$; **(b)** 2000: about 43.62 sec; 2010: about 43.25 sec **34.** C **35.** A **36.** $f(x) = 3.09x + 3.75$

Understanding Through Discussion and Writing

1. A line's x- and y-intercepts are the same only when the line passes through the origin. The equation for such a line is of the form $y = mx$. **2.** The concept of slope is useful in describing how a line slants. A line with positive slope slants up from left to right. A line with negative slope slants down from left to right. The larger the absolute value of the slope, the steeper the slant. **3.** Find the slope–intercept form of the equation:

$$4x + 5y = 12$$
$$5y = -4x + 12$$
$$y = -\frac{4}{5}x + \frac{12}{5}.$$

This form of the equation indicates that the line has a negative slope and thus should slant down from left to right. The student may have graphed $y = \frac{4}{5}x + \frac{12}{5}$. **4.** For $R(t) = 50t + 35$, $m = 50$ and $b = 35$; 50 signifies that the cost per hour of a repair is $50; 35 signifies that the minimum cost of a repair job is $35.

5. $m = \dfrac{\text{change in } y}{\text{change in } x}$

As we move from one point to another on a vertical line, the y-coordinate changes but the x-coordinate does not. Thus the change in x is 0. Since division by 0 is not defined, the slope of a vertical line is not defined. As we move from one point to another on a horizontal line, the y-coordinate does not change but the x-coordinate does. Thus the change in y is 0 whereas the change in x is a nonzero number, so the slope is 0. **6.** Using algebra, we find that the slope–intercept form of the equation is $y = \frac{5}{2}x - \frac{3}{2}$. This indicates that the y-intercept is $(0, -\frac{3}{2})$, so a mistake has been made. It appears that the student graphed $y = \frac{5}{2}x + \frac{3}{2}$.

Test: Chapter 2, p. 196

1. [2.2a] Yes **2.** [2.2a] No **3.** [2.2b] $-4; 2$ **4.** [2.2b] 7; 8 **5.** [2.2b] $-6; -6$ **6.** [2.2b] 3; 0

7. [2.1c]

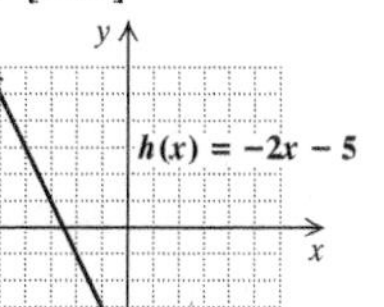

8. [2.2c] **9.** [2.2c]

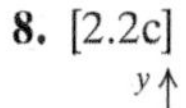
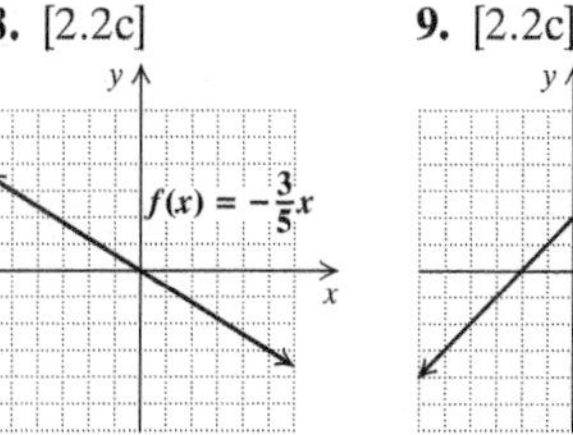

10. [2.2c]

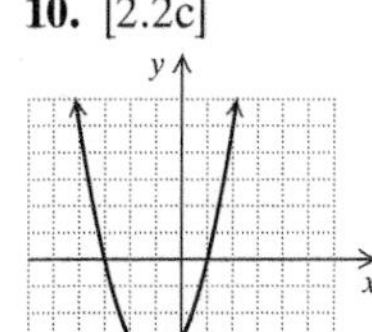

11. [2.5c] **12.** [2.5c]

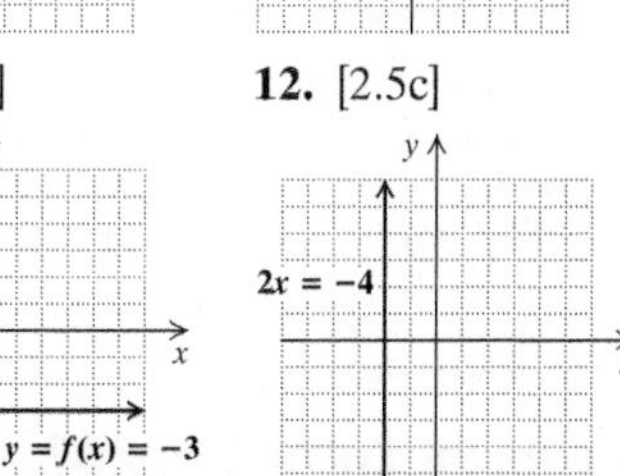

13. [2.2e] **(a)** About 10.3 years; **(b)** 2016 **14.** [2.2d] Yes **15.** [2.2d] No **16.** [2.3a] $\{x \mid x$ is a real number *and* $x \neq -\frac{3}{2}\}$, or $(-\infty, -\frac{3}{2}) \cup (-\frac{3}{2}, \infty)$ **17.** [2.3a] All real numbers **18.** [2.3a] **(a)** 1; **(b)** $[-3, 4]$; **(c)** -3; **(d)** $[-1, 2]$ **19.** [2.4b] Slope: $-\frac{3}{5}$; y-intercept: $(0, 12)$ **20.** [2.4b] Slope: $-\frac{2}{5}$; y-intercept: $(0, -\frac{7}{5})$ **21.** [2.4b] $\frac{5}{8}$ **22.** [2.4b] 0 **23.** [2.4c] $\frac{4}{5}$ km/min

24. [2.5a]

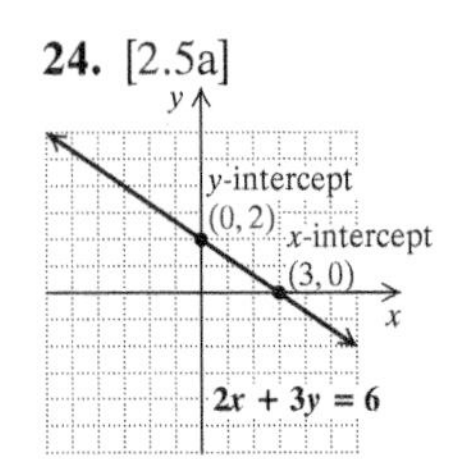

25. [2.5b]

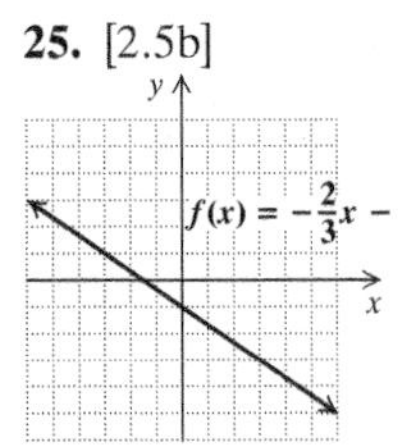

26. [2.5d] Parallel **27.** [2.5d] Perpendicular **28.** [2.6a] $y = -3x + 4.8$ **29.** [2.6a] $f(x) = 5.2x - \frac{5}{8}$ **30.** [2.6b] $y = -4x + 2$ **31.** [2.6c] $y = -\frac{3}{2}x$ **32.** [2.6d] $y = \frac{1}{2}x - 3$ **33.** [2.6d] $y = 3x - 1$ **34.** [2.6e] **(a)** $A(x) = 0.133x + 23.2$; **(b)** 28.25 years; 29.72 years **35.** [2.6b] B **36.** [2.5d] $\frac{24}{5}$ **37.** [2.2b] $f(x) = 3$; answers may vary

Cumulative Review: Chapters 1–2, p. 199

1. [2.6e] **(a)** $R(x) = -0.006x + 3.85$; **(b)** 3.50 min; 3.49 min **2.** [2.3a] **(a)** 6; **(b)** $[0, 30]$; **(c)** 25; **(d)** $[0, 15]$ **3.** [1.1b] -22 **4.** [1.1d] $\frac{15}{88}$ **5.** [1.1c] 20 **6.** [1.1d] $-\frac{21}{4}$ **7.** [1.1d] -5 **8.** [1.1d] No solution **9.** [1.2a] $x = \dfrac{W - By}{A}$ **10.** [1.2a] $A = \dfrac{M}{1 + 4B}$ **11.** [1.4c] $\{y \mid y \le 7\}$, or $(-\infty, 7]$ **12.** [1.4c] $\{x \mid x < -\frac{3}{2}\}$, or $(-\infty, -\frac{3}{2})$ **13.** [1.4c] $\{x \mid x > -\frac{1}{11}\}$, or $(-\frac{1}{11}, \infty)$ **14.** [1.5b] All real numbers **15.** [1.5a] $\{x \mid -7 < x \le 4\}$, or $(-7, 4]$ **16.** [1.5a] $\{x \mid -2 \le x \le \frac{3}{2}\}$, or $[-2, \frac{3}{2}]$ **17.** [1.6c] $\{-8, 8\}$ **18.** [1.6e] $\{y \mid y < -4 \text{ or } y > 4\}$, or $(-\infty, -4) \cup (4, \infty)$ **19.** [1.6e] $\{x \mid -\frac{3}{2} \le x \le 2\}$, or $[-\frac{3}{2}, 2]$ **20.** [2.6d] $y = -4x - 22$ **21.** [2.6d] $y = \frac{1}{4}x - 5$

22. [2.1c]

23. [2.5a]

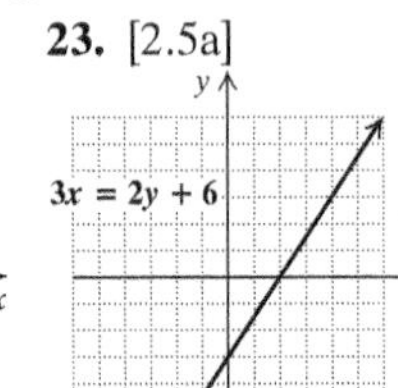

24. [2.5c]

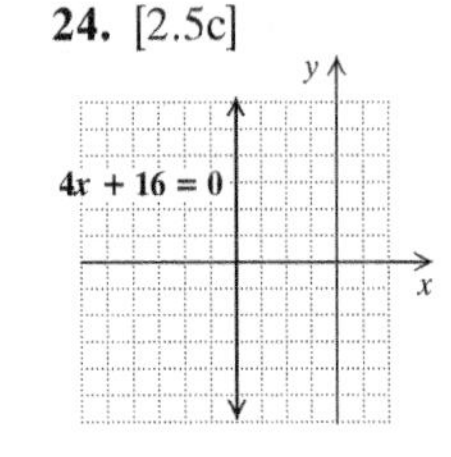

25. [2.5c]

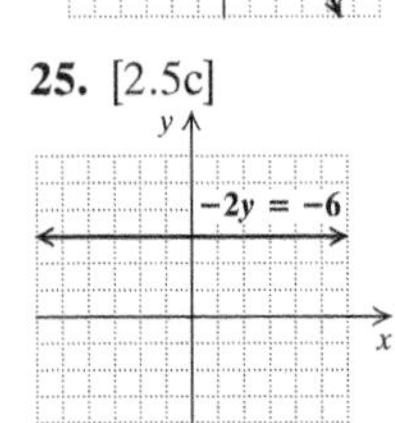

26. [2.2c]

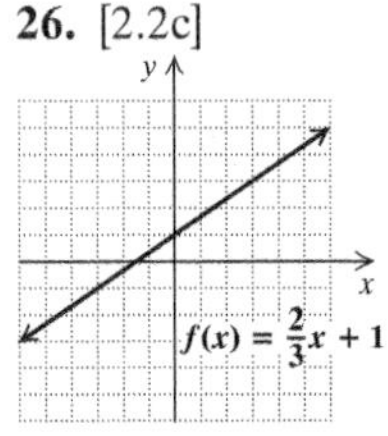

27. [2.2c]

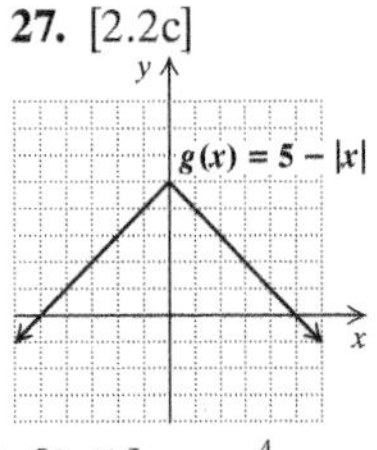

28. [2.4b] Slope: $\frac{9}{4}$; y-intercept: $(0, -3)$ **29.** [2.4b] $m = \frac{4}{3}$ **30.** [2.6b] $y = -3x - 5$ **31.** [2.6c] $y = -\frac{1}{10}x + \frac{12}{5}$ **32.** [1.3a] $w = 17$ m, $l = 23$ m **33.** [1.3a] \$22,500 **34.** [2.5d] (1), (4) **35.** [2.6e] \$151,000 **36.** [1.5a] $\{x \mid 6 < x \le 10\}$, or $(6, 10]$

CHAPTER 3

Calculator Corner, p. 207

1. $(2, 3)$ **2.** $(-4, -1)$ **3.** $(-1, 5)$ **4.** $(3, -1)$

Exercise Set 3.1, p. 207

RC1. False **RC2.** True **RC3.** True **RC4.** True **CC1.** Yes **CC2.** Yes **CC3.** No **1.** $(3, 1)$; consistent; independent **3.** $(1, -2)$; consistent; independent **5.** $(4, -2)$; consistent; independent **7.** $(2, 1)$; consistent; independent **9.** $(\frac{5}{2}, -2)$; consistent; independent **11.** $(3, -2)$; consistent; independent **13.** No solution; inconsistent; independent **15.** Infinitely many solutions; consistent; dependent **17.** $(4, -5)$; consistent; independent **19.** $(2, -3)$; consistent; independent **21.** Consistent; independent; F **23.** Consistent; dependent; B **25.** Inconsistent; independent; D **27.** -3 **28.** -20 **29.** $\frac{9}{20}$ **30.** -38 **31.** $(2.23, 1.14)$ **33.** $(3, 3), (-5, 5)$ **35.** 1

Exercise Set 3.2, p. 214

RC1. True **RC2.** True **RC3.** False **RC4.** False **CC1.** $b = 4$; $(8, 4)$ **CC2.** $x = 19$; $(19, \frac{47}{2})$ **CC3.** $w = -\frac{1}{2}$; $(-\frac{19}{2}, -\frac{1}{2})$ **1.** $(2, -3)$ **3.** $(\frac{21}{5}, \frac{12}{5})$ **5.** $(2, -2)$ **7.** $(-2, -6)$ **9.** $(-2, 1)$ **11.** No solution **13.** $(\frac{19}{8}, \frac{1}{8})$ **15.** Infinitely many solutions **17.** $(\frac{1}{2}, \frac{1}{2})$ **19.** Length: 25 m; width: 5 m **21.** 48° and 132° **23.** Wins: 23; ties: 14 **25.** 1.3 **26.** $-15y - 39$ **27.** $p = \dfrac{7A}{q}$ **28.** $\frac{7}{3}$ **29.** -23 **30.** $\frac{29}{22}$ **31.** $m = -\frac{1}{2}$; $b = \frac{5}{2}$ **33.** Length: 57.6 in.; width: 20.4 in.

Exercise Set 3.3, p. 223

RC1. Consistent **RC2.** Inconsistent **RC3.** Consistent **RC4.** Dependent **RC5.** Inconsistent **RC6.** Independent **CC1.** -3 **CC2.** 2 **CC3.** -1 **1.** $(1, 2)$ **3.** $(-1, 3)$ **5.** $(-1, -2)$ **7.** $(5, 2)$ **9.** Infinitely many solutions **11.** $(\frac{1}{2}, -\frac{1}{2})$ **13.** $(4, 6)$ **15.** No solution **17.** $(10, -8)$ **19.** $(12, 15)$ **21.** $(10, 8)$ **23.** $(-4, 6)$ **25.** $(10, -5)$ **27.** $(140, 60)$ **29.** 36 and 27 **31.** 18 and -15 **33.** 48° and 42° **35.** Two-point shots: 21; three-point shots: 6 **37.** 3-credit courses: 25; 4-credit courses: 8 **39.** 1 **40.** 5 **41.** 15 **42.** $12a^2 - 2a + 1$ **43.** $\{x \mid x \text{ is a real number } and\ x \ne -7\}$, or $(-\infty, -7) \cup (-7, \infty)$ **44.** Domain: all real numbers; range: $\{y \mid y \le 5\}$, or $(-\infty, 5]$ **45.** $y = -\frac{3}{5}x - 7$ **46.** $y = x + 12$ **47.** $(23.12, -12.04)$ **49.** $A = 2, B = 4$ **51.** $p = 2, q = -\frac{1}{3}$

Translating for Success, p. 234

1. G **2.** E **3.** D **4.** A **5.** J **6.** B **7.** C **8.** I **9.** F **10.** H

Exercise Set 3.4, p. 235

RC1. 10 **RC2.** 15 **RC3.** $0.15y$ **RC4.** 2 **1.** Developed: 2322 locations; published: 526 locations **3.** Books: 45; games: 23 **5.** Olive oil: $22\frac{1}{2}$ oz; vinegar: $7\frac{1}{2}$ oz **7.** 5 lb of each **9.** 25%-acid: 4 L; 50%-acid: 6 L **11.** Sweet-pepper packets: 11; hot-pepper packets: 5 **13.** \$7500 at 6%; \$4500 at 3% **15.** Whole milk: $169\frac{3}{13}$ lb; cream: $30\frac{10}{13}$ lb **17.** \$1800 at 5.5%; \$1400 at 4% **19.** \$5 bills: 7; \$1 bills: 15 **21.** 30-sec commercials: 45; 1-min commercials: 17 **23.** 375 mi **25.** 14 km/h **27.** Headwind: 30 mph; plane: 120 mph **29.** $1\frac{1}{3}$ hr **31.** About 1489 mi **33.** $\{6, 8, 10\}$ **34.** $\{2, 4, 6, 7, 8, 9, 10\}$ **35.** $3|a|$ **36.** $7x^2$ **37.** $\dfrac{3}{|y|}$ **38.** $\dfrac{a^4}{|c|}$ **39.** $4\frac{4}{7}$ L **41.** City: 261 mi; highway: 204 mi

Mid-Chapter Review: Chapter 3, p. 239

1. False **2.** False **3.** True **4.** True

5.
$$\begin{aligned} x + 2(x - 6) &= 3 \\ x + 2x - 12 &= 3 \\ 3x - 12 &= 3 \\ 3x &= 15 \\ x &= 5 \end{aligned}$$
$$y = 5 - 6$$
$$y = -1$$
The solution is $(5, -1)$.

6. $\begin{aligned} 6x - 4y &= 10 \\ 2x + 4y &= 14 \\ \hline 8x \quad &= 24 \\ x &= 3 \\ 2 \cdot 3 + 4y &= 14 \\ 6 + 4y &= 14 \\ 4y &= 8 \\ y &= 2 \end{aligned}$

The solution is $(3, 2)$.
7. $(5, -1)$; consistent; independent **8.** $(0, 3)$; consistent; independent **9.** Infinitely many solutions; consistent; dependent **10.** No solution; inconsistent; independent **11.** $(8, 6)$ **12.** $(2, -3)$ **13.** $(-3, 5)$ **14.** $(-1, -2)$ **15.** $(2, -2)$ **16.** $(5, -4)$ **17.** $(-1, -2)$ **18.** $(3, 1)$ **19.** No solution **20.** Infinitely many solutions **21.** $(10, -12)$ **22.** $(-9, 8)$ **23.** Length: 12 ft; width: 10 ft **24.** \$2100 at 2%; \$2900 at 3% **25.** 20% acid: 56 L; 50% acid: 28 L **26.** 26 mph
27. *Graphically*: **1.** Graph $y = \frac{3}{4}x + 2$ and $y = \frac{2}{5}x - 5$ and find the point of intersection. The first coordinate of this point is the solution of the original equation. **2.** Rewrite the equation as $\frac{7}{20}x + 7 = 0$. Then graph $y = \frac{7}{20}x + 7$ and find the x-intercept. The first coordinate of this point is the solution of the original equation. *Algebraically*: **1.** Use the addition and multiplication principles for equations. **2.** Multiply by 20 to clear the fractions and then use the addition and multiplication principles for equations.
28. (a) Answers may vary.
$x + y = 1,$
$x - y = 7$
(b) Answers may vary.
$x + 2y = 5,$
$3x + 6y = 10$
(c) Answers may vary.
$x - 2y = 3,$
$3x - 6y = 9$
29. Answers may vary. Form a linear expression in two variables and set it equal to two different constants. See Exercises 10 and 19 in this review for examples. **30.** Answers may vary. Let any linear equation be one equation in the system. Multiply by a constant on both sides of that equation to get the second equation in the system. See Exercises 9 and 20 in this review for examples.

Exercise Set 3.5, p. 245

RC1. (b) **RC2.** (c) **RC3.** (a) **RC4.** (a) **CC1.** Yes **CC2.** No **CC3.** Yes
1. $(1, 2, -1)$ **3.** $(2, 0, 1)$ **5.** $(3, 1, 2)$ **7.** $(-3, -4, 2)$ **9.** $(2, 4, 1)$ **11.** $(-3, 0, 4)$ **13.** $(2, 2, 4)$ **15.** $\left(\frac{1}{2}, 4, -6\right)$ **17.** $(-2, 3, -1)$ **19.** $\left(\frac{1}{2}, \frac{1}{3}, \frac{1}{6}\right)$ **21.** $(3, -5, 8)$ **23.** $(15, 33, 9)$ **25.** $(4, 1, -2)$ **27.** $(17, 9, 79)$ **28.** $a = \frac{F}{3b}$ **29.** $a = \frac{Q - 4b}{4}$, or $\frac{Q}{4} - b$ **30.** $d = \frac{tc - 2F}{t}$, or $c - \frac{2F}{t}$ **31.** $c = \frac{2F + td}{t}$, or $\frac{2F}{t} + d$ **32.** $y = \frac{c - Ax}{B}$ **33.** $y = \frac{Ax - c}{B}$ **34.** Slope: $-\frac{2}{3}$; y-intercept: $\left(0, -\frac{5}{4}\right)$ **35.** Slope: -4; y-intercept: $(0, 5)$ **36.** Slope: $\frac{2}{5}$; y-intercept: $(0, -2)$ **37.** Slope: 1.09375; y-intercept: $(0, -3.125)$ **39.** $(1, -2, 4, -1)$

Exercise Set 3.6, p. 251

RC1. (c) **RC2.** (d) **RC3.** (a) **RC4.** (b)
1. Chinese: 13,000 exams; French: 22,000 exams; German: 5000 exams **3.** 32°, 96°, 52° **5.** $-7, 20, 42$ **7.** Small: 5 dozen; medium: 6 dozen; large: 11 dozen **9.** Egg: 274 mg; cupcake: 19 mg; pizza: 9 mg **11.** Automatic transmission: \$865; power door locks: \$520; air conditioning: \$375 **13.** Birds: 20 million; cats: 94 million; dogs: 90 million **15.** Roast beef: 2; baked potato: 1; broccoli: 2 **17.** First fund: \$45,000; second fund: \$10,000; third fund: \$25,000 **19.** Par-3: 6 holes; par-4: 8 holes; par-5: 4 holes **21.** A: 1500 lenses; B: 1900 lenses; C: 2300 lenses

23. $f(x) = 2x - 3$

24.

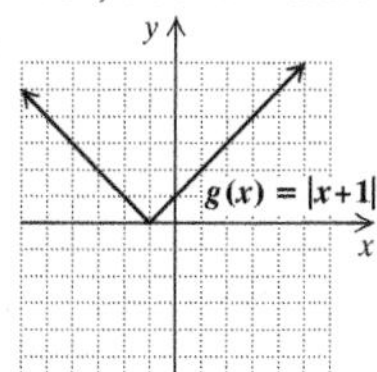

25. $h(x) = x^2 - 2$

26. No **27.** Yes **28.** Yes **29.** 180° **31.** 464 **33.** E: 701 games; E10+: 298 games; T: 313 games; M: 164 games

Visualizing for Success, p. 263

1. D **2.** B **3.** E **4.** C **5.** I **6.** G **7.** F **8.** H **9.** A **10.** J

Exercise Set 3.7, p. 264

RC1. Graph **RC2.** Inequality **RC3.** Half-plane **RC4.** Solution **RC5.** Equation **RC6.** Test **CC1.** Dashed **CC2.** Solid **CC3.** Solid **CC4.** Yes **CC5.** No **CC6.** Yes
1. Yes **3.** Yes

5.

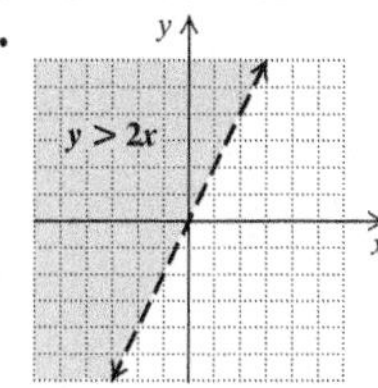

7.

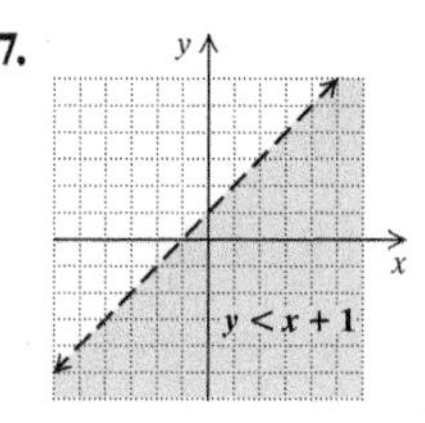

9.

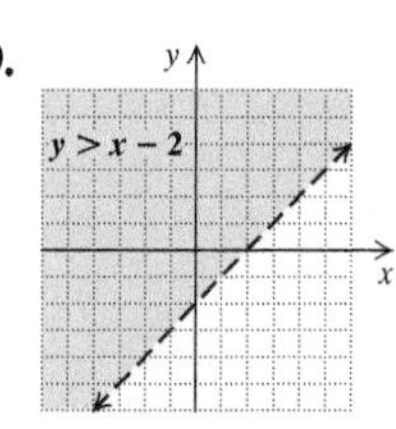

11.

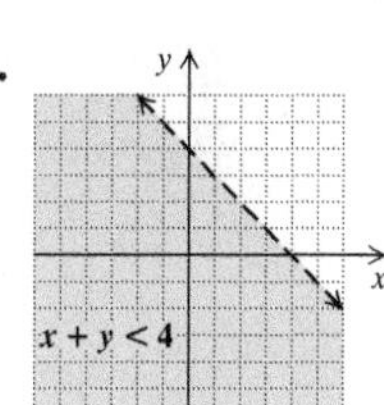

13.

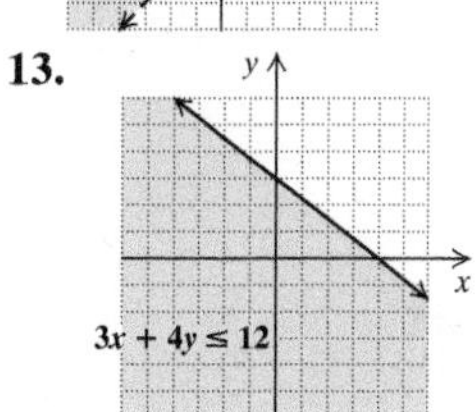

15.

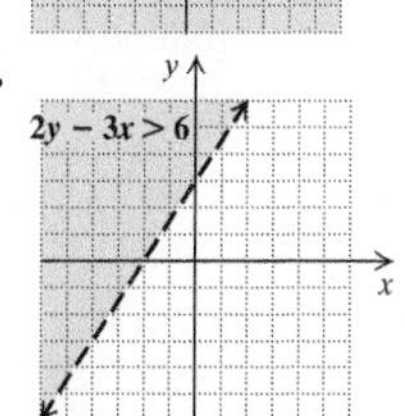

17.

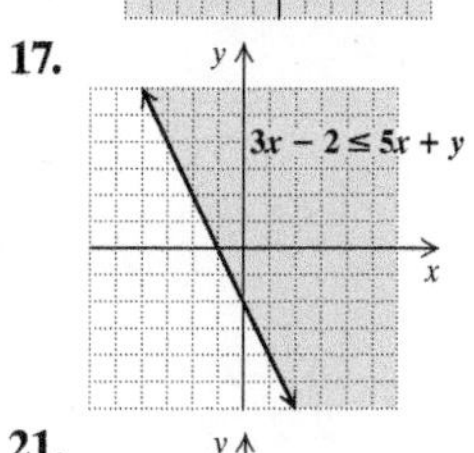

19.

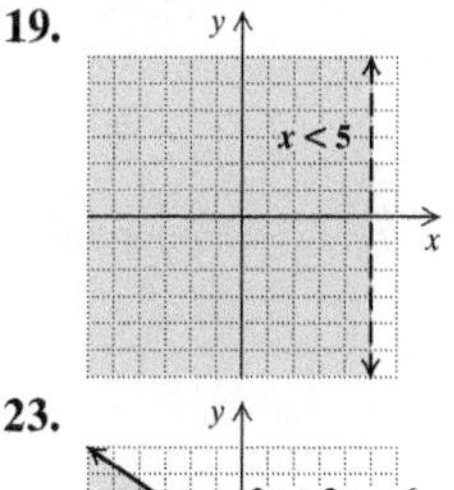

21.

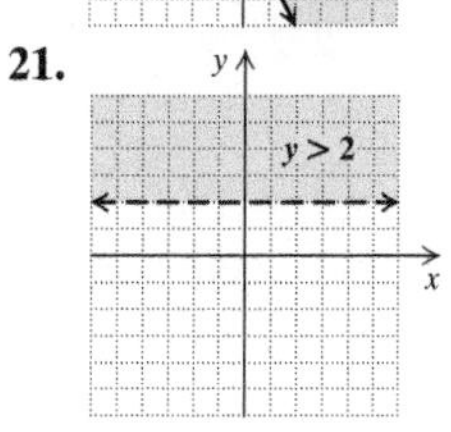

23.

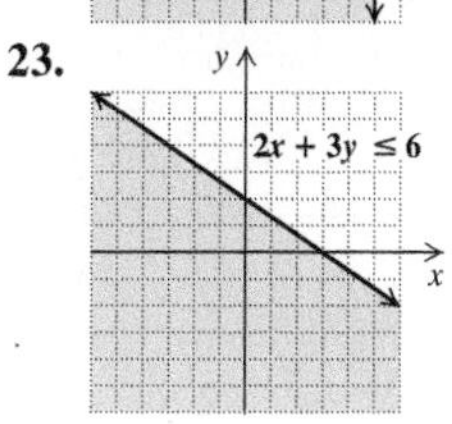

25. F **27.** B **29.** C

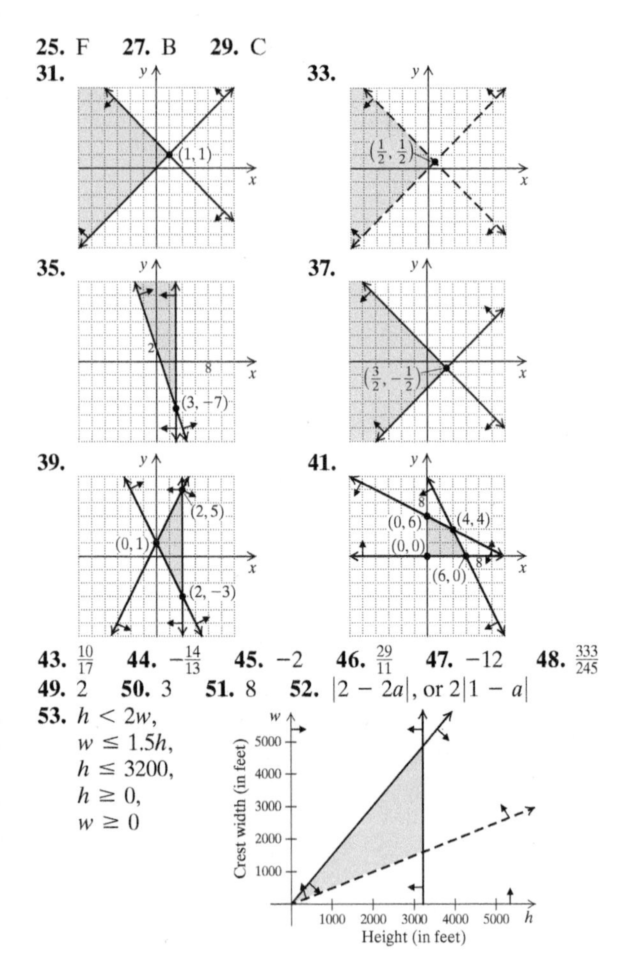

43. $\frac{10}{17}$ **44.** $-\frac{14}{13}$ **45.** -2 **46.** $\frac{29}{11}$ **47.** -12 **48.** $\frac{333}{245}$
49. 2 **50.** 3 **51.** 8 **52.** $|2 - 2a|$, or $2|1 - a|$
53. $h < 2w$,
$w \leq 1.5h$,
$h \leq 3200$,
$h \geq 0$,
$w \geq 0$

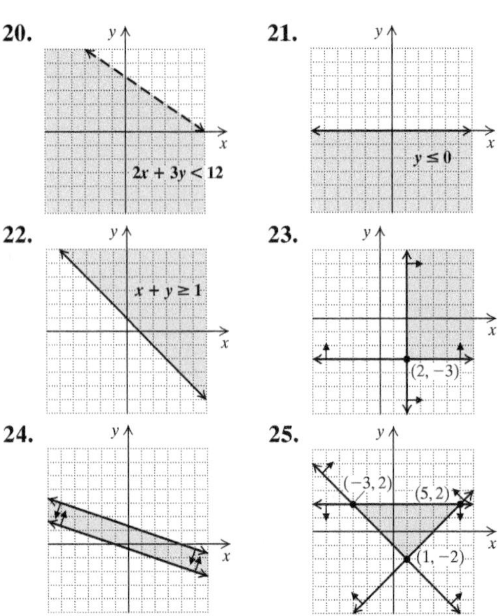

26. C **27.** A **28.** (0, 2) and (1, 3)

Understanding Through Discussion and Writing

1. Answers may vary. One day, a florist sold a total of 23 hanging baskets and flats of petunias. Hanging baskets cost \$10.95 each and flats of petunias cost \$12.95 each. The sales totaled \$269.85. How many of each were sold? **2.** We know that machines A, B, and C can polish 5700 lenses in one week when working together. We also know that A and B together can polish 3400 lenses in one week, so C can polish 5700 − 3400, or 2300, lenses in one week alone. We also know that B and C together can polish 4200 lenses in one week, so A can polish 5700 − 4200, or 1500, lenses in one week alone. Also, B can polish 4200 − 2300, or 1900, lenses in one week alone. **3.** Let x = the number of adults in the audience, y = the number of senior citizens, and z = the number of children. The total attendance is 100, so we have equation (1), $x + y + z = 100$. The amount taken in was \$100, so equation (2) is $10x + 3y + 0.5z = 100$. There is no other information that can be translated to an equation. Clearing decimals in equation (2) and then eliminating z gives us equation (3), $95x + 25y = 500$. Dividing by 5 on both sides, we have equation (4), $19x + 5y = 100$. Since we have only two equations, it is not possible to eliminate z from another pair of equations. However, in $19x + 5y = 100$, note that 5 is a factor of both $5y$ and 100. Therefore, 5 must also be a factor of $19x$, and hence of x, since 5 is not a factor of 19. Then for some positive integer n, $x = 5n$. (We require n to be positive, since the number of adults clearly cannot be negative and must also be nonzero since the exercise states that the audience consists of adults, senior citizens, and children.) We have

$$19 \cdot 5n + 5y = 100$$
$$19n + y = 20. \quad \text{Dividing by 5}$$

Since n and y must both be positive, $n = 1$. (If $n > 1$, then $19n + y > 20$.) Then $x = 5 \cdot 1$, or 5.

$$19 \cdot 5 + 5y = 100 \quad \text{Substituting in (4)}$$
$$y = 1$$
$$5 + 1 + z = 100 \quad \text{Substituting in (1)}$$
$$z = 94$$

There were 5 adults, 1 senior citizen, and 94 children in the audience. **4.** No; the symbol $\geq$ does not always yield a graph in which the half-plane above the line is shaded. For the inequality $-y \geq 3$, for example, the half-plane below the line $y = -3$ is shaded.

Summary and Review: Chapter 3, p. 268

Vocabulary Reinforcement

1. Pair **2.** Consistent **3.** Algebraic **4.** Triple
5. Independent **6.** Half-plane

Concept Reinforcement

1. False **2.** True **3.** True **4.** False

Study Guide

1. $(4, -1)$; consistent; independent **2.** $(-1, 4)$ **3.** $(-2, 3)$
4. \$8700 at 6%; \$14,300 at 5% **5.** $(3, -5, 1)$
6.

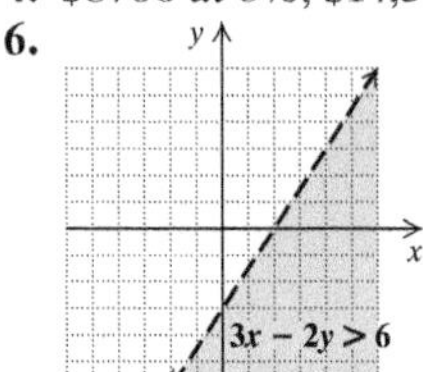

7.

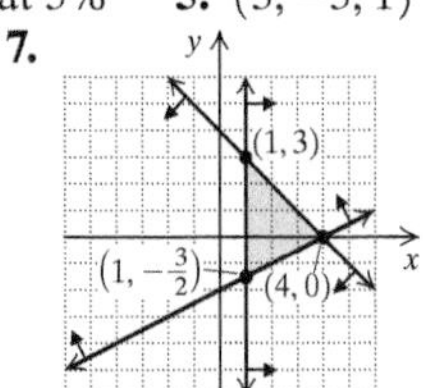

Review Exercises

1. $(-2, 1)$; consistent; independent
2. Infinitely many solutions; consistent; dependent
3. No solution; inconsistent; independent **4.** $(1, -1)$
5. No solution **6.** $\left(\frac{2}{5}, -\frac{4}{5}\right)$ **7.** $(6, -3)$ **8.** $(2, 2)$
9. $(5, -3)$ **10.** Infinitely many solutions
11. 32 brushes at \$8.50; 13 brushes at \$9.75
12. 5 L of each **13.** $5\frac{1}{2}$ hr **14.** $(10, 4, -8)$ **15.** $(-1, 3, -2)$
16. $(2, 0, 4)$ **17.** $\left(2, \frac{1}{3}, -\frac{2}{3}\right)$ **18.** $90°, 67\frac{1}{2}°, 22\frac{1}{2}°$ **19.** Caramel nut crunch: \$30; plain: \$5; mocha choco latte: \$14

Test: Chapter 3, p. 275

1. [3.1a] $(-2, 1)$; consistent; independent **2.** [3.1a] No solution; inconsistent; independent **3.** [3.1a] Infinitely many solutions; consistent; dependent **4.** [3.2a] $(2, -3)$ **5.** [3.2a] Infinitely many solutions **6.** [3.2a] $(-4, 5)$ **7.** [3.3a] $(-1, 1)$ **8.** [3.3a] $\left(-\frac{3}{2}, -\frac{1}{2}\right)$ **9.** [3.3a] No solution **10.** [3.2b] Length: 93 ft; width: 51 ft **11.** [3.4b] 120 km/h **12.** [3.3b], [3.4a] Buckets: 17; dinners: 11 **13.** [3.4a] 20% solution: 12 L; 45% solution: 8 L **14.** [3.5a] $\left(2, -\frac{1}{2}, -1\right)$ **15.** [3.6a] 3.5 hr

16. [3.7b]

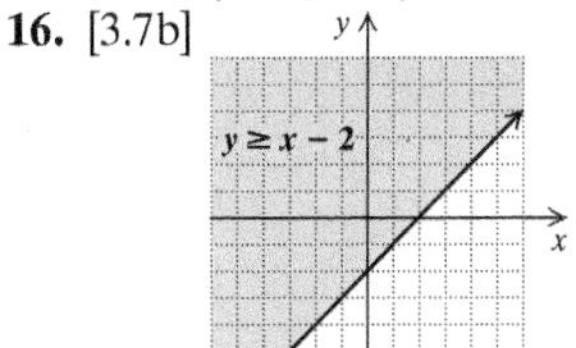

17. [3.7b]

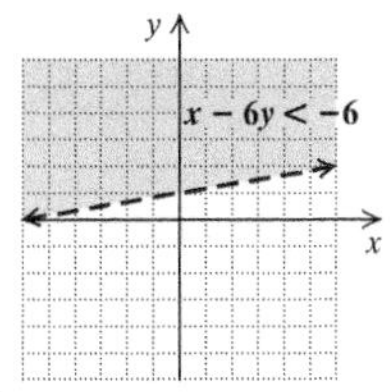

18. [3.7c]

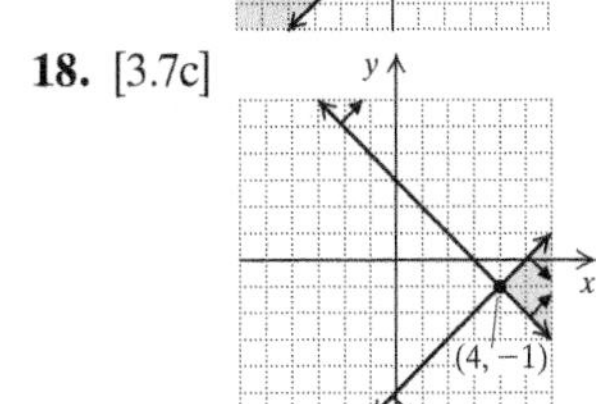

19. [3.7c]

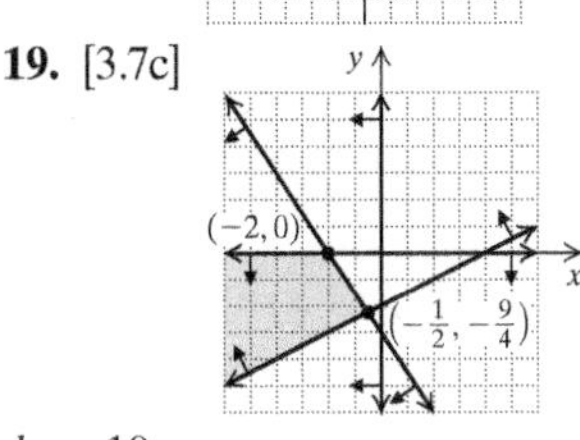

20. [3.6a] B **21.** [3.3a] $m = 7$; $b = 10$

Cumulative Review: Chapters 1–3, p. 277

1. [1.1d] $\frac{10}{9}$ **2.** [1.1d] 6 **3.** [1.2a] $h = \dfrac{A}{\pi r^2}$

4. [1.2a] $p = \dfrac{3L}{m} - k$, or $\dfrac{3L - km}{m}$ **5.** [1.4c] $\{x | x > -1\}$, or $(-1, \infty)$ **6.** [1.5a] $\{x | \frac{1}{3} < x \leq \frac{13}{3}\}$, or $\left(\frac{1}{3}, \frac{13}{3}\right]$

7. [1.5b] $\{x | x \leq 3 \text{ or } x \geq 7\}$, or $(-\infty, 3] \cup [7, \infty)$

8. [1.6c] $\{-5, 3\}$ **9.** [1.6e] $\{y | y \leq -\frac{3}{2} \text{ or } y \geq \frac{9}{4}\}$, or $\left(-\infty, -\frac{3}{2}\right] \cup \left[\frac{9}{4}, \infty\right)$ **10.** [1.6d] $\{-5, 1\}$ **11.** [1.6b] 11

12. [2.5c]

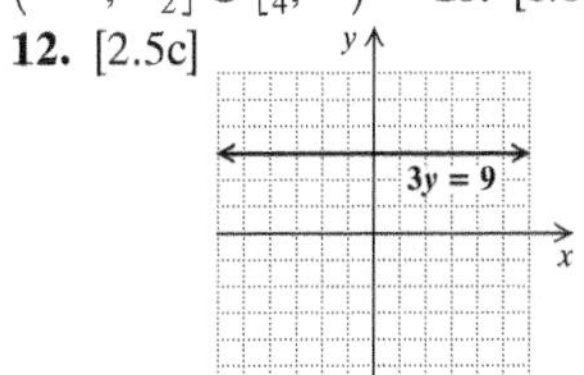

13. [2.2c]

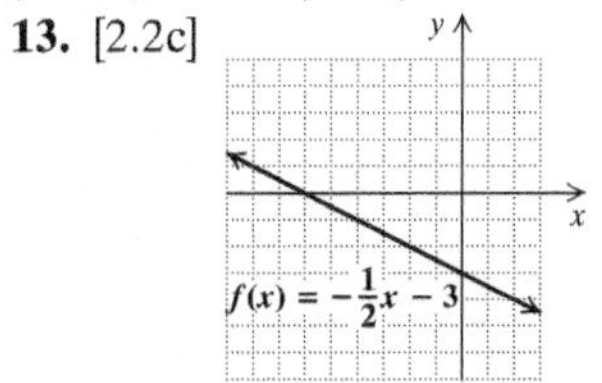

14. [2.5b]

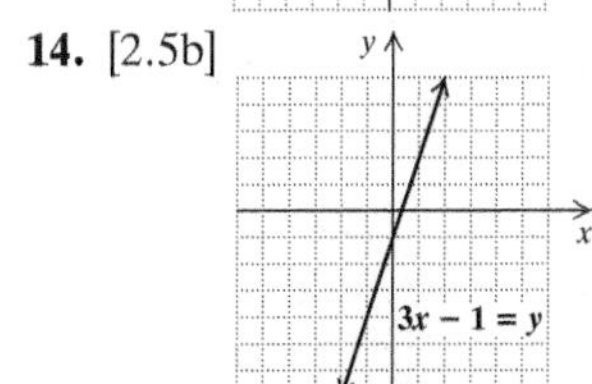

15. [2.5a]

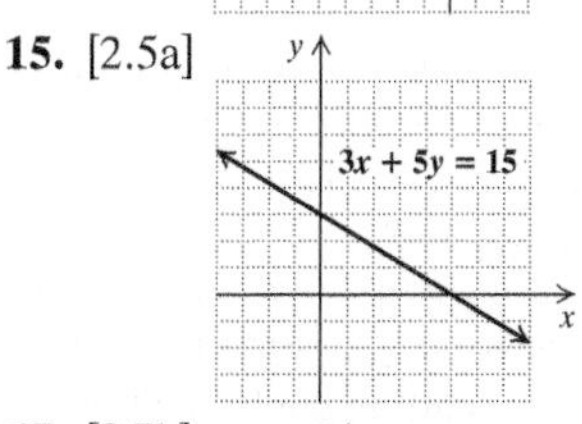

16. [3.7b]

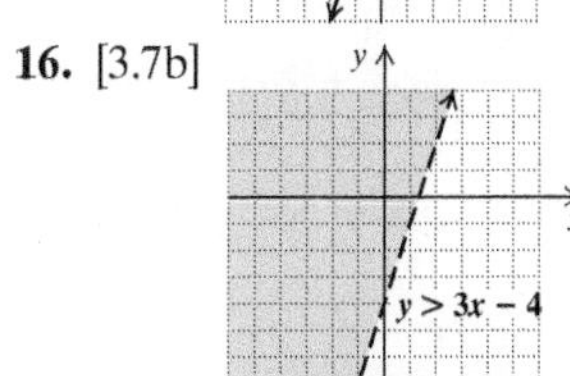

17. [3.7b]

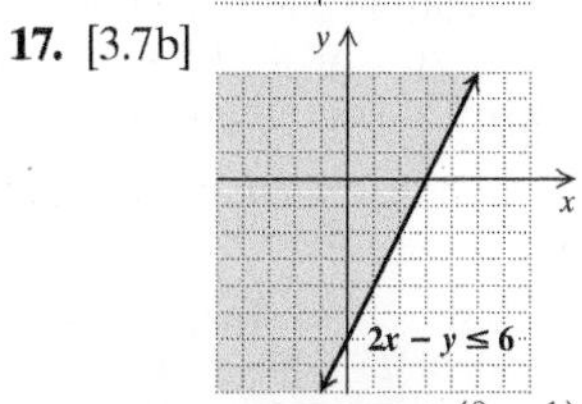

18. [3.1a] $(3, -1)$; consistent; independent **19.** [3.2a] $\left(\frac{8}{5}, -\frac{1}{5}\right)$ **20.** [3.3a] $(1, -1)$ **21.** [3.3a] $(-1, 3)$ **22.** [3.5a] $(2, 0, -1)$

23. [3.7c]

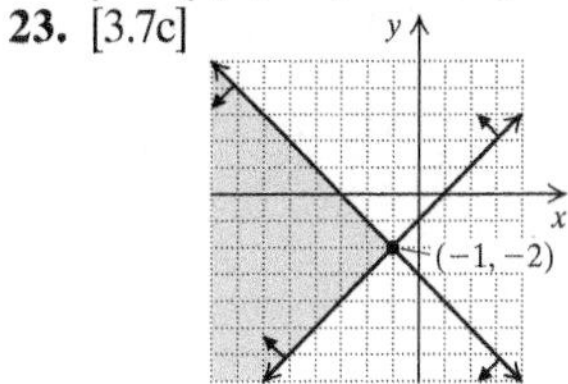

24. [3.7c]

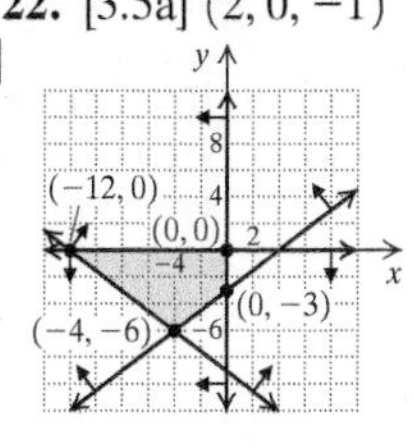

25. [2.3a] **(a)** $\{-5, -3, -1, 1, 3\}$; **(b)** $\{-3, -2, 1, 4, 5\}$; **(c)** -2; **(d)** 3 **26.** [2.3a] $\{x | x \text{ is a real number } and\ x \neq \frac{1}{2}\}$, or $\left(-\infty, \frac{1}{2}\right) \cup \left(\frac{1}{2}, \infty\right)$ **27.** [2.2b] -1; 1; -17 **28.** [2.4b] Slope: $\frac{4}{5}$; y-intercept: $(0, 4)$ **29.** [2.6b] $y = -3x + 17$ **30.** [2.6c] $y = -4x - 7$ **31.** [2.5d] Perpendicular **32.** [2.6d] $y = \frac{1}{3}x + 4$ **33.** [1.3a] 4 m; 6 m **34.** [1.4d] $\{S | S \geq 88\}$ **35.** [3.4a] Scientific: 18; graphing: 27 **36.** [3.4a] 15%: 21 L; 25%: 9 L **37.** [3.4b] 720 km **38.** [3.6a] \$120 **39.** [2.6e] \$151,000 **40.** [3.3a] $m = -\frac{5}{9}$; $b = -\frac{2}{9}$

CHAPTER 4

Exercise Set 4.1, p. 286

RC1. (c) **RC2.** (d) **RC3.** (a) **RC4.** (b) **RC5.** (h) **RC6.** (f) **RC7.** (g) **RC8.** (e)

CC1. $4x$ and $-3x$; $7x^2$ and $-x^2$; $2x^3$ and $5x^3$
CC2. $2x^2y$ and $7x^2y$; $-3xy^2$ and $4xy^2$
CC3. $2a^2$, a^2, and $8a^2$; -7 and 3 **CC4.** $-5x$ and $8x$

1. $-9x^4, -x^3, 7x^2, 6x, -8$; 4, 3, 2, 1, 0; 4; $-9x^4$; -9; -8
3. $t^3, 4t^7, s^2t^4, -2$; 3, 7, 6, 0; 7; $4t^7$; 4; -2
5. $u^7, 8u^2v^6, 3uv, 4u, -1$; 7, 8, 2, 1, 0; 8; $8u^2v^6$; 8; -1
7. $-4y^3 - 6y^2 + 7y + 23$ **9.** $-xy^3 + x^2y^2 + x^3y + 1$
11. $-9b^5y^5 - 8b^2y^3 + 2by$ **13.** $5 + 12x - 4x^3 + 8x^5$
15. $3xy^3 + x^2y^2 - 9x^3y + 2x^4$ **17.** $-7ab + 4ax - 7ax^2 + 4x^6$
19. 45; 21; 5 **21.** -168; -9; 4; $-7\frac{7}{8}$ **23.** About 288 watts
25. **(a)** About 12.7 million citizens; **(b)** about 11.25 million citizens; **(c)** $f(7) \approx 11.9$ million; **(d)** $f(4) \approx 12.8$ million **27.** **(a)** \$10,750; **(b)** \$18,287.50 **29.** $P(x) = -x^2 + 280x - 7000$
31. 17 **33.** $2x^2$ **35.** $3x + 4y$ **37.** $7a + 14$
39. $-6a^2b - 2b^2$ **41.** $9x^2 + 2xy + 15y^2$
43. $-x^2y + 4y + 9xy^2$ **45.** $5x^2 + 2y^2 + 5$ **47.** $6a + b + c$
49. $-4a^2 - b^2 + 3c^2$ **51.** $-3x^2 + 2x + xy - 1$
53. $5x^2y - 4xy^2 + 5xy$ **55.** $9x^2 - y^2 + 7xy - 9y$
57. $9r^2 + 9r - 9$ **59.** $-\frac{2}{15}xy + \frac{19}{12}xy^2 + 1.7x^2y$
61. $-(5x^3 - 7x^2 + 3x - 6)$; $-5x^3 + 7x^2 - 3x + 6$
63. $-(-13y^2 + 6ay^4 - 5by^2)$; $13y^2 - 6ay^4 + 5by^2$
65. $11x - 7$ **67.** $-4x^2 - 3x + 13$ **69.** $2a + 3c - 4b$
71. $-2x^2 + 6x$ **73.** $-4a^2 + 8ab - 5b^2$
75. $16ab + 8a^2b + 3ab^2$ **77.** $0.06y^4 + 0.032y^3 - 0.94y^2 + 0.93$
79. $x^4 - x^2 - 1$ **81.** $-\frac{1}{2}$ **82.** 3 **83.** No solution
84. $\{y | y \geq 2\}$, or $[2, \infty)$

85.

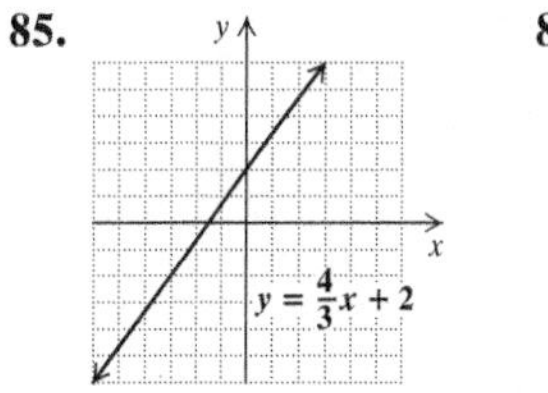

86.

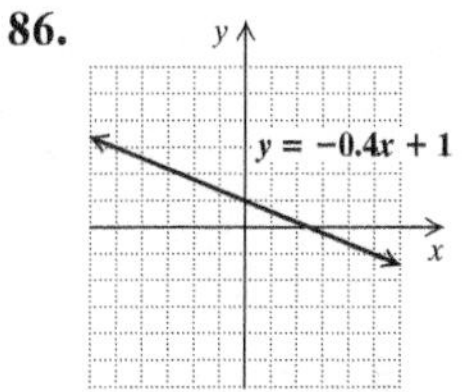

87.

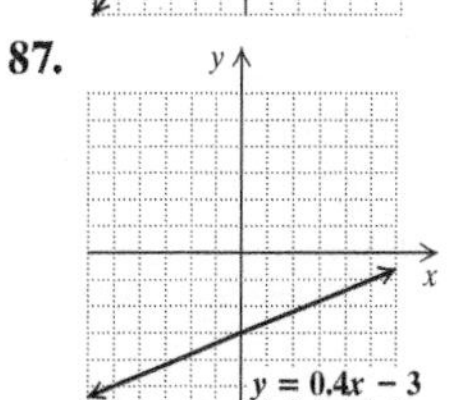

88.

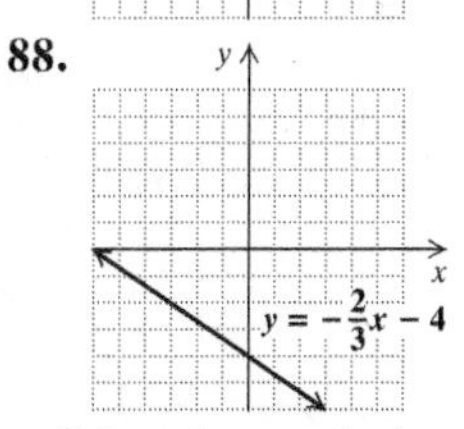

89. 494.55 cm^3 **91.** $f(0) = 5$, but the graph shows that $f(0) = -5$ **93.** $47x^{4a} + 40x^{3a} + 30x^{2a} + x^a + 4$

Calculator Corner, p. 296

1. Correct **2.** Incorrect **3.** Correct **4.** Incorrect **5.** Incorrect **6.** Correct **7.** Correct **8.** Incorrect **9.** Correct **10.** Incorrect

Exercise Set 4.2, p. 298

RC1. True **RC2.** False **RC3.** True **RC4.** True
CC1. (c) **CC2.** (d) **CC3.** (b) **CC4.** (f) **CC5.** (h)
CC6. (g)
1. $24y^3$ **3.** $-20x^3y$ **5.** $-10x^6y^7$ **7.** $14z - 2zx$
9. $6a^2b + 6ab^2$ **11.** $15c^3d^2 - 25c^2d^3$ **13.** $15x^2 + x - 2$
15. $s^2 - 9t^2$ **17.** $x^2 - 2xy + y^2$ **19.** $x^6 + 3x^3 - 40$
21. $a^4 - 5a^2b^2 + 6b^4$ **23.** $x^3 - 64$ **25.** $x^3 + y^3$
27. $a^4 + 5a^3 - 2a^2 - 9a + 5$
29. $4a^3b^2 + 4a^3b - 10a^2b^2 - 2a^2b + 3ab^3 + 7ab^2 - 6b^3$
31. $x^2 + \frac{1}{2}x + \frac{1}{16}$ **33.** $\frac{1}{8}x^2 - \frac{2}{9}$ **35.** $3.25x^2 - 0.9xy - 28y^2$
37. $a^2 + 13a + 40$ **39.** $y^2 + 3y - 28$ **41.** $9a^2 + 3a + \frac{1}{4}$
43. $x^2 - 4xy + 4y^2$ **45.** $b^2 - \frac{5}{6}b + \frac{1}{6}$ **47.** $2x^2 + 13x + 18$
49. $400a^2 - 6.4ab + 0.0256b^2$ **51.** $4x^2 - 4xy - 3y^2$
53. $x^6 + 4x^3 + 4$ **55.** $4x^4 - 12x^2y^2 + 9y^4$
57. $a^6b^4 + 2a^3b^2 + 1$ **59.** $0.01a^4 - a^2b + 25b^2$
61. $A = P + 2Pi + Pi^2$ **63.** $d^2 - 64$ **65.** $4c^2 - 9$
67. $36m^2 - 25n^2$ **69.** $x^4 - y^2z^2$ **71.** $m^4 - m^2n^2$
73. $16p^4 - 9p^2t^2$ **75.** $\frac{1}{4}p^2 - \frac{4}{9}n^2$ **77.** $x^4 - 1$
79. $a^4 - 2a^2b^2 + b^4$ **81.** $a^2 + 2ab + b^2 - 1$
83. $4x^2 + 12xy + 9y^2 - 16$
85. $t^2 + 3t - 4, p^2 + 7p + 6, h^2 + 2ah + 5h,$
$t^2 + t - 6 + c, a^2 + 5a + 5$ **87.** $3t^2 - 13t + 18,$
$3p^2 - p + 4, 3h^2 + 6ah - 7h, 3t^2 - 19t + 34 + c, 3a^2 - 7a + 13$
89. $-t^2 + 7t - 6, -p^2 + 3p + 4, -h^2 - 2ah + 5h,$
$-t^2 + 9t - 14 + c, -a^2 + 5a + 5$
91. $-t^2 + 5t, -p^2 + p + 6, -h^2 - 2ah + 3h,$
$-t^2 + 7t - 6 + c, -a^2 + 3a + 9$
93.

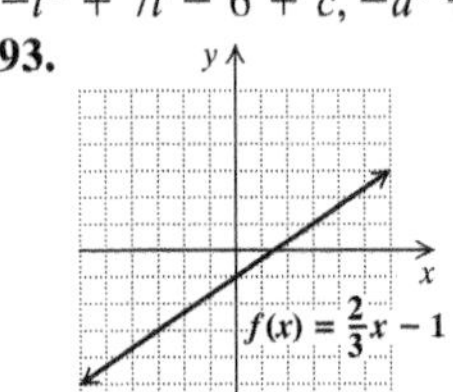

94.

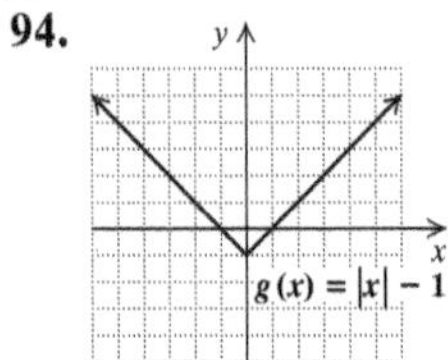

95.

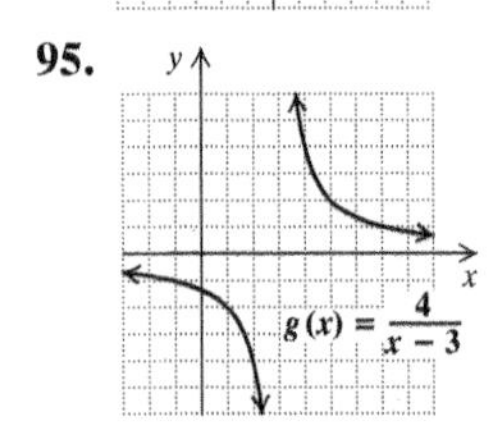

96.

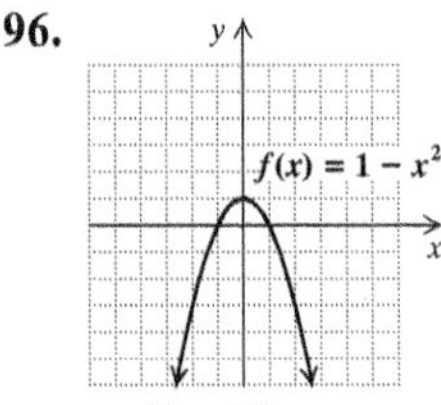

97. 5.5 hr **98.** 180 mph **99.** $\left(\frac{4}{3}, -\frac{14}{27}\right)$ **100.** $(1, 3)$
101. Infinitely many solutions **102.** $\left(\frac{10}{21}, \frac{11}{14}\right)$
103. Left to the student **105.** z^{5n^5} **107.** $r^8 - 2r^4s^4 + s^8$
109. $9x^{10} - \frac{30}{11}x^5 + \frac{25}{121}$ **111.** $x^{4a} - y^{4b}$ **113.** $x^6 - 1$

Exercise Set 4.3, p. 305

RC1. Product **RC2.** Factors **RC3.** Factorization
RC4. Prime **RC5.** Common **RC6.** Binomial
CC1. 4 **CC2.** x^3 **CC3.** $6a^3$ **CC4.** $4a^2bc^4$
1. $3a(2a + 1)$ **3.** $x^2(x + 9)$ **5.** $4x^2(2 - x^2)$
7. $4xy(x - 3y)$ **9.** $3(y^2 - y - 3)$ **11.** $2a(2b - 3c + 6d)$
13. $5(2a^4 + 3a^2 - 5a - 6)$ **15.** $3x^2y^4z^3(5y - 4x^2z^4)$
17. $7a^3b^3c^3(2ac^2 + 3b^2c - 5ab)$ **19.** $-5(x + 9)$
21. $-6(a + 14)$ **23.** $-2(x^2 - x + 12)$ **25.** $-3y(y - 8)$
27. $-(a^4 - 2a^3 + 13a^2 + 1)$ **29.** $-3(y^3 - 4y^2 + 5y - 8)$
31. $\pi r^2(h + \frac{4}{3}r)$, or $\frac{1}{3}\pi r^2(3h + 4r)$
33. **(a)** $h(t) = -8t(2t - 9)$; **(b)** $h(2) = 80$ in each
35. $R(x) = 0.4x(700 + x)$ **37.** $(b - 2)(a + c)$
39. $(x - 2)(2x + 13)$ **41.** $(y - 7)(y^7 + 1)$
43. $(c + d)(a + b)$ **45.** $(b - 1)(b^2 + 2)$
47. $(y + 8)(y^2 - 5)$ **49.** $12(x^2 + 3)(2x - 3)$
51. $a(a^3 - a^2 + a + 1)$ **53.** $(y^2 + 3)(2y^2 - 5)$ **55.** $-7, 13$
56. $-8, -\frac{2}{5}$ **57.** $\{x \mid -10 \le x \le 14\}$, or $[-10, 14]$
58. $\{y \mid y < \frac{1}{3}$ *or* $y > \frac{13}{3}\}$, or $\left(-\infty, \frac{1}{3}\right) \cup \left(\frac{13}{3}, \infty\right)$
59. $\{x \mid 15 \le x \le 17\}$, or $[15, 17]$ **60.** $\{x \mid \frac{1}{3} < x < 1\}$, or $\left(\frac{1}{3}, 1\right)$
61. $\{x \mid x < \frac{1}{3}$ *or* $x > \frac{13}{2}\}$, or $\left(-\infty, \frac{1}{3}\right) \cup \left(\frac{13}{2}, \infty\right)$
62. All real numbers, or $(-\infty, \infty)$
63. $x^5y^4 + x^4y^6 = x^3y(x^2y^3 + xy^5)$ **65.** $(x^2 - x + 5)(r + s)$
67. $(x^4 + x^2 + 5)(a^4 + a^2 + 5)$ **69.** $x^{1/3}(1 - 7x)$
71. $x^{1/3}(1 - 5x^{1/6} + 3x^{5/12})$ **73.** $3a^n(a + 2 - 5a^2)$
75. $y^{a+b}(7y^a - 5 + 3y^b)$

Exercise Set 4.4, p. 311

RC1. (c) **RC2.** (d) **RC3.** (a) **RC4.** (b)
CC1. 1, 36; 2, 18; 3, 12; 4, 9; 6, 6 **CC2.** 1, 18; 2, 9; 3, 6
CC3. 1, 54; 2, 27; 3, 18; 6, 9 **CC4.** 1, 72; 2, 36; 3, 24; 4, 18; 6, 12; 8, 9 **CC5.** 1, 80; 2, 40; 4, 20; 5, 16; 8, 10
CC6. 1, 84; 2, 42; 3, 28; 4, 21; 6, 14; 7, 12
1. $(x + 4)(x + 9)$ **3.** $(t - 5)(t - 3)$
5. $(x - 11)(x + 3)$ **7.** $2(y - 4)(y - 4)$ **9.** $(p + 9)(p - 6)$
11. $(x + 3)(x + 9)$ **13.** $\left(y - \frac{1}{3}\right)\left(y - \frac{1}{3}\right)$ **15.** $(t - 3)(t - 1)$
17. $(x + 7)(x - 2)$ **19.** $(x + 2)(x + 3)$
21. $-1(x - 8)(x + 7)$, or $(-x + 8)(x + 7)$, or $(x - 8)(-x - 7)$
23. $-y(y - 8)(y + 4)$, or $y(-y + 8)(y + 4)$, or $y(y - 8)(-y - 4)$ **25.** $(x^2 + 16)(x^2 - 5)$ **27.** Not factorable
29. $(x + 9y)(x + 3y)$ **31.** $2(x - 9)(x + 5)$
33. $-1(z + 12)(z - 3)$, or $(-z - 12)(z - 3)$, or $(z + 12)(-z + 3)$ **35.** $(x^2 + 49)(x^2 + 1)$ **37.** $(x^3 + 9)(x^3 + 2)$
39. $(x^4 - 3)(x^4 - 8)$ **41.** $(y - 0.4)(y - 0.4)$
43. $(4 + b^{10})(3 - b^{10})$, or $-1(b^{10} + 4)(b^{10} - 3)$
45. Countryside: $9\frac{3}{8}$ lb; Mystic: $15\frac{5}{8}$ lb **46.** 8 weekdays
47. Yes **48.** No **49.** No **50.** Yes **51.** All real numbers
52. All real numbers **53.** $\{x \mid x$ is a real number *and* $x \neq \frac{7}{4}\}$, or $\left(-\infty, \frac{7}{4}\right) \cup \left(\frac{7}{4}, \infty\right)$ **54.** All real numbers
55. 76, −76, 28, −28, 20, −20 **57.** $x - 365$

Mid-Chapter Review: Chapter 4, p. 314

1. True **2.** False **3.** True **4.** False **5.** True
6.

$$\begin{aligned} & \qquad\qquad\quad \text{F} \qquad\qquad \text{O} \qquad\qquad \text{I} \qquad\qquad \text{L} \\ (8w - 3)(w - 5) &= (8w)(w) + (8w)(-5) + (-3)(w) + (-3)(-5) \\ &= 8w^2 - 40w - 3w + 15 \\ &= 8w^2 - 43w + 15 \end{aligned}$$

7.
$$\begin{aligned} c^3 - 8c^2 - 48c &= c \cdot c^2 - c \cdot 8c - c \cdot 48 \\ &= c(c^2 - 8c - 48) = c(c + 4)(c - 12) \end{aligned}$$

8.
$$\begin{aligned} x^{10} + 8x^5 - 9 &= (x^5)^2 + 8(x^5) - 9 \\ &= (x^5 + 9)(x^5 - 1) \end{aligned}$$

9.
$$\begin{aligned} 5y^3 + 20y^2 - y - 4 &= 5y^2(y + 4) - 1(y + 4) \\ &= (y + 4)(5y^2 - 1) \end{aligned}$$

10. Terms: $-a^7, a^4, -a, 8$; degree of each term: 7, 4, 1, 0; degree of polynomial: 7; leading term: $-a^7$; leading coefficient: −1; constant term: 8 **11.** Terms: $3x^4, 2x^3w^5, -12x^2w, 4x^2, -1$; degree of each term: 4, 8, 3, 2, 0; degree of polynomial: 8; leading term: $2x^3w^5$; leading coefficient: 2; constant term: −1
12. $5 - 2y - y^3 - 2y^4 + y^9$ **13.** $2x^5 - 4qx^2 + 2qx - 9qr$
14. $h(0) = 5; h(-2) = 21; h\left(\frac{1}{2}\right) = 2\frac{7}{8}$, or $\frac{23}{8}$ **15.** $f(-1) = 1\frac{1}{2}$, or $\frac{3}{2}$; $f(1) = -\frac{1}{2}$; $f(0) = 0$ **16.** $f(a - 2) = a^2 - 2a - 9$; $f(a + h) - f(a) = 2ah + h^2 + 2h$
17. $-2a^2 - 3b - 4ab - 1$ **18.** $11x^2 + 7x - 8$
19. $b^2 - 11b - 12$ **20.** $3c^4 - c^5$ **21.** $y^8 - 3y^4 - 18$
22. $4y^3 + 6y^2 - 2y$ **23.** $9x - 12$ **24.** $16x^2 - 40x + 25$
25. $4x^2 + 20x + 25$ **26.** $0.11x - 3y$ **27.** $-130x^3y$
28. $x^3 - x^2y + xy^2 + 3y^3$ **29.** $10x^2 + 31x - 63$
30. $81x^2 - 16$ **31.** $h(5h + 7)$ **32.** $(x + 10)(x - 2)$
33. $-(b + 7)(b - 3)$, or $(7 + b)(3 - b)$ **34.** $\left(m + \frac{1}{7}\right)^2$
35. $(2 - x)(xy + 5)$ **36.** $3(w - 1)^2$ **37.** $(t + 3)(t^2 + 1)$
38. $8xy^3z(3y^3z^3 - 2x^3)$ **39.** Not factorable **40.** One explanation is as follows. The expression $-(a - b)$ is the opposite of $a - b$. Since $(a - b) + (b - a) = 0$, then $-(a - b) = b - a$. **41.** No; if the coefficients of at least one pair of like terms are opposites, then the sum is a monomial.

For example, $(2x + 3) + (-2x + 1) = 4$, a monomial. **42.** No; consider the polynomial $3x^{11} + 5x^7$. All the coefficients and exponents are prime numbers, yet the polynomial can be factored so it is not prime. **43.** When coefficients and/or exponents are large, a polynomial is more easily evaluated after it has been factored. **44. (a)** The middle term, $2 \cdot a \cdot 3$, is missing from the right-hand side.

$$(a + 3)^2 = a^2 + 6a + 9$$

(b) The middle term, $-2ab$, is missing from the right-hand side and the sign preceding b^2 is incorrect.

$$(a - b)(a - b) = a^2 - 2ab + b^2$$

(c) The product of the outside terms and the product of the inside terms are missing from the right-hand side.

$$(x + 3)(x - 4) = x^2 - x - 12$$

(d) There should be a minus sign between the terms of the product.

$$(p + 7)(p - 7) = p^2 - 49$$

(e) The middle term, $-2 \cdot t \cdot 3$, is missing from the right-hand side and the sign preceding 9 is incorrect.

$$(t - 3)^2 = t^2 - 6t + 9$$

45. Answers may vary. For the polynomial $4a^3 - 12a$, an incorrect factorization is $4a(a - 3)$. Evaluating both the polynomial and the factorization for $a = 0$, we get 0 in each case. Thus the evaluation does not catch the mistake.

Exercise Set 4.5, p. 321

CC1. $2x^2$ **CC2.** 1 **CC3.** $-3x$ **CC4.** Negative **CC5.** Leading; constant **CC6.** Product; sum **CC7.** $21x$ **CC8.** Factor **RC1.** True **RC2.** True **RC3.** False **RC4.** False **RC5.** True
1. $(3x + 1)(x - 5)$ **3.** $y(5y - 7)(2y + 3)$ **5.** $(3c - 8)(c - 4)$ **7.** $(5y + 2)(7y + 4)$ **9.** $2(5t - 3)(t + 1)$ **11.** $4(2x + 1)(x - 4)$ **13.** $3(3a - 1)(2a - 5)$ **15.** $5(3t + 1)(2t + 5)$ **17.** $x(3x - 4)(4x - 5)$ **19.** $x^2(7x + 1)(2x - 3)$ **21.** $(3a - 4)(a + 1)$ **23.** $(3x + 1)(3x + 4)$ **25.** $-1(z - 3)(12z + 1)$, or $(-z + 3)(12z + 1)$, or $(z - 3)(-12z - 1)$ **27.** $-1(2t - 3)(2t + 5)$, or $(-2t + 3)(2t + 5)$, or $(2t - 3)(-2t - 5)$ **29.** $x(3x + 1)(x - 2)$ **31.** $(24x + 1)(x - 2)$ **33.** $-2t(2t + 5)(2t - 3)$ **35.** $-x(24x + 1)(x - 2)$ **37.** $(7x + 3)(3x + 4)$ **39.** $4(10x^4 + 4x^2 - 3)$ **41.** $(4a - 3b)(3a - 2b)$ **43.** $(2x - 3y)(x + 2y)$ **45.** $2(3x - 4y)(2x - 7y)$ **47.** $(3x - 5y)(3x - 5y)$ **49.** $(3x^3 - 2)(x^3 + 2)$ **51.** $2x(8x^2 + 6x - 3)$ **53. (a)** 224 ft; 288 ft; 320 ft; 288 ft; 128 ft; **(b)** $h(t) = -16(t - 7)(t + 2)$ **55.** $(2, -1, 0)$ **56.** $\left(\frac{3}{2}, -4, 3\right)$ **57.** $(1, -1, 2)$ **58.** $(2, 4, 1)$ **59.** Parallel **60.** Parallel **61.** Neither **62.** Perpendicular **63.** $y = -\frac{1}{7}x - \frac{23}{7}$ **64.** $y = -\frac{1}{3}x - \frac{7}{3}$ **65.** $y = -\frac{7}{17}x - \frac{19}{17}$ **66.** $y = -\frac{5}{2}x - \frac{2}{3}$ **67.** Left to the student **69.** $(7a + 6)(ab + 1)$ **71.** $(9xy - 4)(xy + 1)$ **73.** $(x^a + 8)(x^a - 3)$

Visualizing for Success, p. 330

1. A, E **2.** F, J **3.** G, K **4.** L, S **5.** P, Q **6.** C, I **7.** D, H **8.** M, O **9.** N, T **10.** B, R

Exercise Set 4.6, p. 371

RC1. Difference of squares **RC2.** Trinomial square **RC3.** Difference of cubes **RC4.** None of these **RC5.** Trinomial square **RC6.** None of these **RC7.** Sum of cubes **RC8.** Difference of squares **CC1.** Yes **CC2.** No **CC3.** No **CC4.** Yes **CC5.** Yes **CC6.** Yes

1. $(x - 2)^2$ **3.** $(y + 9)^2$ **5.** $(x + 1)^2$ **7.** $(3y + 2)^2$ **9.** $y(y - 9)^2$ **11.** $3(2a + 3)^2$ **13.** $2(x - 10)^2$ **15.** $(1 - 4d)^2$, or $(4d - 1)^2$ **17.** $3a(a - 1)^2$ **19.** $(0.5x + 0.3)^2$ **21.** $(p - t)^2$ **23.** $(a + 2b)^2$ **25.** $(5a - 3b)^2$ **27.** $(y^3 + 13)^2$ **29.** $(4x^5 - 1)^2$ **31.** $(x^2 + y^2)^2$ **33.** $-5(2x - 3)^2$ **35.** $-2a^4(3 - 5a)^2$, or $-2a^4(5a - 3)^2$ **37.** $(p + 7)(p - 7)$ **39.** $(y + 2)^2(y - 2)^2$ **41.** $(pq + 5)(pq - 5)$ **43.** $6(x + y)(x - y)$ **45.** $4x(y^2 + z^2)(y + z)(y - z)$ **47.** $a(2a + 7)(2a - 7)$ **49.** $3(x^4 + y^4)(x^2 + y^2)(x + y)(x - y)$ **51.** $a^2(3a + 5b^2)(3a - 5b^2)$ **53.** $\left(\frac{1}{6} + z\right)\left(\frac{1}{6} - z\right)$ **55.** $(0.2x + 0.3y)(0.2x - 0.3y)$ **57.** $(m - 7)(m + 2)(m - 2)$ **59.** $(a - 2)(a + b)(a - b)$ **61.** $(a + b + 10)(a + b - 10)$ **63.** $(12 - p + 8)(12 + p - 8)$, or $(20 - p)(4 + p)$ **65.** $(a + b + 3)(a + b - 3)$ **67.** $(r - 1 + 2s)(r - 1 - 2s)$ **69.** $2(m + n + 5b)(m + n - 5b)$ **71.** $[3 + (a + b)][3 - (a + b)]$, or $(3 + a + b)(3 - a - b)$ **73.** $(z + 3)(z^2 - 3z + 9)$ **75.** $(x - 1)(x^2 + x + 1)$ **77.** $(2 - 3b)(4 + 6b + 9b^2)$ **79.** $(2a + 1)(4a^2 - 2a + 1)$ **81.** $(2x + 3)(4x^2 - 6x + 9)$ **83.** $(a - b)(a^2 + ab + b^2)$ **85.** $\left(a + \frac{1}{2}\right)\left(a^2 - \frac{1}{2}a + \frac{1}{4}\right)$ **87.** $(x + 0.1)(x^2 - 0.1x + 0.01)$ **89.** $2(y - 4)(y^2 + 4y + 16)$ **91.** $3(2a + 1)(4a^2 - 2a + 1)$ **93.** $r(s + 4)(s^2 - 4s + 16)$ **95.** $5x^2(x - 2z)(x^2 + 2xz + 4z^2)$ **97.** $8(2x^2 - t^2)(4x^4 + 2x^2t^2 + t^4)$ **99.** $(z - 1)(z^2 + z + 1)(z + 1)(z^2 - z + 1)$ **101.** $(t^2 + 4y^2)(t^4 - 4t^2y^2 + 16y^4)$ **103.** $(2w^3 - z^3)(4w^6 + 2w^3z^3 + z^6)$ **105.** $(x - 2y)(x^2 + 2xy + 4y^2)(x + 2y)(x^2 - 2xy + 4y^2)$ **107.** $(3w - 2z^2)(9w^2 + 6wz^2 + 4z^4)(3w + 2z^2)(9w^2 - 6wz^2 + 4z^4)$ **109.** $\left(\frac{1}{2}c + d\right)\left(\frac{1}{4}c^2 - \frac{1}{2}cd + d^2\right)$ **111.** $(0.1x - 0.2y)(0.01x^2 + 0.02xy + 0.04y^2)$ **113.** $\left(-\frac{41}{53}, \frac{148}{53}\right)$ **114.** $\left(-\frac{26}{7}, -\frac{134}{7}\right)$ **115.** $(1, 13)$ **116.** No solution

117.

118.

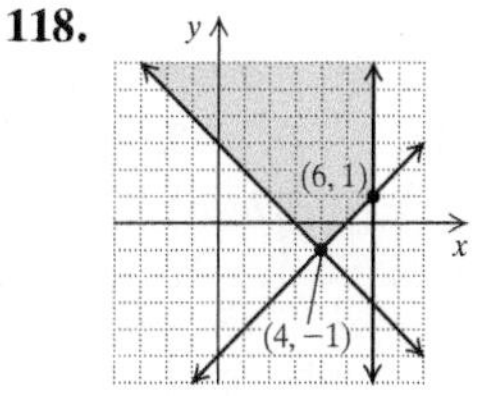

119.

120.

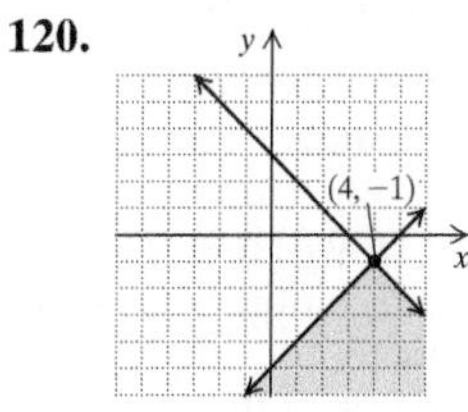

121. $y = x - 2$; $y = -x - 6$ **122.** $y = \frac{2}{3}x - \frac{23}{3}$; $y = -\frac{3}{2}x - \frac{11}{2}$ **123.** $y = -\frac{1}{2}x + 7$; $y = 2x - 3$ **124.** $y = \frac{1}{4}x - \frac{3}{2}$; $y = -4x + 24$ **125.** $h(3a^2 + 3ah + h^2)$ **127. (a)** $\pi h(R + r)(R - r)$; **(b)** 3,014,400 cm^3, or 3.0144 m^3 **129.** $5(c^{50} + 4d^{50})(c^{25} + 2d^{25})(c^{25} - 2d^{25})$ **131.** $(x^{2a} + y^b)(x^{4a} - x^{2a}y^b + y^{2b})$ **133.** $3(x^a + 2y^b)(x^{2a} - 2x^ay^b + 4y^{2b})$ **135.** $\frac{1}{3}\left(\frac{1}{2}xy + z\right)\left(\frac{1}{4}x^2y^2 - \frac{1}{2}xyz + z^2\right)$ **137.** $y(3x^2 + 3xy + y^2)$ **139.** $4(3a^2 + 4)$

Exercise Set 4.7, p. 337

RC1. Common **RC2.** Difference **RC3.** Square **RC4.** Grouping **RC5.** Completely **RC6.** Check **CC1.** (b) **CC2.** (d) **CC3.** (f) **CC4.** (a) **CC5.** (c) **CC6.** (a) **CC7.** (e) **CC8.** (f)

1. $(y + 15)(y - 15)$ **3.** $(2x + 3)(x + 4)$
5. $5(x^2 + 2)(x^2 - 2)$ **7.** $(p + 6)^2$ **9.** $2(x - 11)(x + 6)$
11. $(3x + 5y)(3x - 5y)$ **13.** $4(m^2 + 5)(m^2 - 5)$
15. $6(w - 1)(w + 3)$ **17.** $2x(y + 5)(y - 5)$
19. $(18 - a)(12 + a)$ **21.** $6x^3(x^2 + 9)$
23. $(m + 1)(m^2 - m + 1)(m - 1)(m^2 + m + 1)$
25. $(x + 3 + y)(x + 3 - y)$
27. $2(5x - 4y)(25x^2 + 20xy + 16y^2)$
29. $(m^3 + 10)(m^3 - 2)$ **31.** $(a + d)(c - b)$
33. $(5b - a)(10b + a)$ **35.** $5(a + 1)(2 - ac)$
37. $(2x - 7)(x^2 + 2)$ **39.** $2(x + 3)(x + 2)(x - 2)$
41. $2(2x + 3y)(4x^2 - 6xy + 9y^2)$
43. $-3y(5x + 2)(4x - 1)$, or $3y(-5x - 2)(4x - 1)$, or $3y(5x + 2)(-4x + 1)$ **45.** $(a^4 + b^4)(a^2 + b^2)(a + b)(a - b)$
47. $ab(a + 4b)(a - 4b)$ **49.** $(\frac{1}{4}x - \frac{1}{3}y^2)^2$
51. $5(x - y)^2(x + y)$ **53.** $(9ab + 2)(3ab + 4)$
55. $y(2y - 5)(4y^2 + 10y + 25)$
57. $4x(x + 3)(2x + 5)$ **59.** $3a(a^2b^2 + 5)(a^2b^2 - 5)$
61. $(a - b - 3)(a + b + 3)$ **63.** $(q - 5 + r)(q - 5 - r)$
65. Correct answers: 55; incorrect answers: 20 **66.** $\frac{80}{7}$
67. $(6y^2 - 5x)(5y^2 - 12x)$ **69.** $5(x - \frac{1}{3})(x^2 + \frac{1}{3}x + \frac{1}{9})$
71. $x(x - 2p)$ **73.** $y(y - 1)^2(y - 2)$
75. $(2x + y - r + 3s)(2x + y + r - 3s)$
77. $c(c^w + 1)^2$ **79.** $3x(x + 5)$ **81.** $(x - 1)^3(x^2 + 1)(x + 1)$
83. $y(y^4 + 1)(y^2 + 1)(y + 1)(y - 1)$

Calculator Corner, p. 344

1. Left to the student

Translating for Success, p. 347

1. Q **2.** F **3.** B **4.** A **5.** P **6.** D **7.** O **8.** H
9. I **10.** N

Exercise Set 4.8, p. 348

RC1. False **RC2.** False **RC3.** False **RC4.** False
CC1. 1, 3 **CC2.** −9, −4 **CC3.** $-\frac{3}{2}, \frac{8}{5}$ **CC4.** −8, 0
CC5. $0, \frac{1}{3}$ **CC6.** −5, 2, 6
1. −7, 4 **3.** 3 **5.** −10 **7.** −5, −4 **9.** 0, −8 **11.** −5, 5
13. −12, 12 **15.** 7, −9 **17.** −4, 8 **19.** $-2, -\frac{2}{3}$ **21.** $\frac{1}{2}, \frac{3}{4}$
23. 0, 6 **25.** $\frac{2}{3}, -\frac{3}{4}$ **27.** −1, 1 **29.** $\frac{2}{3}, -\frac{5}{7}$ **31.** $0, \frac{1}{5}$
33. 7, −2 **35.** 0, −2, 3 **37.** 0, −8, 8 **39.** 5, −5, 1, −1
41. −6, 6 **43.** $-\frac{7}{4}, \frac{4}{3}$ **45.** −8, −4 **47.** $-4, \frac{3}{2}$ **49.** −9, −3
51. $\{x | x$ is a real number *and* $x \neq -1$ *and* $x \neq 5\}$
53. $\{x | x$ is a real number *and* $x \neq -3$ *and* $x \neq 3\}$
55. $\{x | x$ is a real number *and* $x \neq \frac{1}{5}\}$
57. $\{x | x$ is a real number *and* $x \neq 0$ *and* $x \neq 2$ *and* $x \neq 5\}$
59. x-intercepts: $(-5, 0)$ and $(9, 0)$; solutions: −5, 9
61. x-intercepts: $(-4, 0)$ and $(8, 0)$; solutions: −4, 8
63. Length: 12 cm; width: 7 cm **65.** Height: 6 ft; base: 4 ft
67. 6 cm **69.** 16, 18, 20 **71.** 3 cm **73.** Length: 12 ft; width: 8 ft **75.** $d = 12$ ft; $h = 16$ ft **77.** 41 ft **79.** 6, 8, 10
81. 11 sec **83.** 1 **84.** 1.3 **85.** $\frac{19}{15}$ **86.** 1023
87. $y = \frac{11}{6}x + \frac{32}{3}$ **88.** $y = -\frac{11}{10}x + \frac{24}{5}$
89. $y = -\frac{3}{10}x + \frac{32}{5}$ **90.** $y = \frac{26}{31}x + \frac{934}{31}$
91. $\{-3, 1\}$; $\{x | -4 \leq x \leq 2\}$, or $[-4, 2]$
93. (a) 1.2522305, 3.1578935; **(b)** −0.3027756, 0, 3.3027756; **(c)** 2.1387475, 2.7238657; **(d)** −0.7462555, 3.3276509

Summary and Review: Chapter 4, p. 353

Vocabulary Reinforcement

1. Ascending **2.** Factor **3.** Factor **4.** Factorization
5. Grouping **6.** Binomial **7.** Zero **8.** Difference

Concept Reinforcement

1. False **2.** True **3.** False

Study Guide

1. Terms: $-6x^4, 5x^3, -x^2, 10x, -1$; degree of each term: 4, 3, 2, 1, 0; degree of polynomial: 4; leading term: $-6x^4$; leading coefficient: −6; constant term: −1 **2.** $2y^3 + 2y^2 + 17y - 8$
3. $3x^2 + xy - 10y^2$ **4.** $4y^2 + 28y + 49$ **5.** $25d^2 - 100$
6. $f(x + 1) = 3x^2 + 5x + 4$; $f(a + h) - f(a) = 3h^2 + 6ah - h$
7. $(y + 3)(y^2 - 8)$ **8.** $(3x - 8)(x + 9)$ **9.** $(2x - 7)(5x + 1)$
10. $(9x - 4)^2$ **11.** $(10t + 1)(10t - 1)$
12. $(6x + 1)(36x^2 - 6x + 1)$ **13.** $(10y - 3)(100y^2 + 30y + 9)$
14. $-2, \frac{7}{3}$

Review Exercises

1. (a) 7, 11, 3, 2; 11; **(b)** $-7x^8y^3$; −7; **(c)** $-3y^2 + 2x^3 + 3x^6y - 7x^8y^3$; **(d)** $-7x^8y^3 - 3y^2 + 3x^6y + 2x^3$
2. 0; −6 **3.** 4; −31 **4.** $-7x + 23y$ **5.** $ab + 12ab^2 + 4$
6. \$45,000 **7.** $-x^3 + 2x^2 + 5x + 2$ **8.** $x^3 + 6x^2 - x - 4$
9. $13x^2y - 8xy^2 + 4xy$ **10.** $9x - 7$ **11.** $-2a + 6b + 7c$
12. $16p^2 - 8p$ **13.** $4x^2 - 7xy + 3y^2$ **14.** $-18x^3y^4$
15. $x^8 - x^6 + 5x^2 - 3$ **16.** $8a^2b^2 + 2abc - 3c^2$ **17.** $4x^2 - 25y^2$
18. $4x^2 - 20xy + 25y^2$ **19.** $20x^4 - 18x^3 - 47x^2 + 69x - 27$
20. $x^4 + 8x^2y^3 + 16y^6$ **21.** $x^3 - 125$ **22.** $x^2 - \frac{1}{2}x + \frac{1}{18}$
23. $a^2 - 4a - 4$; $2ah + h^2 - 2h$ **24.** $3y^2(3y^2 - 1)$
25. $3x(5x^3 - 6x^2 + 7x - 3)$ **26.** $(a - 9)(a - 3)$
27. $(3m + 2)(m + 4)$ **28.** $(5x + 2)^2$ **29.** $4(y + 2)(y - 2)$
30. $(a + 2b)(x - y)$ **31.** $4(x^4 + x^2 + 5)$
32. $(3x - 2)(9x^2 + 6x + 4)$
33. $(0.4b - 0.5c)(0.16b^2 + 0.2bc + 0.25c^2)$
34. $y(y^2 + 1)(y + 1)(y - 1)$ **35.** $2z^6(z^2 - 8)$
36. $2y(3x^2 - 1)(9x^4 + 3x^2 + 1)$ **37.** $(1 + a)(1 - a + a^2)$
38. $4(3x - 5)^2$ **39.** $(3t + p)(2t + 5p)$
40. $(x + 2)(x + 3)(x - 3)$ **41.** $(a - b + 2t)(a - b - 2t)$
42. 10 **43.** $\frac{2}{3}, \frac{3}{2}$ **44.** $0, \frac{7}{4}$ **45.** −4, 4 **46.** −4, 11
47. $\{x | x$ is a real number *and* $x \neq \frac{2}{3}$ *and* $x \neq -7\}$
48. Length: 8 in.; width: 5 in.
49. −7, −5, −3; 3, 5, 7 **50.** 7 **51.** A **52.** C
53. $2(2x + y)(4x^2 - 2xy + y^2)(2x - y)(4x^2 + 2xy + y^2)$
54. $2(3x^2 + 1)$ **55.** $a^3 - (b - 1)^3$ **56.** $0, \frac{1}{8}, -\frac{1}{8}$

Understanding Through Discussion and Writing

1. A sum of two squares can be factored when there is a common factor that is a perfect square. For example, consider $4 + 4x^2$:

$$4 + 4x^2 = 4(1 + x^2).$$

2. See the procedure on p. 319 of the text. **3.** Add the opposite of the polynomial being subtracted. **4.** To solve $P(x) = 0$, find the first coordinate(s) of the x-intercept(s) of $y = P(x)$. To solve $P(x) = 4$, find the first coordinate(s) of the points of intersection of the graphs of $y_1 = P(x)$ and $y_2 = 4$.
5. To use factoring, write $x^3 - 8 = (x - 2)(x^2 + 2x + 4)$ and $(x - 2)^3 = (x - 2)(x - 2)(x - 2)$. Since $(x - 2)(x^2 + 2x + 4) \neq (x - 2)(x - 2)(x - 2)$, then $x^3 - 8 \neq (x - 2)^3$. To use graphing, graph $y_1 = x^3 - 8$ and $y_2 = (x - 2)^3$, and show that the graphs are different.
6. Both are correct. The factorizations are equivalent:

$$\begin{aligned}(a - b)(x - y) &= -1(b - a)(-1)(y - x)\\ &= (-1)(-1)(b - a)(y - x)\\ &= (b - a)(y - x)\end{aligned}$$

7. $x = 5$ *or* $x = -3$
$x - 5 = 0$ *or* $x + 3 = 0$
$(x - 5)(x + 3) = 0$
$x^2 - 2x - 15 = 0;$
No; there cannot be more than two solutions of a quadratic equation. This is because a quadratic equation is factorable into at most two different linear factors. Each of these has one solution when set equal to zero as required by the principle of zero products. **8.** The discussion could include the following points: **(a)** We can now solve certain polynomial equations. **(b)** Whereas most linear equations have exactly one solution, nonlinear polynomial equations can have more than one solution. **(c)** We used factoring and the principle of zero products to solve polynomial equations.

Test: Chapter 4, p. 359

1. [4.1a] **(a)** 4, 3, 9, 5; 9; **(b)** $5x^5y^4$; 5; **(c)** $3xy^3 - 4x^2y - 2x^4y + 5x^5y^4$; **(d)** $5x^5y^4 + 3xy^3 - 4x^2y - 2x^4y$, or $5x^5y^4 + 3xy^3 - 2x^4y - 4x^2y$ **2.** [4.1b] 4; 2 **3.** [4.1b] About 12% **4.** [4.1c] $3xy + 3xy^2$ **5.** [4.1c] $-3x^3 + 3x^2 - 6y - 7y^2$ **6.** [4.1c] $7a^3 - 6a^2 + 3a - 3$ **7.** [4.1c] $7m^3 + 2m^2n + 3mn^2 - 7n^3$ **8.** [4.1d] $6a - 8b$ **9.** [4.1d] $7x^2 - 7x + 13$ **10.** [4.1d] $2y^2 + 5y + y^3$ **11.** [4.2a] $64x^3y^3$ **12.** [4.2a] $12a^2 - 4ab - 5b^2$ **13.** [4.2a] $x^3 - 2x^2y + y^3$ **14.** [4.2a] $-3m^4 - 13m^3 + 5m^2 + 26m - 10$ **15.** [4.2c] $16y^2 - 72y + 81$ **16.** [4.2d] $x^2 - 4y^2$ **17.** [4.2e] $a^2 + 15a + 50$; $2ah + h^2 - 5h$ **18.** [4.3a] $x(9x + 7)$ **19.** [4.3a] $8y^2(3y + 2)$ **20.** [4.6c] $(y + 5)(y + 2)(y - 2)$ **21.** [4.4a] $(p - 14)(p + 2)$ **22.** [4.5a, b] $(6m + 1)(2m + 3)$ **23.** [4.6b] $(3y + 5)(3y - 5)$ **24.** [4.6d] $3(r - 1)(r^2 + r + 1)$ **25.** [4.6a] $(3x - 5)^2$ **26.** [4.6b] $(z + 1 + b)(z + 1 - b)$ **27.** [4.6b] $(x^4 + y^4)(x^2 + y^2)(x + y)(x - y)$ **28.** [4.6c] $(y + 4 + 10t)(y + 4 - 10t)$ **29.** [4.6b] $5(2a + b)(2a - b)$ **30.** [4.5a, b] $2(4x - 1)(3x - 5)$ **31.** [4.6d] $2ab(2a^2 + 3b^2)(4a^4 - 6a^2b^2 + 9b^4)$ **32.** [4.8a] $-3, 6$ **33.** [4.8a] $-5, 5$ **34.** [4.8a] $-\frac{3}{2}, -7$ **35.** [4.8a] 0, 5 **36.** [4.8a] $\{x|x$ is a real number *and* $x \neq -1\}$ **37.** [4.8b] Length: 8 cm; width: 5 cm **38.** [4.8b] 13 ft **39.** [4.3a] $f(n) = \frac{1}{2}n(n - 1)$ **40.** [4.6d] C **41.** [4.7a] $(3x^n + 4)(2x^n - 5)$ **42.** [4.2c] 19

Cumulative Review: Chapters 1–4, p. 361

1. [4.1c] $-2x^2 + x - xy - 1$ **2.** [4.1d] $-2x^2 + 6x$ **3.** [4.2a] $a^4 + a^3 - 8a^2 - 3a + 9$ **4.** [4.2b] $x^2 + 13x + 36$ **5.** [1.1d] 2 **6.** [1.1d] 13 **7.** [1.2a] $b = \dfrac{2A - ha}{h}$, or $\dfrac{2A}{h} - a$ **8.** [1.4c] $\{x|x \geq -\frac{7}{9}\}$, or $[-\frac{7}{9}, \infty)$ **9.** [1.5b] $\{x|x < \frac{5}{4}$ *or* $x > 4\}$, or $(-\infty, \frac{5}{4}) \cup (4, \infty)$ **10.** [1.6e] $\{x|-2 < x < 5\}$, or $(-2, 5)$ **11.** [3.5a] $(1, 3, -9)$ **12.** [3.3a] $(4, -2)$ **13.** [3.3a] $(\frac{19}{8}, \frac{1}{8})$ **14.** [3.5a] $(-1, 0, -1)$ **15.** [4.8a] $-3, -8$ **16.** [4.8a] $\frac{1}{2}, 7$ **17.** [4.8a] $\frac{2}{3}, -2$ **18.** [4.8a] $\{x|x$ is a real number *and* $x \neq 5$ *and* $x \neq -3\}$ **19.** [4.3a] $3x^2(x - 4)$ **20.** [4.3b], [4.6d] $(2x + 1)(x + 1)(x^2 - x + 1)$ **21.** [4.4a] $(x - 2)(x + 7)$ **22.** [4.5a, b] $(4a - 3)(5a - 2)$ **23.** [4.6b] $(2x + 5)(2x - 5)$ **24.** [4.6a] $2(x - 7)^2$ **25.** [4.6d] $(a + 10)(a^2 - 10a + 100)$ **26.** [4.6d] $(4x - 1)(16x^2 + 4x + 1)$ **27.** [4.4a] $(a^3 + 6)(a^3 - 2)$ **28.** [4.6b] $x^2y^2(2x + y)(2x - y)$ **29.** [3.6a] 8 × 10: \$9.20; 11 × 14: \$23; 24 × 36: \$69 **30.** [1.5b]

0 1 2

31. [2.1c]

y = −2x

32. [2.5c]

6y + 24 = 0

33. [3.7b]

y > x + 6

34. [2.2c]

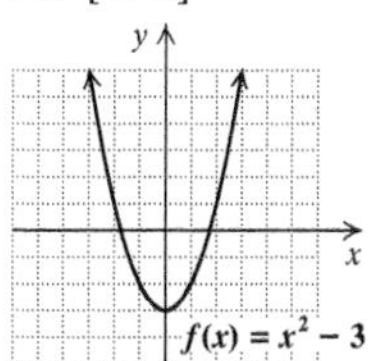

35. [2.2c]

g(x) = 4 − |x|

36. [3.7c]

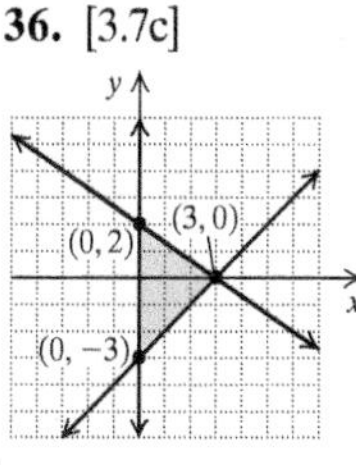

37. [2.6d] $y = -\frac{1}{2}x + \frac{17}{2}$ **38.** [2.6d] $y = \frac{4}{3}x - 6$ **39.** [2.6c] $y = 4x + 8$ **40.** [2.6b] $y = -3x + 7$ **41.** [2.4c] 1.12 million gal per year **42.** **(a)** [4.1b] 30 games; **(b)** [4.8b] 9 teams **43.** [4.8b] 11 cm by 10 cm **44.** [1.6e] $\{x|x \leq 1\}$, or $(-\infty, 1]$

CHAPTER 5

Calculator Corner, p. 368

1. Correct **2.** Correct **3.** Incorrect **4.** Incorrect **5.** Correct

Exercise Set 5.1, p. 371

RC1. Quotient **RC2.** Factors **RC3.** Equivalent **RC4.** Denominator **CC1.** (c) **CC2.** (g) **CC3.** (e) **CC4.** (a) **CC5.** (b) **CC6.** (h) **CC7.** (f) **CC8.** (d) **1.** $-\frac{17}{3}$ **3.** $-7, -5$ **5.** $\{x|x$ is a real number *and* $x \neq -7\}$, or $(-\infty, -7) \cup (-7, \infty)$ **7.** $\{x|x$ is a real number *and* $x \neq 0$ *and* $x \neq 3\}$, or $(-\infty, 0) \cup (0, 3) \cup (3, \infty)$ **9.** $\{x|x$ is a real number *and* $x \neq -\frac{17}{3}\}$, or $(-\infty, -\frac{17}{3}) \cup (-\frac{17}{3}, \infty)$ **11.** $\{x|x$ is a real number *and* $x \neq -7$ *and* $x \neq -5\}$, or $(-\infty, -7) \cup (-7, -5) \cup (-5, \infty)$ **13.** $\dfrac{(x + 2)(7x)}{(x + 8)(7x)}$ **15.** $\dfrac{(q - 5)(q + 5)}{(q + 3)(q + 5)}$ **17.** $3y$ **19.** $\dfrac{2}{3p^4}$ **21.** $a - 3$ **23.** $\dfrac{4x - 5}{7}$ **25.** $\dfrac{y - 3}{y + 3}$ **27.** $\dfrac{t + 4}{t - 4}$ **29.** $\dfrac{x - 8}{x + 4}$ **31.** $\dfrac{w^2 + wz + z^2}{w + z}$ **33.** $\dfrac{1}{3x^3}$ **35.** $\dfrac{(x - 4)(x + 4)}{x(x + 3)}$ **37.** $\dfrac{y + 4}{2}$ **39.** $\dfrac{(2x + 3)(x + 5)}{7x}$ **41.** $c - 2$ **43.** $\dfrac{1}{x + y}$ **45.** $\dfrac{3x^5}{2y^3}$ **47.** 3 **49.** $\dfrac{(y - 3)(y + 2)}{y}$ **51.** $\dfrac{2a + 1}{a + 2}$ **53.** $\dfrac{(x + 4)(x + 2)}{3(x - 5)}$ **55.** $\dfrac{y(y^2 + 3)}{(y + 3)(y - 2)}$ **57.** $\dfrac{x^2 + 4x + 16}{(x + 4)(x + 4)}$ **59.** $\dfrac{4y^2 - 6y + 9}{(4y - 1)(2y - 3)}$ **61.** $\dfrac{2s}{r + 2s}$ **63.** $\dfrac{y^5}{(y + 2)^3(y + 4)}$ **65.** Domain = $\{-4, -2, 0, 2, 4, 6\}$; range = $\{-3, -2, 0, 1, 3, 4\}$ **66.** Domain = $[-4, 5]$; range = $[-3, 2]$ **67.** Domain = $[-5, 5]$; range = $[-4, 4]$ **68.** Domain = $[-4, 5]$; range = $[0, 2]$ **69.** $10(x - 7)(x - 1)$ **70.** $(5x - 4)(2x - 1)$ **71.** $2x(x - 11)(x + 3)$ **72.** $10(y + 13)(y - 5)$

73. $y = -\frac{2}{3}x - 5$ **74.** $y = -\frac{2}{7}x + \frac{48}{7}$ **75.** $\frac{m - t}{m + t + 1}$

77. $\frac{13}{19}$; -3; not defined; $\frac{2a + 2h + 3}{4a + 4h - 1}$

Calculator Corner, p. 382

Left to the student

Exercise Set 5.2, p. 382

RC1. Common **RC2.** Multiple **RC3.** Denominator **RC4.** Greatest **RC5.** $2 \cdot 3 \cdot 3 \cdot 3 \cdot 3 \cdot 5$ **CC1. (a)** $3x + 5$; **(b)** $10x - 3x - 5$; **(c)** $7x - 5$ **CC2. (a)** $4 - 9a$; **(b)** $7 - 4 + 9a$; **(c)** $3 + 9a$

1. 120 **3.** 144 **5.** 210 **7.** 45 **9.** $\frac{11}{10}$ **11.** $\frac{17}{72}$ **13.** $\frac{251}{240}$ **15.** $21x^2y$ **17.** $10(y - 10)(y + 10)$ **19.** $30a^3b^2$ **21.** $5(y - 3)^2$ **23.** $(y + 5)(y - 5)$, or $(y + 5)(5 - y)$, or $-(y + 5)(y - 5)$ **25.** $(2r + 3)(r - 4)(3r - 1)(r + 4)$ **27.** $x^3(x - 2)^2(x^2 + 4)$ **29.** $10x^3(x - 1)^2(x + 1)(x^2 + 1)$ **31.** $\frac{2x + 7y}{x + y}$ **33.** $\frac{3y + 5}{y - 2}$ **35.** $a + b$ **37.** $\frac{13}{y}$ **39.** $\frac{1}{a + 7}$ **41.** $a^2 + ab + b^2$ **43.** $\frac{2(y^2 + 11)}{(y + 4)(y - 5)}$ **45.** $\frac{x + y}{x - y}$ **47.** $\frac{3x - 4}{(x - 2)(x - 1)}$ **49.** $\frac{8x + 1}{(x + 1)(x - 1)}$ **51.** $\frac{2(x - 7)}{15(x + 5)}$ **53.** $\frac{-a^2 + 7ab - b^2}{(a + b)(a - b)}$ **55.** $\frac{y}{(y - 2)(y - 3)}$ **57.** $\frac{3y - 10}{(y - 5)(y + 4)}$ **59.** $\frac{3y^2 - 3y - 29}{(y + 8)(y - 3)(y - 4)}$ **61.** $\frac{2x^2 - 13x + 7}{(x + 3)(x - 1)(x - 3)}$ **63.** 0 **65.** $\frac{4y - 11}{(y + 4)(y - 4)}$ **67.** $\frac{-2y - 3}{(y + 4)(y - 4)}$ **69.** $\frac{-3x^2 - 3x - 4}{(x + 1)(x - 1)}$ **71.** $\frac{-2}{x - y}$, or $\frac{2}{y - x}$ **73.** $\frac{1}{x - 3}$

75.

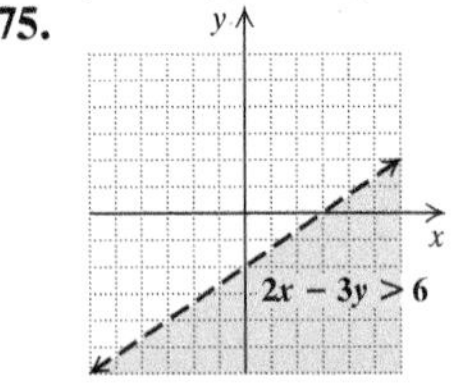

76.

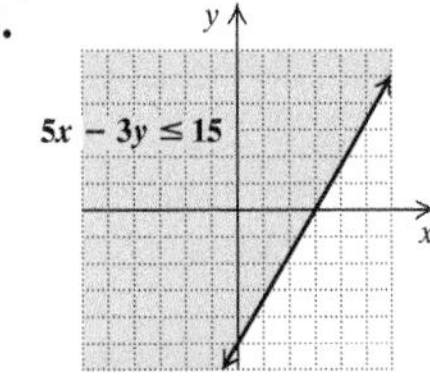

77. $23x(x + 1)(x^2 - x + 1)$ **78.** $(4a - 3b)(16a^2 + 12ab + 9b^2)$ **79.** $y = -\frac{8}{3}x + \frac{14}{3}$ **80.** $y = \frac{5}{4}x + \frac{11}{2}$ **81.** Domain $= (-\infty, 2) \cup (2, \infty)$; range $= (-\infty, 0) \cup (0, \infty)$ **83.** $x^4(x + 1)(x - 1)(x^2 + 1)(x^2 + x + 1)(x^2 - x + 1)$ **85.** -1

Exercise Set 5.3, p. 392

RC1. 4 **RC2.** -6 **RC3.** 7 **RC4.** 6 **RC5.** -4 **RC6.** -7

1. $4x^4 + 3x^3 - 6$ **3.** $9y^5 - 4y^2 + 3$ **5.** $16a^2b^2 + 7ab - 11$ **7.** $x + 7$ **9.** $a - 12$, R 32, or $a - 12 + \frac{32}{a + 4}$ **11.** $x + 2$, R 4, or $x + 2 + \frac{4}{x + 5}$ **13.** $2y^2 - y + 2$, R 6, or $2y^2 - y + 2 + \frac{6}{2y + 4}$ **15.** $2y^2 + 2y - 1$, R 8, or $2y^2 + 2y - 1 + \frac{8}{5y - 2}$ **17.** $2x^2 - x - 9$, R$(3x + 12)$, or $2x^2 - x - 9 + \frac{3x + 12}{x^2 + 2}$ **19.** $2x^3 + 5x^2 + 17x + 51$, R $152x$, or $2x^3 + 5x^2 + 17x + 51 + \frac{152x}{x^2 - 3x}$ **21.** $x^2 - x + 1$, R -4, or $x^2 - x + 1 + \frac{-4}{x - 1}$ **23.** $a + 7$, R -47, or $a + 7 + \frac{-47}{a + 4}$ **25.** $x^2 - 5x - 23$, R -43, or $x^2 - 5x - 23 + \frac{-43}{x - 2}$ **27.** $3x^2 - 2x + 2$, R -3, or $3x^2 - 2x + 2 + \frac{-3}{x + 3}$ **29.** $y^2 + 2y + 1$, R 12, or $y^2 + 2y + 1 + \frac{12}{y - 2}$ **31.** $3x^3 + 9x^2 + 2x + 6$ **33.** $x^2 + 2x + 4$ **35.** $y^3 + 2y^2 + 4y + 8$ **37.** $y^7 - y^6 + y^5 - y^4 + y^3 - y^2 + y - 1$

39.

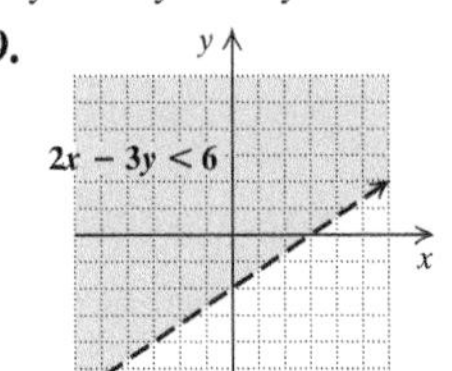

40.

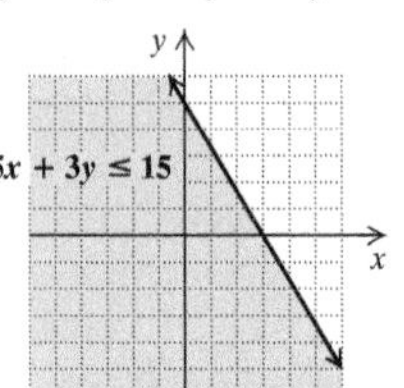

41.

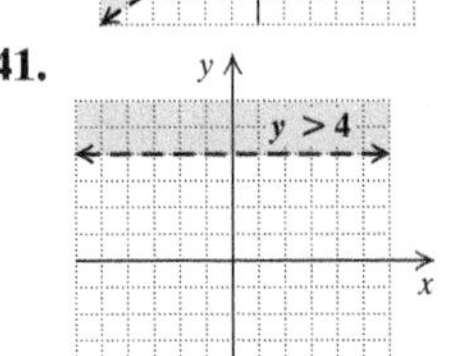

42.

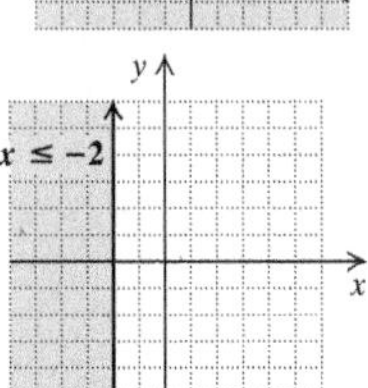

43.

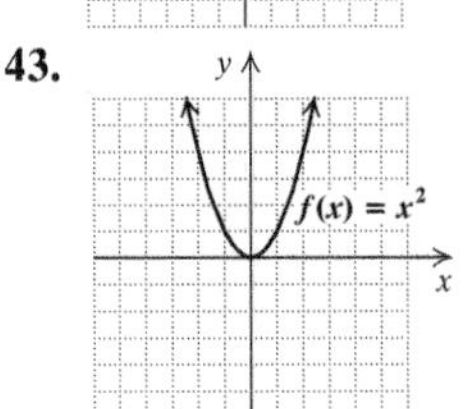

44.

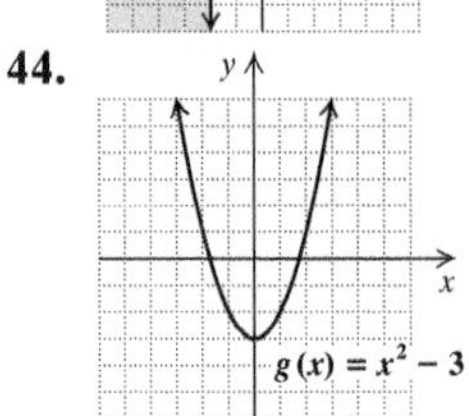

45.

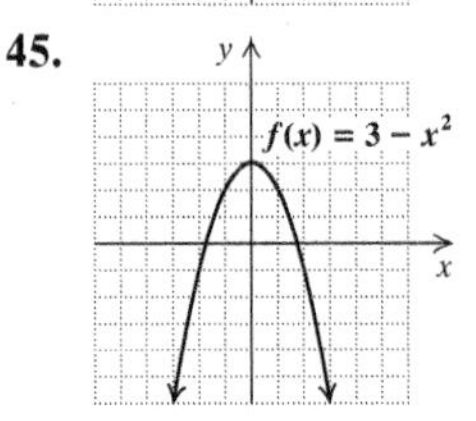

46.

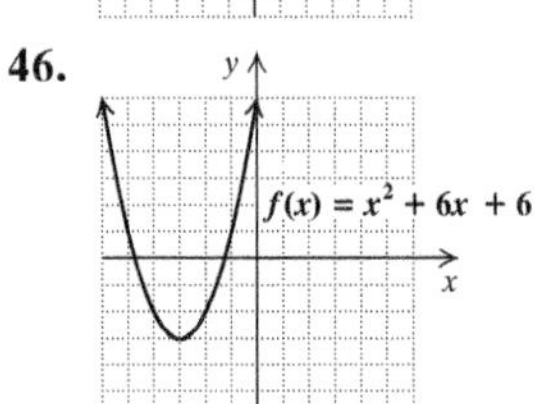

47. 0, 5 **48.** $-\frac{8}{5}, \frac{8}{5}$ **49.** $-\frac{1}{4}, \frac{5}{3}$ **50.** $-\frac{1}{4}, -\frac{2}{3}$ **51.** 0; -3, $-\frac{5}{2}, \frac{3}{2}$ **53.** $-\frac{3}{2}$ **55.** $a^2 + ab$

Exercise Set 5.4, p. 399

RC1. Complex **RC2.** Numerator **RC3.** Least common denominator **RC4.** Reciprocal **CC1.** 40; 70; 280 **CC2.** $15a$; $24a$; $120a$

1. $\frac{26}{35}$ **3.** $\frac{88}{15}$ **5.** $\frac{x^3}{y^5}$ **7.** $\frac{3x + y}{x}$ **9.** $\frac{1 + 2a}{1 - a}$ **11.** $\frac{x^2 - 1}{x^2 + 1}$ **13.** $\frac{3y + 4x}{4y - 3x}$ **15.** $\frac{a^2(b - 3)}{b^2(a - 1)}$ **17.** $\frac{1}{a - b}$ **19.** $\frac{-1}{x(x + h)}$ **21.** $\frac{(x - 4)(x - 7)}{(x - 5)(x + 6)}$ **23.** $\frac{x + 1}{5 - x}$ **25.** $\frac{5x - 16}{4x + 1}$ **27.** $\frac{zw(w - z)}{w^2 - wz + z^2}$ **29.** $\frac{2x^2 - 11x - 27}{2x^2 + 21x + 13}$ **31.** $2x(2x^2 + 10x + 3)$ **32.** $(y + 2)(y^2 - 2y + 4)$ **33.** $(y - 2)(y^2 + 2y + 4)$ **34.** $2x(x - 9)(x - 7)$ **35.** $(10x + 1)(100x^2 - 10x + 1)$ **36.** $(1 - 10a)(1 + 10a + 100a^2)$ **37.** $(y - 4x)(y^2 + 4xy + 16x^2)$ **38.** $(\frac{1}{2}a - 7)(\frac{1}{4}a^2 + \frac{7}{2}a + 49)$ **39.** $s = 3T - r$

40.

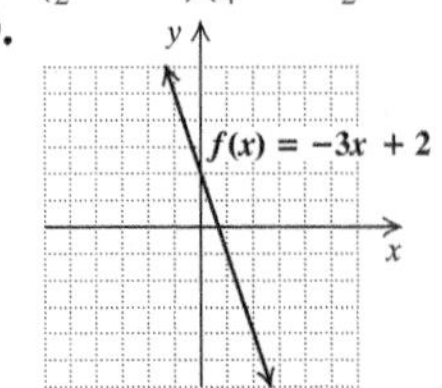

41. 22 **42.** −1, 6 **43.** $\frac{-3(2a + h)}{a^2(a + h)^2}$ **45.** $\frac{1}{(1 - a - h)(1 - a)}$

47. $\frac{5}{6}$

Mid-Chapter Review: Chapter 5, p. 402

1. True **2.** False **3.** False

4. $\frac{7x - 2}{x - 4} - \frac{x + 1}{x + 3} = \frac{7x - 2}{x - 4} \cdot \frac{x + 3}{x + 3} - \frac{x + 1}{x + 3} \cdot \frac{x - 4}{x - 4}$
$= \frac{7x^2 + 19x - 6}{(x - 4)(x + 3)} - \frac{x^2 - 3x - 4}{(x + 3)(x - 4)}$
$= \frac{7x^2 + 19x - 6 - x^2 + 3x + 4}{(x - 4)(x + 3)}$
$= \frac{6x^2 + 22x - 2}{(x - 4)(x + 3)}$

5. $\frac{\frac{1}{m} + 3}{\frac{1}{m} - 5} = \frac{\frac{1}{m} + 3}{\frac{1}{m} - 5} \cdot \frac{m}{m} = \frac{1 + 3m}{1 - 5m}$

6. $\{x | x$ is a real number *and* $x \neq -10$ *and* $x \neq 10\}$, or $(-\infty, -10) \cup (-10, 10) \cup (10, \infty)$ **7.** $\{x | x$ is a real number *and* $x \neq 7\}$, or $(-\infty, 7) \cup (7, \infty)$ **8.** $\{x | x$ is a real number *and* $x \neq -9$ *and* $x \neq 1\}$, or $(-\infty, -9) \cup (-9, 1) \cup (1, \infty)$ **9.** $\frac{2}{3p^7}$

10. $\frac{14y - 1}{11}$ **11.** $\frac{x - y}{x^2 - xy + y^2}$ **12.** $\frac{x + 5}{x + 2}$ **13.** $\frac{a - 2}{a + 2}$

14. $\frac{-1}{t + 2}$ **15.** $70x^4y^5$ **16.** $(x - 5)^2(x + 5)(x + 8)$

17. $\frac{45}{(x + 1)^2}$ **18.** $\frac{4x^2 - x + 2}{(x + 6)(x - 2)}$ **19.** $\frac{-3q - 2}{q(q + 2)}$

20. $\frac{-y^2 - 6y - 3}{(y - 1)(y + 3)(y + 2)}$ **21.** $\frac{b}{1 + b}$ **22.** $\frac{w - z}{5}$

23. $(t - 1)(t^2 + 2t + 4)$ **24.** $\frac{25c^2 + 6a}{15c}$ **25.** $\frac{x + 2}{x - 4}$

26. $3x + 2$, R 17, or $3x + 2 + \frac{17}{2x - 3}$ **27.** $x^3 - x^2 + x - 1$

28. $2x^2 - 5x + 15$, R -34, or $2x^2 - 5x + 15 + \frac{-34}{x + 2}$

29. $x + 2$ **30.** $x^3 - 3x^2 + 6x - 18$, R 56, or $x^3 - 3x^2 + 6x - 18 + \frac{56}{x + 3}$ **31.** $3x - 1$, R 7, or $3x - 1 + \frac{7}{5x + 1}$

32. A remainder of 0 indicates that $x - a$ is a factor of the polynomial. The quotient is a polynomial of one degree less than the original polynomial and can be factored further, if possible, using synthetic division again or another factoring method. **33.** Addition, subtraction, and multiplication of polynomials always result in a polynomial, because each is defined in terms of addition, subtraction, or multiplication of monomials, and the sum, difference, and product of monomials is a monomial. Division of polynomials does not always result in a polynomial, because the quotient is not always a monomial or a sum of monomials. Example 1 in Section 5.3 in the text illustrates this. **34.** No; when we simplify a rational expression by removing a factor of 1, we are actually reversing the multiplication process. **35.** Janine's answer was correct. It is equivalent to the answer at the back of the book:

$\frac{3 - x}{x - 5} = \frac{-x + 3}{x - 5} = \frac{-1(-x + 3)}{-1(x - 5)} = \frac{x - 3}{-x + 5} = \frac{x - 3}{5 - x}$.

36. Nancy's misconception is that x is a factor of the numerator. $\left(\frac{x + 2}{x} = 3 \text{ only for } x = 1.\right)$ **37.** Most would agree that it is easier to find the LCM of all the denominators, bd, and then to multiply by $bd/(bd)$ than it is to add in the numerator, subtract in the denominator, and then divide the numerator by the denominator.

Calculator Corner, p. 407

1. Left to the student **2.** Left to the student

Exercise Set 5.5, p. 409

RC1. Rational expression **RC2.** Solutions
RC3. Rational expression **RC4.** Rational expression
RC5. Solutions **RC6.** Solutions
RC7. Rational expression **RC8.** Solutions
CC1. True **CC2.** True
1. $\frac{31}{4}$ **3.** $-\frac{12}{7}$ **5.** 144 **7.** −1, −8 **9.** 2 **11.** 11 **13.** 11 **15.** No solution **17.** 2 **19.** 5 **21.** −145 **23.** $-\frac{10}{3}$ **25.** −3 **27.** $\frac{31}{5}$ **29.** $\frac{85}{12}$ **31.** −6, 5 **33.** No solution **35.** 2 **37.** No solution **39.** −1, 0 **41.** $-\frac{3}{2}, 2$ **43.** $\frac{17}{4}$ **45.** $\frac{3}{5}$ **46.** $4(t + 5)(t^2 - 5t + 25)$ **47.** $(1 - t)(1 + t + t^2)(1 + t)(1 - t + t^2)$ **48.** $(a + 2b)(a^2 - 2ab + 4b^2)$ **49.** $(a - 2b)(a^2 + 2ab + 4b^2)$ **50.** 3 **51.** −4, 3 **52.** −7, 7 **53.** $\frac{1}{4}, \frac{2}{3}$ **54.** About \$67.2 billion per year **55.** A decrease of about 2.5 billion pieces of first-class mail per year **57. (a)** $(-3.5, 1.\overline{3})$; **(b), (c)** left to the student

Translating for Success, p. 420

1. N **2.** B **3.** A **4.** C **5.** E **6.** G **7.** I **8.** K **9.** M **10.** O

Exercise Set 5.6, p. 421

RC1. Quotient **RC2.** Proportion
RC3. Rate **RC4.** Distance
RC5. Distance **RC6.** Proportional
1. $15\frac{3}{4}$ hr **3.** $13\frac{1}{3}$ hr **5.** Machine A: 2 hr; machine B: 6 hr **7.** 4.375 hr, or $4\frac{3}{8}$ hr **9.** Cole: 8 min; Jim: 24 min **11.** About 49 three-point field goals **13.** 7.5 in. **15.** 28.8 lb **17.** 1160 trees **19.** About 14.5 tons of grapes **21.** 287 trout **23.** $10\frac{1}{2}$ ft; $17\frac{1}{2}$ ft **25.** About 20,658 kg **27.** 35 mph **29.** 7 mph **31.** Bus: 60 mph; trolley: 45 mph **33.** 5.2 ft/sec **35.** Domain: $[-5, 5]$; range: $[-4, 3]$ **36.** Domain: $\{-4, -2, 0, 1, 2, 4\}$; range: $\{-2, 0, 2, 4, 5\}$ **37.** Domain: $[-5, 5]$; range: $[-5, 3]$ **38.** Domain: $[-5, 5]$; range: $[-5, 0]$

39.

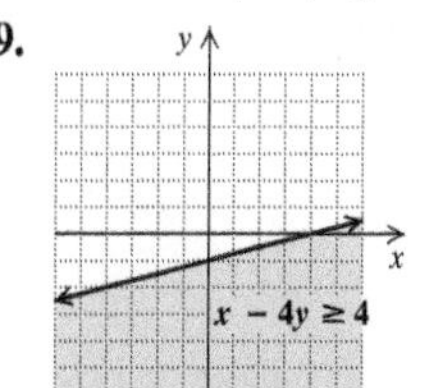

40.

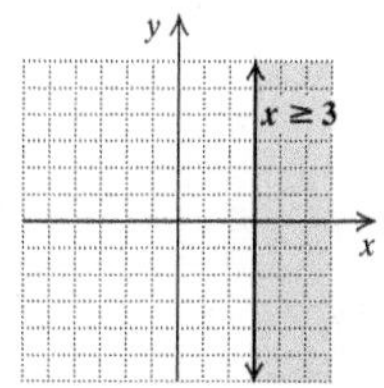

41.

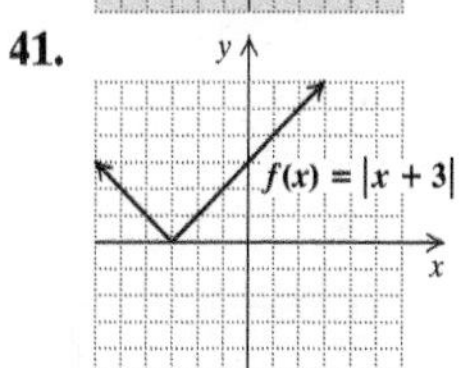

42.

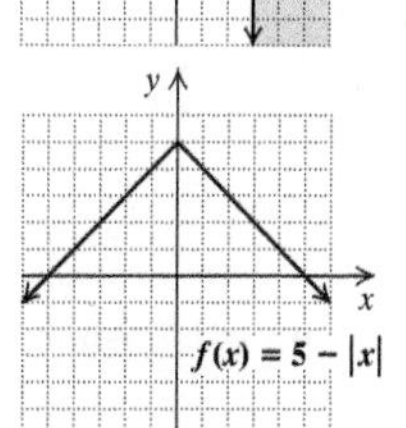

43. $t = \frac{2}{3}$ hr **45.** City: 261 mi; highway: 204 mi

Exercise Set 5.7, p. 428

RC1. (c) **RC2.** (a) **RC3.** (e) **RC4.** (d)

1. $W_2 = \frac{d_2W_1}{d_1}$ **3.** $r_2 = \frac{Rr_1}{r_1 - R}$ **5.** $t = \frac{2s}{v_1 + v_2}$

7. $s = \frac{Rg}{g - R}$ **9.** $p = \frac{qf}{q - f}$ **11.** $a = \frac{bt}{b - t}$
13. $E = \frac{Inr}{n - I}$ **15.** $H^2 = \frac{704.5W}{I}$ **17.** $r = \frac{eR}{E - e}$
19. $R = \frac{3V + \pi h^3}{3\pi h^2}$ **21.** $h = \frac{S - 2\pi r^2}{2\pi r}$ **23.** $t_2 = \frac{d_2 - d_1 + t_1 v}{v}$
25. Dimes: 2 rolls; nickels: 5 rolls; quarters: 5 rolls
26. -6 **27.** 6 **28.** 0 **29.** $8a^3 - 2a$ **30.** $-\frac{4}{5}$
31. $y = -\frac{4}{5}x + \frac{17}{5}$

Exercise Set 5.8, p. 437

RC1. Inverse **RC2.** Direct
CC1. (f) **CC2.** (d) **CC3.** (h) **CC4.** (i)
CC5. (c) **CC6.** (a) **CC7.** (g) **CC8.** (b)
1. 5; $y = 5x$ **3.** $\frac{2}{15}$; $y = \frac{2}{15}x$ **5.** $\frac{9}{4}$; $y = \frac{9}{4}x$
7. 175 semi trucks **9.** 90 g **11.** 40 kg
13. 76,361,280 cans **15.** 98; $y = \frac{98}{x}$ **17.** 36; $y = \frac{36}{x}$
19. 0.05; $y = \frac{0.05}{x}$ **21.** 3.5 hr **23.** $\frac{2}{9}$ ampere **25.** 960 lb
27. $5\frac{5}{7}$ hr **29.** $y = 15x^2$ **31.** $y = \frac{0.0015}{x^2}$ **33.** $y = xz$
35. $y = \frac{3}{10}xz^2$ **37.** $y = \frac{xz}{5wp}$ **39.** 36 mph **41.** **(a)** $W = kd^3$; **(b)** $102\frac{2}{5}$ lb **43.** 39 earned runs **45.** 729 gal **47.** $[-8, \infty)$
48. $(-\infty, \infty)$ **49.** $[-5, 15)$ **50.** $(-\infty, -4) \cup [0, \infty)$
51. $\left(-\infty, -\frac{1}{2}\right)$ **52.** $\left(-\infty, -\frac{5}{8}\right] \cup (2, \infty)$ **53.** $(2, -1)$
54. $(4, 6)$ **55.** **(a)** Inversely; **(b)** neither; **(c)** directly; **(d)** directly

Summary and Review: Chapter 5, p. 441

Vocabulary Reinforcement

1. Proportion **2.** Rational **3.** Rational
4. Positive, inverse **5.** Complex **6.** Positive, direct, constant

Concept Reinforcement

1. True **2.** True

Study Guide

1. $\{x \mid x \text{ is a real number } and\ x \neq -9\ and\ x \neq 6\}$, or $(-\infty, -9) \cup (-9, 6) \cup (6, \infty)$ **2.** $\frac{b - 3}{b - 8}$
3. $\frac{(w - 5)(w^2 + 5w + 25)}{w + 3}$
4. $x^4(x - 3)(x + 3)(2x + 5)$ **5.** $\frac{r^2 - 3rs - 9s^2}{(r + 2s)(r - s)(r + s)}$
6. $y - 4$, R 5, or $y - 4 + \frac{5}{y - 1}$
7. $x^2 - 8x + 24$, R -73, or $x^2 - 8x + 24 + \frac{-73}{x + 3}$
8. $\frac{b + 4a}{4b - a}$ **9.** $-\frac{33}{2}$ **10.** $k = 93$; $y = 93x$ **11.** $k = \frac{9}{2}$; $y = \frac{9}{2x}$

Review Exercises

1. $-3, 3$ **2.** $\{x \mid x \text{ is a real number } and\ x \neq -3 \text{ and } x \neq 3\}$, or $(-\infty, -3) \cup (-3, 3) \cup (3, \infty)$ **3.** $\frac{x - 2}{3x + 2}$ **4.** $\frac{1}{a - 2}$
5. $48x^3$ **6.** $(x - 7)(x + 7)(3x + 1)$
7. $(x + 5)(x - 4)(x - 2)$ **8.** $\frac{y - 8}{2}$ **9.** $\frac{(x - 2)(x + 5)}{x - 5}$
10. $\frac{3a - 1}{a - 3}$ **11.** $\frac{(x^2 + 4x + 16)(x - 6)}{(x + 4)(x + 2)}$
12. $\frac{x - 3}{(x + 1)(x + 3)}$ **13.** $\frac{2x^3 + 2x^2y + 2xy^2 - 2y^3}{(x - y)(x + y)}$
14. $\frac{-y}{(y + 4)(y - 1)}$ **15.** $4b^2c - \frac{5}{2}bc^2 + 3abc$ **16.** $y - 14$, R -20; or $y - 14 + \frac{-20}{y - 6}$ **17.** $6x^2 - 9$, R $(5x + 22)$, or $6x^2 - 9 + \frac{5x + 22}{x^2 + 2}$ **18.** $x^2 + 9x + 40$, R 153, or $x^2 + 9x + 40 + \frac{153}{x - 4}$ **19.** $3x^3 - 8x^2 + 8x - 6$, R -1, or $3x^3 - 8x^2 + 8x - 6 + \frac{-1}{x + 1}$ **20.** $\frac{3}{4}$ **21.** $\frac{a^2b^2}{2(a^2 - ab + b^2)}$
22. $\frac{(x - 9)(x - 6)}{(x - 3)(x + 6)}$ **23.** $\frac{2(2x^2 - 7x + 1)}{3x^2 + 7x - 11}$
24. $\frac{28}{11}$ **25.** 6 **26.** No solution **27.** 3 **28.** $-\frac{11}{3}$
29. 2 **30.** $5\frac{1}{7}$ hr
31.

	Distance	Speed	Time
Downstream	50 mi	$x + 6$	t
Upstream	30 mi	$x - 6$	t

24 mph
32. 4000 mi **33.** $d = \frac{Wc}{c - W}$; $c = \frac{Wd}{d - W}$
34. $b = \frac{ta}{Sa - p}$; $t = \frac{Sab - pb}{a}$ **35.** $y = 4x$ **36.** $y = \frac{2500}{x}$
37. 20 min **38.** 75 **39.** 500 watts **40.** B **41.** C
42. $a^2 + ab + b^2$ **43.** All real numbers except 0 and 13

Understanding Through Discussion and Writing

1. When adding or subtracting rational expressions, we use the LCM of the denominators (the LCD). When solving a rational equation or when solving a formula for a given letter, we multiply by the LCM of all the denominators to clear fractions. When simplifying a complex rational expression, we can use the LCM in either of two ways. We can multiply by a/a, where a is the LCM of all the denominators occurring in the expression. Or we can use the LCM to add or subtract as necessary in the numerator and in the denominator. **2.** Rational equations differ from those previously studied because they contain variables in denominators. Because of this, possible solutions must be checked in the original equation to avoid division by 0. **3.** Assuming all algebraic procedures have been performed correctly, a possible solution of a rational equation would fail to be an actual solution only if it were not in the domain of one of the rational expressions in the equation. This occurs when the number in question makes a denominator 0. **4.** Let $y = k_1 x$ and $x = \frac{k_2}{z}$. Then $y = k_1 \cdot \frac{k_2}{z}$, or $y = \frac{k_1 k_2}{z}$, so y varies inversely as z. **5.** Answers may vary. From Example 4 of Section 5.5, we see that one form of such an equation is $\frac{x^2}{x - a} = \frac{a^2}{x - a}$. **6.** Answers may vary. Many would probably argue that it is easier to solve $\frac{1}{a} + \frac{1}{b} = \frac{1}{x}$ since it is easier to multiply a and b than 38 and 47. Others might argue that it is easier to solve $\frac{1}{38} + \frac{1}{47} = \frac{1}{x}$ since it is easier to work with constants than with variables.

Test: Chapter 5, p. 447

1. [5.1a] 1, 2 **2.** [5.1a] $\{x|x$ is a real number *and* $x \neq 1$ *and* $x \neq 2\}$, or $(-\infty, 1) \cup (1, 2) \cup (2, \infty)$ **3.** [5.1c] $\frac{3x+2}{x-2}$
4. [5.1c] $\frac{p^2-p+1}{p-2}$ **5.** [5.2a] $(x+3)(x-2)(x+5)$
6. [5.1d] $\frac{2(x+5)}{x-2}$ **7.** [5.2b] $\frac{x-6}{(x+4)(x+6)}$
8. [5.1e] $\frac{y+4}{2}$ **9.** [5.2b] $x+y$ **10.** [5.2c] $\frac{3x}{(x-1)(x+1)}$
11. [5.2c] $\frac{a^3+a^2b+ab^2+ab-b^2-2}{(a-b)(a^2+ab+b^2)}$
12. [5.3a] $4s^2+3s-2rs^2$ **13.** [5.3b] $y^2-5y+25$
14. [5.3b] $4x^2+3x-4$, R $(-8x+2)$, or $4x^2+3x-4+\frac{-8x+2}{x^2+1}$ **15.** [5.3c] $x^2+6x+20$, R 54, or $x^2+6x+20+\frac{54}{x-3}$ **16.** [5.3c] $3x^2+10x-40$
17. [5.4a] $\frac{x+1}{x}$ **18.** [5.4a] $\frac{b^2-ab+a^2}{a^2b^2}$ **19.** [5.5a] $-1, 4$
20. [5.5a] 9 **21.** [5.5a] No solution **22.** [5.5a] $-\frac{7}{2}, 5$
23. [5.5a] $\frac{17}{8}$ **24.** [5.6a] 2 hr **25.** [5.6c] $3\frac{3}{11}$ mph
26. [5.6b] $14\frac{2}{17}$ gal **27.** [5.7a] $a = \frac{Tb}{T-b}$; $b = \frac{Ta}{a+T}$
28. [5.7a] $a = \frac{2b}{Qb+t}$ **29.** [5.8e] $Q = \frac{5}{2}xy$
30. [5.8c] $y = \frac{250}{x}$ **31.** [5.8b] \$990 **32.** [5.8d] $7\frac{1}{2}$ hr
33. [5.8f] 615.44 cm^2 **34.** [5.2a] D
35. [5.5a] All real numbers except 0 and 15
36. [5.4a], [5.5a] x-intercept: $(11, 0)$; y-intercept: $(0, -\frac{33}{5})$

Cumulative Review: Chapters 1–5, p. 449

1. [2.1c]

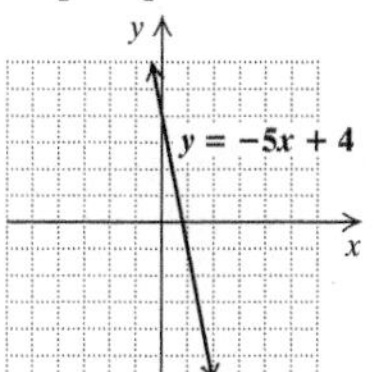

2. [2.5c]

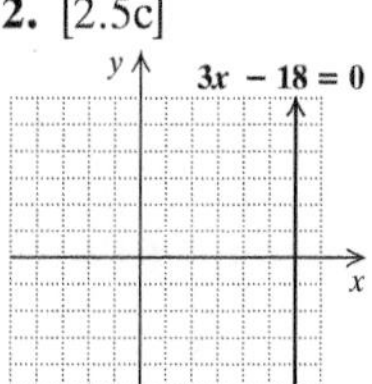

3. [3.7b]

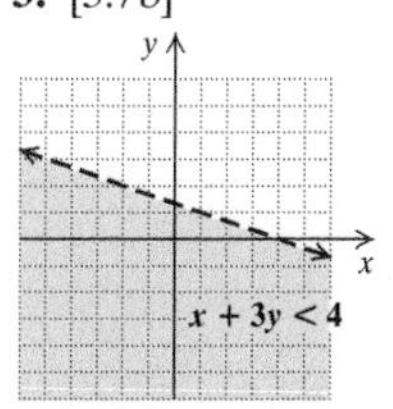

4. [3.7c]

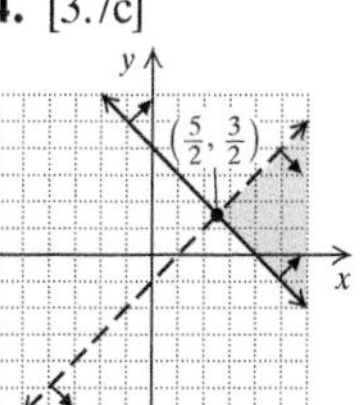

5. [2.2b] 11 **6.** [5.1a] $\{x|x$ is a real number *and* $x \neq -5$ *and* $x \neq 5\}$, or $(-\infty, -5) \cup (-5, 5) \cup (5, \infty)$ **7.** [2.3a] Domain: $(-5, 5)$; range: $[-2, 4]$ **8.** [4.2c] $36m^2-12mn+n^2$
9. [4.2b] $15a^2-14ab-8b^2$ **10.** [5.1d] $\frac{y-2}{3}$
11. [5.1e] $\frac{3x-5}{x+4}$ **12.** [5.2b] $\frac{6x+13}{20(x-3)}$ **13.** [5.4a] $\frac{y^3-2y}{y^3-1}$
14. [4.1d] $16p^2-8p$ **15.** [5.2c] $\frac{4x+1}{(x+2)(x-2)}$
16. [5.3b] $2x^2-11x+23+\frac{-49}{x+2}$ or $2x^2-11x+23+\frac{-49}{x+2}$
17. [1.1d] $\frac{15}{2}$ **18.** [1.5a] $\{x|-3 < x < -\frac{3}{2}\}$, or $(-3, -\frac{3}{2})$
19. [5.5a] No solution **20.** [5.7a] $a = \frac{bP}{3-P}$

21. [1.2a] $C = \frac{5}{9}(F-32)$ **22.** [1.6e] $\{x|x \leq -2.1$ *or* $x \geq 2.1\}$, or $(-\infty, -2.1] \cup [2.1, \infty)$ **23.** [5.5a] -1 **24.** [4.8a] $\frac{1}{4}$
25. [4.8a] $-2, \frac{7}{2}$ **26.** [3.2a], [3.3a] Infinite number of solutions
27. [3.2a], [3.3a] $(-2, 1)$ **28.** [3.5a] $(3, 2, -1)$
29. [3.5a] $(\frac{5}{8}, \frac{1}{16}, -\frac{3}{4})$ **30.** [4.3a] $2x^2(2x+9)$
31. [4.3b] $(2a-1)(4a^2-3)$ **32.** [4.4a] $(x-6)(x+14)$
33. [4.5a, b] $(2x+5)(3x-2)$ **34.** [4.6b] $(4y+9)(4y-9)$
35. [4.6a] $(t-8)^2$ **36.** [4.6d] $8(2x+1)(4x^2-2x+1)$
37. [4.6b] $(0.3b-0.2c)(0.09b^2+0.06bc+0.04c^2)$
38. [4.7a] $x^2(x^2+1)(x+1)(x-1)$
39. [4.5a, b] $(4x-1)(5x+3)$ **40.** [2.6b] $y = -\frac{1}{2}x - 1$
41. [2.6d] $y = \frac{1}{2}x - \frac{5}{2}$ **42.** [3.6a] Win: 44 games; lose: 14 games; tie: 5 games **43.** [5.8b] About 202.3 lb **44.** [5.5a] A
45. [4.8a] C **46.** [5.6a] B **47.** [3.6a] $a = 1, b = -5, c = 6$
48. [4.8a] $0, \frac{1}{4}, -\frac{1}{4}$ **49.** [5.5a] All real numbers except 9 and -5

CHAPTER 6

Exercise Set 6.1, p. 459

RC1. True **RC2.** False **RC3.** False **RC4.** True
RC5. True **RC6.** False **CC1.** (j) **CC2.** (b)
CC3. (h) **CC4.** (i) **CC5.** (i) **CC6.** (g)
1. $4, -4$ **3.** $12, -12$ **5.** $20, -20$ **7.** $-\frac{7}{6}$ **9.** 14 **11.** 0.06
13. Does not exist as a real number **15.** 18.628 **17.** 1.962
19. y^2+16 **21.** $\frac{x}{y-1}$ **23.** $\sqrt{20} \approx 4.472$; 0; does not exist as a real number; does not exist as a real number
25. $\sqrt{11} \approx 3.317$; does not exist as a real number; $\sqrt{11} \approx 3.317$; 12
27. $\{x|x \geq 2\}$, or $[2, \infty)$ **29.** 21 spaces; 25 spaces
31.

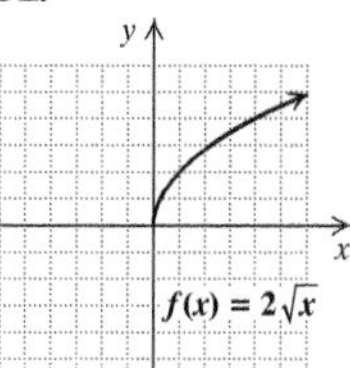

33.
f(x) = −3√x

35.

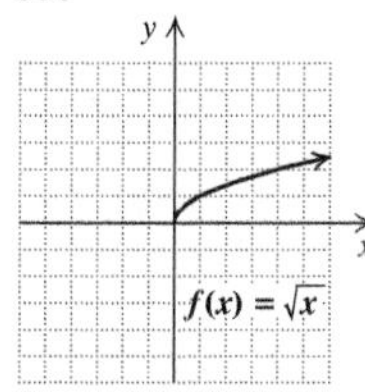

37.
f(x) = √(x − 2)

39.

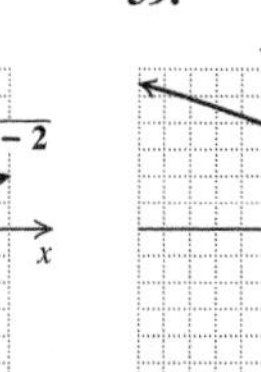

41.

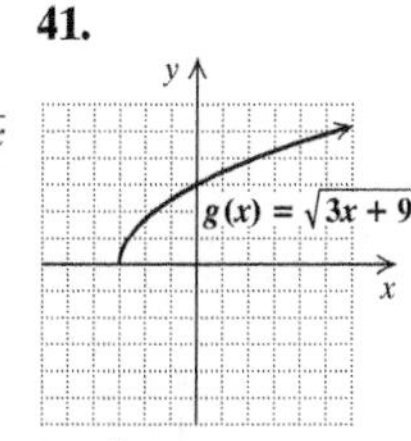

43. $4|x|$ **45.** $12|c|$ **47.** $|p+3|$ **49.** $|x-2|$ **51.** 3
53. $-4x$ **55.** -6 **57.** $0.7(x+1)$ **59.** $2; 3; -2; -4$
61. $-1; -\sqrt[3]{-20} \approx 2.714; -4; -10$ **63.** -5 **65.** -1
67. $-\frac{2}{3}$ **69.** $|x|$ **71.** $5|a|$ **73.** 6 **75.** $|a+b|$ **77.** y
79. $x-2$ **81.** $-\frac{7}{2}, \frac{7}{2}$ **82.** 4, 9 **83.** $0, \frac{5}{2}$ **84.** $\frac{5}{2}$ **85.** $a^9b^6c^{15}$
86. $10a^{10}b^9$ **87.** $\{x|-3 \leq x < 2\}$, or $[-3, 2)$
89. **(a)** Domain: $(-\infty, \infty)$; range: $(-\infty, \infty)$;
(b) domain: $(-\infty, \infty)$; range: $(-\infty, \infty)$;
(c) domain: $[-3, \infty)$; range: $(-\infty, 2]$; **(d)** domain: $[0, \infty)$; range: $[0, \infty)$; **(e)** domain: $[3, \infty)$; range: $[0, \infty)$

Calculator Corner, p. 465

1. 3.344 **2.** 3.281 **3.** 0.283 **4.** 11.053 **5.** 5.527×10^{-5}
6. 2

Exercise Set 6.2, p. 467

RC1. (h) **RC2.** (b) **RC3.** (c) **RC4.** (g) **RC5.** (e)
RC6. (d) **RC7.** (f) **RC8.** (a)
1. $\sqrt[7]{y}$ **3.** 2 **5.** $\sqrt[5]{a^3b^3}$ **7.** 8 **9.** 343 **11.** $17^{1/2}$
13. $18^{1/3}$ **15.** $(xy^2z)^{1/5}$ **17.** $(3mn)^{3/2}$ **19.** $(8x^2y)^{5/7}$

21. $\frac{1}{3}$ 23. $\frac{1}{1000}$ 25. $\frac{3}{x^{1/4}}$ 27. $\frac{1}{(2rs)^{3/4}}$ 29. $\frac{2a^{3/4}c^{2/3}}{b^{1/2}}$
31. $\left(\frac{8yz}{7x}\right)^{3/5}$ 33. $x^{2/3}$ 35. $\frac{x^4}{2^{1/3}y^{2/7}}$ 37. $\frac{7x}{z^{1/3}}$ 39. $\frac{5ac^{1/2}}{3}$
41. $5^{7/8}$ 43. $7^{1/4}$ 45. $4.9^{1/2}$ 47. $6^{3/28}$ 49. $a^{23/12}$
51. $a^{8/3}b^{5/2}$ 53. $\frac{1}{x^{2/7}}$ 55. $\frac{y^{1/3}}{x^{1/2}}$ 57. $m^{3/5}n^2$ 59. $\sqrt[3]{a}$
61. x^5 63. $\frac{1}{x^3}$ 65. a^5b^5 67. $\sqrt{2}$ 69. $\sqrt[3]{2x}$ 71. x^2y^3
73. $2c^2d^3$ 75. $\sqrt[12]{7^4 \cdot 5^3}$ 77. $\sqrt[20]{5^5 \cdot 7^4}$ 79. $\sqrt[6]{4x^5}$
81. a^6b^{12} 83. $\sqrt[18]{m}$ 85. $\sqrt[12]{x^4y^3z^2}$ 87. $\sqrt[30]{\frac{d^{35}}{c^{99}}}$
89. $\{-\frac{4}{7}, 2\}$ 90. $\{-40, 40\}$ 91. $\{-\frac{15}{2}, \frac{5}{2}\}$ 92. $\{-\frac{11}{8}, \frac{3}{8}\}$
93. Left to the student

Exercise Set 6.3, p. 475

RC1. True **RC2.** False **RC3.** True **RC4.** True
CC1. No **CC2.** No **CC3.** Yes **CC4.** No **CC5.** Yes
CC6. No
1. $2\sqrt{6}$ 3. $3\sqrt{10}$ 5. $5\sqrt[3]{2}$ 7. $6x^2\sqrt{5}$ 9. $3x^2\sqrt[3]{2x^2}$
11. $2t^2\sqrt[3]{10t^2}$ 13. $2\sqrt[4]{5}$ 15. $4a\sqrt{2b}$ 17. $3x^2y^2\sqrt[4]{3y^2}$
19. $2xy^3\sqrt[5]{3x^2}$ 21. $5\sqrt{2}$ 23. $3\sqrt{10}$ 25. 2 27. $30\sqrt{3}$
29. $3x^4\sqrt{2}$ 31. $5bc^2\sqrt{2b}$ 33. $a\sqrt[3]{10}$ 35. $2y^3\sqrt[3]{2}$
37. $4\sqrt[4]{4}$ 39. $4a^3b\sqrt{6ab}$ 41. $\sqrt[6]{200}$ 43. $\sqrt[4]{12}$ 45. $a\sqrt[4]{a}$
47. $b\sqrt[10]{b^9}$ 49. $xy\sqrt[6]{xy^5}$ 51. $2ab\sqrt[4]{2a^3}$ 53. $3\sqrt{2}$ 55. $\sqrt{5}$
57. 3 59. $y\sqrt{7y}$ 61. $2\sqrt[3]{a^2b}$ 63. $4\sqrt{xy}$ 65. $2x^2y^2$
67. $\frac{1}{\sqrt[6]{a}}$ 69. $\sqrt[12]{a^5}$ 71. $\sqrt[12]{x^2y^5}$ 73. $\frac{5}{6}$ 75. $\frac{4}{7}$ 77. $\frac{5}{3}$
79. $\frac{7}{y}$ 81. $\frac{5y\sqrt{y}}{x^2}$ 83. $\frac{3y\sqrt[3]{3y^2}}{4}$ 85. $\frac{3a\sqrt[3]{a}}{2b}$ 87. $\frac{3x}{2}$
89. $\frac{2a^3}{bc^4}$ 91. $\frac{2x\sqrt[5]{x^3}}{y^2}$ 93. $\frac{w\sqrt[5]{w^2}}{z^2}$ 95. $\frac{x^2\sqrt[6]{x}}{yz^2}$ 97. $-10, 9$
98. Height: 4 in.; base: 6 in. 99. 8 100. No solution
101. **(a)** 1.62 sec; **(b)** 1.99 sec; **(c)** 2.20 sec 103. $2yz\sqrt{2z}$

Exercise Set 6.4, p. 481

RC1. False **RC2.** True **RC3.** True **RC4.** True
CC1. Yes **CC2.** No **CC3.** No **CC4.** Yes
CC5. Yes **CC6.** No **CC7.** No **CC8.** Yes
1. $11\sqrt{5}$ 3. $\sqrt[3]{7}$ 5. $13\sqrt[3]{y}$ 7. $-8\sqrt{6}$ 9. $6\sqrt[3]{3}$
11. $21\sqrt{3}$ 13. $38\sqrt{5}$ 15. $122\sqrt{2}$ 17. $9\sqrt[3]{2}$ 19. $29\sqrt{2}$
21. $(1 + 6a)\sqrt{5a}$ 23. $(2 - x)\sqrt[3]{3x}$ 25. $(21x + 1)\sqrt{3x}$
27. $2 + 3\sqrt{2}$ 29. $15\sqrt[3]{4}$ 31. $(x + 1)\sqrt[3]{6x}$ 33. $3\sqrt{a - 1}$
35. $(x + 3)\sqrt{x - 1}$ 37. $4\sqrt{5} - 10$ 39. $\sqrt{6} - \sqrt{21}$
41. $-12 + 6\sqrt{3}$ 43. $2\sqrt{15} - 6\sqrt{3}$ 45. -6
47. $6y - 12\sqrt[3]{y^2}$ 49. $3a\sqrt[3]{2}$ 51. 1 53. -12 55. 44
57. 1 59. 3 61. -19 63. $c - d$ 65. $1 + \sqrt{5}$
67. $7 + 3\sqrt{3}$ 69. -6 71. $a + \sqrt{3a} + \sqrt{2a} + \sqrt{6}$
73. $2\sqrt[3]{9} - 3\sqrt[3]{6} - 2\sqrt[3]{4}$ 75. $7 + 4\sqrt{3}$
77. $\sqrt[5]{72} + 3 - \sqrt[5]{24} - \sqrt[5]{81}$ 79. $a - 2$ 80. $\frac{(a + 2)(a + 4)}{a}$
81. $\frac{4(3x - 1)}{3(4x + 1)}$ 82. $\frac{x}{x + 1}$ 83. $\frac{pq}{q + p}$ 84. $\frac{a^2b^2}{b^2 - ab + a^2}$
85. $-\frac{29}{3}, 5$ 86. $\{x \mid -\frac{29}{3} < x < 5\}$, or $(-\frac{29}{3}, 5)$
87. $\{x \mid x \le -\frac{29}{3}$ *or* $x \ge 5\}$, or $(-\infty, -\frac{29}{3}] \cup [5, \infty)$
88. $-12, -\frac{2}{5}$ 89. $(-\infty, \infty)$ 91. 6
93. $14 + 2\sqrt{15} - 6\sqrt{2} - 2\sqrt{30}$ 95. $3\sqrt[3]{3} + 2\sqrt[3]{9} - 8$

Mid-Chapter Review: Chapter 6, p. 485

1. False 2. True 3. False 4. True
5. $\sqrt{6}\sqrt{10} = \sqrt{6 \cdot 10} = \sqrt{2 \cdot 3 \cdot 2 \cdot 5} = 2\sqrt{15}$
6. $5\sqrt{32} - 3\sqrt{18} = 5\sqrt{16 \cdot 2} - 3\sqrt{9 \cdot 2} = 5 \cdot 4\sqrt{2} - 3 \cdot 3\sqrt{2} = 20\sqrt{2} - 9\sqrt{2} = 11\sqrt{2}$ 7. 9 8. -12 9. $\frac{4}{5}$
10. Does not exist as a real number 11. 3; does not exist as a real number 12. $\{x \mid x \le 4\}$, or $(-\infty, 4]$
13.

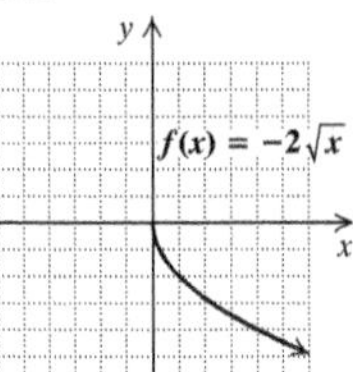

14.

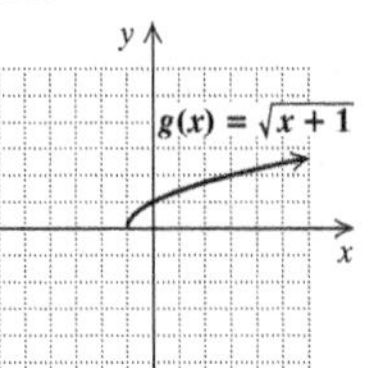

15. $6|z|$ 16. $|x - 4|$ 17. -4 18. $-3a$ 19. 2
20. $|y|$ 21. 5 22. $\sqrt[4]{a^3b}$ 23. $16^{1/5}$ 24. $(6m^2n)^{1/3}$
25. $\frac{1}{3^{3/8}}$ 26. $7^{4/5}$ 27. $\frac{x^{3/2}}{y^{4/3}}$ 28. $\frac{1}{n^{3/4}}$ 29. $\sqrt[3]{4}$ 30. $\sqrt{ab}$
31. $\sqrt[6]{y^5}$ 32. $\sqrt[15]{a^{10}b^9}$ 33. $5\sqrt{3}$ 34. $2xy\sqrt[3]{3y^2}$
35. $2\sqrt[3]{5}$ 36. $\frac{7a^2\sqrt{a}}{b^4}$ 37. $11\sqrt{7}$ 38. $(9x - 24)\sqrt{2x}$
39. $2\sqrt{3} - 15$ 40. $3 - 4\sqrt{x} + x$ 41. $m - n$
42. $11 + 4\sqrt{7}$ 43. $-42 + \sqrt{15}$ 44. Yes; since x^2 is nonnegative for any value of x, the nth root of x^2 exists regardless of whether n is even or odd. Thus the nth root of x^2 always exists. 45. Formulate an expression containing a radical term with an even index and a radicand R such that the solution of the inequality $R \ge 0$ is $\{x \mid x \le 5\}$. One expression is $\sqrt{5 - x}$. Other expressions could be formulated as $a\sqrt[k]{b(5 - x)} + c$, where $a \ne 0$, $b > 0$, and k is an even integer.
46. Since $x^6 \ge 0$ and $x^2 \ge 0$ for any value of x, then $\sqrt[3]{x^6} = x^2$. However, $x^3 \ge 0$ only for $x \ge 0$, so $\sqrt{x^6} = x^3$ only when $x \ge 0$.
47. No; for example, $\frac{\sqrt{8}}{\sqrt{2}} = \sqrt{\frac{8}{2}} = \sqrt{4} = 2$.

Exercise Set 6.5, p. 490

RC1. No **RC2.** Yes **RC3.** Yes **RC4.** No **CC1.** (g)
CC2. (c) **CC3.** (e) **CC4.** (h) **CC5.** (c) **CC6.** (d)
1. $\frac{\sqrt{15}}{3}$ 3. $\frac{\sqrt{22}}{2}$ 5. $\frac{2\sqrt{15}}{35}$ 7. $\frac{2\sqrt[3]{6}}{3}$ 9. $\frac{\sqrt[3]{75ac^2}}{5c}$
11. $\frac{y\sqrt[3]{9yx^2}}{3x^2}$ 13. $\frac{\sqrt[4]{s^3t^3}}{st}$ 15. $\frac{\sqrt{15x}}{10}$ 17. $\frac{\sqrt[3]{100xy}}{5x^2y}$
19. $\frac{\sqrt[4]{2xy}}{2x^2y}$ 21. $\frac{54 + 9\sqrt{10}}{26}$ 23. $-2\sqrt{35} + 2\sqrt{21}$
25. $\frac{18\sqrt{6} + 6\sqrt{15}}{13}$ 27. $\frac{3\sqrt{2} - 3\sqrt{5} + \sqrt{10} - 5}{-3}$
29. $\frac{3 + \sqrt{21} - \sqrt{6} - \sqrt{14}}{-4}$ 31. $\frac{\sqrt{15} + 20 - 6\sqrt{2} - 8\sqrt{30}}{-77}$
33. $\frac{6 - 5\sqrt{a} + a}{9 - a}$ 35. $\frac{6 + 5\sqrt{x} - 6x}{9 - 4x}$ 37. $\frac{3\sqrt{6} + 4}{2}$
39. $\frac{x - 2\sqrt{xy} + y}{x - y}$ 41. 30 42. $-\frac{19}{5}$ 43. 1 44. $\frac{x - 2}{x + 3}$
45. $-\frac{3\sqrt{a^2 - 3}}{a^2 - 3}$

Calculator Corner, p. 494

1. Left to the student 2. Left to the student

Exercise Set 6.6, p. 498

RC1. Radical **RC2.** Powers **RC3.** Isolate **RC4.** Radicands **RC5.** Even **CC1.** 1 **CC2.** 1 **CC3.** 2 **CC4.** 2

1. $\frac{19}{2}$ **3.** $\frac{49}{6}$ **5.** 57 **7.** $\frac{92}{5}$ **9.** -1 **11.** No solution **13.** 3 **15.** 19 **17.** -6 **19.** $\frac{1}{64}$ **21.** 9 **23.** 15 **25.** 2, 5 **27.** 6 **29.** 5 **31.** 9 **33.** 7 **35.** $\frac{80}{9}$ **37.** 2, 6 **39.** -1 **41.** No solution **43.** 3 **45.** About 54.7 mi **47.** About 680 ft **49.** About 642 ft **51.** 151.25 ft; 281.25 ft **53.** About 80.2 pounds per square inch **55.** About 0.81 ft **57.** About 3.9 ft **59.** 2808 mi **60.** Jeff: $1\frac{1}{3}$ hr; Grace: 4 hr **61.** $0, -2.8$ **62.** $0, \frac{5}{3}$ **63.** $-8, 8$ **64.** $-3, \frac{7}{2}$ **65.** $2ah + h^2$ **66.** $2ah + h^2 - h$ **67.** $4ah + 2h^2 - 3h$ **68.** $4ah + 2h^2 + 3h$ **69.** 0 **71.** $-6, -3$ **73.** $0, \frac{125}{4}$ **75.** 3

Translating for Success, p. 505

1. J **2.** B **3.** O **4.** M **5.** K **6.** I **7.** G **8.** E **9.** F **10.** A

Exercise Set 6.7, p. 506

RC1. Pythagorean **RC2.** Legs **RC3.** Hypotenuse **CC1.** (e) **CC2.** (c) **CC3.** (g) **CC4.** (d)

1. $\sqrt{34}$; 5.831 **3.** $\sqrt{450}$; 21.213 **5.** 5 **7.** $\sqrt{43}$; 6.557 **9.** $\sqrt{12}$; 3.464 **11.** $\sqrt{n-1}$ **13.** About 7.066 ft **15.** $\sqrt{26.35}$ in.; about 5.133 in. **17.** $\sqrt{4959}$ ft; about 70.420 ft **19.** $(3, 0), (-3, 0)$ **21.** $\sqrt{10{,}561}$ ft; 102.767 ft **23.** $\sqrt{14{,}824}$ ft; 121.754 ft **25.** $\sqrt{420.125}$ in.; 20.497 in. **27.** $\sqrt{181}$ cm; 13.454 cm **29.** $s + s\sqrt{2}$ **31.** Flash: $67\frac{2}{3}$ mph; Crawler: $53\frac{2}{3}$ mph **32.** $3\frac{3}{4}$ mph **33.** $-7, \frac{3}{2}$ **34.** 3, 8 **35.** 1 **36.** $-2, 2$ **37.** 13 **38.** 7 **39.** 26 packets

Calculator Corner, p. 515

1. $-2 - 9i$ **2.** $20 + 17i$ **3.** $-47 - 161i$ **4.** $-\frac{151}{290} + \frac{73}{290}i$ **5.** -20 **6.** -28.373 **7.** $-\frac{16}{25} - \frac{1}{50}i$ **8.** 81 **9.** $117 + 118i$

Exercise Set 6.8, p. 516

RC1. True **RC2.** True **RC3.** False **RC4.** False **RC5.** True **RC6.** False **RC7.** True **RC8.** True

1. $i\sqrt{35}$, or $\sqrt{35}i$ **3.** $4i$ **5.** $-2i\sqrt{3}$, or $-2\sqrt{3}i$ **7.** $i\sqrt{3}$, or $\sqrt{3}i$ **9.** $9i$ **11.** $7i\sqrt{2}$, or $7\sqrt{2}i$ **13.** $-7i$ **15.** $4 - 2i\sqrt{15}$, or $4 - 2\sqrt{15}i$ **17.** $(2 + 2\sqrt{3})i$ **19.** $12 - 4i$ **21.** $9 - 5i$ **23.** $7 + 4i$ **25.** $-4 - 4i$ **27.** $-1 + i$ **29.** $11 + 6i$ **31.** -18 **33.** $-\sqrt{14}$ **35.** 21 **37.** $-6 + 24i$ **39.** $1 + 5i$ **41.** $18 + 14i$ **43.** $38 + 9i$ **45.** $2 - 46i$ **47.** $5 - 12i$ **49.** $-24 + 10i$ **51.** $-5 - 12i$ **53.** $-i$ **55.** 1 **57.** -1 **59.** i **61.** -1 **63.** $-125i$ **65.** 8 **67.** $1 - 23i$ **69.** 0 **71.** 0 **73.** 1 **75.** $5 - 8i$ **77.** $2 - \frac{\sqrt{6}}{2}i$ **79.** $\frac{9}{10} + \frac{13}{10}i$ **81.** $-i$ **83.** $-\frac{3}{7} - \frac{8}{7}i$ **85.** $\frac{6}{5} - \frac{2}{5}i$ **87.** $-\frac{8}{41} + \frac{10}{41}i$ **89.** $-\frac{4}{3}i$ **91.** $-\frac{1}{2} - \frac{1}{4}i$ **93.** $\frac{3}{5} + \frac{4}{5}i$

95.

$x^2 - 2x + 5 = 0$	
$(1 - 2i)^2 - 2(1 - 2i) + 5$	? 0
$1 - 4i + 4i^2 - 2 + 4i + 5$	
$1 - 4i - 4 - 2 + 4i + 5$	
0	TRUE

Yes

97.

$x^2 - 4x - 5 = 0$	
$(2 + i)^2 - 4(2 + i) - 5$	? 0
$4 + 4i + i^2 - 8 - 4i - 5$	
$4 + 4i - 1 - 8 - 4i - 5$	
-10	FALSE

No

99. x-intercept: $(-15, 0)$, y-intercept: $(0, 30)$ **100.** x-intercept: $(-24, 0)$, y-intercept: $(0, 18)$ **101.** x-intercept: $(2, 0)$, y-intercept: $(0, 5)$ **102.** $-\frac{7}{3}$ **103.** $\frac{1}{3}$ **104.** Not defined **105.** 3.0×10^3 **106.** 2.665×10^5 **107.** $-4 - 8i; -2 + 4i; 8 - 6i$ **109.** $-3 - 4i$ **111.** $-88i$ **113.** 8 **115.** $\frac{3}{5} + \frac{9}{5}i$ **117.** 1

Summary and Review: Chapter 6, p. 521

Vocabulary Reinforcement

1. Cube **2.** Complex **3.** Principal **4.** Rationalizing **5.** Radical **6.** Imaginary **7.** Square **8.** Radicand **9.** Conjugate **10.** Index

Concept Reinforcement

1. True **2.** False **3.** False **4.** False **5.** True **6.** True

Study Guide

1. $6|y|$ **2.** $|a + 2|$ **3.** $\sqrt[5]{z^3}$ **4.** $(6ab)^{5/2}$ **5.** $\frac{1}{9^{3/2}} = \frac{1}{27}$ **6.** $\sqrt[4]{a^3b}$ **7.** $5y\sqrt{6}$ **8.** $2\sqrt{a}$ **9.** $2\sqrt{3}$ **10.** $25 - 10\sqrt{x} + x$ **11.** 5 **12.** 6 **13.** $-21 - 20i$ **14.** $\frac{4}{5} - \frac{7}{5}i$

Review Exercises

1. 27.893 **2.** 6.378 **3.** $f(0), f(-1)$, and $f(1)$ do not exist as real numbers; $f\left(\frac{41}{3}\right) = 5$ **4.** $\left\{x \middle| x \geq \frac{16}{3}\right\}$, or $\left[\frac{16}{3}, \infty\right)$ **5.** $9|a|$ **6.** $7|z|$ **7.** $|6 - b|$ **8.** $|x + 3|$ **9.** -10 **10.** $-\frac{1}{3}$ **11.** $2; -2; 3$ **12.** $|x|$ **13.** 3 **14.** $\sqrt[5]{a}$ **15.** 512 **16.** $31^{1/2}$ **17.** $(a^2b^3)^{1/5}$ **18.** $\frac{1}{7}$ **19.** $\frac{1}{4x^{2/3}y^{2/3}}$ **20.** $\frac{5b^{1/2}}{a^{3/4}c^{2/3}}$ **21.** $\frac{3a}{t^{1/4}}$ **22.** $\frac{1}{x^{2/5}}$ **23.** $7^{1/6}$ **24.** x^7 **25.** $3x^2$ **26.** $\sqrt[12]{x^4y^3}$ **27.** $\sqrt[12]{x^7}$ **28.** $7\sqrt{5}$ **29.** $-3\sqrt[3]{4}$ **30.** $5b^2\sqrt[3]{2a^2}$ **31.** $\frac{7}{6}$ **32.** $\frac{4x^2}{3}$ **33.** $\frac{2x^2}{3y^3}$ **34.** $\sqrt{15xy}$ **35.** $3a\sqrt[3]{a^2b^2}$ **36.** $\sqrt[15]{a^5b^9}$ **37.** $y\sqrt[3]{6}$ **38.** $\frac{5}{2}\sqrt{x}$ **39.** $\sqrt[12]{x^5}$ **40.** $7\sqrt[3]{x}$ **41.** $3\sqrt{3}$ **42.** $15\sqrt{2}$ **43.** $(2x + y^2)\sqrt[3]{x}$ **44.** $-43 - 2\sqrt{10}$ **45.** $8 - 2\sqrt{7}$ **46.** $9 - \sqrt[3]{4}$ **47.** $\frac{2\sqrt{6}}{3}$ **48.** $\frac{2\sqrt{a} - 2\sqrt{b}}{a - b}$ **49.** 4 **50.** 13 **51.** 1 **52.** About 4166 rpm **53.** 4480 rpm **54.** 9 cm **55.** $\sqrt{24}$ ft; 4.899 ft **56.** 25 **57.** $\sqrt{46}$; 6.782 **58.** $(5 + 2\sqrt{2})i$ **59.** $-2 - 9i$ **60.** $1 + i$ **61.** 29 **62.** i **63.** $9 - 12i$ **64.** $\frac{2}{5} + \frac{3}{5}i$ **65.** $\frac{1}{10} - \frac{7}{10}i$

66.

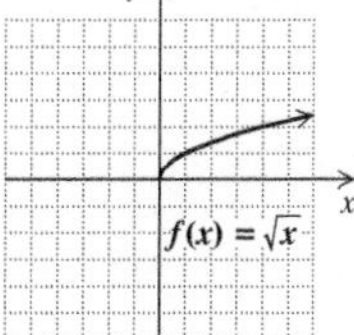

67. D **68.** -1 **69.** 3

Understanding Through Discussion and Writing

1. $f(x) = (x + 5)^{1/2}(x + 7)^{-1/2}$. Consider $(x + 5)^{1/2}$. Since the exponent is $\frac{1}{2}$, $x + 5$ must be nonnegative. Then $x + 5 \geq 0$, or $x \geq -5$. Consider $(x + 7)^{-1/2}$. Since the exponent is $-\frac{1}{2}$, $x + 7$ must be positive. Then $x + 7 > 0$, or $x > -7$. Then the domain of $f = \{x | x \geq -5 \textit{ and } x > -7\}$, or $\{x | x \geq -5\}$.
2. Since $\sqrt{x}$ exists only for $\{x | x \geq 0\}$, this is the domain of $y = \sqrt{x} \cdot \sqrt{x}$. **3.** The distributive law is used to collect radical expressions with the same indices and radicands just as it is used to collect monomials with the same variables and exponents.
4. No; when n is odd, it is true that if $a^n = b^n$, then $a = b$.

5. Use a calculator to show that $\frac{5 + \sqrt{2}}{\sqrt{18}} \neq 2$. Explain that we multiply by 1 to rationalize a denominator. In this case, we would write 1 as $\sqrt{2}/\sqrt{2}$. **6.** When two radical expressions are conjugates, their product contains no radicals. Similarly, the product of a complex number and its conjugate does not contain i.

Test: Chapter 6, p. 527

1. [6.1a] 12.166 **2.** [6.1a] 2; does not exist as a real number **3.** [6.1a] $\{x \mid x \le 2\}$, or $(-\infty, 2]$ **4.** [6.1b] $3|q|$ **5.** [6.1b] $|x + 5|$ **6.** [6.1c] $-\frac{1}{10}$ **7.** [6.1d] x **8.** [6.1d] 4 **9.** [6.2a] $\sqrt[3]{a^2}$ **10.** [6.2a] 8 **11.** [6.2a] $37^{1/2}$ **12.** [6.2a] $(5xy^2)^{5/2}$ **13.** [6.2b] $\frac{1}{10}$ **14.** [6.2b] $\frac{8a^{3/4}}{b^{3/2}c^{2/5}}$ **15.** [6.2c] $\frac{x^{8/5}}{y^{9/5}}$ **16.** [6.2c] $\frac{1}{2.9^{31/24}}$ **17.** [6.2d] $\sqrt[4]{x}$ **18.** [6.2d] $2x\sqrt{x}$ **19.** [6.2d] $\sqrt[15]{a^6b^5}$ **20.** [6.2d] $\sqrt[12]{8y^7}$ **21.** [6.3a] $2\sqrt{37}$ **22.** [6.3a] $2\sqrt[4]{5}$ **23.** [6.3a] $2a^3b^4\sqrt[3]{3a^2b}$ **24.** [6.3b] $\frac{2x\sqrt[3]{2x^2}}{y^2}$ **25.** [6.3b] $\frac{5x}{6y^2}$ **26.** [6.3a] $\sqrt[3]{10xy^2}$ **27.** [6.3a] $xy\sqrt[4]{x}$ **28.** [6.3b] $\sqrt[5]{x^2y^2}$ **29.** [6.3b] $2\sqrt{a}$ **30.** [6.4a] $38\sqrt{2}$ **31.** [6.4b] -20 **32.** [6.4b] $9 + 6\sqrt{x} + x$ **33.** [6.5b] $\frac{13 + 8\sqrt{2}}{-41}$ **34.** [6.6a] 35 **35.** [6.6b] 7 **36.** [6.6a] 5 **37.** [6.7a] 7 ft **38.** [6.6c] 3600 ft **39.** [6.7a] $\sqrt{98}$; 9.899 **40.** [6.7a] 2 **41.** [6.8a] $11i$ **42.** [6.8b] $7 + 5i$ **43.** [6.8c] $37 + 9i$ **44.** [6.8d] $-i$ **45.** [6.8e] $-\frac{77}{50} + \frac{7}{25}i$ **46.** [6.8f] No **47.** [6.6a] A **48.** [6.8c, e] $-\frac{17}{4}i$ **49.** [6.6b] 3

Cumulative Review: Chapters 1–6, p. 529

1. [4.1c] $-3x^3 + 9x^2 + 3x - 3$ **2.** [4.2c] $4x^4 - 4x^2y + y^2$ **3.** [4.2a] $15x^4 - x^3 - 9x^2 + 5x - 2$ **4.** [5.1d] $\frac{(x + 4)(x - 7)}{x + 7}$ **5.** [5.4a] $\frac{y - 6}{y - 9}$ **6.** [5.2c] $\frac{-2x + 4}{(x + 2)(x - 3)}$, or $\frac{-2(x - 2)}{(x + 2)(x - 3)}$ **7.** [5.3b, c] $y^2 + y - 2 + \frac{-1}{y + 2}$ **8.** [6.1c] $-2x$ **9.** [6.1b], [6.3a] $4(x - 1)$ **10.** [6.4a] $57\sqrt{3}$ **11.** [6.3a] $4xy^2\sqrt{y}$ **12.** [6.5b] $\sqrt{30} + \sqrt{15}$ **13.** [6.1d], [6.3b] $\frac{m^2n^4}{2}$ **14.** [6.2c] $6^{8/9}$ **15.** [6.8b] $3 + 5i$ **16.** [6.8e] $\frac{7}{61} - \frac{16}{61}i$ **17.** [1.1d] 2 **18.** [1.2a] $c = 8M + 3$ **19.** [1.4c] $\{a \mid a > -7\}$, or $(-7, \infty)$ **20.** [1.5a] $\{x \mid -10 < x < 13\}$, or $(-10, 13)$ **21.** [1.6c] $\frac{4}{3}, \frac{8}{3}$ **22.** [4.8a] $\frac{25}{7}, -\frac{25}{7}$ **23.** [3.3a] $(5, 3)$ **24.** [3.5a] $(-1, 0, 4)$ **25.** [5.5a] -5 **26.** [5.5a] $\frac{1}{3}$ **27.** [1.2a], [5.7a] $R = \frac{nE - nrI}{I}$ **28.** [6.6a] 6 **29.** [6.6b] $-\frac{1}{4}$ **30.** [6.6a] 5

31. [2.2c]

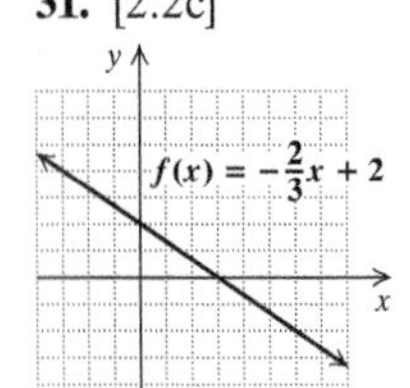

32. [2.5a]

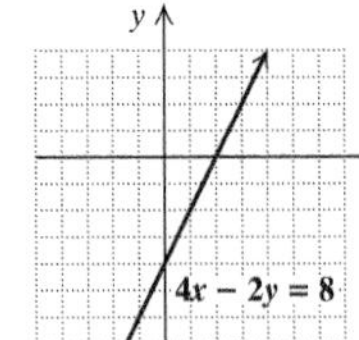

33. [3.7b]

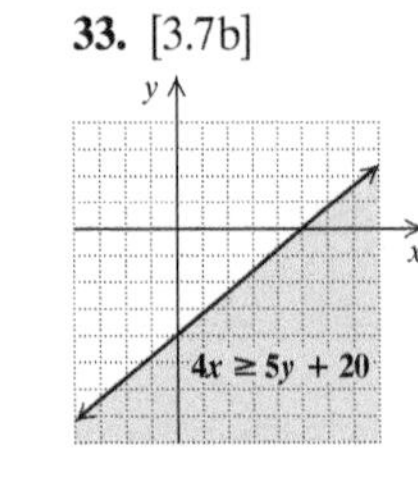

34. [3.7c]

(−3, −3)

35. [2.2c]

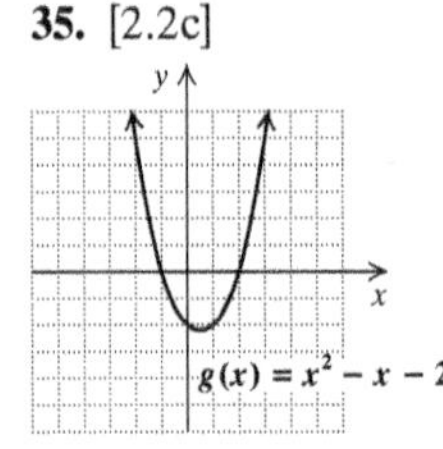

36. [2.2c]

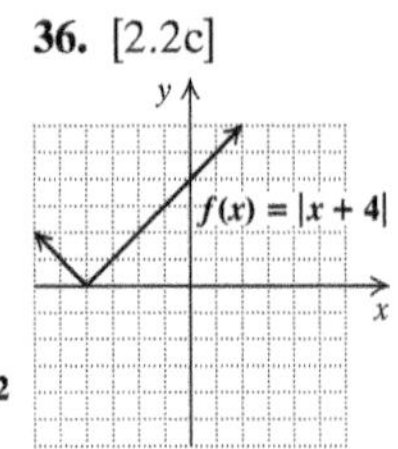

37. [2.2c]

$g(x) = \frac{4}{x - 3}$

38. [2.2c], [6.1a]

$f(x) = 2 - \sqrt{x}$

39. [4.3a] $6xy^2(2x - 5y)$ **40.** [4.5a, b], $(3x + 4)(x - 7)$ **41.** [4.4a] $(y + 11)(y - 12)$ **42.** [4.6d] $(3y + 2)(9y^2 - 6y + 4)$ **43.** [4.6b] $(2x + 25)(2x - 25)$ **44.** [2.3a] Domain: $[-5, 5]$; range: $[-3, 4]$ **45.** [2.3a] Domain: $(-\infty, \infty)$; range: $[-5, \infty)$ **46.** [2.4b] Slope: $\frac{3}{2}$; y-intercept: $(0, -4)$ **47.** [2.6d] $y = -\frac{1}{3}x + \frac{13}{3}$ **48.** [5.8d] 125 ft; 1000 ft^2 **49.** [5.6a] 1 hr **50.** [5.8f] 64 L **51.** [6.2a] D **52.** [5.6a] A **53.** [5.3b, c] A **54.** [6.6a] B **55.** [6.6b] $-\frac{8}{9}$

CHAPTER 7

Calculator Corner, p. 536

The calculator returns an ERROR message because the graph of $y = 4x^2 + 9$ has no x-intercepts. This indicates that the equation $4x^2 + 9 = 0$ has no real-number solutions.

Exercise Set 7.1, p. 541

RC1. True **RC2.** True **RC3.** True **RC4.** True **RC5.** False **RC6.** False **CC1.** 25; 5 **CC2.** 36; 6 **CC3.** $\frac{25}{4}; \frac{5}{2}$ **CC4.** $\frac{1}{4}; \frac{1}{2}$ **CC5.** $\frac{1}{25}; \frac{1}{5}$ **CC6.** $\frac{9}{16}; \frac{3}{4}$
1. **(a)** $\sqrt{5}, -\sqrt{5}$, or $\pm\sqrt{5}$; **(b)** $(-\sqrt{5}, 0), (\sqrt{5}, 0)$
3. **(a)** $\frac{5}{3}i, -\frac{5}{3}i$, or $\pm\frac{5}{3}i$; **(b)** no x-intercepts
5. $\pm\frac{\sqrt{6}}{2}$; ± 1.225 **7.** $5, -9$ **9.** $8, 0$
11. $11 \pm \sqrt{7}$; 13.646, 8.354 **13.** $7 \pm 2i$ **15.** $18, 0$
17. $\frac{3}{2} \pm \frac{\sqrt{14}}{2}$; 3.371, −0.371 **19.** $5, -11$ **21.** $9, 5$
23. $-2 \pm \sqrt{6}$ **25.** $11 \pm 2\sqrt{33}$ **27.** $-\frac{1}{2} \pm \frac{\sqrt{5}}{2}$
29. $\frac{5}{2} \pm \frac{\sqrt{53}}{2}$ **31.** $-\frac{3}{4} \pm \frac{\sqrt{57}}{4}$ **33.** $\frac{9}{4} \pm \frac{\sqrt{105}}{4}$
35. $2, -8$ **37.** $-11 \pm \sqrt{19}$ **39.** $5 \pm \sqrt{29}$
41. **(a)** $-\frac{7}{2} \pm \frac{\sqrt{57}}{2}$; **(b)** $\left(-\frac{7}{2} - \frac{\sqrt{57}}{2}, 0\right), \left(-\frac{7}{2} + \frac{\sqrt{57}}{2}, 0\right)$
43. **(a)** $\frac{5}{4} \pm \frac{\sqrt{39}}{4}i$; **(b)** no x-intercepts
45. $\frac{3}{4} \pm \frac{\sqrt{17}}{4}$ **47.** $\frac{3}{4} \pm \frac{\sqrt{145}}{4}$ **49.** $\frac{2}{3} \pm \frac{\sqrt{7}}{3}$
51. $-\frac{1}{2} \pm \frac{\sqrt{7}}{2}i$ **53.** $2 \pm 3i$ **55.** About 0.890 sec
57. About 5.3 sec **59.** About 4.9 sec **61.** About 5.9 sec

63. **(a)** $R(t) = -\frac{2}{15}t + 128$, where t is the number of years since 1981; **(b)** about 122.8 min; **(c)** 2026

64.

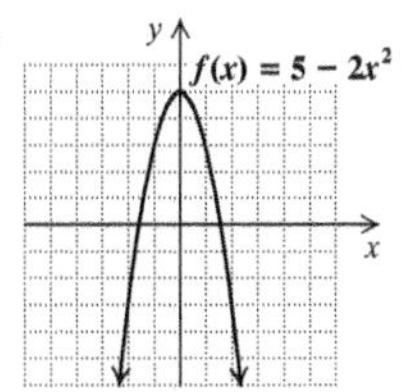

65.

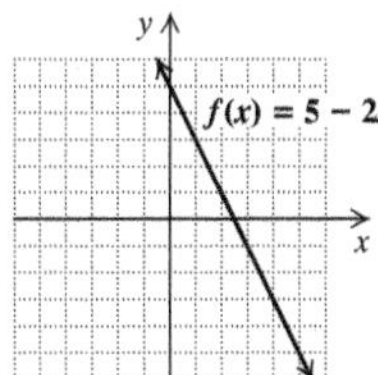

66.

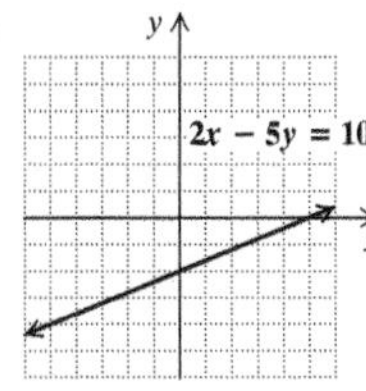

67.

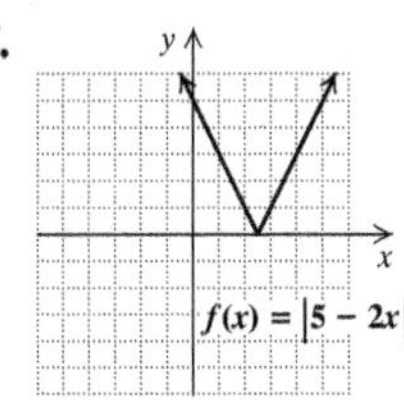

68. $2\sqrt{22}$ **69.** $2x$ **70.** $\frac{t^2 + 2t + 4}{t - 3}$ **71.** $\frac{2(2x - 5)}{3(x - 1)}$ **72.** 6

73. $\frac{t^2}{(t + 1)^2(t - 1)}$ **75.** Left to the student

77. 16, −16 **79.** $0, \frac{7}{2}, -8, -\frac{10}{3}$

Calculator Corner, p. 547

1. −3, 0.8 **2.** −1.5, 5 **3.** 3, 8 **4.** 2, 4

Calculator Corner, p. 548

1.–3. Left to the student

Exercise Set 7.2, p. 550

RC1. True **RC2.** False **RC3.** False **RC4.** True
CC1. 3 **CC2.** −1 **CC3.** $5x^2 + x - 9$ **CC4.** 0

1. $-4 \pm \sqrt{14}$ **3.** $\frac{-4 \pm \sqrt{13}}{3}$ **5.** $\frac{1}{2} \pm \frac{\sqrt{3}}{2}i$

7. $2 \pm 3i$ **9.** $\frac{-3 \pm \sqrt{41}}{2}$ **11.** $-1 \pm 2i$

13. **(a)** 0, −1; **(b)** (0, 0), (−1, 0)

15. **(a)** $\frac{3 \pm \sqrt{229}}{22}$; **(b)** $\left(\frac{3 + \sqrt{229}}{22}, 0\right), \left(\frac{3 - \sqrt{229}}{22}, 0\right)$

17. **(a)** $\frac{2}{5}$; **(b)** $\left(\frac{2}{5}, 0\right)$ **19.** −1, −2 **21.** 5, 10

23. $\frac{17 \pm \sqrt{249}}{10}$ **25.** $2 \pm i$ **27.** $\frac{2}{3}, \frac{3}{2}$ **29.** $2 \pm \sqrt{10}$

31. $\frac{3}{4}, -2$ **33.** $\frac{1}{2} \pm \frac{3}{2}i$ **35.** $1, -\frac{1}{2} \pm \frac{\sqrt{3}}{2}i$

37. $-3 \pm \sqrt{5}$; −0.764, −5.236 **39.** $3 \pm \sqrt{5}$; 5.236, 0.764

41. $\frac{3 \pm \sqrt{65}}{4}$; 2.766, −1.266 **43.** $\frac{4 \pm \sqrt{31}}{5}$; 1.914, −0.314

45. 2 **46.** 8 **47.** No solution **48.** $\frac{15}{4}$ **49.** $-1, \frac{3}{2}$ **50.** $\frac{1}{10}$

51. −2 **52.** $-\frac{10}{13}$ **53.** Left to the student; −0.797, 0.570

55. $\frac{1 \pm \sqrt{1 + 8\sqrt{5}}}{4}$ **57.** $\frac{-i \pm i\sqrt{1 + 4i}}{2}$

59. $\frac{-1 \pm 3\sqrt{5}}{6}$ **61.** $3 \pm \sqrt{13}$

Translating for Success, p. 559

1. B **2.** G **3.** F **4.** L **5.** N
6. C **7.** J **8.** E **9.** K **10.** A

Exercise Set 7.3, p. 560

RC1. (f) **RC2.** (e) **RC3.** (b)
RC4. (a) **RC5.** (d) **RC6.** (c)

1. Length: 9 ft; width: 2 ft **3.** Length: 18 yd; width: 9 yd

5. Base: 15 mi; height: 8 mi

7. Length: $\frac{51 + \sqrt{122{,}399}}{2}$ ft; width: $\frac{\sqrt{122{,}399} - 51}{2}$ ft

9. 2 in. **11.** 6 ft, 8 ft **13.** 23 and 24

15. Length: $(2 + \sqrt{14})$ ft $\approx$ 5.742 ft; width: $(\sqrt{14} - 2)$ ft $\approx$ 1.742 ft

17. $\frac{17 - \sqrt{109}}{2}$ in. $\approx$ 3.280 in.

19. $(7 + \sqrt{239})$ ft $\approx$ 22.460 ft; $(\sqrt{239} - 7)$ ft $\approx$ 8.460 ft

21. First part: 60 mph; second part: 50 mph

23. 6 km/h **25.** Cessna: 150 mph; Beechcraft: 200 mph; or Cessna: 200 mph; Beechcraft: 250 mph

27. To Hillsboro: 10 mph; return trip: 4 mph

29. About 11 mph **31.** $s = \sqrt{\frac{A}{6}}$ **33.** $r = \sqrt{\frac{Gm_1m_2}{F}}$

35. $c = \sqrt{\frac{E}{m}}$ **37.** $b = \sqrt{c^2 - a^2}$ **39.** $k = \frac{3 + \sqrt{9 + 8N}}{2}$

41. $r = \frac{-\pi h + \sqrt{\pi^2h^2 + 2\pi A}}{2\pi}$ **43.** $g = \frac{4\pi^2L}{T^2}$

45. $H = \sqrt{\frac{703W}{I}}$ **47.** $v = \frac{c\sqrt{m^2 - (m_0)^2}}{m}$ **49.** $\frac{1}{x - 2}$

50. $\frac{(x + 1)(x^2 + 2)}{(x - 1)(x^2 + x + 1)}$ **51.** $\frac{-x}{(x + 3)(x - 1)}$ **52.** $3x^2\sqrt{x}$

53. $2\sqrt{5}\,i$ **55.** $\pm\sqrt{2}$ **57.** $A(S) = \frac{\pi S}{6}$ **59.** $l = \frac{w + w\sqrt{5}}{2}$

Exercise Set 7.4, p. 570

RC1. (b) **RC2.** (a) **RC3.** (c)
RC4. (b) **RC5.** (b) **RC6.** (c)
CC1. $\sqrt{x}$ **CC2.** p^2 **CC3.** x^{-1} **CC4.** $2y + 7$

1. One real **3.** Two nonreal **5.** Two real **7.** One real
9. Two nonreal **11.** Two real **13.** Two real **15.** One real

17. $x^2 - 16 = 0$ **19.** $x^2 + 16 = 0$ **21.** $x^2 - 16x + 64 = 0$

23. $25x^2 - 20x - 12 = 0$ **25.** $12x^2 - (4k + 3m)x + km = 0$

27. $x^2 - \sqrt{3}x - 6 = 0$ **29.** $x^2 + 36 = 0$ **31.** $\pm\sqrt{3}$

33. 1, 81 **35.** −1, 1, 5, 7 **37.** $-\frac{1}{4}, \frac{1}{9}$ **39.** 1 **41.** −1, 1, 4, 6

43. ±2, ±5 **45.** −1, 2 **47.** $\pm\frac{\sqrt{15}}{3}, \pm\frac{\sqrt{6}}{2}$

49. −1, 125 **51.** $-\frac{11}{6}, -\frac{1}{6}$ **53.** $-\frac{3}{2}$ **55.** $\frac{9 \pm \sqrt{89}}{2}, -1 \pm \sqrt{3}$

57. $\left(\frac{4}{25}, 0\right)$ **59.** $(4, 0), (-1, 0), \left(\frac{3 + \sqrt{33}}{2}, 0\right), \left(\frac{3 - \sqrt{33}}{2}, 0\right)$

61. (−8, 0), (1, 0) **63.** Kenyan: 30 lb; Peruvian: 20 lb

64. Solution A: 4 L; solution B: 8 L **65.** $4x$ **66.** $3x^2$

67. $3a\sqrt[4]{2a}$ **68.** 4

69. f(x) = −(3/5)x + 4

70.

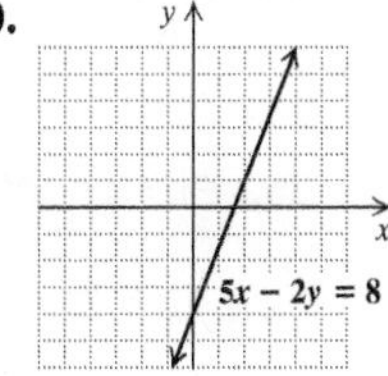

71.

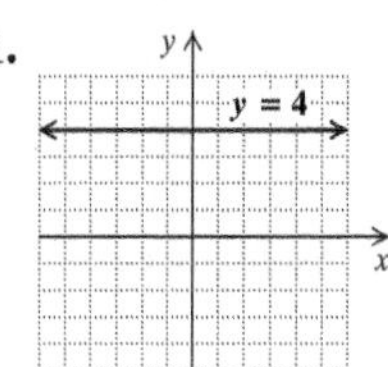

72.

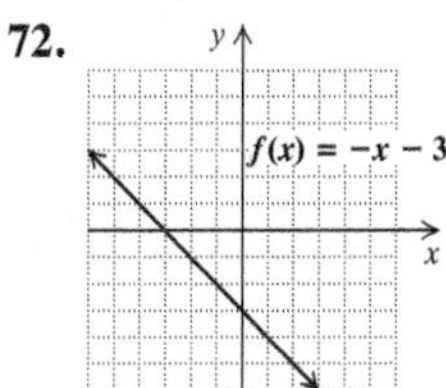

73. Left to the student **75. (a)** $-\frac{3}{5}$; **(b)** $-\frac{1}{3}$
77. $x^2 - \sqrt{3}x + 8 = 0$ **79.** $a = 1, b = 2, c = -3$
81. $\frac{100}{99}$ **83.** 259 **85.** $1, 3, -\frac{1}{2} \pm \frac{\sqrt{3}}{2}i, -\frac{3}{2} \pm \frac{3\sqrt{3}}{2}i$

Mid-Chapter Review: Chapter 7, p. 574

1. False **2.** True **3.** True **4.** False

5.
$$5x^2 + 3x = 4$$
$$\frac{1}{5}(5x^2 + 3x) = \frac{1}{5}\cdot 4$$
$$x^2 + \frac{3}{5}x = \frac{4}{5}$$
$$x^2 + \frac{3}{5}x + \frac{9}{100} = \frac{4}{5} + \frac{9}{100}$$
$$\left(x + \frac{3}{10}\right)^2 = \frac{89}{100}$$
$$x + \frac{3}{10} = \sqrt{\frac{89}{100}} \quad or \quad x + \frac{3}{10} = -\sqrt{\frac{89}{100}}$$
$$x + \frac{3}{10} = \frac{\sqrt{89}}{10} \quad or \quad x + \frac{3}{10} = -\frac{\sqrt{89}}{10}$$
$$x = -\frac{3}{10} + \frac{\sqrt{89}}{10} \quad or \quad x = -\frac{3}{10} - \frac{\sqrt{89}}{10}$$
The solutions are $-\frac{3}{10} \pm \frac{\sqrt{89}}{10}$.

6.
$$5x^2 + 3x = 4$$
$$5x^2 + 3x - 4 = 0$$
$$a = 5, \quad b = 3, \quad c = -4$$
$$x = \frac{-b \pm \sqrt{b^2 - 4ac}}{2a}$$
$$x = \frac{-3 \pm \sqrt{3^2 - 4\cdot 5\cdot(-4)}}{2\cdot 5}$$
$$x = \frac{-3 \pm \sqrt{9 + 80}}{10}$$
$$x = \frac{-3 \pm \sqrt{89}}{10}$$
$$x = -\frac{3}{10} \pm \frac{\sqrt{89}}{10}$$

7. $-2 \pm \sqrt{3}$ **8.** $-3, \frac{1}{2}$ **9.** $-5 \pm \sqrt{31}$ **10.** $\frac{1}{2} \pm \frac{\sqrt{21}}{2}$
11. One real solution; one x-intercept
12. Two real solutions; two x-intercepts
13. Two nonreal solutions; no x-intercepts
14. Two nonreal solutions; no x-intercepts
15. Two real solutions; two x-intercepts
16. Two real solutions; two x-intercepts
17. $x^2 - 9x - 10 = 0$ **18.** $x^2 - 169 = 0$
19. $x^2 - 2\sqrt{5}x - 15 = 0$ **20.** $x^2 + 16 = 0$
21. $x^2 + 12x + 36 = 0$ **22.** $21x^2 + 22x - 8 = 0$
23. 60 mph **24.** $s = \sqrt{\frac{R}{a}}$ **25.** $-\frac{4}{3}, 1$ **26.** $\pm\sqrt{3}, \pm\sqrt{5}$
27. $\frac{15 \pm \sqrt{145}}{8}$ **28.** $-1, -\frac{2}{7}$ **29.** $-1, 0$ **30.** $-11, 5$
31. $\pm\frac{4}{7}i$ **32.** $\pm\sqrt{6}, \pm 2i$ **33.** $\frac{-5 \pm \sqrt{73}}{2}$ **34.** $-6 \pm i$
35. $\frac{5}{2} \pm \frac{\sqrt{11}}{2}$ **36.** $\frac{7 \pm \sqrt{13}}{6}$ **37.** $\frac{1 \pm \sqrt{2}}{2}$ **38.** $-1 \pm 4i$
39. $8 \pm \sqrt{3}$ **40.** $3 \pm \sqrt{10}i$ **41.** $4 \pm \sqrt{26}$ **42.** 9
43. Given the solutions of a quadratic equation, it is possible to find an equation equivalent to the original equation but not necessarily expressed in the same form as the original equation. For example, we can find a quadratic equation with solutions -2 and 4:
$$[x - (-2)](x - 4) = 0$$
$$(x + 2)(x - 4) = 0$$
$$x^2 - 2x - 8 = 0.$$
Now $x^2 - 2x - 8 = 0$ has solutions -2 and 4. However, the original equation might have been in another form, such as $2x(x - 3) - x(x - 4) = 8$.
44. Given the quadratic equation $ax^2 + bx + c = 0$, we find $x = \frac{-b + \sqrt{b^2 - 4ac}}{2a}$ or $x = \frac{-b - \sqrt{b^2 - 4ac}}{2a}$ using the quadratic formula.
Then we have $ax^2 + bx + c =$
$$\left(x - \frac{-b + \sqrt{b^2 - 4ac}}{2a}\right)\left(x - \frac{-b - \sqrt{b^2 - 4ac}}{2a}\right).$$
Consider $5x^2 + 8x - 3$. First, we use the quadratic formula to solve $5x^2 + 8x - 3 = 0$:
$$x = \frac{-8 \pm \sqrt{8^2 - 4\cdot 5\cdot(-3)}}{2\cdot 5}$$
$$x = \frac{-8 \pm \sqrt{124}}{10} = \frac{-8 \pm 2\sqrt{31}}{10}$$
$$x = \frac{-4 \pm \sqrt{31}}{5}.$$
Then $5x^2 + 8x - 3 = \left(x - \frac{-4 - \sqrt{31}}{5}\right)\left(x - \frac{-4 + \sqrt{31}}{5}\right)$.
45. Set the product $(x - 1)(x - 2)(x - 3)(x - 4)(x - 5)(x - 6)(x - 7)$ equal to 0. **46.** Write an equation of the form $a(3x^2 + 1)^2 + b(3x^2 + 1) + c = 0$, where $a \neq 0$. To ensure that this equation has real-number solutions, select a, b, and c so that $b^2 - 4ac \geq 0$ and $3x^2 + 1 \geq 0$, which is true for $x \in \mathbb{R}$.

Exercise Set 7.5, p. 581

RC1. False **RC2.** True **RC3.** True **RC4.** False
CC1. Down **CC2.** Up **CC3.** Up **CC4.** Down

1.

x	$f(x)$
0	0
1	4
2	16
−1	4
−2	16

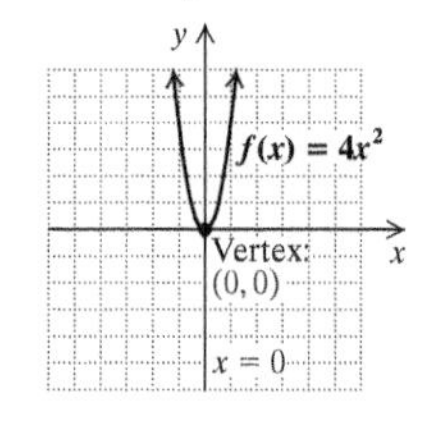

3.

x	$f(x)$
0	0
1	$\frac{1}{3}$
2	$\frac{4}{3}$
−1	$\frac{1}{3}$
−2	$\frac{4}{3}$

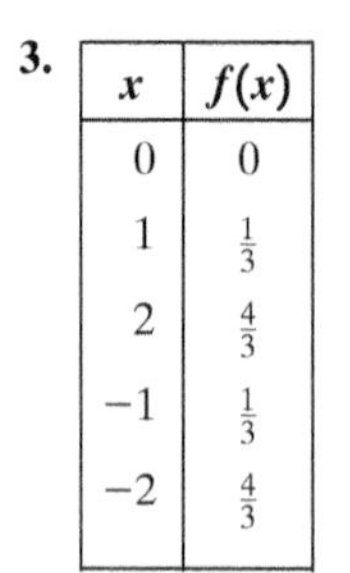
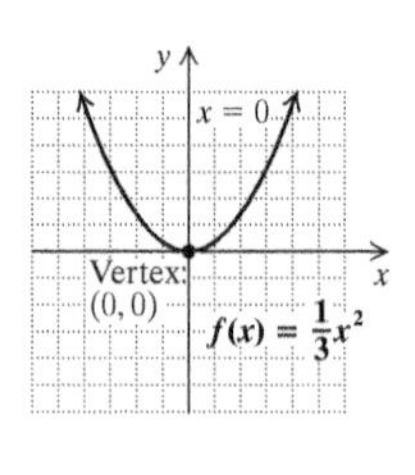

5.

x	$f(x)$
−3	0
−2	1
−1	4
−4	1
−5	4

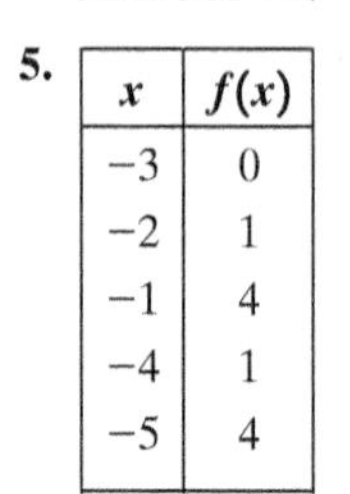
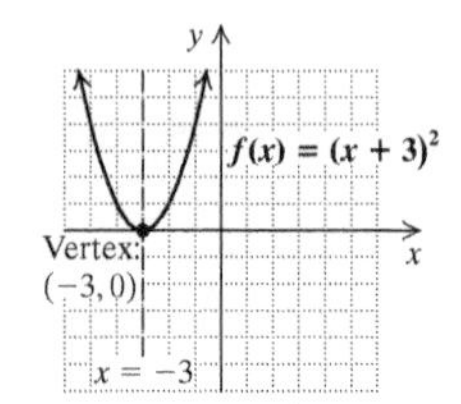

7.

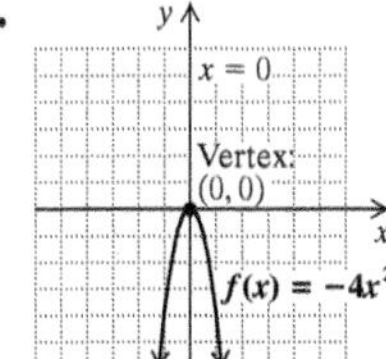

9.

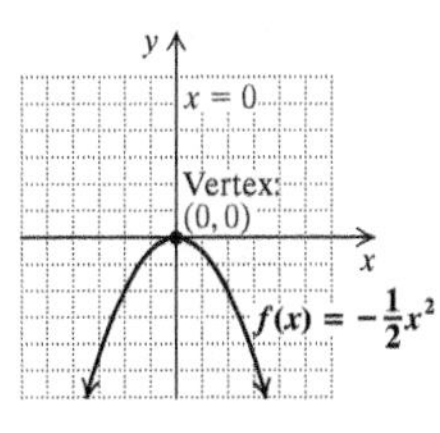

11.

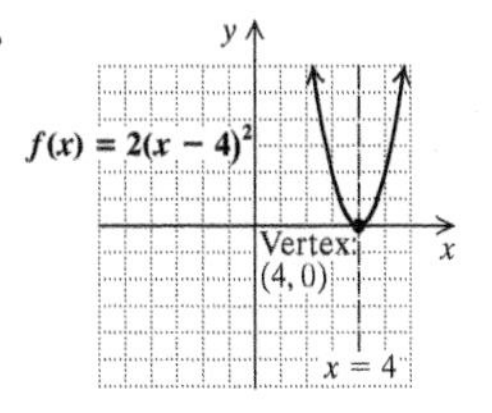

13.

x	$f(x)$
−2	0
−3	−2
−1	−2
−4	−8
0	−8

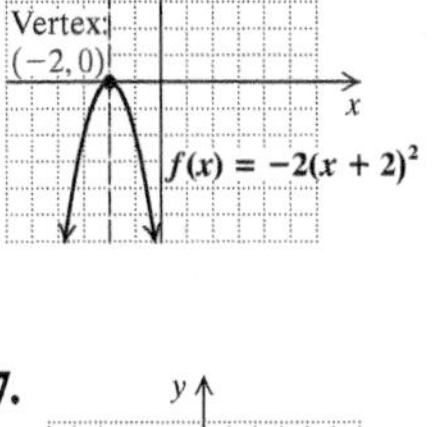

15.

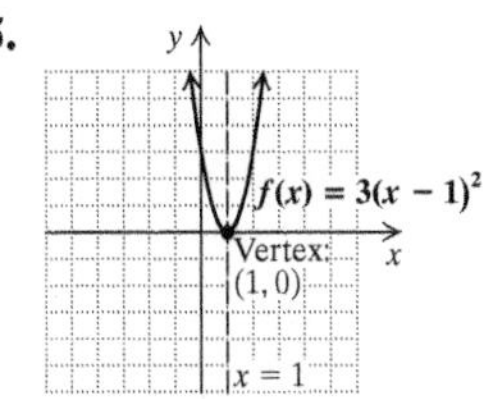

17.

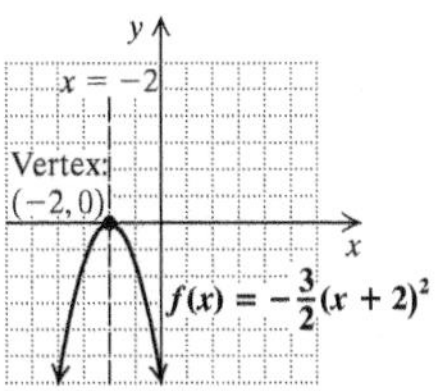

19.

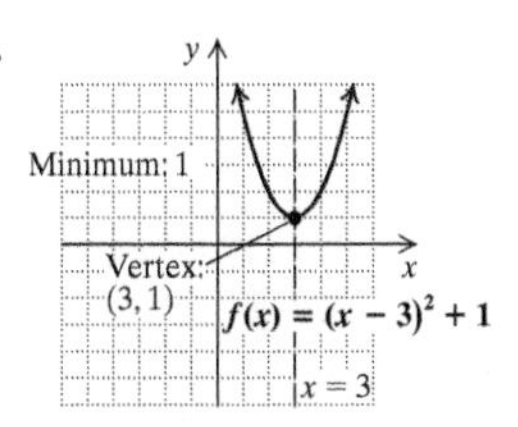

21.

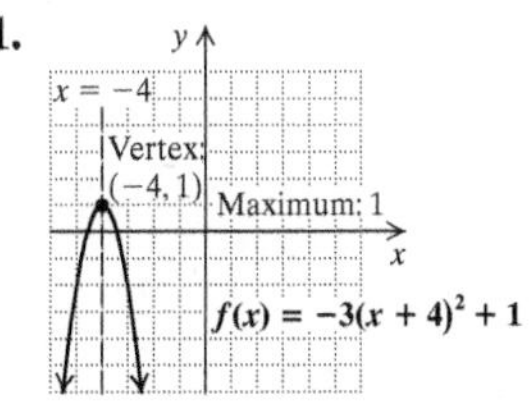

23.

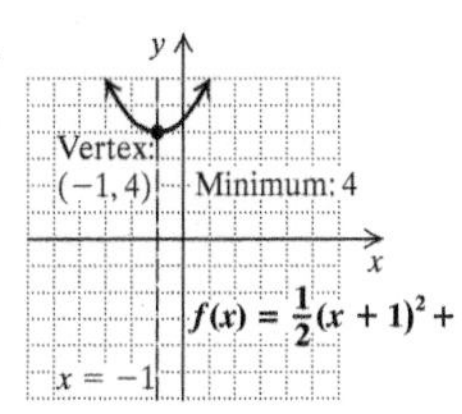

25.

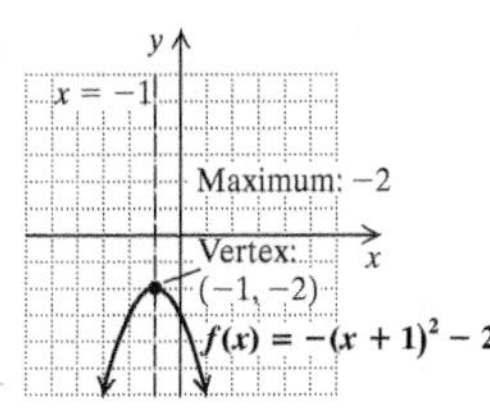

27. $-\frac{2}{5}$ **28.** 0

Visualizing for Success, p. 590

1. F **2.** H **3.** A **4.** I **5.** C **6.** J **7.** G **8.** B **9.** E **10.** D

Exercise Set 7.6, p. 591

RC1. Upward **RC2.** Minimum **RC3.** Vertex **RC4.** y-intercept

1. (a) $(1, -4)$; **(b)** $x = 1$; **(c)** minimum: -4;
(d)

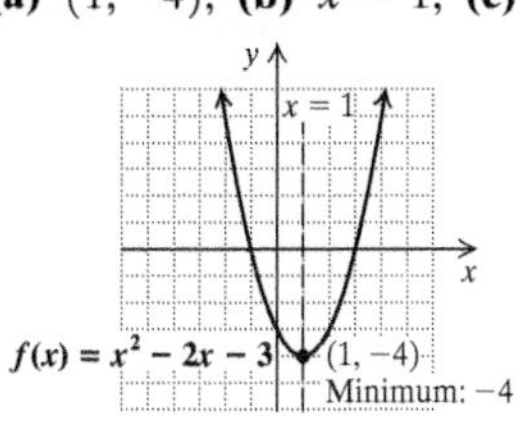

3. (a) $(-2, 2)$; **(b)** $x = -2$; **(c)** maximum: 2;
(d)

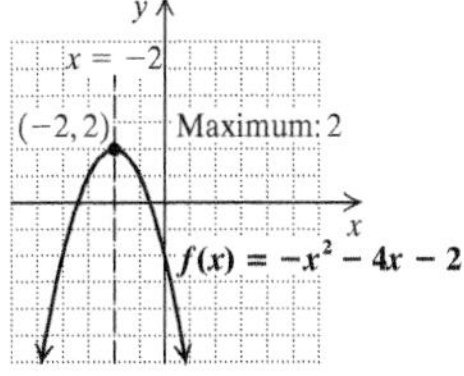

5. (a) $(4, 2)$; **(b)** $x = 4$; **(c)** minimum: 2;
(d)

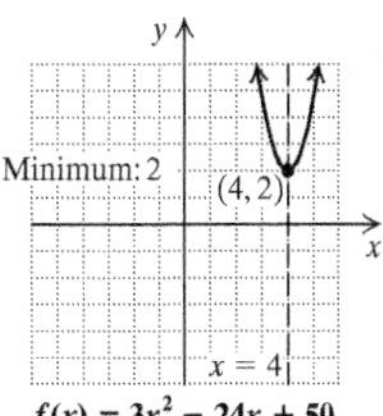

7. (a) $\left(-\frac{1}{2}, \frac{7}{2}\right)$; **(b)** $x = -\frac{1}{2}$; **(c)** maximum: $\frac{7}{2}$;
(d)

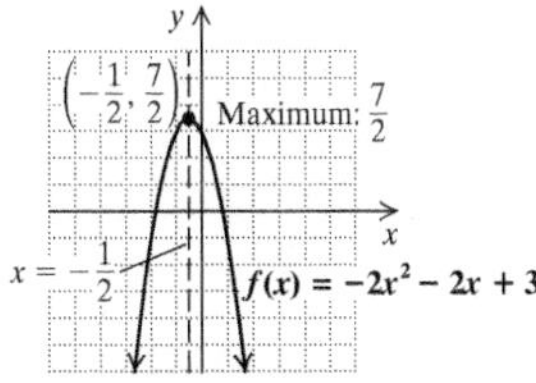

9. (a) $(0, 5)$; **(b)** $x = 0$; **(c)** maximum: 5;
(d)

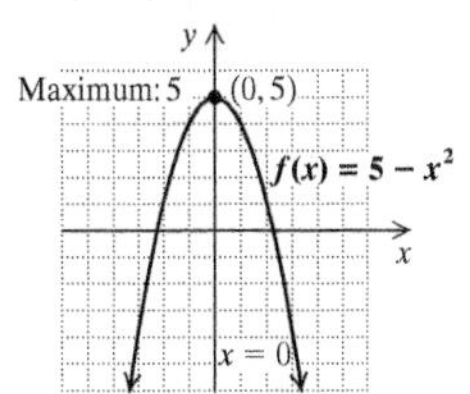

11. (a) $\left(-\frac{5}{4}, -\frac{41}{8}\right)$; **(b)** $x = -\frac{5}{4}$; **(c)** minimum: $-\frac{41}{8}$;
(d)

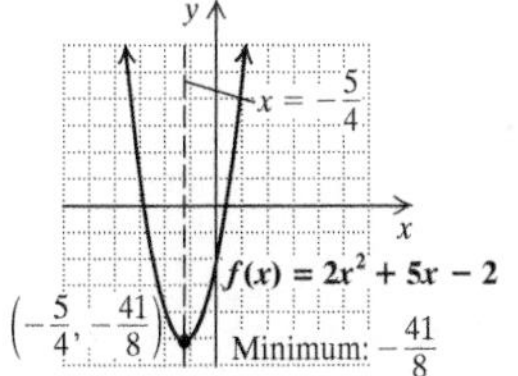

13. y-intercept: $(0, 1)$; x-intercepts: $(3 + 2\sqrt{2}, 0)$, $(3 - 2\sqrt{2}, 0)$
15. y-intercept: $(0, 20)$; x-intercepts: $(5, 0)$, $(-4, 0)$
17. y-intercept: $(0, 9)$; x-intercept: $\left(-\frac{3}{2}, 0\right)$
19. y-intercept: $(0, 8)$; x-intercepts: none
21. $D = 15w$ **22.** $C = \frac{89}{6}t$ **23.** 250; $y = \frac{250}{x}$
24. 250; $y = \frac{250}{x}$ **25.** $\frac{125}{2}$; $y = \frac{125}{2}x$ **26.** $\frac{2}{125}$; $y = \frac{2}{125}x$
27. (a) Minimum: −6.954; **(b)** maximum: 7.014

29.

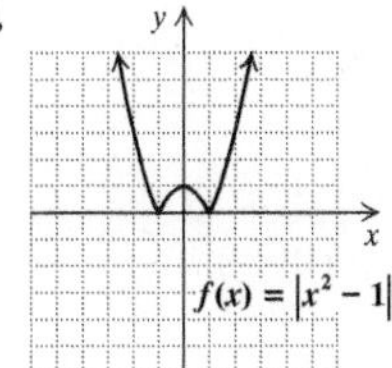

31.

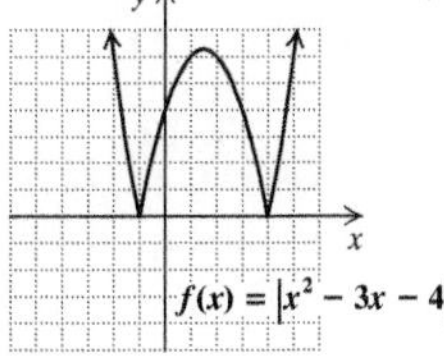

33. $f(x) = \frac{5}{16}x^2 - \frac{15}{8}x - \frac{35}{16}$, or $f(x) = \frac{5}{16}(x - 3)^2 - 5$

35.

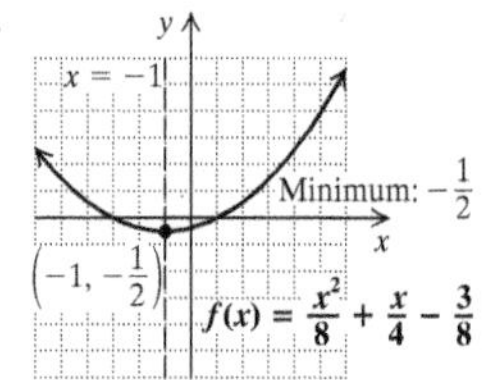

Calculator Corner, p. 596

1. Minimum: 1 **2.** Minimum: 4.875
3. Maximum: 6 **4.** Maximum: 0.5625

Exercise Set 7.7, p. 600

RC1. True **RC2.** False **RC3.** True **RC4.** False
CC1. $4 = 4a + 2b + c$ **CC2.** $0 = 25a + 5b + c$
CC3. $10 = 36a - 6b + c$
1. 180 ft by 180 ft **3.** 3.5 in. **5.** 3.5 hundred, or 350 bicycles
7. 200 ft^2; 10 ft by 20 ft **9.** 11 days after the concert was announced; about 62 tickets **11.** $P(x) = -x^2 + 980x - 3000$; \$237,100 at $x = 490$ **13.** 121; 11 and 11 **15.** −4; 2 and −2
17. 36; −6 and −6 **19.** $f(x) = ax^2 + bx + c, a < 0$
21. $f(x) = ax^2 + bx + c, a > 0$ **23.** Polynomial, neither quadratic nor linear **25.** $f(x) = mx + b$
27. $f(x) = 2x^2 + 3x - 1$ **29.** $f(x) = -\frac{1}{4}x^2 + 3x - 5$
31. (a) $A(s) = \frac{3}{16}s^2 - \frac{135}{4}s + 1750$; **(b)** about 531 per 200,000,000 kilometers driven **33.** $D(x) = -0.008x^2 + 0.8x$; 15 ft **35.** $H(x) = -\frac{1}{32}x^2 + \frac{5}{8}x + 65$, where x is the number of years after 1995; 68.125% **37.** $-2x^2 + 2x - 9$ **38.** $-4m^2n$
39. $c^4d^2 - 4y^2$ **40.** $(10t + 9)(10t - 9)$ **41.** $3x(2x - 5)^2$
42. $(3y - 4)(2y + 3)$

Exercise Set 7.8, p. 610

RC1. Negative **RC2.** Positive
RC3. Positive **RC4.** Negative
CC1. Yes **CC2.** No **CC3.** No **CC4.** Yes
1. $\{x \mid x < -2 \text{ or } x > 6\}$, or $(-\infty, -2) \cup (6, \infty)$
3. $\{x \mid -2 \le x \le 2\}$, or $[-2, 2]$ **5.** $\{x \mid -1 \le x \le 4\}$, or $[-1, 4]$
7. $\{x \mid -1 < x < 2\}$, or $(-1, 2)$ **9.** All real numbers, or $(-\infty, \infty)$ **11.** $\{x \mid 2 < x < 4\}$, or $(2, 4)$
13. $\{x \mid x < -2 \text{ or } 0 < x < 2\}$, or $(-\infty, -2) \cup (0, 2)$
15. $\{x \mid -9 < x < -1 \text{ or } x > 4\}$, or $(-9, -1) \cup (4, \infty)$
17. $\{x \mid x < -3 \text{ or } -2 < x < 1\}$, or $(-\infty, -3) \cup (-2, 1)$
19. $\{x \mid x < 6\}$, or $(-\infty, 6)$ **21.** $\{x \mid x < -1 \text{ or } x > 3\}$, or $(-\infty, -1) \cup (3, \infty)$ **23.** $\{x \mid -\frac{2}{3} \le x < 3\}$, or $[-\frac{2}{3}, 3)$
25. $\{x \mid 2 < x < \frac{5}{2}\}$, or $(2, \frac{5}{2})$ **27.** $\{x \mid x < -1 \text{ or } 2 < x < 5\}$, or $(-\infty, -1) \cup (2, 5)$ **29.** $\{x \mid -3 \le x < 0\}$, or $[-3, 0)$
31. $\{x \mid 1 < x < 2\}$, or $(1, 2)$ **33.** $\{x \mid x < -4 \text{ or } 1 < x < 3\}$, or $(-\infty, -4) \cup (1, 3)$ **35.** $\{x \mid 0 < x < \frac{1}{3}\}$, or $(0, \frac{1}{3})$
37. $\{x \mid x < -3 \text{ or } -2 < x < 1 \text{ or } x > 4\}$, or $(-\infty, -3) \cup (-2, 1) \cup (4, \infty)$ **39.** $\frac{5}{3}$ **40.** $\frac{5}{2a}$ **41.** $\frac{4a}{b^2}\sqrt{a}$
42. $\frac{3c}{7d}\sqrt[3]{c^2}$ **43.** $\sqrt{2}$ **44.** $17\sqrt{5}$ **45.** $(10a + 7)\sqrt[3]{2a}$
46. $3\sqrt{10} - 4\sqrt{5}$ **47.** Left to the student
49. $\{x \mid 1 - \sqrt{3} \le x \le 1 + \sqrt{3}\}$, or $[1 - \sqrt{3}, 1 + \sqrt{3}]$
51. All real numbers except 0, or $(-\infty, 0) \cup (0, \infty)$
53. $\{x \mid x < \frac{1}{4} \text{ or } x > \frac{5}{2}\}$, or $(-\infty, \frac{1}{4}) \cup (\frac{5}{2}, \infty)$
55. (a) $\{t \mid 0 < t < 2\}$, or $(0, 2)$; **(b)** $\{t \mid t > 10\}$, or $(10, \infty)$

Summary and Review: Chapter 7, p. 613

Vocabulary Reinforcement

1. Quadratic **2.** Rational **3.** Complete **4.** Discriminant
5. Quadratic **6.** Parabola **7.** Symmetry **8.** Vertex

Concept Reinforcement

1. False **2.** True **3.** False

Study Guide

1. $2 \pm 3i$ **2.** $6 \pm \sqrt{5}$ **3.** $5 \pm \sqrt{2}$, or 6.414 and 3.586
4. (a) Two real solutions; **(b)** two nonreal solutions
5. $5x^2 - 13x - 6 = 0$ **6.** $\pm\sqrt{2}, \pm 3$ **7.** Vertex: $(-1, -2)$; line of symmetry: $x = -1$; maximum: −2;

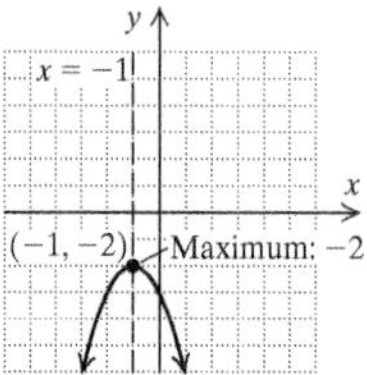

8. y-intercept: $(0, 4)$; x-intercepts: $(3 - \sqrt{5}, 0)$, $(3 + \sqrt{5}, 0)$
9. $\{x \mid x < 4 \text{ or } x > 10\}$, or $(-\infty, 4) \cup (10, \infty)$
10. $\{x \mid 5 < x \le 11\}$, or $(5, 11]$

Review Exercises

1. (a) $\pm\frac{\sqrt{14}}{2}$; **(b)** $\left(-\frac{\sqrt{14}}{2}, 0\right), \left(\frac{\sqrt{14}}{2}, 0\right)$ **2.** $0, -\frac{5}{14}$
3. 3, 9 **4.** $\frac{7}{2} \pm \frac{\sqrt{3}}{2}i$ **5.** $\frac{-3 \pm \sqrt{13}}{4}$ **6.** 3, 5
7. $-2 \pm \sqrt{3}$; −0.268, −3.732 **8.** 4, −2 **9.** $4 \pm 4\sqrt{2}$
10. $\frac{1 \pm \sqrt{481}}{15}$ **11.** $-3 \pm \sqrt{7}$ **12.** 0.901 sec
13. Length: 14 cm; width: 9 cm **14.** 1 in. **15.** First part: 50 mph; second part: 40 mph **16.** Two real **17.** Two nonreal
18. $25x^2 + 10x - 3 = 0$ **19.** $x^2 + 8x + 16 = 0$
20. $p = \frac{9\pi^2}{N^2}$ **21.** $T = \sqrt{\frac{3B}{2A}}$ **22.** 2, −2, 3, −3
23. 3, −5 **24.** $\pm\sqrt{7}, \pm\sqrt{2}$ **25.** 81, 16
26. (a) $(1, 3)$; **(b)** $x = 1$; **(c)** maximum: 3;
(d)

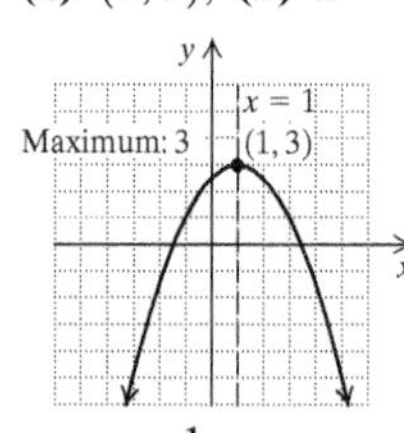

27. (a) $\left(\frac{1}{2}, \frac{23}{4}\right)$; **(b)** $x = \frac{1}{2}$; **(c)** minimum: $\frac{23}{4}$;
(d)

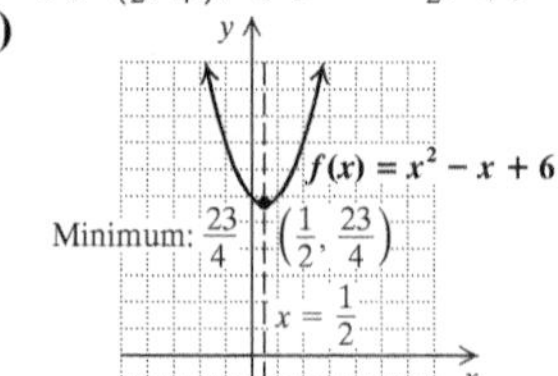

28. **(a)** $(-2, 4)$; **(b)** $x = -2$; **(c)** maximum: 4;
(d)

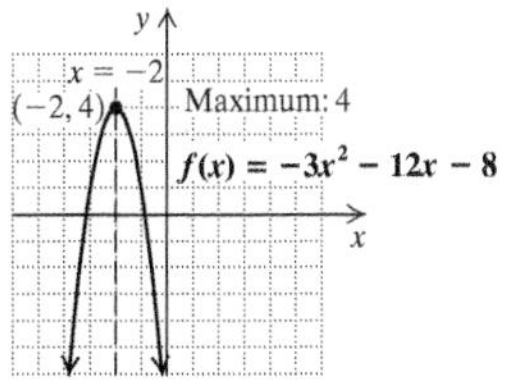

29. y-intercept: $(0, 14)$; x-intercepts: $(2, 0)$, $(7, 0)$
30. y-intercept: $(0, -3)$; x-intercepts: $(2 - \sqrt{7}, 0)$ and $(2 + \sqrt{7}, 0)$ **31.** -121; 11 and -11
32. $f(x) = -x^2 + 6x - 2$ **33.** $\{x|-2 < x < 1 \text{ or } x > 2\}$, or $(-2, 1) \cup (2, \infty)$ **34.** $\{x|x < -4 \text{ or } -2 < x < 1\}$, or $(-\infty, -4) \cup (-2, 1)$ **35.** **(a)** $f(x) = \frac{37}{12}x^2 + \frac{121}{6}x + 170$;
(b) \$462.25 million **36.** B **37.** D
38. $b = 19$ cm, $h = 19$ cm; $A = 180.5\text{ cm}^2$ **39.** 18 and 324

Understanding Through Discussion and Writing

1. Yes; for any quadratic function $f(x) = ax^2 + bx + c$, $f(0) = c$, so the graph of every quadratic function has a y-intercept, $(0, c)$. **2.** If the leading coefficient is positive, the graph of the function opens up and hence has a minimum value. If the leading coefficient is negative, the graph of the function opens down and hence has a maximum value. **3.** When an input of $y = (x + 3)^2$ is 3 less than (or 3 units to the left of) an input of $y = x^2$, the outputs are the same. In addition, for any input, the output of $f(x) = (x + 3)^2 - 4$ is 4 less than (or 4 units down from) the output of $f(x) = (x + 3)^2$. Thus the graph of $f(x) = (x + 3)^2 - 4$ looks like the graph of $f(x) = x^2$ translated 3 units to the left and 4 units down. **4.** Find a quadratic function $f(x)$ whose graph lies entirely above the x-axis or a quadratic function $g(x)$ whose graph lies entirely below the x-axis. Then write $f(x) < 0, f(x) \le 0, g(x) > 0$, or $g(x) \ge 0$. For example, the quadratic inequalities $x^2 + 1 < 0$ and $-x^2 - 5 \ge 0$ have no solution. **5.** No; if the vertex is off the x-axis, then due to symmetry, the graph has either no x-intercept or two x-intercepts. **6.** The x-coordinate of the vertex lies halfway between the x-coordinates of the x-intercepts. The function must be evaluated for this value of x in order to determine the y-coordinate of the vertex.

Test: Chapter 7, p. 619

1. [7.1a] **(a)** $\pm\frac{2\sqrt{3}}{3}$; **(b)** $\left(\frac{2\sqrt{3}}{3}, 0\right), \left(-\frac{2\sqrt{3}}{3}, 0\right)$
2. [7.2a] $-\frac{1}{2} \pm \frac{\sqrt{3}}{2}i$ **3.** [7.4c] 49, 1 **4.** [7.2a] 9, 2
5. [7.4c] $\pm\frac{\sqrt{5}}{2}, \pm\sqrt{3}$ **6.** [7.2a] $-2 \pm \sqrt{6}$; 0.449, -4.449
7. [7.2a] 0, 2 **8.** [7.1b] $2 \pm \sqrt{3}$ **9.** [7.1c] About 6.7 sec
10. [7.3a] About 2.89 mph **11.** [7.7a] 7 cm by 7 cm
12. [7.1c] About 0.946 sec **13.** [7.4a] Two nonreal
14. [7.4b] $x^2 - 4\sqrt{3}x + 9 = 0$ **15.** [7.3b] $T = \sqrt{\frac{V}{48}}$, or $\frac{\sqrt{3V}}{12}$
16. [7.6a] **(a)** $(-1, 1)$; **(b)** $x = -1$; **(c)** maximum: 1;
(d)

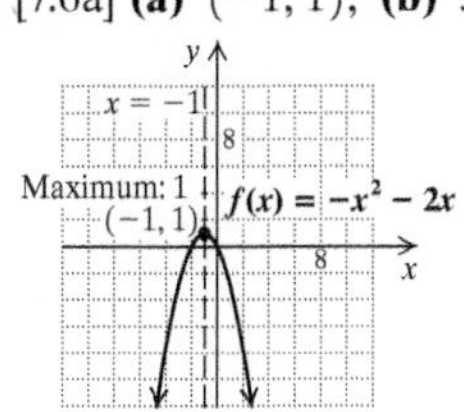

17. [7.6a] **(a)** $(3, 5)$; **(b)** $x = 3$; **(c)** minimum: 5;
(d)

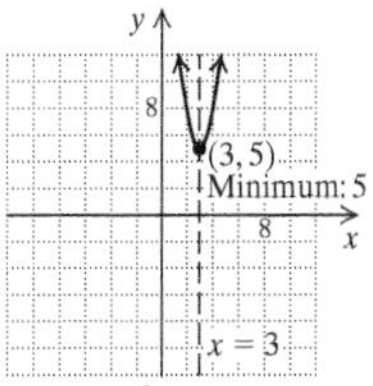

18. [7.6b] y-intercept: $(0, -1)$; x-intercepts: $(2 - \sqrt{3}, 0)$, $(2 + \sqrt{3}, 0)$ **19.** [7.7a] -16; 4 and -4
20. [7.7b] $f(x) = \frac{1}{5}x^2 - \frac{3}{5}x$
21. [7.7b] **(a)** $P(d) = \frac{1}{64}d^2 + \frac{5}{16}d + \frac{13}{2}$; **(b)** \$13.94
22. [7.8a] $\{x|-1 < x < 7\}$, or $(-1, 7)$
23. [7.8b] $\{x|-3 < x < 5\}$, or $(-3, 5)$
24. [7.8b] $\{x|-3 < x < 1 \text{ or } x \ge 2\}$, or $(-3, 1) \cup [2, \infty)$
25. [7.4b] C
26. [7.6a, b] $f(x) = -\frac{4}{7}x^2 + \frac{20}{7}x + 8$; maximum: $\frac{81}{7}$ **27.** [7.2a] $\frac{1}{2}$

Cumulative Review: Chapters 1–7, p. 621

1. [6.7a] About 422 yd **2.** [4.1d] $10x^2 - 8x + 6$
3. [4.2a] $2x^3 - 9x^2 + 7x - 12$ **4.** [5.1d] $\frac{2(a - 4)}{5}$
5. [5.1e] $\frac{1}{y(y + 6)}$ **6.** [5.2c] $\frac{-(m - 3)(m - 2)}{(m + 1)(m - 5)}$
7. [5.3b, c] $9x^2 - 13x + 26 + \frac{-50}{x + 2}$ **8.** [5.4a] $\frac{y - x}{xy(x + y)}$
9. [6.1b] 0.6 **10.** [6.1b] $3(x - 2)$
11. [6.4a] $12\sqrt{5}$ **12.** [6.5b] $\frac{\sqrt{6} + 9\sqrt{2} - 12\sqrt{3} - 4}{-26}$
13. [6.2d] 256 **14.** [6.8c] $17 + 7i$ **15.** [6.8e] $-\frac{2}{3} - 2i$
16. [4.5a, b] $(2t + 5)(t - 6)$ **17.** [4.4a] $(a + 9)(a - 6)$
18. [4.3a] $-3a^2(a - 4)$ **19.** [4.6d] $(8a + 3b)(8a - 3b)$
20. [4.6a] $3(a - 6)^2$ **21.** [4.6d] $\left(\frac{1}{3}a - 1\right)\left(\frac{1}{9}a^2 + \frac{1}{3}a + 1\right)$
22. [4.3b] $(4a + 3)(6a^2 - 5)$ **23.** [4.3a] $(x + 1)(2x + 1)$
24. [1.1d] $\frac{11}{13}$ **25.** [1.2a] $r = \frac{mv^2}{F}$
26. [1.4c] $\{x|x \ge \frac{5}{14}\}$, or $\left[\frac{5}{14}, \infty\right)$ **27.** [1.5b] $\{x|x < -\frac{4}{3} \text{ or } x > 6\}$, or $\left(-\infty, -\frac{4}{3}\right) \cup (6, \infty)$ **28.** [1.6e] $\{x|-\frac{13}{4} \le x \le \frac{15}{4}\}$, or $\left[-\frac{13}{4}, \frac{15}{4}\right]$
29. [3.3a] $(-4, 1)$ **30.** [3.5a] $\left(\frac{1}{2}, 3, -5\right)$ **31.** [4.8a] $\frac{1}{5}, -3$
32. [5.5a] $-\frac{5}{3}$ **33.** [5.5a] 3 **34.** [5.7a] $m = \frac{aA}{h - A}$
35. [6.6a] $\frac{37}{2}$ **36.** [6.6b] 11 **37.** [4.8a] 4
38. [7.2a] $\frac{3}{2} \pm \frac{\sqrt{55}}{2}i$ **39.** [7.2a] $\frac{17 \pm \sqrt{145}}{2}$
40. [7.3b] $a = \sqrt{P^2 + b^2}$
41. [7.8b] $\{x|-3 < x < -2 \text{ or } -1 < x < 1\}$, or $(-3, -2) \cup (-1, 1)$
42. [7.8a] $\{x|x < -\frac{5}{2} \text{ or } x > \frac{5}{2}\}$, or $\left(-\infty, -\frac{5}{2}\right) \cup \left(\frac{5}{2}, \infty\right)$
43. [2.5a] **44.** [3.7b]

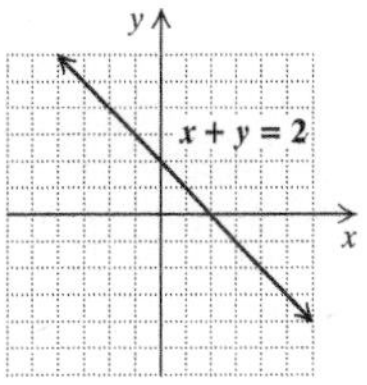

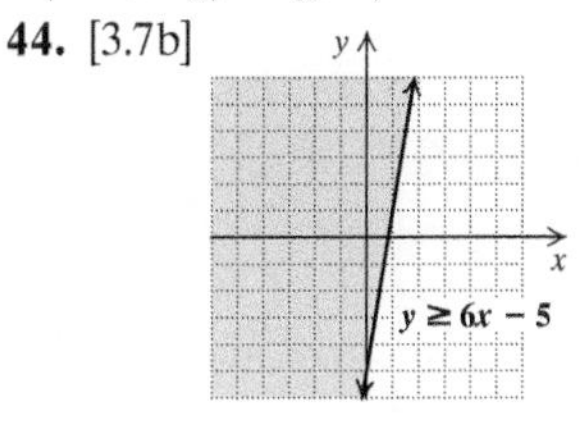

45. [3.7b]

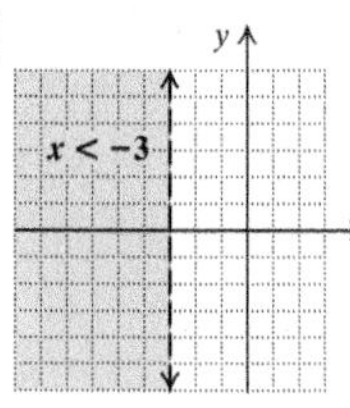

46. [3.7c]

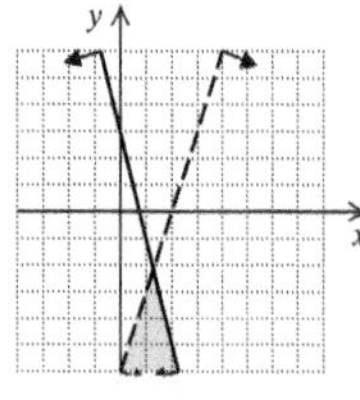

47. [7.6a]

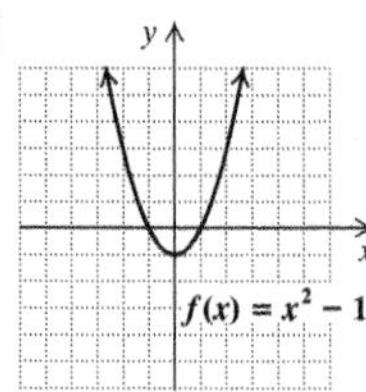

48. [7.6a]

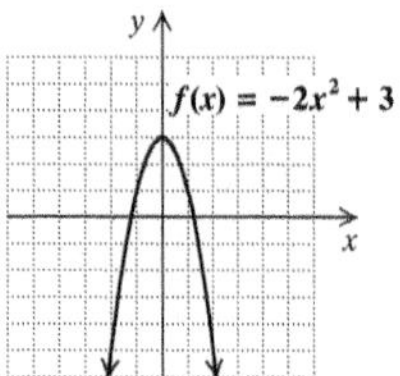

49. [2.6b] $y = \frac{1}{2}x + 4$ **50.** [2.6d] $y = -3x + 1$
51. [7.3a] 16 km/h **52.** [7.7a] 14 ft by 14 ft; 196 ft^2
53. [3.2b], [3.3b] 36 **54.** [5.6a] 2 hr **55.** [7.1b] A
56. [7.4c] B **57.** [7.4c] $\frac{2}{51 + 7\sqrt{61}}$, or $\frac{-51 + 7\sqrt{61}}{194}$

58. [4.6d] $\left(\frac{a}{2} + \frac{2b}{9}\right)\left(\frac{a^2}{4} - \frac{ab}{9} + \frac{4b^2}{81}\right)$

CHAPTER 8

Calculator Corner, p. 628

1. Left to the student

Exercise Set 8.1, p. 632

RC1. True **RC2.** False **RC3.** False **RC4.** True
CC1. $a > 1$ **CC2.** $0 < a < 1$ **CC3.** $0 < a < 1$
CC4. $a > 1$

1.

x	$f(x)$
0	1
1	2
2	4
3	8
−1	$\frac{1}{2}$
−2	$\frac{1}{4}$
−3	$\frac{1}{8}$

3.

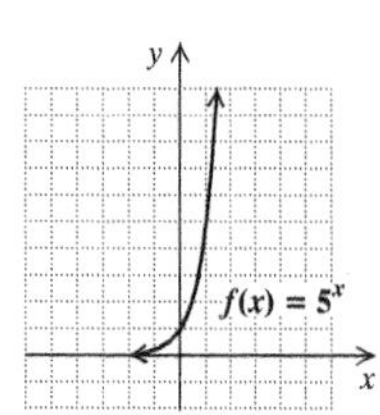

5.

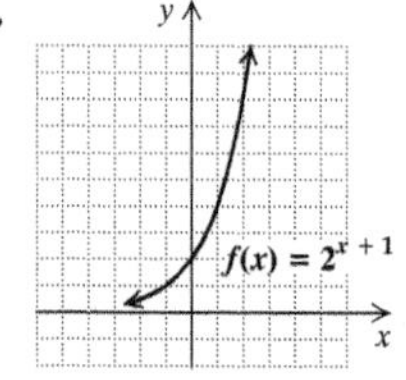

7.

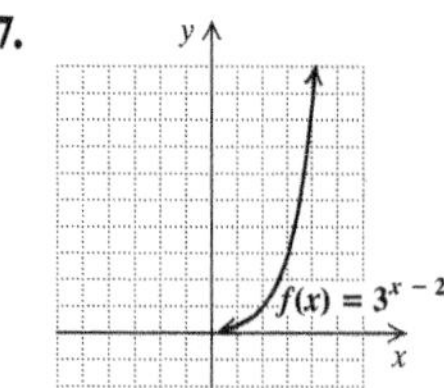

9.

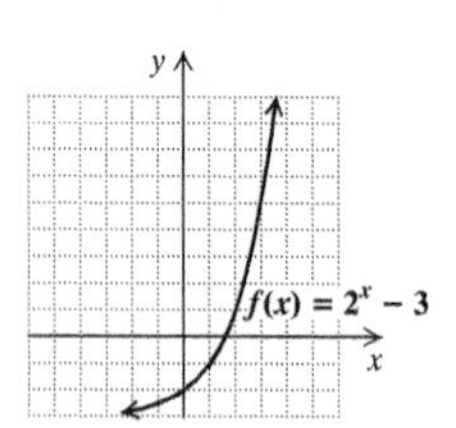

11.

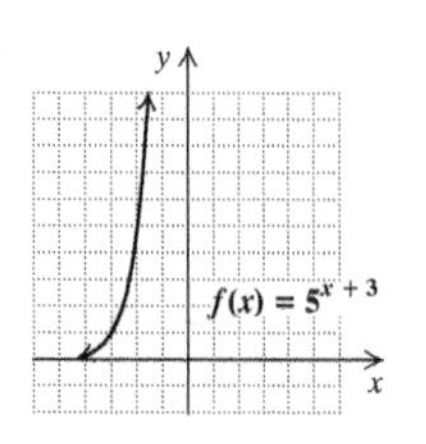

13.

x	$f(x)$
0	1
1	$\frac{1}{2}$
2	$\frac{1}{4}$
3	$\frac{1}{8}$
−1	2
−2	4
−3	8

15.

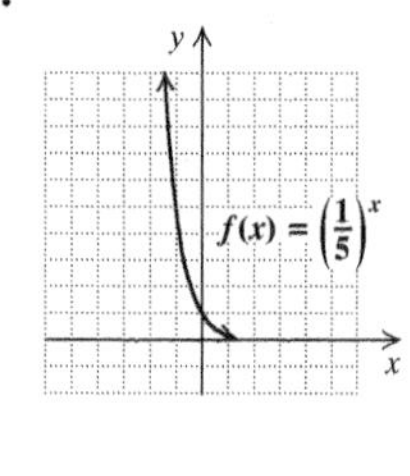

17.

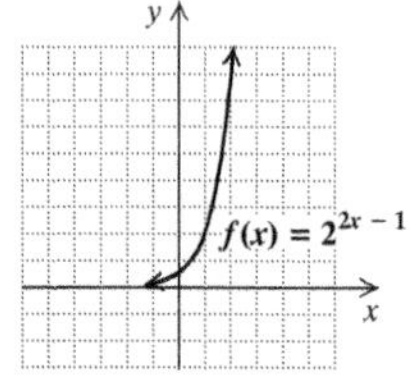

19.

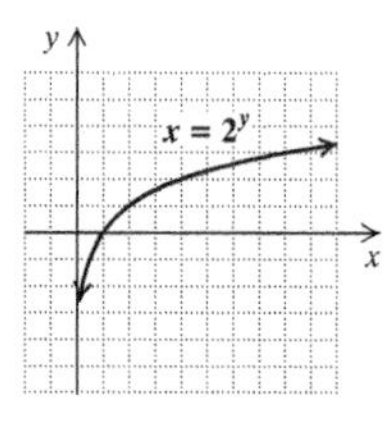

21.

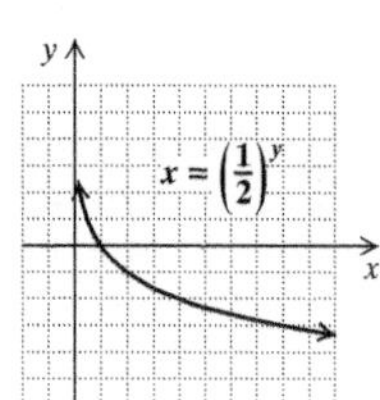

23.

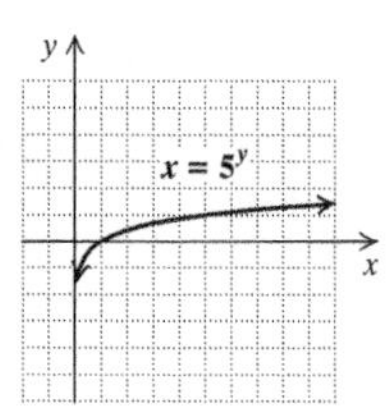

25.

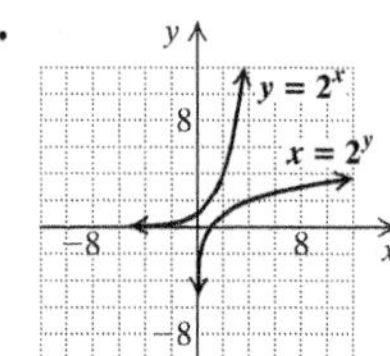

27. **(a)** $A(t) = \$50{,}000(1.02)^t$; **(b)** \$50,000; \$51,000; \$52,020; \$54,121.61; \$58,582.97; \$60,949.72; \$74,297.37;
(c)

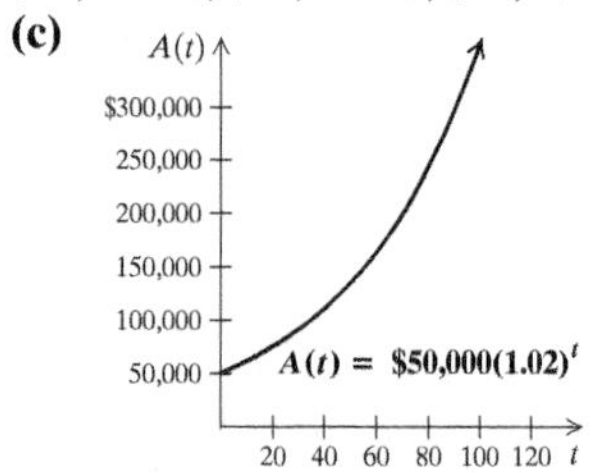

29. \$2161.16 **31.** \$5287.54
33. **(a)** 374 nene; about 1812 nene; about 3065 nene;
(b)

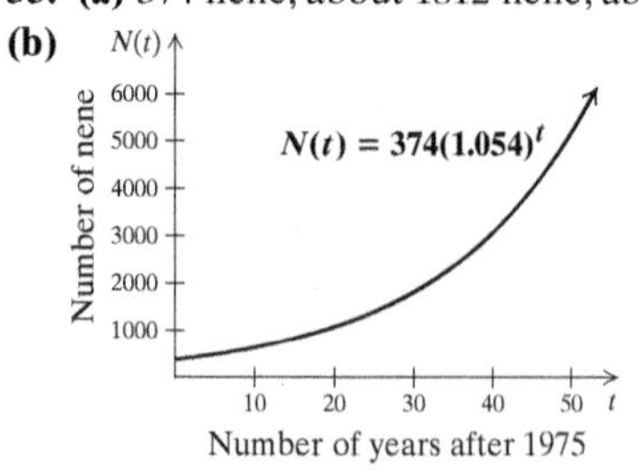

35. **(a)** \$6553; about \$2454; about \$919;
(b)

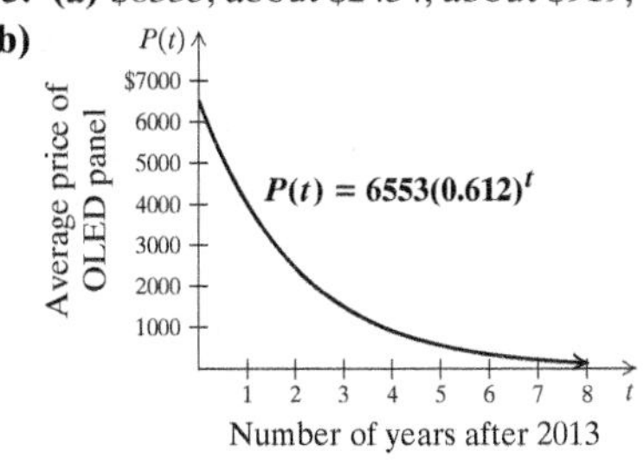

37. **(a)** 335,000 cans; 150,382 cans; 30,304 cans;
(b)

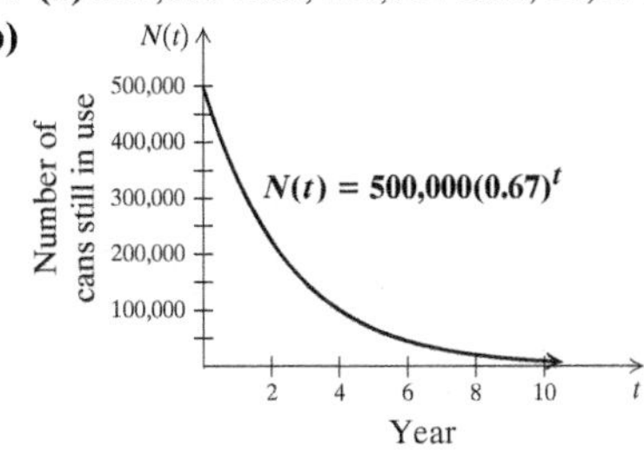

39. $\frac{1}{x^2}$ **40.** $\frac{1}{x^{12}}$ **41.** 1 **42.** 1 **43.** $\frac{2}{3}$ **44.** 2.7

45. $\frac{1}{x^7}$ **46.** $\frac{1}{x^{10}}$ **47.** x **48.** x **49.** 5^4, or 625

51.

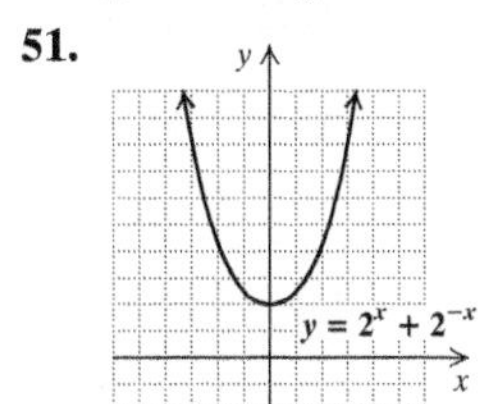

53.

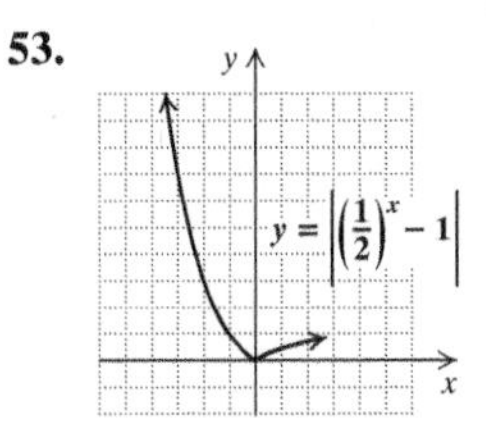

55.

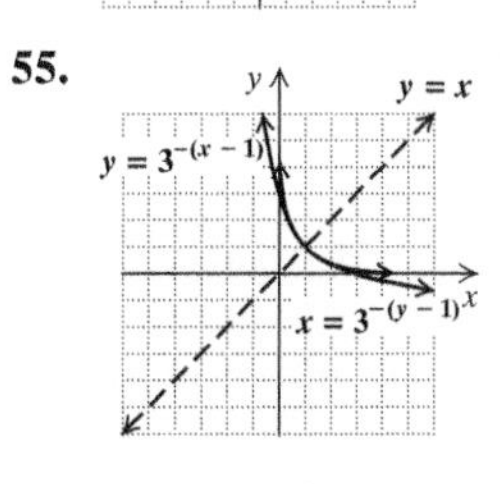

57. Left to the student

Calculator Corner, p. 645

1.

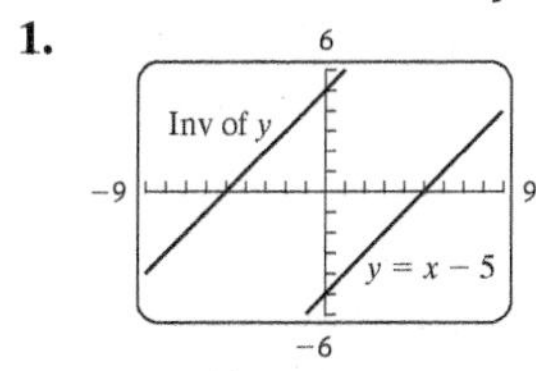

2.

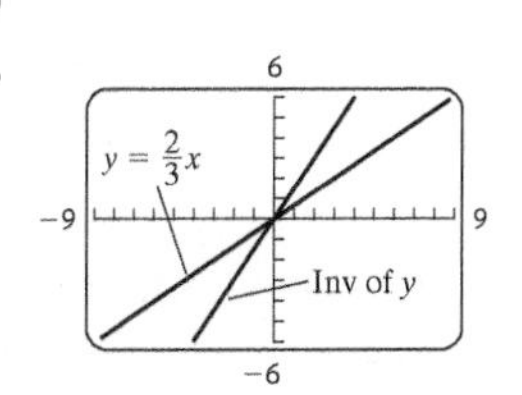

3.

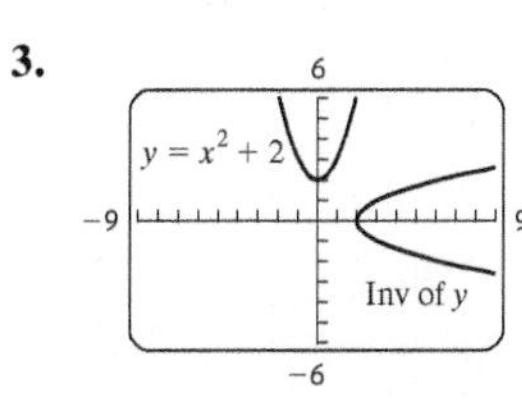

4.

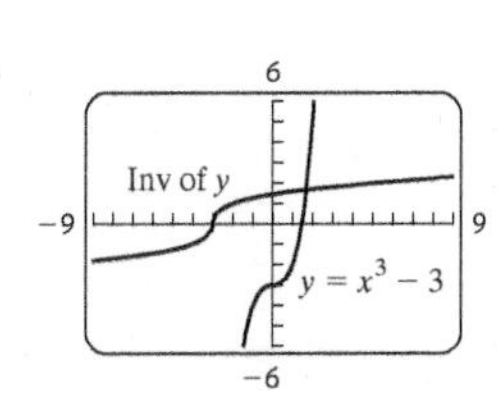

Exercise Set 8.2, p. 649

RC1. Relation **RC2.** Inverse **RC3.** Horizontal
RC4. One-to-one **RC5.** Inverse **RC6.** Composition
CC1. 3 **CC2.** 24 **CC3.** 1 **CC4.** −2 **CC5.** 3
CC6. Not defined
1. $-8x + 9; -8x + 18$ **3.** $12x^2 - 12x + 5; 6x^2 + 3$

5. $\frac{16}{x^2} - 1; \frac{2}{4x^2 - 1}$ **7.** $x^4 - 10x^2 + 30; x^4 + 10x^2 + 20$

9. $f(x) = x^2, g(x) = 5 - 3x$ **11.** $f(x) = \sqrt{x}, g(x) = 5x + 2$

13. $f(x) = \frac{1}{x}, g(x) = x - 1$ **15.** $f(x) = \frac{1}{\sqrt{x}}, g(x) = 7x + 2$

17. $f(x) = x^4, g(x) = \sqrt{x} + 5$

19. Inverse: $\{(2, 1), (-3, 6), (-5, -3)\}$

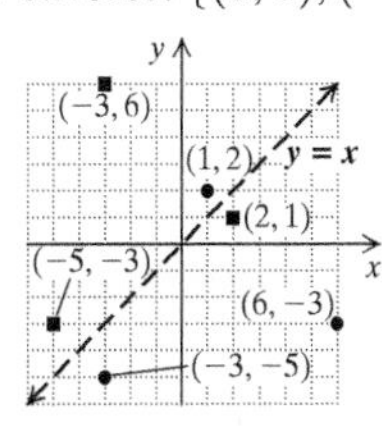

21. Inverse: $x = 2y + 6$

x	y
4	−1
6	0
8	1
10	2
12	3

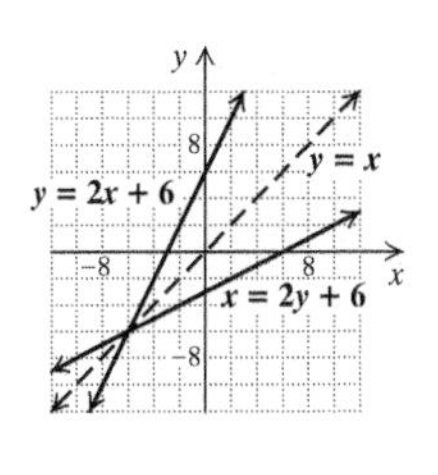

23. Yes **25.** No **27.** No **29.** Yes **31.** $f^{-1}(x) = \frac{x + 2}{5}$

33. $f^{-1}(x) = \frac{-2}{x}$ **35.** $f^{-1}(x) = \frac{3}{4}(x - 7)$ **37.** $f^{-1}(x) = \frac{2}{x} - 5$

39. Not one-to-one **41.** $f^{-1}(x) = \frac{1 - 3x}{5x - 2}$

43. $f^{-1}(x) = \sqrt[3]{x + 1}$ **45.** $f^{-1}(x) = x^3$

47. $f^{-1}(x) = 2x + 6$

x	$f(x)$
−4	−5
0	−3
2	−2
4	−1

x	$f^{-1}(x)$
−5	−4
−3	0
−2	2
−1	4

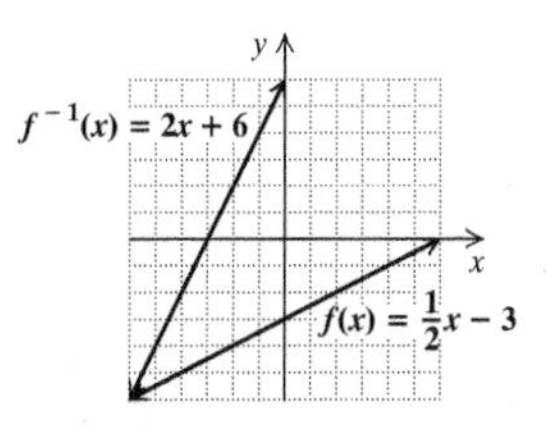

49. $f^{-1}(x) = \sqrt[3]{x}$

x	$f(x)$
0	0
1	1
2	8
3	27
−1	−1
−2	−8
−3	−27

x	$f^{-1}(x)$
0	0
1	1
8	2
27	3
−1	−1
−8	−2
−27	−3

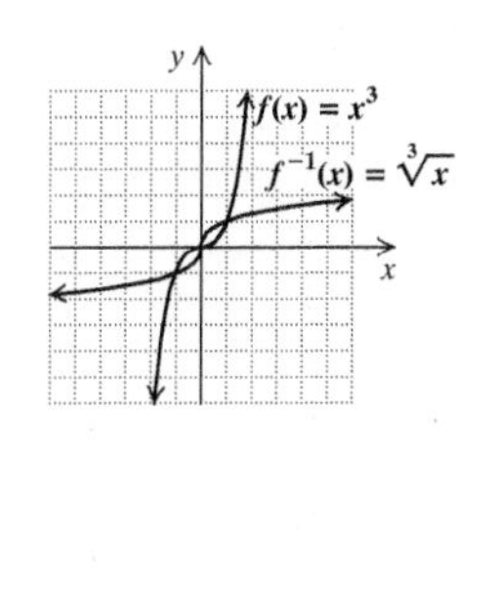

51. $(f^{-1} \circ f)(x) = f^{-1}(f(x)) = f^{-1}\left(\frac{4}{5}x\right) = \frac{5}{4}\left(\frac{4}{5}x\right) = x;$
$(f \circ f^{-1})(x) = f(f^{-1}(x)) = f\left(\frac{5}{4}x\right) = \frac{4}{5}\left(\frac{5}{4}x\right) = x$

53. $(f^{-1} \circ f)(x) = f^{-1}(f(x)) = f^{-1}\left(\frac{x + 7}{2}\right)$
$= 2\left(\frac{x + 7}{2}\right) - 7 = x + 7 - 7 = x;$
$(f \circ f^{-1})(x) = f(f^{-1}(x)) = f(2x - 7)$
$= \frac{2x - 7 + 7}{2} = \frac{2x}{2} = x$

55. $(f^{-1} \circ f)(x) = f^{-1}(f(x)) = f^{-1}\left(\frac{1-x}{x}\right)$

$$= \frac{1}{\frac{1-x}{x}+1} = \frac{1}{\frac{1}{x}} = x;$$

$(f \circ f^{-1})(x) = f(f^{-1}(x)) = f\left(\frac{1}{x+1}\right)$

$$= \frac{1 - \frac{1}{x+1}}{\frac{1}{x+1}} = \frac{\frac{x}{x+1}}{\frac{1}{x+1}} = x$$

57. $f^{-1}(x) = \frac{1}{3}x$ **59.** $f^{-1}(x) = -x$ **61.** $f^{-1}(x) = x^3 + 5$
63. **(a)** 40, 42, 46, 50; **(b)** $f^{-1}(x) = x - 32$; **(c)** 8, 10, 14, 18
65. $\sqrt[3]{a}$ **66.** $\sqrt{3}$ **67.** $\sqrt{2xy}$ **68.** $3a^2b^2$ **69.** $-i$
70. $\frac{y^{4/3}}{25\,x^2z^4}$ **71.** $20\sqrt{6}$ **72.** $x^3 + x^2 - 7x - 7$
73. $4y^2 - 20y + 25$ **74.** $49a^4 - c^2$ **75.** No **77.** Yes
79. (1) C; (2) A; (3) B; (4) D
81.

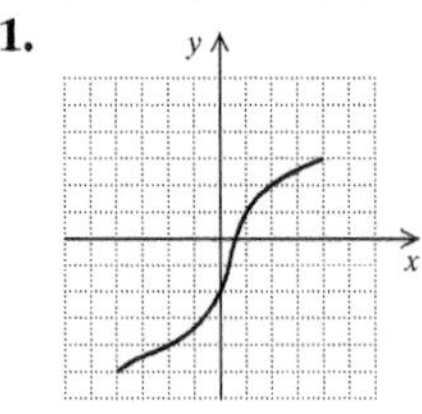

83. $f(x) = \frac{1}{2}x + 3$; $g(x) = 2x - 6$; yes

Exercise Set 8.3, p. 662

RC1. False **RC2.** True **RC3.** True **RC4.** True
RC5. False **RC6.** True **RC7.** True **RC8.** True
CC1. 2 **CC2.** 2 **CC3.** 3 **CC4.** 10
1. $x = 2^y$

x, or 2^y	y
1	0
2	1
4	2
8	3
$\frac{1}{2}$	-1
$\frac{1}{4}$	-2
$\frac{1}{8}$	-3

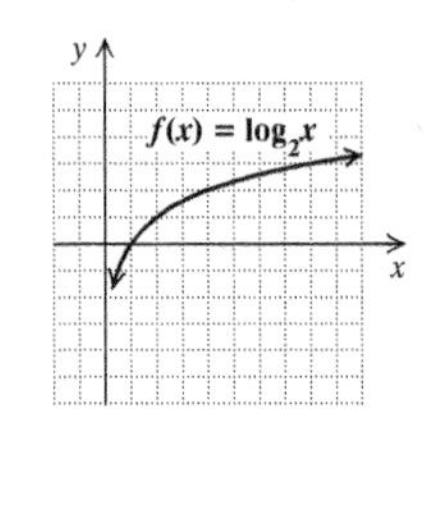

3.

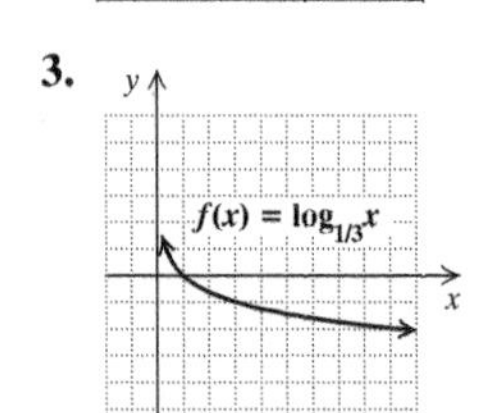

5.

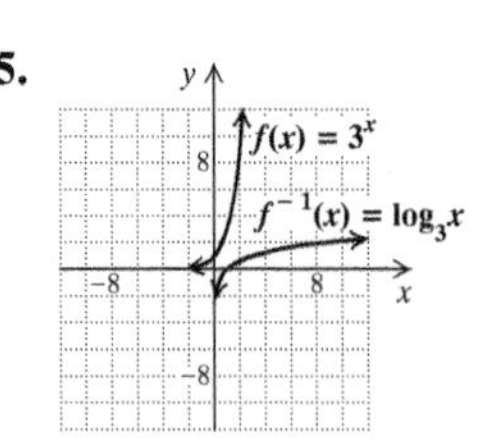

7. $3 = \log_{10} 1000$ **9.** $-3 = \log_5 \frac{1}{125}$ **11.** $\frac{1}{3} = \log_8 2$
13. $0.3010 = \log_{10} 2$ **15.** $2 = \log_e t$ **17.** $t = \log_Q x$
19. $2 = \log_e 7.3891$ **21.** $-2 = \log_e 0.1353$ **23.** $4^w = 10$
25. $6^2 = 36$ **27.** $10^{-2} = 0.01$ **29.** $10^{0.9031} = 8$
31. $e^{4.6052} = 100$ **33.** $t^k = Q$ **35.** 9 **37.** 4 **39.** 4
41. 3 **43.** 25 **45.** 1 **47.** $\frac{1}{2}$ **49.** 2 **51.** 2 **53.** -1
55. 0 **57.** 4 **59.** 2 **61.** 3 **63.** -2 **65.** 0 **67.** 1
69. $\frac{2}{3}$ **71.** 4.8970 **73.** -0.1739 **75.** Does not exist as a real number **77.** 0.9464 **79.** $6 = 10^{0.7782}$; $84 = 10^{1.9243}$; $987{,}606 = 10^{5.9946}$; $0.00987606 = 10^{-2.0054}$; $98{,}760.6 = 10^{4.9946}$; $70{,}000{,}000 = 10^{7.8451}$; $7000 = 10^{3.8451}$ **81.** $\frac{t-3}{t-2}$ **82.** $c + p$
83. $\frac{5}{(x-2)(2x+1)}$ **84.** $(10x + 3)(10x - 3)$
85. $3x(x - 4)^2$ **86.** $(4a - y)(16a^2 + 4ay + y^2)$
87. $(2x - 3)(x^2 + 1)$ **88.** $(5t^3 - 3)(2t^3 + 5)$
89.

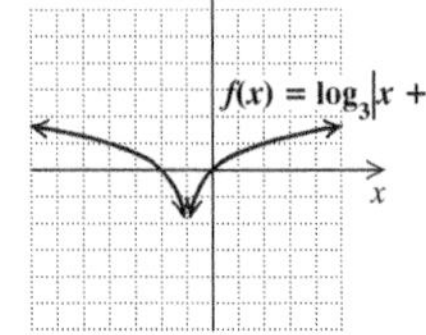

91. 25 **93.** 32 **95.** $-\frac{7}{16}$ **97.** 3 **99.** 0 **101.** -2

Calculator Corner, p. 667

1. Not correct **2.** Correct **3.** Not correct **4.** Correct
5. Not correct **6.** Correct **7.** Not correct **8.** Not correct

Exercise Set 8.4, p. 670

RC1. (b) **RC2.** (c) **RC3.** (d) **RC4.** (a) **RC5.** (e)
RC6. (f) **CC1.** (f) **CC2.** (d) **CC3.** (b) **CC4.** (e)
1. $\log_2 32 + \log_2 8$ **3.** $\log_4 64 + \log_4 16$ **5.** $\log_a Q + \log_a x$
7. $\log_b 252$ **9.** $\log_c Ky$ **11.** $4 \log_c y$ **13.** $6 \log_b t$
15. $-3 \log_b C$ **17.** $\log_a 67 - \log_a 5$ **19.** $\log_b 2 - \log_b 5$
21. $\log_c \frac{22}{3}$ **23.** $2 \log_a x + 3 \log_a y + \log_a z$
25. $\log_b x + 2 \log_b y - 3 \log_b z$ **27.** $\frac{4}{3} \log_c x - \log_c y - \frac{2}{3} \log_c z$
29. $2 \log_a m + 3 \log_a n - \frac{3}{4} - \frac{5}{4} \log_a b$ **31.** $\log_a \frac{x^{2/3}}{y^{1/2}}$, or $\log_a \frac{\sqrt[3]{x^2}}{\sqrt{y}}$
33. $\log_a \frac{2x^4}{y^3}$ **35.** $\log_a \frac{\sqrt{a}}{x}$ **37.** 2.708 **39.** 0.51 **41.** -1.609
43. $\frac{1}{2}$ **45.** 2.609 **47.** Cannot be found using the properties of logarithms **49.** t **51.** 5 **53.** 7 **55.** -7 **57.** i
58. -1 **59.** 5 **60.** $\frac{3}{5} + \frac{4}{5}i$ **61.** $23 - 18i$ **62.** $10i$
63. $-34 - 31i$ **64.** $3 - 4i$ **65.** False **67.** True **69.** False

Mid-Chapter Review: Chapter 8, p. 672

1. False **2.** True **3.** False **4.** True
5. $\log_5 x = 3$
$5^3 = x$
$125 = x$
6. $\log_a 18 = \log_a (2 \cdot 9) = \log_a 2 + \log_a 9$
$= 0.648 + 2.046 = 2.694$
7.

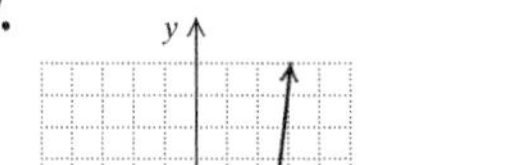

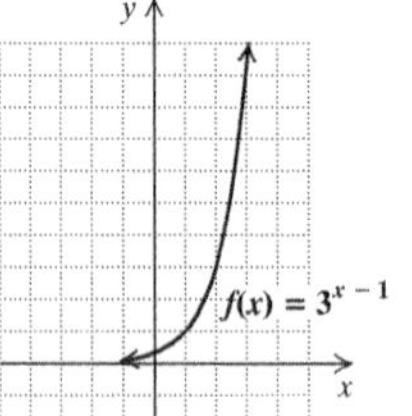

8.

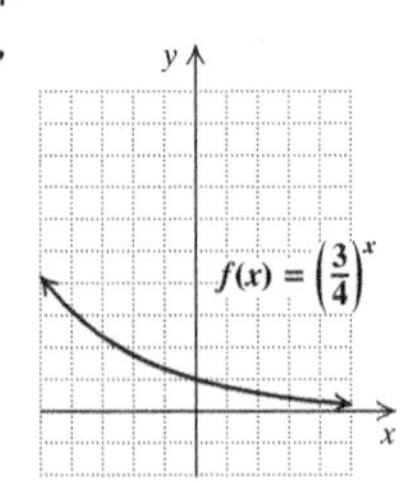

9.

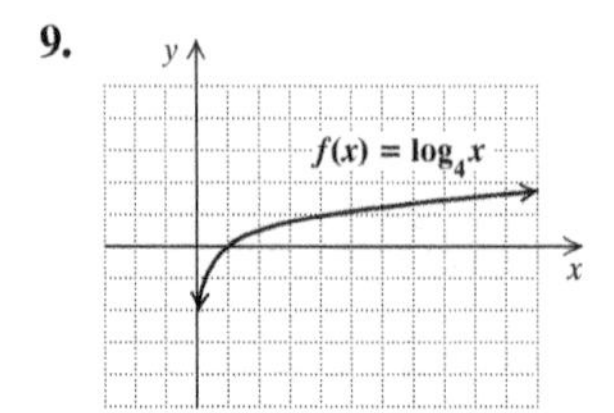

10.

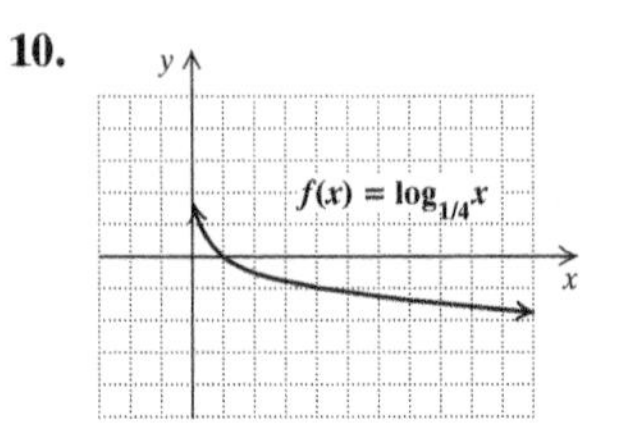

11. **(a)** $A(t) = \$500(1.04)^t$; **(b)** \$500; \$584.93; \$740.12
12. \$1580.49 **13.** $f^{-1}(x) = \dfrac{x-1}{3}$ **14.** $f^{-1}(x) = \sqrt[3]{x-2}$
15. $1 - 2x; 8 - 2x$ **16.** $9x^2 - 6x + 2; 3x^2 + 2$
17. $f(x) = \dfrac{3}{x}; g(x) = x + 4$ **18.** $f(x) = \sqrt{x}; g(x) = 6x - 7$

19. $(f^{-1} \circ f)(x) = f^{-1}(f(x)) = f^{-1}\left(\dfrac{x}{3}\right) = 3\left(\dfrac{x}{3}\right) = x;$

$(f \circ f^{-1})(x) = f(f^{-1}(x)) = f(3x) = \dfrac{3x}{3} = x$

20. $(f^{-1} \circ f)(x) = f^{-1}(f(x)) = f^{-1}(\sqrt[3]{x+4})$
$= (\sqrt[3]{x+4})^3 - 4 = x + 4 - 4 = x;$
$(f \circ f^{-1})(x) = f(f^{-1}(x)) = f(x^3 - 4)$
$= \sqrt[3]{x^3 - 4 + 4} = \sqrt[3]{x^3} = x$

21. $3 = \log_7 343$ **22.** $-4 = \log_3 \dfrac{1}{81}$ **23.** $6^t = 12$ **24.** $n^m = T$
25. 3 **26.** 2 **27.** 2 **28.** 5 **29.** 2.3869 **30.** −0.6383
31. $\log_b 2 + \log_b x + 2\log_b y - 3\log_b z$
32. $\frac{2}{3}\log_a x + \frac{5}{3}\log_a y - \frac{4}{3}\log_a z$ **33.** $\log_a \dfrac{x\sqrt{z}}{y^2}$
34. $\log_m (b - 4)$ **35.** 0 **36.** 1 **37.** −3 **38.** 5
39. $V^{-1}(t)$ could be used to predict when the value of the stamp will be t, where $V^{-1}(t)$ is the number of years after 2010.
40. $\log_a b$ is the number to which a is raised to get b. Since $\log_a b = c$, then $a^c = b$. **41.** Express $\dfrac{x}{5}$ as $x \cdot 5^{-1}$ and then use the product rule and the power rule to get
$\log_a\left(\dfrac{x}{5}\right) = \log_a(x \cdot 5^{-1}) = \log_a x + \log_a 5^{-1} =$
$\log_a x + (-1)\log_a 5 = \log_a x - \log_a 5.$
42. The student didn't subtract the logarithm of the entire denominator after using the quotient rule. The correct procedure is as follows:

$$\begin{aligned}\log_b \frac{1}{x} &= \log_b \frac{x}{xx}\\ &= \log_b x - \log_b xx\\ &= \log_b x - (\log_b x + \log_b x)\\ &= \log_b x - \log_b x - \log_b x\\ &= -\log_b x.\end{aligned}$$

(Note that $-\log_b x$ is equivalent to $\log_b 1 - \log_b x$.)

Calculator Corner, p. 678

1.

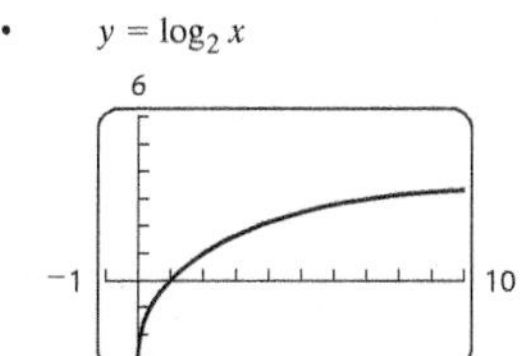

2.

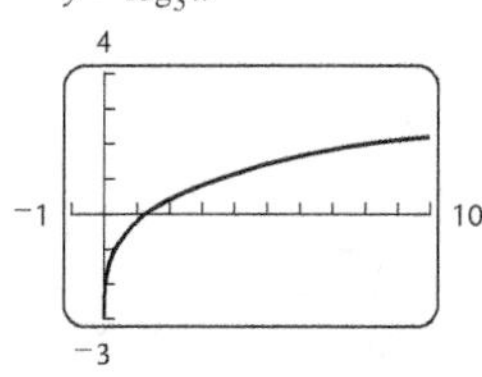

3.

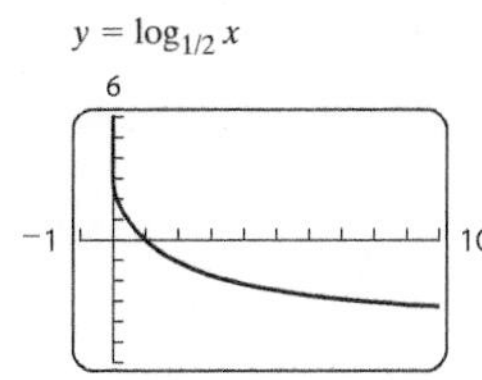

4.

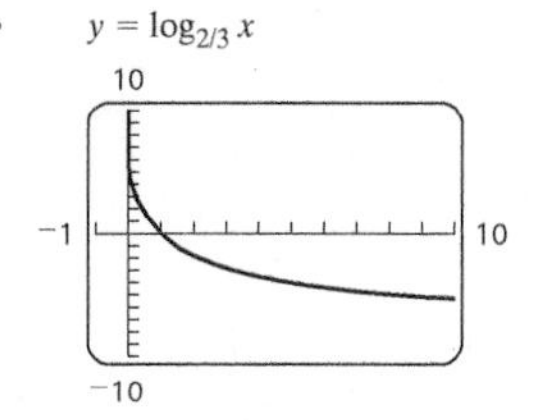

Visualizing for Success, p. 679

1. J **2.** B **3.** O **4.** G **5.** N **6.** F **7.** A **8.** H **9.** I **10.** K

Exercise Set 8.5, p. 680

RC1. Base 10 **RC2.** Base e **RC3.** Base 10 **RC4.** Base e
CC1. (b) **CC2.** (a) **CC3.** (b) **CC4.** (a)
1. 0.6931 **3.** 4.1271 **5.** 8.3814 **7.** −5.0832 **9.** −1.6094
11. Does not exist **13.** −1.7455 **15.** 1 **17.** 15.0293
19. 0.0305 **21.** 109.9472 **23.** 5 **25.** 2.5702 **27.** 6.6439
29. 2.1452 **31.** −2.3219 **33.** −2.3219 **35.** 4.6284

37.

x	$f(x)$
0	1
1	2.7
2	7.4
−1	0.4
−2	0.1

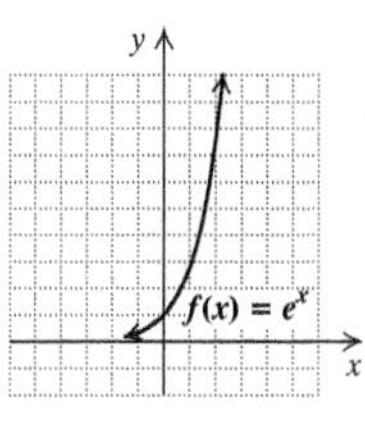

39. $f(x) = e^{-0.5x}$

41.

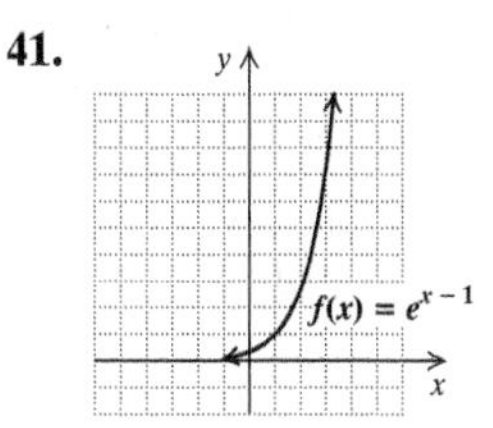

43. $f(x) = e^{x+2}$

45.

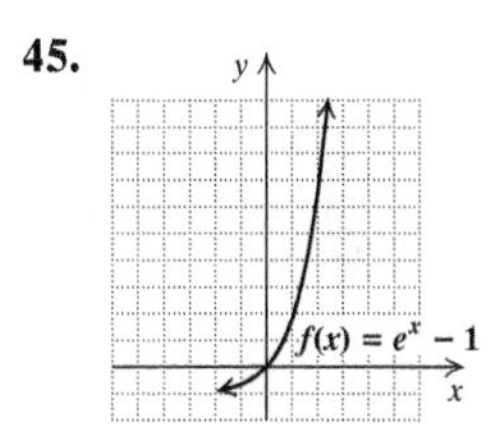

47.

x	$f(x)$
0	0.7
1	1.1
2	1.4
3	1.6
−0.5	0.4
−1	0
−1.5	−0.7

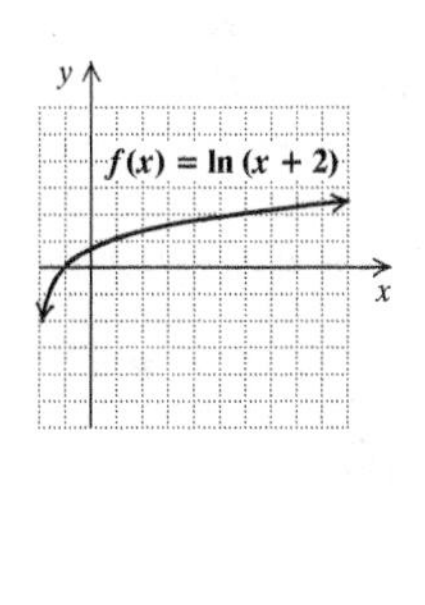

49. $f(x) = \ln(x - 3)$

51.

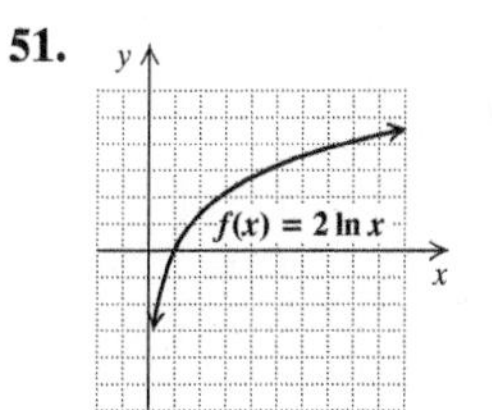

53. $f(x) = \frac{1}{2}\ln x + 1$

55.

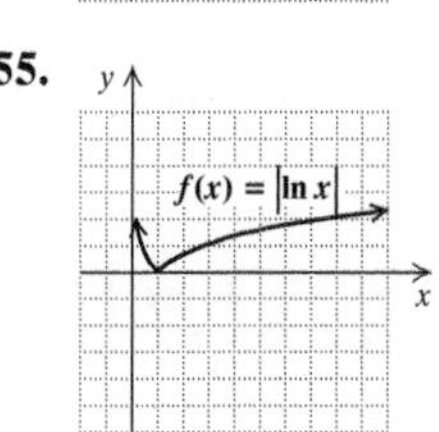

57. 16, 256 **58.** $\frac{1}{4}$, 9 **59.** 49, 121 **60.** $\pm 3, \pm 4$
61. Domain: $(-\infty, \infty)$; range: $[0, \infty)$ **63.** Domain: $(-\infty, \infty)$; range: $(-\infty, 100)$ **65.** $(\frac{5}{2}, \infty)$

Exercise Set 8.6, p. 687

RC1. True **RC2.** True **RC3.** True **RC4.** False
CC1. False **CC2.** True **CC3.** False **CC4.** True
CC5. True **CC6.** True

1. 3 **3.** 4 **5.** $\frac{5}{2}$ **7.** $\frac{3}{5}$ **9.** 3.4594 **11.** 5.4263 **13.** $\frac{5}{2}$
15. $-3, -1$ **17.** $\frac{3}{2}$ **19.** 4.6052 **21.** 2.3026 **23.** 140.6705
25. 2.7095 **27.** 3.2220 **29.** 256 **31.** $\frac{1}{32}$ **33.** 10 **35.** $\frac{1}{100}$
37. $e^2 \approx 7.3891$ **39.** $\frac{1}{e} \approx 0.3679$ **41.** 121 **43.** 10
45. $\frac{1}{3}$ **47.** 3 **49.** $\frac{2}{5}$ **51.** 5 **53.** No solution
55. $\{x \mid 9 \le x < 16\}$, or $[9, 16)$
56. $\{x \mid x < 1 \text{ or } x > 4\}$, or $(-\infty, 1) \cup (4, \infty)$ **57.** $-3, 4$
58. $\{x \mid -3 \le x \le 4\}$, or $[-3, 4]$ **59.** 65 **60.** 7 **61.** 1
63. **(a)** 0.3770; **(b)** -1.9617; **(c)** 0.9036; **(d)** -1.5318
65. 3, 4 **67.** -4 **69.** 2 **71.** $\pm\sqrt{34}$ **73.** $10^{100,000}$
75. 1, 100 **77.** $3, -7$ **79.** $1, \frac{\log 5}{\log 3} \approx 1.465$

Translating for Success, p. 698

1. D **2.** M **3.** I **4.** A **5.** E **6.** H **7.** C **8.** G
9. N **10.** B

Exercise Set 8.7, p. 699

RC1. (b) **RC2.** (d) **RC3.** (c) **RC4.** (a) **CC1.** 550
CC2. 0.13 **CC3.** 2 **CC4.** 713
1. 188 dB **3.** $10^{-7.5}$ W/m², or about 3.2×10^{-8} W/m²; 10^{-6} W/m² **5.** About 6.8 **7.** 1.58×10^{-8} moles per liter
9. 2.25 ft/sec **11.** 3.40 ft/sec **13.** **(a)** About 1752 riverbanks; **(b)** 2002; **(c)** 9.7 years **15.** **(a)** About 1811 deaths; **(b)** 2016; **(c)** 4.0 years **17.** **(a)** $P(t) = P_0 e^{0.03t}$; **(b)** \$5152.27; \$5309.18; \$6749.29; **(c)** in 23.1 years
19. **(a)** $P(t) = 7.5e^{0.0106t}$; **(b)** 8.2 billion; **(c)** 2082; **(d)** 65.4 years
21. **(a)** $k \approx 0.340$; $P(t) = 56e^{0.340t}$; **(b)** 605 thousand employees; **(c)** about 8 years after 2011, or in 2019 **23.** About 2103 years
25. About 7.2 days **27.** 69.3% per year **29.** **(a)** $k \approx 0.149$; $A(t) = 451e^{-0.149t}$; **(b)** about 65 million albums; **(c)** 2010
31. **(a)** $k \approx 0.002$; $P(t) = 396{,}000e^{-0.002t}$; **(b)** about 387,000; **(c)** 2031 **33.** **(a)** $k \approx 0.263$; $V(t) = 19{,}000e^{0.263t}$; **(b)** about \$246 million; **(c)** 2.6 years; **(d)** 41.3 years **35.** $\left(-1, \frac{1}{2}\right)$
36. $(2, -1, 1)$ **37.** $-1 \pm \sqrt{2}i$ **38.** $-\frac{6}{5}, 4$ **39.** $\frac{9}{2}$
40. $\pm\sqrt{3}i$ **41.** $-0.937, 1.078, 58.770$ **43.** $-0.767, 2, 4$
45. About 80,922 years

Summary and Review: Chapter 8, p. 704

Vocabulary Reinforcement

1. Exponential **2.** Inverse **3.** Compound **4.** Common
5. Exponent **6.** Half-life

Concept Reinforcement

1. True **2.** False **3.** True **4.** True **5.** False **6.** True

Study Guide

1.

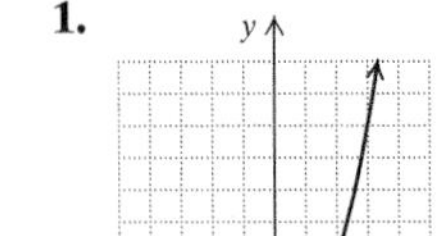

2. $8x + 2$; $8x + 1$

3. $f(x) = \frac{1}{x}$, $g(x) = 3x + 2$; answers may vary

4. Yes **5.** $g^{-1}(x) = 4 - x$

6.

7.

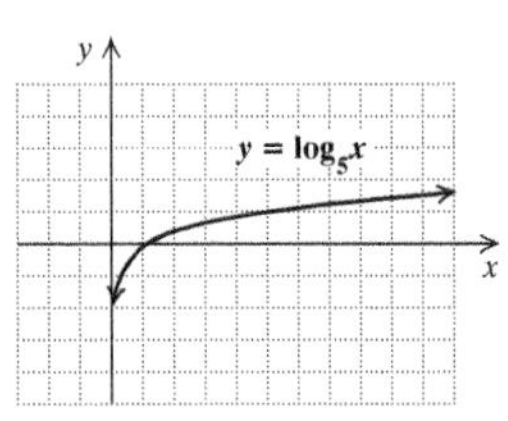

8. $\frac{3}{5}\log_a x - \frac{2}{5}\log_a y$ **9.** $\log_a \frac{\sqrt{x}}{y^3}$, or $\log_a \frac{x^{1/2}}{y^3}$

10.

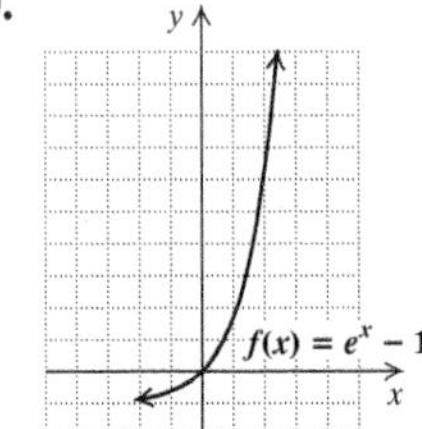

11.

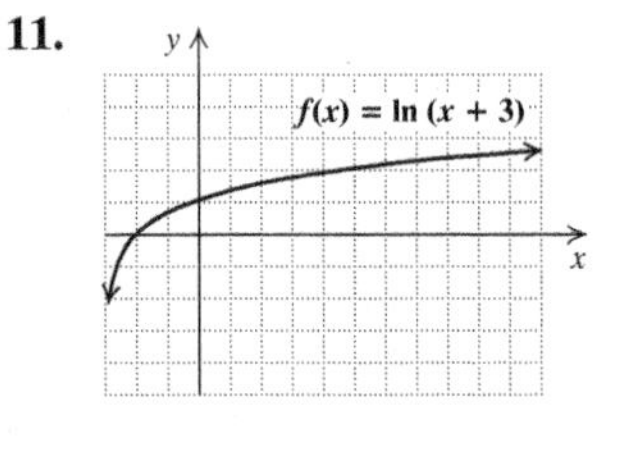

12. $\frac{4}{3}$ **13.** 3

Review Exercises

1. $\{(2, -4), (-7, 5), (-2, -1), (11, 10)\}$ **2.** Not one-to-one
3. $g^{-1}(x) = \frac{7x + 3}{2}$ **4.** $f^{-1}(x) = \frac{1}{2}\sqrt[3]{x}$ **5.** $f^{-1}(x) = \frac{3x - 4}{2x}$

6.

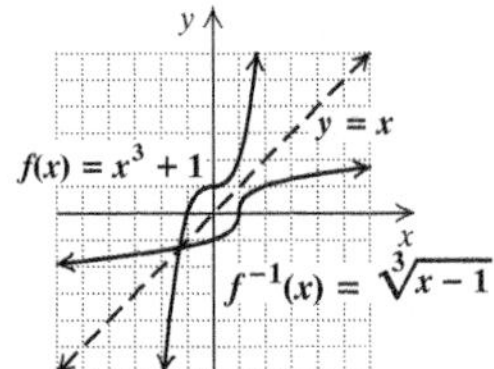

7.

x	$f(x)$
0	$\frac{1}{3}$
1	1
2	3
3	9
-1	$\frac{1}{9}$
-2	$\frac{1}{27}$
-3	$\frac{1}{81}$

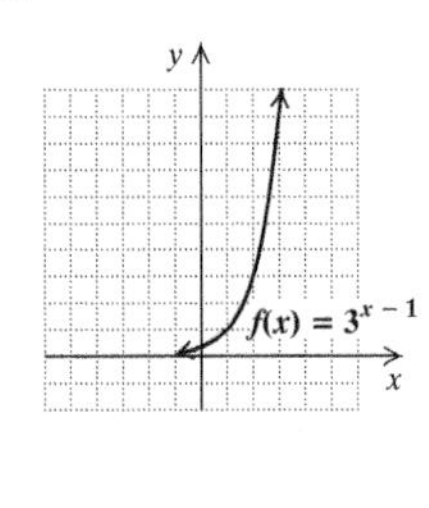

8. 3^y

x, or $3y$	y
1	0
3	1
9	2
27	3
$\frac{1}{3}$	-1
$\frac{1}{9}$	-2
$\frac{1}{27}$	-3

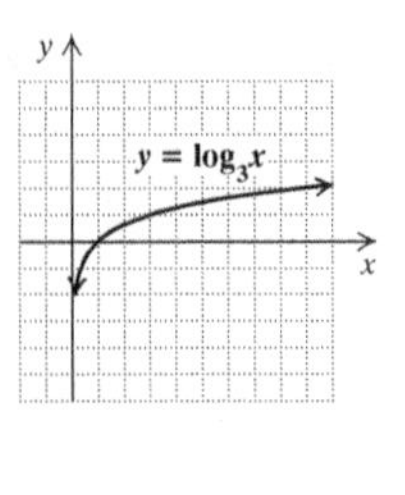

9.

x	$f(x)$
0	2.7
1	7.4
2	20.1
3	54.6
−1	1
−2	0.4
−3	0.1

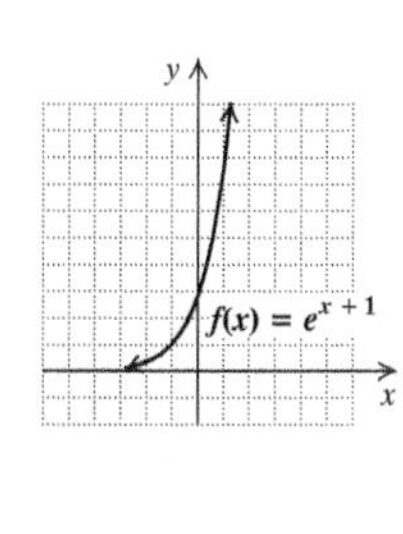

10.

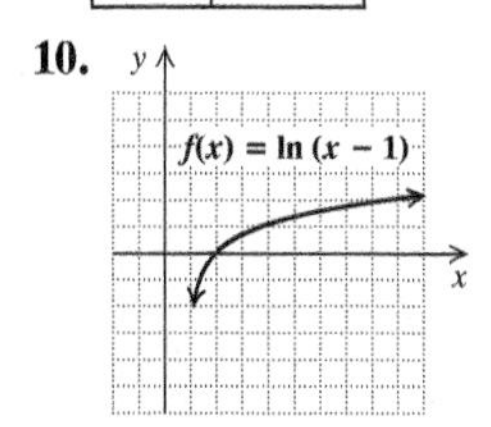

11. $(f \circ g)(x) = 9x^2 - 30x + 25$; $(g \circ f)(x) = 3x^2 - 5$
12. $f(x) = \sqrt{x}$, $g(x) = 4 - 7x$; answers may vary
13. $4 = \log 10{,}000$ **14.** $\frac{1}{2} = \log_{25} 5$ **15.** $4^x = 16$
16. $\left(\frac{1}{2}\right)^{-3} = 8$ **17.** 2 **18.** −1 **19.** 1 **20.** 0
21. −2.7425 **22.** Does not exist as a real number
23. $4 \log_a x + 2 \log_a y + 3 \log_a z$ **24.** $\frac{1}{2} \log z - \frac{3}{4} \log x - \frac{1}{4} \log y$
25. $\log_a 120$ **26.** $\log \dfrac{a^{1/2}}{bc^2}$, or $\log \dfrac{\sqrt{a}}{bc^2}$ **27.** 17 **28.** −7
29. 8.7601 **30.** 3.2698 **31.** 2.54995 **32.** −3.6602
33. −2.6921 **34.** 0.3753 **35.** 18.3568 **36.** 0 **37.** Does not exist **38.** 1 **39.** 0.4307 **40.** 1.7097 **41.** $\frac{1}{9}$ **42.** 2
43. $\frac{1}{10{,}000}$ **44.** $e^{-2} \approx 0.1353$ **45.** $\frac{7}{2}$ **46.** 1, −5
47. $\dfrac{\log 8.3}{\log 4} \approx 1.5266$ **48.** $\dfrac{\ln 0.03}{-0.1} \approx 35.0656$
49. 2 **50.** 8 **51.** $\frac{17}{5}$ **52.** $\sqrt{43}$ **53.** 137 dB
54. **(a)** 5204 PB; 35,721 PB; 77,190 PB; **(b)** 2020;
(c) about 1.8 years;
(d)

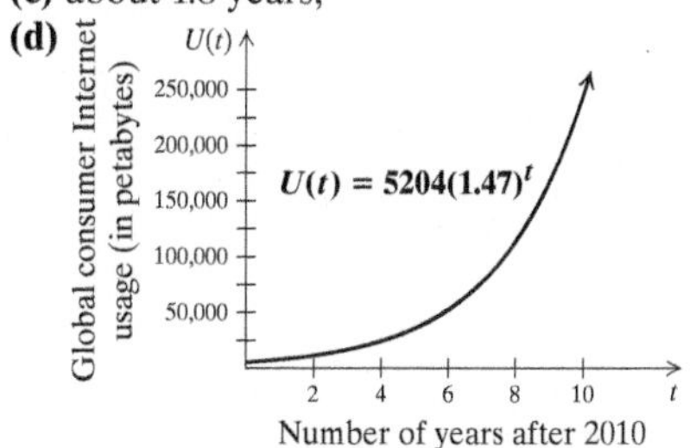

55. **(a)** $k \approx 0.094$; $V(t) = 40{,}000e^{0.094t}$; **(b)** \$102,399; **(c)** 2023
56. $k \approx 0.231$ **57.** About 20.4 years **58.** About 3463 years
59. C **60.** D **61.** e^{e^3} **62.** $\left(\frac{8}{3}, -\frac{2}{3}\right)$

Understanding Through Discussion and Writing

1. Reflect the graph of $f(x) = e^x$ across the line $y = x$ and then translate it up one unit. **2.** Christina mistakenly thinks that, because negative numbers do not have logarithms, negative numbers cannot be solutions of logarithmic equations.

3.

$$C(x) = \frac{100 + 5x}{x}$$

$$y = \frac{100 + 5x}{x} \quad \text{Replace } C(x) \text{ with } y.$$

$$x = \frac{100 + 5y}{y} \quad \text{Interchange variables.}$$

$$y = \frac{100}{x - 5}; \quad \text{Solve for } y.$$

$$C^{-1}(x) = \frac{100}{x - 5} \quad \text{Replace } y \text{ with } C^{-1}(x)$$

$C^{-1}(x)$ gives the number of people in the group, where x is the cost per person, in dollars.
4. To solve $\ln x = 3$, graph $f(x) = \ln x$ and $g(x) = 3$ on the same set of axes. The solution is the first coordinate of the point of intersection of the two graphs. **5.** You cannot take the logarithm of a negative number because logarithm bases are positive and there is no real-number power to which a positive number can be raised to yield a negative number. **6.** Answers will vary.

Test: Chapter 8, p. 712

1. [8.1a]

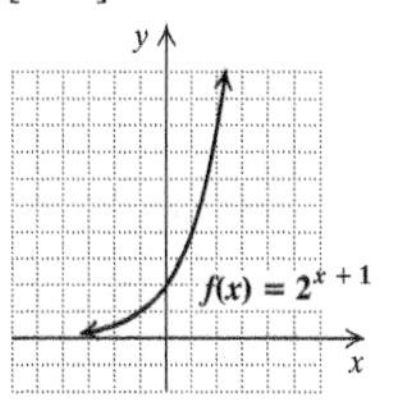

2. [8.3a]

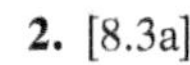

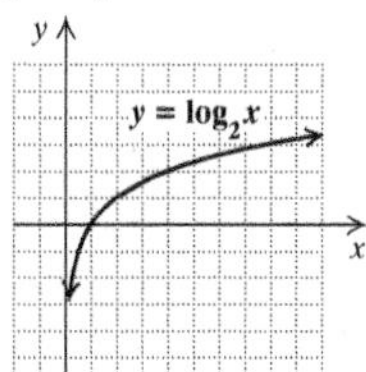

3. [8.5c]

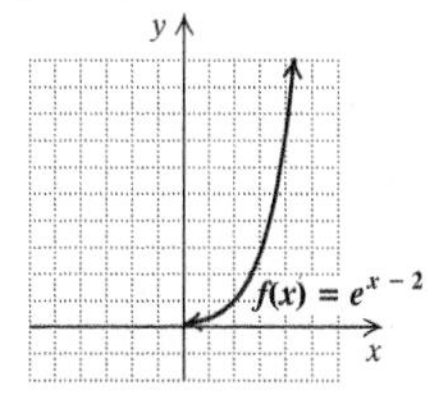

4. [8.5c]

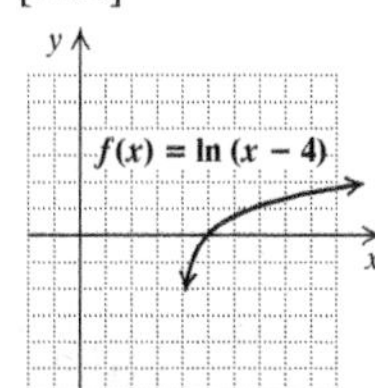

5. [8.2a] $(f \circ g)(x) = 25x^2 - 15x + 2$, $(g \circ f)(x) = 5x^2 + 5x - 2$
6. [8.2b] $\{(3, -4), (-8, 5), (-3, -1), (12, 10)\}$
7. [8.2c, d] $f^{-1}(x) = \dfrac{x + 3}{4}$ **8.** [8.2c, d] $f^{-1}(x) = \sqrt[3]{x} - 1$
9. [8.2c] Not one-to-one **10.** [8.3b] $\log_{256} 16 = \frac{1}{2}$
11. [8.3b] $7^m = 49$ **12.** [8.3c] 3 **13.** [8.4e] 23 **14.** [8.3c] 0
15. [8.3d] −1.9101 **16.** [8.3d] Does not exist as a real number
17. [8.4d] $3 \log a + \frac{1}{2} \log b - 2 \log c$ **18.** [8.4d] $\log_a \dfrac{x^{1/3}z^2}{y^3}$
19. [8.4d] −0.544 **20.** [8.4d] 1.079 **21.** [8.5a] 6.6938
22. [8.5a] 107.7701 **23.** [8.5a] 0 **24.** [8.5b] 1.1881
25. [8.6b] 5 **26.** [8.6b] 2 **27.** [8.6b] 10,000
28. [8.6b] $e^{1/4} \approx 1.2840$ **29.** [8.6a] $\dfrac{\log 1.2}{\log 7} \approx 0.0937$
30. [8.6b] 9 **31.** [8.6b] 1 **32.** [8.7a] 4.2
33. [8.7b] **(a)** 4860 people; **(b)** 3.6 days; **(c)** 0.63 day
34. [8.7b] **(a)** $k \approx 0.028$, or 2.8%; $P(t) = 1000e^{0.028t}$; **(b)** \$1251.07; **(c)** after 13 years; **(d)** about 24.8 years
35. [8.7b] About 3% **36.** [8.7b] About 4684 years
37. [8.6b] B **38.** [8.6b] 44, −37 **39.** [8.4d] 2

Cumulative Review: Chapters 1–8, p. 715

1. [1.1d] $\frac{11}{2}$ **2.** [4.8a] −2, 5 **3.** [3.3a] (3, −1)
4. [3.5a] (1, −2, 0) **5.** [5.5a] $\frac{9}{2}$ **6.** [6.6b] 5 **7.** [7.4c] 9,25
8. [7.4c] ±2, ±3 **9.** [8.6b] 8 **10.** [8.6a] $\dfrac{\log 7}{5 \log 3} \approx 0.3542$
11. [8.6b] $\frac{80}{9}$ **12.** [7.8a] $\{x \mid x < -5 \text{ or } x > 1\}$, or $(-\infty, -5) \cup (1, \infty)$ **13.** [1.6e] $\{x \mid x \le -3 \text{ or } x \ge 6\}$, or $(-\infty, -3] \cup [6, \infty)$ **14.** [7.2a] $-3 \pm 2\sqrt{5}$
15. [5.7a] $a = \dfrac{Db}{b - D}$ **16.** [5.7a] $q = \dfrac{pf}{p - f}$
17. [5.1a] $\left(-\infty, -\frac{1}{3}\right) \cup \left(-\frac{1}{3}, 2\right) \cup (2, \infty)$
18. [5.6a] $\frac{60}{11}$ min, or $5\frac{5}{11}$ min **19.** [8.7a] **(a)** 78; **(b)** 67.5
20. [3.4a] Swim Clean: 60 L; Pure Swim: 40 L

21. [5.6c] $2\frac{7}{9}$ km/h **22.** [8.7b] **(a)** $P(t) = 49e^{0.0084t}$, where $P(t)$ is in millions and t is the number of years after 2016; **(b)** about 50.7 million, about 52.8 million; **(c)** about 82.5 years
23. [4.8b] 10 ft **24.** [5.8e] 18
25. [2.5a]

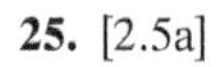

26. [3.7b]

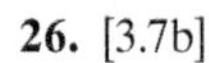

27. [7.6a]

28. [8.1a]

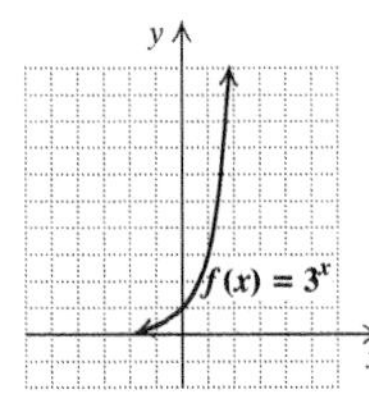

29. [8.3a]

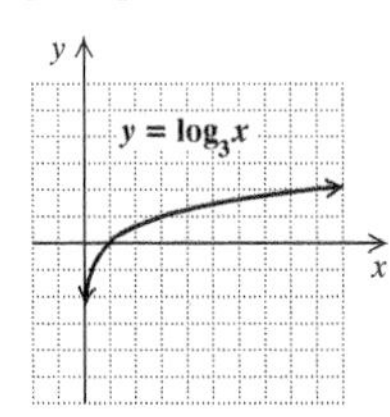

30. [4.1d] $8x^2 - 11x - 1$ **31.** [4.2c] $9x^4 - 12x^2y + 4y^2$
32. [4.2b] $10a^2 - 9ab - 9b^2$ **33.** [5.1e] $\frac{(x + 4)(x - 3)}{2(x - 1)}$
34. [5.4a] $\frac{1}{x - 4}$ **35.** [5.2c] $\frac{7x + 4}{(x + 6)(x - 6)}$
36. [4.6d] $(1 - 5x)(1 + 5x + 25x^2)$
37. [4.3a], [4.5a,b] $2(3x - 2y)(x + 2y)$
38. [4.3b] $(x^3 + 7)(x - 4)$ **39.** [4.3a], [4.6a] $2(m + 3n)^2$
40. [4.6b] $(x - 2y)(x + 2y)(x^2 + 4y^2)$ **41.** [2.2b] -12
42. [5.3b,c] $x^3 - 2x^2 - 4x - 12 + \frac{-42}{x - 3}$ **43.** [6.3a] $14xy^2\sqrt{x}$
44. [6.3b] $2y^2\sqrt[3]{y}$ **45.** [6.5b] $\frac{6 + \sqrt{y} - y}{4 - y}$
46. [6.8c] $12 + 4\sqrt{3}i$ **47.** [8.2d] $f^{-1}(x) = \frac{x - 7}{-2}$, or $\frac{7 - x}{2}$
48. [2.6d] $y = \frac{1}{2}x + \frac{13}{2}$ **49.** [8.4d] $\log\left(\frac{x^3}{y^{1/2}z^2}\right)$
50. [8.3b] $a^x = 5$ **51.** [8.3d] -1.2545 **52.** [8.3d] 776.2471
53. [8.5a] 2.5479 **54.** [8.5a] 0.2466 **55.** [7.6a] D
56. [7.3b] D **57.** [5.5a] All real numbers except 1 and -2
58. [8.6b] $\frac{1}{3}, \frac{10{,}000}{3}$ **59.** [5.6c] 35 mph

CHAPTER 9

Calculator Corner, p. 721

1. $x = y^2 + 4y + 7$

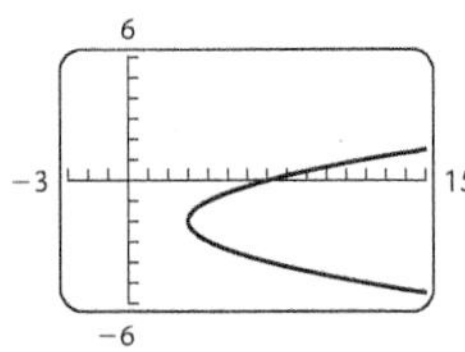

2. $x = -2y^2 + 10y - 7$

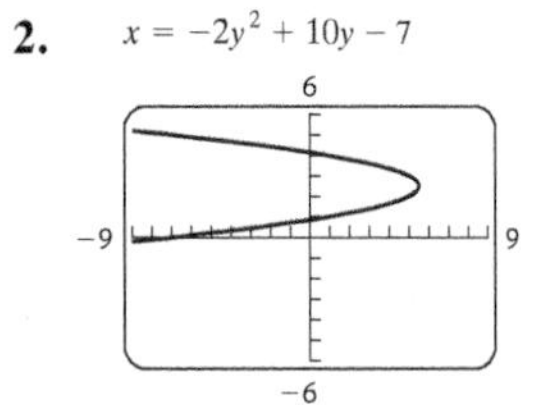

3. $x = 4y^2 - 12y + 5$

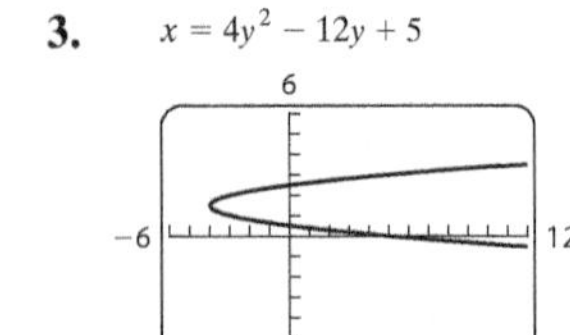

Calculator Corner, p. 726

1. $(x - 1)^2 + (y + 2)^2 = 4$

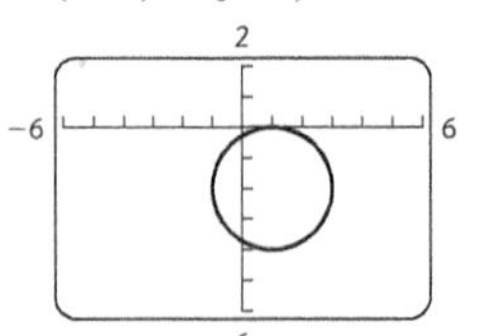

2. $(x + 2)^2 + (y - 2)^2 = 25$

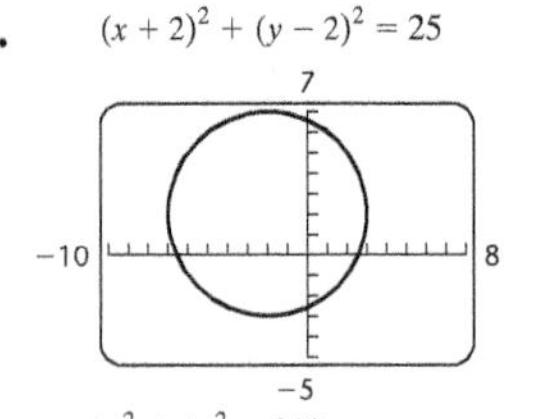

3. $x^2 + y^2 - 16 = 0$

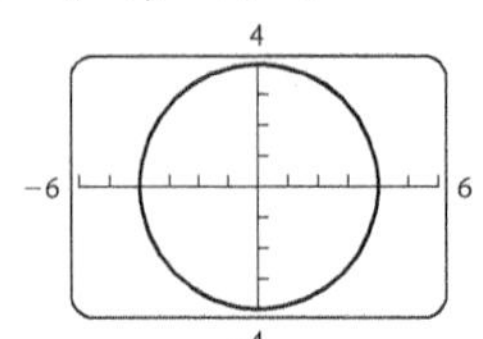

4. $4x^2 + 4y^2 = 100$

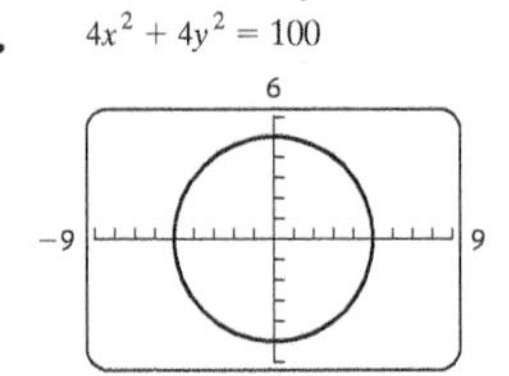

5. $x^2 + y^2 - 10x - 11 = 0$

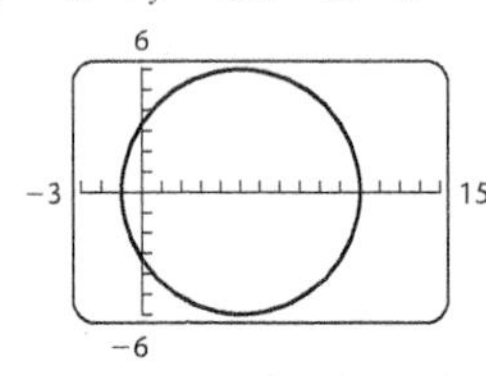

Exercise Set 9.1, p. 727

RC1. True **RC2.** False **RC3.** True **RC4.** True
RC5. True **RC6.** False **RC7.** False **RC8.** True

1.

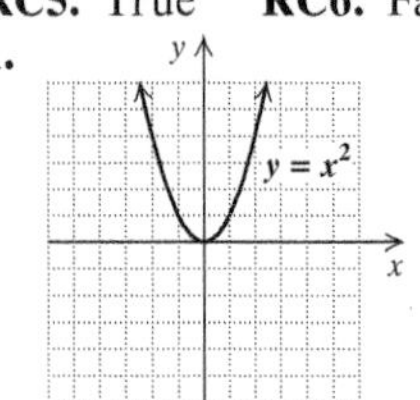

3.

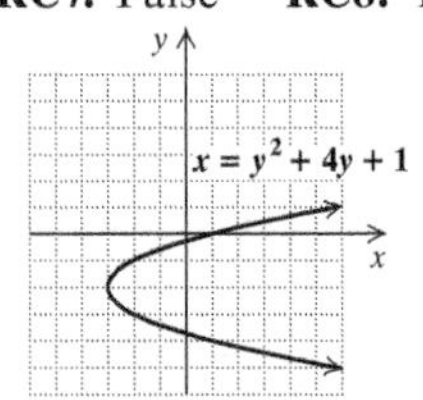

5.

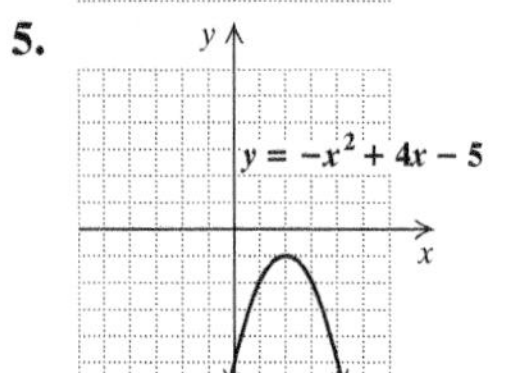

7.

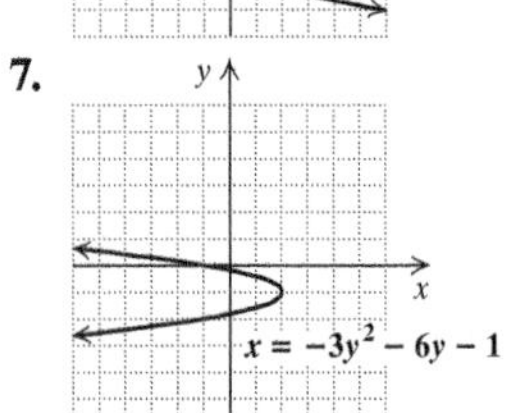

9. 5 **11.** $\sqrt{29} \approx 5.385$ **13.** $\sqrt{648} \approx 25.456$
15. 7.1 **17.** $\frac{\sqrt{41}}{7} \approx 0.915$ **19.** $\sqrt{6970} \approx 83.487$
21. $\sqrt{a^2 + b^2}$ **23.** $\sqrt{17 + 2\sqrt{14} + 2\sqrt{15}} \approx 5.677$
25. $\sqrt{9{,}672{,}400} \approx 3110.048$ **27.** $\left(\frac{3}{2}, \frac{7}{2}\right)$ **29.** $\left(0, \frac{11}{2}\right)$
31. $\left(-1, -\frac{17}{2}\right)$ **33.** $(-0.25, -0.3)$ **35.** $\left(-\frac{1}{12}, \frac{1}{24}\right)$
37. $\left(\frac{\sqrt{2} + \sqrt{3}}{2}, \frac{3}{2}\right)$ **39.**

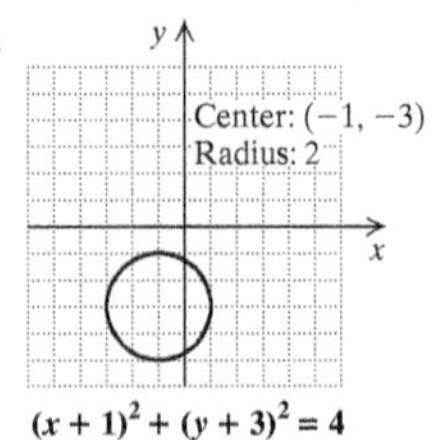

41.

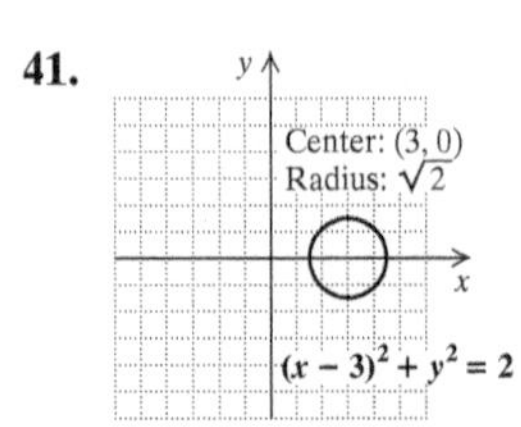

43.

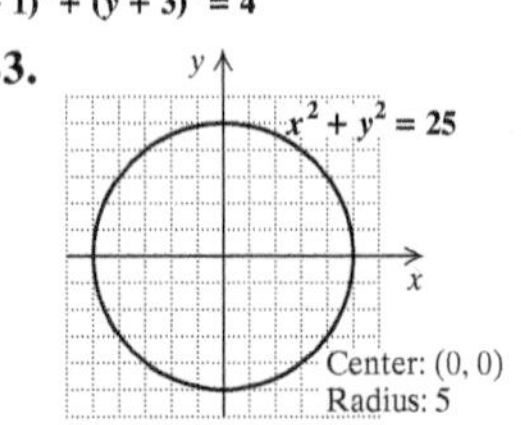

45. $x^2 + y^2 = 49$ **47.** $(x + 5)^2 + (y - 3)^2 = 7$
49. $(-4, 3), r = 2\sqrt{10}$ **51.** $(4, -1), r = 2$ **53.** $(2, 0), r = 2$
55. $(9, 2)$ **56.** $(-8, 16)$ **57.** $\left(-\frac{21}{5}, -\frac{73}{5}\right)$ **58.** $(1, 2)$
59. No solution **60.** $(2a + b)(2a - b)$ **61.** $(x - 4)(x + 4)$
62. $(a - 3b)(a + 3b)$ **63.** $(8p - 9q)(8p + 9q)$
64. $25(4cd - 3)(4cd + 3)$ **65.** $(x + 3)^2 + (y + 2)^2 = 9$
67. $8\sqrt{m^2 + n^2}$ **69.** Yes **71.** **(a)** $(0, -8467.8)$; **(b)** 8487.3 mm

Calculator Corner, p. 734

1. $y_1 = \sqrt{\frac{36 - 4x^2}{9}}$, $y_2 = -\sqrt{\frac{36 - 4x^2}{9}}$

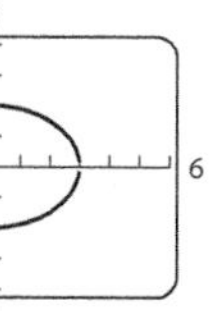

2. $y_1 = \sqrt{\frac{144 - 16x^2}{9}}$, $y_2 = -\sqrt{\frac{144 - 16x^2}{9}}$

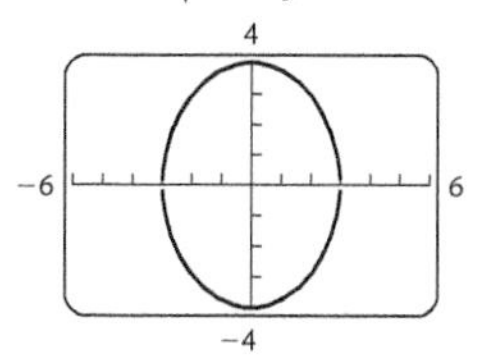

3. 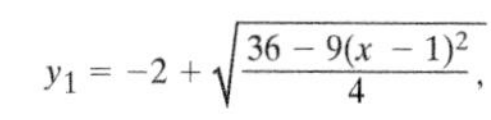
$y_1 = -2 + \sqrt{\frac{36 - 9(x - 1)^2}{4}}$,

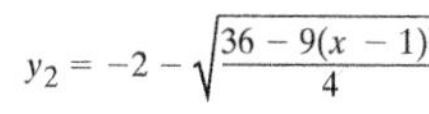
$y_2 = -2 - \sqrt{\frac{36 - 9(x - 1)^2}{4}}$

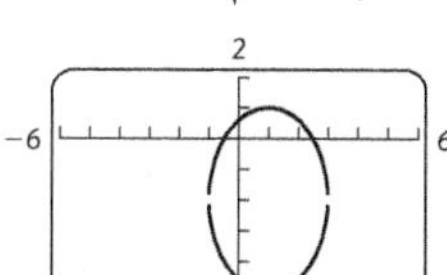

4. $y_1 = 3 + \sqrt{\frac{144 - 9(x + 2)^2}{16}}$, $y_2 = 3 - \sqrt{\frac{144 - 9(x + 2)^2}{16}}$

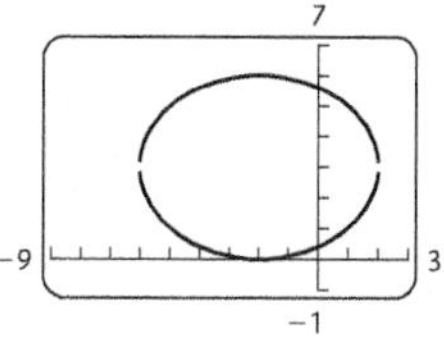

Exercise Set 9.2, p. 736

RC1. A **RC2.** C **RC3.** B **RC4.** D
CC1. x-intercepts: $(-8, 0)$ and $(8, 0)$; **CC2.** Center: $(-5, 2)$; vertices: $(-5, 12)$ and $(-5, -8)$

1.

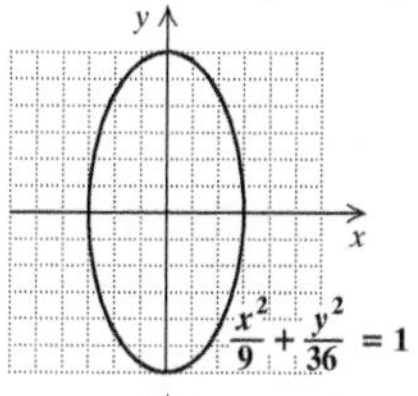

3.

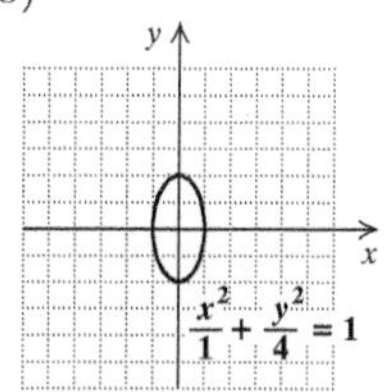

5.

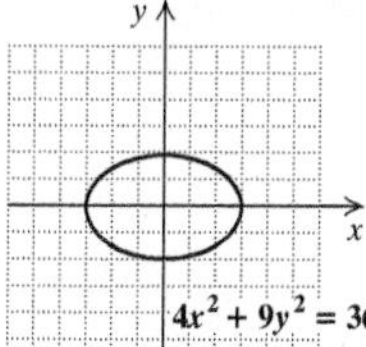

7.

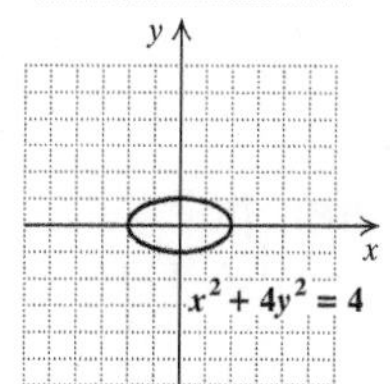

9.

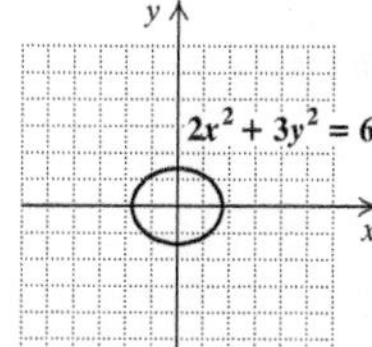

11.

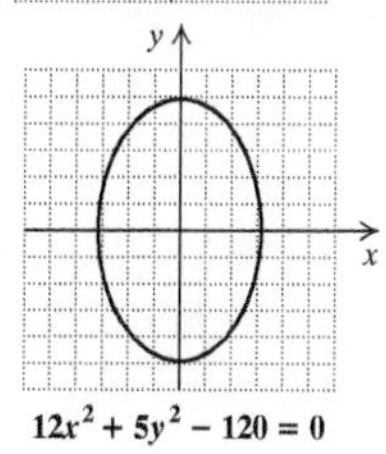

$12x^2 + 5y^2 - 120 = 0$

13.

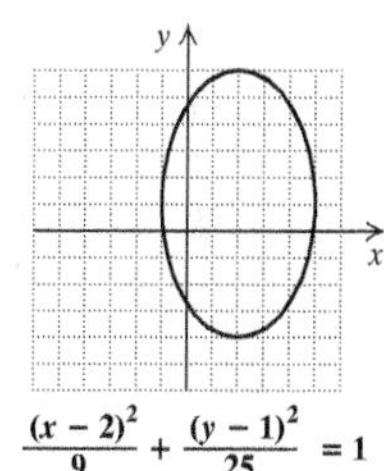

$\frac{(x-2)^2}{9} + \frac{(y-1)^2}{25} = 1$

15.

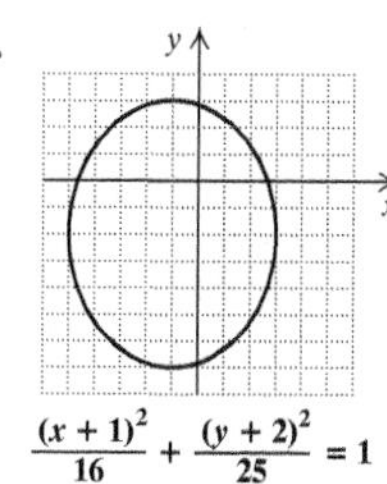

$\frac{(x+1)^2}{16} + \frac{(y+2)^2}{25} = 1$

17.

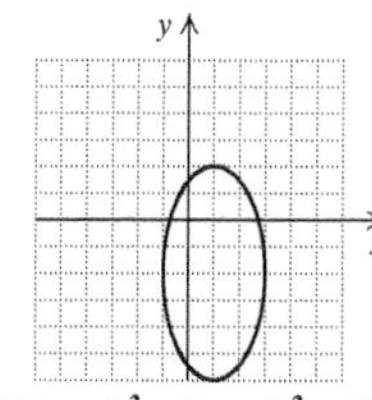

$12(x - 1)^2 + 3(y + 2)^2 = 48$

19.

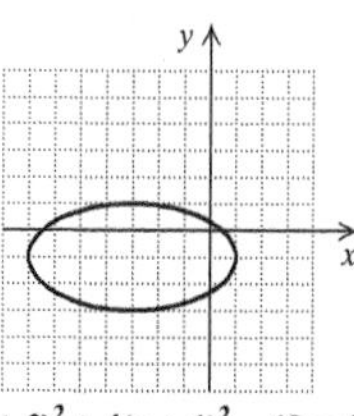

$(x + 3)^2 + 4(y + 1)^2 - 10 = 6$

21. $\frac{1 \pm 2i\sqrt{5}}{3}$ **22.** $\frac{6 \pm \sqrt{15}}{3}$ **23.** $\frac{-1 \pm i\sqrt{7}}{2}$
24. $-1 \pm \sqrt{11}$ **25.** $-1 \pm 3\sqrt{2}$; 3.2, −5.2
26. $1 \pm \sqrt{11}$; 4.3, −2.3 **27.** $\frac{17 \pm \sqrt{337}}{8}$; 4.4, −0.2
28. $\frac{-3 \pm \sqrt{41}}{4}$; 0.9, −2.4 **29.** $\log_a b = -t$
30. $\log_8 17 = a$ **31.** $e^{3.1781} = 24$ **32.** $e^p = W$
33. $\frac{x^2}{49} + \frac{y^2}{25} = 1$ **35.** $\frac{x^2}{9} + \frac{y^2}{25} = 1$

Mid-Chapter Review: Chapter 9, p. 739

1. True **2.** True **3.** True **4.** False
5. **(a)** $d = \sqrt{(x_2 - x_1)^2 + (y_2 - y_1)^2} = \sqrt{(4 - (-6))^2 + (-1 - 2)^2} = \sqrt{(10)^2 + (-3)^2} = \sqrt{100 + 9} = \sqrt{109} \approx 10.440$; **(b)** $\left(\frac{x_1 + x_2}{2}, \frac{y_1 + y_2}{2}\right) = \left(\frac{-6 + 4}{2}, \frac{2 + (-1)}{2}\right) = \left(\frac{-2}{2}, \frac{1}{2}\right) = \left(-1, \frac{1}{2}\right)$

6.
$$x^2 - 20x + y^2 + 4y = -79$$
$$x^2 - 20x + 100 + y^2 + 4y + 4 = -79 + 100 + 4$$
$$(x - 10)^2 + (y + 2)^2 = 25$$
$$(x - 10)^2 + (y - (-2))^2 = 5^2$$
Center: $(10, -2)$; radius: 5

7. $3\sqrt{2} \approx 4.243$ **8.** $\sqrt{120.53} \approx 10.979$ **9.** $\sqrt{11} \approx 3.317$
10. $\left(-\frac{19}{2}, \frac{15}{2}\right)$ **11.** $\left(-\frac{1}{6}, \frac{3}{8}\right)$ **12.** $(-1.5, -3.9)$ **13.** Center: $(0, 0)$; radius: 11 **14.** Center: $(13, -9)$; radius: $\sqrt{109}$ **15.** Center: $(0, 5)$; radius: $\sqrt{14}$ **16.** Center: $(-3, 7)$; radius: 4
17. $x^2 + y^2 = 1$ **18.** $\left(x + \frac{1}{2}\right)^2 + \left(y - \frac{3}{4}\right)^2 = \frac{81}{4}$
19. $(x + 8)^2 + (y - 6)^2 = 17$ **20.** $(x - 3)^2 + (y + 5)^2 = 20$

21.

$\frac{x^2}{4} + \frac{y^2}{36} = 1$

22.

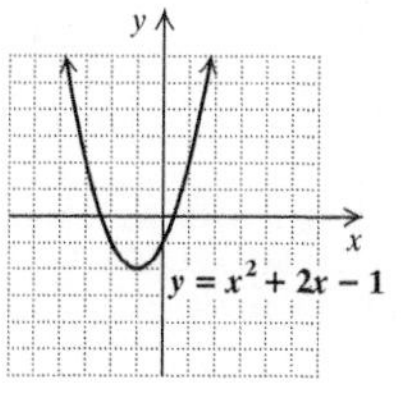

23.

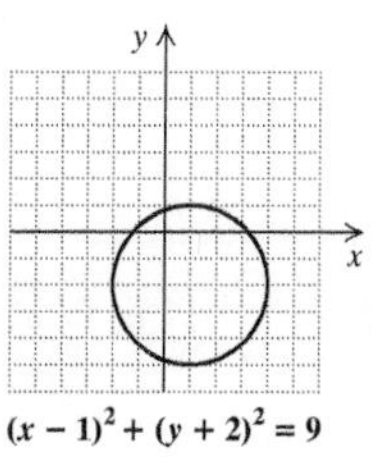

$(x - 1)^2 + (y + 2)^2 = 9$

24.

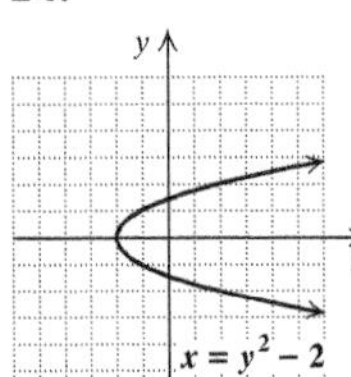

25.

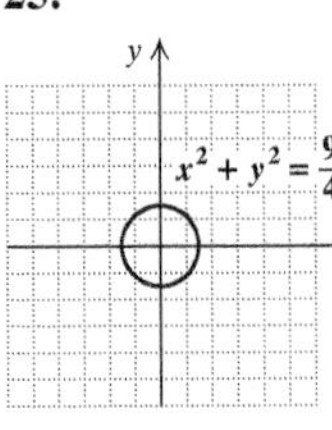

26.

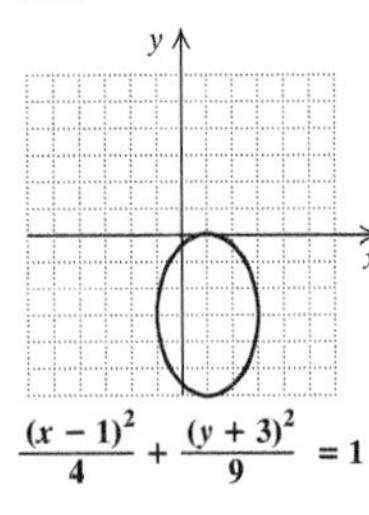

27.

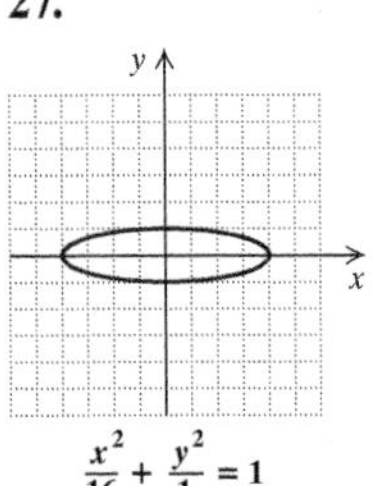

28.

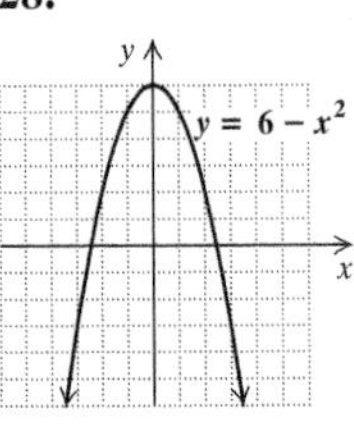

29. One method is to graph $y = ax^2 + bx + c$ and then use the DrawInv feature to graph the inverse relation, $x = ay^2 + by + c$. Another method is to use the quadratic formula to solve $x = ay^2 + by + c$, or $ay^2 + by + c - x = 0$. The solutions are $\frac{-b \pm \sqrt{b^2 - 4a(c - x)}}{2a}$. Then graph

$$y_1 = \frac{-b + \sqrt{b^2 - 4a(c - x)}}{2a} \text{ and}$$

$$y_2 = \frac{-b - \sqrt{b^2 - 4a(c - x)}}{2a} \text{ on the same screen.}$$

30. No; a circle is defined to be the set of points in a plane that are a fixed distance from the center. Thus, unless $r = 0$ and the "circle" is one point, the center is not part of the circle.
31. Bank shots originating at one focus (the tiny dot) are deflected to the other focus (the hole).
32. (a)

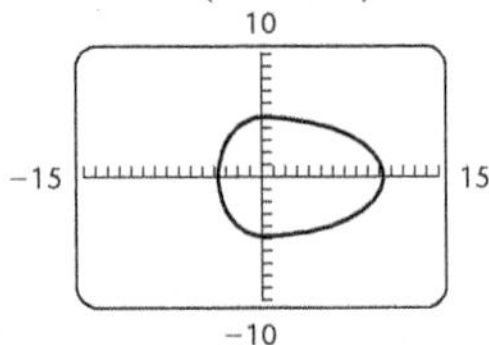

(b) Some other factors are the wind speed, the amount of rainfall in the preceding months, and the composition of the forest.

Calculator Corner, p. 744

1. $y_1 = \sqrt{\frac{15x^2 - 240}{4}}$, $y_2 = -\sqrt{\frac{15x^2 - 240}{4}}$

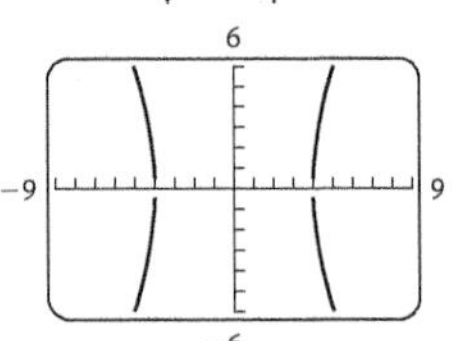

2. $y_1 = \sqrt{\frac{16x^2 - 320}{5}}$, $y_2 = -\sqrt{\frac{16x^2 - 320}{5}}$

6
−9 9
−6

3. $y_1 = \sqrt{\frac{16x^2 - 48}{3}}$, $y_2 = -\sqrt{\frac{16x^2 - 48}{3}}$

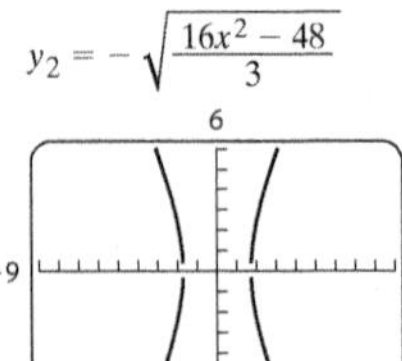

4. $y_1 = \sqrt{\frac{45x^2 - 405}{9}}$, $y_2 = -\sqrt{\frac{45x^2 - 405}{9}}$

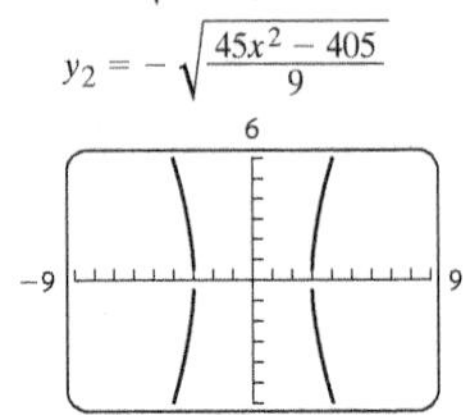

Visualizing for Success, p. 745

1. C **2.** E **3.** G **4.** J **5.** B **6.** F **7.** A **8.** H **9.** I **10.** D

Exercise Set 9.3, p. 746

RC1. B **RC2.** E **RC3.** A **RC4.** D **RC5.** F **RC6.** C **CC1.** True **CC2.** True **CC3.** True **CC4.** False **CC5.** False

1.

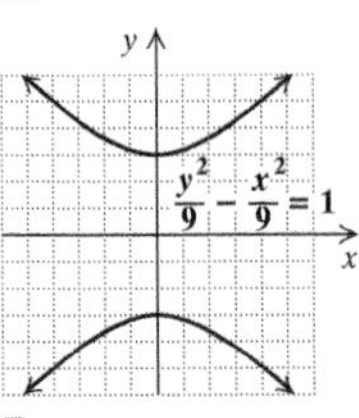

3.

$\frac{x^2}{4} - \frac{y^2}{25} = 1$

5.

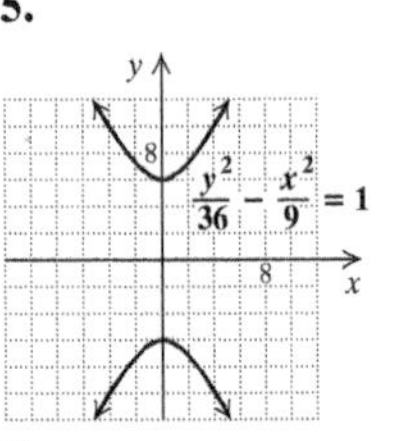

7.

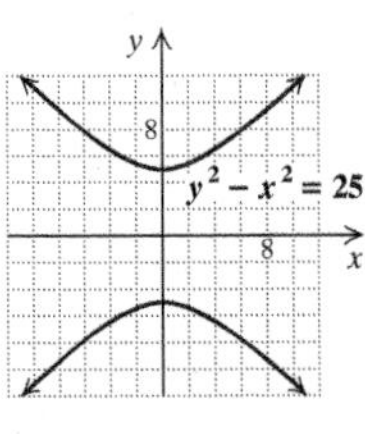

9.

$x^2 = 1 + y^2$

11.

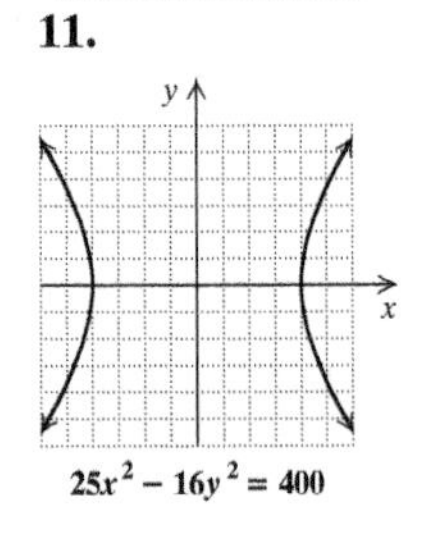

13.

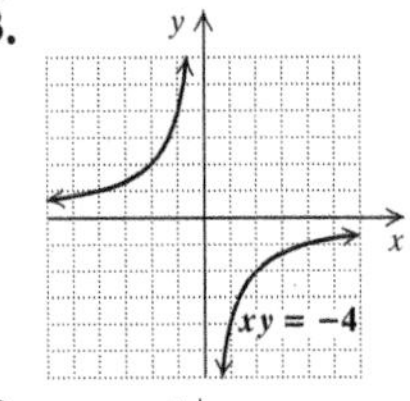

15.

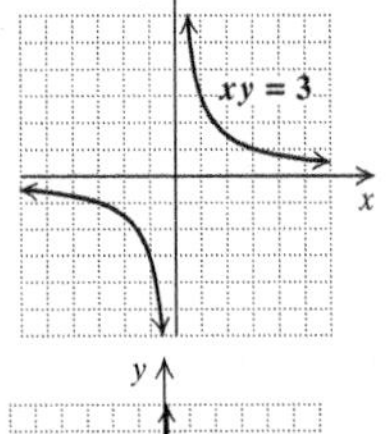

17.

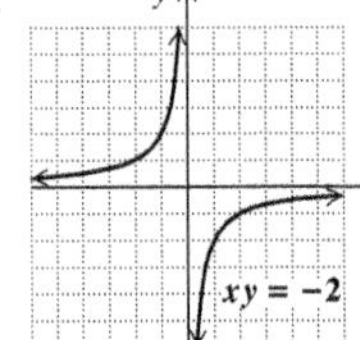

19.

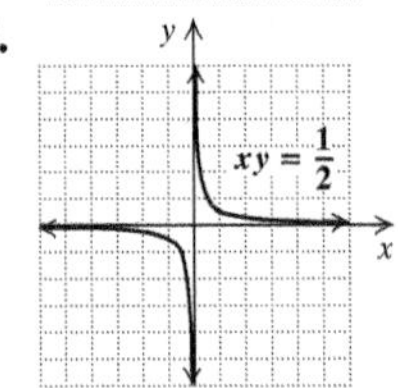

21. $\{x \mid x \geq 0\}$, or $[0, \infty)$ **22.** 6.75 **23.** $\{y \mid 3 < y < 9\}$, or $(3, 9)$ **24.** $\{x \mid x < -7 \text{ or } x > 7\}$, or $(-\infty, -7) \cup (7, \infty)$ **25.** $-3, \frac{1}{2}$ **26.** $\frac{17}{5}$ **27.** 4 **28.** 3 **29.** Left to the student **31.** Circle **33.** Parabola **35.** Ellipse **37.** Circle **39.** Hyperbola **41.** Circle

Calculator Corner, p. 749

1. Left to the student **2.** Left to the student

Exercise Set 9.4, p. 755

RC1. True **RC2.** False **RC3.** False **RC4.** True **RC5.** True

1. $(-8, -6), (6, 8)$ **3.** $(2, 0), (0, 3)$ **5.** $(-2, 1)$

7. $\left(\frac{5 + \sqrt{70}}{3}, \frac{-1 + \sqrt{70}}{3}\right), \left(\frac{5 - \sqrt{70}}{3}, \frac{-1 - \sqrt{70}}{3}\right)$

9. $\left(\frac{7}{3}, \frac{1}{3}\right), (1, -1)$ **11.** $(-7, 1), (1, -7)$ **13.** $(3, -5), (-1, 3)$

15. $\left(\frac{8 + 3i\sqrt{6}}{2}, \frac{-8 + 3i\sqrt{6}}{2}\right), \left(\frac{8 - 3i\sqrt{6}}{2}, \frac{-8 - 3i\sqrt{6}}{2}\right)$

17. $(-5, 0), (4, 3), (4, -3)$ **19.** $(3, 0), (-3, 0)$
21. $(2, 4), (-2, -4), (4, 2), (-4, -2)$
23. $(2, 3), (-2, -3), (3, 2), (-3, -2)$ **25.** $(2, 1), (-2, -1)$
27. $(5, 2), (-5, 2), \left(2, -\frac{4}{5}\right), \left(-2, -\frac{4}{5}\right)$
29. $(\sqrt{2}, -\sqrt{2}), (-\sqrt{2}, \sqrt{2})$

31. $\left(\frac{8i\sqrt{5}}{5}, \frac{3\sqrt{105}}{5}\right), \left(-\frac{8i\sqrt{5}}{5}, \frac{3\sqrt{105}}{5}\right),$ $\left(\frac{8i\sqrt{5}}{5}, -\frac{3\sqrt{105}}{5}\right), \left(-\frac{8i\sqrt{5}}{5}, -\frac{3\sqrt{105}}{5}\right)$
33. Length: 8 cm; width: 6 cm **35.** Length: 7 in.; width: 2 in.
37. Length: 12 ft; width: 5 ft **39.** Length: $\sqrt{2}$ m; width: 1 m
41. 24 ft; 16 ft **43.** Length: 24.8 cm; height: 18.6 cm
45. $f^{-1}(x) = \log x$ **46.** $f^{-1}(x) = x^3 - 2$ **47.** Does not exist
48. $f^{-1}(x) = \frac{7x + 3}{2x}$ **49.** $f^{-1}(x) = \frac{3x + 2}{1 - x}$ **50.** $f^{-1}(x) = e^x$
51. $\left(\frac{1}{2}, \frac{1}{3}\right), \left(\frac{1}{3}, \frac{1}{2}\right)$ **53.** One piece: $38\frac{12}{25}$ cm; other piece: $61\frac{13}{25}$ cm
55. 30 units

Summary and Review: Chapter 9, p. 759

Vocabulary Reinforcement

1. Circle **2.** Hyperbola **3.** Horizontal **4.** Vertex
5. Center **6.** Center

Concept Reinforcement

1. False **2.** True **3.** False **4.** True

Study Guide

1.

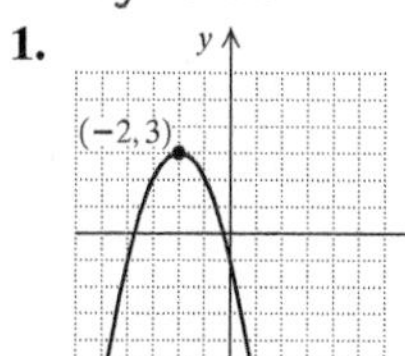

2. $\sqrt{10} \approx 3.162$ **3.** $(4, -8)$

4. Center: $(2, -1)$; radius: 4;

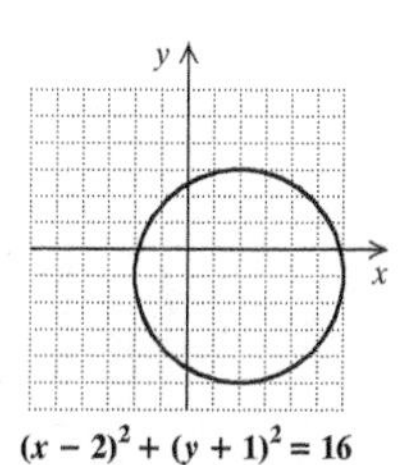

5. $x^2 + (y - 3)^2 = 36$

6.

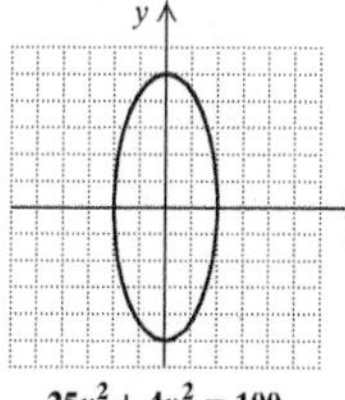

7.

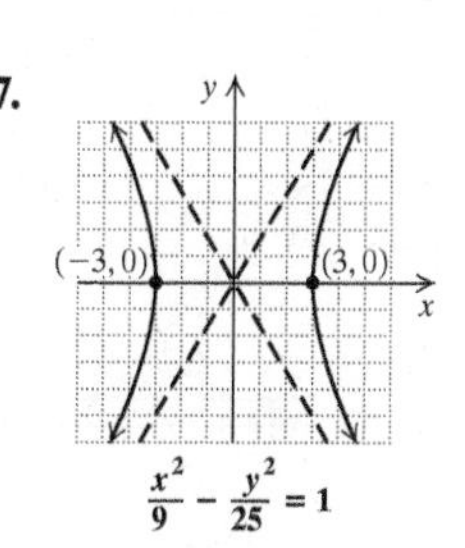

8. $(-6, 0)$ and $(0, 2)$

Review Exercises

1. 4 **2.** 5 **3.** $\sqrt{90.1} \approx 9.492$ **4.** $\sqrt{9 + 4a^2}$ **5.** $(4, 6)$
6. $\left(-3, \frac{5}{2}\right)$ **7.** $\left(\frac{3}{4}, \frac{\sqrt{3} - \sqrt{2}}{2}\right)$ **8.** $\left(\frac{1}{2}, 2a\right)$ **9.** $(-2, 3), \sqrt{2}$
10. $(5, 0), 7$ **11.** $(3, 1), 3$ **12.** $(-4, 3), \sqrt{35}$
13. $(x + 4)^2 + (y - 3)^2 = 48$ **14.** $(x - 7)^2 + (y + 2)^2 = 20$

15.

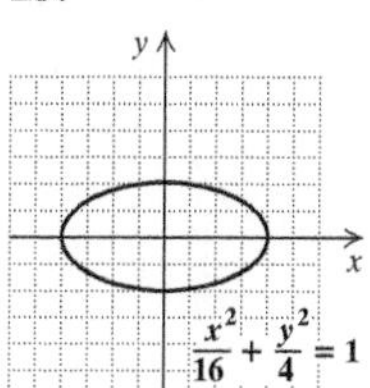

16.

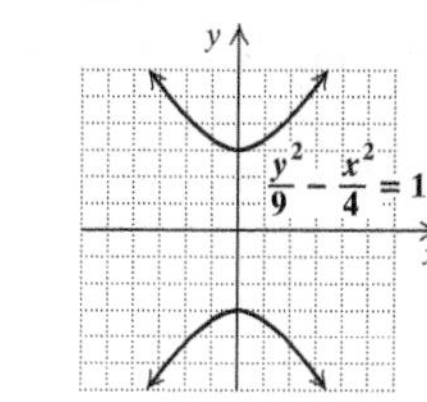

17.

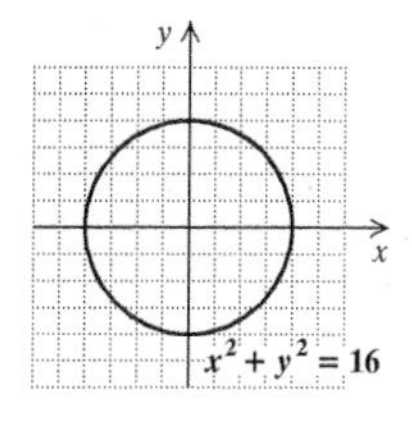

18.

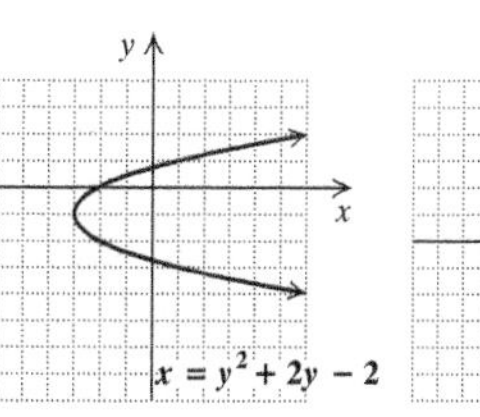

19.

$y = -2x^2 - 2x + 3$

20.

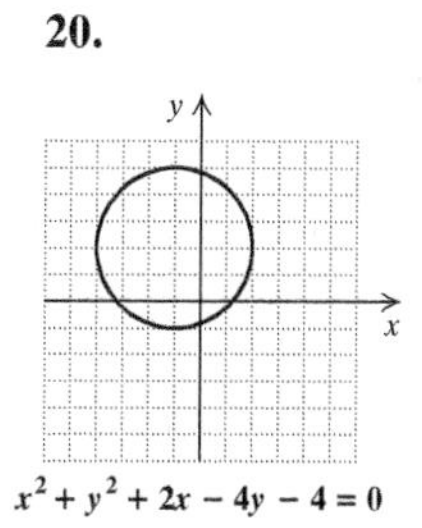

21.

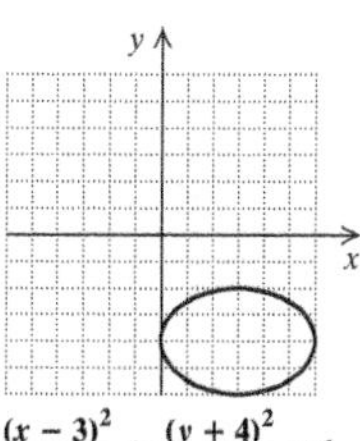

22.

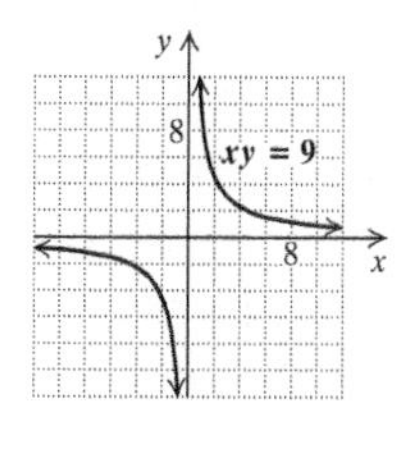

23.

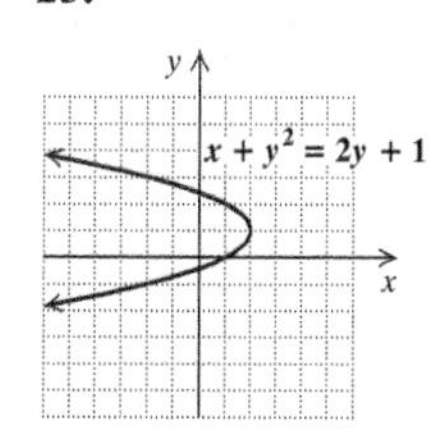

24.

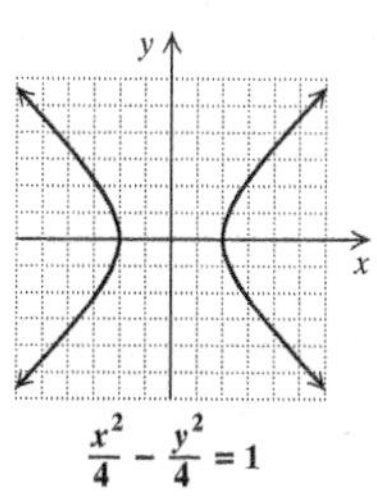

25. $(7, 4)$ **26.** $(2, 2), \left(\frac{32}{9}, -\frac{10}{9}\right)$ **27.** $(0, -3), (2, 1)$
28. $(4, 3), (4, -3), (-4, 3), (-4, -3)$
29. $(2, 1), (\sqrt{3}, 0), (-2, 1), (-\sqrt{3}, 0)$
30. $(3, -3), \left(-\frac{3}{5}, \frac{21}{5}\right)$ **31.** $(6, 8), (6, -8), (-6, 8), (-6, -8)$
32. $(2, 2), (-2, -2), (2\sqrt{2}, \sqrt{2}), (-2\sqrt{2}, -\sqrt{2})$
33. Length: 4 in.; width: 3 in. **34.** 11 ft, 3 ft **35.** 4 and 8
36. Length: 12 m; width: 7 m **37.** B **38.** D
39. $(-5, -4\sqrt{2}), (-5, 4\sqrt{2}), (3, -2\sqrt{2}), (3, 2\sqrt{2})$
40. $(x - 2)^2 + (y + 1)^2 = 25$ **41.** $\frac{x^2}{49} + \frac{y^2}{9} = 1$ **42.** $\left(\frac{9}{4}, 0\right)$
43. Parabola **44.** Hyperbola **45.** Circle **46.** Ellipse
47. Circle **48.** Hyperbola

Understanding Through Discussion and Writing

1. In Chapter 3, we studied systems of linear equations. In this chapter, we studied systems of two equations in which at least one equation is of second degree. **2.** Parabolas of the form $y = ax^2 + bx + c$ and hyperbolas of the form $xy = c$ pass the vertical-line test, so they are functions. Circles, ellipses, parabolas of the form $x = ay^2 + by + c$, and hyperbolas of the form $\frac{x^2}{a^2} - \frac{y^2}{b^2} = 1$ or $\frac{y^2}{b^2} - \frac{x^2}{a^2} = 1$ fail the vertical-line test, and hence are not functions. **3.** The graph of a parabola has one branch whereas the graph of a hyperbola has two branches. A hyperbola has asymptotes, but a parabola does not. **4.** The asymptotes are $y = x$ and $y = -x$, because for $a = b$, $\pm\frac{b}{a} = \pm 1$.

Test: Chapter 9, p. 765

1. [9.1b] $\sqrt{180} \approx 13.416$ **2.** [9.1b] $\sqrt{36 + 4a^2}$, or $2\sqrt{9 + a^2}$
3. [9.1c] $(0, 5)$ **4.** [9.1c] $(0, 0)$ **5.** [9.1d] Center: $(-2, 3)$; radius: 8 **6.** [9.1d] Center: $(-2, 3)$; radius: 3
7. [9.1d] $(x + 2)^2 + (y + 5)^2 = 18$

8. [9.1a]

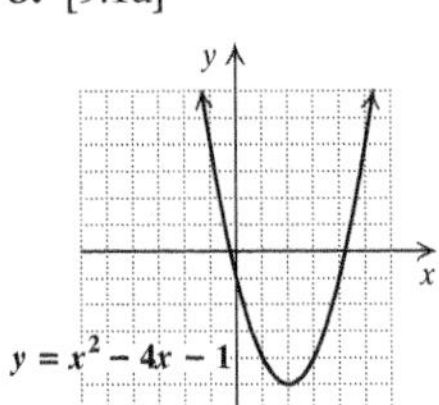

9. [9.1d]

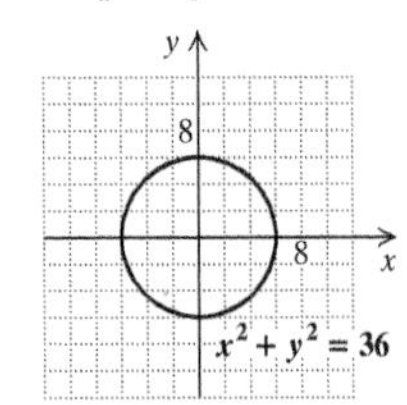

10. [9.3a]

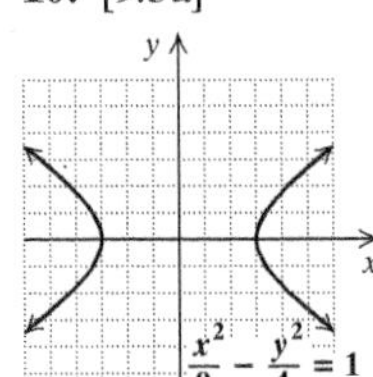

11. [9.2a]

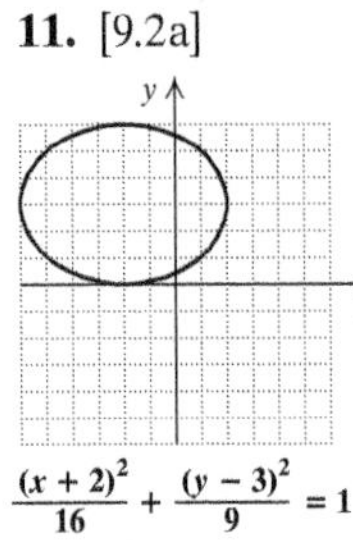

12. [9.1d]

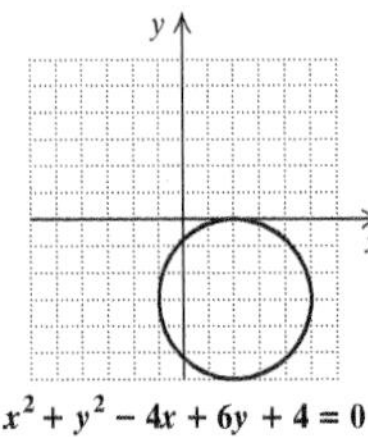

13. [9.2a]

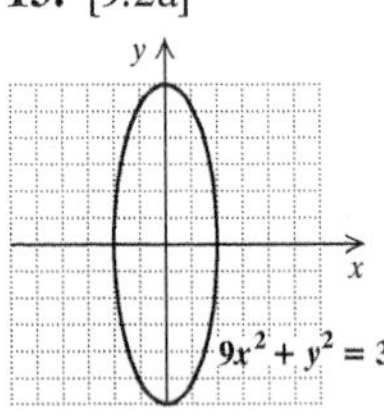

14. [9.3b]

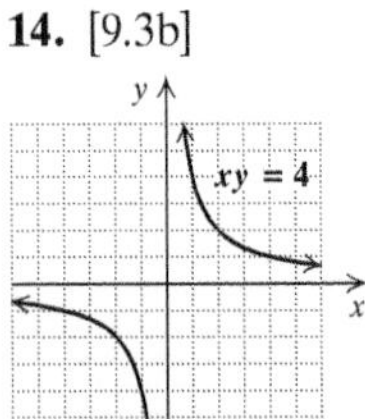

15. [9.1a]

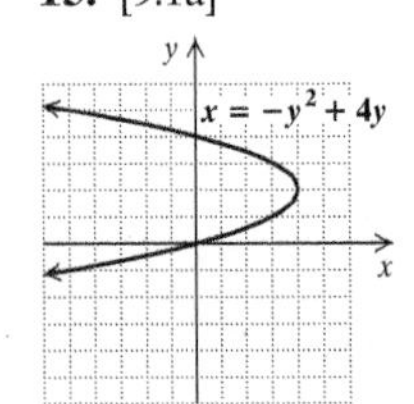

16. [9.4a] $(0, 3), (4, 0)$ **17.** [9.4a] $(4, 0), (-4, 0)$
18. [9.4b] 16 ft by 12 ft **19.** [9.4b] \$1200, 6%
20. [9.4b] 11 yd by 2 yd **21.** [9.4b] $\sqrt{5}$ m, $\sqrt{3}$ m **22.** [9.4a] B
23. [9.2a] $\frac{(x-6)^2}{25} + \frac{(y-3)^2}{9} = 1$
24. [9.1d] $\{(x, y) | (x-8)^2 + y^2 = 100\}$
25. [9.4b] 9 **26.** [9.1b] $\left(0, -\frac{31}{4}\right)$

Cumulative Review: Chapters 1–9, p. 767

1. [1.4c] $\{x | x \geq -1\}$, or $[-1, \infty)$
2. [1.6e] $\{x | x < -6.4 \text{ or } x > 6.4\}$, or $(-\infty, -6.4) \cup (6.4, \infty)$
3. [1.5a] $\{x | -1 \leq x < 6\}$, or $[-1, 6)$ **4.** [3.2a], [3.3a] $\left(-\frac{1}{3}, 5\right)$
5. [7.4c] $-3, -2, 2, 3$ **6.** [4.8a] $-1, \frac{3}{2}$ **7.** [5.5a] $-\frac{2}{3}, 3$ **8.** [6.6a] 4
9. [4.8a] $-7, -3$ **10.** [7.2a] $-\frac{1}{4} \pm i\frac{\sqrt{7}}{4}$ **11.** [8.6a] 1.748
12. [7.8b] $\{x | x < -1 \text{ or } x > 2\}$, or $(-\infty, -1) \cup (2, \infty)$
13. [8.6b] 9 **14.** [7.8a] $\{x | x \leq -1 \text{ or } x \geq 1\}$, or $(-\infty, -1] \cup [1, \infty)$
15. [8.6b] 1 **16.** [5.7a] $p = \frac{qf}{q - f}$ **17.** [3.5a] $(-1, 2, 3)$
18. [9.4a] $(1, 2), (-1, 2), (1, -2,), (-1, -2)$ **19.** [5.5a] -16
20. [1.2a] $N = \frac{4P - 3M}{6}$ **21.** [8.1c], [8.7b] **(a)** About 104 million barrels per day; about 123 million barrels per day; **(b)** about 37 years;
(c)

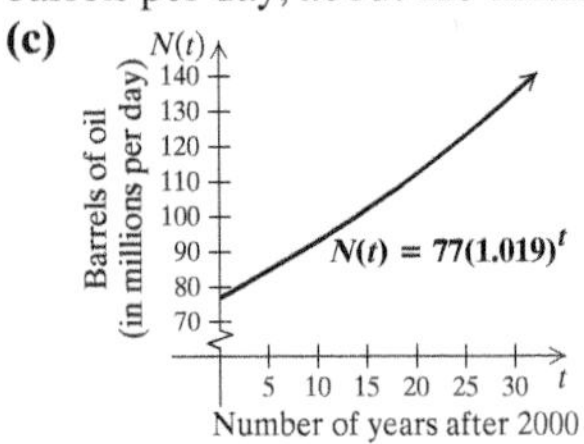

22. [8.7b] **(a)** $A(t) = \$50{,}000(1.04)^t$; **(b)** \$50,000; \$58,492.93; \$68,428.45; \$74,012.21;
(c)

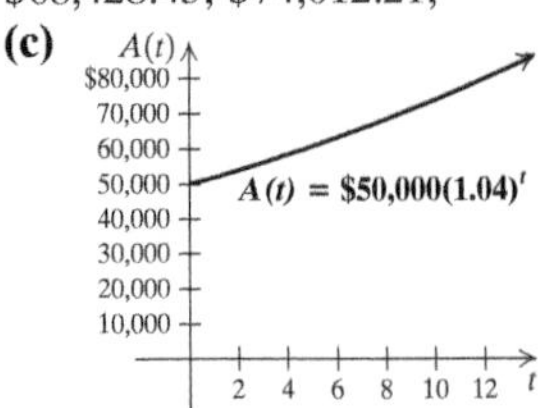

23. [4.2a] $2x^3 - x^2 - 8x - 3$ **24.** [4.1d] $-x^3 + 3x^2 - x - 6$
25. [5.1d] $\frac{2m - 1}{m + 1}$ **26.** [5.2c] $\frac{x + 2}{x + 1}$ **27.** [5.4a] $\frac{1}{x + 1}$
28. [5.3b, c] $x^3 + 2x^2 - 2x + 1 + \frac{3}{x + 1}$
29. [6.3b] $5x^2\sqrt{y}$ **30.** [6.4a] $11\sqrt{2}$ **31.** [6.2d] 8
32. [6.8c] $16 + i\sqrt{2}$ **33.** [6.8e] $\frac{3}{10} + \frac{11}{10}i$
34. [2.4c], [2.6e] **(a)** 2018: 74.75 million; 2021: 77.9 million; 2025: 82.1 million;
(b)

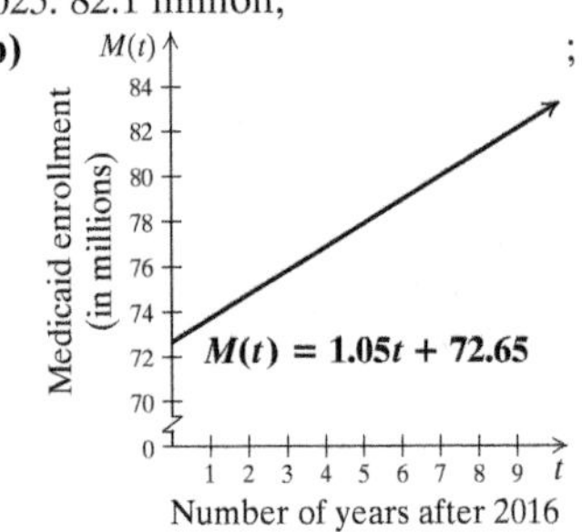

;
(c) $(0, 72.65)$; **(d)** 1.05; **(e)** an increase of 1.05 million per year
35. [2.6c] $y = 2x + 2$ **36.** [2.6d] $y = -\frac{1}{2}x + \frac{5}{2}$

37. [2.5a]

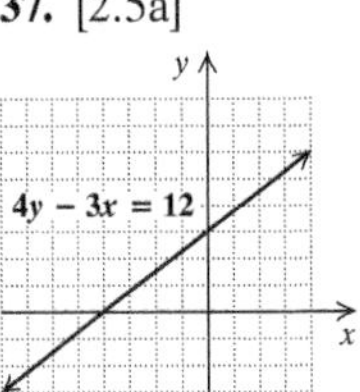

38. [3.7b]

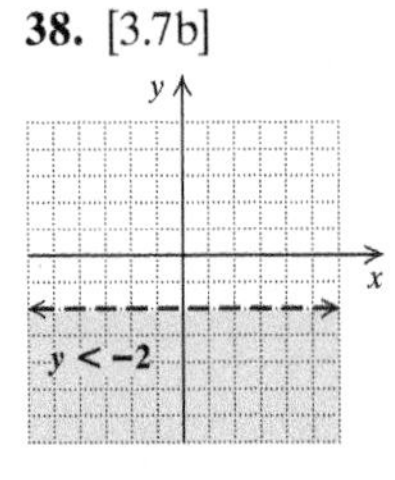

39. [9.2a]

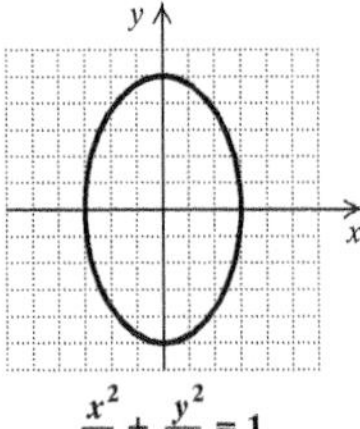

40. [9.1d]

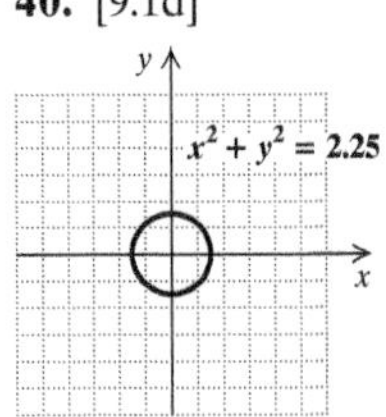

41. [3.7c]

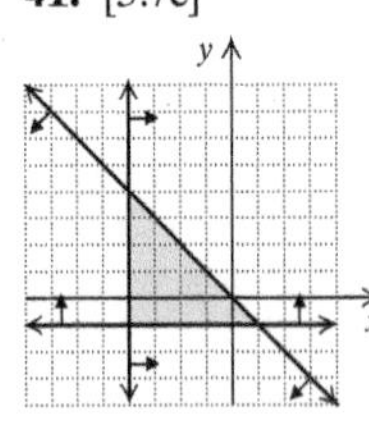

42. [9.3a]

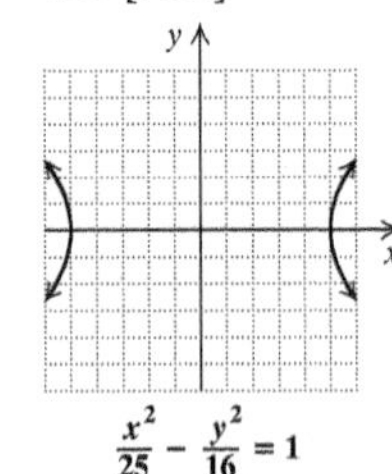

43. [9.1d]

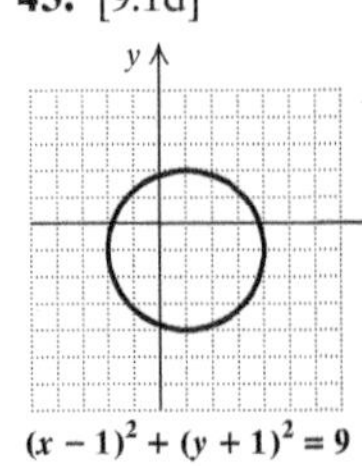

44. [7.6a]

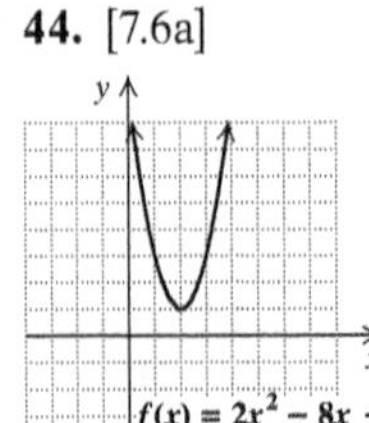

45. [2.5c]

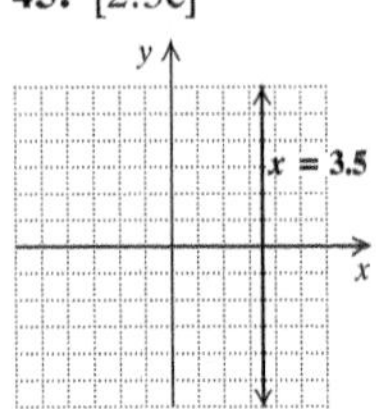

46. [9.1a]

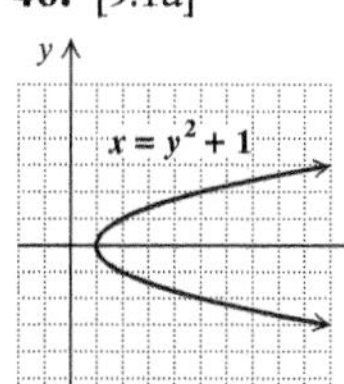

47. [8.5c]

48. [8.3a]

49. [4.3b] $(x - 6)(2x^3 + 1)$ **50.** [4.4a] $3(a - 9b)(a + 5b)$
51. [4.4a] $(x - 8)(x - 9)$
52. [4.6b] $(9m^2 + n^2)(3m + n)(3m - n)$
53. [4.6a] $4(2x - 1)^2$ **54.** [4.6d] $3(3a - 2)(9a^2 + 6a + 4)$
55. [4.5a, b] $2(5x - 2)(x + 7)$ **56.** [4.5a, b] $3x(2x - 1)(x + 5)$
57. [9.1d] Center: $(8, -3)$; radius: $\sqrt{5}$
58. [8.2d] $f^{-1}(x) = \frac{1}{2}(x + 3)$ **59.** [5.8e] $\frac{4}{5}$ **60.** [2.2b] -10
61. [9.1b] 10 **62.** [9.1c] $\left(1, -\frac{3}{2}\right)$ **63.** [6.5b] $\dfrac{15 + 8\sqrt{a} + a}{9 - a}$
64. [5.1a] $\left(-\infty, -\frac{1}{3}\right) \cup \left(-\frac{1}{3}, 0\right) \cup (0, \infty)$ **65.** [4.8a] $-1, \frac{2}{3}$
66. [1.4d] More than 4 **67.** [3.4b] 612 mi **68.** [1.3a] $11\frac{3}{7}$
69. [3.4a] 24 L of A; 56 L of B **70.** [5.6c] 350 mph
71. [5.6a] $8\frac{2}{5}$ min **72.** [5.8f] 20 **73.** [9.4b] 5 ft by 12 ft
74. [9.4b] Length: 20 ft; width: 15 ft **75.** [7.7a] 1250 ft^2
76. [8.7b] 2397 years **77.** [5.8d] 3360 kg
78. [2.6c] $f(x) = -\frac{1}{3}x - \frac{7}{3}$ **79.** [7.7b] $f(x) = -\frac{17}{18}x^2 - \frac{59}{18}x + \frac{11}{9}$
80. [8.3b] $\log r = 6$ **81.** [8.3b] $3^x = Q$
82. [8.4d] $\log_b\left(\dfrac{x^7}{yz^8}\right)^{1/5}$, or $\log_b \dfrac{x^{7/5}}{y^{1/5}z^{8/5}}$
83. [8.4d] $-6 \log_b x - 30 \log_b y + 6 \log_b z$ **84.** [7.7a] 169
85. [8.2c] No **86.** [2.3a] **(a)** -5; **(b)** $(-\infty, \infty)$;
(c) $-2, -1, 1, 2$; **(d)** $[-7, \infty)$
87. [8.7b] **(a)** $P(t) = 1{,}998{,}257e^{0.025t}$; **(b)** 3,294,569;
(c) about 2028 **88.** [5.5a] All real numbers except 0 and -12
89. [8.6b] 81 **90.** [9.1d] Circle centered at the origin with radius $|a|$ **91.** [1.3a] 84 years

APPENDIXES

Exercise Set A, p. 782

1. $\frac{9}{12}$ **3.** $\frac{60}{100}$ **5.** $\frac{21}{24}$ **7.** $\frac{2}{3}$ **9.** 4 **11.** $\frac{1}{7}$ **13.** 8 **15.** $\frac{1}{4}$
17. 5 **19.** $\frac{17}{21}$ **21.** $\frac{13}{7}$ **23.** $\frac{4}{3}$ **25.** $\frac{1}{12}$ **27.** $\frac{45}{16}$ **29.** $\frac{7}{6}$
31. $\frac{5}{6}$ **33.** $\frac{1}{2}$ **35.** $\frac{13}{24}$ **37.** $\frac{35}{18}$ **39.** $\frac{10}{3}$ **41.** $\frac{5}{36}$ **43.** 500
45. $\frac{3}{40}$ **47.** $\frac{99{,}999}{100}$

Exercise Set B, p. 787

1. 10 **3.** 0 **5.** -48 **7.** 0 **9.** -10 **11.** -3 **13.** 5
15. 0 **17.** $(2, 0)$ **19.** $\left(-\frac{25}{2}, -\frac{11}{2}\right)$ **21.** $\left(\frac{3}{2}, \frac{5}{2}\right)$ **23.** $(-4, 3)$
25. $(2, -1, 4)$ **27.** $(1, 2, 3)$ **29.** $\left(\frac{3}{2}, -4, 3\right)$ **31.** $(2, -2, 1)$

Exercise Set C, p. 791

1. $\left(\frac{3}{2}, \frac{5}{2}\right)$ **3.** $(-4, 3)$ **5.** $\left(\frac{1}{2}, \frac{3}{2}\right)$ **7.** $(10, -10)$ **9.** $\left(\frac{3}{2}, -4, 3\right)$
11. $(2, -2, 1)$ **13.** $(0, 2, 1)$ **15.** $\left(4, \frac{1}{2}, -\frac{1}{2}\right)$

Exercise Set D, p. 794

1. 1 **3.** -41 **5.** 12 **7.** $\frac{13}{18}$ **9.** 5 **11.** 2 **13.** $x^2 - x + 1$
15. 21 **17.** 5 **19.** $-x^3 + 4x^2 + 3x - 12$ **21.** 42 **23.** $-\frac{3}{4}$
25. $\frac{1}{6}$ **27.** $x^2 + 3x - 4$; $x^2 - 3x + 4$; $3x^3 - 4x^2$; $\dfrac{x^2}{3x - 4}$
29. $\dfrac{1}{x - 2} + 4x^3$; $\dfrac{1}{x - 2} - 4x^3$; $\dfrac{4x^3}{x - 2}$; $\dfrac{1}{4x^3(x - 2)}$
31. $\dfrac{3}{x - 2} + \dfrac{5}{4 - x}$; $\dfrac{3}{x - 2} - \dfrac{5}{4 - x}$; $\dfrac{15}{(x - 2)(4 - x)}$; $\dfrac{3(4 - x)}{5(x - 2)}$

Guided Solutions

CHAPTER 1

Section 1.1

10.
$$x + \frac{1}{4} = -\frac{3}{5}$$
$$x + \frac{1}{4} + \left(-\frac{1}{4}\right) = -\frac{3}{5} + \left(-\frac{1}{4}\right)$$
$$x + \frac{1}{4} - \frac{1}{4} = -\frac{3}{5} - \frac{1}{4}$$
$$x + 0 = -\frac{3}{5}\cdot\frac{4}{4} - \frac{1}{4}\cdot\frac{5}{5}$$
$$x = -\frac{12}{20} - \frac{5}{20}$$
$$x = -\frac{17}{20}$$

15.
$$-4x = -\frac{6}{7}$$
$$-\frac{1}{4}\cdot(-4x) = -\frac{1}{4}\cdot\left(-\frac{6}{7}\right)$$
$$x = \frac{6}{28}$$
$$x = \frac{3}{14}$$

20.
$$\frac{2}{3} - \frac{5}{6}y = \frac{1}{3}$$
LCD = 6
$$6\left(\frac{2}{3} - \frac{5}{6}y\right) = 6\cdot\frac{1}{3}$$
$$6\cdot\frac{2}{3} - 6\cdot\frac{5}{6}y = 2$$
$$4 - 5y = 2$$
$$-4 + 4 - 5y = -4 + 2$$
$$-5y = -2$$
$$-\frac{1}{5}\cdot(-5y) = -\frac{1}{5}\cdot(-2)$$
$$1\cdot y = \frac{2}{5}$$
$$y = \frac{2}{5}$$

Section 1.2

4.
$$P = \frac{3}{5}(c + 10)$$
$$5\cdot P = 5\cdot\frac{3}{5}(c + 10)$$
$$5P = 3(c + 10)$$
$$5P = 3c + 30$$
$$5P - 30 = 3c$$
$$\frac{5P - 30}{3} = c, \text{ or}$$
$$c = \frac{5}{3}P - 10$$

Section 1.4

11.
$$2x - 3 \ge 3x - 1$$
$$2x - 3 - 2x \ge 3x - 1 - 2x$$
$$-3 \ge x - 1$$
$$-3 + 1 \ge x - 1 + 1$$
$$-2 \ge x, \text{ or}$$
$$x \le -2$$
The solution set is $\{x|x \le -2\}$, or $(-\infty, -2]$.

15.
$$6 - 5y \ge 7$$
$$6 - 5y - 6 \ge 7 - 6$$
$$-5y \ge 1$$
$$\frac{-5y}{-5} \le \frac{1}{-5}$$
$$y \le -\frac{1}{5}$$
The solution set is $\left\{y|y \le -\frac{1}{5}\right\}$, or $\left(-\infty, \frac{1}{5}\right]$.

Section 1.5

7.
$$-4 \le 8 - 2x \le 4$$
$$-4 - 8 \le 8 - 2x - 8 \le 4 - 8$$
$$-12 \le -2x \le -4$$
$$\frac{-12}{-2} \ge \frac{-2x}{-2} \ge \frac{-4}{-2}$$
$$6 \ge x \ge 2, \text{ or}$$
$$2 \le x \le 6$$
The solution set is $\{x|2 \le x \le 6\}$, or $[2, 6]$.

13.
$$-3x - 7 < -1 \quad or \quad x + 4 < -1$$
$$-3x < 6 \quad or \quad x < -5$$
$$\frac{-3x}{-3} > \frac{6}{-3} \quad or \quad x < -5$$
$$x > -2 \quad or \quad x < -5$$
The solution set is $\{x|x < -5 \text{ or } x > -2\}$, or $(-\infty, -5) \cup (-2, \infty)$.

Section 1.6

6. $|-6-(-35)| = |-6+35|$
$= |29| = 29$

16. $|3x-4| = 17$

$3x - 4 = -17$ *or* $3x - 4 = 17$

$3x = -13$ *or* $3x = 21$

$x = -\frac{13}{3}$ *or* $x = 7$

The solution set is $\left\{-\frac{13}{3}, 7\right\}$.

24. $|7-3x| \le 4$

$-4 \le 7 - 3x \le 4$

$-11 \le -3x \le -3$

$\frac{-11}{-3} \ge \frac{-3x}{-3} \ge \frac{-3}{-3}$

$\frac{11}{3} \ge x \ge 1$

The solution set is $\left\{x \,|\, 1 \le x \le \frac{11}{3}\right\}$, or $\left[1, \frac{11}{3}\right]$.

25. $|3x+2| \ge 5$

$3x + 2 \le -5$ *or* $3x + 2 \ge 5$

$3x \le -7$ *or* $3x \ge 3$

$x \le -\frac{7}{3}$ *or* $x \ge 1$

The solution set is $\left\{x \,|\, x \le -\frac{7}{3} \text{ or } x \ge 1\right\}$, or $\left(-\infty, -\frac{7}{3}\right] \cup [1, \infty)$.

CHAPTER 2

Section 2.1

17. $y = \frac{1}{2}x$

x	y	(x, y)
4	2	(4, 2)
2	1	(2, 1)
0	0	(0, 0)
−2	−1	(−2, −1)
−4	−2	(−4, −2)

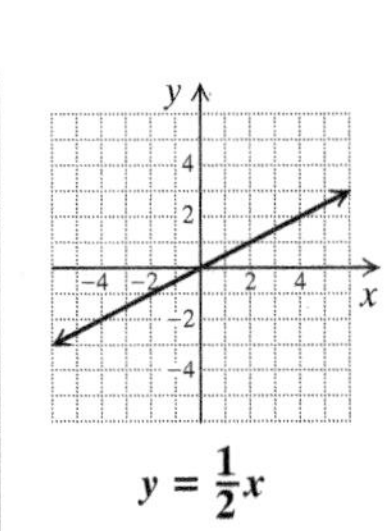

$y = \frac{1}{2}x$

20. Graph: $4y - 3x = -8$.

We first solve for y:

$4y - 3x = -8$

$4y = 3x - 8$

$y = \frac{3}{4}x - 2$

x	y	(x, y)
0	−2	(0, −2)
4	1	(4, 1)
−4	−5	(−4, −5)

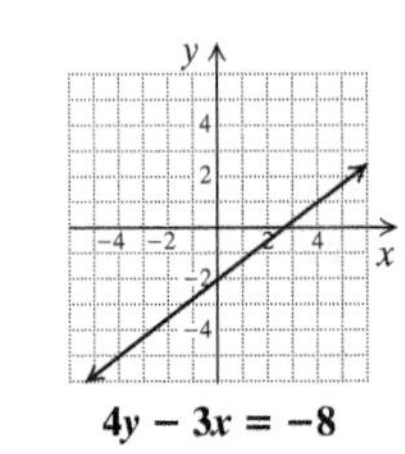

$4y - 3x = -8$

Section 2.2

8. a) $f(x) = 2x^2 + 3x - 4$

$f(8) = 2 \cdot 8^2 + 3 \cdot 8 - 4$
$= 2 \cdot 64 + 24 - 4$
$= 128 + 24 - 4$
$= 152 - 4$
$= 148$

Section 2.3

6. Find the domain of $f(x) = \frac{4}{3x+2}$.

Set the denominator equal to 0 and solve for x:

$3x + 2 = 0$

$3x = -2$

$x = -\frac{2}{3}$.

Thus, $-\frac{2}{3}$ is not in the domain of $f(x)$; all other real numbers are.

Domain $= \left\{x \,|\, x \text{ is a real number } and\ x \ne -\frac{2}{3}\right\}$, or $\left(-\infty, -\frac{2}{3}\right) \cup \left(-\frac{2}{3}, \infty\right)$

Section 2.4

6. Graph the line through $(-1, -1)$ and $(2, -4)$ and find its slope.

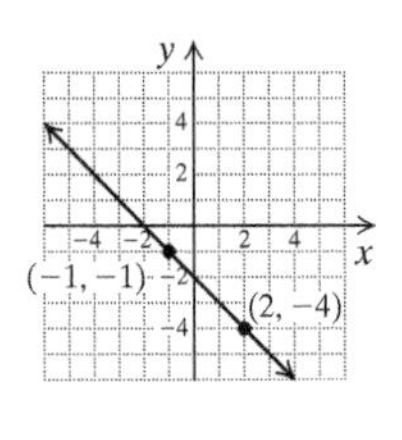

$m = \frac{-4-(-1)}{2-(-1)}$

$= \frac{-3}{3}$

$= -1$

10. Find the slope and the y-intercept of $5x - 10y = 25$.

First solve for y:

$5x - 10y = 25$

$-10y = -5x + 25$

$y = \frac{-5x + 25}{-10}$

$y = \frac{1}{2}x - \frac{5}{2}$.

Slope is $\frac{1}{2}$; y-intercept is $\left(0, -\frac{5}{2}\right)$.

Section 2.5

1. Find the x-intercepts of $4y - 12 = -6x$ and then graph the line.

To find the y-intercept, set $x = 0$ and solve for y:

$4y - 12 = -6 \cdot 0$

$4y - 12 = 0$

$4y = 12$

$y = 3$.

The y-intercept is $(0, 3)$.

To find the x-intercept, set $y = 0$ and solve for x:

$4 \cdot 0 - 12 = -6x$

$-12 = -6x$

$2 = x$.

The x-intercept is $(2, 0)$.

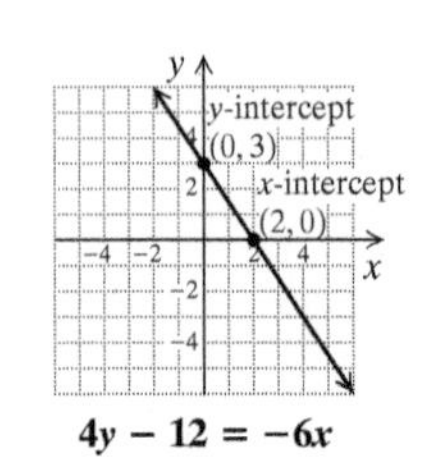

$4y - 12 = -6x$

11. Determine whether the graphs of the lines $x + 4 = y$ and $y - x = -3$ are parallel.

Write each equation in the form $y = mx + b$:

$$x + 4 = y \rightarrow y = x + 4;$$
$$y - x = -3 \rightarrow y = x - 3.$$

The slope of each line is 1, and the y-intercepts, $(0, 4)$ and $(0, -3)$, are different. Thus the lines ___are___ parallel.

14. Determine whether the graphs of the lines $2y - x = 2$ and $y + 2x = 4$ are perpendicular.

Write each equation in the form $y = mx + b$:

$$2y - x = 2 \rightarrow y = \frac{1}{2}x + 1;$$
$$y + 2x = 4 \rightarrow y = -2x + 4.$$

The slopes of these lines are $\frac{1}{2}$ and -2. The product of the slopes is $\frac{1}{2} \cdot (-2) = -1$. Thus the lines ___are___ perpendicular.

Section 2.6

6. Find an equation of the line containing the points $(4, -3)$ and $(1, 2)$.

First, find the slope:

$$m = \frac{2 - (-3)}{1 - 4} = \frac{5}{-3} = -\frac{5}{3}.$$

Using the point–slope equation,

$$y - y_1 = m(x - x_1),$$

substitute 4 for x_1, -3 for y_1, and $-\frac{5}{3}$ for m:

$$y - (-3) = -\frac{5}{3}(x - 4)$$
$$y + 3 = -\frac{5}{3}x + \frac{20}{3}$$
$$y = -\frac{5}{3}x + \frac{20}{3} - 3$$
$$y = -\frac{5}{3}x + \frac{20}{3} - \frac{9}{3}$$
$$y = -\frac{5}{3}x + \frac{11}{3}.$$

8. Find an equation of the line containing the point $(2, -1)$ and parallel to the line $9x - 3y = 5$.

Find the slope of the given line:

$$9x - 3y = 5$$
$$-3y = -9x + 5$$
$$y = 3x - \frac{5}{3}.$$

The slope is 3.
The line parallel to $9x - 3y = 5$ must have slope 3.

Using the slope–intercept equation,

$$y = mx + b,$$

substitute 3 for m, 2 for x, and -1 for y, and solve for b:

$$-1 = 3 \cdot 2 + b$$
$$-1 = 6 + b$$
$$-7 = b.$$

Substitute 3 for m and -7 for b in $y = mx + b$:

$$y = 3x + (-7)$$
$$y = 3x - 7.$$

9. Find an equation of the line containing the point $(5, 4)$ and perpendicular to the line $2x - 4y = 9$.

Find the slope of the given line:

$$2x - 4y = 9$$
$$-4y = -2x + 9$$
$$y = \frac{1}{2}x - \frac{9}{4}.$$

The slope is $\frac{1}{2}$.
The slope of a line perpendicular to $2x - 4y = 9$ is the opposite of the reciprocal of $\frac{1}{2}$, or -2.

Using the point–slope equation,

$$y - y_1 = m(x - x_1),$$

substitute -2 for m, 5 for x_1, and 4 for y_1:

$$y - 4 = -2(x - 5)$$
$$y - 4 = -2x + 10$$
$$y = -2x + 14.$$

CHAPTER 3

Section 3.1

4. Classify each of the systems in Margin Exercises 1–3 as consistent or inconsistent.

The system in Margin Exercise 1 has a solution, so it is consistent.

The system in Margin Exercise 2 has a solution, so it is consistent.

The system in Margin Exercise 3 does not have a solution, so it is inconsistent.

6. Classify the equations in Margin Exercises 1, 2, 3, and 5 as dependent or independent.

In Margin Exercise 1, the graphs are different, so the equations are independent.

In Margin Exercise 2, the graphs are different, so the equations are independent.

In Margin Exercise 3, the graphs are different, so the equations are independent.

In Margin Exercise 5, the graphs are the same, so the equations are dependent.

Section 3.2

4. $8x - 5y = 12,$ **(1)**

$x - y = 3$ **(2)**

Solve for x in equation (2):

$$x - y = 3$$
$$x = y + 3. \quad \textbf{(3)}$$

Substitute $y + 3$ for x in equation (1) and solve for y:

$$8x - 5y = 12$$
$$8(y + 3) - 5y = 12$$
$$8y + 24 - 5y = 12$$
$$3y + 24 = 12$$
$$3y = -12$$
$$y = -4.$$

Substitute -4 for y in equation (3) and solve for x:

$$x = y + 3$$
$$= -4 + 3$$
$$= -1.$$

The ordered pair checks in both equations.
The solution is $(-1, -4)$.

Section 3.3

3. Solve by the elimination method:

$$2y + 3x = 12, \quad (1)$$
$$-4y + 5x = -2. \quad (2)$$

Multiply by 2 on both sides of equation (1) and add:

$$\begin{aligned} 4y + 6x &= 24 \\ -4y + 5x &= -2 \\ \hline 0 + 11x &= 22 \\ 11x &= 22 \\ x &= 2. \end{aligned}$$

Substitute 2 for x in equation (1) and solve for y:

$$\begin{aligned} 2y + 3x &= 12 \\ 2y + 3(2) &= 12 \\ 2y + 6 &= 12 \\ 2y &= 6 \\ y &= 3. \end{aligned}$$

The ordered pair checks in both equations, so the solution is (2, 3).

8. Solve by the elimination method:

$$y + 2x = 3,$$
$$y + 2x = -1.$$

Multiply the second equation by -1 and add:

$$\begin{aligned} y + 2x &= 3 \\ -y - 2x &= 1 \\ \hline 0 &= 4. \end{aligned}$$

The equation is $\frac{\text{false}}{\text{true/false}}$ so the system has no solution.

Section 3.4

1.

White	Red	Total	
w	r	30	→ $w + r = 30$
\$18.95	\$19.50		
$18.95w$	$19.50r$	572.90	→ $18.95w + 19.50r = 572.90$

3.

First Investment	Second Investment	Total	
x	y	\$3700	→ $x + y = 3700$
7%	9%		
1 year	1 year		
$0.07x$	$0.09y$	\$297	→ $0.07x + 0.09y = 297$

4.

Attack	Blast	Mixture	
a	b	60	→ $a + b = 60$
2%	6%	5%	
$0.02a$	$0.06b$	0.05×60, or 3	→ $0.02a + 0.06b = 3$

5.

Distance	Rate	Time	
d	35 km/h	t	→ $d = 35t$
d	40 km/h	$t - 1$	→ $d = 40(t - 1)$

6.

Distance	Rate	Time	
d	$r + 20$	4 hr	→ $d = 4(r + 20)$
d	$r - 20$	5 hr	→ $d = 5(r - 20)$

Section 3.5

3. Solve. Don't forget to check.

$$x + y + z = 100, \quad (1)$$
$$x - y = -10, \quad (2)$$
$$x - z = -30 \quad (3)$$

Add equations (1) and (3):

$$\begin{aligned} x + y + z &= 100 \quad (1) \\ x - z &= -30 \quad (3) \\ \hline 2x + y &= 70. \quad (4) \end{aligned}$$

Add equations (2) and (4) and solve for x:

$$\begin{aligned} x - y &= -10 \quad (2) \\ 2x + y &= 70 \quad (4) \\ \hline 3x &= 60 \\ x &= 20. \end{aligned}$$

Substitute 20 for x in equation (4) and solve for y:

$$\begin{aligned} 2(20) + y &= 70 \\ y &= 30. \end{aligned}$$

Substitute 20 for x and 30 for y in equation (1) and solve for z:

$$\begin{aligned} 20 + 30 + z &= 100 \\ z &= 50. \end{aligned}$$

The numbers check. The solution is (20, 30, 50).

Section 3.6

2.

	First Investment	Second Investment	Third Investment	Total
Principal, P	x	y	z	\$25,000
Rate of interest, r	3%	4%	5%	
Time, t	1 year	1 year	1 year	
Interest, I	$0.03x$	$0.04y$	$0.05z$	\$1120

Section 3.7

4. $4x + 3y \geq 12$

1. Graph the related equation $4x + 3y = 12$.
2. Since the inequality symbol is $\geq$, draw a $\frac{\text{solid}}{\text{dashed/solid}}$ line.
3. Try the test point (0, 0).

$$\begin{array}{c|c} 4x + 3y & \geq 12 \\ \hline 4(0) + 3(0) & ?\ 12 \\ 0 & 12 \end{array} \quad \text{FALSE}$$

(0, 0) is not a solution, so we shade the opposite half-plane.

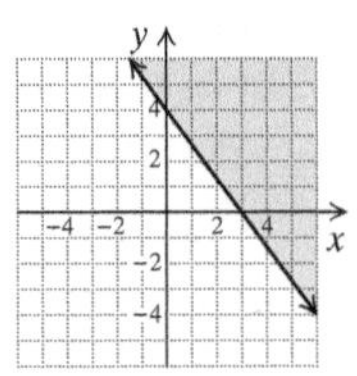

$4x + 3y \geq 12$

CHAPTER 4

Section 4.1

10. Collect like terms:

$3xy^3 + 2x^3y + 5xy^3 - 8x + 15 - 3x^2y - 6x^2y + 11x - 8.$

$3xy^3$ and $5xy^3$ are like terms.

$-3x^2y$ and $-6x^2y$ are like terms.

$-8x$ and $11x$ are like terms.

15 and -8 are like terms.

Rearranging:

$2x^3y + 3xy^3 + 5xy^3 - 3x^2y - 6x^2y - 8x + 11x + 15 - 8.$

Collecting like terms:

$2x^3y + 8xy^3 - 9x^2y + 3x + 7.$

14. Write two equivalent expressions for the opposite of $4x^3 - 5x^2 + \frac{1}{4}x - 10$.

One expression uses an inverse sign in front:
$-(4x^3 - 5x^2 + \frac{1}{4}x - 10)$.

For a second expression, write the opposite of each term:
$-4x^3 + 5x^2 - \frac{1}{4}x + 10$.

Section 4.2

19. $\left(m^3 + \frac{1}{4}n\right)^2$

$= (m^3)^2 + 2(m^3)\left(\frac{1}{4}n\right) + \left(\frac{1}{4}n\right)^2$

$= m^6 + \frac{1}{2}m^3n + \frac{1}{16}n^2$

23. $\left(3w - \frac{3}{5}q^2\right)\left(3w + \frac{3}{5}q^2\right)$

$= (3w)^2 - \left(\frac{3}{5}q^2\right)^2$

$= 9w^2 - \frac{9}{25}q^4$

Section 4.3

10. Factor: $(y + 3)(y - 21) + (y + 3)(y + 10)$.

The common binomial factor is $y + 3$.

$(y + 3)(y - 21) + (y + 3)(y + 10)$

$= (y + 3)(y - 21 + y + 10)$

$= (y + 3)(2y - 11)$

11. $5y^3 + 2y^2 - 10y - 4$

$= y^2(5y + 2) - 2(5y + 2)$

$= (5y + 2)(y^2 - 2)$

Section 4.4

2. Factor: $y^2 + 7y + 10$.

Because both 7 and 10 are positive, we need consider only positive factors of 10.

Pairs of Factors	Sums of Factors
1, 10	11
2, 5	7

The numbers we need are 2 and 5.

Thus, $y^2 + 7y + 10 = (y + 2)(y + 5)$.

15. Factor: $y^6 + y^3 - 6$.

We look for numbers p and q such that

$y^6 + y^3 - 6 = (y^3 + p)(y^3 + q)$.

The product of p and q is -6.
The sum of p and q is 1.
The factors we want are -2 and 3.
The factorization is $(y^3 - 2)(y^3 + 3)$.

Section 4.5

4. Factor: $20x^5 - 46x^4 + 24x^3$.

First, factor out the largest common factor:
$2x^3(10x^2 - 23x + 12)$.
$10x^2$ can be factored as $2x \cdot 5x$ or as $10x \cdot x$.
Since -23 is negative, the factors of 12 must be negative:
12 can be factored as $(-1)(-12)$, $(-2)(-6)$, or $(-3)(-4)$.
Next, list all possibilities with $2x$ and $5x$ as the first terms.
If none of these is correct, we will check $10x$ and x.

$(2x - 1)(5x - 12)$,

$(2x - 12)(5x - 1)$,

$(2x - 2)(5x - 6)$,

$(2x - 6)(5x - 2)$,

$(2x - 3)(5x - 4)$,

$(2x - 4)(5x - 3)$

Only two possibilities have no common factor:

$(2x - 1)(5x - 12) = 10x^2 - 29x + 24$,

$(2x - 3)(5x - 4) = 10x^2 - 23x + 12$.

The factorization of $20x^5 - 46x^4 + 24x^3$ is
$2x^3(2x - 3)(5x - 4)$.

10. Factor: $4x^2 + 37x + 9$

1. There is no common factor.
2. Multiply the leading coefficient and the constant: $4(9) = 36$.
3. Look for a pair of factors of 36 whose sum is 37. Both factors will be positive.

Pairs of Factors	Sums of Factors
1, 36	37
2, 18	20
3, 12	15
4, 9	13
6, 6	12

4. Split the middle term, $37x$: $37x = x + 36x$.
5. Factor by grouping:

   $4x^2 + x + 36x + 9$

   $= x(4x + 1) + 9(4x + 1)$

   $= (4x + 1)(x + 9)$.

6. *Check:* $(4x + 1)(x + 9) = 4x^2 + 37x + 9$.

Section 4.6

5. Factor as a trinomial square:

$16x^4 - 40x^2y^3 + 25y^6$

$= (4x^2)^2 - 2 \cdot 4x^2 \cdot 5y^3 + (5y^3)^2$

$= (4x^2 - 5y^3)^2$.

10. Factor as a difference of squares:

$m^2 - \frac{1}{9}$

$= m^2 - \left(\frac{1}{3}\right)^2$

$= \left(m + \frac{1}{3}\right)\left(m - \frac{1}{3}\right)$.

Section 4.7

1. Factor completely: $3y^3 - 12x^2y$.
   a) Factor out the largest common factor:
   $3y^3 - 12x^2y = 3y(y^2 - 4x^2)$.
   b) There are two terms inside the parentheses. The expression is a difference of squares. Factor the difference of squares:
   $3y(y^2 - 4x^2) = 3y(y + 2x)(y - 2x)$.
   c) We have factored completely.
   d) *Check:* $3y(y + 2x)(y - 2x)$
   $= 3y(y^2 - 4x^2)$
   $= 3y^3 - 12x^2y$.

Section 4.8

1.
$$x^2 + 8 = 6x$$
$$x^2 - 6x + 8 = 0$$
$$(x - 2)(x - 4) = 0$$
$$x - 2 = 0 \quad or \quad x - 4 = 0$$
$$x = 2 \quad or \quad x = 4$$
Both numbers check. The solutions are 2 and 4.

10. 1. *Familiarize.* Let x and $x + 2$ represent the lengths of the sides of the triangle.

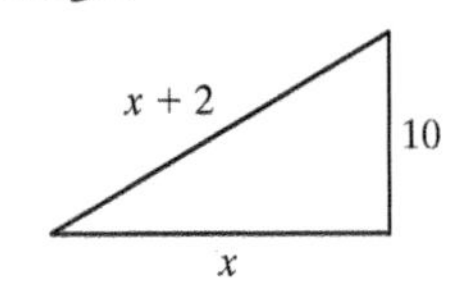

2. *Translate.*
$$a^2 + b^2 = c^2$$
$$x^2 + 10^2 = (x + 2)^2$$
3. *Solve.*
$$x^2 + 100 = x^2 + 4x + 4$$
$$100 = 4x + 4$$
$$96 = 4x$$
$$24 = x$$
4. *Check.* If $x = 24$, then $x + 2 = 26$.
Then $10^2 + 24^2 = 100 + 576 = 676 = 26^2$.
5. *State.* The lengths are 24 cm and 26 cm.

CHAPTER 5

Section 5.1

14.
$$\frac{y-3}{3-y} = \frac{y-3}{-(y-3)}$$
$$= \frac{1(y-3)}{-1(y-3)}$$
$$= \frac{1}{-1} \cdot \frac{y-3}{y-3}$$
$$= -1 \cdot 1$$
$$= -1$$

22.
$$\frac{x^2 + 7x + 10}{2x - 4} \div \frac{x^2 - 3x - 10}{x - 2}$$
$$= \frac{x^2 + 7x + 10}{2x - 4} \cdot \frac{x - 2}{x^2 - 3x - 10}$$
$$= \frac{(x + 5)(x + 2)(x - 2)}{2(x - 2)(x - 5)(x + 2)}$$
$$= \frac{(x + 5)\cancel{(x + 2)}\cancel{(x - 2)}}{2\cancel{(x - 2)}(x - 5)\cancel{(x + 2)}}$$
$$= \frac{x + 5}{2(x - 5)}$$

Section 5.2

13.
$$\frac{3x}{7} + \frac{4y}{3x}$$
$$= \frac{3x}{7} \cdot \frac{3x}{3x} + \frac{4y}{3x} \cdot \frac{7}{7}$$
$$= \frac{9x^2}{21x} + \frac{28y}{21x}$$
$$= \frac{9x^2 + 28y}{21x}$$

20.
$$\frac{4x^2}{2x - y} - \frac{7x^2}{y - 2x}$$
$$= \frac{4x^2}{2x - y} - \frac{7x^2}{y - 2x} \cdot \frac{-1}{-1}$$
$$= \frac{4x^2}{2x - y} - \frac{-7x^2}{2x - y}$$
$$= \frac{4x^2 - (-7x^2)}{2x - y}$$
$$= \frac{11x^2}{2x - y}$$

Section 5.3

4.
$$\begin{array}{r} x + 5 \\ x + 2 \overline{)x^2 + 7x + 9} \\ \underline{x^2 + 2x} \\ 5x + 9 \\ \underline{5x + 10} \\ -1 \end{array}$$
The answer is $x + 5$, R -1; or $x + 5 + \frac{-1}{x + 2}$.

11. $(2x^3 - 4x^2 + 8x - 8) \div (x - 3)$
$$\begin{array}{r|rrrr} 3 & 2 & -4 & 8 & -8 \\ & & 6 & 6 & 42 \\ \hline & 2 & 2 & 14 & 34 \end{array}$$
The answer is $2x^2 + 2x + 14$, R 34; or $2x^2 + 2x + 14 + \frac{34}{x - 3}$.

Section 5.4

3.
$$\frac{\frac{1}{a} + \frac{1}{b}}{\frac{1}{a} - \frac{1}{b}}$$
$$= \frac{\frac{1}{a} + \frac{1}{b}}{\frac{1}{a} - \frac{1}{b}} \cdot \frac{ab}{ab}$$
$$= \frac{\left(\frac{1}{a} + \frac{1}{b}\right) \cdot ab}{\left(\frac{1}{a} - \frac{1}{b}\right) \cdot ab}$$

$$= \frac{\frac{1}{a}\cdot ab + \frac{1}{b}\cdot ab}{\frac{1}{a}\cdot ab - \frac{1}{b}\cdot ab}$$

$$= \frac{b + a}{b - a}$$

6. $$\frac{1 - \frac{1}{x}}{1 - \frac{1}{x^2}}$$

$$= \frac{1\cdot\frac{x}{x} - \frac{1}{x}}{1\cdot\frac{x^2}{x^2} - \frac{1}{x^2}}$$

$$= \frac{\frac{x}{x} - \frac{1}{x}}{\frac{x^2}{x^2} - \frac{1}{x^2}}$$

$$= \frac{\frac{x-1}{x}}{\frac{x^2-1}{x^2}}$$

$$= \frac{x-1}{x}\cdot\frac{x^2}{x^2-1}$$

$$= \frac{(x-1)\cdot x\cdot x}{x(x+1)(x-1)}$$

$$= \frac{\cancel{(x-1)}\cdot\cancel{x}\cdot x}{\cancel{x}(x+1)\cancel{(x-1)}}$$

$$= \frac{x}{x+1}$$

Section 5.5

1. Solve: $\frac{2}{3} + \frac{5}{6} = \frac{1}{x}$.

LCM $= 6x$, or $2\cdot 3\cdot x$

$$2\cdot 3\cdot x\left(\frac{2}{3} + \frac{5}{6}\right) = 2\cdot 3\cdot x\cdot\frac{1}{x}$$

$$2\cdot 3\cdot x\cdot\frac{2}{3} + 2\cdot 3\cdot x\cdot\frac{5}{6} = 2\cdot 3\cdot x\cdot\frac{1}{x}$$

$$2\cdot x\cdot 2 + x\cdot 5 = 2\cdot 3$$

$$4x + 5x = 6$$

$$9x = 6$$

$$x = \frac{6}{9}$$

$$x = \frac{2}{3}$$

The number $\frac{2}{3}$ checks and is the solution.

5. Solve: $\frac{2}{x-1} = \frac{3}{x+2}$.

LCM $= (x-1)(x+2)$

$$(x-1)(x+2)\cdot\frac{2}{x-1} = (x-1)(x+2)\cdot\frac{3}{x+2}$$

$$2(x+2) = 3(x-1)$$

$$2x + 4 = 3x - 3$$

$$7 = x$$

The number 7 checks and is the solution.

Section 5.6

3. Let c = the number of calories burned in 35 min.

$$\frac{20}{110} = \frac{35}{c}$$

$$20c = 110\cdot 35$$

$$20c = 3850$$

$$c = 192.5$$

In 35 min, Mia will burn 192.5 calories.

Section 5.7

2. $$I = \frac{pT}{M + pn}$$

$$I(M + pn) = pT$$

$$IM + Ipn = pT$$

$$Ipn = pT - IM$$

$$n = \frac{pT - IM}{Ip}$$

Section 5.8

1. $y = kx$

$$8 = k\cdot 20$$

$$\frac{8}{20} = k$$

$$\frac{2}{5} = k$$

The variation constant is $\frac{2}{5}$. The equation of variation is $y = \frac{2}{5}\cdot x$.

5. $$y = \frac{k}{x}$$

$$0.012 = \frac{k}{50}$$

$$0.012\cdot 50 = k$$

$$0.6 = k$$

The variation constant is 0.6. The equation of variation is $y = \frac{0.6}{x}$.

7. $y = kx^2$

$$175 = k\cdot 5^2$$

$$175 = k\cdot 25$$

$$7 = k$$

The equation of variation is $y = 7x^2$.

9. $y = kxz$

$$65 = k\cdot 10\cdot 13$$

$$65 = k\cdot 130$$

$$\frac{65}{130} = k$$

$$\frac{1}{2} = k$$

The equation of variation is $y = \frac{1}{2}xz$.

CHAPTER 6

Section 6.1

20. $g(x) = \sqrt{6x + 4}$; $g(0)$, $g(3)$, and $g(-5)$

$$g(0) = \sqrt{6\cdot 0 + 4}$$

$$= \sqrt{0 + 4}$$

$$= \sqrt{4} = 2$$

$$g(3) = \sqrt{6\cdot 3 + 4} = \sqrt{18 + 4} = \sqrt{22}$$

$$g(-5) = \sqrt{6(-5) + 4} = \sqrt{-30 + 4} = \sqrt{-26}$$

-26 is a negative radicand. No real-number function value exists.

33. $\sqrt{x^2 - 6x + 9} = \sqrt{(x-3)^2} = |x - 3|$

Section 6.2

17. $81^{-3/4} = \dfrac{1}{81^{3/4}} = \dfrac{1}{(\sqrt[4]{81})^3} = \dfrac{1}{3^3} = \dfrac{1}{27}$

32. $\dfrac{a^{1/2}b^{3/8}}{a^{1/4}b^{1/8}} = a^{1/2-1/4}\cdot b^{3/8-1/8} = a^{1/4}\cdot b^{1/4} = \sqrt[4]{ab}$

Section 6.3

13. $\sqrt{12ab^3c^2}$
$= \sqrt{4\cdot 3\cdot a\cdot b^2\cdot b\cdot c^2}$
$= \sqrt{4}\cdot\sqrt{b^2}\cdot\sqrt{c^2}\cdot\sqrt{3ab}$
$= 2\cdot b\cdot c\sqrt{3ab}$

17. $\sqrt[3]{3x^2y}\,\sqrt[3]{36x}$
$= \sqrt[3]{3x^2y\cdot 36x}$
$= \sqrt[3]{3\cdot x\cdot x\cdot y\cdot 2\cdot 2\cdot 3\cdot 3\cdot x}$
$= 3\cdot x\sqrt[3]{4y}$

Section 6.4

5. $\sqrt{25x - 25} - \sqrt{9x - 9}$
$= \sqrt{25(x-1)} - \sqrt{9(x-1)}$
$= 5\sqrt{x-1} - 3\sqrt{x-1}$
$= 2\sqrt{x-1}$

10. $(\sqrt{2} + \sqrt{5})(\sqrt{2} - \sqrt{5})$
$= (\sqrt{2})^2 - (\sqrt{5})^2$
$= 2 - 5$
$= -3$

Section 6.5

3. $\sqrt{\dfrac{4a}{3b}} = \dfrac{\sqrt{4a}}{\sqrt{3b}}\cdot\dfrac{\sqrt{3b}}{\sqrt{3b}}$
$= \dfrac{\sqrt{12ab}}{\sqrt{3^2\cdot b^2}}$
$= \dfrac{2\sqrt{3ab}}{3b}$

11. $\dfrac{14}{3 + \sqrt{2}}$
$= \dfrac{14}{3+\sqrt{2}}\cdot\dfrac{3-\sqrt{2}}{3-\sqrt{2}}$
$= \dfrac{14(3-\sqrt{2})}{9 - 2}$
$= \dfrac{14(3-\sqrt{2})}{7}$
$= 2(3 - \sqrt{2})$
$= 6 - 2\sqrt{2}$

Section 6.6

3.
$x + 2 = \sqrt{2x + 7}$
$(x+2)^2 = (\sqrt{2x+7})^2$
$x^2 + 4x + 4 = 2x + 7$
$x^2 + 2x - 3 = 0$
$(x+3)(x-1) = 0$
$x + 3 = 0$ *or* $x - 1 = 0$
$x = -3$ *or* $x = 1$

The number -3 does not check, but the number 1 does check. The solution is 1.

6. $\sqrt[4]{x-1} - 2 = 0$
$\sqrt[4]{x-1} = 2$
$(\sqrt[4]{x-1})^4 = 2^4$
$x - 1 = 16$
$x = 17$

Section 6.8

5. $\sqrt{-54} = \sqrt{-1\cdot 54}$
$= \sqrt{-1}\cdot\sqrt{54} = i\sqrt{9\cdot 6}$
$= i\cdot 3\sqrt{6}$
$= 3\cdot i\sqrt{6}$, or $3\sqrt{6}i$

15. $(1 + 3i)(1 + 5i)$
$= 1 + 5i + 3i + 15i^2$
$= 1 + 8i + 15(-1)$
$= 1 + 8i - 15$
$= -14 + 8i$

25. $6i^{11} + 7i^{14}$
$= 6\cdot i^{10}\cdot i + 7\cdot i^{14}$
$= 6(i^2)^5\cdot i + 7(i^2)^7$
$= 6(-1)^5\cdot i + 7(-1)^7$
$= 6(-1)i + 7(-1)$
$= -6i - 7$
$= -7 - 6i$

CHAPTER 7

Section 7.1

5.
$5x^2 = 15$
$x^2 = 3$
$x = \sqrt{3}$ *or* $x = -\sqrt{3}$

The solutions can also be written $\pm\sqrt{3}$. If we round to three decimal places, the solutions are ± 1.732.

9.
$x^2 + 16x + 64 = 11$
$(x+8)^2 = 11$
$x + 8 = \sqrt{11}$ *or* $x + 8 = -\sqrt{11}$
$x = -8 + \sqrt{11}$ *or* $x = -8 - \sqrt{11}$

The solutions can also be written $-8 \pm \sqrt{11}$.

11. $x^2 - 8x - 20 = 0$

$$x^2 - 8x = 20$$
$$x^2 - 8x + 16 = 20 + 16$$
$$(x-4)^2 = 36$$
$$x - 4 = 6 \quad or \quad x - 4 = -6$$
$$x = 10 \quad or \quad x = -2$$

Section 7.2

2. Write the equation in standard form.

$$3x^2 + 2x - 7 = 0$$
$$a = 3, \quad b = 2, \quad c = -7$$
$$x = \frac{-2 \pm \sqrt{2^2 - 4 \cdot 3 \cdot (-7)}}{2 \cdot 3}$$
$$x = \frac{-2 \pm \sqrt{88}}{6}$$
$$x = \frac{-2 \pm 2\sqrt{22}}{6}$$
$$x = \frac{2(-1 \pm \sqrt{22})}{2 \cdot 3}$$
$$x = \frac{-1 \pm \sqrt{22}}{3}$$

Approximate the solutions and round to three decimal places.

$$\frac{-1 + \sqrt{22}}{3} \approx 1.230$$
$$\frac{-1 - \sqrt{22}}{3} \approx -1.897$$

Section 7.3

4.

	Distance	Speed	Time
Faster Ship	3000	$r + 10$	$t - 50$
Slower Ship	3000	r	t

6. $V = \pi r^2 h$

$$\frac{V}{\pi h} = r^2$$
$$\sqrt{\frac{V}{\pi h}} = r$$

Section 7.4

3. Determine the nature of the solutions of $3x^2 - 2x + 1 = 0$.

$$a = 3, \quad b = -2, \quad c = 1$$
$$b^2 - 4ac = (-2)^2 - 4 \cdot 3 \cdot 1 = -8$$

Since the discriminant is negative, there are 2 nonreal solutions.

9. Solve: $x^4 - 10x^2 + 9 = 0$.

Let $u = x^2$. Then $u^2 = x^4$.

$$x^4 - 10x^2 + 9 = 0$$
$$u^2 - 10u + 9 = 0$$
$$(u - 1)(u - 9) = 0$$
$$u - 1 = 0 \quad or \quad u - 9 = 0$$
$$u = 1 \quad or \quad u = 9$$
$$x^2 = 1 \quad or \quad x^2 = 9$$
$$x = \pm 1 \quad or \quad x = \pm 3$$

All four numbers check. The solutions are $-1, 1, -3$, and 3.

Section 7.5

4. Compute $g(x)$ for each value of x shown.

x	$g(x) = (x + 2)^2$
-2	0
-1	1
0	4
1	9
-3	1
-4	4
-5	9

Plot the points and draw the curve.

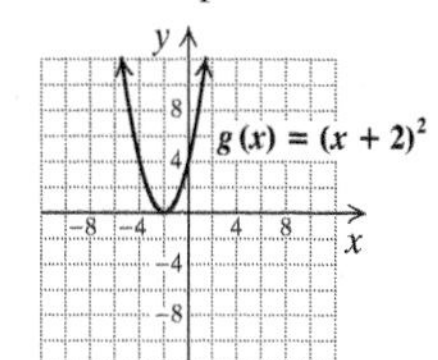

7. Compute $g(x)$ for each value of x shown.

x	$g(x) = (x - 1)^2 - 3$
1	-3
2	-2
3	1
4	6
0	-2
-1	1
-2	6

The vertex is $(1, -3)$.

The line of symmetry is $x = 1$, and the minimum function value is -3.

Plot the points and draw the curve.

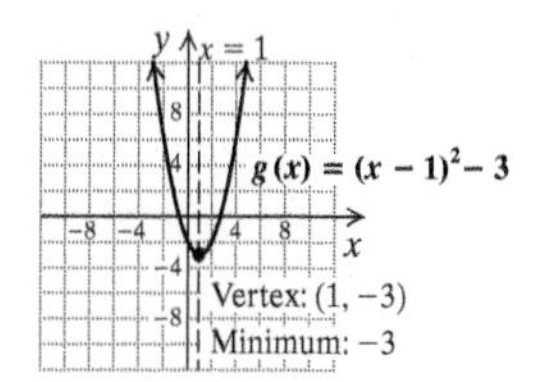

Section 7.6

6. Find the vertex of the parabola given by $f(x) = -4x^2 + 12x - 5$.

The x-coordinate of the vertex is $-\frac{b}{2a} = -\frac{12}{2(-4)} = \frac{3}{2}$.

The second coordinate of the vertex is $f\left(\frac{3}{2}\right)$.

$$f\left(\frac{3}{2}\right) = -4\left(\frac{3}{2}\right)^2 + 12\left(\frac{3}{2}\right) - 5$$
$$= -4\left(\frac{9}{4}\right) + 18 - 5$$
$$= -9 + 13$$
$$= 4$$

The vertex is $\left(\frac{3}{2}, 4\right)$.

7. Find the intercepts of $f(x) = x^2 + 2x - 3$.

The y-intercept is $(0, f(0))$.

$f(0) = 0^2 + 2 \cdot 0 - 3 = -3$

The y-intercept is $(0, -3)$. To find the x-intercepts, we solve $0 = x^2 + 2x - 3$:

$$0 = x^2 + 2x - 3$$
$$0 = (x - 1)(x + 3)$$
$$x - 1 = 0 \quad or \quad x + 3 = 0$$
$$x = 1 \quad or \quad x = -3.$$

The x-intercepts are $(1, 0)$ and $(-3, 0)$.

Section 7.8

6. Solve: $6x(x + 1)(x - 1) < 0$.

The solutions of $6x(x + 1)(x - 1) = 0$ are 0, -1, and 1. Divide the number line into four intervals and test values of $f(x) = 6x(x + 1)(x - 1)$.

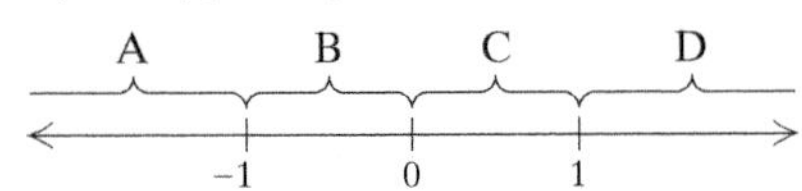

A: Test -2.

$$f(-2) = 6(-2)(-2 + 1)(-2 - 1)$$
$$= -36$$

B: Test $-\frac{1}{2}$.

$$f\left(-\frac{1}{2}\right) = 6\left(-\frac{1}{2}\right)\left(-\frac{1}{2} + 1\right)\left(-\frac{1}{2} - 1\right)$$
$$= \frac{9}{4}$$

C: Test $\frac{1}{2}$.

$$f\left(\frac{1}{2}\right) = 6\left(\frac{1}{2}\right)\left(\frac{1}{2} + 1\right)\left(\frac{1}{2} - 1\right)$$
$$= -\frac{9}{4}$$

D: Test 2.

$$f(2) = 6(2)(2 + 1)(2 - 1)$$
$$= 36$$

The expression is negative for values of x in intervals A and C. The solution set is $\{x \mid x < -1 \text{ or } 0 < x < 1\}$, or $(-\infty, -1) \cup (0, 1)$.

CHAPTER 8

Section 8.1

8. Amount after 1 year:

$\$40{,}000 + \$40{,}000 \times 0.02$
$= \$40{,}000(1.02)$
$= \$40{,}800$

Amount after 2 years:

$\$40{,}800 + \$40{,}800 \times 0.02$
$= \$40{,}800(1.02)$
$= \$41{,}616$

10. $$A = P \cdot \left(1 + \frac{r}{n}\right)^{n \cdot t}$$
$$= 7000 \cdot \left(1 + \frac{3.4\%}{4}\right)^{4 \cdot \frac{11}{2}}$$
$$= 7000 \cdot \left(1 + \frac{0.034}{4}\right)^{22}$$
$$= 7000(1.0085)^{22}$$
$$\approx \$8432.72$$

Section 8.2

2. $$(f \circ g)(x) = f(g(x))$$
$$= f(\sqrt[3]{x})$$
$$= 4(\sqrt[3]{x}) + 5$$
$$(g \circ f)(x) = g(f(x))$$
$$= g(4x + 5)$$
$$= \sqrt[3]{4x + 5}$$

Section 8.3

11. $$\log_{10} x = 4$$
$$10^4 = x$$
$$10{,}000 = x$$

14. Let $\log_{10} 10{,}000 = x$.
$$10^x = 10{,}000$$
$$10^x = 10^4$$
$$x = 4$$

Section 8.4

11. $$\log_a \frac{x^3 y^4}{z^5 w^9} = \log_a x^3 y^4 - \log_a z^5 w^9$$
$$= (\log_a x^3 + \log_a y^4) - (\log_a z^5 + \log_a w^9)$$
$$= \log_a x^3 + \log_a y^4 - \log_a z^5 - \log_a w^9$$
$$= 3 \log_a x + 4 \log_a y - 5 \log_a z - 9 \log_a w$$

22. $$\log_a 16 = \log_a 2^4$$
$$= 4 \log_a 2$$
$$= 4(0.301)$$
$$= 1.204$$

Section 8.5

8. **(a)** $$\log_6 7 = \frac{\log 7}{\log 6};$$
$$\approx 1.0860$$

(b) $$\log_6 7 = \frac{\ln 7}{\ln 6}$$
$$\approx 1.0860$$

Section 8.6

3. $$7^x = 20$$
$$\log 7^x = \log 20$$
$$x \cdot \log 7 = \log 20$$
$$x = \frac{\log 20}{\log 7}$$
$$x \approx 1.5395$$

8. $\log_3(2x - 1) - \log_3(x - 4) = 2$

$$\log_3 \frac{2x - 1}{x - 4} = 2$$

$$\frac{2x - 1}{x - 4} = 3^2$$

$$\frac{2x - 1}{x - 4} = 9$$

$$2x - 1 = 9(x - 4)$$

$$2x - 1 = 9x - 36$$

$$35 = 7x$$

$$5 = x$$

Section 8.7

1. $L = 10 \cdot \log \frac{I}{I_0}$

$= 10 \cdot \log \frac{3.2 \times 10^{-6}}{10^{-12}}$

$= 10 \cdot \log(3.2 \times 10^6)$

$= 10 \cdot (\log 3.2 + \log 10^6)$

$\approx 10 \cdot (0.5051 + 6)$

≈ 65

The sound level is about 65 decibels.

4. $\text{pH} = -\log[\text{H}^+]$

$7 = -\log[\text{H}^+]$

$-7 = \log[\text{H}^+]$

$10^{-7} = [\text{H}^+]$

CHAPTER 9

Section 9.1

5. $d = \sqrt{(x_2 - x_1)^2 + (y_2 - y_1)^2}$

$= \sqrt{(4 - (-2))^2 + (2 - 1)^2}$

$= \sqrt{6^2 + 1^2}$

$= \sqrt{37}$

≈ 6.083

10. $(x - h)^2 + (y - k)^2 = r^2$

$(x - (-3))^2 + (y - 1)^2 = 6^2$

$(x + 3)^2 + (y - 1)^2 = 36$

Section 9.2

1. $\frac{x^2}{9} + \frac{y^2}{4} = \frac{x^2}{3^2} + \frac{y^2}{2^2}$

$a = 3,\ b = 2$

The x-intercepts are $(-3, 0)$ and $(3, 0)$.
The y-intercepts are $(0, -2)$ and $(0, 2)$.

We plot the intercepts and draw the ellipse.

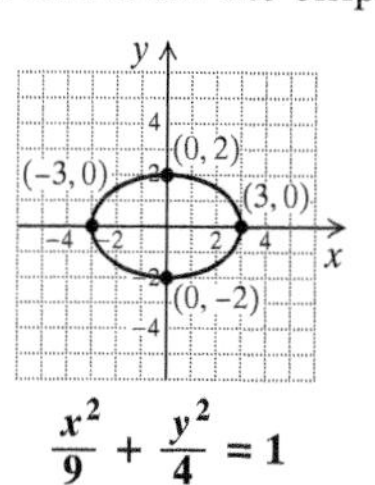

$\frac{x^2}{9} + \frac{y^2}{4} = 1$

3. $16x^2 + 9y^2 = 144$

$$\frac{1}{144}(16x^2 + 9y^2) = \frac{1}{144}(144)$$

$$\frac{x^2}{9} + \frac{y^2}{16} = 1$$

$$\frac{x^2}{3^2} + \frac{y^2}{4^2} = 1$$

$a = 3,\ b = 4$

The x-intercepts are $(-3, 0)$ and $(3, 0)$.
The y-intercepts are $(0, -4)$ and $(0, 4)$.
We draw the ellipse.

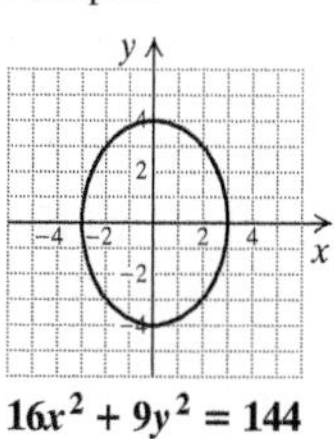

$16x^2 + 9y^2 = 144$

Section 9.3

1. $\frac{x^2}{16} - \frac{y^2}{25} = \frac{x^2}{4^2} - \frac{y^2}{5^2}$

$a = 4, b = 5$

The asymptotes are $y = \frac{5}{4}x$ and $y = -\frac{5}{4}x$.

The hyperbola is horizontal.

The x-intercepts are $(-4, 0)$ and $(4, 0)$.

We sketch the asymptotes, plot the intercepts, and draw the hyperbola.

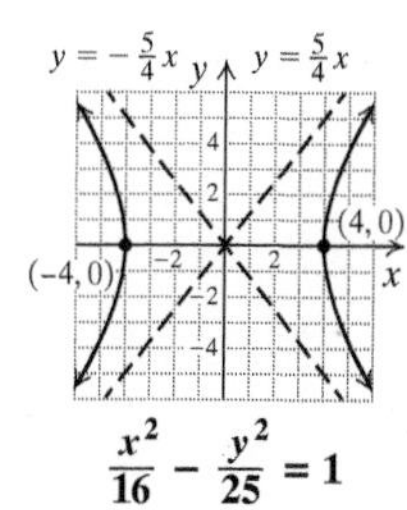

$\frac{x^2}{16} - \frac{y^2}{25} = 1$

Section 9.4

4. $9x^2 - 4y^2 = 36,$

$5x + 2y = 0$

Solve $5x + 2y = 0$ for x:

$$x = -\frac{2}{5}y.$$

Substitute $-\frac{2}{5}y$ for x in the first equation and solve for y:

$$9\left(-\frac{2}{5}y\right)^2 - 4y^2 = 36$$

$$9\left(\frac{4}{25}y^2\right) - 4y^2 = 36$$

$$\frac{36}{25}y^2 - \frac{100}{25}y^2 = 36$$

$$-\frac{64}{25}y^2 = 36$$

$$y^2 = -\frac{225}{16}$$

$$y = \pm\sqrt{-\frac{225}{16}} = \pm\frac{15}{4}i.$$

If $y = \frac{15}{4}i$, then $x = -\frac{2}{5}y = -\frac{2}{5}\cdot\frac{15}{4}i = -\frac{3}{2}i.$

If $y = -\frac{15}{4}i$, then $x = -\frac{2}{5}y = -\frac{2}{5}\cdot\left(-\frac{15}{4}i\right) = \frac{3}{2}i.$

The solutions are $\left(-\frac{3}{2}i, \frac{15}{4}i\right)$ and $\left(\frac{3}{2}i, -\frac{15}{4}i\right)$.

5. $2y^2 - 3x^2 = 6,$

$5y^2 + 2x^2 = 53$

To eliminate the x^2-terms, multiply the first equation by 2 and the second equation by 3:

$$\begin{array}{l} 4y^2 - 6x^2 = 12 \\ 15y^2 + 6x^2 = 159 \\ \hline 19y^2 \qquad\quad = 171 \\ \qquad y^2 = 9 \\ \qquad y = \pm 3. \end{array}$$

Substitute 3 for y in $5y^2 + 2x^2 = 53$:

$$5(3)^2 + 2x^2 = 53$$

$$45 + 2x^2 = 53$$

$$2x^2 = 8$$

$$x^2 = 4$$

$$x = \pm 2.$$

We get the same result when $y = -3$. The solutions are $(2, 3)$ $(2, -3)$, $(-2, 3)$, and $(-2, -3)$.

Glossary

A

Abscissa The first coordinate in an ordered pair

Absolute value The distance that a number is from 0 on the number line

***ac*-method** A method for factoring trinomials of the type $ax^2 + bx + c, a \neq 1$, involving the product, ac, of the leading coefficient a and the last term c

Additive identity The number 0

Additive inverse A number's opposite; two numbers are additive inverses of each other if their sum is 0

Algebraic expression An expression consisting of variables, constants, numerals, operation signs, and/or grouping symbols

Ascending order When a polynomial in one variable is arranged so that the exponents increase from left to right, it is said to be in ascending order.

Associative law of addition The statement that when three numbers are added, regrouping the addends gives the same sum

Associative law of multiplication The statement that when three numbers are multiplied, regrouping the factors gives the same product

Asymptote A line that a graph approaches more and more closely as x increases or as x decreases

Axes Two perpendicular number lines used to locate points in a plane

Axis of symmetry A line that can be drawn through a graph such that the part of the graph on one side of the line is an exact reflection of the part on the opposite side; also called *line of symmetry*

B

Base In exponential notation, the number being raised to a power

Binomial A polynomial composed of two terms

Break-even point In business, the point of intersection of the revenue function and the cost function

C

Circle The set of all points in a plane that are a fixed distance r, called the radius, from a fixed point (h, k), called the center

Circumference The distance around a circle

Coefficient The numerical multiplier of a variable

Common logarithm A logarithm with base 10

Commutative law of addition The statement that when two numbers are added, changing the order in which the numbers are added does not affect the sum

Commutative law of multiplication The statement that when two numbers are multiplied, changing the order in which the numbers are multiplied does not affect the product

Complementary angles Angles whose sum is 90°

Completing the square Adding a particular constant to an expression so that the resulting sum is a perfect square

Complex number Any number that can be named $a + bi$, where a and b are any real numbers

Complex number *i* The square root of -1; that is, $i = \sqrt{-1}$ and $i^2 = -1$

Complex rational expression A rational expression that contains rational expressions within its numerator and/or denominator

Complex-number system A number system that contains the real-number system and is designed so that negative numbers have defined square roots

Composite function A function in which a quantity depends on a variable that, in turn, depends on another variable

Compound inequality A statement in which two or more inequalities are joined by the word *and* or the word *or*

Compound interest Interest computed on the sum of an original principal and the interest previously accrued by that principal

Conic section A curve formed by the intersection of a plane and a cone

Conjugate of a complex number The conjugate of a complex number $a + bi$ is $a - bi$ and the conjugate of $a - bi$ is $a + bi$.

Conjugate of a radical expression Pairs of radical terms, like $\sqrt{a} + \sqrt{b}$ and $\sqrt{a} - \sqrt{b}$ or $c + \sqrt{d}$ and $c - \sqrt{d}$, are called conjugates.

Conjunction A statement in which two or more sentences are joined by the word *and*

Consecutive even integers Even integers that are two units apart

Consecutive integers Integers that are one unit apart

Consecutive odd integers Odd integers that are two units apart

Consistent system of equations A system of equations that has at least one solution

Constant A known number

Constant function A function given by an equation of the form $y = b$, or $f(x) = b$, where b is a real number

Constant of proportionality The constant in an equation of direct variation or inverse variation

Coordinates The numbers in an ordered pair

Cube root The number c is the cube root of a, written $\sqrt[3]{a}$, if the third power of c is a

D

Degree of a polynomial The degree of the term of highest degree in a polynomial

Degree of a term The sum of the exponents of the variables

Dependent equations The equations in a system are dependent if one equation can be removed without changing the solution set.

Dependent variable The variable that represents the output of a function

Descending order When a polynomial is arranged so that the exponents decrease from left to right, it is said to be in descending order.

Diameter A segment that passes through the center of a circle and has its endpoints on the circle

Difference of cubes Any expression that can be written in the form $A^3 - B^3$

Difference of squares Any expression that can be written in the form $A^2 - B^2$

Direct variation A situation that gives rise to a linear function $f(x) = kx$, or $y = kx$, where k is a positive constant

Discriminant The expression $b^2 - 4ac$ from the quadratic formula

Disjoint sets Two sets with an empty intersection

Disjunction A statement in which two or more sentences are joined by the word *or*

Distributive law of multiplication over addition The statement that multiplying a factor by the sum of two numbers gives the same result as multiplying the factor by each of the two numbers and then adding

Distributive law of multiplication over subtraction The statement that multiplying a factor by the difference of two numbers gives the same result as multiplying the factor by each of the two numbers and then subtracting

Domain The set of all first coordinates of the ordered pairs in a function

Doubling time The time necessary for a population to double in size

E

Elimination method An algebraic method that uses the addition principle to solve a system of equations

Ellipse The set of all points in a plane for which the sum of the distances from two fixed points F_1 and F_2 is constant

Empty set The set without members

Equation A number sentence that says that the expressions on either side of the equals sign, =, represent the same number

Equation of direct variation An equation described by $y = kx$, with k a positive constant, used to represent direct variation

Equation of inverse variation An equation described by $y = k/x$, with k a positive constant, used to represent inverse variation

Equivalent equations Equations with the same solutions

Equivalent expressions Expressions that have the same value for all allowable replacements

Equivalent inequalities Inequalities that have the same solution set

Evaluate To substitute a value for each occurrence of a variable in an expression and carry out the operations

Even root When the number k in $\sqrt[k]{\ }$ is an even number, we say that we are taking an even root.

Exponent In expressions of the form a^n, the number n is an exponent.

Exponential decay model A decrease in quantity over time that can be modeled by an exponential function of the form $P(t) = P_0e^{-kt}$, $k > 0$

Exponential decay rate The variable k in the exponential decay model $P(t) = P_0e^{-kt}$

Exponential equation An equation in which a variable appears as an exponent

Exponential function The function $f(x) = a^x$, where a is a positive constant different from 1

Exponential growth model An increase in quantity over time that can be modeled by an exponential function of the form $P(t) = P_0e^{kt}$, $k > 0$

Exponential growth rate The variable k in the exponential growth model $P(t) = P_0e^{kt}$

Exponential notation A representation of a number using a base raised to a power

F

Factor *Verb*: To write an equivalent expression that is a product. *Noun*: A multiplier

Factorization of a polynomial An expression that names the polynomial as a product of factors

Focus One of two fixed points that determine the points of an ellipse

FOIL To multiply two binomials by multiplying the First terms, the Outside terms, the Inside terms, and then the Last terms

Formula An equation that uses numbers or letters to represent a relationship between two or more quantities

Fraction equation An equation containing one or more rational expressions; also called a *rational equation*

Function A correspondence that assigns to each member of a set called the domain *exactly one* member of a set called the range

G

Grade The measure of a road's steepness

Graph A picture or a diagram of the data in a table; a line, a curve, or a collection of points that represents all the solutions of an equation or an inequality

Greatest common factor (GCF) The common factor of a polynomial with the largest possible coefficient and the largest possible exponent(s)

H

Half-life The amount of time necessary for half of a quantity to decay

Hyperbola The set of all points in a plane for which the difference of the distances from two fixed points F_1 and F_2 is constant

Hypotenuse In a right triangle, the side opposite the right angle

I

Identity property of 1 The statement that the product of a number and 1 is always the original number

Identity property of 0 The statement that the sum of a number and 0 is always the original number

Imaginary number A number that can be named bi, where b is some real number and $b \neq 0$
Inconsistent system of equations A system of equations for which there is no solution
Independent equations Equations that are not dependent
Independent variable The variable that represents the input of a function
Index In the expression $\sqrt[k]{a}$, the number k is called the index.
Inequality A mathematical sentence using $<, >, \leq, \geq$, or $\neq$
Input A member of the domain of a function
Integers The whole numbers and their opposites
Intercept The point at which a graph intersects the x-axis or the y-axis
Intersection of sets *A* and *B* The set of all members that are common to A and B
Interval notation The use of a pair of numbers inside parentheses and brackets to represent the set of all numbers between those two numbers
Inverse relation The relation formed by interchanging the coordinates of the ordered pairs in a relation
Inverse variation A situation that gives rise to a function $f(x) = k/x$, or $y = k/x$, where k is a positive constant
Irrational number A real number whose decimal notation neither terminates nor has a repeating block of digits. An irrational number cannot be represented as a quotient of two integers.

J

Joint variation A situation that gives rise to an equation of the form $y = kxz$, where k is a positive constant

L

Leading coefficient The coefficient of the term of highest degree in a polynomial
Leading term The term of highest degree in a polynomial
Least common denominator (LCD) The least common multiple of the denominators
Least common multiple (LCM) The smallest number that is a multiple of two or more numbers
Legs In a right triangle, the two sides that form the right angle
Like radicals Radicals having the same index and radicand
Like terms Terms that have exactly the same variable factors; also called *similar terms*
Line of symmetry A line that can be drawn through a graph such that the part of the graph on one side of the line is an exact reflection of the part on the opposite side; also called *axis of symmetry*
Linear equation in two variables Any equation that can be written in the form $y = mx + b$ or $Ax + By = C$, where x and y are variables
Linear equation in three variables An equation equivalent to one of the type $Ax + By + Cz = D$, where x and y are variables
Linear function A function that can be described by an equation of the form $y = mx + b$, where x and y are variables
Linear inequality An inequality whose related equation is a linear equation
Logarithmic equation An equation containing a logarithmic expression
Logarithmic function, base *a* The inverse of an exponential function $f(x) = a^x$

M

Mathematical model A model in which the essential parts of a problem are described in mathematical language
Matrix A rectangular array of numbers
Maximum The largest function value (output) achieved by a function
Minimum The smallest function value (output) achieved by a function
Monomial A constant, or a constant times a variable or variables raised to powers that are nonnegative integers
Motion formula The formula
Distance = Rate (or Speed) · Time
Multiplication property of 0 The statement that the product of 0 and any real number is 0
Multiplicative identity The number 1
Multiplicative inverses Reciprocals; two numbers whose product is 1

N

Natural logarithm A logarithm with base e
Natural numbers The counting numbers: 1, 2, 3, 4, 5, . . .
Negative integers The integers to the left of zero on the number line
Nonlinear equation An equation whose graph is not a straight line
Nonlinear function A function whose graph is not a straight line

O

Odd root When the number k in $\sqrt[k]{}$ is an odd number, we say that we are taking an odd root.
One-to-one function A function for which different inputs have different outputs
Opposite The opposite, or additive inverse, of a number a is denoted $-a$. Opposites are the same distance from 0 on the number line but on different sides of 0.
Opposite of a polynomial To find the opposite of a polynomial, replace each term with its opposite—that is, change the sign of every term.
Ordered pair A pair of numbers of the form (h, k) for which the order in which the numbers are listed is important
Ordinate The second coordinate in an ordered pair
Origin The point on a graph where the two axes intersect
Output A member of the range of a function

P

Parabola A graph of a quadratic function
Parallel lines Lines in the same plane that never intersect. Two nonvertical lines are parallel if they have the same slope and different y-intercepts.
Perfect square A rational number p for which there exists a number a for which $a^2 = p$
Perfect-square trinomial A trinomial that is the square of a binomial

Perimeter The sum of the lengths of the sides of a polygon
Perpendicular lines Lines that form a right angle. Two lines are perpendicular if the product of their slopes is -1 or if one line is vertical and the other is horizontal.
Pi (π) The number that results when the circumference of a circle is divided by its diameter; $\pi \approx 3.14$, or $22/7$
Point–slope equation An equation of the form $y - y_1 = m(x - x_1)$, where m is the slope and (x_1, y_1) is a point on the line
Polynomial A monomial or a combination of sums and/or differences of monomials
Polynomial equation An equation in which two polynomials are set equal to each other
Positive integers The natural numbers or the integers to the right of zero on the number line
Prime polynomial A polynomial that cannot be factored using only integer coefficients
Principal square root The nonnegative square root of a number
Principle of zero products The statement that an equation $ab = 0$ is true if and only if $a = 0$ is true or $b = 0$ is true, or both are true
Proportion An equation stating that two ratios are equal
Proportional numbers Two pairs of numbers having the same ratio
Pythagorean theorem In any right triangle, if a and b are the lengths of the legs and c is the length of the hypotenuse, then $a^2 + b^2 = c^2$.

Q

Quadrants The four regions into which the axes divide a plane
Quadratic equation An equation of the type $ax^2 + bx + c = 0$, where a, b, and c are real-number constants and $a > 0$
Quadratic formula The solutions of $ax^2 + bx + c = 0$ are given by the equation $x = \dfrac{-b \pm \sqrt{b^2 - 4ac}}{2a}$.
Quadratic inequality A second-degree polynomial inequality in one variable

R

Radical The symbol $\sqrt{}$
Radical equation An equation in which a variable appears in one or more radicands
Radical expression An algebraic expression written with a radical
Radicand The expression written under the radical
Radius A segment with one endpoint on the center of a circle and the other endpoint on the circle
Range The set of all second coordinates of the ordered pairs in a function
Rate The ratio of two different kinds of measure
Ratio Any rational expression a/b
Rational equation An equation containing one or more rational expressions; also called a *fraction equation*
Rational expression A quotient of two polynomials
Rational inequality An inequality containing a rational expression
Rational number A number that can be written in the form p/q, where p and q are integers and $q \neq 0$
Rationalizing the denominator A procedure for finding an equivalent expression without a radical in the denominator
Real numbers All rational and irrational numbers; the set of all numbers corresponding to points on the number line
Reciprocal A multiplicative inverse; two numbers are reciprocals if their product is 1.
Rectangle A four-sided polygon with four right angles
Relation A correspondence between a first set, called the domain, and a second set, called the range, such that each member of the domain corresponds to *at least one* member of the range
Repeating decimal A decimal in which a block of digits repeats indefinitely
Right triangle A triangle that includes a 90° angle
Rise The change in the second coordinate between two points on a line
Roster method A way of naming sets by listing all the elements in the set
Run The change in the first coordinate between two points on a line

S

Scientific notation A representation of a number of the form $M \times 10^n$, where n is an integer, $1 \leq M < 10$, and M is expressed in decimal notation
Set A collection of objects
Set-builder notation The naming of a set by describing basic characteristics of the elements in the set
Similar terms Terms that have exactly the same variable factors; also called *like terms*
Simplify To rewrite an expression in an equivalent, abbreviated, form
Slope The ratio of the rise to the run for any two points on a line
Slope–intercept equation An equation of the form $y = mx + b$, where x and y are variables, the slope is m, and the y-intercept is $(0, b)$
Solution A replacement for the variable that makes an equation or an inequality true
Solution of a system of linear inequalities An ordered pair (x, y) that is a solution of *all* inequalities
Solution of a system of three equations An ordered triple (x, y, z) that makes *all three* equations true
Solution of a system of two equations An ordered pair (x, y) that makes *both* equations true
Solution set The set of all solutions of an equation, an inequality, or a system of equations or inequalities
Solve To find all solutions of an equation, an inequality, or a system of equations or inequalities; to find the solution(s) of a problem
Square A four-sided polygon with four right angles and all sides of equal length
Square of a number A number multiplied by itself
Square root The number c is a square root of a if $c^2 = a$.
Standard form of a quadratic equation A quadratic equation in the form $ax^2 + bx + c = 0$, where $a \neq 0$
Subsets Sets that are contained within other sets
Substitute To replace a variable with a number or an equivalent expression
Substitution method A nongraphical method for solving systems of equations

Sum of cubes An expression that can be written in the form $A^3 + B^3$

Sum of squares An expression that can be written in the form $A^2 + B^2$

Supplementary angles Angles whose sum is 180°

Synthetic division A simplified process for dividing a polynomial by a binomial of the type $x - a$

System of equations A set of two or more equations that are to be solved simultaneously

System of linear inequalities A set of two or more inequalities that are to be solved simultaneously

T

Term A number, a variable, or a product or a quotient of numbers and/or variables

Terminating decimal A decimal that can be written using a finite number of decimal places

Trinomial A polynomial that is composed of three terms

Trinomial square The square of a binomial expressed as a polynomial with three terms

U

Union of sets *A* and *B* The set of all elements belonging to A and/or B

V

Value The numerical result after a number has been substituted into an expression and the operations have been carried out

Variable A letter that represents an unknown number

Variation constant The constant in an equation of direct variation or inverse variation

Vertex The point at which the graph of a quadratic equation crosses its axis of symmetry

Vertical-line test If it is possible for a vertical line to cross a graph more than once, then the graph is *not* the graph of a function.

W

Whole numbers The natural numbers and 0: 0, 1, 2, 3, . . .

X

***x*-intercept** The point at which a graph crosses the x-axis

Y

***y*-intercept** The point at which a graph crosses the y-axis

Photo Credits

JUST-IN-TIME REVIEW: 1, Michal Adamczyk/123RF; VintageMind/Shutterstock **CHAPTER 1: 31,** rdonar/Shutterstock; Candus Camera/Shutterstock **46,** Sho Tamura/AFLO SPORT/Alamy Live News/Alamy Stock Photo **47,** Xinhua/Alamy Stock Photo **53,** GagliardiImages/Shutterstock **54,** Darren Brode/Fotolia **57,** degist/Fotolia **58,** Comstock/Jupiterimages/Getty Images **59,** goodluz/Fotolia **64 (left),** Simon Eeman/123RF **64 (right),** Kondoruk/Shutterstock **66,** Miroslav Beneda/123RF **77,** Vitaliy Vodolazsky/123RF **78,** gwimages/Fotolia **84,** Bob Orsillo/Shutterstock **85,** Icatnews/Shutterstock **86,** John Kwan/Shutterstock **93,** Monkey Business/Shutterstock **96,** Andrey N. Bannov/Shutterstock **CHAPTER 2: 117,** miroslav110/123RF; Softulka/Shutterstock; arcady31/123RF **133,** Mesut Dogan/123RF **138,** Geri Lavrov/Getty Images **141 (left),** WENN Ltd/Alamy Stock Photo **141 (right),** Graham Hardy USA/Alamy Stock Photo **142,** pstedrak/123RF **181,** Andrey Cherkasov/123RF **182,** Thampapon/Shutterstock **186 (left),** Cathleen A. Clapper/Shutterstock **186 (right),** Mali lucky/Shutterstock **199,** Natursports/Shutterstock **CHAPTER 3: 201,** dotshock/Shutterstock; red mango/Shutterstock; Iculig/123RF **202,** Cathy Yeulet/123RF **215,** StephanScherhag/Shutterstock **222,** John David Mercer/Mobile Press-Register **225 (left),** Michael Dwyer/Alamy Stock Photo **225 (right),** V. J. Matthew/Shutterstock **226,** Tyler Olson/123RF **227,** Vladimir Wrangel/Shutterstock **235 (left),** REDPIXEL.PL/Shutterstock **235 (right),** dizanna/123RF **249,** ShutterOK/Shutterstock **251 (left),** smolaw/Shutterstock **251 (right),** Anton Estrada/123RF **252 (left),** Alena Ozerova/123RF **252 (right),** Bram Reusen/Shutterstock **267,** Leksele/Shutterstock **CHAPTER 4: 279,** Shao-Chun Wang/123RF; Stuart Miles/123RF **287 (left),** qingwa/Fotolia **287 (right),** Original Pyramid Stacker™ **303,** Mark Scheuern/Alamy Stock Photo **345,** steheap/Fotolia **CHAPTER 5: 363,** Mark Winfrey/123RF; Josep Jorba/123RF **412,** The Sandbagger Corporation **413,** ALCE/Fotolia **414,** Andriy Blokhin/Shutterstock **415,** EuToch/Fotolia **416,** milan2099/Shutterstock **417,** Chris Curtis/Shutterstock **419,** Hansenn/Fotolia **421,** Umat34/Shutterstock **422 (top left),** Elena Elisseeva/Shutterstock **422 (top right),** John Kwan/Shutterstock **422 (bottom left),** HodagMedia/Shutterstock **422 (bottom right),** Tom Zeman/Shutterstock **423 (left),** Kaband/Shutterstock **423 (right),** Emaria/Shutterstock **424 (top left),** Elenathewise/Fotolia **424 (top right),** Marv Bittinger **424 (bottom),** Eric Gavaert/Fotolia **425 (left),** Pavel Losevsky/Fotolia **425 (right),** Jay Crihfield/Shutterstock **429,** Paul Yates/Fotolia **430,** Lisa F. Young/Fotolia **432,** Ikonoklast_hh/Fotolia **436,** 578foot/123RF **437 (left and right),** S_oleg/Shutterstock **438,** Barry Blackburn/123RF **450 (bottom left),** Kathy Libby/123RF **450 (top right),** 1000 Words/Shutterstock **CHAPTER 6: 451,** Blue Jean Images/Alamy Stock Photo; Testing/Shutterstock **460,** ESB Professional/Shutterstock **498,** NurPhoto/Contributor/Getty Images **501,** jivan child/Fotolia **502 (left),** Akhararat Wathanasing/123RF **502 (right),** Kara Grubis/Shutterstock **507 (left),** Mo Peerbacus/Alamy Stock Photo **507 (right),** Lasse Kristensen/Fotolia **525,** Hildegard Williams/123RF **530,** Jochen Tack/Alamy Stock Photo **CHAPTER 7: 531,** Hanna Kuprevich/Shutterstock; Dirk Ercken/123RF **544 (left),** Ingo Oeland/Alamy Stock Photo **544 (right),** Xinhua/Alamy Stock Photo **557,** ARENA Creative/Shutterstock **560,** Bill Bachman/Alamy Stock Photo **562 (left),** Rob Kints/Shutterstock **562 (right),** Stefan Schurr/Shutterstock **599,** Bob Cheung/Fotolia **CHAPTER 8: 623,** Vaclav Volrab/123RF; Jason Winter/Shutterstock **636 (left),** Artush/Shutterstock **636 (right),** senrakawa/Shutterstock **637 (left),** Jenny Thompson/Fotolia **637 (right),** Deco Images II/Alamy Stock Photo **638,** Alex Van Wyhe and Shane Kimberlin of Tea Mug Collective **690,** Aaron Amat/Shutterstock **693,** Roman Fedin/123RF **696,** Rancz Andrei/Shutterstock **697,** Reprinted with permission of the Baltimore Sun Media Group. All rights reserved. **699 (left),** Michael Rosskothen/123RF **699 (right),** Fotolia RAW/Fotolia **700 (left),** Nicky Rhodes/Shutterstock **700 (right),** rawpixel/123RF **702 (left),** Trial/Shutterstock **702 (right),** pmstephens/Fotolia **703 (left),** ukartpics/Alamy Stock Photo **703 (right),** David Chedgy/Alamy Stock Photo **710,** Victor Korchenko/Alamy Stock Photo **711,** rawpixel/123RF **CHAPTER 9: 717,** Aflo Co. Ltd./Alamy Stock Photo; Zoonar GmbH/Alamy Stock Photo **719,** Bob Richardson/Alamy Stock Photo **758 (left),** Andrey Popov/Shutterstock **758 (right),** alexskopje/Shutterstock **767,** xiaoliangge/fotolia

Index

T

Geometric Formulas

PLANE GEOMETRY

Rectangle

Area: $A = l \cdot w$
Perimeter: $P = 2 \cdot l + 2 \cdot w$

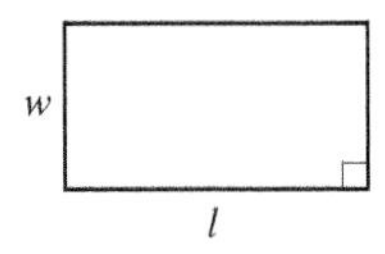

Square

Area: $A = s^2$
Perimeter: $P = 4 \cdot s$

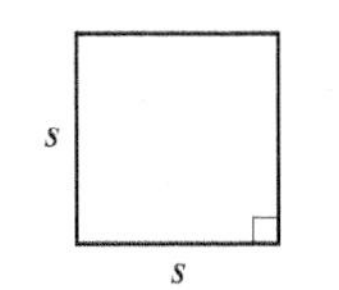

Triangle

Area: $A = \frac{1}{2} \cdot b \cdot h$

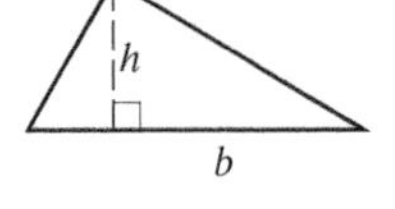

Sum of Angle Measures

$A + B + C = 180°$

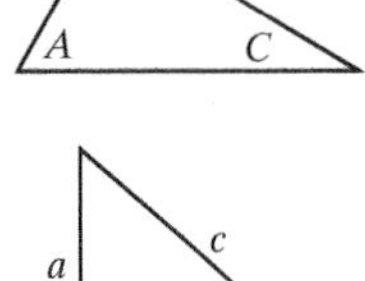

Right Triangle

Pythagorean Theorem:
$a^2 + b^2 = c^2$

Parallelogram

Area: $A = b \cdot h$

Trapezoid

Area: $A = \frac{1}{2} \cdot h \cdot (a + b)$

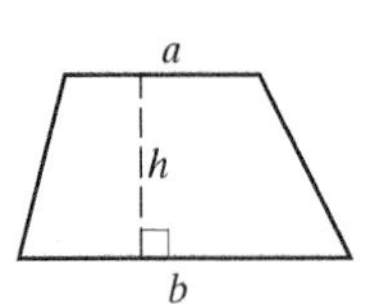

Circle

Area: $A = \pi \cdot r^2$
Circumference:
$C = \pi \cdot d = 2 \cdot \pi \cdot r$ ($\frac{22}{7}$ and 3.14 are different approximations for π)

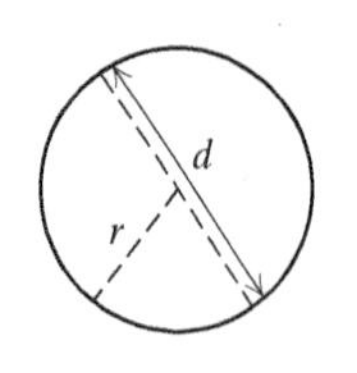

SOLID GEOMETRY

Rectangular Solid

Volume: $V = l \cdot w \cdot h$

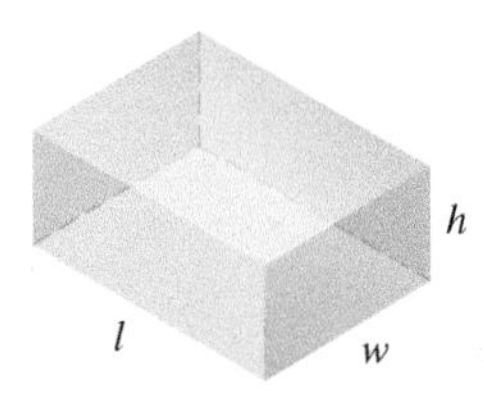

Cube

Volume: $V = s^3$

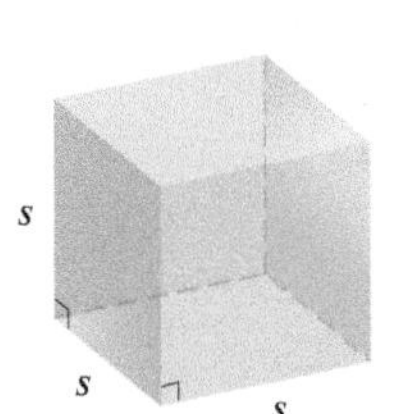

Right Circular Cylinder

Volume: $V = \pi \cdot r^2 \cdot h$
Surface Area:
$S = 2 \cdot \pi \cdot r \cdot h + 2 \cdot \pi \cdot r^2$

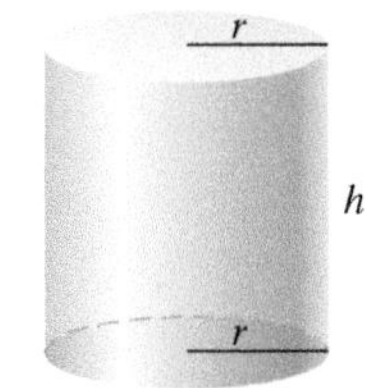

Right Circular Cone

Volume: $V = \frac{1}{3} \cdot \pi \cdot r^2 \cdot h$
Surface Area:
$S = \pi \cdot r^2 + \pi \cdot r \cdot s$

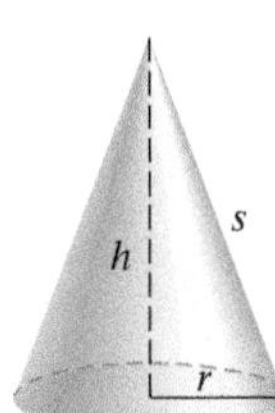

Sphere

Volume: $V = \frac{4}{3} \cdot \pi \cdot r^3$
Surface Area: $S = 4 \cdot \pi \cdot r^2$

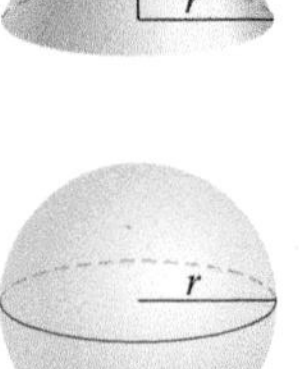